LANDOLT-BÖRNSTEIN

Numerical Data and Functional Relationships in Science and Technology

New Series

Editor in Chief: K.-H. Hellwege

Group III: Crystal and Solid State Physics

Volume 4
Magnetic and Other Properties of Oxides and Related Compounds

Part b

D. Bonnenberg · E. L. Boyd · B. A. Calhoun · V. J. Folen · W. Gräper
A. P. Greifer · C. J. Kriessman · R. A. Lefever · T. R. Mc Guire
M. Paulus · G. H. Stauss · R. Vautier · H. P. J. Wijn

Editors: K.-H. Hellwege and A. M. Hellwege

Springer-Verlag Berlin · Heidelberg · New York 1970

LANDOLT-BÖRNSTEIN

Zahlenwerte und Funktionen
aus Naturwissenschaften und Technik

Neue Serie

Gesamtherausgabe: K.-H. Hellwege

Gruppe III: Kristall- und Festkörperphysik

Band 4

Magnetische und andere Eigenschaften
von Oxiden und verwandten Verbindungen

Teil b

D. Bonnenberg · E. L. Boyd · B. A. Calhoun · V. J. Folen · W. Gräper
A. P. Greifer · C. J. Kriessman · R. A. Lefever · T. R. Mc Guire
M. Paulus · G. H. Stauss · R. Vautier · H. P. J. Wijn

Herausgeber: K.-H. Hellwege und A. M. Hellwege

Springer-Verlag Berlin · Heidelberg · New York 1970

ISBN 3-540-05176-7 Springer-Verlag Berlin-Heidelberg-New York
ISBN 0-387-05176-7 Springer-Verlag New York-Heidelberg-Berlin

Library of Congress Catalog Card Number 62-53136

Vorwort

Nachdem Teil a schon vor einem halben Jahr herausgegeben wurde, erscheint jetzt mit Teil b der Rest des Bandes III/4 der Neuen Serie des Landolt-Börnstein. Wie schon im ersten Teil ausgeführt, handelt es sich um eine Neubearbeitung eines Kapitels des 1962 erschienenen Bandes II/9 der 6. Auflage (siehe das Vorwort zu Teilband III/4a).

Auch diesem Teilband ist eine Liste der Symbole und Abkürzungen vorangestellt. Das alphabetische Substanzenverzeichnis für alle Substanzen aus beiden Teilbänden befindet sich am Schluß des vorliegenden Teilbandes.

Den Autoren möchten wir für ihre schnelle und sorgfältige Arbeit danken. Dem Verlag gebührt besonderer Dank für die Aufgeschlossenheit, mit der er die bei der Herstellung auftauchenden Probleme bewältigte.

Darmstadt, September 1970 **Die Herausgeber**

Preface

Subsequent to part 4a which was published six months ago, part 4b is now available, thus completing Vol. III/4 in the Landolt-Börnstein New Series. It has already been explained in part 4a that Vol. III/4 is a revision of a chapter of Vol. II/9 of the 6th edition published in 1962 (see the preface to Vol. III/4a).

As in part 4a, there is a preceding list of symbols and abbreviations. The alphabetical index of substances for all the compounds in both parts is placed at the end of this part.

We would like to thank the authors for their prompt and careful work. The publishers deserve special thanks for their readiness to cope with the problems which arose during publication.

Darmstadt, September 1970 **The Editors**

Survey

Part a

Part b

Übersicht

Teil a

Teil b

Table of contents — Inhaltsverzeichnis

5 Non-iron garnets

By W. GRÄPER, Technische Hochschule Darmstadt, Germany

6 Spinels

7 Hexagonal ferrites

By H. P. J. WIJN, Philips Research Laboratories, Eindhoven, Netherlands

Alphabetical list of abbreviations and symbols

(for specific symbols and their definitions see special lists in the respective chapters)

a) Abbreviations and indices

A	anisotropy
A	per gram atom; per atom
AF	antiferromagnetic
a	applied
$A, B; a, c, d; 1, 2, \cdots$	for sublattices or sites $A, B; a, c, d; 1, 2, \cdots$
ac	alternating current
anis	anisotropic
calc	calculated
crit	critical
cub	cubic
dc	direct current
dia	diamagnetic
dod	dodecahedral
eff	effective
ex	exchange
exp	experimental
ext	external
f	ferromagnetic
g	per gram
H (lower index)	at field strength H
hex	hexagonal
i	initial
int	internal
is	isotropic
lat	lattice
liq	liquid
loc	local
ME	magnetoelastic
Me	metal

m	per mole; per molecule
magn	magnetic
max	maximum
min	minimum
mon	monoclinic
n	nuclear; per nucleus
o, orth	orthorhombic
oct	octahedral
p	paramagnetic
p	planar
RT	room temperature
rem	remanent
rev	reversal
res	resonance
resid	residual
rh	rhombohedral
s	saturation
ss, SS	solid solution
T (upper index)	at temperature T
tetr	tetragonal
tot	total
vap	vapor
0 (lower index)	spontaneous
0 (upper index)	at 0 °K
$\parallel, \perp$	parallel, perpendicular
[ ]	(octahedral) B sites } in chemical formulae of spinels
( )	(tetrahedral) A sites } in chemical formulae of spinels
□	ion vacancy

b) Symbols

A [cm^{-1}]	absorption (coefficient)
A, B [cm^{-1}]	hyperfine (nuclear spin) splitting parameters in the spin Hamiltonian
a [erg/Gauss^2 mole; emu/mole]	temperature-independent paramagnetic term of the susceptibility
a, b, c [Å]	lattice parameters
B [Gauss]	induction
B_{rem} [Gauss]	remanent induction
B_{resid} [Gauss]	residual induction
B_s [Gauss]	saturation induction
B_V [Gauss]	induction at the Villari reversal point
$(BH)_{max}$ [Gauss Oe]	maximum energy product of permanent magnet materials
C [cm^{-1} Gauss^{-2}]	Cotton-Mouton constant
C [cal mole^{-1} deg^{-1}]	specific heat per mole, molar heat capacity
C_A [cm^3 °K/g-atom]	atomic Curie constant
C_g [cm^3 °K/g]	Curie constant per gram
C_{lat} [cal mole^{-1} deg^{-1}]	lattice contribution to specific heat
C_m [cm^3 °K/mole]	Curie constant per mole
C_{magn} [cal mole^{-1}deg^{-1}]	magnetic heat per mole
c_{ij} [dyn/cm^2]	elastic constants
c_{magn} [cal °K^{-1} g^{-1}]	magnetic heat per gram
c_n [cal °K g^{-1}]	nuclear contribution to c_{magn}
c_p [cal °K^{-1} g^{-1}]	specific heat at constant pressure
D	stiffness parameter
D, E, F [cm^{-1}]	fine structure (crystal field) parameters in the spin Hamiltonian
d [m]	diameter

E [dyn/cm³]	Young's modulus
E_k [erg/cm³]	magneto-crystalline energy per unit volume
ΔE	shift of center of gravity of Mössbauer absorption lines
ΔE [eV]	energy difference between two levels
e	electron charge
e [V/°K]	thermoelectric power
eqQ/h [Mc/s]	quadrupole coupling constant
f [Hz, c/s]	frequency
f_s [c/s]	spin resonance frequency
f, F [erg/cm³] g, G [erg/cm³]	induced anisotropy constants
g	spectroscopic splitting factor
g'	magnetomechanical ratio
$g_{\parallel,\perp}$	g-value parallel or perpendicular to the principal magnetic axis
g_1, g_2, g_3	principal values of the g-tensor
$g_{A,B'}$; $g_{a,b,d}$	sublattice g-values
g_{app}	"apparent" g-factor
g_{eff}	effective g-value for two- or three-sublattice system
g_{exp}	experimentally observed g-factor
g_n	nuclear g-factor
H^A [Oe]	magneto-crystalline anisotropy field
H_a [Oe]	applied (or external) magnetic field
H_A, H_B [Oe]	hyperfine field for the A and B sites*)
$H_{a,d}$ [Oe]	molecular field at a, d sites*)
H_{bias} [Oe]	biasing (magnetic) field
H_c [Oe]	coercive field
${}_MH_C$ [Oe]	coercive field for magnetization
H_{const} [Oe]	constant external magnetic field
H_{crit}, H_a^{crit} [Oe]	critical applied field, e.g. for antiferromagnetic-ferromagnetic transition or spinflop transition
H_{drive} [Oe]	magnetic drive field
H_{eff} [Oe]	effective magnetic field (e.g. at the nucleus)*)
H_{ext} [Oe]	external magnetic field
H_{ext}^{crit} [Oe]	critical external field
H_{int} [Oe]	internal magnetic field*)
H_0 [Oe]	magnetic field for antiferromagnetic resonance or: threshold field for switching
H_Q [Oe]	quenching field
H_{res} [Oe]	magnetic field for resonance
H_T	hyperfine magnetic field
H_T	annealing field
ΔH [Oe]	full linewitdh of magnetic resonance absorption (distance between the inflection points of the absorption curve)
$\Delta H_{1/2}$ [Oe]	full half-width of magnetic resonance absorption (distance between the points of the absorption curve with half the maximum value)
ΔH_{exp} [Oe]	measured full linewidth
ΔH_k [Oe]	spin-wave linewidth
ΔH_{slow} [Oe]	slow relaxation linewidth
ΔH_{theor} [Oe]	calculated theoretical full linewidth
δH [Oe]	shift of resonance field
$h_{a,c,d}$ [Oe]	local magnetic field for sublattice a, c, d
h_c [Oe]	critical switching field
h_{crit} [Oe]	critical amplitude of the r.f. field for nonlinear effects in the ferromagnetic resonance
h_{thr} [Oe]	threshold field
J	exchange energy (J/hc in [cm⁻¹], J/k in [°K])
K [deg Gauss⁻¹ cm⁻¹]	magnetooptical Kerr constant
K_1/M, K_2/M [Oe]	first-, second-order anisotropy field
K_i [erg/cm³]	anisotropy coefficients
K_u [erg/cm³]	uniaxial induced anisotropy constant
k [erg/°K]	Boltzmann constant
k [cm⁻¹]	wave number (of spin waves]
$\mathbf{k}$ [cm⁻¹]	wave vector

*) For special definition, see original paper.

k	magnetomechanical coupling factor
k_i	coefficient of the linear magnetostriction; see p. 549
L [erg/cm³]	torque
$M = P/\mathrm{cm}^3$ [Gauss = emu/cm³]	magnetic moment per unit volume = magnetization
(M_A) [g/mole]	molar mass of the substance A
M_a, M_b [Gauss]	sublattice magnetization
M_0	spontaneous magnetization ($H = 0$)
M_R, M_{rem} [Gauss]	remanent magnetization
M_s [Gauss]	saturation magnetization
m [g]	electron mass
N (N_x, N_y, N_z)	demagnetization factor
$N_L = 6.02 \cdot 10^{23}$ mole⁻¹	Avogadro (Loschmidt) number
n	index of refraction
$\Delta n_{CM} = n_\parallel - n_\perp$	Cotton-Mouton birefringence
$\Delta n_F = n_r - n_l$	Faraday birefringence
P [Gauss cm³ = emu]	total magnetic moment
P [cm⁻¹]	quadrupole interaction constant in the spin Hamiltonian
p [at]	pressure
p_A [μ_B]	magnetic moment per atom (or ion)
$(p_A)_{a,c,d}$ [μ_B]	magnetic moment per ion of sublattice a, c, d
p_{eff} [μ_B]; $(p_m)_{eff} = 2.84 \sqrt{C_m}$ [μ_B]	effective magnetic moment per molecule
p_m [μ_B]	magnetic moment per molecule
p_{ms} [μ_B]	saturation magnetic moment per molecule
p_n [μ_n]	nuclear magnetic moment
p_0, (p_{0m}) [μ_B]	spontaneous magnetic moment (per molecule)
Q [cal °K⁻¹]	heat capacity
Q	quality factor
Q^{-1}	attenuation
Q_{trans} ($Q_{\alpha,\beta}\cdots$)	heat capacity due to (α, $\beta\cdots$) transformation
q [eV]	activation energy for conductivity
R [cm sec⁻¹ Oe⁻¹]	domain wall mobility
R [Ω]	resistance
R [V m A⁻¹ Gauss⁻¹]	ordinary Hall constant
R_0	spontaneous Hall constant
R'	extraordinary Hall constant
R_s	squareness ratio; see p. 249
$r = 1/\mu_i$	reluctivity
S [V °K⁻¹]	Seebeck coefficient
$\mathbf{S}$	effective spin
$\mathbf{S}_i$	spin of ion i
S_{magn} [cal °K⁻¹]	magnetic contribution to entropy
S/R	relative entropy
ΔS	entropy change
$(\Delta S)_{lat}$	lattice contribution to entropy change
S_w [Oe μsec]	switching coefficient
T [°K]	temperature
T_1, τ_1 [sec]	spin-lattice relaxation time, T or τ, according to experimental method
T_2, τ_2 [sec]	spin-spin relaxation time, T or τ, according to experimental method
$T_{1,2}$ [sec]	cross relaxation time
T_C [°K]	ferroelectric Curie temperature
T_{comp} [°K]	compensation temperature ($M_a + M_b = 0$ in ferrimagnetic substances)
T_{crit} [°K]	critical temperature
T_{max}, $T_{max}(C)$ [°K]	temperature associated with the maximum in the specific heat C
T_{melt} [°K]	melting temperature
T_N [°K]	magnetic transition temperature
T_Q [°K]	quenching temperature (temperature from which the sample is quenched)
T_s [°K]	effective temperature of spin system
T_{trans}, $T_{\alpha,\beta}$, ... [°K]	transition temperature
t [sec]	time
u	oxygen parameter; anion parameter

V [deg cm^{-1} Oe^{-1}]	Verdet constant
V [cm^3]	volume
V_0, v [Å^3]	volume of unit cell
V, V_c	crystal field potential (of rare earth sites)
v [cm sec^{-1}]	domain wall velocity; velocity; source velocity of the Mössbauer effect
v	valency
v_{app} [cm sec^{-1}]	apparent domain wall velocity
$v_{T,L}$ [cm sec^{-1}]	velocity of sound, transversal or longitudinal
$\Delta v/v$	relative variation of sound velocity with magnetic field
W [erg]	energy
ΔW	energy of the crystal field components of the ground state above the lowest component ($\Delta W/hc$ in [cm^{-1}], $\Delta W/k$ in [°K])
w_K [erg/cm^3]	magneto-crystalline anisotropy energy density
w_U [erg/cm^3]	anneal-induced uniaxial magneto-crystalline anisotropy energy density
x, y, z	symmetry axes, rectangular coordinates
x_O, y_O, z_O	oxygen parameters
Z	number of molecules in the unit cell; counting rate (in arbitrary scale) in Mössbauer spectrum
α [emu/mole	temperature-independent paramagnetic susceptibility
α [°K^{-1}]	coefficient of thermal expansion
α [Gauss cm V^{-1}]	magnetoelectric susceptibility
α [deg/cm]	specific rotation of the plane of polarization (Faraday rotation)
α, β, γ [°]	angles between crystallographic axes
$\alpha_1, \alpha_2, \alpha_3; \beta_1, \beta_2, \cdots$ $\alpha_x, \alpha_y, \alpha_z$	direction cosines
β [cm^2/g, cm^2/dyn]	compressibility
Γ	half-width of Mössbauer absorption lines at half absorption
γ	gryomagnetic ratio
Δ	energy splitting of an exicted state
δ [m sec^{-1}]	isomer shift in Mössbauer spectrum
$\tan \delta_m = \mu''/\mu'$	magnetic loss tangent
ε	dielectric constant
ε; Σ' [mm sec^{-1}; eV]	quadrupole splitting in Mössbauer spectrum
Θ [°K, °C]	transition temperature
Θ_a	asymptotic Curie temperature
Θ_C [°K]	Curie temperature
Θ_c [°K]	transition temperature for superconductivity
Θ_{comp} [°K]	compensation temperature (see T_{comp})
Θ_D [°K]	Debye temperature
Θ_N [°K]	Néel temperature
Θ_{ord} [°K]	ordering temperature
Θ_p [°K]	paramagnetic Curie temperature
Θ_r [°K]	temperature of spin reorientation
Θ_{trans} [°K]	crystallographic transition temperature
θ [°, deg]	Faraday rotation angle angle between measurement direction and magnetic field
θ_0 [°, deg]	angle for the preferred direction of the magnetization with respect to the c-axis of the crystal
$\theta_{\boldsymbol{k}}$ [°, deg]	angle between dc field and wave vector $\boldsymbol{k}$
$\varkappa$ [cal sec^{-1} °K^{-1} cm^{-1}]	thermal conductivity
$\varkappa_H$ [cal sec^{-1} °K^{-1} cm^{-1}]	thermal conductivity in magnetic field H
$\varkappa_0$ [cal sec^{-1} °K^{-1} cm^{-1}]	thermal conductivity in zero field
λ [Å, nm]	wavelength
$\lambda = \Delta l/l$	linear magnetostriction
λ'	dynamic magnetostriction constant
λ_s; λ_s^0	saturation magnetostriction; $\lambda_s^0 = \lambda_s$ at 0 °K
μ [cm^2/V sec]	mobility of the charge carriers
$\mu = \mu' - i\mu''$	complex magnetic permeability
μ_B	Bohr magneton
μ_i	initial permeability
μ_n	nuclear magneton
ν [c/s]	frequency
$\tilde{\nu}$ [cm^{-1}]	wave number
ϱ [Ω cm]	resistivity

ϱ [deg cm^{-1}]	specific magneto-optical rotation, Faraday rotation
ϱ [g/cm^3]	density
ϱ_B [g/cm^3]	bulk density
ϱ_P [g/cm^3]	pycnometric density
ϱ_X [g/cm^3]	X-ray density
$\Delta\varrho_{magn}$ [Ω cm]	magnetoresistivity = change of resistivity due to an applied magnetic field
$\Delta\varrho_{magn}/\varrho_0$	relative magnetoresistivity
σ [Ω^{-1} cm^{-1}]	electric conductivity
σ [emu/g = Gauss cm^3/g]	specific magnetization = magnetic moment per gram, $\sigma = M/\varrho$
σ_a [emu/g]	specific magnetization of sublattice *a*
σ_A [emu/g-atom]	magnetic moment per gram atom
σ_f [emu/g]	ferromagnetic moment per gram
σ_m [emu/mole]	magnetic moment per mole
σ_{mf} [emu/mole]	ferromagnetic moment per mole
σ_0, σ_{sp} [emu/g]	spontaneous specific magnetization
σ_s [emu/g]	specific saturation magnetization
σ_s^0	$\sigma_s^0 = \sigma_s$ at 0 °K
τ [sec]	ferromagnetic resonance relaxation time
τ_1 [sec]	spin lattice relaxation time, see T_1
τ_2 [sec]	spin-spin relaxation time, see T_2
τ_d [sec]	pulse duration
τ_e [sec]	elastic relaxation time
τ_r [sec]	pulse rise time
τ_s [sec]	switching time
ϕ [deg]	magneto-optical Kerr effect (rotation angle)
$\chi = \chi' - i\chi''$	complex magnetic susceptibility
$\chi_\parallel$, $\chi_\perp$	magnetic susceptibility parallel and perpendicular to a principal crystallographic direction
χ_A [emu/g-atom]	susceptibility per g-atom
χ_g [emu/g = cm^3/g]	specific paramagnetic susceptibility, susceptibility per gram
χ_{gD}, $\chi_{g\,diam}$ [emu/g]	diamagnetic susceptibility per gram
χ_i	initial volume susceptibility
χ_m [emu/mole = cm^3/mole]	molar paramagnetic susceptibility, susceptibility per mole
$\overline{\chi_m}$	mean molar susceptibility, determined by appropriate averaging of $\chi_\parallel$ and $\chi_\perp$
χ_{mD}, $\chi_{m\,diam}$ [emu/mole]	diamagnetic susceptibility per mole
χ_v [emu/cm^3 = 1]	susceptibility per cm^3
ω [rad sec^{-1}]	angular frequency
ω_{res} [rad sec^{-1}]	ferromagnetic angular resonance frequency

Errata in Vol. III/4b

p. 110 below, after Fig. 10 insert: Dependence on oxygen content in $Mn_xFe_{3-x}O_{4\pm\gamma}$: $p_{O_2} = f(\gamma)$

p. 112 above: insert the text (8 lines) after Tab. 4, p. 113, instead of the line "$Mn_xFe_{3-x}O_{4+\gamma}$: Dependence..."

p. 113, Tab. 7: omit the second line in the heading of column 3

p. 116, Fig. 18, reference: instead of [*62 P*(3)] **read** [*62 P*(2)]
Fig. 19, 20, reference: instead of [*58 B*] **read** [*58 B*(2)]

p. 121, Fig. 36, second line of the subscript: instead of Fig. 7 **read** Fig. 6
Fig. 37, 38, reference: instead of [*65 B*] **read** [*65 B*(1)]

p. 122, Fig. 40, 41, reference: instead of [*60 B*] **read** [*60 B*(2)]

p. 123, Fig. 44, 46, reference: instaed [*65 B*] **read** [*65 B*(1)]
Fig. 47, reference: instead of [*62 P*] **read** [*62 P* (1)]

Berichtigungen zu Band III/4a

S. 177, letzte Zeile: statt Ba_2TdPaO_6 **lies** Ba_2TbPaO_6

S. 219, Zeile 16 von unten: statt $KMg_{1-x}Ni_xFe_3$ **lies** $KMg_{1-x}Ni_xF_3$

S. 252, Zeile 26 von oben (Überschrift): statt $Sr_3Fe_3UO_9$ **lies** $Sr_3Fe_2UO_9$

Errata in Vol. III/4a

p. 177, bottom line: instead of Ba_2TdPaO_6 **read** Ba_2TbPaO_6

p. 219, line 16 from the bottom: instead of $KMg_{1-x}Ni_xFe_3$ **read** $KMg_{1-x}Ni_xF_3$

p. 252, line 26 from above (headline): instead of $Sr_3Fe_3UO_9$ **read** $Sr_3Fe_2UO_9$

5 Non-iron garnets

5.0 Introduction — Einleitung

During the last few years considerable attention has been paid to magnetic, calorimetric, and spectroscopic investigation of trivalent rare earth ions in non-magnetic garnet host lattices. The most important non-iron garnets are rare earth gallium and rare earth aluminium garnets of the general formula $R_3Ga_5O_{12}$ and $R_3Al_5O_{12}$.

The understanding of the crystal field interaction in these garnets is interesting in itself as an example of large and very asymmetric crystal fields. Most interesting results are those obtained for $Dy_3Al_5O_{12}$ (DyAlG) which undergoes a transition to an antiferromagnetic state at 2.53 °K. The *g* values of the magnetic Dy^{3+} ions in this compound are so highly anisotropic as to form a good approximation to an Ising system. None of the other rare earth garnets has a transition at so high a temperature.

The rare earth aluminium and gallium garnets were initially investigated because they are isomorphous with the magnetically important ferrimagnetic rare earth iron garnets $R_3Fe_5O_{12}$ (RIG). To shed light on the effects of anisotropy and magnetization of the iron garnets, it is generally customary to study the properties of the individual rare earth ions in a non-magnetic environment having a structure as near as possible to that of the ordered compound. In the case of the garnets, however, extensive experiments have shown that in several cases the magnetic properties of the rare earth ions are extremely sensitive to the exact diamagnetic garnet environment, and it is therefore necessary to investigate in more detail the nature of the crystal field at the rare earth site.

For further remarks about garnet systems, especially iron garnets, we refer to the introduction of 4.1 in part a of this volume.

The references are listed in two categories: the primary references, denoted by the last two digits of the year of publication and the author's initials, give the original papers from which the information appearing in the text was obtained. Additional relevant secondary references are given running numbers. The literature has been considered up to August 1969. More recent references will be found in the annual digest "Magnetism and Magnetic Materials" (Academic Press, New York and London).

Während der letzten Jahre wurde den magnetischen, kalorimetrischen und spektroskopischen Untersuchungen an dreiwertigen Selten-Erd-Ionen in nicht-magnetischen Granaten als Wirtsgittern erhebliche Aufmerksamkeit geschenkt. Die Selten-Erd-Gallium- und -Aluminium-Granate sind die wichtigsten Nichteisengranate; sie haben die allgemeine Formel $R_3Ga_5O_{12}$ und $R_3Al_5O_{12}$.

Es ist schon an sich interessant, die Kristallfeld-Wechselwirkung in diesen Granaten als Beispiel für große und sehr asymmetrische Kristallfelder zu verstehen. Besonders interessant sind die am $Dy_3Al_5O_{12}$ gewonnenen Ergebnisse: DyAlG geht bei 2.53 °K in einen antiferromagnetischen Zustand über. Wegen der hohen Anisotropie der *g*-Werte der magnetischen Dy^{3+}-Ionen ist diese Verbindung in guter Näherung ein Ising System. Kein anderer Selten-Erd-Granat hat eine so hohe Übergangstemperatur.

Zunächst wurden die Selten-Erd-Aluminium- und -Gallium-Granate untersucht, weil sie mit den magnetisch bedeutsamen ferrimagnetischen Selten-Erd-Eisen-Granaten (RIG) isomorph sind. Um die Erscheinung der Anisotropie und der Magnetisierung der Eisengranate zu verstehen, ist es allgemein üblich, die Eigenschaften der einzelnen Selten-Erd-Ionen in einer nicht-magnetischen Umgebung zu ermitteln, deren Struktur der des geordneten Systems möglichst nahe kommt. Im Fall der Granate jedoch haben umfassende Experimente ergeben, daß die magnetischen Eigenschaften der Selten-Erd-Ionen in einigen Fällen sehr empfindlich gegen die exakte diamagnetische Umgebung im Granat sind. Daher ist es nötig, die Natur des Kristallfeldes am Selten-Erd-Gitterplatz im einzelnen genauer zu erforschen.

Weitere Bemerkungen über Granatsysteme, besonders Eisengranate, sind in 4.1 im Teil a dieses Bandes zu finden.

Die Literatur ist in zwei Kategorien aufgeführt. Die für die Tabellen benutzte Originalliteratur ist durch die letzten beiden Stellen der Jahreszahl des Erscheinungsjahres und die Anfangsbuchstaben der Autorennamen gekennzeichnet. Weitere einschlägige Literatur ist zusätzlich mit laufenden Nummern aufgeführt. Die Literatur wurde bis August 1969 erfaßt. Spätere Literatur kann man in der jährlichen Zusammenstellung "Magnetism and Magnetic Materials" (Academic Press, New York and London) finden.

Gräper

5.1 Magnetic susceptibility data

A. Paramagnetic susceptibility above the ordering temperature

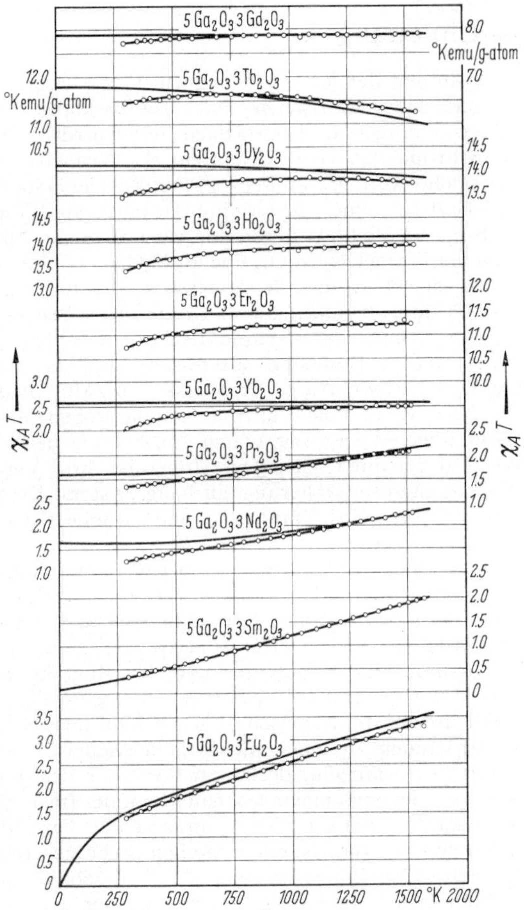

Fig. 1. Gallium garnets. Paramagnetic behavior at high temperatures [*62AP*]. The solid lines represent theoretical curves from VanVleck's theory [*3*]. The experimental data are corrected for diamagnetism.

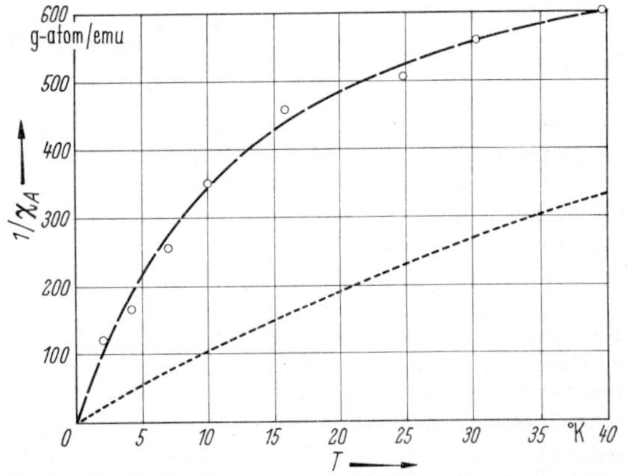

Fig. 2. SmGaG. Inverse susceptibility, χ_A^{-1} vs. temperature for a spherical single crystal. Dashed line: experimental data; dotted line: theoretical curve for free Sm^{3+}ion ($^6H_{5/2}$). At room temperature the susceptibility is $1.08 \cdot 10^{-3}$ emu/g-atom. No anisotropy is found over the range of temperature (2.1° ··· 300 °K) and field (up to 8 kOe) [*65SH*].

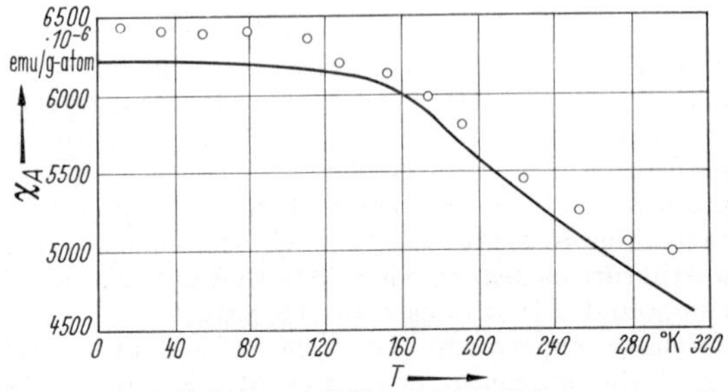

Fig. 3. EuGaG. Susceptibility, χ_A, vs. temperature for a sphere of a single crystal. No anisotropy is detected over the range of temperature (2.1 ··· 300 °K) and fields (up to 8 kOe). The solid line represents VanVleck's formula for the free Eu^{3+}ion ($^7F_0$) with a separation of 480 °K between $J = 0$ and $J = 1$ levels [*65SH*].

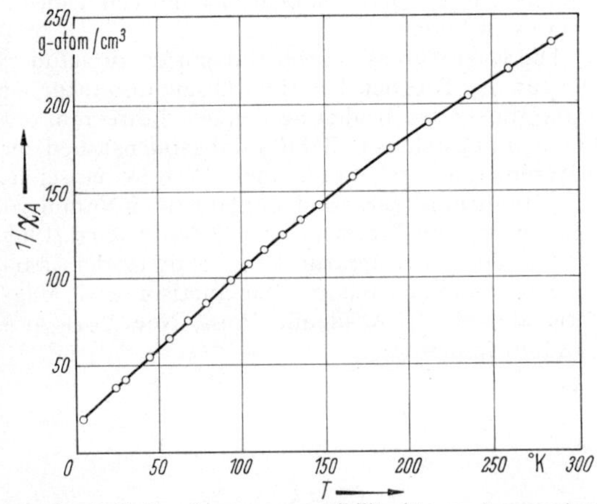

Fig. 4. PrGaG. Inverse susceptibility $1/\chi_A$ vs. temperature [*59CD*]. The data obey Curie laws with Curie constants $C_A = 1.56$ for ordinary temperatures, and $C_A = 1.04$ °K emu/g-atom for low temperatures.

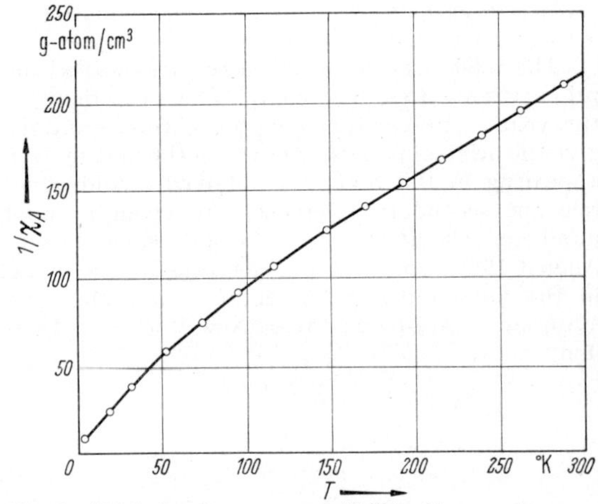

Fig. 5. NdGaG. Inverse susceptibility, $1/\chi_A$, vs. temperature [*59CD*]. The data fit Curie law with $C_A = 1.64$ for ordinary temperatures, $C_A = 0.92$ °K emu/g-atom for low temperatures.

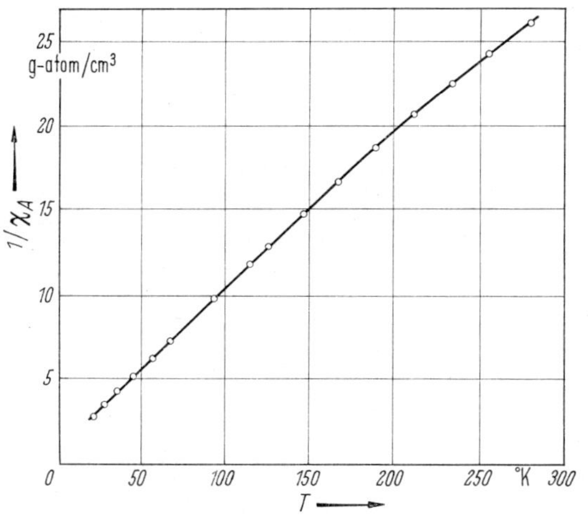

Fig. 6. ErGaG. Inverse susceptibility, $1/\chi_A$, vs. temperature [*59CD*].

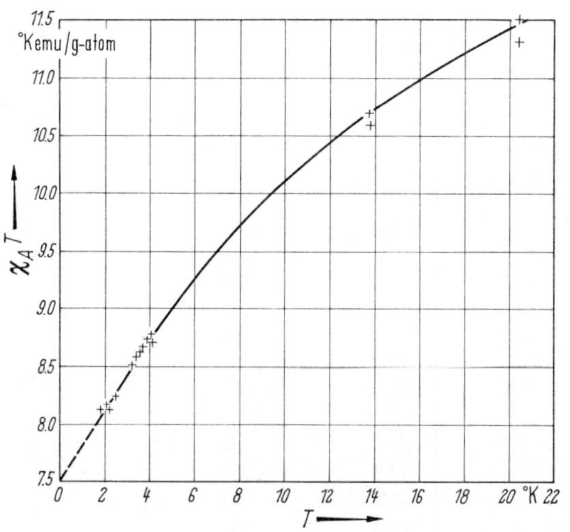

Fig. 8. Dy^{3+} in DyGaG. Paramagnetic behavior ($\chi_A T$) vs. T [*66AMR*]. $T = 4 \cdots 20$ °K. For $T < 5$ °K $\chi_A T = 7.5 + 0.30 \cdot T$ represents the data. Compare Fig. 9, 10.

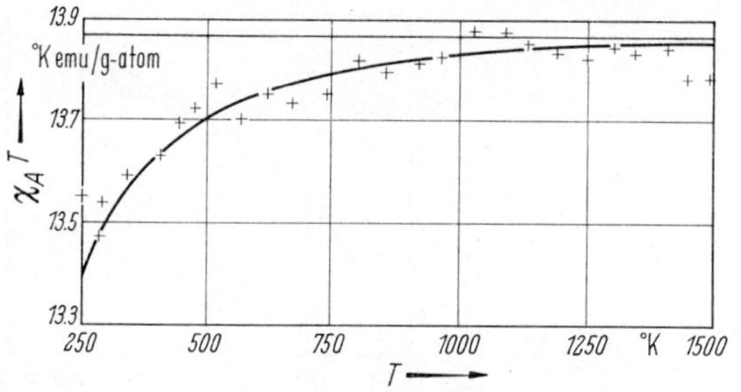

Fig. 10. Dy^{3+} in DyGaG. ($\chi_A T$) vs. T [*66AMR*]. $T = 250 \cdots 1500$ °K [*66AMR, 62AP*]. Compare Fig. 8, 9.

→

Fig. 12. $5Ga_2O_3 \cdot 3Er_2O_3$. ($\chi_A T$) vs. T [*64ARV*]. *1*: experimental data [*62AP*]; *2, 3*: calculations with energy differences $\Delta W/k = 720$ °K (2) and $\Delta W/k = 580$ °K (3) between the ground state multiplet [*63ST*] and the first excited multiplet.

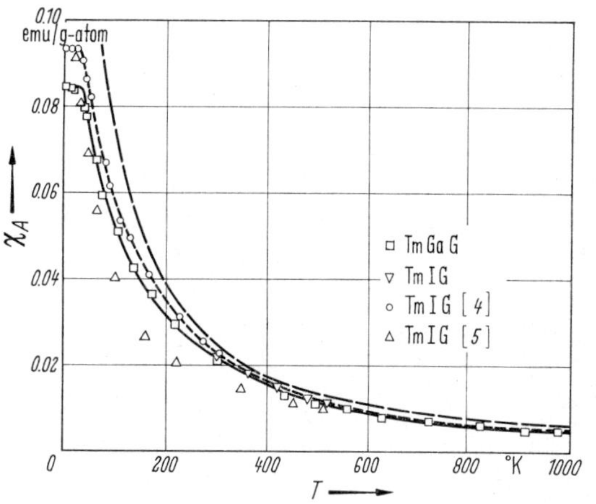

Fig. 7. TmGaG, TmIG. Magnetic susceptibilities per g-atom of Tm vs. temperature directly measured as $\chi = (\partial M/\partial H)$ [*66SLV*].

—— —— free ion curve, $\chi = 7.14/T$,

– – – – cubic field with $\Delta W/k = 150$ °K.

———— cubic field with $\Delta W/k = 160$ °K.

The temperature independent susceptibility in TmGaG is confirmed at $T = 0$ °K in [*62WBHLW*]. See Fig. 17.

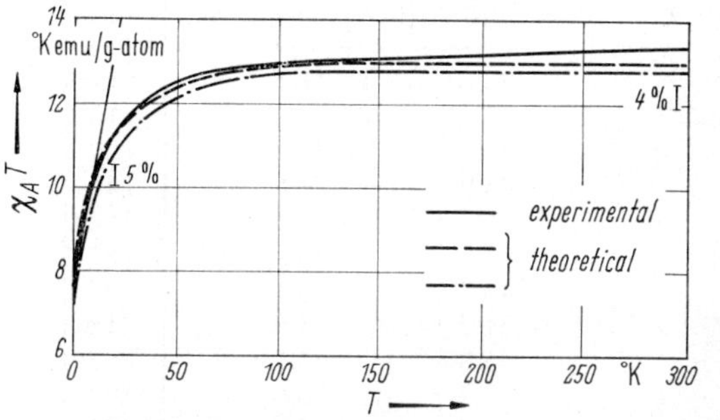

Fig. 9. Dy^{3+} in DyGaG. ($\chi_A T$) vs. T [*66AMR*]. $T \leq 300$ °K. Theoretical calculations with VanVleck's formula neglecting energy levels higher than 150 cm^{-1}. Compare Fig. 8, 10.

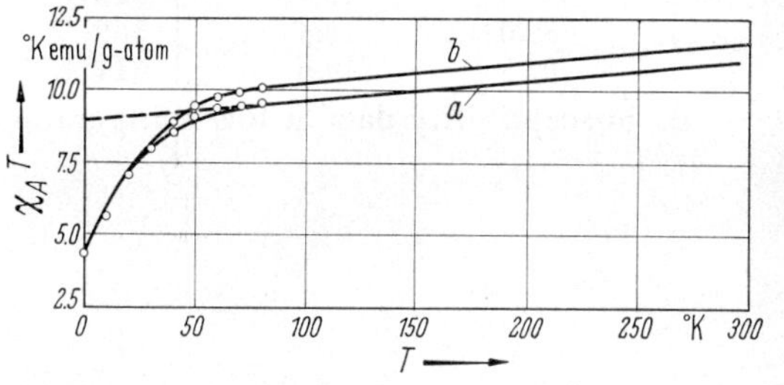

Fig. 11. $5Ga_2O_3 \cdot 3Er_2O_3$. ($\chi_A T$) vs. temperature [*64ARV*].
a experimental data from [*8*] and [*62AP*].
b theoretical curve with VanVleck's formula [*64ARV*] using distances of energy levels measured by [*63ST*].

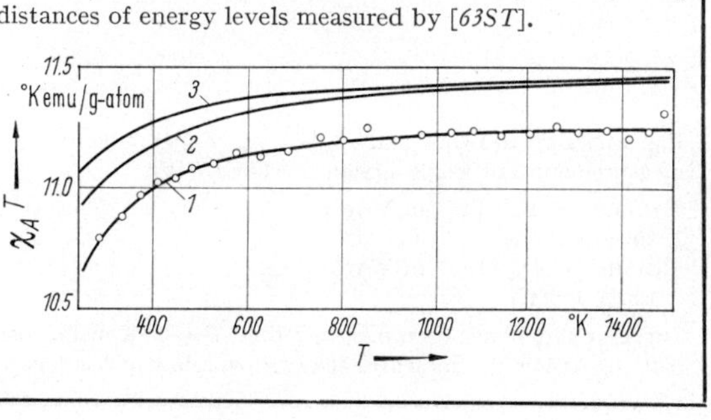

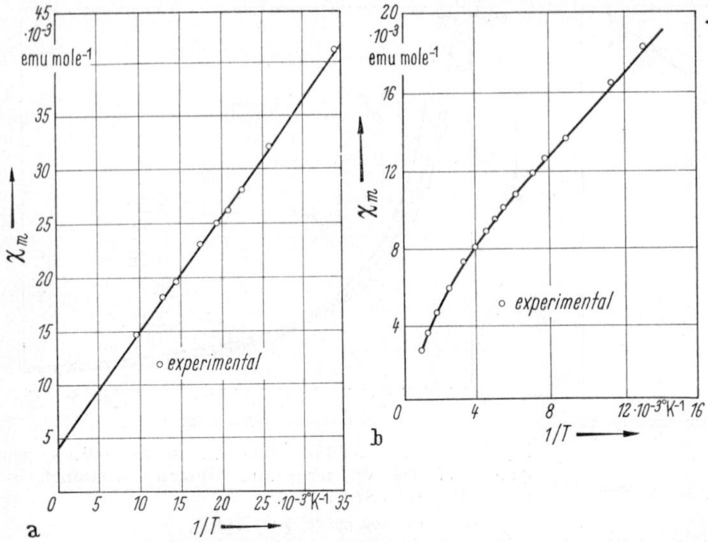

← Fig. 13a, b. YbGaG. Magnetic susceptibility vs. $1/T$ [*63BLV*].

a) $T = 30 \cdots 100$°K. Solid curve: theoretical values according to $\chi_m = 1.082/T + 0.00397$.

b) $T = 77 \cdots 1000$ °K. Solid curve: theoretical values using $E(\Gamma_8) - E(\Gamma_7) = 515 \text{cm}^{-1}$ and $E(\Gamma_6) - E(\Gamma_7) = 800 \text{ cm}^{-1}$ for the ground manifold (Yb^{3+}: $^2F_{7/2}$) splitting in the crystalline field.

Tab. 1. Magnetic susceptibilities (emu/g-atom of Yb^{3+}) of YbGaG. The diamagnetic correction made was taken as $-91 \cdot 10^{-6}$ emu/g-atom of Yb^{3+} [*63BLV*]

T °K	χ_A 10^{-3} emu/g-atom	T °K	χ_A 10^{-3} emu/g-atom	T °K	χ_A 10^{-3} emu/g-atom
29.3	41.1	111	13.7	639	3.92
38.6	32.2	113	13.5	658	3.81
44.5	28.1	129	12.5	684	3.67
45.9	27.8	141	11.8	724	3.50
46.3	27.4	162	10.6	776	3.28
47.9	26.2	181	9.92	814	3.15
48.3	26.1	197	9.36	866	2.98
51.2	24.9	218	8.70	907	2.86
53.8	24.1	252	7.94	928	2.83
57.2	23.0	295	7.11	967	2.69
68.7	19.6	326	6.59	998	2.60
69.8	19.3	384	5.82	1038	2.50
70.8	19.2	452	5.12	1069	2.45
77.3	18.1	526	4.59	1109	2.37
89.0	16.4	568	4.28		
102.5	14.6	611	4.10		

B. Susceptibility data at low temperatures in the vicinity of an ordering temperature

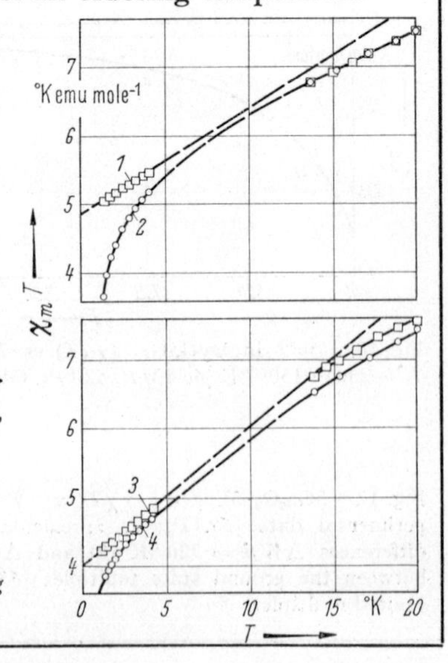

→

Fig. 14. Er^{3+} in YGaG and YAlG. $\chi_m \cdot T$ (per mole Er^{3+}) vs. absolute temperature, measurements on single crystals [*61BGLRW*].

1 dilute (≈ 1%) Er^{3+} in YGaG.

2 100% ErGaG.

3 dilute (≈ 1%) Er^{3+} in YAlG.

4 100% ErAlG.

Curves *1* and *3* have been normalized at $T = 0$ °K, using paramagnetic resonance g values. At low temperatures they follow a law of the form $\chi_m = C_m/T + \alpha$.

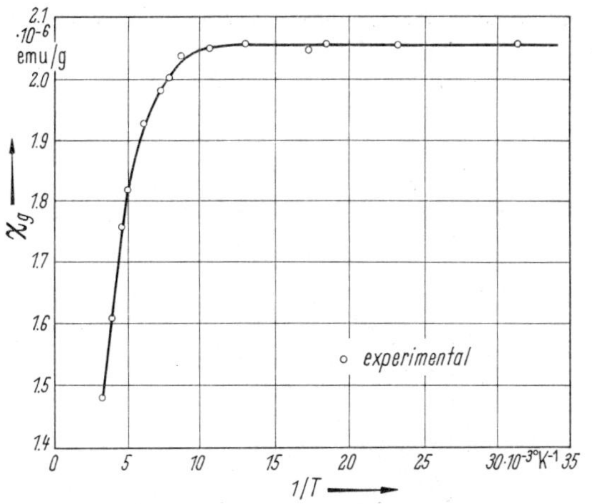

Fig. 15. Eu^{3+} in YGaG. Paramagnetic susceptibilities per gram sample vs. T for the interval 30° ··· 300 °K. Solid curve: calculated susceptibility assuming the crystal field to be very small compared to spin-orbit coupling constant. The best fit was found to be $\Delta_1 = 325$ cm^{-1}, where Δ_1 gives the center of gravity of the $^7F_1$ states of Eu^{3+} in YGaG [*68BM*].

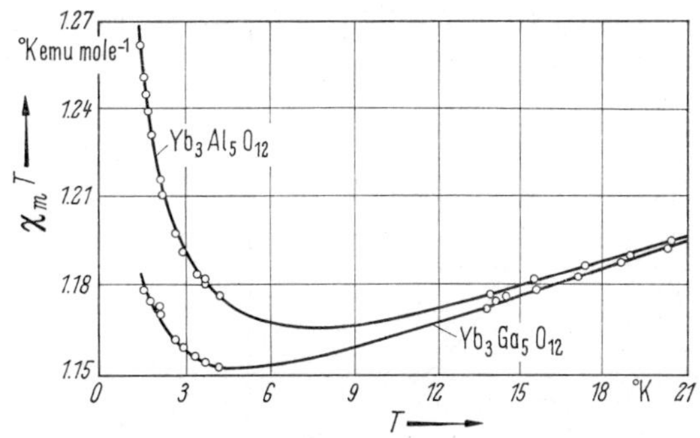

Fig. 16. YbGaG and YbAlG. Susceptibility of single crystals [*60BGLW*]. The results can be fitted to
$\chi_m = 1.116/(T - 0.048\ °K) + 0.00358$ for YbGaG and
$\chi_m = 1.111/(T - 0.139\ °K) + 0.00359$ for YbAlG (corrected to spherical samples).

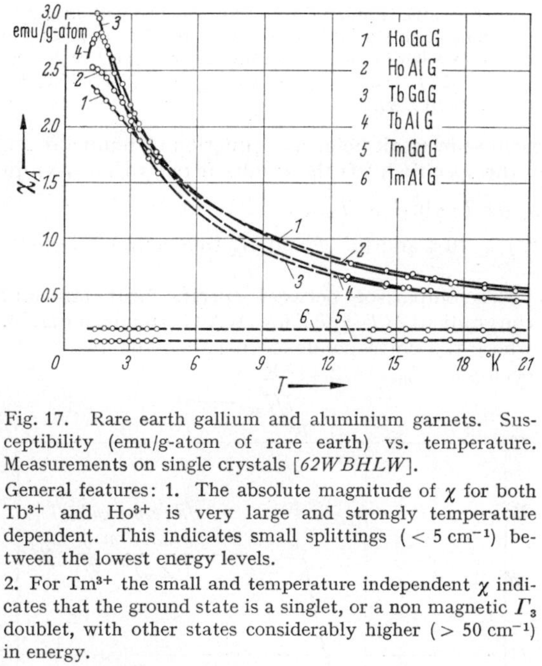

Fig. 17. Rare earth gallium and aluminium garnets. Susceptibility (emu/g-atom of rare earth) vs. temperature. Measurements on single crystals [*62WBHLW*].

General features: 1. The absolute magnitude of χ for both Tb^{3+} and Ho^{3+} is very large and strongly temperature dependent. This indicates small splittings (< 5 cm^{-1}) between the lowest energy levels.

2. For Tm^{3+} the small and temperature independent χ indicates that the ground state is a singlet, or a non magnetic Γ_3 doublet, with other states considerably higher (> 50 cm^{-1}) in energy.

3. The susceptibility of TbAlG has a maximum at T = 1.5 °K, indicating magnetic ordering.

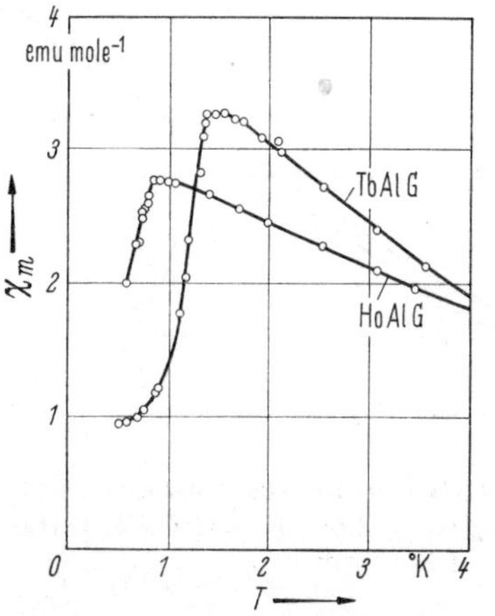

Fig. 18. TbAlG and HoAlG. Susceptibility (in units of emu mole^{-1} of rare earth) vs. temperature. The specimens consisted of small randomly oriented single crystals, the results are presented without demagnetizing corrections [*67CTW*]. Transition temperatures to antiferromagnetism: 1.35 °K (TbAlG) and 0.85 °K (HoAlG).

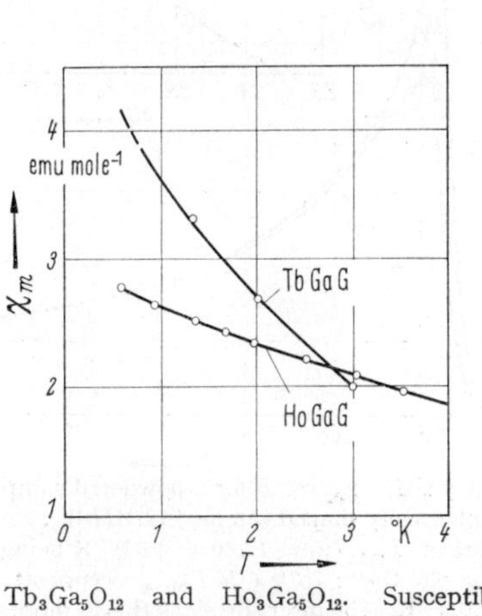

Fig. 19. $Tb_3Ga_5O_{12}$ and $Ho_3Ga_5O_{12}$. Susceptibility in emu/(mole of rare earth) vs. T. The results are presented without demagnetizing corrections. There is no sign of an approach to a magnetic transition or to temperature-independent paramagnetism between 0.6 °K and 4.2 °K [*67CTW*].

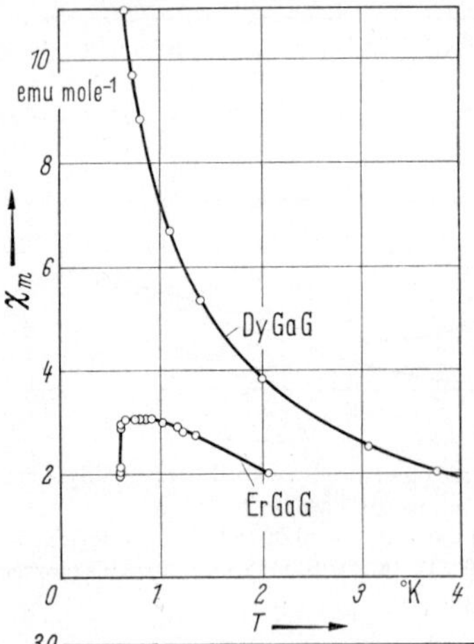

← Fig. 20. DyGaG and ErGaG. Susceptibility in emu/(mole of rare earth), uncorrected for demagnetizing effects [67CTW].

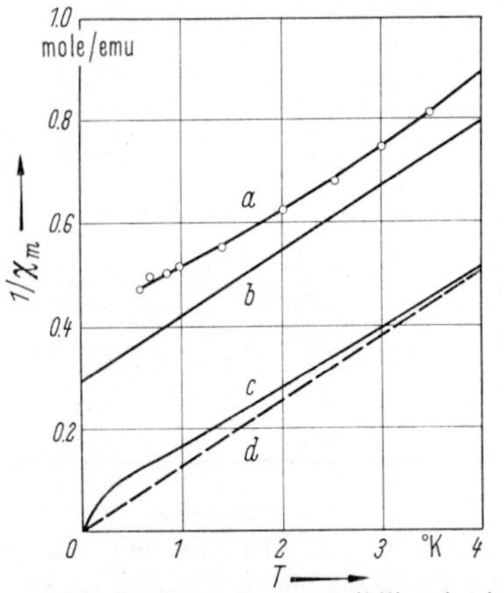

Fig. 22. GdGaG. Inverse susceptibility in (emu/mole of rare earth)$^{-1}$ vs. temperature [67CTW]. (a) experimental results [67CTW] on single crystals, (b) extrapolation of Curie-Weiss law $\chi_m^{-1} = (T + 2.3)/7.82$ [62WBHLW], (c) computed variation for isolated Gd^{3+} ions with cubic field splitting, giving a ground-state doublet, a quartet state at 1.42 cm^{-1} and a doublet at 4.40 cm^{-1} [7], (d) variation for $^8S_{7/2}$ ions with no ground-state splitting [67CTW].

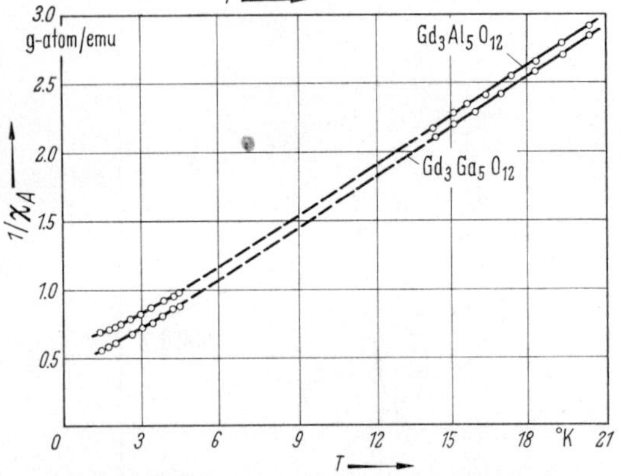

← Fig. 21. GdGaG and GdAlG. Inverse susceptibility, χ_A^{-1}, in (emu/g-atom Gd^{3+})$^{-1}$ vs. temperature. The data fit the Curie-Weiss law in the temperature region from 4 ⋯ 20 °K. $\chi_A = 7.82/(T + 2.3)$ for GdGaG, $\chi_A = 7.82/(T + 3.0)$ for GdAlG [62WBHLW].

In [64BWWa] the susceptibility of DyAlG is remeasured in weak fields as a function of temperature. Near $T_{max}(c) = 2.49$ °K [63BLWW] (the maximum in the specific heat) the results fitted to an asymptotic law of the form

$$\chi T = (\chi T)_\infty \{A - B_\pm(1 - T_{max}(c)/T) \cdot \ln|1 - T_{max}(c)/T|\}$$

as suggested in [9]. For $(\chi T)_\infty$ they took $\frac{1}{3}(N g_z^2 \cdot \mu_B^2/4k) = 10.4$ and for A and $B_\pm$ they found $A = 0.76$, $B_+ = 0.42(T > 2.49$ °K), $B_- = 1.34(T < 2.49$ °K).

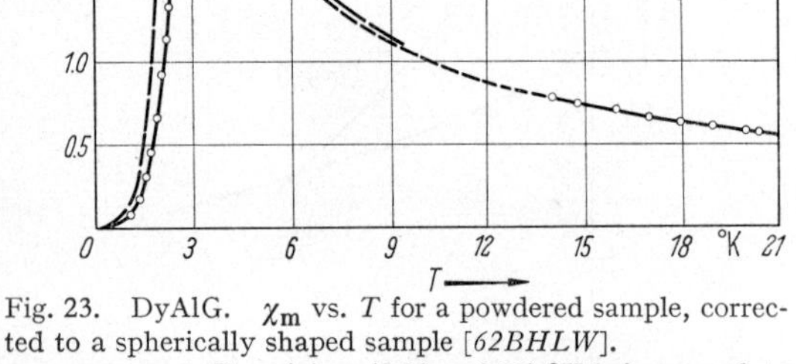

Fig. 23. DyAlG. χ_m vs. T for a powdered sample, corrected to a spherically shaped sample [62BHLW].
χ_{max} occurs at $T_{max}(\chi) = (2.80 \pm 0.02)$ °K being equal to 2.25 emu/mole Dy^{3+}; $[\partial(\chi T)/\partial T]_{max}$ occurs at $T_N(\chi) = (2.55 \pm 0.01)$ °K. The insert displays this in more detail, the triangles and circles referring to two separate samples. The larger dashed line is the theoretical variation calculated for magnetic dipole interaction only. The estimated Néel temperature is $\Theta_N = 2.68$ °K [62BHLW].

A comparison between specific heat data and the gradient $\partial(T \cdot \chi)/\partial T$ for DyAlG is made in Fig. 24, where the correspondence between the two peaks occurs at $T_{max} = 2.49$ °K.

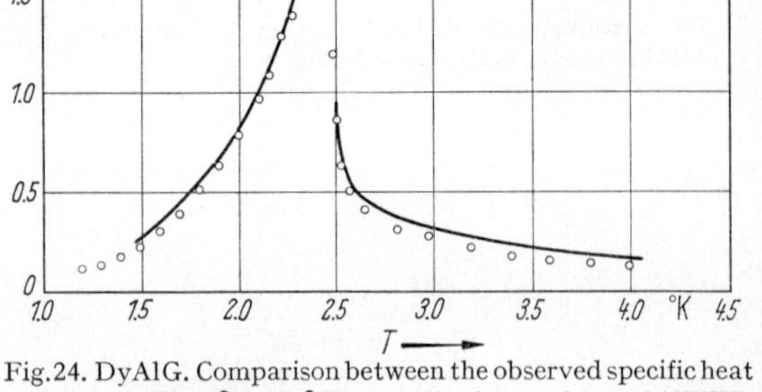

Fig. 24. DyAlG. Comparison between the observed specific heat and the gradient $\partial(\chi T)/\partial T$ normalized by a factor [64WW].

C. Summary of susceptibility parameters and transition temperatures

Tab. 2. Paramagnetic susceptibility of several Ga and Al garnets. Theoretical and experimental constants to fit the equation $\chi_m = C_m/(T - \Theta_p) + \alpha$

garnet	C_m °K emu mole^{-1}	Θ_p °K	α emu mole^{-1}	temperature range	Ref.
PrGaG	1.56 (theor.: 1.62)			ordinary temperature	*59CD*
	1.04 (theor.: 1.03)			low temperature	
NdGaG	1.64 (theor.: 1.63)			ordinary temperature	*59CD*
	0.92 (theor.: 1.21)			low temperature	
	1.03 ± 0.02		$0.11 \cdot 10^{-2}$	180···300 °K	*61T*
	0.6		$0.55 \cdot 10^{-2}$	<180 °K	
GdGaG	7.82	−2.3		5···20 °K	*62WBHLW, 67CTW*
GdAlG	7.82	−3.0		5···20 °K	*62WBHLW*
DyGaG	7.55	−0.05	0.14	He-temperature	*67CTW*
	or: 8.00	−0.1	0.1	He-temperature	
	7.80	−0.436	0.212		*67CTW*, value from Wolf (private communication)
Dy^{3+} (5%) in YAlG	C		$0.006 \cdot C$	He-temperature	*62BHLW*
ErGaG	8.85		0.0072	>70 °K	*61ATCD*
YbGaG	2.35 (theor.: 2.51)			ordinary temperature	*59CD*
	1.07 (theor.: 1.09)			low temperature	
	1.09		0.003	20···150 °K	*61ATCD*
	1.116 ± 0.005	−0.048 ± 0.10	0.00358 ± 0.00002	1··· 20 °K	*60BGLW*
	1.082		0.00397	<150 °K	*63BLV*
	1.089		0.00345	theoretical	*67PHWB*
YbAlG	1.111 ± 0.005	−0.139 ± 0.010	0.00359 ± 0.00002	1··· 20 °K	*60BGLW*
	1.100		0.00272	theoretical	*67PHWB*

Tab. 3. Summary of transition temperatures [°K] for external field $H_a = 0$, indicating the onset of long range order and the transition to an antiferromagnetic state

garnet	$T_{max}(C_{magn})$ [1]	$T_{max}(\chi)$ [2]	$T_N(C_{magn})$ [3]	$T_N(n)$ [4]	$T_N(\chi)$ [5]	Θ_N [6] (theory)	Ref.
NdGaG	0.514		0.516				*67OMR*
SmGaG *)	0.960		0.967				*67OMR*
	0.918		0.921				
TbAlG	1.35				1.35		*67CTW*
		1.5					*62WBHLW*
DyGaG	0.35		≲0.36				*67OMR*
DyAlG			2.537				*67KLW*
			2.535 ± 0.01				*66KLSW*
						2.52	*65C*
		2.8			2.49		*64WW*
				2.54 ± 0.02			*64HM*
			2.49 ± 0.01				*63BLWW*
		2.80 ± 0.02			2.55 ± 0.01	2.68	*62BHLW*
HoAlG					0.85		*67CTW*
ErGaG				0.789			*68Hb*
	0.786		0.790				*67OMR*
						0.82	*65C*

*) Two different samples.
[1] Temperature, where the specific heat has its maximum. [2] Temperature, where the susceptibility has its maximum.
[3] Transition temperature, deduced from a plot of $\ln(C_{magn})$ or C_{magn} vs. $\ln|(T - T_N)/T_N|$.
[4] Transition temperature, indicated by neutron diffraction experiments.
[5] Transition temperature from a sharp fall of χ vs. T or from the maximum of $\partial(\chi \cdot T)/\partial T$.
[6] Estimated or predicted Néel temperature.

5.2 Magnetization

(Critical magnetic fields, magnetic anisotropy, magnetic phase diagram)

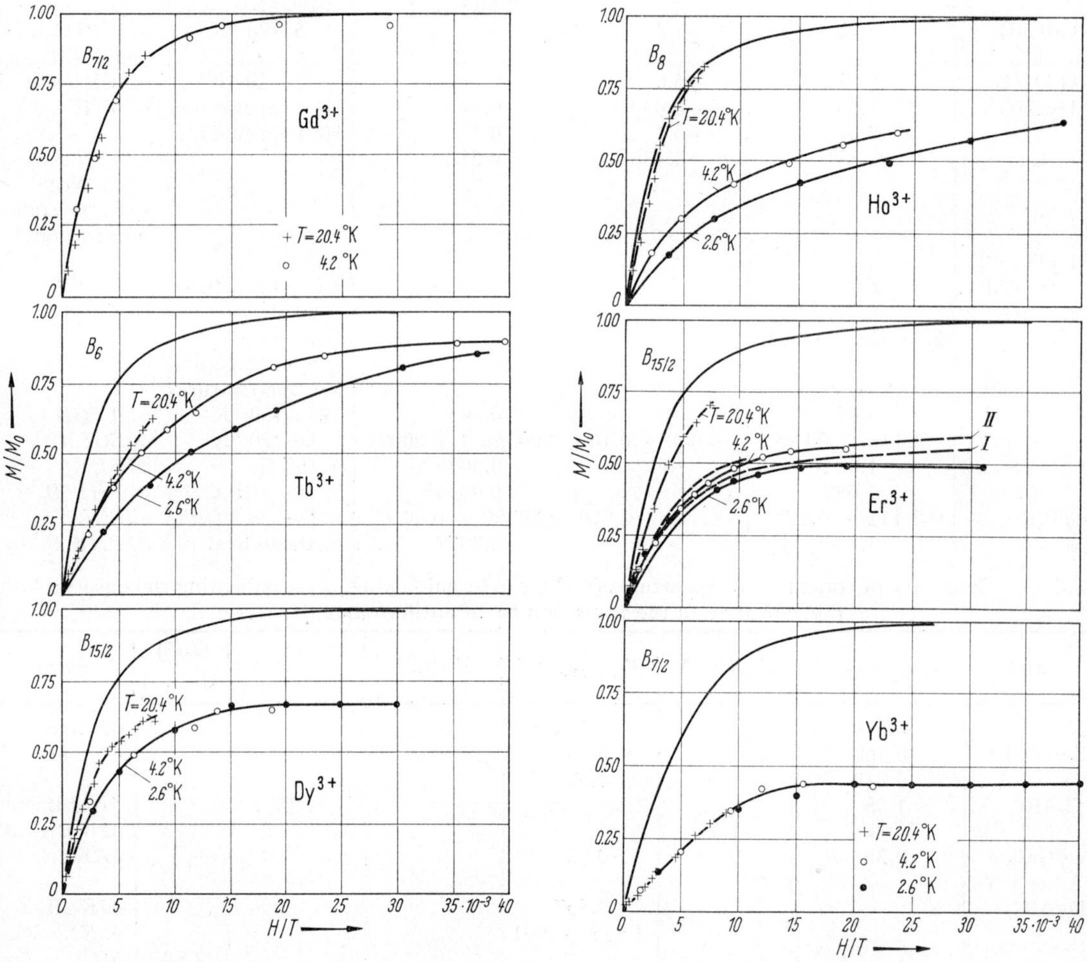

Fig. 25. $5Ga_2O_3 \cdot 3Me_2O_3$. Me = Gd, Dy, Er, Tb, Ho, and Yb. Normalized magnetization (M/M_0) vs. (H/T). Solid curves (Brillouin functions) correspond to free ion calculations. The quenching of the rare-earth ion moment is clearly shown from the fact that the magnetization laws for ions other than Gd^{3+} (^8S ground state) differ from Brillouin's free ion law and are dependent on temperature except for Yb^{3+} [*64GP a, 65GP*]. The results relating to the Yb^{3+} ion are in good agreement with the theory [*8*]. The curves I and II for ErGaG are calculated with crystalline field model [*65A BGR*]. The measurements were performed in pulsed fields [*64GP b*] up to 200 kOe and for temperatures between 2.5 and 20.4 °K.

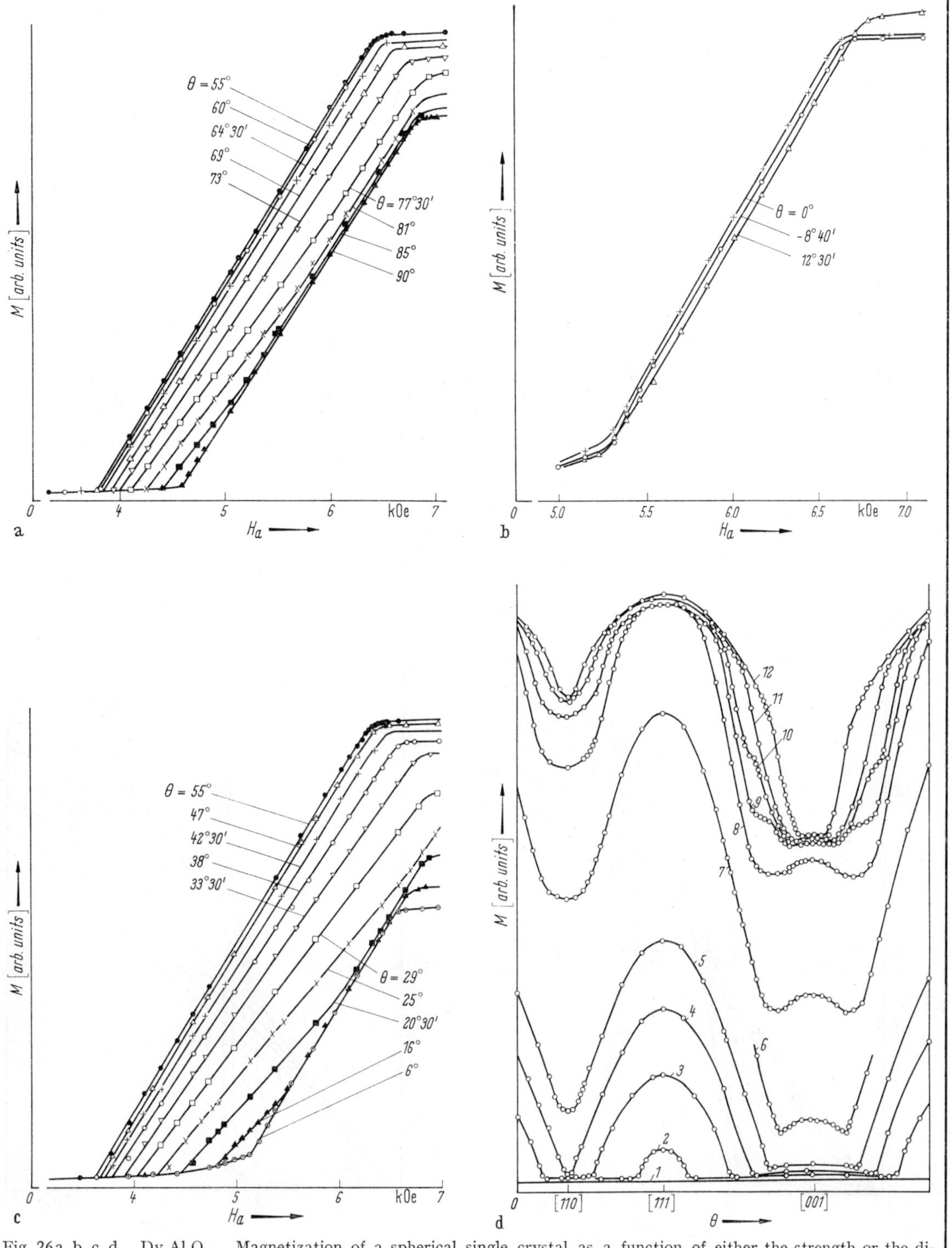

Fig. 26a, b, c, d. $Dy_3Al_5O_{12}$. Magnetization of a spherical single crystal as a function of either the strength or the direction of the applied field. $T = 0.37$ °K [*68BCV*]. The field moves in the ($1\bar{1}0$) plane, θ measures the angle between the [001] direction and the field direction.

a) Magnetization curves, when the field direction is turning from [110] to [111] direction.
b) $M(H)$ in the vicinity of the [001] direction.
c) $M(H)$, field direction is turning from [111] near to [001] direction.
d) Magnetization $M(\theta)$ for different fields: 1: 3640 Oe, 2: 3950 Oe, 3: 4265 Oe, 4: 4575 Oe, 5: 4900 Oe, 6: 5370 Oe, 7: 5920 Oe, 8: 6555 Oe, 9: 6790 Oe, 10: 7030 Oe, 11: 7900 Oe, 12: 12 630 Oe.

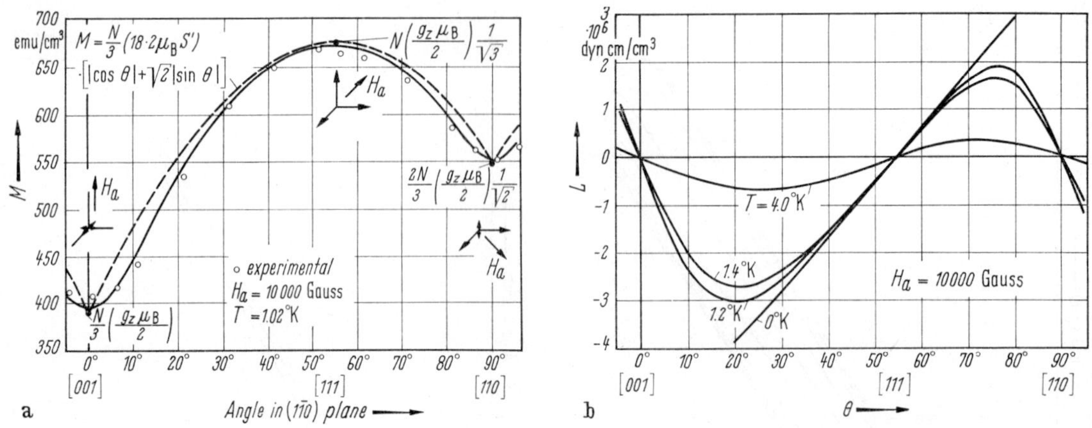

Fig. 27a, b. DyAlG. Magnetic anisotropy. The field H_a is applied in the (1$\bar{1}$0) plane. θ: angle between H_a and [001].
a) Magnetization in the direction of H_a. The broken curve is obtained from asymptotic theory, the full curve from exact theory, neglecting interactions [*64W*]. N: total number of Dy^{3+} ions per unit volume.
b) Corresponding to the anisotropy in the magnetization one can predict the torque L in a large applied field. Evaluating simply $\Gamma = M \times H_a$, one finds the shown figure [*64W*].

Fig. 28, 29 see next page

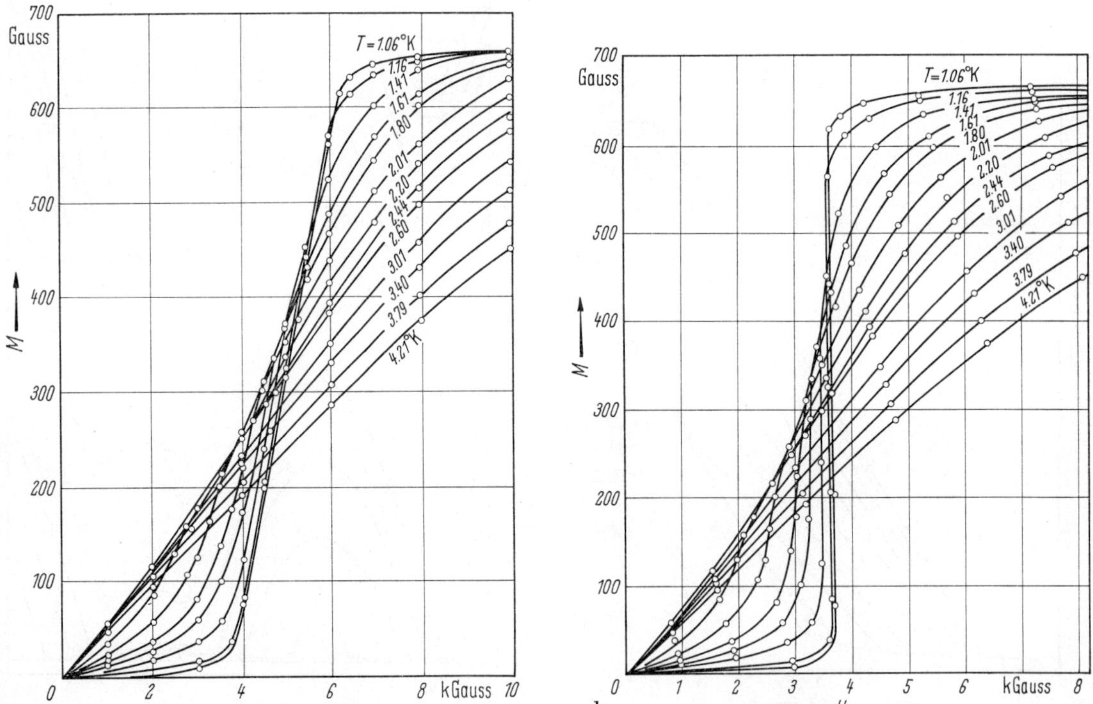

Fig. 30a, b. DyAlG. Magnetization vs. applied field. Spherical crystal at temperatures between 4.2 and 1.06 °K, with the applied field along [111] [*64BWWb*].
a) M vs. applied field H_a: For $T \lesssim 1.4$ °K (0.55 T_N) the curves have a region which is approximately linear. This region becomes larger as T decreases, while $\partial M/\partial H_a$ remains practically constant. The gradient of the linear portion is equal to $1/N$ where N is the demagnetizing factor. [*68W*] envisaged that the spherical sample exists in a mixed state with well-defined regions of parallel and antiparallel phases.
b) M vs. internal field, corrected for demagnetization effects.

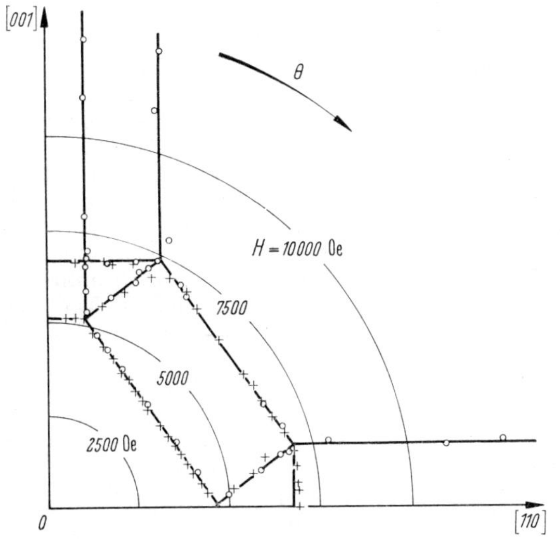

Fig. 28. DyAlG. Experimental phase diagram of a spherical single crystal at 0.37 °K [*68BCV*]. The results agree well with theoretical predictions. For details see [*68BCV*].

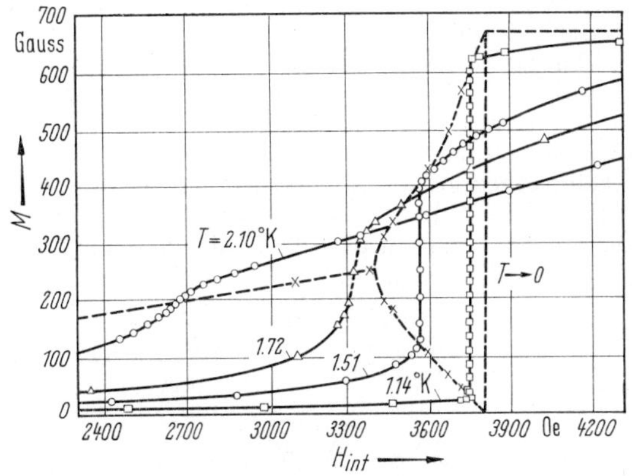

Fig. 31. DyAlG. Isothermal magnetization vs. internal field, $H_{int} = H_{ext} - N \cdot M$, along [111]. High resolution measurements on a spherical sample [*66KLSW*].

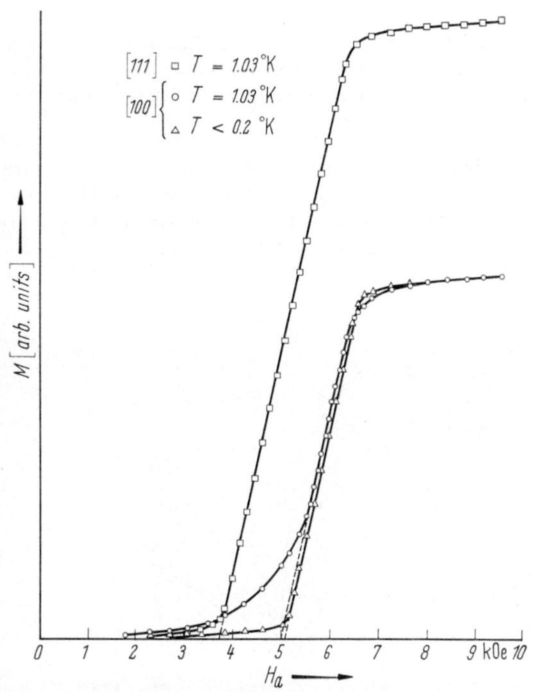

Fig. 29. DyAlG. Magnetization of a single crystal vs. applied field. The transition from the antiferromagnetic state to a mixed phase occurs at H_{crit} [100] = (5040 ± 75) Oe (field in [100] direction) and at H_{crit} [111] = (3770 ± 60) Oe (field in [111] direction) [*67BCVa*].

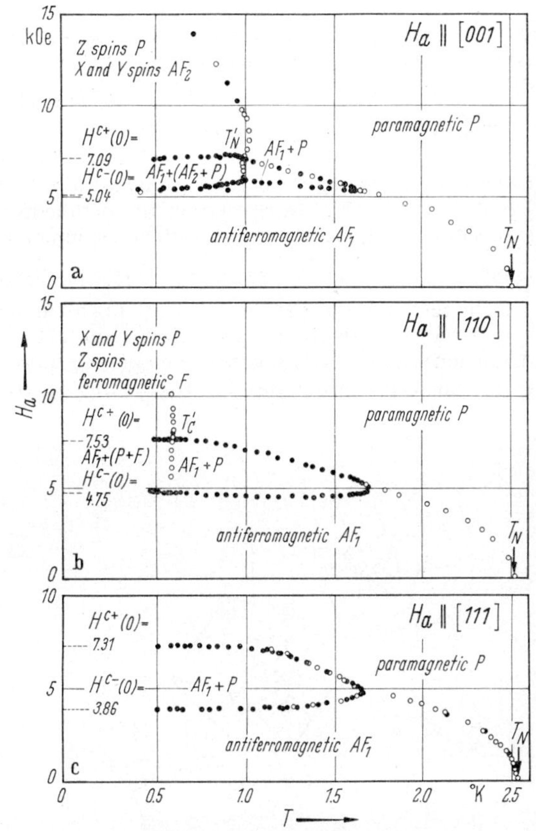

Fig. 32. DyAlG. Magnetic phase bounderies as measured on an ellipsoid of revolution with demagnetizing factor $N_{\perp} = 5.35$. The direction of the applied magnetic field is a) ∥ [001], b) ∥ [110], c) ∥ [111]. The regions lying between the lines marked H^{c+} and H^{c-} correspond to mixed phases of the upper and lower states. (●) determined from $(\partial S/\partial H)_T$ vs. H; (○) determined from $C_H(T)$. [*66KLW*].

5.3 Crystallographic and structural data

A. Location and environment of ions in the unit cell of the garnet lattice

Informations can be found in the references [*69SMHR, 68AT, 68MO, 67Ka, 67PHWB, 66BGB, 66CC, 66HWW, 66RK, 66SMR, 66SZV, 65EB, 64HW, 64HWW, 64R, 64RBa, 61CW, 61G*, et al.].

The garnet crystals have the general chemical formula $A_3B_5O_{12}$ with eight of such molecules contained in the unit cell.

The space group is O_h^{10}-Ia3d with an over all symmetry being cubic [*65EB*]. The A ions (in this paper: rare earth or Y) occupy 24 dodecahedral *c* sites (D_2 symmetry), with eightfold oxygen coordination. The 40 B ions (in this paper: Ga, Al, Ge, In) are at 24 tetrahedral *d* sites (S_4 point symmetry), with fourfold oxygen coordination and at 16 octahedral *a* sites (C_{3i} point symmetry), with sixfold oxygen coordination. Each oxygen ion lies at a vertex that is common to four polyhedra of oxygen ions, one octahedron, one tetrahedron and two dodecahedra (Fig. 33). The locations of the sites *a*, *d*, and *c* are given in Tab. 5 of section 4.1, p. 327 in part A of this volume.

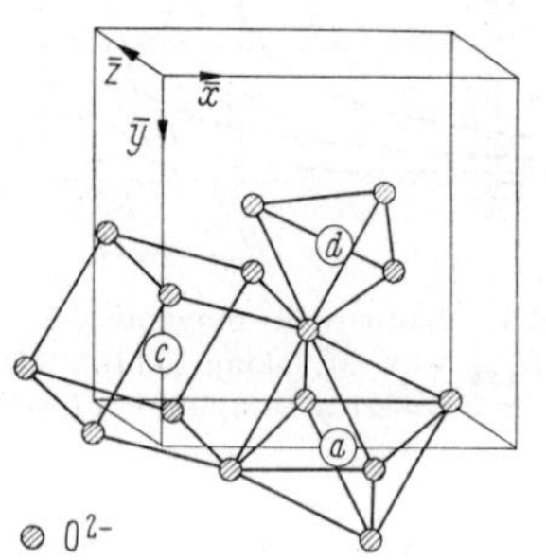

Fig. 33. Unit cell of a garnet with the different sites *a*, *c*, *d* for ions and their oxygen environment [*69O*].

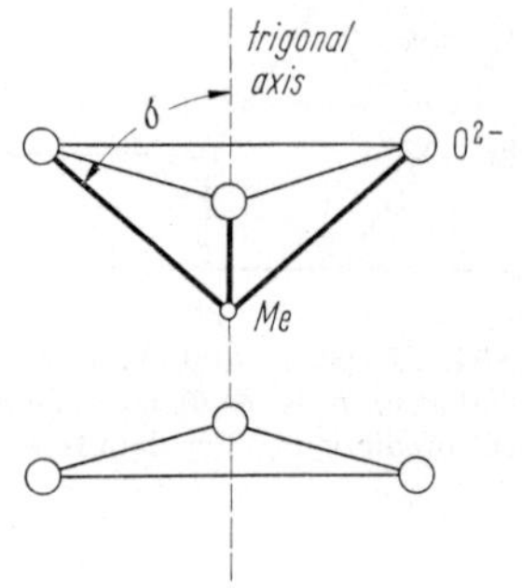

Fig. 34. The polar angle σ which characterizes the trigonal distortion of the octahedral (*a*) site [*68MO*].

I. Octahedral *a* sites

The O^{2-} ions lie at the centers of six faces of a cube. The six oxygens can be considered to form two equilateral triangles centered on a [111] axis of the crystal as shown in Fig. 34. In an undistorted octahedron the polar angle σ between a ligand direction (parallel to a cube edge) and a threefold symmetry axis [111] is 54.75°. In the garnet the octahedron is distorted by being streched along the [111] axis, proceeding a trigonal distortion with site symmetry C_{3i}. Fig. 35 shows two nonequivalent octahedral sites.

II. Tetrahedral *d* sites

The tetrahedra are distorted by pulling along one of their fourfold rotary-inversion axes and have site symmetry S_4, with the tetragonal axes along any one of the [100] directions of the crystal. The axial distortion may be fairly severe. Even among those sites having a common axis of distortion there are two types which are distinguished from each other by a rotation in opposite directions about the axis. See Fig. 36.

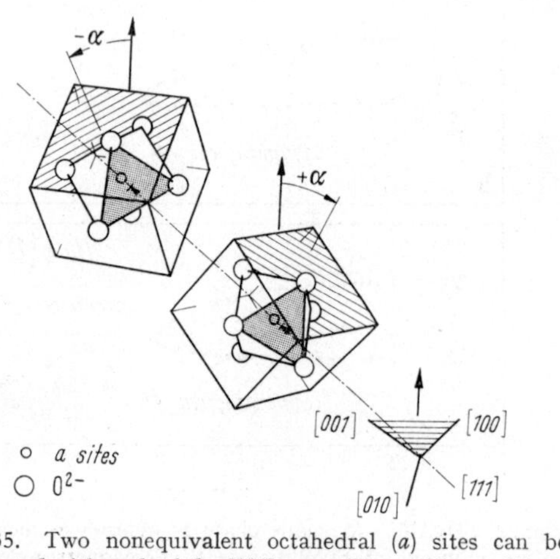

Fig. 35. Two nonequivalent octahedral (*a*) sites can be associated with each of the [111] axes. The positions of these octahedra may be visualized by starting from a pair of regular octahedra inscribed on cubes whose edges are parallel to the crystal axes, [100]. The octahedra are first distorted by stretching along a [111] direction. They are then rotated about this crystal [111] direction through opposite angle α [*61G*].

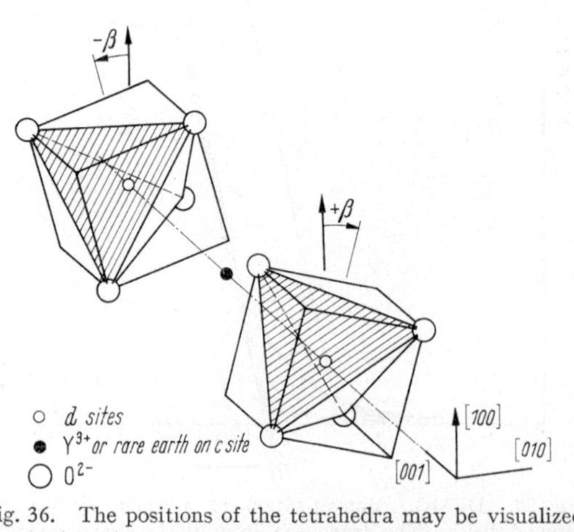

Fig. 36. The positions of the tetrahedra may be visualized by starting with regular tetrahedra inscribed in a cube whose axes are coincident with the unit cell edges. The cube is then stretched along a unit cell edge, [001], and then rotated about this axis through opposite angles β giving rise to two types of sites for this common axis of distortion [*61G*].

III. Dodecahedral *c* sites

These sites are occupied by Y^{3+} or rare earth ions in pure garnets. Each ion on a *c* site is surrounded by 8 oxygen ions, roughly at the corners of a cube (the so-called "pseudocube"). The magnetic anisotropy is dominant by the local symmetry at the *c* site, which is only D_2. The unit cell contains 6 magnetically inequivalent *c* sites which are simply related by 90° rotations and reflections (Fig. 37).

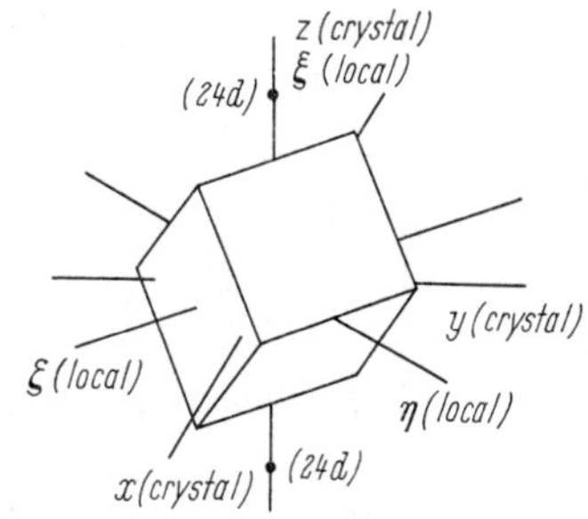

Fig. 37. Orientation of local symmetry axes (ξ, η, ζ) relative to cubic crystal axes (x, y, z) at a type 1 rare earth site. Axes for the five other types of sites are given in the table. The corners of the cube show the approximate positions of the 8 nearest O^{2-} neighbors and the points marked (24d) the positions of the two nearest metal ions, which are only slightly further away [*64W, 68J*]. ←

Inequivalent sites:

local axes	sites: 1	2	3	4	5	6	g- and χ-value axes
ξ	001	001	100	100	010	010	g_z; χ_1
η	$\bar{1}\bar{1}0$	$1\bar{1}0$	011	$0\bar{1}1$	$\bar{1}01$	101	g_x; χ_2 or g_y; χ_3
ζ	$0\bar{1}0$	110	$01\bar{1}$	011	101	$10\bar{1}$	g_y; χ_3 or g_x; χ_2

B. Oxygen coordinates, interionic spacings and angles of compounds with garnet structure

The definition of the polar coordinates and interatomic distances is given in Fig. 38.

The oxygen structure parameters x_0, y_0, z_0 and p are given as fractions of the lattice parameter, which is also given in Tab. 4. The relationship between some structure parameters and the atomic numbers of the rare earths are shown in Fig. 39 and Fig. 40. For comparison, the data of some iron garnets are tabled as well.

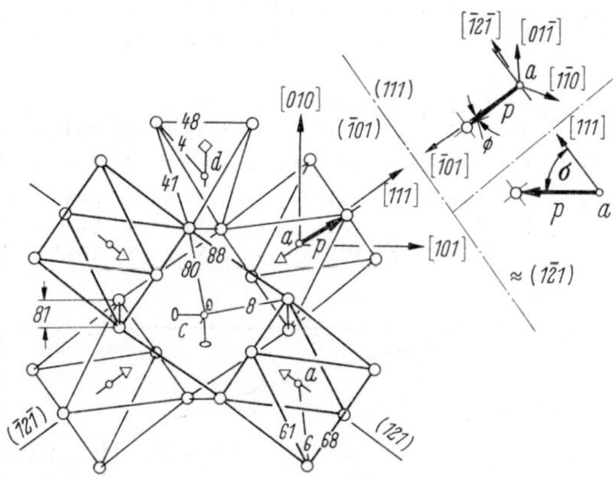

Fig. 38. Arrangement of the anion polyhedra (CN 4, 6, 8) in garnets, ($\bar{1}01$) projection; cation sites (*d*), (*a*), (*c*); point symmetries $\bar{4}$, $\bar{3}$, 222. Interionic distances d_i (i: 4, 6, 8, 80; 41, 61, 81; 48, 68, 88; the last three pertain to shared edges of neighboring polyhedra). Cell origin at *a* site (intersection of [010] and [101]); vector $\boldsymbol{p}$, pointing from there to anion position x, y, z (solid circle), is specified by magnititude p, angles σ and φ, shown in projections ≈ ($1\bar{2}1$) and (111). The octahedra are regular if $\sigma = \sigma_0 = \cos^{-1}(1/\sqrt{3}) = 54.75°$. Right trigonal prisms are found between two octahedra if $\varphi = 0$, associated with degenerate dodecahedra bound by two parallel rectangles ($d_8 = d_{80}$); their most degenerate form ($d_{88} = d_{68} = d_{61}$ and $d_{48} = d_{81}$) requires, in addition, $p = 0.1686$. The tetrahedra are regular if $p = 0.1470$ (with $\sigma = \sigma_0$ and $\varphi = 0$) [*65EB*].

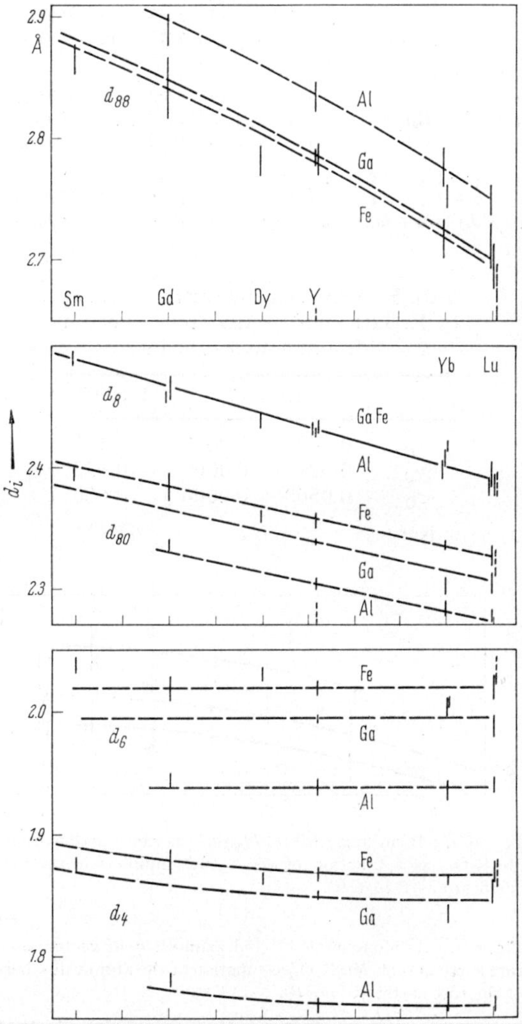

→

Fig. 39. Selected interionic distances, d_i, (see Fig. 38) in rare earth garnets $Me_5R_3O_{12}$ as function of the atomic numbers of the rare earths R [*65EB*]. Where data fall close together, Me reads from left to right: Al, Ga, Fe.

Tab. 4. Structure data of garnets $Me_5R_3O_{12}$ from X-ray experiments. The definition of the coordinates and distances are given in Fig. 38 [*65EB*]

Me (3 in *d*, 2 in *a* site) R (3 in *c* site)	Si, Al Mg, Fe, Ca	Al Lu	Al Yb	Al Y	Al Gd	Ga Lu	Ga Yb	Ga Y
Oxygen parameters								
$-x_O$	339 ± 5	294 ± 4	296 ± 5	306 ± 4	311 ± 4	252 ± 5	259 ± 5	272 ± 2
y_O	491 ± 6	537 ± 5	529 ± 6	512 ± 4	509 ± 4	570 ± 8	563 ± 5	558 ± 3
z_O	1535 ± 6	1509 ± 6	1504 ± 7	1500 ± 5	1490 ± 5	1506 ± 8	1519 ± 7	1501 ± 3
Polar coordinates								
radial p	1647 ± 6	1628 ± 6	1622 ± 7	1614 ± 5	1605 ± 5	1630 ± 8	1641 ± 7	1624 ± 3
cone angle σ	537 ± 2	516 ± 2	518 ± 2	524 ± 1	526 ± 1	498 ± 2	501 ± 2	506 ± 1
azimuth φ	38 ± 2	26 ± 2	28 ± 2	31 ± 2	30 ± 2	21 ± 3	25 ± 2	21 ± 1
Distances								
cell edge a	11556 ± 4	11906 ± 4	11931 ± 4	12000 ± 4	12113 ± 4	12188 ± 4	12204 ± 4	12280 ± 4
cation-anion								
$d - h$ (tetr.) d_4	1635 ± 7	1760 ± 6	1762 ± 8	1761 ± 5	1781 ± 5	1852 ± 9	1835 ± 7	1849 ± 3
$a - h$ (oct.) d_6	1903 ± 7	1939 ± 7	1935 ± 8	1937 ± 6	1944 ± 6	1987 ± 10	2002 ± 8	1995 ± 4
$c - h$ (dod.) 1) d_8	2377 ± 7	2383 ± 6	2397 ± 7	2432 ± 5	2458 ± 5	2393 ± 10	2407 ± 7	2428 ± 3
2) d_{80}	2222 ± 6	2276 ± 6	2283 ± 7	2303 ± 5	2335 ± 5	2303 ± 8	2302 ± 7	2338 ± 3
anion-anion ($h - h$)								
shared (tetr., dod.) d_{48}	2502 ± 14	2684 ± 13	2691 ± 16	2696 ± 11	2740 ± 11	2793 ± 19	2761 ± 15	2810 ± 7
(oct., dod.) d_{68}	2658 ± 11	2632 ± 11	2633 ± 13	2658 ± 9	2675 ± 8	2626 ± 14	2660 ± 13	2668 ± 6
(dod., dod.) d_{88}	2760 ± 15	2750 ± 13	2777 ± 16	2837 ± 11	2889 ± 12	2713 ± 20	2719 ± 15	2786 ± 7
unshared (tetr.) d_{41}	2750 ± 11	2964 ± 10	2967 ± 12	2961 ± 9	2988 ± 8	3134 ± 14	3108 ± 11	3117 ± 6
(oct.) d_{61}	2724 ± 13	2848 ± 12	2835 ± 15	2818 ± 10	2821 ± 11	2981 ± 19	2993 ± 14	2966 ± 7
(dod.) d_{81}	2808 ± 11	2877 ± 10	2891 ± 11	2917 ± 8	2936 ± 8	2959 ± 16	2970 ± 13	2965 ± 6
Bond angles								
$a - h - d$	1317 ± 4	1282 ± 4	1288 ± 4	1301 ± 3	1307 ± 3	1251 ± 5	1254 ± 4	1265 ± 2
$a - h - (d_{80}) - c$	1208 ± 2	1040 ± 2	1042 ± 2	1042 ± 2	1042 ± 2	1049 ± 3	1046 ± 3	1045 ± 1
$a - h - (d_8) - c$	974 ± 3	1002 ± 3	1001 ± 3	997 ± 2	999 ± 2	1017 ± 4	1010 ± 3	1014 ± 1
$d - h - (d_{80}) - c$	958 ± 3	942 ± 3	941 ± 3	942 ± 2	938 ± 2	937 ± 4	944 ± 3	936 ± 1
$d - h - (d_8) - c$	1226 ± 3	1226 ± 3	1221 ± 3	1216 ± 2	1213 ± 2	1226 ± 4	1229 ± 4	1225 ± 2
$c - (d_8) - h - (d_{80}) - c$	1005 ± 3	1029 ± 2	1026 ± 3	1017 ± 2	1014 ± 2	1053 ± 4	1050 ± 3	1042 ± 1

Tab. 5. Oxygen parameters x_O, y_O, z_O of ErGaG and DyAlG from neutron diffraction experiments

	ErGaG	DyAlG
x_O	0.1498 ± 0.0015	0.151
y_O	0.0285 ± 0.0008	−0.032
z_O	0.0568 ± 0.0008	0.056
Ref.	*68Hb*	*64HM*

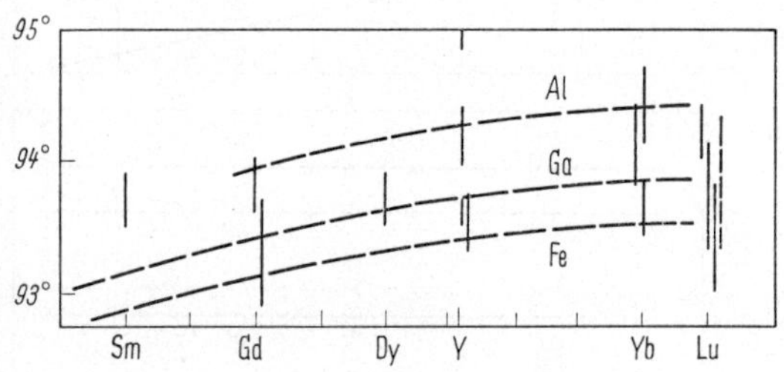

Fig. 40a. Bond angle (d)-(h)-(d_{80})-(c) in rare earth garnets $Me_5R_3O_{12}$ as a function of the atomic numbers of the rare earth atoms R [*65EB*].

→

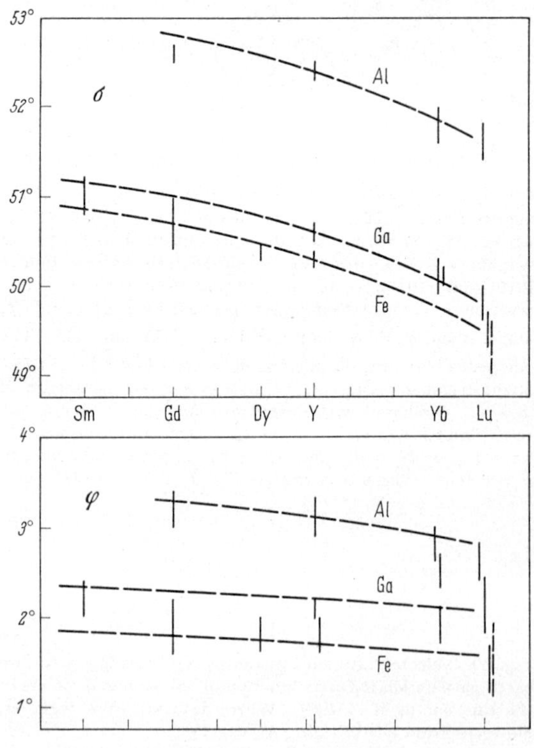

Fig. 40b, c. "Cone angle" σ and azimuth φ of vector $\boldsymbol{p}$ in rare earth garnets $Me_5R_3O_{12}$ as function of the atomic numbers of the rare earths R [*65EB*].

Tab. 4. (continued)

Me(3 in *d*, 2 in *a* site) R(3 in *c* site)	Fe Lu	Fe Lu	Fe Yb	Fe Y	Fe Dy	Fe Sm	
Oxygen parameters							
$-x_O$	251 ± 8	255 ± 5	263 ± 3	270 ± 4	272 ± 4	282 ± 4	10^{-4}
y_O	595 ± 8	592 ± 8	570 ± 5	569 ± 5	570 ± 5	552 ± 5	10^{-4}
z_O	1523 ± 10	1514 ± 8	1506 ± 5	1505 ± 5	1511 ± 5	1503 ± 5	10^{-4}
Polar coordinates							
radial p	1654 ± 10	1645 ± 8	1632 ± 4	1631 ± 5	1638 ± 5	1626 ± 5	10^{-4}
cone angle σ	493 ± 3	495 ± 2	501 ± 1	503 ± 1	504 ± 1	510 ± 2	10^{-1} deg
azimuth φ	15 ± 4	14 ± 3	19 ± 2	18 ± 2	18 ± 2	22 ± 2	10^{-1} deg
Distances							
cell edge a	12290 ± 4	12278 ± 4	12302 ± 1	12376 ± 4	12405 ± 1	12540 ± 4	10^{-3} Å
cation-anion							
$d - h$ (tetr.) d_4	1867 ± 11	1868 ± 8	1861 ± 4	1866 ± 5	1864 ± 5	1875 ± 5	10^{-3} Å
$a - h$ (oct.) d_6	2033 ± 12	2019 ± 10	2007 ± 5	2019 ± 6	2031 ± 6	2039 ± 6	10^{-3} Å
$c - h$ (dod.) 1) d_8	2385 ± 9	2385 ± 10	2417 ± 5	2434 ± 6	2439 ± 6	2489 ± 6	10^{-3} Å
2) d_{80}	2319 ± 11	2326 ± 8	2335 ± 4	2356 ± 5	2360 ± 5	2394 ± 5	10^{-3} Å
anion-anion ($h - h$)							
shared (tetr., dod.) d_{48}	2812 ± 23	2826 ± 19	2820 ± 10	2837 ± 12	2832 ± 12	2858 ± 12	10^{-3} Å
(oct., dod.) d_{68}	2671 ± 19	2659 ± 13	2667 ± 7	2692 ± 9	2710 ± 9	2743 ± 9	10^{-3} Å
(dod., dod.) d_{88}	2674 ± 23	2695 ± 19	2754 ± 10	2783 ± 12	2782 ± 12	2865 ± 13	10^{-3} Å
unshared (tetr.) d_{41}	3160 ± 19	3157 ± 13	3142 ± 7	3148 ± 9	3146 ± 9	3159 ± 9	10^{-3} Å
(oct.) d_{61}	3066 ± 22	3039 ± 18	3001 ± 9	3010 ± 11	3027 ± 11	3017 ± 12	10^{-3} Å
(dod.) d_{81}	2952 ± 19	2942 ± 16	2968 ± 8	2975 ± 10	2980 ± 9	3022 ± 10	10^{-3} Å
Bond angles							
$a - h - d$	1234 ± 5	1239 ± 5	1255 ± 3	1258 ± 3	1257 ± 3	1271 ± 3	10^{-1} deg
$a - h - (d_{80}) - c$	1040 ± 4	1041 ± 3	1045 ± 2	1042 ± 2	1040 ± 2	1042 ± 2	10^{-1} deg
$a - h - (d_8) - c$	1018 ± 5	1020 ± 4	1016 ± 2	1015 ± 2	1013 ± 2	1010 ± 2	10^{-1} deg
$d - h - (d_{80}) - c$	938 ± 5	934 ± 4	936 ± 2	935 ± 2	937 ± 2	937 ± 2	10^{-1} deg
$d - h - (d_8) - c$	1241 ± 5	1238 ± 4	1229 ± 2	1230 ± 3	1233 ± 3	1226 ± 3	10^{-1} deg
$c - (d_8) - h - (d_{80}) - c$	1062 ± 4	1058 ± 3	1049 ± 2	1046 ± 2	1046 ± 2	1037 ± 2	10^{-1} deg

C. Lattice parameters and density of garnet systems

Tab. 6. Lattice parameters and density of some gallium and aluminium garnets

	ϱ [g/cm³]		a [Å]			
Ref.	*56BF*	*64RBa*	*56BF*	*65EB*	*64RBa*	
RGaG						
R = Pr	6.40		12.57			
Nd	6.57		12.50			
Sm	6.58		12.42			12.431 [*65SH*]
Eu					12.407	
Gd	7.02		12.39			
Dy	7.25		12.32			
Er	7.49		12.25		12.255	12.255 [*68Hb*]
Tm					12.23	
Yb				12.204		
Lu				12.188		
Y	5.73		12.30	12.280		
RAlG						
R = Gd	5.95		12.11	12.113		
Tb		6.063			12.074	
Dy	6.15	6.193	12.06		12.042	
Ho		6.297			12.011	
Er	6.36	6.397	11.98		11.981	
Tm		6.476			11.957	
Yb				11.931		
Lu				11.906		
Y	4.50		12.02	12.000		12.01 ± 0.02 [*68AT*]

Tab. 7. Lattice constants of several germanate garnets [*64R*]

	a [Å]		a [Å]
$Ca_3Zr_2Ge_2CoO_{12}$	12.62	$Ca_3ZrNiGe_3O_{12}$ *)	12.49_5
$Ca_3Sn_2Ge_2CoO_{12}$	12.52	$Ca_3SnNiGe_3O_{12}$	12.43
$Ca_3ZrMgGe_3O_{12}$ *)	12.50_5	$Ca_3TiNiGe_3O_{12}$ *)	12.335
$Ca_3SnMgGe_3O_{12}$ *)	12.44	$Y_2Mg_3Ge_3O_{12}$	12.21
$Ca_3TiMgGe_3O_{12}$ *)	12.35	$Y_2MgCo_2Ge_3O_{12}$	12.23_5
$Ca_3ZrCoGe_3O_{12}$ *)	12.52_5	$Y_2Co_3Ge_3O_{12}$ *)	12.26
$Ca_3SnCoGe_3O_{12}$ *)	12.46	$Y_2CaMg_2Ge_3O_{12}$	12.31_5
$Ca_3TiCoGe_3O_{12}$ *)	12.35_5	$Y_2CaCo_2Ge_3O_{12}$	12.35_5
		$GdMn_2Ge_2GaO_{12}$ **)	12.550

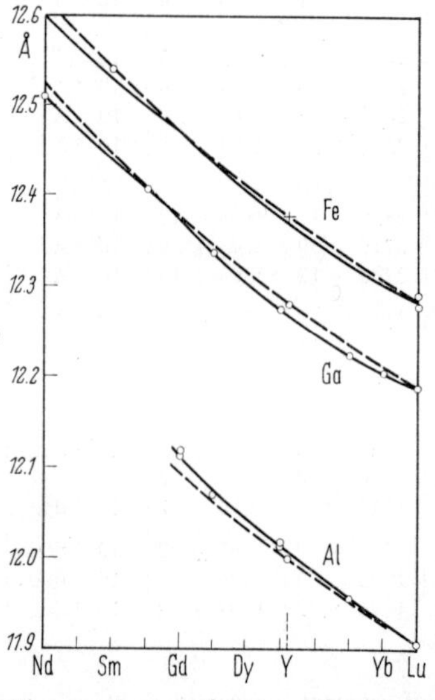

Fig. 41. Lattice constant a of rare earth garnets $Me_5R_3O_{12}$ as function of the atomic numbers of the rare earths R. Circles indicate measurements from [*65EB*].

Tab. 8. Lattice parameters for some Y (Ga, In, Cr) garnets [*64SM*]

$Y_3Ga_{4-x}In_xCrO_{12}$		$Y_3Ga_3Cr_{2-y}In_yO_{12}$	
x	a [Å]	y	a [Å]
0	12.275	0	12.271
0.2	12.309	1.0	12.417
0.4	12.337	1.5	12.480
0.6	12.365	1.8	12.513
1.0	12.417	2.0	12.543

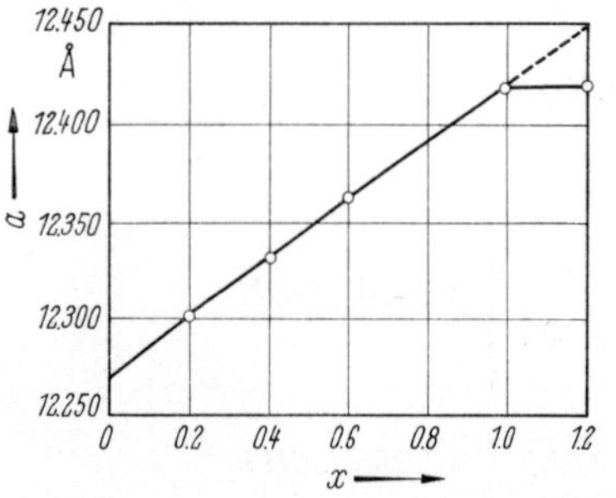

Fig. 42. Lattice parameter of the mixed garnets $Y_3Ga_{4-x}In_xCrO_{12}$ as a function of x [*64SM*].

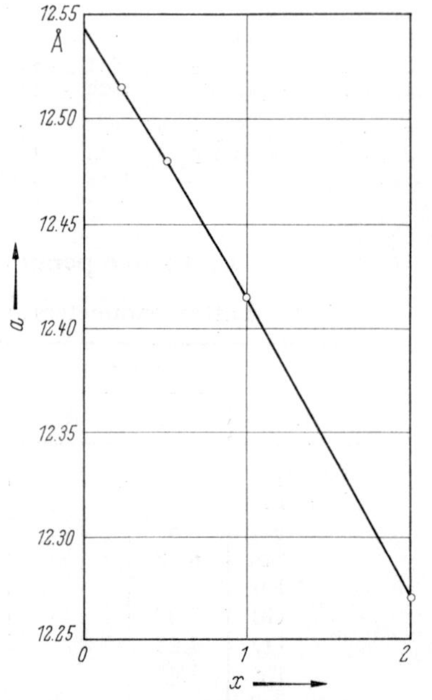

Fig. 43. Lattice parameter of the mixed garnets $Y_3Ga_3In_{2-x}Cr_xO_{12}$ vs. x [*64SM*].

D. Magnetic structure of DyAlG and ErGaG

I. DyAlG

The magnetic structure was first presented in [*62BHLW*] (Fig. 44b). [*65C*] gives a theoretical solution of a molecular field with six sublattices, which was verified by neutron diffraction [*64HM, 65HCW*]. For a molecular field treatment involving 24 sublattices see [*68BV*]. Below the Néel temperature the sublattice structure consists of antiferromagnetic chains parallel to each of the cubic unit cell axes, the main interaction being inter- rather than intra-chain [*62BHLW*]. The Dy^{3+} ions are in the positions 24 *c*.

*) Can be found in reference [*50*] as well. **) From reference [*59GG*].

Tab. 9. Magnetic sublattice structure of DyAlG below the Néel temperature in zero magnetic field [*64HM*]

Sub-lattice	coordinates	direction of magnetization
1	($^1/_8$, 0, $^1/_4$); ($^7/_8$, 0, $^3/_4$)	100
2	($^1/_4$, $^1/_8$, 0); ($^3/_4$, $^7/_8$, 0)	010
3	(0, $^1/_4$, $^1/_8$); (0, $^3/_4$, $^7/_8$)	001
4	($^5/_8$, 0, $^1/_4$); ($^3/_8$, 0, $^3/_4$)	$\bar{1}00$
5	($^1/_4$, $^5/_8$, 0); ($^3/_4$, $^3/_8$, 0)	$0\bar{1}0$
6	(0, $^1/_4$, $^5/_8$); (0, $^3/_4$, $^3/_8$)	$00\bar{1}$

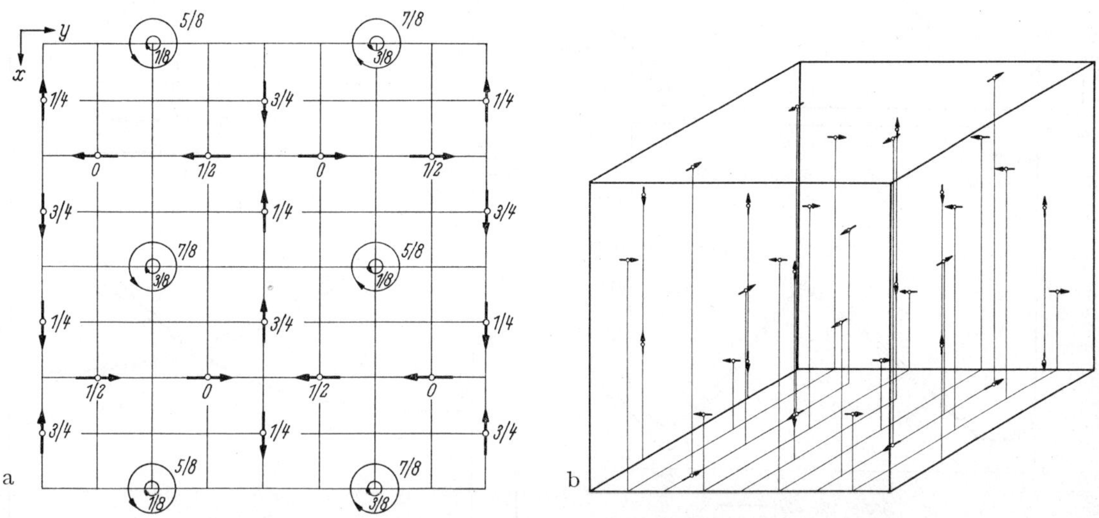

Fig. 44a, b. DyAlG. Zero field antiferromagnetic spin arrangement below Θ_N.
a) Spin arrangement shown in a [001] projection of the unit cell. The numbers in the figure give the height above the $z = 0$ plane. Spins pointing along the z axis are indicated as current loops in a righthand sense [*65HCW*].
b) Cubic unit cell with the spin arrangement [*62BHLW*].

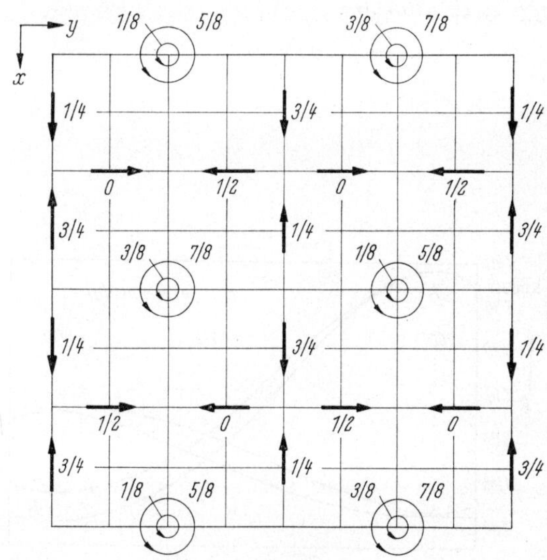

Fig. 45. DyAlG. Proposed antiferrimagnetic spin arrangement of the X and Y spins when Z spins are aligned by a strong magnetic field along [001]. [*66KLW*]. For comparison see the phase diagram in Fig. 32. →

II. ErGaG

Neutron diffraction experiments for ErGaG give the zero field spin arrangement shown in Fig. 44a for DyAlG [*68Hb*]

5.4 Results of magnetic resonance experiments

(Gyromagnetic ratio, hyperfine constant, spin-Hamiltonian parameters)

A. Paramagnetic resonance of rare earth impurities in diamagnetic garnets

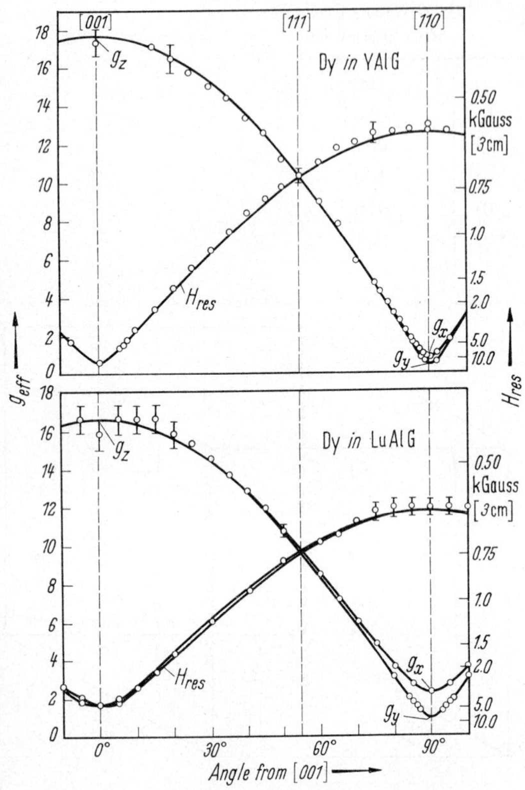

Fig. 46. Dy^{3+} (1%) in LuAlG and YAlG. Resonance field as a function of angle in the (1$\bar{1}$0) plane, showing the variation for the four inequivalent oriented ions in this plane. Both spectra are seen to have essentially the same form with $g_z \gg g_x, g_y$ [*62BHLW*].

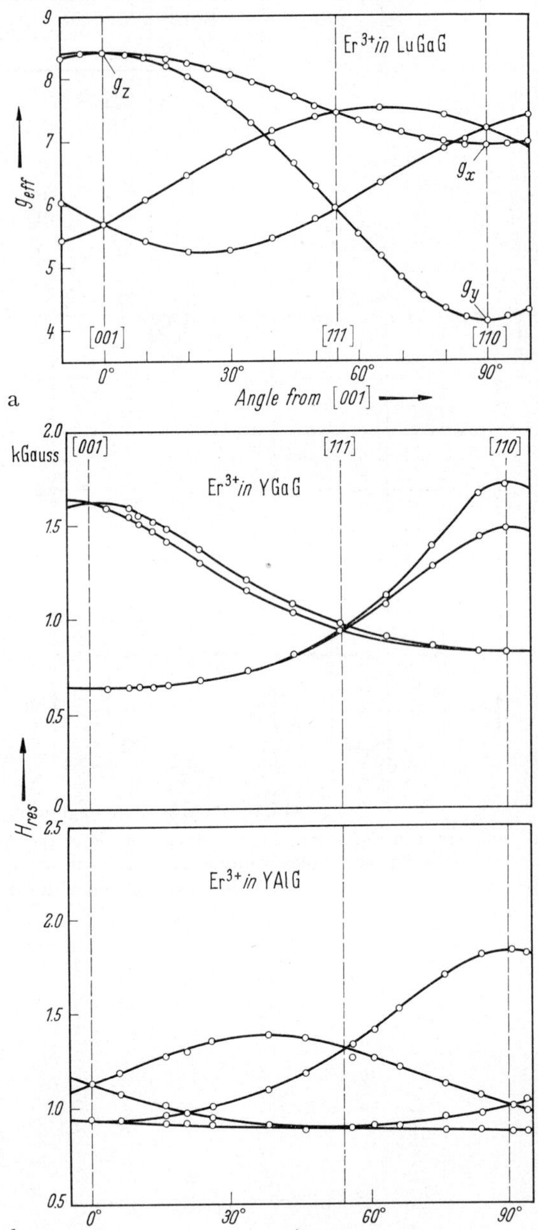

Fig. 47a, b. Er^{3+} ($\lesssim$ 1%) in some diamagnetic host garnets. Paramagnetic resonance at 4° K. Angular variation of the four resonances observed when a magnetic field is applied in a [1$\bar{1}$0) plane. The four curves correspond to the four types of ions which are inequivalently oriented with respect to a magnetic field in that plane.

Circles: experimental results

Full lines: calculated variation from the three principal *g* values chosen to give the best fit (Tab. 10).

a) Angular variation of the effective *g* values for Er^{3+} in LuGaG [*62WBHLW*].

b) Variation of the resonance field for Er^{3+} in YGaG and in YAlG ($f = 9$ Gc/s) [*61BGLRW*].

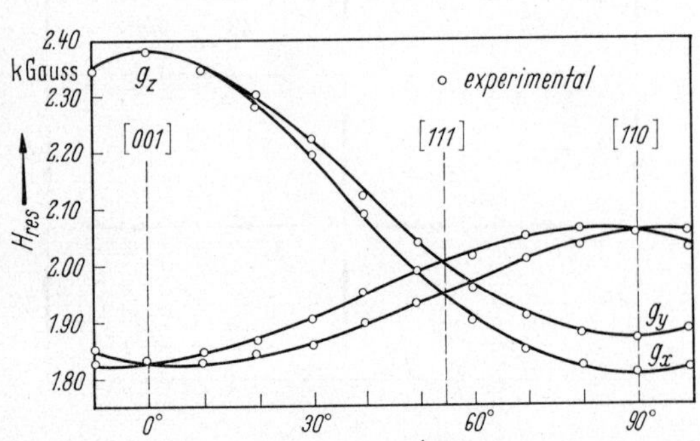

Fig. 48. Yb^{3+} ($\approx$ 1%) in YGaG. Variation of resonance field with angle in the (1$\bar{1}$0) plane at 9.677 Gc/s. Full lines: curves calculated from *g* values given in Tab. 11 [*59BGRW*].

Tab. 10. Representation of the resonance data by a g tensor with the principal values g_x, g_y, g_z (Fig. 37) ($T = 4.2$ °K)

impurity ($\lesssim 1\%$)	dia-magnetic host garnet	g_x	g_y	g_z	f Gc/s	Ref.
Ce^{3+}	YAlG	1.872	0.9098	2.738	9.4	*66L*
		1.872	0.9092	2.738	9.6	
		1.85	0.90	2.74	theor. values	
Nd^{3+}	LuGaG	2.083 ± 0.007	1.323 ± 0.007	3.550 ± 0.007	≈9	*62WBHLW*
	YGaG	2.027 ± 0.008	1.251 ± 0.008	3.667 ± 0.018		*18*
	LuAlG	1.789 ± 0.005	1.237 ± 0.006	3.834 ± 0.005		*62WBHLW*
	YAlG	1.733 ± 0.002	1.179 ± 0.002	3.915 ± 0.026		*18*
		1.739 ± 0.002	1.158 ± 0.001	3.908 ± 0.002		*69BDM*
Dy^{3+}	LuGaG	13.45 ± 0.10	0.57 ± 0.10	3.41 ± 0.03	≈9	*62WBHLW*
	YGaG	11.07 ± 0.20	1.07 ± 0.14	7.85 ± 0.10	9	*61BGLRW*
	LuAlG	2.29 ± 0.03	0.91 ± 0.05	16.6 ± 0.3		*62BHLW, 62WBHLW*
	YAlG	0.73 ± 0.15	0.4 ± 0.2	17.7 ± 0.7	9	*62BHLW*
Dy^{3+} (100%)	DyAlG*)	0.7	0.4	18.2		*64W, 68BCV, 65HCW*
Er^{3+}	LuGaG	3.183 ± 0.015	3.183 ± 0.015	12.62 ± 0.10		*62WBHLW*
	YGaG	4.69 ± 0.03	4.03 ± 0.02	10.73 ± 0.05		*61BGLRW*
	LuAlG	6.93 ± 0.02	4.12 ± 0.03	8.43 ± 0.04		*62WBHLW*
	YAlG	7.75 ± 0.09	3.71 ± 0.02	7.35 ± 0.08		*61BGLRW*
Yb^{3+}	LuGaG	3.653 ± 0.013	3.559 ± 0.013	2.994 ± 0.011		*62WBHLW*
	YGaG	3.73 ± 0.02	3.60 ± 0.02	2.85 ± 0.02	9.677	*59BGRW*
		3.74	3.60	2.85	34.2	*60CW*
		3.78	3.62	2.86	theor. values	*60AT*
	LuAlG	3.842 ± 0.005	3.738 ± 0.007	2.594 ± 0.04		*62WBHLW*
	YAlG	3.87	3.78	2.47		*60CW*

Tab. 11. Spin-Hamiltonian parameters of Gd^{3+} in diamagnetic garnets [*62RM*]. The interaction of the ground state with the crystal field and the effective magnetic field $\boldsymbol{H}$ may be described phenomenologically by a spin-Hamiltonian [*20*]

$$\mathscr{H}(\boldsymbol{S}) = g\mu_B \boldsymbol{H}\cdot\boldsymbol{S} + \sum_{l,m} b_{l,m} Y_{l,m}(\boldsymbol{S}). \quad (1)$$

The spectroscopic splitting tensor may be taken as isotropic for S state ions.
The measurements were performed at a frequency of 36 Gc/s.

	Gd^{3+} in YGaG		Gd^{3+} in YAlG		
T	300	4.2	300	4.2	°K
g	1.992 ± 0.003	1.992 ± 0.003	1.990 ± 0.003	1.990 ± 0.003	
$b_{2,0}$	441.1 ± 0.5	448.5 ± 0.5	777.7 ± 0.5	776.4 ± 0.5	10^{-4} cm^{-1}
$b_{2,2}$	−269.7 ± 0.5	−283.8 ± 0.5	−85.1 ± 0.5	−96.9 ± 0.5	10^{-4} cm^{-1}
$b_{4,0}$	− 42.2 ± 0.5	− 45.2 ± 0.5	−46.9 ± 0.5	−48.6 ± 0.5	10^{-4} cm^{-1}
$b_{4,4}$	21.1 ± 0.5	22.1 ± 0.5	23.2 ± 0.5	23.5 ± 0.5	10^{-4} cm^{-1}
$b_{6,0}$	+0.6 ± 0.5	+4.15 ± 0.5	+3.77 ± 0.5	4.62 ± 0.5	10^{-4} cm^{-1}
$b_{6,4}$	0.1 ± 0.5	1.28 ± 0.5	2.43 ± 0.5	4.29 ± 0.5	10^{-4} cm^{-1}

Rare earth ions as impurities in diamagnetic garnets occupy the 24 pseudocubic c sites with D_2 symmetry. The spin-Hamiltonian, referred to the D_2 symmetry axes, may be written according to [*66HWW*] as

$$\mathscr{H} = \mu_B(g_x H_x S_x + g_y H_y S_y + g_z H_z S_z) + B_2^0 O_2^0 + B_2^2 O_2^2 + B_4^0 O_4^0 + B_4^2 O_4^2 + B_4^4 O_4^4 + B_6^0 O_6^0 + B_6^2 O_6^2 + B_6^4 O_6^4 + B_6^6 O_6^6 \quad (2)$$

where the O_n^m are the standard spin operators [*14, 15*]. Here x is chosen to lie along the local D_2 axis parallel to a crystallographic [001] axis, which is different from the convention used in Tab. 10 and Fig. 37.

Tab. 12 shows some experimental results fitted to this spin-Hamiltonian (2).

*) From magnetization measurements [*64W*] and neutron diffraction experiments [*68BCV, 65HCW*].

Tab. 12. Experimental spin-Hamiltonian parameters for Gd^{3+} on c sites in some diamagnetic host garnets [*66HWW*]

T	300	4.2	4.2	300	300	°K
Host lattice	YGaG	LuGaG	YAlG	LuAlG	LuGaG	
lattice constant a	12.280	12.188	12.000	11.906		Å
Ref.	[*63OGFC*]	[*63OGFC*]	[*62RM*]	[*63OGFC*]	[*63OGFC*]	
$g_x = g_y = g_z$	1.991	1.99	1.990	1.989	1.99	
B_2^0	146.9	93	258.8	190.5	93	10^{-4} cm^{-1}
B_2^2	216.1	238	96.9	112.4	238	10^{-4} cm^{-1}
B_4^0	−0.720	−0.788	−0.81	−0.84	−0.789	10^{-4} cm^{-1}
B_4^2	0.05					10^{-4} cm^{-1}
B_4^4	3.01	3.61	3.92	3.37	3.61	10^{-4} cm^{-1}
B_6^0	0.0002					10^{-4} cm^{-1}
B_6^2	0.0007					10^{-4} cm^{-1}
B_6^4	−0.012					10^{-4} cm^{-1}
B_6^6	0.011					10^{-4} cm^{-1}

B. Paramagnetic resonance of impurities with *d* electrons in diamagnetic garnets

The spectrum of impurities in garnets can be analysed in terms of the following spin-Hamiltonian [*61G*] given in [*13*]

$$\begin{aligned}\mathscr{H} = & g\,\mu_B \boldsymbol{H}\cdot\boldsymbol{S} + D[S_z^2 - (1/3)\,S(S+1)] \\ & + (1/6)\,a[S_\xi^4 + S_\eta^4 + S_\zeta^4 - (1/5)\,S(S+1)(3S^2+3S-1)] \\ & + (1/180)\,F[35S_z^4 - 30S(S+1)\,S_z^2 + 25S_z^2 - 6S(S+1) + 3S^2(S+1)^2] + A\,\boldsymbol{S}\cdot\boldsymbol{I}\end{aligned} \tag{3}$$

D and F correspond to axial fields of the second and fourth degree, respectively, with z direction chosen along the direction of the axial crystal field. a is the cubic crystal field splitting parameter; ξ, η, ζ are the axes of the cubic crystalline field. A is the hyperfine splitting constant.

It is convenient to rewrite the spin-Hamiltonian relative to a set of axes which more specifically express the point symmetry of octahedral and tetrahedral site [*61G*].

Octahedral a site (C_{3i}) (zero nuclear spin):

$$\begin{aligned}\mathscr{H} = & g_\parallel \mu_B H S_z \cos\theta + (1/2)\,g_\perp \mu_B H \sin\theta\,(S_+ + S_-) + D[S_z^2 - (1/3)\,S(S+1)] \\ & - (1/180)\,(a - F)\,\{35S_z^4 - 30S(S+1)\,S_z^2 + 25S_z^2 - 6S(S+1) + 3S^2(S+1)^2\} \\ & + (a\sqrt{2}/36)\,[S_z(S_+^3 e^{i3(\varphi\mp\alpha)} + S_-^3 e^{-i3(\varphi\mp\alpha)}) + (S_+^3 e^{i3(\varphi\mp\alpha)} + S_-^3 e^{-i3(\varphi\mp\alpha)})S_z]\end{aligned} \tag{4}$$

Here the z axis is along the axis of distortion of the octahedron (a [111] direction of the crystal), and the x axis is chosen to coincide with the projection of $\boldsymbol{H}$ upon a plane perpendicular to z. θ is the angle between $\boldsymbol{H}$ and z. $S_+ = S_x + iS_y$; $S_- = S_x - iS_y$. α is the angle through which the octahedra are rotated about the z axis (Fig. 35), φ is the angle between $H_\perp$ and the projection of one oft he unit-cell edges onto the (111) plane.

Tetrahedral d site (S_4) (zero nuclear spin):

$$\begin{aligned}\mathscr{H} = & g_\parallel \mu_B H S_z \cos\theta + (1/2)\,g_\perp \mu_B H \sin\theta\,(S_+ + S_-) + D[S_z^2 - (1/3)\,S(S+1)] \\ & + (1/120)\,(a + 2F/3)\,[35S_z^4 - 30S(S+1)\,S_z^2 + 25S_z^2 - 6S(S+1) + 3S^2(S+1)^2] \\ & + (1/48)\,a\,[S_+^4 e^{i4(\varphi\pm\beta)} + S_-^4 e^{-i4(\varphi\pm\beta)}]\end{aligned} \tag{5}$$

Here φ is the angle between $H_\perp$ and the unit-cell edges, β is the angle of rotation of the cubic axes of the tetrahedra about their z axes (Fig. 36) [*61G*].

Tab. 13. Ground-state crystal field splitting parameters of Fe^{3+} in YGaG. The paramagnetic resonance data were fitted to the above spin-Hamiltonians (3, 4, 5) [*61G*]. (Lattice parameter 12.280 Å)

	octahedral (a site)		tetrahedral (d site)		
T	295	4.2	295	4.2	°K
D	−0.1295 ± 0.0003	−0.1320 ± 0.0004	−0.0885 ± 0.0005	−0.0880 ± 0.0006	cm^{-1}
a	+0.0185 ± 0.0003	+0.0189 ± 0.0007	+0.0062 ± 0.0003	+0.0062 ± 0.0004	cm^{-1}
F	+0.0026 ± 0.0004	+0.0034 ± 0.0007	−0.0037 ± 0.0004	−0.0038 ± 0.0005	cm^{-1}
$g_\parallel = g_\perp$	2.003 ± 0.001		2.0047 ± 0.0005		
α	28				deg
β			16.5		deg

The sign of the coefficient c in the cubic crystalline potential $V = c(x^4 + y^4 + z^4 - 3r^4/5)$ at the center of an octahedron of point charges is opposite to the sign of c at the center of a tetrahedron of point charges of the same polarity. The finding of a positive in both types of sites indicates that in the mechanism responsible for this splitting terms proportional to even powers of V are dominant.

Tab. 14. Experimental spin-Hamiltonian parameters for Fe^{3+} in some diamagnetic garnets on different sites [*64RBa, 66RK*]

Host lattice	a Å	site	$g_\perp = g_\parallel$	D cm^{-1}	a cm^{-1}	F cm^{-1}	α deg	β deg	T °K	f Gc/s
YAlG	12.00	*a*	2.0015	−0.1053	0.0205	0.0027	27.5		4.2	16; 35
	12.00	*d*	2.004	−0.1028	0.0075	−0.0110		16	4.2	35
LuAlG	11.906	*a*	2.001	−0.0935	0.0292	0.0064	27.5			
	11.906	*d*	2.004	−0.1249	0.0084	−0.0104		16.5		
LuGaG	12.172	*a*	2.001	−0.1290	0.0174	−0.0033	28			
	12.172	*d*	2.003	−0.1131	0.0065	−0.0047		16		

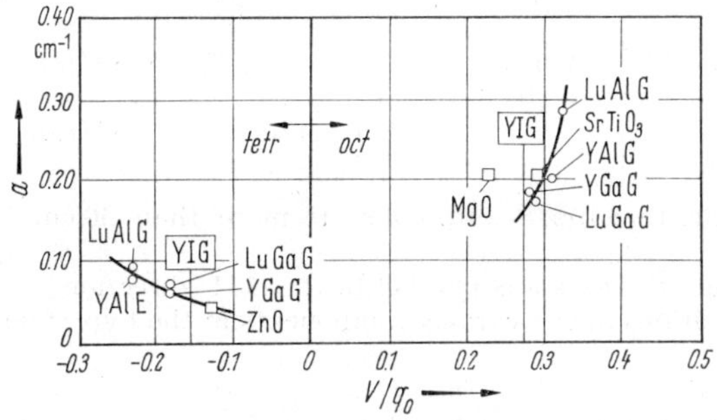

Fig. 49. The cubic field splitting parameter a for the ground state of Fe^{3+} in various host garnets vs. V/q_0 proportional to the point charge potential ($q_0 = 6$ or 4 for octahedral or tetrahedral sites) [*66RK*].

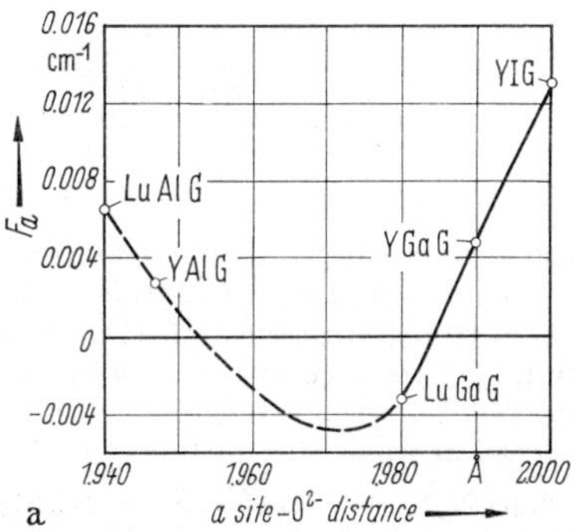

a

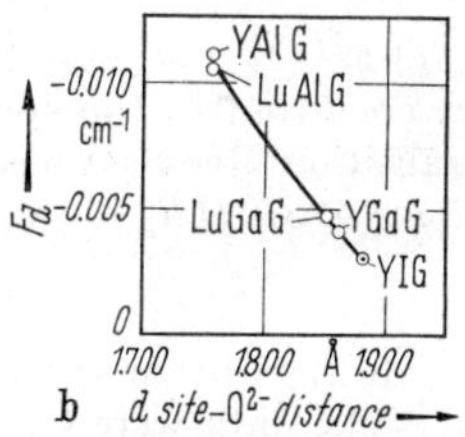

b

Fig. 50a, b. Variation of the fourth-order axial-field parameter F of the Fe^{3+} with the ion oxygen separation in various garnet hosts [*66RK*].
(a) octahedral sites,
(b) tetrahedral sites.
Linear extrapolation to a diamagnetic lattice equivalent to YIG:
$F_a = 0.013$ cm^{-1}
$F_d = -0.003$ cm^{-1}

Tab. 15 see next page

Tab. 16. Resonance spectrum of Cr^{3+} in YGaG and YAlG at 35 Gc/s. The Cr^{3+} enters predominantly *a* sites. The trigonal distortion gives rise to the zero-field splitting $2D$ of the ground state spin quartet into two doublets [*61CW*]. The data are fitted to the parameters of the spin-Hamiltonian (7) (see Tab. 15, footnote c); they are compatible with the theoretical results in [*23*]

Host garnet	$g_\parallel = g_\perp$	$2D$	T °K	Ref.
YGaG	1.98	21.0 Gc/s	300	*61CW*
	1.98	20.9 Gc/s	77	
	1.968 ± 0.002	0.1868 cm^{-1}		*66SZV*
YAlG	1.98	15.3 Gc/s	300	*61CW*
	1.98	15.7 Gc/s	77	
	1.977 ± 0.002	5527 ± 1.0 Gauss		*66SC*

Tab. 15. Spin resonance of Mn and Co ions in the diamagnetic YGaG host lattice

ion	site	S_{eff}	I	T °K	$g_\parallel$	$g_\perp$	D cm^{-1}	$\|a\|$, $(a-F)$ cm^{-1}	A, $\|A\|$ 10^{-4} cm^{-1}	$\|B\|$ 10^{-4} cm^{-1}		Ref.
								$a-F$	A			
Mn^{2+} (0.1%)	a	5/2		300	2.001 ±0.001	2.001 ±0.001	(−897 ± 0.1)·10^{-4}	(28.3 ± 0.5)·10^{-4}	−90.6 ± 0.5		a)	*66CC*
									$\|A\|$			
Co^{2+} (0.1%)	a	1/2	7/2	4.2	7.01 ±0.01	2.69 ±0.02			308 ± 3	15 ± 5	b)	*66CC*
(0.3%)	a	1/2	7/2	4.2	7.035 ±0.005	2.675 ±0.002			306 ± 3	17 ± 3	b)	*66SMR*
				4.2	7.027 ±0.002	2.665 ±0.002			307.5 ± 2	15 ± 3	b)	*69SMHR*
*)	d	3/2	7/2	77	2.42 ±0.01	2.19 ±0.01	−18		<40	<30	c)	*69SMHR*
								$\|a\|$				
Co^{3+} (0.13%)	d	2	7/2	4.2, 1.2	2.168 ±0.003		≈ −14	0.660 ±0.003	<4		d)	*69SMHR*
Co^{4+}	a	1/2	7/2	1.5	3.3385 ±0.001	0.84 ±0.03			367 ± 1	170 ±1	b)	*69SMHR*

*) The spectrum disappears on cooling to 20 °K, showing that it arises from a state more than 30 cm^{-1} above the ground state [*69SMHR*].

a) Results were fitted to the spin-Hamiltonian (3), where the axis was parallel to the [111] direction. The sign of A relative to D was inferred from the positions of the various components in the hyperfine pattern [*66CC*].

b) The spectrum was fitted to the spin-Hamiltonian [*20*]

$$\mathcal{H} = g_\parallel \mu_B H_z S_z + g_\perp \mu_B (H_x S_x + H_y S_y) + A S_z I_z + B(S_x I_x + S_y I_y) \tag{6}$$

with an effective spin S_{eff} and where the axis coincided with the [111] directions in the crystal.

c) Spectrum is consistent with the spin-Hamiltonian

$$\mathcal{H} = g_\parallel \mu_B H_z S_z + g_\perp \mu_B (H_x S_x + H_y S_y) + D[S_z^2 - (1/3)\, S\,(S+1)]. \tag{7}$$

d) Data were fitted to the spin-Hamiltonian (3), neglecting the small fourth order axial term ($F \approx 0$). The transition observed has $\Delta S_z = 4$; mechanisms by which such a transition is observed are discussed in [*69SMHR*].

The ESR-absorption spectrum of Pt^{3+} at octahedral sites in YAlG could be observed below 120 °K [*66HSM*]. The data were fitted to the spin-Hamiltonian

$$\mathcal{H} = \mu_B \boldsymbol{H g S} + \boldsymbol{I A S},$$

with $S = 1/2$ and $I = 1/2$.

The g and A tensor have orthorhombic symmetry with principal components given in Tab. 17. The orientation of these tensors are shown in Fig. 51.

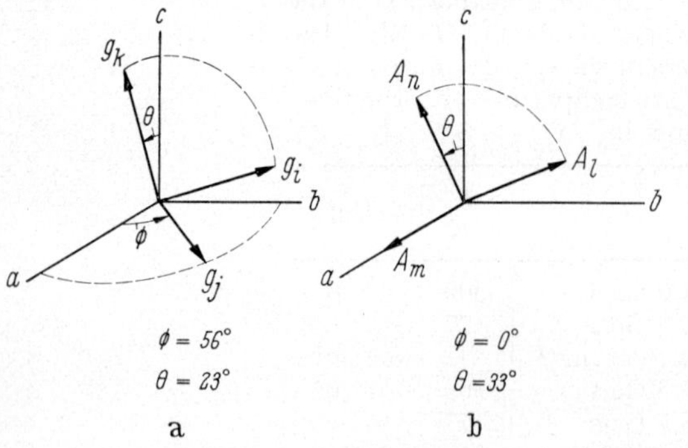

Fig. 51. Orientations of (a) the g tensor components and (b) the A tensor components for Pt^{3+} at one octahedral site in YAlG shown relative to a set of orthogonal axes a, b, c. The 24 orientations of the g tensor arise from the three static Jahn-Teller distortions at each of the eight inequivalent octahedral sites per unit cell [*66HSM*].

Tab. 17. Spin-Hamiltonian g- and A-tensor principal values for Pt in YAlG [*66HSM*].

	principal component values	principal component directions[a])
g_i	1.962	[0.763, 0.515, 0.391]
g_j	2.361	[0.559, 0.829, 0]
g_k	2.424	[0.324, 0.219, 0.921]
A_l	99 Oe	[0, 0.84, 0.55]
A_m	88 Oe	[1.00, 0, 0]
A_n	30 Oe	[0, 0.55, 0.84]

Tab. 18. Electron spin resonance of Ru^{3+} on octahedral a sites in two diamagnetic host garnets at 4.2 °K [*68MO*]. The data were fitted to the axial spin-Hamiltonian (6) (see Tab. 15, footnote b)

Host garnet	S_{eff}	I	$g_{\parallel}$	$g_{\perp}$	A 10^{-4} cm^{-1}	B 10^{-4} cm^{-1}	H_c [b]) k Gauss/ unpaired spin
YGaG	1/2	5/2	3.113 ± 0.003	1.148 ± 0.002	63.3 ± 0.8	38 ± 5	−220
YAlG	1/2	5/2	2.88 ± 0.02	1.300 ± 0.004	56.4 ± 0.7	41 ± 2	−250

C. Paramagnetic resonance of impurities in paramagnetic garnets

(Shifts in the g and D values.)

I Spin-Hamiltonian parameters:

Tab. 19. Experimental spin-Hamiltonian (Eq. (3), for a sites Eq. (4)) parameters for Fe^{3+} as impurity in some paramagnetic garnets. Tm garnets [*64RBa*], Eu garnet [*66HWW*]
y_0: oxygen parameter (for explanation see section 5.3 B)

Host lattice	site	lattice constant Å	$y_0 \cdot 10^4$	$g_{\parallel}$	$g_{\perp}$	D cm^{-1}	$a-F$ cm^{-1}	$\lvert a \rvert$ cm^{-1}	α [d]) deg $a>0$	α [d]) deg $a<0$	ν Gc/s
TmGaG	a	12.23	564	2.175	2.123	−0.1572	0.0185	0.0239	30	30	16, 35
TmAlG	a	11.96	523	1.985	1.931	−0.1287	0.0214	0.0227	29	31	16, 35
	d	11.96		1.235	1.755	(0.2)					35
				1.23	1.72						16
EuGaG (77 °K)	a	12.407		2.012 ± 0.001	2.001 ± 0.001	− 0.012 ± 0.0001	0.0161 [c]) ± 0.0028	$a =$ + 0.018 [c]) ± 0.002	26.8 [c]) (± 1.0)		25.981

Tab. 20. Experimental spin-Hamiltonian (2) parameters for Gd^{3+} on c sites in EuGaG (lattice constant 12.407 Å) measured at 77 °K and 25.995 Gc/s [*66HWW*]

g_x	g_y	g_z	B_2^0	B_2^2	B_4^0	B_4^2	B_4^4	B_6^0	B_6^4	$B_6^6 + B_6^2$
			10^{-4} cm^{-1}							
1.984 ± 0.001	1.984 ± 0.001	1.987 ± 0.001	198.3 ± 0.3	159.5 ± 0.6	−0.687 ±0.003	0.039 ± 0.020	3.03 ± 0.02	0.000 ±0.0001	−0.0017 ±0.0020	0.0004 ± 0.0020

Cr^{3+} in EuGaG: In [*66SZV*] the resonance data are described by the standard axial spin-Hamiltonian with the parameters $D = +0.0934$ cm^{-1}, $g = 1.968 \pm 0.002$.

II. Shifts:

By comparing the spin-Hamiltonian parameters found for paramagnetic garnets doped with impurities (subsection C) with those for other diamagnetic garnets (subsections A and B), shifts due to the interaction of the substituted ions with the rare earth ions in the host lattice can be determined. A theory for shifts of S-state ion resonance in europium compounds is given in [*66HWW*]. Exchange contribution to the D-parameter-shifts of Fe^{3+} and Cr^{3+} in EuGaG are discussed by [*67TS*].

a) These components are given with respect to the three cubic directions of the YAlG structure.
b) H_c is the effective magnetic hyperfine field due to the core polarization.
c) Assuming that a has the same sign as in YGaG.
d) As discussed in [*61G*], the resonance data yield the absolute value of a and two possible values for the angle α; $a \gtrless 0$ according to $\alpha \gtrless 30°$.

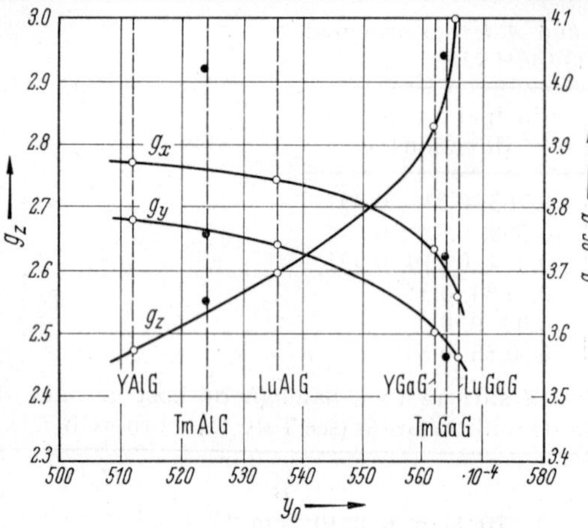

← Fig. 52. Variation of the principal g values with oxygen parameter y_O for Yb^{3+} in four diamagnetic garnets (Tab. 10) YGaG, YAlG, LuGaG, and LuAlG (open circles) and two paramagnetic garnets TmAlG and TmGaG (closed circles). A similar variation is found with the parameter x_O. The parameters x_O and y_O define the positions of the O^{2-} ions that dominate the crystal field [*63HW*] (See section 5.3 B).

Tab. 21. Estimated shift in the spin-Hamiltonian parameters. Shifts are defined as $\Delta X = X$ (experimental in paramagnetic host garnet) $-$ X (extrapolated from several diamagnetic garnets).

Tab. 21a. Fe^{3+} and Cr^{3+} in Al and Ga garnets

impurity	paramagnetic host garnet	site	$\Delta g_\parallel$	$\Delta g_\perp$	ΔD cm^{-1}	Δa cm^{-1}	Ref.
Fe^{3+}	EuGaG*)	*a*	$+0.0095 \pm 0.0020$	-0.0015 ± 0.0020	$+0.115 \pm 0.010$	-0.002 ± 0.005	*66HWW*
Fe^{3+}	TmGaG	*a*	$+0.175$	$+0.122$	-0.030	0.005	*64RBa*
Fe^{3+}	TmAlG	*a*	-0.016	-0.069	-0.025		*64RBa*
Fe^{3+}		*d*	-0.770	-0.245	(≈ -0.1)		*64RBa*
Cr^{3+}	EuGaG	*a*	≈ -0.01		≈ -0.26		*66SZV*

Tab. 21b. Estimated shift in the spin-Hamiltonian (2) parameters for Gd^{3+} on pseudocubic c sites in EuGaG [*66HWW*]. The three shifts in the g tensor are well explained by an isotropic exchange interaction $J_{is} = +0.07$ cm^{-1} and the magnetic dipole interaction calculated from the structure, $J_d = -0.047$ cm^{-1}.

Δg_x	Δg_y	Δg_z	ΔB_2^0	ΔB_2^2	ΔB_4^0	ΔB_4^4
			10^{-4} cm^{-1}			
-0.007 ± 0.002	-0.007 ± 0.002	0.004 ± 0.002	-20 ± 80	-40 ± 60	0 ± 0.2	-0.3 ± 0.5

D. Nuclear magnetic resonance experiments

^{169}Tm in TmAlG and TmGaG, ^{27}Al in TmAlG.

The field for resonance is $H_{res} = H_{res}^{ext} + H_{hf}$, where H_{res}^{ext} is the applied magnetic field and H_{hf} is the hyperfine field, in terms of the atomic susceptibility tensor $\{\chi\}$

$$H_{hf} = -(A/\gamma \hbar g_J \mu_B) \cdot H \{\chi\} \qquad [69JS].$$

From a detailed study of the angular dependence of the NMR spectra (Fig. 53) and comparison with measured susceptibility values [*62WBHLW*], the components of the orthorhombic susceptibility tensor (Fig. 37) and the strength of the ^{169}Tm hyperfine interaction $A\,\boldsymbol{I}\langle\boldsymbol{J}\rangle$ can be determined [*68J* and *69JS*]:

$$A/h = -388.8 \text{ Mc/s}$$

Tab. 22. Components of the orthorhombic susceptibility tensor

	^{169}Tm in TmGaG	^{169}Tm in TmAlG	
χ_1	0.0571	0.014	emu/g-atom
χ_2	0.142	0.551	emu/g-atom
χ_3	0.080	0.030	emu/g-atom
Ref.	*68J*	*69JS*	

*) From the shift in the Fe^{3+} g values, the isotropic part of Eu^{3+} (c) $-$ Fe^{3+} (a) interaction is found to be small and ferromagnetic in sign, $J_{is} \approx -0.016$ cm^{-1} (where $H = +J_{is} S^1 S^2$). The anisotropy of Δg is partly accounted for by the magnetic dipole interactions which predict $\Delta g_\parallel = 0.002$ and $\Delta g_\perp = -0.001$ for the calculated $J_d = -0.062$ cm^{-1}. The discrepancy for $\Delta g_\parallel$ indicates the importance of some other anisotropic interactions [*66HWW*]. In [*68HV*] orbitally anisotropic exchange interaction are introduced to explain the observed shifts.

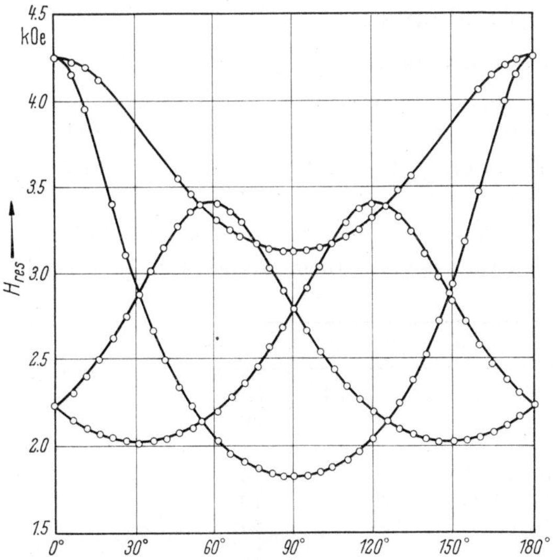

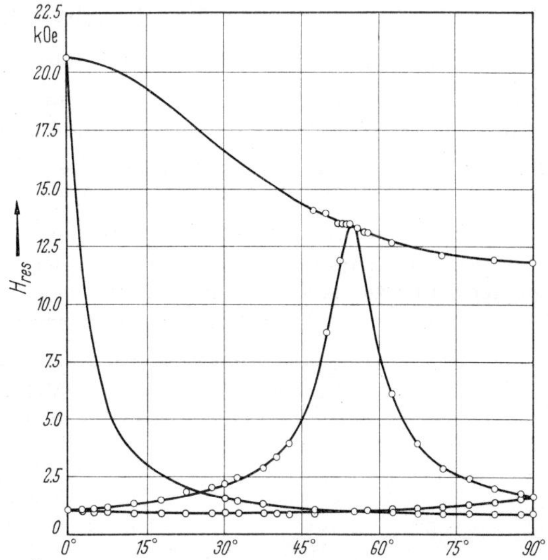

Fig. 53a, b. The angular dependence of ^{169}Tm NMR fields H_{res} in the (110) plane. The [100] and [110] directions are given by $\phi = 0$ and $\phi = 90°$, respectively. The open circles are the measured NMR data and the solid lines are the calculated values.

a) ^{169}Tm in TmGaG; $f = 16.000$ Mc/s, $T = 4$ °K [68J]. The three NMR fields for the (ξ, η, ζ) axes are respectively 4.262, 1.816, and 3.124 kOe.

b) ^{169}Tm in TmAlG; $f = 25$ Mc/s, $T = 1.5$ °K [69JS]. The three NMR fields for the (ξ, η, ζ) axes are respectively 20.658, 0.750, and 11.644 kOe.

(The orientations of ξ, η, ζ and χ_1, χ_2, χ_3 are given in Fig. 37).

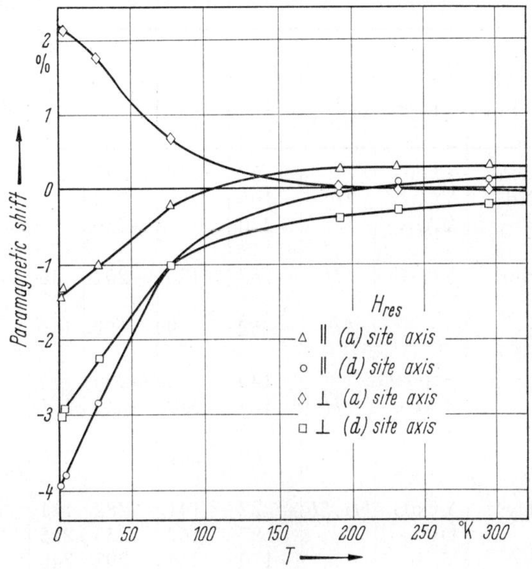

→

Fig. 54. TmAlG. The temperature dependence of the ^{27}Al NMR frequency shift. [69JS].

Tab. 23. ^{27}Al in TmAlG. Paramagnetic frequency shifts for H_{res}^{ext} parallel and perpendicular to site axes. The shifts, calculated from Tm point dipole sums at 300 and 1.5 °K, are based on the free Tm^{3+} ion and the 1.5 °K susceptibility tensor, respectively [69JS]

T °K	site	H_{res}^{ext}, direction	measured (%)	calculated (%)
297	*a*	parallel ∥	+0.28 ± 0.01	+0.392
	a	perpendicular ⊥	−0.02 ± 0.02	−0.196
	d	∥	+0.012 ± 0.01	+0.404
	d	⊥	−0.22 ± 0.01	−0.202
1.5	*a*	∥	−1.41 ± 0.20	−1.59
	a	⊥	+2.16 ± 0.10	+1.77
	d	∥	−3.95 ± 0.10	−3.68
	d	⊥	−3.01 ± 0.30	−2.83

5.5 Results from optical spectra

A. Crystal field analysis and parameters

The measured crystal-field energy levels of one or more multiplets of rare earth ions in garnets can be fitted to the perturbing Hamiltonian appropriate for the D_2 site symmetry of the ions [*69GHOS*]

$$\mathcal{H}_c = A_2^0\langle r^2\rangle V_2^0 + A_2^2\langle r^2\rangle V_2^2 + A_4^0\langle r^4\rangle V_4^0 + A_4^2\langle r^4\rangle V_4^2 + A_4^4\langle r^4\rangle V_4^4 + A_6^0\langle r^6\rangle V_6^0 + A_6^2\langle r^6\rangle V_6^2 + A_6^4\langle r^6\rangle V_6^4 + A_6^6\langle r^6\rangle V_6^6$$

to find the best set of the crystal-field parameters $A_k^q\langle r^k\rangle$. The tensor operators V_k^q are defined in [*62AD*].

Crystal-field parameters are also found from EPR-spectra. Some of them are listed in the preceding chapter 5.4 for the following substances:

Ion	host	Tab.
Gd^{3+}	YGaG, YAlG, LuGaG, LuAlG	11, 12
	EuGaG	20
Fe^{3+}	YGaG	13
	YAlG, LuGaG, LuAlG	14
	TmGaG, TmAlG	19
Cr^{3+}	YGaG, YAlG	16
Mn^{2+}, Co^{2+}, Co^{3+}, Co^{4+}	YGaG	15

I. Kramers ions

Tab. 24. Crystal-field parameters $A_k^q\langle r^k\rangle$ for Kramers ions.
If the fit is made for the states of special multiplets this is indicated in the third column.

Ion	host	multiplet [1]	$A_2^0\langle r^2\rangle$	$A_2^2\langle r^2\rangle$	$A_4^0\langle r^4\rangle$	$A_4^2\langle r^4\rangle$	$A_4^4\langle r^4\rangle$	$A_6^0\langle r^6\rangle$	$A_6^2\langle r^6\rangle$	$A_6^4\langle r^6\rangle$	$A_6^6\langle r^6\rangle$	Ref.	
			$A_k^q\langle r^k\rangle$ in [cm⁻¹]										
Ce^{3+}	YGaG		−52	178	−347	286	963	101	−321	1841	−326	*66HPWB*	
Nd^{3+}	YGaG	^4F	105		−270		−1255	80		−1092		*67Kb, (66Ka)*	
	YAlG	^4F	270		−250		−1250	92		−965		*67Kb, (66Ka, 64KGa)*	
Sm^{3+}	YGaG	^6H	−52	126	−231	46	1283	56	−21	949	88	*69Ga, b; 69GHOS*	
		^6F	52	−9	−189	148	661	22	25	656	−233	*69Ga, b; 69GHOS*	
	SmGaG		140		−200			40				*67VD*	
	YAlG	^6H	−184	319	−246	165	1373	45	−164	1038	41	*69Ga, b; 69GHOS*	
		^6F	68	667	−154	345	1201	33	129	638	357	*69Ga, b; 69GHOS*	
Dy^{3+}	YGaG	^6H, ^6F	−22	141	−282	181	1040	36	−108	725	−57	*69Gb, 69GHOS*	[2]
	DyGaG		87	162	−235	215	945	67	−10	902	−200	*67VD*	[2][3]
	YAlG		−170	334	−292	240	1046	41	−142	747	−77	*69Gb, 69GHOS*	
Er^{3+}	YGaG		−14	89	−238	255	920	33	−58	645	−70	*67CM*	[4]
	LuGaG		32	58	−245	175	948	338	−83	664	−48	*67CM*	
	ErGaG		6	90	−241	245	893	324	−49	641	−51	*69O*	[5][6]
	YAlG		260		−160		−800	45		−710		*67Kb, (64KGb)*	
	ErAlG		−163	262	−237	244	1057	367	−122	655	−92	*69O*	[5][7]
Yb^{3+}	YGaG		−70	288	−177	221	621	49	−462	822	−233	*67BWPH*	[8][9]
	YAlG		−12	352	−151	102	692	107	−767	1009	−230	*67BWPH*	[9]

[1]) Notation after [*62W*].
[2]) A fit to a pseudo-cubic Hamiltonian will be found in [*67GHH*]: $A_4^0\langle r^4\rangle = -290$ cm⁻¹; $A_6^0\langle r^6\rangle = +45$ cm⁻¹.
[3]) Four parameters are given in [*66AMR*]: $A_2^0\langle r^2\rangle = 147$ cm⁻¹, $A_4^0\langle r^4\rangle = -207$ cm⁻¹, $|A_2^2\langle r^2\rangle| = 250$ cm⁻¹, $A_6^0\langle r^6\rangle = 58$ cm⁻¹.
[4]) Crystal field calculation also in [*63P*].
[5]) Slightly different parameters are given in [*69O*] (second analysis) and [*69OH*].
[6]) Earlier results from [*65DVV*] are given in [*69O*]: $A_4^0\langle r^4\rangle = -260$ cm⁻¹, $A_4^4\langle r^4\rangle = 920$ cm⁻¹, $A_6^0\langle r^6\rangle = 40$ cm⁻¹, $A_6^4\langle r^6\rangle = 595$ cm⁻¹. From [*67VD*]: $A_2^0\langle r^2\rangle = 100$ cm⁻¹, $A_4^0\langle r^4\rangle = -260$ cm⁻¹, $A_6^0\langle r^6\rangle = 40$ cm⁻¹.
[7]) These parameters differ markedly from those given in [*66HHSS*].
[8]) Slightly different parameters were found from a fit to experimental data and point-charge ratios in [*64HW*]. See also [*59AT, 60AT, 61ATCD, 63W*].
[9]) Earlier crystal-field analysis is given in [*65Kb, 60PW, 67Kb*].

II. Non-Kramers ions

Tab. 25. Crystal-field parameters $A_k^q \langle r^k \rangle$ in [cm^{-1}] for non-Kramers ions

Ion	host	$A_2^0\langle r^2\rangle$	$A_2^2\langle r^2\rangle$	$A_4^0\langle r^4\rangle$	$A_4^2\langle r^4\rangle$	$A_4^4\langle r^4\rangle$	$A_6^0\langle r^6\rangle$	$A_6^2\langle r^6\rangle$	$A_6^4\langle r^6\rangle$	Ref.
Eu^{3+}	YGaG	105	95	−170	−100	850	50	150	− 945	*67Kb*, (*65Ka*)
	YAlG	268	51	−190		− 950	94		−1150	*67Kb*, (*64K*)
Tb^{3+}	YAlG	266	50	−184		− 920	93		−1115	*67Kb*, (*64K*)
Tm^{3+}	YAlG	270	95	−170	−410	−1020	30	115	− 475	*67Kb*, (*66Kb*)

In [*69BPH*] evidence is reported for the operation of selection rules in the absorption spectrum of Tm^{3+} in YGaG and YAlG. These selection rules result from the rare earth site symmetry and are expected to influence the spectrum of every non-Kramers ion occupying the rare-earth ion site in the garnet structure. No previous spectral analysis of any non-Kramers rare earth ion in garnets has considered selection rules. Therefore all previous crystalline field calculations made from spectra of non-Kramers rare earth ions in garnets may be meaningless [*69BPH*].

B. Summary of references for some results from optical measurements

Ion	host	Key for references concerning: spectrum		*g*-factors	crystal field		References				
		absorption	emission		levels	parameters	a	b	c	d	e
Ce^{3+}	YGaG				a	a	*66HPWB*				
Pr^{3+}	YGaG	a	a, b		a		*66H*	*61KP*			
	YAlG	a, b	a, b		a		*66H*	*63FPC*			
Nd^{3+}	YGaG	a			a	a, b, c	*66Ka*	*65Ka*	*67Kb*		
	YAlG	a, b	b		a, b, c	a, b, c	*66Ka*	*64KGa*	*67Kb*		
Sm^{3+}	YGaG	a, b	a, b, c		a, b, d	a, b, d	*69Ga*	*69Ga*	*61KP*	*69GHOS*	
	SmGaG	a			a	a	*67VD*				
	YAlG	a, b	a, b		a, b, c	a, b, c	*69Ga*	*69Gb*	*69GHOS*		
Eu^{3+}	YGaG	a, c, d	a, e		a, b	a, b, c	*65Ka*	*67Kb*	*65Kc*	*64K*	*61KP*
	YAlG	a, b	a		a, c	a, b, c	*64K*	*65Kc*	*67Kb*		
Tb^{3+}	YGaG		a				*61KP*				
	YAlG	a	a		a	a, b	*64K*	*67Kb*			
Dy^{3+}	YGaG	a	a, c	a	a, b	a, b	*69Gb*	*69GHOS*	*61KP*		
	DyGaG	a, b	a		b, c	b, c	*69Gb*	*67VD*	*67GHH*		
	YAlG	a, d	a	a	a, b, c	a, b	*69Gb*	*69GHOS*	*67GHH*	*65HSS*	
	DyAlG	a	a				*69Gb*				
Ho^{3+}	YGaG	b	a, b		b		*61KP*	*69JDR*			
Er^{3+}	YGaG	a	c		a	a, b, d	*63P*	*67Kb*	*61KP*	*67CM*	
	LuGaG					a	*67CM*				
	ErGaG	a, c		a, b	a, b, c	a, b, c, d	*69O*	*69OH*	*65DVV*	*67VD*	
	YAlG	a			a, b	a, b	*64KGb*	*67Kb*			
	ErAlG	a, c		a, b	a, b, c	a, b, c	*69O*	*69OH*	*66HHSS*		
Tm^{3+}	YGaG	a	b				*69Gb*	*61KP*			
	TmGaG	a					*69Gb*				
	YAlG	a, b			b	b, c	*69Gb*	*66Kb*	*67Kb*		
	TmAlG	a					*69Gb*				
Yb^{3+}	YGaG	a, e	a	b	a, b, c	a, b, c, d	*67BWPH*, *66HPWB*	*64HW*	*65Kb*	*60PW*, *67Kb*	*63W*
	LuGaG			b				*64HW*			
	YAlG	a, c, d	a, c, d	b	a, c	a, c, e	*67BWPH*	*64HW*	*65Kb*	*63W*	*67Kb*
	LuAlG			b				*64HW*			
Co^{2+}, Co^{3+}	YAlG, YGaG	a					*67WR*				
Cr^{3+}			a				*66BGB*				

Electronic Raman spectra of Yb, Eu, Nd in YGaG have been observed [*67Ka*].

5.6 Spin-lattice relaxation of paramagnetic ions in garnets

It has been shown by several authors [*34, 35, 36, 37*] that the dominant spin-lattice interaction is through the thermal modulation of the crystalline Stark field.

According to these authors, the energy transfer between the spin system and the lattice vibration gives rise to two processes:

(1) a one-phonon process in which a phonon is adsorbed or emitted accompanied by a quantum jump of a spin between two energy levels. This process will determine the relaxation at very low temperatures, where the spin-lattice relaxation time, T_1, is proportional to the reciprocal temperature.

(2) a two-phonon (Raman) process in which a phonon is scattered by a spin and another phonon with different energy is emitted accompanied by a quantum jump of a spin. The theory predicts $T_1^{-1} \approx T^7$ for non-Kramers doublets and $T_1^{-1} \approx T^9$ for Kramers doublets.

Relaxation in the rare earths has been considered in [*27*], basing on

(3) a process involving two phonons in resonance with the first excited state (two phonon resonance or Orbach process). There is a rapid exponential temperature dependence $T_1^{-1} \approx \exp(-\Delta/\mathrm{k}T)$, where Δ is the energy splitting of the first excited state of ions in garnets.

A Raman process often is not observed because of the dominance of the exponential term of the Orbach process.

A. Temperature dependence of the spin-lattice relaxation time T_1

I. Rare earth ions in diamagnetic host garnets

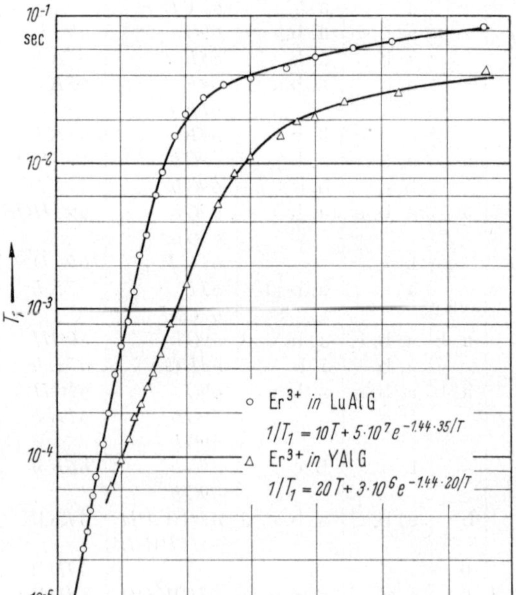

Fig. 55. Er^{3+} (1%) in YAlG and LuAlG. Relaxation time, T_1, vs. reciprocal temperature [*63SS*].

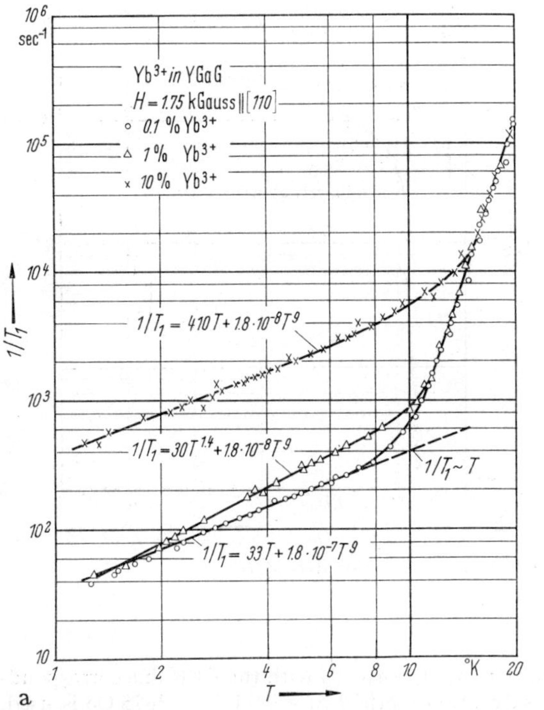

a

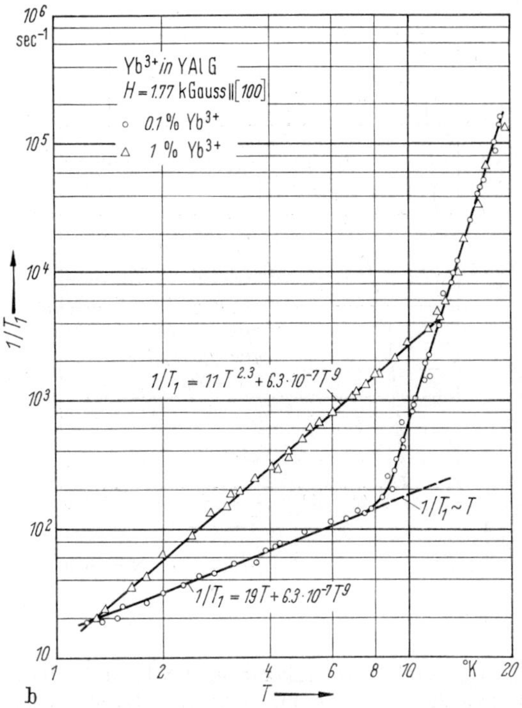

b

Fig. 56a, b, c. Yb^{3+} in some garnets. Relaxation rate, T_1^{-1}, vs. temperature [*65H*].

a) Yb^{3+} in YGaG. The deviation from $T_1^{-1} \sim T$ in the low temperature region for samples with 1% Yb^{3+} can be interpreted as arising from an additional contribution to the relaxation rate from exchange coupled pairs [*28*]. This is supported by the results for samples with 10% Yb^{3+}.

b) Yb^{3+} in YAlG. The $T^{2.3}$ term for 1% Yb^{3+} again may arise from exchange coupled pairs. The data are in reasonable agreement with the data at 4.2 °K in Tab. 26.

c) Yb^{3+} in LuGaG. The first term $T^{1.7}$ is again interpreted as arising from exchange coupled pairs.

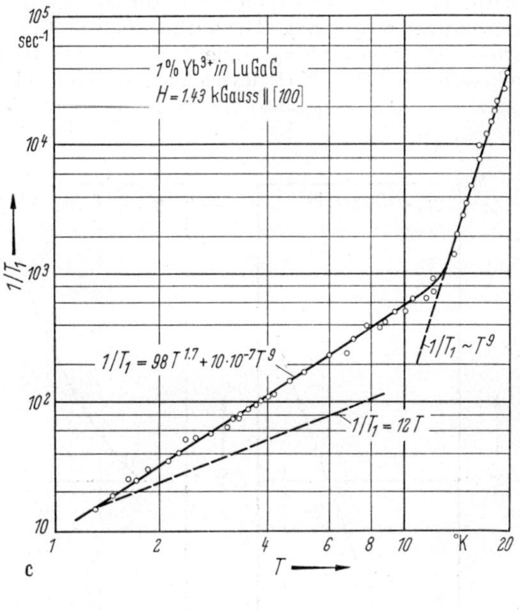

c

Tab. 26. Longitudinal relaxation time, T_1, in YAlG as a function of Yb^{3+} concentration at 4.2 °K. T_1 at 1.8 °K was 2⋯4 times longer than at 4.2 °K. The data were taken from measurements of the line width as a function of microwave power [*60CW*]

Yb^{3+} %	line width Gauss	T_1 sec
2	98	0.001
1	61	0.005
0.75	43	0.006
0.5	32	0.01
0.25	30	0.01
0.1	20	0.02

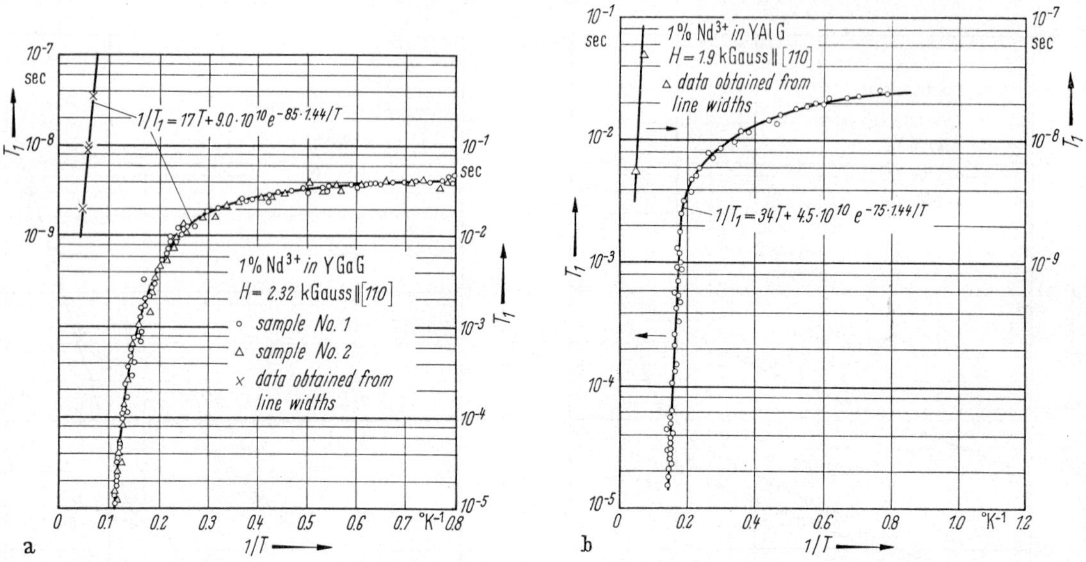

Fig. 57a, b. 1% Nd^{3+} in a) YGaG and b) YAlG. Relaxation data [*65H*].

In [*68BDM*] the relaxation of Nd^{3+} in YAlG was investigated at 4.2 °K with the EPR line corresponding to the g-tensor component $g_2 = 1.158$. A pulsed saturation method in a field $H = 3875$ Oe is used. The observed relaxation curves were of the double-exponential type. The characteristic times τ_1 and τ_2 of the fast and slow exponentials turned out to depend strongly on the temperature and on the Nd^{3+} concentration (Fig. 58). The authors believe that some anomalies (temperature dependence of the strong cross relaxation time τ_1, anomalous concentration dependence of τ_2) are connected with effects of cross relaxation between Nd^{3+} levels and excited levels of extraneous impurity ions.

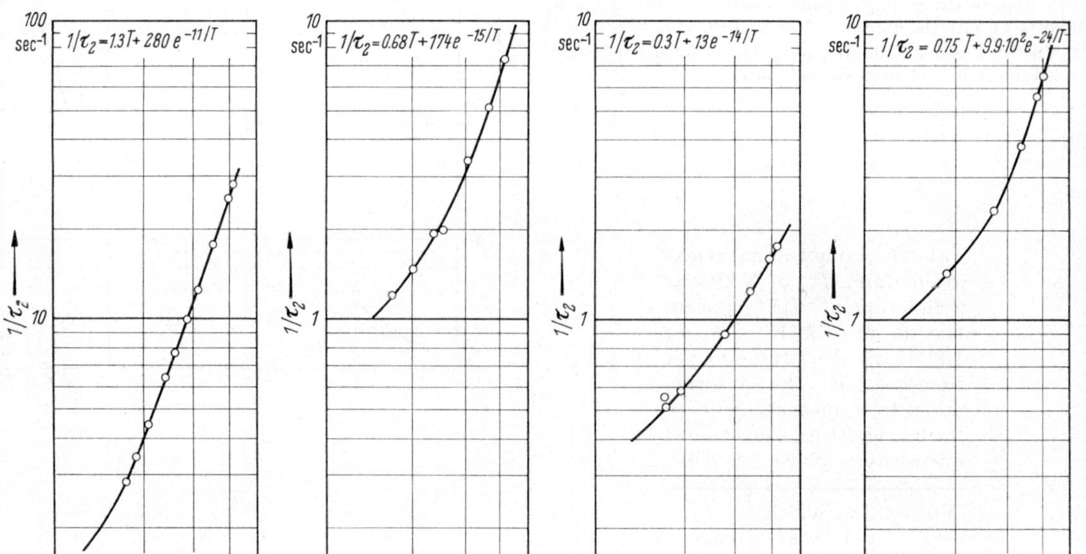

Fig. 58. Nd^{3+} in YAlG. Temperature dependence of the relaxation rates τ_2^{-1} for samples with different Nd^{3+} concentrations [*68BDM*].

Tab. 27. Comparison of measured data and theoretical [a]) estimates of the spin-lattice relaxation rates at temperatures $T \lesssim 30$ °K for rare-earth ions in diamagnetic garnets
Δ: energy splitting of the first excited state

ion	concentration	host garnet	H_a		direct process	Orbach process	Raman process	Δ [d])	Ref.
	%		kGauss		T^{-1} in sec^{-1}			cm^{-1}	
Nd^{3+}	1	YGaG	2.32 [110]	exp.	17 T	$9 \cdot 10^{10} \exp(-85 \cdot 1.44/T)$		85	*65H*
				theor.	10 T	$10^{12} \exp(-85 \cdot 1.44/T)$			
	1	YGaG		exp.	17 T	$9 \cdot 10^{10} \exp(-85 \cdot 1.44/T)$		85	*63SS*
	1	YAlG	1.9 [110]	exp.	34 T	$4.5 \cdot 10^{10} \exp(-75 \cdot 1.44/T)$			*65H*
				theor.	10 T	$10^{11} \exp(-75 \cdot 1.44/T)$			
	0.18	YAlG	3.875	exp.	1.3 T	$280 \exp(-11/T)$			*68BDM*
	0.41		3.875	exp.	0.68 T	$174 \exp(-15/T)$			
	0.61		3.875	exp.	0.3 T	$13 \exp(-14/T)$			
	2.5		3.875	exp.	0.75 T	$990 \exp(-24/T)$			
Dy^{3+}	1	YAlG [c])		exp.	10 T	$2.4 \cdot 10^{9} \exp(-53 \cdot 1.44/T)$		53	*63SS*
	1	YGaG		exp.	440 T	$1.1 \cdot 10^{7} \exp(-14 \cdot 1.44/T)$		14	*63SS*
Er^{3+}	1	YAlG		exp.	20 T	$3.2 \cdot 10^{6} \exp(-20 \cdot 1.44/T)$		20	*63SS*
	1	LuAlG		exp.	10 T	$6 \cdot 10^{7} \exp(-36 \cdot 1.44/T)$		36	*63SS*
	1	YGaG		exp.	70 T	$1.1 \cdot 10^{9} \exp(-46 \cdot 1.44/T)$		46	*63SS*
	1	LuGaG		exp.	50 T	$1.4 \cdot 10^{9} \exp(-47 \cdot 1.44/T)$		47	*63SS*
Yb^{3+}	0.1	YGaG	1.75 [100]	exp.	33 T		$1.8 \cdot 10^{-7} T^9$		*63SS, 65H*
				theor.	4.2 T		$5.7 \cdot 10^{-7} T^9$		*65H*
	1	YGaG		exp.	30 $T^{1.4}$ [b])		$1.8 \cdot 10^{-7} T^9$		*63SS, 65H*
	10	YGaG		exp.	410 T		$1.8 \cdot 10^{-7} T^9$		*65H*
	0.1	YAlG	1.77 [100]	exp.	15 T		$6.3 \cdot 10^{-7} T^9$		*65H*
				theor.	5.3 T		$9.0 \cdot 10^{-7} T^9$		
	1	YAlG		exp.	11 $T^{2.3}$ [b])		$6.3 \cdot 10^{-7} T^9$		*65H*
	1	LuGaG	1.43 [100]	exp.	9.8 $T^{1.7}$ [b])		$1.0 \cdot 10^{-7} T^9$		*65H*
				theor.	3.1 T		$3.2 \cdot 10^{-7} T^9$		

[a]) [*65H*] gives theoretical expressions for the spin-lattice relaxation rate for Kramers salts of XY_8 symmetry for the direct process, the Raman process and the Orbach process.
[b]) The deviation from $T_1^{-1} \approx T$ is interpreted as arising from an additional contribution to the relaxation from exchange coupled pairs.
[c]) In [*65NW*] "effective" relaxation times for DyAlG are determined from Mössbauer experiments: $\geq 10^{-8}$ sec at 4 °K; $\approx 3 \cdot 10^{-9}$ sec at 20 °K.
[d]) Determined from the Orbach process.

II. Rare earth ions in paramagnetic garnets

Yb^{3+} in TmAlG and TmGaG [*65BR*]

A sufficiently strong exchange contribution from the Tm^{3+} impurity magnetic-exchange interaction can cause a large change in both the magnitude and the temperature dependence of the relaxation rate, compared to that found in similar diamagnetic garnets.

The best fit for the high-temperature portion of the relaxation rates of Yb^{3+} in TmAlG and TmGaG is given by an exponential temperature dependence (Orbach process); this is in contrast with the results obtained for Yb^{3+} in YGaG and YAlG (Tab. 27) where $T_1^{-1} \approx T^9$, indicating a Raman process.

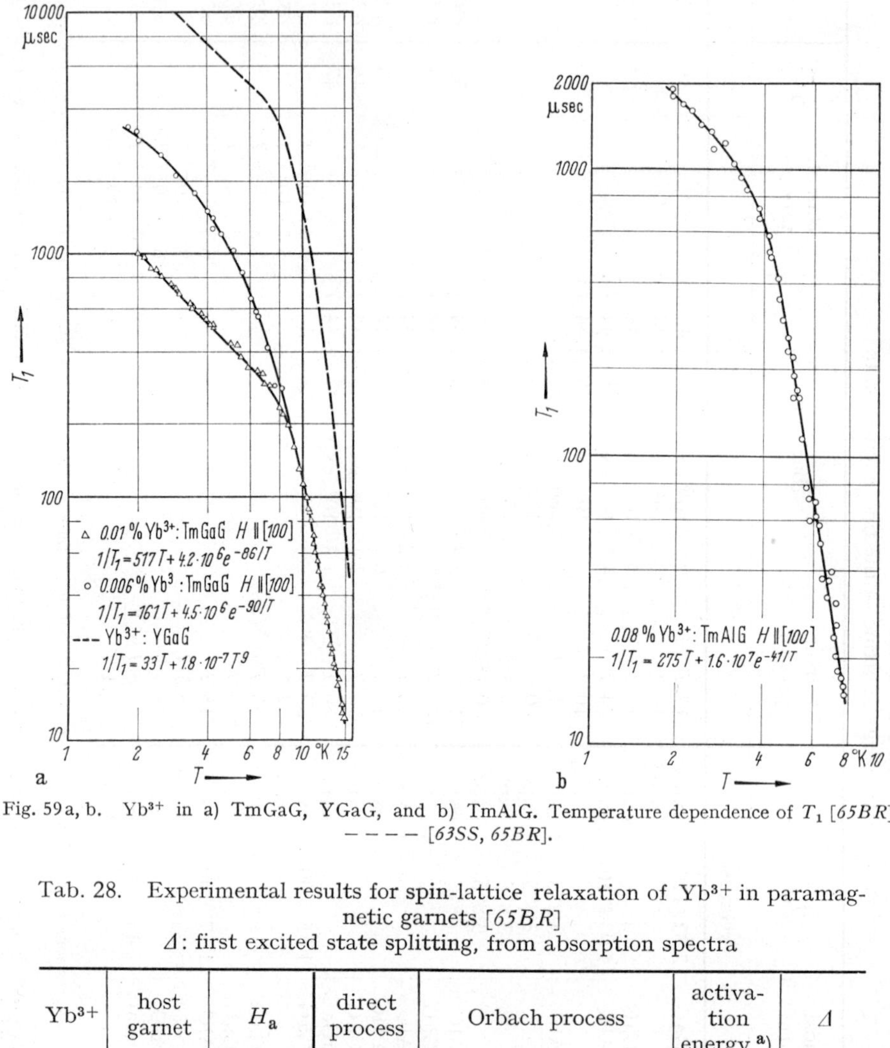

Fig. 59a, b. Yb^{3+} in a) TmGaG, YGaG, and b) TmAlG. Temperature dependence of T_1 [*65BR*].
– – – – [*63SS, 65BR*].

Tab. 28. Experimental results for spin-lattice relaxation of Yb^{3+} in paramagnetic garnets [*65BR*]
Δ: first excited state splitting, from absorption spectra

Yb^{3+}	host garnet	H_a	direct process	Orbach process	activation energy a)	Δ
%			T_1^{-1} in sec^{-1}		cm^{-1}	cm^{-1}
0.01	TmGaG	∥ [100]	550 T	$8.9 \cdot 10^6 \exp(-75/T)$	60 ± 10	63 ± 5
0.006	TmGaG	∥ [100]	161 T	$4.5 \cdot 10^6 \exp(-90/T)$	60 ± 10	
0.01	TmAlG		200 T	$2 \cdot 10^7 \exp(-45/T)$	30 ± 6	35 ± 3
0.08	TmAlG		275 T	$1.6 \cdot 10^7 \exp(-41/T)$	30 ± 6	

III. Fe^{3+} ions in diamagnetic and paramagnetic host garnets

For Fe^{3+} in YGaG and LuAlG, only a direct process in the indicated temperature region was found. The measurements on Fe^{3+} in TmAlG and TmGaG indicate the presence of Fe^{3+}-Tm^{3+} interaction effects [*64RBa*].

a) Determined from the Orbach process.

Tab. 29. Expressions for T_1^{-1} obtained by fitting the experimental data of the temperature dependence of the spin-lattice relaxation time for Fe^{3+} in various host garnets [*64RBa*]
ΔW: splitting between the ground and first excited state; from relaxation data

Fe^{3+}	site	host garnet	transition	direct process	Orbach process	T	ΔW/hc
%				T_1^{-1} in [sec^{-1}]		°K	cm^{-1}
0.1	*a*	TmAlG	5/2 → 3/2	$5 \cdot 10^2\,T$	$10^{10}\exp(-55/T)$	$1.9 \leq T \leq 5$	39 ± 5
0.1	*d*	TmAlG	1/2 → −1/2	$1.96 \cdot 10^2\,T$	$6.5 \cdot 10^8\exp(-57/T)$	$1.9 \leq T \leq 8$	39 ± 5
0.1	*a*	TmGaG	5/2 → 3/2	$7.8 \cdot 10^2\,T$	$3.47 \cdot 10^8\exp(-82/T)$	$3 \leq T \leq 10$	58 ± 5
0.1	*a*	TmGaG	3/2 → 1/2	$0.574 \cdot 10^2\,T$	$2.2 \cdot 10^8\exp(-84/T)$	$3 \leq T \leq 11$	58 ± 5
≈0.1	*a*	YGaG	5/2 → 3/2	$1.4 \cdot 10^4\,T$		$1.9 \leq T \leq 15$	
<0.1	*a*	LuAlG	5/2 → 3/2	$1.08 \cdot 10^2\,T$		$1.9 \leq T \leq 30$	

IV. High temperature electronic relaxation at the rare earth ions in garnets [*66CR*]

The relaxation time in this case was measured by the attenuation of the angular correlation of a nuclear gamma-gamma cascade [*38*]. The temperature dependence of the relaxation time of the Tb^{3+} ion in LuGaG, LuAlG (nonmagnetic) and LuIG (magnetic) is shown in Fig. 60 in terms of $\lambda \cdot \tau$, where τ is the life time of the intermediate nuclear state. The relaxation times ($\approx 10^{-11}$ sec) do not vary with temperature within the statistical error of the experiment.

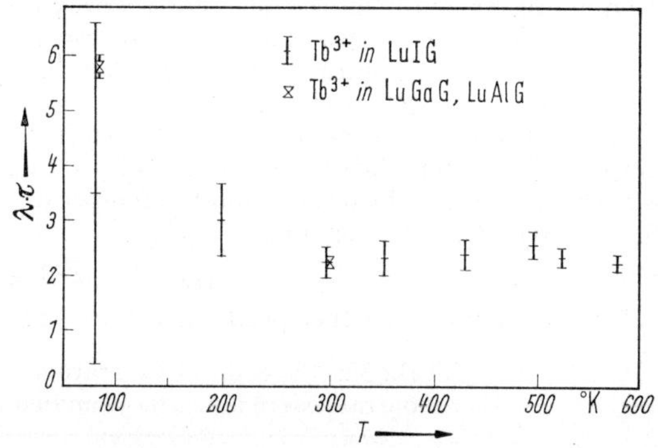

Fig. 60. Tb^{3+} in magnetic LuIG and in nonmagnetic garnets, LuGaG and LuAlG. Electronic relaxation time, $\lambda \cdot \tau$, vs. temperature, T [*66CR*].

B. The dependence of the spin-lattice relaxation time on magnetic field

In [*66RBS*] it is shown that for a number of Kramers doublet rare earth ions, at frequencies above 70 Gc/s and at liquid Helium temperatures (1.8⋯4.2 °K), the only important relaxation mechanism is the direct spin-phonon interaction due to modulation of the crystalline field by lattice vibration [31].

There are theoretical grounds [30] to expect the transition rate T_1^{-1} for a Kramers doublet do depend on magnetic field as

$$T_1^{-1} = k_5 \cdot H^5 \cdot \coth\left(\frac{g\,\mu_B H}{2kT}\right).$$

In the samples examined this field dependence was found (Tab. 30). Below 10 kOe, as in lower frequency measurements, there is evidence that concentration dependent mechanisms with a different field dependence dominate the spin-lattice relaxation [32, 39].

Tab. 30. Experimental values for the relaxation coefficient, k_5, of rare earth doublets from pulsed field EPR spectra. Experimental range: $1.8\ °K \leq T \leq 4.2\ °K$, $10\ kOe \leq H_a \leq 60\ kOe$ [*66RBS*]
T_1(exp.): measured at the X-band; 10 Gc/s; 1.85 °K

ion	host	g [a]	k_5 10^{-3} sec^{-1} kOe^{-5}	T_1^{-1}(calc.) sec^{-1}	T_1^{-1}(exp.)
Yb^{3+}	YGaG	3.7	0.25	0.0065	≈ 61 [*63SS*]
		3.26	0.20	0.10	
	YAlG	3.85	0.33	0.007	≈ 26 [*65H*]
		3.2	0.29	0.016	
Er^{3+}	YAlG	7.3	(9.0	0.008	≈ 37 [*63SS*]
		4.0	(6.0)	0.1	

There is a large discrepancy in the results for T_1^{-1}. According to [*66RBS*] it may be possible that in previous measurements at the X-band, relaxation rates were mashed by complicating effects related to cross relaxation or spin-diffusion processes.

a) $H \parallel [110]$.

Tab. 31. Using experimental values for k_5 one can obtain values for the matrix elements combinations of V_+ (dynamical crystal field) and J_H (component of the angular magnetic moment parallel to the magnetic field) from

$$k_5 = 1.553 \cdot 10^{22}\, \varrho^{-1} V^{-1} g^3 H^5 \wedge^2 \sum_i \Delta_i^{-2} |\langle 0 - | J_H | i + \rangle \times \langle i + | V | 0 + \rangle|^2 \coth (g \mu_B H/2kT)$$

$$= 1.553 \cdot 10^{22}\, \varrho^{-1} V^{-1} g^3 H^5 \wedge^2 \sum_i (\Delta_i^{-2} M_i^2) \coth (g \mu_B H/2kT) \quad [66RBS]$$

host	ϱ *) g cm^{-3}	$\langle v^5 \rangle^{1/5}$ *) 10^5 cm sec^{-1}	ion	g	$\Sigma (M_i^2 \Delta_i^{-2})^{1/2}$
YGaG	5.79	4	Yb^{3+}	3.7	3.9
				3.26	4
YAlG	4.55	5	Yb^{3+}	3.85	2
				3.2	2.5
			Er^{3+}	7.3	4
				4.0	8

5.7 Thermal properties near the transition temperature

A. Specific heat of rare earth gallium garnets [*67OMR*]

Calorimetric studies from 0.35 to 4.2 °K show rounded specific heat maxima of a few millidegree for all samples. There are indications that the way of preparation of the garnets effects the properties of the sample. Smoothed values of the data are presented in Tab. 32. The magnetic specific heat C_{magn} is obtained by subtracting the lattice contribution $C_{lat} = B \cdot T^3$ with $B = 2.8 \cdot 10^{-5} R$ °K^{-3}. **)

I. SmGaG, ErGaG, and NdGaG.

The transition temperature T_N is defined in such a way that the specific heat plotted against the log of the relative temperature difference $|(T - T_N)/T_N|$ gives two "parallel" straight lines over a certain temperature region which correspond to temperatures above and below T_N. Such a plot is shown in Fig. 61 for SmGaG. The parameters of the plot

$$C_{magn}/R = a \cdot \log_{10} |1 - T/T_N| + \Delta_\pm$$

and T_{max} (maximum of the specific heat) are presented in Tab. 33. See also Tab. 3.

Tab. 32. Specific heat of rare earth gallium garnets expressed in C/R per mole rare earth ion. The Roman numerals are sample numbers [*67OMR*]

T	C/R						
°K	Nd	Sm (II)	Gd (II)	Er	Dy	Ho	Yb (II)
0.35		0.0196	0.54	0.0621	0.551	0.0634	
0.4	0.583	0.0329	0.606	0.1105	0.364	0.0493	
0.45	0.911	0.0534	0.668	0.1770	0.270	0.0383	0.136
0.5	1.565	0.0868	0.734	0.2636	0.241	0.0313	0.1205
0.55	0.408	0.1340	0.784	0.385	0.225	0.0271	0.1060
0.6	0.327	0.2005	0.826	0.535	0.215	0.0248	0.0930
0.65	0.287	0.293	0.854	0.736	0.210	0.0243	0.0815
0.7	0.257	0.413	0.873	1.009	0.209	0.0253	0.0715
0.75	0.233	0.563	0.881	1.43	0.208	0.0281	0.0634
0.8	0.214	0.754	0.880	0.678	0.207	0.0323	0.0570
0.85	0.198	1.014	0.871	0.412	0.203	0.0376	0.0515
0.9	0.185	1.386	0.858	0.341	0.200	0.0439	0.0467
0.95	0.177	2.073	0.840	0.306	0.195	0.0516	0.0425
1.0	0.152	0.642	0.820	0.281	0.190	0.0662	0.0391
1.2	0.109	0.320	0.720	0.213	0.170	0.105	0.0289
1.5	0.071	0.205	0.567	0.150	0.143	0.182	0.0201
2.0	0.0402	0.116	0.385	0.0942	0.103	0.311	0.01320
2.5	0.0261	0.075	0.272	0.0641	0.0782	0.387	0.00955
3	0.0183	0.0527	0.202	0.0459	0.0648	0.420	0.00738
3.5	0.0139	0.0395	0.156	0.0357	0.0603	0.415	0.00622
4.0	0.0115	0.0308	0.126	0.0289	0.0676	0.395	0.0055
4.2	0.0109	0.0286	0.118	0.0271		0.387	0.0054

*) Experimental values of density, ϱ, and sound velocity, v, from [*33*].

**) The contribution of the nuclear specific heat C_n resulting from electron-nucleus interaction was calculated to be negligible in comparison with C_{magn} above 0.35 °K, except for HoGaG. For ErGaG, for example, $C_n T^2/R \approx 10^{-3}$ °K^2 is estimated.

Fig. 61. SmGaG. Magnetic specific heat of two samples plotted as C_{magn}/R vs. $\log_{10}|1 - T/T_N|$. SmGaG (I) was sintered, SmGaG (II) consisted of several single crystals [67OMR].

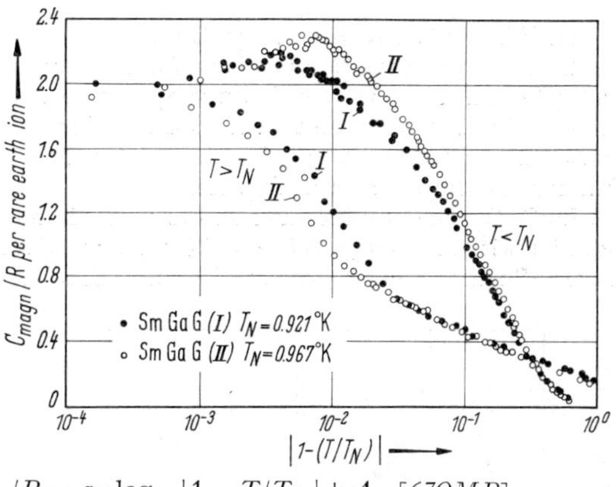

Tab. 33. Summary of λ-point analysis $C_{magn}/R = a \cdot \log_{10}|1 - T/T_N| + \Delta_{\pm}$ [67OMR]
T_N and T_{max} correspond to T_N (C_{magn}) and T_{max} (C_{magn}) of Tab. 3
$\Delta_{\pm}$: values for fitting the data above and below T_N

Sample	a	Δ_+	Δ_-	$\Delta_+ - \Delta_-$	T_N [°K]	T_{max} [°K]
SmGaG (I)	−1.2	−1.2	−0.2	1.0	0.921	0.918
SmGaG (II)	−1.5	−2.2	−0.5	1.7	0.967	0.960
ErGaG	−1.25	−1.8	−0.2	1.6	0.790	0.786
NdGaG	−1.2	−1.8	−0.2	1.6	0.516	0.514

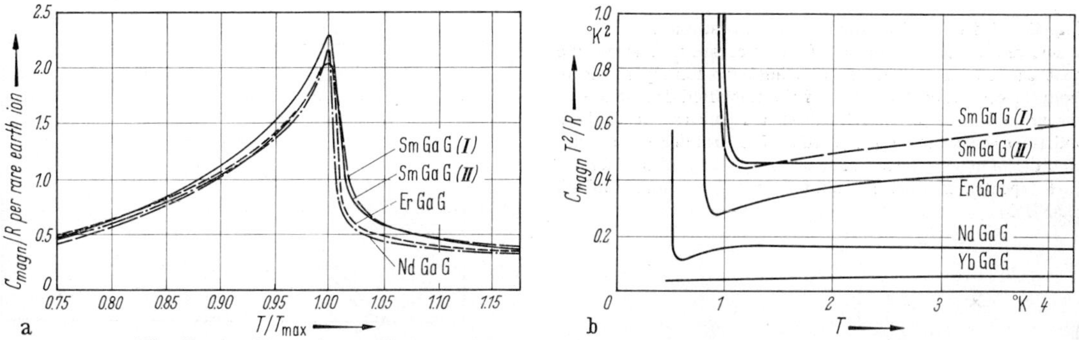

Fig. 62a, b. Magnetic specific heat of SmGaG, ErGaG, NdGaG, and YbGaG. [67OMR].
a) C_{magn}/R vs. T/T_{max}. b) $C_{magn}T^2/R$ vs. T.

Tab. 34. The specific heat for $T/T_N < 0.5$ follows closely a T^3 dependence. So it is extrapolated from $T = 0.35$ °K to $T = 0$ °K assuming such a dependence. At high temperatures, the specific heat is taken to be proportional to T^{-2} and the limiting entropy S_∞ is found to be close to $R \cdot \ln 2$, the value for a system with spin $\frac{1}{2}$.
The entropy and magnetic energy for the three compounds were computed *) for an Ising spin $\frac{1}{2}$ system and are also shown for a Heisenberg model spin $\frac{1}{2}$ [67OMR]

Sample	$\frac{S_\infty - S_0}{R}$	$\frac{S_\infty - S_c}{R}$	$\frac{S_c - S_0}{R}$	$\frac{S_\infty - S_c}{S_c - S_0}$	$\frac{E_\infty - E_0}{R T_N}$	$\frac{E_\infty - E_c}{R T_N}$	$\frac{E_c - E_0}{R T_N}$	$\frac{E_\infty - E_c}{E_c - E_0}$
NdGaG	0.72	0.264	0.46	0.58	0.91	0.554	0.36	1.58
SmGaG (I)	0.699	0.304	0.395	0.77	0.944	0.625	0.319	1.96
SmGaG (II)	0.689	0.272	0.417	0.65	0.869	0.526	0.343	1.53
ErGaG	0.700	0.271	0.429	0.63	0.923	0.582	0.341	1.70
Heisenberg (fcc) ferromagnet $S = \frac{1}{2}$ a)	0.693	0.265	0.428	0.62	0.736	0.439	0.297	1.48
as above but corrected as antiferromagnet $S = \frac{1}{2}$	0.693				0.785			
Ising (fcc) ferromagnet $S = \frac{1}{2}$ a)	0.693	0.107	0.586	0.18	0.629	0.169	0.460	0.37
DyAlG b)	0.693	0.204	0.489	0.42	0.612	0.213	0.399	0.53

a) [1]; b) [63BLWW].
*) Using the notation of [49]. For comparison, the calculated values for the entropy, S_c, and energy, E_c, at the Néel temperature.

II. RGaG with R = Yb, Dy, Gd, and Ho.

YbGaG: No transition has been found above 0.35 °K. The magnetic specific heat rises monotonically as the temperature is lowered. At temperature above 2.5 °K one obtains $C_{magn} T^2/R = 0.061$ °K². The constancy of this ratio confirms the absence of any excited level for energies of less than about 100 cm⁻¹ as shown by susceptibility [*60BGLW*] and optical data [*60PW*]. The ratio $C_{magn}\,T^2/R$ is plotted as a function of temperature in Fig. 62b.

DyGaG: The specific heat shows a maximum at about 0.35 °K (Fig. 64). The effect of an excited doublet with an energy $\Delta E/k = 34$ °K was substracted from the total C_{magn}. Possibly impurities may have influenced the experimental results [*67OMR*].

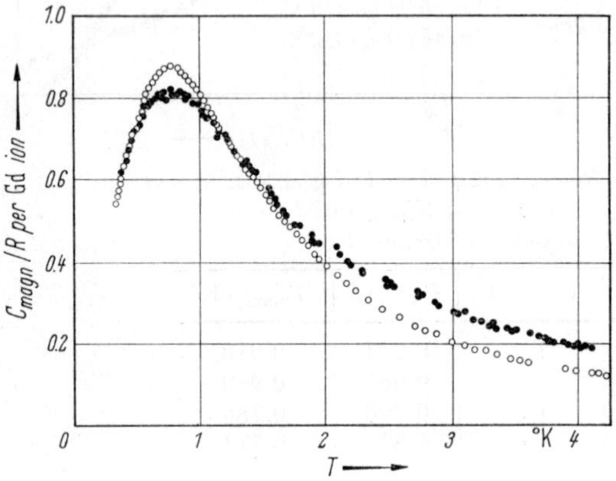

Fig. 63. GdGaG. Specific heat of two samples. • sintered sample, ○ cluster of single crystals. The specific heat can be interpreted as a Schottky anomaly resulting from the crystal field splitting of the eight-fold degeneracy of the ground state [*67OMR*]. From a rough estimate, these splittings appear to be of the order of several degrees Kelvin, an order of magnitude larger than the splittings obtained from ESR spectra in GdGaG diluted in YGaG and LuGaG [*63OGFC*].

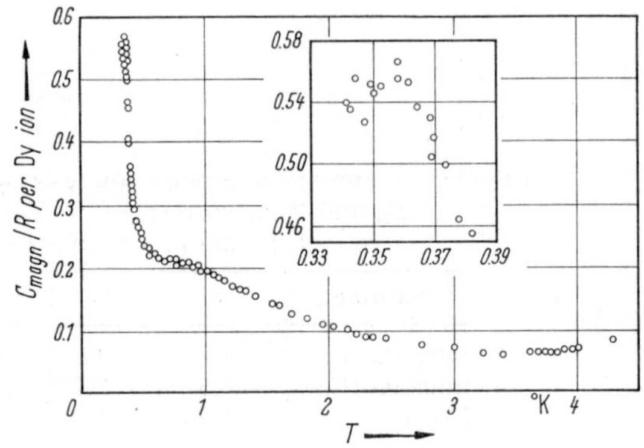

Fig. 64. DyGaG. Specific heat vs. T [*67OMR*].

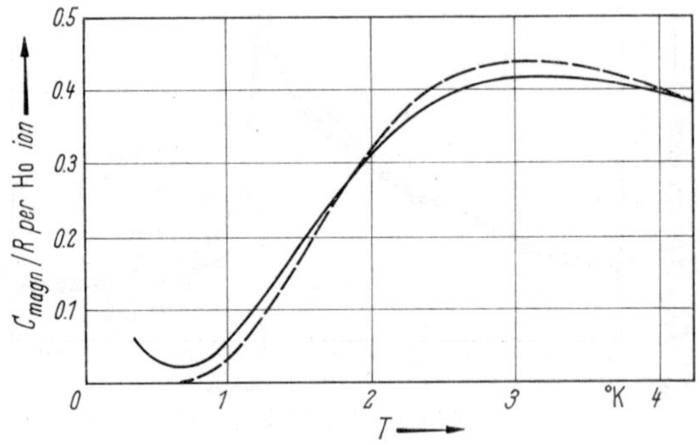

Fig. 65. HoGaG. Specific heat C_{magn}. The solid line is a smoothed reproduction of the data, which scatter around this line by about 1%. The dashed line represents a calculated Schottky anomaly of two electronic singlet levels separated by an overage energy of $\Delta E/k = 7.4$ °K. There might be a certain distribution of splittings through the lattice since the fit of a simple Schottky anomaly to the experimental points is not perfect [*67OMR*].

B. Specific heat of two rare earth aluminium garnets

I. DyAlG.

A sharp λ peak confirms the onset of long range order at (2.49 ± 0.01) °K [*63BLWW*]. The results for specific heat measurements between 1.2 and 4 °K are shown in Fig. 66. Measurements were made on a 21 g polycrystalline block sintered to a high density.

The temperature of the maximum (2.49 °K) is to be compared with the point at which $\partial(\chi T)/\partial T$ is a maximum. A comparison between experiments on the same sample is given in Fig. 24.

High-resolution measurements [*67KLW*] of the specific heat of three samples of DyAlG in the critical point region of 2.5 °K, using large optical-quality single crystals and heating steps down to 50 μdeg, in all cases show rounded maxima 2 to 3 mdeg wide, with peak values between 4.2 and 5.1 R. Below these maxima, the results can be fitted to simple laws of the form

$$C = A \cdot \log|T - T_N| + B$$

(Fig. 67), but above the maxima a better fit can be obtained using a law of the form

$$C = A'(T - T_N)^{-\alpha}$$

with $T_N = 2.537$ °K and $\alpha = 0.31 \pm 0.02$ (Fig. 68).

Fig. 66. $Dy_3Al_5O_{12}$. Specific heat vs. temperature. Curves (a) and (b) were calculated using the molecular field approximation, with the constant for (a) estimated on the basis of pure magnetic dipole-dipole interaction, and for (b) by fitting the low-temperature experimental points. The small bump at 3.4 °K is due to a small (≈ 1%) "impurity" of the perovskite phase $DyAlO_3$ [63BLWW]. →

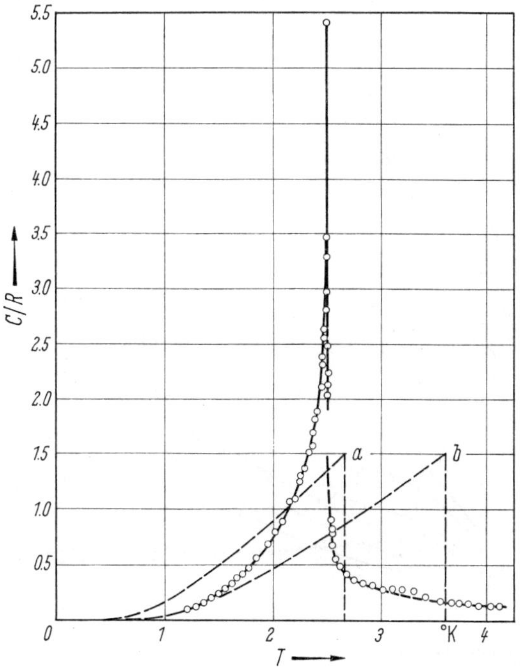

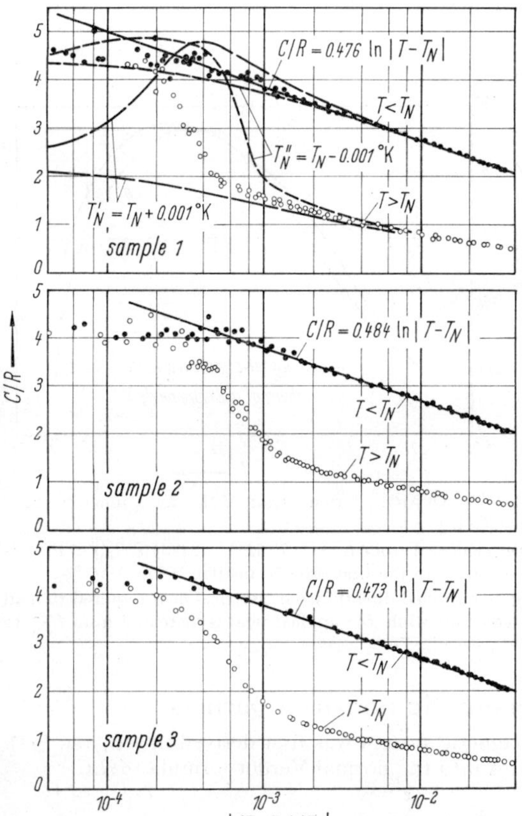

Fig. 67. DyAlG. C/R vs. $\log|(T - T_N)/T_N|$ with T_N = 2.537 °K. The broken lines in the top graph show the effect of changing T_N by ±0.001 °K [67KLW].

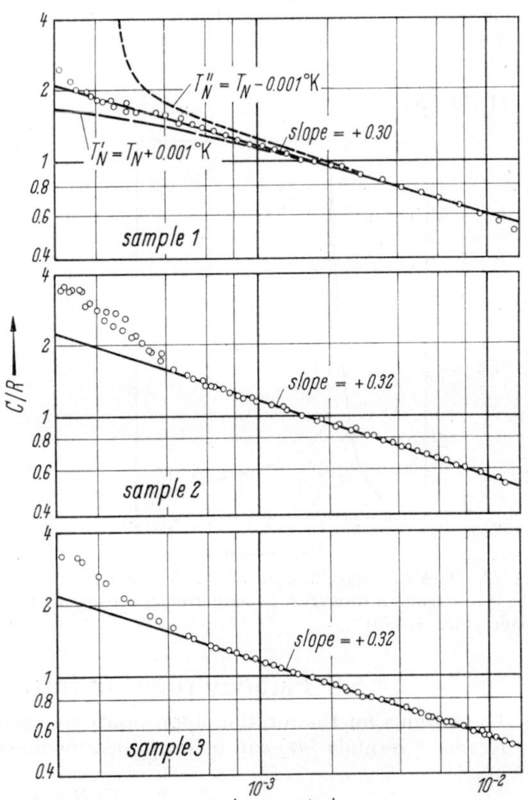

Fig. 68. DyAlG. Plot of $\log(C/R)$ vs. $\log|(T - T_N)/T_N|$ for $T > T_N$ = 2.537 °K. The broken lines in the top graph show the effect of changing T_N by ±0.001 °K [67KLW].

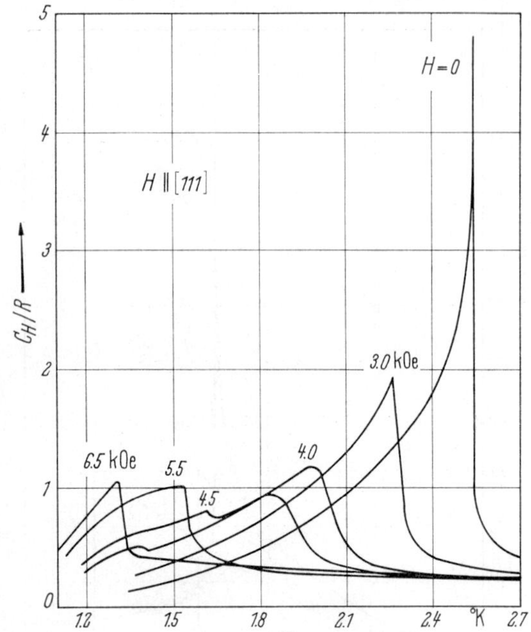

Fig. 69. DyAlG. Specific heat, C_H, as a function of T in different constant external magnetic fields H along [111] axis [*66KLSW*]. At temperatures above 1.66 °K slightly rounded λ-type peaks were found which indicate the paramagnetic antiferromagnetic transition. Below 1.66 °K, broad peaks with fairly sharp high-temperature edges indicate the transition into the mixed antiferromagnetic-paramagnetic region (Fig. 70).

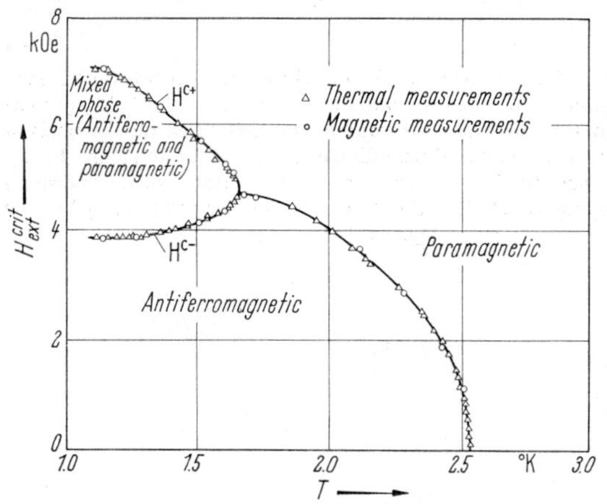

Fig. 70. DyAlG. Magnetic phase diagram for H along a [111] axis. The lines marked H^{c+} and H^{c-} show the phase boundaries calculated from the internal field results for the case of a demagnetizing factor $N = 5.35$, which corresponds to the ellipsoidal sample shape [*66KLSW*]. Close to T_N the specific-heat peaks determine the phase boundary rather accurately, the latter was fitted to $H_{int}^{crit} = A(1 - T/T_N)^n$ with $n = 0.50 \pm 0.02$, $A = 6.52$ kOe and $T_N = (2.53_5 \pm 0.01)$ °K. See also Fig. 32.

II. TbAlG.

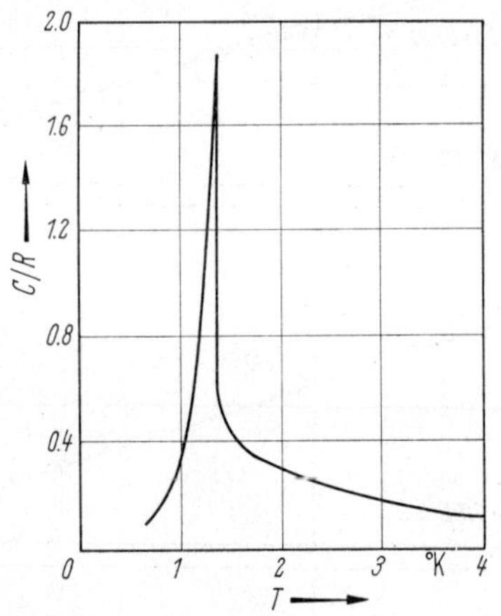

Fig. 72. TbAlG. Specific heat vs. T. The λ anomaly at 1.35 °K indicates a transition to antiferromagnetism at this temperature [*67CTW*].

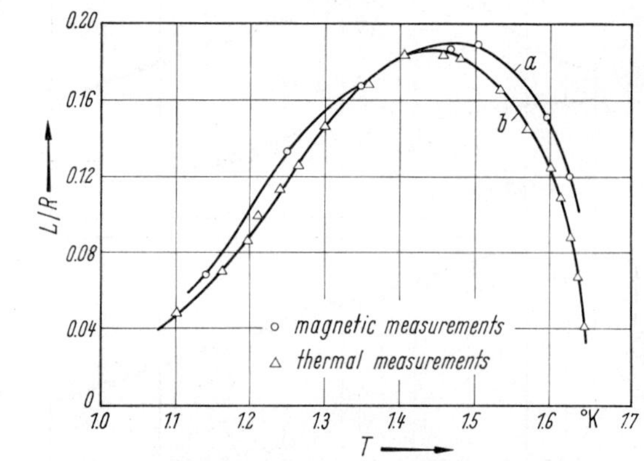

Fig. 71. DyAlG. Latent heat, L/R, as a function of T [*66KLSW*].
(a) Calculated from $L = -T(M^{c+} - M^{c-})(\partial H_{int}^{crit}/\partial T)$ corresponding to the first order transition below 1.66 °K.
(b) Determined directly from electrical heat required to hold T constant while the applied field is increased from H^{c-} to H^{c+}; see Fig. 70.

5.8 Faraday rotation in paramagnetic non-iron garnets

The formula for the rotation appropriate to a paramagnetic system was first derived in [*40*] and [*41*]. Kramers' formula [*41*] can be simplified for most cases into the normal Verdet formula [*68LC*]

$$\theta = V\, l\, \boldsymbol{H}\, \boldsymbol{e} \text{ with } V = K\,\frac{\nu^2}{\nu^2 - \nu_{ng}^2}$$

where θ is the rotation angle, ν the frequency of the incident radiation, ν_{ng} the frequency for the transition between the ground state g and an intermediate state n, l is the length of the medium, $\boldsymbol{H}$ is the strength of the applied field, $\boldsymbol{e}$ the unit vector in direction of propagation, V is the Verdet constant.

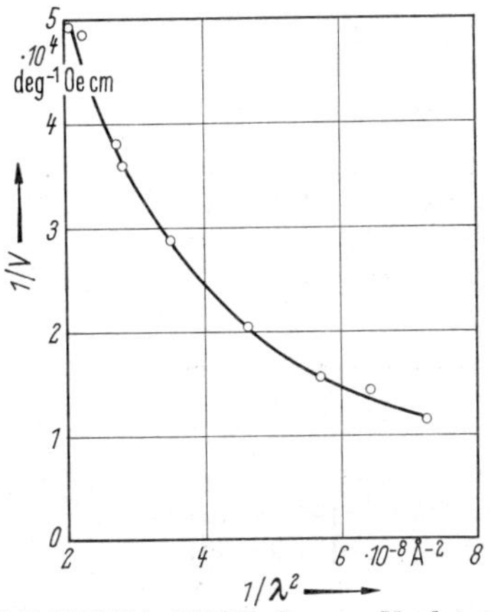

Fig. 73. Er^{3+} (9.7%) in YAlG. Inverse Verdet constant vs. $1/\lambda^2$ at about 20 °K. Magnetic fields up to 7 kGauss and wavelengths 3000 ··· 8000 Å were used. $\nu_{ng} = (62 \pm 4) \cdot 10^3$ cm^{-1} is determined from the plot [68LC].

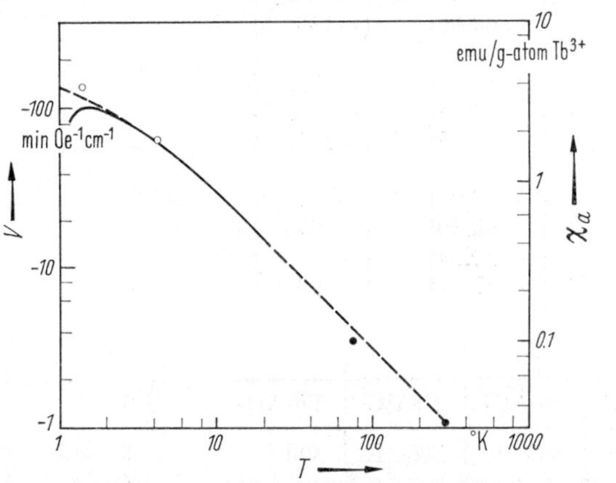

Fig. 75. TbAlG. Verdet constant as a function of temperature showing the results of [64RUG] (●) and [67S] (○) for the wavelength $\lambda = 5200$ Å.
The solid curve gives magnetic susceptibility data for TbAlG [62WBHLW]. The dashed curve is a Curie-Weiss law with a Curie constant of $C_A = 9.1$ °K emu/g-atom and $\Theta_p = -1.3$ °K. The Verdet constant exhibits the same temperature dependence as the magnetic susceptibility, as predicted in [42].

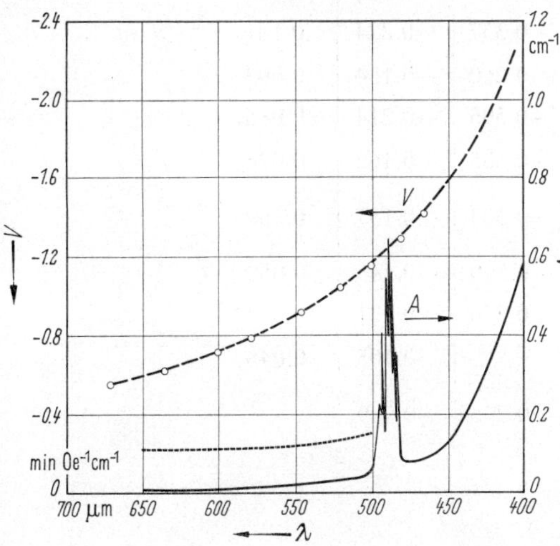

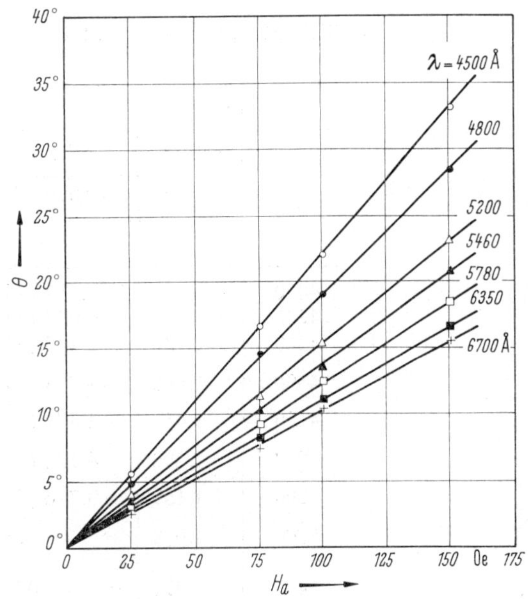

Fig. 74. TbAlG. The rotation of the plane of polarisation as a function of applied transverse magnetic field. T = 1.45 °K. Parameter: wavelengths of light in the visible range. "Effective" light path thickness equals approximately 0.14 cm [67S].

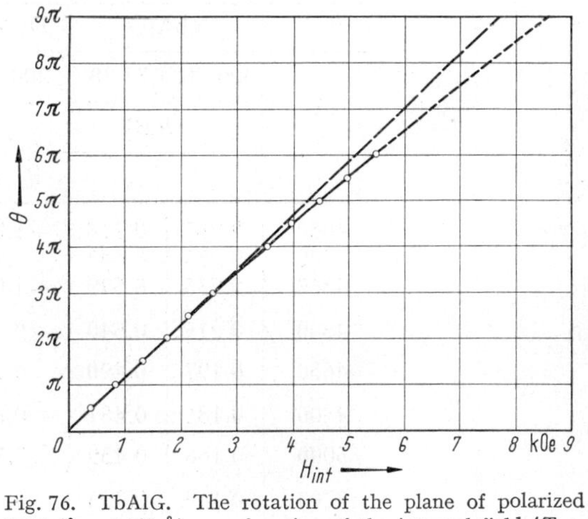

Fig. 76. TbAlG. The rotation of the plane of polarized light (λ = 5460 Å) as a function of the internal field (T = 1.45 °K), showing the onset of magnetic saturation [67S].

←

Fig. 77. TbAlG. Absorption coefficient, A, and Verdet constant, V, at room temperature as a function of wavelength, λ. The dotted curve represents the possible upper limit of error for the absorption measurements [64RUG].

Tab. 35. Measured Verdet constant of several rare earth aluminium garnets

λ Å	TbAlG 300 °K [64RUG]	TbAlG 77 °K [64RUG]	TbAlG 4.22 °K [67S]	TbAlG 1.45 °K [67S]
	V [min Oe^{-1} cm^{-1}]			
4050	−2.256	−6.717 (86 °K)		
4200	−1.928	−6.044 (83 °K)		
4360	−1.730	−5.440 (83 °K)		
4500	−1.565	−4.974 (80 °K)	{− 65.25, −102.16} *)	{− 97.10, −200.95} *)
4650	−1.417	−4.471 (80 °K)		
4800	−1.290	−4.121 (79 °K)	{− 53.30, − 83.45}	{− 83.37, −172.52}
5000	−1.159	−3.716 (79 °K)		
5200	−1.039	−3.425	{− 41.39, − 64.80}	{− 67.31, −139.28}
5460	−0.921	−3.051	{− 37.26, − 58.35}	{− 60.44, −125.07}
5780	−0.787	−2.603	{− 34.34, − 53.77}	{− 53.77, −111.27}
6000	−0.714	−2.349		
6350	−0.620	−2.00	{− 30.91, − 48.39}	{− 47.10, − 97.47}
6700	−0.542	−1.815	{− 30.04, − 47.04}	{− 45.15, − 93.42}

λ Å	YbAlG 300 °K [65RB]	YbAlG 77 °K [65RB]	DyAlG 300 °K [64RUG]	HoAlG 300 °K [64RUG]	ErAlG 300 °K [64RUG]	TmAlG 300 °K [64RUG]
	V [min Oe^{-1} cm^{-1}]					
4050	0.287	0.718	−1.241	−0.709	−0.189	0.151
4200	0.256	0.643	−1.127		−0.283	0.134
4360	0.233	0.579	−1.019	−0.582	−0.251	0.120
4500	0.215	0.540	−0.942	−0.320	−0.240	0.103
4650	0.197	0.490	−0.866	−0.337	−0.224	0.101
4800	0.186	0.481	−0.803	−0.260	−0.154	0.093
5000	0.166	0.439	−0.732	−0.395	−0.214	0.086_7
5200	0.149	0.393	−0.667	−0.335	−0.162	0.076_5
5460	0.133	0.342	−0.592	−0.304	−0.157	0.069
5780	0.116	0.302	−0.518	−0.299	−0.145	0.059
6000	0.107	0.272	−0.473	−0.276	−0.131	0.054
6350	0.094	0.239	−0.411		−0.105	0.048_9
6700			−0.359	−0.206	-0.089_9	

*) { $V_{uncorr.}$, based on the values of the applied external field
$V_{corr.}$, based on the values of the internal field }

5.9 Magnetomechanical properties

A. Magnetostriction of rare earth gallates and germanates with a garnet structure

The magnetostriction is caused chiefly by the interaction of the anisotropic electron cloud (which is associated with the orbital momentum of the rare earth ion) with the crystal field of the lattice [*69BSH*].

The high values for λ in the paramagnetic garnets may be explained by the approximity of the ordering point as suggested from the susceptibility data [*66BS*].

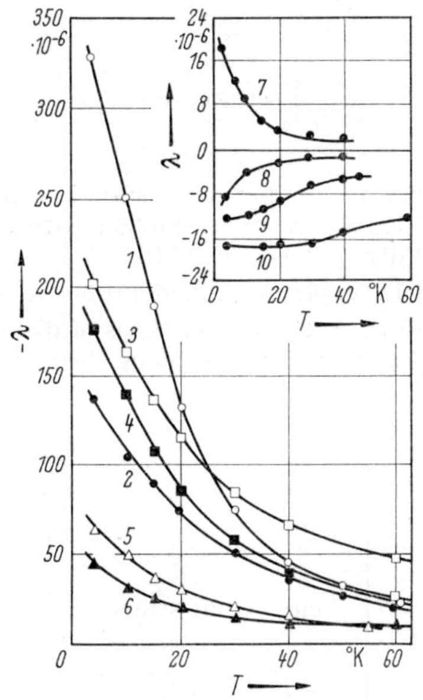

Fig. 78. Rare earth gallates and germanates with a garnet structure. Temperature dependence of the magnetostriction. (H = 50 kOe). [*69BSH*].

1) $Sr_3Ho_2Ge_3O_{12}$;
2) $Ho_3Ga_5O_{12}$;
3) $Ca_3Dy_2Ge_3O_{12}$:
4) $Dy_3Ga_5O_{12}$;
5) $Sr_3Er_2Ge_3O_{12}$;
6) $Er_3Ga_5O_{12}$;
7) $Sr_3Yb_2Ge_3O_{12}$;
8) $Yb_3Ga_5O_{12}$;
9) $Sr_3Tm_2Ge_3O_{12}$;
10) $Tm_3Ga_5O_{12}$.

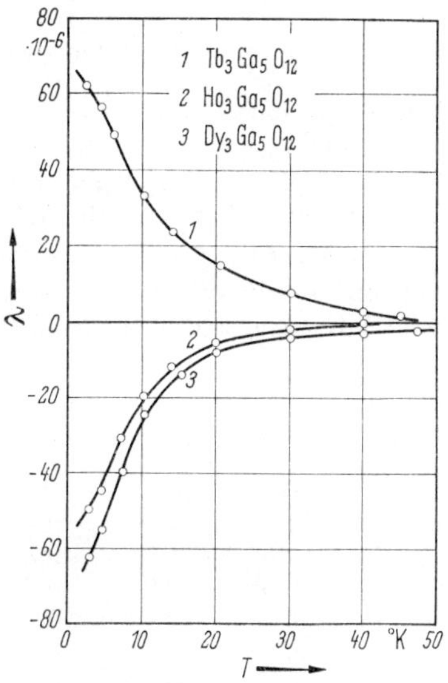

Fig. 79. Tb, Dy and Ho gallate garnets. Magnetostriction vs. temperature (H = 22 kOe) [*66BS*].

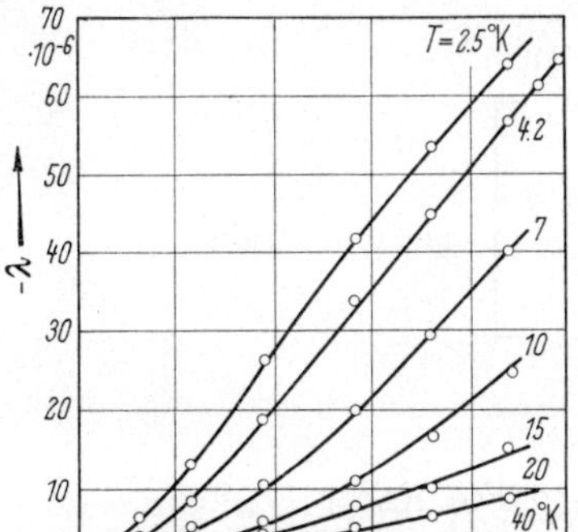

Fig. 80. DyGaG. Isotherms of the magnetostriction, which show a quadratic growth of magnetostriction with magnetic field characteristic of ferromagnets. Deviation from this relation takes places at 4.2 °K and 2.5 °K [*66BS*].

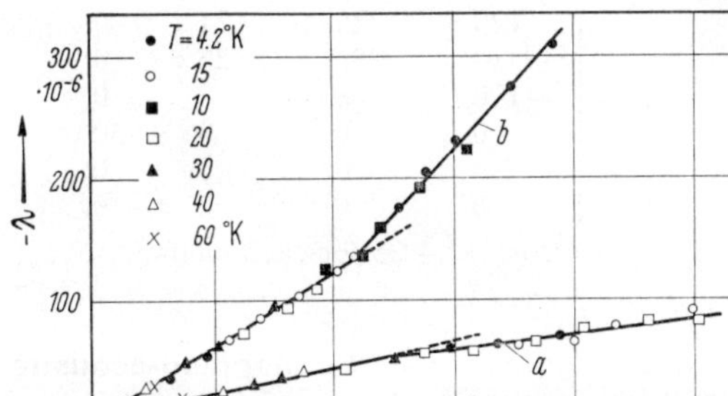

Fig. 81. $Ho_3Ga_5O_{12}$ (a), $Sr_3Ho_2Ge_3O_{12}$ (b). Dependence of the magnetostriction on the square of the specific magnetization. The proportionality coefficients differ for the temperature ranges above and belov 25 °K [*69BSH*].

B. Microscopic magnetoelastic tensors for paramagnetic rare earth ions in diamagnetic garnet hosts [*67PW*]

For the lowest-lying manifold of states the magnetic portion of the single ion spin-Hamiltonian may be written as

$$\mathcal{H}_{\mathrm{M}} = \mu_{\mathrm{B}} \boldsymbol{H}^0 \mathbf{g} \boldsymbol{S} + \boldsymbol{S}\mathbf{D}\boldsymbol{S} + \text{terms in } S^4,\, S^6.$$

The magnetoelastic tensor is given by the strain dependence of $\mathcal{H}_{\mathrm{M}}$ and is

$$\mathcal{H}_{\mathrm{ME}} = \mu_{\mathrm{B}} \sum_{ijkl} \left(\frac{\partial g_{ij}}{\partial \varepsilon_{kl}}\right)_0 H_i^0 S_j \varepsilon_{kl} + \sum_{ijkl} \left(\frac{\partial D_{ij}}{\partial \varepsilon_{kl}}\right)_0 S_i S_j \varepsilon_{kl} + \text{(higher order terms in the spin coordinates)},$$

where ε_{kl} is the strain tensor in Love's notation.

$F_{ijkl} = \left(\frac{\partial g_{ij}}{\partial \varepsilon_{kl}}\right)_0$ is the strain dependence of the **g** tensor, and $G_{ijkl} = \left(\frac{\partial D_{ij}}{\partial \varepsilon_{kl}}\right)_0$ is the strain dependence of the crystal field components.

For the S-state ions, g is taken to be isotropic and strain-independent; for non S-state ions the spin-Hamiltonian is simply $\mathcal{H} = \mu_{\mathrm{B}} \boldsymbol{H} \mathbf{g} \boldsymbol{S}'$, if the ground state of the ion is a Kramers doublet with effective spin of $\frac{1}{2}$. For exact definition of the notations and for detailed discussions, see [*67PW*].

The components of the **G** and **F** tensor were determined from EPR experiments under pressure [*67PW*].

Experimental results of some rare earth ions substituted into the orthorhombic sites of diamagnetic YGaG and YAlG, expressed in the local site axis system, where the major axes lie along the cube ⟨001⟩, ⟨110⟩ and ⟨1$\bar{1}$0⟩.

a) Gd^{3+} in YGaG (S ground state, electronic spin $S = 7/2$)

G = (× cm⁻¹/(unit stress))

	ε_{xx}	ε_{yy}	ε_{zz}	ε_{yz}	ε_{zx}	ε_{xy}
α_x^2	1.1	−1.3	0.1	0	0	0
α_y^2	−1.0	1.5	−0.8	0	0	0
α_z^2	0.0	−0.2	0.7	0	0	0
$\alpha_y\alpha_z$	0	0	0	2.5	0	0
$\alpha_z\alpha_x$	0	0	0	0	3.2	0
$\alpha_x\alpha_y$	0	0	0	0	0	0.1

error: ≈ ±20%

b) Yb^{3+} in YGaG (Kramers doublet ground state, effective spin $\frac{1}{2}$)

F = (per unit strain)

113	85	42	0	0	0
−81	−153	−55	0	0	0
−32	68	13	0	0	0
0	0	0	6	0	0
0	0	0	0	37	0
0	0	0	0	0	47

error: ≈ ±6 for all elements

c) Er^{3+} in YAlG (Kramers doublet ground state, effective spin $\frac{1}{2}$)

F = (per unit strain)

120	225	21	0	0	0
− 10	−285	−23	0	0	0
−110	58	2	0	0	0
0	0	0	30	0	0
0	0	0	0	−126	0
0	0	0	0	0	25

error: ≈ ±25 for each element

C. Magneto-acoustic properties [*65FGCRSU*]

The resonant frequency and Q of the acoustic modes of YAlG may be changed by substituting paramagnetic impurities. For the samples with Mn or Ni additions, the resonant frequency $f_0 \sim (1/d)\,(C_{44}/\varrho)^{1/2}$ of the shear modes has a maximum as a function of temperature (Fig. 82), but not so with Nd, Sm and Pr additions. (d = diameter of the spheric sample; ϱ = density; C_{44} = the appropriate elastic constant).

The observed effects are attributed to Jahn-Teller distortions due to the orbitally degenerated ground state of Mn^{3+} [*65FGCRSU*].

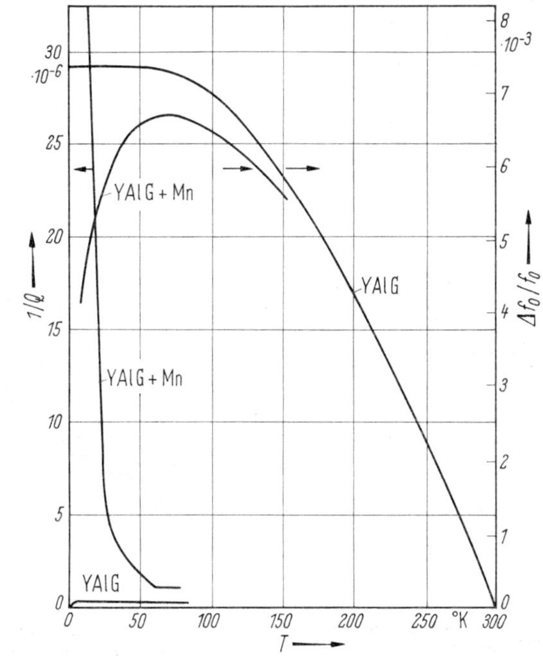

Fig. 82. YAlG, and YAlG with 0.02% Mn. The attenuation Q^{-1}, and resonant frequency of the acoustic modes of spherical samples. The ordinate $\Delta f_0/f_0$ is defined as the difference of the resonant frequency at temperature T and at 300 °K divided by the resonant frequency at 300 °K [*65FGCRSU*].

5.10 Neutron diffraction results near and below the magnetic ordering point

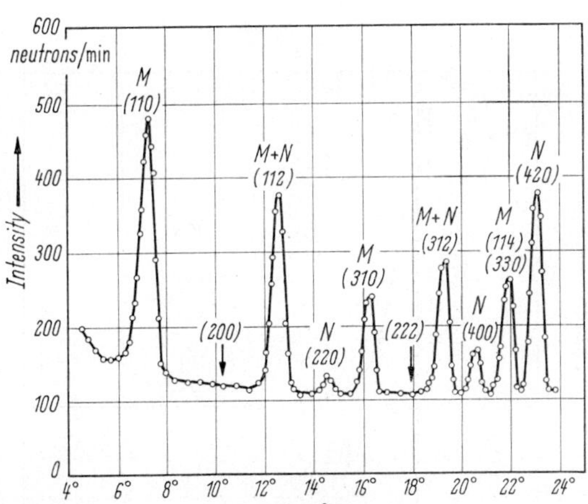

Fig. 83. $Dy_3Al_5O_{12}$. Neutron diffraction pattern at 1.35 °K. Magnetic and nuclear reflections are indicated by the symbols M and N [*65HCW*]. Intensity calculations confirm the magnetic structure (Fig. 44b) assumed in computing the susceptibility [*62BHLW*] below the Néel temperature.

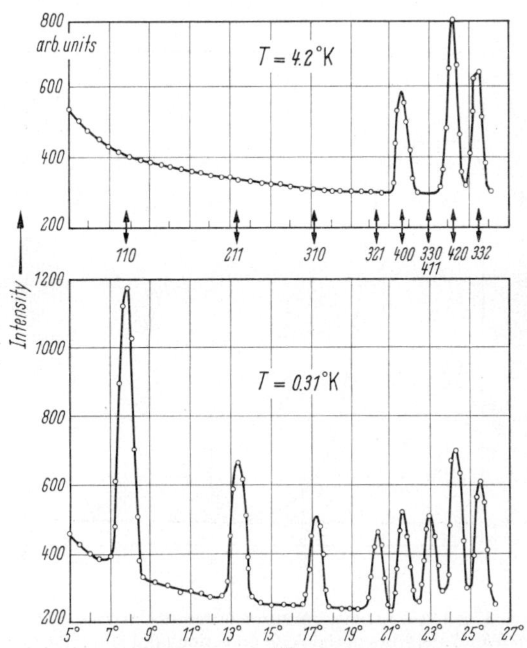

Fig. 84. ErGaG. Neutron diffraction pattern at $T = 4.2$ °K and $T = 0.31$ °K [*68Hb*]. One finds that ErGaG has the same magnetic structure (in zero field) as DyAlG (Fig. 44).

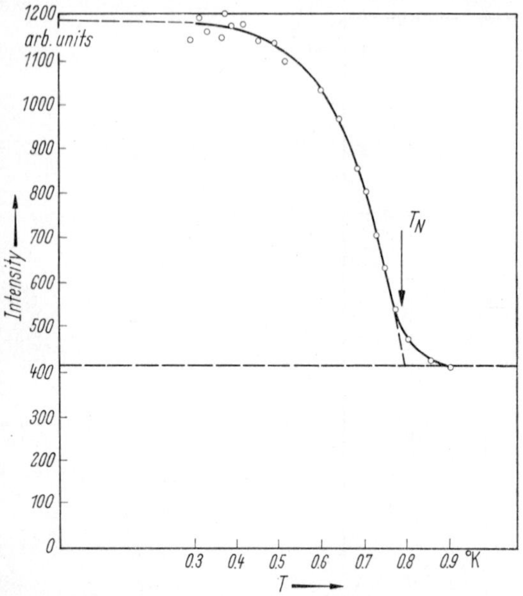

Fig. 85. ErGaG. Intensity of the (110) reflection vs. temperature [68Hb]. A transition temperature of T_N (n) = 0.789 °K is deduced. Extrapolating to 0 °K one finds the magnetic moment of the Er^{3+} ion to be $p_A^0 \approx p_A^{0.31\,°K} \approx 5.9\,\mu_B$.

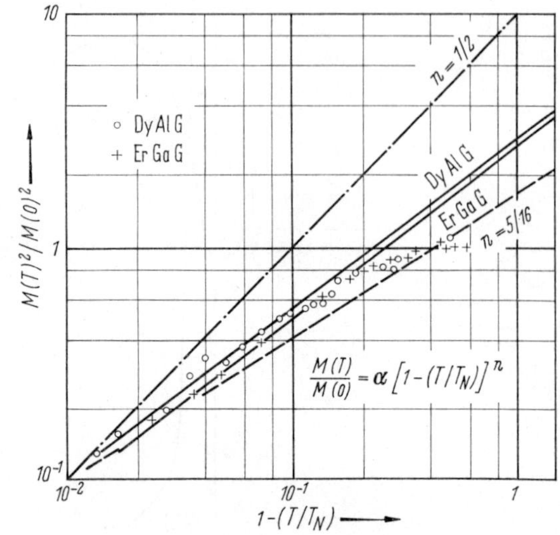

Fig. 86. ErGaG and DyAlG. Comparison of the temperature variation of the magnetic moment in the vicinity of the transition temperature T_N with theoretical models: molecular field approximation ($n = 1/2$) and Ising model ($n = 5/16$) [68Hb].

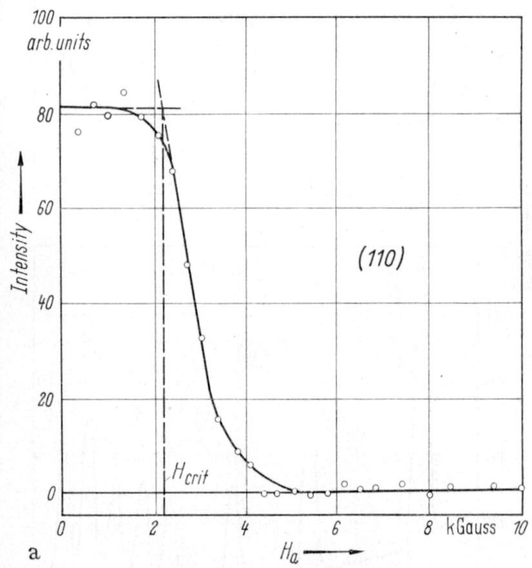

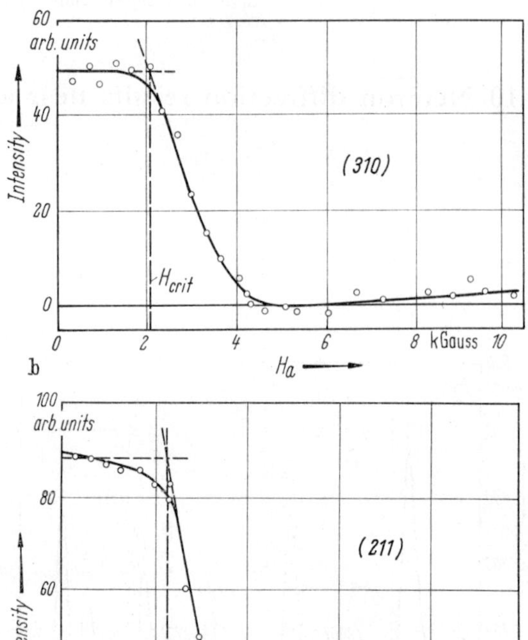

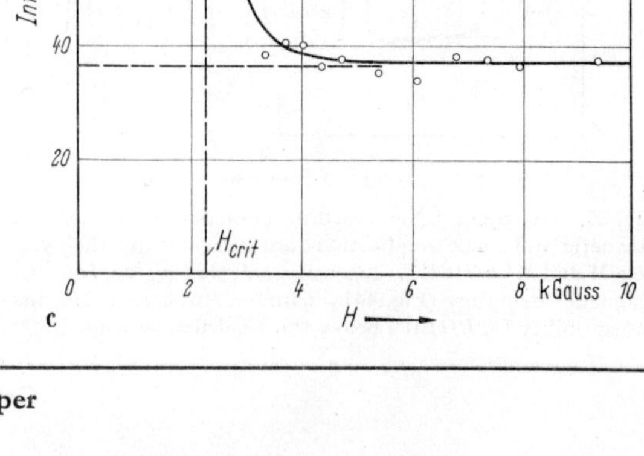

Fig. 87a, b, c. ErGaG. Neutron diffraction at $T \approx 0.31$ °K [68Hb]. Intensities of the (110) reflection (a), (310) reflection (b) and (211) reflection (c) vs. applied magnetic field parallel to the diffusion vector. One deduces the critical fields for the transitions from antiferromagnetism to a mixed phase:

$H_{crit}(110) = 2.2$ kGauss
$H_{crit}(310) = 2.1$ kGauss
$H_{crit}(211) = 2.3$ kGauss.

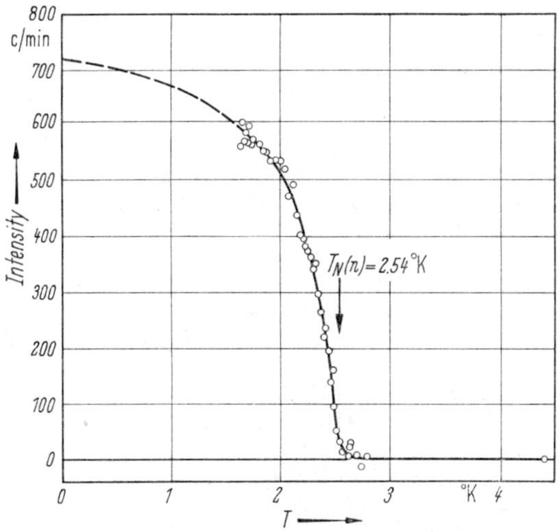

Fig. 88. DyAlG. Temperature variation of the intensity of the (110) reflection for the neutron pattern. From this plot a transition temperature $T_N(n) = (2.54 \pm 0.02)$ °K is found [*64HM*]. From extrapolation one obtains for the Dy^{3+} magnetic moment at 0 °K: $p_A^0 \approx 9.5\,\mu_B$.

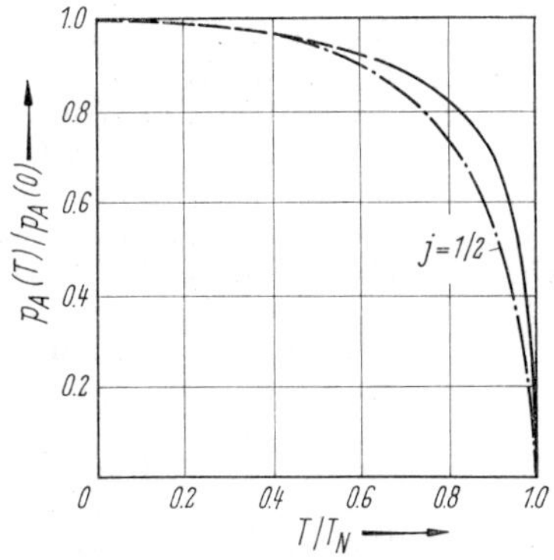

Fig. 89. Variation of the magnetic moment of Dy^{3+} (solid line) (from intensity variation from the (110) reflection) compared with the calculated variation of an Ising model ($J = 1/2$) [*64HM*].

5.11 References

A. Primary references

56BF	Bertaut, F., and F. Forret: Compt. Rend. **243** (1956) 1219.
59AT	Ayant, Y., and J. Thomas: Compt. Rend. **248** (1959) 387.
59BGRW	Boakes, D., G. Garton, D. Ryan, and W. P. Wolf: Proc. Phys. Soc. (London) **74** (1959) 663.
59CD	Cohen, J., and J. Ducloz: J. Phys. Radium **20** (1959) 402.
59GG	Gilleo, M. A., and S. Geller: J. Appl. Phys. **30** (1959) 297 S.
59Pa	Pauthenet, R.: J. Phys. Radium **20** (1959) 388.
59Pb	Pauthenet, R.: J. Appl. Phys. **30** (1959) 290 S.
60AT	Ayant, Y., and J. Thomas: Compt. Rend. **250** (1960) 2688.
60BGLW	Ball, M., G. Garton, M. J. M. Leask, and W. P. Wolf: Proc. 7th Intern. Conf. Low Temp. Phys., Toronto, Ont. **1960**, 128.
60CW	Carson, J. W., and R. L. White: J. Appl. Phys. **31** (1960) 53 S.
60PW	Pappalardo, R., and D. L. Wood: J. Chem. Phys. **33** (1960) 1734.
61ATCD	Ayant, Y., J. Thomas, J. Cohen and J. Ducloz: J. Phys. Radium **22** (1961) 63 S.
61BGLRW	Ball, M., G. Garton, M. J. M. Leask, D. Ryan, and W. P. Wolf: J. Appl. Phys. **32** (1961) 267 S.
61CW	Carson, J. W., and R. L. White: J. Appl. Phys. **32** (1961) 1787.
61G	Geschwind, S.: Phys. Rev. **121** (1961) 363.
61KP	Keller, S. P., and G. D. Pettit: Phys. Rev. **121** (1961) 1639.
61T	Thomas, J.: Compt. Rend. **253** (1961) 1772.
62AD	Axe, J. D., and G. H. Dieke: J. Chem. Phys. **37** (1962) 2364.
62AP	Aléonard, R., and R. Pauthenet: Compt. Rend. **255** (1962) 1727.
62BHLW	Ball, M., M. T. Hutchings, M. J. M. Leask, and W. P. Wolf: Proc. 8th Intern. Conf. Low Temp. Phys., London **1962**, 248.
62BW	Bierig, R. W., and M. J. Weber: Phys. Rev. **126** (1962) 923.
62RM	Rimai, L., and G. A. DeMars: J. Appl. Phys. **33** (1962) 1254.
62W	Wybourne, B. G.: J. Chem. Phys. **36** (1962) 2301.
62WBHLW	Wolf, W. P., M. Ball, M. T. Hutchings, M. J. M. Leask, and A. F. G. Wyatt: J. Phys. Soc. Japan **17**, Suppl. B-I (1962) 443.
62WW	Wickersheim, K. A., and R. L. White: Phys. Rev. Letters **8** (1962) 483.
63BLV	Brumage, W. H., C. C. Lin, and J. H. VanVleck: Phys. Rev. **132** (1963) 608.
63BLWW	Ball, M., M. J. M. Leask, W. P. Wolf, and A. P. G. Wyatt: J. Appl. Phys. **34** (1963) 1104.
63FPC	Fairchild, C. F., A. B. Prag, and K. C. Clark: J. Chem. Phys. **39** (1963) 794.
63HW	Hutchings, M. T., and W. P. Wolf: Phys. Rev. Letters **11** (1963) 187.
63OGFC	Overmeyer, J., E. A. Giess, M. J. Freiser, and B. A. Calhoun: Paramagnetic Resonance, Academic Press. Inc., New York, Vol. I, 1963, 224.

63P	Pappalardo, R.: Z. Physik **173** (1963) 374.
63SS	Svare, J., and G. Seidel: Paramagnetic Resonance, Academic Press, Inc., New York Vol. II, 1963, p. 430.
63ST	Sievers, A. J., and M. Tinkham: J. Phys. Rev. **129** (1963) 1995.
63W	Wood, D. L.: J. Chem. Phys. **39** (1963) 1671.
64ARV	Ayant, Y., J. Rosset, and M. Veyssie: Compt. Rend. **259** (1964) 1698.
64BWWa	Ball, M., W. P. Wolf, and A. F. G. Wyatt: J. Appl. Phys. **35** (1964) 937.
64BWWb	Ball, M., W. P. Wolf, and A. F. G. Wyatt: Phys. Letters **10** (1964) 7.
64C	Capel, H. W.: Phys. Letters **11** (1964) 7.
64GPa	Guillot, M., and R. Pauthenet: Compt. Rend. **259** (1964) 1303.
64GPb	Guillot, M., and R. Pauthenet: Compt. Rend. **258** (1964) 3242.
64HHSS	Hellwege, K. H., S. Hüfner, M. Schinkmann, and H. Schmidt: Phys. Letters **12** (1964) 107.
64HM	Herpin, A., and P. Meriel: Compt. Rend. **259** (1964) 2416.
64HW	Hutchings, M. T., and W. P. Wolf: J. Chem. Phys. **41** (1964) 617.
64K	Koningstein, J. A.: Phys. Rev. **136** (1964) A 717.
64KGa	Koningstein, J. A., and J. E. Geusic: Phys. Rev. **136** (1964) A 711.
64KGb	Koningstein, J. A., and J. E. Geusic: Phys. Rev. **136** (1964) A 726.
64R	Reinen, D.: Z. Anorg. Allgem. Chem. **327** (1964) 238.
64RBa	Rimai, L., and R. W. Bierig: Phys. Rev. Letters **12** (1964) 284.
64RBb	Rubinstein, C. B., and R. L. Berns: Am. Mineralogist **49** (1964) 1489.
64RUG	Rubinstein, C. B., L. G. VanUitert, and W. H. Grodkiewicz: J. Appl. Phys. **35** (1964) 3069.
64SM	Schmitz-DuMont, O., and N. Moulin: Z. Anorg. Allgem. Chem. **330** (1964) 259.
64W	Wolf, W. P.: Proc. Intern. Conf. Magnetism, Nottingham 1964, 555.
64WW	Wolf, W. P., and A. F. G. Wyatt: Phys. Rev. Letters **13** (1964) 368.
65ABGR	Ayant, Y., E. Belorizky, M. Guillot, and J. Rosset: J. Phys. (Paris) **26** (1965) 385.
65BR	Bierig, R. W., and L. Rimai: J. Appl. Phys. **36** (1965) 1199.
65C	Capel, H. W.: Physica **31** (1965) 1152.
65CGLST	Cooke, A. H., K. A. Gehring, M. J. M. Leask, D. Smith, and J. H. M. Thornley: Phys. Rev. Letters **14** (1965) 685.
65DVV	Dreyfus, B., J. Verdone, and M. Veyssie: J. Phys. Chem. Solids **26** (1965) 107.
65EB	Euler, F., and J. A. Bruce: Acta Cryst. **19** (1965) 971.
65FGCRSU	Fraser, D. B., E. M. Gyorgy, R. C. LeCraw, J. P. Remeika, F. J. Schmettler, and L. G. VanUitert: J. Appl. Phys. **36** (1965) 1016.
65GP	Guillot, M., and R. Pauthenet: J. Appl. Phys. **36** (1965) 1003.
65HCW	Hastings, J. M., L. M. Corliss, and C. G. Windsor: Phys. Rev. **138** (1965) A 176.
65H	Huang, C. Y.: Phys. Rev. **139** (1965) A 241.
65HSS	Hüfner, S., M. Schinkmann, and H. Schmidt: Physik Kondensierten Materie **4** (1965) 108.
65Ka	Koningstein, J. A.: J. Chem. Phys. **42** (1965) 3195.
65Kb	Koningstein, J. A.: Theoret. Chim. Acta **3** (1965) 271.
65Kc	Koningstein, J. A.: J. Chem. Phys. **42** (1965) 1423.
65NW	Nowik, J., and H. H. Wickmann: Phys. Rev. **140** (1965) A 869.
65RB	Rubinstein, C. B., and S. B. Berger: J. Appl. Phys. **36** (1965) 3951.
65SH	Schieber, M., and L. Holmes: J. Appl. Phys. **36** (1965) 1159.
66AMR	Ayant, Y., P. Mollard, and J. Rosset: J. Phys. (Paris) **27** (1966) 536.
66BGB	Burns, G., E. A. Giess, and K. W. Blazey: J. Appl. Phys. **37** (1966) 1301.
66BS	Belov, K. P., and V. I. Sokolov: JETP Letters (English Transl.) **4** (1966) 127.
66CC	Chamberlain, J. R., and R. W. Cooper: Proc. Phys. Soc. (London) **87** (1966) 967.
66CR	Caspary, M. E., and F. D. Ray: J. Appl. Phys. **37** (1966) 1326.
66H	Hooge, F. N.: J. Chem. Phys. **45** (1966) 4504.
66HHSS	Hellwege, K. H., S. Hüfner, M. Schinkmann, and H. Schmidt: Physik Kondensierten Materie **4** (1966) 397.
66HPWB	Herrmann, G. F., J. J. Pearson, K. A. Wickersheim, and R. A. Buchanan: J. Appl. Phys. **37** (1966) 1312.
66HSM	Hodges, J. A., R. A. Serway, and S. A. Marshall: Phys. Rev. **151** (1966) 196.
66HWW	Hutchings, M. T., C. G. Windsor, and W. P. Wolf: Phys. Rev. **148** (1966) 444.
66KLSW	Keen, B. E., D. Landau, B. Schneider, and W. P. Wolf: J. Appl. Phys. **37** (1966) 1120.
66KLW	Keen, B. E., D. P. Landau, and W. P. Wolf: Phys. Letters **23** (1966) 202.
66Ka	Koningstein, J. A.: J. Chem. Phys. **44** (1966) 3957.
66Kb	Koningstein, J. A.: Theoret. Chim. Acta **5** (1966) 327.
66L	Lewis, H. R.: J. Appl. Phys. **37** (1966) 739.

66RBS	RIMAI, L., R. B. BIERIG, and B. D. SILVERMAN: Phys. Rev. **146** (1966) 222.
66RK	RIMAI, L., and T. KUSHIDA: Phys. Rev. **143** (1966) 161.
66SC	SCHWEE, L. J., and J. R. CUNNINGHAM, Jr.: J. Appl. Phys. **37** (1966) 449.
66SLKW	SCHNEIDER, B., D. P. LANDAU, B. E. KEEN, and W. P. WOLF: Phys. Letters **23** (1966) 210.
66SLV	SCHIEBER, M. M., C. C. LIN, and J. H. VANVLECK: J. Phys. Chem. Solids **27** (1966) 1041.
66SMR	STURGE, M. D., F. R. MERRIT, and J. P. REMEIKA: Appl. Phys. Letters **9** (1966) 63.
66SZV	ŠROUBEK, Z., K. ŽDÁNSKÝ, and M. VICHR: Phys. Letters **21** (1966) 264.
66V	VANVLECK, J. H.: J. Phys. Chem. Solids **27** (1966) 1047.
67BCVa	BIDAUX, R., P. CARRARA, and V. VIVET: J. Phys. (Paris) **28** (1967) 187.
67BCVb	BIDAUX, R., P. CARRARA, and B. VIVET: Phys. Letters **25A** (1967) 332.
67BWPH	BUCHANAN, R. A., K. A. WICKERSHEIM, J. J. PEARSON, and G. F. HERRMANN: Phys. Rev. **159** (1967) 245.
67CM	CROSSWHITE, H. M., and H. W. MOOS: Optical properties of ions in crystals, Interscience Publishers, Inc., New York, 1967, p. 3.
67CTW	COOKE, A.H., T.L.THORP, and M.R.WELLS: Proc. Phys. Soc. (London) **92** (1967) 400.
67GHH	GRÜNBERG, P., K. H. HELLWEGE, and S. HÜFNER: Physik Kondensierten Materie **6** (1967) 95.
67Ka	KONINGSTEIN, J. A.: J. Chem. Phys. **46** (1967) 2811.
67Kb	KONINGSTEIN, J. A.: In H. M. CROSSWHITE, H. W. MOOS (Editors): Optical properties of ions in crystals, Interscience Publishers, Inc., New York (1967) 105.
67KLW	KEEN, B. E., D. P. LANDAU, and W. P. WOLF: J. Appl. Phys. **38** (1967) 967.
67OMR	ONN, D. G., H. MEYER, and J. P. REMEIKA: Phys. Rev. **156** (1967) 663.
67PHWB	PEARSON, J. J., G. J. HERRMANN, K. A. WICKERSHEIM, and R. A. BUCHANAN: Phys. Rev. **159** (1967) 251.
67PW	PHILLIPS, T. G., and R. J. WHITE: Phys. Rev. **160** (1967) 316.
67S	DESORBO, W.: Phys. Rev. **158** (1967) 839.
67TS	TACHIKI, M., and Z. ŠROUBEK: Solid State Commun. **5** (1967) 361.
67VD	VEYSSIE, M., and B. DREYFUS: J. Phys. Chem. Solids **28** (1967) 499.
67WR	WOOD, D. L., and J. P. REMEIKA: J. Chem. Phys. **46** (1967) 3595.
68AT	ASHBEE, K. H. G., and G. THOMAS: J. Appl. Phys. **39** (1968) 3778.
68BCV	BIDAUX, P., P. CARRARA, and B. VIVET: J. Phys. (Paris) **29** (1968) 357.
68BDM	BAGDASAROV, KH. S., D. M. DARASELIYA, and A. A. MANENKOV: JETP Letters (English Transl.) **8** (1968) 323.
68BM	BRUMAGE, W. H., and J. P. MAHONEY: J. Chem. Phys. **48** (1968) 5733.
68BV	BIDAUX, R., and B. VIVET: J. Phys. (Paris) **29** (1968) 57.
68Ha	HAMMANN, J.: Phys. Letters **26** (1968) A 263.
68Hb	HAMMANN, J.: J. Phys. (Paris) **29** (1968) 495.
68HV	HUANG, N. L., and J. H. VANVLECK: Solid State Commun. **6** (1968) 557.
68J	JONES, E. D.: J. Phys. Chem. Solids **29** (1968) 1305.
68LC	LUNG, W. Y., and J. R. CHAMBERLAIN: Phys. Letters **27A** (1968) 365.
68MO	MILLER, I. A., and E. I. OFFENBACHER: Phys. Rev. **166** (1968) 269.
68W	WYATT, A. F. G.: J. Phys. C (Proc. Phys. Soc.) **1** (1968) 684.
69BPH	BUCHANAN, R. A., J. J. PEARSON, and G. F. HERRMANN: Solid State Commun. **7** (1969) 195.
69BSH	BELOV, K. P., V. I. SOKOLOV, and THAN DYK HIEN: Soviet Phys.-Solid State (English Transl.) **10** (1969) 2946.
69Ga	GRÜNBERG, P.: Z. Physik **225** (1969) 376.
69Gb	GRÜNBERG, P.: Dissertation, Darmstadt 1969.
69GHOS	GRÜNBERG, P., S. HÜFNER, E. ORLICH, and J. SCHMITT: J. Appl. Phys. **40** (1969) 1501.
69JDR	JOHNSON, L. F., J. F. DILLON Jr., and J. P. REMEIKA: J. Appl. Phys. **40** (1969) 1499.
69JS	JONES, E. D., and V. H. SCHMIDT: J. Appl. Phys. **40** (1969) 1406.
69O	ORLICH, E.: Dissertation, Darmstadt 1969, 1. analysis.
69OH	ORLICH, E., and S. HÜFNER: J. Appl. Phys. **40** (1969) 1503.
69SMHR	STURGE, M. D., F. R. MERRIT, J. C. HENSEL, and J. P. REMEIKA: Phys. Rev. **180** (1969) 402.
69SOCR	SLACK, G. A., D. W. OLIVER, R. M. CHRENKO, and S. ROBERTS: Phys. Rev. **177** (1969) 1308.

B. Supplementary references

1	DOMB, C.: In "Magnetism", edited by SUHL, H., and G. T. RADO, Academic Press Inc., New York, 1965, Vol. II A.
2	FISHER, M. E.: Phil. Mag. **7** (1962) 1731.

3 VanVleck, J. H.: Electric and magnetic susceptibilities, Oxford 1932.
4 Pappalardo, R., G. T. Wood, and R. C. Linares: J. Chem. Phys. **35** (1961) 2041.
5 Pauthenet, R.: Ann. Phys. (Paris) **3** (1958) 424.
6 Geller, R., J. P. Remeika, R. C. Sherwood, H. J. Williams, and E. G. Espinosa: Phys. Rev. **137** (1965) 1034A.
7 Onn, D. G., H. Meyer, and J. P. Remeika: Ann. Acad. Sci. Fennicae **210** (1966) 218.
8 Thomas, J.: Thèse, Grenoble (1962).
9 Fisher, M. E., and M. F. Sykes: Physica **28** (1962) 939.
10 Caola, M., and J. Thomas: Compt. Rend. **258** (1964) 3234.
11 Wolf, W. P.: Proc. Phys. Soc. (London) **74** (1959) 665.
12 Tanabe, Y., and S. Sugano: J. Phys. Soc. Japan **9** (1954) 766.
13 Bleaney, B., and R. S. Trenam: Proc. Roy. Soc. (London) **A 223** (1954) 1.
14 Jones, D. A., J. M. Baker, and D. F. D. Pope: Proc. Phys. Soc. (London) **74** (1959) 249.
15 Weiler, M. H., and D. W. Wylie: Proc. Phys. Soc. (London) **85** (1965) 608.
16 Jones, E. D., and V. H. Schmidt: J. Appl. Phys. **40** (1969) 1406.
17 Watson, R. E., and A. J. Freeman: Hyperfine interactions, Academic Press Inc., New York, 1967.
18 Ryan, D.: Thesis, unpublished; from *62WBHLW*.
19 Bertaut, F., and F. Forrat: Compt. Rend. **244** (1957) 96.
20 Bleaney, B., and K. W. H. Stevens: Rept. Progr. Phys. **16** (1953) 108.
21 Pappalardo, R., G. T. Wood, and R. C. Linares: J. Chem. Phys. **35** (1961) 1460.
22 Ball, M., W. P. Wolf, and A. F. G. Wyatt: Clarendon Lab. Techn. Report [AF61(052)-125], 1963, unpublished.
23 Davis, C. F., Jr., and M. W. D. Strandberg: Phys. Rev. **105** (1957) 447.
24 Levy, P. M.: Phys. Rev. **135** (1964) A 155.
25 Bierig, R. W., and M. J. Weber: Phys. Rev. **134** (1964) A 1492.
26 VanVleck, J. H.: J. Chem. Phys. **7** (1939), 72; Phys. Rev. **51** (1940) 526.
27 Orbach, R.: Proc. Roy. Soc. (London) **A 264** (1961) 458.
28 Orbach, R.: Paramagnetic Resonance, Academic Press. Inc., New York, Vol. II, 1963, A 456.
29 Finn, B. P., R. Orbach, and W. P. Wolf: Proc. Phys. Soc. (London) **77** (1961) 261.
30 Davids, D. A., and P. E. Wagner: Phys. Rev. Letters **12** (1964) 141.
31 Orbach, R.: Proc. Phys. Soc. (London) **76** (1960) 821.
32 Seidel, G., and I. Svare: Paramagnetic Resonance, Academic Press Inc., New York, Vol. II, 1963, 468.
33 American Institute of Physics, Handbook, McGraw-Hill Book Company, Inc., New York, **1957**.
34 Heitler, W., and E. Teller: Proc. Roy. Soc. (London) **A 155** (1936) 629.
35 Fierz, M.: Physica **5** (1938) 433.
36 Kronig, R. M. L.: Physica **6** (1939) 33.
37 VanVleck, J. H.: Phys. Rev. **57** (1940) 426.
38 Caspari, M. E., S. Frankel, and G. T. Wood: Phys. Rev. **127** (1962) 1519.
39 Bierig, R. W., M. J. Weber, and S. I. Warshaw: Phys. Rev. **134** (1964) A 1504.
40 Rosenfeldt, L.: Z. Physik **57** (1930) 837.
41 Kramers, H. A.: Proc. Acad. Sci. Amsterdam **33** (1930) 959.
42 VanVleck, J. H., and M. H. Hebb: Phys. Rev. **46** (1934) 17.
43 Geller, S., and M. A. Gilleo: J. Phys. Chem. Solids **3** (1957) 30.
44 Menzer, G.: Z. Krist. **69** (1929) 300.
45 Baker, J. M., B. Bleaney, and W. Hayes: Proc. Roy. Soc. (London) **A 247** (1958) 141.
46 Gruber, J. B., and J. G. Conway: J. Chem. Phys. **32** (1960) 1178.
47 Freeman, A. J., and R. E. Watson: Phys. Rev. **127** (1962) 2058.
48 Fisher, M. E.: Proc. Roy. Soc. (London) **A 254** (1960) 66.
49 Domb, C., and A. R. Miedema: Progress in Low Temperature Physics, edited by C. J. Gorter, John Wiley and Sons, Inc., New York, **1964** Vol. 4.
50 Geller, S., and C. E. Miller: Acta Cryst. **13** (1960) 179.
51 Johnson, L. F., J. F. Dillon, Jr., and J. P. Remeika: J. Appl. Phys. **40** (1969) 1499.
52 Gehring, K. H., M. J. M. Leask, and J. H. M. Thomley: J. Physics C (Solid State) **2** (1969) 484.
53 Belyaeva, A. I., V. V. Eremenko, V. N. Pavlov, and A. V. Antonov: Opt. Spectry. (USSR) (English Transl.) **25** (1968) 55.
54 Schuchert, H., S. Hüfner, and R. Faulhaber: Z. Physik **222** (1969) 105.
55 Holloway, W. W., Jr., and M. Kestigian: J. Opt. Soc. Am. **59** (1969) 60.
56 Zverev, G. M., G. Ya. Kolodnyi, and A. M. Onishchenko: Soviet Phys. JETP (English Transl.) **28** (1969) 21.
57 Slack, G. H., D. W. Oliver, R. M. Chrenko, and S. Roberts: Phys. Rev. **177** (1969) 1308.

58 Birgenau, R. J., M. T. Hutchings, J. M. Baker, and J. D. Riley: J. Appl. Phys. **40** (1969) 1070.

59 Phillips, T. G., and R. L. White: J. Appl. Phys. **38** (1967) 1222.

60 Skorobogatov, B. S., and V. S. Konevskii: Opt. Spectry. (USSR) (English Transl.) **24** (1968) 544.

61 Geschwind, S.: Phys. Rev. Letters **3** (1959) 207.

62 Wolf, W. P.: Proc. Phys. Soc. (London) **74** (1959) 665.

63 Timofeeva, V. A., T. V. Lebedeva, and T. S. Kon'kova: Soviet Phys. Cryst. (English Transl.) **10** (1965) 72.

64 Inoue, M.: Phys. Rev. Letters **11** (1963) 196.

65 Phillips, T. G., and R. L. White: Phys. Rev. Letters **16** (1966) 650.

66 Nowik, J., and H. H. Wickman: Phys. Rev. **140** (1965) A 869.

67 Grant, P. M., and W. Ruppel: Solid State Commun. **5** (1967) 543.

68 Faulhaber, R., and S. Hüfner: Solid State Commun. **7** (1969) 389.

69 Sychugov, V. A., and G. P. Shipulo: Soviet Phys.-Solid State (English Transl.) **10** (1969) 2224.

70 Yudin, V. M., S. A. Poltinnikov, O. B. Proskuryakov, and N. N. Parfenova: Phys. Letters **28** (1969) A 483.

71 Marezio, M., J. P. Remeika, and P. D. Dernier: Acta Cryst. **B 24** (1968) 1670.

72 Skorobogatov, B. S., and V. S. Konevskii: Opt. Spectry. (USSR) (English Transl.) **24** (1968) 544.

73 Kushida, T., and J. E. Geusic: Phys. Letters **21** (1968) 1172.

74 Haubenreisser, W.: Phys. Status Solidi **27** (1968) K 145.

75 Ashbee, K. H. G., and G. Thomas: J. Appl. Phys. **39** (1968) 3778.

76 Cockayne, B., D. B. Gasson, D. Findlay, D. W. Goodwin, and R. A. Clay: J. Phys. Chem. Solids **29** (1968) 905.

77 Suzuishi, M., H. Ito, and G. Yamaguchi: Japan. J. Appl. Phys. **7** (1968) 183.

78 Grant, P. M.: Appl. Phys. Letters **11** (1967) 177.

79 Weaver, L. H., I. Liberman, and C. H. Church: J. Appl. Phys. **38** (1967) 5405.

80 Timofeeva, V. A.: J. Phys. Chem. Solids, Suppl. Nr. **1** (1967) 445.

81 Puttbach, R. C., R. R. Monchamp, and J. W. Nielson: J. Phys. Chem. Solids, Suppl. Nr. **1** (1967) 569.

82 Monchamp, R. R., R. Belt, and J. W. Nielson: J. Phys. Chem. Solids, Suppl. No. **1** (1967) 463.

83 Kestigian, M., and W. W. Holloway, Jr.: J. Phys. Chem. Solids Suppl. No. **1** (1967) 451.

84 Bagdasarov, Kh. S., F. M. Zubkowa, A. A. Kaminskii, A. M. Kevorkov, and V. Ya. Khaimov-Mal'kov: Soviet Phys.-Cryst. (English Transl.) **12** (1967).

85 Grant, P. M., and W. Ruppel: Solid State Commun. **5** (1967) 543.

86 Koningstein, J. A.: J. Opt. Soc. Am. **56** (1966) 1405.

87 Veyssie, M., A. Hadni, and P. Strimer: J. Phys. (Paris) **27** (1966) 43.

88 Low, W., and E. L. Offenbacher: Solid State Phys. **17** (1965) 135.

89 Pauthenet, R.: Proc. Intern. Conf. on Magnetism, London **1965**, p. 659.

90 Wolf, W. P.: Proc. Intern. Conf. on Magnetism, London **1965**, p. 555.

91 Bayer, G.: J. Am. Ceram. Soc. **48** (1965) 600.

92 Feofilov, P. P., V. A. Timofeeva, M. N. Tolstoi, and L. M. Belyaev: Opt. Spectry. (USSR) (English Transl.) **19** (1965) 451.

93 Linares, R. C.: Solid State Commun. **2** (1964) 229.

94 Néel, L., R. Pauthenet, and B. Dreyfus: Prog. Low Temp. Phys. **4** (1964) 344.

95 Paulus, M.: Z. Angew. Phys. **17** (1964) 216.

96 Geusic, I. E., H. M. Marcos, and L. G. VanUitert: Appl. Phys. Letters **4** (1964) 182.

97 Keen, B. E., D. Landau, B. Schneider, and W. P. Wolf: J. Appl. Phys. **37** (1966) 1120.

98 Keen, B. E., D. Landau, and W. P. Wolf: Phys. Letters **23** (1966) 202.

99 Belov, K. P., and S. A. Nikitin: Phys. Status Solidi **12** (1965) 453.

100 Nowik, I., and H. H. Wickman: Phys. Rev. **140** (1965) A 869.

101 Schneider, B., D. Landau, B. E. Keen, and W. P. Wolf: Phys. Letters **23** (1966) 210.

102 Burns, G., E. A. Gress, and K. W. Beazey: J. Appl. Phys. **37** (1966) 1301.

103 Herrmann, G. F., J. J. Pearson, K. A. Wickersheim, and R. A. Buchanan: J. Appl. Phys. **37** (1966) 1312.

104 Rimai, L., and T. Kushida: Phys. Rev. **143** (1965) 160.

105 Ayant, Y., É. Belorizky, J. Rosset, and M. Veyssie: Compt. Rend. **263** (1966) 246.

106 Caspary, M. E., and F. D. Ray: J. Appl. Phys. **37** (1966) 1326.

107 Burns, G.: J. Chem. Phys. **42** (1965) 377.

108 Caspary, M. E., A. Koicki, S. Koicki, and G. T. Wood: Phys. Letters **11** (1964) 195.

6 Spinels

Arrangement

Because of the variety and complexity of possible compositions of the spinels, $Me'Me''_2X_4$, a perfect chemical-systematical arrangement was not aimed at. Each compound, however, may easily be found by using the alphabetical index of substances at the end of the volume.

Some overlapping has been tolerated in order to avoid possible gaps.

Sequence of the spinels $Me'Me''X_4$ in the sections 6.1···6.3 (each subsection includes the respective compounds with substitutions):

Anordnung

Wegen der Vielfalt und der Variationsmöglichkeit bei der Zusammensetzung der Spinelle $Me'Me''_2X_4$ wurde auf eine vollkommene chemisch-systematische Anordnung verzichtet. Jede Substanz kann aber leicht mit Hilfe des alphabetischen Substanzenverzeichnisses am Ende des Bandes gefunden werden.

Gelegentliche Überlappungen wurden in Kauf genommen, da sie die Möglichkeit von Lücken verringern.

Reihenfolge der Spinelle $Me'Me''X_4$ in den Abschnitten 6.1···6.3 (jeder Unterabschnitt enthält auch die entsprechenden Verbindungen mit Substitutionen):

6.0 Introduction

6.1 **Fe^{3+} spinels** (cubic ferrites, Me" = Fe^{3+}; X = O, S, Se, Te)

6.1.1···6.1.8: **Me' = Fe^{2+}, Ni, Mn, Mg, Zn, Li, Cu, Co**

6.2 **Cr spinels** (chromites, Me" = Cr^{3+})

6.2.1, 6.2.2: **X = O; S, Se, Te**

6.3 **Other spinels** (X = O, S, Se, Te)

6.3.1···6.3.21: **Me" = V, Ge, Rh, Al, Ga, In, Co, Ni, Te, Mn, Ti, Mo, W, Nb, Zr, Sb, Sn, Si**

6.0 Introduction — Einleitung

6.0.1 General remarks — Allgemeine Bemerkungen

In this section, all spinels are included for which crystallographic and/or magnetic data are known. Though not all of the $Me'Me''_2X_4$ compounds (Me', Me'' = transition elements, X = O, S, Se, Te) with the spinel structure are magnetic or thoroughly investigated, a systematic compilation of the available data on spinels seemed to be justified by the importance of many of these compounds. Most of the investigations of the magnetic properties of spinels have dealt with Fe^{3+} spinels (cubic ferrites) and the discussion of the magnetic behavior in 6.1.0 is similarly concerned primarily with ferrites although it also applies to other spinels. The exchange interactions responsible for the magnetic ordering in spinels are much stronger when Fe^{3+} ions are involved and consequently magnetic ordering occurs at relatively low temperatures in spinels other than ferrites.

In diesem Kapitel sind alle Spinelle aufgenommen, für die kristallographische und/oder magnetische Daten bekannt sind. Obwohl nicht alle $Me'Me''_2X_4$-Verbindungen (Me', Me'' = Übergangselemente, X = O, S, Se, Te) mit Spinellstruktur magnetisch sind oder gründlich untersucht sind, schien eine systematische Zusammenstellung aller über Spinelle vorliegenden Daten durch die Wichtigkeit vieler dieser Verbindungen gerechtfertigt. Die meisten Untersuchungen magnetischer Eigenschaften von Spinellen sind an Fe^{3+}-Spinellen (kubische Ferrite) durchgeführt. Daher betrifft die Diskussion des magnetischen Verhaltens in 6.1.0 in erster Linie die Ferrite, obwohl sie auch für die anderen Spinelle gültig ist. Die Austausch-Wechselwirkungen, die für die magnetische Ordnung in Spinellen verantwortlich sind, sind viel stärker, wenn Fe^{3+}-Ionen beteiligt sind; folglich tritt die magnetische Ordnung bei den anderen Spinellen im Vergleich zu den Ferriten bei verhältnismäßig niedrigen Temperaturen auf.

6.0.2 Structure — Struktur

The spinel structure, named for the mineral spinel $MgAl_2O_4$, is facecentered cubic with a large unit cell containing eight formula units. It can be described as a cubic close-packed array of anions with the metal ions located in interstitial positions in this array. Two types of interstices are available, one formed by four anions at the vertices of a tetrahedron and the other by six anions at the vertices of an octahedron; these interstices are commonly called the tetrahedral or *A* sites and the octahedral or *B* sites, respectively. Each unit cell contains 8 occupied *A* sites and 16 occupied *B* sites. The positions of the metal ions are fixed by the symmetry of the structure but the anion positions are variable and are specified by the parameter u. The value $u = \frac{3}{8}$ corresponds to perfect close-packing of the anions; observed values of u range from 0.38 to 0.39. A schematic representation of two octants of the unit cell of the spinel structure is shown in Fig. 1. If the *A* sites are occupied by Me' ions (in the formula $Me'Me''_2O_4$) and the *B* sites by Me'' ions, the structure is *normal*. If the *A* sites are

Die nach dem Mineral Spinell, $MgAl_2O_4$, genannte Spinell-Struktur ist kubisch flächenzentriert mit einer großen, 8 Formeleinheiten enthaltenden Einheitszelle. Sie läßt sich als kubisch dichteste Kugelpackung von Anionen auffassen, in deren Lücken sich Metall-Ionen befinden. Es gibt zwei Arten solcher Lücken; die eine wird von vier Anionen an den Eckpunkten eines Tetraeders und die anderen von sechs Anionen an den Eckpunkten eines Oktaeders gebildet; diese Lücken werden gewöhnlich die Tetraeder- oder *A*-Gitterplätze bzw. die Oktaeder- oder *B*-Gitterplätze genannt. Jede Gitterzelle enthält 8 besetzte *A*- und 16 besetzte *B*-Gitterplätze. Die Lage der Metall-Ionen ist durch die Symmetrie der Struktur festgelegt, während die Lage der Anionen variabel ist und durch den Parameter u bezeichnet wird. Der Wert $u = \frac{3}{8}$ entspricht einer vollständig dichten Packung der Anionen; die beobachteten Werte von u liegen zwischen 0.38 und 0.39. In Fig. 1 wird eine schematische Darstellung zweier Oktanten einer Einheitszelle der Spinell-Struktur gezeigt. Wenn die *A*-Gitterplätze von Me'-Ionen und die *B*-Gitterplätze von Me''-Ionen (in der Formel $Me'Me''_2O_4$) besetzt sind, bezeichnet man die Struktur als *normal*. Wenn die

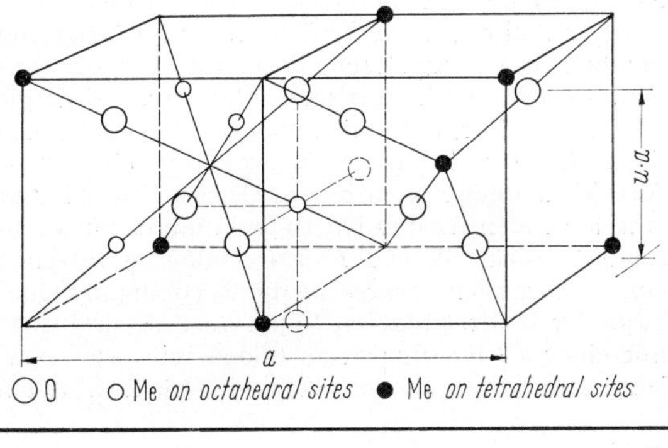

Fig. 1. Two octants of the spinel structure. (From SMART: Landolt-Börnstein, 6. ed., vol. II/9).

occupied by Me" ions and the B sites by a random arrangement of Me' and Me" ions, the structure is *inverse*. In the oxide spinels, intermediate arrangements are also possible. A convenient way of indicating the structure is the use of square brackets around the metal ions occupying the B sites; thus, the formula for a normal spinel would be written Me'[Me_2'']X_4 and for an inverse spinel Me"[Me'Me"]X_4.

The principal difference between the oxides and the other chalcogenide spinels results from the decreased electronegativity of S, Se, and Te compared to oxygen. The oxide spinels are ionic crystals whereas the sulfides and selenides have much less ionic character and, indeed, some of them are metallic. In the oxide spinels, two main combinations of valence states of the cations occur: the 2-3 spinels in which Me' is divalent and Me" is trivalent and the 4-2 spinels in which Me' is quadrivalent and Me" is divalent. Additional combinations of valence states of the metal ions are possible since, for example, the divalent cations can be replaced by an equiatomic mixture of mono- and trivalent cations. Lithium ferrite ($Li_{1/2}Fe_{5/2}O_4$ or $LiFe_5O_8$) is an important example of this. The oxide spinels will also tolerate a large number of metal ion vacancies; an extreme case of this is γ-Fe_2O_3 with $2\frac{2}{3}$ vacancies per unit cell. The existence of these metal ion vacancies frequently makes it difficult to prepare spinels with the stoichiometric metal:oxygen ratio.

Oxide spinels can be formed with cations whose charges vary from +1 to +6 and whose radii range from less than 0.5 Å to almost 1.0 Å. In addition some ions which are too large (such as Ca^{2+}) or too small (such as Si^{4+}) to form the spinel structure, can be substituted to a limited extent into other spinels. Since there is generally complete miscibility between different oxide spinels, the range of oxide compositions which occur with the spinel structure is very large.

The distribution of the metal ions between A and B sites is determined by a number of effects. The A sites are smaller than the B sites; thus, smaller ions which usually have a higher charge tend to occupy A sites. This effect becomes less important as the u parameter increases and the sizes of the two sites become nearly equal. — The electrostatic (Madelung) energy is more favorable when the cations with the higher valence occupy the B sites where they are coordinated by a large number of anions. This effect depends on the inter-ionic distances and increases with increasing u. — The electronic configurations of some cations cause them to prefer a particular site. Ions with a d^{10} configuration such as Zn^{2+} and Ga^{3+} form partially covalent bonds in the tetrahedral interstices. Ions with a d^3 (Cr^{3+}) or a d^8 (Ni^{2+}) configuration in octahedral interstices have their energy substantial-

A-Gitterplätze von Me"-Ionen und die B-Gitterplätze von einer zufälligen Anordnung von Me'- und Me"-Ionen besetzt sind, bezeichnet man die Struktur als *invers*. In den Oxid-Spinellen sind gemischte Anordnungen ebenfalls möglich. Eine geeignete Methode zur Kennzeichnung der Struktur ist das Setzen eckiger Klammern um die auf den B-Gitterplätzen befindlichen Metall-Ionen; auf diese Weise wird die Formel eines normalen Spinells Me'[Me_2'']X_4 und eines inversen Spinells Me"[Me'Me"]X_4 geschrieben.

Der hauptsächliche Unterschied zwischen den Oxid- und den anderen Chalcogenid-Spinellen beruht auf der verglichen mit O verminderten Elektronegativität von S, Se und Te. Die Oxid-Spinelle sind Ionen-Kristalle, während die Sulfide und Selenide einen weit weniger ionischen Charakter besitzen und einige von ihnen sogar metallisch sind. In den Oxid-Spinellen treten zwei hauptsächliche Kombinationen der Wertigkeitsstufen auf: die 2-3 Spinelle, in denen Me' zwei- und Me" dreiwertig, und die 4-2 Spinelle, in denen Me' vier- und Me" zweiwertig ist. Weitere Kombinationen von Wertigkeitsstufen der Metallionen sind möglich, da z.B. die zweiwertigen Kationen durch eine Mischung ein- und dreiwertiger Kationen des gleichen Elements ersetzt werden kann. Lithiumferrit ($Li_{1/2}Fe_{5/2}O_4$ oder $LiFe_5O_8$) ist hierfür ein wichtiges Beispiel. Die Oxid-Spinelle lassen auch eine große Anzahl von Leerstellen an Stelle von Metall-Ionen zu; hierfür ist γ-Fe_2O_3 ein besonders stark ausgeprägtes Beispiel mit $2\frac{2}{3}$ Leerstellen pro Einheitszelle. Die Existenz dieser Leerstellen an Stelle von Metall-Ionen ist häufig der Grund für die schwierige Herstellung von Spinellen mit stöchiometrischem Verhältnis zwischen Metall und Sauerstoff.

Oxid-Spinelle können aus Kationen gebildet werden, deren Ladungen von +1 bis +6 variieren und deren Radien einen Wert zwischen weniger als 0.5 bis fast 1.0 Å haben. Außerdem können einige Ionen, die für die Bildung der Spinell-Struktur zu groß (wie Ca^{2+}) oder zu klein (wie Si^{4+}) sind, in begrenztem Ausmaß in andere Spinelle eingebaut werden. Da gewöhnlich die verschiedenen Oxid-Spinelle vollständig mischbar sind, ist der Bereich der Oxid-Verbindungen mit Spinell-Struktur sehr groß.

Die Verteilung der Metall-Ionen auf A- und B-Gitterplätze wird durch verschiedene Effekte bestimmt. Die A-Plätze sind kleiner als die B-Plätze; daher streben die kleineren Ionen, die gewöhnlich eine höhere Ladung haben, danach, die A-Gitterplätze zu besetzen. Dieser Effekt wird weniger wichtig, wenn der Parameter u wächst und die beiden Gitterplätze nahezu gleich groß werden. — Die elektrostatische (Madelung-) Energie wird begünstigt, wenn die höher-wertigen Kationen die B-Gitterplätze besetzen, wo ihnen eine große Anzahl Anionen zugeordnet ist. Dieser Effekt hängt von den Ionen-Abständen ab und nimmt mit wachsendem u zu. — Die Elektronenkonfigurationen einiger Kationen verursacht die Bevorzugung eines besonderen Gitterplatzes. Ionen wie Zn^{2+} und Ga^{3+} mit einer d^{10}-Konfiguration bilden teilweise kovalente Bindungen in den Tetraeder-Lücken. Ionen

ly lowered by crystal field effects. — With so many different effects involved, the site occupancy of any cation generally depends on the other cations present. In [3] the "octahedral-site preference energies" for twenty-one cations in 2-3 oxide spinels have been calculated. The chromites $Me'Cr_2O_4$ are all normal because of the very great preference of Cr^{3+} ions for the *B* sites. Fe, Co, Ni and Li ferrites have the inverse structure; Zn ferrites are normal; and Mn, Mg and Cu ferrites have mixed structures. In the latter two ferrites, the ionic distribution varies substantially with the heat treatment of the sample. — [1] gives a more extensive description of the structure of oxide spinels with emphasis on those features of interest in connection with the magnetic properties of the ferrites.

The spinel structure becomes less stable as the electronegativity of the anion decreases. The number of sulfide spinels is small and the number of selenide spinels is even smaller. Only two telluride spinels $CuCr_2Te_4$ and $ZnMn_2Te_4$, have been reported. Many sulfides and selenides, e.g. $Fe_{1-x}S$, $Fe_{1-x}Se$ and Cr_3S_4, form the nickel arsenide structure rather than spinel. There is apparently little difference in the energies of these two structures and, in fact, the sulfide spinels $FeCr_2S_4$ and $CoCr_2S_4$ can be transformed to a NiAs-type structure by the application of a pressure of 65 kbar [26]. Most of the reported sulfide spinels belong to one of the series $Me'Cr_2S_4$ (with Me' = Mn, Fe, Co, Cu, Zn, Cd and Hg) and $CuMe''_2S_4$ (with Me'' = Ti, V, Cr, Co and Rh). The analogous selenide spinels, with Me' limited to Cu, Zn, Cd and Hg and Me'' limited to Cr and Rh, exist. In addition, a few spinels which can be regarded as substitutions in Co_3S_4 and the complete set of chalcogenides, $ZnMn_2X_4$ with X = S, Se and Te, have been reported. In [27] the magnetic and electrical properties of these sulfide and selenide spinels have been tabulated.

Almost all sulfide and selenide spinels are normal 2-3 spinels. In contrast to the oxides, where there is generally complete miscibility, only very limited miscibility was found between some sulfide and selenide spinels [27]. Another difference from the oxide spinels is the possibility of anion substitutions. The mixed sulfide—selenide systems, $CdCr_2S_{4-x}Se_x$ [28] and $CuCr_2S_{4-x}Se_x$ [29], have been investigated. Halogen substitutions, i.e. Cl for S, Br for Se, and I for Te, in $CuCr_2$ chalcogenides have been made by several investigators [30, 31, 32].

Interest in the metallic $CuMe''_2X_4$ chalcogenide spinels was stimulated by the discovery [7] that the compounds $CuCr_2X_4$ (X = S, Se and Te) are ferromagnetic with Curie temperatures in the neighborhood of 400 °K (see 6.1.0.2). Two different models have been proposed to account for the observed properties of these compounds. According

wie Cr^{3+} mit einer d^3- oder Ni^{2+} mit einer d^8-Konfiguration haben auf Oktaeder-Lücken eine durch Kristallfeld-Effekte erheblich erniedrigte Energie.— Es sind also viele verschiedene Effekte beteiligt. Daher hängt die Besetzung eines Gitterplatzes durch irgendein Anion gewöhnlich von den anderen vorhandenen Kationen ab. In [3] sind die Energien für die Bevorzugung von Oktaeder-Gitterplätzen für 21 Kationen in 2-3 Oxid-Spinellen berechnet. Alle Chromite $Me'Cr_2O_4$ haben normale Spinellstruktur, weil die Cr^{3+}-Ionen die *B*-Gitterplätze bevorzugen. Fe-, Co-, Ni- und Li-Ferrite haben die inverse Struktur; Zn-Ferrite haben normale, Mg- und Cu-Ferrite gemischte Strukturen. Bei den beiden letztgenannten Ferriten ändert sich die Ionen-Verteilung erheblich mit der Wärmebehandlung der Probe. — In [1] ist eine ausführlichere Beschreibung der Struktur der Oxid-Spinelle enthalten, in der diejenigen Merkmale betont werden, die in Zusammenhang mit den magnetischen Eigenschaften der Ferrite wichtig sind.

Die Spinell-Struktur wird weniger stabil, wenn die Elektronegativität der Anionen abnimmt. Die Anzahl der Sulfid-Spinelle ist klein und die der Selenid-Spinelle noch kleiner. Tellurid-Spinelle sind bisher nur in zwei Verbindungen, $CuCr_2Te_4$ und $ZnMn_2Te_4$, bekannt geworden. Viele Sulfide und Selenide, z.B. $Fe_{1-x}S$, $Fe_{1-x}Se$ und Cr_3S_4, bilden eher die Nickelarsenid-Struktur als die Spinell-Struktur. Anscheinend ist der Unterschied in den Energien dieser beiden Strukturen gering; tatsächlich können auch die beiden Sulfid-Spinelle $FeCr_2S_4$ und $CoCr_2S_4$ bei Anwendung eines Drucks von 65 kbar die Struktur vom NiAs-Typ annehmen [26]. Die meisten der bekannten Sulfid-Spinelle gehören zu den Reihen $Me'Cr_2S_4$ (mit Me' = Mn, Fe, Co, Cu, Zn, Cd und Hg) und $CuMe''_2S_4$ (mit Me'' = Ti, V, Cr, Co und Rh). Die entsprechenden Selenid-Spinelle sind auf Me' = Cu, Zn, Cd und Hg bzw. Me'' = Cr, Rh beschränkt. Außerdem sind einige wenige Spinelle, die als Substitutionen in Co_3S_4 angesehen werden können, und die vollständige Reihe der Chalcogenid-Verbindungen $ZnMn_2X_4$ mit X = S, Se und Te, bekannt. In [27] sind die magnetischen und elektrischen Eigenschaften dieser Sulfid- und Selenid-Spinelle zusammengestellt.

Beinahe alle Sulfid- und Selenid-Spinelle sind normale 2-3 Spinelle. Im Gegensatz zu den Oxiden, die gewöhnlich vollständig mischbar sind, ist die Mischbarkeit zwischen einigen Sulfid- und Selenid-Spinellen sehr begrenzt [27]. Ein anderer Unterschied zu den Oxid-Spinellen ist die Möglichkeit der Anionen-Substitutionen. Die gemischten Sulfid-Selenid-Systeme $CdCr_2S_{4-x}Se_x$ [28] und $CuCr_2S_{4-x}Se_x$ [29] sind untersucht worden. In $CuCr_2$-Chalcogeniden wurden Halogene substituiert, z.B. Cl für S, Br für Se und J für Te [30, 31, 32].

Das Interesse an den metallischen Chalcogenid-Spinellen $CuMe''_2X_4$ wurde durch die Entdeckung [7] verstärkt, daß die Verbindungen $CuCr_2X_4$ (X = S, Se, Te) ferromagnetisch sind und ihre Curie-Temperaturen bei etwa 400 °K liegen (siehe 6.1.0.2). Um die beobachteten Eigenschaften dieser Verbindungen verstehen zu können, wurden zwei

to one model [33], the electronic configuration is $Cu^{+}Cr^{3+}Cr^{4+}X_4$ and the band responsible for the metallic conduction is associated with chromium in the *B* sites. The alternative model [34] postulates the configuration $Cu^{2+}Cr_2^{3+}X_4$. According to this model, the band responsible for the metallic behavior is associated with copper on the *A* sites. In an effort to resolve this question, other copper chalcogenide spinels, $CuTi_2S_4$, CuV_2S_4, $CuCo_2S_4$, $CuRh_2S_4$ and $CuRh_2Se_4$, and mixed compounds with $CuCr_2S_4$ and $CuCr_2Se_4$ have been investigated [27, 33]. Both models are apparently capable of explaining the observed behavior of these compounds.

A surprising result of this work was the discovery [35, 36] that three of these compounds, CuV_2S_4, $CuRh_2S_4$ and $CuRh_2Se_4$, are superconductors with transition temperatures in the range 3.5···4.8 °K. This undoubtedly presages an increase in the investigation of these chalcogenide spinels.

verschiedene Modelle vorgeschlagen. In dem einen Modell [33] ist die Elektronenkonfiguration $Cu^{+}Cr^{3+}Cr^{4+}X_4$, und das für die metallische Leitung verantwortliche Elektronenband gehört zu Cr auf den *B*-Gitterplätzen. Das andere Modell [34] setzt die Konfiguration $Cu^{2+}Cr_2^{3+}X_4$ voraus. Nach diesem Modell gehört das für die metallische Leitung verantwortliche Elektronenband zu Cu auf den *A*-Gitterplätzen. Um diese Frage zu lösen, wurden andere Cu Chalcogenid-Spinelle untersucht: $CuTi_2S_4$, CuV_2S_4, $CuCo_2S_4$, $CuRh_2S_4$ und $CuRh_2Se_4$, sowie die Mischungen von $CuCr_2S_4$ und $CuCr_2Se_4$ [27, 33]. Beide Modelle scheinen das beobachtete Verhalten dieser Verbindungen erklären zu können.

Ein überraschendes Ergebnis dieser Arbeit war die Entdeckung [35, 36], daß drei dieser Verbindungen, CuV_2S_4, $CuRh_2S_4$ und $CuRh_2Se_4$, Supraleiter mit Übergangstemperaturen zwischen 3.5··· 4.8 °K sind. Dies wird ohne Zweifel die Untersuchungen an den Chalcogenid-Spinellen stark zunehmen lassen.

6.1 Fe^{3+} spinels (cubic ferrites)

6.1.0 Ferrimagnetism; magnetic behaviour of ferrites – Ferrimagnetismus; magnetisches Verhalten der Ferrite

6.1.0.1 *A-B* exchange interaction; magnetization — (*A-B*)-Austauschwechselwirkung; Magnetisierung

The first clear understanding of the magnetic properties of the ferrites was provided by Néel [2]. In his original paper, Néel proposed the concept of ferrimagnetism, devised a simple "molecular field" theory to predict the properties of such materials, and showed that the ferrites were probably ferrimagnetic instead of ferromagnetic (as had been previously supposed). According to Néel, there are two necessary conditions for ferrimagnetism: 1) the crystal structure should contain two non-equivalent sets of sites (*A* and *B*) for the magnetic ions, and 2) the principal exchange interaction should be a negative interaction between nearest neighbor *A-B* pairs. If these conditions are fulfilled, a spontaneously ordered arrangement of the magnetic moments occurs at low temperatures. This ordered arrangement resembles antiferromagnetism in that M_a and M_b are aligned antiparallel but differs from it in that a net magnetization appears because the two sublattices are not equivalent.

Because of the ferrimagnetic nature of ferrites, the interpretation of their magnetic data differs considerably from the interpretation of corresponding data on ferromagnetic systems. Beginning with the static magnetic properties, we have

Das erste klare Verständnis der magnetischen Eigenschaften der Ferrite verdankt man Néel [2]. In seiner Arbeit schlug Néel den Begriff des Ferrimagnetismus vor, stellte eine einfache „Molekularfeld"-Theorie auf, um die Eigenschaften solcher Substanzen vorauszusagen, und zeigte, daß die Ferrite wahrscheinlich ferrimagnetisch statt ferromagnetisch sind, wie man bislang angenommen hatte. Nach Néel gibt es zwei notwendige Bedingungen für den Ferrimagnetismus: 1) die Kristallstruktur muß zwei nicht gleichwertige Untergitter (*A* und *B*) für die magnetischen Ionen enthalten, und 2) die vorherrschende Austauschwechselwirkung muß eine negative Wechselwirkung zwischen den nächstbenachbarten (*A-B*)-Paaren sein. Wenn diese Bedingungen erfüllt sind, erfolgt bei niedrigen Temperaturen spontan eine Ausrichtung der magnetischen Momente. Dieser geordnete Zustand ähnelt dem Antiferromagnetismus darin, daß M_a und M_b antiparallel ausgerichtet sind, unterscheidet sich aber von ihm dadurch, daß eine resultierende Magnetisierung auftritt, weil die beiden Untergitter nicht gleichwertig sind.

Wegen der ferrimagnetischen Natur der Ferrite weicht die Deutung ihrer magnetischen Meßwerte beträchtlich von der Deutung der entsprechenden Meßwerte für ferromagnetische Systeme ab. Betrachtet man zunächst die statischen magnetischen Eigenschaften, so gilt

$$M = M_b - M_a \qquad (1)$$

where M is the observed spontaneous magnetization and M_a and M_b are the magnitudes of the spontaneous magnetizations of the *A* and *B* sublattices, respectively. This arrangement is shown schematically in Fig. 2a. As a corollary of Eq. (1),

wobei M die beobachtete spontane Magnetisierung ist und M_a und M_b die Beträge der spontanen Magnetisierung der *A*- bzw. *B*-Untergitter sind. Dieser Zusammenhang ist in Abb. 2a schematisch dargestellt. Parallel zu Gl. (1) folgt für das aus der

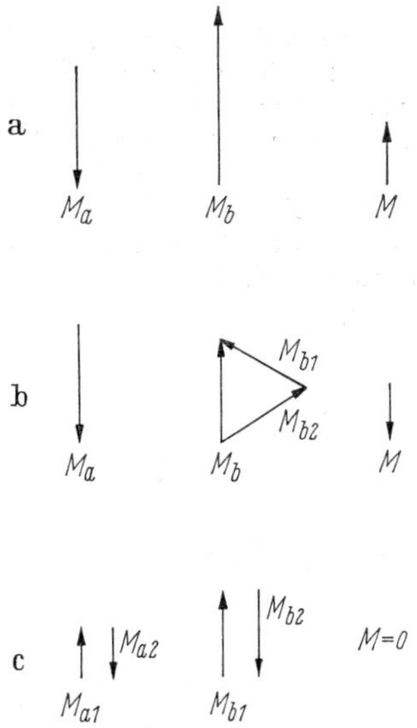

Fig. 2. Arrangements of magnetic moments in ferrites. (From SMART: Landolt-Börnstein, 6. ed., vol. II/9).

the magnetic moment per unit $MeFe_2O_4$, as determined from the saturation magnetization at low temperatures, is given by

Sättigungsmagnetisierung bei tiefen Temperaturen bestimmte magnetische Moment pro Formeleinheit $MeFe_2O_4$

$$p_m^0 = (2n_b - n_a)\,\mu_B \qquad (2)$$

where μ_B is the Bohr magneton, n_a is the number of Bohr magnetons per A ion, and n_b is the number of Bohr magnetons per B ion. If a given kind of site is occupied by more than one kind of ion, the average values of n_b and/or n_a should be used. For example, for inverse ferrites, we have

wobei μ_B das Bohrsche Magneton, n_a die Anzahl der Bohrschen Magnetonen je A-Ion und n_b die Anzahl der Bohrschen Magnetonen je B-Ion ist. Wenn eine bestimmte Art von Gitterplätzen von mehr als einer Ionensorte besetzt ist, sind dabei die Mittelwerte von n_b und/oder n_a einzusetzen. So ergibt sich z. B. für inverse Ferrite

$$p_m^0 = (n_{Fe^{3+}} + n_{Me^{2+}} - n_{Fe^{3+}})\,\mu_B = n_{Me^{2+}}\,\mu_B\,. \qquad (2a)$$

Corresponding results for solid solutions of two or more ferrites can be obtained by obvious extensions of the rule. It is clear from Eq. (2) that the distribution of metal ions over the A and B sites can have profound effects on the magnetic properties of the ferrites. GOODENOUGH [3] has extensively discussed the site occupancy preferences of the cations which are incorporated in ferrites.

An example of the type of experimental result which was not understood prior to Néel's work is the data on the saturation magnetization of solid solutions of nickel ferrite and zinc ferrite. When small quantities of non-magnetic zinc ferrite are added to nickel ferrite, the saturation magnetization is observed to increase. In a ferromagnetic model, it is difficult to see how substitution of a non-magnetic constituent could increase the saturation moment; using the Néel model, one sees that, if the diamagnetic zinc ions go to A sites, M_a will decrease as expected but the difference $M = M_b - M_a$ will increase. Experimental magnetization and structure investigations have shown this explanation to be correct.

As another example, the reciprocal susceptibility vs. temperature curve for ferrites is not approximately linear, as for ferromagnets and antiferromagnets, but is instead approximately hyperbolic

Durch sinngemäße Erweiterung der Regel kann man entsprechende Resultate für Mischkristalle aus zwei (oder mehr) Ferriten erhalten. Aus Gl. (2) sieht man sofort, daß die Verteilung der Metall-Ionen auf die A- und B-Gitterplätze die magnetischen Eigenschaften der Ferrite erheblich beeinflussen kann. GOODENOUGH [3] hat die Bevorzugung bestimmter Gitterplätze durch die in die Ferrite eingebauten Kationen ausführlich erörtert.

Ein Beispiel für diese Art experimenteller Ergebnisse, die man vor der Néelschen Arbeit nicht verstand, sind die Daten über die Sättigungsmagnetisierung der Mischkristalle von Nickel- und Zinkferrit. Fügt man kleine Mengen nichtmagnetischen Zinkferrits zu Nickelferrit, so beobachtet man eine Zunahme der Sättigungsmagnetisierung. Es ist schwer einzusehen, wie in einem ferromagnetischen Modell die Substitution eines nichtmagnetischen Bestandteils das Sättigungsmoment erhöhen könnte; im Néelschen Modell nimmt M_a zwar wie erwartet ab, falls die diamagnetischen Zink-Ionen die A-Plätze besetzen, jedoch die Differenz $M = M_b - M_a$ nimmt zu. Experimentelle Magnetisierungs- und Strukturuntersuchungen haben gezeigt, daß diese Erklärung richtig ist.

Ein weiteres Beispiel: die Temperaturabhängigkeit der reziproken Suszeptibilität ist für Ferrite nicht angenähert linear wie für die Ferromagnetika und Antiferromagnetika, sondern angenähert hy-

and is concave toward the T-axis. Such behavior is predicted by the Néel theory. Fig. 3 shows a typical $1/\chi$ vs. T curve for a ferrite. The T-intercept of the linear extrapolation of the high temperature portion of the curve is called paramagnetic (or asymptotic) Curie (or Néel) temperature Θ_p (or Θ_a); it is analogous to Θ_p in the Curie-Weiss law for an antiferromagnet.

perbolisch und konkav gegen die T-Achse. Ein solches Verhalten wird von der Néelschen Theorie vorhergesagt. Abb. 3 zeigt den typischen Verlauf von $1/\chi$ als Funktion von T für einen Ferrit. Den Schnittpunkt der linearen Extrapolation des für hohe Temperaturen gültigen Teils der Kurve mit der T-Achse nennt man die paramagnetische (oder auch die asymptotische) Curie- (oder Néel-) Temperatur Θ_p (oder Θ_a); sie entspricht Θ_p im Curie-Weißschen Gesetz für Antiferromagnetika.

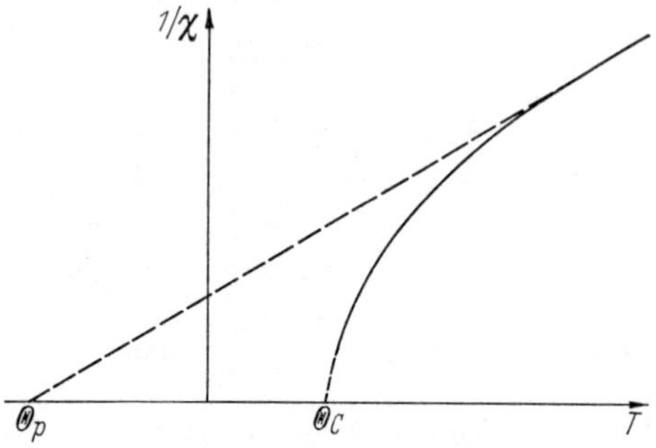

Fig. 3. Typical $1/\chi$ vs. T-curve for a ferrite. (From SMART: Landolt-Börnstein, 6. ed., vol. II/9.)

6.1.0.2 *A-A*, *B-B* exchange interaction; magnetic order — (*A-A*), (*B-B*) Austauschwechselwirkung; magnetische Ordnung

In addition to the A-B exchange interaction there are also exchange interactions between pairs of ions on the same kind of site (A-A and B-B). These intrasublattice interactions are usually much weaker than the intersublattice A-B interaction but, as Néel showed, they can profoundly affect the temperature dependence of the magnetization. Although both sublattice magnetizations, M_a and M_b, are monotonically decreasing functions of temperature, the observed spontaneous magnetization M (Eq. 1) can vary with temperature in a variety of ways. If M_a decreases more rapidly than M_b, due to an antiferromagnetic A-A interaction, M can increase with temperature and have a maximum below the Curie point. Conversely, if M_b decreases more rapidly than M_a, M can become zero at some temperature below the Curie point. The experimental demonstration of the existence of such a "compensation Point" by GORTER and SCHULKES [4] provided direct evidence for the essential correctness of Néel's theory.

Außer der (A-B)-Austauschwechselwirkung gibt es auch Austauschwechselwirkungen zwischen Ionenpaaren auf Plätzen desselben Untergitters, (A-A und B-B). Diese Wechselwirkungen innerhalb *eines* Untergitters sind gewöhnlich viel schwächer als die (A-B)-Wechselwirkung zwischen verschiedenen Untergittern; sie können aber, wie Néel gezeigt hat, die Temperaturabhängigkeit der Magnetisierung erheblich beeinflussen. Obwohl die beiden Untergitter-Magnetisierungen M_a und M_b monoton abnehmende Funktionen der Temperatur sind, kann die beobachtete spontane Magnetisierung M (Gl. 1) mit der Temperatur auf die verschiedenste Weise variieren. Wenn M_a infolge einer antiferromagnetischen (A-A)-Wechselwirkung schneller abnimmt als M_b, kann M mit der Temperatur zunehmen und ein Maximum unterhalb des Curie-Punktes haben. Umgekehrt kann M, wenn M_b schneller als M_a abnimmt, bei einer bestimmten Temperatur unterhalb des Curie-Punktes verschwinden. Der experimentelle Nachweis durch GORTER und SCHULKES [4] für das Vorhandensein eines solchen „Kompensationspunktes" lieferte einen direkten Beweis für die grundsätzliche Richtigkeit der Néelschen Theorie.

As the A-A or B-B interactions become sufficiently large compared to the A-B interaction, the magnetic ordering at low temperature no longer corresponds to a simple anti-parallel arrangement of M_a and M_b (Fig. 2a, p. 55). YAFET and KITTEL [5] extended Néel's treatment by allowing for the formation of more than one magnetic sublattice on both the A and B sites. They obtained the triangular and doubly antiferromagnetic arrangements, shown schematically in Fig. 2b and 2c (p. 55), as the description of the magnetic ordering at low temperature when the A-A or B-B interactions are sufficiently large. More recent work by KAPLAN et al. [6] indicated that, while these triangular configurations have a lower energy than the Néel arrangement, they do not represent the magnetic ground

Wenn die (A-A)- oder (B-B)-Wechselwirkungen im Vergleich zur (A-B)-Wechselwirkung genügend groß werden, entspricht die magnetische Ordnung nicht mehr einer einfachen antiparallelen Anordnung von M_a und M_b (Abb. 2a, S. 55). YAFET und KITTEL [5] erweiterten Néels Theorie, indem sie die Möglichkeit der Ausbildung von mehr als einem magnetischen Untergitter sowohl auf den A- als auch auf den B-Plätzen berücksichtigten. Sie erhielten die Dreiecks-Anordnung und die doppelt-antiferromagnetische Anordnung, die schematisch in Abb. 2b und 2c (S. 55), dargestellt sind, als Beschreibung der magnetischen Ordnung bei niedriger Temperatur, wenn die (A-A)- oder (B-B)-Wechselwirkung genügend groß ist. Eine neuere Arbeit von KAPLAN et. al. [6] weist darauf hin, daß diese Dreiecks-Anord-

state. Instead, the ground state is a spiral configuration which cannot be described in terms of sublattices.

nungen nicht den magnetischen Grundzustand darstellen, obwohl sie eine geringere Energie als die Néelsche Anordnung haben. Der Grundzustand ist statt dessen durch eine spiralförmige Anordnung gegeben, die sich nicht durch Untergitter beschreiben läßt.

The recent discoveries [*7, 8*] that several chromium sulfide and selenide spinels are ferromagnetic have stimulated renewed interest in the *B-B* interactions. Some of these compounds ($CdCr_2S_4$, $CdCr_2Se_4$, $HgCr_2S_4$ and $HgCr_2Se_4$ [*8*]) are semiconductors with nonmagnetic ions in the *A* sites. The exchange interaction between Cr^{3+} ions in nearest-neighbor *B* sites becomes more ferromagnetic as the anion becomes less electronegative and as Cr-Cr separation, i.e. the unit cell edge, becomes larger [*8, 37*]. The exchange interactions between Cr^{3+} ions in more widely separated *B* sites are weaker than those between nearest neighbors but these distant-neighbor interactions play a significant role in determining the magnetic properties of the chromium chalcogenide spinels [*8, 27, 38*].

Die jüngsten Entdeckungen [*7, 8*], die gezeigt haben, daß mehrere Chrom-Sulfid- und Selenid-Spinelle ferromagnetisch sind, haben erneutes Interesse an den (*B-B*)-Wechselwirkungen hervorgerufen. Einige dieser Verbindungen ($CdCr_2S_4$, $CdCr_2Se_4$, $HgCr_2S_4$ und $HgCr_2Se_4$ [*8*]) sind Halbleiter mit nicht-magnetischen Ionen auf den *A*-Gitterplätzen. Die Austausch-Wechselwirkung zwischen Cr^{3+}-Ionen auf nächstgelegenen *B*-Gitterplätzen wird um so stärker ferromagnetisch je weniger elektronegativ die Anionen werden, und je größer der Cr-Cr-Abstand, d.h. die Größe der Einheitszelle, wird [*8, 37*]. Die Austausch-Wechselwirkungen zwischen Cr^{3+}-Ionen auf weiter entfernten *B*-Gitterplätzen sind schwächer als die zwischen nächsten Nachbarn, aber diese Wechselwirkungen mit entfernten Nachbarn spielen eine wichtige Rolle für die Festlegung der magnetischen Eigenschaften der Cr-Chalcogenid-Spinelle [*8, 27, 38*].

The other ferromagnetic chromium spinels ($CuCr_2X_4$ with X = S, Se and Te) are metallic conductors. Both models of the electronic structure of these compounds, discussed in 6.0.2 indicate that there would be no localized magnetic moments associated with the *A* site copper ions and this has been confirmed by neutron diffraction studies [*39, 40*]. The ferromagnetic character of these compounds is due to a combination of *B-B* interactions, similar to those in the ferromagnetic semiconductors discussed above, and a long range coupling via the "delocalized electrons". The presence of this additional coupling is indicated by the much higher Curie temperatures of the conducting ferromagnets (390···433 °K) as compared to the semiconducting ones (36···130 °K).

Die anderen ferromagnetischen Cr Spinelle ($CuCr_2X_4$ mit X = S, Se, Te) sind metallische Leiter. Beide in 6.0.2 besprochenen Modelle der Elektronenstruktur dieser Verbindungen zeigen, daß keine lokalisierten magnetischen Momente zu den Cu-Ionen auf *A*-Gitterplätzen gehören können; das ist durch Untersuchungen mit Neutronenstreuung bestätigt worden [*39, 40*]. Der ferromagnetische Charakter dieser Verbindungen entspricht einer Kombination von (*B-B*)-Wechselwirkungen, ähnlich den oben besprochenen bei ferromagnetischen Halbleitern, und einer weit reichenden Kopplung über die ,,delokalisierten Elektronen". Das Vorhandensein dieser zusätzlichen Kopplung zeigt sich in den viel höheren Curie-Temperaturen der leitenden Ferromagnetika (390···433 °K) im Vergleich zu den halbleitenden (36···130 °K).

6.1.0.3 Hyperfine interactions (Mössbauer and NMR methods) — Hyperfeinwechselwirkungen (Mössbauer- und NMR-Verfahren)

The investigation of the hyperfine interactions of various nuclei in ferrites is providing significant new information about these materials. This is an active, rapidly developing field, and in this brief discussion, we can only indicate the significant contributions being made to our knowledge of ferrites. Freeman and Watson [*10*] have discussed the origins of the hyperfine interactions in magnetic materials and reviewed several of the experimental methods available to measure them. The interest in using these hyperfine interactions to investigate the electronic magnetism of many different materials has resulted from the recent discovery of two effective techniques, the Mössbauer effect (ME, the recoilfree emission and absorption of γ-rays) and nuclear magnetic resonance (NMR), for measuring these interactions. There are two hyperfine interactions which split the energy levels of the nucleus of a magnetic ion; one is proportional to

Die Untersuchung der Hyperfeinwechselwirkungen verschiedener Kerne in Ferriten liefert wichtige neue Informationen über diese Substanzen. Dieses Gebiet befindet sich in rascher Entwicklung; in der vorliegenden kurzen Erörterung können wir deshalb nur auf die wichtigsten Beiträge, die zu unseren Kenntnissen über die Ferrite beitragen, hinweisen. Freeman und Watson [*10*] haben die Ursache der Hyperfeinwechselwirkungen in magnetischen Substanzen diskutiert und einige experimentelle Methoden zu ihrer Messung erörtert. Das Interesse an der Anwendung dieser Hyperfeinwechselwirkungen zur Untersuchung des elektronischen Magnetismus vieler verschiedener Substanzen ergab sich kürzlich aus der Entdeckung zweier wirkungsvoller Verfahren zur Messung dieser Wechselwirkungen, des Mössbauer-Effektes (ME, der rückstoßfreien Emission und Absorption von γ-Strahlen) und der magnetischen Kernresonanz

the magnetic moment and the other to the electric quadrupole moment of the nucleus. In addition, ME experiments permit the measurement of "isomer shifts", the small changes of the nuclear energy levels which depend on the electron density at the nucleus.

In many cases, the magnetic hyperfine interaction can be represented in the Hamiltonian by the term $A\,I \cdot S$ where I is the nuclear spin, S the electronic spin and A a coefficient which is usually independent of temperature. There are a few cases in which the anisotropy of this hyperfine interaction or the explicit temperature dependence of this coefficient must be considered. The coefficient A is primarily a property of the nuclear species involved but it also depends on the environment of the ion. In ferrites, A is about five or six percent larger if the ion is in an octahedral site than if it is in a tetrahedral site. Alternatively, the hyperfine interaction can be written in the form

(NMR). Es gibt zwei Hyperfeinwechselwirkungen, welche die Kernenergieniveaus eines magnetischen Ions aufspalten; die eine ist dem magnetischen Moment und die andere dem elektrischen Quadupolmoment des Kerns proportional. Außerdem ermöglichen die ME-Versuche die Messung von „Isomerie-Verschiebungen", jenen Änderungen der Kernenergieniveaus, die von der Elektronendichte am Kern abhängen.

In vielen Fällen kann die magnetische Hyperfeinwechselwirkung im Hamilton-Operator durch den Ausdruck $A\,I \cdot S$ dargestellt werden, wobei I der Kernspin, S der Elektronenspin und A ein Koeffizient ist, der gewöhnlich nicht von der Temperatur abhängt. Es gibt einige wenige Fälle, in denen die Anisotropie dieser Hyperfeinwechselwirkung oder die Temperaturabhängigkeit dieses Koeffizienten berücksichtigt werden muß. Der Koeffizient A ist charakteristisch für die betreffende Kernart, hängt aber auch von der Umgebung des Ions ab. In den Ferriten ist A für ein Ion auf einem Oktaeder-Gitterplatz um etwa fünf oder sechs Prozent größer als für ein Ion auf einem Tetraeder-Gitterplatz. Andererseits kann die Hyperfeinwechselwirkung auch in der folgenden Form geschrieben werden

$$\mathcal{H} = -\boldsymbol{p}_{\mathrm{n}} \cdot \boldsymbol{H}_{\mathrm{n}} = -g_{\mathrm{n}} \mu_{\mathrm{n}} \boldsymbol{I} \cdot \boldsymbol{H}_{\mathrm{n}}$$

where $\boldsymbol{p}_{\mathrm{n}}$ is the nuclear magnetic moment, g_{n} is the nuclear spectroscopic splitting factor, and μ_{n} is the nuclear magneton. This interaction splits the nuclear energy level into $2I + 1$ equally spaced levels. The energy separation between these levels, ΔE, is proportional to the average value of the electronic spin S which in turn is proportional to the magnetization of the sublattice on which the ion is located. The measurement of this splitting ΔE yields the magnetization of the individual sublattice and also the spin S and, hence, the valence state of the ion involved.

The NMR technique for measuring the hyperfine interaction is conceptually very simple. An rf magnetic field of frequency $\nu = \Delta E/h$ will induce transitions between the nuclear energy levels and this is observed as an absorption of rf power at this frequency. The only restriction on the nuclear species is that the ground level must have a spin $I \neq 0$. All the elements from Cr to Cu have at least one stable isotope suitable for NMR measurements. The amount of power absorbed in an NMR experiment is very small and the sample must satisfy stringent conditions on chemical purity and physical form in order to make the experiment possible. PORTIS and LINDQUIST [11] have described the use of this technique for investigating ferro- and ferrimagnetic materials and have summarized the results of some studies of ferrites.

The nuclear resonant absorption of γ-rays was discovered in 1958. MÖSSBAUER [12] realized that nuclei bound in solids cannot recoil arbitrarily but, instead, their recoil energy is quantized. When the average recoil energy is less than the phonon energy

wobei $\boldsymbol{p}_{\mathrm{n}}$ das kernmagnetische Moment, g_{n} der kernspektroskopische Aufspaltungsfaktor und μ_{n} das Kernmagneton ist. Diese Wechselwirkung spaltet das Kernenergieniveau in $2I + 1$ Niveaus im gleichen Energieabstand voneinander auf. Die Energiedifferenz ΔE zwischen diesen Niveaus ist dem Durchschnittswert des Elektronenspins S proportional, der wiederum der Magnetisierung des Untergitters, in dem sich das Ion befindet, proportional ist. Aus der Messung dieser Aufspaltung ΔE ergibt sich die Magnetisierung des einzelnen Untergitters und auch der Spin S und daraus die Wertigkeit des betreffenden Ions.

Die NMR-Methode zur Messung der Hyperfeinwechselwirkung ist begrifflich sehr einfach. Ein hochfrequentes Magnetfeld der Frequenz $\nu = \Delta E/h$ induziert Übergänge zwischen den Kernenergieniveaus, und dies wird als Absorption der Hochfrequenzleistung bei dieser Frequenz beobachtet. Die einzige Einschränkung bei den Kernen ist, daß das Grundniveau einen Spin $I \neq 0$ haben muß. Alle Elemente von Cr bis Cu haben wenigstens ein stabiles, für NMR-Messungen geeignetes Isotop. Die in einem NMR-Versuch absorbierte Leistung ist sehr gering und die Probe muß in chemischer und physikalischer Hinsicht strengen Bedingungen genügen, um das Experiment zu ermöglichen. PORTIS und LINDQUIST [11] haben die Anwendung dieser Methode zur Untersuchung ferro- und ferrimagnetischer Substanzen beschrieben und die Ergebnisse einiger Untersuchungen an Ferriten zusammengefaßt.

Die Kernresonanzabsorption von γ-Strahlen wurde im Jahre 1958 entdeckt. MÖSSBAUER [12] erkannte, daß bei Kernen, die in Festkörpern eingebaut sind, die Rückstoßenergie nicht beliebige Werte annehmen kann, sondern gequantelt ist.

associated with the nuclear recoil, a fraction of the γ-rays are emitted without the creation of a phonon. Effectively, the recoil momentum is absorbed by the entire solid and, because of the large mass involved, the recoil energy is negligibly small. The importance of the ME results from the fact that the energy of these "recoil-free" γ-rays is extremely sharply defined.

When the emitting and absorbing nuclei are incorporated in different materials, the different hyperfine interactions will destroy the resonant absorption provided the energy shifts which result from these interactions are larger than the width of the energy distribution of the emitted γ-rays. The resonant absorption can be restored by modulating the γ-ray energy using the Doppler effect; a relative velocity v between the source and absorber will produce a fractional change v/c, where c is the velocity of light, in the γ-ray energy. The nuclide Fe^{57} is one of the few with a suitable lowlying, long-lived excited state and has been used in practically all of the ME studies of ferrites. Since the excited state of Fe^{57}, at an energy of 14.4 keV above the ground state, has a nuclear spin $I = \frac{3}{2}$ and the ground state has $I = \frac{1}{2}$, a magnetic hyperfine interaction will split these states into four and two levels, respectively. A total of six transitions are allowed by the usual selection rule for magnetic dipole transitions. A small quadrupole splitting of the excited state of Fe^{57} on the B sites of a ferrite is possible but this is usually significant only as a source of line broadening. The differences in the magnetic hyperfine interactions and isomer shifts of iron ions on the A and B sites result in a twelve-line ME spectrum for ferrites. The identification of the lines is frequently facilitated by investigating the changes produced in the ME spectrum by an external magnetic field. If the ferrite sample is magnetized parallel to the direction of propagation of the γ-rays, the number of allowed transitions for each site is reduced from six to four. An external field of approximately 20 kOe will produce measurable changes in the total field acting on the iron nuclei in a ferrite. Assuming, as is almost always the case, that M_b is greater than M_a, the total field at the A sites is increased and that at the B sites is decreased.

Wenn die mittlere Rückstoßenergie geringer als die mit dem Kernrückstoß verbundene Phononenenergie ist, wird ein Teil der γ-Strahlen emittiert, ohne ein Phonon zu erzeugen. Tatsächlich wird dann der Rückstoßimpuls vom gesamten Festkörper aufgenommen, und wegen der großen betroffenen Masse ist die Rückstoßenergie vernachlässigbar gering. Die Bedeutung der ME-Ergebnisse beruht auf der Tatsache, daß die Energie dieser „rückstoßfreien" γ-Strahlen äußerst genau definiert ist.

Wenn die emittierenden und absorbierenden Kerne in verschiedenartige Substanzen eingebaut sind, verschwindet die Resonanzabsorption auf Grund der unterschiedlichen Hyperfeinwechselwirkungen, vorausgesetzt, daß die Energieverschiebungen, die aus diesen Wechselwirkungen folgen, größer sind als die Breite der Energieverteilung der emittierten γ-Strahlen. Die Resonanzabsorption läßt sich durch Modulation der Energie der γ-Strahlen mit Hilfe des Doppler-Effekts wiederherstellen; eine relative Geschwindigkeit v zwischen der Strahlenquelle und dem Absorber wird die γ-Strahlen-Energie um den Bruchteil v/c ändern (c = Lichtgeschwindigkeit). Fe^{57} ist eines der wenigen Nuklide mit einem geeigneten niedrigen, langlebigen Anregungszustand und wurde in praktisch allen ME-Untersuchungen an Ferriten benutzt. Da der angeregte Zustand von Fe^{57} mit einer Energie von 14.4 keV über dem Grundzustand einen Kernspin $I = \frac{3}{2}$ und der Grundzustand einen Kernspin $I = \frac{1}{2}$ hat, spaltet die magnetische Hyperfeinwechselwirkung diese Zustände in vier bzw. zwei Niveaus auf. Insgesamt sind sechs Übergänge durch die übliche Auswahlregel für magnetische Dipolübergänge erlaubt. Eine kleine Quadrupolaufspaltung des angeregten Zustands von Fe^{57} auf den B-Gitterplätzen eines Ferrits ist möglich, sie hat jedoch im allgemeinen nur Bedeutung als Ursache der Linienverbreiterung. Die Unterschiede zwischen den magnetischen Hyperfeinwechselwirkungen und den Isomerie-Verschiebungen der Eisenionen auf den A- und B-Gitterplätzen ergeben ein ME-Spektrum von zwölf Linien. Die Identifizierung der Linien wird häufig durch die Untersuchung der im ME-Spektrum durch ein äußeres Magnetfeld erzeugten Änderungen erleichtert. Wenn die Ferrit-Probe parallel zur Ausbreitungsrichtung der γ-Strahlen magnetisiert ist, reduziert sich die Zahl der erlaubten Übergänge für jeden Gitterplatz von sechs auf vier. Ein äußeres Feld von ungefähr 20 kOe ruft bereits meßbare Änderungen des Gesamtfeldes am Ort der Eisenkerne im Ferrit hervor. Falls, wie fast immer der Fall, M_b größer ist als M_a, wird das Gesamtfeld an den A-Gitterplätzen dann größer und das an den B-Gitterplätzen kleiner.

6.1.0.4 Anisotropy — Anisotropie

Thus far, we have emphasized those properties of ferrites which are strongly affected by the ferrimagnetic character of these materials. Other properties such as the crystalline anisotropy and magnetostriction depend on the symmetry of the crystal structure and the existence of a spontaneous magnetization but are not explicitly dependent on

Bis jetzt haben wir diejenigen Eigenschaften der Ferrite hervorgehoben, die stark durch den ferrimagnetischen Charakter dieser Substanzen beeinflußt werden. Andere Eigenschaften wie die Kristall-Anisotropie und die Magnetostriktion hängen von der Symmetrie der Kristallstruktur und der Existenz einer spontanen Magnetisierung ab, sind aber

the detailed structure of this magnetization. Thus, phenomenological descriptions of these properties are similar to those of ferromagnetic materials. The atomic mechanisms responsible for these properties are characteristic of the ferrites so that their magnitudes and temperature dependences are, in many cases, quite different from those usually found in ferromagnets. The crystalline anisotropy of the ferrites has been extensively studied, both theoretically and experimentically. YOSIDA and TACHIKI [13] have analyzed the various mechanisms which could contribute to the anisotropy and concluded that, in most cases, "single ion" effects predominate. These "single ion" effects result from the anisotropic terms in the Hamiltonian of an isolated magnetic ion in a non-magnetic crystal lattice and are functions of both the specific ion and the site it occupies in the spinel lattice. FOLEN and RADO [14] found that the anisotropy of magnesium ferrites was due to additive contributions of ferric ions in the A and B sites. These "single ion" anisotropies have been extensively studied by WOLF [15]. They depend strongly on the magnetization of the relevant sublattice and, thus, implicitly on the temperature. Because the sublattice magnetizations of a ferrite depend differently on temperature, the total anisotropy can exhibit a wide variety of temperature dependences.

nicht explizit von der magnetischen Struktur im einzelnen abhängig. So ähnelt die phänomenologische Beschreibung dieser Eigenschaften derjenigen der ferromagnetischen Substanzen. Die verantwortlichen atomaren Mechanismen sind jedoch für die Ferrite charakteristisch, so daß diese Eigenschaften in vielen Fällen eine ganz andere Größe und Temperaturabhängigkeit aufweisen als sie in ferromagnetischen Substanzen gefunden werden. Die Kristall-Anisotropie der Ferrite ist sowohl theoretisch wie auch experimentell ausführlich untersucht worden. YOSIDA und TACHIKI [*13*] haben die verschiedenen Mechanismen analysiert, die zur Anisotropie beitragen könnten, und folgerten, daß in den meisten Fällen „Einzel-Ion"-Effekte vorherrschen. Diese „Einzel-Ion"-Effekte ergeben sich aus den anisotropen Gliedern des Hamilton-Operators für ein isoliertes magnetisches Ion in einem nichtmagnetischen Kristallgitter und sind Funktionen sowohl des spezifischen Ions als auch des Gitterplatzes, den es im Spinell-Gitter einnimmt. FOLEN und RADO [*14*] fanden, daß die Anisotropie der Magnesium-Ferrite auf additive Beiträge der Eisen-Ionen an den A- und B-Gitterplätzen zurückzuführen ist. Diese „Einzel-Ion"-Anisotropien wurden von WOLF [*15*] eingehend untersucht. Sie hängen stark von der Magnetisierung des entsprechenden Untergitters und damit von der Temperatur ab. Da die Untergittermagnetisierungen eines Ferrites in unterschiedlicher Weise von der Temperatur abhängen, kann die Gesamtanisotropie die verschiedenartigsten Temperaturabhängigkeiten aufweisen.

6.1.0.5 Ferrimagnetic resonance — Ferrimagnetische Resonanz

a) g-factor

In the absence of demagnetizing, damping and anisotropy effects, the magnetization in a sample subjected to a steady field H will obey the equation

a) g-Faktor

Beim Fehlen von Entmagnetisierungs-, Dämpfungs- und Anisotropie-Effekten folgt die Magnetisierung einer Probe in einem statischen Feld H der Gleichung

$$\dot{\boldsymbol{M}} = -\gamma(\boldsymbol{M} \times \boldsymbol{H}) \tag{4}$$

and the resulting motion of the magnetization will be a precession about the field H at the Larmor frequency

und führt eine Präzessionsbewegung um das Feld H aus mit der Larmor-Frequenz

$$\omega = \gamma H = \frac{g\,e}{2mc} H\,. \tag{5}$$

If an rf field is applied at right angles to the steady field H, there will be a transfer of power from this field to the precessing magnetization when the angular frequency of the alternating field equals ω. KITTEL [16] analyzed the resonance behaviour in ferrite samples and showed that, for spheres, the field H used in Eq. 4 and 5 should be the applied dc field. If the sample is an ellipsoid with demagnetizing factors N_x, N_y and N_z and the rf and dc fields are applied along the x- and z-axes respectively, the field H is replaced by the expression

Wenn ein Wechselfeld senkrecht zum statischen Feld H angelegt wird, erfolgt eine Energieübertragung von diesem Feld zur präzedierenden Magnetisierung, wenn die Winkelfrequenz des Wechselfelds gleich ω ist. KITTEL [*16*] hat das Resonanzverhalten in Ferrit-Proben analysiert und gezeigt, daß bei kugelförmigen Proben in den Gln. 4 und 5 für das Feld H das anglegte statische Feld einzusetzen ist. Wenn die Probe ein Ellipsoid mit den Entmagnetisierungsfaktoren N_x, N_y und N_z ist und das Hochfrequenzfeld und das statische Feld längs der x- bzw. z-Achse anglegt werden, ist das Feld H durch den folgenden Ausdruck zu ersetzen

$$\{[H + (N_y - N_z) M_s]\,[H + (N_x - N_z) M_s]\}^{1/2}\,.$$

TSUYA [*17*] and WANGSNESS [*18*] have shown that, with the usual assumptions of a sublattice coupled

TSUYA [*17*] und WANGSNESS [*18*] haben gezeigt, daß bei der üblichen Annahme eines durch ein

by a molecular field, the observed g-factor should be given by

$$g_{eff} = \frac{M_b - M_a}{M_b g_b^{-1} - M_a g_a^{-1}} \quad (6)$$

where g_a and g_b are the average spectroscopic splitting factors for the A and B sites, respectively. For most $3d$ transition metal ions (Co^{2+} is the best known exception) in ferrites, the quenching of the orbital angular momentum is nearly complete and, hence, the g-factors are fairly close to the "spin only" value of 2.00. Thus, the averages g_a and g_b and also g_{eff} are fairly close to 2.00. However, there are certain ferrites which have a "compensation point", i. e. the spontaneous magnetization vanishes at some temperature below the Curie point. Since g_a will not exactly equal g_b, the numerator and denominator of Eq. 6 will vanish at different temperatures. Thus at temperatures near the "compensation point", g_{eff} will take on a wide range of both positive and negative values. This has been observed experimentally in ferrites [*19*] and also in rare earth iron garnets .

b) Anisotropy

When the sample is a single crystal, the crystalline anisotropy produces a variation of the magnetic field for the resonance with crystal orientation. This has become a very common method of measuring the anisotropy constants of ferrites because of the small size of the sample and the relative ease of performing the experiment. The ferrimagnetic character of the ferrites does not play an essential role in this measurement and the equations usually used to interpret the data are those originally developed by Kittel for ferromagnets. This treatment is limited to cases where the anisotropy field, K_1/M, is much smaller than the applied field. Artman [*20*] has developed a more general method which can be used to handle very large anisotropies such as are encountered in $CoFe_2O_4$. It has recently been shown that, by observing the changes produced by known stresses in the fields for resonance, the magnetostriction constants can be measured [*21*].

c) Resonance line width

Ferrimagnetic resonance experiments provide the major part of our knowledge about the processes by which the magnetization relaxes to its equilibrium value. To discuss the relaxation an appropriate damping term must be included in the equation of motion of the magnetization (Eq. 4). The first phenomenological equation was proposed by Landau and Lifshitz and has the form

Molekularfeld gekoppelten Untergitters der beobachtete g-Faktor durch

wiedergegeben werden muß, wobei g_a und g_b die mittleren spektroskopischen Aufspaltungsfaktoren für die A- bzw. B-Gitterplätze sind. Für die meisten $3d$-Übergangsmetallionen (Co^{2+} ist die bekannteste Ausnahme) in Ferriten ist die Unterdrückung des Bahndrehimpulses beinahe vollständig und infolgedessen liegen die g-Faktoren ziemlich dicht beim Wert des reinen Spins, 2.00. Ebenso liegen die Mittelwerte g_a und g_b und auch g_{eff} ziemlich dicht bei 2.00. Es gibt jedoch bestimmte Ferrite, die einen „Kompensationspunkt" haben, d. h. die spontane Magnetisierung verschwindet bei einer bestimmten Temperatur unterhalb des Curie-Punkts. Da g_a nicht genau gleich g_b sein wird, werden der Zähler und der Nenner der Gl. 6 bei verschiedenen Temperaturen gleich Null. So kann g_{eff} bei Temperaturen nahe dem „Kompensationspunkt" innerhalb weiter Grenzen sowohl positive als auch negative Werte annehmen. Dies wurde experimentell bei Ferriten [*19*] und auch bei Selten-Erd-Eisengranaten beobachtet.

b) Anisotropie

Wenn die Probe ein Einkristall ist, erzeugt die Kristall-Anisotropie eine Änderung des Resonanzfeldes mit der Kristallorientierung. Darauf beruht ein Verfahren zur Messung der Anisotropiekonstanten der Ferrite, das wegen der benötigten geringen Größen der Probe und der relativ leichten Ausführbarkeit des Versuchs häufig verwendet wird. Der ferrimagnetische Charakter der Ferrite spielt bei dieser Messung keine wesentliche Rolle und die zur Erklärung der Meßwerte im allgemeinen benutzten Gleichungen sind dieselben, die ursprünglich von Kittel für die Ferromagnetika entwickelt wurden. Die Anwendbarkeit dieser Gleichungen beschränkt sich auf jene Fälle, in denen das Anisotropiefeld K_1/M viel kleiner als das angelegte Feld ist. Artman [*20*] hat ein allgemeineres Verfahren entwickelt, das man bei sehr großen Anisotropien anwenden kann, wie man sie in $CoFe_2O_4$ findet. Kürzlich wurde gezeigt, daß man die Konstanten der Magnetostriktion messen kann, indem man die Änderung der Resonanzfeldstärke unter dem Einfluß einer bekannten mechanischen Spannung beobachtet [*21*].

c) Resonanzlinienbreite

Unsere Kenntnisse über die Prozesse, durch welche die Magnetisierung zu ihrem Gleichgewichtszustand zurückkehrt, beruhen im wesentlichen auf ferrimagnetischen Resonanzversuchen. Um diese Relaxation diskutieren zu können, muß man ein geeignetes Dämpfungsglied in die Bewegungsgleichung der Magnetisierung (Gl. 4) einführen. Die erste phänomenologische Gleichung wurde von Landau und Lifshitz vorgeschlagen und hat die folgende Form

$$\dot{\boldsymbol{M}} = -\gamma(\boldsymbol{M}\times\boldsymbol{H}) - \lambda[\boldsymbol{M}\times(\boldsymbol{M}\times\boldsymbol{H})]. \quad (7)$$

The damping parameter λ is related to the width of the observed resonance by the equation

$$\Delta H_{1/2} = 2\gamma^{-1}\lambda M H_0 \tag{8}$$

where the line width, $\Delta H_{1/2}$, is the difference between the values of H at which the absorption equals one-half of its maximum value. In examining the data on resonance line widths, the reader must exercise considerable care because of the multiplicity of definitions of this quantity. Two other definitions in common use are the "half-width" equal to $\frac{1}{2}(\Delta H_{1/2})$ and the width ΔH_{ms} between the points of steepest slope on the sides of the absorption line. Several alternative forms for the damping term in Eq. 7 have been suggested but they are all limited by the fact that they describe the relaxation processes in terms of one or two parameters.

The microscopic description of the relaxation processes, recently summarized by Haas and Callen [*22*], is based on describing the deviation of the magnetization from complete alignment in terms of spin waves which are quantized and treated like particles ("magnons") of energy $\hbar\omega$ and momentum $\hbar\boldsymbol{k}$, where $|k| = 2\pi/\lambda$ is the wave number. In a resonance experiment, the rf field excites a uniform precession of the magnetization, i. e. it creates $k = 0$ magnons. The principal relaxation in ferrites involves the conversion of these $k = 0$ spin waves into $k \neq 0$ spin waves primarily due to "scattering" by static deviations from perfect periodicity in the spinel lattice. These deviations can be such things as local variations in the distribution of magnetic ions, voids, surface irregularities, etc. The relaxation process is completed by the conversion of these $k \neq 0$ spin waves into phonons (quantized lattice vibrations).

This relaxation behavior cannot be adequately described by a single damping parameter or a line width. Callen [*23*] has formulated an equation of motion in which the probabilities of occurence of the various magnon-magnon and magnon-phonon processes involved in the microscopic description of the relaxation appear explicitly. The presentation of the experimental results in terms of a single quantity, the line width, is limited in a way similar to the use of an equation of motion with a single damping parameter. Much progress has been made by measuring line widths under conditions such that one particular mechanism dominated the overall relaxation behavior. Another fruitful area has been the study of the non-linear phenomena which occur as the rf field is increased. Many aspects of this non-linear behavior can be associated with the generation of specific $k \neq 0$ spin waves. Hence,

Der Dämpfungsparameter λ ist mit der beobachteten Resonanzlinienbreite durch die Gleichung (8) verknüpft, wobei die Linienbreite (Halbwertbreite) $\Delta H_{1/2}$ den Abstand zwischen den Werten von H darstellt, bei denen die Absorption auf die Hälfte des Maximalwertes abgesunken ist. Die Angaben über die Resonanzlinienbreiten sind vom Leser wegen der verschiedenen Definitionen dieser Größe besonders kritisch aufzunehmen. Zwei weitere oft gebrauchte Begriffe sind „half-width" gleich $\frac{1}{2}(\Delta H_{1/2})$ und die Breite ΔH_{ms}, die zwischen den Punkten größter Steigung der Absorptionslinie gemessen wird. Abweichende Formen für das Dämpfungsglied in Gl. 7 sind vorgeschlagen worden; sie unterliegen jedoch der Einschränkung, daß zur Beschreibung der Relaxationsprozesse ein oder zwei Parameter eingeführt werden.

Die atomistische Beschreibung der Relaxationsprozesse, wie sie kürzlich von Haas und Callen [*22*] zusammengefaßt wurde, beruht auf der Beschreibung der Abweichung der Magnetisierung von der vollständigen Ausrichtung in Form von Spinwellen, die gequantelt sind und wie Teilchen („Magnonen") mit der Energie $\hbar\omega$ und dem Impuls $\hbar\boldsymbol{k}$ behandelt werden, wobei $|k| = 2\pi/\lambda$ die Wellenzahl ist. Im Resonanzversuch ruft das Hochfrequenzfeld eine einheitliche Präzession der Magnetisierung hervor, d. h. es erzeugt $k = 0$-Magnonen. Der wichtigste Relaxationsprozeß in den Ferriten hängt mit der Umwandlung dieser $k = 0$-Spinwellen in $k \neq 0$-Spinwellen zusammen, die in erster Linie auf die „Streuung" durch statische Abweichungen von der vollkommenen Periodizität des Spinell-Gitters zurückzuführen ist. Diese Abweichungen können örtliche Schwankungen in der Verteilung der magnetischen Ionen, Leerstellen, Unregelmäßigkeiten in der Oberfläche usw. sein. Der Relaxationsprozeß wird durch die Umwandlung dieser $k \neq 0$-Spinwellen in Phononen (gequantelte Gitterschwingungen) abgeschlossen.

Dieses Relaxationsverhalten kann durch einen einzelnen Dämpfungsparameter oder eine Linienbreite nicht hinreichend beschrieben werden. Callen [*23*] hat eine Bewegungsgleichung aufgestellt, in der die Wahrscheinlichkeiten der verschiedenen Magnon-Magnon- und Magnon-Phonon-Prozesse, die in der atomistischen Beschreibung der Relaxation vorkommen, explizit erscheinen. Die Darstellung der experimentellen Ergebnisse in Form einer einzigen Größe, der Linienbreite, ist ähnlich wie bei Verwendung einer Bewegungsgleichung mit einem einzigen Dämpfungsparameter nur begrenzt möglich. Durch die Messung der Linienbreiten unter Versuchsbedingungen, bei denen ein spezieller Mechanismus die gesamte Relaxation beherrschte, wurden große Fortschritte erzielt. Ein anderes lohnendes Gebiet ist durch die Untersuchung der nichtlinearen Phänomene erschlossen worden, die bei Erhöhung der Hochfrequenzfeldstärke auftreten. Viele Aspekte dieses nichtlinearen Verhaltens können mit der Erzeugung von spezifischen $k \neq 0$-Spinwellen in Zusammenhang gebracht werden. Folglich sind diese Untersuchungen geeignet,

these studies provide information about some features of the relaxation processes in ferrites which cannot be obtained from studies of their linear behavior.

Informationen über einige wesentliche Punkte der Relaxationsprozesse in Ferriten zu liefern, die man aus Untersuchungen ihres linearen Verhaltens nicht entnehmen kann.

d) Polycrystalline samples

In polycrystalline samples, two additional effects come into play. The different orientations of the individual grains in a polycrystalline sample lead to a variation of resonance fields in the different grains. However, the dipolar forces between the moments of the individual grains couple their motions so that the resultant line width is significantly less than would be expected for a collection of isolated grains. In addition, the pores in polycrystalline ferrites produce free magnetic poles inside the sample. SCHLOMANN *[24]* has shown that, in polycrystals, the peak of the resonance line is shifted and the line is broadened and becomes asymmetrical in shape. There is, in the first order approximation, no shift in the centroid of the resonance line. For a spherical sample, the peak of the resonance line is given by the equation

d) Polykristalline Proben

Bei polykristallinen Proben spielen zwei zusätzliche Effekte eine Rolle. Die verschiedenen Orientierungen der einzelnen Körner einer polykristallinen Probe führen zu unterschiedlichen Resonanzfeldern in den verschiedenen Körnern. Die Dipol-Kräfte zwischen den Momenten der einzelnen Körner koppeln jedoch ihre Bewegungen so, daß die sich ergebende Linienbreite bedeutend geringer ist als für eine Summe von isolierten Körnern zu erwarten wäre. Außerdem bilden die Poren in polykristallinen Ferriten freie Magnetpole im Innern der Probe. SCHLOMANN *[24]* hat gezeigt, daß sich in Polykristallen das Maximum der Resonanzlinie verschiebt, die Linie sich verbreitert und eine asymmetrische Form annimmt. In erster Näherung tritt dabei keine Verschiebung des Flächenschwerpunktes der Resonanzlinie auf. Für eine kugelförmige Probe ist das Maximum der Resonanzlinie durch die folgende Gleichung gegeben

$$\omega = \gamma (H + H_{\text{int}}) \tag{9}$$

where the internal field

wobei das innere Feld

$$H_{\text{int}} = -\frac{K_1}{2M_s} + \frac{4\pi M_s}{3} p$$

and p is the fractional volume of the sample occupied by pores.

und p das von den Poren eingenommene Teilvolumen der Probe ist.

6.1.0.6 Accuracy — Genauigkeit

In conclusion, we wish to caution the reader in his use of the tables that there is an unusually strong likelihood of error due to uncertainties in the composition and the physical state of the samples on which the measurements were made. The various original samples which provide the data are of varying quality; in some cases the composition specified is that of the starting ingredients while in others, it is a result of chemical analysis of the finished product. The chemical analysis of ferrites is often inadequate to specify the amounts and states of oxidation of the ingredients to any but the crudest degree of precision. Important factors which depend on the thermal history of a sample and which may vary among nominally identical samples are the oxygen content, the distribution of cations between A and B sites, and the porosity. Two factors must be kept in mind: certain physical properties such as permeability, electrical conductivity and ferrimagnetic resonance line width are extremely sensitive to variations in the samples and certain elements, especially cobalt, can have profound effects on the properties of ferrites even when present in trace amounts. The original literature should be consulted whenever the details of sample preparation or treatment may be important.

Schließlich möchten wir dem Leser beim Gebrauch der Tabellen zur Vorsicht raten, da Fehler wegen der Unsicherheit in der Zusammensetzung und dem physikalischen Zustand der untersuchten Proben sehr leicht möglich sind. Die verschiedenen Proben, von denen die tabellierten Werte stammen, sind von unterschiedlicher Qualität; in einigen Fällen entspricht die angegebene Zusammensetzung derjenigen in der Ausgangssubstanz, während sie in anderen Fällen ein Ergebnis der chemischen Analyse des Endproduktes ist. Die chemische Analyse von Ferriten kann die Menge und den Oxidationszustand der einzelnen Bestandteile jedoch bestenfalls nur sehr ungenau bestimmen. Wichtige Faktoren, die von der Wärmebehandlung der Probe abhängen und die auch unter nominell identischen Proben variieren können, sind der Sauerstoffgehalt, die Verteilung der Kationen auf die A- und B-Gitterplätze und die Porosität. Man muß stets die folgenden beiden Punkte im Auge behalten: einmal reagieren bestimmte physikalische Eigenschaften wie Permeabilität, elektrische Leitfähigkeit und ferrimagnetische Resonanzlinienbreite äußerst empfindlich auf Änderungen in den Proben und zum anderen können bestimmte Elemente, insbesondere Kobalt, die Eigenschaften der Ferrite sehr stark beeinflussen, selbst wenn sie nur in Spuren vorhanden sind. Man sollte immer auf die Originalliteratur zurückgreifen, sobald die Einzelheiten der Probenherstellung oder -behandlung von Bedeutung sind.

6.1.0.7 References for 6.0 and 6.1.0

1 GORTER, E. W.: Philips Res. Rept. **9** (1954) 295.
2 NÉEL, L.: Ann. Phys. (Paris) **3** (1948) 137.
3 GOODENOUGH, J. B.: Magnetism, Volume III, Chapter 1, edited by G. T. RADO and H. SUHL, Academic Press, New York (1963).
4 GORTER, E. W., and J. A. SCHULKES: Phys. Rev. **90** (1953) 487.
5 YAFET, Y., and C. KITTEL: Phys. Rev. **87** (1952) 290.
6 KAPLAN, T. A., K. DWIGHT, D. LYONS, and N. MENYUK: J. Appl. Phys. **32** (1961) 13S.
7 LOTGERING, F. K.: Proc. Intern. Conf. Magnetism, Nottingham, England, 1964. Inst. Phys. and Phys. Soc., London (1965) 533.
8 BALTZER, P. K., P. J. WOJTOWICZ, M. ROBBINS, and E. LOPATIN: Phys. Rev. **151** (1966) 367.
10 FREEMAN, A. J., and R. E. WATSON: Magnetism, Volume IIA, Chapter 4, edited by G. T. RADO and H. SUHL, Academic Press, New York (1965).
11 PORTIS, A. M., and R. H. LINDQUIST: Magnetism. Volume IIA, Chapter 6, edited by G. T. RADO and H. SUHL, Academic Press, New York (1965).
12 MÖSSBAUER, R. L.: Z. Physik **151** (1958) 124.
13 YOSIDA, K., and M. TACHIKI: Progr. Theoret. Phys. **17** (1957) 331.
14 FOLEN, V. J., and G. T. RADO: J. Appl. Phys. **29** (1958) 438S.
15 WOLF, W. P.: Phys. Rev. **108** (1957) 1152.
16 KITTEL, C.: Phys. Rev. **71** (1947) 270; **73** (1948) 155.
17 TSUYA, N.: Progr. Theoret. Phys. **7** (1952) 263.
18 WANGSNESS, R. K.: Phys. Rev. **91** (1954) 1085.
19 VAN WIERINGEN, J. S.: Phys. Rev. **90** (1953) 488.
20 ARTMAN, J. O.: Phys. Rev. **105** (1957) 62.
21 SMITH, A. B., and R. V. JONES: J. Appl. Phys. **34** (1963) 1283.
22 HAAS, C. W., and H. B. CALLEN: Magnetism, Volume I, Chapter 10, edited by G. T. RADO and H. SUHL, Academic Press, New York (1963).
23 CALLEN, H. B.: Phys. Chem. Solids **4** (1958) 256.
24 SCHLOMANN, E.: Conf. Magnetism Magnetic Materials, Boston, 1956 (1957) 600.
26 BOUCHARD, R. J.: Mat. Res. Bull. **2** (1967) 459.
27 LOTGERING, F. K.: Phys. Chem. Solids **29** (1968) 699.
28 WOJTOWICZ, P. J., P. K. BALTZER, and M. ROBBINS: Phys. Chem. Solids **28** (1967) 2423.
29 OHBAYASHI, K., Y. TOMINAGA, and S. IIDA: J. Phys. Soc. Japan **24** (1968) 1173.
30 SLEIGHT, A. W., H. S. JARRETT: Phys. Chem. Solids **29** (1968) 868.
31 ROBBINS, M., P. K. BALTZER, and E. LOPATIN: J. Appl. Phys. **39** (1968) 662.
32 WHITE, J. G., and M. ROBBINS: J. Appl. Phys. **39** (1968) 664.
33 LOTGERING, F. K., and R. P. VAN STAPELE: Solid State Commun. **5** (1967) 143; J. Appl. Phys. **39** (1968) 417.
34 GOODENOUGH, J. B.: Solid State Commun. **5** (1967) 577.
35 VAN MAAREN, N. H., G. M. SCHAEFFER, and F. K. LOTGERING: Phys. Letters **25A** (1967) 238.
36 ROBBINS, M., R. H. WILLENS, and R. C. MILLER: Solid State Commun. **5** (1967) 933.
37 HUANG, N. L., and R. ORBACH: J. Appl. Phys. **39** (1968) 426.
38 DWIGHT, K., and N. MENYUK: J. Appl. Phys. **39** (1968) 660.
39 COLOMINAS, C.: Phys. Rev. **153** (1967) 558.
40 ROBBINS, M., H. W. LEHMANN, and J. G. WHITE: Phys. Chem. Solids **28** (1967) 897.

6.1.1 Fe^{2+}-Fe^{3+} spinels (magnetite) and Fe^{2+}-Fe^{3+} spinels with substitutions

6.1.1.1 Tables and figures

General references 1 ··· 5 see page 74.

Tab. 1. General properties of stoichiometric magnetite, Fe_3O_4. Room temperature unless otherwise specified.

Crystal type	Inverse spinel—cubic above 119 °K	*51SWK*
	Orthorhombic below 119 °K	*47VH*
Structure type	$MgAl_2O_4$	
Lattice constant	$a = 8.397$ Å	*65S(2)*
Volume of unit cell	$592.1 \cdot 10^{-24}$ cm³ (containing 8 molecules)	
Molecular weight	231.6 g/mole	
X-ray density	$\varrho_X = 5.238$ g/cm³	*55W*
Melting point	T_m = See Fig. 36	
Thermal conductivity	$\varkappa \cong 1.5 \cdot 10^{-2}$ cal sec⁻¹ cm⁻¹ deg⁻¹	*59SW(a)*
Specific heat	$c_p \cong 0.2$ cal g⁻¹ deg⁻¹	*59SW(a)*
Electrical conductivity	$\sigma \cong 250\ \Omega^{-1}$ cm⁻¹	*54C*
Young's modulus	$E \cong 10^{12}$ dyne/cm²	*59SW(a)*
Compressibility	$\beta = 0.5643 \cdot 10^{-12}$ cm²/dyne	*55W*
Elastic constants		
calculated	$C_{11} = C_{12} = 1.772 \cdot 10^{12}$ dyne/cm²	*55W*
experimental	$C_{11} = 2.73 \cdot 10^{12}$ dyne/cm²	*58H, 47D*
	$C_{12} = 1.06 \cdot 10^{12}$ dyne/cm²	
	$C_{44} = 0.97 \cdot 10^{12}$ dyne/cm²	

Lefever

Tab. 2. Magnetic properties of stoichiometric magnetite.
Room temperature unless otherwise specified.

Curie temperature	$\Theta_C = 585$ °C	*55G*
Magnetic ordering temperature (octahedral site)	$T = 119$ °K	*47VH*
Saturation magnetization	$M_s = 471$ Gauss	*55G*
Saturation induction at 0 °K	$4\pi M_s = 6000$ Gauss	*59SW(b)*
Saturation magnetic moment/molecule at 0 °K	$p^0_{ms} = 4.1\ \mu_B$	*59SW(c)*
Saturation magnetic moment/g at 0 °K	$\sigma^0_s = 98$ Gauss cm³/g	*59SW(b)*
Initial permeability, single crystal	$\mu_i \cong 100$	*55BPS*
Anisotropy constant, first order	$K_1 = (-1.10 \pm 0.08) \cdot 10^5$ erg/cm³	*50B(b)*
Magnetostriction at saturation, polycrystalline	$\lambda_s \cong 35 \cdot 10^{-6}$	*55BPS*
single crystal	$\lambda_{100} = -19.5 \cdot 10^{-6}$	*55BPS*
	$\lambda_{110} = 57.1 \cdot 10^{-6}$	*55BPS*
	$\lambda_{111} = 77.6 \cdot 10^{-6}$	*55BPS*
Effective spectroscopic splitting factor	$g = 2.12 \pm 0.04$	*50B(b)*
Magnetomechanical ratio	$g' = 1.93$	*56K*

Tab. 3. Summary of data from Mössbauer experiments [*63IOI*].
See also Fig. 29 and [*61BCMOS*].

T °K	Internal field H_{eff} kOe			Quadrupole splitting e^2qQ mm/sec			Isomer shift δ mm/sec		
	$Fe^{2+}(B)$	$Fe^{3+}(B)$	$Fe^{3+}(A)$	$Fe^{2+}(B)$	$Fe^{3+}(B)$	$Fe^{3+}(A)$	$Fe^{2+}(B)$	$Fe^{3+}(B)$	$Fe^{3+}(A)$
300	470 ± 20		495 ± 20	−0.1 ± 0.2		0.1 ± 0.2	0.87 ± 0.10		0.46 ± 0.10
200	485 ± 20		510 ± 20						
85	$470\ ^{+30}_{-20}$	510 ± 20		−1.9 ± 0.2	−0.2 ± 0.2		0.88 ± 0.15	0.60 ± 0.10	
30	$470\ ^{+30}_{-20}$	510 ± 20		−1.9 ± 0.2	−0.2 ± 0.2				

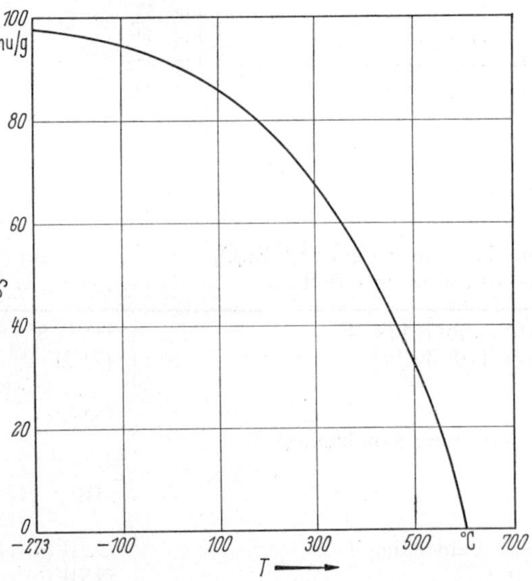

Fig. 1. Fe_3O_4. Specific saturation magnetization vs. temperature [*59SW*].

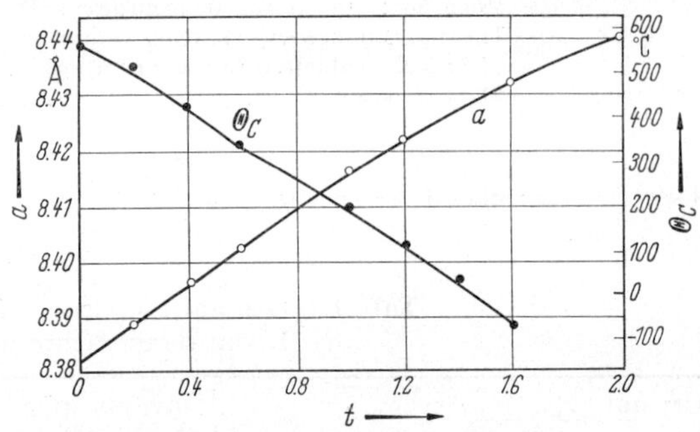

Fig. 2. $Fe^{2+}Fe^{3+}_{2-t}V^{3+}_tO_4$. Lattice constant, *a*, and Curie temperature, Θ_C, vs. *t* [*57LM*].

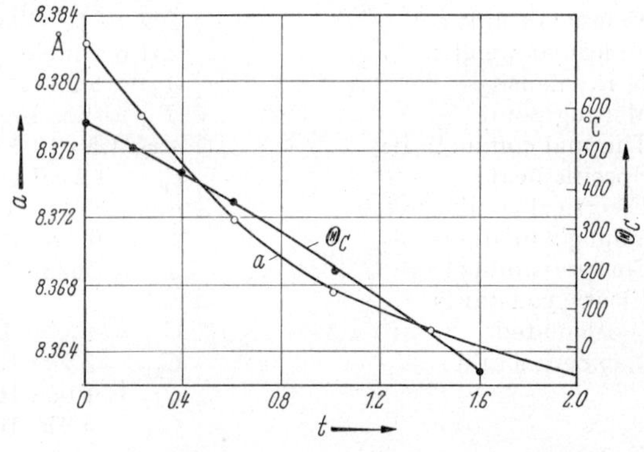

→

Fig. 3. $Fe^{2+}Fe^{3+}_{2-t}Ga^{3+}_tO_4$. Lattice constant, *a*, and Curie temperature, Θ_C, vs. *t* [*57LM*].

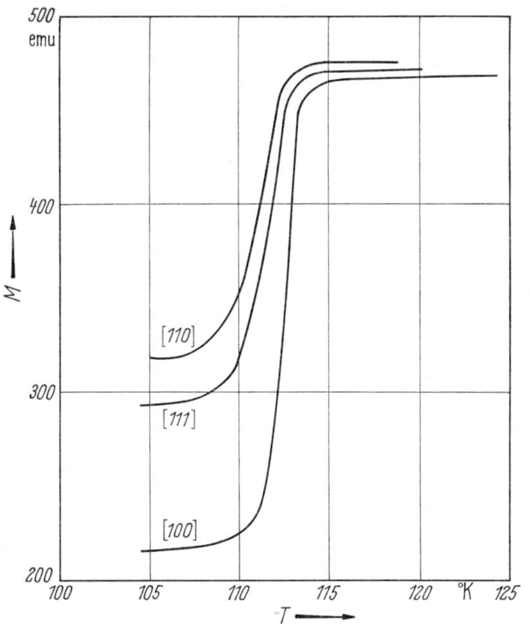

Fig. 4. Fe_3O_4. Magnetization vs. temperature (warming curves) parallel to the three principal crystallographic directions, natural magnetite single crystal. Relatively weak magnetic fields must be used for this measurement as strong fields enhance the magnetization below the transition temperature and remove the discontinuity. Remanent magnetization must be avoided. The detailed behavior is complex and the original literature should be consulted [*50D*].

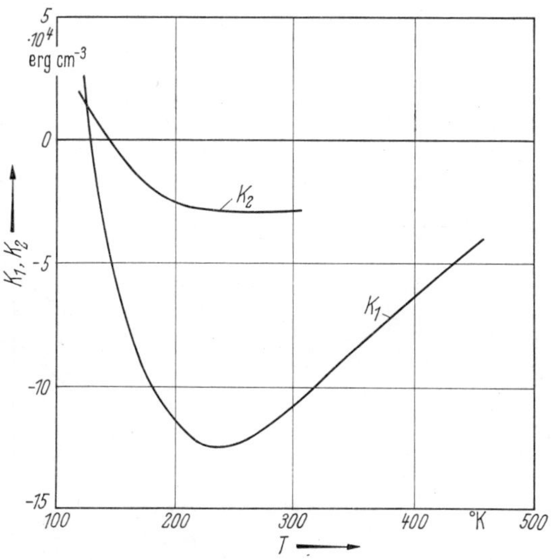

Fig. 6. Fe_3O_4. First- and second-order anisotropy constants vs. temperature as measured by the torque method [*57BBP*]. See also [*65S(1)*] and [*66GO*] for effect of oxygen stoichiometry.

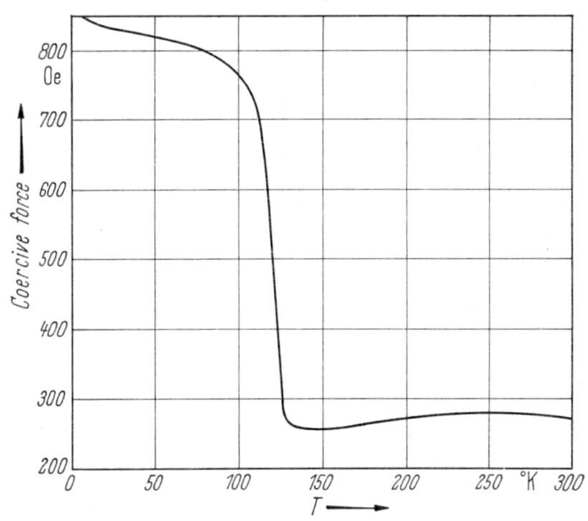

Fig. 5. Fe_3O_4. Coercive force vs. temperature for single-domain magnetite particles with a mean axial ratio of 6.5 [*61BMW*].

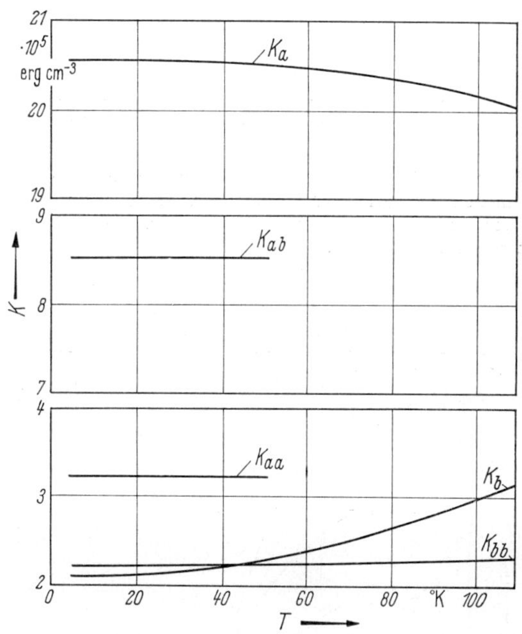

Fig. 7. Fe_3O_4. Anisotropy constants vs. temperature for the orthorhombic phase. Data from torque measurements on single crystal sphere. $E_a = K_a\alpha_a^2 + K_b\alpha_b^2 + K_{aa}\alpha_a^4 + K_{ab}\alpha_a^2\alpha_b^2 + K_{bb}\alpha_b^4$ [*63P*]. See also [*61BMW*] and [*54C*].

→

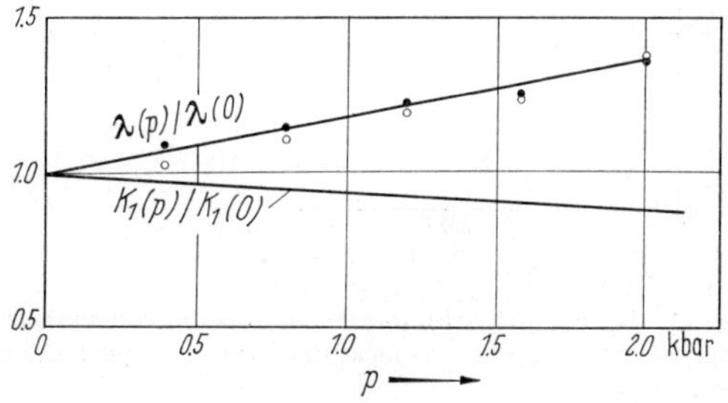

Fig. 8. Fe_3O_4. Dependence of saturation magnetostriction constants and crystalline anisotropy constant, K_1, on hydrostatic pressure: open circles, λ_{111}; closed circles, λ_{100}. The pressure dependence of K_2 is approximately the same as that of K_1 [*67NK*].

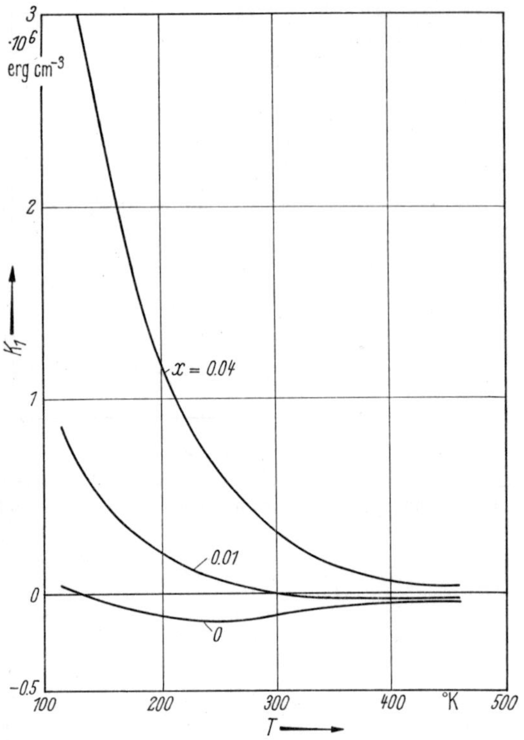

Fig. 9. $Co_xFe_{3-x}O_4$. First-order anisotropy constant as a function of temperature. Parameter x. These measurements were made on carefully prepared single crystals using the torque method [*57BBP*].

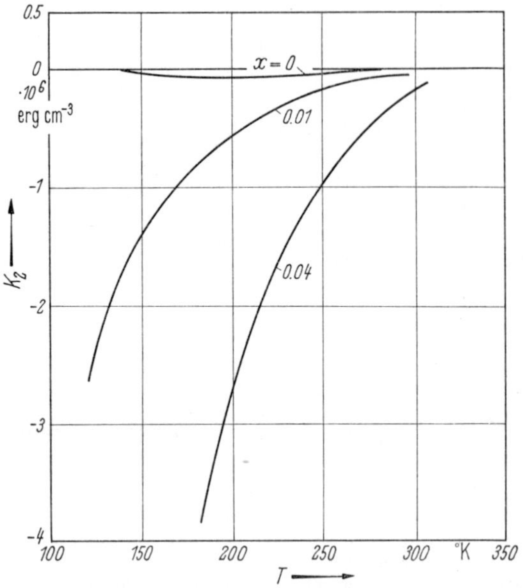

Fig. 10. $Co_xFe_{3-x}O_4$. Second-order anisotropy constant as a function of temperature, as measured by the torque method. Parameter x [*57BBP*].

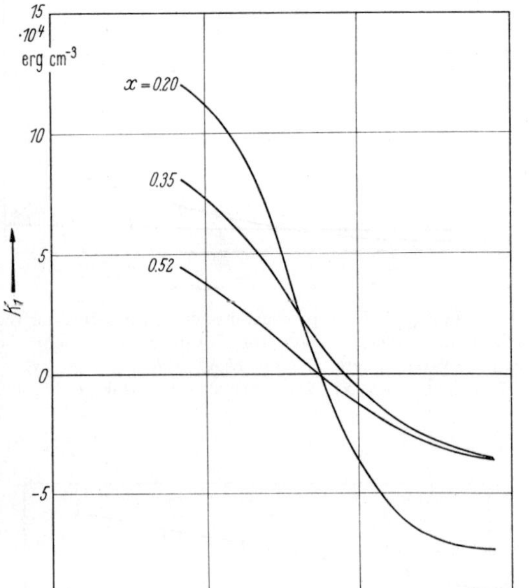

Fig. 11. $Ni_xFe_{3-x}O_4$. First-order anisotropy constant as a function of temperature, as measured by the torque method. Parameter x [*61M*].

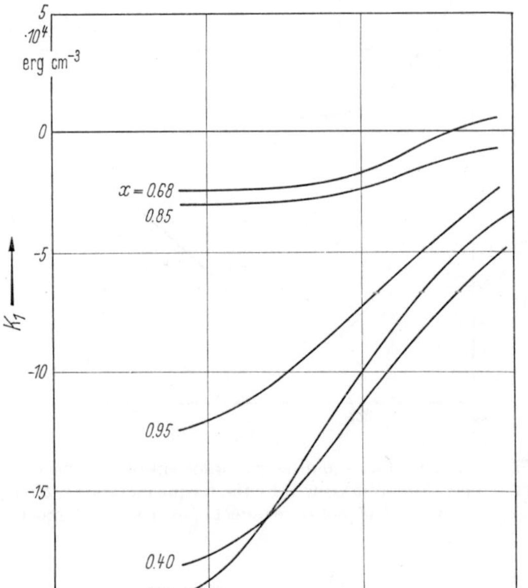

Fig. 12. $Mn_xFe_{3-x}O_4$. First-order anisotropy constant as a function of temperature, as measured by the torque method. Parameter x [*61M*].

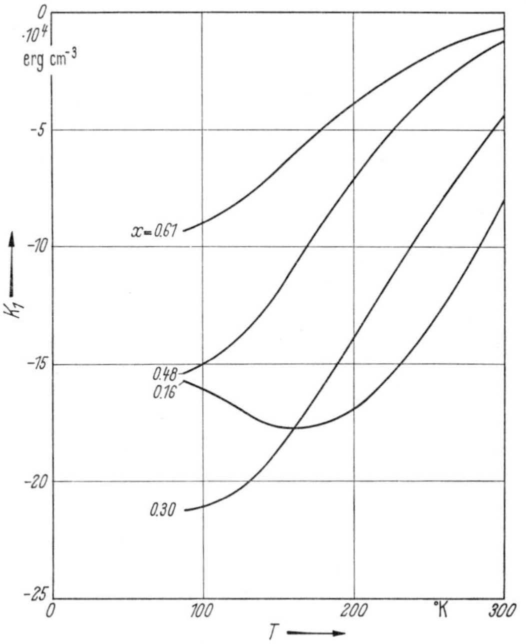

Fig. 13. $Zn_xFe_{3-x}O_4$. First-order anisotropy constant as a function of temperature, as measured by the torque method. Parameter x [*61M(a)*].

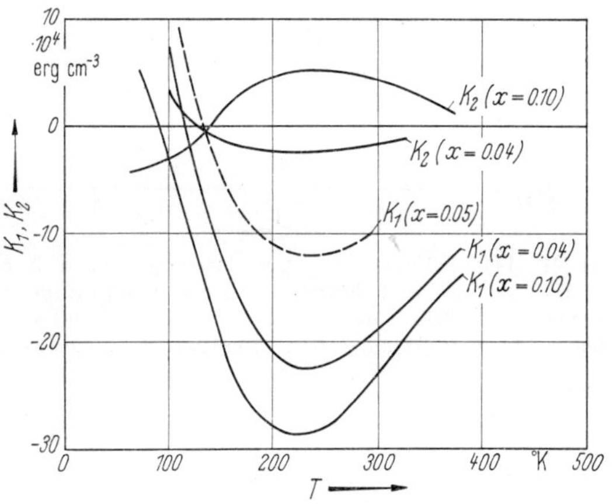

Fig. 15. $xFe_2TiO_4 \cdot (1-x)Fe_3O_4$. First- and second-order anisotropy constants vs. temperature as measured by the torque method. Solid curves from [*63SI(1)*]. Dotted curve was obtained from a crystal containing excess oxygen [*63MKI*].

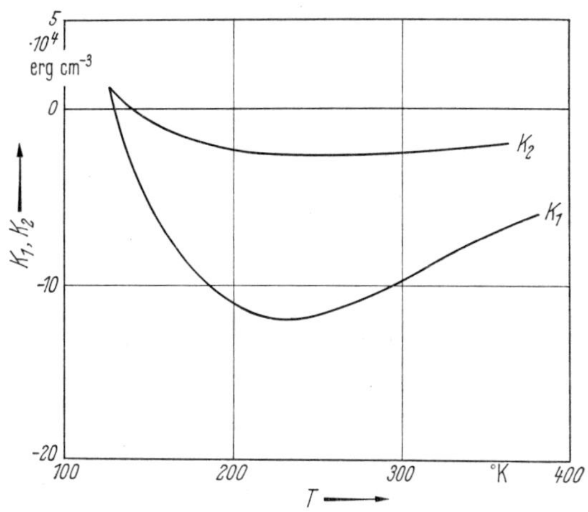

Fig. 14. $Li_{0.05}Fe_{2.95}O_4$. First- and second-order anisotropy constants vs. temperature as measured by the torque method [*63SI(3)*].

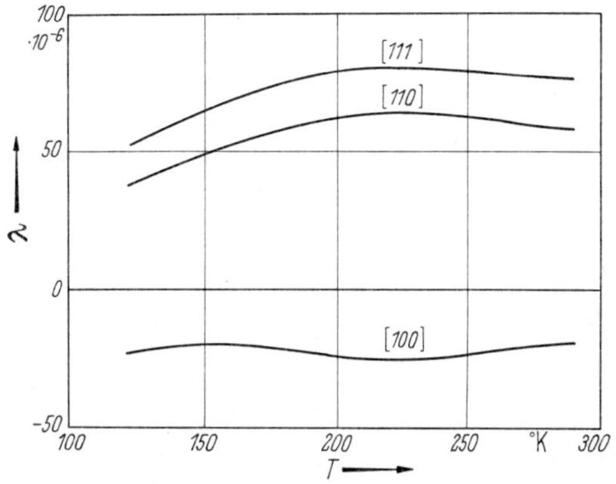

Fig. 16. Fe_3O_4. Magnetostriction constants as a function of temperature for the three principal directions of a synthetic single crystal [*55BPS*]. See also [*50D(a)*].

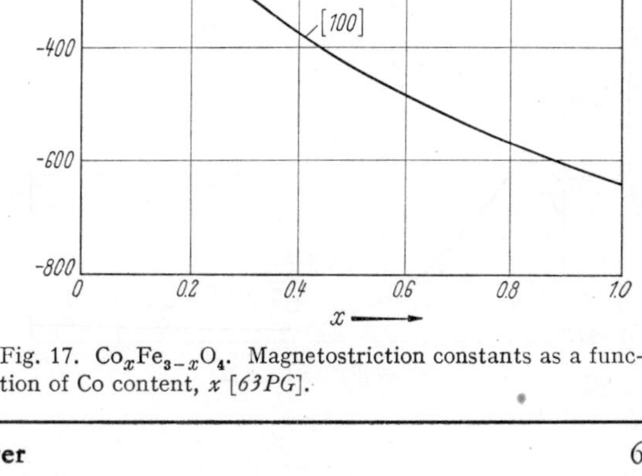

Fig. 17. $Co_xFe_{3-x}O_4$. Magnetostriction constants as a function of Co content, x [*63PG*].

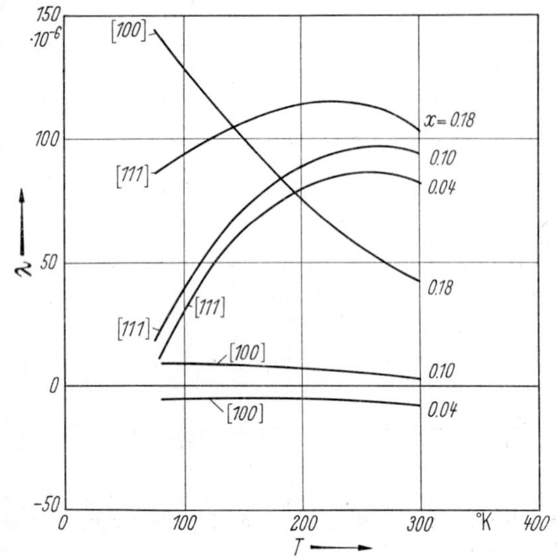

Fig. 18. $x Fe_2TiO_4 \cdot (1-x) Fe_3O_4$. Magnetostriction constants as a function of temperature. Parameter x [*63SI(2)*].

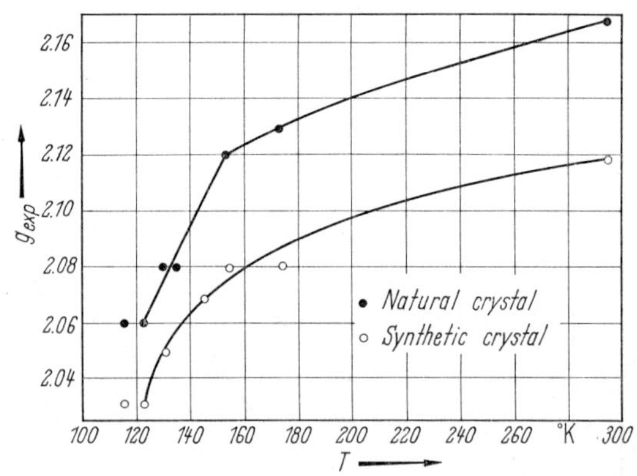

Fig. 19. Fe_3O_4. Values of g_{exp} vs. temperature for natural and synthetic crystals [*50B*].

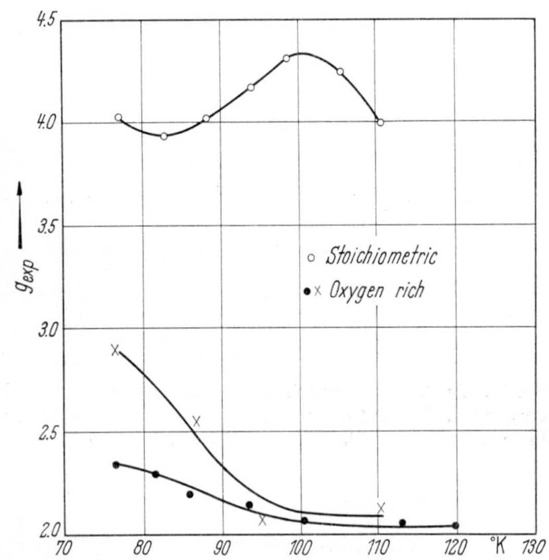

Fig. 20. Fe_3O_4. Values of g_{exp} vs. temperature for stoichiometric and oxygen-rich samples [*65S(2)*].

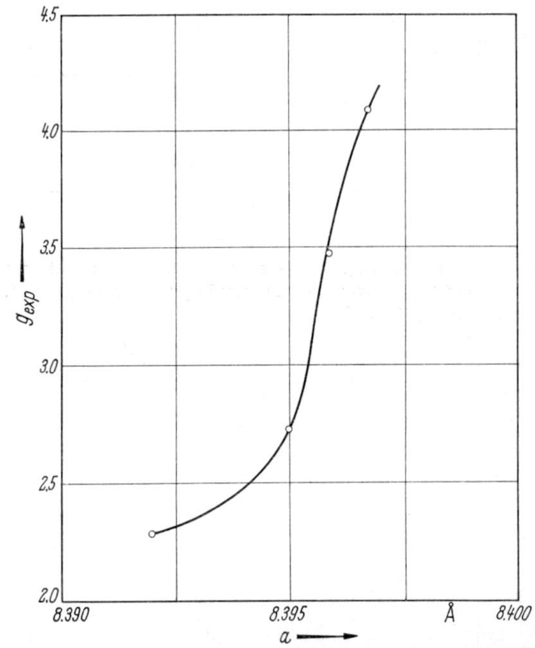

Fig. 21. Fe_3O_4. Values of g_{exp} vs. lattice parameter, a, at 77 °K. Samples were heated at various temperatures in oxygen. The lattice parameter of 8.397 Å, corresponding to the stoichiometric composition, decreased with increasing oxygen content [*65S(2a)*].

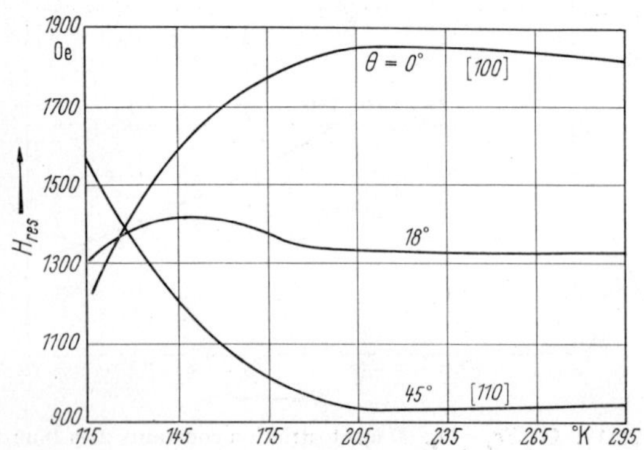

←

Fig. 22. Fe_3O_4. Field for ferromagnetic resonance vs. temperature for fields in (100) plane of synthetic crystal at 8960 Mc/s (points at 295 °K correspond to a frequency of 8930 Mc/s). Measurements were made on a thin disc-shaped specimen for which the difference in demagnetizing factors parallel and perpendicular to the plane of the disc, $N_y - N_z$, was 0.935. The resonance absorption lines are broad, highly asymmetric and change drastically in appearance as one passes below the transition temperature range of about 188 °K [*50B(a)*].

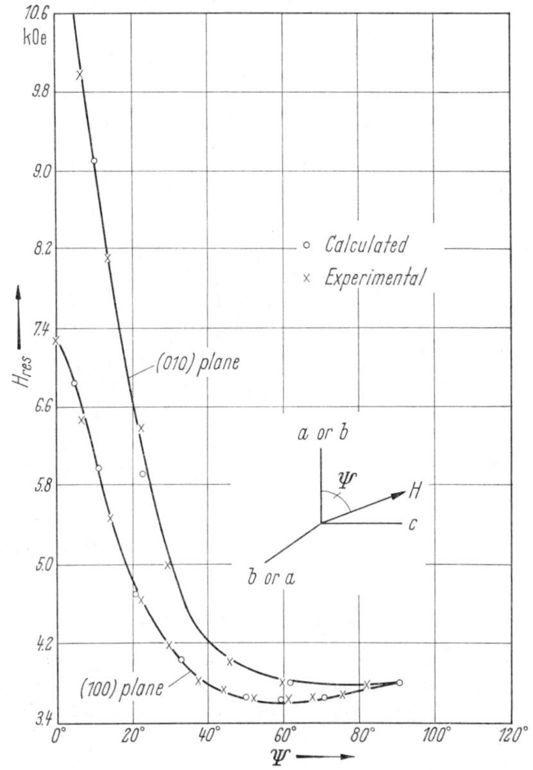

Fig. 23. Fe_3O_4. Ferromagnetic resonance field vs. angle between field direction and a or b axes at 4.2 °K and 24 Gc/s [61BMW].

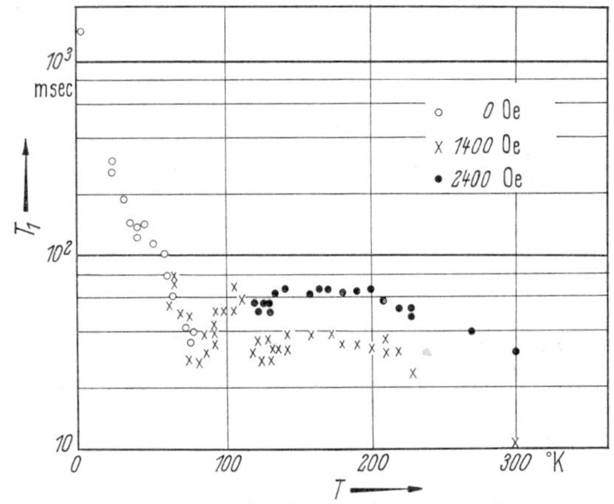

Fig. 24. Fe_3O_4. Longitudinal nuclear magnetic resonance relaxation time T_1 vs. temperature for Fe^{57} on the A site [66MI]. See also [65MT].

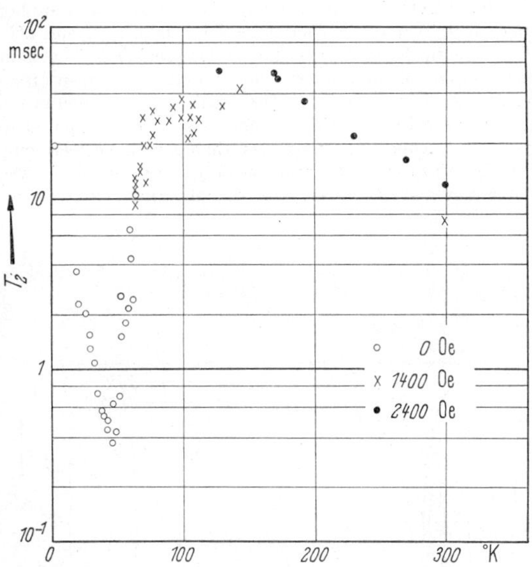

Fig. 25. Fe_3O_4. Transverse nuclear magnetic resonance relaxation time T_2 vs. temperature for Fe^{57} on the A site [66MI(a)]. See also [65MT].

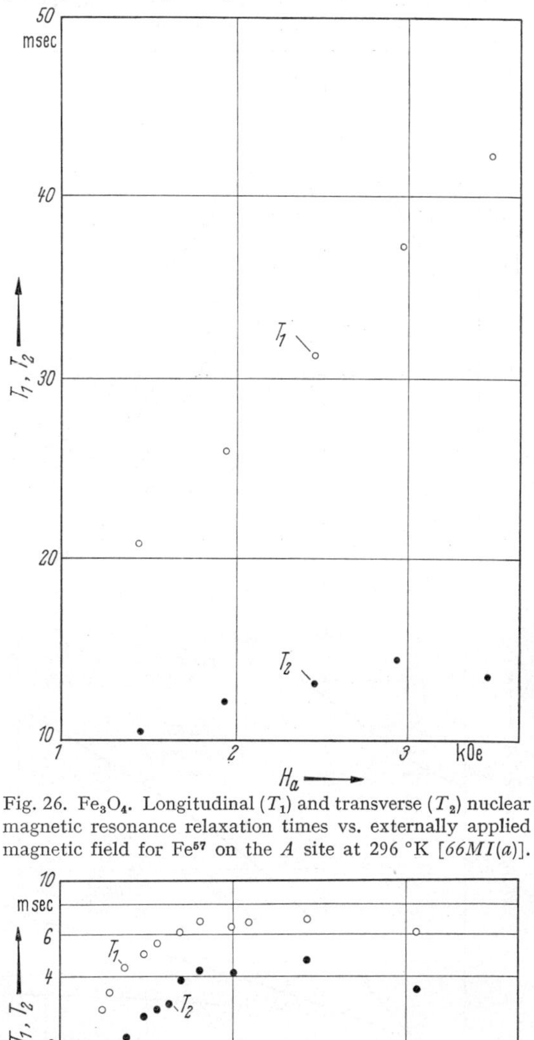

Fig. 26. Fe_3O_4. Longitudinal (T_1) and transverse (T_2) nuclear magnetic resonance relaxation times vs. externally applied magnetic field for Fe^{57} on the A site at 296 °K [66MI(a)].

→

Fig. 27. Fe_3O_4. Longitudinal (T_1) and transverse (T_2) nuclear magnetic resonance relaxation times vs. temperature for Fe^{57} on the B site [66MI(a)].

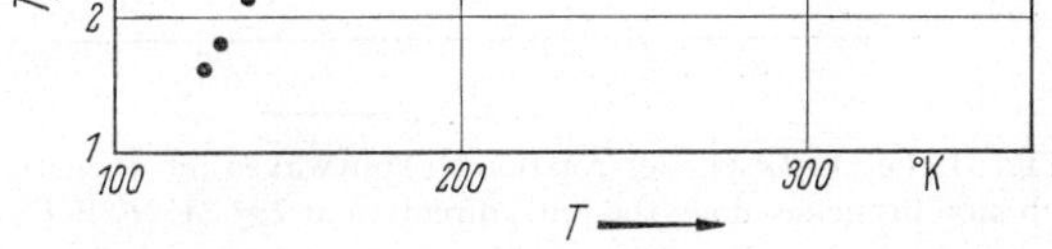

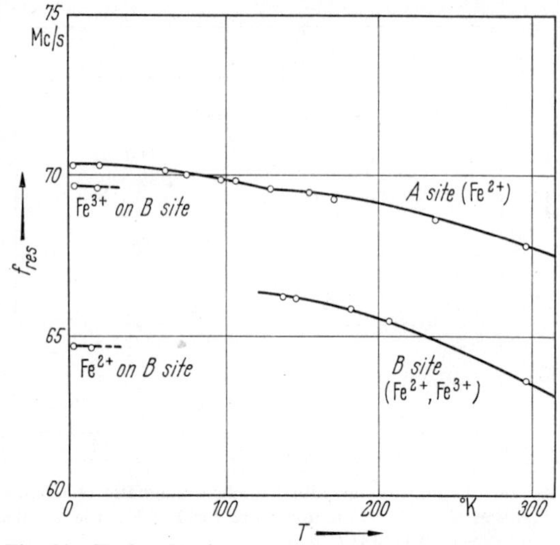

Fig. 28. Fe_3O_4. Nuclear magnetic resonance frequency vs. temperature for Fe^{57} [*66MI*]. See also [*63B*].

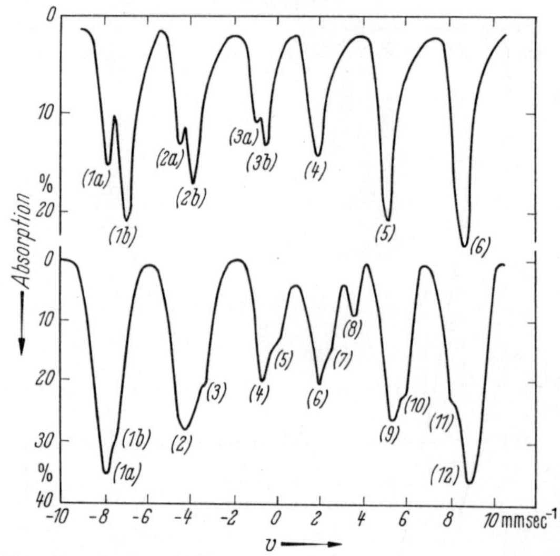

Fig. 29. Fe_3O_4. Mössbauer absorption spectra at 300 °K (upper curve) and 85 °K (lower curve) for Fe^{57}-enriched sample of polycrystalline Fe_3O_4. In the upper curve, absorption peaks *1a, 2a, 3a, 4, 5* and *6* are assigned to iron on *A* sites, while peaks *1b, 2b, 3b, 4, 5* and *6* are assigned to iron on the *B* sites. On the basis of single crystal data, the absorption peaks in the lower curve are assigned as follows: *1b, 3, 5, 7, 10* and *11* correspond to ferrous ions and *1a, 2, 4, 6, 9* and *12* correspond to ferric ions on both *A* and *B* sites [*63IOI*]. See also [*61BCMOS*].

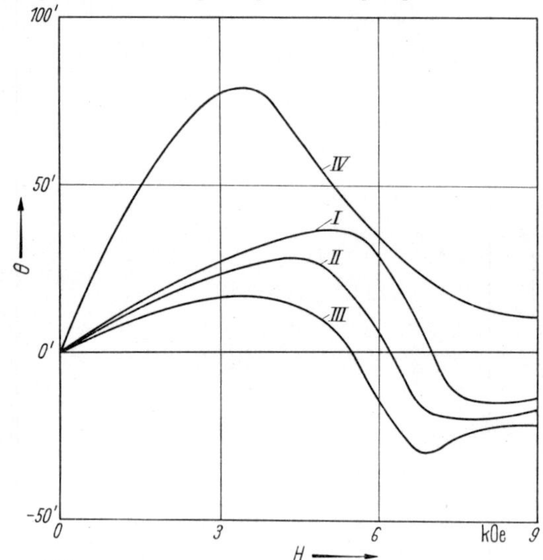

←

Fig. 30. Fe_3O_4. The magnetic Kerr effect and the Faraday effect at 9375 Mc/s: rotation angle θ vs. field. *I, II* and *III*: the magnetic Kerr effect obtained for specimens 23 mm in diameter and 3.6 mm, 6.2 mm and 4.8 mm thick, respectively. *IV*: the Faraday effect. The depth of penetration of the microwave fields into the specimen was small compared to specimen thickness. Nevertheless, there may have been some Faraday effect from reflections from the wall of the sample. There may also have been other complicating factors [*57N*].

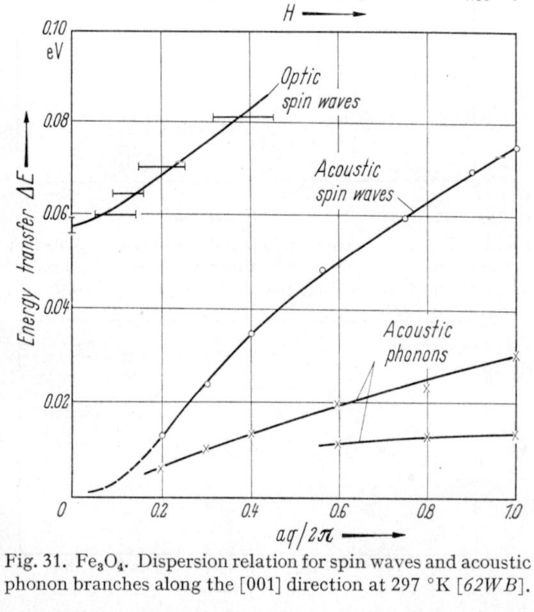

Fig. 31. Fe_3O_4. Dispersion relation for spin waves and acoustic phonon branches along the [001] direction at 297 °K [*62WB*].

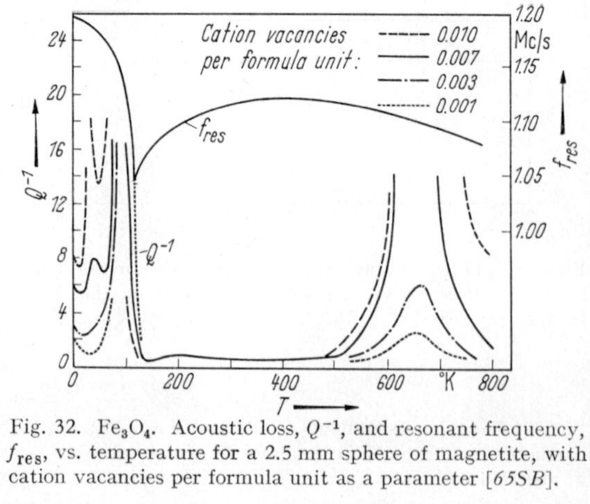

Fig. 32. Fe_3O_4. Acoustic loss, Q^{-1}, and resonant frequency, f_{res}, vs. temperature for a 2.5 mm sphere of magnetite, with cation vacancies per formula unit as a parameter [*65SB*].

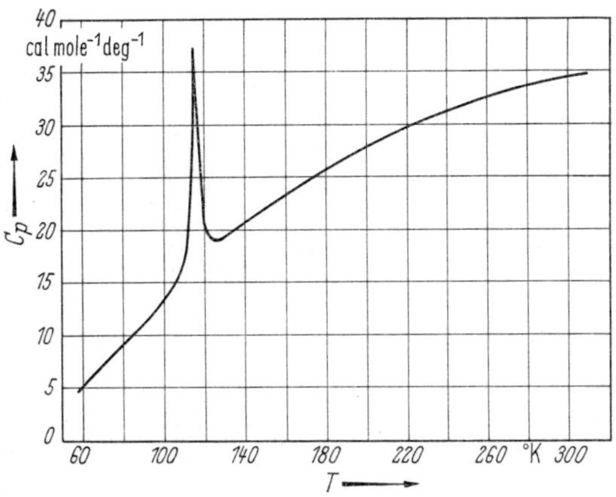

Fig. 33. Fe_3O_4. Molar heat capacity C_p vs. temperature. Analysis of sample: Fe_3O_4, 99%; Fe_2O_3, 0.65%; and impurities, largely quartz, 0.3% [*29M*].

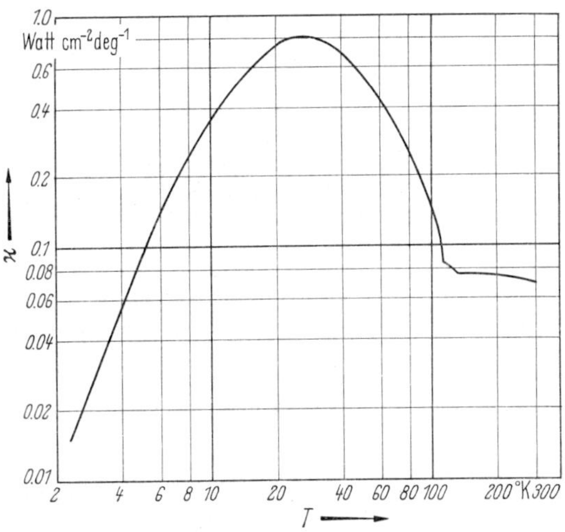

Fig. 34. Fe_3O_4. Thermal conductivity, $\varkappa$, of a synthetic single crystal vs. temperature. These measurements were made on a nearly stoichiometric crystal in a zero magnetic field [*62S*].

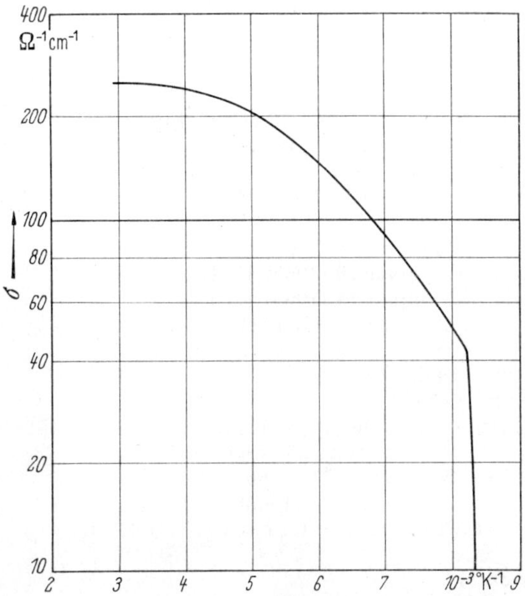

Fig. 35. Fe_3O_4. Electrical conductivity, σ, as a function of temperature for synthetic crystals [*54C*]. See also [*61H*].

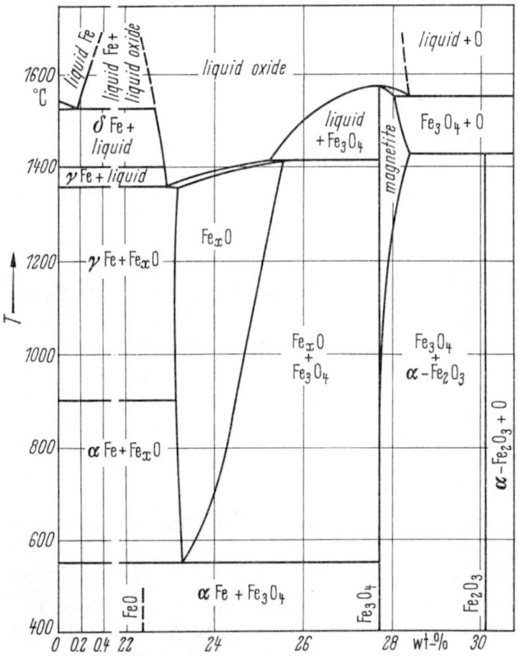

Fig. 36. Iron-oxygen system at a total pressure of 1 atmosphere [*46DG*]. See also [*57S*], [*61S*].

6.1.1.2 References for 6.1.1.1

29M	MILLAR, R. W.: J. Am. Chem. Soc. **51** (1929) 216.
46DG	DARKEN, L. S., and R. W. GURRY: J. Am. Chem. Soc. **68** (1946) 799.
47D	DORAISWAMI, M. S.: Proc. Indian Acad. Sci. Sect. A **25** (1947) 413.
47VH	VERWEY, E. J. W., and E. L. HEILMANN: J. Chem. Phys. **15** (1947) 174.
50B	BICKFORD Jr., L. R.: Phys. Rev. **78** (1950) 449.
50B(a)	BICKFORD Jr., L. R.: Phys. Rev. **78** (1950) 454.
50B(b)	BICKFORD Jr., L. R.: Phys. Rev. **78** (1950) 452.
50D	DOMENICALI, C. A.: Phys. Rev. **78** (1950) 462.
50D(a)	DOMENICALI, C. A.: Phys. Rev. **78** (1950) 461.
51SWK	SHULL, C. G., E. O. WOLLEN, and W. C. KOHLER: Phys. Rev. **84** (1951) 912.

54C	Calhoun, B. A.: Phys. Rev. **94** (1954) 1582.
55BPS	Bickford Jr., L. R., J. Pappis, and J. L. Stull: Phys. Rev. **99** (1955) 1211.
55G	Gorter, E. W.: Proc. IRE **43** (1955) 245.
55W	Waldron, R. D.: Phys. Rev. **99** (1955) 1727.
56K	Kittel, C.: Introduction to Solid State Physics, Wiley, New York, (1956) 410.
57BBP	Bickford Jr., L. R., J. M. Brownlow, and R. F. Penoyer: Proc. Inst. Elec. Engrs. (London) **104 B** Suppl. No. 5 (1957) 238.
57LM	Lensen, M., and A. Michel: Congr. Intern. Chim. Pure Appl. 16^e Paris 1957.
57N	Neprimerov, N. N.: Bull. Acad. Sci. USSR, Phys. Ser. (English Transl.) **21** (1957) 1276.
57S	Smiltens, J.: J. Am. Chem. Soc. **79** (1957) 1877.
58H	Huntington, H. B.: The Elastic Constants of Crystals. New York: Academic Press (1958).
59SW	Smit, J., and H. P. J. Wijn: Ferrites. New York: John Wiley and Sons 1959, p. 156.
59SW(a)	Smit, J., and H. P. J. Wijn: Ferrites. New York: John Wiley and Sons 1959, p. 225.
59SW(b)	Smit, J., and H. P. J. Wijn: Ferrites. New York: John Wiley and Sons 1959, p. 157.
59SW(c)	Smit, J., and H. P. J. Wijn: Ferrites. New York: John Wiley and Sons 1959, p. 149.
61BCMOS	Bauminger, R., S. G. Cowan, A. Marinov, S. Ofer, and E. Segal: Phys. Rev. **122** (1961) 1449.
61BMW	Bonstrom, D. B., A. H. Morrish, and L. A. K. Watt: J. Appl. Phys. **32** (1961) 273S.
61H	Haubenreisser, W.: Phys. Status Solidi **1** (1961) 619.
61M	Miyata, N.: J. Phys. Soc. Japan **16** (1961) 1293.
61M(a)	Miyata, N.: J. Phys. Soc. Japan **16** (1961) 1295.
61S	Salmon, O. N.: J. Phys. Chem. **65** (1961) 550.
62S	Slack, G. A.: Phys. Rev. **126** (1962) 438.
62WB	Watanabe, H., and B. N. Brockhouse: Phys. Letters **1** (1962) 190.
63B	Boyd, E. L.: Phys. Rev. **129** (1963) 1962.
63IOI	Ito, A., K. Ôno, and Y. Ishikawa: J. Phys. Soc. Japan **18** (1963) 1468.
63MKI	Miyata, N., T. Kobayashi, and K. Itoh: J. Phys. Soc. Japan **18** (1963) 1351.
63P	Palmer, P.: Phys. Rev. **131** (1963) 1060.
63PG	Pines, B. Ya., and N. M. Gumen: Fiz. Tverd. Tela **5** (1963) 3486. Soviet Phys.-Solid State (English Transl.) **5** (1964) 2558.
63SI(1)	Syono, I., and Y. Ishikawa: J. Phys. Soc. Japan **18** (1963) 1230.
63SI(2)	Syono, I., and Y. Ishikawa: J. Phys. Soc. Japan **18** (1963) 1231.
63SI(3)	Syono, I., and Y. Ishikawa: J. Phys. Soc. Japan **18** (1963) 1704.
65MT	Mizoguchi, T., and Y. Tomono: J. Appl. Phys. **36** (1965) 1021.
65S(1)	Sharma, V. N.: Phys. Letters **19** (1965) 463.
65S(2)	Sharma, V. N.: J. Appl. Phys. **36** (1965) 1452.
65S(2a)	Sharma, V. N.: J. Appl. Phys. **36** (1965) 1451.
65SB	Suiter Jr., W. B., and R. Blair: J. Appl. Phys. **36** (1965) 1157.
66GO	Gyorgy, E. M., and H. M. O'Bryan Jr.: Phys. Letters **23** (1966) 513.
66MI	Mizouguchi, T., and M. Inoue: J. Phys. Soc. Japan **21** (1966) 1312.
66MI(a)	Mizouguchi, T., and M. Inoue: J. Phys. Soc. Japan **21** (1966) 1313.
67NK	Nagata, T., and H. Kinoshita: Phys. Earth Planet. Interiors **1** (1967) 46, 48.

General References

1	Domenicali, C. A.: Phys. Rev. **78** (1950) 458.
2	Calhoun, B. A.: Phys. Rev. **94** (1954) 1577.
3	Smit, J., and H. P. J. Wijn: Ferrites. John Wiley and Sons, New York (1959).
4	Chikazumi, S.: Physics of Magnetism, Wiley, New York (1964).
5	Callen, E.: Phys. Rev **150** (1966) 367.

6.1.2 Ni-Fe^{3+} spinels (Ni ferrites) and Ni-Fe^{3+} spinels with substitutions

6.1.2.0 Introduction

This chapter contains a survey of the published experimental values of the physical properties of nickel ferrite and nickel ferrite substituted with many additives. Since the optical properties and the small signal permeabilities are strong functions of sample preparation and purity those figures concerned with the properties are presented as indications of what has been achieved, not as intrinsic properties of the nickel spinels. Readers are referred to the original publication for detail of sample preparation. Readers should not be surprised at disagreements between various authors on this account.

6.1.2.1 Ni ferrite $NiFe_2O_4$

Tab. 1. General properties of stoichiometric Ni ferrite

Chemical formula: $NiFe_2O_4$. Appearance: In sintered solid form—opacity very high—black appearance. Room temperature unless otherwise specified.

Certain parameters are very sensitive to the porosity of the structure and values given must be approximate.

		Ref.
Crystal type	Inverse spinel—cubic	*47VH*
Structure type	$MgAl_2O_4$	
Space group	O_h^7	
Lattice constant	$a = 8.337$ Å	*54G*
Volume of unit cell	$579.472 \cdot 10^{-24}$ cm³ (containing 8 molecules)	
Molecular weight (mass of unit cell)	234.39 amu · $1.6595 \cdot 10^{-24}$ g/amu · 8 = $3111.76 \cdot 10^{-24}$ g	
X-ray density	$\varrho_X = 5.37$ g/cm³	
Melting point*)	$T_{melt} = 1690 \cdots 1750$ °C (see Fig. 26e to 26i)	
Sintering temperature	$1200 \cdots 1450$ °C	
Thermal conductivity	$\varkappa \cong 9 \cdot 10^{-3}$ cal sec^{-1} cm^{-1} deg^{-1}	*63K*
Specific heat	$c_p \cong 0.2$ cal g^{-1} deg^{-1}	*54SW(1)*
Young's modulus	$E \cong 1.6 \cdot 10^{12}$ dyne/cm²	
Compressibility	$\beta = 0.5144 \cdot 10^{-12}$ cm²/dyne	*55W*
Elastic constants	$C_{11} = C_{12} = 1/\beta = 1.944 \cdot 10^{12}$ dyne/cm²	*55W*
Velocity of sound	6010 m/sec	*55W*
Dielectric constant	$\varepsilon = 19$ (single crystal); $\varepsilon = 21$ (polycrystal)	*50GMR*

*) Nickel ferrite decomposes when melted in air to a spinel of $Ni_{1.78}Fe_{2.22}O_4$ and NiO.

Tab. 2. Magnetic properties of stoichiometric nickel ferrite. Room temperature unless otherwise specified.

		Ref.
Curie temperature	$\Theta_C = 585$ °C	*54G*
Saturation magnetization = magnetic moment per unit volume	$M_s = 300$ emu/cm³	
Saturation induction	$4\pi M_s = 3770$ Gauss	
Saturation magnetic moment/molecule (0 °K)	$p^0_{sm} = 2.2\ \mu_B$	*59SW(1)*
Saturation magnetic moment/g at 0 °K	$\sigma^0_s = 56$ emu/g	*59SW(1)*
Initial permeability	$\mu_i \cong 80$ (single crystal)	*50GMR*
	$\cong 39$ (polycrystal)	*57E*
Anisotropy constant, first order (microwave) 20 °C	$K_1 \cong -6.8 \cdot 10^4$ erg/cm³	*55YGM*
(magnetization energy)	$\cong -7.0 \cdot 10^4$ erg/cm³	*62E*
Magnetostriction at saturation, (polycrystalline)	$\lambda_s = -26 \cdot 10^{-6}$,	*35R*
(single crystal)	$\lambda_{100} = -46 \cdot 10^{-6}$	*66SJ*
	$\lambda_{111} = -21 \cdot 10^{-6}$	*66SJ*
Effective spectroscopic splitting factor	$g = 2.198$	*55YGM*
Magnetomechanical ratio	$g' = 1.94 \pm 0.04$	*35R*
Distribution of magnetic ions		*52WG*
A site (tetrahedral)	Fe^{3+} 5 μ_B per ion	
B site (octahedral)	Ni^{2+} 2.2 μ_B per ion, Fe^{3+} 5 μ_B per ion	

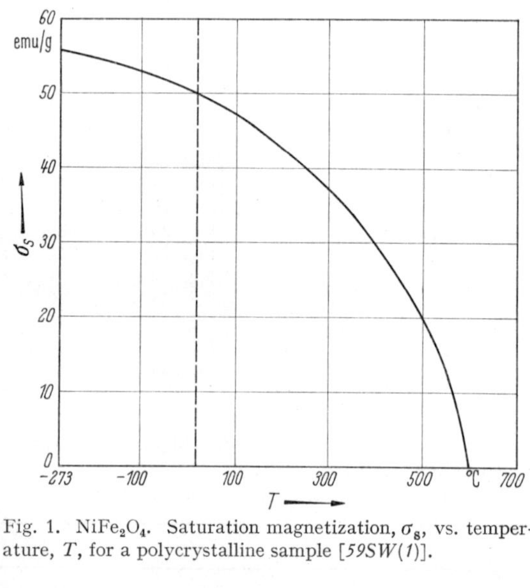

Fig. 1. $NiFe_2O_4$. Saturation magnetization, σ_s, vs. temperature, T, for a polycrystalline sample [*59SW(1)*].

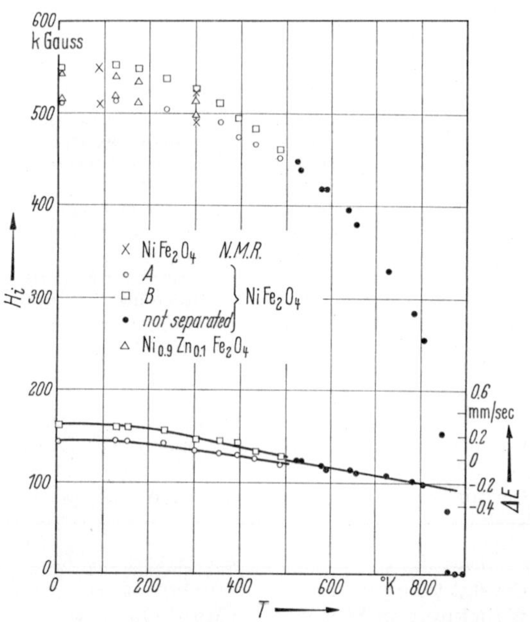

Fig. 2. $NiFe_2O_4$. Hyperfine field and isomer shift for the Fe^{3+} ions on *A* and *B* sites as a function of temperature, *T* [*67Mo*].

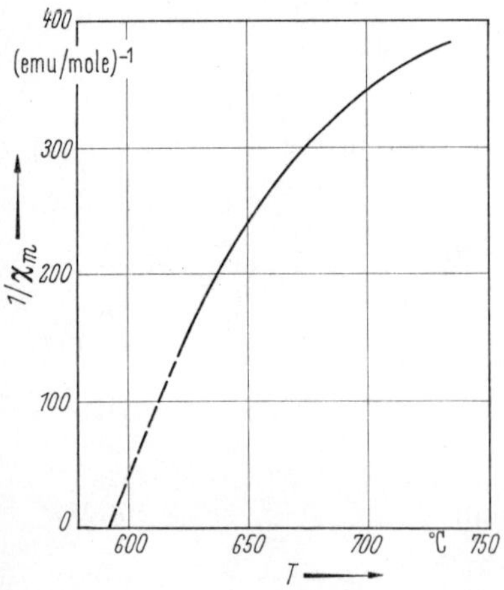

Fig. 3. $NiFe_2O_4$. Reciprocal magnetic susceptibility, $1/\chi_m$, vs. temperature, *T* [*32S*].

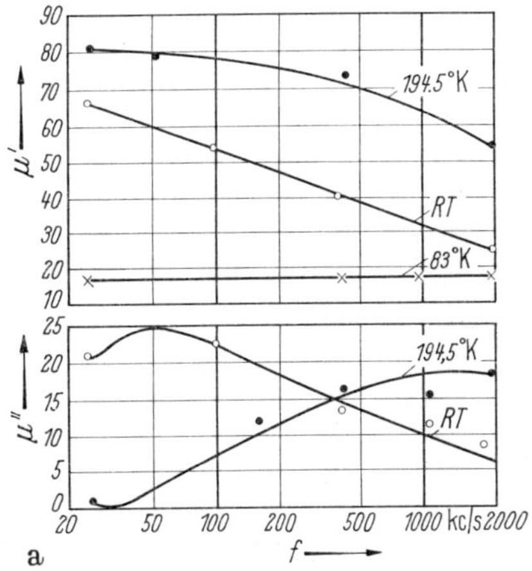

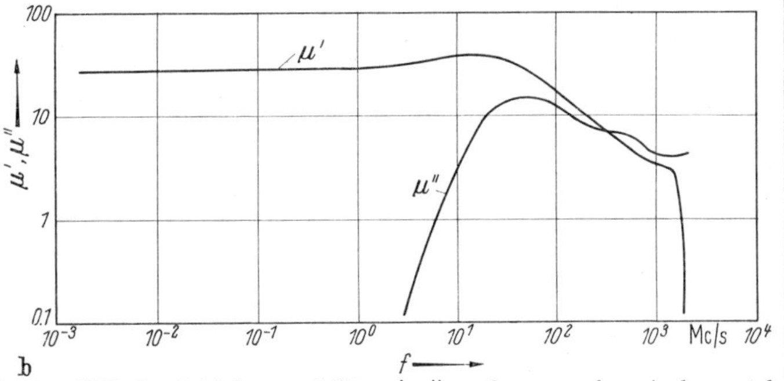

Fig. 4a. $NiFe_2O_4$. Initial permeability, μ', μ'', vs. frequency for a single-crystal ring in [111] plane; parameter: T [*50GMR*]. μ' and μ'' change by a considerable amount depending on sample preparation. See [*58PH*].

Fig. 4b. $NiFe_2O_4$. Real and imaginary parts of initial permeability, μ', μ'', of a polycrystalline sample vs. frequency, f, at RT [*60F*].

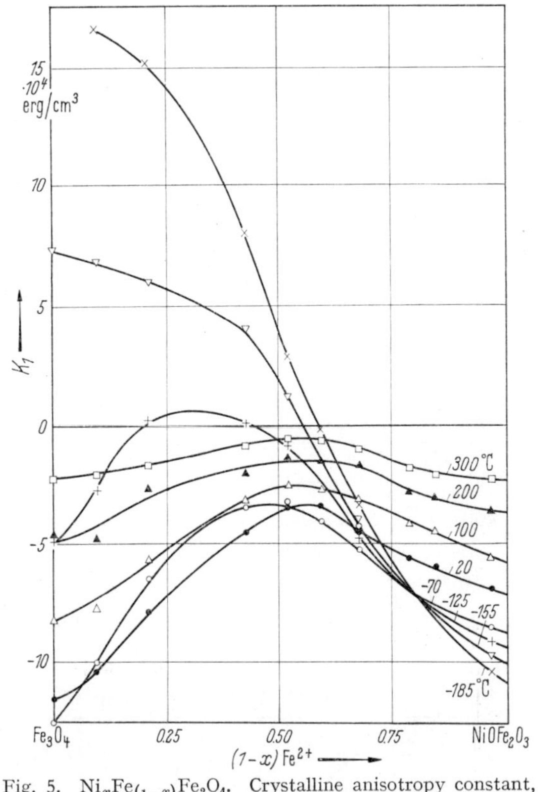

Fig. 5. $Ni_xFe_{(1-x)}Fe_2O_4$. Crystalline anisotropy constant, K_1, plotted for several temperatures, as a function of composition in the solid solution where $Fe_{(1-x)}$ is divalent iron. Measurements were made by static torque method [*62E*].

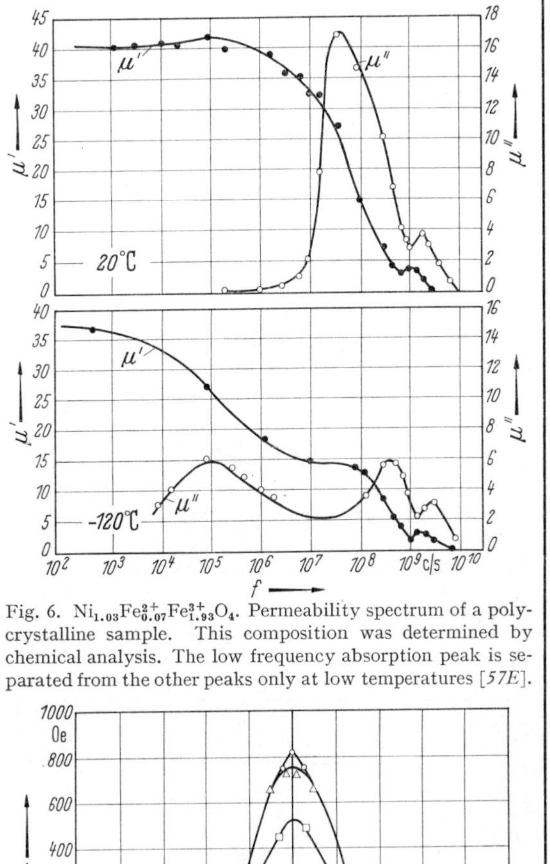

Fig. 6. $Ni_{1.03}Fe^{2+}_{0.07}Fe^{3+}_{1.93}O_4$. Permeability spectrum of a polycrystalline sample. This composition was determined by chemical analysis. The low frequency absorption peak is separated from the other peaks only at low temperatures [*57E*].

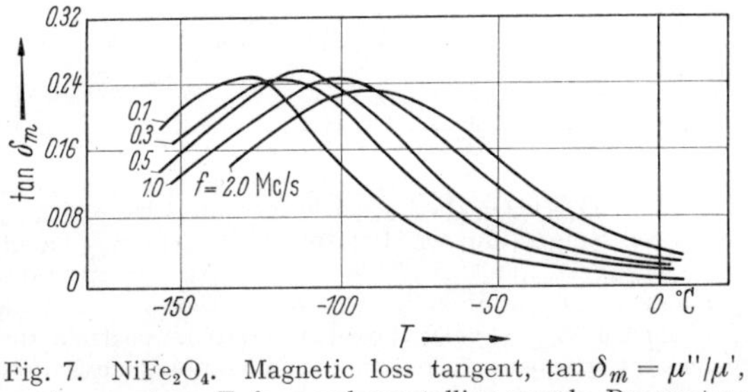

Fig. 7. $NiFe_2O_4$. Magnetic loss tangent, $\tan \delta_m = \mu''/\mu'$, vs. temperature, T, for a polycrystalline sample. Parameter: frequency f [*57E*].

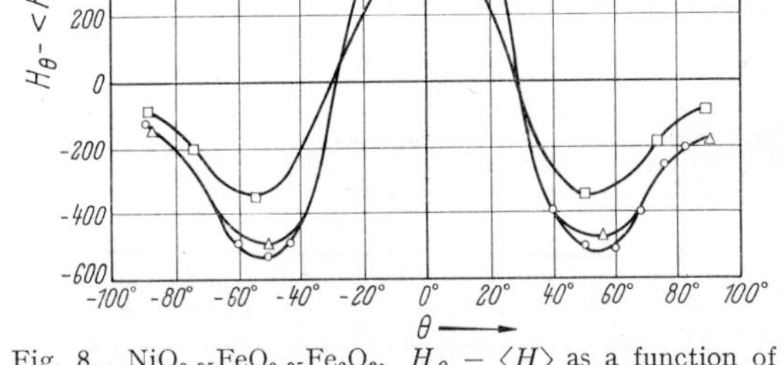

Fig. 8. $NiO_{0.95}FeO_{0.05}Fe_2O_3$. $H_\theta - \langle H \rangle$ as a function of angle, θ, between M and [100] direction in (110) plane for a single crystal. At the angle θ, resonance occurs at the field H_θ; the mean value of H_θ averaged over 180° is $\langle H \rangle$. ○ 4.2 °K; △ 85 °K; □ 300 °K [*55YGM*].

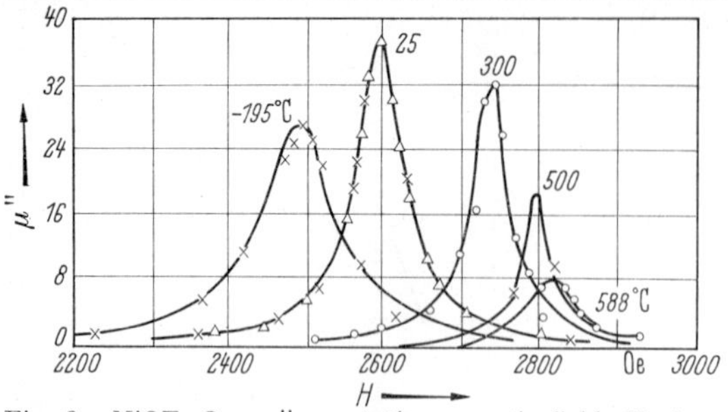

Fig. 9. $NiOFe_2O_3$. μ'' vs. static magnetic field, H, for a single-crystal sphere. Parameter: various temperatures [52H]. Sphere was 0.019 inches in diameter. Frequency was about 9000 Mc/s and field was in [111] direction.

×, ○ experimental points. Solid lines are theoretical curves fitted in experimental values at the peak of μ''.

△: theoretical curve drawn using Yager's form of damping constant (room temperature only).

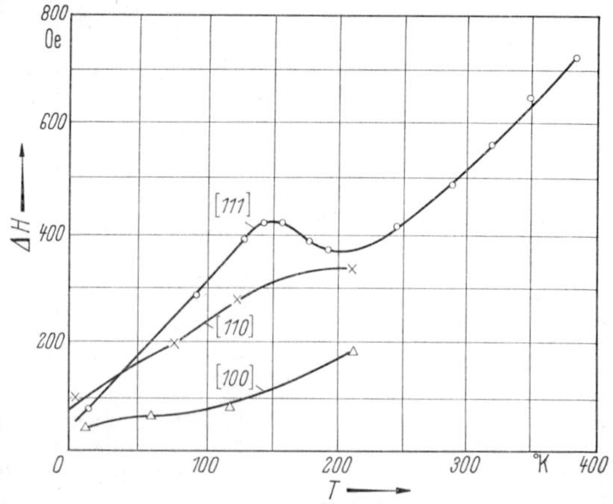

Fig. 10. $(NiO)_{0.75}(FeO)_{0.25}Fe_2O_3$. Ferromagnetic resonance line width, ΔH, vs. temperature, T, for a single crystal. Data obtained from 0.009 inches diameter sphere at a frequency of 24 000 Mc/s with the applied magnetic field in the principle crystal directions as marked [55YGM].

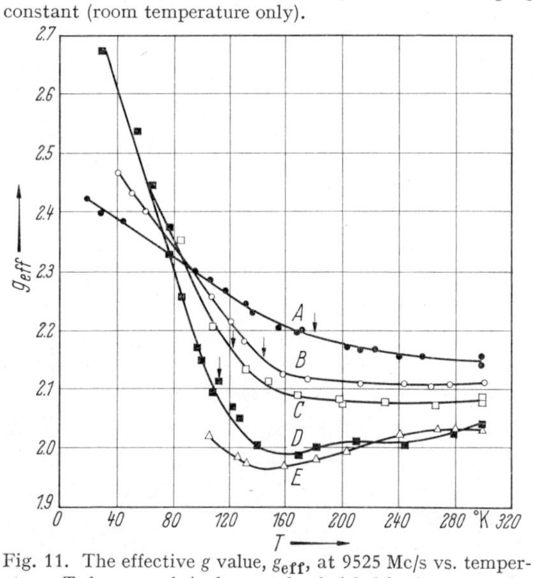

Fig. 11. The effective g value, g_{eff}, at 9525 Mc/s vs. temperature, T, for several single crystals of nickel ferrite-magnetite solid solutions having the following compositions [56MM]:

A $Ni_{0.78}Fe^{2+}_{0.16}Fe^{3+}_{2.02}O_4$

B $Ni_{0.62}Fe_{0.29}Fe_{2.09}O_4$

C $Ni_{0.51}Fe_{0.46}Fe_{1.95}O_4$

D $Ni_{0.28}Fe_{0.61}Fe_{1.93}O_4$

E $Ni_{0.13}Fe_{0.85}Fe_{1.97}O_4$

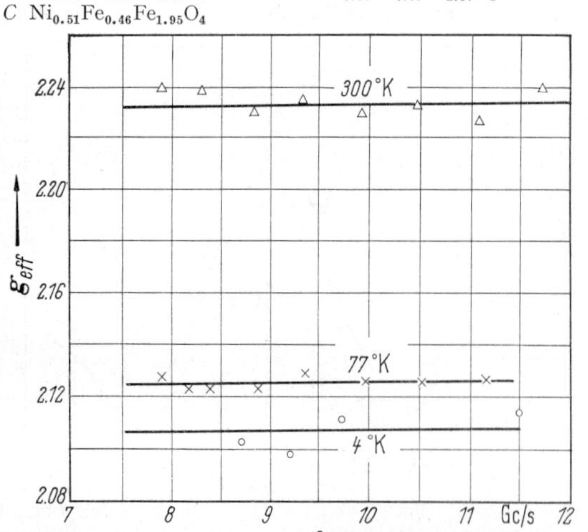

Fig. 13. The g value corrected for internal field in the manner described in the introduction. The internal field $H_i = 336$ Oe at 4 °K, 279 Oe at 77 °K and 87 Oe at 300 °K [56HJ].

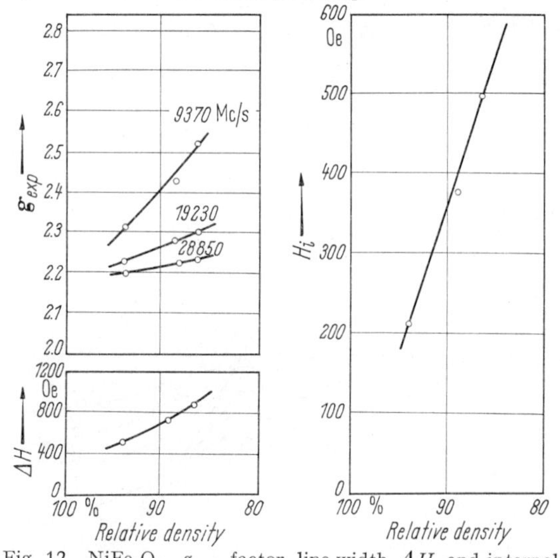

Fig. 12. $NiFe_2O_4$. g_{exp} factor, line width, ΔH, and internal field, H_i, of a polycrystalline sample vs. density. The internal field, H_i, is a method of representation of the effects of porosity in producing a frequency dependence of g_{exp} [54K].

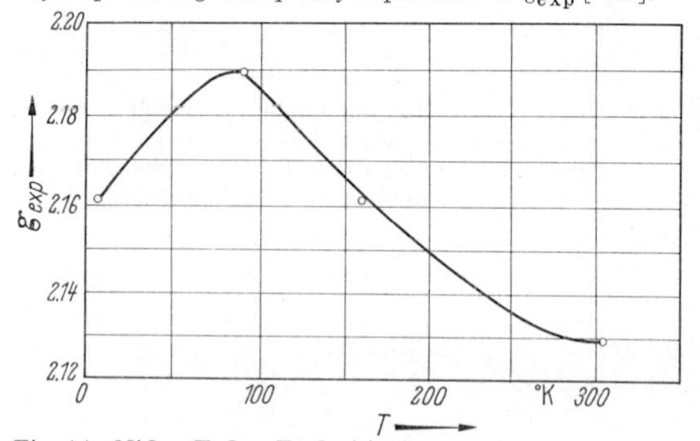

Fig. 14. $NiO_{0.75}FeO_{0.25}Fe_2O_3$ (single crystal). Observed g_{exp} factor vs. temperature for [100] direction. The g_{exp} factor for $NiO_{0.95}FeO_{0.05}Fe_2O_3$ is 2.198 and is not temperature dependent. The variation in g_{exp} factor shown cannot be explained as a size effect; no such variation occurs in the value of the resonance field in the [111] direction, for example, among the various sphere sizes used in the experiments. The frequency for these measurements was 24 000 Mc/s [55YGM].

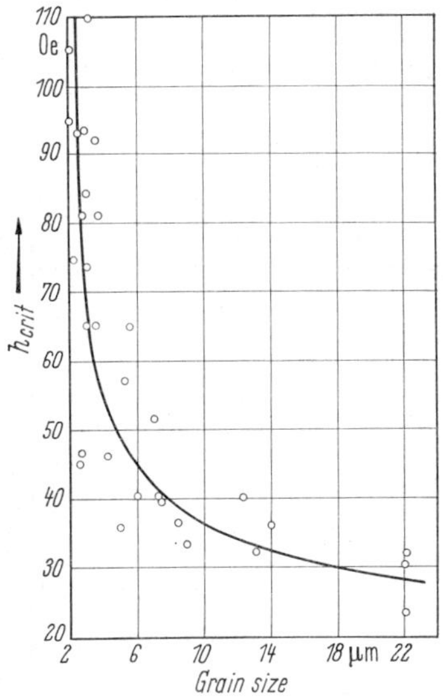

Fig. 15. Dependence of h_{crit}, the microwave field at which saturation occurs, upon grain size for Ni ferrite and Ni ferrite with Mn additions. The frequency is 16.5 Gc/s [*66PWGB*].

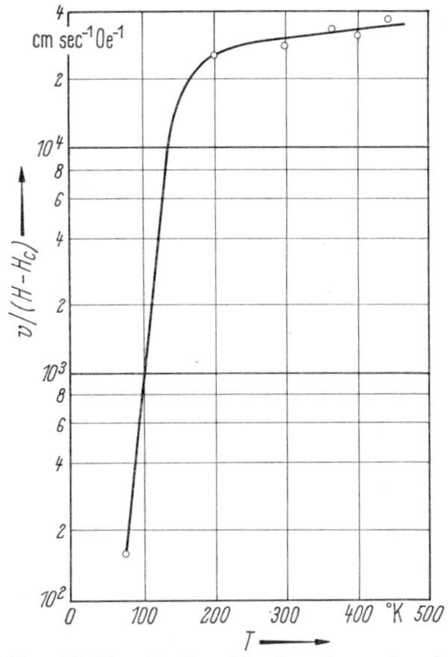

Fig. 16. $(NiO)_{0.75}(FeO)_{0.25}Fe_2O_3$. The domain wall velocity coefficient vs. temperature, T, in single crystals. When a field H in excess of the coercive field H_c (about 0.075 Oe for this sample) is applied, the domain wall will move with a velocity v linearly proportional to the excess field up to excess fields of about 1 Oe at 77 °K where the proportionality becomes non-linear [*54Ga*].

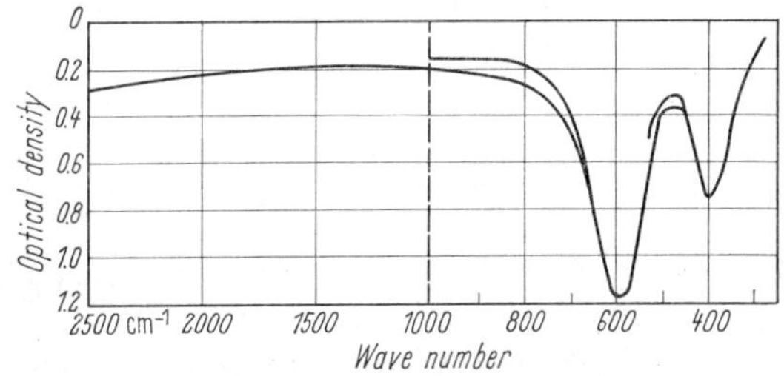

Fig. 17. $NiFe_2O_4$. Optical density in the infrared spectrum region as plotted in wave numbers. These measurements were made by vacuum pressing into a solid clear disc a small percentage of ferrite powder mixed into powdered KBr. This is then compared to a standard pure KBr disc. The threshold energy of the electronic absorption band is (2700 ± 600) cm⁻¹ or 0.33 eV [*55W*].

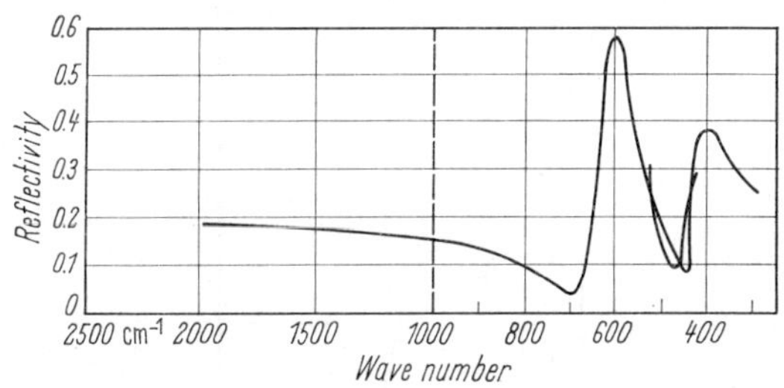

Fig. 18. $NiFe_2O_4$. Reflection spectrum of single crystals (normal incidence (110) face) [*55W*].

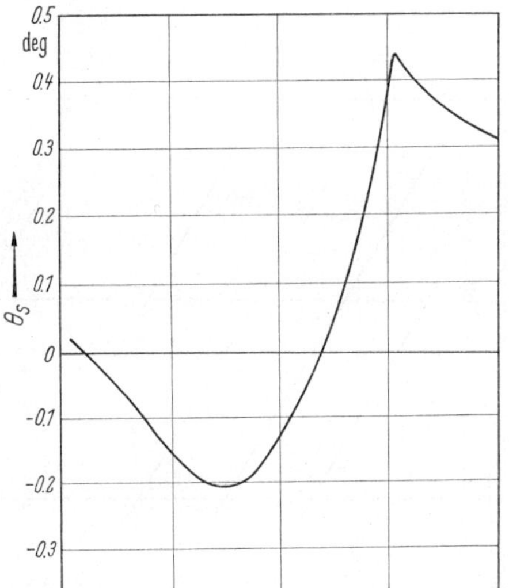

Fig. 19. $NiFe_2O_4$. Saturation Faraday rotation, θ_s, at RT vs. wave number, $\tilde{\nu}$, (λ = 2700···7000 Å) for a nickel ferrite film, 0.16 microns thick. The specific magneto-optical rotation (ϱ) is obtained by multiplying the degrees rotation by $0.624 \cdot 10^5$ [*64CF*].

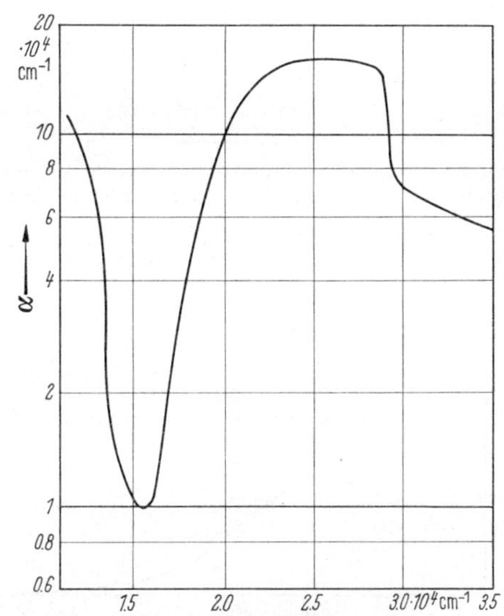

Fig. 20. $NiFe_2O_4$. Absorption coefficient, α, at RT for nickel ferrite film as a function of wave number, $\tilde{\nu}$ [*64CF*].

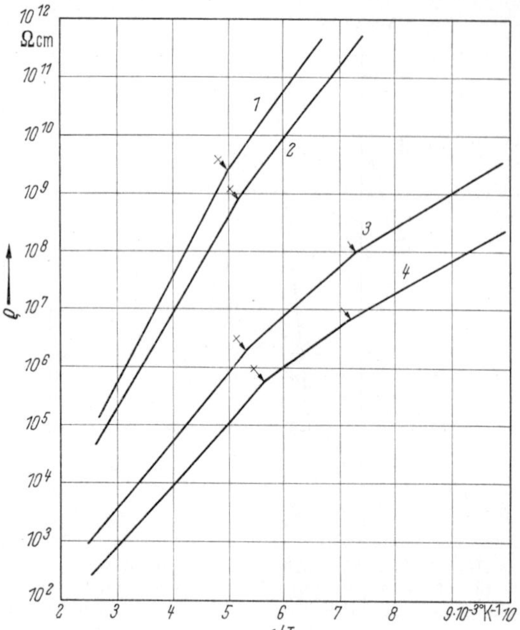

Fig. 21. $Ni_{1.0-}Fe_{1.9}O_{4\pm}$. Resistivity, ϱ, vs. reciprocal temperature, $1/T$. The samples were fired for 10 hours at various temperatures, as follows: curve *1*: 1200 °C; curve *2*: 1250 °C; curve *3*: 1300 °C; curve *4*: 1350 °C [*56U(1)*].

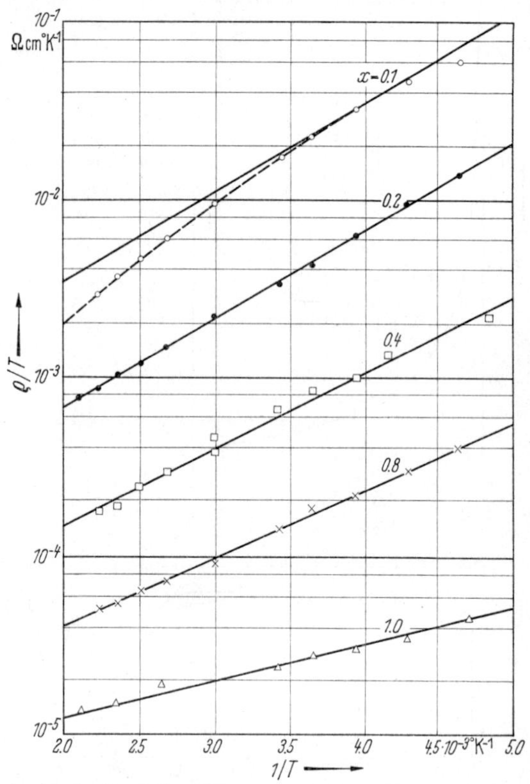

Fig. 22. $Ni^{2+}_{1-x}Fe^{2+}_xFe^{3+}_2O_4$. Resistivity, ϱ/T, vs. reciprocal temperature, $1/T$, for various x values [*62KS*].

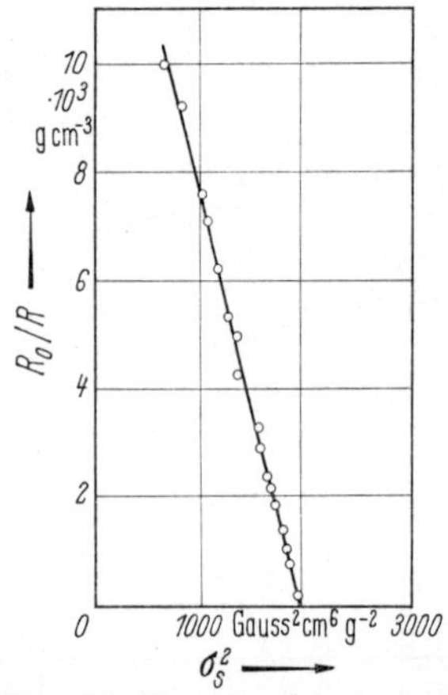

Fig. 23. $Ni_{0.69}Fe_{2.31}O_4$. The ratio of the anomalous to normal Hall coefficients, R_0/R, vs. σ_s^2 for a single crystal [*66SMM*].

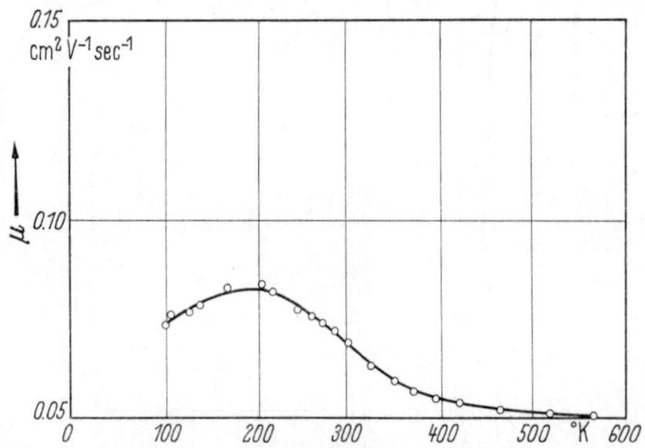

Fig. 24. $Ni_{0.69}Fe_{2.31}O_4$. A plot of Hall mobility, μ, vs. temperature, T, for a single crystal [*66SMM*].

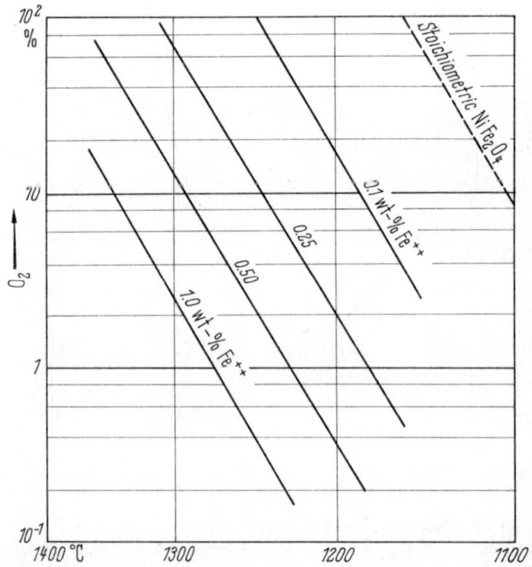

Fig. 25. Atmosphere-temperature diagram for $NiFe_2O_4$ reduction [*67Ma*].

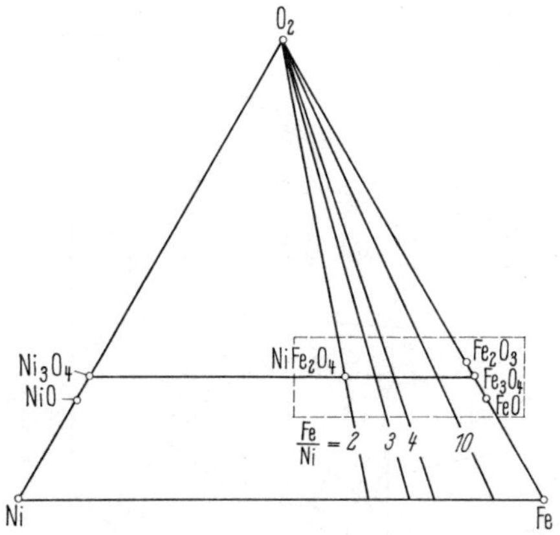

Fig. 26a. A compositional plot for the ternary system Ni – Fe – O. The coordinates are wt-% of the respective materials. The straight lines originating from the oxygen apex show Fe/Ni ratios. The line connecting Fe_3O_4 and Ni_3O_4 shows where the metal to oxygen ratio is three to four. For example, a $NiFe_2O_4$ composition would be where the Fe/Ni = 2.1 line intersects the Fe_3O_4-Ni_3O_4 line [*61Sh*].

←

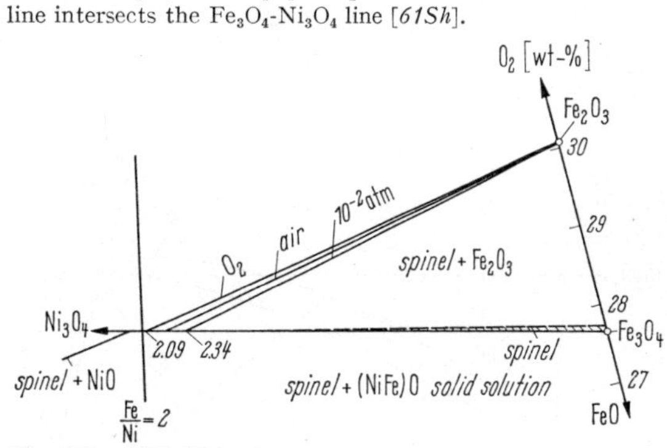

Fig. 26b. 1000 °C isotherm.

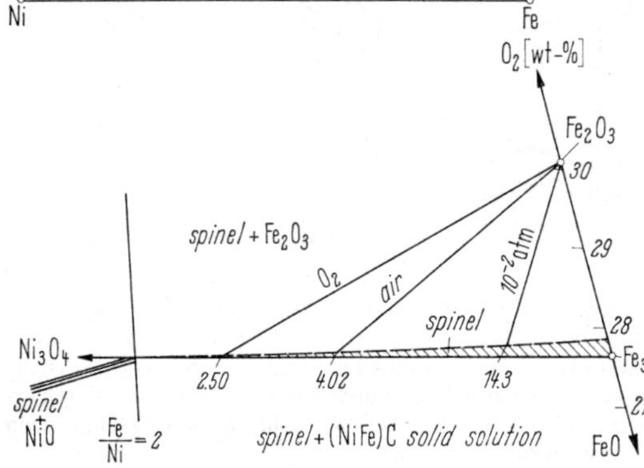

Fig. 26c. 1100 °C isotherm.

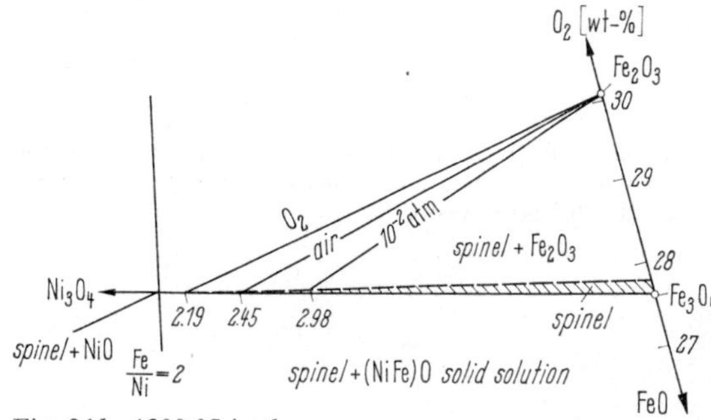

Fig. 26d. 1200 °C isotherm.

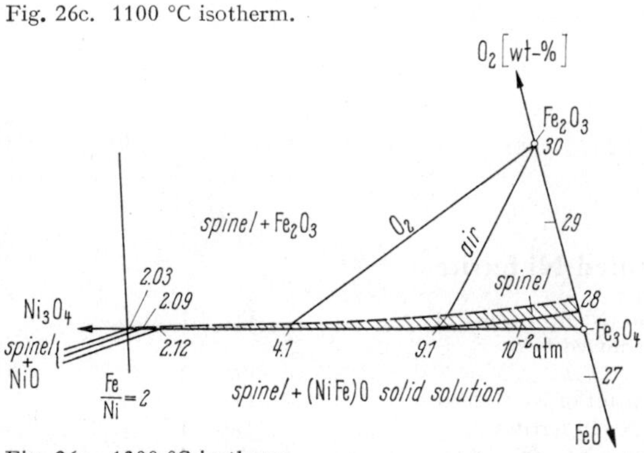

Fig. 26e. 1300 °C isotherm.

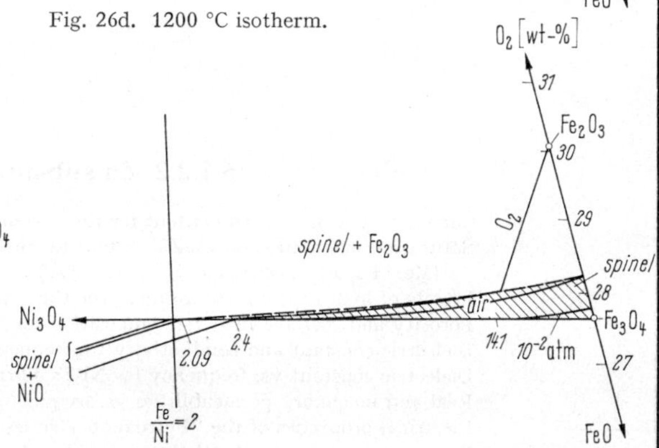

Fig. 26f. 1400 °C isotherm.

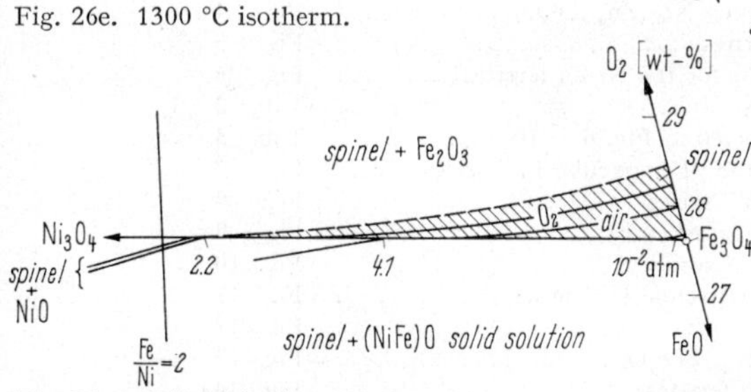

Fig. 26g. 1500 °C isotherm.

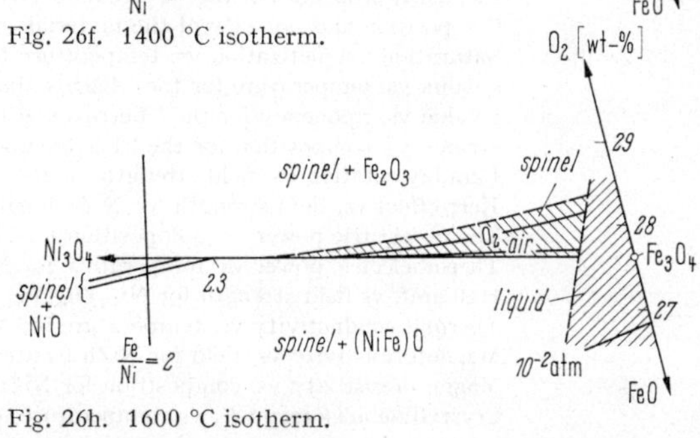

Fig. 26h. 1600 °C isotherm.

Fig. 26b to 26h. A part of the ternary system Ni – Fe – O showing the nickel ferrite region. (That section which is outlined by the dashed lines in Fig. a). The isobars are shown as solid lines labeled "Oxygen" to indicate 1 atmosphere of oxygen, "air" to indicate 10^{-2} atmospheres of oxygen. The area of single phase spinel (nickel ferrites) is the shaded area. The Fe/Ni ratios are given where the isobars enter and leave the spinel field. Figs. b, c, d, e from data in [*59P*]; f, g, h from [*61Sh*]; a disagreeing phase diagram was obtained by PALMER [*54P*].

Note: It is only in the spinel field where the nickel ferrites exist as single phase material. On the oxygen side of the Fe_3O_4-Ni_3O_4 join, two solid phases coexist, spinel and αFe_2O_3. On the metal side two solid solutions with the rock salt structure coexist. (Cf. Fig. 36 in 6.1.1.)

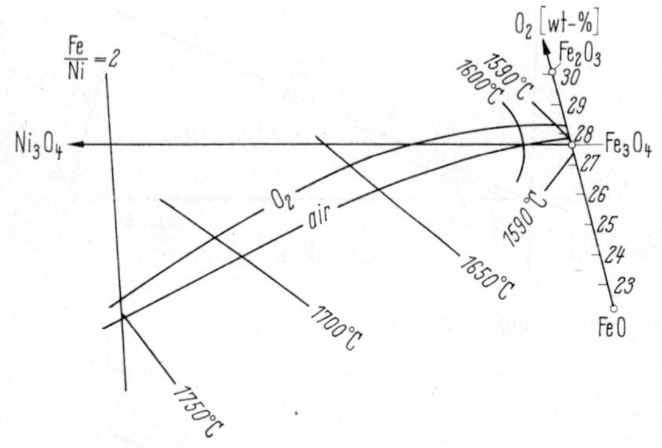

Fig. 26i. Liquidus projection showing compositions for the oxygen and air isobars. The isotherms are indicated by their respective temperatures [*61Sh*].

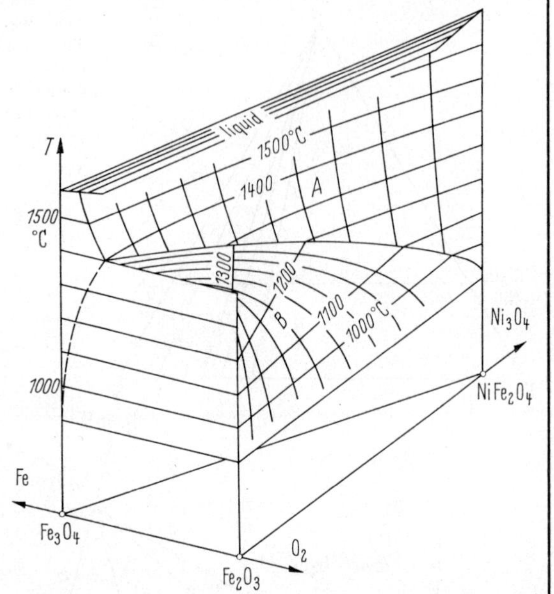

Fig. 26j. Perspective view of the equilibrium composition surface in air. The volume enclosed by surface *A* is a single phase of spinel while the volume underneath surface *B* is the two phase region where spinel and Fe_2O_3 coexist. This perspective view neglects the relationship between composition and oxygen pressure [*61Sh*].

6.1.2.2 Zn substituted Ni ferrite

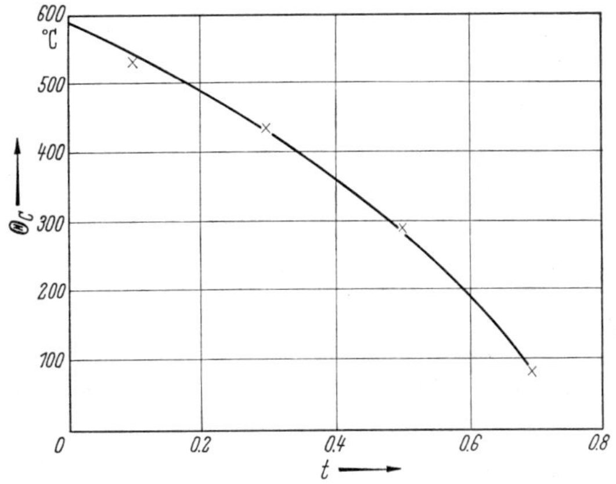

Fig. 1. $(NiO)_{1-t}(ZnO)_t Fe_2O_3$. Curie temperature, Θ_C, vs. zinc content, t [52WG]. See also Fig. 21 in 6.1.2.2.

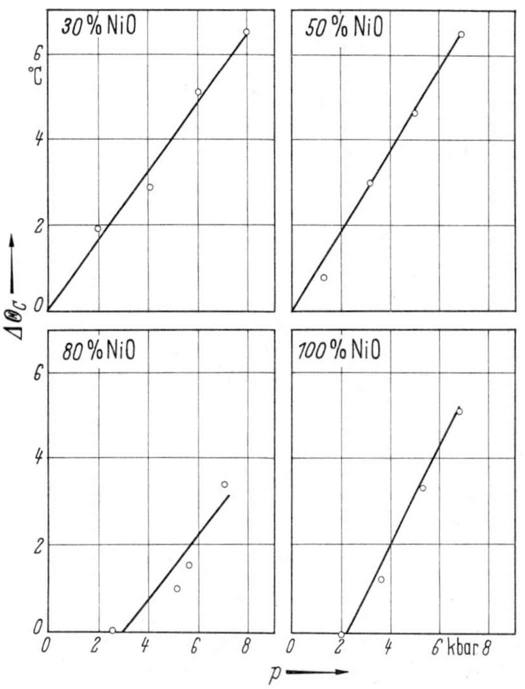

Fig. 3. $(NiO)_x(ZnO)_{1-x}Fe_2O_3$. Plots of Curie point changes, $\Delta\Theta_C$, for the Ni-Zn series vs. pressure, p. For the 30% and 50% NiO samples the changes are relative to the Curie points at atmospheric pressure. For the 80% and 100% NiO samples the changes are relative to the Curie points at 2.53 and 2.10 kbar, respectively [65FT].

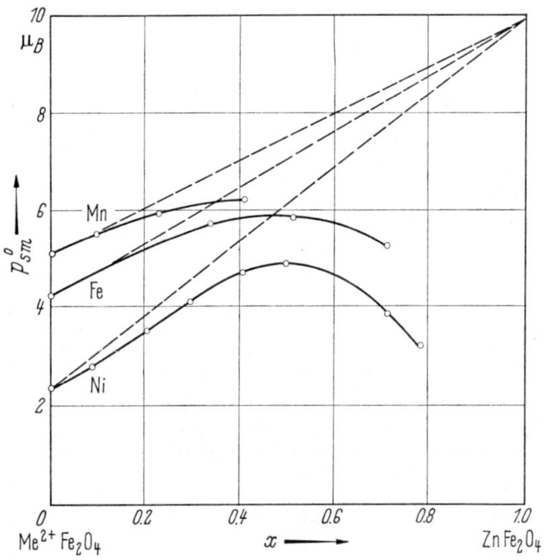

←

Fig. 2. $(Me^{2+}Fe_2O_4)_{1-x}(ZnFe_2O_4)_x$. The saturation magnetization per molecule, p^0_{sm}, of mixed ferrites as a function of the zinc ferrite content, x. The broken lines represent the relationship which would be obtained if the ions in the tetrahedral sites were always able to direct the magnetic moments of the ions in the octahedral sites [52WG].

Tab. 1. Relation between porosity and coercive field of NiZn ferrites in ferroxcube series [54SW(2)].

Ferroxcube IV		Composition (balance Fe_2O_3)		Porosity*) %	M_s at RT emu/cm^3	μ_i	H_c (measured) Oe
Grade	Firing temperature °C	NiO mole-%	ZnO mole-%				
A	1250	17.5	33.2	8.9	292	650	0.4
B	1250	24.9	24.9	15.4	332	230	1.4
C	1250	31.7	16.5	22.5	321	90	4.0
D	1250	39.0	9.4	24.3	283	45	6.8
E	1250	48.2	0.7	24.8	196	17	13.7
A	1450			9.5	290	470	0.3
B	1450			3.2	380	312	0.4
C	1450	See above		8.0	381	86	1.1
D	1450			8.9	341	63	1.7
E	1450			9.9	235	42	3.2

*) Porosity is the proportion of pores in the ferrite that can be inferred from a comparison of apparent density with the X-ray density.

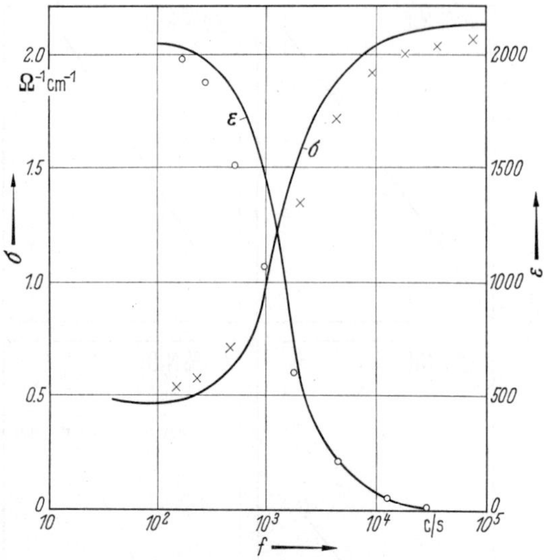

Fig. 4. $Ni_{0.4}Zn_{0.6}Fe_2O_4$. Dispersion of dielectric constant, ε, and specific conductivity, σ, vs. frequency, f. The circles and crosses are the results of measurements made at RT on a ceramic disc fired in air at 1280 °C and cooled slowly in air. The solid lines are the result of a theoretical model which assumes that the specimen is composed of well-conducting grains separated by layers of lower conductivity [*51K*].

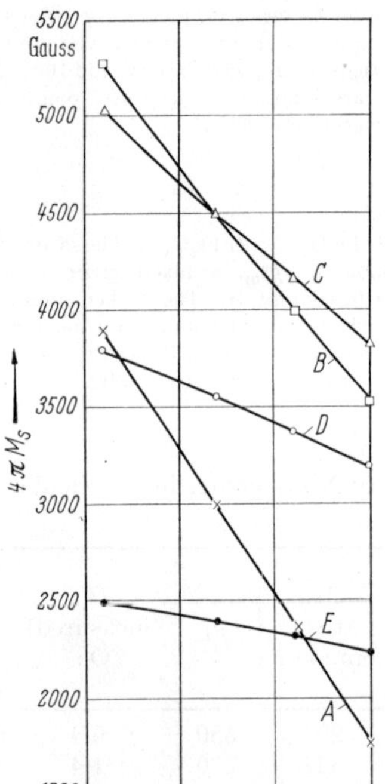

Fig. 7. The saturation magnetization, $4\pi M_s$, for the ferroxcube IV series of Ni-Zn ferrites of Tab. 3 as a function of temperature, T [*58S*].

→

Fig. 8. Observed g value, g_{exp}, vs. temperature, T, for polycrystalline Ni-Zn ferrites (Philips Ferroxcube Series, IV, A, B, C, D, E). g values corrected for porosity for these samples can be seen in Fig. 10. These values were measured at a wavelength of 3.2 cm using spheres of radius 0.02 cm [*57S*].

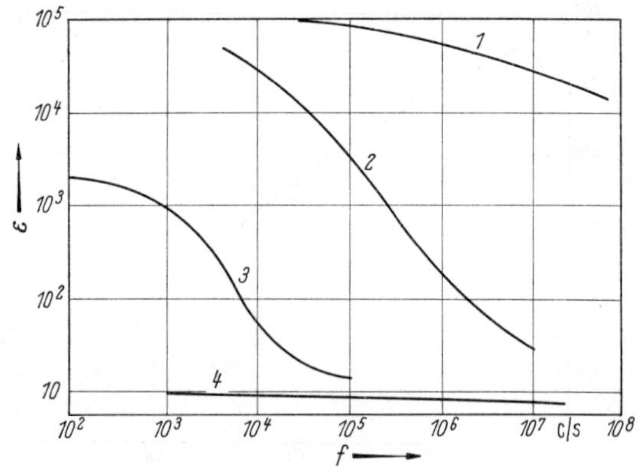

Fig. 5. Parallel dielectric constants, ε, vs. frequency, f, for the ferrites listed below [*56U(2)*].

Curve Nr.	Ferrite	ϱ at 10^3 c/s Ωcm	ϱ at x c/s Ωcm	x c/s
1	$Mn_{0.5}$-$Zn_{0.5}$	130	10	10^8
2	Cu-Zn	$1 \cdot 10^3$	400	10^7
3	$Ni_{0.4}$-$Zn_{0.6}$	$2 \cdot 10^6$	$5 \cdot 10^5$**)	10^5
4	Ferramic-A*)	$2 \cdot 10^9$	$1 \cdot 10^7$	10^7

*) Trade name for an iron-magnesium ferrite purchased from the General Ceramics and Steatite Corporation 45% Fe, 14% Mg, 3% Mn, 2% Ca, 1% Zn; the balance is presumably oxygen [*50RWE*].

**) Would be ≈$5 \cdot 10^4$ at 10^7 c/s.

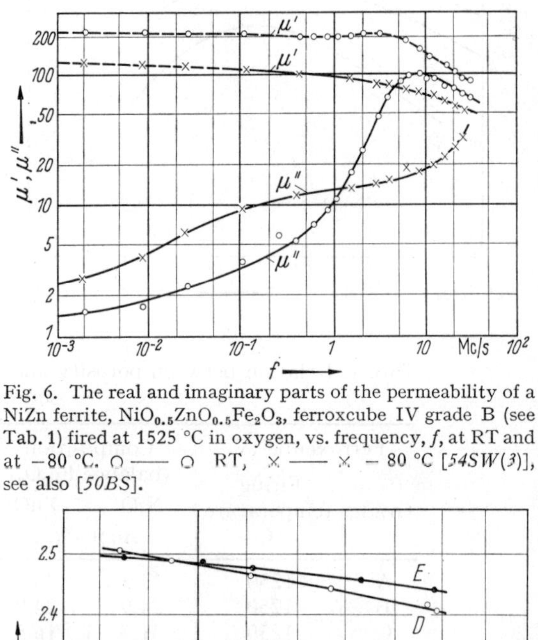

Fig. 6. The real and imaginary parts of the permeability of a NiZn ferrite, $NiO_{0.5}ZnO_{0.5}Fe_2O_3$, ferroxcube IV grade B (see Tab. 1) fired at 1525 °C in oxygen, vs. frequency, f, at RT and at −80 °C. ○——○ RT, ×——× −80 °C [*54SW(3)*], see also [*50BS*].

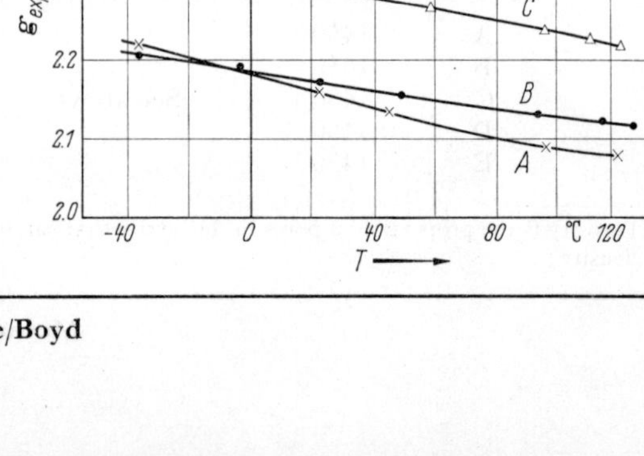

Tab. 2. Electrical and magnetic properties of NiZn ferrites in ferroxcube*) series**).

Ferroxcube IV		Grade A	B	C	D	E
Composition in mole-% (balance Fe_2O_3)	NiO	17.5	24.9	31.7	39.0	48.2
	ZnO	33.2	24.9	16.5	9.4	0.7
Frequency range (approx.) for tuned applications		1 ··· 500 kc/s	0.5 ··· 2 Mc/s	2 ··· 5 Mc/s	5 ··· 10 Mc/s	10 ··· 50 Mc/s
Frequency range (approx.) for non-tuned applications		1 kc/s ··· 20 Mc/s	1 ··· 50 Mc/s	1 ··· 70 Mc/s	1 ··· 100 Mc/s	1 ··· 200 Mc/s
Initial permeability μ_i at induction $\leqq$ 7.5 Gauss		650	200	125	57	20
Loss factor $\frac{\tan\delta}{\mu}$ at	60 kc/s	$22 \cdot 10^{-6}$				
	250 kc/s	$38 \cdot 10^{-6}$				
	450 kc/s	$53 \cdot 10^{-6}$				
	0.5 Mc/s		$90 \cdot 10^{-6}$			
	1 Mc/s		$100 \cdot 10^{-6}$	$95 \cdot 10^{-6}$		
	2 Mc/s		$135 \cdot 10^{-6}$	$130 \cdot 10^{-6}$		
	5 Mc/s			$150 \cdot 10^{-6}$	$170 \cdot 10^{-6}$	
	8 Mc/s				$180 \cdot 10^{-6}$	
	10 Mc/s				$200 \cdot 10^{-6}$	$350 \cdot 10^{-6}$
	20 Mc/s					$380 \cdot 10^{-6}$
	30 Mc/s					$700 \cdot 10^{-6}$
	40 Mc/s					$1\,700 \cdot 10^{-6}$
	50 Mc/s					$2\,500 \cdot 10^{-6}$
Temperature factor of μ_i between 20 and 50 °C $\frac{1}{\mu^2} \cdot \frac{d\mu_i}{dT}$	in $(°C)^{-1}$	$8 \cdot 10^{-6}$	$14 \cdot 10^{-6}$	$15 \cdot 10^{-6}$	$20 \cdot 10^{-6}$	$30 \cdot 10^{-6}$
Magnetizing force H	in Oe	10	20	30	60	80
B_{sat} at 20 °C	in Gauss	2 900	3 200	2 700	2 500	1 900
B_{sat} at 100 °C	in Gauss	1 800	2 650	2 450	2 300	1 750
Resistivity, ϱ	in Ωcm	10^5	10^5	10^5	10^5	10^5
Curie point, Θ_C	in °C	135	250	380	460	550
Specific gravity		4.8	4.6	4.4	4.2	3.7

Tab. 3. Materials used to obtain properties shown in Fig. 7 ··· 11 [*58S*].

Ferrox-cube	Composition in mole-% NiO	ZnO	Specific gravity	Porosity %	μ_i at 20 °C
IV A	17.5	33.2	4.45	16.6	342
IV B	24.9	24.9	4.80	11.0	286
IV C	31.7	16.5	4.52	16.2	138
IV D	39.0	9.4	3.98	26.2	54.1
IV E	48.2	0.7	3.80	29.5	16.6

A comparison of the parameters in this table with those listed in Tab. 2 will indicate the order of disagreement to be found among specimens of the same material.

Fig. 7, 8 see p. 84

*) "Ferroxcube" is the trademark of N. V. Philips' Gloeilampenfabrieken.
**) Ferroxcube, Tech. Publications Dept. for Mullard Overseas Limited, London, Issue 2, page 6.

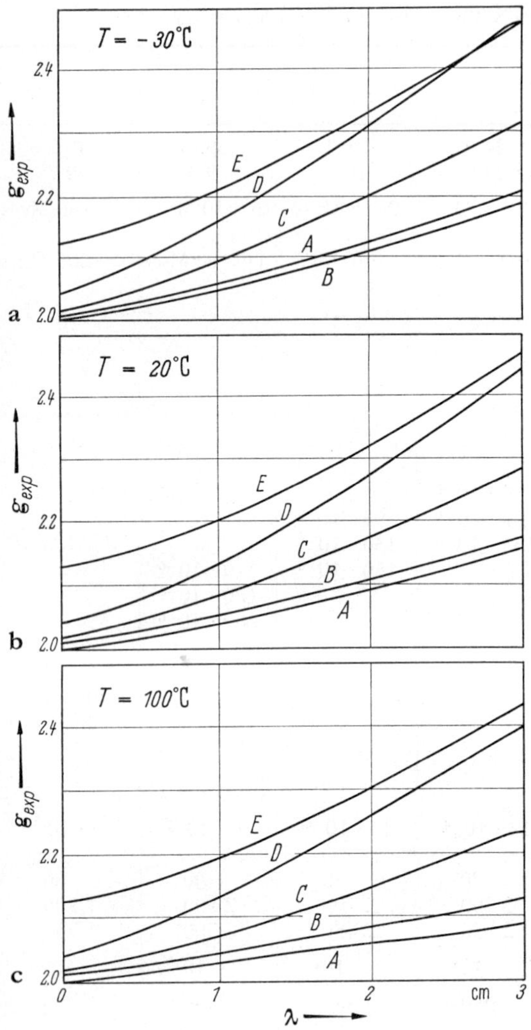

Fig. 9a, b, c. The observed g value for the ferroxcube IV series of Ni-Zn ferrites of Tab. 3 vs. wavelength, λ, for three temperatures. These curves represent the internal field formula (see Introduction); result fitted to experimental data at $\lambda = 3.2$ cm and $\lambda = 1.6$ cm and also, in Fig. 9b, at $\lambda = 1.25$ cm [58S].

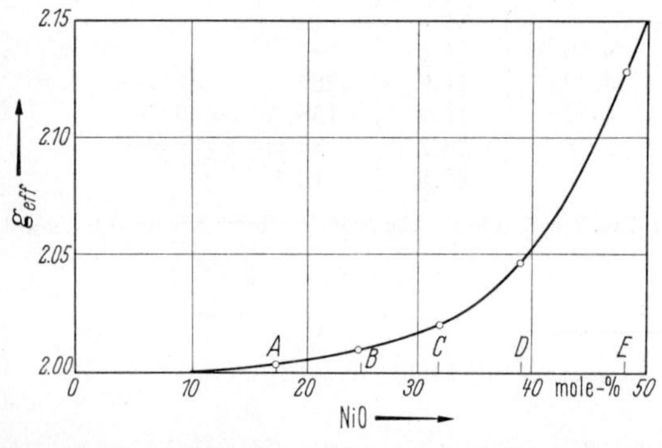

Fig. 10. The g value corrected for porosity in the ferroxcube series of Ni-Zn ferrites of Tab. 3. These g_{eff} values are independent of temperature and wavelength [58S].

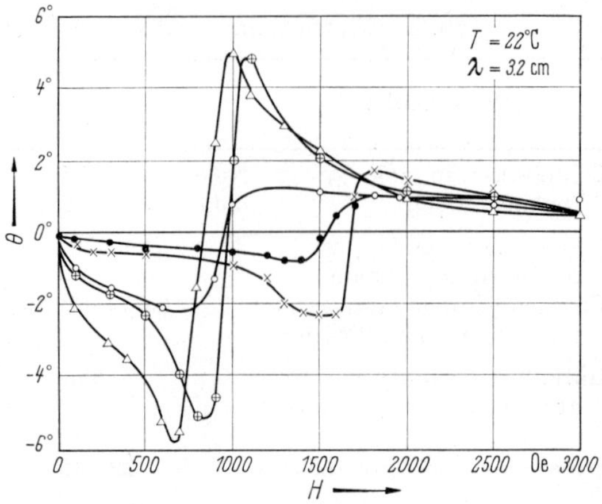

Fig. 11. The Faraday rotation angle, θ, vs. the static magnetic field strength, H, for rods of the ferrites A (×), B (⊗), C (△), D (○), E (●) of Tab. 3. Length of the rods was 3.5 cm, radius 0.05 cm for A, B, D, E, and 0.06 cm for C [58S].

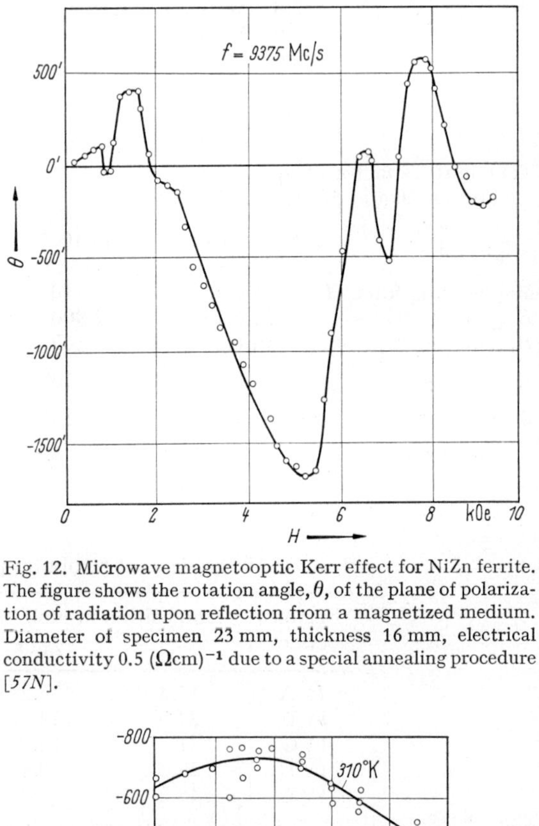

Fig. 12. Microwave magnetooptic Kerr effect for NiZn ferrite. The figure shows the rotation angle, θ, of the plane of polarization of radiation upon reflection from a magnetized medium. Diameter of specimen 23 mm, thickness 16 mm, electrical conductivity 0.5 $(\Omega\text{cm})^{-1}$ due to a special annealing procedure [57N].

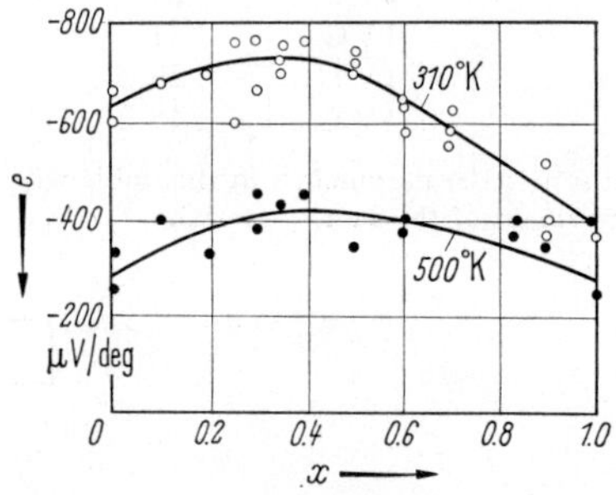

Fig. 13. $Ni_xZn_{1-x}Fe_2O_4$. Thermoelectric power, e, at two temperatures vs. x [63RSF].

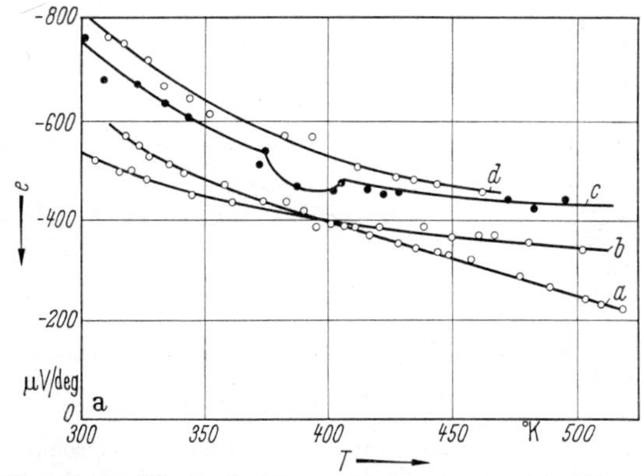

Fig. 14a, b. Thermoelectric power, e, vs. temperature, T.
Fig. 14a. a $ZnFe_2O_4$; b $Ni_{0.9}Zn_{0.1}Fe_2O_4$; c $Ni_{0.35}Zn_{0.65}Fe_2O_4$; d $Ni_{0.30}Zn_{0.70}Fe_2O_4$ [64SR].

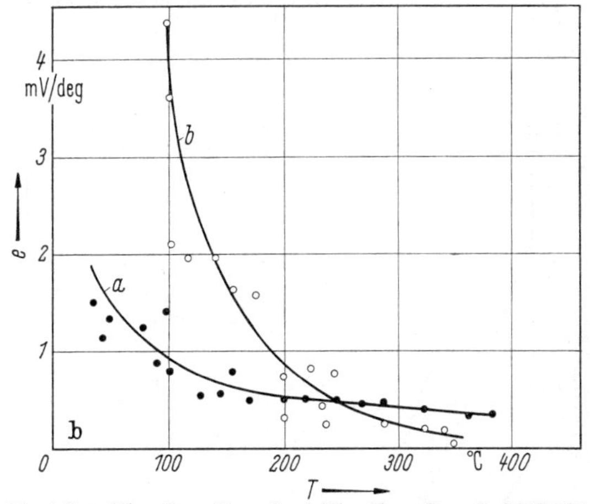

Fig. 14b. a $Ni_{0.70}Zn_{0.40}Fe_{1.90}O_4$; b $Ni_{0.50}Zn_{0.60}Fe_{1.90}O_4$ [63RSF].

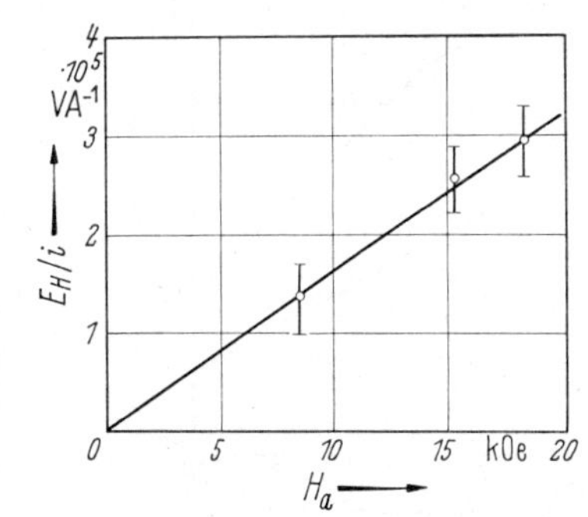

Fig. 15. Dependence of the Hall emf on applied magnetic field, H_a, at 20 °C for material $Ni_{0.70}Zn_{0.40}Fe_{1.90}O_4$. Sample dimensions were $1.994 \cdot 0.580 \cdot 0.103\ cm^3$ [64SR].

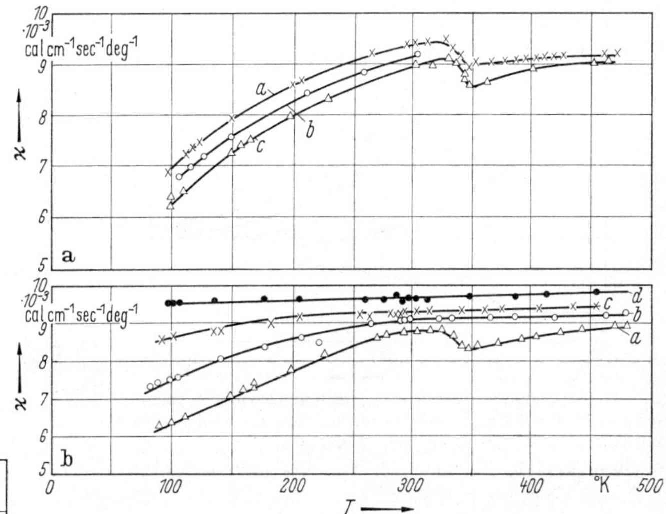

Fig. 16a, b. $Ni_{0.3}Zn_{0.7}Fe_2O_4$. Thermal conductivity, $\varkappa$, vs. temperature, T [63K].

Fig. 16a. Samples having three densities ranging from 4.3 g/cm³ with small grains (plot a) to 5.1 g/cm³ with large grains (plot c). Plot b represents an intermediate density.

Fig. 16b. Samples having the following compositions: a $Ni_{0.3}Zn_{0.7}Fe_2O_4$; b $Ni_{0.25}Zn_{0.75}Fe_2O_4$; c $Ni_{0.10}Zn_{0.90}Fe_2O_4$; d $ZnFe_2O_4$.

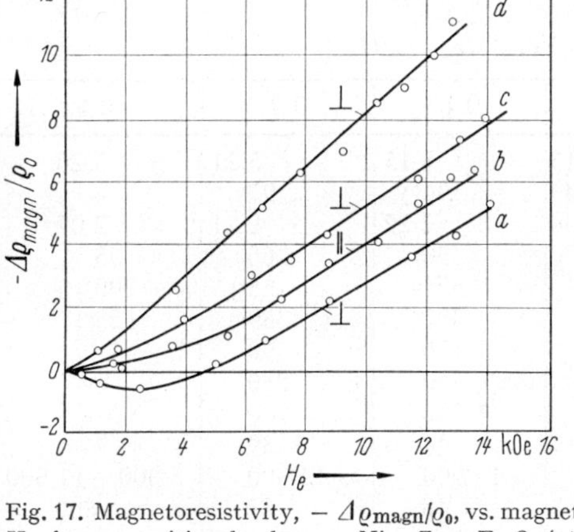

Fig. 17. Magnetoresistivity, $-\Delta\varrho_{magn}/\varrho_0$, vs. magnetic field, H_e, for compositional values: a $Ni_{0.20}Zn_{0.80}Fe_2O_4$ (at 33 °C); b $Ni_{0.20}Zn_{0.80}Fe_2O_4$ (at 20 °C); c $Ni_{0.40}Zn_{0.60}Fe_2O_4$ (at 20 °C); d $Ni_{0.30}Zn_{0.70}Fe_2O_4$ (at 20 °C).
Plot a, c, and d magnetic field is perpendicular to current flow while plot b field is parallel to current flow [63RSF]. $\varrho_0 : \varrho$ at $H_e = 0$.

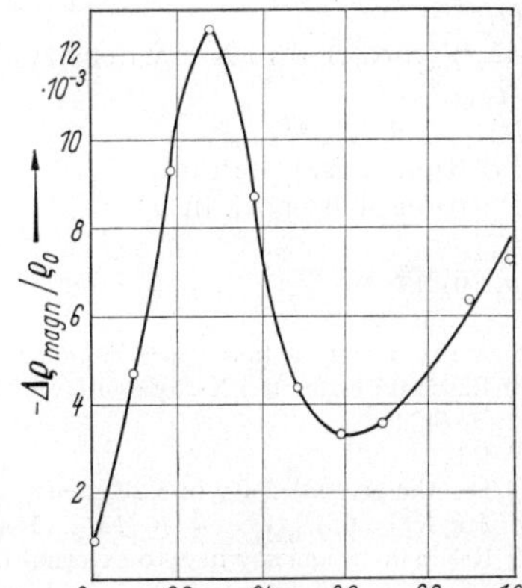

Fig. 18. $Ni_xZn_{(1-x)}Fe_2O_4$. Magnetoresistivity, $-\Delta\varrho_{magn}/\varrho_0$, at 20 °C in a field of 14 000 Oe vs. x [63RSF]. $\varrho_0 : \varrho$ at $x = 0$.

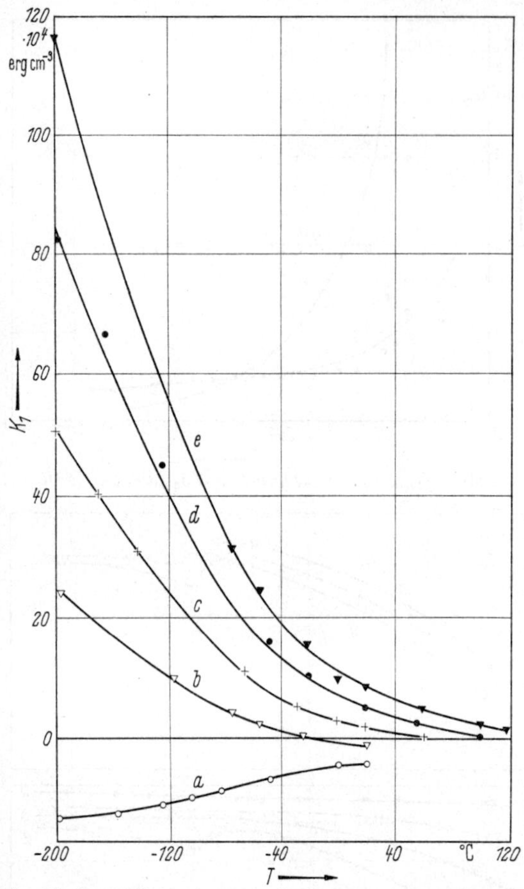

Fig. 19. Crystalline anisotropy, K_1, vs. temperature, T, measured at 10 000 Oe using static torque method for compositional values: *a* $Ni_{0.46}Zn_{0.29}Fe_{2.25}O_4$; *b* $Ni_{0.45}Co_{0.01}Zn_{0.29}Fe_{2.25}O_4$; *c* $Ni_{0.441}Co_{0.019}Zn_{0.29}Fe_{2.25}O_4$; *d* $Ni_{0.432}Co_{0.028}Zn_{0.29}Fe_{2.25}O_4$; *e* $Ni_{0.423}Co_{0.037}Zn_{0.29}Fe_{2.25}O_4$ [*65M*].

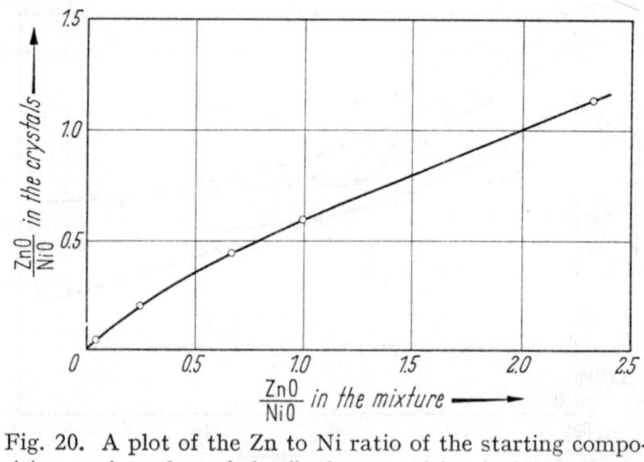

Fig. 20. A plot of the Zn to Ni ratio of the starting composition against that of the final composition in Ni-Zn ferrite single crystals. A PbO-PbF$_2$ flux was used [*67Mak*].

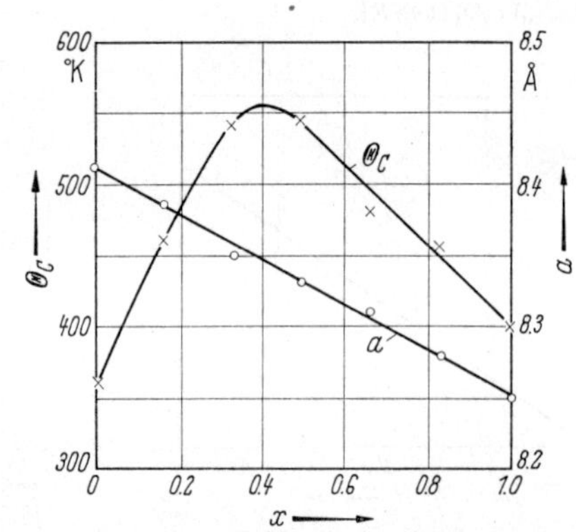

Fig. 21. $Ni_{0.25+1.5x}Zn_{0.75-0.75x}Fe_{2-1.5x}Ge_{0.75x}O_4$. Curie temperature, Θ_C, and cell edge, a, vs. composition, x [*65B*]. Compare Fig. 1.

Fig. 22, 23 see p. 89

Tab. 4. Properties of $Ni_{1-t}Zn_tMn_{0.02}Fe_{1.9}O_{4\pm}$ [*59PH*].

Zn content t		0.0	0.1	0.2	0.6
Density ϱ a)	in g/cm³	5.13	5.15	5.21	5.20
	% of maximum	95.4	96 b)	97 b)	—
g_{exp}		2.34	2.22	2.21	2.07
Resonance line width ΔH	in Oe	460	330	265	95
Θ_C	in °C	585	525 c)	455 c)	200 c)
M_s d) (corrected to X-ray density)	in emu/cm³	257	392 b)	389 b)	323 e)
H_{anis} f)	in Oe	260	≈ 150	≈ 90	—
f g) $= 1.4\, g_{exp} H_{anis}$	in Mc/s	850	≈ 460	≈ 280	—
Position of 2nd peak in μ''	in Mc/s	1 700	1 200	≈ 900	—
Position of 1st peak in μ''	in Mc/s	100	44	39	7
f_{max} h)	in Mc/s	12 000	≈ 13 000	≈ 14 000	8 000 ··· 11 000
$\mu_i'(0)$ i)		28	48	82	≈ 800

a) X-ray density is taken as 5.38 g/cm³ for small Co content, see Tab. 2 in 6.1.2.4.
b) Based on estimated X-ray density of 5.4 g/cm³ for nickel-zinc ferrites with low zinc content.
c) [*56Be*].
d) RT.
e) For the given density of 5.20 g/cm³.
f) For $K_1 < 0$, $H_{anis} = \frac{4}{3}|K_1|/M_s$. For $K_1 > 0$, $H_{anis} = 2K_1/M_s$.
g) Resonant frequency in zero external field for a spherical sample.
h) The highest resonant frequency which can occur in zero external field.
i) Initial real permeability at low frequency.

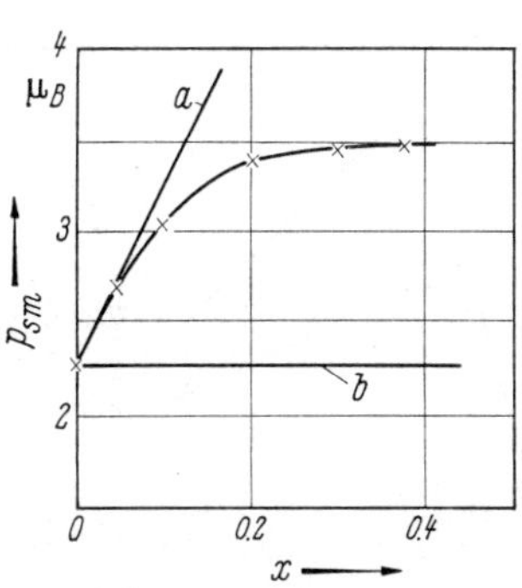

Fig. 22. $NiZn_xFe_{2-2x}Ge_xO_4$. Saturation moment, p_{sm}, vs. x. a calculated for all Zn^{2+} and Ge^{4+} ions on A sites; b calculated for equal amounts of diamagnetic ions on A sites and B sites (e. g. Zn^{2+} on A sites and Ge^{4+} on B sites) [65B]. Compare Fig. 2.

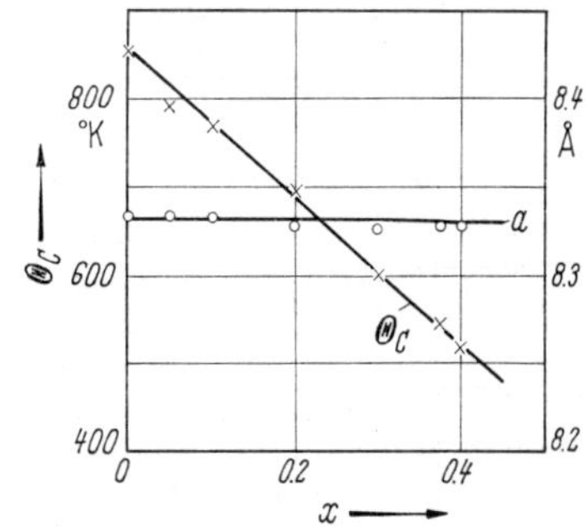

Fig. 23. $NiZn_xFe_{2-2x}Ge_xO_4$. Curie temperature, Θ_C, and cell edge, a [65B]. Compare Fig. 1.

6.1.2.3 Ni ferrite with Co substitutions

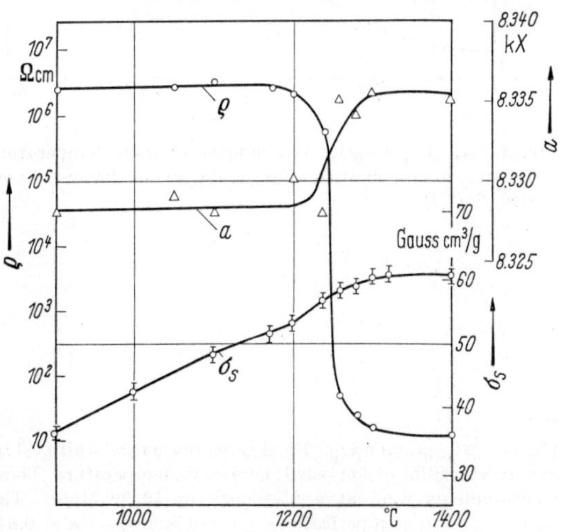

Fig. 1. Saturation magnetization, σ_s, volume resistivity, ϱ, and lattice parameter, a, vs. sintering temperature, T_s, for a starting mixture $(NiO)_{0.41}(ZnO)_{0.03}(CoO)_{0.01}(Fe_2O_3)_{0.55}$ [65SF].

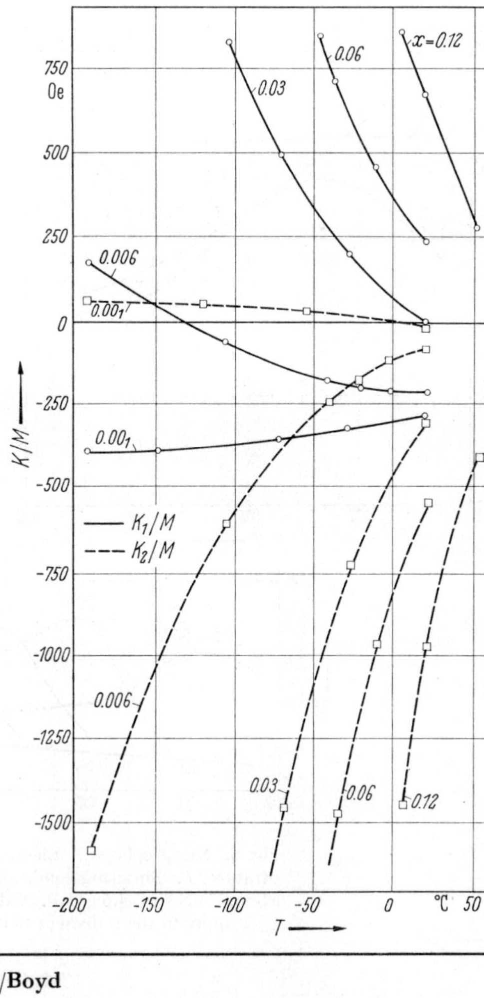

→

Fig. 2. $Co_xNi_{1-x}Fe_2O_4$. Magnetic anisotropy constants, K_1/M (solid lines), K_2/M (dashed lines), vs. temperature, T. Parameter x. These values were determined on single crystals from microwave resonant absorption at 9.27 Gc/s [65MTI].

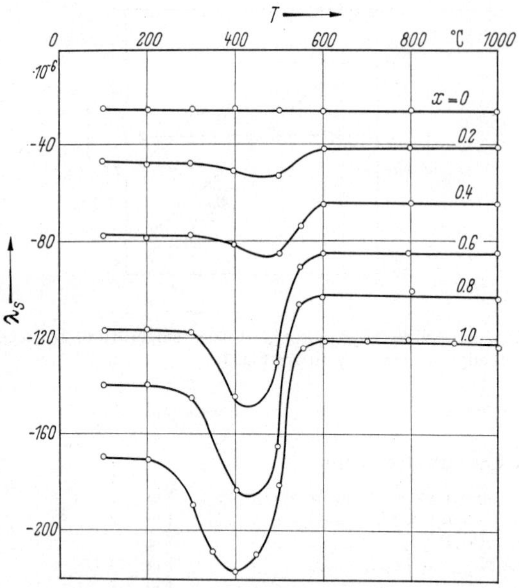

Fig. 3. $Ni_{1-x}Co_xFe_2O_4$. Saturation magnetostriction, λ_s, vs. temperature, T. Parameter x [*65PG*].

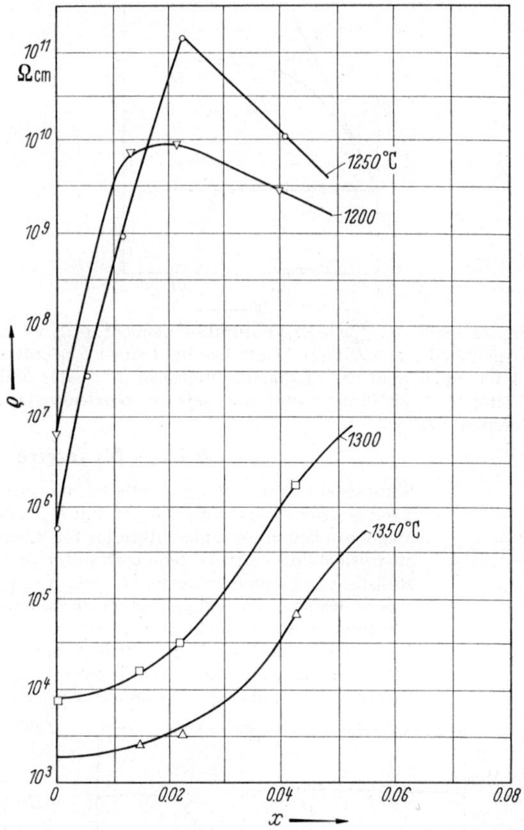

Fig. 5. $Ni_{1.0}Fe_{1.9}Co_xO_4$. Dependence of room temperature resistivity, ϱ, on cobalt content, x, for several firing temperatures [*56U(1)*].

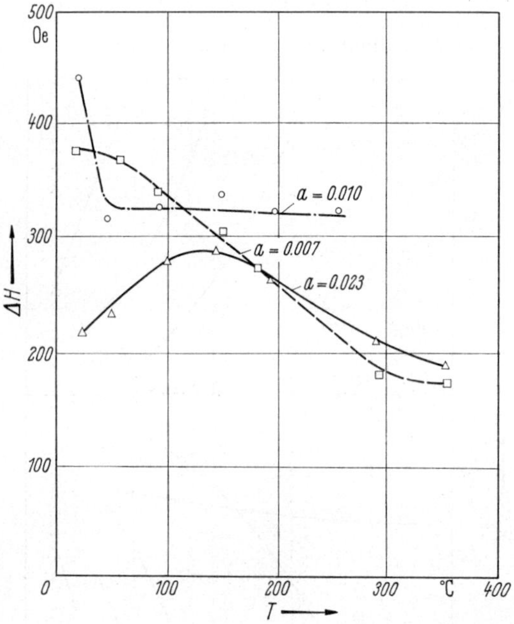

←

Fig. 4. $Ni_{1-a}Co_aFe_2O_4$. Electronic resonance width, ΔH, in polycrystalline nickel-cobalt ferrites vs. temperature. These measurements were at a frequency of 10 000 Mc/s. The effects of ceramic imperfections are not known. ○ $a = 0.04$; □ $a = 0.007$; △ $a = 0.023$ [*56SS*].

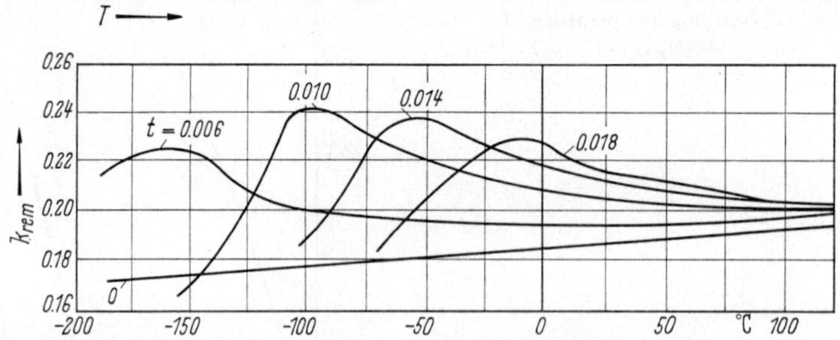

Fig. 6. $Ni_{1-t}Co_tFe_2O_4$. Electromechanical coupling coefficient vs. temperature, T. Sintering conditions: 2 hours at 1260 °C in O_2 atmosphere; average density 4.6 g/cm³. Values of t are shown on curves. All data k_{rem} apply to the remanent field state of the transducer [*56B*].

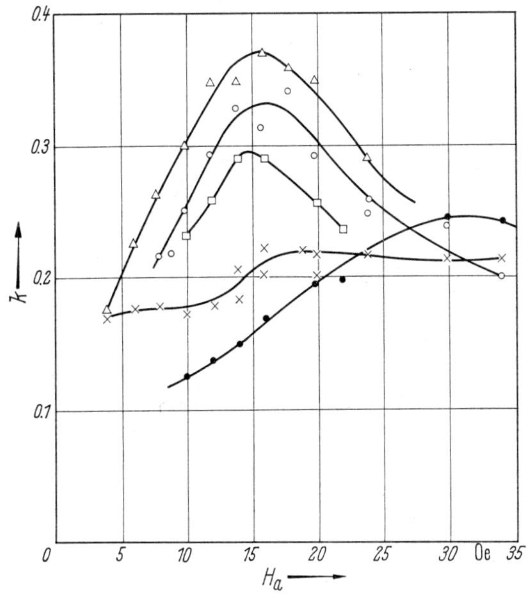

Fig. 7. $(NiO)_{0.975-x}(CoO)_{0.025}(FeO)_xFe_2O_3$. Electromechanical coupling coefficient, k, vs. applied magnetic field, H_a, for various mole percentage of FeO [*58DF*].
× 0.00 FeO, ○ 0.01 FeO, △ 0.06 FeO, □ 0.07 FeO, ● 0.08 FeO.

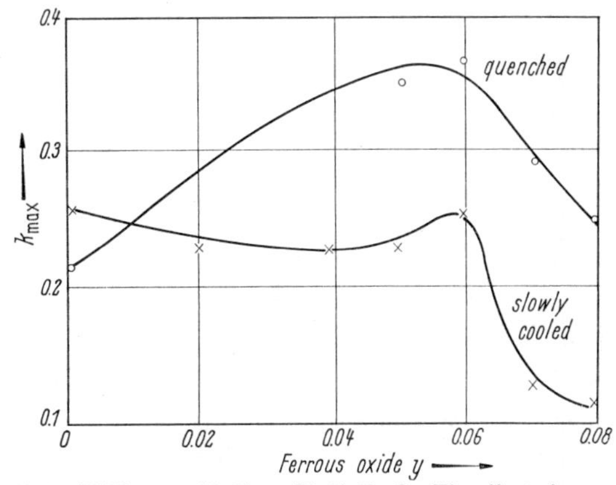

Fig. 8. $(NiO)_{0.973-y}(CoO)_{0.027}(FeO)_yFe_2O_3$. The effect of excess iron on the electromechanical coupling coefficient, k_{max}, of polycrystalline cobalt-substituted nickel ferrite. k_{max} refers to the maximum value of k with an optimum biasing magnetic field. The quenched specimens were cooled by removal from the furnace at 1400 °C while the slowly cooled specimens were cooled at 1 °C per minute in the furnace [*58FD*].

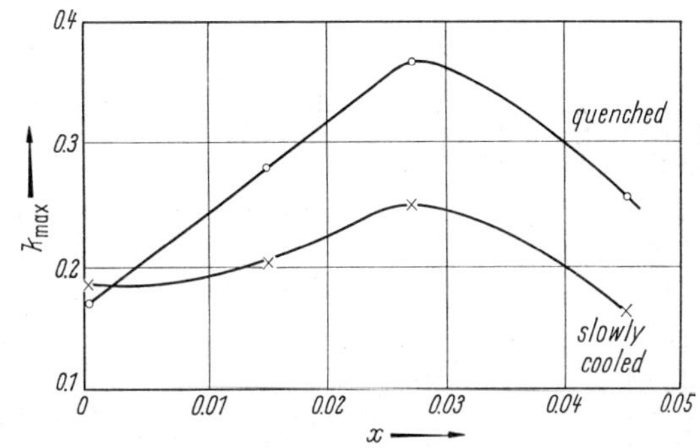

Fig. 9. $(NiO)_{0.94-x}(CoO)_x(FeO)_{0.06}Fe_2O_3$. The effect of cobalt on the electromechanical coupling coefficient, k_{max}, of polycrystalline nickel ferrous ferrite. Quenched and slowly cooled have the same meaning as in Fig. 8 [*58FD*].

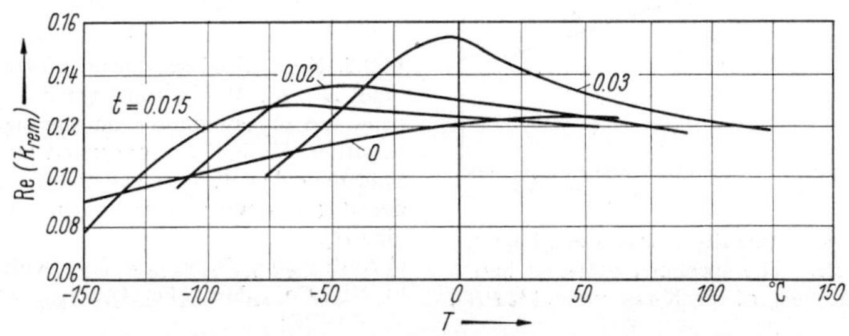

Fig. 10. $(Ni_{0.5}Zn_{0.5})_{1-t}Co_tFe_2O_4$. The real part of the electromechanical coupling coefficient, Re(k_{rem}), vs. temperature, T, for samples sintered for 2 hours at 1260 °C in oxygen. Parameter Co content t. The phase angle is ordinarily negligible for ferrite transducers [*57B*].

6.1.2.4 Dilute Mn and Co substitutions in Ni ferrite and Ni ferrite aluminate

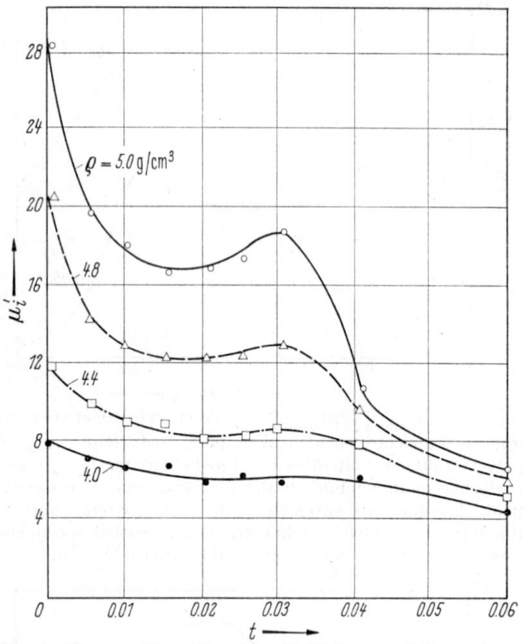

Fig. 1. $Ni_{1-t}Co_tMn_{0.02}Fe_{1.9}O_{4\pm}$. Initial permeability, μ_i', vs. Co content, t, at RT and a frequency of 10 Mc/s. Parameter: specimen densities [*59PT*].

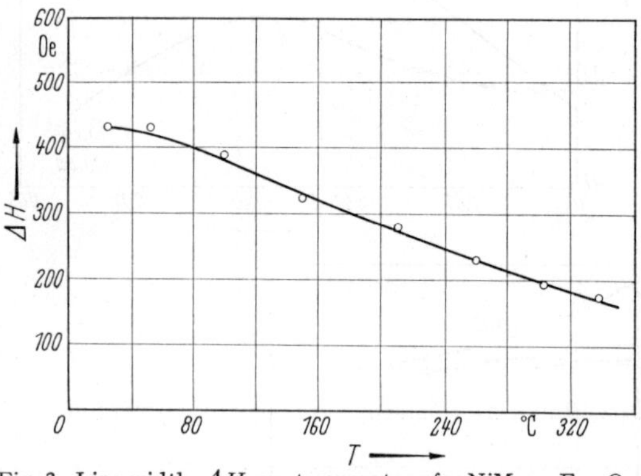

Fig. 3. Line width, ΔH, vs. temperature for $NiMn_{0.02}Fe_{1.9}O_{4\pm}$ at frequency of 9208 Mc/s. The specimen material had a density between 95% and 96% of the X-ray value [*58PH*].

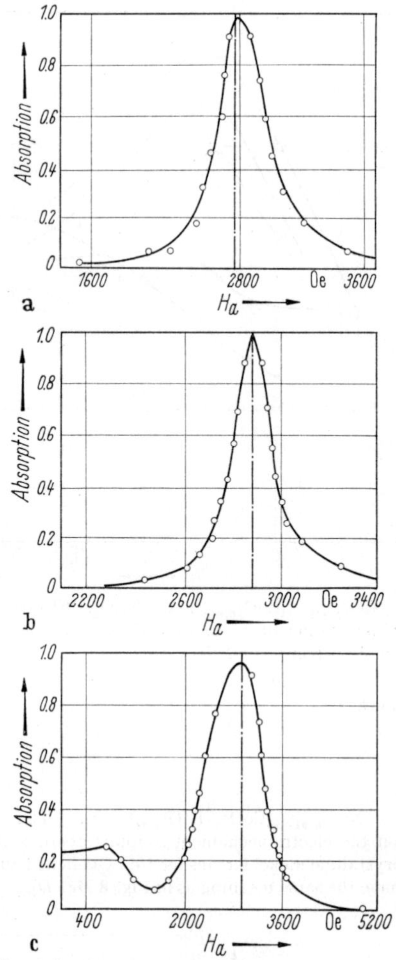

Fig. 2. Normalized resonance absorption vs. applied magnetic field, H_a, for Mn-Co substituted Ni ferrites illustrating dependence of asymmetry on the sign of K_1. For a, K_1 is negative; for b, K_1 is presumed zero; for c, K_1 is positive. Specimen densities were between 95% and 96% of X-ray density; resonance measurements were made at 9208 Mc/s [*58PH*].

a) $NiMn_{0.02}Fe_{1.9}O_4$; $g_{exp} = 2.34$; width = 460 Oe; $T = 25$ °C;
b) $Ni_{0.97}Co_{0.027}Mn_{0.02}Fe_{1.9}O_4$; $g_{exp} = 2.28$; width = 170 Oe; $T = 19$ °C;
c) $Ni_{0.9}Co_{0.09}Mn_{0.02}Fe_{1.9}O_4$; $g_{exp} = 2.24$; width = 1040 Oe; $T = 25$ °C.

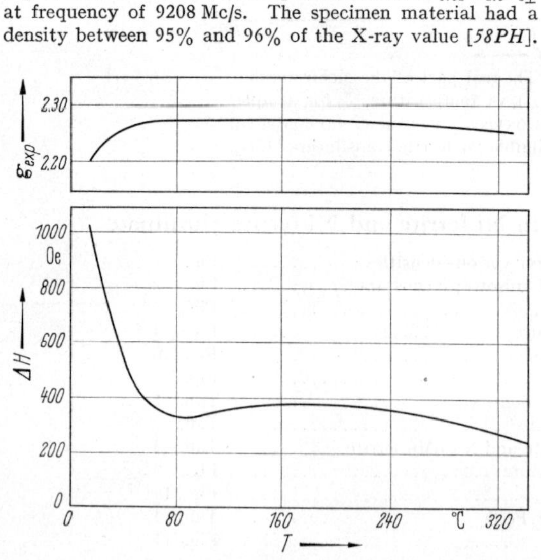

←

Fig. 4. $Ni_{0.9}Co_{0.09}Mn_{0.02}Fe_{1.9}O_{4\pm}$. Line width, ΔH, and observed g factor, g_{exp}, vs. temperature, T; density between 95% and 96% of X-ray density. These resonance measurements were made at 9208 Mc/s [*58PH*].

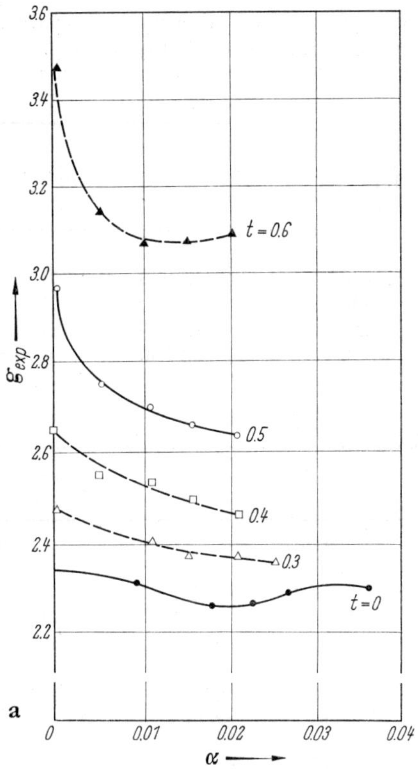

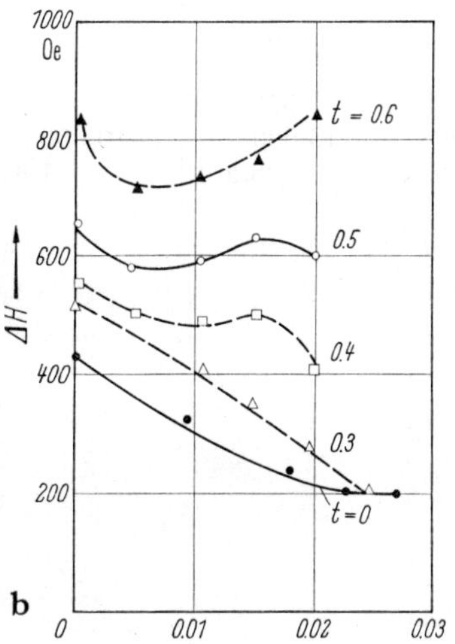

Fig. 6a, b. $Ni_{1-\alpha}Co_{\alpha}Mn_{0.02}Fe_{2-t}Al_tO_{4\pm}$. Line width, ΔH, and observed g factor, g_{exp}, vs. Co content, α, at RT [58PH].

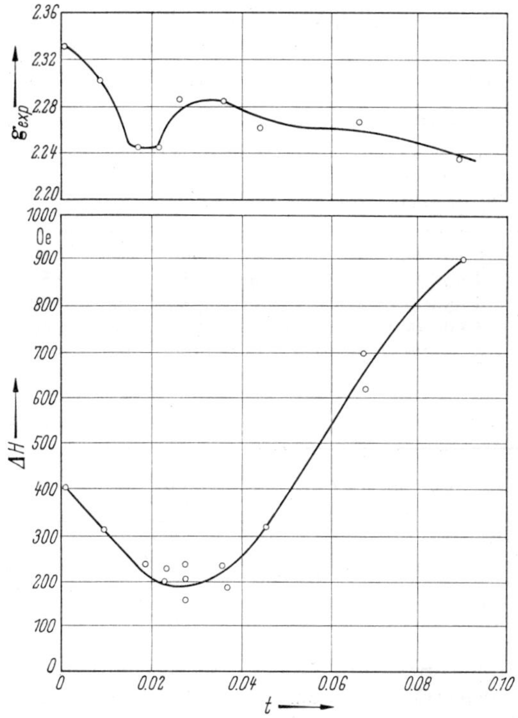

Fig. 5. $Ni_{1-t}Co_tMn_{0.02}Fe_{1.9}O_{4\pm}$. Line width, ΔH, and observed g factor, g_{exp}, vs. Co content, t, at RT [58PH].

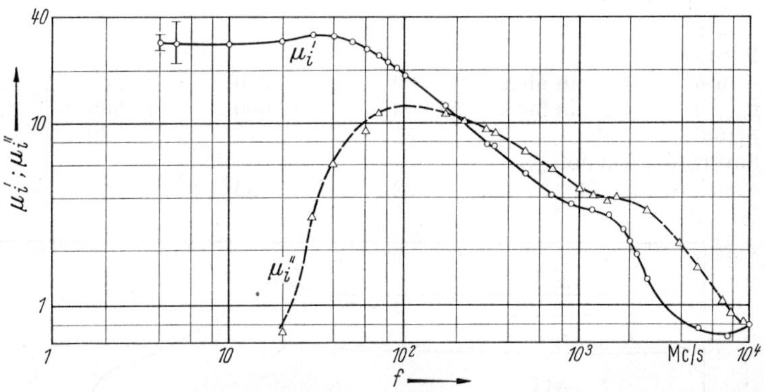

Fig. 7. $NiMn_{0.02}Fe_{1.9}O_4$. Real and imaginary parts of the initial permeability, μ_i, vs. frequency, f, for polycrystalline material of density 5.13 g/cm³ at RT [59PH].

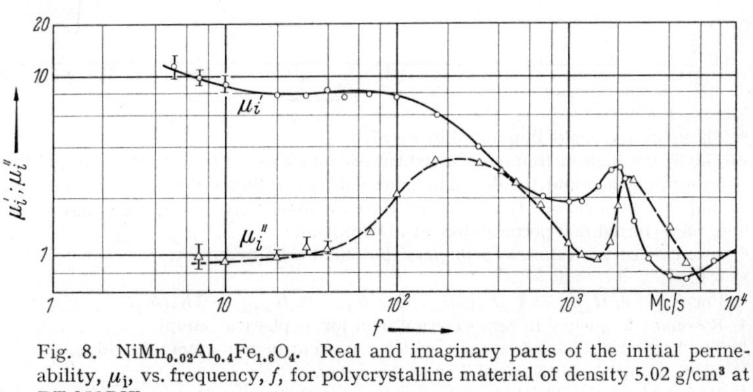

Fig. 8. $NiMn_{0.02}Al_{0.4}Fe_{1.6}O_4$. Real and imaginary parts of the initial permeability, μ_i, vs. frequency, f, for polycrystalline material of density 5.02 g/cm³ at RT [59PH].

Tab. 1. Some microwave properties of the ferrites of composition $NiMn_{0.02}Fe_{1.9}O_{4\pm}$ with density as a variable. $\varrho_X = 5.38$ g/cm³; $K_1 = -5 \cdot 10^4$ erg/cm³ and $\Theta_C = 585$ °C. $4\pi M_s \approx 3230$ Oe for X-ray density. $H_{anis} = \frac{4}{3}|K_1|/M_s$; $f = 1.4\, g_{exp} H_{anis}$ is the resonant frequency in a zero external field for a spherical sample [*59PH*].

ϱ g/cm³ (% of maximum)	g_{exp}	ΔH Oe	H_{anis} Oe	f Mc/s	μ_i'' a) 2nd peak Mc/s	μ_i'' a) 1st peak Mc/s	f_{max} b) Mc/s	$\mu_i'(0)$ c)
5.13 (95.4%)	2.34	460	260	850	1700	100	12 000	28
4.96 (92.3%)	2.36	515	260	850	1600	170	≈ 12 000	20.5
4.85 (90.2%)	2.35	550	260	850	—	330	—	17
4.59 (85.3%)	2.39	770	260	850	—	350	≈ 12 000	13
3.81 (70.8%)	2.43	1320	260	850	—	650	≈ 12 000	6.5

Tab. 2. Properties of $Ni_{1-t}Co_tMn_{0.02}Fe_{1.9}O_{4\pm}$ [*58PH*].

Co content t	0.00	0.01	0.027	0.036	0.068	$Ni_{0.7}Co_{0.27}$
ϱ d) in g/cm³	5.13	5.11	5.12	5.08	4.83	5.14
% of maximum	95.4	95.0	95.2	94.5	89.8	95.5
g_{exp}	2.34	2.31	2.28	2.29	2.26	—
ΔH in Oe	460	330	170	200	700	1 440
Θ_C in °C	585	585	585	585	585	585
M_s e) (corrected to X-ray density) in emu/cm³	257	261	264	271	277	321
H_{anis} f) in Oe	260	155	small	160	620	high
f g) $= 1.4\, g_{exp} H_{anis}$ in Mc/s	850	500	small	510	1 950	high
Position of 2nd peak in μ'' in Mc/s	1 700	1 400	—	—	2 500	—
Position of 1st peak in μ'' in Mc/s	100	≈ 270	280	≈ 500	550	—
f_{max} h) in Mc/s	12 000	≈ 12 000	≈ 12 000	12 000	13 000	> 10 000
$\mu_i'(0)$ i)	28	17	20	16	8.5	1.8
Saturation magnetostriction $\lambda_s \cdot 10^6$	−22	−33 j)	—	—	—	—

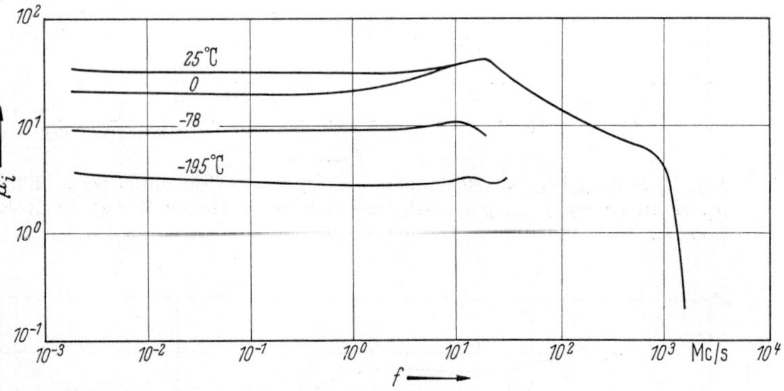

←

Fig. 9. $Ni_{0.925}Co_{0.025}Fe_{0.050}Fe_2O_4$. Real part of initial permeability, μ_i', of a polycrystalline sample vs. frequency, f. Parameter temperature, T [*60F*].

a) These are the peaks illustrated in Fig. 7.

b) This is the highest frequency at which resonance can occur in zero external magnetic field. It occurs when the effective anisotropy field and the demagnetizing field are aiding and have their maximum values due to favorable orientation and shape of the domain. It is given by the expression $f_{max} = 1.4\, g_{exp}(H_{anis} + 4\pi M_s)$.

c) $\mu_i'(0)$ = initial real permeability at low frequency.

d) X-ray density is taken as 5.38 g/cm³ for small cobalt content.

e) RT.

f) For $K_1 < 0$, $H_{anis} = \frac{4}{3}|K_1|/M_s$. For $K_1 > 0$, $H_{anis} = 2K_1/M_s$.

g) Resonant frequency in zero external field for a spherical sample.

h) The highest resonant frequency which can occur in zero external field.

i) $\mu_i'(0)$ = initial real permeability at low frequency.

j) Hoffman, P. O.: The Lionel Corporation Final Progress Report Contract Nr. 729 (60) (1955).

Tab. 3. Properties of $NiMn_{0.02}Fe_{2-t}Al_tO_{4\pm}$ and of $Ni_{0.975}Co_{0.025}Mn_{0.02}Fe_{1.7}Al_{0.3}O_{4\pm}$ [*59PH*].

Al content t	0	0.2	0.4	0.5	0.65	0.75	1.20	1.30	$Ni_{0.975}Co_{0.025}$
ϱ in g/cm³	5.13	5.08	5.02	4.98	4.90	4.74	4.27	4.98	5.15
% of maximum	95.4	95.6	96.0	96.0	95.5	93.2	87.4	94.5	97.7
g_{exp}	2.34	2.38	2.65	2.97	4.40	—	1.38	2.47	2.35
ΔH in Oe	460	480	550	660	1 180	large	150	530	200
Θ_C in °C	585	500	420	375	310	275	80	460	460
M_s a) (corrected to X-ray density) in emu/cm³	257	174	119	83	37	20	27	157	158
H_{anis} in Oe	260	450	590	800	1 500	large	125	480	$<$ 170
f b) $= 1.4 g_{exp} H_{anis}$ in Mc/s	850	1 500	2 190	3 320	9 200	—	240	1 660	$<$560
Position of 2nd peak in μ'' c) in Mc/s	1 700	1 800	2 300	3 000	≈ 8 000	—	—	2 100	—
Position of 1st peak in μ'' c) in Mc/s	100	100	230	350	≈ 800	—	≈ 250	—	—
f_{max} d) in Mc/s	12 000	≈ 9 000	≈ 7 000	≈ 6 500	≈ 10 000	> 10 000	≈ 1 200	≈ 7 000	≈ 7 000
$\mu'_i(0)$ e)	28	20	≈ 10	≈ 6	≈ 2	—	≈ 3	—	—
μ'_i in "flat" region between dispersions	—	2.5	2.0	1.5	1.2	—	—	—	—

a) RT.
b) Resonant frequency for spherical specimen in zero external field.
c) See Fig. 8 for illustration of these peaks.
d) The highest resonant frequency which can occur in zero external field.
e) $\mu'_i(0)$ = Initial real permeability at low frequency.

Tab. 4. Other substitutions for divalent Ni in $Ni_{1-x}Me_xFe_2O_4$.

Me	Ref.
Cd	*51WTT, 54T*
Li	*54WGEG, 57WH*
Mg	*39FV, 51W, 54WGEG, 57WH, 51OT, 55ST*
MgZn	*56F*
MnZn	*53Mo*
MnMg	*54WGEG*
MnMgZn	*52A—S*

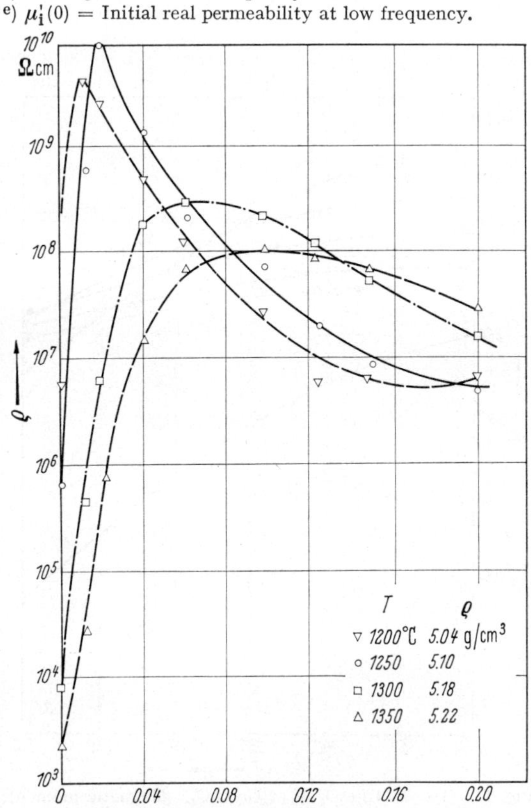

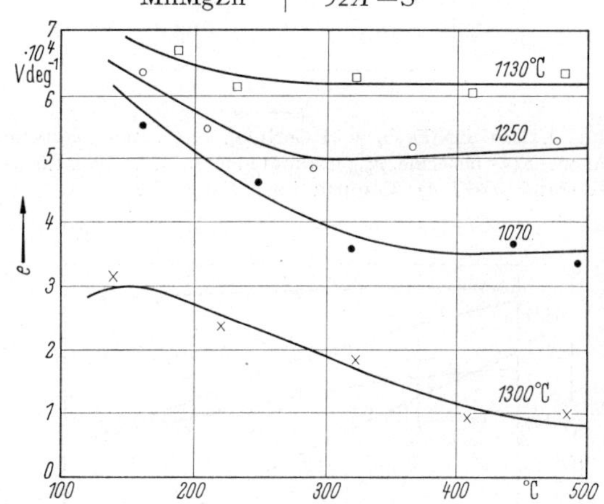

Fig. 11. $NiMn_{0.02}Fe_{1.98}O_4$. Thermoelectric power, e, vs. temperature, T, for material prepared at four different temperatures in air as given in the figure. Gold electrodes were used [*64H*].

Fig. 10. $Ni_{1.0}Fe_{1.9}Mn_tO_4$. Room temperature resistivity, ϱ, vs. Mn content, t, for various firing temperatures [*56U(1)*].

6.1.2.5 Ge substituted Ni ferrite

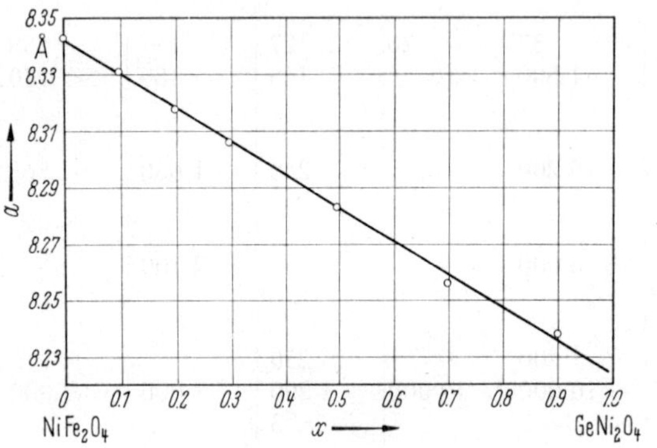

Fig. 1. $(1-x)NiFe_2O_4 + (x)GeNi_2O_4$. Lattice constant, a, at RT vs. $GeNi_2O_4$ content, x, in the solid solution system [*65WHA*]. Compare Fig. 21 in 6.1.2.2.

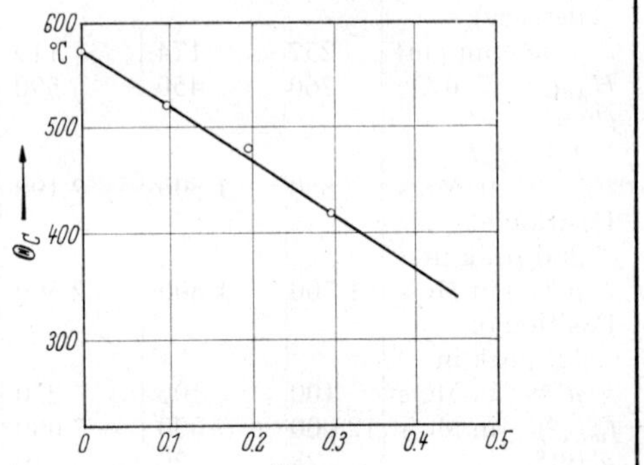

Fig. 2. $(1-x)NiFe_2O_4 + (x)GeNi_2O_4$. Curie temperature, Θ_C, vs. $GeNi_2O_4$ content, x [*65WHA*]. Compare Fig. 21, 23 in 6.1.2.2.

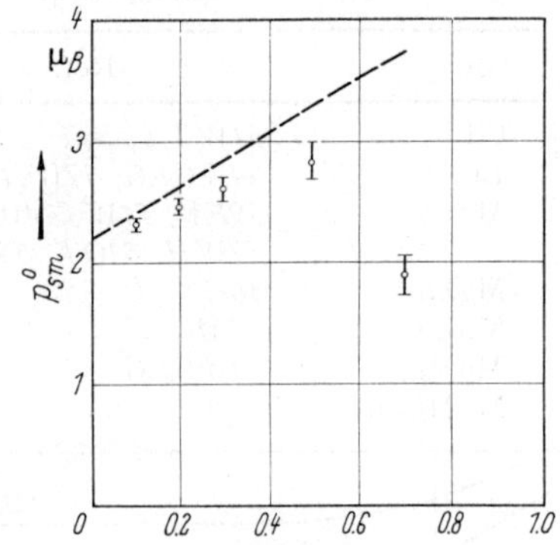

Fig. 3. $(1-x)NiFe_2O_4 + (x)GeNi_2O_4$. Saturation magnetic moment per molecule, p^0_{sm} (estimated for 0 °K in the saturated state) [*65WHA*]. Compare Fig. 22 in 6.1.2.2.

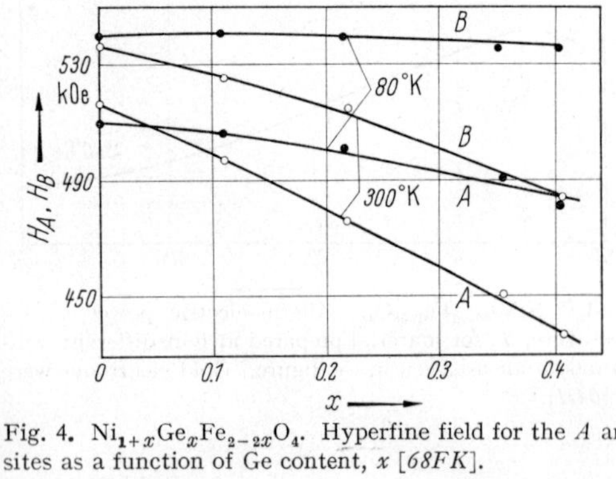

Fig. 4. $Ni_{1+x}Ge_xFe_{2-2x}O_4$. Hyperfine field for the A and B sites as a function of Ge content, x [*68FK*].

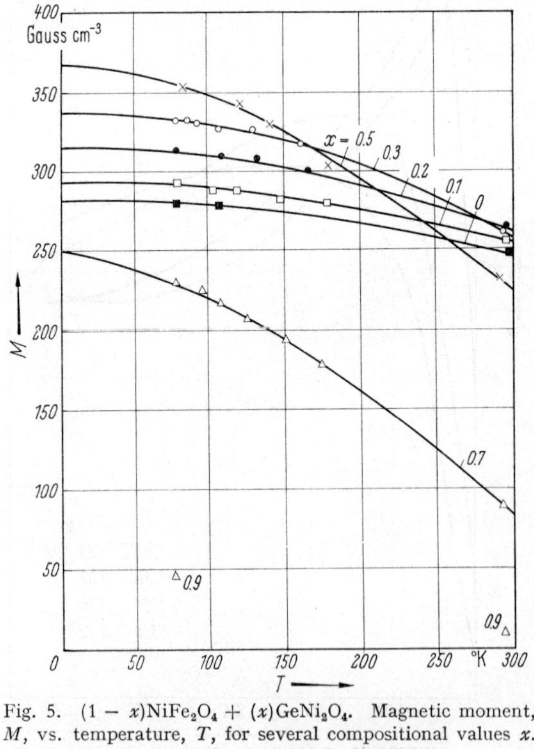

Fig. 5. $(1-x)NiFe_2O_4 + (x)GeNi_2O_4$. Magnetic moment, M, vs. temperature, T, for several compositional values x. Measurements were made at 13 300 Oe [*65WHA*].

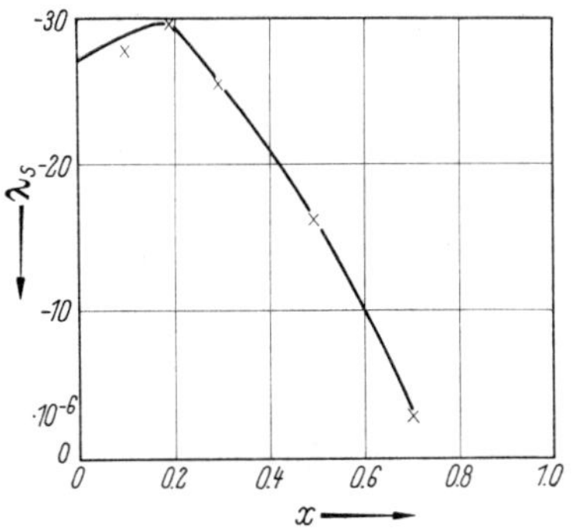

Fig. 6. $(1-x)NiFe_2O_4 + (x)GeNi_2O_4$. Saturation magnetostriction, λ_s, at RT vs. $GeNi_2O_4$ content, x. Measurements were made at 13 300 Oe [*65WHA*].

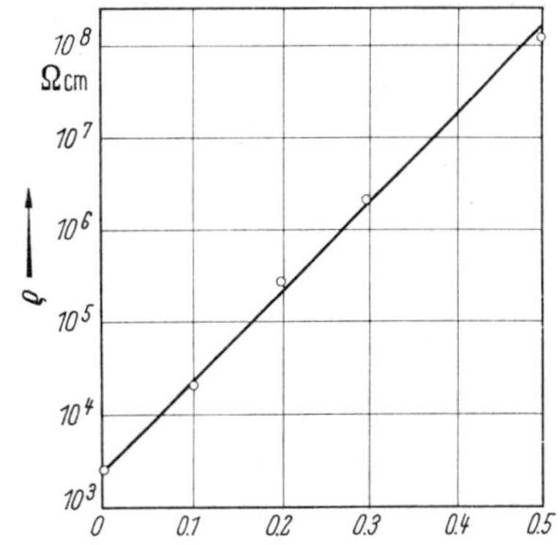

Fig. 7. $(1-x)NiFe_2O_4 + (x)GeNi_2O_4$. Specific electrical resistance, ϱ, at RT vs. $GeNi_2O_4$ content, x [*65WHA*].

6.1.2.6 Ni ferrite chromite

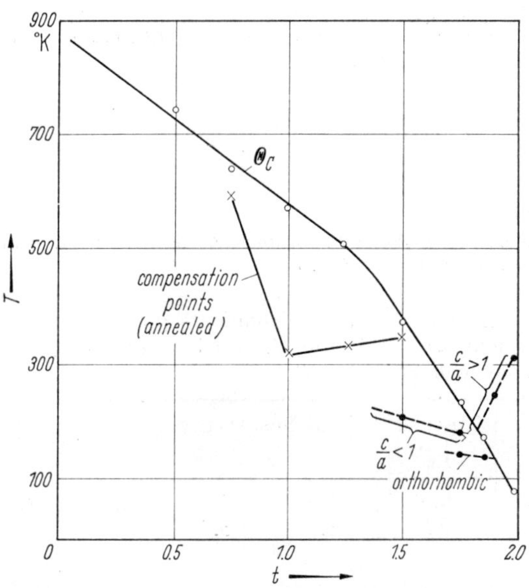

Fig. 1. $NiCr_tFe_{2-t}O_4$. Transition temperatures, T, vs. Cr content, t. The solid lines are magnetic transitions while the dotted lines are crystallographic phase changes. The latter only occur for $t > 1.25$. The cubic-tetragonal structure change has a c/a value greater than or less than one on either side of the minimum temperature as indicated. The orthorhombic phase occurs at lower temperatures but over a very small t range [*60MG*].

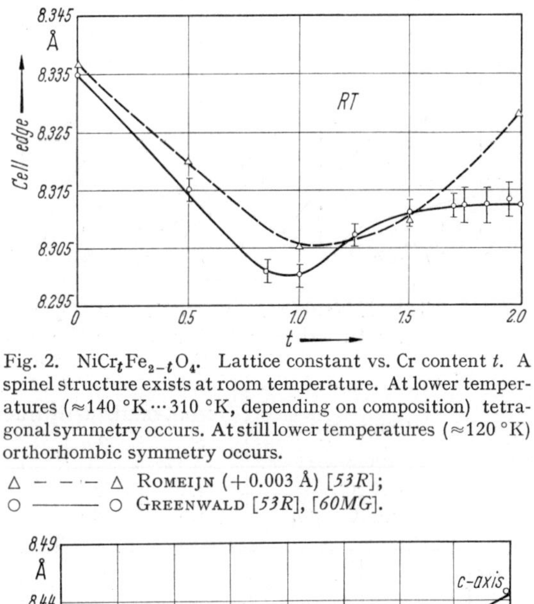

Fig. 2. $NiCr_tFe_{2-t}O_4$. Lattice constant vs. Cr content t. A spinel structure exists at room temperature. At lower temperatures (≈140 °K···310 °K, depending on composition) tetragonal symmetry occurs. At still lower temperatures (≈120 °K) orthorhombic symmetry occurs.
△ — — — △ Romeijn (+0.003 Å) [*53R*];
○ ——— ○ Greenwald [*53R*], [*60MG*].

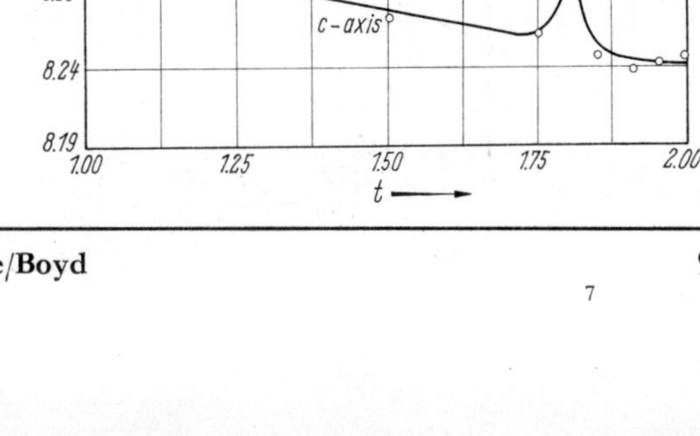

→

Fig. 3. $NiCr_tFe_{2-t}O_4$. Lattice constants vs. Cr content, t, 15 °K below the tetragonal transition temperature shown in Fig. 1 [*60MG*].

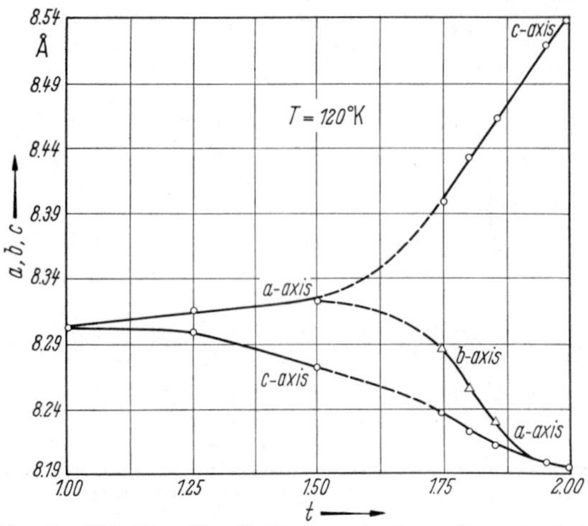

Fig. 4. $NiCr_tFe_{2-t}O_4$. Lattice constants vs. Cr content, t, at 120 °K. The orthorhombic transition has already taken place at this temperature [*60MG*].

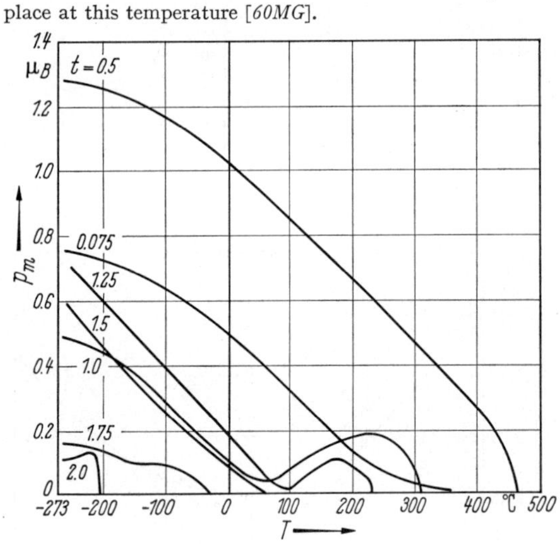

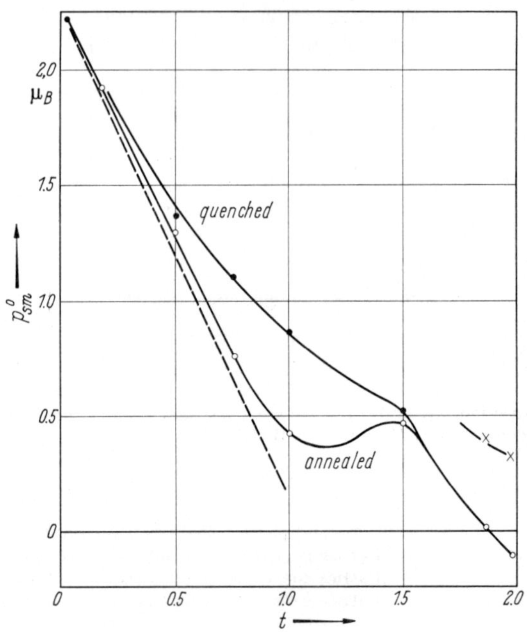

Fig. 6. $NiCr_tFe_{2-t}O_4$. Estimated saturation magnetic moments per formula unit for 0 °K vs. Cr content, t. Measurements made in fields up to 6800 Gauss. The dashed line is calculated on the assumption that both Ni and Cr are on B sites [*60MG*].
crosses: at $t = 1.85$ and $t = 2$: from [*60J*], measurements up to fields of 140 000 Gauss.

←

Fig. 5. $NiOCr_tFe_{2-t}O_3$. Thermomagnetization curves. Parameter Cr content, t [*60MG*].

Tab. 1. $NiFeCrO_4$ and $NiFe_{0.5}Cr_{1.5}O_4$. Comparison of the sublattice moments p_A and p_B, with values calculated from the ionic moment and with the observed magnetization. All numbers are Bohr magnetons [*58P*]

Composition	from neutron diffraction			Observed	from Néel theory		
	p_A	$2p_B$	p_m^0	p_m^0	p_A	$2p_B$	
$NiFeCrO_4$	4.5	5.2	0.7	0.8	5.0	5.2	μ_B
$NiFe_{0.5}Cr_{1.5}O_4$	3.6	4.0	0.4	0.5	3.6	5.6	μ_B

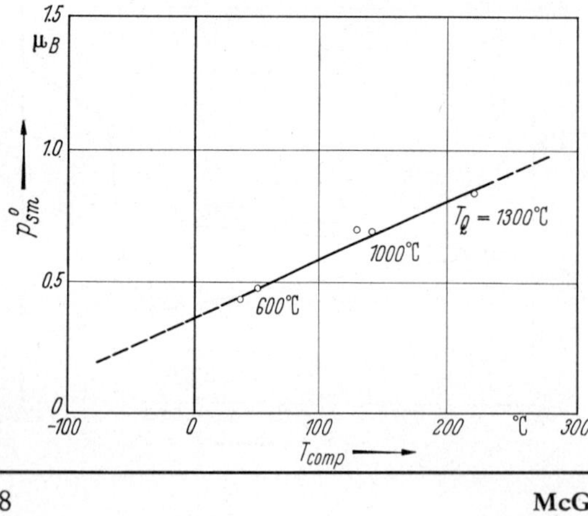

←

Fig. 7. $NiCrFeO_4$. Compensation temperature, T_{comp}, and saturation moments (estimated for 0 °K) for different quenching temperatures, T_Q [*60MG*].

6.1.2.7 Ni ferrite vanadate, $NiFe_{2-x}V_xO_4$

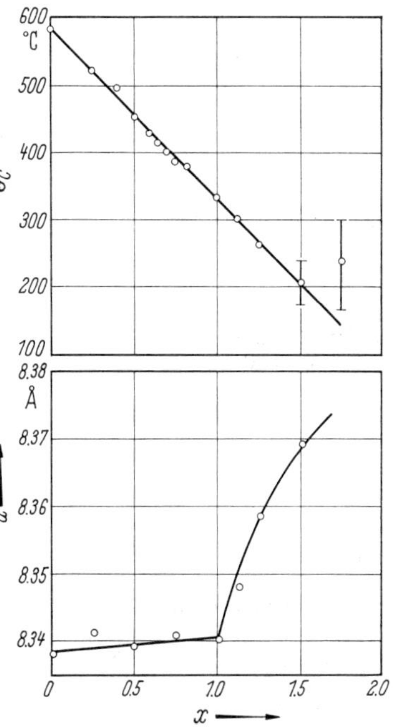

Fig. 1. $NiFe_{2-x}V_xO_4$. Curie temperatures, Θ_C, and cell edges, *a*, vs. V content, *x* [*62BG*].

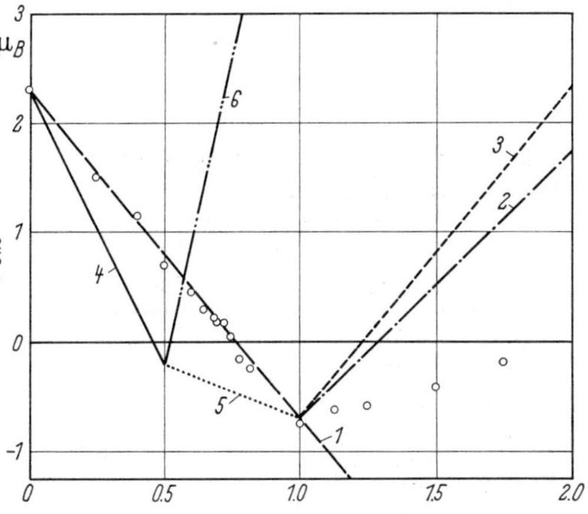

Fig. 2. $NiFe_{2-x}V_xO_4$. Saturation moments, p_{sm} vs. *x* [*62BG*].
○ experimental points.
1 calculated for V^{3+} and Ni^{2+} ions both in octahedral sites ($x \leq 1$).
2 calculated for all V^{3+} ions in octahecdral sites ($x > 1$).
3 calculated for all Ni^{2+} ions in octahedral sites ($x > 1$).
4 calculated for Fe^{2+} and V^{4+} ions, both in octahedral sites ($0 < x < 0.5$).
5 calculated for all V^{4+} in octahedral sites ($0.5 < x < 1$).
6 calculated for all Fe^{2+} in octahedral sites ($0.5 < x < 1$).

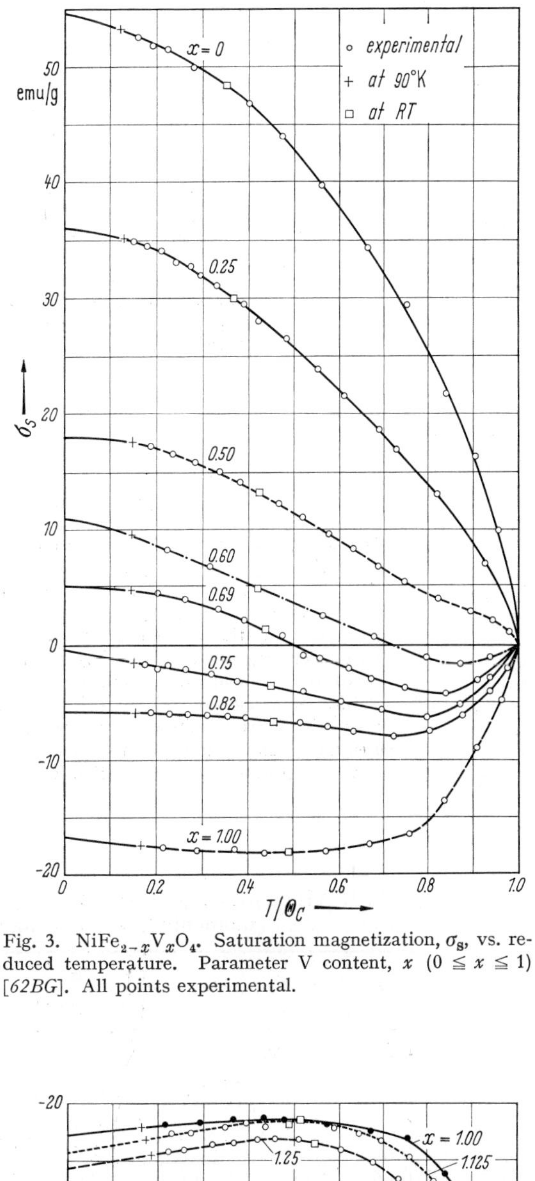

Fig. 3. $NiFe_{2-x}V_xO_4$. Saturation magnetization, σ_s, vs. reduced temperature. Parameter V content, *x* ($0 \leq x \leq 1$) [*62BG*]. All points experimental.

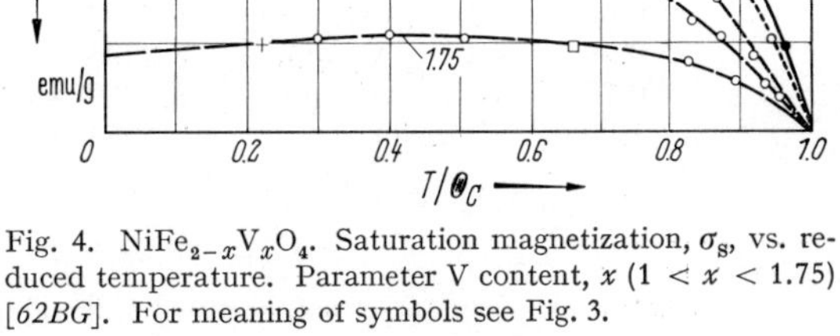

Fig. 4. $NiFe_{2-x}V_xO_4$. Saturation magnetization, σ_s, vs. reduced temperature. Parameter V content, *x* ($1 < x < 1.75$) [*62BG*]. For meaning of symbols see Fig. 3.

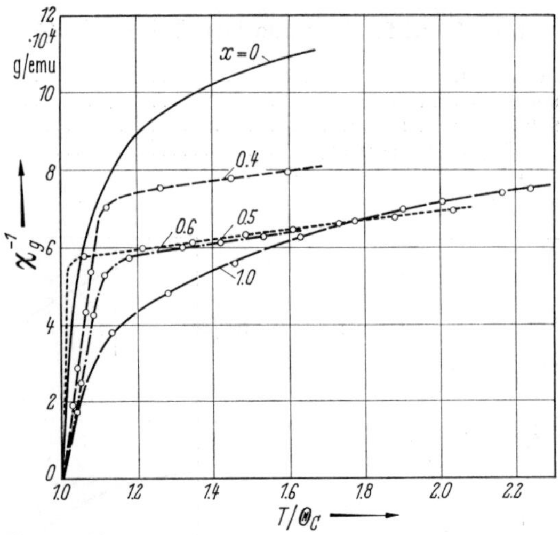

Fig. 5. $NiFe_{2-x}V_xO_4$. Reciprocal susceptibility per gram, $1/\chi_g$, vs. reduced temperature. Parameter V content, x [62BG].

6.1.2.8 Ni ferrite aluminate

Tab. 1. $NiFe_{2-t}Al_tO_4$ ($0 < t < 1$). Cell edges, a, Curie temperatures, Θ_C, saturation moments, p^0_{sm}, [54G] and observed g factors, g_{exp} [53M].

t	Formula	a [Å] annealed	Θ_C °C	p^0_{sm} [μ_B] annealed	quenched*)	g_{exp} annealed
0.00	$NiFe_2O_4$	8.3370 ± 4	580	2.29	2.29	2.3
0.25	$NiFe_{1.75}Al_{0.25}O_4$	8.3062 ± 7	506	1.30	1.59	2.7
0.45	$NiFe_{1.55}Al_{0.45}O_4$	8.2769 ± 4	465	0.61	1.19	—
0.50	$NiFe_{1.50}Al_{0.50}O_4$	8.2705 ± 6	430	0.44	0.99	6.9
0.625	$NiFe_{1.375}Al_{0.625}O_4$	8.2521 ± 7	360	0···0.045	—	3.8
0.66	$NiFe_{1.34}Al_{0.66}O_4$	8.2420 ± 20	368	**)	—	—
0.68	$NiFe_{1.32}Al_{0.68}O_4$	8.2485 ± 1	356	**)	—	—
0.686	$NiFe_{1.314}Al_{0.686}O_4$	8.2388 ± 4	} ≅ 340	**)	—	—
0.688	$NiFe_{1.312}Al_{0.688}O_4$	8.2387 ± 2	} ≅ 340	**)	—	—
0.70	$NiFe_{1.30}Al_{0.70}O_4$	8.2385 ± 7	} ≅ 340	**)	—	—
0.75	$NiFe_{1.25}Al_{0.75}O_4$	8.2329 ± 7	294	0.38	0.58	1.5
1.00	$NiFeAlO_4$	8.1951 ± 7	198	0.64	0.42	—

*) Quenched from 1350 °C.
**) Extrapolation too uncertain, or not measured at liquid hydrogen temperature.

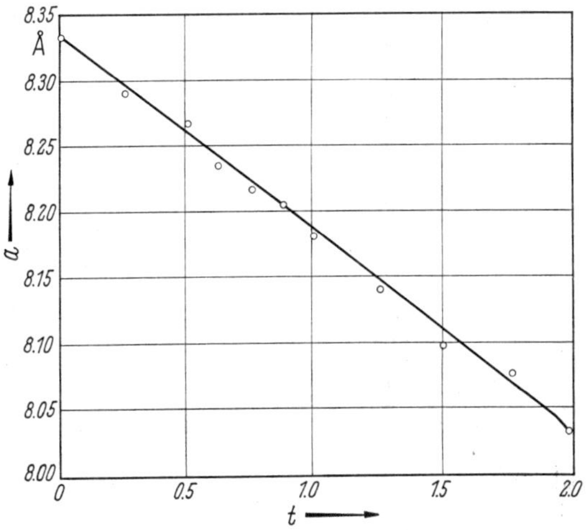

Fig. 1. $NiFe_{2-t}Al_tO_3$. Unit cell size, *a*, vs. Al content, *t* [*53MP*].

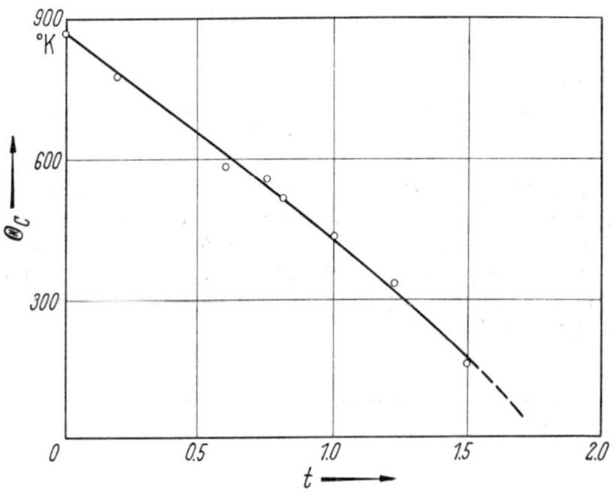

Fig. 2. $NiFe_{2-t}Al_tO_3$. Curie temperature, Θ_C, vs. Al content, *t* [*53MP*].

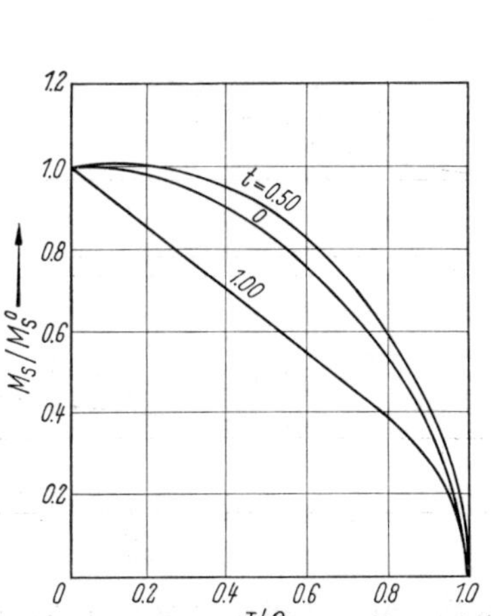

Fig. 3. $NiFe_{2-t}Al_tO_3$. Reduced saturation magnetization vs. temperature. Parameter Al content, *t*. Specimens annealed at 1°/min [*53MP*].

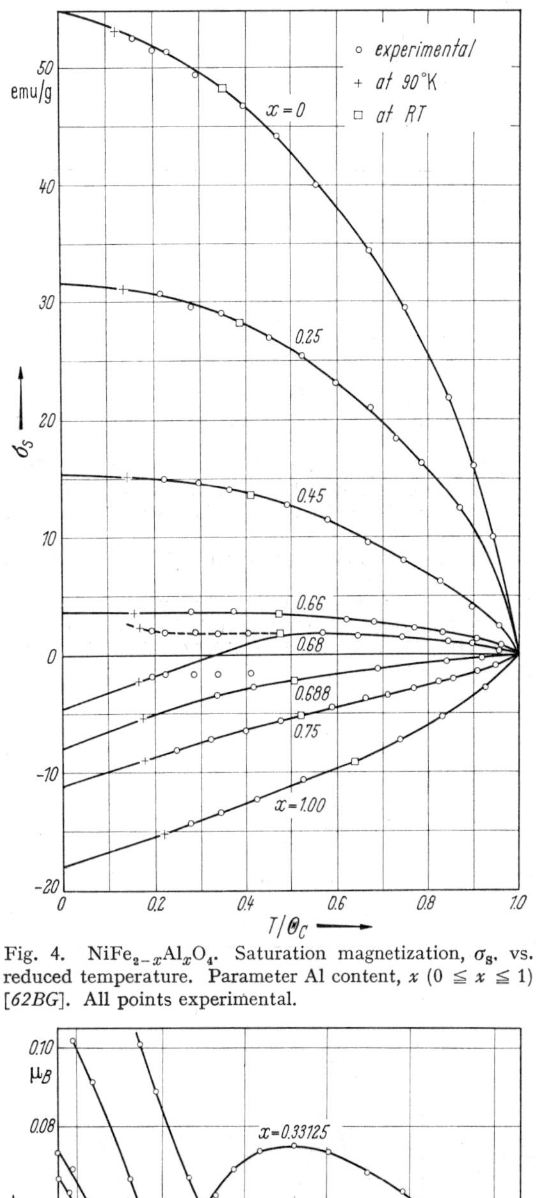

Fig. 4. $NiFe_{2-x}Al_xO_4$. Saturation magnetization, σ_s, vs. reduced temperature. Parameter Al content, *x* ($0 \leq x \leq 1$) [*62BG*]. All points experimental.

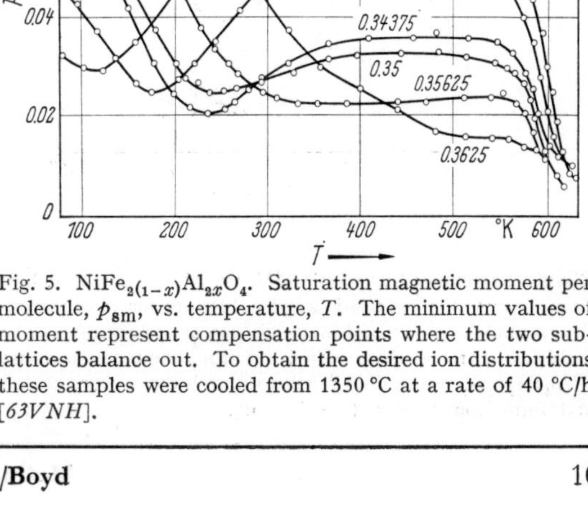

Fig. 5. $NiFe_{2(1-x)}Al_{2x}O_4$. Saturation magnetic moment per molecule, p_{sm}, vs. temperature, *T*. The minimum values of moment represent compensation points where the two sublattices balance out. To obtain the desired ion distributions these samples were cooled from 1350 °C at a rate of 40 °C/h [*63VNH*].

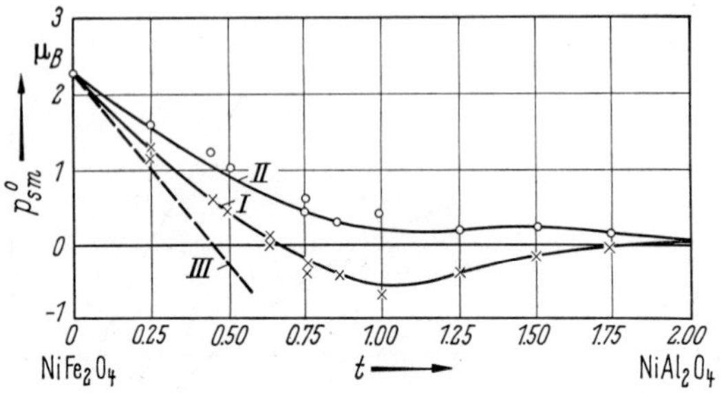

Fig. 6. $NiFe_{2-t}Al_tO_4$. Saturation moments per molecule at absolute zero, p^0_{sm}, vs. Al content, t, between $t = 0$ and $t = 2$. A negative value of p^0_{sm} indicates that the A-sites predominate [*53MP*], [*54G*].
Curve I: Experimental values for annealed samples: [*53MP*], [*54G*].
Curve II: Experimental values for quenched samples: quenched from 1420 °C [*53MP*], quenched from 1350 °C [*54G*].
Curve III: Values calculated for all Ni^{2+} and all Al^{3+} ions in the octahedral position.

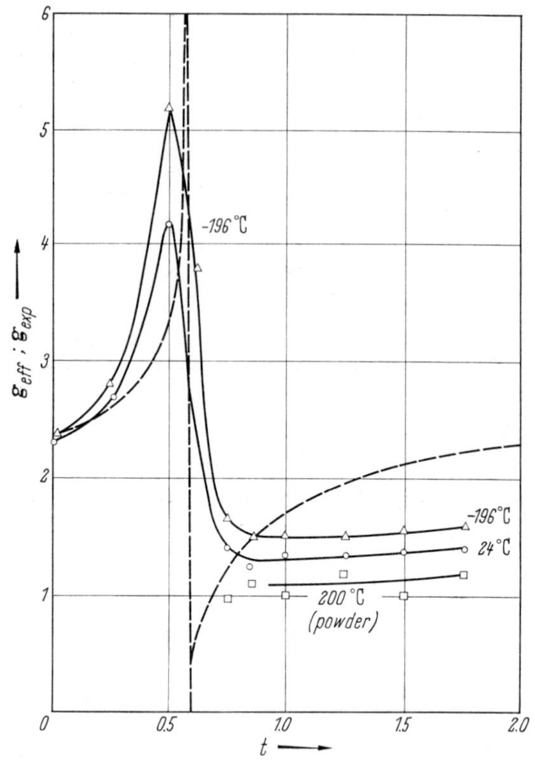

Fig. 7. $NiFe_{2-t}Al_tO_4$. g_{exp} vs. composition, t. Parameter temperature, T. All specimens were cooled from 1350 °C at 1°/min to RT. Points shown are for powder samples. Measurements were made at 9300 Mc/s. The dotted line is a theoretical curve of g_{eff} taking the compensation point into account [*54M*], [*53W*].

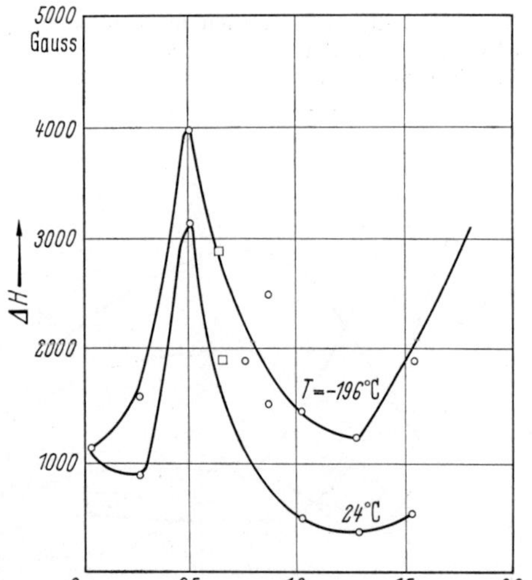

Fig. 8. Resonance line width, ΔH, vs. Al content, t, at half power points for annealed samples. Points for two compositions made on powdered materials are also shown (□). The materials can be represented by the chemical formula $NiAl_tFe_{2-t}O_4$. Annealing refers to samples that were cooled at 1°/min from 1350 °C to RT [*54M*].

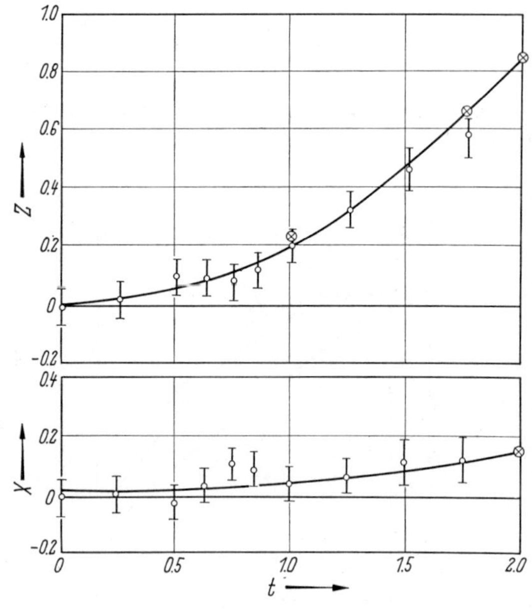

Fig. 9. $NiOFe_{2-t}Al_tO_3$. Cation distribution vs. Al content, t. Z and X are the fractional number of trivalent and divalent ions on the A sites, respectively, as calculated (○) from experimental (⊗) values of magnetization and g-values [*54S*], [*54GPG*].

6.1.2.9 Ni gallate ferrite

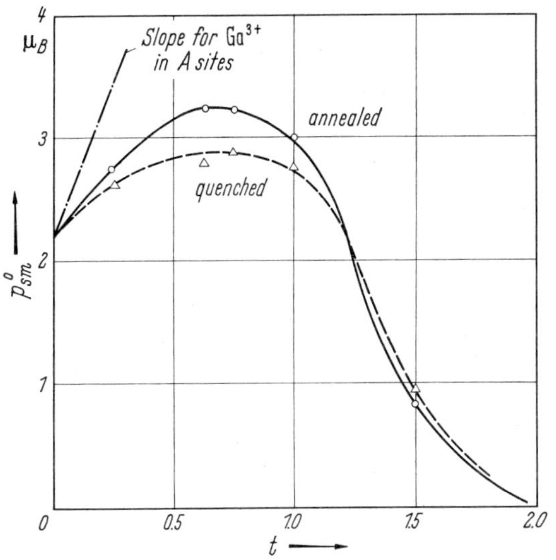

Fig. 1. $NiFe_{2-t}Ga_tO_4$. Saturation magnetic moment, p^0_{sm}, vs. Ga content, t. Annealed at 1°/min; quenching carried out from 1410 °C. Points shown are extrapolated to absolute zero [*53MP*].

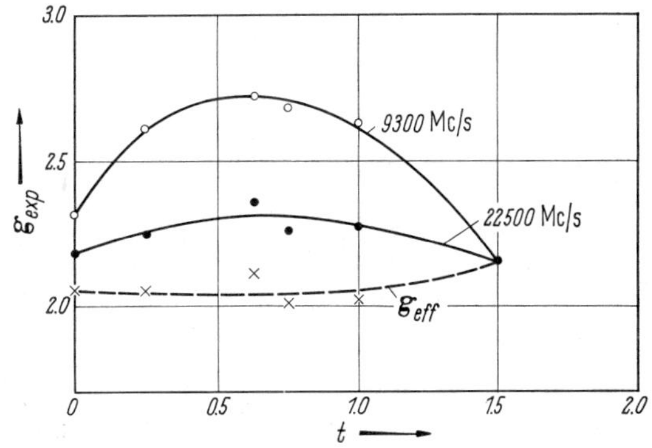

Fig. 2. $NiGa_tFe_{2-t}O_4$. g_{exp} values at 77 °K vs. Ga content, t, at two frequencies. The dotted curve is corrected for the frequency dependence and represents g_{eff}. It was found from these measurements that the internal field, H_{int}, is proportional to the magnetic moment of the sample [*55M*].

6.1.2.10 References for 6.1.2

30P Passevini, L.: Gazz. Chim. Ital. **60** (1930) 754.
31B Buschendorf, F.: Z. Physik. Chem. (Leipzig) (B) **14** (1931) 297.
32BP Barth, T. F. W., and E. Posnjak: Z. Krist. **82** (1932) 325.
32M Machatschki, F.: Z. Krist. **82** (1932) 348.
32S Serres, A.: Ann. Phys. (Paris) **17** (1932) 5.
35R Raychaudhuri, D. P.: Indian J. Phys. **9** (1935) 383.
36VB Verwey, E. J. M., and J. H. de Boer: Rec. Trav. Chim. **55** (1936) 531.
39FV Forrester, H., and M. Vetter: Compt. Rend. **209** (1939) 164.
46D Darken, L. S., and R. W. Gurry: J. Am. Chem. Soc. **68** (1946) 799.
47M Mason, B.: Am. Mineralogist **32** (1947) 426.
47RR Rudorf, W., and B. Reuter: Z. Anorg. Allgem. Chem. **253** (1947) 177, 194.
47VH Verwey, E. J. W., and E. L. Heilmann: J. Chem. Phys. **15** (1947) 174.
50B Bickford Jr., L. R.: Phys. Rev. **78** (1950) 449.
50BS Belgers, H. G., and J. L. Snock: Philips Tech. Rev. **11** (1950) 314.
50D Domenicali, C. A.: Phys. Rev. **78** (1950) 462.
50G Gorter, E. W.: Nature **165** (1950) 799.
50GMR Galt, J. K., B. T. Matthias, and J. P. Remeika: Phys. Rev. **79** (1950) 391.
50MSM McMurdie, H. F., M. Sullivan, and F. A. Mauer: J. Res. Natl. Bur. Std. **45** (1950) 35.
50P Pauthenet, R.: Compt. Rend. **230** (1950) 1842.
50RWE Rado, G. T., R. W. Wright, and W. H. Emerson: Phys. Rev. **80** (1950) 273.
50W Wells, A. F.: Structural Inorganic Chemistry. London: 1950.
50YGMW Yager, W. A., J. K. Galt, F. R. Merritt, and E. A. Wood: Phys. Rev. **80** (1950) 748.
51B Bertaut, E. F.: J. Phys. Radium **12** (1951) 252.
51K Koops, C. G.: Phys. Rev. **83** (1951) 121.
51L Lovell, G. H. B.: Trans. Brit. Ceram. Soc. **50** (1951) 315.
51OT Okamura, T., and Y. Torizuka: Nature **167** (1951) 986.
51W Wagenknecht, F.: Frequenz **5** (1951) 145.
51WTT Watanabe, H., N. Tsuya, and T. Takahashi: Sci. Rep. Res. Inst. Tohoku Univ. **A 3** (1951) 62.

52A-S	ALBERS-SCHÖNBERG, E.: Ceramic Age **59** (1952) 30.
52BW	BOZORTH, R. M., and J. G. WALKER: Phys. Rev. **88** (1952) 1209.
52H	HEALY Jr., D. W.: Phys. Rev. **86** (1952) 1009.
52MHS(1)	MCGUIRE, T. R., L. N. HOWARD, and J. S. SMART: Ceramic Age **60** (1952) 22.
52MHS(2)	MCGUIRE, T. R., L. N. HOWARD, and J. S. SMART: Phys. Rev. **86** (1952) 599.
52WG	WENT, J. J., and E. W. GORTER: Philips Tech. Rev. **13** (1952) 185.
53AC	ABRAHAMS, S. C., and B. A. CALHOUN: Acta Cryst. **6** (1953) 105.
53HC	HASTINGS, J. M., and L. M. CORLISS: Rev. Mod. Phys. **25** (1953) 114.
53M	MCGUIRE, T. R.: Phys. Rev. **91** (1953) 206.
53Mo	MORRIS, A. L.: "Soft Magnetic Materials for Telecommunication H", London 1953, p. 278.
53MP	MAXWELL, L. R., and S. J. PICKART: Phys. Rev. **92** (1953) 1120.
53R	ROMEIJN, F. C.: Philips Res. Rept. **8** (1953) 337.
53R(a)	ROMEIJN, F. C.: Philips Res. Rept. **8** (1953) 304.
53W	WANGSNESS, R. K.: Phys. Rev. **91** (1953) 1088.
54C	CALHOUN, B. A.: Phys. Rev. **94** (1954) 1580.
54C(a)	CALHOUN, B. A.: Phys. Rev. **94** (1954) 1577.
54G	GORTER, E. W.: Philips Res. Rept. **9** (1954) 295.
54Ga	GALT, J. K.: Bell System Tech. J. **33** (1954) 1023.
54GPG	GREENWALD, S., S. J. PICKART, and F. H. GRANNIS: J. Chem. Phys. **22** (1954) 1597.
54K	KOJIMA, Y.: Sci. Rept. Res. Inst. Tohoku Univ. **A 6** (1954) 620.
54M	MCGUIRE, T. R.: Phys. Rev. **93** (1954) 682.
54Mi	MILES, P. A.: Nature **174** (1954) 178.
54O	OKAMURA, T.: Sci. Rept. Res. Inst. Tohoku Univ. **A 6** (1954) 93.
54P	PALMER, G. G.: Interim Report on the Study of Nonmetallic Ferromagnetic Materials No. 6, Wright Air Development Center, Dayton, Ohio, prepared by General Electric Co.
54S	SMART, J. S.: Phys. Rev. **94** (1954) 847.
54SW(1)	SMIT, J., and H. P. J. WIJN: Advances in Electronics and Electron Physics, Vol. **6**, 83. New York: Academic Press 1954.
54SW(2)	SMIT, J., and H. P. J. WIJN: Ibid., p. 100.
54SW(3)	SMIT, J., and H. P. J. WIJN: Ibid., p. 115.
54T	TSUYA, N.: J. Phys. Soc. Japan **9** (1954) 644.
54WGEG	WIJN, H. P. J., E. W. GORTER, C. J. ESVELDT, and P. GELDERMANS: Philips Tech. Rev. **16** (1954) 49.
55BPS	BICKFORD Jr., L. R., J. PAPPIS, and J. L. STULL: Phys. Rev. **99** (1955) 1210.
55BTW	BOZORTH, R. M., E. F. TILDEN, and A. J. WILLIAMS: Phys. Rev. **99** (1955) 1792.
55D	DELORME, C.: Compt. Rend. **241** (1955) 1588.
55GL	GOODENOUGH, J. B., and A. L. LOEB: Phys. Rev. **98** (1955) 391.
55M	MCGUIRE, T. R.: Pittsburgh Conference on Magnetism and Magnetic Materials, A. I. E. E. **43** (1955).
55ST	SMOL'KOV, N. A., and E. V. TALALAEVA: Fiz. Metal i Metalloved. **1** (1955) 417.
55W	WALDRON, R. D.: Phys. Rev. **99** (1955) 1727.
55YGM	YAGER, W. A., J. K. GALT, and F. R. MERRITT: Phys. Rev. **99** (1955) 1203.
56Be	BELGERS, H. G.: Philips Tech. Rev. **18** (1956) 82.
56B	BURGT, C. M. VAN DER: J. Acoust. Soc. Am. **28** (1956) 1031.
56F	FOMENKO, L. A.: Izv. Akad. Nauk USSR Ser. Fiz. **20** (1956) 1348.
56HJ	HEALY Jr., D. W., and R. A. JOHNSON: Phys. Rev. **104** (1956) 635.
56L	LOTGERING, F. K.: Philips Res. Rept. **11** (1956) 221.
56MM	MCGUIRE, T. R. and L. R. MAXWELL: Bull. Am. Phys. Soc. Ser. II, **1** (1956) 23.
56MO	MIYAHARA. S., and H. OHNISHI: J. Phys. Soc. Japan **11** (1956) 1296.
56SS	SIRVETZ, M. H., and J. H. SAUNDERS: Phys. Rev. **102** (1956) 366.
56U(1)	UITERT, L. G. VAN: J. Chem. Phys. **24** (1956) 306.
56U(2)	UITERT, L. G. VAN: Proc. Inst. Radio Engrs. **44** (1956) 1294.
57B	BURGT, C. M. VAN DER: Philips Res. Rept. **12** (1957) 97.
57BBP	BICKFORD, L. R., Jr., J. M. BROWNLOW, and R. F. PENOYER: Proc. Inst. Elec. Engrs. (London) Pt. **B 104** Suppl. No. 5 (1957) 238.
57E	EPSTEIN, D. J.: Boston Conference on Magnetism and Magnetic Materials, A. I. E. E. (1957) 498.
57F	FRANCOMBE, M. H.: Phys. Chem. Solids **3** (1957) 37.
57MP	MAXWELL, L. R., and S. J. PICKART: Proc. Inst. Elec. Engrs. (London) Pt. **B 104**, Suppl. No. 5, (1957) 152.
57N	NEPRIMEROV, N. N.: Bull. Acad. Sci. USSR, Phys. Ser. (English Transl.) **21** (1957) 1276.
57Sh	SHENKER, H.: Phys. Rev. **107** (1957) 1248.

57S	Snieder, J.: Appl. Sci. Res. **B 6** (1957) 306.
57Sm	Smiltens, J.: J. Am. Chem. Soc. **79** (1957) 1877.
57WH	Wijn, H. P. J., and H. van der Heide: Proc. Inst. Elect. Engrs. (London) Pt. **B 104** Suppl. No. 5 (1957) 422.
58DF	Davis, Jr., C. M., and S. F. Ferebee: U. S. Navy Journal of Underwater Acoustics **8** (1958) 491.
58DM	Dwight, K., and N. Menyuk: Bull. Am. Phys. Soc., Ser. II, **3** (1958) 41.
58E	Edwards, P. I.: Bull. Am. Phys. Soc., Ser. II, **3** (1958) 43.
58FD	Ferebee, S. F., and C. M. Davis Jr.: J. Acoust. Soc. Am. **30** (1958) 747.
58P	Pickart, S. J.: "A Neutron Diffraction Study of Chromium Substituted Ferrite", NAVORD Report 6101, U. S. Naval Ordnance Laboratory, White Oak, Silver Spring, Md., USA (1958).
58MT	Miyahara, S., and T. Tsushima: J. Phys. Soc. Japan **13** (1958) 758.
58PH	Pippin, J. E., and C. L. Hogan: I. R. E. Trans. on Microwave Theory Tech. Vol. MTT **6** (1958) 79.
58S	Snieder, J.: Appl. Sci. Res. **B 7** (1958) 185.
59E	Edwards, P. L.: Phys. Rev. **116** (1959) 294.
59PH	Pippin, J. E., and C. L. Hogan: Scientific Report No. 1, Gordon McKay Lab. of Applied Science, Harvard Univ. (1959).
59P	Paladino, A. E., Jr.: J. Am. Ceram. Soc. **42** (1959) 168.
59PT	Pippin, J. E., and G. H. Thiess: J. Appl. Phys. **30** (1959) 81S.
59SW(1)	Smit, J., and H. P. J. Wijn: Ferrites. John Wiley and Sons, New York (1959) p. 156.
59SW(2)	Smit, J., and H. P. J. Wijn: Ferrites. John Wiley and Sons, New York (1959), p. 222.
60F	Ferebee, S. F.: U. S. Naval Ordnance Laboratory, White Oak, Silver Spring, Md., USA, unpublished private communication.
60J	Jacobs, I. S.: J. Phys. Chem. Solids **15** (1960) 54.
60MG	McGuire, T. R., and S. Greenwald: Proceedings of an International Conference on Solid State Physics, Brussels 1958; London and New York: Academic Press (1960), Vol. 3.
60Sh	Shafer, M. W.: J. Phys. Chem. **65** (1961) 2055.
61S	Salmon, O. N.: J. Phys. Chem. **65** (1961) 550.
62BG	Blasse, G., and E. W. Gorter: J. Phys. Soc. Japan **17** B-1 (1962) Suppl.
62E	Elbinger, G.: Z. Angew. Phys. **4** (1962) 274.
62KS	Kohane, T., and B. D. Silverman: J. Phys. Soc. Japan **17** (1962) 249.
63K	Kamilov, I. K.: Soviet Phys. — Solid State (English Transl.) **4** (1963) 1693.
63RSF	Rustamov, A. G., A. A. Samokhvalov, and I. G. Fakidov: Soviet Phys. — Solid State (English Transl.) **5** (1963) 751.
63VNH	Vassiliev, A., J. Nicholas, and M. Hildebrandt: Compt. Rend. **254** (1963) 1235.
64CMP	Chapman, F. W., R. E. Michel, and R. Poplawsky: J. Appl. Phys. **35** (1964) 1063.
64CF	Coren, R. L., and M. H. Francombe: J. Phys. (Paris) **25** (1964) 233.
64H	Hughes, D. O.: Brit. J. Appl. Phys. **15** (1964) 1250.
64SR	Samokhvalov, A. A., and A. G. Rustamov: Soviet Phys. — Solid State (English Transl.) **6** (1964) 749.
65B	Blasse, G.: Philips Res. Rept. **20** (1964) 528.
65FT	Foiles, C. L., and C. T. Tomizuka: J. Appl. Phys. **36** (1965) 3839.
65M	Michalowsky, L.: Phys. Status Solidi **8** (1965) 543.
65MTI	Miyamoto, S., N. Tanaka, and S. Iida: J. Phys. Soc. Japan **20** (1965) 753.
65PG	Pines, B. Yu., and N. M. Gumen: Phys. Status Solidi **8** (1965) 751.
65SF	Sokolov, A. D., and M. N. Fedorova: Fiz. Metal. i. Metalloved. **22** (1965) 631.
65WHA	Watanabe, M., T. Hirasawa, and Y. Aiyama: Bull. Electrotech. Lab. (Tokyo) **29** (1965) 22.
66PWGB	Paladino, A. E., J. S. Waught, J. J. Green, and A. E. Booth: J. Appl. Phys. **37** (1966) 3371.
66SJ	Smith, A. B., and R. V. Jones: J. Appl. Phys. **37** (1966) 1001.
66SMM	Svirina, E. P., O. A. Malikova, and M. A. Malikova: Fiz. Tverd. Tela **8** (1966) 1599.
67Ma	Maklen, E. D.: Czech. J. Phys. **B 17** (1967) 376.
67Mak	Makrow, H.: Czech. J. Phys. **B 17** (1967) 387.
67Mo	Morel, J. P.: J. Phys. Chem. Solids **28** (1967) 629.
68FK	Fatseos, G. A., and R. Krishnan: J. Appl. Phys. **39** (1968) 1256.

6.1.3 Mn-Fe^{3+} spinels (Mn ferrites) and Mn-Fe^{3+} spinels with substitutions

6.1.3.0 Introduction

Though the central compound in this article is Mn ferrite $\mathbf{MnFe_2O_4}$, we are in fact concerned primarily with the spinel part of the entire Fe-Mn-O system. In addition we will deal with the properties of the Mn-Me-Fe oxides having the spinel structure: $Mn_{1-x}Me'_xFe_2O_4$ with Me' = Ni, Mg, Zn, Li, Cu, Co and $MnFe_{2-y}Me''_yO_4$ with Me" = Cr, V, Rh, Al, Ga. The mixed Mn-Fe oxides which we call "ferrites" can thus be seen to extend over a wide range of compositions. It is, in fact, generally considered to center on compositions in which the cation/oxygen atom ratio is 3/4, that is in $\mathbf{Mn_xFe_{3-x}O_4}$. Deviations from this so-called stoichiometry can occur, as indicated above. In some instances we shall give compositions as $\mathbf{Mn_xFe_yO_4}$, and it will be seen that $x + y \neq 3$. On other occasions where it is desirable to emphasize the slight deviations from the 3/4 ratio, compositions may be given as $\mathbf{Mn_xFe_{3-x}O_{4\pm\delta}}$. It will be seen in the discussion of ionic distribution and magnetization that the formula may be used to indicate the site occupied. For example, if square brackets are used for the ions on octahedral sites we would interpret $Mn^{++}[Fe_2^{+++}]O_4$ as indicating a normal spinel in which the divalent Mn ions occupy the tetrahedral sites (*A* sites) and trivalent Fe ions occupy the octahedral sites (*B* sites). More complicated examples will be given later if necessary.

General References: [*47S*], [*48N*], [*59SW*].

Part of this compilation is the reproduction of the compilation of Dr. J. F. DILLON Jr. in volume II/9, Magnetic properties I, of the sixth edition of Landolt-Börnstein (Springer, Berlin-Göttingen-Heidelberg, 1962) p. **2**—41 ··· **2**—80. We express our sincere acknowledgment to Dr. DILLON.

6.1.3.1 Phase equilibria and structure

6.1.3.1.1 Phase equilibria

Introduction

The close interaction between the magnetic and electric properties of spinel ferrites and their chemical transformations during synthesis and heat treatment has stimulated the interest in the study of the phase diagrams of the basic ferrite-forming systems, especially, the Fe-Mn-O and Fe-Mn-O substituted systems. The phase diagrams have been constructed only in the most general outlines. The main reason for this is that the partial pressure of oxygen and its change with temperature has to be considered.

a) Phase diagrams Mn-Fe-O

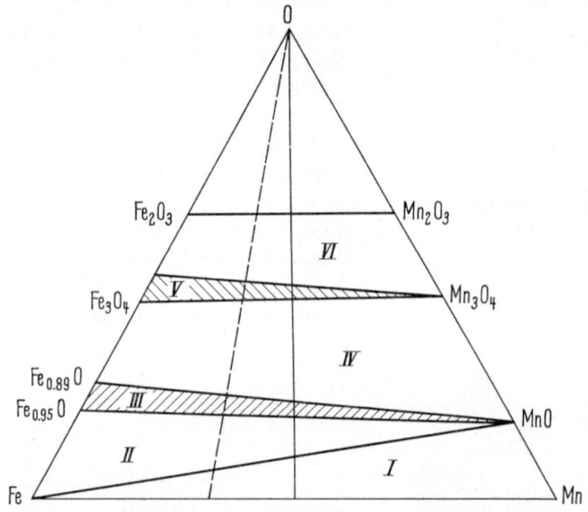

Fig. 1. Schematic depiction of the diagram for rather high temperatures (> 900 °C). Region I corresponds to a two-phase (without considering the gas phase) equilibrium of manganous oxide and a metallic alloy of Fe-Mn, region II corresponds to equilibrium of Fe and a solid solution of the wustite type (Mn, Fe)O, region III corresponds to a lower oxide phase with variable oxygen content, region IV corresponds to a two-phase equilibrium of a wustite-type solid solution and spinel, region V corresponds to a one-phase spinel, in the structure of which the ferrite crystallizes, and region VI corresponds to equilibrium of spinel with solid solutions of the sesquioxides (the shapes and dimensions of the individual phase regions are shown in an arbitrary scale). From [*65TSG*].

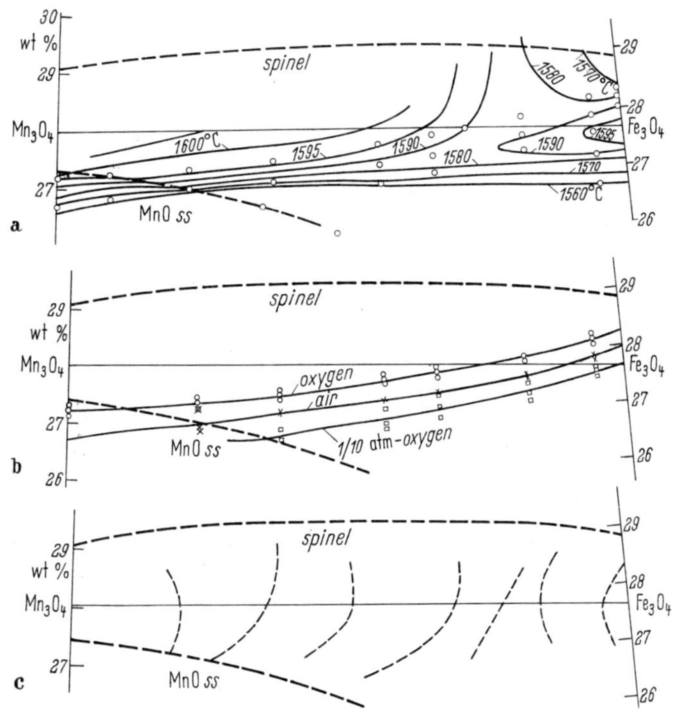

Fig. 2. Phase diagram for the ferrite region of the Mn-Fe-O system in three parts [*58Sh*].

a) Isotherms showing the relationship between liquid composition and temperature. The open circles represent the results of iron determinations. These isotherms were obtained by heating mixtures with known Mn/Fe ratios to successively higher temperatures, in a known atmosphere, until they were completely liquid. They are then quenched and their composition determined by chemical analysis. The heavy dashed line only shows approximately the boundaries of the spinel region.

b) Relationship between oxygen pressure and liquid composition. These were determined in the same manner as the isotherms (Fig. 2a).

c) Relationship of compositions of liquid and crystalline phases in equilibrium involves the fractionation curves. These fractionation curves are constructed from the straight line segments which join the composition of a liquid with the composition of the solid with which it is in equilibrium (conjugation lines). That is to say the conjugation lines are the tangents to the fractionation curve. [*58Sh*] uses these data to discuss the possible paths of crystallization.

→

Fig. 3. Fe_3O_4-Mn_3O_4-O. Isothermal section of the diagram at 1400 °C. The samples were quenched in water. Isobars p_{O_2}: △ 10^{-3} atm, ○ 0.21 atm, □ 1 atm. From [*65KOST*]. There are no significant differences from the data of the isothermal section of the diagram Fe_3O_4-$MnFe_2O_4$-Fe_2O_3 at 1390 °C and p_{O_2} = 0.21 atm and 1 atm given by [*65TO*].

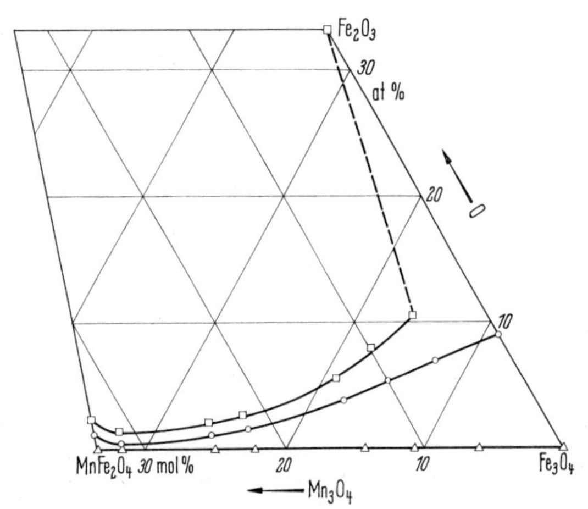

Fig. 4. Fe_3O_4-Mn_3O_4-MnO-FeO Phase diagram at 1000 °C obtained by the method of electromotive forces and X-ray diffraction study [*65TSG*]. It is assumed from literature data [*58Sh*] that at a temperature of the order of 900 ··· 1000 °C, the region of spinel phase is virtually limited by the stoichiometric compositions with respect to oxygen, i.e. those compositions lying on the straight line Fe_3O_4-Mn_3O_4. The lines in the two-phases region give the relative composition of spinel and wustite and the equilibrium partial pressure of oxygen.

↓

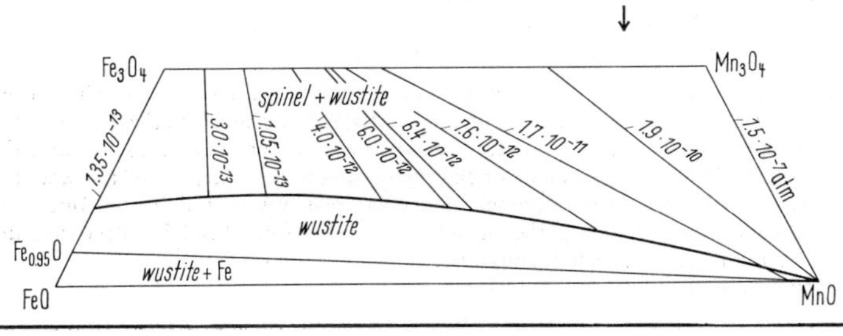

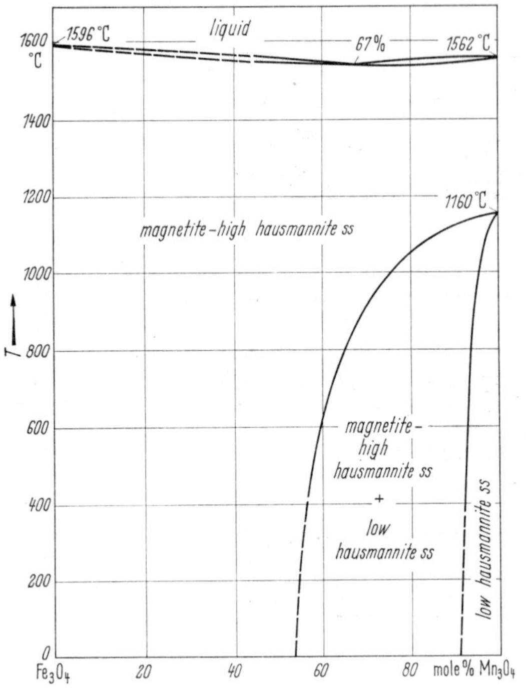

←
Fig. 5. Fe_3O_4-Mn_3O_4 system. After [58HK]. Compare with Fig. 7; some differences are obvious at high temperatures.

b) Phase diagrams Zn-Mn-Fe-O

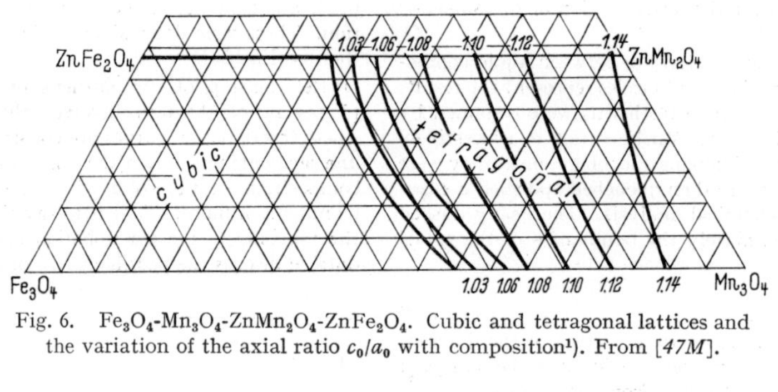

Fig. 6. Fe_3O_4-Mn_3O_4-$ZnMn_2O_4$-$ZnFe_2O_4$. Cubic and tetragonal lattices and the variation of the axial ratio c_0/a_0 with composition[1]). From [47M].

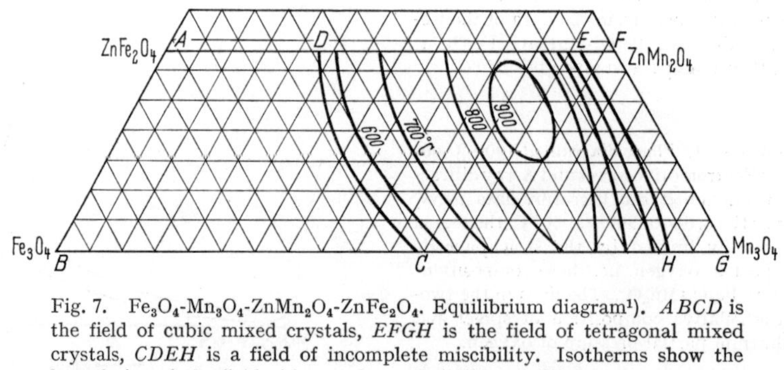

Fig. 7. Fe_3O_4-Mn_3O_4-$ZnMn_2O_4$-$ZnFe_2O_4$. Equilibrium diagram[1]). *ABCD* is the field of cubic mixed crystals, *EFGH* is the field of tetragonal mixed crystals, *CDEH* is a field of incomplete miscibility. Isotherms show the boundaries of the field of incomplete miscibility at different temperatures. This disappears between 900 and 1000 °C. From [47M].

[1]) Compounds were prepared by [47M] in the system Fe_3O_4-Mn_3O_4-$ZnMn_2O_4$-$ZnFe_2O_4$ by firing powder at temperatures up to 1300 °C for times up to 24 h. The data were plotted on part of the plane within the Fe-Mn-Zn-O phase tetrahedron defined by the four compounds mentioned above. The compositions were determined from the ratios of the starting materials, and in some cases by analyses before the final firing. No mention was made of the possibility of oxygen rich and oxygen poor compositions. There seem to have been no analyses of any mixtures containing Zn, the most volatile constituent. About twenty-five compounds around the periphery of the phase area were made, as well as six within the polygon. Most samples were fired in air, but some along the Mn_3O_4-Fe_3O_4 join were fired in evacuated and sealed ampoules.

c) Phase diagram Cu-Mn-Fe-O.

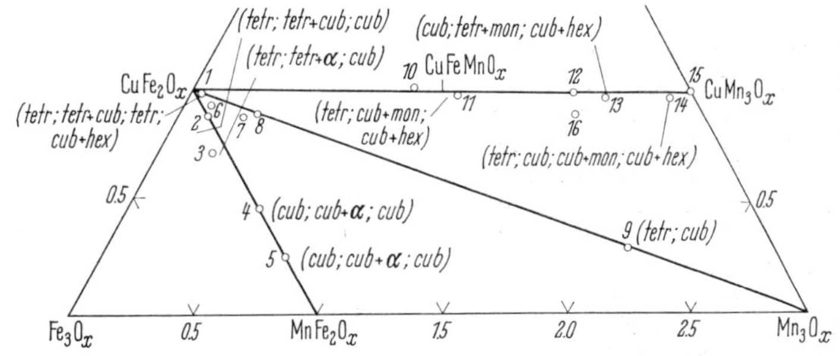

Fig. 8. Mn-Fe-Cu-O system.

Nr.	1	2	3	4	5	6	7	8	9	10	11	12	13	14	15	16
Cu	0.99	0.874	0.72	0.47	0.27	0.94	0.88	0.9	0.34	1.02	0.94	1.01	0.96	1.0	1.0	0.92
Mn	0.03	0.126	0.22	0.53	0.73	0.10	0.26	0.3	2.1	0.88	1.10	1.52	1.69	1.92	2.0	1.59
Fe	1.98	2.00	2.06	2.00	2.00	1.96	1.86	1.8	0.66	1.86	1.06	0.47	0.35	0.1		0.59

Compositions under investigation and phases found (in the order of rising temperature): cub = cubic spinel; tetr = tetragonally distorted spinel; mon = monoclinic phase, analogous to tenorite (CuO); hex = hexagonal phase, analogous to delafossite ($CuFeO_2$); α = solid solutions with the structure of hematite (α-Fe_2O_3) [*65BSK*].

Temperature of precipitation:

Nr.		
1	tetr $\xrightarrow{1050\ °C}$ cub + hex	
5	cub $\xrightarrow{\simeq 650\ °C}$ cub + α	
9	tetr $\xrightarrow{450\cdots600\ °C}$ cub	
12	tetr $\xrightarrow{450\cdots600\ °C}$ cub	
14	tetr $\xrightarrow{450\cdots600\ °C}$ mon + cub,	$\xrightarrow{950\cdots1000\ °C}$ cub
	cub $\xrightarrow{1000\cdots1100\ °C}$ hex	

These results are obtained by means of thermogravimetric, dilatometric, and X-ray measurements in air at temperatures between 20 ··· 1200 °C.

6.1.3.1.2 Equilibrium oxygen partial pressure

Introduction

The reproducibility of the magnetic and electrical properties of spinel ferrites is often quite poor, possibly because of failure to remember that the ferrites are phases of variable composition. A system of a gas and a condensed phase should, according to the phase rule, show one degree of variance, regardless of the number of components. That is to say the maintenance of a fixed composition corresponding to a predetermined relative number of defects is possible only under independent variation of just one of the two variables, temperature and pressure, i.e., with $p_{O_2} = f(T)$. Thus any alteration in the sintering temperature during the synthesis of a ferrite of fixed composition must be accompanied by a corresponding alteration in the partial pressure of gaseous oxygen to the equilibrium value for the ferrite in question. $p_{O_2} = f(T)$ function varies with the type of the ferrite and the relative number of defects. The plot of $\log p_{O_2}$ vs. reciprocal absolute temperature is generally a straight line, with the line slope almost independent of the composition of the spinel, and the nature and the valency of the cations.

$$\log p_{O_2} = A - B/T.$$

Note: there are great differences in the results, especially between those from the electromotive forces and those from thermogravimetric measurements.

a) System Mn-Fe-O

Dependence on Mn content x in $\mathbf{Mn_xFe_{3-x}O_4}$: $p_{O_2} = f(x)$ (T = 900, 1000, 1150, 1300 °C):

Tab. 1. Equilibrium pressure of oxygen, p_{O_2}, above the spinel $Mn_xFe_{3-x}O_4$ at 1000 °C from investigation by the method of electromotive forces [*65TSG*]. These pressures allow the coexistence of the wustite phase. See Fig. 4 for wustite compositions and [*65T* (2)], [*64GTK*] for other results

Mn content x in spinel	Equilibrium p_{O_2} [atm]
0.000	$1.35 \cdot 10^{-13}$
0.225	$3.0 \cdot 10^{-13}$
0.450	$1.05 \cdot 10^{-12}$
0.705	$4.0 \cdot 10^{-12}$
0.870	$6.0 \cdot 10^{-12}$
0.915	$6.4 \cdot 10^{-12}$
1.000	$7.6 \cdot 10^{-12}$
1.200	$1.7 \cdot 10^{-11}$
2.130	$1.9 \cdot 10^{-10}$
3.000	$1.5 \cdot 10^{-7}$

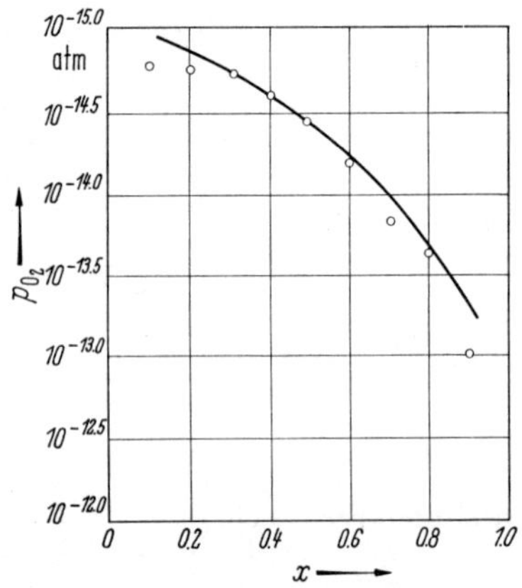

Fig. 9. $x\,MnFe_2O_4 \cdot (1-x)\,Fe_3O_4$. Concentration dependence of the partial pressure of oxygen at 900 °C. From [*66BZMSA*].

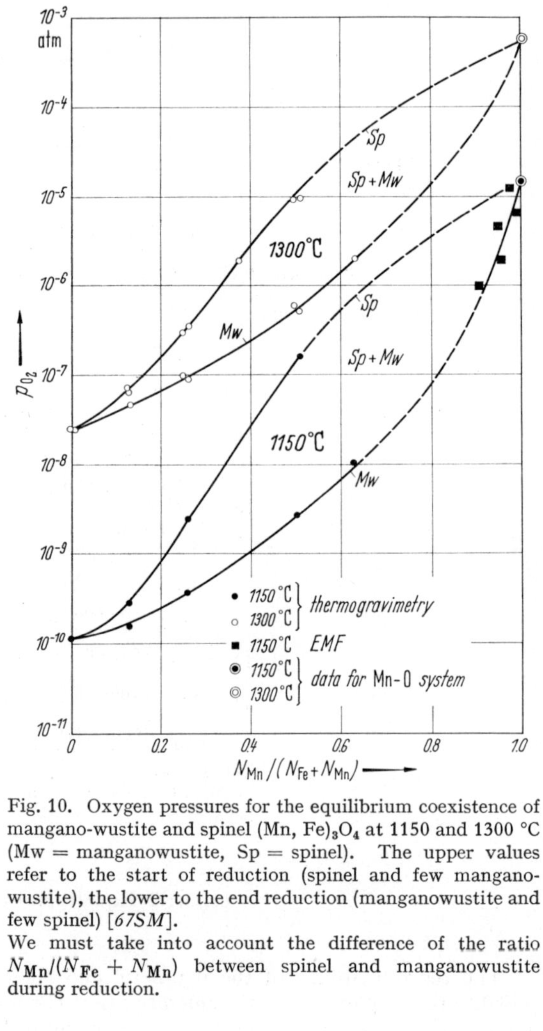

Fig. 10. Oxygen pressures for the equilibrium coexistence of mangano-wustite and spinel $(Mn, Fe)_3O_4$ at 1150 and 1300 °C (Mw = manganowustite, Sp = spinel). The upper values refer to the start of reduction (spinel and few mangano-wustite), the lower to the end reduction (manganowustite and few spinel) [*67SM*].

We must take into account the difference of the ratio $N_{Mn}/(N_{Fe}+N_{Mn})$ between spinel and manganowustite during reduction.

$MnFe_{1.932}O_x$:

The dependence of $\log(p_{H_2O}/p_{H_2})$, and $\log p_{O_2}$ [atm] on $1/T$ for the ferrite $\mathbf{MnFe_{1.932}O_{3.897}}$ which from the requirement of electrical neutrality contains only Mn^{2+} and Fe^{3+}, is represented by the linear equations [*66BZR*]:

$\log(p_{H_2O}/p_{H_2}) = 6.64 - 3.83 \cdot 10^3\,T^{-1}$ (1050 ··· 1450 °K) compare Fig. 11;

$\log p_{O_2} = 19.02 - 33.39 \cdot 10^3\,T^{-1}$ (1050 ··· 1450 °K) compare Tab. 2.

→

Fig. 11. $MnFe_{1.932}O_x$. Equilibrium with H_2-H_2O mixtures at 1) 777 °C 2) 877 °C 3) 977 °C 4) 1077 °C 5) 1177 °C and for p_{H_2O}/p_{H_2} ratios: a) ≧ 1.00 b) 0.01 ··· 1.00.

The ferrite exists as a pure phase of variable composition over the range 3.897 ≧ x ≧ 3.860. The solid solution of lower oxides exists over the range 3.1 > x > 2.8. The composition $MnFe_{1.932}O_{1.07-0.97}$ consists of MnO and Fe. From [*66BZR*].

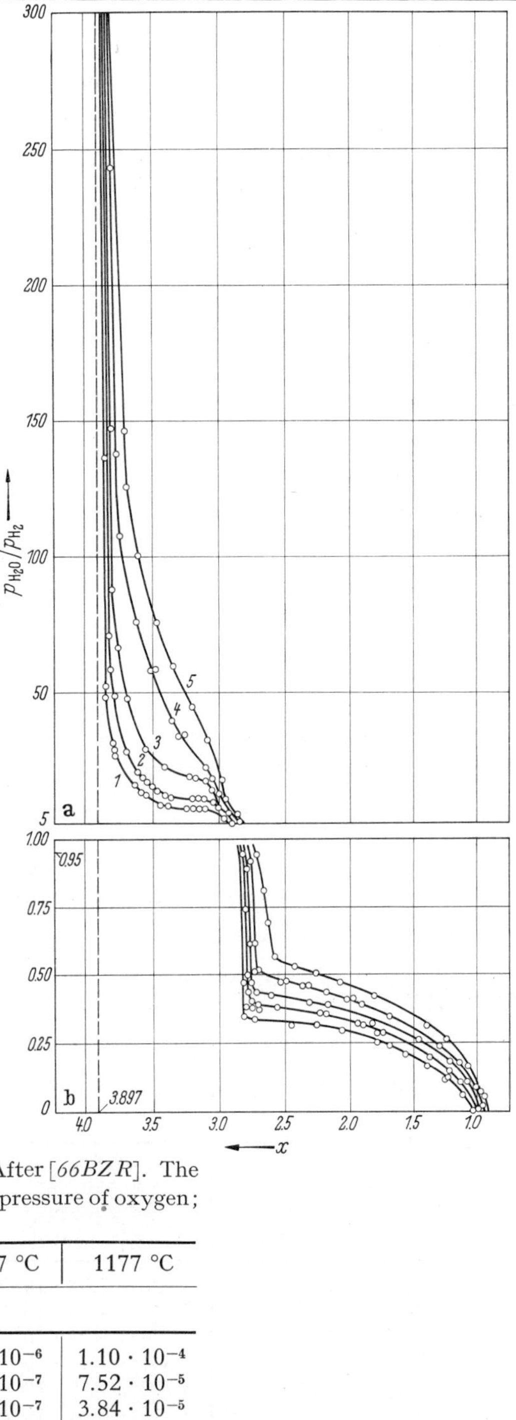

Tab. 2. $MnFe_{1.932}O_x$. Partial pressure of oxygen, p_{O_2}. After [*66BZR*]. The corresponding numerical data for the calculated partial pressure of oxygen; see Fig. 11

T	777 °C	877 °C	977 °C	1077 °C	1177 °C
x	p_{O_2} [atm]				
3.897	$1.37 \cdot 10^{-13}$	$1.41 \cdot 10^{-10}$	$1.69 \cdot 10^{-8}$	$1.55 \cdot 10^{-6}$	$1.10 \cdot 10^{-4}$
3.89	$1.31 \cdot 10^{-15}$	$2.15 \cdot 10^{-11}$	$8.87 \cdot 10^{-9}$	$9.86 \cdot 10^{-7}$	$7.52 \cdot 10^{-5}$
3.88	$3.93 \cdot 10^{-16}$	$8.98 \cdot 10^{-13}$	$1.70 \cdot 10^{-9}$	$7.04 \cdot 10^{-7}$	$3.84 \cdot 10^{-5}$
3.87	$2.91 \cdot 10^{-16}$	$2.02 \cdot 10^{-13}$	$9.66 \cdot 10^{-11}$	$6.00 \cdot 10^{-8}$	$1.22 \cdot 10^{-5}$
3.86	$2.49 \cdot 10^{-16}$	$1.24 \cdot 10^{-13}$	$3.64 \cdot 10^{-11}$	$3.56 \cdot 10^{-9}$	$9.38 \cdot 10^{-7}$
3.85	$2.08 \cdot 10^{-16}$	$9.55 \cdot 10^{-14}$	$1.33 \cdot 10^{-11}$	$2.68 \cdot 10^{-9}$	$7.62 \cdot 10^{-8}$
3.84	$1.82 \cdot 10^{-16}$	$7.43 \cdot 10^{-14}$	$1.25 \cdot 10^{-11}$	$1.94 \cdot 10^{-9}$	$1.30 \cdot 10^{-8}$
3.83	$1.59 \cdot 10^{-16}$	$5.79 \cdot 10^{-14}$	$8.48 \cdot 10^{-12}$	$1.36 \cdot 10^{-9}$	$5.56 \cdot 10^{-8}$
3.82	$1.39 \cdot 10^{-16}$	$4.89 \cdot 10^{-14}$	$7.41 \cdot 10^{-12}$	$9.16 \cdot 10^{-10}$	$4.72 \cdot 10^{-8}$
3.81	$1.24 \cdot 10^{-16}$	$4.23 \cdot 10^{-14}$	$6.56 \cdot 10^{-12}$	$7.36 \cdot 10^{-10}$	$4.09 \cdot 10^{-8}$
3.80	$1.13 \cdot 10^{-16}$	$3.66 \cdot 10^{-14}$	$5.86 \cdot 10^{-12}$	$6.34 \cdot 10^{-10}$	$3.61 \cdot 10^{-8}$
3.75	$7.40 \cdot 10^{-17}$	$2.04 \cdot 10^{-14}$	$3.76 \cdot 10^{-12}$	$4.25 \cdot 10^{-10}$	$2.03 \cdot 10^{-8}$
3.70	$5.01 \cdot 10^{-17}$	$1.33 \cdot 10^{-14}$	$2.40 \cdot 10^{-12}$	$3.26 \cdot 10^{-10}$	$1.38 \cdot 10^{-8}$
3.60	$2.22 \cdot 10^{-17}$	$6.48 \cdot 10^{-15}$	$1.23 \cdot 10^{-12}$	$2.06 \cdot 10^{-10}$	$7.67 \cdot 10^{-9}$

$MnFe_{2.065}O_x$:

The change in the lattice parameters indicates that the lower oxide phase formed during reduction of spinel is markedly enriched with manganous oxide. With increasing reduction of spinel the lower-oxide phase is enriched with wustite, which leads to a decrease in the lattice parameter. From the data obtained for $\mathbf{MnFe_{2.065}O_{4.098}}$, linear equations for the gas-phase composition in terms of $1/T$ in the range 1090···1198 °K are derived as follows:

$\log (p_{H_2O}/p_{H_2}) = 5.57 - 4.13 \cdot 10^3 T^{-1}$ (See Fig. 12)
$\log p_{O_2}$ [atm] $= 17.34 - 34.80 \cdot 10^3 T^{-1}$.

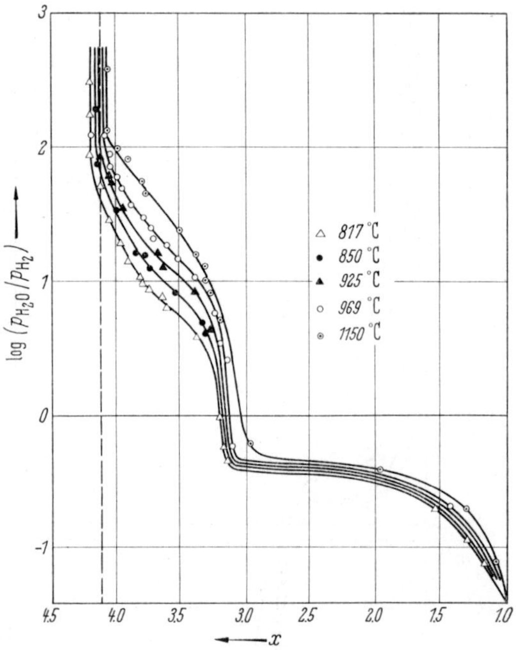

Fig. 12. $MnFe_{2.065}O_x$. The equilibrium with H_2-H_2O mixtures at different temperatures [67RBZ].

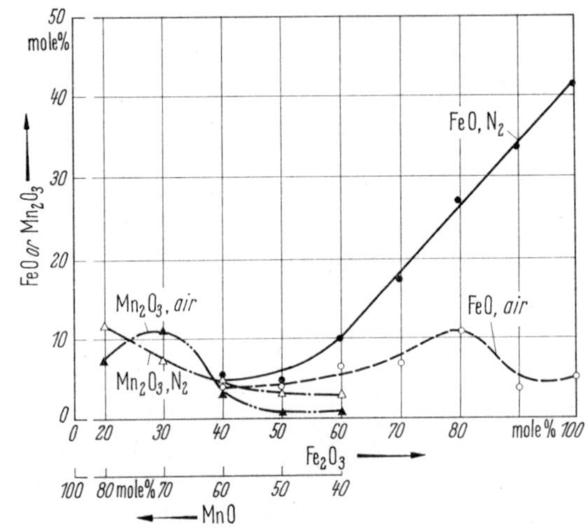

Fig. 13. Results of chemical analysis for the MnO-Fe_2O_3 system. Moulded samples were heated in air for 2 h at 1400 °C and cooled to 800 °C at a cooling rate of 400 °C per hour and quenched in air from 800 °C or heated in nitrogen gas in the same way, except the specimens were slowly cooled in the furnace to RT. From [61S].

Tab. 3. $MnFe_{2.065}O_x$. The corresponding calculated partial pressure of oxygen at different temperatures, see Fig. 12. [67RBZ]

817 °C		850 °C		969 °C		1150 °C	
x	p_{O_2} atm	x	p_{O_2} atm	x	p_{O_2} atm	x	p_{O_2} atm
4.164	$1.00 \cdot 10^{-13}$	4.126	$1.78 \cdot 10^{-13}$	4.097	$1.35 \cdot 10^{-10}$	4.080	$3.47 \cdot 10^{-8}$
4.177	$2.95 \cdot 10^{-14}$	4.068	$2.95 \cdot 10^{-14}$	4.089	$1.02 \cdot 10^{-10}$	4.101	$6.46 \cdot 10^{-9}$
4.156	$1.45 \cdot 10^{-14}$	3.978	$5.25 \cdot 10^{-15}$	4.091	$7.76 \cdot 10^{-11}$	3.998	$3.80 \cdot 10^{-9}$
4.147	$6.92 \cdot 10^{-15}$	3.841	$1.15 \cdot 10^{-15}$	4.082	$1.10 \cdot 10^{-11}$	3.922	$2.63 \cdot 10^{-9}$
4.142	$5.01 \cdot 10^{-15}$	3.777	$1.10 \cdot 10^{-15}$	4.039	$6.17 \cdot 10^{-12}$	3.811	$1.26 \cdot 10^{-9}$
4.096	$2.45 \cdot 10^{-15}$	3.747	$6.61 \cdot 10^{-16}$	4.044	$3.72 \cdot 10^{-12}$	3.775	$7.94 \cdot 10^{-10}$
4.034	$7.59 \cdot 10^{-16}$	3.551	$2.95 \cdot 10^{-16}$	3.984	$3.17 \cdot 10^{-12}$	3.528	$2.40 \cdot 10^{-10}$
3.954	$3.32 \cdot 10^{-16}$	4.102	$2.04 \cdot 10^{-12}$	4.001	$3.09 \cdot 10^{-12}$	3.413	$9.33 \cdot 10^{-11}$
3.891	$1.78 \cdot 10^{-16}$	4.019	$5.01 \cdot 10^{-13}$	3.931	$1.21 \cdot 10^{-12}$	3.351	$6.92 \cdot 10^{-11}$
3.792	$1.00 \cdot 10^{-16}$	4.004	$4.37 \cdot 10^{-13}$	3.871	$1.10 \cdot 10^{-12}$	3.336	$4.07 \cdot 10^{-11}$
3.795	$8.32 \cdot 10^{-17}$	3.911	$1.58 \cdot 10^{-13}$	3.779	$6.31 \cdot 10^{-13}$	3.279	$2.63 \cdot 10^{-11}$
3.735	$6.46 \cdot 10^{-17}$	3.677	$3.47 \cdot 10^{-14}$	3.729	$4.79 \cdot 10^{-13}$		
3.645	$5.25 \cdot 10^{-17}$	3.645	$2.14 \cdot 10^{-14}$	3.706	$3.39 \cdot 10^{-13}$		
3.608	$3.89 \cdot 10^{-17}$	3.402	$8.91 \cdot 10^{-15}$	3.600	$2.57 \cdot 10^{-13}$		
				3.498	$1.74 \cdot 10^{-13}$		
				3.396	$9.12 \cdot 10^{-14}$		

Tab. 4. $MnFe_{2.065}O_x$. X-ray analysis data for $MnFe_{2.065}O_x$ specimens in equilibrium with different gas compositions at 969 °C [*67RBZ*]

$\log \frac{p_{H_2O}}{p_{H_2}}$	x	Phase		Lattice parameter [Å]	
		basic	excess	spinel	lower-oxide phase
2.060	4.082	spinel	lower-oxide phase	8.50	4.42
1.438	3.779	spinel	lower-oxide phase	8.47	4.41
0.954	3.352	lower-oxide phase	—	—	4.35

$Mn_xFe_{3-x}O_{4+\gamma}$: Dependence on oxygen content in $\mathbf{Mn_xFe_{3-x}O_{4\pm\gamma}}$: $p_{O_2} = f(\gamma)$.

Tab. 5. Composition of condensed phases $Mn_xFe_{3-x}O_{4+\gamma}$. γ = oxygen excess content. Sample was heated 4 h to 1400 °C in equilibrium with gas mixtures with different partial pressure of oxygen, p_{O_2}, and then subjected to abrupt quenching by dropping into cold water [*65KOST*]

p_{O_2} = 0.001 atm		p_{O_2} = 0.21 atm		p_{O_2} = 1 atm	
x	γ	x	γ	x	γ
0	0	0	0.0955	0	0.5000
0.1788	0.000	0.1788	0.0732	0.1788	0.1180
0.3172	0.000	0.3172	0.0566	0.3172	0.0862
0.4302	0.002	0.4302	0.0386	0.4302	0.0573
0.6678	0.001	0.6678	0.0141	0.6678	0.0257
0.7514	0.000	0.7514	0.0124	0.7514	0.0193
0.9480	0.000	0.9480	0.00276	0.9480	0.0135
0.9612	0.0121	0.9612	0.0159	0.9612	0.01726
0.9920	−0.0011	0.9920	0.0105	0.9920	0.01006
1.0309	−0.00405	1.0309	−0.0036	1.0309	0.0045
1.2375	−0.02667	1.2375	−0.0160	1.2375	−0.0070
1.5650	−0.0350	1.5650	−0.0259	1.5650	−0.0140
1.8725	−0.0455	1.8725	−0.0284	1.8725	−0.0160
2.2510	−0.0701	2.2510	−0.0329	2.2510	−0.0220
2.6065	−0.1013	2.6065	−0.0380	2.6065	−0.0376
2.7767	−0.1480	2.7767	−0.0668	2.7767	−0.0526
3.0000	−0.1865	3.000	−0.1156	3.000	−0.072

Dependence on temperature in $\mathbf{Mn_xFe_{3-x}O_{4\pm\gamma}}$: $p_{O_2} = f(T)$.

Tab. 6. $Mn_xFe_{3-x}O_4$ spinel. Equations for the partial pressure of oxygen, p_{O_2} as a function of temperature, after [*65TSGZG*]. The dissociation pressure was determined by the EMF method with a solid electrolyte between 900 ··· 1200 °C. See also [*66G*]

$\log p_{O_2} = 13.45 - 334 \cdot 10^2 \cdot T^{-1}$ for Fe_3O_4
$\log p_{O_2} = 12.41 - 317 \cdot 10^2 \cdot T^{-1}$ for $Mn_{0.3}Fe_{2.7}O_4$
$\log p_{O_2} = 13.57 - 318 \cdot 10^2 \cdot T^{-1}$ for $Mn_{0.6}Fe_{2.4}O_4$
$\log p_{O_2} = 13.89 - 319 \cdot 10^2 \cdot T^{-1}$ for $Mn_{0.9}Fe_{2.1}O_4$
$\log p_{O_2} = 14.52 - 324 \cdot 10^2 \cdot T^{-1}$ for $MnFe_2O_4$
$\log p_{O_2} = 13.32 - 307 \cdot 10^2 \cdot T^{-1}$ for $Mn_{1.2}Fe_{1.8}O_4$
$\log p_{O_2} = 13.51 - 303 \cdot 10^2 \cdot T^{-1}$ for $Mn_{1.5}Fe_{1.5}O_4$
$\log p_{O_2} = 12.61 - 286 \cdot 10^2 \cdot T^{-1}$ for $Mn_{1.8}Fe_{1.2}O_4$
$\log p_{O_2} = 12.49 - 280 \cdot 10^2 \cdot T^{-1}$ for $Mn_{2.1}Fe_{0.9}O_4$
$\log p_{O_2} = 11.73 - 273 \cdot 10^2 \cdot T^{-1}$ for $Mn_{2.4}Fe_{0.6}O_4$
$\log p_{O_2} = 11.65 - 267 \cdot 10^2 \cdot T^{-1}$ for $Mn_{2.7}Fe_{0.3}O_4$
$\log p_{O_2} = 9.89 - 236 \cdot 10^2 \cdot T^{-1}$ for Mn_3O_4.

Tab. 7. $MnFe_2O_n$. Coefficients of the equation for the oxygen partial pressure as a function of temperature: $\log p_{O_2} = A_1 + B_1 \cdot T^{-1}$ (p in atm)

n	A_1	$-B_1 \cdot 10^{-3}$	Ref.
4.30	10.69	14.91	*67RL*
4.21	11.50	16.98	*67RL*
4.06	12.77	19.49	*67RL*
4.02	13.03	19.82	*67RL*
3.99	11.70	28.15	*66RL*
3.98	11.71	28.38	*66RL*
3.96	11.93	28.83	*66RL*
3.77	13.28	31.42	*66RL*
3.72	13.72	32.10	*66RL*
3.25	9.38	28.07	*67RL*
3.02	8.54	28.05	*67RL*
2.84	6.49	27.64	*66RL*
2.27	6.48	27.71	*66RL*
1.92	6.17	27.43	*66RL*
1.32	6.03	27.54	*66RL*

Vautier/Paulus

b) System Al-Mn-Fe-O

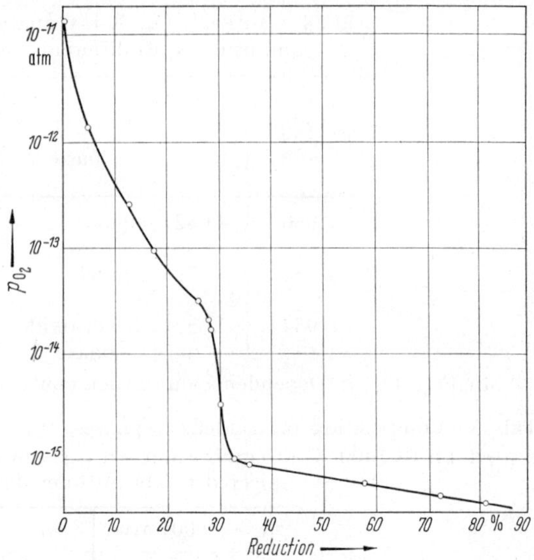

Fig. 14. $MnAl_{0.25}Fe_{1.75}O_4$. Equilibrium oxygen pressure, p_{O_2}, as a function of the degree of reduction at 1000 °C. 100% corresponding to reduction of $MnAl_{0.25}Fe_{1.75}O_4$ to $MnAl_2O_4$, MnO and Fe.
The vapor pressure of oxygen was calculated according to the experimental values p_{H_2O}/p_{H_2} from the equation

$$\log p_{O_2} = 2 \log \left(\frac{p_{H_2O}}{p_{H_2}} \cdot K_{H_2O} \right)$$

where K_{H_2O} is the equilibrium constant of the dissociation of water vapor (at 1000 °C, $\log K_{H_2O} = -7.2878$). After [*65TBC*].

Tab. 8. $MnAl_{0.25}Fe_{1.75}O_4$. The compositions of equilibrium solid and gas phases at $p_{H_2O} = 4.579$ mmHg and 1000 °C. See also Tab. 12, Fig. 30 for the variation of lattice parameters. From [*65TBC*]

Degree of reduction %	Spinel phase	Lower oxide phase	Metal	p_{H_2} mm Hg
0	$(MnAl_2O_4)_{0.125}(MnFe_2O_4)_{0.875}$	—	—	—
0.2	$MnAl_{0.25}Fe_{1.75}O_4$	present	—	0.06
5.8	$0.849\,(MnAl_2O_4)_{0.15}(MnFe_2O_4)_{0.59}(Fe_3O_4)_{0.26}$	$0.453\,Mn_{0.82}Fe_{0.18}O$	—	0.21
12.7	$0.667\,(MnAl_2O_4)_{0.19}(MnFe_2O_4)_{0.43}(Fe_3O_4)_{0.38}$	$0.999\,Mn_{0.59}Fe_{0.41}O$	—	0.45
17.9	$0.530\,(MnAl_2O_4)_{0.24}(MnFe_2O_4)_{0.34}(Fe_3O_4)_{0.42}$	$1.410\,Mn_{0.49}Fe_{0.51}O$	—	0.76
26.2	$0.312\,(MnAl_2O_4)_{0.40}(MnFe_2O_4)_{0.18}(Fe_3O_4)_{0.42}$	$2.064\,Mn_{0.40}Fe_{0.60}O$	—	1.34
28.0	$0.265\,(MnAl_2O_4)_{0.48}(MnFe_2O_4)_{0.13}(Fe_3O_4)_{0.39}$	$2.205\,Mn_{0.38}Fe_{0.62}O$	—	1.66
28.6	$0.249\,(MnAl_2O_4)_{0.50}(MnFe_2O_4)_{0.12}(Fe_3O_4)_{0.38}$	$2.253\,Mn_{0.37}Fe_{0.63}O$	—	1.79
30.5	$0.199\,(MnAl_2O_4)_{0.63}(MnFe_2O_4)_{0.08}(Fe_3O_4)_{0.29}$	$2.403\,Mn_{0.36}Fe_{0.64}O$	—	4.17
33.1	$0.131\,(MnAl_2O_4)_{0.95}(Fe_3O_4)_{0.05}$	$2.607\,Mn_{0.33}Fe_{0.67}O$	—	7.61
35.9	$0.125\ MnAl_2O_4$	$2.558\,Mn_{0.34}Fe_{0.66}O$	0.067 Fe	7.92
57.7	$0.125\ MnAl_2O_4$	$1.987\,Mn_{0.44}Fe_{0.56}O$	0.638 Fe	9.68
72.1	$0.125\ MnAl_2O_4$	$1.607\,Mn_{0.54}Fe_{0.46}O$	1.018 Fe	11.49
80.7	$0.125\ MnAl_2O_4$	$1.382\,Mn_{0.63}Fe_{0.37}O$	1.243 Fe	12.62

c) System Cr-Al-Mn-Fe-O

Tab. 9. $MnAl_{0.25}Cr_{0.25}Fe_{1.50}O_4$. The amount and the composition of equilibrium solid and gas phases at $p_{H_2O} = 4.579$ mmHg for different reduction stages at 1000 °C. Complete reduction of Fe cations from solution to the metallic state was taken as 100%. After [*66TCB*].
The oxygen pressure above the solid solutions may be calculated according to the equation

$$\log p_{O_2} = 2\,(\log\,(p_{H_2O}/p_{H_2}) + \log K_{H_2O}); \text{ (at 1000 °C, } \log K_{H_2O} = 7.2878)$$

reduction %	Spinel phase	Lower oxide phase	Metal	p_{H_2} mm Hg
0	$(MnAlCrO_4)_{0.25}(MnFe_2O_4)_{0.75}$	—	—	—
3.00	$0.932\,(MnAlCrO_4)_{0.26}(MnFe_2O_4)_{0.64}(Fe_3O_4)_{0.10}$	$0.202\,Mn_{0.76}Fe_{0.24}O$	—	0.45
10.90	$0.755\,(MnAlCrO_4)_{0.34}(MnFe_2O_4)_{0.39}(Fe_3O_4)_{0.27}$	$0.736\,Mn_{0.62}Fe_{0.38}O$	—	0.49
12.67	$0.715\,(MnAlCrO_4)_{0.35}(MnFe_2O_4)_{0.35}(Fe_3O_4)_{0.30}$	$0.855\,Mn_{0.59}Fe_{0.41}O$	—	0.56
17.52	$0.606\,(MnAlCrO_4)_{0.41}(MnFe_2O_4)_{0.22}(Fe_3O_4)_{0.37}$	$1.182\,Mn_{0.52}Fe_{0.48}O$	—	0.76
25.50	$0.426\,(MnAlCrO_4)_{0.58}(MnFe_2O_4)_{0.13}(Fe_3O_4)_{0.29}$	$1.721\,Mn_{0.41}Fe_{0.59}O$	—	1.59
30.00	$0.325\,(MnAlCrO_4)_{0.76}(MnFe_2O_4)_{0.01}(Fe_3O_4)_{0.20}$	$2.025\,Mn_{0.36}Fe_{0.64}O$	—	4.00
34.20	$0.25\ MnAlCrO_4$	$2.230\,Mn_{0.34}Fe_{0.66}O$	0.021 Fe	8.50
50.10	$0.25\ MnAlCrO_4$	$1.875\,Mn_{0.40}Fe_{0.60}O$	0.375 Fe	10.30
63.4	$0.25\ MnAlCrO_4$	$1.573\,Mn_{0.48}Fe_{0.52}O$	0.677 Fe	12.20
96.2	$0.25\ MnAlCrO_4$	$0.836\,Mn_{0.90}Fe_{0.10}O$	1.414 Fe	14.50

d) System Cr-Mn-Fe-O

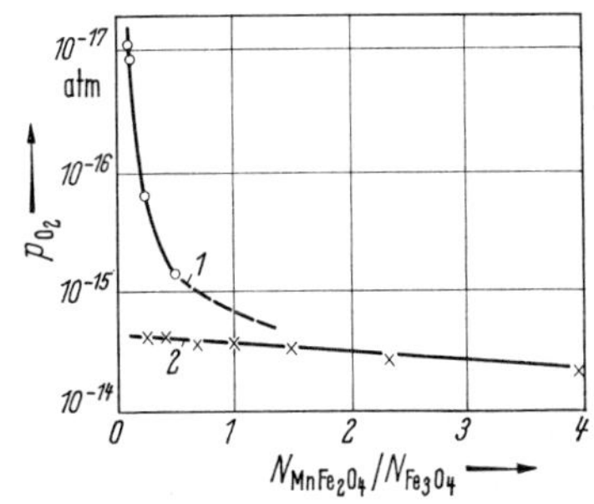

Fig. 15. Equilibrium pressure of oxygen vs. ratio $MnFe_2O_4/Fe_3O_4$ for the spinel solution components:
1) $MnFe_2O_4$-$MnCr_2O_4$-Fe_3O_4 2) $MnFe_2O_4$-Fe_3O_4.
An increase in the Mn chromite content in the solid solution of Mn ferrite and magnetite reduces the equilibrium pressure of oxygen. From [*67TABC*].

e) System Zn-Ca-Mg-Mn-Fe-O

Tab. 10. Coefficients A and B of the equation $\log p_{O_2} = A + B \cdot T^{-1}$ (p in atm; T in °K) for the temperature dependence between 900 ··· 1200 °C of the partial pressure of oxygen for Mn-Mg, Mn-Mg-Ca and Mn-Mg-Zn-Ca ferrites [*66G*] and [*65TSGZG*]

Composition mole %	A	$-B \cdot 10^{-3}$
Fe_2O_3—43; MnO—43; MgO—14	13.54	32.3
Fe_2O_3—39.26; MnO—22.09; MgO—38.67	14.33	31.9
Fe_2O_3—41.76; MnO—41.57; MgO—14.84; CaO—1.83	13.63	31.9
Fe_2O_3—37, MnO—38; MgO—10; CaO—3; ZnO—12	13.84	32.4

f) System Zn-Mn-Fe-O

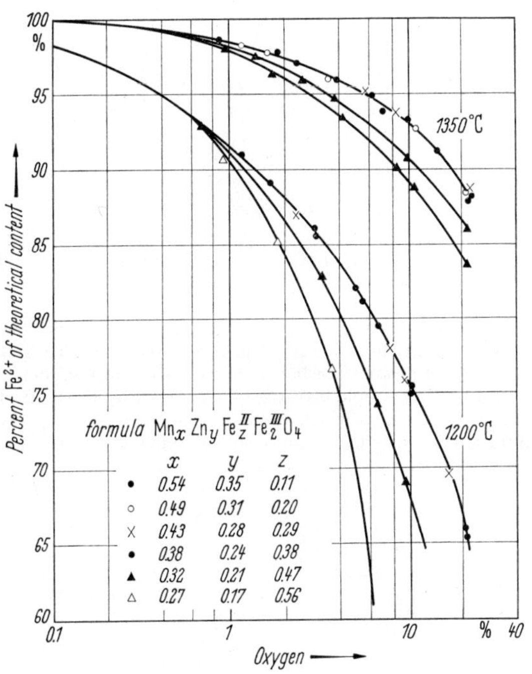

Fig. 16. Percent of Fe^{2+} vs. oxygen content of the furnace atmosphere for different compositions of MnZn ferrites, after [*65MJ*]. Although samples with different Mn/Zn/Fe ratios at fixed temperature and oxygen content possessed widely different Fe^{2+} contents, these values were almost the same when expressed as a percentage of the optimum Fe^{2+} content, i.e. the Fe^{2+} content if the excess iron was present as Fe_3O_4. This rule is valid only for $Fe^{2+} < 10$ wt %; see Fig. 17.

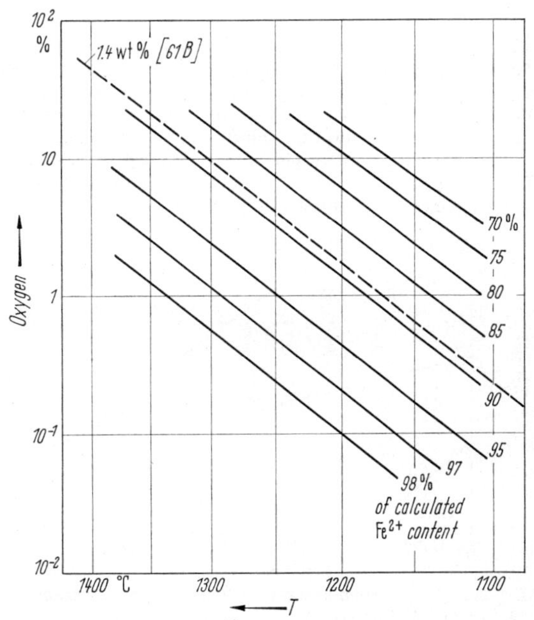

Fig. 17. Equilibrium diagram, percent of oxygen of the furnace atmosphere vs. temperature for MnZn ferrite containing < 10 wt % Fe^{2+} [*65MJ*]. Composition lines are for constant Fe^{2+} content given in percent of the theoretical value, i.e. if the excess iron is present as Fe_3O_4. This diagram is essentially the same as the universal one proposed by [*61B*]. At a given temperature a straight linear relationship is observed between log of the oxygen percent of the atmosphere and log of Fe^{2+} content in the MnZn ferrite.

Tab. 11. Oxygen content of the atmosphere in the furnace for sintering MnZn ferrites with constant degree of oxidation. From [*67AP*]

53.5 Fe_2O_3, 28.5 MnO, 18.0 ZnO		53.0 Fe_2O_3, 25.0 MnO, 22.0 ZnO	
T °C	Oxygen %	*T* °C	Oxygen %
1275	99	1300	98
1270	20	1280	50
1250	5	1260	20
1220	1	1240	1
1150	0.1	1220	0.5
		1200	0.2
		1150	0.06

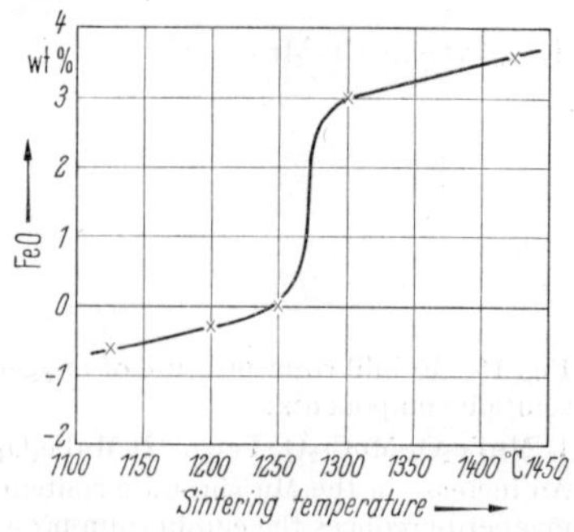

Fig. 18. FeO content (in wt %) in 52.5 Fe_2O_3, 28.3 MnO, 19.2 ZnO (in mol%) sintered 4 h at various temperatures; p_{O_2} = 0.01 atm. From [*62P*(3)].

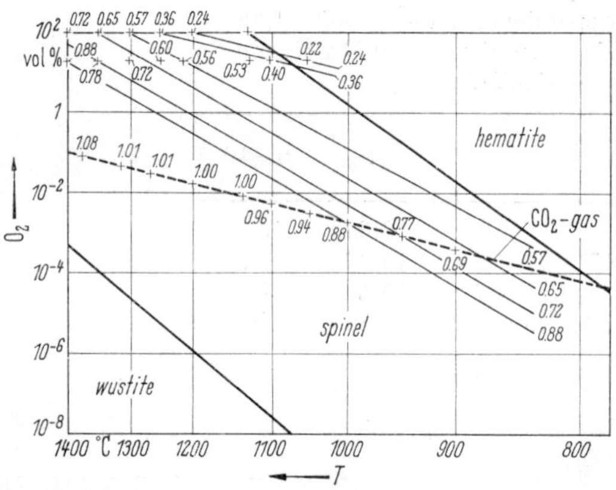

Fig. 19. Equilibrium atmospheres and phase diagram for 52.7 Fe_2O_3, 30.5 MnO_2, 16.8 ZnO (in mole%). The heavy lines are solid phase boundaries. Light lines give the temperature dependence of oxygen content on the firing atmosphere for constant Fe^{2+} content in the spinel phase. Numerical values are the Fe^{2+} wt % of the spinel phase obtained after quenching of the samples. The dotted line gives the equivalent percentage of oxygen for a CO_2 atmosphere. After [*58B*].

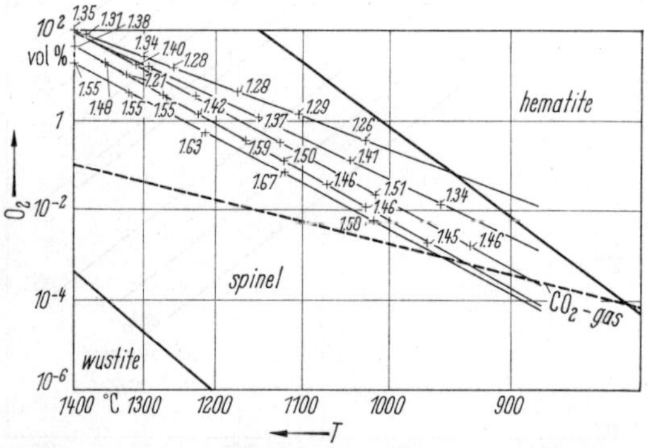

Fig. 20. Equilibrium atmospheres and phase diagram for 52.7 Fe_2O_3, 29.3 MnO_2, 16.1 ZnO (in mole), i.e. 53.7 % Fe_2O_3, 29.9 mol % MnO_2, 16.4 mol% ZnO. Same conditions as for Fig. 19. From [*58B*].

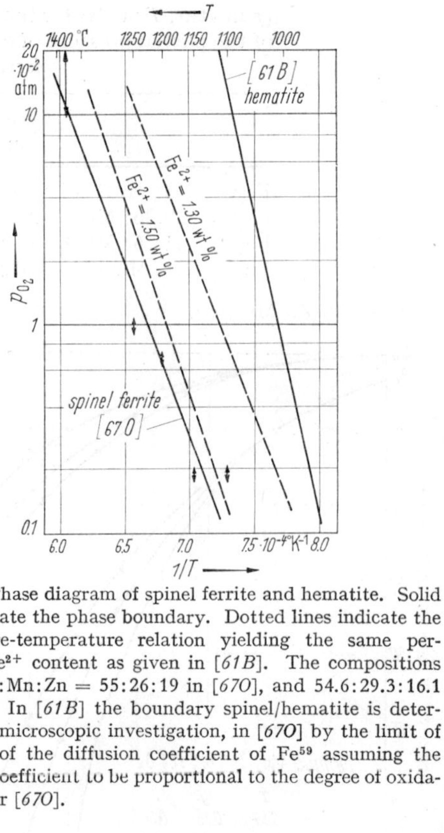

Fig. 21. Phase diagram of spinel ferrite and hematite. Solid lines indicate the phase boundary. Dotted lines indicate the atmosphere-temperature relation yielding the same percentage Fe^{2+} content as given in [*61B*]. The compositions were Fe/2:Mn:Zn = 55:26:19 in [*67O*], and 54.6:29.3:16.1 in [*61B*]. In [*61B*] the boundary spinel/hematite is determined by microscopic investigation, in [*67O*] by the limit of variation of the diffusion coefficient of Fe^{59} assuming the diffusion coefficient to be proportional to the degree of oxidation. After [*67O*].

g) System V-Mn-Fe-O

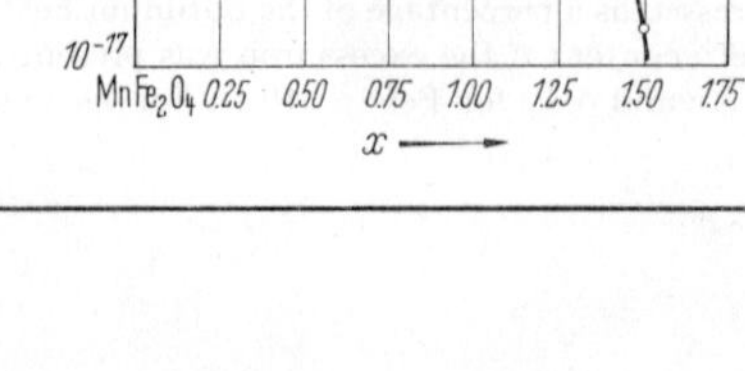

Fig. 22. $MnFe_{2-x}V_xO_4$. Equilibrium oxygen pressure, p_{O_2}, vs. V content, x, at 1000 °C. From [*67LBSC*].

6.1.3.1.3 Lattice constants

a) System Mn-Fe-O Mn/Fe ratio dependence of the cell constant:

Tab. 12. $Mn_xFe_{3-x}O_4$. Variation of the lattice constant [*65TSGZG*]. See also [*66C*]. Probably the irregular variation of the lattice constant in the system $Mn_xFe_{3-x}O_4$ is due to the fact that, when Fe in magnetite is replaced by Mn, the latter enters the lattice primarily in the form of Mn^{2+} up to the composition $MnFe_2O_4$, and then in the form of Mn^{3+}. This irregular variation is not obvious in Fig. 24

Molar fractions		Lattice constants [Å]			Phase composition
Mn_3O_4	Fe_3O_4	*a*	*c*	*c*/*a*	
—	1.0	8.396			spinel
0.1	0.9	8.432			spinel
0.2	0.8	8.472			spinel
0.3	0.7	8.503			spinel
0.33	0.67	8.516			spinel
0.4	0.6	8.519			spinel
0.5	0.5	8.523			spinel
0.6	0.4	8.525		1.05 ··· 1.06	spinel + a little of a tetragonal phase
0.67	0.33	8.525		1.05 ··· 1.06	spinel + a little of a tetragonal phase
0.7	0.3	8.525		1.05 ··· 1.06	spinel + tetragonal phase
0.8	0.2	8.169	9.397	1.1503	tetragonal phase + a little spinel
0.9	0.1	8.160	9.425	1.1551	tetragonal phase + very little spinel
1.0	—	8.145	9.464	1.1619	tetragonal phase

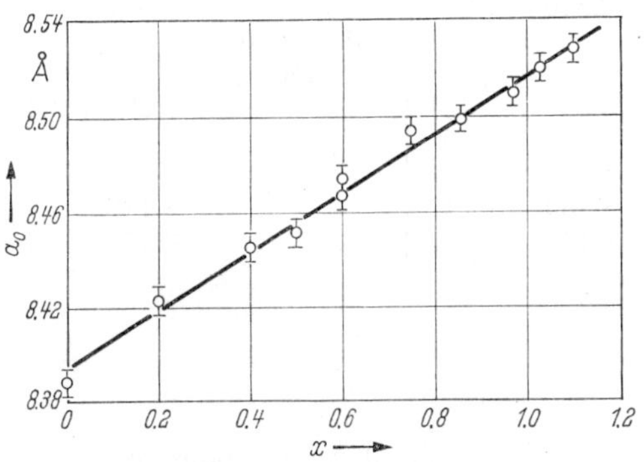

Fig. 23. $Mn_xFe_{3-x}O_4$. Lattice constant, a_0, vs. Mn content, x. From [*59FMU* (2)].

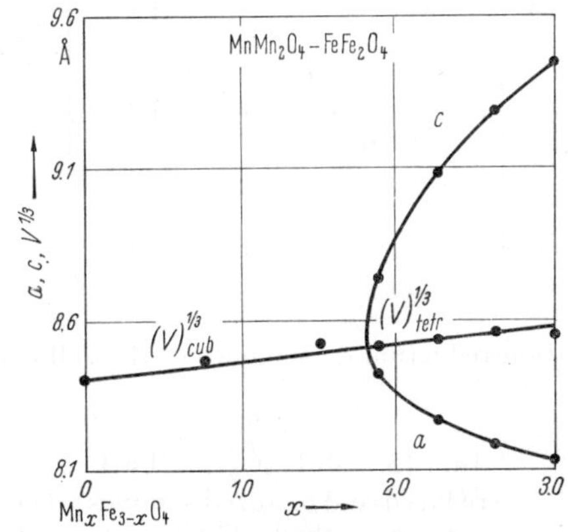

Fig. 24. $Mn_xFe_{3-x}O_4$. Cell dimensions vs. Mn content, x [*57FSS*]. This includes some data from [*47M*]. The transition occurs at the unit cell composition $Fe^{3+}_{3.2}Mn^{2+}_{4.8}[Mn^{3+}_{9.6}Fe^{2+}_{3.2}Fe^{3+}_{3.2}]O^{2-}_{32}$.

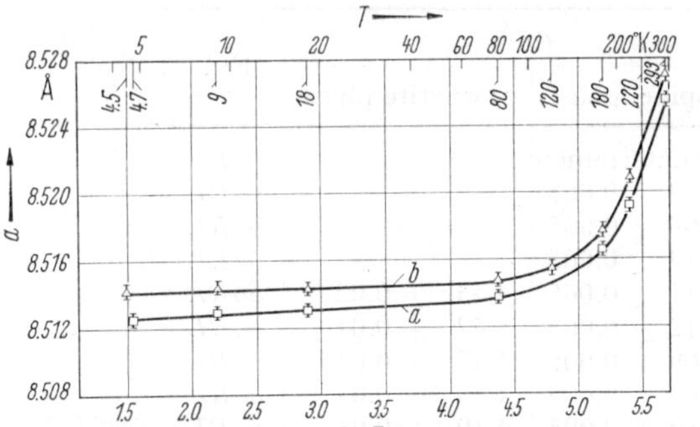

Fig. 25. $Mn_xFe_{3-x}O_4$. The lattice parameter, a, of two samples from the Mn ferrite system vs. ln T [*68CV*]. T in °K. a) x = 1.66; polycrystalline sample; sintered at 1230 °C for 6 h at an air pressure of about 2 mm Hg and cooled slowly from 1200 °C in Ar having a pressure of about 10 mm Hg. b) x = 1.88; prepared by the Verneuil method at 1580 °C and cooled slowly from approximately 1200 °C.

Tab. 13. $Mn_xFe_{3-x}O_4$. Lattice parameters of spinel phase coexisting with the wustite phase; the samples were quenched from 1000 °C [*65TSG*]. See also equilibrium pressure of oxygen, Tab. 1

x	a Å
0.000	8.396
0.225	8.423_2
0.450	8.451
0.750	8.482
0.870	8.501
0.915	8.506
1.000	8.515
1.200	8.518
2.130	a = 8.526 c/a = 1.1619
3.00	a = 8.145 c/a = 1.1619

Lattice parameters as a function of p_{O_2}:

Tab. 14. $Mn_xFe_{3-x}O_{4+\gamma}$. Lattice parameters and phases for samples heated 4 h at 1400 °C in different atmospheres of equilibrium oxygen pressure, p_{O_2}, and then subjected to abrupt quenching by dropping into cold water [*65KOST*]

p_{O_2}	0.001 atm			0.21 atm			1 atm		
x	Phase composition	a Å	c/a	Phase composition	a Å	c/a	Phase composition	a Å	c/a
0.000	spinel	8.397	1	spinel	8.389	1	hematite	—	
0.1788		8.417	1		8.413	1	spinel	8.413	1
0.3172		8.436	1		8.432	1		8.432	1
0.4302		8.459	1		8.448	1		8.446	1
0.6678		8.483	1		8.483	1		8.482	1
0.7514		8.490	1		8.489	1		8.487	1
0.9480		8.513	1		8.511	1		8.509	1
0.9612		8.514	1		8.514	1		8.514	1
0.9920		8.516	1		8.516	1		8.516	1
1.0309		8.518	1		8.518	1		8.518	1
1.2375		8.522	1		8.522	1		8.522	1
1.5650		8.525	1		8.525	1		8.525	1
1.8725		8.528	1		8.528	1		8.529	1
2.2510		8.528	1	spinel + tetragonal phase	8.310	1.084	tetragonally distorted spinel	8.301	1.084
2.6065	tetragonally distorted spinel	8.196	1.126	tetragonally distorted spinel	8.197	1.124		8.208	1.124
2.7767					8.159	1.142		8.148	1.144
3.0000		8.138	1.160		8.136	1.160		8.136	1.160

Oxidation-reduction dependence of the cell constant:

Tab. 15. $MnFe_2O_{4+\gamma}$. Lattice constants, phases, and chemical compositions of reduced and oxidized samples. The initial Mn ferrite $MnFe_2O_{4.02}$ was produced by a ceramic method. The samples were sintered in air at 1350 °C for 5 h and quenched in a stream of Ar from 1150 °C. The amount of oxygen in the samples was determined according to the weight loss after reduction with hydrogen to a mixture of Fe and manganous oxide

Sample	Phase composition	a [Å] of spinel phase	a [Å] of wustite phase	Ref.
$MnFe_2O_{4.30}$	spinel + Fe_2O_3, Mn_2O_3	not determined	—	*67RL*
$MnFe_2O_{4.21}$	spinel + Fe_2O_3	8.512 ± 0.005	—	*67RL*
$MnFe_2O_{4.06}$	spinel	8.509 ± 0.005	—	*67RL*
$MnFe_2O_{4.02}$	spinel	8.513 ± 0.005	—	*67RL*
$MnFe_2O_{3.99}$	spinel + wustite	8.511 ± 0.005	4.43 ± 0.02	*66RL*
$MnFe_2O_{3.98}$	spinel + wustite	8.512 ± 0.005	4.42 ± 0.01	*66RL*
$MnFe_2O_{3.96}$	spinel + wustite	8.509 ± 0.005	4.42 ± 0.01	*66RL*
$MnFe_2O_{3.77}$	spinel + wustite	8.486 ± 0.005	4.40 ± 0.01	*66RL*
$MnFe_2O_{3.72}$	spinel + wustite	8.480 ± 0.005	4.39 ± 0.01	*66RL*
$MnFe_2O_{3.25}$	wustite	—	4.342 ± 0.005	*67RL*
$MnFe_2O_{3.02}$	wustite	—	4.362 ± 0.005	*67RL*
$MnFe_2O_{2.84}$	wustite + Fe	—	4.366 ± 0.0005	*66RL*
$MnFe_2O_{2.27}$	wustite + Fe	—	4.380 ± 0.0005	*66RL*
$MnFe_2O_{1.92}$	wustite + Fe	—	4.393 ± 0.0005	*66RL*
$MnFe_2O_{1.32}$	wustite + Fe	—	4.416 ± 0.0005	*66RL*

Temperature variation of the lattice parameter:

Tab. 16. $Mn_xFe_{3-x}O_{4+\gamma}$. Comparison of lattice constants, a, measured at high temperature, and at RT after annealing and hardening. The differences in the slope below and above $x = 1$ is attributed to the fact that for $x < 1$ Mn^{2+} ions ($r = 0.91$ Å) are substituted with growing x for Fe^{2+} ions ($r = 0.83$ Å) while at $x > 1$ the substitution mechanism takes place between Fe^{3+} ($r = 0.67$ Å) and Mn^{3+} ions ($r = 0.70$ Å). The differences in slope and values between high temperature results and the results for quenched samples at the same annealing temperature are ascribed to changes in the distribution of cations, Jahn-Teller distortion, and the exchange of electrons $Mn^{2+} + Fe^{3+} \rightleftarrows Mn^{3+} + Fe^{2+}$. From [*63B*].
For influence of calcining parameters and oxygen partial pressure on the redox-equilibrium and on the unit cell, see [*59B* (1)] and [*59B* (2)]

x	High temp. X-ray at		Hardened samples at RT; annealing temperature:		
	1210 °C	1250 °C	1210 °C	1260 °C	1310 °C
	a [Å] (±0.003 Å)				
0.522	—	—	8.455	8.458	8.452
0.780	8.591	8.601	8.488	8.491	8.491
			8.488		
0.852	8.602	8.603	8.503	8.502	8.501
0.953	8.605	8.609	8.516	8.514	8.516
		8.610			
1.000	8.611	8.615	8.513	8.510	8.511
			8.512	8.510	8.510
			8.513		
1.107	8.611	6.618	8.514	8.510	
			8.514		8.516
1.119	8.614	8.617	8.512	8.517	8.516
				8.518	
1.200	8.614	8.617	8.511	8.516	8.521
					8.524
1.368	8.616	8.617	8.514	8.523	8.525
1.577	8.618	8.622	8.517	8.523	8.523
				8.524	8.523
1.967	8.641	8.626	8.530	8.543	8.525
			8.529		

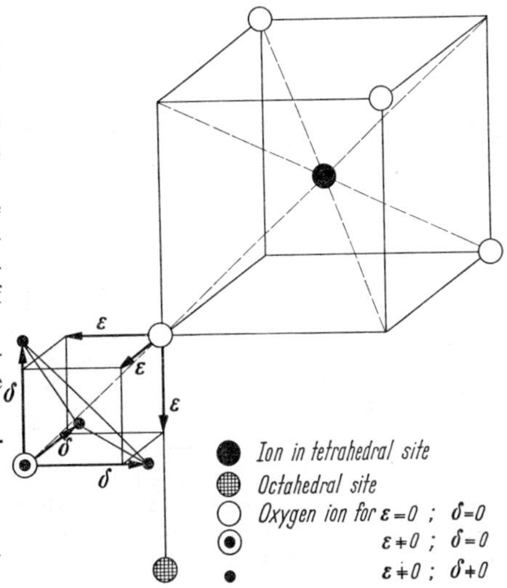

Fig. 27. $Mn_{1.66}Fe_{1.33}O_4$. The ε and δ parameters [*65C* (3)]. In this case of Mn ferrite the ε parameter in $u = 0.375 + \varepsilon$ is not sufficient for the description of the lattice. Another parameter δ was introduced as defined in this figure. This δ parameter is assumed to have a close connection with the local Jahn-Teller distortion of $Mn^{3+}O_6^{2-}$ octahedron. The value of δ enables not only the estimation of the number of distorted octahedrons in the elementary cell, but also their distortion.

Distortion of the lattice:

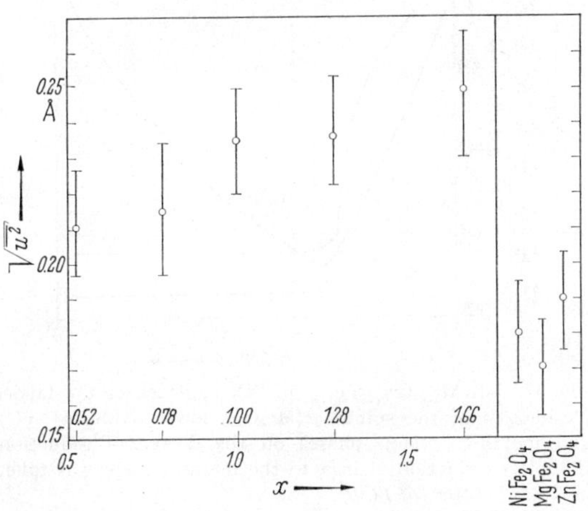

Fig. 26. $Mn_xFe_{3-x}O_4$. Square root of the mean square amplitude of lattice vibrations measured in the cubic part of the Mn-Fe-O system as a function of Mn per formula unit [*65C* (1)]. Comparison of the values of $\overline{u^2}$ measured in the Mn ferrite system with the values obtained on other ferrospinels (Ni, Mg, Zn) enables the separation of the displacement component $\overline{u_D^2}$ from the thermal component $\overline{u_T^2}$. The existence of the component $\overline{u_D^2}$ is explained by the tendency of Mn^{3+} ions in octahedral sites to distort their immediate neighbourhood in the lattice by the Jahn-Teller effect. Same measurement on single crystal and for $x = 1.66$. [*65C* (2)] gives lower value of $\overline{u^2}$. In the same way, the use of a monochromator [*62C*] or of a counter changes $\overline{u^2}$, but the displacement component $\overline{u_D^2}$ remains the same. $\overline{u_D^2}$ may also be obtained from the difference between $\overline{u^2}$ for reflections which do not depend on the atomic factor of oxygen at all and $\overline{u^2}$ for those which are a function of all three atomic factors: (f Mn, f O, f Fe).

b) System Mn-Fe-O-F

The unit cell dimension of fluorine containing spinel type ferrite $\mathbf{Mn_{0.50}Fe^{2+}_{0.15}Fe^{3+}_{2.00}O_{3.3}F_{0.7}}$ fired between 1100 ··· 1300 °C under CO_2 is 8.462 Å. From [*64S2*].

c) System Al-Mn-Fe-O

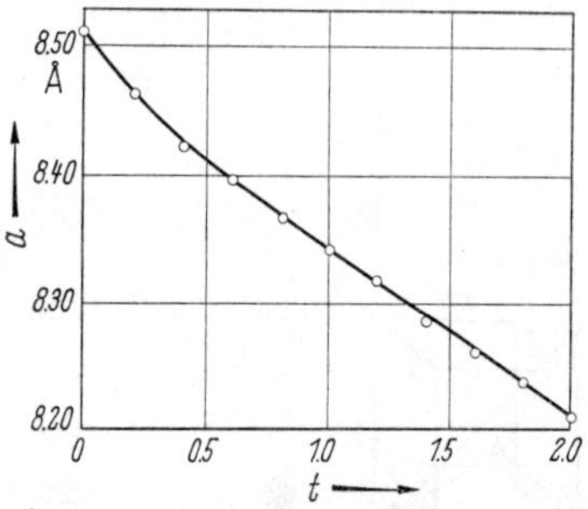

Fig. 28. $MnAl_tFe_{2-t}O_4$. Cell edges, *a*, vs. Al content, *t*. Materials prepared starting from $MnCO_3$, $Al(OH)_3$, Fe_2O_3. Heat treatment 4 h, 1300° C in nitrogen. Curie temperature, Θ_C, see Fig. 34 in 6.1.3.5. From [*65B*].

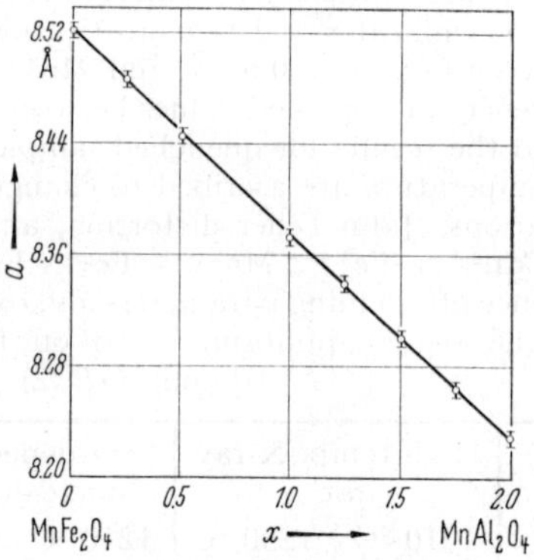

Fig. 29. $MnAl_xFe_{2-x}O_4$. Lattice parameter, *a*, vs. Al content, *x*, of the solid solutions prepared by sintering mixtures of the oxides MnO, Fe_2O_3 and $Al(OH)_3$ pressed into tablets at 1275 °C for 15 h in an atmosphere of purified nitrogen, followed by cooling in the same atmosphere [*65TBC*].

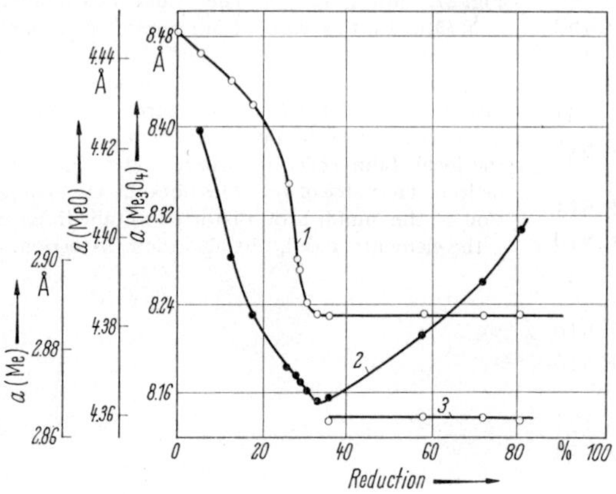

←

Fig. 30. $MnAl_{0.25}Fe_{1.75}O_4$. Variation of the lattice parameter, *a*, of the phases 1) spinel (Me_3O_4) 2) lower oxide (MeO) 3) metallic (Me) as a function of the degree of reduction [*65TBC*]. 100% correspond to reduction of $MnAl_{0.25}Fe_{1.76}O_4$ to $MnAl_2O_4$, MnO, and Fe. See also Tab. 8, 9 for the equilibrium oxygen partial pressure and phase composition.

d) System Cr-Al-Mn-Fe-O

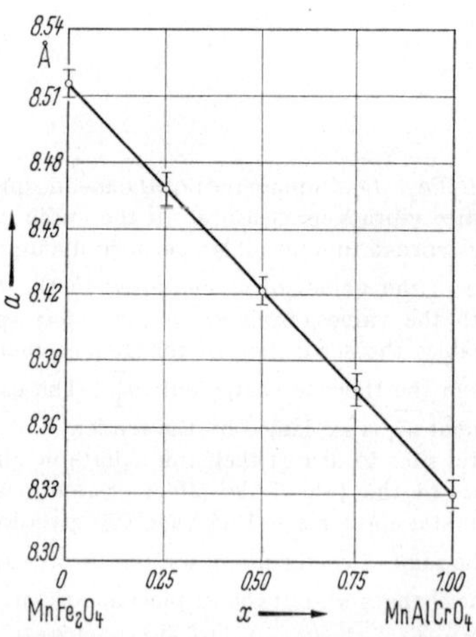

Fig. 31. $MnAl_xCr_xFe_{2-2x}O_4$. Lattice parameter, *a*, vs. *x* in solid solutions. The samples were synthesized by sintering for 15 h at 1275 °C in nitrogen and cooled together with the furnace in an inert atmosphere. From [*66TCB*].

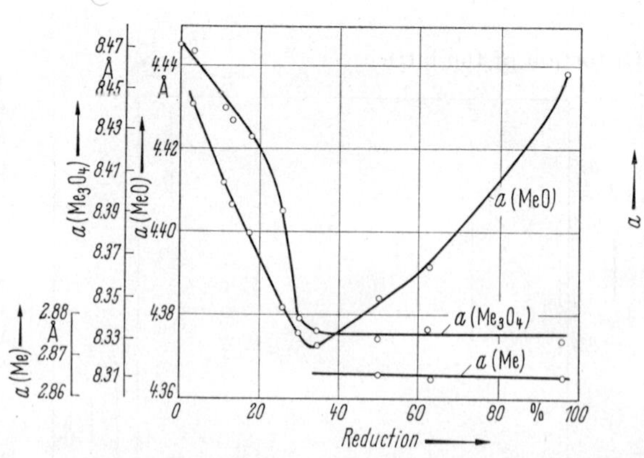

Fig. 32. $MnAl_{0.25}Cr_{0.25}Fe_{1.50}O_4$. Dependence of the lattice parameters of the spinel, $a(Me_3O_4)$, lower oxide, a(MeO), and metallic, a(Me), phases on the degree of reduction. Complete reduction of iron to the metallic state was taken as 100%. After [*66TCB*].

e) System Co-Mn-Fe-O

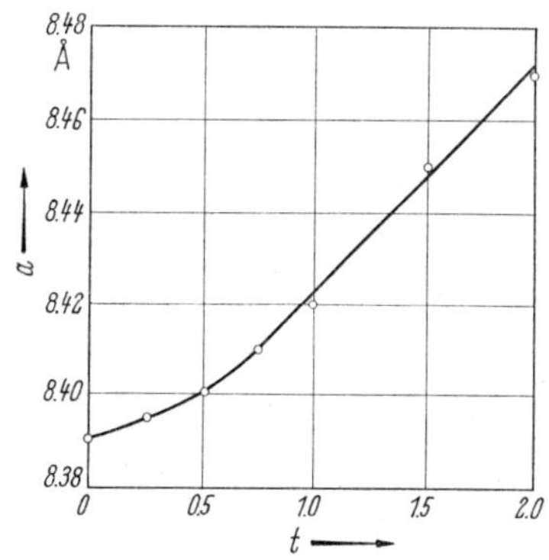

→

Fig. 33. $CoMn_tFe_{2-t}O_4$. Cell edges, a, vs. Mn content, t. For $t \geqq 1.5$, the value of $(a^2c)^{1/3}$ has been plotted. Materials prepared starting from $CoCO_3$, $MnCO_3$, Fe_2O_3. Heat treatment: 4h, 1200 °C in air. Curie temperature, Θ_C, see Fig. 29 in 6.1.3.5. From [*65B*].

f) System Cu-Mn-Fe-O

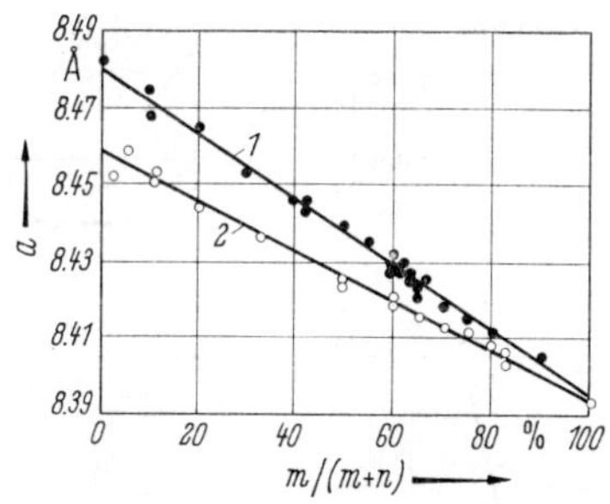

Fig. 34. $m(CuOFe_2O_3)\cdot n(MnOFe_2O_3)$. Crystal lattice parameter, a, of Cu-Mn ferrites vs. composition. From [*65AMY*]. Curve 1 is for specimens prepared on a base of Cu_2O, Fe_2O_3, MnO_2 first annealed at 1000 °C and then at 1200 °C.
Curve 2 is for cupromanganese ferrites produced on a base of CuO, Fe_2O_3, MnO_2 after roasting at 1200 °C. Even with slight departures from the processes used, the spread of results was considerable.

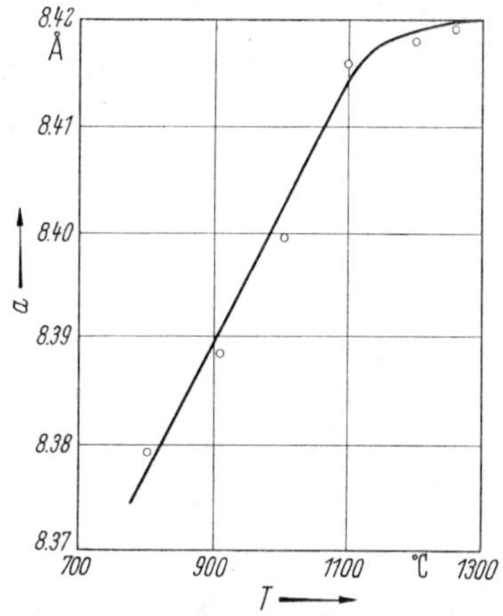

Fig. 35. $3(CuOFe_2O_3)\cdot 2(MnOFe_2O_3)$. Crystal lattice parameter, a, as a function of annealing temperature. After [*65AMY*].

g) System Li-Mn-Fe-O

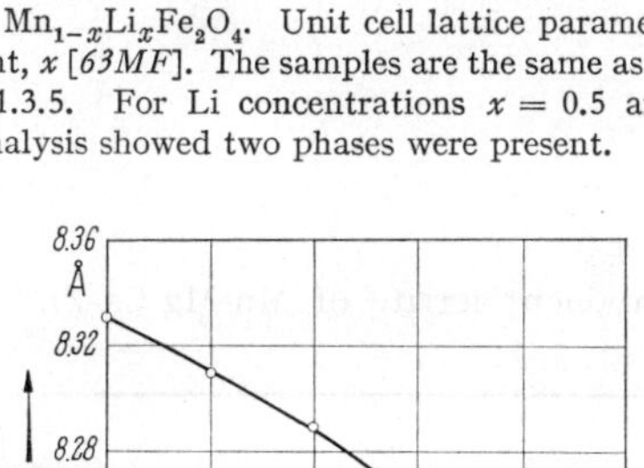

Fig. 36. $Mn_{1-x}Li_xFe_2O_4$. Unit cell lattice parameter, a, vs. Li content, x [*63MF*]. The samples are the same as for Fig. 7, 16 in 6.1.3.5. For Li concentrations $x = 0.5$ and above, X-ray analysis showed two phases were present.

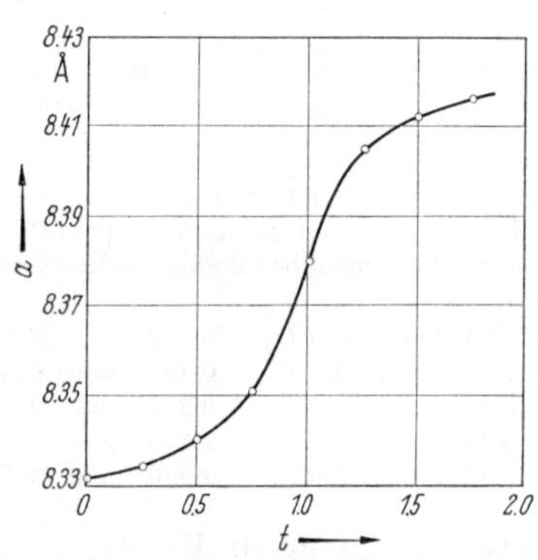

Fig. 38. $Li_{0.5}Mn_tFe_{2.5-t}O_4$. Cell edges, a, vs. Mn content, t. Materials were prepared starting from Li_2CO_3, $MnCO_3$, Fe_2O_3. Heat treatment 6h, 1100 °C in air. Curie temperature, Θ_C, see Fig. 33 in 6.1.3.5. After [*65B*].

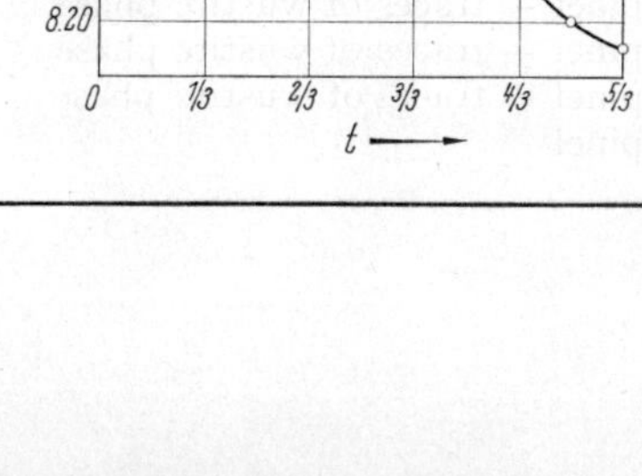

←

Fig. 37. $Li_{0.5+0.5t}Fe_{2.5-1.5t}Mn_tO_4$. Cell edges, a, vs. composition. Materials with the above formula were prepared starting from Li_2CO_3, Fe_2O_3, $MnCO_3$; heat treatment: 24 h, 750 °C, O_2. Ionic distribution see Fig. 51 in this section. From [*65B*].

h) System Ni-Mn-Fe-O

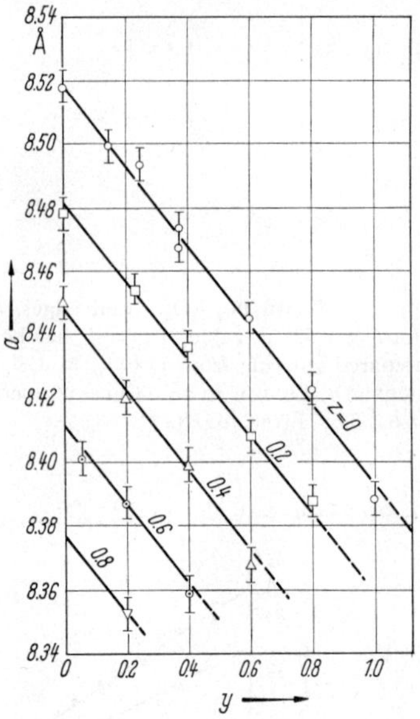

Fig. 39. $(Mn_xFe_yNi_z)Fe_2O_4$. Lattice constants, a, vs. composition of the solid solution. (The lattice constant of NaCl, $a = 5.640$ Å, is chosen as the standard). Lines show $a = (8.518 - 0.125y - 0.176z)$ Å, in which z is taken as a parameter [*61M*]. →

i) System Mg-Mn-Fe-O

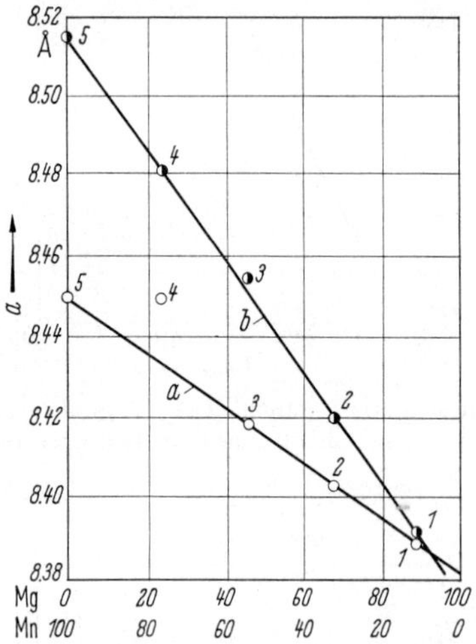

Fig. 40. Lattice constant, a, plotted against the molecular ratio Mg/Mn of the spinel ferrite with an amount of 0.444 mole of Fe_2O_3. Curve (a) after annealing in air at 1180 °C and (b) at 1480 °C [*60B*]. Care must be taken about the total number of molecules.

Plots Nr. 1 0.0692 mole $MnCO_3$ – 0.582 mole $MgCO_3$
2 0.186 mole $MnCO_3$ – 0.387 mole $MgCO_3$
3 0.304 mole $MnCO_3$ – 0.2512 mole $MgCO_3$
4 0.423 mole $MnCO_3$ – 0.1307 mole $MgCO_3$
5 0.556 mole $MnCO_3$ – 0.000 mole $MgCO_3$

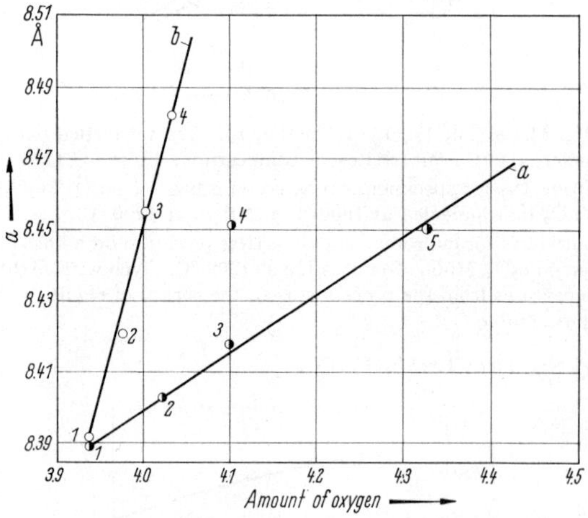

Fig. 41. Lattice parameter, a, vs. the total amount of oxygen for the same compositions and annealing temperatures as for Fig. 40. This total amount of oxygen and the valence of Mn-ion increase with the ratio Mg/Mn. From [*60B*].

j) System Zn-Ca-Mg-Mn-Fe-O

Tab. 17. Lattice constants and phases for polycomponent ferrite of Mn-Mg-Ca-Zn [*65TSGZG*]

Composition according to initial oxides mol %					a Å	Phase composition
Fe_2O_3	MnO	MgO	CaO	ZnO		
43	43	14	—	—	8.485	spinel + traces of wustite phase
41.76	41.57	14.84	1.83	—	8.495	spinel + traces of wustite phase
37	38	10	3	12	8.493	spinel + traces of wustite phase
39.26	22.09	38.67	—	—	8.448	spinel

k) System Zn-Mn-Fe-O

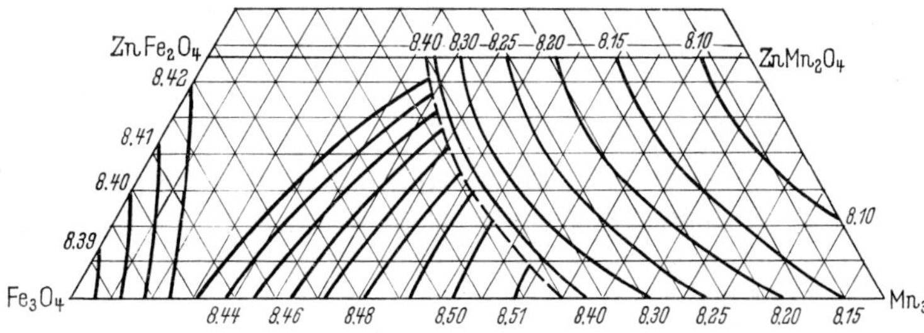

→
Fig. 42. Relationship between composition and lattice constant a_0 for cubic and tetragonal lattices. From [*47M*]. See also Fig. 6

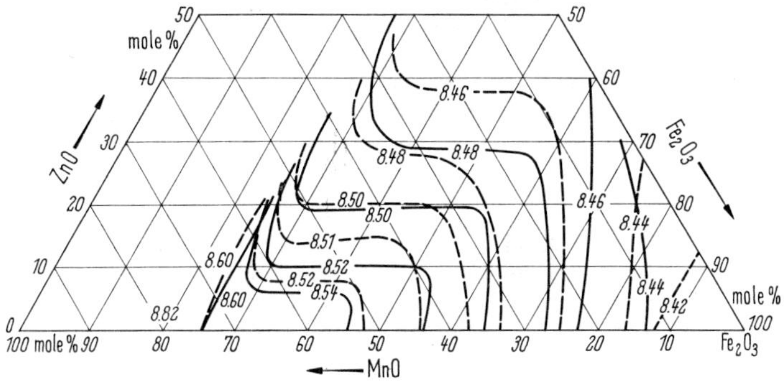

←
Fig. 43. Contours of the lattice constants for Mn-Zn-Fe ferrites. The solid lines represent the data for the specimens sintered in nitrogen and the broken lines data for the specimens sintered in air. The moulded samples were heated in air for 2h at 1400 °C and cooled to 800 °C at a cooling rate of 400 °C per hour and then quenched in air from 800 °C (broken lines) or heated in nitrogen gas in the same way, except the specimens were slowly cooled in the furnace to RT (solid lines). From [*61S*].

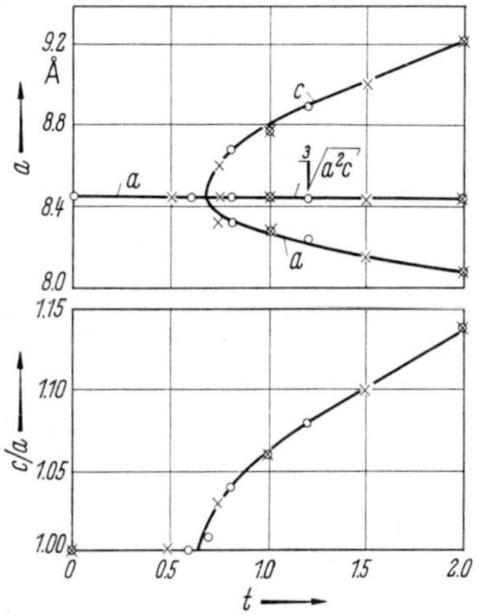

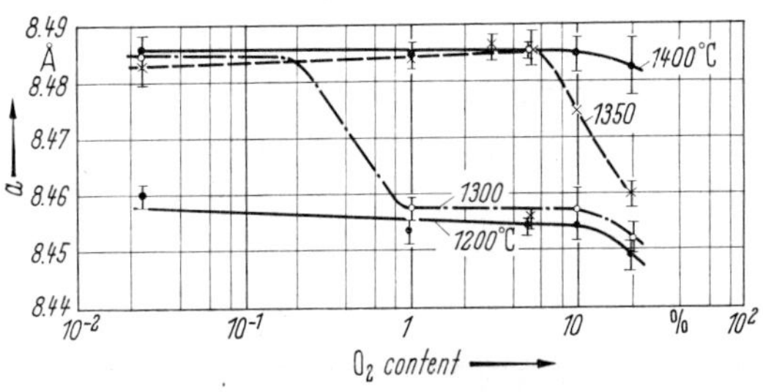

Fig. 45. Lattice constant, *a*, of MnZn ferrite with a starting composition containing 54.3 mole % Fe_2O_3, 27.9 mole % MnO, 17.8 mole % ZnO as a function of the oxygen content of the firing atmosphere. The samples were maintained at the peak temperature for 4 h, then air quenched [*64MJ*].

←
Fig. 44. $ZnMn_tFe_{2-t}O_4$. Cell edges and *c*/*a* ratio vs. Mn content, *t*. Materials were prepared starting from ZnO, $MnCO_3$, Fe_2O_3. Heat treatment 4h, 1200 °C in air. Crosses from [*65B*], circles after [*61O*].

l) System Rh-Mn-Fe-O

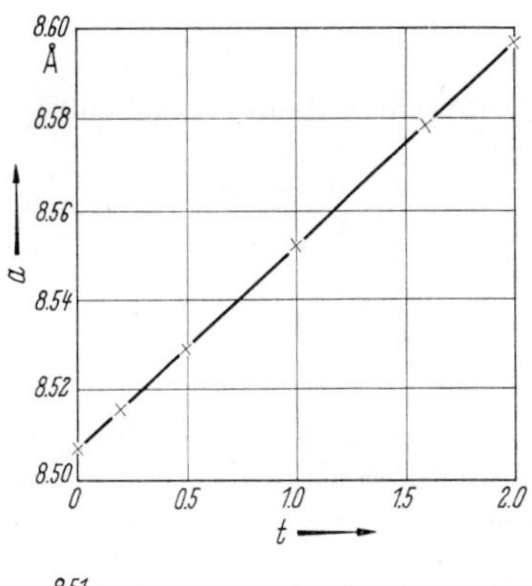

→
Fig. 46. $MnRh_tFe_{2-t}O_4$. Cell edges, *a*, vs. Rh content, *t*. Materials prepared starting from $MnCO_3$, Fe_2O_3, Rh_2O_3. Heat treatment 24 h, 1150 °C in air. Curie temperature, Θ_C, see Fig. 31 in 6.1.3.5. After [*65B*].

m) System V-Mn-Fe-O

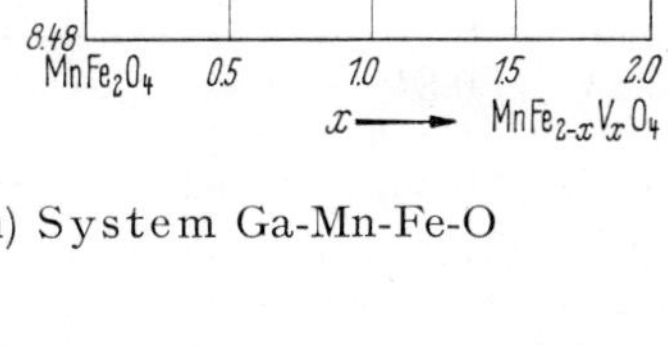

←
Fig. 47. $MnFe_{2-x}V_xO_4$. Lattice parameter, *a*, vs. V content, *x*. After [*67LBSC*]. Vegard's rule is satisfied by the almost uniform distribution of cations in vanadite and ferrite lattice; MnV_2O_4 has a normal structure [*54RBR*], [*64APPZL*], while $MnFe_2O_4$ is no more than 20% inverse. [*62P*]. See also [*57LM*], [*59L(2)*].

n) System Ga-Mn-Fe-O

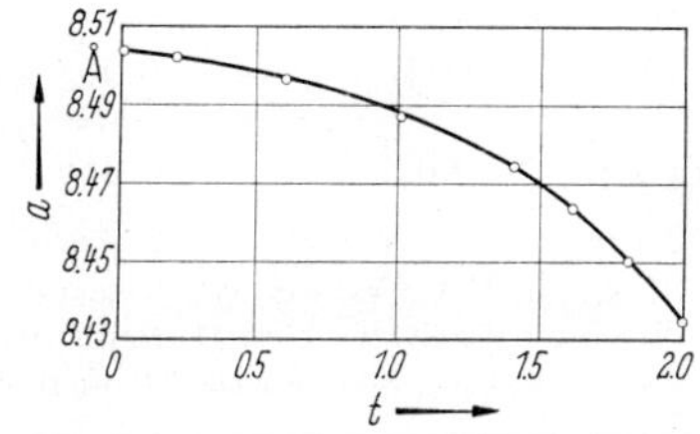

→
Fig. 48. $MnFe_{2-t}Ga_tO_4$. Lattice constant, *a*, vs. Ga content, *t*. Curie temperature, Θ_C, see Fig. 32 in 6.1.3.5. [*57LM*].

6.1.3.1.4 Ionic distribution

a) System Mn-Fe-O

Though $\mathbf{MnFe_2O_4}$ has the spinel structure, some members of the series $\mathbf{Mn_xFe_{3-x}O_4}$ near $x = 3$ are tetragonally distorted. Except for this and the determination of the oxygen parameter the fundamental problem which dominates all others with regard to the $\mathbf{Mn_xFe_{3-x}O_4}$ has been that of ionic distribution and spin arrangement.

Neutron diffraction:

The distribution of ions of $\mathbf{MnFe_2O_4}$ produced from aqueous solution was found by neutron diffraction [*67SS*] to be $\mathbf{Mn_{0.33}Fe_{0.67}(Mn_{0.67}Fe_{1.33})O_4}$ (round brackets refer to ions in tetrahedral *A* sites) where the possible error is 0.03. The ε oxygen parameter (see Fig. 42) is 0.003 ± 0.003 and the Debye-Waller factor $(1.1 \pm 0.5) \cdot 10^{-16}\ cm^2$. These results support the marked difference found by using NMR techniques [*66Y, 67YHSKBT*] for the ionic distribution of Mn ferrite prepared from water solution from that of sintered samples and confirm the belief that ionic distributions of Mn ferrite samples differ according to the methods of preparation of samples.

Tab. 18. $MnFe_2O_4$. Results of the neutron diffraction experiments on a number of samples at 300 °K and 4.2 °K [*56HC*]. Values for *z*, the fraction of *A* sites occupied by Mn^{2+}, and *u*, the oxygen space group parameter, were determined from the ratios of the intensities of various pairs of reflections. The values of magnetization, *z* and *u*, are essentially the same for the three samples though somewhat different preparative procedures were used. A quantitative analysis of sample C shows 23.36% Mn and 48.65% Fe compared with the stoichiometric composition of 23.82 %Mn and 48.44 %Fe. The analysis corresponds to $Mn_{0.972}Fe_{1.992}O_4$

Sample		A 1)	B 2)	C 3)	
Firing conditions		4 h 1200 °C N_2 atmosphere	6 h 1300 °C 2000 mil/min N_2 5 mil/min O_2 slow cooling in N_2	24 h 1400 °C air atmosphere water quench	
Lattice constant				8.517	Å
Magnetization:	*T*				
p_A, *A* lattice	4.2 °K			4.60	μ_B
	300 °K			4.33	μ_B
p_A, *B* lattice	4.2 °K			4.60	μ_B
	300 °K			3.78	μ_B
p_m (neutron diffraction)	4.2 °K	4.6	4.6	4.60 ± 0.2	μ_B
	300 °K			3.23 ± 0.3	μ_B
p_m (ballistic)	4.2 °K			4.60	μ_B
	300 °K			3.27	μ_B
ξ (fraction of *A* sites occupied by Mn^{++})		0.79	0.82	0.81 ± 0.03	
u (oxygen parameter)		0.385	0.386	0.3846 ± 0.0003	

Thus we may conclude from these experiments that $MnFe_2O_4$ is 80 or 90% normal spinel.

Tab. 19. $Mn_xFe_{3-x}O_4$. Occupation of the tetrahedral sites by Mn, ξ, as a function of the chemical composition, *x*, from a neutronographic analysis [*62YBN*]. Note the unusual values of the *u* parameter

x	0.43	0.53	0.84	1.03	1.06	1.10	1.28	1.50	1.58
ξ ($\Delta\xi = \pm 0.04$)	0.00	0.00	0.78	0.80	0.81	0.83	0.84	0.89	0.92
u ($\Delta u = \pm 0.02$)	0.256	0.258	0.258	0.260	0.260	0.261	0.261	0.262	0.263

1) From Dr. V. C. Wilson, GE Research Laboratory.
2) From Dr. L. G. van Uitert, Bell Telephone Laboratories.
3) From Dr. C. J. Kriessman, Remington Rand Corp.

Mn^{55} NMR in $MnFe_2O_4$:

From [*63HGB*]: the nuclear magnetic resonance frequency in a sphere of $\mathbf{MnFe_2O_4}$ at 4.2 °K in an external field of 2730 Oe was 586.0 Mc/s. The identification of the 586.0 Mc/s line as being associated with the Mn^{2+} ion was made by comparing the observed frequency with measurements of the hyperfine constant obtained from paramagnetic resonance. The absence of quadrupole splitting or broadening of the observed line indicated the Mn^{2+} ions to be on the *A* sites.

From [*63YAMH*]: the spin echoes on Mn^{55} in $\mathbf{MnFe_2O_4}$ were observed at about 582 Mc/s and are considered to be due to Mn^{2+} ion located at *A* sites. From [*66Y*]: in $\mathbf{MnFe_2O_4}$ prepared by usual ceramic method, the only signal observed is identified as Mn^{55} NMR associated with Mn^{2+} ions located at *A* sites. In $\mathbf{MnFe_2O_4}$ prepared by precipitation from ferric and manganous liquid solution, the Mn^{55} NMR is identified with Mn^{2+} ions located at *B* sites.

Mössbauer absorption spectra of $\mathbf{MnFe_2O_4}$: the ratio of the number of Fe ions in *A* and *B* sites was found to be 0.12 ± 0.02 [*67SVM*], in agreement with the cation distribution determined from neutron diffraction experiments, as $\mathbf{(Mn_{0.8}Fe_{0.2})Mn_{0.2}Fe_{1.8}O_4}$, ($0.12 \simeq 0.2/1.8$).

According to the Jahn-Teller effect Mn^{3+} is stabilized in octahedral sites with a stabilization energy of about 0.25 eV. Similarly, Fe^{2+} tends to lower the symmetry in octahedral positions and thus acquires an extra-stabilization (of some 0.1 eV). One may expect that if the Mn^{3+} and Fe^{2+} ions are situated in the nearest neighbouring octahedral sites the two distortions might support each other. The resulting stabilization of the coupled pair Mn^{3+}-Fe^{2+} may considerably exceed the simple sum of the stabilization energies of the individual ions [*65S*].

Other results:

Composition	ξ	u	Ref.
$Mn_{1.01}Fe_{1.99}O_{3.999}$	0.81 ± 0.03	0.3848 ± 0.0008	*65KWKP*
$Mn_{1.30}Fe_{1.97}O_4$	0.80 ± 0.04	0.390 ± 0.03	*61NY*

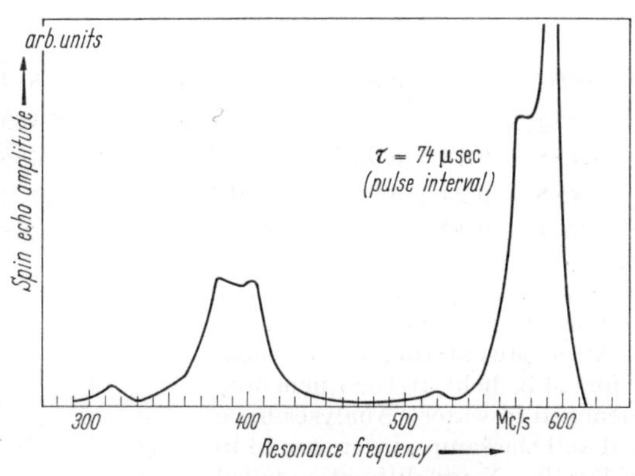

Fig. 49. $MnFe_2O_4$. Spin echo spectra of Mn ferrite at 1.6 °K. The strongest signal between 570 Mc/s and 591 Mc/s was due to Mn^{55} NMR associated with Mn^{2+} ions at *A* sites. The next strong signal around 400 Mc/s was considered to be due to the Mn^{55} NMR associated with Mn^{3+} ions at *B* sites [*67KTAH*].

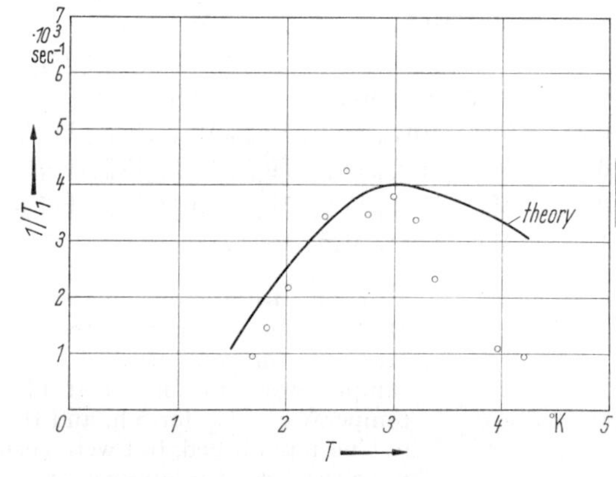

Fig. 50. $MnFe_2O_4$. Temperature dependence of nuclear spin-lattice relaxation time for Mn^{55}. The solid curve is a plot of the theoretical expression for $1/T_1$ using the slow relaxation theory with no adjustable parameters and using the same values for ΔE and $1/\tau$ as Fig. 10 in 6.1.3.9 [*64HBG*].

b) System Zn-Mn-Fe-O

The distribution of cations of sintered $\mathbf{Mn_{0.547}Zn_{0.372}Fe^{2+}_{0.081}Fe^{3+}_{2.000}O_{4.000}}$ between tetrahedral and octahedral sites was determined by neutron diffraction at 575 °K (Curie temperature 150 °C). With the assumption, Zn ions are only in tetrahedral sites, the distribution is:
A sites: 54.7% Mn^{2+}, 37.2% Zn^{2+}, 8.1% Fe^{2+} or Fe^{3+}
B sites: 96···100% Fe^{3+}, 4···0% Fe^{2+}.

The oxygen parameter is $u = 0.385 \pm 0.002$ in agreement with [*56HC*] but [*63KS*] found for the composition $\mathbf{Mn_{0.65}Zn_{0.29}Fe_{2.07}O_4}$ and $u = 0.3881 \pm 0.0008$ an inversion parameter $z = 0.89$.

From the patterns below the Curie point, the following magnetic moments for the ions at *A* and *B* sites were obtained: $p_A = p_B = 2.6\ \mu_B$ (at 295 °K) and $p_A = 2.97\ \mu_B$; $p_B = 4.26\ \mu_B$ (at 80 °K). From [*67K*].

Tab. 20. $Mn_xZn_{1-x}Fe_2O_4$. Cation distribution. Calculations are based on the geometry of the spinel crystal and the lattice constant, *a*. From [*64SS*]. R_A, R_B: Experimental oxygen to cation length

x	a Å	R_A Å	R_B Å	Calculated cation distribution
0	8.443	1.974	2.026	$Zn^{2+}[Fe_2^{3+}]O_4$
0.3	8.466	1.979	2.035	$Zn_{0.70}^{2+}Mn_{0.24}^{2+}Fe_{0.06}^{3+}[Mn_{0.06}^{3+}Fe_{0.06}^{2+}Fe_{1.88}^{3+}]O_4$
0.5	8.4843	1.984	2.040	$Zn_{0.50}^{2+}Mn_{0.41}^{2+}Fe_{0.09}^{3+}[Mn_{0.09}^{3+}Fe_{0.09}^{2+}Fe_{1.82}^{3+}]O_4$
0.7	8.4995	1.987	2.043	$Zn_{0.30}^{2+}Mn_{0.57}^{2+}Fe_{0.13}^{3+}[Mn_{0.13}^{3+}Fe_{0.13}^{2+}Fe_{1.74}^{3+}]O_4$
0.9	8.5144	1.991	2.047	$Zn_{0.10}^{2+}Mn_{0.73}^{2+}Fe_{0.17}^{3+}[Mn_{0.17}^{3+}Fe_{0.17}^{2+}Fe_{1.66}^{3+}]O_4$
1.0	8.5210	1.992	2.048	$Mn_{0.81}^{2+}Fe_{0.19}^{3+}[Mn_{0.19}^{3+}Fe_{0.19}^{2+}Fe_{1.62}^{3+}]O_4$
*) 1.0	8.517	2.018	2.126	$Mn_{0.8}^{2+}Fe_{0.2}^{3+}[Mn_{0.2}^{3+}Fe_{0.2}^{+}Fe_{1.6}^{3+}]O_4$

c) System Cr-Mn-Fe-O and Mg-Mn-Fe-O

Tab. 21. Cation distribution for Cr-Mn ferrites and Mg-Mn ferrites, calculated from the geometry of the spinel crystal and the lattice constant.
ξ = Mn cation distribution in tetrahedral site; η = Mg cation distribution in tetrahedral site

t	Composition and cation distribution	Ref.	Experiment			
			η	ξ	u	a Å
	$Mn_{\xi}^{2+}Fe_{1-\xi}^{3+}[Mn_{1-\xi}^{3+}Fe_{1-\xi}^{2+}Fe_{2\xi-t}^{3+}Cr_t^{3+}]O_4$ **)					
0.5	$Mn_{0.84}^{2+}Fe_{0.16}^{3+}[Mn_{0.16}^{3+}Fe_{0.16}^{2+}Fe_{1.18}^{3+}Cr_{0.50}^{3+}]O_4$	*64SS*		0.84	0.3867 ± 0.0003	8.49
1	$Mn_{0.87}^{2+}Fe_{0.13}^{3+}[Mn_{0.13}^{3+}Fe_{0.13}^{2+}Fe_{0.74}^{3+}Cr_{1.00}^{3+}]O_4$	*64SS*		0.87	0.3883 ± 0.0003	8.46
1.5	$Mn_{0.98}^{2+}Fe_{0.02}^{3+}[Mn_{0.02}^{3+}Fe_{0.02}^{2+}Fe_{0.46}^{3+}Cr_{1.50}^{3+}]O_4$	*64SS*		0.98	0.3892 ± 0.0003	8.45
	$Mg_{\eta t}^{2+}Mn_{\varepsilon(1-t)}^{2+}Fe_{1-\eta t-\varepsilon(1-t)}^{3+}[Mg_{(1-\eta)t}^{2+}$ $Mn_{(1-t)(1-\varepsilon)}^{3+}Fe_{(1-t)(1-\varepsilon)}^{2+}Fe_{2\varepsilon(1-t)+t(\eta+2)-1}^{3+}]O_4$					
0	$Mn_{0.8}^{2+}Fe_{0.2}^{3+}[Mn_{0.2}^{3+}Fe_{0.2}^{2+}Fe_{1.6}^{3+}]O_4$	*56HC*		0.81	0.385	8.517
0.25	$Mg_{0.03}^{2+}Mn_{0.67}^{2+}Fe_{0.30}^{3+}[Mg_{0.22}^{2+}Mn_{0.08}^{3+}Fe_{0.08}^{2+}Fe_{1.62}^{3+}]O_4$	*64SS*	0.12	0.89	0.385	8.485
0.5	$Mg_{0.09}^{2+}Mn_{0.46}^{2+}Fe_{0.45}^{3+}[Mg_{0.41}^{2+}Mn_{0.04}^{3+}Fe_{0.04}^{2+}Fe_{1.51}^{3+}]O_4$	*64SS*	0.18	0.92	0.3839	8.455
0.75	$Mg_{0.18}^{2+}Mn_{0.22}^{2+}Fe_{0.60}^{3+}[Mg_{0.57}^{2+}Mn_{0.03}^{3+}Fe_{0.03}^{2+}Fe_{1.37}^{3+}]O_4$	*64SS*	0.24	0.90	0.3834	8.424
1	$Mg_{0.12}^{2+}Fe_{0.88}^{3+}[Mg_{0.88}^{2+}Fe_{1.12}^{3+}]O_4$	*53CH*	0.12		0.381 ± 0.001	8.395

Tab. 22. Results of neutron diffraction measurements on a series of $Mg_xMn_{1-x}Fe_2O_4$. For all the compositions studied about 90% of the Mn ions were on tetrahedral sites. After presintering twice, these samples were fired in air at 1400 °C for 24 h, held at the quenching temperature, T_Q, for 5 h, and then quenched in water. Analyses have not been published, but were performed and the samples were found to have the correct composition to within 1 ··· 2%. X-ray diffraction failed to disclose the presence of phases other than spinel. From [*57NPHK*]
ξ = fraction of Mn on A sites; η = fraction of Mg on A sites

Sample	T_Q °C	ξ	u ±0.0003	η	a Å
$Mg_{0.25}Mn_{0.75}Fe_2O_4$	1400	0.89 ± 0.05	0.3850	0.12	8.485
	1000	0.78 ± 0.05	0.3848	0.09	8.479
$Mg_{0.5}Mn_{0.5}Fe_2O_4$	1400	0.92 ± 0.05	0.3839	0.18	8.455
	1000	0.82 ± 0.05	0.3836	0.14	8.450
$Mg_{0.75}Mn_{0.25}Fe_2O_4$	1400	0.90 ± 0.07	0.3834	0.24	8.424
	1000	0.98 ± 0.07	0.3831	0.21	8.421
$Mg_{0.9}Mn_{0.1}Fe_2O_4$	1400	1.0 ± 0.15	0.3831	0.25	8.406
	1000	0.90 ± 0.15	0.3821	0.21	8.403

*) Experimental value [*56HC*].
**) The possibility of the existence of Mn^{3+}-Fe^{2+} in octahedral positions according to [*57HOT*] and [*59M*] was taken into account in the Mn ferrite.

Tab. 23. Cation distribution estimated from neutron and X-ray diffraction experiments [*65WK*], [*65KWKP*].
η = amount of Mg ions in tetrahedral sites, ξ = amount of Mn ions in tetrahedral sites.
The values from [*57NPHK*] are from saturation magnetization measurements

	$Mn_{0.57}Mg_{0.74}Fe_{1.70}O_{3.99}$	$Mn_{0.51}Mg_{0.52}Fe_{1.97}O_{4.00}$	$Mn_{0.5}Mg_{0.5}Fe_2O_4$ [*57NPHK*]
u	0.3830 ± 0.0007	0.3824 ± 0.0008	0.3836 ± 0.0003
η	0.21 ± 0.02	0.13 ± 0.02	0.07
ξ	0.26 ± 0.02	0.34 ± 0.02	0.41 ± 0.03

d) System Li-Mn-Fe-O

Kubo et al. [*66KYH*] have observed two NMR signals around 583 Mc/s and 310 Mc/s in $\mathbf{Li_{0.5}Fe_{1.75}Mn_{0.75}O_4}$ at 90 °K. The 583 Mc/s signal should be identified as associated with Mn^{2+} ions. The NMR signal around 310 Mc/s can be identified without contradiction as associated with Mn^{55} NMR of Mn^{4+} ions at *B* sites.

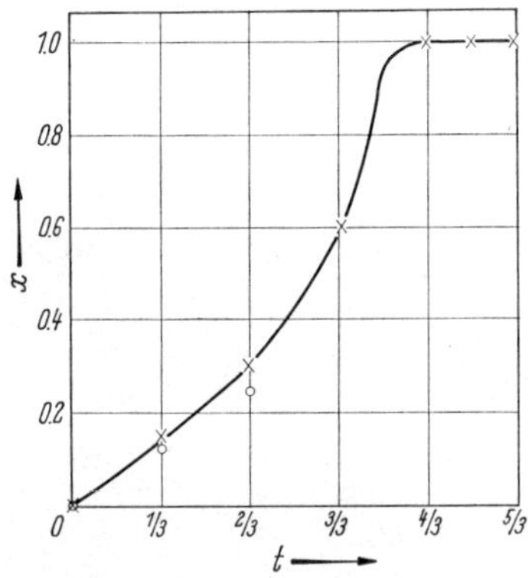

Fig. 51. $Li_{0.5+0.5t}Fe_{2.5-1.5t}Mn_tO_4$. Amount of Li^+ ions on *A* sites, *x*, vs. composition. (Crosses from X-ray intensities; circles from magnetic moments). [*65B*]. For material, see Fig. 37.

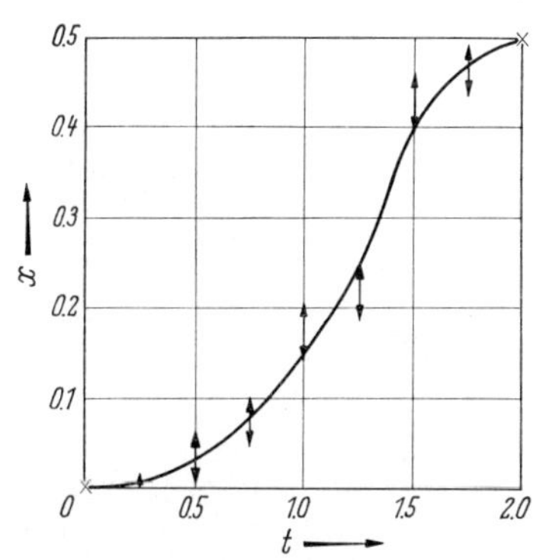

Fig. 52. $Li_{0.5}Mn_tFe_{2.5-t}O_4$. Amount of Li^+ ions on *A* sites, *x*, vs. Mn content, *t* [*65B*]. For materials, see Fig. 38.

6.1.3.1.5 Spin arrangements

The results of Mössbauer investigation on $\mathbf{MnFe_2O_4}$, $\mathbf{Mg_{0.5}Mn_{0.5}Fe_2O_4}$, and $\mathbf{Mg_{0.75}Mn_{0.5}Fe_{1.75}O_4}$ at RT and liquid nitrogen temperature are explained as being due to deviations from collinear Néel spin-configurations on octahedral sites [*66WMK* and *68WPMK*].

Tab. 24. Mössbauer effect in Mn- and Mg-Mn ferrites [*64BDZMST*]
H_{loc} = local magnetic field at the Fe^{57} nuclei; δ = chemical shift

Composition	H_{loc} kOe	δ mm/sec	M_s Gauss/cm³
Fe_3O_4	473	0.6 ± 0.1	450
$Mn_{0.43}Fe_{2.57}O_4$	472	0.5 ± 0.1	
$Mn_{0.59}Fe_{2.41}O_4$	477	0.5 ± 0.1	
$Mn_{0.61}Fe_{2.39}O_4$	470	0.5 ± 0.1	
$Mn_{0.84}Fe_{2.16}O_4$	468	0.5 ± 0.1	470
33.4MgO, 16.6MnO, 50Fe_2O_3 [wt %]	436	0.3 ± 0.1	240
37.8MgO, 12.2MnO, 50Fe_2O_3 [wt %]	427	0.4 ± 0.1	197
3.1MgO, 54MnO, 42.9Fe_2O_3 [wt %]	373	0.5 ± 0.1	120
96.93$MnFe_2O_4$, 3.07Mn_3O_4 [wt %]	432	0.4 ± 0.1	420
87.2$MnFe_2O_4$, 12.8Mn_3O_4 [wt %]	450	0.4 ± 0.1	350
81.55$MnFe_2O_4$, 18.45Mn_3O_4 [wt %]	460	0.4 ± 0.1	280

6.1.3.2 Thermal properties

6.1.3.2.1 Specific heat

a) System Mn-Fe-O

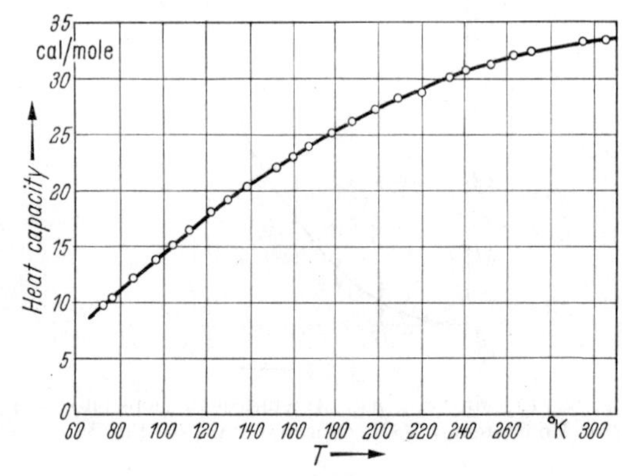

Fig. 1. Mn_3O_4. The molar heat capacity [*28M*]. There is undoubtedly an anomaly associated with the Curie point near 43 °K, as well as with the tetragonal to cubic phase transition at about 1445 °K [*49K* (2)]. See also [*58ISB*].

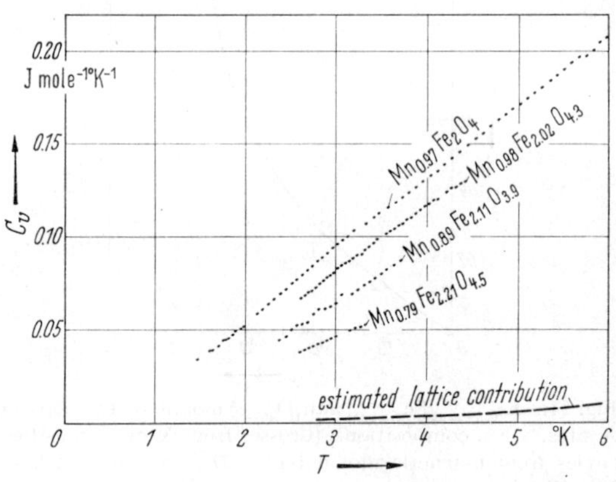

Fig. 2. Specific heat, C_v, vs. temperature for polycrystalline Mn ferrites of various compositions [*63L*]. The following conclusions are evident from the results: (a) transitions involving Mn ions occur in Mn ferrites at low temperatures, (b) the energies of the transitions correspond to temperatures of the order of 20 °K, (c) the associated specific heat is more dependent on Mn concentration near the concentration $MnFe_2O_4$ than at lower Mn concentrations.

b) System Zn-Mn-Fe-O

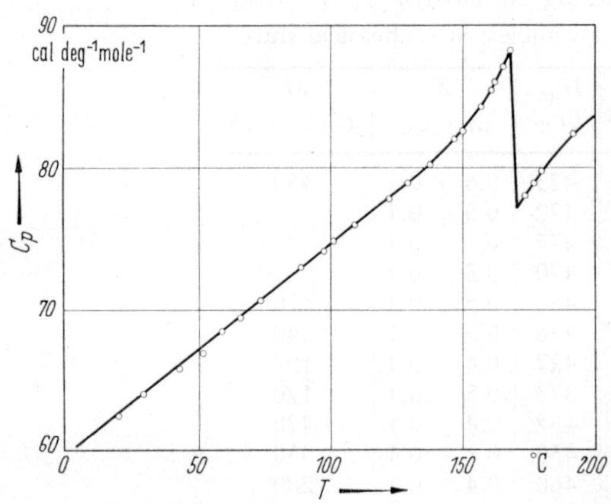

←

Fig. 3. $Mn_{1-x}Zn_xFe_2O_4$ ($x \cong 0.38$). Specific heat, C_p, vs. temperature. After [*60VRB*]. The curve was measured with rising and falling temperature, no thermal hysteresis being observed. The magnitude of the jump at the Curie temperature is in accordance with the relation $\Delta C_p/n_B = 1.7$ as previously found for Ni-Zn and Mg-Zn ferrites.
For discontinuity of specific heat at the Curie temperature see also [*66ZO*].

6.1.3.2.2 Thermal conductivity

a) System Mn-Fe-O

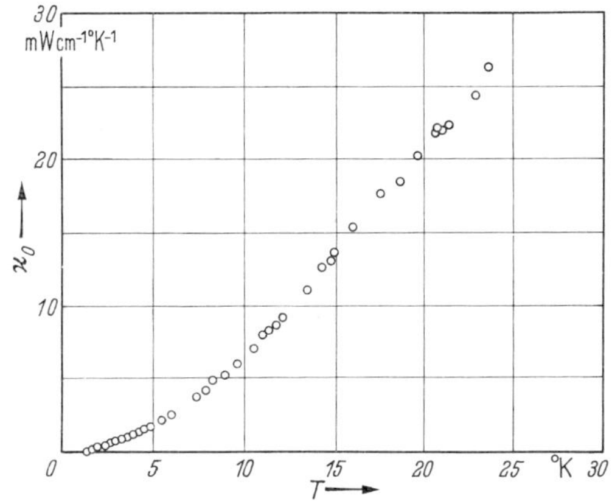

Fig. 4. The thermal conductivity in zero magnetic field, $\varkappa_0$, of a single crystal whose analysis corresponds to $Mn_{1.51}Fe_{1.52}O_4$. The crystal was grown by the Verneuil process. Its long axis lay 45° from the [111] axis. The Curie temperature was 573 °K. From [58DF].

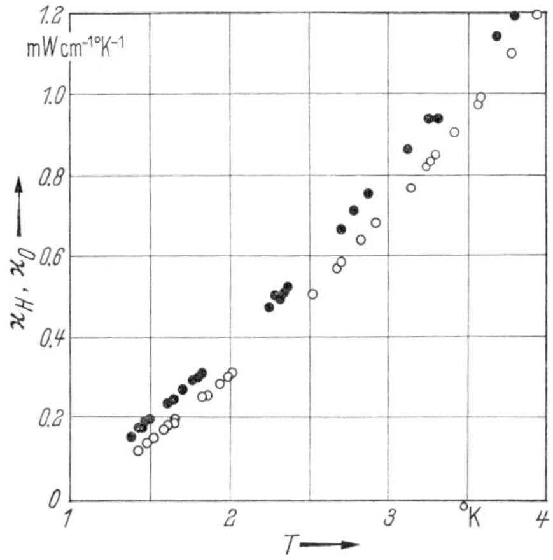

Fig. 5. The thermal conductivity in zero and maximum available magnetic field, $\varkappa_0$ and $\varkappa_H$, of the crystal described in caption of Fig. 4 in this section. From [58DF]. ● H = 9400 Gauss, ○ H = 0.

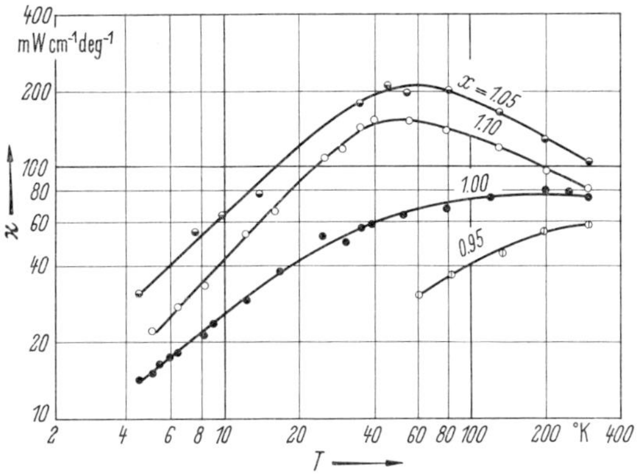

Fig. 6. $Mn_xFe_{3-x}O_4$. Thermal conductivity vs. temperature of Mn ferrites. Parameter x. After [66S (3)]. Measurements on single crystals grown in air by a Bridgman technique.

b) System Zn-Mn-Fe-O

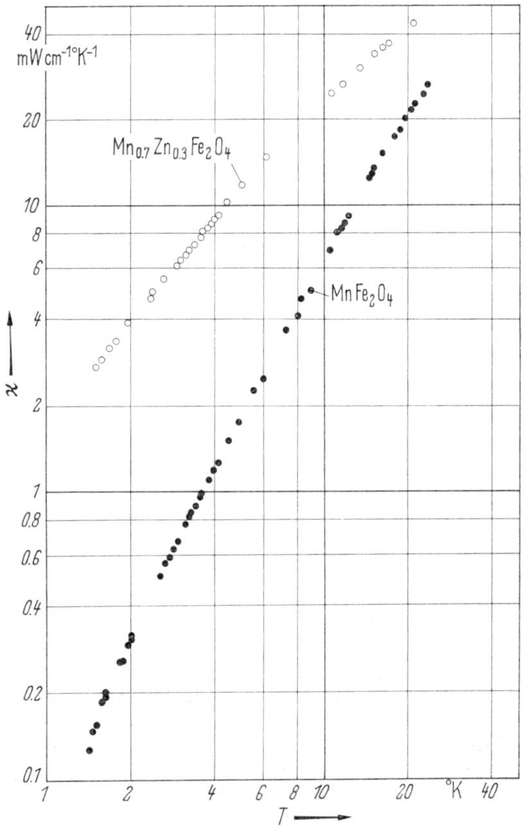

Fig. 7. $MnFe_2O_4$, $Mn_{0.7}Zn_{0.3}Fe_2O_4$. Log-log plot of temperature variation of the thermal conductivities of two ferrite specimens between 1.4 °K and 25 °K. The samples are single crystals. Compositions are only close estimates. After [61DF].

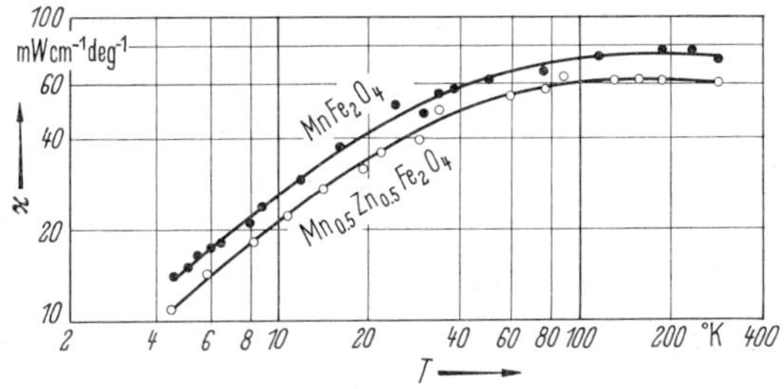

Fig. 8. $MnFe_2O_4$, $Mn_{0.5}Zn_{0.5}Fe_2O_4$. Thermal conductivity vs. temperature. After [66S (3)]. Measurements on single crystals grown in air by a Bridgman technique.

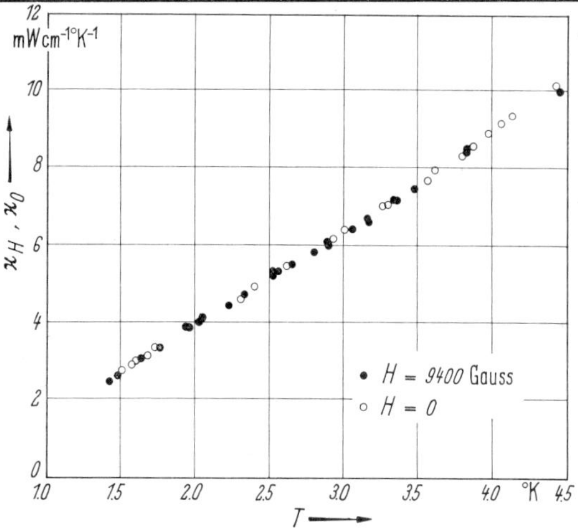

← Fig. 9. $Mn_{0.7}Zn_{0.3}Fe_2O_4$. The temperature variation of the thermal conductivity in both zero and 9400 Gauss external longitudinal field. The sample is a single crystal. The composition is only a close estimate. From [*61DF*].

c) System Ni-Mn-Fe-O

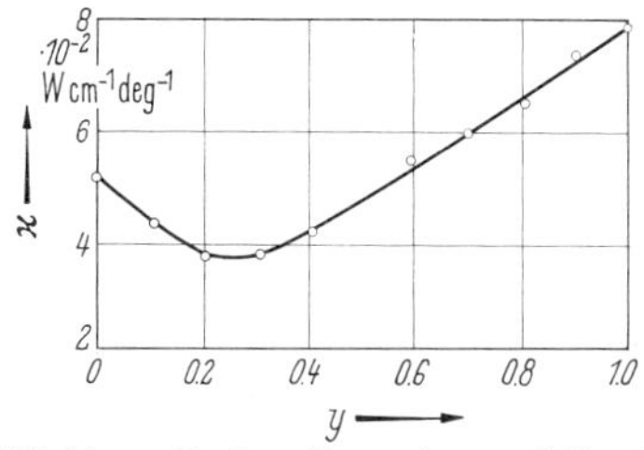

Fig. 10. $Ni_yMn_{1-y}Fe_2O_4$. Dependence of the thermal conductivity, $\varkappa$, on the composition of the solid solution. From [*67B* (1)].

6.1.3.2.3 Thermal expansion

a) System Ni-Mn-Fe-O

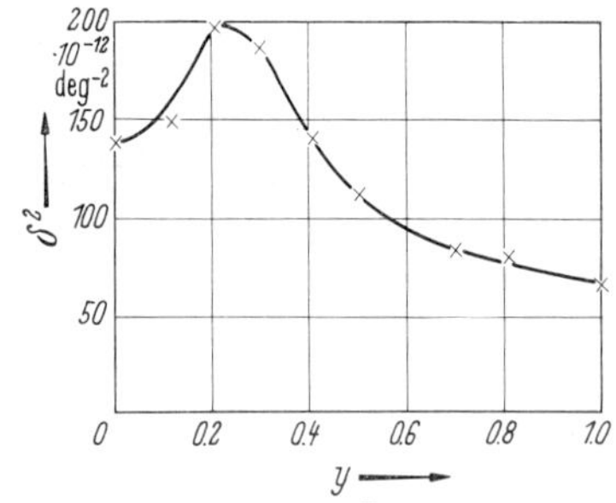

Fig. 11. $Ni_yMn_{1-y}Fe_2O_4$. Dependence of the thermal expansion, δ, on the composition of the solid solution. From [*67B* (1)]. →

6.1.3.2.4 Diffusion and oxidation coefficients

a) System Zn-Mn-Fe-O

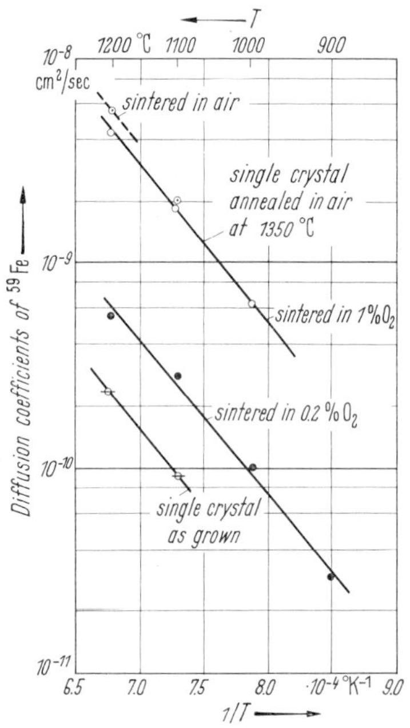

Fig. 12

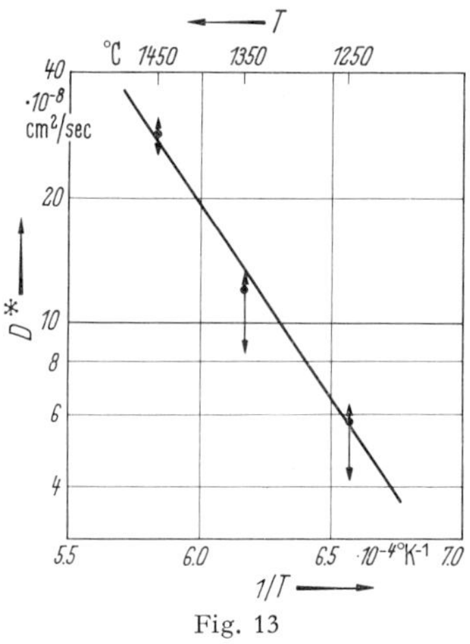

Fig. 13

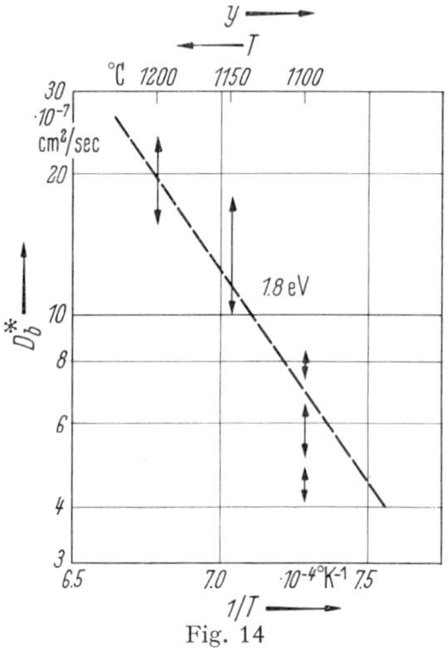

Fig. 14

Fig. 12. Temperature dependence of the diffusion coefficient of ^{59}Fe in MnZn ferrite $(Fe_2O_3)_{55}$, $(MnO)_{26}$, $(ZnO)_{19}$ sintered in various atmospheres at 1250 °C for 4 h and cooled slowly in nitrogen gas. A surface layer of 0.2 mm thickness was removed from the sample after the sintering process. Single crystals of MnZn ferrites $(Fe_2O_{3-x})_{54.8}$, $MnO_{28.6}$, $ZnO_{16.6}$ (x being undetermined) were grown at 1650 °C in air using the Bridgman method. After [*67ON*]. The volume diffusion coefficient of ^{59}Fe varies with oxygen content, i.e. the density of cation vacancies in the sample. The activation energy of the diffusion was (1.50 ± 0.05) eV, being independent of the vacancy density. This means that the activation energy is due only to the migration energy and not to the formation of thermal vacancies as in Zn ferrite (3.5 eV) and α-Fe_2O_3 (4.9 eV).

Fig. 13. $(Fe_2O_3)_{53}$, $(MnO)_{31}$, $(ZnO)_{16}$. Temperature dependence of the coefficient D^* of oxidation in air for a single crystal [*67O*]. The single crystal is grown at 1650 °C in air by means of the Bridgman method, so that there is not any oxygen excess. This state is one of the boundary conditions for applying this coefficient D^* the other being the constant oxygen excess C_0 at the surface. The oxidation rate is expressed by the coefficient D^* having the same dimension as a diffusion coefficient. It can be used in the usual equation: $C = C_0 \left[1 - \Theta\left(x/2\sqrt{D^* t}\right)\right]$ t = oxidation time, C = oxygen excess at the distance x from the surface, Θ = error function.
The activation energy of oxidation is (1.8 ± 0.2) eV.

Fig. 14. $(Fe_2O_3)_{55}$, $(MnO)_{26}$, $(ZnO)_{19}$. Temperature dependence of the coefficient D_b^* of oxidation in air for a polycrystalline sample [*67O*]. The sample was sintered at 1200 °C in pure nitrogen gas. Same remarks as for Fig. 13. The D_b^* coefficient cannot be considered as a boundary oxidation coefficient.

b) System Co-Zn-Mn-Fe-O

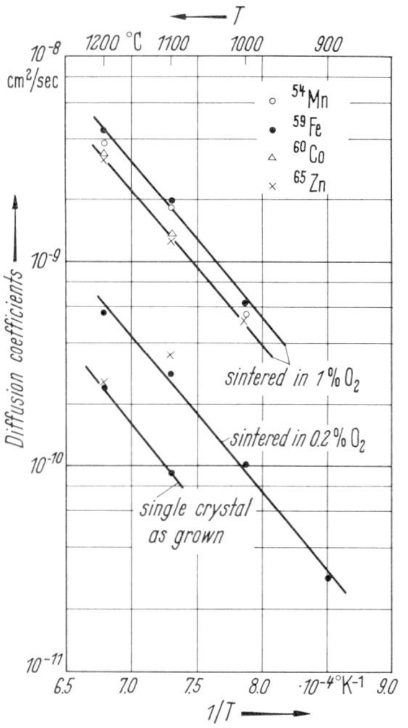

Fig. 15. Temperature dependence of the diffusion coefficients of ^{54}Mn, ^{59}Fe, ^{60}Co, ^{65}Zn in the same sample of MnZn-ferrite as for Fig. 12. From [*67ON*]. The volume diffusion coefficients obtained are the same for the 4 cations, but slightly different for ^{45}Ca. See Fig. 16 and 17.

c) System Ca-Zn-Mn-Fe-O

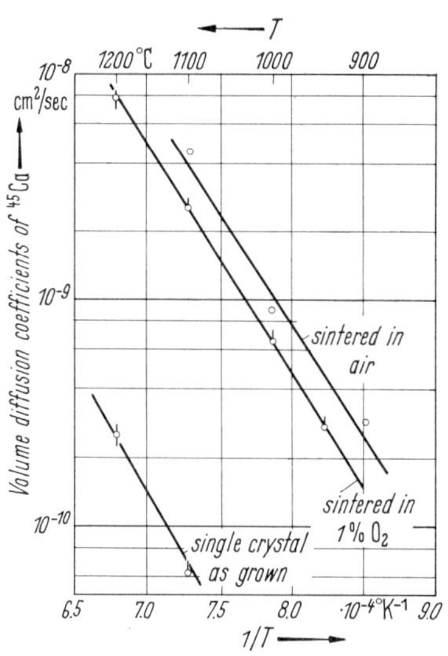

Fig. 16. Temperature dependence of the volume diffusion coefficient of ^{45}Ca in the same sample of MnZn-ferrite as for Fig. 12. From [*67ON*].

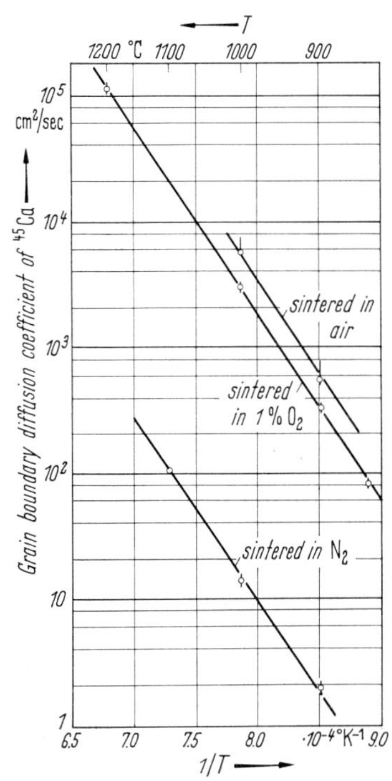

Fig. 17. Temperature dependence of the grain boundary diffusion coefficients of ^{45}Ca in the same sample of MnZn-ferrite as for Fig. 12. After [*67ON*].

6.1.3.2.5 Ferrite thermodynamics

Introduction

Studies of ferrite thermodynamics are stimulated by the importance of their knowledge for the preparation of ferrites with predetermined and strictly reproducible properties.

Considering the formation $x\text{Me} + O_2 = \text{Me}_xO_2$, the free energy of formation of the oxide from the elements is:

$$\Delta G_T = \Delta G_T^0 + \text{RT} \cdot \log \frac{a_{\text{Me}_xO_2}}{a_{\text{Me}}^x \cdot p_{O_2}},$$

where ΔG_T^0 is the free standard energy of formation, $a_{\text{Me}_xO_2}$ the activity of the oxide, and a_{Me} the activity of the metal.

Measurements of the reversible emf's of galvanic cells of the type

$$\text{Fe, Fe}_{0.95}\text{O} \,|\, 0.15\,\text{CaO} - 0.85\,\text{ZrO}_2 \,|_{\text{s.p., w.p.}}$$

(s.p. designates a spinel ferrite or solid solution, and w.p. a wustite phase) have made it possible to determine $\Delta\bar{G}_{O_2}$, the partial molar free energy, $\Delta\bar{S}_{O_2} = -\partial\,\Delta\bar{G}_{O_2}/\partial T$, the partial molar entropy, and $\Delta\bar{H}_{O_2} = [\Delta\bar{G}_{O_2} + T\,\Delta\bar{S}_{O_2}]$, the partial molar enthalpy of the oxygen in the spinel phase, as well as $\log p_{O_2}\,(\text{atm}) = \Delta\bar{G}_{O_2}/\text{RT}$, the equilibrium dissociation pressure of the ferrite or the solid solution.

a) System Mn-Fe-O

Tab. 1. Values of the free energy of formation of spinel from magnetite, Fe_3O_4, and hausmannite, Mn_3O_4, ΔG^{Free} [*65TSG*]. The values given here are free of errors related to the assumption of strict quasi-binary character of the system Fe_3O_4-Mn_3O_4-MnO-FeO and to the use of literature data on the values of ΔG of the formation of the individual oxides as for [*64GT*]

Composition of the spinel phase	ΔG^{Free} kcal/mole
Fe_3O_4	0
$Mn_{0.225}Fe_{2.775}O_4$	− 2.95
$Mn_{0.45}Fe_{2.55}O_4$	− 4.53
$Mn_{0.705}Fe_{2.295}O_4$	− 6.46
$Mn_{0.87}Fe_{2.13}O_4$	− 7.91
$Mn_{0.915}Fe_{2.085}O_4$	− 8.50
$MnFe_2O_4$	−11.0
$Mn_{1.2}Fe_{1.8}O_4$	−17.7
$Mn_{2.13}Fe_{0.87}O_4$	−18.24
Mn_3O_4	0

Tab. 3. Free molar energy (isobaric potential) of the reaction for the formation of $MnFe_2O_n$ from MnO, Fe and O_2. After [*60RK*]. ΔG_T^0 is the free standard energy of formation

n	ΔG_T^0 [kcal/mole] 1073 °K	1173 °K	1273 °K	1373 °K
1.30	14.591	14.160	13.710	13.293
1.50	24.207	23.509	22.777	22.092
1.70	33.771	32.804	31.796	30.840
1.90	43.304	42.060	40.774	39.541
2.10	52.819	51.288	49.716	48.196
2.40	67.067	65.092	63.076	61.113
2.60	76.548	74.274	71.958	69.694
2.94	92.613	89.768	86.987	84.243
3.02	96.143	93.103	90.192	87.319
3.25	105.835	102.337	98.924	95.246
3.40	112.008	108.102	104.258	100.270
3.60	120.173	115.634	111.130	106.500
3.72	124.992	120.069	115.180	110.161
3.77	126.968	121.892	116.847	111.674
3.90	131.980	126.520	121.100	115.541
3.96	134.211	128.586	123.001	117.274
3.98	134.940	129.260	123.621	117.840
3.99	135.299	129.592	123.927	118.119
4.00	135.331	129.625	123.959	118.149

Tab. 2. Free energies of formation of $MnFe_{1.932}O_x$ from the elements [*66BZ*]. ΔG_f^0 is the free standard energy of formation from the elements. The temperature variation of ΔG_f^0 is described by the linear equation $\Delta G_f^0 = a + bT$. The a and b values are given by [*66BZ*]

T [°K] / x	ΔG_f^0 [kcal/mole] 1050 °K	1150 °K	1250 °K	1350 °K	1450 °K
3.897	−205.63	−198.44	−190.94	−182.74	−174.58
3.89	−205.43	−198.31	−190.84	−182.68	−174.57
3.88	−205.07	−198.01	−190.61	−182.49	−174.43
3.87	−204.70	−197.68	−190.34	−182.28	−174.27
3.86	−204.33	−197.34	−190.04	−182.03	−174.09
3.85	−203.94	−197.00	−189.74	−181.77	−173.87
3.84	−203.57	−196.66	−189.43	−181.50	−173.64
3.83	−203.19	−196.31	−189.12	−181.23	−173.40
3.82	−202.81	−195.96	−188.80	−180.96	−173.16
3.81	−202.43	−195.61	−188.48	−180.68	−172.91
3.80	−202.04	−195.26	−188.16	−180.39	−172.67
3.75	−200.12	−193.44	−186.54	−178.96	−171.41
3.70	−198.18	−191.66	−184.89	−177.50	−170.12
3.60	−194.22	−187.97	−181.52	−174.54	−167.46

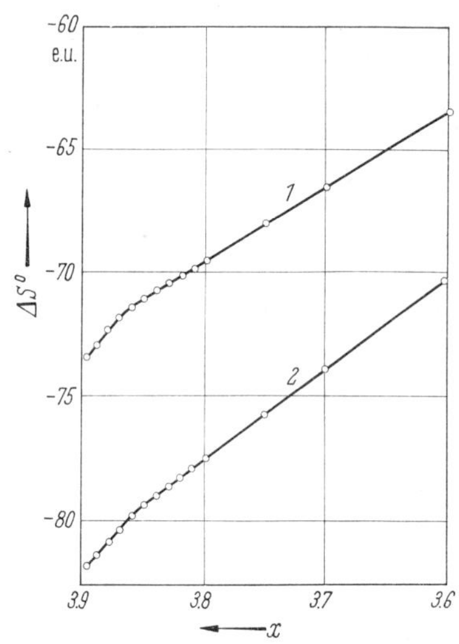

Fig. 18. $MnFe_{1.932}O_x$. Entropy of formation, ΔS^0, (in entropy units) vs. x. 1: 1050 ⋯ 1250 °K, 2: 1250 ⋯ 1450 °K. [*66ZB*].

Tab. 4. Molar values of enthalpy, ΔH_{298}, and entropy, ΔS_{298}^0, of $MnFe_2O_4$ formation at 298 °K

Reaction of formation	ΔS_{298}^0 e.u.	ΔH_{298} kcal/mole	Ref.	Method
$Mn + 2Fe + 2O_2 \rightarrow MnFe_2O_4$	−8.16	−292.8	*60RK*	E.M.F.
		−293.5 ± 1.0	*66F*	E.M.F.
$MnO + Fe_2O_3 \rightarrow MnFe_2O_4$	−1.2	− 3.9	*60RK*	E.M.F.
		− 4.0	*60RK*	calorimetric
		− 7.7	*54L*	
		− 4.8 ± 0.5	*66F*	E.M.F.

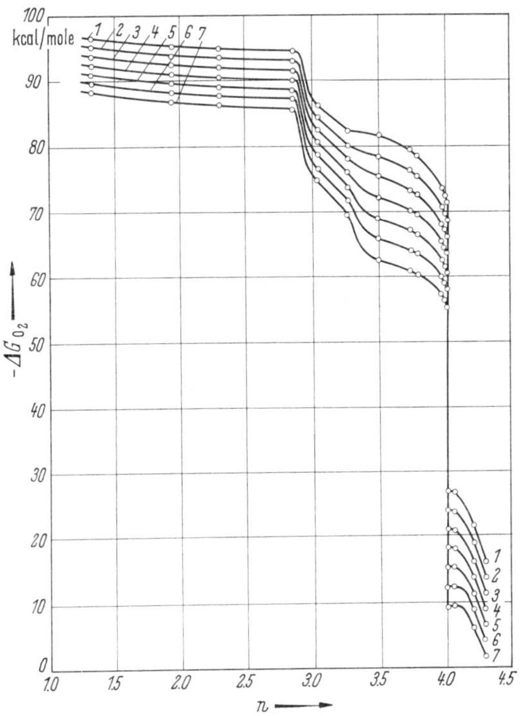

Tab. 5. Enthalpy of transfer of oxygen from the gaseous phase to the spinel phase for Mn ferrites, $\Delta\bar{H}_{O_2}$, between 900···1200 °C [*66G*]

Composition	Mean value of $\Delta\bar{H}_{O_2}$ kcal/mole
$Mn_{0.3}Fe_{2.7}O_4$	−145
$Mn_{0.6}Fe_{2.4}O_4$	−146
$Mn_{0.9}Fe_{2.1}O_4$	−147
$Mn_{1.0}Fe_{2.0}O_4$	−148

←
Fig. 19. $MnFe_2O_n$. Dependence of the partial molar free energy of oxidation on the oxygen content, *n*, for temperatures 1) 1073 °K 2) 1123 °K 3) 1173 °K 4) 1223 °K 5) 1273 °K 6) 1323 °K 7) 1373. $\Delta\bar{G}_{O_2}$ is the partial molar energy of oxidation for the reaction $(O_2) \rightarrow 2[O]$ in $MnFe_2O_n$. Data obtained from measurements of the electromotive force of a galvanic cell [*67RL*].

b) System Ca-Mg-Zn-Mn-Fe-O

Tab. 6. Enthalpy of transfer of oxygen from the gaseous to the spinel phase for Mn-Mg, Mn-Mg-Ca and Mn-Mg-Ca-Zn ferrites, $\Delta\bar{H}_{O_2}$, between 900···1200 °C [*66G*]

Composition	Mean valuc of $\Delta\bar{H}_{O_2}$
43 Fe_2O_3, 43 MnO, 14 MgO	−148
39.26 Fe_2O_3, 22.09 MnO, 38.67 MgO	−146
41.76 Fe_2O_3, 41.57 MnO, 14.84 MgO, 1.83 CaO	−146
37 Fe_2O_3, 38 MnO, 10 MgO, 3 CaO, 12 ZnO	−148

Activities

About the problem of the activities of magnetite and hausmannite in the spinel phase, see [*65T* (1)], [*65T* (2)], [*65TO*], [*65TSG*], [*65KOST*], [*66G*], [*67BZMAC*], [*66BZMSA*], [*67SM*].

6.1.3.2.6 Solid crystal growth

Introduction

The permeability is strongly dependent on the grain size of the ferrite samples. Therefore, a great interest has been devoted to the grain growth in polycrystalline ferrite.

It was shown in [*63P* (1)] that all ferrites with the spinel structure have approximately the same value of the activation energy for grain boundary migration, irrespective of their composition, provided migration of grain boundaries is not hindered by voids or inclusions. Furthermore, the mean free energy of the boundaries (as calculated from the kinetics of crystal growth) doubles when a reducing atmosphere is replaced by an oxidizing one. Thus, concerning the grain growth, special attention must be paid to: crystal defects (grinding, impurity, atmosphere), particle size distribution, inclusions and pores.

a) Normal crystal growth of Zn-Mn-Fe-O, Mn-Fe-O, Cd-Mn-Fe-O

The effect of powder parameters on grain growth in MnZn-ferrite was studied in [*64Y*]: not only the particle size and distribution but also the calcination temperature were shown to be critical factors responsible for the growth behavior of ferrite.

The rate of crystal growth may be increased by oxidation or reduction during the sintering process. With this technique single crystals 1 mm in diameter were obtained for MnZn- and Mn-ferrites [*63MV*]. The facies of crystal growth in different atmospheres for Mn- and MnCd-ferrites were studied in [*66P*].

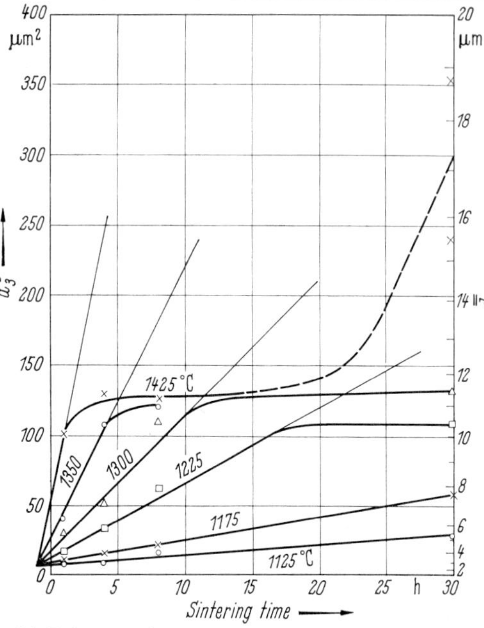

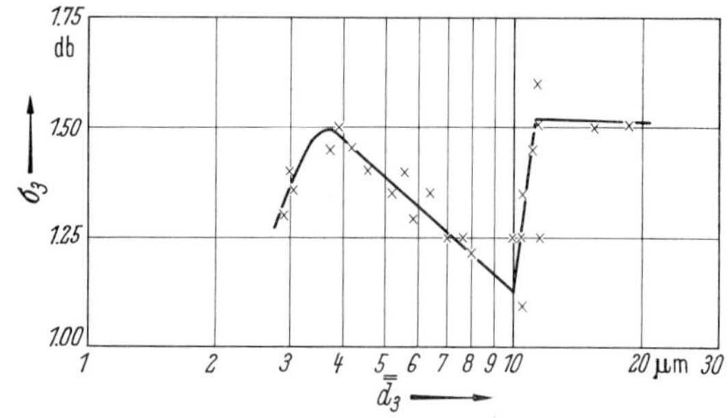

Fig. 21. Change of the standard deviation (in db) of the median crystal diameter of ZnMn-ferrite, approaching the limit value of crystal diameter ($\cong$ 10 μm). This value depends on porosity. Same sample as for Fig. 20 [*63P* (5)].

←

Fig. 20. Square of the median crystal diameter $\bar{\bar{d}}_3^2$ of the ferrite 53 Fe_2O_3, 28 MnO, 19 ZnO vs. time of sintering for different temperatures. The wet-mixed starting oxides were pressed in toroids at 5000 kg/cm², and then subjected to the heat treatment in an atmosphere of nitrogen with 1% oxygen. The limiting grain size $\cong$ 10 μm for crystal growth is due to the presence of voids [*62P* (2)].

b) Discontinuous crystal growth of Zn-Mn-Fe-O, Mn-Fe-O

Controlled discontinuous grain growth of ferrite may be used to grow single crystals from polycrystalline sample. After [*63P* (1)].

After [*64K* (2)] the anisotropic exaggerated grain growth, observed in **$MnFe_2O_4$** without additions, sintered for 3 h at 1370 °C in nitrogen, is due to the presence of a liquid phase at the grain boundaries. The rate of exaggerated grain growth is then determined by a solution-precipitation process rather than by grain boundary migration.

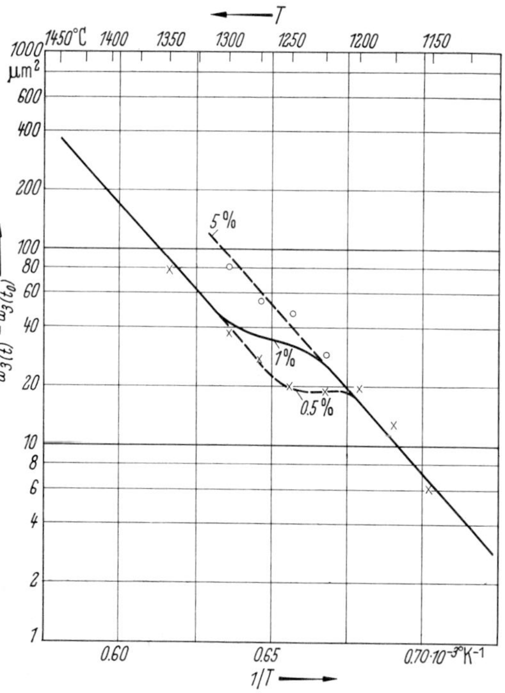

Fig. 22. Difference of the square median crystal diameter of ZnMn-ferrites for a sintering-annealing time of 4 h and different oxygen content in the heat treatment atmosphere, as a function of the inverse of temperature. The value of the activation energy $Q \cong 83\,000$ cal/mole for crystal growth should be considered as an activation energy of the self-diffusion of oxygen ions along the grain boundaries between 1125 and 1425 °C. Same samples as for Fig. 20 [*62P* (4)].

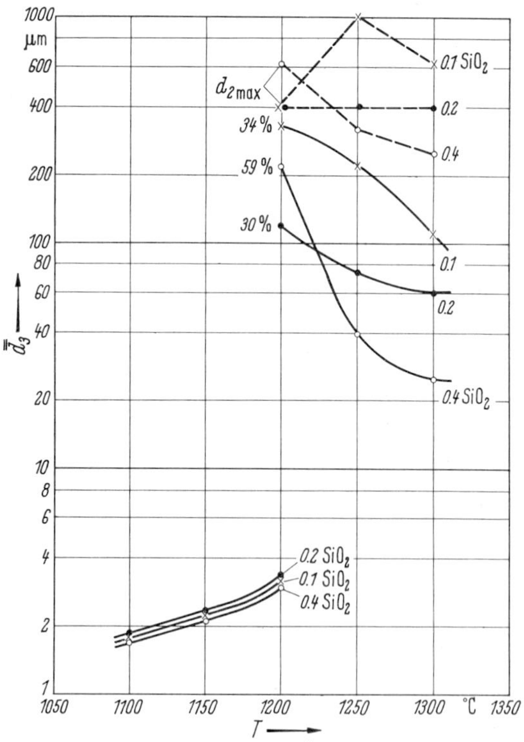

Fig. 23. Median crystal diameter $\bar{\bar{d}}_3$ of the ferrite 53 Fe_2O_3, 28 MnO, 19 ZnO, x% SiO_2 vs. annealing-sintering temperature. Heat treatment of 4 h in nitrogen +1% oxygen. Same preparation as for Fig. 20. From [*62P* (5)]. The discontinuous crystal growth is due to the dissolution of SiO_2 inclusions at 1200 °C. — — — maximum crystal size.

→

Fig. 24. $MnFe_2O_4$. Crystal diameter, d, vs. temperature, T, of annealing and addition of Ba ions. Heat treatment 48 h in a 5% O_2 atmosphere. Single crystals of 1 cm in diameter and containing less than 30 ppM ($30 \cdot 10^{-6}$) of Ba are obtained. The discontinuous crystal growth is due to a thin liquid film at certain grain boundaries. From [*68HP*].

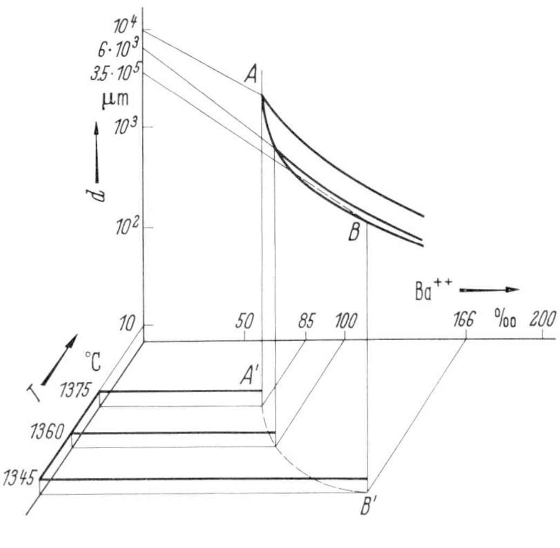

6.1.3.2.7 Reactivity and sintering

a) System Zn-Mn-Fe-O

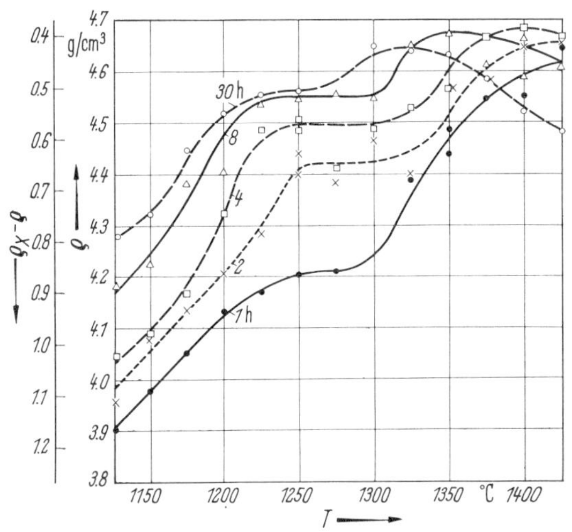

←

Fig. 25. Density, ϱ, vs. sintering temperature, T, for 53 Fe_2O_3, 28 MnO, 19 ZnO. Sintering time is given at the respective curve. Same sample preparation as for Fig. 20. Furnace atmosphere: nitrogen + 1% oxygen. It is shown that the sintering depends on the rate of crystal growth. Compare with Fig. 20 [*63P* (1)].

b) System Ca-Zn-Mn-Fe-O

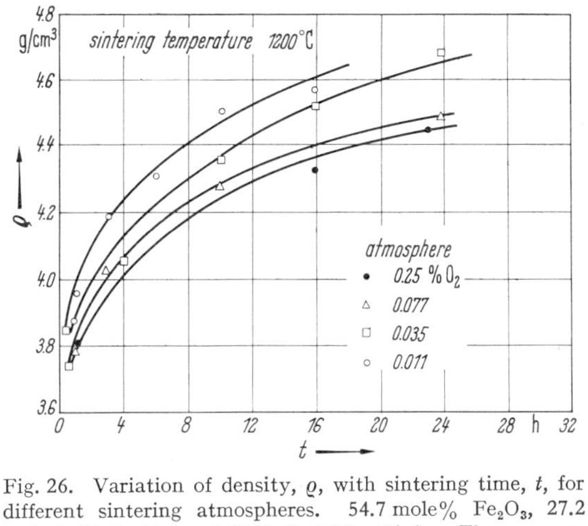

Fig. 26. Variation of density, ϱ, with sintering time, t, for different sintering atmospheres. 54.7 mole% Fe_2O_3, 27.2 mole% MnO, 18.1 mole% ZnO, 0.06 wt% Ca. The raw materials were mixed and then calcined at 950 °C for $1\frac{1}{2}$ h [*66SB*].

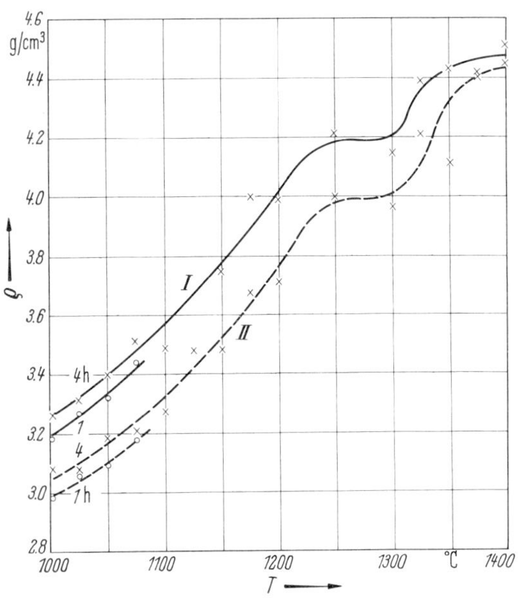

Fig. 27. Temperature dependence of density, ϱ, for 53 Fe_2O_3, 28 MnO, 19 ZnO, x CaO sintered in a nitrogen + 1% oxygen atmosphere. Sintering time is given at the respective curve. I: 0.00% CaO, II: 0.12% CaO. Same sample preparation as for Fig. 20, except for the pellet volume. Small CaO additions inhibit the earlier sintering stage of ZnMn-ferrites. [*63P*(1)].

Despite a great deal of work on reactivity and sintering of ferrites, there are not yet well established laws and results. The main reasons are the raw material and sample volume dependence of the results and the lack of knowledge on the main parameters of reactivity.

For further information about reactivity and sintering of Mn- and ZnMn-ferrites, see also [*60Y*], [*62K* (1)], [*64S* (4)], [*66PM*], [*66S* (4)].

6.1.3.3 Mechanical properties

(Acoustic relaxation measurements: a) Mn_3O_4; b) Mn-Fe-O; c) Mn-Zn-Fe-O.)

Acoustic relaxation experiments were reported by [*58G*] for single crystals in the Mn and MnZn ferrite. systems. These crystals had been grown by the flame fusion technique. The measurements were made on rod shaped samples. A field of 2300 Gauss parallel to the specimen axis was applied in order to saturate the specimen, and thus "freeze" domain wall movement which would otherwise provide an additional source for energy loss. In all cases the longitudinal acoustic mode was excited with a strain amplitude $\approx 10^{-7}$ and frequency in the range 100 ··· 150 kc/s.

a) Mn_3O_4

In [*57FC*] acoustic relaxation is studied in sintered polycrystalline samples of nominally $\mathbf{Mn_3O_4}$. They found broad peaks in internal friction which shifted with frequency at a rate implying an activation energy of 0.4 eV per process.

b) System Mn-Fe-O

In both of the curves for $\mathbf{Mn_{1.37}Fe_{1.41}O_4}$ (Fig. 2) and $\mathbf{Mn_{1.18}Fe_{1.6}O_4}$ (Fig. 1) there are two loss peaks, one near 40 °K and the other near 260 °K. Both peaks behave as typical stress induced relaxation. That is to say the temperature shifts with frequency and there is an inflection in the modulus curve at that temperature. The simple Arrhenius relation $\tau = \tau_0 e^{\varepsilon/kT}$ can be used to arrive at an activation energy ε. The half widths of the relaxation peaks observed are too broad by a factor of between three and ten to be accounted for by a process involving a simple relaxation time. This is not entirely unexpected since one would expect defects and localized inhomogeneities to occur in crystals grown by the flame fusion technique, and these would presumably perturb any relaxation process. Note also the steep rise in both modulus and internal friction below 10 °K.

The exact activation energy of this peak at about 260 °K in Fig. 3 varies with composition and/or homogeneity in the range 0.27 ··· 0.34 eV; and τ_0^{-1} is 10^{12} c/s. The peak near 40 °K (see Fig. 1 and 2) has an activation energy ε of (0.030 ± 0.004) eV per unit process and τ_0^{-1} of 10^{10} c/s. Measurements of these quantities in magnetite and Ni ferrite show a loss peak near 40 °K with the same activation energy. It is thought that the stress induced relaxation near 40 °K with an activation of 0.03 eV per electron jump and common to all the ferrites containing di- and trivalent ferrous ions on the octahedral sites is fairly well established as a stress induced electron migration between octahedral sites. The mechanism of the loss near 260 °K with an activation energy of about 0.3 eV is not at all certain. After [*58G*] it might represent the reaction

$$Mn_1^{2+} + Mn_2^{4+} + E \rightleftarrows Mn_1^{3+} + Mn_2^{3+}.$$

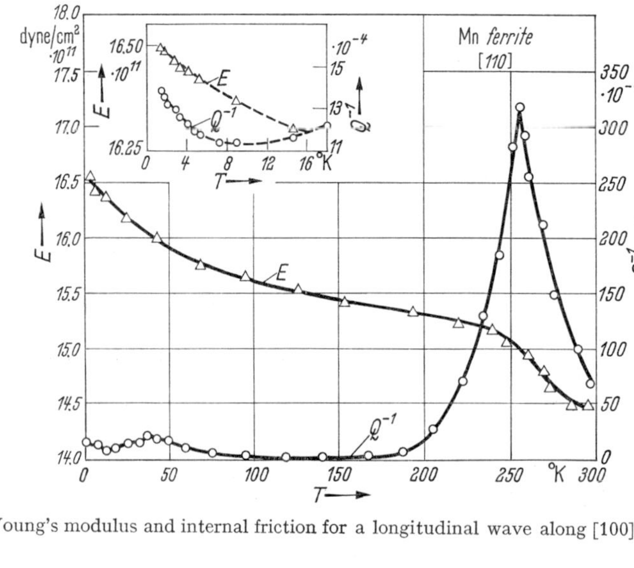

Fig. 1. $Mn_{1.18}Fe_{1.6}O_4$. Young's modulus and internal friction for a longitudinal wave along [100]. $f \approx 150$ kc/s. From [*58G*].

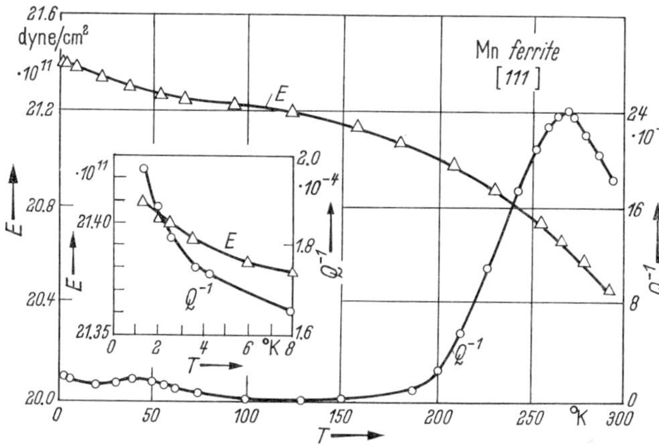

Fig. 2. $Mn_{1.37}Fe_{1.41}O_4$. Young's modulus and internal friction for a longitudinal wave along [111]. f = 110 kc/s. From [58G].

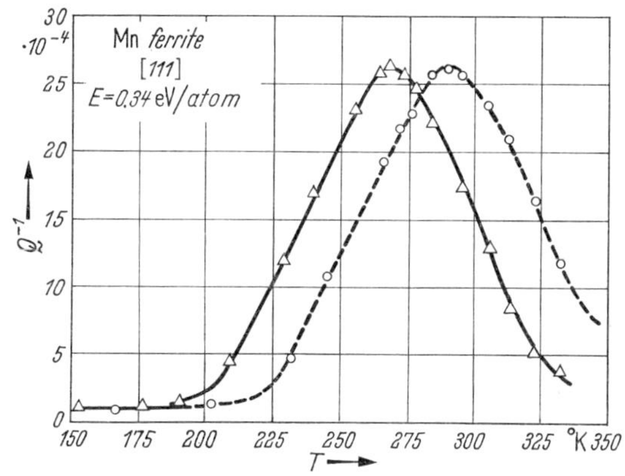

Fig. 3. $Mn_{1.37}Fe_{1.14}O_4$. Details of the stress induced relaxation of Mn ferrite occurring near 260 °K. [58G]. This shows the internal friction vs. temperature for a longitudinal wave along [111]. △ 11.35 kc/s, ○ 329.30 kc/s.

c) System Mn-Zn-Fe-O

The acoustic measurements in MnZn ferrite $\mathbf{Mn_{0.716}Zn_{0.182}Fe_{1.89}O_4}$ and $\mathbf{Mn_{0.766}Zn_{0.090}Fe_{2.03}O_4}$ give very different results [58G]; Fig. 4 and 5. The stress induced loss peaks near 40 °K and 260 °K are not to be seen at all.

There is a loss peak below 15 °K. The temperature at which this occurs is invariant with frequency. It is located at 11.4 °K for the crystal of Fig. 4, and at 14.0 °K for the crystal of Fig. 5. The modulus curve shows a sharp discontinuity just above the temperature of the maximum in the internal friction curve. These are characteristics of a transformation. Note also the decrease in the magnetization below 14 °K. In Fig. 1 and 2 there was evidence of a peak in internal friction below 1.25 °K. It is thought that this is another manifestation of the peaks in the Mn-Zn crystals. Apparently there is an electronic ordering process taking place. Also there seems to exist a critical concentration of Zn for which the transition temperature is a maximum, the transition temperature decreasing on either side. Below the transformation there is a decrease in the magnetization.

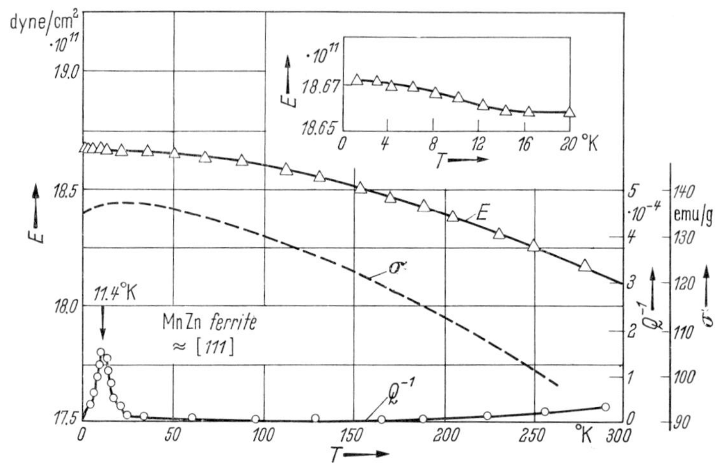

Fig. 4. $Mn_{0.716}Zn_{0.182}Fe_{1.89}O_4$. Young's modulus, internal friction, and intensity of magnetization per unit mass for a longitudinal wave along a direction approximately 20° from [111]. f = 100 kc/s. From [*58G*].

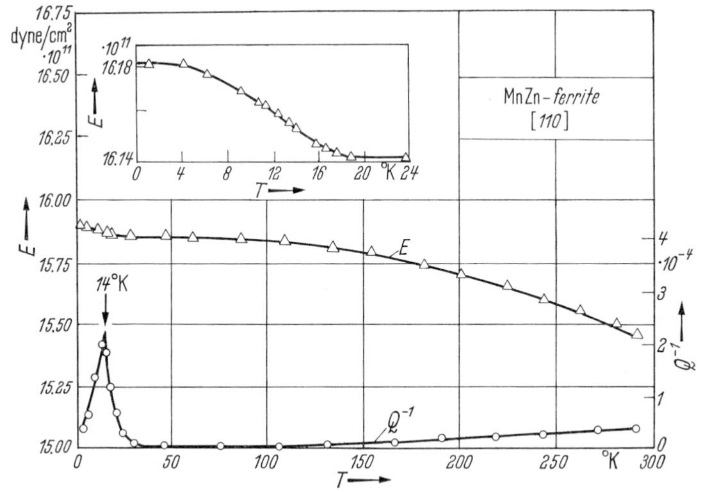

Fig. 5. $Mn_{0.766}Zn_{0.090}Fe_{2.03}O_4$. Young's modulus and internal friction for a longitudinal wave along [110]. f = 100 kc/s. From [*58G*].

Tab. 1. Physical parameters for two MnZn ferrites from magnetostrictively excited acoustic modes in toroidal samples of a number of polycrystalline ferrites [*53B*]

	$Mn_{0.495}Zn_{0.565}Fe_{1.98}O_4$	$Mn_{0.524}Zn_{0.389}Fe_{2.12}O_4$	
Apparent density, ϱ'	4.85	4.93	g/cm³
Magnetomechanical coupling coefficient, k	−0.027	−0.0015	number
Young's modulus, E	1.61	1.64	10^{12} dyne/cm²
Elastic quality in air	1970	3300	number
Poisson's coefficient of lateral contraction	—	0.33	number
Velocity of acoustic wave propagation $(E/p^1)^{1/2}$	$5.77 \cdot 10^5$	$5.77 \cdot 10^5$	cm/sec

There is not enough data here to justify an exposition of the various quantities which enter into the magnetostrictive excitation of vibrations in ferrites. The parameter k^2*) is equal to the fractional conversion of elastic energy into magnetic energy by an elastically driven sample or receiver operating below resonance. Conversely it is also equal to the fractional conversion of magnetic energy into elastic energy by a magnetically driven sample or radiator operating below resonance. The measured parameters are given in this table. These were obtained from extensional vibration in the uniform radial mode of toroidal samples in the frequency range 60···75 kc/s.

References for section 6.1.3.3: [*53B*], [*57BB*], [*57FC*], [*58G*].

*) $k^2 = 4\pi\lambda'\mu_r/E$, where λ' = dynamic magnetostrictive constant and E = Young's modulus.

6.1.3.4 Electrical properties

6.1.3.4.1 Electrical resistivity

a) Mn_3O_4

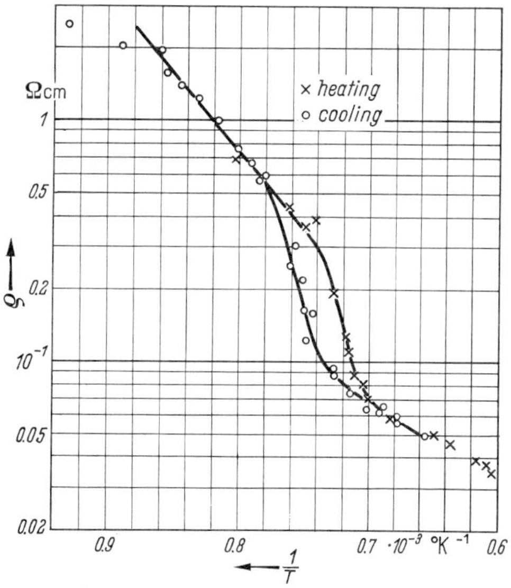

→ Fig. 1. Mn_3O_4. Resistivity, ϱ, vs. reciprocal temperature. Up to about 1350 °K the resistivity is described rather well by $\varrho = \varrho_\infty \exp(q/kT)$, $q = 1.3$ eV. From about 1350 to about 1425 °K the resistivity drops sharply to about $\frac{1}{3}$ of its former value, and then follows an exponential law with $q = 0.75$ eV. The same phenomenon is observed on cooling with a marked hysteresis. Note that the behavior here is very different from that observed in the case of Fe_3O_4, where the change in resistivity at the transition is very much larger than that here. The discontinuity is concluded to be caused only by the change in symmetry rather than an electron ordering [*53Ro*].

b) System Mn-Fe-O

Resistivity measurements were made on a series of samples $\mathbf{Mn_xFe_{3-x}O_4}$ where x ran from 0 to 1.14 [*59FMU* (1)]. log ϱ depends linearly on $1/T$ between 120 °K and 200 °K for $x \geq 0.1$, and a break is observed at about 120 °K for $0.2 \leq x \leq 0.6$. Thus the resistivity may be expressed by $\varrho = \varrho_0 \exp(q/kT)$. The variation of the two constants q and ϱ_0 with x is given in Fig. 2 and 3. The authors used these data to draw conclusions about the distribution of Mn ions.

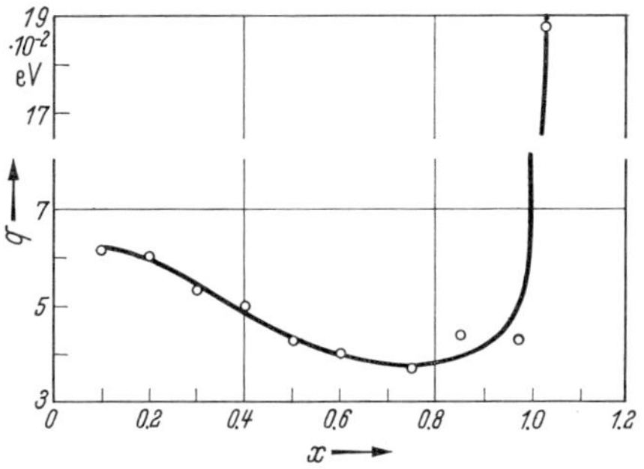

← Fig. 2. $Mn_xFe_{3-x}O_4$. Resistivity activation energy, q, vs. x. See text. From [*59FMU* (1)].

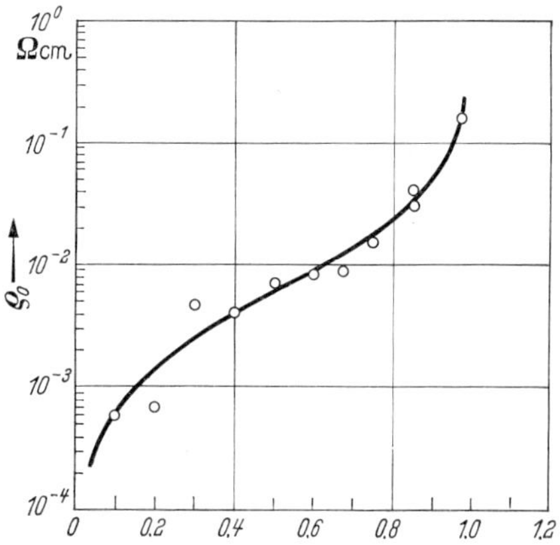

Fig. 3. $Mn_xFe_{3-x}O_4$. ϱ_0 vs. x [*59FMU* (1)]. ○: experimental values; curve: $\varrho_0 = 6 \cdot 10^{-3}(x/(1-x))$.

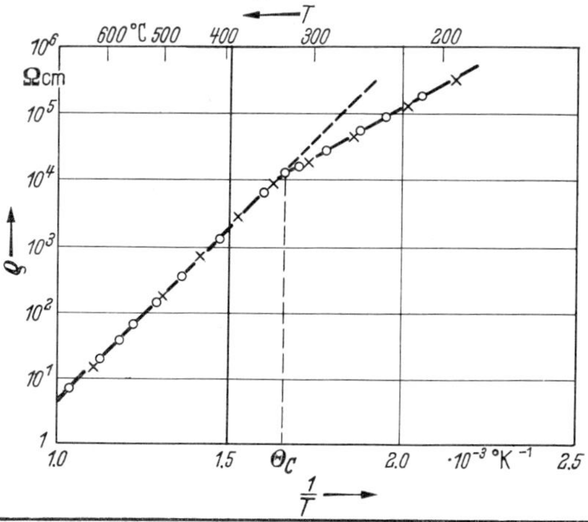

← Fig. 4. Resistivity, ϱ, vs. T^{-1} for Mn ferrite [*54KK*]. The starting mixture appears to have been MnO and Fe_2O_3 in the ratio 45:50, but no analysis of the samples is given. There is a break in the curve near the Curie point. The break is estimated to be at (335 ± 5) °C whereas the independently measured Curie temperature is 315 °C. This type of behavior was found also for specimens of Co, Ni, and NiZn ferrites.

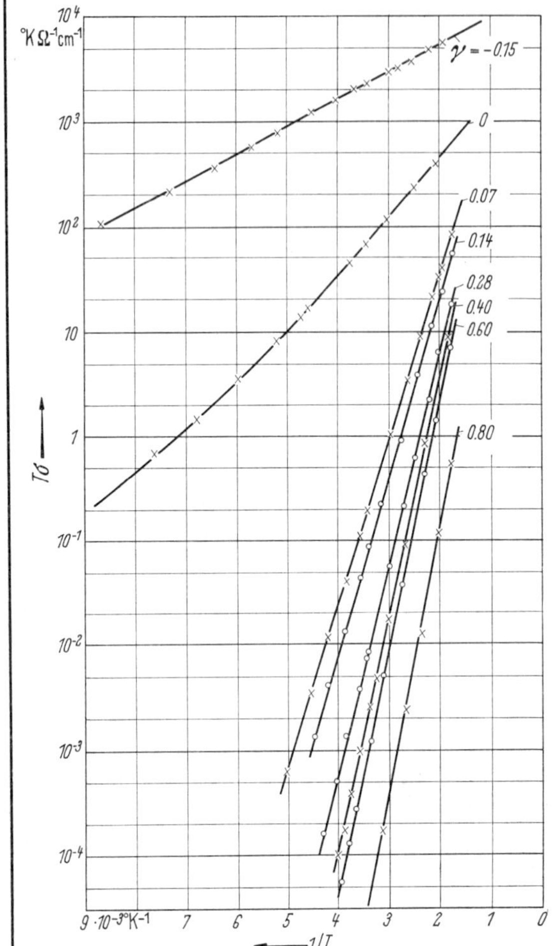

Fig. 5. $Mn_{1+\gamma}Fe_{2-\gamma}O_4$. $T\sigma$ vs. T^{-1}. Measurements on single crystals, $\gamma = -0.15$, 0, 0.07, 0.28 and 0.8, and on polycrystalline samples, $\gamma = 0.4$ and 0.6 [64L].

Tab. 1. $MnFe_2O_4$. Activation energies of a single crystal grown by the flame fusion method. q_ϱ = the activation energy of the electrical conduction. q_θ = the activation energy in the thermoelectric power. From [64RNB]

T °C	q_ϱ eV	q_θ eV
$T > 380$	1.19	0.572
$350 > T > 300$	0.758	0.182
Θ_C	—	—
$280 > T$	0.594	0.154

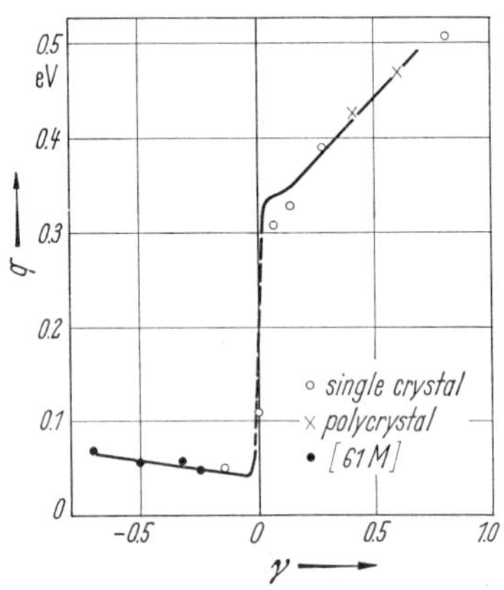

Fig. 6. $Mn_{1+\gamma}Fe_{2-\gamma}O_4$. Activation energy q, vs. composition, corresponding to the slopes of the graphs in Fig. 5 [64L].

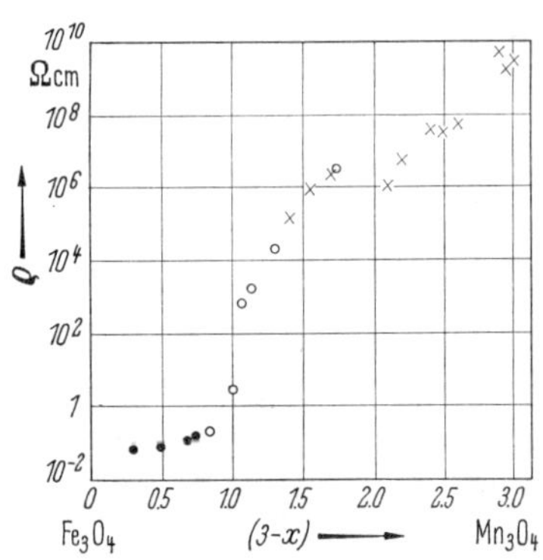

Fig. 9. $Fe_xMn_{3-x}O_4$. Resistivity, ϱ, as a function of composition [66RNB]. • [61M], ○ [64L], × [66RNB].

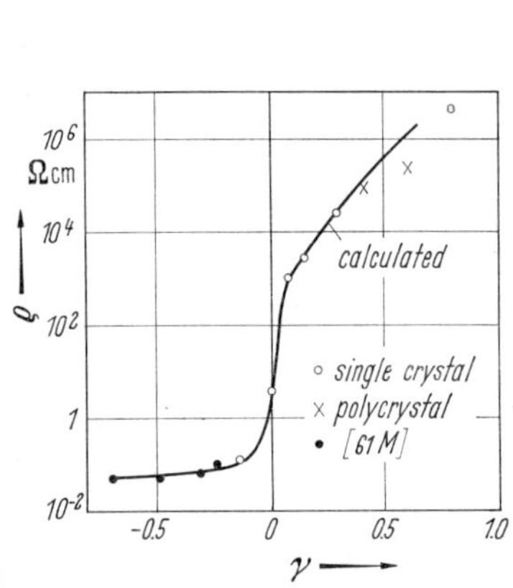

Fig. 7. $Mn_{1+\gamma}Fe_{2-\gamma}O_4$. Resistivity, ϱ, at 300 °K as a function of composition, γ [64L].

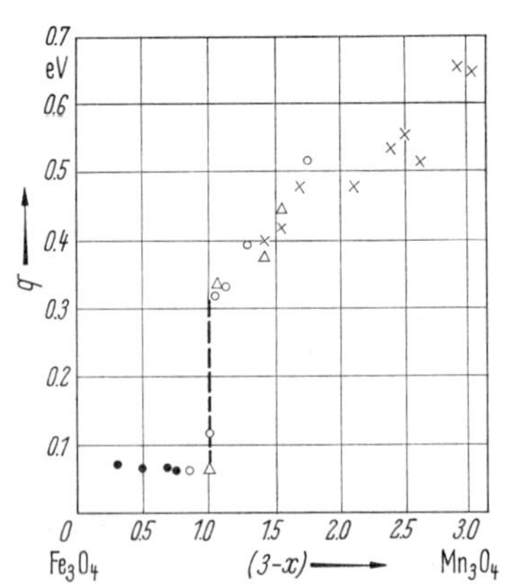

Fig. 8. $Fe_xMn_{3-x}O_4$. Activation energy, q, as a function of composition [66RNB]. • [61M], ○ [64L], △ [59BKZ], × [66RNB].

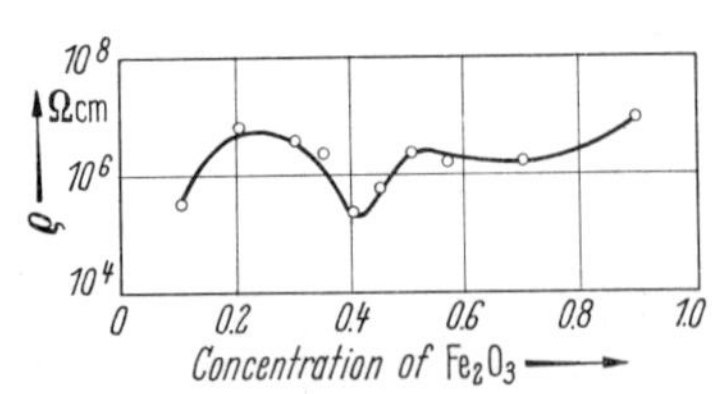

Fig. 10. $Mn_{1-x}Fe_{2+x}O_4$. Resistivity, ϱ, as a function of composition [67SR (2)].

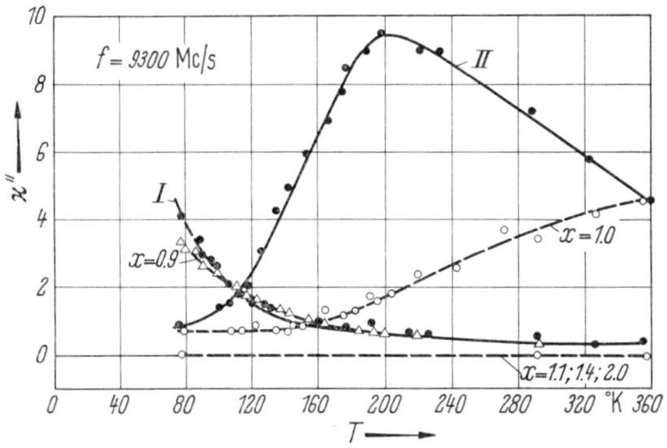

Fig. 11. $Mn_xFe_{3-x}O_4$. Temperature variation of dielectric losses, $\varkappa''$, at 9300 Mc/s of single and polycrystals. Solid lines show the results of single crystals with $x \cong 1$. Sample I is more conductive than sample II, indicating $x_I < x_{II} \cong 1$. Dotted lines show the results of polycrystals with $x = 0.9$, 1.0, 1.1, 1.4, 2.0 [67MI].

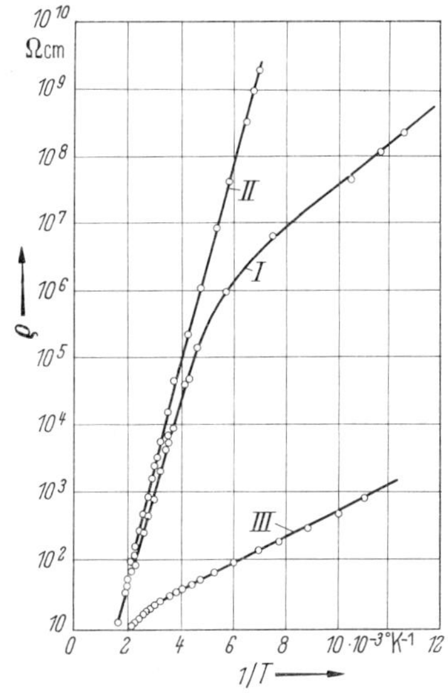

Fig. 12. $Mn_{1.16}Fe_{1.84}O_4$. Temperature dependence of electric resistivity of Mn ferrite single crystal. I) before annealing, $q = 0.61$ eV; II) annealing at 800 °C in air for 5 h, $q = 0.29$ eV; III) annealing at 800 °C in vacuo for 10 h, $q = 0.04$ eV [62ZSM].

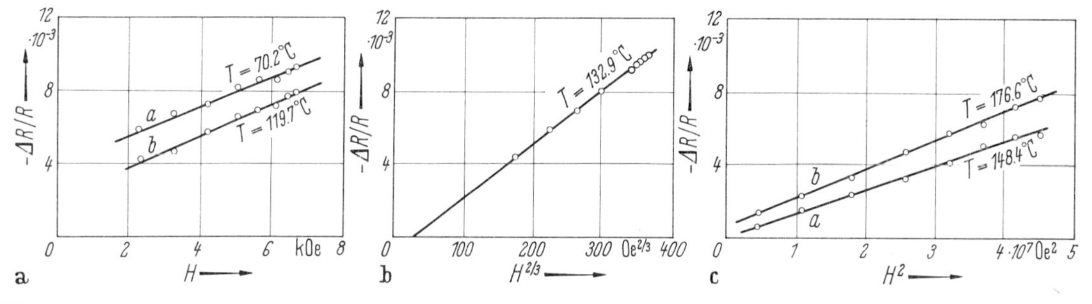

Fig. 13. Mn ferrite (composition not specified) [66PRM.] The variation of resistance is proportional to H below the Curie temperature (Fig. a), proportional to $H^{2/3}$ in the region of the Curie point $\Theta_C \cong 133$ °C (Fig. b), and proportional to H^2 above the Curie temperature (Fig. c).

c) System Zn-Mn-Fe-O

Tab. 2: see p. 142

Tab. 3. Mn-Zn ferrites. Electrical conductivity [61Z]

Composition	Θ_C °C	$T_{\Delta q}$ °C	Δq eV	q eV	$\log \varrho_0$	$\log \varrho_{\Theta_C}$
$Mn_{0.56}Zn_{0.49}Fe_{1.95}O_{4.04}$	83.1			0.37_8	−0.91	4.4_5
$Mn_{0.56}Zn_{0.49}Fe_{1.95}O_{4.05}$	70.5			0.43_9	−0.66	5.7_7
$Mn_{0.64}Zn_{0.48}Fe_{1.88}O_{4.01}$	94.5			0.31_6	−0.75	3.5_8
$Mn_{0.64}Zn_{0.48}Fe_{1.88}O_{4.01}$	81.0	77.0	0.02_4	0.40_9	−1.30	4.5_0
$Mn_{0.64}Zn_{0.48}Fe_{1.88}O_{4.02}$	73.0	74.0	0	0.44_5	−0.85	5.5_7
$Mn_{0.64}Zn_{0.48}Fe_{1.88}O_{4.02}$	73.8			0.42_1	−0.83	5.2_9
$Mn_{0.74}Zn_{0.44}Fe_{1.82}O_{4.01}$	71.2	69.0	0.01_8	0.43_6	−1.20	5.1_8
$Mn_{0.74}Zn_{0.44}Fe_{1.82}O_{4.01}$	71.5	69.5	0.02_6	0.44_6	−1.06	5.4_6

The point of inflexion of the curve $\mu(T)$ was taken for the Curie temperature Θ_C.

$T_{\Delta q}$ = the temperature corresponding to the eventual break or shift of the linear plot $\log \varrho(1/T)$.
Δq = the magnitude of the change in the activation energy, q, from the formula $\varrho = \varrho_0 \exp(q/kT)$.
ϱ_{Θ_C} = the resistivity at the transition point.

Tab. 2. Mn ferrites, MnZn ferrite. Electrical resistivity, ϱ, and activation energy, q, above the Curie temperatures. Single crystals grown in air by Brigdman technique. From [*66S* (3)]. See thermal conductivity results

Sample	ϱ (300 °K) Ωcm	q above Θ_C eV
$Mn_{0.95}Fe_{2.05}O_4$	0.075	0.074
$Mn_{1.00}Fe_{2.00}O_4$	0.22	0.075
$Mn_{1.05}Fe_{1.95}O_4$	2.5	0.211
$Mn_{1.10}Fe_{1.90}O_4$	58	0.354
$Mn_{0.5}Zn_{0.5}Fe_2O_4$	4.8	0.177

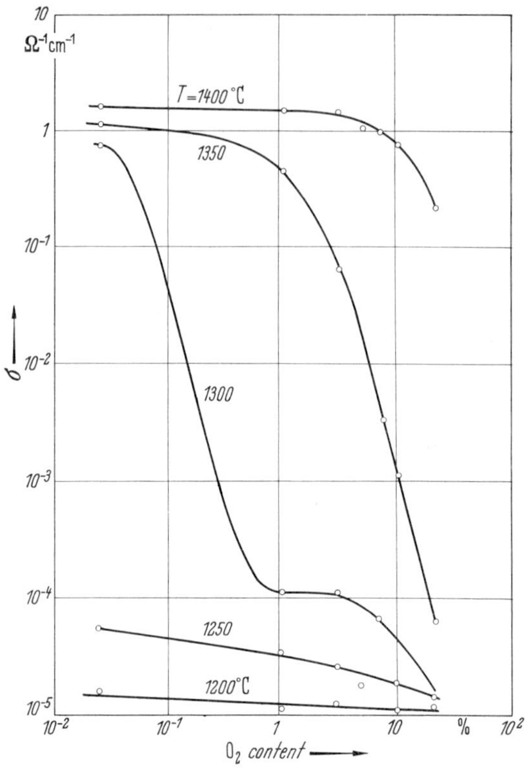

→

Fig. 14. MnZn ferrite. Conductivity, σ, as a function of the oxygen content of the firing atmosphere. Starting composition containing 54.3 mol% Fe_2O_3, 27.9 mol% MnO, 17.8 mol% ZnO. The samples are maintained at the peak temperature for 4 h, then air quenched [*64MJ*].

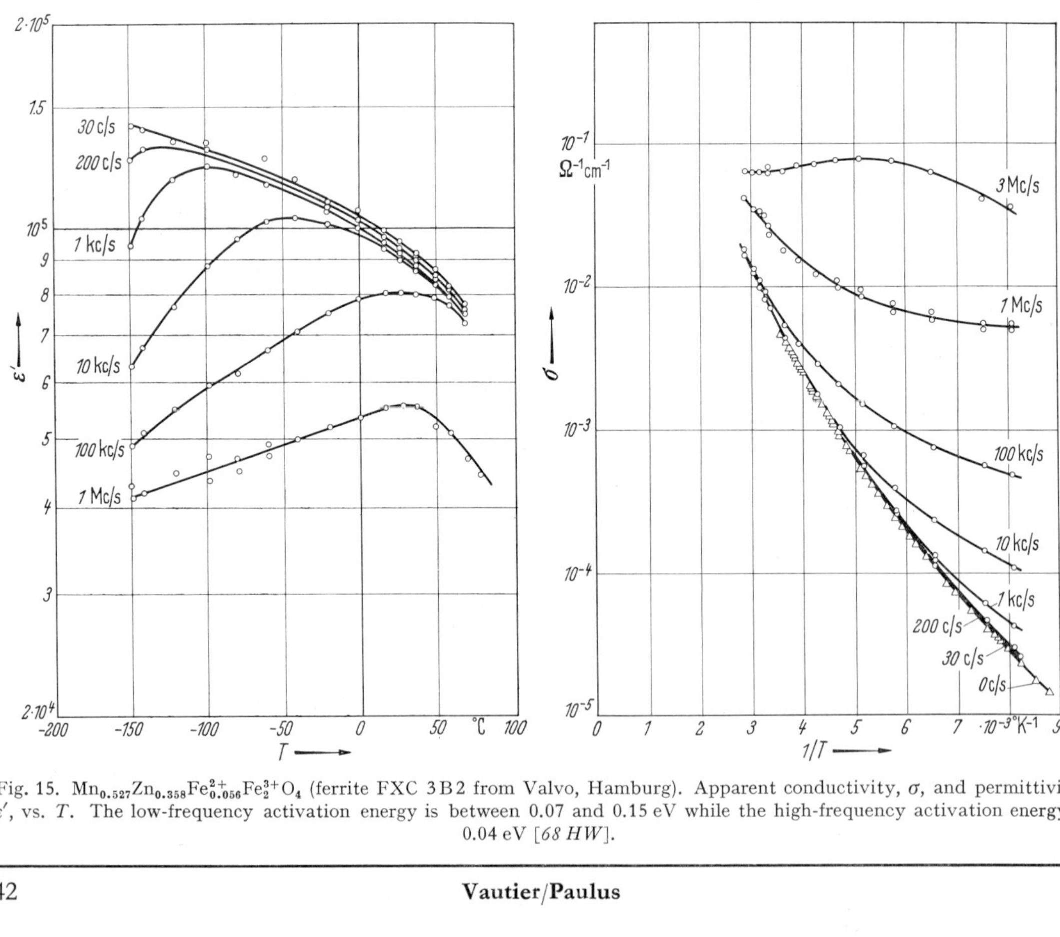

Fig. 15. $Mn_{0.527}Zn_{0.358}Fe^{2+}_{0.056}Fe^{3+}_2O_4$ (ferrite FXC 3B2 from Valvo, Hamburg). Apparent conductivity, σ, and permittivity, ε', vs. T. The low-frequency activation energy is between 0.07 and 0.15 eV while the high-frequency activation energy is 0.04 eV [*68 HW*].

d) System Mg-Mn-Fe-O

Tab. 4. Physical parameters of solid solutions of Mg and Mn ferrites [*67SR* (1)]. $\bar{d}$: mean crystallite size, q_c = conduction-electron activation energy

Composition	$\bar{d}$ μm	ϱ g/cm³	μ_i	q_c eV	a Å	ϱ Ωcm
$MgFe_2O_4$	7	3.73	28	0.33	8.352	$1.34 \cdot 10^6$
$Mn_{0.2}Mg_{0.8}Fe_2O_4$	6	3.80	46	0.56	8.375	$5.67 \cdot 10^8$
$Mn_{0.4}Mg_{0.6}Fe_2O_4$	50	3.83	53	0.54	—	$1.87 \cdot 10^8$
$Mn_{0.6}Mg_{0.4}Fe_2O_4$	90	3.96	60	0.53	8.391	$1.4 \cdot 10^8$
$Mn_{0.6}Mg_{0.4}Fe_2O_4$	130	4.06	70	0.49	—	$2.78 \cdot 10^7$
$Mn_{0.8}Mg_{0.2}Fe_2O_4$	170	—	77	0.46	8.413	$6.15 \cdot 10^6$
$MnFe_2O_4$	150	4.11	84	0.45	8.425	$3.03 \cdot 10^6$

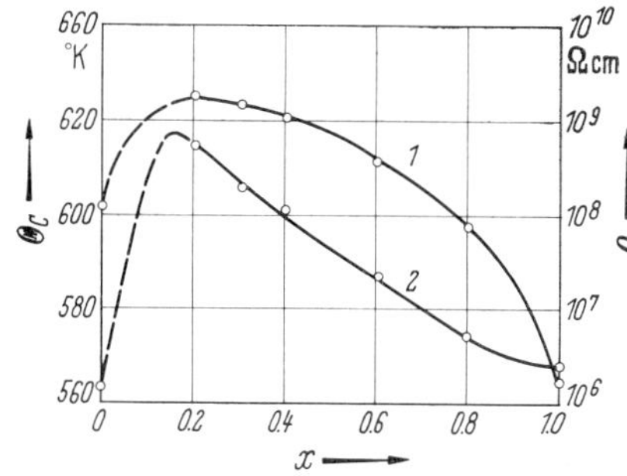

Fig. 16. $Mn_xMg_{1-x}Fe_2O_4$. Curie temperature, Θ_C, (curve 1) and the resistivity, ϱ, (curve 2) as a function of Mn concentration, x. [*67SR* (1)].

e) System Li-Mn-Fe-O

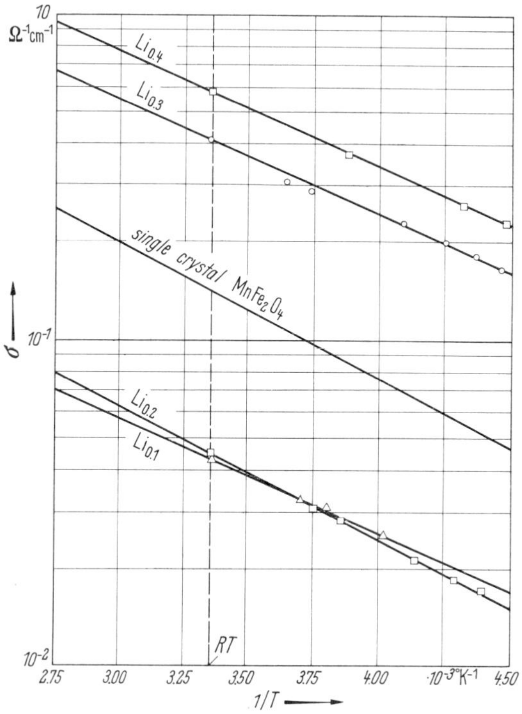

←

Fig. 17. MnLi ferrite. Electrical conductivity, σ, as a function of temperature for different compositions [*63MF*]. The samples are the same as for Fig. 15 in 6.1.3.5.

f) System Cu-Mn-Fe-O

Tab. 5. Cu Mn ferrites. Room temperature resistivity, ϱ, values of the activation energies, q_1 and q_2, below and above the temperature, T_1, at which a change of the slope in log R vs. $1/T$ curve occurs, and Curie temperature, Θ_C, for different compositions. All samples were quenched to room temperature [*66RNB*]

Composition	Sintering temperature °C	ϱ (300 °K) Ωcm	q_1 eV	q_2 eV	T_1 °C	Θ_C °C
$Cu_{0.14}Mn_{0.81}Fe_{2.05}O_4$	1250	$2.9 \cdot 10^5$	0.4	0.5	325	325
$Cu_{0.28}Mn_{0.69}Fe_{2.03}O_4$	1250	$9.1 \cdot 10^5$	0.39	0.53	333	338
$Cu_{0.49}Mn_{0.45}Fe_{2.06}O_4$	1200	—	0.52	0.56	352	378

g) CaO and SiO_2 or GeO_2 additions

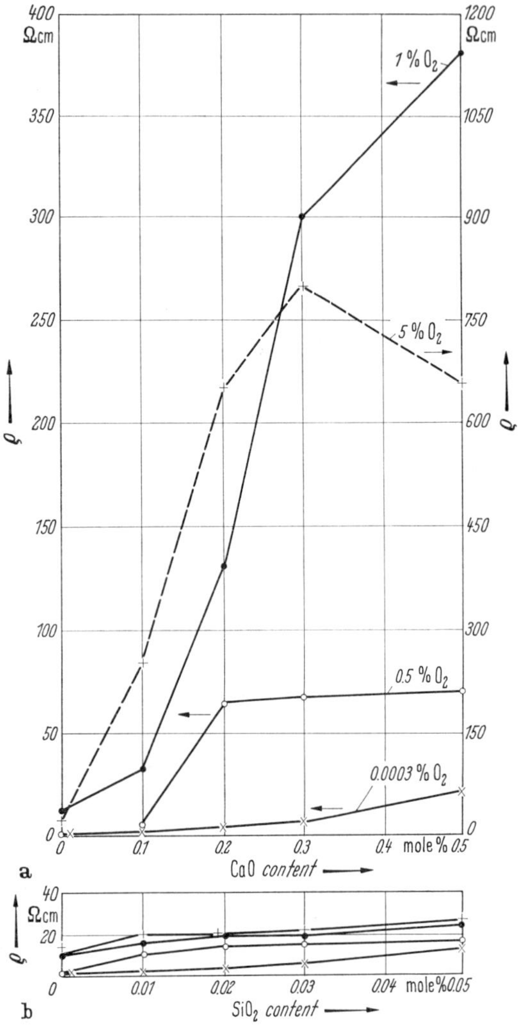

Fig. 18. Changes in resistivity of Mn ferrites by the addition of CaO (a), or SiO_2 (b) [*66A*]. The Mn ferrites were fired at 1250 °C in varying atmospheres. Values at the right hand ordinate and dotted line are used for the case oxygen content is 5%). Signs in (b) indicate the same oxygen contents as in (a).

Fig. 19 see next page

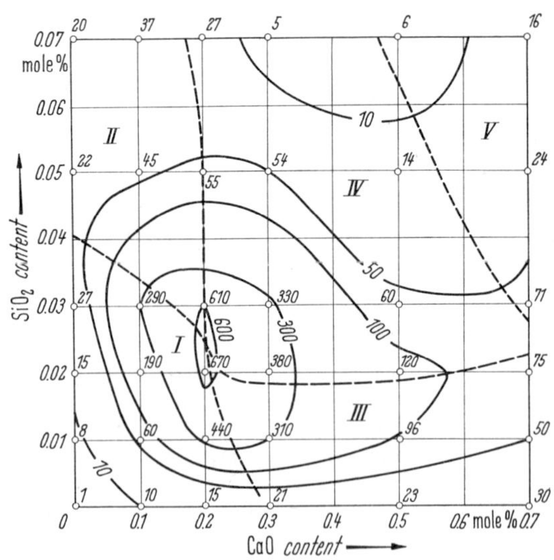

Fig. 20. Specific resistivity (in Ω cm) as a function of the CaO and SiO_2 contents in $Mn_{0.68}Zn_{0.21}Fe_{2.11}O_{4+\delta}$. Dots with a numerical value represent the resistivity measured at the corresponding amounts of impurities. All the ferrite specimens were fired at 1200 °C for 4 h in an atmosphere of N_2 [*61A*].

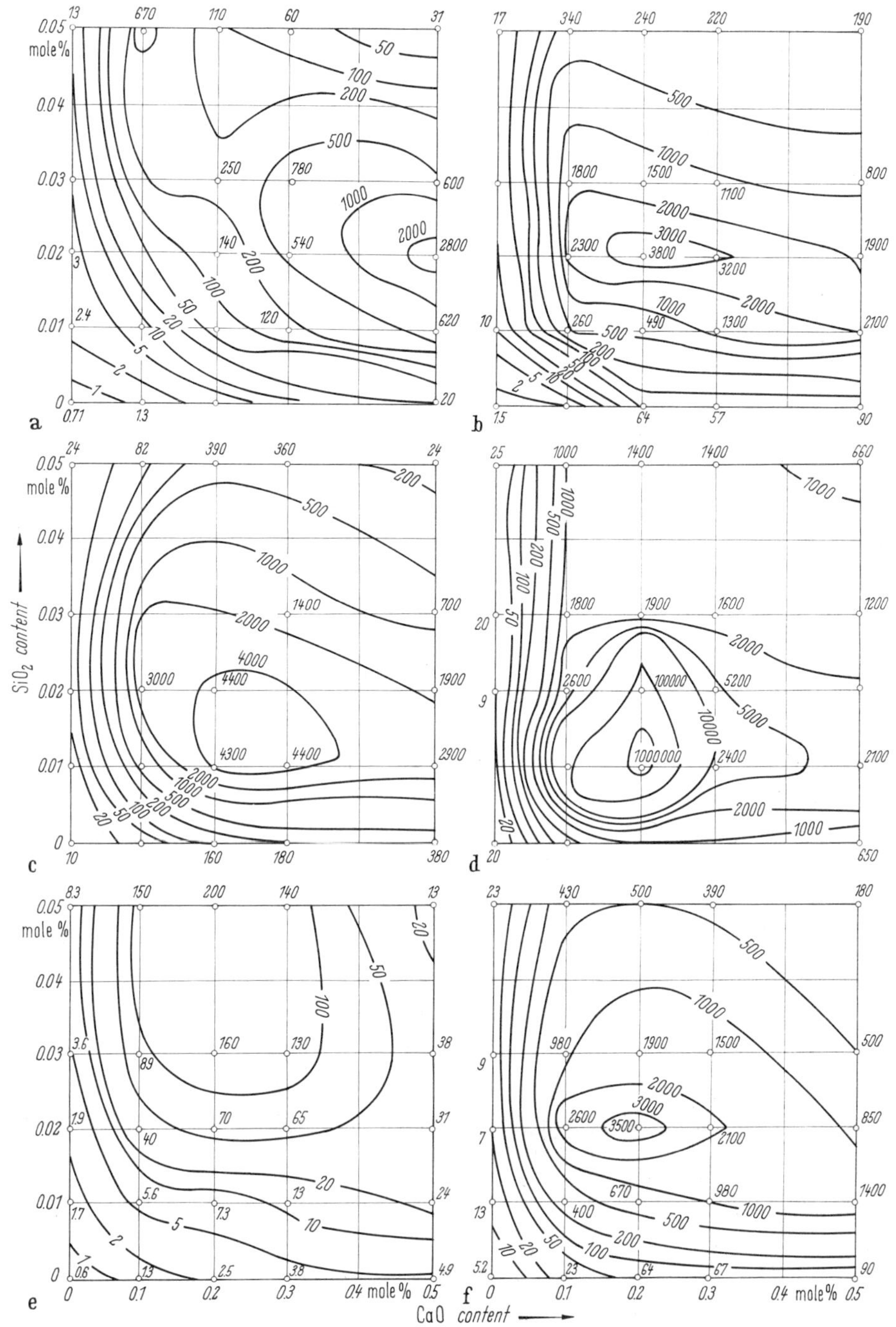

Fig. 19. Mn ferrites. Changes in resistivity by simultaneous additions of CaO and SiO_2. Numerical values in the diagrams indicate the resistivities in Ωcm. The six diagrams differ in the firing conditions [*66A*].

Diagram	Firing temperature °C	O_2 content in atmosphere %
a	1250	0.0003
b	1250	0.5
c	1250	1
d	1250	5
e	1300	0.0003
f	1300	1

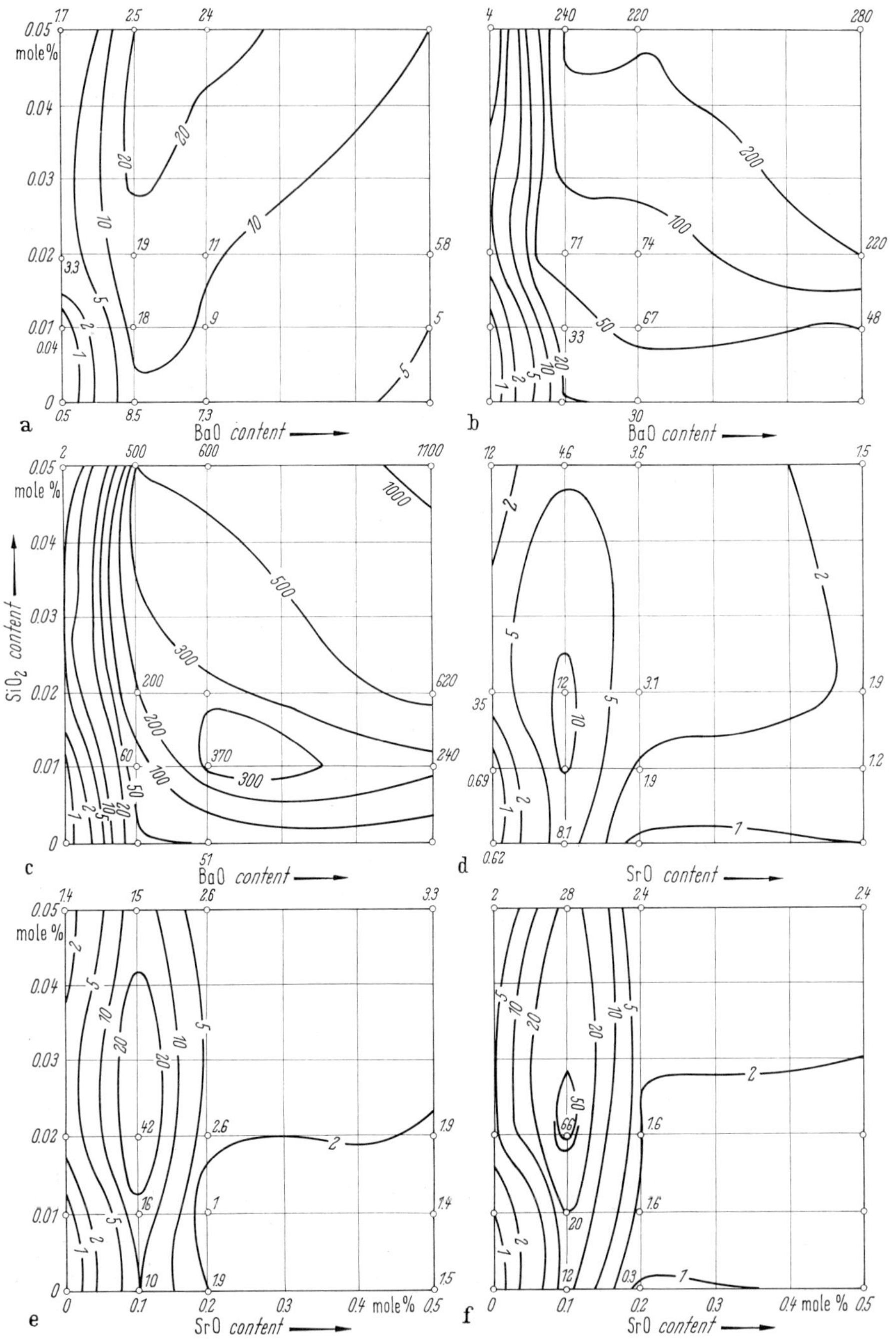

Fig. 21. Changes in resistivity of $(Mn_{0.59}Zn_{0.33}Fe_{2.08}O_{4+\delta})_{1-x-y}(MeO)_x(SiO_2)_y$ by simultaneous additions of MeO and SiO_2. The six diagrams differ in the kind of Me and also in the firing conditions [*66A*].

Diagram	Me	Firing temperature °C	O_2 content in atmosphere %
a	Ba	1200	0.0003
b	Ba	1200	0.5
c	Ba	1200	1
d	Sr	1200	0.0003
e	Sr	1200	0.5
f	Sr	1200	1

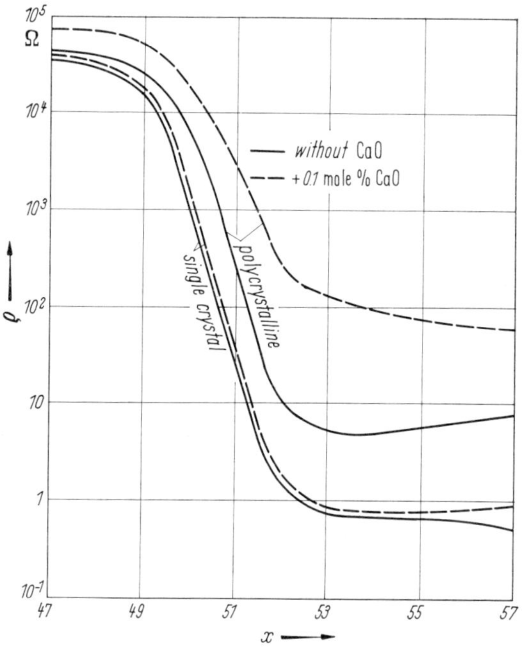

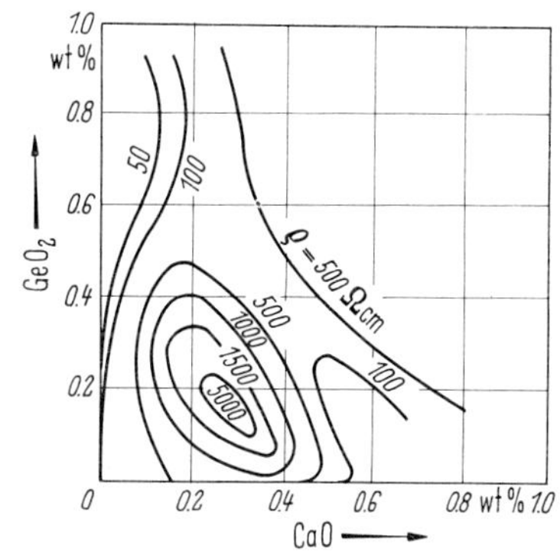

Fig. 23. Mn-Zn ferrite (57 mol% Fe_2O_3, 29.5 mol% MnO, 13.5 mol% ZnO) with CaO and GeO_2. Resistivity, ϱ, as a function of added amounts of CaO and GeO_2. The ferrites were prepared by sintering at 1250 °C for 2 h in air and furnace-cooling in N_2 + 0.2% O_2 [66H].

←

Fig. 22. MnZn ferrites (28 MnO-(72 − x) ZnO). Resistivity, ϱ, of the polycrystalline samples and inside the crystals as a function of the Fe_2O_3 content, x, without and with 0.1 mol% CaO [62PG].

References for section 6.1.3.4.1: [41VH], [50GB (1)], [53Ro], [54KK], [56B], [56GPV], [57BB], [59FMU (1)].

6.1.3.4.2 Magnetoresistance and thermomagnetic effect

System Mn-Fe-O

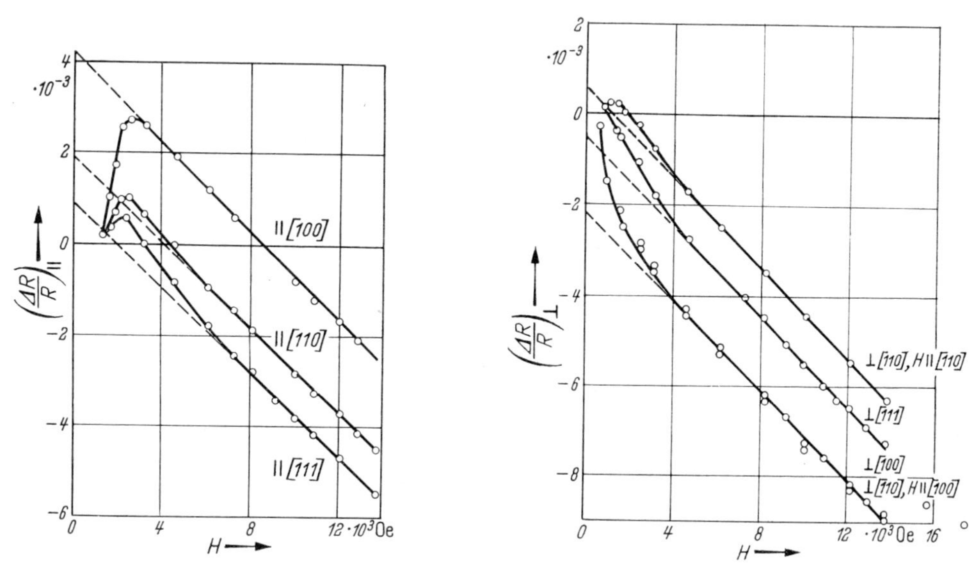

Fig. 24a. The main longitudinal magnetoresistance effect. Fig. 24b. The main transverse magnetoresistance effects.

Fig. 24. $Mn_{1.12}Fe_{1.88}O_4$. Magnetoresistance effects on a single crystal [59Z]. Circular disks with their major plane parallel to each of the principal crystallographic planes; diameter 0.6 cm, thickness 0.276 cm. Silver contacts were fired on to the flat surfaces. To measure the longitudinal magnetoresistance effect the field was normal to these faces. To measure the transverse effect, the field was in the plane of the surface. It was found that the expression

$$\Delta R/R = 3/2\, h_{100}(\alpha_1^2\beta_1^2 + \alpha_2^2\beta_2^2 + \alpha_3^2\beta_3^2 - 1/3) + 3 h_{111}(\alpha_1\alpha_2\beta_1\beta_2 + \alpha_2\alpha_3\beta_2\beta_3 + \alpha_3\alpha_1\beta_3\beta_1)$$

from [39A] describes the effects observed. The constants were found to be $h_{100} = 4.25\cdot 10^{-3}$ and $h_{111} = 0.9\cdot 10^{-3}$.

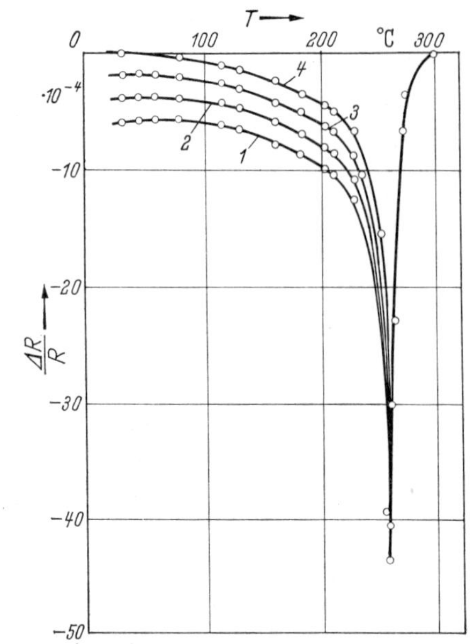

Fig. 25. $Mn_{1.0}Fe_{2.0}O_4$. Temperature dependence of the longitudinal magnetoresistance for a single crystal. The sample was a cylinder 5.5 cm long and 0.28 cm in diameter. The [111] direction was very nearly parallel to the axis of the rod. [*58BPT*].

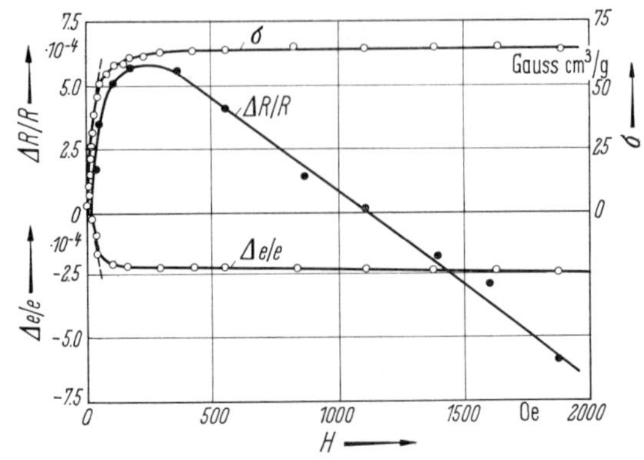

Fig. 26. $Mn_{1.13}Fe_{1.87}O_4$. The dependence of the thermomagnetic effect, $\Delta e/e$, of magnetoresistance, $\Delta R/R$, and of the specific magnetization, σ, on magnetic field for a ferrite single crystal [*65BTK*]. e = thermo emf.

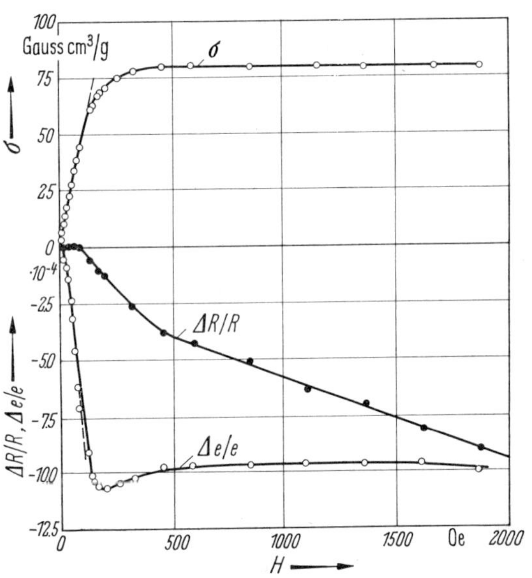

Fig. 27. $Mn_{0.87}Fe_{2.13}O_4$. The dependence of the thermomagnetic effect, $\Delta e/e$, of magnetoresistance, $\Delta R/R$, and the specific magnetization, σ, on magnetic field for a ferrite single crystal [*65BTK*].

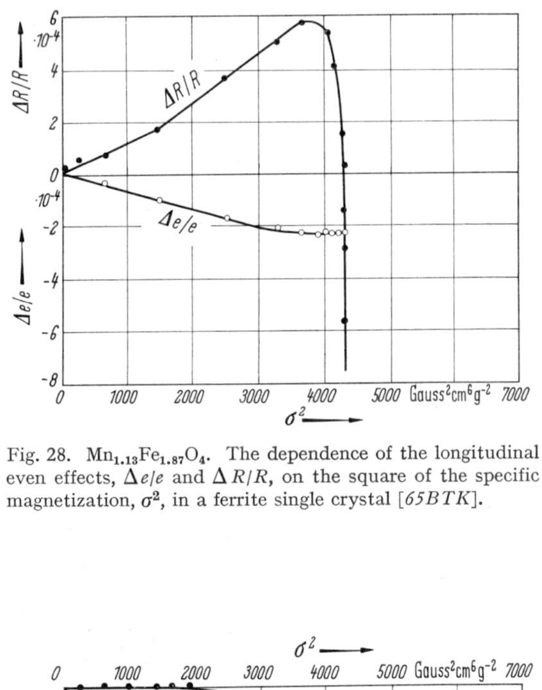

Fig. 28. $Mn_{1.13}Fe_{1.87}O_4$. The dependence of the longitudinal even effects, $\Delta e/e$ and $\Delta R/R$, on the square of the specific magnetization, σ^2, in a ferrite single crystal [*65BTK*].

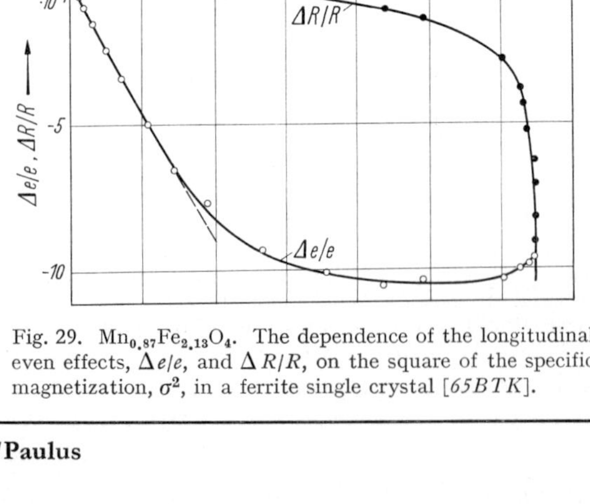

Fig. 29. $Mn_{0.87}Fe_{2.13}O_4$. The dependence of the longitudinal even effects, $\Delta e/e$, and $\Delta R/R$, on the square of the specific magnetization, σ^2, in a ferrite single crystal [*65BTK*].

6.1.3.4.3 Hall effect

System Mn-Fe-O

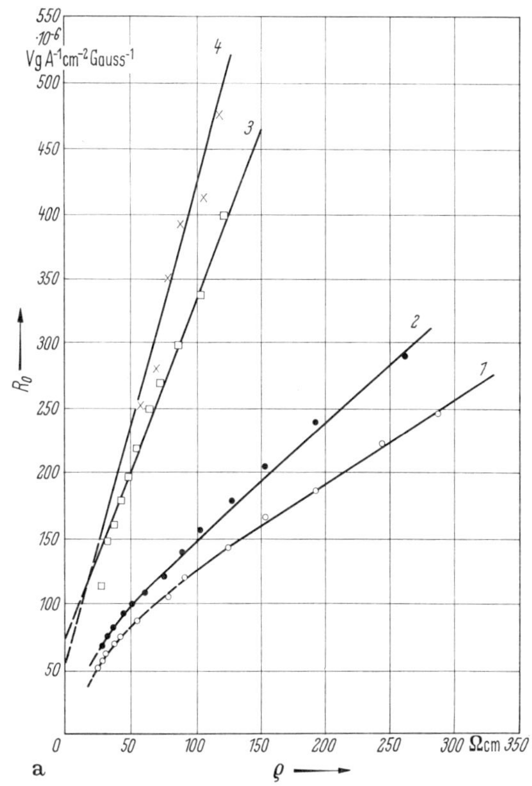

E_{Hall} = $R \cdot H + R_0\sigma_0 + R_i\sigma_i$ (above saturation)
R = classical Hall coefficient
R_0 = spontaneous Hall coefficient
R_i = Hall coefficient due to the intrinsic magnetization σ_i
σ_0 = spontaneous magnetization
σ_i = increase of the intrinsic magnetization (magnetization of the paraprocess).

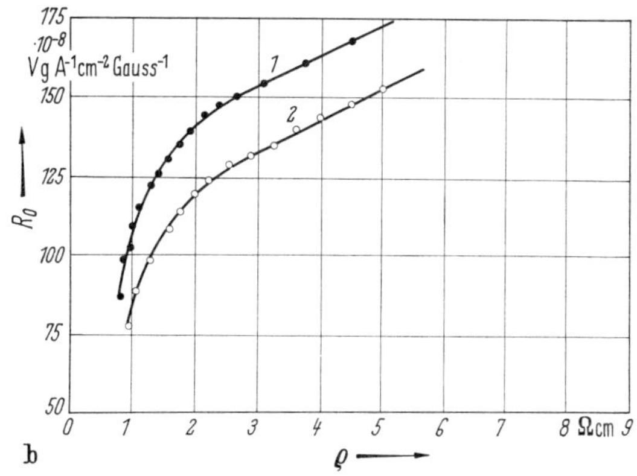

Fig. 30. Dependence of the spontaneous Hall coefficient, R_0, on the electrical resistivity of Mn ferrite single crystals: a) With a deficiency of Fe ions (1) $Mn_{1.03}Fe_{1.97}O_4$; (2) $Mn_{1.06}Fe_{1.94}O_4$; (3) $Mn_{1.16}Fe_{1.84}O_4$; (4) $Mn_{1.28}Fe_{1.72}O_4$); b) with an excess of iron ions (1) $Mn_{0.87}Fe_{2.13}O_4$; (2) $Mn_{0.84}Fe_{2.16}O_4$) [*64S*]. R_0 was defined by the relation $E_{Hall} = RH + R_0\sigma_0 + R_i\sigma_i$ valid above technical saturation [*61BS*].

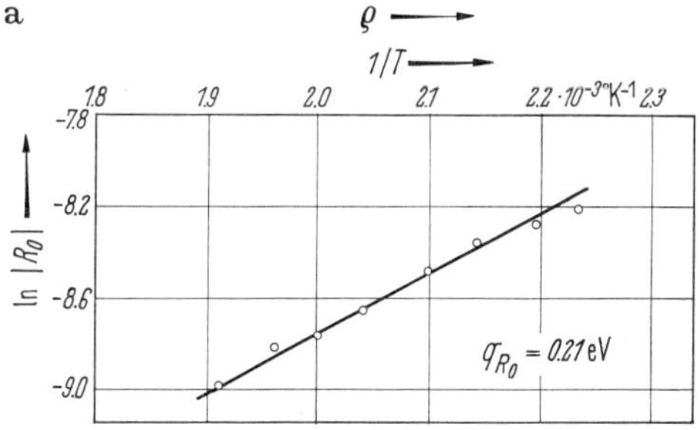

Fig. 31. $Mn_{1.16}Fe_{1.84}O_4$. Dependence of lnR_0 (R_0 = spontaneous Hall coefficient) on $1/T$ for single crystal Mn ferrite with a deficiency of Fe ions [*64S* (5)]. q = activation energy.

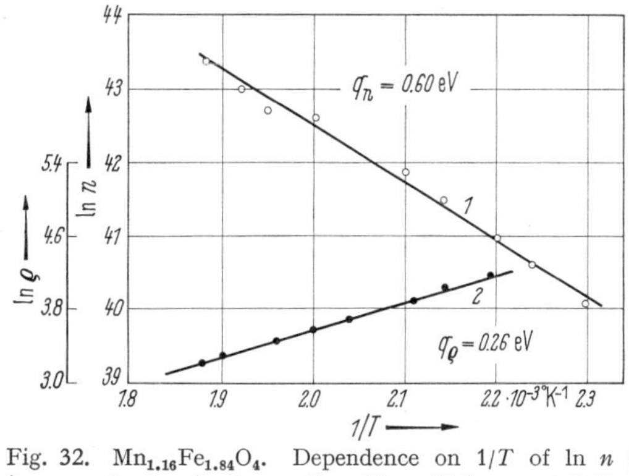

Fig. 32. $Mn_{1.16}Fe_{1.84}O_4$. Dependence on $1/T$ of ln n (1) (n = carrier density in cm^{-3}) and ln ϱ (2) (ϱ = resistivity in Ωcm) for a single crystal of Mn ferrite with a deficiency of Fe ions [*64S* (5)]. q = activation energy.

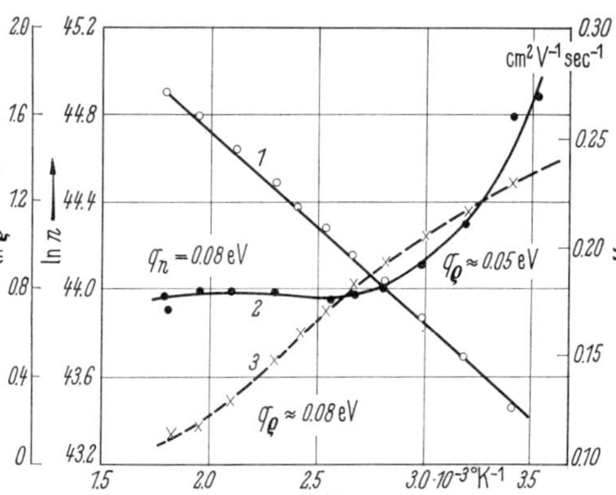

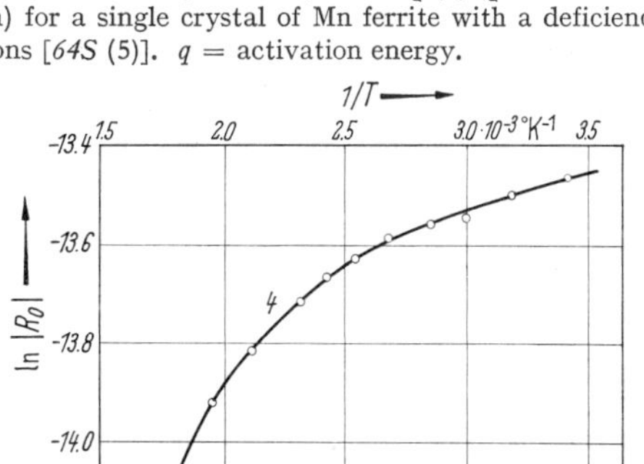

Fig. 33. $Mn_{0.84}Fe_{2.16}O_4$. Dependence on $1/T$ of (*1*) ln n (n = carrier density in cm^{-3}); (*2*) μ (μ = carrier mobility in $cm^2V^{-1}sec^{-1}$); (*3*) ln ϱ (ϱ = resistivity in Ωcm); (*4*) ln R_0 (R_0 = spontaneous Hall coefficient in $VgA^{-1}cm^{-2}Gauss^{-1}$) for single crystal Mn ferrite with an excess of Fe ions [*64S* (5)]. The respective activation energies are given in the figure.

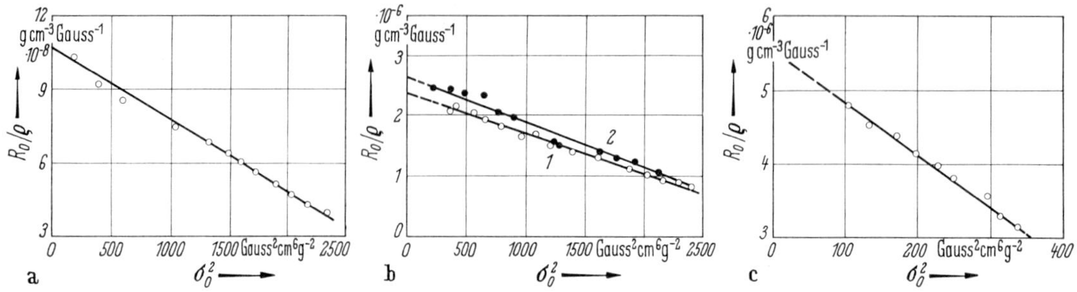

Fig. 34. Dependence of the ratio R_0/ϱ on the square of the spontaneous magnetization, σ_0^2, for Mn ferrites of various compositions [*64S* (5)].

a) $Mn_{0.84}Fe_{2.16}O_4$, b) (1) $Mn_{1.03}Fe_{1.97}O_4$, (2) $Mn_{1.06}Fe_{1.94}O_4$, c) $Mn_{1.16}Fe_{1.84}O_4$.

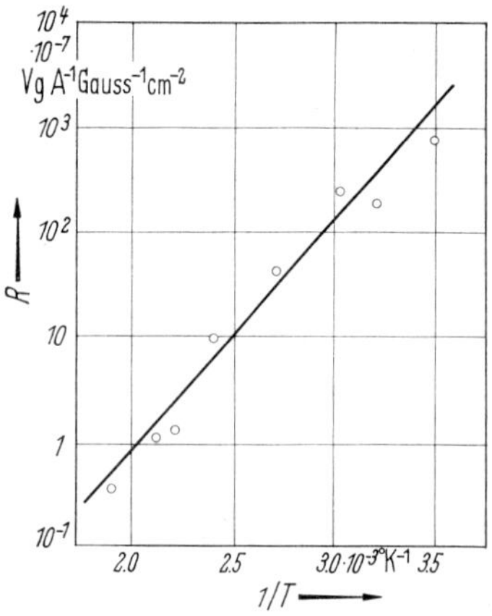

Fig. 35. $Mn_{1.16}Fe_{1.84}O_4$. Temperature dependence of classical Hall constant, R, for a Mn ferrite single crystal [*62ZSM*].

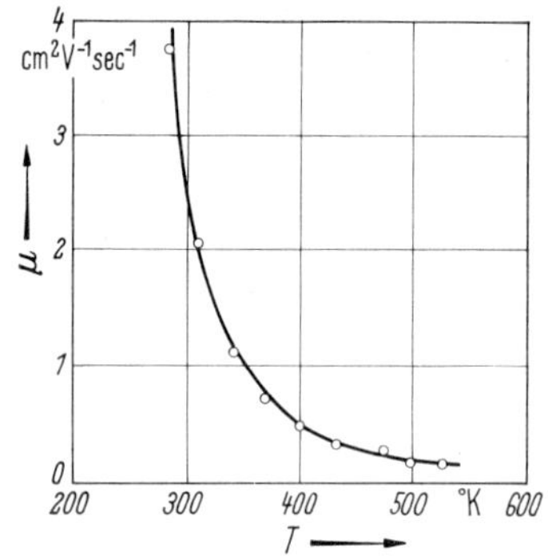

Fig. 36. $Mn_{1.16}Fe_{1.84}O_4$. Temperature dependence of current-carrier mobility, μ, in Mn ferrite single crystal [*62ZSM*].

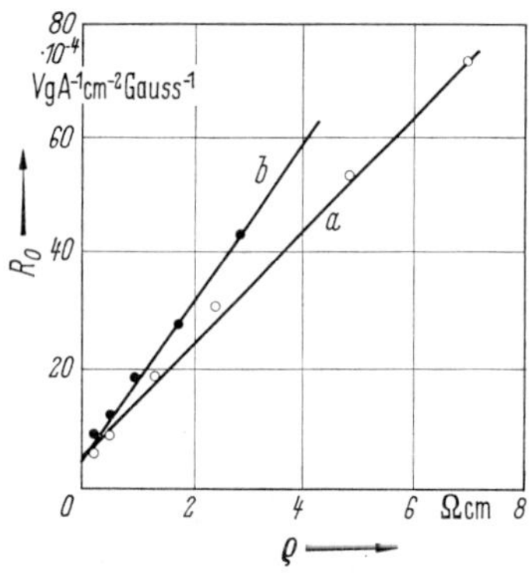

Fig. 37. $Mn_{1.16}Fe_{1.84}O_4$. Dependence of the spontaneous Hall coefficient, R_0, on the resistivity, ϱ, of Mn ferrite single crystal [*62ZSM*].

a) before annealing, b) annealing at 800 °C in air for 5h.

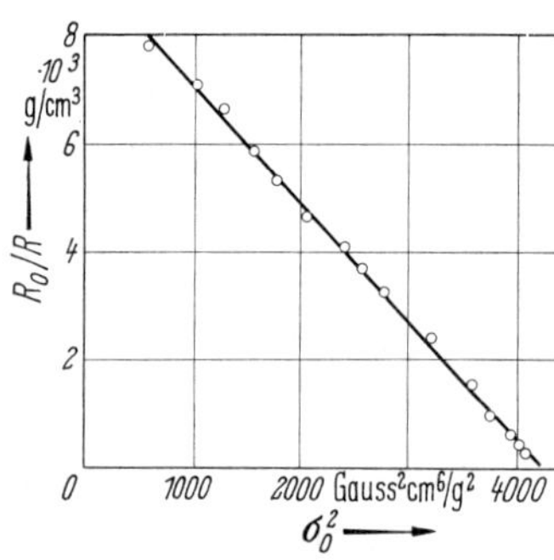

Fig. 38. $Mn_{0.74}Fe_{2.26}O_4$. Dependence of the ratio R_0/R on the square of the spontaneous magnetization, σ, for a single crystal of Mn ferrite [*66SMM*].

R = classical Hall coefficient; R_0 = spontaneous Hall coefficient. For definition see Fig. 30.

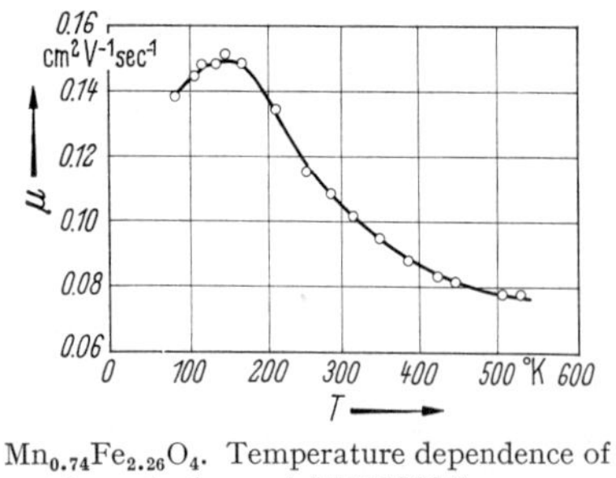

Fig. 39. $Mn_{0.74}Fe_{2.26}O_4$. Temperature dependence of the Hall mobility, μ, for a single crystal [*66SMM*].

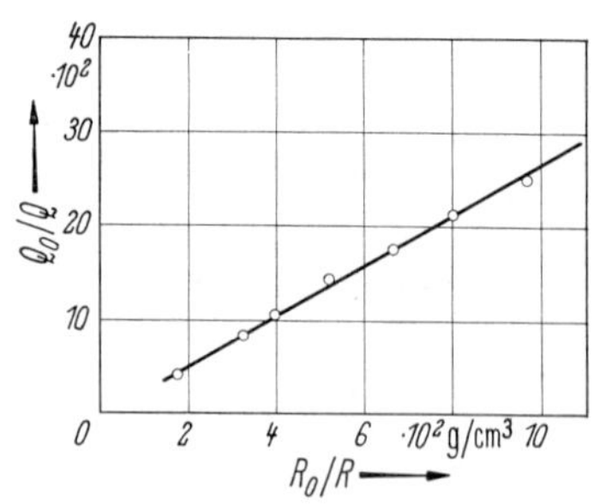

Fig. 40. $Mn_{0.74}Fe_{2.26}O_4$. Dependence of the ratio Q_0/Q on R_0/R for a single crystal [*66SMM*]. Q_0 = spontaneous Nernst-Ettingshausen coefficient, Q = classical Nernst-Ettingshausen coefficient, R_0 = spontaneous Hall coefficient, R = classical Hall coefficient. For definition see Fig. 30.

6.1.3.4.4 Seebeck effect

a) System Mn-Fe-O

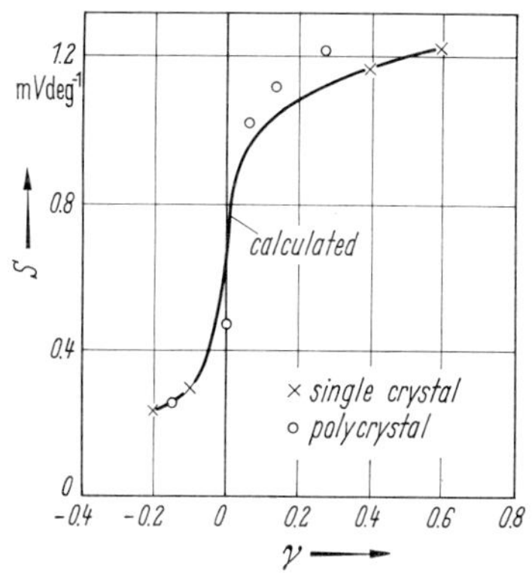

Fig. 41. $Mn_{1+\gamma}Fe_{2-\gamma}O_4$. Seebeck coefficient, S, at about 300 °K as a function of the composition [*64L*].

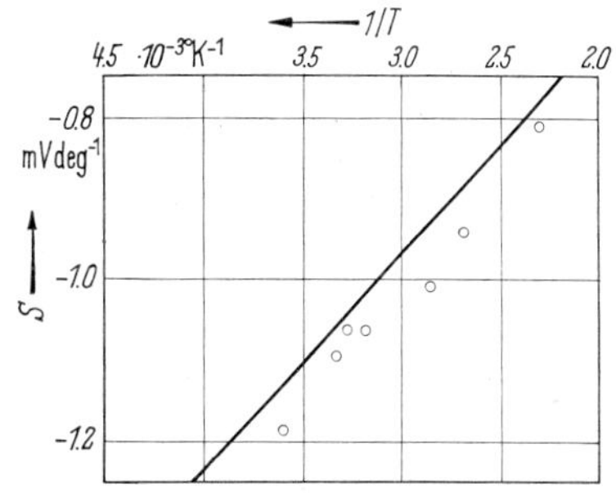

Fig. 42. $Mn_{1.14}Fe_{1.86}O_4$, single crystal. Seebeck coefficient, S, as a function of T^{-1} [*64L*]. The straight curve is calculated by means of the equation

$$S = \frac{k}{e} \ln \frac{[Mn^{2+}]_0}{[Mn^{3+}]_0} - \frac{E_g + \alpha}{eT}$$

putting $E = 0.30$ eV and $\alpha = 0.013$ eV. E_g = energy required for the reaction $Mn^{2+} + Fe^{3+} \rightarrow Mn^{3+} + Fe^{2+}$;
α = parameter for taking into account a possible lattice energy transport by the hopping electron.

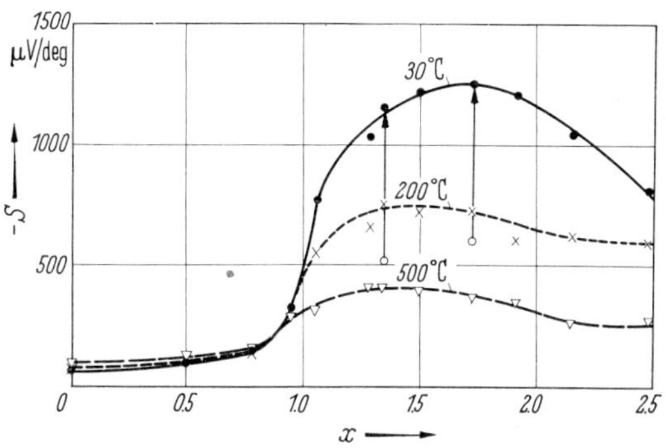

Fig. 43. $Mn_xFe_{3-x}O_{4+\gamma}$. Dependence of Seebeck coefficient, S, on chemical composition, x, in single crystals for temperatures of 30, 200 and 500 °C.

○ Values of S before thermal treatment.

● Values of S after annealing (4 h at 1280 °C in air for single crystals prepared by the Bridgman method or 34 h at 1350 °C in Ar for single crystals prepared by the Verneuil method, and rapid cooling) [*66GSV*].

▽ [*57HOT*]

× [*61M*]

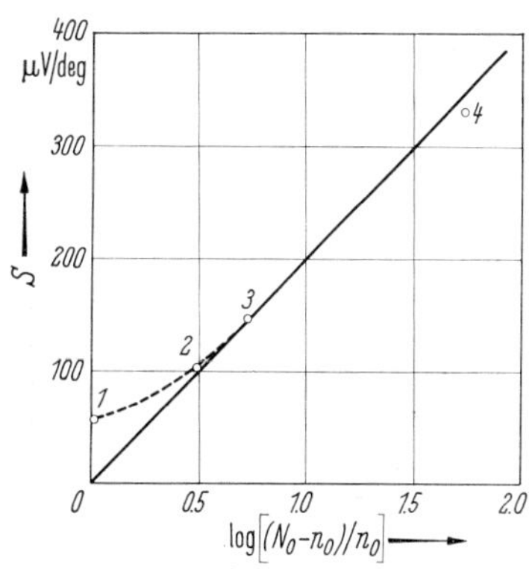

Fig. 44. $Mn_xFe_{3-x}O_{4+\gamma}$. Theoretical and experimental values of Seebeck coefficient, S, at RT of 4 samples

Sample	x	γ
1	0	0.010
2	0.50	0.010
3	0.78	−0.045
4	0.95	0.010

The theoretical curve was calculated on the basis of the equation

$$S = -\frac{k}{e}(\eta + a) = 198.5 \log \frac{n_0}{N_0 - n_0} - \frac{a k}{e}$$

where k is the Boltzmann constant, η the reduced Fermi energy, a the kinetic term (= 0 in this case), N_0 the number of states on the Fe^{2+} levels and n_0 the number of the free electrons on the Fe^{2+} levels at 0 °K. These results were given as a proof in favour of the model where the pairs $Mn^{3+} + Fe^{2+}$ are more stable than the pairs $Mn^{2+} + Fe^{3+}$ as a consequence of the Jahn-Teller effect. [*66S* (1)]. For $x > 1$, the disagreement between experiment and theory can be explained by the formation of clusters of Mn ions [*67S*].

b) System Cu-Mn-Fe-O

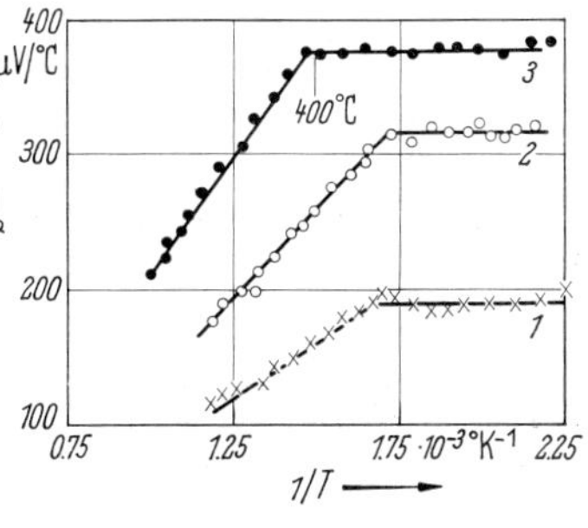

Fig. 45. Seebeck coefficient, S, vs. T^{-1} for the compositions
1) $Cu_{0.14}Mn_{0.81}Fe_{2.05}O_4$
2) $Cu_{0.28}Mn_{0.69}Fe_{2.03}O_4$
3) $Cu_{0.49}Mn_{0.45}Fe_{2.06}O_4$
from [*66RNB*].

6.1.3.5 Spontaneous magnetization

6.1.3.5.1 Saturation moment at 0 °K

a) System Mn-Fe-O

Tab. 1. Saturation moment, p_{sm}, at 0 °K of Mn ferrite samples. The magnetization was measured at room temperature and at liquid nitrogen temperature between 5 000 and 13 000 Oe and extrapolated back to zero field, then to absolute zero using a $T^{3/2}$ relation with the coefficient of $5.5 \cdot 10^{-5}$ (°K^{-1}) [*63BB*]. x gives the composition by the "formula" for Mn ferrite: $(Mn^{2+}_{1-x}Fe^{3+}_x)[Mn^{3+}_xFe^{2+}_xFe^{3+}_{2-2x}]O_4$

T °C	Firing conditions: gas mixture	p_{O_2} [atm]	a Å	p_{sm} μ_B	x	
1330	$CO_2/H_2 = 100$	$4.8 \cdot 10^{-6}$	8.509 ± 0.001	4.68	0.16	more reducing conditions
1330	CO_2	$4 \cdot 10^{-4}$	8.510	4.72	0.14	
1250	CO_2	$2 \cdot 10^{-4}$	8.513	4.70	0.15	
1200	CO_2	$1 \cdot 10^{-4}$	8.514	4.72	0.14	
1330	air	0.21	8.510	4.88	0.06	intermediate conditions
1250	air	0.21	8.509	4.88	0.06	
1330	O_2	1.0	8.508	4.88	0.06	
1200	air	0.21	8.510	4.76	0.12	more oxidizing conditions
1250	O_2	1.0	8.504	4.68	0.16	
1200	O_2	1.0	8.501	4.55	0.23	

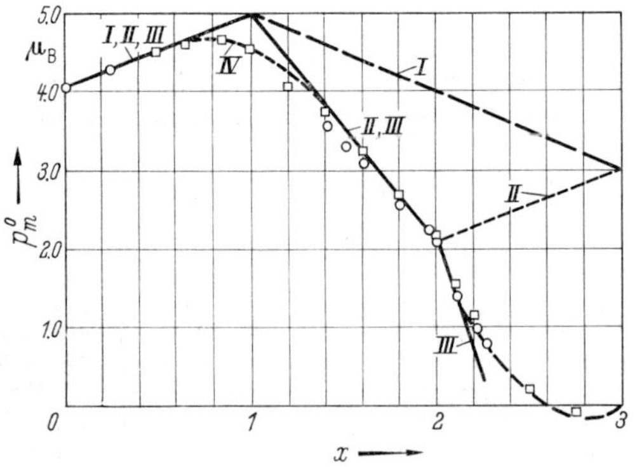

Fig. 1. $Mn_xFe_{3-x}O_4$. The magnetic moments at absolute zero, $p^0_m = n^0_m\,\mu_B$, vs. Fe content, x [*58Es*].
The measurements were made by determination of the magnetic moment at a number of temperatures in the range from 70 °K···300 °K. Except for compositions very close to Fe_3O_4 it was found that the magnetization fell off from 0 °K as $T^{3/2}$, that is in $p_m = p^0_m(1 - T^{3/2})$. (See Fig. 13). For the most Fe rich samples a T^2 term was more appropriate. n^0 determinations were made by extrapolating back from the measurements just mentioned, and in a number of cases these were checked by actual determinations at liquid helium temperatures. The samples were made by very careful ceramic techniques. The firing temperature was 1400 °C, and samples of each composition were fired in two different atmospheres: (1) air, and (2) that pressure between 10^{-6} and 10^{-2} atmospheres of oxygen that had been previously determined to yield the best oxygen stoichiometry for the particular composition. All the samples were single phase though those with a Mn to Fe ratio of greater than 2.0 were found to have extensive spinel twinning. Note that at $MnFe_2O_4$ the moment is stable with respect to atmosphere. Samples fired in oxygen at pressures from $2 \cdot 10^{-4} \cdots 2 \cdot 10^{-1}$ gave the same value, $p_m = (4.58 \pm 0.01)\,\mu_B$. In the region of Mn_2FeO_4 the moment is quite sensitive to oxygen pressure during firing.
○ Air fired, □ controlled atmosphere. I, II, III, IV: theoretical curves.

Tab. 2. Temperature variation of magnetization in the system $Mn_xFe_{3-x}O_4$, $0 \leqq x \leqq 2$. The determination of the magnetic moment at 0 °K as given in Fig. 1 of this section required the measurements of the temperature dependence of the magnetization, and extrapolation to 0 °K. The appropriate form of the temperature variation is given in this table [60E]

Range of x	$p_m(T) = n_m(T) \cdot \mu_B$	C_n [°K$^{-3/2}$]
$x = 0$	$n_m = n_m^0(1 - C_n T^2)$	—
$0.2 \leqq x \leqq 1.0$	$n_m = n_m^0(1 - C_n T^{3/2})$	$C_n = (5.4x + 0.5)$
$1.0 \leqq x \leqq 2.0$	$n_m = n_m^0(1 - C_n T^{3/2})$	$C_n = (6.1x - 0.2)$

b) System Zn-Mn-Fe-O

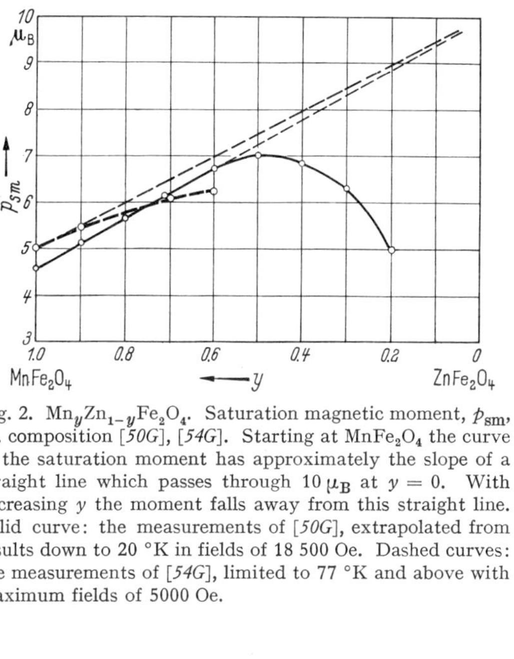

Fig. 2. $Mn_yZn_{1-y}Fe_2O_4$. Saturation magnetic moment, p_{sm}, vs. composition [50G], [54G]. Starting at $MnFe_2O_4$ the curve of the saturation moment has approximately the slope of a straight line which passes through 10 μ_B at $y = 0$. With decreasing y the moment falls away from this straight line. Solid curve: the measurements of [50G], extrapolated from results down to 20 °K in fields of 18 500 Oe. Dashed curves: the measurements of [54G], limited to 77 °K and above with maximum fields of 5000 Oe.

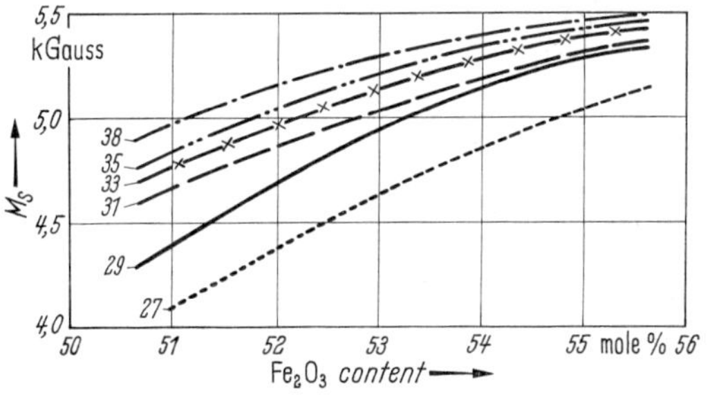

Fig. 3. Saturation magnetization, M_s, of a number of $MnZnFe^{2+}$ ferrites in the composition range of commercial interest [57Gu]. The compositions are in mole % and represent the percentage of the oxides in the starting mixtures. The abscissa gives the Fe_2O_3 fraction, and the remainder is ZnO. Parameter mole % MnO.

c) System Li-Mn-Fe-O

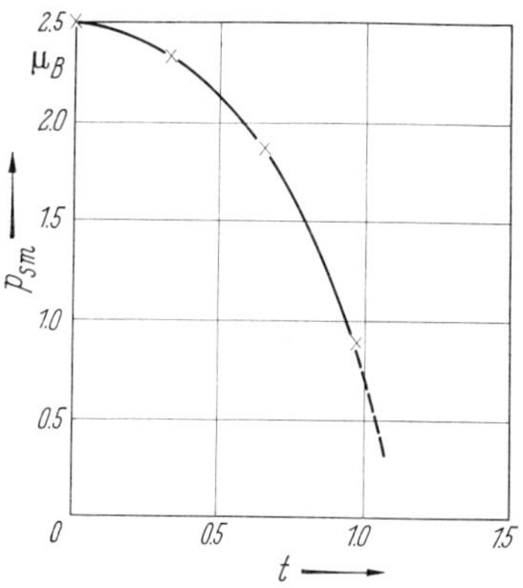

Fig. 4. $Li_{0.5+0.5t}Fe_{2.5-1.5t}Mn_tO_4$. Saturation moments, p_{sm}, vs. Li and Mn content [65B]. For lattice constants see Fig. 37 of 6.1.3.1.

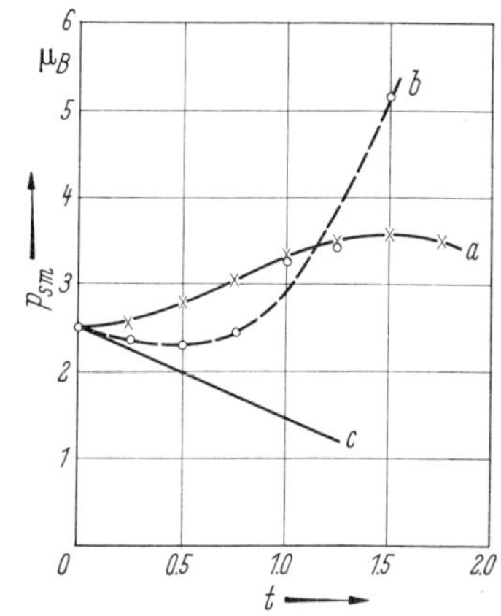

Fig. 5. $Li_{0.5}Mn_tFe_{2.5-t}O_4$. Saturation moments, p_{sm}, vs. Mn content, t [65B]. (a) experimental (crosses); (b) calculated from the cation distribution found from X-ray-reflexion intensities; (c) calculated for all Li^+ and Mn^{3+} ions on B sites. For lattice constants see Fig. 38 of 6.1.3.1.

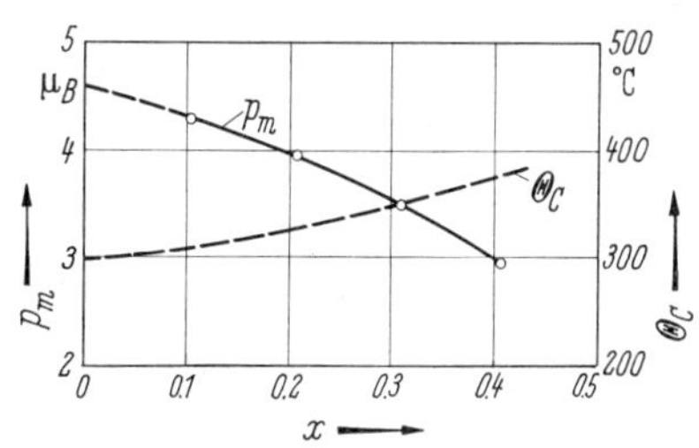

Fig. 6. $Mn_{1-x}Li_xFe_2O_4$. Magnetic moments and Curie temperatures vs. Li substitution, x [63MF]. In preparing the samples the oxides were mixed by ball milling for 8 h in a tungsten carbide mill containing ethyl alcohol. After drying, the temperature was raised over a 5 h period and then held at 1150 °C for 2½ h. The samples were again ball milled for 8 h and 3 wt% of binder added. Pellets of ½ inch diameter were pressed at 13 000 psi followed by sintering at 1400 °C for 1½ h. All heating was carried out in a helium atmosphere. See Fig. 16 and for lattice constants see Fig. 36 of 6.1.3.1.

d) System Mn-Fe-Cr-O, Zn-Mn-Fe-Cr-O

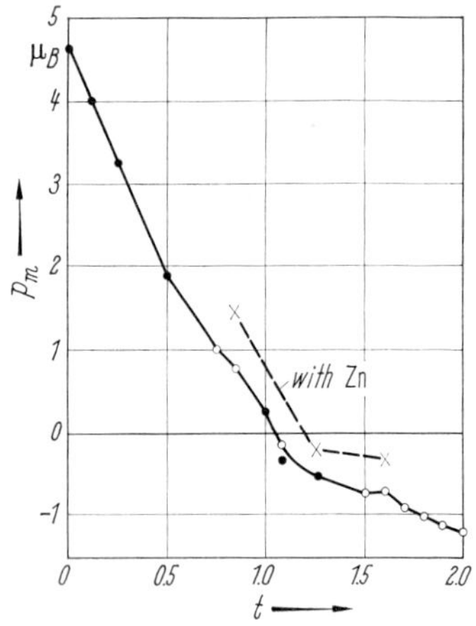

← Fig. 7. $MnFe_{2-t}Cr_tO_4$ (circles), $Mn_{0.8}Zn_{0.2}Fe_{2-t}Cr_tO_4$ (crosses). Magnetic moments vs. composition at 4.2 °K. ● [54G], ○, × [63MP].

e) System Mn-Fe-Al-O

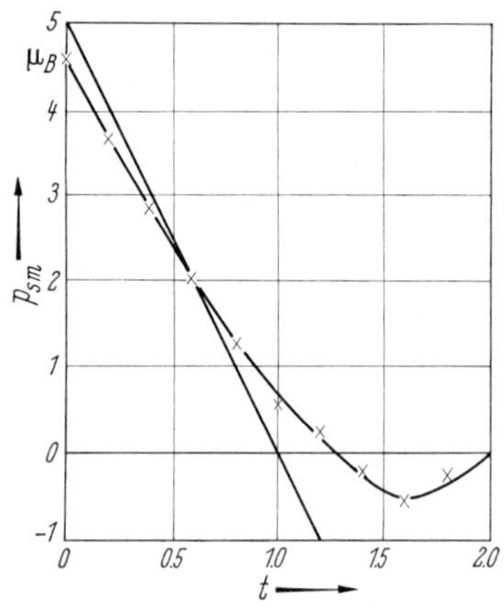

Fig. 8. $MnAl_tFe_{2-t}O_4$. Saturation moments, p_{sm}, vs. Al content, t [65B]. Straight line calculated according to the Néel-coupling scheme assuming all Al^{3+} ions to be on B sites.

f) System Mn-Fe-Rh-O

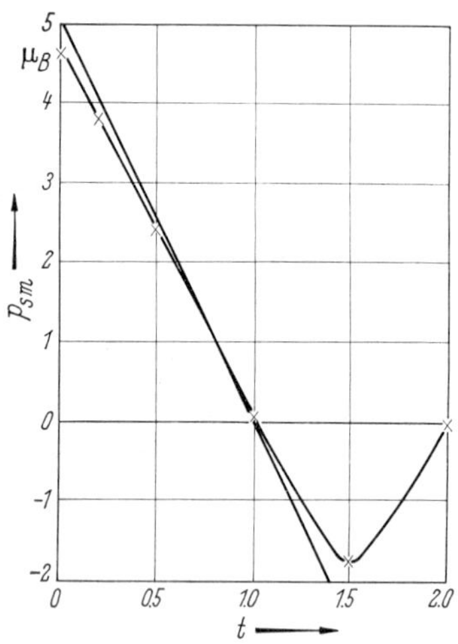

← Fig. 9. $MnFe_{2-t}Rh_tO_4$. Saturation moments, p_{sm}, vs. Rh content, t [65B]. Straight line calculated according to the Néel-coupling scheme assuming all Rh^{3+} ions to be on B sites.

g) System Mn-Fe-V-O, Mn-Fe-Ga-O

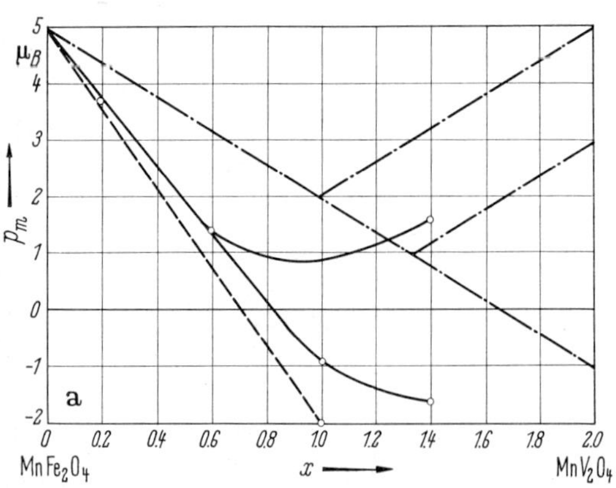

Fig. 10a. $MnFe_{2-x}V_xO_4$. p_m vs. x [59L].
Solid curves: experimental values; dashed curves: calculated values, see [59L].

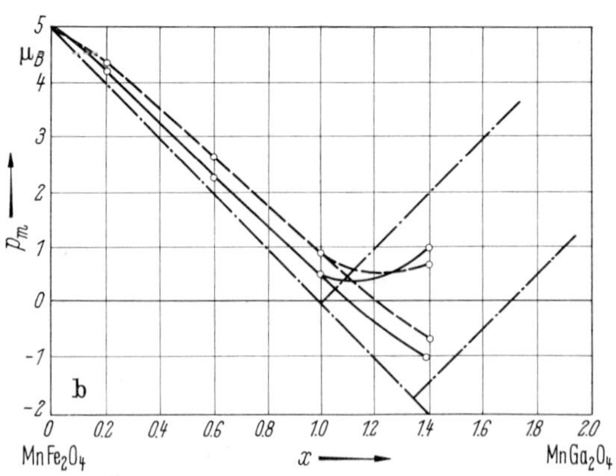

Fig. 10b. $MnFe_{2-x}Ga_xO_4$. p_m vs. x [59L]. Curves as in Fig. 10a.

6.1.3.5.2 Magnetization as a function of temperature

a) Mn_3O_4

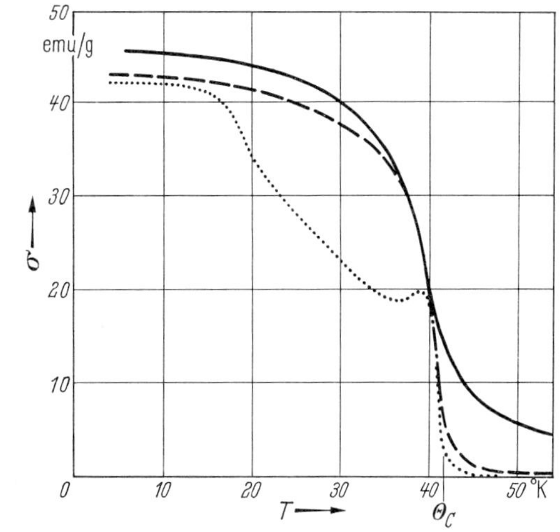

Fig. 11. Mn_3O_4. Magnetization, σ, vs. temperature for single crystals [*60DM*]. The curve is appreciably squarer than predicted by a Brillouin function.
—— cooling in 10 000 Oe; — — — cooling in 1000 Oe; · · · · warming in 250 Oe.

b) System Mn-Fe-O

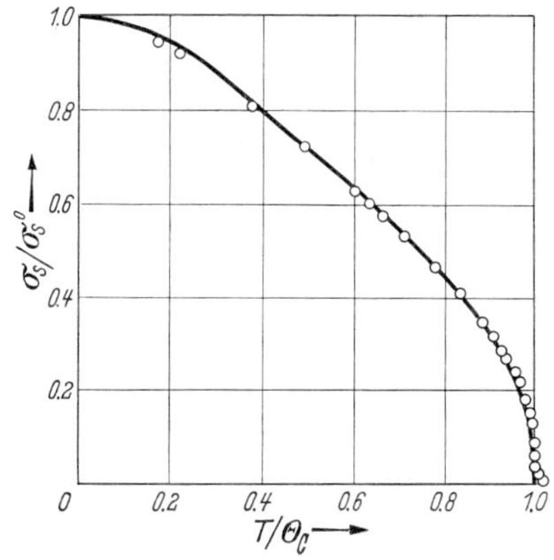

Fig. 12. $MnFe_2O_4$. The reduced magnetization, σ_S/σ_S^0, vs. reduced temperature. The solid line is a theoretical curve obtained from the molecular field theory with constants derived from the high temperature susceptibility data [*54CS*]. The force as determined with a Sucksmith balance was used to measure the magnetization. In the region of the Curie point two more sophisticated methods were used[1]).

[1]) In one method the magnetic isothermals were drawn on a magnetization vs. magnetic field graph. If a long high field section of one of these isothermals was linear, and of small slope, that section was used to extrapolate back to the zero field axis, thus giving the saturation magnetization at zero field at some temperature. In some cases the temperature rise due to the magnetocaloric effect was measured and plotted as a function of (magnetization)^2. It can be shown that if molecular field theory applies the change in temperature, ΔT, varies as σ^2 below the Curie point once the hysteresis effects are passed. Also in the region above Θ_C it can be shown that $\Delta T \propto \sigma^2$. The straight line section of the ΔT vs. σ^2 curves can be used to extrapolate back to find σ at a temperature.

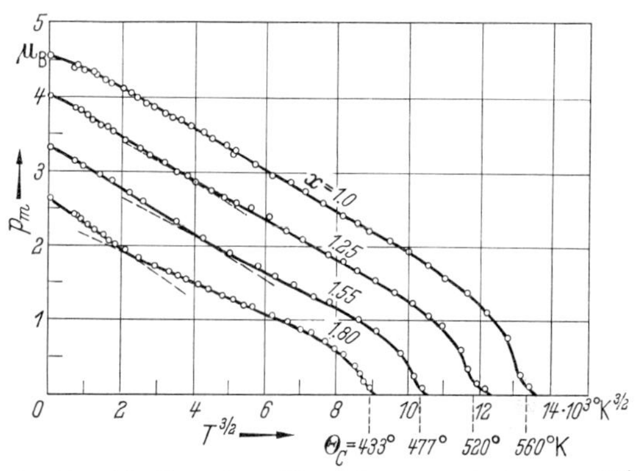

Fig. 13. $Mn_xFe_{3-x}O_4$. Magnetic moment, p_m, vs. $T^{3/2}$. Parameter x. The samples here were single crystals grown by the Stockbarger method. The curves for $x = 1.25$, 1.55, and 1.80 show distinct breaks at 231 °K, 256 °K, and 172 °K, respectively. These data are from [*61M* (1)], [*61M* (2)]. He suggests that the breaks probably represent transitions from a triangular arrangement to an antiparallel Néel spin arrangement. No break was seen in the magnetization curve of Mn_3O_4.

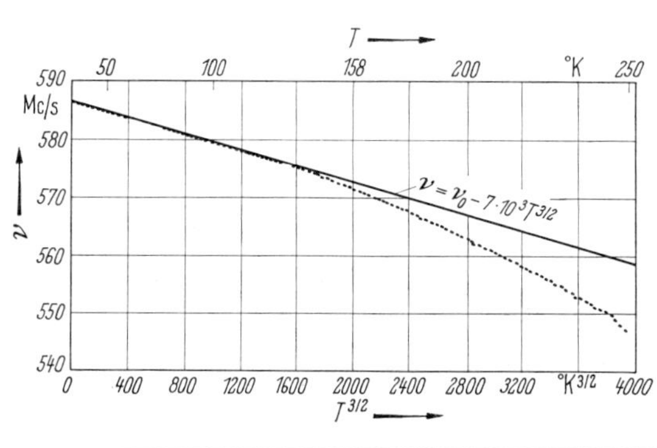

←

Fig. 14. $MnFe_2O_4$. Temperature dependence of the Mn^{55} NMR resonance frequency. The data points and the solid line show the initial deviation from saturation at the lower temperatures. The initial deviation from saturation is proportional to $T^{3/2}$, but no $T^{5/2}$ term is observed. These results are found in agreement with a ferrimagnetic spin-wave calculation for the spinel lattice. Two $T^{5/2}$ terms of opposite sign originate from the k^4 terms of the dispersion relation and the k-dependent transformation coefficients from single sublattice spin-deviation operators to the diagonal spin-wave variables. These two terms cancel, for the spin lattice, to within 1% [*64HH*].

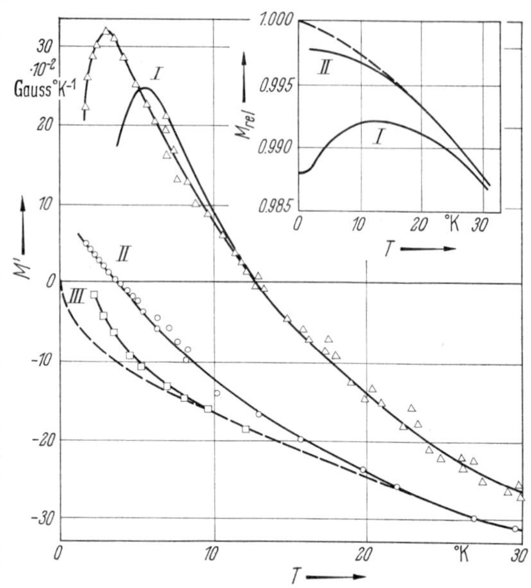

Fig. 15. Single crystals of Mn ferrite grown by the Verneuil method. Temperature dependence of $M' = \mathrm{d}M/\mathrm{d}T$ for samples of compositions $Mn_{0.84}Fe_{2.16}O_4$ (I and II); $Mn_{0.9}Fe_{2.1}O_4$ (III).
The insert shows the results of the integration of the $M'(T)$ curves for samples I and II (the dashed lines in the figure represent Bloch's law) [*64TZ*].

c) System Li-Mn-Fe-O

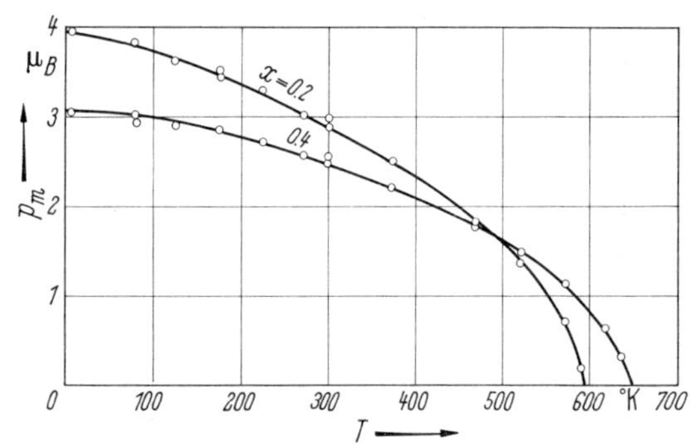

Fig. 16. $Mn_{1-x}Li_xFe_2O_4$. Magnetic moment, p_m, vs. temperature [*63MF*]. See Fig. 6, 7.

d) System Mg-Mn-Fe-O

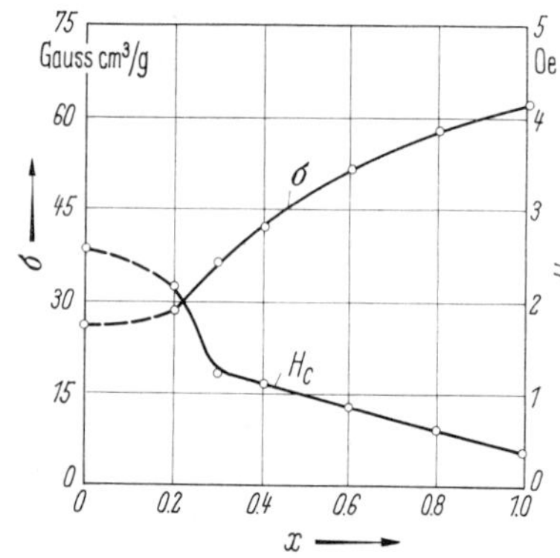

→

Fig. 17. $Mn_xMg_{1-x}Fe_2O_4$. Magnetization, σ, and coercive force, H_c, vs. Mn concentration, x [*67SR* (1)].

6.1.3.5.3 Magnetization in the region of the Curie point

System Zn-Mn-Fe-O

Though the temperature dependence of the spontaneous magnetization of a magnetic material shows an abrupt fall to zero at the Curie point, the magnetization actually measured in a magnetic field shows a "tail" extending out considerably above the Curie temperature. The true magnetization as contrasted with the spontaneous magnetization is termed the "paraprocess" and it has been studied in some detail for a number of samples of Mn Zn ferrites $\mathbf{Mn_yZn_{1-y}Fe_2O_4}$ [*57BBE*].

The experiments show that the paraprocess is closely characterized by the equation

$$\alpha + \beta\sigma^2 = H/\sigma$$

where $\sigma = M/\varrho'$ is the specific magnetization (M is the magnetization and ϱ' is the density) and α and β are thermodynamic coefficients depending on temperature and elastic stresses.

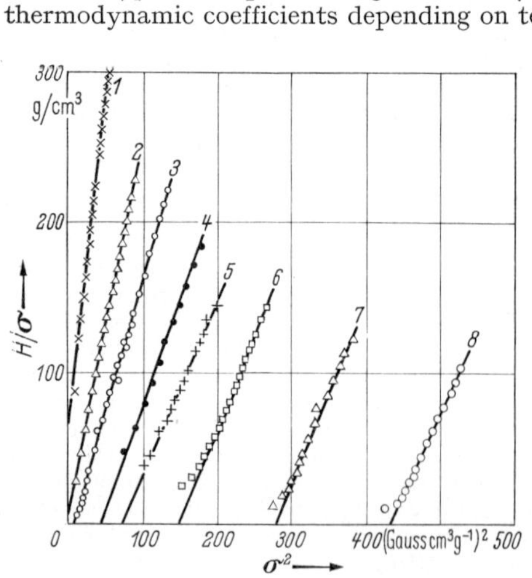

←

Fig. 18. Variation of H/σ with σ^2 near the Curie point for a ferrite made of 35.5% MnO + 15% ZnO + 49.5% Fe_2O_3 at (1) 458 °K, (2) 453 °K, (3) 451 °K, (4) 448 °K, (5) 446 °K, (6) 443 °K, (7) 438 °K, and (8) 433 °K. Note that above the region of low magnetization, the points fall on straight lines, and the slope and intercepts of these can be used to determine α and β as in the equation above. From [*57BBE*].

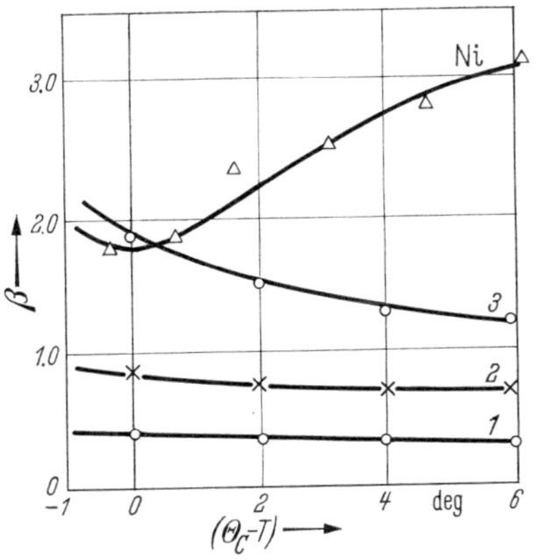

Fig. 19. Temperature variation of the coefficient β for several MnZn ferrites. Approximate compositions:
(1) $Mn_{0.46}Zn_{0.54}Fe_2O_4$,
(2) $Mn_{0.56}Zn_{0.44}Fe_2O_4$,
(3) $Mn_{0.71}Zn_{0.29}Fe_2O_4$.
The curve for pure Ni is also shown for comparison. Note that at $T = T_0$ when $\alpha = 0, \sigma = \frac{1}{3\sqrt{\beta}} H^{1/3}$. Thus the value of $\beta^{-1/3}$ is a measure of the intensity of the paraprocess since it represents the slope of the magnetization vs. field relation. This seems to be no particular correlation between the magnitude of β and the position of the Curie point. Experiments on the ferromagnetic metals have shown that β is dependent on the thermal history of the sample, and no search was made for such an effect in the work reported here. From [*57BBE*].

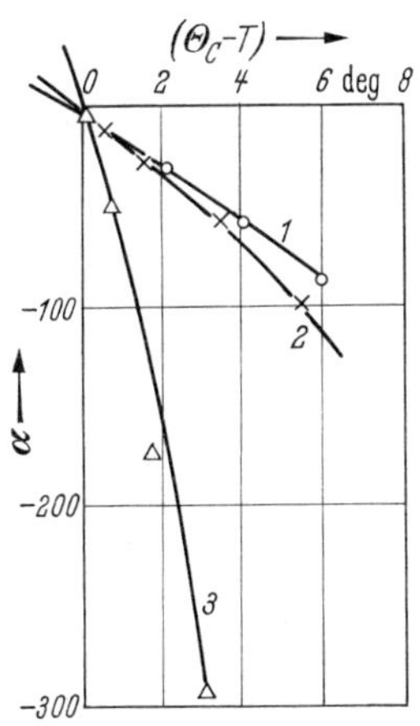

Fig. 20. The temperature dependence of the coefficient α for (1) approximately $Mn_{0.46}Zn_{0.54}Fe_2O_4$, (2) approximately $Mn_{0.56}Zn_{0.44}Fe_2O_4$, and (3) pure Ni. From [*57BBE*].

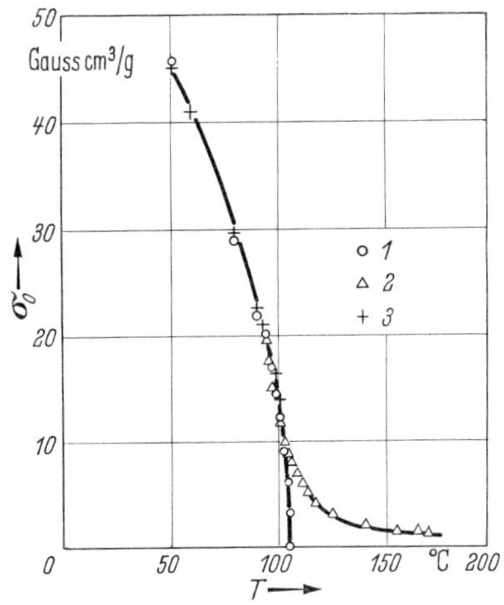

Fig. 21. The temperature dependence of the spontaneous magnetization, σ_0, near the Curie point for a ferrite made of 28% MnO + 21.5% ZnO + 50.5% Fe_2O_3. (1) points determined by the method of thermodynamic coefficients just discussed. (2) values determined by the "lines of equal magnetization method" proposed by Weiss and Forrer, and (3) by extrapolation of the true magnetization isotherms to $H = 0$. Note that more than say 5° below the Curie point, the three methods agree. The thermodynamic coefficient procedure gives a series of points which fall off steeply to zero at the Curie temperature. On other hand the lines of equal magnetization method give a long tail above the Curie point. It was not possible to determine the spontaneous magnetization from the magnetization isotherms at temperatures higher than about 10° below the Curie point since the isotherms become nonlinear as the Curie point is approached. Note the length of the tail here is perhaps 70° whereas that given in Fig. 11 of this section for a simple ferrite is very much shorter. From [*57BBE*].

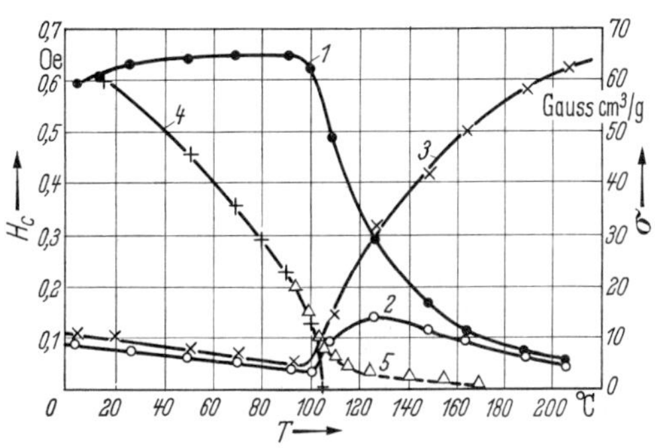

Fig. 22. Temperature dependences of the residual magnetization and coercivity in a ferrite made of 28% MnO, 22.5% ZnO, 49.5% Fe_2O_3 in the region near the Curie point. (1) magnetization in a 0.8 Oe field, (2) residual magnetization, field was 190 Oe, (3) coercive force, (4) saturation magnetization calculated by the method of thermodynamic coefficients, (5) calculated from the equal magnetization curve. In these curves both the residual magnetization and coercive force increase above some point very close to the Curie point. This behavior was found in a number of MnZn ferrites from at least two sources. It was not found in samples of CoZn ferrites. From [*57BBE*].

References for 6.1.3.5.3: [*57BBE*], [*58B*].

6.1.3.5.4 Curie temperatures

a) System Mn-Fe-O

Tab. 3. Curie temperature, Θ_C, saturation magnetization σ_s per gram at 4.2 °K and corresponding moment per formula unit, p_m, measured on two polycrystalline samples and a single crystal of $Mn^{2+}_{1-\varepsilon}Fe^{3+}_{\varepsilon}[Mn^{2+}_{\varepsilon}Fe^{3+}_{2-\varepsilon}]O_4$. From [65L]

Sample	Heat treatment	Θ_C °K	σ_s Gauss cm³ g⁻¹	p_m μ_B	ε
polycrystal	1150 °C, quenched	610	108.5 ± 0.2	4.49 ± 0.01	0.11···0.12
	1150 °C, slowly cooled and annealed at 1000, 900 and 700 °C	555	111.8 ± 0.2	4.61 ± 0.01	0.05···0.08
single crystal	rapidly cooled from melting point	585	110.4 ± 0.2	4.56 ± 0.01	—
	1100···1200 °C under pressure of 60 000 bars, rapidly cooled	620	106.8 ± 0.2	4.42 ± 0.01	—

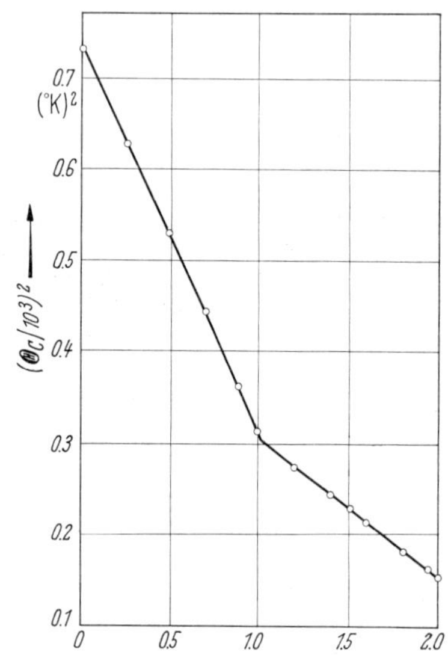

a

Fig. 23a. $Mn_xFe_{3-x}O_4$. The variation of the Curie temperature with composition in the range $0 \leqq x \leqq 2$ [60E]. In the range $0 \leqq x \leqq 1$ the curve exactly fits the theoretically derived relation

$$\left(\frac{\Theta_C}{10^3}\right)^2 = 0.735 - 0.377x - 0.042x^2.$$

In the range $1.1 \leqq x \leqq 2.0$ it is found that the appropriate relation is

$$\left(\frac{\Theta_C}{10^3}\right)^2 = 0.459 - 0.154x.$$

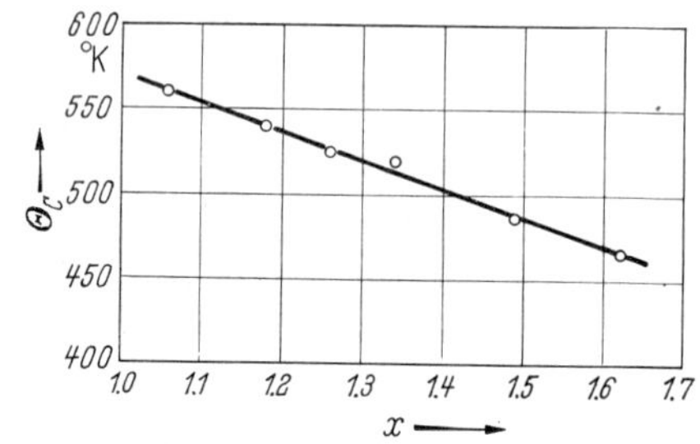

Fig. 23b. $Mn_xFe_{3-x}O_4$, $x = 1.1 \cdots 1.6$. Θ_C vs. Mn content, x [58HCP]. Coincidence with Fig. 23a only for $x = 1$. It is not possible to choose one set of measurements as being more nearly correct.

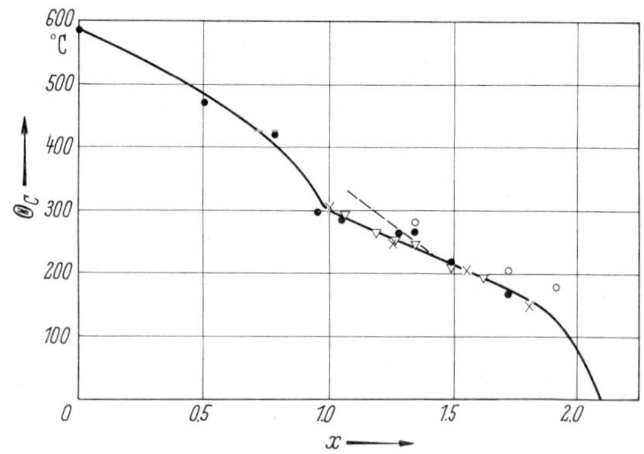

Fig. 24. $Mn_xFe_{3-x}O_{4+\gamma}$. Dependence of the Curie temperature, Θ_C, on chemical composition, x, in single crystals. ○ values of Θ_C before thermal treatment, ● values of Θ_C after annealing (4 h at 1280 °C in air for single crystals prepared by the Bridgman method or 34 h at 1350 °C in Ar for single crystals prepared by the Verneuil method, and rapid cooling) [66GSV]. × [61M], – – – [60B], ▽ [57HOT].

b) System Zn-Mn-Fe-O

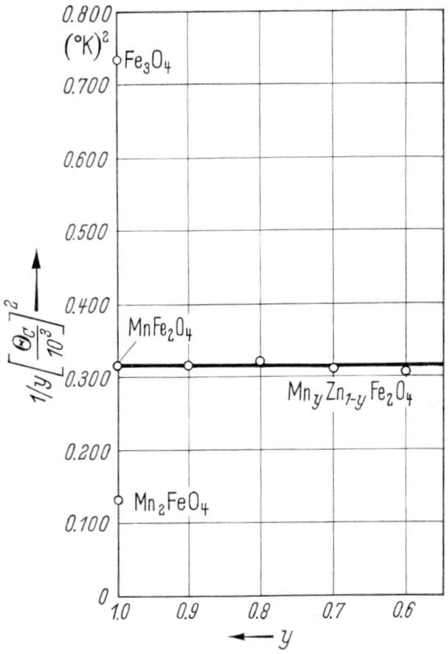

Fig. 25. $Mn_yZn_{1-y}Fe_2O_4$. The Curie temperatures for a series of MnZn ferrites in the range $0.6 \leq y \leq 1.0$ [*60E*]. In this plot the quantity

$$\frac{1}{y}\left[\frac{\Theta_C\,[°K]}{10^3}\right]^2$$

is plotted against y in order to point up the relationship

$$\frac{1}{y}\left(\frac{\Theta_C}{10^3}\right)^2 = 0.316 \pm 0.006.$$

The primary interest in this curve resides in the question of ionic distribution. For if $MnFe_2O_4$ is a normal spinel, substitution of Zn for Mn should result in a Θ_C^2 which is proportional to the number of $Mn^{2+} - Fe^{3+}$ pairs. The quantity 0.316 is just twice the $Mn^{2+} - Fe^{3+}$ interaction constant.

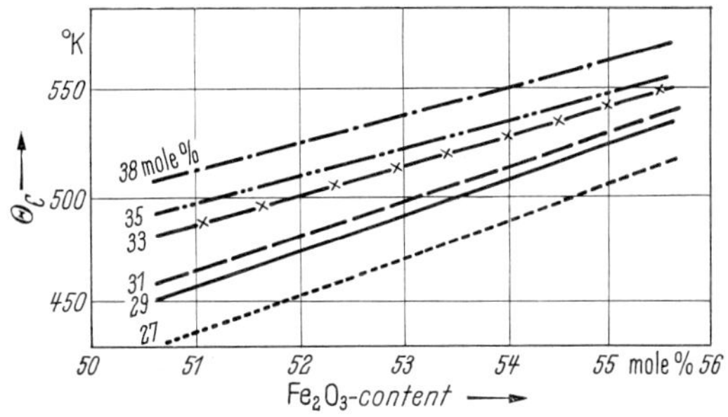

Fig. 26. The Curie points for a number of MnZnFe ferrites in the range of technological interest [*57Gu*]. Parameter MnO content in mole %. The abscissa gives the Fe_2O_3 content. The remainder is ZnO.

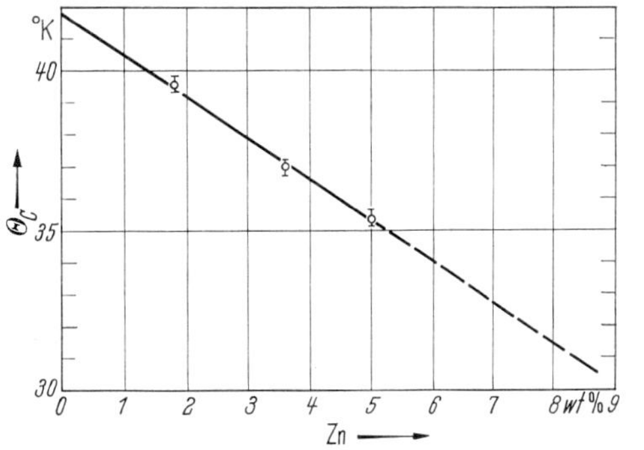

Fig. 27. $Zn_yMn^{2+}_{1-y}Mn^{3+}_2O_4$. The variation of the ferrimagnetic Curie temperature with Zn concentration in wt%. From [*60DM*].

Pressure effect on the Curie temperature:

For $Mn_{0.5}Zn_{0.5}Fe_2O_4$:

$\Theta_C = 90$ °C and $(\Delta\Theta_C/\Delta p) = 0.9 \cdot 10^{-3}$ deg/atm [*54P*].
$\Theta_C = 96$ °C and $(\Delta\Theta_C/\Delta p) = 0.7 \cdot 10^{-3}$ deg/atm [*64TNT*].

→

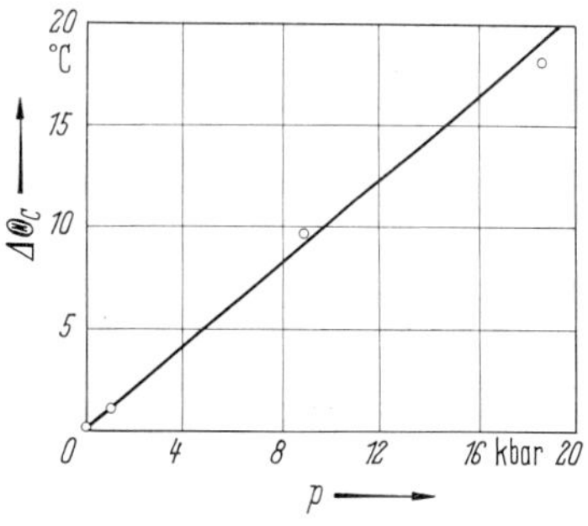

Fig. 28. $Mn_{0.48}Zn_{0.48}Fe_{2.08}O_{4.08}$. Shift of Curie point, $\Delta\Theta_C$, with the application of pressure. Curie point under atmospheric condition was 86.7 °C [*65EKKOH*].

c) System Co-Mn-Fe-O

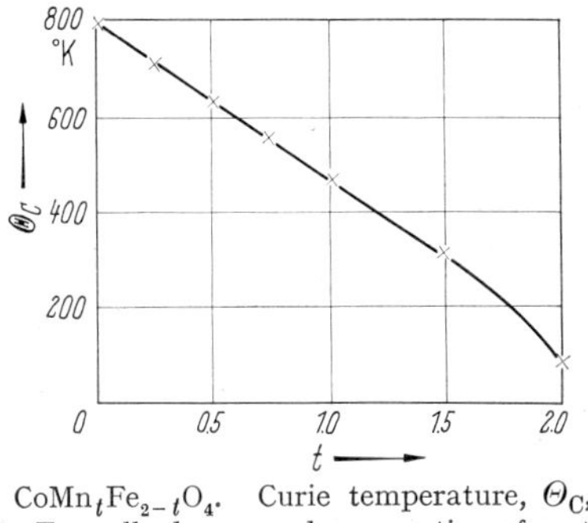

Fig. 29. $CoMn_tFe_{2-t}O_4$. Curie temperature, Θ_C, vs. Mn content, t. For cell edge, a, and preparation of material, see Fig. 33 in 6.1.3.1 [*65B*].

d) System Mn-Fe-V-O

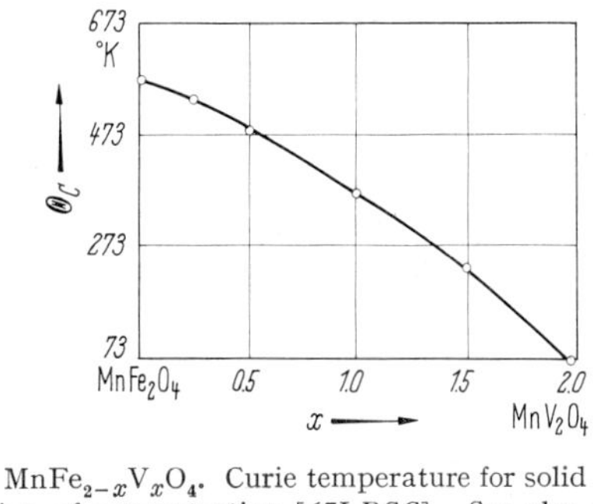

Fig. 30. $MnFe_{2-x}V_xO_4$. Curie temperature for solid solution as function of concentration [*67LBSC*]. See also [*57LM*], [*59L*].

e) System Mn-Fe-Rh-O

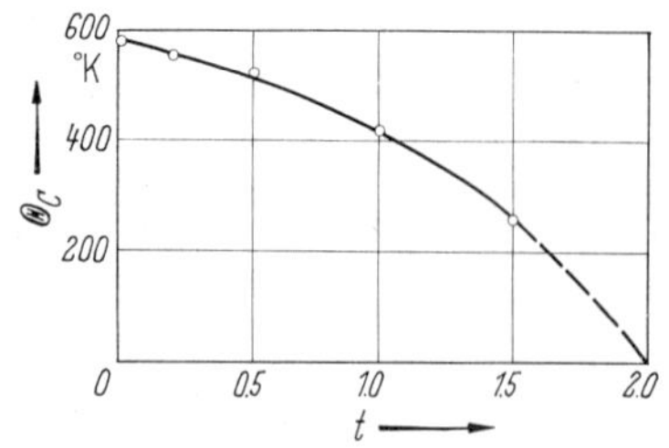

Fig. 31. $MnRh_tFe_{2-t}O_4$. Curie temperature, Θ_C, vs. Rh content, t. For cell edge and preparation of material, see Fig. 46 in 6.1.3.1 [*65B*].

f) System Mn-Fe-Ga-O

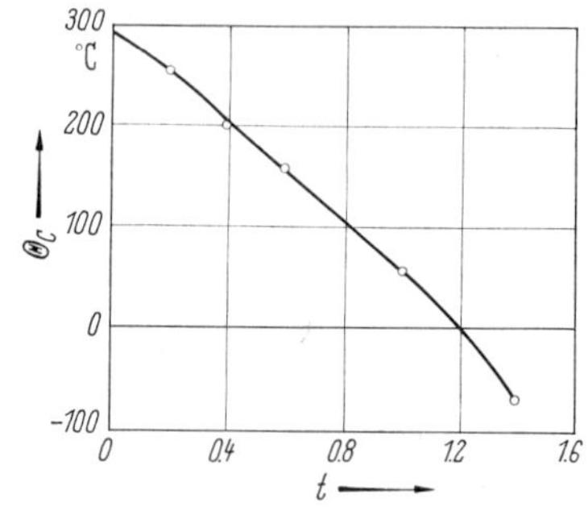

Fig. 32. $MnFe_{2-t}Ga_tO_4$. Curie temperature, Θ_C, vs. Ga content, t [*59L*]. For cell edge, a, see Fig. 48 in 6.1.3.1.

g) System Li-Fe-Mn-O

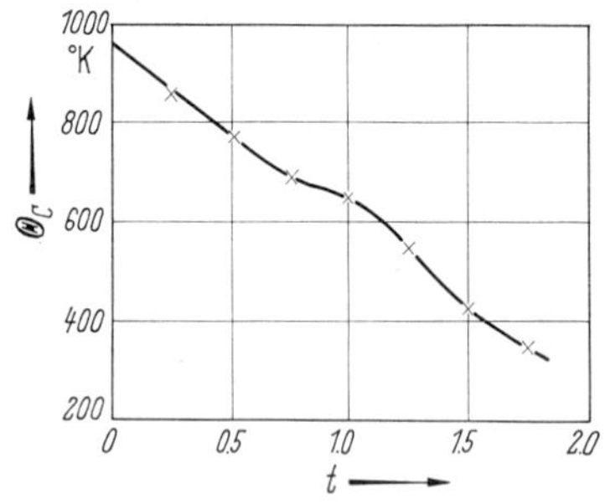

Fig. 33. $Li_{0.5}Mn_tFe_{2.5-t}O_4$. Curie temperature, Θ_C, vs. t. Cell edge, a, see Fig. 38 in 6.1.3.1 [*65B*].

h) System Mn-Fe-Al-O

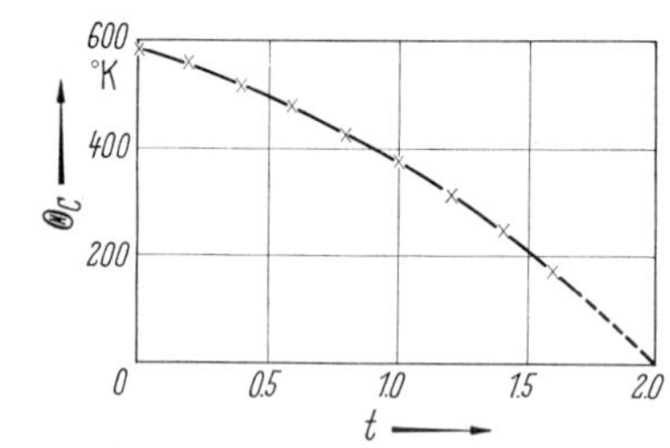

Fig. 34. $MnAl_tFe_{2-t}O_4$. Curie temperature, Θ_C, vs. t. Cell edge, a, see Fig. 28 in 6.1.3.1 [*65B*].

References for section 6.1.3.5.4: [*50GC*], [*57Gu*], [*58HKP*], [*60DM*], [*60E*].

6.1.3.6 Magnetocrystalline anisotropy

6.1.3.6.1 Magnetocrystalline anisotropy in Mn ferrite

The magnetocrystalline anisotropy energy of a cubic substance is customarily expressed by means of the coefficients in the series

$$E_{\text{anis}} = K_1(\alpha_1^2\alpha_2^2 + \alpha_2^2\alpha_3^2 + \alpha_3^2\alpha_1^2) + K_2\alpha_1^2\alpha_2^2\alpha_3^2 + \text{terms of higher order in } \alpha_1\alpha_2\alpha_3.$$

In this, α_1, α_2, and α_3 are the direction cosines of the magnetization relative to the four-fold axes of the crystal.

This crystalline anisotropy can be measured by a torque method, by magnetic resonance experiments, and by the comparison of the magnetization curves obtained with the applied field along several different crystal directions. Though unquestioned results can only be obtained by using single crystal samples with each of these methods considerably cruder estimates of anisotropy can be made using polycrystalline specimens.

Ferrimagnetic resonance experiments on sound spherical samples of crystals with a sufficiently narrow line width give a measure of anisotropy fields, the ratios K_1/M_s and K_2/M_s. If $M(T)$ is measured over the same range as the anisotropy fields have been measured then K_1 and K_2 can be obtained. In most cases the necessary magnetization measurements are not available, and so we give K_2/M_s for most of the resonance measurements. The torque and resonance measurements complement each other. It is

of special interest to study cases in which the anisotropy torque shows a frequency variation. Obviously the torque determination is a dc value, and resonance measurements can in principle be made over a wide frequency range.

a) Single crystals of $Mn_xFe_{3-x}O_4$:

The crystals of Fig. 6 were grown by the Bridgman Stockbarger [*52S*] technique at about 1° per hour, and when completely crystallized were quenched to RT at about 100° per hour. Chemical analysis showed that the actual composition were within about less than one per cent of those given by the formulas. The cubic to tetragonal phase transformation caused crystals with x greater than 1.90 to shatter on cooling. The interesting features observed are as follows:

(1) Small Mn concentrations in magnetite ($x = 0$) are shown to decrease the temperature of the isotropy point known to exist in magnetite at low temperatures.

(2) An unexpected variation in the temperature dependence and magnitude of the anisotropy energy in the $x = 0.4$ to 1.0 composition region is shown.

$Mn_{1.26}Fe_{1.74}O_4$. A plot of $(-K_1/M_s)$ vs. temperature in the range from about 130 °K to about 450 °K for a crystal is given in Fig. 3 of 6.1.3.9.1 [*58HBK*]. These authors found that they could express the variation of $K_1(T)$:

$$\frac{K_1(T)}{K_1(0)} = \left\{\frac{M_s(T)}{M_s(0)}\right\}^5.$$

b) Polycrystalline samples of $Mn_xFe_{3-x}O_4$:

Reference to the variation of permeability with frequency as shown in the section on permeability shows that a polycrystalline sample of one of the Mn ferrites usually has three peaks in permeability, see Fig. 16 in 6.1.3.8. Two of these are apparently associated with domain walls, but that appearing at the highest field is thought to be associated [*58HKP*] with the resonance of the electron spins in the internal magnetocrystalline anisotropy field. It was found that for a series $\mathbf{Mn_xFe_{3-x}O_4}$ where x varied from 1.0 to a little over 1.6 the frequency of this high frequency peak varied linearly with composition. This is to be seen in Tab. 2 of 6.1.3.8. If this frequency is converted to a field by means of the spectroscopic splitting factor g it presumably represents some effective anisotropy field. On dividing this by the magnetization we obtain an effective anisotropy constant K_{eff}. It was found that within the experimental error, $K_{eff} = 2K_1$.

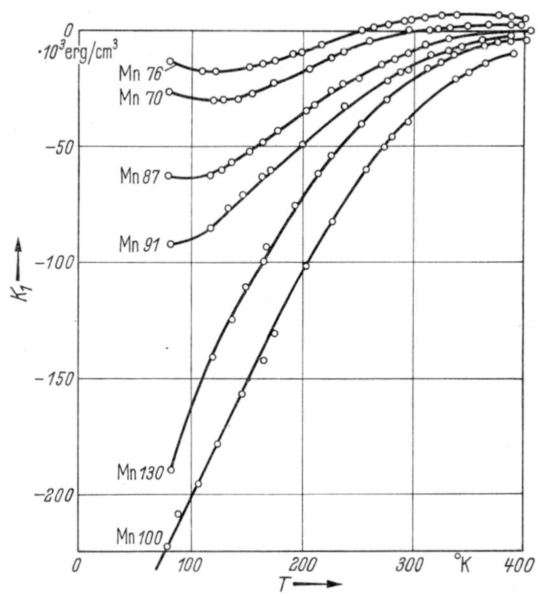

Fig. 1. $Mn_xFe_{3-x}O_4$, $-0.6 \leq x \leq 1.3$. First order anisotropy constant K_1 determined for single crystals over a temperature range from 77 °K to about 400 °K [*60P* (1)]. The composition approximating stoichiometric Mn ferrite most closely lies lowest on this plot. For several compositions the anisotropy actually changes sign as the temperature is increased. See Fig. 13.

Mn 130: $Mn_{1.32}Fe_{1.68}O_4$ Mn 87: $Mn_{0.819}Fe_{2.135}O_4$
Mn 100: $Mn_{0.98}Fe_{1.99}O_4$ Mn 76: $Mn_{0.747}Fe_{2.241}O_4$
Mn 91: $Mn_{0.891}Fe_{2.093}O_4$ Mn 70: $Mn_{0.697}Fe_{2.291}O_4$

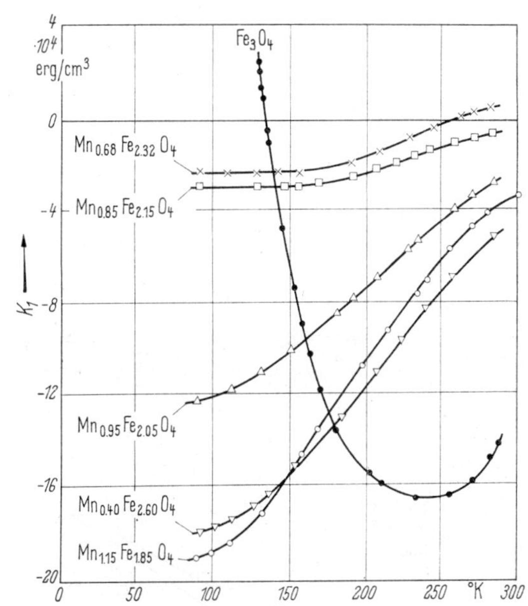

Fig. 2. $MnFe_2O_4$-Fe_3O_4 system. Temperature dependence of K_1. From [*61M* (2)]. Measurements on single-crystal sphere by the torque method. Crystal grown by the Bridgman technique in air.

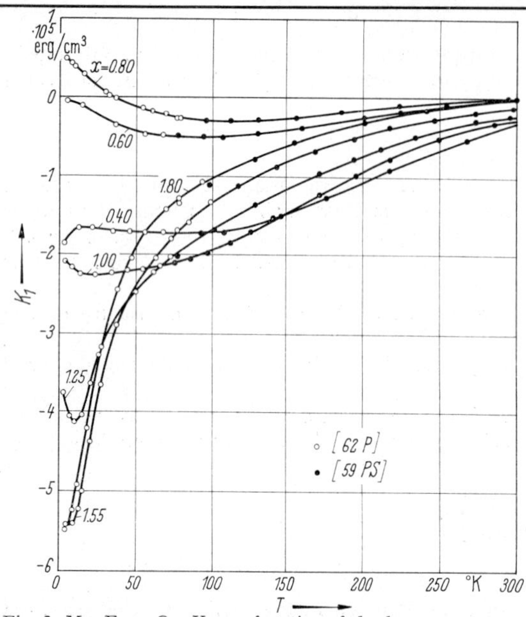

Fig. 3. $Mn_xFe_{3-x}O_4$. K_1 as a function of absolute temperature for various ferrites. From [*62P* (1), *59PS*]. The samples are the same as in Fig. 6. In [*63K*] a model is proposed in which the large negative anisotropy at low temperatures in the composition range $1 < x < 1.8$ is ascribed to Mn^{3+} ions in octahedral positions.

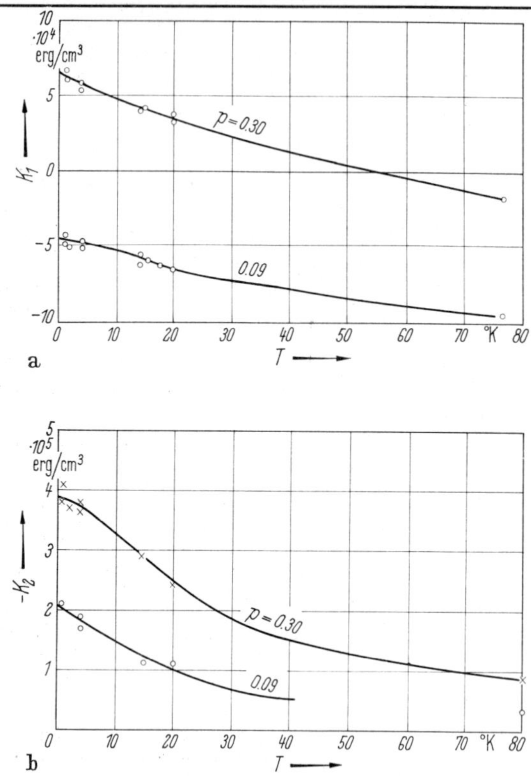

Fig. 4. $Mn_{1-p}Fe_{2+p}O_4$. Anisotropy constants K_1 and K_2 as a function of temperature for two single crystals, grown by the flame fusion method. From [*67BP*].

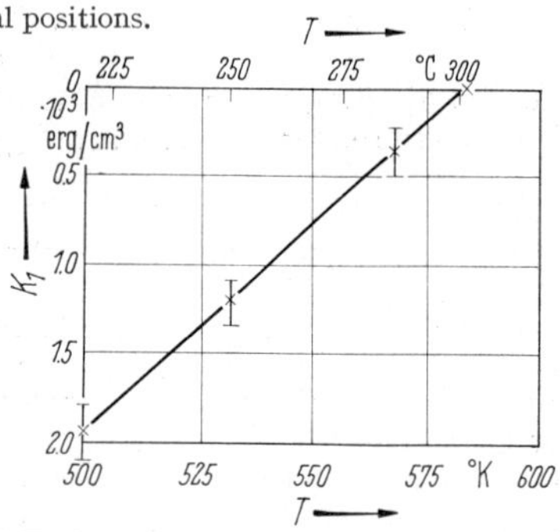

Fig. 5. $MnFe_2O_4$. Magnetic anisotropy constant K_1 vs. T in a single crystal sphere near the Curie point. Determined by the torque method [*66DS* (1)].

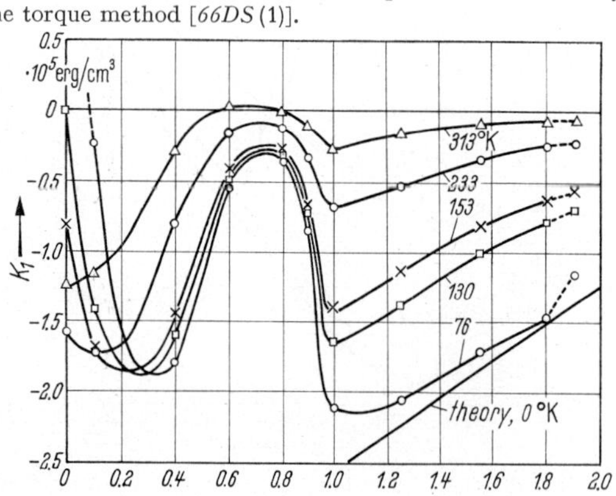

Fig. 6. $Mn_xFe_{3-x}O_4$, $0 \leq x \leq 1.90$. The magnetic anisotropy constant K_1 vs. x in single crystals. Parameter: temperatures. Torque method. From [*59PS*]. Results from [*64GV*] are in good agreement with those of [*59PS*], although the measurements were performed on samples prepared by the Verneuil method [*64GV*], while [*59PS*] used the Bridgman-Stockbarger technique, moreover, in both cases, the oxygen parameter γ remained undetermined.

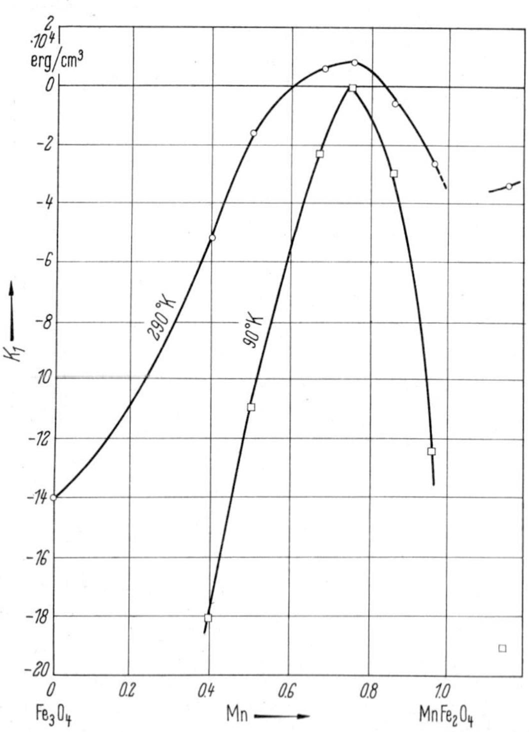

Fig. 7. $MnFe_2O_4$-Fe_3O_4 system. Concentration dependence of K_1. From [*61M*]. Measurements on single-crystal sphere by the torque method. Crystal grown by the Bridgman technique in air.

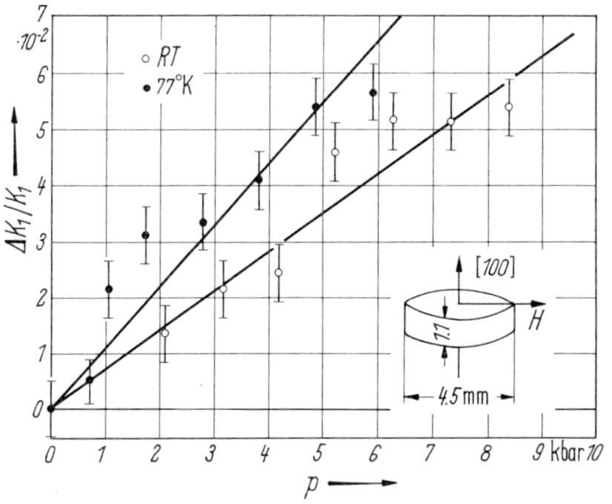

Fig. 8. $MnFe_2O_4$. Hydrostatic pressure dependence of magnetic first order anisotropy constant K_1 of Mn ferrite at RT and 77 °K [*67SKK*]. The specimen was annealed before the measurements.

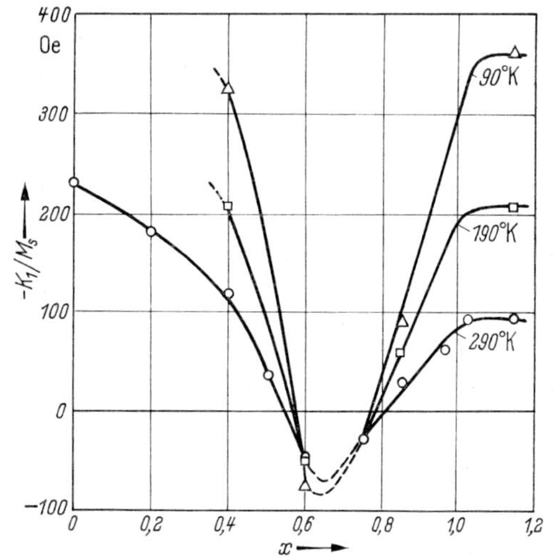

Fig. 10. $Mn_xFe_{3-x}O_4$. $-K_1/M_s$ vs. composition [*59FMU* (2)]. These data are to be associated with Fig. 2 in 6.1.3.9.

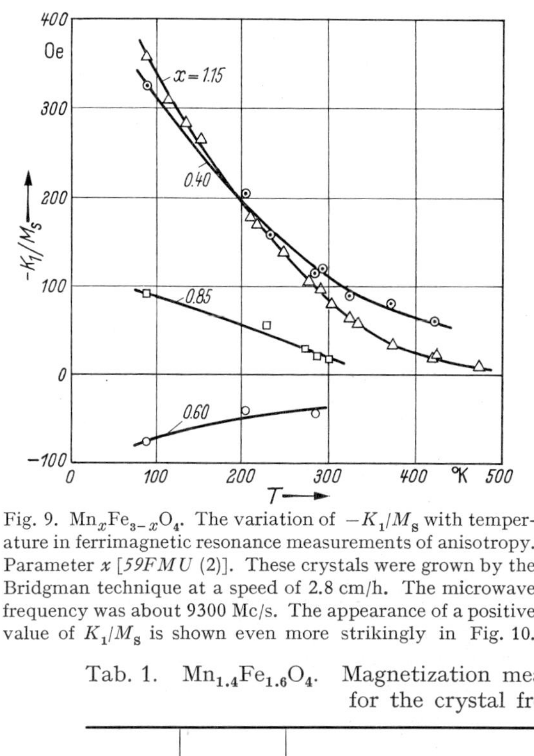

Fig. 9. $Mn_xFe_{3-x}O_4$. The variation of $-K_1/M_s$ with temperature in ferrimagnetic resonance measurements of anisotropy. Parameter x [*59FMU* (2)]. These crystals were grown by the Bridgman technique at a speed of 2.8 cm/h. The microwave frequency was about 9300 Mc/s. The appearance of a positive value of K_1/M_s is shown even more strikingly in Fig. 10.

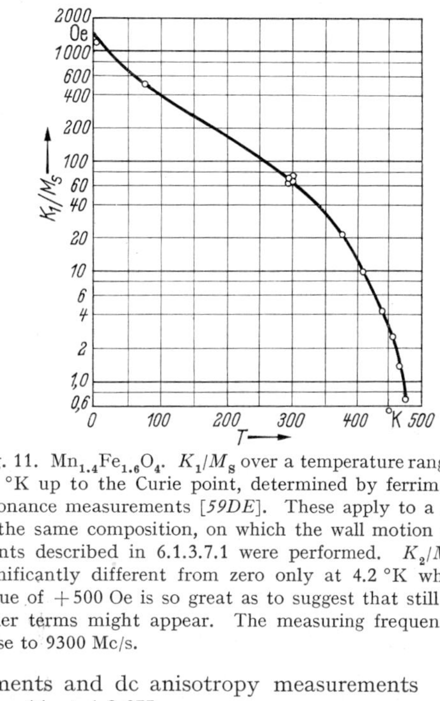

Fig. 11. $Mn_{1.4}Fe_{1.6}O_4$. K_1/M_s over a temperature range from 4.2 °K up to the Curie point, determined by ferrimagnetic resonance measurements [*59DE*]. These apply to a crystal of the same composition, on which the wall motion experiments described in 6.1.3.7.1 were performed. K_2/M_s was significantly different from zero only at 4.2 °K where its value of +500 Oe is so great as to suggest that still higher order terms might appear. The measuring frequency was close to 9300 Mc/s.

Tab. 1. $Mn_{1.4}Fe_{1.6}O_4$. Magnetization measurements and dc anisotropy measurements for the crystal from Fig. 11 at 4.2 °K

T °K	$4\pi M_s$ Gauss	dc				rf	
		K_1 erg/cm³	K_2 erg/cm³	K_1/M_s Oe	K_2/M_s Oe	K_1/M_s Oe	K_2/M_s Oe
1.4	5120 a)	$-4.0 \cdot 10^5$ a)	$-3.0 \cdot 10^5$ a)	−1000 a)	−750 a)	−1255	−287
4.2	5120 a)						
45	4980 a)					− 499	0
77	4850 a)					− 140	0
195	4240 b)					− 72	0
294	3220 b)						
479	0						

a) Data obtained by Bozorth and Walsh [*58 BW*].
b) Magnetizations obtained by normalizing the B values for B_{max} obtained with the Cioffi Fluxmeter to the low temperature data.

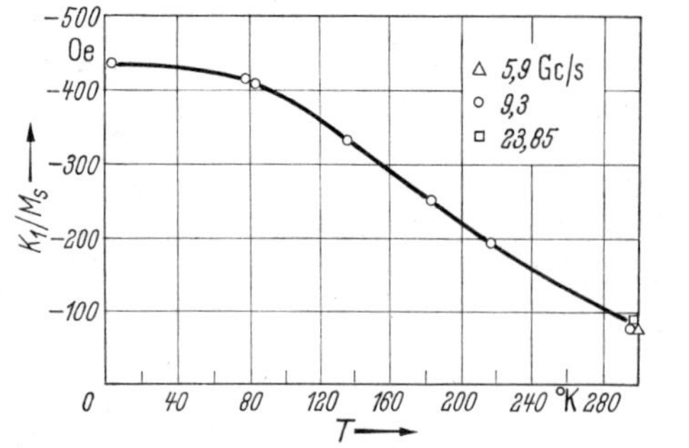

Fig. 12. $Mn_{0.98}Fe_{1.86}O_4$. The variation of K_1/M_s with temperature as measured in ferrimagnetic resonance experiments [*55DGJ*]. At about 300 °K, values taken over a wide frequency range coincide.

Tab. 2. $Mn_{0.98}Fe_{1.86}O_4$. Values of K_1/M_s and K_2/M_s with their errors as measured in a 0.0105 inch diameter sphere at 9300 Mc/sec [*55DGJ*]. The errors were estimated on the basis of reproducibility

T °K	K_1/M_s Oe	K_2/M_s Oe	M Gauss
300	− 79 ± 3	6 ± 10	351
77	−412 ± 5	− 5 ± 20	493
4.2	−433 ± 15	−200	

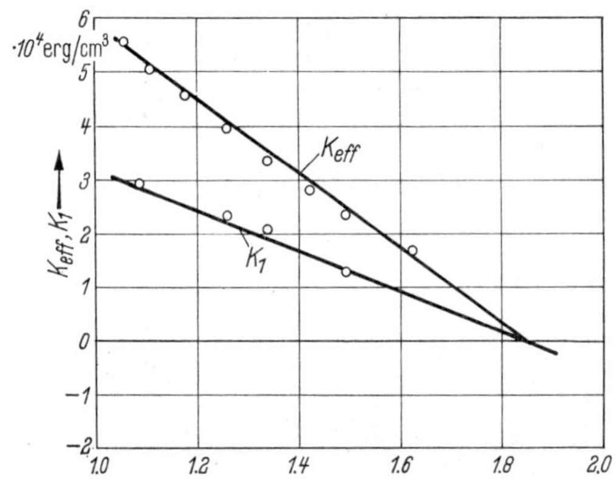

Fig. 13. $Mn_xFe_{3-x}O_4$. Effective anisotropy constant K_{eff} calculated from the natural spin resonance observed in initial permeability measurements [*58HKP*] in polycrystalline samples, and the real anisotropy constant K_1 obtained from microwave resonance experiments in single crystals vs. Mn content x. RT. The crystals were grown by sintering discs of pressed ferrites for long periods of time.

6.1.3.6.2 Magnetocrystalline anisotropy in Mn ferrites with substitutions

a) System Co-Mn-Fe-O

The anisotropy of a series of Mn ferrite crystals containing from 1···25% Co ferrite was measured over the temperature range from 160 °K···300 °K [*59P* (1)]. The crystals were grown by the flame fusion method. The analyses are given in Tab. 3 along with a designation which is used to identify the curves in Fig. 14.

On some of these crystals a discrepancy was noted between anisotropy constants derived from resonance measurements and those derived from torque experiments [*60PT*]. Tab. 4 below gives their results for crystals of $\mathbf{Co_xMn_{1-x}Fe_2O_4}$ for which x is 0.0, 0.01, and 0.25.

It will be seen that for $x = 0$, K_1 obtained by the two methods is not significantly different. For $x = 0.25$, however, the difference is (48 000 ± 13 000) erg cm^{-3}, definitely larger than the experimental error. For the crystal in which $x = 0.01$ at the lower temperature there is a very great difference. However, the appearence of higher order anisotropy terms complicates the comparison. [*59AD*] suggested that microwave and static measurements measure the anisotropy at constant strain and constant stress, respectively. [*49K* (1)] showed that the difference ΔK_1 is given by

$$\Delta K_1 = 9/4\,(c_{11} - c_{12})\,\lambda_{100}^2 - 2c_{44}\,\lambda_{111}^2$$

where c_{11}, c_{12}, and c_{44} are the elastic constants of a cubic crystal and λ_{100} and λ_{111} are the magnetostriction coefficients. For the $\mathbf{Co_{0.25}Mn_{0.75}Fe_2O_4}$ crystal [*60PT*] estimate $c_{11} - c_{12} = c_{44} = 10^{12}$ erg cm^{-3} and using the measured values of the magnetostriction for 290 °K, $\lambda_{100} = (-111 \pm 6) \cdot 10^{-6}$, and $\lambda_{111} = (+19 \pm 1) \cdot 10^{-6}$, they obtain $\Delta K_1 = (28\,000 \pm 3\,000)$ erg cm^{-3}.

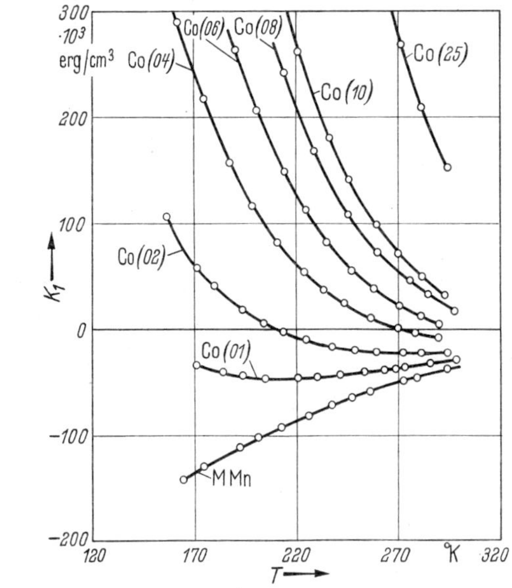

Fig. 14. Values of K_1 plotted against temperature for Co substituted Mn ferrite crystals. The analyses corresponding to each of the designations are given in Tab. 3 [*59P* (1)].

Tab. 3. Chemical analyses of the Co-Mn ferrites in Fig. 14 [*59P* (1)]

Chemical analysis	Designation
$Mn_{0.98}Fe_{1.99}O_4$	MMn
$Co_{0.009}Mn_{0.99}Fe_{1.98}O_4$	Co (01)
$Co_{0.019}Mn_{1.03}Fe_{1.94}O_4$	Co (02)
$Co_{0.038}Mn_{0.98}Fe_{1.97}O_4$	Co (04)
$Co_{0.056}Mn_{0.96}Fe_{1.97}O_4$	Co (06)
$Co_{0.077}Mn_{0.92}Fe_{1.99}O_4$	Co (08)
$Co_{0.094}Mn_{0.91}Fe_{1.97}O_4$	Co (10)
$Co_{0.245}Mn_{0.75}Fe_{1.99}O_4$	Co (25)

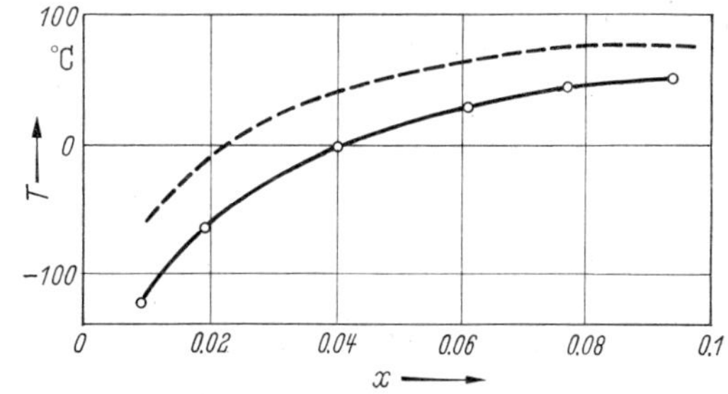

Fig. 15. It can be seen from the data in Fig. 14 that a number of the compositions $Co_xMn_{1-x}Fe_2O_4$ pass through a point in temperature at which the first order anisotropy is zero. This figure is a plot of the temperature at which this isotropy point occurs as a function of Co concentration, x: $-K_1 = 0$. [*57Bu*] working with polycrystalline materials of similar composition found a peak in initial permeability vs. temperature curves which moved with additions of Co. The position of this peak is plotted as the dotted line in the figure. From [*59P* (1)].

Tab. 4. $Co_xMn_{1-x}Fe_2O_4$. Magnetic anisotropy constants [*60PT*]

At room temperature:

x	K_1 [10^3 erg cm^{-3}] Torque	K_1 [10^3 erg cm^{-3}] Resonance	T °K
0	-37.9 ± 1.5	-36.0 ± 1.5	298
0.25	$+164.0 \pm 3.0$	$+116.0 \pm 10.0$	291

At $T = 82$ °K:

x	Torque K_1	Torque K_2	Torque K_3	Resonance K_1	Resonance K_2	Resonance K_3	
0	-221.0 ± 3	—	—	-229.0 ± 20	-47.0 ± 30.0	—	10^3 erg cm^{-3}
0.01	$+198.0$	-1040	-67.0	$+258.0$	-1230	-126.0	10^3 erg cm^{-3}

x	Nominal composition	Chemical analysis
0	$MnFe_2O_4$	$Mn_{0.99}Fe_{1.99}O_4$
0.01	$Co_{0.01}Mn_{0.99}Fe_2O_4$	$Co_{0.009}Mn_{0.99}Fe_{1.98}O_4$
0.25	$Co_{0.25}Mn_{0.75}Fe_2O_4$	$Co_{0.245}Mn_{0.747}Fe_{1.99}O_4$

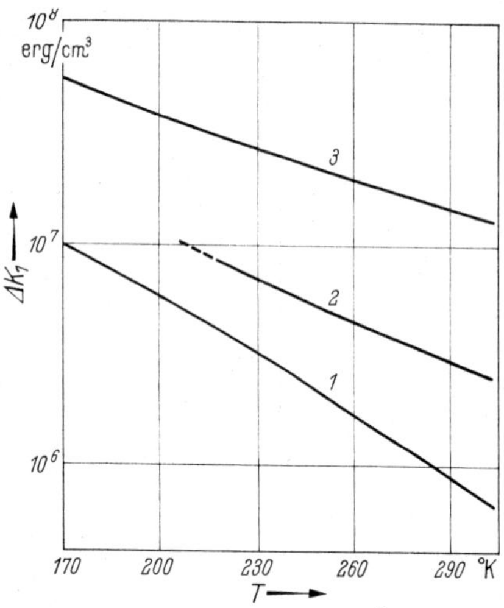

Fig. 16. Anisotropy contribution ΔK_1 of Co^{++} ion (extrapolated to 100% substitution) against temperature in (1) $MnFe_2O_4$, (2) $Mn_{0.7}Fe_{2.3}O_4$, (3) Fe_3O_4 (from Bickford). From [60P (2)].

Tab. 5. Anisotropy measurements in Co substituted $MnFe^{2+}$ ferrites for two temperatures. The contribution of the Co ion increases rapidly as the concentration of divalent iron goes up [60P (2)]

Nominal composition	Chemical analysis	K_1 [10^3 erg cm^{-3}] 290 °K	200 °K
$Mn_{0.91}Fe_{2.09}O_4$	$Mn_{0.891}Fe_{2.093}O_4$	− 16.8	− 48.5
$Co_{0.04}Mn_{0.87}Fe_{2.09}O_4$	$Co_{0.041}Mn_{0.878}Fe_{2.068}O_4$	+ 15.9	+172.9
$Mn_{0.87}Fe_{2.73}O_4$	$Mn_{0.819}Fe_{2.135}O_4$	− 9.6	− 35.5
$Co_{0.04}Mn_{0.83}Fe_{2.13}O_4$	$Co_{0.0358}Mn_{0.820}Fe_{2.135}O_4$	+ 44.3	+224.0
$Mn_{0.76}Fe_{2.24}O_4$	$Mn_{0.747}Fe_{2.241}O_4$	+ 4.9	− 8.1
$Co_{0.04}Mn_{0.72}Fe_{2.24}O_4$	$Co_{0.0371}Mn_{0.722}Fe_{2.234}O_4$	+ 93.6	+328.8
$Mn_{0.70}Fe_{2.3}O_4$	$Mn_{0.697}Fe_{2.291}O_4$	+ 0.6	− 17.2
$Co_{0.04}Mn_{0.66}Fe_{2.3}O_4$	$Co_{0.0372}Mn_{0.647}Fe_{2.31}O_4$	+118.1	—

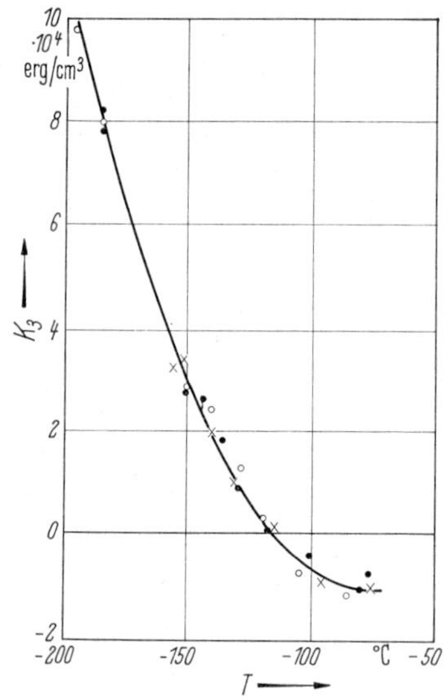

→

Fig. 17. $Co_{0.0067}Mn_{1.015}Fe_{1.98}O_{3.99}$. Temperature dependence of K_3 for a single crystal grown by the Verneuil process. From [64E].

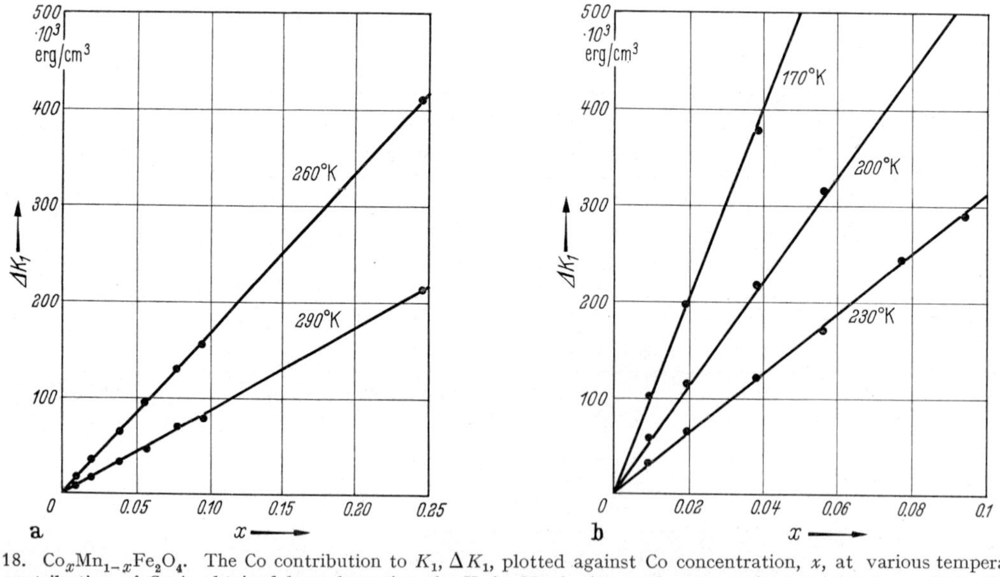

Fig. 18. $Co_xMn_{1-x}Fe_2O_4$. The Co contribution to K_1, ΔK_1, plotted against Co concentration, x, at various temperatures. The contribution of Co is obtained by subtracting the K_1 for Mn ferrite at the appropriate temperature. Note that the contribution of the Co ions is linear over this range of concentration [59P (1)].

b) System Co-Mg-Mn-Fe-O

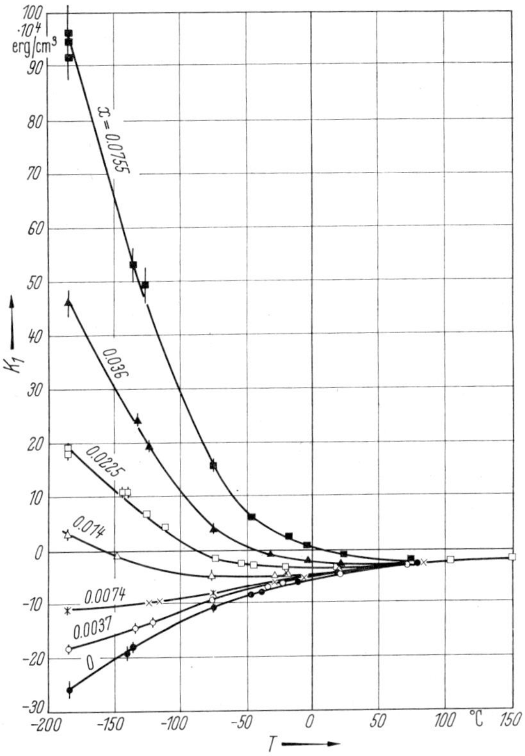

Fig. 19. $Co_xMg_{0.75}Mn_{0.53}Fe_{1.71}O_{3.96+x}$ $(0 \leq x \leq 0.0755)$. Temperature dependence of K_1. Parameter x. Measurement in the plane (100) by the torque method. Crystal grown by the Verneuil process. After [*65E*].

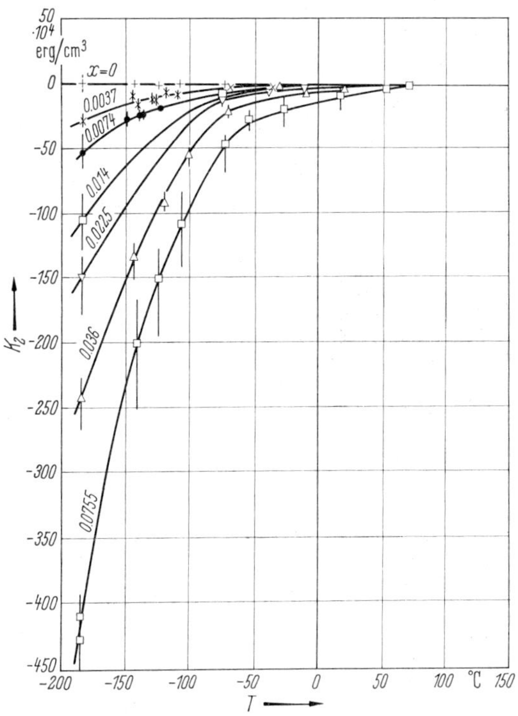

Fig. 20. $Co_xMg_{0.75}Mn_{0.53}Fe_{1.71}O_{3.96+x}$ $(0 \leq x \leq 0.0755)$. Temperature dependence of K_2. Parameter x. Measurement in the plane (110). Small differences in the plane (111). From [*65E*].

Tab. 6. Numerical values of K_3 measured at −185 °C [*64E*]

	K_3 [10^4 erg/cm³]
$Co_{0.014}Mg_{0.75}Mn_{0.53}Fe_{1.71}O_{3.97}$	11.8 ± 2.2
$Co_{0.012}Mg_{0.75}Mn_{0.53}Fe_{1.71}O_{3.97}$	9.1 ± 1.8
$Co_{0.010}Mg_{0.75}Mn_{0.53}Fe_{1.71}O_{3.97}$	7.45 ± 1.5

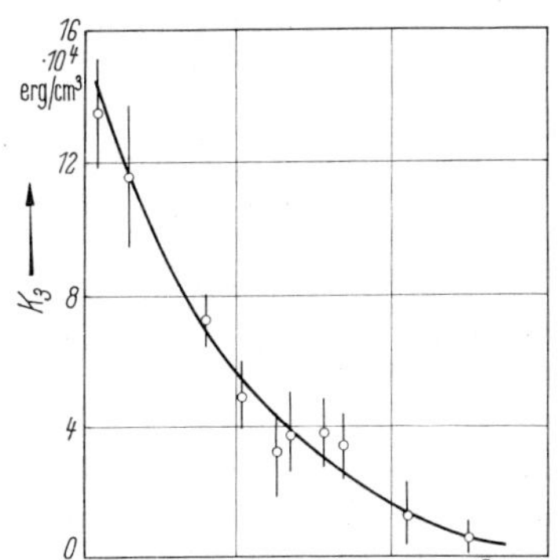

Fig. 21. $Co_{0.014}Mg_{0.75}Mn_{0.53}Fe_{1.71}O_{3.97}$. Thermal change of K_3. Measurement by the torque method. Crystal grown by the Verneuil process [*64E*].

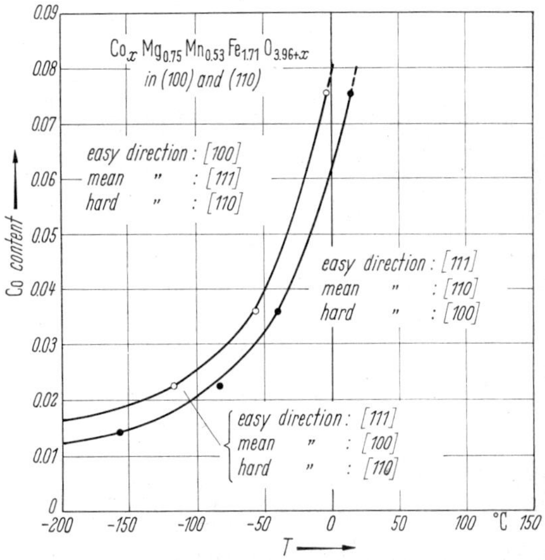

Fig. 22. $Co_xMg_{0.75}Mn_{0.53}Fe_{1.71}O_{3.96+x}$ $(0 \leq x \leq 0.0755)$. Easy, mean and hard directions at different temperatures. From [*65E*].

c) System Mg-Mn-Fe-O

Tab. 8. Magnetocrystalline anisotropy constant K_1 for MnMg ferrites at $T = 300$ °K. The values of K_1 were determined from measurements of the variation of magnetic and magnetostrictive susceptibilities as functions of magnetic field strength in the region approaching saturation. From [67URBS]

Composition mole %	K_1 10^3 erg/cm³	$4\pi M_s$ Gauss
57 MnO 43 Fe_2O_3	−14.7	2450
35 MnO, 22 MgO, 43 Fe_2O_3	−10.8	1600
27 MnO, 30 MgO, 43 Fe_2O_3	− 9.8	1380
20 MnO, 37 MgO, 43 Fe_2O_3	− 8.7	1130
13 MnO, 44 MgO, 43 Fe_2O_3	− 8.0	1030
7 MnO, 50 MgO, 43 Fe_2O_3	− 7.5	920
57 MgO, 43 Fe_2O_3	− 6.7	750

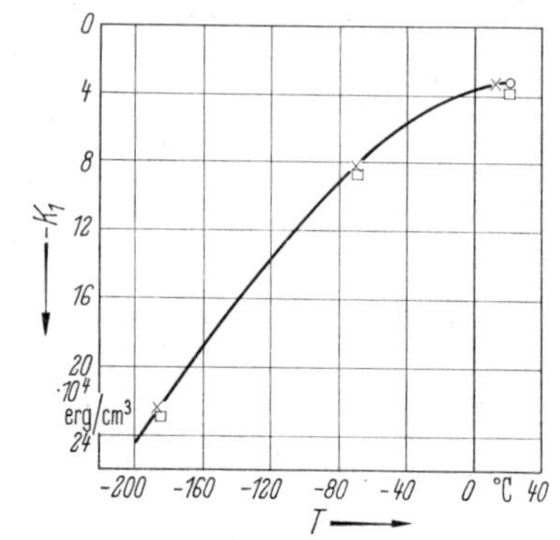

Fig. 23. $K_1 = f(T)$ for a single crystal 25% MnO, 35% MgO, 40% Fe_2O_3.
× = dynamic determination, ○ = static determination at RT, □ = values from Elbinger (private communication). From [63MV].

Tab. 7. Effective anisotropy constants K_1 for MnMg ferrites derived from the slope of the straight part of the magnetization, $M = f(1/H^2)$. Polycrystalline sintered sphere shaped specimens were used. From [62S]

Molar ratio of oxides			K_1 10^5 erg/cm³		M_∞ abs. units	
MnO	MgO	Fe_2O_3	RT	Liquid N_2	RT	Liquid N_2
1.30	—	1	6.71	13.34	356	500
0.426	0.13	0.444	5.00	9.22	302	406
0.304	0.251	0.444	3.79	6.85	266.5	350
0.176	0.38	0.444	2.39	4.21	211.5	279
0.055	0.5	0.444	2.34	4.27	202.5	273

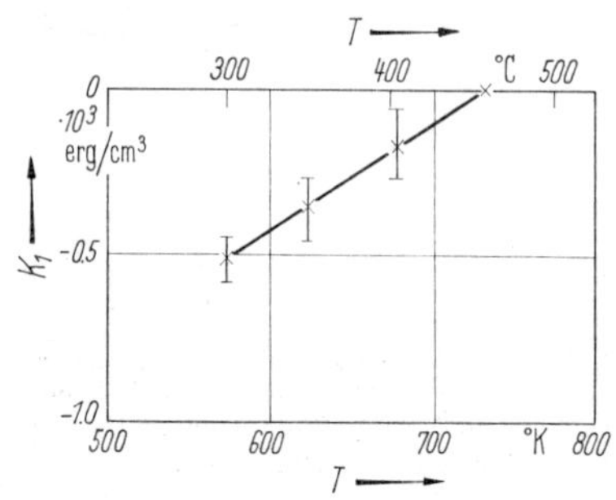

Fig. 24. $Mg_{0.5}$, $Mn_{0.5}$, Fe_2O_4. The $K_1(T)$ dependence of a single-crystal sphere on approaching the Curie point. Determined by the torque method. [66DS (1)].

d) System Ni-Mn-Fe-O

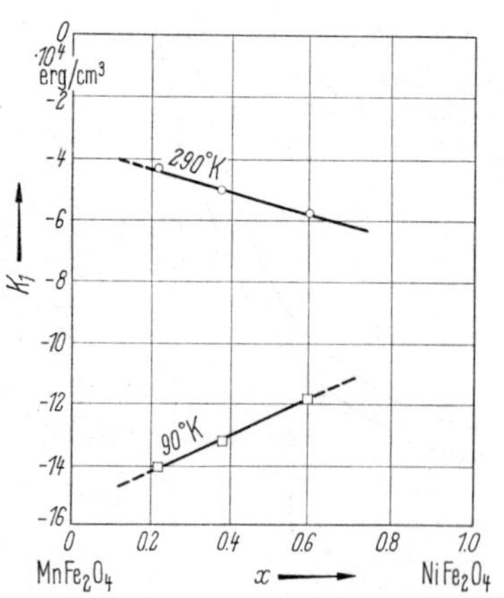

Fig. 25. $MnFe_2O_4$-$NiFe_2O_4$ system. Concentration dependence of K_1. Measurements on single-crystal spheres by the torque method. Crystals grown by the Bridgman technique in air [61M].

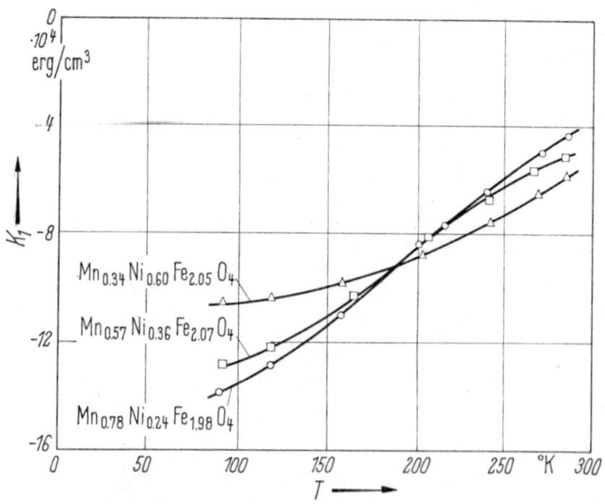

Fig. 26. $MnFe_2O_4$-$NiFe_2O_4$ system. Temperature dependence of K_1. Measurements on single-crystal spheres by the torque method. Crystals grown by the Bridgman technique in air [61M].

e) System Zn-Mn-Fe-O

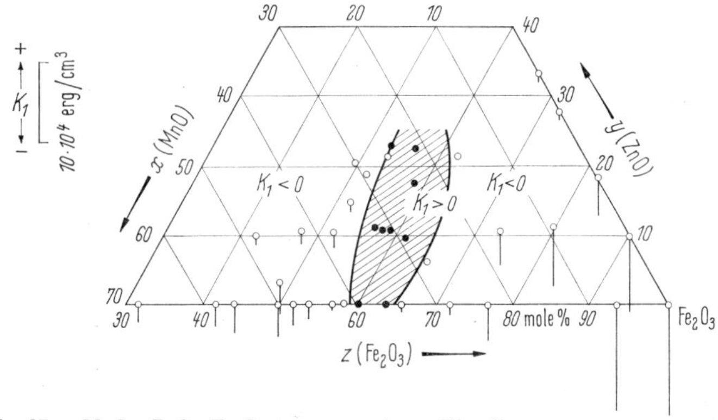

Fig. 27. xMnO-yZnO-$z$$Fe_2O_3$ ternary system. The diagram of K_1 at 20 °C. Compositions of monocrystal specimens are represented by the parameter x, y, and z (e.g. z = 100% is corresponding to Fe_3O_4). Open and closed circles are negative and positive K_1, respectively. From [63O], with data of [59PS] and [61M (2)].

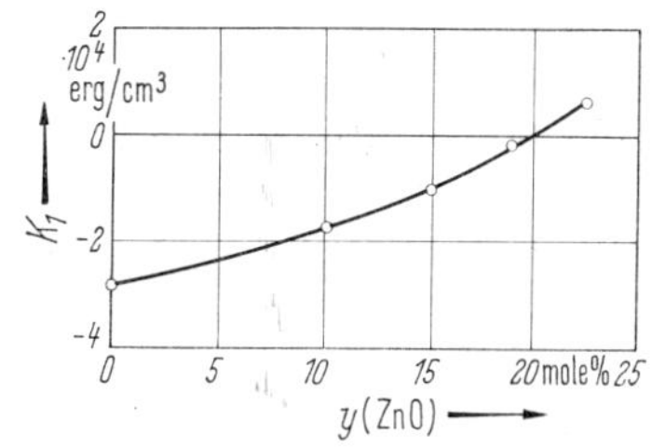

Fig. 28. xMnO-yZnO-52Fe_2O_3. K_1 vs. parameter y at 20 °C. From [63O].

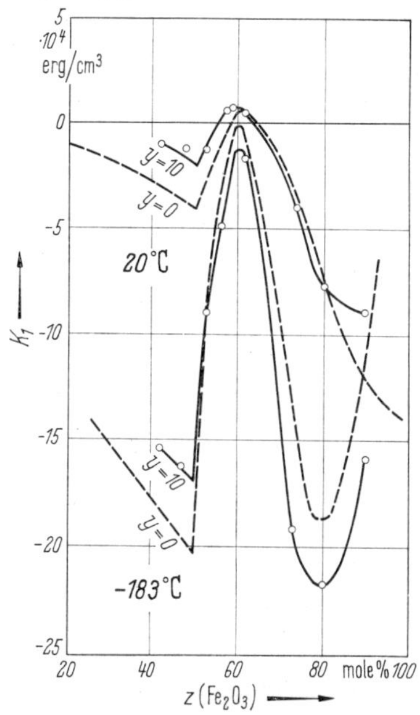

Fig. 29. xMnO-10ZnO-$z$$Fe_2O_3$. K_1 vs. the parameter z at 20 °C and liquid oxygen temperature [63O]. Broken curves are the results of [59PS] and [61M (2)].

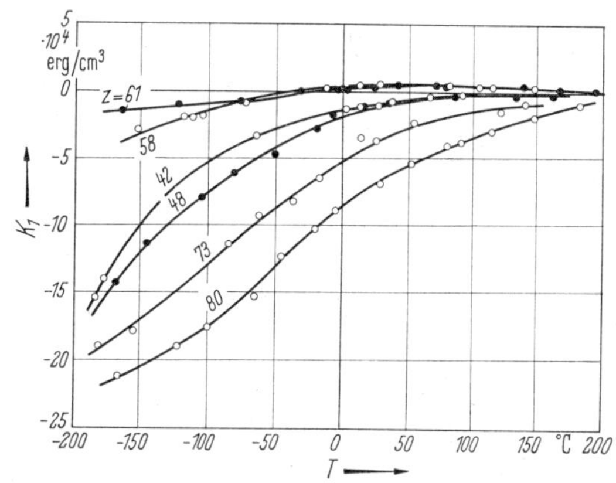

Fig. 30. xMnO-10ZnO-$z$$Fe_2O_3$. K_1 vs. T. Parameter z. From [63O].

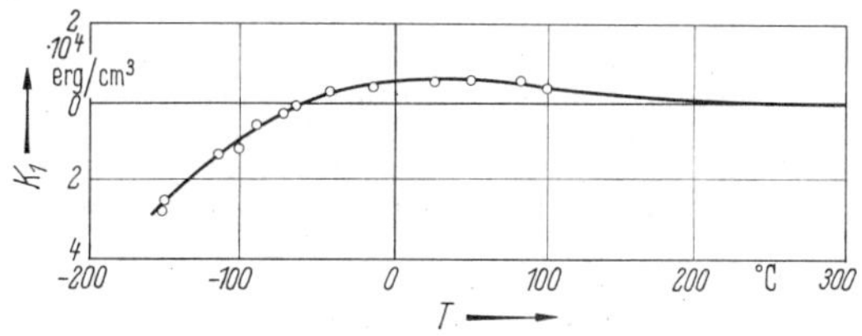

Fig. 32. 31MnO-11ZnO-58Fe_2O_3. K_1 vs. T. From [63O].

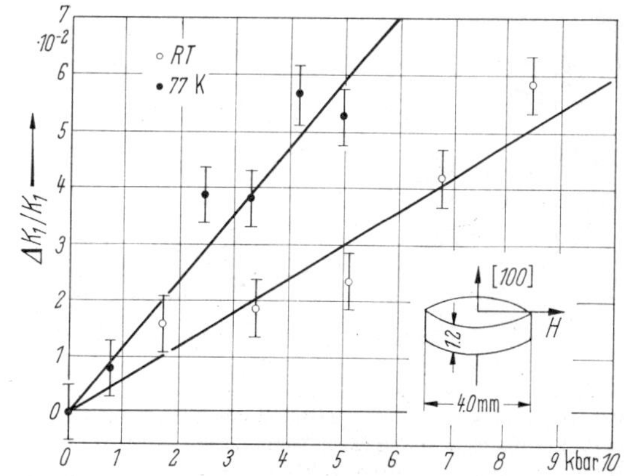

Fig. 33. $Mn_{0.8}Zn_{0.2}Fe_2O_4$. Hydrostatic pressure dependence of magnetic first order anisotropy constant K_1 of MnZn ferrite at RT (○) and 77 °K (●). The specimen was annealed before the measurements. From [67SKK].

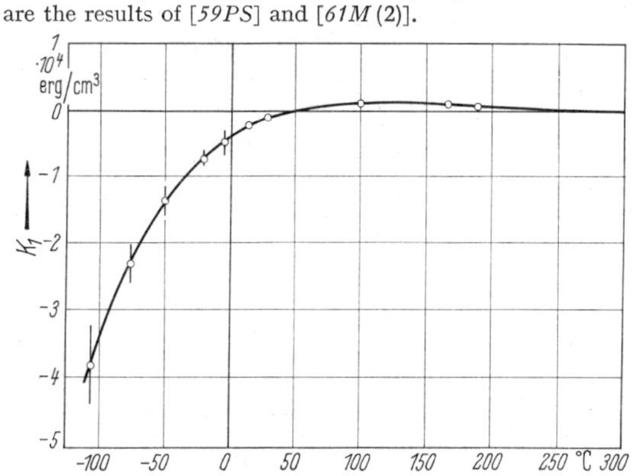

Fig. 31. 17MnO-21ZnO-62Fe_2O_3. K_1 vs. T. From [63O].

f) System Ti-Mn-Fe-O

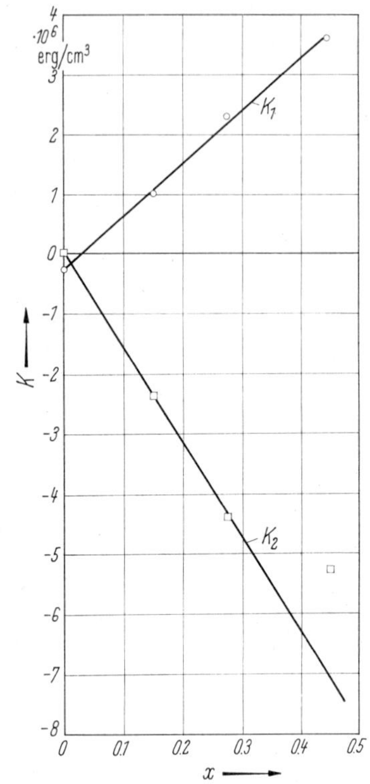

Fig. 34. $MnTi_xFe_{2-x}O_4$. K_1 and K_2 at $T = 4.2$ °K as a function of x. Single crystals made by a zone melting method and annealed afterwards. Measurements in the (110) plane and in a field of 32 kOe using an automatic torque balance [*62SLS*].

6.1.3.6.3 Linear magnetostriction

a) System Mn-Fe-O

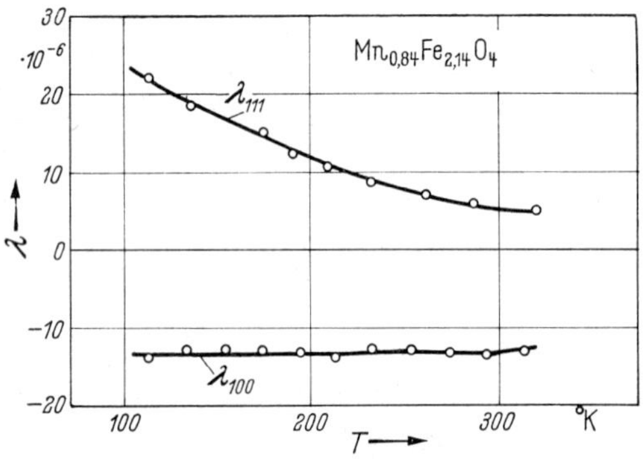

Fig. 35. $Mn_{0.84}Fe_{2.14}O_4$. The temperature dependence of λ_{111} and λ_{100} [*58En* (2)]. The sample was a (110) circular disc, and the values were deduced from measurements with the disc demagnetized and magnetized. In one case a more exact procedure was used in which the change in length with the disc magnetized along a range of crystal directions was plotted. This made it possible to extract a value of the magnetostriction coefficients which did not depend on the uncertainty in the initial state of the crystal. The RT values were $\lambda_{111} = (+5.3 \pm 0.2)\cdot 10^{-6}$; $\lambda_{100} = (-13.9 \pm 0.2)\cdot 10^{-6}$. These values agree very well with those measured in the conventional way.

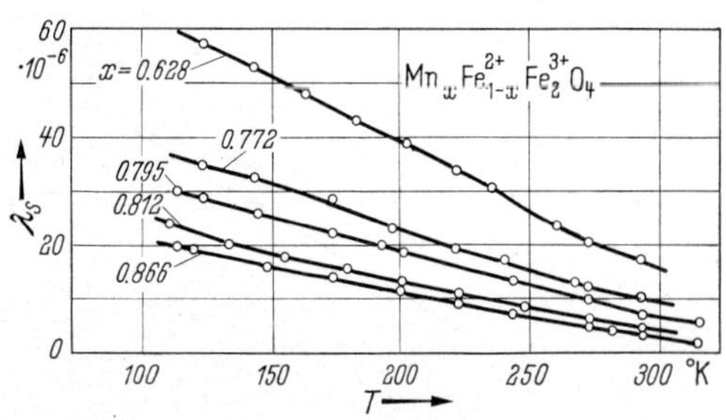

Fig. 36. $Mn_xFe_{3-x}O_4$. The saturation magnetostriction λ of polycrystalline $MnFe^{2+}$ ferrite as dependent on temperature. Parameter x [*58En* (2)]. An increasing content of Fe^{++} causes an increased positive value of the saturation magnetostriction, and a higher rate of change with temperature. Comparing this curve with Fig. 35, it may be concluded that the main influence of the bivalent iron is on the magnetostriction in the preferred direction, λ_{111}; and that λ_{100} is not much affected by it.

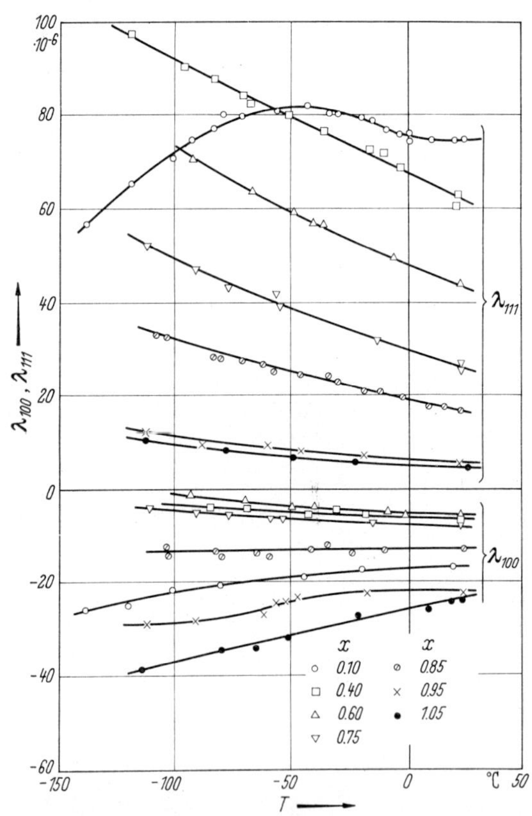

Fig. 37. $Mn_xFe_{3-x}O_4$. Temperature dependence of magnetostriction constants. Parameter x. From [*62MF*].

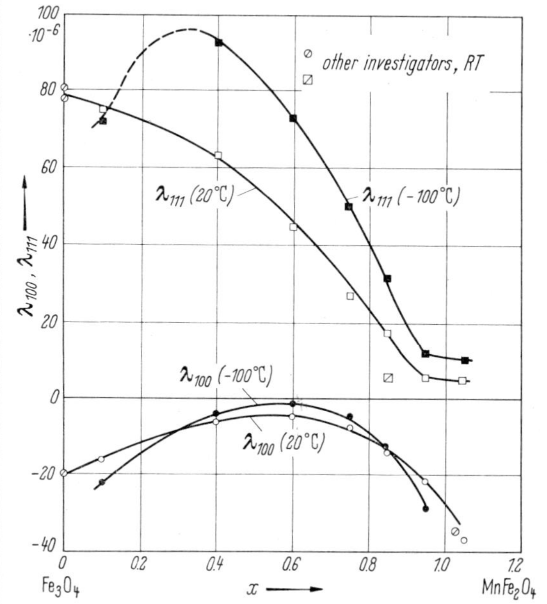

←

Fig. 38. $Mn_xFe_{3-x}O_4$. Concentration dependence of magnetostriction constants. From [*62MF*].

b) System Mn-Zn-Fe-O

→

Fig. 39. (Mn, Zn)Fe_2O_4. Saturation magnetostriction λ_s and permeability μ_i vs. δ. It has been found that the permeability of MnZn ferrites can often be improved by making a ternary ferrite with Fe_3O_4. This is thought to be associated with a minimum in the magnetostrictive energy which occurs for a small percentage of divalent Fe in the ferrite.
This figure shows data [*59L*] taken on a series of ferrites made by sintering in nitrogen for two hours at 1290 °C oxide mixtures consisting of 18 mole% ZnO, (50 + δ) mole% Fe_2O_3, and the balance MnO. Curve (a) shows the divalent iron content expressed in mole% FeO. Curve (b) shows the saturation magnetostriction, λ_s, and (c) the initial permeability, μ_i, of the samples. The maximum in the permeability of the samples occurs near the point where the magnetostriction constant is zero.

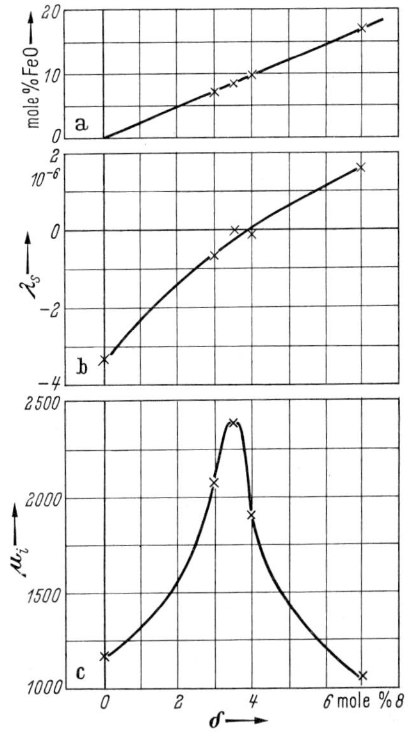

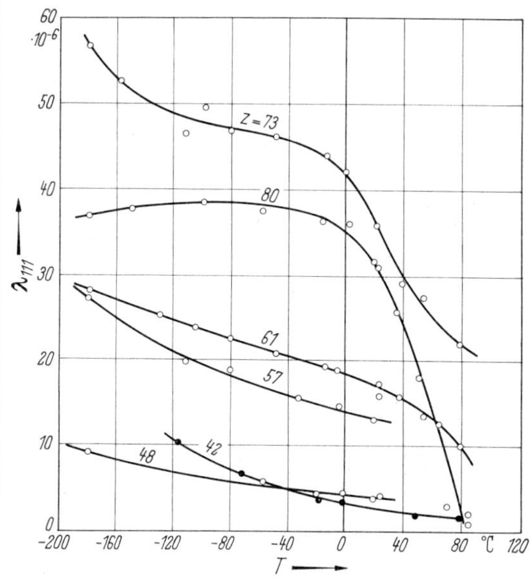

Fig. 40. xMnO-yZnO-$z$$Fe_2O_3$. Thermal change of λ_{111} of specimens with $y \approx 10$ mole%. Parameter z. [*640K*].

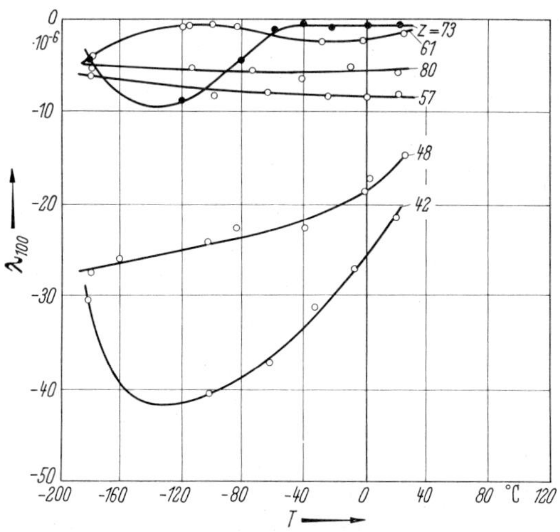

Fig. 41. xMnO-yZnO-$z$$Fe_2O_3$. Thermal change of λ_{100} of specimens with $y \approx 10$ mole%. Parameter z. [*640K*].

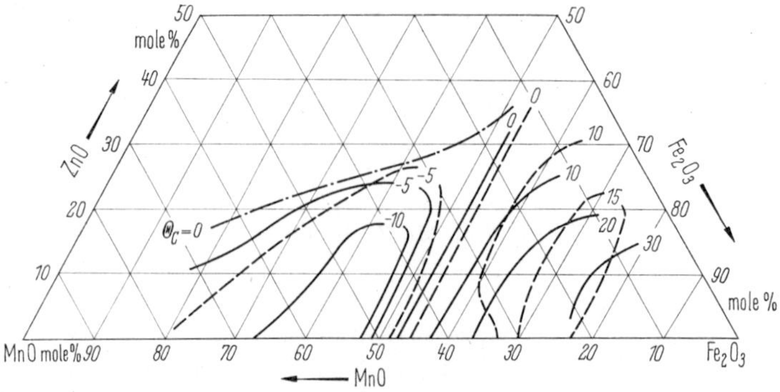

Fig. 42. Mn-Zn-Fe ternary ferrite system. Diagram of the magnetostriction λ_s. The solid lines represent the data of the specimens sintered in nitrogen and the broken lines the data of the specimens sintered in air [*61S*].

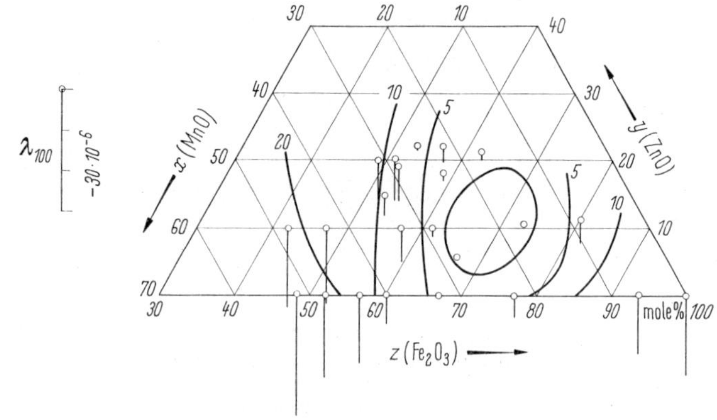

→

Fig. 43. Mn-Zn-Fe ternary ferrite system. Diagram of λ_{100} at 20 °C [*640K*]. Compositions of monocrystalline specimens are represented by xMnO-yZnO-$z Fe_2O_3$ (e.g. $z = 100$ mole% is corresponding to Fe_3O_4). Data of [*62MF*] are included.

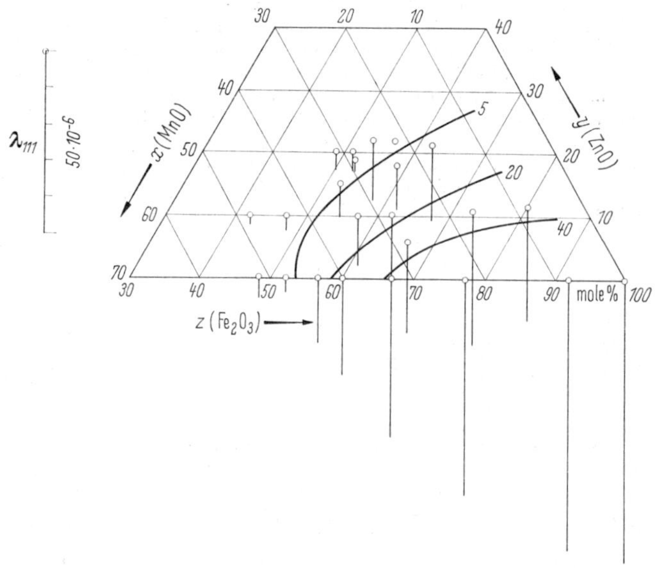

→

Fig. 44. Mn-Zn-Fe ternary ferrite system. Diagram of λ_{111} at 20 °C [*640K*].

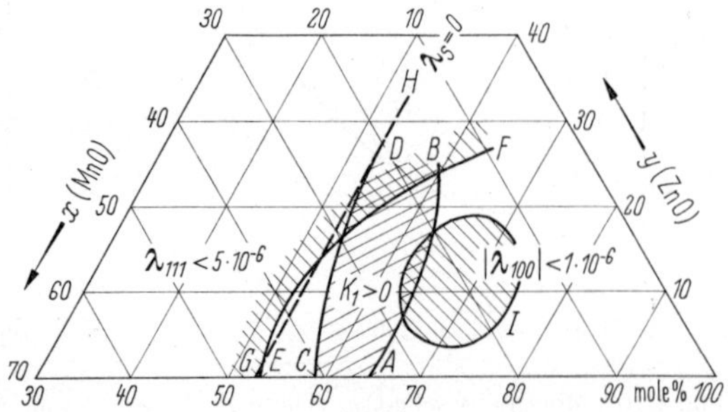

←

Fig. 45. Mn-Zn-Fe ternary ferrite system. Relation among magnetostriction constants and crystalline anisotropy constant at 20 °C [*640K*].

Line AB and CD: $K_1 = 0$,
Line GH: $\lambda_s = 0$,
Left side of the line EF: small λ_{111},
In the circle I: small λ_{100}.

c) System Me-Mn-Fe-O (Mn = Zn, Co, Mg)

Tab. 9. Magnetostriction constants and magnetoelastic constants

Composition	T °K	$\lambda_{100} \cdot 10^{-6}$	$\lambda_{111} \cdot 10^{-6}$	$\lambda_s \cdot 10^{-6}$	$4\pi M_s$ Gauss	B_1 *) cm⁻¹ per molecule	B_2 *)	Ref.
$MnFe_2O_4$	300	− 25	+ 5	− 7		+ 15	− 4	*62MF*
	150	− 35	+15	− 5		+ 21	−13	
$Mn_{0.98}Fe_{1.86}O_4$	300	− 35	− 1	−15				*55BTW*
$Mn_{0.6}Zn_{0.1}Fe_{2.1}O_4$	300	− 14	+14	+ 3		+ 8	−12	*55BTW*
$Fe_{0.15}Mn_{0.85}Fe_2O_4$	300	− 14	+16	+ 4		+ 8	−14	*62MF*
	77	− 15	+32	+13		+ 9	−28	
$Co_{0.05}Mn_{0.95}Fe_2O_4$	300	− 27				+ 16		*64LR*
	100	− 51				+ 30		
$Co_{0.3}Mn_{0.4}Fe_2O_4$	300	−200	+65	−40		+120	−58	*55BTW*
$Co_{0.5}Mn_{0.5}Fe_2O_4$	300	−230				+140		*64LR*
	77	−440				+260		
57MnO, 43Fe_2O_3	300	− 43.0	−12.6	−24.8	2450			*67URBS***)
35MnO, 22MgO, 43Fe_2O_3	300	− 35.6	−11.4	−20.8	1600			*67URBS***)
27MnO, 30MgO, 43Fe_2O_3	300	− 34.6	−13.5	−21.9	1380			*67URBS***)
20MnO, 37MgO, 43Fe_2O_3	300	− 31.6	−12.6	−20.2	1130			*67URBS***)
13MnO, 44MgO, 43Fe_2O_3	300	− 32.0	−13.6	−21.0	1030			*67URBS***)
7MnO, 50MgO, 43Fe_2O_3	300	− 29.2	−12.7	−19.3	920			*67URBS***)
57MgO, 43Fe_2O_3	300	− 25.8	−11.3	−17.1	750			*67URBS***)

*) The magnetoelastic constants B_1 and B_2 are estimated values using approximate values of the cubic elastic constants and unit cell sizes [*66J*].

**) The values of λ_{100} and λ_{111} were determined from measurements of the variation of magnetic and magnetostrictive susceptibilities as functions of magnetic field strength in the region approaching saturation.

6.1.3.6.4 Magnetic annealing

a) System Mn-Fe-O

For the composition $\mathbf{Mn_{0.4}Fe_{2.6}O_4}$ a small magnetic annealing effect was observed by [*59PS*][1]). A spectroscopic analysis for Co on this particular crystal showed that the Co concentration was less than 10 parts per million and could not account for the annealing effect.

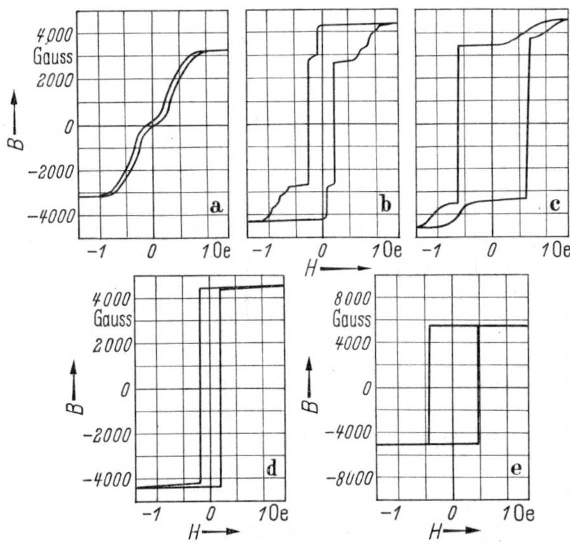

Fig. 46. $Mn_{1.4}Fe_{1.6}O_4$. A series of hysteresis loops showing a very clean cut example of a magnetic annealing. These were taken for one of the cores used in the domain wall motion experiments described in 6.1.3.7. The traces (a), (b) and (c) were taken at 300 °K, 195 °K, and 77 °K, respectively, and the cooling accomplished without biasing field. When the core was brought back to room temperature, a loop identical with (a) was obtained. The traces (d) and (e) were obtained at 195 °K and 77 °K after cooling from room temperature with a biasing field of 2.3 Oe [*59DE*].

[1]) The experimental procedure used for measuring this effect consisted of holding the temperature of the crystal at 370 °K for a period of time in the presence of a saturating magnetic field which was applied along either a cube edge or a face diagonal direction and then the crystal was quenched to RT where the annealing effect was essentially "frozen in". A quantitative measure of the effect was obtained by comparing torque curves record at RT before and after the heat treatment. The induced energy in the (100) plane could be described by $K_u \sin^2\theta$ where K_u is the anisotropy constant and θ is the angle between the magnetization and the [100] axis. The magnitude of K_u is dependent on the direction of applied field during the anneal such that it is a maximum if the field is parallel to the [100] axis and zero if parallel to the [110] axis. K_u reached a maximum value of $-2.0 \cdot 10^3$ erg/cm³ for an annealing time of the order of 10 minutes at 370 °K with torque measurements at 300 °K and the annealing field parallel to the [100] axis.

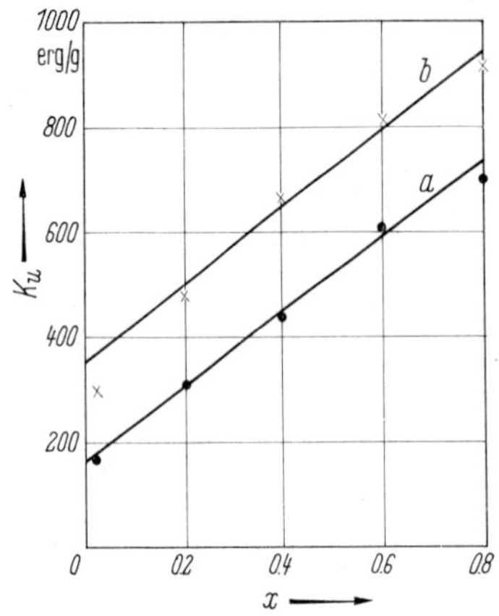

←
Fig. 47. $Mn_{1+x}Fe_{2-x}O_{4+\gamma}$. Induced uniaxial anisotropy constant K_u of Mn ferrites measured at 88 °K as a function of x and γ. The samples are fired:
Curve a: at 1200 °C in air, $p = 760$ Torr.
Curve b: at 1200 °C in air, $p = 10^{-2}$ Torr.
From [*66GMMS*]. The corresponding values of γ are given in Fig. 48.

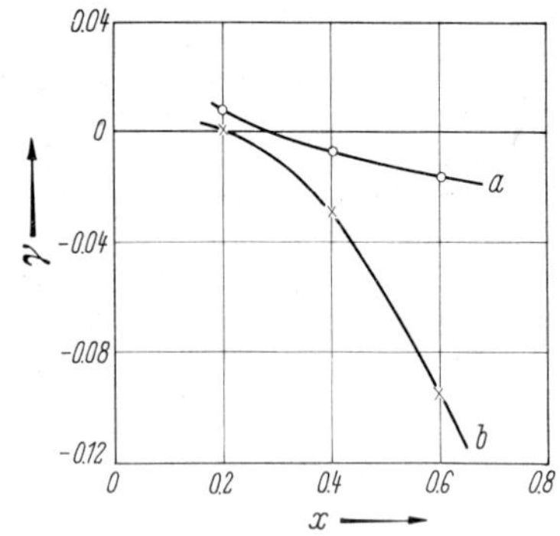

Fig. 48. $Mn_{1+x}Fe_{2-x}O_{4+\gamma}$. Oxygen content for different Mn ferrites [*66GMMS*]. Same samples as in Fig. 47.

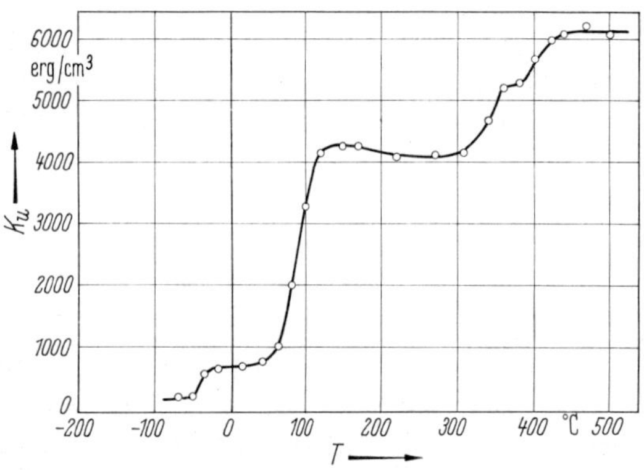

Fig. 49. $Mn_{0.2}Fe_{2.8}O_{4.014}$. Induced uniaxial anisotropy constant, K_u, of a polycrystalline Mn ferrite measured at 77 °K vs. magnetic annealing temperature, T. The Fe^{2+} content is 0.77 and the vacancies concentration 0.011 . A field strength of 10 000 Oe is applied 15 min during the magnetic annealing. From [*65MMP*].

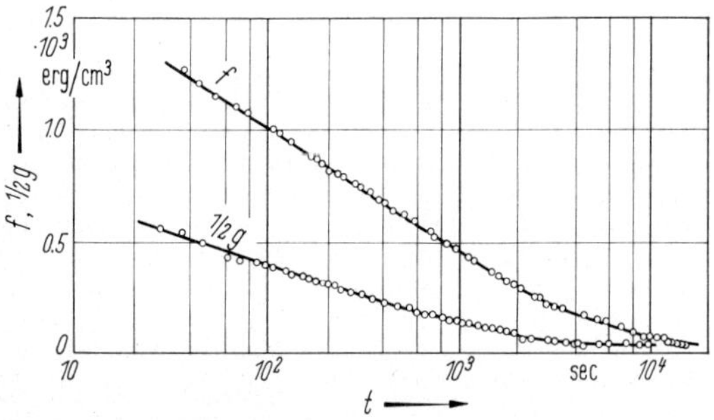

Fig. 50. $Mn_{0.4}Fe_{2.6}O_{4+\gamma}$. Induced anisotropy constants, f and $\frac{1}{2}g$, as functions of time. Measured on a single crystal in (001) plane, in [110] and [100] directions, after previous annealing in magnetic field in [100] and [110] directions, respectively. Annealing temperature equals measuring temperature of +393 °K. From [*64GV*]. For polycrystalline samples K_u is connected with f and g by the relation $K_u = 4/10\, f + 3/10\, g$. For the same Mn ferrite with an annealing temperature of 370 °K and a symmetry axis [100]: $-g = 0$ and $-f = 2 \cdot 10^3$ erg/cm³ [*59PS*].

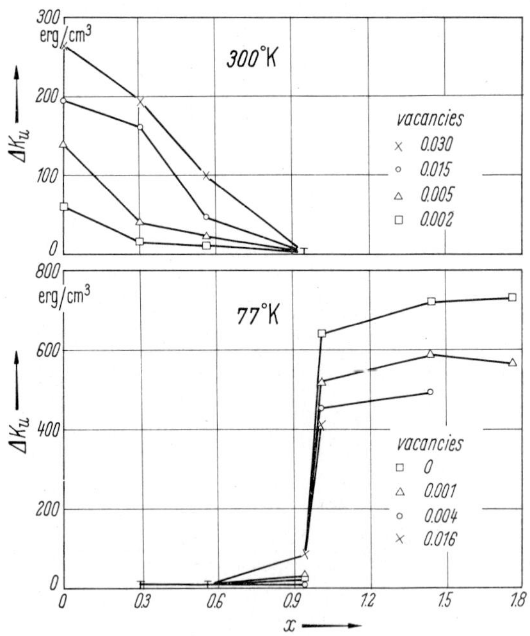

Fig. 51. $Mn_xFe_{3-x}O_{4+\gamma}$. Induced anisotropy as a function of Mn content, x, for various vacancy levels. The induced anisotropy was measured by annealing the sample in a magnetic field and then quickly rotating the magnet for 45 deg. The decay of the induced anisotropy was recorded from the time at the end of the rotation. ΔK_u is the decay between 5 and 500 seconds. [*67BG*].

b) System Mn-Co-Fe-O

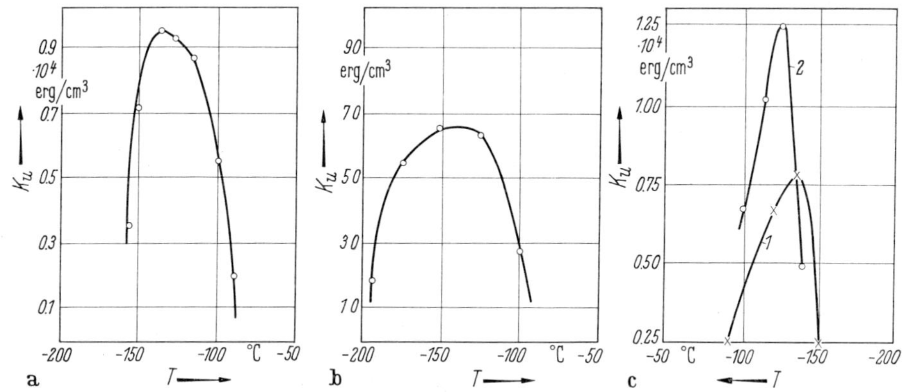

Fig. 52. Maximum uniaxial anisotropy constant, K_u, as a function of magnetic annealing temperature determined from the curve showing the time build-up of magnetic intensity:

a) Ferrite $Mn_{1.75}Co_{0.05}Fe_{1.2}O_4$ [*64BY*].
b) Ferrite $Mn_{1.8}Fe_{1.2}O_4$ [*64BY*].
c) *1* Ferrite $Mn_{1.75}Co_{0.05}Fe_{1.2}O_4$ [*66BA*].
2 Ferrite $Mn_{1.70}Co_{0.1}Fe_{1.2}O_4$ [*66BA*].

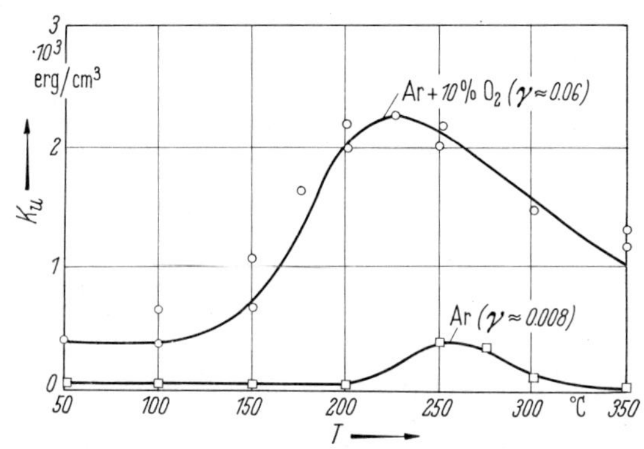

Fig. 53. Dependence of K_u constant of induced anisotropy on magnetic annealing temperature and degree of oxidation of the sample. Initial composition of the sample: 58 mole % Fe_2O_3, 41 mole % MnO, 1 mole % CoO, sintered at 1250 °C in Ar or Ar + 10% O_2. From [*64K3*]. γ = oxygen excess.

c) System Me-Mn-Fe-O (Me = Zn, Ni, Mg)

Fig. 54. $Me_zMn_{1.4-z}Fe_{1.6}O_{4+\gamma}$. Induced uniaxial anisotropy constant, K_u, measured at 88 °K as a function of x.

The samples are fired:
curve a: at 1200 °C in air, p = 760 Torr,
curve b: at 1200 °C in air, p = 10^{-2} Torr.
From [*66GMMS*].

→

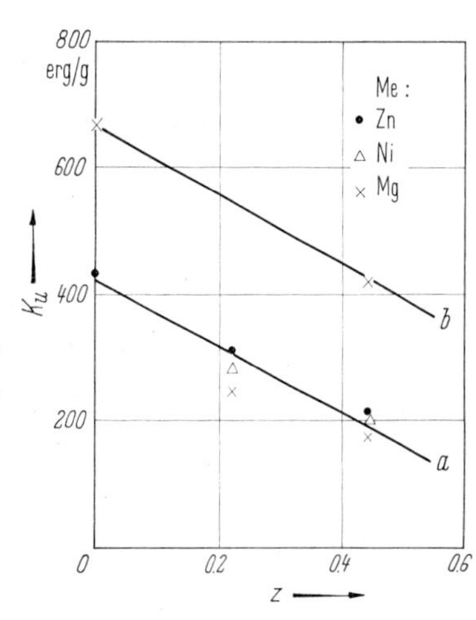

6.1.3.7 Domains and domain walls

To a large extent the low field and low frequency properties of magnetic materials, in particular their response to applied fields, depend on details of domain structure and the properties of the domain walls. These are structure dependent characteristics in a very sensitive way. In polycrystalline samples the influence of the random orientation of crystallites, of voids, grain boundaries, precipitates, etc., is so important that it alters completely the properties as observed on sound single crystal samples.

6.1.3.7.1 Domain wall motion experiments

a) System Mn-Fe-O

The wall motion experiments reported by *[59DE]* were performed on a crystal of $\mathbf{Mn_{1.4}Fe_{1.6}O_4}$ which has been grown by the flame fusion process. So-called "picture frame" samples were cut in such a way that the magnetization could form a closed loop around the central hole, and the magnetization could always be parallel to an easy direction.

There were two separate windings. The current through one is proportional to magnetic field. If the magnetization reversal is effected by the motion of a single wall, the induced voltage on the second winding yields a wall velocity. With the particular samples used it was found necessary to cool the cores to a temperature below 200 °K with an applied field in order to observe magnetization reversal by a single wall system. See Fig. 1 and 2.

In studying the switching behavior of the cores, rectangular current pulses of both signs were applied in sequence at some convenient repetition rate. The output voltage was viewed on the screen of an oscilloscope. The simplest sequence of pulses consisted first of a pulse capable of driving the core to saturation followed by a test pulse of opposite polarity which might or might not switch the core completely.

Let us take as the equation of motion of the magnetization

$$\boldsymbol{M} = \gamma \boldsymbol{M} \cdot (\boldsymbol{H}_n + \boldsymbol{H}_{ex} + \boldsymbol{H}_{an}) + (\alpha/M)\, \boldsymbol{M} \cdot \boldsymbol{M}.$$

In this the first term on the right represents the effect of the net field, applied plus demagnetizing. The second term $\gamma \boldsymbol{M} \cdot \boldsymbol{H}_{ex}$ gives the torque on the magnetic moment at a point arising from the fact that the moments in nearby regions are along slightly different directions. The gyromagnetic ratio is γ. The structure of the anisotropy energy surface introduces torques which are indicated by $\gamma \boldsymbol{M} \cdot \boldsymbol{H}_{an}$. These may be written as derivatives of that energy with respect to angle evaluated at the direction in question. Finally, the damping is introduced by the term $(\alpha/M)\boldsymbol{M} \cdot \boldsymbol{M}$.

A value for the damping constant α can be derived from the wall motion data, though the procedure is open to question. It is also possible to obtain a value of the damping parameter from ferrimagnetic resonance experiments. This is given by $\alpha = (\frac{1}{2})\,(\Delta H/H_{eff})$, where ΔH is the line width and H_{eff} the field for resonance after anisotropy effects have been taken out. In addition to H_c and R, Tab. 1 gives the values of α_{dw} derived from domain wall motion experiments alongside corresponding damping constants, $\alpha_{res}^{(9)}$ and $\alpha_{res}^{(24)}$, obtained from resonance experiments at 9.3 Mc/s and 24.0 Mc/s, respectively, on samples from the same crystal.

Frequency of observed nuclear and domain wall resonance for $\mathbf{MnFe_2O_4}$ *[62BBBB]*.

Fe^{57} nuclear	58.78 ··· 54.18 Mc/s for single crystal
wall	70 ··· 50 Mc/s for single crystal
wall	≈70 Mc/s for ceramic.

In measurements of the magnetization distribution in the domain boundaries of FeMn ferrite by means of the Kerr effect, it is found that on the same plane and in the same crystal the boundaries may have different polarities and different widths. From *[66DS* (2)].

Tab. 1. $Mn_{1.4}Fe_{1.6}O_4$. Average value of the internal coercive force, H_c, and the wall mobility, R, for various temperatures at which the wall motion experiment has been performed. Values of the damping parameter α suitable to the Landau-Lifschitz equation as derived from wall motion experiments are compared with those derived from ferrimagnetic resonance experiments (see text) *[59DE]*

T °K	H_c Oe	R cm sec^{-1} Oe^{-1}	α_{dw}	$\alpha_{res}^{(9)}$	$\alpha_{res}^{(24)}$
4.2	7.5	9.5	1.96		0.055
20	1.4	94	0.22		
65	0.175	630	0.037	0.034	
77	0.06	870	0.035	0.026	0.016
87	0.035	1040	0.030		
195	0.03	2120	0.029	0.0092	
294				0.0046	0.0035

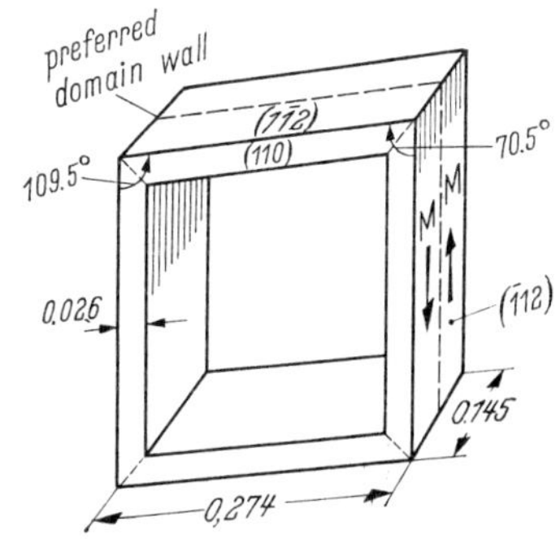

→ Fig. 1. The dimensions and crystallographic orientation of one of the cores. It was found to be important that the cross section of the legs of the core should not be square or nearly square. For if they are, the area of two possible domain walls becomes equal, and the domain structure actually assumed by the magnetization is not well defined. The difficulty in the experiment resides mostly in the selection and preparation of the sample. All the outer surfaces are polished. The planes are accurate to within a few minutes of arc [*59DE*].

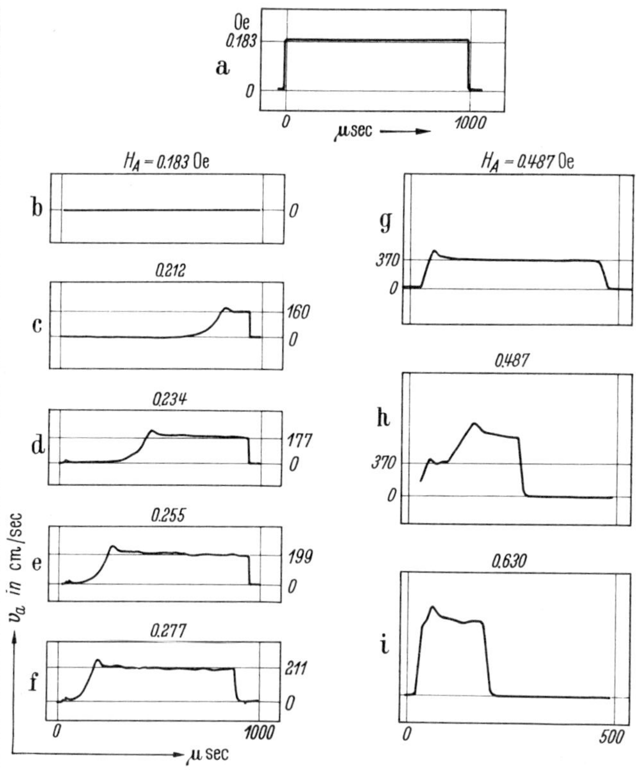

← Fig. 2. The response to the test pulse in the low field region. In Fig. 2a is shown the shape of the driving pulse as read across a resistor in series with the input winding. The point is that the current pulse is rectangular on this time scale. Fig. 2b shows the output signal corresponding to a current pulse of 0.183 Oe; there is none. However, in (c) we see that after a long period of time, about 800 μsec, the rate of change of magnetization builds up at an increasing rate and finally reaches a constant value which it keeps till the end of the drive pulse. With increasing fields this initial period decreases in duration till in (f) the magnetization stops changing before the end of the driving pulse. Fig. 2g and 2h were taken at the same field, but it is a field at which a discontinuity occurs. These traces are interpreted as representing the nucleation of a single domain wall, and its motion at a constant velocity across the sample. In (f) the wall reaches the opposite side of the crystal, it is thus saturated. No further magnetization change is possible. In (h), however, a wall is thought to be nucleated at the opposite face, whence it is driven to meet the wall which started first. Since the wall area is doubled, the rate of change of magnetization also is doubled. Note that there seem to be two different nucleation times associated with the walls starting from two surfaces [*59DE*].

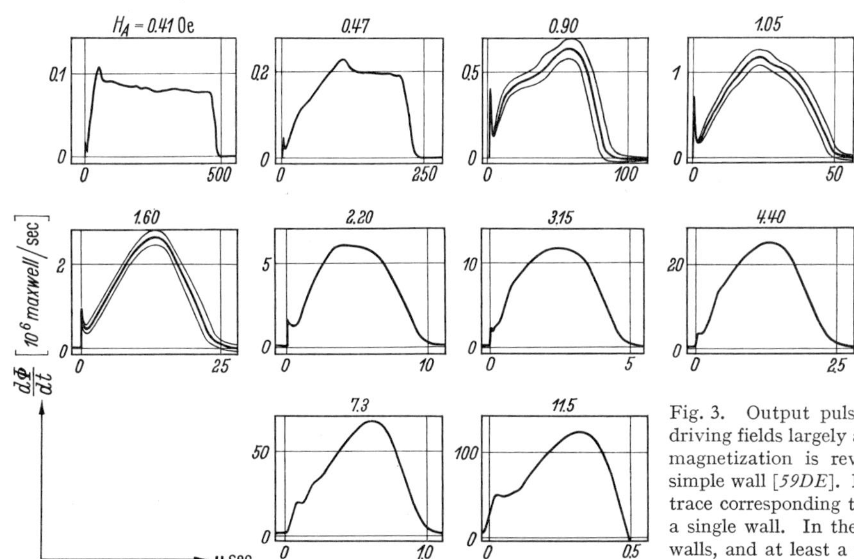

Fig. 3. Output pulses obtained for a series of driving fields largely above the level at which the magnetization is reversed by the motion of a simple wall [*59DE*]. In the first figure is shown a trace corresponding to magnetization reversal by a single wall. In the next frame there are two walls, and at least a part of the output is constant in time. Note that in this sequence of traces the horizontal and vertical scales have been altered reciprocally so that area, i.e., flux, is conserved. The next few traces show a cluster of discrete traces rather than a single line as is observed at higher fields. This presumably represents a field region in which the magnetization reversal is accomplished by the motion of relatively simple wall structures. At still higher fields the curve looks more and more like the response of a polycrystalline core.

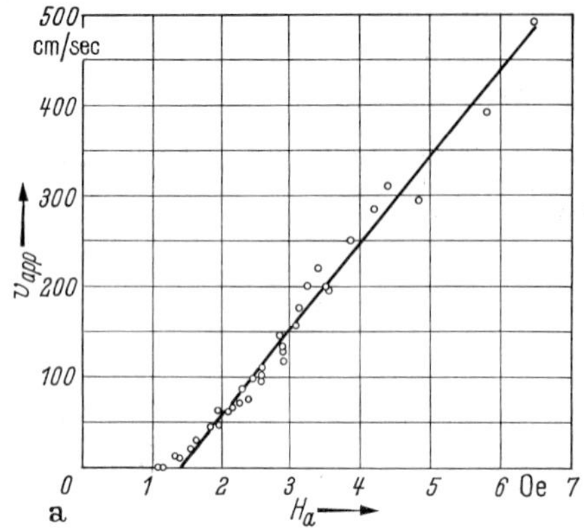

a

Fig. 4. $Mn_{1.4}Fe_{2.6}O_4$. Apparent wall velocity v_{app} vs. applied field H_a. It seems possible to account for the data with a straight line of the form $v_{app} = R(H_a - H_c)$ where v_{app} is the apparent wall velocity, H_a is the applied field, and H_c is some sort of internal coercive field. R is a constant which we might call domain wall mobility. It represents the slope of the line [59DE].

Fig. 4a. A core immersed in liquid hydrogen. The low field points were obtained by using a three pulse sequence in which the core is first saturated, then a wall is nucleated and driven part way across the crystal. The test pulse which follows does not have to nucleate a wall; the established wall merely moves under the influence of the test pulse. $H_c = 1.4$ Oe; $R = 94$ cm sec^{-1} Oe^{-1}.

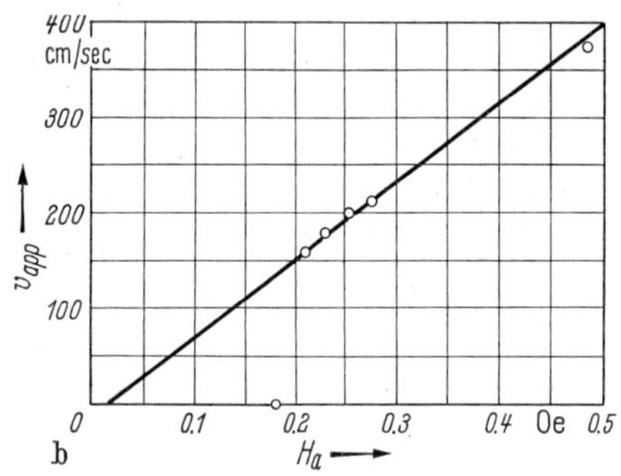

b

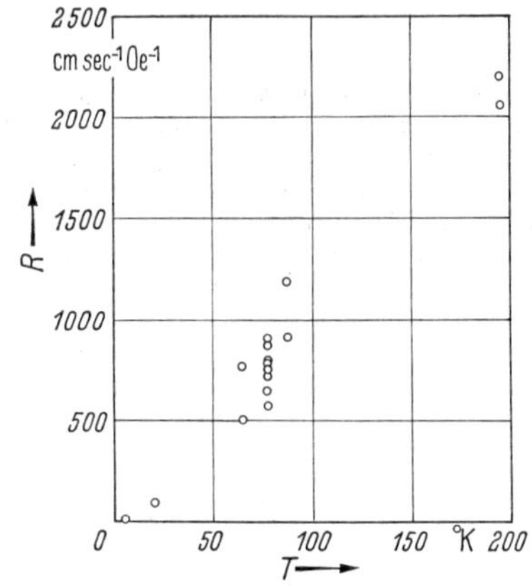

Fig. 5. $Mn_{1.4}Fe_{2.6}O_4$. R = domain wall velocity/net driving field H_n vs. temperature [59DE].

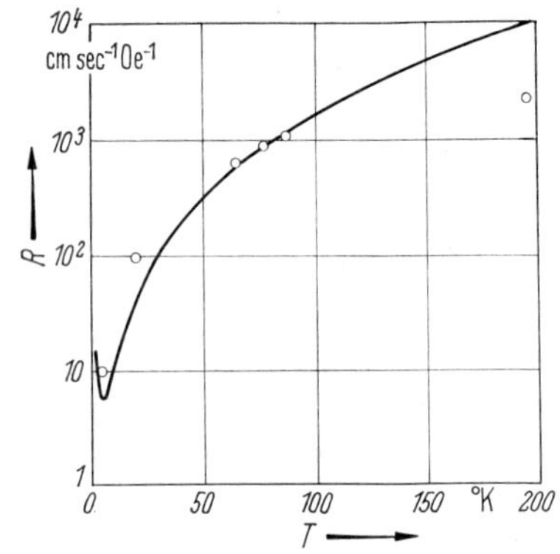

Fig. 6. Magnetic domain wall mobility, R, in Mn rich Mn ferrite ($\approx Mn_{1.4}Fe_{1.6}O_4$) as a function of temperature. From [67HT]. The curve is deduced from microwave resonance measurements at 9.3 Gc/s and 24 Gc/s; the points are mean measured values from [59DE].

←

Fig. 4b. Data taken from the traces of Fig. 1, 2, only in the field region where the magnetization reversal is effected by a single wall. $H_c = 0.015$ Oe; $R = 805$ cm sec^{-1} Oe^{-1}.

Tab. 2. Domain wall mobility, R, and damping constant, α, for different Mn, MnMg, and MnMgZn ferrites. After [66S (2)]. A = exchange energy constant.

Composition	T °C	M Gauss	K_1 10^4 erg cm^{-3}	K_u erg cm^{-3}	A 10^{-7} erg cm^{-1}	R cm sec^{-1} Oe^{-1}	α
$Mn_{0.5}Fe_{2.5}O_4$	−78	490	−6.5	$2.0 \cdot 10^3$	2.9	2000…3300	
$Mn_{1.02}Fe_{1.98}O_4$	−185	532	−20		2.1	1500…2300	≈ 0.020
$Mn_{1.2}Fe_{1.8}O_4$	−185	462	−20		1.7	1350…2050	≈ 0.021
			−6.8			(2120)*)	(0.029)*)
$Mn_{1.4}Fe_{1.6}O_4$	−78	348			1.5	{1200…1800	{≈ 0.023
	−185	418	−19			(1040)*)	(0.030)*)
$Mn_{1.6}Fe_{1.4}O_4$	−78	285	−5.2		1.4	2100…3100	≈ 0.024
	−185	360	−14.5			1050…1550	≈ 0.028
$Mn_{1.8}Fe_{1.2}O_4$	−78	196	−4.0		1.3	2000…3000	≈ 0.028
	−185	270	−12.5			1000…1500	≈ 0.031
$Mg_{0.76}Mn_{0.54}Fe_{1.7}O_4$	−78	317	−10.8		1.7	1700…2000	0.028
	−185	367	−26.0			850…1250	0.030
$Mg_{0.5}Mn_{0.51}Zn_{0.42}Fe_{1.57}O_4$	−78	362	−1.7		1.3	2200…3400	0.036
	−185	474	−9.0			1000…1600	0.036

*) From [59DE].

b) Systems Mg-Mn-Fe-O, Zn-Mg-Mn-Fe-O

See Tab. 2, p. 178

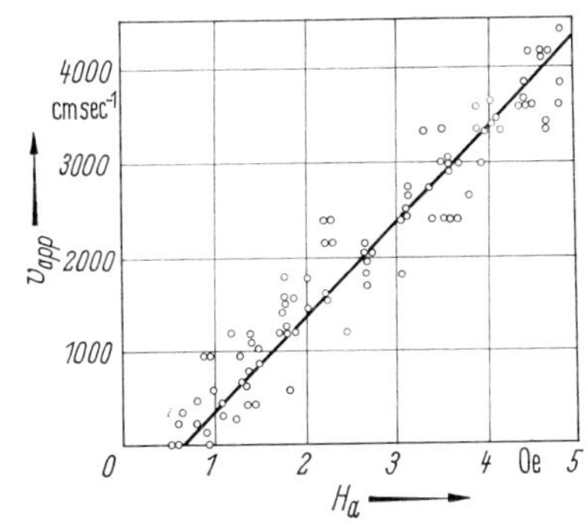

Fig. 7. $Mn_{0.47}Mg_{0.84}Fe_{1.69}O_4$. The variation of apparent wall velocity, v_{app}, with applied field, H_a, for a "square loop" ferrite. The motion of the wall was shown to be impeded by Néel spike domain structures around the pores. From [*61K*]. The full line represents $v_{app} = 985\,(H_a - 0.62)$ cm sec^{-1}.

6.1.3.7.2 Domain structure

Systems Mn-Fe-O, Mg-Mn-Fe-O, Zn-Mn-Fe-O

See also: [*57P, 59CG*].

Tab. 3

Substance		Observed structures	Ref.
$MnFe_2O_4$	crystals grown by Verneuil process [*04V*]	large fern type structure and, quite distinct from that, a very fine band	*58BCG*
$Mn_xFe_{3-x}O_4$	polished crystals	very simple structures and labyrinth structures	*59DE* *63HM*
$Mn_{0.7}Zn_{0.3}Fe_2O_4$	polycrystalline density 86% of X-ray density	no bulk domain walls	*58BCCG*
MnMg-ferrite type M.M. 7	crystal grown on a Verneuil apparatus	plane (100): the structure is like fern ice crystal plane (110): domains with very twisted boundaries and drop-like closed regions	*63DD*

6.1.3.8 Response of magnetization to a field

Permeability.

Important though it may be technically, the permeability of a sample of a magnetic material is very far removed from being an intrinsic property associated with a particular composition. It is a property extremely sensitive to structure, and in the case of sintered ferrites this means that all of the possible variables of fabrication have a bearing on the permeability of the finished sample. Thus permeability measurements only have real meaning if they are part of a series in which one parameter has been varied.

6.1.3.8.1 High temperature susceptibility

The remarks just made above do not apply in full to measurements of the susceptibility in the paramagnetic region. Here, however, it must be noted that the experiments reported to date are on polycrystalline samples, and thus represent a sort of average over all crystal directions.

$MnFe_2O_4$

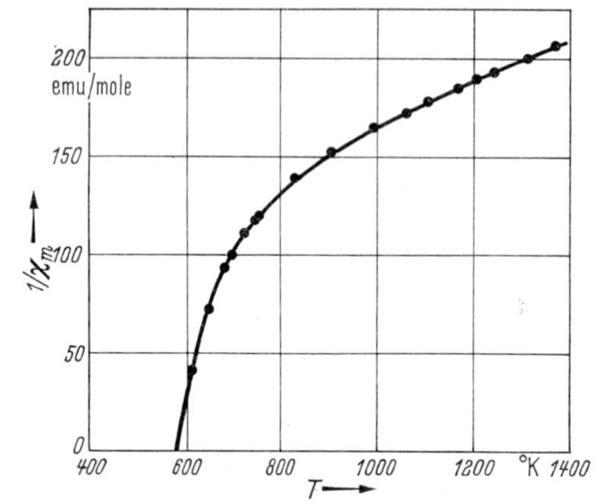

← Fig. 1. $MnFe_2O_4$. The inverse of the susceptibility per gram molecule as a function of the absolute temperature [*54CS*]. The line was obtained by fitting Néel's [*48N*] equation

$$\frac{1}{\chi} = \frac{T}{C} + \frac{1}{\chi_0} - \frac{\sigma}{T - \Theta}$$

to the points. In this, C is the Curie constant and $1/\chi_0$, σ, and Θ are parameters defined in terms of the Curie constant, the molecular field constants, and the fractions of magnetic ions on A and B sites, respectively.

System Zn-Mn-Fe-O

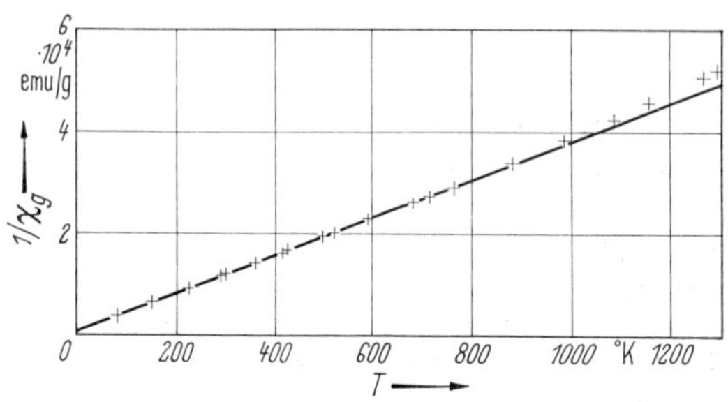

← Fig. 2. $ZnMn_{0.5}Fe_{1.5}O_4$. Reciprocal susceptibility per gram vs. temperature. Heat treatment 4 h 1200 °C in air. From [*65B* (1)].

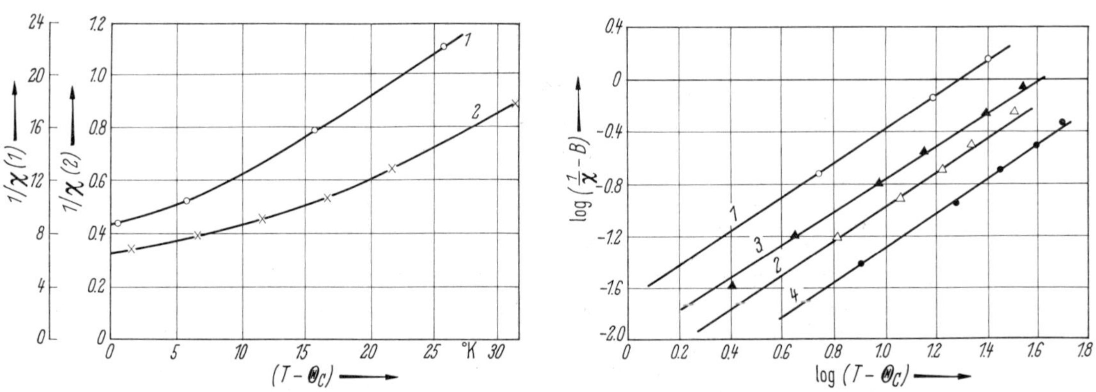

Fig. 3a, b. MnZn ferrites. Inverse of initial susceptibility, χ*), near the Curie point, Θ_C, described by $1/\chi = A(T - \Theta_C)^\gamma + B$. A is a constant depending on the type of lattice and on the value of the spin, $\gamma \cong 4/3$ for these samples, and B is given by the intercept on the ordinate of Fig. 3a. The samples are:

Nr.	compositions	Θ_C[°C]
1	MnO – 35.5%, ZnO – 15%, Fe_2O_3 – 49.5%	194.5
2	MnO – 30%, ZnO – 20%, Fe_2O_3 – 50%	128.5
3	MnO – 28%, ZnO – 22.5%, Fe_2O_3 – 49.5%	114
4	MnO – 26%, ZnO – 24%, Fe_2O_3 – 50%	95.5

(a) $1/\chi$ vs. $(T - \Theta_C)$, (b) $\log(1/\chi - B)$ vs. $\log(T - \Theta_C)$. From [*66BE*].

Further references for section 6.1.3.8.1: [*48N*], [*54CS*], [*54GDM*], [*57Bo*], [*59C*].

*) According to the compiler's inquiries, χ is the susceptibility per gram.

6.1.3.8.2 High field susceptibility in the ferrimagnetic state

System Mn-Fe-O

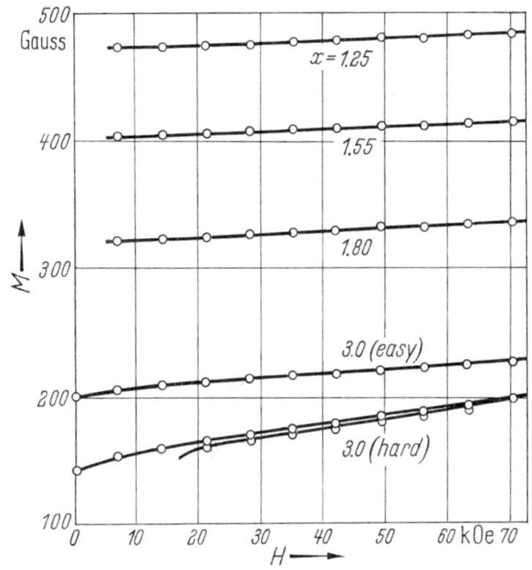

Fig. 4. $Mn_xFe_{3-x}O_4$. Magnetization vs. field. Parameter x. Pulsed fields of up to 75 kOe and about 1.5 msec duration. The susceptibilities obtained from these data are

x in $Mn_xFe_{3-x}O_4$	χ 10^{-4} Gauss/Oe
1.00	0.7 ± 0.5
1.25	1.4
1.55	1.7
1.80	2.3
3.00	3.5

[*61M* (3)].

6.1.3.8.3 Permeability variation with composition, impurities, irradiation, and grain size

a) Composition, impurities, and irradiation

System Zn-Mn-Fe-O

Tab. 1. The characteristic properties of the typical high-permeability MnZn ferrites obtained by the vacuum reduction method. Firing was performed in a vacuum of 10^{-4} Torr at 1250 °C. The temperature was then raised to 1330 °C and nitrogen gas containing 1.0 vol% of oxygen was introduced into the furnace. Finally, the samples were cooled in a vacuum of 10^{-3} Torr [*64SAT*].

Composition and properties		Sample *a*	Sample *b*
Main components (mole-%)	Fe_2O_3	51	51
	MnO	24	27
	ZnO	25	22
B_{10}, flux density at 10 Oe [Gauss]		3 700	4 250
H_c, coercive force		0.013	0.030
$\mu_{0.01}$, permeability at 0.01 Oe		35 000	22 000
μ_{max}, maximum permeability		40 000	32 000
Θ_C, Curie point [°C]		93	125
ϱ, density [g/cm³]		5.15	5.10
ϱ, electrical resistivity [Ω · cm]		1.1	1.2
μ, permeability at 1 kc/s		22 000	18 000
disaccommodation		small	small

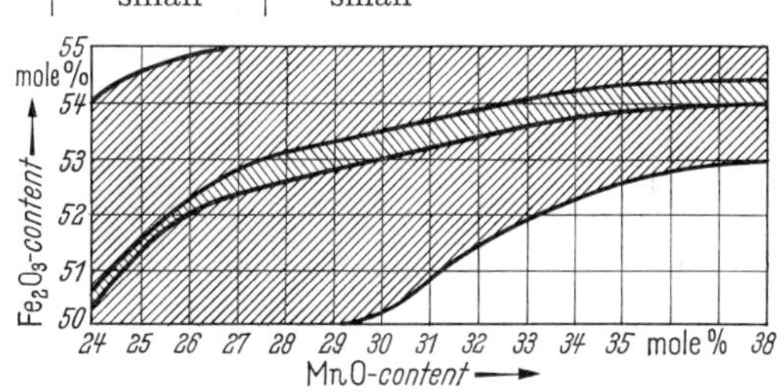

Fig. 5. $MnZnFe^{2+}$ ferrites. [*57Gu*] shows by shading the composition region in which permeabilities exceeding 2000 have been measured. The narrow band in the center shows the range for permeabilities exceeding 5000. It must be emphasized that permeability depends on a great many factors of which composition is merely one. The remaining constituent in these composition is, of course, ZnO.

Permeability is strongly dependent on preparation parameters: raw material origin, grinding-mixing, calcination, thermochemical cycles. The four following figures are given, more to point out the effect of the sintering-annealing atmosphere, than for the intrinsic value of the results.

For further information, see the following references: [*60LP*], [*61HW*], [*64P*], [*65PG*], [*65IMS*], [*66BS* (2)], [*66DFI*], [*66IMS*], [*66SB*].

[*57Gu*] showed that the temperature coefficient of permeability near room temperature could be adjusted in manganese-zinc-ferrous ferrite **(Mn, Zn)Fe_2O_4** by the substitution of Al^{3+} or Cr^{3+} for Fe^{3+}. For a ferrite whose starting materials consisted of 26 mole-% MnO, 6 mole-% Al_2O_3, 47.5 mole-% Fe_2O_3, and 20.5 mole-% ZnO the permeability was essentially constant from 230···350 °K. A mixture containing 6 mole-% Cr_2O_3 gave a similar result.

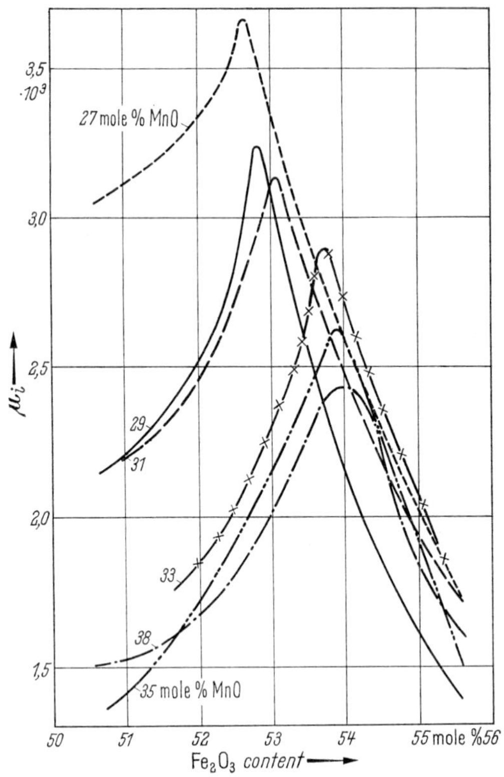

← Fig. 6. MnZnFe^{2+} ferrites. The variation in initial permeability, μ_i, for several series. Parameter mole% MnO. Special care was taken to compare samples which were nearly stochiometric and showed equal grain size, as well as having been given similar heat treatment. These were presented by [*57Gu*] who pointed out that for each percentage of Mn in the ferrite there is a specific FeO content which yields the maximum permeability. This is apparently due to the zero in the magnetotriction at about the same composition. The remainder of the starting mixture was ZnO. See also Fig. 37 in 6.1.3.6.

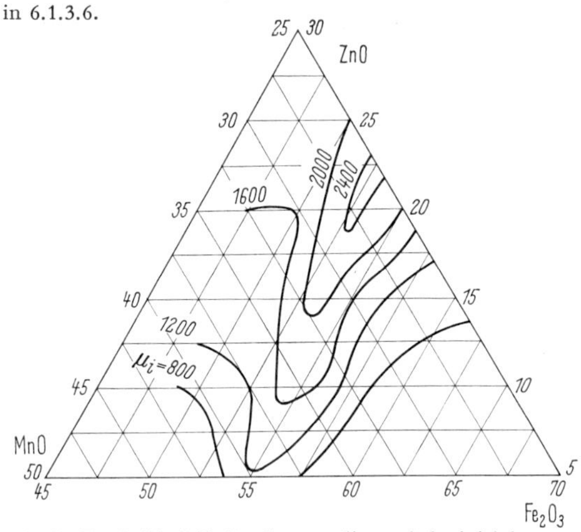

Fig. 8. Fe_2O_3-MnO-ZnO. Contour lines of the initial permeability, μ_i, Thermal treatment: 1240 °C in nitrogen gas with 1% oxygen. From [*60LP*].

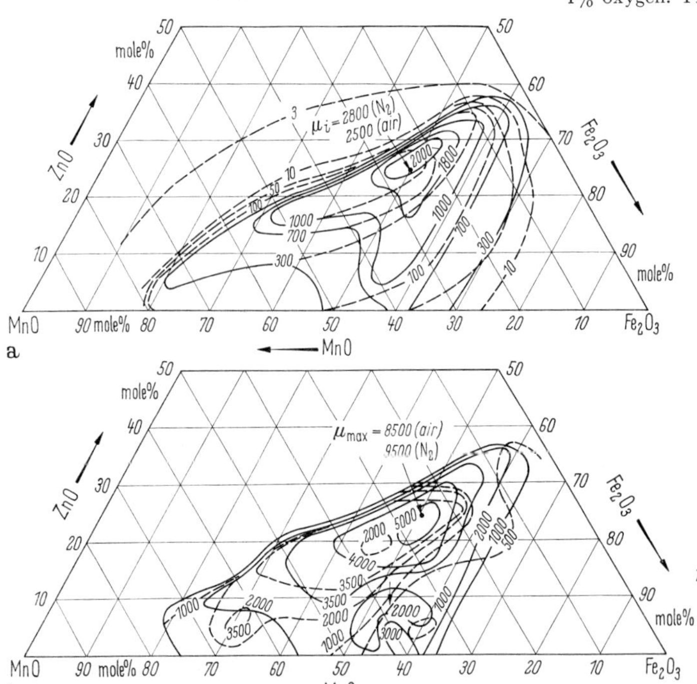

Fig. 7 a, b. MnO-ZnO-Fe_2O_3. (a) Contour lines of the initial permeability (μ_i), (b) contour lines of the maximum permeability (μ_{max}). The solid lines represent the data for the specimens sintered in nitrogen and the broken lines the data for the specimens sintered in air. Moulded samples were heated in air for 2 h at 1400 °C and cooled to 800 °C at a cooling rate of 400 °C per hour and then quenched in air from 800 °C (broken lines) or heated in nitrogen gas in the same way, except the specimens were slowly cooled in the furnace to room temperature (solid lines). From [*61S*].

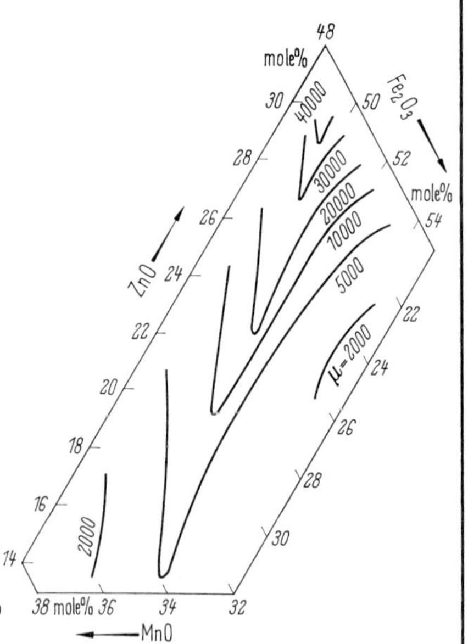

Fig. 9. MnO-ZnO-Fe_2O_3. Contour curves of the permeability, μ, at 1 kc/s of MnZn ferrite obtained by the vacuum reduction method and containing 0.6 wt% of In and 0.01 wt% of Ca oxide. Firing is performed in a vacuum of 10^{-4} Torr at 1250 °C. The temperature is then raised to 1330 °C in nitrogen gas containing 1% O_2. The samples are cooled in a vacuum of 10^{-3} Torr to room temperature. From [*64SAT*].

System Zn-Mn-Li-Fe-O

→

Fig. 10. ZnMnLi ferrites. Dependence of the maximum variation of permeability, μ, on neutrons irridiation for two samples. The initial composition is the same for both, but the properties before irradiation are different:

a: $\mu_{max} = 13\,000$,

b: $\mu_{max} = 6\,000$.

The μ decrease for sample (a) is due to the production of defects. The μ increase for sample (b) is attributed to the shatter of the big crystals and to a better chemical homogeneity. From [*64LAM*].

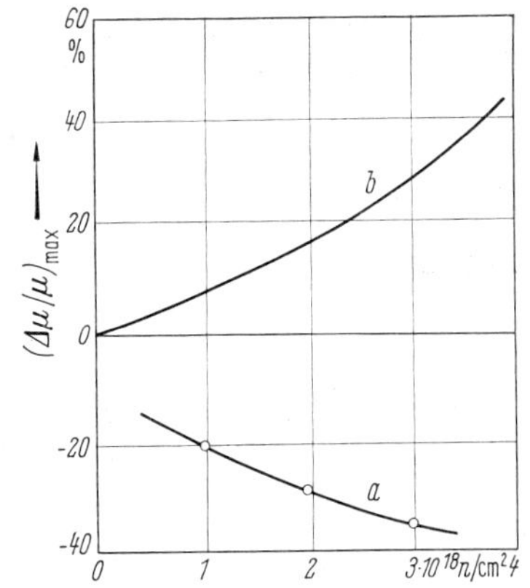

System Zn-Mn-Fe-O

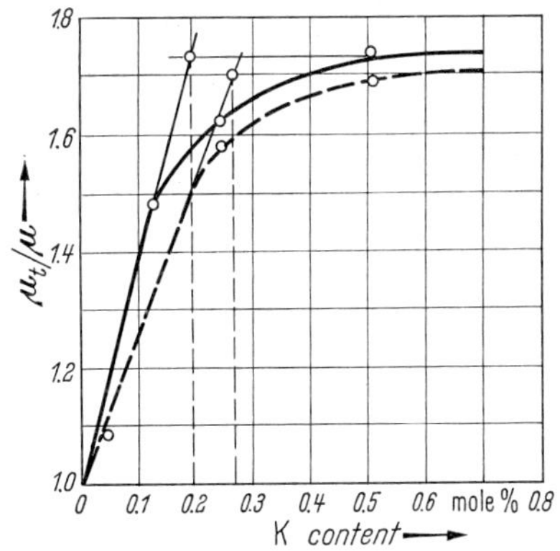

Fig. 11. MnO-ZnO-FeO-Fe_2O_3. The effect of impurities in the initial permeability of sintered $MnZnFe^{2+}$ ferrite [*57Gu*]. The samples were made from an initial oxide mixture consisting of 52.5 mole% Fe_2O_3, 19 mole% ZnO, and 28.5 mole% MnO, and the impurities were introduced by coprecipitation or by the addition to the mixture of a salt which would easily decompose on heating. Care was taken in the heat treatment to produce samples having the same composition, grain size, and grain structure. – The ordinate is the ratio μ_t/μ of the permeability of an undoped specimen to that of a specimen containing the impurity in question. At the lowest levels the permeability changes linearly with impurity concentration. In this range it is thought to be entering the lattice, thus the limit of this initial slope represents the limit of solubility of the ion in question. Above this level the curve levels off, and in this region the impurity is thought to be rejected by the lattice, and located in the grain boundaries where it no longer has much effect on permeability. The two curves represent the spread of observations at the various impurity contents.

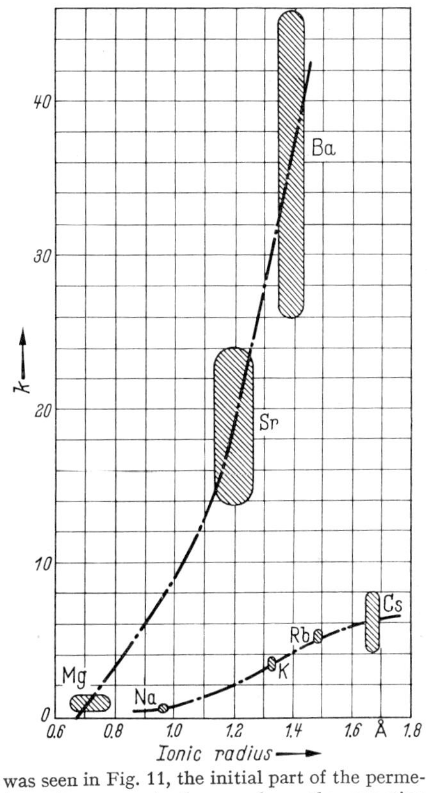

Fig. 12. As was seen in Fig. 11, the initial part of the permeability vs. impurity concentration η obeys the equation $[(\mu_t/\mu) - 1]/\eta = k$. In this figure [*57Gu*] the values of k for some alkali metal and alkaline earth ions are given against ionic size. Note the separate courses of the divalent and monovalent curves. Guillaud interprets these data as indicating that the impurities affect the permeability only through changing the internal stresses in the lattice, since he was not able to observe any effect on the magnetostriction or anisotropy energies. He postulates that the stresses associated with an impurity ion increase with ionic radius and for equal radius the lattice will support large stresses when the charge is higher.

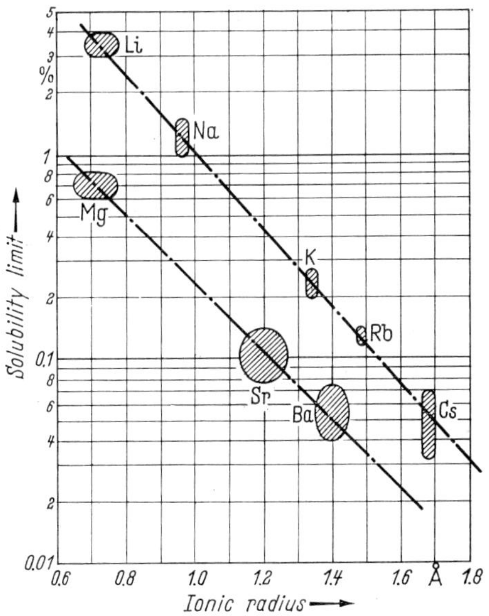

←

Fig. 13. From the impurity level at which the curves of Fig. 11 flatten out, it is possible to judge the solubility of various impurities in the $MnZnFe^{2+}$ ferrite prepared as described under Fig. 11. Here these are given against the ionic radius of the ions. Note again that the divalent and monovalent ions fall along quite separate curves [*57Gu*].

For the effect of Ca and Si on initial permeability see [*61A*] and [*66H*].

b) Granular structure

[*56GP*] prepared a series of MnZn ferrites **(Mn, Zn)Fe_2O_4** which so far as possible differed from one another only in the average size of the grains. This was accomplished by making all the samples from the same starting materials, 52.5 mole-% Fe_2O_3, 28.3 mole-% MnO, and 19.2 mole-% ZnO. The heat treatment was arranged so that the Fe_2O_3 content of the final samples lay between 49.7 and 50.6 mole-%. A set of similar samples was chosen by measurement of the anisotropy and determining the anisotropy coefficient K from magnetization curves. The samples thus had the same magnetocrystalline and the same magnetoelastic energies, and the only variable was grain size, as determined by microscopic examination of a cross section which has been polished and etched. The curve in Fig. 14 shows the relation between the initial permeability and the mean grain diameter. This curve is interpreted as indicating that the permeability process for grains less than about $5 \cdot 10^{-4}$ cm is a rotational process leading to a permeability of about 500, whereas for larger grains there is a wall displacement process leading to a wall displacement permeability of about 4000. It was found that samples with grains larger than about $20 \cdot 10^{-4}$ cm tended to have pores within the grains, which disturbed the wall displacement mechanism, and lowered the initial permeability. [*64RHM*] prepared samples with nearly the same composition: 52.5 mole-% Fe_2O_3, 25.5 mole-% MnO, and 22 mole-% ZnO, but with a lower porosity. The grain size dependence of permeability is linear for grains larger than 10^{-4} cm (Fig. 15). This supports the point of view of [*56GP*] about the initial permeability decrease due to the hindrance of the wall displacement by pores within large grains.

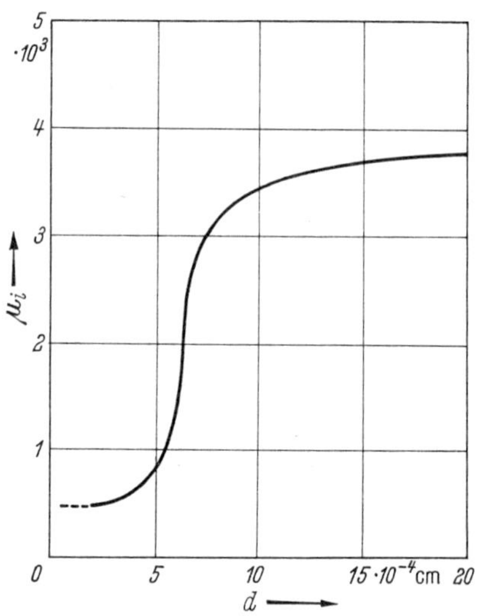

Fig. 14. MnO-ZnO-FeO-Fe_2O_3. Initial permeability, μ_i, vs. grain size, d, for samples with pores within the large grains. (See text above). From [*56GP*].

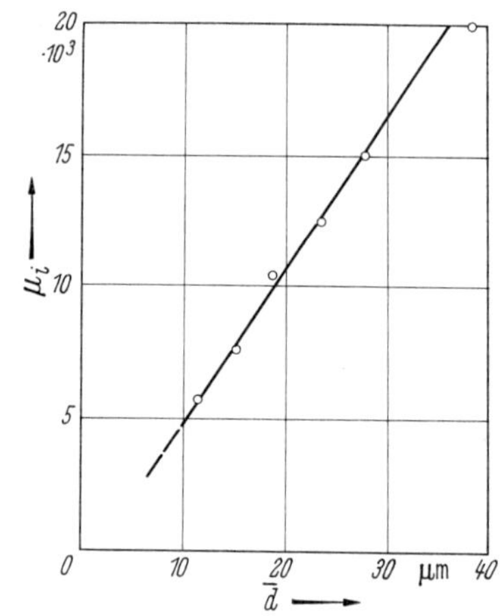

Fig. 15. MnO-ZnO-FeO-Fe_2O_3. Initial permeability, μ_i, vs. mean grain size, $\bar{d}$, for samples with very few pores within the large grains. (See text above). From [*64RHM*].

6.1.3.8.4 Permeability vs. temperature

System Mn-Fe-O

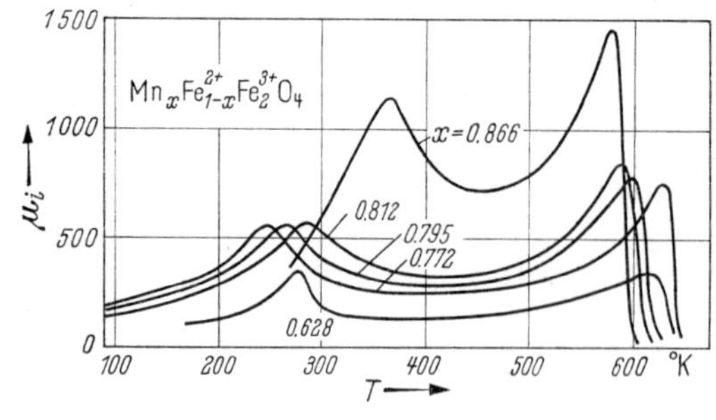

Fig. 16. $Mn_xFe_{3-x}O_4$. Initial permeability, μ_i, vs. temperature. Parameter x [*58En* (2)]. Certain $MnFe^{2+}$ ferrites show an unusual temperature dependence of permeability. The permeability shows two peaks, one just below the Curie point, and a secondary peak at a lower temperature, often near 270 °K. These samples were fired in a reducing atmosphere of nitrogen in such a way as to form a stoichiometric ferrite, and the composition was checked by chemical analysis. The analysis assumed that only bivalent Mn was present. (See section 6.1.3.1.2).
The secondary peak is found to be an effect of disaccomodation [*58En* (2)]. Additional curves of initial permeability vs. temperature for a single crystal of $Mn_{0.84}Fe_{2.14}O_4$ are given in the discussion of disaccomodation, section 6.1.3.8.1, 6.1.3.8.5, Fig. 38. In [*63O*] it is pointed out that the second peak occurs at the temperature at which the anisotropy constant reduces to zero.

System Zn-Mn-Fe-O

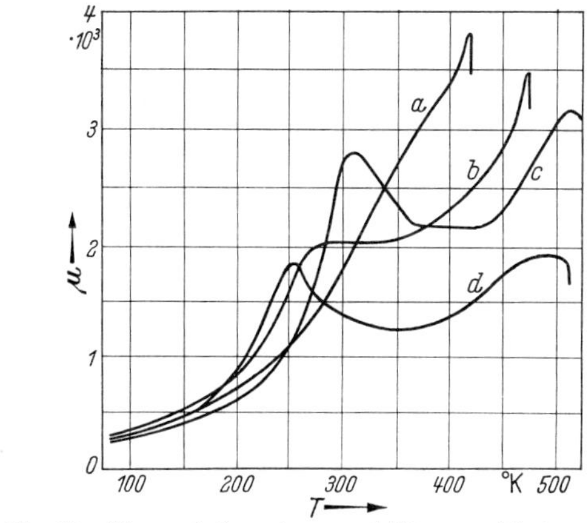

Fig. 17. The variation of permeability, μ, with temperature and composition for a number of $MnZnFe^{2+}$ ferrites in the range of commercial interest [*57Gu*]. These are mole% of the starting mixtures:
a) 50.6% Fe_2O_3, 31% MnO
b) 53.6% 31%
c) 55.6% 38%
d) 55.6% 31%
See also [*63H*].

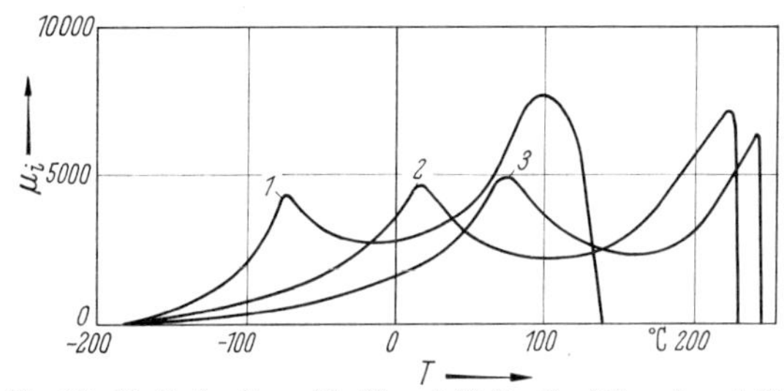

Fig. 19. MnZn ferrites with 55 mole% Fe_2O_3, $(45 - x)$ mole% MnO, x mole% ZnO. μ_i vs. T.
(1) $x = 24$ mole%
(2) 12 mole%
(3) 8 mole%
From [*62R*].

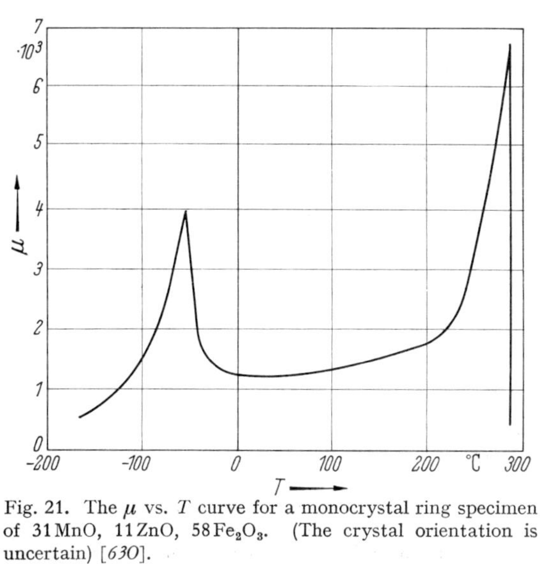

Fig. 21. The μ vs. T curve for a monocrystal ring specimen of 31 MnO, 11 ZnO, 58 Fe_2O_3. (The crystal orientation is uncertain) [*63O*].

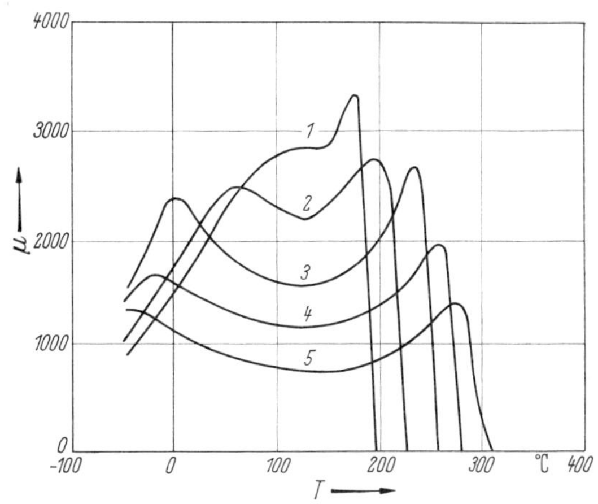

Fig. 18. MnZn ferrites with a constant MnO content of 32.5 mole%. The variation of permeability with temperature and composition. 1) 53 mole% Fe_2O_3, 2) 54 mole% Fe_2O_3, 3) 55 mole% Fe_2O_3. 4) 56 mole% Fe_2O_3, 5) 57 mole% Fe_2O_3. From [*60LP*].

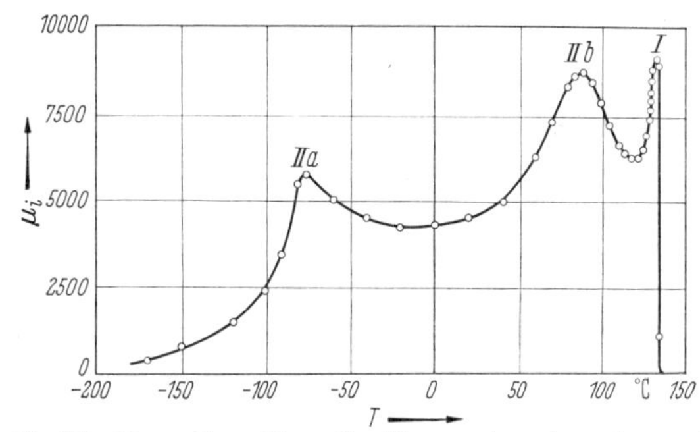

Fig. 20. $Mn_{0.412}Zn_{0.481}Fe_{2.107}O_4$. Temperature dependence of initial permeability, μ_i, for

$f = 200$ c/s, $H_a = 1$ mOe.

In a narrow range of compositions a number of MnZn ferrites show a third peak beside the peak at the Curie temperature and the known secondary one in the permeability vs. temperature curve. A distinct difference between this peak and the Curie peak can only be observed at small field strengths and low frequencies. The third peak in analogy to the known secondary peak is due to a zero-axis crossing of the crystalline anisotropy. Thus the [100]-direction is shown to be the preferred direction of spontaneous magnetization only between the two anomalous peaks. From [*66R*]. See also [*64SAT*].

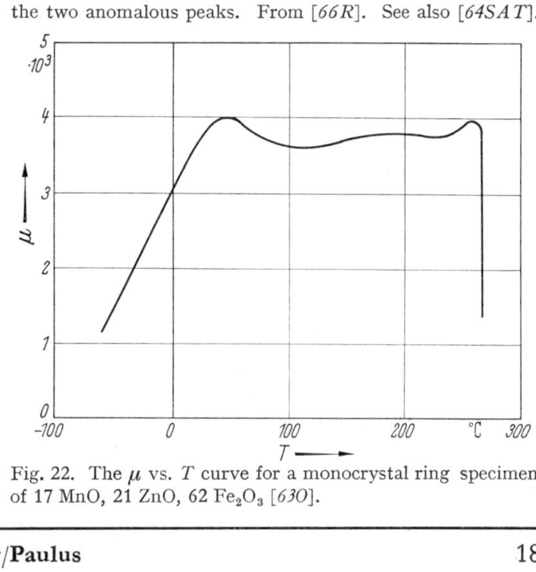

Fig. 22. The μ vs. T curve for a monocrystal ring specimen of 17 MnO, 21 ZnO, 62 Fe_2O_3 [*63O*].

System Co-Mn-Fe-O

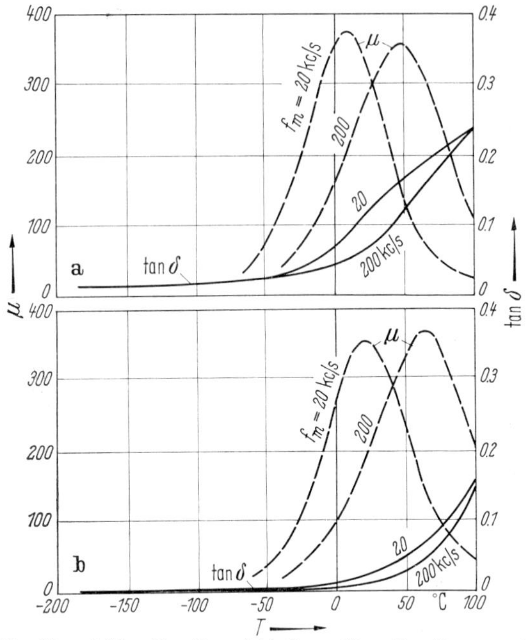

Fig. 23. a) $Mn_{1.5}Fe_{1.5}O_4$ and b) $Co_{0.05}$ $Mn_{1.45}Fe_{1.5}O_4$. μ and tan δ vs. temperature T. Parameter measuring frequency, f_m. Field strength 3 mOe. From [*58K* (1)].

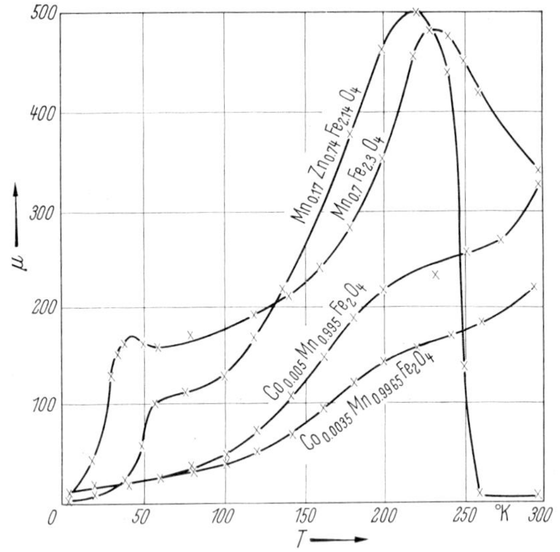

Fig. 24. μ vs. T for several polycrystalline spinel ferrites measured at 1.6 kc/s. From [*67CP*].

For the influence of additions on the temperature variation of ZnMn-ferrites, approaching room temperature, see: Al_2O_3 [*57Gu*], [*60LP*], [*64VK*]; Cr_2O_3 [*57Gu*], [*64VK*]; CuO [*60LP*]; CdO [*60CP*]; MoO [*60LP*]; NiO [*64VK*].

6.1.3.8.5 Permeability vs. pressure

System Zn-Mn-Fe-O

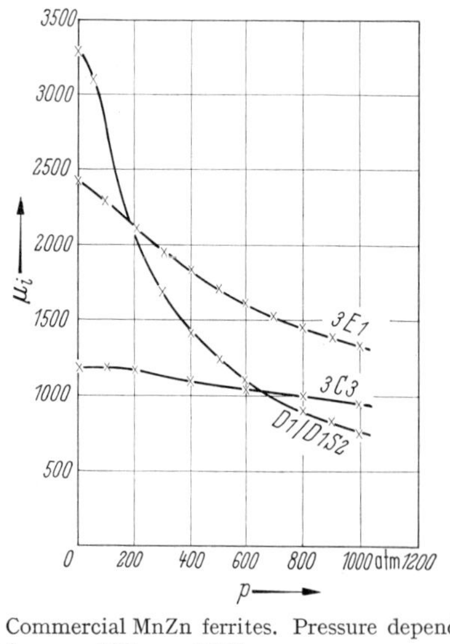

Fig. 25. Commercial MnZn ferrites. Pressure dependence of the initial permeability, μ_i [*63G*].
3 El Ferroxcube (Valvo)
3 C 3 Ferroxcube (Valvo)
D 1/D 1 S 2 Hyperox (Krupp)

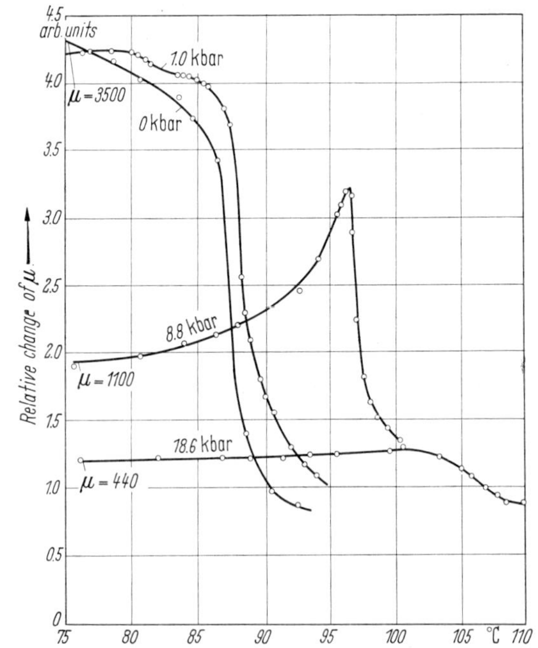

Fig. 26. $Mn_{0.48}Zn_{0.48}Fe_{2.08}O_{4.08}$. Curve μ vs. T under pressure of different magnitudes. The absolute values of permeability in the figure are those at room temperature [*65EKKOH*].

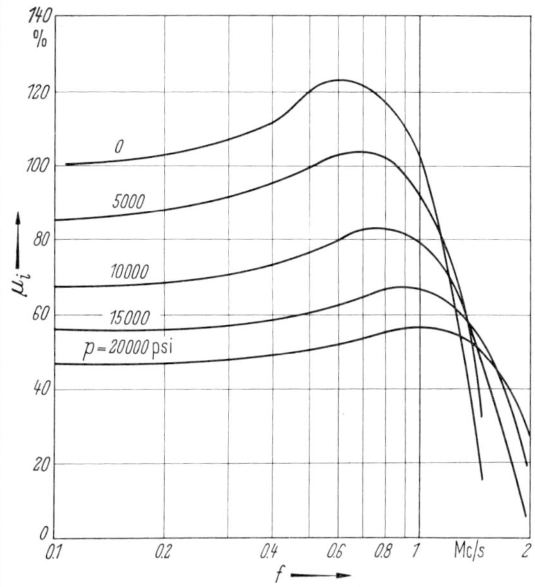

Fig. 27. MnZn ferrite. Initial permeability, μ_i, vs. frequency, f, in % of the value of μ_i at 100 kc/s and $p = 0$. "100%" permeability = 2440. Parameter pressure, p. The composition was not specified. From [60AL].
(5 000 psi = 351 500 g/cm² ≅ 350 kg/cm²
10 000 psi ≅ 700 kg/cm²
15 000 psi ≅ 1050 kg/cm²
20 000 psi ≅ 1400 kg/cm²)

6.1.3.8.6 Permeability vs. frequency

System Mn-Fe-O

Tab. 2. The frequencies at which the three resonances of $Mn_xFe_{3-x}O_4$ occur for the various compositions [58HKP]

x	f [Mc/s]		
	f_1	f_2	f_3
1.06	29	93	570
1.11	25	80, 150	540
1.18	(25)	110	510
1.26	(40)	105	480
1.34	30	150	425
1.42	29	125	400
1.49	33	110	375
1.62	25	70, 140	300

→

Fig. 29. $Mn_{1.42}Fe_{1.58}O_4$. The real ($\mu' - 1$) and imaginary (μ'') parts of the complex initial permeability as a function of frequency (20 ··· 2000 Mc/s) [58HKP]. Here as in most of the samples of the series of $Mn_xFe_{3-x}O_4$, x from 1.06 to 1.62, three resonances are observed. The magnitudes of the two lower peaks are considerably reduced if the sample is placed in the remanent state as compared with the demagnetized state. The high frequency peak is relatively unaffected by the change in magnetization. The two low frequency peaks were observed to disappear if the sintered samples were broken up into powder in which the particle size was approximately single domain. Since the lower frequency peaks are substantially altered when the magnetic state is changed from the unmagnetized to the remanent condition, it is assumed that these peaks are due to domain wall resonance of electron spins in the internal magnetocrystalline anisotropy field. For the influence of composition see Tab. 2.

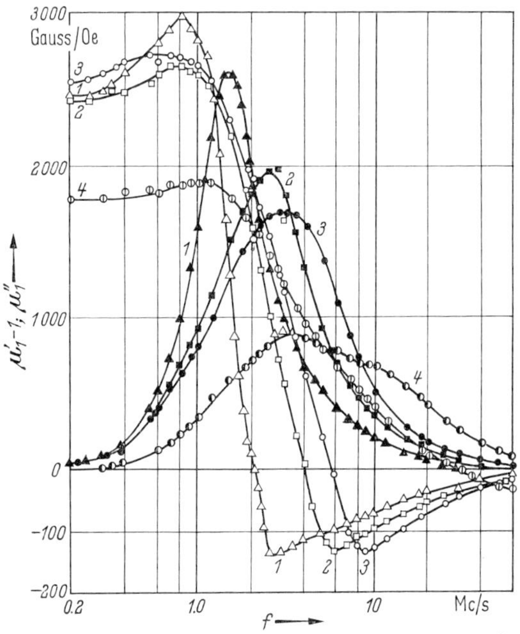

Fig. 28. MnZn ferrite. The real and imaginary parts of the apparent permeability have been measured for a number of cores of different sizes made from the same ferrite [59F (1)]. The open points are $\mu_1 - 1$, solid points μ_1''. The material was the Russian Oksifer M-2000 (initial composition in mole% 53.8 Fe_2O_3, 13.8 ZnO, 32.4 MnO). The dimensions of the toroids were as follows (o.d. and i.d.: outer and inner diameter; h = height).

Sample	o.d. cm	i.d. cm	h cm	V cm³
1	6.80	4.30	1.28	27.9
2	3.94	2.49	0.712	5.18
3	2.14	1.15	2.05	5.29
4	0.742	0.400	0.229	0.70

Note that as the size of the toroid decreases the characteristic frequencies increase, the heights decrease. In the smallest sample there seem to be two maxima.

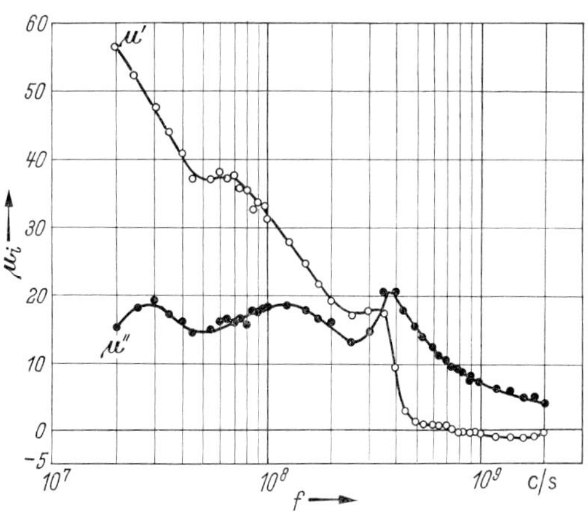

In measurements of the initial permeability of ferrite samples care must be taken to insure that the wavelength of the radiation inside the material remains much longer than the sample dimensions. At low frequencies it is sufficient to obtain data from the impedance of ring shaped cores. A procedure that is generally quite useful at higher frequencies consists of inserting thin toroidal shaped cores of rectangular cross section adjacent to a short circuit termination in coaxial line [*50R, 50RWE, 52RWE, 53Ra*].

Several cases in which the size was such that one or more wavelengths of the radiation occured within the samples are reported. These dimensional resonances are fundamentally not of much interest, however they do become of some importance in the design of certain devices [*49BDS, 50BDS*].

[*49BDS, 50BDS*] first reported the appearence of dimensional resonances in specimens of this size at frequencies of one or two Mc/s. See Fig. 28. In [*50BDS*] it was shown that it implied very high apparent dielectric constants. It also means that measurements of dielectric constant must take into account the existence of the resonances. See also [*59F* (1)].

Further references for section 6.1.3.8.6: [*49BDS, 50BDS, 50BS, 50GB* (2), *50GC, 50P, 50R, 50RWE, 52RWE, 53AS, 53H, 53Ra, 53TR, 55E, 56RFE, 57H, 57MWH, 57St* (1), *58HKP, 59F* (1), *59F* (2), *59RN*].

6.1.3.8.7 After-effect, disaccommodation, and viscosity

a) Relaxation measurements

System Mn-Fe-O

A relaxation ($\mu - T$) type loss was found in Mn ferrites ranging in composition approximately from $\mathbf{MnFe_2O_4}$ to $\mathbf{Mn_{1.4}Fe_{1.6}O_4}$ [*57KV*]. The samples used were prepared from a mixture of 1 mole $Fe_2O_3 + (1 + x)$ mole $MnCO_3$ by sintering at 1200 °C. Firing was done in air of normal or lowered pressure to control the degree of oxidation. The oxygen content of the samples was determined by chemical analysis.

Tab. 3. The compositions of the samples tested, and in a separate column their oxygen excess, along with the quantities measured by these experiments, q, τ_∞, $\tan\delta_{max}$, and T_{max}. Note that the activation energies, q, monotonically increase with Mn content. There seems to be little correlation between oxygen content and q. From [*57KV*]

	Composition	Oxygen excess [1])	q eV	τ_∞ 10^{-12} sec	$\tan\delta_{max}$ at 100 kc/s	T_{max} [°K] for 100 kc/s
1	$MnFe_2O_4$	0.0	0.30	0.9	0.25	214.2
2	$Mn_{0.983}Fe_{1.965}O_4$	+0.07	0.35	0.5	0.25	232.7
3	$Mn_{1.067}Fe_{1.940}O_4$	−0.008	0.36	0.45	0.48	251.2
4	$Mn_{1.142}Fe_{1.844}O_4$	+0.019	0.39	0.2	0.24	258.7
5	$Mn_{1.142}Fe_{1.844}O_4$	+0.019	0.41	0.25	0.40	269.8
6	$Mn_{1.142}Fe_{1.842}O_4$	+0.019	0.38	0.3	0.22	256.2
7	$Mn_{1.35}Fe_{1.830}O_4$	+0.055	0.40	0.2	0.21	268.2
8	$Mn_{1.35}Fe_{1.830}O_4$	+0.055	0.41	0.2	0.24	269.8
9	$Mn_{1.357}Fe_{1.635}O_4$	+0.027	0.49	0.03	0.31	292.2
10	$Mn_{1.372}Fe_{1.644}O_4$	−0.026	0.42	0.5	0.47	293.2

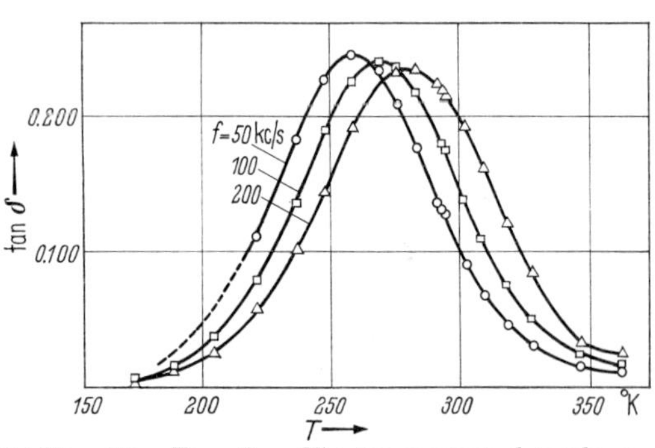

Fig. 30. $Mn_{1.35}Fe_{1.830}O_4$. The temperature dependence of $\tan\delta = \mu''/\mu'$ for sample 8 in Tab. 3 with a slight oxygen excess for three different measuring frequencies. [*57KV*].

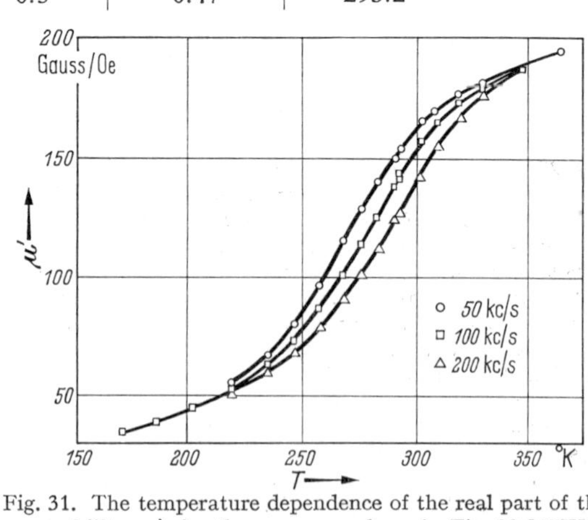

Fig. 31. The temperature dependence of the real part of the permeability, μ', for the same sample as in Fig. 30 [*57KV*].

[1]) If the formula is written $MnFe_2O_4 \cdot \frac{1}{3}x\ Mn_3O_4 \cdot y\ O$, the oxygen excess is y.

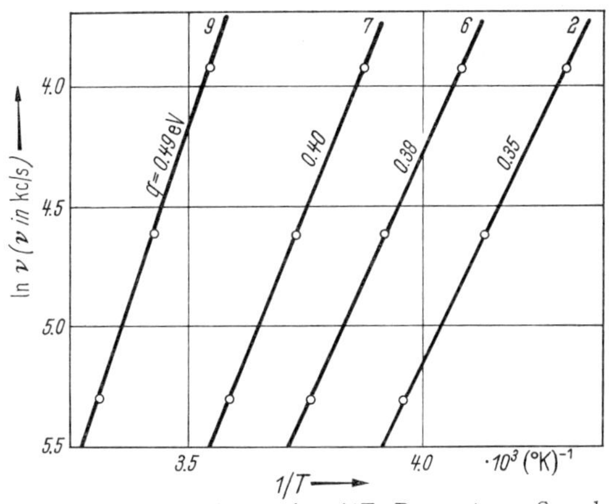

Fig. 32. $Mn_xFe_zO_4$. ln ν against $1/T$. Parameter q. Samples 2, 6, 7, 9 of Tab. 3. Data from [*57KV*]. For a simple relaxation type loss the relaxation time obeys the equation $\tau = \tau_\infty e^{-q/kT}$, and on a temperature plot the maximum loss for some frequency ν occurs when $q = 1/kT$. If we take the temperatures of the loss maxima for the three different frequencies at which measurements were made, and plot ln ν_{max} against $1/T$, we find that the points define a line, and from its slope we can measure q.

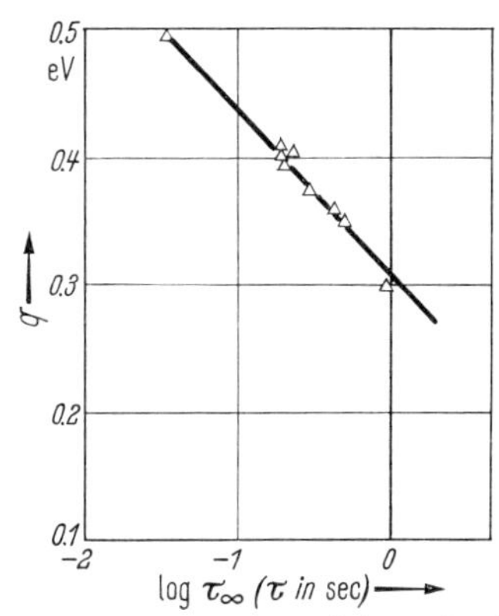

Fig. 33. $Mn_xFe_{3-x}O_4$. The values of the activation energy q plotted against the corresponding values of log τ_∞ for the MnFe oxides of Tab. 3 define a straight line, that is they fall on a curve $q = \log \tau_\infty + B$. From [*58K* (2)].

The three ionic-migration after-effects observed in iron-rich ferrites containing vacancies can be ascribed to the Fe^{2+} anisotropic interaction with the spontaneous magnetization in the spinel lattice by using Slonezewski's theory. It is believed that this interaction causes migration of Fe^{2+} ions through intermediary of vacancies (vacancy = □).

Tab. 4. Some characteristic data on three ionic-migration after-effects in $Mn_{0.6}Fe_{2.4}O_{4+\gamma}$, $0 < \varrho < 0.05$. From [*65B* (2)]

Effect		III	II	I
Temperature range	T [°K]	243 ··· 273	418 ··· 510	553 ··· 653
Typical temperature of measurement	T_0 [°K]	263	418	573
Mean relaxation time constant	$\Theta_0(T_0)$ [sec]	$10^2 \cdots 10^3$	$10^2 \cdots 10^5$	$10^2 \cdots 10^4$
Relaxation time dependence on the cation vacancies concentration ϱ	$\Theta_0(\varrho)$	$\Theta_0 \simeq$ constant	$\Theta_0 \varrho = a$ a decreases when ϱ increases	$\Theta_0 \varrho =$ constant
Relaxation time bandwidth in logarithmic approximation	$\bar{k} = \sqrt{\frac{\Theta_2}{\Theta_1}}$	30 ± 5	1 ··· 100	<10
Constant of the Arrhenius-equation	$\Theta_{0\infty}$ [sec]	$10^{-14} \cdots 10^{-13}$	$10^{-8} \cdots 10^{-11}$	$10^{-15} \cdots 10^{-13}$
Activation energy	q [eV]	0.8 ± 0.2	≈1	≈2
Maximum stabilization energy	$W_0(T_0)$ $\left[\frac{erg}{cm^3}\right]$	<50	<300	<30
Stabilization energy dependence on ϱ	$W_0(\varrho)$ $\left[\frac{erg}{cm^3}\right]$	$10^3\, \varrho$		
Elementary interaction energy of Fe^{2+}, [*64T*]	$w_\vartheta\,(T_0)$ [erg]	$3.5 \cdot 10^{-15}$	$3.1 \cdot 10^{-15}$	$2.6 \cdot 10^{-15}$
Elementary interaction energy of □, [*62Y*]	$w_\square$ [erg]	$0.3 \cdot 10^{-15}$ (300 °K)	non-estimated, lower than $0.3 \cdot 10^{-15}$	
Elementary interaction energy, related to □, experiment	w_ϱ [erg]	$0.24 \cdot 10^{-15}$	$1.1 \cdots 1.5 \cdot 10^{-15}$	$0.5 \cdots 0.9 \cdot 10^{-15}$
Local interaction symmetry axis		[111]	[110]	[100]

Tab. 5. $Mn_{1+x}Fe_{2-x}O_4$. Magnetic diffusion after-effect with $0.1 \leqq x \leqq 0.8$. To each tan δ_{max} corresponds a D_{max}, the position of which has been calculated and is shown in the table. After [*67MM*]

x	tan δ_{max} (f = 5000 c/s)	T [°K] of tan δ_{max} (f = 5000 c/s)	q eV	T [°K] of D_{max} calculated
0.1	0.21	213	0.50	127
0.2	0.18	220	0.40	117
0.3	0.24	233	0.46	129
0.4	0.23	239	0.43	127
0.5	0.27	243	0.49	136
0.6	0.28	257	0.50	142
0.7	0.31	257	0.52	144
0.8	0.06	256	0.56	148

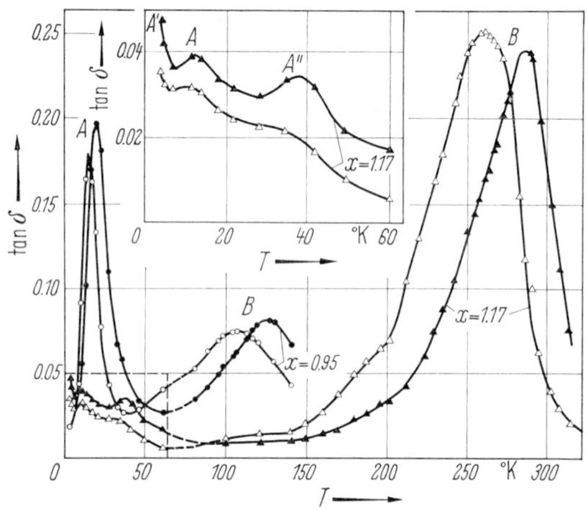

←
Fig. 34. $Mn_xFe_{3-x}O_4$. Temperature dependence of tan δ. Open symbols for 75 kc/s, full for 600 kc/s. From [*66ZTZ*]. The activation energies for the processes A and B are:

x	q_A[eV]	q_B[eV]
0.95	0.012	0.11
1.06	≃0.01	0.42
1.17	≃0.01	0.49

System Zn-Mn-Fe-O

MnZnFe^{2+} ferrites were first reported to show a Richter type after-effect [*53WH*]. The material had the composition $\mathbf{Mn_{0.56}Zn_{0.38}Fe_{2.06}O_4}$ and the loss factor was measured as a function of temperature at three frequencies, 2, 8, and 100 kc/s. The activation energy, q, was 0.1 eV and was attributed to the electron transfer between Fe^{2+} and Fe^{3+}. Another Richter type after-effect also has been observed in Mn ferrite containing Mn ions in excess of stoichiometric ratio, whose activation energy was about 0.4 eV [*57KV*]. The origin of this after-effect was regarded as the electron transfer between Mn ions having different valencies.

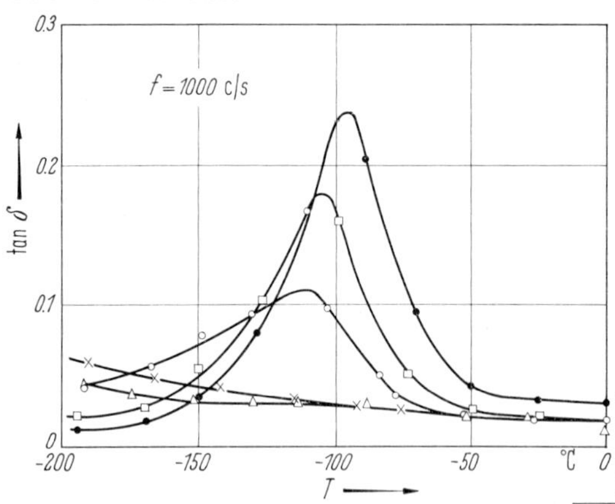

Fig. 35. tan δ vs. temperature for MnZn ferrites sintered at 1250 °C, 4 h in an atmosphere with 0.5% oxygen. Measurements at 1000 c/s and 10 mOe. From [*64M*]. Characteristics of the samples:

	Initial composition in mole%			FeO mole%	q_T eV	q_σ eV	τ_∞ sec	σ $\Omega^{-1}cm^{-1}$
	Fe_2O_3	MnO	ZnO					
●	49	30	21	oxide	0.38	0.20	$2\cdot10^{-13}$	$3.78\cdot10^{-8}$
□	49.5	30	20.5	—	0.27	0.18	$6\cdot10^{-12}$	$8.61\cdot10^{-7}$
○	50	30	20	0.18	0.21	0.12	10^{-10}	$1.17\cdot10^{-7}$
×	51	30	19	1.28	—	0.08	—	$3.02\cdot10^{-5}$
△	52	30	18	2.54	—	0.09	—	$3.87\cdot10^{-3}$

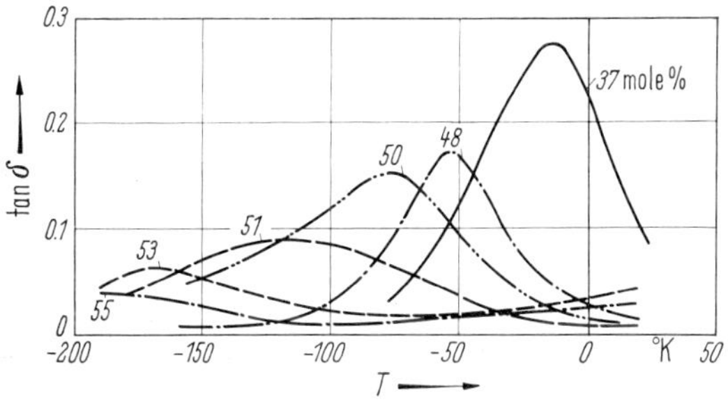

←

Fig. 36. MnZn ferrites. tan δ vs. temperature measured at 50 kc/s. Parameter mole% Fe_2O_3. In all specimens, mixing mole ratio of MnO to ZnO is 30:17. Specimens were sintered in oxygen-nitrogen mixture ($p_{O_2} = 2\cdot10^{-3}$ atm) at 1200 °C. From [65OA]. See also [61AO].

System Mg-Mn-Fe-O

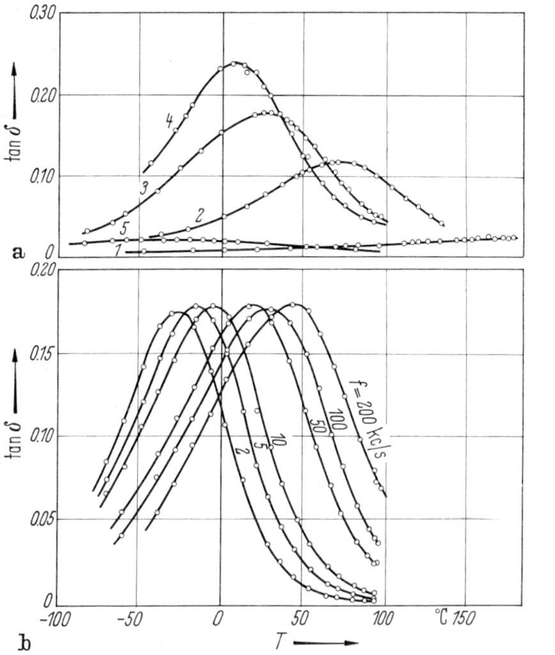

←

Fig. 37. a) Temperature dependence of tan δ for 100 kc/s in different MgMn-ferrite samples.
Composition: *1* $Mg_{1.154}Mn_{0.137}Fe_{1.760}O_4$
2 $Mg_{0.795}Mn_{0.382}Fe_{1.825}O_4$
3 $Mg_{0.522}Mn_{0.632}Fe_{1.845}O_4$
4 $Mg_{0.270}Mn_{0.870}Fe_{1.833}O_4$
5 $Mg_{1.003}Fe_{2.006}O_4$
b) $Mg_{0.522}Mn_{0.632}Fe_{1.845}O_4$. Temperature dependence of tan δ at different frequencies. From [60K (2)].

b) Disaccommodation

System Mn-Fe-O

MnFe^{2+} ferrites have been found to exhibit a phenomenon known as disaccommodation [58En (2)]. This consists in a decrease of the permeability with time after the specimen has been brought into an initial state by demagnetization.

The phenomenon of disaccommodation is also found in MnZn ferrites containing an excess of Fe_2O_3 which have been fired in an oxidizing atmosphere. Though similar effects had been seen before in metals, disaccommodation was first observed in such a MnZn ferrite [47S]. It appears [47S], [58En (2)] to arise from a lowering of the energy of a domain wall by an electron diffusion mechanism. Thus the potential well in which a piece of wall rests for a time grows deeper, and that part of the permeability which arises from wall motion is drastically reduced. [58En (2)] found that at any temperature the time variation of the permeability closely obeys the relation

$$\frac{1}{\mu(t,T)-\mu_{\text{rot}}} = \frac{1}{\mu_{\text{f}}} + \frac{1-e^{-t/T}}{\mu_{\text{b}}}.$$

In this, μ_{rot} is the part of the permeability which is not time dependent, it depends on domain rotation. μ_{f} is the free permeability before disaccommodation, and μ_{b} is the blocked wall permeability for $t \to \infty$. The relaxation time is characteristic of the electron diffusion process, and is very different from the time characteristic of the decay of the permeability.

Disaccommodation has been studied in the system $\mathbf{Mn}_x\mathbf{Fe}_{3-x}\mathbf{O}_{4+y}$ [60C] where $x = 0.5$, 0.7, 0.85, and 0.9. The stoichiometric samples ($y = 0$) were fired to maintain the metal oxygen ratio at 3 : 4, while the nonstoichiometric samples were fired to produce the maximum number of vacancies, $y \simeq 0.025$. The permeabilities were measured at 10 kc/s with a driving field of 0.002 Oe. Demagnetization was done at 60 c/s, with a peak driving field ten times the coercive force.

It was found that the decrease of permeability with time could not be represented by a single relaxation time. If the quantity dealt with is the reciprocal of the permeability, the reluctivity r, it was found that the following equation is valid:

$$r = r_0 + \Sigma_i r_i (1 - e^{-t/\tau_i}).$$

In this τ_i and r_i are characteristic relaxation times and reluctivity associated with the i-th term. r_∞ is the value of the reluctivity measured at $t = \infty$. r_0 is obtained by extrapolating a curve of ln $(r_\infty - r)$ back to $t = 0$.

It was found that within the experimental error, the relaxation times observed could be represented by an equation of the form $\tau_i = \tau_{0i} \exp(Q/kT)$. The longest relaxation times observed, τ_i's for the stoichiometric samples with $x = 0.7$, 0.85, 0.9, showed the least scatter. It was found in this case that the equation representing the data was

$$\tau_i = \tau_{0i} \exp(Q/kT);\; Q = (0.46 \pm 0.06)\ \text{eV}.$$

The shorter times showed a larger scatter, but within the error the activation energies for all the compositions were the same, $Q = (0.4 \pm 0.1)$ eV. The relaxation times for the nonstoichiometric samples were consistently shorter and had a larger scatter than the corresponding times for stoichiometric compounds.

The parameter A serves to normalize the part of the total reluctivity which relaxes, $A = (r_\infty - r_0)/r_\infty$. Tab. 6 gives A for various Mn/Fe ratios in $\mathbf{Mn_xFe_{3-x}O_{4-y}}$.

We see in this that there is a general increase in A as the Mn/Fe ratio increases. Also A is essentially independent of temperature. See Tab. 7.

Tab. 6 $Mn_xFe_{3-x}O_{4-y}$. Parameter A (see text). [*60C*]

x	Mn/Fe	A at 298 °K	A at 273 °K
0.9	0.428	0.02	—
0.85	0.395	0.06	0.05
0.70	0.304	0.13	0.11
0.50	0.200	0.23	0.25

Tab. 7. Comparison of the relaxation times for the composition of Fig. 42, $Mn_{0.5}Fe_{2.5}O_{4+y}$ from [*60C*]

T	$y = 0$			$y \simeq 0.025$		
°K	τ_1 min	τ_2 min	τ_3 min	τ_1 min	τ_2 min	τ_3 min
298	12.5	1.3	0.16	3.6	0.15	≈0.03
273	37.6	4.65	0.54	5.8	0.72	0.08

The activation energies observed suggest that the mechanism responsible for the pinning of the domain walls in an electron diffusion process. It must be noted that the energy is much greater than that associated with the Fe^{2+}, Fe^{3+} exchange in octahedral sites as deduced from resistivity measurements.

Further references for section 6.1.3.8.7: [*58En* (2)], [*60C*], [*41VH*].

c) Temperature dependence of disaccommodation

System Mn-Fe-O

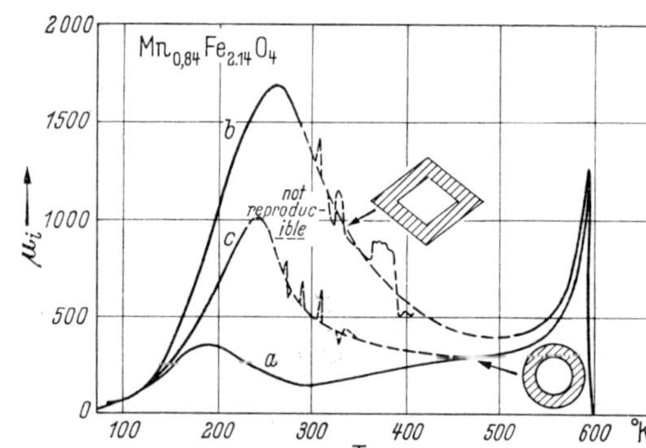

→ Fig. 38. $Mn_{0.84}Fe_{2.14}O_4$. The temperature variation of initial permeability for two samples cut from a single crystal [*58En* (2)]. The picture frame sample was cut with the same orientations as those discussed in the section on domain wall motion, 6.1.3.7, Fig. 1, however there was no effort to keep one dimension very small compared to the others. The toroidal sample was cut out of a slice parallel to the (110) plane. The curve (a) was measured with the crystal in a thermal bath with a long rest between demagnetization and measurement. Curves (b) and (c) were measured with a fast automatic $\mu - T$ recorder with particular care taken to keep the measuring field very low in order to measure the true initial permeability. In the dashed regions between about 240° and 330°K the curve shows various little peaks and irregularities. The effect always occurs on both heating and cooling, but cannot be reproduced in detail. The high peak in (b) and (c) parallels that in $MnFe^{2+}$ ferrites shown in Fig. 16. The domain structure in the picture frame presumably has more 180° domain walls than in the toroid.

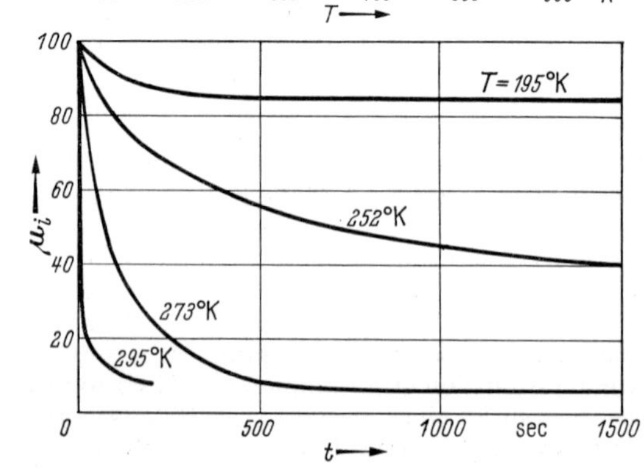

→ Fig. 39. $Mn_{0.84}Fe_{2.14}O_4$. Disaccommodation in a single crystal. The disaccommodation was measured directly in a 10 kc/s measuring field of $3 \cdot 10^{-3}$ Oe. The discharge of a condenser-inductance circuit demagnetized the core. The output of a third winding was recorded, and the trace was thus a plot of initial permeability vs. time. The curves have been normalized so that $\mu = 100$ at time $t = 0$ [*58En* (2)].

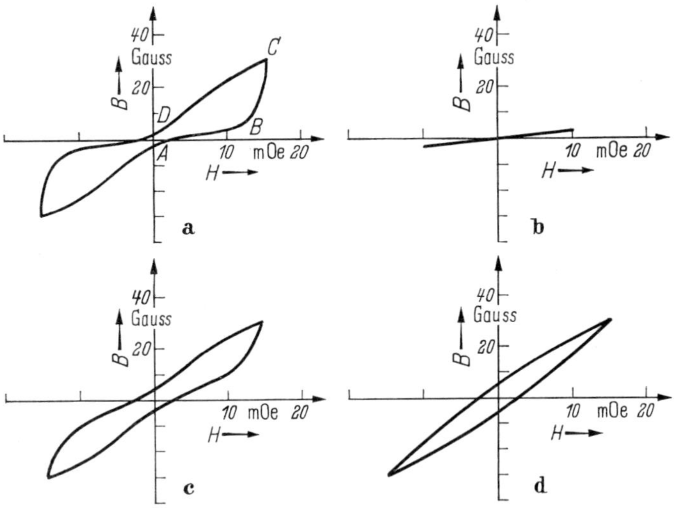

Fig. 40. $Mn_{0.84}Fe_{2.14}O_4$. Constricted hysteresis loops are another manifestation of disaccommodation, observed reproducibly in the single crystal core [*58En* (2)]. The sample was first demagnetized. The sample was allowed to age for about one minute at room temperature. A minor hysteresis loop is then found by applying an alternating current field of $15 \cdot 10^{-3}$ Oe at 10 kc/s. This loop is shown in (a). On starting from zero applied field a very flat loop is first found, (b). Above some critical alternating field the constricted loop appears suddenly. The constriction shown in (a) gradually disappears with time. (c) was taken after about 30 sec. (d) was taken just after the demagnetization with a capacitor discharge through a damped resonant circuit.

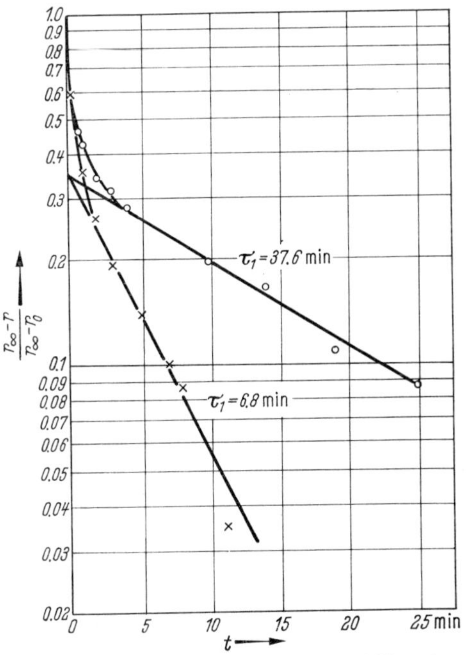

Fig. 41. The time decrease of initial permeability of stoichiometric ($y = 0$) (×) and nonstoichiometric ($y = 0.025$) (○) $Mn_{0.5}Fe_{2.5}O_{4+y}$ at 273 °K [*60C*]. The longest relaxation time is indicated on the graph, and its relative reluctivity $(r_\infty - r)/(r_\infty - r_0)$. The procedure is to subtract from the data the contribution of the component characterized by τ_i and r_i/Σ_i, then to repeat the process to find the next relaxation time and relative reluctivity.

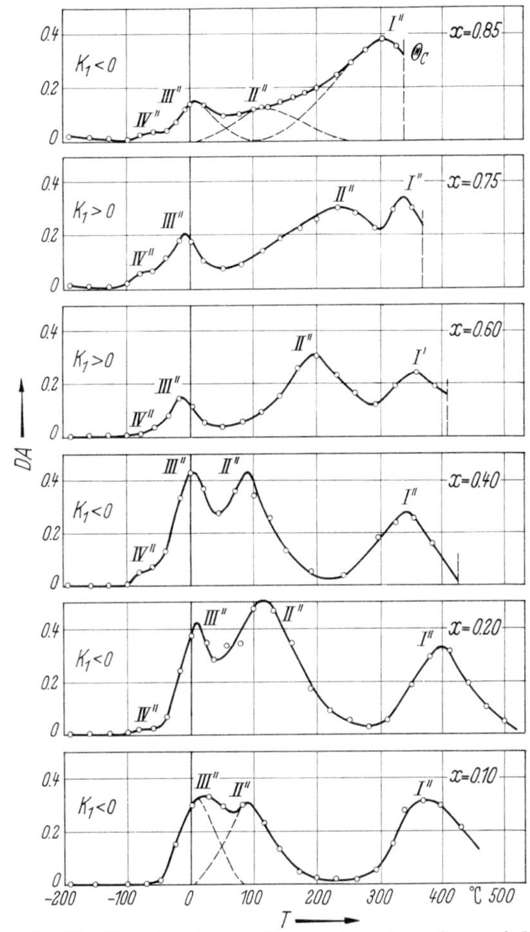

Fig. 42. $Mn_xFe_{3-x}O_{4+\gamma}$. Temperature dependence of disaccommodation $DA = (\mu_{t_1} - \mu_{t_2})/\mu_{t_1}$ for $x = 0.10$, 0.20, 0.40, 0.60, 0.75, 0.85; $\gamma \simeq 0.03 = C^{te}$, $\mu_{t_1} = \mu$ (1 sec after demagnetization) and $\mu_{t_2} = \mu$ (30 min after demagnetization). From [*64M*].

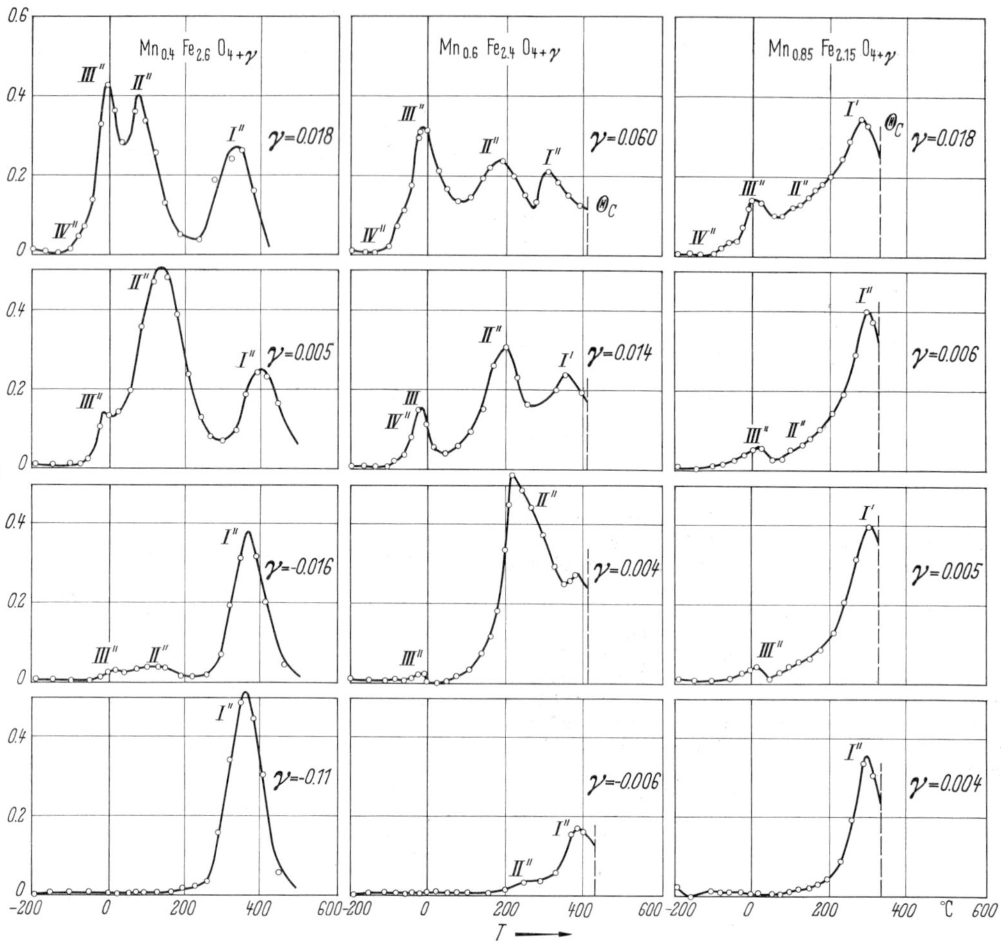

Fig. 43. $Mn_xFe_{3-x}O_{4+\gamma}$. Temperature dependence of disaccommodation with different γ and x values. $t_1 = 1$ sec, $t_2 = 30$ min. After [64M].

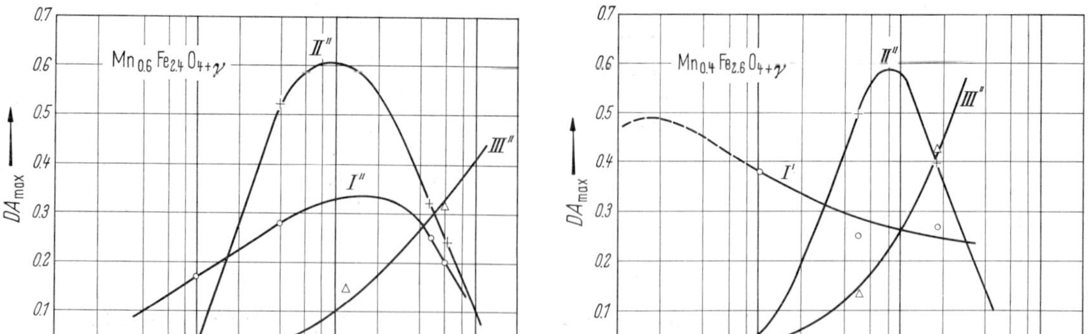

Fig. 44a and b. Maximum value of disaccommodation, DA, of the three effects as a function of γ. Same samples as for Fig. 41 and 42. From [64M]. The three disaccommodation effects seem to be connected with the different geometrical arrangements of Fe^{2+} in presence of vacancies.

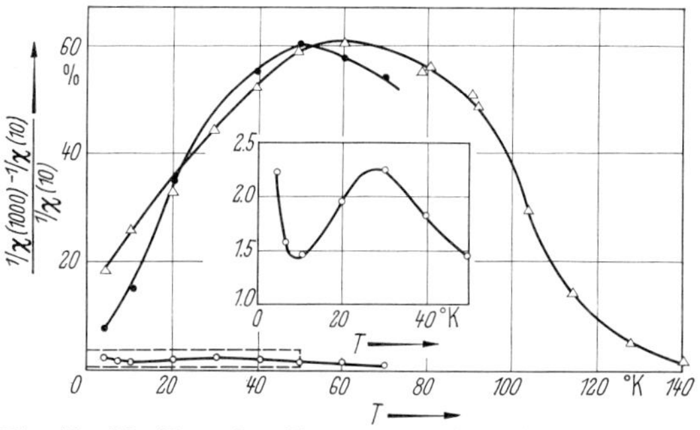

Fig. 45. $Mn_xFe_{3-x}O_4$. Temperature dependence of permeability disaccommodation. $x = 0.95$ (○), $x = 1.06$ (●), $x = 1.17$ (△). After [66ZTZ].

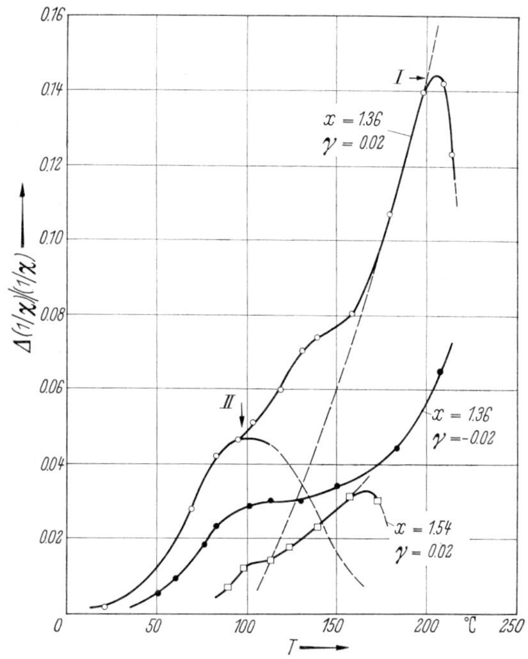

Fig. 46. Disaccommodation spectra of $Mn_xFe_{3-x}O_{4+\gamma}$ ferrites with $x > 1$. ($t_1 = 5$ sec, $t_2 = 500$ sec). The disaccommodation maximum near room temperature is subject to the presence of Fe^{2+} ions. The high temperature process depends only on the presence of vacancies. From [64K (1)].

For Mn ferrites, see also [66BS (1)], [61KV], [62K (2)], [62BMM], [65MMP].

System Zn-Mn-Fe-O

$Mn_{0.597}Zn_{0.265}Fe_{2.07}O_4$, $Mn_{0.584}Zn_{0.353}Fe_{2.05}O_4$. Disaccommodation of the complex susceptibility of two MnZn ferrite materials was studied by [65R] in response to demagnetizing frequency and rate, temperature, frequencies of ≈20 kc/s··· ≈ 200 kc/s, and rf magnetic field intensities of ≈10^{-4} Oe and ≈10^{-2} Oe. Precise susceptibility spectra near −10.0 °C show that the resonance maxima shift to higher frequency and decrease in magnitude with increasing time after demagnetization and that the disaccommodation rate varies considerably and changes sign with frequency. Activation energies of the order of 0.8 eV were computed at 100 kc/s.

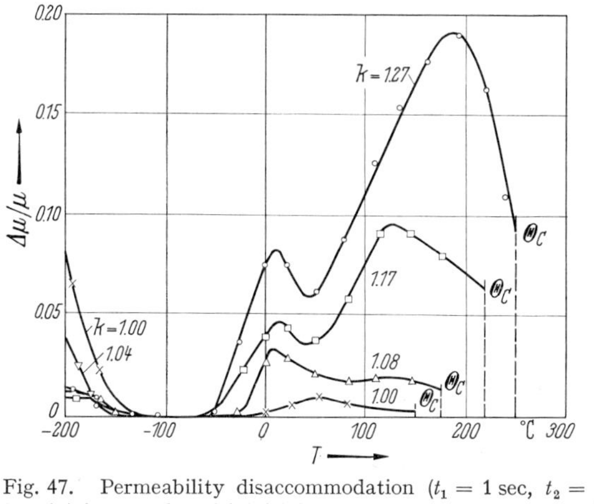

Fig. 47. Permeability disaccommodation ($t_1 = 1$ sec, $t_2 = 30$ min) for samples with initial composition x ZnO, 30 MnO, $(70 - x)$ Fe_2O_3 sintered 4 h at 1200 °C in an atmosphere containing 0.5% oxygen. Parameter $k = (70 - x)/(x + 30)$. From [61BMM]. See also [64M].

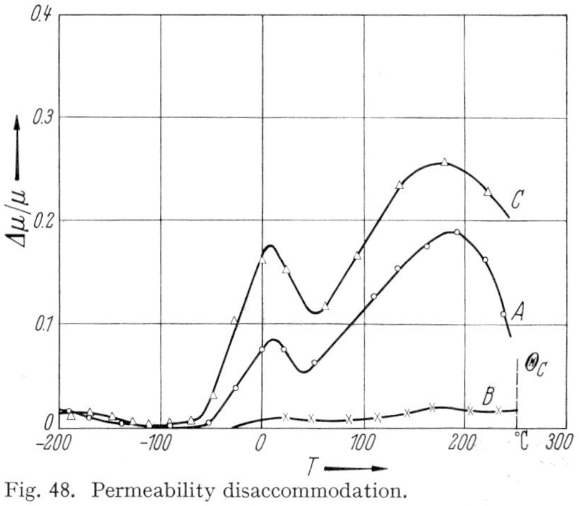

Fig. 48. Permeability disaccommodation.
A = sample of Fig. 48 for $k = 1.27$ sintered 4 h at 1250 °C in 0.5% O_2.
B = sample A reduced in N_2 at 1250 °C for 4 h.
C = sample B oxidized in N_2 + 5% O_2 at 1250 °C.
From [61BMM].

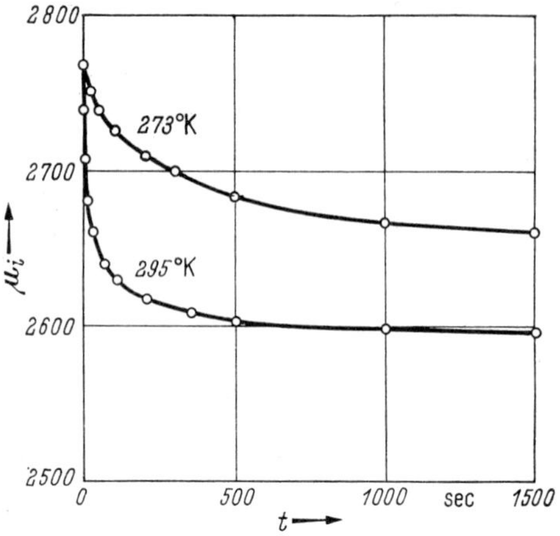

← Fig. 49. The disaccommodation of a polycrystalline sample whose composition might be represented by $Mn_{0.514}Zn_{0.369}Fe_{2.09}O_4$ [*58En* (2)]. For the method see Fig. 38.

System Ti-Mn-Fe-O

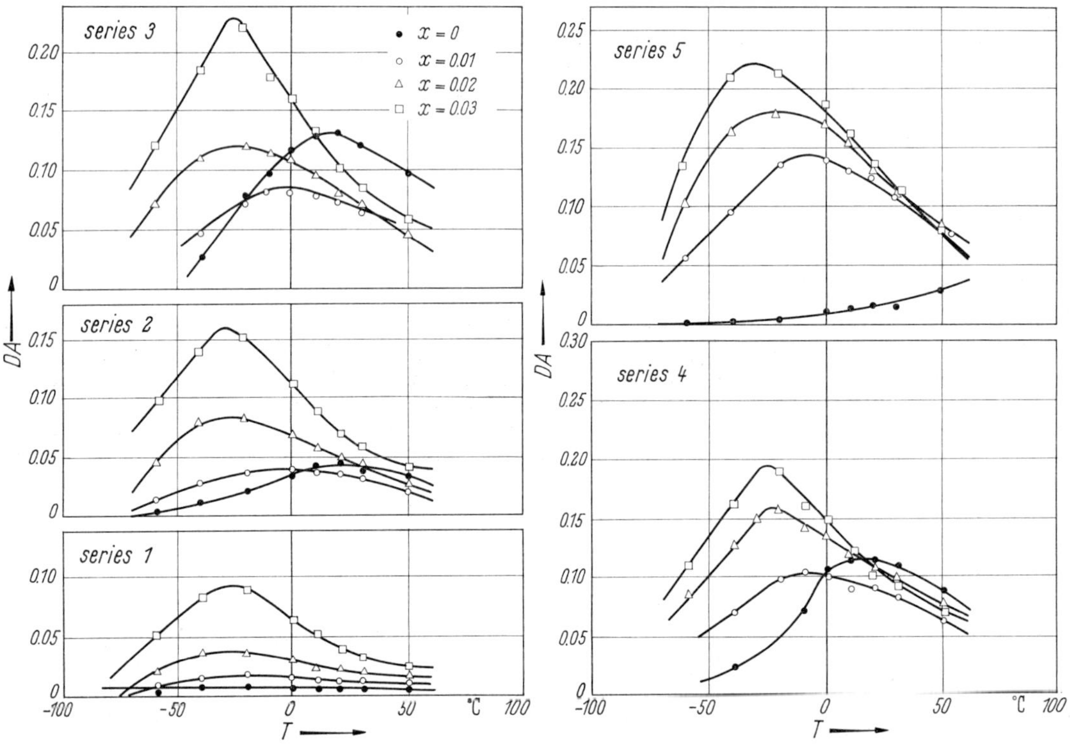

Fig. 50. $(Fe_2O_3)_{0.53}(MnO)_{0.28}(ZnO)_{0.19-x}(TiO_2)_x$. Disaccommodation, DA, of MnZn ferrites with Ti substituted to Zn vs. temperature. Samples were sintered for 4 h at 1250 °C in different atmospheres.

Series 1: N_2 + 0.5% O_2
2: N_2 + 1.0% O_2
3: N_2 + 2.0% O_2
4: N_2 + 5.0% O_2
5: N_2 + 10.0% O_2

The table below gives the FeO content for each sample in mole% FeO.

FeO content

x	1 0.5 mole% O_2	2 1 mole% O_2	3 2 mole% O_2	4 5 mole% O_2	5 10 mole% O_2
0.00	5.10	3.80	2.96	0.52	–
0.01	6.56	5.78	5.53	4.74	4.07
0.02	8.21	7.45	7.20	5.77	5.56
0.03	10.07	8.97	9.15	7.00	5.93

From this it is concluded that the disaccommodation vanishes at RT with vanishing number of Fe^{2+} or of the vacancies, since the disaccommodation varies like N(Fe^{2+})·N (vacancies), where N gives the number. From [*63MM* (2)].

System Li-Mn-Fe-O

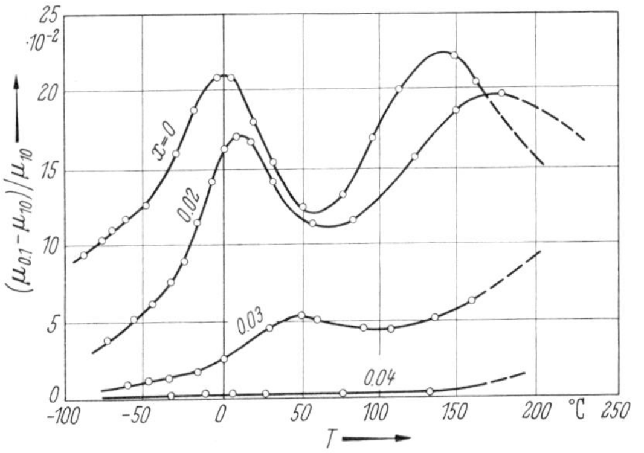

Fig. 51. $(MnO)_{0.4}(Li_2O)_x(Fe_2O_3)_{0.6-x}$. Temperature spectra of disaccommodation. Parameter x. $t_1 = 0.1$ min, $t_2 = 10$ min. From [62OYM].

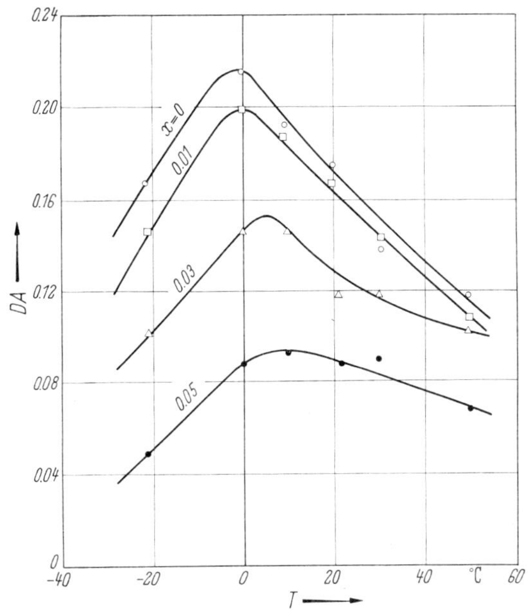

Fig. 52. $(Fe_2O_3)_{0.580}(MnO)_{0.325}(LiO_{1/2})_x(ZnO)_{0.095-x}$. Temperature dependence of disaccommodation for MnZn ferrites with Li substituted to Zn.

Samples were sintered for 4 h at 1250 °C in a 10% oxygen atmosphere. Three other series sintered in different atmospheres (5%, 2.5%, 1% O_2) have been studied. DA $= f(T)$ is of the same kind as for the series sintered in 10% O_2. It is deduced that addition of Li to MnZn ferrites decreases the Fe^{2+} content without changing the concentration of vacancies. Therefore disaccommodation decreases strongly at room temperature.
After [63MM (1)].

d) Pressure after-effect

System Zn-Mn-Fe-O

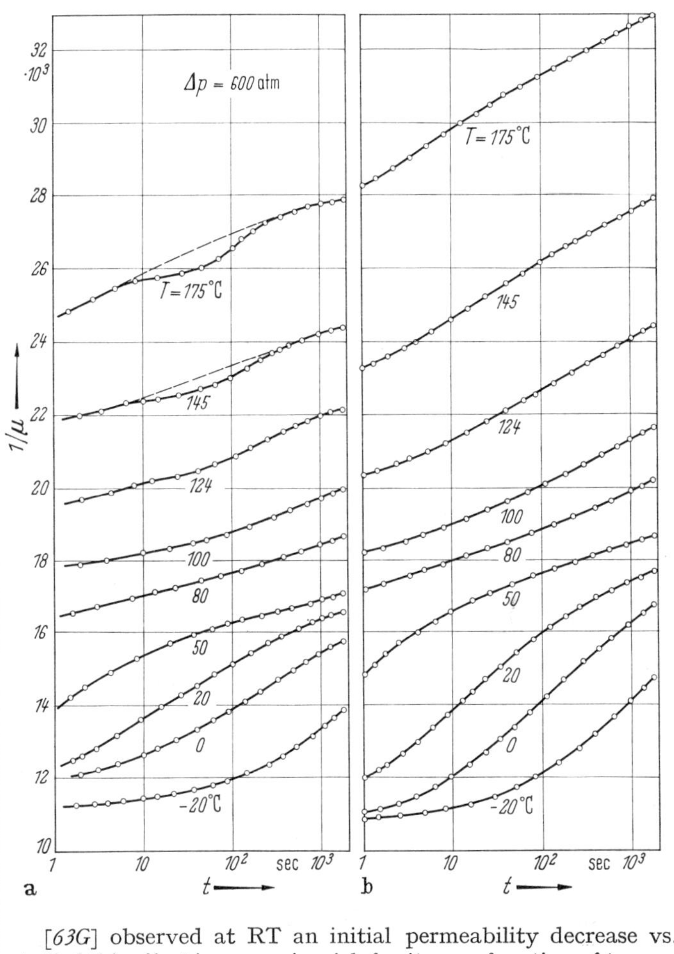

←
Fig. 53. 58% Fe_2O_3, 26% MnO, 16% ZnO, sintered in air at 1300 °C.

a) $1/\mu = f(t)$ after pressure change $\Delta p = 600$ atm
b) $1/\mu = f(t)$ after demagnetization

at different temperatures of measurement. The time dependence of the two $1/\mu$ variations is the same, except for $T > 100$ °C. It is shown, that this difference is due to the thermal effect during the pressure change. The identity of disaccommodation and pressure after-effect is well understood by the fact that pressure change acts as a demagnetization by the build-up of a new Bloch wall distribution due to a new distribution of stresses. From [65SSD].

[63G] observed at RT an initial permeability decrease vs. time after pressure changes. [65SSD] studied this effect in vacancies rich ferrite as a function of temperature and called it "pressure after-effect".

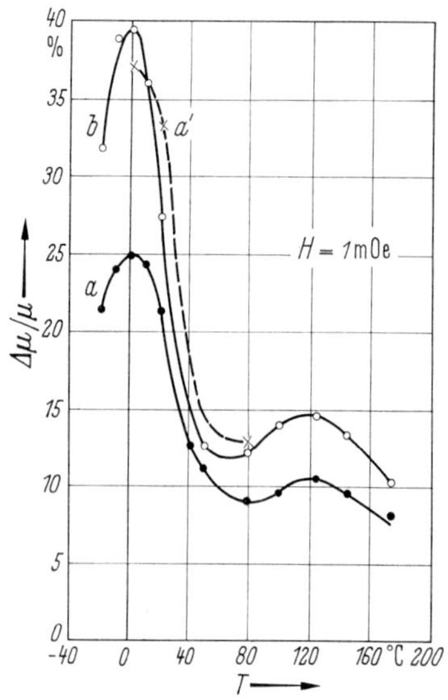

Fig. 54. 58% Fe_2O_3, 26% MnO, 16% ZnO. Temperature dependence of initial permeability disaccommodation. After [*65SSD*].

a) after Δp = 600 atm – pressure after-effect
a') after Δp = 3000 atm – pressure after-effect
b) after demagnetization – permeability disaccommodation

$$\frac{\Delta\mu}{\mu} = \frac{\mu\,(10\text{ sec}) - \mu\,(1800\text{ sec})}{\mu\,(1800\text{ sec})}$$

See caption of Fig. 53.

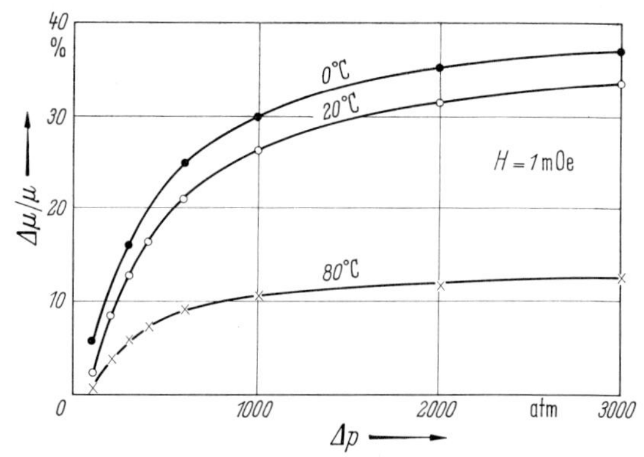

Fig. 55. 58% Fe_2O_3, 26% MnO, 16% ZnO. Pressure after-effect vs. pressure difference Δp by decompression. Parameter *T*. From [*65SSD*]. See captions of Fig. 53, 54.

e) Viscosity

System Mn-Fe-O

→

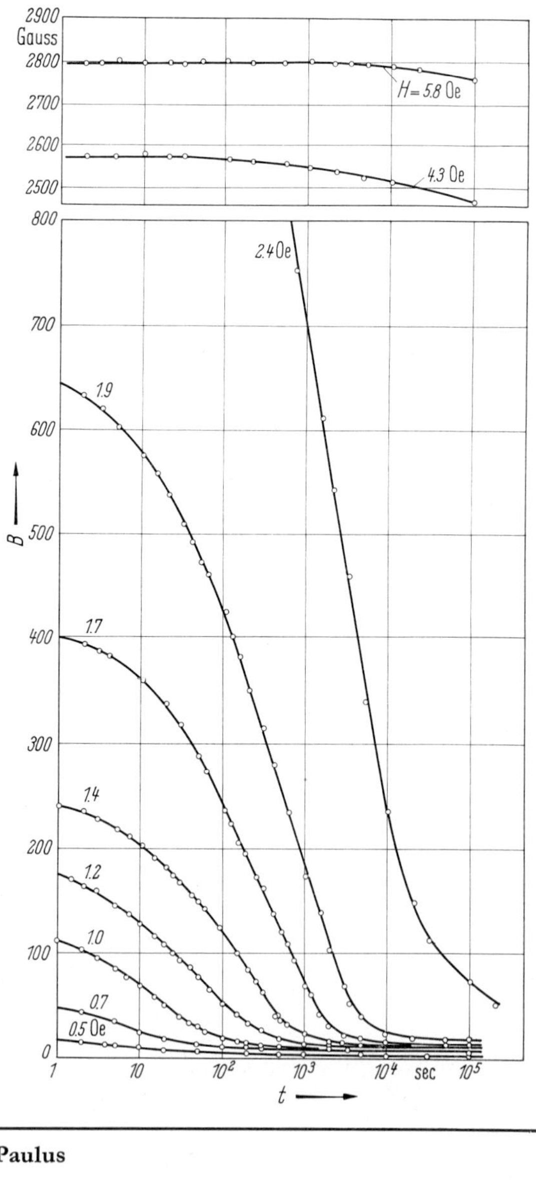

Fig. 56. $Mn_{1.36}Fe_{1.64}O_4$. Disaccommodation curves *B* vs. *t*. Parameter *H*. $T = -196$ °C. From [*65K*(1)]. The decrease in magnetic induction *B* in a dc magnetic field *H* as a function of the time after demagnetization was measured by the ballistic method on a polycrystalline sample of the Mn ferrite $Mn_{1.36}Fe_{1.64}O_4$ for which a strong magnetic after-effect at liquid nitrogen temperature had already been observed by [*57KV*]. The preparation of the sample was as follows: after reaching a temperature $T \approx 1000$ °C the air pressure was lowered to 0.1 mm Hg. The technical vacuum was maintained throughout firing. The maximum temperature of firing was $T = 1200$ °C. See also [*66B*].

System Cu-Mn-Fe-O

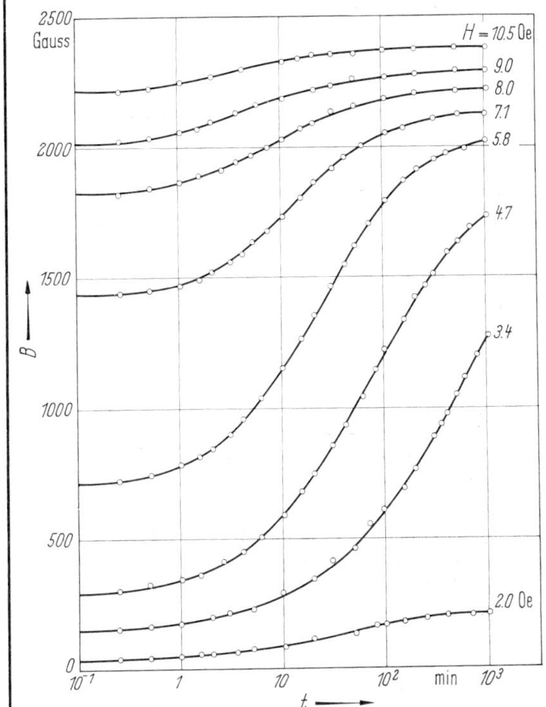

←
Fig. 57. $Cu_{0.47}Mn_{0.53}Fe_2O_4$. Curves of development of magnetic induction B as a function of time t after switching on magnetic field H; $T = -195$ °C. The sample was demagnetized at RT. The demagnetized state of the sample was stabilized by cooling the sample to the temperature of liquid nitrogen. The samples were prepared as follows: they were annealed for 8 h at a temperature of 1200 °C (in air). Cooling took place in three stages: 1200°···900 °C slowly, 900 °···400 °C quickly, 400°···20 °C slowly. From [*65K* (2)]. The same ferrite was also studied with the simultaneous action of a pulse and dc field. See [*66ZS*].

Tab. 8. Viscosity field $(H_t)_{max}$ defined as the maximum horizontal shift between the magnetization curves, one of which has been taken immediately after demagnetization and the other after a considerable period, when the process of stabilization of the demagnetized state is already completed [*64BY*]

T °C	$Mn_{1.75}Co_{0.05}Fe_{1.2}O_4$ $(H_t)_{max}$ Oe	$Mn_{1.8}Fe_{1.2}O_4$ $(H_t)_{max}$ Oe
−100	23	0.1
−125	45	—
−150	—	0.3
−157	12	—
−180	—	0.2

System Co-Mn-Fe-O

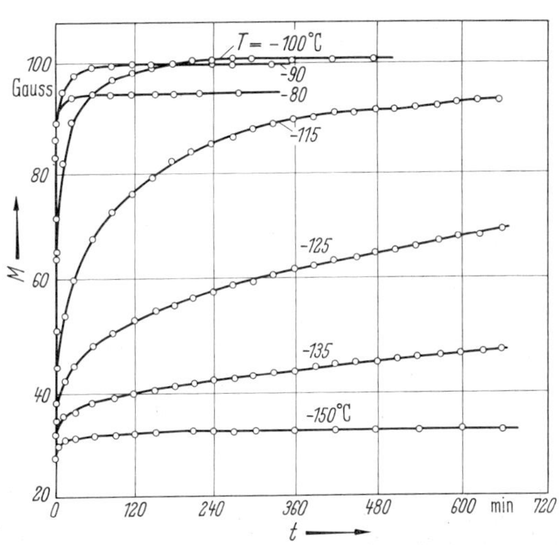

Fig. 58. $Mn_{1.7}Co_{0.1}Fe_{1.2}O_4$. Graph of intensity of magnetization against time (viscosity curves). Single crystal in an H = 130 Oe field at various temperatures, T. The specimen was demagnetized at RT, cooled to the temperature of measurement, held at this temperature for 30 min, and then the field was switched on and the increase in the intensity of magnetization was measured by a ballistic method over a period of time. From [*66BA*].

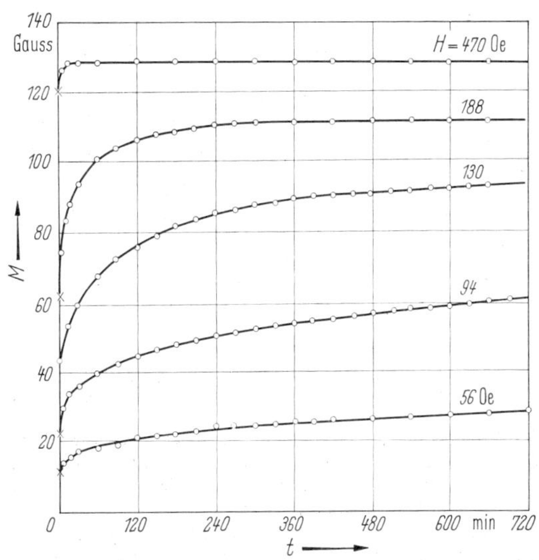

Fig. 59. $Mn_{1.7}Co_{0.1}Fe_{1.2}O_4$. Viscosity curves at −115 °C for various magnetic field strengths, H. Same measurement method as in Fig. 58. After [*66BA*].

System Mg-Mn-Fe-O

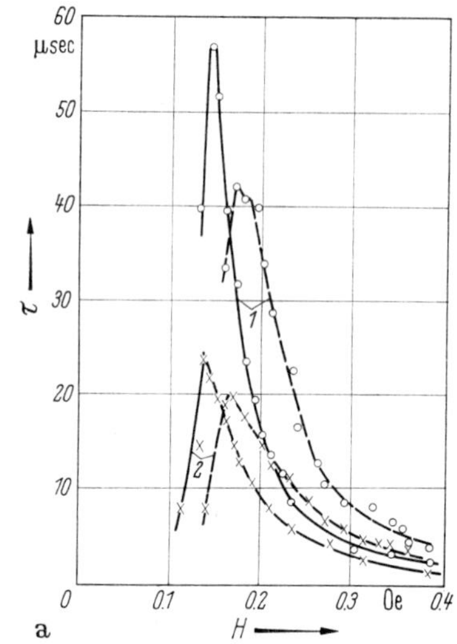

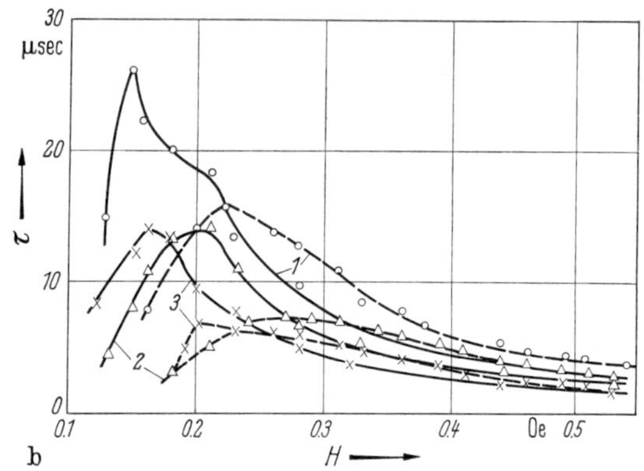

Fig. 60a, b. Magnetic viscosity τ vs. the magnetizing field H for MgMn ferrite [*65TM*]. As the measure of magnetic viscosity, they took τ, the magnetic reversal time of the specimen, i.e. the width of the secondary e.m.f. pulse generated in one of its coils, measured at 0.1 of the amplitude.

a) 24 MgO, 32 MnO, 44 Fe_2O_3
1 crystallographic direction [111], $H_c = 0.18$ Oe
2 crystallographic direction [100], $H_c = 0.178$ Oe
Solid lines were taken 7½ months after the dashed ones.

b) 10 MgO, 40 MnO, 50 Fe_2O_3
1 crystallographic direction [111], $H_c = 0.19$ Oe
2 crystallographic direction [110], $H_c = 0.18$ Oe
3 crystallographic direction [100], $H_c = 0.175$ Oe

Solid lines were taken 6 months after the dashed ones.

6.1.3.9 Microwave properties

6.1.3.9.1 Resonance line width

a) Mn-Fe-O

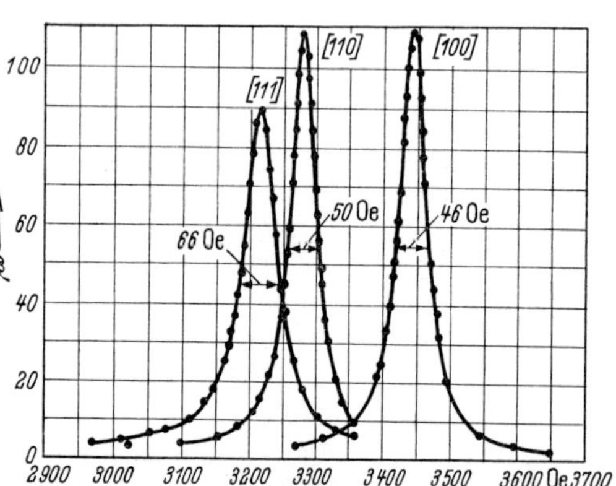

Fig. 1. $Mn_{0.98}Fe_{1.86}O_4$. Ferrimagnetic resonance line shapes [*55DGJ*] for samples measured with the steady magnetic field along the principal crystallographic directions. Highly polished 0.026 cm diameter sphere, RT, $f = 9.3$ Gc/s. The shape of such symmetric curves as these is often indicated merely by giving the line width ΔH. This we will define for our use here as the field interval between points of half the maximum absorption. These curves were found to be very closely Lorentz shaped over the measured region.

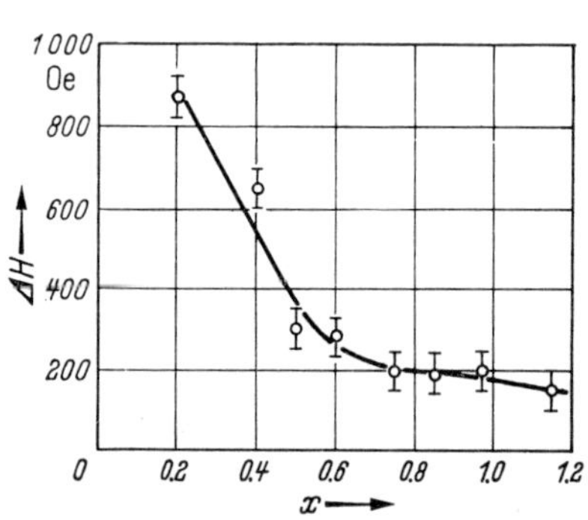

Fig. 2. Measurements of ΔH for various compositions $Mn_xFe_{3-x}O_4$, $0.2 \leq x \leq 2$. These measurements were at RT, and represent the average of ΔH for the steady field along the three principal directions [100], [110] and [111]. Unfortunately we do not have corresponding measurements of the Curie temperatures for these crystals, so that it is difficult to compare them. There is a plot of the room temperature anisotropy field K_1/M_s for these samples in Fig. 10 in 6.1.3.6. There K_1/M_s increases on both sides of $x = 0.6$ whereas in this plot of line width $x = 0.6$ represents the point at which the ΔH levels off [*59FMU* (2)].

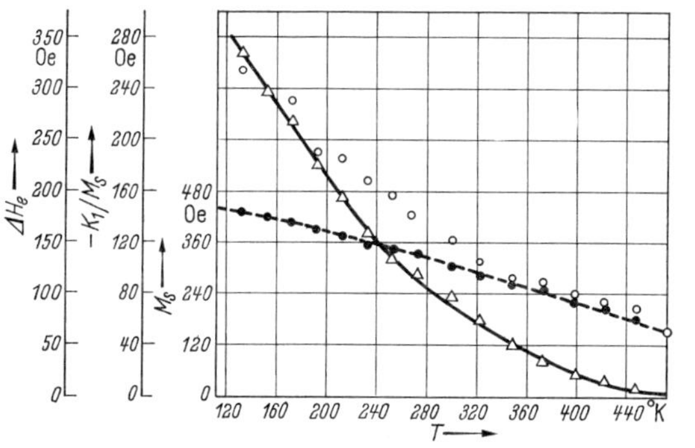

Fig. 3. $Mn_{1.26}Fe_{1.74}O_4$. The relative ferrimagnetic resonance line widths in single crystal and polycrystal [*58HBK*]. The figure shows the temperature variation of the anisotropy field $-K_1/M_s$ (△), the magnetic moment M_s (●), and the "excess line width" ΔH_e (○). This last is defined as the line width observed in the polycrystalline sample less the line width observed in the single crystal sample. The curves show that in the region above RT ΔH_e varies with the magnetization, and somewhat below RT ΔH_e varies with K_1/M_s. The authors conclude that for Mn rich Mn ferrites the determining factor as to whether the anisotropy field or the magnetic moment makes the chief contribution to ΔH_e appears to depend on the ratio K_1/M_s^2 being greater or less than 0.25, respectively.

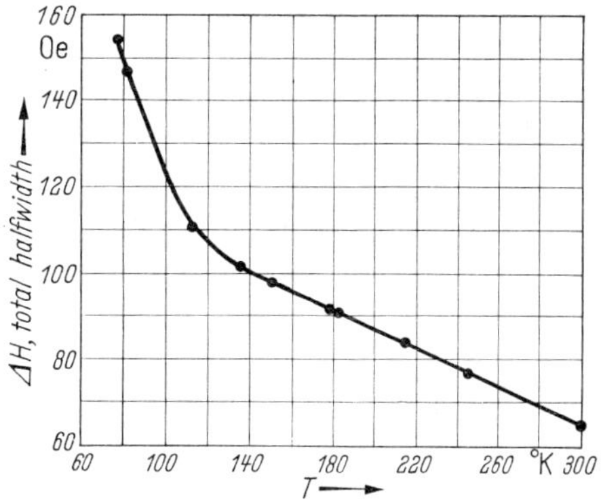

Fig. 4. Single crystal $Mn_{0.98}Fe_{1.96}O_4$. The variation of line width with temperature [*55DGJ*]. The steady field was along the [111] axis. f = 9.3 Gc/s.

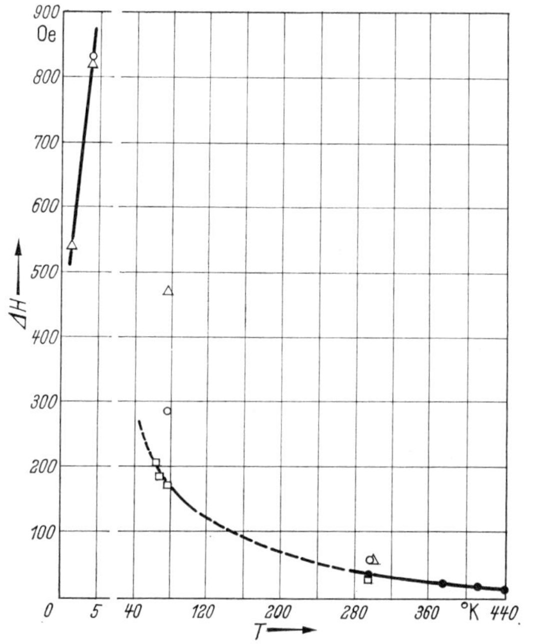

Fig. 5. $Mn_{1.4}Fe_{1.6}O_4$. Resonance line width vs. temperature. It was necessary to introduce points for a disc shaped sample at the lowest temperatures in order to have a sample which could be saturated in the fields available. There is a maximum in ΔH between liquid nitrogen and liquid helium temperatures. Some of these data are reduced to values of the damping parameter in the equation of motion in Tab. 1 in 6.1.3.7 [*59DE*].
○ ΔH_1, 24 Gc/s; spheres ≈ 0.03 cm diameter; △ ΔH_2, 24 Gc/s; 0.10 · 0.0033 cm disc; field in plane; □ ΔH_1, 9.1 Gc/s; spheres ≈ 0.03 cm diameter; ● ΔH_1, 6.4 Gc/s; spheres ≈ 0.03 cm diameter.

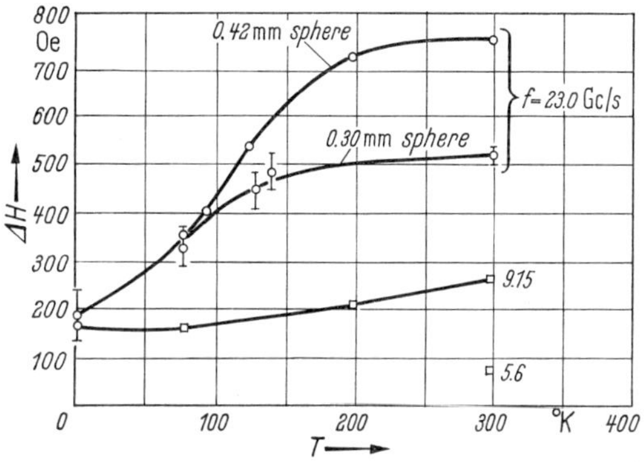

Fig. 6. MnFe spinel. The temperature dependence of the line width in ferrimagnetic resonance measurements on single crystals. No analyses were performed on the crystals but the experiments were performed at several frequencies and attempts were made to fit the line shape in detail. Note the larger line width at 25 Gc/s and the variation of ΔH with sample size. No size effect was apparent at 9.15 Gc/s [*55T*].

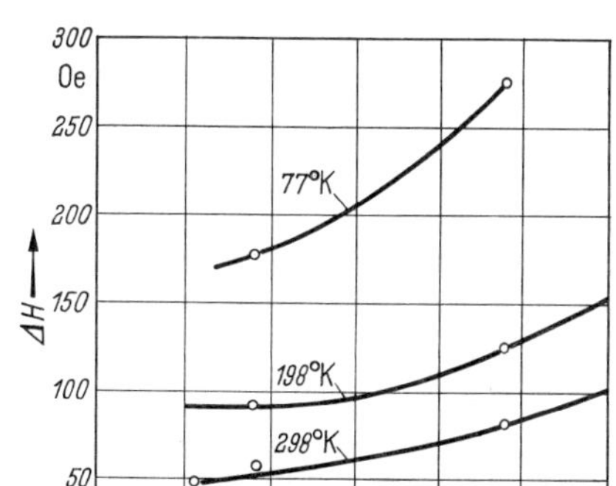

Fig. 7. Mn ferrite. The frequency variation of line width at several temperatures [*55T*].

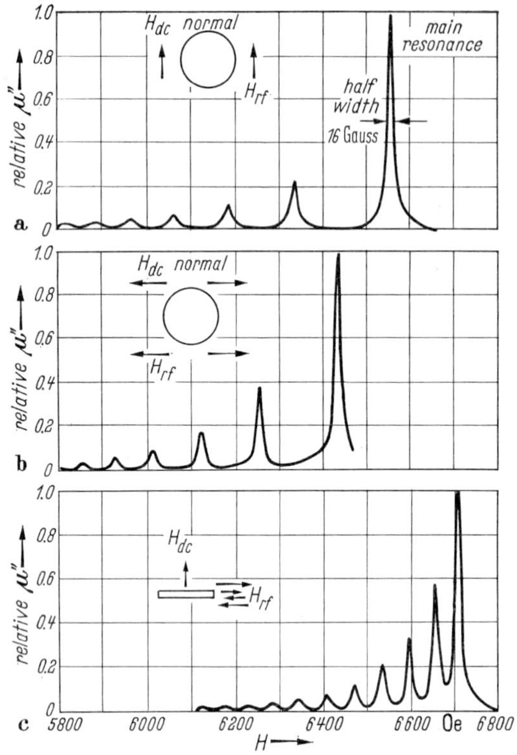

Fig. 8. Absorption curves obtained with a 0.0125 cm thick by 0.25 cm diameter circular disc of Mn ferrite with three different transverse rf magnetic field configurations. The frequencies were close to 9226 Mc/s. This multiplicity of absorption maxima results from the excitation of the so-called "magnetostatic modes", in which the phase and amplitude of the spin precession varies throughout the sample. These were first observed in Mn ferrite single crystals, and must be continually reckoned with in any ferrimagnetic resonance experiments on narrow line width materials [*60D*].

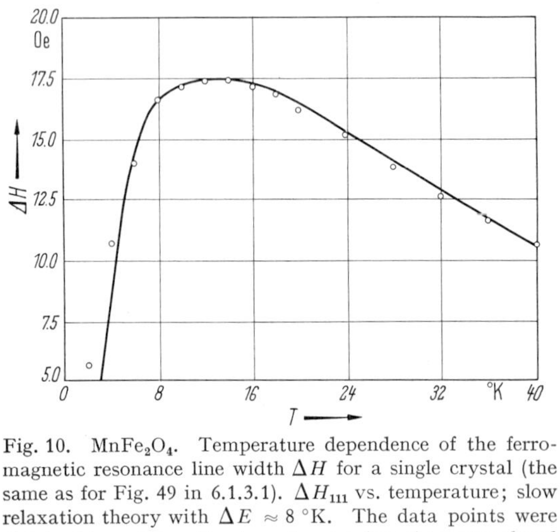

Fig. 10. $MnFe_2O_4$. Temperature dependence of the ferromagnetic resonance line width ΔH for a single crystal (the same as for Fig. 49 in 6.1.3.1). ΔH_{111} vs. temperature; slow relaxation theory with $\Delta E \approx 8$ °K. The data points were obtained by substracting from the raw data the extrapolated contribution from a second peak in ΔH occurring at higher temperatures. The peak in ΔH at 14 °K occurs when $\omega_{res} \cdot \tau \cong 1$ where τ is the relaxation time associated with the Fe^{2+} impurities and ω_{res} is the ferromagnetic resonance frequency [*64HBG*].

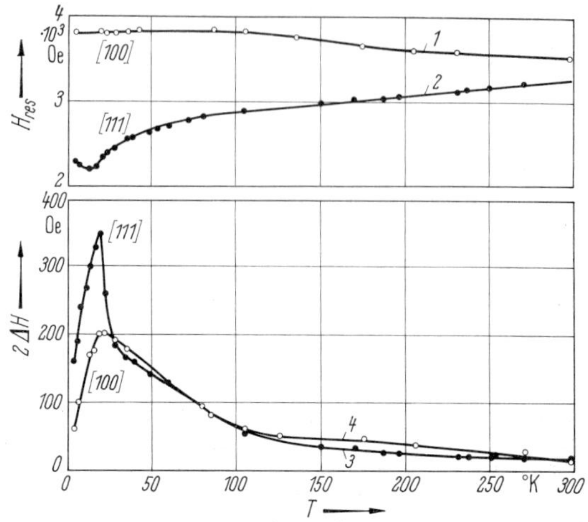

Fig. 9. Single crystal $Mn_{1.03}Fe_{1.95}O_4$. Temperature dependence of the resonance field H_{res} (1, 2) and the resonance curve width $2\Delta H$ (3, 4) for the [100] and [111] axes. $\lambda = 3.2$ cm. Spherical sample $d = 0.6$ mm [*64S* (3)].

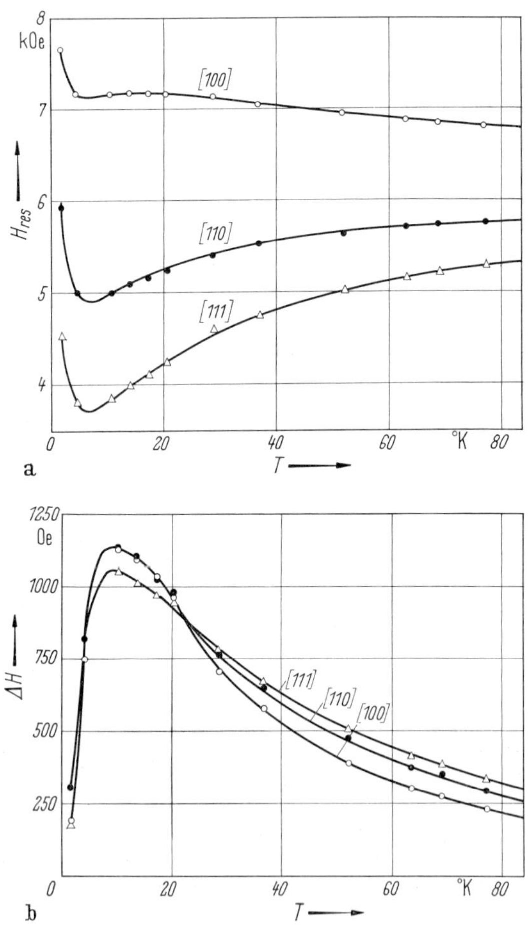

Fig. 11. $Mn_{1.318}Fe_{1.682}O_4$. Variation with temperature of (a) applied fields for resonance, H_{res}, and of (b) line widths, ΔH, in principal directions in a (110) plane at 16.85 Gc/s [*64CT*].

Tab. 1. Properties of materials investigated in Fig. 12, 13, 14 [*67LPJ*]

	Composition mole-%	$4\pi M_s$ Gauss	ΔH Oe	ϱ_X g/cm³	ϱ Ωcm	Grain size μm
Monocrystalline	1) 50 MnO · 50 Fe_2O_3	5000	42	4.997	5…10	—
	2) 63.5 MnO · 36.5 Fe_2O_3	3200	19	5.16	10^4	—
	3) 70 MnO · 30 Fe_2O_3	2500	28	—	—	—
	4) 80 MnO · 20 Fe_2O_3	1300	50	—	—	—
Polycrystalline	1) 50 MnO · 50 Fe_2O_3	4850	340	4.99	$4.3 \cdot 10^3$	50
	2) 60 MnO · 40 Fe_2O_3	3900	—	5.11	$8.8 \cdot 10^5$	—
	3) 70 MnO · 30 Fe_2O_3	2550	270	5.26	$1.0 \cdot 10^6$	25
	4) 80 MnO · 20 Fe_2O_3	2300	268	5.47	$3.2 \cdot 10^6$	8…12

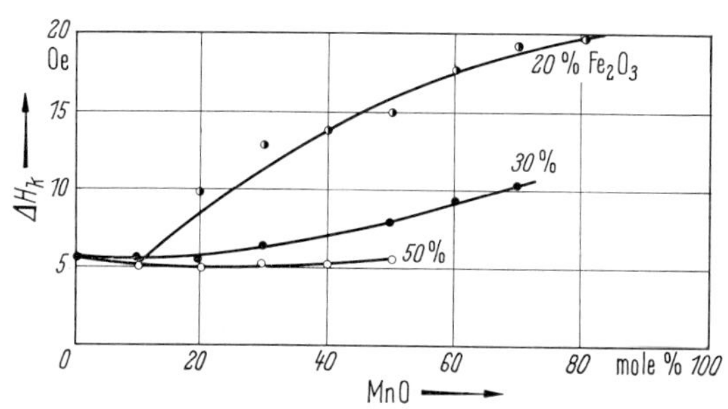

Fig. 12. Graph of spin wave loss factor, ΔH_k, at 9200 Mc/s against composition for polycrystalline MgMn ferrites [*67LPJ*]. See Tab. 1.

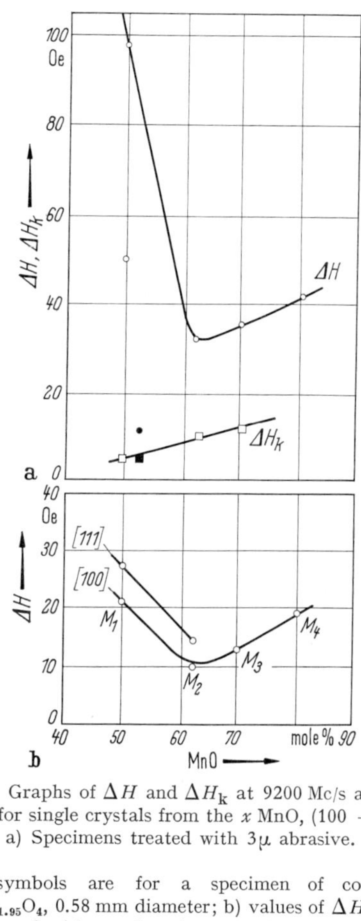

Fig. 13. Graphs of ΔH and ΔH_k at 9200 Mc/s against Mn content for single crystals from the x MnO, $(100 - x)$ Fe_2O_3 system. a) Specimens treated with 3μ abrasive. Direction [111].
Black symbols are for a specimen of composition $Mn_{1.03}Fe_{1.95}O_4$, 0.58 mm diameter; b) values of ΔH for specimens treated with powder of 1μm grain size [*67LPJ*]. See Tab. 1.

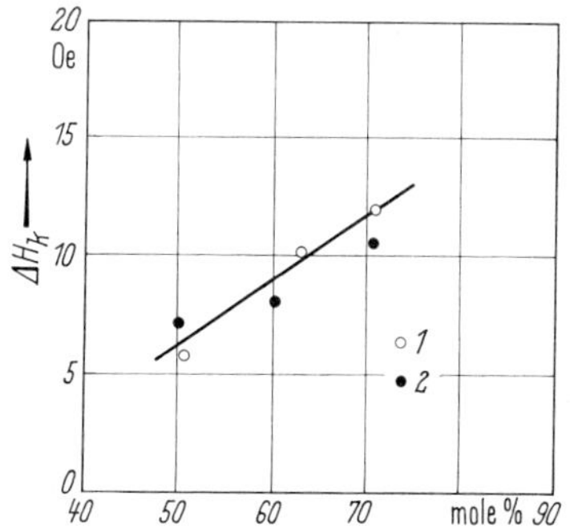

Fig. 14. Comparison of ΔH_k for monocrystalline (1) and polycrystalline (2) Mn ferrites of the same composition. In the single crystals measurements were made along the [111] axis [*67LPJ*]. See Tab. 1.

b) Zn-Mn-Fe-O

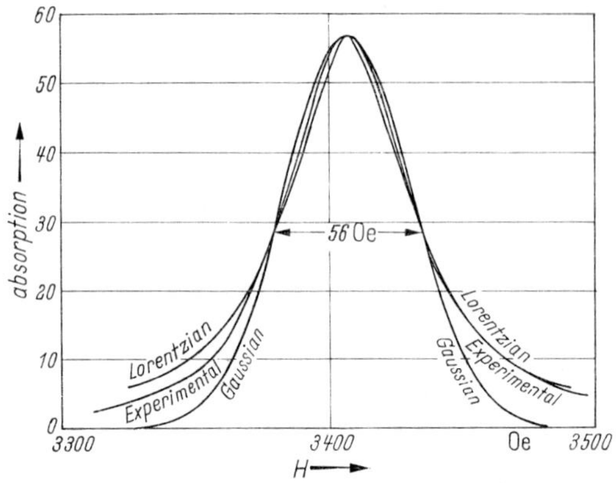

Fig. 15. MnZn ferrite. The line shape of resonance absorption at 9.1 Gc/s with the steady field along [100]. For comparison the ideal Lorentzian and Gaussian shapes are shown for lines of the same half width [*55T*].

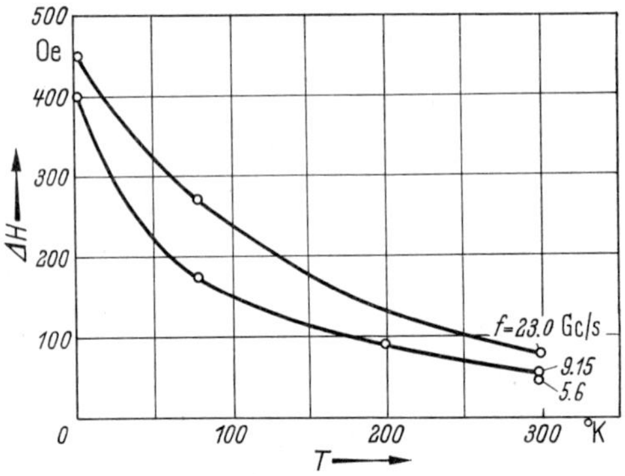

Fig. 16. MnZn ferrite. ΔH vs. T in ferrimagnetic resonance measurements on single crystals. See Fig. 6. [*55T*].

c) Co-Mn-Fe-O

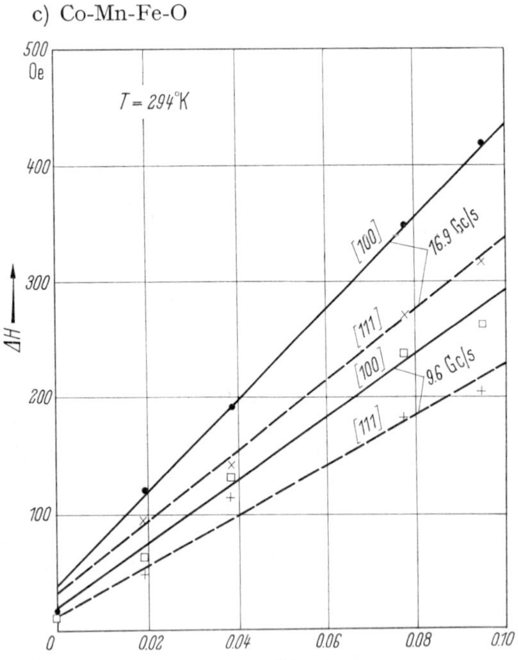

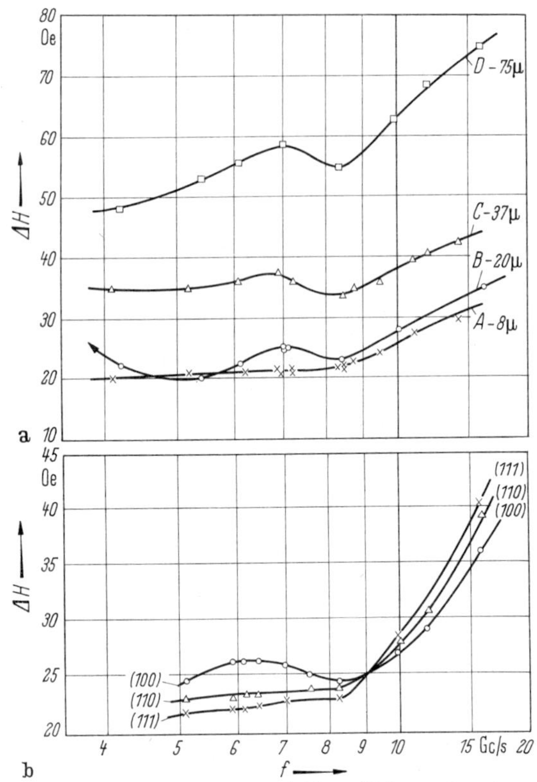

Fig. 17. MnZn ferrite [*60B*]. a) Line width as a function of frequency and surface condition. Note: curves labeled according to surface roughness. b) Line width anisotropy as a function of frequency. The composition was not specified.

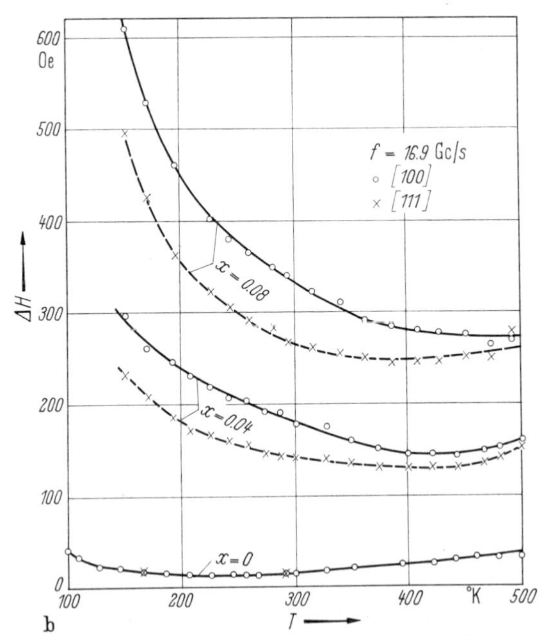

Fig. 18. $Co_xMn_{1-x}Fe_2O_4$. Microwave resonance line width $2\Delta H$. (a) $2\Delta H$ vs. x. (b) $2\Delta H$ vs. temperature. The exact composition of each sample is given in Tab. 3 in 6.1.3.6. The Curie temperature of the crystals $Co_xMn_{1-x}Fe_2O_4$ vary linearly with x from 550 °K for $x = 0$ to 575 °K for $x = 0.10$ [*62T*].

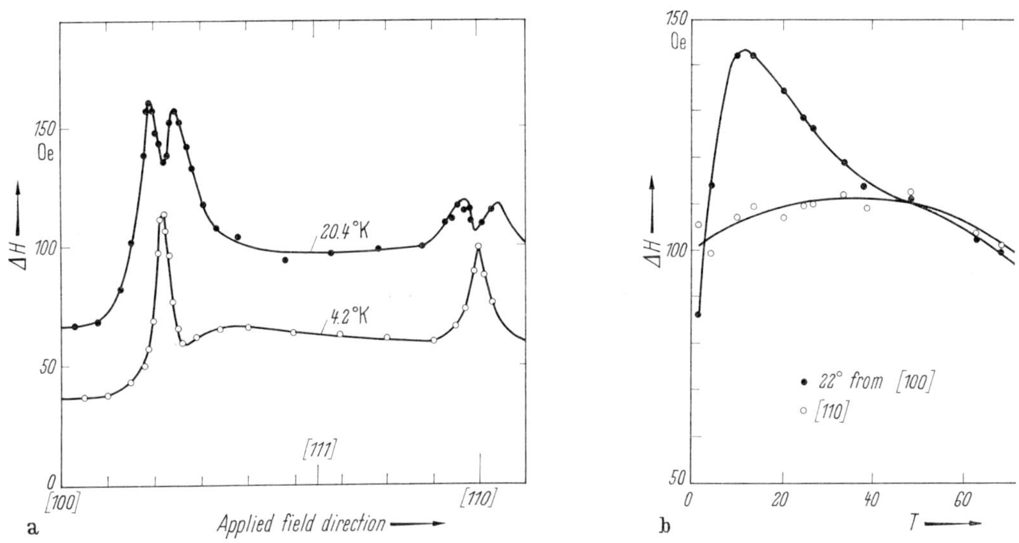

Fig. 19. $Co_{0.005}Mn_{0.995}Fe_{1.99}O_4$. The crystal was grown by the flame fusion method. (a) Line width vs. orientation of the applied field. (b) Line width vs. temperature.
At 4.2 °K and 1.5 °K sharp peaks were found in H_{res} measured as a function of the orientation of this field in (110), (100), and (210) planes. In a (110) plane sharp maxima were observed at 22° from [100] and at [110]. In a (100) plane there was a sharp peak along [110], and in a (210) plane they occurred at 23.5° and 84.5° from [100]. These field orientations are consistent with their lying on cones with their axes along [111] directions and semivertical angle 33.5° ± 1°. Sharp maxima in ΔH as a function of the crystallographic orientation of the applied field were observed at 4.2 °K and 1.5 °K for all of the orientations for which sharp maxima in H_{res} occurred and for no other orientations. These localized maxima in H_{res} and ΔH are attributed to the Co doping. These effects resemble those associated with the doping of YIG with certain rare earth ions [*63TC*].

d) Mg-Mn-Fe-O

Tab. 2. MnMg ferrites. Width of resonant absorption line, ΔH, saturation magnetization, $4\pi M_s$, and electric resistivity, ϱ, in single crystals [*60BBP*]. Single crystals were grown by the Verneuil method. The surfaces of the spherical samples were polished with 10μm grains. The measurements were made at 9470 Mc/s. ΔH varies lightly with the crystal axis

Composition (calculated) wt%	ΔH Oe	$4\pi M_s$ Gauss	ϱ Ωcm
9.4MgO; 16.5MnO; 74.1FeO	18	3320	800
8.4MgO; 23.9MnO; 67.5FeO	12	2950	$1.6 \cdot 10^5$
8.0MgO; 28.3MnO; 63.7FeO	16	2740	10^6
6.9MgO; 37.3MnO; 55.9FeO	18	2480	$4.6 \cdot 10^6$

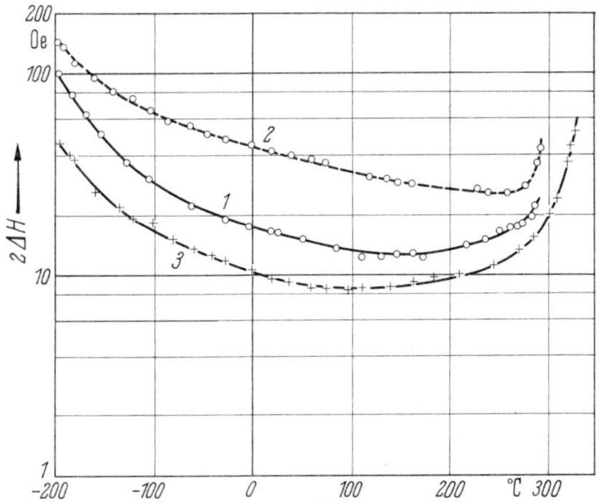

Fig. 20. Temperature dependences of $2\Delta H$. Frequency 9100 Mc/s. The numbers near the curves are the numbers of specimens (for the dimensions and characteristics of the specimens see table). The ordinate axis has a logarithmic scale [*61GGT*].

Nr. of specimens	*1*	*2*	*3*
Composition	$Mn_{1.03}Fe_{1.95}O_4$		$Mg_{0.525}Mn_{0.665}Fe_{1.81}O_4$
Diameter of spheres [mm]	0.60	0.58	0.76
Dimension of abrasive grain (abrasive cloth), with which the spheres were finally treated [μm]	1···3	60	1···3
Intensity of magnetization at 20 °C. $H = 3250$ Oe	320	320	230
Curie point, Θ_C [°C]	300	300	333

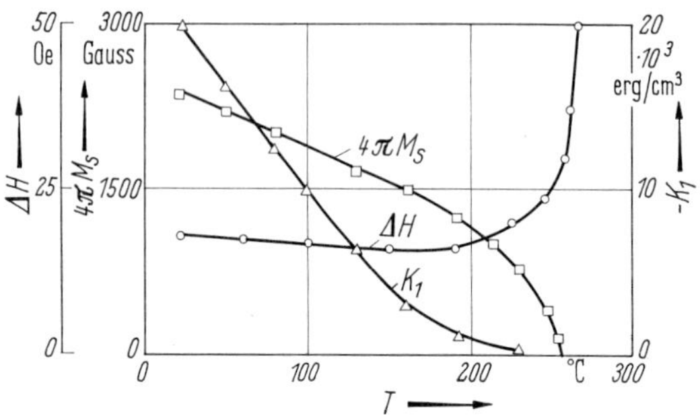

Fig. 21. 6.9 MgO, 37.3 MnO, 55.9 Fe_2O_3. Temperature dependence of the width, ΔH, of the line of ferromagnetic resonance absorption in the [110] direction, of the saturation magnetization, $4\pi M_S$, and of the constant of magnetic anisotropy, K_1 [*60BBP*].

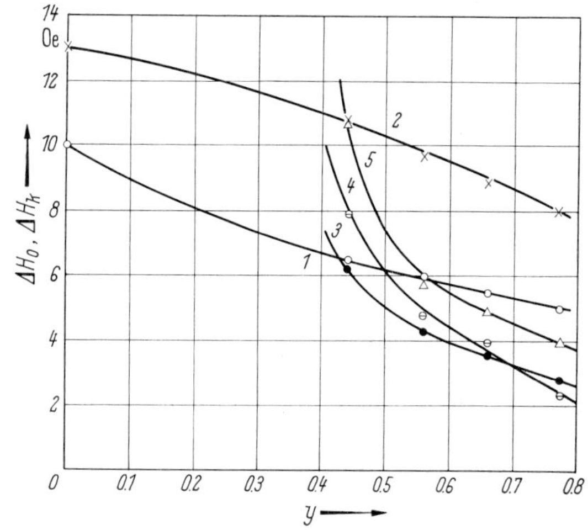

Fig. 22. $Mg_yMn_{1.3-y}Fe_{1.7}O_{4+y}$. Width of resonance curve, ΔH_0, and width of spin wave curve, ΔH_k, vs. composition y. Single crystals grown by the Verneuil technique [*66MY*]. $\lambda \cong 3$ cm.

Compositions of the crystals
1 ΔH_0 [111]
2 ΔH_0 [100]
3 ΔH_k for $H_0 =$ 400 Oe
4 ΔH_k for $H_0 =$ 600 Oe
5 ΔH_k for $H_0 =$ 1000 Oe

Further references for 6.1.3.9.1: [*50BS, 52OFD* (1), *52OFD* (2), *52OFD* (3), *54T, 55AT, 55DGJ, 55T, 55Th, 56AVM, 56D, 56W, 56WS, 56WSM, 57D, 57K, 58HBK, 58Su, 58W, 59DE, 59FMU* (2), *59SF, 60D*].

6.1.3.9.2 Spectroscopic splitting factor

a) Mn-Fe-O

Tab. 3. g-value determinations for $Mn_xFe_{3-x}O_4$

Composition	T °K	f Mc/s	g_{exp} [1])	Ref.
$Mn_{1.4}Fe_{1.6}O_4$	4.2	23 900	2.26	*59DE*
	77		2.017	
	295		2.0047	
	409		2.0033	
	498		2.004	
$Mn_{0.98}Fe_{1.86}O_4$	4.2	9 300	2.060	*55DGJ*
	77		2.019 ± 0.003	
	300		2.044 ± 0.002	
$Mn_{1.26}Fe_{1.74}O_4$	133	—	2.062	*58HBK*
	473		2.028	

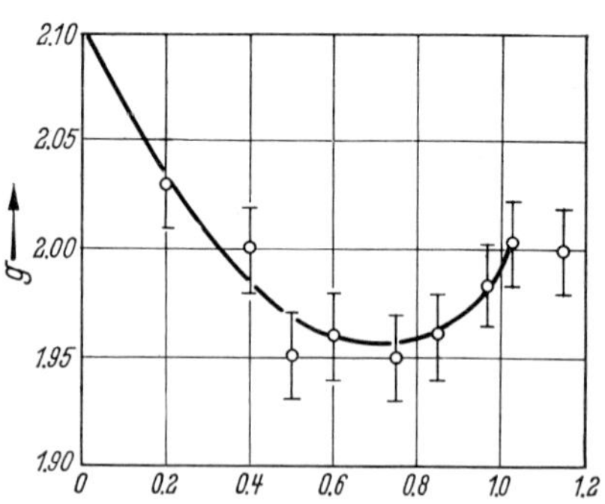

Fig. 23. $Mn_xFe_{3-x}O_4$. g_{exp} at RT vs. composition parameter, x. From [*59FMU* (2)]. The experimental splitting factor g_{exp} is defined by $h\nu_{res} = g_{exp}\mu_B H_{res}$.

1) The experimental splitting factor, g_{exp}, is defined by $h\nu_{res} = g_{exp}\mu_B H_{res}$.

b) Cr-Mn-Fe-O

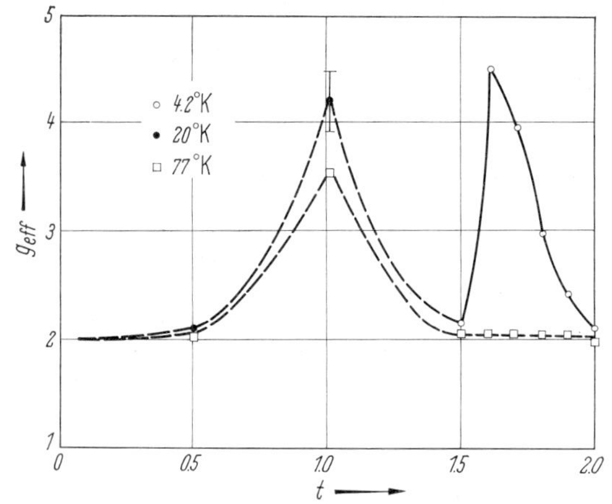

Fig. 24. $MnFe_{2-t}Cr_tO_4$. g_{eff} vs. composition parameter, t, at 9200 Mc/s at various temperatures [*63MP*].

Further references for 6.1.3.9.2: [*55DGJ*, *56D*, *56W*, *56WS*, *56WSM*, *57D*, *57K*, *57L*, *57W*, *58HBK*, *59DE*, *59FMU* (2), *59P* (2), *60D*].

6.1.3.10 Optical properties (absorption)

a) $MnFe_2O_4$

The absorption spectra in the infrared have been measured for a number of ferrites [*55W*] including Mn ferrites $\mathbf{MnFe_2O_4}$. The ferrite was made by precipitation and thermal decomposition of the oxalates. Transparent samples were prepared pressing the ferrite powder with KBr powder in a vacuum to form clear discs. These transmitted out to about 280 cm^{-1}.

There are absorption bands at 392 and 550 cm^{-1}. Also there seem to be shoulders at about 320 and 625 cm^{-1}. [*55W*] reads his data as indicating the threshold of an electronic absorption band at (930 ± 50) cm^{-1}. This corresponds to an energy of 0.12 eV. The vibrational bands did not become sharper when the absorption was measured at 77 °K. The original paper includes an actual assignment of the vibrations associated with the deep bands noted above as well as the force constants derivable from them.

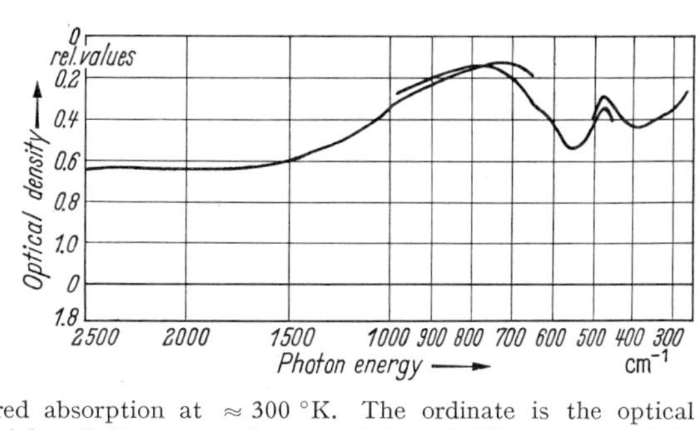

Fig. 1. $MnFe_2O_4$. The infrared absorption at ≈ 300 °K. The ordinate is the optical density measured relative to a standard KBr disc. Sample weight: 2.50 mg; sample area: 3.14 cm^2. The overlap of the lines indicates a prism change.

6.1.3.11 References for 6.1.3

38KN	Kussmann, A., and H. Nitka: Physik. Z. **39** (1938) 208.
39A	Akulov, N. S.: Ferromagnetism, ONTI (1939).
41VH	Verwey, E. J. W., and P. W. Haayman: Physica **8** (1941) 979.
47M	Mason, B.: Am. Mineralogist **32** (1947) 426.
47S	Snoek, J. L.: New Developments in Ferromagnetic Materials, Elsevier, New York, Amsterdam (1947).
48MG	McMurdie, H. F., and E. Golovato: J. Res. Natl. Bur. Std. **41** (1948) 589.
48N	Néel, L.: Ann. Phys. **3** (1948) 137.
49BDS	Brockman, F. G., P. H. Dowling, and W. G. Steneck: Phys. Rev. **75** (1949) 1440.
49K (1)	Kittel, C.: Rev. Mod. Phys. **21** (1949) 541.
49K (2)	Kelly, K. K.: U. S. Bur. Mines Bull. **1949** 476.
50BDS	Brockman, F. G., P. H. Dowling, and W. G. Steneck: Phys. Rev. **77** (1950) 85.
50BS	Beljers, H. G., and J. L. Snoek: Philips Tech. Rev. **11** (1950) 313.
50G	Guillaud, C.: J. Rech. Centre Natl. Rech. Sci. Lab. Bellevue (Paris) **3** (1950) 113.
50GB (1)	Guillaud, C., and R. Bertrand: J. Rech. Centre Natl. Rech. Sci. Lab. Bellevue (Paris) **3** (1950) 73.
50GB (2)	Guillaud, C., and A. Barbezat: J. Rech. Centre Natl. Rech. Sci. Lab. Bellevue (Paris) **3** (1950) 83.
50GC	Guillaud, C., and H. Creveaux: Compt. Rend. **230** (1950) 1256.
50HHL	Harvey, R. L., I. J. Hegyi, and H. W. Leverenz: RCA Rev. **11** (1950) 321.
50P	Polder, D.: Proc. Inst. Elec. Engrs. (London) **97** (1950) 246.
50R	Rado, G. T.: Advances in Electronics **2**, 251, Academic Press, New York (1950).
50RWE	Rado, G. T., R. W. Wright, and W. H. Emerson: Phys. Rev. **80** (1950) 273.
50W	Wells, A. F.: Structural Inorganic Chemistry, 2nd ed., Oxford University Press, London (1950) 381.
51K	Kornetzki, M.: Z. Angew. Phys. **3** (1951) 5.
51OKT	Okamura, T., Y. Kojima, and Y. Torizuka: Sci. Rept. Res. Inst. Tohoku Univ. Ser. **A 3** (1951) 725.
51W	Weis, A.: Elektrotechnik (Berlin) **5** (1951) 213.
51WW (1)	Went, J. J., and H. P. J. Wijn: Phys. Rev. **82** (1951) 269.
51WW (2)	Wijn, H. P. J., and J. J. Went: Physica **17** (1951) 976.
52A	Ahrens, L. H.: Geochim. Cosmochim. Acta **2** (1952) 155.
52KBFG	Kornetzki, M., J. Brackman, J. Frey, and W. Gieseke: Z. Angew. Phys. **4** (1952) 371.
52M	Möltgen, G.: Z. Angew. Phys. **4** (1952) 216.
52OFD (1)	Okamura, T., T. Fujimura, and M. Date: Sci. Rept. Res. Inst. Tohoku Univ. Ser. **A 4** (1952) 191.
52OFD (2)	Okamura, T., T. Fujimura, and M. Date: Phys. Rev. **85** (1952) 1041.
52OFD (3)	Okamura, T., T. Fujimura, and M. Date: Phys. Rev. **88** (1952) 1435.
52P	Pauthenet, R.: Ann. Phys. **7** (1952) 710.
52PB	Prache, P. M., and H. Billottet: Cables Transmission **6** (1952) 317.
52RWE	Rado, G. T., R. W. Wright, and W. H. Emerson: Phys. Rev. **88** (1952) 909.
52S	Smiltens, J.: J. Chem. Phys. **20** (1952) 990.
52W	Wangsness, R. K.: Phys. Rev. **86** (1952) 263.
52YK	Yafet, Y., and C. Kittel: Phys. Rev. **87** (1952) 290.
53AS	Alley, R. E., and F. J. Schnettler: J. Appl. Phys. **24** (1953) 1524.
53B	Burgt, C. M. van der: Philips Res. Rept. **8** (1953) 91.
53BBP	Blewett, J. P., M. H. Blewett, and M. Plotkin: Rev. Sci. Instr. **24** (1953) 800.
53CHB	Corliss, L. M., J. M. Hastings, and F. G. Brockman: Phys. Rev. **90** (1953) 1013.
53H	Heister, W.: Arch. Elektrotech. Berlin **41** (1953) 142.
53Ra	Rado, G. T.: Rev. Mod. Phys. **25** (1953) 81.
53Ro	Romeijn, F. C.: Philips Res. Rept. **8** (1953) 304.
53S	Shimoizaka, J.: J. Chem. Soc. Japan, Pure Chem. Sect. **74** (1953) 728.
53TRFE	Toropov, N. A., L. I. Rabkin, E. Z. Friedenfeld, and B. S. Epstein: Zh. Prikl. Khim. **26** (1953) 982.
53WH	Wijn, H. P. J., and H. van der Heide: Rev. Mod. Phys. **25** (1953) 98.
54CS	Clark, C. A., and W. Sucksmith: Proc. Roy. Soc. (London) Ser. **A 225** (1954) 147.
54G	Gorter, E. W.: Philips Res. Rept. **9** (1954) 295, 321, 403.
54GDM	Galperin, F., G. Dmitrakova, and L. Molodtsova: Dokl. Akad. Nauk. SSSR **94** (1954) 833.
54KK	Komar, A. P., and V. V. Kliushin: Izv. Akad. Nauk. SSSR Ser. Fiz. **18** (1954) 400. — Bull. Acad. Sci. USSR Phys. Ser. (English Transl.) **18** (1954) 96.

54L Lewin, B. E.: Nachr. Akad. Wiss. USSR Physik. **18** (1954) 519.
54P Patrick, L.: Phys. Rev. **93** (1954) 384.
54RBR Richardson, H. M., F. Ball, and C. R. Rigby: Trans. Brit. Ceram. Soc. **53** (1954) 376.
54SW Smit, J., and H. P. J. Wijn: V. 6 of Advances in Electronics and Electronic Physics, Academic Press, N. Y. (1954) 70.
54T Tannenwald, P. E.: Phys. Rev. **96** (1954) 846.
54W Wijn, H. P. J.: Onde Élec. **34** (1954) 418.
55AT Artman, J. O., and P. E. Tannenwald: J. Appl. Phys. **26** (1955) 1124.
55B Birks, J. B.: Phys. Rev. **99** (1955) 1821.
55BCGMY Bozorth, R. M., B. B. Cetlin, J. K. Galt, F. R. Merritt, and W. A. Yager: Phys. Rev. **99** (1955) 1898.
55BTW Bozorth, R. M., E. F. Tilden, and A. J. Williams: Phys. Rev. **99** (1955) 1788.
55DGJ Dillon, J. F., Jr., S. Geschwind, and V. Jaccarino: Phys. Rev. **100** (1955) 750.
55E Economos, G.: J. Am. Ceram. Soc. **38** (1955) 292.
55G Gorter, E. W.: Proc. I. R. E. **43** (1955) 1945.
55GL Goodenough, J. B., and A. L. Loeb: Phys. Rev. **98** (1955) 391.
55H Halem, J.: Elektron. Rundschau **9** (1955) 98.
55Hi Hipskind, G. S.: Elec. Mfg. **56** (1955) 108.
55T Tannenwald, P. E.: Phys. Rev. **100** (1955) 1713.
55Th Thompson, G. H. B.: Nature **175** (1955) 1135.
55W Waldron, R. D.: Phys. Rev. **99** (1955) 1727.
56AVM Ayres, W. P., P. H. Vartanian, and J. L. Melchor: J. Appl. Phys. **27** (1956) 188.
56B Broz, J.: Czech. J. Phys. **6** (1956) 328.
56CHK Callen, H. B., S. E. Harrison, and C. J. Kriessman: Phys. Rev. **103** (1956) 851.
56D Dillon, J. F., Jr.: Bull. Am. Phys. Soc. Ser. II **1** (1956) 125.
56GP Guillaud, C., and M. Paulus: Compt. Rend. **242** (1956) 2525.
56GPV Guillaud, C., M. Paulus, and R. Vautier: Compt. Rend. **242** (1956) 2712.
56HC Hastings, J. M., and L. M. Corliss: Phys. Rev. **104** (1956) 328.
56KH Kriessman, C. J., and S. E. Harrison: Phys. Rev. **103** (1956) 857.
56KT Kedesdy, H. H., and A. Tauber: J. Am. Ceram. Soc. **39** (1956) 425.
56L Lotgering, F. K.: Philips Res. Rept. **11** (1956) 190.
56O Osmond, W. P.: Proc. Phys. Soc. (London) **B 69** (1956) 1319.
56RFE Rado, G. T., V. J. Folen, and W. H. Emerson: Proc. Inst. Elec. Engrs. (London) **B 104**, Suppl. 5 (1956) 198.
56V Vautier, R.: Compt. Rend. **242** (1956) 2814.
56W Walker, L. R.: Bull. Am. Phys. Soc. Ser. II **1** (1956) 125.
56WS White, R. L., and I. H. Solt: Phys. Rev. **104** (1956) 56.
56WSM White, R. L., I. H. Solt, and J. E. Mercereau: Bull. Am. Phys. Soc. Ser. II **1** (1956) 12.
57Ba Baltzer, P. K.: Phys. Rev. **108** (1957) 580.
57BB Bergstein, A., and A. Bohun: Czech. J. Phys. **7** (1957) 489.
57BBP Bickford, L. R., Jr., M. S. Brownlow, and R. F. Penoyer: Proc. Inst. Elec. Engrs. (London) **B 104**, Suppl. 5 (1957) 238.
57BC Bates, L. F., and D. A. Christoffel: Proc. Inst. Elec. Engrs. (London) **B 104**, Suppl. 5 (1957) 231, 245.
57Bo Bongers, P. F.: Ph. D. Thesis, University of Leiden (1957).
57BT Belov, K. P., and E. V. Talalaeva: Zh. Eksperim. i. Teor. Fiz. **33** (1957) 6.
57Bu Burgt, C. M. van der: Philips Res. Repts. **12** (1957) 97.
57BBE Belov, K. P., K. M. Bol'shova, and T. A. Elkina: Izv. Akad. Nauk SSSR, Ser. Fiz. **21** (1957) 1047. — Bull. Acad. Sci. USSR. Phys. Ser. (English Transl.) **21** (1957) 1051.
57BBKVZ Brož, J., A. Bergstein, S. Krupička, J. Vintera, and K. Závěta: Czech. J. Phys. **7** (1957) 66.
57BZ Brož, J., and K. Závěta: Czech. J. Phys. **7** (1957) 242.
57D Dillon, J. F., Jr.: Bull. Am. Phys. Soc. Ser. II **2** (1957) 128.
57DO Dunitz, J. D., and L. E. Orgel: Phys. Chem. Solids **3** (1957) 20, 318.
57FC Fine, M. E., and C. Chiou: Phys. Rev. **105** (1957) 121.
57FSS Finch, G. I., A. P. B. Sinha, and K. P. Sinha: Proc. Roy. Soc. (London) **A 242** (1957) 28.
57Go Gorter, E. J.: Proc. Inst. Elec. Engrs. (London) **B 104**, Suppl. 5 (1957) 255.
57Gu Guillaud, C.: Proc. Inst. Elec. Engrs. (London) **B 104**, Suppl. 5 (1957) 165.
57H Heister, W.: Tech. Mitt. Krupp **15** (1957) 130.
57HOT Harrison, F. W., W. P. Osmand, and R. W. Teale: Phys. Rev. **106** (1957) 865.
57K Krupička, S.: Czech. J. Phys. **7** (1957) 344.
57K (2) Krupička, S.: Czech. J. Phys. **7** (1957) 769.

57KT	KEDESDY, H., and A. TAUBER: J. Metals **9** (1957) 1140.
57KV	KRUPIČKA, S., and F. VILÍM: Czech. J. Phys. **7** (1957) 723.
57L	LJUNG, P. E.: Proc. Inst. Elec. Engrs. (London) **B 104**, Suppl. 5 (1957) 154.
57LM	LENSEN, M., and A. MICHEL: Congr. Intern. Chim. Pure Appl. 16e Paris (1959) in Struct. Rept. **23** (1959) 381.
57M	MCCLURE, D. S.: Phys. Chem. Solids **3** (1957) 311.
57MWH	MILES, P. A., W. B. WESTPHAL, and A. VON HIPPEL: Rev. Mod. Phys. **29** (1957) 279.
57NPHK	NATHANS, R., S. J. PICKART, S. E. HARRISON, and C. J. KRIESSMAN: Proc. Inst. Elec. Engrs. (London) **B 104**, Suppl. 5 (1957) 217.
57P	PEARSON, R. F.: Proc. Phys. Soc. (London) **B 70** (1957) 441.
57S	SHENKER, H.: Phys. Rev. **107** (1957) 1246.
57SSB	SINHA, A. P. B., H. R. SAJANA, and A. B. BISWAS: Acta Cryst. **10** (1957) 439.
57St (1)	ŠTERNBERK, J.: Czech. J. Phys. **4** (1957) 372.
57St (2)	ŠTERNBERK, J.: Czech. J. Phys. **7** (1957) 339.
57W	WALKER, L. R.: Phys. Rev. **105** (1957) 390.
57We	WEISZ, R. S.: Am. Ceram. Soc. Bull. **40** (1957) 139.
57YT	YOSIDA, K., and M. TACHIKI: Progr. Theoret. Phys. (Kyoto) **17** (1957) 331.
58B (1)	BELOV, K. P.: Usp. Fiz. Nauk. **63** (1958) 207.
58B (2)	BLANK, J. M.: German Patent No. 1 158 482, 6. Oct., 1958.
58BCCG	BATES, L. F., H. CLOW, D. J. CRAIK, and P. M. GRIFFITHS: Proc. Phys. Soc. (London) **72** (1958) 224.
58BCG	BATES, L. F., D. J. CRAIK, and P. M. GRIFFITHS: Proc. Phys. Soc. (London) **71** (1958) 789.
58BPT	BELOV, K. P., A. A. POPOVA, and J. E. V. TALALAEVA: Kristallografiya **3** (1958) 733.
58Br	BROWNLOW, J. M.: J. Appl. Phys. **29** (1958) 373.
58BW	BOZORTH, R. M., and D. WALSH: Private communication.
58DF	DOUTHETT, D., and S. A. FRIEDBERG: Carnegie Inst. Technol., ONR Contract NONR-760(05) Technical Report No. 3, June 10, 1958.
58En (1)	ENZ, U.: Physica **24** (1958) 68.
58 En (2)	ENZ, U.: Physica **24** (1958) 609.
58Es	ESCHENFELDER, A. H.: J. Appl. Phys. **29** (1958) 378.
58G	GIBBONS, D. F.: J. Appl. Phys. **28** (1957) 810.
58HBK	HARRISON, S. E., H. S. BELSON, and C. J. KRIESSMAN: J. Appl. Phys. **29** (1958) 337.
58HK	VAN HOOK, H. J., and M. L. KEITH: Am. Mineralogist **43** (1958) 80.
58HKP	HARRISON, S. E., C. J. KRIESSMAN, and S. R. POLLAK: Phys. Rev. **110** (1958) 844.
58ISB	IRANI, K. S., A. P. B. SINHA, and A. B. BISWAS: Proc. Phys. Soc. (London) **71** (1958) 270.
58JL	JACOBS, I. S., and P. E. LAWRENCE: Rev. Sci. Instr. **29** (1958) 713.
58K (1)	KIENLIN, A. VON: Z. Angew. Phys. **10** (1958) 562.
58K (2)	KRUPIČKA, S.: Czechoslov. J. Phys. **7** (1958) 769.
58KBF	KORNETZKI, M., J. BRACKMANN, and J. FREY: Siemens-Z. **32** (1958) 412.
58S	SHAFER, M. W.: I. B. M. J. Res. Develop. **2** (1958) 193.
58Su	SUHL, H.: J. Appl. Phys. **29** (1958) 416.
58W	WALKER, L. R.: J. Appl. Phys. **29** (1958) 318.
58WC	WICKHAM, D. G., and W. J. CROFT: Phys. Chem. Solids **7** (1958) 351.
58Z	ZITKA, B. H.: Czech. J. Phys. **8** (1958) 375.
59AD	ANDERSON, J. C., and B. DONOVAN: Proc. Phys. Soc. (London) **73** (1959) 593.
59B (1)	BERGSTEIN, A.: Collection Czech. Chem. Commun. **29** (1959) 3362.
59B (2)	BERGSTEIN, A.: Collection Czech. Chem. Commun. **24** (1959) 885.
59BKZ	BROŽ, I., S. KRUPIČKA, and K. ZÁVĚTA: Czech. J. Phys. **9** (1959) 481.
59C	CHECHERNIKOV, V. I.: Vestn. Mosk. Univ. Ser. Mat. Mekhan. Astron. Fiz. Khim. **3** (1959) 93.
59CG	CRAIK, D. J., and P. M. GRIFFITHS: Proc. Phys. Soc. (London) **73** (1959) 1.
59DE	DILLON, J. F., Jr., and H. E. EARL: J. Appl. Phys. **30** (1959) 202.
59E	EDWARDS, P. L.: Phys. Rev. **116** (1959) 294.
59F (1)	FOMENKO, L. A.: Izv. Akad. Nauk SSSR Ser. Fiz. **23** (1959) 329.
59F (2)	FOMENKO, L. A.: Izv. Akad. Nauk. SSSR Ser. Fiz. **23** (1959) 336.
59FMU (1)	FUNATOGAWA, Z., H. MIYATA, and S. USAMI: J. Phys. Soc. Japan **14** (1959) 854.
59FMU (2)	FUNATOGAWA, Z., H. MIYATA, and S. USAMI: J. Phys. Soc. Japan **14** (1959) 1583.
59Go	GOODENOUGH, J. B.: J. Phys. Radium **20** (1959) 153.
59J	JACOBS, I. S.: Phys. Chem. Solids **11** (1959) 1.
59Kp	KAPLAN, T. A.: Quarterly Progress Report, Solid State Research. Lincoln Laboratory, (Oct. 15, 1959) 91.
59Ks	KASPER, J. S.: Bull. Am. Phys. Soc. **4** (1959) 178.
59KZ	KRUPIČKA, S., and K. ZÁVĚTA: Czech. J. Phys. **9** (1959) 324.

59L (1) Louwerse, M.: Unpublished data quoted in [*59SW*], 256.
59L (2) Lensen, M.: Ann. Chim. (Paris) **4** (1959) 891.
59M Miller, A.: Phys. Rev. **116** (1959) 1481.
59MD (1) Menyuk, N., and K. Dwight Jr.: Quarterly Progress Report, Solid State Research, Lincoln Lab., (Apr. 15, 1959) 56.
59MD (2) Menyuk, N., and K. Dwight Jr.: Quarterly Progress Report, Solid State Research, Lincoln Lab. (July 31, 1959) 91.
59MD (3) Menyuk, N., and K. Dwight Jr.: Quarterly Progress Report, Solid State Research, Lincoln Lab. (Oct. 15, 1959) 88.
59P (1) Pearson, R. F.: Proc. Phys. Soc. (London) **74** (1959) 505.
59P (2) Picherit, F.: Compt. Rend. **249** (1959) 69.
59PN Pickart, S. J., and R. Nathans: Phys. Rev. **116** (1959) 317.
59PS Penoyer, R. F., and M. W. Shafer: J. Appl. Phys. **30** (1959) 315S.
59RN Rabkin, L. I., and Z. I. Novikova: Izv. Akad. Nauk SSSR Ser. Fiz. **23** (1959) 388.
59SF Smol'kov, N. A., and E. I. Fomenko: Izv. Akad. Nauk SSSR Ser. Fiz. **23** (1959) 377.
59St Šternberk, J.: Czech. J. Phys. **9** (1959) 408.
59SW Smit, J., and H. P. J. Wijn: Ferrites, John Wiley, New York/Eindhoven (1959).
59Z Zalesskii, A. V.: Kristallografiya **4** (1959) 867. – Soviet Phys. Cryst. (English Transl.) **4** (1960) 822.
60AL Alley, R. E., and V. E. Legg: J. Appl. Phys. Suppl. **31** (1960) 239S.
60AP Alperin, M. A., and S. J. Pickart: Bull. Am. Phys. Soc. Ser. II, **5** (1960) 458.
60B (1) Belov, K. P. et al.: Ferrity. Doklady 3. Vsesojuznogo soveščanija, Minsk 1960 (Ferrites. Proceedings of the 3rd Conference) Izv. Akad. Nauk Belorusk. SSR, **10** (1960) 83.
60B (2) Bergstein, A.: Collection Czech. Chem. Commun. **25** (1960) 2274.
60B (3) Buffler, C. R.: J. Appl. Phys. Suppl. **31** (1960) 222S.
60BBP Belov, K. P., V. F. Belov, and A. A. Popova: Zh. Eksperim. i Teor. Fiz. **38** (1960) 1908.
60C Crapo, W. A.: J. Appl. Phys. Suppl. **31** (1960) 267S.
60Ca (1) Callen, E. R.: Bull. Am. Phys. Soc. **II 5** (1960) 458.
60Ca (2) Callen, E. R.: Private Communication.
60D Dillon, J. F., Jr.: J. Appl. Phys. **31** (1960) 1605.
60DM Dwight, K., and H. Menyuk: Phys. Rev. **119** (1960) 1470.
60E Eschenfelder, A. H.: Private Communication.
60J Jacobs, I. S.: J. Appl. Phys. **31** (1960) 263S.
60K (1) Kaplan, T. A.: J. Appl. Phys. Suppl. **31** (1960) 364S.
60K (2) Krupička, S.: Czech. J. Phys. **B 10** (1960) 782.
60K (3) Kaplan, T. A.: Phys. Rev. **119** (1960) 1460.
60LP Lescroel, V., and A. Pierrot: Căbles Transmission **14** (1960) 220.
60M Miller, A.: J. Appl. Phys. Suppl. **31** (1960) 261S.
60P (1) Pearson, R. F.: Private Communication.
60P (2) Pearson, R. F.: J. Appl. Phys. Suppl. **31** (1960) 160S.
60PT Pearson, R. F., and R. W. Teale: Proc. Phys. Soc. (London) **75** (1960) 314.
60RK Reznitskii, L. A., and K. G. Khomyakov: Vestn. Mosk. Univ., Ser. II, Khim. **6** (1960) 24.
60VRB Verhaeghe, J. C., G. C. Robbrecht, and W. N. Bruynooghe: Appl. Sci. Res., Sect. B, **8** (1960) 128.
60Y Yamaguchi, T.: Sci. Papers Inst. Phys. Chem. Res. (Tokyo) **54** (1960) 394.
60Z Zaveta, K.: Fiz. Tverd. Tela **2** (1960) 106.
61A Akashi, T.: Trans. Jap. Inst. Metals **2** (1961) 171.
61AO Akashi, T., and T. Okada: J. Phys. Soc. Japan **16** (1961) 1641.
61B Blank, J. M.: J. Appl. Phys. **32** (1961) 378S.
61BMM Braginski, A., A. Marais, and Th. Merceron: Compt. Rend. **253** (1961) 239.
61BS Belov, K. P., and E. P. Svirina: Fiz. Tverd. Tela **3** (1961) 2495; Soviet Phys. Solid State (English Transl.) **3** (1962) 1814.
61DF Douthett, D., and S. A. Friedberg: Phys. Rev. **121** (1961) 1662.
61GGT Gurevich, A. G., I. E. Gubler, and A. G. Titova: Fiz. Tverd. Tela **3** (1961) 19. – Soviet Phys. Solid State (English Transl.) **3** (1961) 13.
61HC Hastings, J. H., and L. M. Corliss: Phys. Rev. **104** (1961) 328.
61HW Heck, C., and J. Weber: Ceram. Ind. **77** (1961) 74, 94.
61K Knowles, J. E.: Proc. Phys. Soc. (London) **78** (1961) 233.
61KV Krupička, S., and F. Vilím: Czech. J. Phys. **B 11** (1961) 10.
61M Moruzzi, V. L.: Private Communication.
61M (1) Miyata, N.: J. Phys. Soc. Japan **16** (1961) 206.
61 M(2) Miyata, N.: J. Phys. Soc. Japan **16** (1961) 1291.

61M (3)	MORUZZI, V. L.: J. Appl. Phys. Suppl. **32** (1961) 59 S.
61O	O'KEEFFE, M.: J. Phys. Chem. Solids **21** (1961) 172.
61S	SHICHIJO, Y.: Trans. Jap. Inst. Metals **2** (1961) 204.
61Z	ZÁVĚTA, K.: Czech. J. Phys. **11** (1961) 376.
62BBBB	BOYD, E. L., J. I. BUDNICK, L. J. BRUNER, and R. J. BLUME: J. Appl. Phys. **33** (1962) 2484.
62BMM	BRAGINSKI, A., A. MARAIS, and TH. MERCERON: J. Phys. Soc. Japan **17**, Suppl. B-I (1962) 300.
62C	CERVINKA, L.: Czech. J. Phys. **12** (1962) 490.
62K (1)	KIMURA, Y.: Mem. Defense Acad. Math. Phys. Chem. Eng. (Yokosuka, Japan) **2** (1962) 203.
62K (2)	KRUPIČKA, S.: J. Phys. Soc. Japan **17**, Suppl. B-I (1962) 304.
62MF	MIYATA, N., and Z. FUNATOGAWA: J. Phys. Soc. Japan **17**, Suppl. B-I (1962) 279.
62OYM	OKADA, T., T. YAMADAYA, and S. MIYAHARA: J. Phys. Soc. Japan **17** (1962) 1799.
62P (1)	PALMER, W.: J. Appl. Phys. Suppl. **33** (1962) 1201.
62P (2)	PAULUS, M.: Phys. Status Solidi **2** (1962) 1181.
62P (3)	PLUMIER, M. P.: Compt. Rend. **255** (1962) 2244.
62P (4)	PAULUS, M.: Compt. Rend. **255** (1962) 871.
62P (5)	PAULUS, M.: Phys. Status Solidi **2** (1962) 1325.
62R	RÖSS, E.: Z. Angew. Phys. **14** (1962) 495.
62S	ŠTERNBERK, J.: Czech. J. Phys. **B 12** (1962) 83.
62SLS	SMIT, J., F. K. LOTGERING, and R. P. VAN STAPELE: J. Phys. Soc. Japan **17**, Suppl. B-I (1962) 268.
62T	TEALE, R. W.: J. Appl. Phys. Suppl. **33** (1962) 1295.
62Y	YANASE, A.: J. Phys. Soc. Japan **17** (1962) 1005.
62YBN	YAMZIN, I. I., N. V. BELOV, and YU. Z. NOZIK: J. Phys. Soc. Japan **17**, Suppl. B-III (1962) 55.
62ZSM	ZÁVĚTA, K., E. SVIRINA, and O. MALIKOVA: Fiz. Tverd. Tela **4** (1962) 3593. — Soviet Phys. Solid State (English Transl.) **4** (1963) 2628.
63B	BERGSTEIN, A.: Czech. J. Phys. **B 13** (1963) 613.
63BB	BUTLER, S. R., and W. R. BUESSEM: J. Appl. Phys. **34** (1963) 1754.
63DD	DROKIN, A. I., and V. D. DYLGERON: Fiz. Metal. Metalloved. **15** (1963) 128. — Phys. Metals Metallog. USSR (English Transl.) **15** (1963) 116.
63G	GROH, G.: Z. Angew. Phys. **16** (1963) 172.
63H	HEISTER, W.: Tech. Mitt. Krupp **21** (1963) 15.
63HGB	HEEGER, A. J., S. K. GHOSH, and T. G. BLOCKER, III: J. Appl. Phys. **34** (1963) 1034.
63HM	HANKE, I., and W. METZDORF: Z. Angew. Phys. **15** (1963) 191.
63K	KRUPIČKA, S.: Phys. Status Solidi **3** (1963) K118.
63KS	KANOKAVITCH, Y. Y., and Y. G. SAKSONOV: Kristallografiya **8** (1963) 25.
63L	LOW, B.: J. Appl. Phys. **34** (1963) 1250.
63MF	MCGUIRE, T. R., and F. S. FEREBEE: J. Appl. Phys. **34** (1963) 1821.
63MM (1)	MARAIS, A., and TH. MERCERON: Compt. Rend. **256** (1963) 2560.
63MM (2)	MARAIS, A., and TH. MERCERON: Compt. Rend. **257** (1963) 1760.
63MP	MCGUIRE, T. R., and S. J. PICKART: J. Phys. Chem. Solids **24** (1963) 1531.
63MV	MICHALOWSKY, L., and G. VOGLER: Phys. Status Solidi **3** (1963) K313.
63O	OHTA, K.: J. Phys. Soc. Japan **18** (1963) 685.
63P (1)	PAULUS, M.: J. Rech. Centre Natl. Rech. Sci. Lab. Bellevue (Paris) No. **63** (1963) 141.
63P (2)	PAULUS, M.: Compt. Rend. **256** (1963) 638.
63TC	TEALE, R. W., and B. H. CLARKE: J. Appl. Phys. **34** (1963) 1248.
64APPZL	AMIROVA, S. A., V. V. PECHKOVSKII, V. G. PROKHOROVA, T. V. ZHEVELEVA, and A. A. LEZHNEVA: Zh. Fiz. Khim. **38** (1964) 108.
64BDZMST	BELOV, V. F., N. M. DEVISHEVA, I. S. ZHELUDEV, E. F. MAKAROV, R. A. STUKAN, and V. A. TRUKHTANOV: Fiz. Tverd. Tela **6** (1964) 3435. — Soviet Phys. Solid State (English Transl.) **6** (1965) 2747.
64BY	BOL'SHOVA, K. M., and T. A. YELKINA: Fiz. Metal. Metalloved. **17** (1964) 819. — Phys. Metals Metallog. USSR (English Transl.) **17** (1964) 21.
64CT	CLARKE, B. H., and R. W. TEALE: J. Appl. Phys. **35** (1964) 892.
64E	ELBINGER, G.: Phys. Status Solidi **5** (1964) 87.
64GT	GORDEEV, I. V., and YU. D. TRET'YAKOV: Zh. Neorgan. Khim. **9** (1964) 164.
64GTK	GORDEEV, I. V., YU. D. TRET'YAKOV, and K. G. KHOMYAKOV: Zh. Neorgan. Khim. **9** (1964) 164.
64GV	GERBER, R., and M. VICHR: Czech. J. Phys. **14** (1964) 888.
64HBG	HEEGER, A. J., T. G. BLOCKER III, and S. K. GHOSH: J. Appl. Phys. **35** (1964) 840.

64HH	Heeger, A. J., and T. Houston: J. Appl. Phys. **35** (1964) 836.
64K (1)	Krupička, S.: Czech. J. Phys. **14** (1964) 29.
64K (2)	Kooy, C.: Science of Ceramics **1** (1964) 21. – Published by the Academic Presse London and New York for the British Ceramic Society.
64K (3)	Kulikowski, J.: Bull. Acad. Polon. Sci. Sér. Sci. Tech. **12** (1964) 81.
64L	Lotgering, F. K.: J. Phys. Chem. Solids **25** (1964) 95.
64LAM	Labusca, E., N. Andreescu, and C. Motoc: Rev. Phys. Acad. Rep. Populaire Roumaine **9** (1964) 419.
64LR	Lee, E. W., and J. A. Robey: Proc. Intern. Magnetism Conf. Nottingham 1964, 642.
64M	Merceron, Th.: Thèse Faculté des Sciences de Paris (1964).
64MJ	Macklen, E. D., and P. Johns: Meeting of Basic Science Brighton 1964.
64OK	Ohta, K., and N. Kobayashi: Japan. J. Appl. Phys. **3** (1964) 576.
64P	Proske, J. W.: Thesis Pennsylvania State University (1964).
64RHM	Röss, E., I. Hanke, and E. Moser: Z. Angew. Phys. **17** (1964) 504.
64RNB	Rosenberg, M., P. Nicolau, and I. Bunget: Phys. Status Solidi **4** (1964) K125.
64S (1)	Sakurai, J.: J. Phys. Soc. Japan **19** (1964) 311.
64S (2)	Schieber, M.: J. Appl. Phys. **35** (1964) 1072.
64S (3)	Solov'ev, V. I.: Fiz. Tverd. Tela **6** (1964) 2524. – Soviet Phys. Solid State (English Transl.) **6** (1965) 2005.
64S (4)	Stuijts, A. L.: Proc. Brit. Ceram. Soc. **12** (1964) 73.
64S (5)	Svirina, E. P.: Fiz. Tverd. Tela **6** (1964) 3378. – Soviet Phys. Solid State (English Transl.) **6** (1965) 2702.
64SAT	Shichijo, Y., G. Asano, and E. Takama: J. Appl. Phys. **35** (1964) 1646.
64SS	Saksonov, Yu. G., and V. A. Somenkov: Fiz. Metal. Metalloved. **18** (1964) 853. – Phys. Metals Metallog. USSR (English Transl.) **18** (1964) 52.
64T	Tweedale, K.: Unpublished results, Mullard Res. Lab. 1964.
64TNT	Tsuboi, T., T. Nakajima, and H. Takaki: J. Phys. Soc. Japan **19** (1964) 768.
64TZ	Tsarev, V. A., and N. V. Zavaritskiĭ: Zh. Eksperim. i Teor. Fiz. **47** (1964) 2102. – Soviet Phys. JETP (English Transl.) **20** (1965) 1412.
64VK	Vorob'ev, A. S., and V. V. Kolchin: Izv. Vysshikh Uchebn. Zavedenii, Fiz. No. 4, **1964** 180.
64Y	Yamaguchi, T.: J. Am. Ceram. Soc. **47** (1964) 131.
65AMY	Andriyeuski, A. I., G. F. Mocharnyuk, and Yu. G. Yuskevich: Fiz. Metal. Metalloved. **20** (1965) 216. – Phys. Metals Metallog. USSR (English Transl.) **20** (1965) 47.
65B (1)	Blasse, G.: Philips Res. Rept. **20** (1965) 528.
65B (2)	Braginski, A.: Phys. Status Solidi **11** (1965) 603.
65BSK	Bergstein, A., J. Sestak, and E. Kitzinger: Am. Mineralogist **50** (1965) 762.
65BTK	Belov, K. P., E. V. Talalaeva, and T. V. Kudryavtseva: Fiz. Tverd. Tela **7** (1965) 981. – Soviet Phys. Solid State (English Transl.) **7** (1965) 789.
65C (1)	Cervinka, L.: Czech. J. Phys. **15** (1965) 747.
65C (2)	Cervinka, L.: Czech. J. Phys. **15** (1965) 425.
65C (3)	Cervinka, L.: J. Phys. Chem. Solids **26** (1965) 1917.
65E	Elbinger, G.: Phys. Status Solidi **9** (1965) 843.
65EKKOH	Endo, S., S. Kume, M. Koizumi, C. Okazaki, and E. Hirota: Phys. Letters **18** (1965) 232.
65IMS	Isaka, A., Y. Matsumoto, and M. Sugihara: J. Japan Inst. Metals (Sendai) **29** (1965) 1066.
65K (1)	Kratochvilova, E.: Czech. J. Phys. **15** (1965) 718.
65K (2)	Kratochvilova, E.: Czech. J. Phys. **15** (1965) 912.
65KOST	Komarov, V. F., N. N. Oleinikov, Yu. G. Saksonov, and Yu. O. Tret'yakov: Izv. Akad. Nauk SSSR, Neorgan. Materialy **1** (1965) 395.
65KWKP	Kleinstück, K., E. Wieser, P. Kleinert, and R. Perthel: Phys. Status Solidi **8** (1965) 271.
65L	Lotgering, F. K.: Philips Res. Rept. **20** (1965) 320.
65MJ	Macklen, E. D., and P. Johns: J. Appl. Phys. **36** (1965) 1022.
65MMP	Marais, A., Th. Merceron, and M. Porte: Compt. Rend. **261** (1965) 2188.
65OA	Okada, T., and T. Akashi: J. Phys. Soc. Japan **20** (1965) 639.
65PG	Paulus, M., and C. Guillaud: Bull. Soc. Chim. France **1965** 1175.
65R	Rasmussen, A. L.: J. Res. Natl. Bur. Std. **69A** (1965) 181.
65S	Šimša, Z.: Czech. J. Phys. **15** (1965) 435.
65SSD	Schäfer, D., H. Ch. Semmelhack, and G. Dietzmann: Phys. Status Solidi **10** (1965) K25.
65T (1)	Tret'yakov, Yu. D.: Izv. Akad. Nauk SSSR, Neorgan. Materialy **1** (1965) 405; Inorg. Mater. (English Transl.) **1** (1965) 374.

65T (2)	TRET'YAKOV, YU. O.: Izv. Akad. Nauk SSSR, Neorgan. Materialy **1** (1965) 240; Inorg. Mater. **1** (1965) 221.
65TBC	TAKENOV, T. D., V. F. BALAKIREV, and G. I. CHUFAROV: Izv. Akad. Nauk SSSR, Neorgan. Materialy **1** (1965) 587 Inorg. Mater. (English Transl.) **1** (1965) 536.
65TM	TELESNIN, R. V., and K. T. MAKAROV: Fiz. Metal. Metalloved. **20** (1965) 349; Phys. Metals Metallog. USSR (English Transl.) **20** (1965) 349.
65TO	TRET'YAKOV, YU. B., and N. N. OLEINIKOV: Izv. Akad. Nauk SSSR Neorgan. Materialy **1** (1965) 254.
65TSG	TRET'YAKOV, YU. O., YU. G. SAKSONOV, and I. V. GORDEEV: Izv. Akad. Nauk SSSR, Neorgan. Materialy **1** (1965) 413; Inorg. Mater. (English Transl.) **1** (1965) 382.
65TSGZG	TRET'YAKOV, YU. D., YU. G. SAKSONOV, I. V. GORDEEV, A. M. ZAJONCHKOSKIJ, and A. M. GORDINA: Izv. Akad. Nauk SSSR, Neorgan. Materialy **1** (1965) 408; Inorg. Mater. (English Transl.) **1** (1965) 378.
65V	VEITH, H.: Z. Angew. Phys. **19** (1965) 414.
65WK	WIESER, E., and K. KLEINSTÜCK: Z. Krist. **121** (1956) 14.
66A	AKASHI, T.: N. E. C. Research and Development **1966** 89.
66B	BRAGINSKI, A.: Arch. Elektrotec. (Poland) **15** (1966) 29.
66BA	BOL'SHOVA, K. M., and T. B. ANDREEVA: Fiz. Tverd. Tela **8** (1966) 2313; Soviet Phys. Solid State (English Transl.) **8** (1966) 1846.
66BE	BOL'SHOVA, K. M., and T. A. ELKINA: Fiz. Tverd. Tela **8** (1966) 2811.
66BS (1)	BRAGINSKI, A., and Z. SIERMINSKI: Phys. Status Solidi **17** (1966) K211.
66BS (2)	BEER, A., and J. SCHWARZ: IEEE Trans. Magnetics **2** (1966) 470.
66BZ	BULGAKOVA, T. I., and O. S. ZAITSEV: Zh. Fiz. Khim. **40** (1966) 1938; Russ. J. Phys. Chem. (English Transl.) **40** (1966) 1039.
66BZMSA	BOGOLOVSKI, V. N., M. G. ZHURAVLEVA, A. N. MEN', M. N. STAFEEVAN, and V. D. AVERBUKH: Zh. Fiz. Khim. **40** (1966) 3000.
66BZR	BULGAKOVA, T. I., O. S. ZAITSEV, and A. G. ROZANOV: Zh. Fiz. Khim. **40** (1966) 1098; Russ. J. Phys. Chem. (English Transl.) **40** (1966) 590.
66C	CERVINKA, L.: Dissertation — Institute of Solid State Physics. Czech. Acad. Sci. — Prague 1966.
66CDFI	CRUCEANU, E., B. DAVID, A. FRIEDMAN, and V. IVAȘCU: Phys. Status Solidi **14** (1966) K111.
66DS (1)	DROKIN, A. I., and L. I. SLOBODSKOI: Zh. Eksperim. i Teor. Fiz. **51** (1966) 1023; Soviet Phys. JETP (English Transl.) **24** (1967) 682.
66DS (2)	DROKIN, A. I., and V. I. SINEGUBOV: Soviet Phys. JETP (English Transl.) **22** (1966) 493.
66F	FISCHER, M.: Z. Anorg. Allg. Chem. **345** (1966) 134.
66G	GORDEEV, I. V.: Izv. Akad. Nauk SSSR, Neorgan. Materialy **2** (1966) 281; Inorg. Mater. (English Transl.) **2** (1966) 241.
66GMMS	GERBER, R., L. MICHALOWSKY, K. MOTZKE, and E. STEINBEISS: Phys. Status Solidi **16** (1966) 795.
66GSV	GERBER, R., Z. ŠIMŠA, and M. VICHR: Czech. J. Phys. **16** (1966) 913.
66H	HIROTA, E.: Japan. J. Appl. Phys. **5** (1966) 1125.
66IMS	ISAKA, A., Y. MATSUMOTO, and M. SUGIHARA: J. Japan Inst. Metals (Sendai) **30** (1966) 990.
66J	JONES, R. V.: IEEE Trans. Sonics Ultrasonics **13** (1966) 86.
66KYH	KUBA, T., H. YASUOKA, and A. HIRAI: J. Phys. Soc. Japan **21** (1966) 812.
66MY	MUKHIN, V. P., and E. K. YUR'EVA: Fiz. Tverd. Tela **8** (1966) 2011; Soviet Phys. Solid State (English Transl.) **8** (1967) 1599.
66P	PILLOT, J.: Thèse Doct.-Ing. Strasbourg (1966).
66PM	PAULUS, M., and D. MERCIER: Conf. Intern. sur la Métallurgie des Poudres à Stary Smokovec (Haut Tatra) (Tchécoslovaquie) (1966).
66PRM	PETRESCU, V., N. REZLESCU, and G. MAXIM: Rev. Phys. Acad. Rep. Populaire Roumaine **11** (1966) 697.
66R	RÖSS, D.: Z. Angew. Phys. **21** (1966) 391.
66RL	ROSCHUPKIN, V. I., and V. I. LAVRENT'EV: Izv. Akad. Nauk SSSR, Neorgan. Materialy **2** (1966) 712; Inorg. Mater. (English Transl.) **2** (1966) 612.
66RNB (1)	ROSENBERG, M., P. NICOLAU, and I. BUNGET: Phys. Status Solidi **14** (1966) K65.
66RNB (2)	ROSENBERG, M., P. NICOLAU, and I. BUNGET: Phys. Status Solidi **15** (1966) 521.
66S (1)	ŠIMŠA, Z.: Czech. J. Phys. **16** (1966) 919.
66S (2)	STEINBEISS, E.: Phys. Status Solidi **18** (1966) 241.
66S (3)	SUEMUNE, Y.: Japan J. Appl. Phys. **5** (1966) 455.
66S (4)	SUGIMOTO, M.: Sci. Papers Inst. Phys. Chem. Res. (Tokyo) **60** (1966) 47.
66SB	SLICK, P. I., and H. BASSECHES: IEEE Trans. Magnetics **2** (1966) 603.

66SMM	SVIRINA, E. P., O. A. MALIKOVA, and M. A. MALIKOVA: Fiz. Tverd. Tela **8** (1966) 1599; Soviet Phys. Solid State (English Transl.) **8** (1966) 1269.
66TCB	TAKENOV, T. D., G. I. CHUFAROV, and V. F. BALAKIREV: Ivz. Akad. Nauk SSSR, Neorgan. Materialy **2** (1966) 161; Inorg. Mater. (English Transl.) **2** (1966) 136.
66WMK	WIESER, E., W. MEISEL, and K. KLEINSTÜCK: Phys. Status Solidi **16** (1966) 127.
66ZB	ZAITSEV, O. S., and T. I. BULKAKOVA: Zh. Fiz. Khim. **40** (1966) 2623.
66ZO	ZAZLIEV, S., and A. ORAZSAKHATOV: Izv. Akad. Nauk Turkm. SSR, Ser. Fiz. Tekhn. Khim. i Geol. Nauk No. 2, **1966**, 27.
66ZS	ZITKA, B., and J. ŠTERNBERK: Czech. J. Phys. **16** (1966) 536.
66ZTZ	ZAVETA, K., E. I. TRINKLER, and F. ZOUNOVA: Phys. Status Solidi **14** (1966) K9.
66Y	YASUOKA, H.: J. Phys. Soc. Japan **21** (1966) 393.
67 B(1)	BEKKER, YA. M.: Izv. Akad. Nauk SSSR, Neorgan. Materialy **3** (1967) 196; Inorg. Mater. (English Transl.) **3** (1967) 168.
67B (2)	BROESE VAN GROENOU, A.: J. Appl. Phys. **38** (1967) 3317.
67BG	BABA, P. D., and E. M. GYORGY: J. Appl. Phys. **38** (1967) 909.
67BP	BROESE VAN GROENOU, A., and R. F. PEARSON: J. Phys. Chem. Solids **28** (1967) 1027.
67BZMAC	BOGOSLOVSKII, V. N., M. G. ZHURAVLEVA, A. N. MEN', V. D. AVERBUKH, and G. I. CHUFAROV: Inorg. Mater. (English Transl.) **2** (1967) 902.
67CP	COSIER, J. P., and R. F. PEARSON: Brit. J. Appl. Phys. **18** (1967) 615.
67HT	HARPER, H., and R. W. TEALE: Solid State Commun. **5** (1967) 327.
67K	KÖNIG, U.: Z. Angew. Phys. **22** (1967) 103.
67KTAH	KUBO, T., T. TSUDA, H. ABE, and A. HIRAI: J. Phys. Soc. Japan **22** (1967) 679.
67AP	Austrian Patent No. 252 592 (1967).
67LBSC	LEONT'EVA, I. A., V. F. BALAKIREV, A. D. SOKOLOV, and G. I. CHUFAROV: Izv. Akad. Nauk SSSR, Neorgan. Materialy **3** (1967); Inorg. Mater. (English Transl.) **3** (1967) 1142.
67LPJ	LEBEDEVA, E. V., A. I. PIL'SHCHIKOV, and E. K. JUR'EVA: Fiz. Tverd. Tela **9** (1967) 317; Soviet Phys. Solid State (English Transl.) **9** (1967) 237.
67MI	MIZUSHIMA, K., and S. IIDA: J. Phys. Soc. Japan **22** (1967) 1300.
67MM	MARAIS, A., and TH. MERCERON: Phys. Status Solidi **24** (1967) 635.
67O	OGAWA, S.: Japan. J. Appl. Phys. **6** (1967) 1427.
67ON	OGAWA, S., and Y. NAKAGAWA: J. Phys. Soc. Japan **23** (1967) 179.
67RBZ	ROZANOV, A. G., T. I. BULGAKOVA, and O. S. ZAITSEV: Izv. Akad. Nauk SSSR, Neorgan. Materialy **3** (1967) 192; Inorg. Mater. (English Transl.) **3** (1967) 165.
67RL	ROSCHUPKIN, V. I., and V. I. LAVRENT'EV: Izv. Akad. Nauk SSSR, Neorgan. Materialy **3** (1967) 551; Inorg. Mater. (English Transl.) **3** (1967) 490.
67S	ŠIMŠA, Z.: J. Phys. Chem. Solids **28** (1967) 2435.
67SKK	SAWAOKA, A., N. KAWAI, and S. KIKUCHI: Phys. Status Solidi **24** (1967) K83.
67SM	SCHWERDTFEGER, K., and A. MUAN: Trans. Met. Soc. AIME, **239** (1967) 1114.
67SR (1)	SMOLIN, R. P., and L. I. RYABINKINA: Zh. Tekhn. Fiz. **37** (1967) 572; Soviet Phys. Tech. Phys. (English Transl.) **12** (1967) 411.
67SR (2)	SMOLIN, R. P., and L. I. RYABINKINA: Zh. Tekhn. Fiz. **37** (1967) 575; Soviet Phys. Tech. Phys. (English Transl.) **12** (1967) 414.
67SS	SAKURAI, J., and T. SHINJO: J. Phys. Soc. Japan **23** (1967) 1426.
67SWM	SAWATZKY, G. A., F. VAN DER WOUDE, and A. H. MORRISH: Phys. Letters, A, **25** (1967) 147.
67TABC	TAKENOV, T. D., B. D. AVERBUKH, V. F. BALAKIREV, and G. I. CHUFAROV: Izv. Akad. Nauk SSSR, Neorgan. Materialy **3** (1967) 557; Inorg. Mater. (English Transl.) **3** (1967) 494.
67URBS	USATOV, V. U., V. P. RYABININ, K. K. BURMAKINA, and R. P. SMOLIN: Zh. Tekhn. Fiz. **37** (1967) 1701; Soviet Phys. — Techn. Phys. (English Transl.) **12** (1968) 1245.
67YHSKBT	YASUOKA, H., A. HIRAI, T. SHINJO, M. KIYAMA, Y. BANDO, and T. TAKADA: J. Phys. Soc. Japan **22** (1967) 174.
68CV	CERVINKA, L., and D. VETTERKIND: J. Phys. Chem. Solids **29** (1968) 171.
68HP	HAMELIN, A., and M. PAULUS: Compt. Rend. **266** (1968) 1422.
68HW	HABEREY, F., and H. P. J. WIJN: Phys. Status Solidi **26** (1968) 231.
68WPMK	WIESER, E., V. A. POVITSKII, E. F. MAKAROV, and K. KLEINSTÜCK: Phys. Status Solidi **25** (1968) 607.

6.1.4 Mg-Fe^{3+} spinels (Mg ferrites) and Mg-Fe^{3+} spinels with substitutions

6.1.4.1 Magnesium ferrite $MgFe_2O_4$

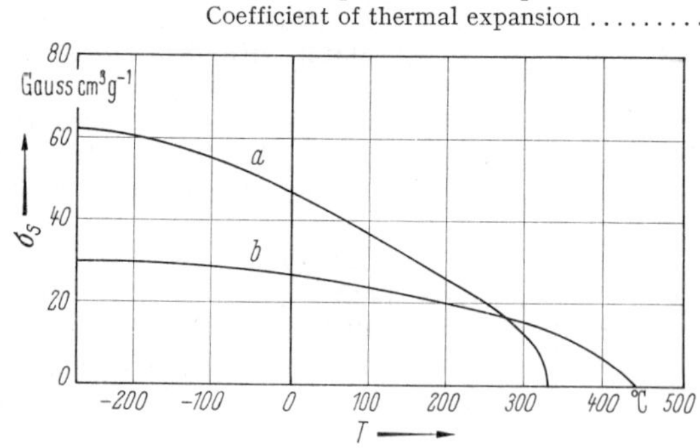

Fig. 1. $MgFe_2O_4$. Typical saturation magnetization, σ_s, vs. temperature, T, for two samples with different distributions of Mg ions.
Sample a [*56KH*] has 0.23 Mg ions on A sites (Θ_C = 331 °C, σ_s^0 = 62 Gauss cm^3g^{-1}). Sample b [*59SW*] has 0.11 Mg ions on A sites (Θ_C = 440 °C, σ_s^0 = 31 Gauss cm^3g^{-1}).

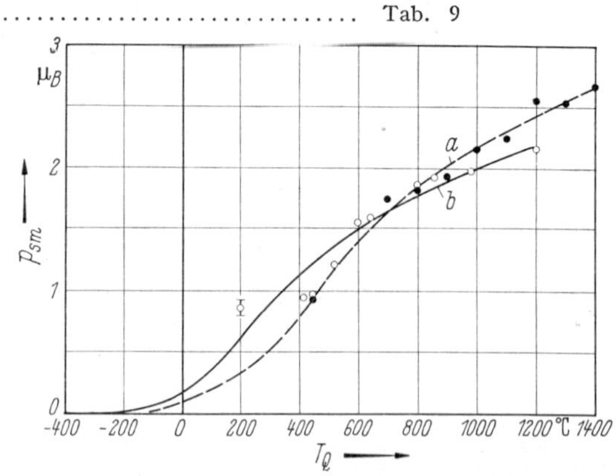

Fig. 2. $MgFe_2O_4$. Saturation magnetic moment, p_{sm}, vs. quenching temperature, T_Q. a) [*56KH*]. b) [*51PB*]. See also [*58EF*].

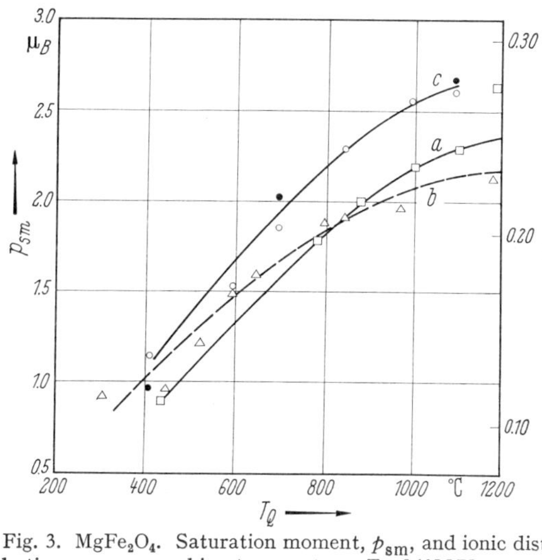

Fig. 3. $MgFe_2O_4$. Saturation moment, p_{sm}, and ionic distribution, η, vs. quenching temperature, T_Q [63MP].
a) [56KH], b) [51PB], and c) [63MP]. Originally, [51PB] represented the ionic distribution by

$$\eta(1+\eta)/(1-\eta)^2 = \exp(-\Theta/T)$$

η = fraction of Mg ions on A sites; Θ = 1200 °K; T in °K. Later measurements have shown the distribution to be better represented by

$$\Theta = \Theta_0 - \Theta_1 \cdot \eta$$

where Θ depends on the fraction of Mg ions on A sites. Θ_0, Θ_1 = constants (given in Tab. 1).

Tab. 1. $MgFe_2O_4$. Values of Θ_0 and Θ_1 in °K.

Θ_0	Θ_1	Ref.
2020	4350	*56KH*
1920	4200	*58EF*
1850	4480	*63MP*

Values of Θ_0 and Θ_1 derived from experimental distribution data of $MgFe_2O_4$. The physical significance of Θ_0 and Θ_1 may be seen from $(1/R)\,(\partial u_1/\partial\eta) = \Theta_0 - \Theta_1\eta$ where Θ_1 is the non-thermal contribution to the energy (Madelung constant, covalent constant, Born repulsive constant, etc.). Thus, Θ_0 is the non-thermal work required for the transfer of the first Mg ion to a tetrahedral site, and Θ_1 is the difference in work required for the transfer of the first and last Mg ions.

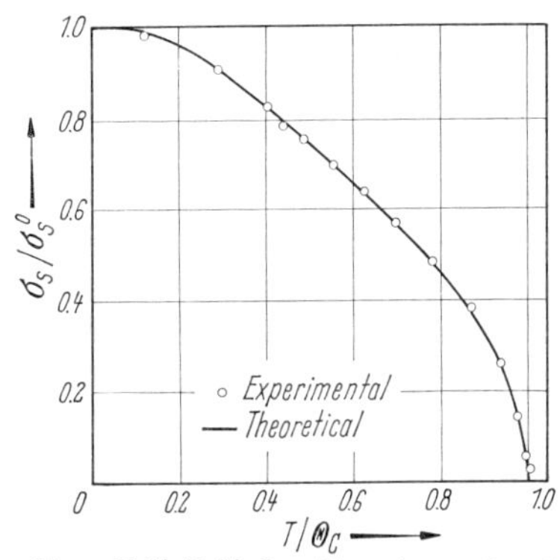

Fig. 4. $Mg_{0.71}Fe^{3+}_{2.15}Fe^{2+}_{0.06}O_4$. Experimental and theoretical data for σ_s/σ_s^0 vs. T/Θ_C [60RF].
The value of σ_s^0 is 68.3 Gauss cm³ g⁻¹ for this case. The Néel molecular field coefficients which give the best agreement between the Néel theory and the experimental data are presented below for two different ionic distributions in this single-crystal ferrite:

σ_s^0 Gauss cm³ g⁻¹	p^0_{sm} μ_B	Θ_C °K	n mole cm⁻³	α	β
68.3	2.51	692	276	−0.72	−0.30
51.1	1.87	701	256	−0.63	−0.27

n = molecular field coefficient referring to the strengths of the AB interaction; $n\alpha$, $n\beta$, referring to the AA, BB interactions.
p^0_{sm} = spontaneous magnetic moment at T = 0 °K.

Tab. 2. $MgFe_2O_4$. Curie temperatures for samples with different ionic distributions [57K].

η	Θ_C [°C]
0.27	325
0.23	331
0.19	339
0.18	355
0.16	372
0.09	410

η = fraction of Mg^{2+} ions on A sites.
Also, a linear relationship between Θ_C and η has been reported in the literature [58EF]:
$\Theta_C = (690 - 490\,\eta)$ °K.

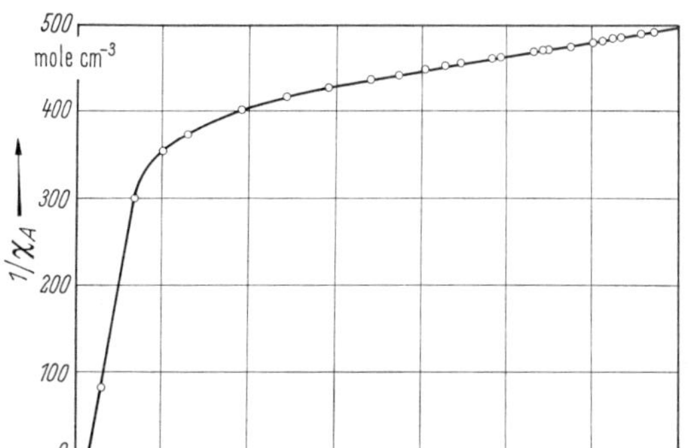

Fig. 5. $MgFe_2O_4$. Reciprocal magnetic susceptibility, $1/\chi_A$, vs. temperature, T [50FS].

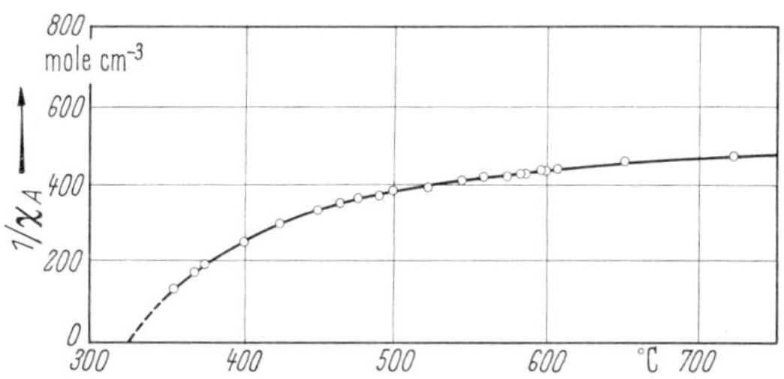

Fig. 6. $MgFe_2O_4$. Reciprocal susceptibility, $1/\chi_A$, vs. temperature, T [32S].
Sample had a Curie temperature of 339 °C.

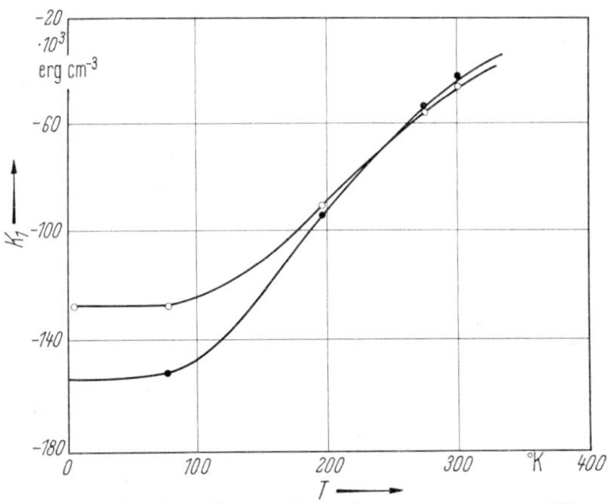

Fig. 7. $MgFe_2O_4$. First order anisotropy constant, K_1, vs. temperature, T, for a single crystal which contained 6 mole-% of Fe_3O_4 [*58FR*].
The data depend on the ionic distribution and results are shown for two different distributions.
——— theoretical,
● high M_s, ○ low M_s: experimental, referring to ionic distributions corresponding to relatively high and relatively low saturation magnetization.

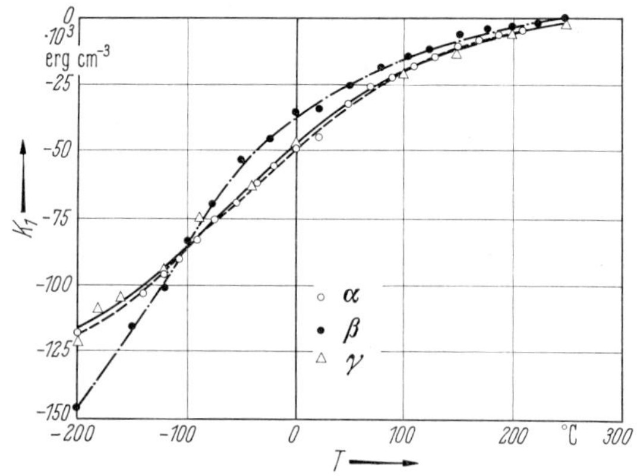

Fig. 8. $MgFe_2O_4$. First-order magnetocrystalline anisotropy constant, K_1, vs. temperature, T, in a single crystal [*59BK*]. α, β, and γ refer to heat treatments where the saturation magnetizations are 1.28, 2.23, and 1.41 μ_B, respectively. See Fig. 12.

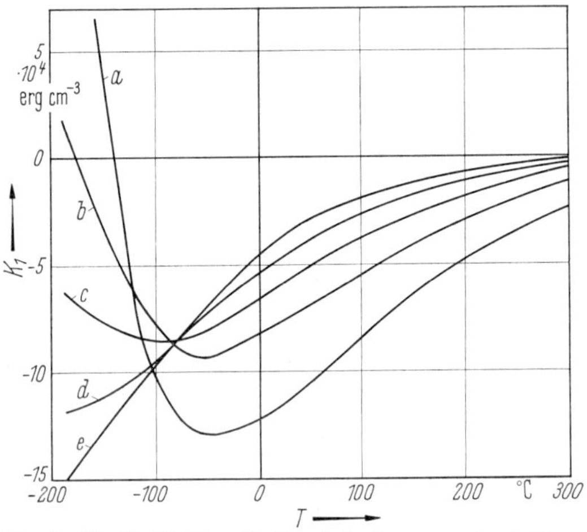

Fig. 9. $(MgO)_x(FeO)_{1-x}Fe_2O_3$. Anisotropy constant, K_1, vs. temperature, T, for various MgFe ferrites [*61E*].
Single crystals with MgO content, x: a) $x = 0$, b) $x = 0.25$, c) $x = 0.5$, d) $x = 0.75$, e) $x = 1$.

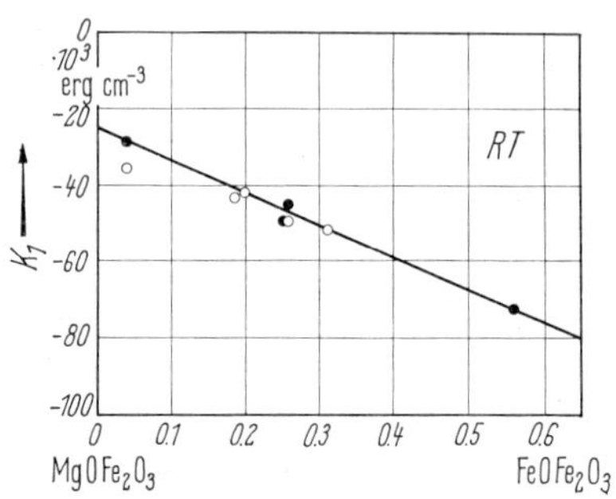

Fig. 10. $Mg_xFe_{3-x}O_4$. First-order magneto-crystalline anisotropy constant, K_1, vs. composition at RT [*57RFE*]. Open and closed circles refer to samples with different ionic distributions. ○ low M_s; ● high M_s (see Fig. 7).

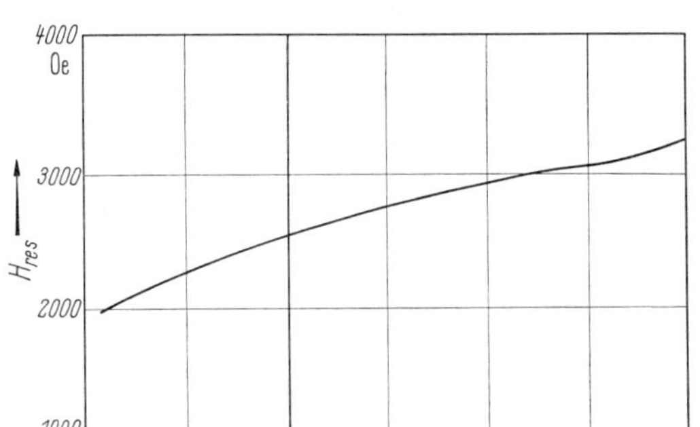

Fig. 11. $MgFe_2O_4$. Resonance field, H_{res}, vs. temperature, T, for a polycrystalline sample sintered for 3 hours at 1200 °C and quenched in air (frequency 9.310 Gc/s) [*51OKT*].

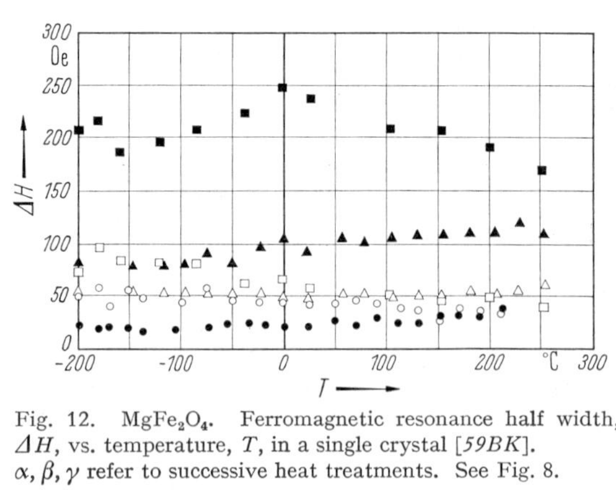

Fig. 12. $MgFe_2O_4$. Ferromagnetic resonance half width, ΔH, vs. temperature, T, in a single crystal [*59BK*].
α, β, γ refer to successive heat treatments. See Fig. 8.

○ Easy direction, ● Hard direction: α
△ Easy direction, ▲ Hard direction: β
□ Easy direction, ■ Hard direction: γ

Tab. 3. $MgFe_2O_4$. Dependence of resonance half width, ΔH, on temperature, T, for a polycrystalline sample sintered at 1200 °C for 3 h and quenched in air [51OKT]

T	−190	−130	−110	−82	−50	+16	+51	+79	+124	+164	+204	+242	+281	+320	+336	+390	+398	°C
ΔH	840	680	640	588	502	435	388	338	316	270	240	200	160	157	116	98	97	Oe

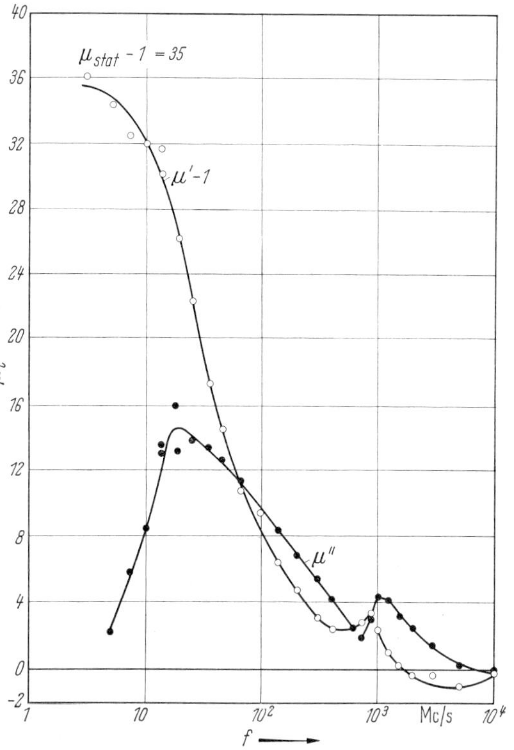

Fig. 13. $(MgO)_{0.97}(FeO)_{0.03}Fe_2O_3$. Initial permeability, μ_i, vs. frequency, f [57RFE].
Polycrystalline sample with $\varrho = 4.26\ gcm^{-3}$ and σ_s (26 °C) = 40.3 Gauss cm^3g^{-1}. See also 6.1.4.3, Fig. 45.

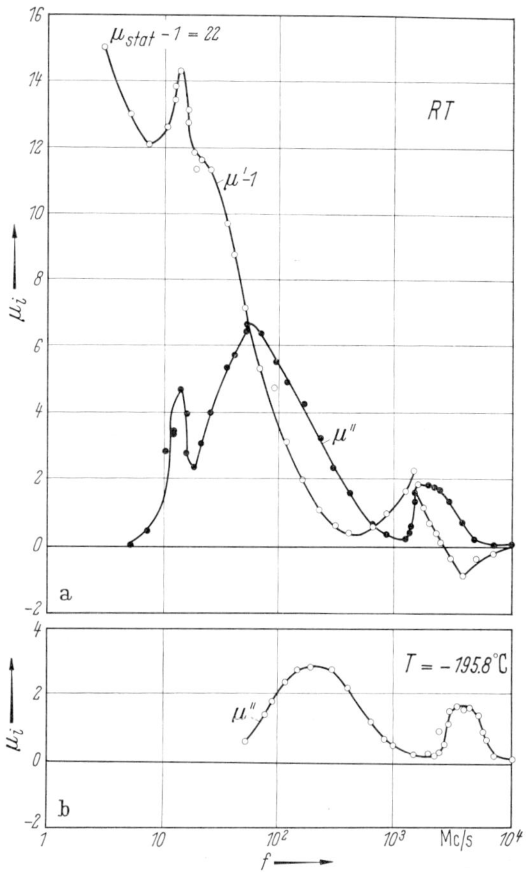

↑ Fig. 14a and b. $(MgO)_{0.81}(FeO)_{0.19}Fe_2O_3$. Initial permeability, μ_i, at RT (a) and at liquid nitrogen temperature (b), where only μ'' is shown [57RFE].
Sample was polycrystalline with $\varrho = 3.68\ gcm^{-3}$ and σ_s(26 °C) = 26.4 Gauss cm^3g^{-1}.

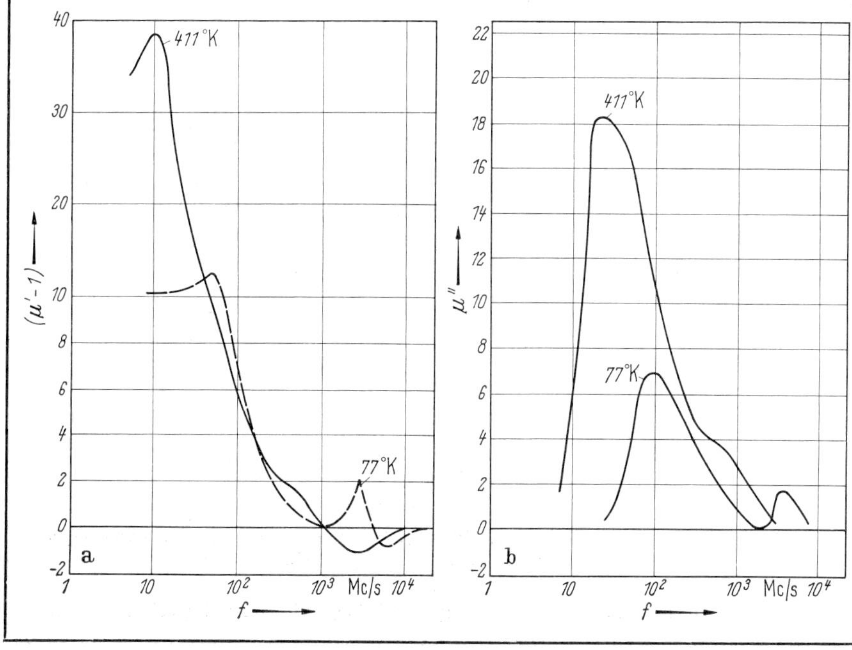

← Fig. 15a and b. Ferramic A (a commercial ferrite). Components of complex initial permeability, μ', and μ'', vs. frequency, f, at 411 °K and 77 °K [52RWE].

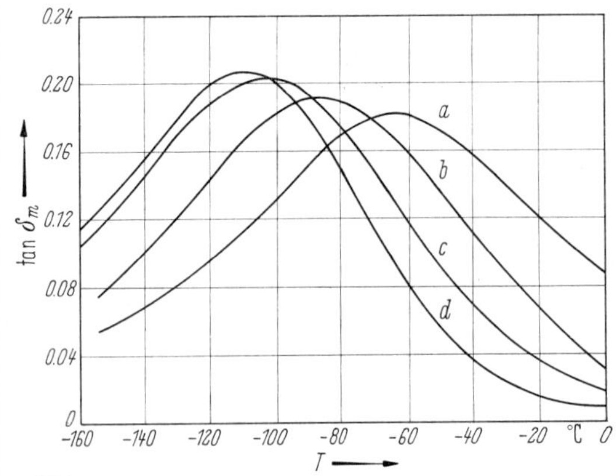

← Fig. 16. $MgFe_2O_4$. Magnetic loss tangent, δ_m, vs. temperature, T. Parameter frequency, f [*56E*].
Sample was polycrystalline and quenched from 1000 °C. *a*) 1 Mc/s, *b*) 300 kc/s, *c*) 100 kc/s, *d*) 50 kc/s.

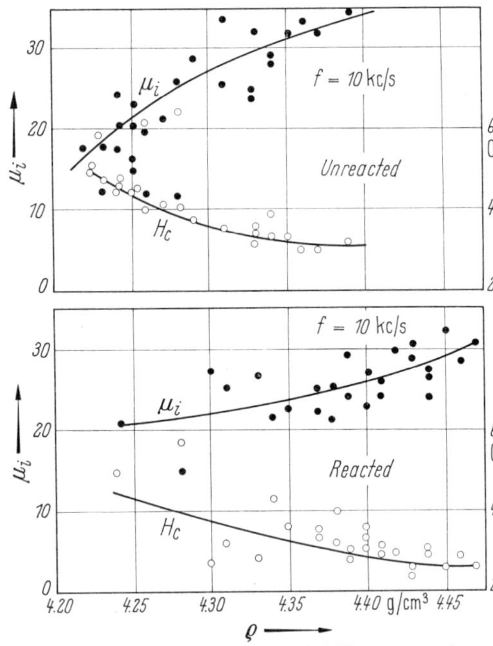

Fig. 17. $MgFe_2O_4$. Initial permeability, μ_i, and coercive force, H_c, vs. density, ϱ, at 10 kc/s [*55E(1)*].

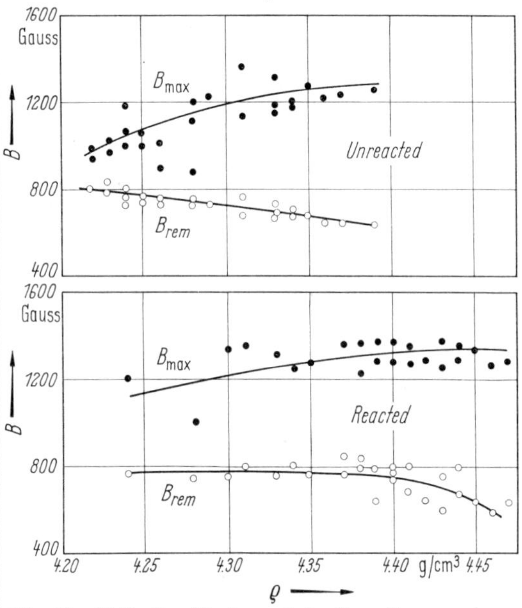

Fig. 18. $MgFe_2O_4$. Maximum induction, B_{max}, and remanence, B_{rem}, vs. density, ϱ [*55E(1)*].
Unreacted: before treatment, powders are mixed and heated at 500 °C; reacted: powders heated at 1050 °C, some spinel formed.

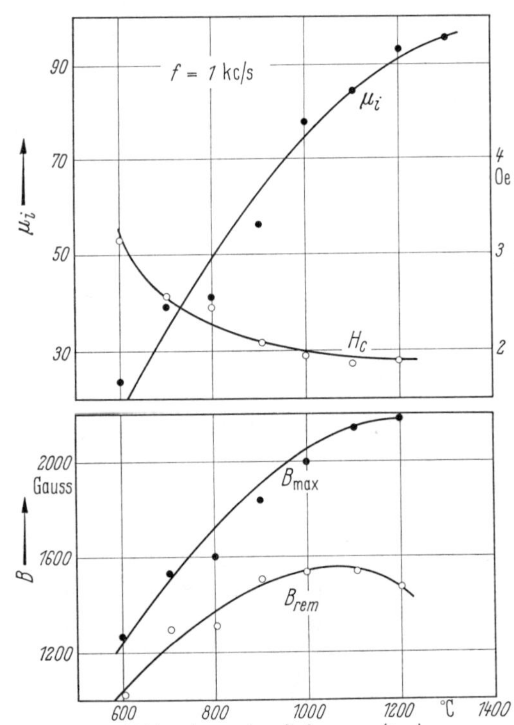

Fig. 19. $MgFe_2O_4$. Effect of quenching treatment on magnetic parameters, μ_i, H_c, B_{max}, B_{rem} [*55E(1)*].

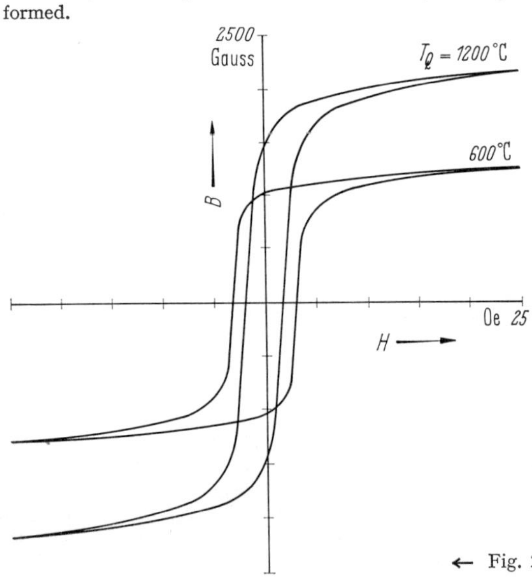

← Fig. 20. $MgFe_2O_4$. Hysteresis loops of quenched specimens [*55E(3)*].

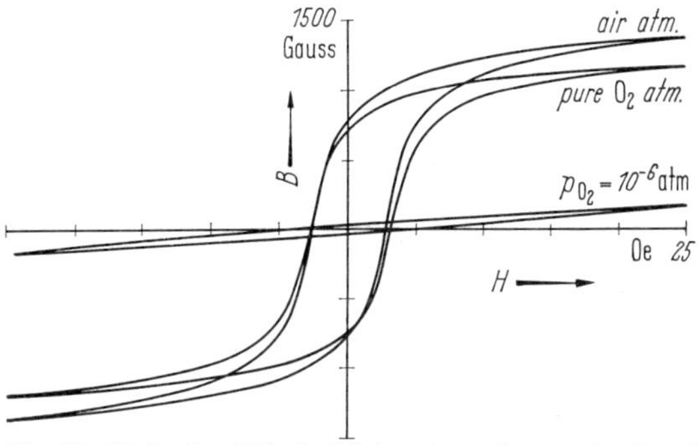

Fig. 21. $MgFe_2O_4$. Effect of firing atmosphere on hysteresis loops [*55E(3)*].

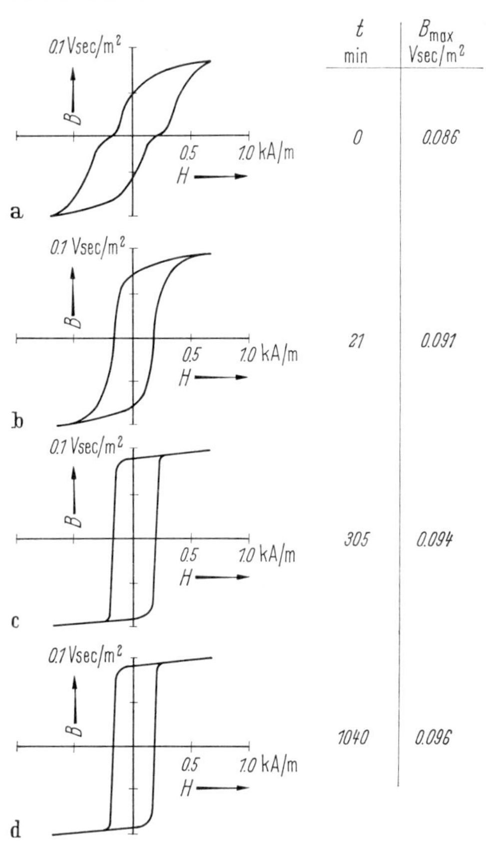

Fig. 22. $MgO \cdot 2FeO \cdot 4Fe_2O_3$. Transformation of hysteresis loops at 95 °C [*64HV*]. →
Square hysteresis loops are achieved by applying a field of 450 ··· 600 A/m oscillating at 50 c/s. The activation energy is about 1.3 eV, and the transformation is assumed to occur through an ordering process consisting of a diffusion of vacancies. t = time after switching.

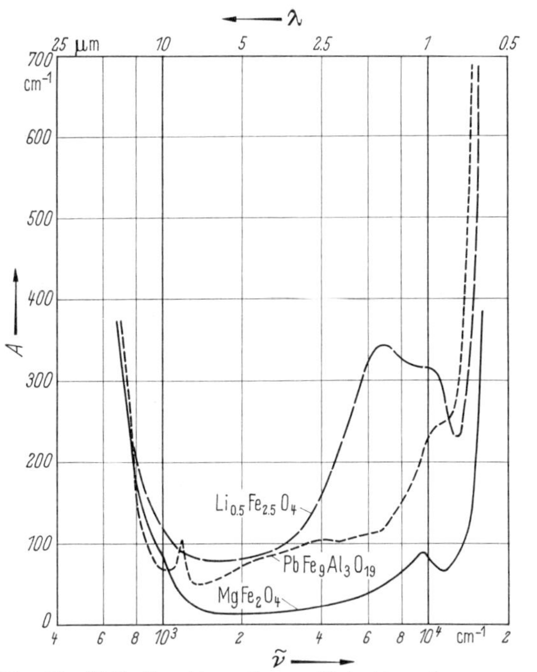

Fig. 24. $MgFe_2O_4$. Absorption coefficient/cm, A, vs. wave number, $\tilde{\nu}$, and wavelength, λ, for a single crystal.
Crystal was grown from PbO flux. The observed Faraday rotation was ϱ = 1100 deg cm⁻¹ [*59SRW*].
$I/I_0 = 10^{-A \cdot t}$, t = thickness [cm].

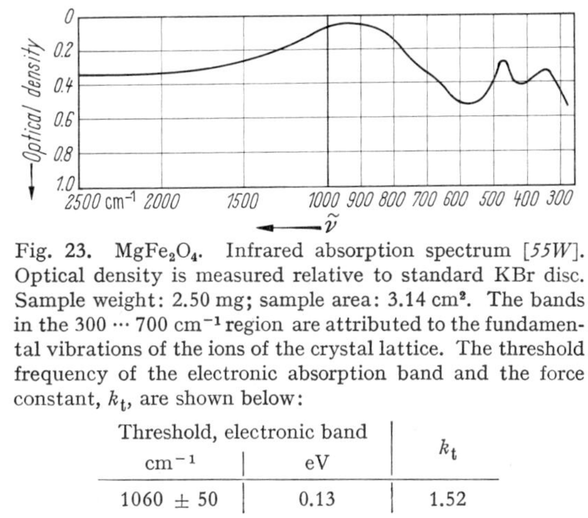

Fig. 23. $MgFe_2O_4$. Infrared absorption spectrum [*55W*]. Optical density is measured relative to standard KBr disc. Sample weight: 2.50 mg; sample area: 3.14 cm². The bands in the 300 ··· 700 cm⁻¹ region are attributed to the fundamental vibrations of the ions of the crystal lattice. The threshold frequency of the electronic absorption band and the force constant, k_t, are shown below:

Threshold, electronic band		k_t
cm⁻¹	eV	
1060 ± 50	0.13	1.52

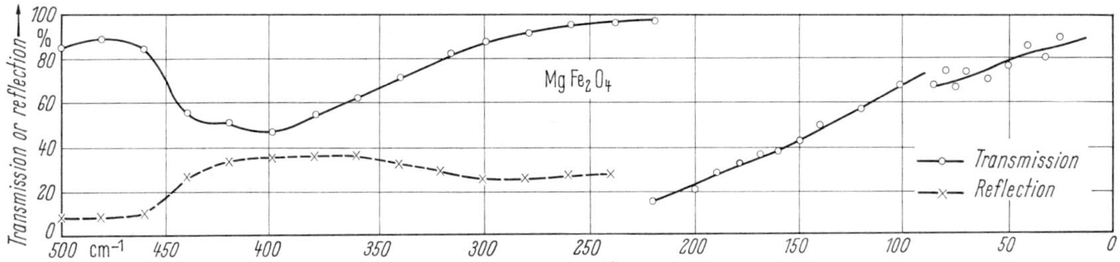

Fig. 25. $MgFe_2O_4$. Far infrared absorption and reflection spectrum [*58MYF*].

Tab. 4. $MgFe_2O_4$. Values of dielectric constant and permeability measured at $\lambda = 6.6$ cm for samples sintered at 1200 °C for 3 h. Samples were discs 1.5 mm thick and 20 mm in diameter [*52OFD*].

sample	ε'	ε''	$\tan\delta$	μ'	μ''	$\tan\delta_m$
quenched	9.66	0.174	0.018	0.80	0.974	0.812
slowly cooled	8.53	0.132	0.016	−0.84	0.341	0.120

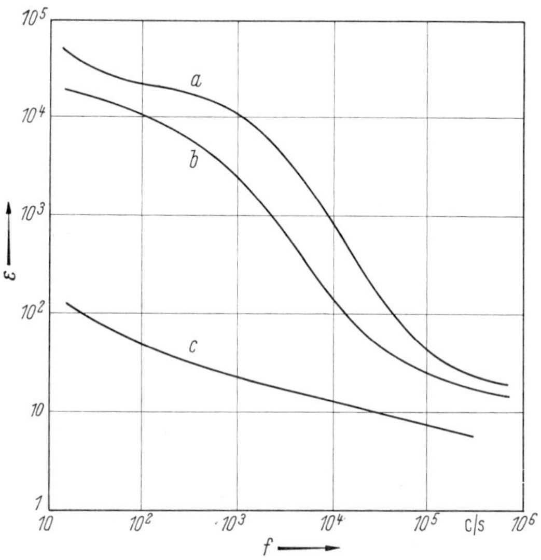

Fig. 26. $MgFe_2O_4$. Dielectric constant, ε, vs. frequency, f [*51FF*].
a) $MgAl_{0.4}Fe_{1.6}O_4$, *b*) $MgFe_2O_4$, *c*) $MgAlFeO_4$.

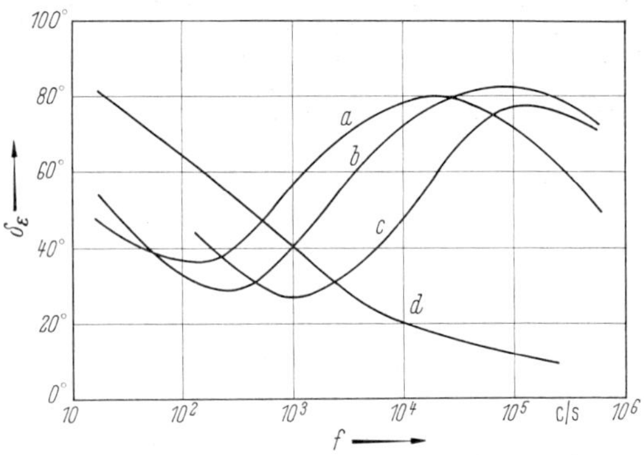

Fig. 27. Dielectric loss angle, δ_ε, vs. frequency, f [*51FF*].
a) $MgFe_2O_4$ at 20 °C, *b*) $MgAl_{0.4}Fe_{1.6}O_4$ at 20 °C, *c*) $MgAl_{0.4}Fe_{1.6}O_4$ at 44 °C, *d*) $MgAlFeO_4$ at 20 °C.

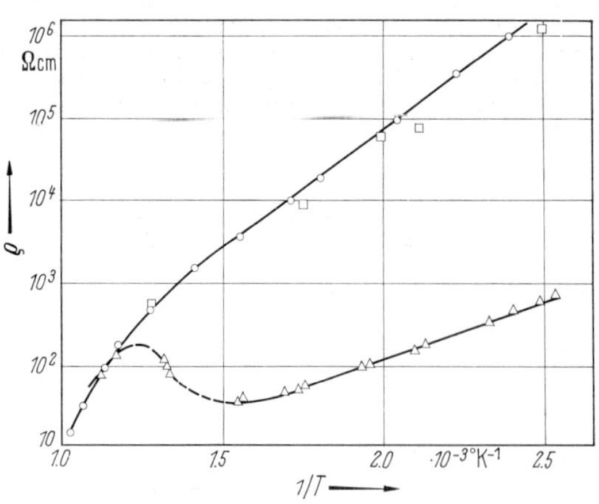

Fig. 28. $MgFe_2O_4$. Resistivity, ϱ, vs. inverse temperature, $1/T$ [*51B*].
Samples were sintered and had a density of 4 g/cm³.
○ Slowly cooled; △ quenched from 700 °K; □ [*50GB*].

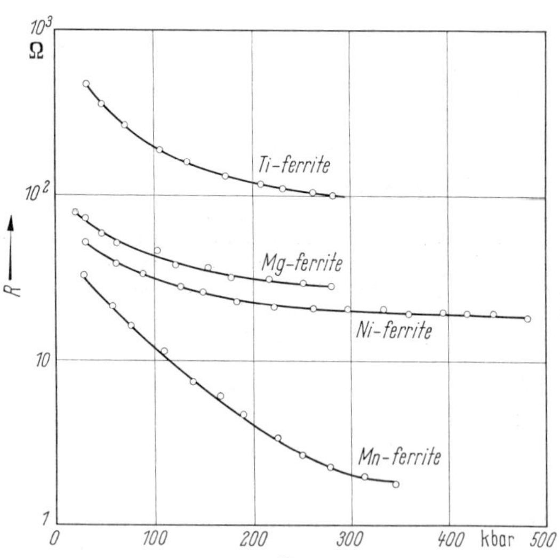

Fig. 29. $MgFe_2O_4$. Electrical resistance, R, vs. pressure, p [*63MD*].
For comparison: Ti-, Ni-, Mn-ferrite.

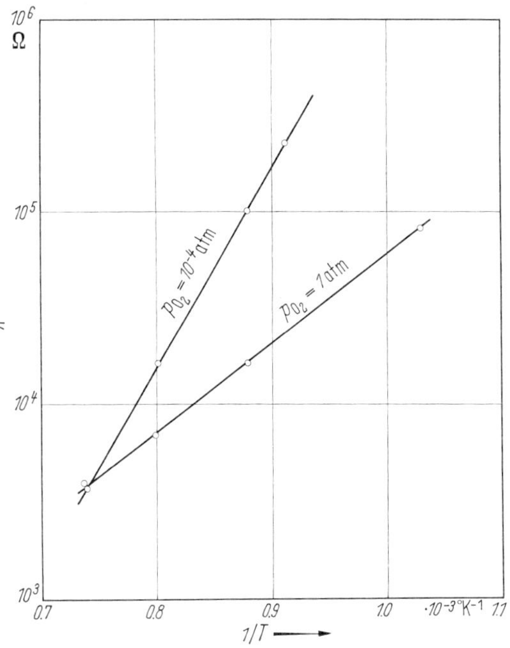

Fig. 30. $MgFe_2O_4$. Electrical resistance, R, vs. inverse temperature, $1/T$ [*63WM*].
Parameter O_2 pressure, p_{O_2}.

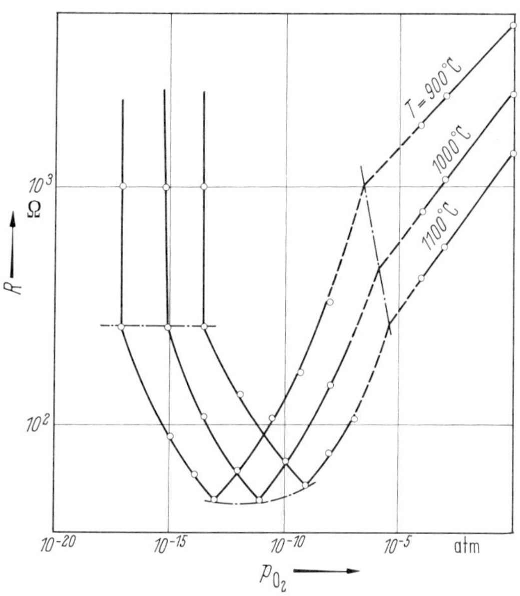

Fig. 31. $MgFe_2O_4$. Electrical resistance, R, vs. O_2 pressure, p_{O_2}. Parameter temperature, T [*63WM*].

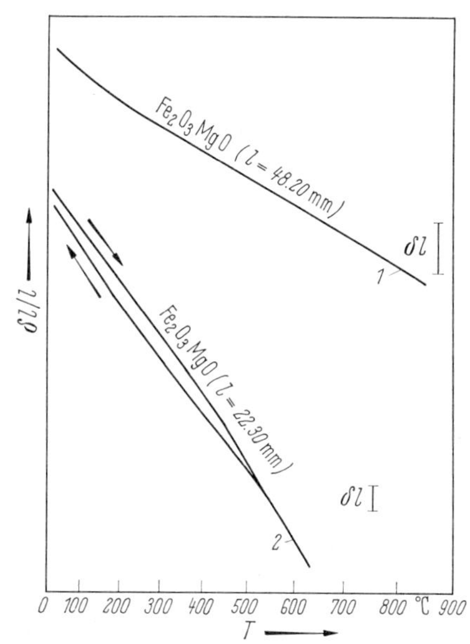

Fig. 32. $MgFe_2O_4$. Relative change in length, $\delta l/l$, vs. temperature, T, for two samples; one slowly cooled, the other quenched [*51W*].
$\delta l = 1\permil$ is given in the figure for both samples.

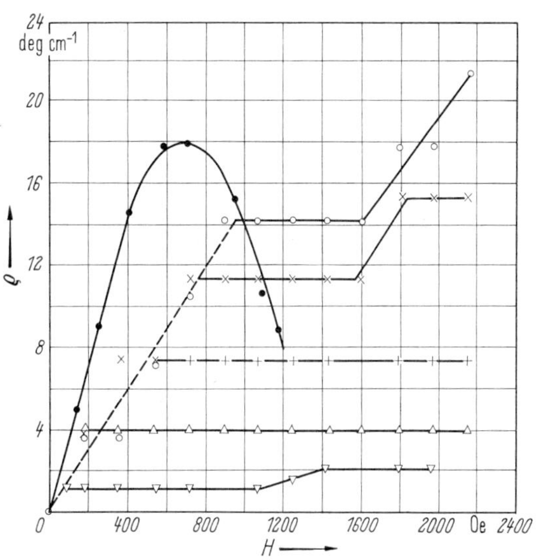

Fig. 33. $MgFe_2O_4$ and Mg ferrite aluminates. Faraday rotation, ϱ, vs. applied axial field, H [*51R(2)*].

- ○ $MgOFe_2O_3$
- × $MgO(Al_2O_3)_{0.1}(Fe_2O_3)_{0.9}$
- \+ $MgO(Al_2O_3)_{0.2}(Fe_2O_3)_{0.8}$ — at $\lambda = 3.2$ cm
- △ $MgO(Al_2O_3)_{0.3}(Fe_2O_3)_{0.7}$
- ▽ $MgO(Al_2O_3)_{0.4}(Fe_2O_3)_{0.6}$
- ● $MgOFe_2O_3$ — at $\lambda = 10$ cm

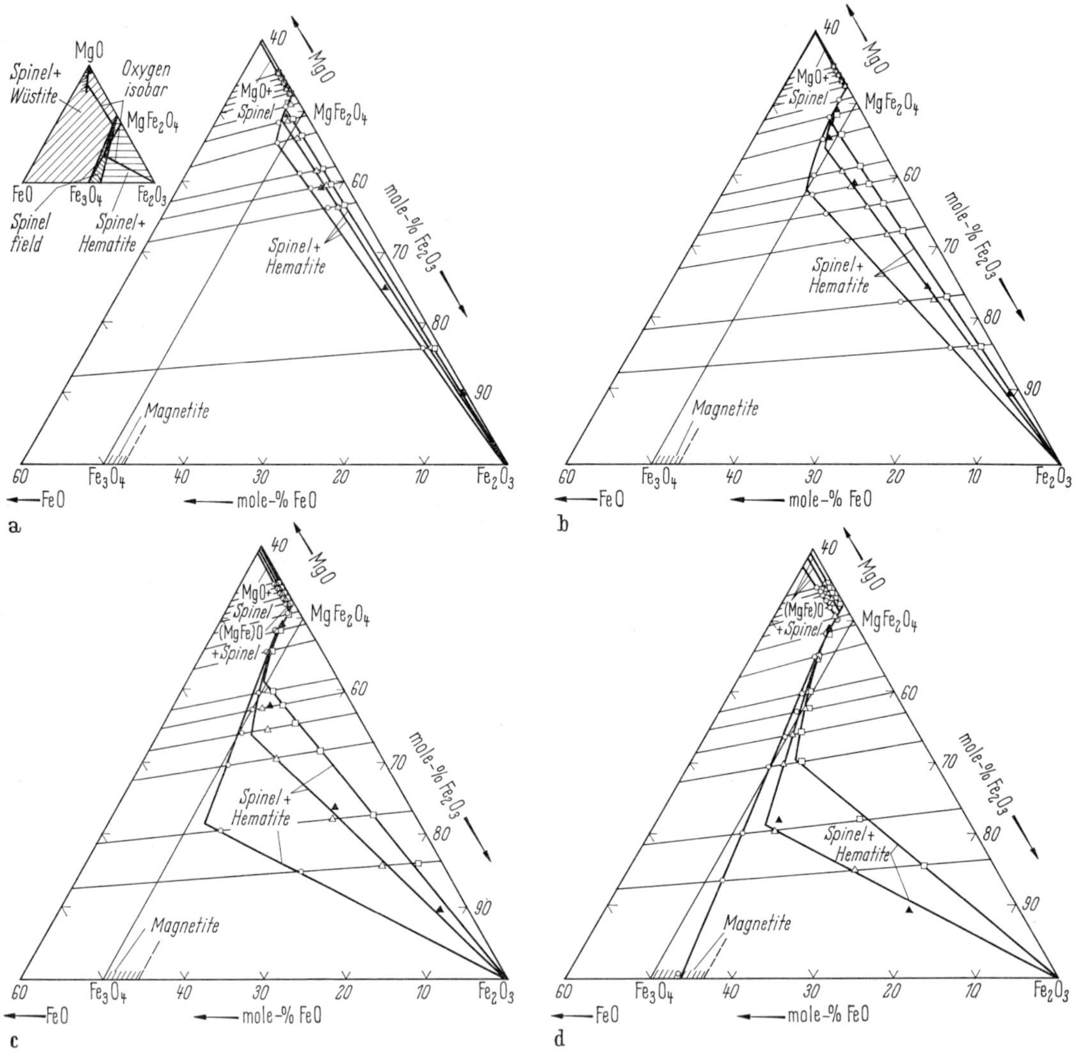

Fig. 34a…d. FeO-MgO-Fe_2O_3. Phase equilibria relations in the ternary system.

a) at 1000 °C, c) 1200 °C,
b) at 1100 °C, d) 1300 °C.

Isobars:
□ 1.0 atm O_2 [*60P*], ○ 0.01 atm O_2 [*60P*],
△ 0.2 atm O_2 [*60P*], ▲ air data [*55WW*].

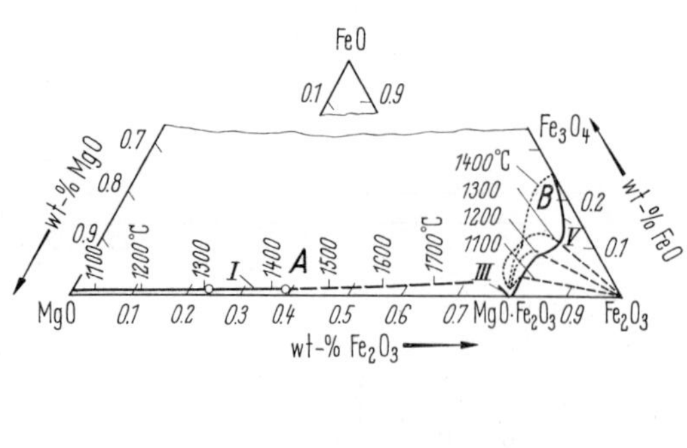

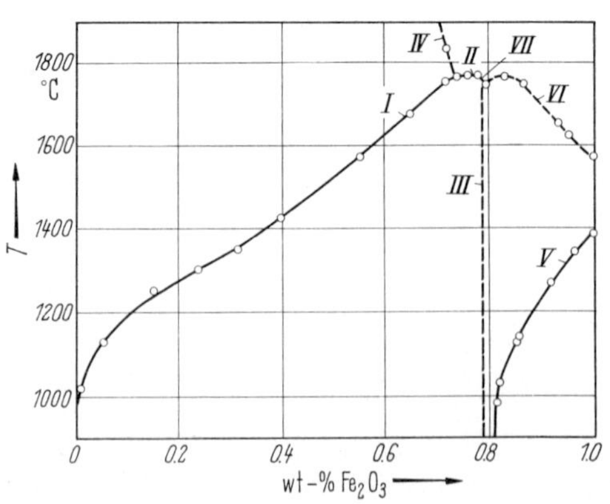

Fig. 35. FeO-MgO-Fe_2O_3. Phase relations in the ternary system [*31RM*]. See Tab. 5. Compare Fig. 37 p. 226.

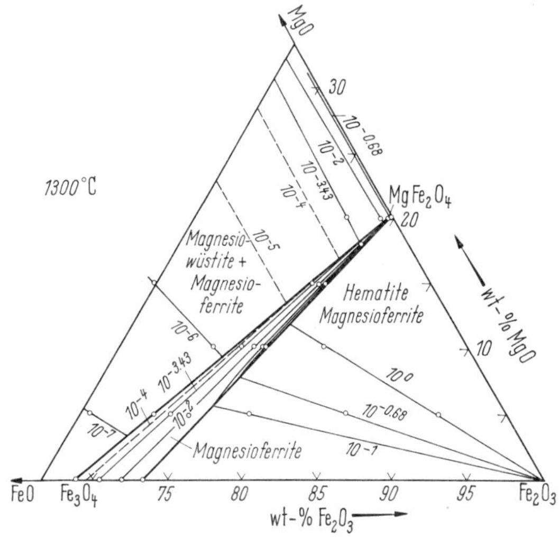

Fig. 36. FeO-MgO-Fe_2O_3. Ferrite region of the ternary system at 1300 °C [*67S*].
Heavy lines are phase boundaries, light lines are oxygen isobars, dashed were approximated.
This conflicts with data in Fig. 34d on the composition of the Mg end member of the spinel solid solution $MgFe_2O_4$. Data in Fig. 34d gives the Mg end member as $[MgO]_xMgFe_2O_4$, where $x = 0.092$, which represents a cation to anion ratio of 0.773 . [*67S*] and [*65R*] give $x = 0$ (cation to anion ratio of 0.750). See also [*65KK*] and [*61PSM*].

Fig. 37 see next page

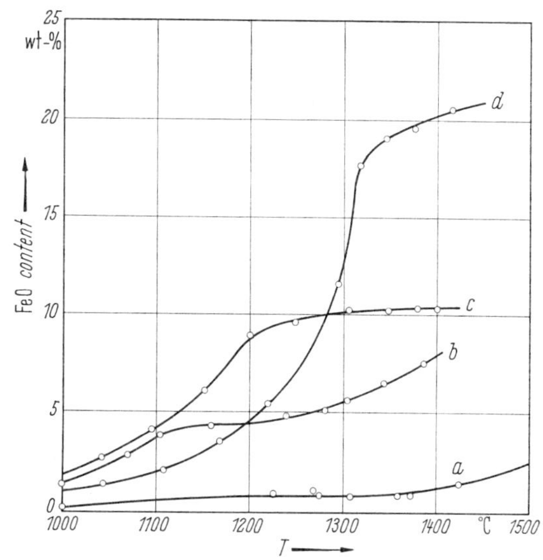

Fig. 38. FeO content vs. temperature, T, for four compositions in solid solution field B of Fig. 35 [*31RM*].
Initial compositions (in wt-%): *a*) 80.0% Fe_2O_3, 20% MgO
b) 82.9% Fe_2O_3, 17.1% MgO
c) 86.8% Fe_2O_3, 13.2% MgO
d) 94.1% Fe_2O_3, 5.9% MgO.

Tab. 5. Solid solution limits [*31RM*] (see Fig. 35).

Initial composition, wt-% Fe_2O_3	MgO	T °C	Time of heating	Second phase	Equilibrium: % iron oxide as Fe_2O_3	$T \pm 10$ °C
Curve *I* in Fig. 35. Excess magnesioferrite in MgO solid solutions.						
1	99	1000	70 h	trace	1	1.020
same		1040	16 h	none	1	1.020
5	95	1112	20 h	2 phases		
same		1152	16 h	none	5	1.132
14	86	1250	140 h	none		
16	84	1250	140 h	ca. 5%	15	1.250
25	75	1302	700 h	trace	23.5	1.302
30	70	1347	700 h	none		
35	65	1347	700 h	ca. 5%	31	1.347
40	60	1426	50 h	trace	39.5	1.426
55	45	1550	10 min	several %		
same		1575	10 min	trace	55	1.575
65	35	1660	10 min	small amount		
same		1700	15 min	none	65	1.680
72	28	1755	10 min	trace	72	1.755
Curve *V* in Fig. 35. Excess hematite in magnesioferrite solid solutions.						
81.4	18.6	984	17 h	trace		
same		1040	16 h	none	81.4	990
82.9	17.1	1040	16 h	ca. 5%		
same		1085	17 h	none	82.0	1.040
86.9	13.1	1134	16 h	ca. 10%	85.5	1.134
same		1146	16 h	ca. 5%	86.2	1.146
94.1	5.9	1275	16 h	ca. 20%	92.6	1.275
98.2	1.8	1334	16 h	ca. 50%	96.4	1.334
same		1346	16 h	ca. 25%	97.3	1.346

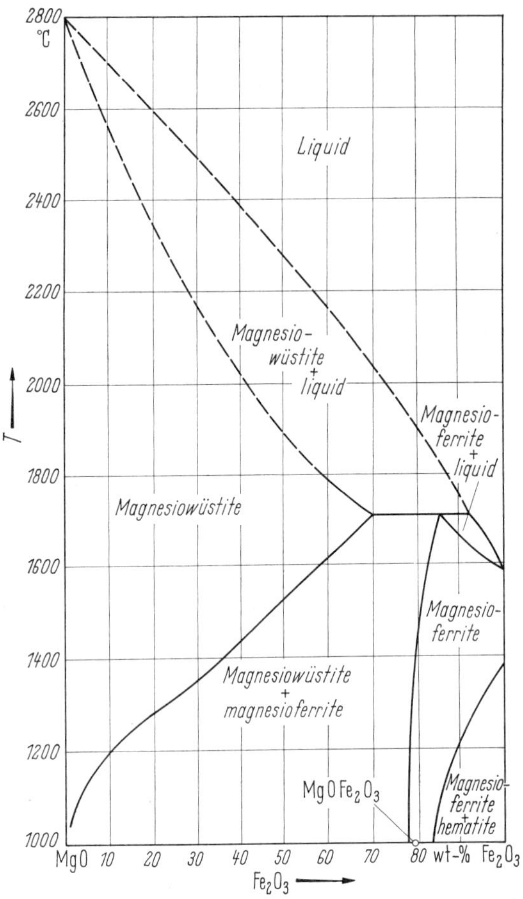

Fig. 37. MgO-Fe_2O_3. Phase relations in air [*61PSM*].

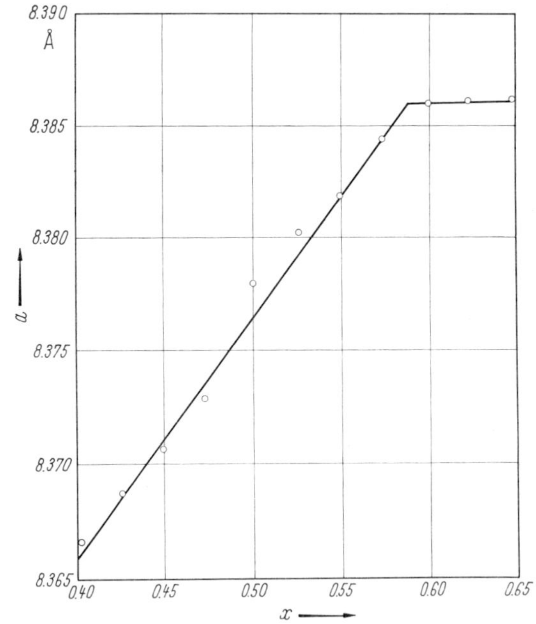

Fig. 39. $(MgO)_x(Fe_2O_3)_{1-x}$. Unit cell dimension, a, vs. MgO content, x [*56F*].
All samples were fired at 1350 °C and cooled at 50 °C per hour. See 6.1.4.2, Fig. 1.

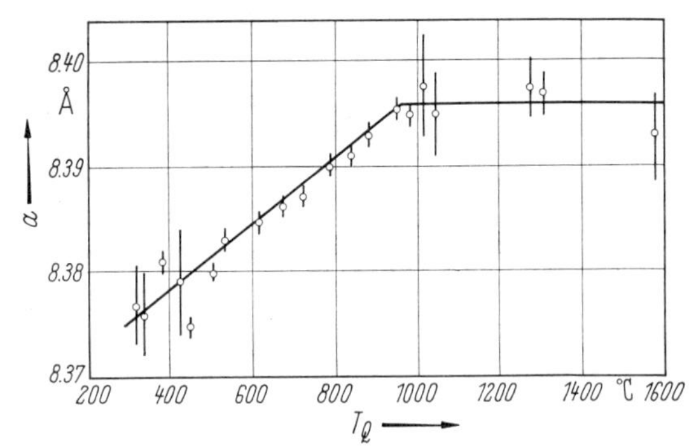

Fig. 40. $MgFe_2O_4$. Lattice parameter, a, vs. quenching temperature, T_Q [*66A*].

Tab. 6. $MgFe_2O_4$. Summary of properties [*66A*].

Property	High-temperature form	Low-temperature form	Literature values	Ref.
a [Å]	8.397 ± 0.005	8.375 ± 0.005	8.379···8.402	*56KH* *58B*
ϱ_{calc} [$g \cdot cm^{-3}$]	4.486	4.522	4.518	*30P*
Color, bulk	Black	Red		
Color under microscope	Brown	Brown		
n_{calc}	2.34	2.35	2.34 2.39	*30P* *31RM*
η	0.33	0.18	0.09 0.27	*56KH* *56KH*

n = refractive index, η = fraction of Mg ions on A sites.

Tab. 7. $MgFe_2O_4$. Neutron diffraction intensities [53BR].

	Calculated intensities			Experimental intensity
	Normal $u = 0.385$	Inverse $u = 0.382$	Incompletely inverted $u = 0.382$	
A sites	Mg	Fe	$Mg_{0.1}Fe_{0.9}$	
B sites	Fe_2	MgFe	$Mg_{0.9}Fe_{1.1}$	
(111)	126	10	16	18 ± 3
(220)	39	127	114	112 ± 4
(311, 222)	320	400	385	374 ± 10
(400)	448	256	274	279 ± 6
(511, 333)	270	297	292	274 ± 7
(440, 531)	820	758	762	770 ± 30

Calculated and experimental reflections are shown for a temperature of 400 °C. Only nuclear reflections are considered since no coherent magnetic scattering is expected at 400 °C. Agreement is best for the incompletely inverted structure with O_2 parameter, $u = 0.382$. A powdered sample with a Curie point of 300 °C was used.

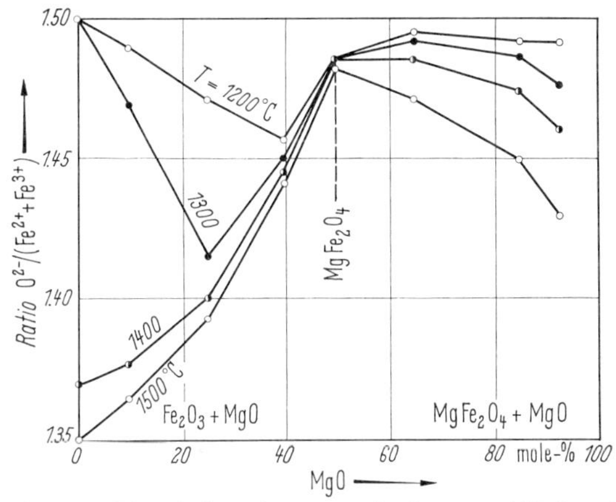

Fig. 41. Dissociation of prereacted mixtures of MgO and Fe_2O_3 in air.
The data are from [54RW] and have been presented in this rearranged form in [59Bla]. Further corroboration is found in [55WW]. The extent of the dissociation, measured by the loss of weight of the samples, is expressed by the ratio $O^{2-}/(Fe^{2+} + Fe^{3+})$, which refers to the iron oxide phase only:

$$Fe_2O_3 \rightleftarrows Fe_x^{2+}Fe_{2-x}^{3+}O_{3-x/2}^{2-} + O_{x/2}^{2-}.$$

Thus, a ratio of 1.5 indicates a stable compound, as in Fe_2O_3 at 1200 and 1300 °C. $MgFe_2O_4$ is only slightly dissociated at all temperatures.

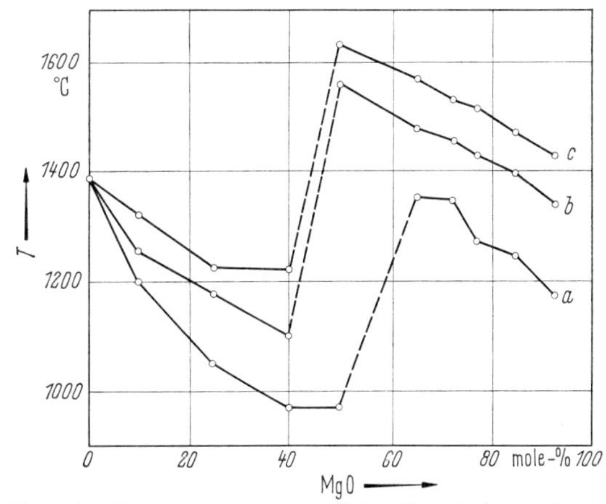

Fig. 42. Temperatures required for dissociation ratios of *a*) 1.49, *b*) 1.47, *c*) 1.45 [59B]. Curve *c* is taken from [54RW]. See Fig. 41.

Tab. 8. Heat, ΔH, and free energy, ΔF^0, of formation for MgO(c) + Fe_2O_3(c) = $MgFe_2O_4$ [61KBK].

T °K	ΔH kcal/mole	ΔF^0 kcal/mole	*T* °K	ΔH kcal/mole	ΔF^0 kcal/mole
298.15	−0.6	−1.25	1050 [3])	−1.55	−3.15
400	−0.45	−1.5	1050	−1.55	−3.15
500	−0.5	−1.75	1100	−1.55	−3.3
600	−0.4	−2.0	1200	−1.65	−3.45
665 [1])	0.0	−2.2	1230 [4])	−1.7	−3.5
665	0.0	−2.2	1230	−1.35	−3.5
700	0.0	−2.35	1300	−1.55	−3.6
800	−0.25	−2.65	1400	−1.85	−3.75
900	−0.75	−2.9	1500	−2.0	−3.85
950 [2])	−1.1	−2.95	1600	−2.05	−4.0
950	−1.3	−2.95	1700	−2.05	−4.1
1000	−1.4	−3.1	1800	−2.15	−4.2

[1]) Transition point of $MgFe_2O_4$ ($\alpha \rightarrow \beta$).
[2]) Transition point of Fe_2O_3 ($\alpha \rightarrow \beta$).
[3]) Transition point of Fe_2O_3 ($\beta \rightarrow \gamma$).
[4]) Transition point of $MgFe_2O_4$ ($\beta \rightarrow \gamma$).

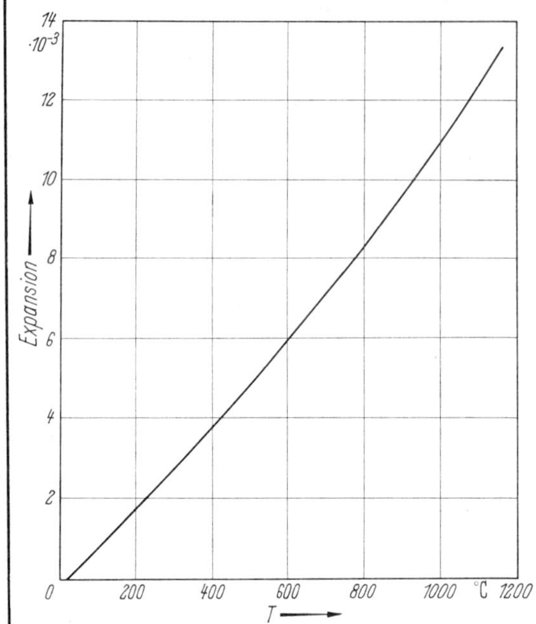

Fig. 43. $MgFe_2O_4$. Thermal expansion $\delta l/l$**) vs. temperature, T [*59C*].

Tab. 9. Coefficient of thermal expansion for $MgFe_2O_4$ [*59C*].

T	200	400	600	800	1000	Ref.
*)	10.4	10.3	10.6	11.0	11.5	*59C*
	8.9	10.4	11.5	12.3	13.0	*46RLG*
					12.8 (850 °C)	*32PA*

6.1.4.2 MgMn ferrites and MgMn ferrites with substitutions

6.1.4.2.0 Introductory remarks

The tables and figures in this section are grouped according to subject matter. At the same time, an attempt was made to keep together each author's contribution on a topic of research so that his final results and conclusions could be indicated wherever possible. The sequence of the presentation is from physical measurements (structural data, resistivity, dielectric constant etc.) and magnetization data to anisotropy constants, losses and, finally, to measurements of properties of technical importance (square loop and microwave applications).

The ferrites of this section find their greatest application as memory elements by virtue of their square-loop characteristic and in microwave devices. Within each topic the unsubstituted MgMn ferrites are offered first, in as far as possible, followed by the more complex compositions.

Although many of the Russian journals are available in English translation, the reference date listed is for the Russian publication; the reference for the translation is given after the Russian entry.

6.1.4.2.1 Crystallographic properties, physical properties, phase data, magnetic moments

*) Coefficient was not defined in original paper; probably $\frac{\delta l}{l} \cdot \frac{1}{T}$ in °C^{-1}.
**) Not defined in original paper.

Tab. 1. Distribution of magnetic ions and oxygen parameters, *u*, for MgMn ferrites [57NPHK].

Sample	T_Q °C	η	u*)	ξ
$Mg_{0.25}Mn_{0.75}Fe_2O_4$	1400	0.12	0.3850	0.89 ± 0.05
	1000	0.09	0.3848	0.78 ± 0.05
$Mg_{0.5}Mn_{0.5}Fe_2O_4$	1400	0.18	0.3839	0.92 ± 0.05
	1000	0.14	0.3836	0.82 ± 0.05
$Mg_{0.75}Mn_{0.25}Fe_2O_4$	1400	0.24	0.3834	0.90 ± 0.07
	1000	0.21	0.3831	0.98 ± 0.07
$Mg_{0.9}Mn_{0.1}Fe_2O_4$	1400	0.25	0.3831	1.0 ± 0.15
	1000	0.21	0.3823	0.90 ± 0.15

T_Q = Quenching temperature, η = fraction of Mg^{2+} ions on *A* sites. ξ = fraction of Mn^{2+} ions on *A* sites. Also see Fig. 25 in 6.1.4.3.

Tab. 2. Tetrahedral cation-anion distances in $Mg_yMn_{1-y}Fe_2O_4$ [57NPHK].

y	T_Q °C	a_0 Å	d Å	Mn-O distance Å
0	1400	8.517	1.99	2.01
0.25	1400	8.485	1.99	2.03
	1000	8.479	1.98	2.05
0.50	1400	8.455	1.96	2.02
	1000	8.450	1.96	2.03
0.75	1400	8.424	1.95	2.01
	1000	8.421	1.94	2.00
0.90	1400	8.406	1.94	1.99
	1000	8.403	1.93	1.97
1.0	—	8.380	1.90	—

T_Q = Quenching temperature, d = tetrahedral metal-oxygen distance = $\sqrt{3}\,(u - \frac{1}{4})\,a_0$.
The Mn-O distance is calculated from d and from the Fe-O, Mg-O bond lengths for $MgFe_2O_4$ weighted according to ξ and η. The fact that the Mn-O distance so calculated is only 0.9 of the value obtained from the Goldschmidt radii is accepted as evidence for covalent bond formation and taken to account for the preference of Mn^{2+} for the *A* site.

*) All values have a tolerance of ± 0.0003.

Tab. 3. Crystallographic data for some MgMn ferrites

Structural formula	η	ξ	u	a_0 Å	Ref.
$Mg_{0.03}Mn^{2+}_{0.67}Fe^{3+}_{0.30}[Mg_{0.22}Mn^{3+}_{0.08}Fe^{2+}_{0.08}Fe^{3+}_{1.62}]O_4$	0.12	0.89	0.385	8.485	*56HC*
$Mg_{0.09}Mn^{2+}_{0.46}Fe^{3+}_{0.45}[Mg_{0.41}Mn^{3+}_{0.04}Fe^{2+}_{0.04}Fe^{3+}_{1.51}]O_4$	0.18	0.92	0.3839	8.455	*56HC*
$Mg_{0.18}Mn^{2+}_{0.22}Fe^{3+}_{0.60}[Mg_{0.57}Mn^{3+}_{0.03}Fe^{2+}_{0.03}Fe^{3+}_{1.37}]O_4$	0.24	0.88	0.3834	8.424	*56HC*
$Mg_{0.12}Fe^{3+}_{0.88}[Mg_{0.88}Fe^{3+}_{1.12}]O_4$	0.12	—	0.381 ± 0.001	8.395	*53CHB*

η = fraction of Mg on A site; ξ = fraction of Mn on A sites; u = oxygen parameter.
For a relationship between a_0, u, η, ξ see [*64SS*].

Tab. 4. Structural parameters for MgMn ferrites by neutron and X-ray diffraction [*65KWKP*].

	$Mg_{0.52}Mn_{0.51}Fe_{1.97}O_{4.001}$	$Mg_{0.73}Mn_{0.57}Fe_{1.70}O_{3.986}$
u [Å]	0.3824 ± 0.0008	0.3827 ± 0.0007
η	0.25	0.29
ξ	0.67	0.44
$(p_A)_a$ [μ_B]	3.93 ± 0.14	3.52 ± 0.11
$(p_A)_b$ [μ_B]	3.16 ± 0.14	2.73 ± 0.11
Structural formula	$(Mg^{2+}_{0.13}Mn^{2+}_{0.34}Fe^{3+}_{0.53})[Mg^{2+}_{0.39}Mn^{3+}_{0.17}Fe^{2+}_{0.14}Fe^{3+}_{1.30}]O_{4.001}$	$(Mg^{2+}_{0.21}Mn^{2+}_{0.25}Fe^{3+}_{0.54})[Mg^{2+}_{0.52}Mn^{3+}_{0.32}Fe^{2+}_{0.05}Fe^{3+}_{1.11}]O_{3.986}$

u = oxygen parameter, η = fraction of Mg on A site, ξ = fraction of Mn on A site, $(p_A)_a$ = magnetic moment per tetrahedral ion, $(p_A)_b$ = magnetic moment per octahedral ion.
Compositions including O_2 content were determined by chemical analysis.
Materials were prepared by precipitation of mixed oxalates, prefiring for one hour in air at 600 °C, firing for one hour in air at 1100 °C. See also [*65WK*].

Tab. 5. Magnetic moment measurements for the MgMn ferrites of Tab. 4 [*65KWKP*].

Composition	σ_s (290 °K) emu/g	σ_s^0 (0 °K) emu/g	p_m (290 °K) μ_B	p_m^0 (0 °K) μ_B	Θ_C °K
$Mn_{1.01}Fe_{1.99}O_{3.999}$	78.8 ± 2.2	111.0 ± 3.5	3.25 ± 0.09	4.58 ± 0.14	587 ± 5
$Mg_{0.52}Mn_{0.51}Fe_{1.97}O_{4.001}$	62.1 ± 1.7	83.3 ± 2.5	2.40 ± 0.07	3.22 ± 0.10	621 ± 5
$Mg_{0.73}Mn_{0.57}Fe_{1.70}O_{3.986}$	52.2 ± 1.3	77.3 ± 2.5	1.94 ± 0.05	2.87 ± 0.10	553 ± 5

Measurements were made on a magnetic balance from RT to 90 °K and extrapolated to 0 °K. Mn ferrite was included as a point of reference.
p_m = magnetic moment per molecule.

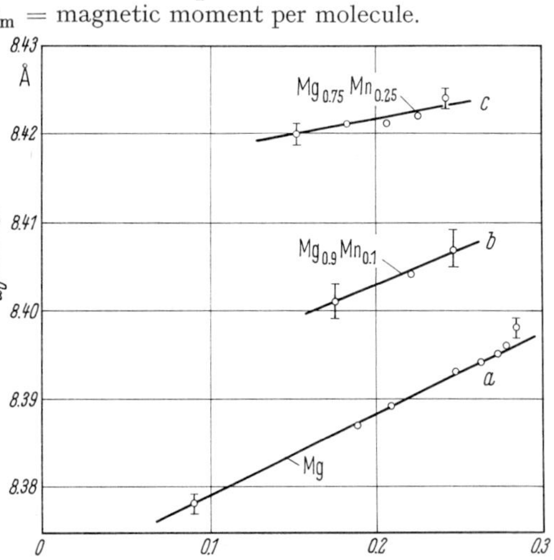

Fig. 1. Lattice constant, a_0, vs. η, the fraction of Mg ions on A sites [*56KH*].
a) $MgFe_2O_4$; *b*) $Mg_{0.9}Mn_{0.1}Fe_2O_4$; *c*) $Mg_{0.75}Mn_{0.25}Fe_2O_4$.

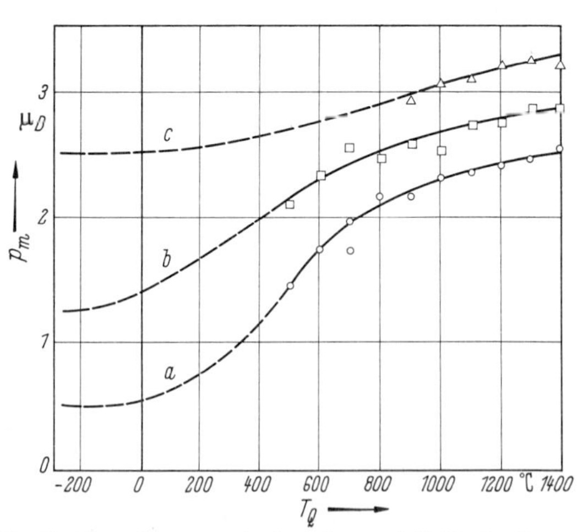

Fig. 2. Magnetic moments, p_m, of several MgMn ferrites vs. quenching temperature, T_Q [*56KH*].
a) $Mn_{0.1}Mg_{0.9}Fe_2O_4$; *b*) $Mn_{0.25}Mg_{0.75}Fe_2O_4$; *c*) $Mn_{0.5}Mg_{0.5}Fe_2O_4$.

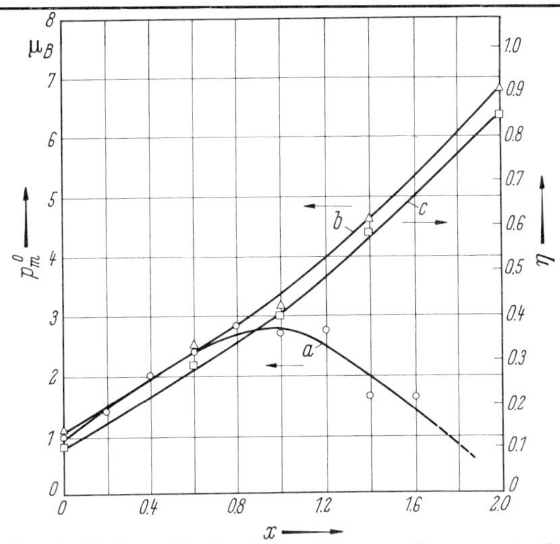

Fig. 3. $MgFe_{2-x}Mn_xO_4$. Molecular magnetic moment, p_m^0, and ionic distribution, η, vs. Mn content, x [60MM].
a) Experimental, p_m^0 vs. x. Magnetization was measured by Sucksmith's ring balance in a field of 1300 Oe. Powder samples were fired in air at 1200 °C, annealed for 24 hours in air at 900 °C, cooled to RT at 0.5 °C/min (presumably in an inert atmosphere). Measurements were made from Curie temperature to 77 °K.
b) Theoretical, p_m^0 vs. x. Calculated from Néel's theory using cation distribution given by curve *c*.
c) Fraction of the Mg ions located on *A* site, η, as determined from intensity ratios of (220) and (422) X-ray diffraction lines to that of (400) line for which all Mn ions were assumed to be located on *B* sites. Structure is cubic for compositions from $x = 0$ to 1.2 and tetragonal from $x = 1.8$ to 2.0. Intermediate compositions contain both phases.

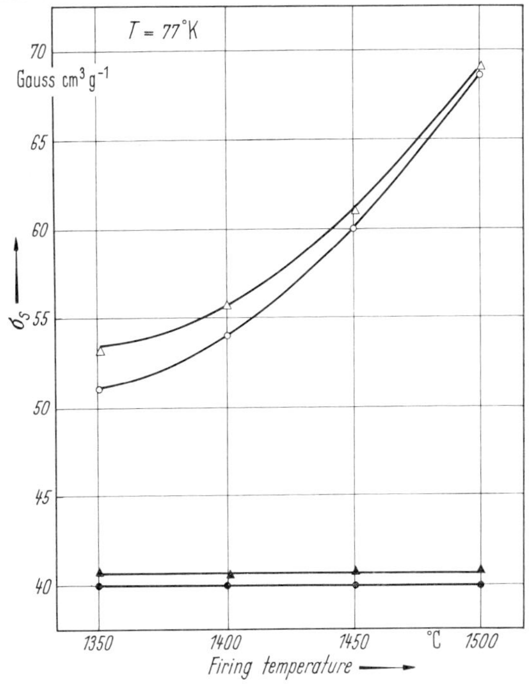

Fig. 4. Fe-deficient Mg and MgMn ferrites. Specific saturation magnetization, σ_s, vs. firing temperature [58B].
A MgO phase was present in all the materials.
○ $MgFe_{1.6}O_{3.4\pm}$; air quenched after firing for 4 hours in O_2.
△ $Mg_{0.9}Mn_{0.1}Fe_{1.6}O_{3.4\pm}$; air quenched after firing for 4 hours in O_2.
● $MgFe_{1.6}O_{3.4\pm}$; slowly cooled to 700 °C, annealed for 16 hours at 700 °C.
▲ $Mg_{0.9}Mn_{0.1}Fe_{1.6}O_{3.4\pm}$; slowly cooled to 700 °C, annealed for 16 hours at 700 °C.
The milled powders were calcined at 1000 °C, remilled and pressed into rods 0.125 in. in diameter. Bulk density varied from 3.7 $g cm^{-3}$ (18% porosity) for the 1350 °C firing to 4.0 $g cm^{-3}$ (11% porosity) for the 1500° C firing. The magnetization measurements were made by a ballistic method at 77 °K in a field of 6000 Oe. Precision of measurement was ±0.1%; reproducibility from sample to sample of the same material was ±1%.

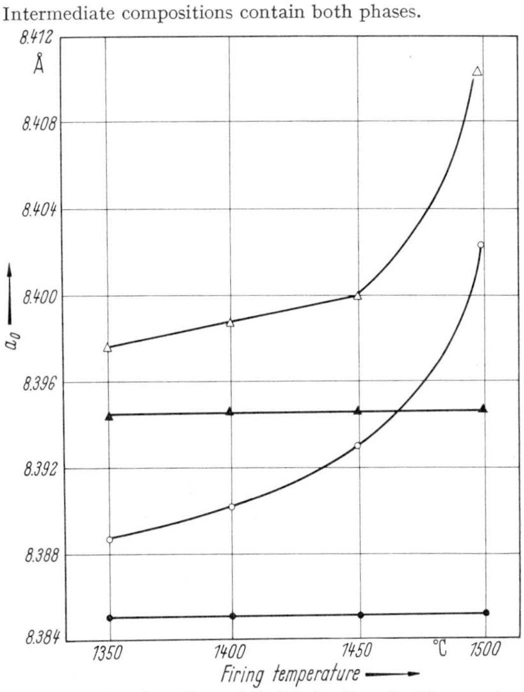

Fig. 5. Fe-deficient Mg and MgMn ferrites. Lattice constant, a_0, vs. firing temperature [58B].
Refer to Fig. 4 for explanation of symbols.

→

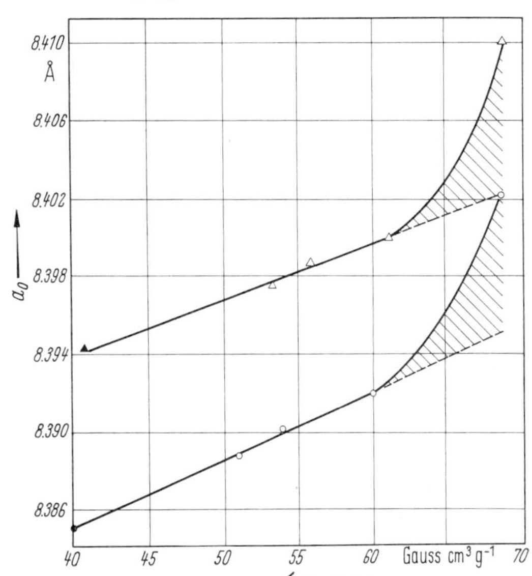

Fig. 6. Fe-deficient Mg and MgMn ferrites. Lattice constant, a_0, vs. specific saturation magnetization, σ_s [58B].
Refer to Fig. 4 for explanation of symbols. Shaded areas for firings above 1450 °C represent the region and degree of dissociation.

Fig. 8. Fe-deficient Mg and MgMn ferrites. Resistivity, ϱ, → vs. firing temperature [*58B*]. Refer to Fig. 4 for explanation of symbols. A two-contact method was used. Contacts were prepared by rubbing graphite on the rod faces.

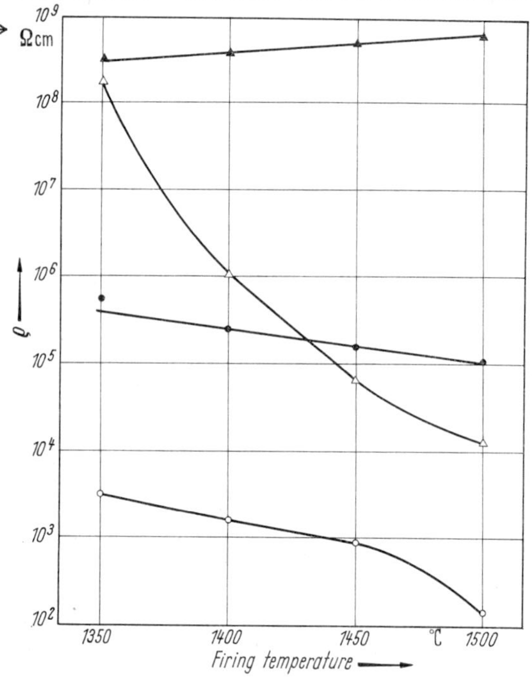

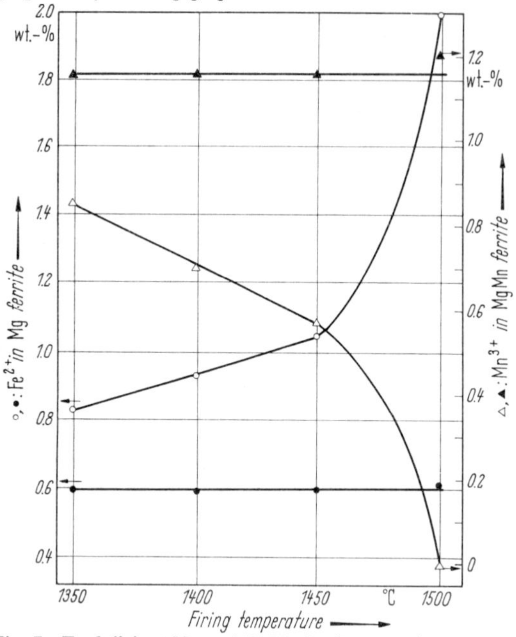

Fig. 7. Fe-deficient Mg and MgMn ferrites. Fe^{2+} and Mn^{3+} content vs. firing temperature [*58B*].
Refer to Fig. 4 for explanation of symbols. Data obtained from chemical analysis. Data for Mn^{3+} assumes absence of Fe^{2+} in the ferrite since any Fe^{2+}, if present, would be oxidized to Fe^{3+} by Mn^{3+} in preparing the sample for analysis. The increase in Fe^{2+} for quenched Mg ferrite and the corresponding decrease in Mn^{3+} for quenched MgMn ferrite coupled with the observation that their sums total (1.69 ± 0.08)% for firings below 1500 °C indicates that Fe^{2+} is formed at high temperatures in MgMn ferrite. Resistivity data (Fig. 8) suggest that at least some of the Fe^{2+} thus formed does coexist with Mn^{3+} in the solid. For the 1500 °C quenched sample (0.0% Mn^{3+}), 0.2% Fe^{2+} was found. All the Fe^{2+} is contained in the spinel phase.

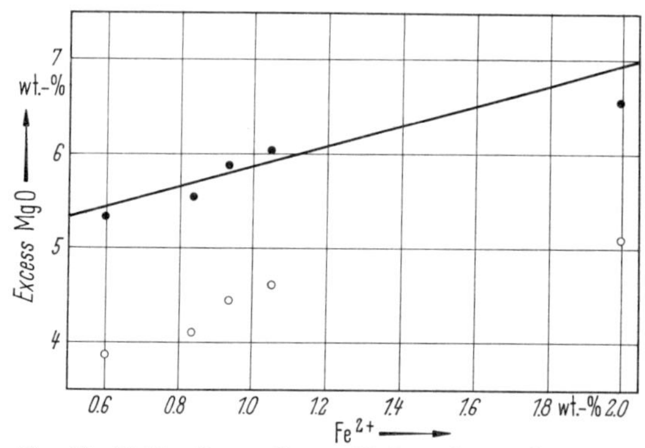

Fig. 10. $MgFe_{1.6}O_{3.4\pm}$. Excess MgO vs. ferrous ion content [*58B*].
Material was quenched or annealed. See Fig. 4 for details.
○ Amount of MgO phase.
● Amount of MgO phase + amount of extra MgO dissolved in spinel phase (1.4%).
— Theoretical relationship according to x vs. $(0.25 + 3x/2)$ in $[Mg_{1-3x/2}Fe^{2+}_{x}Fe^{3+}_{2-x}O_{4-2x}] - (MgO)_{0.25+3x/2}$.

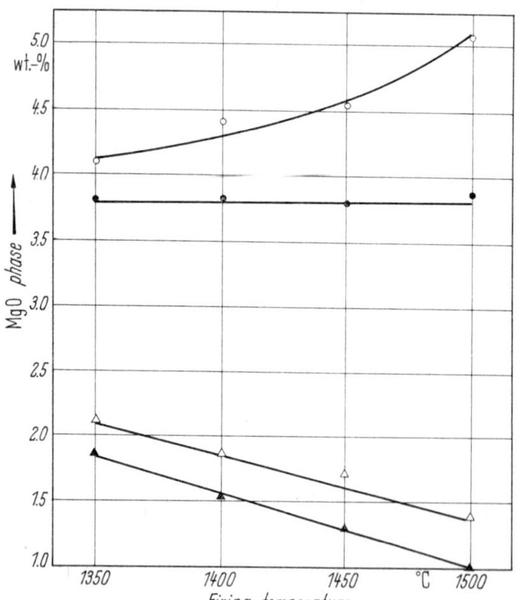

← Fig. 9. Fe-deficient Mg and MgMn ferrites. Amount of MgO phase vs. firing temperature [*58B*].
Refer to Fig. 4 for explanation of symbols.
The MgO present as second phase was determined by extraction in 0.01 *N* HCl. These data do not include the extra MgO present in the spinel part as solid solution. For pure (Fe^{2+} absent), stoichiometric Mg ferrite, this amounts to 0.07 moles MgO per mole $MgFe_2O_4$ or 1.4 w.-% of the spinel phase. Compare with 6.1.4.1, Fig. 34 [*60P*]. For the Fe-deficient Mg ferrite of this research, 0.18 moles of a MgO phase per mole of spinel phase should be present in the absence of Fe^{2+} (3.45 wt.-% MgO phase). The additional amount found, as given in the graph, was attributed to the presence of Fe^{2+} in the spinel phase (Fig. 7) thereby forcing an equivalent amount of bound MgO into the MgO phase according to $[Mg_{1-3x/2}Fe^{2+}_{x}Fe^{3+}_{2-x}O_{4-2x}]\cdot(MgO)_{0.07}+(MgO)_{0.18+3x/2}$. (See Fig. 10). The data for the MgMn ferrite were not understood.

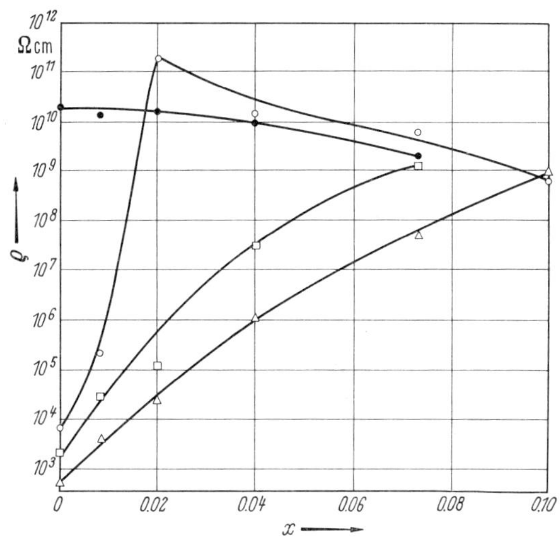

Fig. 11. $Mg_{1.0}Fe_{1.36}Mn_xO_4$. Room-temperature resistivity, ϱ, vs. Mn content, x, for various firing temperatures, T, and firing times, t, and densities, ϱ [*56vU(1)*].

	T °C	t h	ϱ $g cm^{-3}$
●	1250	10	3.6
○	1250	24	4.15
□	1300	10	4.3
△	1350	10	4.4

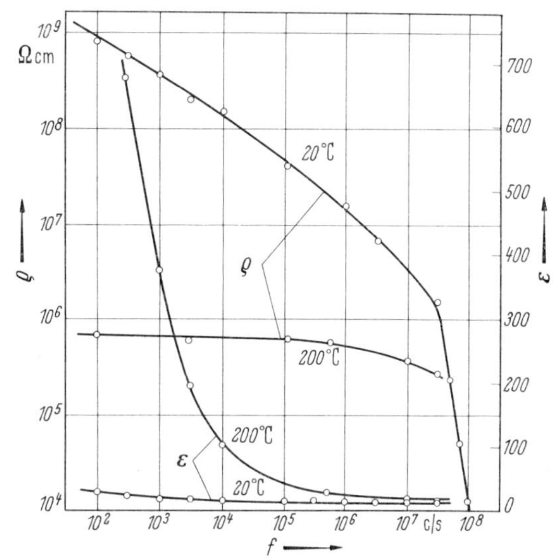

Fig. 12. $Mg_{0.9}Mn_{0.1}Fe_{1.6}O_{3.4}$ (approximate). Effective resistivity, ϱ, and dielectric constant, ε, vs. frequency, f, at 20 °C and 200 °C [*58PS*]. See note in Fig. 14.

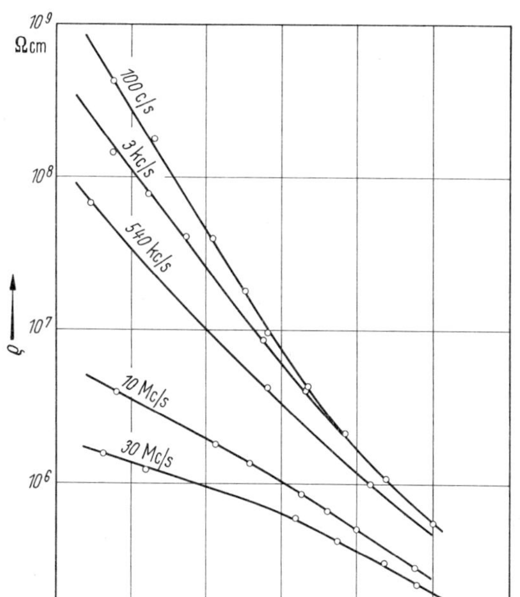

Fig. 13. $Mg_{0.9}Mn_{0.1}Fe_{1.6}O_{3.4}$ (approximate). Resistivity, ϱ, vs. temperature, T, at frequencies from 100 c/s to 30 Mc/s [*58PS*]. See note in Fig. 14.

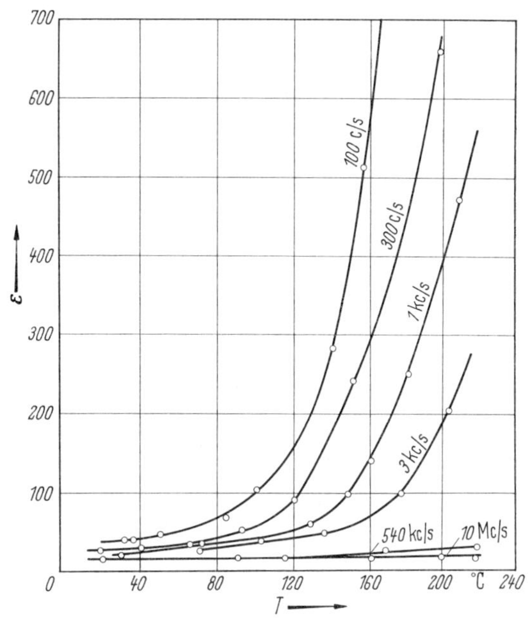

Fig. 14. $Mg_{0.9}Mn_{0.1}Fe_{1.6}O_{3.4}$ (approximate). Dielectric constant, ε, vs. temperature, T, at frequencies from 100 c/s to 10 Mc/s [*58PS*].

Note for Fig. 12···14:

Measurements were made by a substitution method or a Schering bridge at the lower frequencies and by a Wayne-Kerr admittance bridge at the higher frequencies. Samples were 1 mm thick and 2.5 cm in diameter. Contacts were formed by rubbing carbon or evaporating Ag or Au to the ground surfaces and clamping on brass discs to which leads had been soldered. The density of the material was about 92% of the X-ray value. Ceramic samples.

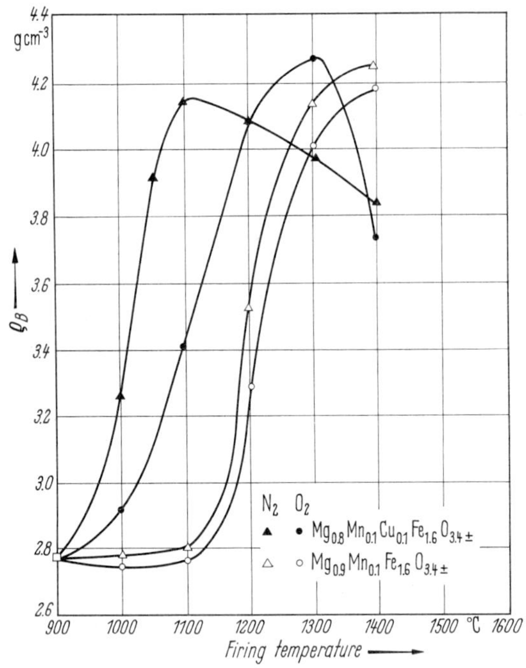

Fig. 15. MgMn ferrites. Bulk density, ϱ_B, vs. firing temperature as a function of firing atmosphere and Cu addition [*57B*]. The addition of Cu permits firing at lower temperatures and reduces possibility of Fe^{2+} formation.

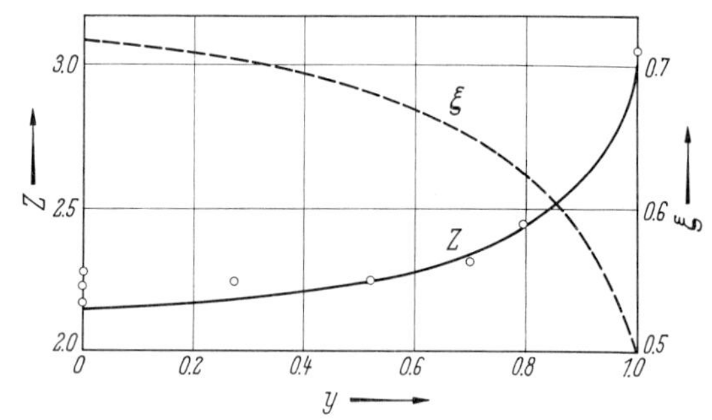

Fig. 16. $Mn_xMg_yFe_{3-(x+y)}O_{4+\gamma}$. Effective average valence of Mn, Z, and its distribution coefficients, ξ, vs. Mg content, y [*60Br*].
The solid line is the plot of Z vs. y from $Z = 3 - (1-y)/(a-y)$ in which $a = x + y \cong 1.15$. Points denote experimental values from Tab. 6. All Fe is assumed to be trivalent. Fraction of Mn on A sites, ξ, is plotted vs. y from $\xi = \alpha\,[3 - (1-y)/(a-y)]^{-1} + \beta$ where α, β are constants of value 1.60 and −0.0346, respectively. This equation is derived assuming a linear relationship between ξ and $1/Z$.

Tab. 6. Properties of some Fe-deficient MgMn ferrites as a function of Mg content [*60Br*].

$Mn_xMg_yFe_{3-(x+y)}O_{4+\gamma}$				Average valency of Mn	Saturation magnetic moment	
x	y	$3-(x+y)$	γ	Z	$(p_m^0)'$ μ_B	p_m^0 μ_B
0.144	1.000	1.856	$+0.004_3$	3.06	2.14	2.14
0.382	0.795	1.823	-0.002_4	2.45	2.28_5	2.28_5
0.455	0.700	1.845	-0.004_8	2.32	2.37	2.37
0.632	0.522	1.846	$+0.002_3$	2.25	2.81	2.82
0.880	0.272	1.848	$+0.028_6$	2.24	3.39	3.48
1.155	0	1.845	$+0.078_9$	2.27	3.98	4.24
1.15	0	1.85	$+0.050_9$	2.22	4.05	4.21
1.15	0	1.85	$+0.017_6$	2.16	4.15	4.21

For all compositions $x + y \simeq 1.15$ and Fe content $\simeq 1.84$. See Fig. 16.
Materials were prepared by prefiring mixed powders for 2 hours at 950 °C, firing in air for 3 hours at 1400 °C, and cooling to RT in air for $\simeq$ 3 hours. X-ray diffraction established the presence of a single phase except for the sample having $y = 1.000$ which contained a small percentage of MgO. Compositions, including O_2 content, γ, were established by chemical analysis. From γ, an average *effective* valency of Mn, Z, was calculated by assuming all Fe ions are trivalent. The author's model postulates the presence of Mn^{2+}, Mn^{4+}, Fe^{2+} as well as Mn^{3+}, Fe^{3+}.
The magnetic moment was measured by a ballistic method using spheres of 2···4 mm diameter. $(p_m^0)'$ was determined by extrapolating p_m data from 77···293 °K to 0 °K. p_m^0 represents the moments corrected to $\gamma = 0$ by using $p_m^0 = (p_m^0)' + 3.3\,\gamma$. See source for more details.

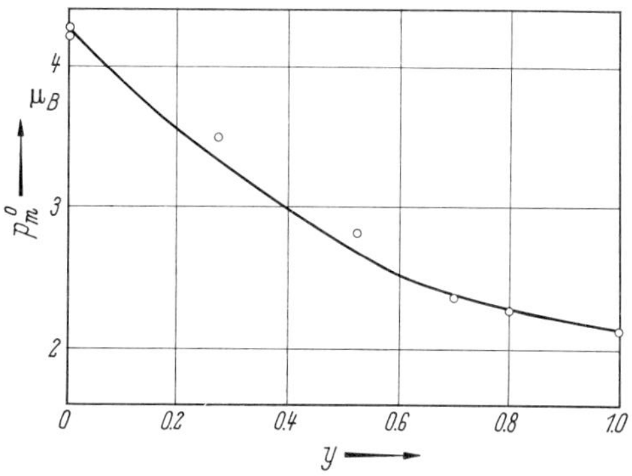

Fig. 17. $Mn_xMg_yFe_{3-(x+y)}O_{4+\gamma}$. Magnetic moment, p_m^0, vs. Mg content, y [*60Br*].
Solid line calculated from $p_m^0 = 0.935 + 4.62\,\xi - (3 + 2\,\xi - 10\,\eta)\,y$ for $y < 0.61$, and from $p_m^0 = 0.535 + 4.62\,\xi - (1 + 4\,\xi - 10\,\eta)\,y$ for $y > 0.61$. ξ, the fraction of Mn occupying *A* sites, is determined from Fig. 16 and η, the fraction of Mg occupying *A* sites, is calculated from $\eta = 0.75\,[1 - \xi(a - y)]/(2 + y)$ where $a = x + y \cong 1.15$. See source for details. Points denote experimental values of p_m^0 as given in Tab. 6.

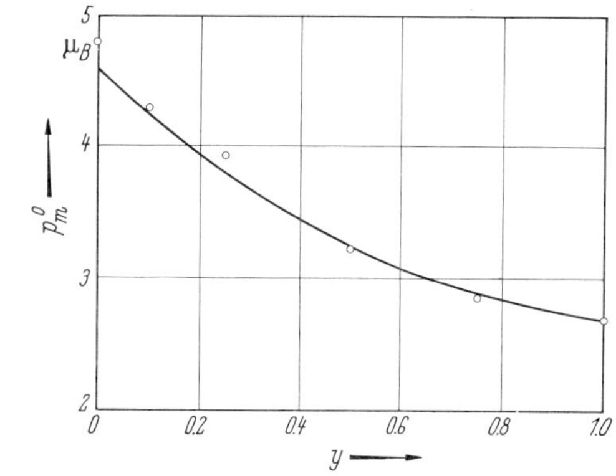

Fig. 18. $Mn_{1-y}Mg_yFe_2O_4$. Magnetic moment, p_m^0, vs. Mg content, y [*60Br*].
Solid line calculated from $p_m^0 = 4.6 - (4.6 - 10\eta)\,y$. η, the fraction of Mg on *A* sites, is found from $\eta = 0.8\,[1 - 0.8\,(1 - y)]/(2 + y)$. Points denote experimental values from [*56KH*]. (Some of these values are given in Fig. 2.) See source for details.

Tab. 7. MgMn ferrites. Average valency of Mn and amount of excess Mn^{3+} (or Mn^{2+}) vs. firing atmosphere and composition [*61BeBo*].

Composition	Average valency of Mn after firing in:			Excess Mn^{3+} or Mn^{2+} after firing in:			
	Oxygen	Air	"Vacuum"	Oxygen	Air	"Vacuum"	"Vacuum"
				Mn^{3+}	Mn^{3+}	Mn^{3+}	Mn^{2+}
$Mg_{1.10}Mn_{0.13}Fe_{1.77}O_{\approx 4}$	3.356	2.840	2.360	—	0.0328	—	0.02895
$Mg_{0.780}Mn_{0.383}Fe_{1.84}O_{\approx 4}$	2.506	2.530	2.355	0.039	0.041	—	0.0225
$Mg_{0.525}Mn_{0.633}Fe_{1.85}O_{\approx 4}$	2.295	2.292	2.226	0.048	0.0435	—	0.0087
$Mg_{0.275}Mn_{0.883}Fe_{1.85}O_{\approx 4}$	2.32	2.290	2.222	0.153	0.089	0.039	—
$Mn_{1.15}Fe_{1.85}O_{\approx 4}$	2.354	2.317	2.222	0.280	0.238	0.132	—

Oxygen firing: $3\frac{1}{2}$ hours in air at 1400 °C, 2 hours in O_2 at 1400 °C, cooled in O_2.
Air firing: $5\frac{1}{2}$ hours in air at 1400 °C, cooled in air. "Vacuum" firing: $5\frac{1}{2}$ hours in air at 1400 °C, 2 hours at 1250 °C in a vacuum of $\approx$ 0.8 mm Hg.
Compositions, including O_2 content, were determined by chemical analysis. Data were calculated with the assumption that all Fe ions are trivalent. For the first entry, a MgO phase was found and some of the Mn is Mn^{4+}. Excess Mn ions represent the amount per mole beyond the requirement for stoichiometry and disregards presence of 2nd phase for the first entry. In addition to the observed increase in valence of Mn with increase in Mg/Mn, a decrease in O_2 content was observed as reflected in the decrease of excess Mn^{3+}. See Fig. 19.

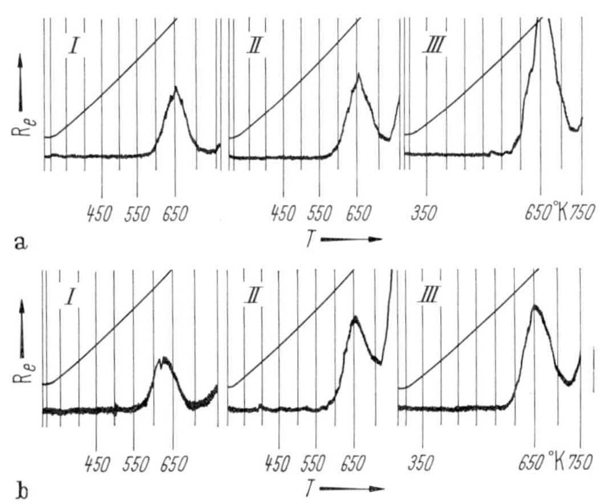

Fig. 19a, b. MgMn ferrites. Typical curves of thermionic emission vs. temperature, T [*61BeBo*].
R_e is the relative rate of electron emission. Experiments were carried out in air. Rate of temperature rise (straight lines) was 2°/sec. *I*) Oxygen-fired, *II*) Air-fired, *III*) "Vacuum"-fired. See Tab. 7 for details of *I, II, III*. All samples showed a peak at 650 °K. Curves, together with data of Tab. 7, are interpreted to indicate the liberation of electrons bound to oxygen vacancies. The characteristic temperature, 650 °K, indicates the bonding energy while the height of the peak indicates the density of vacancies which is a function of Mg content and O_2 partial pressure during firing.
a) $Mg_{1.10}Mn_{0.13}Fe_{1.77}O_{\approx 4}$, b) $Mg_{0.525}Mn_{0.633}Fe_{1.85}O_{\approx 4}$.

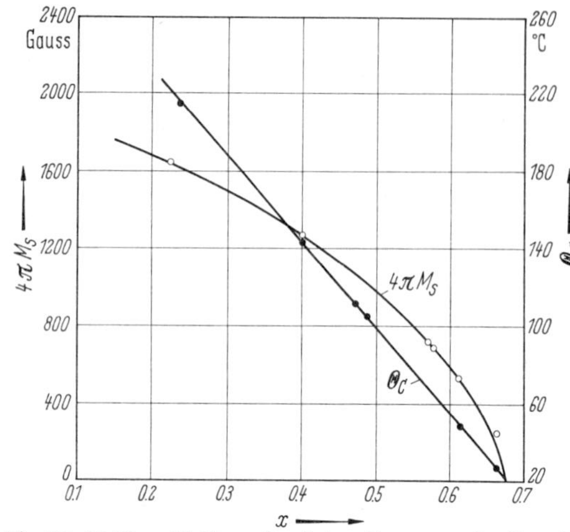

Fig. 20. $MgMn_{0.1}Al_xFe_{2-x}O_4$. Saturation magnetization, M_s, and Curie temperature, Θ_C, vs. Al content, x [55vU].
$4\pi M_s$: At RT and uncorrected for porosity. Maximal density attained by firing at 1400 °C, ϱ varying monotonically from 4.36 to 4.14 $g cm^{-3}$ for x values between 0.229 and 0.676.

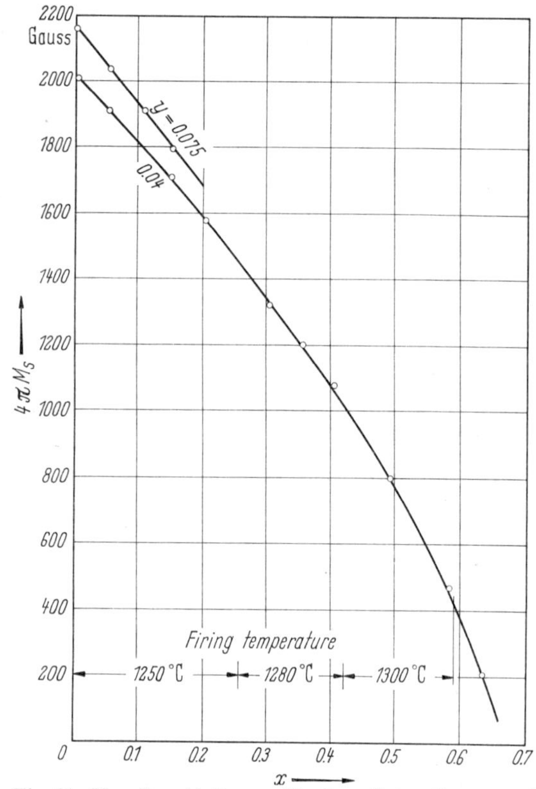

Fig. 21. $Mg_{0.9}Cu_{0.1}Al_xFe_{1.75-x}Mn_yO_{4\pm}$. Saturation magnetization, M_s, vs. Al content, x [57vU].
Data for 28 °C and uncorrected for porosity. Samples were fired in O_2 at the temperatures indicated for 20 hours. By the addition of Cu, the required firing temperature can be reduced from 1350 ··· 1450 °C to 1200 ··· 1300 °C. Θ_C is given by: $\Theta_C = (340 - 460x)$°C. The bulk density, ϱ_B, varied linearly with Al content, x, according to: $\varrho_B = (4.55 - 0.56x)g\ cm^{-3}$.

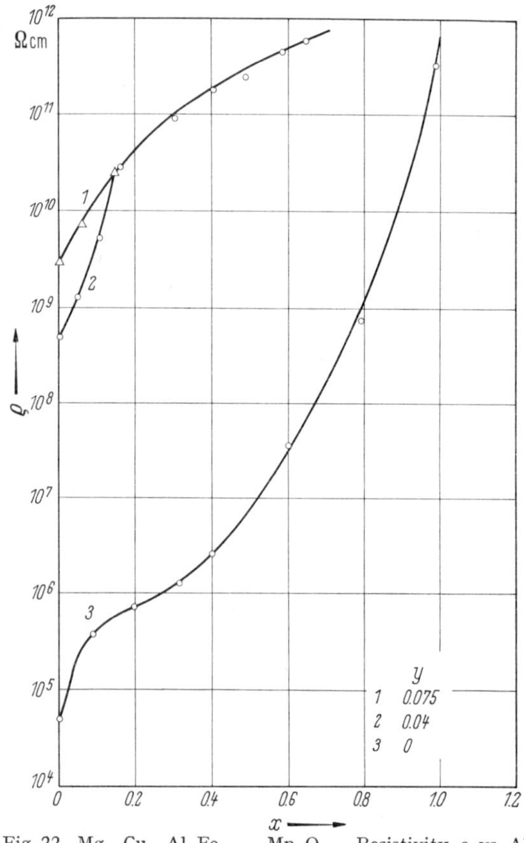

Fig. 22. $Mg_{0.9}Cu_{0.1}Al_xFe_{1.75-x}Mn_yO_{4\pm}$. Resistivity, ϱ, vs. Al content, x [57vU].
Fired in O_2 at 1250 °C for 20 hours.
Mn increased resistivity; increase of resistivity with x is attributed to a dilution effect inhibiting the Fe^{2+}-Fe^{3+} conduction process.

→

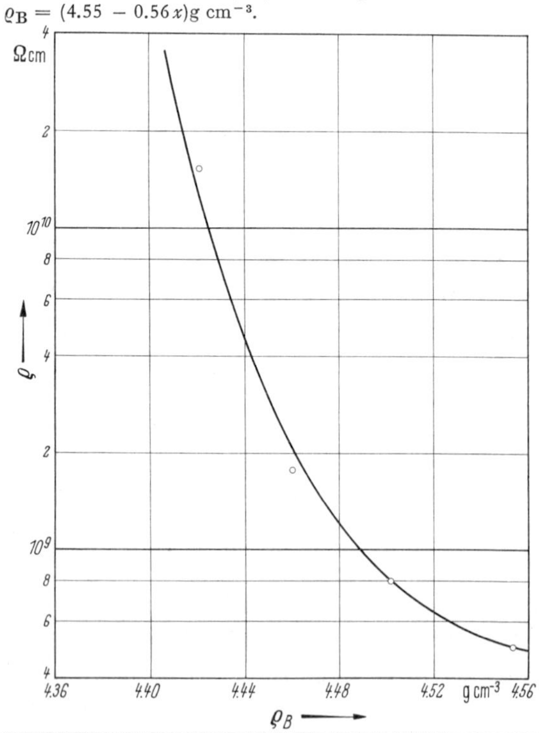

Fig. 23. $Mg_{0.9}Cu_{0.1}Mn_{0.04}Fe_{1.75}O_{4\pm}$. Resistivity, ϱ, vs. bulk density, ϱ_B [57vU].
Samples were furnace cooled.

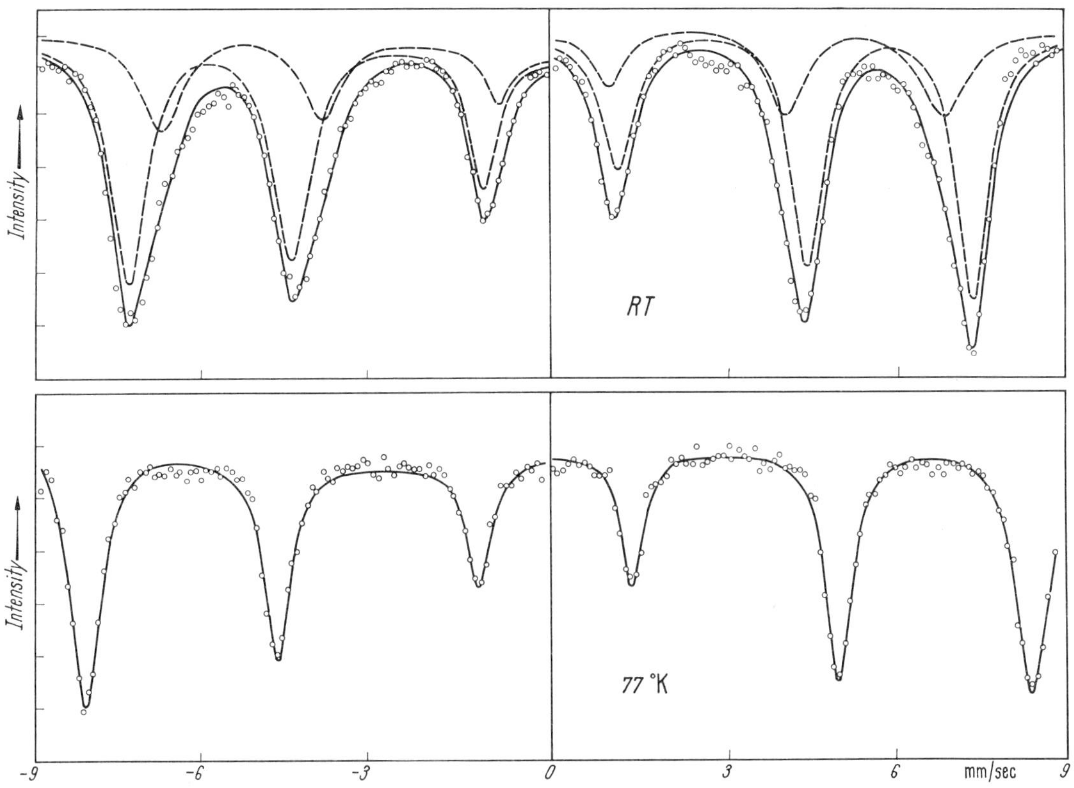

Fig. 24. $Mg_{0.5}Mn_{0.5}Fe_2O_4$. Mössbauer spectrum [*66WMK*]. See Tab. 8, 9 p. 238.

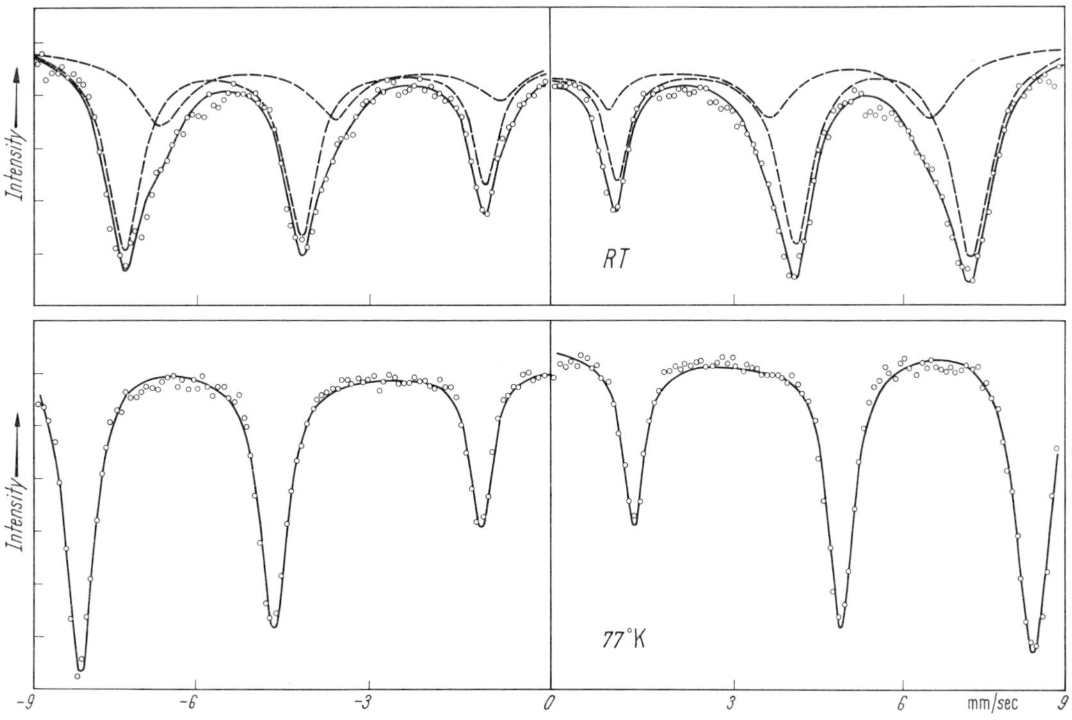

Fig. 25. $Mg_{0.75}Mn_{0.5}Fe_{1.75}O_4$. Mössbauer spectrum [*66WMK*]. See Tab. 8, 9 p. 238.

Tab. 8. Mössbauer effect in MgMn ferrites [*64BDZMST*].

Composition	M_s Gauss	H_{loc} 10^5 Oe	δ mm/sec
$Mg_{0.668}Mn_{0.332}Fe_2O_4$	240	4.36	0.3 ± 0.1
$Mg_{0.756}Mn_{0.244}Fe_2O_4$	197	4.27	0.4 ± 0.1
$Mg_{0.065}Mn_{1.134}Fe_{1.801}O_4$	120	3.73	0.5 ± 0.1

H_{loc} is the local magnetic field at the Fe^{57} nuclei. δ is the isomer shift which depends on density of *s* electrons at the nucleus. See source for data on Mn ferrites.
The powdered ferrites, pressed in a disc with paraffin, were used as an absorber. Cross sections were 10 mg cm^{-2}. Source was Co^{57} in stainless steel. Compare with Tab. 9.

Tab. 9. MgMn ferrites. Mössbauer effect at two temperatures [*66WMK*].

Composition	H_{loc_1} 10^5 Oe	H_{loc_2} 10^5 Oe	δ_1 mm/sec	δ_2 mm/sec	Intensity V	T °K
$Mg_{0.5}Mn_{0.5}Fe_2O_4$	4.75 ±0.25	4.40 ±0.25	−0.02 ±0.15	0.04 ±0.15	2.8 ± 1	293
	5.10 ±0.20	—	0.12 ±0.10	—	—	77
$Mg_{0.75}Mn_{0.5}Fe_{1.75}O_4$	4.50 ±0.25	4.0 ±0.25	−0.03 ±0.15	−0.01 ±0.15	3.1 ± 1	293
	5.10 ±0.20	—	0.10 ±0.10	—	—	77

See Tab. 8 for definitions of parameters and data on other MgMn ferrites. Two values are listed for H_{loc} and δ at RT corresponding to the two superimposed spectra observed at RT only (see Fig. 24, 25). V gives the intensity dependence for the two spectra. Cross sections were 35 mg cm^{-2}. Co^{57} in Pt was used for the γ-ray source. Cation distributions for these compositions are given in Tab. 4. Data were explained on basis of deviations from collinear Néel spin configurations on *B* sites.

Tab. 10. *K*-edge absorption of Fe in MgMn and NiZn ferrites [*66Ki*].

Material	Absorption intensity eV	a_0 Å
Fe	0 ± 0.3	
Fe_3O_4	5 ± 0.5	
Fe_2O_3	7.3 ± 0.5	
$MgFe_2O_4$	6.3 ± 0.4	8.36
$Mg_{0.8}Mn_{0.2}Fe_2O_4$	7.0 ± 0.5	8.38
$Mg_{0.7}Mn_{0.3}Fe_2O_4$	7.1 ± 0.4	8.40
$Mg_{0.6}Mn_{0.4}Fe_2O_4$	7.4 ± 0.5	8.41
$Mg_{0.5}Mn_{0.5}Fe_2O_4$	7.6 ± 0.4	8.46
$Mg_{0.4}Mn_{0.6}Fe_2O_4$	7.2 ± 0.7	8.45
$Mg_{0.3}Mn_{0.7}Fe_2O_4$	7.4 ± 0.6	8.45
$Mg_{0.2}Mn_{0.8}Fe_2O_4$	7.8 ± 0.6	8.48
$Mg_{0.1}Mn_{0.9}Fe_2O_4$	8.0 ± 0.6	8.49
$MnFe_2O_4$	8.3 ± 0.5	8.51
$NiFe_2O_4$	6.0 ± 0.4	8.34
$Ni_{0.7}Zn_{0.3}Fe_2O_4$	6.2 ± 0.4	8.37
$Ni_{0.5}Zn_{0.5}Fe_2O_4$	6.6 ± 0.5	8.39
$Ni_{0.1}Zn_{0.9}Fe_2O_4$	7.4 ± 0.4	8.41
$ZnFe_2O_4$	7.3 ± 0.6	8.44

Tab. 11. *K*-edge absorption of Mn in MgMn ferrites [*66Ki*].

Material	Absorption intensity eV
MnO	0 ± 0.3
MnO_2	4.6 ± 0.4
$KMnO_4$	8.7 ± 0.5
$Mg_{0.8}Mn_{0.2}Fe_2O_4$	1.0 ± 0.4
$Mg_{0.7}Mn_{0.3}Fe_2O_4$	0.6 ± 0.5
$Mg_{0.5}Mn_{0.5}Fe_2O_4$	0.5 ± 0.6
$Mg_{0.3}Mn_{0.7}Fe_2O_4$	1.1 ± 0.5
$Mg_{0.2}Mn_{0.8}Fe_2O_4$	0.8 ± 0.5
$MnFe_2O_4$	1.4 ± 0.5
50% MnO, 50% MnO_2	2.4 ± 0.5

Data indicate a correlation between shift in K-edge absorption of Fe and change in lattice parameter, a_0, and also with the isomer shift in Mössbauer experiments (see Tab. 8).
Fe and Mn appear to be present in ferrites as Fe^{3+} and Mn^{2+} from comparison with data for Fe_2O_3 and MnO, respectively.

Tab. 12. MgMn (Zn, Al) ferrites. Properties of some typical commercial materials [*63 AES*].

Designation	Composition	B at H Gauss	Oe	Θ_C °C	μ_i	$\mu_i Q$ at f $\cdot 10^3$	Mc/s	μ_{max}	H_c Oe
H 1	0.20 MgO-0.22 ZnO-0.04 NiO-0.04 ZnO-0.50 Fe_2O_3	2800	25	125	550	2.5	1	3800	0.35
C	0.05 MgO-0.15 ZnO-0.20 NiO-0.10MnO-0.50 Fe_2O_3	4200	25	330	250	16	1	1100	2.1
R-1	MgO-MnO-Fe_2O_3	2250	250	290	45	0.75	1	320	2.1
R-4	MgO-MnO-Fe_2O_3	2200	250	320	40	0.56	1	470	1.4
R-5	MgO-MnO-Al_2O_3-Fe_2O_3	1500	250	140	65	3.2	1	420	0.8
R-6	MgO-MnO-Al_2O_3-Fe_2O_3	1100	250	100	70	2.8	1	480	0.7
4373-A	MgO-MnO-Fe_2O_3	3500	25	170	1000	250	0.1	1900	0.3
C-2	MgO-MnO-ZnO-Fe_2O_3	3700	25	210	430	4.0	1	4000	0.36
S-1	0.34 MgO-0.23 MnO-0.42 Fe_2O_3	1780	25	265	40	—	1	515	1.5
S-3	0.32 MgO-0.27 MnO-0.41 Fe_2O_3	2000	25	265	45	—	1	1800	0.6
S-6	MgO-MnO-ZnO-CaO-Fe_2O_3	3450	25	200	110	1.0	1	6000	0.2

The 'R' numbers are microwave materials; the 'S' numbers and C-2 are square-loop materials. R-1 is similar in composition to Ferramic A. μ_i measured at f given for $\mu_i Q$ measurement. $Q = 1/\tan\delta_m$. (Materials are manufactured by Electronics Division of Indiana General Corporation USA).

6.1.4.2.2 Anisotropy

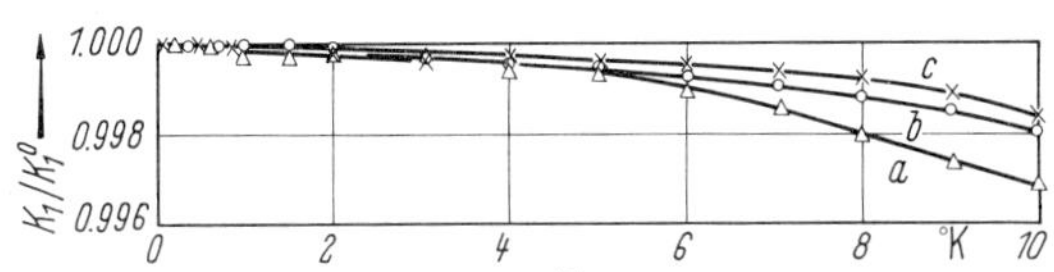

Fig. 26. $Mg_{0.5}Mn_{0.5}Fe_2O_4$. Temperature dependence of the magnetocrystalline anisotropy constant, K_1, from 0 °K to 10 °K for a single crystal [*66DS*].
a) Calculated from the empirical law of Bryukhatov and Kirenskii, $K_1 = K_1^0\, e^{-\alpha T^2}$, where K_1^0 is the anisotropy at 0 °K obtained by extrapolation, $K_1^0 = -20.7 \cdot 10^4$ erg cm^{-3}, and $\alpha = 3.2 \cdot 10^{-5}$ deg^{-2} (see [*63D*]). This relationship is valid up to $\Theta_C/3$ and is used to test the Turov-Mitsek formula (curves *b*, *c*) derived phenomenologically from spin wave theory. Direct measurements of K_1 near 0 °K are difficult to make.
b) Calculated from: $\frac{K_1}{K_1^0} \approx 1 - \frac{10}{(\Theta_C/T)^2 - (\Theta_C/T) + 1}$, which is the Turov-Mitsek formula modified by the Frenkel-Heisenberg equation. $\Theta_C = 733$ °K for the crystal used in curve *a*.
c) Calculated from: $K_1/K_1^0 = 1 - 10\beta(T/\Theta_C)^{3/2}$, which is the Turov-Mitsek formula modified by the Bloch equation. $\beta = 0.1174$. Similar data are presented for Li and Mn ferrites.

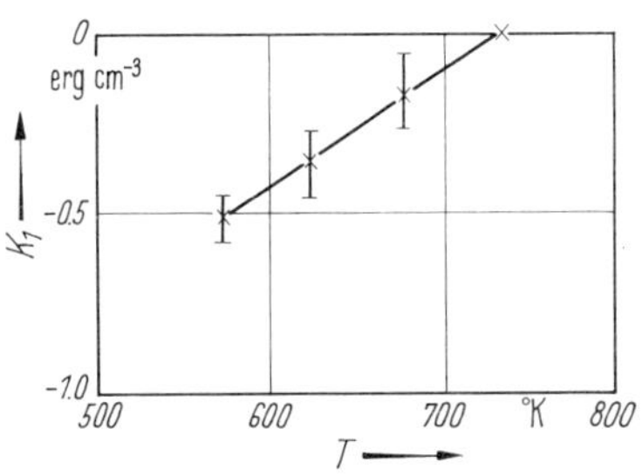

Fig. 27. $Mg_{0.5}Mn_{0.5}Fe_2O_4$. Temperature dependence of the magneto-crystalline anisotropy constant, K_1, near the Curie temperature for a single crystal [*66DS*].
Measurements were made by torque method on a spherical sample having $\Theta_C = 733$ °K. Error limits are indicated for each point. Data check the linear relationship predicted by Vonsovskii on thermodynamic grounds for temperatures approaching Θ_C: $(K_1)_{T\to\Theta_C} = C'(\Theta_C - T)$. C' is a constant independent of temperature. Data also presented for other ferrites.

Tab. 13. Composition and physical properties of samples for Fig. 28 ··· 30 and Tab. 14 [*62S*].

Sample Nr.	Diameter mm	Composition	Porosity by X-ray %	Porosity by powder method %
1	2.273	$Mg_{0.608}Mn_{0.547}Fe_{1.845}O_4$	3.9	1.2
2	2.378	$Mg_{0.700}Mn_{0.455}Fe_{1.845}O_{4.028}$	12.5	10.3
3	2.193	$Mg_{0.700}Mn_{0.455}Fe_{1.845}O_{4.073}$	31.3	31.2

Spherical, polycrystalline samples deviated less than 1% from roundness. Compositions, including O_2 content, are determined on the sintered product.

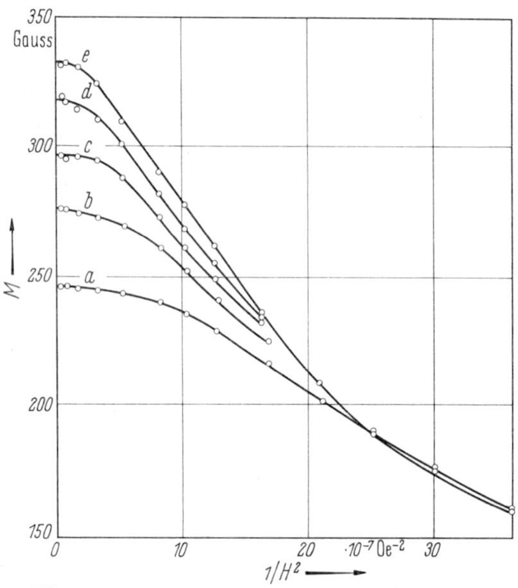

←

Fig. 28. $Mg_{0.608}Mn_{0.547}Fe_{1.845}O_4$. Magnetization, M, vs. $1/H^2$ for a Fe-deficient MgMn ferrite [*62S*].
Material is sample 1 of Tab. 13. Measurements were made over the descendent part of the hysteresis loop from 5000 Oe to 500 Oe using a ballistic method. *a*) 295 °K, *b*) 241 °K, *c*) 200 °K, *d*) 145 °K, *e*) 77 °K. Temperature control to ±2 °K. See Tab. 14.

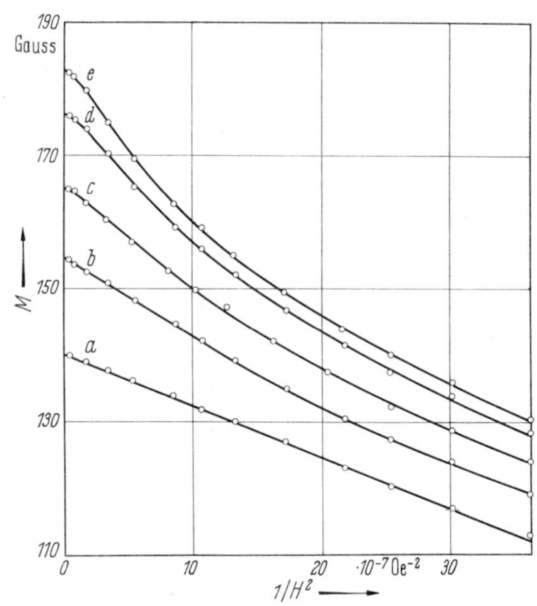

Fig. 30. $Mg_{0.700}Mn_{0.455}Fe_{1.845}O_{4.073}$. Magnetization, M, vs. $1/H^2$ for a Fe-deficient MgMn ferrite [*62S*].
Material is sample 3 of Tab. 13. *a*) 293.5 °K, *b*) 242.5 °K, *c*) 203 °K, *d*) 145.5 °K, *e*) 77 °K. Refer to notes under Fig. 28. See Tab. 14.

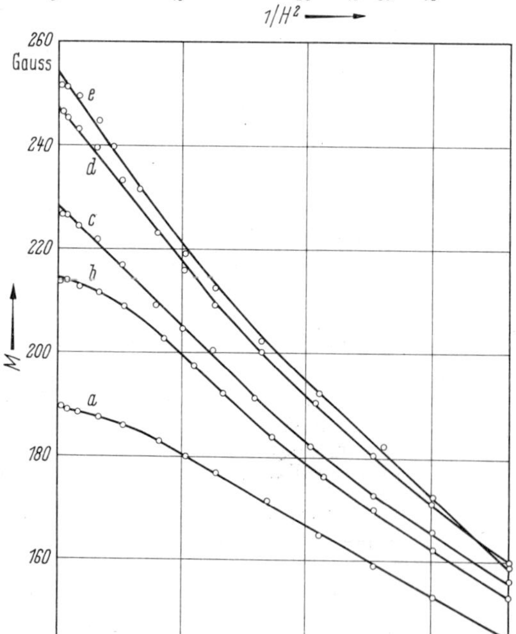

←

Fig. 29. $Mg_{0.700}Mn_{0.455}Fe_{1.845}O_{4.028}$. Magnetization, M, vs. $1/H^2$ for a Fe-deficient MgMn ferrite [*62S*].
Material is sample 2 of Tab. 13. *a*) 297 °K, *b*) 240 °K, *c*) 200 °K, *d*) 145.5 °K, *e*) 77 °K. Refer to notes under Fig. 28. See Tab. 14.

Tab. 14. Effective anisotropy constants of Fe-deficient MgMn ferrites [62S].

Sample Nr.	T °K	$\lvert K_1\rvert$ 10^5 erg/cm³	$\lvert K_u\rvert$ 10^5 erg/cm³	$r = m/M_\infty^3$
1	295	3.26	1.75	1.60
	241	4.06	2.17	1.64
	200	4.57	2.44	1.64
	145	4.95	2.65	1.62
	77	5.53	2.96	1.64
2	297	1.98	1.06	2.02
	240	2.47	1.32	2.00
	200	2.64	1.41	1.97
	145.5	3.05	1.63	1.90
	77	3.33	1.78	2.03
3	293.5	1.21	0.64	2.87
	242.5	1.54	0.82	3.17
	203	1.87	1.00	3.56
	145.5	2.21	1.18	3.79
	77	2.56	1.37	4.37

Values of K_1, magnetocrystalline anisotropy, and K_u, uniaxial anisotropy, can be calculated from Fig. 28···30 and compared to determine which anisotropy is applicable, according to the law of approach to saturation, $M = -b \cdot M_s/H^2$. Therefore, the slope, m, of the linear portion of the curve gives $-b \cdot M_s$ and the extrapolation of the linear segment to infinite H gives M_∞ which is taken as the value of M_s for the calculations performed. For a ceramic spinel, $b = 0.0762\ K_1^2/M_s^2$ if the controlling anisotropy is magnetocrystalline, and $b = 0.267\ K_u^2/M_s^2$ if the controlling anisotropy is uniaxial. It is concluded that the effective anisotropy must be a uniaxial-shape anisotropy arising from demagnetizing effects of the constituent grains since the above values calculated for K_1 are not in agreement with the accepted values ($\approx 10^4$ erg cm^{-3}). See source for significance of r.

Tab. 15. Effective anisotropy constants of Fe-deficient MgMn ferrites [62S].

Molar ratio of oxides			$\lvert K_1\rvert$ 10^5 erg/cm³		M_∞ Gauss		$\lvert K_1\rvert/M_\infty^2$	
Fe_2O_3	MnO	MgO	RT	77 °K	RT	77 °K	RT	77 °K
1	1.30	—	6.71	13.34	356	500	5.3	5.4
0.444	0.426	0.13	5.00	9.22	302	406	5.5	5.6
0.444	0.304	0.251	3.79	6.85	266.₅	350	5.3	5.6
0.444	0.176	0.38	2.39	4.21	211.₅	279	5.3	5.4
0.444	0.055	0.5	2.34	4.27	202.₅	273	5.7	5.7

Refer to notes for Tab. 14. For these materials K_u was not calculated. K_1 cannot be applicable because the above calculated values are too high. As concluded for materials in Tab. 14, a uniaxial-shape anisotropy must also be the controlling anisotropy for these materials.

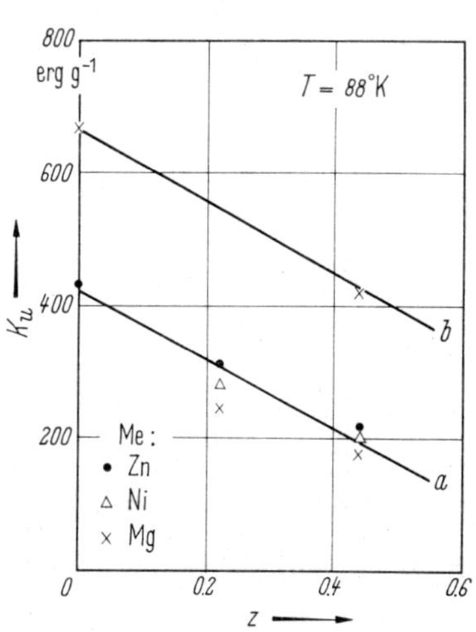

Fig. 31. $Me_zMn_{1.4-z}Fe_{1.6}O_{4+\gamma}$. Induced anisotropy, K_u, vs. Mg, Ni, or Zn content, z, at 88 °K [66GMMS]. Ceramic samples were sintered at 1400 °C in air and then refired for 10 hours at 1200 °C under 760 Torr (curve a) or 10^{-2} Torr (curve b). Measurements were made by torque method.

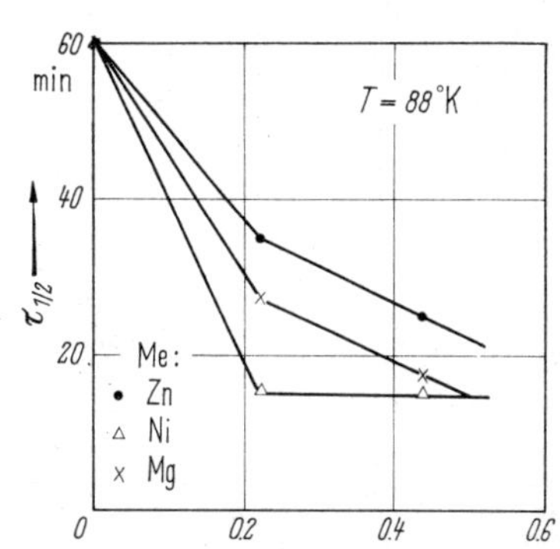

Fig. 32. $Me_zMn_{1.4-z}Fe_{1.6}O_{4+\gamma}$. Half the relaxation time, $\tau_{1/2}$, vs. Mg, Ni, or Zn content, z, at 88 °K [66GMMS]. Same notation as in Fig. 31 for the refiring at 760 Torr.

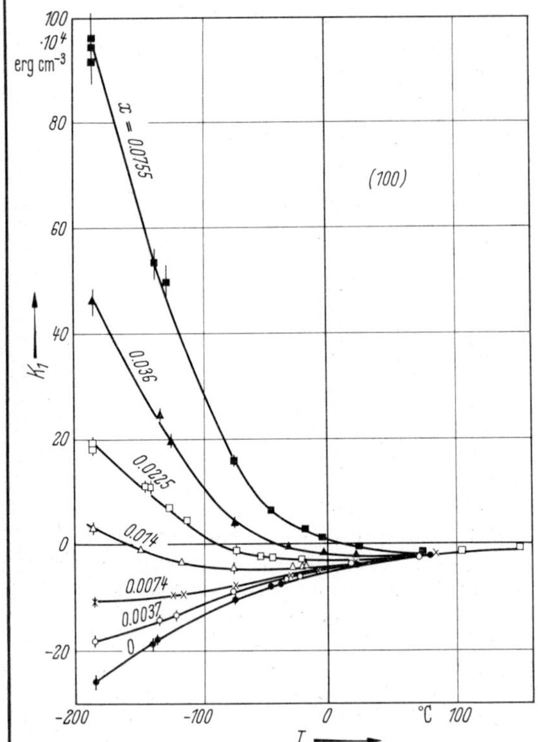

Fig. 33. $Co_xMg_{0.75}Mn_{0.53}Fe_{1.71}O_{3.96+x}$. First order anisotropy constant, K_1, vs. temperature, T. Parameter x [*65E*]. Measurements made in (100) plane by torque method.

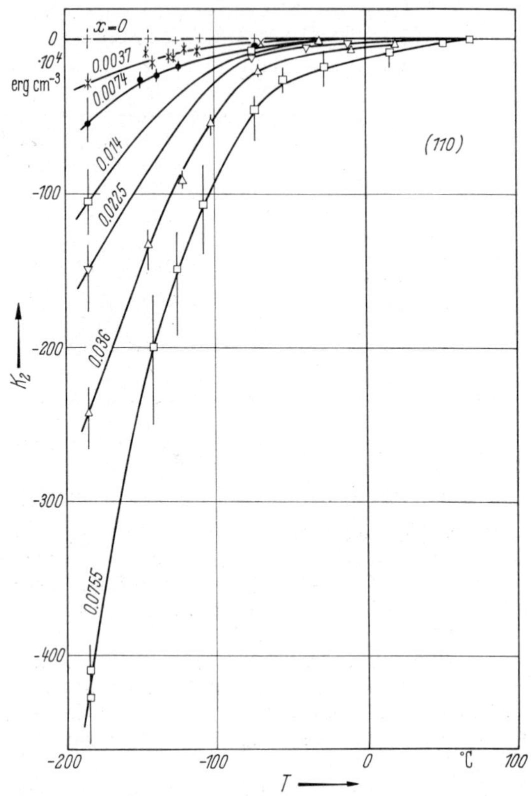

Fig. 34. $Co_xMg_{0.75}Mn_{0.53}Fe_{1.71}O_{3.96+x}$. Second order anisotropy constant, K_2, vs. temperature, T. Parameter x [*65E*]. Measurements in (110) plane.

Tab. 16. $Co_xMg_{0.75}Mn_{0.53}Fe_{1.71}O_{3.96+x}$. K_1 vs. Co content, x, and temperature, T [*65E*].

T [°C]	−185	−75	+20
x	K_1 [10^4 erg cm^{-3}]		
0	−26.0	−10.45	−4.0
0.0037	−18.4	− 9.1	−4.0
0.0074	−10.9	− 8.05	−4.0
0.014	+ 2.85	− 4.6	−3.5
0.0225	+18.9	− 1.1	−2.8
0.036	+46.0	+ 4.2	−2.0
0.0755	+94.0	+15.8	−0.7

→

Fig. 35. $Co_xMg_{0.75}Mn_{0.53}Fe_{1.71}O_{3.96+x}$. Co content, x, for zero anisotropy vs. temperature, T [*65E*].
Measurements made in (100) and (110) planes. Easy, hard, and middle directions are indicated for the three regions of the diagram.

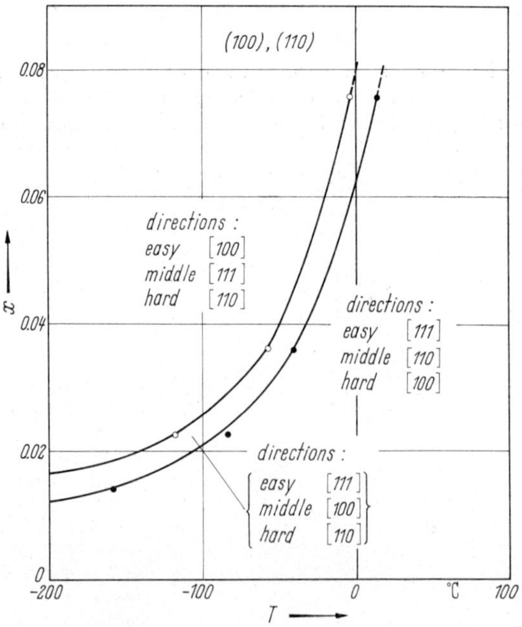

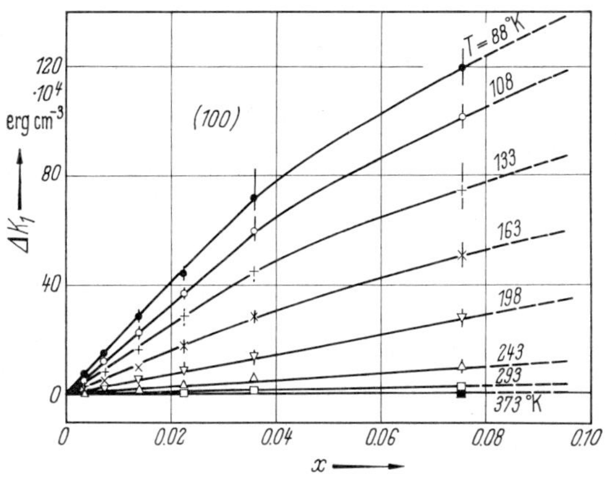

Fig. 36. $Co_xMg_{0.75}Mn_{0.53}Fe_{1.71}O_{3.96+x}$. Change in K_1 vs. Co content, x. Parameter T [*65E*].
Measured in (100) plane.

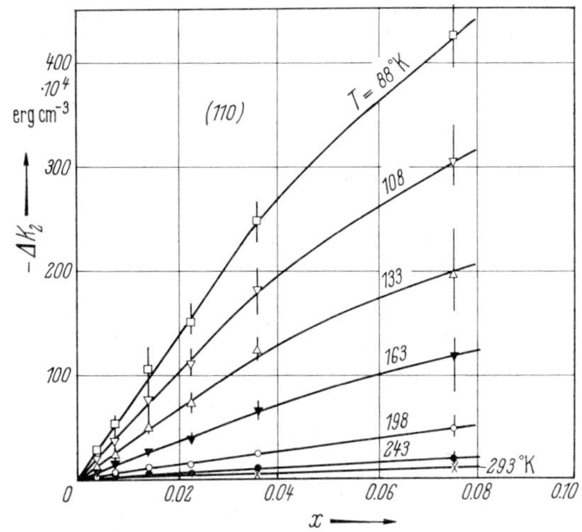

Fig. 37. $Co_xMg_{0.75}Mn_{0.53}Fe_{1.71}O_{3.96+x}$. Change in K_2 vs. Co content, x. Parameter T [*65E*].
Measurements in (110) plane.

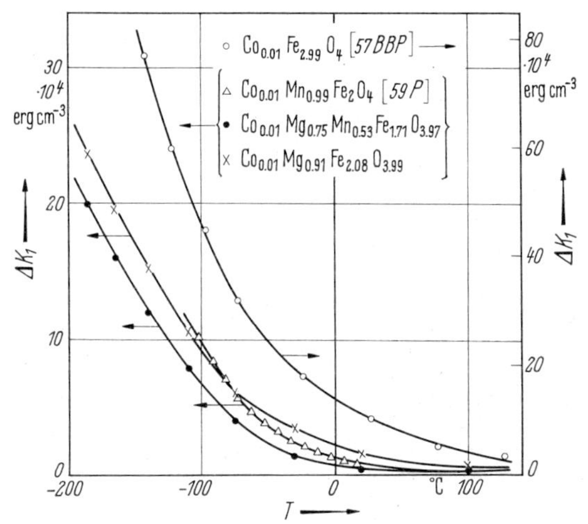

Fig. 38. Comparative effect of 1 mole-% Co ferrite on K_1 for various ferrites [*65E*].
The ordinate is the change in K_1, ΔK_1, caused by the introduction of 1 mole-% Co ferrite into each ferrite. See [*57BBP*], [*59P*].

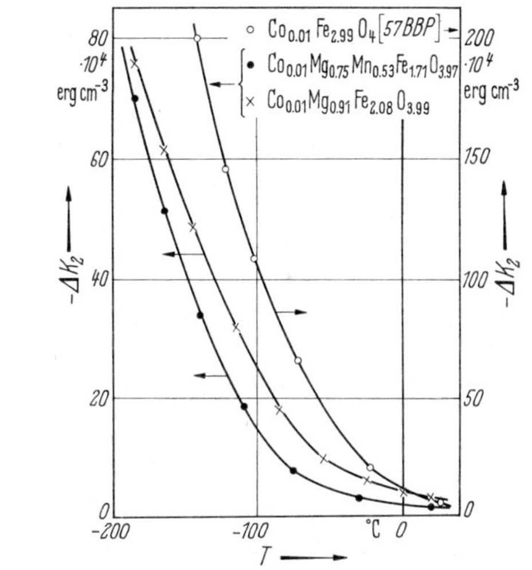

Fig. 39. Comparative effect of 1 mole-% Co ferrite on K_2 for various ferrites [*65E*]. See notation of Fig. 38. See also [*57BBP*].

6.1.4.2.3 Losses

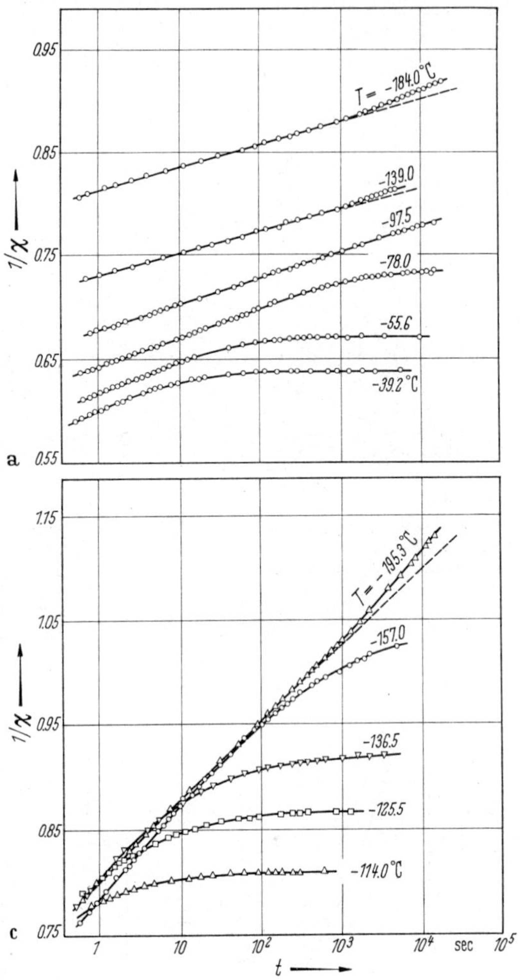

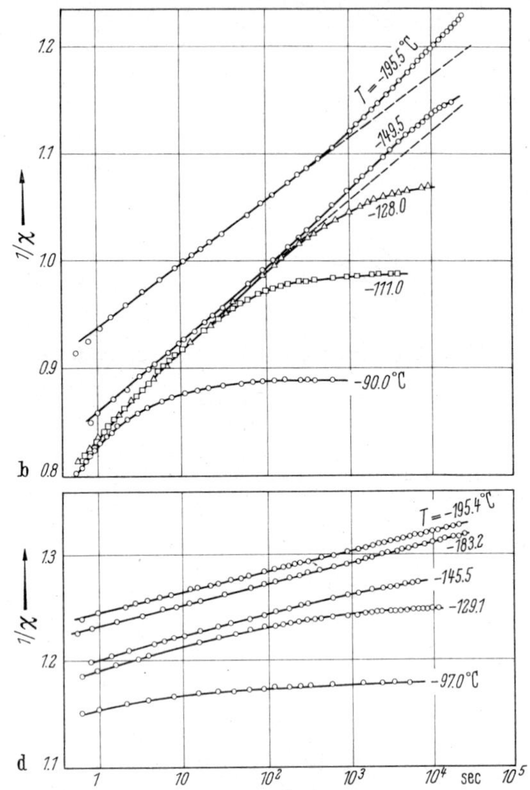

Fig. 40a ··· d. Permeability disaccomodation in MgMn ferrites: reciprocal susceptibility, $1/\chi$, vs. time, t. Parameter T [*60K*]. See notes under Tab. 17.

χ = complex volume susceptibility, t = time passed after demagnetization. Data were taken at frequencies of 100 ··· 200 kc/s using a resonance bridge. The amplitude of the ac field was kept at ≈1 mOe. Measurements were initiated 0.6 sec after demagnetization and corrections were made for time to achieve demagnetization (0.2 sec). Temperature was maintained to ±0.05 °C.

a) $Mg_{1.154}Mn_{0.137}Fe_{1.760}O_4$ (sample 1),
b) $Mg_{0.795}Mn_{0.382}Fe_{1.825}O_4$ (sample 2),
c) $Mg_{0.522}Mn_{0.632}Fe_{1.845}O_4$ (sample 3),
d) $Mg_{1.003}Fe_{2.006}O_4$ (sample 5).

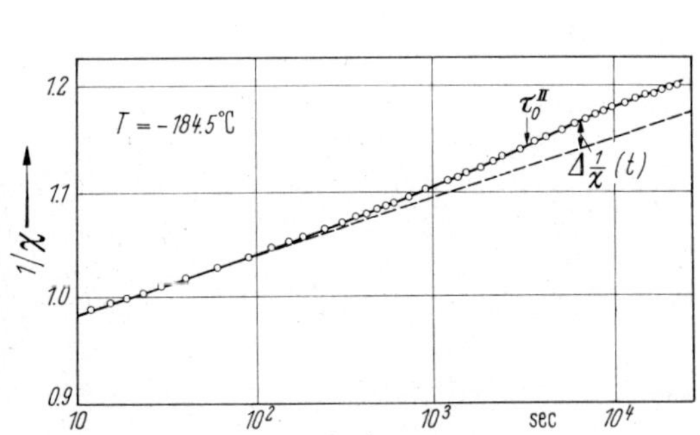

Fig. 41. $Mg_{0.795}Mn_{0.382}Fe_{1.825}O_4$. Main and superimposed disaccomodation processes [*60K*].

Experimental disaccomodation curve of sample 2 of Tab. 17 at −184.5 °C (see Fig. 40b). Deviation from linearity indicates that the main relaxation process is overlapped by another relaxation process. Subtraction of the extrapolated linear part of the $1/\chi$ vs. t curve corresponding to I yields the disaccomodation curve, $\Delta 1/\chi(t)$, of the superimposed relaxation (II) in Fig. 42. I and II are not mutually independent, though for this calculation they are treated as such. τ_0^{II} = the mean time constant for process II; τ_0^{II} can be determined from the displacement of the inflection point. For denotations see Tab. 17.

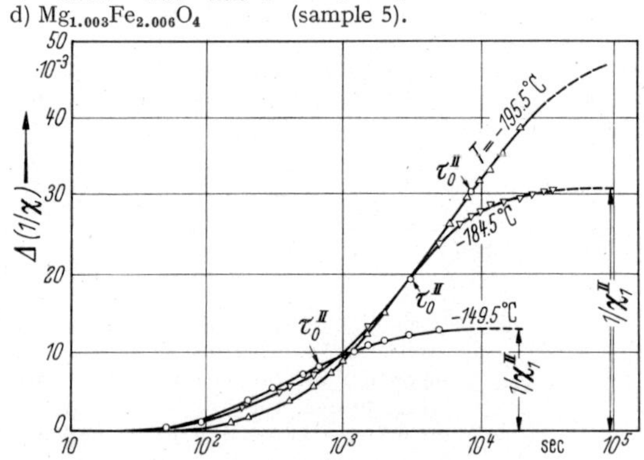

Fig. 42. $Mg_{0.795}Mn_{0.382}Fe_{1.825}O_4$. The superimposed disaccomodation process [*60K*].

Material is sample 2 of Tab. 17. Points were obtained from the smoothed experimental curve for the total disaccomodation by the method indicated in Fig. 41. Arrows indicate mean time constants, τ_0^{II}, at each temperature. Magnitudes of the relaxation for process II, $1/\chi_1^{II}$, are indicated. For denotations see Tab. 17.

Tab. 17. Composition and properties of materials used in study of magnetic after-effect (Fig. 40 ··· 45, Tab. 18) [*60K*].

Nr. of sample	Chemical formula	Mean grain size 10^{-3} mm	V_{rel} %	M_s [Gauss] 77 °K	M_s [Gauss] 300 °K
1	$Mg_{1.154}Mn_{0.137}Fe_{1.760}O_4$	3.1	8	156	122
2	$Mg_{0.795}Mn_{0.382}Fe_{1.825}O_4$	4.5	8	250	180
3	$Mg_{0.522}Mn_{0.632}Fe_{1.845}O_4$	6.2	8	312	217
4	$Mg_{0.270}Mn_{0.870}Fe_{1.833}O_4$	9.5	13	382	265
5	$Mg_{1.003}Fe_{2.006}O_4$	—	11	118	92

V_{rel} = relative volume of pores. Samples were of toroidal shape.
Denotations used in Fig. 40 ··· 45 and Tab 18:

I = main relaxation process;
II = superimposed relaxation process;
III = denotes parameters determined from the temperature dependence of tan δ_m (Fig. 44, 45);
τ_1 = shortest time constant of relaxation process;
τ_2 = longest time constant of relaxation process;
τ_0 = mean time constant;
τ_∞ = stabilized value of τ;
q_1, q_2, q_0 = corresponding activation energies;
χ = volume susceptibility;
χ_0 = unrelaxed susceptibility;
χ_∞ = relaxed susceptibility (stabilized value);

$$\frac{1}{\chi_1} = \frac{1}{\chi_\infty} - \frac{1}{\chi_0}.$$

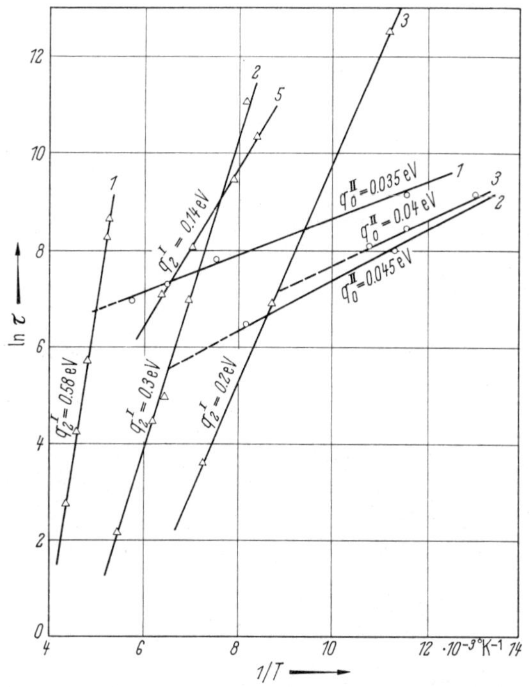

Fig. 43. MgMn ferrites. Activation energies for disaccomodation processes; dependence of ln τ_2^I and ln τ_0^{II} on reciprocal temperature, $1/T$ [*60K*]. τ in sec.
Curve numbers indicate materials as given in Tab. 17. Slope of the plot of the longest time constant for the main process (I), τ_2^I, vs. $1/T$, gives the activation energy, q_2^I, for this relaxation. Likewise, the activation energy, q_0^{II}, corresponding to the mean time constant, τ_0^{II}, for the superimposed process (II) is given by plot of τ_0^{II} vs. $1/T$. The Arrhenius relation, $\tau = \tau_\infty e^{q/kT}$, is obeyed. τ_0^{II} is obtained from the inflection points as shown for sample 2 in Fig. 42. τ_2^I is obtained from $1/\chi_\infty = A + B \ln t$; A and B are constants. For denotations see Tab. 17. Data tabulated in Tab. 18.

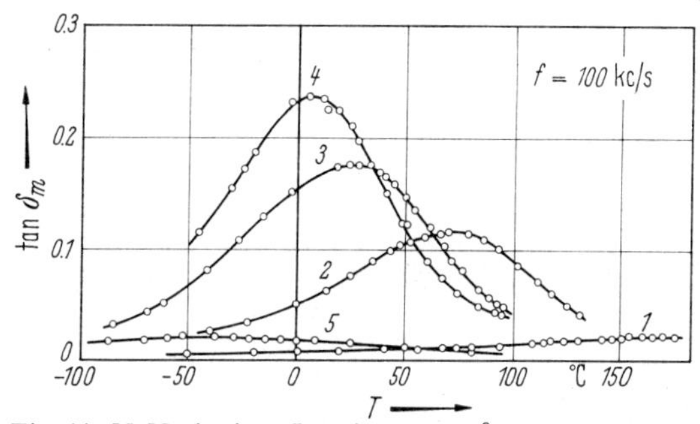

Fig. 44. MgMn ferrites. Loss factor, tan δ_m, vs. temperature, T [*60K*].
Curve numbers indicate compositions given in Tab. 17. Data for Fig. 44, 45 taken at frequencies of 2 ··· 200 kc/s (using the same resonance bridge as for Fig. 40a ··· d; thus time constants $\tau = \omega^{-1}$ lie between $10^{-4} \cdots 5 \cdot 10^{-6}$ sec; ac magnetizing field ≈ 1 mOe. Some of the material is published in [*60DM*]. See also [*61Kr*].

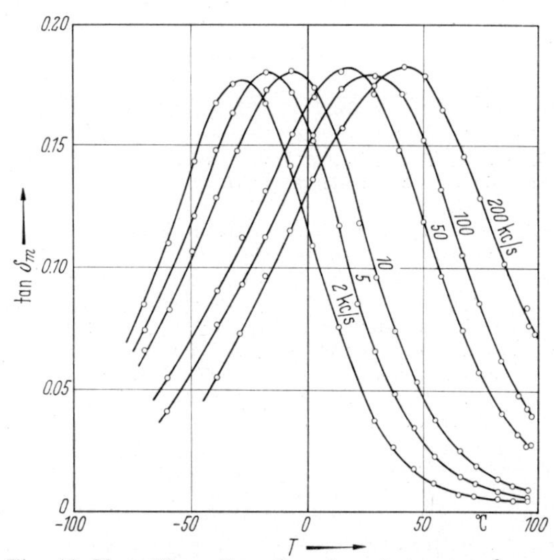

Fig. 45. $Mg_{0.522}Mn_{0.632}Fe_{1.825}O_4$. Loss factor, tan δ_m, vs. temperature, T, for different frequencies [*60K*].
Material is sample 3 of Tab. 17. See notes in Fig. 44.

Tab. 18. MgMn ferrites. Summary of activation energies for relaxation processes [*60K*].

Nr. of sample	Mn content moles	q_2^{I} eV	τ_{200}^{I} sec	q_0^{II} eV	q_0^{III} eV	Δq^{III} eV	τ_{∞}^{III} sec
1	0.137	0.58	$1 \cdot 10^{-12}$	0.035	—	0.3	—
2	0.382	0.30	$1 \cdot 10^{-7}$	0.045	0.55	0.16	$1 \cdot 10^{-14}$
3	0.632	0.20	$2 \cdot 10^{-6}$	0.04	0.45	0.14	$4 \cdot 10^{-14}$
4	0.870	—	—	—	0.39	0.14	$1 \cdot 10^{-13}$
5	0	0.14	$1 \cdot 10^{-6}$	—	0.34	0.3	$3 \cdot 10^{-14}$

Exact compositions are given in Tab. 17.

q_2^{I} = activation energy for the main disaccomodation process;
q_0^{II} = activation energy for the superimposed disaccomodation process;
q_0^{III} = mean activation energy in the frequency-temperature behavior of the complex susceptibility;
τ_{∞}^{III} = stabilization time for process III;
Δq^{III} = the permissible variation from q_0^{III} such that all activation energies are contained within its limits.

For denotations see also Tab. 17.

It is concluded that electron diffusion involving Mn ions on the *B* sites was responsible for the 3 relaxations.

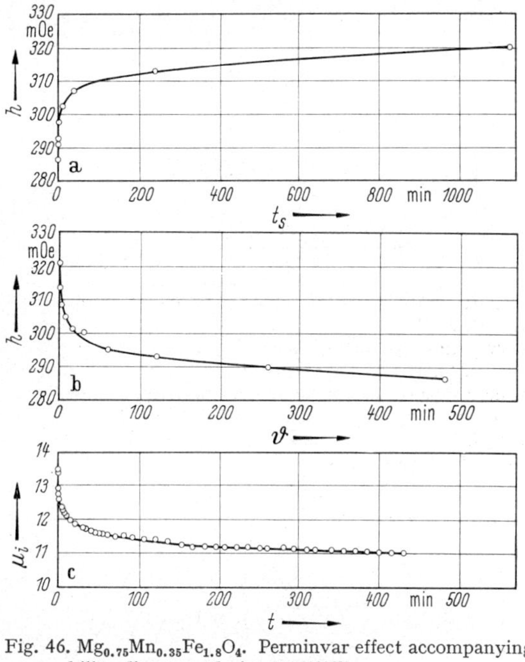

Fig. 46. $Mg_{0.75}Mn_{0.35}Fe_{1.8}O_4$. Perminvar effect accompanying permeability disaccomodation [*59BKZ*].
T = 77 °K.
a) Effective after-effect field, h, vs. stabilization time, t_s. For a perminvar-type hysteresis loop, $h = (H_{max} + H_p)/2$ where H_{max} = applied field and H_p = field at the point of closure. t_s is the time lapse between the end of demagnetizing the toroid and switching on H_{max}. H_{max} = 0.60 Oe at f = 50 c/s. h measured for ϑ = 5 sec. ϑ is time interval between application of H and measurement of h.
b) h vs. ϑ. H_{max} = 0.60 Oe. $t_s = 18\frac{3}{4}$ hours.
c) Initial permeability, μ_i, vs. time after demagnetization, t.

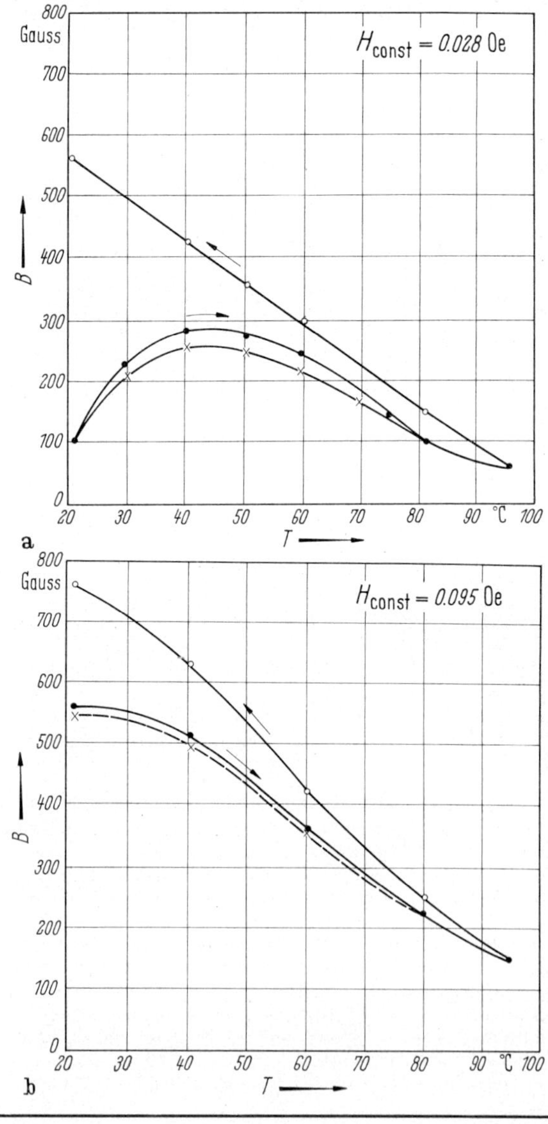

→

Fig. 47. MgMnZn ferrite. Thermomagnetic hysteresis plots. Parameter H_{const} [*64A*], [*66An*].
– × – commutation curve. Composition is not given. For data on μ', μ'' see [*62LAT*].

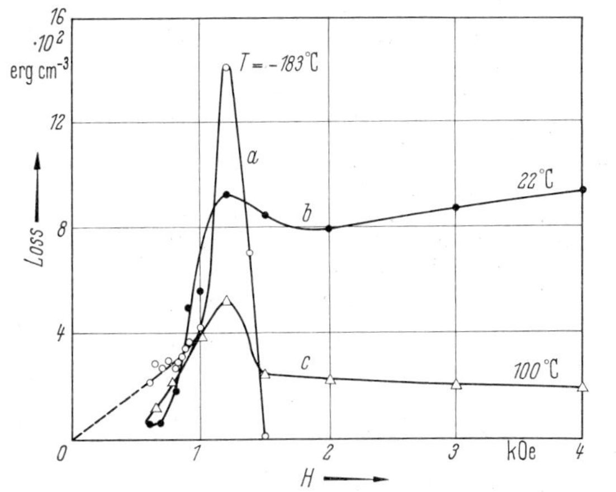

Fig. 48. Hysteresis loss in rotating magnetic fields for a square-loop MgMn ferrite [*62DDSS*].
Data were calculated from measurements (Fig. 49) made in (110) plane of a single-crystal sphere (6 mm diameter) of a stoichiometric composition having a [111] easy direction. (Composition is not given). Only curve *a* conforms to Akulov's theory [*34A*]. Highly distorted domain walls were observed.

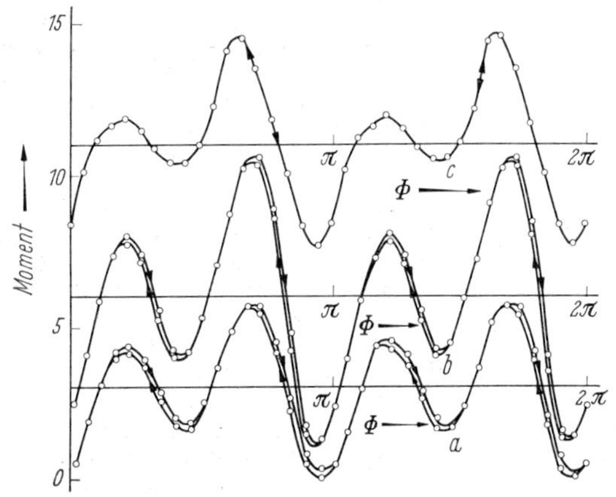

Fig. 49. Rotational mechanical moment vs. field rotation, ϕ, at 1500 Oe [*62DDSS*].

Curve	T °C	Moment per unit scale $erg cm^{-3}$
a	100	520
b	22	1300
c	−183	6050

Measurements made in (110) plane using a torque method. The curves for the forward and reverse processes for 22 °C and 100 °C do not coincide at this field. The loss per cycle at each H and T is determined from the area enclosed between the two processes. Fig. 48 is constructed in this way.

6.1.4.2.4 Square-loop properties

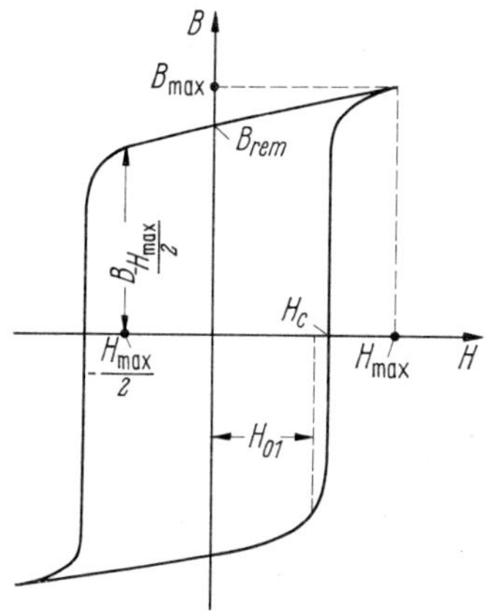

Fig. 50. Definitions of hysteresis loop parameters for square-loop materials.
It is standard practice to define the squareness ratio, R_s, for memory cores used in a 2: 1 coincident current scheme as $R_s = B_{(-H_{max}/2)}/B_{max}$ where H_{max} is the maximum drive field applied and B_{max} is the magnetization at H_{max}. R_s is usually a function of H_{max} and exhibits a maximum, $R_{s(max)}$ (Fig. 52, 104). Since the literature generally uses R_s to denote $R_{s(max)}$, this practice is followed in this section although $R_{s(max)}$ should properly be used. The pulse parameters for a memory core are defined in Fig. 90, 91.
Although the literature as a rule associates the threshold field for switching, H_0, as defined in Fig. 91, with the field at the knee of the maximum squareness hysteresis loop, H_{01}, it is found that $H_0 \ll H_{01}$. However, $H_0 \approx H_{01}$ when the switching curve (Fig. 91) is constructed from data obtained by saturating the core before the application of each reversal field. See [*69G*] for discussion of this subject and for a review of memory materials.

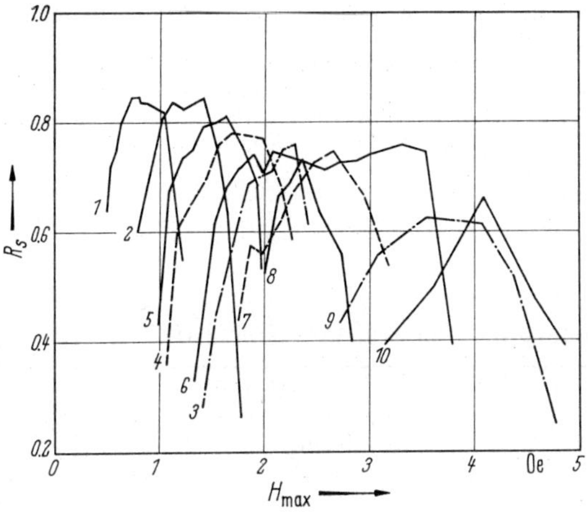

Fig. 52. MgMn ferrites. Squareness ratio, R_s, vs. field, H_{max} [*63Barn*].
Numbers refer to compositions given in Tab. 19.

→

Fig. 53. MgMn ferrites. Field distribution of Barkhausen jumps in hysteresis loop of single crystals [*66DSPS*].
1) Sample 1 in Tab. 20, 4) Sample 4 in Tab. 20.
The rate of change in the number of Barkhausen jumps, dN/dH, is plotted as the magnetizing field is decreased from its maximum positive to maximum negative value. Peaks occurred at the knee of the loops. Loops were rectangular. Samples were rectangular bars $0.2 \times 1.5 \times 10$ mm³, cut lengthwise along the [100] axis with the larger face in the (100) plane. Crystals were grown by the Verneuil method. Measurements were made while reversing magnetization at a rate of 0.001 c/s.

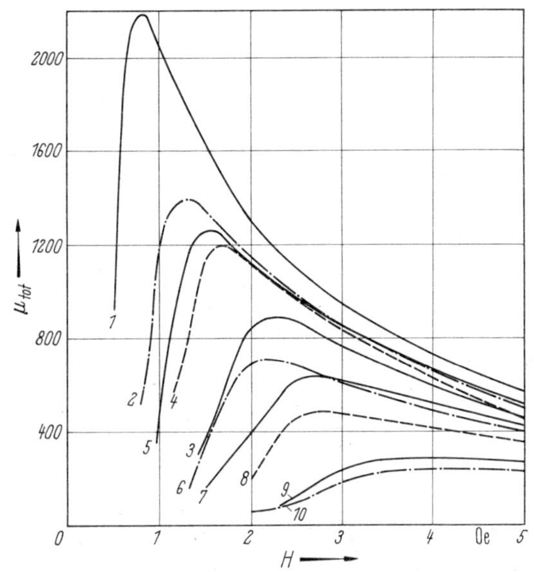

Fig. 51. MgMn ferrites. Total permeability, μ_{tot}, vs. field, H [*63Barn*].
Numbers refer to compositions given in Tab. 19. μ_{tot} is the slope of the line connecting the origin to any point on the initial magnetization curve.

Tab. 19. Compositions for Fig. 51, 52 [*63Barn*].

Sample Nr.	Composition [mole-%]		
	Fe_2O_3	MnO	MgO
1	48.1	45.6	2 ··· 4
2	43.9	45.5	10.58
3	44.8	41.8	13.45
4	45.1	42.0	12.95
5	44.7	32.7	22.60
6	44.4	26.6	29.05
7	44.0	21.8	34.5
8	44.5	16.5	42.0
9	48.8	5.1	46.0
10	43.6	6.55	49.9

Materials were ball milled, calcined at 950 °C, remilled, pressed into cores, fired for 2 hours at 1350 °C and cooled in vacuum. Relations between magnetic properties and microstructure are studied.

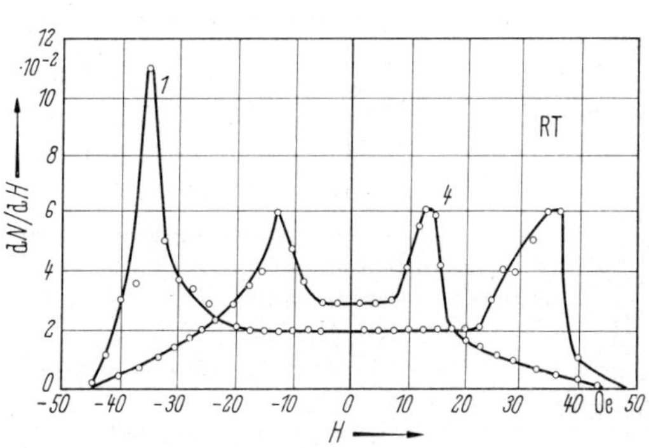

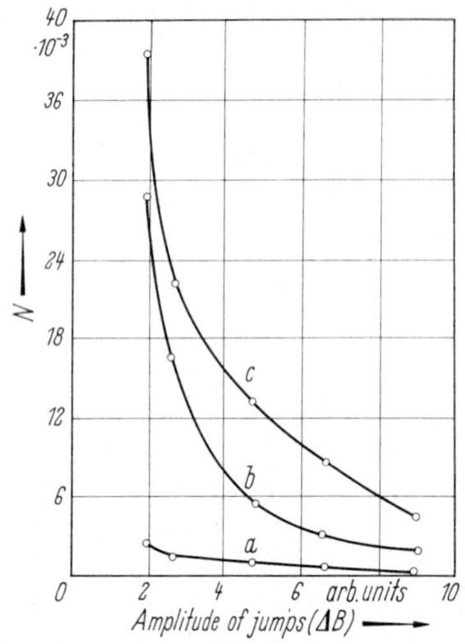

Fig. 54. MgMn ferrite. Amplitude distribution of the number of Barkhausen jumps, N, in a single crystal [*66DSPS*]. Material is sample 4 in Tab. 20.
a) RT for field excursion ±40 Oe,
b) 77 °K for field excursion ±40 Oe,
c) 77 °K for field excursion ±80 Oe.

Tab. 20. Composition of single crystals used in the study of the Barkhausen effect [*66DSPS*].

Sample Nr.	Composition mole-% (calculated)			$\frac{Mn}{Mg}$	ϱ gcm^{-3}
	MgO	MnO	Fe_2O_3		
1	25.0	25.0	50.0	1.0	4.450
2	21.7	34.8	43.5	1.6	4.425
3	20.0	40.0	40.0	2.0	4.400
4	16.5	50.0	33.5	3.0	4.395
5	7.5	55.5	37.0	7.1	4.390
6	20.0	30.0	50.0	1.5	—

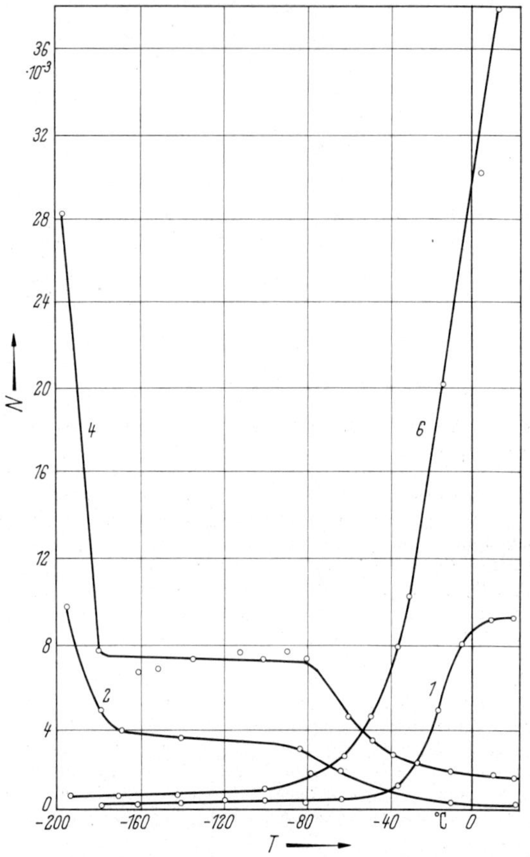

Fig. 55. Number of Barkhausen jumps, N, vs. temperature, T [*66DSPS*].
Numbers refer to compositions given in Tab. 20. Measurements were made while temperature was gradually raised from −195 °C at uniform rate. The type of temperature behaviour is associated with stoichiometry.

All samples except No. 6 had square hysteresis loops. All the specimens contained a small amount of Mn_3O_4 phase.

Tab. 21. Summary of magnetic and switching data for Fig. 56···58 [*65TM*].

Composition mole-%			Sample Nr.	Crystallographic orientation	H_c Oe	B_{rem} Gauss	B_{rem}/B_{max}	H_0 [Oe]		S_w Oe · µsec
MgO	MnO	Fe_2O_3						from $1/\tau_s$ vs. H_{max}	from hysteresis loop	
24	32	44	1	[111]	0.18	3140	0.90	0.18	0.14	1.0
			2	[100]	0.178	2620	0.79	0.16	0.13	0.8
10	40	50	3	[111]	0.19	2100	0.70	0.17	0.15	1.5
			4	[110]	0.18	1620	0.50	0.15	0.15	1.2
			5	[100]	0.175	1600	0.54	0.14	0.13	1.0

Single crystals were grown by the Verneuil method. Picture frames were cut for the magnetization to lie in the directions indicated.
Outer diameter = 4 mm, inner diameter = 2 mm, height = 1 mm. Switching data were measured under pulse conditions. H_c, B_{rem}, B_{rem}/B_{max}, H_0 were measured from hysteresis loop on loop tracer. H_0 is also determined from intercept on H_{max} axis as shown in Fig. 58. S_w is determined from slope of plots as in Fig. 58.
See Fig. 50 for definitions of loop parameters. See source for similar data on $MnFe_2O_4$ and YIG.

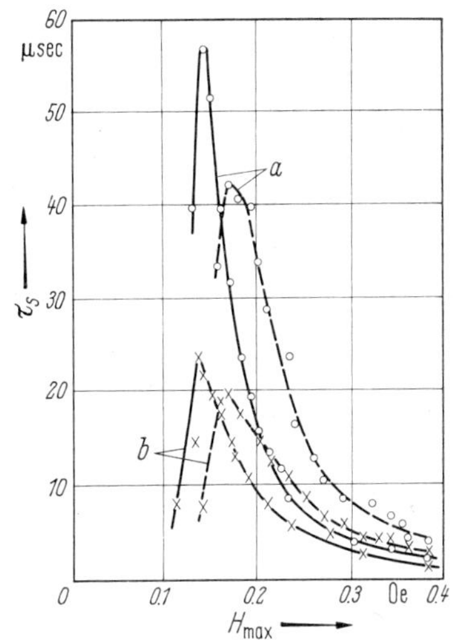

Fig. 56. $(MgO)_{0.24}(MnO)_{0.32}(Fe_2O_3)_{0.44}$. Switching time, τ_s, vs. maximum applied field, H_{max}, as a function of crystalline orientation [65TM].
a) H_{max} in [111] direction, b) H_{max} in [100] direction.
Solid lines: remeasurement after $7\frac{1}{2}$ months. See Tab. 21 for details. H_{max}: see Fig. 50.

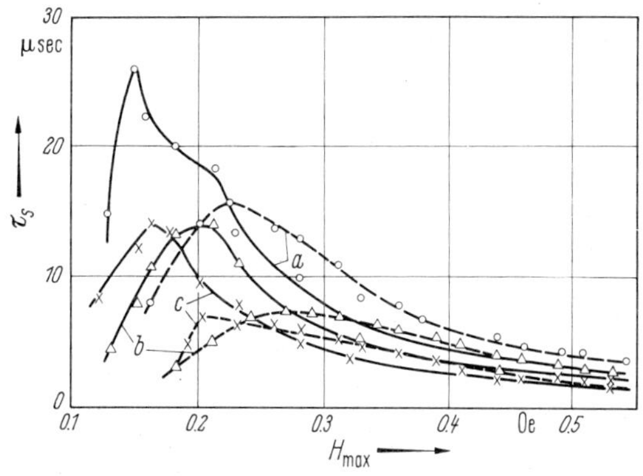

Fig. 57. $(MgO)_{0.1}(MnO)_{0.4}(Fe_2O_3)_{0.5}$. Switching time, τ_s, vs. H_{max} as a function of crystalline orientation [65TM].
a) H_{max} in [111] direction, b) H_{max} in [110] direction, c) H_{max} in [100] direction.
Solid lines: remeasurement after 6 months. See Tab. 21 for details. H_{max}: see Fig. 50.

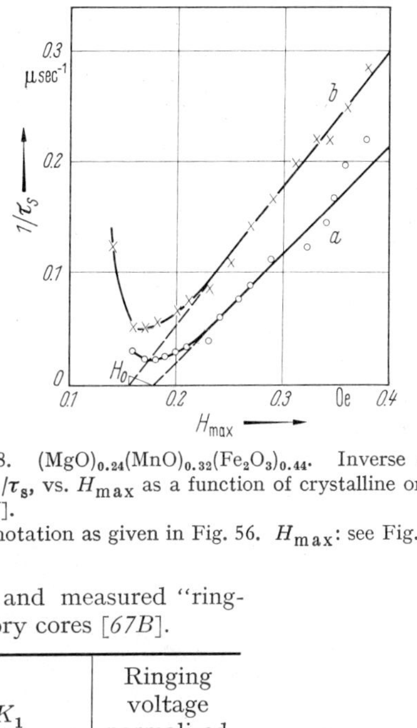

Fig. 58. $(MgO)_{0.24}(MnO)_{0.32}(Fe_2O_3)_{0.44}$. Inverse switching time, $1/\tau_s$, vs. H_{max} as a function of crystalline orientation [65TM].
Same notation as given in Fig. 56. H_{max}: see Fig. 50.

Tab. 22. Comparison of calculated and measured "ringing" voltages for some memory cores [67B].

Material	$\sigma_s \lambda_s / K_1$ emu cm³g⁻¹erg⁻¹	Ringing voltage normalized (calc)	Ringing voltage normalized (exp)
Copper manganese (1)	$\frac{57 \cdot 5 \cdot 10^{-6}}{5 \cdot 10^3}$	1	1
Magnesium manganese	$\frac{57 \cdot 1.3 \cdot 10^{-6}}{10 \cdot 10^3}$	0.13	0.13
Copper manganese (2)	$\frac{57 \cdot 3 \cdot 10^{-6}}{10 \cdot 10^3}$	0.24	0.04
Li ferrite	$\frac{70 \cdot 8 \cdot 10^{-6}}{80 \cdot 10^3}$	0.12	0.10

Assuming that mechanical effects dominate the magnetic effects, the amplitude of the mechanical resonance becomes proportional to $\sigma_s \lambda_s / K_1$. Normalized "ringing" voltages were calculated from this term and compared to normalized values obtained from Fig. 60.

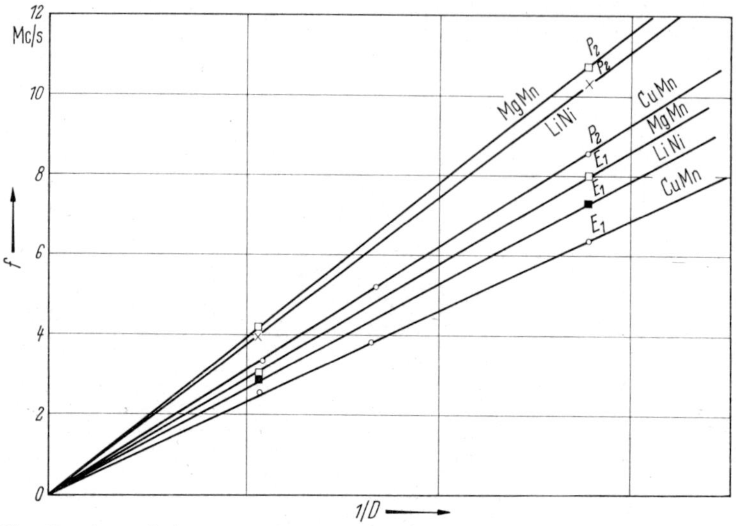

Fig. 59. Acoustical resonant frequency for the E_1 and P_2 modes in memory cores vs. core size [67B].
1/D is the reciprocal of the mean diameter (units are not given. The E_1 and P_2 modes pertain, respectively, to a longitudinal wave having a wave length of one core circumference and to a transverse wave out of the plane of the core having a wave length of ½ the core height. Specific compositions are not given. See also [68B].

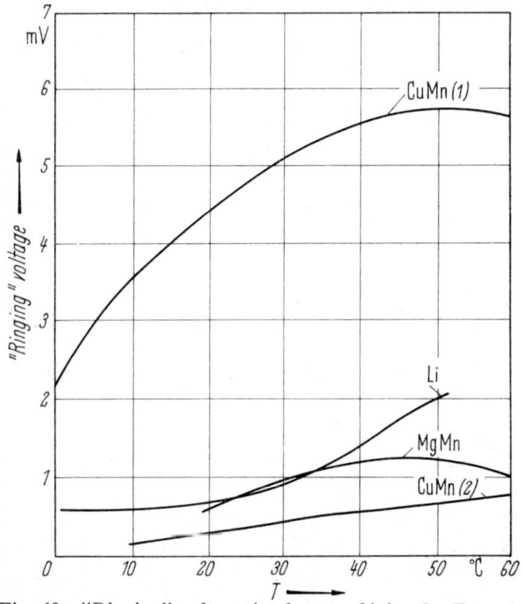

Fig. 60. "Ringing" voltage (peak to peak) for the E_1 mode vs. temperature, T, for various memory core materials [67B]. Voltage is for 100 cores. "Ringing" is the so-called "magnetostrictive ringing" encountered in the operation of some core types. The high mechanical quality factor, Q_{mech}, of core materials (100 to 3000) requires only a weak coupling between H and M to excite mechanical resonance.

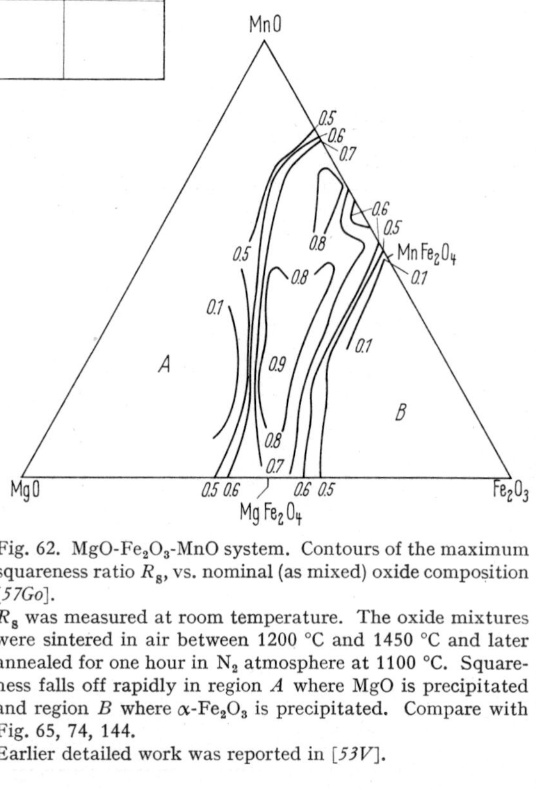

Fig. 62. MgO-Fe_2O_3-MnO system. Contours of the maximum squareness ratio R_s, vs. nominal (as mixed) oxide composition [57Go].
R_s was measured at room temperature. The oxide mixtures were sintered in air between 1200 °C and 1450 °C and later annealed for one hour in N_2 atmosphere at 1100 °C. Squareness falls off rapidly in region A where MgO is precipitated and region B where α-Fe_2O_3 is precipitated. Compare with Fig. 65, 74, 144.
Earlier detailed work was reported in [53V].

Fig. 61. Line width of the mechanical resonance for the E_1 mode vs. temperature, T, for some memory core materials [67B]. →
Line width measured the rate of energy transfer involving coupling of mechanical vibration to motion of charge carriers or vacancies. Mechanism appeared to be related to that for permeability disaccomodation, both processes having activation energies, $q \approx 0.3 \cdots 0.4$ eV [60C].

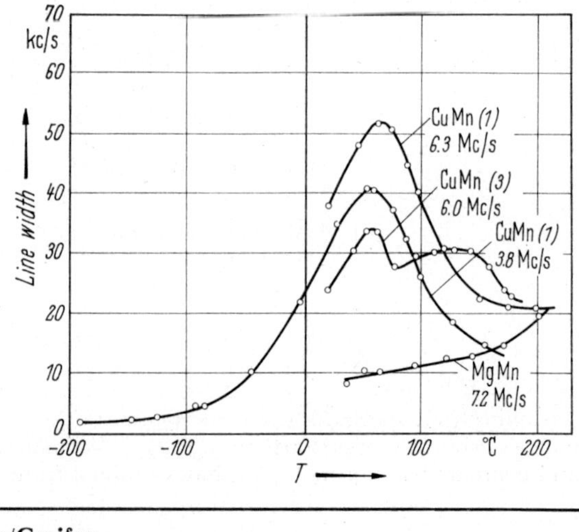

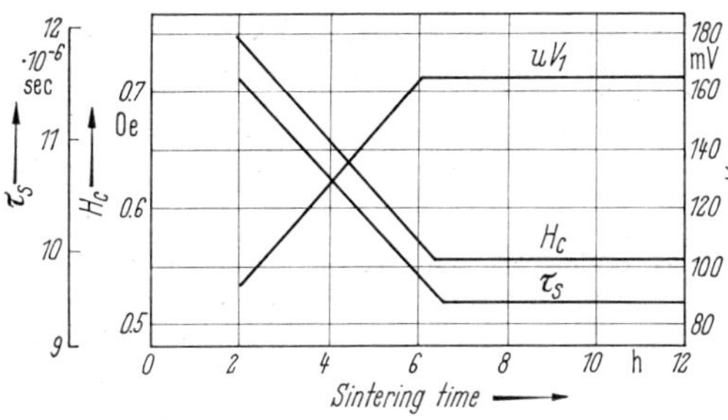

Fig. 63. Switching time, τ_s, coercive force, H_c, and undisturbed output voltage, uV_1, of a memory core vs. sintering time when the sintering temperature is 1350 °C [*57Go*].
The amplitude of the drive pulse is 820 mA. The composition and core size are not given in the reference; a typical composition which gives similar results is $(MgO)_{0.35}(MnO)_{0.30}(Fe_2O_3)_{0.35}$.

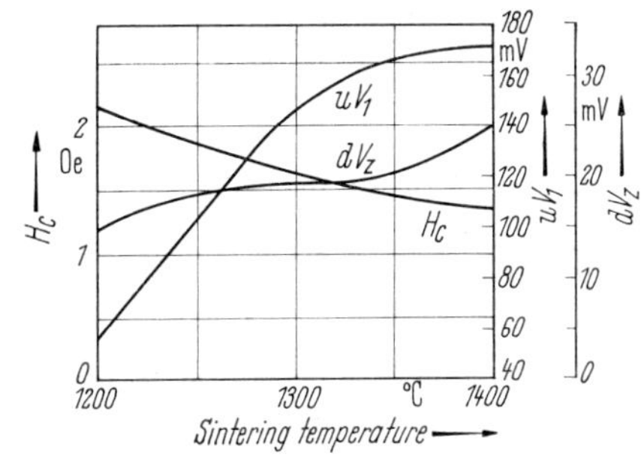

Fig. 64. Memory core parameters, H_c, uV_1, dV_z, vs. sintering temperature [*57Go*].
Soak time in the furnace is 10 hours. The peak output voltage uV_1, is caused by a read pulse of 1000 mA. The noise voltage, dV_z, is that due to a 500 mA pulse. The same material as in Fig. 63.

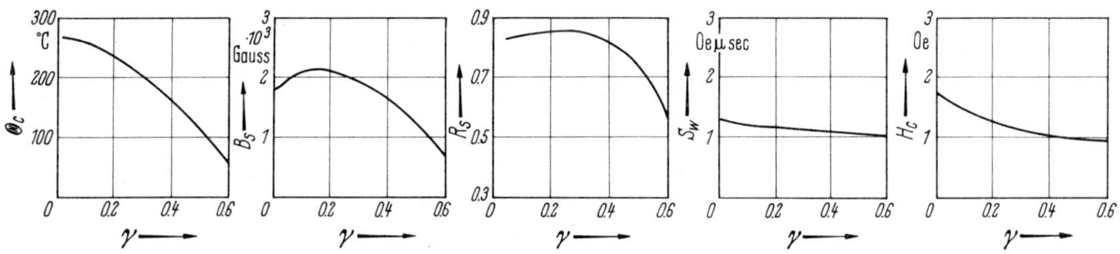

Fig. 65. $(MgFe_2O_4)_{1-\gamma}(Mn_3O_4)_\gamma$ system. Curie temperature, Θ_C, saturation flux density, B_s, maximum squareness, R_s, switching coefficient, S_w, and coercive force, H_c, vs. composition, γ [*53B*], [*55B*]. Compare with Fig. 62, 74, 144.

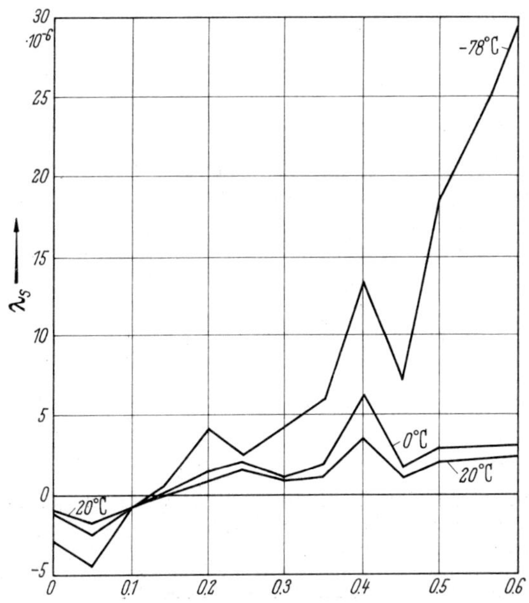

Fig. 66. $(MgFe_2O_4)_{1-\gamma}(Mn_3O_4)_\gamma$ system. Saturation magnetostriction, λ_s, vs. composition, γ [*53B*], [*55B*].

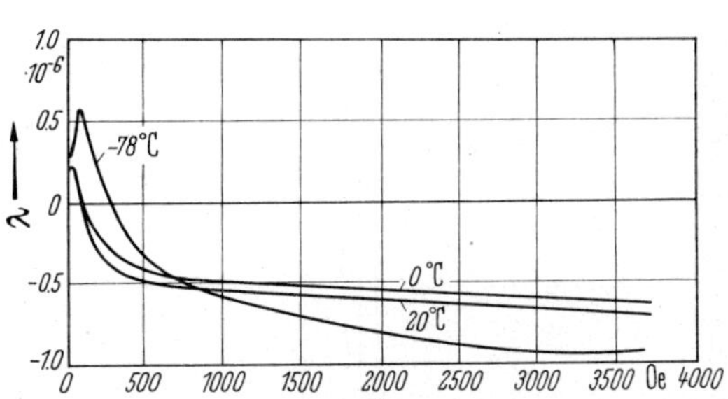

Fig. 67. $(MgFe_2O_4)_{0.9}(Mn_3O_4)_{0.1}$. Magnetostriction, λ, vs. field, H, for a polycrystalline sample [*53B*], [*55B*].

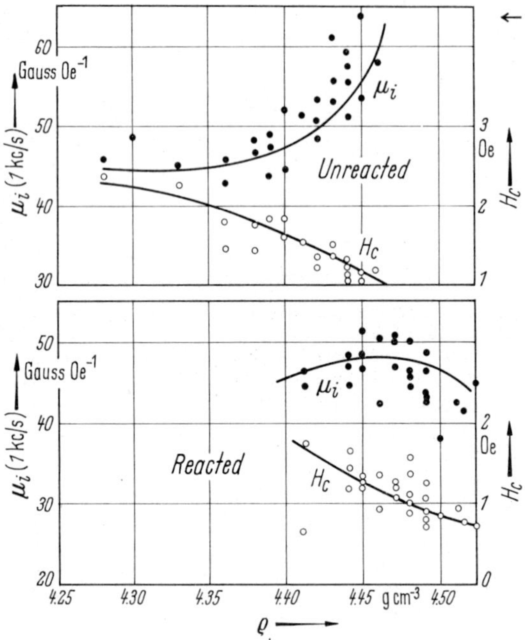

← Fig. 68. $(MgO)_{0.356}(MnO)_{0.341}(Fe_2O_3)_{0.303}$. Initial permeability, μ_i, and coercive force, H_c, vs. density, ϱ [55E(1)].
Unreacted: Powders were dry mixed, calcined at 550 °C for ≈ 15 hours, wet milled for 20 hours (steel balls), dried, binder added, then they were pressed into cores and fired. The pressed bodies consisted of a mixture of oxides.
Reacted: After the 550 °C calcination, the powders were dry milled for 4 hours, recalcined for ≈ 15 hours at 1050 °C, wet milled for 30 hours, dried, binder added, pressed and fired as above. The pressed bodies in this case were in the spinel phase. Differences in fired density were obtained by varying the pressing pressure, firing temperature and firing time.

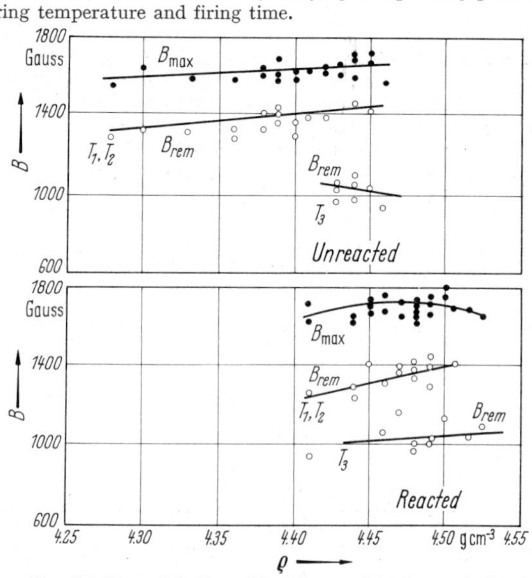

Fig. 69. $(MgO)_{0.356}(MnO)_{0.341}(Fe_2O_3)_{0.303}$. Maximum induction, B_{max}, and remanence, B_{rem} vs. density, ϱ [55E(1)].
Reacted } See Fig. 68.
Unreacted }
Peak firing temperatures: T_1=1350°C, T_2=1400°C, T_3=1450°C.

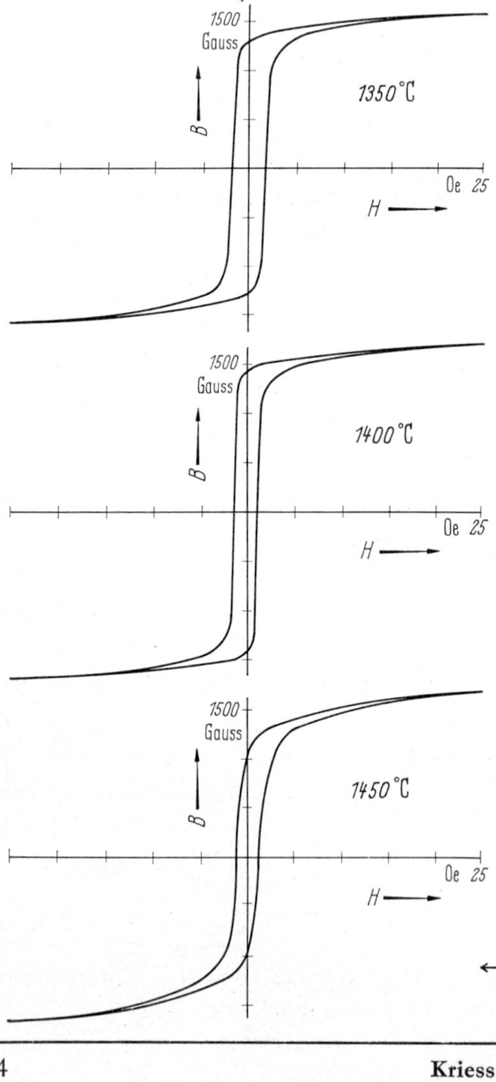

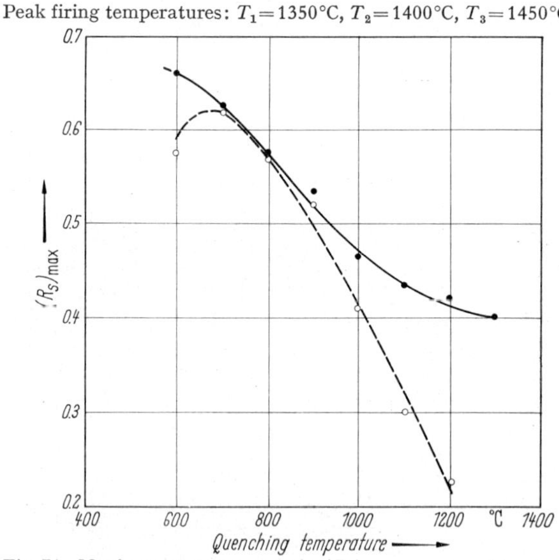

Fig. 71. Maximum squareness ratio, $(R_s)_{max}$, vs. quenching temperature for unreacted specimens [55E(1)].
● $(MgO)_{0.356}(MnO)_{0.341}(Fe_2O_3)_{0.303}$
○ Mg ferrite
Peak firing temperature: 1400 °C. See Fig. 50, 68.

← Fig. 70. $(MgO)_{0.356}(MnO)_{0.341}(Fe_2O_3)_{0.303}$. Effect of firing temperature on hysteresis loop shape [55E(1)].
Material was unreacted: see Fig. 68.

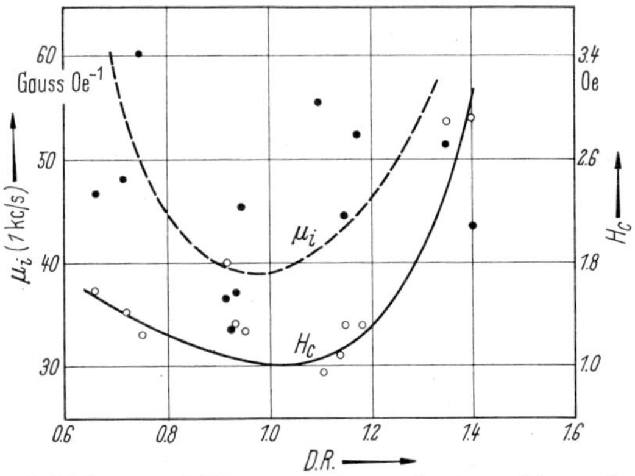

Fig. 72. MgO-MnO-Fe_2O_3. Initial permeability, μ_i, and coercive force, H_c, vs. deviation ratio, *D. R.* [55E(2)].

The numerator of the *D. R.* is the molar difference between the sum of the Mg and the Mn ions and one-half the Fe ions in the hypothetical molecular formulas in the table below, and the denominator is the number of Mn ions in these formulas. Thus,

$$D.\,R. = \frac{[\mathrm{Mg}] + [\mathrm{Mn}] - {}^1/_2[\mathrm{Fe}]}{[\mathrm{Mn}]}$$

Deviation ratio and hypothetical formulas

Code letter	Excess Me over $MeFe_2O_4$ [1]	*D. R.* [2]	Hypothetical formula [3]	a_0 [4] Å
O	0.90	1.15	$Mg_{0.82}Mn_{0.78}Fe_{1.40}O_4$	8.45 ± 0.03
A	0.64	0.93	$Mg_{0.73}Mn_{0.70}Fe_{1.57}O_4$	8.45 ± 0.03
B	0.78	0.94	$Mg_{0.69}Mn_{0.83}Fe_{1.48}O_4$	8.45 ± 0.03
C	1.02	1.14	$Mg_{0.78}Mn_{0.90}Fe_{1.32}O_4$	8.45 ± 0.03
D	1.15	1.37	$Mg_{0.93}Mn_{0.84}Fe_{1.23}O_4$	8.44 ± 0.03
E	1.05	1.40	$Mg_{0.95}Mn_{0.75}Fe_{1.30}O_4$	8.44 ± 0.03
F	0.79	1.18	$Mg_{0.86}Mn_{0.67}Fe_{1.47}O_4$	8.46 ± 0.03
G	0.52	0.92	$Mg_{0.77}Mn_{0.58}Fe_{1.65}O_4$	8.46 ± 0.03
H	0.42	0.67	$Mg_{0.66}Mn_{0.62}Fe_{1.72}O_4$	8.46 ± 0.03
I	0.52	0.72	$Mg_{0.61}Mn_{0.74}Fe_{1.65}O_4$	8.47 ± 0.03
J	0.64	0.75	$Mg_{0.57}Mn_{0.86}Fe_{1.57}O_4$	8.46 ± 0.03
K	0.90	0.95	$Mg_{0.65}Mn_{0.95}Fe_{1.40}O_4$	8.47 ± 0.03
L	1.18	1.12	$Mg_{0.73}Mn_{1.06}Fe_{1.21}O_4$	8.46 ± 0.03

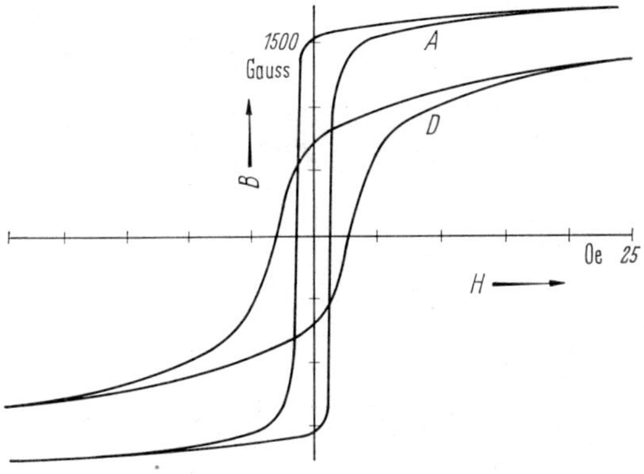

Fig. 73. MgO-MnO-Fe_2O_3. Hysteresis loops [55E(2)]. A and D: see Fig. 72 (Tab. for *D. R.*).

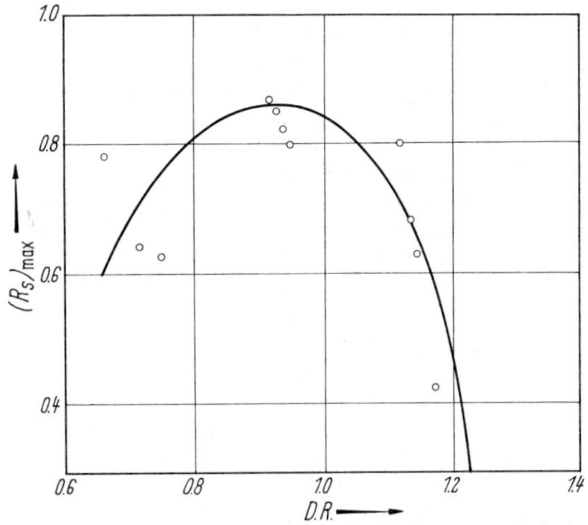

Fig. 74. Maximum squareness ratio, $(R_s)_{max}$, vs. deviation ratio, *D. R.* [55E(2)].
For *D. R.*, see Fig. 72. Compare with Fig. 62, 65, 144.
See Fig. 50.

[1]) Computed from [Mg] + [Mn] − ½[Fe] of the hypothetical formula.
[2]) Ratio of excess Me (above) to the total Mn.
[3]) Spinel formula: $Mg_aMn_bFe_cO_4$ where $a + b + c = 3$.
[4]) Rough measurements made from a fast-sweep X-ray pattern primarily to show that all compositions were spinels with no second phase.

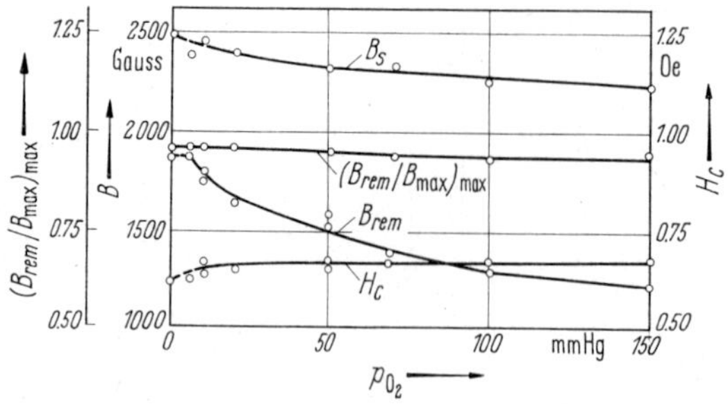

Fig. 75. $Mg_{0.3}Mn_{0.9}Fe_{1.8}O_{4.0}$. Hysteresis loop parameters vs. oxygen partial pressure, p_{O_2}, for annealing at 600 °C [*63BSZSCS*].
Lattice parameter: (8.486 ± 0.003) Å.
See Fig. 50 for definition of loop parameters. $(B_{rem}/B_{max})_{max}$ is obtained from the hysteresis loop having the maximum value of B_{rem}/B_{max}. B_s is the induction measured at 90 Oe. Data are for a toroid annealed for 8 hours at 600 °C at successively higher O_2 pressures. Points on the ordinate are taken for the material as first prepared: fired above 1200 °C, annealed in air at 1200 °C, furnace cooled to 600 °C in CO_2. A weight increase was first detected when p_{O_2} = 50 mm.

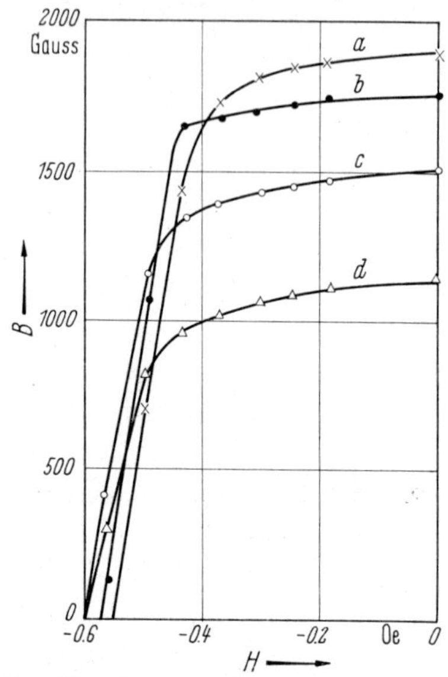

Fig. 76. $Mg_{0.3}Mn_{0.9}Fe_{1.8}O_{4.0}$. Shape of the demagnetization part of the hysteresis loop [*63BSZSCS*].
Parameter p_{O_2} for annealing at 600 °C (see Fig. 75).
a) Initial ferrite (no annealing), *b*) p_{O_2} = 10 mm Hg, *c*) p_{O_2} = 50 mm Hg, *d*) p_{O_2} = 100 mm Hg. These loops are for $(B_{rem}/B_{max})_{max}$.

Tab. 23. $Mg_{0.648}Mn_{0.162}Fe_{2.190}O_{4+\gamma}$. Effect of annealing on hysteresis loop parameters [*64BSSACS*].

Nr. of treatment	Treatment	B_s Gauss	H_c Oe	B_{rem} Gauss	$\left(\frac{B_{rem}}{B_{max}}\right)_{max}$	Δwt. *) g	a_0 Å
1	Original ferrite. Quenching in air from 900 °C	1760	4.8	1220	0.91	—	8.391 ± 0.003
2	After treatment 1: specimen annealed 30 hr at 1280 °C in CO_2 atmosphere. Quenching in air from 900 °C	2080	4.8	550	0.40	−0.0035	8.412
3	After treatment 2: specimen annealed for 4 hr at 400 °C in air	1790	5.3	580	0.47	+0.0004	8.405
4	After treatment 3: specimen annealed for 31 hr at 400 °C in air	1740	4.3	1150	0.81	+0.0034	8.386
5	After treatment 4: specimen annealed for 8 hr at 400 °C in CO_2 atmosphere	1710	4.3	1130	0.82	—	—
6	After treatment 5: specimen annealed for 16 hr at 500 °C in CO_2 atmosphere	1780	4.1	1210	0.86	—	—
7	After treatment 6: specimen annealed for 8 hr at 600 °C in CO_2 atmosphere	1770	4.2	1270	0.87	−0.0001	—
8	After treatment 7: specimen annealed for 8 hr at 500 °C in air	1780	4.2	1210	0.86	+0.0010	—
9	After treatment 8: specimen annealed for 8 hr at 600 °C in air	1800	4.0	1220	0.86	−0.0002	8.380

*) Δwt. = change in weight of 0.5 g sample.
Treatment 1 consisted of pressing a toroid from powder calcined for 4 hours in air at 900 °C, firing in air for 5 hours at 1350 °C, cooling to 900 °C at 100°/hour, and quenching. Compare with Fig. 75 and 76.

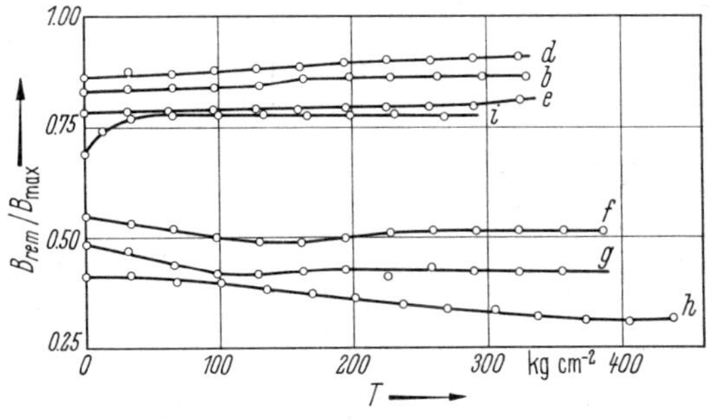

← Fig. 77. MgMn ferrites. Remanence ratio, B_{rem}/B_{max}, vs. compressive stress, T [63C].

	Porosity
b) $Mg_{0.700}Mn_{0.455}Fe_{1.845}O_{3.995}$	20%
d) $Mg_{0.795}Mn_{0.382}Fe_{1.823}O_{3.998}$	–
e) $Mg_{1.000}Mn_{0.144}Fe_{1.856}O_{4.004}$	14%
For comparison with some Mn and MnZn ferrites:	
f) $Mn_{0.95}Fe_{2.05}O_{4+\gamma}$	14%
g) $Mn_{0.85}Fe_{2.15}O_{4+\gamma}$	8%
h) $Mn_{0.5}Fe_{2.5}O_{4+\gamma}$	11%
i) $Mn_{0.64}Zn_{0.47}Fe_{1.89}O_{4+\gamma}$	–

Pressure was applied radially to sintered, toroidal samples in a hydraulic vessel. B_{max} was measured at fields $\approx 2H_c$. H_c for the MgMn samples remained constant with increasing T; for the other materials, H_c increased with increase in T.

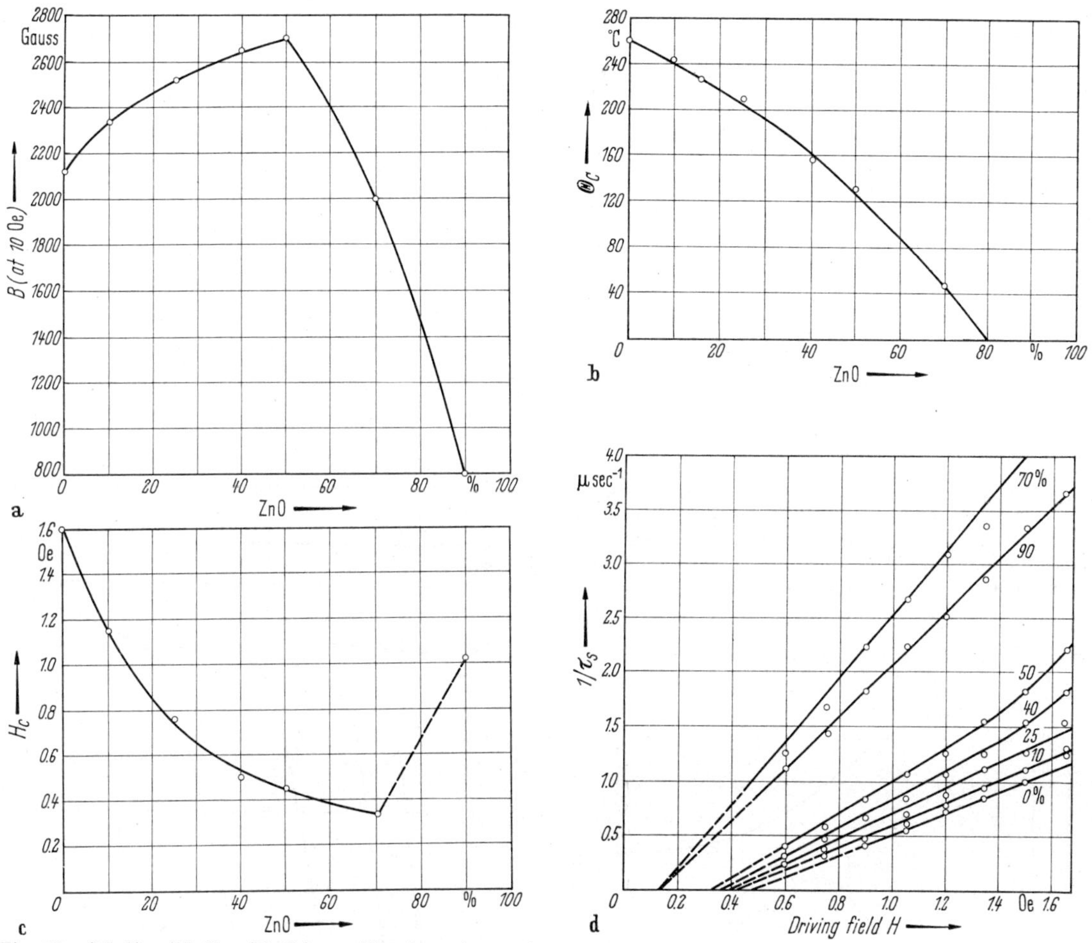

Fig. 78. $(MgO)_{0.35}(MnO)_{0.27}(Fe_2O_3)_{0.38}$. The dependence of a) magnetic induction, B, b) Curie temperature, Θ_C, c) coercive force, H_c, and d) reciprocal switching time, $1/\tau_s$, on the substitution of ZnO for MgO. Percentage of substitution is shown on each curve [57PJS].
Samples were sintered at 1400 °C for 4 hours.

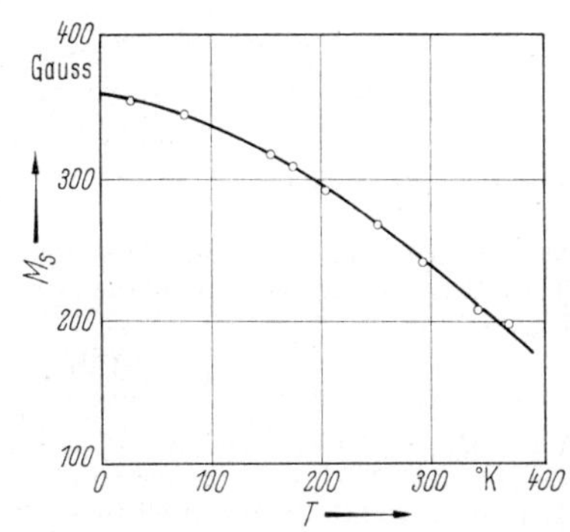

→
Fig. 79. Ferroxcube D1. Temperature dependence of saturation magnetization [61C(1)].
Ferroxcube D1 is a commercial memory-core material of MgMn ferrite. Measurement was made at fields up to 12 kOe using Sucksmith ring-balance.

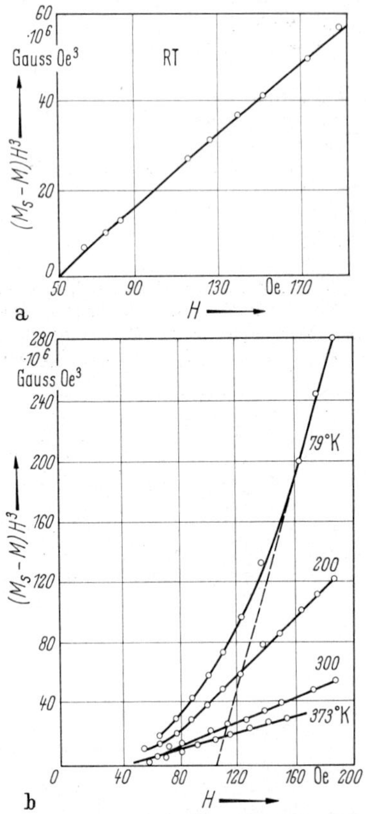

Fig. 80a, b. Ferroxcube D1. The parameter $(M_s - M)H^3$ vs. H [61C(1)].
Data fit the relation $(M_s - M)H^3 = \alpha H + \beta$ where $\alpha = 0.07619\,(K_1^2 M_s)$, $\beta = 0.02362\,K_1^3/M_s^2$, according to Gans' law for approach to saturation [32G]. K_1 is determined either from intercept, β, or slope, α. Tab. 24 gives K_1 data so obtained at various temperatures.

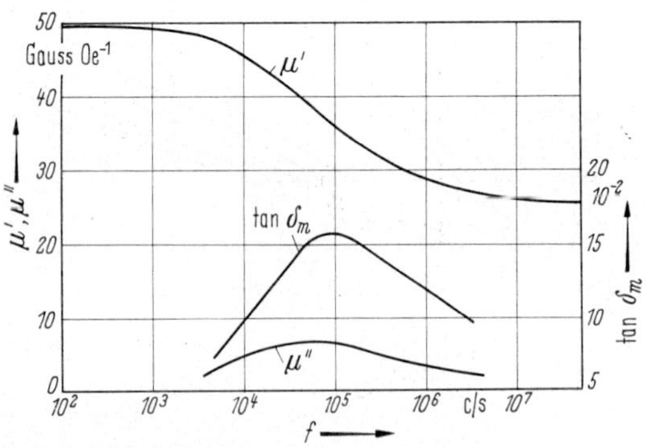

Fig. 83. Ferroxcube D1. Frequency dependence of the complex permeability [61C(2)].
Measurements were made using an impedance bridge method. The rotational contribution to μ' can be calculated by applying

$$\mu_{rot} = 1 + \frac{2\pi M_s^2}{K_1}.$$

With $M_s = 240$ Gauss and $K_1 = 5.03 \cdot 10^4\,\mathrm{erg\,cm^{-3}}$ (from Fig. 79, Tab. 24), we get the rotational permeability, $\mu_{rot} = 8.2$, and the domain wall permeability, $\mu_{dw} = 39.8$.

Tab. 24. Ferroxcube D1. First-order anisotropy constant at various temperatures [61C(1)].

T	79	200	296	373	°K
K_1 (from α)	17.3	8.5	5.03	3.88	10^4 erg cm^{-3}
K_1 (from β)	17.5	8.4	5.1	3.96	10^4 erg cm^{-3}

The values of K_1 obtained by method of Fig. 80 are corrected for the effect of interaction between grains.

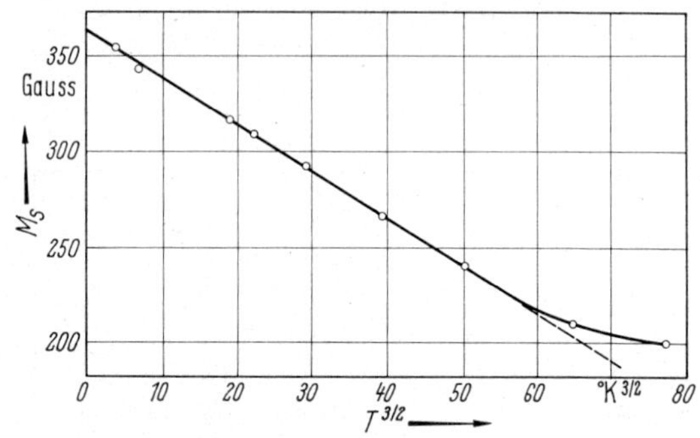

Fig. 81. Ferroxcube D1. M_s vs. $T^{3/2}$ [61C(1)].
Data fit the Bloch equation, $M_s = M_s^0(1 - C\,T^{3/2})$. M_s^0 and C obtained from the plot is inserted into the Herring-Kittel relation [51HK] in order to calculate A, the exchange constant:

$$A = \left(\frac{M_s^0}{2\mu_B}\right)^{1/3} \left(\frac{\mathrm{k}}{13.3\,C^{2/3}}\right)$$

k = Boltzmann's constant. $A = 2.02 \cdot 10^{-7}\,\mathrm{erg\,cm^{-1}}$. From these data, the wall energy, γ, and wall thickness, σ, were calculated:

$$\gamma = 0.20\,\mathrm{erg\,cm^{-2}}, \sigma \simeq 1.6 \cdot 10^{-5}\,\mathrm{cm}.$$

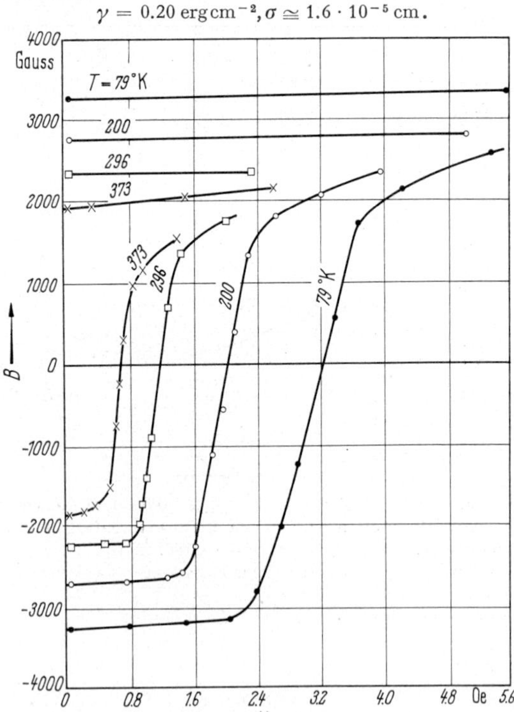

Fig. 82. Ferroxcube D1. Hysteresis loops at various temperatures [61C(1)].
Data (Fig. 79 ··· 82) suggested that the Goodenough theory of domain creation at grain boundaries and the Néel disperse-field theory were not applicable.

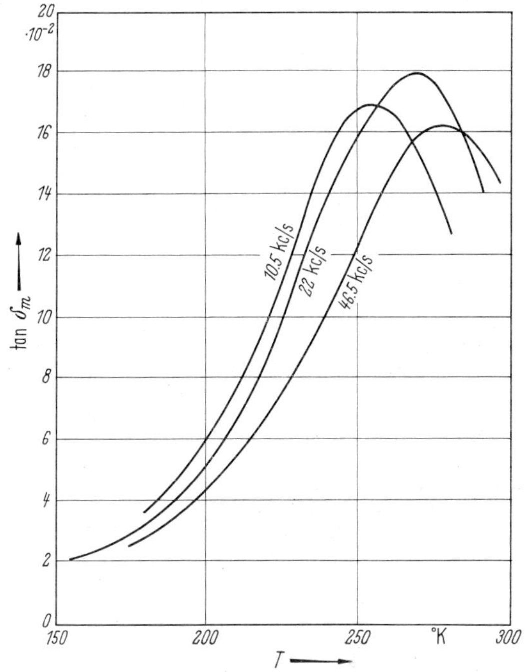

Fig. 84. Ferroxcube D1. Loss tangent, tan δ_m, vs. temperature, T, at three frequencies [*61C(2)*].
Data fit the expression $f = f_\infty e^{-q/kT}$, giving $q = 0.37$ eV and mean time constant, $\tau_0 = 5 \cdot 10^{-12}$ sec. An electron-diffusion relaxation process is indicated.

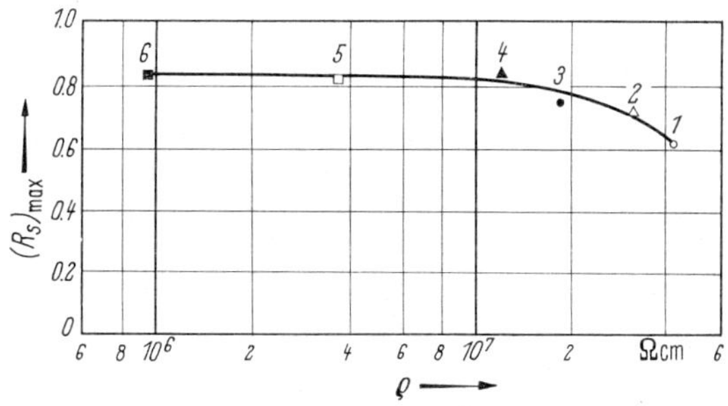

Fig. 85. Ferroxcube D1. Maximum squareness ratio, $(R_s)_{max}$, vs. resistivity, ϱ [*61K*], [*62K*].
R_s attains a maximum value, $(R_s)_{max}$, at a particular drive field, H_{max}. See Fig. 50 for definition of R_s. Material is a commercial square-loop toroid of MgMn ferrite. The samples, numbered in order of decreasing ϱ (and, presumably, increasing Fe^{2+} content), were obtained by firing the material in different atmospheres.

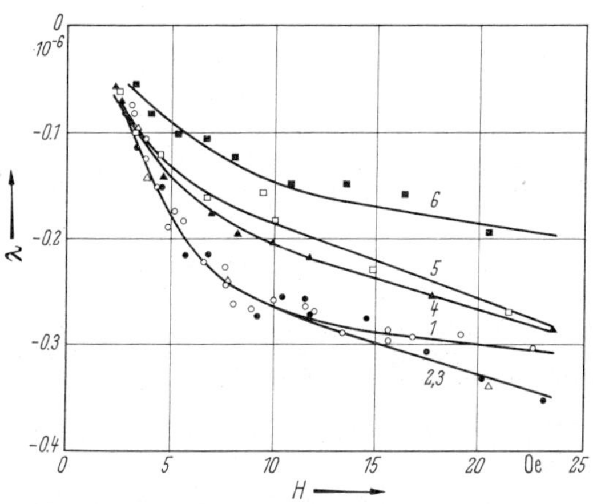

Fig. 86. Ferroxcube D1. Magnetostriction, λ, vs. field, H, as a function of ϱ [*61K*], [*62K*].
Numbers refer to ϱ of the samples as given in Fig. 85. The samples were demagnetized before measurement. $(R_s)_{max}$ appears to depend upon a low $|\lambda|$ which is a function of Fe^{2+} content. A low $|\lambda|$ at low H indicates that λ in the easy direction (in this case, λ_{111}) is approaching zero.

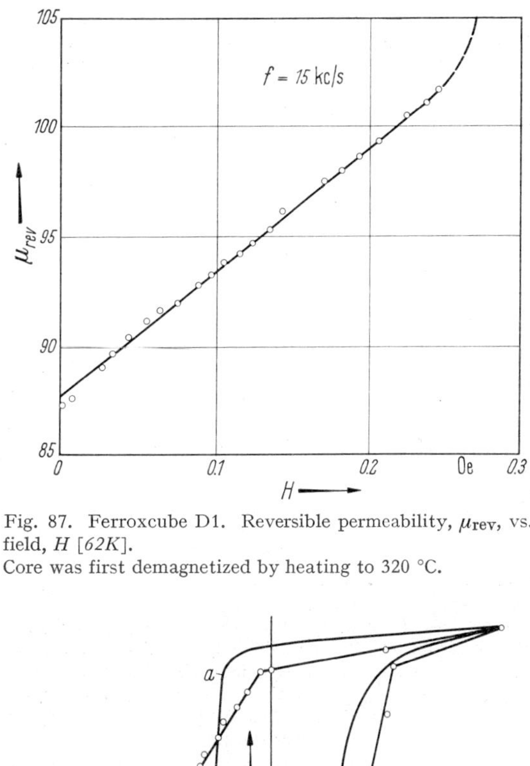

Fig. 87. Ferroxcube D1. Reversible permeability, μ_{rev}, vs. field, H [*62K*].
Core was first demagnetized by heating to 320 °C.

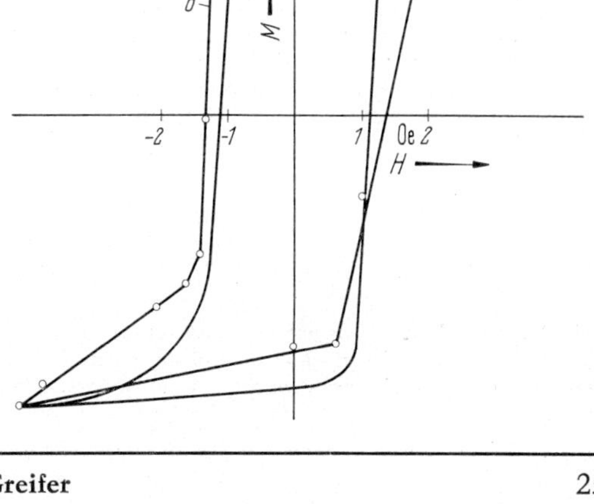

→

Fig. 88. Ferroxcube D1. Hysteresis loops for a memory core and a grain contained in it [*61K*], [*62K*].
a) Loop for the core, *b*) Loop for the grain.
The loop for the grain was deduced from relative areas of domains as a function of H by means of the Bitter technique. The core, fired to obtain large grains, had a low $(R_s)_{max}$ ($\approx$0.70).

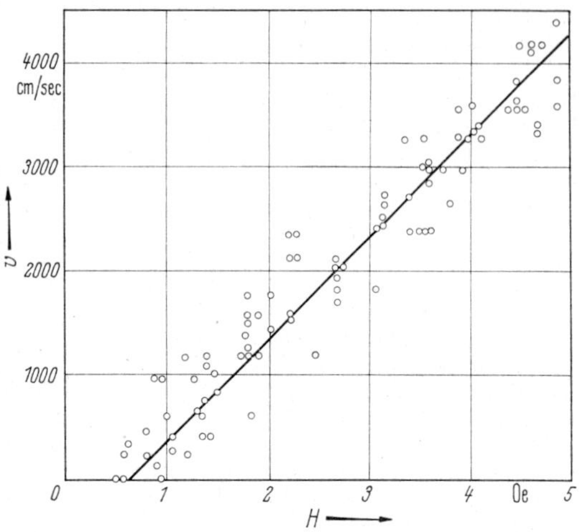

Fig. 89. $Mg_{0.84}Mn_{0.47}Fe_{1.69}O_4$ (nominal). Domain wall velocity, v, vs. field, H [*62K*].
v of a wall in a single grain was estimated by the Bitter technique. Single pulses of controlled height, H, and duration, τ_d, were applied and the distance, d, the domain wall moved, was observed. Then $v \approx d/\tau_d$. v varied by a factor of 3 for different grains. The most consistent set of data indicated that $v = 985\ (H - 0.62)$ [cm/sec].

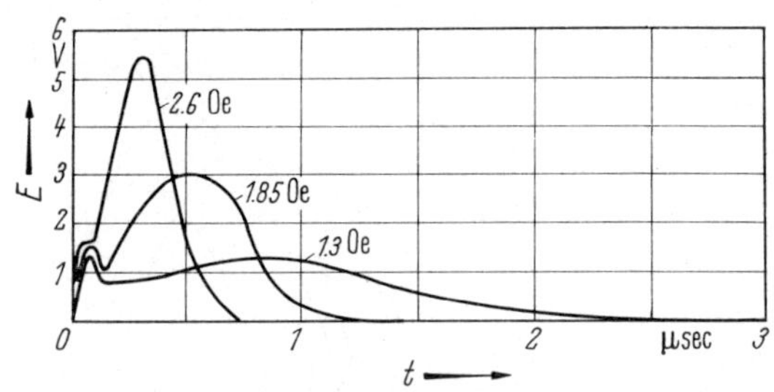

Fig. 90. Ferroxcube 6A. Voltage output pulse, E, as a function of time, t [*57HBW*].
Drive fields are 1.3 Oe, 1.85 Oe, and 2.6 Oe.
These traces are typical of the changes in the form of the output wave as the amplitude of the drive pulse, H_{max}, is increased. The peak voltage gives the undisturbed voltage output, uV_1, for a given H_{max}. If the core had been subject to a disturb pulse prior to the application of H_{max}, a smaller peak output called the disturbed voltage output, dV_1, would be recorded. The parameter dV_z which appears in Tab. 29, 31, 37 refers to a noise voltage arising from a disturb pulse. The switching time, τ_s, of the uV_1 or dV_1, is the duration of the output pulse. It is frequently (but not universally) defined as the interval from the time the drive pulse attains 10% of its maximum amplitude to the time the output voltage drops to 10% of uV_1 or dV_1.

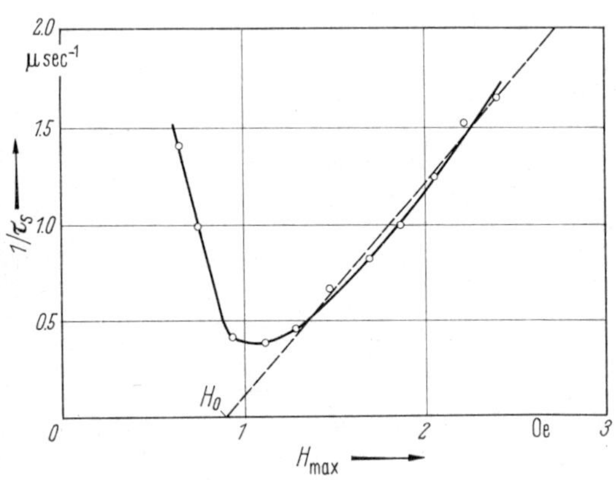

Fig. 91. Ferroxcube 6A. Reciprocal switching time, $1/\tau_s$, vs. field, H_{max} [*57HBW*].
The inverse of the slope of the straight line to the right of the minimum is called the switching constant, S_W. S_W is a merit factor for a core; a low S_W indicates a low τ_s for a given H_{max} or a low H_{max} to attain a given τ_s. The intercept on the H axis is called the threshold field for switching, H_0. H_0 is interpreted theoretically as the minimum field required to reverse the direction of magnetization in a core for the irreversible process that is responsible for the reversal. In practice, the minimum in the curve marks the transition at low fields to another process which applies to the small flux changes incurred over minor hysteresis loops. See source for similar data for Mn ferrite and $CoFe^{2+}$ ferrite.

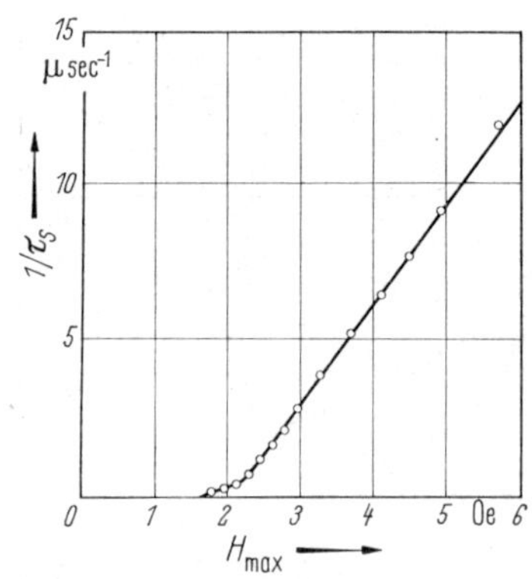

Fig. 92. $Mn_{0.46}Mg_{0.71}Fe_{1.76}O_4$. Reciprocal switching time, $1/\tau_s$, vs. applied field, H_{max}, for a toroid with 0.200 in. outer diameter and 0.180 in. inner diameter [*57Gy*].

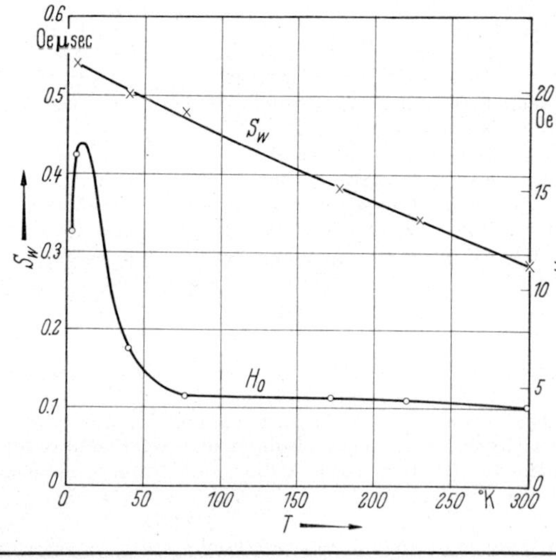

Fig. 93. Switching coefficient, S_W, and threshold field, H_0, vs. temperature, T, for a commercial memory core [*65S*].
For Indiana General core type S-1. Data are taken from Fig. 94.

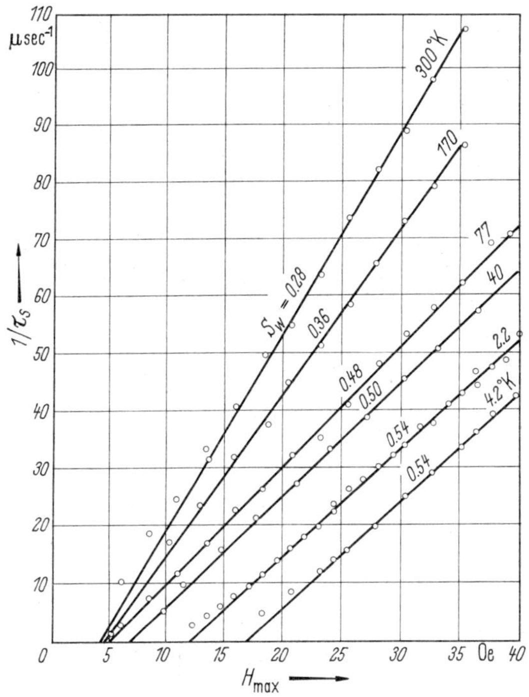

←

Fig. 94. 0.34 MgO · 0.23 MnO · 0.42 Fe_2O_3 (approximate). Reciprocal switching time, $1/\tau_s$, vs. drive field, H_{max}, from 2.2 °K to 300 °K for a commercial memory core [*65S*].

Data for Indiana General core type S-1 (see Tab. 12). τ_s was the time for the switched flux to increase from 10% to 90% of the final value. The core was mounted in an air-dielectric coaxial section of a charged-line switching apparatus. A 30 Oe-field pulse 500 nsec long was applied at the start of each flux-reversal cycle to induce a reproducible initial state. Pulse risetimes were <0.5 nsec. Similar behaviour was found for S-4 type. Switching coefficients, S_w, (the inverse slopes) are indicated for each T. The threshold field, H_0, is the intercept on the H axis. Over the range of fields applied in this experiment switching occurs by a non-uniform rotation mechanism and not by domain wall motion.

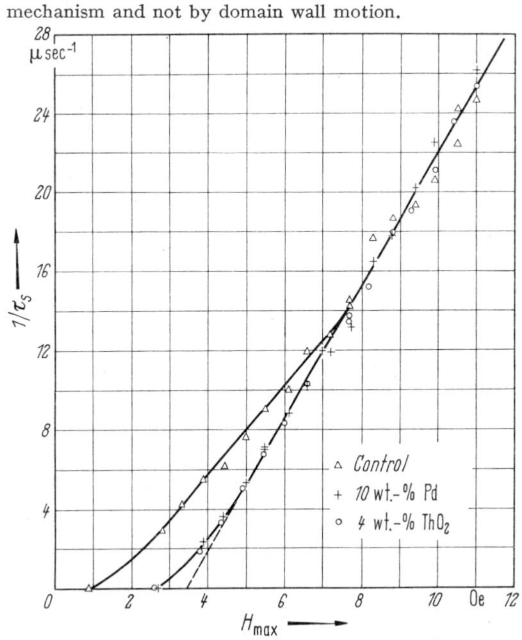

Fig. 95. $Mg_{0.675}Mn_{0.525}Fe_{1.800}O_4$. Reciprocal switching time, $1/\tau_s$, vs. drive field, H_{max}, as a function of Pd and ThO_2 additions [*65Ba*].

τ_s was taken as time for the integrated flux to increase from 10% to 90% of its total value.

Mixed powders were calcined at 900 °C for 16 hours, followed by milling during which the additions (Pd, ThO_2) were made. Toroids were pressed at 35 000 psi, fired at 1400 ··· 1450 °C up to 12 hours in an O_2 atmosphere; cooling was made under N_2. The additives reduced the grain size and increased H_c (Fig. 96) and were found to be located in the grain boundaries and inside the grains. The size of the Pd particles (1000 Å to 2000 Å) was of the order of thickness of the domain wall but Pd was less effective on a percentage basis than ThO_2 (particle size 200 Å ··· 2000 Å) because particles tended to agglomerate. Additives reduced the squareness ratio slightly.

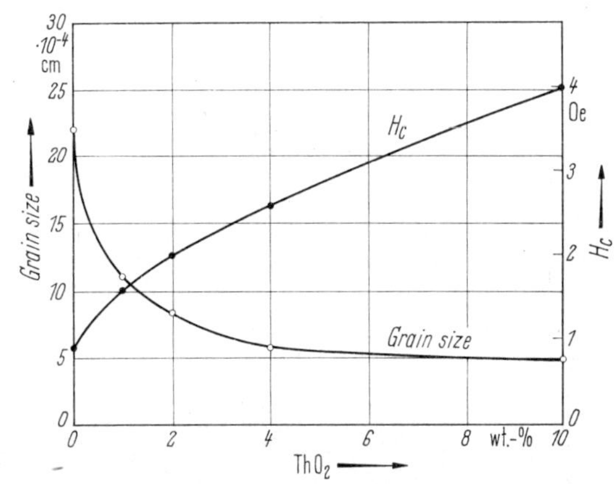

Fig. 96. MgMn ferrite with ThO_2 additions. Grain size and coercive force, H_c, vs. ThO_2 content [*65Ba*]. Basic composition given in Fig. 95.

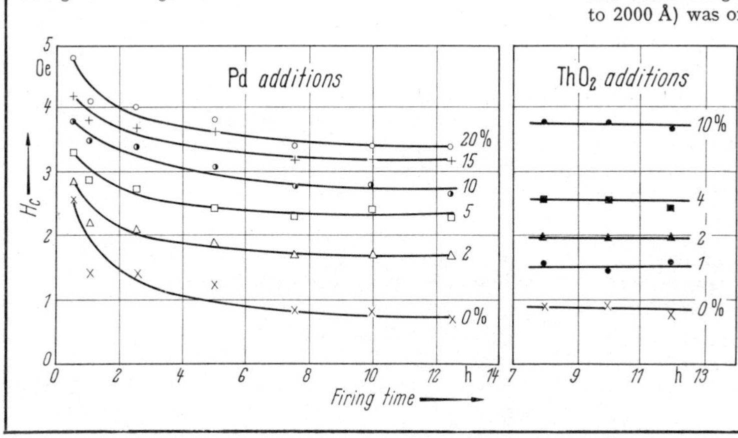

←

Fig. 97. MgMn ferrite with Pd and ThO_2 additions. Coercive force, H_c, vs. firing time (in hours) as a function of Pd and ThO_2 concentration [*65Ba*]. Additions in wt.-%.

Fairly high H_c values can be obtained without additives by firing for short periods but the steepness of the curve made the control difficult. Basic composition given in Fig. 95.

Tab. 25. Reduction of switching time, τ_s, of MgMn ferrite memory cores under coincident current conditions [*60E*].

Compositional and processing changes made to decrease S_w	core size outer diameter in.	core size inner diameter in.	H_c Oe	pulse rise-time, τ_r nsec	drive current mA	disturb current mA	τ_s µsec	S_w Oe · µsec
0.375 MgO · 0.225 MnO · 0.40 Fe_2O_3	0.080	0.050	1.2	200	670	410	1.5	0.95
same as above with a processing change	0.080	0.050	3.5	200	1900	1160	0.5	0.95
same as preceding with τ_r = 100 nsec	0.080	0.050	3.5	100	1800	1100	0.46	0.82
same composition made with o. d. = 0.050 in.	0.050	0.030	3.5	100	1050	640	0.4	0.60
0.325 MgO · 0.225 MnO · 0.35 Fe_2O_3 · 0.05 CaO · 0.05 Cr_2O_3	0.050	0.030	2.8	100	735	445	0.42	0.50

Measurements made at 40 °C.

Tab. 26. Composition of square-loop memory cores containing Sc_2O_3 and Cr_2O_3 [*66B*].

Core type	composition [wt.-%] Fe_2O_3	MgO	MnO	Sc_2O_3	Cr_2O_3
HS-1	64.75	18.98	10.63	5.64	—
HS-2	64.00	19.50	10.30	6.20	—
HS-3	60.50	18.60	10.50	10.40	—
HS-8	57.70	19.00	10.40	8.30	4.60

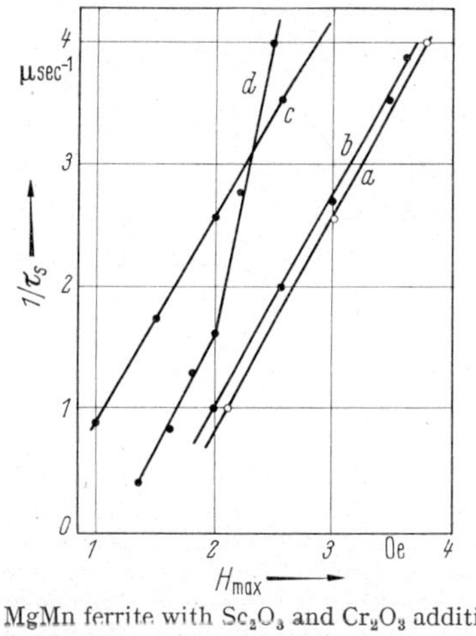

Fig. 98. MgMn ferrite with Sc_2O_3 and Cr_2O_3 additions. Reciprocal switching time, $1/\tau_s$, vs. drive field, H_{max}, for memory cores containing Sc_2O_3 and Cr_2O_3 [*66B*].
a) core type BT-1, *c*) core type HS-8,
b) core type HS-2, *d*) core type HS-2M.
See Tab. 26 for compositions. No additions made to BT-1.

Tab. 28. B_{rem} and temperature coefficient of H_c for cores containing Sc_2O_3 [*64BS*].

Core type	Sc_2O_3 wt.-%	$\Delta H_c/\Delta T$ Oe (°C)$^{-1}$	B_{rem} Gauss	H_c Oe
VT-1	0	0.005	2400	1.4
HS-1	5.64	0.004	900	1.8
HS-2	6.20	0.007	1200	1.20
HS-8	8.30	0.010	800	0.80

See Tab. 26 for compositions.

Tab. 27. Switching coefficient, S_w, coercive force, H_c, and remanent induction, B_{rem}, for memory cores containing Sc_2O_3 and Cr_2O_3 [*66B*].

Core type	S_w Oe·µsec	B_{rem} Gauss	H_c Oe
BT-1	0.50	2350	1.25
HS-2	0.46	1200	1.20
HS-2M	0.28	1100	1.58
HS-8	0.46	800	0.80

BT-1 is a standard composition without Sc_2O_3 included for comparison.
S_w data presumably for domain wall process.
See Tab. 26 for compositions.

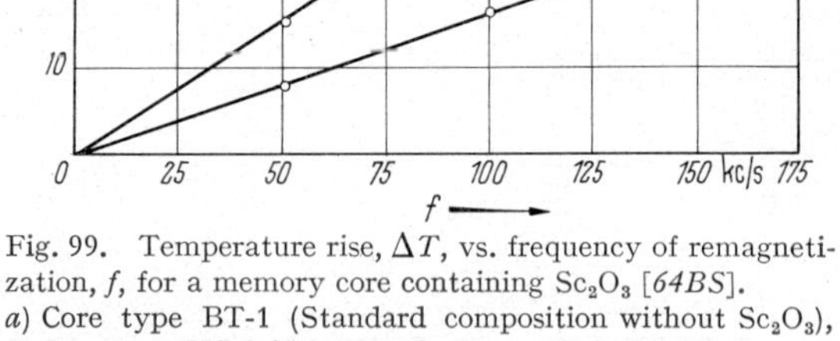

Fig. 99. Temperature rise, ΔT, vs. frequency of remagnetization, *f*, for a memory core containing Sc_2O_3 [*64BS*].
a) Core type BT-1 (Standard composition without Sc_2O_3),
b) Core type HS-2 (Contains Sc_2O_3; see Tab. 26, 27).
Temperature rise estimated by measuring ΔH_c and using $\Delta H_c/\Delta T$ data of Tab. 28. Both core types are of the same size: outer diameter = 1.2 mm, inner diameter = 0.8 mm, height = 0.4 mm.
Self-heating is a function of B_{rem} (Tab. 28).

Tab. 29. Effect of partial replacement of Fe by Sn on pulse properties of square-loop MgMn ferrite [*63WB*].

Composition	drive current mA	u V_1 mV	d V_z mV	Δ_2 mV	τ_s μsec	Θ_C °C	B_s Gauss
0.70 $MgFe_2O_4 \cdot 0.15\, Mg_2SnO_4 \cdot 0.15\, Mn_3O_4$	580	33	4.0	0.3	0.37	160	800
0.85 $MgFe_2O_4 \cdot 0.15\, Mn_3O_4$	800	72	4.0	0.8	0.37	275	2200
0.75 $MgFe_2O_4 \cdot 0.10\, ZnFe_2O_4 \cdot 0.15\, Mn_3O_4$	650	75	6.5	0.9	0.37	260	2250
0.60 $MgFe_2O_4 \cdot 0.25\, ZnFe_2O_4 \cdot 0.15\, Mn_3O_4$	500	80	9.0	1.9	0.37	190	2480

Delta noise, Δ_2, is the voltage difference resulting from half select currents in the magnetizing and demagnetizing directions.
u V_1 = undisturbed one-response voltage.
d V_z = disturbed zero-response voltage.
Pulse risetime and pulse duration were 0.1 and 0.5 μsec, respectively. Core dimensions were: outer diameter = 0.030 in., inner diameter = 0.020 in., height = 0.007 in. Data were obtained at 2 : 1 coincident-current operation. Each type was chosen to have the same τ_s. Δ_2 depends on B_s. Self-heating of a core is a function of $(B_s \cdot H_c)$; reduction of B_s permits operation at high repetition rates and extends the maximum size of the memory, i.e. the total number of cores used in an array. In the system $(0.85 - x)$ $MgFe_2O_4 \cdot x\, Mg_2SnO_4 \cdot 0.15\, Mn_3O_4$, useful properties were found for $x = 0.05$ to 0.15. Similar results were obtained for In substitutions but Al, Ti gave less square hysteresis loops.

Tab. 30. The switching coefficient, S_w, for the system $Mg_xNi_yMn_{1-x-y}Fe_2O_4$ [*64PS(1)*].

Composition x	y	$\frac{1}{\tau_s}$ μsec^{-1}	S_w Oe · μsec	Composition x	y	$\frac{1}{\tau_s}$ μsec^{-1}	S_w Oe · μsec
0.7	0.0	2.46	1.07	0.7	0.2	2.80	1.30
0.6	0.0	1.92	1.25	0.6	0.2	2.10	1.19
0.5	0.0	1.68	1.34	0.5	0.2	1.60	1.31
0.4	0.0	1.36	1.36	0.4	0.2	1.40	1.35
0.3	0.0	0.92	1.54	0.3	0.2	1.36	1.20
0.2	0.0	0.88	1.27	0.2	0.2	1.28	1.35
0.1	0.0	0.58	1.90	0.1	0.2	0.94	1.52
0.0	0.0	0.44	1.72	0.6	0.3	2.60	1.21
0.7	0.1	2.48	1.10	0.5	0.3	1.80	1.23
0.6	0.1	1.90	1.17	0.4	0.3	1.30	1.57
0.5	0.1	1.63	1.25	0.3	0.3	1.20	1.55
0.4	0.1	1.46	1.36	0.2	0.3	0.56	1.80
0.3	0.1	1.30	1.23	0.1	0.3	0.64	2.02
0.2	0.1	1.08	1.01				
0.1	0.1	0.96	1.45				
0.0	0.1	0.86	1.23				

Reciprocal switching time, $1/\tau_s$, is given for drive fields, H_{max}, such that $H_{max} = 2H_0$ where H_0 is the threshold field for switching. Risetime of the drive pulse, τ_r, was 0.1 μsec; pulse duration, τ_d, was 4 μsec. Switching times were measured from the 0.1 points of the disturbed output signal, d V_1. Dimensions of the toroids were: outer diameter = 3 mm, inner diameter = 2 mm, height = 1.5 mm. Consult Fig. 100⋯103 and Tab. 31.

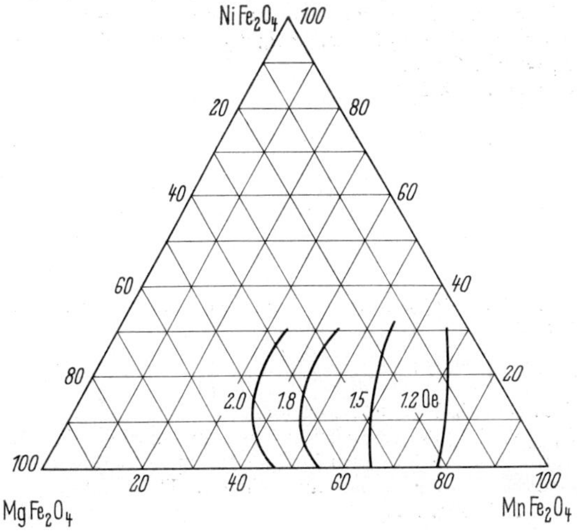

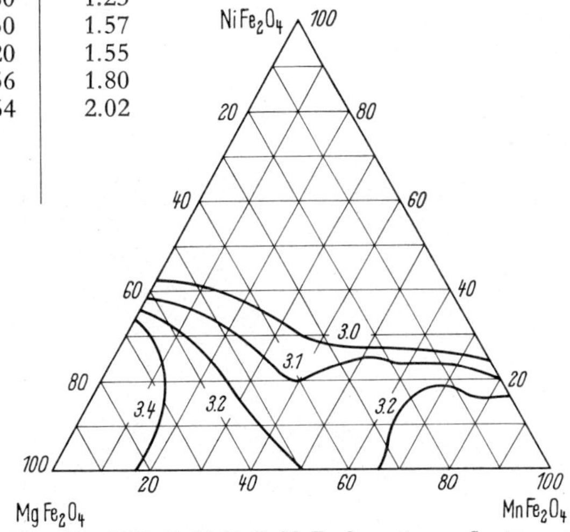

Fig. 101. $NiFe_2O_4$-$MgFe_2O_4$-$MnFe_2O_4$ system. Contours of constant reciprocal switching time, $1/\tau_s$ [*64PS(1)*].
At the curves: $1/\tau_s$ in μsec^{-1} for drive field, H_{max}, of 6 Oe.

←
Fig. 100. $NiFe_2O_4$-$MgFe_2O_4$-$MnFe_2O_4$ system. Contours of constant threshold field, H_0, in the square loop region [*64PS(1)*]. At the curves: H_0 in Oe.

Tab. 31. $Mg_xNi_yMn_{1-x-y}Fe_2O_4$. Output signal, dV_1, and noise/signal ratio, dV_z/dV_1 [*64PS(1)*].

Composition		$H_{max} = 2$ Oe		$H_{max} = 4$ Oe	
x	y	dV_1	dV_z/dV_1	dV_1	dV_z/dV_1
0.7	0.0	0.18	0.56	0.62	0.17
0.6	0.0	0.21	0.44	0.80	0.14
0.5	0.0	0.26	0.42	0.93	0.14
0.4	0.0	0.29	0.40	0.98	0.13
0.3	0.0	0.37	0.36	1.25	0.13
0.2	0.0	0.42	0.35	1.38	0.13
0.1	0.0	0.52	0.35	1.58	0.13
0.0	0.0	0.76	0.32	1.79	0.25
0.7	0.1	0.15	0.64	0.50	0.13
0.6	0.1	0.16	0.55	0.62	0.13
0.5	0.1	0.21	0.48	0.72	0.13
0.4	0.1	0.23	0.45	0.82	0.12
0.3	0.1	0.27	0.42	0.93	0.12
0.2	0.1	0.33	0.40	1.16	0.13
0.1	0.1	0.36	0.39	1.32	0.13
0.0	0.1	0.46	0.37	1.59	0.13
0.7	0.2	0.13	0.72	0.41	0.34
0.6	0.2	0.15	0.63	0.50	0.20
0.5	0.2	0.18	0.55	0.59	0.17
0.4	0.2	0.20	0.50	0.65	0.16
0.3	0.2	0.22	0.48	0.77	0.14
0.2	0.2	0.24	0.45	0.90	0.14
0.1	0.2	0.31	0.43	1.07	0.14
0.0	0.2	0.38	0.41	1.28	0.13
0.6	0.3	0.14	0.71	0.42	0.30
0.5	0.3	0.16	0.65	0.49	0.25
0.4	0.3	0.17	0.59	0.55	0.26
0.3	0.3	0.20	0.56	0.61	0.21
0.2	0.3	0.21	0.54	0.66	0.20
0.1	0.3	0.22	0.52	0.83	0.19
0.0	0.3	0.26	0.52	1.03	0.17

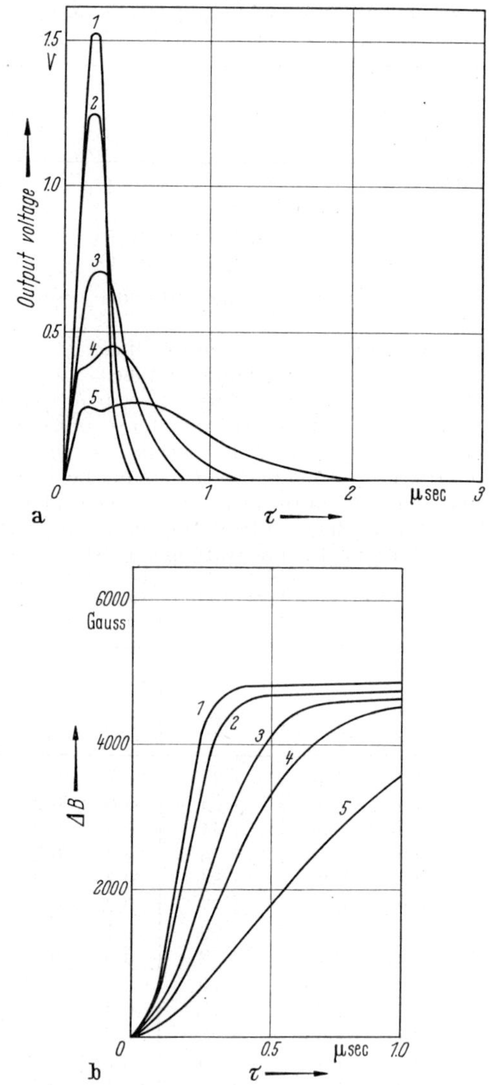

Fig. 102a, b. Square-loop $Mg_{0.3}Mn_{0.4}Ni_{0.3}Fe_2O_4$. Output voltage and induction change, ΔB, vs. time, τ [*64PS(2)*]. Maximum applied field, H_{max}:
1) 6.88 Oe, 2) 5.92 Oe, 3) 4.20 Oe, 4) 3.54 Oe, 5) 2.56 Oe.
Dimensions of toroidal sample: outer diameter = 3 mm, inner diameter = 2 mm, height = 1 mm, H_c = 1.1 Oe, B_s = 2500 Gauss.
Compare with Tab. 30, 31.

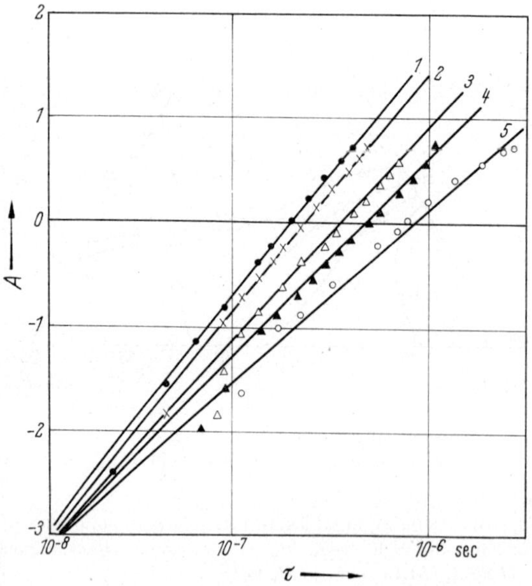

←

Fig. 103. $Mg_{0.3}Mn_{0.4}Ni_{0.3}Fe_2O_4$. The parameter A vs. τ [*64PS(2)*].

$A = \log\left[-\ln\left(1 - \frac{\Delta B}{B_{rem} + B_{max}}\right)\right]$ as calculated from data given in Fig. 102 for each H_{max}.

$B_{rem} + B_{max}$ = area enclosed by each curve in Fig. 102a. The equation was derived from relations used to describe the kinetics of an isothermal phase change. The notation is the same as in Fig. 102.

Tab. 32. Properties of three memory type ferrites [*54WGEG*].

	Composition	μ_i	$\left(\frac{\mu_{rem}}{\mu_i}\right)$	$\left(\frac{B_{rem}}{B_{max}}\right)_{max}$	$(R_s)_{max}$	H_{max} Oe
1	$(MgO)_{0.5}(MnO)_{0.875}Fe_2O_3$	55	0.49	0.96	0.81	1.35
2	$Mn_{0.1}Ni_{0.5}Mg_{0.4}Fe_2O_4$	138	0.48	0.95	0.83	1.7
3	$Mg_{0.4}Ni_{0.6}Fe_2O_4$	28	0.53	0.94	0.84	2.6

See Fig. 50 for definitions.

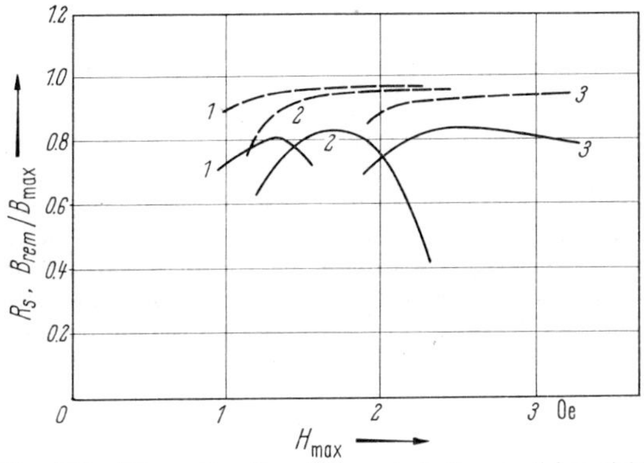

Fig. 104. The ratios B_{rem}/B_{max} and R_s vs. drive field, H_{max}, for the ferrites 1, 2, 3 from Tab. 32 [*54WGEG*]. R_s reaches a peak value and then decreases as H_{max} is increased.
– – – – – B_{rem}/B_{max} ———— R_s
See Fig. 50.

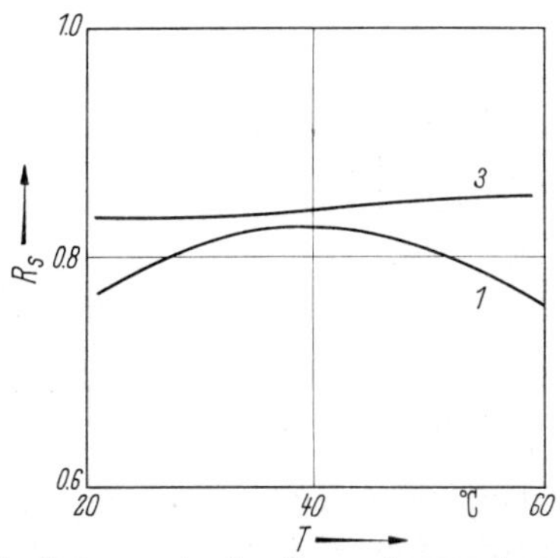

Fig. 105. Squareness ratio, R_s, vs. temperature, T, for the ferrites 1 and 3 of Tab. 32 [*54WGEG*]. See Fig. 50.

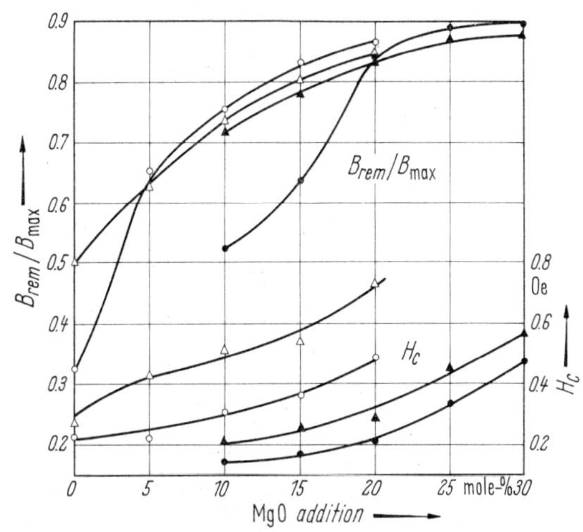

Fig. 106. B_{rem}/B_{max} and H_c vs. mole-% MgO added to $Mn_{0.6}Cd_{0.4}Fe_2O_4$ and $Mn_{0.6}Zn_{0.4}Fe_2O_4$ [*62BCKM*].
○ Cd samples fired at 1280 °C,
● Cd samples fired at 1400 °C,
▵ Zn samples fired at 1280 °C,
▴ Zn samples fired at 1400 °C.
Measurements were made at $H_{max} = 10$ Oe. Cores were sintered for 5 hours in O_2, then cooled in Ar. The compositions given were based on the weighed oxides; it was found that ≈16% of CdO is lost at 1280 °C and ≈24% at 1400 °C regardless of the MgO content.

Tab. 33. Approximate switching coefficient, S_w, and threshold field, H_0, for various ferrite systems [*65Bo*].

Ferrite system	S_w Oe·μsec	H_0 Oe
Mn-Mg	1.7	0.4
Cr-Mn-Fe_2O_4	0.8	0.85
Ni-Mg-Co-Fe_2O_4	2.6	2.5
Mg-Ni-Fe_2O_4	2.2	3.6
Ni-Fe^{2+}-Co-Fe_2O_4	25.0	12.0
Co-Fe^{2+}-Fe_2O_4	105.0	8.0

Compiler's note: S_w for MnMg system appears high. A more realistic range is 0.20 ··· ≈1.0 Oe · μsec, lower values applying to MnMgZn ferrites. See Tab. 35 and 36.

Tab. 34. Effect of CdO on properties of memory cores [*65Bo*].

Core type	H_c Oe	B_s Gauss	S_w Oe · μsec	Θ_C °C	CdO mole-%
S-3	1.3	1950	0.64	280	0.0
DM-1	0.15	1500	0.22	90	20.0
DM-13	0.30	—	0.25	113	16.0
DM-8	0.65	—	0.30	160	10.0

S-3 is the core as listed in Tab. 12. The manufacturer gives 0.6 Oe for H_c in contrast to the value of 1.3 Oe given above. Other data are in general agreement. Presumably, the 3 compositions containing CdO were prepared by using the S-3 composition as base material.

Tab. 35. MnMg-, CaMnMg-, ZnMnMg-ferrites. Switching coefficients, S_w, for various MgMn compositions [58Gy].

Composition	S_{w1} Oe · μsec	S_{w2} Oe · μsec
$Mn_{0.46}Mg_{0.71}Fe_{1.76}O_4$	0.25···0.35	
$Mn_{0.52}Mg_{0.67}Fe_{1.80}O_4$	0.30···0.70	
$Ca_{0.06}Mn_{0.58}Mg_{0.67}Fe_{1.84}O_4$	0.22···0.35 (0.47···0.52)	
$Zn_{0.20}Mn_{0.10}Mg_{0.70}Fe_{1.90}O_4$	0.20	0.42
$Zn_{0.39}Mn_{0.47}Mg_{0.34}Fe_{1.62}O_4$	0.22	
$Zn_{0.92}Mn_{0.46}Mg_{0.79}Fe_{1.62}O_4$	0.21	0.30
$Zn_{0.08}Mn_{0.64}Mg_{0.28}Fe_{1.70}O_4$	0.23	
$Zn_{0.18}Mn_{0.57}Mg_{0.16}Fe_{1.79}O_4$	0.33	

It is found that three linear regions can generally be obtained when $1/\tau_s$ is plotted vs. H_{max}. The low-field region is attributed to a domain wall mechanism, the intermediate-field region to an incoherent rotation, and the high-field region to a coherent rotation mechanism of switching [59HG], [59GH]. For two of the compositions, S_w values are given for two of the regions, the higher value always belonging to the domain wall process. Where a range of values is indicated, it represented a series of measurements made on samples of varying grain size; e.g. $Mn_{0.52}Mg_{0.67}Fe_{1.80}O_4$ has $S_w = 0.30$ Oe · μsec for average grain size = 2μ, and $S_w = 0.70$ Oe · μsec for average grain size = 1500μ. Varying methods of measurement used by different investigators make exact comparisons difficult.

Tab. 36. S_w, R_s for Ferroxcube 6A and other ferrites [57HBW].

Composition	H_0 Oe	S_w Oe · μsec	$(R_s)_{max}$	H_{max} Oe
Ferroxcube 6A	0.9	0.9	0.8	1.5
$(MgO)_{0.5}(MnO)_{0.875}Fe_2O_3$	1.4	0.9	0.75	1.5
$Cu_{0.25}Mn_{0.75}Fe_2O_4$	0.85	0.8		
$Mg_{0.6}Ni_{0.4}Fe_2O_4$	3.6	2.2	0.8	2.2
$Li_{0.47}Ni_{0.06}Fe_{2.47}O_4$	≈ 2.2	≈ 1.0	0.85	
$Co_{0.1}Fe^{II}_{0.9}Fe^{III}_2O_4$	8	105	0.95	10
$MnFe_2O_4$	0.5	3.2	0.8	0.5

Most ferrites have quite similar values of S_w. However, $Co_{0.1}Fe_{2.9}O_4$ has an abnormally large S_w. It is generally found that the "squarer" the hysteresis loop, the slower the switching time. The compound has an exceptionally high R_s due to a special heat treatment in a magnetic field. When magnetization is controlled by a uniaxial or induced anisotropy, as in case of magnetic annealing, relatively few domain walls switch the magnetization. Therefore, τ_s and S_w are large.

$MnFe_2O_4$ also has a long τ_s. Its output voltage curve does not have the second peak that is typical of memory cores.

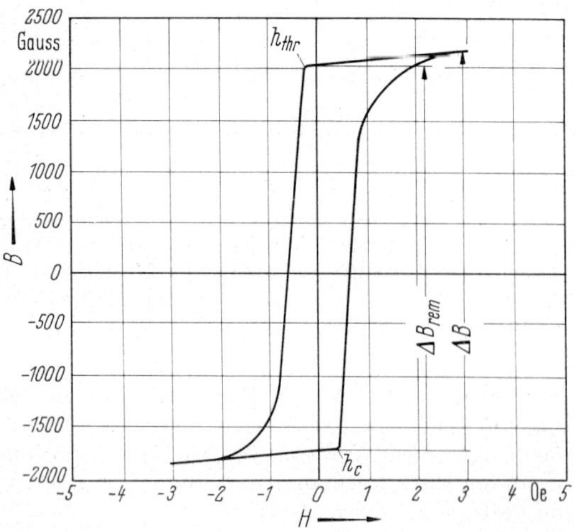

←

Fig. 107. Definitions of square-loop parameters for logic applications [64FW].

The parameters defined here apply to Fig. 108···111 only. The definitions given in Fig. 50 pertain to 2 : 1 coincident current selection applications. The 2 : 1 criterion is not required in logic applications. The portions of the hysteresis loops shown in Fig. 108···111 were obtained from multi-aperture toroids by integrating the voltage outputs as the amplitude of the read pulse was increased. Between each read pulse a block current of constant amplitude (equivalent to a field of 10 h_c) was applied.

$\mu_{max} = (\Delta B/\Delta H)_{max}$ = maximum permeability,
ΔB = magnetic induction switched,
$R_s = \Delta B_{rem}/\Delta B$ = squareness ratio,
h_c = critical switching field,
h_{thr} = threshold field.

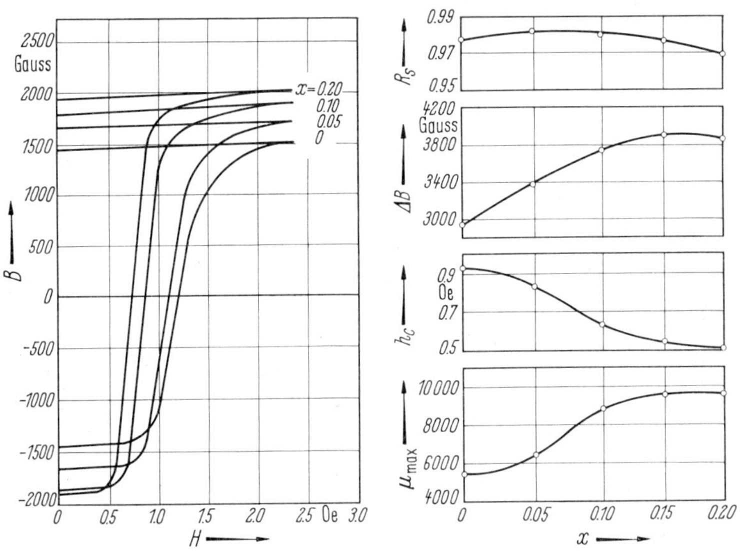

Fig. 108. $Mg_{0.5-x}Mn_{0.8}Zn_xFe_{1.7}O_4$. Effect of Zn content, x, on the square-loop properties [*64FW*]. See Fig. 107 for definition of parameters. Samples were fired for 8 hours in air at 1330 °C, then reheated to 1150 °C and cooled in N_2. Optimum properties were found for $x = 0.15$.

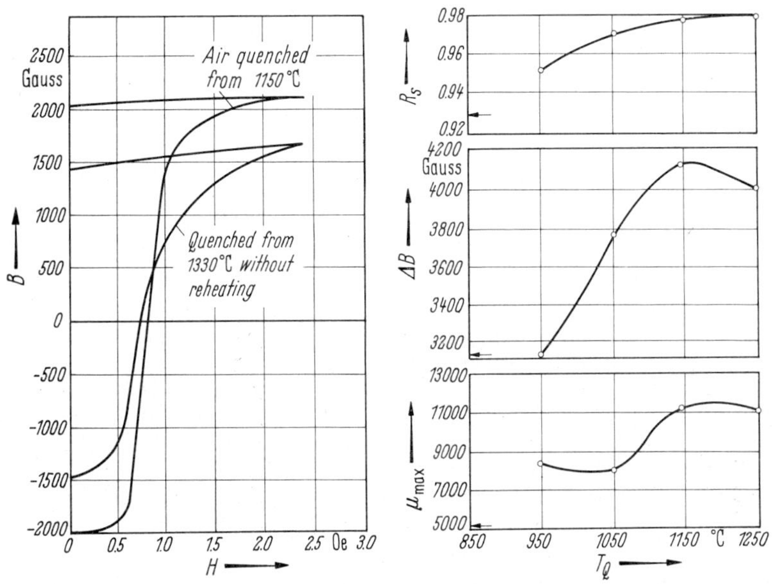

Fig. 109. $Mg_{0.35}Mn_{0.8}Zn_{0.15}Fe_{1.7}O_4$. Effect of quenching temperature, T_Q, on the square-loop properties [*64FW*].
See Fig. 107 for definition of parameters. Samples were fired for 8 hours in air at 1330 °C, then reheated to 950 °C ··· 1250 °C and air quenched. Arrows in right-hand figure indicate values obtained by quenching from 1330 °C without reheating.

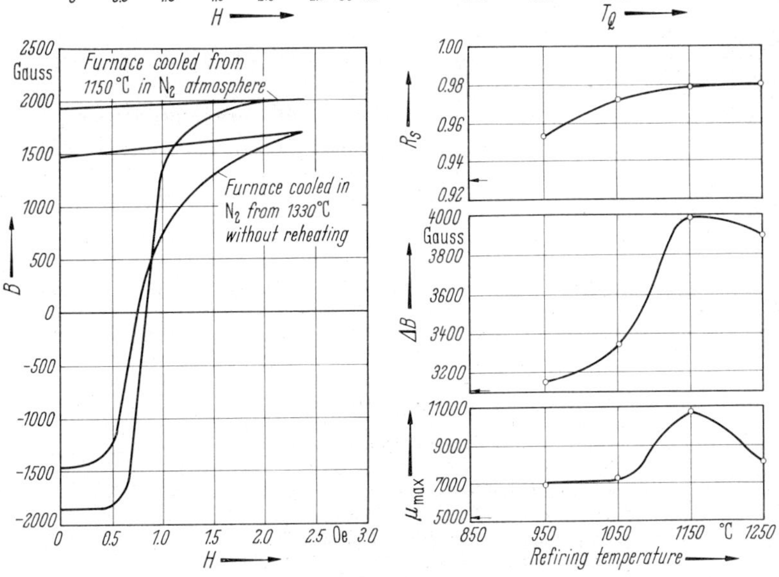

Fig. 110. $Mg_{0.35}Mn_{0.8}Zn_{0.15}Fe_{1.7}O_4$. Effect of refiring temperature on the square-loop properties [*64FW*]. See Fig. 107 for definition of parameters. Samples were fired for 8 hours in air at 1330 °C, then reheated to 950 °C ··· 1250 °C and cooled in N_2. Arrows in right-hand figure indicate values obtained without reheating.

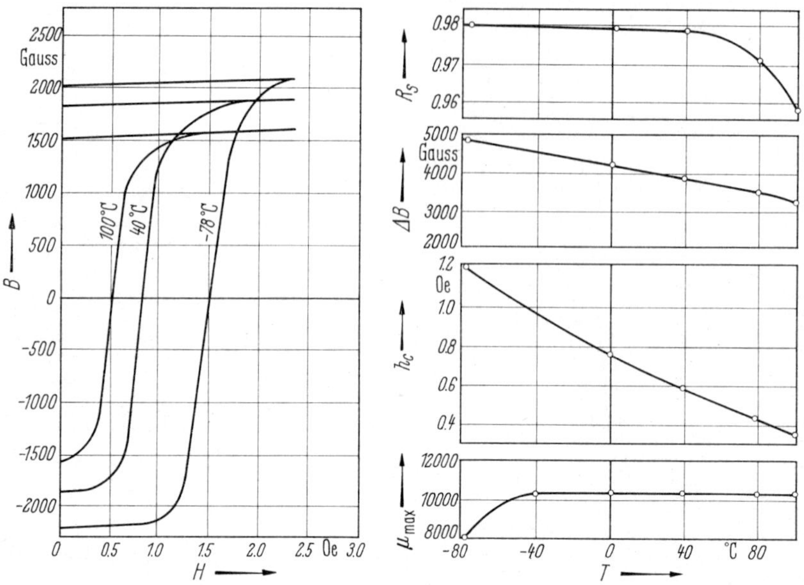

Fig. 111. $Mg_{0.35}Mn_{0.5}Zn_{0.15}Fe_{1.7}O_4$. Effect of ambient temperature on the square-loop properties [*64FW*]. See Fig. 107 for definition of parameters. Sample was fired for 8 hours in air at 1330 °C, then reheated to 1150 °C for 2 hours and air quenched. All materials for Fig. 108 ··· 111 were prepared from uncalcined mixtures of the oxides pressed at 5000 psi (350 kgcm⁻²).

Tab. 37. Pulse properties of some commercial MgMn(Zn) ferrite memory cores [*63AES*].

Designation	System	geometry: outer diameter in.	geometry: inner diameter in.	geometry: height in.	drive currents: full select A	drive currents: partial select A	pulse rise-time τ_r μsec	$(u V_1)_{min}$ mV	$(d V_z)_{max}$ mV	$(\tau_s)_{max}$ μsec
MC 113	MgO-MnO-Fe_2O_3	0.080	0.050	0.025	0.740	0.475	0.20	100	30	1.30
MC 138	MgO-MnO-Fe_2O_3	0.050	0.030	0.015	0.450	0.285	0.20	45	9	1.25
MC 140	MgO-MnO-ZnO-Fe_2O_3	0.050	0.030	0.015	0.400	0.245	0.20	45	11	1.25
MC 142	MgO-MnO-ZnO-Fe_2O_3	0.080	0.050	0.025	0.360	0.220	0.50	42	10	2.90
MC 144	MgO-MnO-Fe_2O_3	0.050	0.030	0.015	0.890	0.540	0.10	100	20	0.50
MC 147	MgO-MnO-ZnO-Fe_2O_3	0.050	0.030	0.015	0.260	0.158	0.50	20	4	2.20
MC 149	MgO-MnO-ZnO-Fe_2O_3	0.050	0.030	0.015	0.320	0.195	0.50	38	6	1.60
MC 151	MgO-MnO-CaO-Fe_2O_3	0.080	0.050	0.025	0.360	0.220	0.50	35	9	4.0
MC 153	MgO-MnO-Fe_2O_3	0.040	0.025	0.015	0.690	0.420	0.10	55	15	0.50
MC 159	MgO-MnO-Fe_2O_3	0.030	0.018	0.009	0.530	0.325	0.10	35	10	0.50

$u V_1$ = undisturbed one-response voltage;
$d V_z$ = disturbed zero-response voltage
(Materials are manufactured by Electronics Division of Indiana General Corporation, USA).

6.1.4.2.5 Microwave properties

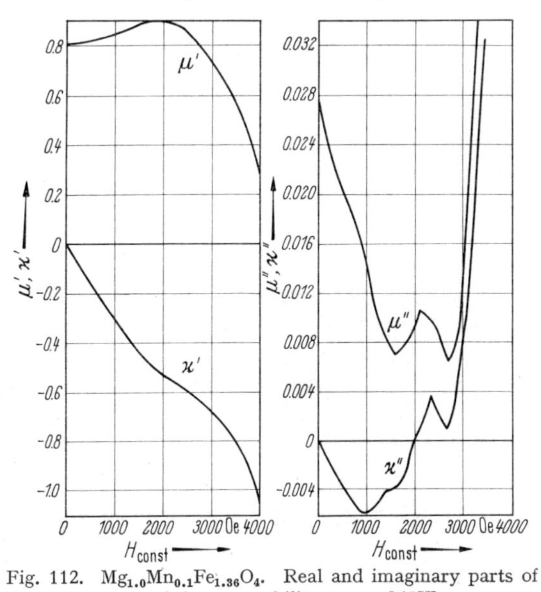

Fig. 112. $Mg_{1.0}Mn_{0.1}Fe_{1.36}O_4$. Real and imaginary parts of the components of the permeability tensor [65K].
$4\pi M_s$ = 2050 Gauss, f = 9200 Mc/s. Data at RT. Thin disc ceramic samples with axial biasing field were used for Fig. 112···117.
The small amount of Mn is introduced into the materials of Fig. 112···117 to increase the resistivity as shown by [56vU(1)], [57vU] (see Fig. 11, 22). For a saturated magnetic medium, a tensor form is required to adequately describe the permeability. μ', μ'', $\varkappa'$, $\varkappa''$ = real and imaginary tensor components of permeability; see Fig. 130.

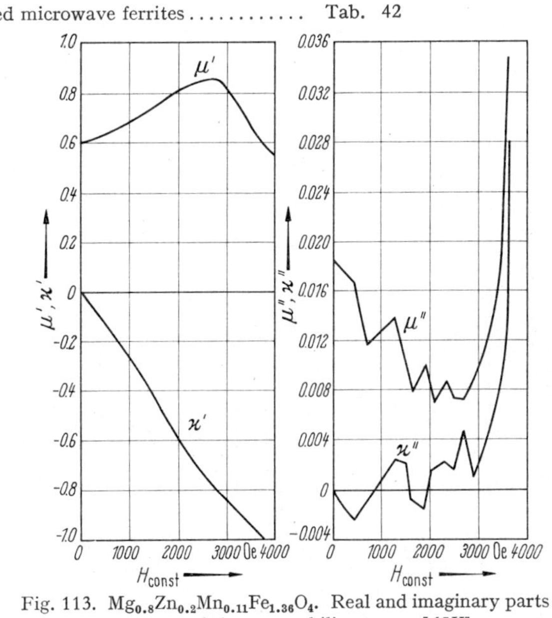

Fig. 113. $Mg_{0.8}Zn_{0.2}Mn_{0.11}Fe_{1.36}O_4$. Real and imaginary parts of the components of the permeability tensor [65K].
$4\pi M_s$ = 2850 Gauss, f = 9200 Mc/s. Data at RT. See Fig. 112.

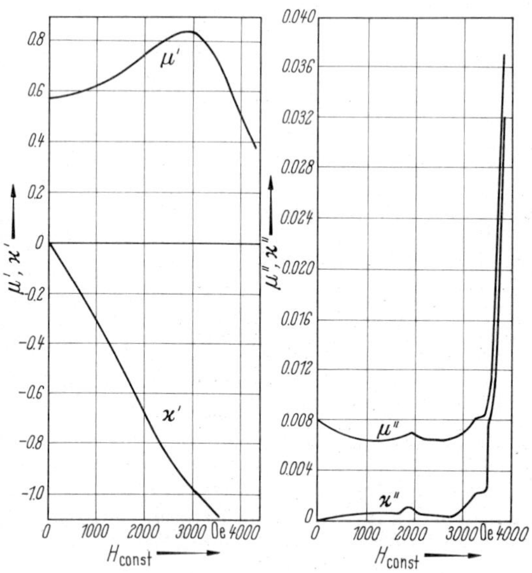

Fig. 114. $Mg_{0.7}Zn_{0.2}Mn_{0.1}Fe_{1.5}O_4$. Real and imaginary parts of the components of the permeability tensor [65K].
$4\pi M_s = 3000$ Gauss, $f = 9200$ Mc/s. Data at RT. See Fig. 112. This material has an extremely small low-field loss. The addition of Zn reduced both μ', μ'' as compared with Mg ferrite.

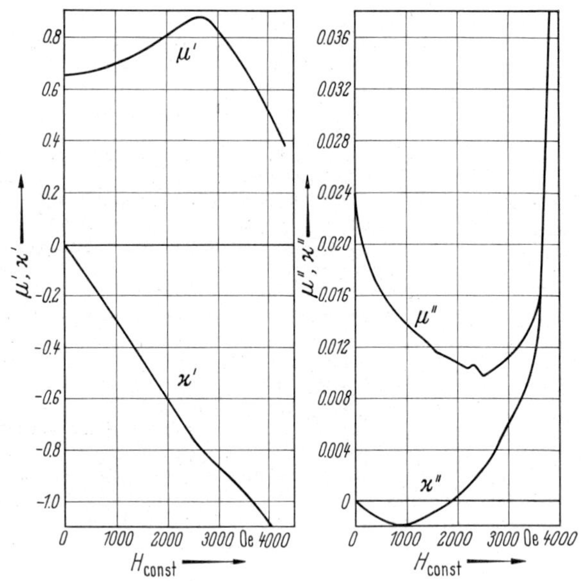

Fig. 115. $Mg_{0.7}Zn_{0.2}Mn_{0.1}Fe_{1.9}O_4$. Real and imaginary parts of the components of the permeability tensor [65K].
$4\pi M_s = 3050$ Gauss, $f = 9200$ Mc/s. Data at RT. See Fig. 112.

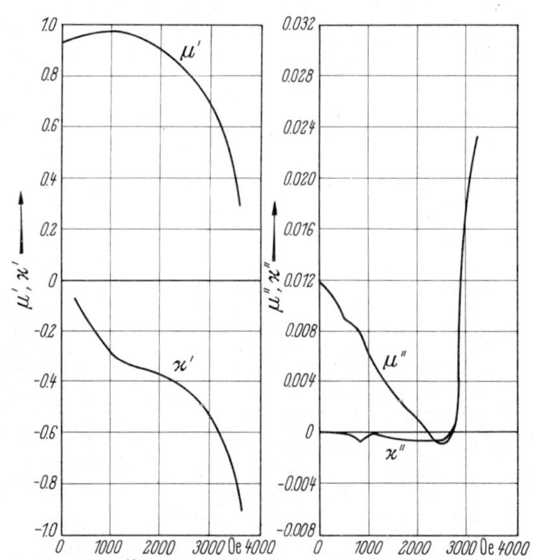

Fig. 116. $Mg_{1.0}Mn_{0.1}Al_{0.2}Fe_{1.70}O_4$. Real and imaginary parts of the components of the permeability tensor [65K].
$4\pi M_s = 1350$ Gauss, $f = 9200$ Mc/s. Data at RT. See Fig. 112. Note the very low loss at ≈2500 Oe. Addition of Al^{3+} not only decreases M_s but further increases resistivity, presumably by inhibiting electron hopping between Fe^{2+} and Fe^{3+} (see Fig. 22 and in 6.1.4.3, Fig. 43, 44, Tab. 7).

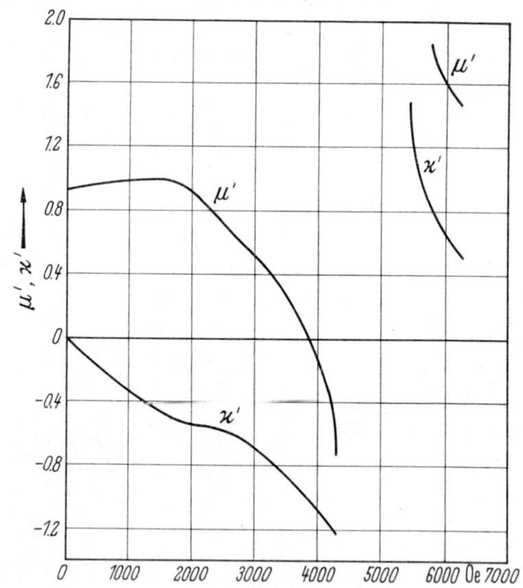

Fig. 117. $Mg_{0.9}Cu_{0.1}Mn_{0.04}Al_{0.15}Fe_{1.6}O_4$. Real parts of the components of the permeability tensor [65K].
$4\pi M_s = 1700$ Gauss, $f = 9200$ Mc/s. Data at RT. See Fig. 112. A further improvement for microwave use is made with a small amount of Cu. Firing temperature is reduced, resistivity further improved, and better uniformity obtained (Fig. 15, 21 ··· 23).

Fig. 118. $(MnO)_{0.63}(MgO)_{0.87}Fe_2O_3$. Resonance curves for a single crystal [*63GV*].
Measurements were made with the field parallel to [110]. Also shown are data for a single crystal of $Ni_{0.5}Fe^{2+}_{0.5}Fe^{3+}_2O_4$ measured with field parallel to [100]. Measurements were made on a recording spectrometer. The results have been explained theoretically in [*63K*].

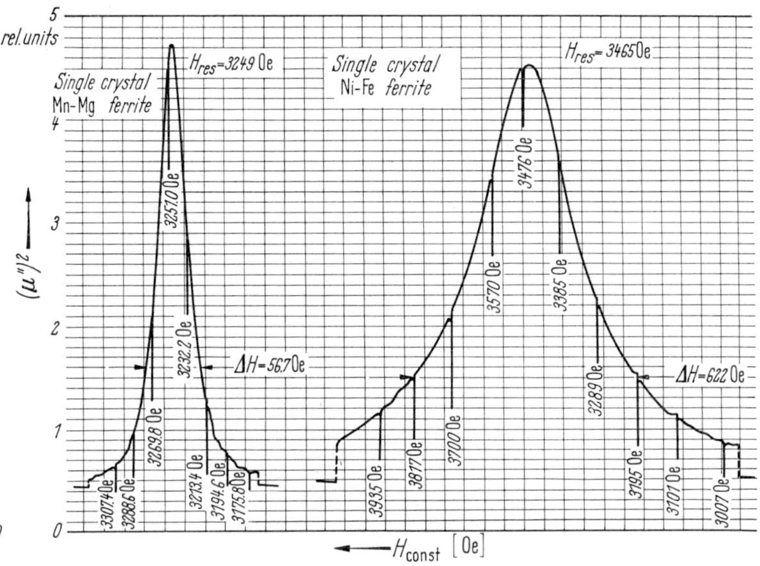

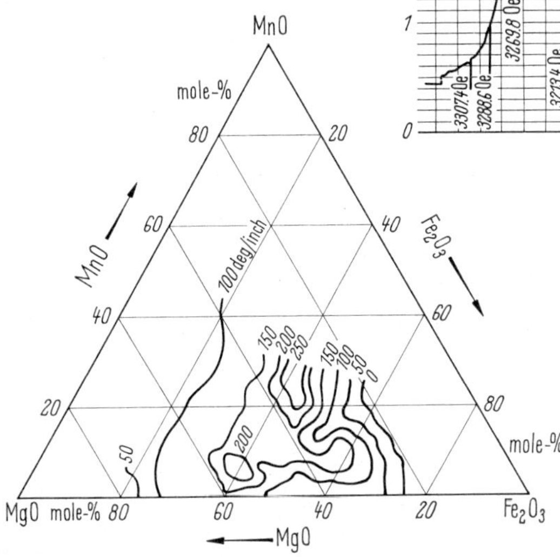

Fig. 119. Specific Faraday rotation, ϱ, in deg/inch at 9400 Mc/s for MgMn ferrites fired at 1350 °C [*56F*]. Compare with Fig. 144.

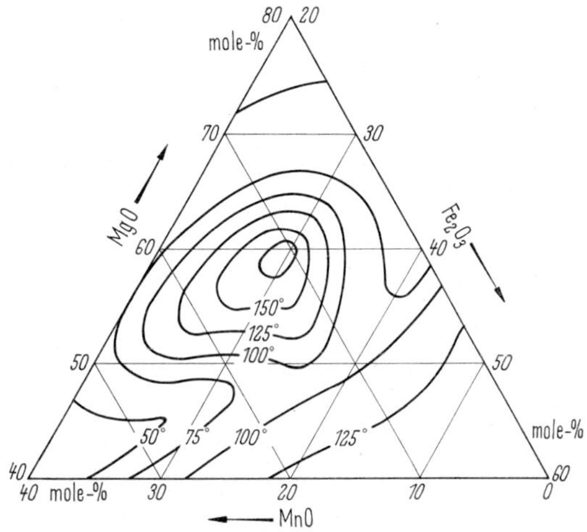

Fig. 120. MgO-MnO-Fe_2O_3. Faraday rotation, θ, at 9525 Mc/s as a function of composition [*57L*]. For Fig. 120···122 measurements were made with a field of 378 Oe applied along the axis of rod specimens having a length of 1.30 in. and a diameter of 0.26 in. The rods were pressed from the milled oxides at 13 tons in.$^{-2}$. They were fired in air at 1300 °C for 105 min and furnace cooled. Compare with Fig. 144.

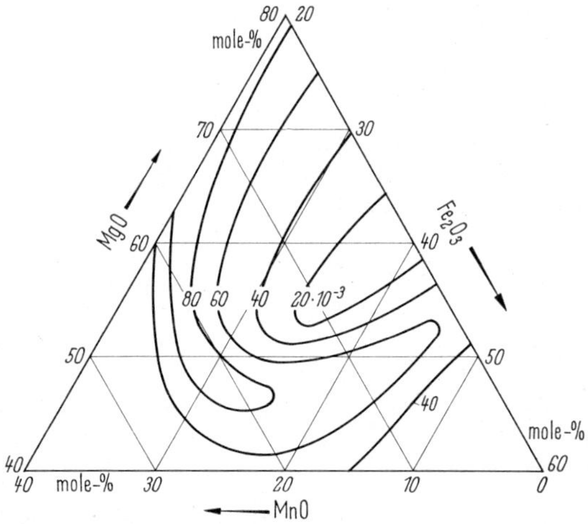

Fig. 121. MgO-MnO-Fe_2O_3. Specific attenuation in db/deg at 4700 Mc/s as a function of composition [*57L*]. See Fig. 120.

→

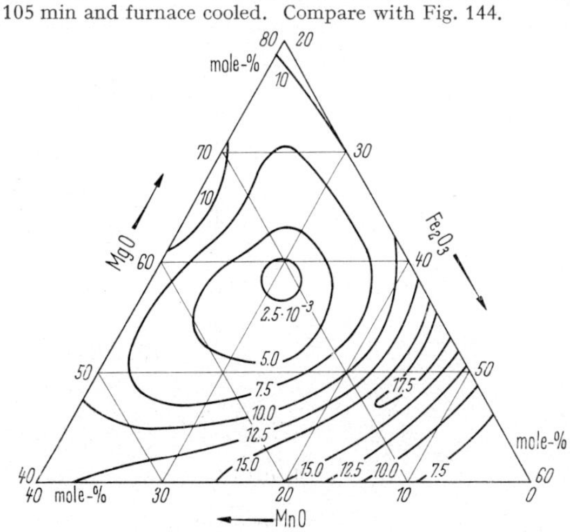

Fig. 122. MgO-MnO-Fe_2O_3. Specific attenuation in db/deg at 9525 Mc/s as a function of composition [*57L*]. Compare with Fig. 144. See Fig. 120.

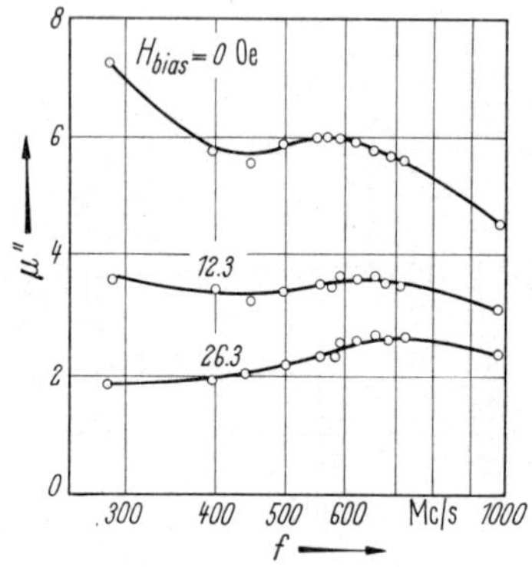

Fig. 123. $(MgO)_{0.356}(MnO)_{0.341}(Fe_2O_3)_{0.303}$. μ'' vs. frequency, f, for different biasing fields [62R].
The material was sintered for 3 hours in air at 1300 °C, quenched from 1200 °C. H_{bias} = biasing field, parallel to hf field. M_s = 185 Gauss, Θ_C = 203 °C, $K_1 = 0.6 \cdot 10^4\ \mathrm{erg\,cm^{-3}}$.

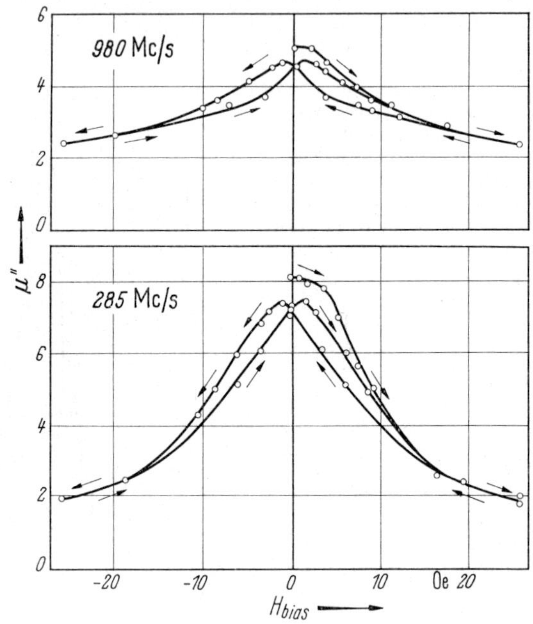

Fig. 124. $(MgO)_{0.356}(MnO)_{0.341}(Fe_2O_3)_{0.303}$. μ'' vs. biasing field, H_{bias}, at two frequencies [62R].
Arrows indicate sequence of measurements. See Fig. 123 for preparation of material.

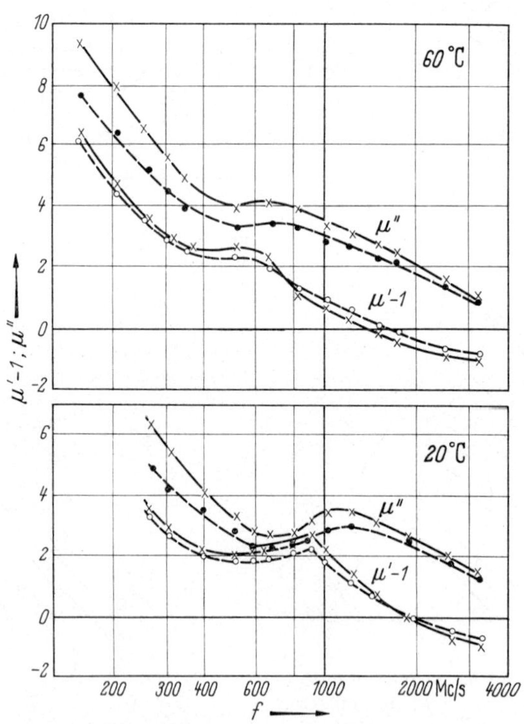

Fig. 125. $(MgO)_{0.356}(MnO)_{0.341}(Fe_2O_3)_{0.303}$. μ' and μ'' vs. frequency, f, at two temperatures [62R].
Same composition as in Fig. 123 but the material was cooled in furnace.
M_s = 160 Gauss, Θ_C = 225 °C, $K_1 = 2.3 \cdot 10^4\ \mathrm{erg\,cm^{-3}}$, H_c = 2.5 Oe, μ_i = 60.
——— after demagnetization
– – – – at remanence.

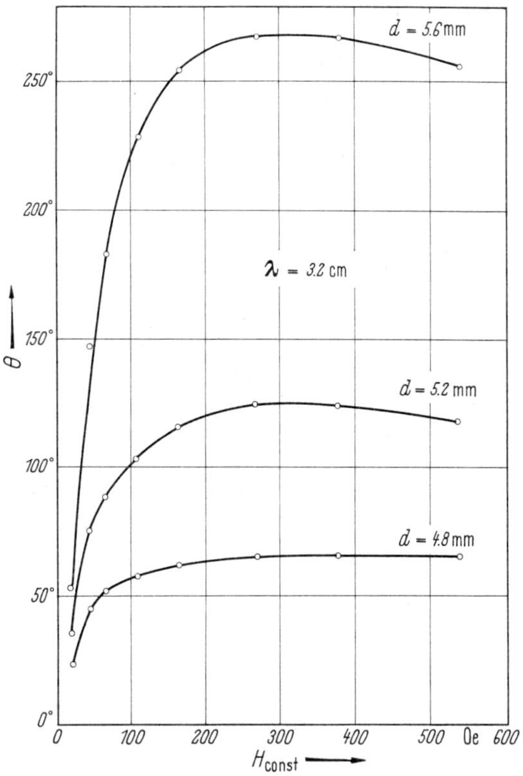

Fig. 126. $(MgO)_{0.35}(MnO)_{0.25}(Fe_2O_3)_{0.40}$. Faraday rotation, θ, vs. constant field, H_{const} [61LV].
$4\pi M_s = 3000$ Gauss, $\Delta H_{1/2} = 250$ Oe, $\varrho = 4.52\ \mathrm{g\,cm^{-3}}$. Measurements were made on cylinders having diameter, d, as indicated at the curves and a length of 50 mm, using a cavity with a radius of 24 mm.

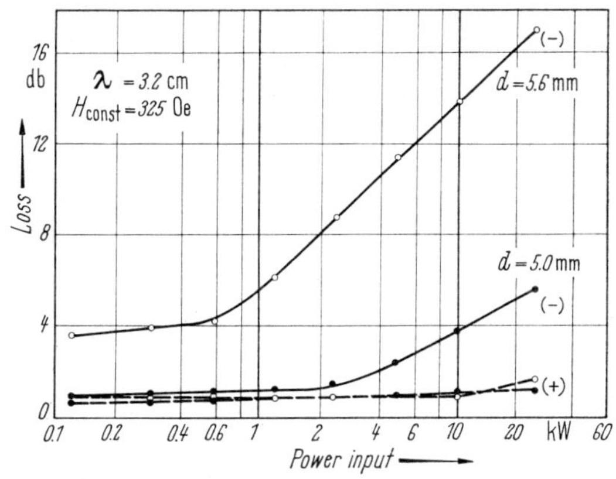

Fig. 127. $(MgO)_{0.35}(MnO)_{0.25}(Fe_2O_3)_{0.40}$. Absorption loss vs. power input for two cylinder diameters, d [61LV].
——— (−) negative circular polarization,
– – – – – (+) positive circular polarization.
Cylinder length = 50 mm. See notes for Fig. 126.

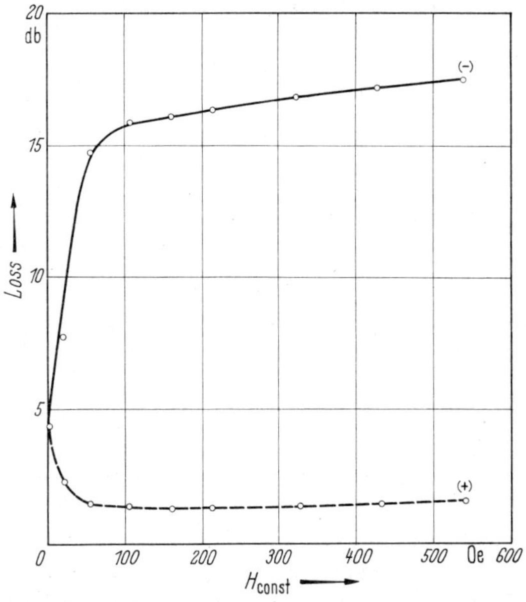

Fig. 128. $(MgO)_{0.35}(MnO)_{0.25}(Fe_2O_3)_{0.40}$. Absorption loss vs. constant field, H_{const}, at a power input of 25 kW [61LV]. Cylinder diameter = 5.6 mm, cylinder length = 50 mm. See notes for Fig. 126, 127.

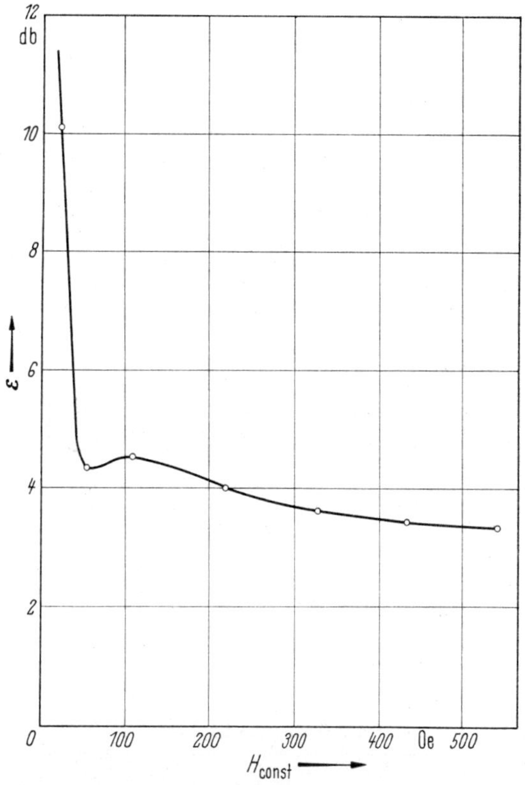

Fig. 129. $(MgO)_{0.35}(MnO)_{0.25}(Fe_2O_3)_{0.40}$. Ellipticity, ε, vs. constant field, H_{const}, at a power input of 25 kW [61LV]. Cylinder diameter = 5.6 mm, cylinder length = 50 mm. ε = ellipticity of circular polarized wave after passing through ferrite. See notes for Fig. 126.

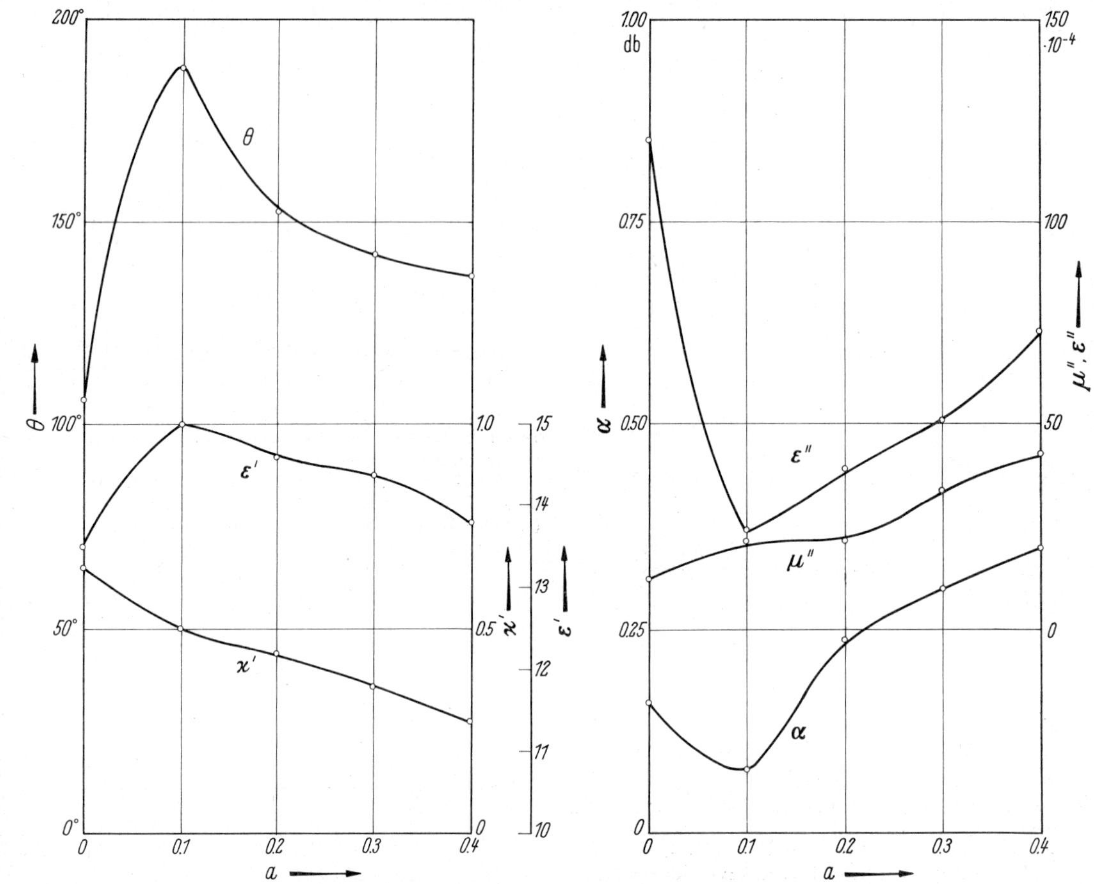

Fig. 130. $Mg_{0.95}Mn_{0.19}Ni^{2+}_aTi^{4+}_aFe_{1.86-2a}O_4$. Properties at X band [*64DC*].
ε', ε'' = real and imaginary parts of the dielectric constant, ε, (scalar form).
μ', μ'', ϰ', and ϰ'' = real and imaginary tensor components of permeability. For all compositions shown, $\mu' \approx 1$ and $\varkappa'' < 3 \cdot 10^{-4}$. θ = Faraday rotation, α = insertion loss. Samples were two cones, 37 mm long, joined at their bases (base diameter = 6.4 mm). Measurements were made at 9575 Mc/s at RT. All samples had been fired at about 1200 °C and had the same density (near the theoretical value).

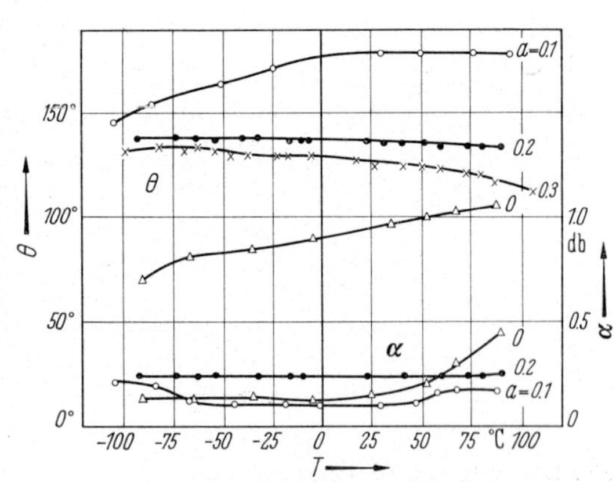

Fig. 131. $Mg_{0.95}Mn_{0.19}Ni^{2+}_aTi^{4+}_aFe_{1.86-2a}O_4$. Faraday rotation, θ, and insertion loss, α, vs. temperature, T, at X band [*64DC*].
Measurements were made at H_{const} = 112 Oe.

Tab. 38. $Mg_{0.95}Mn_{0.19}Ni^{2+}_aTi^{4+}_aFe_{1.86-2a}O_4$. Saturation magnetization, M_s, Curie temperature, Θ_C, and merit factor, θ/α [*64DC*].

a	$4\pi M_s$ Gauss	Θ_C °C	θ/α deg db^{-1}
0	2000	346	665
0.1	1696	291	2350
0.2	1431	266	640
0.3	1132	217	475
0.4	890	158	392

θ = Faraday rotation, α = insertion loss, θ/α and $4\pi M_s$ at RT.
See Fig. 130, 131.

Tab. 39. $Mg_{0.9}Mn_{0.1}Fe_{1.6}O_4$. The dependence of the ferrimagnetic resonance curve on temperature for polycrystalline samples fired at different temperatures [57SP].

Sample Nr.	Firing temperature °C	T °C	$\lambda = 0.866$ cm		$\lambda = 1.244$ cm	
			$\Delta H_{1/2}$ Oe	ΔH_r Oe	$\Delta H_{1/2}$ Oe	ΔH_r Oe
1	1250	20	1070	145	1000	165
		100	810	45	720	55
		150	600	25	600	35
		180	590	25	520	25
		200	570	25	490	25
2	1300	20	900	155	850	160
		100	670	35	610	50
		150	580	25	500	20
		180	510	25	460	15
3	1350	20	550	75	500	90
		100	300	100	310	50
		150	230	35	250	45
		180	220	45	230	45
		200	180	25	210	25
4	1400	20	520	105	450	85
		100	370	80	330	70
		150	350	70	320	45
		180	240	45	250	45
5	1450	20	530	195	500	150
		100	410	95	400	85
		150	365	85	330	50
		180	380	80	330	45
		200	340	80	300	45

All samples were fired in O_2 and were slowly cooled.
ΔH_r = field difference between the observed resonance peak and the peak for which $g = 2.000$.
$\Delta H_{1/2}$ = resonance line half-width.

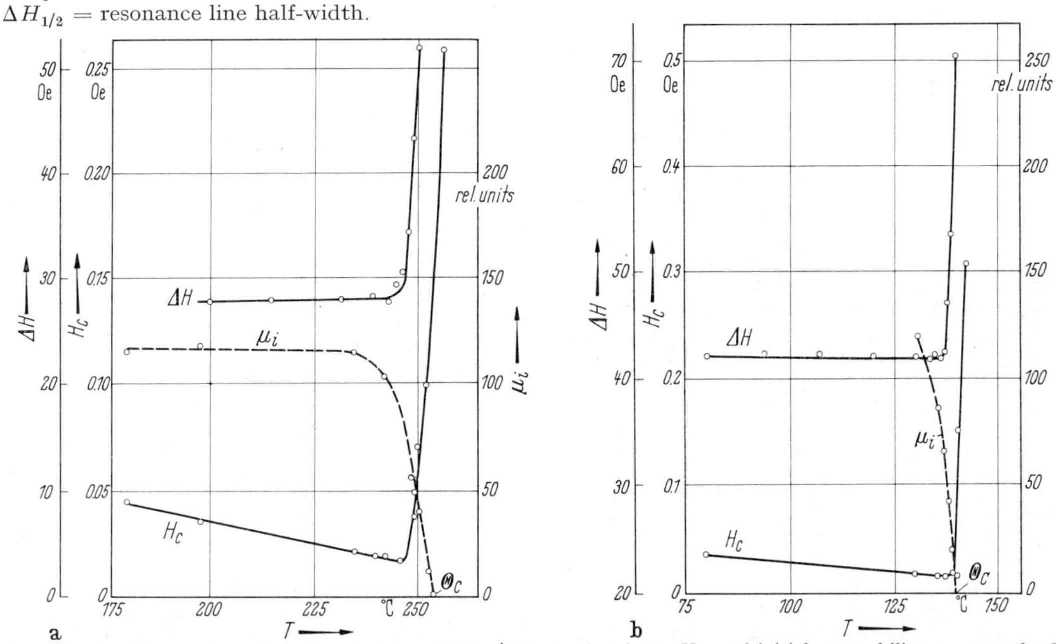

Fig. 132 a, b. Temperature dependence of line width, ΔH, coercive force, H_c, and initial permeability, μ_i, near the Curie point [61BeBe].
a) 1.8 MgO, 32 MnO, 65.7 Fe_2O_3 (wt.-%) – single crystal, b) 2.2 MgO, 54 MnO, 43.6 Fe_2O_3 (wt.-%) – single crystal. ΔH measured at 9470 Mc/s using spheres 0.8 mm diameter. H_c and μ_i measured by a null method with an elastic magnetometer. For μ_i, $H = 0.005$ Oe. The H_c data suggested that the increase in ΔH is due to defects and disorder in cation distribution and not to thermal fluctuations of M_s. Crystals were grown by the Verneuil method. Data were also given for Mn ferrites.

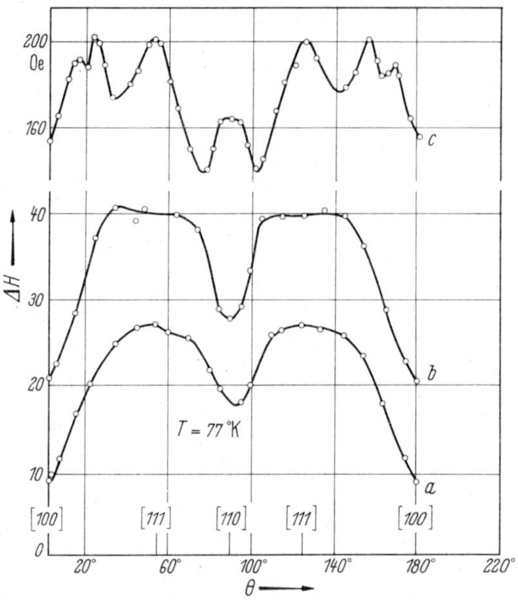

Fig. 133. $Mg_{0.525}Mn_{0.665}Fe_{1.81}O_4$ and YIG. Line width anisotropy for single crystals caused by impurities [*62HSG*].
Measurements were made at 3 cm using polished spheres. Crystallographic directions for various angles, θ, indicated.
a) YIG with about $3 \cdot 10^{-3}$ mole-% rare earth impurities,
b) MgMn ferrite suspected to contain Fe^{2+} as impurity,
c) YIG with about $3 \cdot 10^{-2}$ mole-% rare earth impurities.

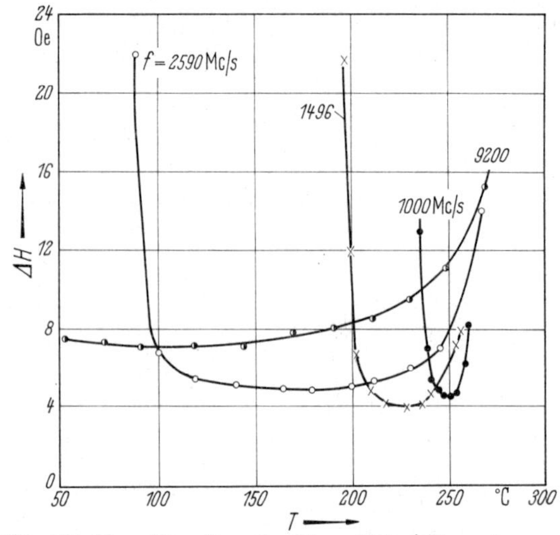

Fig. 134. $Mg_{0.52}Mn_{0.84}Fe_{1.74}O_4$. Line width, ΔH, vs. temperature, T, for a single crystal. Parameter f [*62YL*].
Spherical sample (diameter = 1.4 mm) cut from Verneuil-grown crystal and polished with abrasive powder of 1 μ grains.

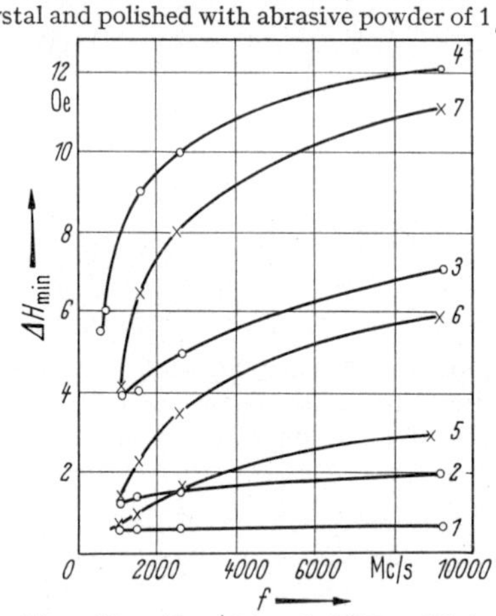

Fig. 135. $Mg_{0.52}Mn_{0.84}Fe_{1.74}O_4$ and YIG. Minimum line width, ΔH_{min}, vs. frequency, f, for single crystals [*62YL*].
1) Polished YIG,
2) Unpolished YIG,
3) $Mg_{0.52}Mn_{0.84}Fe_{1.74}O_4$,
4) Polycrystalline YIG,
5) Polished YIG at 270 °C,
6) Unpolished YIG at 270 °C,
7) $Mg_{0.52}Mn_{0.84}Fe_{1.74}O_4$ at 250 °C.
See source for additional data on YIG.

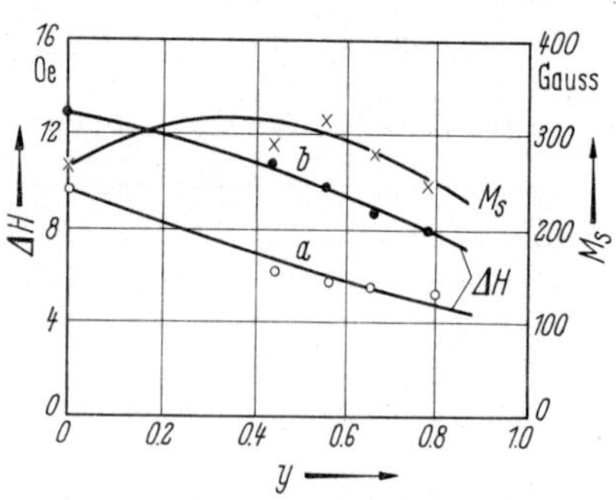

Fig. 136. $Mg_yMn_{1.3-y}Fe_{1.7}O_{4\pm}$. Line width, ΔH, and saturation magnetization, M_s, vs. Mg content, y [*66MY*].
a) ΔH for [111] direction, *b*) ΔH for [100] direction.
Crystals, grown by the Verneuil method, were polished to spheres of 0.5 mm diameter using a 1 μ polishing compound. Measurements were made by parallel pumping in the 3-cm wavelength region for k values (wave numbers) from $0 \cdots \approx 3 \cdot 10^5$ cm^{-1}.

→

Fig. 137. $Mg_yMn_{1.3-y}Fe_{1.7}O_{4\pm}$. Critical power, P_{crit} and wave number, k, vs. constant field, H_{const}, for single crystals [*66MY*].
P_{crit} = threshold power for excitation of spin waves. All measurements in the [111] direction. In Fig. 136 ⋯ 138, the decrease in the relaxation parameters with an increase in Mg is attributed to the increasing order of cation distribution on *A*, *B* sites. For comparison, the lowest curve shows the behavior of a YIG in which mixing of cations does not occur. The effect of the variation of Fe^{2+} content was discounted since the data for samples cut from different portions of the boule, over which Fe^{2+} was expected to vary, were reproducible.

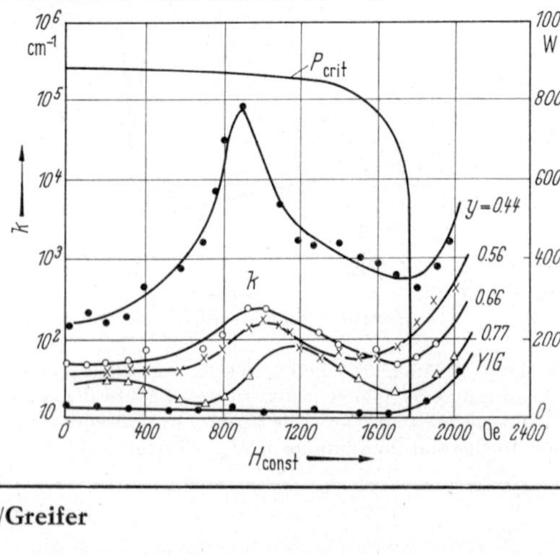

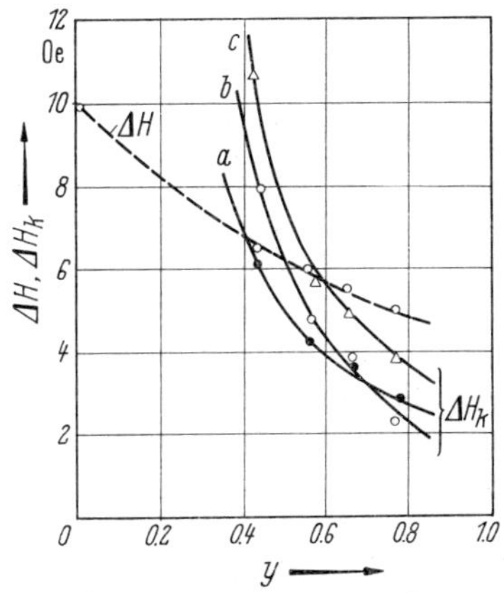

Fig. 138. $Mg_yMn_{1.3-y}Fe_{1.7}O_{4\pm}$. Resonance line width, ΔH, and spin wave line width, ΔH_k, vs. Mg content, y, in single crystals [*66MY*].
a) H_{const} = 400 Oe,
b) H_{const} = 600 Oe,
c) H_{const} = 1000 Oe.
Measurements in [111] direction.

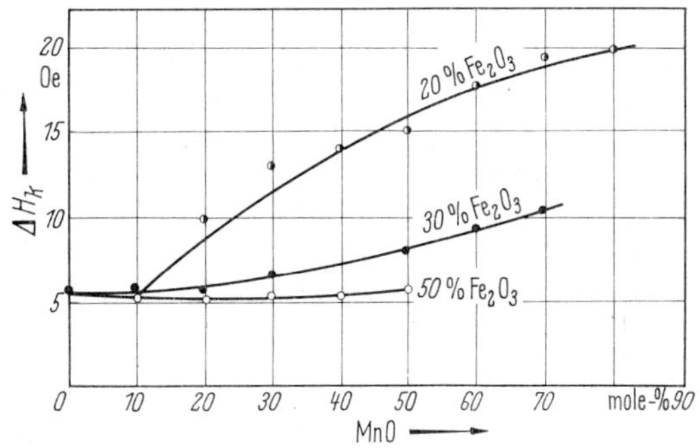

Fig. 139. Polycrystalline MgMn ferrites. Spin wave line width, ΔH_k, vs. mole-% MnO [*67LPY*].
mole-% MgO = 100 − (mole-% Fe_2O_3 + mole-% MnO).
Measurements were made on samples of 0.5 ··· 2 mm diameter by parallel pumping at 9400 Mc/s. ΔH_k was calculated from threshold field minima. All samples with 20 mole-% Fe_2O_3 had the same approximate grain size (8 ··· 12 μ), the same density, and the same value of $4\pi M_s$ (1300 ··· 1500 Gauss) except for the value of 2300 Gauss for the specimen having 80 mole-% MnO. It was concluded that ΔH_k is a function of the Mn^{3+} concentration on the *B* site. See source for additional data on single crystal and polycrystalline Mn ferrites.

Tab. 40. MgMn ferrites. Microwave properties of polycrystalline samples [*65PL*].

Sample Nr.	Composition mole-%			Grain dimensions μ	ϱ g/cm³	$4\pi M_s$ Gauss	ΔH Oe	$(h_{thr})_{min}$ Oe	ΔH_k Oe	Remarks
	Fe_2O_3	MnO	MgO							
1	50	50	0	—	4.67	4570	340	3.2	4.3	Excess hematite phase
2	50	40	10	55 ··· 12	4.58	4360	260	3.5	4.0	
3	50	30	20	34 ··· 10	4.59	3880	280	3.8	4.0	
4	50	20	30	30 ··· 8	4.43	2900	370	4.1	4.5	
5	50	10	40	—	4.61	2460	360	5.5	4.0	
6	50	0	50	4	4.38	1350	400	10.5	4.0	Nonmagnetic MgO phase
7	40	60	0	—	4.61	3520	—	6.7	7.2	
8	40	50	10	30 ··· 10	4.38	3310	180	4.8	4.7	Hausmannite along grain boundaries
9	40	40	20	—	4.55	2850	270	5.8	0.5	
10	40	30	30	—	—	2420	330	5.9	4.3	
11	40	20	40	—	4.36	2130	370	6.4	4.0	
12	40	10	50	—	4.23	2530	380	—	—	
13	40	0	60	4	4.26	1280	450	11.3	4.3	Nonmagnetic MgO phase

ΔH_k = spin wave line width
$(h_{thr})_{min}$ is the minimum microwave threshold field for spin wave excitation by parallel pumping at 9400 Mc/s. Samples were spheres of 2.5 mm diameter. Results were found to be independent of diameter (from 1.7 ··· 3.0 mm) and of degree of surface polishing.

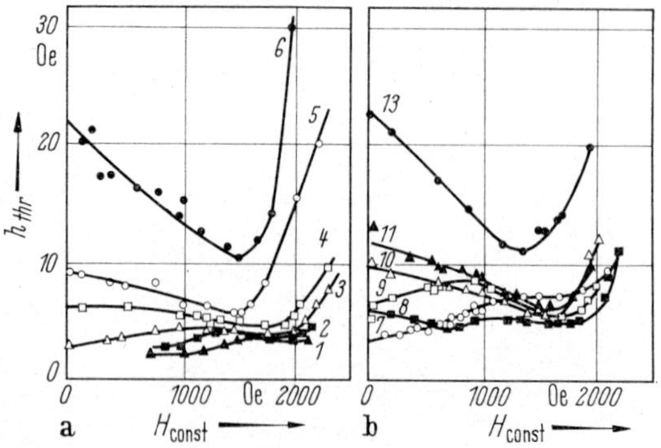

Fig. 140a, b. Polycrystalline MgMn ferrites. Threshold field, h_{thr}, vs. constant field, H_{const} [*65PL*].
Numbers refer to materials listed in Tab. 40.

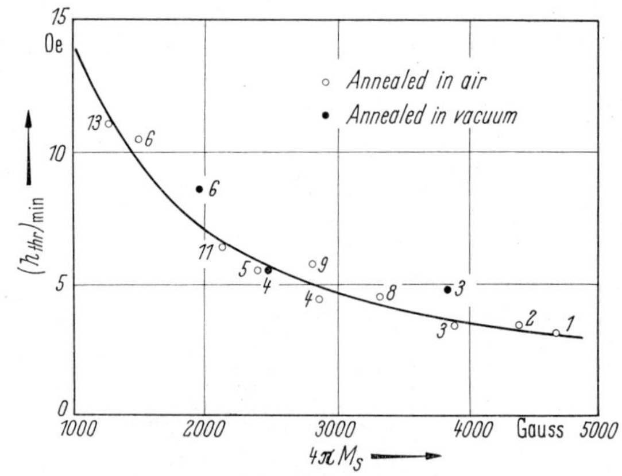

Fig. 141. Polycrystalline MgMn ferrites. Minimum threshold field, $(h_{thr})_{min}$, vs. saturation magnetization, $4\pi M_s$ [*65PL*]. Numbers refer to materials listed in Tab. 40. The curve was calculated on basis of constant ΔH_k; the points are experimental data from Tab. 40. The vacuum annealing reduced $4\pi M_s$ and increased grain size. The data showed that the threshold field can be explained by $4\pi M_s$ regardless of grain sizes as listed in Tab. 40.

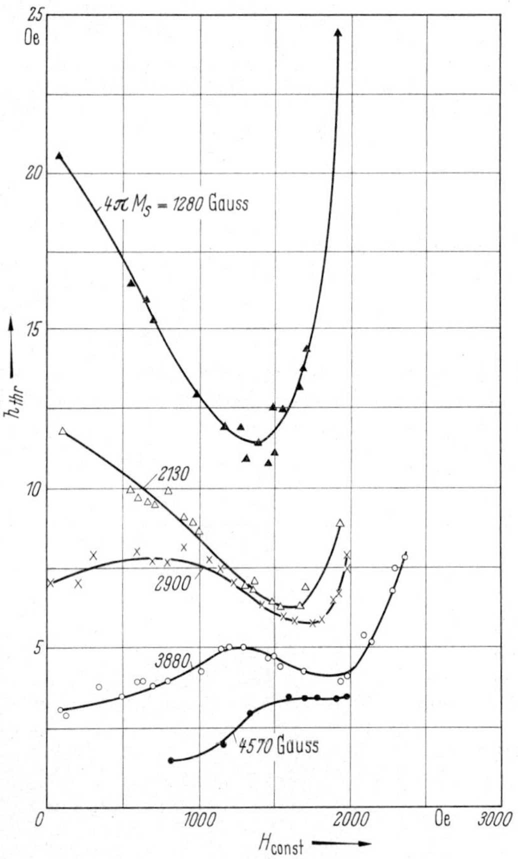

Fig. 143. Threshold field, h_{thr}, vs. constant field, H_{const}, for unsaturated MgMn ferrites as a function of $4\pi M_s$ [*66LPS*]. Ceramic samples of different composition. Measurements were made on spherical samples by parallel pumping at 3 cm. See Fig. 64, 65 in 6.1.4.3 for similar data on MgCr ferrites.

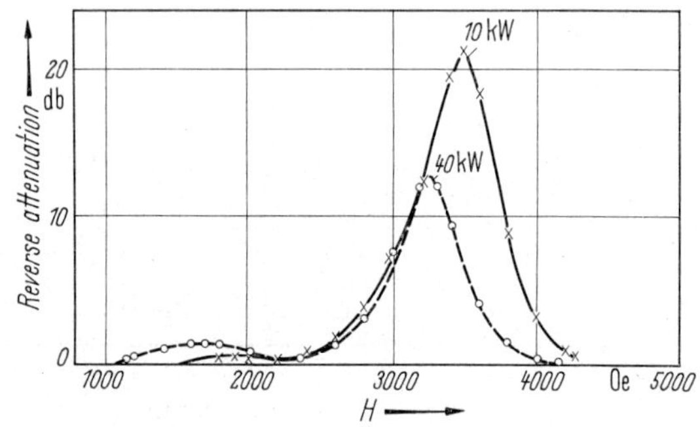

Fig. 142. $Mg_{0.9}Mn_{0.1}Fe_{1.6}O_{3.4\pm}$. Effect of power level on resonance absorption [*60PH*].
Material had the following properties: $4\pi M_s = 2500$ Gauss, $\Theta_C = 290$ °C, hysteresis loss per cycle = 980 erg cm^{-3}.

f	100 kc/s	500 kc/s	1 Mc/s	10 Mc/s
μ_i	30	30	30	42
tan δ	0.05	0.05	0.05	0.05

→

Fig. 144. Composition diagram for the system MgO-MnO-Fe_2O_3 showing optimum areas found for microwave applications (compare with Fig. 119···122) and for square-loop applications (compare with Fig. 62, 65, 74) [*54A*].

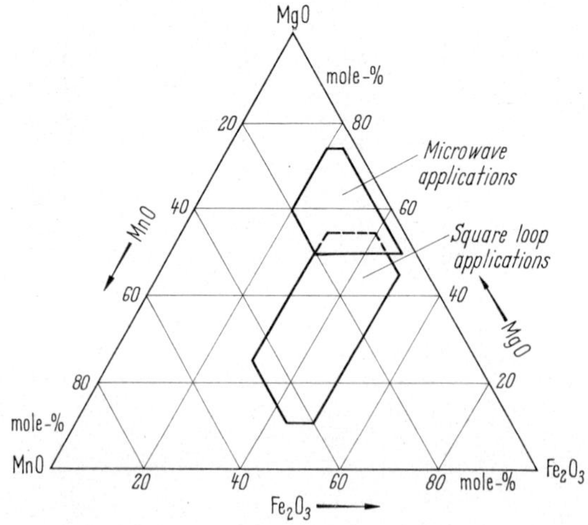

Tab. 41. Dielectric constant, ε, and dielectric loss, tan δ, of some commercial microwave MgMn ferrites [*63BC*].

Sample Nr.	Composition	$4\pi M_s$ Gauss	ε at 25 °C	$\varepsilon_{25°}/\varepsilon_T$				tan δ				
				T [°C]								
				100	150	200	250	25	100	150	200	250
6	MgMn	2300	12.3	1.009	1.018	1.016	—	0.0002	0.0001	0.0003	0.0006	—
7	MgMn	2300	12.7	1.007	1.017	1.013	1.034	0.0001	0.0001	0.0004	0.0012	0.0020
8	MgMn	2000	12.2	1.006	1.019	1.021	1.037	0.0004	0.0002	0.0007	0.0011	0.0032
9	MgMn	1900	12.5	1.007	1.019	1.022	1.032	0.0001	0.0001	0.0002	0.0002	0.0011
11	MgMn	1800	11.9	1.009	1.015	1.021	1.040	0.0004	0.0013	0.0025	0.0034	0.0096
12	MgMn	1800	11.8	1.010	1.023	1.020	1.051	0.0029	0.0076	0.0149	0.0194	0.0393
15	MgMnAl	1500	11.7	1.000	1.014	1.026	—	0.0001	0.0003	0.0007	0.0017	—
18	MgMnAl	1250	11.2	0.994	1.013	—	—	0.0001	0.0002	0.0005	—	—
19	MgMnAl	1200	11.3	1.000	1.000	1.025	—	0.0009	0.0003	0.0006	0.0011	—
20	MgMnAl	1200	10.7	0.988	1.011	—	—	0.0008	0.0016	0.0027	—	—
21	MgMnAl	1100	10.4	1.000	1.018	—	—	0.0009	0.0008	0.0013	—	—
22	MgMnAl	1030	11.2	0.989	1.012	—	—	0.0001	0.0002	0.0005	—	—
23	MgMnAl	950	9.26	1.002	1.010	—	—	0.0002	0.0003	0.0005	—	—
24	MgMnAl	950	11.2	1.002	1.014	—	—	0.0002	0.0001	0.0002	—	—
25	MgMnAl	800	10.8	1.000	1.016	—	—	0.0001	0.0002	0.0004	—	—
27	MgMnAl	700	9.60	1.006	—	—	—	0.0004	0.0003	—	—	—
29	MgMnAl	600	11.2	1.006	—	—	—	0.0001	0.0001	—	—	—
30	MgMnAl	500	10.8	0.997	—	—	—	0.0003	0.0005	—	—	—

Ceramic samples. Sample Nr. refers to manufacturer. Names are not given. Values of $4\pi M_s$ are from the manufacturer. Composition lists principal components. Measurement made at X band (approximately 9400 Mc/s) using rod samples of 0.035 in. diameter, 1.25 in. long and in the manner described by Task Group of ASTM in "Proposed Method of Test for Complex Dielectric Constant of Non-Metallic Magnetic Materials at Microwave Frequencies". Maximum error of ε data is 3%, of tan δ data 0.0005. See the article for similar data on other ferrite materials.

Tab. 42. Properties of commercially available Mg-based microwave ferrites [*65vA*].

Supplier*)	Material code	Composition	$4\pi M_s$ Gauss	Θ_C °K	ΔH Oe	g-factor	H_c Oe	ε	tan δ	ϱ g cm^{-3}
S	83	MgMn	2220	585	500	2.07	—	—	<0.00025	4.20
T	TT1-390	MgMn	2150	595	540	2.10	2.5	13.0	<0.00025	—
R	R-151	MgMn	2100	575	425	2.0	—	12	0.0005	—
A	C8P35	MgMn	2000	555	350	2.0	—	—	—	4.3
K	MGM-11 MW	MgMn	2000	575	400	2.07	—	12	0.0008	4.3
A	C9P28	MgMn	1950	515	150	2.0	—	—	—	4.3
I	R-4	MgMn	1780	595	380	2.06	1.4	12	0.0005	4.15
I	R-1	MgMnAl	1760	565	490	2.11	2.1	12	0.0005	4.10
T	TT1-105	MgMnAl	1700	500	225	2.00	1.0	12.0	<0.00025	—
A	C7P12	MgMn	1650	575	650	2.0	—	—	—	4.3
A	C9P13	MgMnAl	1550	515	375	2.0	—	—	—	4.3
T	TT1-109	MgMnAl	1250	439	155	2.00	0.4	11.5	<0.00025	—
A	C43P11	MgMnAl	1250	475	300	2.0	—	—	—	4.2
S	65	MgMnAl	1140	425	165	2.02	—	—	—	4.10
I	R-5	MgMnAl	1110	415	220	2.04	0.8	12	0.0003	3.95
I	R-6	MgMnAl	730	375	120	2.00	0.7	12	0.0004	3.90
T	TT1-414	MgMnAl	680	375	120	2.02	1.2	11.5	<0.00025	—

Measurements made at X-band. All materials are polycrystalline.

*) Suppliers:

A Airtron Division of Litton Industries Morris Plains, N. J.
I Indiana General Cp. Keasby, N. J.
K Kearfott Division Little Falls, N. J.
R Raytheon Company Waltham, Mass.
S Sperry Microwave Eletronics Co. Clearwater, Fla.
T Trans-Tech. Inc. Gaithersburg, Md.

6.1.4.3 Other ferrite systems containing Mg

6.1.4.3.0 Introductory remarks

This section presents data for Mg ferrites in which substitutions were made for Mg^{2+} and/or Fe^{3+} but does not include materials containing Mn, these latter compositions being the subject of 6.1.4.2. The tables and figures in this section are organized solely on the basis of composition. Divalent ion substitutions are dealt with first (Zn, Cu, Ni, NiCo)[1]), then trivalent ion substitutions (Al, Cr, Al + Cr)[1]), then other substitutions, both trivalent (V^{3+}, Rh^{3+}, Ga^{3+})[1]) and of higher valency. (Ti^{4+}, Sn^{4+}, V^{4+}, Sb^{5+}, V^{5+}, Nb^{5+})[1]) and, finally, spinels in which O^{2-} is partially replaced by F^-, and an identical number of Mg^{2+} ions are replaced by Li^+ to maintain charge balance. Within each material grouping, data pertaining to basic properties (structural parameters, basic magnetic properties) are presented first whenever practicable, following the scheme outlined in 6.1.4.2. Otherwise, remarks made for 6.1.4.2 apply to 6.1.4.3 also.

6.1.4.3.1 Systems containing Zn^{2+}

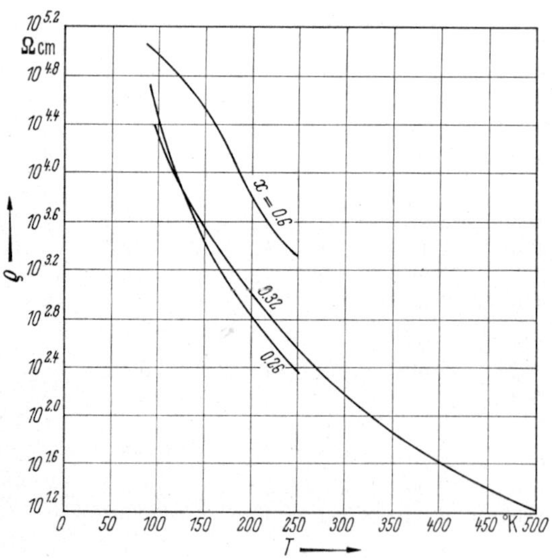

Fig. 1. $Mg_xZn_{1-x}Fe_2O_4$. Resistivity, ϱ, vs. temperature, T, for three compositions [50GB].

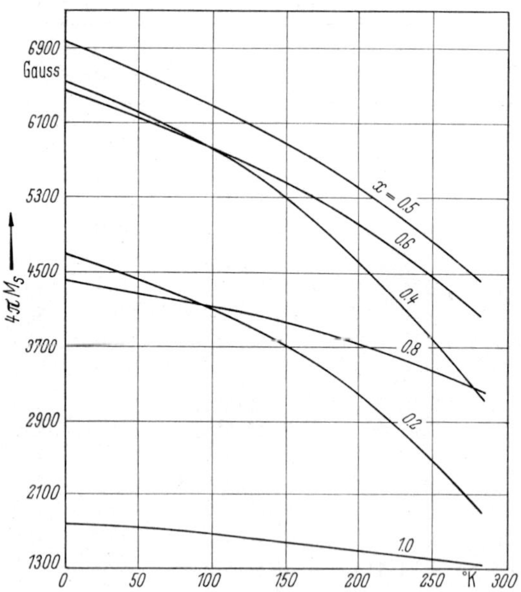

Fig. 2. $Mg_xZn_{1-x}Fe_2O_4$. Saturation magnetization, $4\pi M_s$, vs. temperature, T. Parameter x [51GS]. Data varied with heat treatment.

[1]) See also the respective sections of 6.

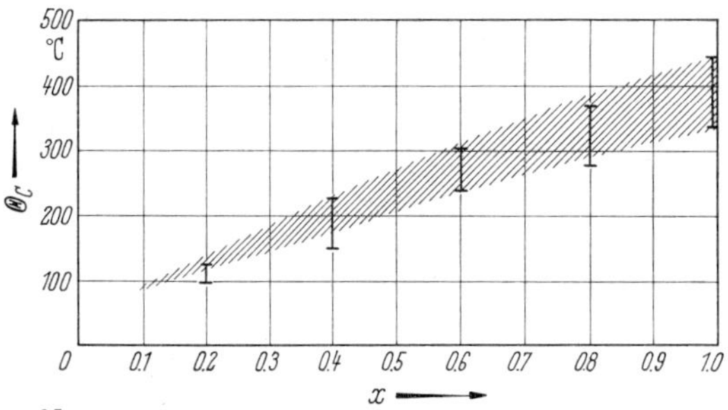

←
Fig. 3. $Mg_xZn_{1-x}Fe_2O_4$. Curie temperature, Θ_C, vs. Mg content, *x* [*51GS*].
(After [*65vA*] p. 285) The increase in bar length with increase in Mg content indicated that samples having high Mg content were sensitive to changes in firing cycle as a consequence of the partial inverse structure of Mg ferrite. The shaded area estimates the range of values to be expected for extreme variations in heat treatment.

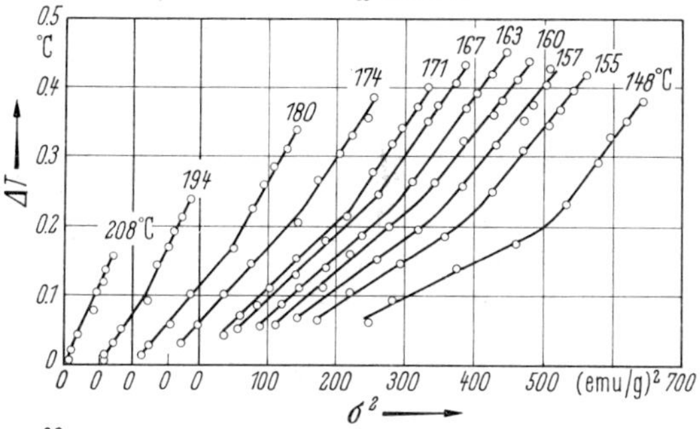

←
Fig. 4. $Mg_{0.5}Zn_{0.5}Fe_2O_4$. Magnetocaloric rise in temperature, ΔT, vs. square of the specific magnetization, σ^2 [*62BC*]. Measurements made from 148 °C to 208 °C as indicated. The existence of two linear portions indicated the existence of two "magnetic phases" with different values of Θ_C, reflecting the partial inverse character of the ionic distribution.

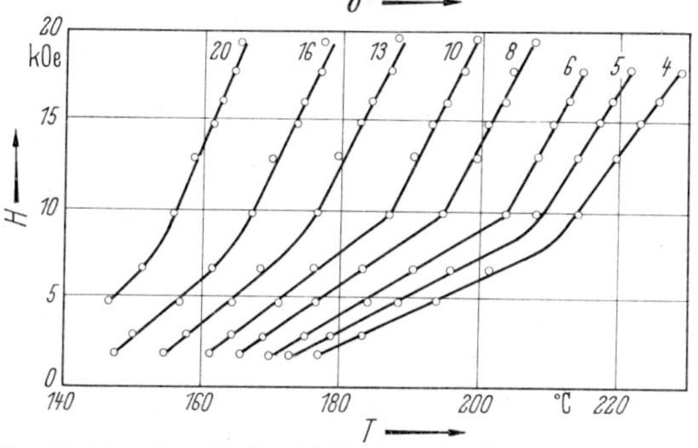

Fig. 5. $Mg_{0.5}Zn_{0.5}Fe_2O_4$. Field strength, *H*, vs. temperature, *T*, to maintain constant magnetization [*62BC*].
The number for each curve is the value of σ in emu/g. The curves are interpreted as discussed in Fig. 4.

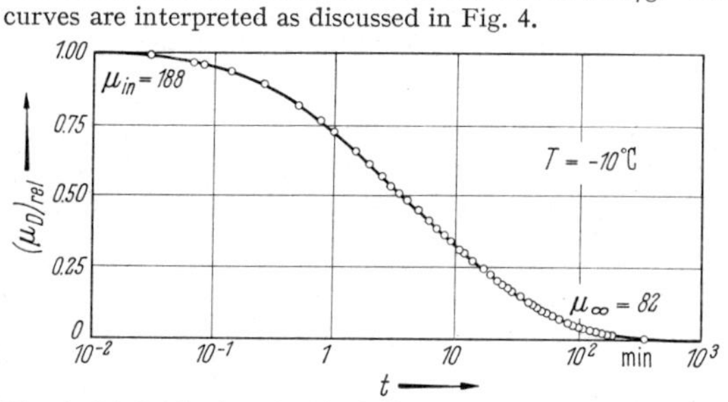

Fig. 6. MgZn ferrite. Typical disaccomodation curve [*61O*].

$$(\mu_D)_{rel} = \frac{\mu_t - \mu_\infty}{\mu_{in} - \mu_\infty}$$

μ_t = initial permeability at any time, *t*,
μ_∞ = relaxed initial permeability at $t = \infty$,
μ_{in} = instantaneous initial permeability at $t = 0$.
The material is 0.4 $[(Mg_{0.6}Zn_{0.4})O] \cdot 0.6[Fe_2O_3]$ for which μ_{in} = 188 and μ_∞ = 82 as obtained by extrapolation. The toroidal sample was first demagnetized by decreasing the amplitude of a 50 c/s field from 10 Oe to zero. Measurements were then made at 1 kc/s in fields <0.05 Oe. Above data are taken at −10 °C. All materials for Fig. 6⋯13 were prepared by standard ceramic procedures: powders calcined at 1000 °C, remilled, pressed, sintered in air for 2 h at 1400 °C, cooled to 1000 °C, then rapidly cooled to RT. The curve obeys Richter's formula [*39BD*]. Compare with Fig. 40, 46 of 6.1.4.2.

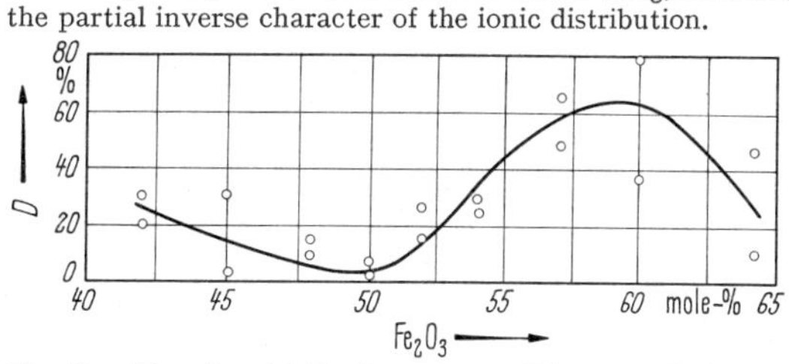

Fig. 7. $(Mg_{0.6}Zn_{0.4})O$-Fe_2O_3 system. Disaccomodation, *D*, vs. mole-% Fe_2O_3 [*61O*].
$D = (\mu_{0.1} - \mu_{10})/\mu_{0.1}$ where $\mu_{0.1}$ and μ_{10} are the initial permeabilities measured 0.1 and 10 minutes after demagnetization, respectively.
The curve for this system is unusual; for other ferrites, *D* values remained ≈0 up to 50 mole-% Fe_2O_3 and then rose steadily. Increase in *D* at low mole-% Fe_2O_3 is believed to occur from interstitial Mg^{2+}.
See also 6.1.4.1, Fig. 34 [*60P*].
Data are also presented for NiZn, MnZn and CuZn ferrites.

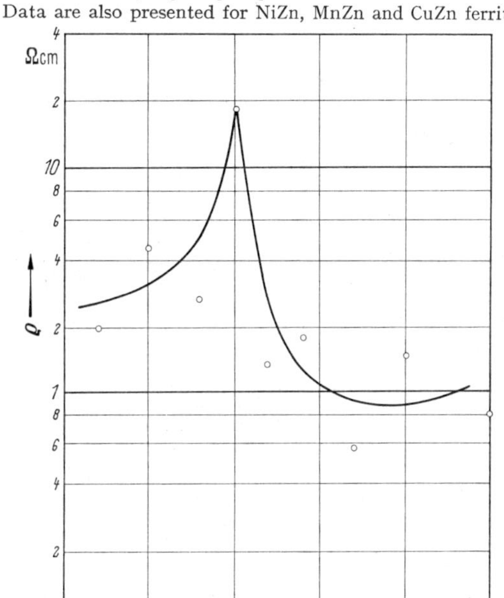

Fig. 8. $(Mg_{0.6}Zn_{0.4})O$-Fe_2O_3 system. Resistivity, ϱ, vs. mole-% Fe_2O_3 [*61O*].
Measurements at RT using 2-point-contact method with dc bridge and indium amalgam electrodes. The changes in ϱ with composition seemed to follow the changes in *D* (Fig. 7).

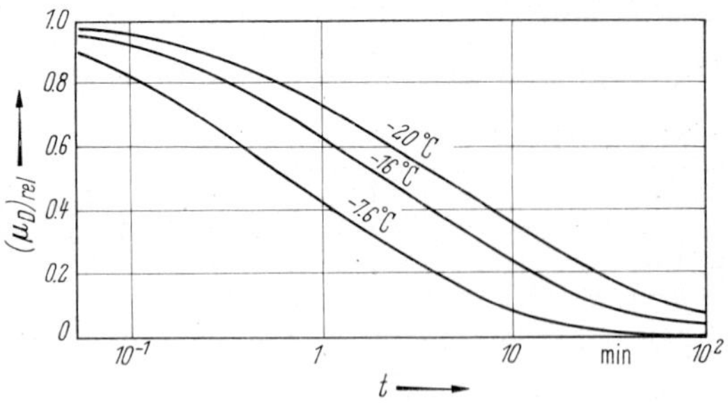

← Fig. 9. $0.58(Mg_{0.6}Zn_{0.4}O) \cdot 0.42Fe_2O_3$. Disaccomodation curves at three temperatures [610].
No disaccomodation is detectable below −80 °C for any of the MgZn compositions studied.

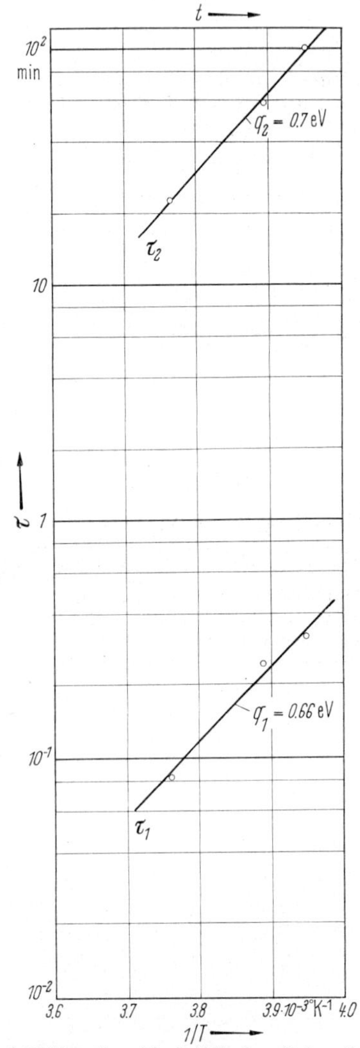

Fig. 10. $0.58(Mg_{0.6}Zn_{0.4}O) \cdot 0.42Fe_2O_3$. Relaxation time, τ, for disaccomodation vs. $1/T$ [610].
τ_1 = minimum relaxation time,
τ_2 = maximum relaxation time.
Values of activation energy, q_1, q_2, are indicated for each plot. τ_1, τ_2 were calculated from Fig. 9.

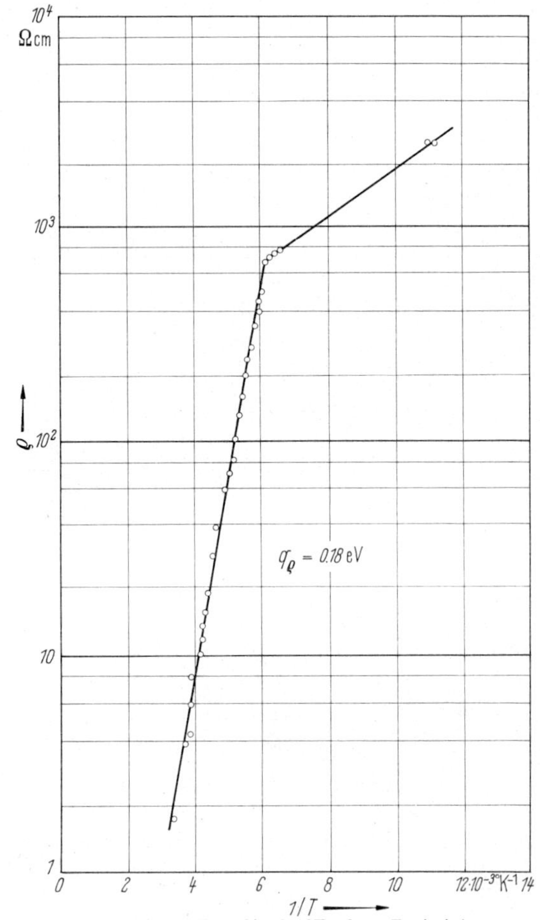

Fig. 11. $0.58(Mg_{0.6}Zn_{0.4}O) \cdot 0.42Fe_2O_3$. Resistivity, ϱ, vs. $1/T$ [610].
Activation energy for conduction in the RT range is $q_\varrho = 0.18$ eV. Comparison of q and q_ϱ values shows that different processes are involved for disaccomodation and conduction:

mole-% Fe_2O_3	42	54	60
q [eV]	0.7	0.5 ··· 0.8	0.5 ··· 0.7
q_ϱ [eV]	0.18	–	0.16

The origin of disaccomodation is concluded to be the diffusion of vacancies or interstitials.

Fig. 12. MgZn ferrite. Dependence of the disaccomodation → on field amplitude, H_{max} [610].

$$D_\infty = \frac{\mu_{in} - \mu_\infty}{\mu_\infty}.$$ See Fig. 6 for notation.

H_{max} = amplitude of the ac field. The material is that of Fig. 6. The shape of the curve is similar to the one predicted by NÉEL for magnetic aftereffect from 180° wall displacements [52N]. Wall motion rather than rotation is involved.

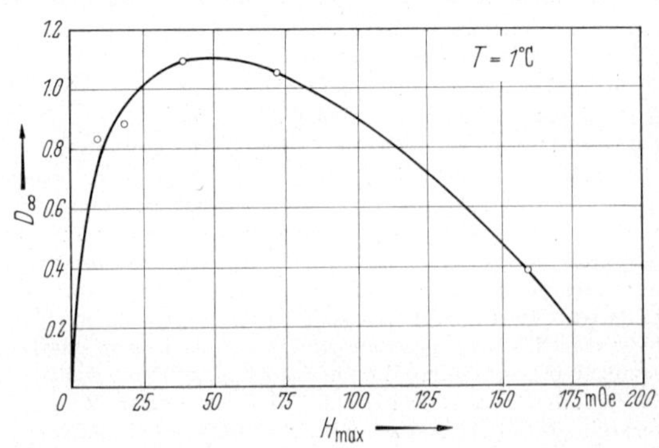

→

Fig. 13. MgZn ferrite. Induction, B, vs. field amplitude, H_{max}, as a function of time after demagnetization [61O]. Same material as in Fig. 6. Curves constructed according to [58Br]. Stabilization field, H_{st}, corresponds to the difference of field between the curves for $t = 0$ and $t = \infty$: $H_{st} \cong 35$ mOe. Assuming a process of ion-vacancy pairing and a vacancy concentration of $2.3 \cdot 10^{20}$ cm^{-3} (from chemical analysis of Fe^{2+} and Fe^{3+} content) the stabilization energy per mole, w, from pairing was calculated: $w = 1.5 \cdot 10^8$ ergmole^{-1}.

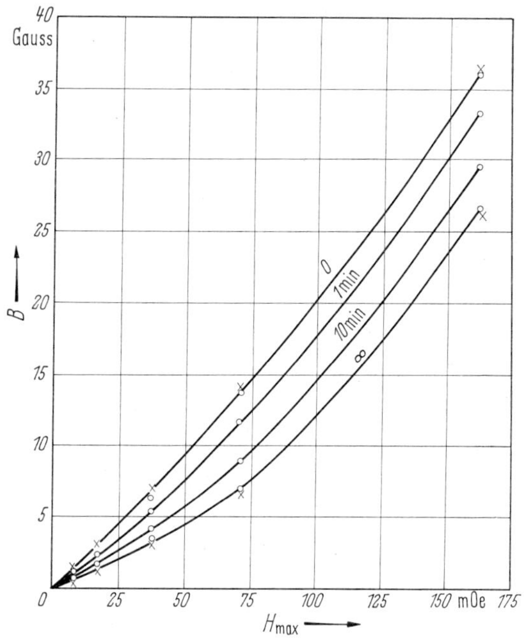

6.1.4.3.2 Systems containing Cu^{2+}

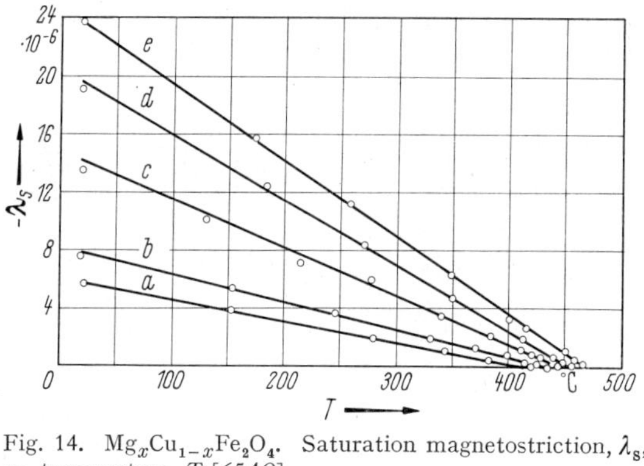

Fig. 14. $Mg_xCu_{1-x}Fe_2O_4$. Saturation magnetostriction, λ_s, vs. temperature, T [65AO].

Curve	x	M_s [Gauss]
a	1.0	110
b	0.75	120
c	0.50	132
d	0.25	141
e	0	147

Akulov's relation is obeyed: $\lambda_s = \lambda_s^0 [1 - (T/\Theta_C)]$ where λ_s^0 is the saturation magnetostriction at 0 °K. Compare with values given in Fig. 16, 33. See source for similar data for $Ni_xCu_{1-x}Fe_2O_4$.

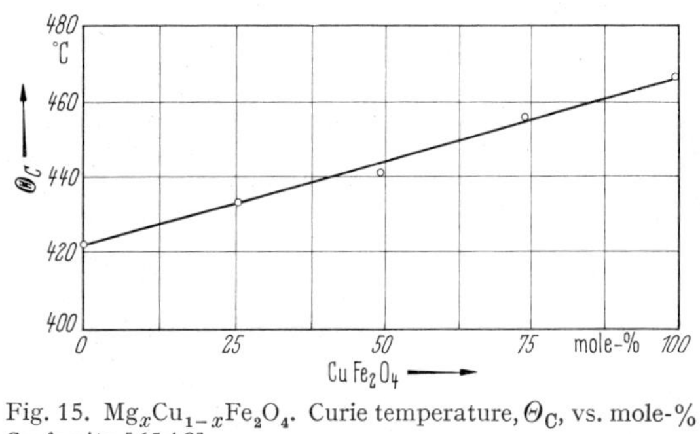

Fig. 15. $Mg_xCu_{1-x}Fe_2O_4$. Curie temperature, Θ_C, vs. mole-% Cu ferrite [65AO].

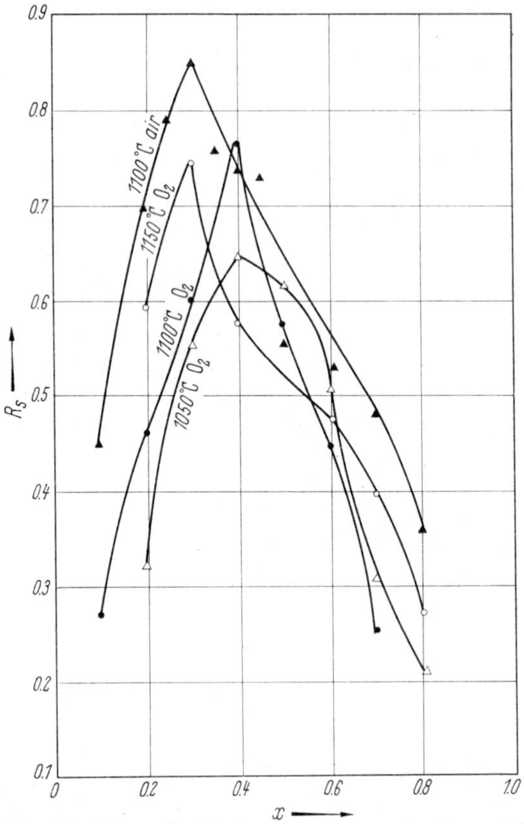

Fig. 16. $Mg_{(1-x)}Cu_xFe_2O_4$. Squareness ratio, R_s, vs. Cu content, x [59GC].

R_s: see definition in Fig. 50 in 6.1.4.2, p. 249.

Cores prepared by standard ceramic practice. X-ray diffraction showed that the composition with $x = 0$ fired to 1050 °C contained a MgO phase which disappeared at 1100 °C firing. For $x = 1$, CuO was present for all firings; for 1150 °C firing, Cu_2O was also present. All other samples were single phase spinels. Annealing at 365 °C converted the compositions from $x = 0.1 \cdots 1.0$ to a homogeneous tetragonal phase. Both quenching and low temperature annealing reduced R_s and H_c and increased B_s.

For $MgFe_2O_4$, $\lambda_s = -5 \cdot 10^{-6}$ and for $x = 0.9$, $\lambda_s = -2 \cdot 10^{-5}$ from strain gauge measurements. Both λ_s and λ_{111} (by calculation) appeared to go through a minimum for the high R_s materials ($-2 \cdot 10^{-6}$ and $-3 \cdot 10^{-6}$, respectively). For these materials, $S_W \approx 3$ Oe · µsec. See Fig. 17···20 for low-temperature behavior.

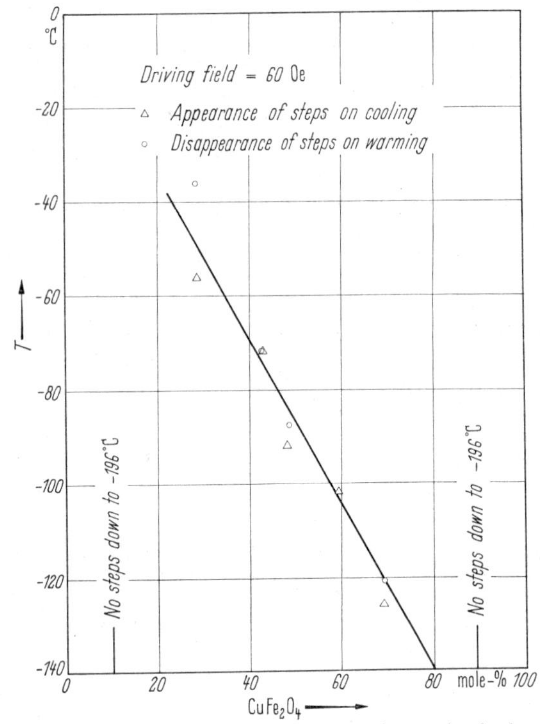

Fig. 17. $Mg_{1-x}Cu_xFe_2O_4$. Appearance of steps in the low temperature hysteresis loops of MgCu ferrites as a function of temperature and composition [60GC].

Measurements were made on a 60 c/s loop tracer. The 60 Oe field was sufficient for saturation. When the steps appeared, the squareness ratio, R_s, increased and approached unity for compositions in the middle range (Tab. 1). Temperature for step formation was also a function of field, steps becoming more pronounced at lower fields. See Fig. 24 for illustration of steps.

Samples were toroids prepared by calcining milled powders at 750···850 °C, remilling, pressing at 20 000 psi, and firing in air or O_2 at 1000···1200 °C. The behavior did not depend on the method of preparation and was found to occur for all ferrite systems containing Cu. Similar observations were reported for Mn ferrites and ferrites with small Co contents.

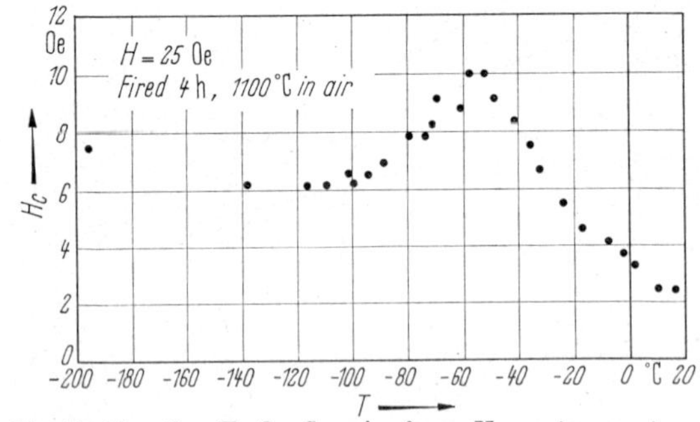

Fig. 18. $Mg_{0.6}Cu_{0.4}Fe_2O_4$. Coercive force, H_c, vs. temperature, T [60GC].

H_c has a maximum value at the temperature at which steps appeared in the hysteresis loop during cooling.

Tab. 1. $Mg_{1-x}Cu_xFe_2O_4$. $(R_s)_{max}$ vs. composition, x, at −195 °C [60GC].

x	$(R_s)_{max}$
0	No loop at 80 Oe
0.1	0.82
0.2	0.89
0.3	0.99
0.4	0.97
0.5	1.0
0.6	1.0
0.7	1.0
0.9	0.70
1.0	No loop at 80 Oe

$(R_s)_{max}$ = maximum squareness ratio as defined for 2 : 1 coincident-current operation. For definition of $(R_s)_{max}$, see Fig. 50 in 6.1.4.2, p. 249.

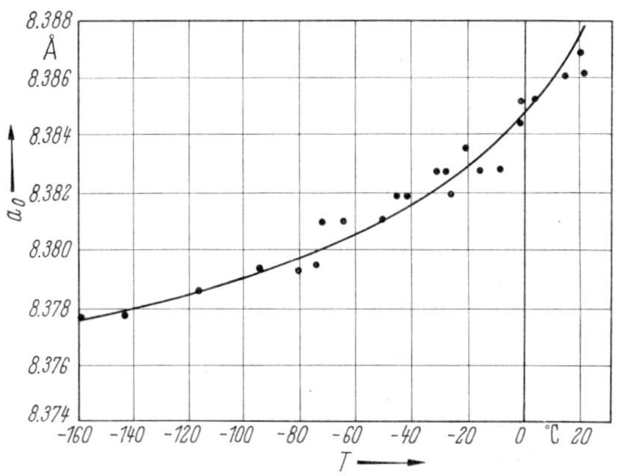

Fig. 19. $Mg_{0.6}Cu_{0.4}Fe_2O_4$. Lattice parameter, a_0, vs. temperature, T [60GC].
No phase transition from spinel was observed down to −195 °C.

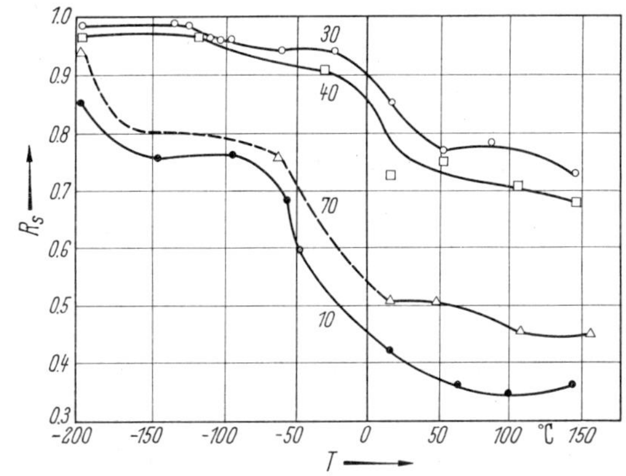

Fig. 20. MgCu ferrites. Squareness ratio, R_s, vs. temperature, T [60GC].
The Cu ferrite content in mole-% is given at the curves.

→

Fig. 21a, b. $Mg_{0.72}Cu_{0.28}Fe_2O_4$. Disaccomodation, D, and remanence ratio, B_{rem}/B_{max}, at low temperatures as a function of oxidation state [65MM], [66MM].

$$D = \frac{\mu_0 - \mu_{30}}{\mu_0}$$

μ_0 = initial permeability measured directly after demagnetization,
μ_{30} = initial permeability measured 30 minutes later.
a) D vs. T curves; b) B_{rem}/B_{max} vs. T curves.
All samples were toroids sintered at 1100 °C in atmospheres of varying O_2 content (balance: N_2):
Sample 1: 1%, 2: 2.5%, 3: 5%, 4: 10%, 5: 20%, 6: 100%.
All samples were single spinel phase with negligible Fe^{2+} but with a Cu^+ content that increased with decreasing percentage furnace oxygen.
Two relaxation processes, A and B, are indicated in (a). Low oxidation was favorable to process A (q = 0.13 eV for sample 1). High oxidation was favorable to process B (q = 0.59 eV for sample 6).
For (b) B_{rem}/B_{max} was measured at $H = 2H_c$ (H_c being the value at saturation). B_{rem}/B_{max} was first measured at 20 °K (indicated by ○ in the Fig.), then a radial dc field of 100 Oe applied for 30 minutes and B_{rem}/B_{max} remeasured (indicated by × in the Fig.). Other points (also ×) were obtained by subsequent magnetic annealing at temperatures indicated in the plot but in every case B_{rem}/B_{max} was always measured at 20 °K. The increase in B_{rem}/B_{max} by annealing shown in (b) appears related to disaccomodation (a).

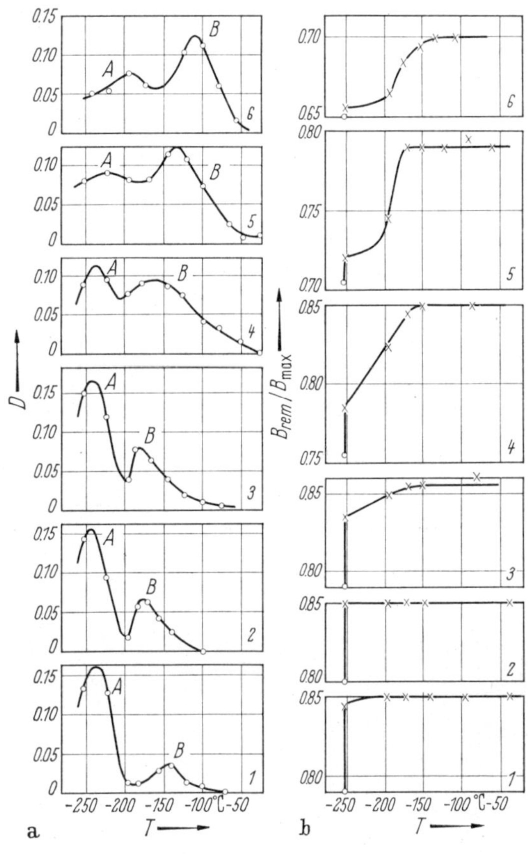

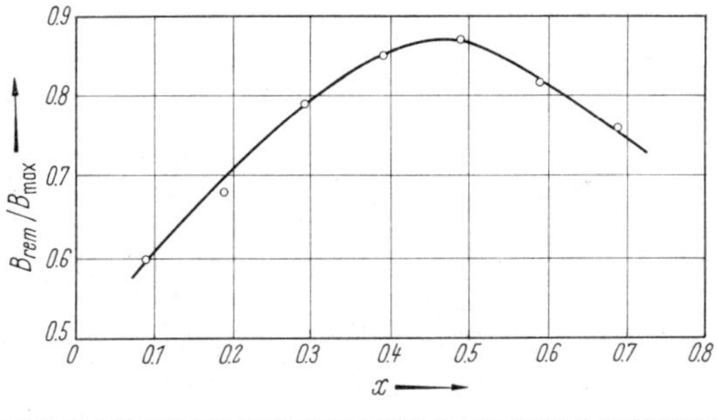

←

Fig. 22. $Mg_{1-x}Cu_xFe_2O_4$. B_{rem}/B_{max} vs. Cu content, x [65MM], [66MM].
B_{rem}/B_{max} measured at 20 °K after dc magnetic annealing at H = 100 Oe at higher temperatures. Before annealing, constricted (perminvar) loops were obtained for x = 0.49, 0.59, 0.69. Samples were sintered 4 h at 1100 °C in 20% O_2.

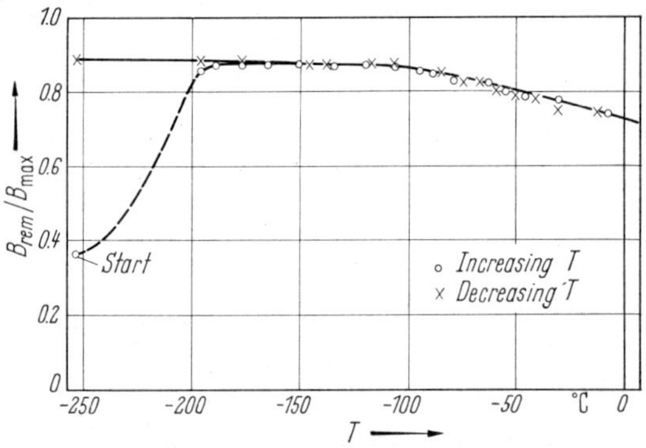

Fig. 23. $Mg_{0.51}Cu_{0.49}Fe_2O_4$. Effect of temperature cycling on B_{rem}/B_{max} *[65MM]*, *[66MM]*.
ac field H = 17.5 Oe.

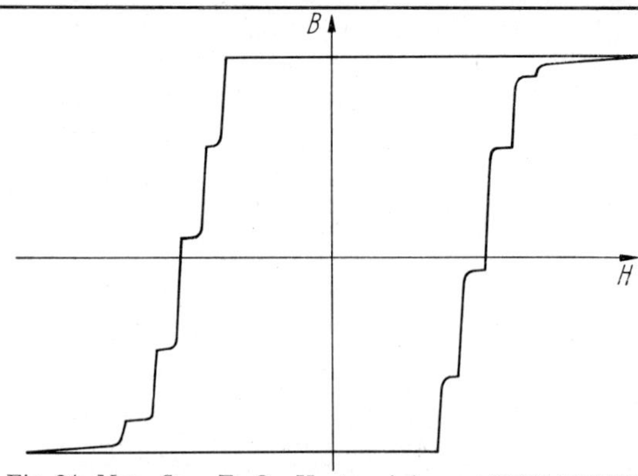

Fig. 24. $Mg_{0.51}Cu_{0.49}Fe_2O_4$. Hysteresis loop at 20 °K *[65MM]*, *[66MM]*.
ac field H = 17.5 Oe, f = 0.005 c/s.
The phenomenon of steps is attributed to directional ordering via electron exchange between Cu^+ and Cu^{++}. The disaccomodation is another manifestation of this ordering. (See Fig. 17.)

6.1.4.3.3 Systems containing Ni^{2+}

Fig. 25 see next page

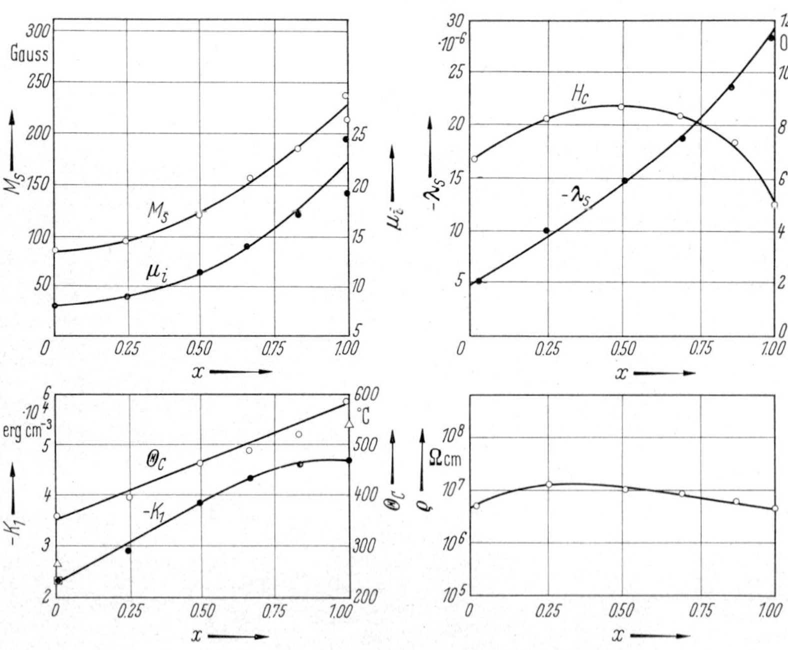

Fig. 26. $Mg_{1-x}Ni_xFe_2O_4$. Magnetic and electrical properties vs. Ni content, x *[62SK]*.
K_1 = first-order magnetocrystalline anisotropy constant. M_s measured by ballistic method. K_1 determined from the law of approach to saturation (see Tab. 14 in 6.1.4.2). For K_1 curve, the △ points are from *[55BTW]* and the □ point is from *[56FR]*.
Samples were prepared by standard ceramic methods and were sintered at 1300 °C. All samples were single phase. The lattice parameter varied linearly from a_0 = 8.36 Å for $MgFe_2O_4$ to a_0 = 8.32 Å for $NiFe_2O_4$. The average diameter of grains for all samples was ≈$4 \cdot 10^{-4}$ cm. The densities were 96 ··· 98% of the X-ray values.

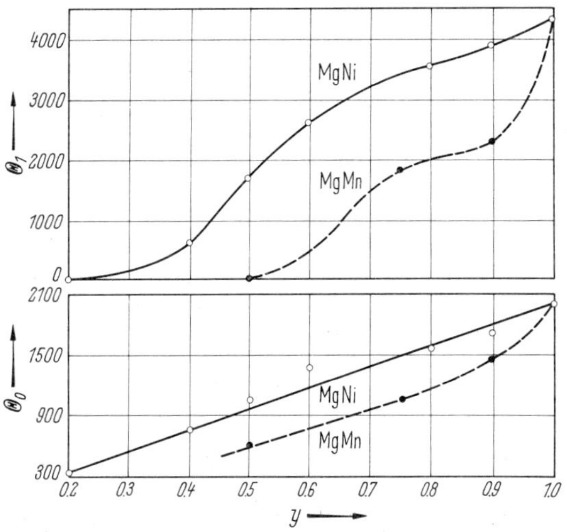

←

Fig. 25. $Mg_yNi_{1-y}Fe_2O_4$ and $Mg_yMn_{1-y}Fe_2O_4$. Variation of Θ_0 and Θ_1 as a function of composition, y [61KH].
Θ_0 and Θ_1 are parameters related to the energies necessary to transport Mg ions from octahedral (B) to tetrahedral (A) sites. [Internal energy $u_1 = u_0 + R(\Theta_0\eta - \frac{1}{2}\Theta_1\eta^2)$. η = fraction of Mg ions on A sites.]

↓ Fig. 27. $Mg_{1-x}Ni_xFe_2O_4$. Complex permeability and complex dielectric constant vs. frequency [62SK].
For μ', μ'', ε' use the vertical scale on the left. For ε'' use the vertical scale on the right.
Measurements at hf made with coaxial lines or coaxial resonators, both methods giving similar results. The two maxima on the μ' vs. f curves correspond to the domain wall resonance frequency, f_d, and the spin resonance frequency, f_s.

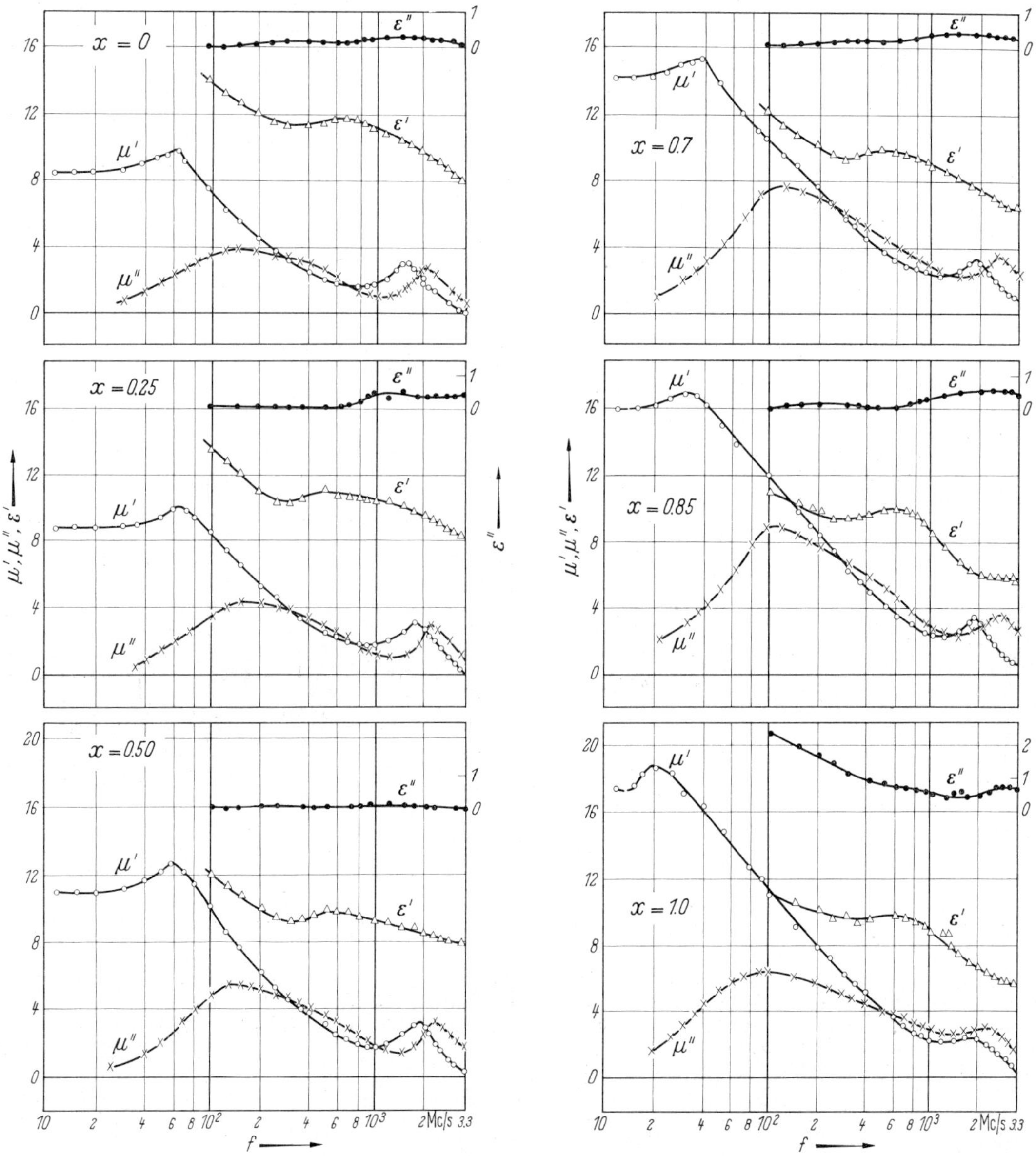

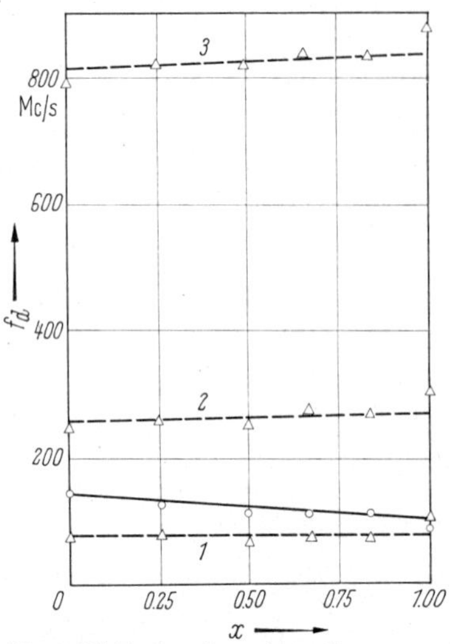

Fig. 28. $Mg_{1-x}Ni_xFe_2O_4$. Domain-wall resonant frequency, f_d, vs. Ni content, x [*62SK*].

○ experimental data from Fig. 27. (The first maximum in the μ'' curve was used.)

△ theoretical data for domain wall resonance calculated from Döring's formula:

$$f_d = \left[50\left(\frac{k}{a_0}\right)^{1/4}\frac{1}{d^{1/2}}\right]\left[\frac{M_s}{\mu_{id}^{1/2}}\left(\frac{\Theta_C}{|K_1|}\right)^{1/4}\right]\ \text{Mc/s}$$

a_0 = lattice parameter in cm; d = transverse linear dimension of a domain in cm; μ_{id} = part of μ_i associated with domain wall displacement; k = Boltzmann constant. f_d is calculated using experimental data given in Fig. 26 and taking for d values 0.1 ··· 10 times the average grain size as has been empirically verified for equiaxial grains that are $\approx 4 \cdot 10^{-4}$ cm in diameter.

Curve *1*: $d = 4 \cdot 10^{-3}$ cm; curve *2*: $d = 4 \cdot 10^{-4}$ cm; curve *3*: $d = 4 \cdot 10^{-5}$ cm. Best fit would be for a d value twice the grain size.

Agreement between the experimental and theoretical curves is taken to verify the domain wall basis for the first peak in the μ', μ'' curves of Fig. 27.

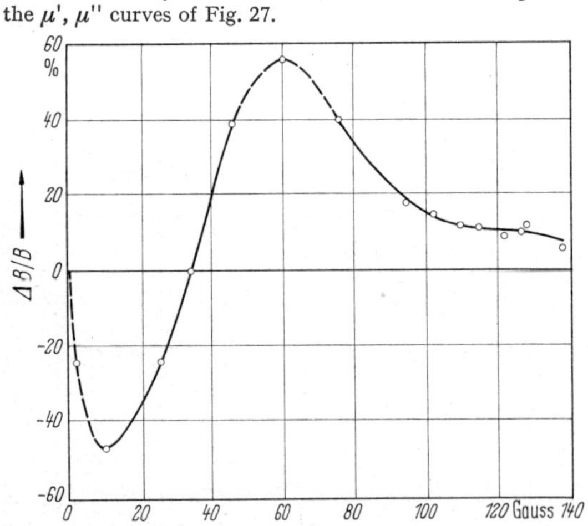

Fig. 30. $Mg_{0.5}Ni_{0.5}Fe_2O_4$. Relative induction change, $\Delta B/B$, vs. induction, B, for a constant compressive stress [*60Be*]. A longitudinal load of 3.2 kg/mm² was applied to the cross section (4 mm · 4 mm) of a ceramic bar, a field was applied and the induction change, ΔB, compared to the unloaded specimen, was noted. (Note: The text is not explicit in stating that the stress was applied prior to biasing field or vice versa).

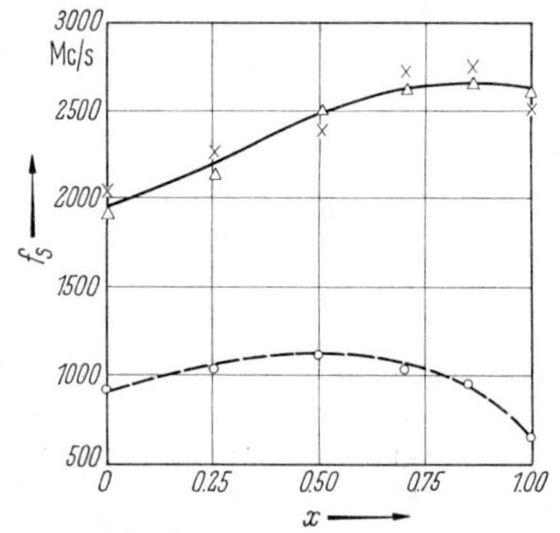

Fig. 29. $Mg_{1-x}Ni_xFe_2O_4$. Spin resonance frequency, f_s, vs. Ni content, x [*62SK*].

× = experimental data from Fig. 27. (The second maximum in the μ'' curve was used.)

○ theoretical points using the relation: $f_s = \gamma/2\pi\,[\frac{3}{4}(|K_1|/M_s)]$. For K_1, M_s, the experimental data of Fig. 26 were used.

△ theoretical points using the same relation as above but adding an additional induced field, of a magnitude up to $4\pi M_s$, that arises from demagnetizing effects due to domain size and shape. The agreement with experimental points is taken to confirm that the second peaks in the permeability spectra result from spin resonance but that the domain structure may raise f_s as proposed by [*54SW*].

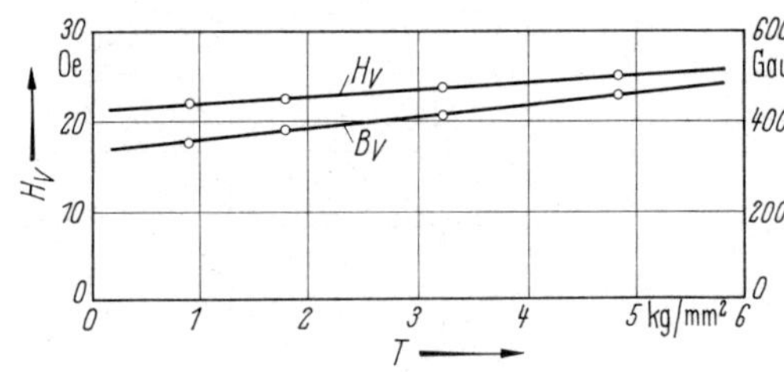

Fig. 31. $Mg_{0.9}Ni_{0.1}Fe_2O_4$. Field, H_V, and induction, B_V, required for Villari reversal vs. compressive stress, T [*60Be*]. At the Villari reversal point the magnetostriction reverses its direction.

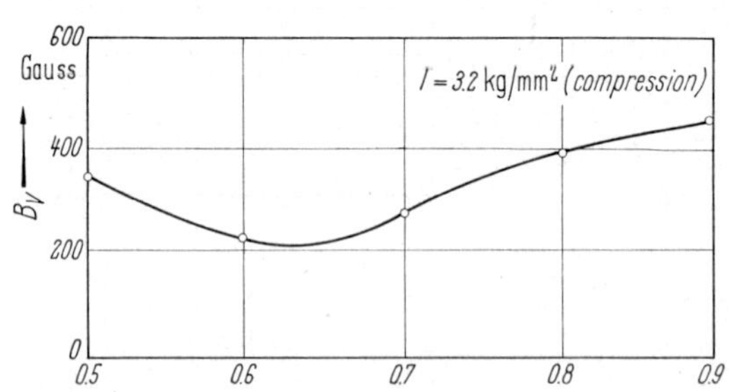

Fig. 32. $Mg_xNi_{1-x}Fe_2O_4$. B_V vs. Mg content, x, under constant stress [*60Be*].

B_V = Induction at which Villari reversal occurred.

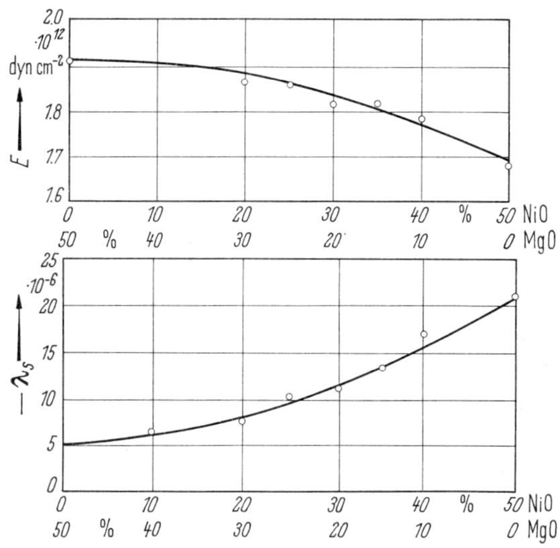

Fig. 33. MgNi ferrites. The variation of Young's modulus, E, and the magnetostriction constant, λ_s, with composition [57H].
Note: Fe_2O_3 content is constant at 50 mole%; thus 50% MgO means $MgFe_2O_4$. Samples were sintered at 1300 °C for one hour. Compare values of λ_s for $MgFe_2O_4$ with those given in Fig. 16, p. 284, and in Fig. 33.

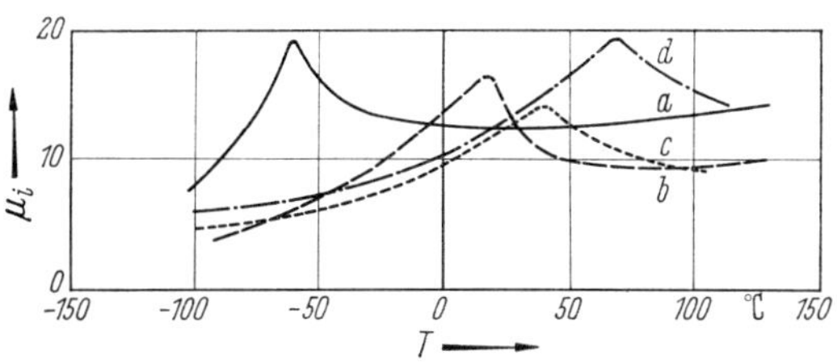

Fig. 34. $(Mg_{0.6}Ni_{0.4})_{1-x}Co_xFe_2O_4$. Initial permeability, μ_i, vs. temperature, T, for MgNiCo ferrites cooled in a magnetic field [57GE].
a) $x = 0.01$; b) $x = 0.02$; c) $x = 0.025$; d) $x = 0.03$.
K_1 is presumed to be zero at T values where μ_i has peaks. The samples were fired in O_2 at 1420 °C. When cooled in a parallel field, square hysteresis loops were obtained; without field, the loops were constricted (Tab. 2).
Samples prepared with $x = 0$ had square loops without magnetic annealing. Samples of Mg ferrite with identical Co contents had neither constricted loops without field nor square loops with field applied during cooling. All samples contained 0.8 ··· 1.1 wt-% FeO.

Tab. 2. $(Mg_{0.6}Ni_{0.4})_{1-x}Co_xFe_2O_4$. Some properties of magnetically annealed MgNiCo ferrites [57GE]

Co content, x	Loop shape before magnetic annealing	After magnetic annealing: R_s	μ_i at peak	T at peak °C ($K_1 \approx 0$)
0.010	$R_s = 0.3$	0.85	19	−60
0.015	Constricted	0.95		
0.020	Constricted	0.95	16	+19
0.025	Constricted	0.85	14	+40
0.030	Constricted	0.70	19	+60
0.035	Constricted	0.60		
0.040	Only slightly constricted	0.55		

See Fig. 34, 35. For definition of R_s, see Fig. 50 in 6.1.4.2, p. 249.

→

Fig. 35. $(Mg_{0.6}Ni_{0.4})_{1-x}Co_xFe_2O_4$. Initial permeability, μ_i, vs. temperature, T, for MgNiCo ferrites as a function of magnetic heat treatment [57GE].

Composition as in Fig. 34.

Curve	x	Cooling treatment	Loop shape
a	0.025	without field	constricted
b	0.025	in field	square
c	0.02	without field	constricted
d	0.02	in field	square
e	0.02	quenched from 600 °C	normal

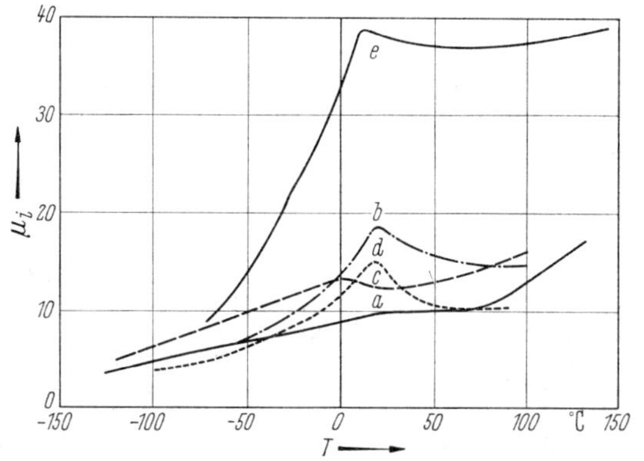

6.1.4.3.4 Systems containing Al^{3+}

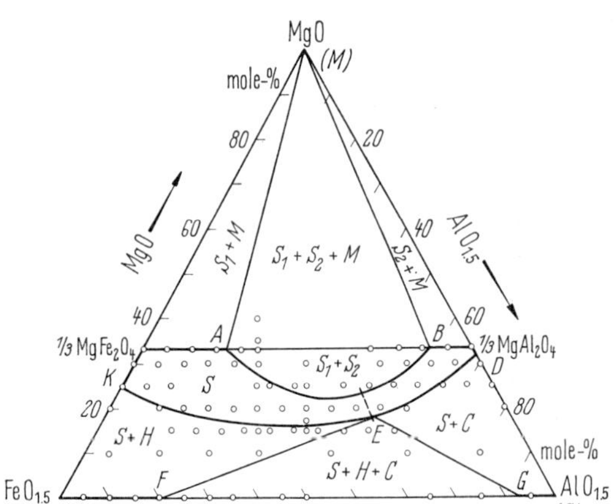

Fig. 36. Phase diagram of the system MgO-Fe_2O_3-Al_2O_3 at 1250 °C in air [*59K*].

Phases:

H = Hematite, S_1 = Spinel ($MgFe_2O_4$, solid solution),
C = Corundum, S_2 = Spinel ($MgAl_2O_4$, solid solution),
M = MgO, S = One-phase spinel band.

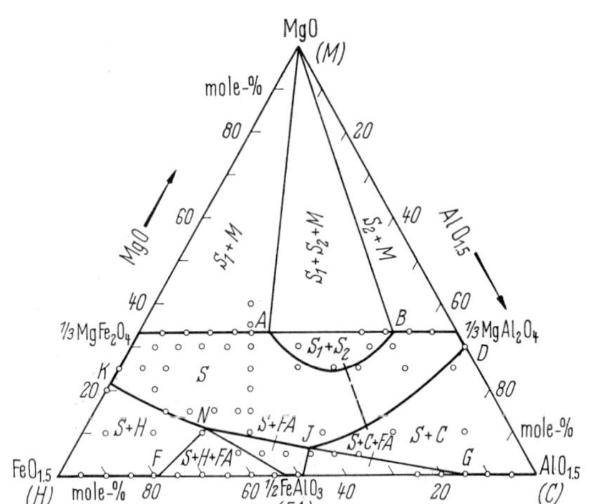

Fig. 37. Phase diagram of the system MgO-Fe_2O_3-Al_2O_3 at 1400 °C in O_2 [*59K*]. O_2 at 1 atm.
FA = $FeAlO_3$ and its solid solutions. $FeAlO_3$ was first reported by [*56MG*].
A detailed study of the system, $MgFe_2O_4$-$MgAl_2O_4$ was carried out by [*66Al*]. See Fig. 66 ··· 68.

The dotted line in the *S* band indicates compositions having $\Theta_C \cong 20$ °C.
The oxides were milled, heated 15 h at 1000 °C, remilled, pressed into discs, fired 2 ··· 18 h, quenched in sand. The phases were determined on powdered samples by X-ray diffraction. The Fe^{2+} content was less than 0.2% and $\varrho \approx 10^7\,\Omega$cm for all samples. The absence of the miscibility gap, AB, in the data of [*52JR*] is attributed to differences in firing and cooling procedures. The data shown here agreed with the results of G. H. Jonker (unpublished) and [*64B*] (Fig. 39). The position of point D is not accurate due to the sluggishness in attaining equilibrium. The data of [*53RRO*] are more accurate for this point. $a_0 = 8.07$ Å for $MgAl_2O_4$ and $a_0 = 8.36$ Å for $MgFe_2O_4$.

Tab. 3. Some physical properties of samples in the system MgO-Fe_2O_3-Al_2O_3 fired at 1250 °C in air [*59K*]

Composition in mole-% MgO	$\frac{1}{2}Fe_2O_3$	$\frac{1}{2}Al_2O_3$	σ_s Gauss cm^3g^{-1}	$\log \varrho$, ϱ in Ω cm	ϱ $g cm^{-3}$
30	70	—	24.3	7.1	4.20
30	65	5	25.4	7.3	4.11
30	60	10	22.3	7.1	4.95
25	75	—	22.1	7.0	4.18
25	70	5	28.8	6.7	4.25
25	65	10	26.5	6.9	4.02
20	70	10	31.5	7.2	4.00
20	65	15	26.7	7.6	4.09
20	60	20	23.6	8.2	3.96
20	55	25	19.3	7.1	3.86

σ_s measured at RT in a field of 6 kOe. See Fig. 36 for material preparation and phase composition.

Tab. 4. Some physical properties of samples in the system MgO-Fe_2O_3-Al_2O_3 fired at 1400 °C in O_2 [*59K*]

Composition in mole-% MgO	$\frac{1}{2}Fe_2O_3$	$\frac{1}{2}Al_2O_3$	σ_s Gauss cm^3g^{-1}	$\log \varrho$, ϱ in Ω cm	ϱ $g cm^{-3}$
$33\frac{1}{3}$	$66\frac{2}{3}$	—	24.9	6.8	4.11
$33\frac{1}{3}$	$61\frac{2}{3}$	5	22.8	7.0	3.86
$33\frac{1}{3}$	$56\frac{2}{3}$	10	19.3	7.4	3.92
25	75	—	31.1	5.6	4.15
25	70	5	27.7	6.8	3.91
25	65	10	24.4	6.7	4.17
20	70	10	31.2	7.0	4.07
20	65	15	27.0	6.6	4.07
20	60	20	23.2	5.8	3.98
15	70	15	28.2	7.1	4.25
15	65	20	27.0	7.1	3.82
15	60	25	23.4	6.7	3.94
10	65	25	22.0	7.4	3.76

O_2 pressure at 1 at.
σ_s measured at RT in a field of 6 kOe.
Compare density data with Fig. 68. For data on a_0, refractive index, n, see Fig. 66, 67, respectively. See Fig. 37 for material preparation and phase composition.

Tab. 5. $MgFeAlO_4$. Neutron diffraction intensities [*53BR*]

	Calculated intensities							Experimental intensity
	a	b	c	d_1	d_2	d'	d''	
u	0.387	0.384	0.388	0.385	0.385	0.385	0.385	
A sites	Mg	Fe	Al	$Mg_{0.5}Fe_{0.5}$	$Al_{0.5}Fe_{0.5}$	$Mg_{0.53}Fe_{0.47}$	$Al_{0.53}Fe_{0.47}$	
B sites	AlFe	MgAl	FeMg	$Mg_{0.5}AlFe_{0.5}$	$MgAl_{0.5}Fe_{0.5}$	$Mg_{0.47}AlFe_{0.53}$	$MgAl_{0.47}Fe_{0.53}$	
(111)	69	3	141	13	26	15	29	17 ± 3
(220)	47	142	24	87	68	85	65	78 ± 4
(311, 222)	232	349	196	284	270	282	266	293 ± 5
(400)	318	164	390	236	271	240	279	246 ± 5

Data are taken at 20 °C. The Curie temperature of the sample is $\Theta_C = -30$ °C. Agreement is best for structure d', although structure d_1 is also very close to experimental data. See Fig. 38···40.

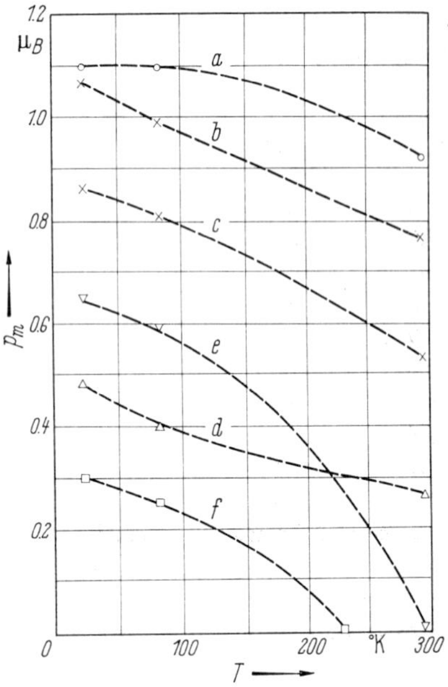

← Fig. 38. Dependence of magnetic moment, p_m, of Mg ferrite and Mg ferrite aluminates on temperature, T [*52JR*].

a $MgFe_2O_4$ *c* $MgAl_{0.4}Fe_{1.6}O_4$ *e* $MgAl_{0.8}Fe_{1.2}O_4$
b $MgAl_{0.2}Fe_{1.8}O_4$ *d* $MgAl_{0.6}Fe_{1.4}O_4$ *f* $MgAlFeO_4$

Since the magnetization decreases as Al replaces Fe, it is hypothesized that Al goes into the *B* sites. See Tab. 5, Fig. 39, 49.

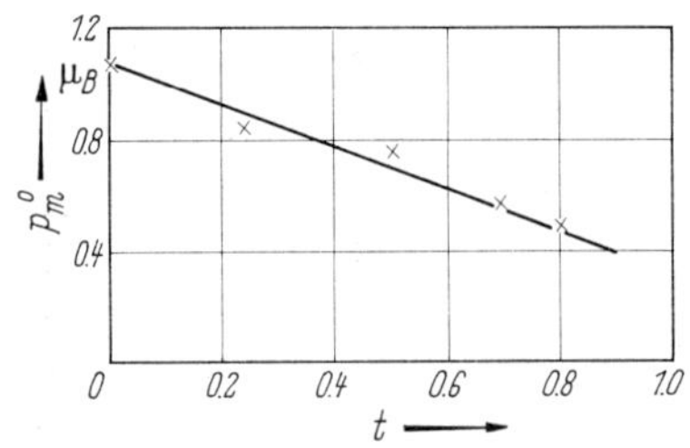

Fig. 39. $MgAl_tFe_{2-t}O_4$ system. Magnetic moment, p_m^0, vs. Al content, t [*64B*].

Materials prepared from $MgCO_3$, Fe_2O_3, and hydrated alumina, similar to procedure given in Fig. 69. Prefiring done at ≈1050 °C in O_2, firing in O_2 at 1250 °C for 4 h, then slowly cooled. Pure single-phase spinels found for $0 \leq t \leq 0.8$ and for $1.7 \leq t \leq 2.0$. Two-spinel phases found for $0.8 < t < 1.7$ as reported by [*59K*], (Fig. 36). Assuming all Al^{3+} on *B* site, the fraction of Mg^{2+} on *A* site, η, is calculated from the moment data:

t	0	0.3	0.5	0.7	0.8
η	0.11	0.24	0.33	0.41	0.45

Values of η are approximate as some Al^{3+} may be on *A* site. Compare with Tab. 5, 8, Fig. 38, 40, 49.

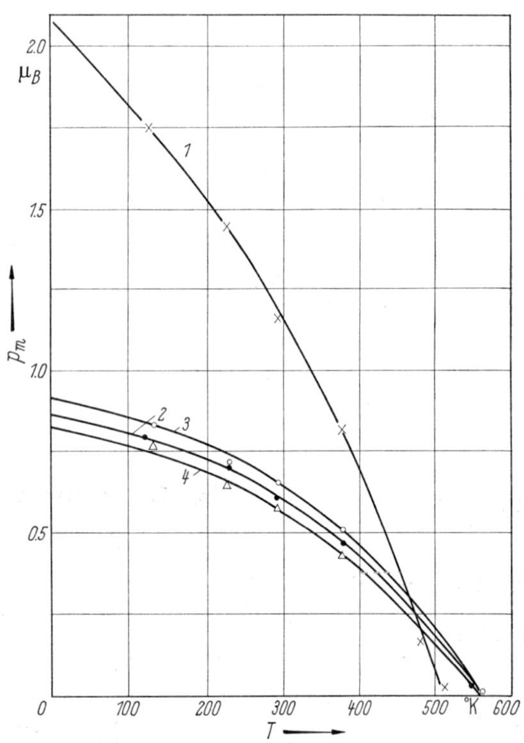

Fig. 40. $MgAl_{0.3}Fe_{1.7}O_4$. Magnetic moment, p_m, vs. temperature, T. The samples were subjected to different heat treatments [*66Kh*].

1) Quenched in water after firing 30 minutes at 1200 °C.
2) Furnace cooled from 1200 °C.
3) Sample of curve 1 annealed 5 h at 500 °C. A sample (3a) slowly cooled from 1200 °C and annealed at 500 °C gave the same curve.
4) Sample of curve 1 annealed 5 h at 800 °C. The same curve was obtained for a sample (4a) that was slowly cooled from 1200 °C ··· 800 °C and annealed.

Relative error of p_m data is 5%.

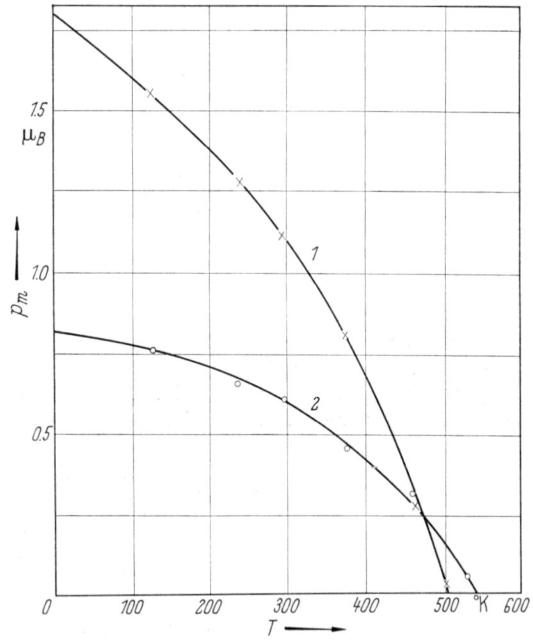

Fig. 41. $MgAl_{0.3}Fe_{1.7}O_4$. Effect of aging on the magnetic moment, p_m [*66Kh*].

Samples 1, 2 of Fig. 40 were remeasured after 4 months. Values for sample 2 had decreased about 6% compared to about 14% for sample 1.

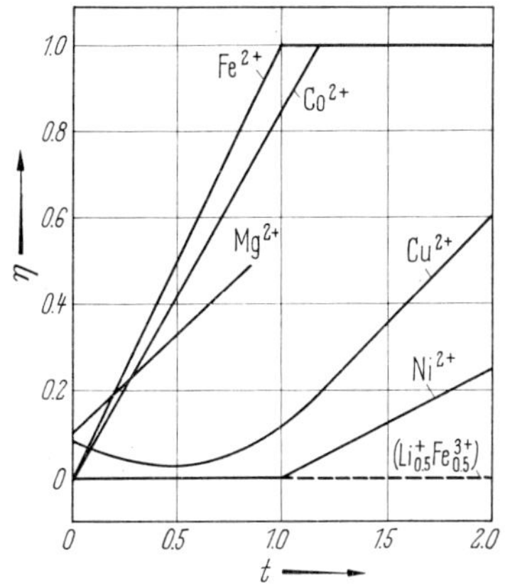

Fig. 42. $Me^{2+}Al_t^{3+}Fe_{2-t}^{3+}O_4$ systems. Approximate cation distribution, η, vs. Al content, t [64B].
η = fraction of Me^{2+} on A site.
Data sources:
($Li_{0.5}^{+}Fe_{0.5}^{3+}$): [63SB], Cu^{2+}: [56D],
Fe^{2+}: [59PT], Co^{2+}: [64B],
Ni^{2+}: [54G], [53MP], [53R], Mg^{2+}: from Fig. 39 [64B].

Tab. 6. $MgAl_{0.3}Fe_{1.7}O_4$. Effect of heat treatment and aging on resistivity, ϱ [66Kh]

Sample	ϱ (initial) $\Omega \cdot$ cm	ϱ (after 4 months) $\Omega \cdot$ cm
1	$2.2 \cdot 10^4$	$2.2 \cdot 10^4$
2	$1.2 \cdot 10^{11}$	$6.8 \cdot 10^{10}$
3	$4.0 \cdot 10^6$	$8.0 \cdot 10^6$
3a	$5.6 \cdot 10^7$	$5.7 \cdot 10^8$
4	$2.6 \cdot 10^7$	$9.7 \cdot 10^7$
4a	$1.4 \cdot 10^7$	$2.0 \cdot 10^8$

Sample numbers refer to Fig. 40. Fe^{2+} content was 0.45% for sample 1 and 0.14% for sample 2.

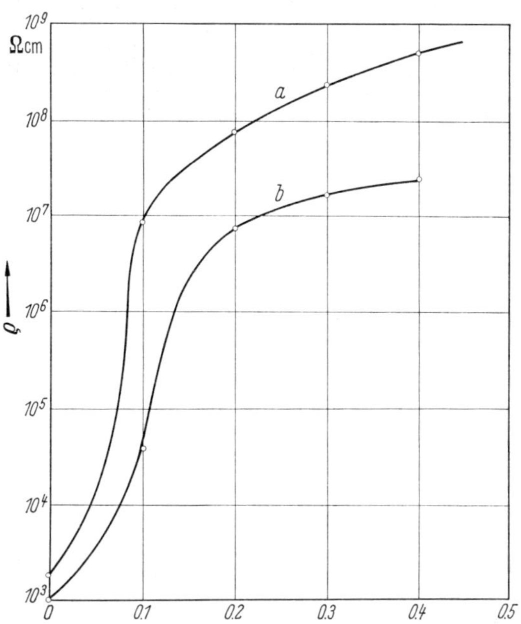

Tab. 7. Resistivity, ϱ, of some MgAl ferrites [66N]

Sample Nr.	Composition (Based on starting materials and assuming $A^{2+}B_2^{3+}O_4$)	ϱ Ωcm
1	$MgFe_2^{3+}O_4$	$7.6 \cdot 10^6$
2	$MgFe_{1.6}^{3+}Al_{0.4}O_4$	$5.2 \cdot 10^7$
3	$MgFe_{1.2}^{3+}Al_{0.8}O_4$	$6.3 \cdot 10^8$
4	$Mg_{0.8}Fe_{0.2}^{2+}Fe_{1.8}^{3+}Al_{0.2}O_4$	$3.3 \cdot 10^7$
5	$Mg_{0.6}Fe_{0.4}^{2+}Fe_{1.6}^{3+}Al_{0.4}O_4$	$3 \cdot 10^8$
6	$Mg_{0.95}Fe_{0.05}^{2+}Fe_{1.6}^{3+}Al_{0.4}O_4$	$4.3 \cdot 10^8$
7	$Mg_{0.95}Fe_{0.05}^{2+}Fe_{1.5}^{3+}Al_{0.5}O_4$	$7.8 \cdot 10^7$
8	$Mg_{0.95}Fe_{0.05}^{2+}Fe_{1.2}^{3+}Al_{0.8}O_4$	$5.7 \cdot 10^6$

All samples were fired 6 h at 1350 °C and were single-phase spinels. See Fig. 44.

Fig. 43. $MgAl_xFe_{2-x}O_4$. Dependence of the resistivity, ϱ, on Al content, x, for two different firing conditions [56vU(2)]. a) 1350 °C in O_2; b) 1375 °C in air.

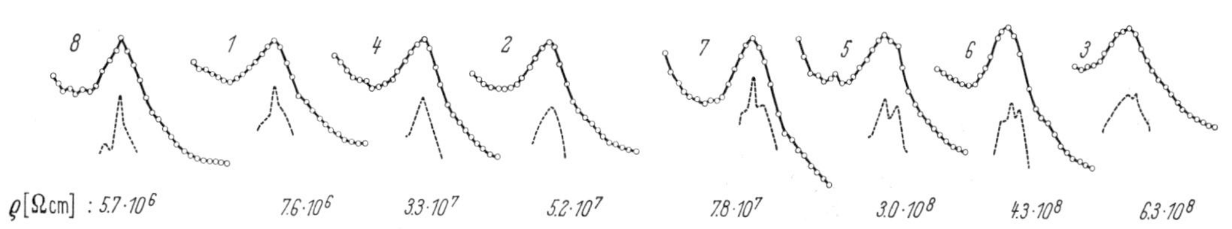

Fig. 44. MgAl ferrites. Effect of resistivity, ϱ, on the shape of the Fe$K\beta_5$ line of the X-ray spectra [66N]. The solid lines are experimental; the dotted lines are corrected curves according to [59B]. Numbers refer to compositions given in Tab. 7.

The peak broadening and the appearance of a break (splitting of the central peak in the corrected curves) with increasing ϱ were interpreted to reflect the inhibition of 3d-electron hopping between A, B-site ions as a result of strong crystal-field splitting of 3d-electron level on A, B-sites. When the splitting is slight (close approach of the threefold degenerate components t_2 and t_{2g}) the peak is sharp and the electrons can hop between A, B-sites. Activation energies range from ≈0.1 eV to ≈4 eV for sample 5. Note that ϱ of sample 8 (Fe^{2+} = 0.05) is lower than that of sample 5 (Fe^{2+} = 0.4). Similar results given for NiZn ferrites.

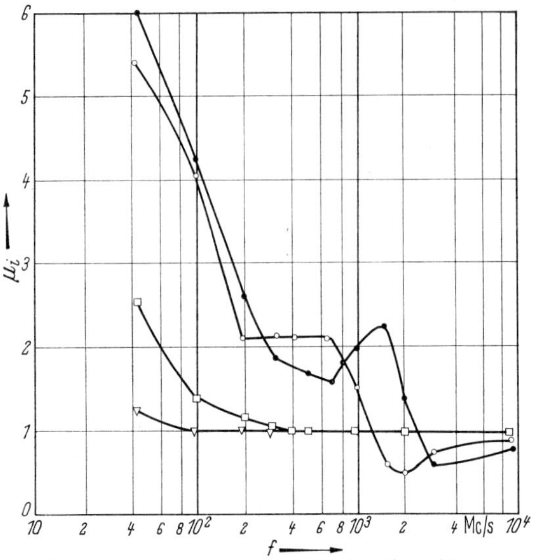

Fig. 45. Initial permeability, μ_i, as a function of frequency, f, for $MgFe_2O_4$ and various MgAl ferrites [51R].

● $MgFe_2O_4$ □ $MgAl_{0.8}Fe_{1.2}O_4$
○ $MgAl_{0.4}Fe_{1.6}O_4$ ▿ $MgAlFeO_4$

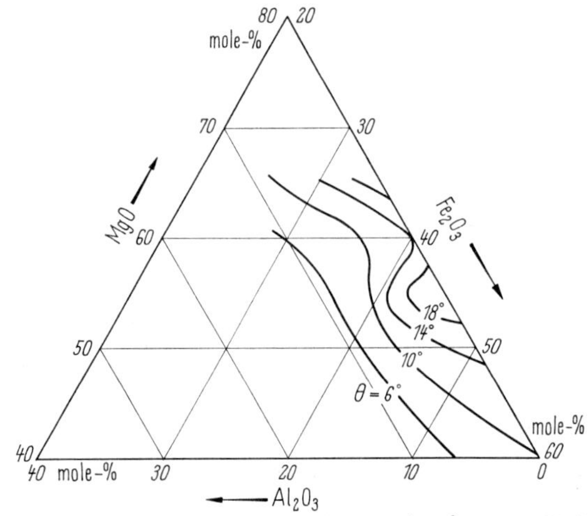

Fig. 46. MgO-Fe_2O_3-Al_2O_3. Faraday rotation, θ, at 4700 Mc/s as a function of composition [57L].

For Fig. 46⋯48 measurements were made with a field of 251 Oe applied along the axis of rod specimens having a length of 2.45 in. and a diameter of 0.33 in. The Faraday rotation data are the values measured on the samples and apparently have not been converted to the specific rotation. The rods were pressed at 25 tons in.$^{-2}$ from powder prepared by milling the oxides, prefiring at 900 °C and remilling. The pressed rods were initially fired to 1000 °C, furnace cooled, refired at 1350 °C for 4 h and cooled at 6 °C/min.

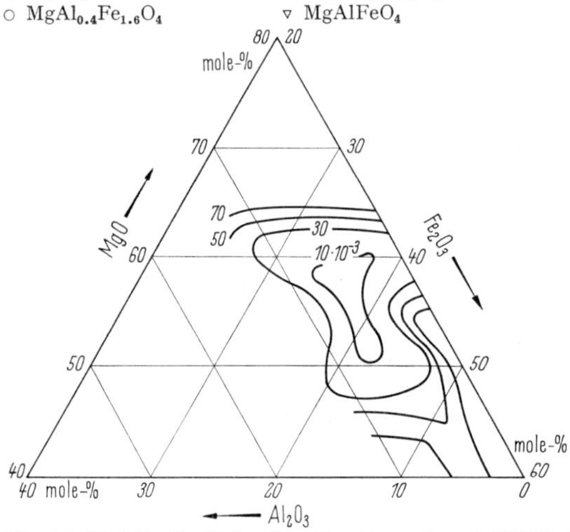

Fig. 47. MgO-Fe_2O_3-Al_2O_3. Specific attenuation at 4700 Mc/s as a function of composition [57L]. See Fig. 46.

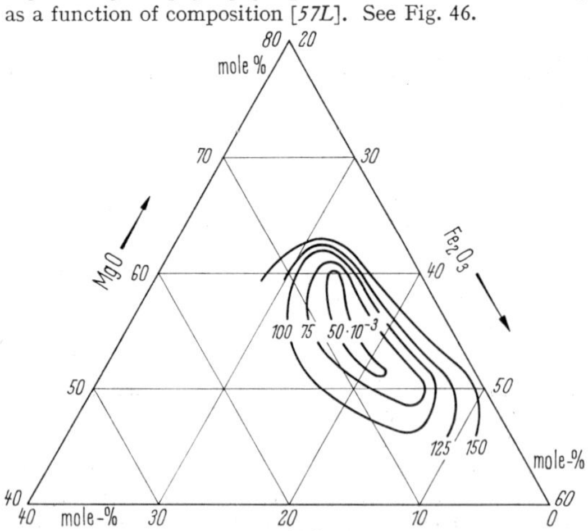

Fig. 48. MgO-Fe_2O_3-Al_2O_3. Specific attenuation at 3000 Mc/s as a function of composition [57L]. See Fig. 46.

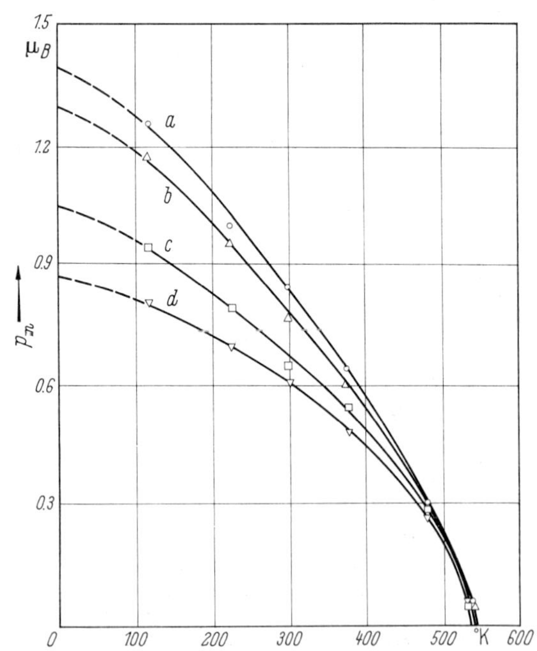

Fig. 49. $MgAl_{0.3}Fe_{1.7}O_4$. Magnetic moment, p_m, vs. temperature, T, for various cooling treatments [63Bart].

a Quenched in water, *c* Furnace cooled in flowing air,
b Quenched in air, *d* Furnace cooled at normal rate.

Tab. 8. $MgAl_{0.3}Fe_{1.7}O_4$. Effect of cooling treatment on magnetic moment, p_m, and fraction of Mg on A site, η (typical data) [*63Bart*]. $\eta = (p_m + 1.5)/10$.

Treatment	p_m at 110 °K μ_B	η
Furnace cooled	0.81	0.23
Furnace cooled in flowing air	0.94	0.24
Air quenched	1.17	0.27
Water quenched	1.27	0.28

Tab. 9. $MgAl_{0.3}Fe_{1.7}O_4$. Effect of cooling treatment on non-reciprocal phase shift, $\Delta\beta$, and merit factor [*63Bart*]

Treatment	p_m at 110 °K μ_B	η	$\Delta\beta$ at 300 Oe deg	Loss db	Merit factor $\frac{\Delta\beta}{\text{Loss}}$ deg · db^{-1}
Furnace cooled	0.91	0.24	79.5 ± 2.5	0.37	215
Furnace cooled in flowing air	1.24	0.27	91 ± 4.5	0.38	240

$\Delta\beta$ = non-reciprocal phase shift in a circular wave guide at a field of 300 Oe.
Samples were in the standard rod form. The cooling characteristics for rods and for the sample shapes used in Tab. 8 were not the same.

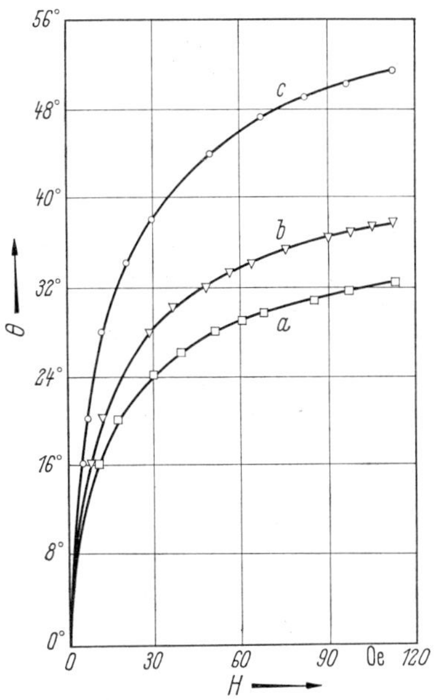

Fig. 50. $MgAl_{0.3}Fe_{1.7}O_4$. Effect of cooling treatment on Faraday rotation, θ [*63Bart*].

a Furnace cooled,
b Furnace cooled in flowing air,
c Air quenched.

Dimensions of samples were not specified.

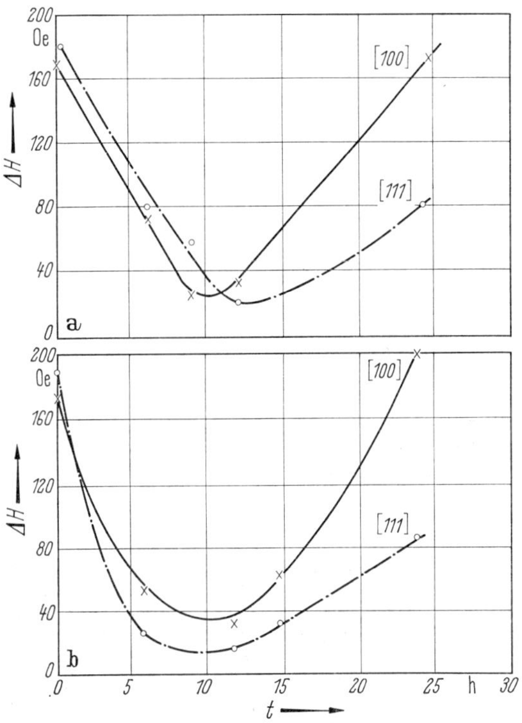

Fig. 51. MgAl ferrites. Line width, ΔH, vs. annealing time, t, of single crystals [*64YT*].

a) $Mg_{0.73}Al_{0.48}Fe^{3+}_{1.52}Fe^{2+}_{0.27}O_4$ } before annealing.
b) $Mg_{0.74}Al_{0.26}Fe^{3+}_{1.75}Fe^{2+}_{0.25}O_4$ }

Crystals, grown by Verneuil method, annealed at 1350 °C in O_2, then cooled at 100 °C/h. Annealing increased ϱ from $\approx 1\,\Omega$cm to $\approx 10^4\,\Omega$cm. Measurements made on polished spheres (using 3 μ abrasive) of 0.5 ··· 1.0 mm diameter, oriented in (110) plane, at 9.2 Gc/s. The change in the sign of the line width anisotropy was associated with the precipitation of α-Fe_2O_3 (to right of minimum). The rise in ΔH to the left of the minimum was associated with the increase in the Fe^{2+} content (see Fig. 52).

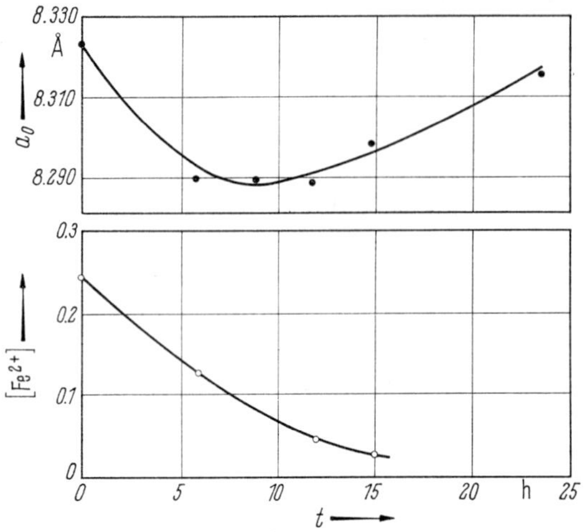

Fig. 52. MgAl ferrite. Fe^{2+} content and cell parameter, a_0, vs. annealing time, t, for a single crystal [64YT].
$[Fe^{2+}]$ = Fe^{2+} content per mole.
X-ray analysis confirmed the presence of α-Fe_2O_3 to the right of the minimum in the a_0 plot. α-Fe_2O_3 grew as single crystal Widmanstätten layers.

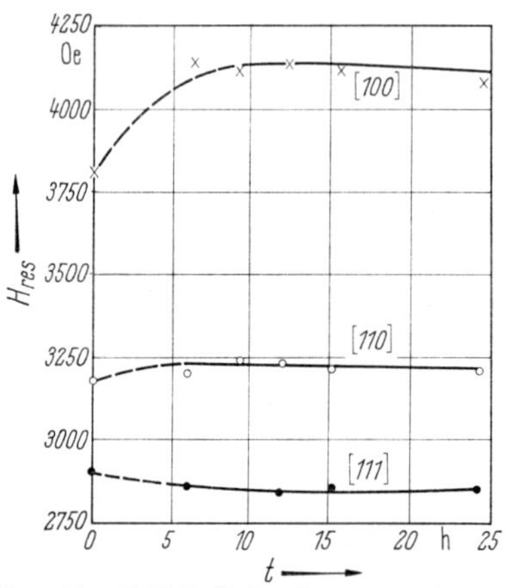

Fig. 53. $Mg_{0.74}Al_{0.26}Fe^{3+}_{1.75}Fe^{2+}_{0.25}O_4$. Resonance field, H_{res}, vs. annealing time, t, for a single crystal [64YT].
Same composition as in Fig. 51b.

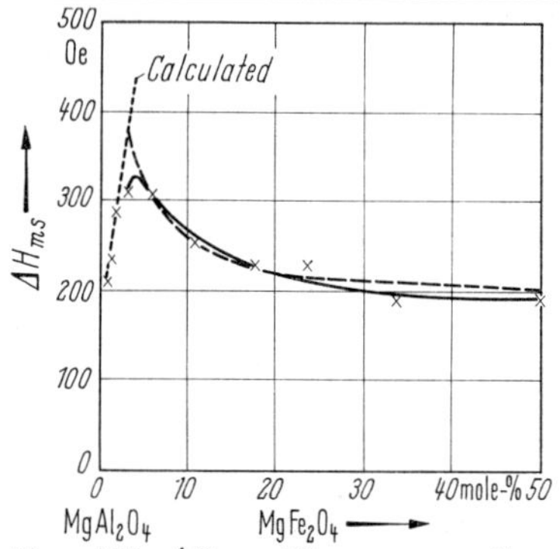

Fig. 54. Line width, ΔH_{ms}, of the paramagnetic resonance absorption of Fe^{3+} in $MgAl_2O_4$ as a function of Fe^{3+} content [60S]. ΔH_{ms} = line width between the points of maximum slope of the absorption curve.

Samples are polycrystalline, prepared from oxides and sintered into a disc in air at 1200 °C. A second unknown phase exists for compositions having concentrations <1% $MgFe_2O_4$. The measurement is made by a standard procedure using a rectangular cavity of TE_{103} mode at $f \cong 24.5$ Gc/s, modulating the field with $f \cong 1$ kc/s. The result becomes the differential of the absorption curves.

Fe^{3+} was shown to occupy A sites at low concentration. The extrapolation of ΔH_{ms} to 0% Fe^{3+}, $\Delta H_{ms} = 190$ Oe, reflects the crystal field environment in $MgAl_2O_4$ at A site. The cubic crystal field splitting parameter for Fe^{3+} in A sites, a_A, was then calculated: $|a_A| = 1.0 \cdot 10^{-2}\,cm^{-1}$.

Similar experiments for Fe^{3+} in $ZnAl_2O_4$ (in which Fe^{3+} seeks B site) gave an extrapolated $\Delta H_{ms} = 580$ Oe and $|a_B| = 3.1 \cdot 10^{-2}\,cm^{-1}$. Signs are indeterminate. The ratio $|a_A|/|a_B|$, is more reliable than separate a-values because effects of the line shape are cancelled out. In MgO, [57Lo] found an a-value of $+2.05 \cdot 10^{-2}\,cm^{-1}$ (octahedral occupancy) and in YGa garnet, $a_A = +0.616 \cdot 10^{-2}\,cm^{-1}$ and $a_B = 1.850 \cdot 10^{-2}\,cm^{-1}$ for Fe^{3+} in A and B sites, respectively. Exchange integrals were estimated from the decrease of line width (due to exchange narrowing) with increasing Fe^{3+} concentration:

$J_{AB}/g\mu_B = 1.32 \cdot 10^5$ Oe, $J_{BB}/g\mu_B = 1.90 \cdot 10^3$ Oe

where J_{AB} is for exchange interaction between Fe^{3+} in A and B sites and J_{BB} between Fe^{3+} in B sites. Θ_C, Θ_N obtained from the exchange integrals agreed with the experimental values for $MgFe_2O_4$ and $ZnFe_2O_4$, respectively. An alternate explanation of the data based on a distribution of Fe^{3+} on both sites at higher concentrations is given in [61S].

6.1.4.3.5 Systems containing Cr^{3+}

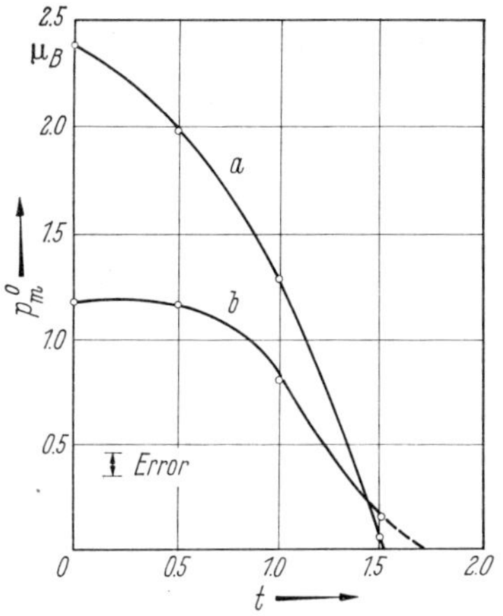

Fig. 55. $MgCr_tFe_{2-t}O_4$. Estimated magnetic moment, p_m^0, vs. Cr content, t [*60MG*].
a Quenched in water from 1350 °C,
b Cooled from 1350 °C ··· 600 °C at 100 °C/h, held at 600 °C for 72 h.
Samples were prepared by standard ceramic procedures. Peak firing temperature extended from 1400 °C up to 1700 °C for samples with high Cr content. Samples were cylinders pressed at 10 000 lbs/in².
The moments were measured by force method above 77 °K and by a ballistic method below 77 °K. See source for data on other compositions.

Tab. 10. MgO-Fe_2O_3-Cr_2O_3 system. Physical and magnetic properties. [*66V*]

Nr.	Open pore porosity %	ϱ g cm^{-3}	M_s Gauss	Θ_C °C
1	3	3.94	87	240
2	8.8	3.84	68	162
3	10.4	3.60	48	100
4	10.7	3.76	29	83
5	3	4.04	146	235
6	1.4	4.17	97	187
7	1.3	3.81	47	231
8	0.04	4.00	147	310
9	12.2	4.21	24	118
10	5.3	4.05	78	172
11	6.7	4.06	56	158
12	9.5	4.08	56	177
13	10.2	4.30	75	203
14	9.9	4.12	70	231
15	0.14	4.07	145	319
16	1.03	4.01	122	229
17	3.0	4.27	89	209
18	6.0	4.22	101	221
19	7.1	4.26	114	226
20	8.1	4.02	105	288
21	0.2	4.15	155	302
22	0.6	4.12	138	307
23	3.3	4.30	108	292
24	5.6	4.29	107	235
25	5.6	4.17	115	282
26	0.04	4.40	183	348
27	2.0	4.38	125	374
28	4.7	4.18	140	338
29	1.0	4.47	182	354
30	4.7	4.32	176	538
31	4.6	4.23	132	408
32	0.2	4.36	159	374
33	1.3	4.26	170	488
34	3.4	4.25	177	444
35	3.8	4.29	206	528
36*)	13.4	4.90	—	—

Numbers refer to compositions in Fig. 57.

*) Composition 36 is stoichiometric $MgO \cdot Cr_2O_3$.

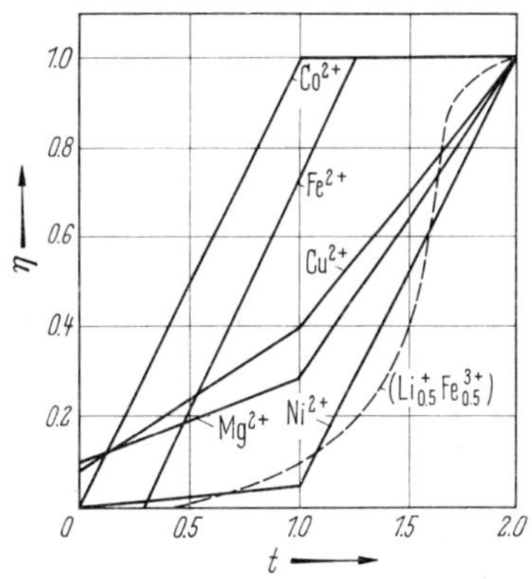

Fig. 56. $Me^{2+}Cr_t^{3+}Fe_{2-t}^{3+}O_4$. Approximate cation distribution, η vs. Cr content, t [*64B*].
η = fraction of Me^{2+} on *A* site.
Data sources:
($Li_{0.5}^{+}Fe_{0.5}^{3+}$): [*54G*],
Mg^{2+} and Ni^{2+}: [*60MG*] (See Fig. 55),
Fe^{2+}: [*54YKL*],
Cu^{2+}: [*61OT*],
Co^{2+}, Ni^{2+}: [*64B*].

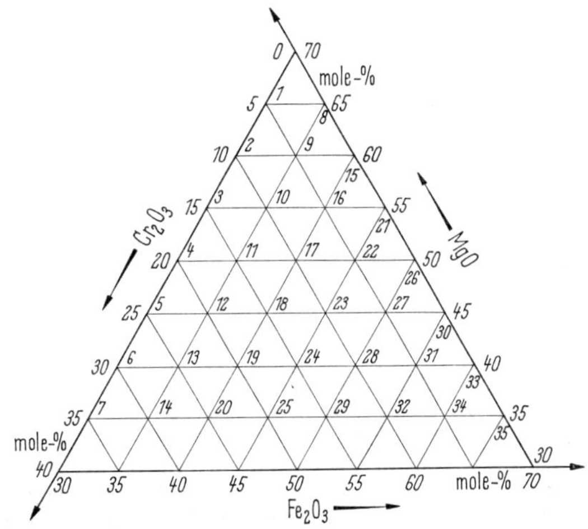

Fig. 57. MgO-Fe_2O_3-Cr_2O_3 system. Composition diagram for Tab. 10, 11 and Fig. 58 [*66V*].
Numbers inside triangle indicate compositions for Fig. 58 and Tab. 10, 11.
Cr_2O_3 was prepared from $(NH_4)_2Cr_2O_7$. The samples were fired 4 h at 1000 °C.

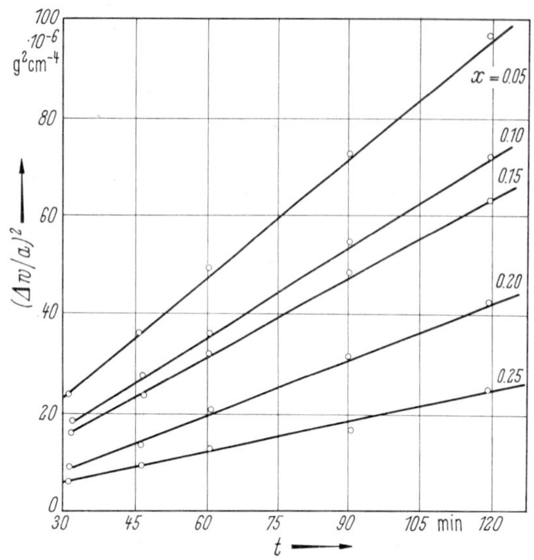

Fig. 58. $0.30Fe_2O_3 \cdot xCr_2O_3 \cdot (0.70 - x)MgO$. Rate of the diffusion contact reaction of Cr_2O_3 into MgCr ferrites at 1200 °C [66V].

Δw = increase in weight,

a = area of a ferrite disc in contact with a disc of Cr_2O_3.

t = time of contact at 1200 °C.

x = initial Cr_2O_3 content in the ferrite disc.

The diffusion follows the parabolic oxidation law, $(\Delta w/a)^2 = k''t$, where k'', the slope, is the rate constant.

The Cr_2O_3 discs were pressed at 600 kg/cm² to a density of 0.9 $g cm^{-3}$. They were fired 2 h at 1200 °C, held at ≈300 °C for 2 h. The discs were 20 mm diameter, 3···5 mm in height and weighed 7···9 g. Any unreacted Cr_2O_3 on the surface of the ferrite discs was removed chemically prior to measurement.

See Fig. 57, Tab. 10 for properties of the ferrite discs.

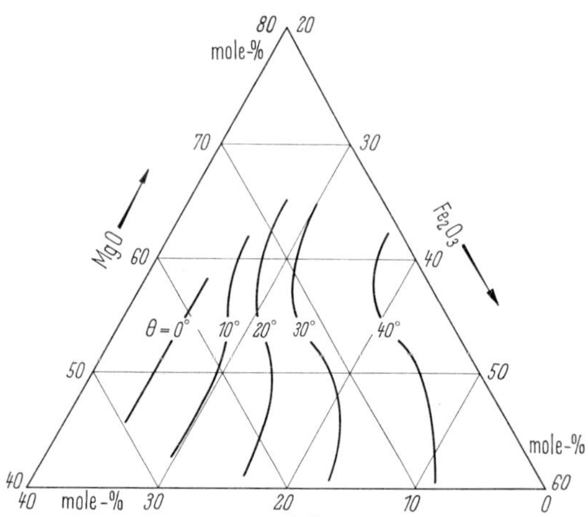

Fig. 60. MgO-Fe_2O_3-Cr_2O_3 system. Faraday rotation, θ, at 3000 Mc/s as a function of composition [57L].

For Fig. 60, 61 measurements were made with a field of 214 Oe applied along the axis of rod specimens having a length of ≈2 in. and a diameter of 0.69 in. The Faraday rotation data are the values measured on the samples and apparently have not been converted to the specific rotation. The rods were prepared by milling the oxides, pressing into blocks, prefiring at 1000 °C, crushing, remilling, pressing at 6 tons in.⁻², firing at 1000 °C, furnace cooling, refiring at 1350 °C for 4 h and cooling at 6 °C/min.

Tab. 11. MgO-Fe_2O_3-Cr_2O_3 system. Reaction rate constant, k'', for the diffusion of Cr_2O_3 into MgCr ferrites at 1300 °C [66V]

Nr.	k'' $10^{-7} g^2 min^{-1} cm^{-4}$	Nr.	k'' $10^{-7} g^2 min^{-1} cm^{-4}$
1	7.8	19	1.8
2	6.0	20	0.7
3	5.0	21	9.5
4	3.7	22	4.8
5	2.0	23	4.7
6	1.3	24	3.8
7	0.3	25	3.5
8	8.3	26	10.3
9	6.5	27	5.5
10	5.0	28	4.2
11	3.7	29	3.8
12	2.0	30	8.3
13	1.6	31	5.0
14	0.3	32	4.0
15	9.0	33	5.3
16	7.0	34	4.7
17	6.0	35	5.0
18	4.0	36	2.5

Numbers refer to compositions in Fig. 57.

k'' depended on density and porosity, as well as Cr_2O_3 content. See Tab. 10.

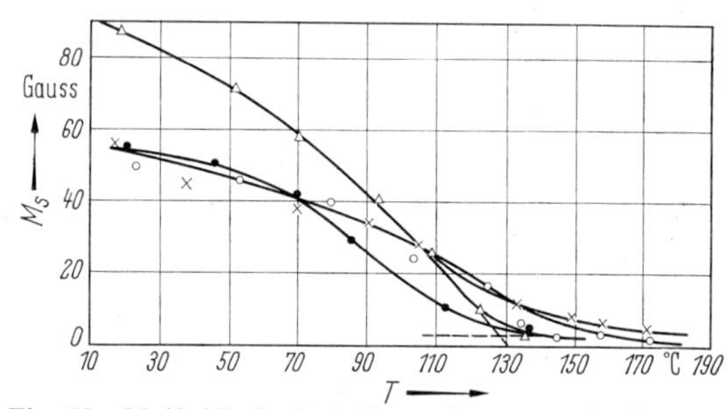

Fig. 59. Mg(Cu)Cr ferrite. Saturation magnetization, M_s, vs. temperature, T [64V].

The material is sample 3 of Tab. 12.

• fired at 1280 °C, ○ fired at 1320 °C,

× fired at 1300 °C, △ fired at 1360 °C.

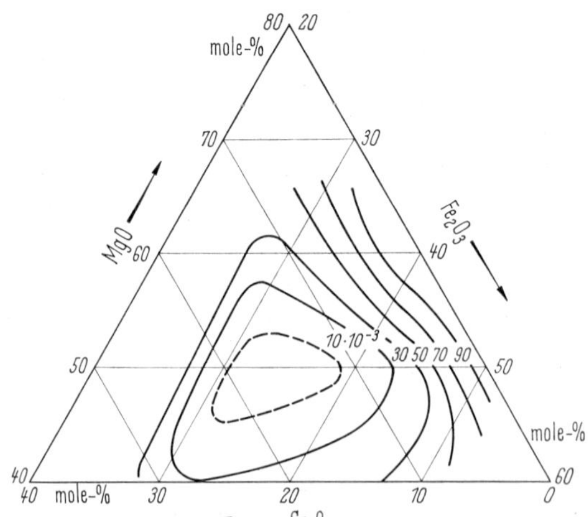

Fig. 61. MgO-Fe_2O_3-Cr_2O_3 system. Specific attenuation at 3000 Mc/s as a function of composition [57L]. See Fig. 60.

Tab. 12. Effect of starting materials on the properties of a Mg(Cu)Cr ferrite [*64V*]

Property	Firing temperature °C	Component	Sample 1	2	3	4	5
			Source of component				
		Mg	High purity MgO		High purity MgO free of S		
		Cr	A	B	A	B	A
		Cu	$CuCO_3$				Aqueous $CuSO_4$
ϱ gm cm^{-3}	1280		2.9	3	3.7	3.1	3.3
	1300		3.4	3.5	4.2	3.9	3.6
	1320		3.7	3.6	4.4	4.1	3.8
	1360		3.8	3.9	4.5	4.4	4.1
open pore porosity %	1280		13.7	13.3	7.1	9.8	11
	1300		7.5	9	2	3.1	6.8
	1320		5.8	6.2	0.13	11.4*)	5.3
	1360		3.8	3.2	0.05	0.05	2.1
M_s Gauss	1280		52	—	52	54	66
	1300		55	31.5	52.5	—	—
	1320		—	39.4	54	50	71
	1360		72.5	74.5	84	80	70
Θ_C °C	1280		170	180	145	145	185
	1300		140	160	145	160	190
	1320		145	160	130	145	170
	1360		180	150	130	135	175
μ_i (at 7.6 MHz)	1280		14	13	25	20	13
	1300		20	18	50	28	20
	1320		21	24	55	30	21
	1360		21	18	55	40	22

A: Cr_2O_3 prepared by decomposition of $(NH_4)_2Cr_2O_7$ at 300 °C. Its surface area was $1.70 \cdot 10^4\,cm^2g^{-1}$.
B: Reagent grade Cr_2O_3. Surface area was $1.03 \cdot 10^4\,cm^2g^{-1}$.
Composition of all samples was the same but not given. "Aqueous" $CuSO_4$ was introduced by wetting the mixture with a given quantity of the standard solution. Cu content (not given) is small.
"Open pore" porosity measures only those pores that are open to the surface of the specimen.
The following data are given for samples 1 and 3 only:

	ε'	ε''	μ'	μ''	
sample 1	9.4	1.82	0.86*)	0.015	Measurements at 3000 Mc/s.
sample 3	11.0	1.64	0.78*)	0.0108	

Tab. 13. $(MgO)_{0.50}(Fe_2O_3)_{0.34}(Cr_2O_3)_{0.16}$. Effect of calcining conditions of mixed sulfates of Mg, Cr, Fe on the properties of the ferrite powder and on the ferromagnetic resonance line widths, ΔH, of ceramic specimens [*65GKL*]

Calcining conditions T °C	t h	Sulfur content of powder, in terms of SO_4^{2-} [%]	ϱ_B g cm^{-3}	Powder surface m^{2}g^{-1}	Ferrite porosity %	ΔH Oe
1000	2	13.00	0.14	1.65	32.3	250
1000	6	1.74	0.27	—	25.6	250
1000	8	1.45	0.24	1.67	20.6	145
1100	2	3.04	—	—	—	210
1200	2	1.30	—	0.62	22.0	170
1300	2	0.74	0.27	—	22.0	180

ϱ_B = bulk density of the powder.
% porosity calculated from difference between X-ray and experimental density. ΔH measured on samples sintered 12 h at 1350 °C. Lattice parameter 8.38 Å. Use of sulfates insures composition uniformity throughout sample and no residue of unreacted oxides. Optimum overall properties obtained with composition (mole-%): 34.5 Fe_2O_3, 15.5 Cr_2O_3, 50 MgO prepared by optimum procedure indicated in this table (8 h at 1000 °C): $\Theta_C = 90$ °C, $M_s = 620$ Gauss, resistivity, $\varrho = 4 \cdot 10^8\,\Omega$cm, density, $\varrho = 3.6$ gcm^{-3}, $\Delta H = 145$ Oe.
For samples of this composition prepared with density $\varrho = 3.75$ gcm^{-3} (porosity = 16%), ΔH was reduced to 95 Oe.

*) Compiler suspects that these values are printing errors.

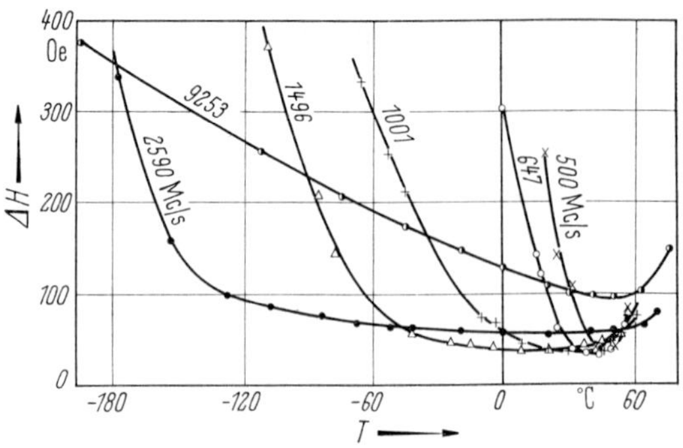

Fig. 62. 0.45MgO · 0.12CuO · $0.38Cr_2O_3$ · $0.76Fe_2O_3$. Resonance line width, ΔH, of a polycrystalline MgCr ferrite as a function of temperature, T, and frequency, f [62LY].
Sample was annealed 12 h in air at 1350 °C. Measurements were made on a single spherical sample starting at 500 Mc/s with a sample diameter of ≈4 mm. As f was increased, the sample was polished down to a smaller diameter. As ΔH increased at low temperatures, the resonance peak became increasingly distorted and the size of the resonance field decreased. At increasing temperatures, the g value decreased monotonically approaching a value of 2.
Note that the mole fractions for the composition as given do not total 1.0.

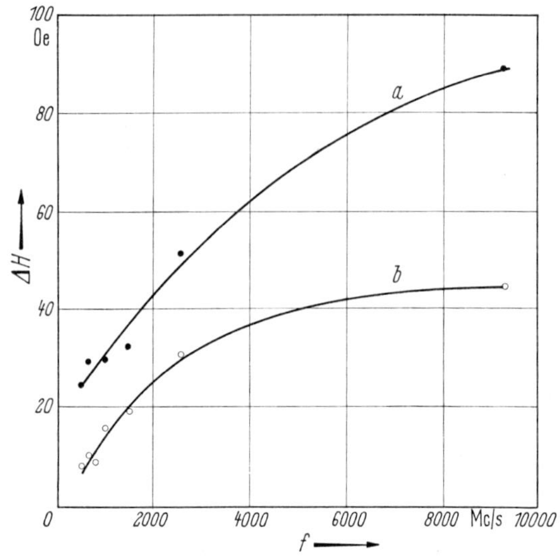

Fig. 63. 0.45MgO · 0.12CuO · $0.38Cr_2O_3$ · $0.76Fe_2O_3$. Line width, ΔH, vs. frequency, f, for a polycrystalline MgCr ferrite (curve a) near its Curie temperature, Θ_C [62LY].
$g \approx 2$ at temperature of measurement.
Curve b: An Al substituted YIG for comparison.
At $f = 500 \cdots 647$ Mc/s ΔH approaches values typical of poorly polished single crystal samples.
See Fig. 62.

Tab. 14. MgO-Cr_2O_3-Fe_2O_3 system. Microwave properties [66PS]

Composition mole-%					M_s Gauss	$K_1 \cdot 10^{-3}$ erg cm^{-3}	Θ_C °C	ΔH ($\lambda = 3$ cm) Oe	ε'
MgO	Cr_2O_3	Fe_2O_3	MnO	CuO					
47.5	20.0	32.5	—	—	30.3	2.0	72	<200	8.2 ··· 10.0
45.0	20	35.0	—	—	40.5	3.8	104	<200	9.6
47.5	15	35	—	—	53.5	4.7	156	200	9.1
47.5	15	35	2	2	—	—	—	—	—
47.5	15	35	3	3	—	—	—	—	—
40	20	40	—	—	62.0	4.0	214	{200 / 300}	{9.4 / 10.3}
40	17.5	42.5	—	—	73.5	12.1	239	300	10.3

Tab. 14. (continued)

Composition mole-%					ε''	$\varrho \cdot 10^{-7}$ Ω · cm	λ_{lim} cm	β ($H = 30$ Oe) deg	Loss ($\lambda = 10$ cm) db
MgO	Cr_2O_3	Fe_2O_3	MnO	CuO					
47.5	20.0	32.5	—	—	0.02	—	>15	63	0.3
45.0	20	35.0	—	—	0.01	2.4	>15	102	0.5
47.5	15	35	—	—	0.01	1	15	155	0.5
47.5	15	35	2	2	—	—	—	240	0.7
47.5	15	35	3	3	—	—	—	380	1.3
40	20	40	—	—	0.02	3.7	>11	130	1.0
40	17.5	42.5	—	—	0.08	1	<11	250	5.0

λ_{lim} = limiting maximum wave length; μ'' remains constant at $\lambda < \lambda_{lim}$
β = phase shift in field of 30 Oe.

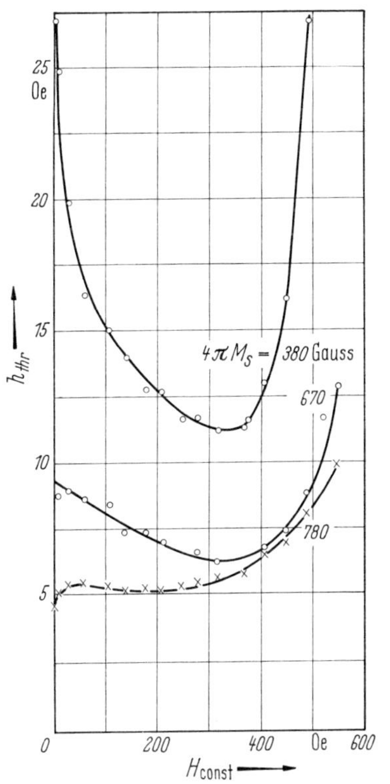

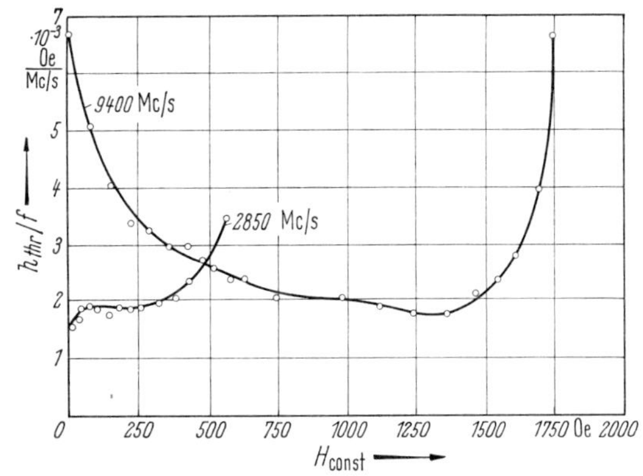

Fig. 65. Threshold field/frequency, h_{thr}/f, vs. constant field, H_{const}, at two frequencies for a polycrystalline MgCr ferrite [*66LPS*].
Material of Fig. 64 having $4\pi M_s$ = 780 Gauss.

Fig. 64. MgCr ferrites. Threshold field, h_{thr}, vs. constant field, H_{const}, for unsaturated polycrystalline MgCr ferrites of various magnetizations, $4\pi M_s$ [*66LPS*].
Measurements made on spherical samples by parallel pumping at 10 cm. See Fig. 143 in 6.1.4.2 for similar data on MgMn ferrites.

6.1.4.3.6 Systems containing Al^{3+} + Cr^{3+}

Lattice parameters in the system $MgAl_2O_4$-$MgCr_2O_4$-$MgFe_2O_4$ .... Fig. 66
Index of refraction in the system $MgAl_2O_4$-$MgCr_2O_4$-$MgFe_2O_4$ .... Fig. 67
Density in the system $MgAl_2O_4$-$MgCr_2O_4$-$MgFe_2O_4$ .... Fig. 68

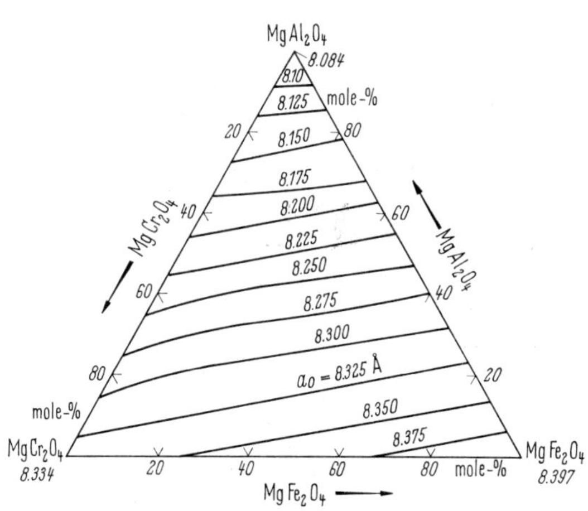

Fig. 66. $MgAl_2O_4$-$MgCr_2O_4$-$MgFe_2O_4$ system. Lattice parameters, a_0 [*66Al*].
Materials quenched from 1600 °C. The system is completely homogeneous at this temperature but below 1000 °C a miscibility gap exists in the system $MgAl_2O_4$-$MgFe_2O_4$ (compare with Fig. 36, 37).
Detailed study showed deviations from Vegard's law in two of the three binary systems but the variation is nearly planar on a gross scale for the ternary system (as shown). Detailed study of system $MgFe_2O_4$-$MgAl_2O_4$ in [*66Al*] (Ph. D. thesis).

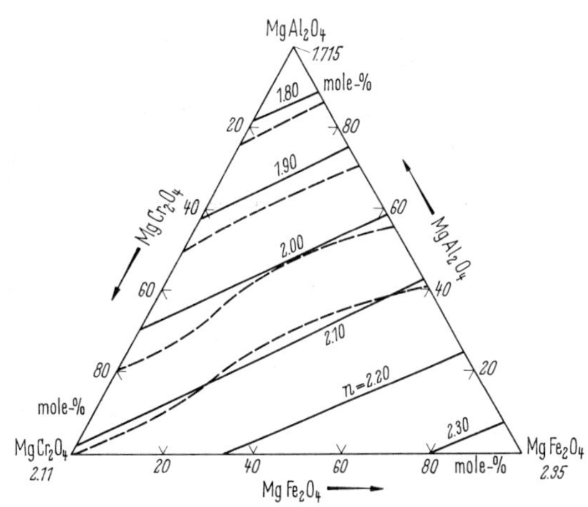

Fig. 67. $MgAl_2O_4$-$MgCr_2O_4$-$MgFe_2O_4$ system. Index of refraction, *n* [*66Al*].
See notes for Fig. 66.
Solid lines: calculated values; dashed lines: measured values.

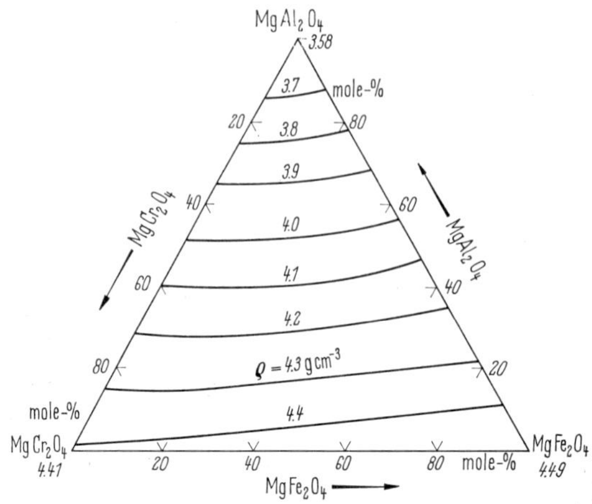

Fig. 68. $MgAl_2O_4$-$MgCr_2O_4$-$MgFe_2O_4$ system. Density, ϱ, calculated from cell size [*66Al*]. See notes for Fig. 66.

6.1.4.3.7 Other systems containing trivalent and higher valency cations

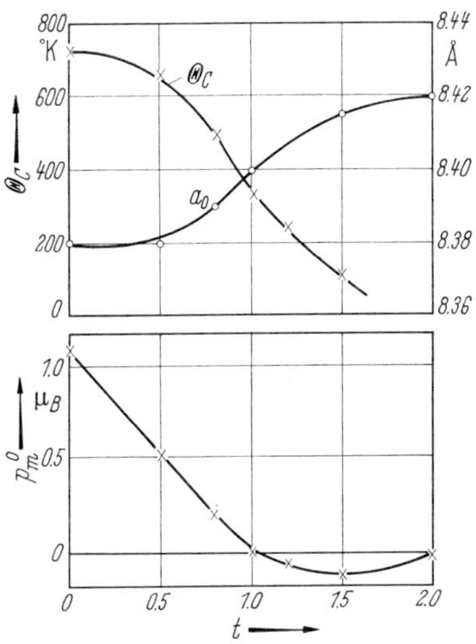

Fig. 69. $MgV_t^{3+}Fe_{2-t}^{3+}O_4$. Curie temperature, Θ_C, lattice parameter, a_0, and magnetic moment, p_m^0, vs. t [64B].

Materials were prepared by milling MgO, V_2O_3, Fe_2O_3 in ethanol, drying, prefiring in vacuum at ≈800 °C, remilling and firing 4 days in vacuum at 1000 °C, and cooling slowly. X-ray analysis showed pure spinel structure but the rather broad reflections indicated materials not completely homogeneous. Therefore, the values of lattice parameters are not very accurate. Assuming validity for the Néel model for $t = 0 \cdots 1$ and presence of V^{3+} on B sites, the structural formula becomes $Mg_\eta Fe_{1-\eta}[Mg_{1-\eta}V_tFe_{1+\eta-t}]O_4$, giving a moment $p_m^0 = 10\eta - 3t$. From measured values of p_m^0, the fraction of Mg on A sites, η, could then be calculated:

t	0	0.5	0.8	1.0
η	0.11	0.20	0.26	0.30

All σ vs. T curves are of normal type, except for $t = 1$ which is of the (Néel)L-type.
For MgV_2O_4 (normal spinel): See 6.3.1.1.

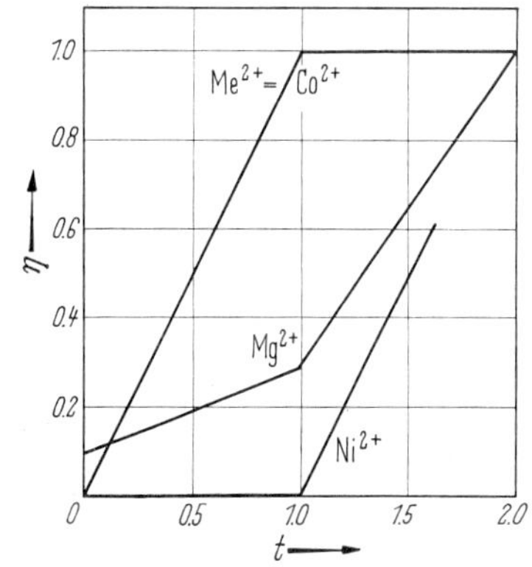

Fig. 70. $Me^{2+}V_t^{3+}Fe_{2-t}^{3+}O_4$. Approximate cation distribution, η, vs. t [64B].
η = fraction of Me^{2+} on A site.

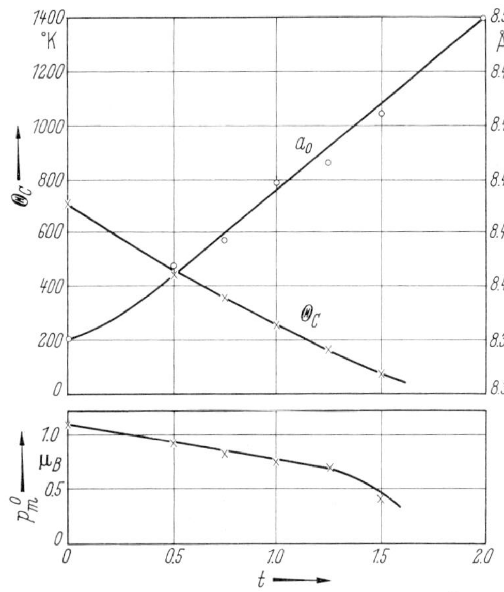

Fig. 71. $MgRh_t^{3+}Fe_{2-t}^{3+}O_4$. Curie temperature, Θ_C, lattice parameter, a_0 and magnetic moment, p_m^0, vs. t [64B].
Materials prepared from $MgCO_3$, Rh_2O_3, Fe_2O_3 similar to the procedure described for Fig. 69. Prefiring done at ≈950 °C in O_2, firing in O_2 at 1150 °C for 24 hours. Fe^{2+} was absent. All samples were spinels. Assuming the Néel model and that Rh^{3+} occupies the B site the structural formula becomes $Mg_\eta Fe_{1-\eta}[Mg_{1-\eta}Rh_tFe_{1+\eta-t}]O_4$, giving a moment $p_m^0 = 10\eta - 5t$.

Tab. 15. $MgRh_t^{3+}Fe_{2-t}^{3+}O_4$. Tabulation of data from Fig. 71 [64B]

t	p_m^0 μ_B	Θ_C °K	a_0 Å
0	1.10	710	8.380
0.5	0.93	445	8.408
0.75	0.82	360	8.417
1	0.76	260	8.439
1.25	0.70	165	8.446
1.5	0.41	75	8.464
2	—	—	8.500

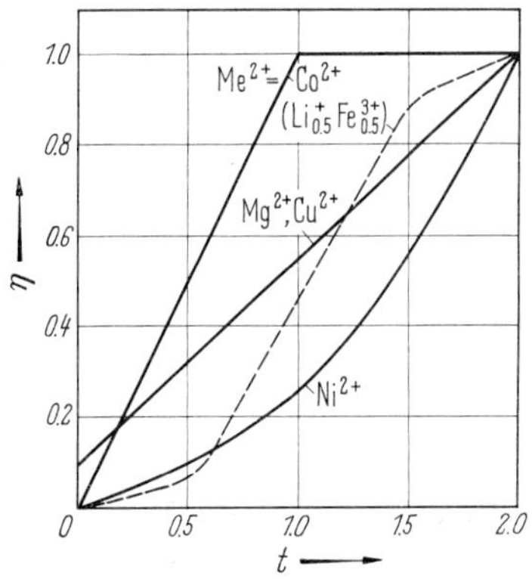

Fig. 72. $Me^{2+}Rh_t^{3+}Fe_{2-t}^{3+}O_4$. Approximate cation distribution, η, vs. t [64B].
η = fraction of Me^{2+} on A site.

Tab. 16. $MgA_t^{3+}Fe_{2-t}^{3+}O_4$ system. The prevailing magnetic interaction as concluded from the sequence of shapes of the σ vs. T curves [*64B*]

A	Sequence of the anomalous σ vs. T curves (Néel shape terminology)	Conclusion		
		interacting ions	sublattices occupied	sign of interaction
V^{3+}	N*, L, P*	V^{3+}-V^{3+}	BB	—
Rh^{3+}	N*	Fe^{3+}-Fe^{3+}	BB	—
Al^{3+}	N*	Fe^{3+}-Fe^{3+}	BB	—

N*, P* mean that N, P curves have not actually been found but the experimental curves approximate these anomalous shapes.

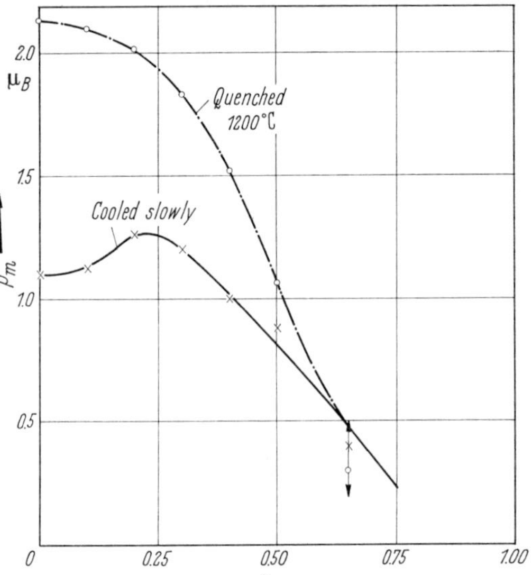

←

Fig. 73. $Mg_{1+a}Fe_{2-2a}^{3+}Ti_a^{4+}O_4$. Magnetic moment, p_m^0, vs. a [*64B*].
Material prepared from $MgCO_3$, Fe_2O_3, TiO_2 similar to the procedure given in Fig. 69. Prefiring done at ≈1000 °C in O_2, firing in O_2 at 1200 °C for 4 h. One set of samples cooled slowly at 2°/min; another set quenched in water. Fe^{2+} was absent. Single spinel phase is present in all samples. Cation distribution is given by:
$Fe_{1-t-s}Mg_tTi_s[Fe_{1+t+s-2a}Mg_{1+a-t}Ti_{a-s}]O_4$ for which the moment is given by the Néel model as $p_m^0 = 10t + 10s - 10a$. Compare Tab. 20. The shape of σ vs. T curves approach N-type Néel anomalous curve indicating the same interactions as given for Al and Rh in Tab. 16.
Compare with Fig. 81.

Tab. 17. $Mg_{1+a}Fe_{2-2a}^{3+}Ti_a^{4+}O_4$. Amount of Fe^{3+} on A and B sites [*64B*]

a	Slowly cooled samples		Quenched samples	
	Fe^{3+} tetrahedral	Fe^{3+} octahedral	Fe^{3+} tetrahedral	Fe^{3+} octahedral
0	0.8_9	1.1_1	0.7_9	1.2_1
0.1	0.7_9	1.0_1	0.6_9	1.1_1
0.2	0.6_7	0.9_3	0.6_0	1.0_0
0.3	0.5_8	0.8_2	0.5_2	0.8_8
0.4	0.5_0	0.7_0	0.4_5	0.7_5
0.5	0.4_1	0.5_9	0.3_9	0.6_1

Calculated from Fig. 73. Amount of Ti^{4+} on A site is expected to be small. Calculation of X-ray intensities assuming Ti^{4+} on B site only, gives good agreement with experimental intensities. Compare with Fig. 78, Tab. 21 ··· 23.

→

Fig. 74. $Mg_{1+a}Fe_{2-2a}^{3+}Ti_a^{4+}O_4$. Curie temperature, Θ_C, and lattice parameter, a_0, vs. a [*64B*].
a_0 data for slowly cooled samples. Compare with Tab. 20.

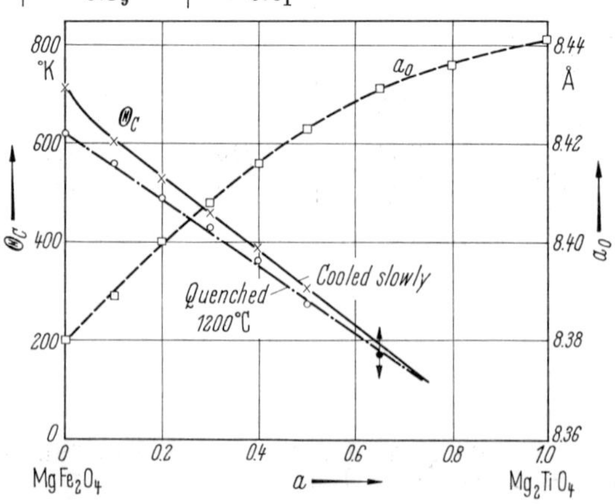

Tab. 18. $Mg_{1+a}Fe^{3+}_{2-2a}Ti^{4+}_{a}O_4$. Tabulation of data from Fig. 73, 74 [*64B*]

a	p^0_m (cooled slowly) μ_B	p^0_m (quenched 1200 °C) μ_B	Θ_C (cooled slowly) °K	Θ_C (quenched 1200 °C) °K	a_0 (cooled slowly) Å
0.00	1.10	2.14	715	620	8.380
0.10	1.12	2.10	605	560	8.389
0.20	1.26	2.02	530	490	8.400
0.30	1.20	1.83	460	430	8.408
0.40	1.00	1.52	390	365	8.416
0.50	0.88	1.06	305	275	8.423
0.65	0.4*	0.3*	170*	170*	8.431
0.80					8.436
1.00	—	—	—	—	8.441

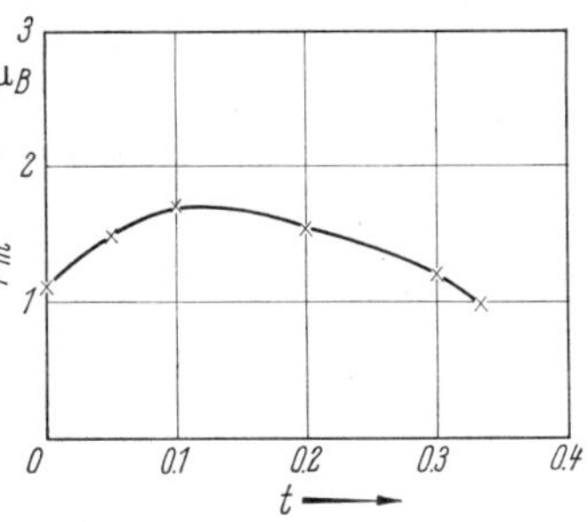

Fig. 75. $Mg_{1+2t}Fe^{3+}_{2-3t}Sb^{5+}_{t}O_4$. Magnetic moment, p^0_m, vs. t [*64B*].

Materials were prepared from $MgCO_3$, Fe_2O_3, Sb_2O_3 similar to procedure given in Fig. 69. Prefiring done at ≈1000 °C in O_2, firing in O_2 at 1200 °C for 4 h. Samples are pure spinels up to $t = 0.33$. Sb^{5+} found on B sites from intensities of X-ray reflections. Then, magnetic moment is given by: $p^0_m = 10x - 15t$ where x = number of Mg^{2+} ions on the A sites. For the distribution $Mg_xFe_{1-x}[Mg_{1+2t-x}Fe_{1-3t+x}Sb_t]O_4$ the experimental moments give the following values of x as function of t:

t	0	0.05	0.10	0.20	0.30	0.33
x	0.11	0.22	0.32	0.45	0.57	0.60

The nature of σ vs. T curves (no anomalies) indicate a weak, negative BB interaction for Fe^{3+}-Fe^{3+}. See Tab. 19a.

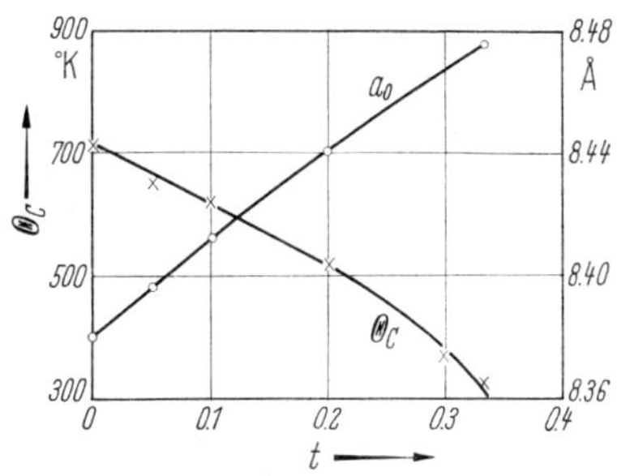

Fig. 76. $Mg_{1+2t}Fe^{3+}_{2-3t}Sb^{5+}_{t}O_4$. Curie temperature, Θ_C, and lattice parameter, a_0, vs. t [*64B*]. See Tab. 19a.

Tab. 19a. $Mg_{1+2t}Fe^{3+}_{2-3t}Sb^{5+}_{t}O_4$. Tabulation of data from Fig. 75, 76 [*64B*]

t	p^0_m μ_B	Θ_C °K	a_0 Å
0.00	1.10	715	8.380
0.05	1.47	650	8.396
0.10	1.69	620	8.412
0.20	1.53	520	8.441
0.30	1.19	370	—
0.33_3	0.97	325	8.477

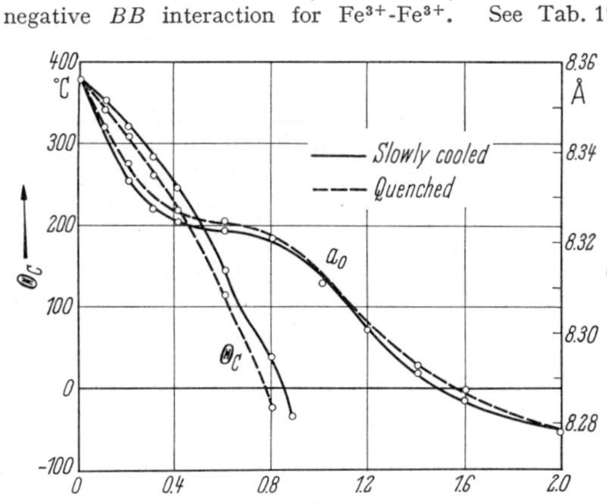

Fig. 77. $MgGa^{3+}_{a}Fe^{3+}_{2-a}O_4$. Lattice parameter, a_0, and Curie temperature, Θ_C, vs. a [*62TL(1)*].

All compositions were spinels. Samples are prepared by firing mixtures of $MgFe_2O_4$ and $MgGa_2O_4$ at 1100 °C for 3 days, regrinding and then repeating this process twice. See Tab. 19b.

Tab. 19b. $MgGa^{3+}_{a}Fe^{3+}_{2-a}O_4$. Tabulation of data from Fig. 77 [*62TL(1)*]

a	Θ_C °C		a_0 Å	
	Slowly cooled	Quenched	Slowly cooled	Quenched
0	380	380	8.356	8.356
0.1	355	345	8.346	8.346
0.2	322	310	8.334	8.338
0.3	287	264	8.328	—
0.4	246	220	8.325	8.326
0.6	148	115	8.323	8.325
0.8	40	−20	8.321	8.322
0.9	−30	—	—	—
1.0	—	—	8.311	8.312
1.2	—	—	8.302	8.302
1.4	—	—	8.291	8.293
1.6	—	—	8.286	8.288
2.0	—	—	8.279	8.279

*) possible error about 20%.

Tab. 20. $Mg_{1+a}Fe^{3+}_{2-2a}Ti^{4+}_{a}O_4$. Curie temperature, Θ_C, lattice parameter, a_0, and magnetic moment, p^0_m [*66TL(1)*]

a	Θ_C °C Slowly cooled from 1200 °C	Θ_C °C Quenched from 900 °C	a_0 Å (±0.001 Å) Slowly cooled from 1200 °C	a_0 Quenched from 1200 °C	a_0 Quenched from 1100 °C	a_0 Quenched from 900 °C	p^0_m μ_B (±0.01 μ_B) Slowly cooled from 1200 °C	p^0_m Quenched from 1200 °C	p^0_m Quenched from 1100 °C	p^0_m Quenched from 900 °C	I_{400}/I_{220} slowly cooled from 1200 °C
0	375	320	8.363	8.376	8.375	8.372	0.72	2.07	1.99	1.80	0.60
0.05	328	280	8.370	8.380	8.379	8.376	0.86	2.08	1.96	1.77	0.65
0.10	285	242	8.375	8.385	8.384	8.380	0.98	2.04	1.92	1.74	0.72
0.15	247	208	8.380	8.388	8.387	8.384	1.03	1.98	1.86	1.68	0.78
0.20	210	174	8.384	8.392	8.391	8.388	1.05	1.90	1.78	1.60	0.85
0.25	174	136	8.389	8.395	8.394	8.391	1.02	1.80	1.66	1.50	0.92
0.30	134	94	8.393	8.398	8.397	8.395	0.96	1.66	1.56	1.40	0.99
0.35	90	48	8.397	8.401	8.400	8.398	0.90	1.50	1.38	1.25	1.05
0.40	46	−5	8.401	8.404	8.403	8.401	0.83	1.31	1.23	1.10	1.13
0.45	−4	−50	8.404	8.407	8.406	8.404	0.72	1.13	1.04	0.94	1.20
0.50	−70	−90	8.407	8.409	8.409	8.407	0.60	0.91	0.86	0.74	1.28
0.55			8.410	8.412	8.412	8.410	0.48	0.70	0.64	0.54	1.36
0.60			8.413	8.415	8.414	8.413	0.35	0.45	0.40	0.36	1.45
0.65			8.416	8.417	8.417	8.416					1.56
0.70			8.418	8.419	8.419	8.418					1.68
0.80			8.425	8.424	8.424	8.424					2.05
0.90			8.433	8.431	8.431	8.431					2.57
1			8.445	8.441	8.441	8.441					3.32

I_{400}/I_{220} = X-ray diffraction intensity ratio corresponding to the (400) and (220) planes.
Cooling rate for slowly cooled samples was 10 °C/h. Identical data obtained for slower cooling rates. Quenched samples were first held at 1200 °C, cooled to T_Q and held at T_Q several hours. Samples were prepared by firing the mixed oxides or the mixed end members in air at 1200 °C, regrinding, refiring until X-ray and moment data remained constant.
The samples were free of Fe^{2+} and were of spinel phase only. The data are consistent with Blasse's results (Fig. 73). Also compare with Fig. 74.

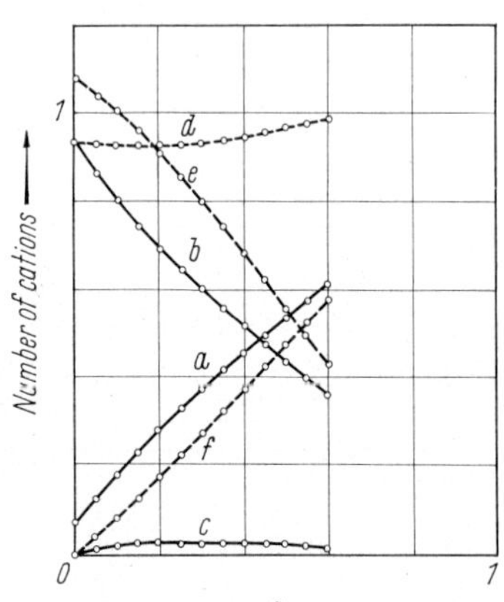

Fig. 78. $Mg_{1+a}Fe^{3+}_{2-2a}Ti^{4+}_{a}O_4$. Cation distribution vs. *a*. [*66TL(1)*].
a) Mg^{2+} on *A* site, *d*) Mg^{2+} on *B* site,
b) Fe^{3+} on *A* site, *e*) Fe^{3+} on *B* site,
c) Ti^{4+} on *A* site, *f*) Ti^{4+} on *B* site.
Data is for the slowly cooled samples of Tab. 20.
The curves were calculated for the best fit from the values of I_{400}/I_{220} and the moment data given in Tab. 20, assuming antiparallel moments for ions in the *A*, *B* sites: $p^0_m = 10t + 10s - 10a$ where t, s = number of Mg^{2+} and Ti^{4+}, respectively, on *A* sites. Compare with Tab. 17.

Tab. 21. $Mg_{1+a}Fe^{3+}_{2-2a}Ti^{4+}_{a}O_4$. Cation distribution (quenched from 900 °C) [*66TL(1)*]

a	I_{400}/I_{220}	Sites *A* Fe^{3+}	Sites *A* Mg^{2+}	Sites *B* Fe^{3+}	Sites *B* Mg^{2+}	Sites *B* Ti^{4+}
0	0.76	0.82	0.18	1.18	0.82	0
0.1	0.88	0.73	0.27	1.07	0.83	0.10
0.2	0.99	0.64	0.36	0.96	0.84	0.20
0.3	1.15	0.56	0.44	0.84	0.86	0.30
0.4	1.23	0.49	0.51	0.71	0.89	0.40
0.5	1.40	0.43	0.57	0.57	0.93	0.50

The distributions were determined in the manner described in Fig. 78.

Tab. 22. $Mg_{1+a}Fe^{3+}_{2-2a}Ti^{4+}_aO_4$. Cation distribution (quenched from 1100 °C) [*66TL(1)*]

a	Sites A		Sites B		
	Fe^{3+}	Mg^{2+}	Fe^{3+}	Mg^{2+}	Ti^{4+}
0	0.80	0.20	1.20	0.80	0
0.05	0.754	0.246	1.146	0.804	0.05
0.10	0.708	0.292	1.092	0.808	0.10
0.15	0.664	0.336	1.036	0.814	0.15
0.20	0.622	0.378	0.978	0.822	0.20
0.25	0.584	0.416	0.916	0.834	0.25
0.30	0.544	0.456	0.856	0.844	0.30
0.35	0.512	0.488	0.788	0.862	0.35
0.40	0.477	0.523	0.723	0.877	0.40
0.45	0.446	0.554	0.654	0.896	0.45
0.50	0.414	0.586	0.586	0.914	0.50
0.55	0.386	0.614	0.514	0.936	0.55
0.60	0.360	0.640	0.440	0.960	0.60

X-ray intensity data not taken. Fe^{3+} distribution calculated from moment data; the Mg^{2+} and Ti^{4+} distributions determined from $t = a + (p^0_m/10)$ assuming $s = 0$ as it is found for 900 °C quenching (Tab. 21). See Fig. 78 for definitions of t, s.

Tab. 23. $Mg_{1+a}Fe^{3+}_{2-2a}Ti^{4+}_aO_4$. Fe^{3+} distribution (quenched from 1200 °C) [*66TL(1)*]

a	Fe^{3+} in A site	Fe^{3+} in B site
0	0.793	1.207
0.05	0.742	1.158
0.10	0.696	1.104
0.15	0.652	1.048
0.20	0.610	0.990
0.25	0.570	0.930
0.30	0.534	0.866
0.35	0.500	0.800
0.40	0.469	0.731
0.45	0.437	0.663
0.50	0.409	0.591
0.55	0.380	0.520
0.60	0.355	0.445

Fe^{3+} distribution is calculated from moment data given in Tab. 20. Compare with Tab. 17.

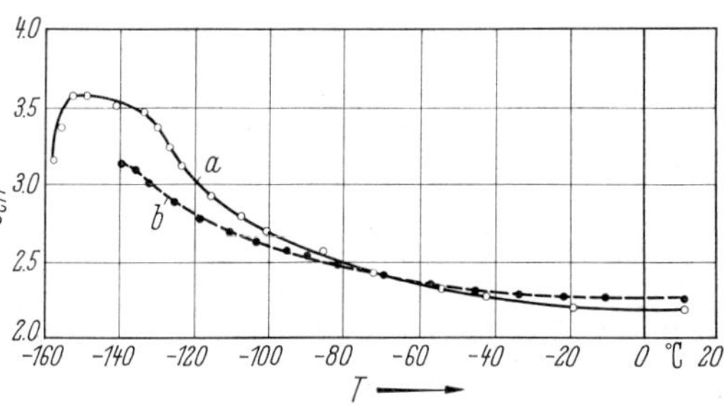

Fig. 79. $Mg_nTi_mFe_{3-m-n}O_4$. Temperature dependence of effective g value, g_{eff} [*59N*].
a) $n = 0.075, m = 0.51, \Theta_C = 125$ °C, $g_{eff} = 2.17 \pm 0.04$ at RT,
b) $n = 0.072, m = 0.50, \Theta_C = 135$ °C, $g_{eff} = 2.23 \pm 0.04$ at RT.
g_{eff} obtained from resonance absorption at 25 700 Mc/s. Materials were naturally occurring single crystals having spinel structure.

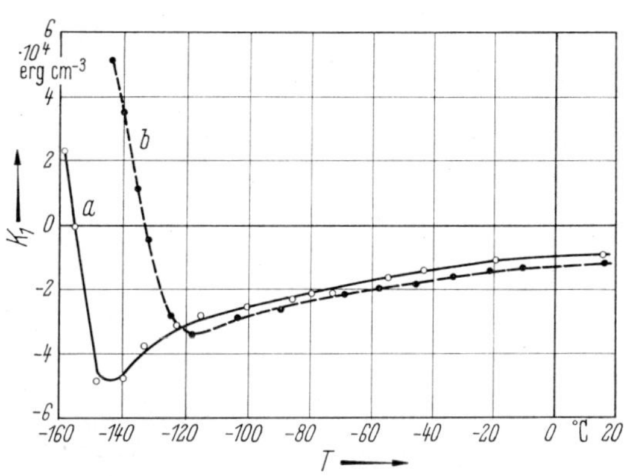

Fig. 80. $Mg_nTi_mFe_{3-m-n}O_4$. Anisotropy constant, K_1, vs. temperature, T [*59N*].
Same notation as Fig. 79.
K_1 obtained from resonance absorption measurements at 25 700 Mc/s. Error <3%. K_2 was within experimental error and neglected.
a) $K_1 = -0.96 \pm 0.15 \cdot 10^4$ erg cm^{-3} at RT,
b) $K_1 = -1.26 \pm 0.19 \cdot 10^4$ erg cm^{-3} at RT.

Fig. 81. $Mg_nTi_mFe_{3-m-n}O_4$. Specific saturation magnetization, σ_s, vs. temperature, T [*59N*]. →
Material is sample *a* of Fig. 79.
Measurements were made on a magnetic torsion balance. The curve shape is similar to those for Néel P-type ferrites such as $Ni_{1.5-a}Mn_aFeTi_{0.5}O_4$ [*54G*] in contrast to Blasse's observation (Fig. 73).

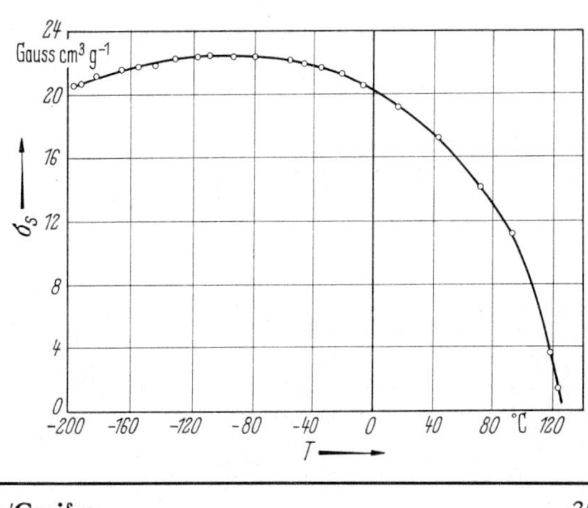

Tab. 24. $Mg_{1+a}Fe^{3+}_{2-2a}Sn^{4+}_{a}O_4$. Curie temperature, Θ_C, lattice parameter, a_0, and magnetic moment, p^0_m [66TL(3)]

a	Θ_C °C		a_0 Å		p^0_m μ_B		I_{400}/I_{220}
	Slowly cooled	Quenched	Slowly cooled	Quenched	Slowly cooled	Quenched	Slowly cooled
0	375	320	8.363	8.372	0.72	1.80	0.60
0.05	333	290	8.373	8.382	0.85	1.87	0.64
0.10	282	242	8.387	8.397	0.99	1.90	0.71
0.15	231	195	8.402	8.411	1.11	1.92	0.81
0.20	179	148	8.418	8.424	1.18	1.86	0.95
0.25	129	100	8.436	8.440	1.23	1.76	1.10
0.30	79	47	8.451	8.453	1.19	1.63	1.29
0.35	20	−10	8.464	8.467	1.08	1.44	1.49
0.40	− 50	−80	8.477	8.480	0.91	1.23	1.71
0.45	−125		8.490	8.492	0.72	0.94	1.92
0.50			8.504	8.504	0.53	0.67	2.18
0.60			8.527	8.528	0.17		2.79
0.70			8.554	8.555			3.59
0.80			8.581	8.582			4.75
0.90			8.609	8.610			6.35
1			8.638	8.638			8.77

Samples were prepared as described in Tab. 20. The quenched samples were quenched from 900 °C. Continuous spinel solid solutions are formed across the system.
See Fig. 82.

Tab. 25. $Mg_{1+a}Fe^{3+}_{2-2a}V^{4+}_{a}O_4$. Curie temperature, Θ_C, lattice parameter, a_0, and magnetic moment, p^0_m [66TL(2)]

a	Θ_C °C		a_0 Å		p^0_m μ_B		I_{400}/I_{220}
	Slowly cooled	Quenched	Slowly cooled	Quenched	Slowly cooled	Quenched	Slowly cooled
0	375	325	8.363	8.372	0.72	1.76	0.60
0.01	340	288	8.372	8.382	1.30	2.30	0.68
0.03	316	274	8.375	8.384	1.40	2.33	0.71
0.05	300	260	8.376	8.385	1.38	2.22	0.73
0.07	286	248	$8.37_7{}^6$	$8.38_6{}^5$	1.35	2.14	0.75
0.10	270	236	$8.37_7{}^6$	8.386	1.29	2.03	0.77
0.15	246	212	$8.37_7{}^6$	8.387	1.19	1.86	0.80
0.20	222	192	8.377	$8.38_8{}^7$	1.08	1.73	0.84
0.25	202	175	8.378	8.388	0.98	1.61	0.88
0.30	183	158	8.379	$8.38_9{}^8$	0.87	1.49	0.92
0.35	162	138	8.380	$8.38_9{}^8$	0.74	1.37	0.96
0.40	130	118	8.381	8.390	0.64	1.24	1.01
0.45	98	90	8.381_5	8.390	0.54	—	1.06
0.50	54	50	8.382	8.390	0.48	0.95	1.14
0.55	8	10	8.383	8.390	0.37	—	1.20
0.60	− 50	− 35	8.383	$8.39_1{}^0$	0.32	0.57	1.30
0.70	−190	−170	8.385	$8.39_1{}^0$			1.49
0.80			8.386	8.391			1.81
0.90			8.387	8.391			2.41
1			8.389	8.391			3.49

I_{400}/I_{220} as in Tab. 20.
Samples were prepared by firing mixed oxides or the mixed end members under vacuum at 850 °C until p^0_m and a_0 attained constant values.
Cooling rate for the "slowly cooled" samples: 10 °C/h. The "quenched" samples were quenched in water from 850 °C. The system forms a continuous solid solution. Fe^{2+} absent.
See Fig. 83.

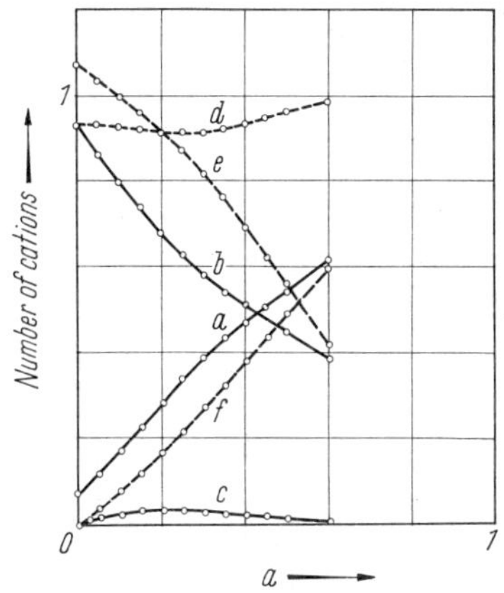

Fig. 82. $Mg_{1+a}Fe^{3+}_{2-2a}Sn^{4+}_{a}O_4$. Cation distribution vs. *a*. [*66TL(3)*].
a) Mg^{2+} on *A* site, *d*) Mg^{2+} on *B* site,
b) Fe^{3+} on *A* site, *e*) Fe^{3+} on *B* site,
c) Sn^{4+} on *A* site, *f*) Sn^{4+} on *B* site.
Data is for the slowly cooled samples of Tab. 24.
The curves were calculated exactly as described for the Ti^{4+} compound in Fig. 78.

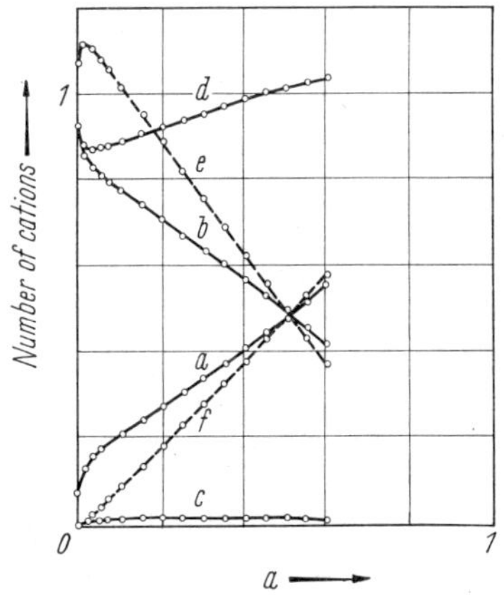

Fig. 83. $Mg_{1+a}Fe^{3+}_{2-2a}V^{4+}_{a}O_4$. Cation distribution vs. *a*. [*66TL(2)*].
a) Mg^{2+} on *A* site, *d*) Mg^{2+} on *B* site,
b) Fe^{3+} on *A* site, *e*) Fe^{3+} on *B* site,
c) V^{4+} on *A* site, *f*) V^{4+} on *B* site.
Data is for the slowly cooled samples of Tab. 25.
The curves were calculated as described in Fig. 78 using $p^0_m = 10t + 8s - 9a$.

Tab. 26. $Mg_{1+(4a/3)}Fe^{3+}_{2-2a}V^{5+}_{2a/3}O_4$. Curie temperature, Θ_C, lattice parameter, a_0, and magnetic moment, p^0_m [*66TL(2)*]

a	Θ_C °C		a_0 Å		p^0_m μ_B		I_{400}/I_{220}
	Slowly cooled	Quenched	Slowly cooled	Quenched	Slowly cooled	Quenched	Slowly cooled
0	375	320	8.363	8.372	0.72	1.80	0.60
0.01	350	295	8.367	8.380	0.88	1.92	0.63
0.02	325	285	8.369	8.384	1.14	2.13	0.68
0.03	318	283	8.370	8.383	1.23	2.08	0.70
0.04	313	282	8.371	8.382	1.28	2.01	0.71
0.05	309	280	8.372	8.380	1.31	1.98	0.72
0.07	305	276	8.373	8.378	1.33	1.94	0.72
0.10	302	274	8.374	8.377	1.38	1.89	0.74
0.15	302	274	8.374				

I_{400}/I_{220} as in Tab. 20.
Samples were prepared from mixed oxides or from $MgFe_2O_4$, MgO, V_2O_5 by firing in air first at 500 °C and then progressively higher up to 1000 °C.
The slowly cooled samples were cooled at 10 °C/h. The quenched samples were cooled to 900 °C, held at 900 °C several hours and quenched in water.
Spinel solid solutions are formed up to $a = 0.1$.
Fe^{2+} and V^{4+} were absent.

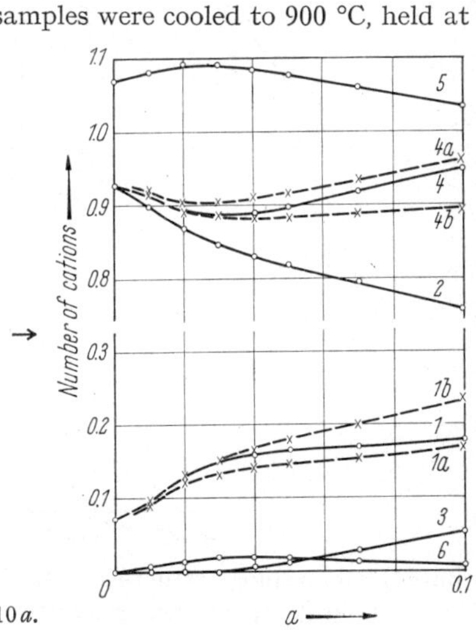

Fig. 84. $Mg_{1+(4a/3)}Fe^{3+}_{2-2a}V^{5+}_{2a/3}O_4$. Cation distribution vs. *a*. [*66TL(2)*].
1) Mg^{2+} on *A* site, *4*) Mg^{2+} on *B* site,
2) Fe^{3+} on *A* site, *5*) Fe^{3+} on *B* site,
3) V^{5+} on *A* site, *6*) V^{5+} on *B* site.
1a) Minimum number of Mg^{2+} on *A* site (all V^{5+} on *A* site),
1b) Maximum number of Mg^{2+} on *A* site (all V^{5+} on *B* site),
4a) Maximum number of Mg^{2+} on *B* site (all V^{5+} on *A* site),
4b) Minimum number of Mg^{2+} on *B* site (all V^{5+} on *B* site).
Data is for she slowly cooled samples of Tab. 26.
The curves were calculated as described in Fig. 78 using $p^0_m = 10t + 10s - 10a$.

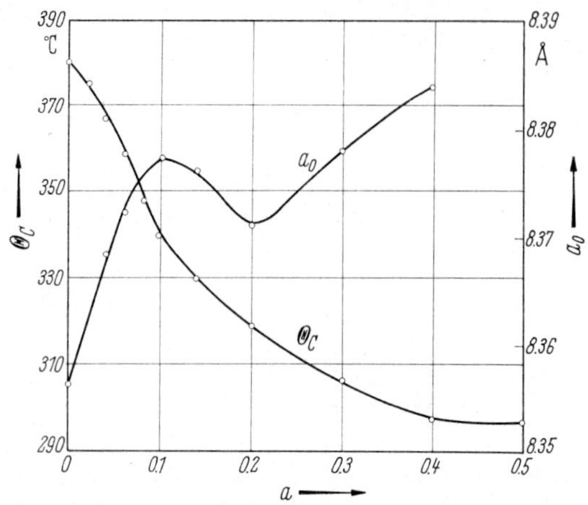

Fig. 85. $Mg_{1+(2a/3)}Fe^{3+}_{2-a}Nb^{5+}_{a/3}O_4$. Lattice parameter, a_0, and Curie temperature, Θ_C, vs. a [*62TL(2)*].
Materials were prepared by firing mixed oxides at 1200 °C for several weeks during which time the powders were reground several times. After the final firing the samples were cooled slowly. A single spinel phase exists for a values up to 0.4. Analogous results stated but not given for Ta^{5+}.

Tab. 27. $Mg_{1+(2a/3)}Fe^{3+}_{2-a}Nb^{5+}_{a/3}O_4$. Tabulation of data from Fig. 85 [*62TL(2)*]

a	Θ_C °C	a_0 Å
0	380	8.356
0.02	375	—
0.04	367	8.368
0.06	359	8.372
0.08	348	—
0.1	340	8.377
0.14	330	8.376
0.2	319	8.371
0.3	307	8.378
0.4	298	8.384
0.5	298	—

6.1.4.3.8 Systems containing F^- and Li^+

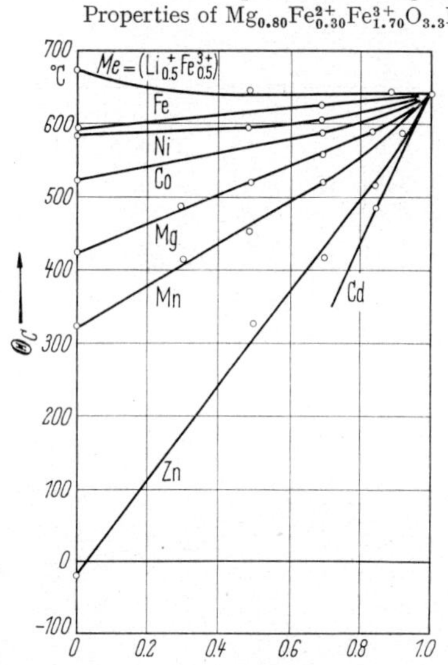

Fig. 86. $(1-x)$MgO · xLiF · Fe_2O_3 and other spinels, $Me_{1-x}Li_xFe_2F_xO_{4-x}$. Curie temperature, Θ_C, vs. LiF content, x [*66OHNON*].

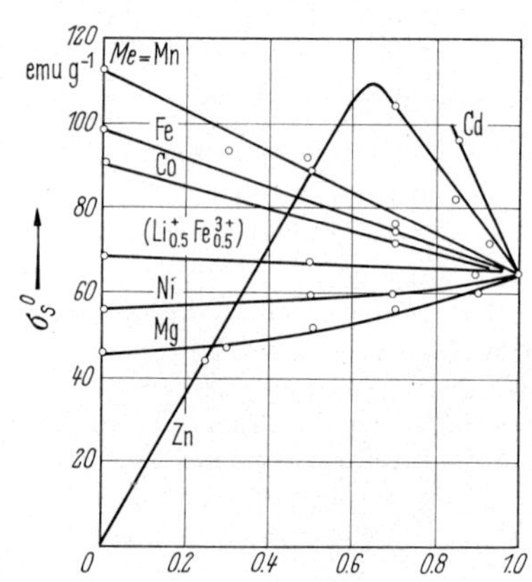

Fig. 87. $(1-x)$MgO · xLiF · Fe_2O_3 and other spinels, $Me_{1-x}Li_xFe_2F_xO_{4-x}$. Specific saturation magnetization at 0 °K, σ_s^0, vs. LiF content, x [*66OHNON*].
Values are estimated from extrapolation of σ_s data measured above 77 °K. Compare with Tab. 28.

Tab. 28. Properties of $Mg_{0.80}Fe^{2+}_{0.30}Fe^{3+}_{1.70}O_{3.3}F_{0.7}$ [*64S*]

a_0 Å	Θ_C °K	σ_s at 90 °K emu/g	Chemical analysis	
			wt-% Fe^{2+}	wt-% F^-
8.367 ± 0.003	693 ± 10	64 ± 2.5%	8.6	8.0

Material had spinel structure. It was prepared from equimolar quantities of MgF_2 and Fe_2O_3 fired 1100···1300 °C under CO_2. Compare with Fig. 86, 87.

6.1.4.4 References for 6.1.4.1 ··· 6.1.4.3

30P	Posnjak, E.: Am. J. Sci. **19** [109] (1930) 61.
31RM	Roberts, H. S., and H. E. Mervin: Am. J. Sci. **21** (1931) 145.
32G	Gans, R.: Ann. Physik (5. Folge) **15** (1932) 28.
32S	Serres, A.: Ann. Phys. **17** (1932) 5.
34A	Akulov, N. S.: Uch. Zap. Mosk. Gos. Univ. (1934) 137.
39BD	Becker, R., and W. Döring: "Ferromagnetismus", p. 247.
50FS	Foëx, G., and A. Serres: Compt. Rend. **230** (1950) 729.
50GB	Guillaud, C., and R. Bertrand: J. Rech. Centre Natl. Rech. Sci. Lab. Bellevue (Paris) **3** (1950) 73.
51B	Bochirol, L.: Compt. Rend. **233** (1951) 736.
51FF	Fairweather, A., and E. J. Frost: Rept. Progr. Phys. **25** (1952) 142; see also Post Office Engineering Department Research Report 1951.
51GS	Guillaud, C., and M. Sage: Compt. Rend. **232** (1951) 944.
51HK	Herring, C., and C. Kittel: Phys. Rev. **81** (1951) 869.
51OKT	Okamura, T., Y. Kojima, and Y. Torizuka: Sci. Rept. Res. Inst. Tohoku Univ. **A 3** (1951) 725.
51PB	Pauthenet, R., and L. Bochirol: J. Phys. Radium **12** (1951) 249.
51R	Roberts, F. F.: J. Phys. Radium **12** (1951) 301.
51R(1)	Roberts, F. F.: J. Phys. Radium **12** (1951) 305.
51W	Weil, L.: J. Phys. Radium **12** (1951) 260.
52JR	Jones, G. O., and F. F. Roberts: Proc. Phys. Soc. (London) **B 65** (1952) 390.
52N	Néel, L.: J. Phys. Radium **13** (1952) 249.
52OFD	Okamura, T., T. Futjimura, and M. Date: Sci. Rept. Res. Inst. Tohoku Univ. **A 4** (1952) 191.
52RWE	Rado, G. T., R. W. Wright, N. H. Emerson, and A. Terris: Phys. Rev. **88** (1952) 909.
53B	Baltzer, P. K.: MIT Digital Computer Lab. Memorandum M-2275 (1953).
53BR	Bacon, G. E., and F. F. Roberts: Acta Cryst. **6** (1953) 57.
53CHB	Corliss, L. M., J. M. Hastings, and F. G. Brockman: Phys. Rev. **90** (1953) 1013.
53MP	Maxwell, L. R., and S. J. Pickart: Phys. Rev. **92** (1953) 1120.
53R	Romeijn, F. C.: Philips Res. Rept. **8** (1953) 304.
53RRO	Roy, D. M., R. Roy, and E. F. Osborne: Am. J. Sci. **251** (1953) 337.
53V	Vinal, F. E.: Memorandum M 2442, Lincoln Laboratory, Lexington, Mass., Sept. 15, **1953**.
54A	Albers-Schoenberg, E.: J. Appl. Phys. **25** (1954) 152.
54G	Gorter, E. W.: Philips Res. Rept. **9** (1954) 403.
54RW	Richards, R. G., and J. White: Trans. Brit. Ceram. Soc. **53** (1954) 422.
54SW	Smit, J., and H. P. J. Wijn: Advan. Electron. Electron Phys. **6** (1954) 91.
54WGEG	Wijn, H. P. J., E. W. Gorter, C. J. Esveldt, and P. Geldermans: Philips Tech. Rev. **16** (1954) 49.
54YKL	Yearian, H. J., J. M. Kortright, and R. H. Langenheim: J. Chem. Phys. **22** (1954) 1196.
55B	Baltzer, P. K.: Proc. Conf. Magnetism and Magn. Materials, Pittsburgh, Pa. (1955) AIEE **T-78** p. 247.
55BTW	Bozorth, R. M., E. F. Tilden, and A. J. Williams: Phys. Rev. **99** (1955) 1788.
55E(1)	Economos, G.: J. Am. Ceram. Soc. **38** (1955) 353.
55E(2)	Economos, G.: J. Am. Ceram. Soc. **38** (1955) 408.
55E(3)	Economos, G.: J. Am. Ceram. Soc. **38** (1955) 335.
55vU	van Uitert, L. G.: J. Appl. Phys. **26** (1955) 1289.
55W	Waldron, R. D.: Phys. Rev. **99** (1955) 1727.
55WW	Woodhouse, D., and J. White: Trans. Brit. Ceram. Soc. **34** (1955) 333.
56D	Delorme, C.: Thesis, Grenoble **1956**.
56E	Epstein, D. J.: Proc. Conf. Magnetism Magnetic Materials, Boston, Am. Inst. Elec. Engrs. (1956) 493.
56F	Fresh, D. L.: Proc. IRE **44** (1956) 1303.
56FR	Folen, V. J., and G. T. Rado: Bull. Am. Phys. Soc. **1** (1956) 132.
56HC	Hastings, J. M., and L. M. Corliss: Phys. Rev. **104** (1956) 328.
56KH	Kriessman, C. J., and S. E. Harrison: Phys. Rev. **103** (1956) 857.
56MG	Muan, A., and C. L. Gee: J. Am. Ceram. Soc. **39** (1956) 207.
56vU(1)	van Uitert, L. G.: J. Chem. Phys. **24** (1956) 306.
56vU(2)	van Uitert, L. G.: Proc. I. R. E. **44** (1956) 1294.
57B	Blackman, L. C. F.: J. Appl. Phys. **28** (1957) 1511.
57BBP	Bickford Jr., L. R., J. M. Brownlow, and R. F. Penoyer: Proc. Inst. Elec. Engrs. (London) **B 104** Suppl. 5 (1957) 238.
57GE	Gorter, E. W., and C. J. Esveldt: Proc. Inst. Elec. Engrs. (London) **B 104** Suppl. 7 (1957) 418.

57Go	GOODENOUGH, J. B.: Proc. Inst. Elec. Engrs. (London) **B 104** Suppl. 7 (1957) 400.
57Gy	GYORGY, E. M.: J. Appl. Phys. **28** (1957) 1011.
57H	HOFFMAN, P. O.: J. Am. Ceram. Soc. **40** (1957) 250.
57HBW	VAN DER HEIDE, H., H. G. BRUIJNING, and H. P. J. WIJN: Philips Tech. Rev. **18** (1957) 336.
57K	KRIESSMAN, C. J.: Unpublished.
57L	LJUNG, P. E.: Proc. Inst. Elec. Engrs. (London) **B 104** Suppl. 5 (1957) 154.
57Lo	LOW, W.: Phys. Rev. **105** (1957) 792.
57NPHK	NATHANS, R., S. J. PICKART, S. E. HARRISON, and C. J. KRIESSMAN: Proc. Inst. Elec. Engrs. (London) **B 104** Suppl. 5 (1957) 217.
57PJS	PALMER, G. G., R. W. JOHNSTON, and R. E. SCHULTZ: J. Am. Ceram. Soc. **40** (1957) 256.
57RFE	RADO, G. T., V. J. FOLEN, and W. H. EMERSON: Proc. Inst. Elec. Engrs. (London) **B 104** Suppl. 5 (1957) 198.
57SP	STANDLEY, K. J., and J. PETERS: Proc. Inst. Elec. Engrs. (London) **B 104** Suppl. 5 (1957) 206.
57vU	VAN UITERT, L. G.: J. Appl. Phys. **28** (1957) 320.
58B	BLACKMAN, L. C. F.: J. Electron. Control **5** (1958) 373.
58Br	BRISSONEAU, P.: J. Phys. Chem. Solids **7** (1958) 22.
58EF	EPSTEIN, D. J., and B. FRACKIEWICZ: J. Appl. Phys. **29** (1958) 376.
58FR	FOLEN, V. J., and G. T. Rado: J. Appl. Phys. **29** (1958) 438.
58Gy	GYORGY, E. M.: J. Appl. Phys. **29** (1958) 283.
58MYF	MITSUISHI, A., H. YISHINAGA, and S. FUJITA: J. Phys. Soc. Japan **13** (1958) 1236.
58PS	PETERS, J., and K. J. STANDLEY: Proc. Phys. Soc. (London) **71** (1958) 131.
59B	BLOKHIN: "Methods of X-ray Spectral Analysis", Fizmatgiz Moscow **1959**.
59Bla	BLACKMAN, L. C. F.: J. Am. Ceram. Soc. **42** (1959) 143.
59BK	BELSON, H. S., and C. J. KRIESSMAN: J. Appl. Phys. **30** (1959) 170S.
59BKZ	BROŽ, J., S. KRUPIČKA, and B. ZITKA: Czech. J. Phys. **9** (1959) 314.
59C	CARTER, R. E.: J. Am. Ceram. Soc. **42** (1959) 143.
59GC	GREIFER, A. P., and W. J. CROFT: J. Appl. Phys. **30** (1959) 34S.
59GH	GYORGY, E. M., and F. B. HAGEDORN: J. Appl. Phys. **30** (1959) 1368.
59HG	HUMPHREY, F. B., and E. M. GYORGY: J. Appl. Phys. **30** (1959) 935.
59K	KWESTROO, W.: J. Inorg. Nucl. Chem. **9** (1959) 65.
59N	NAKAU, T.: Mem. Coll. Sci. Univ. Kyoto, Ser. A, **29** (1959) 241.
59P	PEARSON, R. F.: Proc. Phys. Soc. (London) **74** (1959) 505.
59PT	PICKART, S. J., and A. C. TURNOCK: Phys. Chem. Solids **10** (1959) 242.
59 SRW	SHERWOOD, R. C., J. P. REMEIKA, and H. J. WILLIAMS: J. Appl. Phys. **30** (1959) 217.
59SW	SMIT, J., and H. P. J. WIJN: "Ferrites", Eindhoven and New York 1959.
60Be	BEKKER, YA. M.: Fiz. Tverd. Tela **2** (1960) 1708; Soviet Phys.-Solid State (English Transl.) **2** (1961) 1546.
60Br	BROŽ, J.: Czech. J. Phys. **10** (1960) 428.
60C	CRAPO, W. A.: J. Appl. Phys. **31** (1960) 267S.
60DM	DÉSIRANT, M., and J. L. MICHIELS, editors: Solid State Phys. Electron. Telecommun. Intern. Conf. Brussels 1958, Vol. **3** (1960) 126.
60E	EICHBAUM, B. R.: J. Appl. Phys. **31** (1960) 117S.
60GC	GREIFER, A. P., and W. J. CROFT: J. Appl. Phys. **31** (1960) 85.
60K	KRUPIČKA, S.: Czech. J. Phys. **10** (1960) 782.
60MG	MCGUIRE, T. R., and S. W. GREENWALD: Solid State Phys. Electron. Telecommun. Intern. Conf. Brussels 1958, Vol. **3** (1960) 50.
60MM	MURAMORI, K., and S. MIYAHARA: J. Phys. Soc. Japan **15** (1960) 2354.
60P	PALADINO, A. E.: J. Am. Ceram. Soc. **43** (1960) 183.
60PH	PRINGLE, D. H., and W. D. HEPBURN: Solid State Phys. Electron. Telecommun. Intern. Conf. Brussels 1958, Vol. **3** (1960) 252.
60RF	RADO, G. T., and V. J. FOLEN: J. Appl. Phys. **31** (1960) 62.
60S	SUGIURA, Y.: J. Phys. Soc. Japan **15** (1960) 1217.
61BeBe	BELOV, K. P., and V. J. BELOV: Fiz. Tverd. Tela **3** (1961) 1425; Soviet Phys.-Solid State (English Transl.) **3** (1961) 1034.
61BeBo	BERGSTEIN, A., and A. BOHUN: Collection Czech. Chem. Commun. **26** (1961) 747.
61C(1)	CHAMBERLAIN, J. R.: Proc. Phys. Soc. (London) **78** (1961) 819.
61C(2)	CHAMBERLAIN, J. R.: Proc. Phys. Soc. (London) **78** (1961) 828.
61E	ELBINGER, E.: Naturwiss. **48** (1961) 498; Z. Angew. Phys. **14** (1962) 273.
61K	KNOWLES, J. E.: Proc. Phys. Soc. (London) **77** (1961) 225.
61KBK	KOEHLER, M. F., R. BARANY, and K. K. KELLEY: U. S. Bur. Mines Rep. Invest. (1961) 5711.
61KH	KRIESSMAN, C. J., and S. E. HARRISON: J. Appl. Phys. **32** (1961) 392S.

61Kr	Krupička, S.: Czech. J. Phys. **11** (1961) 457.
61LV	Linzen, D., and F. Voigt: Abhandl. Deut. Akad. Wiss. Berlin Kl. Chem. Geol. Biol. No. 3 **1961**, 49.
61O	Ohta, K.: J. Phys. Soc. Japan **16** (1961) 250.
61OT	Ohnishi, H., and T. Teranishi: J. Phys. Soc. Japan **16** (1961) 35.
61PSM	Phillips, B., S. Somiya, and A. Muan: J. Am. Ceram. Soc. **44** (1961) 167.
61S	Sugiura, Y.: J. Phys. Soc. Japan **16** (1961) 1786.
62BC	Booth, J. G., and J. Crangle: Proc. Phys. Soc. (London) **79** (1962) 1271.
62BCKM	Braginski, A., W. Ciaston, J. Kulikowski, and S. Makolagwa: Proc. Inst. Elec. Engrs. (London) **B 109** (1962) 380.
62DDSS	Drokin, A. I., V. D. Dylgerov, N. I. Sudakov, and I. I. Starostin: Fiz. Tverd. Tela **4** (1962) 2293; Soviet Phys.-Solid State (English Transl.) **4** (1963) 1679.
62HSG	Hsein-Chên, M., S. S. Starobinets, and A. G. Gurevich: Fiz. Tverd. Tela **4** (1962) 1060; Soviet Phys.-Solid State (English Transl.) **4** (1962) 780.
62K	Knowles, J. E.: Philips Tech. Rev. **24** (1962) 242.
62LAT	Lăbuşcă, E., N. Andreescu, and I. Teodorescu: Rev. Roumaine Phys. **7** (1962) 261.
62LY	Lebed', B. M., and Yu. M. Yakolev: Fiz. Tverd. Tela **4** (1962) 1695; Soviet Phys.-Solid State (English Transl.) **4** (1963) 1245.
62R	Rybak, P. M.: Fiz. Metal. i Metalloved. **14** (1962) 778; Phys. Metals Metallog. (USSR) (English Transl.) **14** (1962) 122.
62S	Sternberk, J.: Czech. J. Phys. **12** (1962) 83, 471.
62SK	Shakhov, V. I., and E. I. Kondorskii: Fiz. Tverd. Tela **4** (1962) 29; Soviet Phys.-Solid State (English Transl.) **4** (1962) 20.
62TL(1)	Tellier, J. C., and M. Lensen: Compt. Rend. **255** (1962) 125.
62TL(2)	Tellier, J. C., and M. Lensen: Compt. Rend. **254** (1962) 4174.
62YL	Yakolev, Yu. M., and B. M. Lebed': Fiz. Tverd. Tela **4** (1962) 3654; Soviet Phys.-Solid State (English Transl.) **4** (1963) 2672.
63AES	Albers-Schoenberg, E., M. Eisenberg, and C. L. Snyder: Am. Inst. Phys., Handbook **1963**, p. 5—187. D. Gray, editor, McGraw-Hill, N. Y.
63Barn	Barna, G. I.: Schweiz. Arch. Angew. Wiss. Tech. **29** (1963) 212.
63Bart	Bartashevskii, E. L.: Ukr. Fiz. Zh. **8** (1963) 894.
63BC	Bady, I., and T. Collins: I. E. E. E. Trans. Microwave Theory Tech. **11** (1963) 222.
63BSZSCS	Bogoslovskiy, V. N., I. Ye. Startseva, M. G. Zhuravleva, A. A. Shchepetkin, G. I. Chufarov, and Ya. S. Shur: Fiz. Metal. i Metalloved. **15** (1963) 181; Phys. Metals Metallog. (USSR) (English Transl.) **15** (1963) 19.
63C	Cimpl, Z.: Czech. J. Phys. **13** (1963) 518.
63D	Drokin, A. I.: Izv. Sibirsk. Otd. Akad. Nauk SSSR, Ser. Techn. Nauk **2** (1963) 99.
63GV	Glauche, E., and F. Voigt: Exptl. Tech. Physik **11** (1963) 196.
63K	Klupsch, T.: Monatsber. Deut. Akad. Wiss. Berlin **5** (1963) 125.
63MD	Minomura, S., and H. G. Drickamer: J. Appl. Phys. **34** (1963) 3043.
63MP	Mozzi, R. L., and A. E. Paladino: J. Chem. Phys. **39** (1963) 435.
63SB	Schulkes, J. A., and G. Blasse: Phys. Chem. Solids **24** (1963) 1651.
63WB	Weisz, R. S., and D. L. Brown: J. Appl. Phys. **34** (1963) 1176.
63WM	Wallet, N., and F. Marion: Bull. Soc. Chim. France **1963** 2668; Compt. Rend. **256** (1963) 1290.
64A	Andreescu, N.: Rev. Roumaine Phys. **9** (1964) 315.
64B	Blasse, G.: Philips Res. Rept. Suppl. No. 3 (1964).
64BDZMST	Belov, V. F., M. N. Devisheva, I. S. Zheludev, E. F. Makarov, R. A. Stukan, and V. A. Trukhtanov: Fiz. Tverd. Tela **6** (1964) 3435; Soviet Phys.-Solid State (English Transl.) **6** (1965) 2747.
64BS	Bondarev, D. E., and D. A. Solov'ev: Poroshkovaya Met. Akad. Nauk Ukr. SSR **4** (1964) 97; Soviet Pwdr. Metall. Metal Ceram. (English Transl.) **4** (1964) 342.
64BSSACS	Bogoslovskiy, V. N., A. A. Shchepetkin, I. Ye. Startseva, V. K. Antonov, G. I. Chufarov, and Ya. S. Shur: Fiz. Metal. i Metalloved. **18** (1964) 711; Phys. Metals Metallog. (USSR) (English Transl.) **18** (1964) 72.
64DC	Deschamps, A., and V. Cagan: Compt. Rend. **258** (1964) 3238.
64FW	Fresh, D. L., and R. G. West: J. Am. Ceram. Soc. **47** (1964) 395.
64HV	Heck, C., and G. Vaccari: J. Appl. Phys. **35** (1964) 907.
64PS(1)	Pavlov, V. I., and N. N. Sirota: Dokl. Akad. Nauk Belorussk. SSR **8** (1964) 505.
64PS(2)	Pavlov, V. I., and N. N. Sirota: Fiz. Tverd. Tela **6** (1964) 1267; Soviet Phys.-Solid State (English Transl.) **6** (1964) 990.
64S	Scheiber, M.: J. Appl. Phys. **35** (1964) 1072.
64SS	Saksonov, Yu. G., and V. A. Somenkov: Fiz. Metall. i Metalloved. **18** (1964) 853.
64V	Vorob'ev, A. S.: Izv. Vysshikh Uchebn. Zavedenii, Fiz. **1964** 135.

64YT — Yur'eva, E. K., and O. N. Trubitsyna: Fiz. Tverd. Tela **6** (1964) 247; Soviet Phys.-Solid State (English Transl.) **6** (1964) 196.

65AO — Annaev, R. G., and A. Orazsakhatov: Izv. Akad. Nauk Turkm. SSR, Ser. Fiz. Tekhn. Khim. Geol. Nauk **1965** 11.

65Ba — Baba, P. D.: J. Am. Ceram. Soc. **48** (1965) 305.

65Bo — Bondarev, D. E.: Poroshkovaya Met. Akad. Nauk Ukr. SSR **5** (1965) 93; Soviet Pwdr. Metall. Metal Ceram. (English Transl.) **5** (1965) 422.

65E — Elbinger, E.: Phys. Status Solidi **9** (1965) 843.

65GKL — Grodshtein, A. E., E. M. Kriger, and S. M. Lisitsyn: Poroshkovaya Met. Akad. Nauk Ukr. SSR **5** (1965) 4; Soviet Pwdr. Metall. Metal Ceram. (English Transl.) **5** (1965) 352.

65K — Kankowski, E. F.: in "Handbook of Microwave Ferrite Materials", W. H. von Aulock, editor, Academic Press (1965).

65KK — Katsura, T., and S. Kimura: Bull. Chem. Soc. Japan **38** (1965) 1665.

65KWKP — Kleinstück, K., E. Wieser, P. Kleinert, and R. Perthel: Phys. Status Solidi **8** (1965) 271.

65MM — Marais, A., and Th. Merceron: Compt. Rend. **261** (1965) 901.

65PL — Pil'shchikov, A. I., and E. V. Lebedeva: Fiz. Tverd. Tela **7** (1965) 3320; Soviet Phys.-Solid State (English Transl.) **7** (1966) 2677.

65R — Reynen, P.: "Reactivity of Solids", edited by G. M. Schwab, Elsevier Publishing Co, Amsterdam, Netherlands, 1965, 562—71.

65S — Suiter Jr., W. B.: J. Appl. Phys. **36** (1965) 648.

65TM — Telesnin, R. V., and K. T. Makarov: Fiz. Metal. i Metalloved. **20** (1965) 349; Phys. Metals Metallog. (USSR) (English Transl.) **20** (1965) 29.

65vA — von Aulock, W. H., editor: "Handbook of Microwave Ferrite Materials", Academic Press, N. Y. (1965) xxi.

65WK — Wieser, E., and K. Kleinstück: Z. Krist. **121** (1965) 14.

66A — Allen, W. C.: J. Am. Ceram. Soc. **49** (1966) 251.

66Al — Allen, W. C.: Am. Mineralogist **51** (1966) 239, and Ph. D. Thesis, Rutgers Univ. New Brunswick, N. J.

66An — Andreescu, N.: Rev. Roumaine Phys. **11** (1966) 107.

66B — Bondarev, D. E.: Izv. Akad. Nauk SSSR, Ser. Fiz. **30** (1966) 1076.

66DS — Drokin, A. I., and L. I. Slobodskoi: Zh. Eksperim. Teor. Fiz. **51** (1966) 1023; Soviet Phys. JETP (English Transl.) **24** (1967) 682.

66DSPS — Drokin, A. I., N. M. Salanskii, A. A. Popova, and R. P. Smolin: Zh. Tekh. Fiz. **36** (1966) 521; Soviet Phys.-Tech. Phys. (English Transl.) **11** (1966) 385.

66GMMS — Gerber, R., L. Michalowsky, K. Motzke, and E. Steinbeiss: Phys. Status Solidi **16** (1966) 795.

66Kh — Khitrik, A. S.: Poroshkovaya Met. Akad. Nauk Ukr. SSR **6** (1966) 51; Soviet Pwdr. Metall. Metal Ceram. (English Transl.) **6** (1966) 302.

66Ki — Kirichok, P. P.: Ukr. Fiz. Zh. **11** (1966) 91.

66LPS — Lebedeva, E. V., A. I. Pil'shchikov, and N. S. Sedletskaya: Fiz. Tverd. Tela **8** (1966) 3204; Soviet Phys.-Solid State (English Transl.) **8** (1967) 2564.

66MM — Marais, A., and Th. Merceron: I. E. E. E. Trans. Magnetics **2** (1966) 612.

66MY — Mukhin, V. P., and E. K. Yur'eva: Fiz. Tverd. Tela **8** (1966) 2011; Soviet Phys.-Solid State (English Transl.) **8** (1967) 1599.

66N — Nikol'skii, A. P.: Fiz. Tverd. Tela **8** (1966) 1208; Soviet Phys.-Solid State (English Transl.) **8** (1966) 960.

66OHNON — Okazaki, C., E. Hirota, Y. Neichi, H. Okazaki, and S. Nakajima: J. Phys. Soc. Japan **21** (1966) 199.

66PS — Pil'shchikov, A. I., and N. S. Sedletskaya: Vestn. Mosk. Univ., Ser. III, Fiz., Astron. **21** (1966) 69.

66TL(1) — Tellier, J. C., and M. Lensen: Bull. Soc. Chim. France **1966** 2502.

66TL(2) — Tellier, J. C., and M. Lensen: Bull. Soc. Chim. France **1966** 3939.

66TL(3) — Tellier, J. C., and M. Lensen: Bull. Soc. Chim. France **1966** 3942.

66V — Vorob'ev, A. S.: Izv. Vysshikh Uchebn. Zavedenii, Khim. Tekh. **9** (1966) 290.

66WMK — Wieser, E., W. Meisel, and K. Kleinstück: Phys. Status Solidi **16** (1966) 127.

67B — Boyd, E. L.: J. Appl. Phys. **38** (1967) 2367.

67LPY — Lebedeva, E. V., A. I. Pil'shchikov, and E. K. Yur'eva: Fiz. Tverd. Tela **9** (1967) 317; Soviet Phys.-Solid State (English Transl.) **9** (1967) 237.

67S — Speidel, D. H.: J. Am. Ceram. Soc. **50** (1967) 243.

68B — Boyd, E. L.: J. Appl. Phys. **39** (1968) 1304.

69G — Greifer, A. P.: I. E. E. E. Trans. Magnetics **5** (1969) 774.

6.1.5 Zn-Fe^{3+} spinels (Zn ferrites) and Zn-Fe^{3+} spinels with substitutions

6.1.5.1 Zn, Zn-Fe, Zn-Ti, Zn-Li, Zn-Ti-Li, Zn-Cd, Zn-Mg, Zn-Co and Zn-Cu ferrites

Tab. 1. $ZnFe_2O_4$.

a) General properties (room temperature unless otherwise specified).

		Ref.
Crystal type	Normal spinel — cubic	*47VH* *51B(1)* *53HC*
Structure type	$MgAl_2O_4$	
Space group	O_h^7 — Fd3m	
Lattice constant	$a = 8.440$ Å	*53R*
Volume of unit cell	$601.2 \cdot 10^{-24}$ cm³ (containing 8 molecules)	
Oxygen parameter	$u = 0.385 \pm 0.002$	*53HC*
Molecular weight	241.1 amu	
X-ray density	$\varrho_X = 5.33$ g/cm³	
Melting point	1590 ± 10 °C	*36AVB*
Sintering temperatures	800···1250 °C	
Thermal conductivity	$\varkappa \cong 1.5 \cdot 10^{-2}$ cal sec^{-1} cm^{-1} deg^{-1}	*59SW*
Specific heat	$C_p = 34.3$ cal deg^{-1} mole^{-1}	*58WG*
Young's modulus	$E \approx 10^{12}$ dyne/cm²	*59SW*
Coefficient of linear thermal expansion	$7.5 \cdot 10^{-6}$ °C^{-1}	*50W*
Mohs' scale hardness number	5.5	*32PBB*

b) Magnetic properties. Pure zinc ferrite ($ZnFe_2O_4$) is not ferrimagnetic. It transforms from a paramagnetic state to an antiferrimagnetic state in the vicinity of 15 °K [*66L*]. This transition is accompanied by a λ-type specific heat anomaly [*58WG*]

Néel temperature	$\Theta_N \approx 15$ °K (see Fig. 8)	*66L*
Paramagnetic Curie temperature	$\Theta_p \approx 0$ °K	*66L*

Tab. 2. $ZnFe_2O_4$. Molal heat capacities for quenched Zn ferrite at various temperatures [*58WG*]

T °K	C_p^0 cal deg^{-1} mole^{-1}	T °K	C_p^0 cal deg^{-1} mole^{-1}	T °K	C_p^0 cal deg^{-1} mole^{-1}	T °K	C_p^0 cal deg^{-1} mole^{-1}
10	1.327	70	8.366	170	23.69	270	32.60
15	1.418	80	10.046	180	24.87	280	33.19
20	1.591	90	11.780	190	25.98	290	33.75
25	1.886	100	13.409	200	27.00	300	34.27
30	2.325	110	15.037	210	27.98	350	36.41
						273.15	32.79
35	2.892	120	16.633	220	28.90		
40	3.556	130	18.182	230	29.75	298.15	34.17
45	4.289	140	19.646	240	30.54		
50	5.067	150	21.06	250	31.27		
60	6.693	160	22.41	260	31.96		

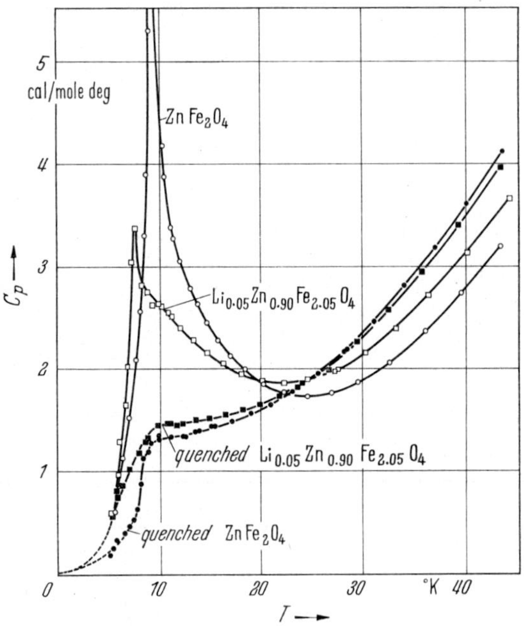

Fig. 1. $ZnFe_2O_4$; $Li_{0.05}Zn_{0.90}Fe_{2.05}O_4$. Molal heat capacities of annealed and quenched samples vs. temperature [*58WG*]. See also Tab. 2 and [*55GW*].

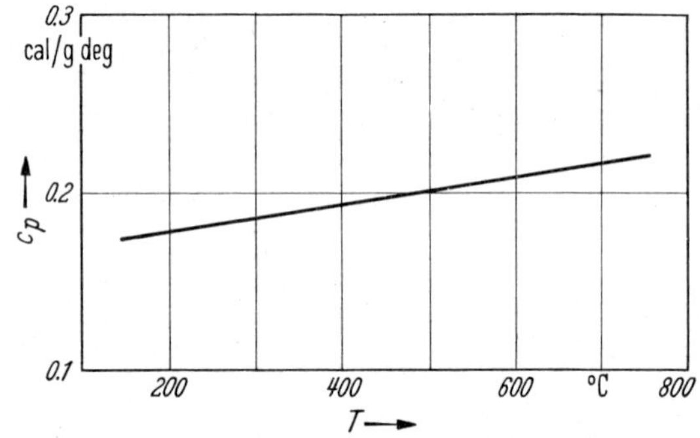

Fig. 2. $ZnFe_2O_4$. Specific heat vs. temperature [*51B2*].

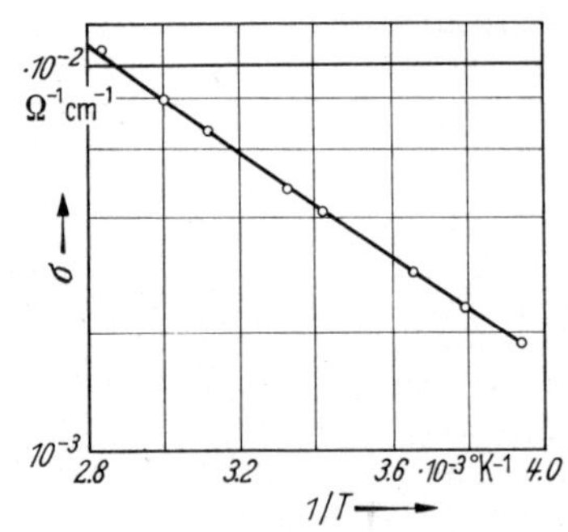

Fig. 4. $ZnFe_2O_4$. Temperature dependence of the dc conductivity, σ [*56E*]. See also [*48BSA*], [*53BR*].

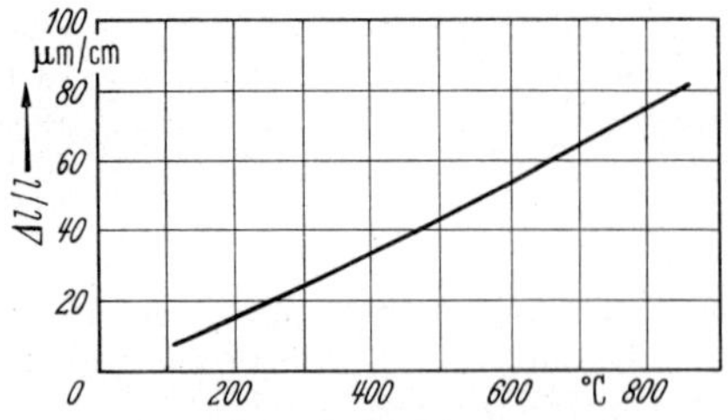

Fig. 3. $ZnFe_2O_4$. Thermal expansion vs. temperature [*32PBB*]; see also [*50W*].

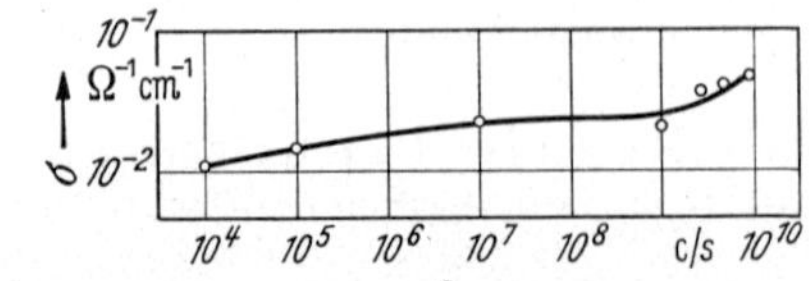

Fig. 5. $ZnFe_2O_4$. Dielectric conductivity, σ, vs. frequency [*57MWH*].

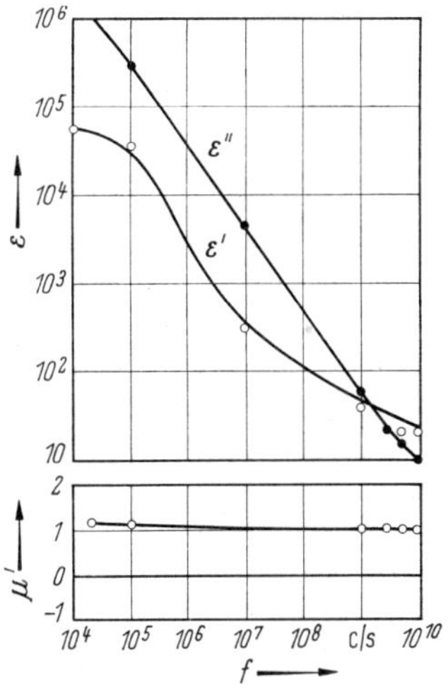

← Fig. 6. $ZnFe_2O_4$. Dielectric constant, ε, and permeability, μ', vs. frequency [*57MWH*].

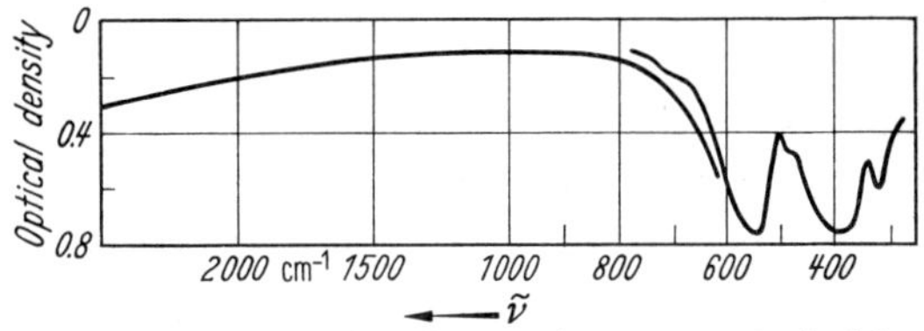

Fig. 7. $ZnFe_2O_4$. Infrared absorption spectra. Optical density measured relative to standard KBr disc. Multiple lines indicate alternate runs or spectrophotometer prism changes. Sample: mixture of 2.5 mg $ZnFe_2O_3$ and 0.8 g of powdered KBr, pressed at 30 t/inch². The threshold energy of the electronic absorption band (obtained from the inflection point of optical density vs. energy) is (2300 ∓ 200) cm⁻¹ [*55W*].

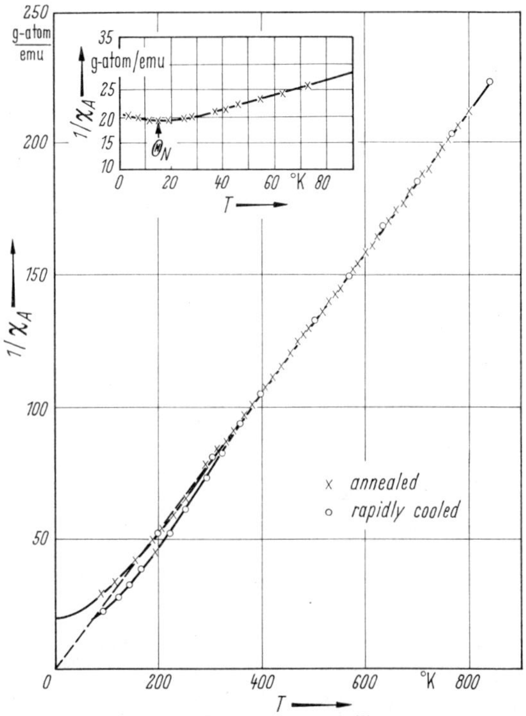

Fig. 8. $ZnFe_2O_4$. Reciprocal susceptibility per gram atom Fe vs. temperature (measured in a field of 9.7 kOe) for two samples. Inset: χ_A^{-1} vs. T curve measured in a field of 27 kOe. $\Theta_N = 15\,°K$ [*66L*].

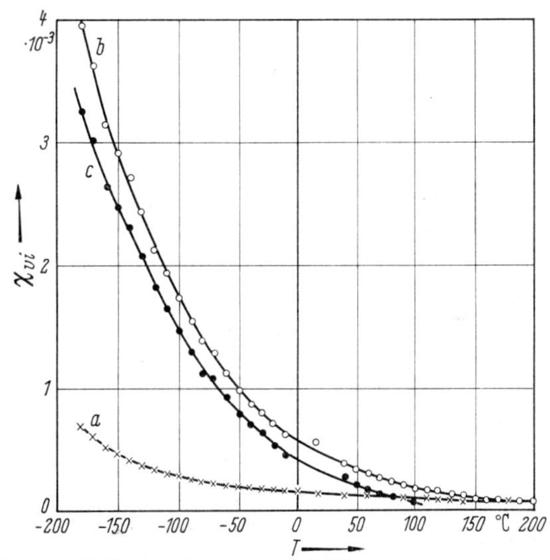

Fig. 9. $ZnFe_2O_4$. Initial susceptibility, χ_{vi}, vs. temperature [*64G*].
(a) at atmospheric pressure; (b) under $3\cdot10^4$ kg/cm² pressure; (c) the difference between curves (b) and (a).

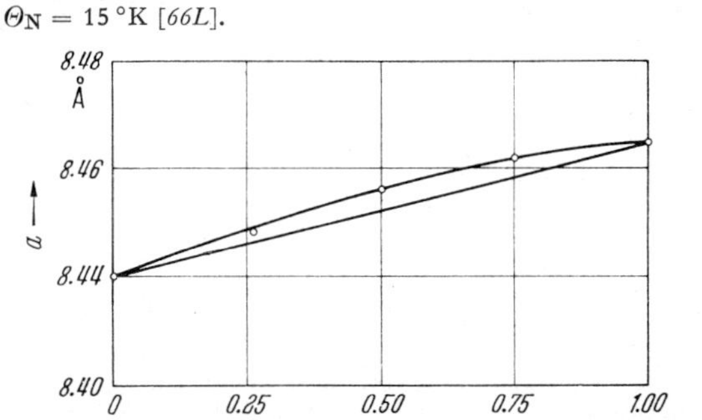

Fig. 10. $ZnFe_2O_4$-Zn(ZnTi)O_4 system. Lattice constant, a, vs. composition [*53R*].

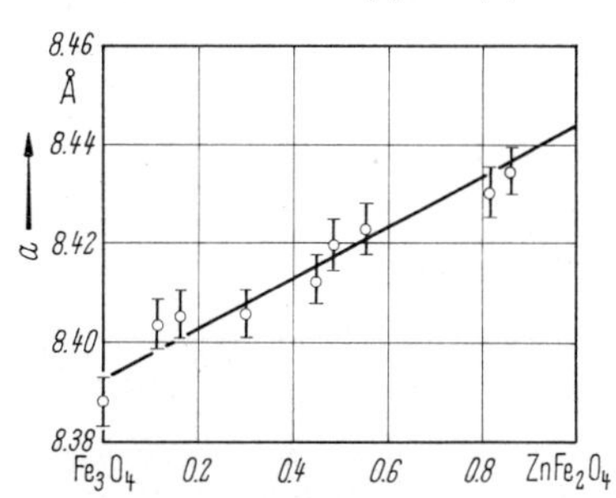

Fig. 11. $ZnFe_2O_4$-Fe_3O_4 system. Lattice constants vs. composition [*61M*].

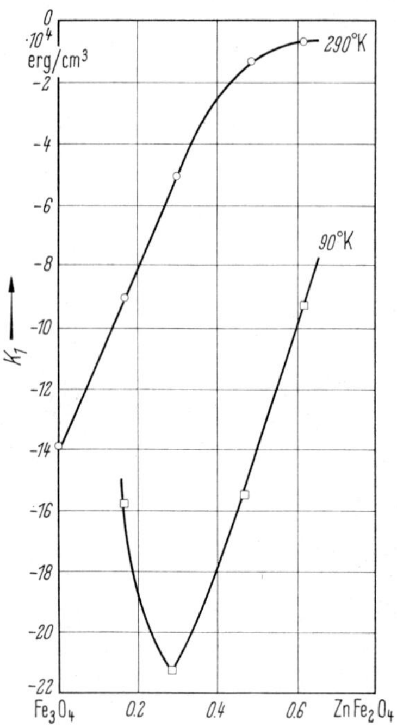

Fig. 12. $ZnFe_2O_4$-Fe_3O_4 system. Compositional dependence of K_1 [*61M*].

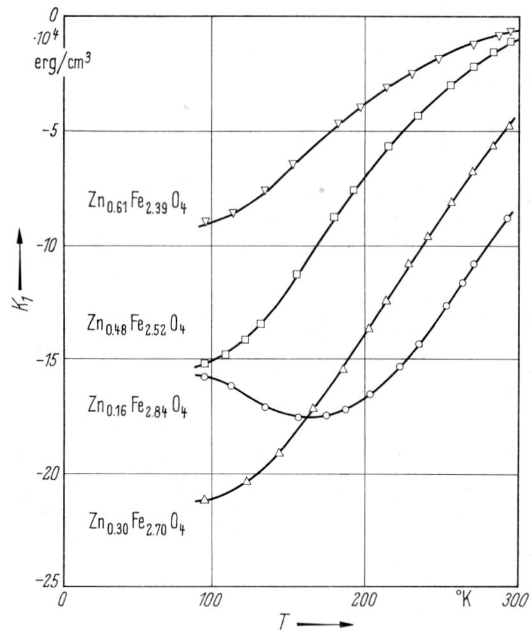

Fig. 13. $ZnFe_2O_4$-Fe_3O_4 system. Temperature dependence of K_1 [*61M*].

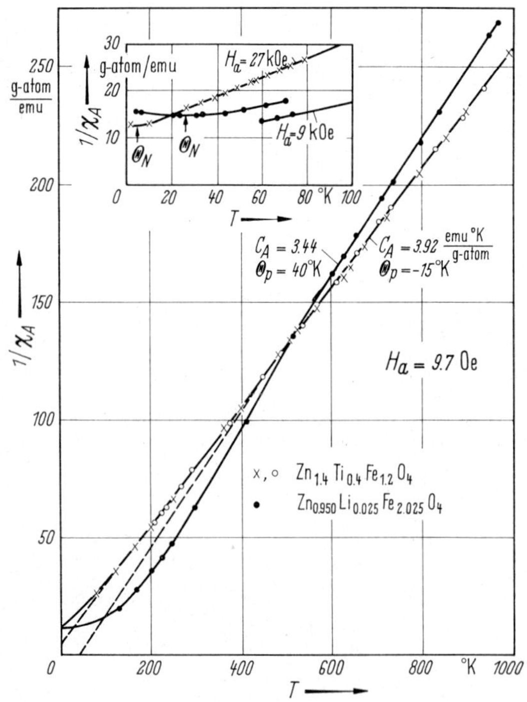

Fig. 14. Reciprocal susceptibility per gram atom Fe vs. temperature (measured in field of 9.7 kOe) for two samples of $Zn_{1.4}Ti_{0.4}Fe_{1.2}O_4$ (Θ_N = 4 °K) and one sample of $Zn_{0.950}Li_{0.025}Fe_{2.025}O_4$ (Θ_N = 26 °K). Inset: χ_A^{-1} vs. T curves measured in fields indicated [*66L*].

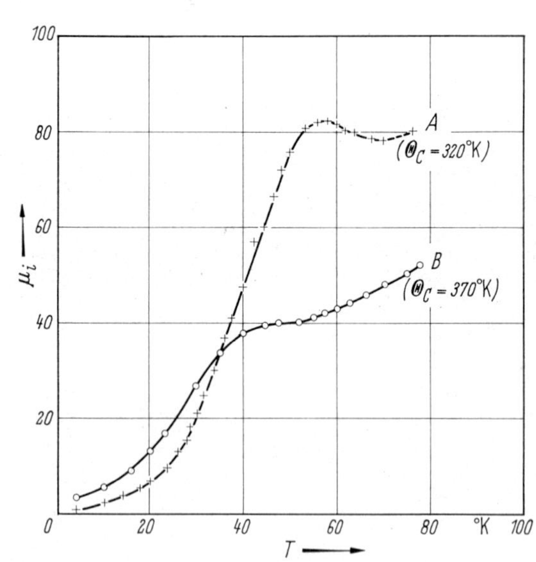

Fig. 15. $Zn_{0.70}Li_{0.15}Fe_{2.15}O_4$. Initial permeability, μ_i, vs. temperature; sample A prepared from Li_2CO_3 and sample B prepared from $LiFeO_2$ [*66B*].

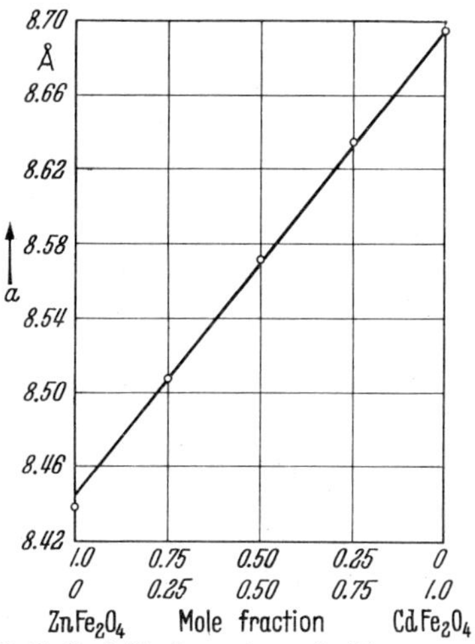

Fig. 16. $ZnFe_2O_4$-$CdFe_2O_4$ system. Lattice constant, *a*, vs. composition [*53R*].

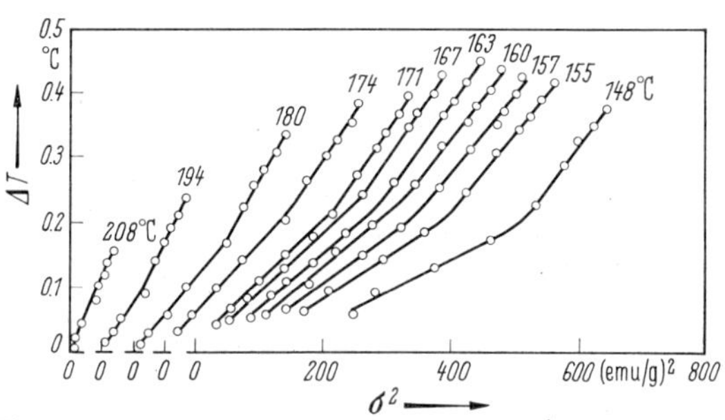

Fig. 17. $Zn_{0.5}Mg_{0.5}Fe_2O_4$. Magnetocaloric (ΔT vs. σ^2) curves [*62BC(1)*].

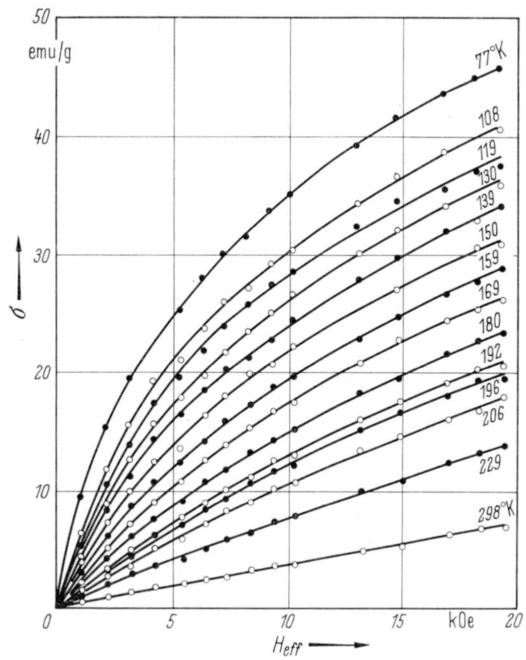

Fig. 19. $(CoO)_{0.18}(ZnO)_{0.75}Fe_2O_3$. Magnetic isothermals from 77 °K to room temperature. Full and open circles are used to distinguish the alternate curves [*62BC(2)*].

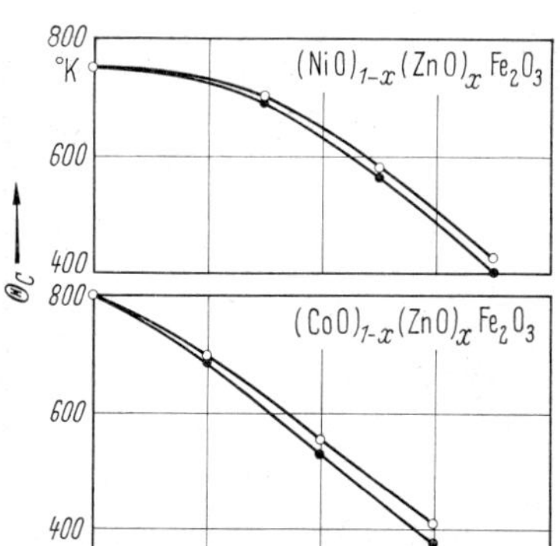

Fig. 18. Ni-Zn and Co-Zn ferrites. The apparent high field (open circles) and low field (full circles) Curie points for various compositions [*62BC(1)*].

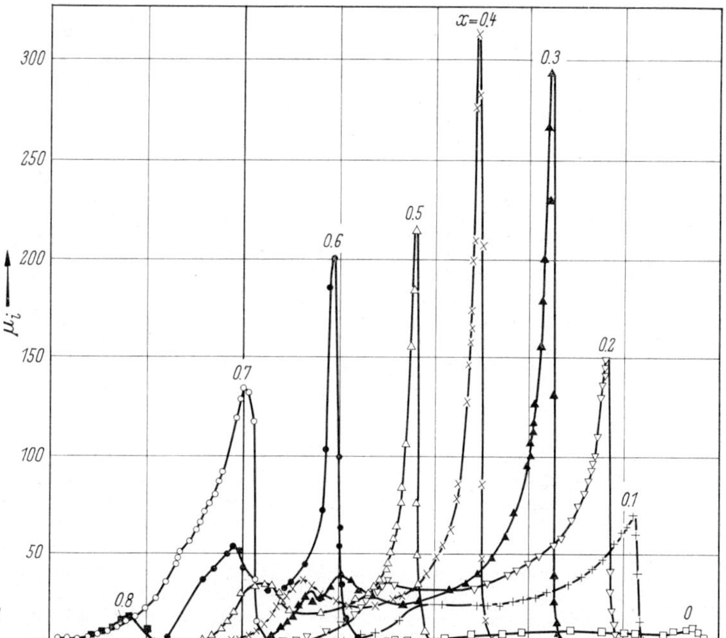

Fig. 20. $Co_{1-x}Zn_xFe_2O_4$. Temperature dependence of the initial permeability, μ_i. [*65PP*].

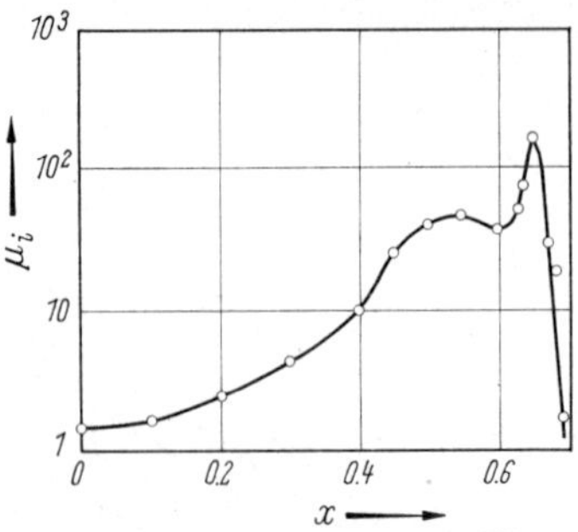

Fig. 21. $Co_{1-x}Zn_xFe_2O_4$. Initial permeability, μ_i, vs. Zn content, x [67PPY].

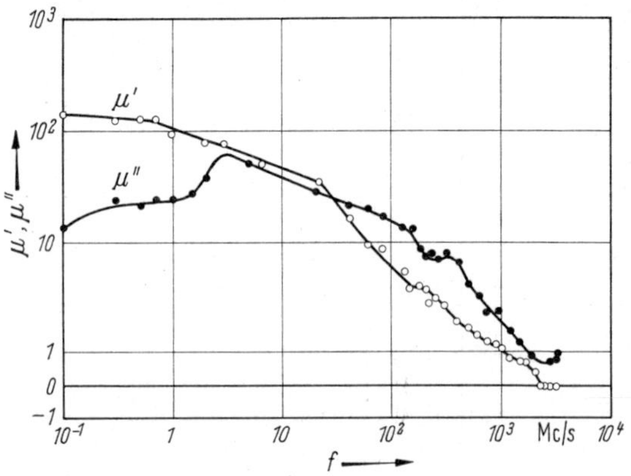

Fig. 22. $Zn_{0.65}Co_{0.35}Fe_2O_4$. Frequency dependence of the permeability at 20 °C [67PPY].

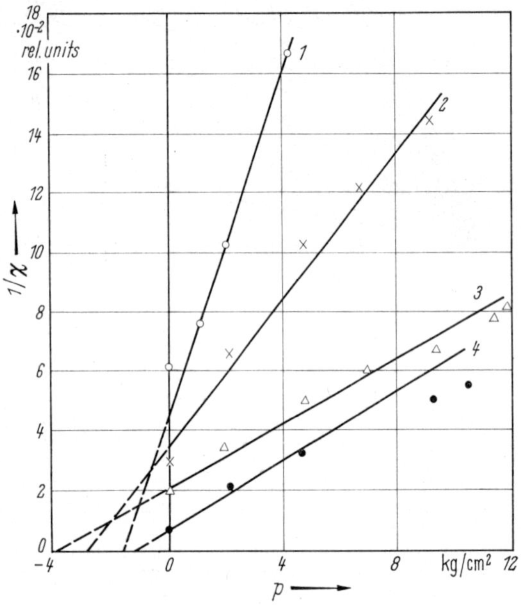

← Fig. 23. Zn-Cu ferrites. Dependence of the inverse magnetic susceptibility on hydrostatic pressure.
1) $Cu_{0.7}Zn_{0.3}Fe_2O_4$; 2) $Cu_{0.6}Zn_{0.4}Fe_2O_4$; 3) $Cu_{0.5}Zn_{0.5}Fe_2O_4$; 4) $Cu_{0.4}Zn_{0.6}Fe_2O_4$ [64SK].

6.1.5.2 Zn-Ni, Zn-Ni-Ge, and Zn-Ni-Sb ferrites

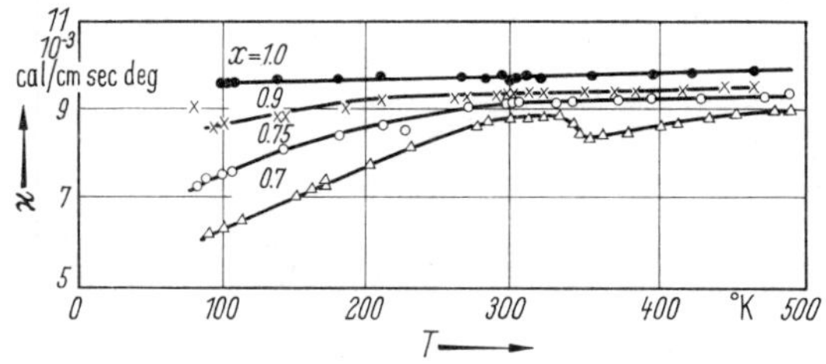

Fig. 1. $Zn_xNi_{1-x}Fe_2O_4$. Dependence of thermal conductivity on temperature [*63K*]. See also [*63FSS*].

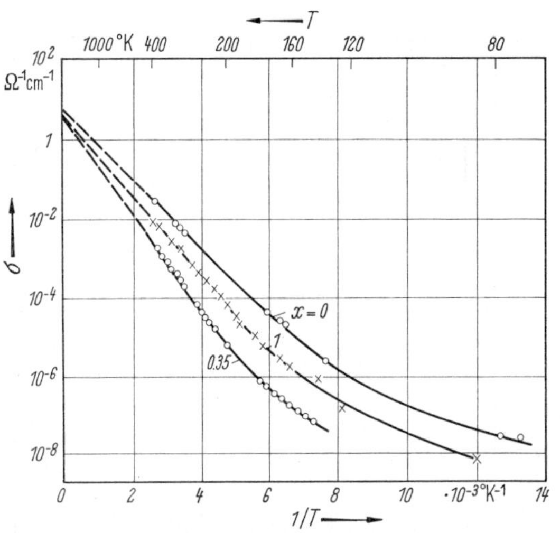

Fig. 3. $Zn_{1-x}Ni_xFe_2O_4$. Temperature dependence of the electrical conductivity [*63RSF*].

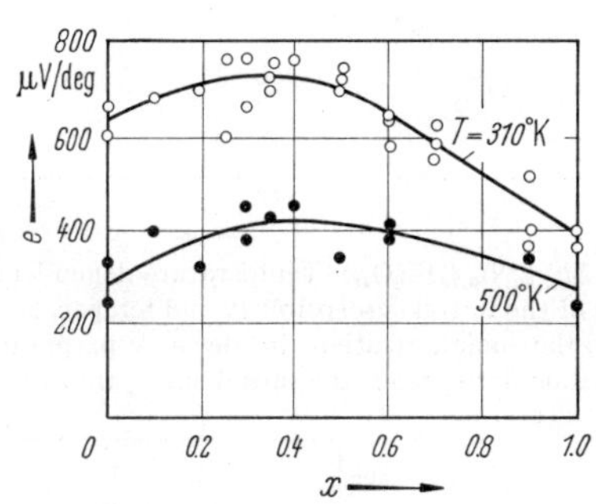

Fig. 5. $Zn_{1-x}Ni_xFe_2O_4$. Compositional dependence of the thermal emf coefficient, *e* [*63RSF*].

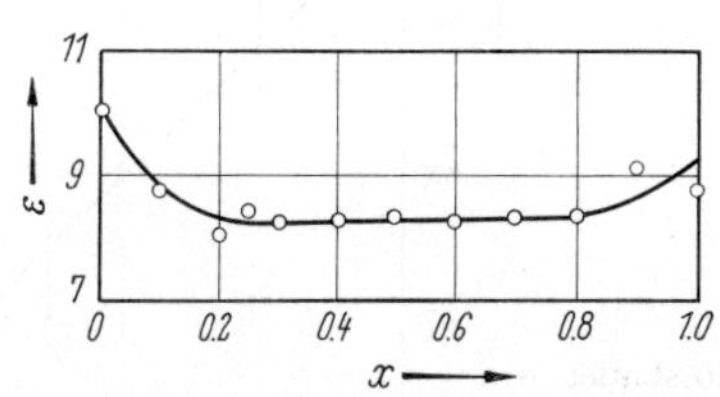

Fig. 7. $Zn_{1-x}Ni_xFe_2O_4$. Compositional dependence of the dielectric constant (measured at 9.5 Gc/s) at 20 °C [*63RSF*].

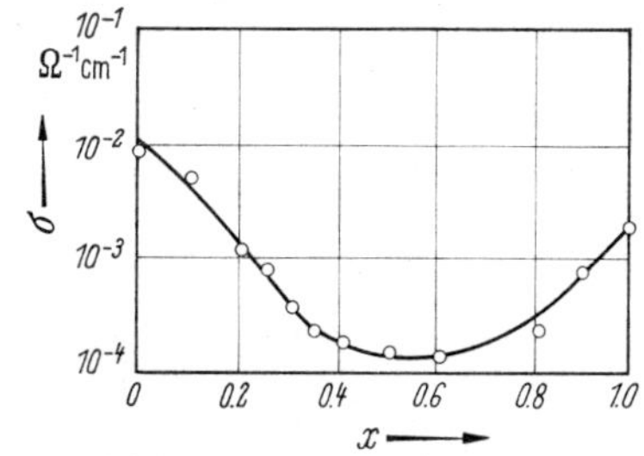

Fig. 2. $Zn_{1-x}Ni_xFe_2O_4$. Compositional dependence of the electrical conductivity at 20 °C [*63RSF*].

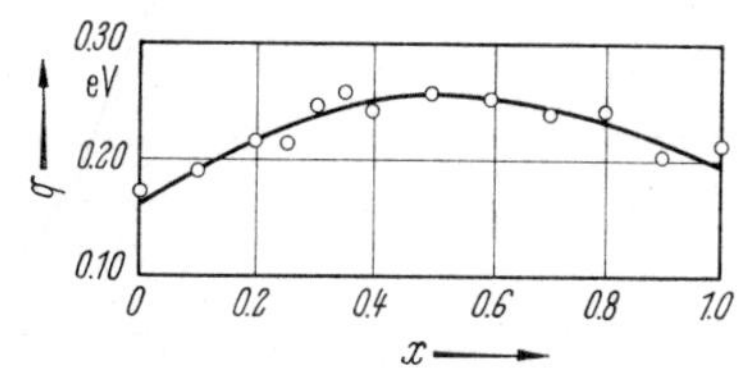

Fig. 4. $Zn_{1-x}Ni_xFe_2O_4$. Compositional dependence of the activation energy, *q*. Results obtained from the temperature dependence of the electrical conductivity in the temperature region 20 ··· 100 °C [*63TSF*].

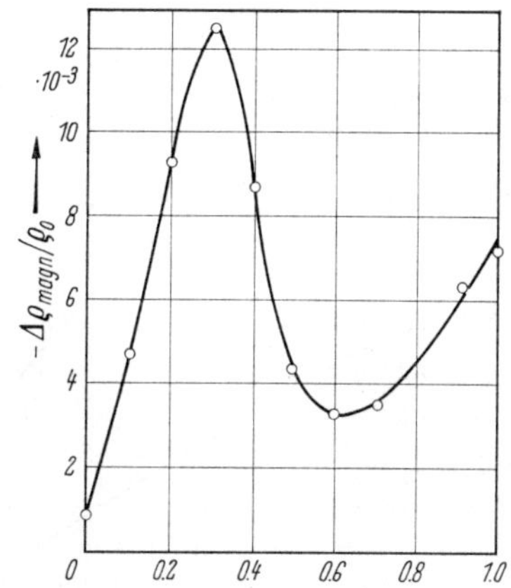

Fig. 6. $Zn_{1-x}Ni_xFe_2O_4$. Compositional dependence of the magnetoresistance at 20 °C [*63RSF*].

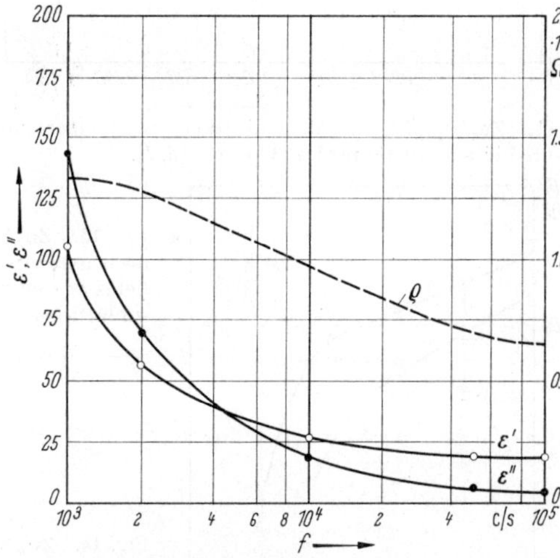

Fig. 8. $Zn_{0.64}Ni_{0.36}Fe_2O_4$. Frequency dependence of the complex dielectric constant $\varepsilon' + i\varepsilon''$ and the resistivity, ϱ, at room temperature [*64V*].

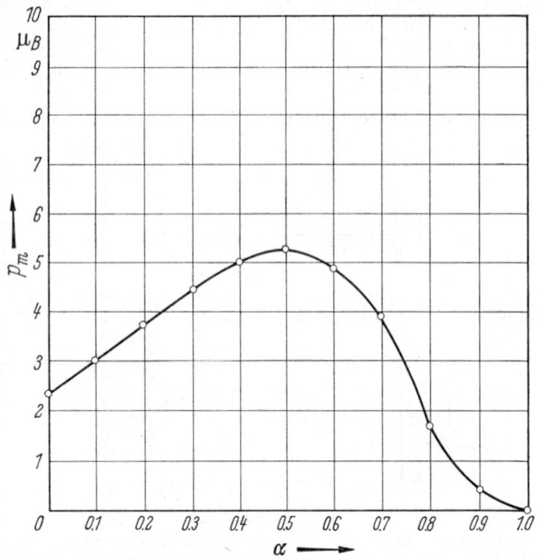

Fig. 9. $Zn_{\alpha}Ni_{1-\alpha}Fe_2O_4$. Compositional dependence of the magnetic moment [*63SV*]. See also [*63RSF*].

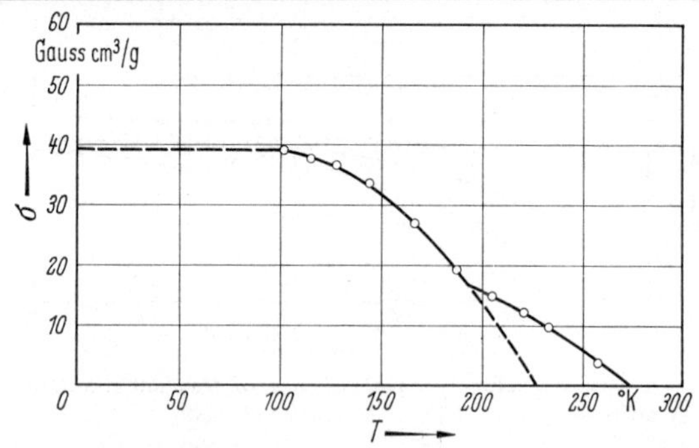

Fig. 10. $Zn_{0.8}Ni_{0.2}Fe_2O_4$. Temperature dependence of the magnetic moment per gram, σ [*63SV*].

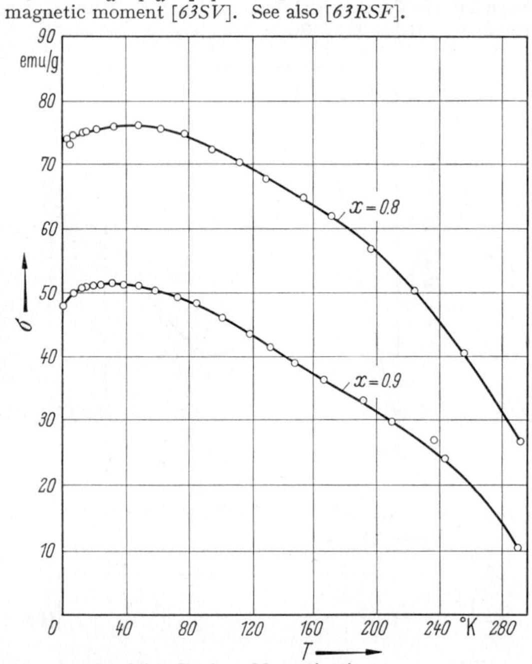

Fig. 11. $Zn_xNi_{1-x}Fe_2O_4$. Magnetization vs. temperature measured in a magnetic field of 30 kOe. [*62I*].

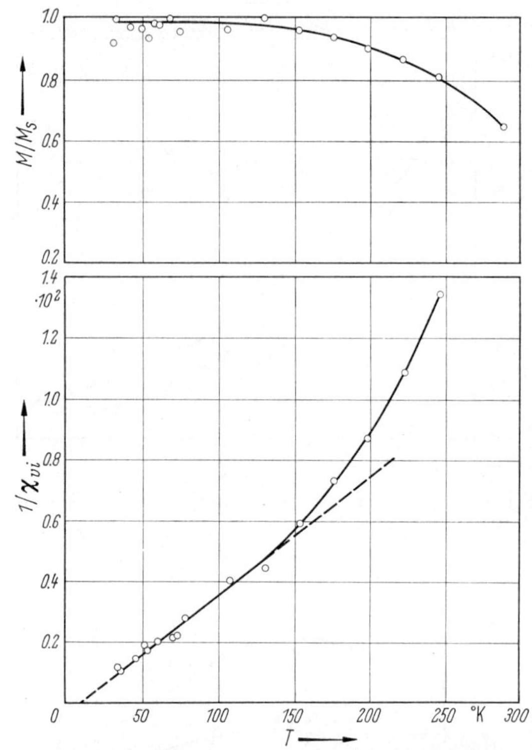

Fig. 12. $Zn_{0.90}Ni_{0.10}Fe_2O_4$. Temperature dependence of the reciprocal of the initial susceptibility and temperature dependence of the magnetization inside a superparamagnetic cluster as calculated from the initial susceptibility [*62I*].

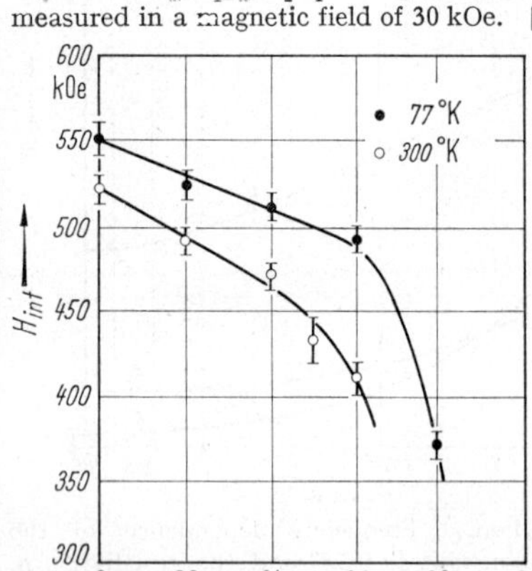

← Fig. 13. $Zn_xNi_{1-x}Fe_2O_4$. Mössbauer measurements of the internal magnetic fields at Fe^{57} nuclei in the octahedral sublattice vs. Zn content [*66GBDT*].

Fig. 14. $Zn_xNi_{1-x}Fe_2O_4$. Mössbauer → measurements of the internal magnetic fields at Fe^{57} nuclei in the tetrahedral sublattice vs. Zn content [*66GBDT*].

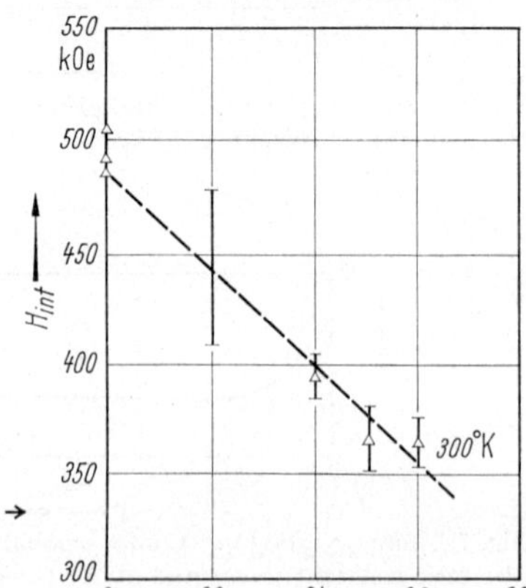

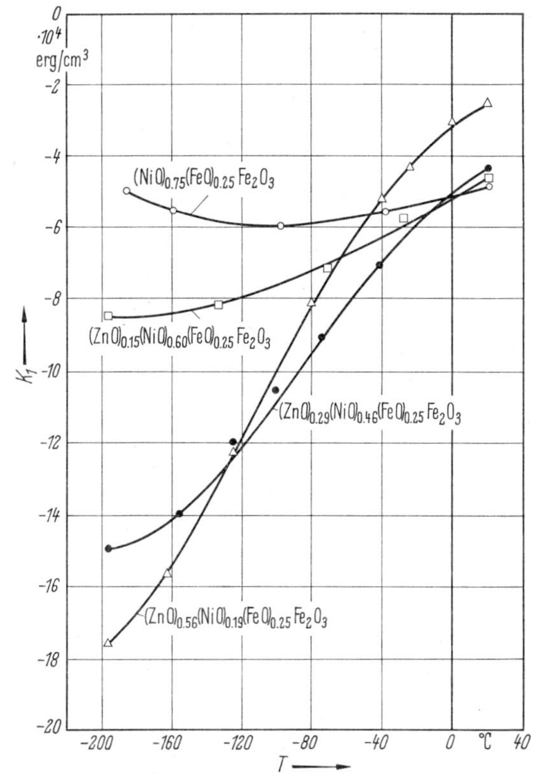

Fig. 15. Zn-Ni-Fe ferrites. Temperature dependence of K_1 [*64M*]. See also [*67BS*] and [*67M*].

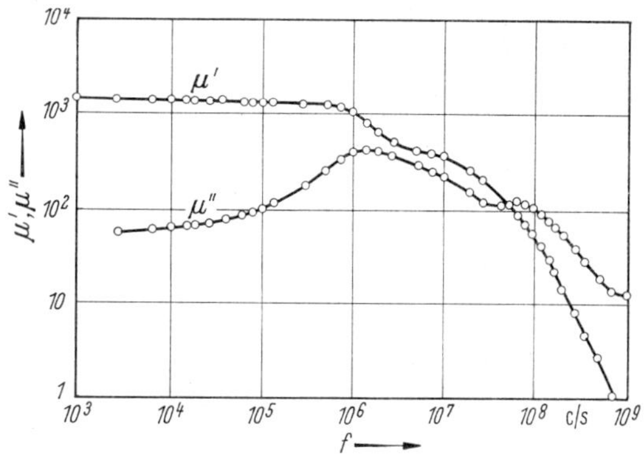

Fig. 16. $Zn_{0.64}Ni_{0.36}Fe_2O_4$. Frequency dependence of the complex initial permeability $\mu_i = \mu' + i\mu''$ at room temperature [*64V*].

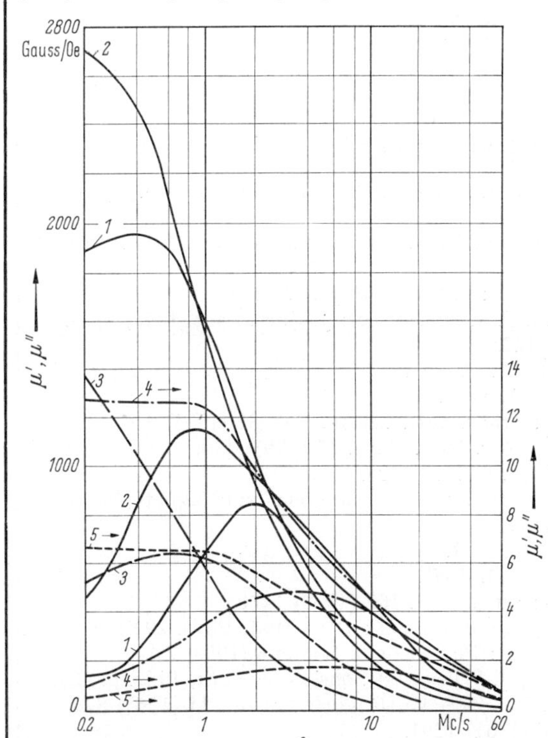

Fig. 17. $Zn_{0.68}Ni_{0.34}Fe_{1.96}O_{3.96}$. Magnetic spectra ($\mu'$, μ'' vs. f) at various temperatures. 1: 20 °C; 2: 81 °C; 3: 113.5 °C; 4: 127.5 °C; 5: 130 °C. Scale for curves 4 and 5 is given on the right [*61F*].

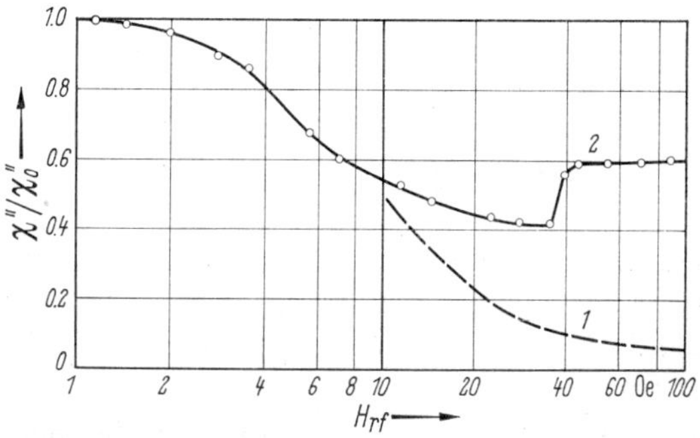

Fig. 18. $Zn_{0.67}Ni_{0.33}Fe_2O_4$. Microwave resonance loss at high power in conventional (1) and fine-grained (2) samples [*62MBS*].

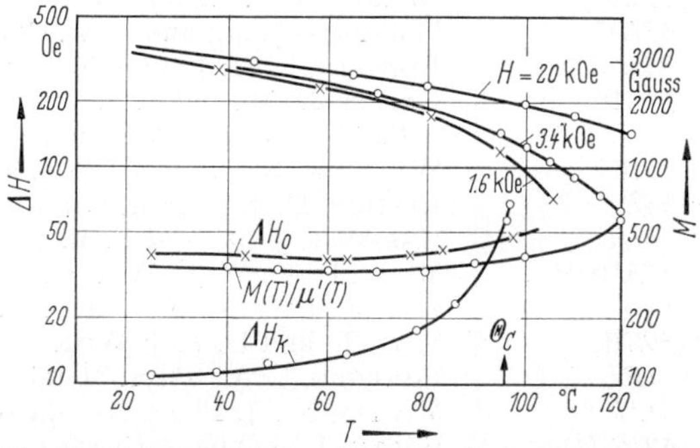

Fig. 19. $Zn_{0.68}Ni_{0.33}Fe_2O_4$. Magnetization in different fields, resonance line width, ΔH_0, and spin-wave line width, ΔH_k, vs. temperature in polycrystalline samples [*65B*].

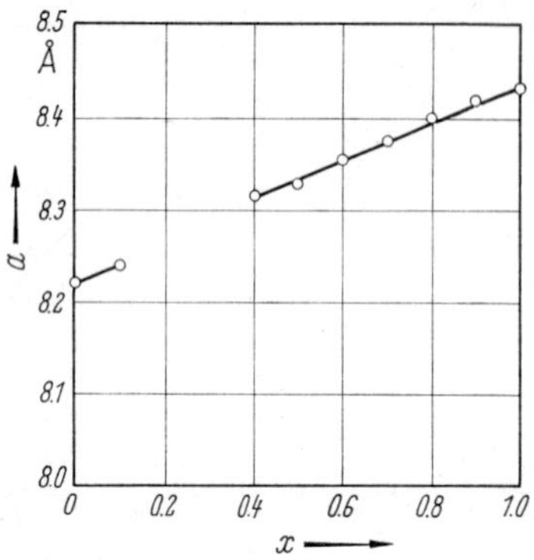

Fig. 20. $Zn_xGe_{1-x}Ni_{2-2x}Fe_{2x}O_4$. Lattice parameter vs. composition [*65R*].

Tab. 1. Magnetic moments of materials in the composition series $Zn_xGe_{1-x}Ni_{2-2x}Fe_{2x}O_4$ [*65R*]

	p_m [μ_B]		
x	29 °K	77 °K	298 °K
0.1	0.39		
0.4	2.28	2.07	0.63
0.5	2.66	2.44	1.00
0.6	2.96	2.78	1.01
0.7	2.97	2.89	1.22
0.8	1.86	1.76	0.47

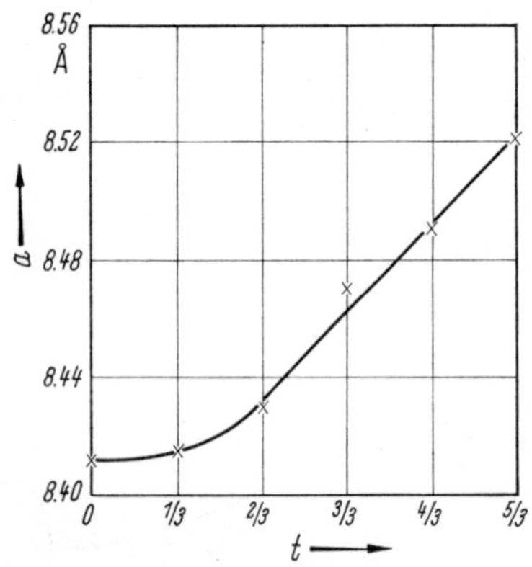

Fig. 21. $Zn_tNi_{5/3-t}FeSb_{1/3}O_4$. Cell edges vs. composition [*64B*].

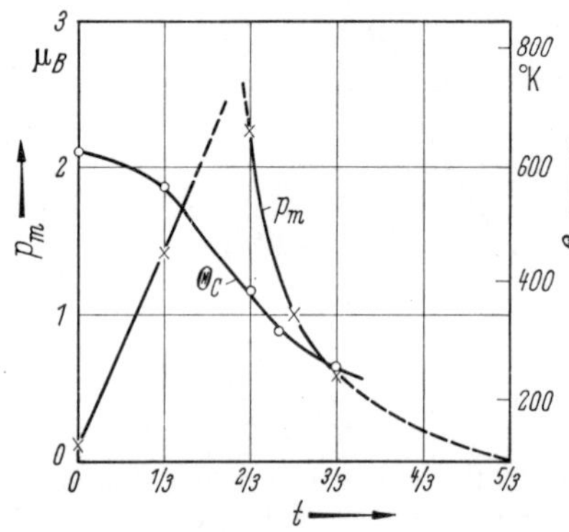

Fig. 22. $Zn_tNi_{5/3-t}FeSb_{1/3}O_4$. Saturation moments, p_m, and Curie temperatures, Θ_C, vs. composition [*64B*].

6.1.5.3 References for 6.1.5.1 and 6.1.5.2

32PBB — PARMELEE, C. W., A. E. BADGER, and G. A. BALLAM: Univ. Illinois Bull. Eng. Exp. Sta. Bull. No., **248** (1932) 7.

36AVB — ARKEL, A. E. VAN, E. J. W. VERWEY, and M. G. VAN BRUGGEN: Rec. Trav. Chim. **55** (1936) 331.

47VH — VERWEY, E. J. W., and E. L. HEILMANN: J. Chem. Phys. **15** (1947) 174.

48BSA — BEVAN, D. J. M., J. P. SHELTON, and J. S. ANDERSON: J. Chem. Soc. **1948**, 1729.

50W — WEIL, L.: Compt. Rend. **231** (1950) 122.

51B(1) — BERTAUT, E. F.: J. Phys. Radium **12** (1951) 252.

51B(2) — BOCHIROL, L.: Compt. Rend. **232** (1951) 1474.

53BR — BRADBURN, T. E., and G. R. RIGBY: Trans. Brit. Ceram. Soc. **52** (1953) 417.

53HC — HASTINGS, J. M., and L. M. CORLISS: Rev. Mod. Phys. **25** (1953) 114.

53R — ROMEIJN, F. C.: Philips Research Rept. **8** (1953) 321.

55GW — GRIMES, D. M., and E. F. WESTRUM: Conference on Magnetism and Magnetic Materials, Pittsburgh, Pa., October 1955, p. 64.

55W — WALDRON, R. D.: Phys. Rev. **99** (1955) 1727.

56E — EPSTEIN, D. J.: Thesis, published in [*57MWH*].

56HC — HASTINGS, J. M., and L. M. CORLISS: Phys. Rev. **102** (1956) 1460.

57MWH — MILES, P. A., W. B. WESTPHAL, and A. VON HIPPEL: Rev. Mod. Phys. **29** (1957) 279.

58WG — WESTRUM, Jr., E. F., and D. M. GRIMES: Phys. Chem. Solids **6** (1958) 280.

59SW — SMIT, J., and H. P. J. WIJN: Ferrites, J. Wiley and Sons, New York, 1959, p. 225.

61F — FOMENKO, L. A.: Phys. Metals Metallog. (USSR) (English Transl.) **11** (1961) 825.

61M — MIYATA, N.: J. Phys. Soc. Japan **16** (1961) 1291.

62BC(1) — BOOTH, J. G., and J. CRANGLE: Proc. Phys. Soc. (London) **79** (1962) 1271.

62BC(2) — BOOTH, J. G., and J. CRANGLE: Proc. Phys. Soc. (London) **79** (1962) 1278.

62I — ISHIKAWA, I.: J. Phys. Soc. Japan **17** (1962) 1877.

62MBS — MALINOFSKY, W. W., R. W. BABBIT, and G. C. SANDS: J. Appl. Phys. **33** (1962) 1206.

63FSS	FOMENKO, L. A., V. A. SHCHELKOTUNOV, and V. L. SOCHIVKO: Soviet Phys. — Solid State (English Transl.) **5** (1963) 643.
63K	KAMILOV, I. K.: Soviet Phys. — Solid State (English Transl.) **4** (1963) 1693.
63RSF	RUSTAMOV, A. G., SAMOKHVALOV, and J. G. FAKIDOV: Soviet Phys. — Solid State (English Transl.) **5** (1963) 751.
63SV	SABOTTA, E. A., and J. VOITLÄNDER: Z. Phys. Chem. **39** (1963) 54.
64B	BLASSE, G.: Philips Res. Rept. Supp. No. 3 (1964).
64G	GOTO, Y.: Japan. J. Appl. Phys. **3** (1964) 309.
64M	MICHALOWSKY, L.: Phys. Status Solidi **6** (1964) K 143.
64SK	SIROTA, N. N., and YU. M. KHACHATRYAN: Soviet Phys. — Solid State (English Transl.) **5** (1964) 2278.
64V	VERWEEL, J.: Philips Res. Rept. **19** (1964) 29.
65B	BELJERS, H. G.: Phys. Letters **18** (1965) 248.
65PP	POLTINNIKOV, S. A., and G. M. PRIKHODKINA: Soviet Phys. — Solid State (English Transl.) **7** (1965) 241.
65R	ROBBINS, M. J.: J. Phys. Chem. Solids **26** (1965) 831.
66B	BROESE VAN GROENOU, A. B.: Philips Res. Rept. **21** (1966) 180.
66GBDT	GOL'DANSKII, V. I., V. F. BELOV, M. N. DEVISHEVA, and V. A. TRUKHTANOV: Soviet Phys. JETP (English Transl.) **22** (1966) 1149.
66L	LOTGERING, F. K.: J. Phys. Chem. Solids **27** (1966) 139.
66O	OKAZAKI, H.: Japan. J. Appl. Phys. **5** (1966) 559.
67BS	BROESE VAN GROENOU, A. B., and J. A. SCHULKES: J. Appl. Phys. **38** (1967) 1133.
67M	MAKRAM, H.: Czech. J. Phys. **17** (1967) 387.
67PPY	POLTINNIKOV, S. A., N. N. PARFENOVA, and T. D. YURCHENKO: Soviet Phys. — Solid State (English Transl.) **9** (1967) 1230.

6.1.6 Li-Fe^{3+} spinels (Li ferrites) and Li-Fe^{3+} spinels with substitutions

6.1.6.1 Li ferrite

Tab. 1. General properties of Li ferrite (room temperature unless otherwise specified)
Chemical formula: $Li_{0.5}Fe_{2.5}O_4$

		Ref.
Crystal type	Inverse spinel — cubic	*52B*
Structure type	$MgAl_2O_4$	*38H*
Space group	O_h^7-Fd3m (quenched from above 755 °C) [1])	*52B*
	O^6-$P4_33$ (or $P4_13$) (annealed below 735 °C) [2])	
Lattice constant	$a = 8.33$ Å [3])	*52B*
Volume of unit cell	$578.0 \cdot 10^{-24}$ cm³ (containing 8 molecules)	
Molecular weight	207.1 amu	
X-ray density	$\varrho_X = 4.75$ g/cm³	
Sintering temperature	700 ··· 1150 °C	
Thermal conductivity	$\varkappa \simeq 1.5 \cdot 10^{-2}$ cal sec^{-1} cm^{-1} deg^{-1}	*54SW*
Specific heat	$c_p \simeq 0.2$ cal g^{-1} deg^{-1}	*54SW*
Elastic constants	$C_{11} = 30.2 \cdot 10^{11}$ dyne/cm²	*64CN*
	$C_{44} = 6.77 \cdot 10^{11}$ dyne/cm³	*64CN*
Longitudinal elastic wave velocity	$v_e = 7.98 \cdot 10^5$ cm/sec [4])	
Sheer elastic wave velocity	$v_s = 3.78 \cdot 10^5$ cm/sec [4]), [5])	

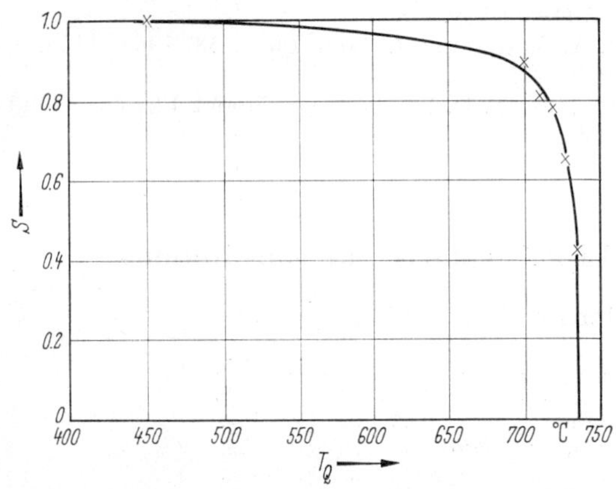

Fig. 1. Li ferrite. Long range order parameter, S, vs. quenching temperature [*64BB*].

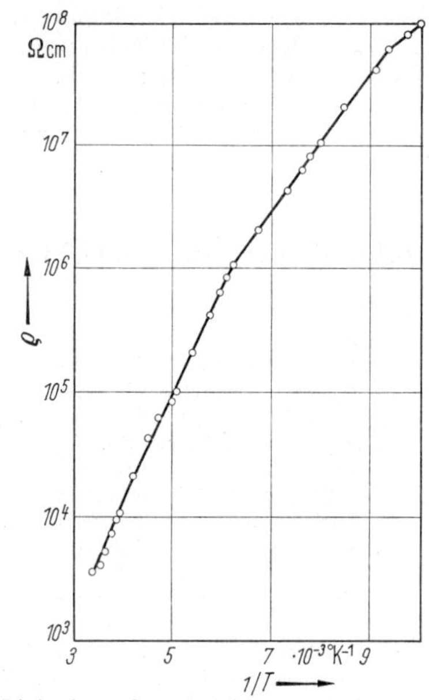

Fig. 3. Li ferrite. dc resistivity, ϱ, of single crystal vs. $1/T$ [*68WGK*].

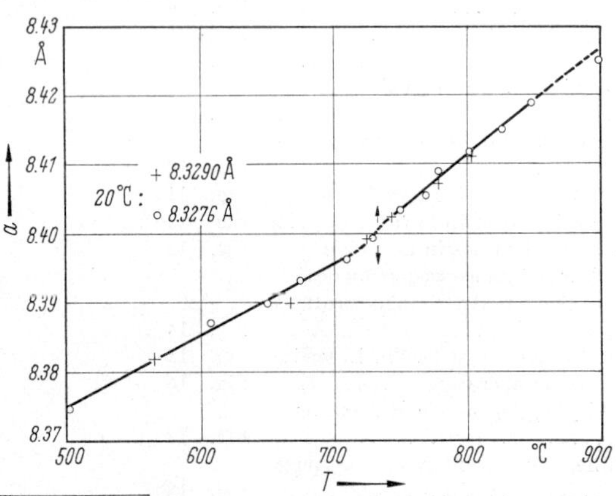

←
Fig. 2. Li ferrite. Temperature dependence of the lattice parameter, a [*64BB*].

---

1) The oxygen parameter, $u = 0.382 \pm 0.005$.

2) $P4_33$ 4 Li at 4 (b)
12 Fe at 12 (d) with $x = 3/8 + \Delta_1$
8 Fe at 8 (c) with $x = 0 + \Delta_2$
24 O at 24 (e) with $x = 1/8 + \Delta_3$, $y = -1/8 - \Delta_4$, $z = 1/8 + \Delta_5$
8 O at 8 (c) with $x = 3/8 + \Delta_6$
with Δ_3, Δ_4, Δ_5, Δ_6 approximately 0.007 and Δ_1, Δ_2 still smaller.

3) See also: [*38H*], [*43B*], [*50CC*], [*56DW*].

4) H. J. McSkimin referred to in [*64CN*].

5) R. C. LeCraw referred to in [*64CN*].

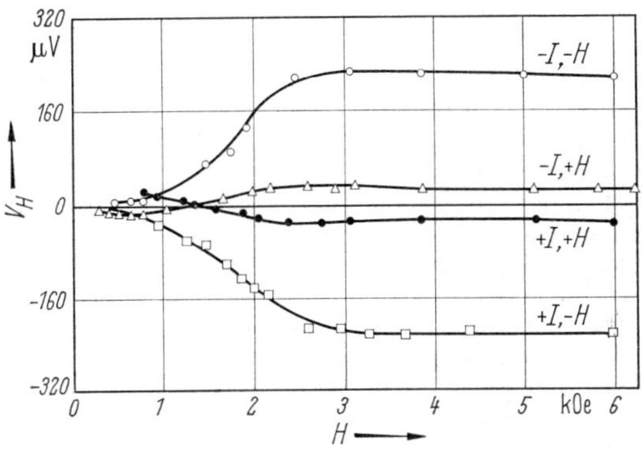

Fig. 4. Li ferrite. Voltage, V_H, from the Hall leads of a single crystal at 295.6 °K vs. magnetic fields for positive and negative electric and magnetic field conditions [*68WGK*].

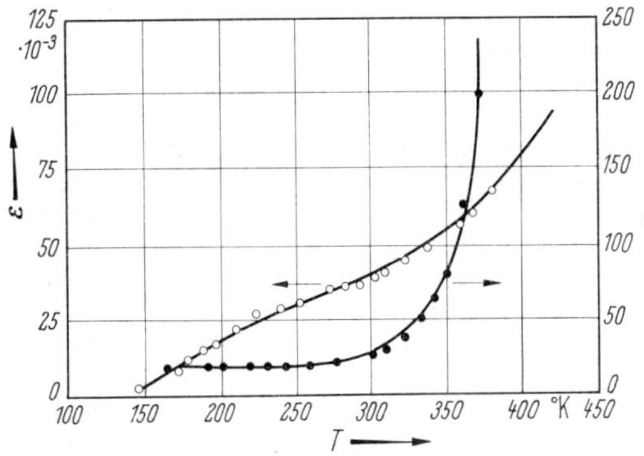

Fig. 5. Li ferrite. Dielectric constant of a polycrystalline sample at 10 kc/s vs. temperature. ○ No heat treatment. ● After annealing for 6 h at 800 °C and cooling at a rate of 0.5 °C/h [*68WGK*].

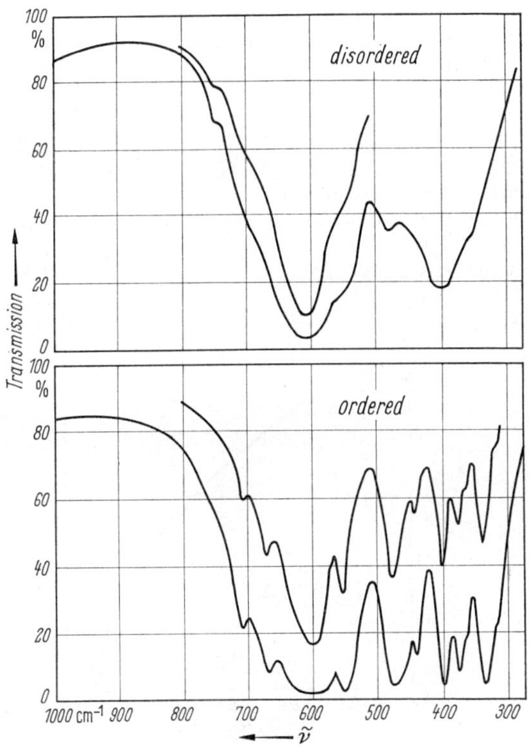

Fig. 6. $LiFe_5O_8$. Infrared spectra in long range disordered and ordered states [*62T*].

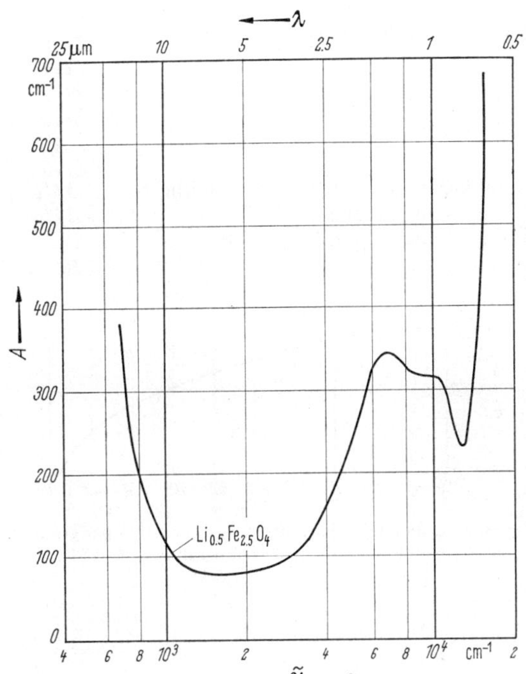

Fig. 7. Li ferrite. Absorption coefficient, A, vs. wave number for a single crystal [*59SRW*]. $I/I_0 = 10^{-At}$, t = thickness in cm.

Tab. 2. Magnetic properties of Li ferrite (room temperature unless otherwise specified)

		Ref.
Ferromagnetic Curie temperature	$\Theta_C = 670$ °C	*59SW(1)*
	$\Theta_C = (620 \pm 2)$ °C	*50CC*
Saturation magnetic moment per molecule (0 °K)	$p^0_{sm} = 2.60\ \mu_B$	*50G(1)*
Saturation magnetic moment per gram	$\sigma_s = 65$ Gauss cm³/g	*59SW(1)*
	$\sigma^0_s = 69$ Gauss cm³/g (0 °K)	*59SW(1)*
Saturation magnetic moment per unit volume	$M_s = 310$ Gauss	*59SW(1)*
	$M^0_s = 330$ Gauss (0 °K)	*59SW(1)*
Saturation induction	$4\pi M_s = 3900$ Gauss (polycrystal)	*59SW(1)*
	$4\pi M_s = 3640$ Gauss (single crystal)	*64CN*
	$4\pi M^0_s = 4200$ Gauss (0 °K)	*59SW(1)*
Saturation magnetostriction	$\lambda_{100} = -14 \cdot 10^{-6}$ (disordered)	*68PS*
	$\lambda_{100} = -42 \cdot 10^{-6}$ (disordered, 77 °K)	*68PS*
	$\lambda_{100} = -\ 8 \cdot 10^{-6}$ (ordered)	*68PS*
	$\lambda_{100} = -28 \cdot 10^{-6}$ (ordered, 77 °K)	*68PS*
	$\lambda_{111} = -\ 1 \cdot 10^{-6}$ (disordered)	*68PS*
	$\lambda_{111} = 4 \cdot 10^{-6}$ (disordered, 77 °K)	*68PS*
	$\lambda_{111} = 1 \cdot 10^{-6}$ (ordered)	*68PS*
	$\lambda_{111} = 8 \cdot 10^{-6}$ (ordered, 77 °K)	*68PS*
First order magnetocrystalline anisotropy constant	$K_1 = -\ 83 \cdot 10^3$ erg/cm³ (ordered) ¹)	*60F*
	$K_1 = -127 \cdot 10^3$ erg/cm³ (ordered, 77 °K) ¹)	*60F*
	$K_1 = -\ 90 \cdot 10^3$ erg/cm³ (disordered) ¹)	*60F*
	$K_1 = -162 \cdot 10^3$ erg/cm³ (disordered, 77 °K) ¹)	*60F*
Second order magnetocrystalline anisotropy constant	$K_2 = -\ 2 \cdot 10^3$ erg/cm³ (ordered) ²)	*60SFR*
	$K_2 = -\ 97 \cdot 10^3$ erg/cm³ (ordered, 77 °K) ²)	*60SFR*
	$K_2 = +0.6 \cdot 10^3$ erg/cm³ (disordered) ²)	*60SFR*
	$K_2 = -\ 32 \cdot 10^3$ erg/cm³ (disordered, 77 °K) ²)	*60SFR*
Experimentally observed g factor	$g_{exp} = 2.003 \pm 0.002$ (ordered) ²)	*60SFR*
	$g_{exp} = 1.958$ (ordered) ³)	*64CN*
	$g_{exp} = 2.012 \pm 0.002$ (ordered, 77 °K) ²)	*60SFR*
	$g_{exp} = 1.962$ (ordered, 4.2 °K) ³)	*64CN*
	$g_{exp} = 2.004 \pm 0.002$ (disordered) ²)	*60SFR*
	$g_{exp} = 2.006 \pm 0.002$ (disordered, 77 °K) ²)	*60SFR*
Spin-wave line width corresponding to long spin wavelengths	$\Delta H_{k\to 0} = 3.52$ Oe ³)⁴)	*64CN*
	$\Delta H_{k\to 0} = 0.060$ Oe (4.2 °K) ⁵)	*68SLN*
Magnetoelastic coupling coefficients	$B_1 = 68.5 \cdot 10^6$ erg/cm³	*64CN*
	$B_2 \ll B_1$	*64CN*

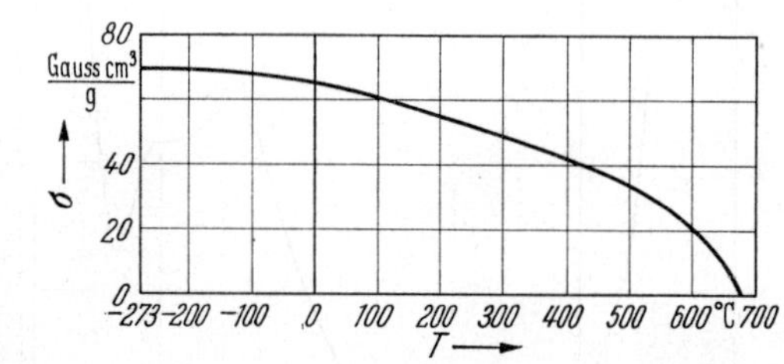

Fig. 8. $Li_{0.5}Fe_{2.5}O_4$. Saturation magnetization per gram, σ, for a polycrystalline sample vs. temperature [*59SW(1)*].

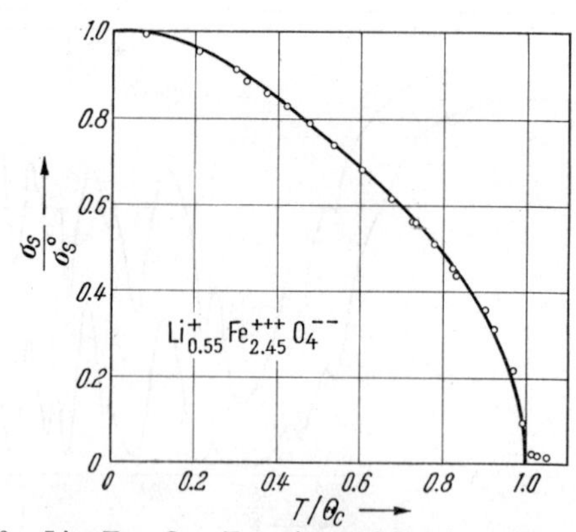

Fig. 9. $Li_{0.55}Fe_{2.45}O_4$. Experimental data (circles) and theoretical curve (full line) for σ_s/σ^0_s vs. T/Θ_C in a "disordered" monocrystal. The value for σ^0_s is 69.6 emu/g. The theoretical curve was based on the Néel [*48N*] theory from which the molecular field constants were determined as follows: $n = 2.63$ mole/cm³, $\alpha = -0.54$ mole/cm³, $\beta = -0.22$ mole/cm³, $\Theta_C = 904$ °K [*60RF*].

¹) Measured by static torque method. Ferromagnetic resonance values of K_1 [*60SFR*] were within 1 percent agreement with those determined statically.
²) Measured at 24 Gc/s.
³) Measured at 17.3 Gc/s with H along [111].
⁴) See also [*68SLN*], [*62DS*], [*64SD*].
⁵) Measured at 10.4 Gc/s with H along [111].

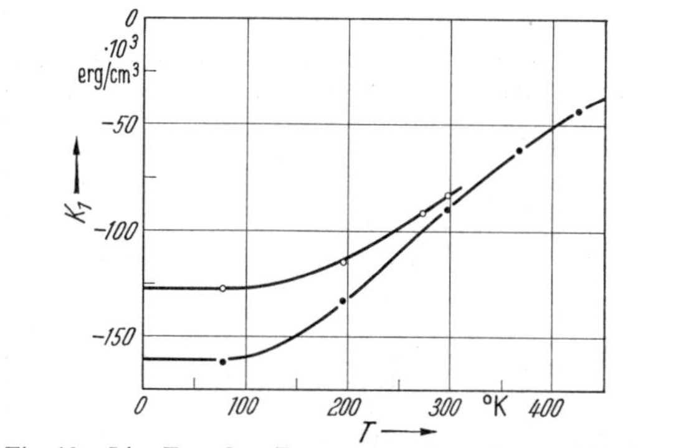

Fig. 10. $Li_{0.55}Fe_{2.45}O_4$. Temperature dependence of K_1 in a monocrystalline sample in "ordered" and "disordered" states [60F]. S is the long range order parameter. The theoretical curves —— are calculated from the Yosida-Tachiki theory [57YT]. ○ $S = 1$; ● $S \approx 0.1$ } experimental.

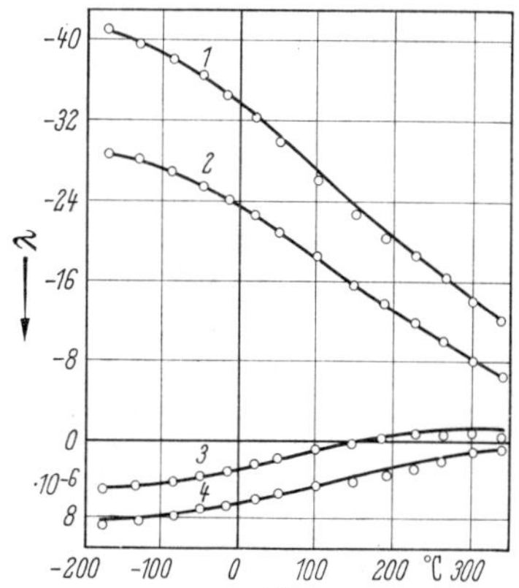

Fig. 11. Li ferrite. Temperature dependence of the magnetostriction constants of a single crystal (points: experimental results; continuous curve: theory developed in [63CCDCC]). Curves 1 and 3 describe λ_{100} and λ_{111}, respectively, for disordered state; curves 2 and 4 describe λ_{100} and λ_{111}, respectively, for ordered state [68PS].

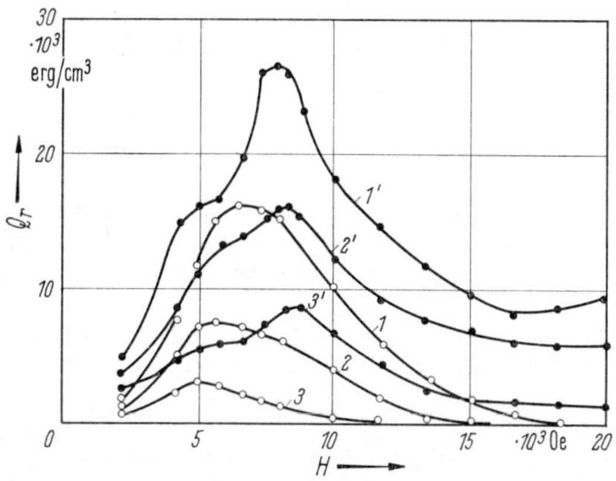

Fig. 12. Li ferrite. Magnetic field dependence of the rotational hysteresis loss, Q_r, in the (010) plane of a single crystal. Curves 1···3 are measured before heat treatment. Curves 1'···3' are measured after quenching in air from 800 °C. Measuring temperatures are: 196 °C for curves 1 and 1'; 18 °C for curves 2 and 2'; and 200 °C for curves 3 and 3'. $Q_r = \phi L$ where L is the torque and ϕ is the angle between the [100] axis and the field H [67DSGY].

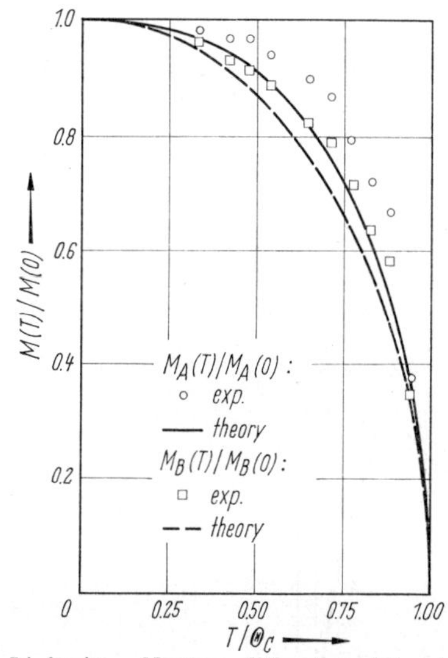

Fig. 13. Li ferrite. Neutron diffraction determination of sublattice magnetizations. The theoretical magnetizations were calculated using the value of molecular field coefficients [60RF] determined from magnetization measurements [64P].

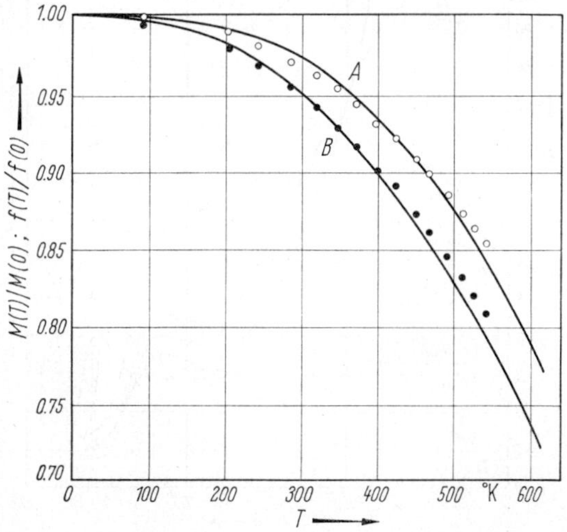

←

Fig. 14. Li ferrite. The experimental temperature dependence of the nuclear resonance frequency of Fe^{57} (see also Fig. 15.) and the theoretical (represented by the solid lines) temperature dependence (calculated from the molecular field coefficients in [60RF]) of the sublattice magnetizations. (A) and (B) correspond to the tetrahedral and octahedral sites, respectively. Open circles and closed circles refer to the experimental data for the tetrahedral and octahedral sites, respectively [62YHMH].

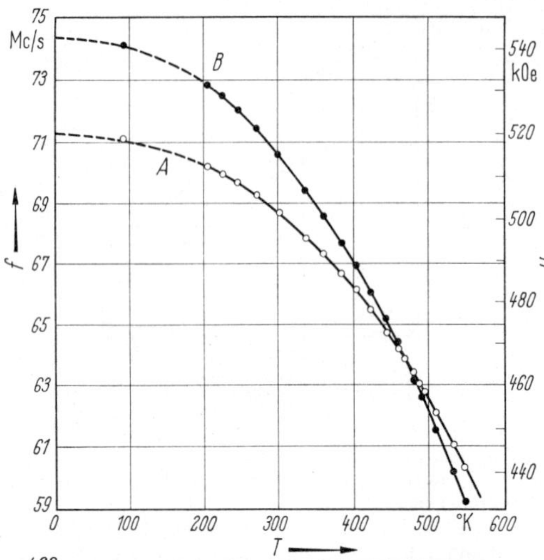

← Fig. 15. Li ferrite. The experimental temperature dependence of the nuclear resonance frequency of Fe^{57} and of the hyperfine field H_{int} (absolute value) at Fe^{57} nuclei. (A) Tetrahedral site, (B) octahedral site [*62YHMH*]. See also [*62KB*] and [*62OMK*].

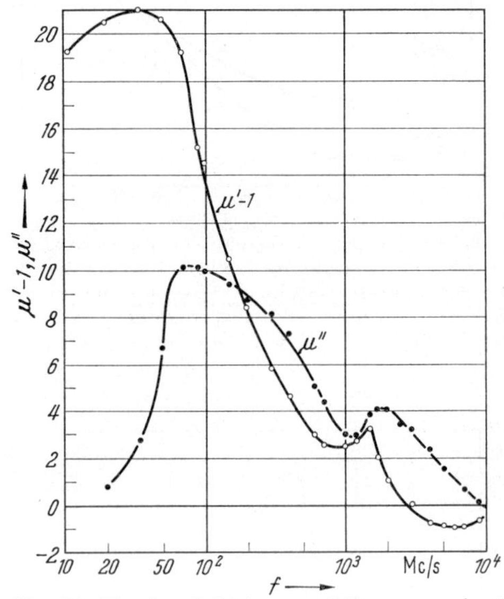

Fig. 17. $Li_{0.5}Fe_{2.5}O_4$. Initial permeability, $\mu_i = \mu' - i\mu''$, vs. frequency of a sintered sample at room temperature. M_S = 258 emu/cm³ and ϱ = 4.15 g/cm³ [*53R*].

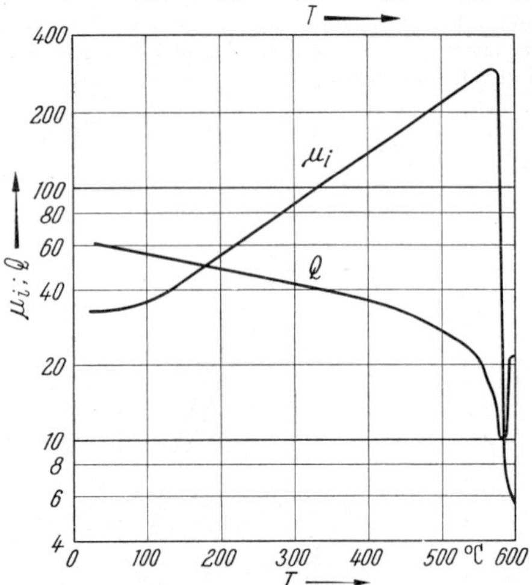

← Fig. 16. $Li_{0.5}Fe_{2.5}O_4$. Initial permeability, μ_i, and relative quality factor, Q, at 500 kc/s vs. temperature. Sample was fired at 1150 °C [*52W(1)*].

↓ Fig. 18a, b. $Li_{0.5}Fe_{2.5}O_4$. Frequency dependence of $\mu' - 1$ in a polycrystalline sample in which various internal transverse magnetic fields, H_{int}, were applied [*58V*].

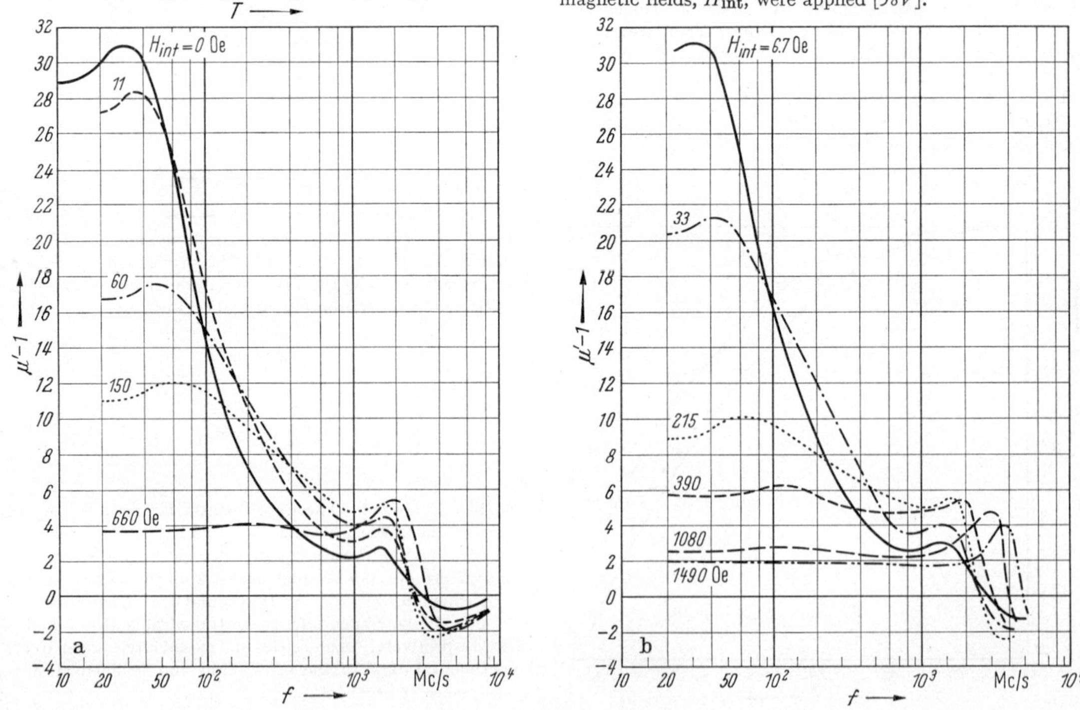

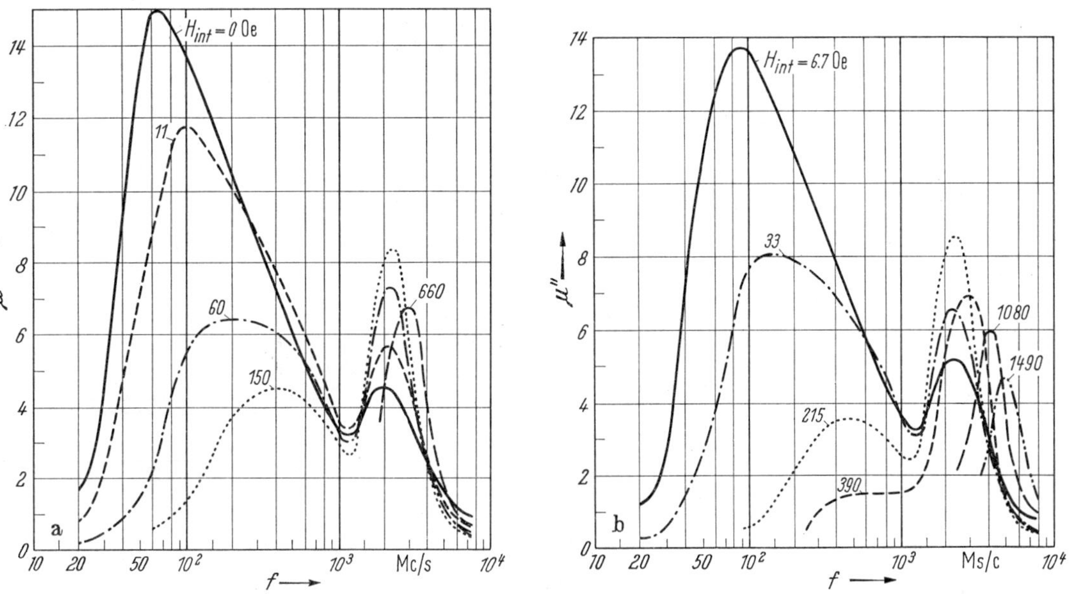

Fig. 19a, b. $Li_{0.5}Fe_{2.5}O_4$. Frequency dependence of μ'' in a polycrystalline sample in which various internal transverse magnetic fields, H_{int}, were applied [*58V*].

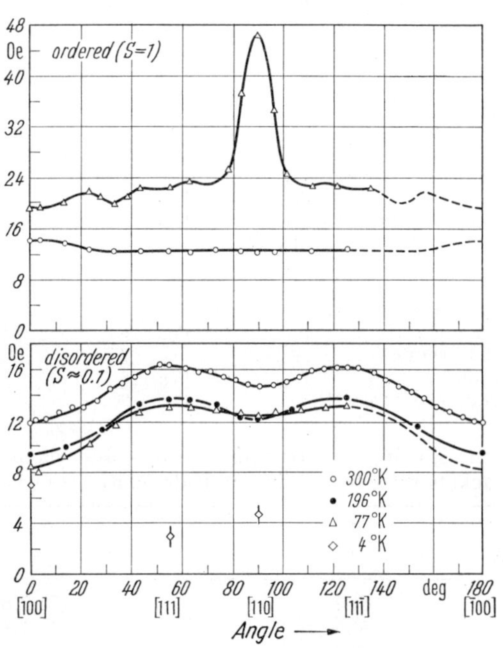

Fig. 20. $Li_{0.55}Fe_{2.45}O_4$. Angular dependence of the resonance line width, ΔH, in the (110) plane for "ordered" and "disordered" monocrystals at 24 Gc/s and various temperatures. *S* is the long range order parameter [*60SFR*].

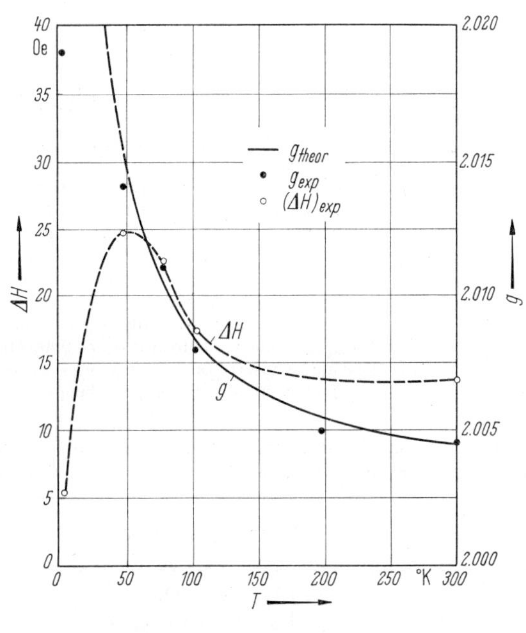

Fig. 21. Li ferrite. Temperature dependence of the theoretical *g* factor, g_{theor} (which is linear in $1/T$), the experimental *g* factor, g_{exp}, and the experimental ferrimagnetic resonance line width, ΔH. Measurements were made at 24 Gc/s. g_{theor} is obtained from the theory of rapidly relaxing impurity ions in [*59K*], [*59DKP*] and [*59W*]. [*62SFR*]. See also Fig. 22 and [*64CR*].

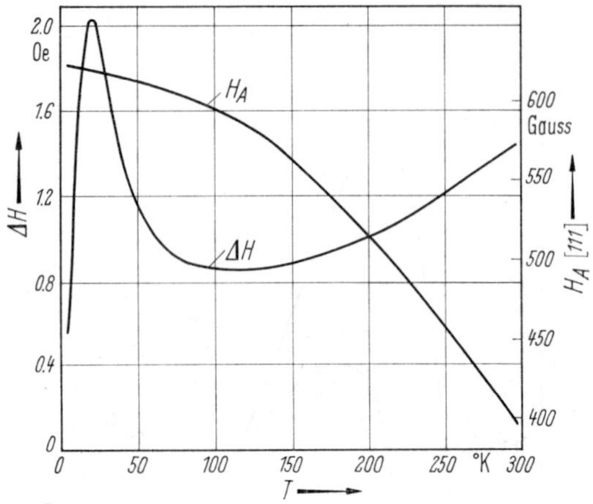

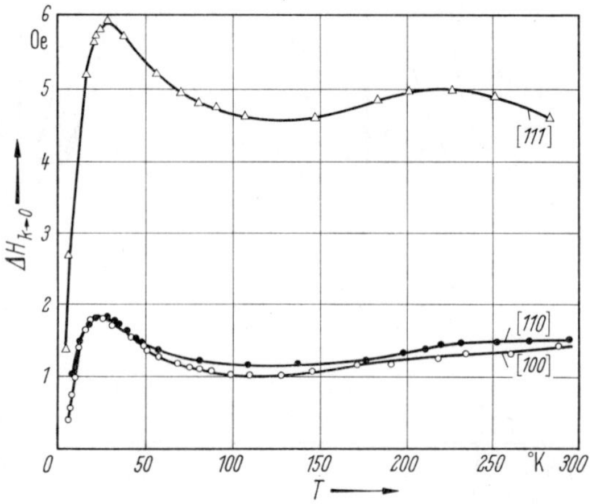

←
Fig. 22. Li ferrite. Ferrimagnetic resonance line width, ΔH, and anisotropy field, H_A, vs. temperature. Measurements were made at 10 Gc/s with H in the [111] direction on a single crystal having near-zero defect concentration [*68SLN*].

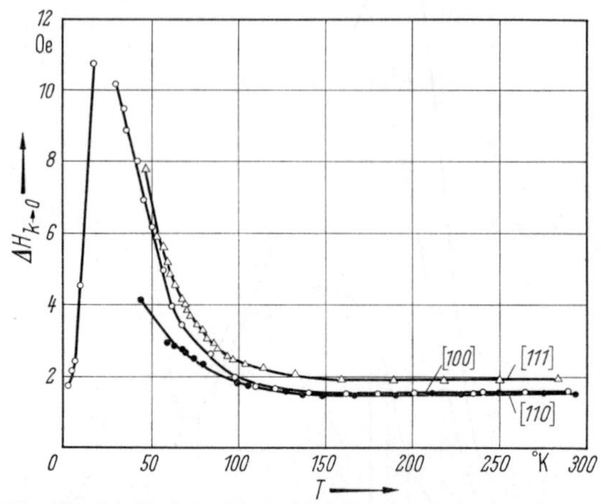

Fig. 23. Li ferrite. Line width for long-wavelength spin waves (obtained by the parallel pump technique) vs. temperature for the dc applied magnetic field along the various principle crystallographic directions in a partially ordered single crystal [*64SD*]. See also [*62DS*] and [*68SLN*].

←
Fig. 24. Li ferrite. Line width for long-wavelength spin waves (obtained by the parallel pump technique) vs. temperature for the dc applied magnetic field along the various principle crystallographic directions in a disordered single crystal [*64SD*].

6.1.6.2 Li-Cr ferrites

Tab. 1. Cation distributions, cell edges, Curie temperatures, compensation temperatures, and magnetic moments for the system $Li_{0.5}Fe_{2.5-a}Cr_aO_4$. Cations in brackets refer to ions in octahedral (B) sites [*54G*].

Cr content a	Ionic distribution	Cell edge Å	Θ_C °C	T_{comp} °C	p_m μ_B
0	$Fe_{1.00}[Li_{0.50}Fe_{1.50}]O_4$	8.331	680		2.47 : 2.60
0.50	$Fe_{1.00}[Li_{0.50}Fe_{1.00}Cr_{0.50}]O_4$	8.306	500		1.62 : 1.50
0.75	$Fe_{0.98}Li_{0.02}[Li_{0.48}Fe_{0.77}Cr_{0.75}]O_4$	8.296	410		1.35
1.00	$Fe_{0.94}Li_{0.06}]Li_{0.44}Fe_{0.56}Cr_{1.00}]O_4$	8.292	315	205	0.84
1.25	$Fe_{0.91}Li_{0.09}[Li_{0.41}Fe_{0.34}Cr_{1.25}]O_4$	8.290	214	+38	0.61
1.50	$Fe_{0.80}Li_{0.20}[Li_{0.30}Fe_{0.20}Cr_{1.50}]O_4$	8.287	119	−16	0.55
1.60	$Fe_{0.64}Li_{0.36}[Li_{0.14}Fe_{0.26}Cr_{1.60}]O_4$	8.288	167	+11	0.42
1.70	$Fe_{0.54}Li_{0.46}[Li_{0.04}Fe_{0.26}Cr_{1.70}]O_4$	8.290	155	+20	0.22
2.00	$Fe_{0.50}Li_{0.50}[Cr_{2.00}]O_4$	8.288	80 ± 16	+37 ± 15 [1])	0.10

[1]) T_{comp} for this sample was obtained from measurements of the remanent magnetization.

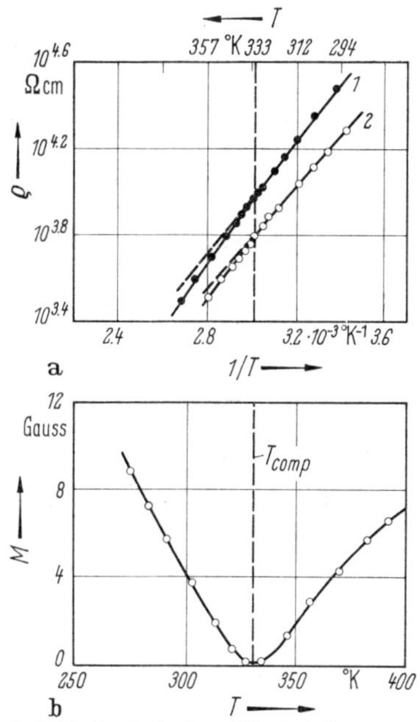

Fig. 1. $Li_2O \cdot 2\frac{1}{2}Fe_2O_3 \cdot 2\frac{1}{2}Cr_2O_3$. a) Resistivity, ϱ, vs. $1/T$ measured with dc (curve 1) and ac (200 kc/s, curve 2). b) Magnetization, M, vs. T at $H = 250$ Oe [*60BGC*]. See also [*64DSR*].

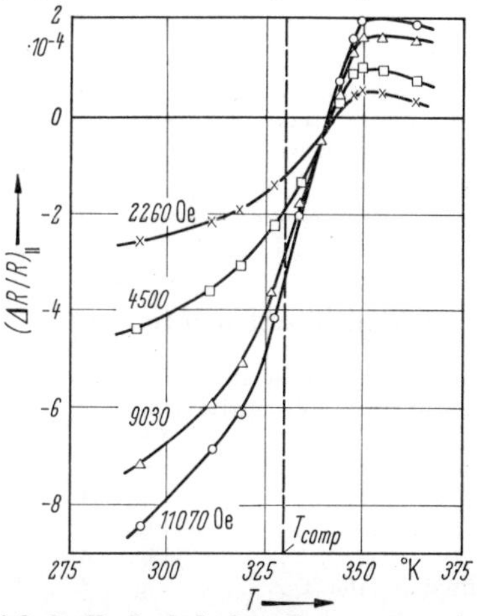

Fig. 2. $Li_2O \cdot 2\frac{1}{2}\ Fe_2O_3 \cdot 2\frac{1}{2}Cr_2O_3$. Temperature dependence of the longitudinal galvanomagnetic effect, $(\Delta R/R)_{\parallel}$, at various applied magnetic fields. T_{comp} is the magnetic compensation temperature [*60BGC*]. See also [*61BGC*].

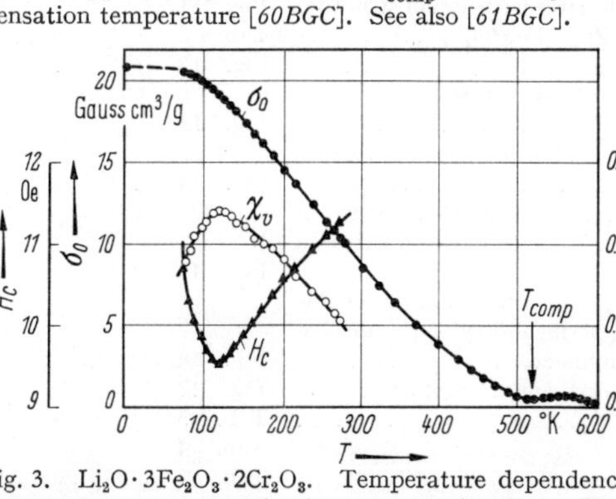

Fig. 3. $Li_2O \cdot 3Fe_2O_3 \cdot 2Cr_2O_3$. Temperature dependences of the spontaneous magnetization, σ_0, coercive force, H_c, and the susceptibility, χ_v (at $H = 10$ Oe). $T_{comp} = 523.5$ °K [*62GC*].

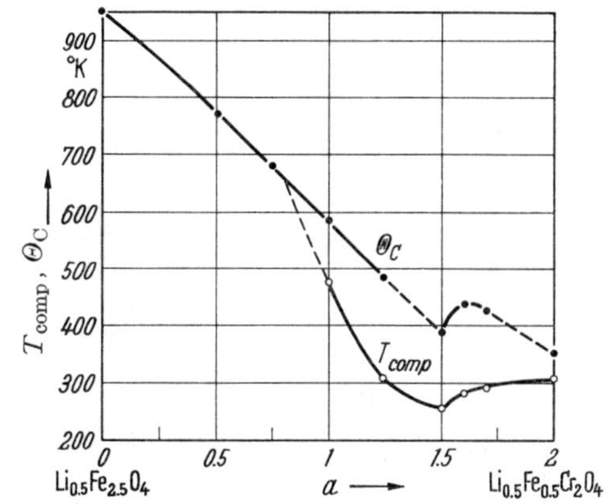

Fig. 4. $Li_{0.5}Fe_{2.5-a}Cr_aO_4$. Curie temperatures, Θ_C, and compensation temperatures, T_{comp}, vs. composition [*54G*].

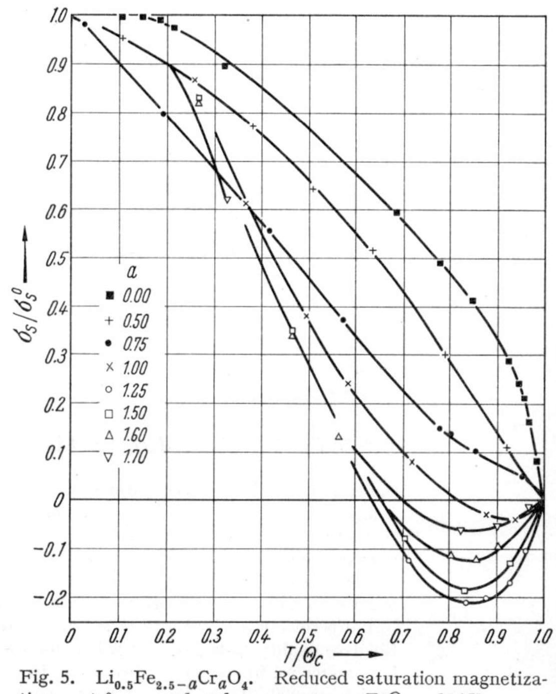

Fig. 5. $Li_{0.5}Fe_{2.5-a}Cr_aO_4$. Reduced saturation magnetization, σ_s/σ_s^0, vs. reduced temperature, T/Θ_C. [*54G*].

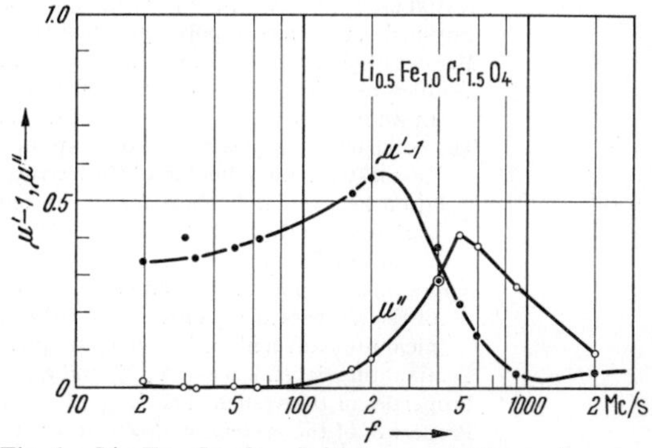

Fig. 6. $Li_{0.5}Fe_{1.0}Cr_{1.5}O_4$. Initial permeability vs. frequency of sintered sample at room temperature. For this sample $\sigma_s = 2.2$ Gauss cm³/g and $\varrho = 4.0$ g/cm³ [*59SW(2)*].

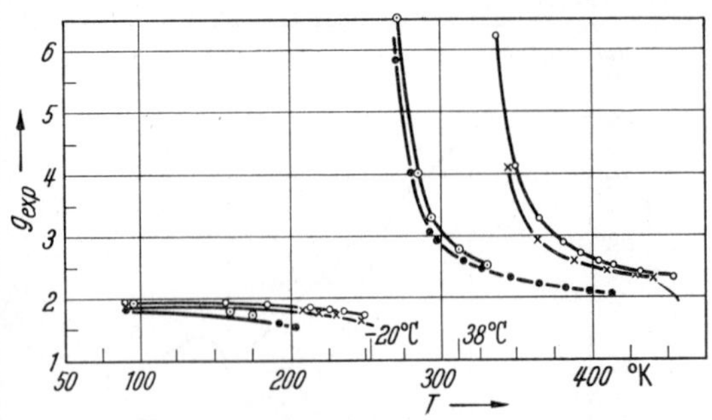

Fig. 7. The g_{exp} factor vs. temperature for $Li_{0.5}Fe_{1.0}Cr_{1.5}O_4$ and $Li_{0.5}Fe_{1.25}Cr_{1.25}O_4$ at 9.5 Gc/s and 24.0 Gc/s [*55W*].
$Li_{0.5}Fe_{1.0}Cr_{1.5}O_4$: ⊙ 9.5 Gc/s
● 24.0 Gc/s
$Li_{0.5}Fe_{1.25}Cr_{1.25}O_4$: ○ 9.5 Gc/s
× 24.0 Gc/s
See also: [*56PD*, *58P*, *53W* and *55M*].

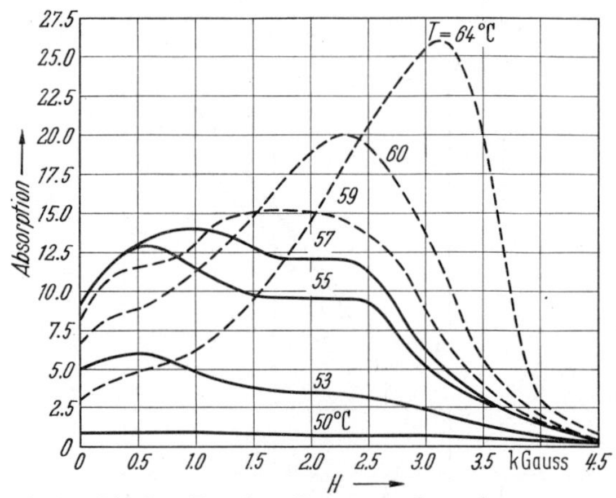

Fig. 8. $Li_{0.5}Cr_{1.25}Fe_{1.25}O_4$. Resonance absorption curves at various temperatures, $f = 22.5$ Gc/s [*55M*].

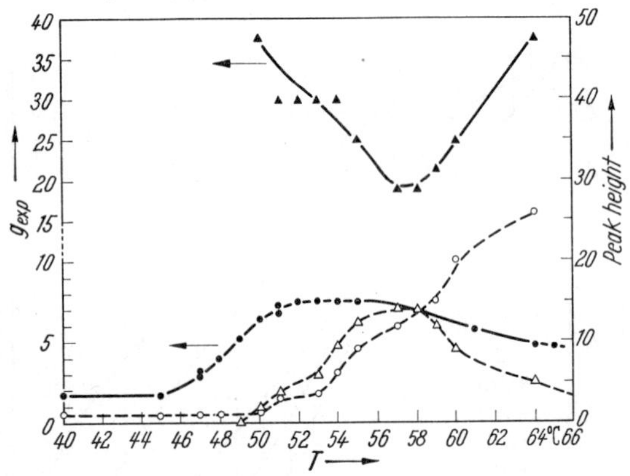

←

Fig. 9. $Li_{0.5}Cr_{1.25}Fe_{1.25}O_4 \cdot g_{exp}$ values and peak heights vs. temperature. The two sets of *g* values were obtained because of the double peaks shown in Fig. 8. $f = 22.5$ Gc/s. Full line: *g*; dashed line: peak height [*55M*].

6.1.6.3 Li-Zn, Li-Zn-Co, Li-Ti, Li-Zn-Ti and Li-Zn-Ti-Al-Mn ferrites

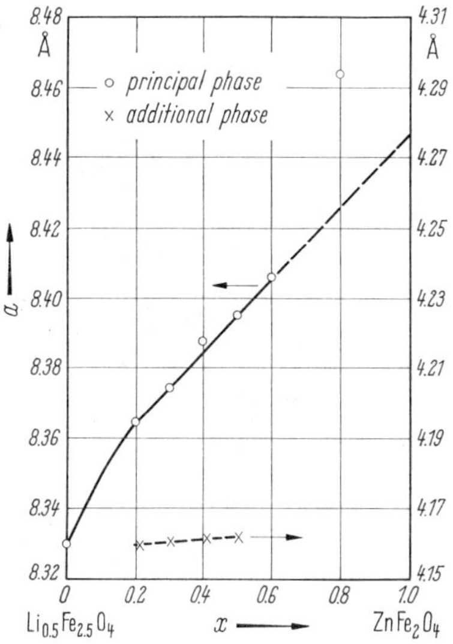

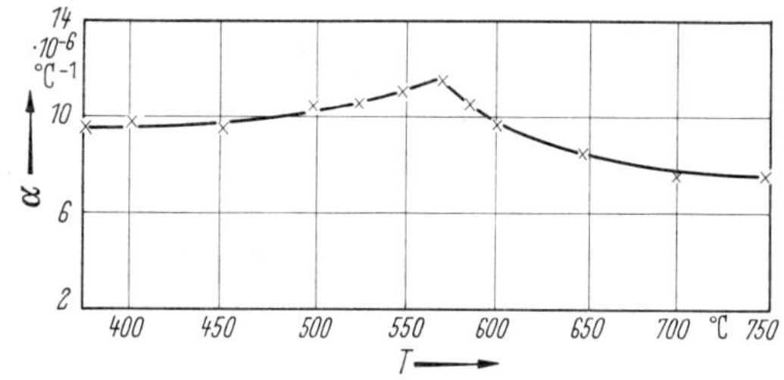

Fig. 2. $3Li_2O(Fe_2O_3)_5 \cdot ZnFe_2O_4$. Temperature dependence of the coefficient of thermal expansion, α [57VDF].

←

Fig. 1. $Li_{0.5-(x/2)}Zn_xFe_{2.5-(x/2)}O_4$. Dependence of the unit cell parameters of solid solutions on composition [65DPB].

Tab. 1. Composition and magnetic properties of certain Li-Zn ferrospinels [52W(2)]

f Mc/s	Molar composition	Firing temperature °C	μ_i	Q	$\mu_i Q$	Θ_C °C	Temperature coefficient of μ_i [°C^{-1}]
0.5	$0.35 LiFe_5O_8 \cdot 0.6 Zn_2Fe_4O_8 \cdot 0.05 Fe_2O_3$ [1]	1100	360	66	23 000	155	+0.36
5	$0.45 LiFe_5O_8 \cdot 0.45 Zn_2Fe_4O_8 \cdot 0.1 Fe_2O_3$ [1]	1050	185	64	11 800	310	+0.42
25	$0.8 LiFe_5O_8 \cdot 0.2 Zn_2Fe_4O_8$ [2]	975	37	90	3 330	470	+0.20
50	$0.147 Li_2O \cdot 0.073 ZnO \cdot 0.78 Fe_2O_3$ [2]	910	20.5	115	2 360	540	−0.22
100	$0.16 Li_2O \cdot 0.036 ZnO \cdot 0.804 Fe_2O_3$ [1]	840	6.9	160	1 100	585	−0.09

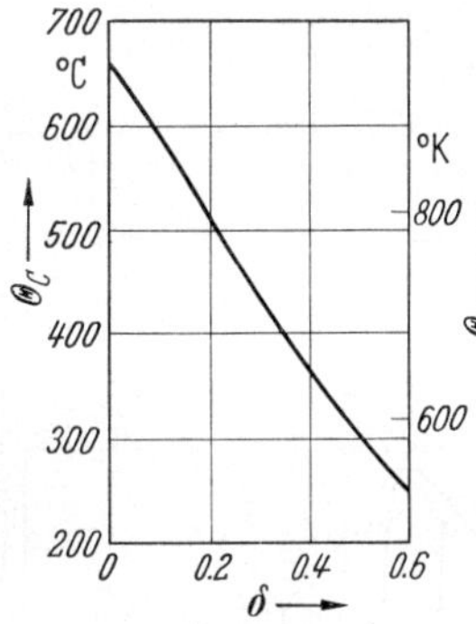

Fig. 3. $(Li_{0.5}Fe_{0.5})_{1-\delta}Zn_\delta Fe_2O_4$. Curie points, Θ_C, of Li-Zn ferrites vs. Zn concentration [59SW(1)]. See also Fig. 14, 15 in 6.1.5.1.

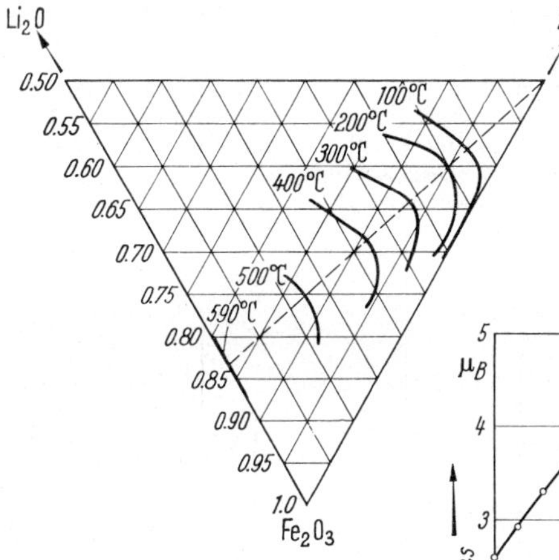

←

Fig. 4. Li-Zn ferrites. Curie temperatures of samples sintered at 1100 °C [52W(1)].

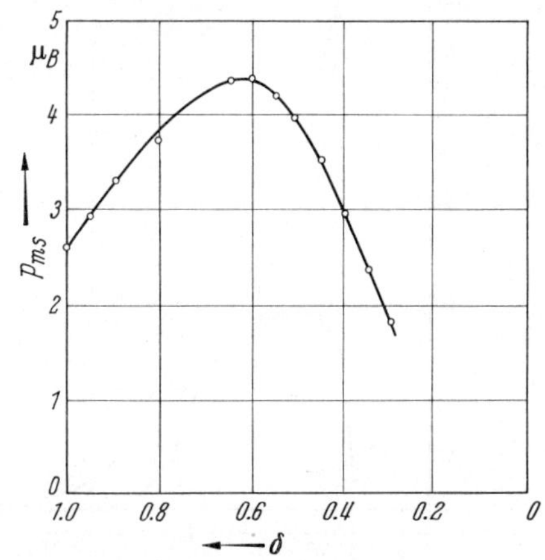

→

Fig. 5. $Zn_{1-\delta}(Li_{0.5}Fe_{0.5})_\delta Fe_2O_4$. Saturation magnetic moment in μ_B at 0 °K for a series of Li-Zn ferrites [50G(2)].

[1] Prepared by dry mixing the constituent Fe and Zn oxides with Li carbonate, pressing and firing.

[2] Prepared by wet mixing the constituents in ball mill with alcohol, presintering 200···300 °C below final firing temperature, regrinding mixing with organic binder and lubricant, firing at 400 °C and at final firing temperature.

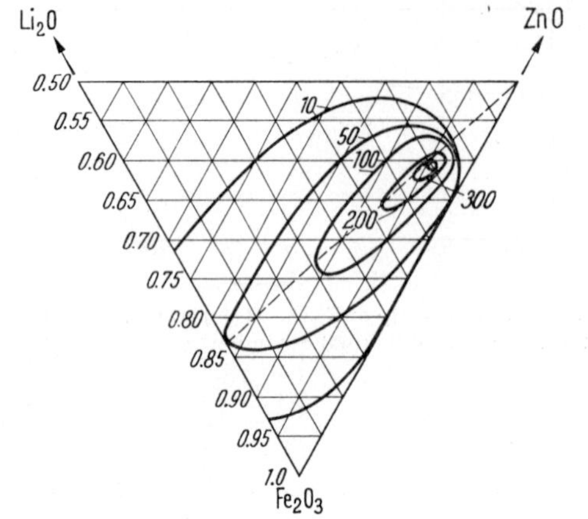

Fig. 6. Li-Zn ferrites. Initial permeability, μ_i. Samples fired at 1100 °C [*52W(1)*]. See also Fig. 15 in 6.1.5.1.

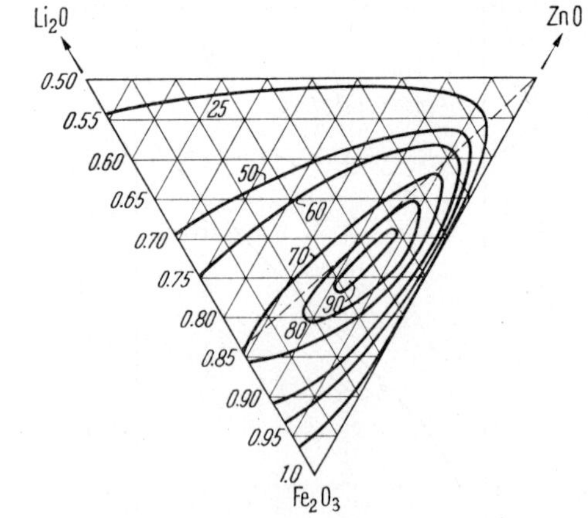

Fig. 7. Li-Zn ferrites. Relative quality factor, Q, at 500 kc/s. Samples fired at 1100 °C [*52W(1)*].

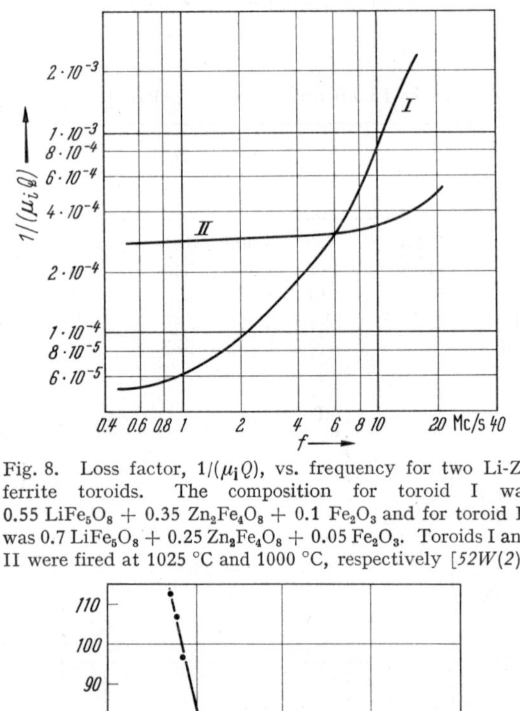

Fig. 8. Loss factor, $1/(\mu_i Q)$, vs. frequency for two Li-Zn ferrite toroids. The composition for toroid I was 0.55 $LiFe_5O_8$ + 0.35 $Zn_2Fe_4O_8$ + 0.1 Fe_2O_3 and for toroid II was 0.7 $LiFe_5O_8$ + 0.25 $Zn_2Fe_4O_8$ + 0.05 Fe_2O_3. Toroids I and II were fired at 1025 °C and 1000 °C, respectively [*52W(2)*].

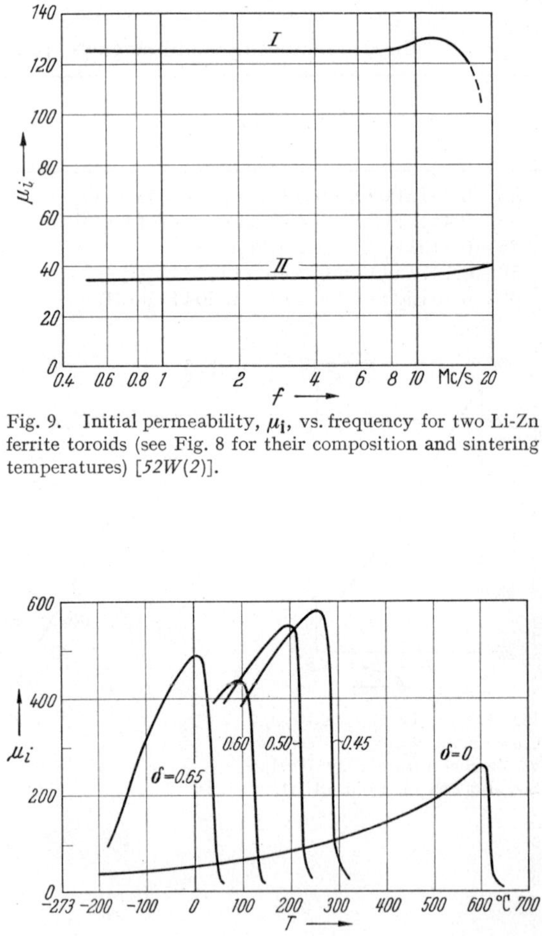

Fig. 9. Initial permeability, μ_i, vs. frequency for two Li-Zn ferrite toroids (see Fig. 8 for their composition and sintering temperatures) [*52W(2)*].

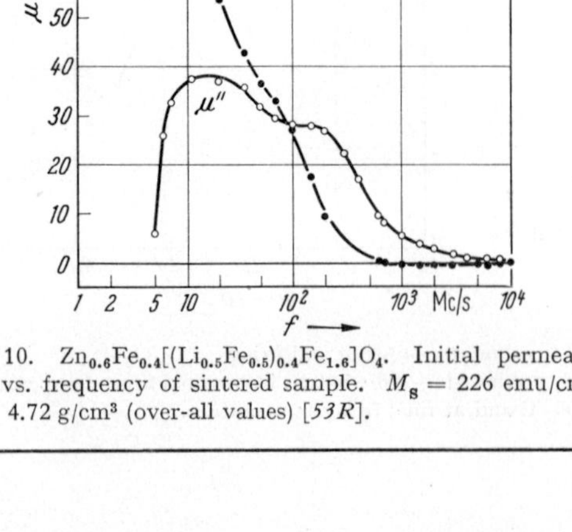

Fig. 10. $Zn_{0.6}Fe_{0.4}[(Li_{0.5}Fe_{0.5})_{0.4}Fe_{1.6}]O_4$. Initial permeability vs. frequency of sintered sample. M_s = 226 emu/cm³; ϱ = 4.72 g/cm³ (over-all values) [*53R*].

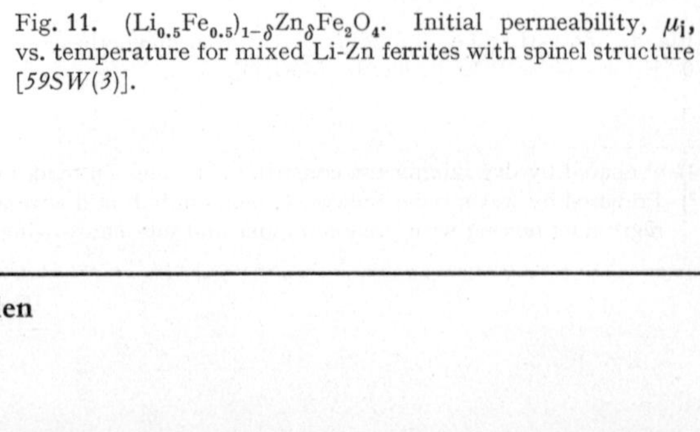

Fig. 11. $(Li_{0.5}Fe_{0.5})_{1-\delta}Zn_\delta Fe_2O_4$. Initial permeability, μ_i, vs. temperature for mixed Li-Zn ferrites with spinel structure [*59SW(3)*].

Tab. 2. Physical parameters of polycrystalline $(Fe^{+3}_{0.5}Li^{+1}_{0.5})_{0.35}Zn^{+2}_{0.65}Fe^{+3}_2O_4$. Sintering times were short so that there was only a slight conversion of Fe^{+3} to Fe^{+2} [53B]. See also Fig. 14 in 6.1.5.1

Measured density	ϱ = 4.50 g/cm³
Initial susceptibility	χ_{vi} = 3.15 · 10^{-5}
Magnetomechanical coupling coefficient [1])	k = −0.013
Young's modulus	E = 1.13 · 10^{12} dyne/cm²

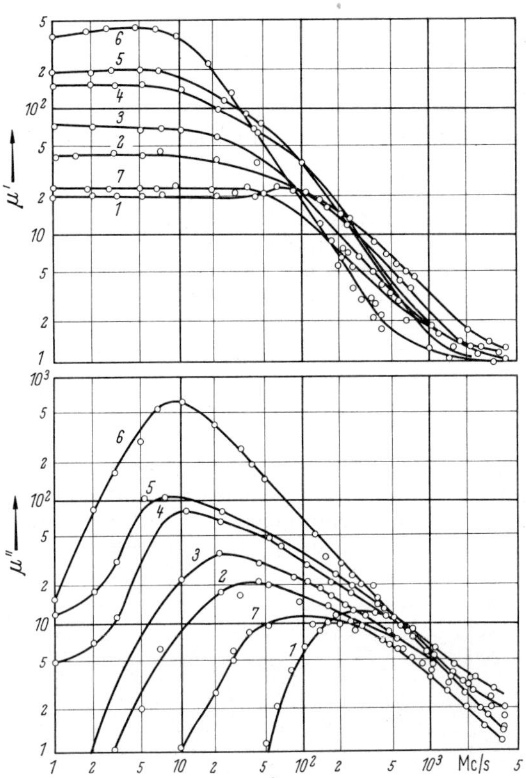

Fig. 12. $Li_{0.5-(x/2)}Zn_xFe_{2.5-(x/2)}O_4$. Magnetic spectra ($\mu'$, μ'' vs. f) [65DPB]. Values of x for the curves are: 1) 0; 2) 0.3; 3) 0.4; 4) 0.5; 5) 0.55; 6) 0.6; 7) 0.65.

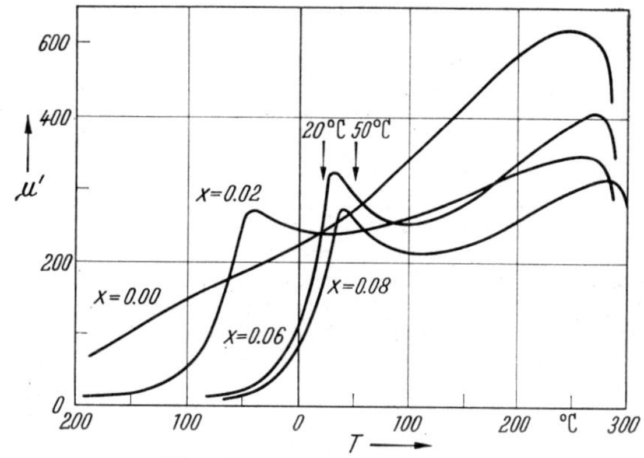

Fig. 13. $[(Li^{1+}_{1/2}Fe^{3+}_{1/2})_{0.5}Zn_{0.5}]_{1-x}Co_xFe_2O_4$. Temperature dependence of the initial permeability, μ', of Co-substituted equimolar Li-Zn ferrites (sintering conditions: 3 h, 1120 °C, air) [57B].

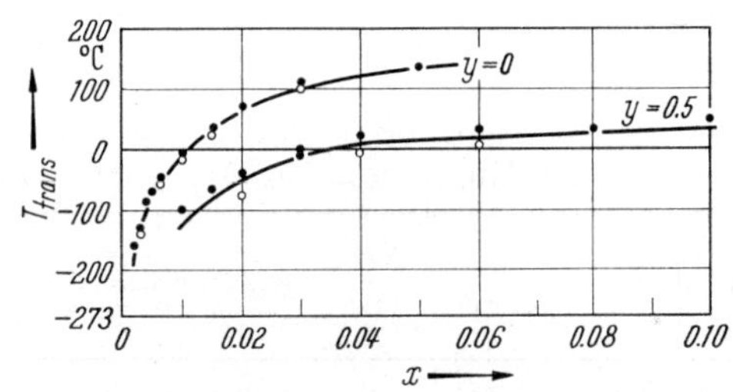

Fig. 15. $[(Li_{1/2}Fe_{1/2})_{1-y}Zn_y]_{1-x}Co_xFe_2O_4$. Relation between the Co-ferrite content, x, and the transition temperature, T_{trans}, as derived from the peaks in the μ' vs. T and k' vs. T curves of Co-substituted Li and Li-Zn ferrites [57B]. Sintering conditions: y = 0: 2 h, 1040 °C, air; y = 0.5; 3 h, 1120 °C, air. ● Temperature at the peak of the μ' vs. T curve, ○ Temperature of the peak of the k' vs. T curve.

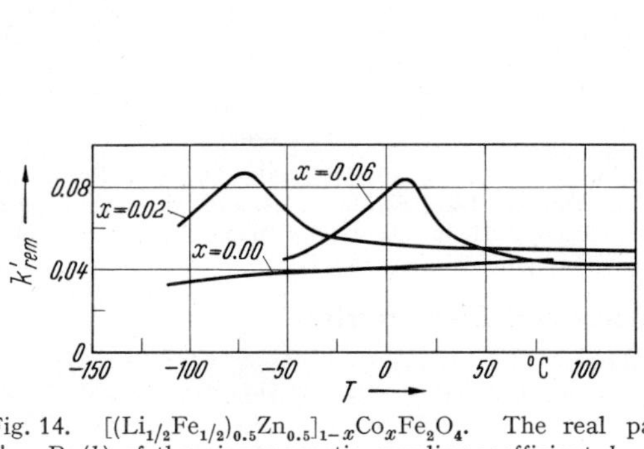

Fig. 14. $[(Li_{1/2}Fe_{1/2})_{0.5}Zn_{0.5}]_{1-x}Co_xFe_2O_4$. The real part k' = Re(k) of the piezomagnetic coupling coefficient, k, vs. temperature for various Co-substituted equimolar Li-Zn ferrites (values at remanence k'_{rem}). Sintering conditions: 3 h, 1120 °C, air [57B].

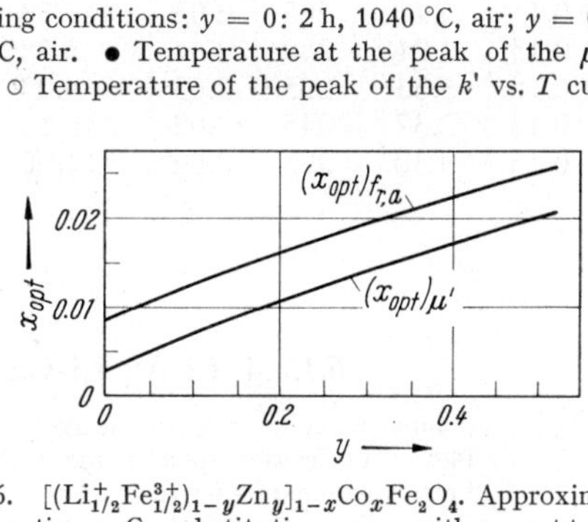

Fig. 16. $[(Li^{+}_{1/2}Fe^{3+}_{1/2})_{1-y}Zn_y]_{1-x}Co_xFe_2O_4$. Approximate values of optimum Co-substitutions, x_{opt}, with respect to a minimum value of the temperature coefficient of either the initial permeability, μ', or the mechanical resonant and antiresonant frequencies, $f_{r,a}$, at remanence. Optimum substitutions are displayed vs. Zn-ferrite content y of Li-Zn ferrites [57B].

[1]) k^2 is defined as the ratio between electromagnetic and elastic energy when only an alternating low frequency stress is applied. Therefore $k^2 = 4\pi\lambda'\mu_r/E$, where λ' is the dynamic magnetostrictive constant, μ_r is the reversible permeability, and E is Young's modulus.

Folen

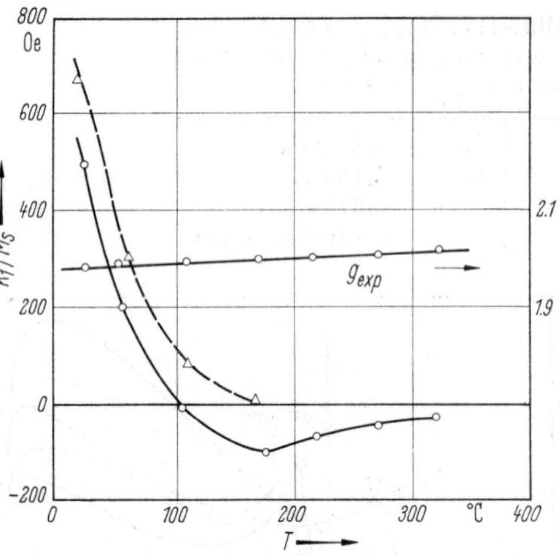

Fig. 17. Temperature dependence of K_1/M_s and g_{exp} for monocrystals of Li-Co ferrite. The monocrystals were flux grown using a mixture of 8.6 Li_2Co_3 + 0.8 Co_2O_3 + 44.6 Fe_2O_3 + 46 PbO mole % [65S].

Tab. 3. Properties of the system $(LiFe)_{1-0.5a}Fe_{2a}Ti_{1-a}O_4$ *) [61JW]

a	ΔH Oe	σ Gauss cm³/g	Θ_C °C	ε'	tan δ	g_{eff}
0.2	420	8.7	231	11	0.005	2.06
0.3	460	13.6	318	10	0.004	2.08
0.4	580	19.3	388			2.06
0.5	680	25.2	451	10	0.004	2.11
0.6	740	33.0	499	11	0.013	2.11
0.7	810	36.8	538	12	0.033	2.21
0.8	820	49.9	602	13	0.129	2.24
0.9	860	57.5	610	60	0.359	2.26
1.0	640	68.4	620	86	0.314	2.21

Tab. 4. Properties of thermally annealed $(LiFe)_{0.6}Zn_{0.4}Fe_{0.8}Ti_{0.6}O_4$.
ΔH refers to the ferromagnetic resonance line width and ΔH_k to the spin-wave line width [63HWS]

Annealing process	ΔH Oe	$4\pi M_s$ Gauss	tan δ_ε	ϱ Ω · cm	ΔH_k Oe
None	110	1630	0.0050	$7.1 \cdot 10^9$	6.10
200 °C, Air	140	1595	0.0047	$2.0 \cdot 10^8$	5.95
400 °C, Air	115	1630	0.0034	$1.7 \cdot 10^7$	6.10
600 °C, Air	100	1400	0.0068	$5.8 \cdot 10^5$	6.40
800 °C, N_2	120	1350	0.0060	$1.1 \cdot 10^4$	8.46
900 °C, N_2	115	1370	0.0425	$3.0 \cdot 10^3$	9.35
1000 °C, N_2	135	1350	1.23	$3.9 \cdot 10^2$	12.30

Tab. 5. Magnetic properties of the system $(LiFe)_{0.85-0.5x}Zn_xMn_{0.02}Fe_{0.6-y}Ti_{0.7}Al_yO_{4\pm}$ **) [61JW]

x	y	ϱ g/cm³	ΔH Oe	g_{eff}	σ_s Gauss cm³/g	Θ_C °C
0.000	0.1	3.20	425	2.08	9.8	247
0.025	0.1	3.27	370	2.05	10.78	228
0.05	0.1	3.33	325	2.05	11.83	228
0.075	0.1	3.36	300	2.04	12.56	207
0.10	0.1	3.21	285	2.03	13.53	199
0.025	0.15	3.21	280	2.04	8.28	188
0.05	0.15	3.33	255	2.03	9.27	190
0.075	0.15	3.32	245	2.02	10.12	177
0.100	0.15	3.38	225	2.02	11.02	172

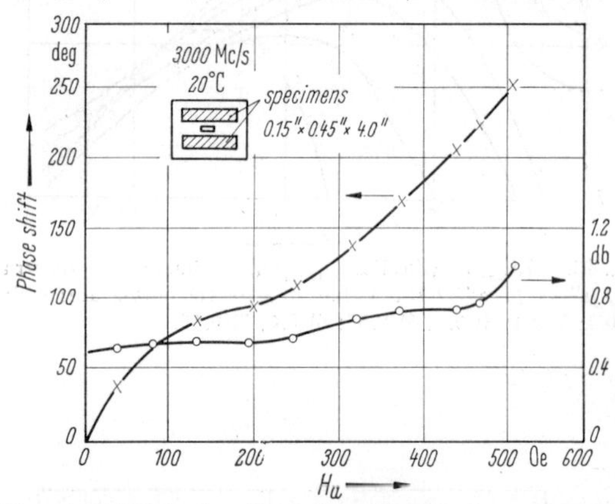

Fig. 18. $(LiFe)_{0.8}Zn_{0.1}Mn_{0.02}Fe_{0.45}Al_{0.15}Ti_{0.7}O_{4\pm}$. Phase shift and microwave absorption losses vs. applied magnetic field. Sample mounted in an axially magnetized, rectangular, coaxial-line phase shifter [61JW].

6.1.6.4 Li-Al, Li-Ga, Li-Cd ferrites and Li-Fe ferrites

*) Sintered at 1200° for 8 h. **) Sintered at 1150° for 5 h.

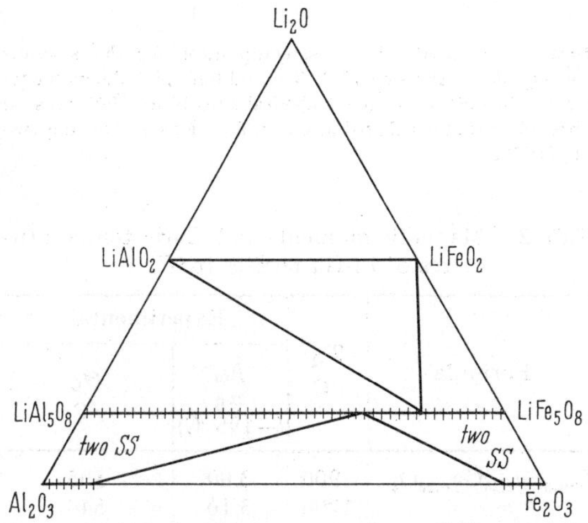

Fig. 1. Li_2O-Fe_2O_3-Al_2O_3 system. Compatibility triangles determined at 1250 °C [*61SR*].

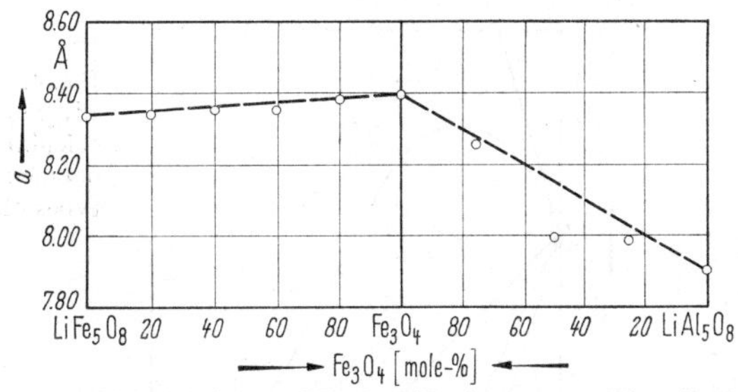

Fig. 2. Fe_3O_4-$LiAl_5O_8$ and Fe_3O_4-$LiFe_5O_8$ systems. Plot of unit cell dimension of spinel phase obtained in sealed platinum tubes at 1400°C [*61SR*].

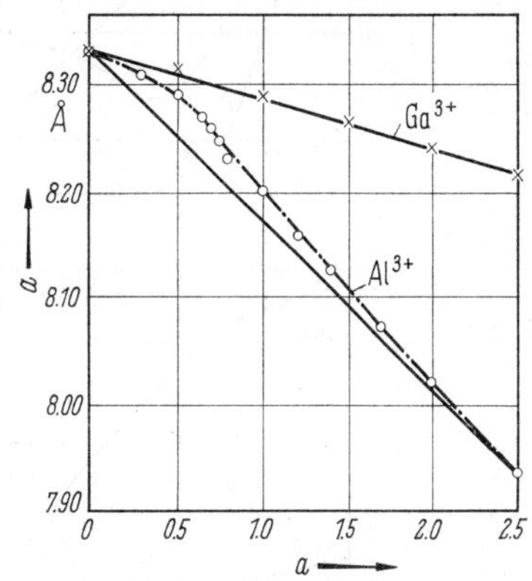

Fig. 3. Lattice parameters vs. composition for the system $Li_{0.5}Fe_{2.5-a}Me_aO_4$ (Me = Al^{3+} or Ga^{3+}). The full lines are straight lines between $a = 0$ and $a = 2.5$ [*63SB*]. See also [*61SR*].

Tab. 1. Cation distributions in Li-Ga ferrites determined from X-ray intensity measurements. Activation energy, q, refers to the energy of redistribution of the trivalent cations among octahedral and tetrahedral sites [*65K*]

Formula	T_Q *) °C	Ga content	% of Ga in tetrahedral sites	q eV
$Fe^{3+}_{1.0}[Li^{1+}_{0.5}Fe^{3+}_{1.5}]O_4$	900 ÷ 1400	—	—	—
$Ga^{3+}_{1.0}[Li^{1+}_{0.5}Ga^{3+}_{1.5}]O_4$	900 ÷ 1400	2.5	40	—
$Ga^{3+}_{0.17}Fe^{3+}_{0.83}[Li^{1+}_{0.5}Ga^{3+}_{0.08}Fe^{3+}_{1.42}]O_4$	900	0.25	68	0.13
$Ga^{3+}_{0.21}Fe^{3+}_{0.79}[Li^{1+}_{0.5}Ga^{3+}_{0.04}Fe^{3+}_{1.45}]O_4$	1000		84	0.25
$Ga^{3+}_{0.22}Fe^{3+}_{0.78}[Li^{1+}_{0.5}Ga^{3+}_{0.03}Fe^{3+}_{1.47}]O_4$	1100		88	0.31
$Ga^{3+}_{0.24}Fe^{3+}_{0.76}[Li^{1+}_{0.5}Ga^{3+}_{0.01}Fe^{3+}_{1.49}]O_4$	1200		96	0.49
$Ga^{3+}_{0.33}Fe^{3+}_{0.67}[Li^{1+}_{0.5}Ga^{3+}_{0.17}Fe^{3+}_{1.33}]O_4$	900	0.50	66	0.13
$Ga^{3+}_{0.36}Fe^{3+}_{0.64}[Li^{1+}_{0.5}Ga^{3+}_{0.14}Fe^{3+}_{1.36}]O_4$	1000		72	0.18
$Ga^{3+}_{0.39}Fe^{3+}_{0.61}[Li^{1+}_{0.5}Ga^{3+}_{0.11}Fe^{3+}_{1.39}]O_4$	1100		78	0.24
$Ga^{3+}_{0.42}Fe^{3+}_{0.58}[Li^{1+}_{0.5}Ga^{3+}_{0.08}Fe^{3+}_{1.42}]O_4$	1200		84	0.32
$Ga^{3+}_{0.40}Fe^{3+}_{0.60}[Li^{1+}_{0.5}Ga^{3+}_{0.35}Fe^{3+}_{1.15}]O_4$	900	0.75	53	0.08
$Ga^{3+}_{0.44}Fe^{3+}_{0.56}[Li^{1+}_{0.5}Ga^{3+}_{0.31}Fe^{3+}_{1.19}]O_4$	1000		58	0.12
$Ga^{3+}_{0.54}Fe^{3+}_{0.46}[Li^{1+}_{0.5}Ga^{3+}_{0.21}Fe^{3+}_{1.29}]O_4$	1100		72	0.23
$Ga^{3+}_{0.64}Fe^{3+}_{0.36}[Li^{1+}_{0.5}Ga^{3+}_{0.11}Fe^{3+}_{1.39}]O_4$	1200		85	0.39
$Ga^{3+}_{0.45}Fe^{3+}_{0.55}[Li^{1+}_{0.5}Ga^{3+}_{0.55}Fe^{3+}_{0.95}]O_4$	900	1.0	45	0.03
$Ga^{3+}_{0.54}Fe^{3+}_{0.46}[Li^{1+}_{0.5}Ga^{3+}_{0.46}Fe^{3+}_{1.04}]O_4$	1000		54	0.11
$Ga^{3+}_{0.56}Fe^{3+}_{0.44}[Li^{1+}_{0.5}Ga^{3+}_{0.44}Fe^{3+}_{1.06}]O_4$	1100		56	0.13
$Ga^{3+}_{0.59}Fe^{3+}_{0.41}[Li^{1+}_{0.5}Ga^{3+}_{0.41}Fe^{3+}_{1.09}]O_4$	1200		59	0.17
$Ga^{3+}_{0.62}Fe^{3+}_{0.38}[Li^{1+}_{0.5}Ga^{3+}_{0.88}Fe^{3+}_{0.62}]O_4$	900	1.5	41	0.06
$Ga^{3+}_{0.76}Fe^{3+}_{0.24}[Li^{1+}_{0.5}Ga^{3+}_{0.74}Fe^{3+}_{0.76}]O_4$	1000		50	0.13
$Ga^{3+}_{0.77}Fe^{3+}_{0.23}[Li^{1+}_{0.5}Ga^{3+}_{0.73}Fe^{3+}_{0.77}]O_4$	1100		51	0.15
$Ga^{3+}_{0.81}Fe^{3+}_{0.19}[Li^{1+}_{0.5}Ga^{3+}_{0.69}Fe^{3+}_{0.81}]O_4$	1200		54	0.20

Note. Time of heat treatment was 5 h in all cases. *) T_Q = quenching temperature.

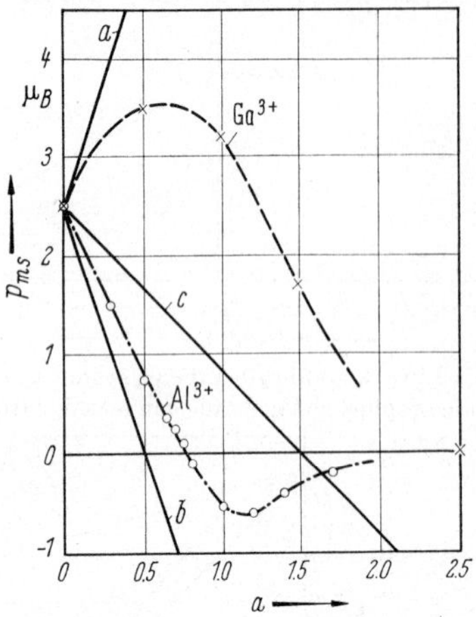

←
Fig. 4. Saturation moments, p_{ms}, vs. composition for the systems $Li_{0.5}Fe_{2.5-a}Me_aO_4$ (Me = Al^{3+} or Ga^{3+}). The full lines give the moments calculated for a) all Me^{3+} ions on tetrahedral sites b) all Me^{3+} ions on octahedral sites c) statistical distribution of Me^{3+} ions among the two types of sites [63SB].

Tab. 2. Magnetic moments and Curie temperatures of Li-Ga ferrites [65K]

Formula	T_Q °C	Experimental p_m μ_B (−195 °C)	Θ_C °C
$Li_{0.5}Fe_{2.25}Ga_{0.25}O_4$	900	3.00	595
	1200	3.10	540
$Li_{0.5}Fe_{2.0}Ga_{0.5}O_4$	900	2.80	420···460
	1200	3.30	410
$Li_{0.5}Fe_{1.75}Ga_{0.75}O_4$	900	2.70	380
	1200	3.10	276
$Li_{0.5}Fe_{1.5}Ga_{1.0}O_4$	900	2.40	285
	1200	2.80	156
$Li_{0.5}Fe_{1.0}Ga_{1.5}O_4$	900	1.40	100···125
	1200	0.80	Below room temperature

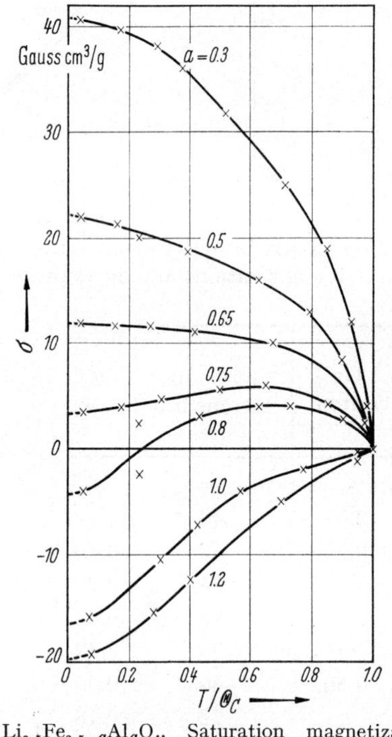

Fig. 5. $Li_{0.5}Fe_{2.5-a}Al_aO_4$. Saturation magnetization per gram, σ, vs. reduced temperature, T/Θ_C. [63SB].

Tab. 3. Magnetic moments per molecule for $LiFe_5O_8$ and its solid solutions with $CdFe_2O_4$ and $LiAl_5O_8$ [57CMW]. See also [54M]

Specimen composition	g_{exp} RT	p^0_{sm} μ_B
$LiFe_5O_8$	1.96(5)	2.38 ± 0.05
$(LiFe_5O_8)_{0.75}(CdFe_2O_4)_{0.25}$	1.97(0)	3.44 ± 0.07
$(LiFe_5O_8)_{0.65}(CdFe_2O_4)_{0.35}$	1.96(9) ± 0.006	4.10 ± 0.08
$(LiFe_5O_8)_{0.50}(CdFe_2O_4)_{0.50}$	1.97(7)	4.70 ± 0.09
$(LiFe_5O_8)_{0.25}(CdFe_2O_4)_{0.75}$	1.99(9)	4.53 ± 0.09
$Li(Fe_{0.9}Al_{0.1})_5O_8$	1.97(1)	1.99 ± 0.04
$Li(Fe_{0.8}Al_{0.2})_5O_8$	1.99(0) ± 0.008	0.62 ± 0.04
$Li(Fe_{0.7}Al_{0.3})_5O_8$	2.01(1)	0.10 ± 0.03
$Li(Fe_{0.6}Al_{0.4})_5O_8$	—	0.13 ± 0.03

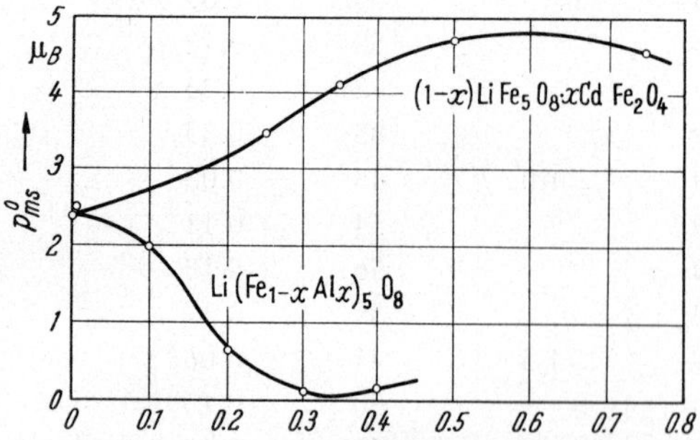

Fig. 6. Variation of the absolute saturation magnetic moment, p^0_{ms}, in μ_B, with composition in solid solutions of Li ferrite with Cd ferrite and Li aluminate [57CMW].

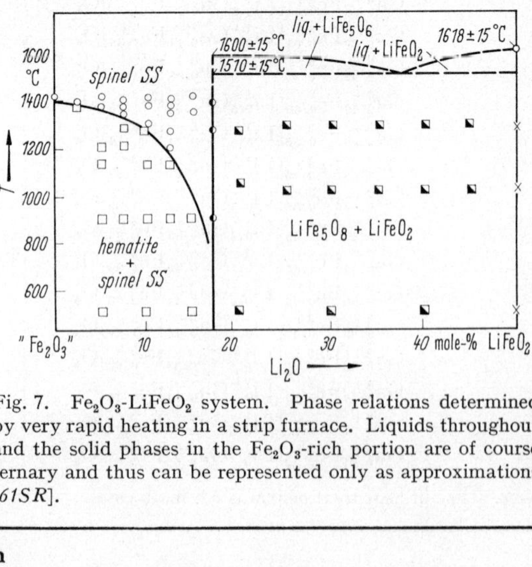

Fig. 7. Fe_2O_3-$LiFeO_2$ system. Phase relations determined by very rapid heating in a strip furnace. Liquids throughout and the solid phases in the Fe_2O_3-rich portion are of course ternary and thus can be represented only as approximations [61SR].

6.1.6.5 Li-Mn, Li-V, and Li-Ge ferrites

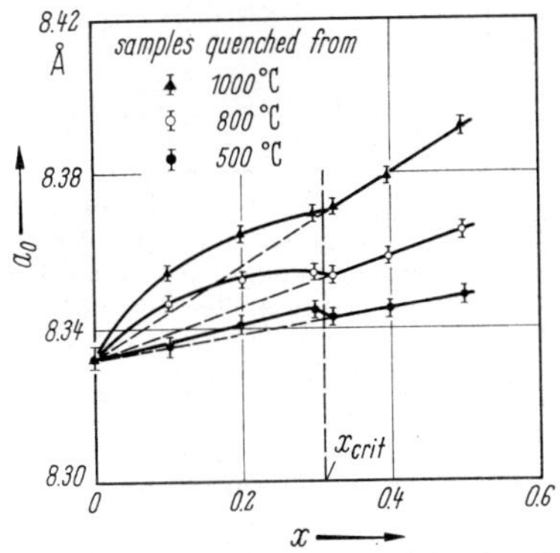

Fig. 1. $Li_{0.5}Fe_{2.5-x}Mn_xO_4$. Compositional and thermal dependence of lattice parameter, a_0 [*65RGA*]. x_{crit} is the critical concentration for removal of octahedral long range order.

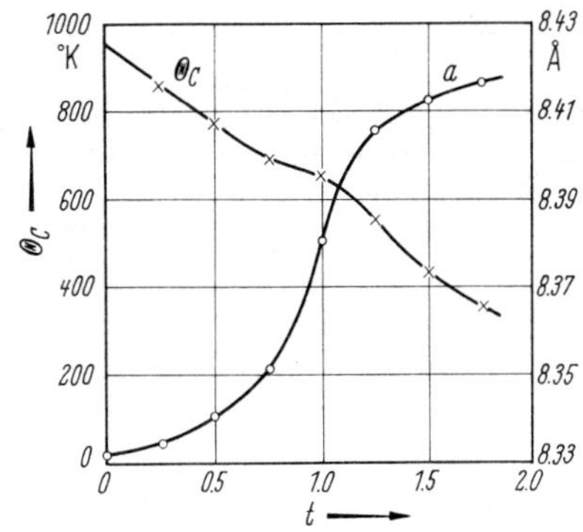

Fig. 2. $Li_{0.5}Mn_tFe_{2.5-t}O_4$. Curie temperature, Θ_C, and cell edge, a, vs. composition [*65B*].

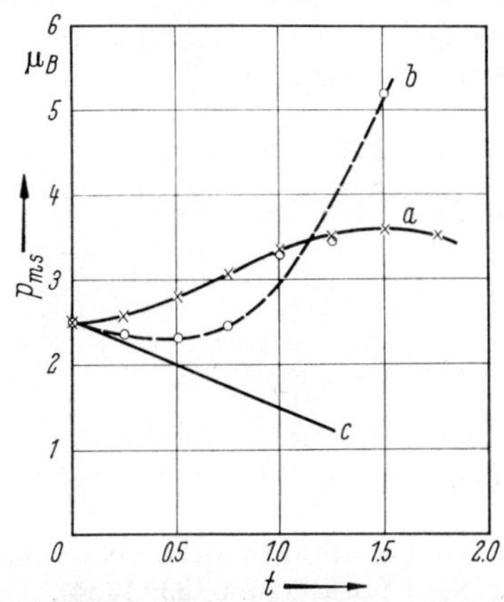

Fig. 3. $Li_{0.5}Mn_tFe_{2.5-t}O_4$. Saturation moments, p_{ms}, vs. composition [*65B*].

a) Experimental ×; b) calculated from cation distribution found from X-ray reflection intensities ○; c) calculated for all Li^+ and Mn^{3+} ions on B-sites.

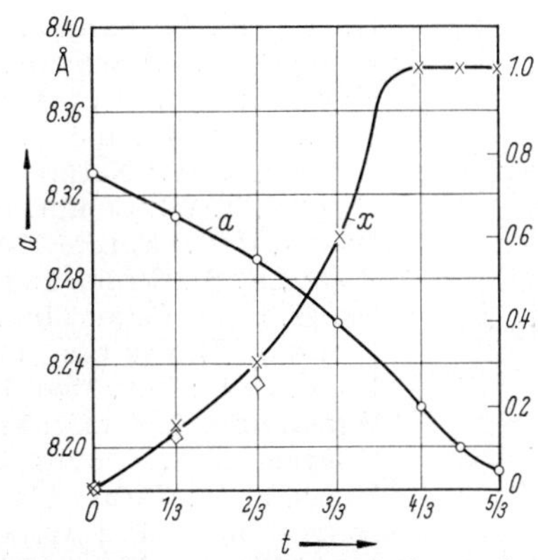

Fig. 4. $Li_{0.5+0.5t}Fe_{2.5-1.5t}Mn_tO_4$. Cell edges and amount of Li^+ ions in A-sites, x, vs. composition (crosses determined from X-ray intensities; squares from magnetic moments) [*65B*].

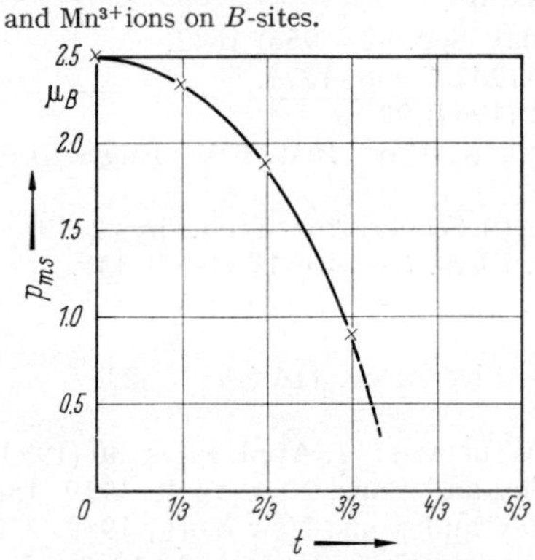

Fig. 5. $Li_{0.5+0.5t}Fe_{2.5-1.5t}Mn_tO_4$. Saturation magnetic moments, p_{ms}, vs. composition [*65B*].

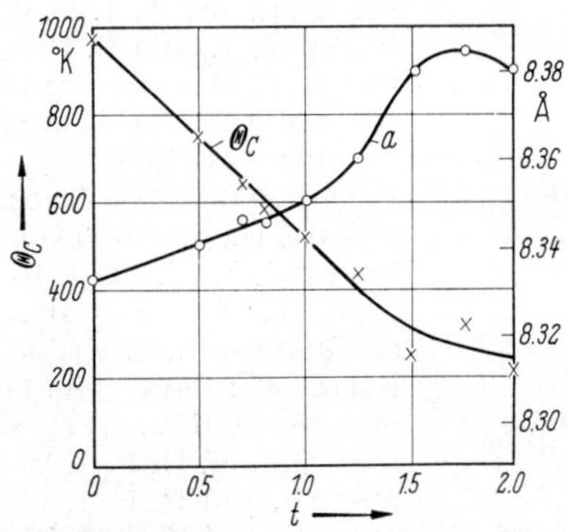

Fig. 6. $Li_{0.5}V_tFe_{2.5-t}O_4$. Curie temperatures and cell edges vs. composition [*65B*].

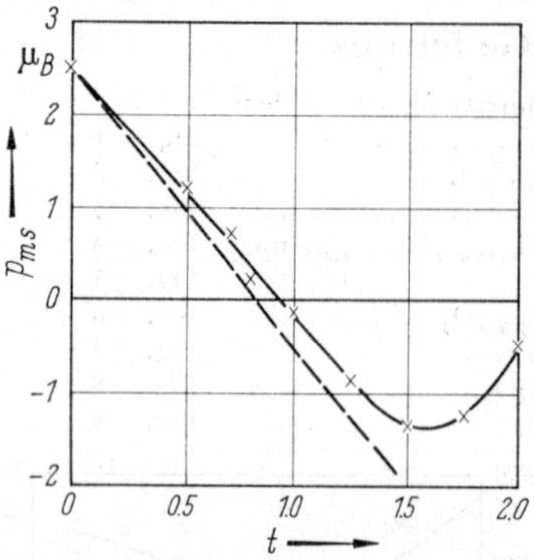

Fig. 7. $Li_{0.5}V_tFe_{2.5-t}O_4$. Saturation moments, p_{ms}, vs. composition. The dashed line is the one calculated for all Li^+ and V^{3+} on B-sites [65B].

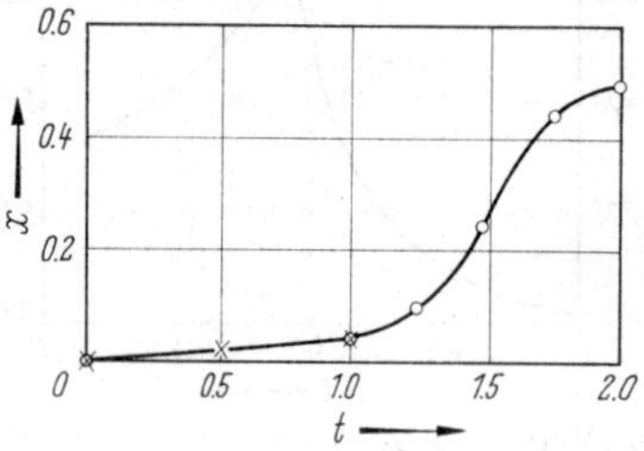

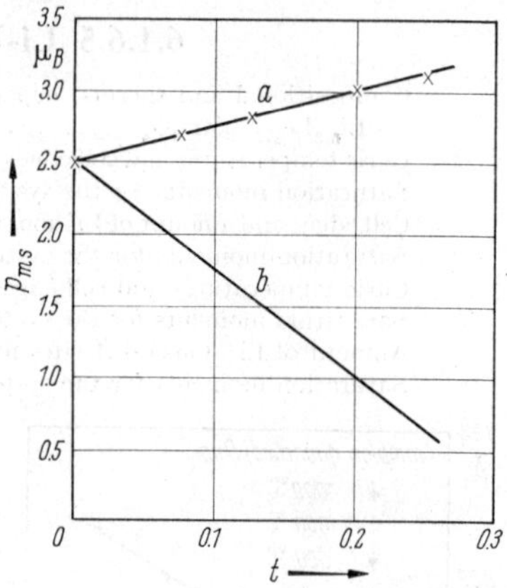

Fig. 9. $Li_{0.5+0.5t}Fe_{2.5-1.5t}Ge_tO_4$. Saturation moments, p_{ms}, vs. composition [65B]. The straight lines are calculated from the Néel-coupling scheme assuming that a) all Li^+ ions are in B-sites and all Ge^{4+} ions are on A-sites; b) all Li^+ and Ge^{4+} ions are on B-sites.

Fig. 8. $Li_{0.5}V_tFe_{2.5-t}O_4$. Amount of Li^+ ions on A-sites, x, vs. composition [65B]. ⊗ from the value of the magnetic moment, ○ from the intensities of X-ray reflections.

6.1.6.6 References for 6.1.6.1 ··· 6.1.6.5

38H Hoffmann, A.: Naturwissenschaften **26** (1938) 431.
43B Barblan, F. F.: Schweiz. Min. Petr. Mitt. **23** (1943) 295.
48N Néel, L.: Ann. Phys. **3** (1948) 137.
50CC Collongues, R., and G. Chaudron: Compt. Rend. **231** (1950) 143.
50G(1) Gorter, E. W.: Nature **165** (1950) 798.
50G(2) Gorter, E. W.: Compt. Rend. **230** (1950) 192.
52B Braun, P. B.: Nature **170** (1952) 1123.
52W(1) Weisz, R. S.: Ceram. Ind. **58** (1952) 130.
52W(2) Weisz, R. S.: Ceram. Ind. **58** (1952) 76.
53B Burgt, C. M. van der: Philips Res. Rept. **8** (1953) 91.
53R Rado, G. T.: Rev. Mod. Phys. **25** (1953) 81.
53W Wieringen, J. S. van: Phys. Rev. **90** (1953) 488.
54G Gorter, E. W.: Philips Res. Rept. **9** (1954) 403.
54M Miles, P. A.: Nature **174** (1954) 177.
54SW Smit, J., and H. P. J. Wijn: "Physical Properties of Ferrites", in Advances in Electronics and Electron Physics (Academic Press Inc., New York, Vol. 6, 83, 1954).
55E Enz, U.: Erzeugung von Ultraschall mit Ferriten, Thesis, Zürich (1955) Cited in [59SW(1)], p. 169.
55M McGuire, T. R.: Phys. Rev. **97** (1955) 831.
55W Wieringen, J. S. van: Cahiers Phys. **62** (1955) 45.
56DW David, I., and A. J. E. Welch: Trans. Faraday Soc. **52** (1956) 1642.
56PD Paulevé, J., and B. Dreyfus: Compt. Rend. **242** (1956) 1273.
57B Burgt, C. M. van der: Philips Res. Rept. **12** (1957) 97.
57CMW Carter, A. E., D. A. Miles, and A. J. E. Welch: Proc. Inst. Elec. Engrs. (London) B-104, **141** (1957).
57VDF Valeev, Kh. S., N. G. Drozdov, and A. L. Frumkin: Soviet. Phys. Tech. Phys. **2** (1957) 2341.
57YT Yosida, K., and M. Tachiki: Progr. Theoret. Phys. (Kyoto) **17** (1957) 331.
58P Paulevé, J.: J. Phys. Radium **19** (1958) 515.
58V Voigt, F.: Ann. Physik **2** (1958) 86.
59DKP de Gennes, P., C. Kittel, and A. M. Portis: Phys. Rev. **116** (1959) 323.
59K Kittel, C.: Phys. Rev. **115** (1959) 1587.
59SRW Sherwood, R. C., J. P. Remeika, and H. J. Williams: J. Appl. Phys. **30** (1959) 217.
59SW(1) Smit, J., and H. P. J. Wijn: Ferrites, J. Wiley and Sons, New York, 1959, 156/157.
59SW(2) Smit, J., and H. P. J. Wijn: Ferrites, J. Wiley and Sons, New York, 1959, 270/271.
59SW(3) Smit, J., and H. P. J. Wijn: Ferrites, J. Wiley and Sons, New York, 1959, 253.
59W White, R. L.: Phys. Rev. Letters **2** (1959) 465.

60BGC	Belov, K. P., A. N. Goryaga, and L. Changta: Soviet Phys. – JETP (English Transl.) **11** (1960) 1376.
60F	Folen, V. J.: J. Appl. Phys. Suppl. **31** (1960) 166 S.
60RF	Rado, G. T., and V. J. Folen: J. Appl. Phys. **31** (1960) 62.
60SFR	Schnitzler, A. D., V. J. Folen, and G. T. Rado: J. Appl. Phys. Suppl. **31** (1960) 348 S.
61BGC	Belov, K. P., A. N. Goryaga, and L. Changta: Soviet Phys. – JETP (English Transl.) **13** (1961) 527.
61JW	Jefferson, C. F., and R. G. West: J. Appl. Phys. **32** (1961) 390 S.
61SR	Strickler, D. W., and R. Roy: J. Am. Ceram. Soc. **44** (1961) 225.
62DS	Denton, R. T., and E. G. Spencer: J. Appl. Phys. **33** (1962) 1300.
62GC	Goryaga, A. N., and L. Changta: Soviet Phys. – JETP (English Transl.) **14** (1962) 502.
62KB	Khoi, L. D., and F. Bertaut: Compt. Rend. **255** (1962) 1211.
62OMK	Ogawa, S., S. Morimoto, and Y. Kimura: J. Phys. Soc. Japan **17** (1962) 1671.
62SFR	Schnitzler, A. D., V. J. Folen, and G. T. Rado: J. Appl. Phys. **33** (1962) 1293.
62T	Tarte, P.: Compt. Rend. **254** (1962) 2008.
62YHMH	Yasuoka, H., A. Hirai, M. Matsuura, and T. Hashi: J. Phys. Soc. Japan **17** (1962) 1071.
63CCDCC	Callen, E. R., A. E. Clark, B. DeSavage, W. Coleman, and H. B. Callen: Phys. Rev. **130** (1963) 1735.
63HWS	Hickernell, F. S., R. G. West, and N. G. Sakiotis: J. Appl. Phys. **34** (1963) 1294.
63SB	Schulkes, J. A., and G. Blasse: J. Phys. Chem. Solids **24** (1963) 1651.
64BB	Brunel, M., and F. de Bergevin: Compt. Rend. **258** (1964) 5628.
64CN	Comstock, R. L., and W. G. Nilsen: Phys. Rev. **136** (1964) A 442.
64CR	Comstock, R. L., and J. P. Remeika: J. Appl. Phys. **35** (1964) 1018.
64DSR	Drokin, A. I., R. P. Smolin, and L. I. Ryabinkina: Soviet Phys. – Solid State (English Transl.) **5** (1964) 1502.
64P	Prince, E.: J. Phys. (Paris) **25** (1964) 503.
64SD	Spencer, E. G., an R. T. Denton: J. Appl. Phys. **35** (1964) 2633.
65B	Blasse, G.: Philips Res. Rept. **20** (1965) 528.
65DPB	Davydov, L. A., S. A. Poltinnikov, and M. F. Bryzhina: Soviet Phys. – Solid State (English Transl.) **1** (1965) 596.
65K	Kapitonova, N. P.: Soviet Phys. – Solid State (English Transl.) **6** (1965) 2111.
65RGA	Rogers, D. B., R. W. Germann, and R. J. Arnett: J. Appl. Phys. **36** (1965) 2338.
65S	Seleznev, V. N.: Izv. Vysshikh Uchebn. Zavedenii, Fiz. **2** (1965) 167.
67DSGY	Drokin, A. I., N. I., Sudakov, S. Sh. Gendeler, and V. K. Yanitskii: Soviet Phys. – Solid State (English Transl.) **8** (1967) 2684.
68PS	Petrakovskii, G. A., and E. M. Smokotin: Soviet Phys. – JETP (English Transl.) **26** (1968) 718.
68SLN	Spencer, E. G., D. A. Lepore, and J. W. Nielsen: J. Appl. Phys. **39** (1968) 732.
68WGK	Wang, F. F. Y., R. L. Gravel, and M. Kestigian: IEEE Trans. on Magnetics **4** (1968) 55.

6.1.7 Cu-Fe^{3+} spinels (Cu ferrites) and Cu-Fe^{3+} spinels with substitutions

6.1.7.1 Cu ferrite $CuFe_2O_4$

Single crystal synthesis

Mixed oxides of Fe_2O_3, CuO and the flux (47 mole % PbF_2 and 53 mole % PbO) were melted at 950 °C in air for several hours. The mole percentages of Fe_2O_3, CuO, and flux were 15, 35, and 50, respectively. The temperature was lowered at a rate of 1···13 °C/h to 500 °C [*64MMM*].

Cu ferrite ($CuFe_2O_4$) has a transformation from cubic (spinel) to a tetragonal (hausmannite) structure. The transformation temperature has been reported to be 760 °C in [*50WBB*], 415 °C in [*50TB*] and 390 °C in [*58II*]. When quenched from above the transition temperature, Cu ferrite possesses a cubic structure. When slowly cooled to room temperature, Cu ferrite possesses a tetragonal structure with a (c/a) ratio between 1.00 and 1.06, depending on the cooling rate [*50WBB*]. Also, the ionic distribution is dependent on the thermal history of Cu ferrite. In the cubic phase, the copper ions are situated both in the tetrahedral and octahedral sites so that cubic Cu ferrite has neither normal or inverse structure [*50WBB*]. In the tetragonal phase, the inverse structure is approached by annealing at temperatures near room temperature [*50WBB*], [*56PT*]

Tab. 1. $CuFe_2O_4$. General properties (room temperature unless otherwise specified).

		Ref.
Crystal type	Partially normal and partially inverse spinel — cubic (quenched)	*50WBB*
	Hausmannite (annealed)	*50WBB*
Structure type	$MgAl_2O_4$ (quenched)	
	Mn_3O_4 (annealed)	
Space group	O_h^7 — Fd3m (quenched)	
	D_{4h}^{19} — I 4_1/amd (annealed)	
Lattice constant	$a = 8.37$ Å (quenched from 960 °C) [1]	*50B*
	$c = 8.70$ Å, $a = 8.22$ Å (annealed) [1]	*56PT*
Volume of unit cell	$586.4 \cdot 10^{-24}$ cm³ (cubic phase, containing 8 molecules)	
Oxygen parameter	$u = 0.380 \pm 0.005$ (cubic phase)	*47VH*
Molecular weight	239.2 amu	
X-ray density	$\varrho_X = 5.42$ g/cm³ (cubic phase)	
Melting point	(1560 ± 10) °C	*36AVB*
Sintering temperatures	700···1200 °C	
Thermal conductivity	$\varkappa \cong 1.5 \cdot 10^{-2}$ cal sec^{-1} cm^{-1} deg^{-1}	*54SW*
Specific heat	$c_p = 0.22$ cal g^{-1} deg^{-1}	*58II*
Young's modulus	$E = \cong 1.5 \cdot 10^{12}$ dyn/cm²	*54SW*
Dielectric constant	$\lvert\varepsilon\rvert = 11.1$ (at 172.5 Mc/s)	*51W*(1)
	$\varepsilon' = 9.29$ (quenched from 1000 °C) [2], 8.65 (slowly cooled)	*52OFD*(1)
	$\varepsilon'' = 0.520$ (quenched from 1000 °C) [2], 0.089 (slowly cooled)	*52OFD*(1)

[1]) See also [*61BC*], [*65AMY*], [*67OI*], and [*66OKI*].
[2]) Measured at 6.6 cm wavelength.

Tab. 2. $CuFe_2O_4$. Structure parameters (determined by neutron diffraction). Space group D_{4h}^{19} — I 4_1/amd. Origin at center of symmetry [56PT]

Atom	x	y	z
Fe (A sites)	0	$\frac{3}{4}$	$\frac{1}{8}$
Fe, Cu (B sites)	0	0	$\frac{1}{2}$
O	0	0.016	0.250

Tab. 3. $CuFe_2O_4$. Fraction of Cu on B sites as a function of atomic fraction of oxygen f_O (f_O = 4/7 for the stoichiometric compound) and quenching temperature T_Q [66OLS]. See also [50WBB]

T_Q °C	f_O 0.5700	0.5704	0.5707	0.5711
900	0.790	0.803	0.814	0.829
700				0.837
500		0.828	0.853	0.880

Tab. 4. $CuFe_2O_4$. Magnetic properties (room temperature unless otherwise specified)

		Ref.
Ferromagnetic Curie temperature	Θ_C = 433···486 °C [1][2]	64RS
Paramagnetic Curie temperature	Θ_p = 407···438 °C [1]	32S
Saturation magnetic moment per molecule (0 °K)	$p_{sm}^0 = 2.3\ \mu_B$ (T_Q = 900 °C)	50G
	$p_{sm}^0 = 1.3\ \mu_B$ (annealed at 360 °C)	50G
Saturation magnetic moment per gram	σ_s = 25 Gauss cm³/g [1]	59SW
	σ_s^0 = 30 Gauss cm³/g (0 °K) [1]	59SW
Saturation magnetic moment per unit volume	M_s = 135 Gauss [1]	59SW
	M_s^0 = 160 Gauss (0 °K) [1]	59SW
Initial permeability	$\mu_i \approx 70$ (quenched)	46S
	μ' = { 0.06 (quenched from 1000 °C) [3]; 0.62 (slowly cooled) [3] }	50OFD(1)
	μ'' = { 1.24 (quenched from 1000 °C) [3]; 0.74 (slowly cooled) [3] }	50OFD(1)
Saturation magnetostriction	$\lambda_s \approx -1.5 \cdot 10^{-5}$ (polycrystal) [4]	54W
Experimentally observed g factor	g_{exp} = 2.09 [5]	65MMM
	g_{exp} = 2.44 (77 °K) [5]	65MMM
Magnetochemical ratio	g' = 1.94	32R
First order magnetocrystalline anisotropy constant	$K_1 = -0.63 \cdot 10^5$ erg/cm³ (15 °C) [6]	52OK
	$K_1 = -2.06 \cdot 10^5$ erg/cm³ (−195 °C) [6]	52OK

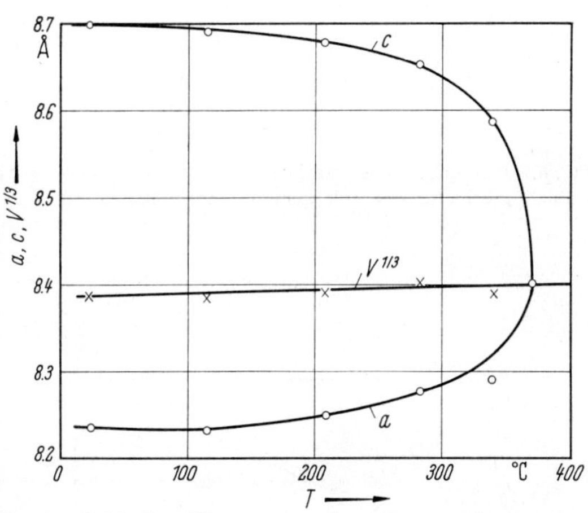

Fig. 1. $CuFe_2O_4$. Temperature dependence of lattice parameters. V is the volume of the unit cell [61OT].

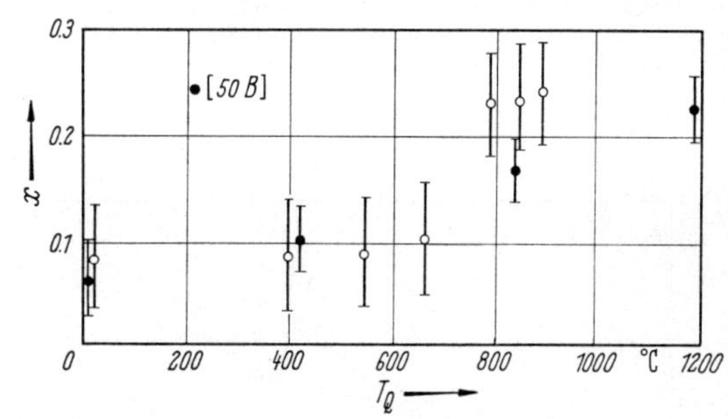

Fig. 2. $CuFe_2O_4$. The cation distribution parameter, x, (see caption of Fig. 10) vs. quenching temperature [61OT]. See also [50B].

[1]) These quantities are dependent on ionic distribution and crystal structure which is a function of thermal treatment before and during measurement.
[2]) See also [59SW] and [51OKT].
[3]) Measurements at 6.6 cm wavelength.
[4]) See also [51S].
[5]) Microwave measurements at 9 Gc/s. See [52OK].
[6]) Microwave measurements at 24 Gc/s. See also [65MMM].

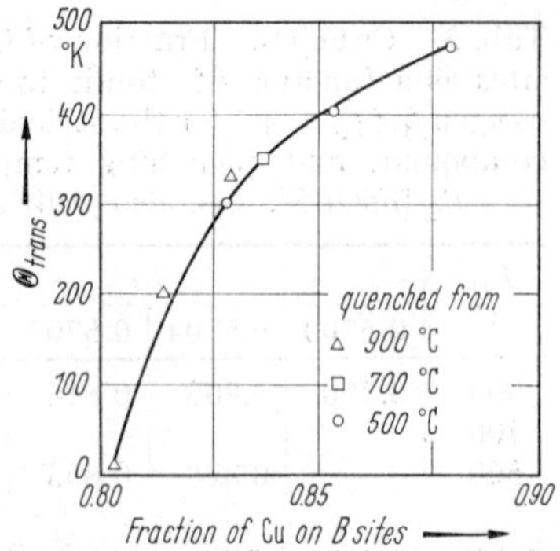

Fig. 3. $CuFe_2O_4$. Cubic to tegragonal transition temperature, Θ_{trans}, vs. atomic fraction of Cu on B sites [*66OIS*]. See also Tab. 3.

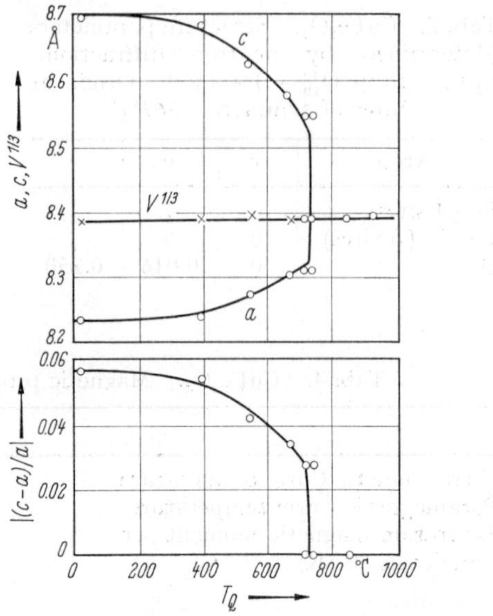

Fig. 4. $CuFe_2O_4$. Lattice parameters and tetragonality vs. quenching temperature. V is the volume of the unit cell [*61OT*].

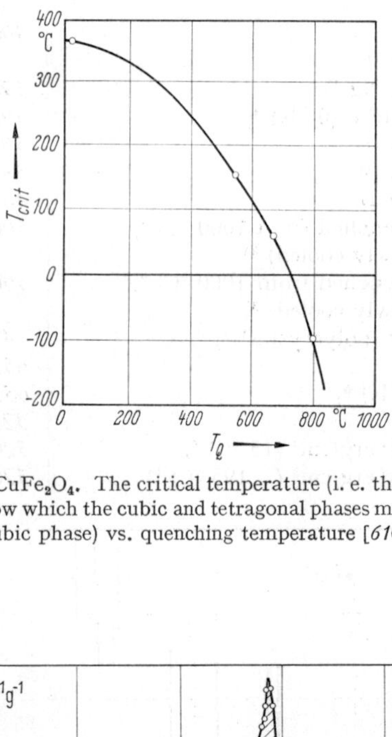

Fig. 5. $CuFe_2O_4$. The critical temperature (i. e. the temperature below which the cubic and tetragonal phases merge into a single cubic phase) vs. quenching temperature [*61OT*].

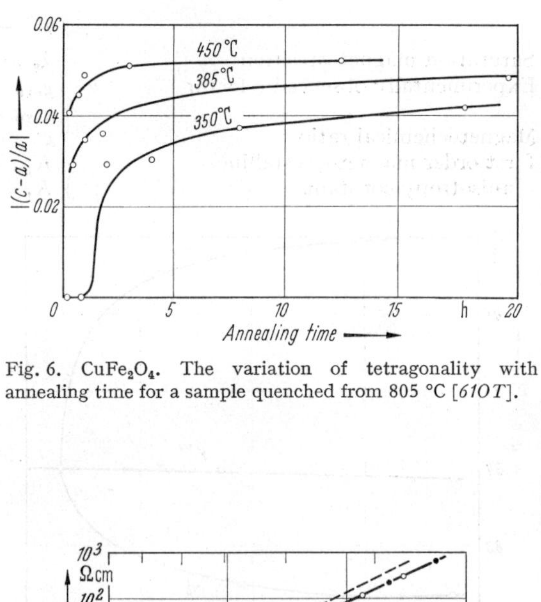

Fig. 6. $CuFe_2O_4$. The variation of tetragonality with annealing time for a sample quenched from 805 °C [*61OT*].

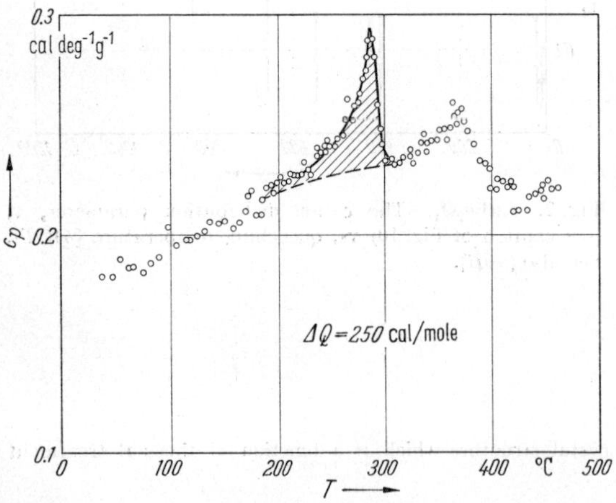

Fig. 7. $CuFe_2O_4$. Specific heat vs. temperature. After rapid cooling in vacuum from 750 °C, the anomaly at 390 °C disappeared. It reappeared after annealing again in air at 900 °C for several hours [*58II*].

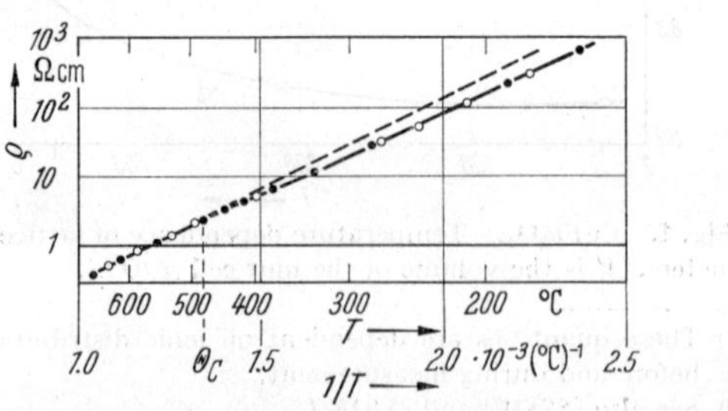

Fig. 8. $CuFe_2O_4$. Temperature dependence of the electrical resistivity. The dashed line indicates a break in the ϱ vs. $1/T$ line near the ferromagnetic Curie point [*54KK*].

Tab. 5. $CuOFe_2O_3$. Dielectric constant and permeability for polycrystalline samples, quenched and slowly cooled, from experiments at 6.6 cm and at room temperature. The specimens were discs 20 mm in diameter and 1.5 mm in thickness [*52OFD*(1)]. See also [*52OFD*(2)].

Ferrite	ε'	ε''	$\tan\delta$	μ'	μ''	$\tan\delta_m$
$CuOFe_2O_3$ quenched from 1000 °C	9.29	0.520	0.056	0.06	1.240	20.6
$CuOFe_2O_3$ slowly cooled	8.65	0.089	0.010	0.62	0.740	1.19

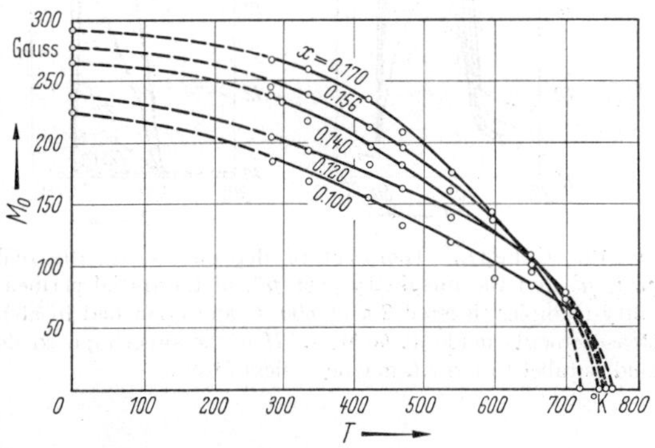

Fig. 9. $CuFe_2O_4$. Spontaneous magnetization, M_0, vs. temperature for various cation distributions x, in $(Cu_xFe_{1-x})[Cu_{1-x}Fe_{1+x}]O_4$ (see Fig. 10) [*64SBR*]. See also [*59SW*], [*52P*], and [*51PB*].

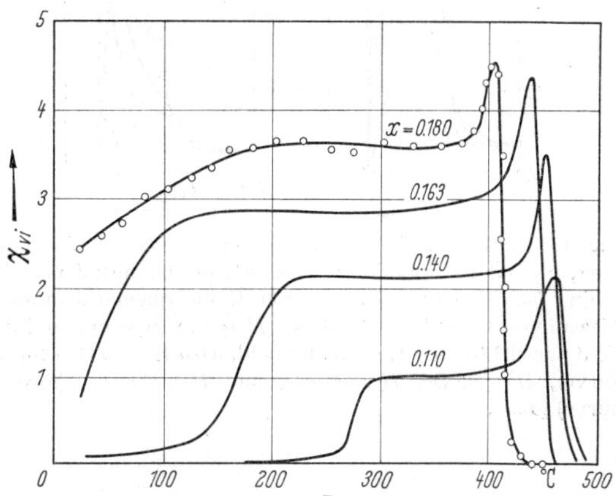

Fig. 10. $CuFe_2O_4$. Initial volume susceptibility, χ_{vi}, vs. temperature for various cation distributions, x, in $(Cu_xFe_{1-x})[Cu_{1-x}Fe_{1+x}]O_4$; the parenthesis and brackets refer to tetrahedral and octahedral site occupancies, respectively [*64SBR*]. See also [*64RS*] and [*29S*].

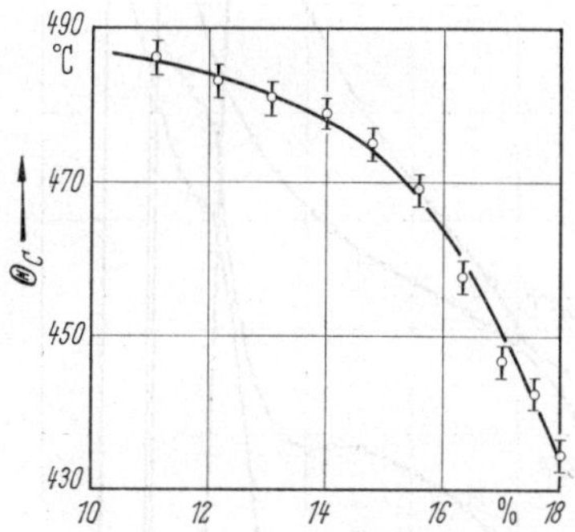

Fig. 11. $CuFe_2O_4$. The dependence of the ferromagnetic Curie temperature, Θ_C, on cation distribution. x is the percent of Cu ions occupying tetrahedral sites [*64RS*].

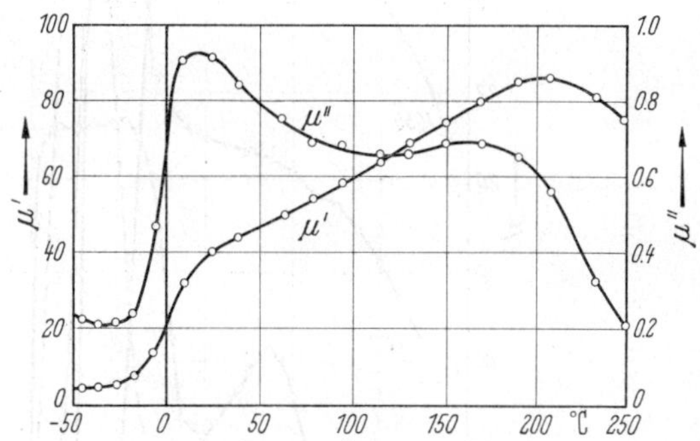

Fig. 12. $CuFe_2O_4$. The real (μ') and imaginary (μ'') parts of the initial permeability measured at 100 kc/s for a sample that transforms from a tetragonal to a cubic structure at 157 °C [*65ISG*].

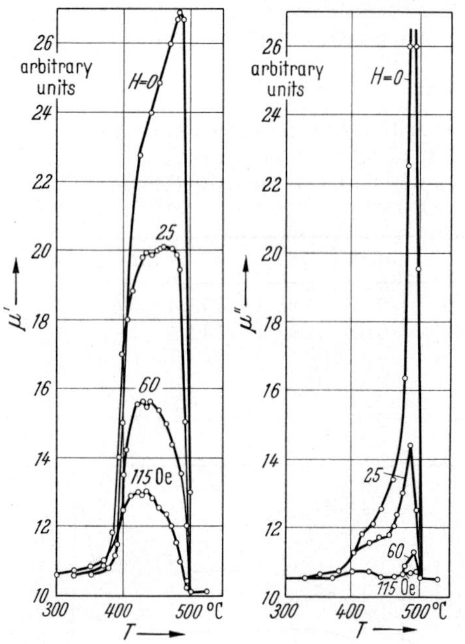

Fig. 13. $CuFe_2O_4$. Temperature dependence of the real part, μ', and the imaginary part, μ'', of the initial permeability (sample sintered 2 h at 800 °C and quenched in air). Measurements made at 13 Mc/s. H is the superimposed dc field parallel to the r.f. magnetic field, with H = 0 Oe (upper curve), H = 25 Oe, H = 60 Oe, and H = 115 Oe (lowest curve) [*56S*].

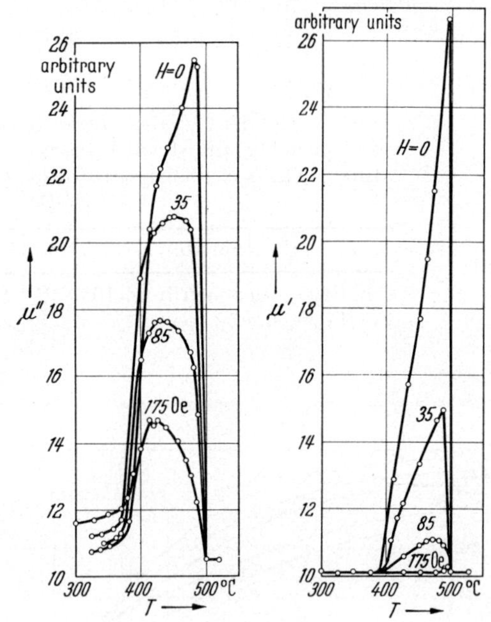

Fig. 15. $CuFe_2O_4$. Temperature dependence of the real part, μ', and the imaginary part, μ'', of the initial permeability (samples sintered 2 h at 800 °C and quenched in air). Measurements made at 60 Mc/s. H is the superimposed dc field parallel to the r.f. magnetic field [*56S*].

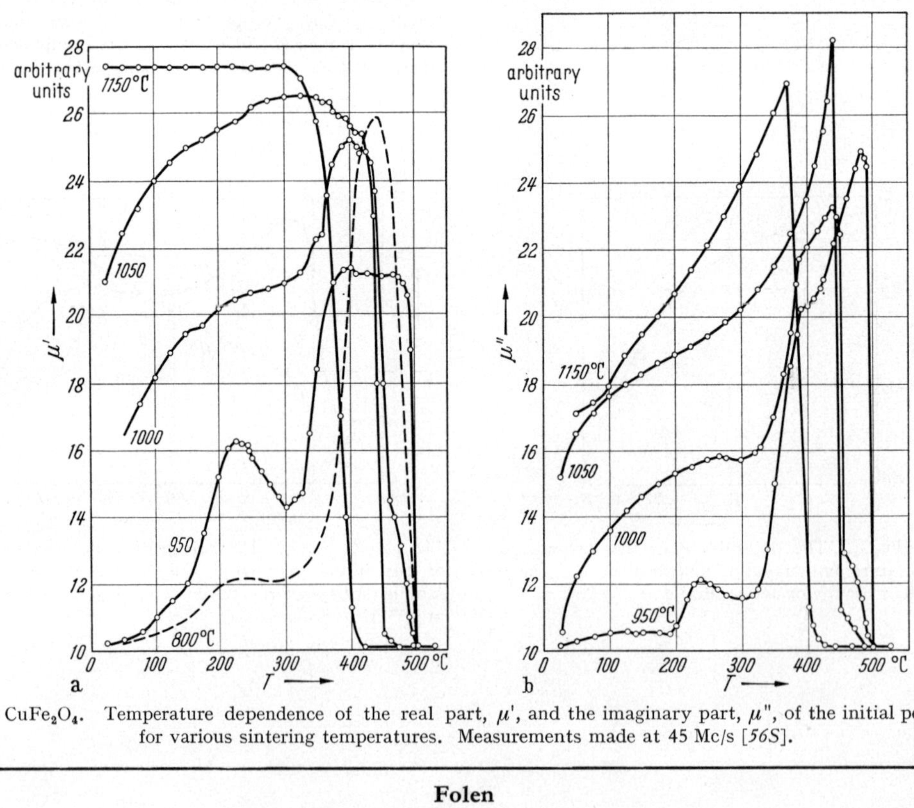

Fig. 14. $CuFe_2O_4$. Temperature dependence of the real part, μ', and the imaginary part, μ'', of the initial permeability for various sintering temperatures. Measurements made at 45 Mc/s [*56S*].

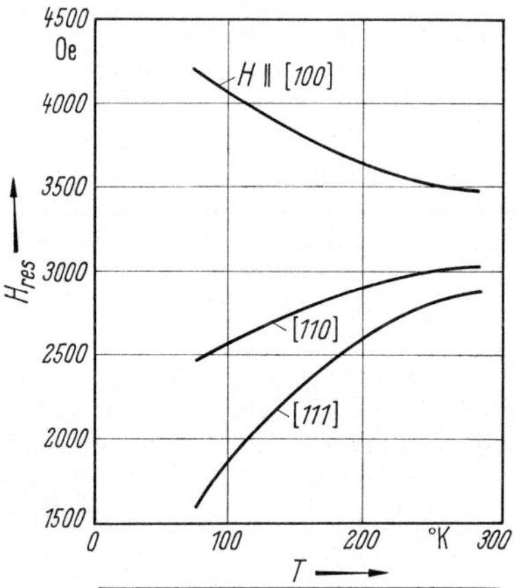

Fig. 18. $CuFe_2O_4$. Temperature dependence of the resonance magnetic fields along the [100], [110], and [111] directions in a cubic sample [*65MMM*]. See also [*51OKT*]. ←

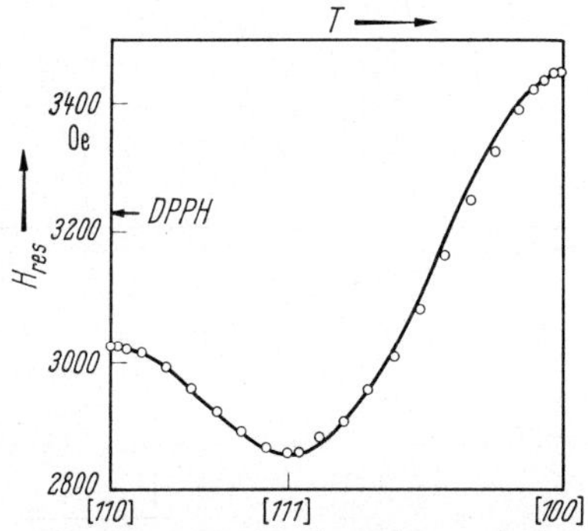

Fig. 16. $CuFe_2O_4$. Angular dependence of the resonance magnetic field in (110) plane at room temperature for a cubic sample. Open circles are observed points and solid curve is calculated with $K_1/M = -172$ Oe, $K_2/M = 0$ and $g_{eff} = 2.09$ [*65MMM*]. See also [*52OK*].

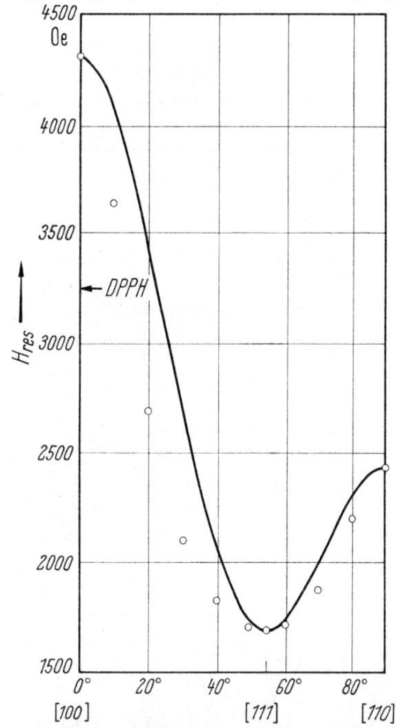

Fig. 17. $CuFe_2O_4$. Angular dependence of the resonance magnetic field in (110) plane at 77 °K for a cubic sample. Open circles are observed points and solid curve is calculated with $K_1/M = -810$ Oe, $K_2/M = +272$ Oe and $g_{eff} = 2.44$ [*65MMM*].

Tab. 6. $CuFe_2O_4$. Temperature dependence of saturation magnetization, anisotropy constant, and g_{exp} factor. g_{exp} and K_1 for single crystal were calculated from the M_s which was determined by measurements on polycrystalline samples. g_{exp} is also given for spherical polycrystalline specimen of diameter 0.8 mm. Although the single crystals were furnace cooled from 900 °C to room temperature, the crystal symmetry was assumed cubic [*52OK*]

T °C	M_s Gauss	K_1 10^4 erg/cm³	g_{exp}	g_{exp} ¹)
−195	315	−2.06	2.20	2.17
−150	310	−1.80	2.13	2.13
−100	305	−1.43	2.09	2.08
− 50	295	−1.01	2.06	2.05
15	280	−0.63	2.04	2.04
50	250	−0.43	2.04	2.05
100	220	−0.24	2.07	2.05
150	180	−0.16	2.01	2.05
200	165	−0.10	2.01	2.05
250	135	−0.07	2.03	2.05
300	110	−0.04	2.04	2.05
350	80	−0.02	2.04	2.06
400	60	−0.01	2.05	2.06
450	30		2.05	2.06

¹) Determined by the resonance experiment on a polycrystalline spherical specimen.

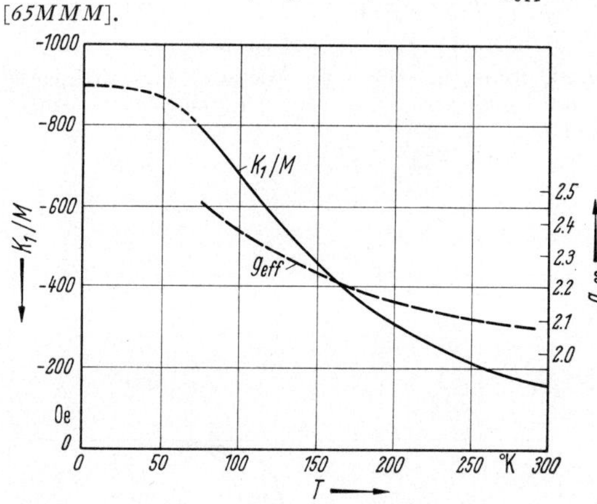

Fig. 19. $CuFe_2O_4$. Temperature dependence of g_{eff} and K_1/M in a cubic sample [*65MMM*].

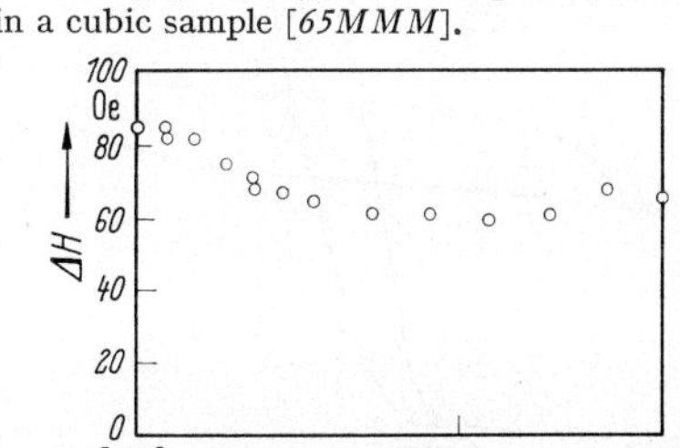

Fig. 20. $CuFe_2O_4$. Angular dependence of the ferromagnetic resonance line width in the (110) plane at room temperature for a cubic sample [*65MMM*]. See also [*51OT*(2)] and [*51OKT*].

Tab. 7. Mössbauer data on $CuFe_2O_4$. *A* and *B* refer to the tetrahedral and octahedral sites, respectively. I and II refer to the magnetically nonequivalent Fe sites and because type I sites exhibit an electric quadrupole interaction $[e^2qQ(3\cos^2\theta-1)]$, type I was assigned to *B* sites. The shift (310 SS) refers to the isomer shift with respect to 310 stainless steel [*66EHK*]. See also Fig. 21

Phase	Tetragonal		Cubic	
Site	*A*	*B*	*A* + *B*	
Pattern	II	I	—	
H_A, H_B (300 °K)	478 ± 5	514 ± 5	473 ± 3	kOe
e^2qQ (3 cos θ − 1)	0.01 ± 0.06	0.69 ± 0.09	0.03 ± 0.05	mm/sec
Shift (310 SS)	0.41 ± 0.02	0.58 ± 0.02	0.40 ± 0.02	mm/sec
Interatomic distance	1.99	2.18	1.89	Å
Intensity ratio II/I	0.65			

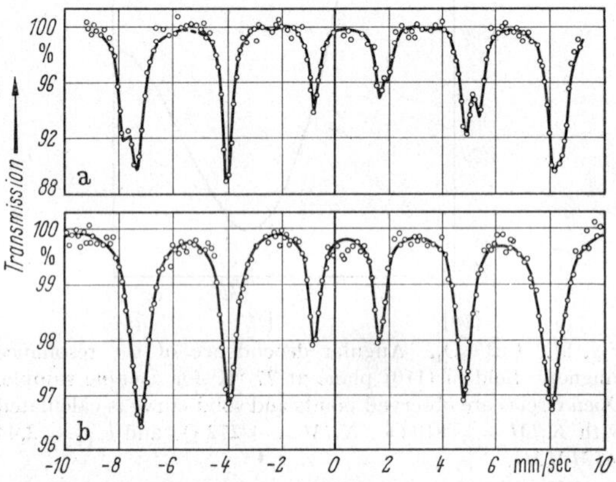

Fig. 21. $CuFe_2O_4$. Mössbauer spectra of tetragonal (curve *a*) and cubic (curve *b*) sample. The source is ^{57}Co in Cr [*66EHK*].

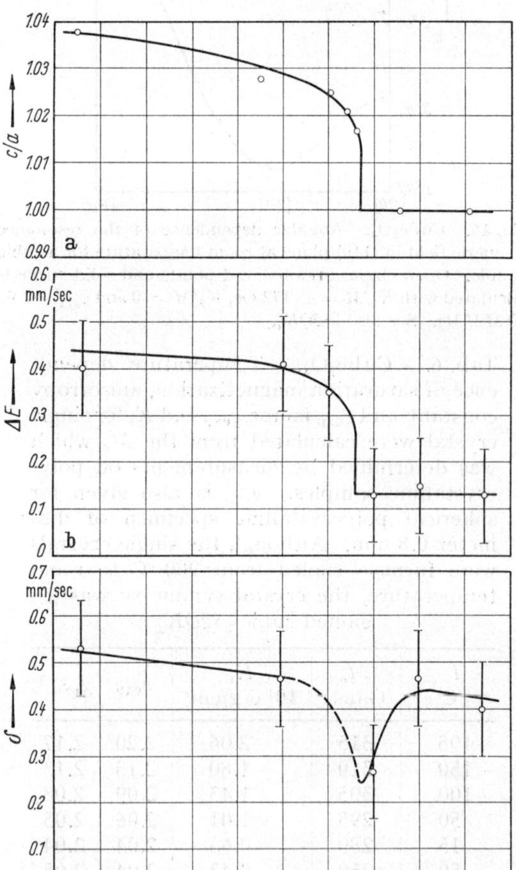

Fig. 22. $CuFe_2O_4$. a) Temperature dependence of the tetragonality (the axial ratio *c*/*a*),

b) temperature dependence of the electric quadrupole interaction, $E = \frac{1}{2}e^2qQ(3\cos^2\theta - 1)$, as determined from Mössbauer measurements (ΔE = shift),

c) temperature dependence of the isomer shift, δ.

The velocity scale was calibrated by using both iron metal and stainless steel absorbers as standards [*62YMO*].

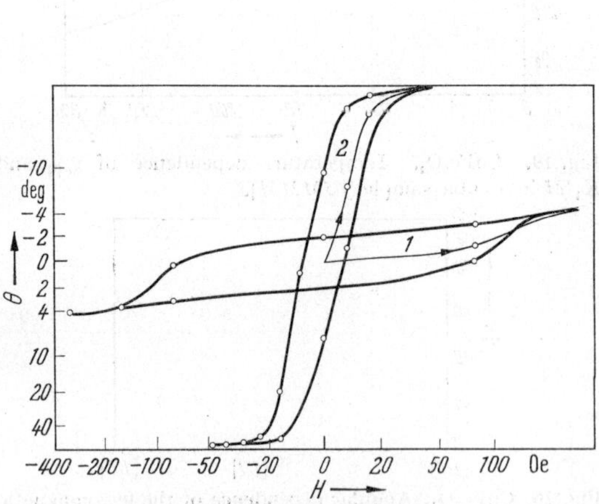

Fig. 23. $CuFe_2O_4$. Faraday rotation angle, θ, in a static longitudinal magnetic field [*56M*]. Curve 1, tetragonal structure. Curve 2, cubic structure. The specimens were ellipsoids having length of 42.5 mm and diameter of 6.7 mm.

6.1.7.2 Cu-Zn and Cu-Cd ferrites

Cu-Zn ferrites containing 5 molar % Zn ferrite or more have been found to possess the cubic spinel structure (O_h^7 — Fd3m). For compositions below this percentage, the Hausmannite structure can exist [*51TB*].

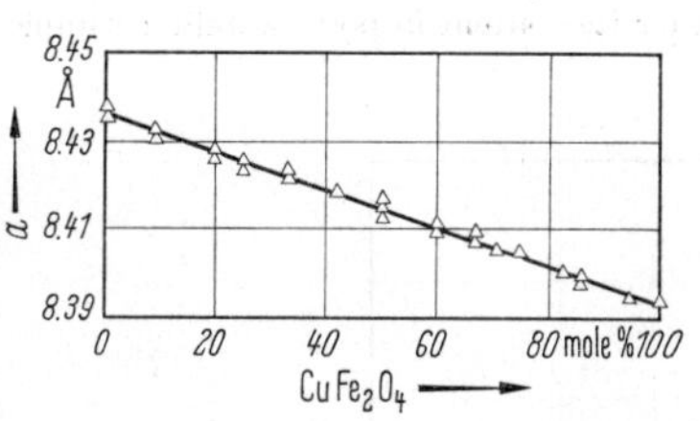

Fig. 1. Cu-Zn ferrites. Lattice parameter vs. composition. Final stage in sample preparation was an annealing at 1100 °C in air followed by a furnace cooling to room temperature in approximately 12 h [*65AMY*].

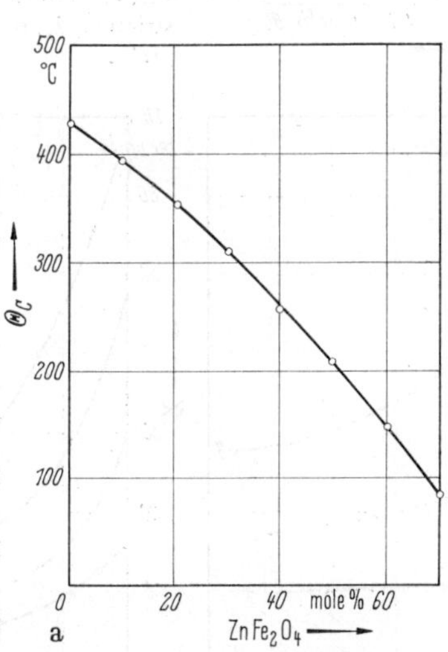

Fig. 2a. Cu-Zn ferrites. Dependence of Curie temperature, Θ_C, on Zn ferrite content in polycrystalline samples [*51S*].

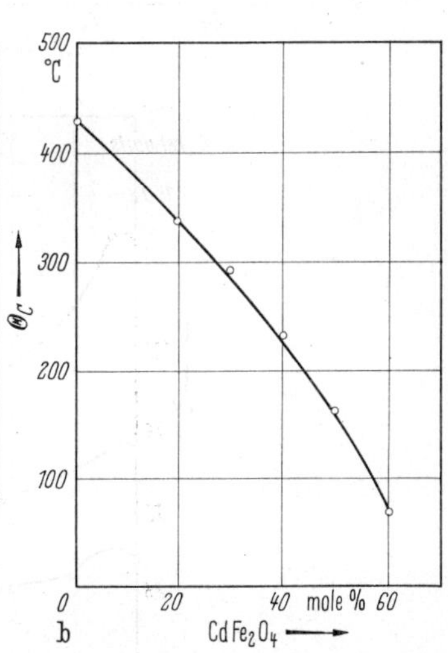

Fig. 2b. Cu-Cd ferrites. Dependence of Curie temperature, Θ_C, on Cd ferrite content in polycrystalline samples [*51S*].

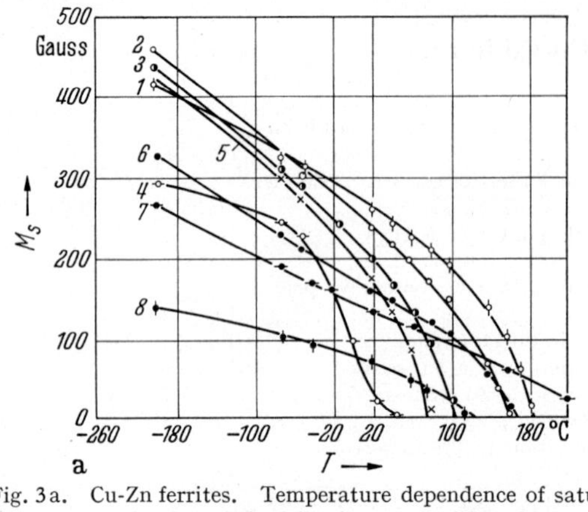

Fig. 3a. Cu-Zn ferrites. Temperature dependence of saturation magnetization of the following compositions [54S]:

	CuO	ZnO	Fe_2O_3	
1	30	20	50	mole%
2	25	25	50	mole%
3	22.5	27.5	50	mole%
4	20.8	31.2	48	mole%
5	19.2	28.8	52	mole%
6	18.4	27.6	54	mole%
7	17.6	26.4	56	mole%
8	12.0	18.0	70	mole%

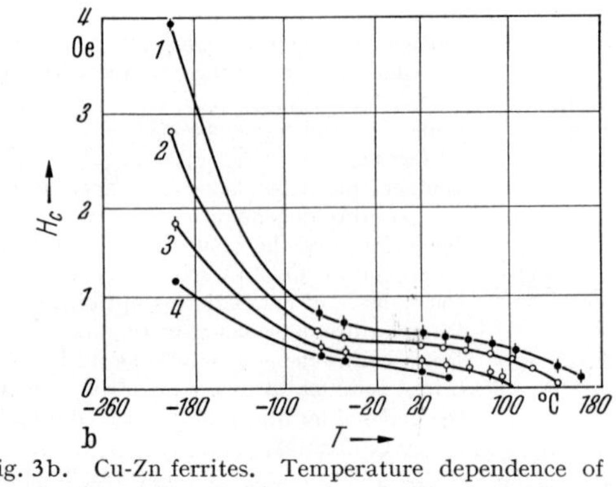

Fig. 3b. Cu-Zn ferrites. Temperature dependence of the coercive force. Compositions given in Fig. 3a [54S].

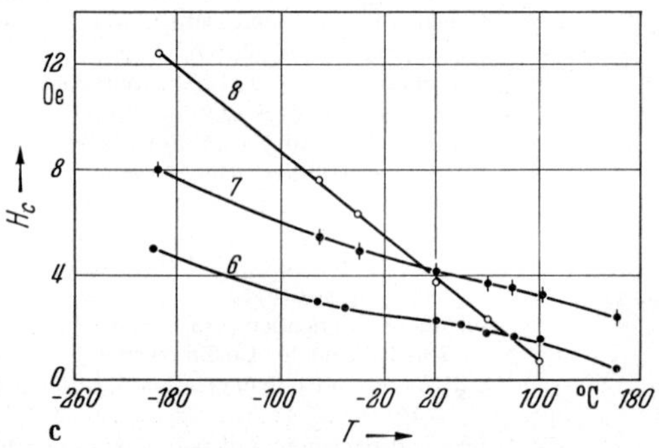

Fig. 3c. Cu-Zn ferrites. Temperature dependence of the coercive force. Compositions given in Fig. 3a [54S].

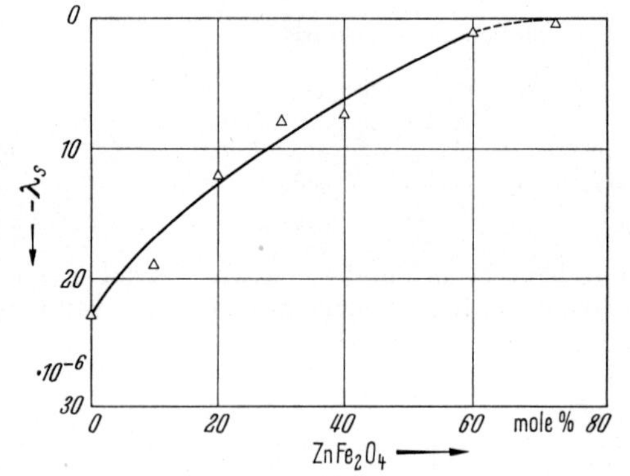

←

Fig. 4. Cu-Zn ferrites. Dependence of saturation magnetostriction, λ_s, on Zn ferrite content in polycrystalline samples [51S].

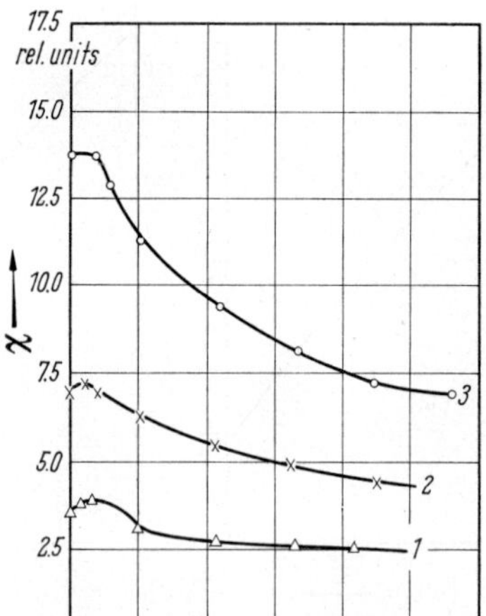

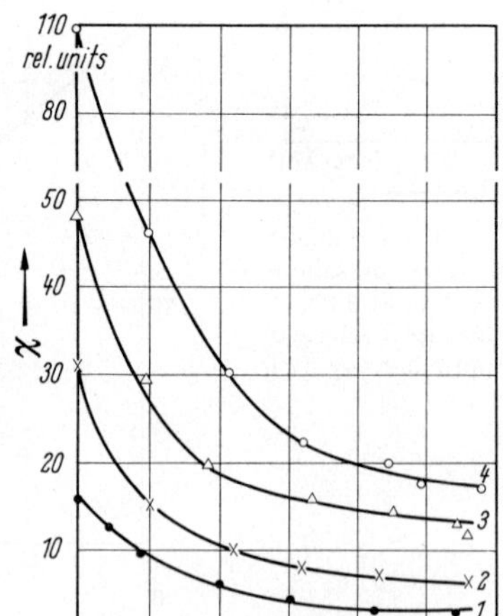

Fig. 5. Cu-Zn ferrites. Dependence of χ on pressure: a) for (1) $CuFe_2O_4$, (2) $Cu_{0.9}Zn_{0.1}Fe_2O_4$, (3) $Cu_{0.8}Zn_{0.2}Fe_2O_4$; b) for (1) $Cu_{0.7}Zn_{0.3}Fe_2O_4$, (2) $Cu_{0.6}Zn_{0.4}Fe_2O_4$, (3) $Cu_{0.5}Zn_{0.5}Fe_2O_4$, (4) $Cu_{0.4}Zn_{0.6}Fe_2O_4$. Measurements at 5 kc/s [64SK].

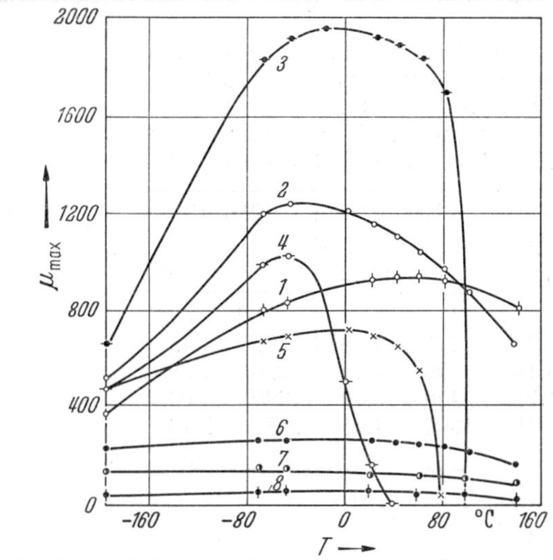

Fig. 6. Cu-Zn ferrites. Temperature dependence of the maximum permeability. Compositions given in Fig. 3a [*54S*].

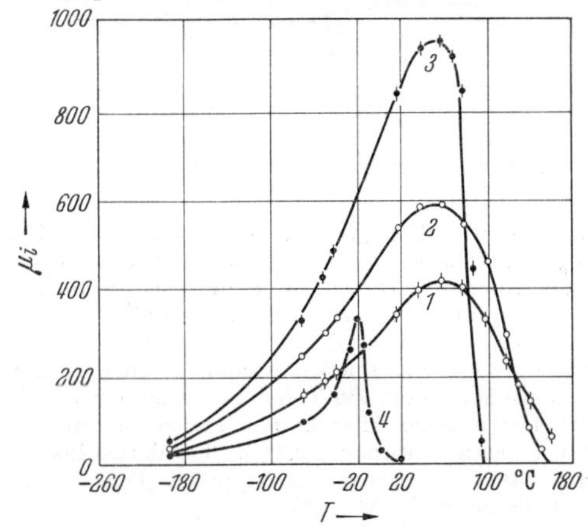

Fig. 7. Cu-Zn ferrites. Temperature dependence of the initial permeability. Compositions given in Fig. 3a [*54S*].

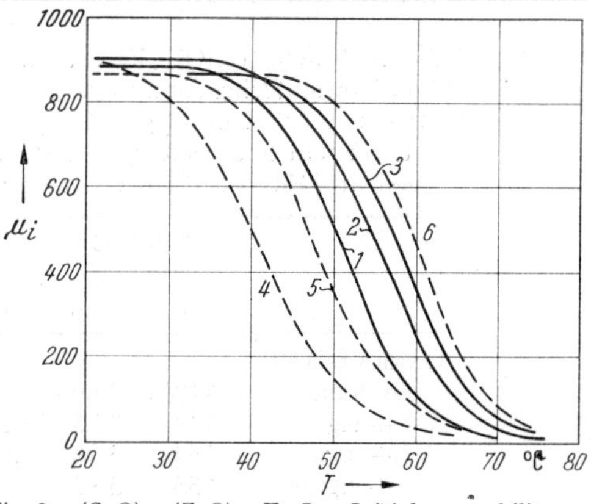

Fig. 8. $(CuO)_{0.40}(ZnO)_{0.60}Fe_2O_3$. Initial permeability, μ_i, vs. temperature [*50Br*].
Curve 1. Sample quenched in air from 550 °C.
Curve 2. Quenched in air from 600 °C.
Curve 3. Quenched in air from 650 °C.
Curve 4. Slowly cooled from 650 °C to room temperature.
Curve 5. Quenched in air from 500 °C.
Curve 6. Quenched in air from 750 °C.
Numbering of curves indicates order in which experiment was performed, same sample being used for all curves.

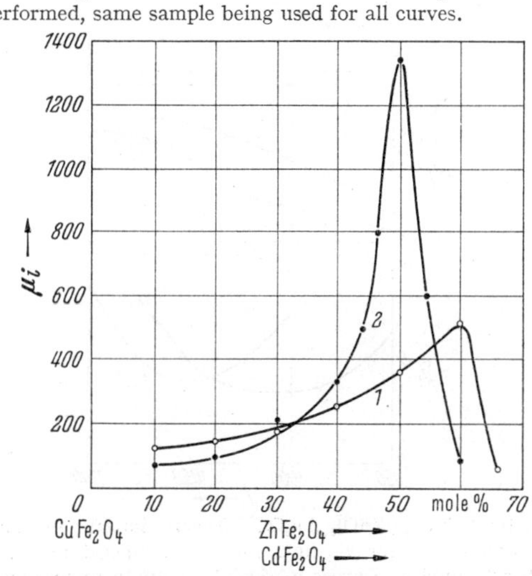

→

Fig. 9. Cu-Zn ferrites (1) and Cu-Cd ferrites (2). Dependence of the initial permeability, μ_i, on composition [*51S*].

Tab. 1. $(CuO)_{0.4}(ZnO)_{0.6}Fe_2O_3$. Initial permeability of polycrystalline samples showing effect of heat treatment. The two samples were prepared from the same mixture and differed only in the heat treatments shown in this table. The samples were sintered in oxygen. The symbols r.c. and s.c. refer to rapid cooling (removed from furnace) and slow cooling (in furnace), respectively. Heat treatments were applied successively, as shown, to each sample [*49S*]

Sample I		μ_i		
Heat treatment	ϱ g/cm³	2 kc/s	100 kc/s	1000 kc/s
6 h 1000 °C s.c.	3.75	112	112	112
+1 h 1200 °C r.s.	5.15	590	590	580
+1 h 1000 °C s.c.	5.15	1145	1052	758
+1 h 950 °C s.c.	5.15	1120	1055	680
Sample II		μ_i		
Heat treatment	ϱ g/cm³	2 kc/s	100 kc/s	1000 kc/s
36 h 1000 °C s.c.	3.90	130	130	130
+1 h 1200 °C s.c.	5.10	1070	990	900
+1 h 1000 °C s.c.	5.10	1040	1015	740
+1 h 950 °C r.s.	5.10	1125	1030	645

Tab. 2. $(CuO)_{0.4}(ZnO)_{0.6}Fe_2O_3$. Dependence of initial permeability, measured at 2 kc/s on sintering conditions. The symbols s.g. in the table stands for sintering for two h at 900 °C in oxygen, followed by grinding for two h in an iron mill. After this treatment had been applied 0, 1, 2 or 3 times, the samples were sintered in oxygen at 1100 °C, 1150 °C, 1200 °C or 1250 °C as shown in the table [*49S*]

	1100 °C		1150 °C		1200 °C		1250 °C	
	μ_i	ϱ g/cm³	μ_i	ϱ g/cm³	μ_i	ϱ g/cm³	μ_i	ϱ g/cm³
Oxides mixed by hand	118	3.60	175	3.75	285	4.05	295	4.45
+ s.g.	148	3.75	260	4.10	390	4.55	1230	4.85
+ s.g.	154	3.90	310	4.25	450	4.80	1290	5.00
+ s.g.	155	3.95	340	4.45	1000	4.95	1580	5.10

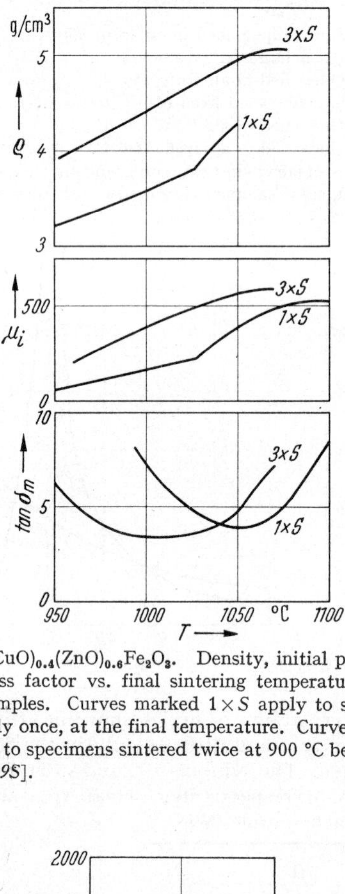

Fig. 10. $(CuO)_{0.4}(ZnO)_{0.6}Fe_2O_3$. Density, initial permeability, and loss factor vs. final sintering temperature, T, in sintered samples. Curves marked $1\times S$ apply to specimens sintered only once, at the final temperature. Curves marked $3\times S$ apply to specimens sintered twice at 900 °C before final sintering [*49S*].

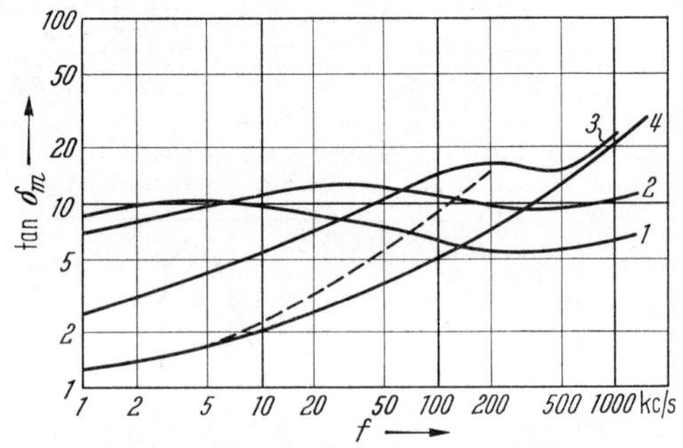

Fig. 11. $(CuO)_{0.4}(ZnO)_{0.6}Fe_2O_3$. The magnetic loss factor, tan δ_m, vs. frequency in porous polycrystalline samples having different oxygen content. Curve 1 applies to the nearly stoichiometric ferrite. Curve 2 is obtained after heat treatment of sample at 1000 °C and cooling in furnace (oxygen deficiency ≈0.01%). Curve 3 is obtained after quenching the sample in air from 950 °C (oxygen deficiency ≈0.03%). Curve 4 is obtained after heating sample in nitrogen at 950 °C (oxygen deficiency ≈0.2%). The dotted line is for Ferroxcube 1a, which is $(CuO)_{0.4}(ZnO)_{0.6}Fe_2O_3$ with an oxygen deficiency ≈0.3% [*49S*].

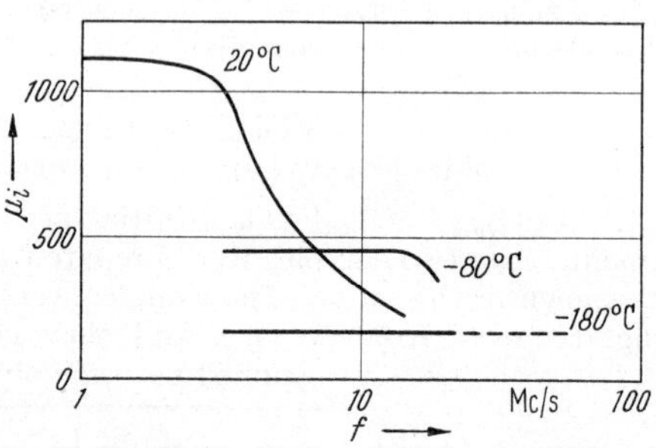

Fig. 13. $(CuO)_{0.4}(ZnO)_{0.6}Fe_2O_3$. Initial permeability of polycrystalline samples vs. frequency for three temperatures [*49S*].

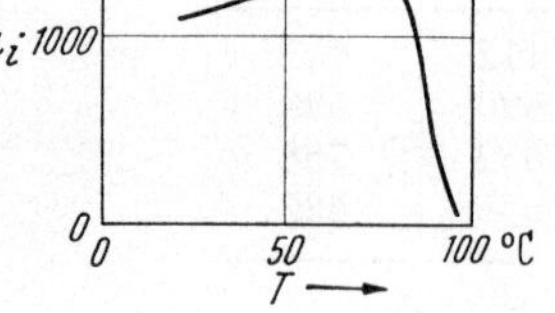

Fig. 12. $(CuO)_{0.4}(ZnO)_{0.6}Fe_2O_3$. Initial permeability vs. temperature for polycrystalline samples [*49S*].

Tab. 3. $(CuO)_{0.4}(ZnO)_{0.6}Fe_2O_3$. Dependence of magnetic loss tangents on frequency in the two polycrystalline samples described in Tab. 1. These measurements were made after the third heat treatment of each sample [*49S*]

f kc/s	tan δ_m		f kc/s	tan δ_m	
	Sample 1	Sample 2		Sample 1	Sample 2
2	0.027	0.030	200	0.173	0.197
16	0.056	0.063	500	0.325	0.386
40	0.080	0.094	750	0.418	0.413
100	0.141	0.158	1000	0.465	0.413

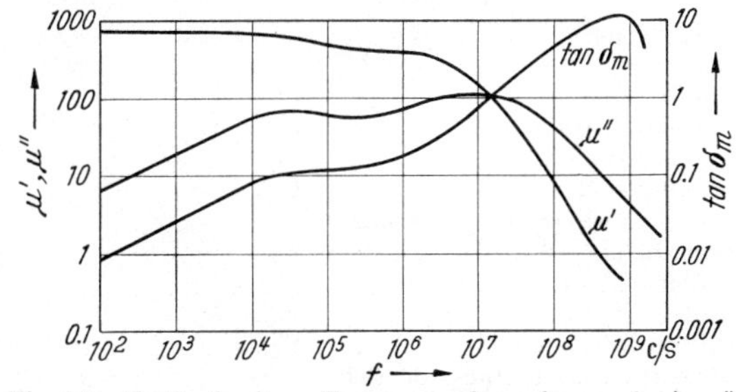

Fig. 14. Cu-Zn ferrite. Frequency dependence of μ', μ'', and $\tan\delta_m$. $\varepsilon' = 8.5$ ($f > 10^8$ c/s), $\sigma \leq 10^{-9}\,\Omega^{-1}\,cm^{-1}$. No molecular formula given [52H]. See also [50F], [53H].

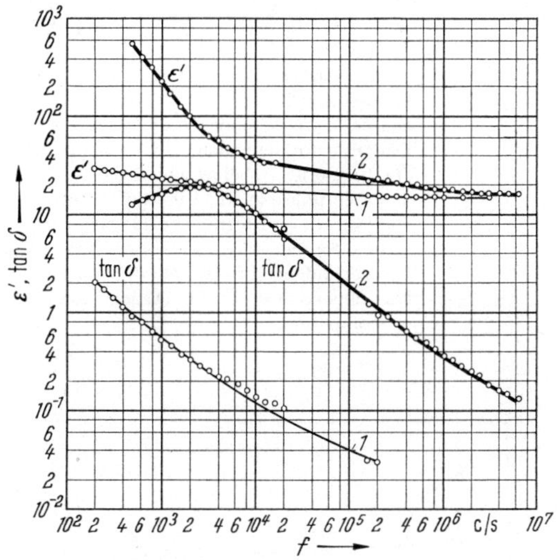

Fig. 15. Cu-Zn ferrites. Frequency dependence of ε' and $\tan\delta$ [52M].
1) $Cu_{0.10}Zn_{0.90}Fe_2O_4$, $\sigma = 7.0 \cdot 10^{-11}\Omega^{-1}cm^{-1}$,
2) $Cu_{0.50}Zn_{0.50}Fe_2O_4$, $\sigma = 1.8 \cdot 10^{-6}\Omega^{-1}cm^{-1}$.

Tab. 4. $Cu_xZn_{1-x}Fe_2O_4$. g_{exp} factors for polycrystalline specimens at 0 °C and −195 °C. Measurements done at 3.2 cm wave length [51OT(1)]

x	g_{exp} (0 °C)	g_{exp} (−195 °C)
0.0	2.09	2.32
0.2	2.10	2.33
0.4	2.21	2.47
0.6	2.11	2.44
0.8	1.95	2.07
1.0	1.93	2.06

Tab. 5. $Cu^{2+}_{0.390}Zn^{2+}_{0.602}Fe^{3+}_2O_4$. Physical parameters of polycrystalline samples. Sintering times were short so that there was only a slight conversion of Fe^{3+} to Fe^{2+} [53B]

Measured density	$\varrho = 4.50\ g/cm^3$
Initial susceptibility	$\chi_{vi} = 4.55 \cdot 10^{-5}$
Magnetomechanical coupling coefficient [1])	$k = -0.006$
Young's modulus	$E = 1.14 \cdot 10^{12}\ dyn/cm^2$

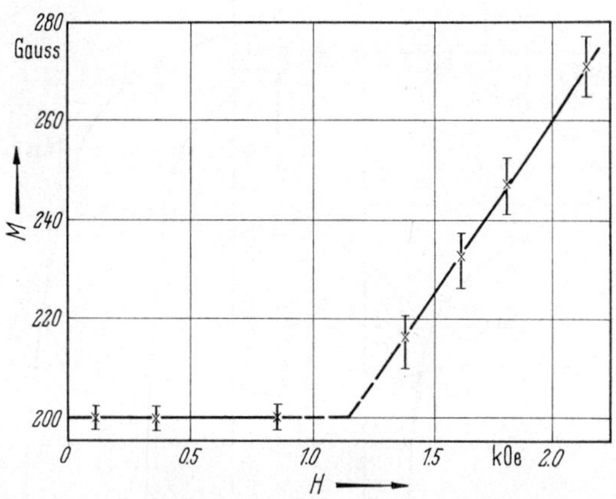

Fig. 16. $Cu_{0.8}Cd_{0.2}Fe_2O_4$. Dependence of magnetization, M, on field, H, indicating the existence of noncollinearity of the magnetic sublattice vectors [64RVK].

[1]) k^2 is defined as the ratio between electromagnetic and elastic energy when only an alternating low frequency stress is applied. Therefore $k^2 = 4\pi\lambda'\mu_r/E$, when λ' is the dynamic magnetostrictive constant, μ_r is the reversible permeability, and E is Young's modulus.

6.1.7.3 Cu-Cr ferrites

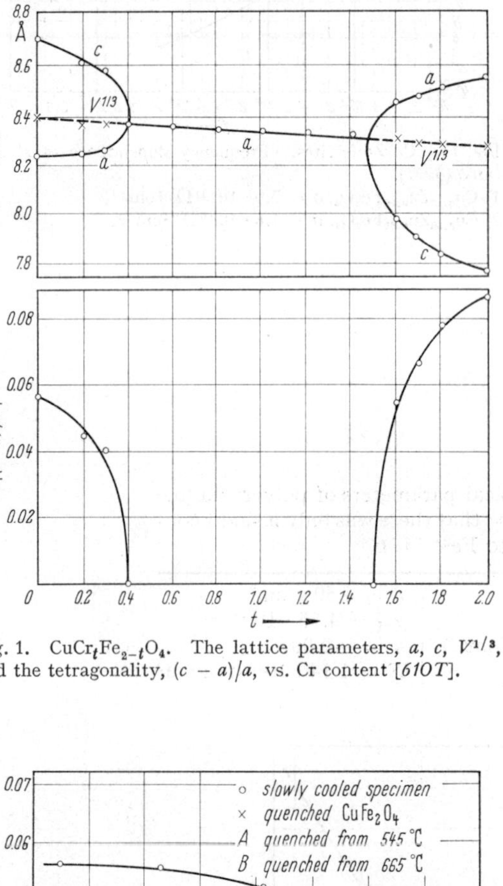

Fig. 1. $CuCr_tFe_{2-t}O_4$. The lattice parameters, a, c, $V^{1/3}$, and the tetragonality, $(c-a)/a$, vs. Cr content [610T].

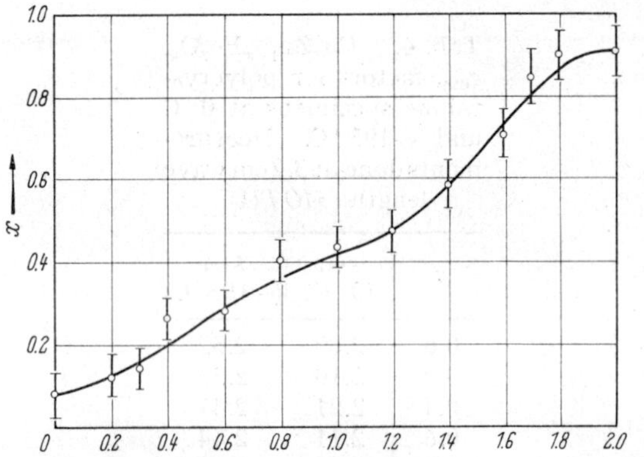

Fig. 2. $(Cu_xFe_{1-x})[Cu_{1-x}Fe_{1+x-t}Cr_t]O_4$. The relation between the cation distribution parameter, x, and the Cr content where parentheses and brackets refer to tetrahedral and octahedral site occupancies, respectively [610T].

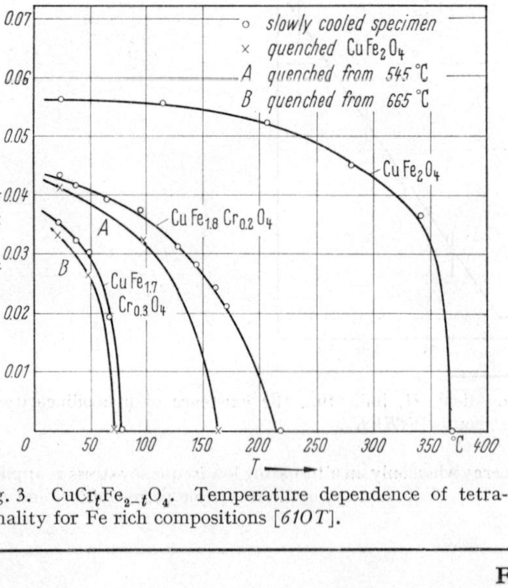

Fig. 3. $CuCr_tFe_{2-t}O_4$. Temperature dependence of tetragonality for Fe rich compositions [610T].

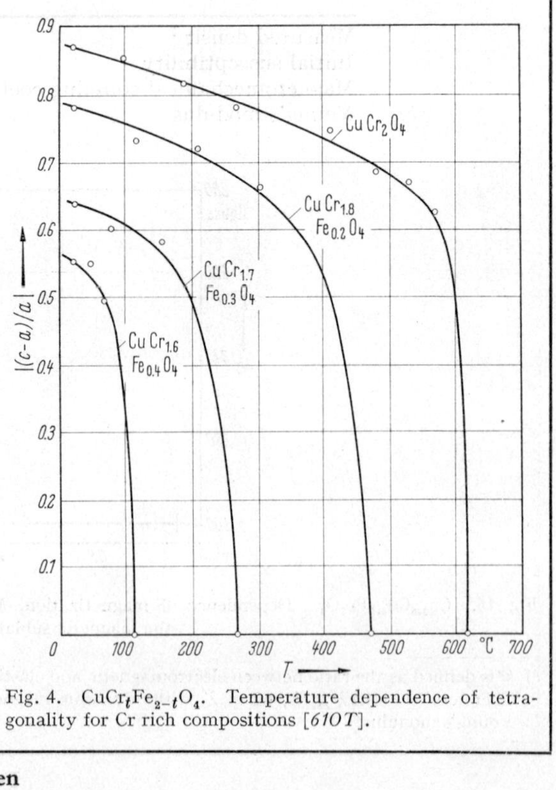

Fig. 4. $CuCr_tFe_{2-t}O_4$. Temperature dependence of tetragonality for Cr rich compositions [610T].

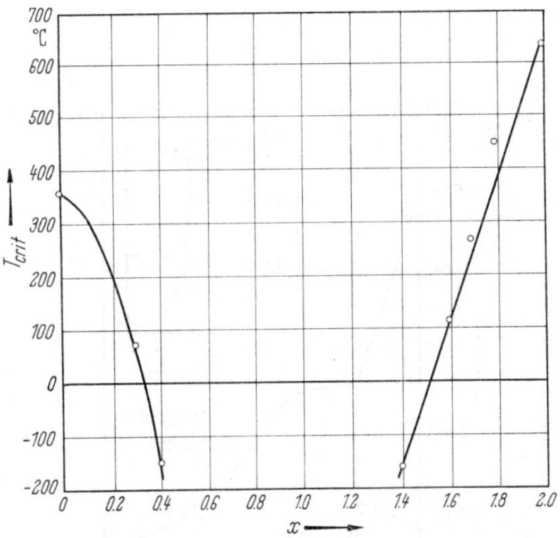

Fig. 5. $CuCr_tFe_{2-t}O_4$. The critical temperature, T_{crit} (i.e. the temperature below which the cubic and tetragonal phases merge into a single cubic phase) vs. Cr content, t [*610T*].

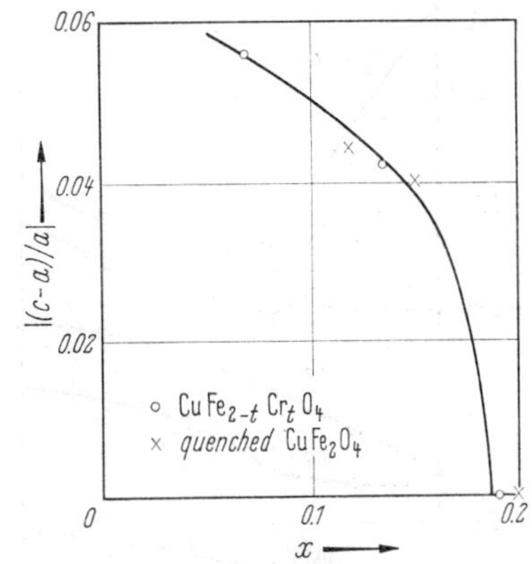

Fig. 6. $CuCr_tFe_{2-t}O_4$. The variation of tetragonality with cation distribution parameter for Fe rich compositions [*610T*].

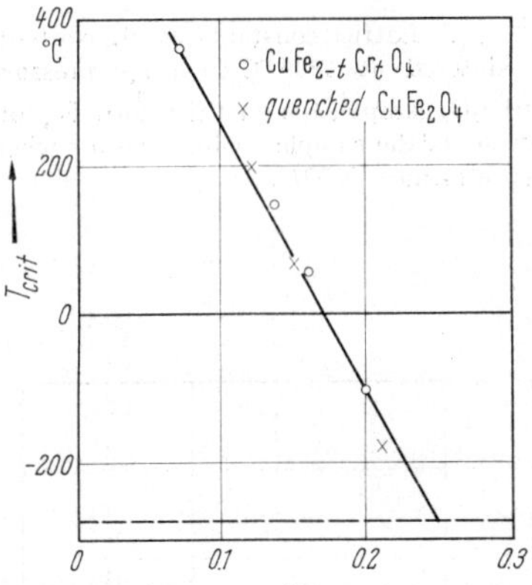

Fig. 7. $CuCr_tFe_{2-t}O_4$. The variation of critical temperature, T_{crit} (see caption of Fig. 5) with cation distribution parameter, x (see caption of Fig. 2) for Fe rich compositions [*610T*].

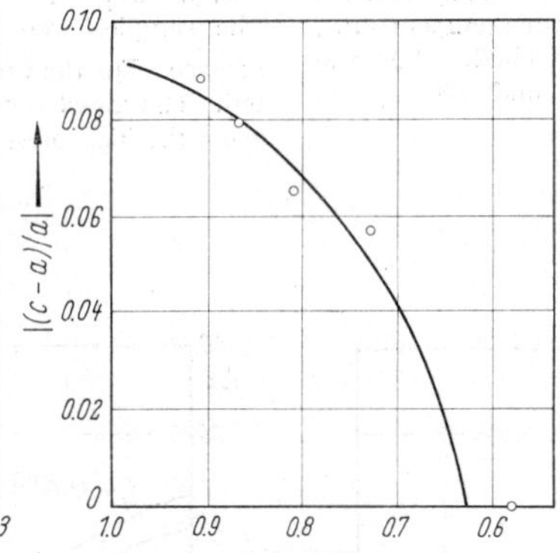

Fig. 8. $CuCr_tFe_{2-t}O_4$. The dependence of tetragonality on cation distribution parameter, x (see caption of Fig. 2) for Cr rich compositions [*610T*].

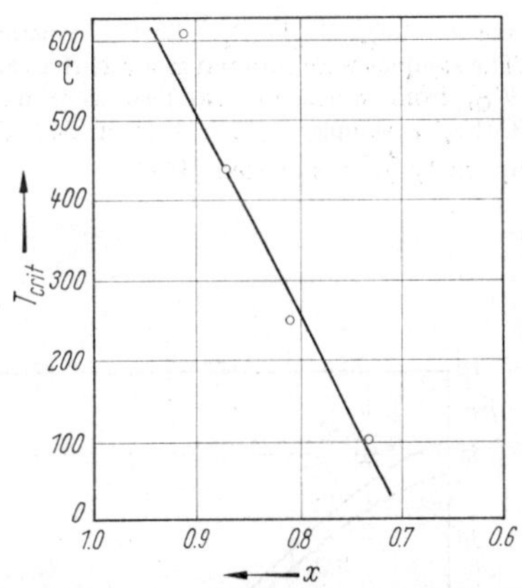

Fig. 9. $CuCr_tFe_{2-t}O_4$. The dependence of critical temperature, T_{crit}, on cation distribution, x, for Cr rich compositions [*610T*]. (See caption of Fig. 5).

6.1.7.4 Cu-Mg ferrites

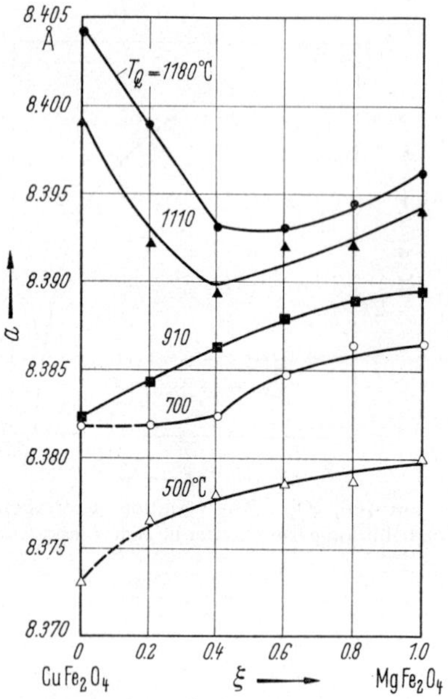

Fig. 1. $Cu_{1-\xi}Mg_{\xi}Fe_2O_4$. Lattice constants vs. Mg content. The samples were sintered in air at the various temperatures, T_Q, from which the samples were quenched. For the $CuFe_2O_4$ sample quenched from 500 °C and 700 °C, the quantity $\sqrt[3]{a^2c}$ is plotted [6701].

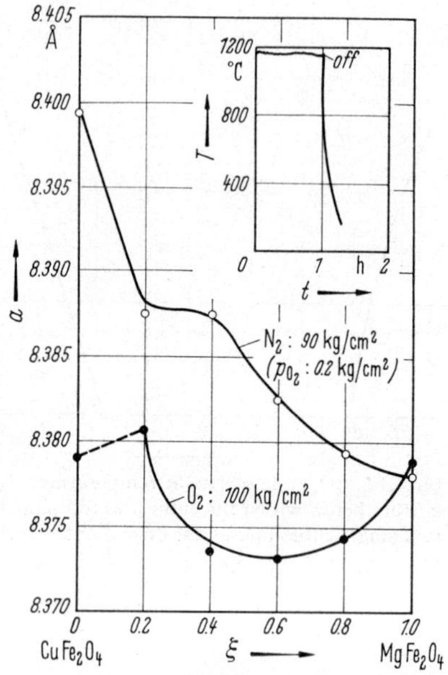

Fig. 2. $Cu_{1-\xi}Mg_{\xi}Fe_2O_4$. Lattice constants vs. Mg content. The samples were sintered at 1200 °C in a high pressure furnace. For the $CuFe_2O_4$ sample, the quantity $\sqrt[3]{a^2c}$ is plotted. The insert represents the sample cooling history when using the high pressure furnace [6701].

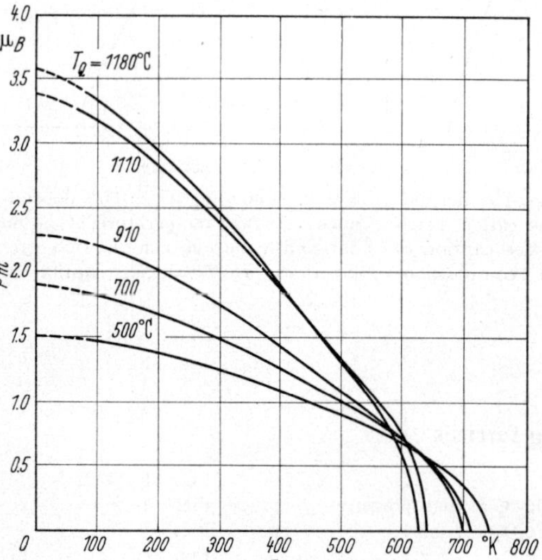

Fig. 3. $Cu_{0.8}Mg_{0.2}Fe_2O_4$. Magnetic moment, p_m, vs. temperature curves of samples fired in air. T_Q denotes the sample quenching temperature [6701].

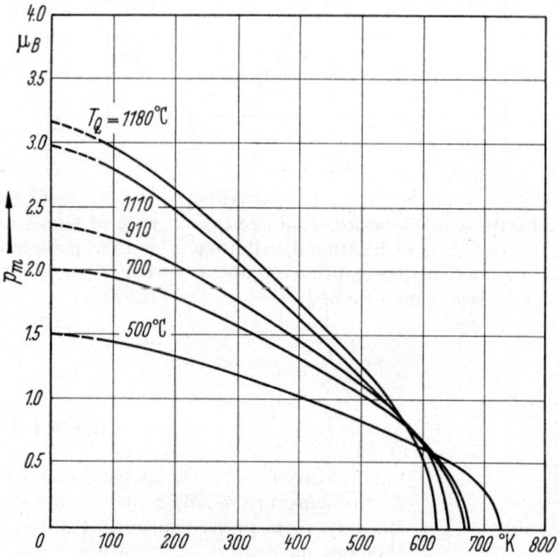

Fig. 4. $Cu_{0.6}Mg_{0.4}Fe_2O_4$. Magnetic moment, p_m, vs. temperature curves of samples fired in air. T_Q denotes the quenching temperature [6701].

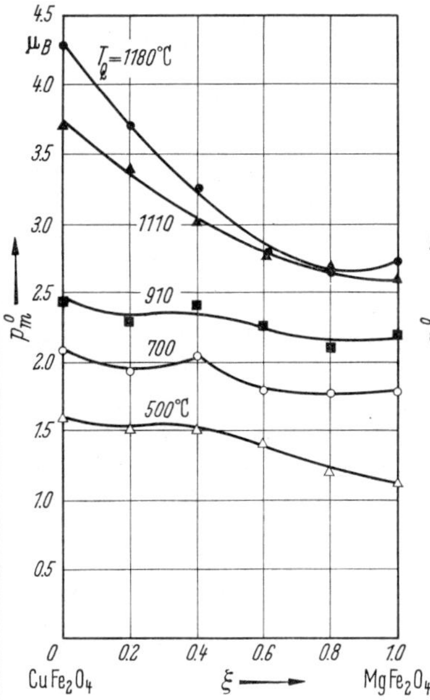

Fig. 5. $Cu_{1-\xi}Mg_{\xi}Fe_2O_4$. Magnetic moment extrapolated to 0 °K, p^0_m, vs. Mg content. The samples were fired in air and T_Q denotes the sample quenching temperature [67OI].

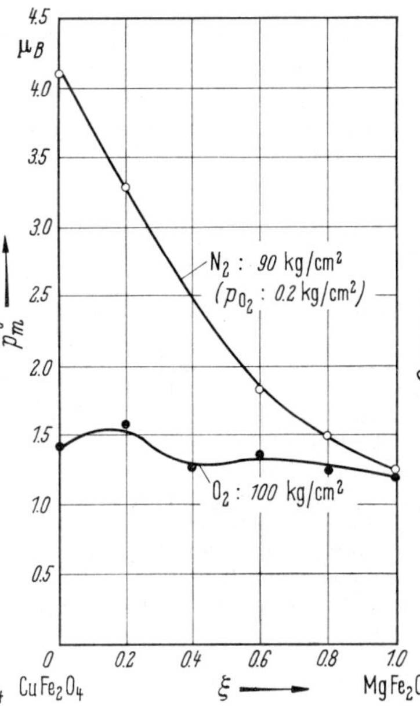

Fig. 6. $Cu_{1-\xi}Mg_{\xi}Fe_2O_4$. Magnetic moment extrapolated to 0 °K, p^0_m, vs. Mg content. The samples were fired at 1200 °C in a high pressure furnace utilizing the indicated atmospheres [67OI].

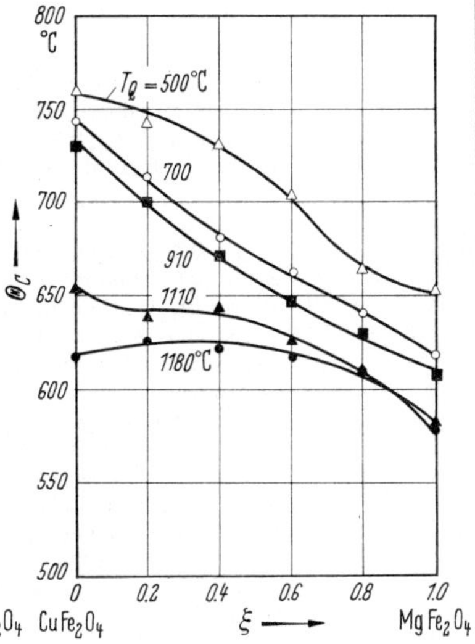

Fig. 7. $Cu_{1-\xi}Mg_{\xi}Fe_2O_4$. Curie temperatures vs. Mg content. The samples were fired in air. T_Q denotes the sample quenching temperature [67OI].

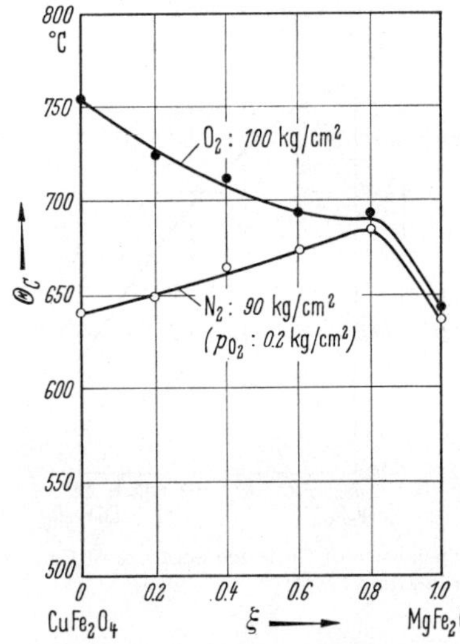

Fig. 8. $Cu_{1-\xi}Mg_{\xi}Fe_2O_4$. Curie temperatures vs. Mg content. The samples were fired at 1200 °C in high pressure furnace utilizing the indicated atmospheres [67OI].

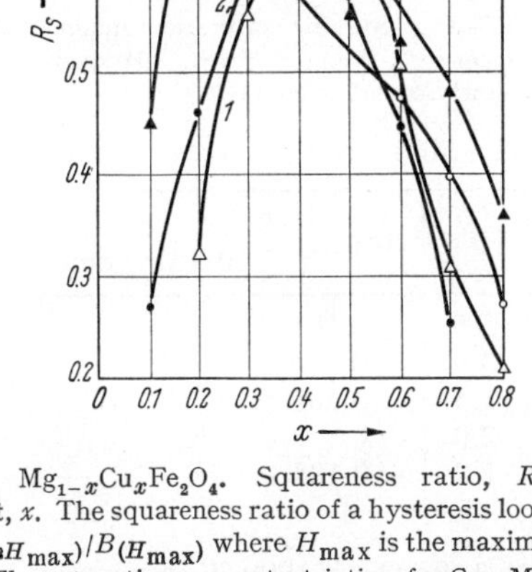

Fig. 9. $Mg_{1-x}Cu_xFe_2O_4$. Squareness ratio, R_s, vs. Cu content, x. The squareness ratio of a hysteresis loop is defined as $B_{(1/2H_{max})}/B_{(H_{max})}$ where H_{max} is the maximum driving field. The saturation magnetostriction for $Cu_{0.9}Mg_{0.1}Fe_2O_4$ is $\lambda_s = -2 \cdot 10^{-5}$ [59GC]. 1) 1050 °C oxygen, 2) 1100 °C oxygen, 3) 1150 °C oxygen, 4) 1100 °C air.

Tab. 1. $Cu_xMg_{1-x}Fe_2O_4$. g_{exp} factors for polycrystalline specimens at 0 °C and −195 °C. Measurements done at 3.2 cm wave length [51OT(1)]

x	g_{exp}		x	g_{exp}	
	(0 °C)	(−195 °C)		(0 °C)	(−195 °C)
0.0	2.09	2.32	0.6	2.11	2.24
0.2	2.09	2.37	0.8	2.11	2.27
0.4	2.11	2.35	1.0	2.04	2.24

6.1.7.5 Other ferrites, oxide compounds, and oxide mixtures containing Cu

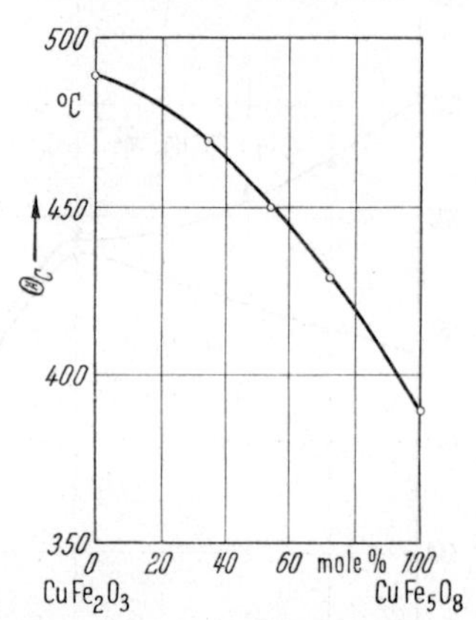

Fig. 1. Dependence of Curie temperature of Cu^{++}-Cu^{+} ferrites on composition [56S].

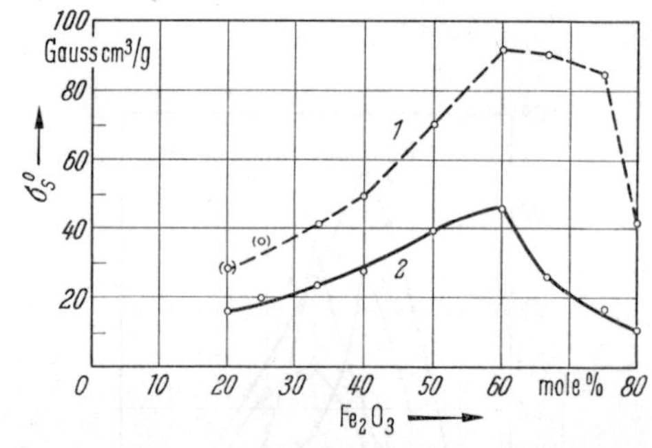

Fig. 2. CuO-Fe_2O_3. Specific saturation magnetization at 0 °K, σ_s^0, vs. mole % Fe_2O_3 [35HS]. 1) quenched from 1100 °C, 2) slowly cooled from 1100 °C.

Tab. 1. Properties of Cu-Fe oxides at 172.5 Mc/s [51W(1)]

Sample	Composition	Fe_2O_3 mole %	$\lvert\varepsilon'\rvert$	tan δ	$\lvert\mu'\rvert$	tan δ_m	σ $10^{-3}\ \Omega^{-1}\ cm^{-1}$
1	$CuO \cdot 1.5Fe_2O_3$	60	7.4	0.087	3.5	0.105	—
2	$CuFe_2O_4$	50	11.1	0.212	3.9	<0.140	2.25
3	$2CuO \cdot Fe_2O_3$	$33\frac{1}{3}$	50.0	0.158	1.5	<0.105	
4	$2CuO \cdot Fe_2O_3$	$33\frac{1}{3}$	9.1	0.158	2.0	0	
5	$CuFe_2O_4 + FeO$	$33\frac{1}{3}$	11.0	0.140	1.2	<0.105	
6	$CuFe_2O_4 + FeO$	$33\frac{1}{3}$	≈15		≈5	+	

Sample 5: Heterogeneous mixture, made by heating $CuO + Fe_2O_3$ to 1000 °C in vacuum and cooling for 10 h to room temperature in N_2.
Sample 6: Solid solution.

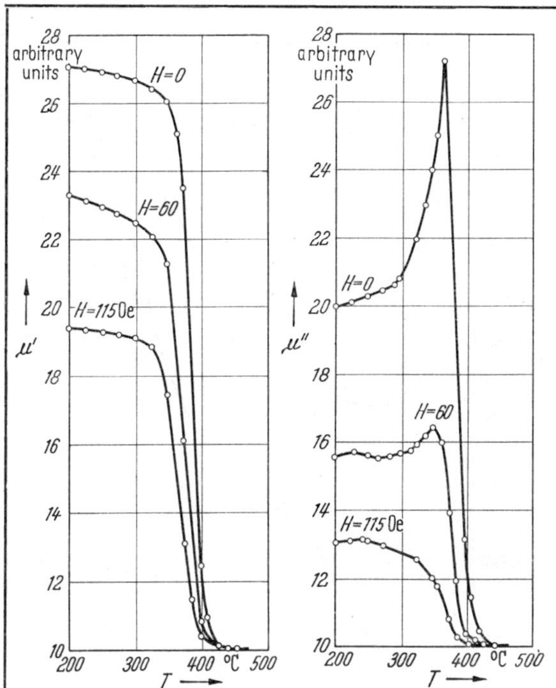

Fig. 3. Temperature dependence of the real part, μ', and the imaginary part, μ'', of the initial permeability (both in arbitrary units) of $Cu^{1+}Fe_5^{3+}O_8$ ($CuO + Fe_2O_3$ sintered 2 h at 1150 °C and quenched in air). H is the superimposed dc field parallel to the r.f. magnetic field [*56S*].

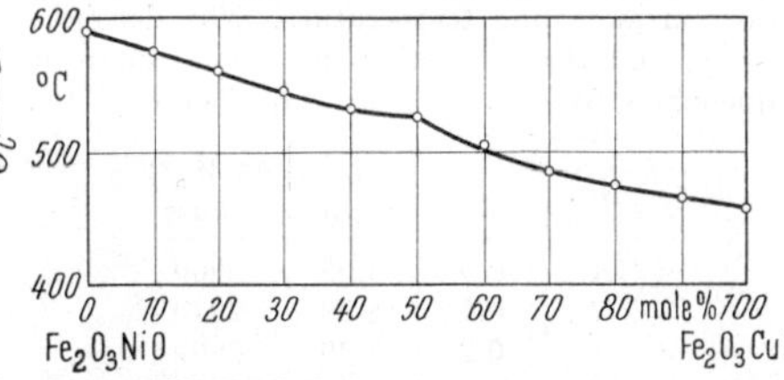

Fig. 5. Fe_2O_3NiO-Fe_2O_3CuO. Ferromagnetic Curie temperature vs. composition [*39FV*].

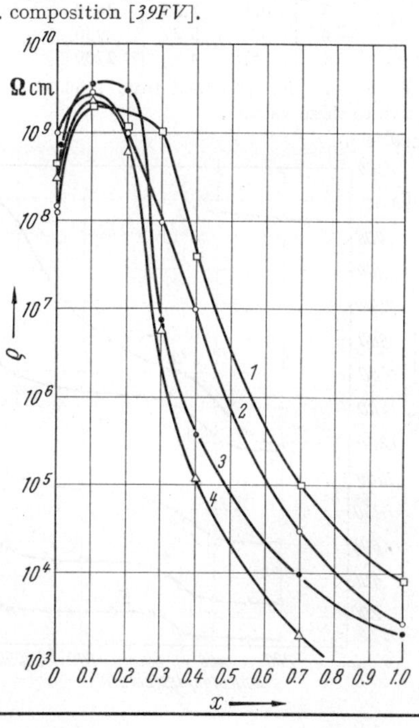

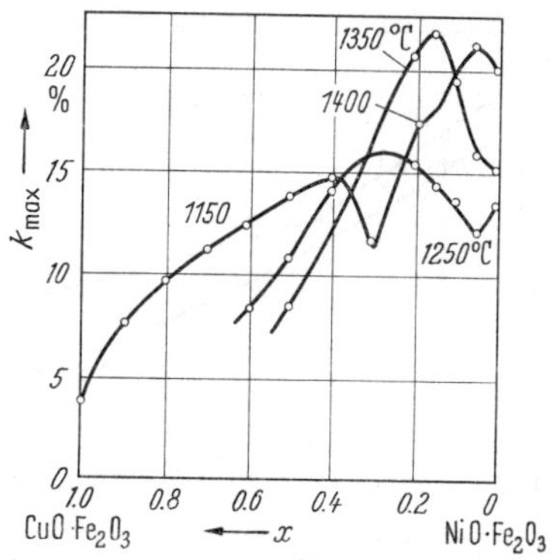

Fig. 4. $(NiO)_{1-x}(CuO)_x(Fe_2O_3)$. Electromechanical coupling coefficient, k_{max}, measured at optimum biasing field vs. composition, x. The temperatures noted are the sintering temperatures [*57K*].

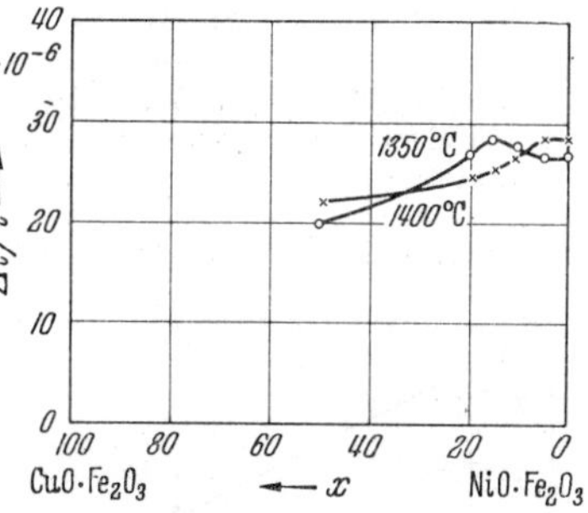

Fig. 6. $(NiO)_{1-x}(CuO)_x(Fe_2O_3)$. Static saturation magnetostriction vs. composition. Sintering temperatures are also shown [*57K*].

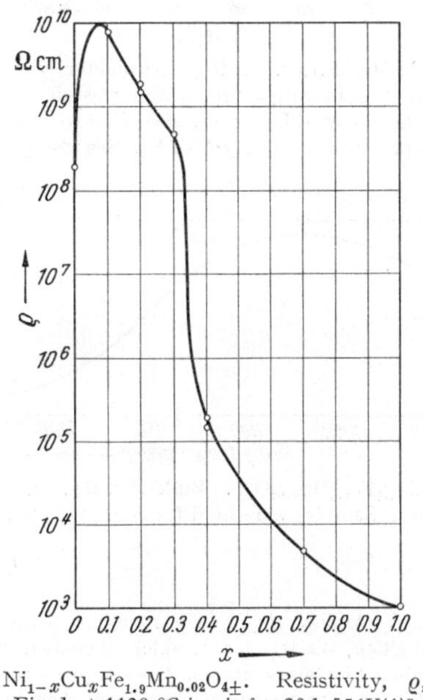

Fig. 8. $Ni_{1-x}Cu_xFe_{1.9}Mn_{0.02}O_{4\pm}$. Resistivity, ϱ, vs. Cu content. Fired at 1130 °C in air for 20 h [*56U*(1)].

←

Fig. 7. $Ni_{1-x}Cu_xFe_{1.9}Mn_{0.02}O_{4\pm}$. Resistivity, ϱ, vs. Cu content [*56U*(1)]. Fired for 20 h in O_2 at various temperatures: 1) 1050 °C, 2) 1100 °C, 3) 1150 °C, 4) 1200 °C.

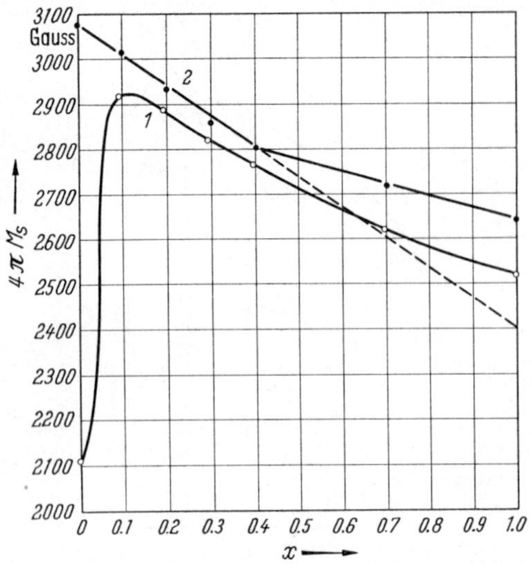

Fig. 9. $Ni_{1-x}Cu_xFe_{1.9}Mn_{0.02}O_{4\pm}$. Saturation induction values fired at 1100 °C for 20 h in O_2 vs. x [*56V*(1)]. Samples were furnace cooled at the rate of about 300 °C/h. 1) measured values, 2) values were corrected for porosity by multiplying by 5.40/bulk density. Dashed line is extrapolated.

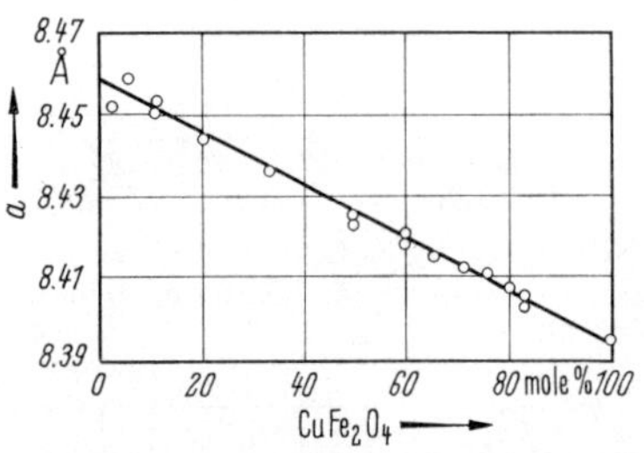

Fig. 11. Cu-Mn ferrites. Lattice parameter vs. composition. Final stage in sample preparation was an annealing at 1200 °C in air followed by a furnace cooling to room temperature in approximately 12 h [*65AMY*]. See also [*61BC*] and [*61BG*].

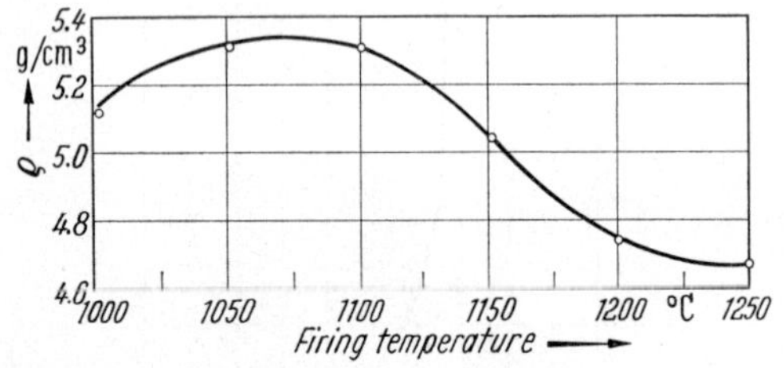

Fig. 12. $Cu_{1.0}Fe_{1.9}Mn_{0.02}O_{4\pm}$. Bulk density, ϱ, vs. firing temperature. Samples were fired for 20 h [*56V*(1)].

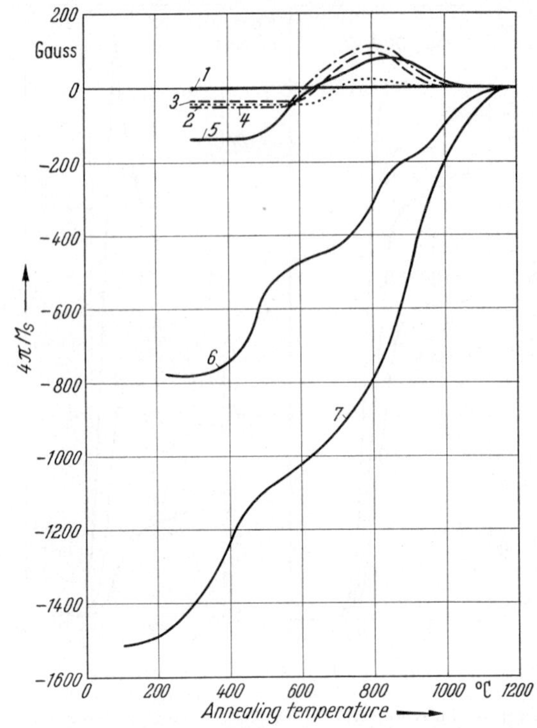

Fig. 10. $Ni_{1-x}Cu_xFe_{1.9}Mn_{0.02}O_{4\pm}$, $Cu_{1.25}Fe_{2.0}O_{4+}$ (sample No. 7). The dependence of the room temperature saturation induction on annealing temperature. The densities of the samples, ϱ, and the saturation induction values measured after quenching from 1150 °C are given below [*57U*].

No.	x	ϱ g/cm³	$4\pi M_s$*) Gauss
1	0.0	5.05	2900
2	0.1	5.29	2970
3	0.2	5.29	2900
4	0.3	5.34	2900
5	0.4	5.35	2850
6	0.7	5.27	2930
7	**)	4.90	2700

*) Values for samples quenched from 1150 °C (curves are relative to these values).

**) $Cu_{1.25}Fe_2O_4$.

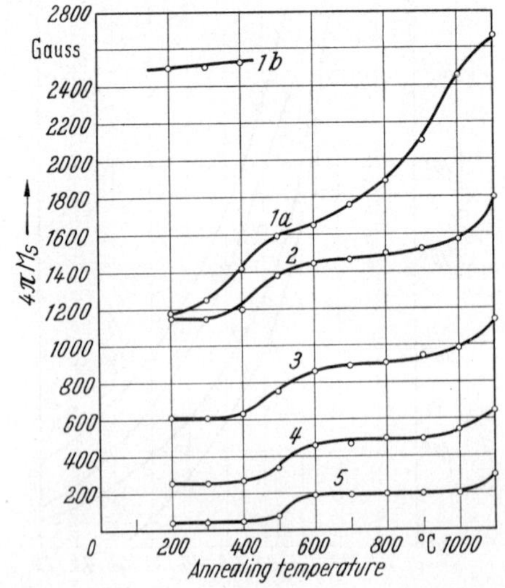

→

Fig. 13. $Cu_{1.25}Al_xFe_{2-x}O_{4+}$. The room temperature saturation induction, $4\pi M_s$, for samples quenched from the indicated annealing temperatures. The samples were initially fired at 1100 °C. The heat treatments were then carried out consistently in direction of decreasing or increasing annealing temperature. Two samples of $Cu_{1.25}Fe_2O_{4+}$ were used (curves 1a and 1b). Sample 1b was heat treated with sample 1a only at 400 °C and lower temperatures. X-ray analysis showed sample 1b to be essentially cubic [*57U*].

No.	1	2	3	4	5
x	0	0.2	0.4	0.6	0.8

Tab. 2a. Cu-Pb-Fe oxides. Electric and magnetic properties at 172.5 Mc/s. Samples quenched from 1000 °C in air [*51W*(1)]. See also [*51W*(2)], [*50F*]

Sample	Composition	$\frac{PbFe_2O_4}{CuFe_2O_4} + Fe_2O_3$	$\|\varepsilon'\|$	tan δ	$\|\mu'\|$	tan δ_m	σ $10^{-3}\ \Omega^{-1}\ cm^{-1}$
1	$CuO \cdot PbO \cdot 3Fe_2O_3$	1.000 + 1.000	15.3	0.185	3.2	0.249	
2	$CuO \cdot PbO \cdot 2Fe_2O_3$	1.000 —	16.6	0.194	3.5	0	
3	$1.5CuO \cdot PbO \cdot 3.8Fe_2O_3$	0.666 + 0.835	13.1	0.158	4.4	0.305	0.20
4	$2CuO \cdot PbO \cdot 3Fe_2O_3$	0.500 —	21.5	0.364	6.2	0.477	
5	$2.8CuO \cdot PbO \cdot 4.2Fe_2O_3$	0.356 + 0.138	35.0	0.370	6.7	0.450	1.30
6	$3CuO \cdot PbO \cdot 5.5Fe_2O_3$	0.333 + 0.500	20.0	0.414	6.1	<0.105	
7	$4CuO \cdot PbO \cdot 7Fe_2O_3$	0.250 + 0.500	17.5	0.435	6.7	0.466	
8	$6CuO \cdot PbO \cdot 10Fe_2O_3$	0.167 + 0.500	11.2	0.296	5.8	0.364	
9	$7CuO \cdot PbO \cdot 9.1Fe_2O_3$	0.142 + 0.153	12.6	0.310	5.6	0.380	0.37
10	$14CuO \cdot PbO \cdot 19.6Fe_2O_3$	0.071 + 0.323	10.0	0.114	4.7	0.296	0.11

Tab. 2b. CuO-PbO-Fe_2O_3. Influence of cooling speed on the electric and magnetic properties. Measurements done at 172.5 Mc/s [*51W*(1)]. See also [*51W*(2)], [*50F*]

Sample	Composition	Quenching	$\|\varepsilon'\|$	tan δ	$\|\mu'\|$	tan δ_m	σ $10^{-3}\ \Omega^{-1}\ cm^{-1}$
9	$7CuO \cdot PbO \cdot 9.1Fe_2O_3$	air	12.6	0.310	5.6	0.380	0.37
9a	$7CuO \cdot PbO \cdot 9.1Fe_2O_3$	water	35.0	0.430	9.4	0.510	1.50
10	$14CuO \cdot PbO \cdot 19.6Fe_2O_3$	air	10.0	0.114	4.7	0.296	0.11
10a	$14CuO \cdot PbO \cdot 19.6Fe_2O_3$	water	28.0	0.480	8.6	0.650	[1])

Tab. 2c. CuO-PbO-Fe_2O_3. Influence of heating duration, sintering temperature, T, and cooling speed on the electric and magnetic properties. Measurements done at 172.5 Mc/s [*51W*(1)]. See also [*51W*(2)], [*50F*]

Sample	Composition	Heating time h	T °C	Quenching	$\|\varepsilon'\|$	tan δ	$\|\mu'\|$	tan δ_m
2	$CuO \cdot PbO \cdot 2Fe_2O_3$	2	1000	air	16.0	0.194	3.5	0
2a	$CuO \cdot PbO \cdot 2Fe_2O_3$	1	1050	water	17.3		2.7	
4	$2CuO \cdot PbO \cdot 3Fe_2O_3$	2	1000	air	21.5	0.364	6.2	0.477
4a	$2CuO \cdot PbO \cdot 3Fe_2O_3$	1	1050	water	51.9		3.8	

Tab. 2d: see next page

Tab. 2e. CuO-PbO-Fe_2O_3. Influence of oxide modifications on the electric and magnetic properties. Measurements done at 172.5 Mc/s [*51W*(1)]. See also [*51W*(2)], [*50F*]

Sample	Composition	CuO		PbO		Quenching	$\|\varepsilon'\|$	tan δ	$\|\mu'\|$	tan δ_m	σ $10^{-3}\ \Omega^{-1}\ cm^{-1}$
		black	red	yellow	red						
11	$4.8CuO \cdot PbO \cdot 7.6Fe_2O_3$	½	½	1	—	air	10.3	0.194	5.7	0.344	0.19
11a	$4.8CuO \cdot PbO \cdot 7.6Fe_2O_3$	1	—	—	1	air	8.3	0.078	2.8	<0.105	
11b	$4.8CuO \cdot PbO \cdot 7.6Fe_2O_3$	½	½	1	—	water	35.0	0.390	9.4	0.480	1.30
11c	$4.8CuO \cdot PbO \cdot 7.6Fe_2O_3$	1	—	—	1	water	35.0	0.330	8.8	0.420	1.10 [2])

[1]) Measurements at 176.5 Mc/s.
[2]) Mixed with 4% paraffin binder.

Tab. 2d. CuO-PbO-Fe_2O_3. Influence of frequency on the electric and magnetic properties. See Tab. 2a, 2b, 2c, and 2e for additional data on samples [*51W*(1)]. See also [*51W*(2)], [*50F*]

Sample	Composition	f Mc/s	$\lvert\varepsilon'\rvert$	$\tan\delta$	$\lvert\mu'\rvert$	$\tan\delta_m$	σ $10^{-3}\ \Omega^{-1}\ cm^{-1}$
2	$CuO \cdot PbO \cdot 2Fe_2O_3$	172.5	16.6	0.194	3.5	0	
		387.0	14.5		2.8		
		3284.8	9.6	⩽1.000	1.1	0.706	
4	$2CuO \cdot PbO \cdot 3Fe_2O_3$	172.5	21.5	0.364	6.2	0.477	
		172.5	15.0	0.325	4.8	0	[1]
		387.0	18.2		4.3		
		2672.2	12.2	0.060	1.35	1.330	
7	$4CuO \cdot PbO \cdot 7Fe_2O_3$	172.5	17.5	0.435	6.7	0.466	
		1010.1	9.0	0.024	1.01	0.253	
11	$4.8CuO \cdot PbO \cdot 7.6Fe_2O_3$	172.5	10.3	0.194	5.7	0.344	0.19
		387.0	10.0	0.140	4.5	0.550	
		771.6	10.0	0.080	3.0	0.740	
11b	$4.8CuO \cdot PbO \cdot 7.6Fe_2O_3$	172.5	35.0	0.390	9.4	0.480	1.30
		387.0	24.0	0.610	6.5	0.700	3.10
		750.6	23.0	0.430	3.9	0.870	4.20
11c	$4.8CuO \cdot PbO \cdot 7.6Fe_2O_3$	172.5	35.0	0.330	8.8	0.420	1.10 [2]
		352.7	22.6	0.340	5.3	1.000	[2]
9a	$7CuO \cdot PbO \cdot 9.1Fe_2O_3$	10^{-3}			7.37		[3]
		172.5	35.0	0.430	9.4	0.510	1.50
		387.0	(27)	(0.40)	6.5	(0.85)	
		750.0	(21)	(0.36)	3.9	(1.13)	
		1172.8	17.1	0.330	3.1	1.230	3.70
		3076.8	15.0	0.170	1.86	1.300	6.40

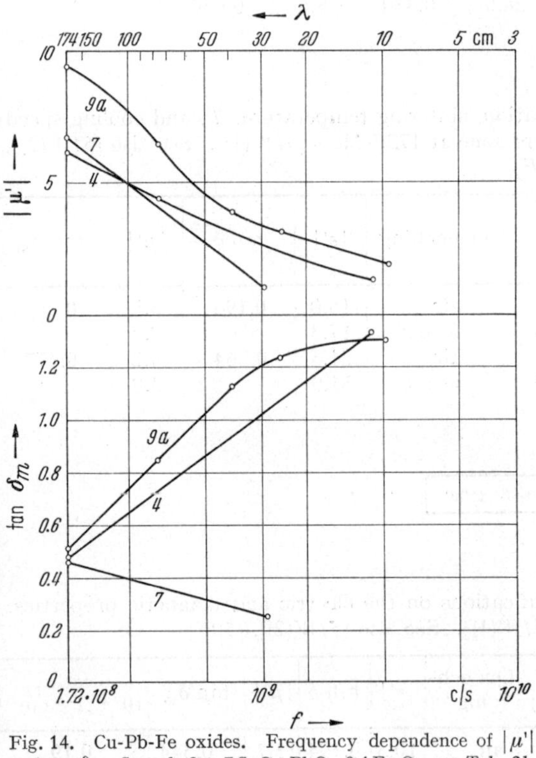

Fig. 14. Cu-Pb-Fe oxides. Frequency dependence of $\lvert\mu'\rvert$ and $\tan\delta_m$. Sample 9a: $7CuO \cdot PbO \cdot 9.1Fe_2O_3$, see Tab. 2b and 2e. Sample 4: $2CuO \cdot PbO \cdot 3Fe_2O_3$, see Tab. 2a, 2c and 2e. Sample 7: $4CuO \cdot PbO \cdot 7Fe_2O_3$, see Tab. 2a and 2e [*51W*(1)]. See also [*50F*].

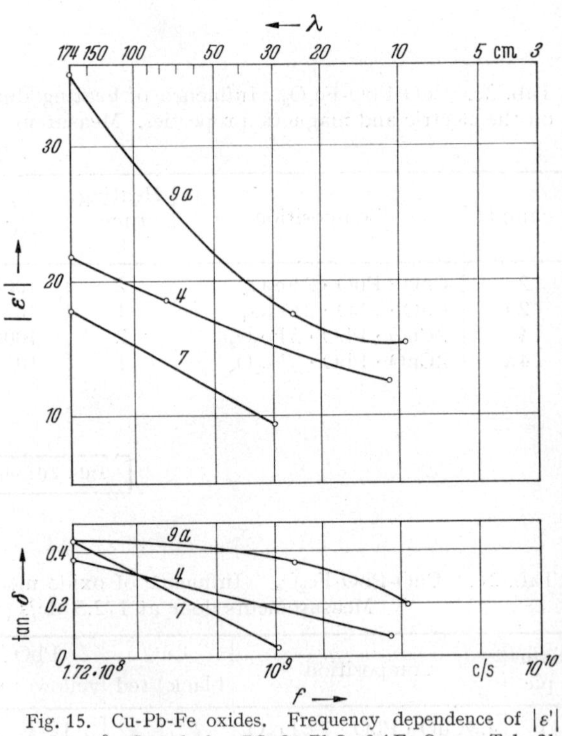

Fig. 15. Cu-Pb-Fe oxides. Frequency dependence of $\lvert\varepsilon'\rvert$ and $\tan\delta$. Sample 9a: $7CuO \cdot PbO \cdot 9.1Fe_2O_3$, see Tab. 2b and 2e. Sample 4: $2CuO \cdot PbO \cdot 3Fe_2O_3$, see Tab. 2a, 2c and 2e. Sample 7: $4CuO \cdot PbO \cdot 7Fe_2O_3$, see Tab. 2a and 2e [*51W*(1)]. See also [*50F*].

[1] Mixed with 10% paraffin.
[2] Mixed with 4% paraffin binder.
[3] 0.11 Oe microwave magnetic field at this frequency.

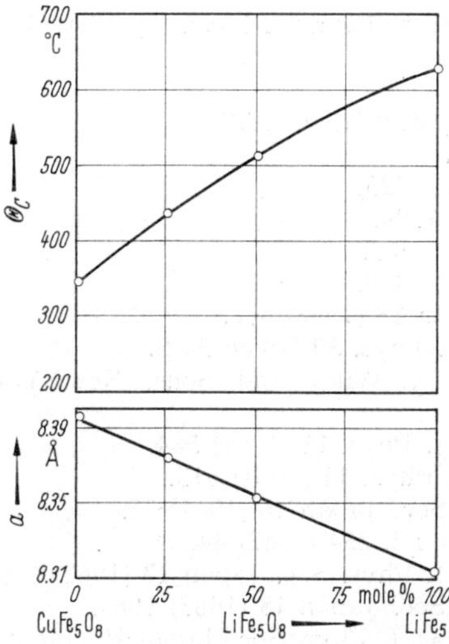

Fig. 16. $CuFe_5O_8$-$LiFe_5O_8$. Lattice constants and Curie temperatures vs. composition [*62TC*].

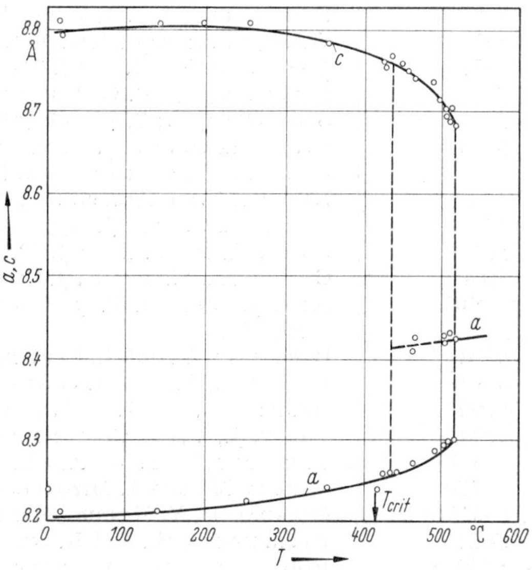

Fig. 17. $Ge_{0.2}Cu_{1.2}Fe_{1.8}O_4$. Temperature dependence of lattice parameters. In this figure, there is a two-phase region between 430 °C and 510 °C [*63TM*].

6.1.7.6 References for 6.1.7.1 ··· 6.1.7.5

29S — Serres, A.: Compt. Rend. **188** (1929) 1239.
32S — Serres, A.: Ann. Phys. **17** (1932) 5.
35HS — Hilpert, R. S., and R. Schweinhagen: Z. Physik. Chem. (Leipzig) (B) **31** (1935) 1.
35R — Raychaudhuri, D. P.: Indian J. Phys. **9** (1935) 383.
36AVB — Arkel, A. E. van, E. J. W. Verwey and M. G. van Bruggen: Rec. Trav. Chim. **55** (1936) 331.
39FV — Forestier, H., and M. Vetter: Compt. Rend. **209** (1939) 164.
46S — Snoek, J. L.: Philips Tech. Rev. **8** (1946) 353.
47VH — Verwey, E. J. W., and E. L. Heilmann: J. Chem. Phys. **15** (1947) 174.
49S — Snoek, J. L.: New Developments in Ferromagnetic Materials (Elsevier Publishing Co., Inc., Amsterdam 1949) p. 78.
50B — Bertaut, F.: Compt. Rend. **230** (1950) 213.
50Br — Brockman, F. G.: Phys. Rev. **77** (1950) 841.
50F — Flegler, E.: Arch. Elektrotechn. **40** Nr. 1 (1950) 4.
50G — Gorter, E. W.: Nature **165** (1950) 798.
50TB — Toropov, N. A., and A. I. Borisenko: J. Appl. Chem. USSR (English Transl.) **23** (1950) 1243.
50WBB — Weil, L., F. Bertaut, and L. Bochirol: J. Phys. Radium **11** (1950) 208.
51PB — Pauthenet, R., and L. Bochirol: J. Phys. Radium **12** (1951) 249.
51OKT — Okamura, T., Y. Kojima, and Y. Torizuka: Sci. Rept. Res. Inst. Tohoku Univ. Ser. **A 3** (1951) 725.
51OT(1) — Okamura, T., and Y. Torizuka: Nature **167** (1951) 986.
51OT(2) — Okamura, T., and Y. Torizuka: Nature **168** (1951) 872.
51S — Smolenskij, G. A.: Dokl. Akad. Nauk SSSR **78** (1951) 921.
51TB — Toropov, N. A., and A. I. Borisenko: Dokl. Akad. Nauk SSSR **76** (1951) 85.
51W(1) — Wagenknecht, F.: Frequenz **5** (1951) 145.
51W(2) — Wagenknecht, F.: Frequenz **5** (1951) 186.
52H — Hersping, A.: Frequenz **6** (1952) 345.
52M — Moltgen, G.: Z. Angew. Phys. **4** (1952) 216.
52OFD(1) — Okamura, T., T. Fujimura, and M. Date: Phys. Rev. **85** (1952) 1041.
52OFD(2) — Okamura, T., T. Fujimura, and M. Date: Phys. Rev. **88** (1952) 1435.
52OK — Okamura, T., and Y. Kojima: Phys. Rev. **86** (1952) 1040.
52P — Pauthenet, R.: Ann. Phys. (Paris) **7** (1952) 710.
53B — Burgt, C. M. van der: Philips Research Rept. **8** (1953) 91.
53H — Heister, W.: Arch. Elektrotech. **41** (1953) 142.
54KK — Komar, A. P., and V. V. Kliushin: Bull. Acad. Sci. USSR, Phys. Ser. (English Transl.) **18** (1954) 99.
54S — Suchkov, A. I.: Zh. Tekhn. Fiz. **24** (1954) 1579.

54SW	SMIT, J., and H. P. J. WIJN: Advan. Electron. Electron Phys. **6** (1954) 83.
54W	WEISZ, R. S.: Phys. Rev. **96** (1954) 800.
56M	MAYER, F.: Compt. Rend. **242** (1956) 81.
56PT	PRINCE, E., and R. G. TRUETING: Acta Cryst. **9** (1956) 1025.
56S	STIERSTADT, K.: Z. Physik **146** (1956) 169.
56U(1)	UITERT, L. G. VAN: J. Appl. Phys. **27** (1956) 723.
56U(2)	UITERT, L. G. VAN: Proc. I.R.E. **44** (1956) 1294.
57K	KIKUCHI, Y.: J. Acoust. Soc. Am. **29** (1957) 569.
57U	UITERT, L. G. VAN: J. Appl. Phys. **28** (1957) 478.
58II	INOUE, T., and S. IIDA: J. Phys. Soc. Japan **13** (1958) 656.
59GC	GREIFER, A. P., and W. J. CROFT: J. Appl. Phys. **30** (1959) 34 S.
59SW	SMIT, J., and H. P. J. WIJN: Ferrites (J. Wiley and Sons, New York, 1959) 156/157.
61BC	BERGSTEIN, A., and L. CERVINKA: Czech. J. Phys. **11** (1961) 584.
61BG	BERGSTEIN, A., and R. GERBER: Czech. J. Phys. **11** (1961) 613.
61OT	OHNISHI, H., and T. TERANISHI: J. Phys. Soc. Japan **16** (1961) 35.
62TC	THERY, J., and R. COLLONGUES: Compt. Rend. **254** (1962) 44.
62YMO	YAMADAYA, T., T. MITUI, and T. OKADA: J. Phys. Soc. Japan **17** (1962) 1897.
63TM	TANAKA, M., and T. MIZOGUCHI: J. Phys. Soc. Japan **18** (1963) 1089.
64MMM	MIYADAI, T., Y. MATSUO, and S. MIYAHARA: J. Phys. Soc. Japan **19** (1964) 1747.
64RS	RECHENBERG, H., and K. STIERSTADT: Z. Angew. Phys. **17** (1964) 242.
64RVK	RODE, V. E., A. V. VEDYAEV, and B. N. KRAINOV: Soviet Phys. — Solid State (English Transl.) **5** (1964) 1277.
64SBR	STIERSTADT, K., H. BENZ, and H. RECHENBERG: Proc. Intern. Conf. Magnetism, Nottingham 1964.
64SK	SIROTA, N. N., and Y. M. KHACHATRYAN: Soviet Phys. — Solid State (English Transl.) **5** (1964) 2278.
65AMY	ANDRIYEVSKIY, A. I., G. F. MOCHARNYUK, and Y. G. YUSKEVICH: Phys. Metals Metallog. (USSR) (English Transl.) **20** (1965) 47.
65LSG	LEVINSTEIN, H. J., F. J. SCHNETTLER, and E. M. GYORGY: J. Appl. Phys. **36** (1965) 1163.
65MMM	MIYADAI, T., S. MIYAHARA, and Y. MATSUO: J. Phys. Soc. Japan **20** (1965) 980.
66EHK	EVANS, B. J., HAFNER, S., and G. M. KALVIUS: Phys. Letters **23** (1966) 24.
66OKI	OHBAYASHI, K., K. KOHN, and S. IIDA: J. Phys. Soc. Japan **21** (1966) 2740.
66OLS	O'BRYAN, H. M., H. J. LEVINSTEIN, and R. C. SHERWOOD: J. Appl. Phys. **37** (1966) 1438.
67OI	OHBAYASHI, K., and S. IIDA: J. Phys. Soc. Japan **23** (1967) 776.
67SK	STIERSTADT, K., and W. A. KAFKA: Angew. Phys. **22** (1967) 91.

6.1.8 Co-Fe^{3+} spinels (Co ferrites) and Co-Fe^{3+} spinels with substitutions

6.1.8.1 Co ferrite

Tab. 1. Co ferrite. General properties (room temperature unless otherwise specified)

		Ref.
Chemical formula	$CoFe_2O_4$	
Crystal type	Inverse spinel — cubic [1],[2]	*56P*
Structure type	$MgAl_2O_4$	
Space group	O_h—Fd3m	
Lattice constant	$a = 8.38$ Å	*54SW*
Volume of unit cell	$588.5 \cdot 10^{-24}$ cm³ (containing 8 molecules)	
Oxygen parameter	$u = 0.381$ [3]	*56P*
Molecular weight	234.6 amu	
X-ray density	$\varrho_X = 5.29$ g/cm³	
Melting point	1570 ± 10 °C	*36AVB*
Sintering temperature	800···1350 °C	
Thermal conductivity	$\varkappa \cong 1.5 \cdot 10^{-2}$ cal sec^{-1} cm^{-1} deg^{-1}	*54SW*
Specific heat	$c_p \cong 0.2$ cal g^{-1} deg^{-1}	*54SW*
Young's modulus	$E \cong 1.5 \cdot 10^{12}$ dyne/cm³	*54SW*
Compressibility	$\beta = 0.5574 \cdot 10^{-12}$ cm²/dyne	*55W*
Elastic constant	$c_{11} = c_{12} = (1/\beta) = 1.794 \cdot 10^{12}$ dyne/cm²	*55W*
Velocity of sound	$v = 5.82 \cdot 10^6$ cm/sec	*55W*
Dielectric constant	$\varepsilon' \approx 10$ (polycrystal) [4]	*52FD*

[1]) Neutron diffraction determination.
[2]) Possesses a tetragonal deformation at 80 °K, with $c/a \approx 0.9987$. See: [*53RW*] see also: [*50GC*(1)].
[3]) Determined for $Co_{0.85}Fe_{2.15}O_4$ by neutron diffraction.
[4]) Measure at 6.6 cm wavelength.

Single crystal synthesis

A 19 to 1 mixture by weight of pure Co ferrite and Na ferrite was heated to 1590 °C in a platinum crucible under an oxygen pressure of 1650 psig. The melt was maintained at this temperature and pressure for two hours and then cooled at the rate of 6 °C/h. The cooling rate was increased to 10 °C/h at 1490 °C but the pressure was maintained constant until the temperature decreased below 1350 °C. The temperature was then decreased at 40 °C/h and the pressure was slowly reduced. At 1000 °C the O_2 pressure was reduced to 0 psig and the furnace power was turned off. [*63FKW*].

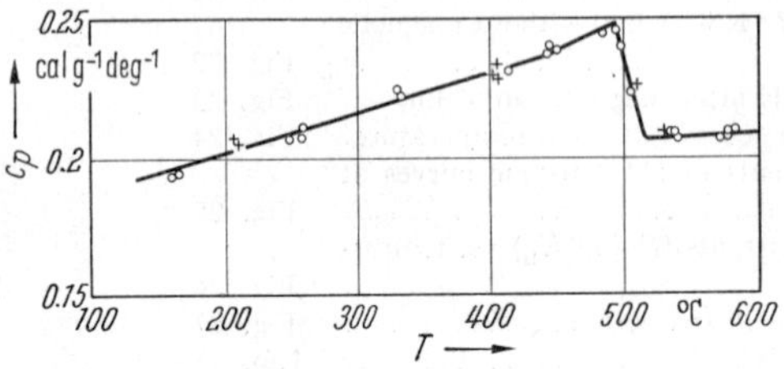

Fig. 1. $CoFe_2O_4$. Specific heat vs. temperature [*51B*].

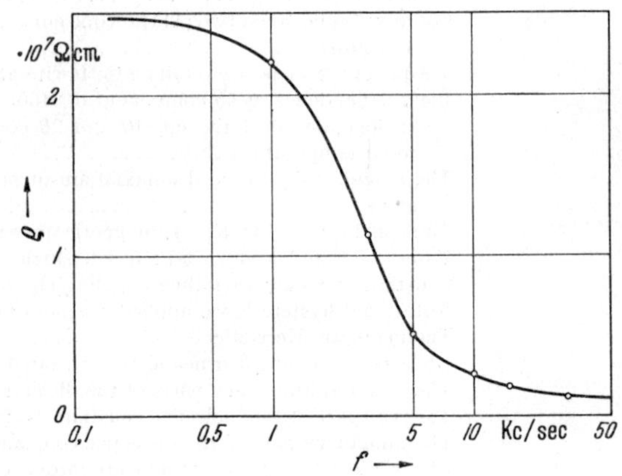

Fig. 2. $CoFe_2O_4$. Specific resistance, ϱ, vs. frequency for a slowly cooled sample at 50 °C [*51K*].

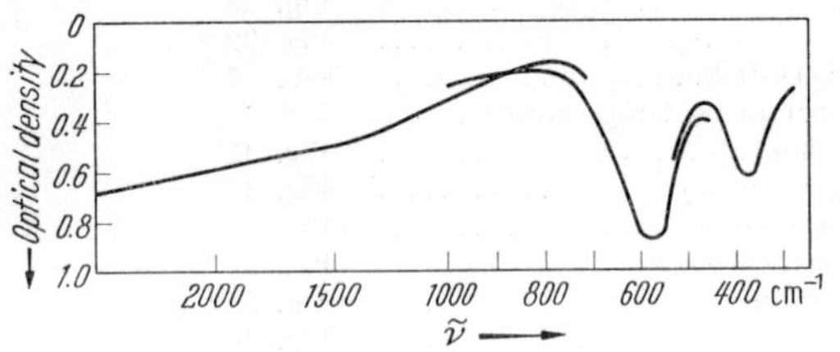

Fig. 3. $CoFe_2O_4$. Infrared absorption spectra at room temperature. Optical density measured relative to standard KBr disc. Sample weight: 2.50 mg; sample area: 3.14 cm². Multiple lines indicate alternate runs or spectrophotometer prism changes [*53W*].

Tab. 2. Co ferrite. Magnetic properties (room temperature unless otherwise specified)

		Ref.
Ferromagnetic Curie temperature	Θ_C = 520 °C	*59SW*(1)
	Θ_C = 496 °C	*50P*
Paramagnetic Curie temperature	Θ_p = 539 °C	*50P*
Saturation magnetic moment per molecule at 0 °K	p^0_{sm} = 3.94 μ_B	*50P*
Saturation magnetic moment per gram	σ_s = 80 Gauss cm³/g	*59SW*(1)
	σ^0_s = 90 Gauss cm³/g (0 °K)	*59SW*(1)
	σ^0_s = 93.9 Gauss cm³/g (0 °K)	*50P*
Saturation magnetization per unit volume	M_s = 425 Gauss	*59SW*(1)
	M^0_s = 475 Gauss (0 °K)	*59SW*(1)
Saturation induction	$4\pi M_s$ = 5300 Gauss	*59SW*(1)
	$4\pi M^0_s$ = 6000 Gauss (0 °K)	*59SW*(1)
Initial permeability	$\mu' \approx 1.3$ (0.45···20 000 Mc/s)	*58PA*
Saturation magnetostriction	$\lambda_s = -110 \cdot 10^{-6}$ (polycrystal)	*53G*
	$\lambda_{100} = -250 \cdot 10^{-6}$ (single crystal) [1]	*55BTW*
Experimentally observed *g* factor	$g_{exp} = 2.7 \pm 0.3$ (90 °C) (single crystal)	*55T*
	g_{exp} = 2.22 (100 °C) (polycrystal)	*51T*
First order magnetocrystalline anisotropy constant	$K_1 = +1.8 \cdot 10^6$ erg/cm³ [2]	*55BTW*
	$K_1 = +2.9 \cdot 10^6$ erg/cm³ [3]	*55BTW*
	$K_1 = +4.4 \cdot 10^6$ erg/cm³ (−196 °C) [3]	*55BTW*
Hyperfine field, tetrahedral (*A*) site	H_A = 510 kOe (90 °K) [4]	*66V*
Hyperfine field, octahedral (*B*) site	H_B = 540 kOe (90 °K) [4]	*66V*
Relative isomer shift	$\delta_A - \delta_B = (-0.06 \pm 0.02)$ mm/sec (80 °K)	*66V*

[1]) The composition of this sample was $Co_{1.1}Fe_{1.9}O_4$; see Tab. 8 of 6.1.8.2 for additional magnetostriction data on Co-Fe ferrites.

[2]) Sample composition was $Co_{1.1}Fe_{1.9}O_4$.

[3]) Sample composition was $Co_{0.9}Fe_{2.1}O_4$. Sample was aged at 150 °C for 3 days.

[4]) Determined from Mössbauer effect measurements.

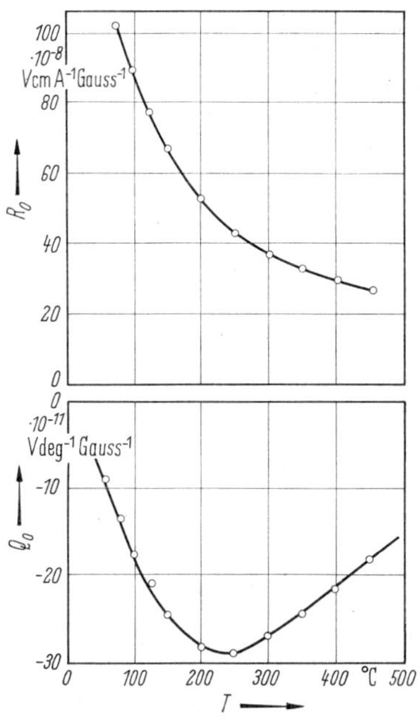

Fig. 4. $Co_{0.94}Fe_{2.06}O_4$. Temperature dependence of the spontaneous Hall (R_0) and Nernst-Ettingshausen (Q_0) coefficients for a single crystal [*65SM*].

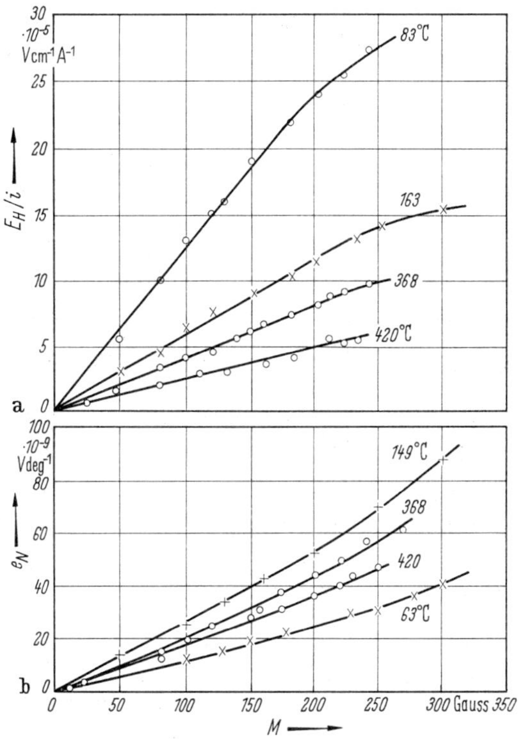

Fig. 5. $Co_{0.94}Fe_{2.06}O_4$. Dependence of the Hall emf, E_H/i, (a) and the Nernst-Ettingshausen emf, e_N, (b) on the magnetization for a single crystal at various temperatures (°C) [*65SM*].

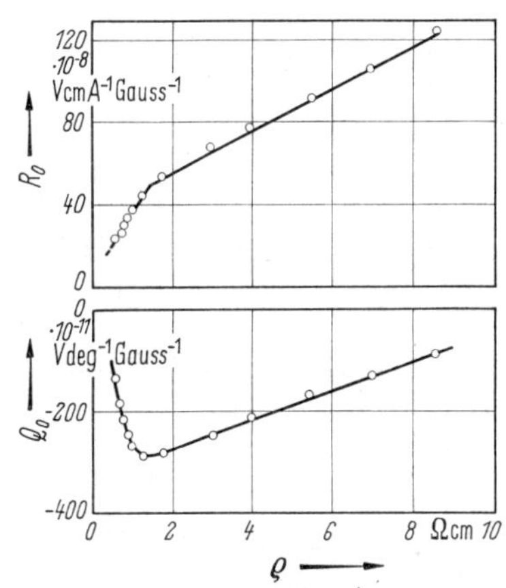

Fig. 6. $Co_{0.94}Fe_{2.06}O_4$. Dependence of the spontaneous Hall (R_0) and Nernst-Ettingshausen (Q_0) coefficients on electrical resistivity, ϱ, in a single crystal [*65SM*].

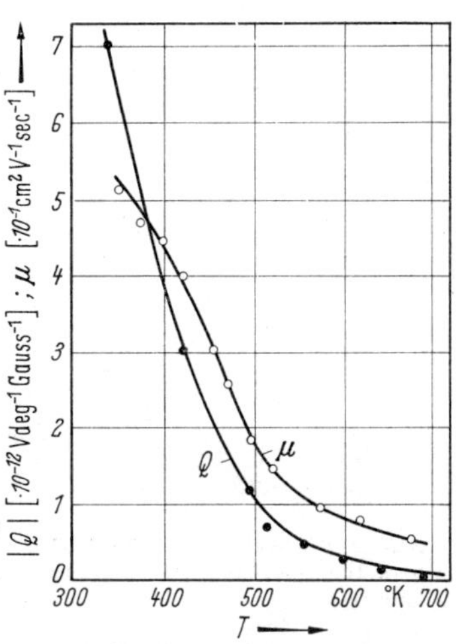

Fig. 7. $Co_{0.94}Fe_{2.06}O_4$. Temperature dependence of the Hall mobility of carrier, μ, and of the classical Nernst-Ettingshausen coefficient, $|Q|$, in a single crystal [*65SM*].

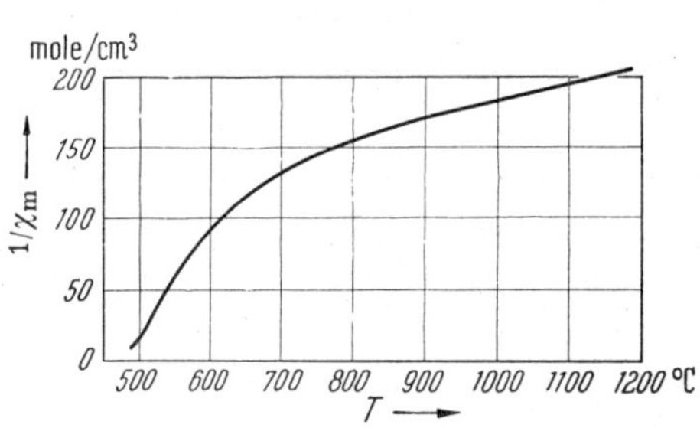

Fig. 8. $CoFe_2O_4$. The reciprocal of the molar susceptibility, $1/\chi_m$, vs. temperature [*51FM*]. See also [*52N*].

→

Fig. 9. $CoFe_2O_4$. Saturation magnetization per gram, σ_s, vs. temperature [*59SW*(1)].

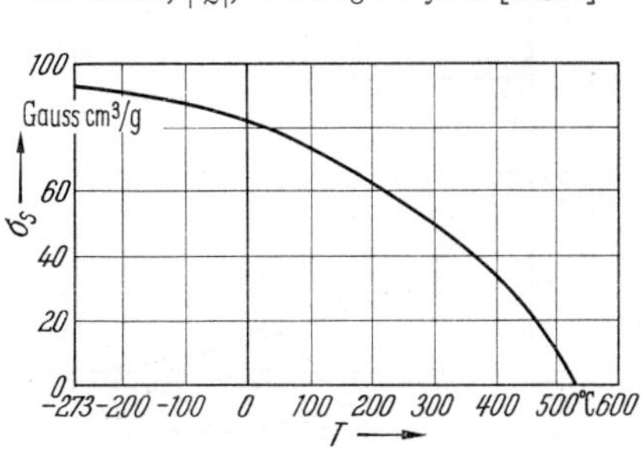

Tab. 3. $CoFe_2O_4$. Dependence of σ_s/σ_s^0 on T/Θ_C [*52P*]

T/Θ_C	σ_s/σ_s^0
Fe_2O_3CoO; $p_{sm}^0 = 3.94\ \mu_B$; $\Theta_C = 769$ °K	
0.369	0.876
0.394	0.861
0.404	0.853
0.417	0.838
0.435	0.821
0.437	0.817
0.457	0.804
0.496	0.77
0.524	0.741
0.581	0.690
0.624	0.648
0.670	0.602
0.700	0.567
0.742	0.517
0.793	0.462
0.838	0.406
0.900	0.314

Tab. 4. $CoFe_2O_4$. Saturation magnetization and coercive field in specimens which differ in density, at 293 °K and 77 °K [*51G*(2)]

M_s Gauss	$T = 293$ °K		$T = 77$ °K		$T = 20$ °K
	H_c Oe	H_cM_s 10^5 erg/cm³	H_c Oe	H_cM_s 10^5 erg/cm³	H_c Oe
198...	3090	6.09	9000	17.82	—
203...	2830	5.74	8770	17.80	17 000
264...	2250	5.93	6800	17.92	—
276...	2110	5.83	6350	17.55	—

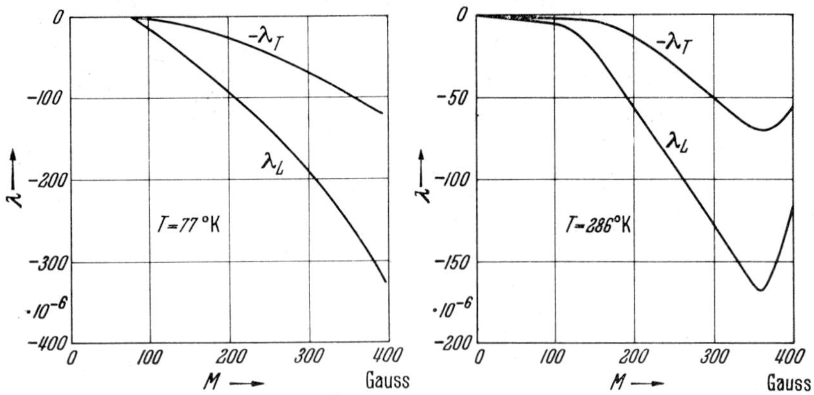

Fig. 11. $CoFe_2O_4$. Longitudinal (λ_L) and transverse (λ_T) magnetostriction vs. magnetization at 77 °K and 286 °K [*50WGP*].

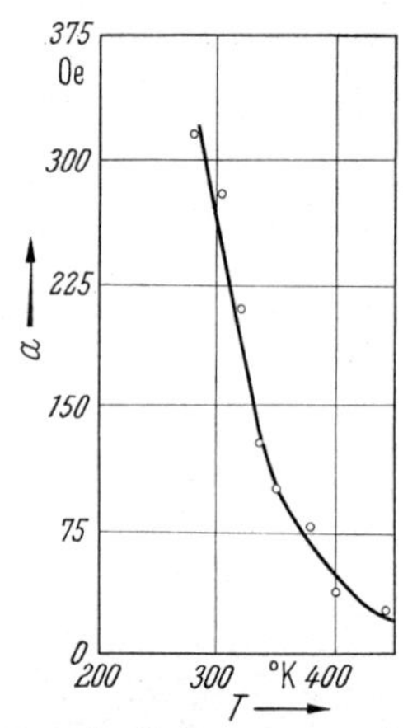

Fig. 10. $CoFe_2O_4$. Coefficient of magnetic hardness, a, vs. temperature where the coefficient of magnetic hardness is defined by the equation:

$\sigma = \sigma_s \left(1 - \frac{a}{H_{int}}\right)$, where H_{int} is the internal field [*52P*].

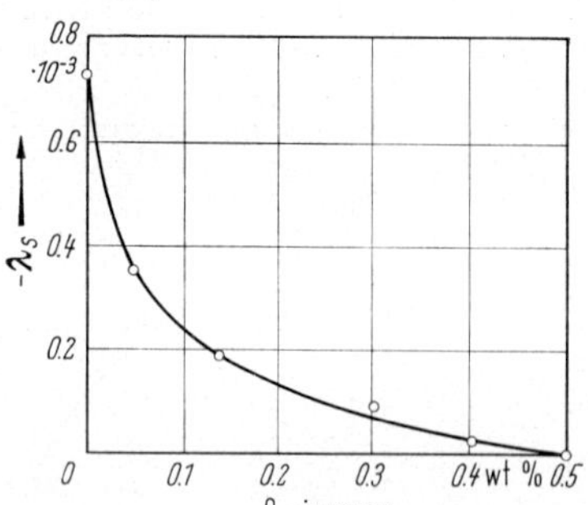

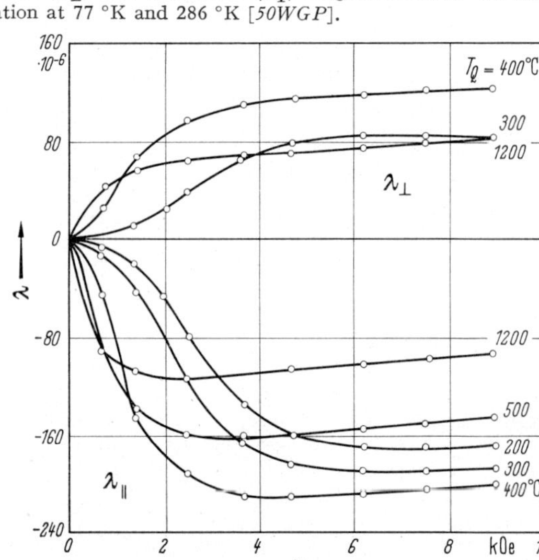

Fig. 13. $CoFe_2O_4$. Room temperature magnetostriction vs. applied magnetic field, H_a, for polycrystalline samples quenched from different temperatures, T_Q. Longitudinal magnetostriction, $\lambda_{\|}$, and transverse magnetostriction, $\lambda_{\perp}$. All heat treatments were in air [*66G*].

←

Fig. 12. $CoFe_2O_4$. Dependence of the saturation magnetostriction of polycrystalline samples on oxidation resulting from thermomagnetic treatment in air. Measurements were made along direction of magnetic annealing. A 7 kOe field was used at a temperature of 300 °C during the 3 h magnetic annealing [*62PG*].

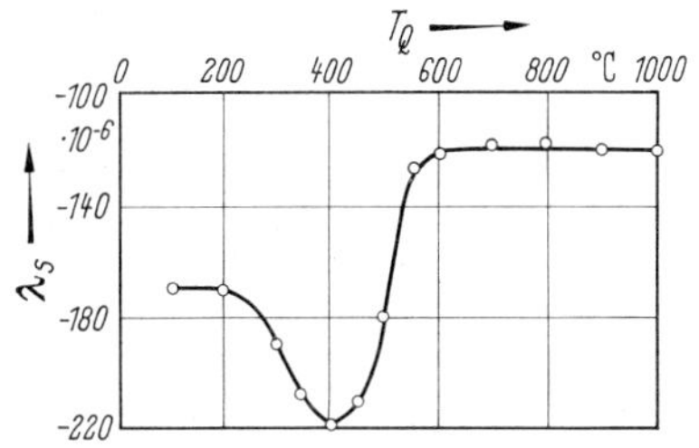

Fig. 14. $CoFe_2O_4$. Room temperature saturation magnetostriction, λ_s, vs. quenching temperature, T_Q. (See Fig. 13) [66G].

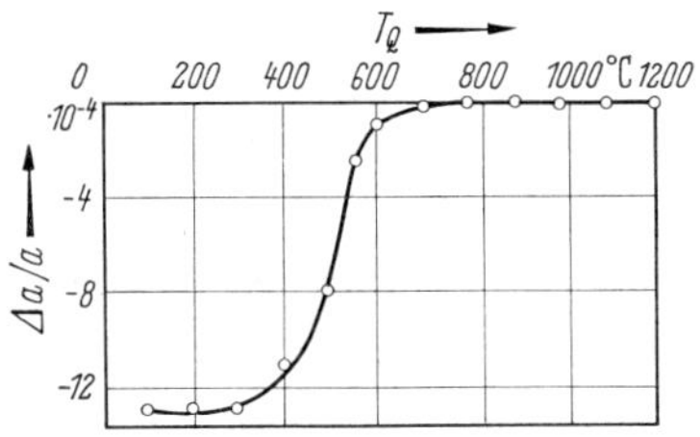

Fig. 15. $CoFe_2O_4$. Relative variation of the lattice constant, $\Delta a/a$, vs. the pre-quenching temperature, T_Q [66G].

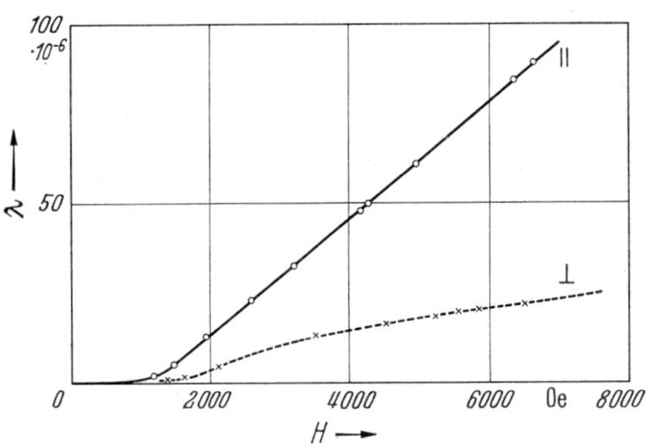

Fig. 16. $CoFe_2O_4$. Transverse magnetostriction parallel and perpendicular to direction of magnetic annealing [54V].

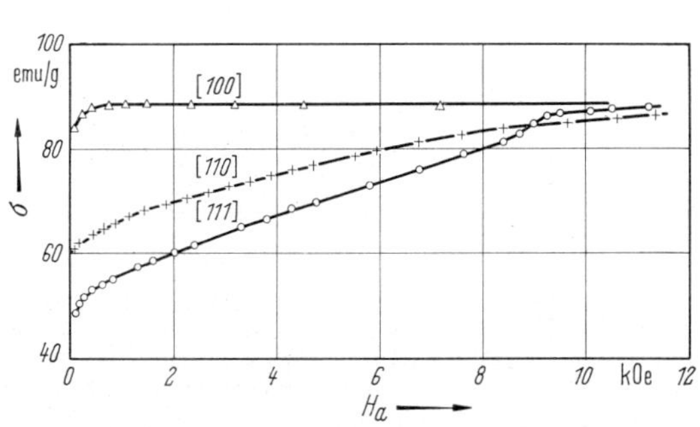

Fig. 17. $CoFe_2O_4$. Specific magnetization vs. applied field for H_a along three crystallographic directions at room temperature. These data yield values of $K_1 = (2.5 \cdots 2.8) \cdot 10^6$ erg cm^{-3} [63PK].

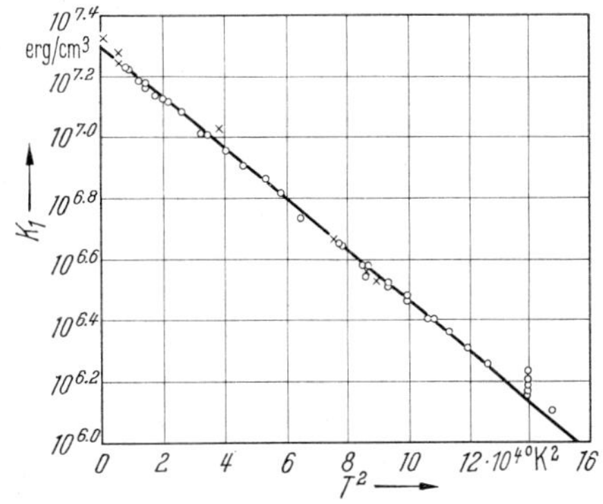

Fig. 18. $Co_{1.01}Fe_{2.00}O_{3.62}$. K_1 as a function of T^2 measured by the static torque method [57S]. $K_1 = A \cdot T^2$, $A = (19.6 \cdot 10^6 \cdot 10^{-8.27 \cdot 10^{-6}})$ erg °K^{-2} cm^{-3}.

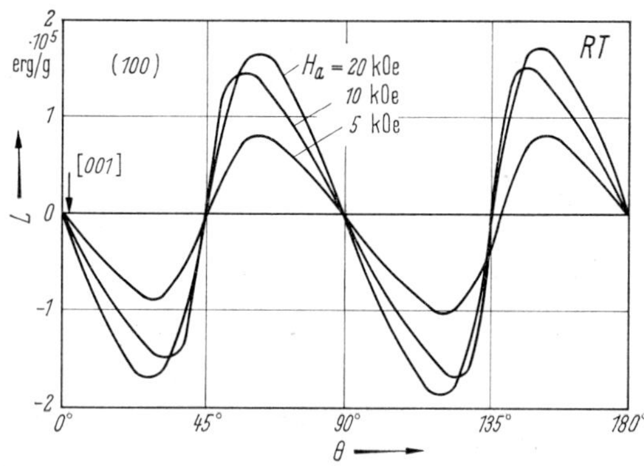

Fig. 19. $Co_{1.1}Fe_{1.9}O_4$. Torque curves at room temperature on a single crystal with magnetic fields H_a in the (100) plane [66II].

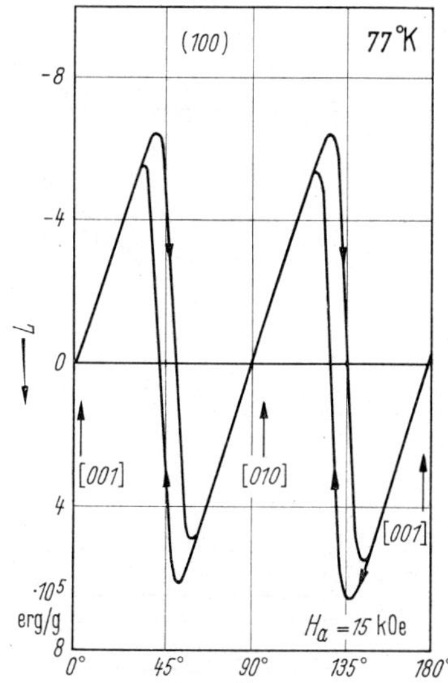

Fig. 20. $Co_{1.1}Fe_{1.9}O_4$. Torque curves at 77 °K on a single crystal with the magnetic field H_a in the (100) plane [66II].

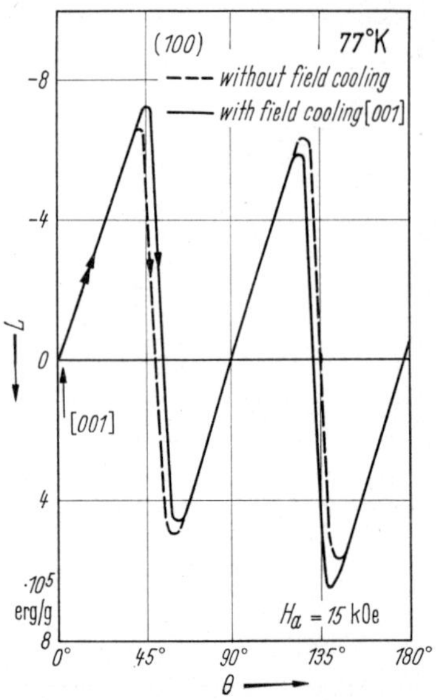

Fig. 21. $Co_{1.1}Fe_{1.9}O_4$. Comparison between two (100) torque curves at 77 °K on a single crystal with and without a magnetic field annealing (H_T = 15 kOe) during cooling [*66II*].

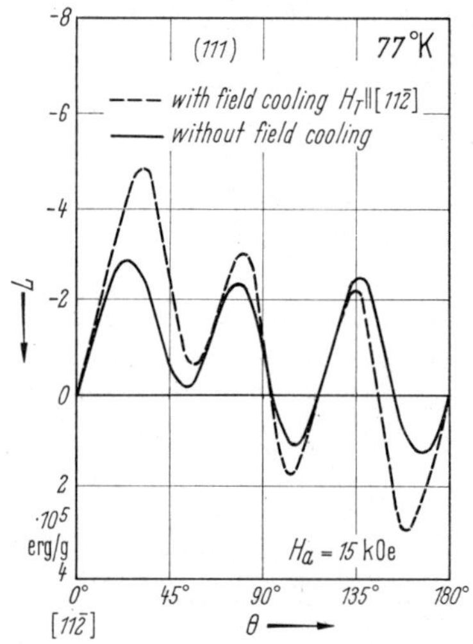

Fig. 22. $Co_{1.1}Fe_{1.9}O_4$. Comparison between two (111) torque curves at 77 °K on a single crystal with and without a magnetic field annealing (H_T = 15 kOe) during cooling [*66II*].

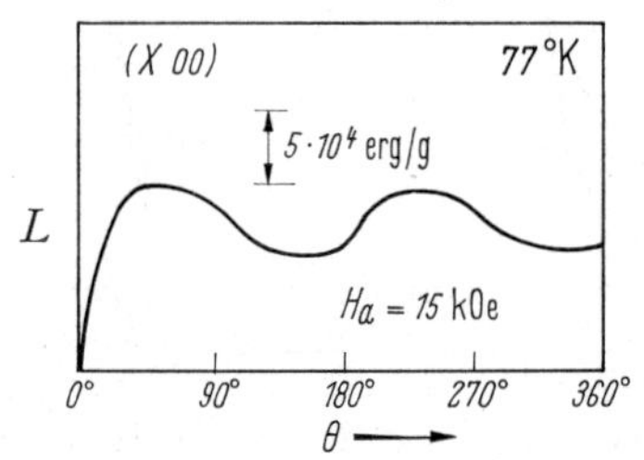

Fig. 23. $Co_{1.0}Fe_{2.0}O_4$. Torque curve of a polycrystalline sample at 77 °K after cooling from room temperature in a magnetic field [*66II*].

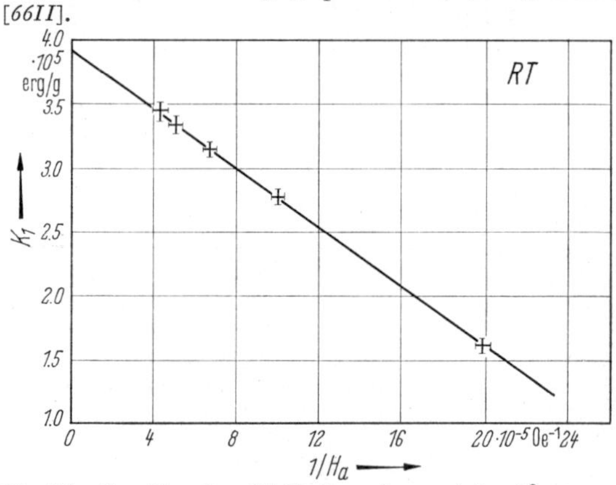

Fig. 24. $Co_{1.1}Fe_{1.9}O_4$. Field dependence of the 4θ component of (100) torque curves for a single crystal at room temperature [*66II*].

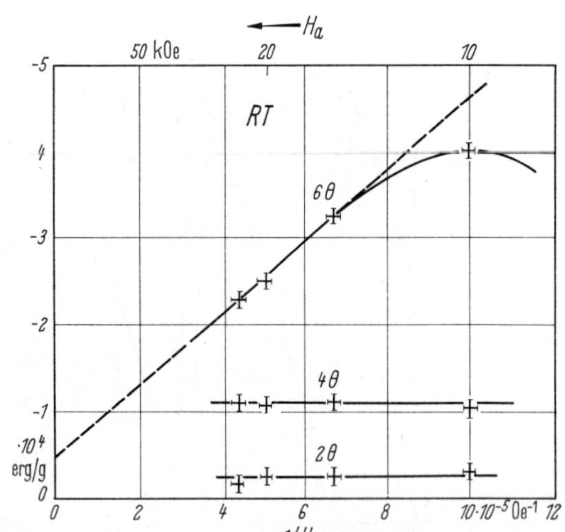

Fig. 25. $Co_{1.1}Fe_{1.9}O_4$. Field dependences of the 6θ, 4θ, and 2θ components of (111) torque curves for a single crystal at room temperature [*66II*].

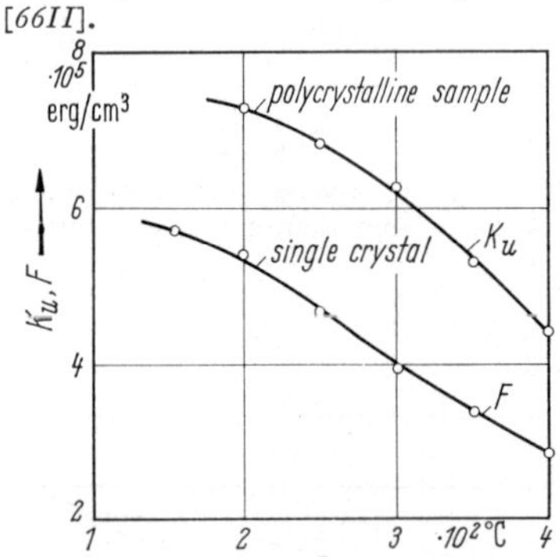

Fig. 26. $Co_{0.9}Fe_{2.1}O_4$. The induced anisotropy constants F (for a single crystal) and K_u (for a polycrystalline specimen) vs. the temperature of annealing in a field H_T = 4 kOe. The constants were obtained at room temperature in a measuring field H_a = 15 kOe. F is defined in the induced anisotropy energy term $-F(\alpha_1^2\beta_1^2 + \alpha_2^2\beta_2^2 + \alpha_3^2\beta_3^2)$ where the α's are direction cosines of the magnetization during the measurement and the β's the corresponding quantities during the magnetic annealing heat treatment. K_u is defined in the relation for polycrystalline induced anisotropy energy $E = -K_u \cos^2 \theta$ where θ is the angle between the magnetic annealing direction and the magnetization during measurement [*64S*].

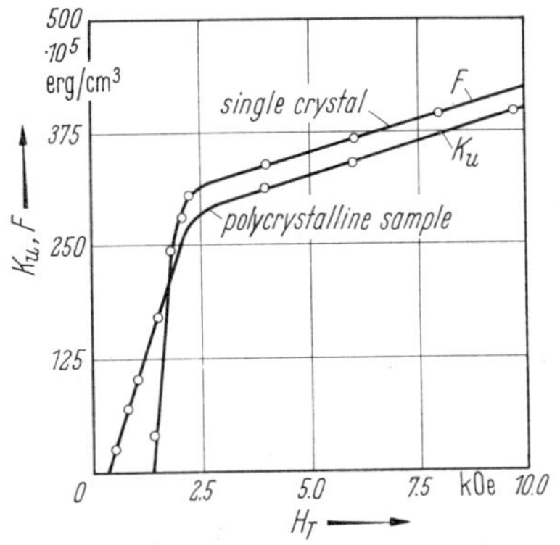

Fig. 27. $Co_{0.9}Fe_{2.1}O_4$. The constants F and K_u (see Fig. 26) vs. the magnetic annealing field strength H_T for a single crystal and polycrystalline specimens at room temperature. Annealing temperature: 207 °C; time of heating: 1.5 h [64S].

Tab. 5. Co-Fe ferrite. Values of induced anisotropy constant, F, and anisotropy constant, K_1, (see Fig. 26 and 27) for single crystals. Room temperature values determined by an oscillation method [64S]

H_a = applied (internal) field during measurement
$H_a = \infty$: extrapolation of different measurements

Composition	$H_a = \infty$		H_a = 15.8 kOe	
	F	K_1	F	K_1
	10^6 erg cm^{-3}			
$Co_{0.9}Fe_{2.1}O_4$	1.6	3.0	0.52	2.1
$Co_1Fe_2O_4$	1.4	2.9	0.36	1.5
$Co_1Fe_2O_4$	0.24	3.8	0.08	1.6

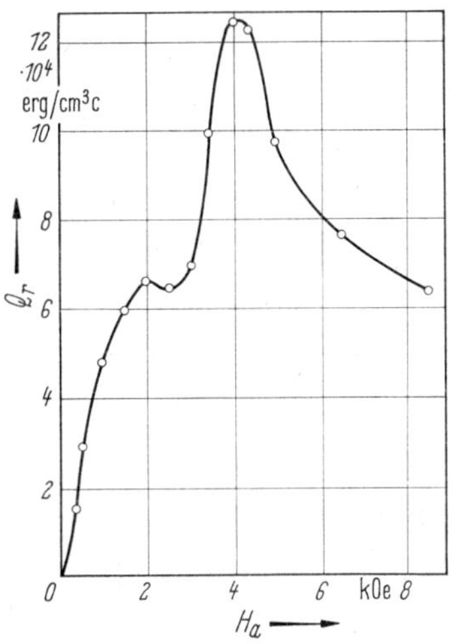

Fig. 28. $Co_{0.94}Fe^{2+}_{0.12}Fe^{3+}_{0.96}O_4$. Rotational hysteresis loss, Q_r, as a function of magnetic field H_a in a single crystal sphere [62DDSZ].

Tab. 6. $Co_{0.9}Fe_{2.1}O_4$. Constants for polycrystalline samples. See Fig. 26 for definition of K_u [64S]

No	Composition	K_u from magnetization curves	K_u by the oscillation method	M_s Gauss	ϱ g · cm^{-3}
		10^6 erg · cm^{-3}			
1	$Cu_{0.9}Fe_{2.1}O_4$	0.44	0.40	432	4.83
2	$Co_{0.9}Fe_{2.1}O_4$	0.85	0.72	429	4.82

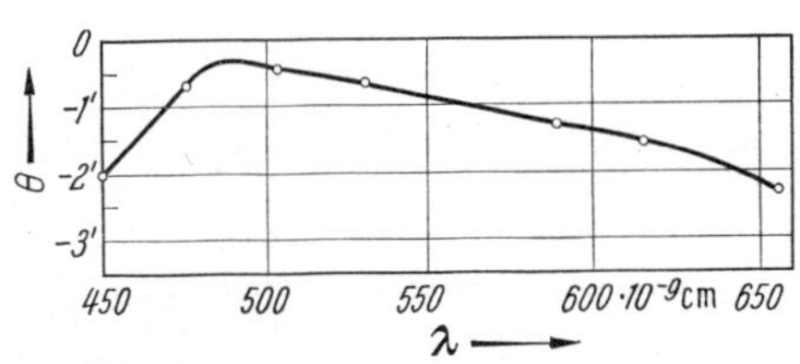

Fig. 29. $CoFe_2O_4$. Rotation angle, θ (in minutes of arc), for the magnetic Kerr effect [18M].

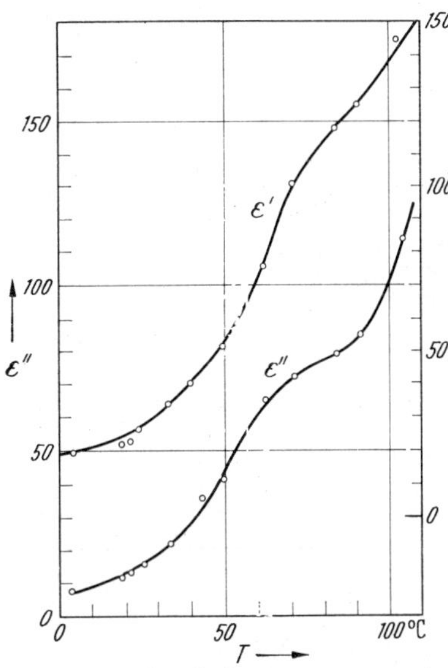

Fig. 30. $CoFe_2O_4$. ε' and ε" vs. temperature for a slowly cooled sample at 30 kc/s [51K].

Tab. 7. $CoOFe_2O_3$. Dielectric constant, permeability, and tan δ in polycrystalline disc specimens [52OFD(1)]. See also [52OFD(2)]

Thickness of disc	Heat treatment	ε'	ε"	tan δ	μ'	μ"
1.5 mm	quenched from 1200 °C	9.49	0.045	0.0047	0.43	0.211
1.5 mm	slowly cooled 1200 °C	9.00	*)	*)	0.10	0.116
1 mm	quenched from 1200 °C	11.23	0.028	0.003	0.43	0.218
1 mm	slowly cooled 1200 °C	9.59	*)	*)	0.47	0.097

*) Too small to be measured.

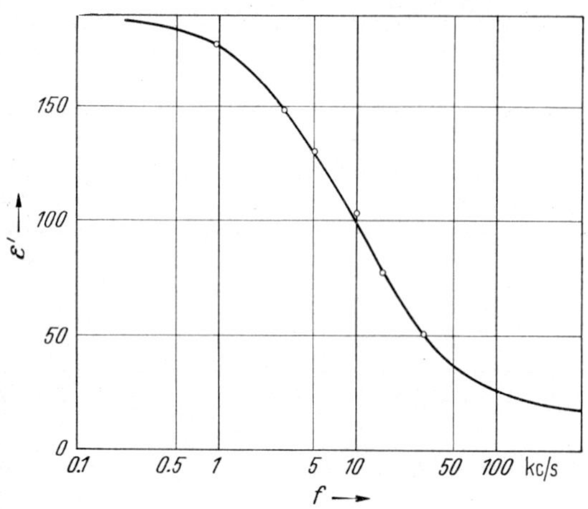

Fig. 31. $CoFe_2O_4$. Dielectric constant, ε', vs. frequency for a slowly cooled sample at 50 °C. $\varepsilon'_0 = 187$, $\varepsilon'_\infty = 18$ [51K].

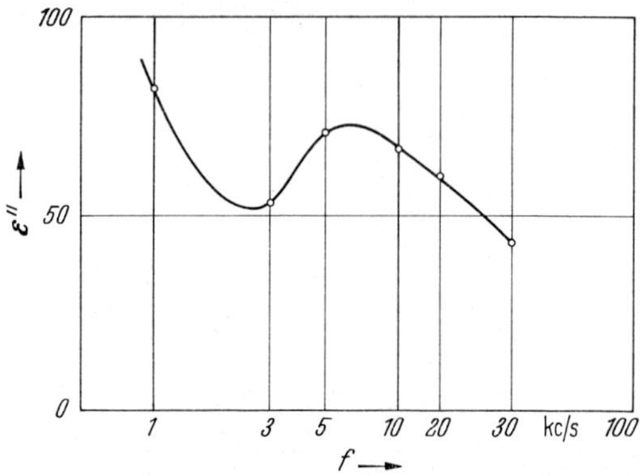

Fig. 32. $CoFe_2O_4$. The imaginary part of the dielectric constant, ε'', vs. frequency for a slowly cooled sample at 50 °C [51K].

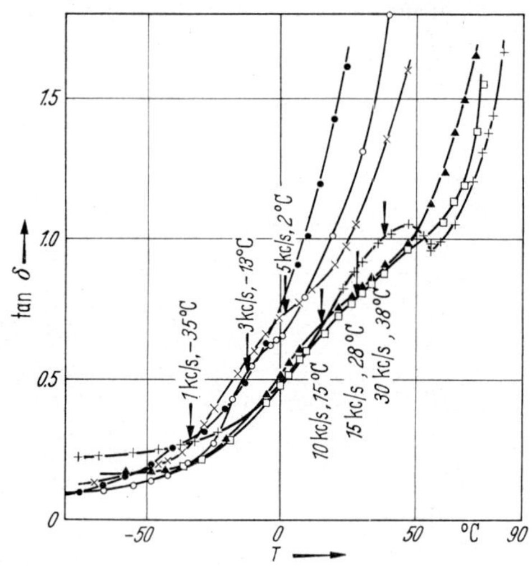

Fig. 33. $CoFe_2O_4$. tan δ vs. temperature at various frequencies for a quenched sample. Sample was prepared by sintering for 3 h at 1200 °C [51K].

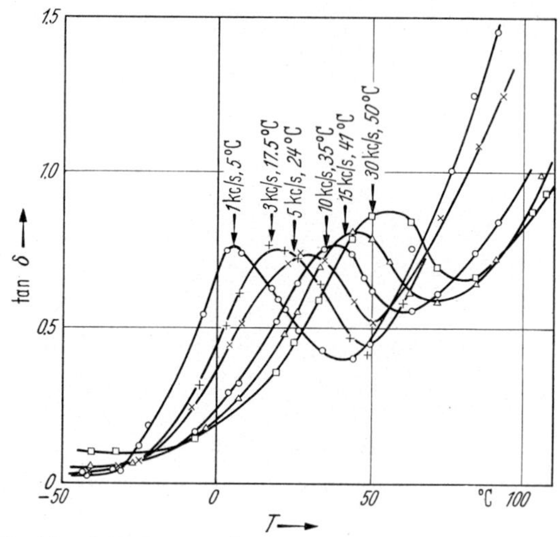

Fig. 34. $CoFe_2O_4$. tan δ vs. temperature at various frequencies for a slowly cooled sample [51K].

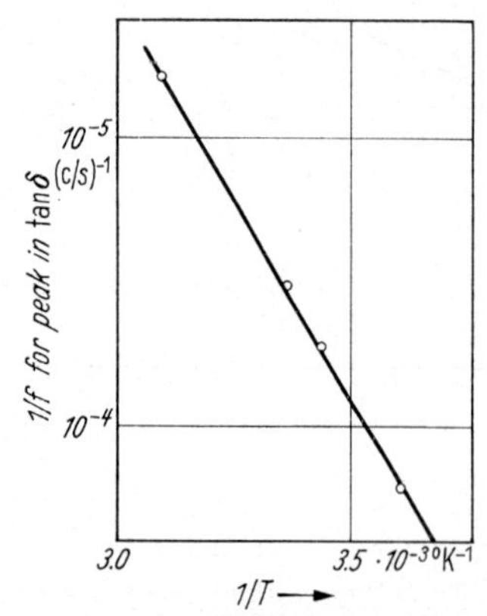

Fig. 35. $CoFe_2O_4$. The dependence of dielectric dispersion on temperature in a slowly cooled sample [56OS]. Cited in [56U].

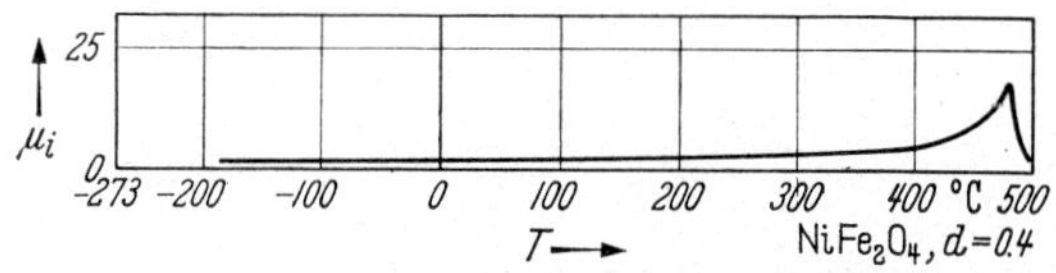

Fig. 36. $CoFe_2O_4$. Initial permeability, μ_i, of a polycrystalline sample vs. temperature. The density of the sample was 4.2 g/cm³ [58SW(2)].

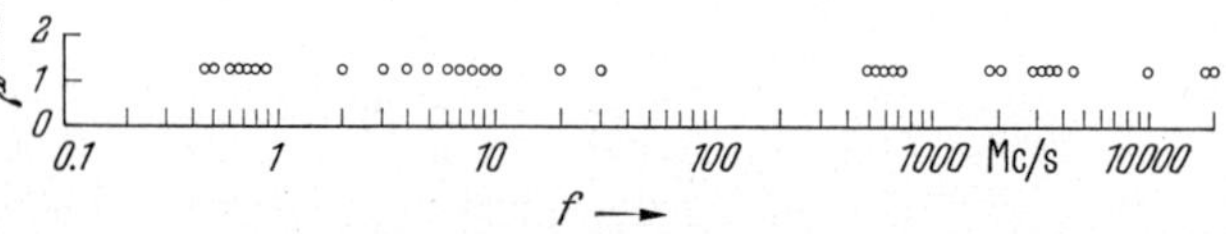

Fig. 37. $CoFe_2O_4$. Initial permeability, μ_i, vs. frequency for a polycrystalline sample [58PA].

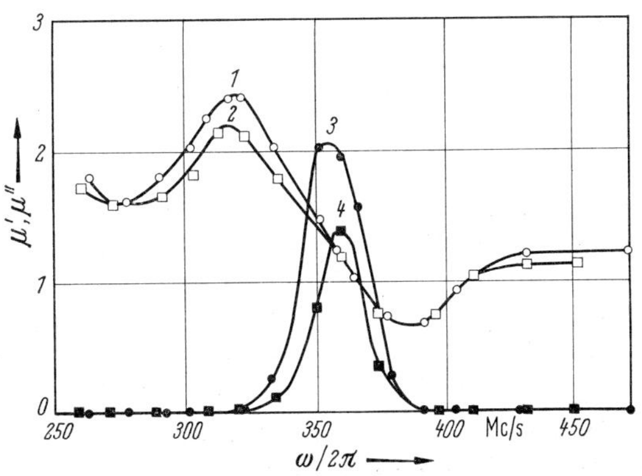

Fig. 38. $Co_{0.94}Fe^{2+}_{0.06}Fe^{3+}_{2.00}O_4$. Frequency dependence of the complex magnetic permeability in a single crystal picture frame sample. The two samples were annealed in oxygen for 6 h and had domain widths from $3.7 \cdot 10^{-3} \cdots 6.4 \cdot 10^{-3}$ cm. Curve 1, μ' for sample 1; curve 2, μ' for sample 2; curve 3, μ'' for sample 1; curve 4, μ'' for sample 2 [61PAS].

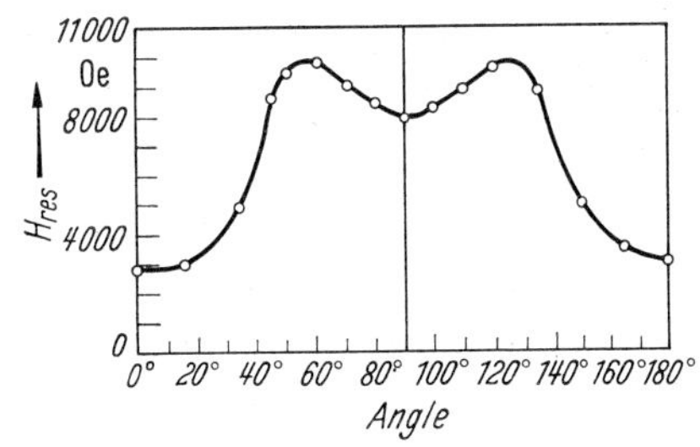

Fig. 39. $CoFe_2O_4$. The resonant fields, H_{res}, at the absorption maxima for a single crystal in (110) plane, at 90 °C and 23800 Mc/s [55T].

Tab. 8. $CoFe_2O_4$. Resonance fields corresponding to maximum absorption, and g factors for a polycrystalline sample [51OTK]. See also [51T]

T °C	H_{res} Oe	g_{exp}
100	1550	2.22
200	1200	2.81
300	1250	2.91
400	1770	2.26
480	2850	2.08

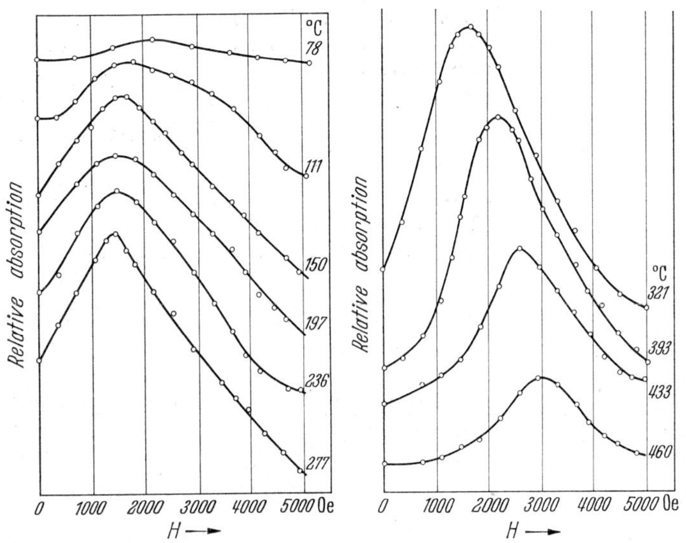

Fig. 40. $CoFe_2O_4$. Resonance absorption vs. dc magnetic field at various high temperatures [51T].

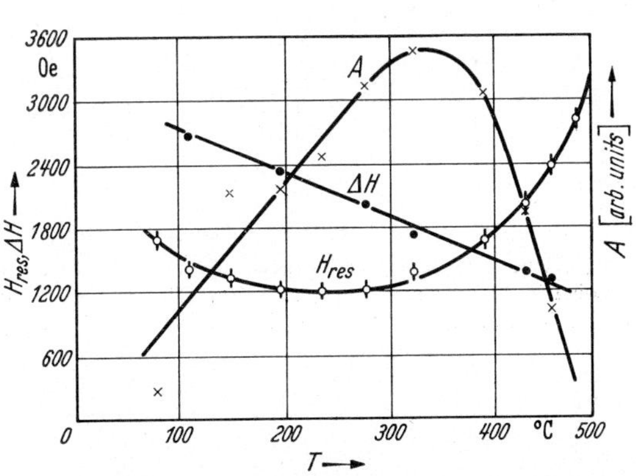

Fig. 41. $CoFe_2O_4$. The resonance field, H_{res}, at the absorption maximum (circles), the amount of the absorption, A, at the maximum (crosses), and the half width, ΔH, of the absorption line (full points) vs. temperature at a wavelength of 3.2 cm. Specimen used was a polycrystalline disc, 3 mm in diameter and 0.22 mm in thickness [51OTK]. See also [51T].

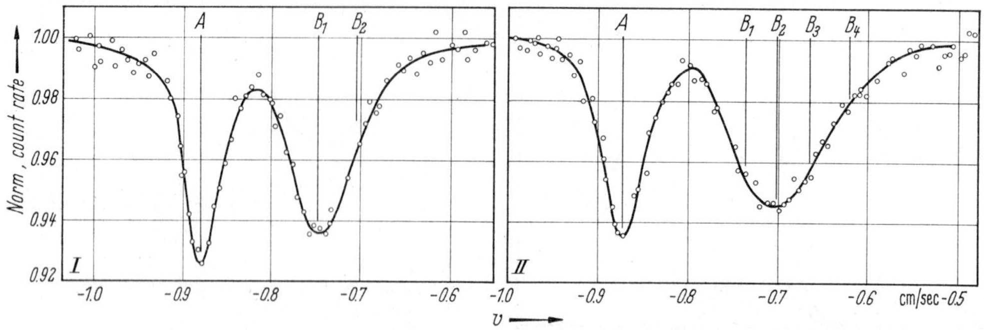

Fig. 42. $CoFe_2O_4$. First few lines of the Mössbauer absorption spectra of ^{57}Fe nuclei in slowly cooled (from 1200 °C) (I) and quenched (from 1200 °C in H_2O) (II) powders in an applied field of 55 kOe. A, B_1, B_2, B_3, B_4 mark the positions of the lines assigned to ^{57}Fe nuclei in tetrahedral and different octahedral sites (i.e. those which have different combinations of Fe and Co as tetrahedral nearest neighbors, see Tab. 9). The length of the drawn lines is proportional to the relative intensities. [*68SVM*]. See also [*62NM*] and [*63DHK*].

Tab. 9. $CoFe_2O_4$. Data for slowly cooled (I) and quenched (II) powders used in Fig. 42 and 43. a_A and a_B are the areas of the tetrahedral and octahedral Mössbauer lines, respectively. (See Fig. 42) [*68SVM*]

	a_A/a_B	Θ_C °K	p_m [μ_B]		Calculated relatived intensities of the B sites [1])			
			calc.	obs.	B_1	B_2	B_3	B_4
I	0.92 ± 0.04	798	3.2	3.4	78	20	2	...
II	0.65 ± 0.03	792	3.8	3.9	24	39	26	9

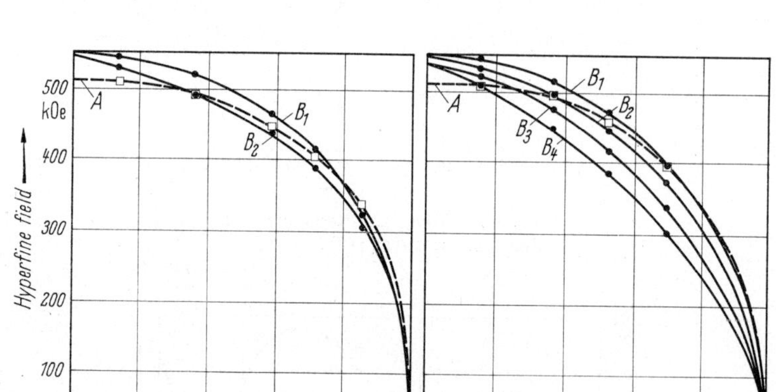

Fig. 43. $CoFe_2O_4$. Temperature dependence of the hyperfine magnetic fields at A site and different B sites (see Fig. 42 and Tab. 9) ^{57}Fe nuclei in slowly cooled (I) and quenched (II) samples. The squares and dashed lines refer to A sites and the dots and solid lines to B sites. The squares and dots are deduced from experiments [*68SVM*].

[1]) B_1, B_2, B_3, and B_4 refer to B ions with the distributions of 6Fe, 5Fe and 1Co, 4Fe and 2Co, and 3Fe and 3Co in the six-nearest A neighbors, respectively.

6.1.8.2 Co ferrite and Co-Fe mixed oxides

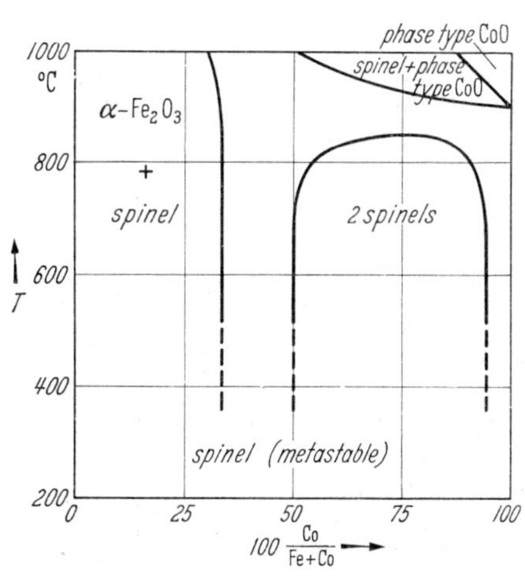

← Fig. 1. γ-Fe_2O_3-Co_3O_4 system. Phase diagram. Below 300 °C γ-Fe_2O_3 (spinel) and Co_3O_4 (spinel) are miscible in all proportions. Above 300 °C γ-Fe_2O_3 is unstable and iron-rich samples decompose into α-Fe_2O_3 (rhombohedral structure) and $CoFe_2O_4$ (spinel). Cobalt-rich samples above 300 °C decompose into two spinels. Between 670 °C and 770 °C these two spinels have a cobalt to iron atomic ratio of 51 : 49 and 90 : 10 [*55R*].

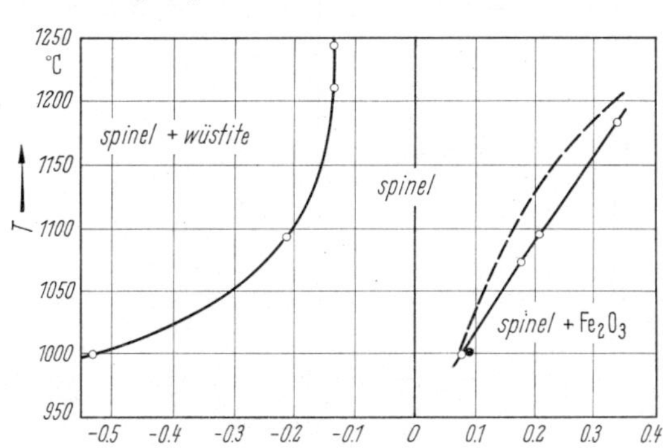

Fig. 2. $Co_{1-x}Fe_{2+x}O_4$. Phase diagram [*64MS*].

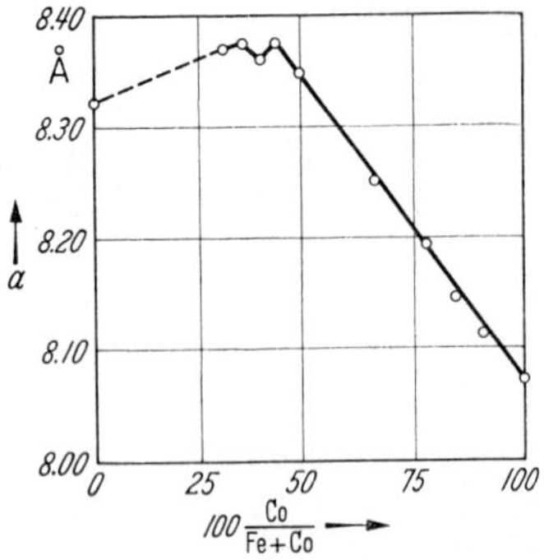

Fig. 3. γ-FeO_3 and Co_3O_4. Lattice parameters, *a*, vs. composition (atomic ratio) for a mixture of the oxides having the spinel structure. The oxygen-cation ratio $\left(\frac{O}{Co + Fe}\right)$ varies with composition from 1.5 for γ-Fe_2O_3 to 1.33 for the mixture containing 33.3% Co, ($CoFe_2O_4$). The oxygen-cation ratio remains constant at 1.33 from $CoFe_2O_4$ to Co_3O_4. The valence states of the cations were not given [*55R*]. See also [*52RB*].

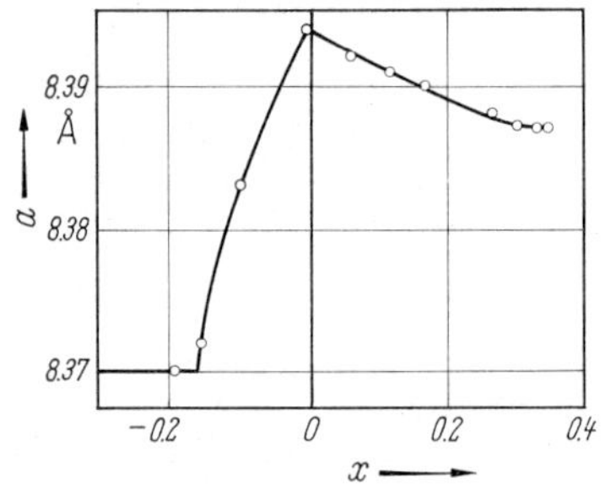

Fig. 4. $Co_{1-x}Fe_{2+x}O_4$. Compositional dependence of the lattice constant. Sample quenched from 1200 °C in 0.21 atm of O_2 [*64MS*]. See Fig. 2.

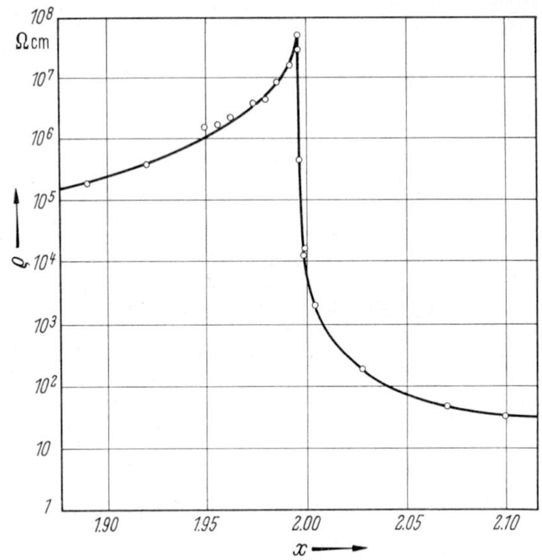

Fig. 5. $Co_{3-x}Fe_xO_4$. Room temperature values of resistivity, ϱ, for a series of mixed crystals near the composition $CoFe_2O_4$ [*59J*].

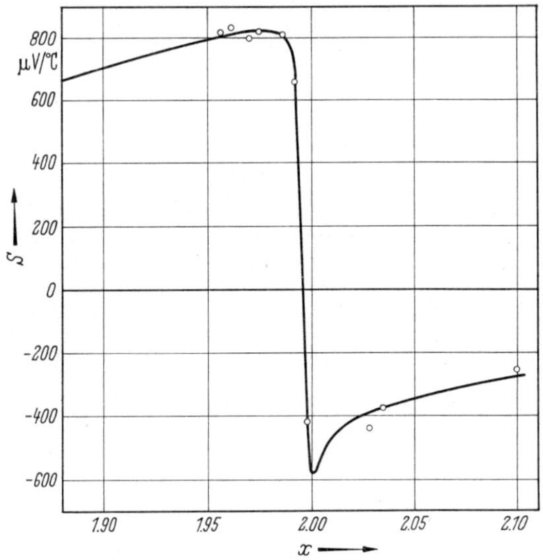

Fig. 7. $Co_{3-x}Fe_xO_4$. Thermoelectric or Seebeck effect vs. composition [*59J*].

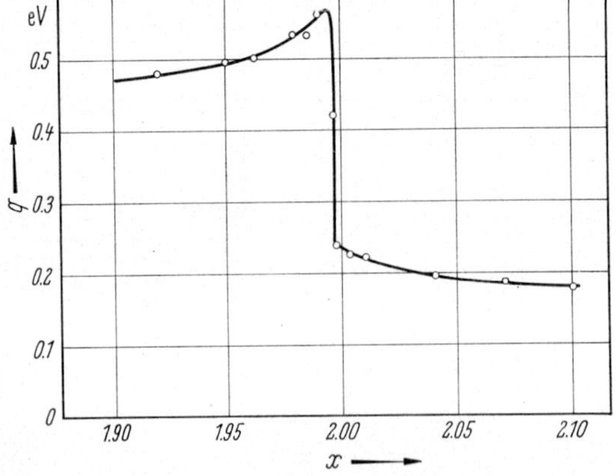

Fig. 6. $Co_{3-x}Fe_xO_4$. Activation energies of the conductivity, *q*, vs. composition [*59J*].

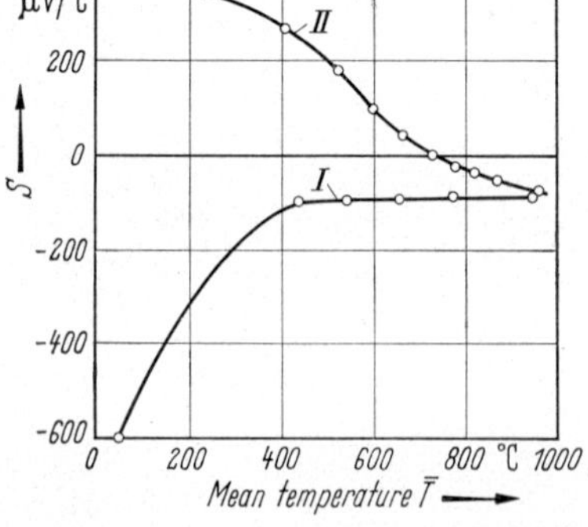

Fig. 8. $CoFe_2O_4$ (I), $Co_{1.01}Fe_{1.99}O_4$ (II). Seebeck effect vs. mean temperature [*59J*].

Compounds in the series $Co_{1-x}Fe_{2+x}O_4$ with $x < 0$, containing an excess of Co, have the Fe^{3+} on octahedral sites replaced by Co^{3+} ions. Compounds for which $x > 0$, containing an excess of Fe, have the Co^{2+} ions on octahedral sites essentially replaced by Fe^{2+} ions [*64MS*].

Tab. 1. $Co_{1-x}Fe_{2+x}O_4$. The self diffusion coefficient, D, of Fe and Co [*64MS*]

	x	−0.21	−0.19	−0.16	−0.11	0.0	0.063	0.19	0.26	0.34
	T °C	D in 10^{-10} cm²/sec								
Co	1180	—	—	2.5	1.6	0.44	1.1	11	—	36
	1130	—	1.5	—	0.5	0.35	—	14	19	—
	1080	0.4	—	—	—	0.27	—	2.7	—	—
Fe	1180	—	—	0.27	—	0.21	—	—	—	25
	1130	—	0.22	—	—	0.1	—	—	20	—
	1080	0.035	—	—	—	—	—	5.8	—	—

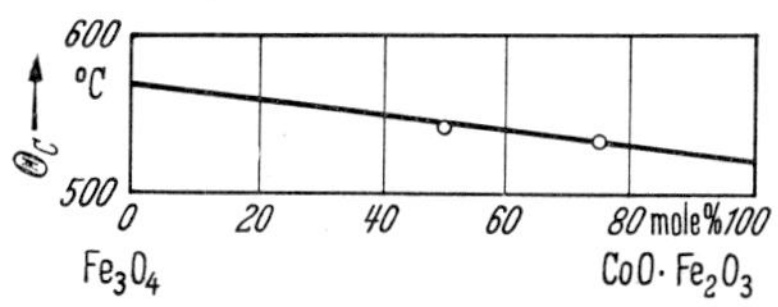

Fig. 9. Fe_3O_4-CoO · Fe_2O_3. Curie temperature as a function of composition for solid solutions [*37BC*].

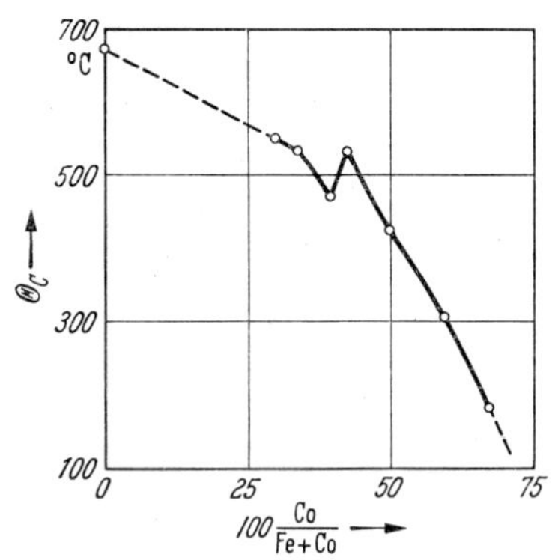

Fig. 10. γ-Fe_2O_3-Co_3O_4. Curie temperatures vs. composition (atomic ratio) in the system of mixed oxides having spinel structure. The valence states of the cations were not given [*55R*]. See also [*53R*].

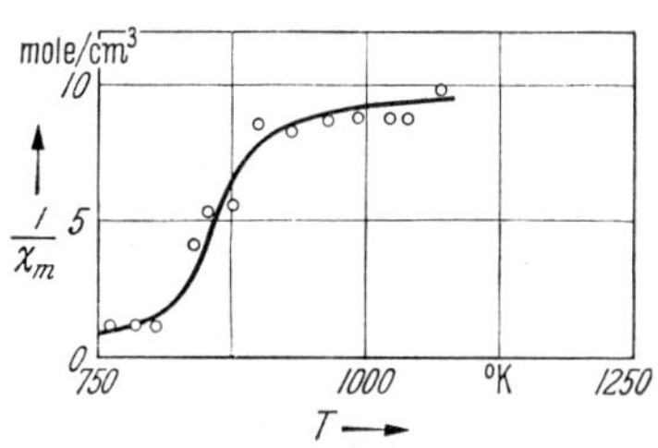

Fig. 11. Vectolite, $0.3Fe_2O_3 \cdot 0.44Fe_3O_4 \cdot 0.26Co_2O_3$. Temperature dependence of the reciprocal susceptibility [*54G*(1)].

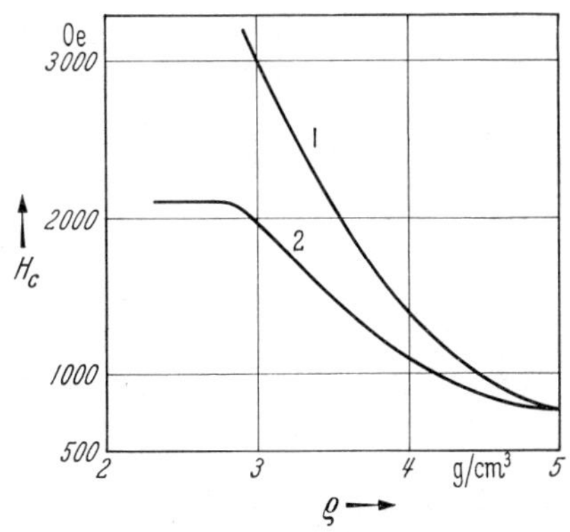

Fig. 12. $xFe_2O_3 \cdot (1-x)CoO$. Coercive field, H_c, vs. density, ϱ, in mixed oxides of Fe and Co. x is the mole fraction varying from 0.5 ⋯ 0.6. Curve 1 is the envelope of all the points made on a great number of samples differing in conditions of preparation. Curve 2 corresponds to a single series of samples with similar conditions of preparation [*53G*].

Tab. 2. $xFe_2O_3(1-x)CoO$. Temperature dependence of the ratio of remanent magnetization, M_{rem}, to saturation magnetization, M_s, along the direction of magnetic annealing in mixed oxides of Fe and Co, x is between 0.5 and 0.65. The conditions of preparation were such that several percent of trivalent iron were transformed into divalent iron [*53G*]

T in [°K]	M_{rem}/M_s	T in [°K]	M_{rem}/M_s
77	0.90	200	0.85
110	0.88	230	0.84
140	0.87	290	0.84

Tab. 3. $Co_xFe^{2+}_{1-x}Fe_2O_4$. Survey of the samples used in Fig. 13 ··· 16

$Co_xFe^{2+}_{1-x}Fe_2O_4$	x	Fe^{2+} content in wt %		firing *) atmosphere cm^3/min	firing temperature °C	ϱ g/cm^3
		analysed	theoretical			
$Co_{0.1}Fe^{2+}_{0.9}Fe^{3+}_2O_4$	0.1	22	21.7	200	1350	5.2
$Co_{0.4}Fe^{2+}_{0.6}Fe^{3+}_2O_4$	0.4	14	14.4	150	1350	5.1
$Co_{0.5}Fe^{2+}_{0.5}Fe^{3+}_2O_4$	0.5	9.5	12.0	70	1350	5.0
$Co_{0.6}Fe^{2+}_{0.4}Fe^{3+}_2O_4$	0.6	8.2	9.6	50	1350	5.1
$Co_{0.8}Fe^{2+}_{0.2}Fe^{3+}_2O_4$	0.8	2.9	4.8	air	1450	5.1

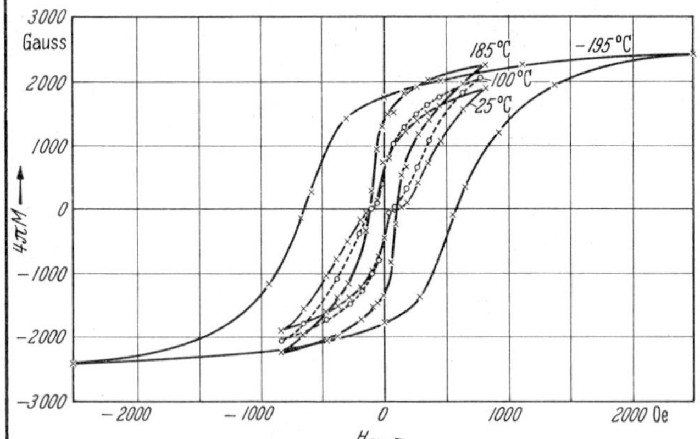

Fig. 14. $Co_{0.6}Fe^{2+}_{0.4}Fe_2O_4$. Minor hysteresis loops measured ballistically at various temperatures; see Tab. 3 [*57WHF*].

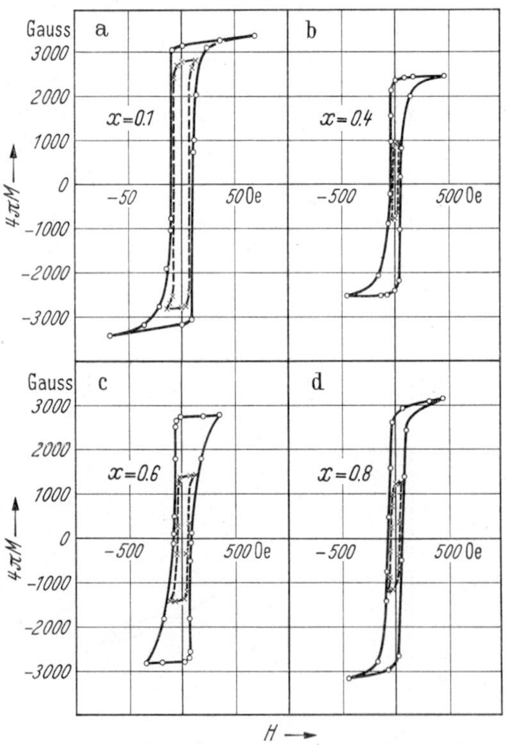

Fig. 15. $Co_xFe^{2+}_{1-x}Fe_2O_4$. Major and minor hysteresis loops for each of the ferrites after magnetic annealing of the samples See Tab. 3 [*57WHF*].

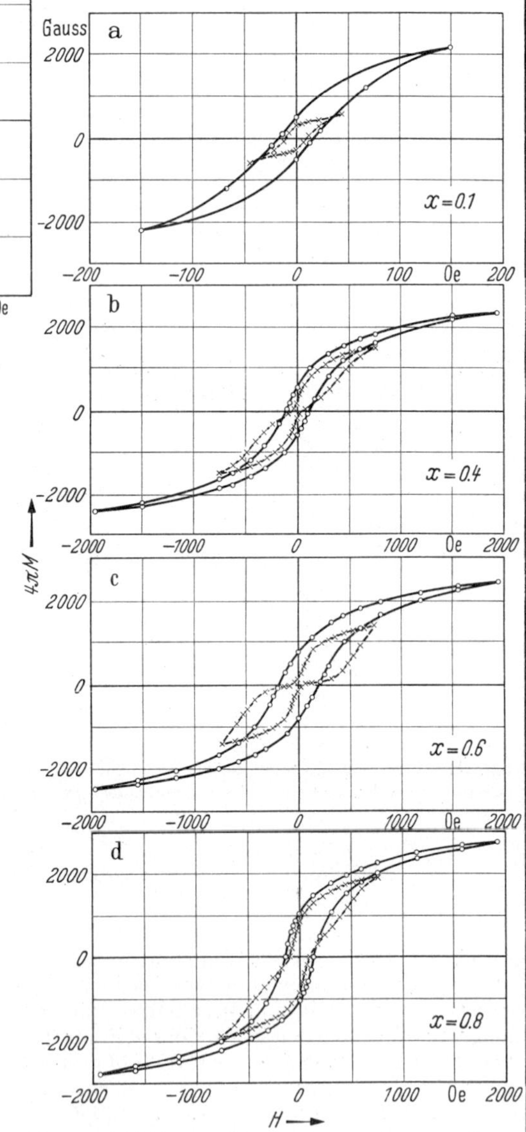

Fig. 13. $Co_xFe^{2+}_{1-x}Fe_2O_4$. Major and minor hysteresis loops for each of the ferrites which have been furnace-cooled slowly without a magnetic field; see Tab. 3 [*57WHF*].

*) A stream of gas passed through the furnace consisting of 600 cm^3/min of CO_2 with an addition of a mixture of 90% nitrogen and 10% hydrogen. The amount of the additions is given in this column [*57WHF*].

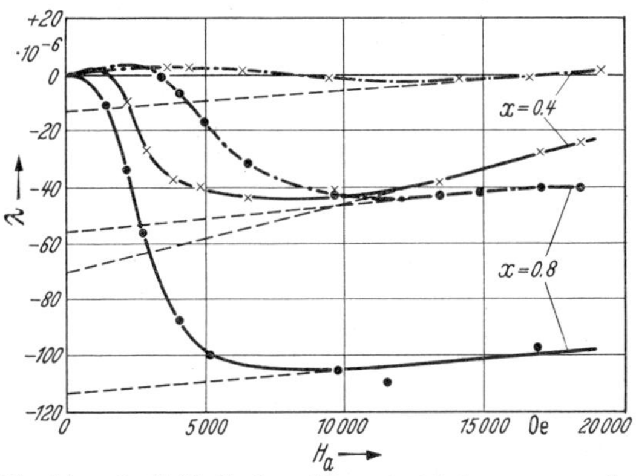

Fig. 16. $Co_xFe^{2+}_{1-x}Fe_2O_4$. Magnetostriction vs. applied field, H_a; $x = 0.4$ and 0.8, see Tab. 3. Solid curves were obtained before magnetic annealing, and broken curves after magnetic annealing [*57WHF*].

Tab. 4. 44CoO · 56Fe_2O_3. Density, magnetostriction at saturation, and coercive field for five specimens [*54V*]. See also [*52V*(1)]

Specimen	ϱ g/cm^3	$\lambda_s \cdot 10^6$	H_c Oe
I	3.81	−110	1040
II	3.64	−110	1260
III	3.56	−110	1490
IV	3.28	−110	1720
V	2.90	−110	2050

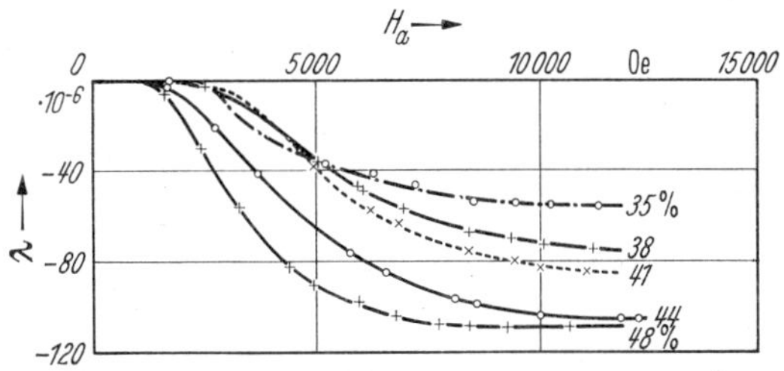

Fig. 17. $(CoO)_x(Fe_2O_3)_{100-x}$. Magnetostriction, λ, vs. applied magnetic field, H_a, for various values of x [*54V*]. See also [*52V*(1)].

Tab. 5. 44CoO · 56Fe_2O_3. Magnetostriction at saturation and coercive field for one of the specimens [*54V*]. See also [*52V*(2)]

T °C	$\lambda_s \cdot 10^6$	H_c Oe
20	−110	2050
− 80	−180	3100
−196	−270	6500

Tab. 6. $x\,Fe_2O_3 \cdot (1-x)CoO$. Dependence of magnetostriction, λ, and coercive field, H_c, on composition in the mixed oxides. The conditions of preparation were such that several percent of trivalent iron were transformed into divalent iron. The applied field was approximately 12 kOe [*53G*]

x	$\lambda \cdot 10^6$	H_c [Oe]
0.52	−110	1500
0.56	−110	2050
0.59	− 90	2000
0.62	− 80	1800
0.65	− 60	1600

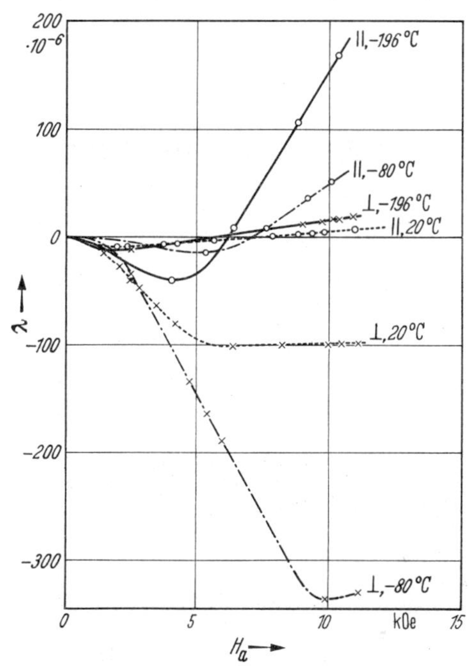

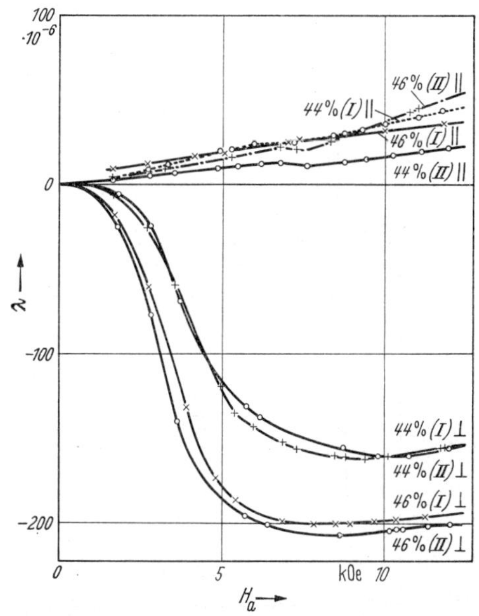

Fig. 18. 44CoO · 56Fe_2O_3; 46CoO · 54Fe_2O_3. Magnetostriction parallel and perpendicular to the direction of magnetic annealing vs. applied magnetic field. Samples I and II in each case differed in unspecified treatment [*54V*]. See also [*52V*(1)].

← Fig. 19. 48CoO · 52Fe_2O_3. Magnetostriction parallel and perpendicular to direction of magnetic annealing vs. applied magnetic field at 20 °C, −80 °C, and −196 °C. The curve shown for magnetostriction parallel to direction of magnetic annealing at −196 °C is obtained after the first cooling. After sample has been warmed and re-cooled several times, the curve changes, and instead shows a continued contraction attaining a value of $\lambda = -145 \cdot 10^{-6}$ in a field of approximately 11 kOe [*54V*]. See also [*52V*(2)].

Tab. 7. Vectolite compound consisting of 44 mole % Fe_3O_4, 30 mole % Fe_2O_3, 26 mole % Co_2O_3. Magnetic properties [*50GVM*]

	T °K	H_c Oe	M_R Gauss	$\lambda \cdot 10^6$
quenched	288	1000	95	−110
quenched	77	3400	153	−125
not quenched	288	1470	86	− 70
not quenched	77	4000	129	− 50
magnetically annealed axis of easy magnetization	288	1200	147	− 70
magnetically annealed axis of hard magnetization	288	1100	52	−200
magnetically annealed axis of easy magnetization	77	6300	154	− 47
magnetically annealed axis of hard magnetization	77	5600	121	− 60

Tab. 8. Co-Fe ferrites. Magnetostriction constants, λ_{100} and λ_{111}, and anisotropy constants, K_1, for some single crystals [*55BTW*]

Composition	$\lambda_{100} \cdot 10^6$	$\lambda_{111} \cdot 10^6$	K_1 at 20 °C 10^{-6} erg/cm³	K_1 at −196 °C 10^{-6} erg/cm³
$Co_{0.8}Fe_{2.2}O_4$ } [2]	−590	120	3.9 [1])	4.4
$Co_{0.8}Fe_{2.2}O_4$ } [2]			2.9 [3])	
$Co_{1.1}Fe_{1.9}O_4$ } [4]	−250		1.8	
$Co_{1.1}Fe_{2.2}O_4$ } [4]			3.8	17.5

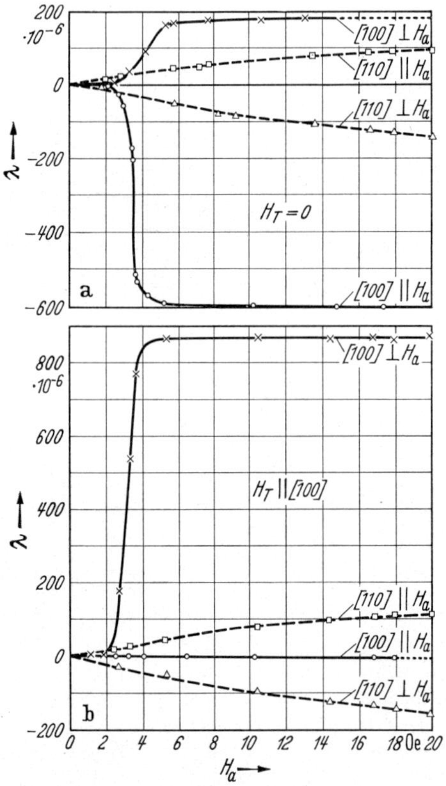

Fig. 20. $Co_{0.8}Fe_{2.2}O_4$. Magnetostriction after treatment in (a) zero magnetic annealing field ($H_T = 0$) and (b) strong magnetic annealing field, H_T, parallel to [100] in the (100) plane [*55BTW*]. H_a is the applied magnetic field.

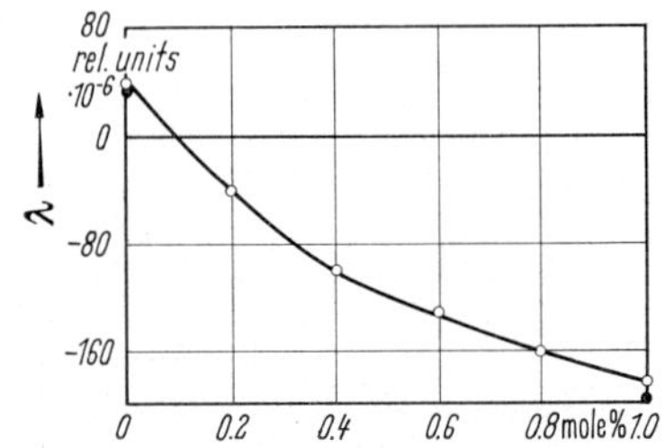

Fig. 21. $Co_xFe_{3-x}O_4$. Concentration dependence of longitudinal saturation magnetostriction in a polycrystalline sample [*64PG*].

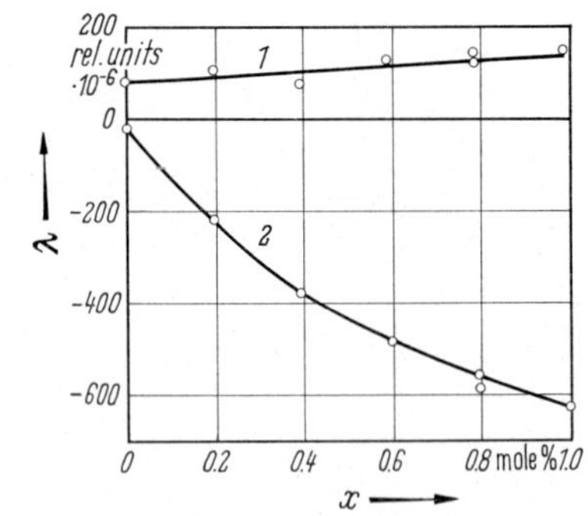

Fig. 22. $Co_xFe_{3-x}O_4$. X-ray determination of concentration dependence of the magnetostriction constants in a polycrystalline sample [*64PG*].

[1]) Quenched in air from 400 °C.
[2]) Brackets indicate that measurements were made on different specimens cut from the same crystal.
[3]) Annealed at 150 °C for three days.
[4]) [*55S, 57S*].

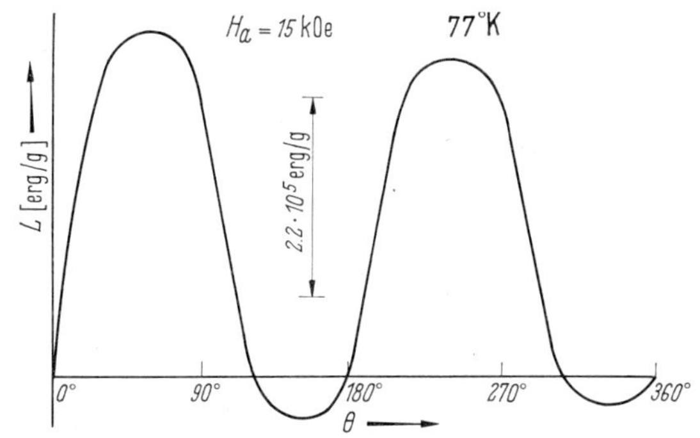

Fig. 23. $Co_{1.4}Fe_{1.6}O_4$. Torque curve of a polycrystalline sample at 77 °K after cooling from room temperature in a magnetic field. [*66II*].

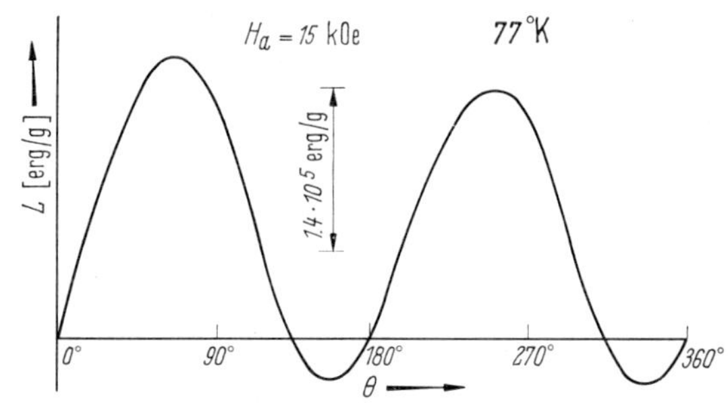

Fig. 24. $Co_{2.0}Fe_{1.0}O_4$. Torque curve of a polycrystalline sample at 77 °K after cooling from room temperature in a magnetic field. [*66II*].

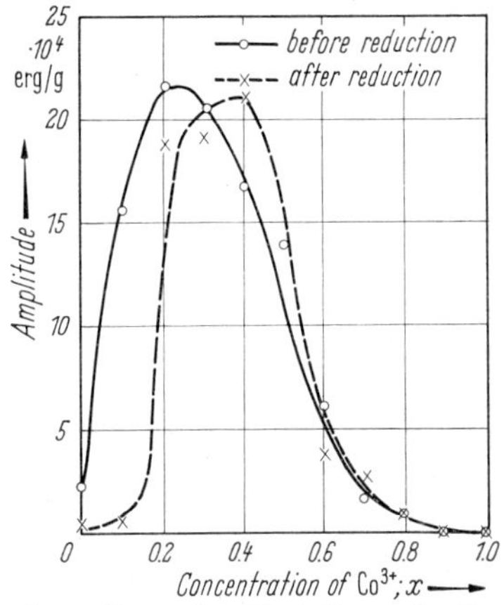

Fig. 25. $Co_{1+2x}Fe_{2-2x}O_4$. Variation of the sinusoidal amplitude of the torque curve of polycrystalline specimens at 77 °K after cooling from room temperature in a magnetic field. Dashed curve represents results for the specimens after reduction of a few percent oxygen [*66II*].

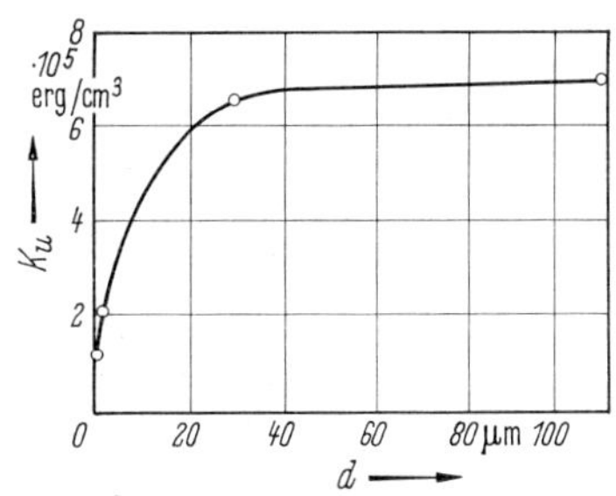

Fig. 26. $Co_{0.7}Fe_{2.3}O_4$. Dependence of the uniaxial anisotropy constant, K_u, on particle size of powders (diameter d) made by crushing single crystals. The powders were subjected to a thermomagnetic treatment consisting of annealing in a 10 kOe field at 400 °C [*67SIS*]. See also [*67ISS*].

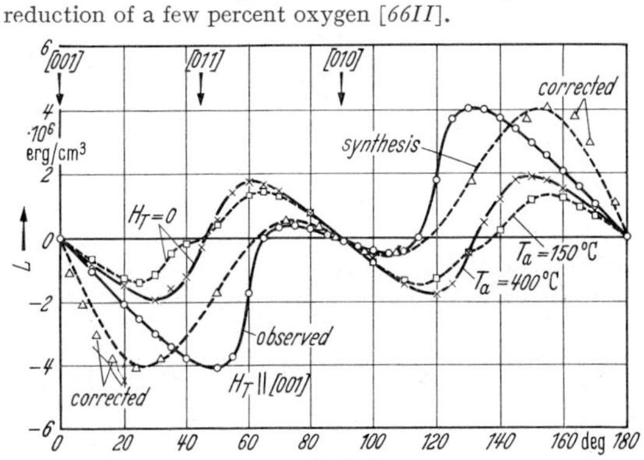

Fig. 27. $Co_{0.73}Fe_{2.27}O_4$. Torque curves for a (100) disc annealed at two temperatures T_a with and without presence of a strong magnetic field, H_T. „Synthesis" curve is drawn for 2θ and 4θ components derived originally from points corrected for lack of saturation (triangles). For this sample $K_1 = 4.3 \cdot 10^6$ erg/cm³ and $K_u = 2.5 \cdot 10^6$ erg/cm³ [*55BTW*].

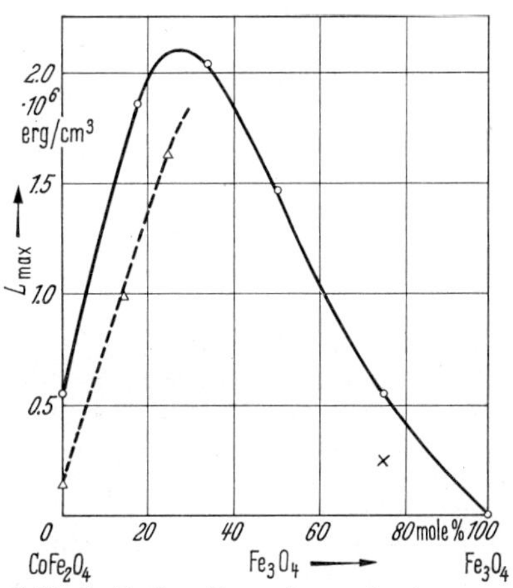

Fig. 28. $CoFe_2O_4$-Fe_3O_4. Dependence of anisotropy, developed in polycrystalline Co-Fe ferrites by magnetic annealing, on composition. Maximum torque, L_{max}, is equal to $\frac{1}{2}K_u$. Experimental points: circles, fired in oxygen; crosses, fired in CO_2; triangles another series, fired in air. Specimens were fired at 1250 °C [*55BTW*]. See also [*57WHF*].

Fig. 29. $Co^{2+}_{0.25}Fe^{2+}_{0.75}Fe^{3+}_2O_4$. Neutron diffraction data giving the temperature variation of the reduced sublattice magnetizations in a single crystal. The square root of the neutron diffraction peak intensities are plotted for the (220) reflection which gives the relative magnetization on tetrahedral A sites and for the (222) reflection which gives the same for octahedral B sites [*61RT*]. →

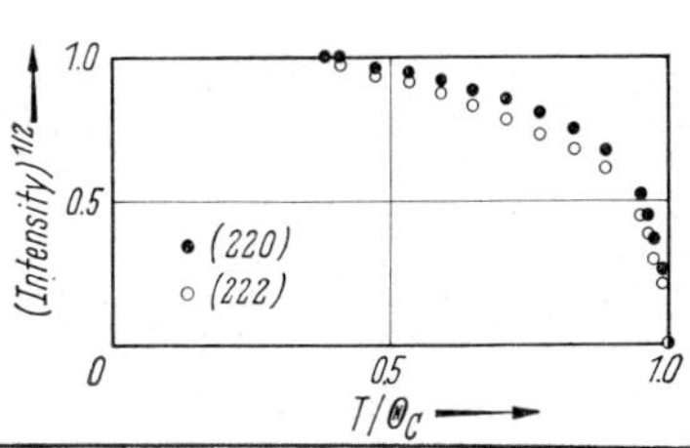

6.1.8.3 Co-Zn ferrites

Tab. 1. Co-Zn ferrite. Physical parameters of polycrystalline $Co^{2+}_{0.5}Zn^{2+}_{0.5}Fe^{3+}_2O_4$. Sintering times were short so that there was only a slight conversion of Fe^{3+} to Fe^{2+} [53B]

Measured density	$\varrho = 5.19$ g/cm³
Initial susceptibility	$\chi_{vi} = 2.25 \cdot 10^{-6}$
Magnetomechanical coupling coefficient [1])	$k = -0.011$
Young's modulus	$E = 1.64 \cdot 10^{12}$ dyn/cm²

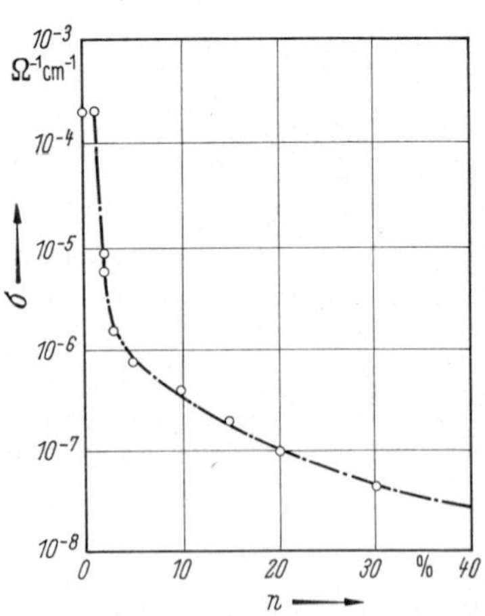

Fig. 1. $Co_{100-n}Zn_nFe_2O_4$. The dependence of electrical conductivity, σ, on concentration in polycrystalline samples [61SA].

[1]) k^2 is defined as the ratio between electromagnetic and elastic energy when only an alternating low frequency stress is applied. Therefore $k^2 = 4\pi\lambda' \mu_r/E$, where λ' is the dynamic magnetostrictive constant, μ_r is the reversible permeability, and E is Young's modulus.

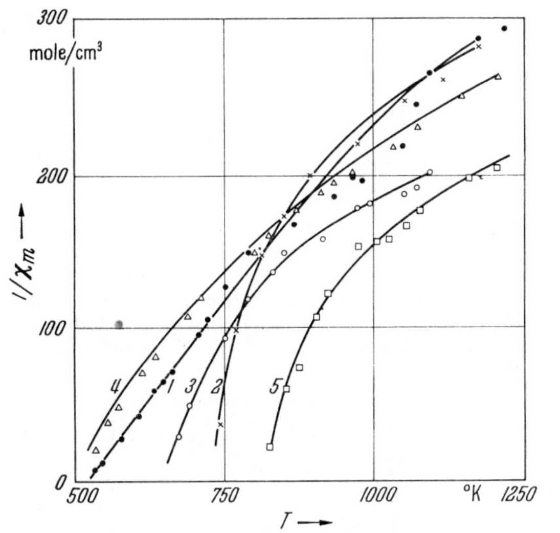

Fig. 2. Co-Zn ferrites. Temperature dependence of the reciprocal molar susceptibility of the following samples [*54D*]:
1 0.5 CoO · 0.5 ZnO · Fe_2O_3
2 0.9 CoO · 0.1 ZnO · Fe_2O_3
3 0.7 CoO · 0.3 ZnO · Fe_2O_3
4 0.3 CoO · 0.7 ZnO · Fe_2O_3
5 CoO · Fe_2O_3.

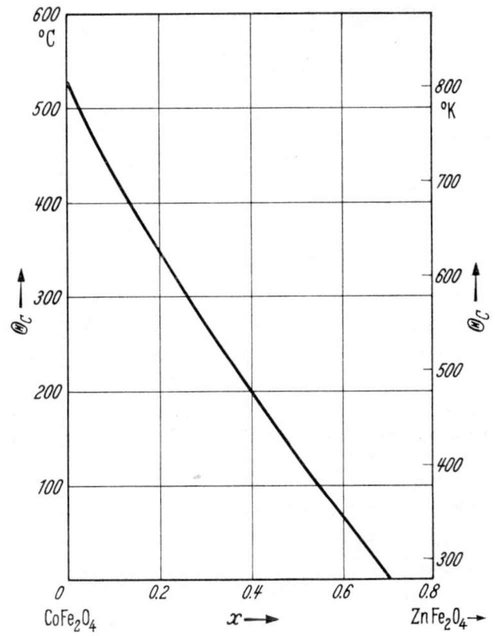

Fig. 3. $Co_{1-x}Zn_xFe_2O_4$. Curie temperature vs. composition [*59SW*(1)]. See also [*51D*].

Tab. 2. Co-Zn ferrites. Specific magnetization, σ, specific saturation magnetization at 0 °K, σ_s^0, magnetic moment per molecule, p_{sm}^0, and Curie temperature, Θ_C, for various compositions [*50G*(1), *50G*(2), *51G*]

CoO	H_{int} = 18 500 Oe σ Gauss cm³/g			$H_{int} \to \infty$ σ_s^0 (0 °K)	p_{sm}^0	Θ_C
(%) *)	(285 °K)	(77 °K)	(20 °K)	Gauss cm³/g	μ_B	°C
10	18.5	62.0	72.6	95	4.06	25
15	39.3	91.0	95.5	114	4.90	82
20	61.0	119.2	126.0	132	5.78	137
30	87.5	128.5	133.2	143.0	6.03	275
40	93.7	110.0	110.0	115.5	4.87	389
45	90.0	93.9	94.1	102.0	4.29	460
50	83.6	83.0	82.6	87.5	3.67	515

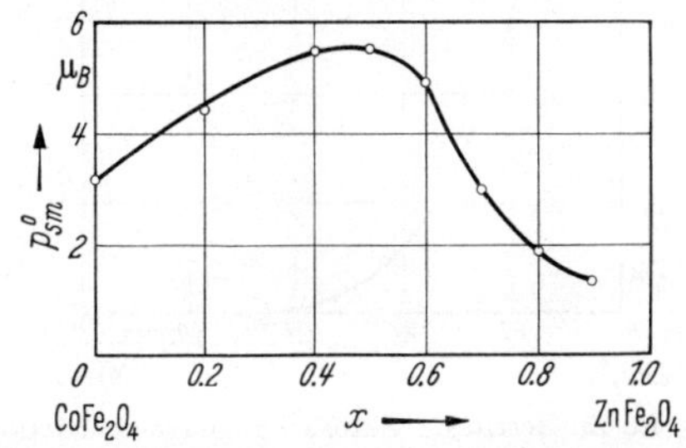

Fig. 4. $Co_{1-x}Zn_xFe_2O_4$. The dependence of the saturation magnetic moment at 0 °K, p_{sm}^0, on the Zn content of Co-Zn ferrites [*57RS*]. See also Fig. 5.

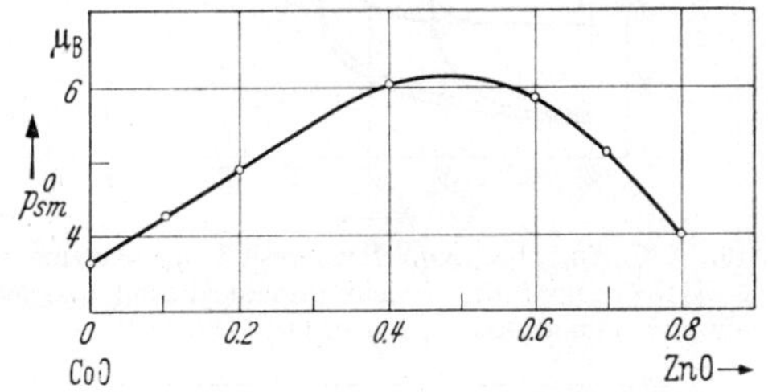

Fig. 5. Co-Zn ferrites. Saturation magnetic moment per molecule at 0 °K, p_{sm}^0, as a function of composition [*51G*(1)]. See also [*50G*(1), *50GC*(2), *50G*(3), *54G*(2), *54D*, and *57RS*].

*) Value for x in x CoO · (50 − x) ZnO · 50 Fe_2O_3.

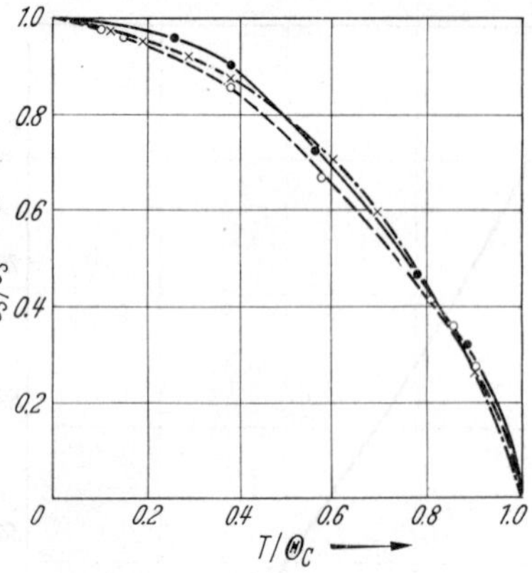

Fig. 6. Co-Zn ferrites. Reduced saturation magnetization, σ_s/σ_s^0, vs. reduced temperature, T/Θ_C, in Co-Zn ferrites [57RS].
○ - - - ○ $(0.7\,CoO\text{-}0.3\,ZnO)Fe_2O_3$
× - · - × $(0.8\,CoO\text{-}0.2\,ZnO)Fe_2O_3$
●——● $(0.9\,CoO\text{-}0.1\,ZnO)Fe_2O_3$.

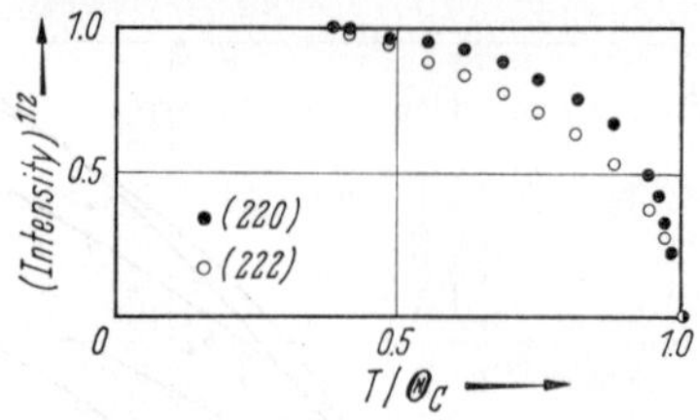

Fig. 7. $Co^{2+}_{0.32}Zn^{2+}_{0.20}Fe^{2+}_{0.48}Fe^{3+}_2O_4$. Neutron diffraction data giving the temperature dependence of the reduced sublattice magnetization in a single crystal. The square of the neutron diffraction peak intensities are plotted for the (220) reflection which gives the relative magnetization on tetrahedral sites and for the (222) reflection which gives the same for the octahedral sites [61RT].

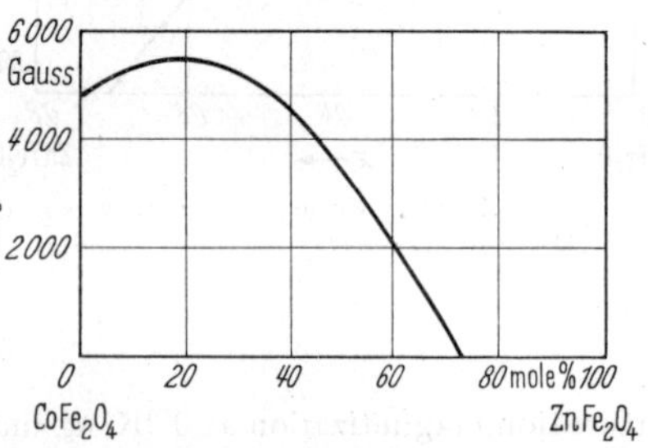

Fig. 8. Co-Zn ferrites. Saturation induction, $4\pi M_s$, vs. mole % Zn ferrite [51D].

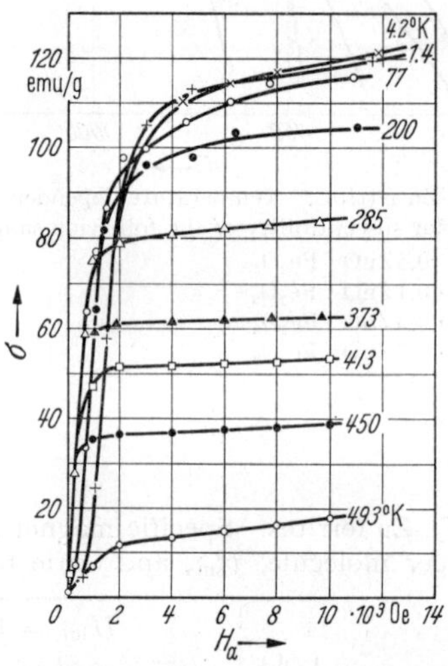

Fig. 9. $Co_{0.7}Zn_{0.3}Fe_2O_4$. Specific magnetization vs. applied magnetic field, H_a, at different temperatures [57RS].

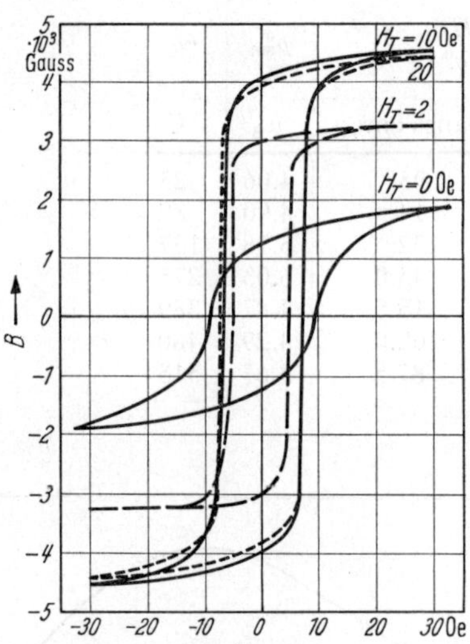

Fig. 10. $Co_{0.32}Zn_{0.24}Fe_{2.18}O_4$. Hysteresis loops showing the effect of the magnitude of field present during magnetic annealing (15 min at 400 °C, H_T in Oe) [55BTW].

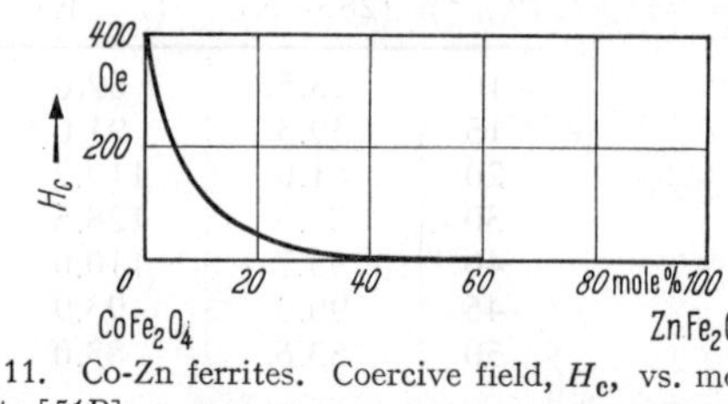

Fig. 11. Co-Zn ferrites. Coercive field, H_c, vs. mole % Zn ferrite [51D].

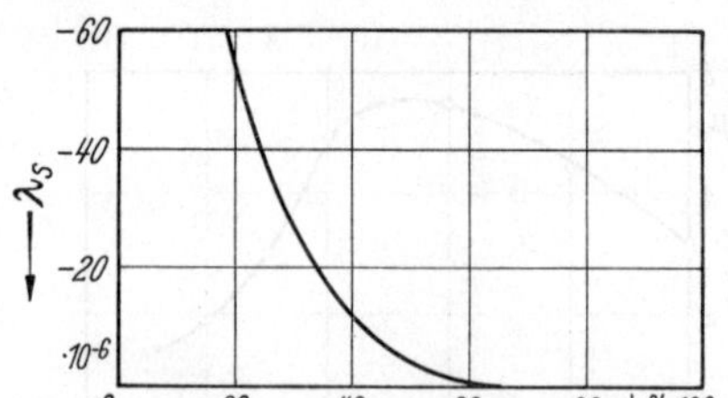

Fig. 13. Co-Zn ferrites. Saturation magnetostriction, λ_s, vs. mole % Zn ferrite [51D].

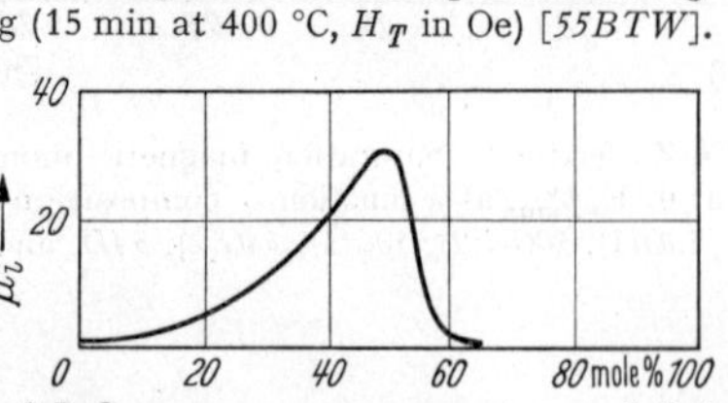

←
Fig. 12. Co-Zn ferrites. Initial permeability, μ_i, vs. mole % Zn ferrite [51D]. See also [65TD].

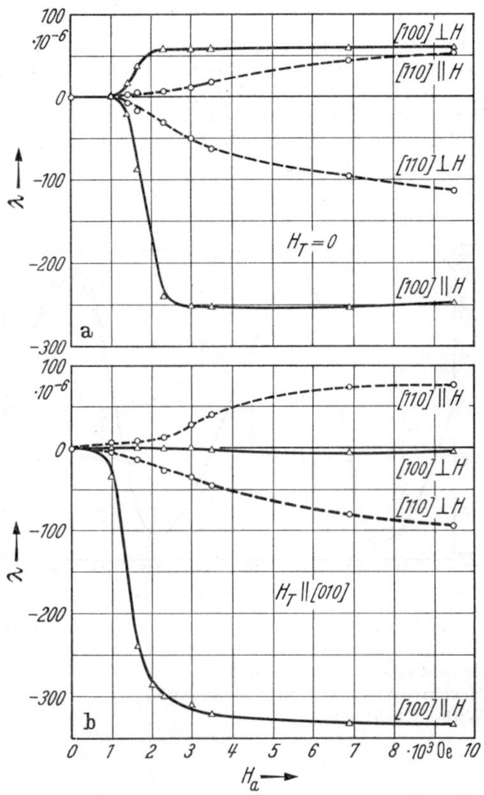

Fig. 14. $Co_{0.3}Zn_{0.2}Fe_{2.2}O_4$. Magnetostriction, λ, vs. H_a. After treatment in (a) zero magnetic annealing field ($H_T = 0$) and (b) strong magnetic annealing field (6 h magnetic annealing at 275 °C) parallel to [010] in the (100) plane. Derived constants: $\lambda_{100} = -210 \cdot 10^{-6}$, $\lambda_{111} = 110 \cdot 10^{-6}$, $\lambda_s = -18 \cdot 10^{-6}$ where $\lambda_s = (2\lambda_{100} + 3\lambda_{111})/5$ [*55BTW*].

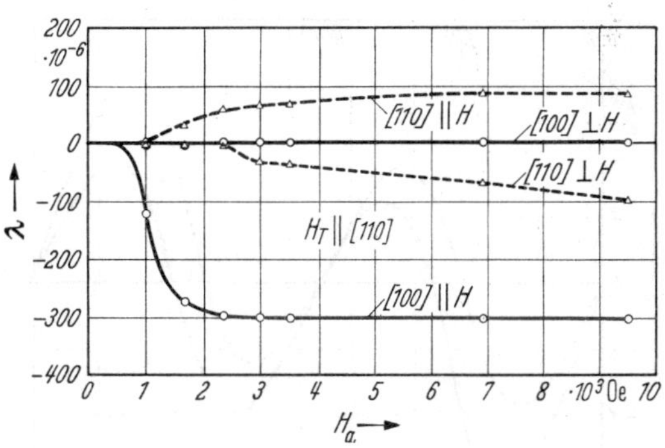

Fig. 15. $Co_{0.32}Zn_{0.24}Fe_{2.18}O_4$. Magnetostriction, λ, in the (100) plane, with annealing field, H_T, parallel to [110] [*55BTW*].

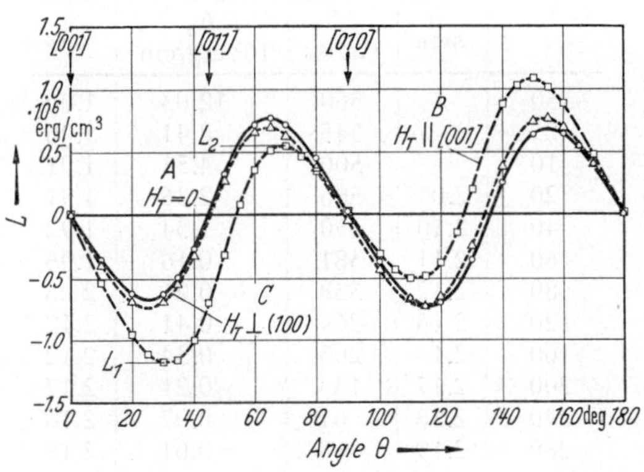

Fig. 16. $Co_{0.32}Zn_{0.24}Fe_{2.18}O_4$. Torque curves, showing effect of magnetic annealing for a (100) disc heated for 3 days at 150 °C. Curve A: no annealing field. Curve B: annealing field, $H_T = 9$ kOe, parallel to [001]. Curve C: $H_T = 9$ kOe perpendicular to (100) plane. For this sample $K_1 = 1.5 \cdot 10^6$ erg/cm³ and $K_u = 0.5 \cdot 10^6$ erg/cm³. K_u is defined by the uniaxial anisotropy energy $E = K_u \sin^2(\theta - \theta_T)$, where θ_T is the angle between the H_T and the nearest crystal axis, and θ is the angle between M_s and [001] [*55BTW*].

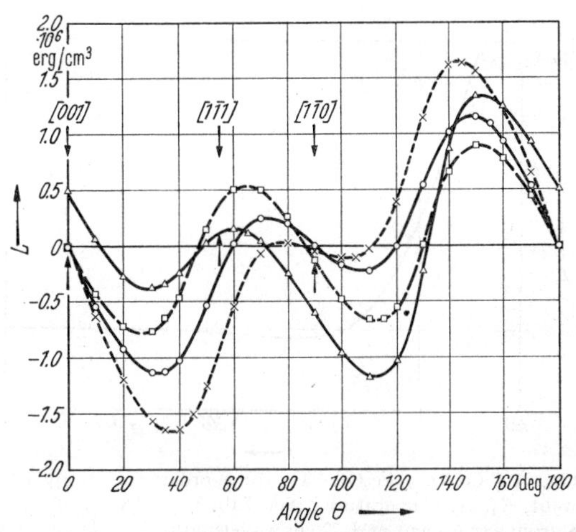

Fig. 17. $Co_{0.32}Zn_{0.24}Fe_{2.18}O_4$. Torque curves for a (110) disc for various directions of the annealing field, H_T. Sample was heated for 3 days at 150 °C in a field of 9 kOe [*55BTW*]. ○ $H_T = 0$, × H_T ∥ [001], △ H_T ∥ [1$\bar{1}$1], □ H_T ∥ [1$\bar{1}$0].

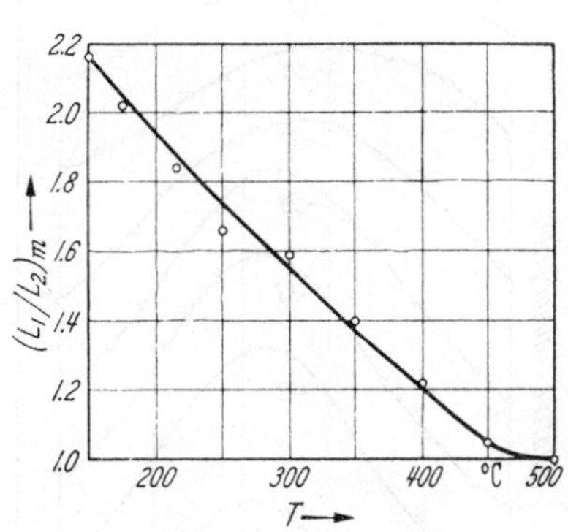

Fig. 18. $Co_{0.32}Zn_{0.24}Fe_{2.18}O_4$. Effect of magnetic annealing at various temperatures, measured by the ratio $(L_1/L_2)_m$ of the two peaks (L_1 and L_2) of the torque curves in (100) plane (see Fig. 16) [*55BTW*].

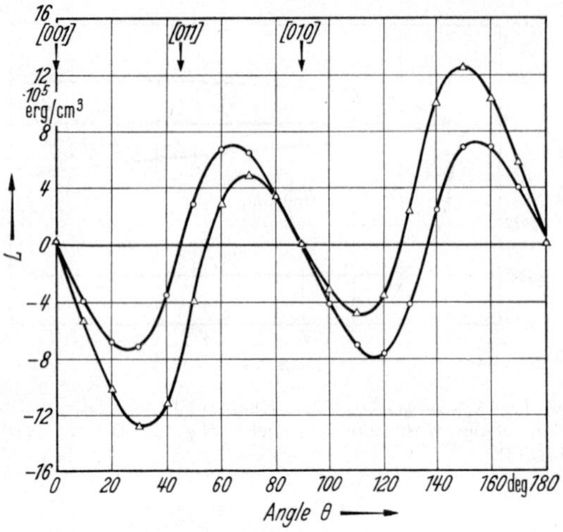

←

Fig. 19. $Co_{0.48}Zn_{0.23}Fe_{2.35}O_4$. Torque curves for a (100) single crystal disc before and after heat treatment in a magnetic field, $H_T = 21$ kOe [*56WHN*].
▵ Heat treatment with field along [001].
○ Without field heat treatment.

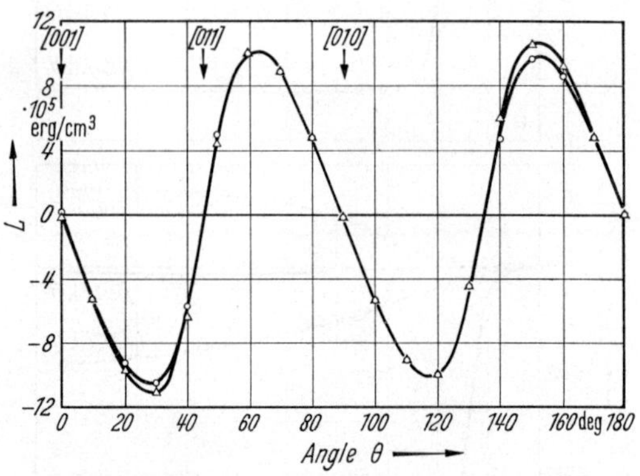

Fig. 20. $Co_{0.84}Zn_{0.18}Fe_{2.16}O_4$. Torque curves for a (100) single crystal disc before and after heat treatment in a magnetic field $H_T = 21$ kOe [*56WHN*].
▵ Heat treatment with field along [001].
○ Without field heat treatment.

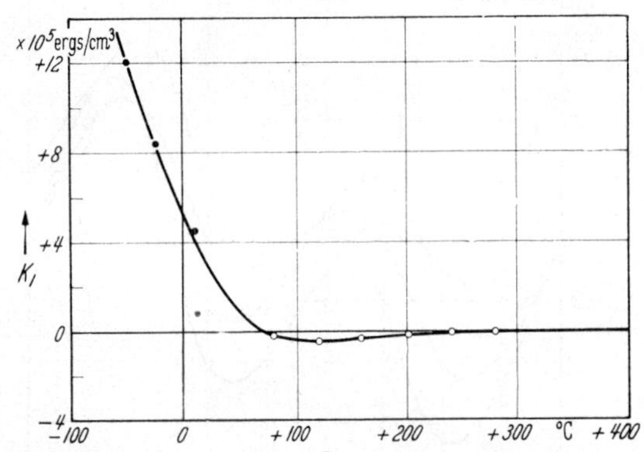

Fig. 21. $Co_{0.7}Zn_{0.3}Fe_2O_4$. The first order anisotropy constant, K_1, vs. temperature. (See Tab. 3) [*52OKT*].
● from experiment at 1.27 cm wavelength.
○ from experiment at 3.22 cm wavelength.

Tab. 3. $Co_{0.7}Zn_{0.3}Fe_2O_4$. g_{exp} factor and M_s for a polycrystalline sample. K_1 and g_{exp} factor for a single crystal. $\Theta_C = 340$ °C. Measurements from −50 °C to 10 °C were made at a wavelength of 1.27 cm. Measurements from 20 °C to 280 °C were made at 3.22 cm [*52OKT*]. See also [*52O*]

T °C	polycrystalline g_{exp}	polycrystalline M_s Gauss	single crystal K_1 10^5 erg/cm³	single crystal g_{exp}
−50	—	560	12.03	1.90
−25	—	545	8.41	1.90
10	—	500	4.51	1.91
20	2.07	506	2.36	1.91
40	2.10	450	1.34	1.92
60	2.11	381	0.16	1.96
80	2.12	338	−0.21	2.06
120	2.15	269	−0.41	2.12
160	2.16	203	−0.34	2.12
200	2.17	134	−0.21	2.17
240	2.18	61	−0.07	2.16
280	2.19	8	−0.01	2.15

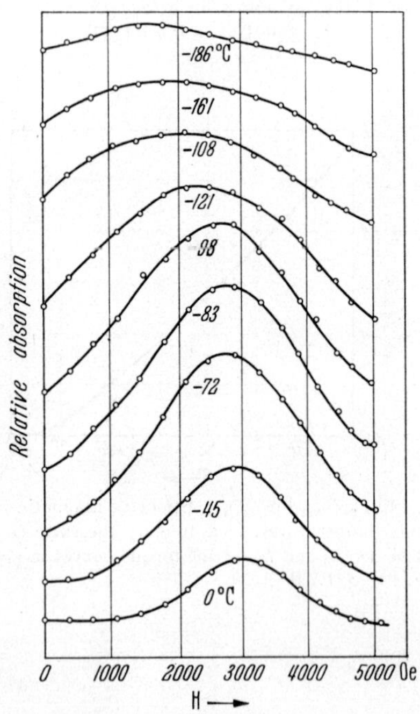

←

Fig. 22. $Co_{0.2}Zn_{0.8}Fe_2O_4$. Resonance absorption vs. dc magnetic field in a polycrystalline sample at various low temperatures [*51T*].

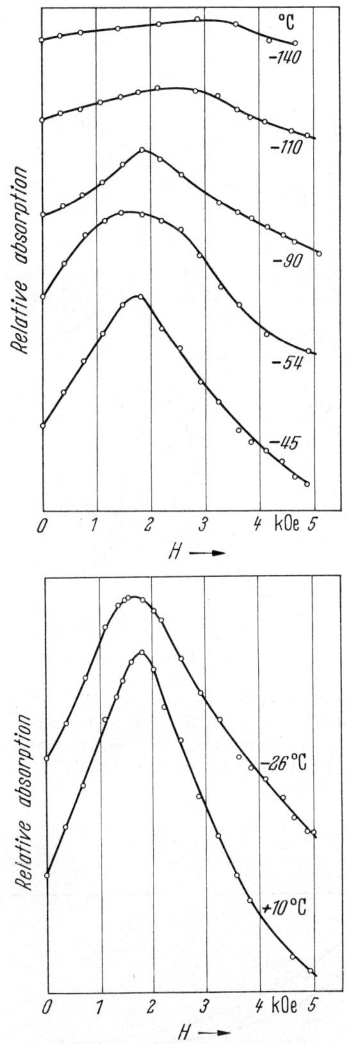

Fig. 23. $Co_{0.6}Zn_{0.4}Fe_2O_4$. Resonance absorption vs. dc magnetic field in polycrystalline samples at various low temperatures [*51T*].

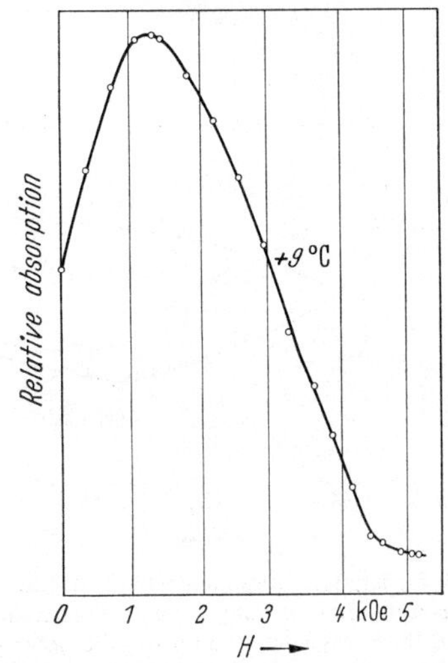

Fig. 24. $Co_{0.8}Zn_{0.2}Fe_2O_4$. Resonance absorption vs. dc magnetic field in a polycrystalline sample at 9 °C. Sample was prepared by sintering at 1300 °C for 3 h [*51T*].

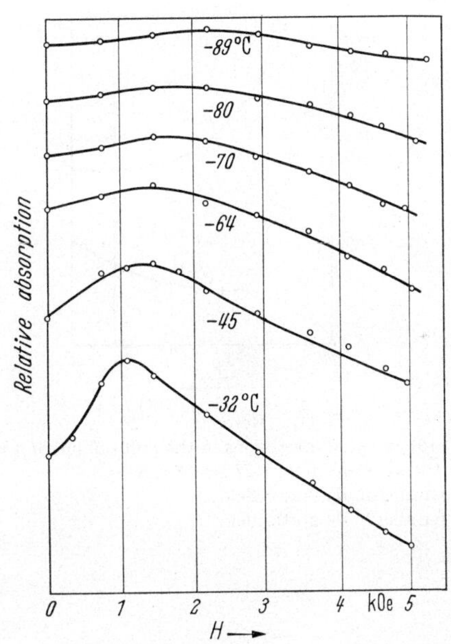

Fig. 25. $Co_{0.8}Zn_{0.2}Fe_2O_4$. Resonance absorption vs. dc magnetic field in a polycrystalline sample at various low temperatures [*51T*].

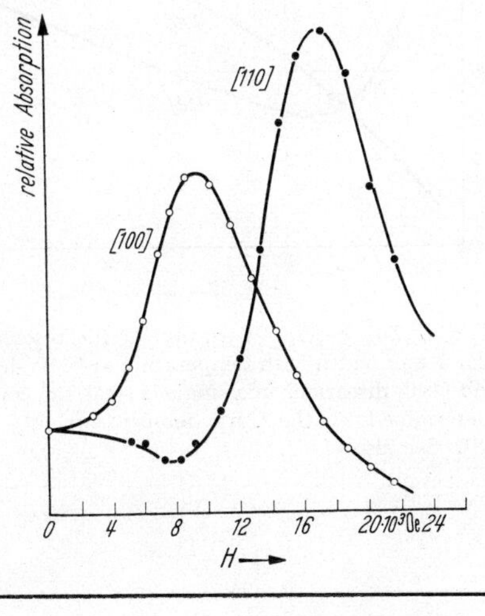

←

Fig. 26. $Co_{0.7}Zn_{0.3}Fe_2O_4$. Typical absorption curves for the [110] and [100] directions in the (100) plane of a single crystal at a wavelength of 1.27 cm.
○ [100] parallel to dc static field.
● [110] parallel to dc static field.
Θ_C = 340 °C [*520KT*].

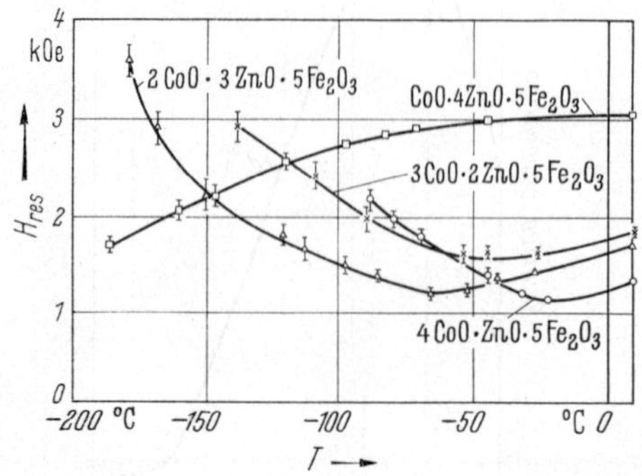

Fig. 27. Co-Zn ferrite. Resonance field at the absorption maximum as a function of temperature. Specimens are discs 0.2⋯0.3 mm thickness and 3⋯5 mm in diameter [*51T*].

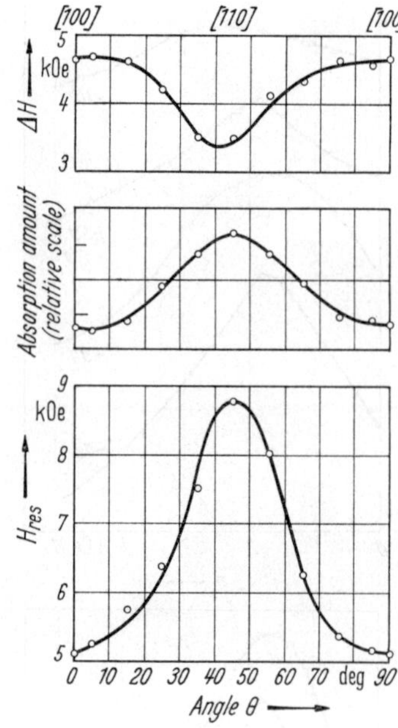

Fig. 28. $Co_{0.7}Zn_{0.3}Fe_2O_4$. Resonance field, H_{res}, relative absorption, and half line width, ΔH, as functions of crystal orientation in a single crystal in the (100) plane at wavelength of 1.27 cm and at room temperature [*520KT*].

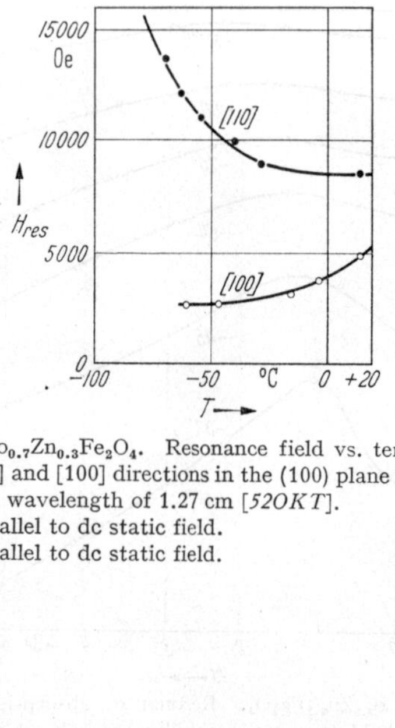

Fig. 29. $Co_{0.7}Zn_{0.3}Fe_2O_4$. Resonance field vs. temperature for the [110] and [100] directions in the (100) plane of a single crystal at a wavelength of 1.27 cm [*520KT*].
○ [100] parallel to dc static field.
● [110] parallel to dc static field.

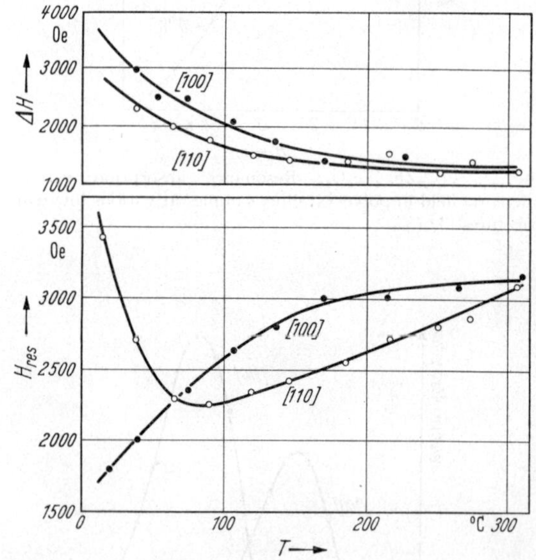

Fig. 30. $Co_{0.7}Zn_{0.3}Fe_2O_4$. Variation of the resonance field and the line half width with temperature at 9310 Mc/s for the [110] and [100] directions in a single crystal, the composition being determined by the Curie temperature, $\Theta_C = 340$ °C [*520KT*]. See also [*520*].

Tab. 4. $Co_{0.7}Zn_{0.3}Fe_2O_4$. Values of g_{exp} of single crystals at room temperature [*540*]

g_{exp}	λ cm	H_{res} Oe
1.92 ± 0.01	0.63	17 100
1.91 ± 0.01	1.26	8 570
1.91 ± 0.01	3.22	3 380

6.1.8.4 Co-Cd, Co-Mg, Co-Ni, and Co-Cr ferrites

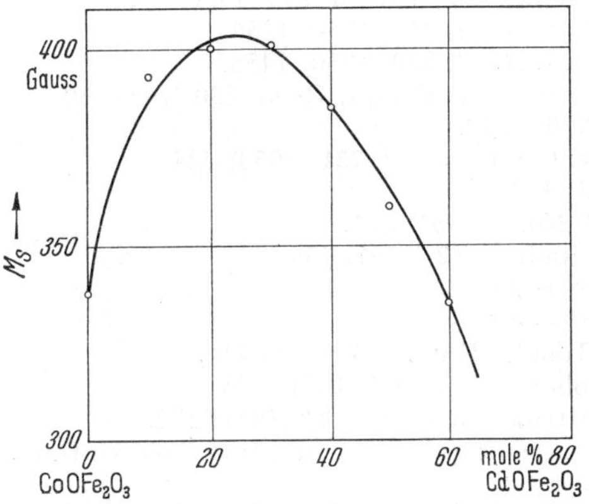

Fig. 1. Co-Cd ferrites. Saturation magnetization, M_s, as a function of composition [*51WTT*].

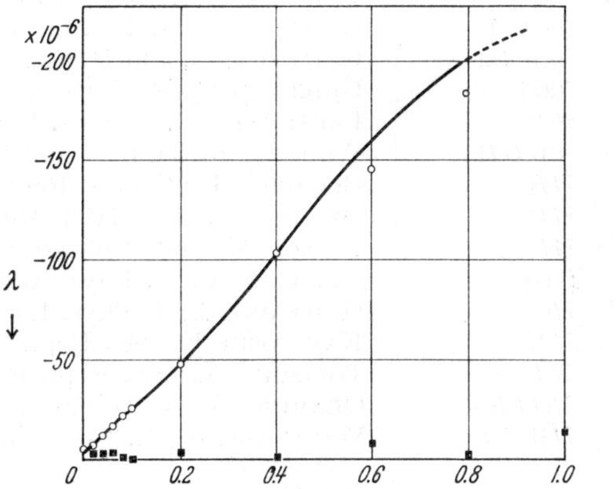

Fig. 2. $Fe_2O_4Co_xMg_{1-x}$. Magnetostriction vs. x measured in a field of 2500 Oe [*52W*].
○ Annealing field, H_T = 12 kOe, perpendicular to field applied during measurements.
■ Annealing field, H_T = 8 kOe, parallel to field applied during measurements.
The solid line represents saturation magnetostriction, determined by extrapolation where necessary.

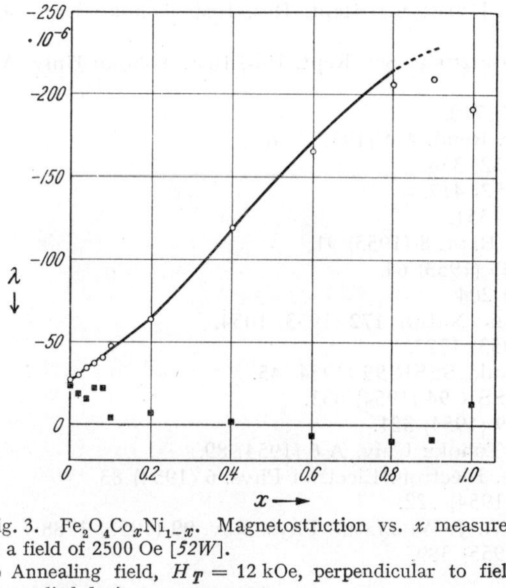

Fig. 3. $Fe_2O_4Co_xNi_{1-x}$. Magnetostriction vs. x measured in a field of 2500 Oe [*52W*].
○ Annealing field, H_T = 12 kOe, perpendicular to field applied during measurements.
■ Annealing field, H_T = 8 kOe, parallel to field applied during measurements.
The solid line represents saturation magnetostriction, determined by extrapolation where necessary.

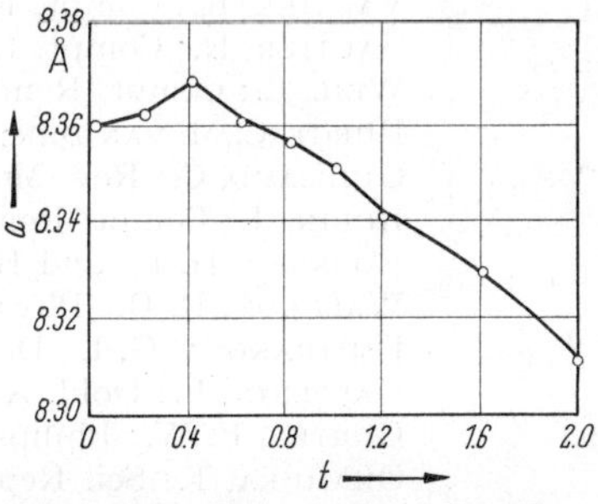

Fig. 4. $CoFe_{2-t}Cr_tO_4$. Compositional dependence of the lattice constants [*63T*].

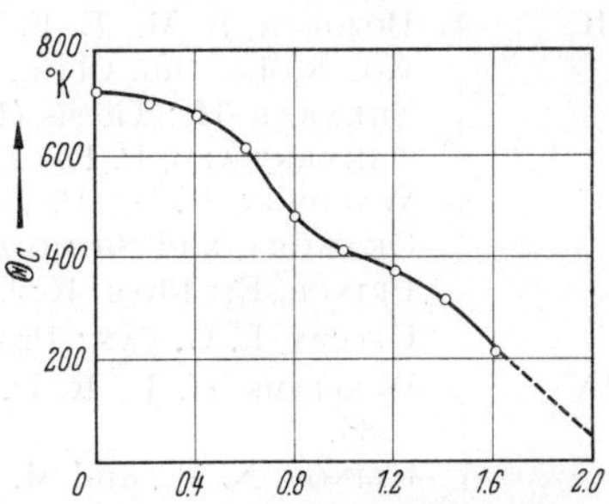

Fig. 5. $CoFe_{2-t}Cr_tO_4$. Compositional dependence of the ferromagnetic Curie temperature [*63T*].

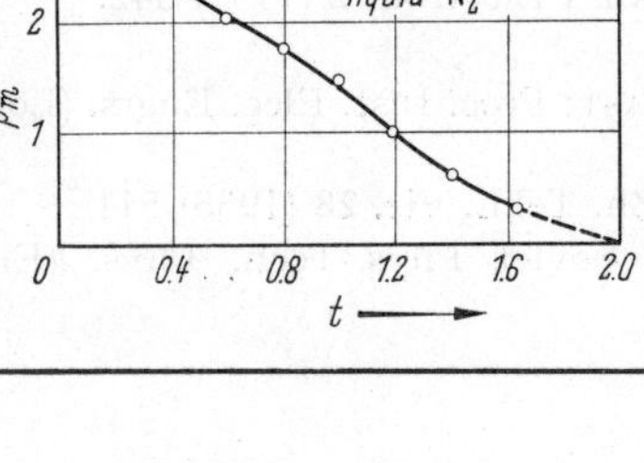

Fig. 6. $CoFe_{2-t}Cr_tO_4$. Compositional dependence of the magnetic moment [*63T*].

6.1.8.5 References for 6.1.8.1 ··· 6.1.8.4

18M	MARTIN, P.: Ann. Phys. **55** (1918) 561.
36AVB	ARKEL, A. E. VAN, E. J. VERWEY, and M. G. VAN BRUGGEN: Rec. Trav. Chim. **55** (1936) 331.
37BC	BERNARD, J., and G. CHAUDRON: Compt. Rend. **204** (1937) 766.
50G(1)	GUILLAUD, C.: J. Rech. Centre Natl. Rech. Sci. Lab. Bellevue (Paris) **12** (1950) 113.
50G(2)	GUILLAUD, C.: Compt. Rend. **230** (1950) 1458.
50G(3)	GORTER, E. W.: Nature **165** (1950) 798 and Compt. Rend. **230** (1950) 192.
50GC(1)	GUILLAUD, C., and H. CREVEAUX: Compt. Rend. **230** (1950) 1256.
50GC(2)	GUILLAUD, C., and H. CREVEAUX: Compt. Rend. **230** (1950) 1458.
50GVM	GUILLAUD, C., R. VAUTIER, and S. MEDVEDIEFF: Compt. Rend. **230** (1950) 60.
50P	PAUTHENET, R.: Compt. Rend. **230** (1950) 1842.
50WGP	WEIL, L., M. GALLAY, and P. POENSIN: Compt. Rend. **231** (1950) 224.
51B	BOCHIROL, L.: Compt. Rend. **232** (1951) 1474.
51D	DIETHELM, C. W.: Tech. Mitt. PTT **29**, No. 8 (1951) 281.
51FM	FALLOT, M., and P. MARONI: J. Phys. Radium **12** (1951) 256.
51G(1)	GUILLAUD, C.: J. Phys. Radium **12** (1951) 239.
51G(2)	GUILLAUD, C.: J. Phys. Radium **12** (1951) 492.
51K	KAMIYOSHI, K.: Sci. Rept. Res. Inst. Tohoku Univ. **A 3** (1951) 716.
51T	TORIZUKA, Y.: Sci. Rept. Res. Inst. Tohoku Univ. **A 3** (1951) 383.
51OTK	OKAMURA, T., Y. TORIZUKA, and Y. KOJIMA: Phys. Rev. **84** (1951) 372.
51WTT	WATANABE, H., N. TSUYA, and T. TAKAHASHI: Sci. Rept. Res. Inst. Tohoku Univ. **A 3** (1951) 62.
52N	NÉEL, L.: Proc. Phys. Soc. (London) **A 65** (1952) 869.
52O	OKAMURA, T.: Phys. Rev. **85** (1952) 690.
52OFD(2)	OKAMURA, T., T. FUJIMURA, and M. DATE: Phys. Rev. **85** (1952) 1041.
52OFD(1)	OKAMURA, T., T. FUJIMURA, and M. DATE: Sci. Rept. Res. Inst. Tohoku Univ. **A 4** (1952) 191.
52OKT	OKAMURA, T., Y. KOJIMA, and Y. TORIZUKA: Sci. Rept. Res. Inst. Tohoku Univ. **A 4** (1952) 72.
52P	PAUTHENET, R.: Ann. Phys. **7** (1952) 710.
52RB	ROBIN, J., and J. BERNARD: Compt. Rend. **234** (1952) 956.
52V(1)	VAUTIER, R.: Compt. Rend. **235** (1952) 356.
52V(2)	VAUTIER, R.: Compt. Rend. **235** (1952) 417.
52W	WEIL, L.: Compt. Rend. **234** (1952) 1351.
53B	BURGT, C. M. VAN DER: Philips Res. Rept. **8** (1953) 91.
53G	GUILLAUD, C.: Rev. Modern Phys. **25** (1953) 64.
53R	ROBIN, J.: Compt. Rend. **236** (1953) 204.
53RW	ROOKSBY, H. P., and B. T. M. WILLIS: Nature **172** (1953) 1054.
53W	WALDRON, R. D.: Phys. Rev. **99** (1953) 1727.
54D	DMITRAKOVA, G. I.: Dokl. Akad. Nauk. SSSR **99** (1954) 45.
54G(1)	GALPERIN, F.: Dokl. Akad. Nauk SSSR **94** (1954) 833.
54G(2)	GORTER, E. W.: Philips Res. Rept. **9** (1954) 321.
54O	OKAMURA, T.: Sci. Rept. Res. Inst. Tohoku Univ. **A 6** (1954) 89.
54SW	SMIT, J., and H. P. J. WIJN: Advan. Electron. Electron Phys. **6** (1954) 83.
54V	VAUTIER, R.: Ann. Phys. (Paris) **9** (1954) 322.
55BTW	BOZORTH, R. M., E. F. TILDEN, and A. J. WILLIAMS: Phys. Rev. **99** (1955) 1788.
55R	ROBIN, J.: Ann. Chim. (Paris) **10** (1955) 389.
55S	SHENKER, H.: Thesis (1955).
55T	TANNENWALD, P. E.: Phys. Rev. **99** (1955) 463.
55W	WALDRON, R. D.: Phys. Rev. **99** (1955) 1727.
56OS	OKAMURA, and SHIMOIZAKA: cited in: [*56U*].
56P	PRINCE, E.: Phys. Rev. **102** (1956) 674.
56U	UITERT, L. G. VAN: Proc. I.R.E. **44** (1956) 1294.
56WHN	WILLIAMS, H. J., R. D. HEIDENREICH, and E. A. NESBITT: J. Appl. Phys. **27** (1956) 85.
57RS	REINOV, N. M., and M. F. STEL'MAKH: Radiotekh. i Electron. **2** (1957) 342.
57S	SHENKER, H.: Phys. Rev. **107** (1957) 1246.
57WHF	WIJN, H. P. J., H. VAN DER HEIDE, and J. F. FAST: Proc. Inst. Elec. Engrs. (London) **104 B**, Suppl. No. 7 (1957) 412.
58PA(1)	PEREKALINA, T. M., and A. A. ASKOCHENSKII: Zh. Tekh. Fiz. **28** (1958) 511.
58PA(2)	PEREKALINA, T. M., and A. A. ASKOCHENSKII: Soviet Phys.-Tech. Phys. (English Transl.) **3** (1958) 480.

59J Jonker, G. H.: J. Phys. Chem. Solids **9** (1959) 165.
59SW(1) Smit, J., and H. P. J. Wijn: "Ferrites", J. Wiley and Sons, N. Y., 1959, p. 156/157.
59SW(2) ibid. p. 251
61PAS Perekalina, T. M., A. A. Askochinskii, and D. G. Sannikov: Soviet Phys. — JETP (English Transl.) **13** (1961) 303.
61RT Riste, T., and L. Tenzer: J. Phys. Chem. Solids **19** (1961) 117.
61SA Sinyakov, E. V., and V. P. Avramenko: Soviet Phys. — Solid State (English Transl.) **3** (1961) 299.
62DDSZ Drokin, A. I., V. D. Dylgerov, N. I. Sudakov, and Y. K. Zayirova: Phys. Met. Metallog. (USSR) (English Transl.) **13** (1962) 150.
62NM Nishikubo, T., and K. Motizuki: J. Phys. Soc. Japan **17** (1962) 871.
62PG Pines, B. Y., and N. M. Gumen: Soviet Phys. — Cryst. (English Transl.) **6** (1962) 727.
63DHK Dézsi, I., A. Z. Hrynkiewicz, and D. S. Kulgawczuk: Acta Physica Polonica **24** (1963) 283.
63FKW Ferreti, A., W. Kunnmann, and A. Wold: J. Appl. Phys. **34** (1963) 388.
63PK Perthel, R., and W. Keilig: Monatsber. Deut. Akad. Wiss. Berlin **5** (1963) 109.
63T Tsushima, T.: J. Phys. Soc. Japan **18** (1963) 1162.
64MS Müller, W., and H. Schmalzried: Ber. d. Bunsengesellschaft **68** (1964) 270.
64PG Pines, B. Y., and N. M. Gumen: Soviet Phys. — Solid State (English Transl.) **5** (1964) 2557.
64S Szydlowski, H.: Acta Physica Polonica **25** (1964) 401.
65PG Pines, B. Y., and N. M. Gumen: Phys. Status Solidi **8** (1965) 751.
65SM Svirina, E. P., and M. A. Malikova: Soviet Phys. — Solid State (English Transl.) **6** (1965) 2901.
65TD Toltinnikov, S. A., and L. A. Davydov: Soviet Phys. — Solid State (English Transl.) **6** (1965) 1730.
66G Gumen, N. M.: Soviet Phys. JETP (English Transl.) **22** (1966) 251.
66II Iizuka, T., and S. Iida: J. Phys. Soc. Japan **21** (1966) 222.
66V van Loef, J. J.: Physica **32** (1966) 2102.
67ISS Ivanov, O. A., Y. S. Shur, and E. V. Shtol'ts: Soviet Phys. — Solid State (English Transl.) **9** (1967) 857.
67SIS Shur, Y. A., O. A. Ivanov, and E. V. Shtol'ts: Soviet Phys. — Solid State (English Transl.) **9** (1967) 30.
68SVM Sawatzky, G. A., F. van der Woude, and A. H. Morrish: J. Appl. Phys. **39** (1968) 1204.

6.2 Cr Spinelle — Cr spinels

6.2.1 Cr Oxid-Spinelle

6.2.1.1 Binäre Cr Oxid-Spinelle

Tab. 1. Kristallografische Eigenschaften binärer Cr^{3+} Oxid-Spinelle

Verbindung	Kationenverteilung	a Å	u	ϱ_X g/cm³	ϱ_P g/cm³	Farbe	Lit.	Fig.	Tab.
γ-Cr_2O_3	$Cr^{3+}[Cr^{3}_{5/3}\square_{1/3}]O_4$	8.36	0.383				*57SS, 64B-1*		
$CdCr_2O_4$		8.60 [1])					*27H-2, 30H*	5	2
		8.57 +0.01 −0.02				lichtgrün	*34KT, 38S*		
	$Cd[Cr_2]O_4$	8.567	0.385 ± 0.01				*47VH, 55GL, 55HFKSS, 64B-1*		
		8.584		5.89			*47···48SR*		
		8.58	0.375 ···0.395				*60H*		
		8.596	0.396				*66BWRL*		
		8.596	0.3896				*66RBW, 66MDAW*		
							54E [2]), *57M, 61VP, 68MRW*		
							68NK [3])		
$CoCr_2O_4$		8.319 ± 0.003		5.24			*27H-1, 27H-2, 31S, 37S*	6, 7, 8, 9, 10	2
		8.31		5.26	5.14		*29NP*		
		8.32				dunkelblaugrün [4])	*34KT, 31EH, 31S, 38S*		
		8.11				dunkelgrün [4])	*34KT*		
		8.32 +0.01 −0.02				blaugrün [4])	*34KT*		
	$Co[Cr_2]O_4$	8.32					*47VH, 55GL, 55HFKSS, 59WG, 60H*		
		8.34		5.20			*47···48SR*		
		8.31		5.24			*53BR, 61H*		
		8.332 ± 0.002					*62WW, 64B-1*		
							54E [2]), *56L, 57B, 57M, 59E, 60MG, 62MWRD, 67P, 67S, 52MHS-1, 64MDW, 63WAWG, 58MT, 68NK* [3]), *68P-2, 68TKF, 52SMHS, 69ILM*		
$CuCr_2O_4$		8.34				lichtgrau [4])	*34KT*	1, 6, 7	2
		8.34				grünlichbraun [4])	*34KT*		
		8.35 +0.00 −0.01				schwarz [4])	*34KT*		

Fußnoten siehe S. 397.

Verbindung	Kationenverteilung	a Å	u	ϱ_X g/cm³	ϱ_P g/cm³	Farbe	Lit.	Fig.	Tab.
$CuCr_2O_4$ (Forts.)		8.37		6.69			*47⋯48SR*		
		[5]					*52MHS-1, 52MHS-2,*		
		[5]					*55GL, 54BD*		
	$Cu[Cr_2]O_4$	8.35 [5]					*55HFKSS, 58D, 57B*		
		8.27 [6]					*61H*		
	$Cu_yCr_{1-y}[Cu_{1-y}Cr_{1+y}]O_4$		0.389 ± 0.003 [7]				*61OT*		
		8.532 ± 0.003 [8]					*62WW, 64B-1*		
	$Cu^{2+}[Cr_2^{3+}]O_4$						*63G-2, 61WWA*		
		[5]					*64BL, 64MH,*		
		[5]					*65SA, 67S,*		
		[5]					*66S, 64AWR,*		
							60MG, 57M,		
							68NK [3]*, 67NK,* [9]		
							52SMHS		
$FeCr_2O_4$		8.357 ± 0.004					*27H-1, 31S*	2, 3, 10, 11, 12, 13	2
		8.344 ± 0.003		5.085			*31CAB, 37S*		
		8.36					*31EH, 31S*		
		8.33					*31E*		
		8.344					*32HN*		
		8.35				grün-braun [4]	*34KT*		
		8.35				dunkel-grau-grün [4]	*34KT, 38S*		
		8.40 ± 0.02				braun-schwarz [4]	*34KT*		
		8.42		4.98			*47⋯48SR*		
	$Fe[Cr_2]O_4$	8.34		5.11			*47VH, 53BR,*		
							55HFKSS, 61H		
		8.376 ± 0.002					*54YKL*		
	$Fe^{2+}[Cr_2^{3+}]O_4$	8.392					*56L, 59WG*		
		8.36	0.388				*60H*		
		8.377 ± 0.002 [10]					*62WW,*		
							64B-1, 65AR		
		8.381 ± 0.005					*65R*		
							51L, 53R, 55GL,		
							58DY, 59G, 57M,		
							54E [2]*, 61N, 65MSY,*		
							66TTA, 66NGS,		
							68NK [3]*, 57F,* [11]		
							62MN [12]*, 67S, 64G,*		
							68HI, 63WAWG,		
							64AWR, 52SMHS		
$InCr_2O_4$							*54E*		
$MgCr_2O_4$		8.290 ± 0.005		4.49			*28P, 29P, 37S*	10, 14, 15, 16	2
		8.32		4.45			*30H*		
		8.305 ± 0.001		4.429			*31CAB, 37S,*		
							32PBB		
		8.30 ± 0.01				schmut-ziggrün	*34KT, 38S*		
		8.329	0.385 ± 0.01	4.42			*47⋯48SR,*		
							47RR		
		8.312	0.385 ± 0.01				*47VH*		

Fußnoten siehe S. 397.

Verbindung	Kationenverteilung	a Å	u	ϱ_x g/cm³	ϱ_p g/cm³	Farbe	Lit.	Fig.	Tab.
$MgCr_2O_4$ (Forts.)	$Mg[Cr_2]O_4$	8.31		4.43			53BR		
		8.32	0.387				53R		
		8.334 ± 0.005	0.385 ± 0.01				54WK		
		8.31	0.375 ⋯ 0.395				60H, 61H		
		8.333 ± 0.002	0.385				62WW, 64B-1, 66BWRL		
		8.334 ± 0.001		4.41		grün	66A, 68A		
		8.322 ± 0.003 [13])					67DR		
							51L, 47VHR, 52MHS-1, 52MHS-2, 53Y, 55GL, 54E, 55HFKSS, 57BC, 57M, 62L, 63BF, 63B-3, 64AMDK, 62MN, [12]) 68P-1, 68NK [3]), 67R, 67KGW, 69GS, 68MRW, 69PS		
$MnCr_2O_4$		8.487 ± 0.006		4.85			27H-1, 27H-2, 31S	7, 8, 17, 18, 19	2
		8.49					31EH, 31S		
		8.436 ± 0.001		4.90			31CAB, 32PBB, 37S		
		8.42				leicht grünlich braun [4])	34KT, 38S		
		8.43				grau-braun [4])	34KT, 60H		
		8.42 + 0.02 − 0.01				grün-grau [4])	34KT		
		8.44		4.83			47⋯48SR		
	$Mn[Cr_2]O_4$	8.49					47VH, 55HFKSS		
		8.426		4.951			54G, 54SR		
		8.425		4.952			54G, 54SR		
		8.437 ± 0.002					62WW, 64B-1		
	$Mn^{2+}[Cr_2^{3+}]O_4$						63G-2		
		8.437	0.3890				66RBW		
		8.432					58E		
							52SMHS, 52MHS-1, 54E [2]), 56L, 60K, 59E, 59WG, 60MG, 62MWRD, 62CH, 57M, 65STT, 64HH, 65SA, 66DMFW, 68TKF, 68P-2, 68NK [3]), 67S, 52MHS-2		
Mn_2CrO_4	$Mn^{2+}[Cr^{3+}Mn^{3+}]O_4$	8.33 [14])					57B, 63G-2, 65B		
$NiCr_2O_4$		8.32					32H	4, 7, 17, 20	2
		8.31				oliv-grün [4])	34KT, 38S, 61H		
		8.31				dunkel-grün [4])	34KT		
		8.31 + 0.02 − 0.01				moos-grün [4])	34KT		
		8.33					47⋯48SR		
	$Ni[Cr_2]O_4$	8.299	0.385				47VH, 64B-1		

Fußnoten siehe S. 397.

Verbindung	Kationenverteilung	a Å	u	ϱ_x g/cm³	ϱ_p g/cm³	Farbe	Lit.	Fig.	Tab.
$NiCr_2O_4$ (Forts.)			0.39 ± 0.01				*47VH*		
		8.328 [15]					*53R, 53SR, 68DR* *57B, 58D, 59W, 65KM, 69PM*		
		8.24 [15]					*59WG*		
		8.31	0.378 ··· 0.398				*60H*		
		8.248 ± 0.002 [16]					*62WW, 64B-1*		
		8.300 ± 0.003 [17]					*67DR*		
							52SMHS, 62T, 52MHS-1, 56LA, 56L, 55GL, 57M, 58MT, 55HFKSS, 60MG, 61P, 63G-2, 63WAWG, 64AWR, 64MH, 65STT 68NK [3], *54E* [2], *67S, 68KKI, 68TKF, 68MRW*		
$ZnCr_2O_4$		8.323 ± 0.004					*27H-1, 27H-2, 31S*	5, 10, 14	2
		8.300 ± 0.005					*29P, 37S*		
		8.280		5.47			*30P, 37S*		
		8.296 ± 0.002		5.393			*31CAB, 31S*		
		8.296					*32PBB*		
		8.32				grün-grau [4]	*34KT, 31EH, 37S*		
		8.32				grau-grün [4]	*34KT, 65SK*		
		8.32 + 0.01 − 0.02				grünlich braun	*34KT*		
	$Zn[Cr_2]O_4$	8.30		5.41			*47VH, 53BR, 55HFKSS, 60H, 64B-1*		
		8.31					*53R, 59WG, 61H*		
		8.327 ± 0.002					*62WW, 64B-1*		
		8.32	0.387				*64BL*		
		8.321	0.3899				*66RBW*		
		8.328	0.387				*66BWRL*		
		8.312 ± 0.002 [18]					*67DR* [19]		

[1]) a = 8.59 Å mit NaCl als Standardsubstanz [*27H-2*]. [2]) Strukturtyp unbekannt [*54E*].
[3]) Thermodynamische Eigenschaften (Entropie, Enthalpie) [*68NK*].
[4]) Unterschiedliche Farben und Gitterparameter bei verschiedenen Herstellungsverfahren [*34KT*].
[5]) Tetragonal verzerrt c/a = 0.92 [*57B*], [*58D*], [*54BD*]; c/a = 0.9 [*66S*]; c/a = 0.91 [*65SA*], [*67S*]; c/a = 0.913 [*64BL*]; c/a = 1.1 [*52MHS-1*].
[6]) $\sqrt[3]{a^2 c}$; c = c/a [*61H*].
[7]) Für y = 0.90 ± 0.06 [*61OT*]. [8]) c = (7.788 ± 0.003) Å [*62WW*].
[9]) Thermodynamische Eigenschaften (Entropie, Enthalpie) [*67NK*].
[10]) Tetragonal unterhalb von −138 °C ± 3 °C; Kubisch oberhalb von −138 °C ± 3 °C [*62WW*].
[11]) Tetragonal unterhalb 180 °K; bei 90 °K c/a = 0.986, a = (8.377 ± 0.001) Å [*57F*].
[12]) Thermodynamische Eigenschaften (Entropie) [*62MN*].
[13]) Hergestellt bei 700 °C und 703 at, abgeschreckt; von 1300 °C abgeschreckt ist a = (8.321 ± 0.001) Å [*67DR*].
[14]) c = 8.75 Å; c/a = 1.05 [*57B*], [*63G-2*].
[15]) Tetragonal verzerrt c/a = 1.025; kubisch oberhalb 36 °C [*57B*]; c = 8.44 Å [*59WG*], c/a = 1.04 (bei tiefen Temperaturen) [*67S*]; c/a > 1 [*58D*]; c/a = 1.04 bei −170 °C (Max); c/a = 1.022 bei 27 °C; Kubisch bei 34 °C [*59W*]; Kubisch bei 28 °C [*65KM*]; c/a = 1.039 bei 92 °K; Kubisch bei 21 °C [*69PM*].
[16]) Übergang von tetragonal zu kubisch bei 37 °C ± 2 °C; c = (8.454 ± 0.002) Å [*62WW*].
[17]) Hergestellt bei 700 °C und 703 at, abgeschreckt; von 1300 °C abgeschreckt beträgt a = (8.328 ± 0.001) Å [*67DR*].
[18]) Hergestellt bei 650 °C und 703 at, abgeschreckt; von 1300 °C abgeschreckt beträgt a = (8.328 ± 0.001) Å [*67DR*].
[19]) Weitere Literatur: [*47RR, 52MHS-1, 56L, 56LA, 57BC, 63B-3, 57M, 63G-1, 61VP, 47VHR, 63G-2, 58H, 63BF, 67S, 68NK* [3]) *68TKF, 68HI, 65KM, 66MDAW*].

Fig. 1…4: Seite 400

Tab. 2. Elektrische und magnetische Eigenschaften binärer Cr^{3+} Oxid-Spinelle

Verbindung	q eV	σ $10^{-3}\ \Omega^{-1}\ cm^{-1}$	Leitungsträger	$\chi_{m\ diam}$ emu/mole	Θ_C °K	Θ_a °K	Θ_N °K	p_{eff} μ_B	p_m μ_B	C_m cm^3 °K/mole	σ Gauss cm^3/g	Lit.	Fig.
$CdCr_2O_4$						− 83	9					*66MDAW, 66BWRL*	5
						− 88	9			3.53		*66MDAW*	
							10)					*61VP*	
$CoCr_2O_4$					100		110		0.2 3)			*52MHS-1, 52SMHS*	6, 7,
	1.05 1)	2.49 2)	*p*		6)	−650				7.65		*53BR, 56L, 58MT*	8, 9,
									0.1			*59WG, 60MG*	10
									0.18 5)			*62WW, 62MWRD*	
					≅ 97 6)				0.14			*64MDW, 67P,*	
												68P-2, 68TKF, 69IIM	
$CuCr_2O_4$					135		135		0.7 3)			*52MHS-1, 52SMHS,*	6, 7
												52MHS-2, 60MG	
									0.72 5)			*62WW*	
									0.5			*63G-2*	
$FeCr_2O_4$	0.76 4)	1.83 2)	*p*							6.87 ± 0.15		*52SMHS, 53BR*	10,
					90 ± 5	−400 ± 30			0.8 ± 0.2	6.87 ± 0.15	19.5 ± 5.5	*56L*	11,
					90				0.8			*59WG*	12,
					88							*61N*	13
									0.84 5)			*62WW*	
												65R, 66TTA, 68HI,	
												57F,58DY, 65MSY	
$MgCr_2O_4$		$<10^{-7}$										*47VHR*	10,
									1.22			*52MHS-2, 52MHS-1*	14,
	0.95 7)	0.971 2)	*p*									*53BR*	15,
									0.15 5)			*62WW*	16
				$-0.04 \cdot 10^{-3}$								*62L*	
						−350	15	3.84				*63BF, 66BWRL*	
						−370						*63B-3, 68P-1, 69PS*	
$MnCr_2O_4$					52.5 ± 2				1.25			*52MHS-2, 58E*	7, 8,
									1.5			*52SMHS*	17,
					55 9)	−310			1.2	8.28		*52MHS-1, 56L*	18,
					40							*54G*	19

Verbindung	q eV	σ $10^{-3}\,\Omega^{-1}\,cm^{-1}$	Leitungsträger	$\chi_{m\,diam}$ emu/mole	Θ_C °K	Θ_a °K	Θ_N °K	p_{eff} μ_B	p_m μ_B	C_m cm³ °K/mole	σ Gauss cm³/g	Lit.	Fig.
$MnCr_2O_4$ (Forts.)					53				1.16			*58E, 59E*	
									1.4			*59WG, 60MG*	
									1.20 [5])			*62WW*	
					≅ 43							*62CH*	
							50		1.20			*63G-2, 62CH*	
									1.2			*64HH*	
						−270 ± 5			1.2	7.25 ± 0.05		*66DMFW,*	
												62MWRD, 68P-2	
												68TKF [8])	
$NiCr_2O_4$									0.3 [3])			*52SMHS*	7,
					50	−570				6.60		*52MHS-1*	17,
					80 ± 10 [9])				0.32 ± 0.04		8 ± 1	*56L*	20
					70				0.1 ··· 0.3			*59WG, 60MG*	
									0.33 [5])			*62WW, 68TKF*	
												62T	
	[13])											*69PM*	
$ZnCr_2O_4$	1.27 [11])	0.103 [2])	p									*53BR*	5,
	1.09 [12])											*53BR*	10,
						−380				3.70		*56L*	14
							15					*58H, 63G-2*	
									0.12 [5])			*62WW*	
						−350	15					*63BF, 63B-3*	
						−283				3.14		*64BL*	
						−392	16					*66BWRL, 63BF*	
						−350	16			3.74		*66MDAW, 52MHS-1*	
							[10])					*68TKF, 47VHR,*	
												61VP	

[1]) Bei 500 ··· 1010 °C [*53BR*].
[2]) $1/R$-Messungen für Proben von 1.5 cm ⌀ und 1.5 cm Höhe bei 900 °C [*63BR*].
[3]) Bei 4.2 °K und 7.4 kGauss [*52SMHS*].
[4]) Bei 655 ··· 1000 °C [*53BR*].
[5]) Bei 4.2 °K und 10 kOe [*62WW*].
[6]) Ferrimagnetisch unterhalb von 97 °K [*64MDW*], [*56L*].
[7]) Zwischen 535 ··· 1000 °C [*53BR*].
[8]) Richtung leichtester Magnetisierbarkeit [110], Richtung schwerer Magnetisierbarkeit [100] [*68TKF*].
[9]) Ferrimagnetisch unterhalb von 55 °K für $MnCr_2O_4$ und unterhalb von 80 °K für $NiCr_2O_4$ [*56L*].
[10]) Paramagnetisch bei RT [*61VP*].
[11]) Bei 800 ··· 1020 °C [*53BR*].
[12]) Bei 480 ··· 800 °C [*53BR*].
[13]) $q_1 = 0.145$ eV bei tiefen Temperaturen; zwischen 300 ··· 450 °K $q_2 = 0.400$ eV [*69PM*].

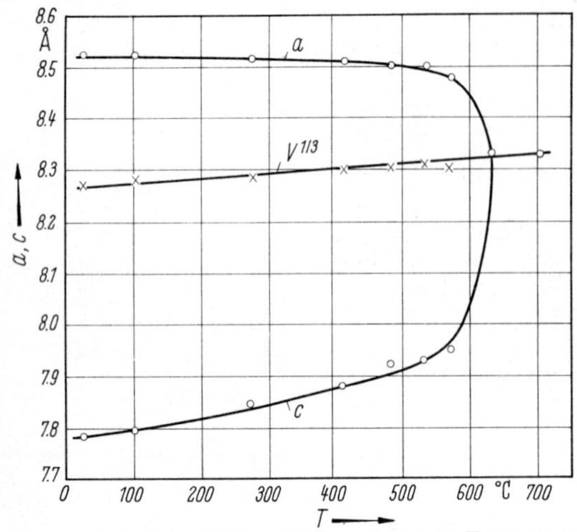

Fig. 1. $CuCr_2O_4$. Gitterparameter *a* und *c* vs. Temperatur *T* [*61OT*].

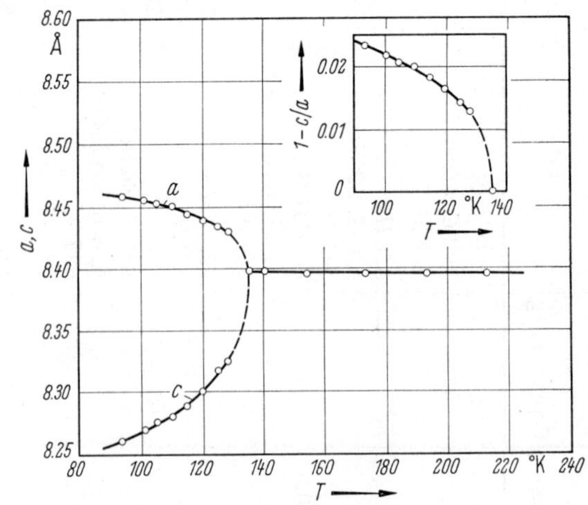

Fig. 3. $FeCr_2O_4$. Gitterparameter *a*, *c* und die Tetragonalität (1 − *c*/*a*) vs. Temperatur *T* [*66TTA*].

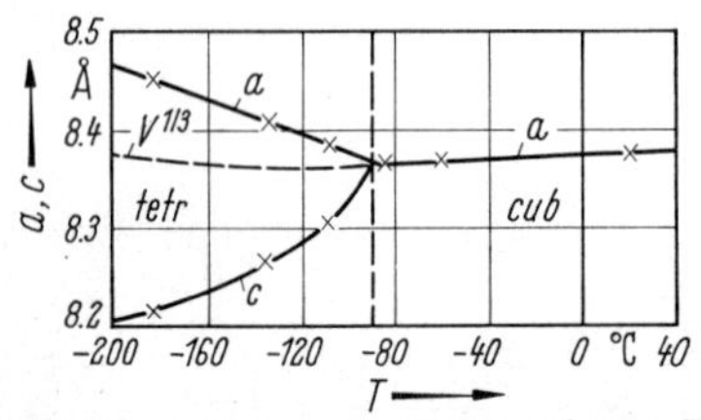

Fig. 2. $FeCr_2O_4$. Gitterparameter *a* und *c* vs. Temperatur *T* [*57F*].

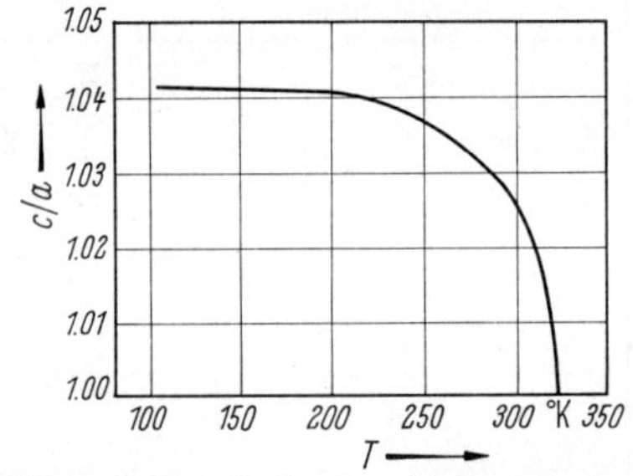

Fig. 4. $NiCr_2O_4$. Verhältnis der Gitterparameter *c*/*a* vs. Temperatur *T* [*65STT*].

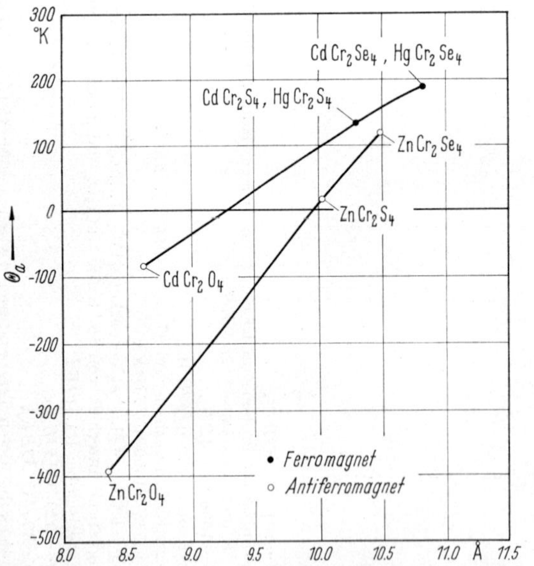

Fig. 5. $CdCr_2O_4$ und $ZnCr_2O_4$ (zum Vergleich: $CdCr_2S_4$, $CdCr_2Se_4$, $ZnCr_2S_4$, $ZnCr_2Se_4$, $HgCr_2S_4$, $HgCr_2Se_4$). Asymptotische Curie-Temperatur Θ_a vs. Gitterparameter *a* [*66BWRL*].

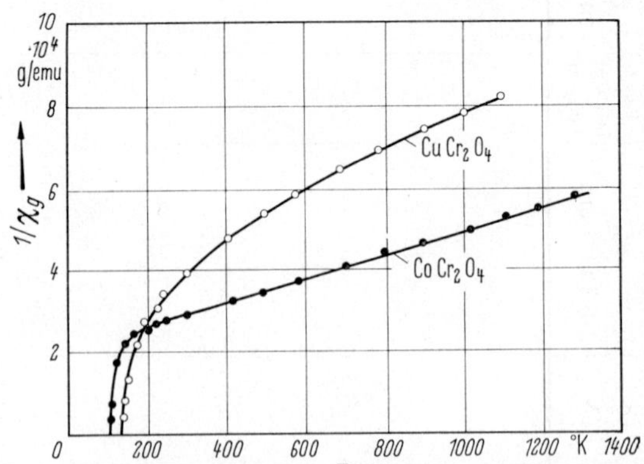

Fig. 6. $CoCr_2O_4$ und $CuCr_2O_4$. Reziproke Suszeptiblität χ_g^{-1} vs. Temperatur *T* [*52MHS-1*].

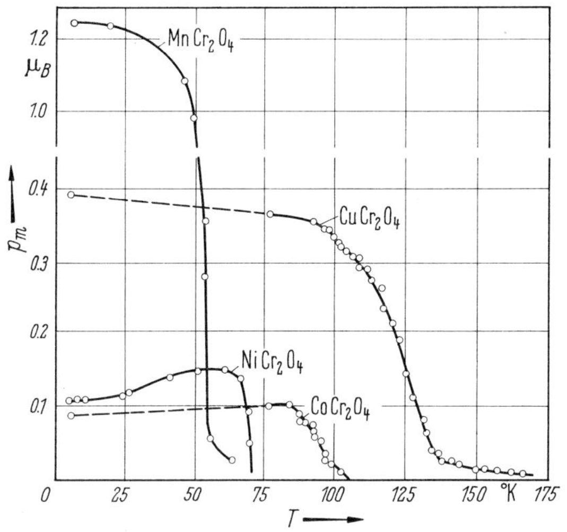

Fig. 7. $CoCr_2O_4$, $CuCr_2O_4$, $NiCr_2O_4$ und $MnCr_2O_4$. Magnetisches Moment p_m vs. Temperatur T [60MG].

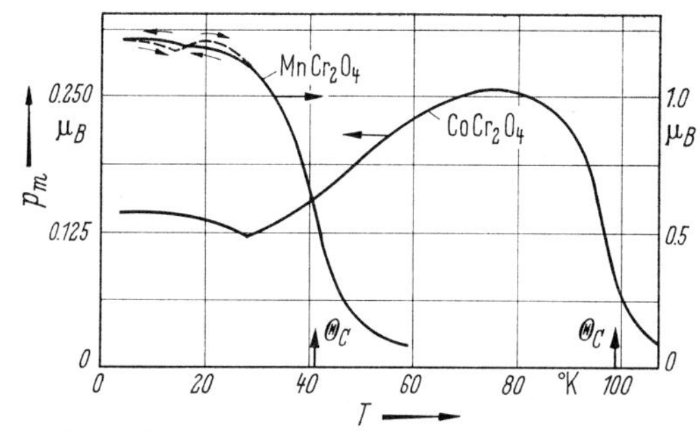

Fig. 8. $CoCr_2O_4$ und $MnCr_2O_4$. Magnetisches Moment p_m vs. Temperatur T bei einem Feld $H_a = 11$ kOe. Die angegebenen Curie-Temperaturen Θ_C sind bei einem Feld $H_a \simeq$ 100 Oe gemessen [62MWRD].

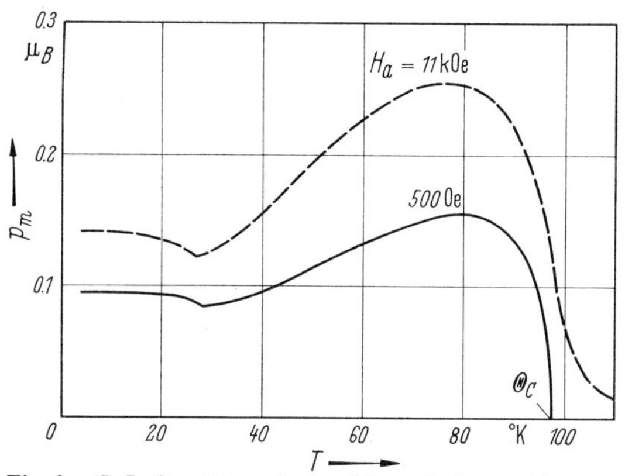

Fig. 9. $CoCr_2O_4$. Magnetisches Moment, p_m vs. Temperatur T bei $H_a = 500$ Oe und $H_a = 11$ kOe [64MDW].

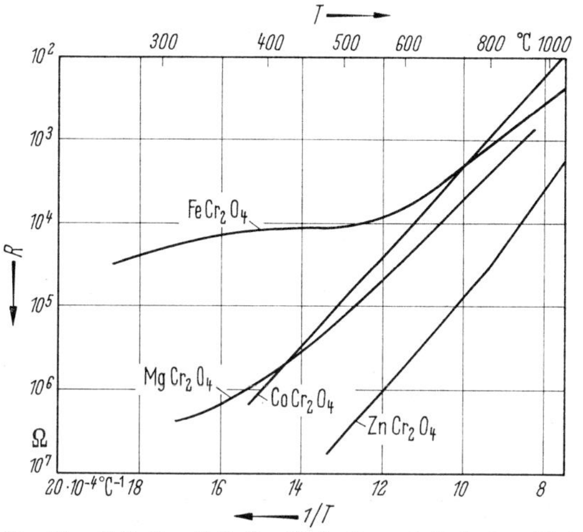

Fig. 10. $CoCr_2O_4$, $FeCr_2O_4$, $MgCr_2O_4$ und $ZnCr_2O_4$. Elektrischer Widerstand R vs. Temperatur T für Proben von 1.5 cm ⌀ und 1.5 cm Höhe [53BR].

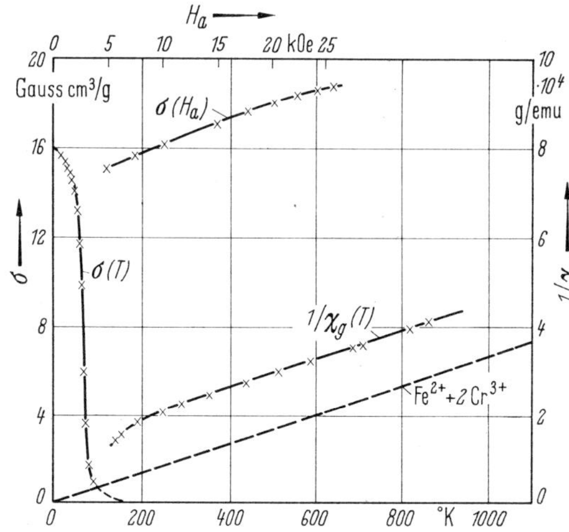

Fig. 11. $FeCr_2O_4$. Magnetisches Moment pro Gramm σ bei $H_a = 8$ kOe vs. Temperatur T; die reziproke Suszeptibilität $1/\chi_g$ vs. Temperatur T und das magnetische Moment pro Gramm σ bei einer Temperatur $T = 20.4$ °K vs. Feldstärke H_a. Die gestrichelte Gerade bezeichnet den Curie-Weiss-Verlauf entsprechend dem „Spin-only" Zustand der zwei Ionen ($Fe^{2+} + 2Cr^{3+}$) [56L].

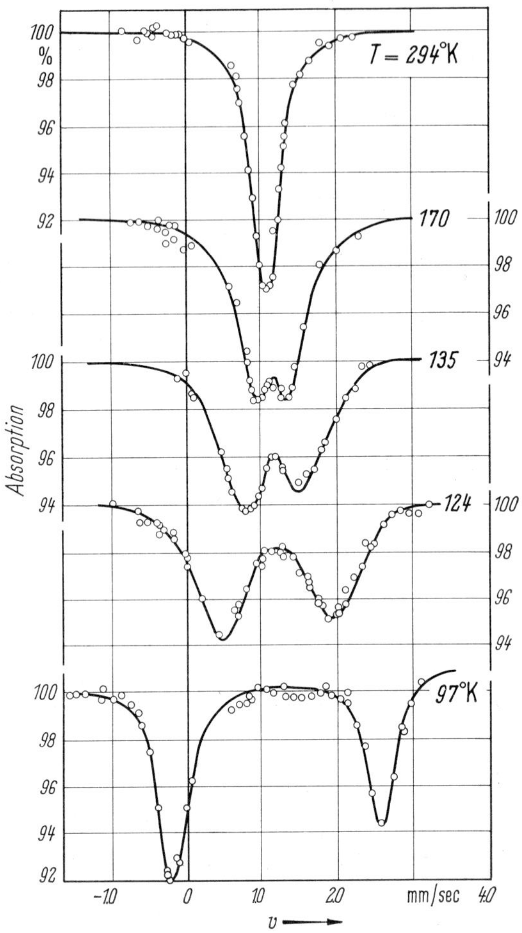

Fig. 12. $FeCr_2O_4$. Mössbauer-Spektren bei verschiedenen Temperaturen [*66TTA*].

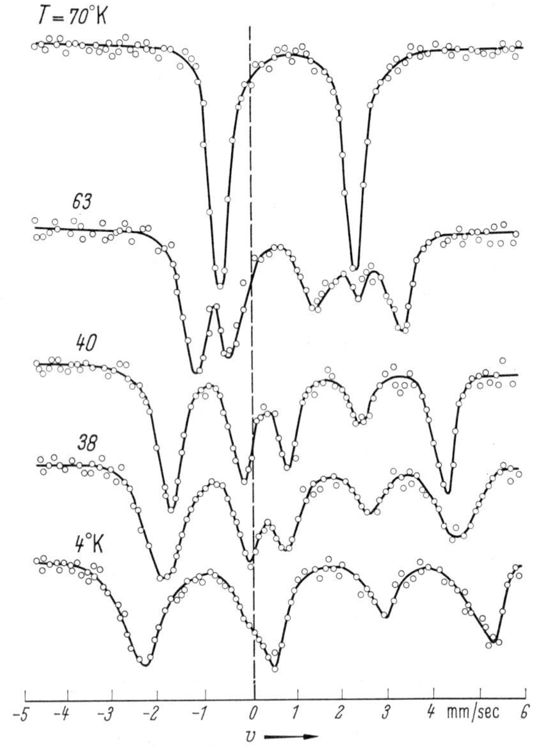

Fig. 13. $FeCr_2O_4$. Mössbauer-Spektren bei verschiedenen Temperaturen [*68HI*].

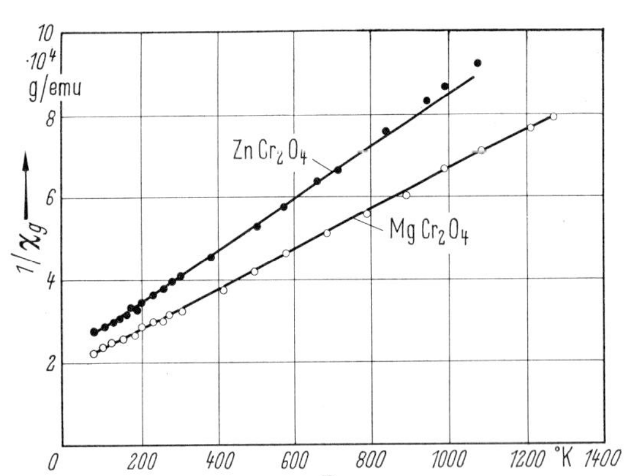

Fig. 14. $MgCr_2O_4$, $ZnCr_2O_4$. Reziproke Suszeptibilität $1/\chi_g$ vs. Temperatur T [*52MHS-1*].

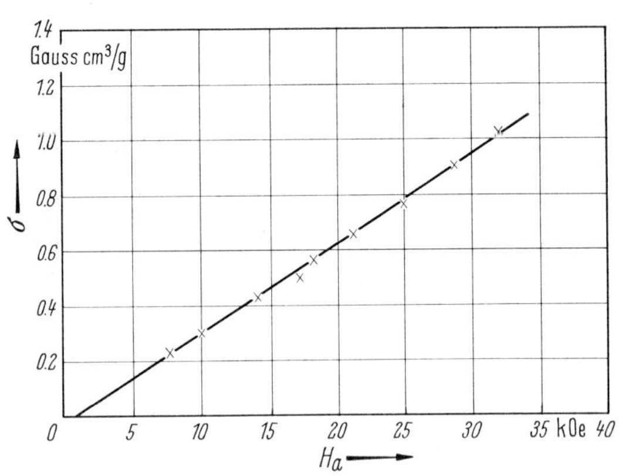

Fig. 15. $MgCr_2O_4$. Magnetisierung σ bei 4.2 °K vs. Feldstärke H_a [*63BF*].

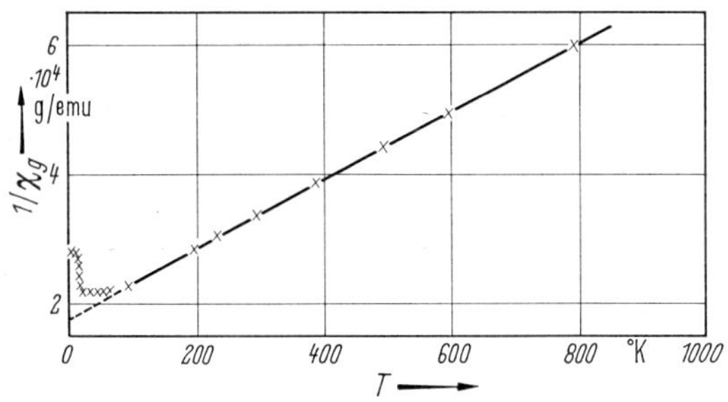

Fig. 16. $MgCr_2O_4$. Reziproke Suszeptibilität $1/\chi_g$ vs. Temperatur *T* [*63BF*].

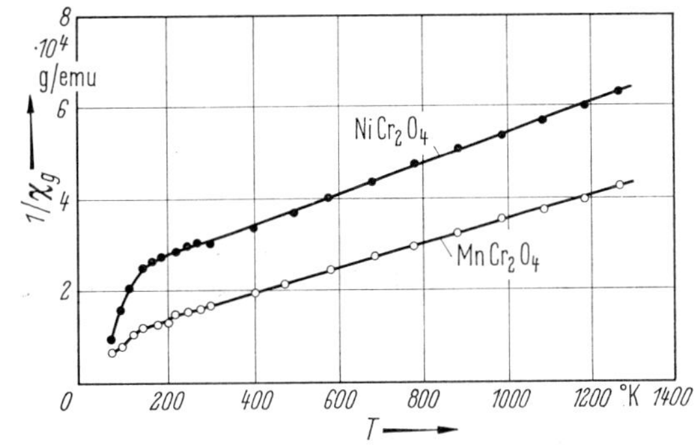

Fig. 17. $MnCr_2O_4$ und $NiCr_2O_4$. Reziproke Suszeptibilität $1/\chi_g$ vs. Temperatur *T* [*52MHS-1*].

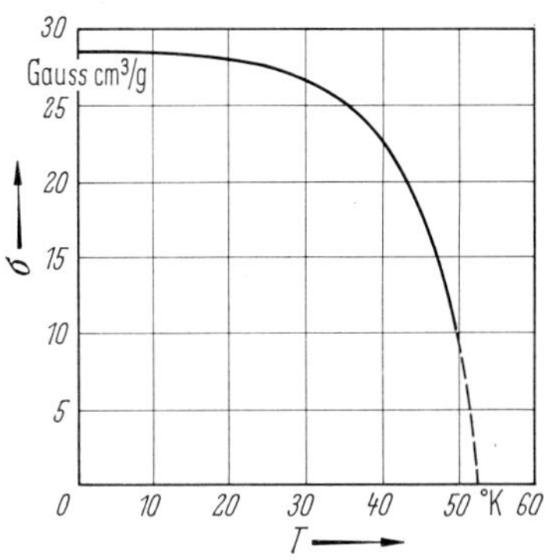

Fig. 18. $MnCr_2O_4$. Magnetisierung σ vs. Temperatur *T* bei einem Feld H_a = 6.6 kOe [*59E*].

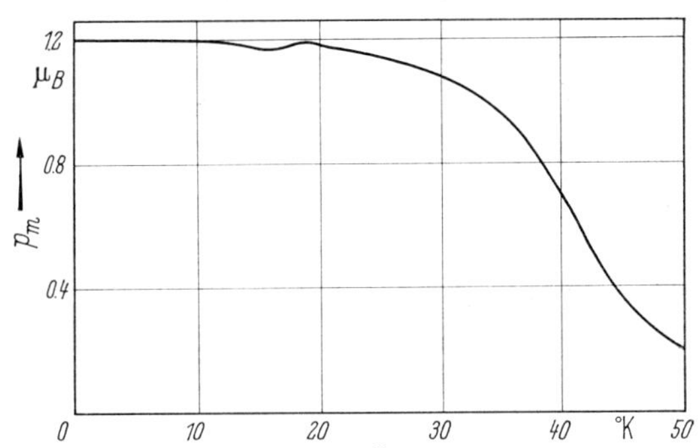

Fig. 19. $MnCr_2O_4$. Magnetisches Moment p_m vs. Temperatur *T* bei H_a = 10 kOe [*66DMFW*].

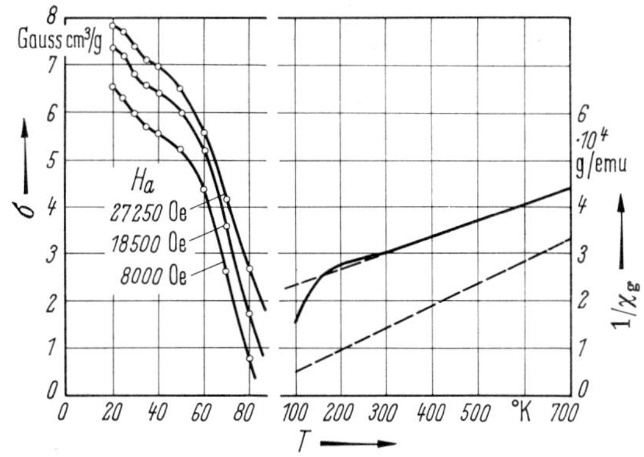

Fig. 20. $NiCr_2O_4$. Magnetisierung σ vs. Temperatur *T* für verschiedene Feldstärken H_a. Reziproke Suszeptibilität $1/\chi_g$ vs. Temperatur *T* (die gestrichelte Gerade vergleiche Fig. 11 nach [*52MHS-1*]) [*56L*].

6.2.1.2 Ternäre und quarternäre Cr Oxid-Spinelle

Tab. 3. Kristallografische Eigenschaften ternärer und quaternärer Cr^{3+} Oxid-Spinelle

Verbindung	x	Kationenverteilung	a Å	c/a	u	ϱ_x g/cm³	Lit.	Fig.	Tab.
$CdMnCrO_4$		$Cd[MnCr]O_4$	8.620				*66BWRL*		5
$CoInCrO_4$ *)							*65SK*		
$CuAl_xCr_{2-x}O_4$							*58D*	21	
$Cu^{2+}Al^{3+}_{2-2x}Cr^{3+}_{2x}O_4$				<1 [1])			*63G-2*		
$Cu_xCd_{1-x}Cr_2O_4$		$Cu^{2+}_xCd^{2+}_{1-x}[Cr^{3+}_2]O_4$		<1 [2])			*63G-2, 62G, 55D*	22	
$Cu_xCo_{1-x}Cr_2O_4$		$Cu^{2+}_xCo^{2+}_{1-x}[Cr^{3+}_2]O_4$		<1 [2]) [10])			*63G-2, 63WAWG, 55D, 62G*	22	
$CuFe_xCr_{2-x}O_4$	0.2	$Cu[Fe_{0.2}Cr_{1.8}]O_4$ $Cu_{0.8}Fe_{0.2}[Cu_{0.2}Cr_{1.8}]O_4$					*58D* *58D*	23,24 25	
$Cu^{2+}Fe^{3+}_{2-2x}Cr^{3+}_{2x}O_4$		$Cu_yFe_{1-y}[Cu_{1-y}Fe_{1+y-2x}Cr_{2x}]O_4$		<1 [3])			*61OT, 63G-2* *56MO, 62G*	42	5
$CuFe_{1+x}Cr_{1-x}O_4$		$Cu^{2+}_yFe^{3+}_{1-y}[Cu^{2+}_{1-y}Fe^{3+}_{x+y}Cr^{3+}_{1-x}]O_4$		>1 [5])			*63G-2, 64B-1*		
$CuGa_xCr_{2-x}O_4$				1 [4])			*58D*		
$CuGa^{3+}_{2-2x}Cr^{3+}_{2x}O_4$				<1 [3])			*63G-2, 61OT*		
$CuGaCrO_4$		$Cu_{0.5}Ga_{0.5}[Cu_{0.5}Ga_{0.5}Cr]O_4$	8.305		0.381		*66RD*		
$Cu_xMg_{1-x}Cr_2O_4$		$Cu^{2+}_xMg^{2+}_{1-x}[Cr^{3+}_2]O_4$		<1 [2])			*63G-2, 62G, 55D*	22	
$CuMn^{3+}_{2-2x}Cr^{3+}_{2x}O_4$				<1 [6])			*63G-2*		
$CuMn_xCr_{2-x}O_4$	1	$Cu^{2+}_{1-x}Cu^{1+}_x[Mn^{4+}_xCr^{3+}_{2-x}]O_4$	8.32		0.385		*64BL*	26 (6.3.12 Fig.55)	5
		$Cu^{2+}[Mn^{3+}_xCr^{3+}_{2-x}]O_4$					*64BL*		
$Cu_xNi_{1-x}Cr_2O_4$		$Cu^{2+}_xNi^{2+}_{1-x}[Cr^{3+}_2]O_4$		<1 [7])			*63G-2, 63WAWG, 55D, 62G*	22, 27	
$Cu_xZn_{1-x}Cr_2O_4$		$Cu^{2+}_xZn^{2+}_{1-x}[Cr^{3+}_2]O_4$		<1 [2])			*63G-2, 62G*	22	
$Cu_{1-x}Zn_xCr_2O_4$							*64BL, 55D*	26	
$Fe_{1-x}Co_xCr_2O_4$				<1 [11])			*64AWR, 64G*		
$FeCoCrO_4$		$Fe^{3+}[Co^{2+}Cr^{3+}]O_4$					*59WG*		5
$Fe_{2-x}CoCr_xO_4$							*58MT*	28	
	1.5						*60MG, 64B-1*	43, 44	

$FeFe_{2-x}Cr_xO_4$	1.28	$Fe^{2+}[Fe^{3+}_{2-x}Cr^{3+}_x]O_4$	8.406 ± 0.002				*54YKL*	29, 30, 31, 45	4, 5
		$Fe^{3+}[Fe^{3+}_{1-x}Fe^{2+}Cr^{3+}_x]O_4$ 8)					*54YKL, 57F, 58DY*		
	1		8.396				*59WG*		
		$Fe^{3+}_{1-y}Fe^{2+}_y[Fe^{2+}_{1-y}Fe^{3+}_{1-x+y}Cr^{3+}_x]O_4$					*64G, 62G*		
$Fe_{3-3x}Mg_xCr_{2x}O_4$	0.5	$Fe^{3+}[Mg^{2+}_{0.5}Fe^{2+}_{0.5}Cr^{3+}]O_4$					*47VHR, 53R*	46, 48	5
	0.75	$Fe^{3+}_{0.5}Mg^{2+}_{0.5}[Fe^{2+}_{0.25}Mg^{2+}_{0.25}Cr_{1.5}]O_4$					*47VHR*		
	0.75	$Fe^{3+}_{0.5}Fe^{2+}_{0.25}Mg^{2+}_{0.25}[Mg^{2+}_{0.5}Cr_{1.5}]O_4$					*47VHR*		
$FeMgCrO_4$							*51KR, 64B-1*		5
$Fe_{3-3x}Zn_xCr_{2x}O_4$		$Fe^{3+}_{1-x}Zn^{2+}_x[Fe^{2+}_{1-x}Fe^{3+}_{1-x}Cr^{3+}_{2x}]O_4$					*47VHR, 53R*	47, 48	5
$Li_{0.5}Al_{2.5-x}Cr_xO_4$							*64B-1*		
$Li_{0.5}Fe_{2.5-x}Cr_xO_4$		$Fe_{0.5}Li_{0.5}[Cr_2]O_4$					*54G, 54SR, 53R, 55GL*	49, 50	5,6
		$Fe[Li_{0.5}Cr_{1.5}]O_4$					*57G*	51, 52	
		$Fe_{0.8}Li_{0.2}[Li_{0.3}Fe_{0.2}Cr_{1.5}]O_4$					*57G*		
	1.25						*53GS, 61N, 64B-1*		
	2						*65H, 66JD*		
$Li_{0.5}Ga_{0.5}Cr_2O_4$			8.243 ± 0.003			4.134	*66JD*		
$LiGeCrO_4$			8.197			4.72	*62D-1, 62D-2, 63B-1*		
		$Li[GeCr]O_4$	8.20				*64B-1*		
$Li_{0.5}In_{0.5}CrO_4$			8.411 ± 0.003			4.347	*66JD*		
$LiMnCrO_4$		$Li^{1+}[Mn^{4+}Cr^{3+}]O_4$	8.19				*63B-2*		
			8.197		0.383		*64B-1, 64BL*		5
$LiSb_{0.5}Cr_{1.5}O_4$		$Li[Sb_{0.5}Cr_{1.5}]O_4$	8.37				*64B-1, 64B-2*		
$LiTiCrO_4$			8.31				*51KR, 51SR*		5
		$Li^{1+}[Ti^{4+}Cr^{3+}]O_4$					*63B-1, 62D-3*		
			8.32				*64B-1*		
			8.32		0.385		*64BL*		
$MgAl_{2-x}Cr_xO_4$							*54WK, 54SR, 67R, 67KS*	32	
$MgAlCrO_4$							*67KBC*		
$MgFe_{2-x}Cr_xO_4$							*60MG*	58	5
$MgInCrO_4$ **)			8.588 ± 0.001				*65SK*		
$MgMn_xCr_{2-x}O_4$	1	$Mg_{0.78}Mn_{0.22}[Mg_{0.22}Mn_{0.78}Cr]O_4$					*67G*	33, (6.3.12 Fig. 52)	

Fußnoten siehe S. 407

Tab. 3. (Fortsetzung)

Verbindung	x	Kationenverteilung	a Å	c/a	u	ϱ_x g/cm³	Lit.	Fig.	Tab.
$MnAl_xCr_{2-x}O_4$		$Mn_{1-yx}Al_{yx}[Mn_{yx}Al_{x(1-y)}Cr_{2-x}]O_4$					*59E, 59SR* *60MG*	53, 54 (6.3.5. Fig. 30, 33)	5
$MnFe_{2-x}Cr_xO_4$							*54G, 54SR, 60MG*		5,6
	1	$Mn^{2+}_{0.9}Fe^{3+}_{0.1}[Mn_{0.1}Fe_{0.9}Cr^{3+}]O_4$			0.3883 ± 0.0003		*59WG, 63G-2, 59PN*	34, 68, 69	
	1.5	$Mn^{2+}[Fe^{3+}_{0.5}Cr^{3+}_{1.5}]O_4$			0.3892 ± 0.0003		*63G-2, 59PN*		
$Mn_{1+x}Ti_xCr_{2-2x}O_4$	0.5 < x < 1	$Mn^{2+}[Mn^{2+}_xCr^{3+}_{2-2x}Ti^{4+}_x]O_4$					*64VLR, 64RLV* *66HGLRV*	(6.3.13 Fig. 16, 32, 33, 34, 35, 36)	5
$Mn^{2+}_{1+x}V^{4+}_xCr^{3+}_{2-2x}O_4$							*66HGLRV*		5
$Ni_xCo_{1-x}Cr_2O_4$							*64AWR*		
$NiFe^{3+}_{2-x}Cr^{3+}_xO_4$							*51KR, 51SR*	28, 35,	5
	1	$Fe[NiCr]O_4$	8.305				*53R, 53SR*	36, 37,	
	1.5	$Ni_{0.5}Fe_{0.5}[Ni_{0.5}Cr_{1.5}]O_4$	8.310				*53R, 53SR*	39, 40,	
	0.5	$Fe[NiCr_{0.5}Fe_{0.5}]O_4$	8.320				*53R, 53SR*	43, 55,	
	1	$Fe_{0.5}Ni_{0.5}[Ni_{0.5}Fe_{0.5}Cr]O_4$					*57G*	56, 57,	
	1.25 < x < 1.80			<1			*60MG, 62BG*	59, 60,	
	1.80 < x < 2.0			>1			*60MG, 65H*	61, 62	
	1		8.301				*59WG*		
							63MT, 62G, 58MT		
							64B-1, 64AWR, 64G		
							67CF		
$Ni_xFe_{1-x}Cr_2O_4$			[12])				*63WAWG, 64AWR, 64G*	38	
$NiInCrO_4$ **)			8.561 ± 0.001				*65SK*		
$Ni_xMg_{1-x}Cr_2O_4$	0.25						*62L*	70	
	0.5						*62L*		5
	0.75						*62L*		
$ZnInCrO_4$ [9])							*65SK*		

$ZnMnCrO_4$			8.25	1.04			*57B*		
		$Zn^{2+}[Mn^{3+}Cr^{3+}]O_4$					*59G, 63G-2*		5
			8.35		0.385		*64BL, 66BWRL*		
$Cd_{0.6}Mg_{0.4}Al_{2-x}Cr_xO_4$	0.5						*67R*		
	1						*67R*		
$MgAl_xFe_{2-2x}Cr_xO_4$							*67KBC*	41	
$Mg_{1-x}Ni_xAl_{2-2x}Cr_{2x}O_4$							*68DR*		
$NiAl_{0.25}Fe_{1.75-x}Cr_xO_4$	1.25			0.999			*60MG*	63, 64	5
	1.5			0.996			*60MG*		
$NiGa_{0.25}Fe_{1.75-x}Cr_xO_4$							*60MG*	65	
$Ni_{0.75}Zn_{0.25}Fe_{2-x}Cr_xO_4$	1.5			1			*60MG*	66, 67	5
	1.75			0.998			*60MG*		

*) schwarz [*65SK*].
**) grün [*65SK*].
1) Für $x > 0.2$ [*63G-2*].
2) Für $x > 0.5$ [*63G-2*].
3) Für $x > 0.7 \cdots 0.75$ [*61OT*].
4) Für $x < 0.3$ [*58D*].
5) Für $x > 0.65$ [*63G-2*].
6) Für $x > 0.8$ [*63G-2*].
7) Für $x > 0.1$ [*63G-2*].
8) Inverse Struktur [*54YKL*].
9) Bei Sinterung von einem stöchiometrischen Gemenge ($2ZnO + Cr_2O_3 + In_2O_3$) wurden 2 Spinellphasen mit $a_1 = (8.42 \pm 0.01)$ Å und $a_2 = (8.70 \pm 0.01)$ Å festgestellt. Es werden die Spinellverbindungen $ZnCr_2O_4$ mit wenig In an Stelle von Cr und $ZnIn_2O_4$ mit wenig Cr an Stelle von In vermutet. [*65SK*].
10) Kubisch bei RT für $0 \leqq x \leqq 0.43$; tetragonal verzerrt $c/a < 1$ für $0.48 \leqq x \leqq 1$ [*63WAWG*].
11) Für $x < 0.3$ [*64AWR*].
12) $c/a < 1$ für $x \leqq 0.2$; $c/a > 1$ für $x \geqq 0.3$ [*64AWR*].

Tab. 4. Kristallografische Eigenschaften der Verbindungsreihe $FeFe_{2-x}Cr_xO_4$ [57F]

x	Kationenverteilung	a [1]) Å	c/a [2])	Fig.
0	$Fe^{3+}[Fe^{2+}Fe^{3+}]O_4$	8.396 ± 0.001	annähernd rhomboedrisch	
0.1···0.8		Fig. 29	kubisch	29
1.0	$Fe^{2+}_{0.6}Fe^{3+}_{0.4}[Fe^{2+}_{0.4}Fe^{3+}_{0.6}Cr^{3+}_{1.0}]O_4$	8.397 ± 0.001	↑	30
1.2		8.405 ± 0.001	tetragonal	
1.3	$Fe^{2+}[Fe^{3+}_{0.7}Cr^{3+}_{1.3}]O_4$	8.406 ± 0.001	>1	
1.4		8.405 ± 0.001	↓	
1.45		8.402 ± 0.001	↑	
1.5	$Fe^{2+}[Fe^{3+}_{0.5}Cr^{3+}_{1.5}]O_4$	8.400 ± 0.001		
1.55		8.398 ± 0.001	orthorhombisch	
1.60		8.395 ± 0.001		
1.70		8.391 ± 0.001		
1.80		8.385 ± 0.001	↓	
2.0	$Fe^{2+}[Cr^{3+}_2]O_4$	8.377 ± 0.001	<1	2

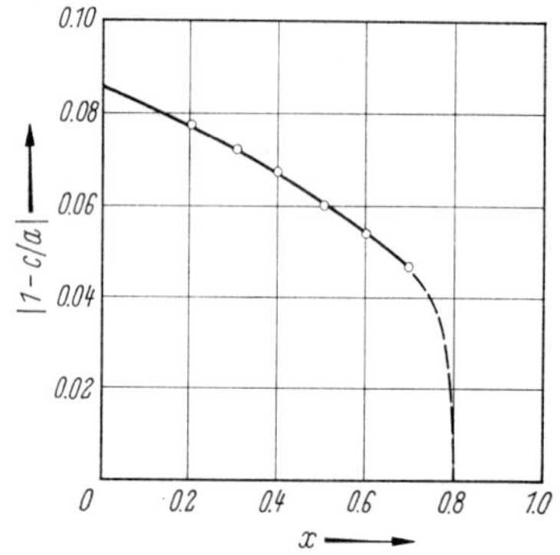

Fig. 21. $CuAl_xCr_{2-x}O_4$. Tetragonale Verzerrung $|1 - c/a|$ vs. x [58D].

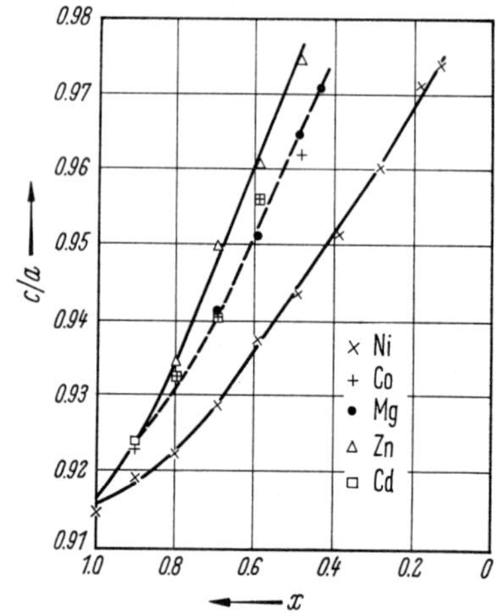

Fig. 22. $Cu_xMe_{1-x}Cr_2O_4$ (Me = Cd, Co, Mg, Ni, Zn). Verhältnis der Gitterparameter c/a vs. x [55D].

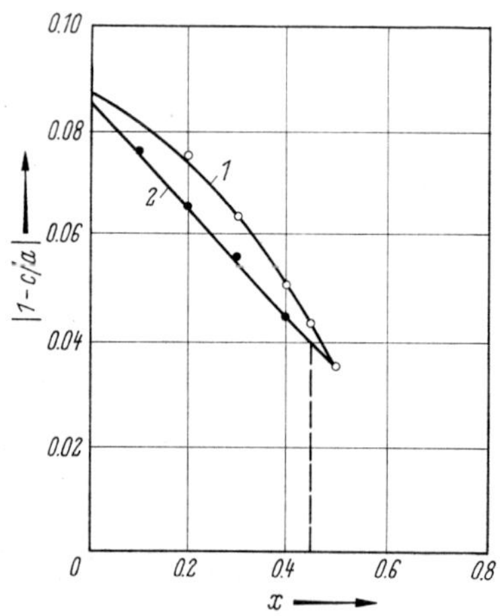

Fig. 23. $CuFe_xCr_{2-x}O_4$. Tetragonale Verzerrung $|1 - c/a|$ vs. x [58D].

1 bei Abkühlung, 2 bei Erwärmung.

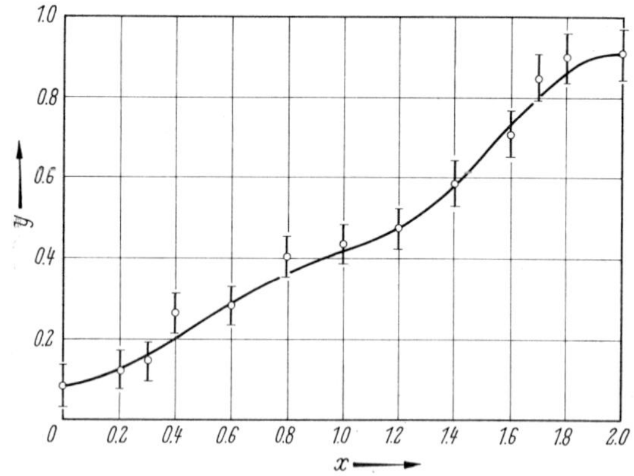

Fig. 24. $CuFe_{2-x}Cr_xO_4$. Kationenverteilung y vs. x [610T].

[1]) Bei 20 °C.

[2]) Struktur bei −183 °C.

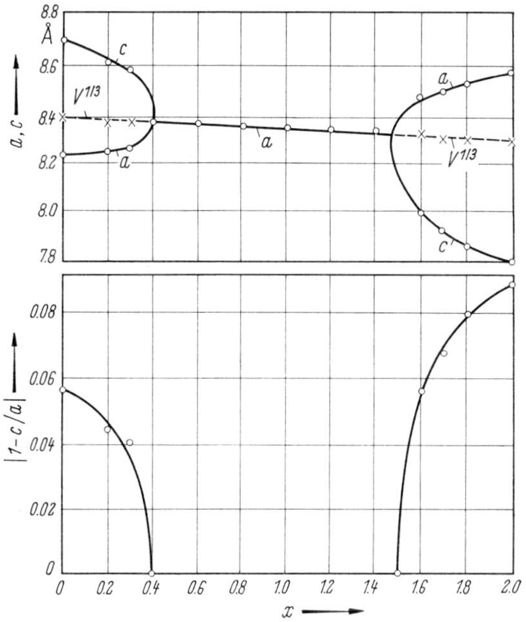

Fig. 25. $CuFe_{2-x}Cr_xO_4$. Gitterparameter *a* und *c* und tetragonale Verzerrung $|1 - c/a|$ vs. *x* [*61OT*].

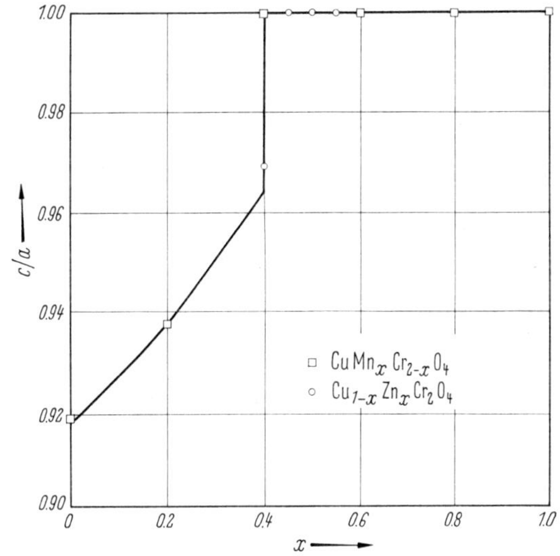

Fig. 26. $CuMn_xCr_{2-x}O_4$ und $Cu_{1-x}Zn_xCr_2O_4$. Verhältnis der Gitterparameter *c*/*a* vs. *x* [*64BL*].

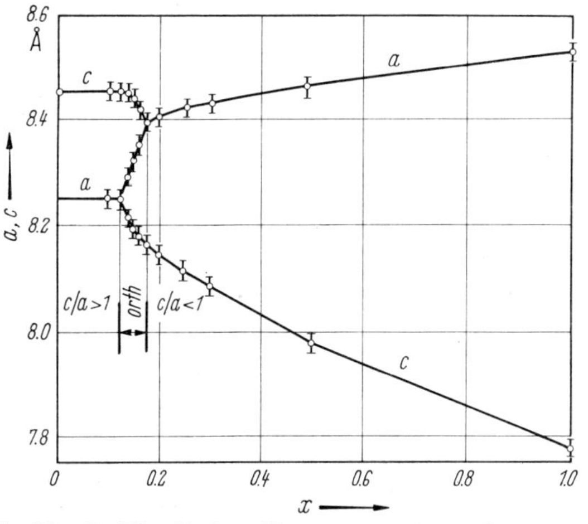

Fig. 27. $Cu_xNi_{1-x}Cr_2O_4$. Gitterparameter *a* und *c* vs. *x* [*63WAWG*].

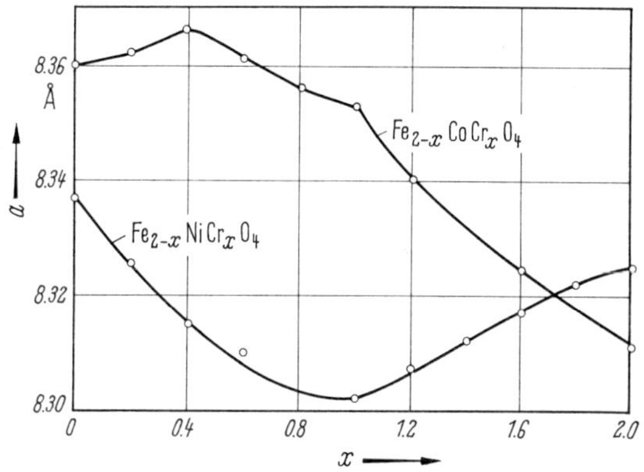

Fig. 28. $Fe_{2-x}CoCr_xO_4$ und $Fe_{2-x}NiCr_xO_4$. Gitterparameter *a* vs. *x* [*58MT*].

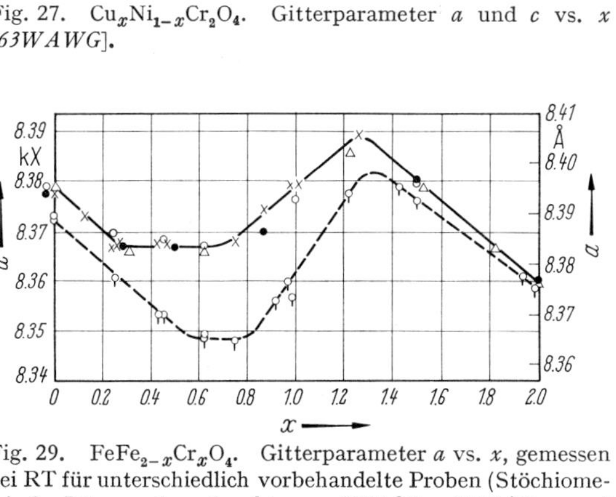

Fig. 29. $FeFe_{2-x}Cr_xO_4$. Gitterparameter *a* vs. *x*, gemessen bei RT für unterschiedlich vorbehandelte Proben (Stöchiometrische Lösung abgeschreckt von: 1180 °C ▵, 1100 °C × und 950 °C ○; metallreiche Lösung ● und oxidreiche Lösung ♀ jeweils von 950 °C abgeschreckt). [*54YKL*].

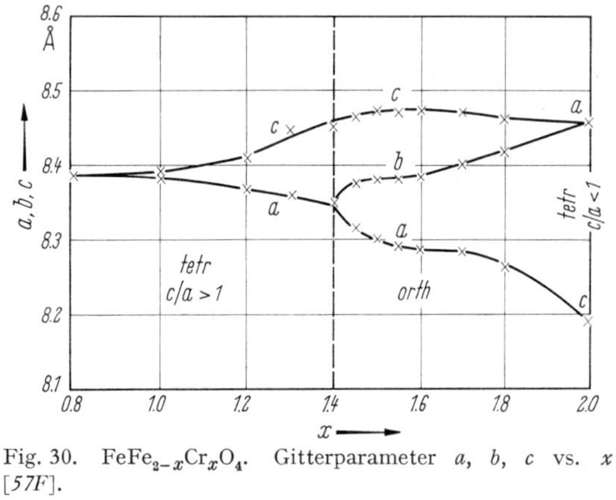

Fig. 30. $FeFe_{2-x}Cr_xO_4$. Gitterparameter *a*, *b*, *c* vs. *x* [*57F*].

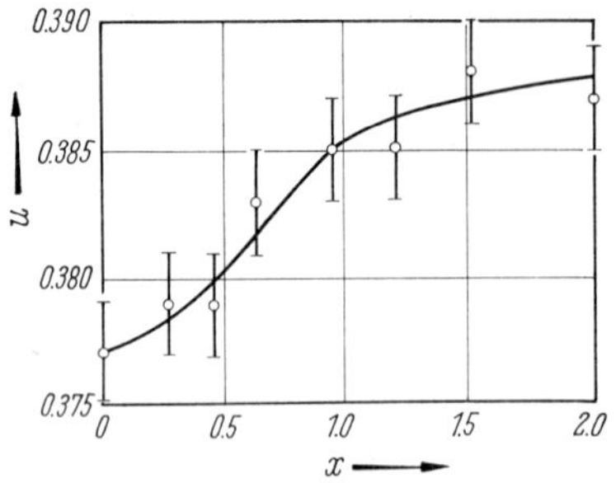

Fig. 31. $FeFe_{2-x}Cr_xO_4$. Anionenparameter u vs. x [*58DY*].

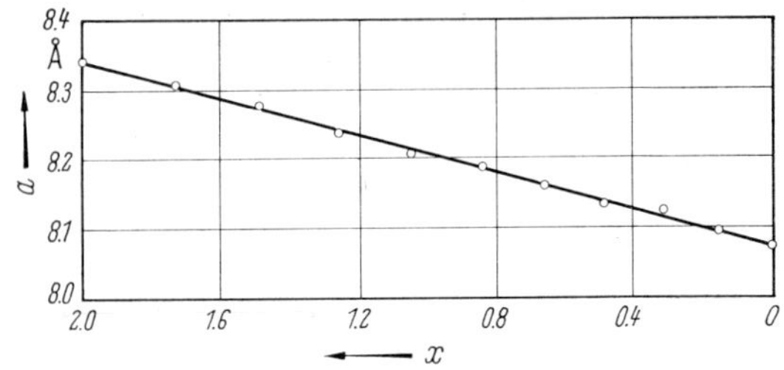

Fig. 32. $MgAl_{2-x}Cr_xO_4$. Gitterparameter a vs. x [*54WK*].

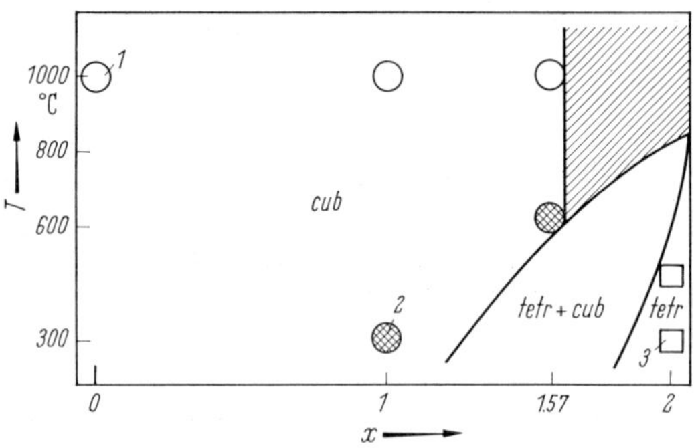

Fig. 33. $MgMn_xCr_{2-x}O_4$. Phasendiagramm (T vs. x).

1 Produkte sind kubisch bei hohen Temperaturen und kubisch nach der Abschreckung.

2 Die Produkte sind kubisch nach dem Glühen oberhalb des Zweiphasengebietes oder bei tiefen Temperaturen.

3 Die Produkte sind tetragonal nach dem Glühen bei tiefen Temperaturen.

Schraffiertes Gebiet: Zone, in der die Verbindungen bei Abschreckung tetragonal sind [*67G*].

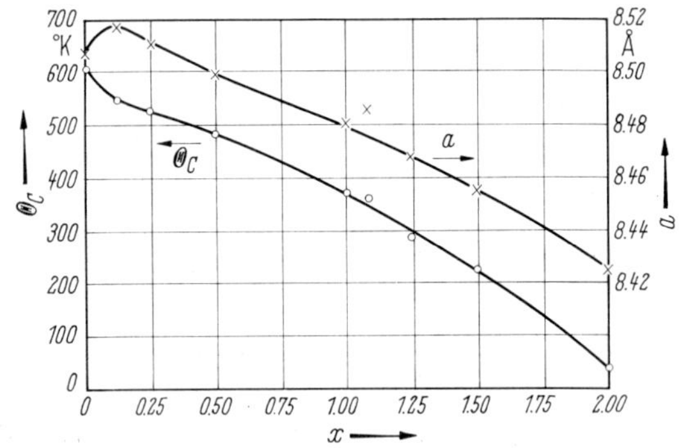

Fig. 34. $MnFe_{2-x}Cr_xO_4$. Curietemperatur Θ_C und Gitterparameter a vs. x [*54G*].

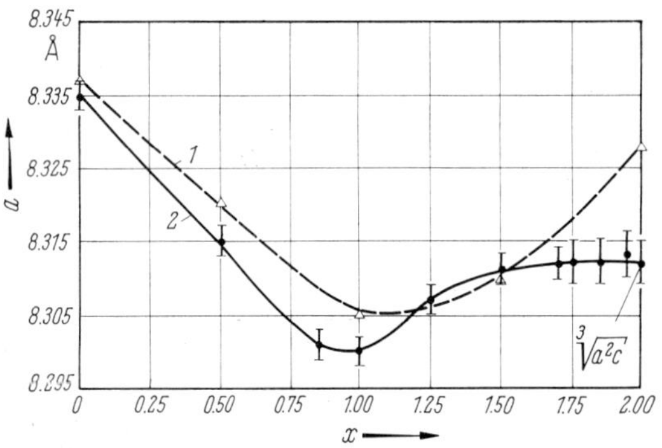

Fig. 35. $Fe_{2-x}NiCr_xO_4$. Gitterparameter a vs. x bei RT. (1 nach [*53R*]) 2 eigene Werte [*60MG*].

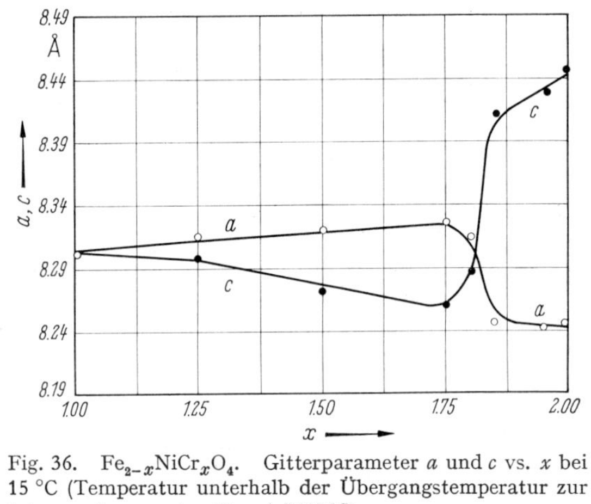

Fig. 36. $Fe_{2-x}NiCr_xO_4$. Gitterparameter a und c vs. x bei 15 °C (Temperatur unterhalb der Übergangstemperatur zur tetragonal verzerrten Form) [*60MG*].

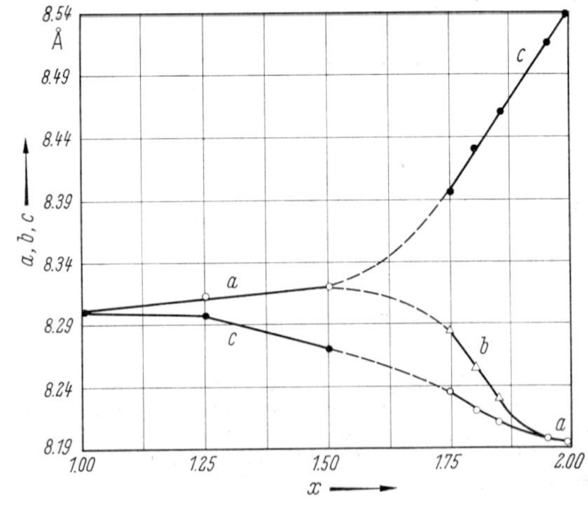

→

Fig. 37. $Fe_{2-x}NiCr_xO_4$. Gitterparameter a, b und c vs. x bei 120 °K [*60MG*].

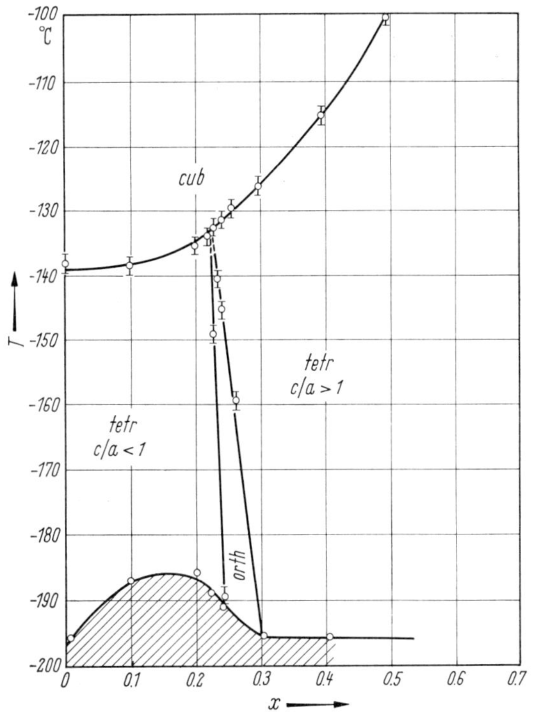

Fig. 38. $Ni_xFe_{1-x}Cr_2O_4$. Phasendiagramm (T vs. x) *63WAWG*].

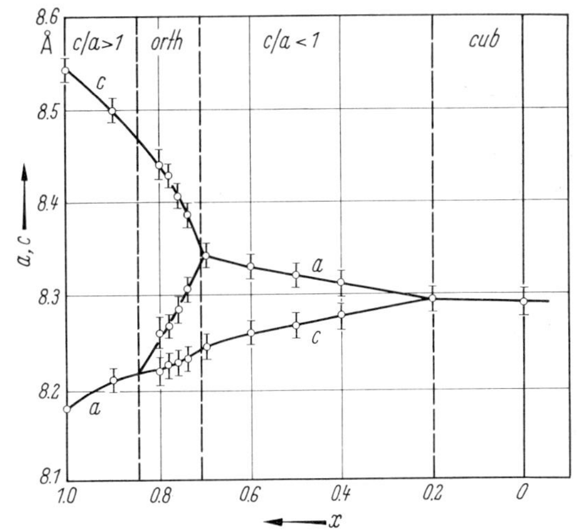

Fig. 39. $NiFe_{2-x}Cr_xO_4$. Gitterparameter a und c vs. x bei (-182 ± 2) °C [*64AWR*].

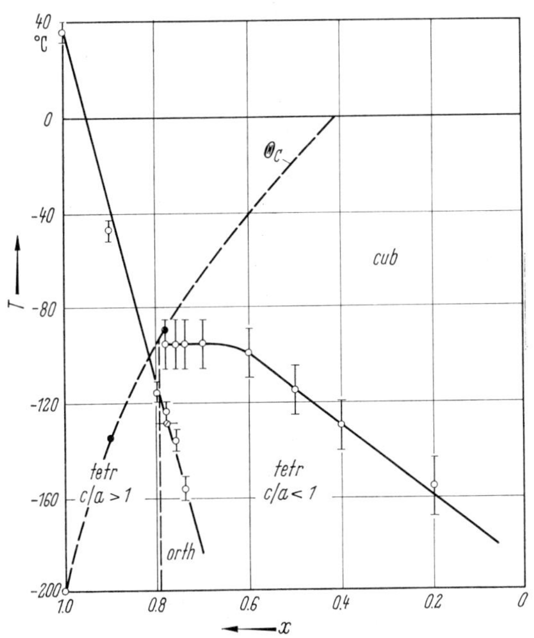

Fig. 40. $NiFe_{2-x}Cr_xO_4$. Phasendiagramm (T vs. x) und Curie-Temperatur Θ_C vs. x [*64AWR*].

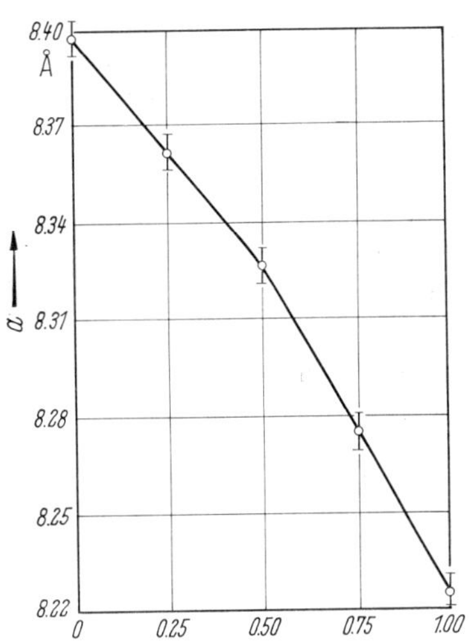

Fig. 41. $MgAl_xFe_{2-2x}Cr_xO_4$. Gitterparameter a vs. x [*67KBC*].

Tab. 5. Magnetische und elektrische Eigenschaften ternärer und quaternärer Cr^{3+} Oxid-Spinelle

Verbindung	x	Θ_C °K	Θ_a °K	C_m cm³ °K/mole	p_m μ_B	$\chi_{g\,diam}$ 10^{-6} emu/g	Lit.	Fig.	Tab.
$CdMnCrO_4$			− 95				*66BWRL*		
$Cu^{2+}Fe^{3+}_{2-2x}Cr^{3+}_{2x}O_4$							*56MO*	42	
$CuMnCrO_4$			−259	4.63			*64BL*	(6.3.12 Fig. 55)	
$Fe_{2-x}CoCr_xO_4$							*58MT*	43, 44	
	1				1.0		*59WG, 60MG*		
$FeFe_{2-x}Cr_xO_4$							*57F, 58DY*	45	
	1	473			2.4		*59WG*		
$Fe_{3-3x}Mg_xCr_{2x}O_4$							*47VHR, 53R*	46, 48	
$FeMgCrO_4$		≈323 ¹)					*51KR*		
$Fe_{3-3x}Zn_xCr_{2x}O_4$							*47VHR, 53R*	47, 48	
$Li_{0.5}Fe_{2.5-x}Cr_xO_4$							*54G, 57G*	49, 50,	6
	1.25						*53GS, 61N*	51, 52	
$LiMnCrO_4$			− 84	2.81			*64BL*		
$LiTiCrO_4$			− 81	1.35			*64BL*		
$MgFe_{2-x}Cr_xO_4$							*60MG*	58	
$MnAl_xCr_{2-x}O_4$							*59E*	53, 54 (6.3.5 Fig. 33)	
$MnFe_{2-x}Cr_xO_4$							*54G*	68, 69,	6
	1	370			0.25		*59WG*	34	
	1	375			0.21		*59PN, 63G-2*		
	1.5	240			0.80		*59PN, 63G-2*		
$Mn_{1+x}Ti_xCr_{2-2x}O_4$							*64RLV*	(6.3.13	
							64VLR	Fig. 32,	
							66HGLRV	33, 34,	
								35, 36)	
$Mn^{2+}_{1+x}V^{4+}_xCr^{3+}_{2-2x}O_4$ ²)							*66HGLRV*		
$NiFe^{3+}_{2-x}Cr^{3+}_xO_4$	1	598 ¹)					*51KR*	43, 55,	
					0.6		*57G*	56, 57,	
		690			0.4		*59WG*	59, 60,	
							60MG, 63MT	61, 62,	
							64AWR, 67CF,	40	
							58MT		
$Ni_xMg_{1-x}Cr_2O_4$						−0.4	*62L*	70	
$ZnMnCrO_4$			−278	4.42			*64BL*		
			−268				*66BWRL*		
$NiAl_{0.25}Fe_{1.75-x}Cr_xO_4$							*60MG*	63, 64	
$NiGa_{0.25}Fe_{1.75-x}Cr_xO_4$							*60MG*	65	
$Ni_{0.75}Zn_{0.25}Fe_{2-x}Cr_xO_4$							*60MG*	66, 67	

¹) Sehr schwach magnetisch bei RT [*51KR*].
²) Ferrimagnetisch [*66HGLRV*].

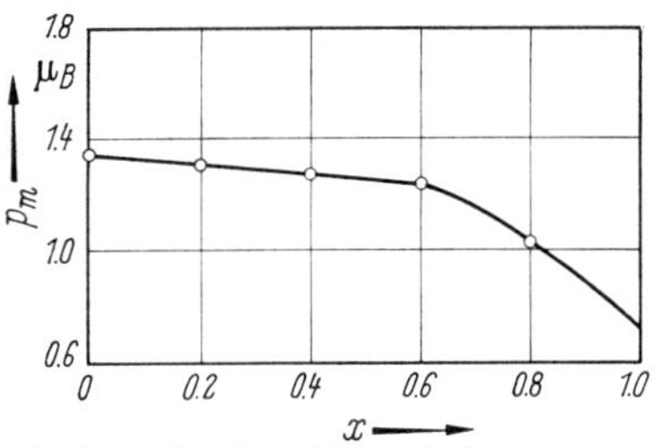

Fig. 42. $CuFe_{2-2x}Cr_{2x}O_4$. Magnetisches Moment p_m vs. x bei einer maximalen Feldstärke $H = 7$ kOe und bei 0° K [56MO].

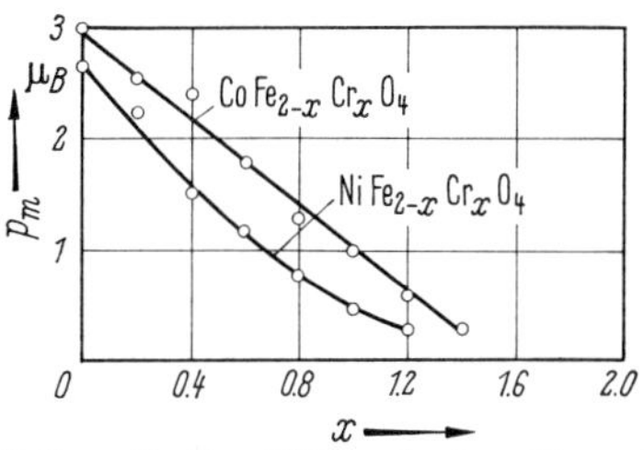

Fig. 43. $CoFe_{2-x}Cr_xO_4$, $NiFe_{2-x}Cr_xO_4$. Magnetisches Moment p_m vs. x bei einer maximalen Feldstärke $H = 8$ kOe und bei 0 °K [58MT].

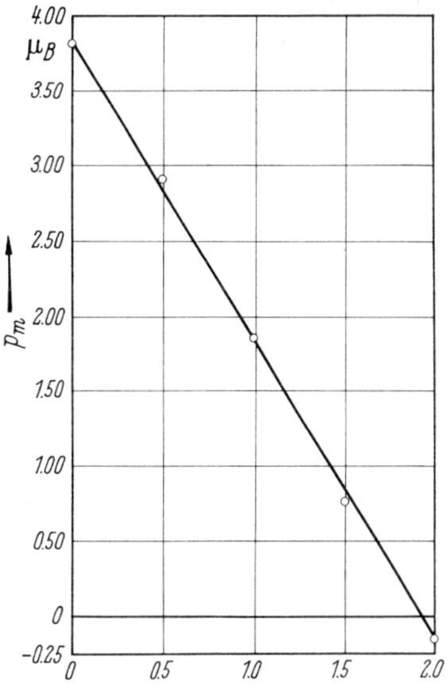

Fig. 44. $Fe_{2-x}CoCr_xO_4$. Magnetisches Moment p_m vs. x bei 0 °K [60MG].

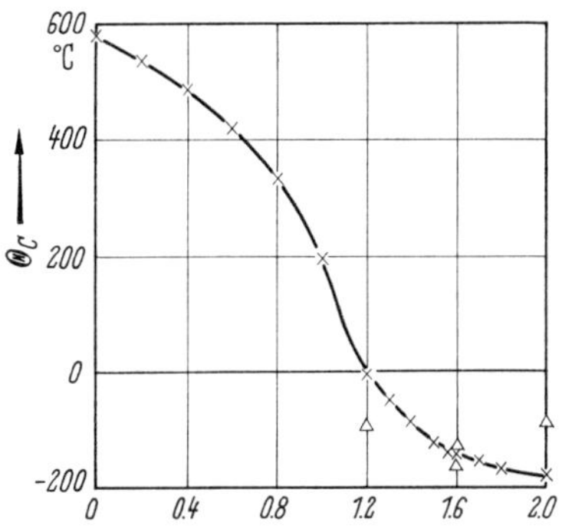

Fig. 45. $FeFe_{2-x}Cr_xO_4$. Curietemperatur Θ_C vs. x (△ = Strukturübergangstemperaturen) [57F].

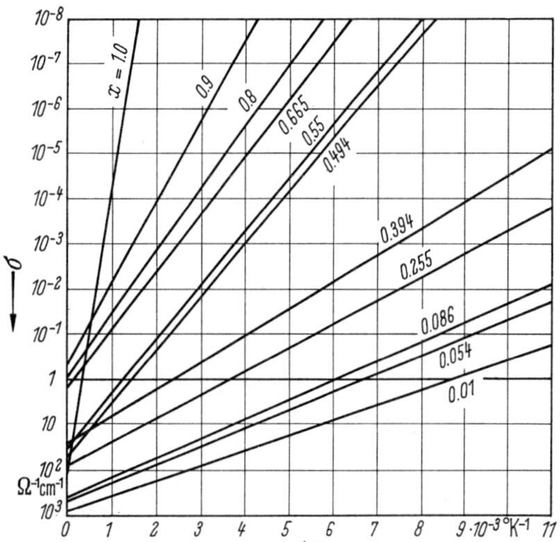

Fig. 46. $Fe_{3-3x}Mg_xCr_{2x}O_4$. Elektrische Leitfähigkeit σ vs. $1/T$. Parameter x [47VHR].

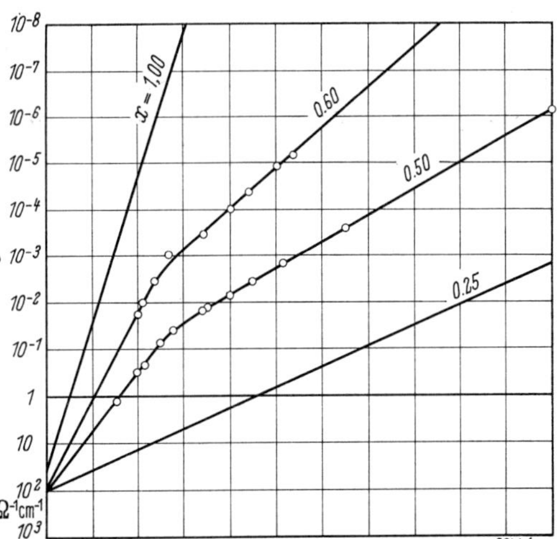

Fig. 47. $Fe_{3-3x}Zn_xCr_{2x}O_4$. Elektrische Leitfähigkeit σ vs. $1/T$. Parameter x [47VHR].

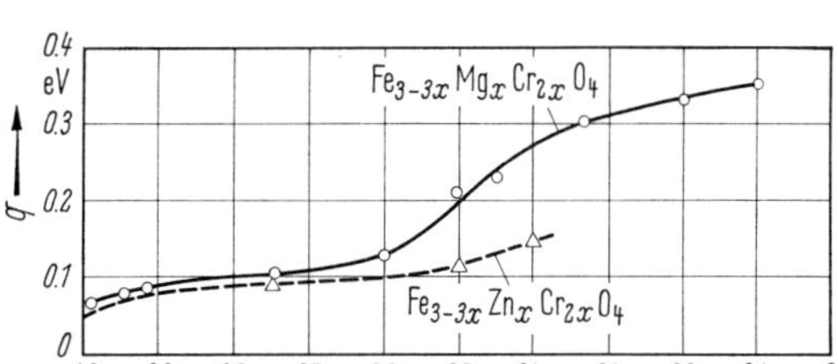

Fig. 48. $Fe_{3-3x}Mg_xCr_{2x}O_4$, $Fe_{3-3x}Zn_xCr_{2x}O_4$. Aktivierungsenergie q für die elektrische Leitfähigkeit vs. x [47VHR].

Tab. 6. Kristallografische und magnetische Eigenschaften der Verbindungsreihen $Li_{0.5}Fe_{2.5-x}Cr_xO_4$ [54G] und $MnFe_{2-x}Cr_xO_4$ [54G]

Verbindung	x	Kationenverteilung	a Å	Θ_C °C	p_m μ_B	Fig.
$Li_{0.5}Fe_{2.5-x}Cr_xO_4$	0	$Fe_{1.0}[Li_{0.5}Fe_{1.5}]O_4$	8.331	680	2.47	49,
	0.50	$Fe_{1.0}[Li_{0.5}Fe_{1.0}Cr_{0.5}]O_4$	8.306	500	1.62	50,
	0.75	$Fe_{0.98}Li_{0.02}[Li_{0.48}Fe_{0.77}Cr_{0.75}]O_4$	8.296	410	1.35	52
	1.0	$Fe_{0.94}Li_{0.06}[Li_{0.44}Fe_{0.56}Cr_{1.0}]O_4$	8.292	315	0.84	
	1.25	$Fe_{0.91}Li_{0.09}[Li_{0.41}Fe_{0.34}Cr_{1.25}]O_4$	8.290	214	0.61	
	1.50	$Fe_{0.80}Li_{0.20}[Li_{0.30}Fe_{0.20}Cr_{1.50}]O_4$	8.287	119	0.55	
	1.60	$Fe_{0.64}Li_{0.36}[Li_{0.14}Fe_{0.26}Cr_{1.60}]O_4$	8.288	167	0.42	
	1.70	$Fe_{0.54}Li_{0.46}[Li_{0.04}Fe_{0.26}Cr_{1.70}]O_4$	8.290	155	0.22	
	2.0	$Fe_{0.5}Li_{0.5}[Cr_{2.0}]O_4$	8.288	80 ± 16	0.10	
$MnFe_{2-x}Cr_xO_4$	0		8.5074 ± 2	330	4.85	34
	0.125		8.5185 ± 1	272	4.04	
	0.25		8.5107 ± 3	247	3.28	
	0.50		8.4977 ± 2	210	1.73	
	1.0		8.4809 ± 2	97	0.25	
	1.085		8.4869 ± 2	88	0.37	
	1.25		8.4680 ± 2	15	0.45	
	1.50		8.4567 ± 1	– 49	0.775	
	2.0		8.426	–233	1.5	19

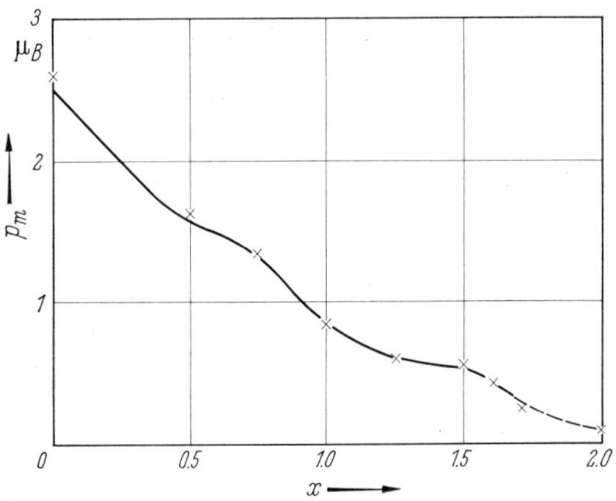

Fig. 49. $Li_{0.5}Fe_{2.5-x}Cr_xO_4$. Magnetisches Moment p_m vs. x [54G].

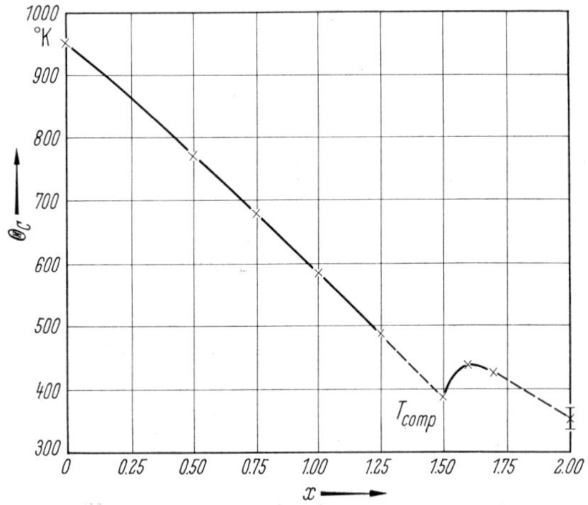

Fig. 50. $Li_{0.5}Fe_{2.5-x}Cr_xO_4$. Curie-Temperatur Θ_C vs. x [54G].

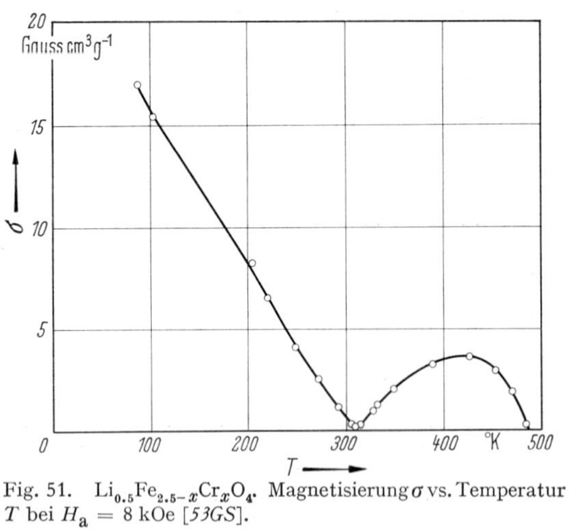

Fig. 51. $Li_{0.5}Fe_{2.5-x}Cr_xO_4$. Magnetisierung σ vs. Temperatur T bei H_a = 8 kOe [53GS].

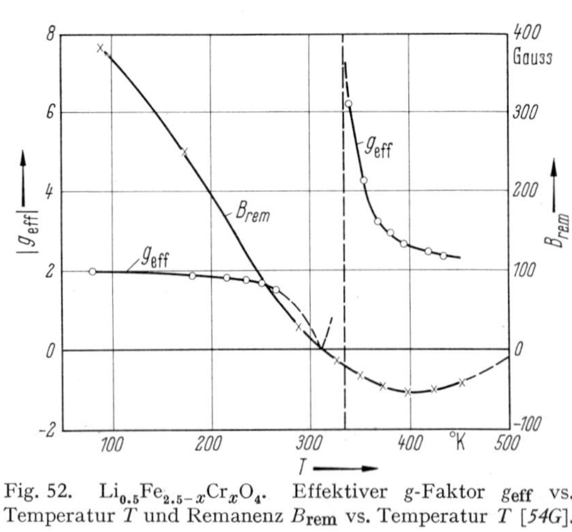

Fig. 52. $Li_{0.5}Fe_{2.5-x}Cr_xO_4$. Effektiver g-Faktor g_{eff} vs. Temperatur T und Remanenz B_{rem} vs. Temperatur T [54G].

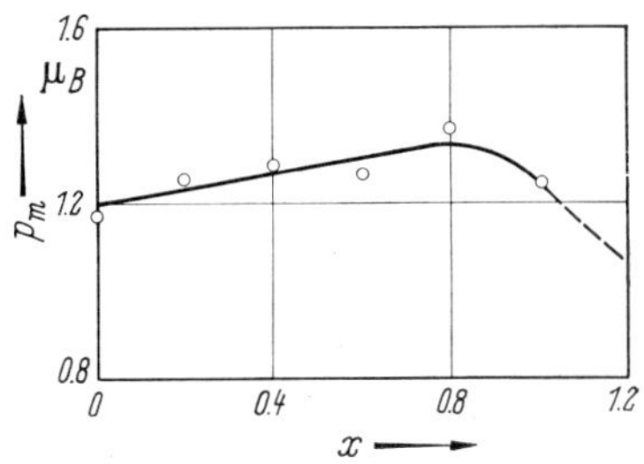

Fig. 53. $MnAl_xCr_{2-x}O_4$. Magnetisches Moment p_m vs. x bei 0° K [*59E*].

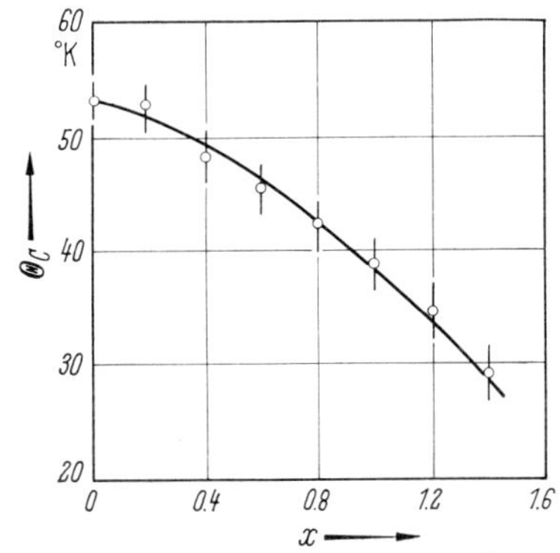

Fig. 54. $MnAl_xCr_{2-x}O_4$. Curie-Temperatur Θ_C vs. x [*59E*].

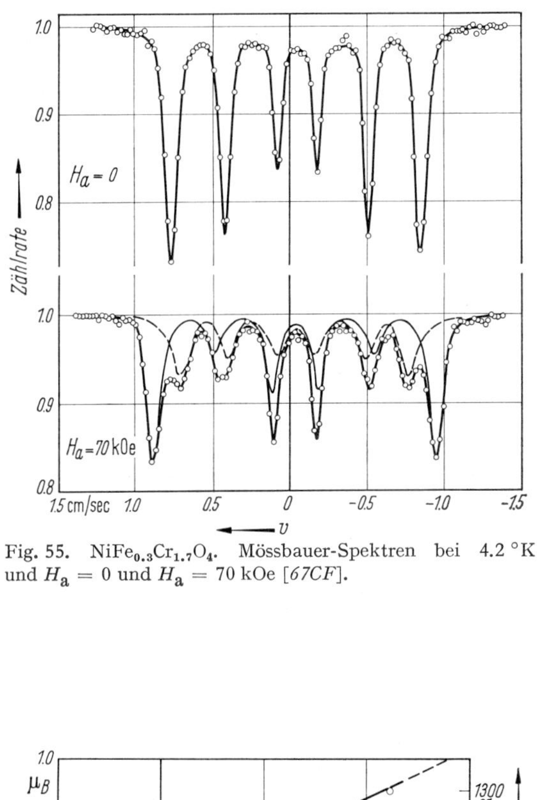

Fig. 55. $NiFe_{0.3}Cr_{1.7}O_4$. Mössbauer-Spektren bei 4.2 °K und $H_a = 0$ und $H_a = 70$ kOe [*67CF*].

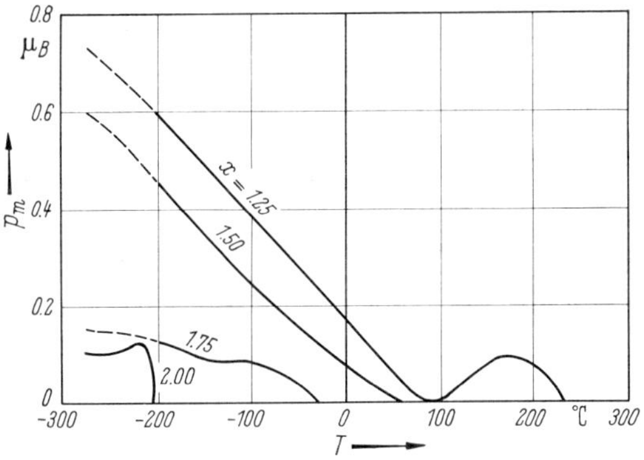

Fig. 56. $NiFe_{2-x}Cr_xO_4$. Magnetisches Moment p_m vs. Temperatur T für getemperte Proben mit $x > 1$ [*60MG*].

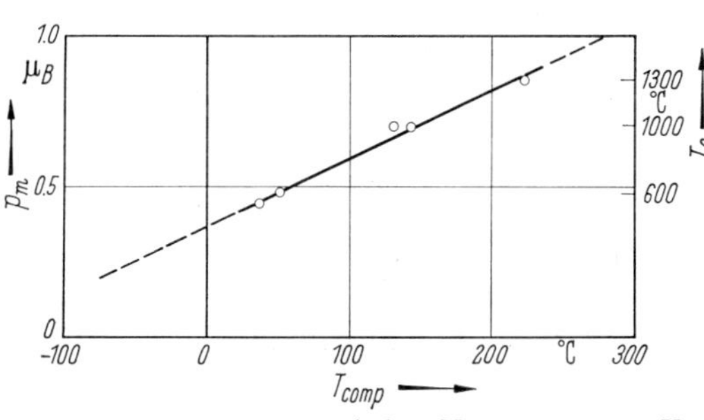

Fig. 57. $NiCrFeO_4$. Magnetisches Moment p_m vs. Kompensationstemperatur T_{comp} für verschiedene Proben (Abschrecktemperaturen T_Q an der rechten Ordinate) [*60MG*].

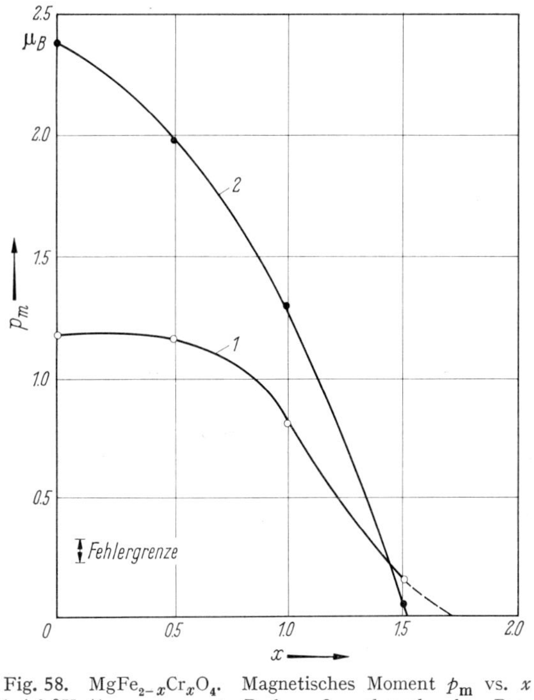

Fig. 58. $MgFe_{2-x}Cr_xO_4$. Magnetisches Moment p_m vs. x bei 0 °K (1 = getemperte Proben, 2 = abgeschreckte Proben) [*60MG*].

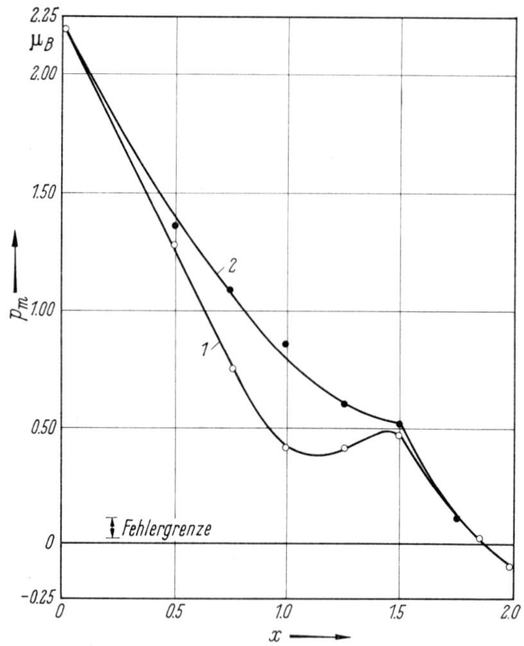

Fig. 59. $NiFe_{2-x}Cr_xO_4$. Magnetisches Moment p_m vs. x bei 0 °K (1 = getemperte Proben, 2 = abgeschreckte Proben) [*60MG*].

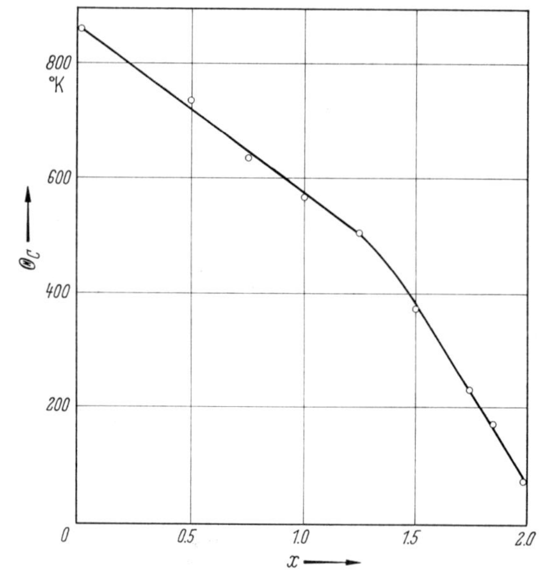

Fig. 60. $NiFe_{2-x}Cr_xO_4$. Curie-Temperatur Θ_C vs. x [*60MG*].

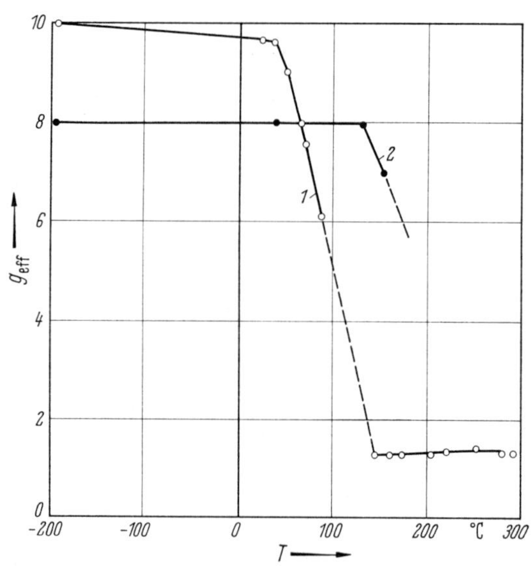

Fig. 61. $NiFeCrO_4$. Effektiver g-Faktor g_{eff} vs. Temperatur T (1 = getemperte Proben, 2 = abgeschreckte Proben) [*60MG*].

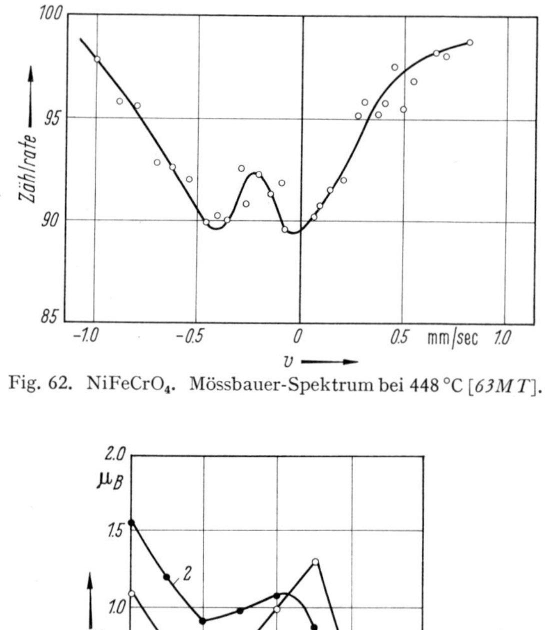

Fig. 62. $NiFeCrO_4$. Mössbauer-Spektrum bei 448 °C [*63MT*].

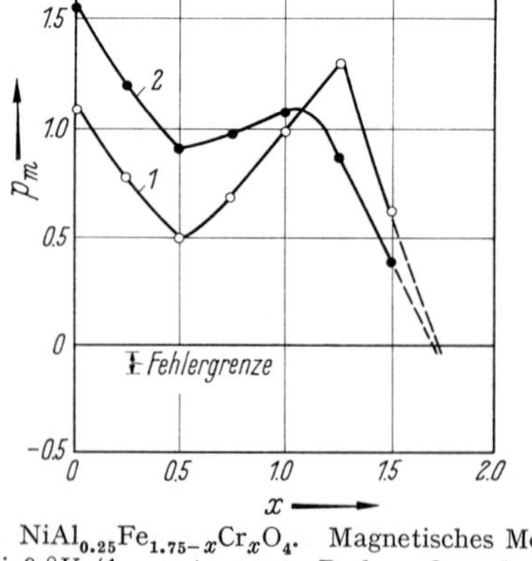

Fig. 63. $NiAl_{0.25}Fe_{1.75-x}Cr_xO_4$. Magnetisches Moment p_m vs. x bei 0 °K (1 = getemperte Proben, 2 = abgeschreckte Proben) [*60MG*].

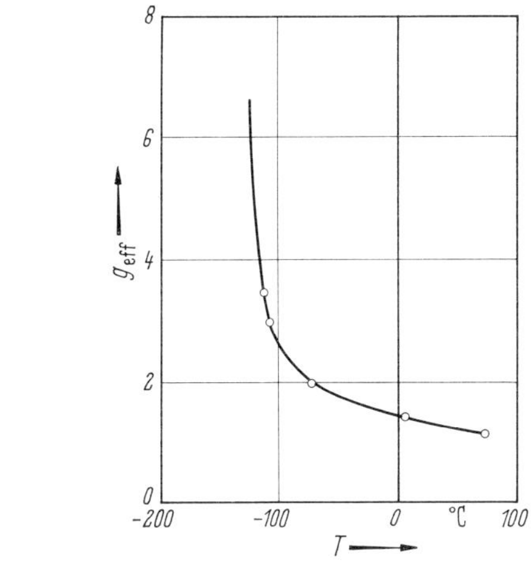

Fig. 64. $NiAl_{0.25}Fe_{0.5}Cr_{1.25}O_4$. Effektiver g-Faktor g_{eff} vs. Temperatur T [60MG].

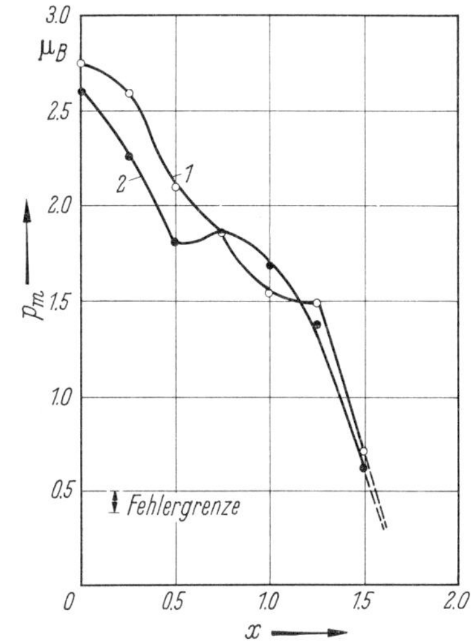

Fig. 65. $NiGa_{0.25}Fe_{1.75-x}Cr_xO_4$. Magnetisches Moment p_m vs. x bei 0 °K (1 = getemperte Proben, 2 = abgeschreckte Proben) [60MG].

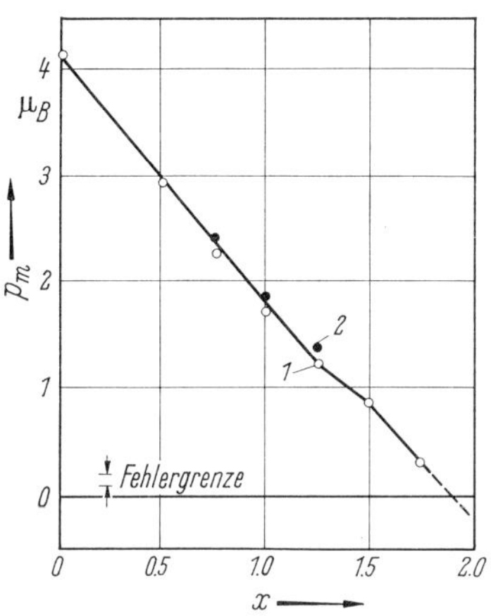

Fig. 66. $Ni_{0.75}Zn_{0.25}Fe_{2-x}Cr_xO_4$. Magnetisches Moment p_m vs. x bei 0 °K (1 = getemperte Proben, 2 = abgeschreckte Proben) [60MG].

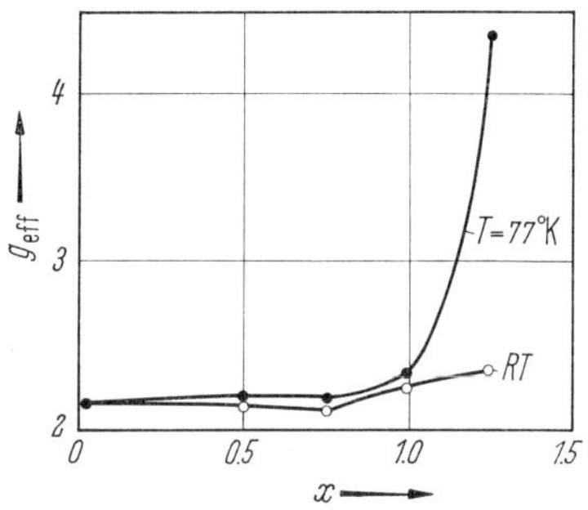

Fig. 67. $Ni_{0.75}Zn_{0.75}Fe_{2-x}Cr_xO_4$. Effektiver g-Faktor g_{eff} vs. x bei RT und 77 °K [60MG].

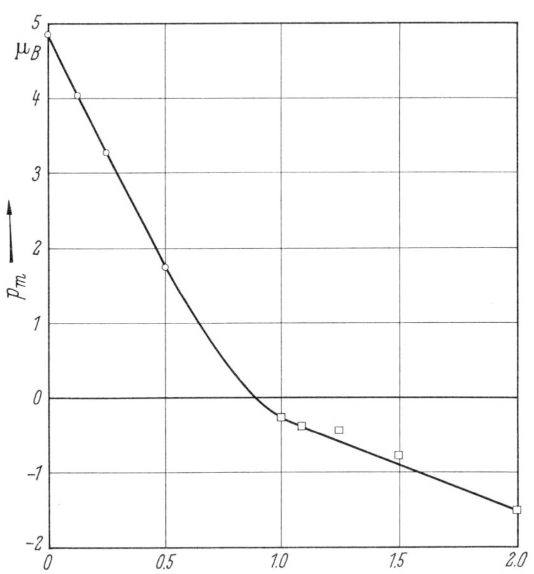

Fig. 68. $MnFe_{2-x}Cr_xO_4$. Magnetisches Moment p_m vs. x [54G].

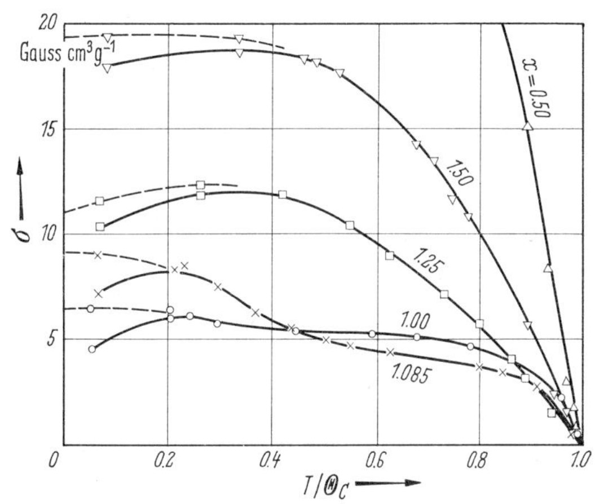

Fig. 69. $MnFe_{2-x}Cr_xO_4$. Magnetisierung σ vs. relative Temperatur T/Θ_C. Parameter x [54G].
($T > 77$ °K bei $H_a \leqq 6$ kOe,
$\left.\begin{array}{l}T = 77\text{ °K}\\ T = 20\text{ °K}\end{array}\right\}$ bei $H_a \leqq 23$ kOe),
– – – $H_a = 23$ kOe,
——— $H_a = 6$ kOe.

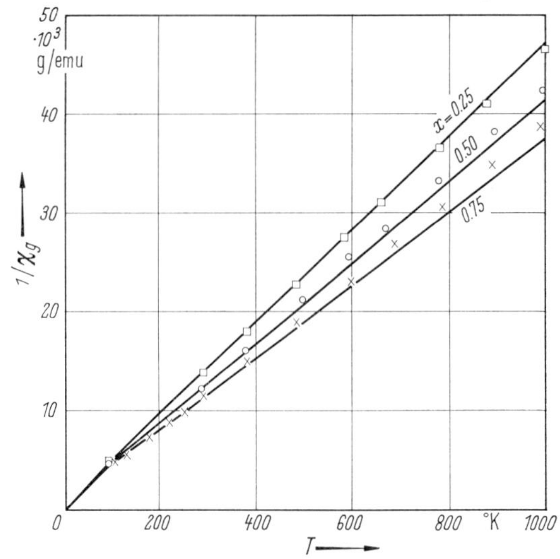

Fig. 70. $Ni_xMg_{1-x}Cr_2O_4$. Reziproke Suszeptibilität $1/\chi_g$ vs. Temperatur T. Parameter x [*62L*].

6.2.1.3 Literatur zu 6.2.1.1 und 6.2.1.2

27H-1	Holgersson, S.: Lunds Univ. Årsskr. N. F., Avd. 2, **23** (1927) Nr. 9.
27H-2	Holgersson, S.: Acta Univ. Lundensis (Leipzig) **23** (1927).
28P	Passerini, L.: Gazz. Chim. Ital. **58** (1928) 775.
29NP	Natta, G., and L. Passerini: Gazz. Chim. Ital. **59** (1929) 280.
29P	Passerini, L.: Rend. Accad. Nazl. Lincei **9** (1926) 338.
30H	Holgersson, S.: Z. Anorg. Allgem. Chem. **192** (1930) 123.
30P	Passerini, L.: Gazz. Chim. Ital. **60** (1930) 389.
31CAB	Clark, G. L., A. Ally, und A. E. Badger: Am. J. Sci., Ser. 5, **22** (1931) 539.
31E	Eilert, W.: Dissertation Universität Münster (1931).
31EH	Ewald, P. P., und C. Hermann: Strukturbericht 1913—26 **1** (1931) 350.
31S	Strukturbericht 1913—26 **1** (1931) 416.
32H	Holgersson, S.: Z. Anorg. Allgem. Chem. **204** (1932) 378.
32HN	Hauptmann, N., und G. Novak: Z. Phys. Chem. **15** (1932) 325.
32PBB	Parmelle, C. W., A. E. Badger, und G. A. Ballam: Univ. Illinois Bull. **29** (1932) Nr. 84.
34KT	Krause, O., und W. Thiel: Ber. Deut. Keram. Ges. **15** (1934) 101.
37S	Strukturbericht 1928—32 **2** (1937) 468.
38S	Strukturbericht 1936 **4** (1938).
47RR	Rüdorff, W., und B. Reuter: Z. Anorg. Allgem. Chem. **253** (1947) 194.
47VH	Verwey, E. J. W., und E. L. Heilmann: J. Chem. Phys. **15** (1947) 174.
47VHR	Verwey, E. J. W., P. W. Haayman, und F. C. Romeijn: J. Chem. Phys. **15** (1947) 181.
47···48SR	Structure Reports **11** (1947—48) 497.
51KR	Kordes, E., und E. Röttig: Z. Anorg. Allgem. Chem. **264** (1951) 34.
51L	Lovell, G. H. B.: Trans. Brit. Ceram. Soc. **50** (1951) 315.
51SR	Structure Reports **14** (1951) 209.
52MHS-1	McGuire, T. R., L. N. Howard, und J. S. Smart: Ceram. Age **60** (1952) 22.
52MHS-2, 52SMHS	Schindler, W. G., T. R. McGuire, L. N. Howard, und J. S. Smart: Phys. Rev. **86** (1952) 599.
53BR	Bradburn, T. E., und G. R. Rigby: Trans. Brit. Ceram. Soc. **52** (1953) 417.
53GS	Gorter, E. W., und J. A. Schulkes: Phys. Rev. **90** (1953) 487.
53R	Romeijn, F. C.: Philips Res. Rept. **8** (1953) 304.
53SR	Structure Reports **17** (1953) 413.
53Y	Yamaguchi, G.: Bull. Chem. Soc. Japan **26** (1953) 204.
54BD	Bertaut, F., und C. Delorme: Compt. Rend. **239** (1954) 504.
54E	Evans, R. C.: Einführung in die Kristallchemie, Barth Verlag, Leipzig 1954.
54G	Gorter, E. W.: Philips Res. Rept. **9** (1954) 295.

54RBR	RICHARDSON, H. M., F. BALL, und G. R. RIGBY: Trans. Brit. Ceram. Soc. **53** (1954) 376.
54SR	Structure Reports **18** (1954) 453.
54WK	WARSHAW, I., und M. L. KEITH: J. Am. Ceram. Soc. **37** (1954) 161.
54YKL	YEARIAN, H. J., J. M. KORTRIGHT, und R. H. LANGENHEIM: J. Chem. Phys. **22** (1954) 1196.
55D	DELORME, C.: Compt. Rend. **241** (1955) 1588.
55GL	GOODENOUGH, J. B., und A. L. LOEB: Phys. Rev. **98** (1955) 391.
55HFKSS	HAHN, H., G. FRANK, W. KLINGLER, A.-D. STÖRGER, und G. STÖRGER: Z. Anorg. Allgem. Chem. **279** (1955) 241.
56L	LOTGERING, F. K.: Philips Res. Rept. **11** (1956) 218.
56LA	LINDNER, R., und Å. ÅKKERSTRÖM: Z. Physik. Chem. (Frankfurt) **6** (1956) 162.
56MO	MIYAHARA, S., und Y. OHNISHI: J. Phys. Soc. Japan **11** (1956) 1296.
57B	BONGERS, P. F.: Thesis, Univ. Leiden, 4. 7. 1957, Verlag Excelsior, 's Gravenhage.
57BC	BEALS, R. J., und R. L. COOK: J. Am. Ceram. Soc. **40** (1957) 279.
57DO	DUNITZ, J. D., und L. E. ORGEL: J. Phys. Chem. Solids **3** (1957) 20.
57F	FRANCOMBE, M. H.: J. Phys. Chem. Solids **3** (1957) 37.
57G	GORTER, E. W.: Advan. Phys. **6** (1957) 336.
57M	MCCLURE, D. S.: J. Phys. Chem. Solids **3** (1957) 311.
57SS	SINHA, K. P., and A. P. B. SINHA: J. Phys. Chem. **61** (1957) 758.
58D	DELORME, C.: Bull. Soc. Franc. Mineral. **81** (1958) 150.
58DY	DERBYSHIRE, W. D., und H. J. YEARIAN: Phys. Rev. **112** (1958) 1603.
58E	EDWARDS, P. L.: Bull. Am. Phys. Soc. **3** (1958) 43.
58H	HANSEN, M.: Constitution of Binary Alloys, McGraw-Hill, New York 1958.
58MT	MIYAHARA, S., und T. TSUSHIMA: J. Phys. Soc. Japan **13** (1958) 758.
58P	PICKART, S. J.: Bull. Am. Phys. Soc. **3** (1958) 42.
58R	RINGWOOD, A. E.: Geochim. Cosmochim. Acta **15** (1958) 195.
59E	EDWARDS, P. L.: Phys. Rev. **116** (1959) 294.
59G	GOODENOUGH, J. B.: J. Phys. Radium **20** (1959) 155.
59PN	PICKART, S. J., und R. NATHANS: Phys. Rev. **116** (1959) 317.
59SR	Structure Reports **23** (1959) 383.
59W	WOJTOWICZ, P. J.: Phys. Rev. **116** (1959) 32.
59WG	WICHHAM, D. G., und J. B. GOODENOUGH: Phys. Rev. **115** (1959) 1156.
60H	HAFNER, S.: Schweiz. Mineral. Petrogr. Mitt. **40** (1960) 33.
60K	KAPLAN, T. A.: J. Appl. Phys. **31** (1960) 364.
60MG	MCGUIRE, T. R., und S. W. GREENWALD: Proc. Intern. Conf. Solid State Physics, Brussels 1958; London and New York: Academic Press (1960), Vol. 3.
61H	HAFNER, S.: Z. Krist. **115** (1961) 65.
61N	NAGATA, T.: Rock Magnetism, 1961 Maruzen Comp. Ltd. Tokyo.
61OT	OHNISHI, H., und T. TERANISHI: J. Phys. Soc. Japan **16** (1961) 35.
61P	PRINCE, E.: J. Appl. Phys. **32** (1961) 68 S.
61VP	VOITLÄNDER, J., und H. PROTSCHKA: Naturwissenschaften **48** (1961) 475.
61WWA	WHIPPLE, E. R., A. WOLD, und R. J. ARNOTT: Solid State Quart. Progr. Repl., Lincoln Laboratory, MIT, Cambridge, Mass., 15. April 1961.
62BG	BLASSE, G., und E. W. GORTER: J. Phys. Soc. Japan **17** Suppl. B 1 (1962) 176.
62CH	CORLISS, L. M., und J. M. HASTINGS: J. Appl. Phys. Suppl. **33** (1962) 1138.
62D-1	DULAC, J.: J. Phys. Radium **23** (1962) 872.
62D-2	DULAC, J.: Compt. Rend. **254** (1962) 1312.
62D-3	DULAC, J.: Bull. Soc. Franc. Mineral. Crist. **85** (1962) 93.
62G	GOODENOUGH, J. B.: J. Phys. Soc. Japan **17** Suppl. B 1 (1962) 185.
62L	LOTGERING, F. K.: J. Phys. Chem. Solids **23** (1962) 1153.
62MN	MOROZOV, A. N., und I. A. NOVOKHARSHI: Russ. Met. Fuels (English Transl.) **6** (1962) 3.
62MWRD	MENYUK, N., A. WOLD, D. ROGERS, und K. DWIGHT: J. Appl. Phys. Suppl. **33** (1962) 1144.
62T	TSUSHIMA, T.: J. Phys. Soc. Japan **17** Suppl. B-1 (1962) 189.
62WW	WHIPPLE, E., und A. WOLD: J. Inorg. Nucl. Chem. **24** (1962) 23.
63B-1	BLASSE, G.: J. Inorg. Nucl. Chem. **25** (1963) 230.
63B-2	BLASSE, G.: J. Inorg. Nucl. Chem. **25** (1963) 743.
63B-3	BLASSE, G.: Philips Res. Rept. **18** (1963) 400.
63BF	BLASSE, G., und J. F. FAST: Philips Res. Rept. **18** (1963) 393.
63G-1	GORTER, E. W.: J. Appl. Phys. **34** (1963) 1253.
63G-2	GOODENOUGH, J. B.: Magnetism and the Chemical Bond. New York: Interscience Publishers 1963.
63MT	MIZOGUCHI, P., und M. TANAKA: J. Phys. Soc. Japan **18** (1963) 1301.

63RNMP Rosenberg, M., P. Nicolau, R. Manaila, und P. Pausescu: J. Phys. Chem. Solids **24** (1963) 1419.

63WAWG Wold, A., R. J. Arnott, E. Whipple, und J. B. Goodenough: J. Appl. Phys. **34** (1963) 1085.

64AWR Arnott, R. J., A. Wold, und D. B. Rogers: J. Phys. Chem. Solids **25** (1964) 161.

64AMDK Alper, A. M., R. N. McNally, R. C. Doman, und F. G. Keihn: J. Am. Ceram. Soc. **47** (1964) 30.

64B-1 Blasse, G.: Philips Res. Rept. Suppl. No. **3** (1964).

64B-2 Blasse, G.: J. Inorg. Nucl. Chem. **26** (1964) 1473.

64BL Baltzer, P. K., und E. Lopatin: Proc. Intern. Conf. Magnetism, Nottingham (1964) 564.

64DMRW Dwight, K., N. Menyuk, D. B. Rogers, und A. Wold: Proc. Intern. Conf. Magnetism, Nottingham (1964) 538.

64G Goodenough, J. B.: J. Phys. Chem. Solids **25** (1964) 151.

64HH Heeger, A., und P. W. Houston: Proc. Intern. Conf. Magnetism., Nottingham (1964) 395.

64MDW Menyuk, N., K. Dwight, und A. Wold: J. Phys. (Paris) **25** (1964) 528.

64MH Miyahara, S., und S. Horiuti: Proc. Intern. Conf. Magnetism, Nottingham (1964) 550.

64RLV Rault, M., A. Lecerf, und G. Villers: Compt. Rend. **258** (1964) 4553.

64VLR Villers, G., A. Lecerf, und M. Rault: Proc. Intern. Conf. Magnetism, Nottingham (1964) 639.

65AR Albers, W., und C. J. M. Rooymans: Solid State Commun. **3** (1965) 417.

65B Blasse, G.: Philips Res. Rept. **20** (1965) 528.

65H Haas, C.: J. Phys. Chem. Solids **26** (1965) 1225.

65KM Kino, Y., und S. Miyahara: J. Phys. Soc. Japan **20** (1965) 1522.

65LRPV Lecerf, A., M. Rault, J. Portier und G. Villers: Bull. Soc. Chim. (Paris) **1965**, 1208.

65MSY Mathur, H. B., A. P. B. Sinha, und C. M. Yagnik: Solid State Commun. **3** (1965) 401.

65R Rossiter, M. J.: J. Phys. Chem. Solids **26** (1965) 775.

65RB O'Reilly, W., und S. K. Banerjee: Phys. Letters **17** (1965) 237.

65RGA Rogers, D. B., R. W. Germann, und R. J. Arnott: J. Appl. Phys. **36** (1965) 2338.

65SA Siratori, K., und Y. Aiyama: J. Phys. Soc. Japan **20** (1965) 1962.

65SK Schmitz-DuMont, O., und H. Kasper: Z. Anorg. Allgem. Chem. **341** (1955) 252.

65STT Siratori, K., A. Tsuchida, und Y. Tomono: J. Appl. Phys. **36** (1965) 1050.

66A Allen, W. C.: Am. Mineralogist **51** (1966) 239.

66B Blasse, G.: J. Phys. Chem. Solids **27** (1966) 383.

66BWRL Baltzer, P. K., P. J. Wojtowicz, M. Robbins, und E. Lopatin: Phys. Rev. **151** (1966) 367.

66DMFW Dwight, K., N. Menyuk, J. Feinleib, und A. Wold: J. Appl. Phys. **37** (1966) 962.

66HGLRV Hagenmuller, P., Ch. Guillaud, A. Lecerf, M. Rault und G. Villers: Bull. Soc. Chim. France **1966**, 2589.

66JD Joubert, J.-C., und A. Durif: Bull. Soc. Franc. Mineral. Crist. **89** (1966) 26.

66MDAW Menyuk, N., K. Dwight, R. J. Arnott, und A. Wold: J. Appl. Phys. **37** (1966) 1387.

66NGS Novokhatskiy, I. A., R. L. Gershman, und A. A. Savinskaya: Russ. Met. Fuels (English Transl.) **1966**, 13.

66R Reinen, D.: Theoret. Chim. Acta **5** (1966) 312.

66RBW Raccah, P. M., R. J. Bouchard, und A. Wold: J. Appl. Phys. **37** (1966) 1436.

66RD Robbins, M., und L. Darey: J. Phys. Chem. Solids **27** (1966) 741.

66S Stone, F. S.: Bull. Soc. Chim. France **1966**, 819.

66TTA Tanaka, M., T. Tokoro, und Y. Aiyama: J. Phys. Soc. Japan **21** (1966) 262.

67CF Chappert, J., und R. B. Frankel: Phys. Rev. Letters **19** (1967) 570.

67DR Datta, R. K., und R. Roy: J. Am. Ceram. Soc. **50** (1967) 578.

67G Grenot, M.: Bull. Soc. Chim. France **1967**, 2043.

67KBC Korneev, Y. A., V. F. Balakirev, und G. I. Chufarov: Russ. Met. Fuels (English Transl.) **1967**, 30.

67KGW Kirchner, H. P., R. M. Gruver, und R. E. Walker: J. Am. Ceram. Soc. **50** (1967) 169.

67KS Kljucarov, J. V., und S. A. Suvorov: Ogneupory. Moskau **32** [5] (1967) 42.

67NK Navrotsky, A., und O. J. Kleppa: J. Inorg. Nucl. Chem. **29** (1967) 2701.

67P Plumier, R.: Compt. Rend. **265 B** (1967) 672.

67R Reinen, D.: Theoret. Chim. Acta **8** (1967) 260.

67S Siratori, K.: J. Phys. Soc. Japan **23** (1967) 948.

68A Allen, W. C.: J. Am. Ceram. Soc. **51** (1968) 485.

68DR	Datta, R. K., und R. Roy: Am. Mineralogist **53** (1968) 1456.
68HI	Hartmann-Boutron, F., und P. Imbert: J. Appl. Phys. **39** (1968) 775.
68KKI	Kanamori, J., M. Kataoka und Y. Itoh: J. Appl. Phys. **39** (1968) 688.
68MRW	Muller, O., R. Roy und W. B. White: J. Am. Ceram. Soc. **51** (1968) 693.
68NK	Navrotsky, A., und O. J. Kleppa: J. Inorg. Nucl. Chem. **30** (1968) 479.
68P-1	Plumier, R.: Compt. Rend. **267 B** (1968) 98.
68P-2	Plumier, R.: J. Appl. Phys. **39** (1968) 635.
68TKF	Tsushima, T., Y. Kino, und S. Funahashi: J. Appl. Phys. **39** (1968) 626.
69GS	Greskovich, C., und V. S. Stubican: J. Phys. Chem. Solids **30** (1969) 909.
69IIM	Isaji, H., K. Iwahashi und Y. Masuda: J. Phys. Soc. Japan **27** (1969) 503.
69PM	Pausescu, P., und R. Manaila: Soviet Phys.-Cryst. (English Transl.) **13** (1969) 533.
69PS	Plumier, R., und M. Sougi: Compt. Rend. **268 B** (1969) 365.

6.2.2 Cr sulfide, selenide, and telluride spinels –

See p. 619

Cr Sulfid-, Selenid- und Tellurid-Spinelle

Siehe S. 619

6.3 Andere Spinelle — Further spinels

For general remarks and special explanations see 6.0 (p. 51) and 6.1.0 (p. 54)

6.3.1 V Spinelle und substituierte V Spinelle

(Literatur siehe 6.3.4, Seite 444)

6.3.1.1 V Spinelle mit V^{3+}

Tab. 1. Kristallografische Eigenschaften binärer V^{3+} Oxid-Spinelle

Verbindung	Kationenverteilung	a [1]) Å	ϱ_x g/cm³	ϱ_p g/cm³	u	Lit.	Fig.	Tab.
CdV_2O_4 3)		8.68 ± 0.01				*60RM*		
		8.695 ± 0.003				*68R-2*		
						63BPM		
CoV_2O_4 4)	$Co^{2+}[V_2^{3+}]O_4$	8.39	5.02	4.93		*49AB, 56RWB*	1, 5	2, 5
		8.393 ± 0.005				*59L, 62MWRD*	6	
						57LM		
		8.407 ± 0.002				*63RAWG, 64B-2*		
						66TS, 63BPM-1		
						64DMRW		
FeV_2O_4 4)	$Fe^{2+}[V_2^{3+}]O_4$	8.44	4.86	4.64		*49AB, 57LM*	2	2
		8.45	4.89			*51L, 54RBR*		
		8.439 ± 0.0005				*59L*		
		8.454 ± 0.002				*63RAWG*		
		8.455 ± 0.002				*66TS*		
		2)				*62WRAM, 67S*		
						65R-2, 66TTA		
						68VBBMC		
LiV_2O_4 6)	$Li[V_2]O_4$					*67RGG*		2
MgV_2O_4 4)	$Mg^{2+}[V_2^{3+}]O_4$	8.394 ± 0.006	4.246	4.23	0.389 ± 0.004	*47RR-1, 47RR-2*	3, 4	2, 3
		8.420	4.29	4.24		*54RBR, 51L*		
		8.413			0.385	*64B-2*		
		8.418 ± 0.002			0.3848 ± 0.0004	*63RAWG, 63PT*		
		8.416 ± 0.002				*60RJ, 63BF, 63G*		
		8.423 ± 0.006				*68C, 66TS*		
MnV_2O_4 5)	$Mn^{2+}[V_2^{3+}]O_4$	8.46	4.76	4.70		*49AB, 47RR-1*	1	2
			4.79	4.76		*51L*	7	
		8.520				*54RBR*		
		8.510 ± 0.0005			0.3883 ± 0.0002	*59L, 62P, 57LM*		
		8.522 ± 0.002				*63RAWG, 66TS*		
		8.522			0.3904	*64DMRW,*		
						62MWRD *)		
ZnV_2O_4 4)	$Zn^{2+}[V_2^{3+}]O_4$	8.393 ± 0.006	5.16	5.10	0.389 ± 0.004	*47RR-1, 47RR-2*		2, 3
			5.17	5.02		*49AB*		
			5.21	5.19		*51L*		
		8.436				*54RBR*		
		8.410 ± 0.002				*63RAWG, 64B-2*		
		8.409 ± 0.002				*66TS* **)		

1) Ausführliche Angaben über die Veränderung der Gitterkonstanten von festen Lösungen dieser Vanadite mit anderen Spinellen (Fe, Cr, Al) finden sich in [*51L*] und [*54RBR*].
2) FeV_2O_4 ist bei hohen Temperaturen kubisch und wird unterhalb 127 °K tetragonal verzerrt mit $c/a = 1.014$ [*63RAWG*].
3) Tiefschwarz [*60RM*].
4) Schwarz [*56RWB*], [*51L*], [*47RR-1*].
5) Dunkelbraun [*51L*].
6) Unter 500 ··· 700 °C bei 3000 at Spinellphase [*67RGG*].
*) Siehe auch [*66VL*], [*68P*].
**) Siehe auch [*63G*], [*69APKD*].

Tab. 2. Magnetische und elektrische Eigenschaften binärer V^{3+} Oxid-Spinelle

Verbindung	q eV	S μV/°K	Θ_C °K	Θ_a °K	Θ_N °K	p_m μ_B	Lit.	Fig.
CoV_2O_4	0.07 ± 0.005	+ 86	145			1.33 (4.2 °K)	*63RAWG* **)	1, 5, 6
FeV_2O_4	0.25 ± 0.01	+400	109[1])			1.95 (4.2 °K)	*63RAWG, 62WRAM* *)	2
LiV_2O_4 [7])							*67RGG*	
MgV_2O_4	0.18 ± 0.01	−304					*63RAWG*	3, 4
				−750	45	3.43 [3])	*63BF, 47RR-1*	
				−600	>80 [2])		*63G, 63PT, 68C*	
MnV_2O_4 [4])	0.37 ± 0.01		56			2.1 (4.2 °K) [5])	*63RAWG, 64DMRW*	1
							62P, 68P **)	
ZnV_2O_4	0.16 ± 0.01	−277					*63RAWG*	
							47RR-1	
				−600	[6])		*63G*	

Tab. 3. Suszeptibilität χ_g und $\chi_{m\,korr.}$ und effektives Moment p_{eff} in Abhängigkeit von der Temperatur T für verschieden hergestellte Proben von MgV_2O_4 und ZnV_2O_4 [*47RR-1*]

Verbindung	T °K	χ_g 10^{-6} emu/g	$\chi_{m\,korr.}$ 10^{-6} emu/mole	p_{eff} [10]) μ_B
ZnV_2O_4 I	293	11.1	1322 [8])	1.77
(10 h 750 °C)	195	12.2	1450	1.51
	90	13.5	1600	1.08
ZnV_2O_4 II	293	12.0	1428	1.84
(48 h 750 °C)	195	12.9	1530	1.55
	90	14.7	1738	1.12
ZnV_2O_4 III	293	11.4	1358	1.79
(144 h 750 °C)	195	12.5	1484	1.53
	90	14.5	1716	1.12
MgV_2O_4 I	293	14.7 [9])	1432	1.84
(5 h mit KCl 900 °C)	195	16.2	1576	1.57
	90	17.6	1707	1.11
MgV_2O_4 II	293	14.4	1404	1.82
(48 h 900 °C)	195	15.2	1480	1.52
	90	17.3	1681	1.11

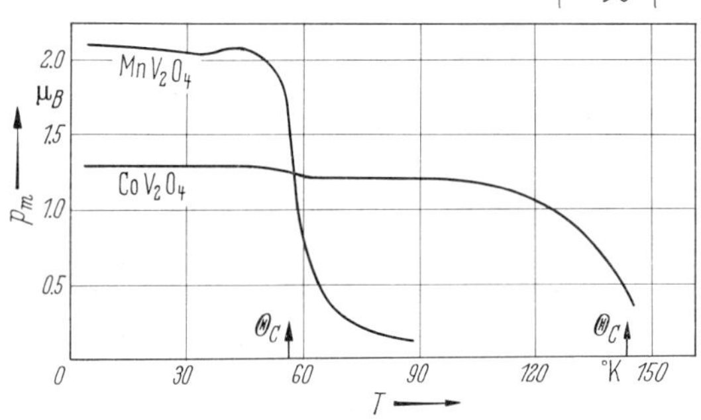

Fig. 1. CoV_2O_4 und MnV_2O_4. Sättigungsmagnetisierung pro Molekül p_m bei H = 11 kOe. Die eingezeichneten Curie-Punkte stützen sich auf Messungen bei einer Feldstärke von 0.1 kOe [*62MWRD*].

[1]) FeV_2O_4 ist unterhalb Θ_C = 109 °K ferrimagnetisch. Bei 70 °K erfolgt ein Übergang der Néel-Konfiguration in eine kompliziertere Konfiguration mit verkanteten Spins (beschrieben in [*62MWRD*]) [*62WRAM*].
[2]) Nach [*63PT*]; nach [*63BF*] unterhalb 45° K antiferromagnetisch.
[3]) p_{eff}, ermittelt ausSuszeptibilitätsmessungen bei hohen Temperaturen [*63BF*].
[4]) MnV_2O_4 ist magnetisch anisotrop. Die schwere Richtung der Magnetisierung ist [111] [*62P*].
[5]) Eine Neutronenbeugungsanalyse führt zu p_m = 3.1 μ_B [*64DMRW*].
[6]) ZnV_2O_4 ist antiferromagnetisch [*63G*].
[7]) Messungen der elektrischen Leitfähigkeit [*67RGG*].
[8]) Die Ioneninkremente zur Berechnung von $\chi_{m\,korr.}$ sind V^{3+}: $-10 \cdot 10^{-6}$; Mg^{2+}: $-3 \cdot 10^{-6}$ u. Zn^{2+}: $-10 \cdot 10^{-6}$ emu/mole [*47RR-1*].
[9]) Fig. 3 [*63BF*].
[10]) Ermittelt aus Suszeptibilitätsmessungen in verschiedenen Temperaturbereichen [*47RR-1*].
*) Siehe auch [*65R-2*].
**) Siehe auch [*62HWRD*].

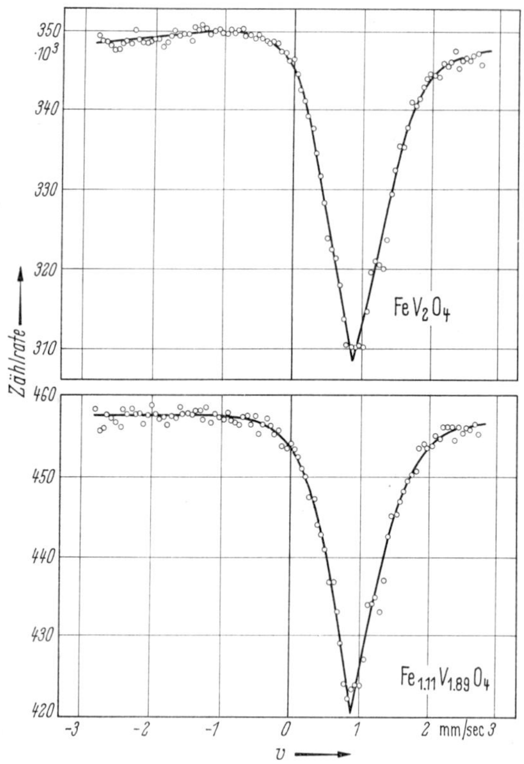

← Fig. 2. FeV_2O_4 und $Fe_{1.11}V_{1.89}O_4$. Mössbauer-Spektren bei Raumtemperatur [*65R-2*].

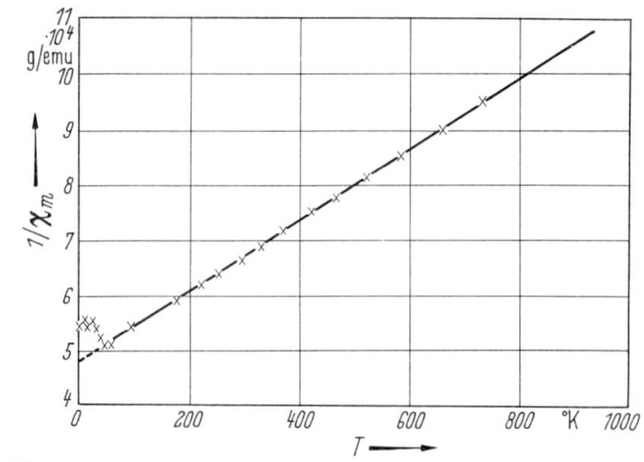

Fig. 3. MgV_2O_4. Reziproke Suszeptibilität pro Gramm, χ_g^{-1}, vs. Temperatur. $\Theta_N = 45$ °K, $p_{eff} = 3.43\,\mu_B$ [*63BF*].

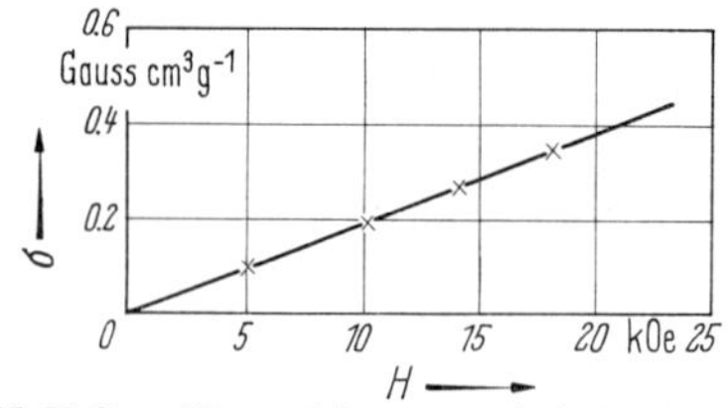

Fig. 4. MgV_2O_4. Magnetisierung σ bei 4.2 °K vs. Feldstärke H [*63BF*].

Tab. 4. Kristallografische Eigenschaften ternärer V^{3+} Oxid-Spinelle

Verbindung	Kationenverteilung	a Å	Lit.	Fig.	Tab.
$Al_xCoV_{2-x}O_4$	$Co^{2+}[Al_x^{3+}V_{2-x}^{3+}]O_4$	Tab. 5	*63RAWG*	5, 6	5
$CoFe_{2-x}V_xO_4$	$Co^{2+}[Fe_{2-x}^{3+}V_x^{3+}]O_4$	8.389 [1])	*59L*	[5])	
$Co_{0.5}Mn_{0.5}V_2O_4$		8.461 ± 0.002	*66TS*		
$FeFe_{2-x}V_xO_4$	$Fe^{2+}[Fe_{2-x}^{3+}V_x^{3+}]O_4$	8.415 [1])	*59L*	[5])	
$FeZnVO_4$ [2])	$Zn^{2+}[Fe^{3+}V^{3+}]O_4$	8.424 ± 0.003	*68R-2*		
$LiTiVO_4$	$Li^{1+}[Ti^{4+}V^{3+}]O_4$	8.35	*64B-2*		
$Li_{0.5}Fe_{2.5-2.5x}V_{2.5x}O_4$	$Li_{0.5}^{1+}[Fe_{2.5-2.5x}^{3+}V_{2.5x}^{3+}]O_4$	Tab. 6	*65L*		6
$MnFe_{2-x}V_xO_4$	$Mn^{2+}[Fe_{2-x}^{3+}V_x^{3+}]O_4$	8.57 [1])	*59L*	[5])	
$Mn_{1+x}V_{2-2x}Ti_xO_4$	$Mn_{1+x}^{2+}[V_{2-2x}^{3+}Ti_x^{4+}]O_4$	Fig. 7	*65VLR*	7···10	
$Mn_xZn_{1-x}V_2O_4$		8.460 ± 0.002 [3]) 8.502 ± 0.002 [4])	*66TS* *66TS*		

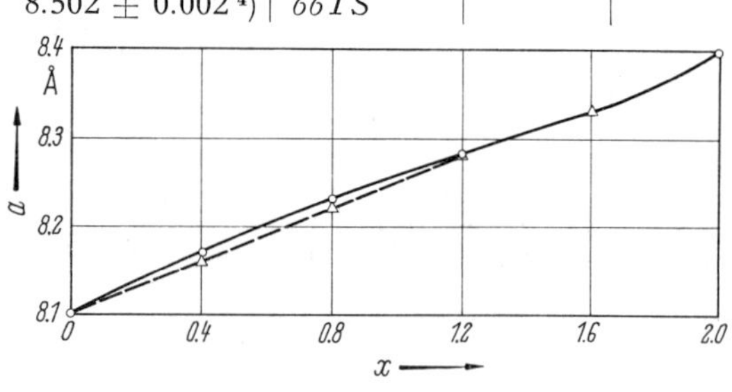

Fig. 5. $Al_xCoV_{2-x}O_4$. Gemessene (○–○) und nach dem Végardschen Gesetz durch Interpolation berechnete (△– –△)-Gitterkonstanten a von polykristallinen Proben vs. Zusammensetzungsparameter x [*63RAWG*].

[1]) für $x = 1$ [*59L*].
[2]) $FeZnVO_4$ hat einen Sauerstoff-Parameter von $u = 0.385 \pm 0.001$ und bildet Mischkristalle mit a) LiV_2O_4 der allgemeinen Zusammensetzung: $Li_{x-y}Zn_{1-x}Fe_y^{3+}[Li_yFe_{1-x-y}^{3+}V_{1.0}^{3+}V_x^{4+}]O_4$ mit $0 \le x \le 1$ und y entsprechend annähernd statistischer Verteilung; b) Fe_3O_4 der allgemeinen Zusammensetzung $Zn_{1-x}Fe_x^{3+}[Fe_x^{2+}Fe_{1.0}^{3+}V_{1-x}^{3+}]O_4$ mit $0 \le x \le 1$ [*68R-2*].
[3]) für $x = 0.5$ [*66TS*].
[4]) für $x = 0.8$ [*66TS*].
[5]) Weitere Angabe: Θ_C vs. x [*57LM*].

Tab. 5. Kristallografische und elektrische Eigenschaften von Verbindungen der Zusammensetzungsreihe $Al_xCoV_{2-x}O_4$ [*63RAWG*]

Verbindung	a [1]) Å	a [2]) Å	q eV
CoV_2O_4	8.401 ± 0.002	—	0.070 ± 0.005
$CoV_{1.5}Al_{0.5}O_4$	8.332 ± 0.002	8.332	0.120 ± 0.005
$CoV_{1.2}Al_{0.8}O_4$	8.286 ± 0.002	8.286	0.182 ± 0.010
$CoV_{0.8}Al_{1.2}O_4$	8.233 ± 0.002	8.228	0.339 ± 0.010
$CoV_{0.4}Al_{1.6}O_4$	8.170 ± 0.002	8.165	0.594 ± 0.015
$CoAl_2O_4$	8.103 ± 0.002	—	—

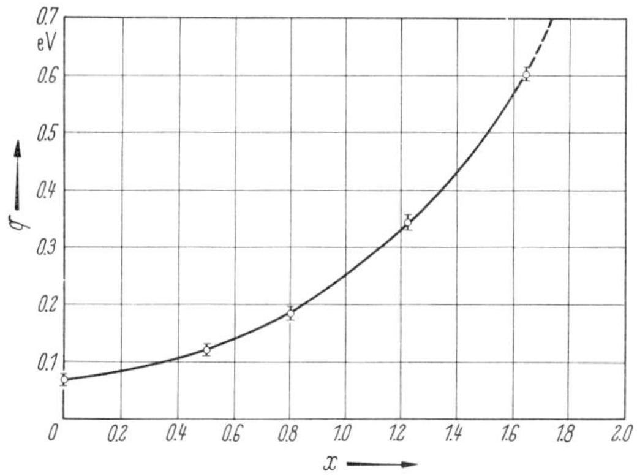

Fig. 6. $Al_xCoV_{2-x}O_4$. Aktivierungsenergie q der Gleichstromleitfähigkeit gemessen an gesinterten Proben vs. Zusammensetzungsparameter x [*63RAWG*].

Tab. 6. Kristallografische und magnetische Eigenschaften von Verbindungen der Zusammensetzungsreihe $Li_{0.5}Fe_{2.5-2.5x}V_{2.5x}O_4$ [*65L*]

Verbindung	Kationenverteilung	x	a Å langsam abgekühlt	a Å abgeschreckt	Θ_C °K	p_{eff} μ_B langsam abgekühlt	p_{eff} μ_B abgeschreckt
$Li_{0.5}Fe_{2.5-2.5x}V_{2.5x}O_4$	$Li^{1+}_{0.5}[Fe^{3+}_{2.5-2.5x}V^{3+}_{2.5x}]O_4$	0	8.313	8.313	620	2.65	2.98
		0.1	8.338	8.338	566	1.70	2.20
		0.2	8.349	8.349	484	1.12	1.62
		0.3	8.352	8.352	368	0.55	0.70
		0.4	8.348	8.346	252	0.46	0.30
		0.5	8.340	8.344	154	1.01	0.84
		0.6	8.326	8.320	56	1.40	1.00
		0.7	8.310	8.302	—	—	—
		0.85	8.280	8.270	—	—	—
		1	8.235	8.235	—	—	—

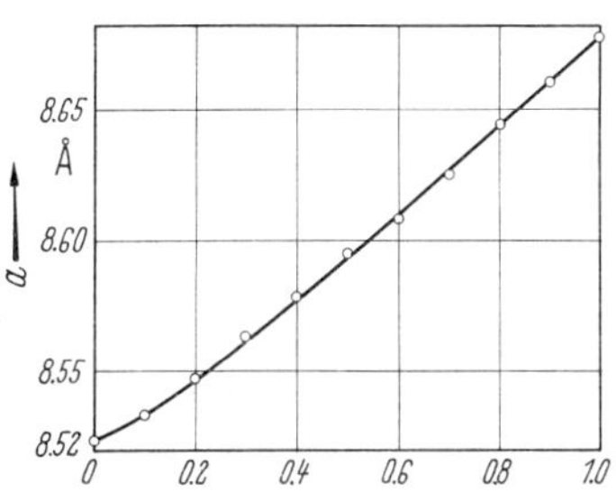

Fig. 7 $Mn_{1+x}V_{2-2x}Ti_xO_4$. Gitterkonstante a vs. Zusammensetzungsparameter x [*65VLR*].

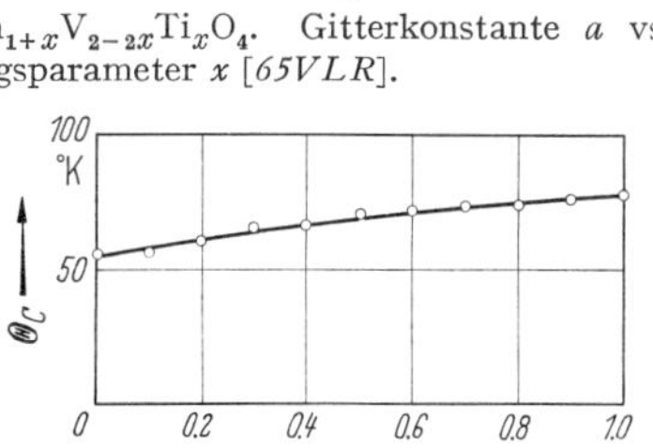

Fig. 8. $Mn_{1+x}V_{2-2x}Ti_xO_4$. Curie-Temperatur Θ_C vs. Zusammensetzungsparameter x [*65VLR*].

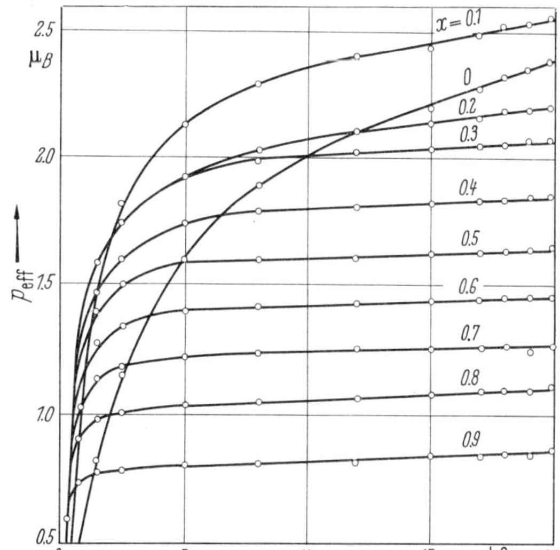

Fig. 9. $Mn_{1+x}V_{2-2x}Ti_xO_4$. Effektives magnetisches Moment pro Molekül, p_{eff}, vs. äußeres Feld H_{ext} bei 20.4 °K für verschiedene Proben aus dieser Zusammensetzungsreihe [*65VLR*].

[1]) gemessene Werte.
[2]) nach dem Végardschen Gesetz berechnete Werte.

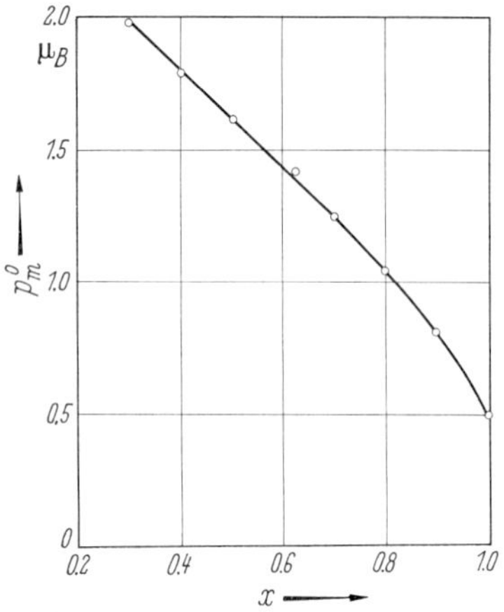

Fig. 10. $Mn_{1+x}V_{2-2x}Ti_xO_4$. Auf 0 °K extrapoliertes molares Moment p_m^0 vs. Zusammensetzungsparameter x [*65VLR*].

Tab. 7. CuV_2S_4. Kristallografische und elektrische Eigenschaften

	CuV_2S_4	Lit.
Kationen-verteilung	$Cu^{2+}[V_2^{3+}]S_4$	*56HLH*
a [Å]	9.804 ± 0.008	*56HLH*
	9.824	*64B-2*
	9.808 ± 0.002	*65BRW*
	9.82	*67MSL*
u	0.384	*56HLH*
ϱ_X [g/cm³]	4.120	*56HLH*
ϱ_p [g/cm³]	4.06	*56HLH*
ϱ [Ω cm] (RT)	$6 \cdot 10^{-4}$ (metallisch)	*65BRW*
S [μV/°C]	+5.2 bei 25 °C	*65BRW*
Θ_c [°K]	4.45 ··· 3.95 für 80 bis 90 Vol% der Probe [1])	*67MSL*

6.3.1.2 V Spinelle mit V^{4+}

Tab. 8. Kristallografische Eigenschaften binärer V^{4+} Oxid-Spinelle

Verbindung	Kationenverteilung	a Å	ϱ_X g/cm³	ϱ_p g/cm³	u	Lit.	Fig.	Tab.
Co_2VO_4	$Co^{2+}[Co^{2+}V^{4+}]O_4$	8.382			0.380 ± 0.003	*63BPM-1*, *62BPM*	11, 12	9
		8.379 ± 0.002				*63RAWG* *56RWB*, *60MDW*		
Fe_2VO_4	$Fe^{2+}[Fe^{2+}V^{4+}]O_4$	8.421 ± 0.002			0.378 ± 0.004	*63RAWG*, *67BP*	13, 14	9
		8.417 ± 0.003				*62WRAM*, *64B-2* *65R-2*, *65R-1*		
Mg_2VO_4 [2])	$Mg^{2+}[Mg^{2+}V^{4+}]O_4$	8.386 ± 0.006	3.66	3.55	0.386 ± 0.004	*47RR-1*, *47RR-2*	15	9, 10
		8.36				*64B-2*		
Mn_2VO_4	$Mn_{0.80}V_{0.20}[Mn_{1.20}V_{0.80}]O_4$	8.575			0.382 ± 0.004	*63BPM-1*, *63BPM-2* *67RR*, *66VL*	16	9
Zn_2VO_4	$Zn^{2+}[Zn^{2+}V^{4+}]O_4$					*64B-2*		
		8.384	5.11	4.73		*47RR-1*, *47RR-2*		
	$Zn_{0.98}V_{0.02}[Zn_{1.02}V_{0.98}]O_4$	8.395			0.375 ± 0.003	*63BPM-1*		

[1]) Scharfer Übergang von Normalleitung zu Supraleitung für 80 ··· 90% der Probe; der restliche Teil wird supraleitend bei Θ_c = 3.95 ··· 3.20 °K. Röntgenaufnahmen zeigen neben reiner Spinellphase Cu_2VS_4 ca. 5 ··· 10 Vol.% Cu_3VS_4 in der Probe [*67MSL*].

[2]) Mg_2VO_4 ist schwarz [*47RR-1*].

Tab. 9. Magnetische und elektrische Eigenschaften binärer V^{4+} Oxid-Spinelle

Verbindung	Θ_C °K	Θ_a °K	C_m cm³ °K/mole	q eV	S μV/°K	p_m μ_B	Lit.
Co_2VO_4	158			0.37 ± 0.01	−167	0.21 (4.2 °K) Fig. 11, 12	*60MDW*, *63RAWG*
Fe_2VO_4	440 [1]) 458 ± 3	−550 ± 20	8.60 ± 0.1	0.27 ± 0.01	− 50	0.72 (4.2 °K) 0.76 (4.2 °K) Fig. 13, 14	*62WRAM* *67BP*
Mg_2VO_4						Tab. 10 Fig. 15	*47RR-1*
Mn_2VO_4	62 ± 3	≈ −310	8.975			Fig. 16	*63BPM-2*

Tab. 10. Suszeptibilität χ_g und $\chi_{m\,korr.}$ und effektives Moment p_{eff} in Abhängigkeit von der Temperatur T für verschieden behandelte Proben der Verbindung Mg_2VO_4 [*47RR-1*]

Verbindung	T °K	χ_g 10⁶ emu/g I	II	Mittel	$\chi_{m\,korr.}$ [2]) 10⁶ emu/mole	p_{eff} μ_B
Mg_2VO_4	293	5.0	5.3	5.15	909	1.47
Mg_2VO_4 48 h 1000 °C nachbehandelt	195	6.0	6.2	6.1	1057	1.29
Mg_2VO_4 72 h 1000 °C nachbehandelt	90	8.8	9.3	9.1	1548	1.06

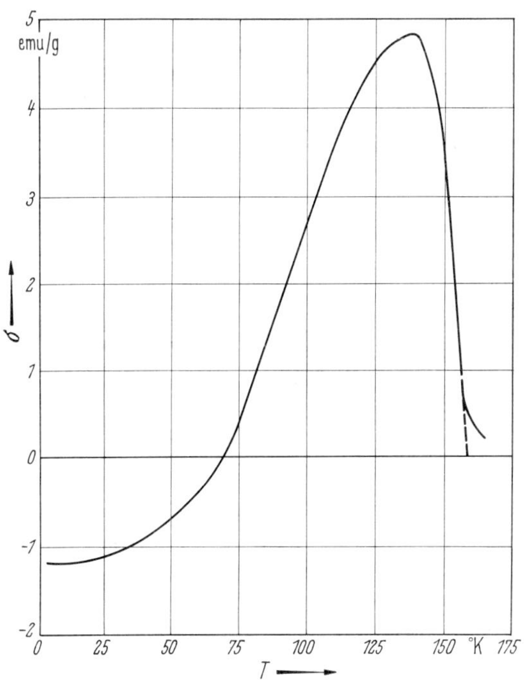

Fig. 11. Co_2VO_4. Magnetisierung σ vs. Temperatur T bei H_{ext} = 700 Oe für eine polykristalline Probe [*60MDW*].

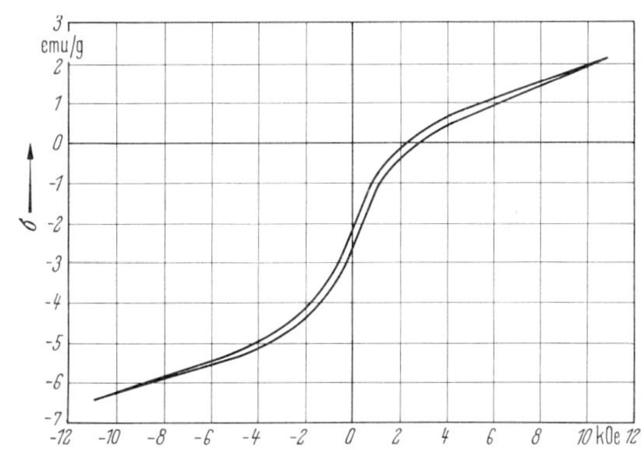

Fig. 12. Co_2VO_4. Hysteresisschleife der Magnetisierung σ einer im Feld abgekühlten Probe bei 4.2 °K. Die Feldstärke bei Abkühlung der Probe lag in positiver Richtung. Die Unsymmetrie ist feldrichtungs- und temperaturabhängig [*60MDW*].

[1]) Fe_2VO_4 ist unterhalb 440 °K ferrimagnetisch. Bei T = 275 °K erfolgt ein Übergang von der Néel-Konfiguration in eine kompliziertere Konfiguration mit verkanteten Spins (beschrieben in [*62MWR*]) [*62MRAM*].

[2]) Die Ioneninkremente zur Berechnung von $\chi_{m\,korr.}$ sind: V^{4+}: $-7 \cdot 10^{-6}$ und Mg^{2+}: $-3 \cdot 10^{-6}$ emu/mole [*47RR-1*].

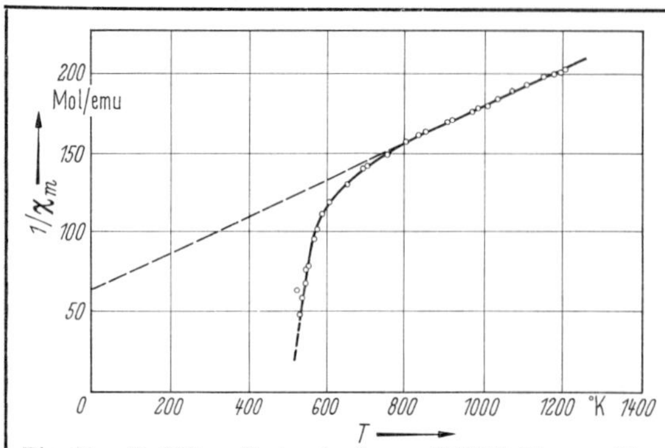

Fig. 13. Fe_2VO_4. Reziproke Suszeptibilität $1/\chi_m$ vs. Temperatur T. Die Kurve ist gezeichnet nach der Gleichung $\frac{1}{\chi_m} = \frac{T}{8.60} + 63 - \frac{930}{T - 530}$, welche der von Néel vorgeschlagenen Form entspricht [*67BP*].

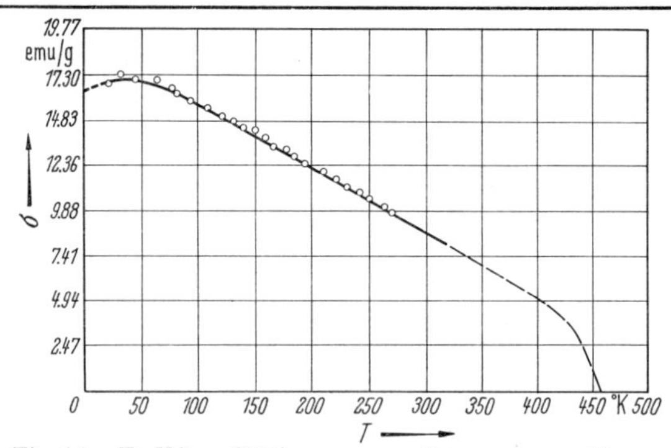

Fig. 14. Fe_2VO_4. Sättigungsmagnetisierung σ vs. Temperatur T gemessen bei einer Feldstärke $H = 22.6$ kOe [*67BP*].

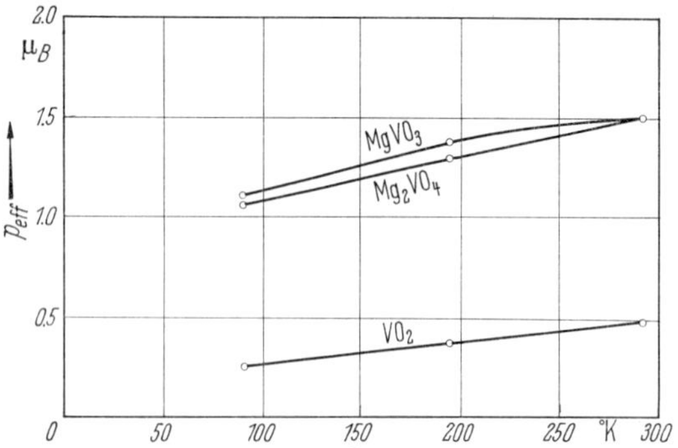

Fig. 15. Mg_2VO_4. Effektives Moment p_{eff} vs. Temperatur T. $MgVO_3$ und VO_2 zum Vergleich. [*47RR-1*].

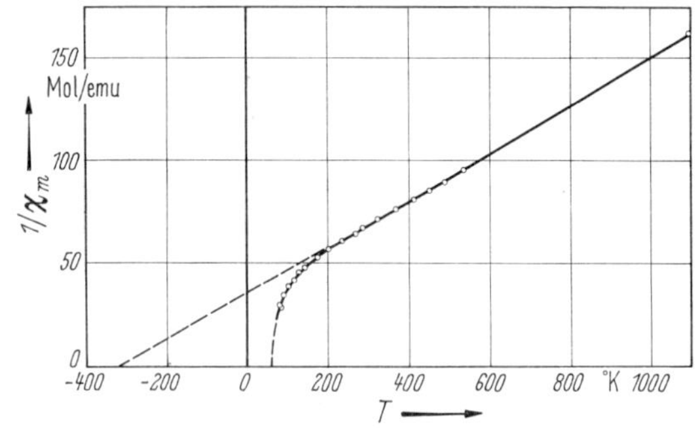

Fig. 16. Mn_2VO_4. Reziproke Suszeptibilität $1/\chi_m$ vs. Temperatur T. Die ausgezogene Kurve entspricht der Gleichung $\frac{1}{\chi_m} = \frac{T}{8.975} + 36.1 - \frac{310}{T - 73}$, welche die von Néel vorgeschlagene Form hat [*63BPM-2*]. $\Theta_a \approx -310$ °K.

Tab. 11. Kristallografische, magnetische und elektrische Eigenschaften ternärer V^{4+} Oxid-Spinelle

Verbindung	Kationenverteilung	x	a Å	Θ_C °K	Θ [1]) °K	$1/\chi_0$ [1]) mole/emu	σ [1]) mole °K/emu	Lit.	Fig.
$Co_{2-2x}Ni_{2x}VO_4$		0	8.382	160				*62BPM*	17, 18
		0.1	8.377	165					
		0.2	8.372	170					
		0.3	8.368	176					
		0.4	8.365	182					
		0.5	8.361	188					
		0.6	8.359	191					
		0.7	8.360	191					
		0.8	8.360						
$Li_{05.+0.83x}Fe_{2.5-2.5x}V_{1.67x}O_4$ [2])		0	8.313					*65L*	
		0.05	8.361						
		0.2	8.330						
		0.4	8.319						
		0.5	8.315						
$Mn_{2-2x}Ni_{2x}VO_4$	$V^{4+}[Mn^{2+}_{2-2x}Ni^{2+}_{2x}]O_4$	0	Fig. 19		73	36.1	310	*65BPM*	19, 20, 21
		0.1			83	36.9	300		
		0.2			98	37.7	280		
		0.3			106	38.6	260		
		0.4			116	40	250		
		0.5			123	41.5	230		
		0.6							
		0.7							

[1]) Formel nach Néel $\frac{1}{\chi_m} = \frac{T}{C_m} - \frac{1}{\chi_0} + \frac{\sigma}{T - \Theta}$ [*65BPM*].

[2]) Weitere Angaben über Θ_C und p_{eff} in Abhängigkeit von der Kationenverteilung im Spinellgitter [*65L*].

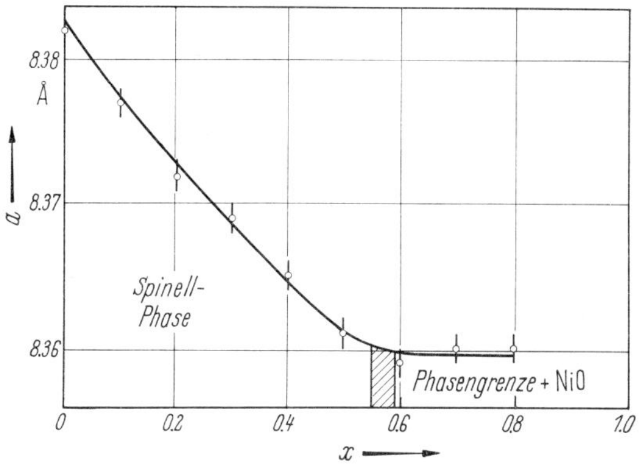

Fig. 17. $Co_{2-2x}Ni_{2x}VO_4$. Gitterkonstanten a und Angabe der Phasengrenze von Verbindungen dieser Reihe. Die Phasengrenze liegt bei $0.55 < x < 0.59$ [*62BPM*].

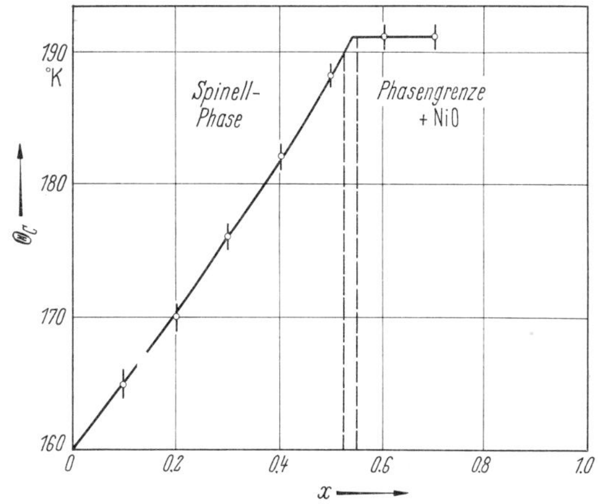

Fig. 18. $Co_{2-2x}Ni_{2x}VO_4$. Curie-Temperaturen und Angabe der Phasengrenze von Verbindungen dieser Reihe. Die Phasengrenze liegt bei $0.52 < x < 0.55$ [*62BPM*].

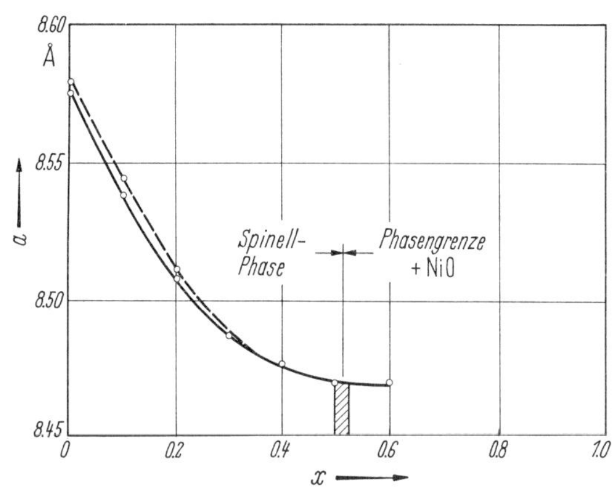

Fig. 19. $Mn_{2-2x}Ni_{2x}VO_4$. Gitterkonstanten a und Angabe der Phasengrenze von Verbindungen dieser Reihe.
Abgeschreckte Probe — — —
Langsam abgekühlte Probe ———
Phasenübergangsbereich ////// [*65BPM*].

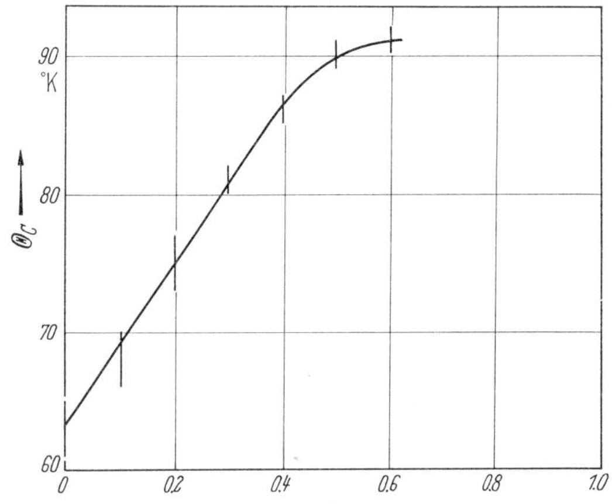

Fig. 20. $Mn_{2-2x}Ni_{2x}VO_4$. Curie-Temperaturen von Verbindungen dieser Reihe [*65BPM*].

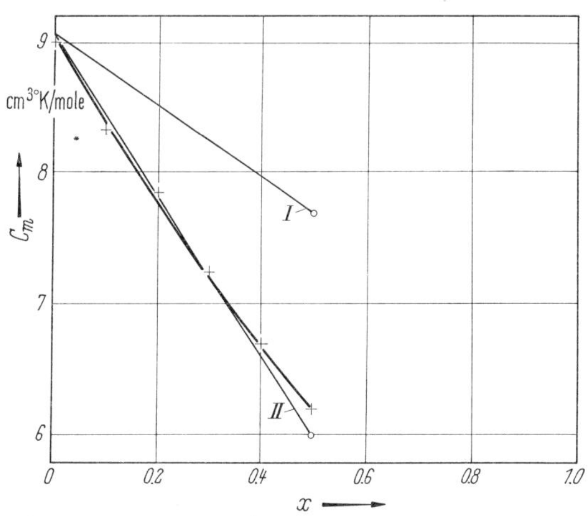

Fig. 21. $Mn_{2-2x}Ni_{2x}VO_4$. Curie-Konstanten C_m (siehe Tab. 11) in Abhängigkeit vom Substitutionsgrad x. Verlauf I: berechnet für die Substitution von Ni auf Tetraederplätzen, Verlauf II: berechnet für die Substitution von Ni auf Oktaederplätzen,
+ gemessene Punkte [*65BPM*].

6.3.1.3 V Spinelle mit V^{5+}

Tab. 12. Kristallografische Eigenschaften ternärer V^{5+} Oxid-Spinelle

Verbindung	Kationen-verteilung	a Å	ϱ_x g/cm³	ϱ_p g/cm³	Lit.	Fig.
$CoLiVO_4$	$V^{5+}[Co^{2+}Li^{1+}]O_4$	8.276 ± 0.001	4.238	4.24 ± 0.01	*61BPM, 64JD* *64B-2*	
$Cr_{0.5}Li_{1.5}VO_4$ ¹)	$V^{5+}[Cr^{3+}_{0.5}Li^{1+}_{1.5}]O_4$	8.19 8.195 ± 0.005	 3.65		*64B-1* *63JD*	
$CuLiVO_4$	$V^{5+}[Cu^{2+}Li^{1+}]O_4$	a = 5.652 ± 0.005 ²) b = 5.810 ± 0.005 c = 8.750 ± 0.005 a = 5.65, b = 5.80, c = 8.74			*65JGD, 66DGJD* *66B*	22
$Ga_{0.5}Li_{0.5}VO_4$	$V^{5+}[Ga_{0.5}Li_{0.5}]O_4$	8.2 ³)			*64B-1*	
$LiMgVO_4$	$V^{5+}[Li^{1+}Mg^{2+}]O_4$	8.27			*64B-2* *63B-4*	
$LiNiVO_4$	$V^{5+}[Li^{1+}Ni^{3+}]O_4$	8.215 ± 0.001	4.328	4.32 ± 0.01	*61BPM*	
$LiZnVO_4$ ⁴)	$V^{5+}[Li^{1+}Zn^{2+}]O_4$	8.309	4.34		*63B-3* *64B-2*	

¹) Weitere Eigenschaften [*63JD*]: Schmelztemperatur (910 ± 10) °C; u = 0.375; Farbe kastanienbraun.
²) $CuLiVO_4$ ist aufgrund des Jahn-Teller Effektes orthorhombisch verzerrt [*65JGD*], [*66B*].
³) $LiMgVO_4$ hat unter Normalbedingungen Olivinstruktur; bei hohem Druck Übergang zur Spinellstruktur [*64B-2*].
⁴) $LiZnVO_4$ hat Phenakitstruktur; bei hohem Druck Übergang zur Spinellstruktur [*63B-3*].

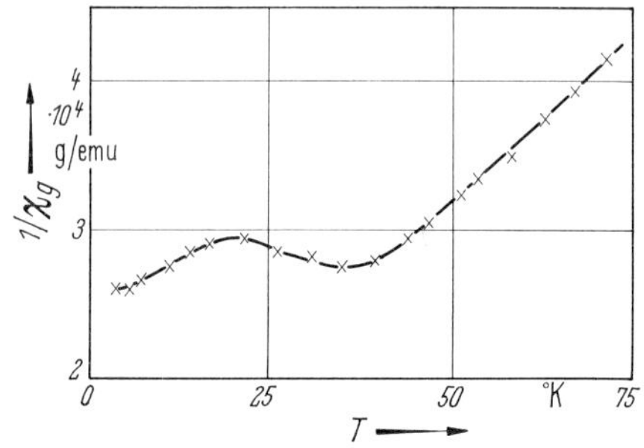

Fig. 22. $CuLiVO_4$. Reziproke Suszeptibilität pro Gramm, $1/\chi_g$, vs. Temperatur T. Oberhalb T = 100 °K gehorcht die Kurve einem Curie-Weiss Verlauf mit C_g = 0.44 und Θ_a = −30 °K [*66B*].

6.3.1.4 V Spinelle mit sowohl V^{3+} als auch V^{4+}

Tab. 13. Kristallografische Eigenschaften binärer V^{3+} V^{4+} Oxid-Spinelle

Verbindung	Kationenverteilung	Obere Phasen-grenze x_1	a Å für $x = 0$ ¹)	a Å für $x = x_1$	Lit.	Fig.	Tab.
$Cd_{1+x}V_{2-x}O_4$	$Cd[Cd_xV^{3+}_{2-2x}V^{4+}_x]O_4$	0.025	8.695	8.700	*65R-1*		14
$Co_{1+x}V_{2-x}O_4$	$Co[Co_xV^{3+}_{2-2x}V^{4+}_x]O_4$	0.9	8.407	8.379	*65R-1, 59RJR* *66RFK*	25, 26, 27, 35, 36	14, 15
$Fe_{1+x}V_{2-x}O_4$	$Fe[Fe_xV^{3+}_{2-2x}V^{4+}_x]O_4$	1.0	8.455	8.419	*65R 1*		14
LiV_2O_4 ²)	$Li^{1+}[V^{3+}V^{4+}]O_4$		8.22	 8.2403 ± 0.0012	*60RJ, 63BPM* *66RJ*		14
$Mg_{1+x}V_{2-x}O_4$	$Mg[Mg_xV^{3+}_{2-2x}V^{4+}]O_4$	0.65	8.418	8.394	*65R-1, 66RLRJ* *59RS, 59RJR*	23, 28, 29, 30, 31, 32, 37, 39, 41, 43	14
$Mn_{1+x}V_{2-x}O_4$	$Mn[Mn_xV^{3+}_{2-2x}V^{4+}_x]O_4$	0.69	8.519	8.570	*65R-1, 67RR* *66VL* ⁴)	24, 33, 34, 38, 40, 42, 44	14
$Zn_{1+x}V_{2-x}O_4$	$Zn[Zn_xV^{3+}_{2-2x}V^{4+}_x]O_4$	0.35	8.410³)	8.410	*65R-1, 68R-2*		14

¹) siehe Tab. 1. ²) u = 0.38 ± 0.01 [*66RJ*].
³) Im Bereich $0 < x < 0.18$ ist die Struktur kubisch mit linearer Änderung von a = (8.4096 ± 0.001) Å bei $x = 0$ auf (8.4065 ± 0.001) Å bei $x = 0.18$. Im Bereich $0.18 < x < 0.5$ (Phasengrenze) sind bei 1000 °C hergestellte Proben kubisch mit linearer Änderung der Gitterkonstanten a = 8.409 Å bei $x = 0.50$; bei 800 °C hergestellte Proben sind dagegen tetragonal mit einer 1 : 1 Ordnung von Zn^{2+} und V^{4+} auf den Oktaederstellen des Gitters [*68R-2*].
⁴) Weitere Angaben: Θ_C vs. x; Auf 0 °K extrapoliertes magnetisches Moment p^0_m vs. x [*66VL*].

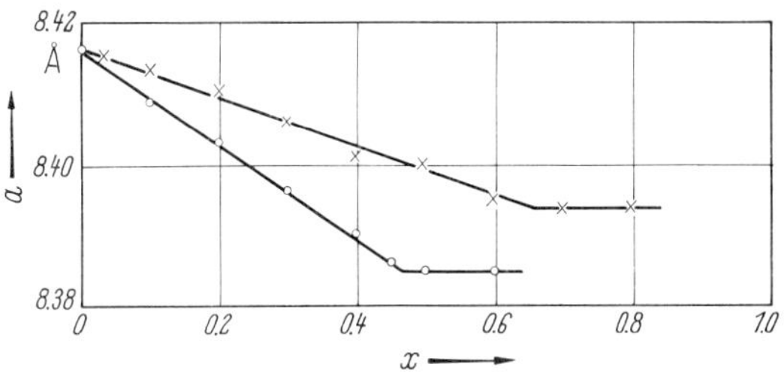

Fig. 23. $Mg_{1+x}V_{2-x}O_4$ (×–×) und $MgNi_xV_{2-x}O_4$ (○–○). Gitterkonstanten a vs. Zusammensetzungsparameter x [*66RLRJ*].

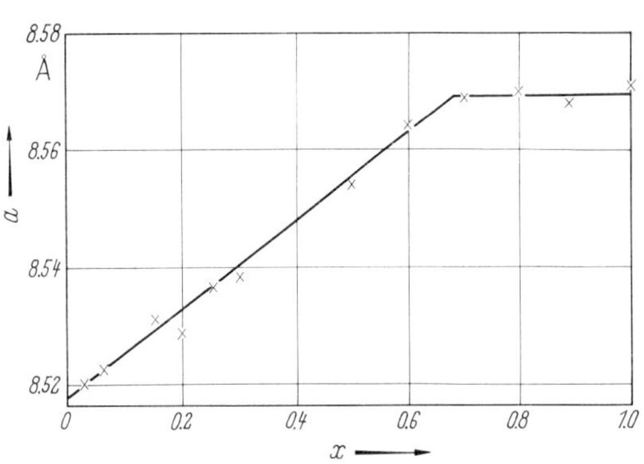

Fig. 24. $Mn_{1+x}V_{2-x}O_4$. Gitterkonstante a vs. x [*67RR*].

Tab. 14. Spezifischer Gleichstromwiderstand ϱ bei 20 °C für einige binäre V^{3+} V^{4+} Oxid-Spinelle [*65R-1*]

Verbindung	$\frac{\varrho}{\Omega cm}$ für $x \approx 0$ [1])	$\frac{\varrho}{\Omega cm}$ für $x = x_1$ [2])	$\frac{\varrho}{\Omega cm}$ für x_{min} [3])	
$Cd_{1+x}V_{2-x}O_4$		950		
$Co_{1+x}V_{2-x}O_4$	0.3	5000	0.3	0
$Fe_{1+x}V_{2-x}O_4$	120	0.31	0.31	1
LiV_2O_4 [4])	0.015 ± 0.002			
$Mg_{1+x}V_{2-x}O_4$	6	3000	1.4	0.16
$Mn_{1+x}V_{2-x}O_4$	60	3500	10	0.19
$Zn_{1+x}V_{2-x}O_4$	7	6.8	5.6	0.18

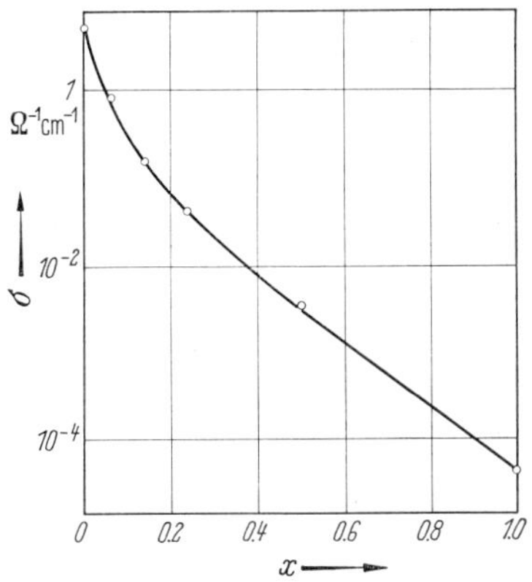

Fig. 25. $Co_{1+x}V^{3+}_{2-2x}V^{4+}_{x}O_4$. Gleichstromleitfähigkeit σ für polykristalline Proben vs. Zusammensetzungsparameter x [*59RJR*].

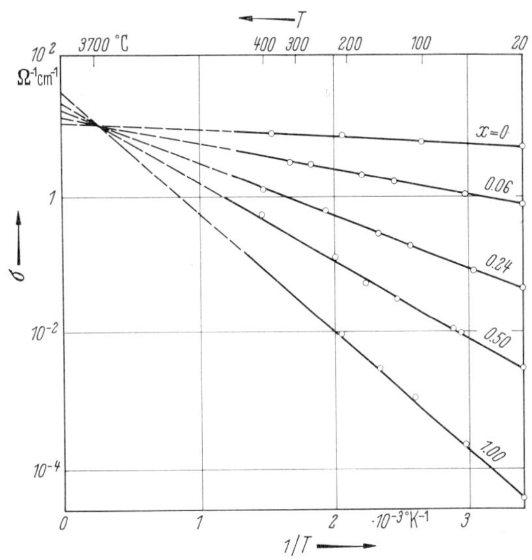

Fig. 26. $Co_{1+x}V^{3+}_{2-2x}V^{4+}_{x}O_4$. Gleichstromleitfähigkeit σ vs. $1/T$ und x [*59RJR*].

---

1) Genauer $x = 0.05$, da in unmittelbarer Nähe des stöchiometrischen Punktes $x = 0$ so starke Abhängigkeiten von ϱ auftreten, daß zuverlässige Angaben nicht möglich sind [*65R-1*].

2) siehe Tab. 13.

3) x_{min}: Die Verbindung hat in der Reihe den niedrigsten spezifischen Gleichstromwiderstand. x_1: Die Verbindung hat eine Zusammensetzung, welche auf der Grenze des Spinellbereiches liegt [*65R-1*].

4) Nach [*66RJ*].

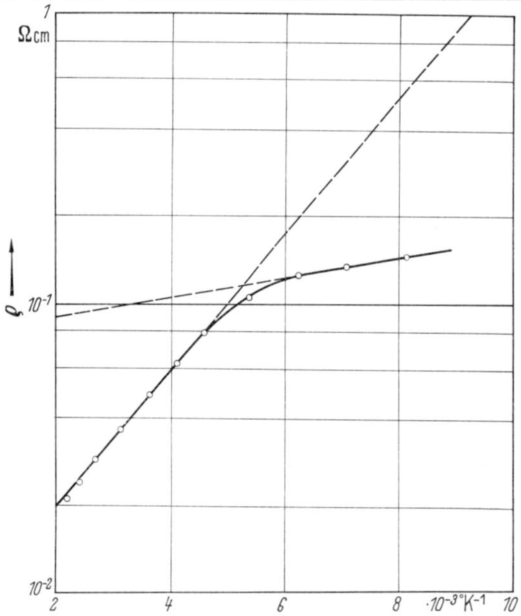

Fig. 27. $Co_{1.56}V_{1.44}O_4$. Spezifischer Gleichstromwiderstand ϱ vs. $1/T$ [66RFK].

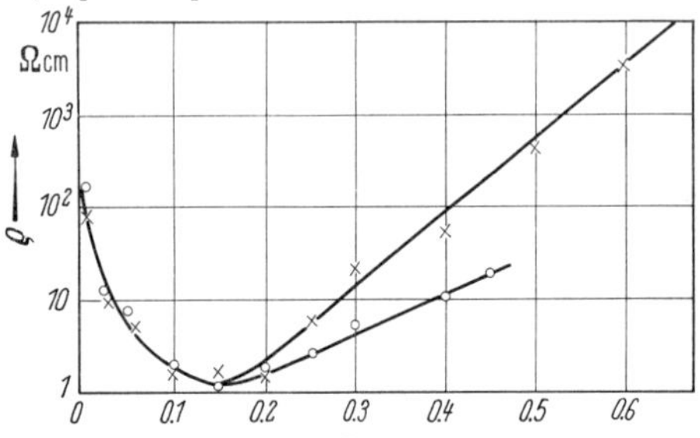

Fig. 28. $Mg_{1+x}V^{3+}_{2-2x}V^{4+}_{x}O_4$ (×-×) und $MgNi_xV^{3+}_{2-2x}V^{4+}_{x}O_4$ (○-○). Spezifischer Gleichstromwiderstand ϱ bei 20 °C vs. Zusammensetzungsparameter x für polykristalline Proben [66RLRJ].

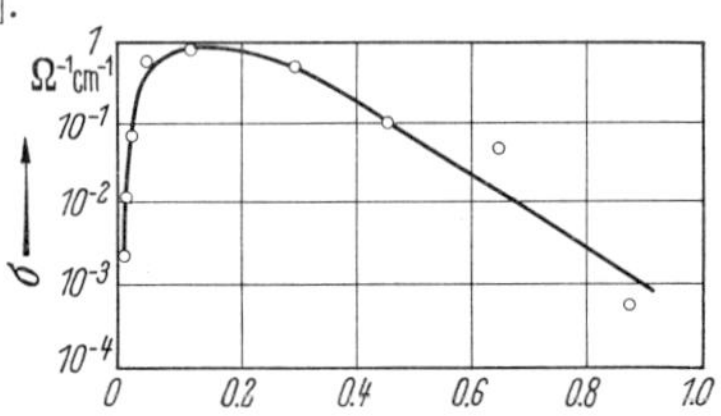

Fig. 29. $Mg_{1+x}V^{3+}_{2-2x}V^{4+}_{x}O_4$. Gleichstromleitfähigkeit σ vs. Zusammensetzungsparameter x [59RJR].

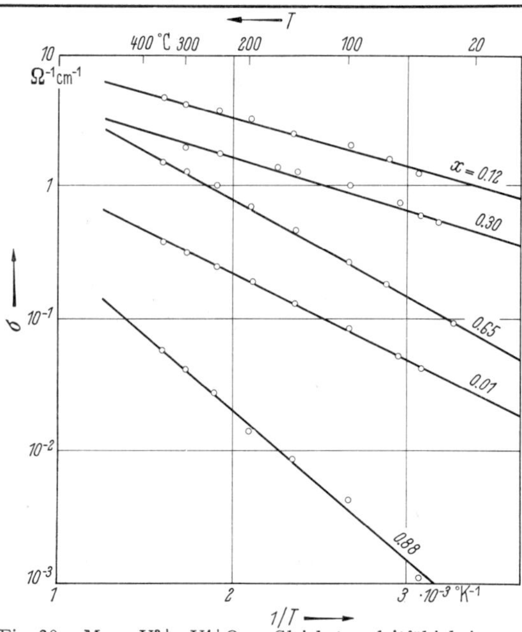

Fig. 30. $Mg_{1+x}V^{3+}_{2-2x}V^{4+}_{x}O_4$. Gleichstromleitfähigkeit σ vs. $1/T$ und x für polykristalline Proben dieser Reihe [59RJR].

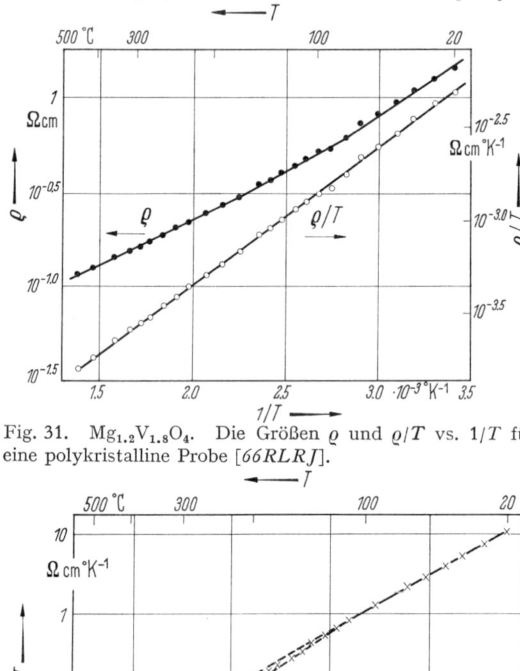

Fig. 31. $Mg_{1.2}V_{1.8}O_4$. Die Größen ϱ und ϱ/T vs. $1/T$ für eine polykristalline Probe [66RLRJ].

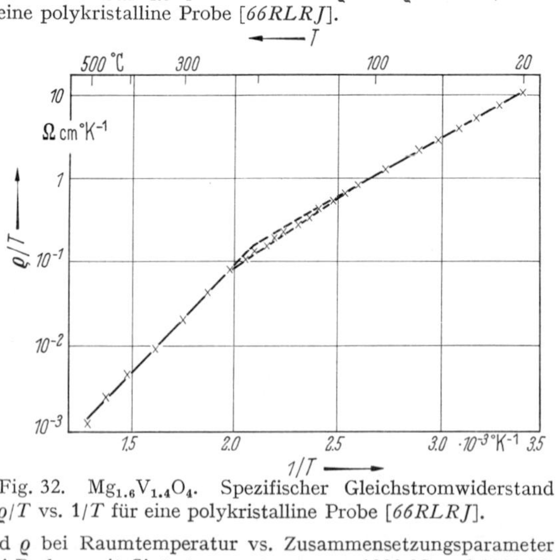

Fig. 32. $Mg_{1.6}V_{1.4}O_4$. Spezifischer Gleichstromwiderstand ϱ/T vs. $1/T$ für eine polykristalline Probe [66RLRJ].

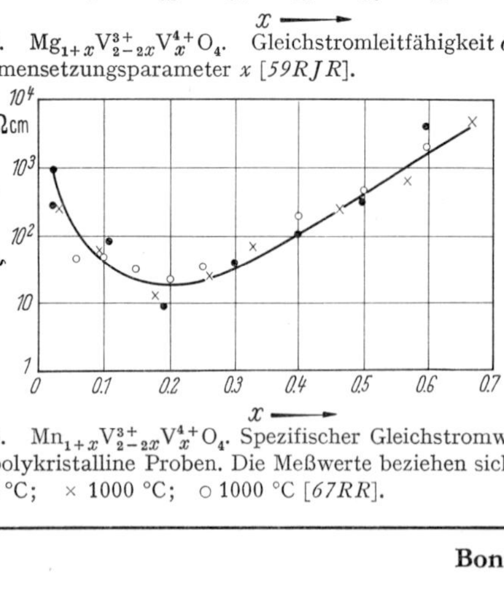

Fig. 33. $Mn_{1+x}V^{3+}_{2-2x}V^{4+}_{x}O_4$. Spezifischer Gleichstromwiderstand ϱ bei Raumtemperatur vs. Zusammensetzungsparameter x für polykristalline Proben. Die Meßwerte beziehen sich auf drei Proben mit Sintertemperaturen von 1000 °C und 1100 °C. ● 1100 °C; × 1000 °C; ○ 1000 °C [67RR].

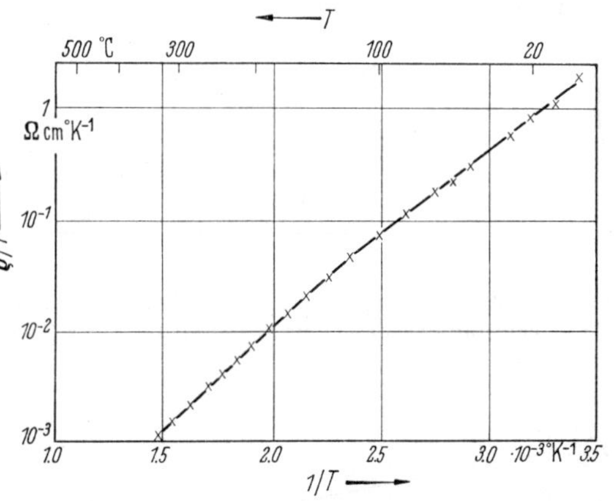

Fig. 34. $Mn_{1.57}V^{3+}_{0.86}V^{4+}_{0.57}O_4$. Spezifischer Gleichstromwiderstand ϱ/T vs. $1/T$ für polykristalline Probe [*67RR*].

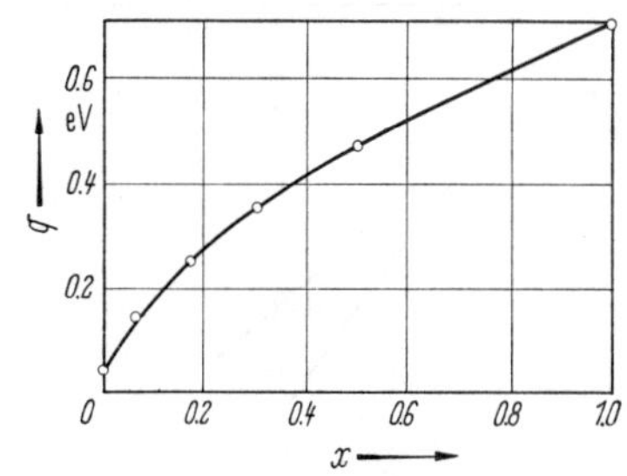

Fig. 35. $Co_{1+x}V^{3+}_{2-2x}V^{4+}_{x}O_4$. Aktivierungsenergie q des Gleichstromwiderstandes vs. Zusammensetzungsparameter x [*59RJR*].

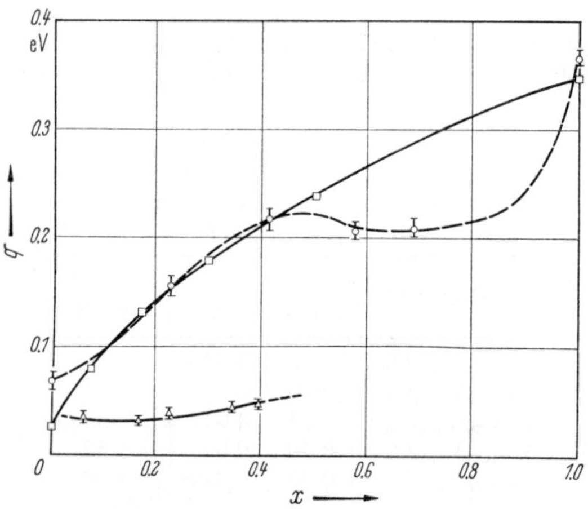

Fig. 36. $Co_{1+x}V^{3+}_{2-2x}V^{4+}_{x}O_4$. Aktivierungsenergie q des Gleichstromwiderstandes vs. Zusammensetzungsparameter x.
□ polykristalline Proben [*59RJR*];
○ polykristalline Proben [*62R*];
△ Einkristallproben [*66RFK*].

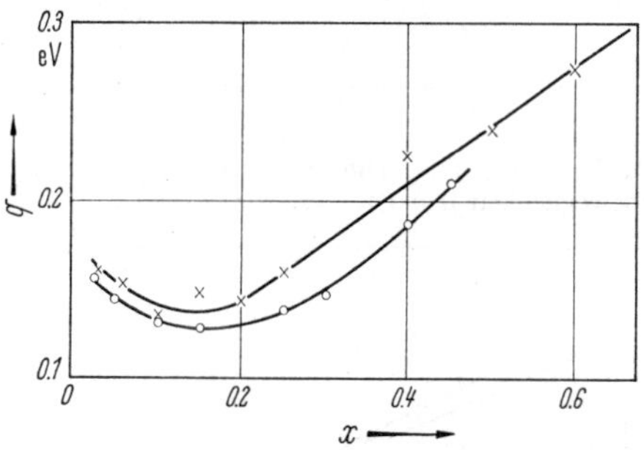

Fig. 37. $Mg_{1+x}V_{2-2x}O_4$ (×-×) und $MgNi_xV_{2-x}O_4$ (○-○). Aktivierungsenergie q des Gleichstromwiderstandes vs. Zusammensetzungsparameter x [*66RLRJ*].

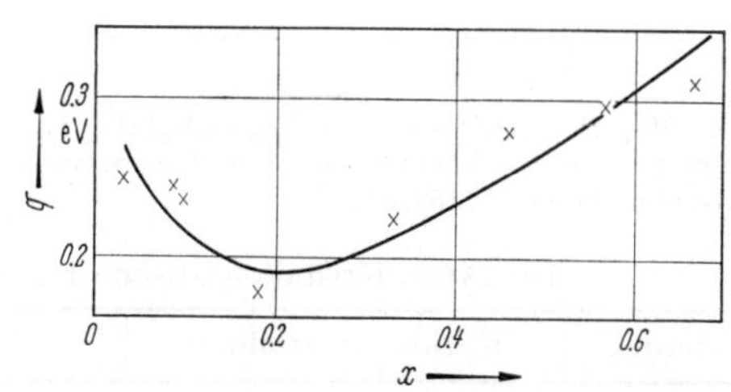

Fig. 38. $Mn_{1+x}V_{2-2x}O_4$. Aktivierungsenergie q des Gleichstromwiderstandes vs. Zusammensetzungsparameter x [*67RR*].

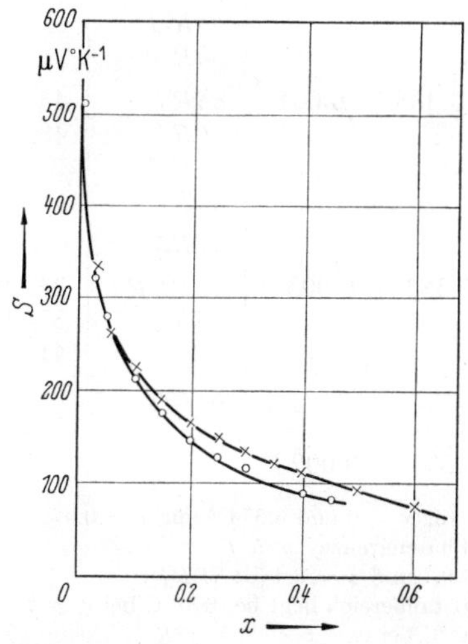

Fig. 39. $Mg_{1+x}V_{2-2x}O_4$ (×-×) und $MgNi_xV_{2-2x}O_4$ (○-○). Seebeck-Koeffizienten S vs. Zusammensetzungsparameter x [*66RLRJ*].

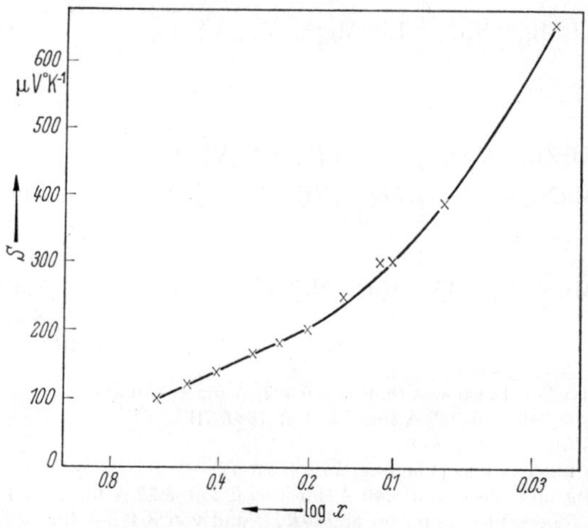

Fig. 40. $Mn_{1+x}V_2O_4$. Seebeck-Koeffizienten S vs. Zusammensetzungsparameter x [*67RR*].

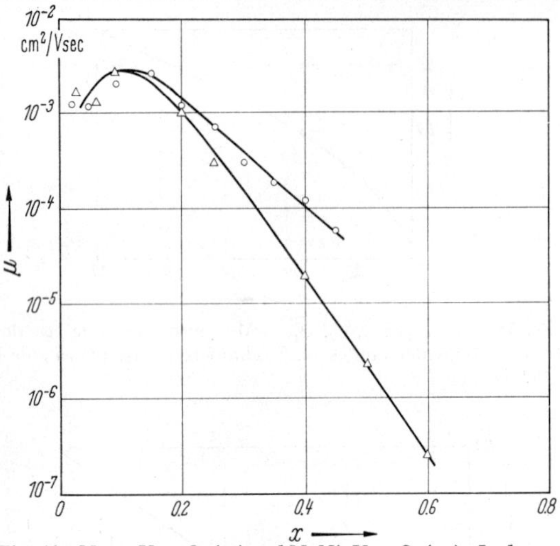

Fig. 41. $Mg_{1+x}V_{2-x}O_4$ (△) und $MgNi_xV_{2-x}O_4$ (○). Ladungsträgerbeweglichkeit μ bei RT vs. x [*66RLRJ*].

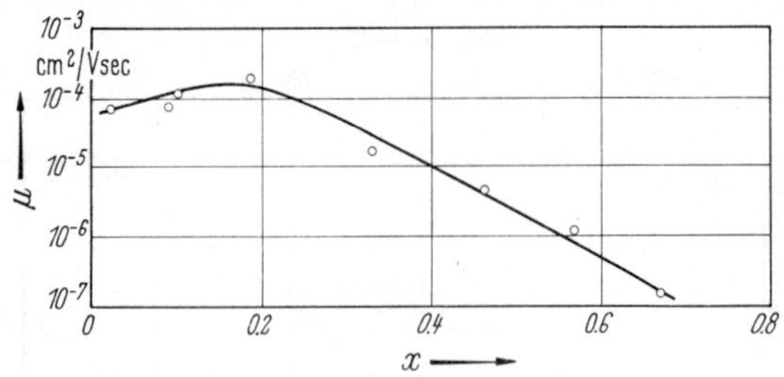

Fig. 42. $Mn_{1+x}V_{2-x}O_4$. Ladungsträgerbeweglichkeit μ bei Raumtemperatur vs. Zusammensetzungsparameter x [*67RR*].

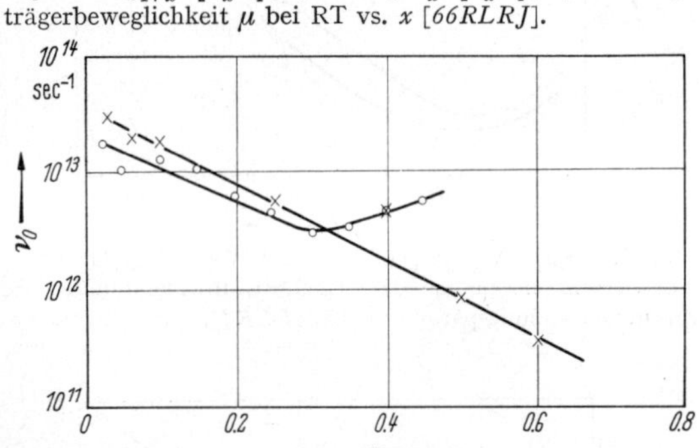

Fig. 43. $Mg_{1+x}V_{2-x}O_4$(×-×) und $MgNi_xV_{2-x}O_4$ (○-○). Sprungfrequenz ν_0 der Ladungsträger für $T = \infty$ vs. x für polykristalline Proben [*66RLRJ*].

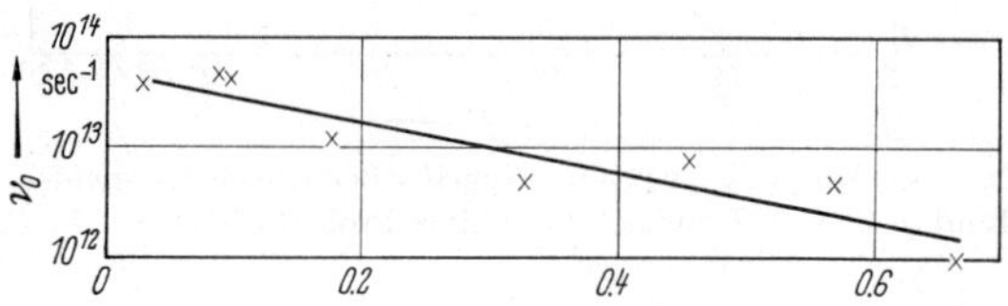

Fig. 44. $Mn_{1+x}V_{2-x}O_4$. Sprungfrequenz ν_0 der Ladungsträger für $T = \infty$ vs. x für polykristalline Proben [*67RR*].

Tab. 15. Elektrische Eigenschaften bei RT für polykristalline Proben der Reihe $Co_{1+x}V_{2-x}O_4$ [*66RFK*]

Verbindung	ϱ Ωcm	q eV	S μV/°K
$Co_{1.11}V_{1.89}O_4$	0.36	0.035	+89
$Co_{1.29}V_{1.71}O_4$	0.40	0.033	+74
$Co_{1.38}V_{1.62}O_4$	0.24	0.042	+58
$Co_{1.51}V_{1.49}O_4$	0.40	0.047	+32
$Co_{1.56}V_{1.44}O_4$	0.40	0.049	+30

Tab. 16. Kristallografische Eigenschaften ternärer V^{3+} V^{4+} Oxid-Spinelle

Verbindung	Kationenverteilung	a [Å]	u	Lit.	Fig.	Tab.
$Co_{1-x}Li_xV_2O_4$		[1]		*64RGW*	47	17
$CoNi_xV_{2-x}O_4$	$Co[Ni_xV_{2-x}]O_4$	8.396 [2]		*65R-1*		17
		8.351 [3]		*65R-1*		
		8.351 ± 0.003		*68R-2*		
$Li_xMg_{1-x}V_2O_4$	$Li^{1+}_xMg^{2+}_{1-x}[V^{3+}_{2-x}V^{4+}_x]O_4$	[4]	0.385 ± 0.001	*66RJ*	45, 46	
				60RJ	48, 49	
				65R-1		
				66J		
$Li_xZn_{1-x}V_2O_4$	$Li^{1+}_xZn^{2+}_{1-x}[V^{3+}_{2-x}V^{4+}_x]O_4$			*67RM*		17
$MgNi_xV_{2-x}O_4$	$Mg^{2+}[Ni^{2+}_xV^{3+}_{2-2x}V^{4+}_x]O_4$	Fig. 23 [5]	0.383 ± 0.003 [6]	*66RLRJ*	23, 28,	
				65R-1	37, 39,	
					41, 43	
$MnNi_xV_{2-x}O_4$ [7]	$Mn^{2+}[Ni^{2+}_xV^{3+}_{2-2x}V^{4+}_x]O_4$	8.478 [8]		*68R-2*		
		8.474 [9]				
		8.487 ± 0.003 [10]	0.394 ± 0.002 [10]			

[1]) a fällt linear von (8.407 ± 0.002) Å für $x = 0$ auf (8.240 ± 0.002) Å für $x = 1$ ab [*64RGW*].
[2]) für $x = 0$ [*65R-1*].
[3]) für $x = 0.89$ (Phasengrenze) [*65R-1*].
[4]) a fällt linear von 8.40 Å für $x = 0$ auf 8.22 Å für $x = 1$ (obere Phasengrenze) ab [*60RJ*]; und von 8.416 Å für $x = 0$ auf 8.240 Å für $x = 1$ [*65R-1*].
[5]) 8.405 Å für $x = 0$ und 8.374 Å für $x = 0.47$ (obere Phasengrenze) [*65R-1*].
[6]) für $x = 0.1$ und $x = 0.4$ [*66RLRJ*].
[7]) Mischkristallbereich liegt bei 800 °C bei $0 < x < 0.6$ und bei 1000 °C bei $0 < x < 0.65$ [*68R-2*].
[8]) für $x = 0.6$ (Phasengrenze) [*68R-2*].
[9]) für $x = 0.65$ (Phasengrenze [*68R-2*].
[10]) für $x = 0.5$ (*68R-2*].

Tab. 17. Elektrische Eigenschaften ternärer V^{3+} V^{4+} Oxid-Spinelle

Verbindung	x	σ_{RT} $\Omega^{-1}cm^{-1}$	q eV	S $\mu V/°K$	Lit.	Fig.
$Co_{1-x}Li_xV_2O_4$	$0 < x < 0.04$	0.5			*64RGW*	47
	$0.05 < x < 0.15$	0.1				
	$0.25 < x < 1.0$	$8 < \sigma < 20$				
$CoNi_xV_{2-x}O_4$	0.0 [1])	1.1			*65R-1*	
	0.13	2 [2])				
	0.89	0.05				
$Li_xZn_{1-x}V_2O_4$	0.05			179	*67RM*	
	0.4		0.2			
	$0.55 < x < 1.0$		0.1	≈ 0		

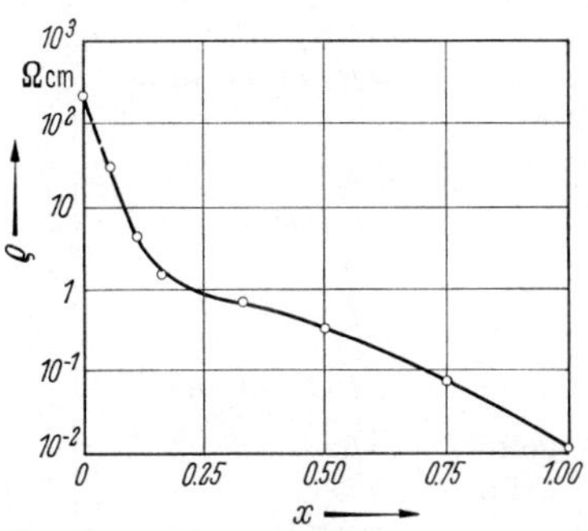

Fig. 45. $Li_xMg_{1-x}V_2O_4$. Spezifischer Gleichstromwiderstand ϱ bei Raumtemperatur vs. Zusammensetzungsparameter x [*66RJ*].

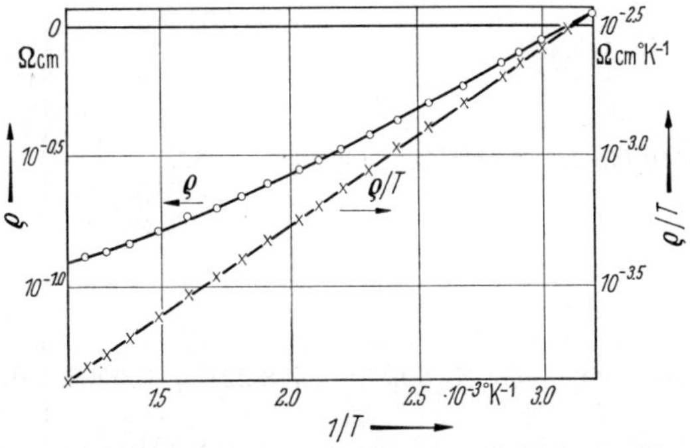

Fig. 46. $Li_{0.17}Mg_{0.83}V_2O_4$. Spezifischer Gleichstromwiderstand ϱ und die Größe ϱ/T vs. $1/T$ für eine polykristalline Probe [*66RJ*].

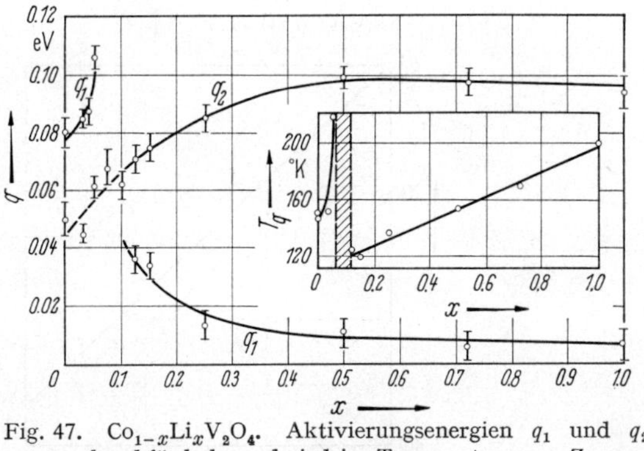

Fig. 47. $Co_{1-x}Li_xV_2O_4$. Aktivierungsenergien q_1 und q_2 entsprechend für hohe und niedrige Temperaturen vs. Zusammensetzungsparameter x. Die Übergangstemperatur, T_q = Differenz zwischen hoher und niedriger Temperatur, vs. x ist in der eingezeichneten Figur gegeben [*64RGW*].

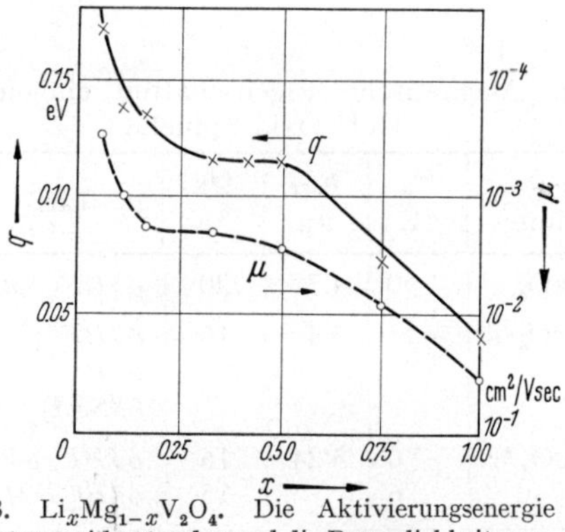

Fig. 48. $Li_xMg_{1-x}V_2O_4$. Die Aktivierungsenergie q des Gleichstromwiderstandes und die Beweglichkeit μ vs. Zusammensetzungsparameter x für polykristalline Proben [*66J*].

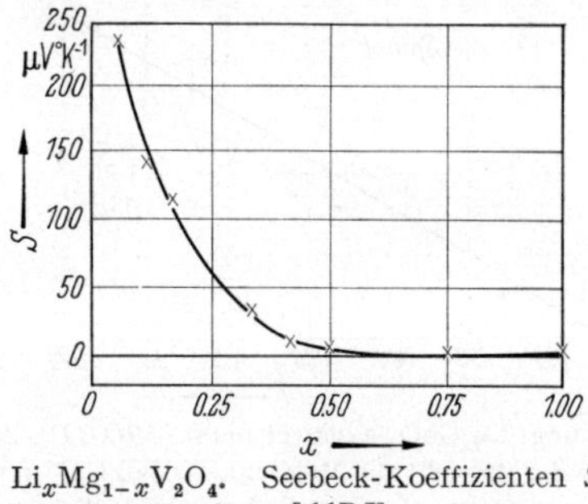

Fig. 49. $Li_xMg_{1-x}V_2O_4$. Seebeck-Koeffizienten S vs. Zusammensetzungsparameter x [*66RJ*].

[1]) Genauer: $x = 0.05$, da in unmittelbarer Nähe des stöchiometrischen Punktes ($x = 0$) so starke x-Abhängigkeiten von σ auftreten, daß zuverlässige Angaben über den stöchiometrischen Punkt selbst nicht gemacht werden können [*65R-1*].

[2]) Maximale Leitfähigkeit der Verbindungsreihe [*65R-1*].

6.3.2 Ge Spinelle und substituierte Ge Spinelle

(Literatur siehe 6.3.4, Seite 444)

Tab. 1. Kristallografische Eigenschaften binärer Ge^{4+} Oxid-Spinelle

Verbindung [1])	Kationenverteilung	a Å	u	Farbe	Lit.	Fig.	Tab.
Co_2GeO_4	$Ge^{4+}[Co_2^{2+}]O_4$	8.3175 ± 0.0005 8.322	0.375 ± 0.003	rosa	*53R* *56DBP* *63BF, 63G* *67P, 68NK*	1, 4, 7	2
Fe_2GeO_4	$Ge^{4+}[Fe_2^{2+}]O_4$ [2])	8.411	 0.375		*56DBP* *64B-2* *58R-2, 58R-3* *63BF, 63G* *64B, 68HI, 66R* *68SS*	1, 5, 7, 8, 9, 10	2
Mg_2GeO_4	$Ge^{4+}[Mg_2^{2+}]O_4$ [3])	8.23 8.255 8.248			*56R* *56DBP* *58R-2* *32JS, 54RR* *56DR, 56R* *58R-1, 60DR*	2, 3	
Ni_2GeO_4	$Ge[Ni_2]O_4$	8.221 ± 0.0005 8.231	0.375 ± 0.003	gelb-grün	*53R* *56DBP* *58R-1, 62RS* *63BF, 63G* *64BQPM, 67DR* *68NK, 65WHA*	1, 6, 7	2

Tab. 2. Magnetische Eigenschaften einiger binärer Ge^{4+} Oxid-Spinelle

Verbindung	Θ_a °K	p_{eff} μ_B	Θ_N °K	Lit.	Fig.
Co_2GeO_4	+90	4.7	20	*63BF, 63G*	4, 7
Fe_2GeO_4	−15	5.4	10	*63BF* *66R* *68SS* [5])	5, 7···10
Ni_2GeO_4 [4])	0 − 6	3.24	16 15	*63BF, 63G* *64BQPM*	6, 7

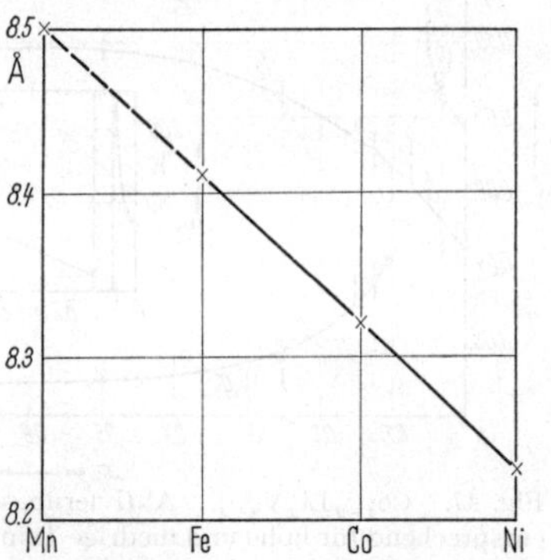

Fig. 1. Vergleich der Gitterkonstanten *a* linearer Ge^{4+} Oxid-Spinelle mit unterschiedlicher Kationenbesetzung auf Oktaederstellen [*56DBP*].

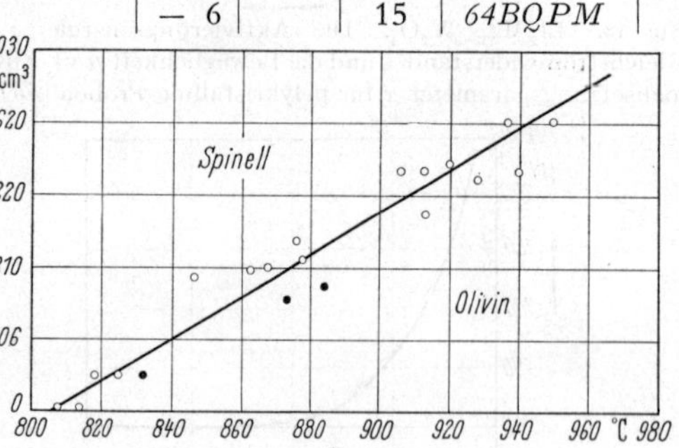

← Fig. 2. Spinell-Olivin-Transformation von Mg_2GeO_4. Die Streuung der Meßpunkte gibt Unregelmäßigkeiten bei der Transformation wieder [*60DR*].

[1]) Anmerkung: Cu_2GeO_4 existiert nicht [*59DBP*], Zn_2GeO_4 hat Phenakitstruktur [*56DBP*].

[2]) Nach [*58R-2*] besteht die Möglichkeit, daß Ge^{4+} teilweise reduziert wird zu Ge^{2+} und entsprechend Fe^{2+} zu Fe^{3+} oxidiert wird.

[3]) Mg_2GeO_4 existiert als Spinell bei tiefen Temperaturen und hohem Druck. Die Strukturtransformation Spinell-Olivin erfolgt bei (1065 ± 10) °C [*32JS*], bei 1005 °C [*54RR*], bei 825 °C, 703 at und ist bei 20 °C, 1000 at aufgehoben [*56DR*]; die Transformation ist nicht reversibel bei 1005 °C [*56DBP*], siehe hierzu auch Fig. 2. Der Schmelzpunkt liegt bei (1855 ± 30) °C [*56DR*].

[4]) Ni_2GeO_4 hat eine Curie-Konstante von $C_g = 1.39$ cm³ °K/g [*64BQPM*]. [5]) Mössbauer Spektrum [*68SS*].

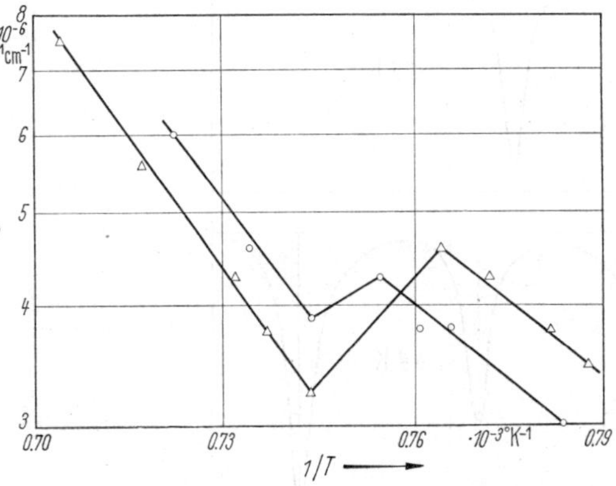

Fig. 3. Mg_2GeO_4. Gleichstromleitfähigkeit σ vs. Kehrwert der Temperatur zweier unterschiedlich hergestellter polykristalliner Proben [32JS].

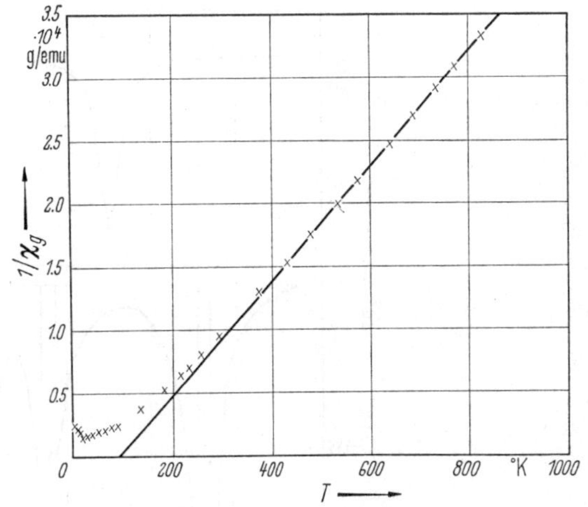

Fig. 4. Co_2GeO_4. Reziproke Suszeptibilität pro Gramm $1/\chi_g$ vs. Temperatur T. Θ_N = 20 °K; p_{eff} = 4.70 μ_B [63BF].

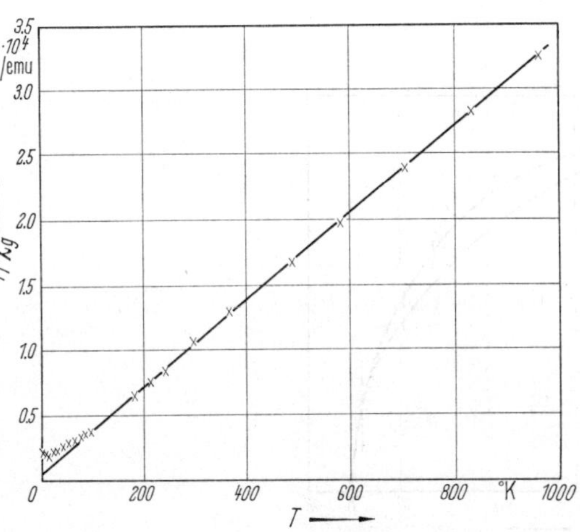

Fig. 5. Fe_2GeO_4. Reziproke Suszeptibilität pro Gramm $1/\chi_g$ vs. Temperatur T. Θ_N = 10 °K; p_{eff} = 5.40 μ_B [63BF].

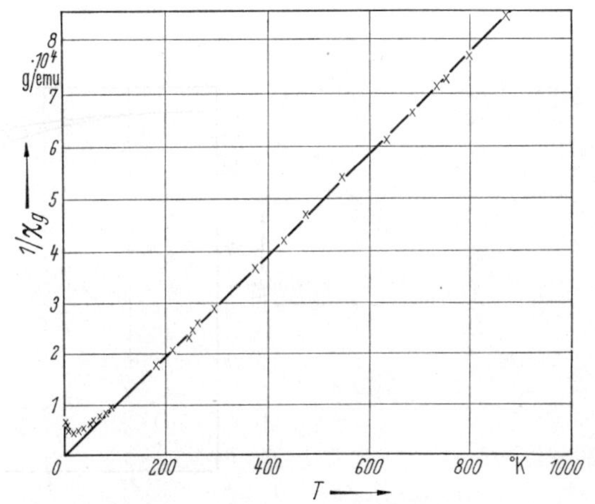

Fig. 6. Ni_2GeO_4. Reziproke Suszeptibilität pro Gramm $1/\chi_g$ vs. Temperatur T. Θ_N = 16 °K; p_{eff} = 3.24 μ_B [63BF

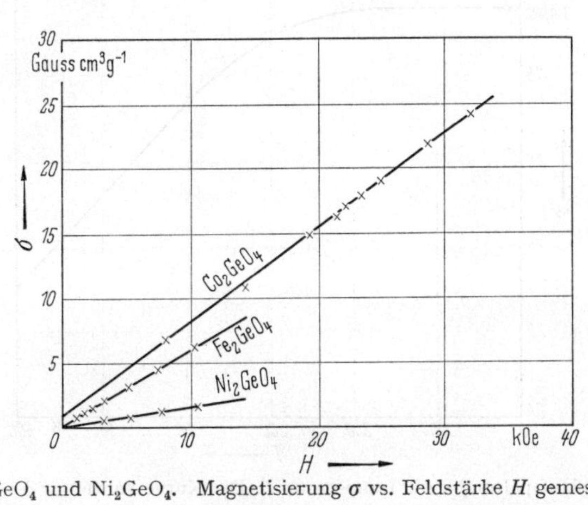

Fig. 7. Co_2GeO_4, Fe_2GeO_4 und Ni_2GeO_4. Magnetisierung σ vs. Feldstärke H gemessen bei 4.2 °K [63BF].

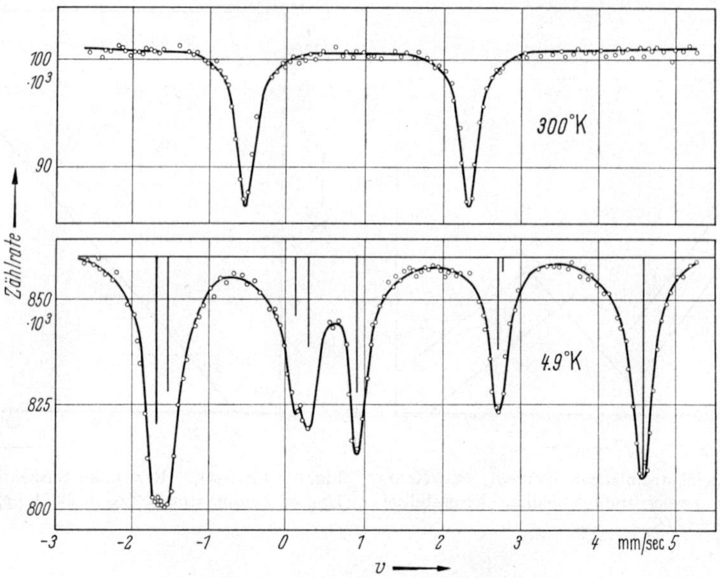

Fig. 8. Fe_2GeO_4. Mössbauer-Spektrum bei 4.9 °K und 300 °K [*68HI*].

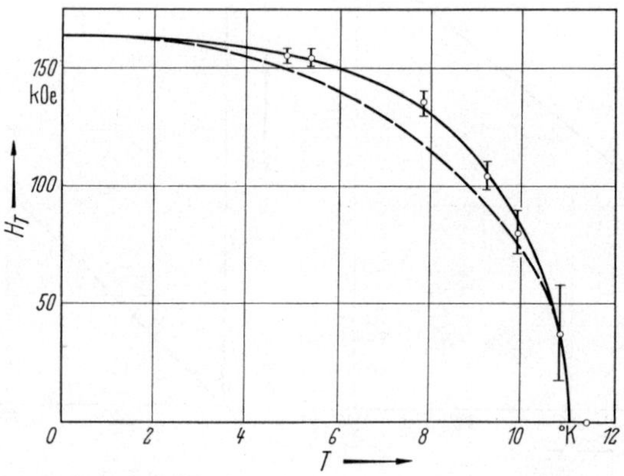

Fig. 9. Fe_2GeO_4. Hyperfeinfeldstärke H_T vs. Temperatur T. ——— Brillouin Kurve für $S = 2$ [*68HI*].

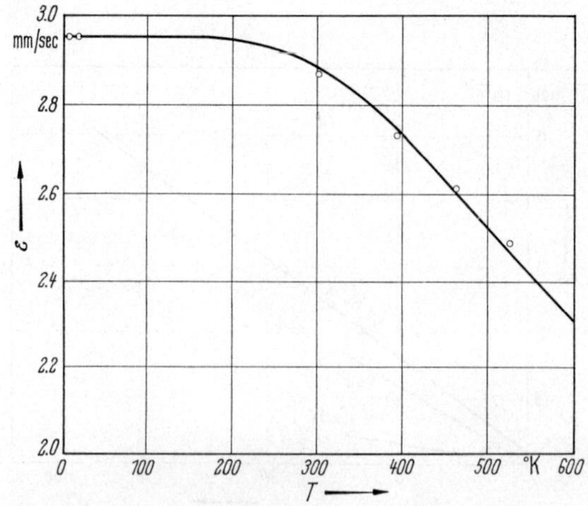

Fig. 10. Fe_2GeO_4. Quadrupolaufspaltung ε vs. Temperatur T. Die Kurve wurde für $\delta = 1020\ cm^{-1}$ berechnet [*68HI*].

Tab. 3. Kristallografische Eigenschaften ternärer Ge^{4+} Oxid-Spinelle

Verbindung	Kationenverteilung	a Å	Struktur-transformation [1])	ϱ_x g/cm³	Farbe	Lit.	Fig.
$Co_{1+x}Ga_{2-2x}Ge_xO_4$ [2])	$Co_y^{2+}Ga_{1-y}^{3+}[Co_{1+x-y}^{2+}Ga_{1-2x+y}^{3+}Ge_x^{4+}]O_4$	Fig. 11			[2])	*68R-1*	11
$Co_{0.5}LiGe_{1.5}O_4$ [3])	$Co_{0.5-0.5x}Li_{0.5-0.5x}[Co_{0.5x}Li_{0.5-0.5x}Ge_{1.5}]O_4$ [4]) $Co_{0.5}Li_{0.5}[Li_{0.5}Ge_{1.5}]O_4$	8.204 ± 0.002 8.20		5.06	dunkelbraun	*62DJ* *64B-2*	
$Co_{2-x}Mg_xGeO_4$ [5])	$Ge^{4+}[Co_{2-2x}^{2+}Mg_x^{2+}]O_4$	Fig. 12, 13	S – O bei $x = 1.5$ S – O bei $x = 1.1$			*56DBP* *68R-1*	12, 13
$CoNiGeO_4$		8.268 ± 0.001			rehbraun	*53R*	
$Co_{2-x}Zn_xGeO_4$	I $Ge[Co_{2-x}Zn_x]O_4$ [6]) II $Co_{1-x}Zn_x[CoGe]O_4$	8.322 [7]) 8.350 [8]) Fig. 14	S – P bei $x = 1$ S – P bei $x = 0.9$			*56DBP* *68R-1*	14
$CrLiGeO_4$	$Li^{1+}[Cr^{3+}Ge^{4+}]O_4$	8.197		4.72	braun	*60SJ, 62D-1*	
$Ga_{2-2x}Ni_{1+x}Ge_xO_4$	$Ga_{1-x+y}Ge_{x-y}[Ni_{1+x}Ga_{1-x-y}Ge_y]O_4$	Fig. 15				*67R, 68R-1*	15
$LiAlGeO_4$ [11])		8.00 7.996 ± 0.001				*65NM* *66GPS*	
$LiGaGeO_4$ [11])		8.16				*65NM*	
$LiNi_{0.5}Ge_{1.5}O_4$	$Li^{1+}[Ni_{0.5}^{2+}Ge_{1.5}^{4+}]O_4$	8.17				*64B-1, 64B-2*	
$LiRhGeO_4$	$Li^{1+}[Ge^{4+}Rh^{3+}]O_4$	8.387		5.55		*62D-1, 62D-2* *64B-2*	
$LiZn_{0.5}Ge_{1.5}O_4$ [9])	$Li_{0.5}^{1+}Zn_{0.5}^{2+}[Li_{0.5}^{1+}Ge_{1.5}^{4+}]O_4$	8.190 ± 0.002		5.11	weiß	*62DJ, 64B-2*	
$Mg_xNi_{2-x}GeO_4$ [10])	$Ge^{4+}[Mg_x^{2+}Ni_{2-x}^{2+}]O_4$	8.23	S – O bei $x = 0.5$ S – O bei $x = 1.4$			*56DBP* *68R-1*	13
$Ni_{2-x}Zn_xGeO_4$	I $Ge[Zn_xNi_{2-x}]O_4$ [6]) II $Zn[Zn_{x-1}Ni_{2-x}Ge]O_4$	Fig. 16	S – O bei $x = 1.3$ Fig. 16			*68R-1*	16

[1]) Transformation Spinell-Olivin S – O; Transformation Spinell – Phenakit S – P.
[2]) In der Reihe $CoGa_2O_4$-Co_2GeO_4 besteht eine Mischungslücke von $x = 0.8 \ldots 1.0$. Die Farbe der ersten Reihe ist blau-grün, die der zweiten rosa [*68R-1*].
[3]) Der Schmelzpunkt liegt bei (980 ± 5) °C [*62DJ*].
[4]) Diese Verteilung gilt für x nahe Null [*62DJ*].
[5]) In der Reihe Co_2GeO_4-Mg_2GeO_4 besteht eine Mischungslücke von $x = 1.1 \cdots 1.6$ [*68R-1*].
[6]) $Co_{2-x}Zn_xGeO_4$ zeigt wie $Ni_{2-x}Zn_xGeO_4$ verschiedene Phasengebiete mit unterschiedlicher Kationenverteilung. Die Normalspinellphase I ist bis zu $x = 0.2$ stabil; für $x > 0.6$ existiert eine Spinellphase II, siehe hierzu auch Fig. 14, 16 [*68R-1*].
[7]) für $x = 0$ [*56DBP*].
[8]) für $x = 1$ [*68R-1*].
[9]) Der Schmelzpunkt liegt bei (995 ± 5) °C [*62DJ*].
[10]) In der Reihe Ni_2GeO_4-Mg_2GeO_4 besteht eine Mischungslücke von $x = 1.4 \cdots 1.8$ [*68R-1*].
[11]) Spinell bei hoher Temperatur und hohem Druck [*65NM*], [*66GPS*].

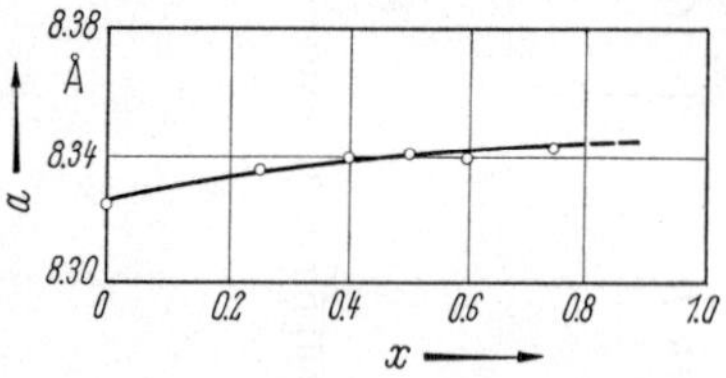

Fig. 11. $Co_{1+x}Ga_{2-2x}Ge_xO_4$. Gitterkonstante a vs. Zusammensetzungsparameter x [*68R-1*].

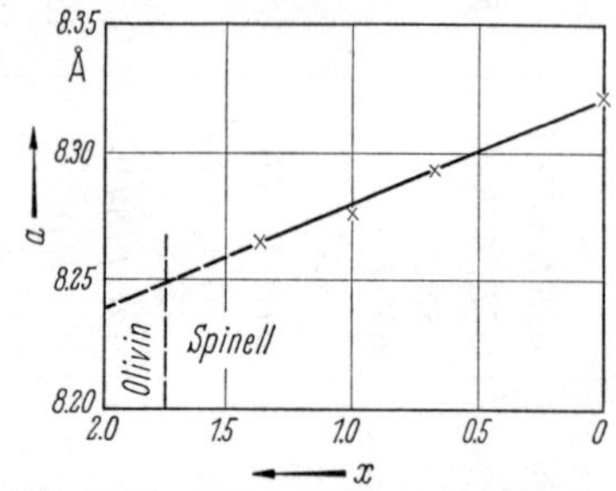

Fig. 12. $Co_{2-x}Mg_xGeO_4$. Gitterkonstante a und Strukturtransformation Olivin-Spinell vs. Zusammensetzungsparameter x [*56DBP*].

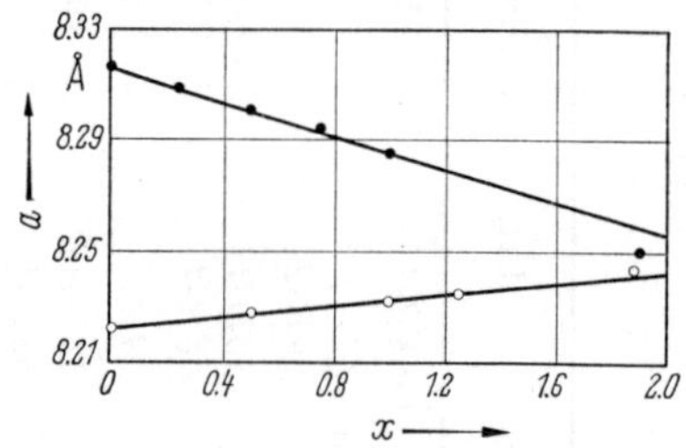

Fig. 13. $Co_{2-x}Mg_xGeO_4$ (●-●) und $Mg_xNi_{2-x}GeO_4$ (○-○). Gitterkonstante a vs. Zusammensetzungsparameter x [*68R-1*].

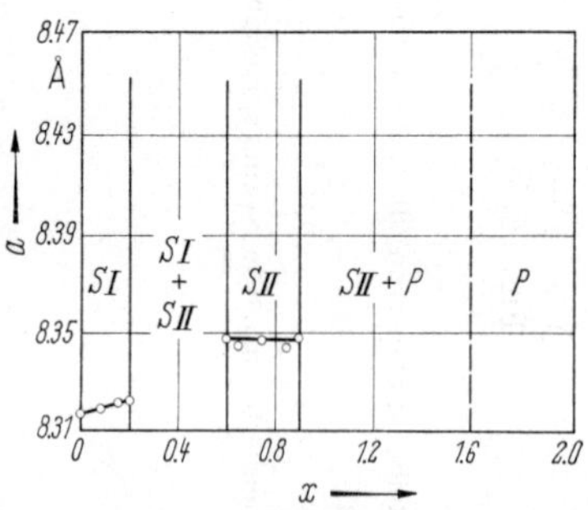

Fig. 14. $Co_{2-x}Zn_xGeO_4$. Gitterkonstante a und Phasenverhältnisse vs. Zusammensetzungsparameter x [*68R-1*].

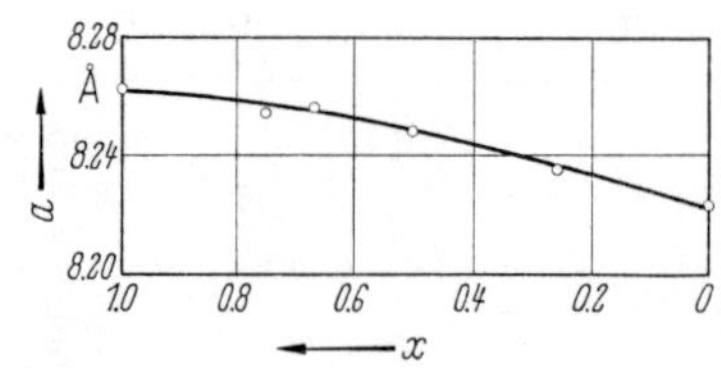

Fig. 15. $Ga_{2x}Ni_{2-x}Ge_{1-x}O_4$. Gitterkonstante a vs. Zusammensetzungsparameter x [*67R*].

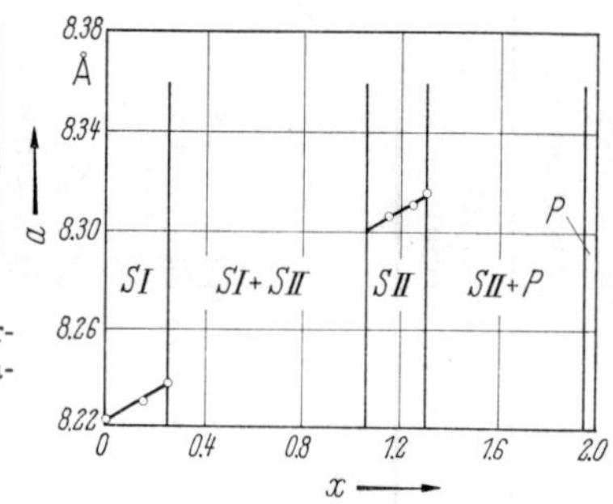

Fig. 16. $Ni_{2-x}Zn_xGeO_4$. Gitterkonstante a und Phasenverhältnisse vs. Zusammensetzungsparameter x [*68R-1*].

Tab. 4. Kristallografische und magnetische Eigenschaften quaternärer Ge^{4+} Oxid-Spinelle

Verbindung	Kationenverteilung	x	a Å	Θ_a °K	Θ_C °K	Lit.	Fig.
$Cu_xMn_{2-2x}ZnGe_xO_4$	$Zn[Cu_xMn_{2-2x}Ge_x]O_4$		Fig. 17			*65RB, 66RD*	17
$Cu_{2x}Mn_{2-2x}Zn_{1-x}Ge_xO_4$	$Zn_{1-x}Cu_x[Cu_xMn_{2-2x}Ge_x]O_4$		Fig. 17			*65RB*	17
$Fe_xLi_xNi_{2-2x}GeO_4$		0.125		+ 45		*63G*	
		0.25		+ 85			
		0.5		+180			
$Ni_{0.25+1.5x}Zn_{0.75-0.75x}Fe_{2-1.5x}Ge_{0.75x}O_4$		1	8.25		8.3	*65B-1*	

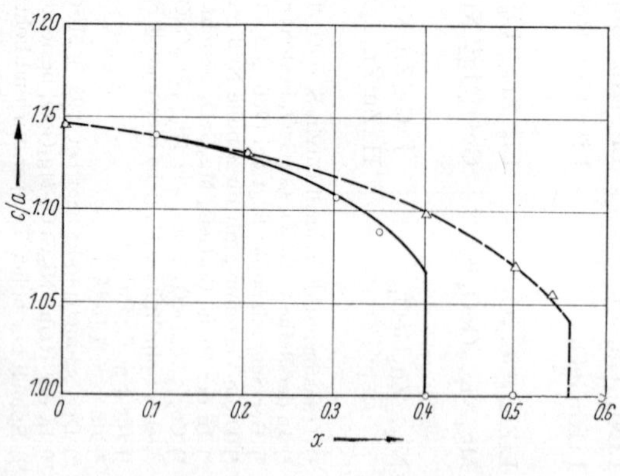

Fig. 17. $Cu_xMn_{2-2x}ZnGe_xO_4$ (○-○) und $Cu_{2x}Mn_{2-2x}Zn_{1-x}Ge_xO_4$ (△-△). Achsenverhältnis c/a vs. Zusammensetzungsparameter x [*65RB*].

6.3.3 Rh Spinelle und substituierte Rh Spinelle

(Literatur siehe 6.3.4, Seite 444)

Tab. 1. Kristallografische Eigenschaften binärer Rh^{3+} Oxid-Spinelle

Verbindung	Kationenverteilung	a Å	Farbe	Lit.	Fig.	Tab.
$CdRh_2O_4$	$Cd^{2+}[Rh_2^{3+}]O_4$	8.781		*59BFD*		
$CoRh_2O_4$	$Co^{2+}[Rh_2^{3+}]O_4$	8.495		*59BFD* *63B-1, 63BS, 64MH*	1	2
$CuRh_2O_4$	$Cu^{2+}[Rh_2^{3+}]O_4$	a = 8.702 [1]) c = 7.914		*63B-1* *59BFD, 64MH*	1	2
$MgRh_2O_4$	$Mg^{2+}[Rh_2^{3+}]O_4$	8.51 ± 0.01 8.53	rostbraun	*47RR-2* *64B-2* *59BFD, 63B-1*		
$MnRh_2O_4$	$Mn^{2+}[Rh_2^{3+}]O_4$	8.613		*59BFD* *63B-1*		2
$NiRh_2O_4$	$Ni^{2+}[Rh_2^{3+}]O_4$	a = 8.36 [2]) c = 8.67		*59BFD* *63BS* *63B-1, 64MH*	1	2
$ZnRh_2O_4$	$Zn^{2+}[Rh_2^{3+}]O_4$	8.52 ± 0.01 8.54	kaffeebraun	*47RR-2* *59BFD*		

[1]) $CuRh_2O_4$ ist tetragonal verzerrt mit c/a = 0.91, es wird kubisch bei 850 °K [*63B-1*].
[2]) $NiRh_2O_4$ ist tetragonal verzerrt mit c/a = 1.038, es wird kubisch bei 400 °K [*63BS*].

Tab. 2. Magnetische Eigenschaften einiger binärer Rh^{3+} Oxid-Spinelle

Verbindung	Θ_N °K	Θ_a °K	p_{eff} μ_B	Lit.	Fig.
$CoRh_2O_4$	27	−30	4.55	*63BS*	1
$CuRh_2O_4$	25	−70	1.87 [1]) 2.2 [2])	*63B-1*	1
$MnRh_2O_4$	15	−25	5.81	*63B-1*	
$NiRh_2O_4$	18	−20	3.38 [3]) 4.08 [4])	*63BS* *63BS, 63B-1*	1

[1]) für einen Meßbereich von 40 °K ≦ T ≦ 800 °K [*63B-1*].
[2]) für einen Meßbereich von 800 °K ≦ T ≦ 1000 °K [*63B-1*].
[3]) für einen Meßbereich von 20 °K ≦ T ≦ 380 °K [*63B-1*].
[4]) für einen Meßbereich von 380 °K ≦ T ≦ 900 °K [*63BS*] bis 1000 °C [*63B-1*].

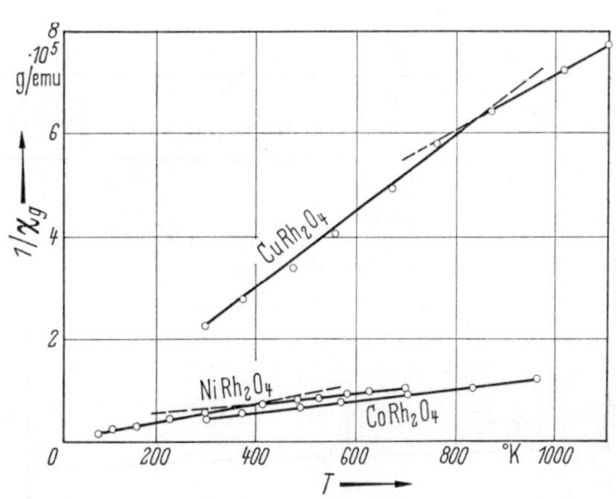

Fig. 1. $CoRh_2O_4$, $CuRh_2O_4$ und $NiRh_2O_4$. Reziproke Suszeptibilität pro Gramm $1/\chi_g$ vs. Temperatur T [*64MH*].

Tab. 3. Kristallografische und magnetische Eigenschaften ternärer Rh^{3+} Oxid-Spinelle

Verbindung	Kationenverteilung	a Å	ϱ_X g/cm³	Θ_N °K	Θ_a °K	p_{eff} μ_B	C_g cm³ °K/g	Lit.
$CoAl_xRh_{2-x}O_4$				24 [1]) 25 [2])	−30	4.5 [1]) 4.55 [2])		*63B-1*
$LiGeRhO_4$	$Li^{1+}[Ge^{4+}Rh^{3+}]O_4$	8.387	5.55					*62D-2, 64B-2* *62D-1*
$Li_{0.5}Ga_{0.5}Rh_2O_4$		8.447 ± 0.005	6.085					*66JD*
$Li_{0.5}In_{0.5}Rh_2O_4$		8.605 ± 0.004	6.225					*66JD*

[1]) für x = 0.4. [2]) für x = 0.2.

Fortsetzung nächste Seite

Tab. 3. Fortsetzung

Verbindung	Kationenverteilung	a Å	ϱ_X g/cm³	Θ_N °K	Θ_a °K	C_m cm³ °K/mole	Lit.
$LiMnRhO_4$	$Li^{1+}[Mn^{4+}Rh^{3+}]O_4$	8.30			−20	1.99	*63B-5*, *63B-2*
$LiSb_{0.5}Rh_{1.5}O_4$	$Li^{1+}[Sb^{5+}_{0.5}Rh^{3+}_{1.5}]O_4$	8.54					*64B-1*
$LiTiRhO_4$	$Li^{1+}[Ti^{4+}Rh^{3+}]O_4$	8.39	4.94				*62D-3*

Tab. 4. Kristallografische, magnetische und elektrische Eigenschaften binärer, ternärer und quaternärer Rh^{3+} Sulfo-Spinelle

Verbindung	Kationenverteilung	a Å	Θ_N °K	Θ_a °K	p_{eff} μ_B	χ_m 10^{-4} emu/mol	Θ_C °K	ϱ_{RT} Ωcm	Lit.	Fig.
$CoRh_2S_4$	$Co^{2+}[Rh^{3+}_2]S_4$	9.71						10	*64B-2**)	
		9.72 1)							*63K*	
		9.74	400	−400	4.3	3)		10^3 4)	*65B-2*	
		9.76 2)							*64BS*	
$CuRh_2S_4$	$Cu^{2+}[Rh^{3+}_2]S_4$	9.72				3.20			*64BS*	2
		9.73 1)							*63K*	
		9.78					6)		*67MSL*	
		9.790					4.35		*67RWM*	
						3.1 ··· 3.6 5)		$1.585 \cdot 10^{-3}$	*68LS*	
							4.37		*68SM**)	
$FeRh_2S_4$		9.80 2)							*64BS*	
$NiRh_2S_4$		9.64 1)							*63K*	
$Co_{1-x}Cu_xRh_2S_4$, $CoCr_xRh_{2-x}S_4$									*69L*, *68L*	
$Fe_{0.5}Cu_{0.5}Rh_2S_4$, $Co_{0.5}Cu_{0.5}Ti_{0.5}Rh_{1.5}S_4$ 10)									*69L*	

Tab. 5. Kristallografische, magnetische und elektrische Eigenschaften der Verbindung $CuRh_2Se_4$

a Å	χ_m 10^{-4} emu/mol	S μV/°K	Θ_c °K	ϱ_{RT} 10^{-3} Ωcm	Lit.	Fig.
10.263 ± 1			3.5		*67RWM*	2
10.34			3.49 8)		*67MSL*	
	0.5 ··· 1.1 7)	7.3		1.257	*68LS*	
			3.48		*68SM*	

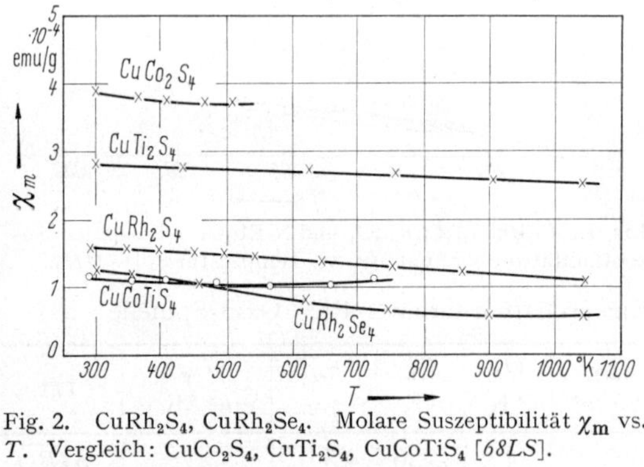

Fig. 2. $CuRh_2S_4$, $CuRh_2Se_4$. Molare Suszeptibilität χ_m vs. T. Vergleich: $CuCo_2S_4$, $CuTi_2S_4$, $CuCoTiS_4$ [*68LS*].

Tab. 6. $Cr_{1+x}CuRh_{1-x}Se_4$. Ferromagnetische Curie-Temperatur Θ_C und spezifischer Gleichstromwiderstand ϱ in Abhängigkeit vom Zusammensetzungsparameter x [*68LS*]

x	Θ_C °K	ϱ_{RT} 10^{-3} Ωcm	Fig.
1.0	460	12.57	3 ··· 8
0.75	425		
0.1		1	
0.0	270 9)	7.95	
−0.25	190	1	
−0.4	160		
−0.55	115		
−0.7	80		
−1.0		1.257	

1) unreine Probe [*63K*]. 2) unreine Probe [*64BS*].
3) Curie-Weiss Verlauf gemessen für $450\ °K < T < 800\ °K$ [*65B-2*].
4) gemessen an einer Kugelprobe mit der Dichte 85% in flüssigem Stickstoff [*65B-2*].
5) Die beobachteten Werte von $\chi_m = 1.1 \cdots 1.6 \cdot 10^{-4}$ emu/mol wurden korrigiert mit $\chi_{m\,diam} = -2 \cdot 10^{-4}$ emu/mol [*68LS*].
6) Keine reinen Proben; für 80 ··· 90 Vol% $\Theta_C = 4.8 \cdots 4.65$ °K [*67MSL*].
7) $\chi_{m\,diam.} = -2.4 \cdot 10^{-4}$ emu/mol und $\chi_{m\,korr.} = (2.9 \cdots 3.5) \cdot 10^{-4}$ emu/mol [*68LS*].
8) Keine reinen Proben, für 80 ··· 90 Vol% 3.49 ··· 3.45 [*67MSL*].
9) $\Theta_C = 255$ °K [*67LS*].
10) Angaben über ϱ [Ω cm], S [μV/°K], χ_m [emu/mol], p_m [μ_B], Θ_N, Θ_C [°K] [*69L*].
*) Siehe auch [*68L*].

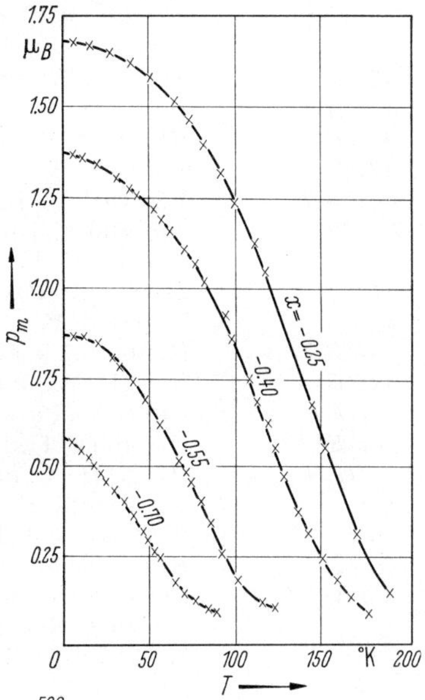

Fig. 4. $Cr_{1+x}CuRh_{1-x}Se_4$. Magnetisches Moment pro Molekül p_m bei $H = 9$ kOe vs. Temperatur T und Zusammensetzungsparameter x [68LS].

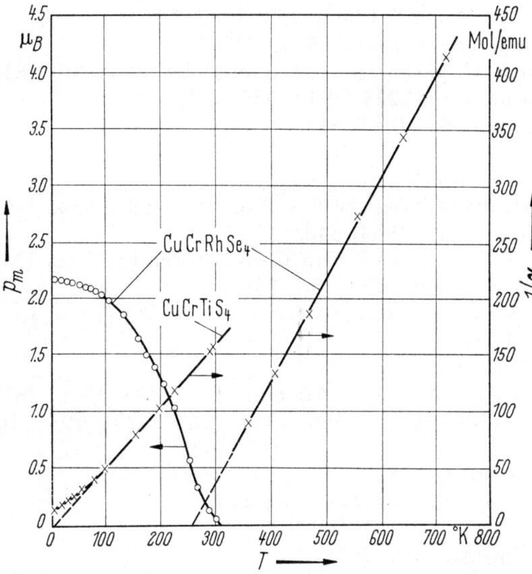

Fig. 3. $CrCuRhSe_4$. Magnetisches Moment pro Molekül p_m bei $H = 9$ kOe und reziproke molare Suszeptibilität $1/\chi_m$ vs. Temperatur T. $CuCrTiS_4$ zum Vergleich [68LS].

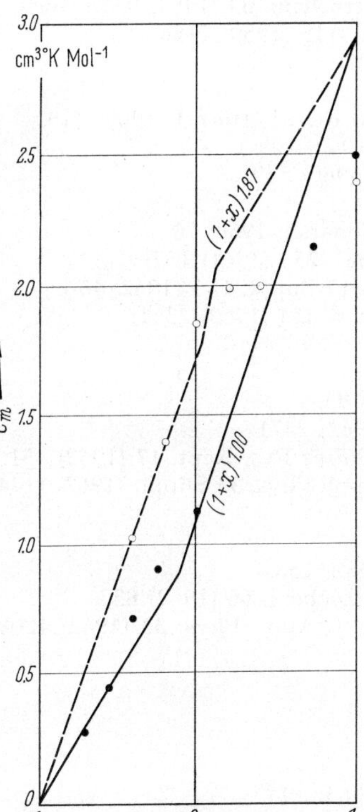

Fig. 5. $Cr_{1+x}CuRh_{1-x}Se_4$. Molare Curie-Konstante C_m vs. Zusammensetzungsparameter x. Der gestrichelte Verlauf bezieht sich auf eine Theorie der Bänderüberschneidung [68LS].
$C_m = 1.1$ bei $x = 0$ [67LS].

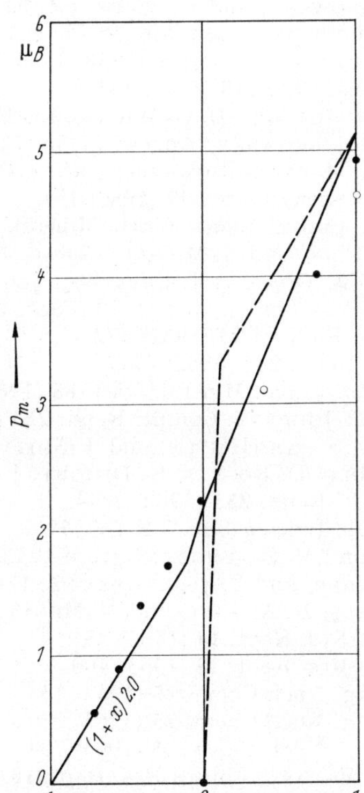

Fig. 6. $Cr_{1+x}CuRh_{1-x}Se_4$. Molares Sättigungsmoment p_m bei 4.5 °K vs. Zusammensetzungsparameter x. Der gestrichelte Verlauf bezieht sich auf eine Theorie der Bänderüberschneidung [68LS].

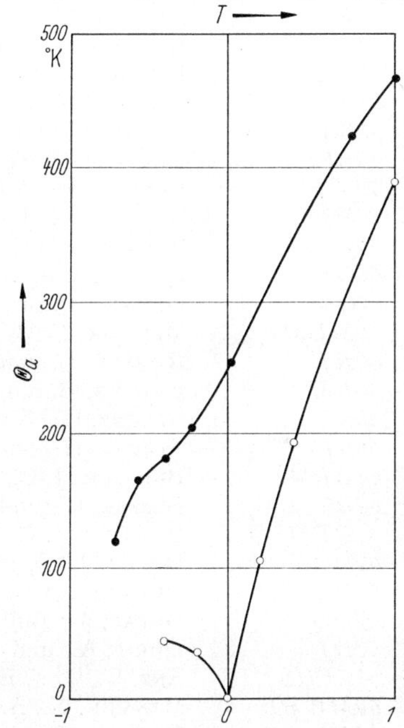

Fig. 7. $Cr_{1+x}CuRh_{1-x}Se_4$ (●); $Cr_{1+x}CuTi_{1-x}S_4$ (○). Asymptotische Curie-Temperatur Θ_a vs. x [68LS].

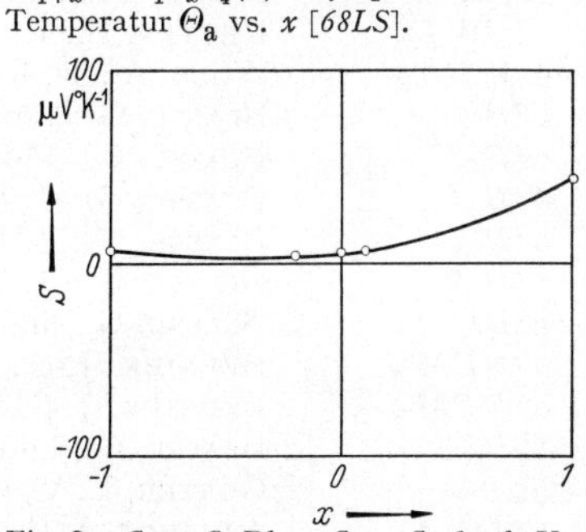

Fig. 8. $Cr_{1+x}CuRh_{1-x}Se_4$. Seebeck-Koeffizienten S vs. Zusammensetzungsparameter x [68LS].

6.3.4 Literatur zu 6.3.1 ··· 6.3.3

32JS	JANDER, W., und W. STAMM: Z. Anorg. Allgem. Chem. **207** (1932) 289.
47RR-1	RÜDORFF, W., und B. REUTER: Z. Anorg. Allgem. Chem. **255** (1947) 177.
47RR-2	RÜDORFF, W., und B. REUTER: Z. Anorg. Allgem. Chem. **253** (1947) 194.
47VH	VERWEY, E. J., und E. L. HEILMANN: J. Chem. Phys. **15** (1947) 174.
47VHR	VERWEY, E. J., P. W. HAAYMAN, und F. C. ROMEIJN: J. Chem. Phys. **15** (1947) 181.
49AB	ANDRIEUX, J.-L., und H. BOZON: Compt. Rend. **228** (1949) 565.
51L	LOVELL, G. H. B.: Trans. Brit. Ceram. Soc. **50** (1951) 315.
51Z	ZENER, C.: Phys. Rev. **82** (1951) 403.
53R	ROMEIJN, F. C.: Philips Res. Rept. **8** (1953) 304.
54RBR	RICHARDSON, H. M., F. BALL, G. R. RIGBY: Trans. Brit. Ceram. Soc. **53** (1954) 376.
54RR	ROY, D. M., und R. ROY: Am. Mineralogist **39** (1954) 340.
56DBP	DURIF-VARAMBON, A., E. F. BERTAUT, R. PAUTHENET: Ann. Chim. (Paris) **13** (1956) 525.
56DR	DACHILLE, F., und R. ROY: Bull. Geol. Soc. Am. **67** (1956) 1682.
56HLH	HAHN, H., C. DE LAURENT, und B. HARDER: Z. Anorg. Allgem. Chem. **283** (1956) 138.
56L	LOTGERING, F. K.: Philips Res. Rept. **11** (1956) 337.
56R	RINGWOOD, A. E.: Am. J. Sci. **254** (1956) 707.
56RWB	RÜDORFF, W., G. WALTER, und H. BECKER: Z. Anorg. Allgem. Chem. **285** (1956) 287.
57LM	LENSEN, M., und A. MICHEL: Congr. Intern. Chim. Pure Appl. 16[e] Paris, **1957**, in Struct. Rep. **23** (1959), 381.
57M	MCCLURE, D. S.: Phys. Chem. Solids **3** (1957) 311.
57RS	REUTER, B., und J. SIEWERT: Naturwissenschaften **44** (1957) 440.
58R-1	RINGWOOD, A. E.: Geochim. Cosmochim. Acta **13** (1958) 303.
58R-2	RINGWOOD, A. E.: Geochim. Cosmochim. Acta **15** (1958) 18.
58R-3	RINGWOOD, A. E.: Bull. Geol. Soc. Am. **69** (1958) 129.
58WC	WICKHAM, D. G., und W. J. CROFT: Phys. Chem. Solids **7** (1958) 351.
59BFD	BERTAUT, E. F., F. FORRAT, und J. DULAC: Compt. Rend. **249** (1959) 726.
59L	LENSEN, M.: Ann. Chim. (Paris) **4** (1959) 891.
59RJR	REUTER, B., J. JASKOWSKY, und E. RIEDEL: Z. Elektrochem. **63** (1959) 937.
59WG	WICKHAM, D. G., und J. B. GOODENOUGH: Phys. Rev. **115** (1959) 1156.
60DR	DACHILLE, F., und F. ROY: Am. J. Sci. **258** (1960) 225.
60G	GENNES, P. DE: Phys. Rev. **118** (1960) 141.
60MDM	MENYUK, N., K. DWIGHT, und D. G. WICKHAM: Phys. Rev. Letters **4** (1960) 119.
60RJ	REUTER, B., und J. JASKOWSKY: Angew. Chem. **72** (1960) 209.
60RM	REUTER, B., und G. MARX: Naturwissenschaften **47** (1960) 539.
60S	STRUNZ, H.: Naturwissenschaften **47** (1960) 154.
60SJ	STRUNZ, H., und P. JAKOB: Neues Jahrb. Mineral. Monatsh. **1960**, 78.
61BPM	BERNIER, J.-C., P. POIX, und A. MICHEL: Compt. Rend. **253** (1961) 1578.
62BG	BLASSE, G., und E. W. GORTER: J. Phys. Soc. Japan **17** Suppl. B-1 (1962) 176.
62BPM	BERNIER, J.-C., P. POIX, und A. MICHEL: Compt. Rend. **254** (1962) 2351.
62D-1	DULAC, J.: J. Phys. Radium **23** (1962) 872.
62D-2	DULAC, J.: Compt. Rend. **254** (1962) 1312.
62D-3	DULAC, J.: Bull. Soc. Franc. Mineral. Crist. **85** (1962) 93.
62DJ	DURIF, A., und J.-C. JOUBERT: Compt. Rend. **255** (1962) 2471.
62EFHS	ENZ, U., J. F. FAST, S. VAN HOUTEN, und J. SMIT: Philips Res. Rept. **17** (1962) 451.
62MWRD	MENYUK, N., A. WOLD, D. ROGERS, K. DWIGHT: J. Appl. Phys. **33** Suppl. (1962) 1144.
62P	PLUMIER, R.: Compt. Rend. **255** (1962) 2244.
62R	ROGERS, D. B.: MIT, Lincoln Lab. TR-269, **1962**.
62RS	RINGWOOD, A. E., und M. SEABROCK: Nature **193** (1962) 158.
62RWJ	REUTER, B., R. WEBER, und J. JASKOWSKY: Z. Elektrochem. **66** (1962) 832.
62WRAM	WOLD, A., D. ROGERS, R. ARNOTT, und N. MENYUK: J. Appl. Phys. **33** (1962) 1208.
63B-1	BLASSE, G.: Philips Res. Rept. **18** (1963) 383.
63B-2	BLASSE, G.: Philips Res. Rept. **18** (1963) 400.
63B-3	BLASSE, G.: J. Inorg. Nucl. Chem. **25** (1963) 136.
63B-4	BLASSE, G.: J. Inorg. Nucl. Chem. **25** (1963) 230.
63B-5	BLASSE, G.: J. Inorg. Nucl. Chem. **25** (1963) 734.
63BF	BLASSE, G., und J. F. FAST: Philips Res. Rept. **18** (1963) 393.
63BPM-1	BERNIER, J.-C., P. POIX, und A. MICHEL: Bull. Soc. Chim. France **1963**, 1724.
63BPM-2	BERNIER, J.-C., P. POIX, und A. MICHEL: Compt. Rend. **256** (1963) 5583.
63BS	BLASSE, G., und D. J. SCHIPPER: Phys. Letters **5** (1963) 300.
63G	GORTER, E. W.: J. Appl. Phys. **34** (1963) 1253.
63JD	JOUBERT, J.-C., und A. DURIF: Bull. Soc. Franc. Mineral. Crist. **86** (1963) 430.
63K	KOERTS, K.: Rec. Trav. Chim. **82** (1963) 1099.
63PT	PLUMIER, R., und A. TARDIEU: Compt. Rend. **257** (1963) 3858.

63RAWG	Rogers, D. B., R. J. Arnott, A. Wold, J. B. Goodenough: J. Phys. Chem. Solids **24** (1963) 347.
64B-1	Blasse, G.: J. Inorg. Nucl. Chem. **26** (1964) 1473.
64B-2	Blasse, G.: Philips Res. Rept. Suppl. (1964).
64BS	Blasse, G., und D. J. Schipper: J. Inorg. Nucl. Chem. **26** (1964) 1467.
64BQPM	Bertaut, E. F., Vu van Qui, R. Pauthenet, und A. Murasik: J. Phys. **25** (1964) 516.
64DMRW	Dwight, K., N. Menyuk, D. B. Rogers, und A. Wold: Proc. Int. Conf. Magnetism, Nottingham **1964**, 538.
64JD	Joubert, J.-C., und A. Durif: Bull. Soc. Franc. Mineral. Crist. **87** (1964) 47.
64MH	Miyahara, S., und S. Horiuti: Proc. Int. Conf. Magnetism, Nottingham, **1964**, 550.
64RGW	Rogers, D. B., J. B. Goodenough, und A. Wold: J. Appl. Phys. **35** (1964) 1069.
65B-1	Blasse, G.: Philips Res. Rept. **20** (1965) 528.
65B-2	Blasse, G.: Phys. Letters **19** (1965) 110.
65BPM	Bernier, J.-C., P. Poix, und A. Michel: Compt. Rend. **261** (1965) 3822.
65BRW	Bouchard, R. J., P. A. Russo, und A. Wold: Inorg. Chem. **4** (1965) 685.
65JGD	Joubert, J.-C., J.-C. Grenier, und A. Durif: Compt. Rend. **260** (1965) 2472.
65L	Lenglet, M.: Rev. Chim. Minerale **2** (1965) 217.
65NM	Neuhaus, A., und H. J. Meyer: Naturwissenschaften **52** (1965) 639.
65R-1	Reuter, B.: Bull. Soc. Chim. France **1965**, 1053.
65R-2	Rossiter, M. J.: J. Phys. Chem. Solids **26** (1965) 775.
65RB	Robbins, M., und R. K. Baltzer: J. Appl. Phys. **36** (1965) 1039.
65VLR	Villers, G., A. Lecerf, und M. Rault: Compt. Rend. **260** (1965) 3017.
65WHA	Watanabe, M., T. Hirasawa, Y. Aiyama: Bull. Electrotech. Lab. (Kyoto) **29** (1965) 22.
66B	Blasse, G.: J. Phys. Chem. Solids **27** (1966) 612.
66DGJD	Durif, A., J.-C. Grenier, J.-C. Joubert, und Tran-Qui Duc: Bull. Soc. Franc. Mineral. Crist. **89** (1966) 407.
66GPS	Gaines, A. M., A. J. Perotta, und D. A. Stephenson: J. Am. Ceram. Soc. **49** (1966) 516.
66I	Imbert, P.: Compt. Rend. **263 B** (1966) 767.
66J	Jaskowsky, J.: Ber. Bunsenges. **70** (1966) 199.
66JD	Joubert, J.-C., und A. Durif: Bull. Soc. Franc. Mineral. Crist. **89** (1966) 26.
66P	Plumier, R.: Compt. Rend. **263** (1966) 173.
66R	Rossiter, M. J.: Phys. Letters **21** (1966) 128.
66RD	Robbins, M., und L. Darcy: J. Phys. Chem. Solids **27** (1966) 741.
66RFK	Rogers, D. B., A. Feretti, und W. Kunnmann: J. Phys. Chem. Solids **27** (1966) 1445.
66RJ	Reuter, B., und J. Jaskowsky: Ber. Bunsenges. **70** (1966) 189.
66RLRJ	Reuter, B., W. Laqua, E. Riedel, und J. Jaskowsky: Ber. Bunsenges. **70** (1966) 194.
66TS	Tourne, G., und J. Schaffner: Compt. Rend. **262C** (1966) 1787.
66TTA	Tanaka, M., T. Tokoro, und Y. Aiyama: J. Phys. Soc. Japan **21** (1966) 262.
66VL	Villers, G., und A. Lecerf: Compt. Rend. **263** (1966) 427.
67BP	Bernier, J.-C., und P. Poix: Ann. Chim. (Paris) **2** (1967) 81.
67DR	Datta, R. K., und R. Rustum: J. Am. Ceram. Soc. **50** (1967) 578.
67LS	Lotgering, F. K., und R. P. van Stapele: Solid State Commun. **5** (1967) 143.
67MSL	Maaren, N. H. van, G. M. Schaeffer, F. K. Lotgering: Phys. Letters **25A** (1967) 238.
67P	Plumier, R.: Compt. Rend. **264 B** (1967) 278.
67R	Reinen, D.: Theoret. Chim. Acta **8** (1967) 260.
67RGG	Rogers, D. B., J. L. Gillson, und T. E. Gier: Solid State Commun. **5** (1967) 263.
67RM	Reuter, B., und K. Müller: Naturwissenschaften **54** (1967) 164.
67RR	Reuter, B., und E. Riedel: Ber. Bunsenges. **71** (1967) 189.
67RWM	Robbins, M., R. H. Willems, und R. C. Miller: Solid State Commun. **5** (1967) 933.
67S	Siratori, K.: J. Phys. Soc. Japan **23** (1967) 948.
68C	Cini, L.: Radex Rundschau **1968**, 102.
68HI	Hartmann-Boutron, F., und P. Imbert: J. Appl. Phys. **39** (1968) 775.
68L	Lotgering, F. K.: J. Phys. Chem. Solids **29** (1968) 699.
68LS	Lotgering, F. K., und R. P. van Stapele: J. Appl. Phys. **39** (1968) 417.
68NK	Navrotsky, A., und O. J. Kleppa: J. Inorg. Nucl. Chem. **30** (1968) 479.
68P	Plumier, R.: Compt. Rend. **267** (1968) 1057.
68R-1	Reinen, D.: Z. Anorg. Allgem. Chem. **356** (1968) 172.
68R-2	Reuter, B.: Private Mitteilung, 1968.
68SM	Schaeffer, G. M., und M. H. van Maaren: Intern. Conf. on Low. Temp. Phys., St. Andrews, **1968**, 2, 1033.
68SS	Sinha, K. P., und A. P. B. Sinha: Indian J. Pure Appl. Phys. **6** (1968), 105.
68VBBMC	Vorob'ev, Yu. P., M. P. Bogdanovich, E. G. Bogacheva, A. N. Men', und G. I. Chufarov: Proc. Acad. Sci. USSR, Phys. Chem. Sect. (English Transl.) **180** (1968) 340.
69APKD	Amirova, S. A., V. G. Prokhorora, V. P. Kudryashor und T. G. Danilova: Russ. J. Inorg. Chem. (English Transl.) **14** (1969) 163.
69L	Lotgering, F. K.: J. Phys. Chem. Solids **30** (1969) 1429.

6.3.5 Al Spinelle und substituierte Al

Tab. 1. Kristallografische Eigenschaften

Verbindung	Kationenverteilung [1]	a Å	u
γ-Al_2O_3		7.985 ± 0.006	
		7.90	
$CdAl_2O_4$	$Cd_{1/3}Al_{2/3}Cd_{2/3}Al_{4/3}O_4$	8.076	0.384
	$CdAl_2O_4$	8.137 [2]	
$CoAl_2O_4$ [20]		8.069	
		8.06	
		8.101	0.390
		8.07 $\begin{cases} +0.02 \\ -0.01 \end{cases}$	
		8.08	
	$Co[Al_2]O_4$	8.105	0.387
	$Co[Al_2]O_4$	8.096	
	$Co[Al_2]O_4$	8.105	0.388
		8.112	
	$Co_{1-y}Al_y[Co_yAl_{2-y}]O_4$ [3]	8.100	
	$Co_{1-y}Al_y[Co_yAl_{2-y}]O_4$	8.106	0.386 ± 0.001
		8.103 ± 0.002	
$Cu_{0.5}Al_{2.5}O_4$ [4]		8.06	
$CuAl_2O_4$		8.072 ± 0.007	
		8.050 ± 0.005	
		8.055 ± 0.005	
		8.064	
		8.074	
		8.07	
		8.08 [6]	
		8.04 [7]	
	$Cu_{1-y}Al_y[Cu_yAl_{2-y}]O_4$ [8]	8.086	
	$Cu_{0.56}Al_{0.44}[Cu_{0.44}Al_{1.56}]O_4$	8.086	
$FeAl_2O_4$ [9][20]		8.146 ± 0.004	
		8.135 ± 0.002	0.390
		8.100	
		8.080 ± 0.002	
		8.13 [7]	
		8.16 [6]	
	$Fe[Al_2]O_4$	8.136	
	$Fe[Al_2]O_4$	8.159 ± 0.002	
		8.139	
		8.125 ± 0.002	
		8.152 ± 0.001	
		8.150 ± 0.002	
	$Fe_{1-y}Al_y[Fe_yAl_{2-y}]O_4$ [11]	8.152 [12]	0.3885 ± 0.001
		8.138 ± 0.001 [10] [13]	
$Li_{0.5}Al_{2.5}O_4$ [16]	$Li_{1/6}Al_{5/6}[Li_{1/6}Al_{5/6}]_2O_4$	7.903 ± 0.004	0.385 ± 0.01
		7.92	
	$Al[Li_{0.5}Al_{1.5}]O_4$	7.911	
		7.910 ± 0.003	
	$Li_{1-y}Al_y[Li_{y-0.5}Al_{2.5-y}]O_4$ [17]	7.925	
	$Li_{0.18}Al_{0.82}[Li_{0.32}Al_{1.68}]O_4$	7.921 ± 0.001	
		7.94	
		7.901 ± 0.002	

Fußnoten siehe S. 450

Spinelle (Literatur siehe 6.3.8, Seite 476)

binärer Al^{3+} Oxid-Spinelle

ϱ_X g/cm³	ϱ_p g/cm³	Farbe	Lit.	Fig.	Tab.
3.646	3.619		*35K*		
			57FS, 57SS, 59YY		
			64L, 68KTHD		
			53H		
5.802	5.75	braun	*55HFKSS*		
			58SBB, 59M		
4.50	4.37	dunkelblau (Thénard-Blau)	*24HH, 27H, 53R*	1, 7···9,	2, 3
4.50	4.57		*29NP*	16	
		Preußisch-Blau	*32BP, 32KT*		
4.44		Preußisch-Blau	*34KT*		
4.44			*53BR-2*		
			53R, 59M, 60CA		
			56C, 55HFKSS, 61H		
			65TS		
			58SBB		
			54GPG		
			61S, 64R, 68R		
			63RAWG		
			64S-2		
			51KR, 69L		
4.60			*27H*		
			29P		
4.625	4.732 [5]		*30P-1*		
			32H-1		
		grünlich-graubraun	*32KT*		
4.55		holzbraun	*34KT, 47SR, 53H*		
			52TK		
4.64			*52TK*		
			54BD, 55HFKSS		
			58D		
			59M, 65ST, 66S		
			27H	10···12,	2
4.392			*31CAB, 32BP*	17, 18	
		hellbraun	*32KT*		
4.34		okergelb	*34KT, 47SR*		
			52TK		
			52TK		
			53R, 55HFKSS		
			65R, 66OIS		
			56FH		
			56HF		
			58AS, 63R		
			58T		
			64R		
4.26		[14]	*64S-2* [15]		
3.6078	3.606		*35K, 47VH, 60DH*	2	2, 4
			51KR, 61KFHSB		
			52B-2, 54G, 59K-2		
			61SR		
			62BH, 62LC-2		
			63DR		
			64B-1		
			65RGA [18]		

(Fortsetzung siehe nächste Seite)

Tab. 1 (Fortsetzung, Fußnoten siehe S. 450)

Verbindung	Kationenverteilung [1]	a Å	u
$MgAl_2O_4$ [19][20]		8.090 ± 0.003	0.384
		8.050 ± 0.005	
	$Mg[Al_2]O_4$	8.086 ± 0.003	0.390 ± 0.006
	$Mg[Al_2]O_4$ [23]	8.0808	0.387 ± 0.001
	$Mg[Al_2]O_4$ [24]	8.089 ± 0.001 [25]	
		$8.08 \begin{cases} +0.03 \\ -0.02 \end{cases}$	0.388 ± 0.005
		8.059 ± 0.004	
		8.073 [22]	
		8.086	
		8.09	
		8.080 ± 0.005	0.387
		8.085	
		8.080	0.386
		8.081	
	$Mg_{1-y}Al_y[Mg_yAl_{2-y}]O_4$ [26]	8.0866 [27]	0.387 ± 0.001
	$Mg_{1-y}Al_y[Mg_yAl_{2-y}]O_4$ [28]	8.0832 ± 0.0005	0.3873 ± 0.0001
		8.0868 [27]	
		8.0843 ± 0.0012	
		8.081	
		8.02	
		8.075	
		8.076 ± 0.02 [42]	
$MnAl_2O_4$ [30]		8.263 ± 0.007	
		8.271 ± 0.002	0.390
	$Mn[Al_2]O_4$	8.269	
	$Mn[Al_2]O_4$	8.277	
		$8.26 \begin{cases} +0.02 \\ -0.01 \end{cases}$	0.390
	$Mn_{1-y}Al_y[Mn_yAl_{2-y}]O_4$ [31]	8.241 [31]	
	$Mn_{1-y}Al_y[Mn_yAl_{2-y}]O_4$ [32]	8.200 [32]	
	$Mn_{1-y}Al_y[Mn_yAl_{2-y}]O_4$ [33]	8.207	0.3907 ± 0.0005
		8.195	
		8.222	
$NiAl_2O_4$ [41]		8.045 ± 0.005	
	$Ni[Al_2]O_4$	8.050	0.390 ± 0.01
	$Ni[Al_2]O_4$	8.046	
		$8.04 \begin{cases} +0.03 \\ -0.01 \end{cases}$	
		8.030	
	$Ni_{0.24\pm0.02}Al_{0.76}[Ni_{0.76}Al_{1.24}]O_4$	8.046	0.381 ± 0.002
	$Ni_{0.15\pm0.03}Al_{0.85}[Ni_{0.85}Al_{1.15}]O_4$ [31]	8.043	
	$Ni_{0.20\pm0.03}Al_{0.80}[Ni_{0.80}Al_{1.20}]O_4$ [32]		
	$Ni_{0.185}Al_{0.815}[Ni_{0.815}Al_{1.185}]O_4$ [34]		
	$Ni_{0.26}Al_{0.74}[Ni_{0.74}Al_{1.26}]O_4$ [35]	8.04	
	$Al_{1.0}[Ni_{1.0}Al_{1.0}]O_4$ [36]		
$SnAl_2O_4$		8.12	
$ZnAl_2O_4$ [38][20]		8.099 ± 0.003	
		8.060 ± 0.005	
		8.050 ± 0.005	
	$Zn[Al_2]O_4$	8.062 ± 0.001	0.390
	$Zn[Al_2]O_4$	8.086	
		8.093 ± 0.004	
		8.087	
		8.09 [6], 8.07 [7]	
		8.06	
		8.088	
		8.086 ± 0.001	0.3889 ± 0.0003
		8.085 ± 0.001 [39]	

Fußnoten siehe S. 450

ϱ_X g/cm³	ϱ_p g/cm³	Farbe	Lit.	Fig.	Tab.
3.58			*15N, 27H*	16, 19	2, 3
3.63		[21]	*29P, 29K, 30P-1*		4
3.548			*31CAB, 32BP, 32PBB*		
3.581			*51SF, 52B-1*		
			60BHHL, 61HL		
		weiß	*31M, 34KT*		
3.588			*32HN*		
3.56			*35K*		
			52TK, 53R		
3.55			*53BR-2*		
3.6	3.54 ± 0.01		*53RRO, 54KFCV, 53WS*		
			58SBB		
			58JS, 60W		
			61LC		
3.58		durchsichtig, leicht rosa	*62S, 64SFHM*		
			67F		
			62S		
			66A		
			66CT		
	3.58		*66MF* [29]		
			69ZK		
			69DR		
4.08			*27H*	13, 14	2
4.031			*31CAB, 32BP, 32PBB*	33	
4.04		braun	*32KT, 53R*		
			55HFKSS, 61H		
		holzbraun	*34KT, 51W*		
			54GPG		
			54GPG		
			64R		
			57B		
			58SBB *)		
4.51			*27H*	3···6,	2, 3
		himmelblau	*32KT, 32BP*	15, 16	
4.50			*43VB*		
		himmelblau	*34KT*		
4.52			*53BR-2, 53MPH*		
		blau	*53R, 66S, 67K*		
			53MP, 54GPG, 62LC-1		
			54GPG, 61S		
			60S		
			60S, 67B, 67DR		
			67DR [37]		
			53SU		
4.59			*27H*	16	2, 3
		weiß	*29P, 34KT*		4
4.68			*30P-1*		
4.615			*31CAB, 32BP, 32PBB*		
4.64			*53BR-2*		
4.565		weiß	*32HN, 32KT*		
4.61			*43VB*		
4.63			*52TK, 55HFKSS*		
			53R, 58SBB, 61H		
			65CT, 66CT, 68KTHD		
			66ANM, 67F		
			67DR [40]		

Fußnoten zu Tab. 1

1) In [*47RR*] ist für die meisten dieser Spinelle das Verhältnis der elektrostatischen Bindungsenergie E_N/E_A für die Anziehung Me-O berechnet, Index N = Normalstruktur, A = Austausch-, inverse Struktur.
2) Durch Extrapolation der Vegardschen Geraden der Verbindungsreihe $Mg_{0.9}Co_{0.1}Al_2O_4$-$Cd_{0.9}Co_{0.1}Al_2O_4$ erhalten [*58SBB*].
3) Von 1400 °C langsam abgekühlt: $y = 0.19$; von 1400 °C abgeschreckt: $y = 0.31$ [*54GPG*].
4) Spinellähnlich, tetragonal: $c/a = 1.08$ [*51KR*].
5) In der Zusammenfassung der Arbeit wird angegeben: 4.532 [*30P-1*].
6) Elektronenbeugung [*52TK*].
7) Röntgenbeugung [*52TK*].
8) Von 1000 °C langsam abgekühlt: $y = 0.44 \pm 0.04$; von 1000 °C abgeschreckt: $y = 0.40 \pm 0.04$ [*54BD*].
9) Schmelzpunkt: 1740 °C [*55FH-2*]; 1780 °C [*56FH*]; Mikrohärte: 1010 ··· 1270 kg/mm² [*56FH*]; Mineral: Hercynit.
10) Mit geringem $MgAl_2O_4$-Gehalt. Berechnet für reinen $FeAl_2O_4$ $a = (8.155 \pm 0.044)$ Å [*64S-2*].
11) Mit $y = 0.144 \pm 0.01$ [*64R*].
12) Neutronenbeugung [*64R*].
13) Einkristall [*64S-2*].
14) Undurchsichtig, grünlich-gelb in dünnen Segmenten [*64S-2*].
15) Weitere Literatur: [*32PBB*], [*55FH-1*], [*55FH-2*], [*57G*], [*59M*], [*59PT*], [*60W*], [*61H*], [*62L*], [*63MT*], [*65RC*], [*66BOGG*], [*66TTA*].
16) Brechungsindex $n = 1.735 \pm 0.001$ [*35K*]; 1.725 [*61SR*]. $Li_{0.5}Al_{2.5}O_4$ tritt in mehreren Phasen auf. Bei 1295 °C findet ein Ordnungs-Unordnungs-Übergang statt. Die Hochtemperaturmodifikation ist ungeordnet und hat Spinellstruktur. Die Tieftemperaturmodifikation hat eine geordnete, einfach kubische Struktur mit kleinerer Gitterkonstanten ($a = 7.908$ Å [*62LC-2*], $a = 7.909$ Å [*59K-2*]). Außerdem existieren verschiedene metastabile Phasen.
17) Mit $y = 0.67 \cdots 1.0$ [*62BH*].
18) Weitere Literatur: [*53Y*], [*55HFKSS*], [*56A*], [*58HSW*], [*61F*], [*61HL*], [*62F*], [*62LC-1*], [*63SB*], [*64L*], [*66ABMP*], [*66K-2*], [*67T*], [*65NM*], [*68DR*], [*69L*].
19) Mineral: Spinell.
20) Weitere Untersuchungen an $MgAl_2O_4$: a) Wärmeausdehnung (auch für $ZnAl_2O_4$) [*57BC*]; b) Wärmekapazität C_p [*55K*]; c) linearer Wärmeausdehnungskoeffizient (auch von $ZnAl_2O_4$) [*32PBB*], [*57BC*], [*52Z*], [*66MF*]; d) Wärmeleitfähigkeit von Einkristallen (auch für $CoAl_2O_4$ und $FeAl_2O_4$) [*64S-2*], [*54KFCV*], [*62S*]; e) Schmelzpunkt, Härte und andere Eigenschaften von $MgAl_2O_4$ [*66MF*], [*32WR*], [*67F*].
21) Synthetisch: rosenrot, natürlich: rubin [*29K*].
22) Synthetischer Kristall [*35K*].
23) Synthetisch: mit „etwas Tendenz zu invers“ [*52B-1*].
24) Die Kationenverteilung ist temperaturabhängig; bei 800 ··· 900 °C Übergang des geordneten natürlichen Spinells in den Strukturzustand des synthetischen Spinells mit weitgehend ungeordneter Verteilung der Mg^{2+} und Al^{3+} Ionen über die Oktaeder- und Tetraederstellen [*60BHHL*], [*61HL*].
25) Natürlicher Kristall; erhitzt: (8.085 ± 0.001) Å [*61HL*].
26) $y = 0.10 \cdots 0.15$ [*64SFHM*].
27) Natürlicher Kristall [*62S*].
28) $y = 0.08 \pm 0.04$ [*67F*].
29) Weitere Literatur: [*28P*], [*28R*], [*30T*], [*32HN*], [*32WR*], [*34CHB*], [*35HS*], [*40K*], [*49KB*], [*51MB*], [*52Z*], [*53O*], [*54SJ*], [*55B*], [*55FH-2*], [*55K*], [*57BC*], [*57J*], [*57SJ*], [*59K-1*], [*59M*], [*59SL*], [*59YY*], [*60CA*], [*60LC*], [*60V*], [*61BLW*], [*61H*], [*61PWL*], [*62A*], [*62BH*], [*62LC-1*], [*63A*], [*64BGS*], [*64L*], [*64S-2*], [*65D-1*], [*65D-2*], [*65S*], [*65SLR*], [*66ANM*], [*66DR*], [*66EW*], [*66NK*], [*66OIS*], [*66VS*], [*66W*], [*67DC*], [*67OCI*], [*67S-1*], [*68D*], [*68L*], [*68KTHD*].
30) Mineral: Galaxit.
31) Von 1400 °C langsam abgekühlt: $y = 0.34 \pm 0.04$ [*54GPG*].
32) Von 1400 °C abgeschreckt: $y = 0.29 \pm 0.04$ [*54GPG*].
33) Genauer Wert des Inversionsgrades: $y = 0.084 \pm 0.044$ [*64R*].
34) Von 900 °C abgeschreckt [*60S*].
35) Von 1500 °C abgeschreckt [*60S*].
36) Von 600 °C abgeschreckt [*67DR*].
37) Weitere Literatur: [*53M*], [*53W*], [*54G*], [*54M-2*], [*55HFKSS*], [*56LÅ*], [*59M*], [*62BH*], [*65SLR*], [*65ST*], [*65TS*], [*67K*], [*68R-1*], [*68KTHD*], [*68R-2*], [*68DR*].
38) Mineral: Gahnit.
39) Bei 600 °C und 703 at oder bei 1200 °C und Normaldruck abgeschreckt [*67DR*].
40) Weitere Literatur: [*31B*], [*32H-2*], [*32H-3*], [*40K*], [*56C*], [*56LÅ*], [*57BC*], [*59M*], [*59SL*], [*60CA*], [*62LC-1*], [*64BGS*], [*64DLW*], [*64EE*], [*64L*], [*64S-1*], [*65ST*], [*68DR*].
41) $(NiO)_{1-x}(Al_2O_3)_x$ für $0.4 \leq x \leq 0.6$ besitzen Spinellstruktur bei abweichender Stöchiometrie [*68R-2*].
42) Bei 1200 °C; 8.084 ± 0.001 Å bei 2100 °C [*68DR*].
*) Siehe auch [*59E*], [*59M*], [*59SL*], [*62BH*], [*62LC-1*], [*64L*], [*66FG*], [*66LN*].

Tab. 2. Magnetische Eigenschaften binärer Al^{3+} Oxid-Spinelle

Verbindung	T °K	p_m μ_B	p_{eff} μ_B	Θ_N °K	Θ_a °K	C_g cm³ °K/g	C_m cm³ °K/mol	$\chi_{g\,diam}$ 10^{-6} emu/g	$\chi_{m\,diam}$ 10^{-6} emu/mol	a 10^{-3} emu/mol	g_A [1]	g_B [1]	Lit.	Fig.	Tab.
$CoAl_2O_4$	< 800					3.07		−0.43	−76		2.0	4.26	*54GPG*	7…9	
	> 800					3.86					2.0	4.30	*54GPG*		
			4.02		− 35		2.023		−76	0.46			*56C, 60CA*		
					− 15 [2])		2.137			0.400			*62L, 63B-3*		
							2.480 [5])		−76	0.522 [3])			*62L*		
			3.12	4	− 19		1.99						*64R*		
$FeAl_2O_4$	77	0 [4])			−144		3.80						*59PT, 62L*	10…12	
							3.245 [3])		−76 [5])	0.261 [3])			*62L*	17, 18	
			4.14	8	− 76		3.18						*64R, 64S-2* *)		
$Li_{0.5}Al_{2.5}O_4$													*66ABMP, 61F*		4
													62F		
$MgAl_2O_4$									−48				*40K* **)	19	
									−57				*60CA*		
									−69.7				*66ANM*		
$MnAl_2O_4$ [6])			5.65			4.08		−0.52			1.94		*54GPG, 57B*	13, 14	
				18	−150								*59E, 66FG*	33	
	4.2	3.58	5.13	(6) [7])	−158 [8])		4.57						*64R*		
$NiAl_2O_4$ [9])	<1000					1.61		−0.43			3.6	2.3	*54GPG*	15	
	>1000					2.04					3.4	2.3	*54GPG* ***)		
$ZnAl_2O_4$									−50				*40K*		
									−76				*56C, 60CA*		
									−74.6				*66ANM*		

[1]) g_A = g-Faktor für die Ionen auf Tetraederstellen; g_B = g-Faktor für die Ionen auf Oktaederstellen.
[2]) und geringer [*63B-3*].
[3]) theoretisch [*62L*].
[4]) Bei 77 °K kein spontanes Moment [*59PT*]; bei Raumtemperatur paramagnetisch [*65R*].
[5]) Es wurde die Suszeptibilität von $ZnAl_2O_4$ angenommen [*62L*].
[6]) Magnetische Struktur wie Co_3O_4 [*63R*], [*64R*].
[7]) Für $\Theta_N > 6.4$ °K [*64R*]; vermutlich $\Theta_N \approx 20$ °K [*66FG*].
[8]) An anderer Stelle bei gleichem Autor: −153 °K [*64R*].
[9]) Paramagnetischer Stoff [*53MP*].
*) [*65R, 63MT, 66ÔIS*].
**) [*67ÔCI*].
***) [*53MP*].

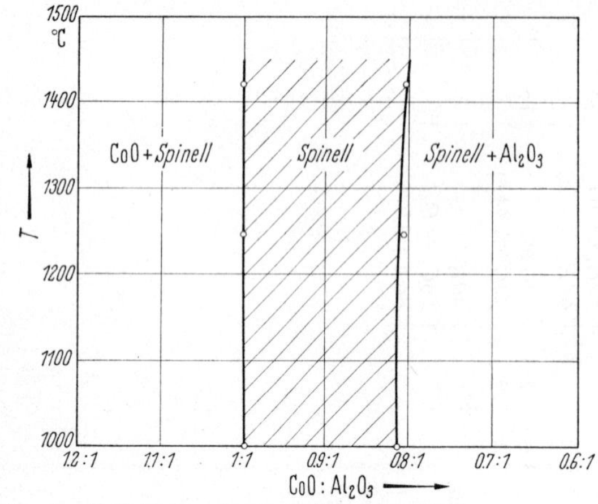

Fig. 1. CoO-Al_2O_3. Zustandsdiagramm [*61S*].

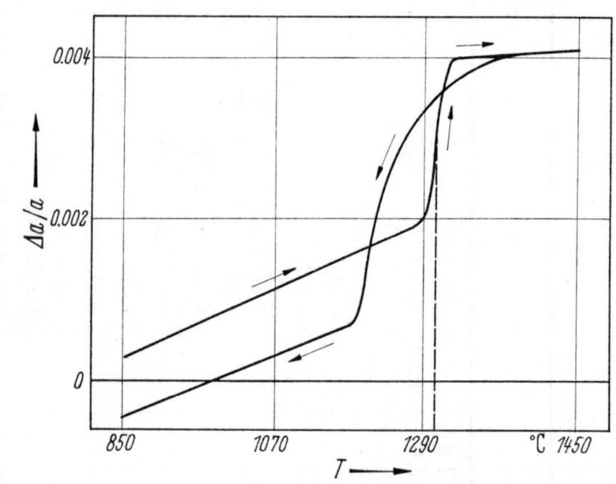

Fig. 2. $Li_{0.5}Al_{2.5}O_4$. Relative Gitterkonstantenänderung $\Delta a/a$ vs. Temperatur T. Bei 1290 °C ist ein Ordnung-Unordnung-Übergang in der Struktur [*62LC-2*], [*64L*].

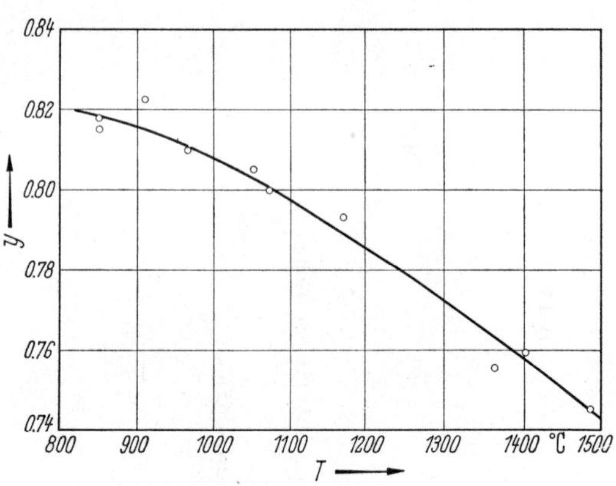

Fig. 3. $NiAl_2O_4$. Kationenverteilung y vs. Temperatur T gemäß der allgemeinen Formel $Ni_{1-y}Al_y[Ni_yAl_{2-y}]O_4$. Die Proben wurden von der jeweiligen Temperatur abgeschreckt [*61S*].

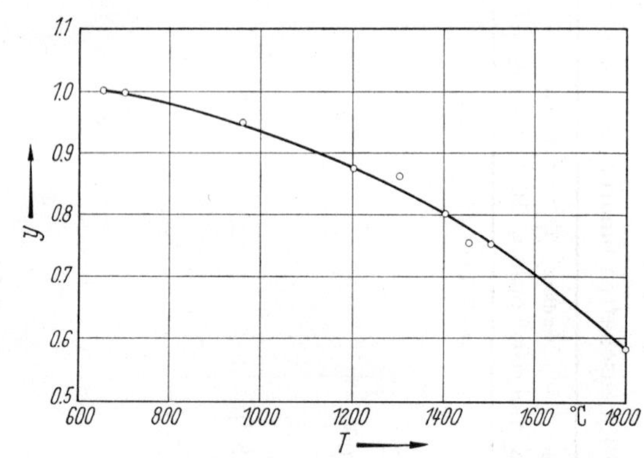

Fig. 4. $NiAl_2O_4$. Kationenverteilung y vs. Temperatur T gemäß der allgemeinen Formel $Ni_{1-y}Al_y[Ni_yAl_{2-y}]O_4$. [*67DR*].

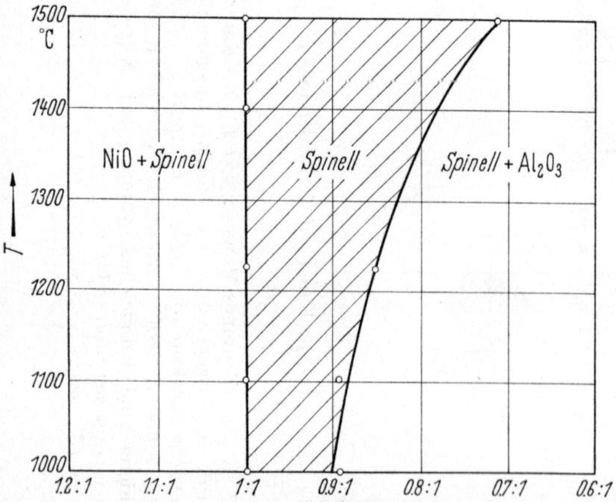

Fig. 5. NiO-Al_2O_3. Zustandsdiagramm [*61S*].

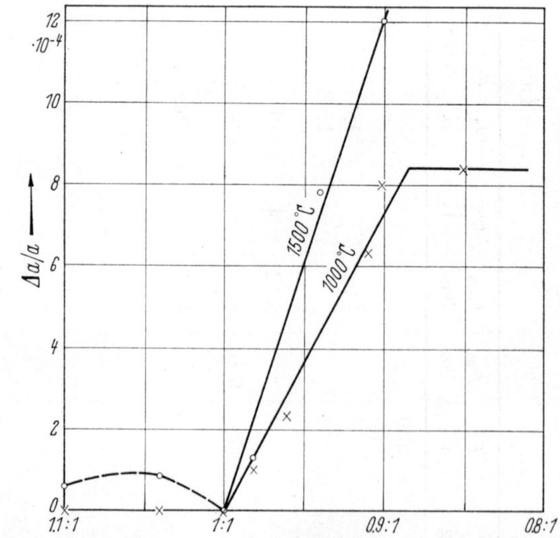

Fig. 6. $NiAl_2O_4$. Relative Gitterkonstantenänderung $\Delta a/a$ in Abhängigkeit vom Verhältnis NiO : Al_2O_3 [*61S*].

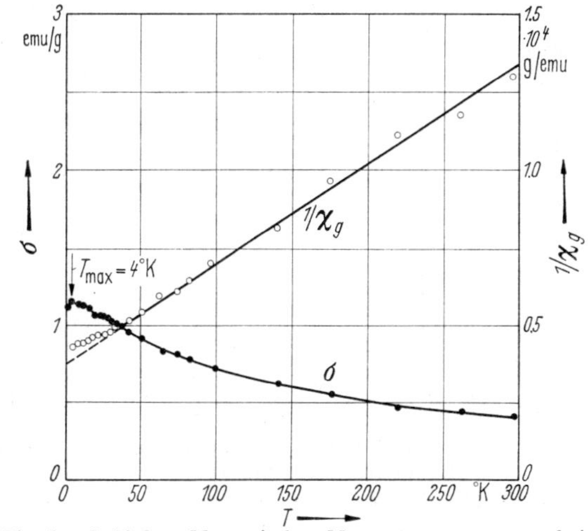

Fig. 7. $CoAl_2O_4$. Magnetisches Moment, σ, gemessen bei einer Feldstärke von 10 kOe, und reziproke Suszeptibilität pro Gramm, $1/\chi_g$, vs. Temperatur T [*64R*].

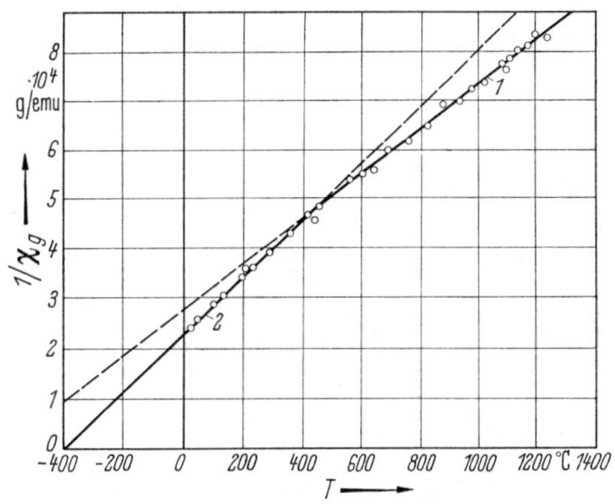

Fig. 8. $CoAl_2O_4$. Reziproke Suszeptibilität pro Gramm, $1/\chi_g$, vs. T. 1: C_g = 3.86 cm³ °K/g, g_A = 2.0; 2: C_g = 3.07 cm³ °K/g, g_B = 4.3 [*54GPG*].

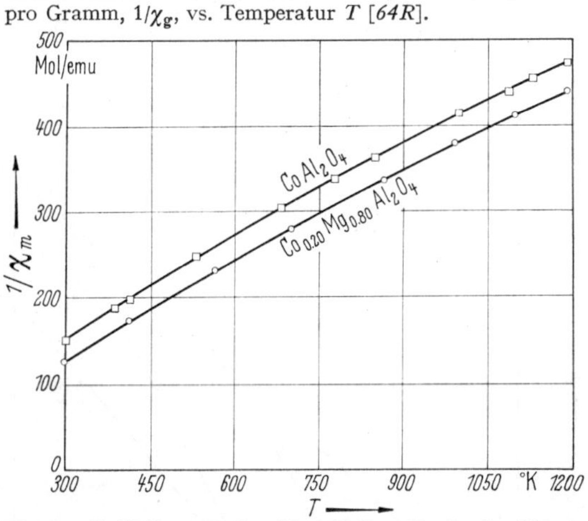

Fig. 9. $CoAl_2O_4$ und $Co_{0.2}Mg_{0.8}Al_2O_4$. Reziproke Suszeptibilität pro Gramm-Molekül, $1/\chi_m$, vs. Temperatur T [*60CA*].

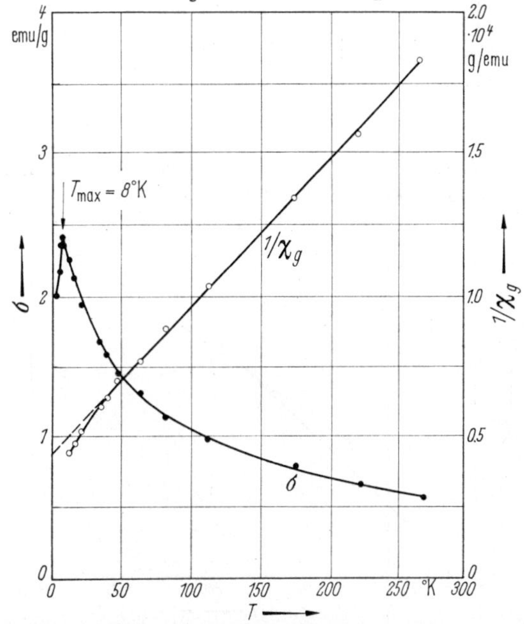

Fig. 10. $FeAl_2O_4$. Magnetisches Moment, σ, gemessen bei einer Feldstärke von 10 kOe, und reziproke Suszeptibilität pro Gramm, $1/\chi_g$, vs. Temperatur T [*64R*].

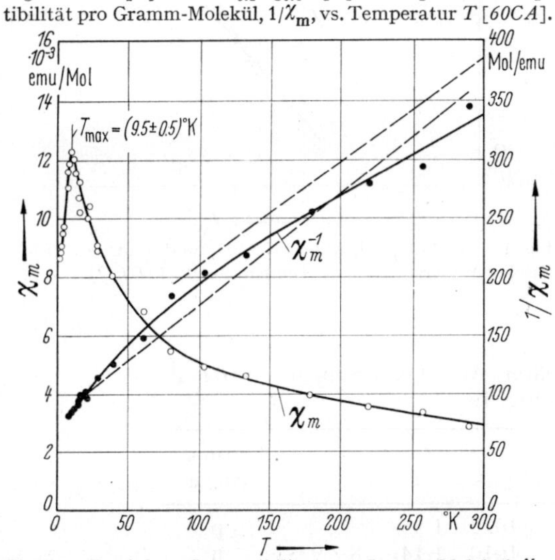

Fig. 11. $FeAl_2O_4$. Suszeptibilität pro Gramm-Molekül, χ_m, und reziproke Suszeptibilität, χ_m^{-1}, vs. Temperatur T. Einkristall [*64S-2*]. Meßergebnisse an Pulverproben anderer Autoren: Untere gestrichelte Kurve: [*64R*], obere gestrichelte Kurve: [*62L*].

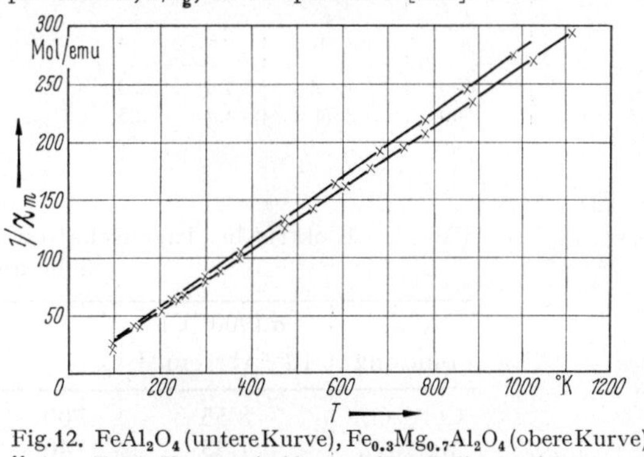

Fig. 12. $FeAl_2O_4$ (untere Kurve), $Fe_{0.3}Mg_{0.7}Al_2O_4$ (obere Kurve). χ_m^{-1}, vs. T. Die Kurven sind in negativer Ordinatenrichtung um den Betrag γ so verschoben, daß sie bei Extrapolation durch den Nullpunkt gehen. Die Werte der Konstanten γ betragen: Untere Kurve: γ = 38.0 mol/emu, (Θ_a = −144 °K); obere Kurve: γ = 14.0 mol/emu, (Θ_a = − 50 °K) [*62L*].

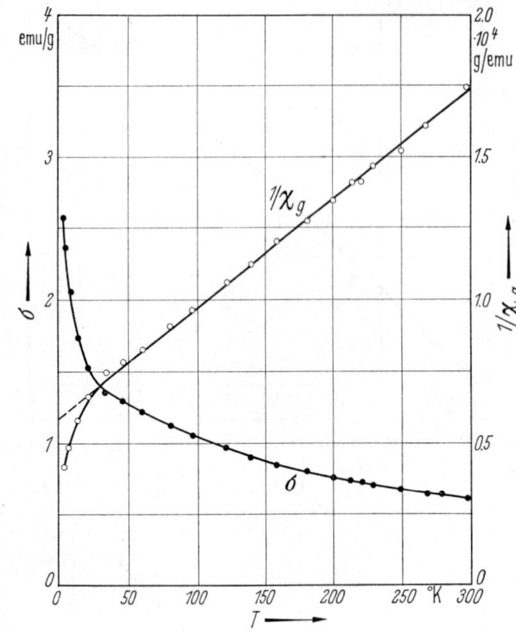

Fig. 13. $MnAl_2O_4$. Magnetisches Moment, σ, gemessen bei einer Feldstärke von 10 kOe, und reziproke Suszeptibilität pro Gramm, $1/\chi_g$, vs. Temperatur T [*64R*].

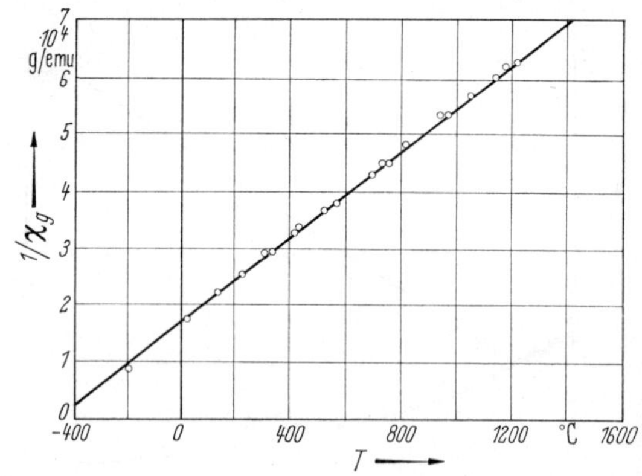

Fig. 14. $MnAl_2O_4$. Reziproke Suszeptibilität pro Gramm, $1/\chi_g$, vs. Temperatur T im Bereich von 100···1600 °C. $C_g = 4.08$ cm³ °K/g; $g = 1.94$ [*54GPG*].

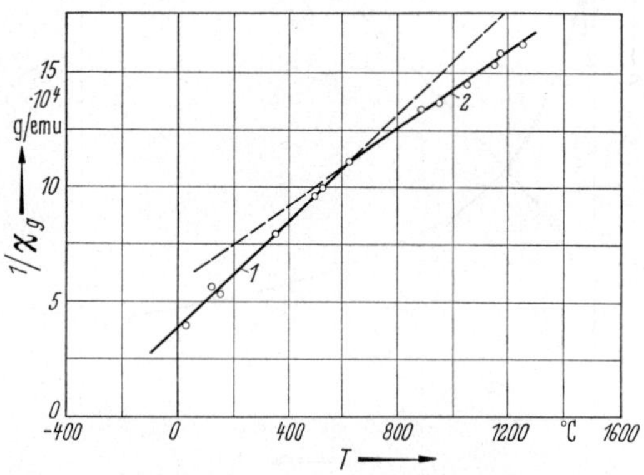

Fig. 15. $NiAl_2O_4$. Reziproke Suszeptibilität pro Gramm, $1/\chi_g$, vs. Temperatur T im Bereich von 300···1600 °C [*54GPG*].

	T °K	C_g cm³ °K/g	g_A	g_B
1	<600	1.61	3.6	2.3
2	>600	2.04	4.4	2.3

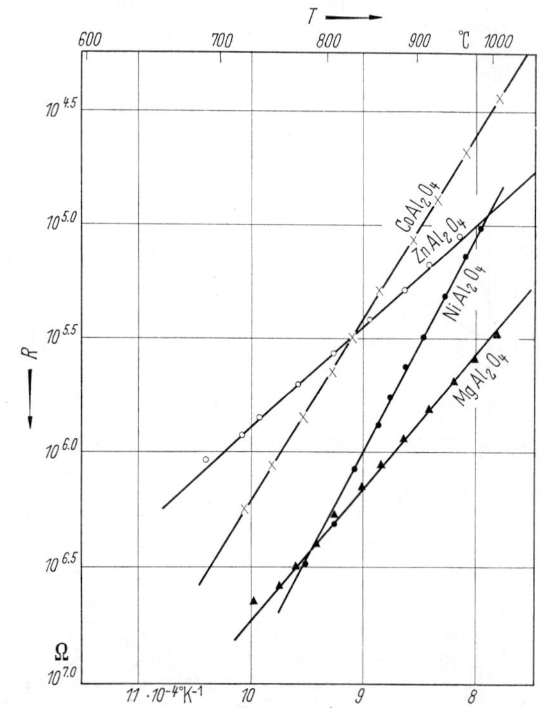

Fig. 16. $CoAl_2O_4$, $MgAl_2O_4$, $NiAl_2O_4$ und $ZnAl_2O_4$. Elektrischer Widerstand R vs. Temperatur T [*53BR-2*].

Tab. 3. Elektrische Eigenschaften binärer Al^{3+} Oxid-Spinelle [*53BR-2*]. Siehe auch Fig. 16

Verbindung	σ (900 °C) $10^{-6}\ \Omega^{-1}cm^{-1}$	q eV		ε	Leitungsträger
$CoAl_2O_4$	8.55	$700 < T < 1020$	1.55		p
$MgAl_2O_4$	1.33	$740 < T < 1010$	1.14	8···9 ¹)	n
$NiAl_2O_4$	2.75	$780 < T < 1020$	1.82		p
$ZnAl_2O_4$	5.38	$640 < T < 980$	0.86		n

¹) Bei 50···1250 MHz [*66MF*].

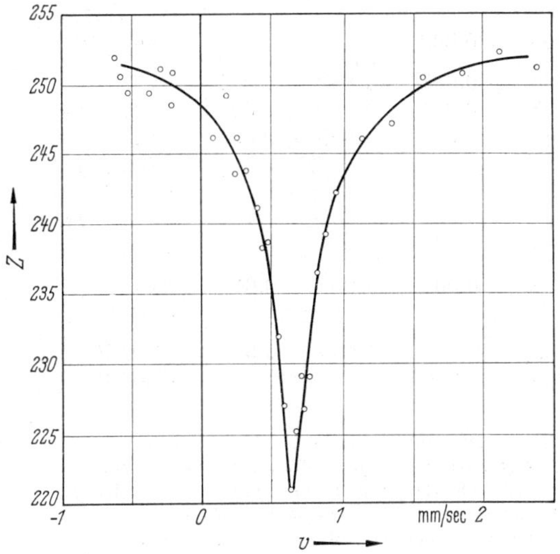

Fig. 17. $FeAl_2O_4$. Mössbauer-Absorptionsspektrum bei RT (paramagnetischer Zustand). Die Zählrate Z ist in willkürlichem Maßstab in Abhängigkeit von der Geschwindigkeit der Quelle v aufgetragen [*63MT*]; siehe auch [*66ÔIS*].

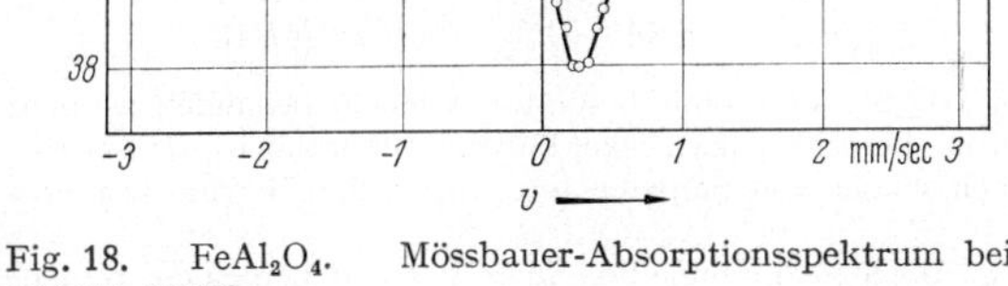

Fig. 18. $FeAl_2O_4$. Mössbauer-Absorptionsspektrum bei 300 °K. [*65R*].

T °K	ε mm/sec	δ mm/sec
77	2.81 ± 0.12	1.28 ± 0.12
295	1.60 ± 0.06	1.08 ± 0.06

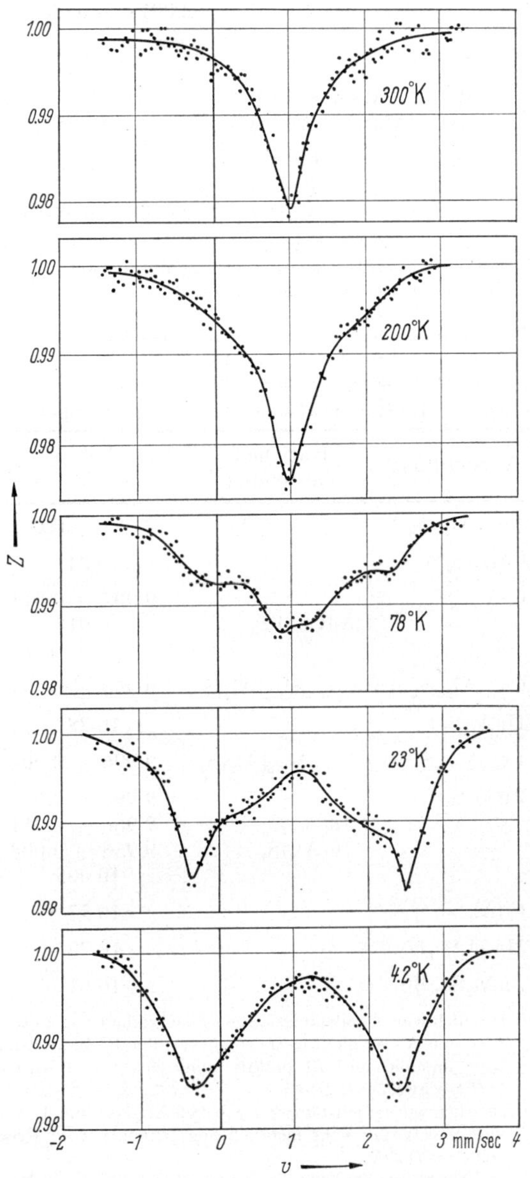

Fig. 19. $MgAl_2O_4$ mit 2.4% Fe^{2+}. Mössbauer-Absorptionsspektren des natürlichen, mit Fe^{2+} dotierten, Spinells bei verschiedenen Temperaturen [*67ÔCI*].

Tab. 4. Dotierte Al^{3+} Oxid-Spinelle. Spektroskopischer Aufspaltungsfaktor paramagnetischer Resonanz- oder Ligandenfeldparameter

Verbindung	g	$g_{\parallel}$	$g_{\perp}$	Lit.
$Li_{0.5}Al_{2.5}O_4$ [1]) : Cr^{3+} [2])	1.97			*66ABMP*
: Fe^{3+} [3])		2.006 ± 0.002		*61F, 62F*
$MgAl_2O_4$: Cr^{3+} [4]) [10])		1.986 ± 0.001	1.989 ± 0.002	*59SL*
: Cr^{3+} [5]) [10])		1.985 ± 0.001	1.980 ± 0.002	*63A, 65D-2*
: Cr^{3+} [6])		1.985 ± 0.001	1.985 ± 0.005	*63A*
: Cr^{3+} [7])				*67DC, 68D*
: Fe^{3+} [8])	2.001 ± 0.007			*61BLW*
: Ni^{2+} [7])				*67DC, 68D*
: V^{3+} [7])				*67DC*
$MgO \cdot 1.1\,Al_2O_3$: Cr^{3+} [6]) [10])		1.985 ± 0.002	1.98 ± 0.01	*63A*
$MgO \cdot 2Al_2O_3$: Co^{2+} [3]) [7])				*66DR*
$ZnAl_2O_4$: Cr^{3+} [9])		1.986 ± 0.001	1.993 ± 0.003	*64DLW*
: Fe^{3+} [9])		2.0018 ± 0.0007		*64DLW*
: Mn^{2+} [10])	2.0002 ± 0.0010			*59SL*

[1]) : Zeichen für die Dotierung.
[2]) Farbe: rosig-weiß bis hellrot je nach Betrag der eingeführten Cr^{3+} Ionen; Einkristall [*66ABMP*].
[3]) Einkristall [*61F*], [*66DR*].
[4]) Einkristall, rubinrot, Cr^{3+} auf Oktaederstellen [*59SL*].
[5]) Natürlicher Kristall, hellrosa [*63A*].
[6]) Einkristall mit 0.6 Gewichts-% Cr_2O_3 [*63A*].
[7]) Dotierung 0.5 Gewichts-% [*67DC*], [*66DR*].
[8]) Natürlicher Kristall, gelbbraun; gemessen bei 8.6···9.8 GHz, 300 °K, bis 10 kGauß [*61BLW*].
[9]) Gemessen bei 33 GHz und 300 °K [*64DLW*].
[10]) Gemessen bei Wellenlängen von 3 cm und anderen Wellenlängen gleicher Größenordnung [*59SL*], [*63A*].

Tab. 5. Kristallografische Eigenschaften binärer Al^{3+} Sulfo-Spinelle und binärer Al^{3+} Seleno-Spinelle

Verbindung	Kationen-verteilung	a Å	u	ϱ_X g/cm³	ϱ_P g/cm³	Lit.
$Ag_{0.5}Al_{2.5}S_4$ [1])	$Ag_{0.5}Al_{0.5}[Al_2]S_4$	10.052 ± 0.004	0.389	3.26	3.19	*60FDO, 61FDGOK*
$CdAl_2S_4$ [6])		10.24		3.65		*68RBW*
$CrAl_2S_4$ [2])		9.904 ± 0.011		3.18	3.09	*51F, 52F*
	$Al[CrAl]S_4$	9.914	0.384 ± 0.004	3.19	3.01···3.09	*61FDGF*
$Cu_{0.5}Al_{2.5}S_4$ [1]) [3])	$Cu_{0.5}Al_{0.5}[Al_2]S_4$	9.956 ± 0.004	0.387 ± 0.002	3.06	3.01	*60FDO, 61FDGOK*
$HgAl_2S_4$ [6])		10.28		4.68		*68RBW*
$Li_{0.5}Al_{2.5}S_4$ [4])	$Li_{0.5}Al_{0.5}[Al_2]S_4$	9.996 ± 0.004	0.386 ± 0.002	2.65	2.49	*60FDO, 61FDGOK*
$ZnAl_2S_4$ [5])		9.99 ± 0.01		3.28	3.25	*51F*
	$Zn[Al_2]S_4$	9.988 ± 0.011		3.28	3.25	*52F*
	$Zn[Al_2]S_4$	9.968 ± 0.008	0.384	3.306	3.30	*52HF, 55HFKSS*
		10.005		3.29		*67S-3*
$CdAl_2Se_4$ [6])		10.77		5.13		*68RBW*
$HgAl_2Se_4$ [6])		10.78		6.06		*68RBW*
$ZnAl_2Se_4$ [6])		10.61		4.84		*68RBW*

[1]) Deutlich widerstandsfähiger gegen Feuchtigkeit und Luft als $Li_{0.5}Al_{2.5}S_4$; kann ohne besondere Vorsicht behandelt werden; wird von Wasser nicht merklich angegriffen. Farbe: Kristalle farblos bei schwacher Dicke, Pulver weiß-gräulich [*61FDGOK*].
[2]) Unveränderlich an Luft, praktisch unlöslich in HCl, wird ziemlich schnell von Salpetersäure angegriffen. Farbe: braunrot [*51F*], schwarz [*61FDGF*].
[3]) Deutlich widerstandsfähiger gegen Feuchtigkeit und Luft als $Li_{0.5}Al_{2.5}S_4$; kann ohne besondere Vorsicht behandelt werden; wird von Wasser nicht merklich angegriffen. Farbe: Kristalle goldgelbe Masse, bei schwacher Dicke transparent, Pulver gelb [*61FDGOK*].
[4]) $Li_{0.5}Al_{2.5}S_4$ ist dimorph. Die zweite Kristallform ist bei Sintertemperaturen höher als 1200 °C oder bei Erwärmen der Verbindung auf 1350 °C vom Gittertyp $MnAl_2S_4$ und läßt sich rhomboedrisch oder hexagonal indizieren.
Verändert sich sehr schnell an Luft, wird von kaltem Wasser angegriffen und sehr schnell von verdünnten Säuren zerstört. Farbe: Kristalle farblos bei schwacher Dicke, Pulver weiß-gräulich [*61FDGOK*].
[5]) Oberhalb 1000 °C (nach [*67S-3*] 1050 °C) geht das Spinellgitter reversibel über in ein Wurtzitgitter mit statistischer Verteilung der beiden Metalle auf die Metallagen des Wurtzitgitters [*55HFKSS*]. Farbe: hellgrau [*52HF*].
[6]) Unter Normalbedingungen Thiogallatstruktur mit tetragonaler Symmetrie; unter Hochdruckbehandlung Umwandlung in Spinellstruktur, die sehr stabil ist; erst nach sehr langer Behandlung bei Temperaturen von 350 °C teilweise Rückumwandlung [*68RBW*].

Tab. 6. Kristallografische Eigenschaften ternärer Al^{3+} Oxid-Spinelle

Verbindung [1])	Kationenverteilung	x	a Å	u	Farbe	Lit.	Fig.	Tab.
$Cd_xMg_{1-x}Al_2O_4$ [2])						*58SBB*	20	
$Co^{2+}Co^{3+}AlO_4$						*63B-3*		7
$CoFe_{2-x}Al_xO_4$	$Co^{2+}[Fe^{3+}Al^{3+}]O_4$ [3])	1.0	8.26			*63B-3, 64B-1*		7
$Co_{2-x}Ge_{1-x}Al_{2x}O_4$ [4])		0.5	8.244			*53R*	21	
		0.75	8.156					
		1.0	8.105					
$Co_xMg_{1-x}Al_2O_4$ [5])		0	8.085			*58BB*	9, 31	7
		0.1	8.087			*58BB, 56C*		
	$Co_{0.2}Mg_{0.8}[Al_2]O_4$	0.2	8.089			*60CA, 56C, 62L*		
		0.503	8.093			*56C*		
		1.0	8.112			*58SBB, 56C*		
$Co_{0.1}Mn_{0.9}Al_2O_4$			8.208			*58SBB*		
$CoRh_{2-x}Al_xO_4$						*63B-3*		
$CoV_{2-x}Al_xO_4$	$Co^{2+}[V^{3+}_{2-x}Al^{3+}_x]O_4$					*63RAWG*	5, 6 in 6.3.1.1	
$Co_{1-x}Zn_xAl_2O_4$	$Co^{2+}_{1-x}Zn^{2+}_x[Al^{3+}_2]O_4$	0	8.105		dunkelblau	*53R, 56C, 60CA*	22, 32	7
		0.25	8.101		↓			
		0.50	8.096					
		0.75	8.091					
		1.0	8.086		hellblau			
$CrZnAlO_4$			8.21 ± 0.01		fleischfarb.	*34KT*		
$CuFe_{2-x}Al_xO_4$ [6])						*58D, 60H-1*		7
						64B-1, 69L [6])		
$CuGaAlO_4$	$Cu^{2+}_{0.5}Ga^{3+}_{0.5}[Cu^{2+}_{0.5}Ga^{3+}_{0.5}Al]O_4$		8.181	0.381		*66RD*		
$CuCr_{2-x}Al_xO_4$						*58D*		
$Fe^{2+}Fe^{3+}_{2-x}Al_xO_4$	$Fe^{2+}_xFe^{3+}_{1-x}[Fe^{2+}_{1-x}Fe^{3+}Al^{3+}_x]O_4$	$0 < x \leq 1$				*59PT, 56HF*	23, 34, 35, 36	7
	$Fe^{2+}[Fe^{3+}_{2-x}Al^{3+}_x]O_4$	$1 < x \leq 2$				*59PT*		
						48MP, 51GM		
		1.0		0.385 ± 0.01		*59P*		
						47VH		
$FeZnAlO_4$			8.21 ± 0.02		hellbraun	*34KT*		

Fußnoten siehe S. 459, 460

Tab. 6. Fortsetzung

Verbindung [1]	Kationenverteilung	x	a Å	u	Farbe	Lit.	Fig.	Tab.
$InNiAlO_4$	$In^{3+}[Ni^{2+}Al^{3+}]O_4$		8.29 ± 0.003		grün	*65SK*		
$Li_{0.5-x}Cu_xAl_{2.5}O_4$						*69L*		
$Li_{0.5}Cr_xAl_{2.5-x}O_4$ [7][8]						*64B-1, 65RGA*	24	
$Li_{0.5}Fe_{2.5-x}Al_xO_4$						*61SR, 69L* *)	25, 26	7
$LiGeAlO_4$			8.00			*65NM*		
$Li_{0.5-0.5x}Mg_xAl_{2.5-0.5x}O_4$						*68DR*		
$Li_{0.5}Mn_xAl_{2.5-x}O_4$ [7][9]	$Al^{3+}[Li_{0.5}Mn_x^{3+}Al_{1.5-x}^{3+}]O_4$	0	7.901 ± 0.002			*65RGA, 65B*		
		0.1	7.948 ± 0.002			*65RGA*		
		0.2	7.970 ± 0.002					
		0.3	7.988 ± 0.002					
		0.4	7.987 ± 0.002					
		0.5	7.987 ± 0.002					
$LiMnAlO_4$	$Li^{+}[Mn^{4+}Al^{3+}]O_4$		8.12			*64B-1*		
$LiTiAlO_4$	$Li^{+}[Ti^{4+}Al^{3+}]O_4$		8.34			*63B-1*		
			8.30			*64B-1*		
						51KR		
$MgCo_{0.35}Al_{1.65}O_{3.65}F_{0.35}$			8.093			*67S-2*		
$MgCr_xAl_{2-x}O_4$						*54WK, 66A*	27, 28, 39	
						67KGW, 67KS		
						67R		
$MgFe_{2-x}Al_xO_4$ [10]	$Fe_{1-y-z}^{3+}Mg_y^{2+}Al_z^{3+}[Fe_{1-x+y+z}^{3+}Mg_{1-y}^{2+}Al_{x-z}^{3+}]O_4$					*52JR, 53BR-1*	27, 28, 39	
						54G, 59K-1		
						64A, 64B-1, 66A		
	$Mg_{0.53}^{2+}Fe_{0.47}^{3+}[Fe_{0.53}^{3+}Mg_{0.47}^{2+}Al^{3+}]O_4$	1				*51KR, 53MP*		
	$Fe_{0.5}^{3+}Al_{0.5}^{3+}[Fe_{0.5}^{3+}Mg^{2+}Al_{0.5}^{3+}]O_4$	1		0.385 ± 0.002		*53BR-1, 54BW*		
		1.9	8.0890 ± 0.0005		rotbraun	*67F*		7
$Mg_{0.73}Fe_{0.33}Al_{1.93}O_4$ [11]			8.117		[11]	*62S*		
$Mg_{0.65}Fe_{0.39}Al_{1.96}O_4$ [12]			8.12		[12]	*53BM*		
$Mg_{0.75}Fe_{0.41}Al_{1.85}O_4$ [11][13]			8.128		[11]	*62S*		
$Mg_{1-x}Fe_xAl_2O_4$		$6 \cdot 10^{-4}$	8.0868		hellrosa	*64S-2*	12	7
		0.02				*66ÔIS*		
		0.024	8.089		[14]	*64S-2*		
		0.19	8.128		[15]	*64S-2*		
		0.27	8.117		[15]	*64S-2*		
		0.30				*62L*		

$MgGa_{2-x}Al_xO_4$ 16)						*58SBB, 61H*		
$MgMn_{0.29}Al_{1.71}O_{3.71}F_{0.29}$			8.152			*67S-2*		
$Mg_{1-x}Mn_xAl_2O_4$ 17)			17)		18)	*58SBB*		
$MgMn_{2x}Al_{2-2x}O_4$ 19)						*60ISB*		
$MgNi_xAl_{2-2x}O_{4-x}F_x$		0.13 0.42	8.081 8.067			*67S-2*		
$Mg_{1-x}Ni_xAl_2O_4$ 17)		0.1	8.085		blaß-blau	*53R, 65SLR* *59SGB, 68DR*	29	
$Mg_{1+x}Si_xAl_{2-2x}O_4$						*65SR, 64C*		
$Mg_{0.235}Zn_{0.79}Al_{1.90}O_4$ 20)	$Mg_{0.21}Zn_{0.79}[Fe_{0.075}Mg_{0.025}Al_{1.90}]O_4$		8.099 ± 0.002	0.384 ± 0.001		*64S-1*		
$MnCr_{2-x}Al_xO_4$	$Mn_{1-0.05x}Al^{3+}_{0.05x}[Cr^{3+}_{2-x}Mn^{2+}_{0.05x}Al^{3+}_{0.95x}]O_4$					*59E*	30, 33	7
$MnFe_{2-x}Al_xO_4$	$Mn^{2+}_{1-y}Fe^{3+}_y[Fe^{3+}_{2-y-x}Mn^{2+}_yAl^{3+}_x]O_4$					*65B*		7
$Mn_{0.5}Zn_{0.5}Al_2O_4$			8.14 ± 0.00		holzbraun	*34KT*		
$NiFe_{2-x}Al_xO_4$	$Ni^{2+}_zFe^{3+}_{1-y-z}Al^{3+}_y[Fe^{3+}_{1-x+y+z}Ni^{2+}_{1-z}Al^{3+}_{x-y}]O_4$					*54M-2* **)	37, 38, 43, 44	7
		1.0	8.1951 ± 0.0007			*54G*		
	$Ni_{0.02}Fe_{0.75}Al_{0.23}[Ni_{0.98}Fe_{0.25}Al_{0.77}]O_4$ 21)	1.0	8.185			*54GPG*		
	$Ni_{0.05}Fe_{0.60}Al_{0.35}[Ni_{0.95}Fe_{0.40}Al_{0.65}]O_4$ 22)	1.0				*54GPG*		
		1.0				*62BG, 63MT,* *51KR*		
	$Ni_{0.08}Fe_{0.25}Al_{0.66}[Ni_{0.92}Al_{1.09}]O_4$ 21)	1.75	8.081			*54GPG, 54M-2*		
	$Ni_{0.18}Fe_{0.18}Al_{0.64}[Ni_{0.82}Fe_{0.07}Al_{1.11}]O_4$ 22)	1.75				*54GPG*		
$Ni_{1+x}Ge_xAl_{2-2x}O_4$ 23)						*68R*		
$Ni_{1-x}Zn_xAl_2O_4$	$Ni^{2+}_{1-x}Zn^{2+}_x[Al^{3+}_2]O_4$	0 0.25 0.50 0.75 1.0	8.046 8.0630 ± 0.001 8.0733 ± 0.001 8.0833 ± 0.001 8.086			*43VB, 53R, 68DR*	22	
$ZnCo_xAl_{2-x}O_4$						*56C*		7

*) Siehe auch [*54M-1, 56CMW, 57CMW, 59K-2, 59K-3, 63SB, 64B-1*].
**) Siehe auch [*53M, 53MP, 53MPH, 53W, 54GPG, 54S, 62NZ, 63VNA, 64B-1, 66K-1, 67B, 67N*].
1) Bei den Verbindungsreihen siehe auch die Endglieder mit $x = 0$ und $x = 2$ oder $x = 2.5$ in Tab. 1.
2) Im Spinell $MgAl_2O_4$ lassen sich nur 70 Atom-% Mg^{2+} durch Cd^{2+} ersetzen, da der Radienunterschied der beiden Kationenarten (0.78 und 1.03 Å) zu groß ist [*58SBB*].
3) Ungefähre Kationenverteilung [*63B-3*].
4) Die Reihe $Ge[Co_2]O_4$-$Co[Al_2]O_4$ ist zweiphasig und bei $0.50 \leq x \leq 0.75$ existiert eine Mischungslücke [*53R*], [*68R*].
5) Das System $MgAl_2O_4$-$CoAl_2O_4$ bildet eine lückenlose Mischkristallreihe. Die Vegardsche Regel ist innerhalb der Fehlergrenzen gültig [*58SBB*].
6) Für kleine Werte von x treten tetragonale Verzerrungen auf [*58D*]. $Cu_{0.5}Fe_{2.5-x}Al_xO_4$ in [*69L*].

Weitere Fußnoten siehe nächste Seite

[7]) Kritische Konzentration x_{crit} von substituierten Ionen zur Aufhebung der 1 : 3-Fernordnung in oktaedrischen Untergitter: $Li_{0.5}Cr_xAl_{2.5-x}O_4$: $x_{crit} < 0.5$ [64B-1]; $Li_{0.5}Mn_xAl_{2.5-x}O_4$: $0.2 < x_{crit} < 0.3$ [65RGA].

[8]) Die Verbindungsreihe hat eine Mischungslücke bei $x > 2$ [64B-1].

[9]) Die Verbindungsreihe ist für $x \leqq 0.25$ einphasig, $0.25 < x < 0.50$ zweiphasig, $0.50 < x \leqq 1.55$ einphasig, $x \geqq 1.75$ nicht mehr einphasig [65B].

[10]) Im Bereich $0.8 < x < 1.7$ besteht eine Mischungslücke [64B-1].

[11]) Pleonast-Einkristall mit Spuren von Titan. Farbe: in dünnen, polierten Segmenten durchsichtig und gelbbraun, als Kristall undurchsichtig [62S].

[12]) Natürlicher Pleonast; Farbe: dunkelgrün bis bräunlichgrün, fast undurchsichtig, $\varrho_X = 3.91$ g/cm³. $\varrho_p = 3.83$ g/cm³; Brechungsindex $n = 1.788 \cdots 1.789$; Härte: 8 Mohs [53BM].

[13]) Wärmeleitfähigkeit: bei 300 °K gemessen: $K = 0.055$ Watt/cm °K,
für 500 °K berechnet: $K = 0.07$ Watt/cm °K [62S].

[14]) Farbe: bläulich-purpurn in massivem Zustand [64S-2].

[15]) Kristalle schwarz undurchsichtig; in dünnen Segmenten gelblich-braun [64S-2].

[16]) $MgAl_2O_4$-$MgGa_2O_4$ bilden eine lückenlose Mischkristallreihe [61H].

[17]) Lückenlose Mischkristallreihe mit linearer Abhängigkeit der Gitterkonstanten a vom Zusammensetzungsparameter x entsprechend der Vegardschen Regel [58SBB], [65SLR].

[18]) Farbe: Sinterung unter O_2-Abschluß: farblos; Sinterung bei Spuren von O_2: mehr oder weniger intensiv braun [58SBB].

[19]) Für $x > x_t = 0.59$ sind die Mischkristalle tetragonal verzerrt, d. h. $c/a > 1$; bei $x = x_t$ ist $a = 8.34$ Å [60ISB].

[20]) Untersuchungen wurden an Einkristallen durchgeführt [64S-1].

[21]) Von 1400 °C langsam abgekühlt [54GPG].

[22]) Von 1400 °C abgeschreckt [54GPG].

[23]) Die Verbindungsreihe $NiAl_2O_4$-Ni_2GeO_4 ist zweiphasig [68R]. Ni_2GeO_4, siehe 6.3.2.

Tab. 7. Magnetische Eigenschaften ternärer Al^{3+} Oxid-Spinelle

Verbindung [1])	x	T °K	p_{eff} μ_B	Θ_a °K	Θ_C °K	C_m cm³ °K/mol	$\chi_{m\,diam}$ 10^{-3} emu/mol	χ_m 10^{-3} emu/mol	a 10^{-3} emu/mol	Lit.	Fig.
$Co^{2+}Co^{3+}AlO_4$ [2])			4.14	− 30						*63B-3*	
$CoFe_{2-x}Al_xO_4$ [3])										*64B-1*	
	1		[4])		420					*63B-3*	
$Co_xMg_{1-x}Al_2O_4$	0.1 [5])		4.28	− 1		2.291	−0.48		0.24	*56C*	9, 31
	0.2 [5])		4.28	− 10		2.292	−0.24		0.29	*56C*	
	0.2		4.28	− 10		2.290			0.350	*60CA, 62L*	
	0.503 [6])		4.31	− 32		2.321	−0.10		0.41	*56C*	
$Co_{1-x}Zn_xAl_2O_4$	0.25		3.88	− 14		1.905	−0.10		0.56	*56C*	32
	0.497		3.83	− 4		1.830	−0.15		0.51		
	0.60		3.82	− 3		1.820	−0.18		0.48		
	0.80 [10])		3.80	+ 12		1.809	−0.38		0.46		
	0.80 [10])		3.94	+ 6		1.938	−0.38		0.38	*56C, 60CA*	
	0.83		4.08	+ 17		2.090	−0.45		0.46	*56C*	
	0.90		4.06	+ 15		2.059	−0.76		0.48		
	0.945						−1.38				
	0.968						−2.34				
	0.967						−2.27				

$CuFe_{2-x}Al_xO_4$											*58D*	
$Cu_{0.5}Fe_{2.5-x}Al_xO_4$											*69L*	
$Fe^{2+}Fe^{3+}_{2-x}Al_xO_4$											*59PT* *)	34, 35, 36
$Li_{0.5}Fe_{2.5-x}Al_xO_4$											*63SB*, *54M-1*	
$MgFe_{2-x}Al_xO_4$ [7])	1.0				370···420						*51KR*	
	1.0				243						*53BR-1* **)	
$Mg_{1-x}Fe_xAl_2O_4$	0.3			− 50		3.58	−0.165 [8])				*62L*	12
$MnCr_{2-x}Al_xO_4$	0			−350							*59E*	33
	1.0			−250								
$MnFe_{2-x}Al_xO_4$											*65B*	
$NiFe_{2-x}Al_xO_4$ [9])	1.0				325						*51KR* *53M*, *53MP*, *53W*, *54G*, *54M-2*, *54S*, *62BG*, *67B*, *67N*, *62NZ*	43, 44
	1.0										*63MT*	37
	1.75										*54M-2*	38
$ZnCo_xAl_{2-x}O_4$	0.3	82					−0.053	0.46			*56C*	
	0.3	287					−0.053	0.16				
	0.5	80					−0.054	0.75				
	0.5	287					−0.054	0.28				
	0.67	80					−0.055	0.81				
	0.67	291					−0.055	0.29				

[1]) Für Verbindungen mit $x = 0$ und $x = 1$ siehe auch Tab. 2.
[2]) $\Theta_N = 30$ °K [*63B-3*].
[3]) $g_A = 2.4$; $g_B = 2.5$ (vergleiche Tab. 2) [*64B-1*].
[4]) $p_m = 0.93\,\mu_B$ bei 0 °K [*63B-3*].
[5]) Von 1100 °C abgeschreckt [*56C*].
[6]) Bei 1100 °C gesintert; von 700 °C abgeschreckt [*56C*].
[7]) Debye-Temperatur von $FeMgAlO_4$: $\Theta_D = 500$ °K [*53BR-1*].
[8]) Berechnet aus den diamagnetischen Suszeptibilitäten der binären Verbindungen [*62L*].
[9]) Bei $x = 1$ ist $p_m = 0.64\,\mu_B$ bei einer geglühten Probe und $p_m = 0.42\,\mu_B$ bei einer von 1350 °C abgeschreckten Probe, gemessen bei H bis zu 9 kOe und T bis 77 °K [*54G*].
[10]) Abweichende Herstellungsbedingungen [*56C*].
*) Siehe auch [*61GH*, *47VH*].
**) Siehe auch [*52JR*, *54G*, *64B-1*].

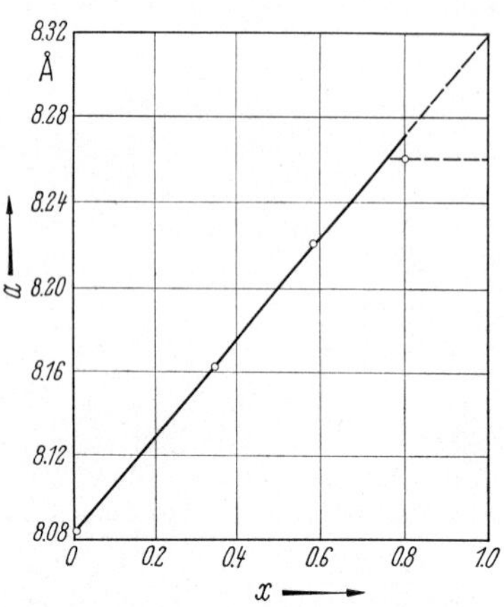

Fig. 20. $Cd_xMg_{1-x}Al_2O_4$. Gitterkonstante a vs. Zusammensetzungsparameter x [*58SBB*].

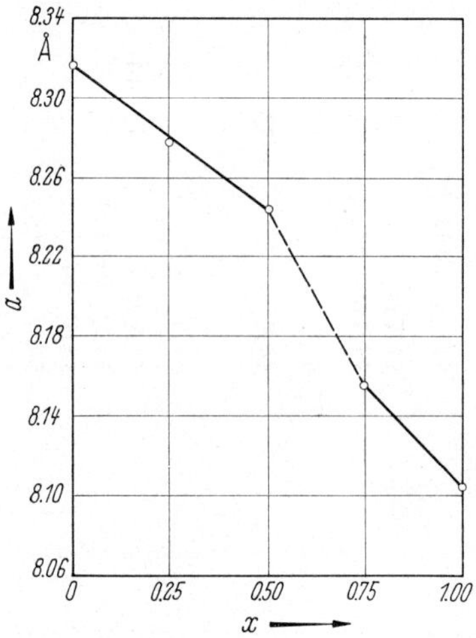

Fig. 21. $Co_{2-x}Ge_{1-x}Al_{2x}O_4$. Gitterkonstante a vs. Zusammensetzungsparameter x [*53R*].

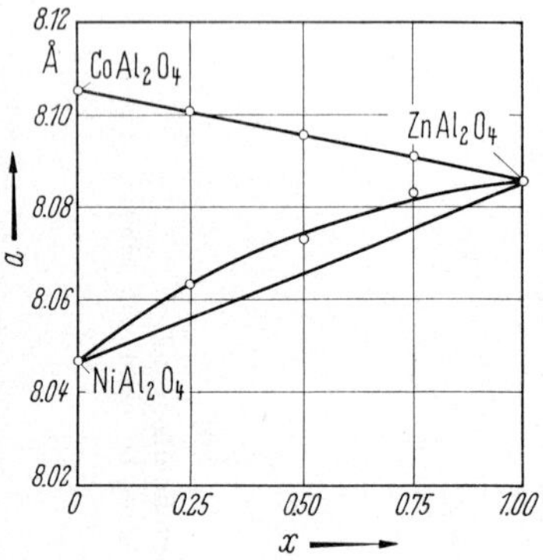

Fig. 22. $Co_{1-x}Zn_xAl_2O_4$ und $Ni_{1-x}Zn_xAl_2O_4$. Gitterkonstante a vs. Zusammensetzungsparameter x [*53R*].

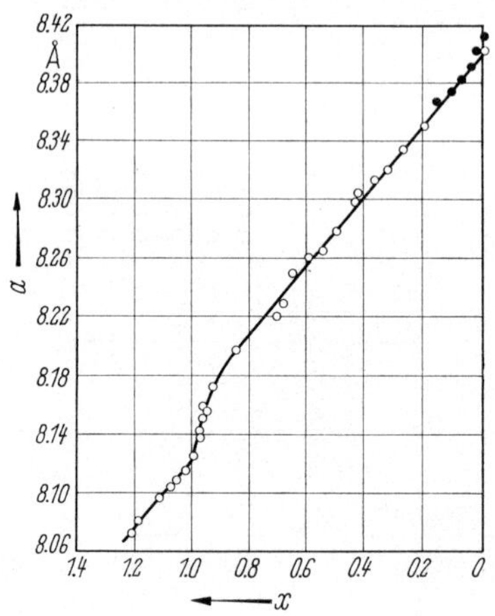

Fig. 23. $Fe^{2+}Fe^{3+}_{2-x}Al_xO_4$. Gitterkonstante a vs. Zusammensetzungsparameter x [*56HF*].

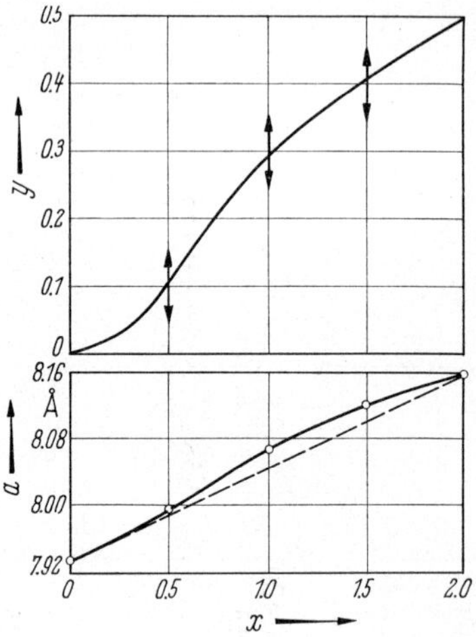

Fig. 24. $Li_{0.5}Cr_xAl_{2.5-x}O_4$. Gitterkonstante a und Anzahl y der Li^+ Ionen pro Molekül auf Tetraederstellen vs. Zusammensetzungsparameter x [*64B-1*].

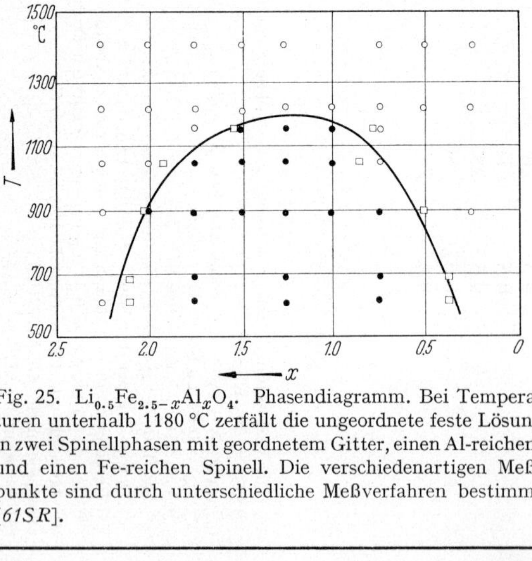

Fig. 25. $Li_{0.5}Fe_{2.5-x}Al_xO_4$. Phasendiagramm. Bei Temperaturen unterhalb 1180 °C zerfällt die ungeordnete feste Lösung in zwei Spinellphasen mit geordnetem Gitter, einen Al-reichen- und einen Fe-reichen Spinell. Die verschiedenartigen Meßpunkte sind durch unterschiedliche Meßverfahren bestimmt [*61SR*].

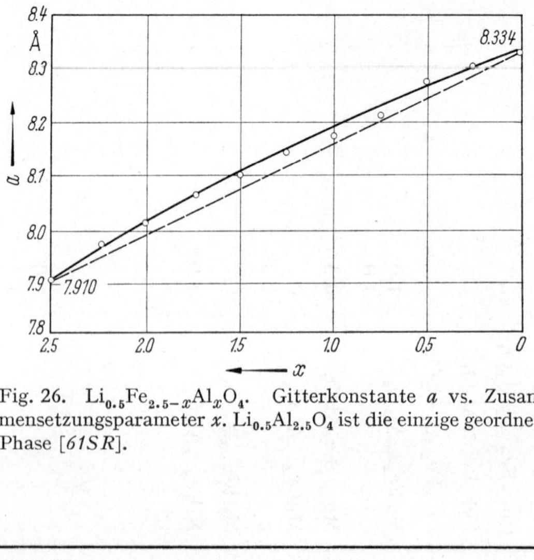

Fig. 26. $Li_{0.5}Fe_{2.5-x}Al_xO_4$. Gitterkonstante a vs. Zusammensetzungsparameter x. $Li_{0.5}Al_{2.5}O_4$ ist die einzige geordnete Phase [*61SR*].

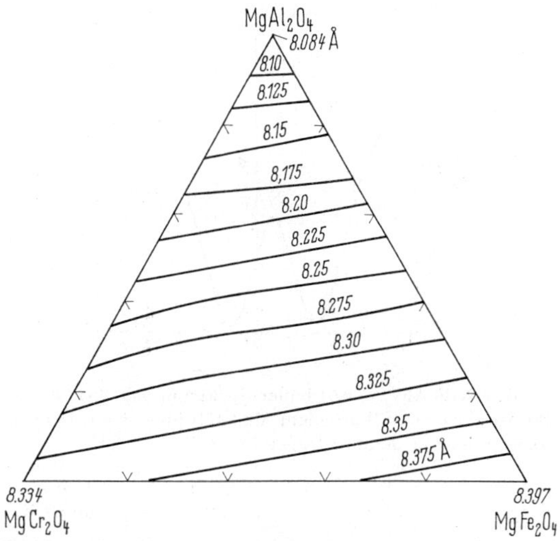

Fig. 27. Gitterkonstante a im Dreistoffsystem $MgAl_2O_4$-$MgCr_2O_4$-$MgFe_2O_4$. Abschrecktemperatur der Proben $T_Q = 1600$ °C [*66A*].

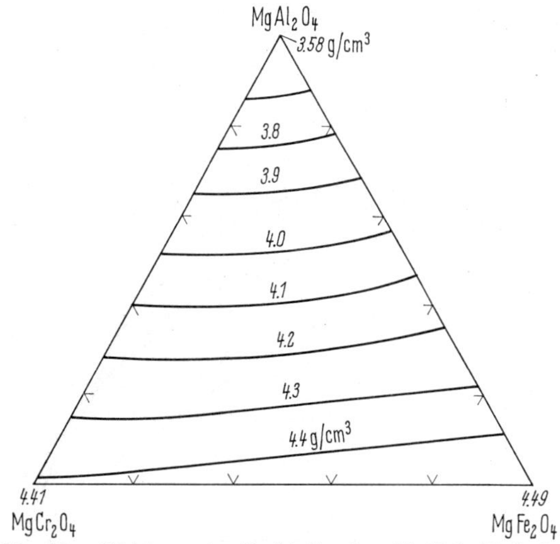

Fig. 28. Dichte ϱ_X im Dreistoffsystem $MgAl_2O_4$-$MgCr_2O_4$-$MgFe_2O_4$. Abschrecktemperatur der Proben $T_Q = 1600$ °C [*66A*].

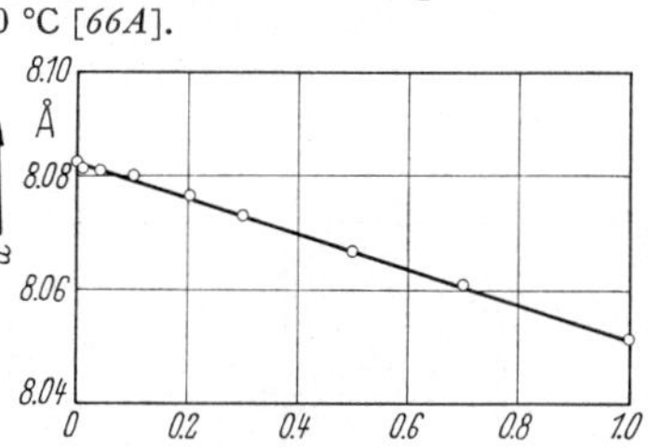

Fig. 29. $Mg_{1-x}Ni_xAl_2O_4$. Gitterkonstante a vs. Zusammensetzungsparameter x [*65SLR*].

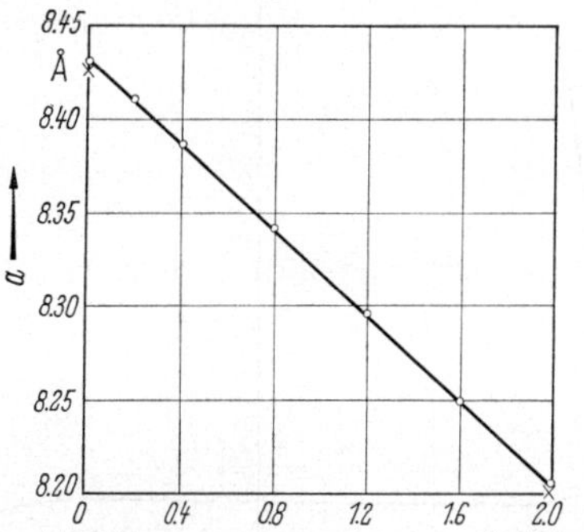

Fig. 30. $MnCr_{2-x}Al_xO_4$. Gitterkonstante a vs. Zusammensetzungsparameter x [*59E*].

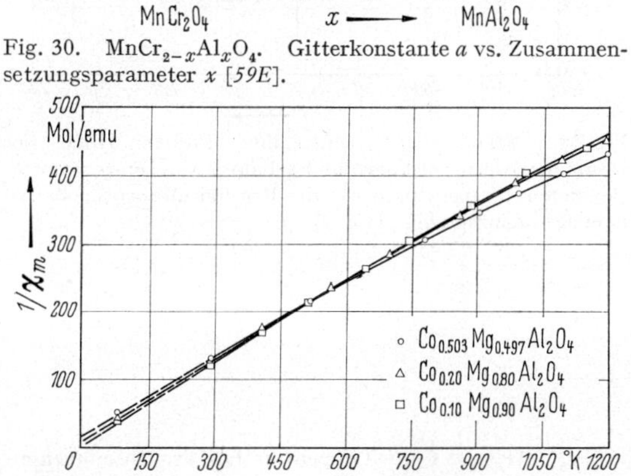

Fig. 31. $Co_xMg_{1-x}Al_2O_4$. Reziproke Suszeptibilität pro Gramm-Molekül, χ_m^{-1}, vs. Temperatur T [*56C*].

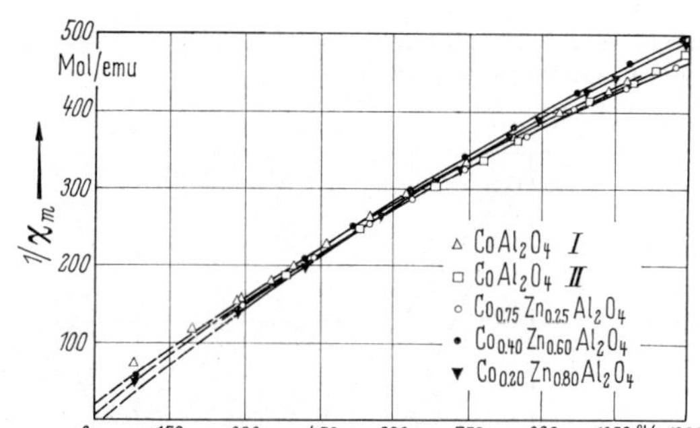

Fig. 32. $Co_{1-x}Zn_xAl_2O_4$. Reziproke Suszeptibilität pro Gramm-Molekül, χ_m^{-1}, vs. Temperatur T [*56C*].

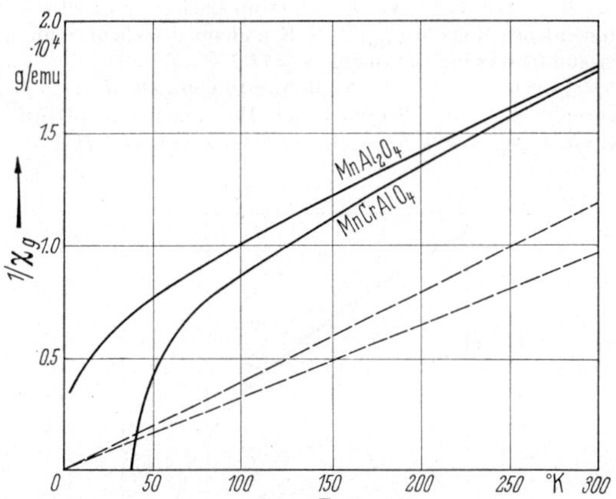

Fig. 33. $MnCrAlO_4$(a) und $MnAl_2O_4$(b). Reziproke Suszeptibilität pro Gramm, χ_g^{-1}, vs. Temperatur T. Die gestrichelten Geraden geben die theoretische Neigung bei hohen Temperaturen für Spins von 5/2 für zweiwertiges Mn und 3/2 für dreiwertiges Cr an [*59E*].

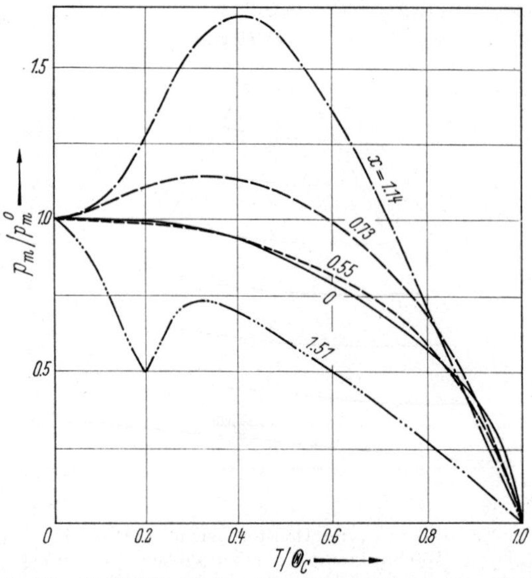

Fig. 34. $Fe^{2+}Fe^{3+}_{2-x}Al_xO_4$. Magnetisches Moment pro Molekül, p_m, bezogen auf das extrapolierte Moment p_m^0 bei 0 °K, in Abhängigkeit von der auf die Curie-Temperatur Θ_C bezogene Temperatur T für verschiedene Zusammensetzungsparameter x. Die Messungen wurden bei einer Feldstärke von 6300 Oe ausgeführt [*59PT*].

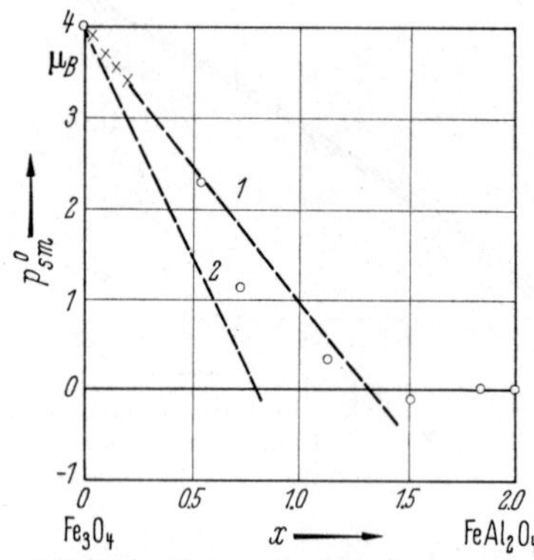

Fig. 35. $Fe^{2+}Fe^{3+}_{2-x}Al_xO_4$. Extrapoliertes magnetisches Moment pro Molekül p_{sm}^0 bei 0 °K und unendlichem Feld vs. Zusammensetzungsparameter x [*59PT*].
○ Meßwerte von [*59PT*], × Meßwerte von [*51GM*].
Kurve 1: $p_m = 4 - 3x$ nach den Messungen von [*51GM*];
Kurve 2: $p_m = 4 - 5x$ nach der Theorie von [*47VH*].

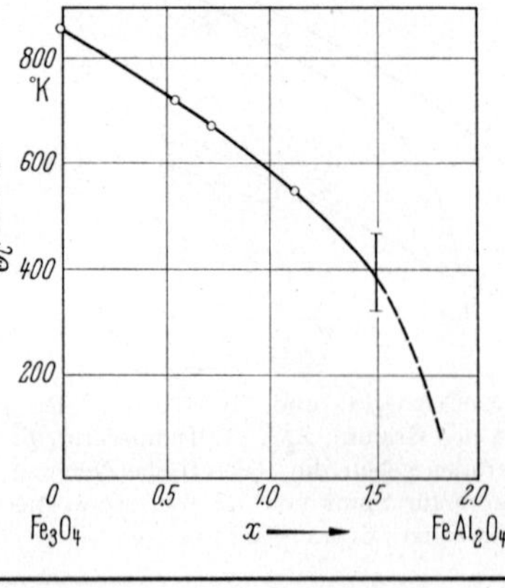

Fig. 36. $Fe^{2+}Fe^{3+}_{2-x}Al_xO_4$. Curie-Temperatur Θ_C vs. Zusammensetzungsparameter x. Der gestrichelte Kurventeil ist extrapoliert, da die Curie-Temperatur bei $x = 1.5$ wegen des äußerst kleinen Momentes nicht bestimmt werden konnte [*59PT*].

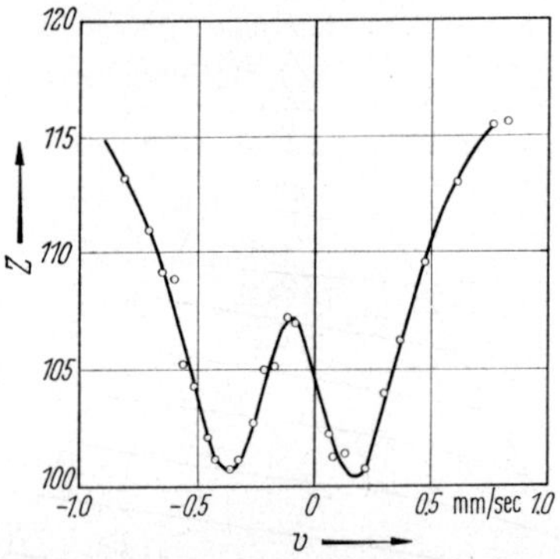

Fig. 37. $NiFeAlO_4$. Mössbauer-Spektrum bei 261 °C. Die Zählrate Z ist in willkürlichem Maßstab über der Geschwindigkeit v der Quelle aufgetragen [*63MT*].

T °C	A mm/sec		ε mm/sec	δ mm/sec
358	0.12 ± 0.01	−0.38 ± 0.01	0.50 ± 0.01	0.21 ± 0.01
261	0.18 ± 0.01	−0.35 ± 0.01	0.53 ± 0.01	0.25 ± 0.01

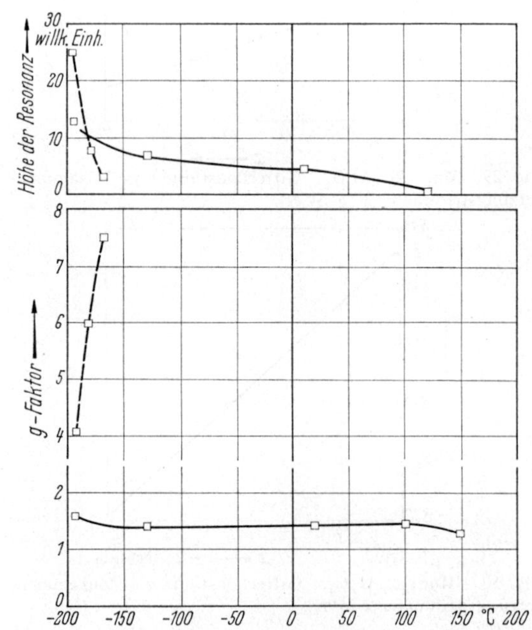

Fig. 38. $NiFe_{0.25}Al_{1.75}O_4$ (geglühte Probe). Höhe der Resonanzabsorptionskurve und g-Faktor vs. Temperatur T. Bei tiefen Temperaturen hat die Resonanzabsorptionskurve zwei Resonanzspitzen [*54M-2*].

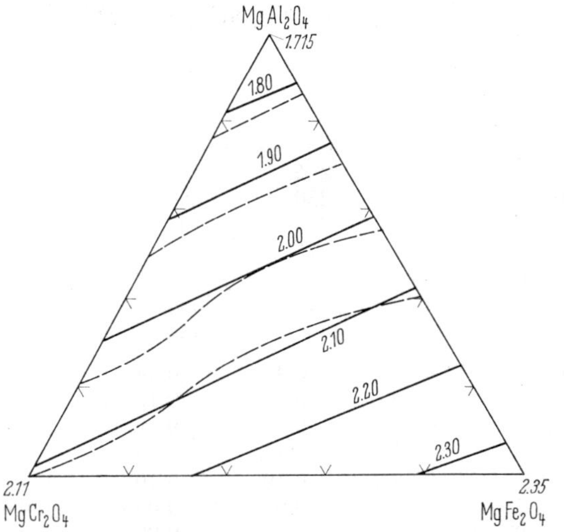

Fig. 39. $MgAl_2O_4$-$MgCr_2O_4$-$MgFe_2O_4$. Brechungsindex n im Dreistoffsystem. —— berechnet, - - - - gemessen [*66A*].

Tab. 8. Kristallografische Eigenschaften des ternären Al^{3+} Sulfo-Spinells $CuSnAlS_4$ [*67HS*], [*68SEH*]

Kationenverteilung	$Cu^+[Sn^{4+}Al^{3+}]S_4$
a	10.24 Å
ϱ_X	4.171 g/cm³
ϱ_P	4.08 g/cm³
u	0.377
Farbe	braun

Tab. 9. Kristallografische Eigenschaften quaternärer Al^{3+} Oxid-Spinelle

Verbindung	a Å	ϱ_X g/cm³	ϱ_P g/cm³	Lit.	Fig.	Tab.
$Cd_{0.7}Co_{0.1}Mg_{0.2}Al_2O_4$	8.249			*58SBB*		
$Cd_{0.6}Mg_{0.4}Cr_xAl_{2-x}O_4$				*67R*		
$Co_{0.1}Mg_{0.9}Ga_xAl_{2-x}O_4$				*58SBB*	40	
$Fe_x^{2+}Fe_{2x}^{3+}Li_{0.5(1-x)}Al_{2.5(1-x)}O_4$ [1]				*61SR*	41	
$Mg_{1-x}^{2+}Mn_x^{2+}Mn_{2x}^{3+}Al_{2-2x}O_4$ [2]				*60ISB*		10
$NiFe_{2-(x+y)}Ga_xAl_yO_4$				*62NZ, 66DP*		
$NiFe_{0.86}Ga_{0.38}Al_{0.76}O_4$		5.25	4.73	*66DP*	42	
$NiFe_{0.905}Ga_{0.337}Al_{0.758}O_4$		5.23	4.39	*66DP*		
$NiFe_{0.95}Ga_{0.30}Al_{0.75}O_4$		5.22	4.67	*66DP*		
$NiFeGa_{0.267}Al_{0.733}O_4$		5.20	4.73	*66DP*		
$NiFe_{1.03}Ga_{0.242}Al_{0.728}O_4$		5.19	4.46	*66DP*	42	
$NiGa_xFe_{2-2x}Al_xO_4$				*62NZ, 66DP*	43, 44, 45	
$NiGa_xFe_{2-3x}Al_{2x}O_4$				*62NZ, 66DP*	43, 44, 45	
$Ni_xMg_{1-x}Cr_{2x}Al_{2-2x}O_4$				*68DR*		

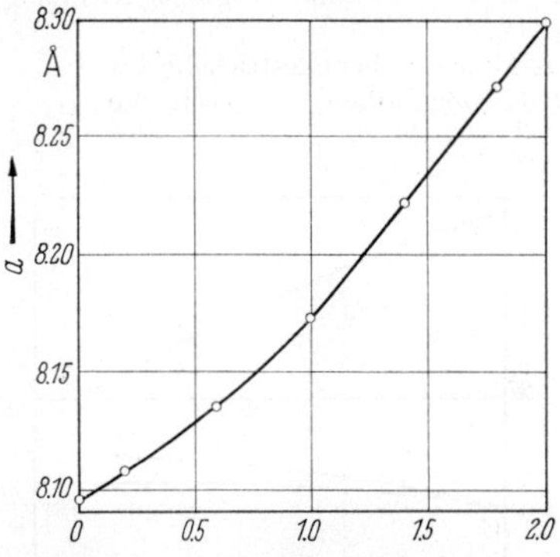

Fig. 40. $Co_{0.1}Mg_{0.9}Ga_xAl_{2-x}O_4$. Gitterkonstante a vs. Zusammensetzungsparameter x [*58SBB*].

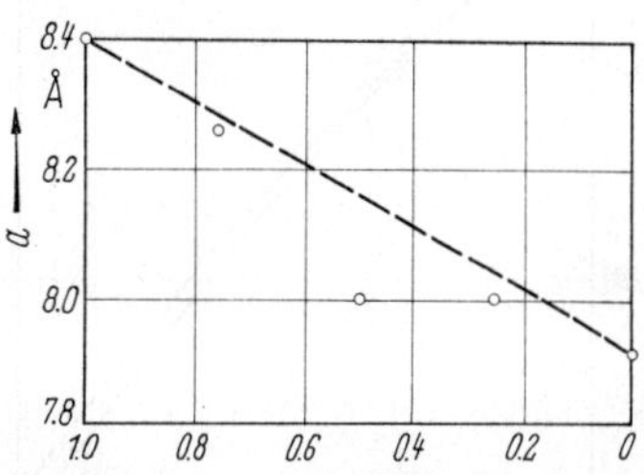

Fig. 41. $Fe_x^{2+}Fe_{2x}^{3+}Li_{0.5(1-x)}Al_{2.5(1-x)}O_4$. Gitterkonstante a vs. Zusammensetzungsparameter x [*61SR*].

[1]) Bei 1400 °C besteht für die Reihe $Fe^{2+}Fe_2^{3+}O_4$-$Li_{0.5}Al_{2.5}O_4$ vollständige Mischbarkeit [*61SR*].

[2]) Bei einem Verhältnis $c/a > 1$ und $x > 0.58$ sind die Mischkristalle tetragonal verzerrt; $a = 8.32$ Å bei $x = 0.58$ [*60ISB*].

Tab. 10. Magnetische Eigenschaften von Verbindungen der Reihe $Fe_{2-(x+y)}Ga_xNiAl_yO_4$ [66DP]. Siehe Fig. 42…45.

x	0.380	0.337	0.300	0.267	0.242
y	0.760	0.758	0.750	0.733	0.728
ΔH bei 9531 MHz [Oe]	150	252	276	315	367
ΔH bei 3001 MHz [Oe]	140	214	230	260	307
$4\pi M_s$ [Gauss]	750	759	737	797	762
Θ_C [°C]	130	150	175	199	209
g-Faktor	2.64	2.69	2.66	2.72	2.84
$\gamma 4\pi M_s$ [MHz]	2765	2865	2745	3034	3020
$\gamma(H^A + 4\pi M_s)$ [MHz]	2915	3000	3050	3450	3390
χ'' ($H = 0$) bei 3001 MHz (Festkörper)	0.16	0.42	0.45	0.64	0.80
χ'' ($H = 0$) bei 3001 MHz (Pulverprobe)		0.12	0.10	0.12	
χ''_{min} bei 3001 MHz	0.08	0.27	0.29	0.35	0.52
χ'' ($H = 0$) bei 3855 MHz	0.03	0.09	0.13	0.14	0.24
χ''_{min} bei 3855 MHz	0.03	0.08	0.10	0.10	0.15
χ'' ($H = 0$) bei 5537 MHz	0.012	0.029	0.030	0.026	0.046
χ'' ($H = 0$) bei 9531 MHz	0.005	0.005	0.005	0.005	0.008

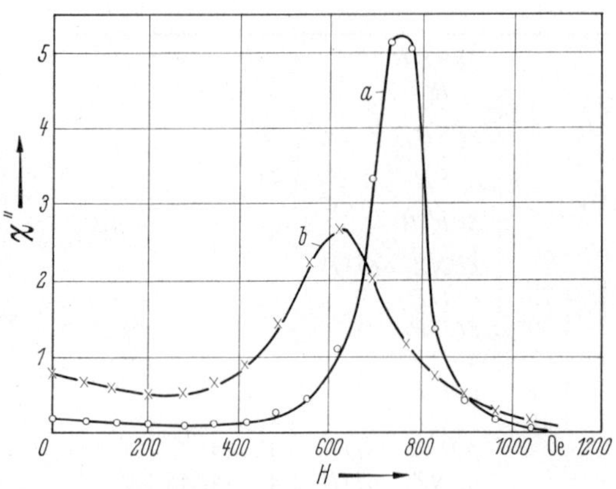

Fig. 42. $NiFe_{0.86}Ga_{0.38}Al_{0.76}O_4$(a) und $NiFe_{1.03}Ga_{0.242}Al_{0.728}O_4$(b). Imaginärteil der Suszeptibilität χ'' bei einer Frequenz von 3001 MHz vs. statisches Magnetfeld H [66DP].

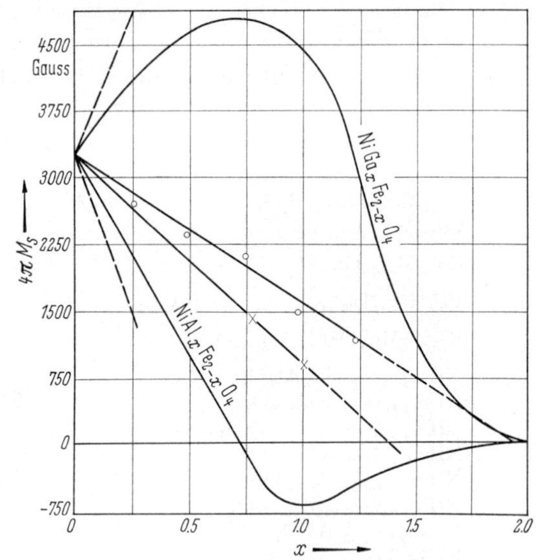

Fig. 43. $NiGa_xFe_{2-2x}Al_xO_4$ (×) und $NiGa_xFe_{2-3x}Al_{2x}O_4$ (○). Sättigungsmagnetisierung M_s vs. Zusammensetzungsparameter x [62NZ]. Zum Vergleich $NiGa_xFe_{2-x}O_4$ und $NiAl_xFe_{2-x}O_4$.
Ga^{3+} auf A-Plätzen: obere gestrichelte Linie.
Al^{3+} auf B-Plätzen: untere gestrichelte Linie.

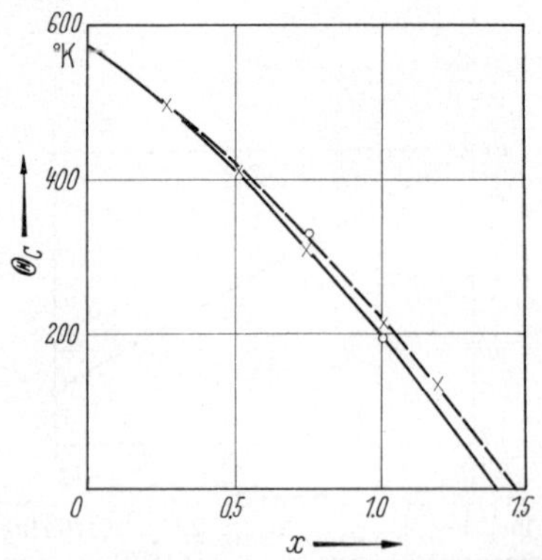

Fig. 44. $NiGa_xFe_{2-2x}Al_xO_4$ (×) und $NiGa_xFe_{2-3x}Al_{2x}O_4$ (○), durchgezogene Kurve $NiGa_xFe_{2-x}O_4$ und $NiAl_xFe_{2-x}O_4$. Curie-Temperatur Θ_C vs. Zusammensetzungsparameter x [62NZ].

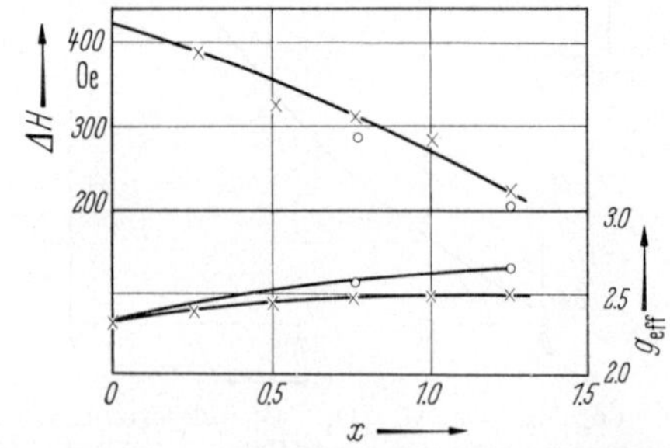

Fig. 45. $NiGa_xFe_{2-2x}Al_xO_4$ (×) und $NiGa_xFe_{2-3x}Al_{2x}O_4$ (○). Linienbreite der ferromagnetischen Resonanz, ΔH, und effektiver g-Faktor, g_{eff}, vs. Zusammensetzungsparameter x bei 9.2 GHz [62NZ].

6.3.6 Ga Spinelle und substituierte Ga Spinelle

(Literatur siehe 6.3.8, Seite 476)

Tab. 1. Kristallografische Eigenschaften binärer Ga^{3+} Oxid-Spinelle (mit Hinweisen auf magnetische Eigenschaften)

Verbindung	Kationenverteilung	a Å	u	ϱ_X g/cm³	Lit.	Fig.
$CdGa_2O_4$ [1][2][3]	$Ga[CdGa]O_4$	8.59 ± 0.01		6.62	*32BP*	
	$Cd_{1/3}Ga_{2/3}[Cd_{2/3}Ga_{4/3}]O_4$	8.57 ± 0.01			*32M, 47RR*	
		8.59 ± 0.05	0.404		*47VH, 53MK*	
		8.39			*58D, 60H-1*	
		8.605			*68SBB*	
					60H-2, 65HB	
$CoGa_2O_4$ [4]	$Co_{0.14}Ga_{0.86}[Co_{0.86}Ga_{1.14}]O_4$	8.307 ± 0.0005			*58LM-1, 59L*	6
	ausgeprägte Fehlordnung				*67NK, 68R*	
$CuGa_2O_4$		8.31 ± 0.05		6.17	*53MK, 53SR*	
		8.39 [5]		6.00	*58D*	
	$Ga[CuGa]O_4$	8.304 [5]	0.381		*65RB, 66RD*	
$FeGa_2O_4$ [4][6]	$Ga[FeGa]O_4$	8.363 ± 0.0005			*58LM-1, 59L*	1
	[7]	8.380			*66R-2*	
$Li_{0.5}Ga_{2.5}O_4$ [8]	$Ga[Li_{0.5}Ga_{1.5}]O_4$	8.21			*64B-1*	Tab. 4
		8.226		5.803	*57HT*	
				ϱ_p		
	$Li_{0.21}Ga_{0.79}[Li_{0.29}Ga_{1.71}]O_4$	8.215		5.78	*57HT, 65L*	
		8.203 ± 0.002			*65RGA*	
					51FS, 61F, 63SB, 66K-2	
				ϱ_X		
$MgGa_2O_4$ [9]	$Ga[MgGa]O_4$	8.26 ± 0.01	0.392 ± 0.006		*32BP, 55HFKSS*	2, 3, 6, 8···10
		8.279 ± 0.006		5.298	*32HN*	
	$Mg_{0.25}Ga_{0.75}[Mg_{0.75}Ga_{1.25}]O_4$	8.30 ± 0.02	(0.375) [10]		*32M, 62BH, 67G*	
		8.29 ± 0.05			*53MK*	
	$Mg_{0.33}Ga_{0.67}[Mg_{0.75}Ga_{1.33}]O_4$ [11]	8.280 ± 0.005	0.379 ± 0.004		*57H*	
		8.280		5.40	*58SBB*	
	[12]		(0.375) [10]		*61S*	
	$Mg_{0.18}Ga_{0.82}[Ga_{1.18}Mo_{0.82}]O_4$ [13]	8.286 ± 0.003	0.3824 ± 0.0003	5.317	*66WSO*	
					31M, 47RR, 59M, 60H-2, 61H, 67R	

Fortsetzung nächste Seite

[1]) In [*47RR*] ist für die meisten dieser Spinelle das Verhältnis der elektrostatischen Bindungsenergie E_N/E_A für die Anziehung Me-O berechnet, Index N = Normalstruktur, A = Austausch-, inverse Struktur.

[2]) In Säure unlöslich; Farbe: weiß [*47RR*].

[3]) Bei etwa 1200 °C scheidet $CdGa_2O_4$ unterschiedliche Mengen von CdO aus, wobei der Spinell seine Struktur behält; im nicht zerstörten Sauerstoffgitter treten in den Oktaederstellen Lücken auf, wie in γ-Fe_2O_3. Es liegt ein reversibles Gleichgewicht vor [*65HB*].

[4]) Nicht ferromagnetisch bei −196 °C [*58LM-1*].

[5]) $CuGa_2O_4$ ist nach [*58D*] kubisch bei allen Temperaturen; Robbins und Baltzer [*65RB*] vermuten bei tiefen Temperaturen tetragonale Jahn-Teller Verzerrungen durch die Cu^{2+}-Ionen.

[6]) Farbe: 24 h im Vakuum bei 1000 °C: graugrün; 48 h im Vakuum bei 1000 °C: schwarz [*66R-2*].

[7]) Aufgrund von Mössbauer-Spektren Tendenz zur normalen Kationenverteilung [*66R-2*].

[8]) $Li_{0.5}Ga_{2.5}O_4$: Fe^{3+} (Einkristall) hat ein $g_\parallel = 2.008 \pm 0.003$ [*61F*].

[9]) [*32BP*], [*32M*], [*57H*] und [*61S*]: Messungen an polykristallinen Proben; [*66WSO*]: Messungen an zwei Einkristallen.

[10]) Angenommener Wert [*61S*].

[11]) Genaue Verteilung der Mg^{2+} Ionen: $Mg_{okt} = 0.67 \pm 0.07$ [*57H*].

[12]) Genaue Verteilung der Mg^{2+} Ionen: $Mg_{okt} = 0.84 \cdots 0.90$ [*61S*].

[13]) Genaue Verteilung der Mg^{2+} Ionen: Röntgenbeugungsdaten: $Mg_{tetr} = 0.16 \pm 0.02$ $Mg_{okt} = 0.81 \pm 0.01$;
Röntgendaten mittels Computer-Kontrolle: $Mg_{tetr} = 0.15 \pm 0.03$ $Mg_{okt} = 0.82 \pm 0.01$;
Röntgenfilmdaten: $Mg_{tetr} = 0.12 \pm 0.05$ $Mg_{okt} = 0.84 \pm 0.03$; [*66WSO*].

Tab. 1 Fortsetzung

Verbindung	Kationenverteilung	a Å	u	ϱ_X g/cm³	Lit.	Fig.
$MnGa_2O_4$ [4])	$Mn_{0.80}Ga_{0.20}[Mn_{0.20}Ga_{1.80}]O_4$	8.435 ± 0.0005			*58LM-1, 59L*	4
		8.461			*58SBB*	
	$Mn_{0.87}Ga_{0.13}[Mn_{0.13}Ga_{1.87}]O_4$ [14])		0.3885 [15])		*66BHO*	
					61GBH	
$NiGa_2O_4$		8.247			*53MP*	5
	$Ga[NiGa]O_4$	8.258	0.387 ± 0.003		*54GPG, 67R, 68R*	
		8.27			*66RV*	
	$Ni_{0.05}Ga_{0.95}[Ni_{0.95}Ga_{1.05}]O_4$	8.26			*66R-1, 66S, 67B*	
					61GBH, 67PB	
$ZnGa_2O_4$ [16])		8.323 ± 0.004		6.1544	*31B*	
		8.37 ± 0.05		6.08	*53MK*	
	$Zn[Ga_2]O_4$				*59M, 61GBH*	
					32B, 47RR, 47VH, 55HFKSS, 62S	

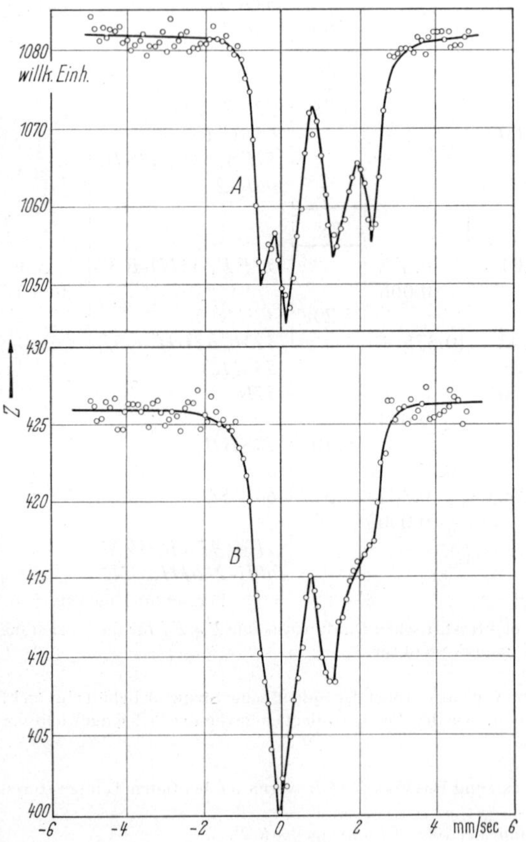

Fig. 1. $FeGa_2O_4$. Mössbauer-Absorptionsspektrum bei Raumtemperatur. Die Zählrate Z ist in willkürlichen Maßstab über der Austrittsgeschwindigkeit v der Quelle aufgetragen [*66R-2*].
A: graugrünes Produkt nach 24 h im Vakuum bei 1000 °C;
B: schwarzes Produkt nach weiteren 24 h im Vakuum bei 1000 °C

	T °K	ε mm/sec	δ mm/sec
A[1])	77	2.7 ± 0.1	0.9 ± 0.1
	300	2.9 ± 0.1 | 1.2 ± 0.1	0.8 ± 0.1 | 0.65 ± 0.1
B	77	2.3	0.9
	300	1.3	0.6

[1]) Das Mössbauer-Spektrum zeigt ein äußeres und ein inneres Quadrupolpaar [*66R-2*].

[4]) Nicht ferromagnetisch bei −196 °C [*58LM-1*].
[14]) Genaue Verteilung: $Mn_{0.87\pm0.02}Ga_{0.13\pm0.02}[Mn_{0.13\pm0.05}Ga_{1.87\pm0.05}]O_4$ [*66BHO*].
[15]) Zuverlässigkeitsfaktor: 4.1% [*66BHO*].
[16]) Farbe: Reinweiß [*31B*]; optisch isotrop; stärkere Lichtbrechung als Methylenjodid, also Brechungsindex $n > 1.74$ [*31B*].

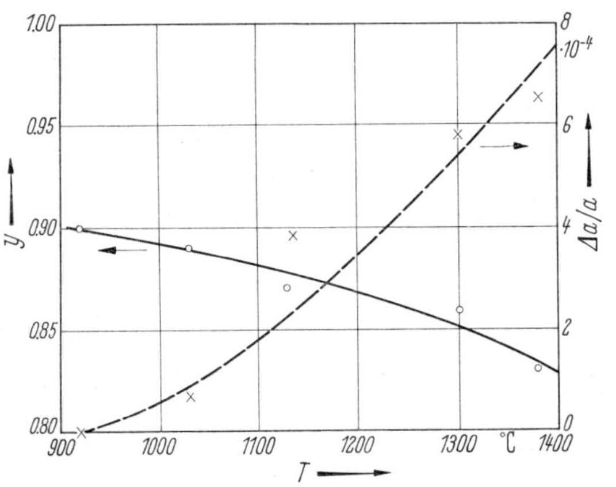

Fig. 2. $MgGa_2O_4$. Relative Gitterkonstantenänderung $\Delta a/a$ und Kationenverteilungsparameter y (gemäß der Formel $Mg_{1-y}Ga_y[Mg_yGa_{2-y}]O_4$) vs. Temperatur T [61S].

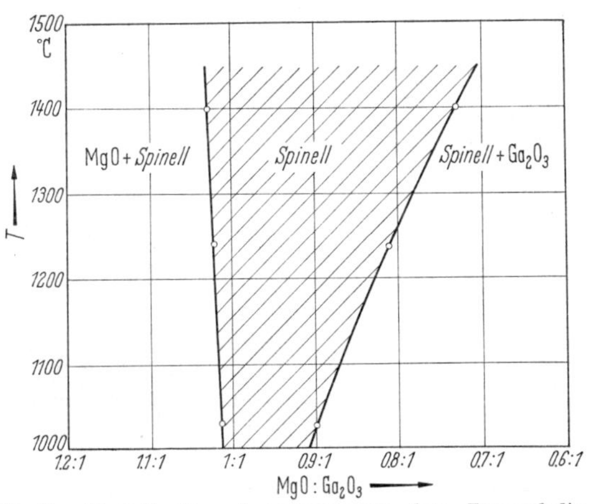

Fig. 3. MgO-Ga_2O_3. Ausschnitt aus dem Zustandsdiagramm [61S].

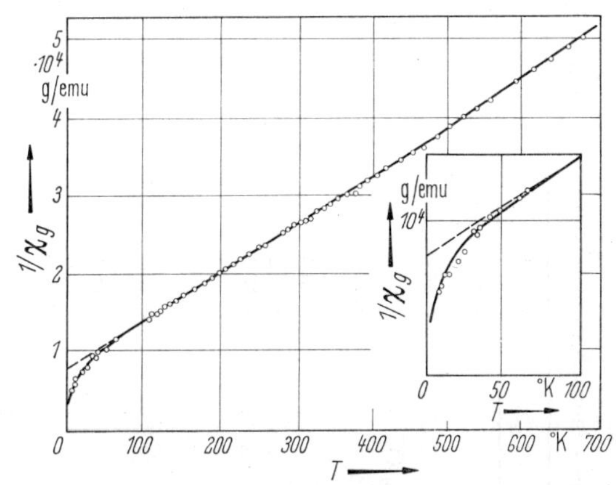

Fig. 4. $MnGa_2O_4$. Experimentelle Werte (○) und berechnete Kurve der reziproken Suszeptibilität pro Gramm χ_g^{-1} vs. Temperatur T.

Die Suszeptibilität wurde nach der Gleichung

$$\chi_g = \frac{C_{gA}}{T - \Theta_a} + \frac{C_{gB}}{T}$$

berechnet, wobei C_{gA} und C_{gB} die Curie-Konstanten pro Gramm für Ionen auf den Tetraeder- oder den Oktaederstellen sind.

$C_{gA} = 0.0163$ cm³ °K/g,
$C_{gB} = 0.0008$ cm³ °K/g,
$\Theta_a = -154$ °K (berechnet) und
-128 °K (experimentell),
$p_{eff} = 3.60\,\mu_B$ und $\Theta_N = 33$ °K,
$K_1 = -4.1 \cdot 10^4$ erg/cm³.

$MnGa_2O_4$ ist in [111]-Richtung antiferromagnetisch [66BHO].

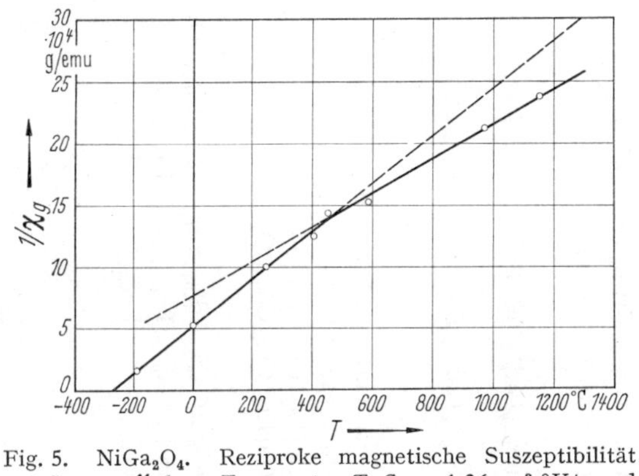

Fig. 5. $NiGa_2O_4$. Reziproke magnetische Suszeptibilität pro Gramm, χ_g^{-1}, vs. Temperatur T. $C_g = 1.36$ cm³ °K/g und g-Faktor = 2.3; $\chi_{g\,diam} = -0.37 \cdot 10^{-6}$ emu/g [54GPG]. Messungen von [67PB] ergaben $C_g = 1.40$ cm³ °K/g und $p_m^0 = 0$.

Tab. 2. Kristallografische Eigenschaften des Ga^{3+} Sulfo-Spinells $CrGa_2S_4$ [61FDGF]

Kationenverteilung	a Å	u	ϱ_X g/cm³	Farbe	chemisches Verhalten
Wegen des nahezu gleichen Streuvermögens von Chrom- und Galliumionen ist eine Bestimmung mittels Röntgenbeugung nicht möglich	9.95	0.387	4.31	schwarz	unveränderlich an Luft, praktisch unlöslich in verdünnter HCl, wird ziemlich schnell von Salpetersäure angegriffen

Tab. 3. Kristallografische Eigenschaften ternärer Ga^{3+} Oxid-Spinelle

Verbindung	x	Kationenverteilung	a [Å]	u	Lit.	Fig.	Tab.
$CdInGa_2O_4$ [1]		$In^{3+}[Cd^{2+}Ga^{3+}]O_4$	8.909 ± 0.002		*65SK*		
$Co_{0.1}Cd_{0.9}Ga_2O_4$			8.581		*58SBB*		
$CoFe_{2-x}Ga_xO_4$	[2]	$Co_z^{2+}Fe_{1-y-z}^{3+}Ga_y^{3+}[Co_{1-z}^{2+}Fe_{1-x+y+z}^{3+}Ga_{x-y}^{3+}]O_4$			*59L, 64B-1, 57LM*		4
	1		8.361		*59L*		
	1.2		8.352				
	1.4		8.343				
	1.6		8.332				
	2		8.307				
$Co_{1+x}Ge_xGa_{2-2x}O_4$	[3]	$Co_y^{2+}Ga_{1-y}^{3+}[Co_{1+x-y}^{2+}Ge_x^{4+}Ga_{1-2x+y}^{3+}]O_4$			*68R*	11 in 6.3.2	
$Co_xMg_{1-x}Ga_2O_4$				0.3805	*61BH, 65HB*	6	
	0.1		8.300		*58SBB*		
$Co_{0.1}Mn_{0.9}Ga_2O_4$			8.450		*58SBB*		
$CoMnGaO_4$		$Ga^{3+}[Co^{2+}Mn^{3+}]O_4$	8.33		*57B*	11	4
$CuAlGaO_4$		$Cu_{0.5}^{2+}Ga_{0.5}^{3+}[Al^{3+}Cu_{0.5}^{2+}Ga_{0.5}^{3+}]O_4$	8.181	0.381 [4]	*66RD*		
$CuCr_{2-x}Ga_xO_4$	1.0	$Cu_{0.5}Ga_{0.5}[CrCu_{0.5}Ga_{0.5}]O_4$ [5]	8.305	0.381 [6]	*58D, 66RD*		
	1.0	$Co_{0.45}Ga_{0.55}[CrCu_{0.55}Ga_{0.45}]O_4$ [7]			*66RD*		
$CuMnGaO_4$		$Cu^{2+}[Mn^{3+}Ga^{3+}]O_4$	8.363	0.383	*64BL*		4
			8.365	0.383 [8]	*66RD*		
$Fe^{2+}Fe_{2-x}^{3+}Ga_xO_4$	[2]	$Fe_{1-y}^{3+}Ga_y^{3+}[Fe^{2+}Fe_{1-x+y}^{3+}Ga_{x-y}^{3+}]O_4$			*59L, 65O, 57LM*		
	1.0		8.368		*59P*		
	1.1	$Fe_{0.36}^{3+}Ga_{0.64}^{3+}[Fe^{2+}Fe_{0.54}^{3+}Ga_{0.46}^{3+}]O_4$	8.377	0.3802 ± 0.0002	*65O*		4
	1.4		8.366		*59L*		
	2		8.363				
$Ge_xNi_{1+x}Ga_{2-2x}O_4$	[2]	$Ge_{x-y}Ga_{1-x+y}[Ni_{1+x}Ge_yGa_{1-x-y}]O_4$ [9]			*67R, 68R*	15 in 6.3.2	
$Li_{0.5}Cr_xGa_{2.5-x}O_4$	[10]	$Ga^{3+}[Li_{0.5}^{+}Cr_x^{3+}Ga_{1.5-x}^{3+}]O_4$			*65RGA*		
	0.3	$Ga[Li_{0.5}Cr_{0.3}Ga_{1.2}]O_4$	8.207 ± 0.002				
	0.5	$Ga[Li_{0.5}Cr_{0.5}Ga_{1.0}]O_4$	8.202 ± 0.002				
$Li_{0.5}Fe_{2.5-x}Ga_xO_4$	[2]				*63SB, 65K*		4
	1.0	$Fe_{0.40}^{3+}Ga_{0.53}^{3+}Li_{0.01}^{1+}[Fe_{1.04}^{3+}Ga_{0.47}^{3+}Li_{0.49}^{1+}]O_4$	8.286 [16]		*65L, 65K*		
	1.25	$Fe_{0.41}^{3+}Ga_{0.59}^{3+}[Fe_{0.84}^{3+}Ga_{0.66}^{3+}Li_{0.5}^{1+}]O_4$	8.282		*65L*		
	1.5	$Fe_{0.38}^{3+}Ga_{0.62}^{3+}[Fe_{0.62}^{3+}Ga_{0.88}^{3+}Li_{0.5}^{1+}]O_4$	8.270 [17]		*65L*		
	1.75		8.262				
	2.0		8.246 [17]		*65L*		
	2.5		8.215				
$LiGeGaO_4$			8.16		*65NM*		

$Li_{0.5}Mn_xGa_{2.5-x}O_4$	10)	$Ga^{3+}[Li^{+}_{0.5}Mn^{3+}_xGa_{1.5-x}]O_4$			*65RGA*	7	
$LiMnGaO_4$		$Li^{+}_{0.5}Ga^{3+}_{0.5}[Li^{+}_{0.5}Ga^{3+}_{0.5}Mn^{4+}]O_4$	8.23		*64B-1*		
$Li_{0.5}Rh_xGa_{2.5-x}O_4$	10)	$Ga^{3+}[Li^{+}_{0.5}Rh^{3+}_xGa^{3+}_{1.5-x}]O_4$			*65RGA*		
	0.3	$Ga[Li_{0.5}Rh_{0.3}Ga_{1.2}]O_4$	8.210		*66JD*		
$LiTiGaO_4$		$Li^{+}_{0.3}Ga^{3+}_{0.7}[Li^{+}_{0.7}Ti^{4+}Ga^{3+}_{0.3}]O_4$	8.28		*63B-2, 64B-1*		
			8.29		*64B-1*		
		$Li^{+}_{1/3}Ga^{3+}_{2/3}[Li^{+}_{2/3}Ti^{4+}Ga^{3+}_{1/3}]O_4$	8.29 11)		*66BP*		
$MgAl_xGa_{2-x}O_4$	2)				*58SBB, 61H*	8	
$Mg_{1-x}Cd_xGa_2O_4$					*60H-2*		
$MgMn_xGa_{2-x}O_4$	12)			13)	*63GH, 63GHM, 65HB, 67G*	9, 10	
	1.0	$Ga^{3+}[Mg^{2+}Mn^{3+}]O_4$	8.35		*57B*	11	4
	1.0	$Mg_{0.4}Mn_{0.1}Ga_{0.5}[Mg_{0.6}Mn_{0.9}Ga_{0.5}]O_4$			*65GH*		
	1.0	$Mg_{0.33}Mn_{0.21}Ga_{0.46}[Mg_{0.67}Mn_{0.79}Ga_{0.54}]O_4$		>0.405	*67G*		
$Mg_{0.9}Ni_{0.1}Ga_2O_4$ 14)			8.280		*59SGB*		
$MnFe_{2-x}Ga_xO_4$	2)	$Fe^{3+}_{1-y-z}Mn^{2+}_zGa^{3+}_y[Fe^{3+}_{1-x+y+z}Mn^{2+}_{1-z}Ga^{3+}_{x-y}]O_4$			*59L, 58LM-2, 57LM*		4
	1.0	$Fe_{0.95}Ga_{0.05}[Fe_{0.05}MnGa_{0.95}]O_4$	8.487		*59L*		
	1.4	$Fe_{0.60}Mn_{0.30}Ga_{0.1}[Mn_{0.7}Ga_{1.3}]O_4$	8.475				
	1.6		8.464				
	1.8		8.450				
	2		8.435				
$NiFe_{2-x}Ga_xO_4$		$Fe^{3+}_{1-y-z}Ni^{2+}_zGa^{3+}_y[Fe^{3+}_{1-x+y+z}Ni^{2+}_{1-z}Ga^{3+}_{x-y}]O_4$			*53MP, 54G*		4
					67B		
$NiInGaO_4$ 15)		$In^{3+}[Ni^{2+}Ga^{3+}]O_4$	8.556 ± 0.001		*65SK*		
$NiMnGaO_4$		$Ga^{3+}[Ni^{2+}Mn^{3+}]O_4$	8.38		*57B, 63G*	11	4
$ZnMn_xGa_{2-x}O_4$	0.5		kubisch		*65B*		
	1.0	$Zn^{2+}[Mn^{3+}Ga^{3+}]O_4$	$c/a = 1.05$		*63G*		4

1) Farbe: weiß [*65SK*].
2) Dieses System bildet eine lückenlose Reihe von Mischkristallen [*59L*], [*65L*], [*67R*], [*58SBB*].
3) In der Verbindungsreihe $CoGa_2O_4$-Co_2GeO_4 besteht eine Mischungslücke für einen Zusammensetzungsparameter von $0.8 < x < 1$. Die Mischkristalle sind intensiv blau oder grün gefärbt; Co_2GeO_4 mit wenig $CoGa_2O_4$ ist rosa [*68R*].
4) Zuverlässigkeitsfaktor 0.027 [*66RD*].
5) Mittels Röntgenbeugung [*66RD*].
6) Minimaler Zuverlässigkeitsfaktor 0.025 [*66RD*].
7) Mittels Neutronenbeugung [*66RD*].
8) Minimaler Zuverlässigkeitsfaktor 0.045 [*66RD*].
9) Von wenigen Ni^{2+} in Tetraederstellen abgesehen [*68R*].
10) Kritische Konzentration x_{crit} von substituierten Ionen für die Beseitigung der 1 : 3 Fernordnung im oktaedrischen Untergitter: $Li_{0.5}Cr_xGa_{2.5-x}O_4$: $x_{crit} > 0.5$; $Li_{0.5}Mn_xGa_{2.5-x}O_4$: $x_{crit} \approx 0.26$; $Li_{0.5}Rh_xGa_{2.5-x}O_4$: $x_{crit} > 0.3$ [*65RGA*].
11) Spinell der Raumgruppe Imma-(D^{28}_{2h}) mit Überstruktur. Das Symbol „:" zeigt eine 1:2-Ordnung auf den Tetraeder- und Oktaederstellen an. Orthorhombisches Gitter: $a' = a/\sqrt{2}$, $b = 3a/\sqrt{2}$, $c = a$, worin a die Gitterkonstante des entsprechenden ungeordneten Spinells ist [*66BP*].
12) Die Gitterkonstante a in Abhängigkeit vom Zusammensetzungsparameter x folgt dem Végardschen Gesetz innerhalb der Meßgenauigkeitsgrenze ($\pm 7 \cdot 10^{-4}$ Å). Beide Oxide $MgGa_2O_4$ und $MgMn_2O_4$ sind vollständig nur oberhalb 850 °C mischbar, vergleiche dazu Fig. 9 [*63GH*], [*63GHM*].
13) Für $x < 1$: $u = 0.407$ für $x \geqq 1$: $u = 0.407 \cdots 0.416$ [*67G*].
14) Farbe: blaßblau [*59SGB*].
15) Farbe: gelbgrün [*65SK*].
16) Langsam abgekühlte Probe; abgeschreckte Proben entsprechend mit höheren Werten [*65L*].
17) Abgeschreckte Proben [*65L*].

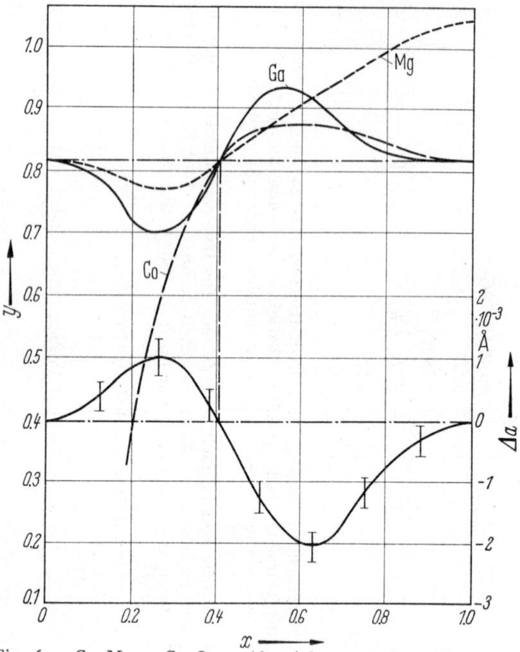

Fig. 6. $Co_xMg_{1-x}Ga_2O_4$. Abweichungen der Gitterkonstanten Δa vom Végardschen Gesetz und Inversionsgrad y vs. x. Für die Kationenverteilung gilt die Formel: $Co_{x(1-y)}Mg_{(1-x)(1-y)}Ga_y[Co_{xy}Mg_{(1-x)y}Ga_{2-y}]O_4$ [65HB].

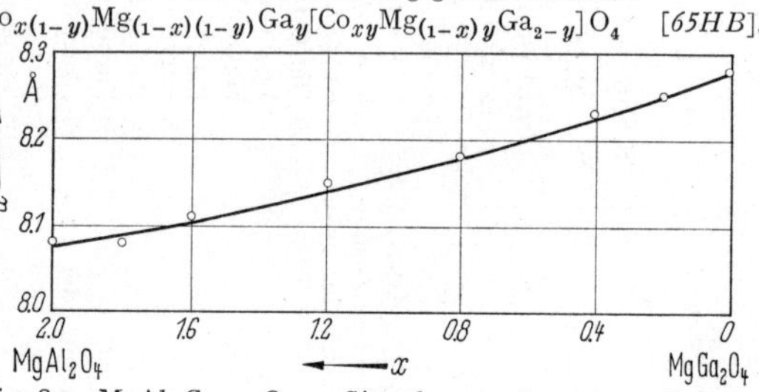

Fig. 8. $MgAl_xGa_{2-x}O_4$. Gitterkonstante a vs. Zusammensetzungsparameter x [61H].

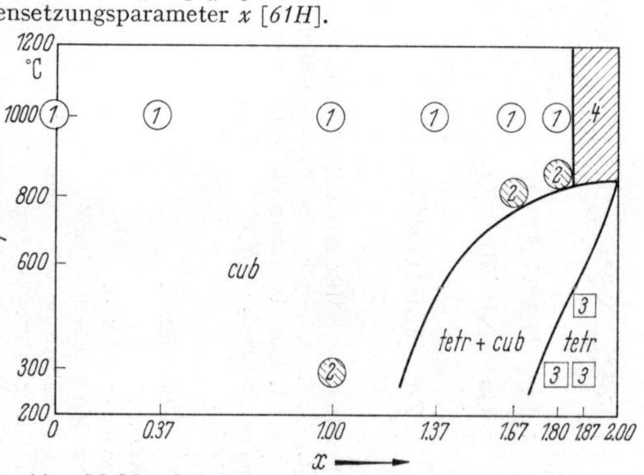

Fig. 10. $MgMn_xGa_{2-x}O_4$. Phasendiagramm [67G].

cub Kubische Phase;

tetr + cub Tetragonal verzerrte Spinell-Mischkristalle;

tetr Tetragonale Phase;

1 Produkte sind kubisch bei hohen Temperaturen und kubisch nach der Abschreckung;

2 Die Produkte sind kubisch nach dem Glühen oberhalb des Zweiphasengebietes oder bei tiefen Temperaturen;

3 Die Produkte sind tetragonal nach dem Glühen bei tiefen Temperaturen;

4 Zone, in der die Verbindungen bei Abschreckung tetragonal sind.

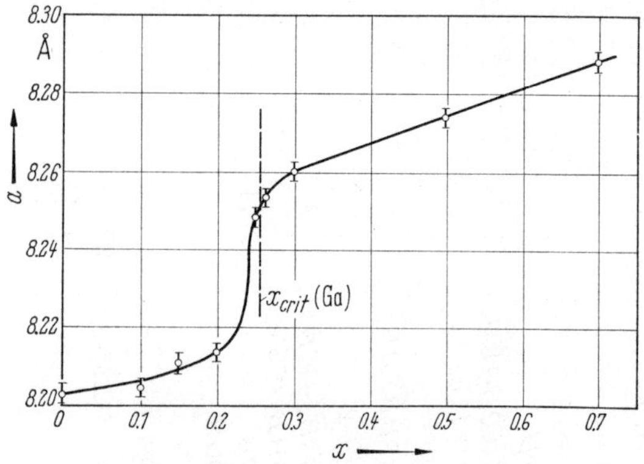

Fig. 7. $Li_{0.5}Mn_xGa_{2.5-x}O_4$. Gitterkonstante a vs. Zusammensetzungsparameter x. Die kritische Konzentration von substituierten Ionen für die Besetzung der Fernordnung im oktaedrischen Untergitter ist $x_{crit} \approx 0.26$ [65RGA].

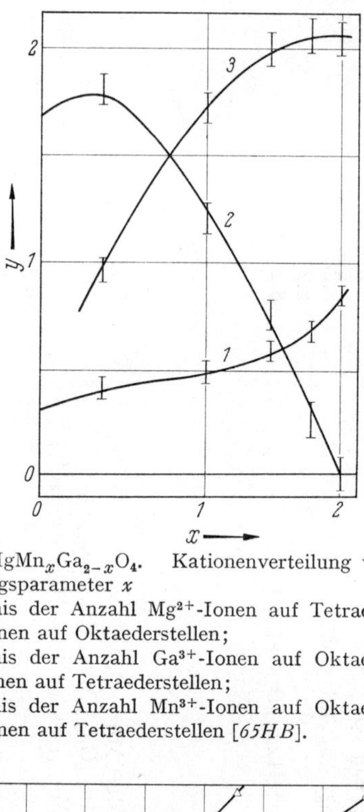

Fig. 9. $MgMn_xGa_{2-x}O_4$. Kationenverteilung vs. Zusammensetzungsparameter x

1 Verhältnis der Anzahl Mg^{2+}-Ionen auf Tetraederstellen/Mg^{2+}-Ionen auf Oktaederstellen;

2 Verhältnis der Anzahl Ga^{3+}-Ionen auf Oktaederstellen/Ga^{3+}-Ionen auf Tetraederstellen;

3 Verhältnis der Anzahl Mn^{3+}-Ionen auf Oktaederstellen/Mn^{3+}-Ionen auf Tetraederstellen [65HB].

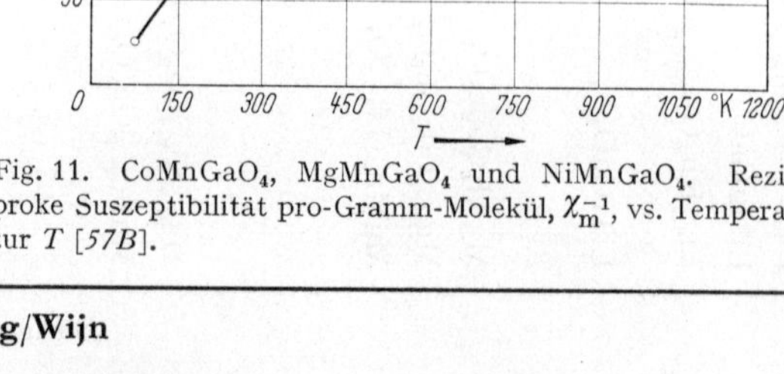

Fig. 11. $CoMnGaO_4$, $MgMnGaO_4$ und $NiMnGaO_4$. Reziproke Suszeptibilität pro-Gramm-Molekül, χ_m^{-1}, vs. Temperatur T [57B].

Tab. 4. Magnetische und elektrische Eigenschaften ternärer und quaternärer Ga^{3+} Oxid Spinelle

Verbindung	x	Θ_C °K	p_m μ_B langsam abgekühlt	p_m μ_B abgeschreckt	p_{eff} μ_B	Θ_a °K	C_m cm³ °K/mol	7) °K	ϱ Ωcm	Lit.	Fig.
$CoFe_{2-x}Ga_xO_4$	1	467	3.85	3.07						*59L*	
	1.2	398	3.17	3.04							
	1.4	315	2.01	2.00							
	1.6	226									
	2										
$CoMnGaO_4$					1)	−264	5.95			*57B*	11
$CuMnGaO_4$						− 63	2.98			*64BL*	
$Fe^{2+}Fe^{3+}_{2-x}Ga_xO_4$	1	473	4.67	4.75						*59L*	
	1.1	470	1.8 ± 0.2 2)							*65O*	
			3.3 ± 0.4 3)							*59L*	
	1.4		3.74	3.75						*59L*	
	1.6	213									
	2										
$Li_{0.5}Fe_{2.5-x}Ga_xO_4$	1	448	2.90	3.28						*65L, 65K*	
	1.25	343	2.13	2.58						*63SB*	
	1.5		1.20	1.46							
$MgMnGaO_4$					4.90 4)	−112	3.00			*57B*	11
$MnFe_{2-x}Ga_xO_4$	1	223	0.55	0.83						*59L*	
	1.4	197	1.00	0.65							
$NiFe_{2-x}Ga_xO_4$										*53MP*	43 in 6.3.5
$NiMnGaO_4$					5)	−150	4.30			*57B*	11
$ZnMn_{0.5}Ga_{1.5}O_4$						− 95	1.50	80 ··· 1300	$10^4 \cdots 10^6$	*65B*	
6)						− 60	1.48	80 ··· 1300	$10^4 \cdots 10^6$		
$Co_{0.1}Mg_{0.9}Al_{2-x}Ga_xO_4$										*58SBB*	40 in 6.3.5
$ZnFe_{0.6}Mn_{0.6}Ga_{0.8}O_4$						−110	4.42	300 ··· 1300	$10^4 \cdots 10^6$	*65B*	

1) $p_{Co^{2+}} = 4.86\ \mu_B$; $p_{Mn^{3+}} = 3.6\ \mu_B$ [*57B*].
2) Magnetisches Moment der Atome auf Tetraederstellen [*65O*].
3) Magnetisches Moment der Atome auf Oktaederstellen [*65O*].
4) $p_{Mn^{3+}} = 4.89\ \mu_B$ [*57B*].
5) $p_{Ni^{2+}} = 3.22\ \mu_B$; $p_{Mn^{3+}} = 4.1\ \mu_B$ [*57B*].
6) Von 1150 °C abgeschreckt [*65B*].
7) Temperaturbereich, in dem Verhalten nach dem Curie-Weiss-Gesetz vorliegt [*65B*].

6.3.7 In Spinelle und substituierte In Spinelle

(Literatur siehe 6.3.8, Seite 476)

Tab. 1. Kristallografische Eigenschaften binärer In^{3+} Oxid-Spinelle

Verbindung	Kationenverteilung	a Å	c Å	c/a	u	ϱ_X g/cm³	Farbe	Lit.
$BaIn_2O_4$ [1])		8.23	8.15	0.99			schmutzig-gelb	*47EV* *30P-2, 64B-1*
$CaIn_2O_4$ [2])[3])		8.770 ± 0.005 8.59	9.822±0.007 9.53	1.12 1.11		5.87	weiß, Stich ins Gelbliche	*30P-2* *47EV* *60G-1, 64B-1* *55GL*
$CdIn_2O_4$ [4])		8.650 ± 0.005	9.875±0.007	1.14		7.30		*30P-2*
						ϱ_p		
	$Cd[In_2]O_4$					7.13		*47EV, 59SDB*
						ϱ_X		
	$In[CdIn]O_4$ [5])	9.115 ± 0.001 9.167 9.165 ± 0.002			0.385 [6])	7.52	gelb bis grün [7])	*59SDB* *64B-1* *65SK* *60G-1* *47VH, 55GL*
$MgIn_2O_4$ [8])	$In[MgIn]O_4$	8.81 ± 0.01 8.81			0.372 ±0.01 0.375			*32BP, 55HFKSS* *47MS, 47RR*
$SrIn_2O_4$ [1])		7.98		1.00			hellgelb	*47EV* *30P-2, 64B-1*

Tab. 2. Kristallografische Eigenschaften ternärer In^{3+} Oxid-Spinelle

Verbindung	Kationenverteilung	a Å	ϱ_X g/cm³	Farbe	Lit.
$CdGaInO_4$	$In^{3+}[Cd^{2+}Ga^{3+}]O_4$	8.909 ± 0.002		weiß	*65SK*
$CoCrInO_4$	$In^{3+}[Co^{2+}Cr^{3+}]O_4$			schwarz	*65SK*
$Li_{0.5}Cr_2In_{0.5}O_4$ [9])		8.411 ± 0.003	4.347		*66JD*
$Li_{0.5}Fe_{2.5-x}In_xO_4$					*65L*
$Li_{0.5}Rh_2In_{0.5}O_4$ [9])		8.605 ± 0.004	6.225		*66JD*
$MgCrInO_4$	$In^{3+}[Cr^{3+}Mg^{2+}]O_4$	8.588 ± 0.001		grün	*65SK*
$NiAlInO_4$	$In^{3+}[Al^{3+}Ni^{2+}]O_4$	8.429 ± 0.003		grün	*65SK*
$NiCrInO_4$	$In^{3+}[Cr^{3+}Ni^{2+}]O_4$	8.561 ± 0.001		grün	*65SK*
$NiGaInO_4$	$In^{3+}[Ga^{3+}Ni^{2+}]O_4$	8.556 ± 0.001		gelbgrün	*65SK*

1) Eine tetragonal deformierte Spinellstruktur nach [*30P-2*] und [*47EV*] ist nach [*64B-1*] zweifelhaft.
2) Die tetragonal deformierte Struktur wurde von [*60G-1*], [*64B-1*] als falsch erklärt; $CaIn_2O_4$ besitzt danach die $CaFe_2O_4$-Struktur.
3) Die Mischkristalle $3CaO \cdot 4In_2O_3$ und $2CaO \cdot 2In_2O_3$ zeigten die gleiche Farbe, der Kristallaufbau blieb bei Aufnahme von weiterem In_2O_3 erhalten [*47EV*].
4) Die tretagonal deformierte Struktur wurde von [*60G-1*], [*64B-1*] als falsch erklärt.
5) Fast vollständig invers [*59SDB*].
6) Zuverlässigkeitsfaktor 0.14 [*59SDB*].
7) Je nach Glühtemperatur: 20 h bei 860 °C: gelb; weitere 15 h bei 1050 °C: gelbgrün; weitere 100 h bei 820 °C: grüngelb [*65SK*].
8) Verhältnis der elektrostatischen Energie berechnet für die Anziehung Me-O: $E_{normal}/E_{invers} = 1.049$ [*47RR*].
9) Die Oxide $Li_{0.5}Rh_2In_{0.5}O_4$ und $Li_{0.5}Cr_2In_{0.5}O_4$ sind definierte Mischungen, die keine Abweichung vom stöchiometrischen Verhältnis erlauben. Dieser Spinelltyp hat eine Kationenordnung von 1 : 1 auf den tetraedrischen Gitterstellen [*66JD*].

Tab. 3. Kristallografische und magnetische Eigenschaften binärer In^{3+} Sulfo-Spinelle

Verbindung [1]	Kationenverteilung	a Å	u	ϱ_X g/cm³	ϱ_p g/cm³	Lit.
β-In_2S_3		10.72				*49HK*
	$\square_{1/3}In_{2/3}[In_2]S_4$	10.774 [2]				*59R, 64B-1*
		10.775				*61FDGF*
						60G-1, 60H-1, 63B-4
$Ag_{0.5}In_{2.5}S_4$ [3]	$Ag_{0.5}In_{0.5}[In_2]S_4$	10.281 ± 0.004	0.389	4.92	4.85	*61FDGOK, 64B-1*
$CaIn_2S_4$ [4]	$Ca[In_2]S_4$	10.774 ± 0.009	<0.393	4.198	4.10	*50HK, 55HFKSS*
$CdIn_2S_4$ [5]	$Cd[In_2]S_4$	10.797 ± 0.007	0.386	4.934	4.93	*50HK, 55HFKSS*
						65BGKP
$CoIn_2S_4$ [7]	$In[CoIn]S_4$	10.559 ± 0.009		4.726	4.60	*50HK, 55HFKSS*
$CrIn_2S_4$ [6][7]	$In[CrIn]S_4$	10.59	0.386±0.004	4.58	4.48	*61FDGF, 50HK*
$Cu_{0.5}In_{2.5}S_4$ [8]	$Cu_{0.5}In_{0.5}[In_2]S_4$	10.698 ± 0.004	0.390	4.85	4.60	*61FDGOK*
$FeIn_2S_4$ [7][12]	$In[FeIn]S_4$	10.598 ± 0.009	0.384	4.671	4.48	*50HK, 55HFKSS*
			<0.387			*64B-1, 67YM*
		10.630 ± 0.002				*67EHS*
$HgIn_2S_4$ [9]	$Hg[In_2]S_4$	10.812 ± 0.007	0.388	5.815	5.79	*50HK, 55HFKSS*
			<0.403			*64B-1*
$MgIn_2S_4$ [10]	$In[MgIn]S_4$	10.687 ± 0.008	<0.387	4.132	4.01	*50HK, 55HFKSS*
$MnIn_2S_4$ [11]	$Mn_{1/3}In_{2/3}[Mn_{2/3}In_{4/3}]S_4$	10.694 ± 0.008	<0.390	4.454	4.45	*50HK, 55HFKSS*
$NiIn_2S_4$ [7]	$In[NiIn]S_4$	10.464 ± 0.009		4.794	4.69	*50HK, 55HFKSS*

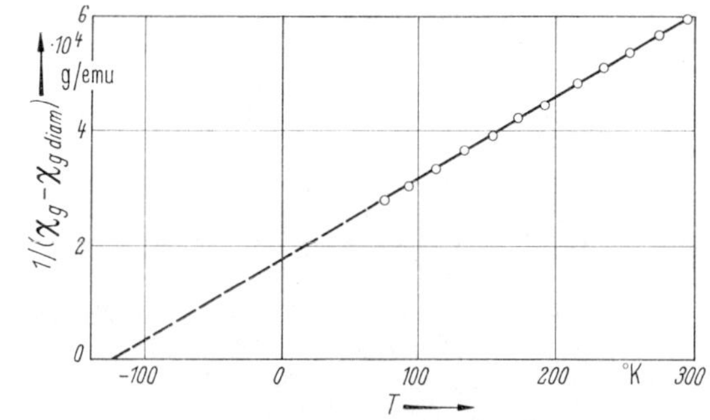

Fig. 1. $FeIn_2S_4$. Reziproke Suszeptibilität pro Gramm $1/(\chi_g - \chi_{g\,diam})$ vs. Temperatur T [*67EHS*].

Tab. 4. Kristallografische Eigenschaften ternärer In^{3+} Sulfo-Spinelle

Verbindung	Kationenverteilung	a Å	u	ϱ_X g/cm³	ϱ_p g/cm³	Farbe	Lit.
$AgSnInS_4$	$Ag^+[Sn^{4+}In^{3+}]S_4$	10.74	0.386	4.923	4.91	dunkelbraun	*67HS, 68SEH*
$CuSnInS_4$	$Cu^+[Sn^{4+}In^{3+}]S_4$	10.50	0.379	4.906	4.86	braun	*67HS, 68SEH*

[1]) In [*50HK*] sind für die meisten dieser Spinelle das Verhältnis der elektrostatischen Bindungsenergie E_N/E_A für die Anziehung Me-S berechnet, wobei Index N = Normalstruktur, A = Austausch-, inverse Struktur.
[2]) Die Struktur von In_2S_3 hat eine tetragonale kubisch-flächenzentrierte Zelle mit $c/a = 3.0$; daneben ist eine kubisch raumzentrierte Indizierung üblich mit $a' = \frac{1}{2} a \sqrt{2}$ und c [*59R*].
[3]) Kristalle schwarz, sehr glänzend; Pulver dunkel-braunrot [*61FDGOK*].
[4]) Hellgelb [*50HK*].
[5]) Orange [*50HK*].
[6]) Die Verbindungen der Reihe CrS-In_2S_3 sind bis zu 50 mol% CrS ebenfalls Spinellphasen:

	a [Å]
In_2S_3	10.775
9 In_2S_3 + CrS	10.74
4 In_2S_3 + CrS	10.70
2 In_2S_3 + CrS	10.65
4 In_2S_3 + 3CrS	10.61
In_2S_3 + CrS	10.59 [*61FDGOK*].

[7]) Schwarz [*50HK*].
[8]) Kristalle schwarz, sehr glänzend; Pulver ziegelrot [*61FDGF*].
[9]) Schwarzbraun [*50HK*].
[10]) Rotbraun [*50HK*].
[11]) Dunkelbraun [*50HK*].
[12]) $\Theta_a = -122 \pm 3$ °K; $C_m = 2.94 \pm 0.02$ cm³ °K/mol; Messungen des Mössbauer-Effektes; Reziproke Suszeptibilität pro Gramm siehe Fig. 1 [*67EHS*].

6.3.8 Literatur zu 6.3.5 ··· 6.3.7

15N	Nishikawa, S.: Proc. Tokyo Math. Phys. Soc. **8** (1915) 199.
24HH	Hill, R., und O. R. Howell: Phil Mag. **48** (1924) 833.
27H	Hogersson, S.: Lunds Univ. Årsskrift, N. F., Avd. 2, **23** (1927) Nr. 9 Acta Universitatis Lundensis, Leipzig, **23** (1927).
28P	Posnjak, E.: Am. J. Sci. **16** (1928) 528.
28R	Rinne, F.: Neues Jahrb. Mineral., Geol., Palaeont. Abt. A, **58** (1928) 43.
29K	Kerr, P. F.: Am. Mineralogist **14** (1929) 259.
29NP	Natta, G., und L. Passerini: Gazz. Chim. Ital. **59** (1929) 280, 620.
29P	Passerini, L.: Rend. Accad. Nazl. Lincei **9** (1929) 338.
30P-1	Passerini, L.: Gazz. Chim. Ital. **60** (1930) 389.
30P-2	Passerini, L.: Gazz. Chim. Ital. **60** (1930) 754.
30T	Taylor, N. W.: Z. Phys. Chem. (B) **9** (1930) 241.
31B	Buschendorf: Z. Phys. Chem. (B) **14** (1931) 297.
31CAB	Clark, G. L., A. Ally, und A. E. Badger: Am. J. Sci. **22** (1931) 539.
31M	Machatschki, F.: Z. Krist. **80** (1931) 416.
32BP	Barth, T. F. W., und E. Posnjak: Z. Krist. **82** (1932) 325.
32H-1	Holgersson, S.: Z. Anorg. Allgem. Chem. **204** (1932) 378.
32H-2	Hild, K.: Mitt. Kais.-Wilh.-Inst. Eisenforschg., Düsseldorf **14** (1932) 69.
32H-3	Hild, K.: Z. Physik. Chem. (Leipzig) (A) **161** (1932) 305.
32HN	Hauptmann, H., und J. Novak: Z. Phys. Chem. (B) **15** (1932) 365.
32KT	Krause, O., und W. Thiel: Z. Anorg. Allgem. Chem. **203** (1932) 120.
32M	Machatschki, M.: Z. Krist. **82** (1932) 348.
32PBB	Parmelee, C. W., A. E. Badger, und G. A. Ballam: Univ. Illinois Bull. Eng. Expt. Sta. Bull. **29** (1932) No. 84 Ceram. Abstr. **13** (1932) [10] (1932) 265.
32WR	Wartenberg, H. v., und H. J. Reusch: Z. Anorg. Allgem. Chem. **207** (1932) 1.
34GHB	Clark, G. L., E. E. Howe, und A. E. Badger: J. Am. Ceram. Soc. **17** (1934) 7.
34KT	Krause, O., und W. Thiel: Ber. Deut. Keram. Ges. **15** (1934) 101, 111.
35HS	Hägg, G., und G. Söderholm: Z. Phys. Chem. (B) **29** (1935) 88.
35K	Kordes, E.: Z. Krist. **91** (1935) 193.
40K	Klemm, W.: Z. Anorg. Allgem. Chem. **244** (1940) 377; Z. Anorg. Allgem. Chem. **246** (1941) 347; Z. Anorg. Allgem. Chem. **250** (1942) 223.
43VB	Vegard, L., und A. Borlaug: Avhandl. Norske Videnskaps-Akad. Oslo. I Mat. Naturv. Kl. No. 5 **1943** 1.
47EV	Ennslin, F., und S. Valentiner: Z. Naturforsch. **2b** (1947) 5.
47MS	Moeller, T., und J. G. Schnizlein: J. Phys. Chem. **51** (1947) 771.
47RR	Rüdorff, W., und B. Reuter: Z. Anorg. Allgem. Chem. **253** (1947) 194.
47SR	Structure Reports **11** (1947) 497–504.
47VH	Verwey, E. J. W., und E. L. Heilmann: J. Phys. Chem. **15** (1947) 174.
48MP	Michel, A., und Pouillard: Compt. Rend. **227** (1948) 194.
49HK	Hahn, H., und W. Klingler: Z. Anorg. Allgem. Chem. **260** (1949) 97.
49KB	Kordes, E., und H. Becker: Z. Anorg. Allgem. Chem. **258** (1949) 227.
50HK	Hahn, H., und W. Klingler: Z. Anorg. Allgem. Chem. **263** (1950) 177.
51B	Bertaut, E.: J. Phys. Radium **12** (1951) 252.
51F	Flahaut, J.: Compt. Rend. **233** (1951) 1279.
51FS	Foster, L. M. , und H. C. Stumph: J. Am. Chem. Soc. **73** (1951) 1590.
51G	Guillaud, C.: J. Phys. Radium **12** (1951) 239.
51GM	Guillaud, C., und A. Michel: J. Phys. Radium **12** (1951) 65.
51KR	Kordes, E., und E. Röttig: Z. Anorg. Allgem. Chem. **264** (1951) 34.
51MB	McCarthy, K. A., und S. S. Ballard: J. Opt. Soc. Am. **41** (1951) 1062.
51SF	Swanson, H. E., und R. K. Fuyat: Natl. Bur. Std. Rep. **1951** No. 1300.
51W	Wykhoff, R. W. G.: Crystal Structures. Interscience Publishers, Inc., New York 1951, Tafel VIII B, 2.
52A	Anderson, H. H.: Sc. D. thesis, Massachusetts Institute of Technology, 1952, zitiert in: [*54KFCV*].
52B-1	Bacon, G. E.: Acta Cryst. **5** (1952) 684.
52B-2	Braun, P. B.: Nature **170** (1952) 1123.
52F	Flahaut, J.: Ann. Chim. (Paris) **7** (1952) 632–696.
52HF	Hahn, H., und G. Frank: Z. Anorg. Allgem. Chem. **269** (1952) 227–231.
52JR	Jones, G. O., und F. F. Roberts: Proc. Phys. Soc. (London) **B 65** (1952) 390.
52TK	Tazaki, H., und S. Kubawara: J. Sci. Hiroshima Univ., Ser. A **15** (1952) 263.
52Z	Zimmermann, W. F.: M. S. thesis, University of Illinois, 1952, zitiert in: [*57BC*].
53BM	Beck, C. W., und C. H. Maxwell: Am. Mineralogist **38** (1953) 329.

53BR-1	Bacon, G. E., und F. F. Roberts: Acta Cryst. **6** (1953) 57.
53BR-2	Bradburn, T. E., und G. R. Rigby: Trans. Brit. Ceram. Soc. **52** (1953) 417.
53H	Honjo, G.: J. Phys. Soc. Japan **8** (1953) 113.
53M	McGuire, T. R.: Phys. Rev. **91** (1953) 206, B 8.
53MK	Moeller, T., und G. L. King: J. Am. Chem. Soc. **75** (1953) 6060.
53MP	Maxwell, L. R., und S. J. Pickart: Phys. Rev. **92** (1953) 1120.
53MPH	Maxwell, L. R., S. J. Pickart, und R. W. Hall: Phys. Rev. **91** (1953) 206, B 7.
53O	Osborn, E. F.: J. Am. Ceram. Soc. **36** (1953) 147.
53R	Romeijn, F. C.: Philips Res. Rept. **8** (1953) 304, 321.
53RRO	Roy, D. M., R. Roy, und E. F. Osborn: Am. J. Sci. **251** (1953) 337.
53SR	Structure Reports for 1953, Vol. **17**, p. 413–416, 418.
53SU	Spandau, H., und T. Ullrich: Z. Anorg. Allgem. Chem. **274** (1953) 271.
53W	Wangsness, R. K.: Phys. Rev. **91** (1953) 1085.
53WS	Weeks, J. L., und R. L. Seifert: Rev. Sci. Instr. **24** (1953) 1054.
53Y	Yamaguchi, G.: Bull. Chem. Soc. Japan **26** (1953) 204.
54BD	Bertaut. F., und C. Delorme: Compt. Rend. **239** (1954) 504.
54BW	Bacon, G. E., und A. J. E. Welch: Acta Cryst. **7** (1954) 361.
54G	Gorter, E. W.: Philips Res. Rept. **9** (1954) 295.
54GPG	Greenwald, S., S. J. Pickart, und F. H. Grannis: J. Chem. Phys. **22** (1954) 1597.
54KFCV	Kingerey, W. D., J. Francl, R. L. Coble, und T. Vasilos: J. Am. Ceram. Soc. **37** (1954) 107.
54M-1	Miles, P. A.: Nature **174** (1954) 177.
54M-2	McGuire, T.: Phys. Rev. **93** (1954) 682.
54S	Smart, J. S.: Phys. Rev. **94** (1954) 847.
54SJ	Saalfeld, H., und H. Jagodzinski: Acta Cryst. **7** (1954) 666.
54WK	Warshaw, I., und M. L. Keith: J. Am. Ceram. Soc. **37** (1954) 161.
55B	Bonnickson, K. R.: J. Phys. Chem. **59** (1955) 220.
55FH-1	Fischer, W. A., und A. Hoffmann: Arch. Eisenhüttenw. **26** (1955) 43.
55FH-2	Fischer, W. A., und A. Hoffmann: Arch. Eisenhüttenw. **26** (1955) 63.
55G	Gorter, E. V.: Usp. Fiz. Nauk **57** (1955) 2, 3.
55GL	Goodenough, J. B., und A. L. Loeb: Phys. Rev. **98** (1955) 391.
55HFKSS	Hahn, H., G. Franck, W. Klingler, A. D. Störger, und G. Störger: Z. Anorg. Allgem. Chem. **279** (1955) 241.
55K	King, E. G.: J. Phys. Chem. **59** (1955) 218.
56A	Anderson, P. W.: Phys. Rev. **102** (1955) 1008.
56C	Cossee, P.: Thesis. Universität Leiden (1956) Verlag Excelsior, 's-Gravenhage.
56CMW	Carter, P. A. Miles, und A. J. E. Welch: Proc. Inst. Electr. Engrs. (London), Part B, Suppl. Convention on Ferrites, Oct.–Nov. **1956**.
56FH	Fischer, W. A., und A. Hoffmann: Arch. Eisenhüttenw. **27** (1956) 343.
56HF	Hoffmann, A., und W. A. Fischer: Z. Physik. Chem. (Frankfurt), **7** (1956) 80.
56LÅ	Lindner, R., und Å. Åkerström: Z. Physik. Chem. (Frankfurt) **6** (1956) 162.
57B	Bongers, P. F.: Thesis, Universität Leiden, 4. 7. 1957 Verlag Excelsior, 's-Gravenhage.
57BC	Beals, R. J., und R. L. Cooks: J. Am. Ceram. Soc. **40** (1957) 279.
57CMW	Carter, P. A. Miles, und A. J. E. Welch: Proc. Inst. Elec. Engrs. (London) **B 104**, Suppl. 5 (1957) 141.
57FS	Finch, G. I., und K. P. Sinha: Proc. Roy. Soc. (London) Ser. **A 239** (1957) 145.
57G	Galakhov, F. Ya.: Bull. Acad. Sci. USSR, Div. Chem. Sci. (English Transl.) **1957** 539.
57H	Huber, M.: Compt. Rend. **244** (1957) 2524.
57HT	Hahn, H., und U. Theune: Naturwiss. **44** (1957) 33.
57J	Jagodzinski, H.: Z. Krist. **109** (1957) 388.
57LM	Lensen, M., und A. Michel: 1957 XVIème Congr. Intern. Chimie; Structure Rep. **23** (1959) 381.
57SJ	Saalfeld, H., und H. Jagodzinski: Z. Krist. **109** (1957) 87.
57SS	Sinha, K. P., und A. P. B. Sinha: J. Phys. Chem. **61** (1957) 758.
58AS	Atlas, L. M., und W. K. Sumida: J. Am. Ceram. Soc. **41** (1958) 150.
58D	Delorme, C.: Bull. Soc. Franc. Mineral. **81** (1958) 79.
58DB	Daunt, J. G., und K. Brugger: Z. Physik. Chem. (Frankfurt) **16** (1958) 203.
58HSW	Hummel, F. H., B. S. R. Sastry, und D. Wotring: J. Am. Ceram. Soc. **41** (1958) 88.
58JS	Jagodzinski, H., und H. Saalfeld: Z. Krist. **110** (1958) 197.
58LM-1	Lensen, M., und A. Michel: Compt. Rend. **246** (1958) 1997.
58LM-2	Lensen, M., und A. Michel: Compt. Rend. **246** (1958) 2766.
58RV	Richardson, J. T., und L. W. Vernon: J. Phys. Chem. **62** (1958) 1153.
58SBB	Schmitz-DuMont, O., H. Brokopf, und K. Burkhardt: Z. Anorg. Allgem. Chem. **295** (1958) 7.

58T	Turnock, A. C.: Carnegie Inst. Wash. Year Book **58** (1958) 134.
59E	Edwards, P. L.: Phys. Rev. **116** (1959) 294.
59K-1	Kwestroo, W.: J. Inorg. Nucl. Chem. **9** (1959) 65.
59K-2	Kato, E.: Bull. Chem. Soc. Japan **32** (1959) 51.
59K-3	Kato, E.: Bull. Chem. Soc. Japan **32** (1959) 585.
59L	Lensen, M.: Ann. Chim. (Paris) **4** (1959) 891.
59M	Miller, A.: J. Appl. Phys. Suppl. **30** (1959) 24S.
59P	Pearson, R. F.: J. Phys. Radium **20** (1959) 409.
59PT	Pickart, S. J., und A. C. Turnock: J. Phys. Chem. Solids **10** (1959) 242.
59R	Rooymans, C. J. M.: J. Inorg. Nucl. Chem. **11** (1959) 78.
59SDB	Skribljak, M., S. Dasgupta, und A. B. Biswas: Acta Cryst. **12** (1959) 1049.
59SGB	Schmitz-DuMont, O., H. Gössling, und H. Brokopf: Z. Anorg. Allgem. Chem. **300** (1959) 159.
59SL	Stahl-Brada, R., und W. Low: Phys. Rev. **116** (1959) 561.
59YY	Yamaguchi, G., und H. Yanagida: Bull. Chem. Soc. Japan **32** (1959) 1264.
60BHHL	Brun, E., S. Hafner, P. Hartmann, und F. Laves: Naturwiss. **47** (1960) 277.
60CA	Cossee, P., und A. E. van Arkel: J. Phys. Chem. Solids **15** (1960) 1.
60DH	Debray, L., und A. Hardy: Compt. Rend. **251** (1960) 725.
60FDO	Flahaut, J., L. Domange, und M. Ourmitchi: Compt. Rend. **250** (1960) 134.
60G-1	Gorter, E. W.: XVII. Internat. Kongr. für Reine und Angew. Chemie München 1959, Band I: Anorg. Chemie. Butterworths, 1960. S. 303–328.
60G-2	Gilleo, M. A.: J. Phys. Chem. Solids **13** (1960) 33.
60H-1	Hafner, S.: Schweiz. Mineral. Petrogr. Mitt. **40** (1960) 207.
60H-2	Huber, M.: J. Chim. Phys. **57** (1960) 202.
60ISB	Irani, K. S., A. P. B. Sinha, und A. B. Biswas: J. Phys. Chem. Solids **17** (1960) 101.
60LC	Lejus, A.-M., und R. Collongues: Compt. Rend. **251** (1960) 959.
60S	Schmalzried, S.: Naturwiss. **47** (1960) 466.
60V	Verma, R. K.: J. Geophys. Res. **65** (1960) 757.
60W	Wykoff, R. W. G.: Crystal Structures. Interscience Publishers. Inc., New York 1948–60.
61BH	Baffier, N., und M. Huber: Compt. Rend. **252** (1961) 2872.
61BLW	Brun, E., K. Loeliger, und F. Waldner: Compt. Rend. Colloque Ampère **10** (1961) 167.
61F	Folen, V. J.: Paramagnetic Resonance, Proc. Intern. Conf., 1st, Jerusalem, 1961. **1**, 68.
61FDGF	Flahaut, J., L. Domange, M. Guittard, und S. Fahrat: Compt. Rend. **253** (1961) 1956.
61FDGOK	Flahaut, J., L. Domange, M. Guittard, M. Ourmitchi, und J. Kam Su Kom: Bull. Soc. Chim. France **1961** 2382.
61GBH	Goffin, J., M. Baffier, und M. Huber: Compt. Rend. **252** (1961) 2744.
61H	Hafner, S.: Z. Krist. **115** (1961) 331.
61HL	Hafner, S., und F. Laves: Z. Krist. **115** (1961) 321.
61KFHSB	Kelly, W. H., V. J. Folen, M. Hass, W. N. Schreiner, und G. B. Beard: Phys. Rev. **124** (1961) 80.
61LC	Lejus, A.-M., und M. R. Collongues: Bull. Soc. Chim. France **1961** 65.
61PWL	Pappalardo, R., D. L. Wood, und R. C. Linares: J. Chem. Phys. **35** (1961) 1460.
61S	Schmalzried, H.: Z. Physik. Chem. (Frankfurt) **28** (1961) 203.
61SR	Strickler, W., und R. Roy: J. Am. Ceram. Soc. **44** (1961) 225.
62A	Arlett, R. H.: J. Am. Ceram. Soc. **45** (1962) 523.
62BG	Blasse, G., und E. W. Gorter: J. Phys. Soc. Japan **17**, Suppl. B 1 (1962) 176.
62BH	Brun, E., und S. Hafner: Z. Krist. **117** (1962) 37.
62F	Folen, V. J.: J. Appl. Phys. **33** (1962) 1084.
62L	Lotgering, F. K.: J. Phys. Chem. Solids **23** (1962) 1153.
62LC-1	Lejus, A.-M., und R. Collongues: Compt. Rend. **254** (1962) 2780.
62LC-2	Lejus, A.-M., und R. Collongues: Compt. Rend. **254** (1962) 2005.
62NZ	Nielsen, J. W., und J. E. Zneimer: J. Appl. Phys. Suppl. **33** (1962) 1370.
62S	Slack, G. A.: Phys. Rev. **126** (1962) 427.
63A	Atsarkin, V. A.: Soviet Phys. JETP (English Transl.) **16** (1963) 593.
63B-1	Blasse, G.: J. Inorg. Nucl. Chem. **25** (1963) 230.
63B-2	Blasse, G.: J. Inorg. Nucl. Chem. **25** (1963) 743.
63B-3	Blasse, G.: Philips Res. Rept. **18** (1963) 383.
63B-4	Blasse, G.: J. Inorg. Nucl. Chem. **25** (1963) 136.
63DR	Datta, R. K., und R. Roy: J. Am. Ceram. Soc. **46** (1963) 388.
63G	Goodenough, J. B.: Magnetism and the Chemical Bond. Interscience Publishers, New York, 1963.

63GH	Grenot, M., und M. Huber: Compt. Rend. **256** (1963) 3074.
63GHM	Grenot, M., M. Huber, und Ch. Mazières: Compt. Rend. **257** (1963) 3372.
63MT	Mizoguchi, T., und M. Tanaka: J. Phys. Soc. Japan **18** (1963) 1301.
63R-1	Roth, W. L.: Bull. Am. Phys. Soc. **8** (1963) 213.
63RAWG	Rogers, D. B., R. J. Arnott, A. Wold, und J. B. Goodenough: J. Phys. Chem. Solids **24** (1963) 347.
63SB	Schulkes, J. A., und G. Blasse: J. Phys. Chem. Solids **24** (1963) 1651.
63VNH	Vassilev, A., J. Nicolas, und M. Hildebrandt: Compt. Rend. **256** (1963) 1235.
64B-1	Blasse, G.: Philips Res. Rept. Suppl. No. 3 (1964).
64B-2	Blasse, G.: J. Inorg. Nucl. Chem. **26** (1964) 1473.
64BGS	Brun, E., S. Ghose, und P. Schindler: Helv. Phys. Acta **37** (1964) 626.
64BL	Baltzer, P. K., und E. Lopatin: Proc. Intern. Conf. Magnetism. Nottingham 1964 The Institute of Physics and the Phys. Soc. in association with the Proc. of the Phys. Soc., S. 564.
64C	Cini, L.: Intern. Keram. Kongr. IX. Bruxelles 1964, 89.
64DLW	Drumheller, J. E., K. Locher, und F. Waldner: Helv. Phys. Acta **37** (1964) 626.
64EE	Ermolenko, N. F., und M. D. Efros: Dokl. Akad. Nauk Belorussk. SSR **8** (1964) 165.
64L	Lejus, A.-M.: Revue Int. Haut. Temp. Refract. **1** (1964) 53.
64R	Roth, W. L.: J. Phys. **25** (1964) 507.
64S-1	Saalfeld, H.: Z. Krist. **120** (1964) 476.
64S-2	Slack, G. A.: Phys. Rev. A **134** (1964) 1268.
64SFHM	Stoll, E., P. Fischer, W. Hälg, und G. Maier: J. Phys. **25** (1964) 447.
65B	Blasse, G.: Philips Res. Rept. **20** (1965) 528.
65BGKP	Baranov, B. V., V. S. Grogor'eva, L. V. Kradinova, und V. D. Prochukhan: Fiz., Dokl. k. XXIII-ei [Dvadtsat Tret'ei] Nauchn. Konf. Leningr. Inzh.- Stroit Inst., Sb., Leningrad **1965**, S. 48.
65CT	Colin, F., und J. Thery: Compt. Rend. **261** (1965) 3141.
65D-1	Drifford, M.: Compt. Rend. **261** (1965) 2205.
65D-2	Dugdale, D. E.: Phys. Letters **16** (1965) 226.
65GH	Grenot, M., und M. Huber: Compt. Rend. **261** (1965) 5124.
65HB	Huber, M., und J. Bénard: Bull. Soc. Chim. France **1965** 1081.
65K	Kapitonova, N. P.: Soviet Phys. — Solid State (English Transl.) **6** (1965) 2111.
65L	Lenglet, M.: Rev. Chim. Minerale **2** (1965) 217.
65NM	Neuhaus, A., und H. J. Meyer: Naturwissenschaften **52** (1965) 639.
65O	Olés, A.: Compt. Rend. **260** (1965) 6075.
65R	Rossiter, M. J.: J. Phys. Chem. Solids **26** (1965) 775.
65RB	Robbins, M., und P. K. Baltzer: J. Appl. Phys. **36** (1965) 1039.
65RC	Rossiter, M. J., und P. T. Clarke: Nature **207** (1965) 402.
65RGA	Rogers, D. B., R. W. Germann, und R. J. Arnott: J. Appl. Phys. **36** (1965) 2338.
65S	Schmitz-DuMont, O.: Bull. Soc. Chim. France **1965** 1099.
65SK	Schmitz-DuMont, O., und H. Kasper: Z. Anorg. Allgem. Chem. **341** (1965) 252.
65SLR	Schmitz-DuMont, O., A. Lulé, und D. Reinen: Ber. Bunsenges. Physik. Chem. **69** (1965) 76.
65SR	Schlaudt, C. M., und D. M. Roy: J. Am. Ceram. Soc. **48** (1965) 248.
65ST	Stone, F. S., und R. J. D. Tilley: 5th Intern. Symposium on the Reactivity of Solids, Munich, Aug. 2—8, 1964. Elsevier Publishing Co., **1965** S. 583.
65TS	Tretjakov, J. D., und H. Schmalzried: Ber. Bunsenges. Physik. Chem. **69** (1965) 396.
66A	Allen, W. C.: Am. Mineralogist **51** (1966) 239.
66ABMP	Ansel'm, L. N., G. L. Bir, I. E. Myl'nikova, und M. P. Petrov: Soviet Phys. — Solid State (English Transl.) **8** (1966) 812.
66ANM	Anishchenko, R. I., A. P. Nikolaev und A. N. Men': Soviet Phys. — Solid State (English Transl.) **8** (1966) 202.
66BHO	Boucher, B., A. G. Herpin, und A. Olés: J. Appl. Phys. **37** (1966) 960.
66BOGG	Banerjee, S. K., W. O'Reilly, T. C. Gibb, und N. N. Greenwood: Phys. Letters **20** (1966) 455.
66BP	Bertaut, E.-F., und G. Patrat: Bull. Soc. Franc. Mineral. Crist. **88** (1966) 586.
66CT	Colin, F., und J. Théry: Rev. Chim. Minerale **3** (1966) 212.
66DP	Demirdjioghlou, S. F., und A. J. Pointon: Proc. Inst. Elec. Engrs. London **113** (1966) 525.
66DR	Drifford, M., und P. Rigny: Compt. Rend. **263** B (1966) 180.
66EW	El-Sharat, R. M., und J. White: Trans. Brit. Ceram. Soc. **65** (1966) 309.
66FG	Friedman, E. A., und A. N. Goland: Phys. Rev. **147** (1966) 457.
66JD	Joubert, J.-C., und A. Durif: Bull. Soc. Franc. Mineral. Crist. **89** (1966) 26.

66K-1	KRISHNAN, R.: Metaux (Corrosion-Ind.) **41** (1966) 28.
66K-2	KOLESOVA, V. A.: Bull. Acad. Sci. USSR, Div. Chem. Sci. (English Transl.) **1966**, 633.
66LN	LENEV, L. M., und I. A. NOVOKHATSKIY: Russ. Met. (English Transl.) **1966**, 40.
66MF	MANASEVIT, H. M., und D. H. FORBES: J. Appl. Phys. **37** (1966) 734.
66NK	NAVROTSKY, A., und O. J. KLEPPA: Inorg. Chem. **5** (1966) 192.
66ÔIS	ÔNO, K., A. ITO, und Y. SYONO: Phys. Letters **19** (1966) 620.
66R-1	REINEN, D.: Theoret. Chim. Acta **5** (1966) 312.
66R-2	ROSSITER, M. J.: Phys. Letters **21** (1966) 128.
66RD	ROBBINS, M., und L. DARCY: J. Phys. Chem. Solids **27** (1966) 741.
66RV	ROOKSBY, H. P., und M. W. VERNON: Brit. J. Appl. Phys. **17** (1966) 1227.
66S	STONE, F. S.: Bull. Soc. Chim. France **1966** 819.
66TTA	TANAKA, M., T. TOKORO, und Y. AIYAMA: J. Phys. Soc. Japan **21** (1966) 262.
66VS	VISHNEVSKII, I. I., und V. N. SKRIPAK: Soviet Phys.-Solid State (English Transl.) **7** (1966) 2373.
66W	WESTBROOK, J. H.: Rev. Hautes Temp. Refract. **3** (1966) 47.
66WSO	WEIDENBORNER, J. E., N. R. STEMPLE, und Y. OKAYA: Acta Cryst. **20** (1966) 761.
67B	BORGHESE, C.: J. Phys. Chem. Solids **28** (1967) 2225.
67DC	DRIFFORD, M., und P. CHARPIN: Compt. Rend. **264** B (1967) 64.
67DR	DATTA, R. K., und R. ROY: J. Am. Ceram. Soc. **50** (1967) 578.
67EHS	EIBSCHÜTZ, M., E. HERMON, und S. SHTRIKMAN: Solid State Commun. **5** (1967) 529.
67F	FISCHER, P.: Z. Krist. **124** (1967) 275.
67G	GRENOT, M.: Bull. Soc. Chim. France **1967** 2043.
67HS	HAHN, H., und G. STRICK: Naturwiss. **54** (1967) 42A.
67KGW	KIRCHNER, H. P., R. M. GRUVER, und R. E. WALKER: J. Am. Ceram. Soc. **50** (1967) 169.
67KS	KLJUČAROV, JA. V., und S. A. SUVOROV: Ogneupory, **32** (1967) 42.
67N	NICOLAS, J.: J. Phys. Chem. Solids **28** (1967) 847.
67NK	NAVROTSKY, A., und O. J. KLEPPA: J. Inorg. Nucl. Chem. **29** (1967) 2701.
67ÔCI	ÔNO, K., L. CHANDLER, und A. ITO: Phys. Letters **24** A (1967) 273.
67PB	POIX, P., und J.-C. BERNIER: Bull. Soc. Chim. France **1967**, 605.
67R	REINEN, D.: Theoret. Chim. Acta **8** (1967) 260.
67S-1	SCHREIBER, E.: J. Appl. Phys. **38** (1967) 2508.
67S-2	SOBEL, A.: J. Phys. Chem. Solids **28** (1967) 185.
67S-3	STEIGMANN, G. A.: Acta Cryst. **23** (1967) 142.
67T	TARTRE, P.: Spectrochim. Acta **23** A (1967) 2127.
67YM	YAGNIK, C. M., und H. B. MATHUR: Solid State Commun. **5** (1967) 841.
68D	DRIFFORD, M.: Compt. Rend. **266** B (1968) 561.
68DR	DATTA, R. K., und R. ROY: Am. Mineralogist **53** (1968) 1456.
68KTHD	KRISCHNER, H., K. TORKAR, R. HOFFMANN, und D. DONNERT: Monatsh. Chem. **99** (1968) 1080.
68L	LEWIS, H.: Phil. Mag. **17** (1968) 481.
68R-1	REINEN, D.: Z. Anorg. Allgem. Chem. **356** (1968) 172.
68R-2	REIJNE, P. J. L.: Science Ceram. **4** (1968) 169.
68RBW	RANGE, K.-J., W. BECKER, und A. WEISS: Z. Naturforsch. B **23** (1968) 1009.
68SEH	STRICK, G.,G. EULENBERGER, und H. HAHN: Z. Anorg. Allgem. Chem. **357** (1968) 338.
69L	LENGLET, M.: Rev. Chim. Mineral. **6** (1969) 703.
69ZK	ZORINA, N. G., und S. S. KVITKA: Soviet. Phys. — Cryst. (English Transl.) **13** (1969) 599.

6.3.9 Co Spinelle und substituierte Co Spinelle

(Literatur siehe 6.3.14, Seite 528)

Tab. 1. Kristallografische Eigenschaften der binären Co Oxid-Spinelle

Verbindung	Kationenverteilung	a Å	ϱ_X g/cm³	u	Lit.	Fig.	Tab.
Co_3O_4 [11]		8.07		≈0.375	*25GBL, 13···26S*	1, 2,	2
		8.04	6.17	≈0.38	*28HA, 28···32S*		
		8.06	6.11 [1]		*29NP-1, 28···32S*	3, 4,	
		8.108			*29HK, 28···32S*	5, 13	
[12]	$CoCo_2O_4$	8.07 [2]			*34KT, 47SR*		
		8.09			*51RB*		
	$CoCo_2^{3+}O_4$	8.070			*52R, 52RB-1,*		
					52RB-2, 52SR		
	$Co^{2+}[Co_2^{3+}]O_4$	8.083			*56C-1, 64B-2, 60H*		
		8.092			*56L*		
		8.065		0.388 [3]	*64R-2*		
		8.110			*62SKT*		
					36VB, 54E, 55R,		
					56C-2, 57KSK, 58C,		
					58RV, 58SS, 58SSK,		
					58WC, 62Sch		
					59SZKT, 61H, 62SZKT,		
					63B-4, 63R, 64AM		
					64R-1, 66JJ		
					66MKKI, 68ABPMKS		
$CuCo_2O_4$ [11]		8.039			*29HK, 28···32S*		
					54E, 56L, 64B-2,		
					62SKT, 68ABPMKS		
$FeCo_2O_4$		8.36	5.34 [4]		*29NP-1, 28···32S*	6, 7	2
$Fe_{0.99}Co_{2.01}O_4$	≈$Co^{2+}[FeCo^{3+}]O^4$	8.254			*56L* [5], *64B-2*		
	$Co^{2+}[FeCo^{3+}]O_4$	8.254			*63B-4*		
					60H		
$MgCo_2O_4$ [11]		8.107			*29HK, 28···32S*		
					52R [6], *52SR, 54E, 56L*		
$MnCo_2O_4$ [13]		8.268			*29HK, 28···32S*	8	2
	$Co^{2+}[Co^{2+}Mn^{4+}]O_4$	8.153			*56L, 64B-2, 63B-4*		
	$Co^{3+}[Mn^{2+}Co^{3+}]O_4$				*57KSK*		
		8.27		0.382	*59SZKT*		
					55ZP, 64VOKL, 66VOK,		
					62SZKT, 66SP		
$NiCo_2O_4$ [11]		8.112			*29HK*	9, 10	2
		8.098	6.015		*52R* [7], *52SR*		
	≈$Co^{2+}[Co^{3+}Ni^{3+}]O_4$	8.121			*56L* [8], *63B-4, 64B-2*		
					51RB, 58SSK, 62SKT,		
					68ABPMKS		
$SnCo_2O_4$ [9]	$Co[SnCo]O_4$				*47VH, 47SR*		
$ZnCo_2O_4$ [11]		8.06			*29NP-2, 28···32S*	1, 2,	2
		8.108			*29HK*	11, 12	
	$Zn[Co_2^{3+}]O_4$	8.047			*56L, 64B-2*		
					52R [10], *54E, 56C-1,*	13	
					56C-2, 58C, 63B-4,		
					66MKKI, 68ABPMKS		

1) $\varrho_p = 6.07$ g/cm³ [*29NP-1*], [*28···32S*].
2) $a = 8.07 \pm {}^{0.02}_{0.03}$ Å [*34KT*]; $a = 8.09$ Å; $\varrho_X = 6.04$ g/cm³ [*47SR*].
3) $u = 0.3881 \pm 0.0004$ durch Neutronenbeugung bestimmt [*64R-2*].
4) $\varrho_p = 5.19$ g/cm³ [*29NP-1*], [*28···32S*].
5) $FeCo_2O_4$ unter 900 °C nicht stabil [*56L*].
6) Kein Beweis für die Existenz eines Spinellgitters gefunden in [*52R*].
7) $NiCo_2O_4$ kann nur unterhalb 450 °C hergestellt werden [*52R*].
8) $NiCo_2O_4$ über 400 °C nicht stabil [*56L*].
9) Inverser Spinell [*47VH*], [*47SR*].
10) $ZnCo_2O_4$ existiert bei tiefen Temperaturen [*52R*].
11) Schwarz [*25GBL*], [*29HK*], [*29NP-2*].
12) Schwarzgrau [*34KT*].
13) Braunschwarz [*29HK*].

Tab. 2. Magnetische und elektrische Eigenschaften der binären Co Oxid-Spinelle

Verbindung	C_A cm³ °K / g-atom	C_m cm³ °K / mol	α 10^{-3} emu/mol	Θ_C °K	Θ_N °K	Θ_a °K	p_{eff} μ_B	p_m μ_B	Lit.	Fig.
Co_3O_4 [1][2][3][4]	≈1.2								*56L*	1, 2,
						−115 [5]			*56C-2*	3, 4,
			0.71				4.14		*58C*	
						−105			*58RV*	5, 13
					46	− 50	4.20		*63B-4*	
					40				*63R*	
		2.14		40		− 53	4.14		*64R-1,*	
									64R-2,	
									56C-1,	
									59SZKT,	
									*62SKT,**)	
									66MKKI	
$FeCo_2O_4$				≈450 [6]				1 [7]	*63B-4*	6, 7
									56L	
$MnCo_2O_4$	1.91			200					*56L*	8
		5.73		≈170				≈0.1 [7]	*63B-4*	
									66SP	
$NiCo_2O_4$ [8][9]				≈350				1.5 [7]	*63B-4*	9, 10
									29HK,	
									52R, 56L,	
									62SKT	
$ZnCo_2O_4$ [10][13]	0.21								*56L*	1, 2,
		0.450 [12]	0.37 [12]			0			*56C-1*	11, 12,
						− 64 [11]			*56C-2*	13
		0	0.2						*58C*	
									66MKKI	

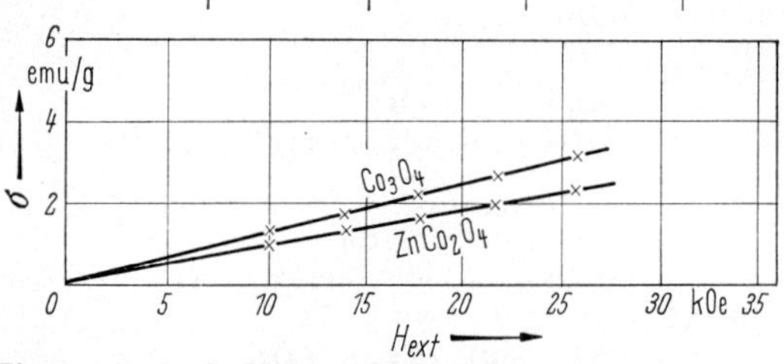

Fig. 1. Co_3O_4, $ZnCo_2O_4$. Magnetisches Moment σ pro Gramm als Funktion von H_{ext} bei 20.4 °K [*56L*].

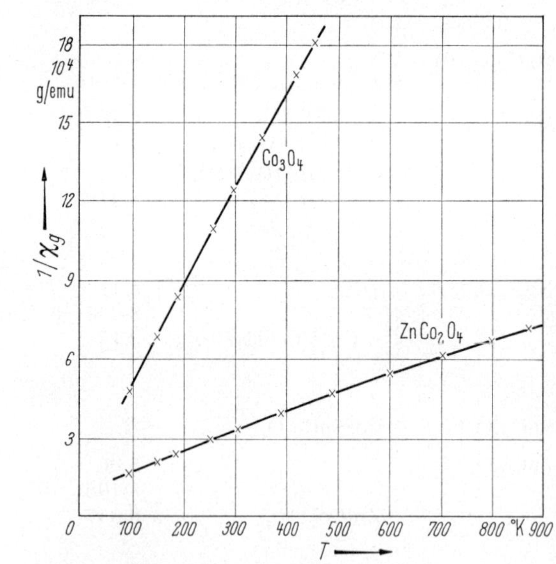

Fig. 2. Co_3O_4, $ZnCo_2O_4$. Reziproke Suszeptibilität pro Gramm $\vec{\chi}_g^{-1}$ vs. Temperatur T [*56L*].

[1]) Paramagnetisch bis 20 °K hinunter [*56L*].
[2]) Maximale Suszeptibilität bei 40 °K; <40 °K antiferromagnetisch [*63R*].
[3]) Angaben über σ, A und q [*59SZKT*], [*62SKT*].
[4]) $\varrho = 8 \cdot 10^4$ Ωcm [*58SS*].
[5]) An anderer Probe $\Theta_a = -109$ °K gemessen [*56C-2*].
[6]) Ferrimagnetisch [*63B-4*].
[7]) Bei 0 °K [*63B-4*].
*) Siehe auch [*58SS*], [*62A-1*], [*62SZKT*].
[8]) Ferrimagnetisch [*29HK*], [*52R*].
[9]) Angaben über σ, A und q [*62SKT*].
[10]) Schwach temperaturabhängiger Paramagnetismus [*56C-2*].
[11]) An anderen Proben $\Theta_a = -76$ und -78 °K gemessen [*56C-2*].
[12]) An anderen Proben weitere Werte für C_m, α gemessen [*56C-1*].
[13]) Geringer, temperaturunabhängiger Paramagnetismus [*58C*], [*65MKKI*].

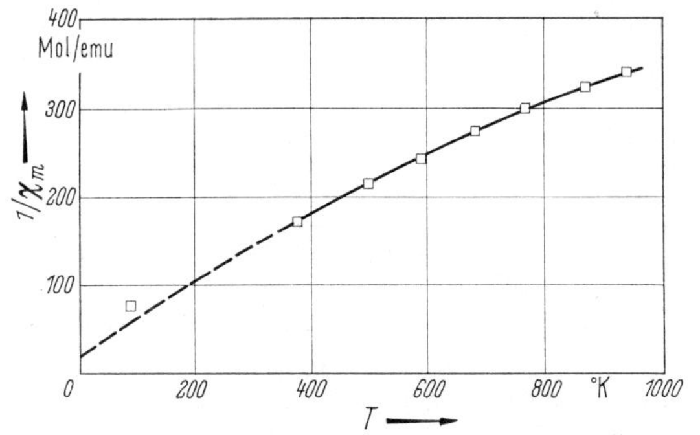

Fig. 3. Co_3O_4. Reziproke Suszeptibilität pro mol χ_m^{-1} vs. Temperatur T [58C].

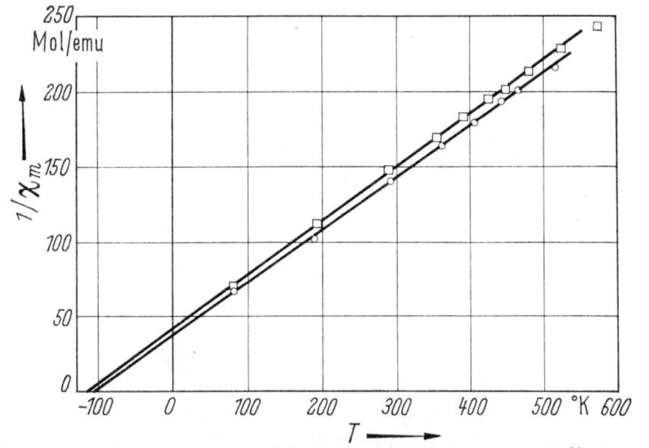

Fig. 4. Co_3O_4. Reziproke Suszeptibilität pro mol χ_m^{-1} vs. Temperatur T für unterschiedliche Proben [56C-2].

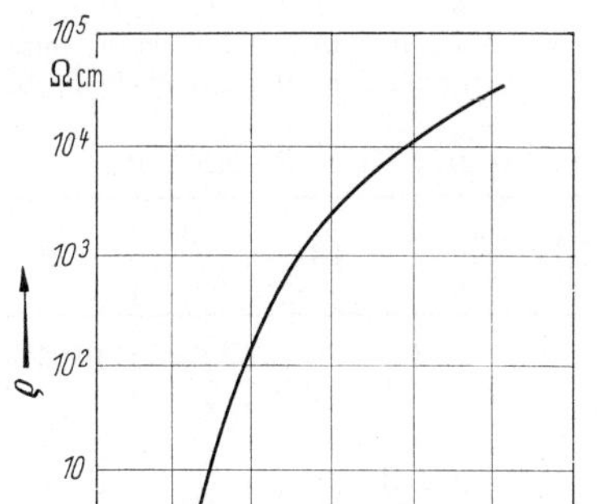

Fig. 5. Co_3O_4. Spezifischer Widerstand ϱ vs. Temperatur T [58SS].

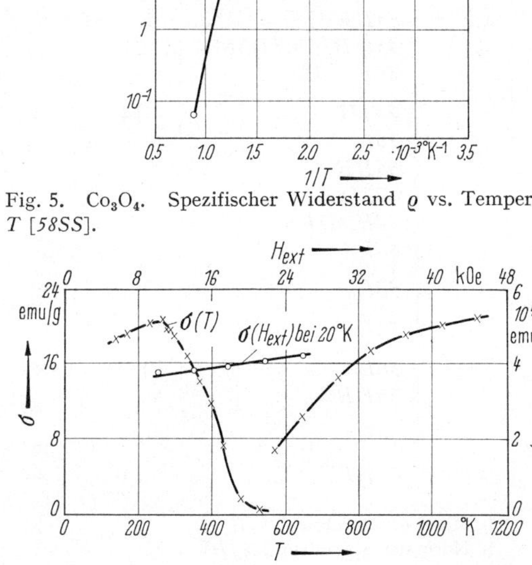

Fig. 6. $FeCo_2O_4$. σ vs. H_{ext} bei 20 °K; σ vs. T für 100 ··· 600 °K bei H_{ext} = 8 kOe; χ_g^{-1} vs. T für 600 ··· 1200 °K [56L].

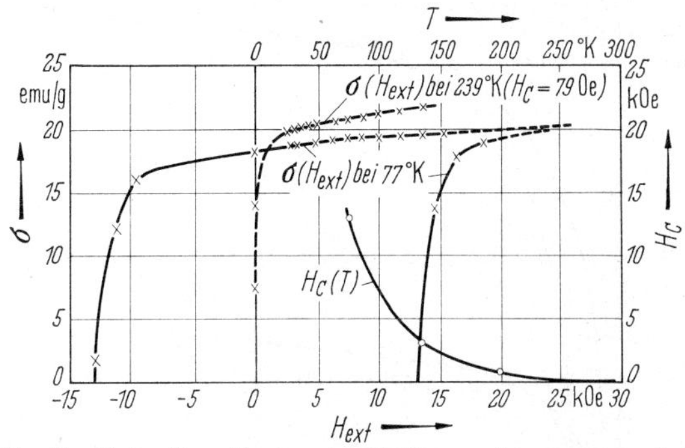

Fig. 7. $FeCo_2O_4$. Hysteresisschleifen aufgenommen bei 77 °K, 293 °K und Koerzitivkraft H_C vs. Temperatur T [56L].

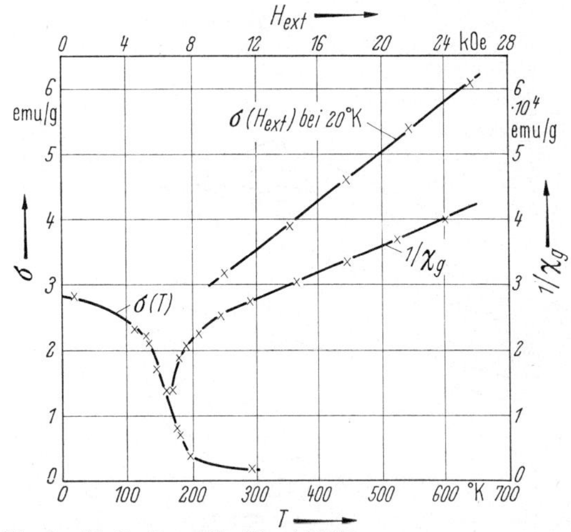

Fig. 8. $MnCo_2O_4$. Wie Fig. 6 [56L].

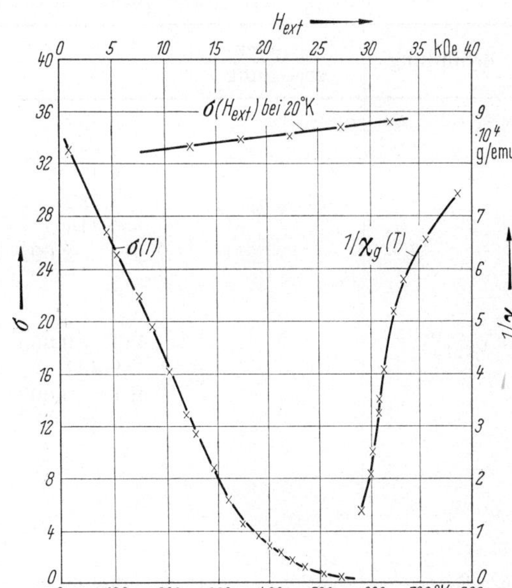

Fig. 9. $NiCo_2O_4$. Wie Fig. 6 [56L]. σ in [emu/g].

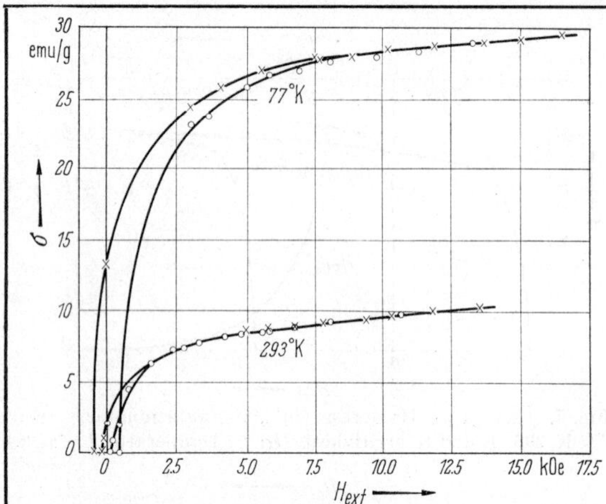

Fig. 10. $NiCo_2O_4$. Hysteresisschleifen aufgenommen bei 77 °K und 293 °K [*56L*].

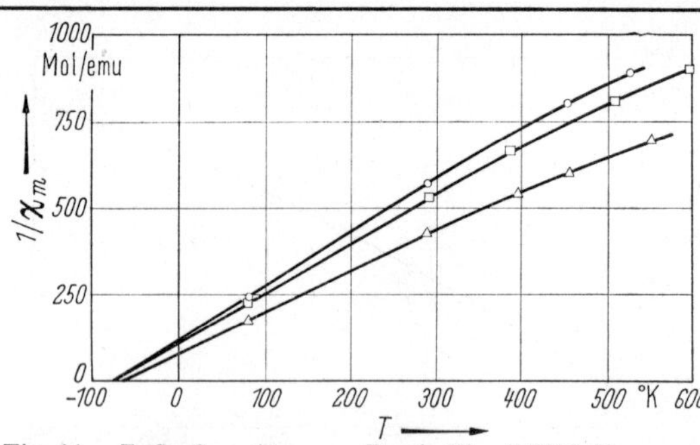

Fig. 11. $ZnCo_2O_4$. $1/\chi_m$ vs. T, wie Fig. 4 [*56C-2*].

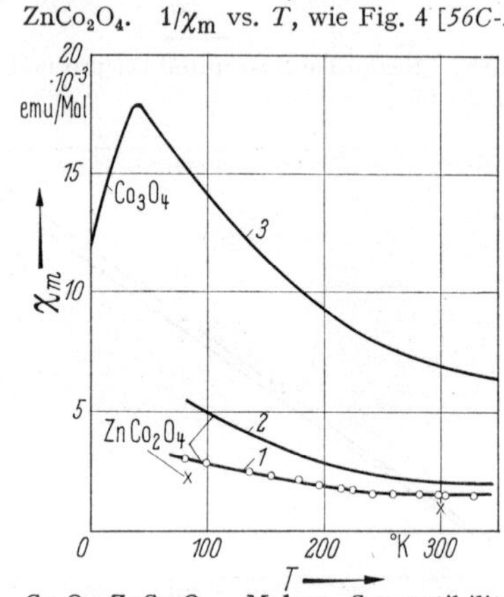

Fig. 13. Co_3O_4, $ZnCo_2O_4$. Molare Suszeptibilität χ_m vs. Temperatur T. 1 [*66MKKI*], 2 [*56L*], × × und 3 [*58C*].

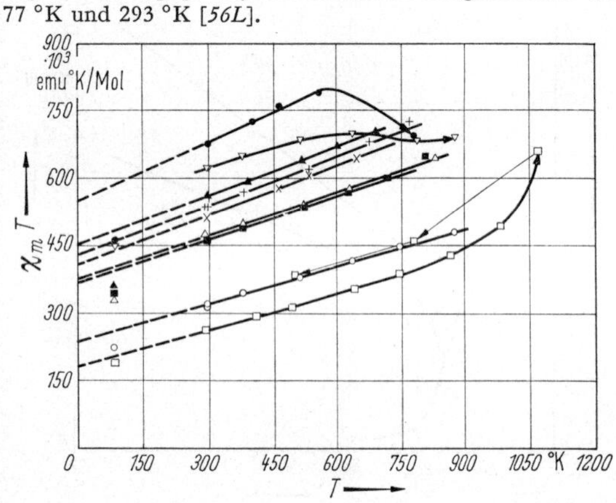

← Fig. 12. $ZnCo_2O_4$ (●, ▲, +, ×, △, ○, □), $Co_{0.125}Zn_{0.875}Co_2O_4$ (▽, ■). Produkt der Temperatur mit der molaren Suszeptibilität ($\chi_m \cdot T$) vs. Temperatur T für unterschiedliche Proben [*56C-1*].

Tab. 3. Kristallografische Eigenschaften binärer Co und Ni Sulfo-Spinelle

Verbindung	Kationen-verteilung	a Å	u	ϱ_X g/cm³	Lit.	Fig.	Tab.
Co_3S_4 [1]		≈ 9.36			*27JW*	14, 15	4
		9.42			*28JH*		
		9.4	0.375	4.87	*31NP, 28···32S*		
	$Co[Co_2]S_4$	9.382			*38LW*		
		9.416			*56L, 64B-2*		
		9.399 ± 0.002			*65BRW*		
					35HB, 39HBM [2]		
					54R, 54E		
$CuCo_2S_4$ [3]	[4]	9.458 ± 0.003	0.375		*28JH*	14	4
		9.482			*56L*		
		9.461 ± 0.002			*65BRW*		
		9.48			*67MSL*		
					54E, 67LS		
					68LS		
$FeCo_2S_4$					*56L*		
$MnCo_2S_4$					*56L*		
$NiCo_2S_4$		9.392			*56L*	14, 15	4
		9.384 ± 0.002			*65BRW*		
$ZnCo_2S_4$					*56L*		
$CoNi_2S_4$		9.418 ± 0.002			*65BRW*	14	4

[1]) Mineral: Linnéit [*27JW*].
[2]) Co_3S_4 geht oberhalb von 680 °C in CoS und CoS_2 über [*39HBM*].
[3]) Mineral: Carrolyte [*28JH*].
[4]) Normaler Spinelltyp [*28JH*].

Tab. 4. Magnetische und elektrische Eigenschaften binärer Co und Ni Sulfo-Spinelle

Verbindung	χ_m emu/mol	C_A $\frac{cm^3\ °K}{g\text{-}atom}$	ϱ_{RT} Ωcm	S μV/°K	Lit.	Fig.
Co_3S_4 [1])		≈0.7	[2]) $3 \cdot 10^{-4}$	4.8 [5])	*56L* *65BRW*	14, 15
$CuCo_2S_4$ [3])[6])	(3.7···3.8) 10^{-4} [4])		$4 \cdot 10^{-4}$	12.7 [5])	*65BRW* *68LS, 67LS* *67MSL*	14
$NiCo_2S_4$ [1])		≈0.3	$4 \cdot 10^{-4}$	−17.7 [5])	*56L* *65BRW*	14, 15
$CoNi_2S_4$			$4 \cdot 10^{-4}$	− 1.8 [5])	*65BRW*	14

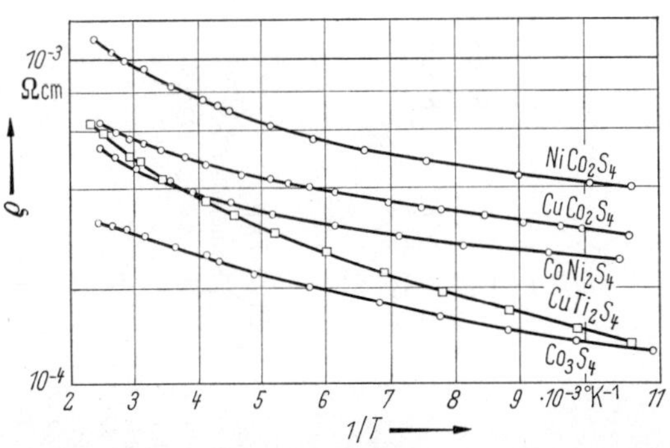

Fig. 14. Co_3S_4, $CuCo_2S_4$, $NiCo_2S_4$, $CoNi_2S_4$, $CuTi_2S_4$. Spezifischer elektrischer Widerstand ϱ vs. Temperatur T [*65BRW*].

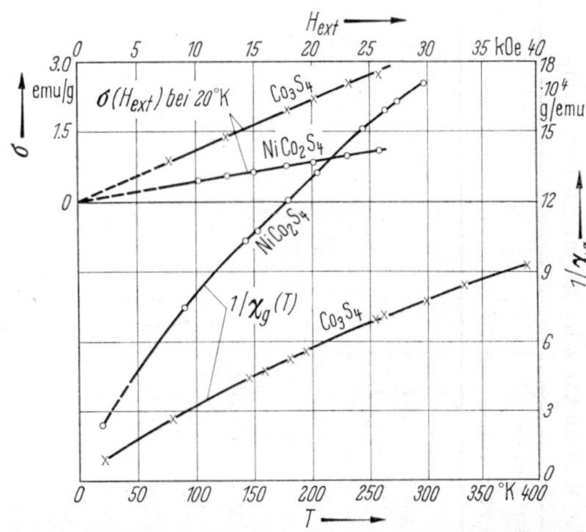

Fig. 15. Co_3S_4, $NiCo_2S_4$. σ vs. H_{ext}; χ_g^{-1} vs. T [*56L*].

[1]) Paramagnetisch bis 20 °K hinunter [*56L*].
[2]) ϱ: metallisch [*65BRW*].
[3]) Halbleiter, *p*-Leitung [*67LS*].
[4]) Korrigiert für Diamagnetismus $\chi_{m\,diam} = (3 \cdots 6)\ 10^{-4}$ emu/mol [*67LS*]; $\chi_{m\,diam} = (5.5 \cdots 5.6)\ 10^{-4}$ emu/mol [*68LS*].
[5]) Bei 25° [*65BRW*].
[6]) Keine Supraleitung oberhalb von 0.05 °K [*67MSL*].

Tab. 5. Kristallografische und elektrische Eigenschaften ternärer und höherer Co, Ni und Te Oxid-Spinelle

Verbindung	Kationen-verteilung	x	a Å	c/a	ϱ_X g/cm³	ϱ_p g/cm³	Lit.	Fig.	Tab.
Co_3O_4-γ-Al_2O_3 1) 2)							*58RV*	16, 17	6
$Co^{2+}AlCo^{3+}O_4$	$Co^{2+}[Al^{3+}Co^{3+}]O_4$						*63B-4*		6
$Co^{2+}Co^{3+}_{2-x}Al_xO_4$	$Co^{2+}[Al^{3+}_{0.5}Co^{3+}_{1.5}]O_4$	0.5					*63B-4*		6
Co_3O_4-γ-Fe_2O_3 2)							*51RB*		
$Co^{2+}Fe^{3+}_{2-2x}Co^{3+}_{2x}O_4$							*52RB-1*, *52RB-2*		
$Mn_xCo_{3-x}O_4$	$Co^{2+}[Mn^{3+}_xCo^{3+}_{2-x}]O_4$		3)				*58WC*, *59G*	18	6
		1.2	8.316						
		2		1.148					
	$Co^{2+}[Mn^{3+}_{1.2}Co^{3+}_{0.8}]O_4$	1.2	8.32				*63G-2*, *57KSK*		
$Ni_xCo^{2+}_{1-x}Co^{3+}_2O_4$ 4)							*52R*		6
$Co_{1-x}Zn_xCo_2O_4$ 5)							*52R*		
		0.875					*56C-1*	12	
CuO-MnO-CoO-O_2 6)							*59SZKT*, *66SP*	19, 20	
CuO-NiO-CoO-O_2 7)							*62SKT*	19, 20	
MnO-NiO-CoO-O_2 8)							*62SZKT*, *66SP*	19, 20	
$Zn_{(2+x)/2}Co_{(3-x)/2}Te_{0.5}O_4$ 9)		0	8.548 ± 0.003		5.88	5.64	*67B*		
$Zn_{(2+x)/2}Ni_{(3-x)/2}Te_{0.5}O_4$ 9)		1	8.505 ± 0.003		6.15	6.20	*67B*		

1) Reihe zwischen Co_3O_4 und γ-Al_2O_3($Co^{2+}_{xm}Al^{3+}_{yn}[\square_t Co^{3+}_{(1-x)m}Al^{3+}_{(1-y)n}]O_4$) [*58RV*].
2) γ-Al_2O_3; γ-Fe_2O_3 = Defekt-Spinelle oder Spinelle mit Leerstellen (□ Spinelle) [*57SS*].
3) Kubischer Spinell für $x \leq 1.2$; tetragonal verzerrter Spinell für $x > 1.2$; $c/a > 1$ [*58WC*], [*63G-2*].
4) Vollständige Mischungsreihe zwischen $NiCo_2O_4$ und Co_3O_4 [*52R*].
5) Vollständige Mischungsreihe zwischen $ZnCo_2O_4$ und Co_3O_4 [*52R*].
6) Angaben über σ [Ω^{-1} cm^{-1}], A [Ω^{-1} cm^{-1}] und q [eV] [*59SZKT*].
7) Angaben über σ [Ω^{-1} cm^{-1}], A [Ω^{-1} cm^{-1}] und q [eV] [*62SKT*].
8) Angaben über σ [Ω^{-1} cm^{-1}], A [Ω^{-1} cm^{-1}] und q [eV] [*62SZKT*].
9) Von [*67B*] erstmalig Te Spinelle untersucht ($Zn_{2+x}Ni_{3-x}TeO_8$; $Zn_{2+x}Co_{3-x}TeO_8$).

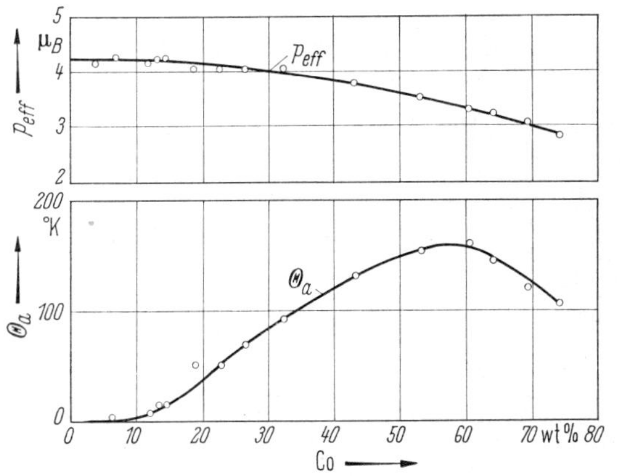

Fig. 16. Co_3O_4-γ-Al_2O_3. Änderung des magnetischen Momentes p_{eff} und der Weiss-Konstanten Θ_a vs. Co-Gehalt [58RV].

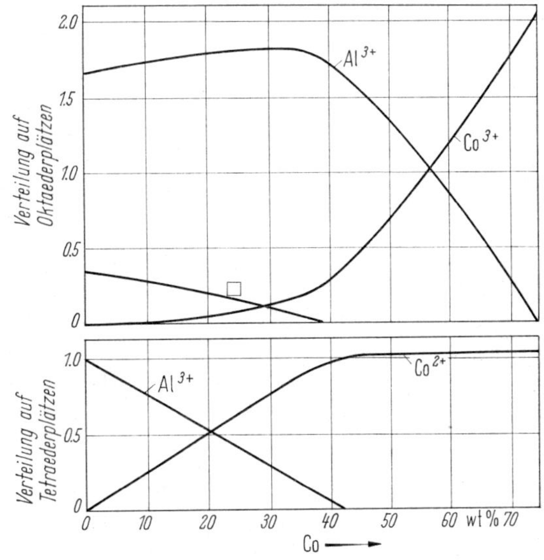

Fig. 17. $Co^{2+}_{xm}Al^{3+}_{yn}[\square_t Co^{3+}_{(1-x)m}Al^{3+}_{(1-y)n}]O_4$. Kationenverteilung vs. Co-Gehalt [58RV].

Tab. 6. Magnetische Eigenschaften ternärer Co Oxid-Spinelle

Verbindung	Θ_N °K	Θ_a °K	p_{eff} μ_B	Lit.	Fig.
Co_3O_4-γ-Al_2O_3 [1])				*58RV*	16
$CoAlCoO_4$	30	−30	4.14	*63B-4*	
$CoAl_{0.5}Co_{1.5}O_4$	38	−40	4.1	*63B-4*	
$Mn_xCo_{3-x}O_4$ [2])				*58WC*	18
$Co_{1-x}Zn_xCo_2O_4$				*56C-1*	12

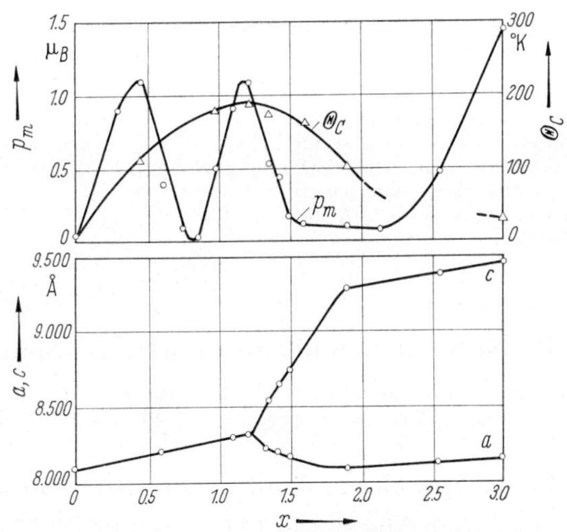

Fig. 18. $Mn_xCo_{3-x}O_4$. Magnetisches Moment p_m pro Molekül bei 4.2 °K, Curie-Temperatur Θ_C und Gitterparameter a und c vs. Zusammensetzungsparameter x [58WC].

[1]) Paramagnetisch unterhalb RT [58RV].
[2]) Maximale Curie-Temperatur Θ_C = 191 °K für x = 1.2 [58WC].

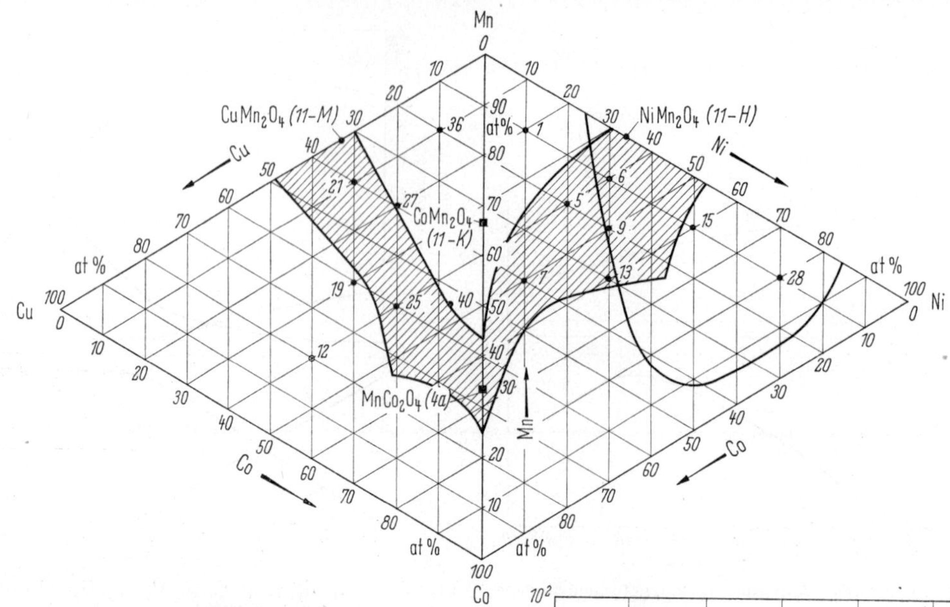

Fig. 19. Kombinierte Dreistoffsysteme MnO-CoO-CuO-O_2 und MnO-CoO-NiO-O_2. Im schraffierten Gebiet liegt allein oder vorwiegend Spinellphase vor.
Die Kurve, die von 25···85% Ni verläuft, trennt das Gebiet der Elektronenhalbleiter ab (300 °K) [*66SP*].

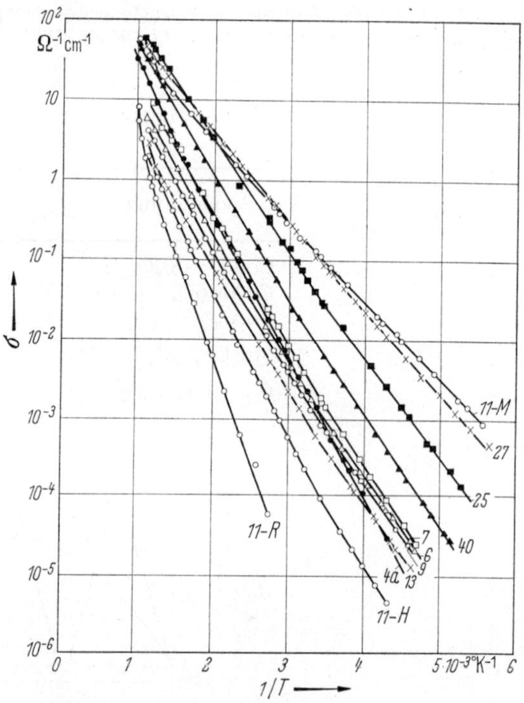

Fig. 20. Elektrische Leitfähigkeit $\vec{\sigma}$ für Halbleiter mit Spinellstruktur in Abhängigkeit von der Temperatur T für Verbindungen der Reihen MnO-CoO-CuO-O_2 and MnO-CoO-NiO-O_2. Die Kurvenbezeichnungen geben die Zusammensetzungen der Verbindungen nach Fig. 19 an, wobei *11-R* mit *11-K* in Fig. 19 übereinstimmt [*66SP*].

6.3.10 Ni Spinelle und substituierte Ni Spinelle

(Literatur siehe 6.3.14, Seite 528)
Siehe Tab. 3···5, Fig. 14, 19, 20 in 6.3.9

6.3.11 Te Spinelle und substituierte Te Spinelle

(Literatur siehe 6.3.14, Seite 528)
Siehe Tab. 5 in 6.3.9

6.3.12 Mn Spinelle und substituierte Mn Spinelle

(Literatur siehe 6.3.14, Seite 528)
Tab. 1 siehe nächste Seite

Tab. 1. Kristallografische Eigenschaften binärer Mn Oxid-Spinelle

Verbindung	Kationenverteilung	x	a Å	c Å	c/a	u	Lit.	Fig.	Tab.
γ-Mn_2O_3					1.16		*55GL, 36VB*	13	3
	$Mn^{3+}[Mn^{3+}_{5/3}\square_{1/3}]O_4$		8.15	9.44	1.16		*57SS, 57BO, 64B-2*		
	$\square_{1/3}Mn_{2/3}Mn_2^{3+}O_4$		8.1	9.4	1.16		*59G, 63G-2*		
							52KPS, 52SR		
Mn_3O_4 [1]			8.157	9.454	1.159		*47M, 48MG, 53H*	1, 2, 14…19	2, 3
	$Mn^{2+}[Mn_2^{3+}]O_4$ [2]		8.14	9.42	1.14 [3]		*53R-1, 50MSM*		
			8.14	9.42	1.16 [3]		*55GL, 59W*		
			8.151	9.450	1.159		*57B, 57BO*		
			8.15	9.44	1.16		*57SS, 57SSB-1, 57SSB-2*		
			Tab. 2				*58HK, 58WC, 59G*		
					1.16		*62ISB, 62SZKT*		
			8.15	9.54	1.17 [4]		*63G-2*		
			8.13	9.43			*64B-2*		
					1.16		*65S-2*		
			8.124	9.439	1.162		*66DR*		
							57FFS, 58D, 59K, 60DM, 58E, 58KHB, 59A, 59J, 60ISB, 61K, 61H, 61N, 63RNMP, 64AM, 64RN, 64VOKL, 65DA, 65H, 66D, 66SP, 66VOK, 67RMS		
$CdMn_2O_4$			8.22	9.87	1.20 [5]		*57SS*		
							59G		
	$Cd^{2+}[Mn_2^{3+}]O_4$					0.400 ± 0.004	*57SSB-1, 57SSB-2, 63G-2, 64B-2*		
			8.074	9.554	1.18		*65DA*	20…29	3
							68NK, 64RN, 66C		
$Cd_xMn_{3-x}O_4$							*64RN, 66D, 65DA*		3

Fußnoten siehe S. 493

Tab. 1 (Fortsetzung)

Verbindung	Kationenverteilung	x	a Å	c Å	c/a	u	Lit.	Fig.	Tab.
$CoMn_2O_4$	$Co^{2+}[Mn_2^{3+}]O_4$		8.10	9.31	1.15 [6])		*57B, 54A*	2, 3,	3
	$Co_{1-y}^{2+}Mn_y^{3+}[Co_y^{3+}Mn_{2-y}^{3+}]O_4$ [8])						*59G, 63G-2*	15, 30	
							58WC, 64B-2		
					1.13		*59A, 62ISB*		
			8.10	9.25	1.14 [9])	0.383	*62ISB*		
			8.157	9.454			*63G-2, 58WC*		
			8.168	8.958	1.097 [7a])		*68BBP*		
			8.137	9.041	1.111 [7b])		*68BBP*		
			8.096	9.133	1.128 [7c])		*68BBP*		
			8.04	9.04	1.12		*57SSB-1,*		
							62SZKT,		
							63VOKL, 64AM,		
							64VOKL, 68NK,		
							65VOK, 66SP,		
							66VOK		
$Co_{3-x}Mn_xO_4$ [10])	$Co^{2+}[Co_{2-x}^{3+}Mn_x^{3+}]O_4$	2.0			1.148		*58WC*	18 in 6.3.9	3
		1.2	8.316				*58WC, 59G,*		
							63G-2		
		2.5			1.15		*59A*		
							62ISB, 62A-1		
							62SZKT, 64AM,		
							64VOKL, 66SP,		
							59J		
$CrMn_2O_4$	$Mn^{2+}[Cr^{3+}Mn^{3+}]O_4$		8.33	8.75	1.05		*57B, 63G-2*		
$CuMn_2O_4$			8.28				*52KPS*	31	3
	$Cu^{2+}[Mn^{2+}Mn^{4+}]O_4$		8.30 [11])				*57B*		
	$Cu^{1+}[Mn^{3+}Mn^{4+}]O_4$		8.33	8.33	1.00	0.389 ± 0.004	*57SSB-1,*		
							64B-2, 59G		
			8.31				*58D*		
			8.33 ± 0.02				*58SSB*		
	$Cu^{2+}[Mn_2^{3+}]O_4$						*61K*		
	[12])		8.28			0.375	*62SZKT,*		
							57KSK,		
							59SZKT		
	$Cu_y^{2+}Mn_{1-y}^{2+}[Cu_{1-y}^{2+}Mn_{1-y}^{4+}Mn_{2y}^{3+}]O_4$ [13])		8.33				*63G-2, 57SSB,*		
							62SZKT		
			8.30				*63ZKZ*		

Fußnoten siehe S. 493

Verbindung	Kationenverteilung	x	a Å	c Å	c/a	u	Lit.	Fig.	Tab.
$CuMn_2O_4$ (Forts.)	$Cu^{2+}[Mn^{3+}]O_4$		8.34			0.392	*64BL*		
			8.373 [14])			0.3884 [15])	*65A-1*		
	$Cu_{0.88}Mn_{0.12}[Cu_{0.12}Mn_{1.88}]O_4$ [16])					0.39	*65A-2*		
			8.33				*66SSB*		
							64RW, 66B, 66SP, 67NK, 68NK		
$Cu_xMn_{3-x}O_4$	$Mn^{2+}[Cu_x^{2+}Mn_x^{4+}Mn_{2-2x}^{3+}]O_4$	$0 \leq x \leq 0.2$					*63RNMP*	4, 32 ··· 34	3
$Cu_{1.5}Mn_{1.5}O_4$	$\approx Cu^{1+}[Cu_{0.5}^{2+}Mn_{1.5}^{4+}]O_4$		8.28				*66B*		3
$FeMn_2O_4$	$Mn_{1-y}^{2+}Fe_y^{3+}[Fe_{1-y}^{2+}Mn_{1+y}^{3+}]O_4$ [17])		8.31	8.85	1.05		*57SSB-1, 59G, 63G-2*		
	$Mn_{1-y}^{2+}Fe_y^{3+}[Fe_y^{2+}Fe_{1-2y}^{3+}Mn_{1+y}^{3+}]O_4$ [18])		8.49				*57FSS, 59G, 63G-2*		
	$Mn^{2+}[Mn^{3+}Fe^{3+}]O_4$						*61K*		
$Fe_{3-3x}Mn_{3x}O_4$	$Mn_{0.54}^{2+}Fe^{3+}[Mn_{0.06}^{3+}Fe_{0.06}^{2+}Fe_{1.88}^{3+}]O_4$	0.2					*57FSS, 67HMST, 60AP, 63G-2*	5, 6	
	$Mn_x^{2+}Fe_{1-x}^{3+}[Fe_{1-x}^{2+}Fe_{1+x}^{3+}]O_4$						*69W, 58HK, 50MSM*		
		0.4	8.49				*59G, 47M*		
	$Mn^{2+}[Mn_{0.8}^{3+}Fe_{1.2}^{3+}]O_4$	0.6			[19])		*58E, 61K, 69W, 66KYH*		
	$Mn_{0.80}^{2+}Fe_{0.20}^{2+}[Fe_{1.00}^{2+}Mn_{1.00}^{4+}]O_4$	0.57					*58WC, 66KYH, 59W*		
$Mn_{1-x}Fe_{2-2x}Mn_{3x}O_4$	$Mn_{0.5}^{2+}Fe_{0.5}^{3+}[Mn^{3+}Mn_{0.5}^{2+}Fe_{0.5}^{3+}]O_4$	0.5					*57FFS*		
$LiMn_2O_4$ [20])	$Li^{1+}[Mn^{3+}Mn^{4+}]O_4$		8.246				*58WC, 59G, 63B-1, 63G-2*	35	3
			8.242			0.383	*64BL*		
			8.25		1		*66B, 65B*		
			8.241 ± 0.005				*66SP*		
$Li_{0.5}Mn_{2.5}O_4$	$Li_{0.5}^{+}Mn_{0.5}^{3+}[Mn_2^{3+}]O_4$		8.132	9.340	1.15		*47M, 58BW, 59G, 63G-2*		
$Li_{4/3}Mn_{5/3}O_4$	$Li[Li_{1/3}Mn_{5/3}^{4+}]O_4$		8.19				*63B-2, 64B-2*		
							63B-3		3
$MgMn_2O_4$			8.07	9.31	1.15		*57B, 59G, 63G-2*	2, 36, 37	3
	$Mg[Mn_2^{3+}]O_4$		8.07	9.28	1.15		*57SS, 57SSB-1, 59G*		

Fußnoten siehe S. 493

Tab. 1 (Fortsetzung)

Verbindung	Kationenverteilung	x	a Å	c Å	c/a	u	Lit.	Fig.	Tab.
$MgMn_2O_4$ (Forts.)	$Mn^{2+}_{1-y}Mn^{3-}_{y}[Mg^{2+}_{y}Mn^{3+}_{2-y}]O_4$ [26]		8.128	9.201	1.13		*47M, 59G*		
			8.08	9.32	1.15		*60ISB*		
						0.385 [27]	*60MM, 64B-2*		
					1.15 [28]		*62ISB*		
							64RN, 59J		
							63GH, 63GHM, 65H, 65MP, 67M, 68NK		
$Mg_xMn_{3-x}O_4$		$0 \leq x \leq 1$					*63RNMP, 59J*	7, 38	
$NiMn_2O_4$			8.37			0.381 ± 0.004	*57SSB-1, 59G, 59A*	15, 35	3
	$Mn[NiMn]O_4$		8.38				*57B*		
			8.39			0.3835	*58BW*		
			8.4028 ± 0.0003			0.386 ± 0.002	*59B, 63G-2*		
			8.37			0.378	*62SZKT*		
	$Mn^{2+}[Ni^{2+}Mn^{4+}]O_4$ [22]						*59L, 61K*		
	$Mn_yNi_{1-y}[Mn_{2-y}Ni_y]O_4$ [24]					0.388	*66BBP, 67BBP*		
$NiMn_{2.02}O_{4.03}$ [23]			8.383 ± 0.005				*66SP*		
							68NK		
$Ni_{0.5}Mn_{2.5}O_4$	$Mn[Ni_{0.5}Mn_{1.5}]O_4$ [21]				[25b]		*59A*		
$Mn_{3x}Ni_{1-x}Mn_{2-2x}O_4$	$Mn^{2+}[Ni^{2+}_{1-x}Mn^{3+}_{2x}Mn^{4+}_{1-x}]O_4$ [25a]						*61K, 58BW 59G*		
$ZnMn_2O_4$	$Zn^{2+}[Mn^{3+}_2]O_4$ [29a]		8.092	9.224	1.142		*47M, 57B*	2, 8, 9, 39, 47…49	3
			8.083 [32]	9.226	1.14 [29b]		*53R-1, 47M, 55GL*		
			8.087	9.228	1.141		*57B*		
			8.10	9.25 [31]	1.14		*57SS, 57SBB-1*		
			8.092 [31]	9.254	1.144		*58WC, 59G*		
					1.14		*62ISB*		
			8.10	9.35	1.15 [29c][30]		*62ISB, 63G-2*		
			8.09	9.23	>1 [29d]		*65B, 64RN*		
					1.14		*65S-2, 63RNMP*		
					1.14		*66A*		
			8.077	9.198			*66DR*		
							60DM, 64B-2, 64BL, 65RB-2, 68NK, 61K		

Fußnoten siehe S. 493

Verbindung	Kationenverteilung	x	a Å	c Å	c/a	u	Lit.	Fig.	Tab.
$Zn_xMn_{3-x}O_4$	$Zn^{2+}_xMn^{2+}_{1-x}[Mn^{3+}_2]O_4$						*60DM, 57B, 59J 63RNMP, 64R, 64RN, 61JK*	10···12 40, 43···46	3
	$Zn^{2+}[Zn^{2+}_{x-1}Mn^{4+}_{x-1}Mn^{3+}_{4-2x}]O_4$	<1			33)		*66DR*		
$Zn_{1+x}Mn_{2-x}O_4$	$Zn^{2+}[Zn^{2+}_xMn^{4+}_xMn^{3+}_{2-2x}]O_4$	$0 \le x \le 0.2$					*63RNMP*	12, 41, 42	3

1) Mineral: Hausmannit.
2) Kationenverteilung $Mn^{2+}[Mn^{2+}Mn^{4+}]O_4$ kann nicht ausgeschlossen werden [*53R-1*].
3) Oberhalb 1150 °C Übergang zur rein kubischen Spinellform [*53R-1*]; oberhalb 1170 °C rein kubische Spinellform mit a = 8.64 Å [*50MSM*], [*55GL*], [*48MG*]; oberhalb 1160 ± 5 °C kubische Spinellform [*58HK*], [*58WC*]; oberhalb 1443 °C kubischer Spinell [*62ISB*]; bei 1130 ± 10 °C Übergang von tetragonal verzerrter zur kubischen Spinellstruktur [*66DR*].
4) Werte für 800 °C nach [*62ISB*], [*63G-2*].
5) Umwandlung in kubisches Gitter bei ≈670 °K [*65DA*].
6) Unterhalb 1173 °K normales Spinellgitter, oberhalb tetragonal verzerrt [*62ISB*].
7a—c) Abgeschreckte Proben von 1250 °C, 1000 °C und 800 °C [*68BBP*].
8) Nach [*66SP*], [*66VOK*], [*64VOKL*] und [*65VOK*] befinden sich auf den Oktaederplätzen neben Mn^{3+}- auch Mn^{2+}- und Mn^{4+}-Ionen.
9) Bei 800 °C [*62ISB*].
10) Kubisch $0 \le x \le 1.2$; tetragonal verzerrt für $x > 1.2$; für $x = 2$, $c/a = 1.148$; die Kationenverteilung $2 \le x \le 3$ ist als $Co^{2+}_{3-x}Mn^{2+}_{x-2}[Mn^{3+}_2]O_4$ angegeben worden [*58WC*].
11) Nach [*57B*] müßte der Spinell tetragonal verzerrt sein, da Cu^{2+}- und Mn^{3+}-Ionen auf Oktaederplätzen sitzen können.
12) Kationenverteilung $Cu_{0.5}Mn_{0.5}[Cu_{0.5}Mn_{1.5}]O_4$ nach [*57KSK*], [*59SZKT*] nach [*62SZKT*] $Cu_{0.33}Mn_{0.67}[Cu_{1.67}Mn_{1.33}]O_4$ bis $Mn[CuMn]O_4$ [*62SZKT*].
13) Für $0 < y < 0.33$ [*63G-2*].
14) Für eine von 1050 °C abgeschreckte Probe; a = 8.335 Å für eine langsam abgekühlte Probe; bei längerer Temperatur-Behandlung Werte bis minimal a = 8.32 Å [*65A-1*].
15) Für $y = 0.42$ bei einer Kationenverteilung von $Mn_yCu_{1-y}[Cu_yMn_{2-y}]O_2$ [*65A-1*].
16) Aus Messungen der elektrischen Leitfähigkeit wird angenommen $Cu^{1+}[Mn^{3+}Mn^{4+}]O_4$ [*66SSB*].
17) $y = 0.5$ [*57SSB-1*].
18) Für $0 \le y \le 0.33$ [*57FSS*].
19) Übergang von kubischer zur tetragonal verzerrten Phase bei $x = 0.6$ [*57FSS*], [*61K*].
20) Farbe: Blauschwarz [*58WC*].
21) Kationenverteilung nach Neutronenbeugungs-Messungen $Mn[Ni_{1.5}Mn_{0.5}]O_4$ [*59A*].
22) Andere Möglichkeit der Kationenverteilung: $Mn^{3+}[Ni^{2+}Mn^{3+}]O_4$ ist jedoch unwahrscheinlicher [*59L*].
23) Formel nach der chemischen Analyse; ϱ_X = 5.22 g/cm³ und ϱ_p = 5.17 g/cm³ [*66SP*].
24) Kationenverteilung $0.74 < y < 0.93$ [*67BBP*].
25a) Andere Kationenverteilung: $Mn^{2+}_{2x}Mn^{3+}_{1-x}[Ni^{2+}_{1-x}Mn^{3+}_{1+x}]O_4$; 25b) Übergang zur tetragonal verzerrten Phase bei $x = 0.42$ [*61K*].
26) Nach der chemischen Analyse bestimmte Formel: $Mg^{2+}_{0.99}Mn^{3+}_{1.85}Mn^{4+}_{0.12}O_4$ und einer Einheitszelle: $Mg^{2+}_{7.92}Mn^{3+}_{14.80}Mn^{4+}_{0.96}\square_{0.32}O^{2-}_{32}$ [*60ISB*].
27) Bei $y = 0.85$ [*60MM*], [*57SSB-1*].
28) Bei 1123 °C Übergang von kubischer zu tetragonal verzerrter Spinellphase [*62ISB*].
29a) Außerdem mögliche Kationenverteilung $Zn^{2+}[Mn^{2+}Mn^{4+}]O_4$; 29b) Oberhalb 1025 °C kubische Form [*53R-1*]; 29c) Übergang bei 1323 °C [*62ISB*]; 29d) Übergang bei ≈1400 °C [*65B*].
30) Werte für 800 °C nach Angabe von [*63G-2*] für [*62ISB*].
31) Nach Angabe von [*63G-2*] c = 9.24 Å für [*57SSB-1*]; a = 8.087 Å für [*58WC*].
32) Nach Angabe von [*64B-2*] a = 8.087 Å für [*53R-1*].
33) Übergangstemperaturen von tetragonal verzerrten – zu kubischen Spinellphasen in Abhängigkeit von x (≈ von 1170··· 860 °C bei $x = 0···1.20$) [*66DR*].

Tab. 2. Gitterparameter *a* und *c* in Abhängigkeit von der Temperatur *T* für die Verbindung Mn_3O_4 [*58HK*]

	T °C	*a* Å	*c* Å
Tetragonale Form	20	8.136	9.422
	250	8.14	9.45
	450	8.15	9.47
	625	8.16	9.49
	800	8.175	9.50
	1000	8.20	9.49
	1160	8.22	9.46
	1162	— erste kubische Reflexe bei Erwärmung	
Kubische Form	1270	8.57	
	1220	8.56	
	1144	— erste tetragonale Reflexe bei Abkühlung	
	1140	8.54	

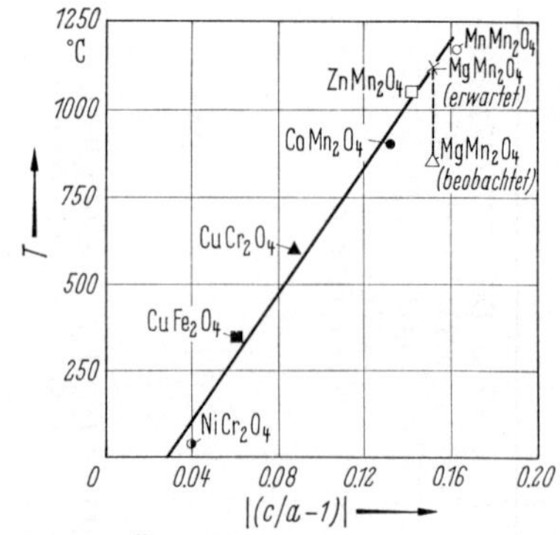

Fig. 2. Übergangstemperatur *T* vom kubischen zum tetragonal verzerrten Spinellgitter vs. Verhältnis der Gitterparameter $|(c/a - 1)|$ dargestellt am Beispiel unterschiedlicher Verbindungen [*62ISB*].

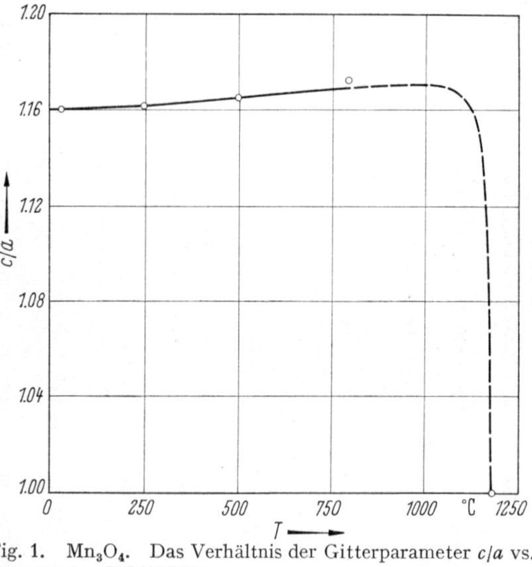

Fig. 1. Mn_3O_4. Das Verhältnis der Gitterparameter c/a vs. Temperatur *T* [*62ISB*].

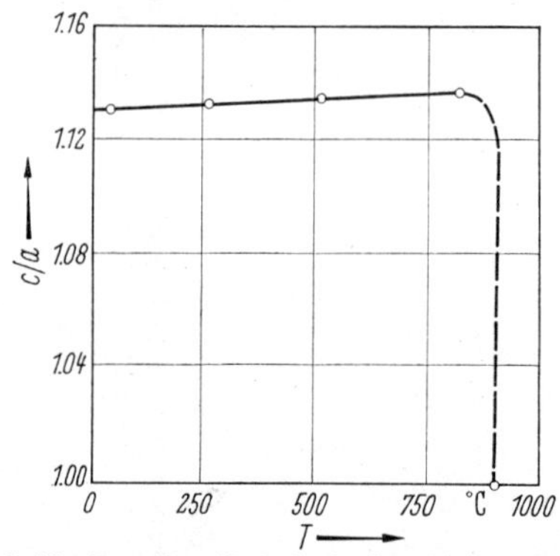

Fig. 3. $CoMn_2O_4$. Das Verhältnis der Gitterparameter c/a vs. Temperatur *T* des tetragonal verzerrten Spinellgitters [*62ISB*].

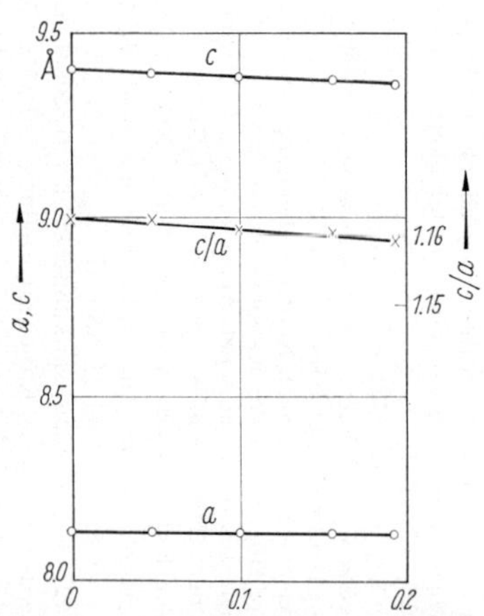

Fig. 4. $Cu_xMn_{3-x}O_4$. Gitterparameter *a* und *c* und das Verhältnis c/a vs. Zusammensetzungsparameter *x* [*63RNMP*].

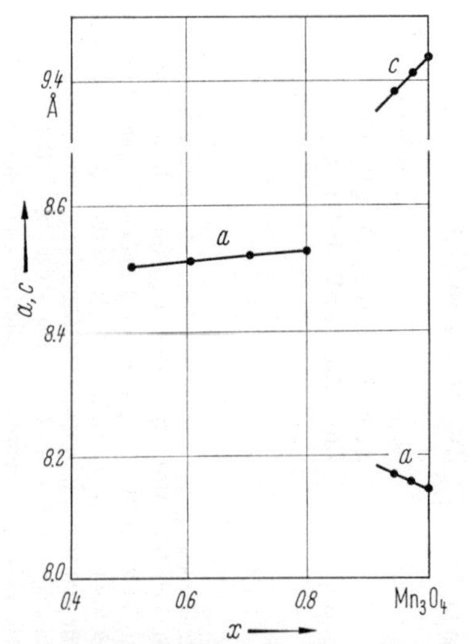

Fig. 5. $Fe_{3-3x}Mn_{3x}O_4$. Gitterparameter *a* und *c* vs. Zusammensetzungsparameter *x*. Bei $x = 0.54$ und $x = 0.91$ existiert die kubische Phase bei von höheren Temperaturen abgeschreckten Proben im metastabilen Zustand [*58HK*].

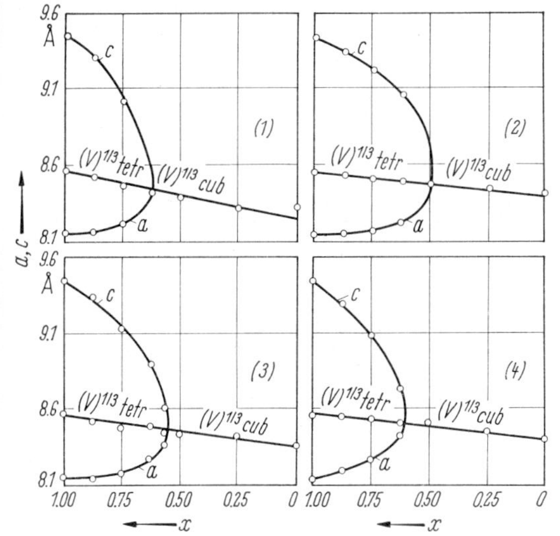

Fig. 6. Gitterparameter a und c vs. Zusammensetzungsparameter x für Verbindungen der Reihen

$Ni_{1-x}Fe_{2-2x}Mn_{3x}O_4$ (1)
$Mn_{1-x}Fe_{2-2x}Mn_{3x}O_4$ (2)
$Co_{1-x}Fe_{2-2x}Mn_{3x}O_4$ (3)
$Fe_{3-3x}Mn_{3x}O_4$ (4)

Der Übergang zur tetragonal verzerrten Spinellphase tritt bei folgenden Zusammensetzungen auf:

(1) $Fe^{3+}_{3/8}Mn^{2+}_{5/8}[Mn^{3+}_{10/8}Ni^{3+}_{3/8}Fe^{3+}_{3/8}]O^{2-}_4$
(2) $Mn^{2+}_{1/2}Fe^{3+}_{1/2}[Mn^{3+}Mn^{2+}_{1/2}Fe^{3+}_{1/2}]O^{2-}_4$
(3) $Fe^{3+}_{3.6/8}Mn^{2+}_{4.4/8}[Mn^{3+}_{1.1}Fe^{3+}_{3.6/8}Co^{2+}_{3.6/8}]O_4{}^{-}$
(4) $Fe^{3+}_{0.4}Mn^{2+}_{0.6}[Mn^{3+}_{1.2}Fe^{2+}_{0.4}Fe^{3+}_{0.4}]O^{2-}_4$ [*57FSS*].

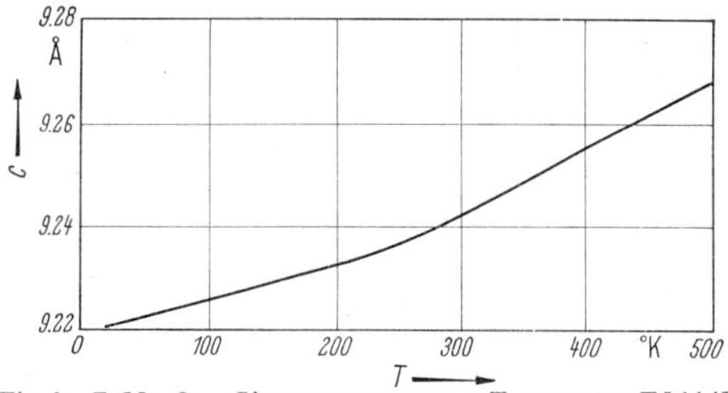

Fig. 9. $ZnMn_2O_4$. Gitterparameter c vs. Temperatur T [*66A*].

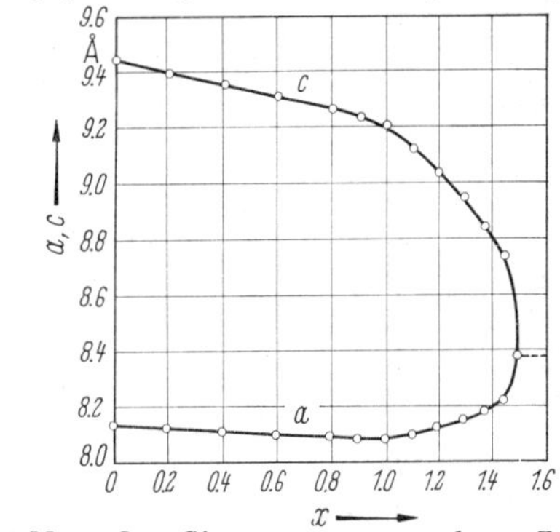

Fig. 10. $Zn_xMn_{3-x}O_4$. Gitterparameter a und c vs. Zusammensetzungsparameter x [*66DR*].

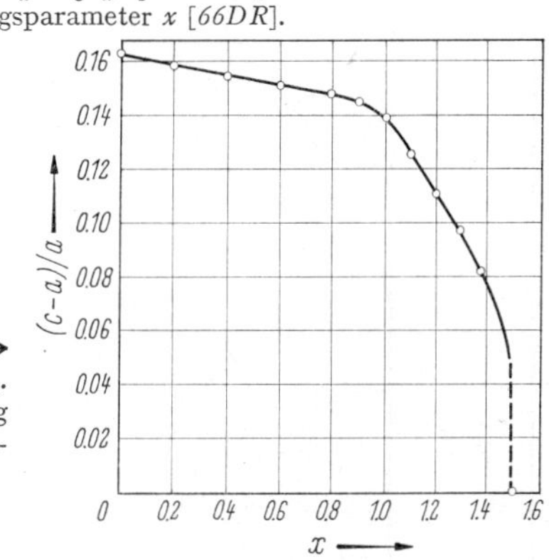

Fig. 11. $Zn_xMn_{3-x}O_4$. Tetragonale Verzerrung $(c-a)/a$ vs. Zusammensetzungsparameter x [*66DR*].

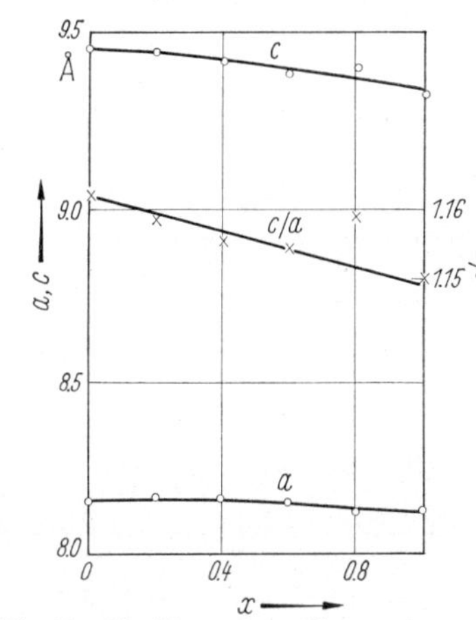

Fig. 7. $Mg_xMn_{3-x}O_4$. Gitterparameter a und c, sowie c/a vs. Zusammensetzungsparameter x [*63RNMP*].

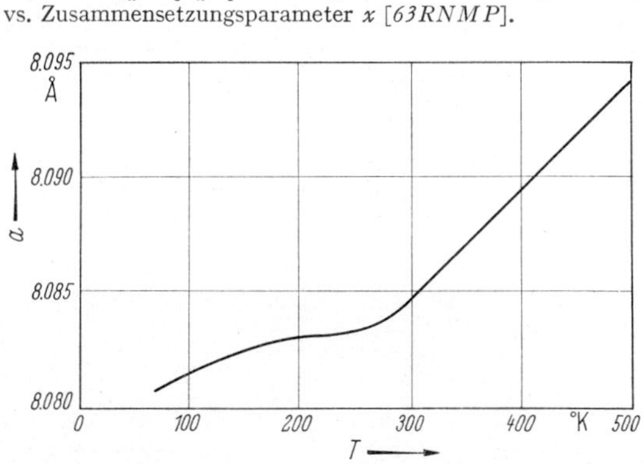

Fig. 8. $ZnMn_2O_4$. Gitterparameter a vs. Temperatur T [*66A*].

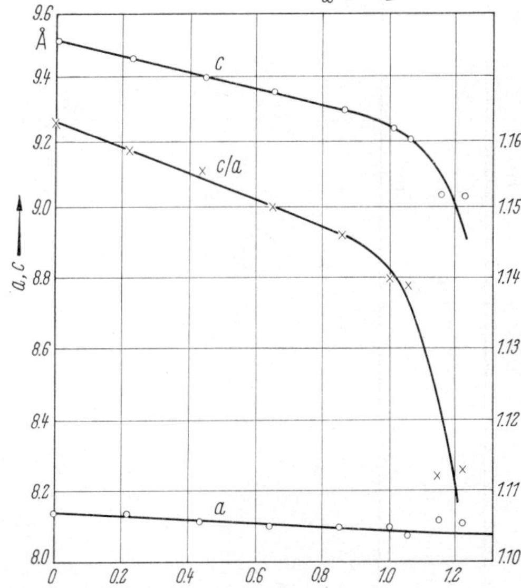

Fig. 12. $Zn_xMn_{3-x}O_4$, $Zn_{1+x}Mn_{2-x}O_4$. Gitterparameter a und c, sowie c/a vs. Zusammensetzungsparameter x [*63RNMP*].

Tab. 3. Magnetische und elektrische Eigenschaften binärer Mn Oxid-Spinelle

Verbindung	x	Θ_a °K	C_m cm³ °K/mol	C_g cm³ °K/g	Θ_C °K	Θ_N °K	p_{eff} μ_B	p_m μ_B	ϱ Ωcm	A $\Omega^{-1}cm^{-1}$	q eV	Lit.	Fig.
γ-Mn_2O_3												*57BO* [1]	13
		≈ −150					4.12					*57SS*	
Mn_3O_4 [2]											1.3 [6]	*53R-1,*	
												55GL	14⋯19
		−530	9.62									*57B*	
			10.4		42.5							*57BO, 61N*	
		≈ −160					3.9					*57SS*	
					30			1.4				*58WC*	
									≈10^{17}			*59A*	
					43			1.6				*59K*	
								1.56 ± 0.04				*59J*	
					41.8 ± 0.1 [4]			1.85 [3]				*60DM*	
									10^7	$3.7 \cdot 10^{-3}$	0.53	*62SZKT*	
						≈ 80		1.85 [5]				*63G-2*	
											3.1	*66DR,*	
												65DA, 64RN,	
												63RNMP,	
												66SP	
												67RMS	
$CdMn_2O_4$												*64RN*	
					< 4							*65DA*	
												66C	
$Cd_xMn_{3-x}O_4$ [7]	0.2	−470		360	< 75							*64RN*	20⋯29
	0.4	−550		340	< 75								
	0.5				< 75								
	0.6	−540		300	< 75								
	0.7												
	0.8	−590		270									
	0.9												
	0.4				< 24.5							*65DA, 66D*	
	0.6				< 15								
	0.8				< 5								
$CoMn_2O_4$ [9]		−730	9.56		80	80⋯94						*57B*	15, 30
									≈10^2			*59A*	
					85			0.1 [8]				*63G-2,*	

$CoMn_2O_4$ [9]) (Forts.)												*58WC*	
		< −550										*68BBP*	
		−550			105 [10])		5.0						
$Co_{3-x}Mn_xO_4$ [11])												*58WC*	18 in 6.3.9
												59J	
$CuMn_2O_4$ [12])		−284	5.49									*64BL*	31 ··· 34
$Cu_xMn_{3-x}O_4$												*63RNMP*	
$Cu_{1.5}Mn_{1.5}O_4$					80		3.35					*66B*	
$LiMn_2O_4$ [13])		−164	3.34									*64BL*	35
		−260	4.77									*66B,*	
												66SP	
$Li_{4/3}Mn_{5/3}O_4$		− 55										*63B-3*	
$MgMn_2O_4$		−450	5.99				4.9					*57B*	36, 37
		−500	5.75									*60MM,*	
								≈0.2 [14])				*59J, 64RN*	
$Mg_xMn_{3-x}O_4$												*64RNMP*	38
$NiMn_2O_4$		−392	7.66									*57B*	15, 35
												66SP	
					≈100							*58BW*	
$y = 0.74$ [15])					164			0.35 } [15])				*66BBP,*	
												67BBP	
$y = 0.93$					116			1.75 }					
$y = 0.80$								0.66				*67BBP*	
$ZnMn_2O_4$											1.0 [16])	*53R-1*	37, 39
												55GL	47 ··· 49
		−450	5.62			≈200	4.74					*57B*	
		−570	5.59									*61K*	
		−560		260								*64RN*	
		−420										*65B*	
		−450	5.62 [17])			≈250						*66A*	
$Zn_xMn_{3-x}O_4$												*60DM*	
												63RNMP	
	0.2	−670		420	[18])							*64RN, 59J*	40
	0.4	−650		370								*64RN*	46
	0.6	−640		330								*64RN*	
	0.8	−600		310								*64RN*	43 ··· 45
$Zn_{1+x}Mn_{2-x}O_4$												*63RNMP*	41, 42

Fußnoten siehe S. 498

Fußnoten zu Tab. 3

[1]) Werte für Mn_2O_3, keine Angabe über Spinellcharakter [*57BO*].

[2]) <80 °K ferrimagnetisch [*57B*].

[3]) $p_m = 1.87 \pm 0.002\ \mu_B$ im Vergleich mit Ni-Standard in einem Feld von 10 kOe gemessen [*60DM*].

[4]) Für einen Einkristall in [001]-Richtung $\Theta_C = (41.95 \pm 0.1)$ °K;
für einen Einkristall in [100]-Richtung $\Theta_C = (41.2\ \pm 0.1)$ °K [*60DM*].

[5]) Bei 4 °K [*63G-2*].

[6]) Bis 1075 °C; zwischen 1075 °C und 1150 °C fällt der elektrische Widerstand ziemlich plötzlich auf etwa 20% ab. Oberhalb 1150 °C $q = 0.75$ eV [*53R-1*].

[7]) Bei $0 < x \leq 0.7$ paramagnetisches Verhalten in einem Temperaturbereich 75 ··· 700 °K [*64RN*].

[8]) Bei 4 °K [*63G-2*].

[9]) Ferrimagnetisch mit sehr schwacher spontaner Magnetisierung bei 4.2 °K [*68BBP*].

[10]) Für eine von 1250 °C abgeschreckte Probe; für von 1000 °C bzw. 800 °C abgeschreckte Proben abweichende Werte für Θ_C, p_{eff} und Θ_a [*68BBP*].

[11]) Magnetisches Moment $\sigma = 0$ für die Konzentration $x = 0.75$ [*58WC*].

[12]) Ferrimagnetisch [*64BL*].

[13]) Paramagnetisch oberhalb 4.2 °K [*66B*].

[14]) Bei 4.2 °K [*59J*].

[15]) Bei 4.2 °K [*66BBP*]. y gibt die Kationenverteilung, siehe Tab. 1.

[16]) Unterhalb 950 °C; zwischen 950 °C und 1025 °C fällt der elektrische Widerstand ziemlich plötzlich auf etwa 20% ab. Bei >1025 °C $q = 0.6$ eV [*53R-1*].

[17]) Für $\frac{1}{2}$ Mol [*66A*].

[18]) Ferrimagnetisch bei niedrigen Temperaturen für $x \leq 0.75$ nach [*64RN*].

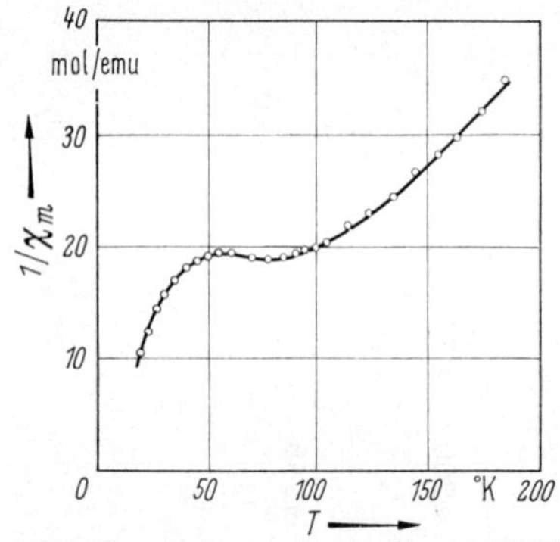

Fig. 13. γ-Mn_2O_3. Molare reziproke Suszeptibilität χ_m^{-1} vs. Temperatur T [*57BO*].

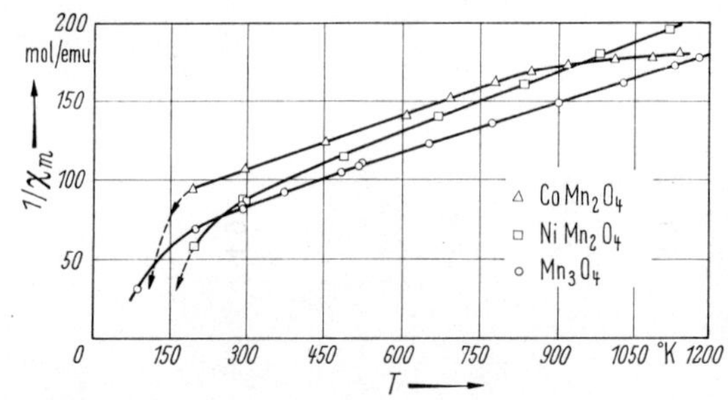

Fig. 15. Mn_3O_4, $NiMn_2O_4$ und $CoMn_2O_4$. Reziproke Suszeptibilität χ_m^{-1} vs. Temperatur T [*57B*].

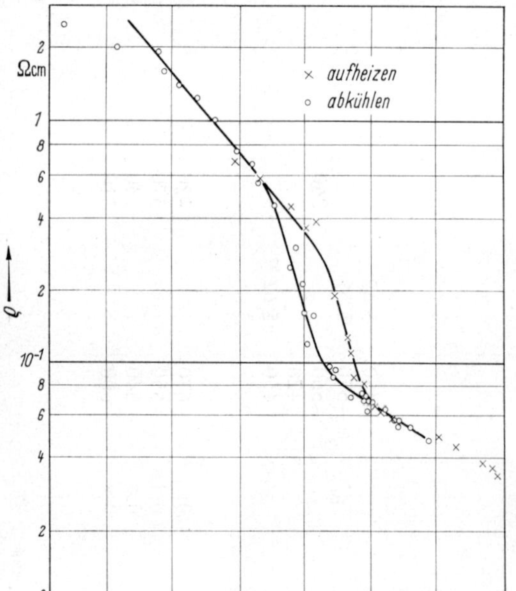

Fig. 14. Mn_3O_4. Spezifischer elektrischer Widerstand ϱ vs. Kehrwert der Temperatur $1/T$. Die Messungen wurden unterhalb und oberhalb der Übergangstemperatur von 1170 °C durchgeführt. Oberhalb 1170 °C kubische Kristallstruktur [*53R-1*].

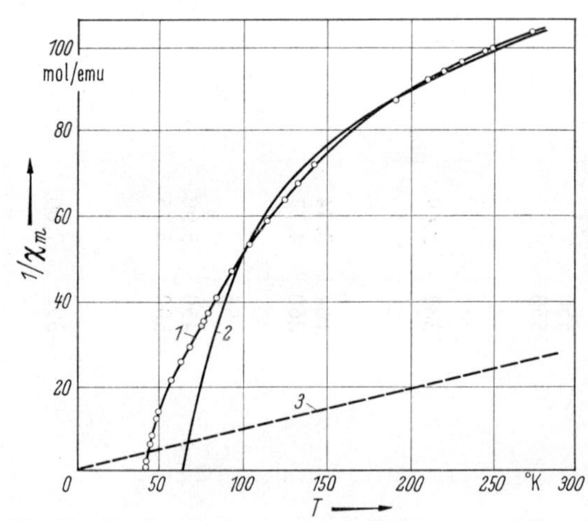

Fig. 16. Mn_3O_4. Reziproke Suszeptibilität pro mol χ_m^{-1} vs. Temperatur T; Kurve 1 = gemessene Werte,
Kurve 2 = berechnete Werte nach der Néel-Formel

$$1/\chi_m = \frac{T}{C_m} + \frac{1}{\chi_0} - \frac{\sigma}{T - \Theta}, \quad \sigma = 3480 \text{ (mol/emu) °K},$$

$1/\chi_0 = 91.3$ mol/emu, $\Theta = 31.1$ °K.
Kurve 3 = nach dem Curie-Weiss-Gesetz [*57BO*].

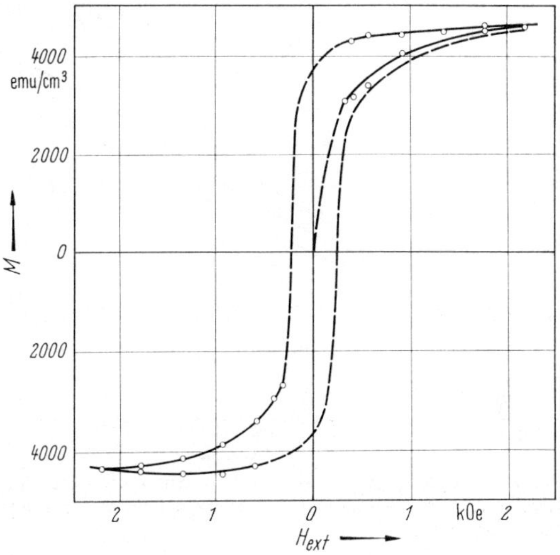

Fig. 17. Mn_3O_4. Magnetisches Moment M vs. angelegte Feldstärke H_{ext} bei T = 20.4 °K [*57BO*].

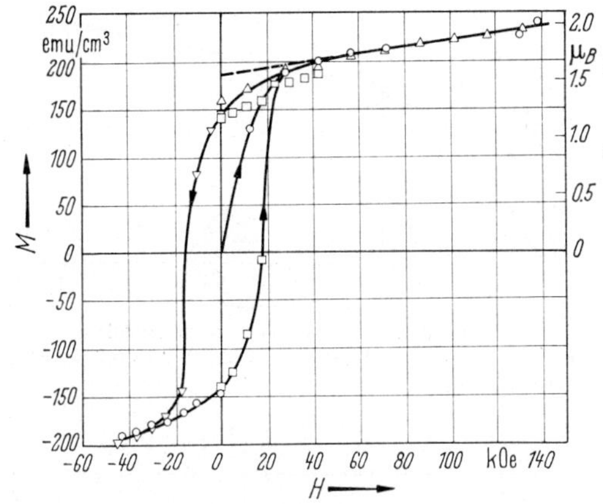

Fig. 18. Mn_3O_4. Magnetisierung M pro Volumeneinheit und magnetisches Moment p_m vs. Feld H für eine pulverisierte Probe bei 4.2 °K [*59J*].

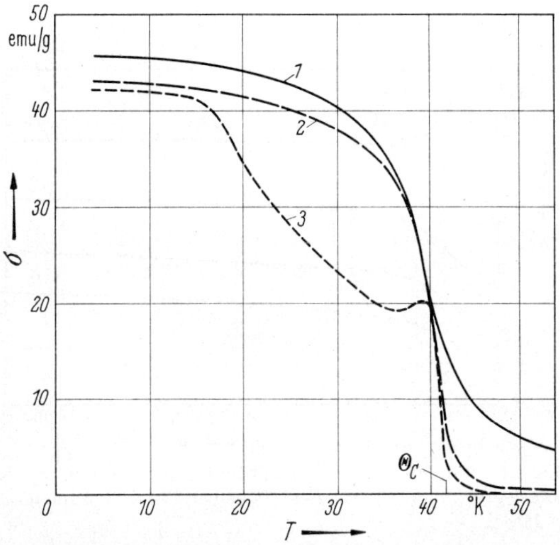

Fig. 19. Mn_3O_4. Magnetisches Moment σ pro Gramm vs. Temperatur T; (1 abgekühlt in einem Feld von 10 kOe, 2 abgekühlt in einem Feld von 1 kOe, 3 erwärmt in einem Feld von 250 Oe) [*60DM*].

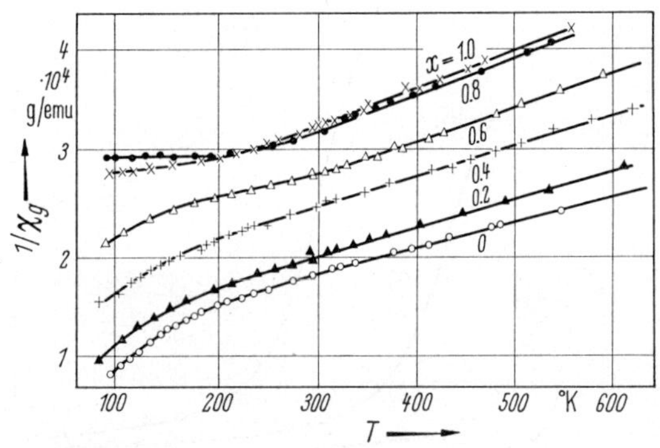

Fig. 20. $Cd_xMn_{3-x}O_4$. Reziproke Suszeptibilität pro Gramm χ_g^{-1} vs. Temperatur T. Parameter x [*64RN*].

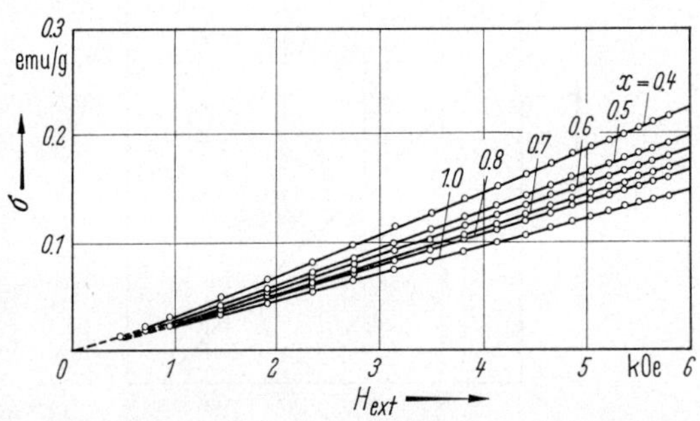

Fig. 21. $Cd_xMn_{3-x}O_4$. Magnetisches Moment σ pro Gramm vs. angelegtes Feld H_{ext} bei Raumtemperatur. Parameter x [*65DA*].

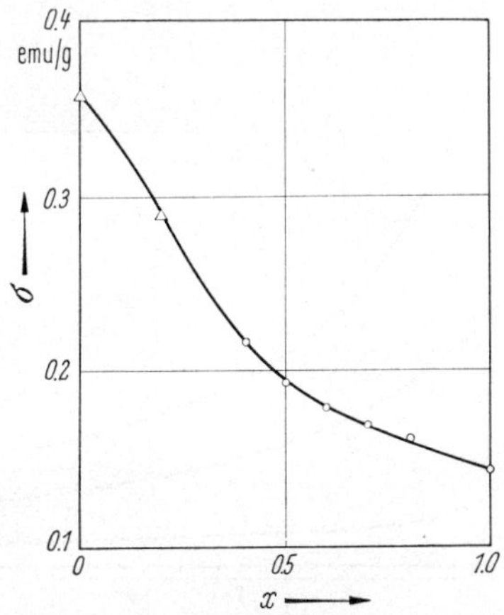

Fig. 22. $Cd_xMn_{3-x}O_4$. Magnetisches Moment σ pro Gramm vs. Zusammensetzungsparameter x bei einem Feld von H_{ext} = 5.8 kOe [*65DA*]; △ = Werte nach [*64RN*].

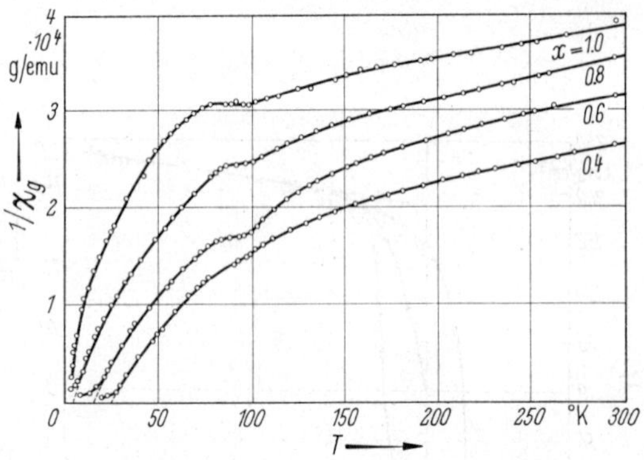

Fig. 23. $Cd_xMn_{3-x}O_4$. Reziproke Suszeptibilität pro Gramm χ_g^{-1} vs. Temperatur T bei einem Feld von 5.2 kOe. Parameter x [*65DA*].

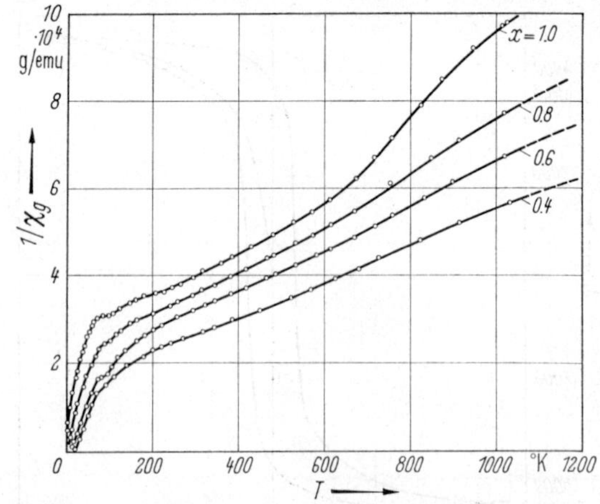

Fig. 24. $Cd_xMn_{3-x}O_4$. Reziproke Suszeptibilität pro Gramm χ_g^{-1} vs. Temperatur T bei einem Feld von 5.2 kOe. Parameter x [*65DA*].

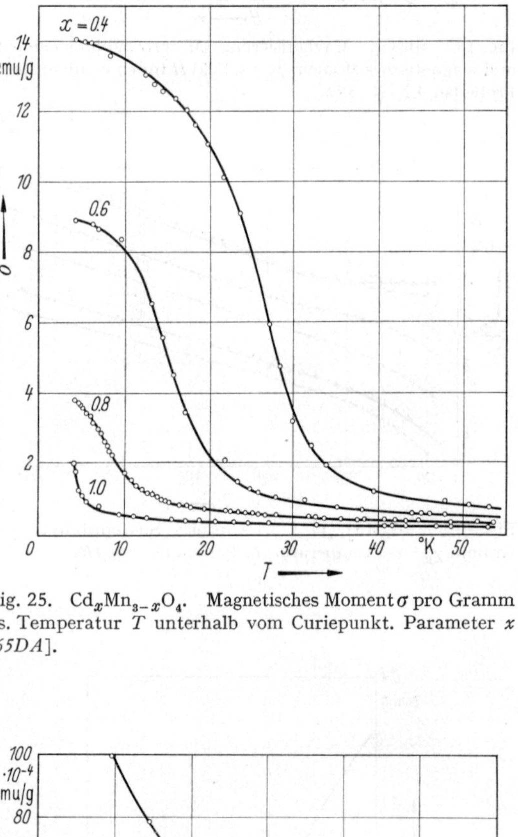

Fig. 25. $Cd_xMn_{3-x}O_4$. Magnetisches Moment σ pro Gramm vs. Temperatur T unterhalb vom Curiepunkt. Parameter x [*65DA*].

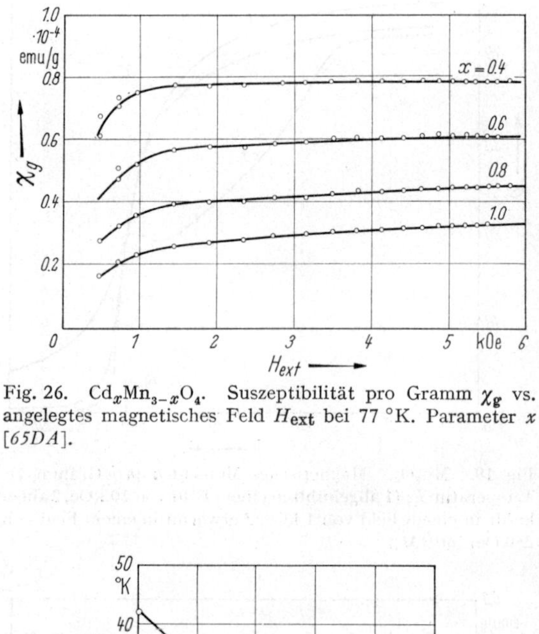

Fig. 26. $Cd_xMn_{3-x}O_4$. Suszeptibilität pro Gramm χ_g vs. angelegtes magnetisches Feld H_{ext} bei 77 °K. Parameter x [*65DA*].

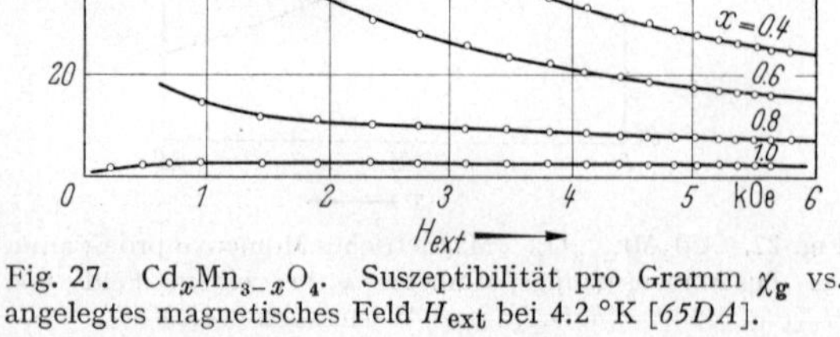

Fig. 27. $Cd_xMn_{3-x}O_4$. Suszeptibilität pro Gramm χ_g vs. angelegtes magnetisches Feld H_{ext} bei 4.2 °K [*65DA*].

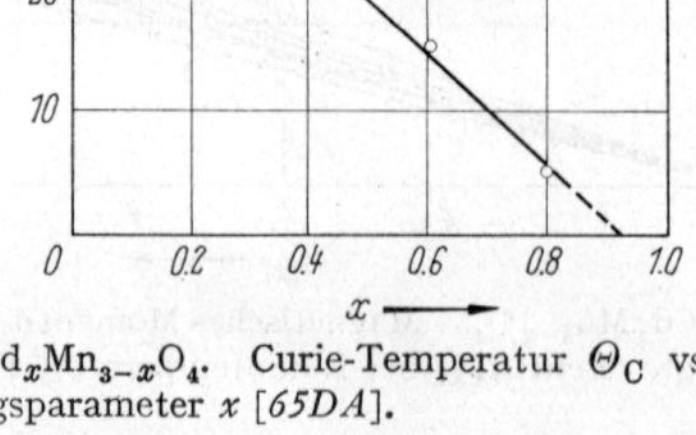

Fig. 28. $Cd_xMn_{3-x}O_4$. Curie-Temperatur Θ_C vs. Zusammensetzungsparameter x [*65DA*].

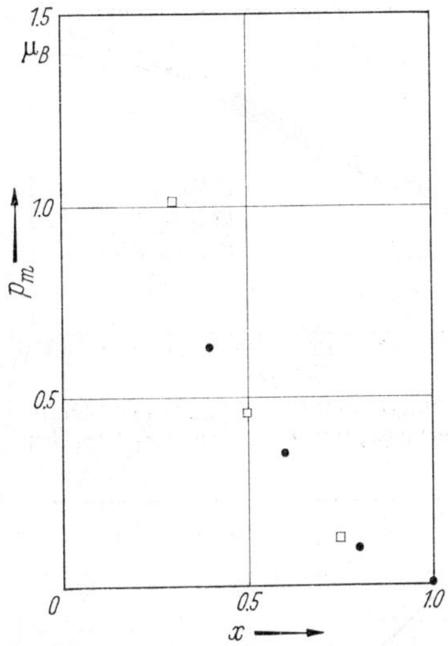

Fig. 29. $Cd_xMn_{3-x}O_4$. Magnetisches Moment p_m vs. Zusammensetzungsparameter x;
Experimentell bestimmte Werte:
• bei Temperaturen zwischen 4.2 ··· 300 °K und einem Feld $H = 0$,
□ bei $H = 5$ kOe und $T = 1.8$ °K [*65DA*].

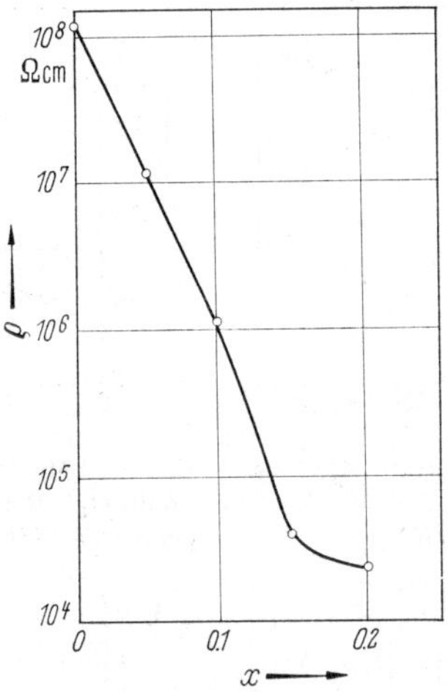

Fig. 32. $Cu_xMn_{3-x}O_4$. Spezifischer elektrischer Widerstand ϱ bei 20 °C vs. Zusammensetzungsparameter x. Die Proben wurden in Luft zwischen 900 ··· 1250 °C gesintert und dann schnell auf Raumtemperatur abgekühlt [*63RNMP*].

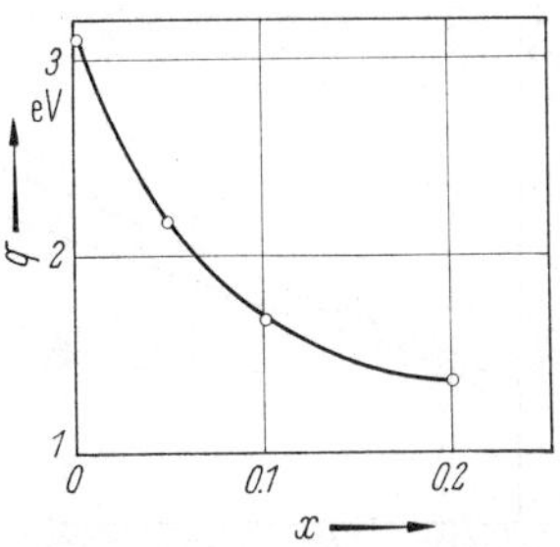

Fig. 33. $Cu_xMn_{3-x}O_4$. Aktivierungsenergie q vs. Zusammensetzungsparameter x nahe unterhalb der Übergangstemperaturen [*63RNMP*]. Herstellung der Proben wie in Fig. 32.

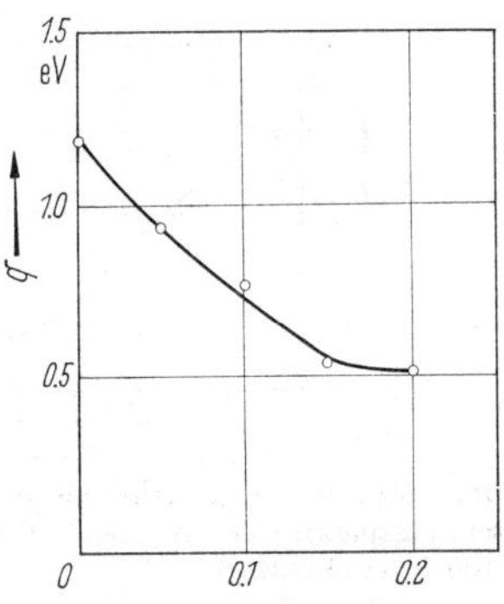

Fig. 34. $Cu_xMn_{3-x}O_4$. Aktivierungsenergie q vs. Zusammensetzungsparameter x bei Temperaturen zwischen 20 ··· 100 °C [*63RNMP*]. Herstellung der Proben wie in Fig. 32.

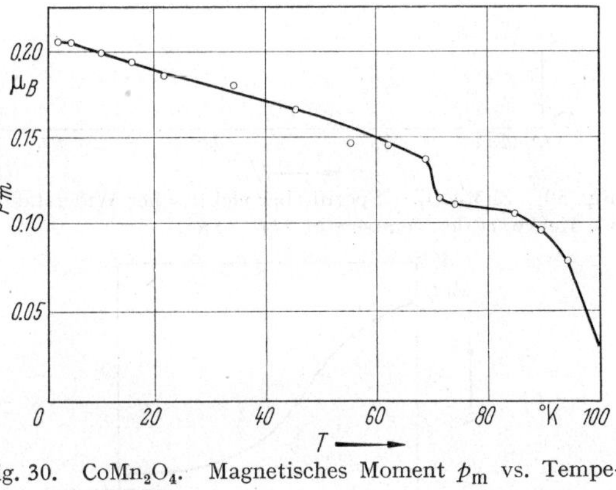

Fig. 30. $CoMn_2O_4$. Magnetisches Moment p_m vs. Temperatur T [*68BBP*].

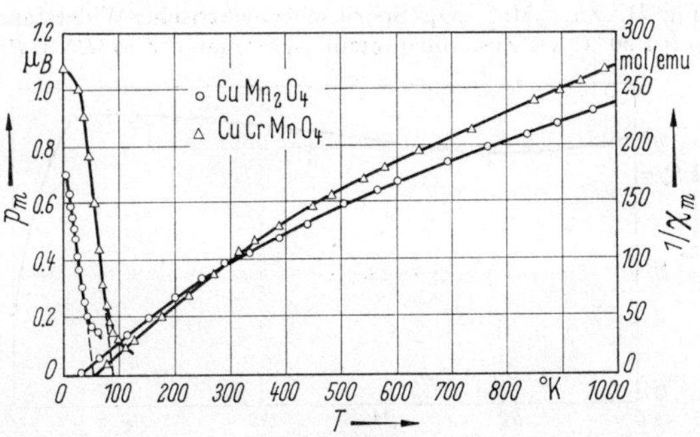

Fig. 31. $CuMn_2O_4$ und $CuCrMnO_4$. Magnetisches Moment p_m und reziproke molare Suszeptibilität χ_m^{-1} vs. Temperatur T bei einem magnetischen Feld H_{ext} von 10 kOe [*64BL*].

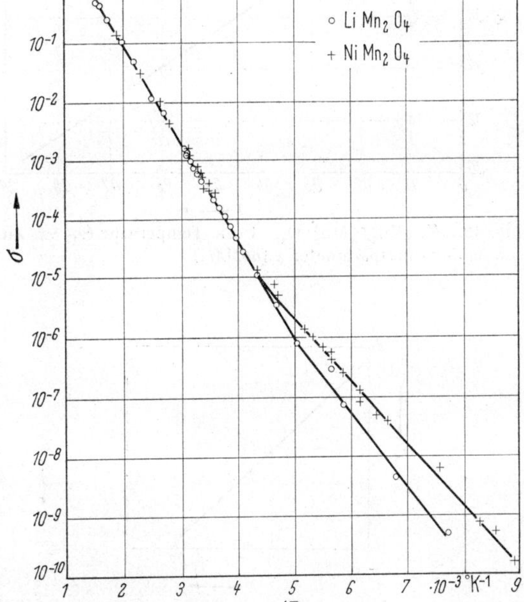

Fig. 35. $LiMn_2O_4$ und $NiMn_2O_4$. Elektrische Leitfähigkeit σ vs. Kehrwert der Temperatur $1/T$ [*66SP*].

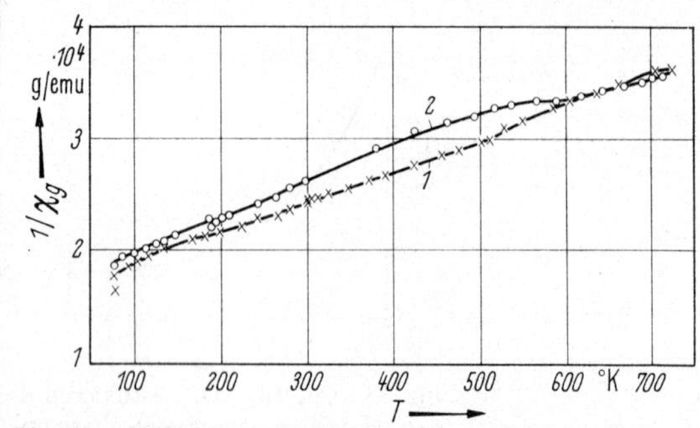

Fig. 36. $MgMn_2O_4$. Reziproke Suszeptibilität χ_g^{-1} vs. Temperatur T
1 Normaler Spinell 2 Inverser Spinell [*64RN*].

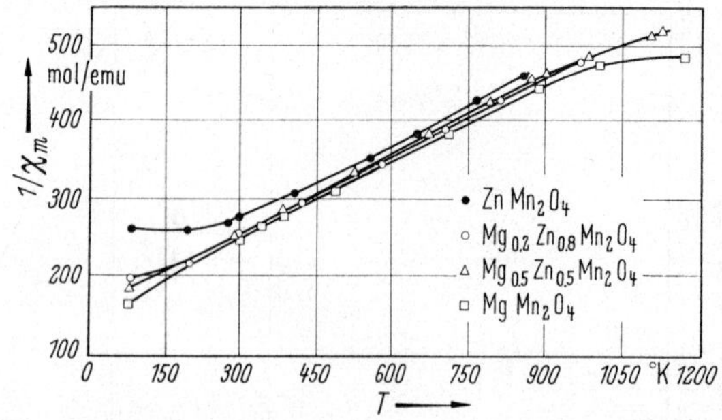

Fig. 37. $MgMn_2O_4$, $Mg_{0.5}Zn_{0.5}Mn_2O_4$, $Mg_{0.2}Zn_{0.8}Mn_2O_4$ und $ZnMn_2O_4$. Reziproke Suszeptibilität pro mol χ_m^{-1} vs. Temperatur T [*57B*].

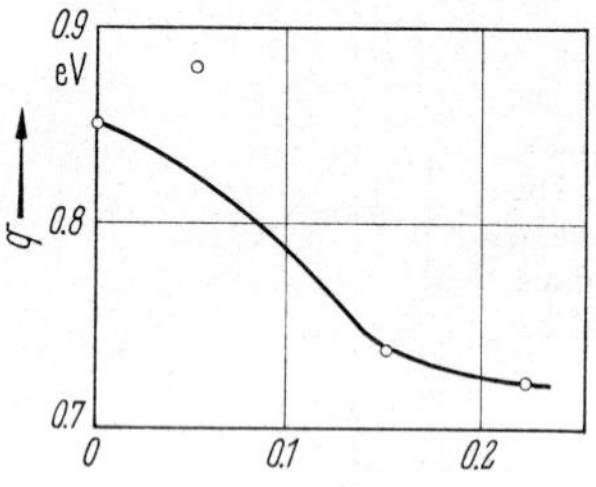

Fig. 38. $Mg_xMn_{3-x}O_4$. Aktivierungsenergie q vs. Zusammensetzungsparameter x bei Temperaturen zwischen 20···100 °C [*63RNMP*].

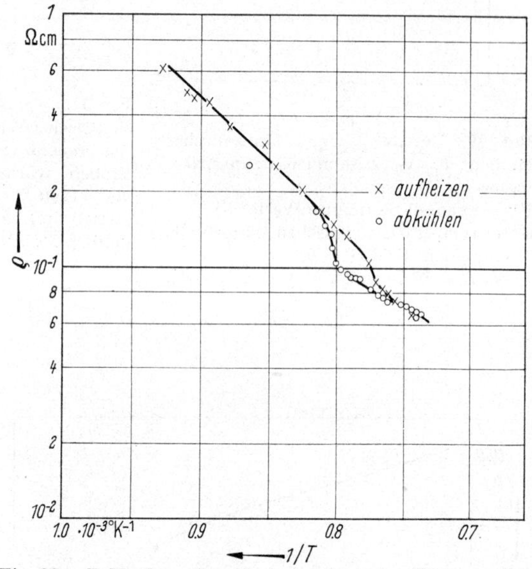

Fig. 39. $ZnMn_2O_4$. Spezifischer elektrischer Widerstand ϱ vs. Kehrwert der Temperatur $1/T$ [*53R-1*].

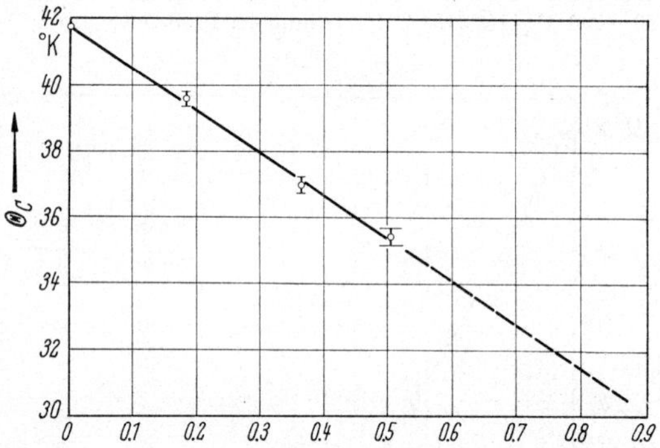

Fig. 40. $Zn_xMn_{1-x}^{2+}Mn_2^{3+}O_4$. Curie-Temperatur Θ_C vs. Zusammensetzungsparameter x [*60DM*].

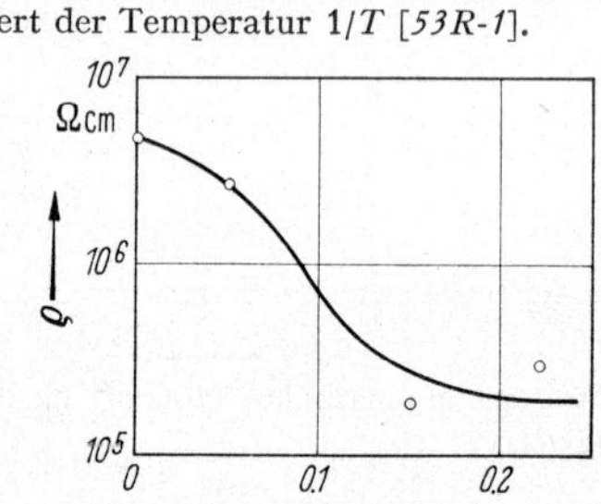

Fig. 41. $Zn_{1+x}Mn_{2-x}O_4$. Spezifischer elektrischer Widerstand ϱ bei 20 °C vs. Zusammensetzungsparameter x [*63RNMP*].

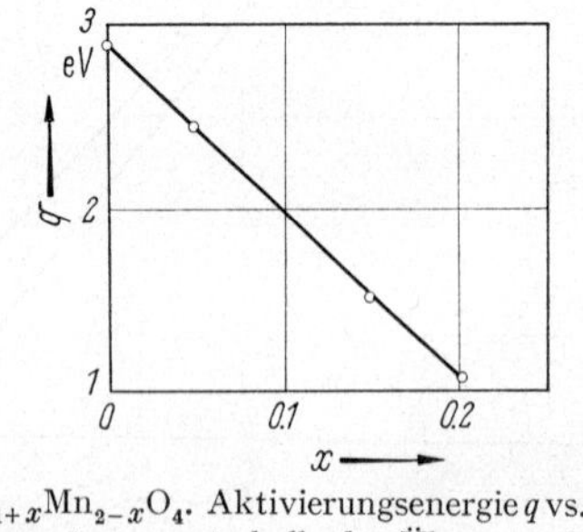

Fig. 42. $Zn_{1+x}Mn_{2-x}O_4$. Aktivierungsenergie q vs. Zusammensetzungsparameter x unterhalb der Übergangstemperaturen [*63RNMP*].

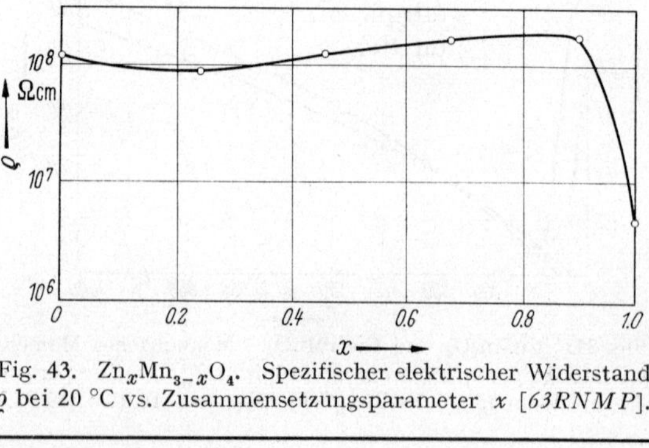

Fig. 43. $Zn_xMn_{3-x}O_4$. Spezifischer elektrischer Widerstand ϱ bei 20 °C vs. Zusammensetzungsparameter x [*63RNMP*].

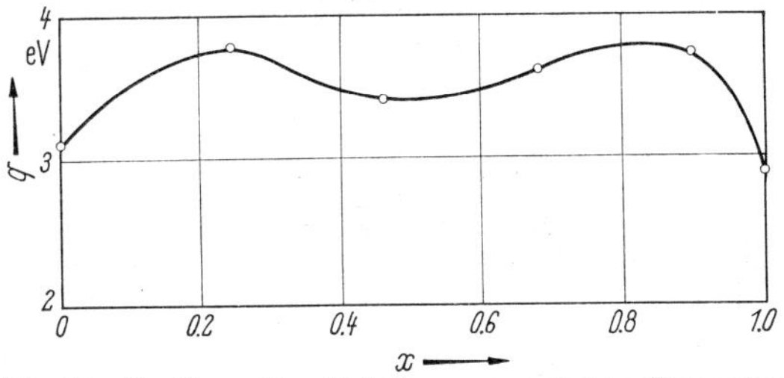

Fig. 44. $Zn_xMn_{3-x}O_4$. Aktivierungsenergie q vs. Zusammensetzungsparameter x nahe unterhalb der Übergangstemperaturen [63RNMP].

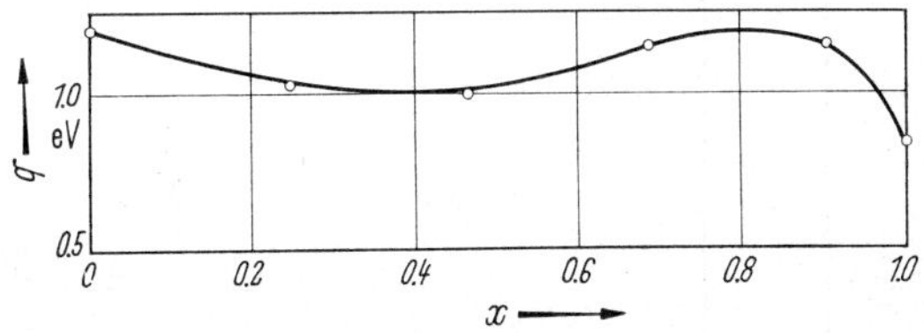

Fig. 45. $Zn_xMn_{3-x}O_4$. Aktivierungsenergie q vs. Zusammensetzungsparameter x bei Temperaturen zwischen 20···100 °C [63RNMP].

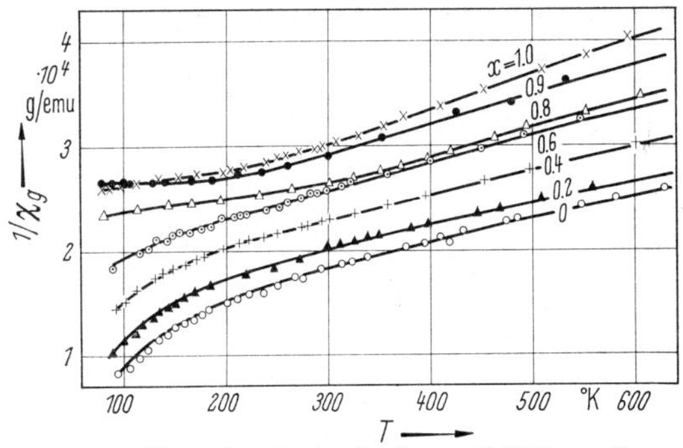

Fig. 46. $Zn_xMn_{3-x}O_4$. Reziproke Suszeptibilität pro Gramm χ_g^{-1} vs. Temperatur T. Parameter x [64RN].

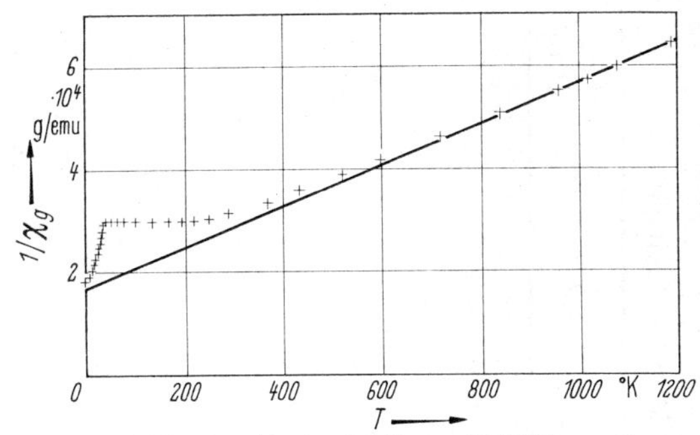

Fig. 47. $ZnMn_2O_4$. Reziproke Suszeptibilität pro Gramm χ_g^{-1} vs. Temperatur T [65B].

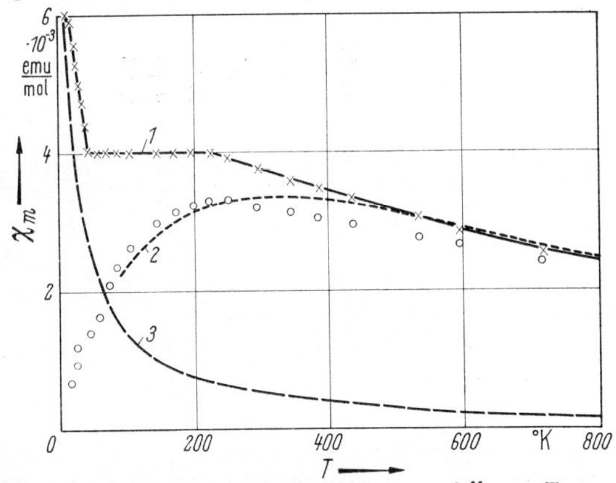

Fig. 48. $ZnMn_2O_4$. Suszeptibilität pro mol χ_m vs. Temperatur T [65B].

1 Experimentell bestimmte Werte ×

2, 3 Theoretische Werte [65B]; Korrigierte Werte: ○.

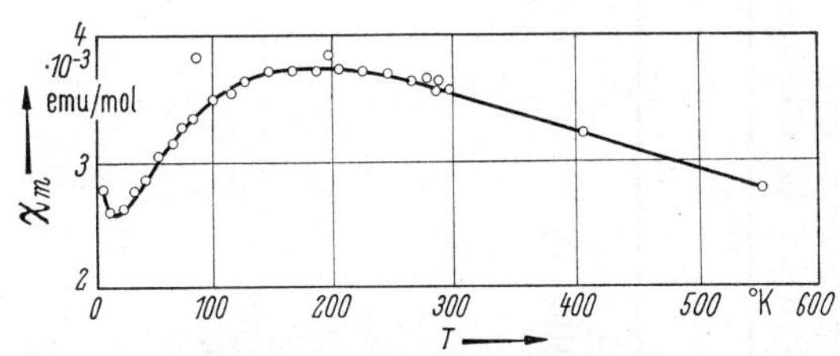

Fig. 49. $ZnMn_2O_4$. Suszeptibilität pro mol χ_m vs. Temperatur T (Werte für ein halbes mol $ZnMn_2O_4$) [66A].

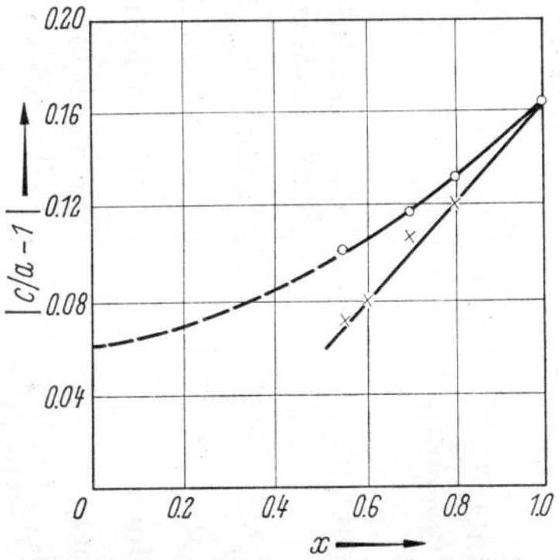

Fig. 50. $Cu_{1-x}Fe_{2-2x}Mn_{3x}O_4$. Das Verhältnis der Gitterparameter $|c/a - 1|$ vs. Zusammensetzungsparameter x für unterschiedlich hergestellte Proben [58D].

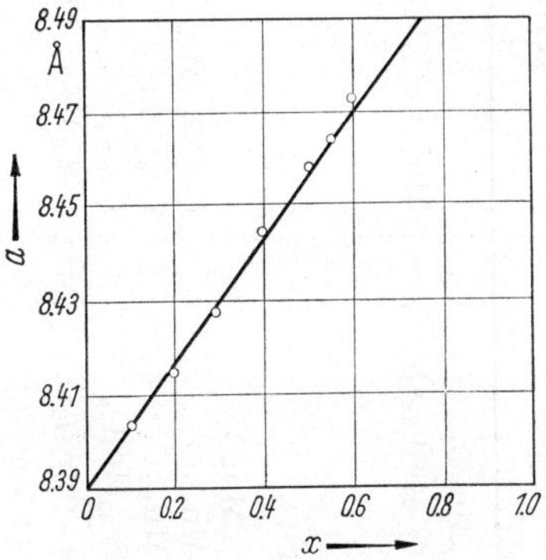

Fig. 51. $Cu_{1-x}Fe_{2-2x}Mn_{3x}O_4$. Gitterparameter a vs. Zusammensetzungsparameter x [58D].

Tab. 4. Kristallografische Eigenschaften ternärer Mn Oxid-Spinelle

Verbindung	Kationenverteilung	x	a Å	c Å	c/a	u	ϱ_X g/cm³	Lit.	Fig.	Tab.
$CoFe_{2-x}Mn_xO_4$		>1.5			>1			*65B*		5
$Co_{1-x}Fe_{2-2x}Mn_{3x}O_4$	$Mn^{2+}_xFe^{3+}_{1-x}[Co^{2+}_{1-x}Fe^{3+}_{1-x}Mn^{3+}_{2x}]O_4$	0.55	8.47		>1			*57FSS, 59G*	6	
		>0.55			>1			*57FSS, 59G, 63G-2*		
$CoGaMnO_4$			8.33					*57B*	11 in 6.3.6	5
$Co_xNi_{1-x}Mn_2O_4$								*65SDYPIDB*		
$CoNi_{0.5}Mn_{1.5}O_4$	$Co[Ni_{0.5}Mn_{1.5}]O_4$							*59A*		
$Co_{0.5}NiMn_{1.5}O_4$	$Co_{0.5}Mn_{0.5}[NiMn]O_4$							*59A*		
$Co_{0.75}Ni_{0.25}Mn_2O_4$	$Co_{0.75}Mn_{0.25}[Ni_{0.25}Mn_{1.75}]O_4$				1.11			*59A*		
$Co_{0.5}Ni_{0.5}Mn_2O_4$	$Co_{0.5}Mn_{0.5}[Ni_{0.5}Mn_{1.5}]O_4$							*59A*		
CuO-CoO-MnO-O_2								*59SZKT*, *66SP*	19, 20 in 6.3.9	5
$Cu_{1-x}Cr_{2-2x}Mn_{3x}O_4$								*58D*		
$CuCrMnO_4$	$Cu[CrMn]O_4$		8.32			0.385		*64BL*	55	5
	$\approx Cu^+[Cr^{3+}Mn^{4+}]O_4$		8.31					*66B*		
	$Cu^{2+}[Cr^{3+}Mn^{3+}]O_4$							*66B*		
$CuCr_{2-x}Mn_xO_4$	$Cu^{2+}_{1-x}Cu^{1+}_x[Cr^{3+}_{2-x}Mn^{4+}_x]O_4$							*64BL, 58D*		
	$Cu^{2+}[Cr^{3+}_{2-x}Mn^{3+}_x]O_4$							*64BL*		
		<0.4			<1			*63G-2*		
$CuFeMnO_x$ 2)								*51KR*		5
$Cu_{1-x}Fe_{2-2x}Mn_{3x}O_4$								*57FSS, 58D*	50, 51	
$CuGaMnO_4$	$Cu^{2+}[GaMn^{3+}]O_4$		8.363		0.383			*64BL*		5
	$Cu^{1+}[GaMn^{4+}]O_4$							*64BL*		
$CuMg_{0.5}Mn_{1.5}O_4$	$\approx Cu^+[Mg_{0.5}Mn^{4+}_{1.5}]O_4$		8.31					*66B*	56	5
$CuNi_{0.5}Mn_{1.5}O_4$	$Cu^+[Ni_{0.5}Mn^{4+}_{1.5}]O_4$		8.31					*66B*	57	5
$CuRhMnO_4$	$\approx Cu^+[RhMn^{4+}]O_4$		8.38					*66B*		5
$LiAlMnO_4$	$Li[AlMn]O_4$		8.12					*64B-2*		
$Li_{1.25}Al_{0.25}Mn_{1.5}O_4$	$Li[Li_{0.25}Al_{0.25}Mn_{1.5}]O_4$		8.15					*64B-1*		
$Li_{0.5}Al_{2.5-x}Mn_xO_4$ 3)								*65B, 63SB*		
$LiCoMnO_4$	$Li[CoMn]O_4$		8.07					*64B-2*		

$LiCo_{0.5}Mn_{1.5}O_4$	$Li[Co_{0.5}Mn_{1.5}]O_4$ 4)		8.16					*64B-1* *64B-2*, *66B*		5
$LiCrMnO_4$	$Li[Cr^{3+}Mn^{4+}]O_4$		8.19					*63B-2*, *64B-2*		5
			8.197		0.383			*64BL*		
$Li_{1.25}Cr_{0.25}Mn_{1.5}O_4$	$Li[Li_{0.25}Cr_{0.25}Mn_{1.5}]O_4$		8.19					*64B-1*		
$LiCu_{0.5}Mn_{1.5}O_4$	$Li[Cu_{0.5}Mn_{1.5}]O_4$ 4)		8.23					*64B-1*, *64B-2*		
$Li_{0.5}Fe_{2.5-x}Mn_xO_4$	5)				0.375			*65B*		
		1	8.366 6)					*65L*, *66KYH*		5
		1.25	8.378 6)					*65L*		
		1.5						*66KYH*		
$Li_{0.5+0.5x}Fe_{2.5-1.5x}Mn_xO_4$								*65B*		5
	$Li[Li_{0.25}Fe_{0.25}Mn_{1.5}]O_4$	1.5	8.20					*64B-1*		
	$Li_{0.6}Fe_{0.4}[Li_{0.4}Fe_{0.6}Mn]O_4$	1	8.26					*64B-2*		
$LiFe_{0.5}Mn_{1.5}O_4$								*65B*		5
$LiGaMnO_4$	$Li_{0.5}Ga_{0.5}[Li_{0.5}Ga_{0.5}Mn]O_4$		8.23					*64B-2*		
$Li_{1.25}Ga_{0.25}Mn_{1.5}O_4$	$Li[Li_{0.25}Ga_{0.25}Mn_{1.5}]O_4$		8.19					*64B-1*		
$LiMg_{0.5}Mn_{1.5}O_4$	$Li[Mg_{0.5}Mn_{1.5}]O_4$ 4)		8.18					*64B-1*, *64B-2* *66B*		5
$LiNi_{0.5}Mn_{1.5}O_4$	$Li[Ni_{0.5}Mn^{4+}_{1.5}]O_4$ 4)		8.17					*64B-1*, *64B-2*, *66B*	58	5
$LiRhMnO_4$	$Li[Rh^{3+}Mn^{4+}]O_4$		8.30					*63B-2*, *63B-3*, *64B-2*		5
$LiZn_{0.5}Mn_{1.5}O_4$ 1)	$Li_{0.5}Zn_{0.5}[Li_{0.5}Mn^{4+}_{1.5}]O_4$ 7)		8.19					*63B-3*, *64B-1*, *64B-2*, *65H*, *66B*	59	5
			8.192 ± 0.003 8)				4.495	*64JD*		
$Li_{1-x}Zn_xMn_2O_4$								*58WC*, *60ISB*		
$Li_{2-2x}Zn_{2x-1}Mn_2O_4$	$Li_{2-2x}Zn^{2+}_{2x-1}[Mn^{3+}_{2x}Mn^{4+}_{2-2x}]O_4$	>0.6		>1				*63G-2*		

Fußnoten siehe S. 507

Tab. 4. (Fortsetzung)

Verbindung	Kationenverteilung	x	a Å	c Å	c/a	u	ϱ_X g/cm³	Lit.	Fig.	Tab.
$MgAlMnO_4$			8.18					*60ISB*		
$MgAl_{2-2x}Mn_{2x}O_4$ [10]	$Mg^{2+}[Mn^{3+}_{2x}Al^{3-}_{2-2x}]$	0.94	8.11	8.94	1.10 [12]			*60ISB*, *63G-2*		
		0.875 [9]	8.32					*60ISB*		
		0.59	8.34					*63G-2*		
$Mg_{1-x}Al_{2-2x}Mn_{3x}O_4$	$Mg^{2+}_{1-x}Mn^{2+}_x[Mn^{3+}_{2x}Al^{3+}_{2-2x}]O_4$	0.875	8.02 [11]	9.15	1.14 [13]			*60ISB*, *63G-2*		
		0.58	8.32					*63G-2*		
$MgCr_{2-x}Mn_xO_4$	$Mg_{0.78}Mn_{0.22}[Mg_{0.22}CrMn_{0.78}]O_4$	1						*67G-1*	52	
$MgGa_{2-x}Mn_xO_4$	$Mg^{2+}_yGa^{3+}_{1-y}[Mg^{2+}_{1-y}Ga^{3+}_yMn^{3+}]O_4$	1	8.35					*57B*, *63G-2*, *59G*	53, 9, 10 in 6.3.(	5
	$Ga^{3+}[Mg^{2+}Mn^{3+}]O_4$	1						*63GH*, *63GHM*, *65HB*, *65GH*, *67G-1*		
$MgIn_{2-x}Mn_xO_4$								*67G-1*		
$Mg_{0.5}Zn_{0.5}Mn_2O_4$								*57B*	37	5
NiO-CoO-MnO-O_2								*62SZKT*, *66SP*	19, 20 in 6.3.9	5
NiO-CuO-MnO-O_2								*62SZKT*		5
$NiFe_{2-2x}Mn_{2x}O_4$								*58BW*	6, 60, 61	5
$Ni_{1-x}Fe_{2-2x}Mn_{3x}O_4$	$Mn^{2+}_xFe^{3+}_{1-x}[Ni^{2+}_{1-x}Fe^{3+}_{1-x}Mn^{3+}_{2x}]O_4$	0.625	8.46		[14]			*57FSS*, *63G-2*, *59G*		
$NiGaMnO_4$	$Ga^{3+}[Ni^{2+}Mn^{3+}]O_4$		8.38					*57B*, *59G*, *63G-2*, *65B*	11 in 6.3.6	5
$ZnAl_{2-x}Mn_xO_4$		1						*57B*		
$ZnCoMnO_4$	$Zn^{2+}[Co^{2+}Mn^{4+}]O_4$		8.25					*57B*, *59G*, *63G-2*	63	5
	$Zn^{2+}[Co^{3+}Mn^{3+}]O_4$							*59G*		

$Zn_xCo_{1-x}Mn_2O_4$								*65SDYPIDB*		
$ZnCrMnO_4$	$Zn^{2+}[Cr^{3+}Mn^{3+}]O_4$		8.25	8.62	1.04			*57B* *59G*, *63G-2*		
$ZnFeMnO_4$	$Zn^{2+}[Fe^{3+}Mn^{3+}]O_4$		8.494					*59M*, *63G-2*		5
$ZnFe_{2-x}Mn_xO_4$								*65B*		
$ZnGaMnO_4$	$Zn^{2+}[Ga^{3+}Mn^{3+}]O_4$		8.23	8.64	1.05			*61O*, *63G-2*, *65B*		
$Zn_xGe_{1-x}Mn_2O_4$ [15][17]	$Zn_x^{2+}Ge_{1-x}^{4+}[Mn_{2-2x}^{2+}Mn_{2x}^{3+}]O_4$	0.6	8.50		[16]			*59G*, *58WC*, *60ISB*, *59W*	54, 62	5
	$Zn_{0.5}^{2+}Ge_{0.5-y}^{4+}Mn_y^{2+}[Ge_y^{4+}Mn_{1-y}^{2+}Mn^{3+}]O_4$	0.5	8.37					*59M*, *63G-2*		
$Zn_{1-x}Mg_xMn_2O_4$	$Zn_{1-x}Mg_x[Mn_2^{3+}]O_4$	0.2	8.091	9.248	1.143			*57B*, *59G*, *63G-2*		
		0.5	8.095	9.278	1.146			*57B*, *59G*		
$ZnNiMnO_4$			8.31					*57B*	63	5
	$Zn^{2+}[Ni^{2+}Mn^{4+}]O_4$ $Zn^{2+}[Ni^{3+}Mn^{3+}]O_4$							*59G*, *63G-2* *59G*		
			8.30					*66B*		

[1]) Farbe: violett [*64JD*].
[2]) Annahme $CuFeMnO_4$ nicht bestätigt. Äußerlich homogene Substanz mit eigenem Curie-Punkt [*51KR*].
[3]) Bei $x \leq 0.25$ einphasige Spinellproben; bei $0.25 < x < 0.50$ Proben mit 2 Spinellphasen; bei $0.5 < x \leq 1.55$ Proben mit einphasiger (kubischer) Spinellphase; bei $x \geq 1.75$ Proben sind nicht einphasig [*65B*].
[4]) Mit Überstruktur auf Oktaederplätzen [*64B-1*].
[5]) Unterschiedliche Kationenverteilungen gefunden [*65L*].
[6]) Für langsam abgekühlte Proben; abgeschreckte Proben bei $x = 1$, $a = 8.368$ Å; $x = 1.25$, $a = 8.381$ Å [*65L*].
[7]) Mit Überstruktur auf Oktaeder- und Tetraederplätzen [*64B-1*].
[8]) Geordnete Struktur. Für ungeordnete Struktur $a = 8.275 \pm 0.003$ Å, $\varrho_X = 4.363$ g/cm³ und Farbe: schwarz; für Unordnung auf Tetraederplätzen $a = 8.193 \pm 0.003$ Å, $\varrho_X = 4.495$ g/cm³ und Farbe: violett [*64JD*].
[9]) Weitere Verbindungen der Reihe $MgMn_2O_4$-$MgAl_2O_4$ mit dem Verhältnis 13 : 3 ($a = 8.33$ Å) und 12 : 4 ($a = 8.28$ Å) [*60ISB*].
[10]) Formeln für die Einheitszelle: (1) $Mg_{7.92}^{2+}Mn_{14.48}^{3+}Mn_{0.48}^{4+}Al_{0.96}^{3+}\square_{0.16}O_{32}$ und (2) $Mg_{7.92}^{2+}Mn_{13.12}^{3+}Mn_{0.72}^{4+}Al_{2.00}^{3+}\square_{0.24}O_{32}$ [*60ISB*].
[11]) Für Verbindungen der Reihe $MgAl_2O_4$-Mn_3O_4 im Verhältnis 14 : 2; für das Verhältnis 12.6 : 3.4: $a = 8.08$ Å, $c = 9.0$ Å; $c/a = 1.11$
Für weitere Proben der Reihe:

Verhältnis	a [Å]
11 : 5	8.34
10 : 6	8.29
9 : 7	8.29
8 : 8	8.26 [*60ISB*].

[12]) $c/a > 1$ für $x > 0.59$ [*63G-2*].
[13]) $c/a > 1$ für $x > 0.58$ [*63G-2*].
[14]) $c/a > 1$ für $x > 0.625$ [*63G-2*].
[15]) Spinellphase nur bei $x > 0.5$ [*58WC*].
[16]) Für $x > 0.6$ $c/a > 1$ [*59G*].
[17]) Farbe: rötlichbraun [*58WC*].

Tab. 5. Magnetische und elektrische Eigenschaften ternärer Mn Oxid-Spinelle

Verbindung	x	C_m $\frac{cm^3\ °K}{mol}$	Θ_a °K	Θ_C °K	p_m μ_B	ϱ Ωcm	Lit.	Fig.
$CoFe_{2-x}Mn_xO_4$	1					10^4	*65B*	
$CoGaMnO_4$							*57B*	11 in 6.3.6
CuO-CoO-MnO-O_2 1)							*59SZKT, 66SP*	20 in 6.3.9
$CuCrMnO_4$		4.63	−259				*64BL*	
				45	1.47 2)3)		*66B*	55
$CuFeMnO_x$				270			*51KR*	
$CuGaMnO_4$		2.98	− 63				*64BL*	
$CuMg_{0.5}Mn_{1.5}O_4$		2.85 10a)	+ 75 10a)	57	3.82 3)10a)		*66B*	56
		2.92 10b)	+ 60 10b)		3.42 10b)			
		3.2 10c)	+ 45 10c)		2.79 10c)			
$CuNi_{0.5}Mn_{1.5}O_4$		3.20	− 20	150	3.15 3)		*66B*	57
$CuRhMnO_4$		1.9	+ 45	35	2.35 3)		*66B*	
$LiCo_{0.5}Mn_{1.5}O_4$		4.14	−240	50	0.33 3)2)		*66B*	
$LiCrMnO_4$		2.81	− 84				*64BL*	
$Li_{0.5}Fe_{2.5-x}Mn_xO_4$	1.25			100			*65L*	
$Li_{0.5+0.5x}Fe_{2.5-1.5x}Mn_xO_4$					9)		*65B*	
$LiFe_{0.5}Mn_{1.5}O_4$		4.55 4)	−105				*65B*	
$LiMg_{0.5}Mn_{1.5}O_4$		2.66	+ 45	38	2.97 2)3)		*66B*	
$LiNi_{0.5}Mn_{1.5}O_4$		3.26	− 60	130	3.28 3)		*66B*	58
$LiRhMnO_4$			− 20				*63B-3*	
$LiZn_{0.5}Mn_{1.5}O_4$			− 55		—		*63B-3*	
		2.74	+ 35	22	4.24 3)		*66B*	59
$MgGaMnO_4$							*57B*	10 in 6.3.6
$Mg_{0.5}Zn_{0.5}Mn_2O_4$							*57B*	37
NiO-CoO-MnO-O_2 6)							*62SZKT, 66SP*	20 in 6.3.9
NiO-CuO-MnO-O_2 6)							*62SZKT*	
$NiFe_{2-2x}Mn_{2x}O_4$							*58BW*	60, 61
$NiGaMnO_4$							*57B*	11 in 6.3.6
$ZnCoMnO_4$		4.5	−260				*57B*	63
$ZnFeMnO_4$		6.54 8)	−135				*65B*	
$ZnFe_{0.5}Mn_{1.5}O_4$		6.27 5)	−285				*65B*	
$ZnGeMnO_4$							*58WC*	62
$ZnNiMnO_4$		3.41	−200				*57B*	
		3.35	−190	90	0.87 3)		*66B*	63

1) Für Angaben über $\sigma[\Omega^{-1}cm^{-1}]$, $A\,[\Omega^{-1}cm^{-1}]$ und q [eV] für verschiedene Verbindungen aus diesem System siehe [*59SZKT*].
2) Wert für $H = 30$ kOe; bei diesem Feld befindet sich die Probe weit entfernt von der Sättigung [*66B*].
3) Bei 4.2 °K [*66B*].
4) Zwischen 80 °K und 800 °K gilt das Curie-Weiss-Gesetz [*65B*].
5) Zwischen 100 °K und 900 °K gilt das Curie-Weiss-Gesetz [*65B*].
6) Für Angaben über $\sigma[\Omega^1 cm^{-1}]$ und q [eV] für verschiedene Verbindungen aus diesem System siehe [*62SZKT*].
8) Zwischen 300 °K und 900 °K gilt das Curie-Weiss-Gesetz [*65B*].
9) Abbildung für $p_m(x)$ siehe Li-Mn-Ferrite, Seite 153.
10a) Werte für eine Probe bei 700 °C in N_2 behandelt und extrem langsam mit 5°/h abgekühlt. 10b, c) Proben bei 900 °C in O_2 behandelt und 10b) extrem langsam 5°/h abgekühlt, 10c) im Ofen abgekühlt [*66B*].

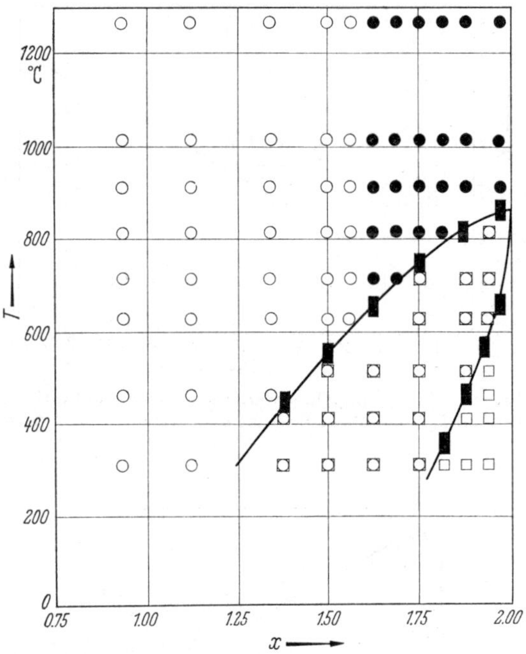

Fig. 52. $MgCr_{2-x}Mn_xO_4$. Phasendiagramm
● Bestätigte kubische Spinellphase
○ Nichtbestätigte kubische Spinellphase
□ Tetragonale Phase
⊡ Tetragonale und kubische Phase [67G-1].

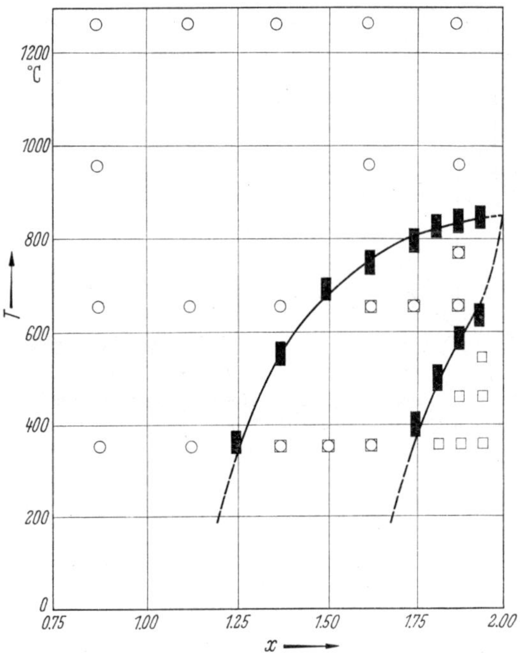

Fig. 53. $MgGa_{2-x}Mn_xO_4$. Phasendiagramm
○ Kubische Spinellphase
⊡ Tetragonale und kubische Phase
□ Tetragonale Phase [63GHM].
Siehe auch Fig. 10, Seite 472.

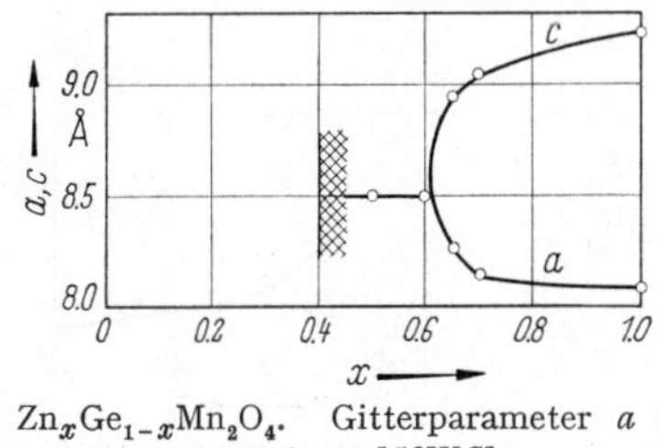

Fig. 54. $Zn_xGe_{1-x}Mn_2O_4$. Gitterparameter a und c vs. Zusammensetzungsparameter x [58WC].

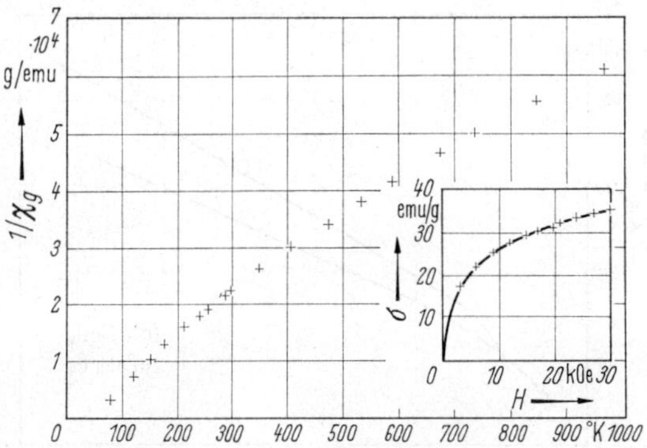

Fig. 55. $CuCrMnO_4$. Reziproke Suszeptibilität pro Gramm χ_g^{-1} vs. Temperatur T, und Magnetisierung σ vs. angelegtes Feld H bei 4.2 °K [66B].

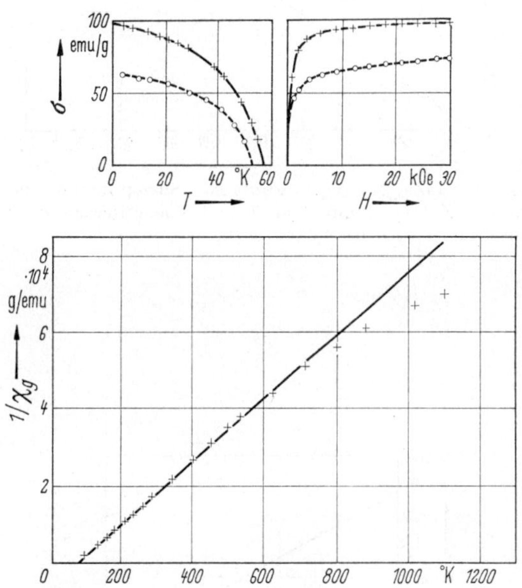

Fig. 56. $CuMg_{0.5}Mn_{1.5}O_4$. Reziproke Suszeptibilität pro Gramm χ_g^{-1} vs. Temperatur T; Magnetisierung σ vs. Temperatur T, und Magnetisierung σ vs. angelegtes Feld H bei 4.2 °K für unterschiedlich behandelte Proben
+ Proben bei 900 °C erhitzt (Ausgangsstoffe Oxide) und im Ofen abgekühlt
○ Proben bei 700 °C erhitzt (Ausgangsstoffe Nitrate) und schnell abgekühlt (5°/h) [66B].

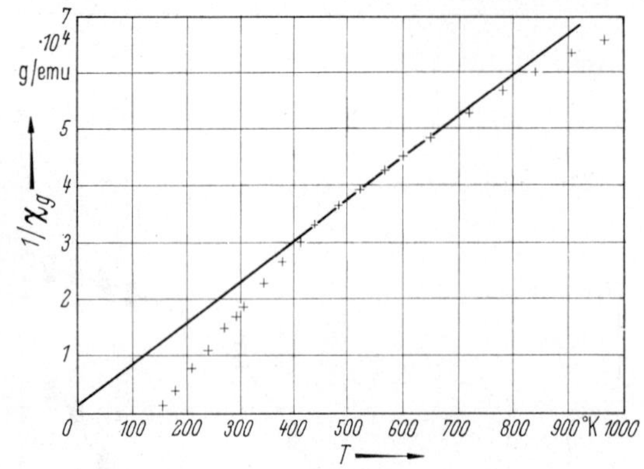

Fig. 57. $CuNi_{0.5}Mn_{1.5}O_4$. Reziproke Suszeptibilität pro Gramm χ_g^{-1} vs. Temperatur T [66B].

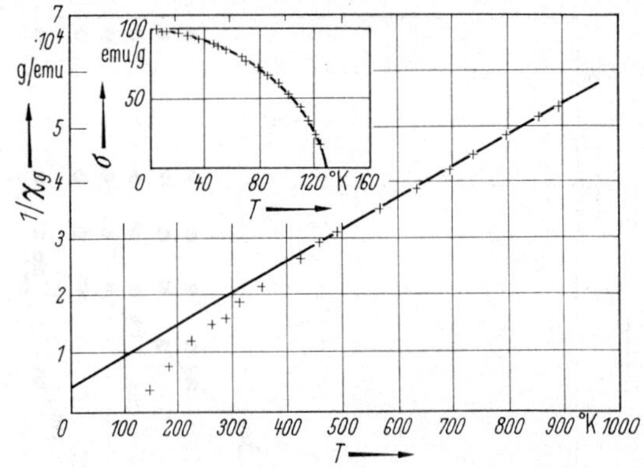

Fig. 58. $LiNi_{0.5}Mn_{1.5}O_4$. Reziproke Suszeptibilität pro Gramm χ_g^{-1} und Magnetisierung σ vs. Temperatur T [66B].

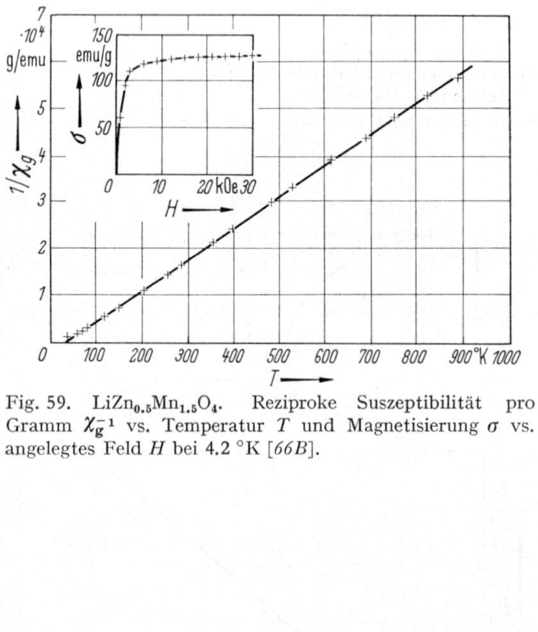

Fig. 59. $LiZn_{0.5}Mn_{1.5}O_4$. Reziproke Suszeptibilität pro Gramm χ_g^{-1} vs. Temperatur T und Magnetisierung σ vs. angelegtes Feld H bei 4.2 °K [66B].

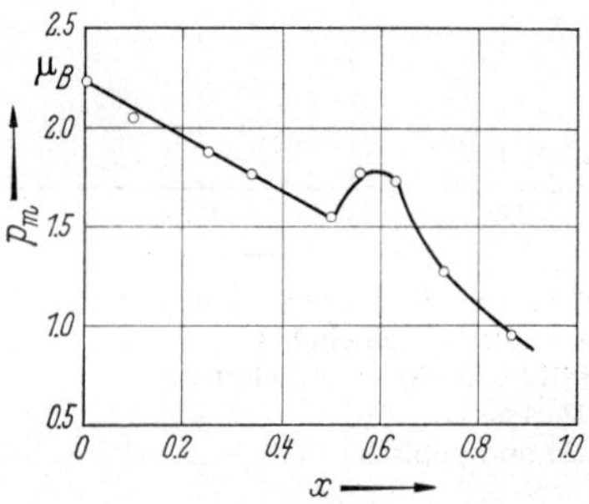

Fig. 60. $NiFe_{2-2x}Mn_{2x}O_4$. Magnetisches Moment p_m vs. Zusammensetzungsparameter x [58BW].

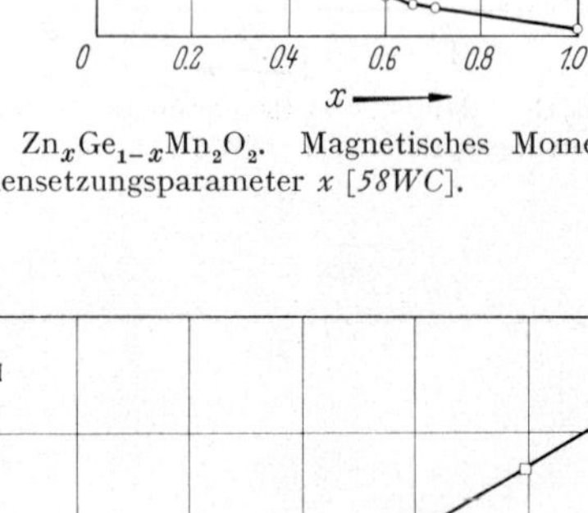

Fig. 62. $Zn_xGe_{1-x}Mn_2O_2$. Magnetisches Moment p_m vs. Zusammensetzungsparameter x [58WC].

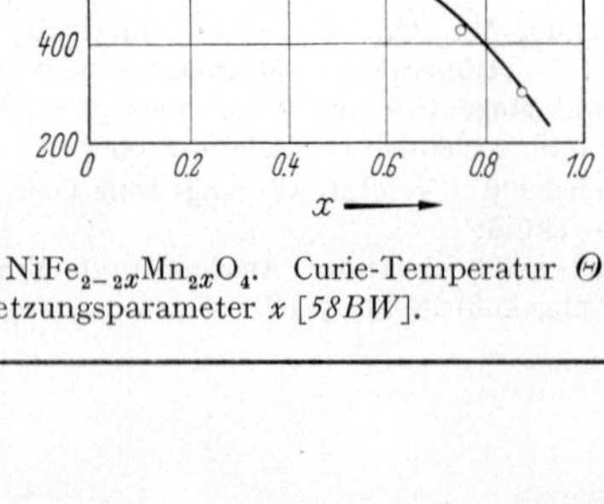

Fig. 61. $NiFe_{2-2x}Mn_{2x}O_4$. Curie-Temperatur Θ_C vs. Zusammensetzungsparameter x [58BW].

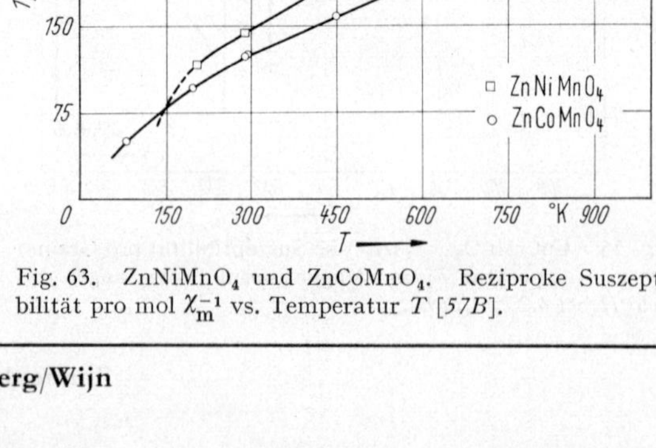

Fig. 63. $ZnNiMnO_4$ und $ZnCoMnO_4$. Reziproke Suszeptibilität pro mol χ_m^{-1} vs. Temperatur T [57B].

Tab. 6. Kristallografische Eigenschaften quaternärer und höherer Mn Oxid-Spinelle

Verbindung	Kationenverteilung	x	a Å	u	Lit.	Fig.	Tab.
$Zn_xGe_{1-x}Co_{2-2x}Mn_{2x}O_4$ [2]	$Zn^{2+}_{0.32}Ge^{4+}_{0.68}[Co^{2+}_{1.36}Mn^{3+}_{0.64}]O_4$	0.32			*58WC, 59G, 63G-2*	66	7
[2]	$Zn^{2+}_{0.65}Ge^{4+}_{0.35}[Co^{2+}_{0.7}Mn^{3+}_{1.30}]O_4$	0.65	8.395		*58WC, 59G, 63G-2*		
[2]	$Zn^{2+}_{0.65}Ge^{4+}_{0.35}[Co^{3+}_{0.7}Mn^{2+}_{0.7}Mn^{3+}_{0.6}]O_4$	0.65	8.395		*58WC, 59G, 59W*		
$Zn_{1-x}Ge_xCu_{2x}Mn_{2-2x}O_4$	$Zn_{0.5}Cu^{1+}_{0.5}[Ge^{4+}_{0.5}Cu^{1+}_{0.5}Mn^{4+}]O_4$	0.5	8.308	0.390	*64BL, 65RB-2*	64	7
	$Zn_{0.5}Cu^{2+}_{0.5}[Ge^{4+}_{0.5}Cu^{2+}_{0.5}Mn^{3+}]O_4$	0.5	8.308	0.390	*64BL*		
$ZnGe_xCu_xMn_{2-2x}O_4$	$Zn[Ge_xCu_xMn_{2-2x}]O_4$				*65RB-2*	64	
$Zn_xGe_{1-x}Fe_{2-2x}Mn_{2x}O_4$	$Zn^{2+}_xGe^{4+}_{1-x}[Fe^{2+}_{2-2x}Mn^{3+}_{2x}]O_4$	0.60	8.37 [1]		*59M, 63G-2*		
		0.73	8.37		*61O*		
$Zn_{1-0.5x}Ge_{0.5x}Fe_xMn_{2-x}O_4$	$Zn_{1-0.5x}Ge_{0.5x}[Fe_xMn_{2-x}]O_4$				*60M*	65	
$ZnMg_{0.5}Ti_{0.5}MnO_4$	$Zn^{2+}[Mg^{2+}_{0.5}Ti^{4+}_{0.5}Mn^{3+}]O_4$		8.397		*61O, 63G-2, 65B*		
$ZnMg_{0.5}Fe_{0.5}Ti_{0.5}Mn_{0.5}O_4$					*65B*		7

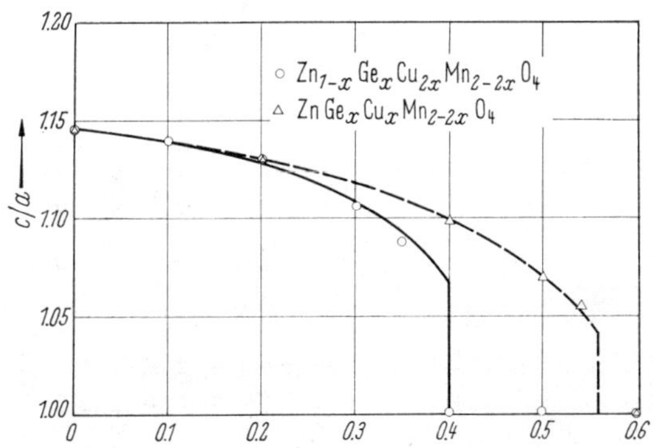

Fig. 64. $Zn_{1-x}Ge_xCu_{2x}Mn_{2-2x}O_4$ und $ZnGe_xCu_xMn_{2-2x}O_4$. Verhältnis der Gitterparameter c/a vs. Zusammensetzungsparameter x [*65RB-2*].

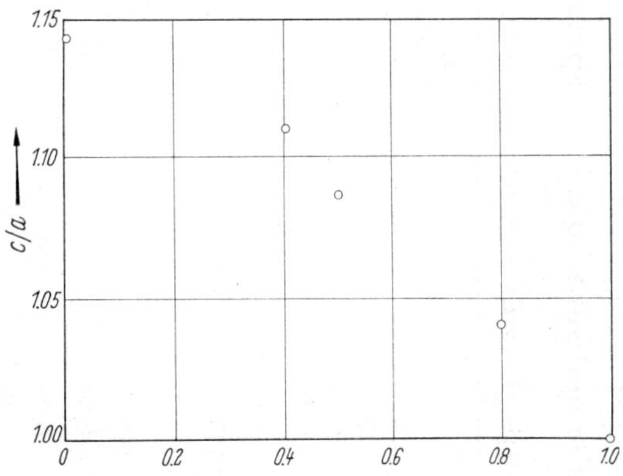

Fig. 65. $Zn_{1-0.5x}Ge_{0.5x}Fe_xMn_{2-x}O_4$. Verhältnis der Gitterparameter c/a vs. Zusammensetzungsparameter x. ○ gemessene Werte [*60M*].

Tab. 7. Magnetische Eigenschaften quaternärer und höherer Mn Oxid-Spinelle. Siehe Fig. 66

Verbindung	x	C_m $\frac{cm^3\ °K}{mol}$	Θ_a °K	p_m μ_B	Lit.
$Zn_xGe_{1-x}Co_{2-2x}Mn_{2x}O_4$	0.65			0.1	*58WC*
$Zn_{1-x}Ge_xCu_{2x}Mn_{2-2x}O_4$	0.5	3	− 84		*64BL*
$ZnMg_{0.5}Fe_{0.5}Ti_{0.5}Mn_{0.5}O_4$		3.41 [3]	−110		*65B*

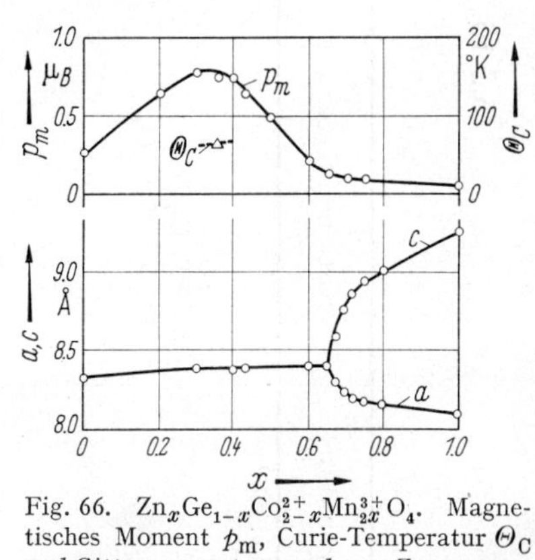

Fig. 66. $Zn_xGe_{1-x}Co^{2+}_{2-x}Mn^{3+}_{2x}O_4$. Magnetisches Moment p_m, Curie-Temperatur Θ_C und Gitterparameter a und c vs. Zusammensetzungsparameter x [*58WC*].

[1]) $c/a > 1$ für $x > 0.6$ [*63G-2*].
[2]) Schwarz [*58WC*].
[3]) Zwischen 300···1300 °K gilt das Curie-Weiss-Gesetz [*65B*].

6.3.13 Ti Spinelle und substituierte Ti Spinelle (Literatur siehe 6.3.14, Seite 528)

Tab. 1. Kristallografische Eigenschaften binärer Ti^{4+} Oxid-Spinelle

Verbindung	Kationenverteilung	a Å	u	ϱ_X g/cm³	ϱ_p g/cm³	Farbe	Lit.	Fig.	Tab.
Co_2TiO_4		8.420		5.12		dunkel-	*30P*	1···3	
		8.431 ± 0.004		5.07		grün	*31HH, 28···32S*		2
	$Co[CoTi]O_4$ 1)	8.445 2)				grün	*53R-1, 53R-2, 64B-2*		
	$Co^{2+}[Co^{2+}Ti^{4+}]O_4$						*57G*	4	
		8.44					*62S*		
		8.445	0.381 ± 0.002				*65P*		
							54E, 67PB, 67NK,		
							68NK		
		8.446					*63BPM-1*		
$CoTi_{1.5}O_4$	$Co[\square_{0.5}Ti_{1.5}]O_4$	8.404 ± 0.005		4.25			*63JD*		
Fe_2TiO_4 3)4)	$Fe[FeTi]O_4$	8.50 ± 0.01	0.390 ± 0.01				*32BP, 64B-2*	5···7	2
							57G		
		8.49 ± 0.010					*43E*		
		8.534					*50P, 55CBM*		
		8.53					*57AKY*		
	$Fe^{2+}[Fe^{2+}Ti^{4+}]O_4$						*61N*		
							67BRJ, 68OC		
							57AKY, 57VWCM, 57SR, 61H,		
							54E, 57W, 64S, 65RB-1, 66BR, 66BZRJB,		
							66BRGG, 66KL, 66LAB, 55A		
$Fe_{4/3}Ti_{4/3}O_4$	$Fe^{2+}_{2/3}\square_{1/3}[Fe^{2+}_{2/3}Ti^{4+}_{4/3}]O_4$						*55CBM*		
$LiTi_{7/4}O_4$	$Li^{1+}_{3/4}\square_{1/4}[Li^{1+}_{1/4}Ti^{4+}_{7/4}]O_4$						*53BD*		
$Li_{4/3}Ti_{5/3}O_4$							*53BD*		
	$Li^{1+}[Li^{1+}_{1/3}Ti^{4+}_{5/3}]O_4$	8.36					*57G, 57J*		
	$Li[Li_{1/3}Ti_{5/3}]O_4$	8.358 5)					*64B-2*		
		8.352					*65L, 56J*		
Mg_2TiO_4		8.439 ± 0.002		3.53	3.52		*31HH, 28···32S*		2
	$Mg[MgTi]O_4$ 4)	8.41 ± 0.01	0.390 ± 0.006				*32BP, 64B-2*		
		8.445 ± 0.005				grün	*53R-1, 53R-2*		
		8.441	0.387				*58SBB*		
		8.445					*62PM-2*		
		8.444	0.386 ± 0.003				*65P*		
		8.445 ± 0.001 6)					*67DR*		
							47VH, 53BD, 53Y, 54E, 54SR, 61H		

Mg_2TiO_4 (Forts.)							*68NK*		
		8.440					*63BPM-1*		
		8.441 ± 0.005	0.3875 ± 0.0005				*66O*		
Mn_2TiO_4		8.674		4.49			*31HH* 7), *28···32S*	8, 9	2
	$Mn[MnTi]O_4$	8.67					*61WMD*, *64B-2*		
		8.679					*64VLR*		
$Mn_2TiO_4\beta$ } 8)9)		8.679 ± 0.001		4.506	4.482		*66HGLRV*		
$Mn_2TiO_4\alpha$ }							*66HGLRV*		
							64HLRV, *54E*, *61LH*, *63LV*, *63VL*, *64RLV*, *65LRV*, *65LRPV*, *65VLR*, *66VJD*, *68NK*		
$MnTi_{1.5}O_4$	$Mn[\square_{0.5}Ti_{1.5}]O_4$	8.48 ± 0.01		4.05			*63JD*		
Zn_2TiO_4		8.410		5.43		weiß	*30P*, *28···32S*		2
		8.462 ± 0.005		5.295			*30T*, *28···32S*		
		8.439 ± 0.003		5.33			*31HH*, *28···32S*		
		8.460		5.12			*38CN*		
		8.445	0.380 ± 0.005	5.32			*47VH*, *47SR*		
		8.55					*53B*, *53SR*		
	$Zn[ZnTi]O_4$	8.465					*53R-1*, *53R-2*		
	$Zn_xTi_{1-x}[Ti_xZn_{2-x}]O_4$ 15)	8.467	≈0.383 ± 0.003				*65P*, *63BPM-1*, *63BP*		
	$Zn[ZnTi]O_4$	8.475 ± 0.005	0.390 ± 0.005				*66O*		
$Zn_2TiO_4\alpha$		8.494 ± 0.003 10)					*65BMPM*, *66VJD*		
		8.494 12)					*67BP*		
$Zn_2TiO_4\beta$ 11)		8.467					*67BP*		
		8.468 ± 0.002 13)					*67DR*		
		8.469 ± 0.001 14)					*67DR*		
							53Y, *54SR*, *61H*, *61PM*, *62BP*, *67NK*, *68NK*, *66BMPM*		
$ZnTi_{1.5}O_4$	$Zn[\square_{0.5}Ti_{1.5}]O_4$	8.395 ± 0.005		4.52			*63JD*		
$Zn_{1.9}Ti_{1.05}O_4$	$Zn_{0.95}\square_{0.05}[Zn_{0.95}Ti_{1.05}]O_4$ 16)						*63BPM-2*		

1) Inverser Spinell [*53R-1*].
2) Angabe von [*64B-2*] $a = 8.465$ Å.
3) Mineral: Ulvöspinell.
4) Inverser Spinell [*32BP*].
5) Geglühte Probe; abgeschreckte Probe $a = 8.359$ Å [*64B-2*].
6) Für eine bei 1500 °C abgeschreckte Probe; $a = 8.446 \pm 0.002$ Å für eine von 650 °C und 703 at abgeschreckte Probe [*67DR*].
7) Deutlich magnetisch [*31HH*], [*28···32S*].
8) Tetragonal verzerrtes Gitter $a = 8.726 \pm 0.001$ Å; $c = 8.567 \pm 0.004$ Å; Übergang von $Mn_2TiO_4\alpha$ in $Mn_2TiO_4\beta$ bei 770 °C [*66HGLRV*].
9) $a = 6.170 \pm 0.003\sqrt{2}$ Å; $c = 8.564 \pm 0.003$ Å; $c/a\sqrt{2} = 0.982$ [*66HGLRV*].
10) Tetragonal verzerrtes Gitter, $c = 8.415 \pm 0.003$ Å [*65BMPM*]; Umwandlung von der verzerrten α-Form in die kubische β-Form bei 500 °C; $a = (6.005 \pm 0.003)\sqrt{2}$, $c = 8.416 \pm 0.003$, $c/a\sqrt{2} = 0.991$ [*66VJD*].
11) Inverser Spinell [*67BP*].
12) Tetragonal verzerrtes Gitter, $c = 8.415$ Å, $c/a = 0.9907$ [*62BP*], [*67BP*], [*63BP*].
13) Bei 600 °C und 703 at hergestellt [*67DR*].
14) Bei 1500 °C hergestellte Probe [*67DR*].
15) Teilweise inverser Spinell [*63BP*].
16) Andere Kationenverteilung Zn $[Zn_{0.9}Ti_{1.05}\square_{0.05}]O_4$ [*63BPM-2*].

Tab. 2. Magnetische Eigenschaften binärer Ti^{4+} Oxid-Spinelle

Verbindung	Θ_a °K	p_m μ_B	p_{eff} μ_B	Θ_C °K	C_m $\frac{cm^3\ °K}{mol}$	Θ_N °K	$\chi_{m\,diam}$ 10^{-6} emu/mol	Lit.	Fig.
Co_2TiO_4								*31HH* 2)	1…4
	−80		4.7	55 1)				*62S*	
	−177				6.17			*67PB*	
								65P	
Fe_2TiO_4 3)				120				*61N*	
				121				*66LAB, 57AKY*	5…7
						120		*66BR*	
		0.36 4)				140		*68OC*	
								67BRJ, 50P	
Mg_2TiO_4							−48	*62PM-2, 65P*	
Mn_2TiO_4		0.45		81				*61WMD*	
								64HLRV, 64VLR	8, 9
Mn_2TiO_4 α				69				*66HGLRV*	
Mn_2TiO_4 β		0.5		78				*66HGLRV*	
Zn_2TiO_4							−63	*61PM*	
							−63 ± 2	*65P*	

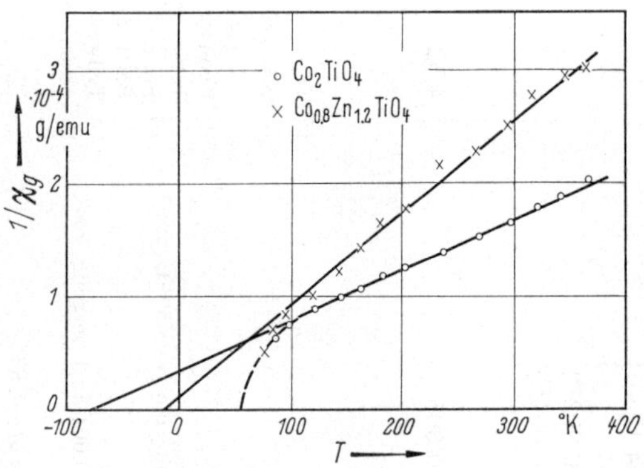

Fig. 1. Co_2TiO_4 und $Co_{0.8}Zn_{1.2}TiO_4$. Reziproke Suszeptibilität pro Gramm χ_g^{-1} vs. Temperatur T [*62S*].

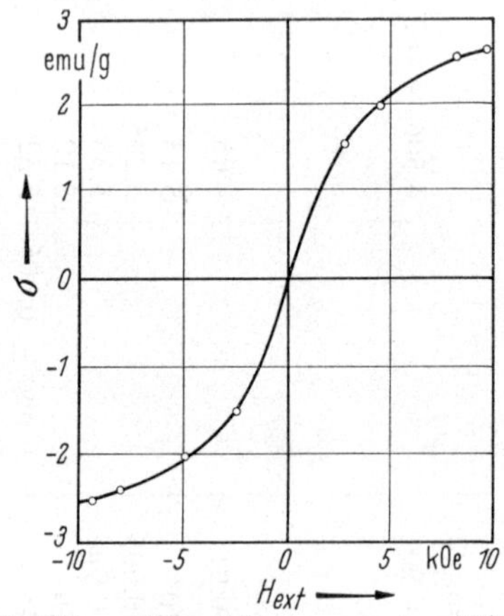

Fig. 2. Co_2TiO_4. Magnetisierung σ in [100]-Richtung vs. angelegtes Feld H_{ext} für T = 45 °K [*62S*].

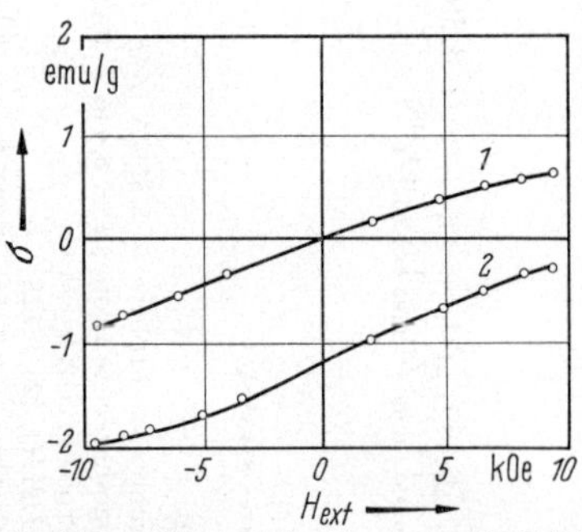

Fig. 3. Co_2TiO_4. Magnetisierung σ in [100]-Richtung vs. angelegtes Feld H_{ext} für T = 4.2 °K [*62S*].

1 ohne Feld abgekühlt

2 nach Abkühlung in einem magnetischen Feld von 8 kOe.

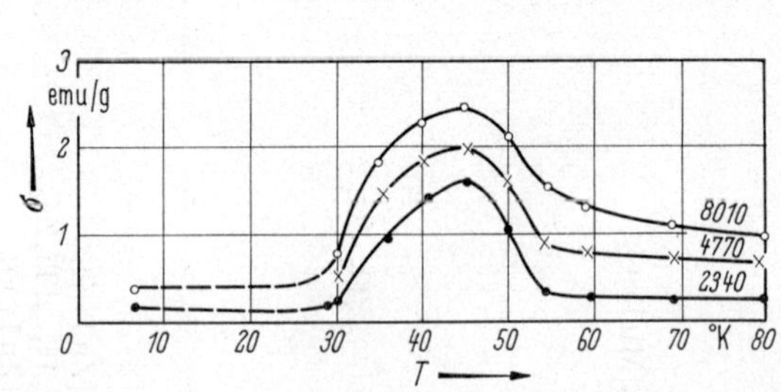

Fig. 4. Co_2TiO_4. Magnetisierung σ vs. Temperatur T bei angelegten Feldstärken von 2340 Oe, 4770 Oe und 8010 Oe [*62S*].

1) Ferrimagnetisch <55 °K [*62S*].

2) Sehr schwache magnetische Eigenschaften [*31HH*].

3) Paramagnetisch bei Raumtemperatur; Inversstruktur bei tieferen Temperaturen antiferromagnetisch [*61N*].

4) Maximale Magnetisierung; unterhalb 115 °K geringe Kristallverzerrung [*68OC*].

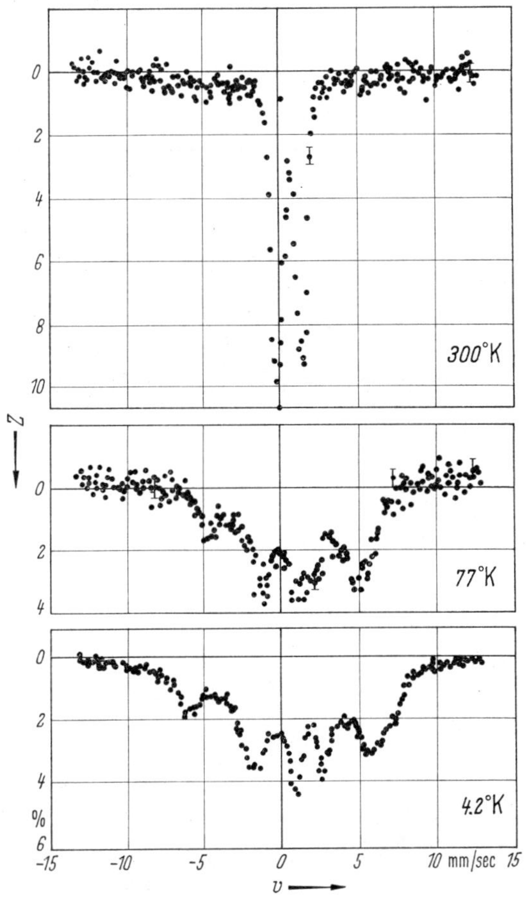

Fig. 5. Fe_2TiO_4. Mössbauerspektren bei 300 °K, 77 °K und 4.2 °K. [*67BRJ*].

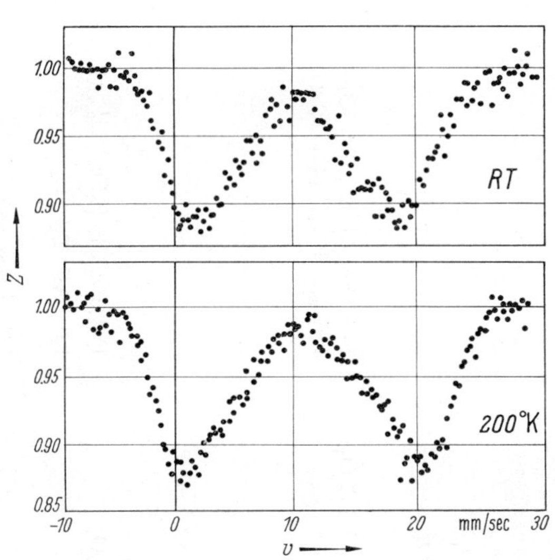

Fig. 6. Fe_2TiO_4. Mössbauerspektren bei Raumtemperatur und 200 °K [*68OC*].

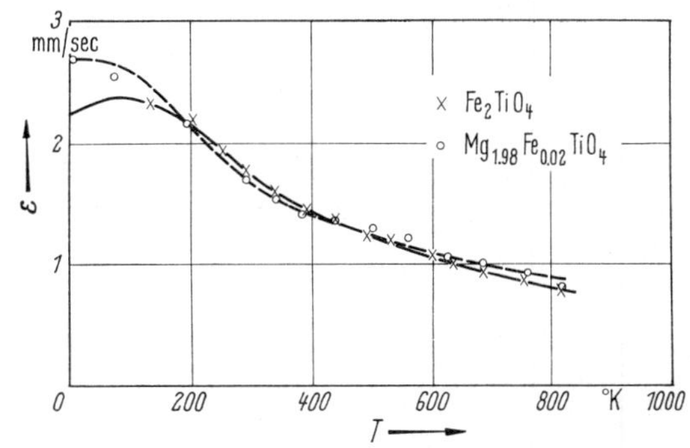

Fig. 7. Fe_2TiO_4, $Mg_{1.98}Fe_{0.02}TiO_4$. Quadrupolaufspaltung ε vs. Temperatur T [*68OC*].

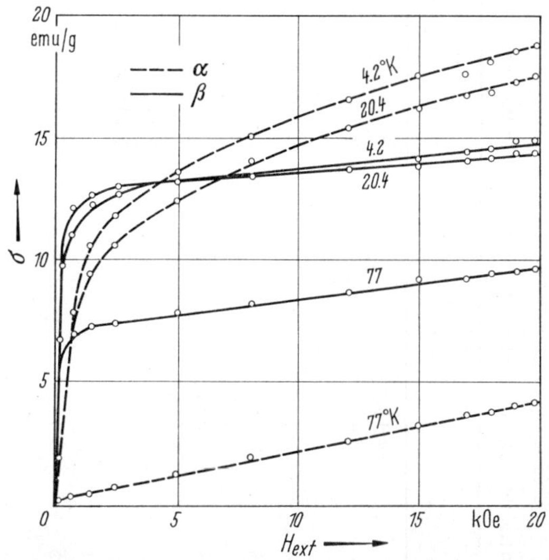

Fig. 8. Mn_2TiO_4. Magnetisierung σ vs. äußeres Feld H_{ext} bei unterschiedlichen Temperaturen für α- und β-Phase [*64VLR*].

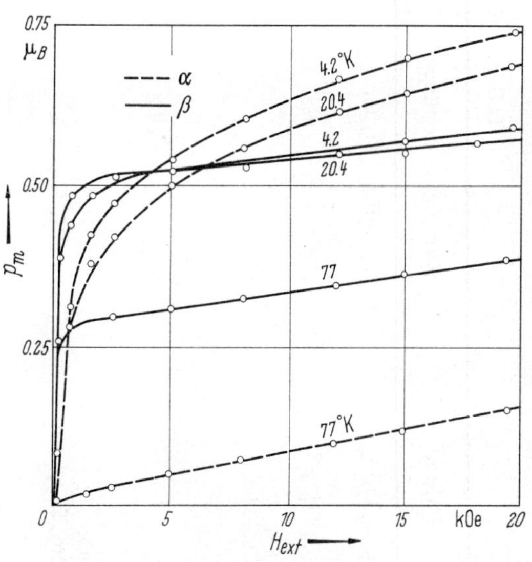

Fig. 9. Mn_2TiO_4. Magnetisches Moment pro Molekül p_m vs. äußeres Feld H_{ext} bei unterschiedlichen Temperaturen für α- und β-Phase [*66HGLRV*].

Tab. 3. Kristallografische Eigenschaften ternärer Ti^{4+} Spinelle

Verbindung	Kationenverteilung	x	a Å	u	Farbe	Lit.	Fig.	Tab.
$Co_{1+x}Fe_{2-2x}Ti_xO_4$	$Co_xFe_{1-x}[CoFe_{1-x}Ti_x]O_4$	0.7	8.436			*64B-2*	19	4
		0.8	8.440			*64B-2*		
$Co^{2+}_{2-x}Fe^{3+}_{2x}Ti^{4+}_{1-x}O_4$						*67ZHRB*		
$CoMgTiO_4$			8.45			*53Y*		
$Co_{2-2x}Mg_{2x}TiO_4$		0.25	8.447			*53R-2*		
		0.5	8.445			*53R-2*	10	
		0.75	8.443			*53R-2*		
		0.9	8.448			*58SBB*		
		0.8			blau	*54SR*		
		0.6			blaugrün	*54SR*		
$CoZnTiO_4$			8.46			*53Y*		4
$Co_{2-2x}Zn_{2x}TiO_4$		0.25	8.445			*53R-2*		
		0.5	8.448		grün	*53R-2*	11, 12	
		0.75	8.456			*53R-2*		
		0.9				*54SR*		
		0.6				*62S*	1	
		0.4		0.382 ± 0.002		*65P*		
						67NK, 68NK		
$CuCoTiO_4$			8.45			*53Y*		
$CuMgTiO_4$			8.45			*53Y*		
$CuZnTiO_4$			8.46			*53Y*		
$Fe^{3+}_{2-2x}Fe^{2+}_{1+x}Ti^{4+}_xO_4$	$Fe^{3+}[Fe^{2+}_{1+x}Fe^{3+}_{1-2x}Ti^{4+}_x]O_4$ [1])	<0.5				*55N, 55CBM*	13,	4
	$Fe^{3+}_{2-2x}Fe^{2+}_{2x-1}[Fe^{2+}_{2-x}Ti^{4+}_x]O_4$	>0.5				*55N, 55A*	20…24	
						57G, 57ANK, 57AKY,		
						61N, 64S,		
	[2])					*65RB-1, 62A-2,*		
						65ISA, 66BR		
$Fe^{3+}_{2x}Fe^{2+}_{2-x}Ti^{4+}_{1-x}O_4$		0.2				*66RB, 67ZHRB*		
		0.4				*66RB*		
						55A		
$Fe_{3-x}Ti_xO_4$	$Fe^{3+}_{0.6}Fe^{2+}_{0.4}[Fe^{2+}_{1.2}Fe^{3+}_{0.2}Ti^{4+}_{0.6}]O_4$	0.6				*67BRJ*	25	4
$LiAlTiO_4$	$Li^{1+}[Al^{3+}Ti^{4+}]O_4$		8.34			*63B-1*		
			8.30			*64B-2*		
$Li_{1+x}Al_xTi_{2-2x}O_4$	$Li^{1+}[Li^{1+}_{0.25}Al^{3+}_{0.25}Ti^{4+}_{1.5}]O_4$	0.25	8.33			*64B-1*		

$LiCd_xTi_{2-x}O_4$ 3)	$Li_{0.5}Cd_{0.5}[Li_{0.5}Ti_{1.5}]O_4$	0.5				*64B-2*		
$LiCo_xTi_{2-x}O_4$	$\approx Li_{0.5}Co_{0.5}[Li_{0.5}Ti_{1.5}]O_4$	0.5	8.36			*63B-2, 64B-2*		
$LiCrTiO_4$			8.31			*51KR*		
	$Li[Cr^{3+}Ti^{4+}]O_4$		8.32			*63B-1, 64B-2*		
			8.32	0.385		*64BL*		
$Li_{1+x}Cr_xTi_{2-2x}O_4$	$Li[Li_{0.25}Cr_{0.25}Ti_{1.5}]O_4$	0.25	8.34			*64B-1*		4
$LiCu_xTi_{2-x}O_4$	$Li_{0.7}Cu_{0.3}[Li_{0.3}Cu_{0.2}Ti_{1.5}]O_4$	0.5	8.39			*63B-2, 64B-2*		
$LiFeTiO_4$			8.342 ± 0.004			*51KR*		
	$Li^{1+}_{0.5}Fe^{3+}_{0.5}[Li^{1+}_{0.5}Fe^{3+}_{0.5}Ti^{4+}]O_4$		8.35			*63B-1*	26	4
			8.356 4)			*64B-2*		
	$Li^{1+}_{0.21}Fe^{3+}_{0.49}Ti^{4+}_{0.3}[Li^{1+}_{0.79}Fe^{3+}_{0.51}Ti^{4+}_{0.7}]O_4$ 7)		8.345			*65L*		
$Li_{0.5+x/2}Fe_{2.5-1.5x}Ti_xO_4$		1.10	8.356 } 5)			*64B-2*	27, 28	4
		1.20	8.357 } 5)					
		1.40	8.356 } 5)					
		1.50	8.357 } 5)					
		1.667	8.358 6)					
$Li_{1.25}Fe_{0.25}Ti_{1.5}O_4$ 8)	$Li_{0.75}Fe_{0.25}[Li_{0.5}Ti_{1.5}]O_4$		8.36			*64B-1*		
$LiGaTiO_4$ 9)	$Li_{0.3}Ga_{0.7}[Li_{0.7}Ga_{0.3}Ti]O_4$		8.28			*63B-2,*		
			8.29			*64B-2*		
	$Li_{1/3}Ga_{2/3}[Li_{2/3}Ga_{1/3}Ti]O_4$		8.29			*66BP*		
$Li_{1.25}Ga_{0.25}Ti_{1.5}O_4$ 8)	$Li_{0.75}Ga_{0.25}[Li_{0.5}Ti_{1.5}]O_4$		8.35			*64B-1*		
$LiMg_xTi_{2-x}O_4$ 3)	$\approx Li_{0.5}Mg_{0.5}[Li_{0.5}Ti_{1.5}]O_4$	0.5	8.37			*63B-2, 64B-2*		
$Li_{0.25}Mg_{1.72}Ti_{1.03}O_{3.8}F_{0.2}$	$Li_{0.25}Mg_{0.75}[Mg_{0.97}Ti_{1.03}]O_4$		8.436 ± 0.005	0.3850 ± 0.0005		*66O*		
$Li_{0.40}Mg_{1.55}Ti_{1.05}O_{3.7}F_{0.3}$	$Li_{0.4}Mg_{0.6}[Mg_{0.95}Ti_{1.05}]O_4$		8.414 ± 0.005	0.3845 ± 0.0005		*66O*		
$LiMnTiO_4$	$Li[Mn^{3+}Ti^{4+}]O_4$		8.30			*63B-2, 64B-2*		
$LiMn_xTi_{2-x}O_4$ 3)	$Li_{0.5}Mn_{0.5}[Li_{0.5}Ti_{1.5}]O_4$	0.5	8.44			*64B-2*		
$LiRhTiO_4$	$Li[Rh^{3+}Ti^{4+}]O_4$		8.36			*63B-1*		
			8.39			*64B-2*		
$Li_{1.5}Sb_{0.5}TiO_4$	$Li[Li_{0.5}TiSb_{0.5}]O_4$		8.35			*64B-1*		
$LiVTiO_4$ 3)	$Li[V^{3+}Ti^{4+}]O_4$		8.35			*63B-1,64B-2*		
$LiZn_xTi_{2-x}O_4$ 3)	$\approx Li_{0.5}Zn_{0.5}[Li_{0.5}Ti_{1.5}]O_4$	0.5	8.37			*63B-2, 64B-2*		
$Mg_{1+x}Fe_{2-2x}Ti_xO_4$		0.8	8.436 10)	0.375		*64B-2, 66TL, 66T*	29, 30	4
		0.65	8.431			*64B-2*		

Fußnoten siehe S. 518

Tab. 3 (Fortsetzung)

Verbindung	Kationenverteilung	x	a [Å]	u	Farbe	Lit.	Fig.	Tab.
$Mg^{2+}Fe^{3+}_{2-2x}Ti_{2x}O_4$						*67T*		
$Mg^{2+}_{2-x}Fe^{3+}_{2x}Ti^{4+}_{1-x}O_4$						*67ZHRB*		
$Mg_2Sn_xTi_{1-x}O_4$		0.5 0.3 0.1	8.548 8.506 8.467	0.384 ± 0.002		*62PM-2, 65P*	14, 15 31	4
$MnCoTiO_4$			8.49			*53Y*		
$Mn^{2+}_{1+x}Cr^{3+}_{2-2x}Ti^{4+}_xO_4$						*64RLV, 64VLR, 65VLR, 66HGLRV*	16, 32··36	4
$MnMgTiO_4$			8.48			*53Y*		
$Mn^{2+}_{1+x}V^{3+}_{2-2x}Ti^{4+}_xO_4$						*65VLR, 66HGLRV*	16,32,33,37	4
$NiCoTiO_4$			8.42			*53Y*		
$Ni_xCo_{2-x}TiO_4$		0.2 [13]	8.438 [13]			*63BPM-1*		
$Ni_{1.5}FeTi_{0.5}O_4$	$Fe_{0.7}Ti_{0.3}[Ni_{1.5}Fe_{0.3}Ti_{0.2}]O_4$ [11]					*54G-1, 54G-2, 57G*		4
$Ni^{2+}_{2-x}Fe^{3+}_{2x}Ti^{4+}_{1-x}O_4$						*67ZHRB*		
$Ni_{0.1}Mg_{1.9}TiO_4$			8.439			*67K*		
$Ni_xMg_{2-x}TiO_4$		0.2 [13]	8.435 [13]			*63BPM-1*		
$Ni_{0.1}Zn_{1.9}TiO_4$			8.460			*67K*		
$NiZnTiO_4$ [12]	$Zn[NiTi]O_4$		8.42		weiß	*53R-2, 66R* *53Y, 50BS, 68NK*		
$Ni_xZn_{2-x}TiO_4$		0.2 [13]	8.454 [13]			*63BPM-1, 63BPM-2*		
$Zn_{0.5}Cd_{0.5}Ti_{1.5}O_4$	$Zn_{0.5}Cd_{0.5}[\square_{0.5}Ti_{1.5}]O_4$		8.394 ± 0.005	4.94		*63JD*		
$Zn_{1.50}Li_{0.40}Ti_{1.10}O_{3.8}F_{0.2}$	$Zn[Li_{0.4}Zn_{0.5}Ti_{1.10}]O_{3.8}F_{0.2}$		8.450 ± 0.005	0.390 ± 0.005		*66O*		
$Zn_{1.30}Li_{0.60}Ti_{1.10}O_{3.6}F_{0.4}$	$Zn[Li_{0.6}Zn_{0.30}Ti_{1.10}]O_{3.6}F_{0.4}$		8.425 ± 0.005	0.385 ± 0.005		*66O*		
$Zn_{0.5}Mn_{0.5}Ti_{1.5}O_4$	$Zn_{0.5}Mn_{0.5}[\square_{0.5}Mn_{1.5}]O_4$		8.449 ± 0.005	4.21		*63JD*		
$Zn_2Sn_xTi_{1-x}O_4$		0.5 0.3 0.1	8.567 8.527 8.487	0.382 ± 0.003		*61PM, 65P*	17, 18 38	4

[1]) Nach [*57G*] Annahme, daß geringe Ti^{4+}-Mengen auf Oktaederplätzen sitzen, später nicht mehr bestätigt [*65ISA*].

[2]) Kationenverteilung: $0 < x < 0.2$ $Fe^{3+}[Fe^{3+}_{1-2x}Fe^{2+}_{1+x}Ti^{4+}_x]O_4$;
$0.2 < x < 0.8$ $Fe^{3+}_{1.2-x}Fe^{2+}_{x-0.2}[Fe^{3+}_{0.8-x}Fe^{2+}_{1.2}Ti^{4+}_x]O_4$;
$0.8 < x < 1.0$ $Fe^{3+}_{2-2x}Fe^{2+}_{2x-1}[Fe^{2+}_{2-x}Ti^{4+}_x]O_4$ [*66BR*].

[3]) Überstruktur [*64B-2*], [*63B-2*].

[4]) Geglühte Probe; $a = 3.59$ Å bei abgeschreckter Probe [*64B-2*].

[5]) Geglühte Probe [*64B-2*].

[6]) Abgeschreckte Probe $a = 8.359$ Å [*64B-2*].

[7]) $Li_{x/2}Fe_{1-x/2}[Li_{0.5}Fe_{1.5-x}Ti_x]O_4$ [*64B-2*].

[8]) Spinell mit Überstruktur auf Oktaederplätzen [*64B-1*].

[9]) Bei langsam abgekühlter Probe schwacher Überstrukturcharakter [*63B-2*].

[10]) Langsam abgekühlte Probe [*64B-2*].

[11]) Allgemeine Kationenverteilung $Fe_{1-x-y}Ni_yTi_x[Ni_{1.5-y}Fe_{x+y}Ti_{0.5-x}]O_4$ [*54G-1*].

[12]) Inverser Spinell [*66R*].

[13]) Weitere Werte für a bei verschiedenen x-Werten siehe [*63BPM-1*].

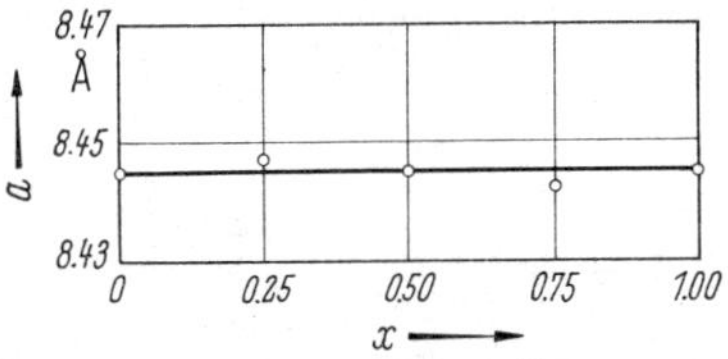

Fig. 10. $Co_{2-2x}Mg_{2x}TiO_4$. Gitterparameter a vs. Zusammensetzungsparameter x [*53R-2*].

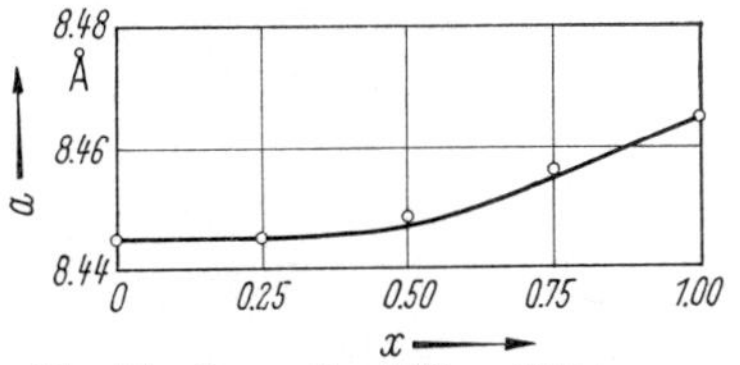

Fig. 11. $Co_{2-2x}Zn_{2x}TiO_4$. Gitterparameter a vs. Zusammensetzungsparameter x [*53R-2*].

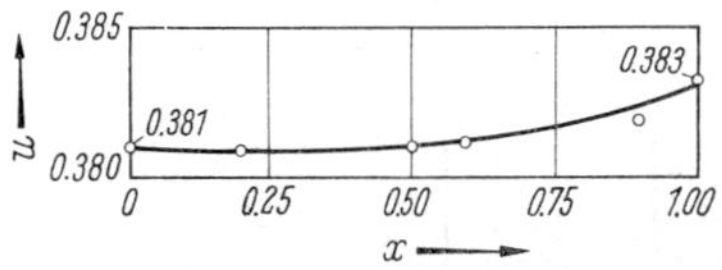

Fig. 12. $Co_{2-2x}Zn_{2x}TiO_4$. Anionenparameter (Sauerstoff) u vs. Zusammensetzungsparameter x [*65P*].

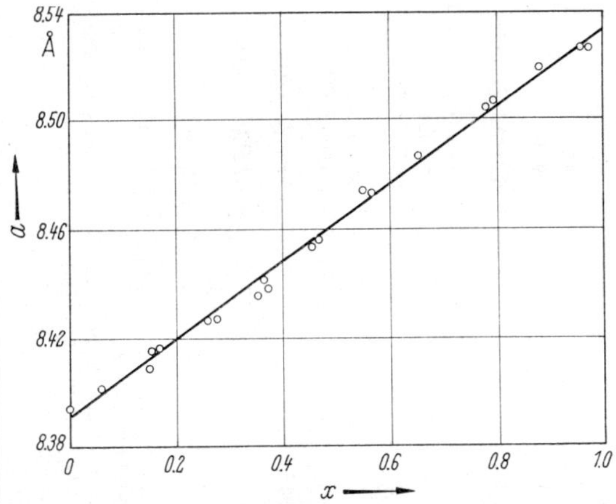

Fig. 13. $Fe^{3+}_{2-2x}Fe^{2+}_{1+x}Ti^{4+}_xO_4$. Gitterparameter a vs. Zusammensetzungsparameter x [*61N*].

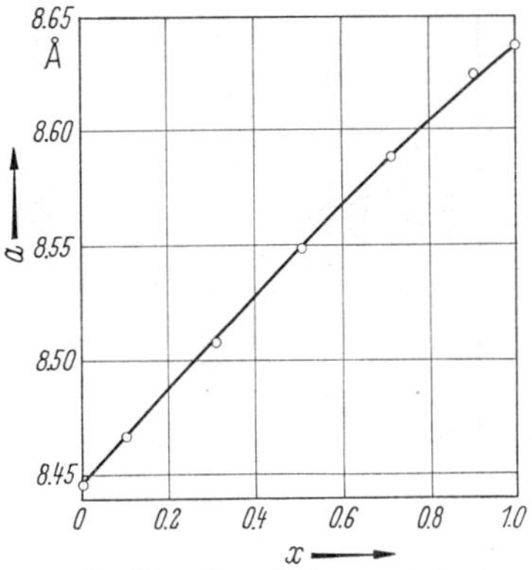

Fig. 14. $Mg_2Sn_xTi_{1-x}O_4$. Gitterparameter a vs. Zusammensetzungsparameter x [*62PM-2*].

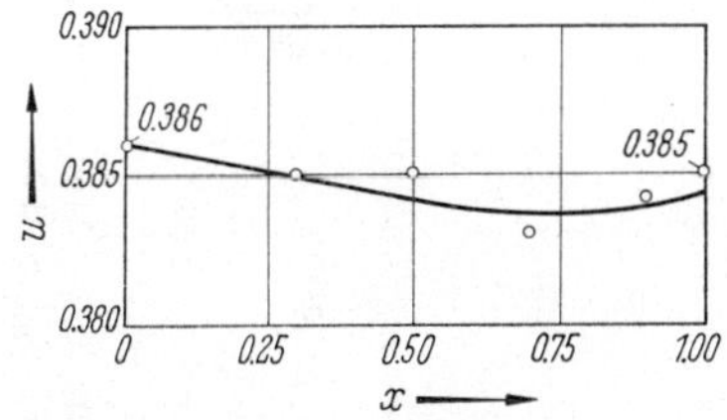

Fig. 15. $Mg_2Sn_xTi_{1-x}O_4$. Anionenparameter (Sauerstoff) u vs. Zusammensetzungsparameter x [*65P*].

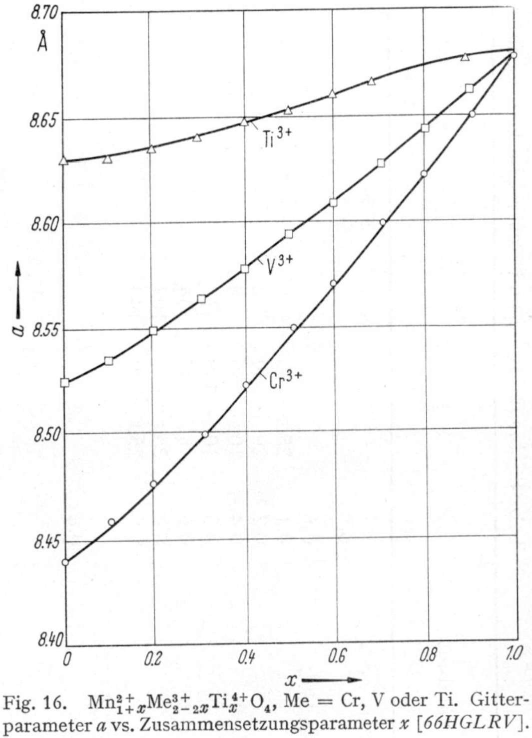

Fig. 16. $Mn^{2+}_{1+x}Me^{3+}_{2-2x}Ti^{4+}_xO_4$, Me = Cr, V oder Ti. Gitterparameter a vs. Zusammensetzungsparameter x [*66HGLRV*].

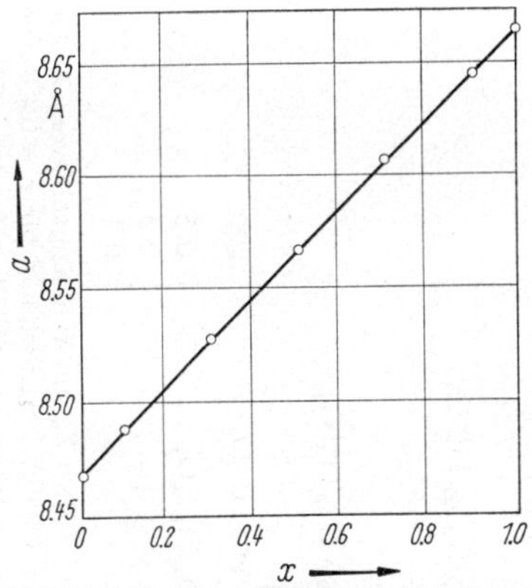

Fig. 17. $Zn_2Sn_xTi_{1-x}O_4$. Gitterparameter a vs. Zusammensetzungsparameter x [*61PM*].

Fig. 18. $Zn_2Sn_xTi_{1-x}O_4$. Anionenparameter (Sauerstoff) u vs. Zusammensetzungsparameter x [*65P*]. →

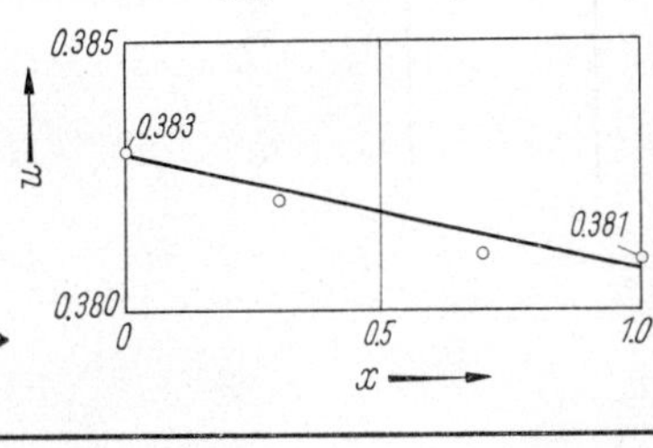

Tab. 4. Magnetische und elektrische Eigenschaften ternärer Ti^{4+} Oxid-Spinelle

Verbindung	x	T °K	Θ_a °K	Θ_C °K	C_m $\frac{cm^3\ °K}{mol}$	p_m μ_B	χ_m 10^{-6} emu/mol	Lit.	Fig.
$Co_{1+x}Fe_{2-2x}Ti_xO_4$	0.7			275		1.31		*64B-2*	19
	0.8			200		0.89		*64B-2*	
$Co_{2-2x}Zn_{2x}TiO_4$	0.6							*62S*	
	0.1	250 ··· 850	≈ 0		0.615		[1]	*65P*	1
	0.2 [1]	200 ··· 850	≈ 2		1.23		[1]		
	0.4 [1]	100 ··· 850	8		2.41				
	0.5	100 ··· 900	17		2.97				
	0.6	80 ··· 900	34		3.50				
	0.8	250 ··· 900	102		4.76				
	0.9	250 ··· 900	140		5.45				
$Fe^{3+}_{2-2x}Fe^{2+}_{1+x}Ti^{4+}_xO_4$						[2]		*65RB-1, 66BR* [2]),	20 ··· 24
$Fe_{3-x}Ti_xO_4$	0.6							*67BRJ*	25
$LiCrTiO_4$			−81		1.35			*64BL*	
$LiFeTiO_4$ [3]				400 [4]		0.08 [4]		*64B-2, 51KR*	26
$Li_{0.5+x/2}Fe_{2.5-1.5x}Ti_xO_4$	1.1			320 [5]		0.25 [5]		*64B-2*	27, 28
	1.2			≈270 [5]		≈0.2 [5]			
	1.4			≈125 [5]					
$Mg_{1+x}Fe_{2-2x}Ti_xO_4$	0.65			170 [6]		0.4 [6]		*64B-2*	29, 30
$Mg_2Sn_xTi_{1-x}O_4$								*62PM-2, 65P*	31
$Mn^{2+}_{1+x}Cr^{3+}_{2-2x}Ti^{4+}_xO_4$								*64RLV, 64VLR* *65VLR* *66HGLRV*	32 ··· 36
$Mn^{2+}_{1+x}V^{3+}_{2-2x}Ti^{4+}_xO_4$								*65VLR* *66HGLRV*	32, 33, 37
$Ni^{2+}_{1.5}Fe^{3+}Ti^{4+}_{0.5}O_4$				≈565		1.45 [7]		*54G-1, 57G*	
$Zn_2Sn_xTi_{1-x}O_4$	0.5						−73	*61PM, 65P*	
	0.3						−69		
	0.1						−60		38

[1]) Für $x = 0.1$ und $x = 0.2$ ist χ_m in Abhängigkeit von T dargestellt; für $x = 0.2$ ist $\Theta_a = 143$ °K, $C_m = 0.98$ cm³ °K/mol und für $x = 0.4$ ist $\Theta_a = 90$ °K, $C_m = 2.21$ cm³ °K/mol für Temperaturen $T > 850$ °K gefunden [*65P*].

[2]) Werte abhängig von der Kationenverteilung [*57G*]. Siehe auch [*55CBH*], [*55N*], [*57G*], [*61N*], [*62A-2*], [*64S*].

[3]) Ferromagnetisch [*51KR*].

[4]) Von 1000 °C abgeschreckte Proben: $p_m = 0.14\ \mu_B$ und $\Theta_C = 380$ °K [*64B-2*].

[5]) Geglühte Proben [*64B-2*].

[6]) Langsam abgekühlte Probe; von 1200 °C abgeschreckte Probe: für $x = 0.65$ ist $p_m = 0.3\ \mu_B$ und $\Theta_C = 170$ °K gefunden [*64B-2*].

[7]) Aus Messungen für $T > 20$ °K [*54G-1*].

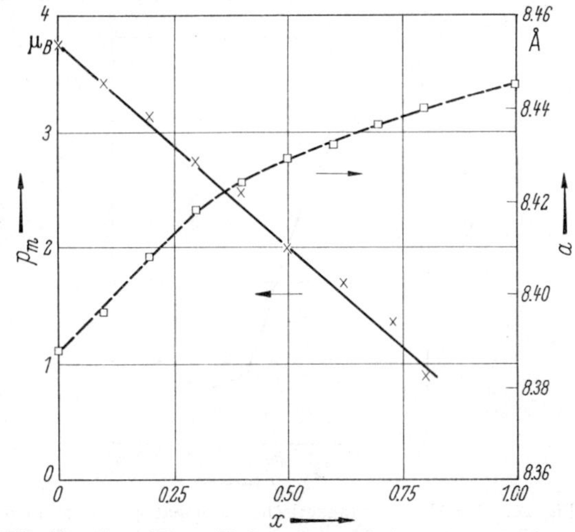

Fig. 19. $Co_{1+x}Fe_{2-2x}Ti_xO_4$. Magnetisches Moment pro Molekül p_m und Gitterparameter a vs. Zusammensetzungsparameter x [64B-2].

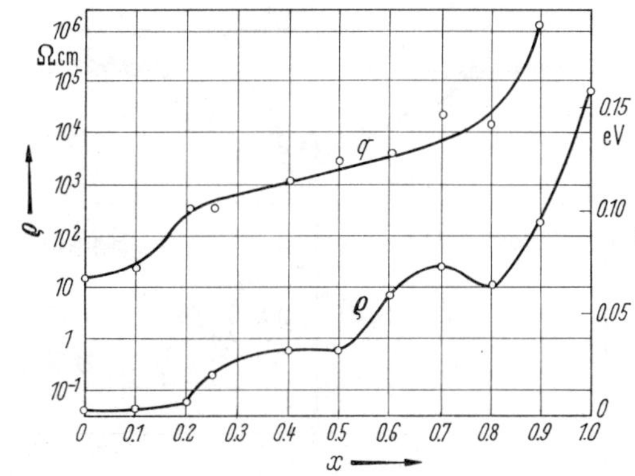

Fig. 20. $Fe^{3+}_{2-2x}Fe^{2+}_{1+x}Ti^{4+}_xO_4$. Der spezifische elektrische Widerstand ϱ bei Raumtemperatur und die Aktivierungsenergie q für die elektrische Leitfähigkeit vs. Zusammensetzungsparameter x [66BR].

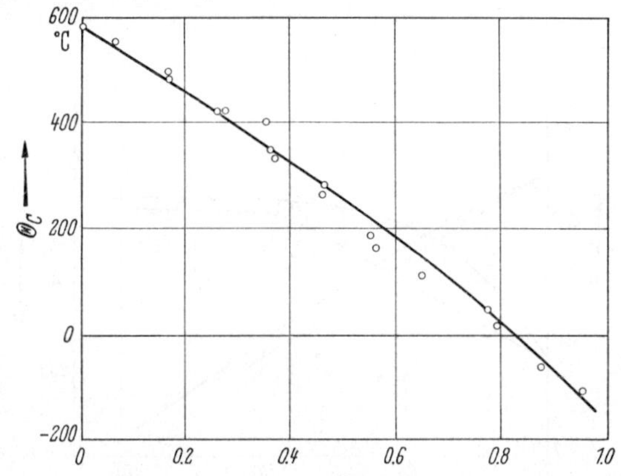

Fig. 21. $Fe^{3+}_{2-2x}Fe^{2+}_{1+x}Ti^{4+}_xO_4$. Curie-Temperatur Θ_C vs. Zusammensetzungsparameter x [61N].

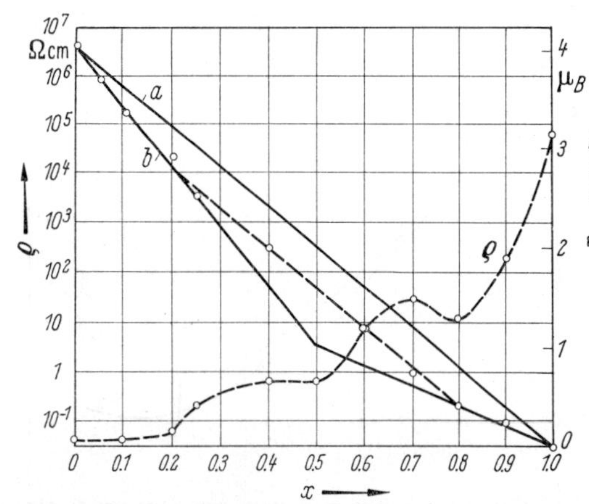

Fig. 22. $Fe^{3+}_{2-2x}Fe^{2+}_{1+x}Ti^{4+}_xO_4$. Magnetisches Moment pro Molekül p_m bei 77 °K und der spezifische elektrische Widerstand ϱ bei Raumtemperatur vs. Zusammensetzungsparameter x [65RB-1]. Die voll ausgezogenen Geraden sind gezeichnet gemäß der Arbeiten von (a) AKIMOTO [62A-2] und (b) NÉEL-CHEVALLIER [55N], [55CBM].

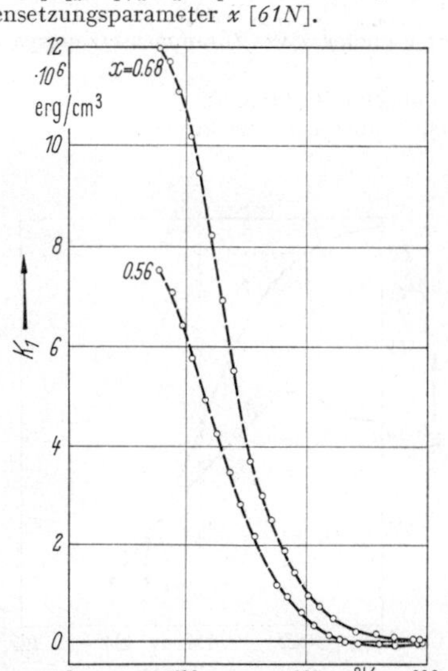

Fig. 23. $Fe^{3+}_{2-2x}Fe^{2+}_{1+x}Ti^{4+}_xO_4$. Kristallanisotropiekonstante K_1 vs. Temperatur T für $x = 0.68$ und $x = 0.56$ ermittelt nach dem Drehmomentverfahren bei einer Meßfeldstärke $H = 15$ kOe [64S].

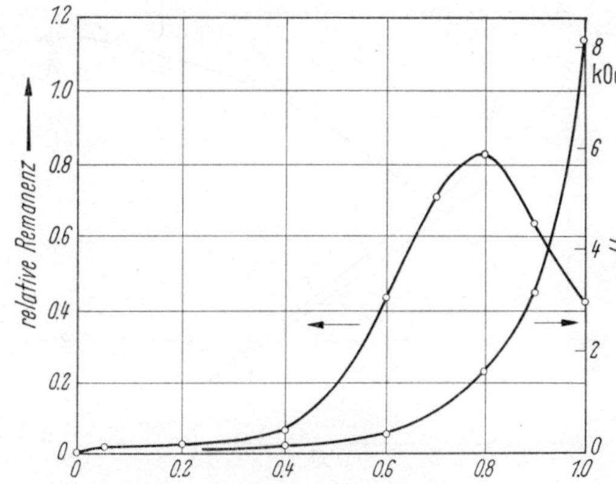

Fig. 24. $Fe^{3+}_{2-2x}Fe^{2+}_{1+x}Ti^{4+}_xO_4$. Koerzitivkraft H_c und relative Remanenz bei 77 °K vs. Zusammensetzungsparameter x. Das Verhältnis zwischen remanenter Magnetisierung und Sättigungsmagnetisierung ist als relative Remanenz bezeichnet [66BR].

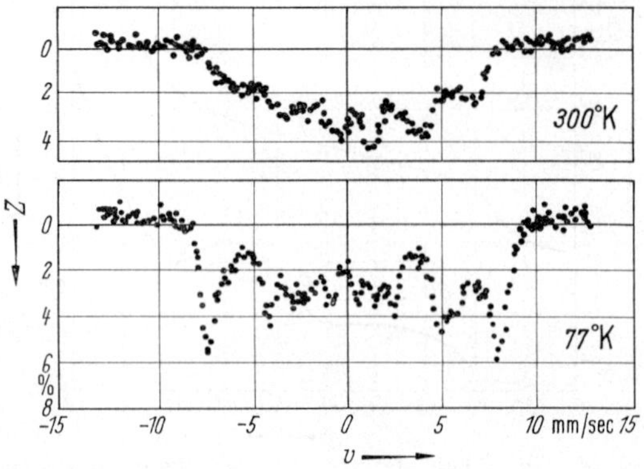

Fig. 25. $Fe_{2.4}Ti_{0.6}O_4$. Mössbauerspektren bei 300 °K und 77 °K [*67BRJ*].

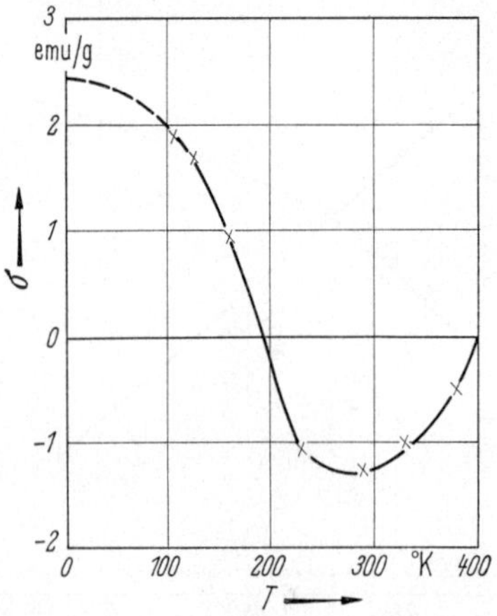

Fig. 26. $LiFeTiO_4$. Magnetisches Moment pro Gramm σ vs. Temperatur T für eine langsam abgekühlte Probe. Nulldurchgang von σ bedeutet gleiche Untergittermagnetisierung [*64B-2*].

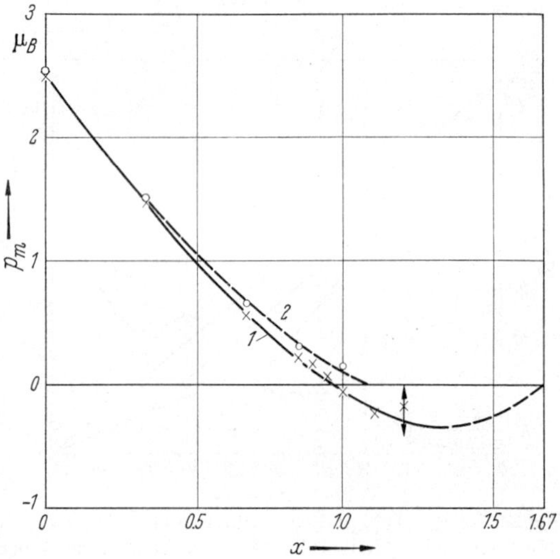

Fig. 27. $Li_{0.5+0.5x}Fe_{2.5-1.5x}Ti_xO_4$. Magnetisches Moment p_m vs. Zusammensetzungsparameter x [*64B-2*].
1 langsam abgekühlte Proben;
2 abgeschreckte Proben.

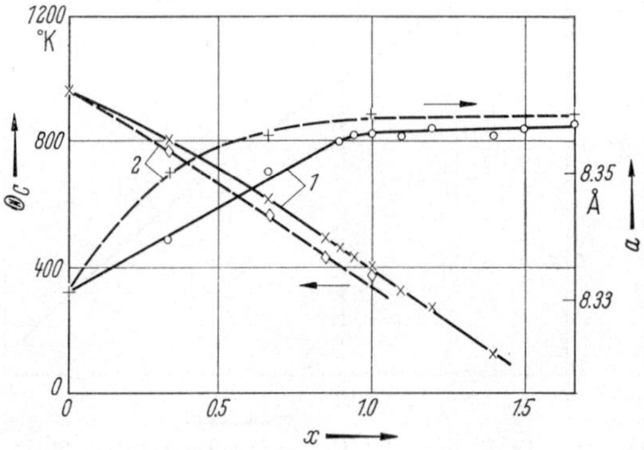

Fig. 28. $Li_{0.5+0.5x}Fe_{2.5-1.5x}Ti_xO_4$. Curie-Temperatur Θ_C und Gitterparameter a vs. Zusammensetzungsparameter x. [*64B-2*].
1 langsam abgekühlte Proben;
2 von 1000 °C abgeschreckte Proben.

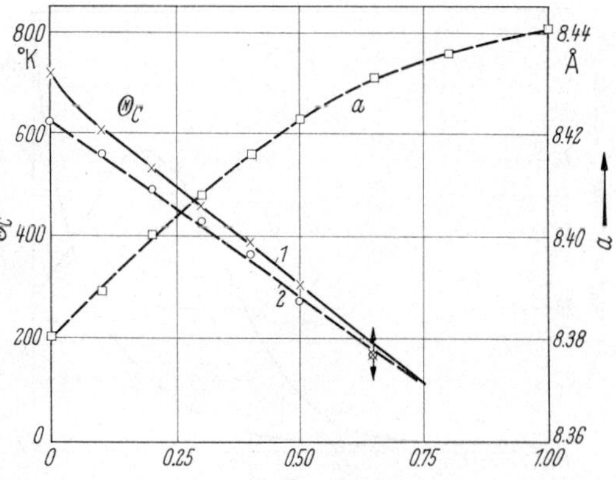

Fig. 29. $Mg_{1+x}Fe_{2-2x}Ti_xO_4$. Curie-Temperatur Θ_C und Gitterparameter a vs. Zusammensetzungsparameter x [*64B-2*].
1 langsam abgekühlte Proben;
2 von 1200 °C abgeschreckte Proben.

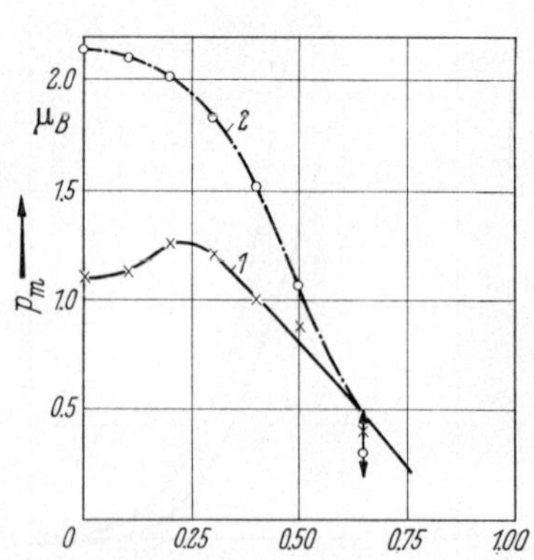

Fig. 30. $Mg_{1+x}Fe_{2-2x}Ti_xO_4$. Magnetisches Moment p_m vs. Zusammensetzungsparameter x [*64B-2*].
1 langsam abgekühlte Proben;
2 von 1200 °C abgeschreckte Proben.

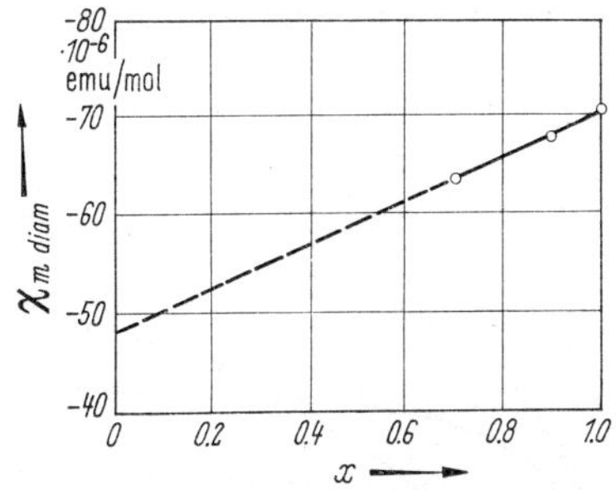

Fig. 31. $Mg_2Sn_xTi_{1-x}O_4$. Molare diamagnetische Suszeptibilität $\chi_{m\,diam}$ vs. Zusammensetzungsparameter x [62PM-2].

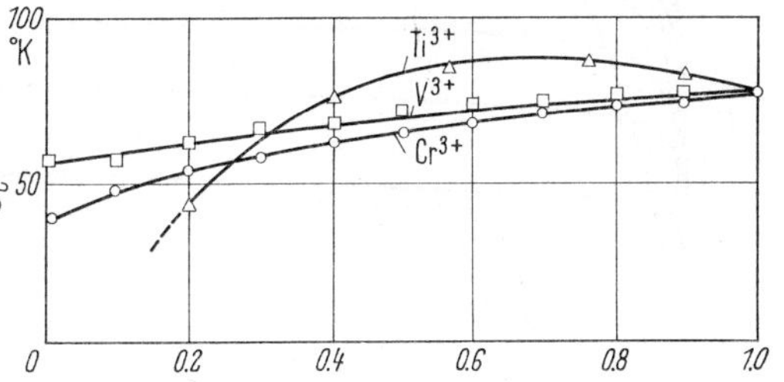

Fig. 32. $Mn^{2+}_{1+x}Me^{3+}_{2-2x}Ti^{4+}_xO_4$, Me = Cr, V, Ti. Curie-Temperatur Θ_C vs. Zusammensetzungsparameter x [66HGLRV].

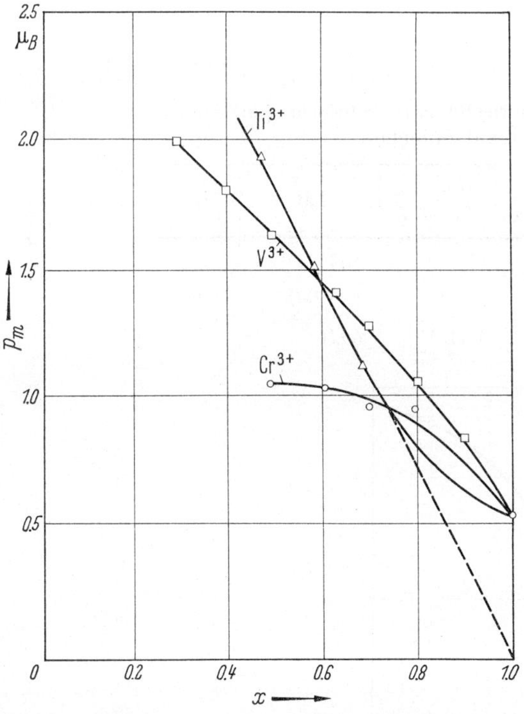

Fig. 33. $Mn^{2+}_{1+x}Me^{3+}_{2-2x}Ti^{4+}_xO_4$, Me = Cr, V, Ti. Magnetisches Moment p_m vs. Zusammensetzungsparameter x [66HGLRV].

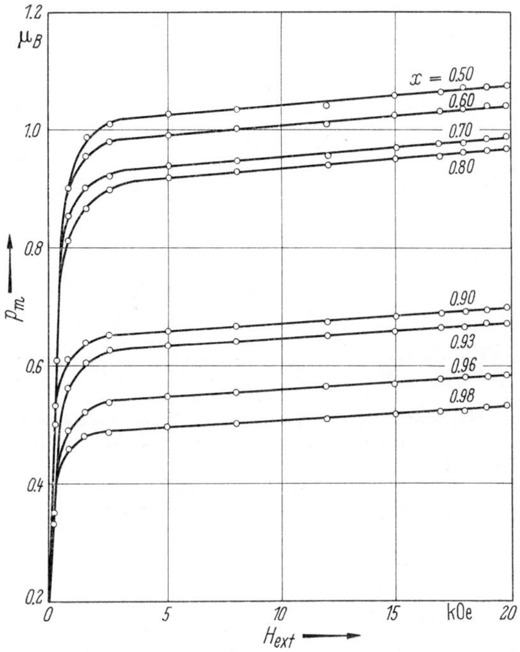

Fig. 34. $Mn^{2+}_{1+x}Cr^{3+}_{2-2x}Ti^{4+}_xO_4$, $0.5 \leq x \leq 0.98$. Magnetisches Moment p_m vs. äußeres Feld H_{ext} bei 20.4 °K. Parameter x [64VLR].

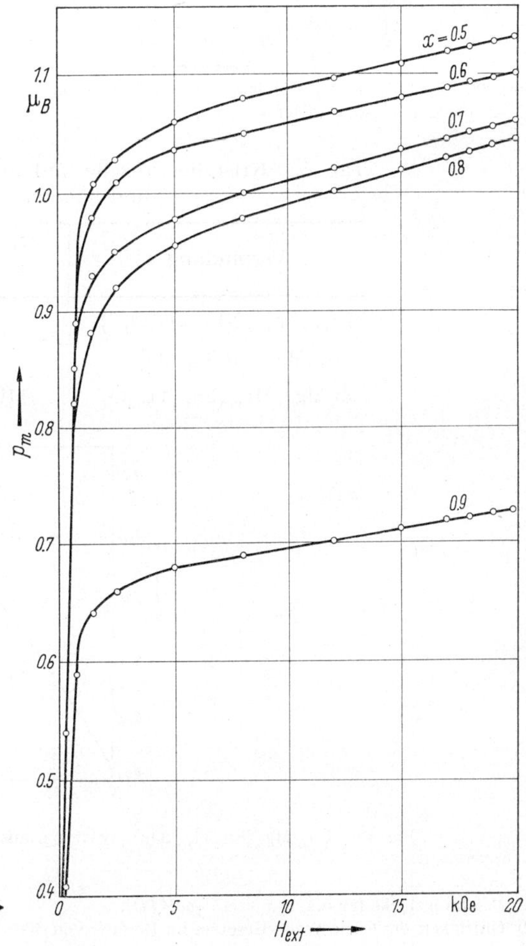

Fig. 35. $Mn^{2+}_{1+x}Cr^{3+}_{2-2x}Ti^{4+}_xO_4$, $0.5 \leq x \leq 0.9$. Magnetisches Moment p_m vs. äußeres Feld H_{ext} bei 4.2 °K [64RLV]. →

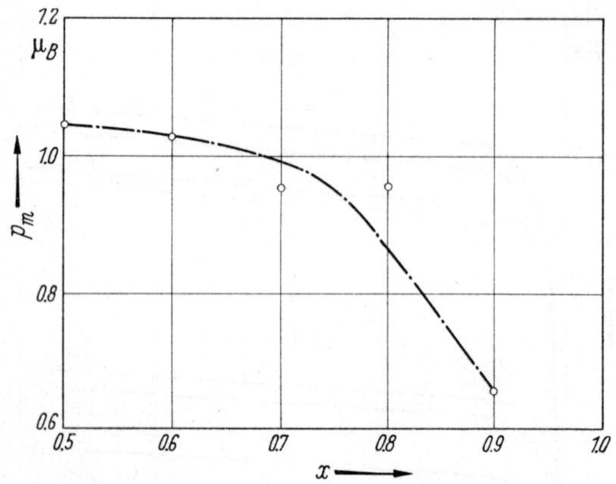

Fig. 36. $Mn^{2+}_{1+x}Cr^{3+}_{2-2x}Ti^{4+}_{x}O_4$. Extrapoliertes magnetisches Moment p_m bei 0 °K vs. Zusammensetzungsparameter x [*64RLV*].

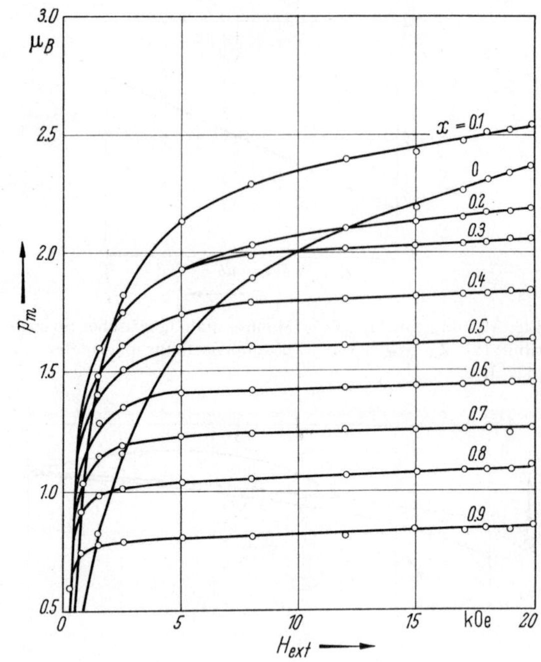

Fig. 37. $Mn^{2+}_{1+x}V^{3+}_{2-2x}Ti^{4+}_{x}O_4$. Magnetisches Moment p_m bei 20.4 °K vs. äußeres Feld H_{ext}. Parameter x [*65VLR*].

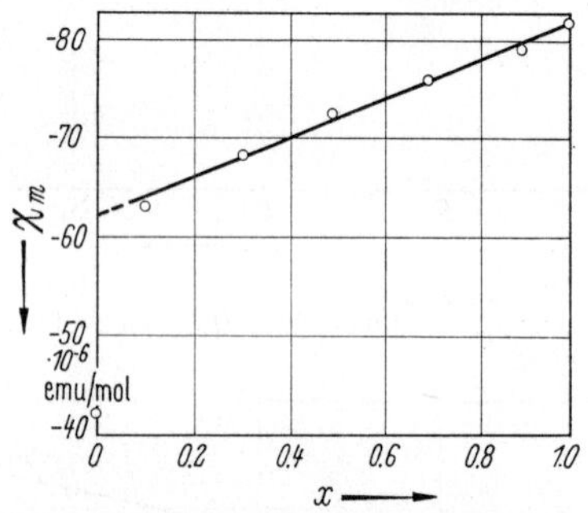

Fig. 38. $Zn_2Sn_xTi_{1-x}O_4$. Molare Suszeptibilität χ_m vs. Zusammensetzungsparameter x [*61PM*].

Tab. 5. Kristallografische und magnetische Eigenschaften quaternärer und höherer Ti^{4+} Oxid-Spinelle

Verbindung	Θ_a °K	C_m cm³ °K/mol	Lit.	Fig.
$Li_{1-x}Zn_{1+x}Nb_{1-x}Ti_xO_4$ [1]			*66VJD*	
$Co_{0.2}Mg_{1.8}Sn_xTi_{1-x}O_4$			*58SBB*	39
$ZnMg_{0.5}Mn_{0.5}Fe_{0.5}Ti_{0.5}O_4$	−110	3.41 [2]	*65B*	

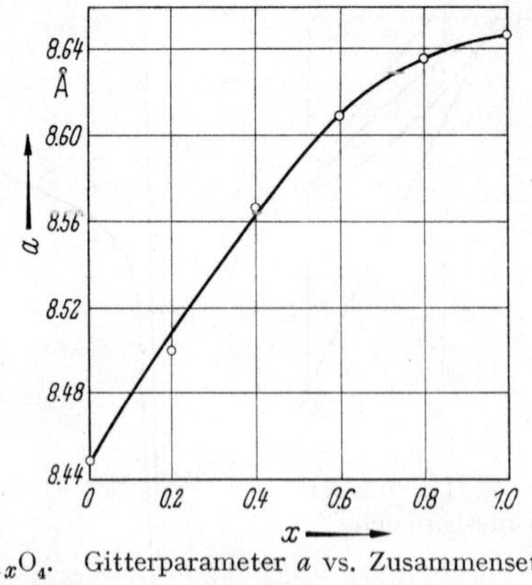

Fig. 39. $Co_{0.2}Mg_{1.8}Sn_xTi_{1-x}O_4$. Gitterparameter a vs. Zusammensetzungsparameter x [*58SBB*].

[1]) Mischungslücke für $0.4 < x < 0.7$ [*66VJD*].
[2]) Gültigkeit des Curie-Weiss-Gesetzes im Bereich von 300···1300 °K [*65B*].

Tab. 6. Kristallografische Eigenschaften von Ti^{3+} und $Ti^{3+}Ti^{4+}$ Oxid-Spinellen

Verbindung	Kationenverteilung	x	a Å	ϱ_p g/cm³	u	Farbe	Lit.	Fig.	Tab.
$MgTi_2O_4$			8.474 ± 0.002	4.05 ± 0.05			*62L*		
$MnTi_2O_4$			8.600 ± 0.002	4.37 ± 0.05			*61LH, 62L*	41	7
							63LV		
	$Mn[Ti_2^{3+}]O_4$						*64VLR*		
							65LRV *)		
$ZnTi_2O_4$							*54E, 56HH*		
$Li_{1.25}Ti_{0.25}^{3+}Ti_{1.5}^{4+}O_4$					0.385	weiß	*53BD*		
$Mn_{1+x}Ti_{2-2x}^{3+}Ti_x^{4+}O_4$							*62L, 65LRPV*	40, 42	
	$Mn_{1-y}^{2+}Ti_y^{4+}[Ti_{2-2x}^{3+}Mn_{x+y}^{2+}Ti_{x-y}^{4+}]O_4$	0.54	8.656				*63VL, 63LV*	43	7
		0.51	8.660				*63VL*		
		0.63	8.660						
		0.67	8.665						
		0.78	8.675						
		0.89	8.681						
	$Mn^{2+}[Mn_x^{2+}Ti_{2-2x}^{3+}Ti_x^{4+}]O_4$						*64VLR, 65LRV, 66HGLRV*		
$Mn_{1+x}^{2+}Cr_{2-2x}^{3+}Ti_x^{4+}O_4$	$Mn^{2+}[Mn_x^{2+}Cr_{2-2x}^{3+}Ti_x^{4+}]O_4$						*66HGLRV* *)		
$Mn_{1+x}^{2+}Mn_{2-2x}^{3+}Ti_x^{4}O_4$							*66HGLRV*		
$Mn_{1+x}^{2+}V_{2-2x}^{3+}Ti_x^{4+}O_4$	$Mn^{2+}[Mn_x^{2+}V_{2-2x}^{3+}Ti_x^{4+}]O_4$						*66HGLRV*		

*) Siehe auch [*65LRPV*].

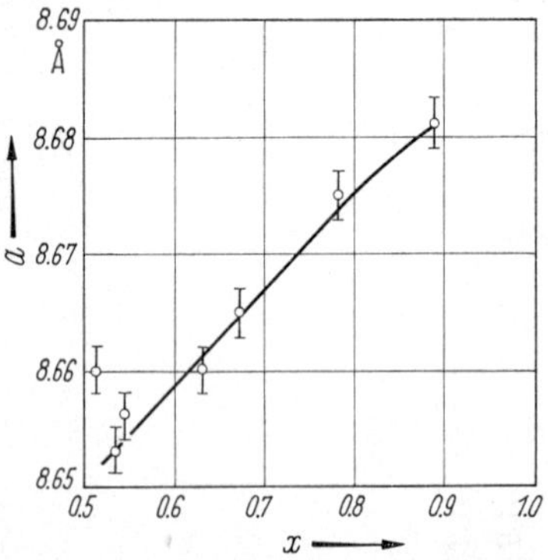

Fig. 40. $Mn^{2+}_{1+x}Ti^{3+}_{2-2x}Ti^{4+}_{x}O_4$. Gitterparameter a vs. Zusammensetzungsparameter x [*63VL*].

Tab. 7. Magnetische Eigenschaften von Ti^{3+} und $Ti^{3+}Ti^{4+}$ Oxid-Spinellen

Verbindung	x	Θ_C °K	p_m μ_B	Lit.	Fig.
$MnTi_2O_4$				*63LV*	41
$Mn^{2+}_{1+x}Ti^{3+}_{2-2x}Ti^{4+}_{x}O_4$	0.54	100		*63LV*	42, 43
	0.51		1.71	*63VL*	
	0.54		1.40		
	0.63		1.40		
	0.67		1.32		
	0.78		0.94		
	0.89		0.62	*64VLR*	
				65LRV, 65LRPV, 66HGLRV	
$Mn^{2+}_{1+x}Cr^{3+}_{2-2x}Ti^{4+}_{x}O_4$				*65LRPV*	

Fig. 41. $MnTi_2O_4$. Magnetisierung σ vs. äußeres Feld H_{ext} → gemessen bei verschiedenen Temperaturen [*63LV*].

2, 3, 4 Proben in 2 h von 1100 °C auf 700 °C abgekühlt, dann normal im Ofen abgekühlt.

1 Probe von 1100 °C abgekühlt, 15 h bei 800 °C und 30 h bei 700 °C gehalten, danach normal im Ofen abgekühlt.

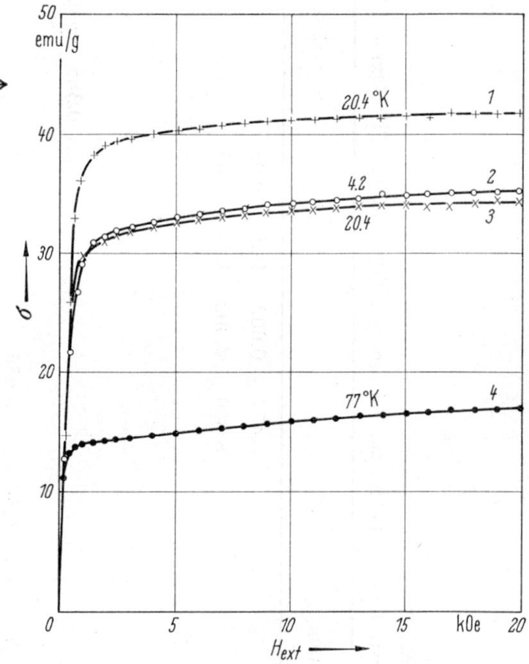

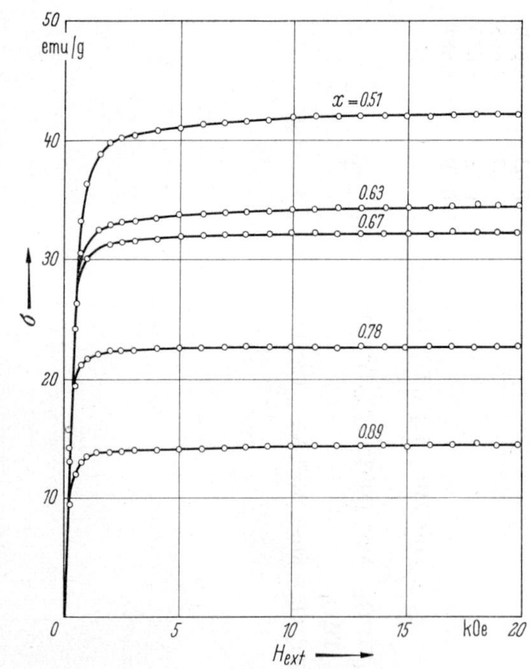

Fig. 42. $Mn^{2+}_{1+x}Ti^{3+}_{2-2x}Ti^{4+}_{x}O_4$. Magnetisierung σ bei 20.4 °K vs. äußeres Feld H_{ext}. Parameter x [*63VL*].

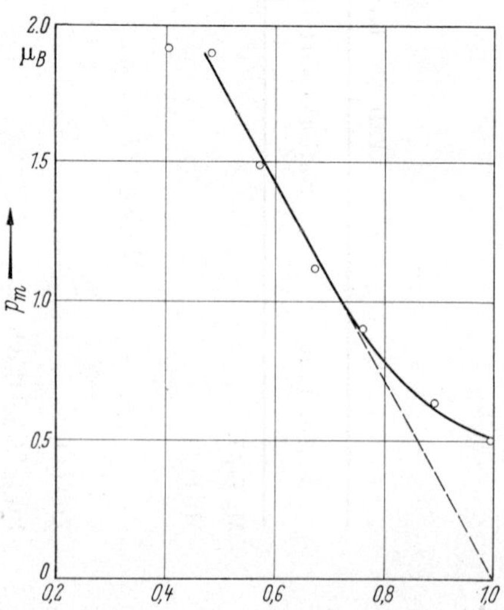

Fig. 43. $Mn^{2+}_{1+x}Ti^{3+}_{2-2x}Ti^{4+}_{x}O_4$. Magnetisches Moment p_m vs. → Zusammensetzungsparameter x [*65LRV*].

Tab. 8. Kristallografische Eigenschaften binärer und ternärer Ti^{3+} und Ti^{4+} Sulfo- und Seleno-Spinelle

Verbindung	Kationen-verteilung	x	a Å	ϱ_X g/cm³	ϱ_p g/cm³	u	Farbe	Lit.	Fig.	Tab.
$CuTi_2S_4$			9.91					*53HH*	14in 6.3.9	
	$Cu[Ti_2^{3+}]S_4$		9.88 ± 0.008	3.953	3.80	0.382		*56HH, 64B-2*		
			9.994 ± 0.002					*65BRW*		
			9.88					*67MSL*		9
								67LS, 68LS *)		
$CuCoTiS_4$	$Cu^{1+}[Co^{3+}Ti^{4+}]S_4$							*68LS*		9
$CuCrTiS_4$	$Cu^{1+}[Cr^{3+}Ti^{4+}]S_4$		9.95	3.930	3.90			*67HS*		
			9.95			0.380	schwarz	*68SEH*		
								67LS, 68LS		9
								68Y, 68S		
$CuCr_{1+x}Ti_{1-x}S_4$		$0 < x \leq 1$						*68LS* **)	44···46	9
$CuCrTiSe_4$			10.45	5.575	5.42			*67HS*		
	$Cu^{1+}[Cr^{3+}Ti^{4+}]Se_4$		10.45	5.572		0.380	schwarz	*68SEH*		

Tab. 9. Magnetische und elektrische Eigenschaften binärer und ternärer Ti^{3+} und Ti^{4+} Sulfo-Spinelle

Verbindung	x	ϱ Ωcm	q eV	S $\frac{\mu V}{°K}$	χ_m $\frac{10^{-3}\,emu}{mol}$	C_m $\frac{cm^3\,°K}{mol}$	Θ_a °K	Lit.	Fig.
$CuTi_2^{3+}S_4$		$4 \cdot 10^{-4}$	[1])	− 11.8 [2])				*65BRW*	14in 6.3.9
					0.3 ···0.6 [3])			*67LS*	
					0.43···0.45 [3])			*68LS*	
					0.25···0.27 [4])			*68LS*	
								67MSL [5]), *68Y*	
$CuCoTiS_4$		0.125		+110	0.28···0.29 [3])			*68LS*	
					0.10···0.11 [4])			*68LS*	
$CuCrTiS_4$ [10])		$5 \cdot 10^{-3}$ [7])				1.97	0 [8])	*67LS*	
		$1 \cdot 10^{-2}$ [9])		−250		1.87	4 [6])	*68LS, 68Y, 68S*	
$CuCr_{1+x}Ti_{1-x}S_4$	−0.2	$0.63 \cdot 10^{-2}$		− 58		1.39	37	*68LS*	44···46
	−0.4	$0.2 \cdot 10^{-2}$		− 30		1.03	45	*68LS* **)	7 in 6.3.3

[1]) Metallisch [*65BRW*].
[2]) Bei 25 °C [*65BRW*].
[3]) Korrigiert für Diamagnetismus [*67LS*],[*68LS*].
[4]) Beobachtete molare Suszeptibilität [*68LS*].
[5]) Bei Temperaturen über 0.05 °K keine Supraleitfähigkeit; *n*-Leiter [*67MSL*].
[6]) Paramagnetisch bis 4.4 °K hinab [*68LS*].
[7]) Temperaturunabhängig, gemessen zwischen 80 und 300 °K [*67LS*].
[8]) An anderer Probe $\Theta_a = 25$ °K gemessen [*67LS*].
[9]) Bei Raumtemperatur [*68LS*].
[10]) Der Spinell $CuCrTiS_4$ ist unmagnetisch und ein Isolator [*68S*].
*) Siehe auch [*68Y*].
**) Siehe auch [*65BRW*].

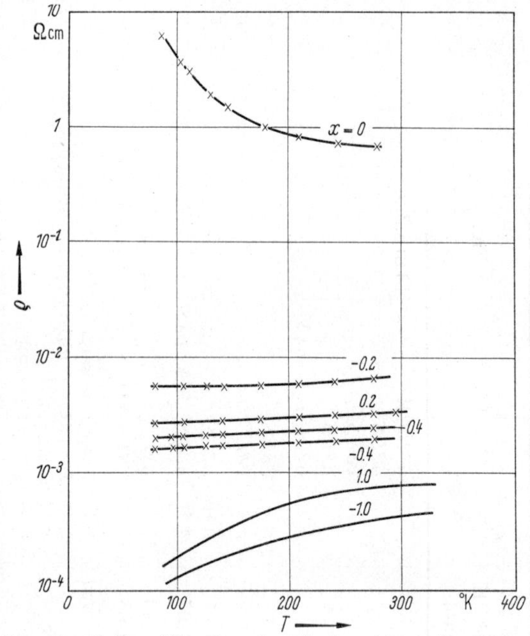

Fig. 44. $CuCr_{1+x}Ti_{1-x}S_4$. Spezifischer elektrischer Widerstand ϱ vs. Temperatur T [*68LS*]; für $x = 1.0$ und -1.0 [*65BRW*].

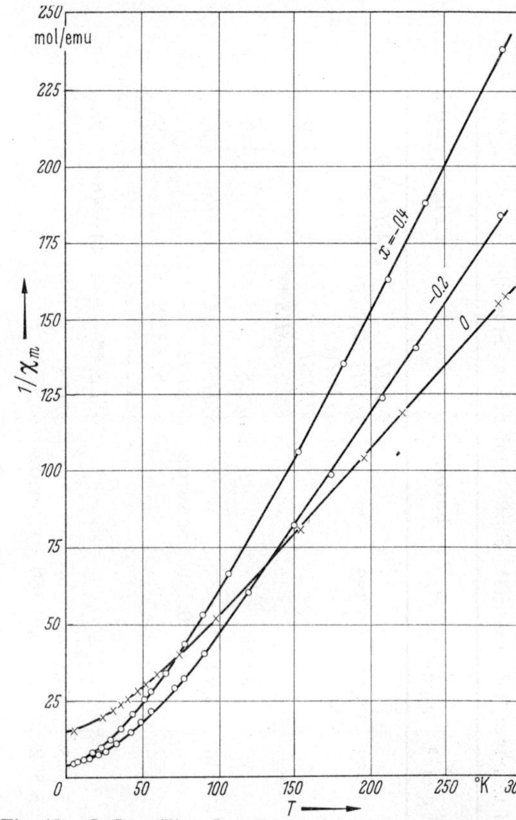

Fig. 45. $CuCr_{1+x}Ti_{1-x}S_4$. Reziproke molare Suszeptibilität χ_m^{-1} vs. Temperatur T. Parameter x [*68LS*].

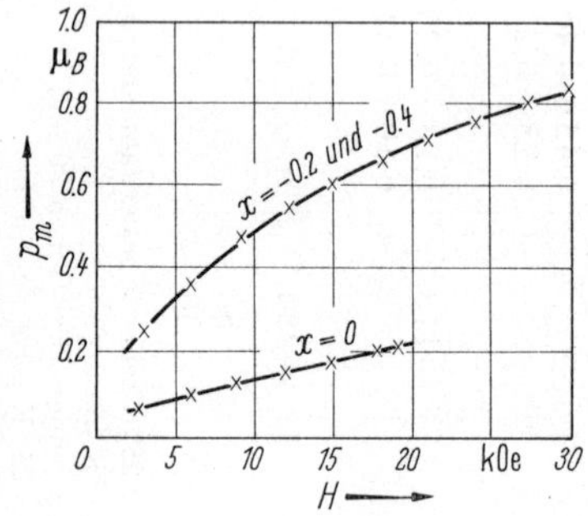

←
Fig. 46. $CuCr_{1+x}Ti_{1-x}S_4$. Magnetisches Moment p_m bei 4.4 °K vs. magnetische Feldstärke H. Parameter x [*68LS*].

6.3.14 Literatur zu 6.3.9 ··· 6.3.13

13···26S	Strukturbericht **1** (1913···26) 420.
25GBL	Goldschmidt, V. M., T. Barth, und G. Lunde: Geochemische Verteilungsgesetze V (1925).
26A	Aminoff, G.: Z. Krist. **64** (1926) 475
27JW	de Jong, W. F., und H. W. V. Willems: Z. Anorg. Allgem. Chem. **161** (1927) 311.
28HA	Hendricks, S. B., und W. H. Albrecht: Chem. Ber. **61** (1928) 2153.
28JH	de Jong, W. F., und A. Hoog: Z. Krist. **66** (1928) 168.
28···32S	Strukturbericht **2** (1928···32) 468, 478, 479.
29BM	le Blanc, M., und E. Möbius: Z. Phys. Chem. **142** (1929) 151.
29HK	Holgersson, S., und A. Karlsson: Z. Anorg. Allgem. Chem. **183** (1929) 384.
29NP-1	Natta, G., und L. Passerini: Gazz. Chim. Ital. **59** (1929) 280.
29NP-2	Natta, G., und L. Passerini: Gazz. Chim. Ital. **59** (1929) 620.
30P	Passerini, L.: Gazz. Chim. Ital. **60** (1930) 957.
30T	Taylor, N. W.: Z. Phys. Chem. **9** (1930) 241.
31HH	Holgersson, S., und A. Herrlin: Z. Anorg. Allgem. Chem. **198** (1931) 69.
31NP	Natta, G., und L. Passerini: Atti Accad. Nazl. Lincei, Rend. **14** (1931) 38.
32BP	Barth, T. F. W., und E. Posnjak: Z. Krist. **82** (1932) 325.
34KT	Krause, J., und W. Thiel: Ber. Deut. Keram. Ges. **15** (1934) 101.
35HB	Hüllsmann, O., und W. Biltz: Z. Anorg. Chem. **224** (1935) 73.
36VB	Verwey, E. J. B., und J. H. de Boer: Rec. Trav. Chim. **55** (1936) 531.
38CN	Cole, S. S., und W. K. Nelson: J. Phys. Chem. **42** (1938) 245.
38LW	Lundquist, D., und A. Westgren: Z. Anorg. Chem. **239** (1938) 85.
39HBM	Heimbrecht, M., W. Biltz, und K. Meisel: Z. Anorg. Chem. **242** (1939) 229.

43E	Ernst, T.: Z. Angew. Min. **4** (1943) 394.
47M	Mason, B.: Am. Mineralogist **32** (1947) 426.
47SR	Structure Rep. **11** (1947) 479.
47VH	Verwey, E. J. W., und E. L. Heilman: J. Chem. Phys. **15** (1947) 174.
48MG	McMurdie, H. F., und E. Golovato: J. Res. Natl. Bur. Std. **41** (1948) 589.
50BS	Birnbaum, H., und R. K. Scott: J. Am. Chem. Soc. **72** (1950) 1398.
50MSM	McMurdie, H . F., B. M. Sullivan, und F. A. Maurer: J. Res. Natl. Bur. Std. **45** (1950) 35.
50P	Pouillard, E.: Ann. Chim. (Paris) **5** (1950) 164.
51KR	Kordes, E., und E. Röttig: Z. Anorg. Allgem. Chem. **264** (1951) 34.
51RB	Robin, J., und C. Bénard: Compt. Rend. **232** (1951) 1830.
52KPS	Kurlina, E. V., V. G. Prokhvatilov, und J. T. Sheftel': Dokl. Akad. Nauk. SSSR **86** (1952) 305.
52R	Robin, J.: Compt. Rend. **235** (1952) 1301.
52RB-1	Robin, J., und J. Bénard: Compt. Rend. **234** (1952) 734.
52RB-2	Robin, J., und J. Bénard: Compt. Rend. **234** (1952) 956.
52SR	Structure Rep. **16** (1952) 240.
53B	Bljumen, L. M.: Jzv. Akad. Nauk Turkm. SSR **5** (1953) 70.
53BD	Bertrant, F., und A. Durif: Compt. Rend. **236** (1953) 212.
53H	Honjo, G.: J. Phys. Soc. Japan **8** (1953) 113.
53HH	Hahn, H., und B. Harder: Angew. Chem. **65** (1953) 325.
53R-1	Romeijn, F. C.: Philips Res. Rept. **8** (1953) 304.
53R-2	Romeijn, F. C.: Philips Res. Rept. **8** (1953) 321.
53Y	Yamaguchi, G.: Bull. Chem. Soc. Japan **26** (1953) 204.
54A	Aoki, I.: J. Coll. Arts and Sci. Chiba Univ. **1** (1954) 150.
54E	Evans, R. C.: Einführung in die Kristallchemie; Barth Verlag, Leipzig 1954.
54G-1	Gorter, E. W.: Nature **173** (1954) 123.
54G-2	Gorter, E. W.: Philips Res. Rept. **9** (1954) 403.
54R	Rosenquist, T.: J. Iron Steel Inst. (London) **176** (1954) 37.
54SR	Schmitz-Dumont, O., und B. Raederscheidt: Naturwissenschaften **41** (1954) 477.
55A	Akimoto, S.: Japan. J. Geophys. **1** [2] (1955) 1.
55CBM	Chevallier, R., J. Bolfa, und S. Mathieu: Bull. Soc. Franc. Mineral. Crist. **78** (1955) 307.
55GL	Goodenough, J. B., und A. L. Loeb: Phys. Rev. **98** (1955) 391.
55N	Néel, L.: Advan. Phys. **4** (1955) 191.
55R	Robin, J.: Ann. Chim. **10** (1955) 389.
55ZP	Zaslavskii, A. I., und V. G. Prokhvatilov: Eighth All-Union Semiconductor Conf. 1955.
56C-1	Cossee, P.: Proefschrift, Universität Leiden (1956) Uitgeverij Excelsior, 's Gravenhage.
56C-2	Cossee, P.: Rec. Trav. Chim. **75** (1956) 1089.
56HH	Hahn, H., und B. Harder: Z. Anorg. Allgem. Chem. **288** (1956) 257.
56J	Jonker, G. H.: Trabajos Reunion Intern. Reactividad Solidos, 3°, Madrid (1956) 17.
56L	Lotgering, F. K.: Philips Res. Rept. **11** (1956) 337.
57ANK	Akimoto, S., T. Nagata, und T. Katsura: Nature **179** (1957) 37.
57AKY	Akimoto, S., T. Katsura, und M. Yoshida: J. Geomagn. Geoelect. (Japan) **9** (1957) 165.
57B	Bongers, P. F.: Proefschrift, Universität Leiden (1957) Uitgeverij Excelsior, 's Gravenhage.
57BO	Borovik-Romanov, A. S., und M. P. Orlova: Soviet Physics — JETP (English Transl.) **5** (1957) 1023.
57FSS	Finch, G. J. F. R. S., A. P. B. Sinha, und K. P. Sinha: Proc. Roy. Soc., (London) **A 242** (1957) 28.
57G	Gorter, E. W.: Advan. Phys. **6** (1957) 336.
57J	Jonker, G. H.: Trabajos Réunion Intern. Reactividad Solidos, 3°, Madrid **1** (1957) 413.
57KSK	Kolomiets, B. T., J. T. Sheftel', und E. V. Kurlina: Soviet Phys.-Tech. Phys. (English Transl.) **2** (1957) 40.
57SS	Sinha, K. P., und A. P. B. Sinha: J. Phys. Chem. **61** (1957) 758.
57SSB-1	Sinha, A. P. B., N. R. Sanjana, und A. B. Biswas: Acta Chryst. **10** (1957) 439.
57SSB-2	Sinha, A. P. B., N. R. Sanjana, und A. B. Biswas: Z. Krist. **109** (1957) 410.
57SR	Structure Rep. **21** (1957) 325.
57VWCM	Vincent, E. A., J. B. Wright, R. Chevalier, und S. Mathieu: Mineral. Mag. **31** (1957) 624.
57W	Wilkinson, J. F. G.: Mineral. Mag. **31** (1957) 443.

58BW	Baltzer, P. K., und J. G. White: J. Appl. Phys. **29** (1958) 445.
58C	Cossee, P.: J. Inorg. Nucl. Chem. **8** (1958) 483.
58D	Delorme, C.: Bull. Soc. Franc. Mineral. Crist. **81** (1958) 79.
58E	Eschenfelder, A. H.: J. Appl. Phys. **29** (1958) 378.
58HK	van Hook, H. J., und M. L. Keith: Am. Mineralogist **43** (1958) 69.
58KHB	Kriessmann, C. J., S. E. Harrison, und H. S. Belson: J. Appl. Phys. **29** (1958) 452.
58RV	Richardson, J. T., und L. W. Vernon: J. Phys. Chem. **62** (1958) 1153.
58SBB	Schmitz-Dumont, O., H. Brokopf, und K. Burkhardt: Z. Anorg. Chem. **295** (1958) 7.
58SS	Sakata, T., und K. Sakata: J. Phys. Soc. Japan **13** (1958) 675.
58SSB	Sinha, A. P. B., N. R. Sanjana, und A. B. Biswas: J. Phys. Chem. **62** (1958) 191.
58SSK	Sakata, T., K. Sakata, und K. Kigusi: J. Phys. Soc. Japan **13** (1958) 675.
58WC	Wickham, D. G., und W. J. Croft: J. Phys. Chem. Solids **7** (1958) 351.
59A	Azároff, L. V.: Z. Krist. **112** (1959) 33.
59B	Boucher, B.: Compt. Rend. **249** (1959) 514.
59G	Goodenough, J. B.: J. Phys. Radium **20** (1959) 155.
59J	Jacobs, I. S.: J. Phys. Chem. Solids **11** (1959) 1.
59K	Kasper, J. S.: Bull. Am. Phys. Soc. **4** (1959) 178.
59L	Larson, E. G.: Lincoln Lab. Quart. Rep. (1959) 96.
59M	Miller, A.: Phys. Rev. **116** (1959) 1481.
59SZKT	Sheftel', J. T., A. J. Zaslavskii, E. V. Kurlina, und G. N. Tekster-Proskuryakova: Soviet Phys.-Solid State (English Transl.) **1** (1959) 203.
59W	Wojtowicz, P. J.: Phys. Rev. **116** (1959) 32.
59WG	Wickham, D. G., und J. B. Goodenough: Phys. Rev. **115** (1959) 1156.
60AP	Alperin, H. A., und S. J. Pickart: Antiferromagnetic Compounds, Landolt-Börnstein Vol. II/9 (1962) 3—143.
60DM	Dwight, K., und N. Menyuk: Phys. Rev. **119** (1960) 1470.
60H	Hafner, S.: Schweiz. Mineral. Petrog. Mitt. **40** (1960) 207.
60ISB	Irani, K. S., A. P. B. Sinha, und A. B. Biswas: J. Phys. Chem. Solids **17** (1960) 101.
60M	Miller, A.: J. Appl. Phys. Suppl. **31** (1960) 261 S.
60MM	Muramori, K., und S. Miyahara: J. Phys. Soc. Japan **15** (1960) 1906.
61H	Hafner, S.: Z. Krist. **115** (1961) 331.
61J	Jacobs, I. S.: J. Phys. Chem. Solids **11** (1961) 1.
61JK	Jacobs, I. S., und J. S. Kouvel: Phys. Rev. **122** (1961) 412.
61K	O'Keeffe, M.: J. Phys. Chem. Solids **21** (1961) 172.
61LH	Lecerf, A., und A. Hardy: Compt. Rend. **252** (1961) 131.
61M	Moruzzi, V. L.: J. Appl. Phys. Suppl. **32** (1961) 59 S.
61N	Nagata, T.: Rock Magnetism; Maruzen Company Ltd. Tokyo 1961.
61PM	Poix, P., und A. Michel: Bull. Soc. Chim. Franc. **1961** 352.
61SKT	Sheftel', J. T., E. V. Kurlina, und G. N. Tekster-Proskuryakova: Soviet Phys.-Solid State (English Transl.) **3** (1961) 1989.
61WMD	Wickham, D. G., N. Menyuk, und K. Dwight: J. Phys. Chem. Solids **20** (1961) 316.
62A-1	Aoki, I.: J. Phys. Soc. Japan **17** (1962) 53.
62A-2	Akimoto, S.: J. Phys. Soc. (Japan) **17** B-1 (1962) 706.
62BP	Billiet, Y., und P. Poix: Bull. Soc. Chim. Franc. **1962** 2031.
62ISB	Irani, K. S., A. P. B. Sinha, und A. B. Biswas: J. Phys. Chem. Solids **23** (1962) 711.
62L	Lecerf, A.: Ann. Chim. (Paris) **7** (1962) 513.
62PM-1	Poix, P., und A. Michel: Compt. Rend. **255** (1962) 2446.
62PM-2	Poix, P., und A. Michel: Bull. Soc. Chim. Franc. **1962** 1010.
62S	Sakamoto, N.: J. Phys. Soc. Japan **17** (1962) 99.
62Sch	Schmalzried, H.: Z. Phys. Chem. **31** (1962) 184.
62SKT	Sheftel', J. T., E. V. Kurlina, und G. N. Tekster-Proskuryakova: Soviet Phys.-Solid State (English Transl.) **3** (1962) 1989.
62SZKT	Sheftel', J. T., A. J. Zaslavskii, E. V. Kurlina, und G. N. Tekster-Proskuryakova: Soviet Phys.-Solid State (English Transl.) **3** (1962) 1979.
62ZKZ	Zaslavskii, A. J., Z. V. Karachentseva, und A. J. Zharinova: Soviet Phys.-Christ. (English Transl.) **7** (1962) 680.
63B-1	Blasse, G.: J. Inorg. Chem. **25** (1963) 230.
63B-2	Blasse, G.: J. Inorg. Chem. **25** (1963) 743.
63B-3	Blasse, G.: Philips Res. Rept. **18** (1963) 400.
63B-4	Blasse, G.: Philips Res. Rept. **18** (1963) 383.
63BP	Billiet, Y., und P. Poix: Bull. Soc. Chim. Franc. **1963** 477.
63BPM-1	Billiet, Y., P. Poix, und A. Michel: Bull. Soc. Chim. Franc. **1963** 2196.
63BPM-2	Billiet, Y., P. Poix, und A. Michel: Compt. Rend. **256** (1963) 4217.
63G-1	Gorter, E. W.: J. Appl. Phys. **34** (1963) 1253.

63G-2	Goodenough, J. B.: Magnetism and the Chemical Bond; John Wiley & Sons, New York — London 1963.
63GH	Grenot, M., und M. Huber: Compt. Rend. **256** (1963) 3074.
63GHM	Grenot, M., M. Huber, und C. Mazières: Compt. Rend. **257** (1963) 3372.
63JD	Joubert, J. C., und A. Durif: Bull. Soc. Franc. Mineral. Crist. **87** (1963) 517.
63LV	Lecerf, A., und G. Villers: Compt. Rend. **256** (1963) 5073.
63R	Roth, W. L.: Bull. Am. Phys. Soc. **8** (1963) 213.
63RNMP	Rosenberg, M., P. Nicolau, R. Manaila, und P. Pausescu: J. Phys. Chem. Solids **24** (1963) 1419.
63SB	Schulkes, J. A., und G. Blasse: J. Phys. Chem. Solids **24** (1963) 1651.
63VL	Villers, G., und A. Lecerf: Compt. Rend. **257** (1963) 1764.
63VOKL	Vainshtein, E. E., R. M. Ovrutskaya, B. I. Kotlyar, und V. R. Linde: Fiz. Tverd. Tela **5** (1963) 2935.
64AM	Aukrust, E., und A. Muan: Trans. AIME **230** (1964) 378.
64B-1	Blasse, G.: J. Inorg. Nucl. Chem. **26** (1964) 1473.
64B-2	Blasse, G.: Philips Res. Rept. Suppl. No. 3 (1964).
64BL	Baltzer, P. K., und E. Lopatin: Proc. Int. Conf. Magnetism Nottingham (1964) 564.
64HLRV	Hardy, A., A. Lecerf, M. Rault, und G. Villers: Compt. Rend. **259** (1964) 3462.
64JD	Joubert, J.-C., und A. Durif: Compt. Rend. **258** (1964) 4482.
64R-1	Roth, W. L.: J. Phys. **25** (1964) 507.
64R-2	Roth, W. L.: J. Phys. Chem. Solids **25** (1964) 1.
64R-3	Rosenberg, M.: Phys. Status Solidi **5** (1964) K 127.
64RLV	Rault, M., A. Lecerf, und G. Villers: Compt. Rend. **258** (1964) 4553.
64RN	Rosenberg, M., und J. Nicolae: Proc. Int. Conf. Magnetism Nottingham (1964) 561.
64RW	Rienäcker, G., und K. Werner: Z. Anorg. Chem. **327** (1964) 275.
64S	Syono, Y.: J. Phys. Soc. Japan **19** (1964) 1752.
64VLR	Villers, G., A. Lecerf, und M. Rault: Proc. Int. Conf. Magnetism Nottingham (1964) 639.
64VOKL	Vainshtein, E. E., R. M. Ovrutskoya, B. J. Kotlyar, und V. R. Linde: Soviet Phys.-Solid State (English Transl.) **5** (1964) 2150.
65A-1	Åsbrink, S.: Acta Chem. Scand. **19** (1965) 1766.
65A-2	Aoki, J.: J. Phys. Soc. Japan **20** (1965) 871.
65B	Blasse, G.: Philips Res. Rept. **20** (1965) 529.
65BC	Bérodias, G., und M. Chevreton: Compt. Rend. **261** (1965) 2202.
65BMPM	Billiet, Y., J. Morgenstern-Baderau, P. Poix, und A. Michel: Compt. Rend. **258** (1965) 4780.
65BRW	Bouchard, R. J., P. A. Russo, und A. Wold: Inorg. Chem. **4** (1965) 685.
65DA	Dey, S. K., und J. C. Anderson: Phil. Mag. **12** (1965) 975.
65GH	Grenot, M., und M. Huber: Compt. Rend. **261** (1965) 5124.
65H	Haas, C.: J. Phys. Chem. Solids **26** (1965) 1225.
65HB	Huber, M., und J. Bénard: Bull. Soc. Chim. Franc. **1965** 1081.
65ISA	Ishikawa, Y., Y. Syono, und S. Akimoto: Ann. Prog. Rept. Rock Magn. Group (Japan) **2** (1965) 14.
65L	Lenglet, M.: Rev. Chim. Minéral. **2** (1965) 217.
65LRPV	Lecerf, A., M. Rault, J. Portier, und G. Villers: Bull. Soc. Chim. (Paris) **1965** 1208.
65LRV	Lecerf, A., M. Rault, und G. Villers: Compt. Rend. **261** (1965) 749.
65MP	Manaila, R., und P. Pausescu: Phys. Status Solidi **9** (1965) 385.
65P	Poix, P.: Ann. Chim. **10** (1965) 49.
65R	Reuter, B.: Bull. Soc. Chim. Franc. **1965** 1053.
65RB-1	O'Reilly, W., und S. K. Banerjee: Phys. Letters **17** (1965) 237.
65RB-2	Robbins, M., und P. K. Baltzer: J. Appl. Phys. **36** (1965) 1039.
65S-1	Schmitz-Dumont, O.: Bull. Soc. Chim. Franc. **1965** 1099.
65S-2	Siratori, K.: J. Phys. Soc. Japan **20** (1965) 1962.
65SDYPIDB	Stefanov, D., I. Danchev, I. Yanachkova, P. Petrov, S. Ivonov, D. Djoglev und L. Bizheva: Izv. Fiz. Inst. s Aneb, Bulgar. Akad. Nauk **13** (1965) 185.
65VLR	Villers, G., A. Lecerf, und M. Rault: Compt. Rend. **260** (1965) 3017.
65VOK	Vainshtein, E. E., R. M. Ovrutskaya, und B. I. Kotlyar: Fiz. Tverd. Tela **7** (1965) 2120.
66A	Aiyama, Y.: J. Phys. Soc. Japan **21** (1966) 1684.
66B	Blasse, G.: J. Phys. Chem. Solids **27** (1966) 383.
66BBP	Boucher, B., R. Buhl, und M. Perrin: Compt. Rend. **263 B** (1966) 344.
66BMPM	Billiet, Y., J. Morgenstern-Badarau, P. Poix, und A. Michel: Compt. Rend. **262C** (1966) 751.

66BP	Bertaut, E.-F., und G. Patrat: Bull. Soc. Franc. Miner. Crist. **88** (1965) 586.
66BR	Banerjee, S. K., und W. O'Reilly: IEEE Trans. Magnetics **2** (1966) 463.
66BRGG	Banerjee, S. K., W. O'Reilly, T. C. Gibb, und N. N. Greenwood: Phys. Letters **20** (1966) 455.
66BZRB	Babkine, J., Cl. Zeller, J.-Cl. Reithler, und J. Bolfa: Compt. Rend. **B 263** (1966) 1135.
66D	Dey, S. K.: Phys. Letters **22** (1966) 375.
66DR	Driessens, F. C. M., und G. D. Rieck: J. Inorg. Nucl. Chem. **28** (1966) 1593.
66HGLRV	Hagenmuller, P., Ch. Guillaud, A. Lecerf, M. Rault, und G. Villers: Bull. Soc. Chim. Franc. **1966** 2589.
66II	Iizuka, T., und S. Iida: J. Phys. Soc. Japan **21** (1966) 222.
66KL	Kiessling, R., und N. Lange: Jernkontorets Ann. **150** (1966) 203.
66KYH	Kubo, T., H. Yasuoka, und A. Hirai: J. Phys. Soc. Japan **21** (1966) 812.
66LAB	Lindsley, D. H., G. E. Andreasen, und J. R. Balsley: Handbook of Physical Constants; The Geological Society of America, Memoir 97 (1966) 543.
66MKKI	Miyatani, K., K. Kohn, H. Kamimura, und S. Iida: J. Phys. Soc. Japan **21** (1966) 464.
66MOKI	Matsumoto, G., K. Ohbayashi, K. Kohn, und S. Iida: J. Phys. Soc. Japan **21** (1966) 2429.
66O	Okazaki, H.: Japan. J. Appl. Phys. **5** (1966) 559.
66R	Reinen, D.: Theor. Chim. Acta **5** (1966) 312.
66RB	Reithler, J.-C., und J. Bolfa: Compt. Rend. **262D** (1966) 537.
66RD	Robbins, M., und L. Darcy: J. Phys. Chem. Solids **27** (1966) 741.
66SP	Sheftel', J. T., und Ya. V. Pavlotskii: Izv. Akad. Nauk SSSR, Neorgan. Materialy **2** (1966) 918.
66SSB	Sabane, C. D., A. P. B. Sinha, und A. B. Biswas: Indian J. Pure Appl. Phys. **4** (1966) 187.
66TL	Tellier, J.-Cl., und M. Lensen: Bull. Soc. Chim. Franc. **8** (1966) 2502.
66VJD	Vincent, H., J.-C. Joubert, und A. Durif: Bull. Soc. Chim. Franc. **1966** 246.
66VOK	Vainshtein, E. E., R. M. Ovrutskaya, und B. I. Kotlyar: Soviet Phys. — Solid State (English Transl.) **7** (1966) 1707.
67B	Bayer, G.: J. Less.-Common Metals **12** (1967) 326.
67BBP	Boucher, B. Y., R. Buhl, und M. Perrin: J. Appl. Phys. **38** (1967) 1109.
67BP	Billiet, Y., und P. Poix: Bull. Soc. Chim. Franc. **1967** 215.
67BRJ	Banerjee, S. K., W. O'Reilley, und C. E. Johnson: J. Appl. Phys. **38** (1967) 1289.
67DR	Datta, R. K., und R. Roy: J. Am. Ceram. Soc. **50** (1967) 578.
67G-1	Grenot, M.: Bull. Soc. Chim. Franc. **1967** 2043.
67HS	Hahn, H., und G. Strick: Naturwissenschaften **54** (1967) 42A.
67K	Kasper, H.: J. Inorg. Nucl. Chem. **29** (1967) 2693.
67LS	Lotgering, F. K., und R. P. van Stapele: Solid State Commun. **5** (1967) 143.
67M	Manaila, R.: Rev. Roumaine Phys. **12** (1967) 497.
67MSL	van Maaren, N. H., G. M. Schaeffer, und F. K. Lotgering: Phys. Letters **25A** (1967) 238.
67NK	Navrotsky, A., und O. J. Kleppa: J. Inorg. Nucl. Chem. **29** (1967) 2701.
67PB	Poix, P., und J.-C. Bernier: Bull. Soc. Chim. Franc. **1967** 605.
67RMS	Rozhdestvenskaya, M. V., V. A. Mokievskii und V. A. Stogova: Soviet Phys. — Cryst. (English Transl.) **11** (1967) 765.
67T	Tellier, J.-Cl.: Rev. Chim. Minéral. **4** (1967) 325.
67ZHRB	Zeller, Cl., J. Hubsch, J.-Cl. Reithler, und J. Bolfa: Compt. Rend. **B 265** (1967) 1335.
68ABPMKS	Andruskevic, T. V., G. K. Boreskov, V. V. Popovskij, V. S. Muzykantov, O. N. Kimchaj, und V. A. Sazanov: Kinetika i Kataliz, Akad. Nauk SSSR, Sb. Statei **9** (1968) 595.
68BBP	Boucher, B., R. Buhl, und M. Perrin: J. Appl. Phys. **39** (1968) 632.
68LS	Lotgering, F. K., und R. P. van Stapele: J. Appl. Phys. **39** (1968) 417.
68NK	Navrotsky, A., und O. J. Kleppa: J. Inorg. Nucl. Chem. **30** (1968) 479.
68OC	Ono, K., und L. Chandler: J. Phys. Soc. Japan **25** (1968) 174.
68S	Smit, J.: Solid State Commun. **6** (1968) 745.
68SEH	Strick, G., G. Eulenberger, und H. Hahn: Z. Anorg. Allgem. Chem. **357** (1968) 338.
68Y	Yokoyama, N. J. H.: Oyo — Butsuri **37** (1968) 171.
69W	Wickham, D. G.: J. Inorg. Nucl. Chem. **31** (1969) 313.

6.3.15 Mo Spinelle und substituierte Mo Spinelle

(Literatur siehe 6.3.22, Seite 545)

Tab. 1. Kristallografische Eigenschaften der Mo^{6+} Oxid-Spinelle

Verbindung	Kationenverteilung	a Å	u	Lit.
Ag_2MoO_4	$Mo[Ag_2]O_4$	9.26	0.37	*22W, 23W*
			0.364 ± 0.002	*47DS*
				51KR, 64B, 66CD
Li_2MoO_4[1])				*26G, 26Z, 28H*
				64B, 51KR, 50L
		8.434 ± 0.003		*67LR*
Na_2MoO_4	$Mo[Na_2]O_4$			*28H*
		8.99		*50L, 60H*
				64B, 66CD
		9.108 [2])		*66P*

6.3.16 W Spinelle und substituierte W Spinelle

(Literatur siehe 6.3.22, Seite 545)

Tab. 1. Kristallografische Eigenschaften der W^{6+} Oxid-Spinelle

Verbindung	Kationenverteilung	a Å	u	Lit.
Li_2WO_4[3])				*26G, 28H*
				64B
Na_2WO_4				*28H*
	$W[Na_2]O_4$	8.99		*50L, 64B, 60H*
		9.13	0.3690 ± 0.0005	*65SS*
				65SK, 66CD
		9.1297 [4])		*66P*

6.3.17 Nb Spinelle und substituierte Nb Spinelle

(Literatur siehe 6.3.22, Seite 545)

Tab. 1. Kristallografische Eigenschaften der Nb^{5+} Oxid-Spinelle

Verbindung	Kationenverteilung	x	a Å	Lit.	Fig.
$Zn_{2.33}Nb_{0.67}O_4$			8.589	*63HD*	
				64B	
$LiZnNbO_4$[5])	$Zn[LiNb]O_4$		a = 6.082 [6])	*63B-1*	
			c = 8.402 [6])	*63B-2, 63B-3*	
				64B, 65H	
				66BMPM	
			a = 6.098 ± 0.003 [7]) c = 8.427 ± 0.003 [7])	*66VJD*	
$LiNi_{0.15}Zn_{0.85}NbO_4$	$Li_{0.15}Zn_{0.85}[Li_{0.85}Ni_{0.15}Nb]O_4$[8])			*66R*	

[1]) Nur die Hochtemperaturphase besitzt Spinellstruktur [*64B*]; unter Normalbedingungen Phenakitstruktur [*50L*], [*26Z*]; Übergang von der Phenakit- zur Spinellstruktur bei 10 at und 400 °C [*67LR*].
[2]) Bei 25 °C; ≦444 °C: sogenanntes Na_2MoO_4 IV hat Spinellstruktur; Umwandlung ab 444 °C und höher in sogenannte Na_2MoO_4 III-I mit orthorhombischer Struktur; Phasendiagramm. Untersuchungen bis ≈46 000 at [*66P*].
[3]) Nur die Hochtemperaturphase besitzt Spinellstruktur [*64B*].
[4]) Bei 25 °C; ≦587.6 °C: sogenanntes Na_2WO_4 III hat Spinellstruktur; Umwandlung oberhalb von 587.6 °C in Na_2WO_4 II und -I mit orthorhombischer Struktur; Phasendiagramm. Untersuchungen bis ≈46 000 at [*66P*].
[5]) Die Zn^{2+}-Ionen können bis zu 20, 50 und 80% durch Co^{2+}-, Mg^{2+}- beziehungsweise Ni^{2+}-Ionen ersetzt werden [*63B-2*].
[6]) Bei leicht tetragonal verzerrter Spinellstruktur a = 8.602 (6.082 · $\sqrt{2}$) Å, c = 8.402 Å; Überstrukturcharakter [*63B-1*].
[7]) $c/a\sqrt{2}$ = 0.977 [*66VJD*].
[8]) Inverser Spinell (auf den Oktaederplätzen geordnet) [*66R*].

Tab. 1 (Fortsetzung)

Verbindung	Kationenverteilung	x	$c/a\sqrt{2}$	Lit.	Fig.
$Li_{1-x}Zn_{1+x}Ti_xNb_{1-x}O_4$		0.1	0.9777	*66VJD*	1
		0.2	0.9800		
		0.3	0.9804		
		0.4 [1]	0.9814		
		0.7	0.9872		
		0.8	0.9880		
		0.9	0.9891		

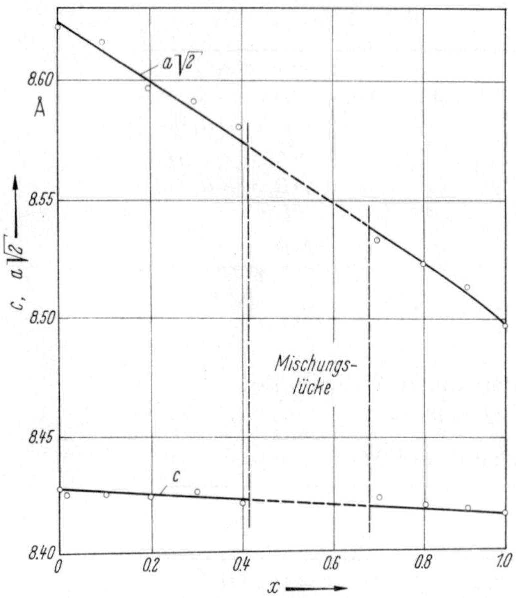

Fig. 1. $Li_{1-x}Zn_{1+x}Ti_xNb_{1-x}O_4$. Gitterkonstanten $a\sqrt{2}$ und c vs. Zusammensetzungsparameter x [*66VJD*].

6.3.18 Zr Spinell

(Literatur siehe 6.3.22, Seite 545)

Tab. 1. Kristallographische Eigenschaften der Zr^{3+} Sulfid-Spinelle

Verbindung	a Å	ϱ_p g/cm³	u	Lit.
Cu_2ZrS_4	10.378 ± 0.003	4.34	0.375	*68TR*

6.3.19 Sb Spinelle und substituierte Sb Spinelle

(Literatur siehe 6.3.22, Seite 545)

Tab. 1. Kristallografische Eigenschaften der Sb^{5+} Oxid-Spinelle

Verbindung	Kationenverteilung	a Å	ϱ_X g/cm³	u	Lit.
$Co_{7/3}Sb_{2/3}O_4$	$Co[Co_{4/3}Sb_{2/3}]O_4$				*41BHM*
		8.523	6.027	0.380 ± 0.005	*60DD, 64B*
$Co_{2.33}Sb_{0.67}O_4$	$Co[Co_{1.33}Sb_{0.67}]O_4$ [2]	8.54 ± 0.005	6.04		*61B*
$Zn_{7/3}Sb_{2/3}O_4$	$Zn[Zn_{4/3}Sb_{2/3}]O_4$				*41BHM*
		8.594	6.230	0.390 ± 0.005	*60DD, 64B*
$Zn_{2.33}Sb_{0.67}O_4$	$Zn[Zn_{1.33}Sb_{0.67}]O_4$ [2]	8.585 ± 0.005	6.24		*61B*
$LiCoSbO_4$	$Li_{1-x}Co_x[Li_xCo_{1-x}Sb]O_4$ [3]	8.562			*63B-2, 64B*
					66BP
$LiZnSbO_4$	$Li_{1-x}Zn_x[Li_xZn_{1-x}Sb]O_4$ [3]	8.55			*63B-2, 64B*
					66BP, 63B-4
$Mg_{1.0}Zn_{1.33}Sb_{0.67}O_4$					*61B*
$LiNi_{0.15}Zn_{0.85}SbO_4$	$Li_{0.65}Zn_{0.35}[Li_{0.35}Zn_{0.5}Ni_{0.15}Sb]O_4$ [4]				*66R*

[1] Zwischen 0.4 < x < 0.7 besteht eine Mischungslücke [*66VJD*].
[2] Inverser Spinelltyp [*61B*].
[3] Spinell mit Überstruktur [*64B*].
[4] Inverser Spinell (auf den Oktaederplätzen geordnet) [*66R*].

6.3.20 Sn Spinelle und substituierte Sn Spinelle

(Literatur siehe 6.3.22, Seite 545)

Tab. 1. Kristallografische Eigenschaften binärer Sn^{4+} Oxid-Spinelle

Verbindung	Kationen-verteilung	a Å	ϱ_X g/cm³	ϱ_p g/cm³	u	Lit.	Fig.	Tab.
Co_2SnO_4	$Co[CoSn]O_4$ [1])	8.605 ± 0.005	6.307	6.108		*29NP, 30T, 47VH, 47/48S*	4, 6	2
		8.658				*58SBB*		
		8.644 ± 0.001			0.375	*59PM, 53Y, 64P*		
		8.644				*61PM-1, 62PM-1, 64B*		
					0.379 ± 0.002	*65P, 66DL, 68GM*		
Fe_2SnO_4	[2])					*66BRGG*		
Mg_2SnO_4	[3])	8.580 ±0.007	4.864	4.738		*29NP, 30T, 52TK*	1, 15	2
		8.590				*53Y*	Fig. 14, 15 in 6.3.13	
		8.636			0.380	*58SBB*		
	$Mg[MgSn]O_4$	8.639 ± 0.001			0.375	*59PM*		
		8.60			0.385	*61WMD, 64B*		
						61PM-1, 62PM-2		
		8.639			0.385 ± 0.002	*65P, 66DL, 64P, 68GM*		
						65HHW		
		8.650 ± 0.005			0.380 ± 0.005	*66O*		
Mn_2SnO_4		8.84 [4])				*59GM-1, 59GM-2*	2, 3, 10, 11, 13	2
	$Mn[MnSn]O_4$	8.865				*61WMD, 64B, 66DL, 68GM*		
		8.879			0.383	*68NP*		
Zn_2SnO_4	$Zn[ZnSn]O_4$	8.650 ± 0.001	6.393			*30T*	4, 6, 16	2
		8.61 ± 0.01			0.390 ± 0.01	*32BP, 60H*	Fig. 17, 18 in 6.3.13	
	[1])	8.667				*47VH, 47/48S, 47RR*		
						54E, 53Y		
		8.665 ± 0.001			0.390	*59PM*		
		8.67				*59GM*		
						61PM-1, 62PM-1, 64B		
						62PM-2, 61PM-2		
		8.665			0.381 ± 0.003	*65P, 64P, 66DL, 68GM*		
		8.65 ± 0.01				*68CDR*		

[1]) Inverser Spinell [*47/48S*].
[2]) Inverser Spinell [*66BRGG*].
[3]) Inverser Spinell [*58SBB*].
[4]) 8.88 Å durch Extrapolation ermittelt bei Annahme der mittleren Wertigkeit 2 von Mn [*59GM-1*].

Tab. 2. Magnetische Eigenschaften binärer Sn^{4+} Oxid-Spinelle

Verbindung	p_m μ_B	$\chi_{m\,diam}$ 10^{-6} emu/mol	C_m cm³ °K/mol	Θ_N °K	Θ_a °K	Θ_C °K	Lit.	Fig.	Tab.
Co_2SnO_4							*59PM*		
		−79					*62PM-1*	6	
		−79 ± 9					*64P*		
			6.132 6)		−180.7 6)		*65P, 68GM*		
Mg_2SnO_4 1)		−70 ± 1					*59PM, 61WMD*	1, 15	
							62PM-2		
		−70					*64P*		
		−71					*65P, 68GM, 65HHW*		
Mn_2SnO_4 2) 3)	0.35					58 ± 3	*59GM-1, 59GM-2*	2, 3, 10,	
	0.41			58		53 ± 3 4)	*61WMD, 68GM*	11, 13	
	0.37 5)						*68NP*		
Zn_2SnO_4 1)		−81 ± 1					*59PM, 61PM-2*	16	
							62PM-1		
		−81					*64P, 65P, 68GM*		

1) Die Orthostannate sind diamagnetisch [*59PM*].
2) Magnetische Werte für eine mittlere Wertigkeit von Mn^{3+} von 2.12 ± 0.01 [*59GM-1*].
3) Ferrimagnetisch [*61WMD*].
4) siehe Fig. 3 [*61WMD*].
5) Für $Mn^{2+}[Mn^{2+}_{1-y/2}Mn^{3+}_{y}Sn^{4+}_{1-y/2}]O_4$ bei $y = 0.255$ [*61WMD*].
6) Für $190 < T < 1200$ °K [*65P*].

Fig. 1 ··· 3, siehe S. 540

Tab. 3. Kristallografische Eigenschaften ternärer und quaternärer Sn^{4+} Oxid-, -Sulfid- und -Seleno-Spinelle

Verbindung	Kationenverteilung	x	a Å	ϱ_X g/cm³	ϱ_p g/cm³	u	Farbe	Lit.	Fig.	Tab.
$Co_{2x}Mg_{2-2x}SnO_4$		0.1	8.640				2)	*61PM-1, 64P*		4
		0.2	8.640 1a)			0.384		*61PM-1, 64P*		
	2a)	0.4	8.642					*61PM-1, 64P, 54SR*		
	$Co_{0.59}Mg_{0.41}[Co_{0.41}Mg_{0.59}Sn]O_4$	0.5	8.642 1b)			0.382		*61PM-1, 64P, 53Y*		
		0.6	8.642					*61PM-1, 64P*		
		0.8	8.643			0.380		*61PM-1, 64P*		
		0.9	8.643					*61PM-1, 64P*		
								58SBB, 66DL, 67PB		

$CoMnSnO_4$			8.68					*53Y*		
		0.5	8.60					*53Y*		
$Co_{2x}Zn_{2-2x}SnO_4$		0.5	8.647					*62PM-1*	4, 6	4
		0.5	8.646			0.381 ± 0.003		*65P*		
$Mg_{1+x}Fe_{2-2x}Sn_xO_4$	$Mg_{0.32}Fe_{0.68}$ $[Mg_{0.78}Fe_{1.12}Sn_{0.1}]O_4$	0.1	8.413			0.378		*68BNPSTY, 67T*	7…9	4
	$Mg_{0.4}Fe_{0.6}[Mg_{0.8}Fe_1Sn_{0.2}]O_4$	0.2	8.447			0.380				
	$Mg_{0.53}Fe_{0.47}$ $[Mg_{0.77}Fe_{0.93}Sn_{0.3}]O_4$	0.3	8.480			0.379				
	$Mg_{0.61}Fe_{0.39}$ $[Mg_{0.79}Fe_{0.81}Sn_{0.4}]O_4$	0.4	8.503			0.378				
	$Mg_{0.67}Fe_{0.33}$ $[Mg_{0.83}Fe_{0.67}Sn_{0.5}]O_4$	0.5	8.531			0.375				
$Mg_{1.5}Li_{0.5}SnO_{3.5}F_{0.5}$	$Mg_{0.75}Li_{0.25}[Li_{0.25}Mg_{0.75}Sn]O_{3.5}F_{0.5}$		8.652 ±0.005			0.335 ± 0.005		*66O*		
$Mg_{2x}Mn_{2-2x}SnO_4$								*61WMD*	5, 10…14	4
		0.5	8.66					*53Y*		
		0.125	8.83[3])					*59GM-1, 59GM-2*		
$Mg_2Ti_{1-x}Sn_xO_4$		0.5	8.548					*62PM-2*	15	4
		0.7	8.589					*65P*	Fig. 14, 15 in 6.3.13	
		0.9	8.628							
$Mn_{1.75}Zn_{0.25}SnO_4$			8.82[4])					*59GM-1, 59GM-2*	10, 13	4
$Mn_2Ge_{0.125}Sn_{0.875}O_4$			8.82[4])					*59GM-1, 59GM-2*	11	4
$NiMgSnO_4$			8.56					*53Y*		
$Ni_{0.5}Mg_{1.5}SnO_4$[5])	$Mg[Ni_{0.5}Mg_{0.5}Sn]O_4$							*66R*		
$NiZnSnO_4$			8.55					*53Y*		
	$Zn[NiSn]O_4$[5])							*66R*		
$Zn_{2-2x}Cd_{2x}SnO_4$	$Zn_{0.8}Cd_{0.2}[ZnSn]O_4$	0.1	8.71 ±0.01					*68CDR*		
	$Zn_{0.67}Cd_{0.33}[Zn_{0.93}Cd_{0.07}Sn]O_4$	0.2	8.76 ±0.01							
	$Zn_{0.50}Cd_{0.50}[Zn_{0.9}Cd_{0.1}Sn]O_4$	0.3	8.82 ±0.01							
	$Zn_{0.36}Cd_{0.64}[Zn_{0.84}Cd_{0.16}Sn]O_4$	0.4	8.88 ±0.01							
	$Zn_{0.27}Cd_{0.73}[Zn_{0.73}Cd_{0.27}Sn]O_4$	0.5	8.94 ±0.01							
Fortsetzung nächste Seite Fußnoten siehe S. 538	$Zn_{0.22}Cd_{0.78}[Zn_{0.58}Cd_{0.42}Sn]O_4$	0.6	8.98 ±0.01							

Tab. 3. (Fortsetzung)

Verbindung	Kationenverteilung	x	a Å	ϱ_X g/cm³	ϱ_p g/cm³	u	Farbe	Lit.	Fig.	Tab.
$Zn_{2-2x}Cd_{2x}SnO_4$ (Fortsetzung)	$Zn_{0.15}Cd_{0.85}[Zn_{0.35}Cd_{0.65}Sn]O_4$	0.75	9.03 ±0.01					*68CDR*		
	$Zn_{0.12}Cd_{0.88}[Zn_{0.28}Cd_{0.72}Sn]O_4$	0.8	9.05 ±0.01							
$Zn_2Ti_{1-x}Sn_xO_4$		0.5 0.7 0.9	8.567 8.607 8.645					*61PM-2, 62PM-1* *65P*	16 Fig. 17, 18 in 6.3.13	4
$AgCrSnS_4$			10.44	4.752	4.67			*67HS*		
	$Ag[CrSn]S_4$					0.387	schwarz	*68SEH*		
$AgInSnS_4$			10.74	4.923	4.91			*67HS*		
	$Ag[InSn]S_4$					0.386	dunkelbraun	*68SEH*		
$CuAlSnS_4$			10.24	4.171	4.08			*67HS*		
	$Cu[AlSn]S_4$					0.377	braun	*68SEH*		
$CuInSnS_4$			10.50	4.906	4.86			*67HS*		
	$Cu[InSn]S_4$					0.379	braun	*68SEH*		
$CuCrSnS_4$			10.16	4.587	4.51			*67HS*		
	$Cu[CrSn]S_4$					0.379	schwarz	*68SEH*		
$AgCrSnSe_4$			10.97	5.978	5.96			*67HS*		
	$Ag[CrSn]Se_4$					0.384	schwarz	*68SEH*		
$CuCrSnSe_4$			10.70	5.962	5.94			*67HS*		
	$Cu[CrSn]Se_4$					0.379	schwarz	*68SEH*		
$Co_{0.2}Mg_{1.8}Ti_{1-x}Sn_xO_4$								*58SBB*		
$Mg_{0.25}Mn_{1.75}Ge_{0.125}Sn_{0.875}O_4$			8.81[6])					*59GM-1, 59GM-2*	10, 11	4

[1a]) 8.646 Å [*58SBB*].
[1b]) 8.59 Å [*53Y*].
[2]) Durch Substitution von Mg durch Co ab $x = 0.4$ beginnende Grünfärbung der anfänglich blauen Spinelle [*54SR*].
[2a]) Ab $x = 0.4$ auch Oktaederlücken besetzt, zuvor nur Tetraederlücken [*54SR*].
[3]) Bei der durchschnittlichen Wertigkeit 2.14 von Mn; 8.81 Å für die durchschnittliche Wertigkeit 2.23 von Mn [*59GM-1*].
[4]) Bei einer durchschnittlichen Wertigkeit 2.1 von Mn [*59GM-1*].
[5]) Inverser Spinell [*66R*].
[6]) Bei einer mittleren Wertigkeit 2.13 von Mn [*59GM-1*].

Tab. 4. Magnetische Eigenschaften ternärer und quaternärer Sn^{4+} Oxid-Spinelle

Verbindung	x	T °K	p_m μ_B	σ Gauss cm³/g	C_m cm³ °K/mol	Θ_a °K	Θ_C °K	Lit.	Fig.
$Co_{2x}Mg_{2-2x}SnO_4$								*64P*	
	0.5				2.87			*67PB*	
$Co_{2x}Zn_{2-2x}SnO_4$								*62PM-1*	6
	0.1	$150 < T < 873$ [1]			0.604	0		*65P*	
	0.2	$120 < T < 1100$ [2]			1.18	−3			
	0.4	$120 < T < 1100$ [3]			2.32	−10			
	0.5	$T > 1000$			2.76	+53			
	0.61	$90 < T < 1200$			3.50	−45			
	0.62	$90 < T < 1200$			3.54	−47			
	0.81	$250 < T < 1200$			4.75	−107			
	0.82	$250 < T < 1200$			4.798	−109			
	0.9	$216 < T < 1208$			5.495	−147			
	1	$190 < T < 1200$			6.132	−180.7			
$Mg_{1+x}Fe_{2-2x}Sn_xO_4$	0.1		2.20	60.6 [8]				*68BNPSTY*	7, 8, 9
	0.2		1.76	47.6 [8]					
	0.3		1.59	42.4 [8]					
	0.4		1.11	29.2 [8]					
	0.5		0.77	20.0 [8]					
$Mg_{2x}Mn_{2-2x}SnO_4$	0.125		0.81 [4]				53	*59GM-1* *59GM-2* *61WMD*	10…14 [7]
$Mg_2Ti_{1-x}Sn_xO_4$								*62PM-2* *65P*	15
$Mn_2Ge_{0.125}Sn_{0.875}O_4$			0.40 [5]					*59GM-1* *59GM-2*	11
$Mn_{1.75}Zn_{0.25}SnO_4$			0.50 [5]				48	*59GM-1* *59GM-2*	10, 13
$Zn_2Ti_{1-x}Sn_xO_4$								*61PM-2* *62PM-1* *65P*	16
$Mg_{0.25}Mn_{1.75}Ge_{0.125}Sn_{0.875}O_4$			0.77 [6]				49	*59GM-1*, *59GM-2*	10, 11

[1] Für 150 °K $< T <$ 840 °K: $C_m = 0.610$ cm³ °K/mol und $\Theta_a = -1.5$ °K; für 820…1200 °K: $C_m = 0.5$ cm³ °K/mol und $\Theta_a = +152$ °K [*65P*].
[2] Für $T >$ 900 °K: $C_m = 1.02$ cm³ °K/mol und $\Theta_a = +128$ °K; für $x = 0.21$ bei 120 °K $< T <$ 900 °K; $C_m = 1.205$ cm³ °K/mol und $\Theta_a = -6$ °K; für $x = 0.22$ bei 120 °K $< T <$ 900 °K: $C_m = 1.190$ cm³ °K/mol und $\Theta_a = -7$ °K [*65P*].
[3] Für $T >$ 900 °K: $C_m = 2.16$ cm³ °K/mol und $\Theta_a = +73$ °K; für $x = 0.41$ bei 90 °K $< T <$ 1000 °K: $C_m = 2.335$ cm³ °K/mol und $\Theta_a = -16$ °K; und für $x = 0.42$ bei 90 °K $< T <$ 1000 °K; $C_m = 2.360$ cm³ °K/mol und $\Theta_a = -17$ °K [*65P*].
[4] Bei der durchschnittlichen Wertigkeit 2.14 von Mn; $p_m = 0.78\ \mu_B$ für Mn mit der Wertigkeit 2.23 [*59GM-1*].
[5] Bei der durchschnittlichen Wertigkeit 2.10 von Mn [*59GM-1*].
[6] Bei der durchschnittlichen Wertigkeit 2.13 von Mn [*59GM-1*].
[7] Für Verbindungen der Reihe $Mn^{2+}[Mn^{2+}_{(1-x-y/2)}Mn^{3+}_yMg^{2+}_xSn^{4+}_{(1-y/2)}]O_4$ [*61WMD*].
[8] Bei 77 °K [*68BNPSTY*].

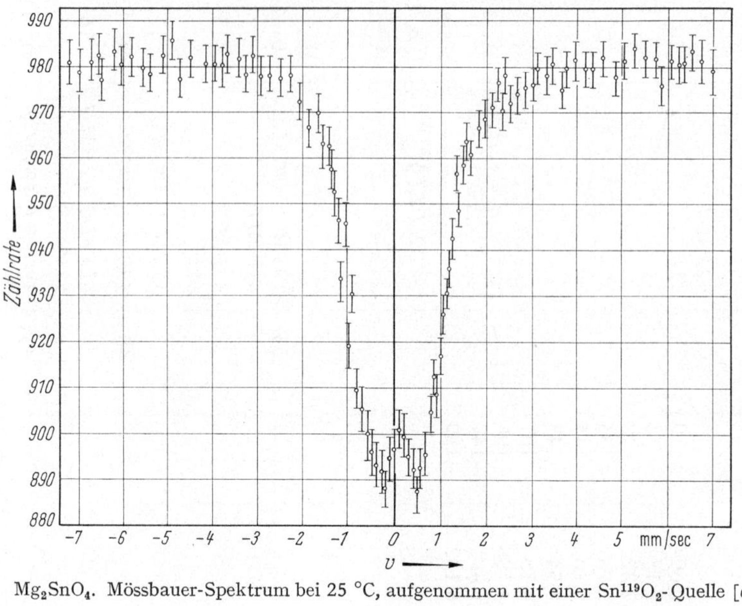

Fig. 1. Mg_2SnO_4. Mössbauer-Spektrum bei 25 °C, aufgenommen mit einer $Sn^{119}O_2$-Quelle [*68GM*].

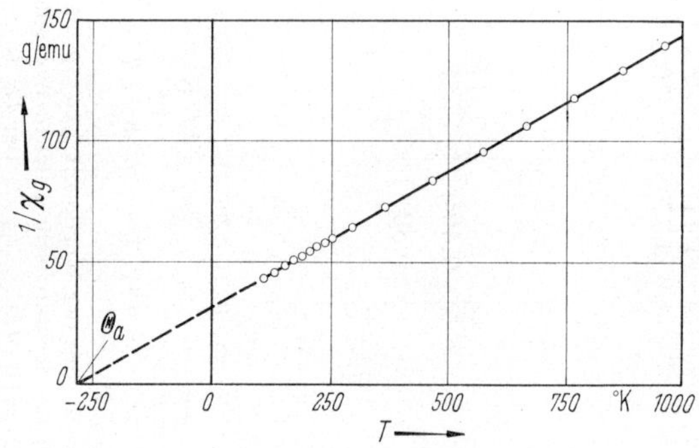

Fig. 2. Mn_2SnO_4. Reziproke Suszeptibilität $1/\chi_g$ vs. Temperatur T [*68NP*].

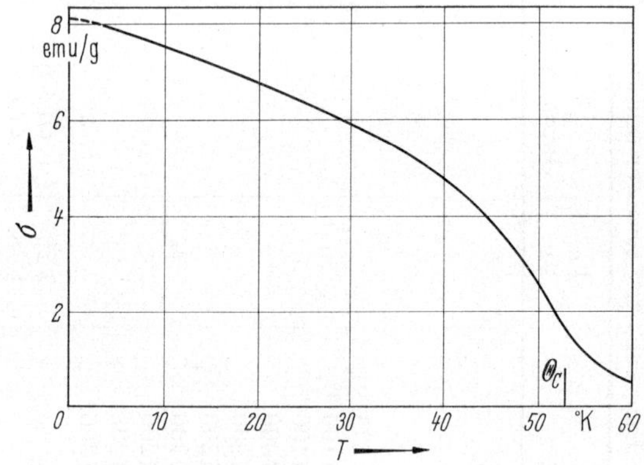

Fig. 3. $Mn^{2+}Mn^{2+}_{(1-y/2)}Mn^{3+}_{y}Sn_{(1-y/2)}O_4$ mit $y = 0.038$. Magnetisierung σ vs. Temperatur T bei einem angelegten Feld von $H_a = 11$ kOe [*61WMD*].

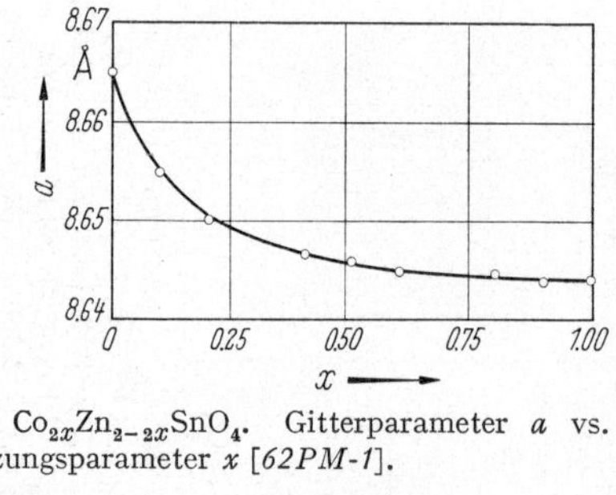

Fig. 4. $Co_{2x}Zn_{2-2x}SnO_4$. Gitterparameter a vs. Zusammensetzungsparameter x [*62PM-1*].

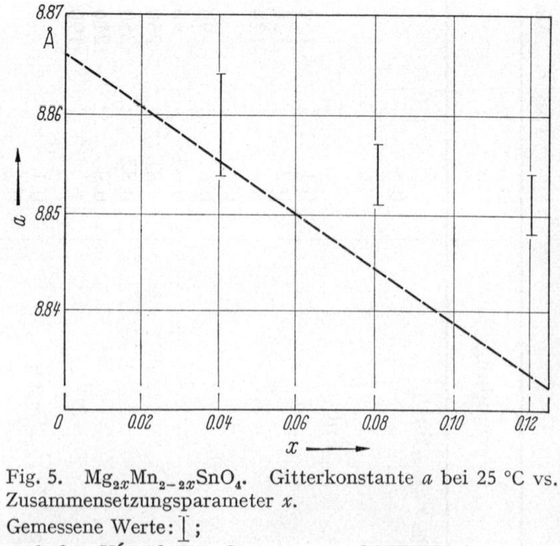

Fig. 5. $Mg_{2x}Mn_{2-2x}SnO_4$. Gitterkonstante a bei 25 °C vs. Zusammensetzungsparameter x.
Gemessene Werte: I ;
nach dem Végardschen Gesetz: - - - - - [*61WMD*].

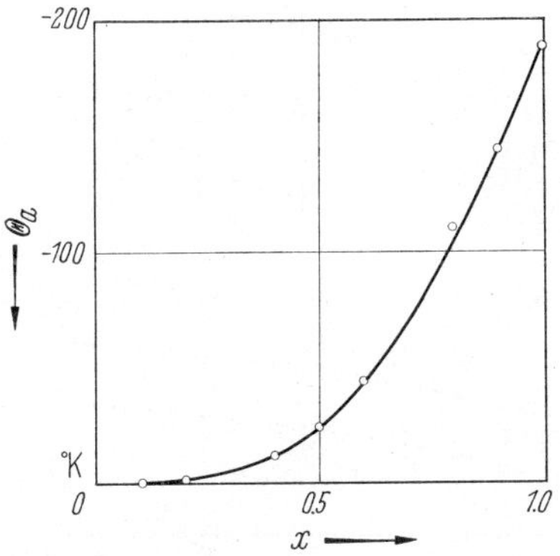

Fig. 6. $Co_{2x}Zn_{2-2x}SnO_4$. Asymptotische Curie-Temperatur Θ_a vs. Zusammensetzungsparameter x [*62PM-1*].

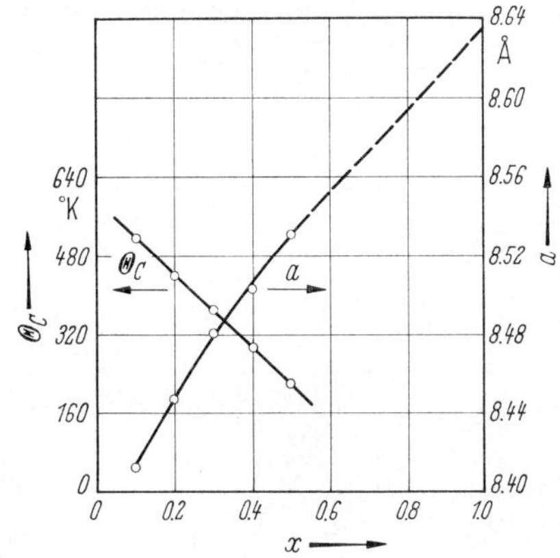

Fig. 7. $Mg_{1+x}Fe_{2-2x}Sn_xO_4$. Gitterparameter a und Curie-Temperatur Θ_C vs. Zusammensetzungsparameter x [*68BNPSTY*].

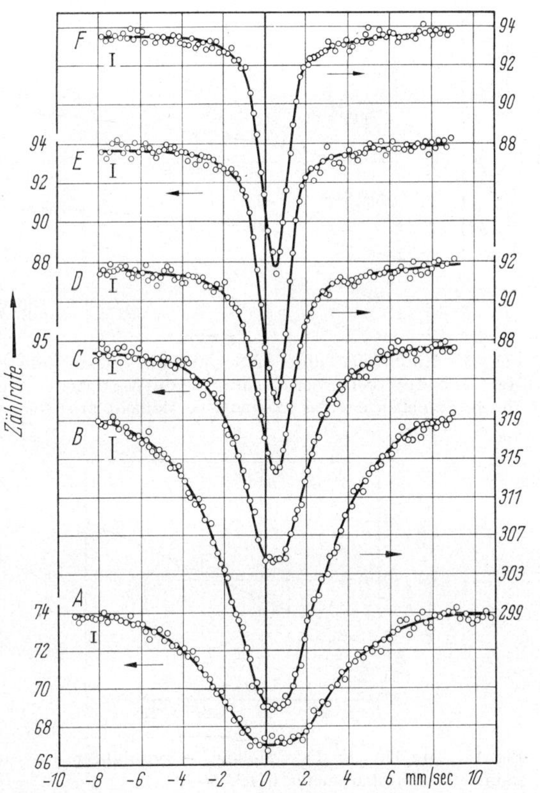

Fig. 8. $Mg_{1+x}Fe_{2-2x}Sn_xO_4$. Mössbauer-Spektren aufgenommen mit einer $Sn^{119m}O_2$-Quelle [*68BNPSTY*].

A: $x = 0.1$ bei $T = 77$ °K; *D*: $x = 0.3$;
B: $x = 0.1$ bei $T = 295$ °K; *E*: $x = 0.4$;
C: $x = 0.2$; *F*: $x = 0.5$.

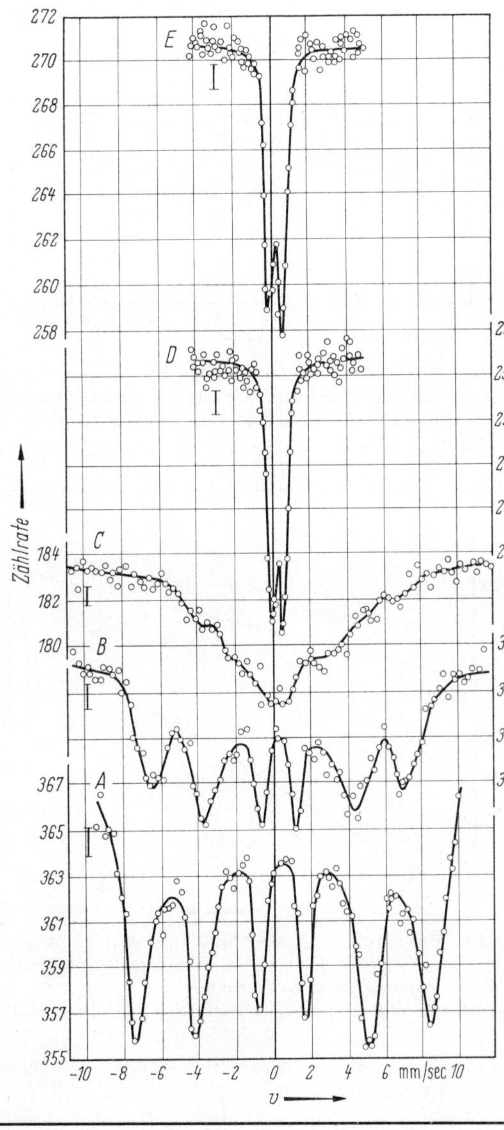

→

Fig. 9. $Mg_{1+x}Fe_{2-2x}Sn_xO_4$. Mössbauer-Spektren bei 295 °K, aufgenommen mit einer $Pd(Co^{57m})$-Quelle [*68BNPSTY*].

A: $x = 0.1$; *D*: $x = 0.4$;
B: $x = 0.2$; *E*: $x = 0.5$.
C: $x = 0.3$:

Fig. 11. Einfluß der teilweisen Substitution von Sn^{4+} durch Ge^{4+} auf das magnetische Moment pro Molekül p_m vs. Feld H_a für Mn-Sn Spinelle bei unterschiedlichen Temperaturen [59GM-1]. →

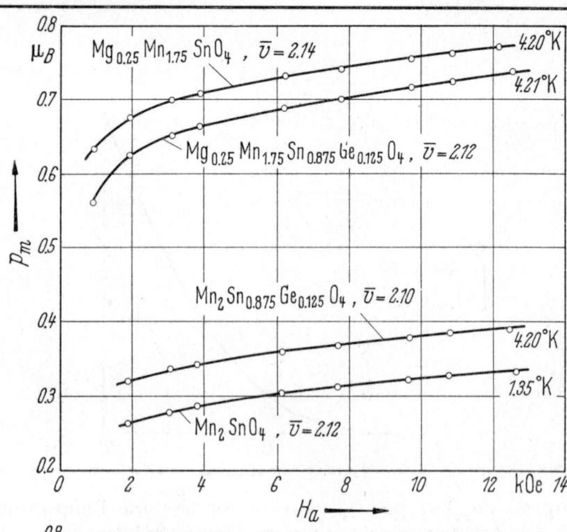

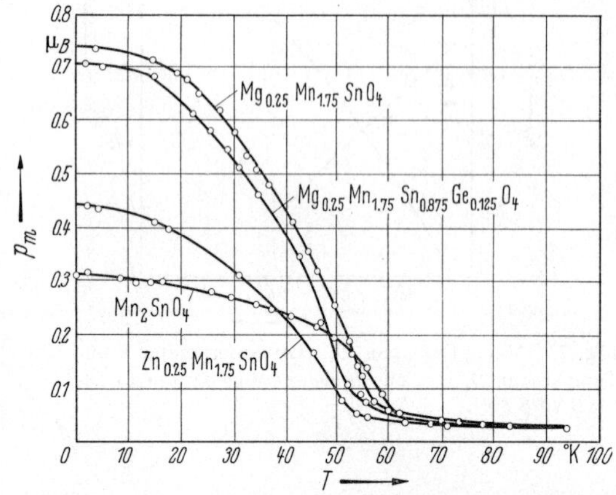

Fig. 10. Magnetisches Moment pro Molekül p_m für binäre, ternäre und quaternäre Mn-Sn Spinelle vs. Temperatur T bei konstantem Feld von $H_a = 7.7$ kOe [59GM-1].

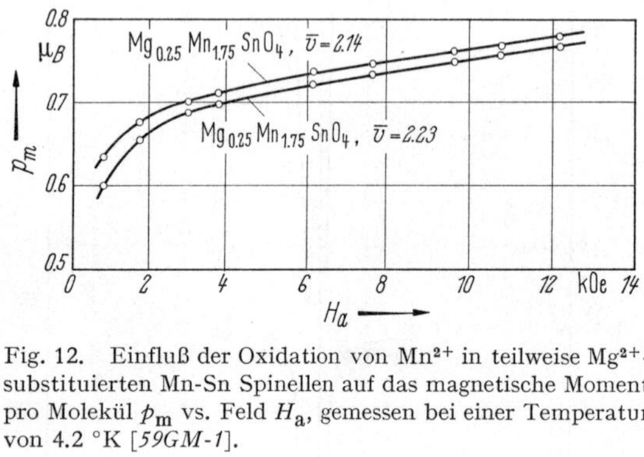

Fig. 12. Einfluß der Oxidation von Mn^{2+} in teilweise Mg^{2+}-substituierten Mn-Sn Spinellen auf das magnetische Moment pro Molekül p_m vs. Feld H_a, gemessen bei einer Temperatur von 4.2 °K [59GM-1].

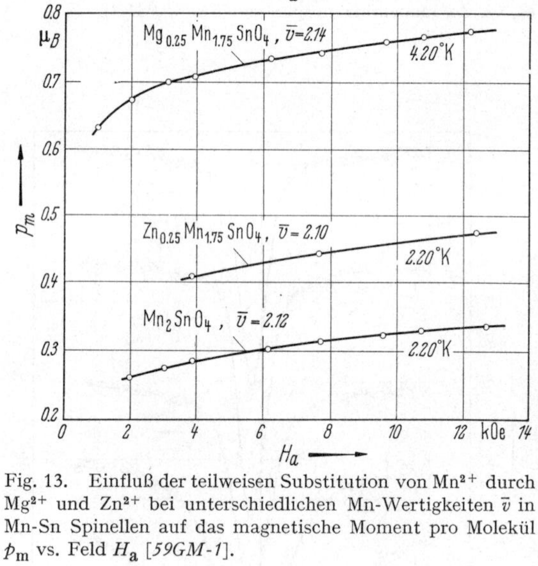

Fig. 13. Einfluß der teilweisen Substitution von Mn^{2+} durch Mg^{2+} und Zn^{2+} bei unterschiedlichen Mn-Wertigkeiten $\bar{v}$ in Mn-Sn Spinellen auf das magnetische Moment pro Molekül p_m vs. Feld H_a [59GM-1].

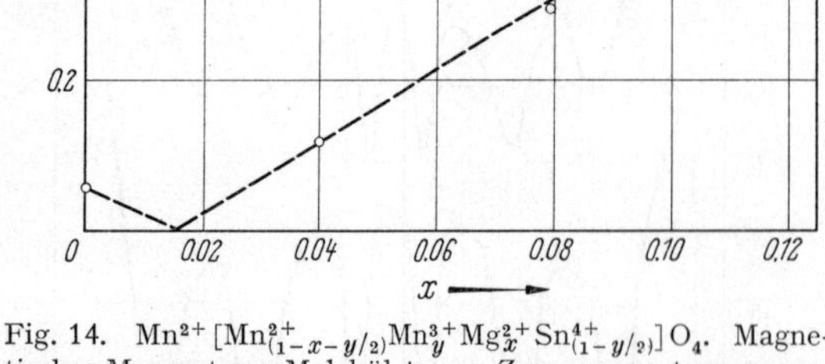

Fig. 14. $Mn^{2+}[Mn^{2+}_{(1-x-y/2)}Mn^{3+}_{y}Mg^{2+}_{x}Sn^{4+}_{(1-y/2)}]O_4$. Magnetisches Moment pro Molekül p_m vs. Zusammensetzungsparameter x für unterschiedliche Werte von y.
Gemessene Werte ——; berechnete Werte - - - - [61WMD].

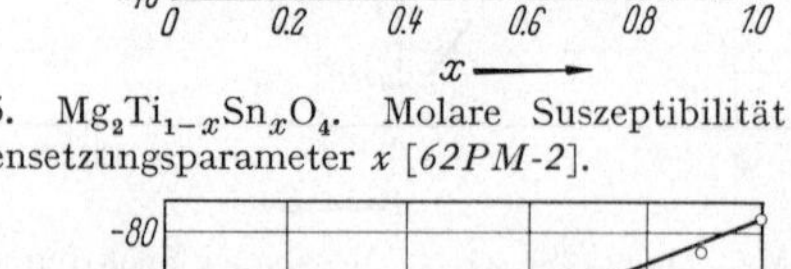

Fig. 15. $Mg_2Ti_{1-x}Sn_xO_4$. Molare Suszeptibilität vs. Zusammensetzungsparameter x [62PM-2].

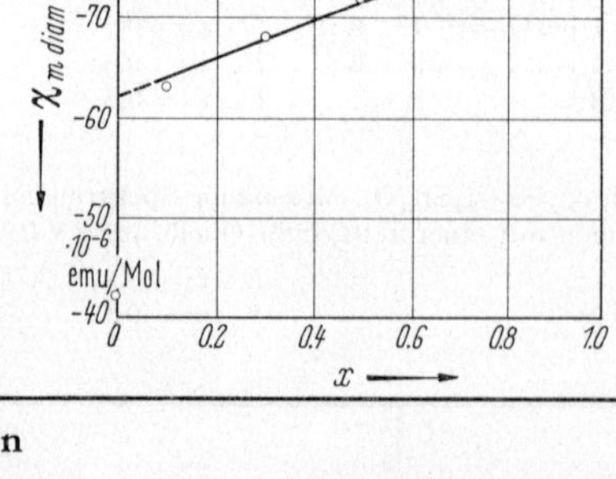

Fig. 16. $Zn_2Ti_{1-x}Sn_xO_4$. Molare Suszeptibilität vs. Zusammensetzungsparameter x [61PM-2]. →

6.3.21 Si Spinelle und substituierte Si Spinelle

(Literatur siehe 6.3.22, Seite 545)

Tab. 1. Kristallografische Eigenschaften binärer, ternärer und quaternärer Si^{4+} Oxid-Spinelle

Verbindung	x	a Å	ϱ_X g/cm³ [1])	Lit.	Fig.
Co_2SiO_4 [2])				*62R*	
		8.140 ± 0.005	9.5%	*63R, 64B*	
		8.137 ± 0.001		*65AKSFK*	
				66RM-1, 66RM-2, 66RM-3, 68NK	
Fe_2SiO_4 [3]) [4])		8.235	10.5 (12)%	*56R-1*	
				58R-3, 58R-2, 58R-1,	2, 3
				59R, 59W, 58MR,	
				63R, 64B,	
				66RM-1, 66RM-2, 66RM-4	
		8.234 ± 0.001		*66AF, 66RM-3, 65AKSFK*	
				68NK, 67EG	
Mg_2SiO_4 [5]) [6])			11 ± 3%	*56R-1, 56R-2, 56DR*	1, 2, 3
		8.07		*58R-2, 58R-3, 58R-4*	
		8.12	8.5 (4.7)%	*60DR, 59R, 59W*	
			>9%	*62RS, 63R, 58MR*	
		8.08	10.6%	*66AI, 66AF, 66SC*	
				66RM-1, 69MS	
		8.09	10.3%	*66RM-2, 68NK*	
		8.085	10.5%	*66RM-3*	
Ni_2SiO_4 [7])		8.026	9%	*58R-2*	
		8.044		*60R, 64B*	
		8.045	9 (8.7)%	*62R, 62RS*	
				62TR,	
				66RM-2,	
				68NK	
$(Mg_{1-x}Fe_x)_2SiO_4$	0.1			*66RM-1*	2, 3
	0.15	8.110 ± 0.002		*66RM-1, 66RM-2*	
	0.2				
	0.25	8.122		*66RM-3*	
	0.3			*66SC*	
	0.5			*58R-4, 58MR*	
	0.65				
	0.7				
	0.8	8.217 ± 0.001 [8])		*66AF*	
	0.83				
	0.9	8.203 ± 0.001 [8])		*60BE*	
$Mg_2(Ge_xSi_{1-x})O_4$	0.5	8.240 [9])		*58R-3, 56DR, 60DR*	
				62RS, 66RM-1	
$(Mg_{1-x}Ni_x)_2(Ge_xSi_{1-x})O_4$	0.45	8.137 [10])		*58R-3*	
	0.47	8.165 ± 0.002 [11])		*62RS*	
	0.37	8.158 ± 0.005 [12])		*62RS*	
	0.3	8.152 ± 0.1 [13])		*62RS*	

[1]) Anstieg der Dichte in % bei O—S (Olivin—Spinell) Transformation [*66RM-2*].
[2]) Unter Normalbedingungen Olivinstruktur. Übergang zur Spinellstruktur bei (700 ± 10) °C und (7.14 ± 2.04) · 10⁴ at [*63R*]; [*62R*].
[3]) Mineral: Fayalit.
[4]) Unter Normalbedingungen Olivinstruktur. Übergang zur Spinellstruktur bei (600 ± 10) °C und (3.87 ± 0.3) · 10⁴ at [*58R-3*]; bei 500 °C und 5.1 · 10⁴ at [*56R-1*].
[5]) Mineral: Forsterit.
[6]) Unter Normalbedingungen Olivinstruktur. Übergang zur stabilen Spinellstruktur bei 1500 °C und (1.785 ± 0.56) · 10⁵ at und bei 500 °C und 1.02 · 10⁵ at [*56R-1*]; bei 685 °C und 9.2 · 10⁴ at [*59R*]; bei 530 °C und (1.02 ± 0.153) · 10⁵ at und bei 820 °C und 0 at [*60DR*]; bei 600 °C und (1.325 ± 0.204) · 10⁵ at [*62RS*] und bei 900 °C und 1.73 · 10⁵ at [*66RM-2*].
[7]) Unter Normalbedingungen Olivinstruktur. Übergang zur Spinellstruktur bei (650 ± 10) °C und (1.84 ± 0.51) · 10⁴ at [*60R*].
[8]) Übergang zur Spinellstruktur für $x = 0.8$ bei (800 ± 10) °C und 8.67 · 10⁴ at; für $x = 0.9$ bei (800 ± 10) °C und 7.35 · 10⁴ at und bei (800 ± 10) °C und 6.72 · 10⁴ at, a = (8.218 ± 0.001) Å [*66AF*].
[9]) Bei 660 °C und 3.06 · 10⁴ at [*58R-3*].
[10]) Bei 580 °C und 7.13 · 10⁴ at, bei 560 °C und 5.61 · 10⁴ at, a = 8.157 Å [*58R-3*].
[11]) Bei (600 ± 10) °C und 6.12 · 10⁴ at, bei 600 °C und 7.12 · 10⁴ at Spinellstruktur mit a = (8.162 ± 0.1) Å [*62RS*].
[12]) Bei 600 °C und 8.12 · 10⁴ at [*62RS*]. [13]) Bei 600 °C und 9.12 · 10⁴ at [*62RS*].

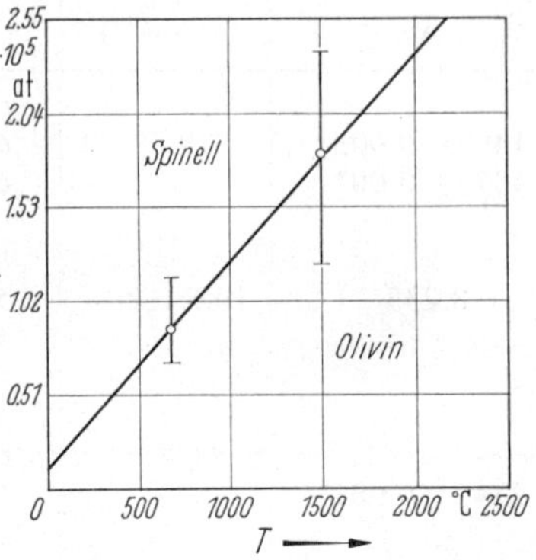

Fig. 1. Mg_2SiO_4. Spinell-Olivin-Übergang ○ gemessene Werte [*58R-3*].

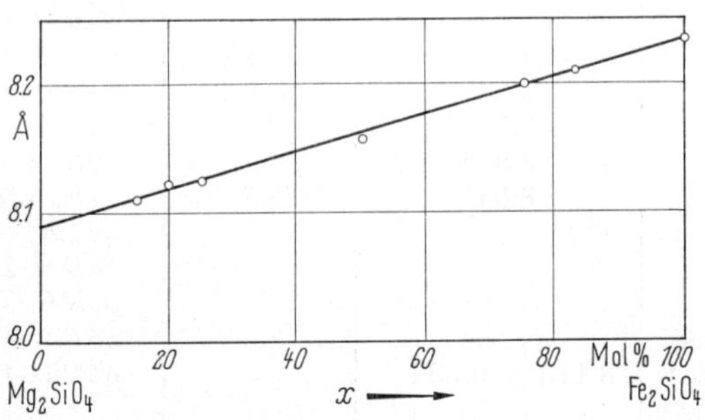

Fig. 2. Mg_2SiO_4—Fe_2SiO_4. Gitterparameter *a* vs. Zusammensetzung [*66RM-2*].

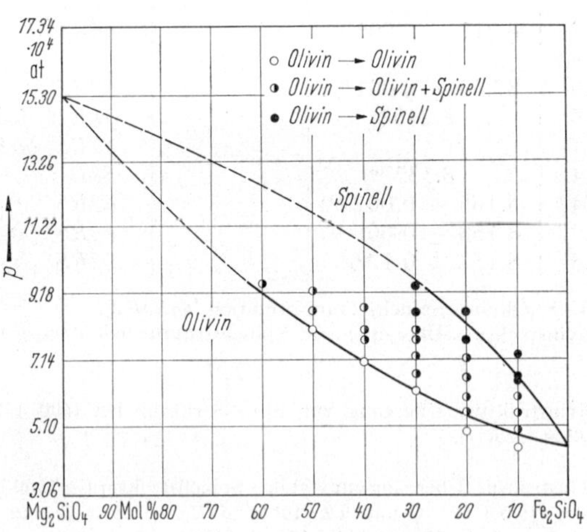

Fig. 3. Mg_2SiO_4—Fe_2SiO_4. Olivin-Spinell-Übergang vs. Zusammensetzung bei 800 °C und $(4.58 \cdots 9.36) \cdot 10^4$ at [*66AF*].

6.3.22 Literatur zu 6.3.15 ··· 6.3.21

22W | Wyckoff, R. W. G.: J. Am. Chem. Soc. **44** (1922) 1994.
23W | Wyckoff, R. W. G.: Chem. Zentr. **1923 I** 26.
26G | Goldschmidt, V. M.: Skr. Akad. Oslo **81** (1926) 107.
26Z | Zachariasen, W.: Norsk. Geol. Tidskr. **9** (1926) 65.
28H | Hoermann, F.: Z. Anorg. Allgem. Chem. **177** (1928) 145.
29NP | Natta, G., und L. Passerini: R. C. Acad. Lincei **9** (1929) 557.
30T | Taylor, N. W.: Z. Physik. Chem., Abt. B, **9** (1930) 241.
32BP | Barth, T. F. W., und E. Posnjak: Z. Krist. **82** (1932) 325.
41BHM | Byström, A., B. Hök, und H. Mason: Ark. Kemi. Min. Geol., **15 B** (1941) 1.
47DS | Donohue, J., und W. Shand, Jr.: J. Am. Chem. Soc. **69** (1947) 222.
47RR | Rüdorff, W., und B. Reuter: Z. Anorg. Allgem. Chem. **253** (1947) 194.
47/48S | Structure Reports **11** (1947—48) 498.
47VH | Verwey, E. J. W., und E. L. Heilmann: J. Chem. Phys. **15** (1947) 174.
50L | Lindquist, I.: Act. Chem. Scand. **4** (1950) 1066.
51KR | Kordes, E., und E. Röttig: Z. Anorg. Allgem. Chem. **264** (1951) 34.
52TK | Tazaki, H., und S. Kuwabara: J. Sci. Hiroshima Univ. A **15** (1952) 263.
53Y | Yamaguchi, G.: Bull. Chem. Soc. Japan **26** (1953) 204.
54E | Evans, R. C.: Einführung in die Kristallchemie, Barth Verlag, Leipzig **1954**.
54SR | Schmitz-Dumont, O., und B. Raederscheidt: Naturwiss. **41** (1954) 477.
56DR | Dachille, F., und R. Roy: Bull. Geol. Soc. Am. **67** (1956) 1682.
56R-1 | Ringwood, A. E.: Nature **178** (1956) 1303.
56R-2 | Ringwood, A. E.: Am. J. Sci. **254** (1956) 707.
58MR | Meijering, J. L., und C. J. M. Rooymans: Proc. Nederl. Ak. B. **61** (1958) 333.
58R-1 | Ringwood, A. E.: Bull. Geol. Soc. Am. **69** (1958) 129.
58R-2 | Ringwood, A. E.: Geochim. Cosmochim. Acta **13** (1958) 303.
58R-3 | Ringwood, A. E.: Geochim. Cosmochim. Acta **15** (1958) 18.
58R-4 | Ringwood, A. E.: Geochim. Cosmochim. Acta **15** (1958) 195.
58SBB | Schmitz-Dumont, O., H. Brokopf, und K. Burghardt: Z. Anorg. Allgem. Chem. **295** (1958) 7.
59GM-1 | Gilleo, M. A., und D. W. Mitchell: J. Phys. Chem. Solids **10** (1959) 182.
59GM-2 | Gilleo, M. A., und D. W. Mitchell: J. Appl. Phys. **30** (1959) 20S.
59PM | Poix, P., und A. Michel: Bull. Soc. Chim. France **1959** 1653.
59R | Ringwood, A. E.: Am. Mineralogist **44** (1959) 659.
59W | Wentorf, R. H.: Nature **183** (1959) 1617.
60BE | Boyd, F. R., und J. L. England: Carnegie Inst. Wash. Year Book **59** (1960) 48.
60DD | Dulac, J., und A. Durif: Compt. Rend. **251** (1960) 747.
60DR | Dachille, F., und R. Roy: Am. J. Sci. **258** (1960) 225.
60H | Hafner, S.: Schweiz. Miner. Petr. Mitt. **40** (1960) 207.
60R | Ringwood, A. E.: Nature **187** (1960) 1019.
61B | Bayer, G.: Naturwiss. **48** (1961) 46.
61PM-1 | Poix, P., und A. Michel: Bull. Soc. Chim. France **1961** 1891.
61PM-2 | Poix, P., und A. Michel: Bull. Soc. Chim. France **1961** 352.
61WMD | Wickham, D. G., N. Menyuk, und K. Dwight: Phys. Chem. Solids **20** (1961) 316.
62PM-1 | Poix, P., und A. Michel: Compt. Rend. **255** (1962) 2446.
62PM-2 | Poix, P., und A. Michel: Bull. Soc. Chim. France **1962** 1010.
62R | Ringwood, A. E.: Geochim. Cosmochim. Acta **26** (1962) 457.
62RS | Ringwood, A. E., und M. Seabrock: J. Geophys. Res. **67** (1962) 1975.
62TR | Tarte, P., und A. E. Ringwood: Nature **193** (1962) 971.
63B-1 | Blasse, G.: J. Inorg. Nucl. Chem. **25** (1963) 136.
63B-2 | Blasse, G.: J. Inorg. Nucl. Chem. **25** (1963) 230.
63B-3 | Blasse, G.: Z. Anorg. Allgem. Chem. **326** (1963) 44.
63B-4 | Blasse, G.: J. Inorg. Nucl. Chem. **25** (1963) 743.
63HD | Harrison, R. W., und E. J. Delgrosso: J. Electrochem. Soc. **110** (1963) 205.
63R | Ringwood, A. E.: Nature **198** (1963) 79.
64B | Blasse, G.: Philips Res. Rept. Suppl. No. 3 (1964).
64P | Poix, P.: Ann. Chim. (Paris) **9** (1964) 261.
65AKSFK | Akimoto, S., T. Katsura, Y. Syono, H. Fujisawa, und E. Komeda: J. Geophys. Res. **70** (1965) 5269.
65H | Haas, C.: J. Phys. Chem. Solids **26** (1965) 1225.
65HHW | Hannaford, P., C. J. Howard, und J. W. G. Wignall: Phys. Letters **19** (1965) 257.
65P | Poix, P.: Ann. Chim. (Paris) **10** (1965) 49.
65SK | Smirnova, J. N., und J. P. Kislyakov: Inorg. Mat. **1**, 2 (1965) 1064.

65SS	Sadikov, G. G., und N. A. Shishakov: Bull. Acad. Sci. USSR, Div. Chem. Sci. (English Transl.) **5** (1965) 1243.
66AF	Akimoto, S., und H. Fujisawa: Earth Planet. Sci. Lett. **1** (1966) 237.
66AI	Akimoto, S., und Y. Ida: Earth Planet. Sci. Lett. **1** (1966) 358.
66BMPM	Billet, Y., J. Morgenstern-Badaran, P. Poix, und A. Michel: Compt. Rend. C **262** (1966) 751.
66BP	Bertaut, E.-F., und G. Patrat: Bull. Soc. Franc. Mineral. Crist. **88** (1966) 586.
66BRGG	Banerjee, S. K., W. O'Reilly, T. G. Gibb, und N. N. Greenwood: Phys. Letters **20** (1966) 455.
66CD	Clark, G. M., und W. P. Doyle: Spectrochim. Acta **22** (1966) 1441.
66DL	Dupius, T., und V. Lorenzelli: Compt. Rend. **262** B (1966) 489.
66O	Okazaki, H.: Japan. J. Appl. Phys. **5** (1966) 559.
66P	Pistorius, C. W. F. T.: J. Chem. Phys. **44** (1966) 4532.
66R	Reinen, D.: Theoret. Chim. Acta **5** (1966) 312.
66RM-1	Ringwood, A. E., und A. Mayor: Earth Planet. Sci. Lett. **1** (1966) 209.
66RM-2	Ringwood, A. E., und A. Mayor: Earth Planet. Sci. Lett. **1** (1966) 241.
66RM-3	Ringwood, A. E., und A. Mayor: J. Geophys. Res. **71** (1966) 4448.
66RM-4	Ringwood, A. E., und A. Mayor: Earth Planet. Sci. Lett. **1** (1966) 135.
66SC	Sclar, C. B., und L. C. Carrison: Trans. Am. Geophys. Union **41** (1966) 207.
66VJD	Vincent, H., J.-C. Joubert, und A. Durif: Bull. Soc. Chim. France **1966** 246.
67EG	Eibschütz, M., und U. Gamiel: Solid State Commun. **5** (1967) 267.
67HMST	Hohmann, H. H., W. Müller, H. Schmalzried und J. D. Tretjakov: Proc. Brit. Ceram. Soc. **8** (1967) 91.
67HS	Hahn, H., und G. Strick: Naturwiss. **54** (1967) 42.
67LR	Liebertz, J., und C. J. M. Rooymans: Solid State Commun. **5** (1967) 405.
67PB	Poix, P., und J.-C. Bernier: Bull. Soc. Chim. France **1967** 605.
67T	Tellier, J.-C.: Rev. Chim. Mineral. **4** (1967) 325.
68BNPSTY	Bokov, V. A., G. V. Novikov, O. B. Proskuryakov, Yu. G. Saksonov, V. A. Trukhtanov, und S. I. Yushchuk: Soviet Phys.-Solid State (English Transl.) **10** (1968) 855.
68CDR	Choisnet, J., A. Deschauvres, und B. Raveau: Compt. Rend. **266** (1968) 543.
68GM	Gupta, M. P., und H. B. Mathur: J. Phys. Chem. Solids **29** (1968) 1479.
68NK	Navrotsky, A., und O. J. Kleppa: J. Inorg. Nucl. Chem. **30** (1968) 479.
68NP	Nogues, M., und P. Poix: Ann. Chim. (Paris) **3** (1968) 335.
68SEH	Strick, G., G. Eulenberger, und H. Hahn: Z. Anorg. Allgem. Chem. **357**(1968) 338.
68TR	Trichet, L., und J. Rouxel: Compt. Rend. **267** (1968) 1322.
69MS	Moore, P. B., und J. V. Smith: Nature **221** (1969) 653.

7 Hexagonal ferrites

7.0 Introduction — Einleitung

Since the discovery in 1951 of the, technically, very important magnetic properties of the ferrite $BaFe_{12}O_{19}$ with hexagonal crystal structure, many related compounds have been synthesized, and the properties of these mostly completely new structures have been investigated. At room temperature the magnetic moments of most of these compounds can be ordered in groups in such a way that the magnetic moments of the ions of one group are mutually parallel oriented, whereas the magnetic moments of the ions of different groups are oriented antiparallel to each other. Such an incompletely compensated antiferromagnetism was called ferrimagnetism by Néel, who was the first to describe this type of magnetism in order to explain the magnetization of ferrites with spinel structure [*48N*]. The technical interest of the oxides with hexagonal crystal structure is given by the fact that some of these materials show a very high uniaxial magnetic anisotropy, so that they can be used as a ceramic permanent magnetic material, which for some applications can compete technically, and economically, with the metallic permanent magnets. Other hexagonal ferrites have interesting properties as magnetic cores at frequencies above about 100 Mc/s. In electronic equipment for microwaves, single crystals and polycrystalline samples of hexagonal ferrites with still other chemical compositions are used successfully, because of their exceptionally high internal magnetic field, or because of the non-linear effects which appear at already relatively low amplitude of the high frequency field. For the h. f. applications it is essential that these semi-conducting ferrites can be prepared in such a way that their conductivity at room temperature is very low.

The chemical compositions and the crystal structures of the hexagonal ferrites show many similarities, and they are therefore reviewed in sections 7.3 and 7.4. Section 7.5 includes the paramagnetic properties of the hexagonal ferrites while sections 7.6 ··· 7.10 deal with the properties of the compounds of each crystal structure separately. Because of the various uncertainties regarding the crystal structures of the calcium ferrites and the substituted calcium ferrites, the properties of these ferrites are given separately in section 7.11.

Seit dem Jahre 1951, als die technisch wichtigen magnetischen Eigenschaften des Bariumferrits $BaFe_{12}O_{19}$ bekannt wurden, sind viele hiermit verwandte Verbindungen mit hexagonaler Kristallstruktur gefunden und insbesondere auf ihre magnetischen Eigenschaften untersucht worden. Die meisten dieser Verbindungen sind bei Zimmertemperatur ferrimagnetisch, d. h. daß man die magnetischen Ionen, die diese Verbindungen enthalten, in mehrere Untergruppen zusammenfassen kann, so daß in jeder Untergruppe die magnetischen Momente der Ionen untereinander parallel gerichtet sind, während die magnetischen Momente von Ionen verschiedener Unterguppen antiparallel zueinander orientiert sind. Dieser nicht völlig kompensierte Antiferromagnetismus, Ferrimagnetismus genannt, wurde zuerst von Néel 1948 für die Deutung der Größe der spontanen Magnetisierung von Ferriten mit Spinellstruktur [*48N*] vorgeschlagen. Die technische Bedeutung einiger dieser Oxide mit hexagonaler Kristallstruktur liegt darin, daß sie eine sehr hohe einachsige magnetische Anisotropie aufweisen, wodurch sie in keramischer Form als Dauermagnetwerkstoff geeignet sind. In einigen Anwendungen können sie dem Stahllegierungsmagnet technisch und wirtschaftlich überlegen sein. Andere hexagonale Ferrite zeigen gute Eigenschaften als Spulenkerne im Frequenzbereich oberhalb von etwa 100 MHz, also dort, wo die Ferrite mit Spinellstruktur wegen zu hoher Verluste nicht mehr eingesetzt werden können. Einkristalline und polykristalline Proben von wieder anderer chemischer Zusammensetzung können in elektronischen Bauelementen der Mikrowellentechnik benutzt werden, und zwar teils wegen ihrer anomal hohen inneren Magnetfelder und teils wegen der leicht anzuregenden nichtlinearen Effekte bei der ferromagnetischen Resonanz. Für die Anwendung in hochfrequenten Wechselfeldern ist es wesentlich, daß diese halbleitenden Stoffe bei Zimmertemperatur einen hohen spezifischen elektrischen Widerstand haben.

Die miteinander eng verwandten chemischen Zusammensetzungen und Kristallstrukturen der hexagonalen Ferrite werden im folgenden zuerst besprochen (7.3, 7.4). Nach den paramagnetischen Eigenschaften der hexagonalen Ferrite (7.5) werden in den Kapiteln 7.6···7.10 jeweils die magnetischen und elektrischen Eigenschaften für die Ferrite vom **M**-, **Z**-, **Y**- und **U**-Typ behandelt. Die Eigenschaften der Calcium-Ferrite und der substituierten Calcium-Ferrite werden wegen der Unklarheiten, die noch hinsichtlich ihrer Struktur bestehen, im letzten Kapitel (7.11) gesondert dargestellt.

7.1 Quantities and units — Größen und Maßsysteme

In literature the magnetic properties are usually discussed in the electromagnetic cgs system, so it is reasonable that this system will also be used in the following tables and figures. The relation

Da in der einschlägigen Literatur die magnetischen Eigenschaften meistens im elektromagnetischen cgs-System gebracht werden, wird dies im folgenden auch benutzt. Es werden dann die In-

between the magnetic induction B, the magnetic field H and the magnetization M is then given by:

duktion B, die magnetische Feldstärke H und die Magnetisierung M durch folgenden Zusammenhang gegeben:

$$B = H + 4\pi M\,. \tag{1}$$

a) Conductivity

The temperature-dependence of the conductivity σ of hexagonal ferrites can usually be described by the equation:

a) Elektrische Leitfähigkeit

Die Temperaturabhängigkeit der elektrischen Leitfähigkeit σ von hexagonalen Ferriten kann im allgemeinen mit folgender Formel beschrieben werden:

$$\sigma = \sigma_\infty\, e^{-q/kT} \tag{2}$$

where the activation energy q is probably a measure for the increase of the mobility of the negative charge carriers with increasing temperature.

wobei q die Aktivierungsenergie ist, die wahrscheinlich die Zunahme der Beweglichkeit der negativen Ladungsträger mit der Temperatur bestimmt.

b) Magnetic moment

Ferrites have mostly been obtained by a sintering process, so that the samples show a certain porosity. The overall density of the sample is then lower than the X-ray density of the compound. The saturation magnetization therefore is often not expressed as a magnetic moment per unit of volume M_s or $4\pi M_s$, but rather as a specific saturation magnetization per gram of the substance, σ_s. The relation between both quantities is given by:

b) Magnetisches Moment

Da Ferrite meist durch Sinterprozesse hergestellt werden, wodurch naturgemäß für die Proben nicht die Röntgendichte erreicht wird, sondern immer noch eine gewisse Porosität vorhanden ist, wird die Sättigungsmagnetisierung oft nicht als magnetisches Moment pro Volumeneinheit M_s oder $4\pi M_s$ angegeben, sondern vielmehr als spezifische Sättigungsmagnetisierung pro Gramm des Stoffes, σ_s. Der Zusammenhang zwischen beiden Größen ist gegeben durch:

$$M_s = \sigma_s \varrho\,, \tag{3}$$

where ϱ denotes the density of the porous sample. The specific saturation magnetization at 0 °K, σ_s^0, is a measure for the magnetic moment p_m per molecule of the substance. The quantity p_m is usually expressed in number of Bohr magnetons according to the equation:

wobei ϱ die Dichte des porösen Werkstoffes darstellt. Die spezifische Sättigungsmagnetisierung beim absoluten Nullpunkt der Temperatur, σ_s^0, ist ein Maß für das magnetische Moment p_m pro Molekül des Stoffes. Der Zusammenhang wird gegeben durch die Formel:

$$p_m = \frac{(M_A)}{N_L \mu_B} \cdot \sigma_s^0 = \frac{(M_A)}{5505}\, \sigma_s^0\,. \tag{4}$$

In this equation (M_A) is the molar mass of the substance A,

Hierin ist p_m ausgedrückt in Bohrschen Magnetonen, (M_A) ist das Molekulargewicht der Verbindung A,

$$N_L = 6.02 \cdot 10^{23}\ \text{mole}^{-1}$$

the number of molecules per mole and

die Loschmidtsche Zahl und

$$\mu_B = 9.27 \cdot 10^{-21}\ \text{erg Gauss}^{-1}$$

the Bohr magneton.

die Größe des Bohrschen Magnetons.

c) Magneto-crystalline anisotropy of hexagonal crystals

Because of the symmetry of the hexagonal crystal lattice, the magneto-crystalline anisotropy energy is given by the equation:

c) Magnetokristalline Anisotropie hexagonaler Kristalle

Aufgrund von Symmetrieüberlegungen kann man allgemein für die magnetokristalline Anisotropie von Verbindungen mit hexagonaler Kristallsymmetrie schreiben:

$$w_K = K_1 \sin^2\theta + K_2 \sin^4\theta + K_3' \sin^6\theta + K_3 \sin^6\theta \cos 6(\phi + \psi)\,. \tag{5}$$

The angles θ and ϕ are polar coordinates and the constants K_i are the coefficients of the magneto-crystalline anisotropy. The phase angle ψ is zero for a particular choice of the axis of the coordinate system. The term with the coefficient K_3' can usually be neglected. In cases where the term

Die Winkel θ und ϕ sind Polarkoordinaten und die Konstanten K_i Kristallanisotropiekoeffizienten. Der Phasenwinkel ψ kann durch richtige Wahl der Bezugsachsen gleich Null gesetzt werden. Der Term mit dem Koeffizienten K_3' wird oft vernachlässigt. Wenn der Term mit K_1 überwiegt, dann ist für

with K_1 is predominant, the spontaneous magnetization is oriented parallel to the c axis for $K_1 > 0$, the crystal has a so-called preferential direction of magnetization. For $K_1 < 0$, the spontaneous magnetization is oriented perpendicular to the c axis, the crystal has a so-called preferential plane of magnetization. In general the angle θ_0 between the direction of the spontaneous magnetization and the c axis is a function of K_1 and K_2, as illustrated in Fig. 1. In cases where $0 < \theta_0 < 90°$ the crystal shows a preferential cone for the spontaneous magnetization with a vertex $2\theta_0$, which is given by the equation:

$K_1 > 0$ die spontane Magnetisierung parallel zur c-Achse des Kristalles orientiert und man spricht von einer Vorzugsrichtung der Magnetisierung im Kristall. Für $K_1 < 0$ ist die spontane Magnetisierung senkrecht zur c-Achse orientiert und man spricht von einer Vorzugsebene der Magnetisierung im Kristall oder auch von planarer Anisotropie. Fig. 1 zeigt den Winkel θ_0, den die spontane Magnetisierung mit der hexagonalen Achse des Kristalls in Abhängigkeit der Koeffizienten K_1 und K_2 bildet. Wenn $0 < \theta_0 < 90°$ ist, spricht man von einem magnetischen Vorzugskegel im Kristall, dessen Öffnungswinkel $2\theta_0$ gegeben ist durch:

$$\sin\theta_0 = \sqrt{-K_1/2K_2}\,. \tag{6}$$

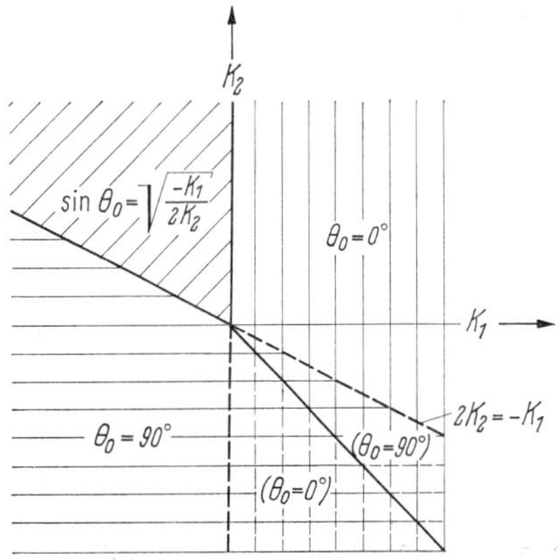

Fig. 1. The relation between the preferential direction of the magnetization vector in a hexagonal crystal and the corresponding values of the magneto-crystalline anisotropy coefficients K_1 and K_2. For $\theta_0 = 0$ the c axis is the preferential direction, for $\theta_0 = 90°$ the basal plane is the preferential plane for the magnetization. In the sector of the diagram for which $\sin\theta_0 = \sqrt{-K_1/2K_2}$ all directions of the magnetization which make an angle of θ_0 with the c axis of the crystal have the lowest energy (preferential cone for the magnetization). In the region $-2K_2 > K_1 > 0$ the spontaneous magnetization has metastable orientations [*59CSEFWGDF*].

The magneto-crystalline anisotropy field H^A is defined as the effective field that causes the same stiffness for a rotation of the magnetization over a small angle out of its preferential direction as the magneto-crystalline anisotropy does. In the case of a rotation with constant angle ϕ we get:

Das Kristallanisotropiefeld H^A ist definiert als die effektive Feldstärke, die die gleiche Steifigkeit wie die Kristallanisotropie ergeben würde, wenn der Magnetisierungsvektor um einen kleinen Winkel aus der Vorzugsrichtung der Magnetisierung herausgedreht wird. Für den Fall, daß bei dieser Drehung der Winkel ϕ konstant bleibt, ergibt sich:

$$H_\theta^A = (1/M_s)\,(\partial^2 w_K/\partial\theta^2)_{\phi\,=\,\text{constant}}\,, \tag{7}$$

so that for

und gilt für

$$\theta_0 = 0: H_\theta^A = H^A = 2K_1/M_s\,, \tag{7a}$$

$$\theta_0 = 90°: H_\theta^A = -(K_1 + 2K_2)/M_s\,, \tag{7b}$$

$$\sin\theta_0 = \sqrt{-K_1/2K_2}: H_\theta^A = 2(K_1/K_2)\,(K_1 + 2K_2)/M_s\,. \tag{7c}$$

The effective anisotropy field for a rotation of the magnetization out of its equilibrium orientation along the surface of the preferential cone is given by:

Das effektive Feld für die Drehung des Magnetisierungsvektors in der Oberfläche des Vorzugskegels ist gegeben durch:

$$H_\phi^A = (1/M_s)\,(\partial^2 w_K/\partial^2\phi^2)_{\substack{\phi=0\\ \theta=\theta_0}} = 36\,|K_3|\sin^4\theta_0\,. \tag{8}$$

d) Linear magnetostriction of a hexagonal crystal

According to [*61G*] the linear magnetostriction of a hexagonal crystal is given by the equation:

d) Lineare Magnetostriktion eines hexagonalen Kristalls

Nach [*61G*] wird die lineare Magnetostriktion eines Kristalls mit hexagonaler Symmetrie gegeben durch die Gleichung:

$$\begin{aligned}\lambda = \Delta l/l = k_0 &+ k_0^*\,(\beta_3^2 - \tfrac{1}{3}) + k_1(\alpha_3^2 - \tfrac{1}{3}) + k_2(\alpha_3^2 - \tfrac{1}{3})\,(\beta_3^2 - \tfrac{1}{3}) + \\ &+ k_3\,\{(\alpha_1\beta_1 + \alpha_2\beta_2)^2 - \tfrac{1}{2}\,(a_1^2 + \alpha_2^2)\,(\beta_1^2 + \beta_2^2)\} + \\ &+ 2k_4\alpha_3\beta_3(\alpha_1\beta_1 + \alpha_2\beta_2) + \cdots + k_0' H + k_0^{*\prime} H\,(\beta_3^2 - \tfrac{1}{3}) + \cdots\end{aligned} \tag{9}$$

where β_i indicate the direction in which the magnetostriction $\Delta l/l$ is measured and α_i are the direction cosines of the magnetization; α_3 and β_3 are the direction cosines with respect to the c axis of the hexagonal crystal.

e) Ferromagnetic resonance in hexagonal crystals

The general equation for the ferromagnetic resonance frequency f_{res} for a magnetization M related to a preferential direction in the polar coordinates (θ and ϕ) is given by:

worin β_i die Richtung angibt, in der die Magnetostriktion $\Delta l/l$ gemessen wird und α_i die Richtungskosinus der Magnetisierung sind; α_3 und β_3 sind die Richtungskosinus in bezug auf die c-Achse des hexagonalen Kristalls.

e) Ferromagnetische Resonanz in hexagonalen Kristallen

Die allgemeine Formel für die ferromagnetische Resonanzfrequenz f_{res} einer Magnetisierung M, die an eine Vorzugsrichtung (gegeben durch die Polarkoordinaten θ und ϕ) gebunden ist, ist nach [*55SB*]:

$$2\pi f_{res} = (\gamma/M \sin\theta)\,(w_{\theta\theta} w_{\phi\phi} - w_{\theta\phi}^2)^{1/2}\,, \tag{10}$$

where $w(\theta, \phi)$ is the direction-dependent energy density of the magnetization [*55SB*]. The gyromagnetic ratio γ is given by

worin $w(\theta, \phi)$ die richtungsabhängige Kristallanisotropieenergiedichte für die Magnetisierung ist. Das gyromagnetische Verhältnis γ ist gegeben durch:

$$\gamma = g_{eff}\, e/2mc \text{ [emu]} \tag{11}$$

where g_{eff} is the effective Landé spectroscopic splitting factor for two or more sublattices and m is the electron mass. For the application of the general equation (10) two cases have to be considered:

I. Hexagonal crystal with preferential direction of magnetization parallel to the c axis.

a) When a d. c.-magnetic field H is applied parallel to the preferential direction eq. (10) yields to the resonance equation first given by Kittel [*48K*]:

worin g_{eff} der effektive Landé-Faktor im Falle mehrerer magnetischer Untergitter und m die Masse eines Elektrons ist. Für die Anwendung der allgemeinen Beziehung (10) lassen sich zwei Sonderfälle unterscheiden:

I. Hexagonaler Kristall mit magnetischer Vorzugsrichtung parallel zur c-Achse.

a) Wenn das Magnetfeld H in Vorzugsrichtung orientiert ist, ergibt (10) die zuerst von Kittel [*48K*] aufgestellte Beziehung:

$$2\pi f_{res} = \gamma\,[(H + H^A) + (N_x - N_z)M]^{1/2} \cdot [(H + H^A) + (N_y - N_z)M]^{1/2}\,. \tag{12}$$

Here N_z is the demagnetization factor in the direction of the c axis and N_x and N_y are the demagnetization factors in the other two directions of the orthogonal coordinate system.

b) For an a. c. magnetic field oriented perpendicular to the c axis, i. e. in the case where $H \perp H^A$ the eq. (10) gives according to [*55SB*]:

Hierin ist N_z der Entmagnetisierungsfaktor in Richtung der c-Achse, N_x und N_y sind Entmagnetisierungsfaktoren in den beiden dazu senkrechten Koordinatenrichtungen.

b) Für den Fall, daß das Magnetfeld senkrecht zur c-Achse orientiert ist, also $H \perp H^A$, ergibt sich nach [*55SB*] aus (10):

$$2\pi f_{res} = \gamma \left[1 - \frac{(N_y - N_x)\,M}{H^A - (N_z - N_y)\,M}\right]^{1/2} \cdot [\{H^A - (N_z - N_y)\,M\}^2 - H^2]^{1/2} \tag{13a}$$

for | für

$$H \leqq H^A - (N_z - N_y)\,M\,,$$
$$H^A > 4\pi M\,,$$

and | und

$$4\pi f_{res} = \gamma[H - (N_y - N_x)\,M]^{1/2} \cdot [H - \{H^A - (N_z - N_y)\,M\}]^{1/2} \tag{13b}$$

for | für

$$H \geqq H^A - (N_z - N_y)\,M\,,$$
$$H^A > 4\pi M\,.$$

Fig. 2 shows the graphs for the eqs. (12), (13a) and (13b) for the cases where demagnetizing fields do not appear.

In Fig. 2 sind die Kurven f_{res} (H) nach den Gleichungen (12), (13a) und (13b) dargestellt für den Fall, daß keine entmagnetisierenden Felder auftreten.

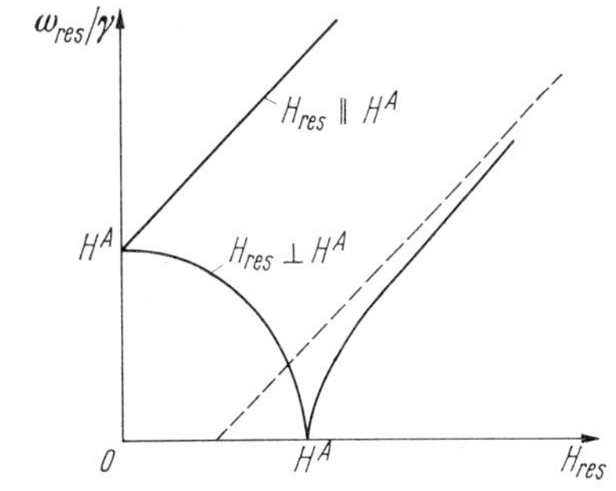

Fig. 2. The angular frequency for ferromagnetic resonance, ω_{res}, as a function of the external magnetic field H_{res} parallel or perpendicular to the preferential direction for the mangetization (c axis) of a crystal with hexagonal symmetry [*59SW*].

II. Hexagonal crystals with planar anisotropy.

a) When the magnetic field H is oriented parallel to the y-direction in the preferential plane, which will also be the xy-plane, of the orthogonal system the resonance frequency given by (10) is as follows:

II. Hexagonaler Kristall mit planarer Anisotropie.

a) Wenn das magnetische Gleichfeld H parallel zur y-Achse in der Vorzugsebene liegt, die gleichzeitig die xy-Ebene des kartesischen Koordinatensystems ist, dann wird gemäß (10)

$$2\pi f_{res} = \gamma[H - (N_y - N_x)M]^{1/2} \cdot [H + H_\theta^A + (N_z - N_y)M]^{1/2}. \tag{14}$$

If only magneto-crystalline energy is considered, according to [*59BK*] this equation reduces to:

Für den Fall, daß nur die Kristallanisotropieenergie berücksichtigt wird, reduziert sich diese Gleichung nach [*59BK*] auf:

$$2\pi f_{res} = \gamma\left[H + \frac{2K_1}{M} - \frac{6K_4}{M}\right]^{1/2} \cdot \left[H \mp 36\frac{K_4}{M}\right]^{1/2}, \tag{14a}$$

where the minus and the plus signs in the second factor correspond to an instable and a stable equilibrium orientation of the magnetization in the plane, respectively.

b) For a magnetic field parallel to the direction for difficult magnetization (c axis) the resonance frequency is given by:

wobei sich das Minus- oder Pluszeichen des zweiten Faktors auf eine instabile bzw. stabile Gleichgewichtslage der Magnetisierung in der Vorzugsebene bezieht.

b) Für den Fall, daß das magnetische Gleichfeld in der schweren Richtung (c-Achse) liegt, ergibt sich:

$$2\pi f_{res} = \gamma[(H - H_\theta^A) - (N_z - N_x)M]^{1/2} \cdot [(H - H_\theta^A) - (N_z - N_y)M]^{1/2} \tag{15}$$

for

für

$$H > H^A + N_z M.$$

In the derivation of eqs. (14) and (15) the anisotropy field strength H_ϕ^A is neglected with respect to H and H_θ^A since it is usually lower by several orders of magnitude. In the special case where only magneto-crystalline energy has to be considered, [*59BK*], the eq. (15) reduces to:

Das Anisotropiefeld H_ϕ^A ist bei der Ableitung der Formeln (14) und (15) nicht berücksichtigt, weil es gegenüber H und H_θ^A meistens um Größenordnungen kleiner ist. Für den Fall, daß keine Formanisotropie berücksichtigt zu werden braucht, ergibt sich nach [*59BK*] anstelle von (15) der Ausdruck:

$$2\pi f_{res} = \gamma H^{1/2}\left(H - \frac{2K_1}{M} - \frac{4K_2}{M} - \frac{6K_3}{M}\right). \tag{15a}$$

7.2 List of symbols — Symbolliste

H [Oe]	Magnetic field	Magnetfeld
H_c	Coercive force for the case of practically no difference between $_MH_c$ and $_BH_c$	Koerzitivkraft für den Fall, daß $_MH_c$ und $_BH_c$ praktisch gleich sind
$_BH_c$	Coercive force for the induction	Koerzitivkraft für die Induktion
$_MH_c$	Coercive force for the magnetization	Koerzitivkraft für die Magnetisierung
H_{res}	Magnetic field for ferromagnetic resonance	Magnetfeld bei der ferromagnetischen Resonanz
H_{ext}	External magnetic field	Äußeres Magnetfeld
H_T	Hyperfine magnetic field	Magnetfeld der Hyperfeinstruktur
H_T^0	Hyperfine magnetic field at 0 °K	Magnetfeld der Hyperfeinstruktur bei 0 °K
H^A	Magneto-crystalline anisotropy field in the case of a preferential direction for the magnetization	Kristallanisotropiefeld für den Fall einer Vorzugsrichtung der Magnetisierung
H_ϕ^A	Magneto-crystalline anisotropy field for rotation of the magnetization in a preferential plane for the magnetization	Kristallanisotropiefeld für Drehung der Magnetisierung in einer Vorzugsebene der Magnetisierung
H_θ^A	Magneto-crystalline anisotropy field for rotation of the magnetization out of a preferential plane for the magnetization	Kristallanisotropiefeld für Drehung der Magnetisierung aus einer Vorzugsebene der Magnetisierung heraus
ΔH	Line width of the magnetic resonance as measured between the two points of maximum slope of the absorption curve	Linienbreite der magnetischen Resonanz gemessen zwischen den Punkten von maximaler Neigung der Absorptionskurve
ΔH_k	Line width for spin waves with wave vector k	Linienbreite für Spinwellen mit dem Wellenvektor k
h_{crit}	Critical amplitude of the r.f. field for non-linear effects in the ferromagnetic resonance	Kritische Amplitude des HF-Feldes für nichtlineare Effekte bei der ferromagnetischen Resonanz
h_{rf}	High-frequency magnetic measuring field	Hochfrequentes magnetisches Meßfeld
B [Gauss]	Induction	Induktion
$(BH)_{max}$ [Gauss · Oe]	Maximum energy product of permanent magnet materials	Maximales Energieprodukt von Dauermagnetwerkstoffen
N	Demagnetization factor	Entmagnetisierungsfaktor
N_x, N_y, N_z	Demagnetization factors in three orthogonal directions	Entmagnetisierungsfaktor in drei orthogonalen Richtungen
M [emu/cm³ = Gauss]	Magnetic moment per unit volume = magnetization	Magnetisches Moment pro Volumeneinheit = Magnetisierung
M_s	Saturation magnetic moment per unit volume	Sättigungsmagnetisierung
M_T	Sublattice magnetization	Untergittermagnetisierung
M_T^0	Sublattice magnetization at 0 °K	Untergittermagnetisierung bei 0 °K
$m_{2\omega}$	Amplitude of second harmonic magnetization	Amplitude der zweiten Harmonischen der Magnetisierung
$\sigma = M/\varrho$ [emu/g] or [Gauss cm³/g]	Magnetic moment per g = specific magnetization	Magnetisches Moment pro Gramm = spezifische Magnetisierung
σ_s	Specific spontaneous magnetization ≡ specific saturation magnetization	Spezifische spontane Magnetisierung ≡ spezifische Sättigungsmagnetisierung
σ_s^0	σ_s at 0 °K	σ_s bei 0 °K
σ_s^{20}, σ_s^{RT}	σ_s at room temperatur	σ_s bei Zimmertemperatur
n_B	number of Bohr magnetons per formula unit	Anzahl der Bohrschen Magnetonen pro Formeleinheit
$\mu_B = 9.27 \cdot 10^{-21}$ erg/Gauss	Bohr magneton	Bohrsches Magneton
p_m [μ_B]	Magnetic moment per molecule	Magnetisches Moment pro Molekül

T [°K; °C]	Temperature	Temperatur
T_Q	Temperature from which the sample is quenched	Abschrecktemperatur
T_C	Curie temperature for the initial permeability (i.e. temperature where μ has decreased to a relatively small part of its room temperature value)	Curie-Temperatur für die Anfangspermeabilität (Temperatur bei der μ nur noch einen Bruchteil seines Wertes bei Raumtemperatur hat)
Θ_p	Paramagnetic Curie temperature	Paramagnetische Curie-Temperatur
Θ_C	Ferromagnetic Curie temperature for the magnetization	Ferromagnetische Curie-Temperatur für die Magnetisierung
$\Theta_{C,1}$	Θ_C for the magnetization perpendicular to the c axis	Θ_C für die Magnetisierung senkrecht zur c-Achse
$\Theta_{C,2}$	Θ_C for the magnetization parallel to the c axis	Θ_C für die Magnetisierung parallel zur c-Achse
C_g [cm³ °K/g]	Curie constant	Curie-Konstante
C_m [cm³ °K/mole]	Molar Curie constant	Molare Curie-Konstante
C_A [cm³ °K/atom]	Atomic Curie constant	Atomare Curie-Konstante
$\mu = \mu' - i\mu''$	Complex initial permeability	Komplexe Anfangspermeabilität
$\tan \delta_m = \mu''/\mu'$	Magnetic loss factor	Magnetischer Verlustfaktor
μ_i	Initial permeability	Anfangspermeabilität
$\chi = \chi' - i\chi''$	Complex reversible susceptibility	Komplexe reversible Suszeptibilität
χ_0''	Imaginary part of the susceptibility at low amplitude of the r.f. field	Imaginärteil der Suszeptibilität bei kleiner Amplitude des HF-Feldes
χ_{res}''	Imaginary part of the susceptibility at ferromagnetic resonance	Imaginärteil der Suszeptibilität bei ferromagnetischer Resonanz
χ_g [emu/g]	Susceptibility per gram	Suszeptibilität pro Gramm
χ_A [emu/atom]	Atomic susceptibility	Suszeptibilität pro g-Atom
χ_m [emu/mole]	Susceptibility per mole	Suszeptibilität pro Mol
x	As an index to a chemical symbol denotes number of these atoms per molecule	Als Index zu einem chemischen Symbol gibt es die Anzahl dieser Atome pro Molekül
x, y, z	Rectangular coordinates	Rechtwinkliges Koordinatensystem
a, c [Å]	Lattice parameters	Gitterparameter
θ, ϕ	Polar angles; θ is the angle with the c axis of the crystal; ϕ is the angle in the basal plane, mostly measured from a vertical mirror symmetry plane (i.e. from the hexagonal [1$\bar{1}$0] direction)	Polkoordinaten; θ ist der Winkel mit der c-Achse; ϕ ist der Winkel in der Basisebene, meistens gerechnet ausgehend von einer vertikalen Spiegelsymmetrieebene (d. h. ausgehend von der hexagonalen [1$\bar{1}$0]-Richtung)
θ_0	Angle for the preferred direction of the magnetization with respect to the c axis of the crystal	Winkel zwischen der Vorzugsrichtung der Magnetisierung und der c-Achse des Kristalls
Z	Number of molecules in the unit cell	Anzahl der Moleküle in der Einheitszelle
( )	In chemical formulae: tetrahedral sites in the unit cell	In chemischen Formeln: Tetraederstellen im Gitter
[ ]	In chemical formulae: octahedral sites in the unit cell	In chemischen Formeln: Oktaederstellen im Gitter
M, X, W, Y, Z, U	Used to identify various hexagonal crystal structures	Zur Bezeichnung verschiedener hexagonaler Kristallstrukturen
Me^{2+}, Me^{3+}, Me^{4+}	Metal ions	Metallionen
□	Not-occupied lattice site, ion vacancy	Unbesetzter Gitterplatz, Leerstelle für ein Ion
(M_A) [g/mole]	Molar mass of the substance A	Molekulargewicht der Verbindung A
$N_L = 6.02 \cdot 10^{23}$ mole^{-1}	Avogadro number	Loschmidtsche Zahl
t [sec]	Time	Zeit
f [c/s, Gc/s]	Frequency	Frequenz

f_{res}	Ferromagnetic resonance frequency	Ferromagnetische Resonanzfrequenz
ω [rad sec^{-1}]	Angular frequency	Kreisfrequenz
ω_{res}	ω for ferromagnetic resonance	ω für ferromagnetische Resonanz
v [cm/sec]	Velocity	Geschwindigkeit
c	Velocity of light	Lichtgeschwindigkeit
d [cm]	Diameter	Durchmesser
ϱ [g/cm^3]	Density	Dichte
ϱ_X	X-ray density	Röntgendichte
$m = 9.1 \cdot 10^{-28}$ g	Mass of the electron	Elektronenmasse
W [erg]	Energy; $W_{\theta\theta} = \partial^2 W/\partial\theta^2$ etc.	Energie; $W_{\theta\theta} = \partial^2 W/\partial\theta^2$ usw.
j [Amp]	Electric current	Elektrischer Strom
R [Ω]	Resistance of a sample	Widerstand einer Probe
ϱ [Ω cm]	Electrical resistivity	Spezifischer Widerstand
σ [Ω^{-1} cm^{-1}]	Electrical conductivity	Spezifische Leitfähigkeit
σ_∞	σ at infinite temperature	σ bei unendlicher Temperatur
$\sigma_\parallel$	σ parallel to the c axis of the crystal	σ parallel zur c-Achse des Kristalls
$\sigma_\perp$	σ perpendicular to the c axis of the crystal	σ senkrecht zur c-Achse des Kristalls
q [eV]	Activation energy for the conductivity	Aktivierungsenergie für die Leitfähigkeit
$q_\parallel$	q for $\sigma_\parallel$	q für $\sigma_\parallel$
$q_\perp$	q for $\sigma_\perp$	q für $\sigma_\perp$
$\varepsilon = \varepsilon' + i\varepsilon''$	Complex dielectric constant	Komplexe Dielektrizitätskonstante
w_K [erg/cm^3]	Magneto-crystalline anisotropy energy density	Kristallanisotropieenergiedichte
w_U	Anneal-induced uniaxial magneto-crystalline anisotropy energy density	Durch Magnetfeldkühlung induzierte einachsige Kristallanisotropieenergiedichte
$K_1, K_2, K_3, K_3', K_4, \ldots$	Magneto-crystalline anisotropy coefficients	Kristallanisotropiekoeffizienten
$S_2, S_4, S_6, S_6', \ldots$	Magneto-crystalline anisotropy coefficients for an expansion of the anisotropy energy in spherical harmonics [62B1]	Kristallanisotropiekoeffizienten für eine Entwicklung der Anisotropieenergie in spherischen Harmonischen [62B1]
$U_1, U_2, \ldots$	Coefficients of induced magneto-crystalline anisotropy	Koeffizienten für die induzierte Kristallanisotropie
$\alpha_1, \alpha_2, \alpha_3, \beta_1, \ldots$	Direction cosines; $\alpha_3, \beta_3, \ldots$ always are direction cosines with the c axis of the hexagonal crystal	Richtungskosinus; $\alpha_3, \beta_3, \ldots$ sind immer die Richtungskosinus mit der c-Achse des hexagonalen Kristalls
$\lambda = \Delta l/l$	Linear magnetostriction	Lineare Magnetostriktion
$k_1, \ldots, k_1^*, \ldots, k_c', \ldots$	Coefficients of λ, see Eq. (9)	Koeffizienten von λ, siehe Gl. (9)
g	Landé spectroscopic splitting factor	Landé-Faktor (spektroskopischer Aufspaltungsfaktor)
g_{eff}	Effective g-value for two or more sublattice systems	Effektiver g-Wert für Systeme mit zwei oder mehr Untergittern
$\gamma = g \cdot \frac{\mu_B}{h}$	Gyromagnetic ratio ($= ge/2mc$ [emu])	Gyromagnetisches Verhältnis
ε [mm/sec]	Quadrupole splitting in Mössbauer spectrum	Quadrupol-Aufspaltung im Mössbauer-Spektrum
δ [mm/sec]	Isomer shift in Mössbauer spectrum	Isomerieverschiebung im Mössbauer-Spektrum
A	Absorption	Absorption

7.3 Chemical compositions and phase diagrams of hexagonal ferrites

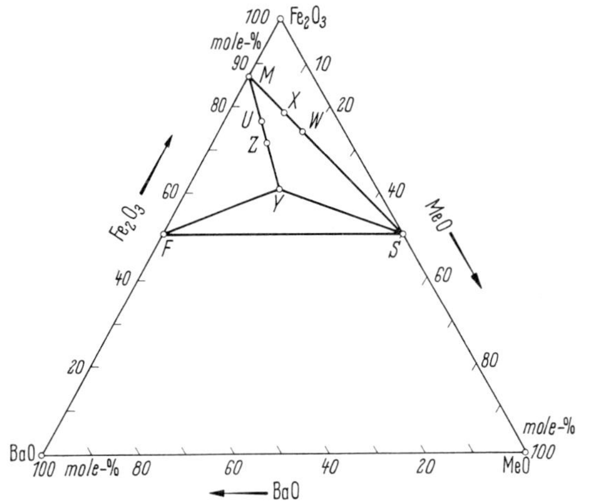

Fig. 3. Composition diagram of ferrimagnetic oxides with a hexagonal crystal structure. The symbol Me represents either a divalent ion of the first transition series, Zn or Mg, or a combination of ions whose valency is two. The symbol **F** represents $BaFe_2O_4$ which is nonferromagnetic; **S** represents a ferrite with the cubic spinel structure.

Symbol	Compound
M	$BaFe_{12}O_{19}$
W = **MS**$_2$	$BaMe_2^{2+}Fe_{16}^{3+}O_{27}$
X = **MS**	$BaMe^{3+}Fe_{14}^{3+}O_{23}$
Y	$BaMe^{2+}Fe_6^{3+}O_{11}$
U = **MY**	$Ba_2Me^{2+}Fe_{18}O_{30}$
Z	$Ba_3Me_2^{2+}Fe_{24}^{3+}O_{41}$
F	$BaFe_2O_4$

see [*52WRGO*], [*52W*], [*52B*], [*57/58SW*], [*59SW*]. For single crystal growth see [*62TSG*], [*64TDS*], [*66AST*], [*67ST*]. Further compositions include the series **M**$_2$**S**, **M**$_3$**S**, the series **MY**$_n$ with n from 1 up to 21 and many compounds of the type **M**$_2$**Y**$_n$, **M**$_3$**Y**$_{14}$, and **M**$_4$**Y**$_{27}$, as well as those given in Tab. 2. [*64KE*], [*65KE1*], [*65KE2*], [*67C*], [*67KEC*]. The various possible chemical substitutions for the different ions are given in the introductory sections for each structure.

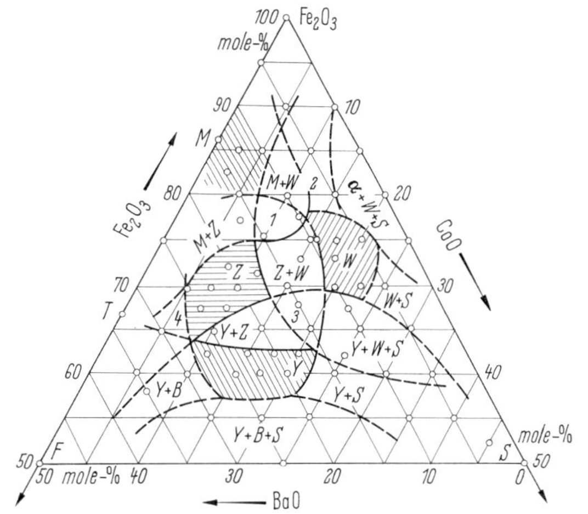

Fig. 4. BaO-CoO-Fe_2O_3 system ($Fe_2O_3 \geq 50\%$). Phase relationships. α-Fe_2O_3; **M** $BaFe_{12}O_{19}$; **F** $BaFe_2O_4$; **S** $CoFe_2O_4$; **Y** or Co_2**Y**, $Ba_2Co_2Fe_{12}O_{22}$; **Z** or Co_2**Z**, $Ba_3Co_2Fe_{24}O_{41}$; **W** or Co_2**W**, $BaCo_2Fe_{16}O_{27}$:
1) **M** + **Z** + **W**; 2) α + **M**; 3) **Y** + **Z** + **W**; 4) **Z** + B. according to [*65V*], where also lattice parameters for solid solutions Co**Y**, Co_2**Z** and Co_2**W** are given.

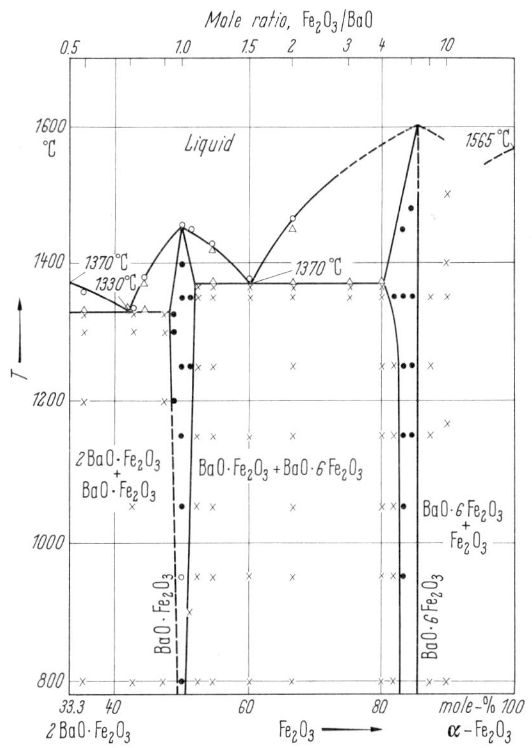

Fig. 5. BaO-Fe_2O_3 system. Phase diagram after [*60GT*]. For the system $BaFeO_{2.5\cdots3.0}$ see e. g. also [*65GMCB*] and [*65MCPSW*].

× two solid phases
● one solid phase
△ liquid and solid phases coexisting
○ one liquid phase.

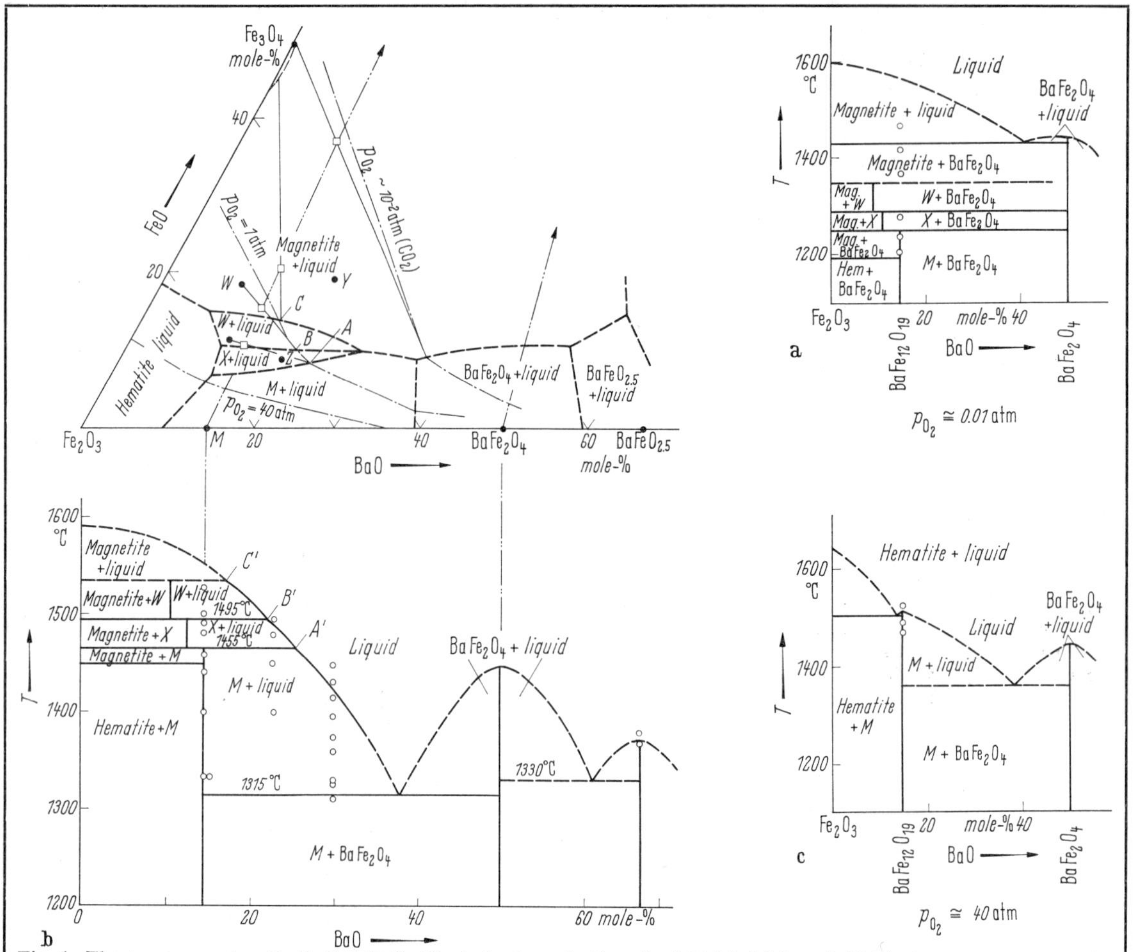

Fig. 6. The ternary system Fe_2O_3-FeO-BaO. The isobaric projections for 0.01 (a), 1 (b), and 40 (c) atm oxygen pressure p_{O_2} are given in the figures a, b and c respectively. These three isobars (dash-dot lines) are shown in the ternary diagram. Dash-double-dot lines are oxygen reaction lines. Square symbols indicate composition of coexisting liquid and crystalline phases joined by tie lines. Heavy dashed lines indicate inferred phase relations [*64vH*].

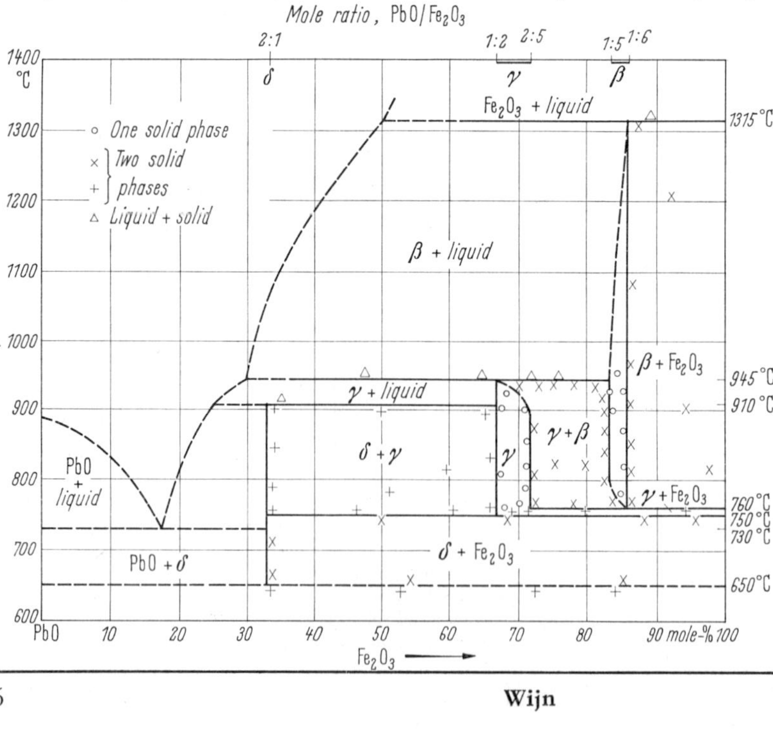

Fig. 7. PbO-Fe_2O_3 system. Phase diagram. The β phase (or **M** phase) ranges from $PbFe_{12}O_{19}$ to $PbFe_{10}O_{16}$ in composition. The authors suggest that in the **M** phase iron and oxygen can be removed from the neighbourhood of the barium ions.
γ phase: $PbFe_4O_7$ ($PbO \cdot 2Fe_2O_3$), $a = 11.86$ Å, $c = 47.14$ Å according to [*28J*]; structure similar to **M** with five formula units per unit cell; presumably antiferromagnetic.
δ phase: $Pb_2Fe_2O_5$ ($2PbO \cdot Fe_2O_3$) is tetragonal, $a = 7.79$ Å, $c = 15.85$ Å, is non-magnetic and has eight formula units per unit cell [*57BP*], [*62MR*].

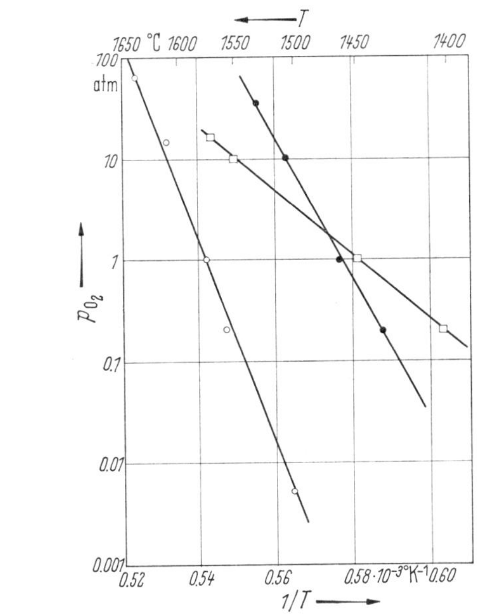

Fig. 8. Pressure-temperature decomposition of barium ferrite compared with hematite and yttrium iron garnet [*64vH*].

- ● $BaFe_{12}O_{19} = BaFe_{15}O_{23}$ + liquid + O_2
- □ $Fe_2O_3 = Fe_3O_{4+y} + O_2$
- ○ $Y_3Fe_5O_{12} = YFeO_3$ + liquid + O_2

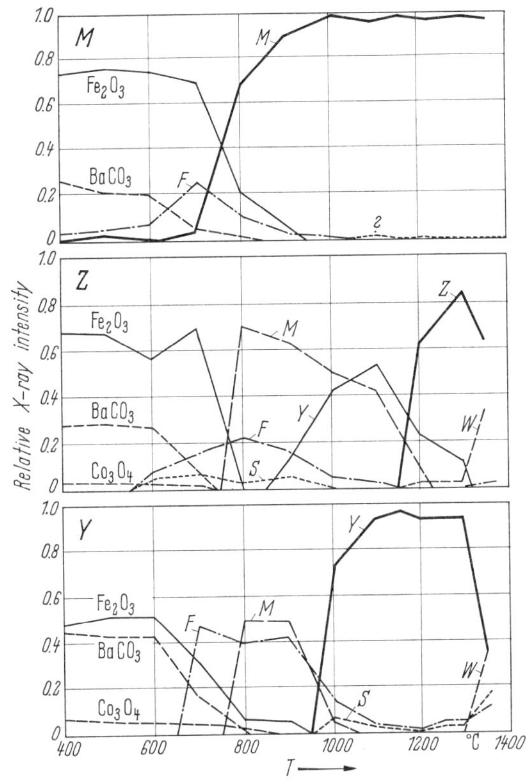

Fig. 9. Formation of the phases **M** ($BaFe_{12}O_{19}$), **Z** ($Ba_3Co_2Fe_{24}O_{41}$) and **Y** ($BaCoFe_6O_{11}$) out of the components $BaCO_3$, Fe_2O_3 and Co_3O_4. The ordinate is a measure of the relative X-ray intensities of the phases which are stable at the firing temperature *T* of the components [*64WT*], [*65W*]. For further information for the formation of **M** and **Y** phases, especially at lower temperature, see [*60ST*], [*65T*] and [*66T*]. For the formation of the **W** phase see [*67VAS*].

7.4 Crystal structures

Tab. 1. Survey of the various crystal structures, see p. 558

Tab. 2. Further crystal structures that have been found for manganese-substituted hexagonal ferrites. **M** = $BaFe_{12}O_{19}$; **Y** = $BaMn_{0.5}Zn_{0.5}Fe_6O_{11}$. [*65KE1*], [*65KE2*], [*67KE*], [*67KEC*]

M : **Y**	Block sequence	Number of anion layers	Structure symbol	Space group	Hexagonal c axis obs. Å	Hexagonal c axis calc. [1]) Å	Ideal stoichiometry
2 : 18	$\mathbf{M(Y)_8M(Y)_{10}}$	3 · 64	192 R_c	$R\bar{3}m$	462.0	462.0	$Ba_{20}(Mn, Zn)_{18}Fe_{132}O_{236}$
2 : 20	$\mathbf{M(Y)_{10}M(Y)_{10}}$	70	70 H_b	$P6_3/mmc$	168.6	168.5	$Ba_{22}(Mn, Zn)_{20}Fe_{144}O_{258}$
2 : 22	$\mathbf{M(Y)_{10}M(Y)_{12}}$	3 · 76	228 R_a	$R\bar{3}m$	549.3	549.2	$Ba_{24}(Mn, Zn)_{22}Fe_{156}O_{280}$
2 : 22	$\mathbf{M(Y)_2M(Y)_{20}}$	76	76 H_a	$P\bar{3}m1$	183.1	183.1	$Ba_{24}(Mn, Zn)_{22}Fe_{156}O_{280}$
2 : 22	$\mathbf{M(Y)_8M(Y)_{14}}$	76	76 H_b	$P\bar{3}m1$	183.1	183.1	$Ba_{24}(Mn, Zn)_{22}Fe_{156}O_{280}$
2 : 22	$\mathbf{M(Y)_{10}M(Y)_{12}}$	3 · 76	228 R_d	$R\bar{3}m$	549.3	549.2	$Ba_{24}(Mn, Zn)_{22}Fe_{156}O_{280}$
2 : 26	$\mathbf{M(Y)_{12}M(Y)_{14}}$	3 · 88	264 R_e	$R\bar{3}m$	636.7	636.4	$Ba_{28}(Mn, Zn)_{26}Fe_{180}O_{324}$

[1]) Calculated for the case that the *c* axis of the **M** compound equals (23.194 ± 0.002) Å and the *c* axis of the **Y** compound equals (21.800 ± 0.004) Å.

Tab. 1. Survey of the various crystal structures of the hexagonal ferrites. [*59 SW*]

Symbol for the compound	Chemical formula	Z	Space group	RT structure data			Crystallographical build-up[2])	Isomorphous with the mineral	Ref.
				ϱ_X[1]) g/cm³	*c* axis[1]) Å	*a* axis[1]) Å			
M	$BaFe_{12}^{3+}O_{19}$	2	$P6_3/mmc$	5.28	23.2	5.9	RSR*S*	magnetoplumbite	*38A, 52WRGO*
W	$BaMe_2^{2+}Fe_{16}^{3+}O_{27}$	2	$P6_3/mmc$	5.31	32.845	5.88	$RS_2R^*S_2^*$		*52W, 52B*
X	$BaMe^{2+}Fe_{14}^{3+}O_{23}$	6	R3 m	5.30	84.11	5.88	$3(RS_2R^*S^*)$		*57B*
Y	$BaMe^{2+}Fe_6^{3+}O_{11}$	6	R3 m	5.39	43.6	5.9	$(ST)_3$		*57B*
Z	$Ba_3Me_2^{2+}Fe_{24}^{3+}O_{41}$	2	$P6_3/mmc$	5.33	52.3	5.9	RSTSR*S*T*S*		*57B*
	$KFe_{11}O_{17}$[3])	2	$P6_3/mmc$	—	23.73	5.932		β-alumina	*38A*
F	$BaFe_2O_4$[4])	2	D_{6h}^4 — $P6_3/mmc$	4.78	8.44	5.51			*61OMK*
	$CaFe_2O_4$[5])								*58PM, 62CT, 66IY, 65H*

[1]) For the case where Me represents a ferrous ion. The *a* axis is practically independent of the type of the metal ion, [*65V*].
[2]) Perspective drawings of the blocks R, S and T are given in Fig. 10. By a rotation of these blocks over an angle of 180° around the *c* axis the blocks indicated by R*, S* and T* are obtained. To build the various crystals these blocks should be extended in all directions perpendicular to the *c* axis and arranged in the direction of the *c* axis in the sequence indicated in this column.
[3]) For magnetic properties of mixed crystals of $KFe_{11}O_{17}$ and **M** see [*64RL*].
[4]) Melting point about 1450°C [*60GT*], for magnetic properties see section 7.5.
[5]) For the properties of calcium ferrites see section 7.11.

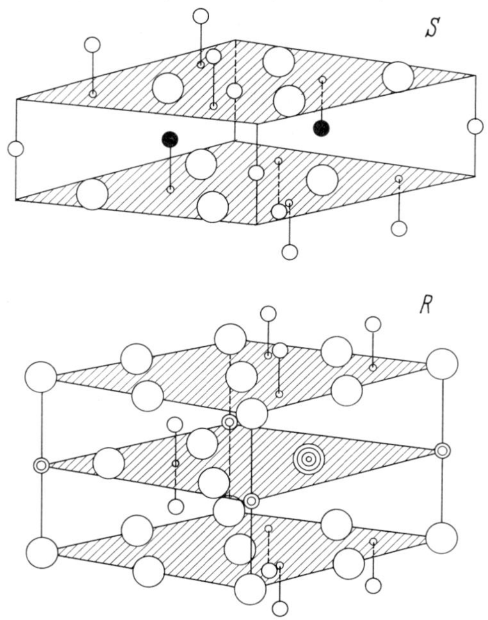

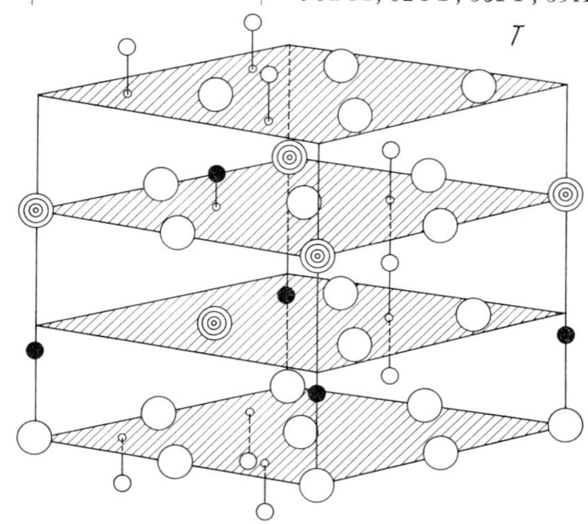

Explanation of the symbols in the Figs. 10 ⋯ 15

- ◯ oxygen ion
- ◎ barium ion (in Fig. 15 potassium ion)
- ↑ relative orientations of the magnetic moments of the metal ions
- X centre of symmetry
- *m* mirror plane perpendicular to the *c* axis

metal ion in

- ○ octahedral surrounding
- ◎ fivefold surrounding
- ● tetrahedral surrounding

Fig. 10. Perspective drawings of the S, R and T blocks with hexagonal symmetry. By extending the blocks according to their symmetry points in all directions perpendicular to the *c* axis, slabs are obtained which exist out of two, three and four oxygen layers, respectively. Various systematical stackings of these slabs lead to the crystal structures of the hexagonal ferrites as indicated in Tab. 1. In the S block the oxygen ions and the interstitial metal ions are arranged in such a way that they form the spinel structure. The R and T blocks do not constitute separately an element of a special crystal structure. They have only been introduced as building blocks for illustration of the relationships between the structures of the various hexagonal ferrites.

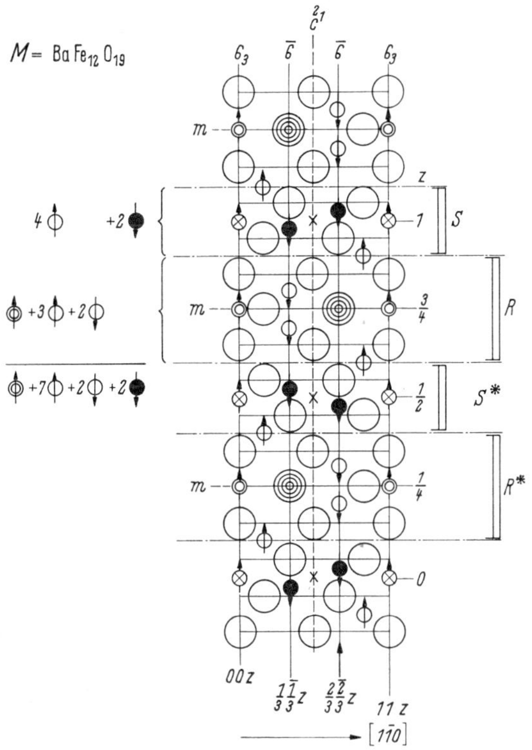

←

Fig. 11. $BaFe_{12}O_{19}$. Crystal structure. The unit cell includes the blocks RSR*S*. Only atoms and symmetry elements in a mirror plane containing the *c* axis are shown. The vertical lines are axes of threefold symmetry [*38A*], [*52WRGO*], [*57B*]. The mutual orientation of the magnetic moments of the ferric ions determines the saturation magnetization of the crystal, see section 7.6.3. The relative orientation of the magnetic moments of the ions with respect to the crystal axes is described in section 7.6.5.

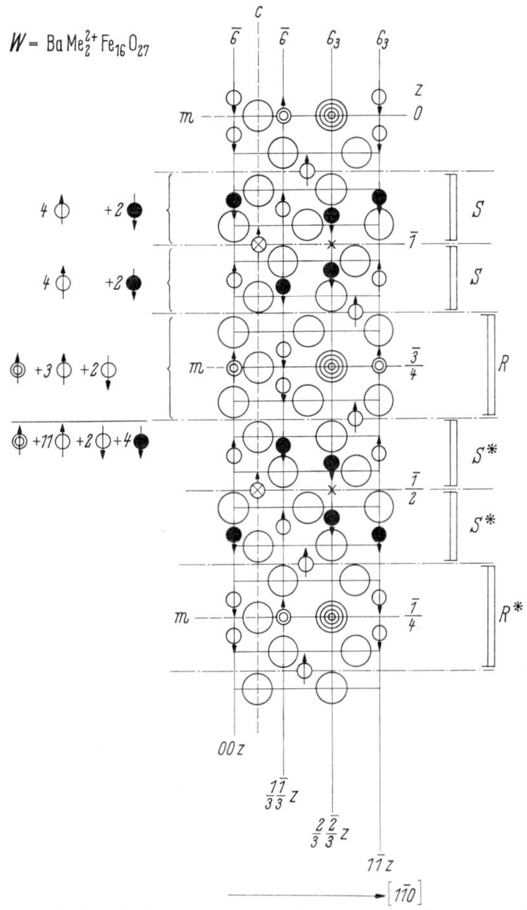

Fig. 12. $BaMe_2^{2+}Fe_{16}O_{27}$. Crystal structure. The unit cell includes the blocks $R^*S_2^*RS_2$ and is obtained from the figure by a displacement in the [$1\bar{1}0$]-direction over a distance of the width of the unit cell. For an explanation of the symbols see Fig. 10 [*52B*], [*52WRGO*].

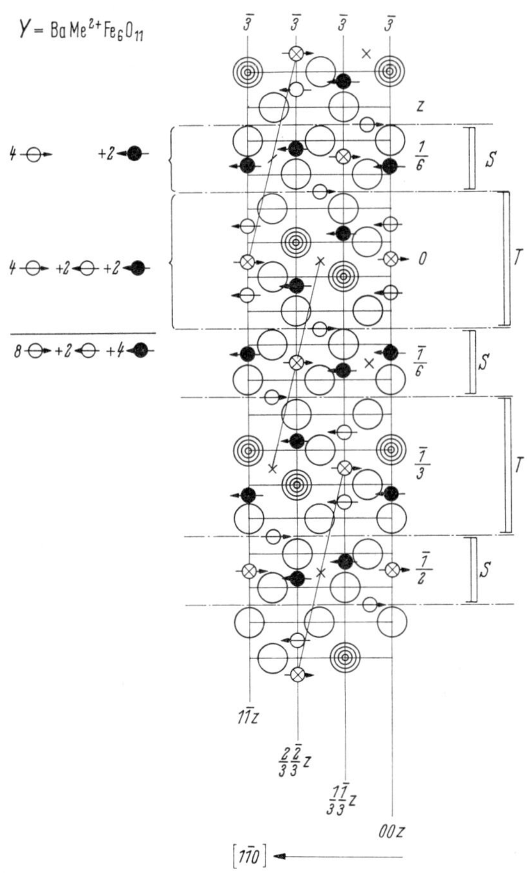

←

Fig. 13. $BaMe^{2+}Fe_6O_{11}$. Crystal structure. Only a part of the hexagonal unit cell is drawn. The complete cell includes the blocks $(ST)_3$. The rhombohedral repeat distance is indicated by the three inclined straight lines. Corresponding to the threefold symmetry of the vertical axes, there are three rhombohedral repeat distances, which could be used to describe this structure. For an explanation of the symbols see Fig. 10 [*56/57JWB*], [*57B*].

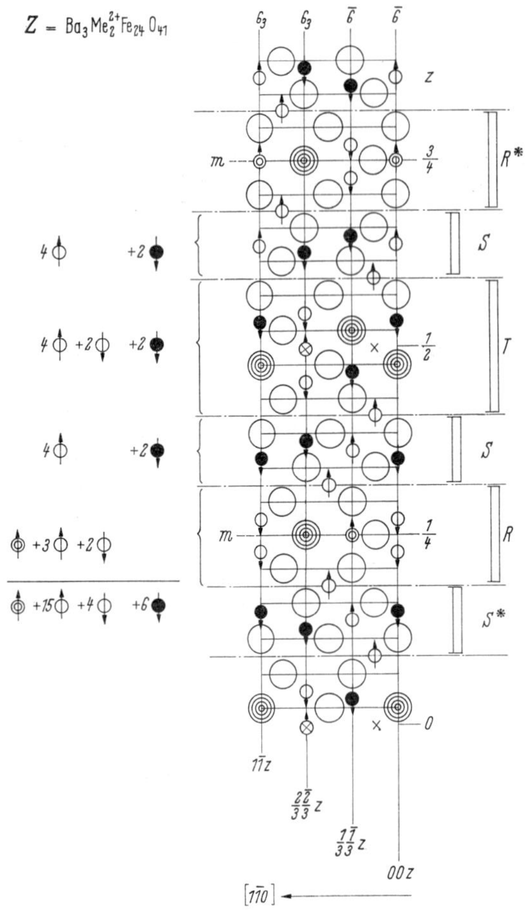

← Fig. 14. $Ba_3Me_2^{2+}Fe_{24}O_{41}$. Crystal structure. Only a part of the hexagonal unit cell is drawn. The complete cell includes the blocks RSTSR*S*T*S*. For an explanation of the symbols see Fig. 10 [*56/57JWB*], [*57B*].

↓ Fig. 15. Comparison between the hexagonal unit cell of $BaFe_{12}O_{19}$ with **M** structure, Fig. 11, and the unit cell of $KFe_{11}O_{17}$ with β-alumina structure, schematically given in the right hand side of this figure. Large circles denote oxygen ions, or K^{1+} or Ba^{2+} ions where indicated; smaller circles represent ferric ions. The ionic positions inside the "spinel blocks" of each structure are not shown: the spinel block contains four planes with each 4 oxygen ions; between these planes nine Fe^{3+} ions are situated, of which seven on the octahedral sites have ionic moments pointing in one direction and two on the tetrahedral sites have ionic moments in the opposite direction.
The difference between the two structures is only determined by the type and the packing of the ions in the layer between the spinel blocks. In $KFe_{11}O_{17}$ the K layer is not close packed, in $BaFe_{12}O_{19}$ the Ba layer is close packed and forms a hexagonal close packed arrangement with the neighbouring close packed oxygen layers [*54G*].

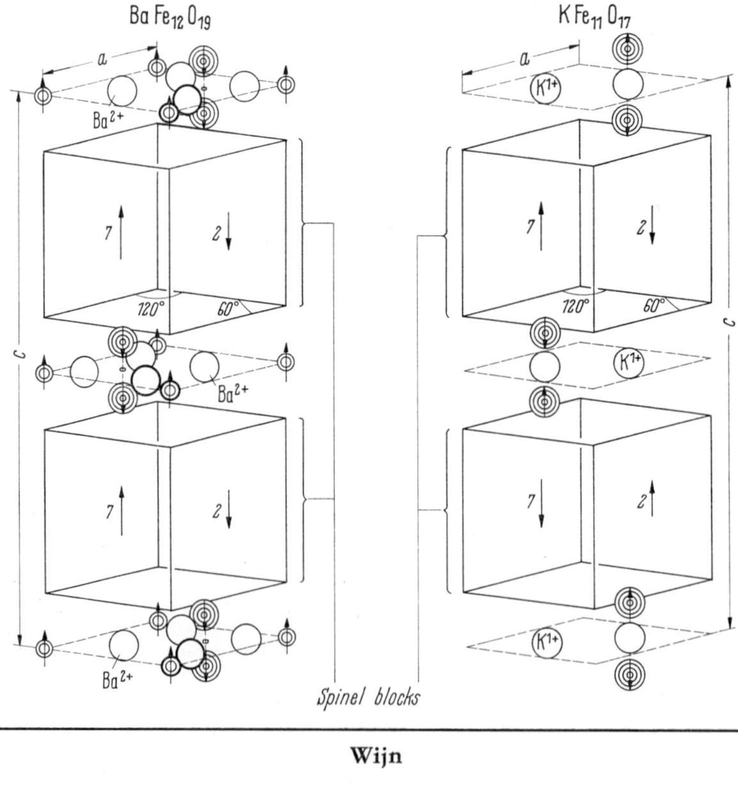

7.5 Paramagnetic properties of ferrites with hexagonal crystal structures

Tab. 3. Néel constants, [*48N*], derived from measurements of the ferromagnetic susceptibility, [*63BM*], according to the formula:

$$\frac{1}{\chi_A} = \frac{1}{\chi_{0,A}} + \frac{T}{C_A} - \frac{\sigma}{T - \Theta}$$

C_A = atomic Curie constant of magneto-active ions
C_m = Curie constant obtained from molar susceptibilities

Compound	C_A cm³ °K/atom	C_m cm³ °K/mole	$1/\chi_{0,A}$ 10^{-4} emu/atom	$\sigma \cdot 10^5$ °K atom/cm³
$SrFe_{12}O_{19}$	0.0475	51	5.8	27
$BaFe_{12}O_{19}$	0.0515	58	5.1	26
$PbFe_{12}O_{19}$	0.0526	61	6.8	31

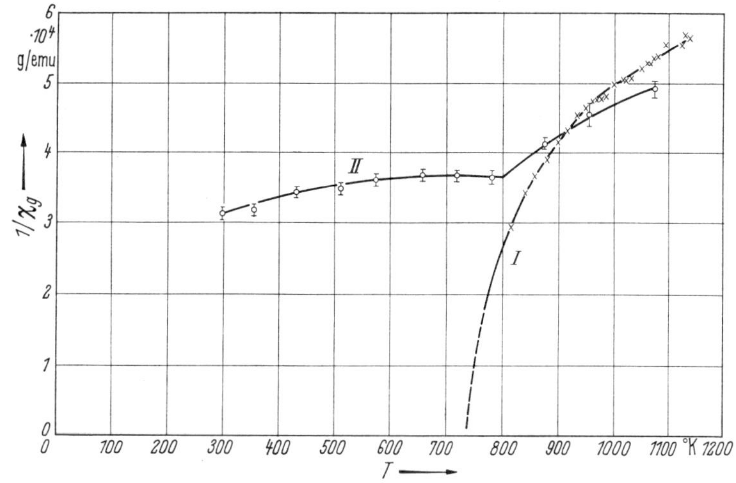

Fig. 16. $BaFe_{12}O_{19}$ and $KFe_{11}O_{17}$. Curve I: the inverse of the susceptibility per gram ($1/\chi_g$) for $BaFe_{12}O_{19}$ above the Curie temperature. The compound $BaFe_{12}O_{19}$ is ferrimagnetic.
Curve II: the inverse of the susceptibility per gram ($1/\chi_g$) for $KFe_{11}O_{17}$. This compound is antiferromagnetic [*54G*].

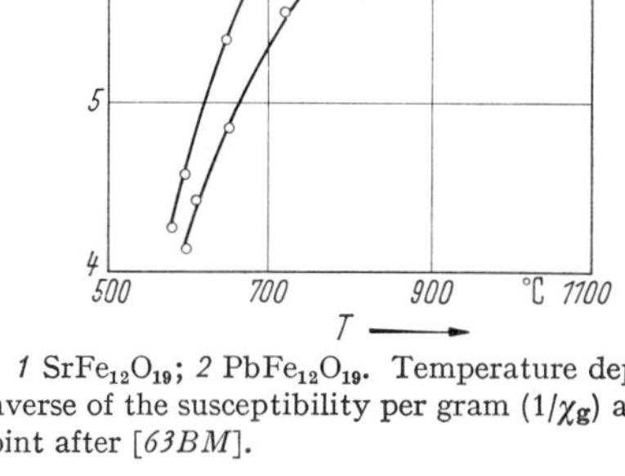

Fig. 17. *1* $SrFe_{12}O_{19}$; *2* $PbFe_{12}O_{19}$. Temperature dependence of the inverse of the susceptibility per gram ($1/\chi_g$) above the Curie point after [*63BM*].

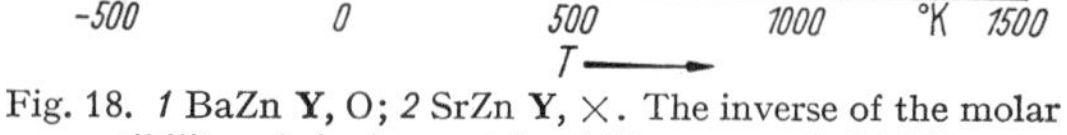

Fig. 18. *1* BaZn **Y**, O; *2* SrZn **Y**, ×. The inverse of the molar susceptibility of single crystals of **Y** compounds [*61E*].

→

Fig. 19. BaO · Fe_2O_3. Curve of inverse susceptibility vs. temperature. The room temperature value of the susceptibility is $\chi_{g,RT} = 7.6 \cdot 10^{-6}$ emu/g. The Curie constant is $C_g = 3.75 \cdot 10^{-3}$ emu°K/g. The compound seems to be antiferromagnetic [*61OMK*]. $\Theta_p = -200$ °K.

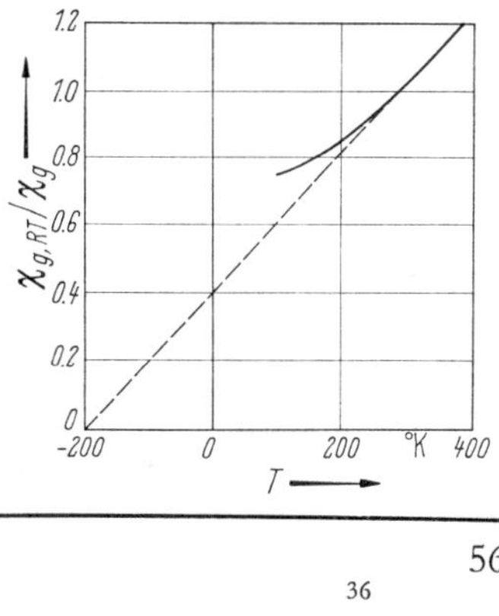

7.6 M (magnetoplumbite)-type ferrites

(general chemical formula $BaFe_{12}O_{19}$)

7.6.1 Survey of the chemical substitutions in the M structure and room temperature lattice constants

Tab. 4. Survey of the chemical substitutions in the **M** structure

Compound	Symbol	c Å	a Å	Ref.	Remarks	Properties: Tables	Properties: Figures
$BaFe_{12}O_{19}$	**M**	23.17 23.12 23.182 23.194 23.2	5.876 5.891 5.889 5.893 5.89	*57SB* *54G* *59B* *63TKS* *63BY*	Microhardness Vickers/ Knoop 1100 kg/mm², [*62CMP*]; domain wall energy [*60KG*], [*62 KG*]	5, 6, 7, 12, 13, 16, 17	24, 25, 30, 31, 32, 33, 34, 35, 36, 37, 62, 63, 64, 78, 79, 80, 81, 82, 83, 84, 85, 86, 89, 90, 91, 92, 93, 96, 98, 100
$SrFe_{12}O_{19}$	Sr**M**	23.031 23.1	5.864 5.85	*59BDPP* *63BY*	$(4\pi M_s)_{20\,°C} = 4652$ Gauss; $\sigma^0 = 20.6$ emu/g; $\sigma^{RT} = 13.69$ emu/g		40, 41, 77
$Ba_{1-x}Sr_xFe_{12}O_{19}$	$Ba_{1-x}Sr_x$**M**			*62BU*			78, 81, 87, 88
$Ba_{0.4}Ca_{0.6}Fe_{12}O_{19}$	$Ba_{0.4}Ca_{0.6}$**M**			*57U*			
$CaFe_{12}O_{19}$	Ca**M**	22.91 ± 0.05	5.877 ± 0.005	*63IK*	$\Theta_C = 718$ °K; $\sigma = 73.4$ emu/g at 77 °K. This ferrite is stabilized with some La_2O_3, see also section 7.11		
$PbFe_{12}O_{19}$	Pb**M**	23.02 23.0	5.877 5.87	*38A, 51B, 56GV, 59PR1, 59PR2, 59V1, 59V2* *63BY*		5	26, 38, 39, 65, 66, 67
$La^{3+}Fe^{2+}Fe^{3+}_{11}O_{19}$	$La^{3+}Fe^{2+}$**M**			*60MS*			42
$Ba_{1-x}Me^{3}_{x}{}^{+}Fe^{2}_{x}{}^{+}Fe^{3+}_{12-x}O_{19}$ Me = La, Pr, Nd, Sm or Eu		Fig. 23	Fig. 22	*57DB*	Me=La, $\sigma \approx 0.1\,\Omega$cm [*60MS2*]. For these substitutions also the arrangement $Ba_{1-3x}Me^{3+}_{2x}\square_x Fe^{3+}_{12}O_{19}$ is proposed, where □ denotes a vacancy at a barium site [*57DB*]	8 for Me = La	
$La_{0.5}Na_{0.5}Fe_{12}O_{19}$ $La_{0.5}K_{0.5}Fe_{12}O_{19}$ $La_{0.5}Rb_{0.5}Fe_{12}O_{19}$	$La_{0.5}Na_{0.5}$**M** $La_{0.5}K_{0.5}$**M** $La_{0.5}Rb_{0.5}$**M**			*57SB* *57SB* *57SB*	$\Theta_C = 470$ °C. Possibly an ordering between Na^{1+} (respectively K^{1+} or Rb^{1+}) and the La^{3+} ions occurs [*57SB*]		43

$La_{0.5}Ag_{0.5}Fe_{12}O_{19}$	$La_{0.5}Ag_{0.5}\mathbf{M}$	22.85	5.85	*63LS*	$\Theta_C = (435 \pm 5)$ °C; ${}_BH_c = 1100$ Oe		
$La_{0.5}Tl_{0.5}Fe_{12}O_{19}$	$LaTl_{0.5}\mathbf{M}$	23.4	5.89	*63L*	$\Theta_C = (430 \pm 5)$ °C		
$Ba_xLa_{1-x}Zn_{1-x}Fe_{11+x}O_{19}$				*58MB*	σ_s^{RT} lower than that of **M**		
$Ba_xLa_{1-x}Mg_{1-x}Fe_{11+x}O_{19}$				*58MB*	σ_s^{RT} lower than that of **M**		
$Ba_{1-x}Bi_xNi_xFe_{12-x}O_{19}$				*61SA*	$x \leqq 0.4$		
$Ba_{1-x}Me_xNi_xFe_{12-x}O_{19}$ Me = La or Pr; $x \leqq 0.8$				*61SA*	$x \leqq 0.8$		44
$Ba_{1-x}Me_xCo_xFe_{12-x}O_{19}$ Me = La or Pr, $x \leqq 0.8$				*61SA*	For $Ba_{0.8}La_{0.2}Ni_{0.2}Ga_{11.8}O_{19}$ and $Ba_{0.8}La_{0.2}Co_{0.2}Ga_{11.8}O_{19}$ see [*63RLB*]		45
$BaMe_x^{3+}Fe_{12-x}O_{19}$ Me = Ga, Cr or Al	$Ga_x\mathbf{M}$; $Cr_x\mathbf{M}$ $Al_x\mathbf{M}$	Fig. 20	Fig. 21	*58MB, 59BDPP* and *59dBdPB*		12, 14, 17	46, 47, 48, 50, 51, 52, 54
$SrAl_xFe_{12-x}O_{19}$	$SrAl_x\mathbf{M}$			*59dBdPB*		12,17	49, 68, 69, 70, 71
$PbAl_xFe_{12-x}O_{19}$	$PbAl_x\mathbf{M}$			*59BK2*			53
$SrCr_2Fe_{10}O_{19}$	$Cr_2\mathbf{M}$	22.94	5.857	*59BDPP*	$(4\pi M_s)_{20°C} = 3228$ Gauss, $\sigma_s^0 = 14.0$ emu/g; $\sigma_s^{RT} = 9.44$ emu/g		46
$SrCr_{3.2}Fe_{8.8}O_{19}$	$Cr_{3.2}\mathbf{M}$			*66BK*		7	27, 28
$SrCr_6Fe_6O_{19}$	$Cr_6\mathbf{M}$	22.77	5.844	*59BDPP*	$(4\pi M_s)_{20°C} = 450$ Gauss, $\sigma_s^0 = 3.2$ emu/g, $\sigma_s = 1.3$ emu/g at 16 °C		
$BaTi_x^{4+}Ni_x^{2+}Fe_{12-2x}O_{19}$ $x < 1.0$	$Ti_xNi_x\mathbf{M}$			*59SW*			72
$BaTi_x^{4+}Co_x^{2+}Fe_{12-2x}O_{19}$ $x < 2.0$	$Ti_xCo_x\mathbf{M}$			*59SW, 61LES*		12, 17	55, 56, 57, 72, 73, 74, 75, 76, 94
$BaTi_x^{4+}Zn_x^{2+}Fe_{12-2x}O_{19}$ $x < 1.2$	$Ti_xZn_x\mathbf{M}$			*59SW*			72
$BaIr_x^{4+}Zn_x^{2+}Fe_{12-2x}O_{19}$ $x < 0.6$	$Ir_xZn_x\mathbf{M}$	23.194 $0 \leqq x \leqq 0.6$	{$x = 0$: 5.893; {$x = 0.6$: 5.920	*63TKS*	$K_1 > 0$ for $x < 0.3$; $K_1 < 0$ for $x > 0.3$	9, 15, 17	58, 77
$BaFe_2^{2+}Fe_{10}^{3+}O_{17}F_2$				*60FSS*	$\varrho = 5.25$ g/cm³; Fe^{2+}-ions presumably in $4f_2$ sites		
$BaNi_xFe_{12-x}O_{19-x}F_x$ $x = 0.5$, 1 and 2				*62BRT*			59
$BaCo_{0.3}Fe_{11.7}O_{18.7}F_{0.3}$		23.13	5.883	*63RB, 63RLB*	For $BaCo_xAl_{12-x}O_{19-x}F_x$ see [*61BR*]	10	60
$BaCo_{0.4}Fe_{11.6}O_{18.6}F_{0.4}$		23.15	5.886	*63RB*		10	60
$BaCo_{0.47}Fe_{11.53}O_{18.53}F_{0.47}$				*63RB*		10	

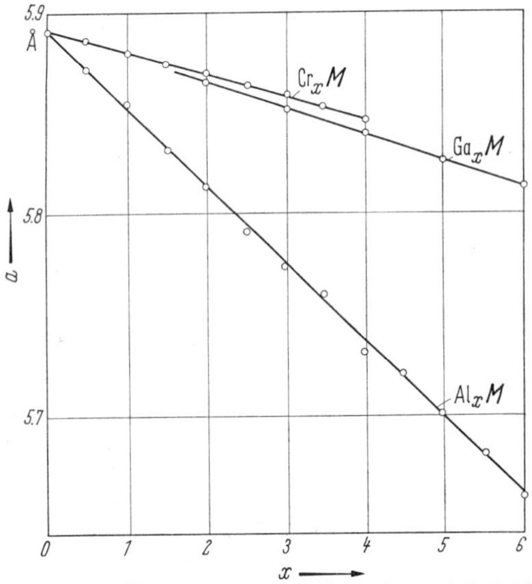

Fig. 20. $BaMe_x^{3+}Fe_{12-x}O_{19}$ system. Dependence of the lattice parameter a on the composition parameter x. Me = Ga, Cr or Al [*59BDPP*], see also [*58MB*].

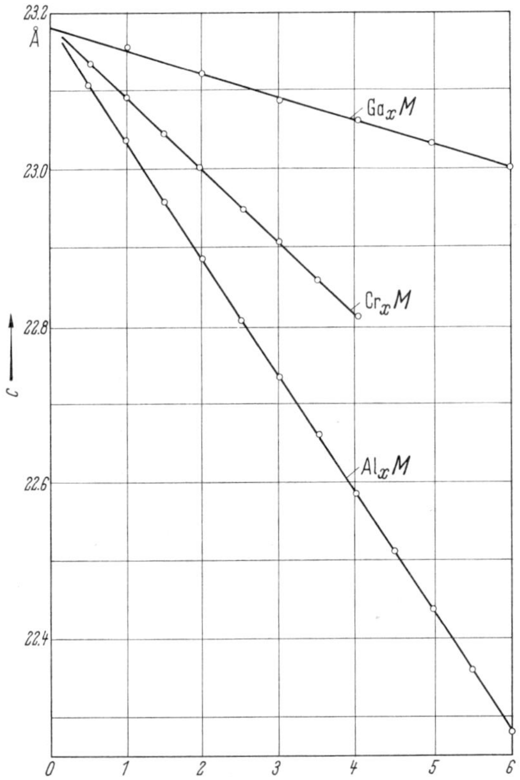

Fig. 21. $BaMe^{3+}Fe_{12-x}O_{19}$ system. Dependence of the lattice parameter c on the composition parameter x. Me = Ga, Cr or Al [*57BDPP*].

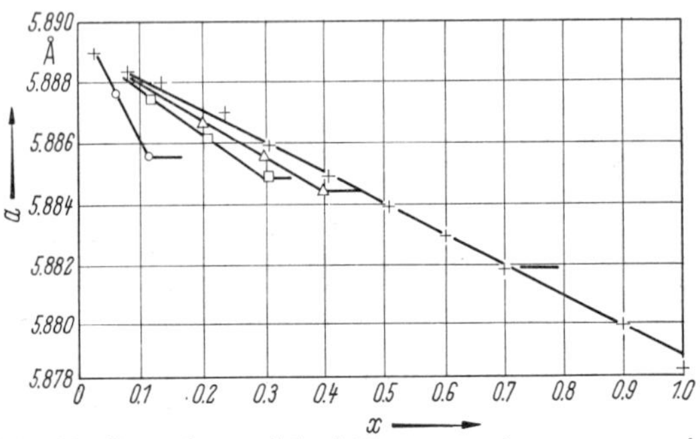

Fig. 22. Dependence of the lattice parameter a on composition x in the systems $Ba_{1-x}Me_x^{3+}Fe_x^{2+}Fe_{12-2x}^{3+}O_{19}$, where Me= La(+), Pr(△), Nd(□), Sm or Eu(○). Solubility limits at 1400 °C are indicated by horizontal lines [*57DB*].

→

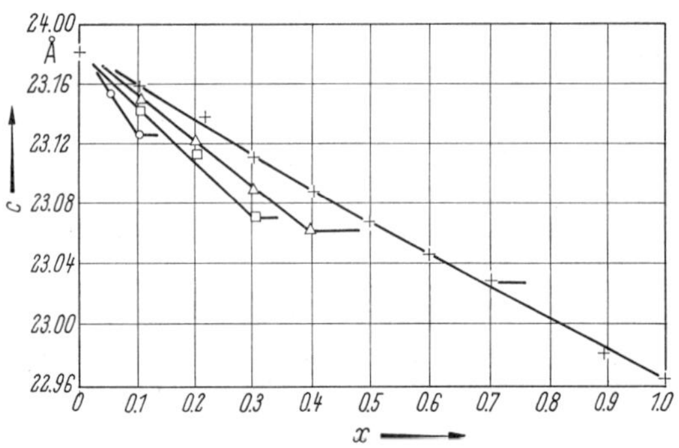

Fig. 23. Dependence of the lattice parameter c on composition x in the systems $Ba_{1-x}Me_x^{3+}Fe_x^{2+}Fe_{12-2x}^{3+}O_{19}$, where Me= La(+), Pr(△), Nd(□), Sm or Eu(○). Solubility limits at 1400 °C are indicated by horizontal lines [*57DB*].

7.6.2 Electric and dielectric properties of ferrites with M structure

Tab. 5. Activation energy q for the d. c. conductivity for some samples of compounds with **M** structure

Compound	Sample	Measuring direction	q eV	Ref.
$BaFe_{12}O_{19}$	single crystal, Fig. 24	∥ c axis	0.09 ···0.15	*63Z*
	single crystal, Fig. 24	⊥ c axis	0.09 ···0.165	*63Z*
$PbFe_{12}O_{19}$	single crystal, Fig. 26	∥ c axis	0.07	*63Z*
	single crystal, Fig. 26	⊥ c axis	0.045···0.065	*63Z*
$BaFe_{12}O_{19}$, stoichiometric	polycrystalline, Fig. 25	—	0.9	*52WRGO*
0.022% by weight of Fe^{2+}	polycrystalline, Fig. 25	—	0.4	*68HW*
0.022% by weight of Fe^{2+}	polycrystalline, Fig. 31	—	0.05*)	*68HW*

*) measured at a frequency of 5 Gc/s.

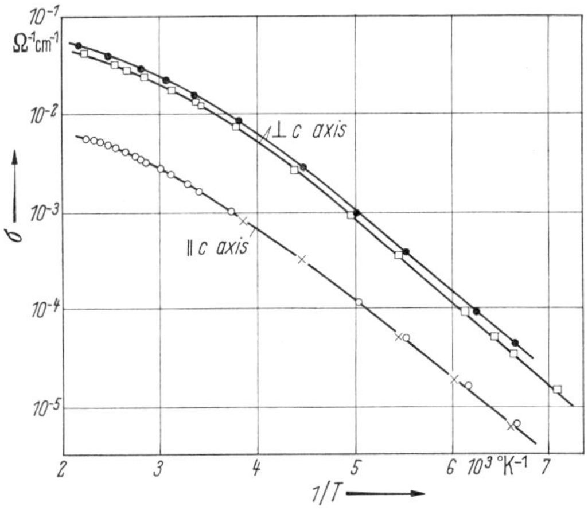

Fig. 24. $BaFe_{12}O_{19}$. Temperature dependence of the electrical d. c. conductivity of a single crystal according to [*63Z*]. ○, × parallel to the *c* axis for two different samples, ●, □ parallel to the basal plane for two different samples.

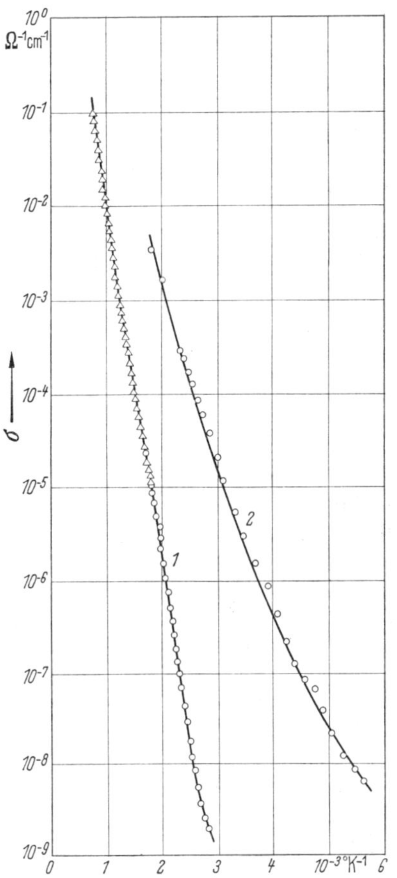

Fig. 25. $BaFe_{12}O_{19}$. Temperature dependence of the electrical d. c. conductivity of polycrystalline samples. Sample *1* is practically stoichiometric and has small crystals ($\leq 1\ \mu m$), sample *2* has an excess of iron ions, or a deficiency of oxygen ions, it contains 0.022% by weight of Fe^{2+} ions and has much larger crystals (about 50 μm in diameter). △ [*52WRGO*], ○ [*68HW*].

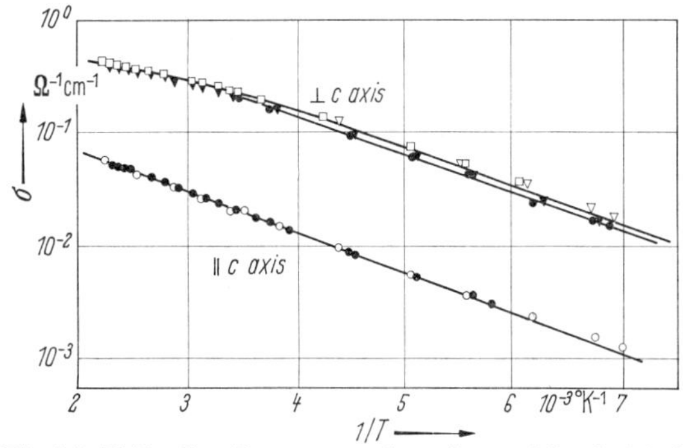

Fig. 26. $PbFe_{12}O_{19}$. Temperature dependence of the electrical d. c. conductivity for a single crystal. The differently indicated measuring points correspond to different samples [*63Z*].

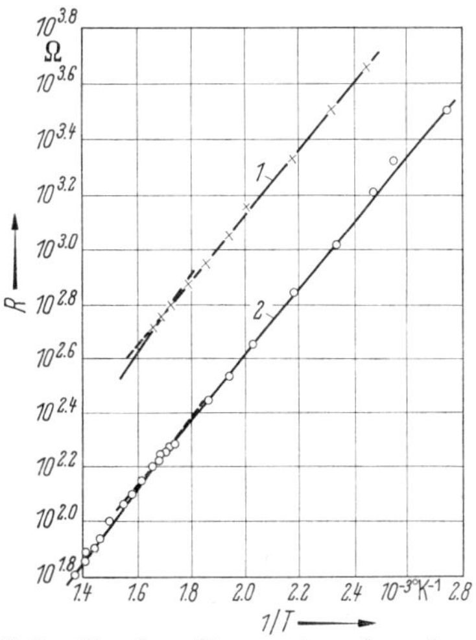

Fig. 27. $SrCr_{3.2}Fe_{8.8}O_{19}$. Temperature dependence of the resistance *R* on $1/T$ for a crystal-oriented specimen in two crystallographic directions: along the *c* axis (*1*) and in the basal plane (*2*). The activation energy is in both cases equal to about 0.23 eV [*66BK*].

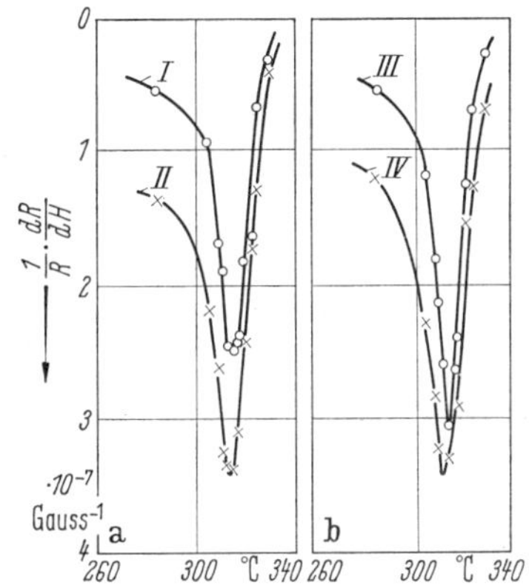

Fig. 28. $SrCr_{3.2}Fe_{8.8}O_{19}$. a) Temperature dependence of the longitudinal paraprocess magnetoresistivity effect $(1/R)$ (dR/dH) for the configurations I and II of Fig. 29; b) temperature dependence of the transverse paraprocess magnetoresistivity effect $(1/R)$ (dR/dH) for the configurations III and IV of Fig. 29. Measurements made on crystal-oriented samples [*66BK*].

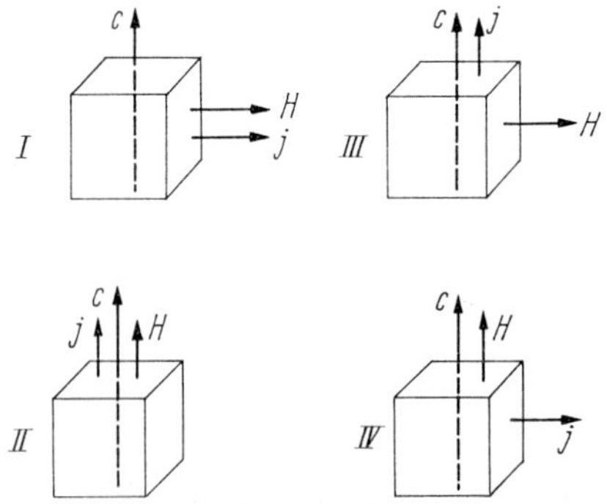

Fig. 29. Relative orientation of the external field H and direction of the current j with respect to the crystallographic c axis for measurement of the magnetoresistivity effect of Fig. 28 [*66BK*].

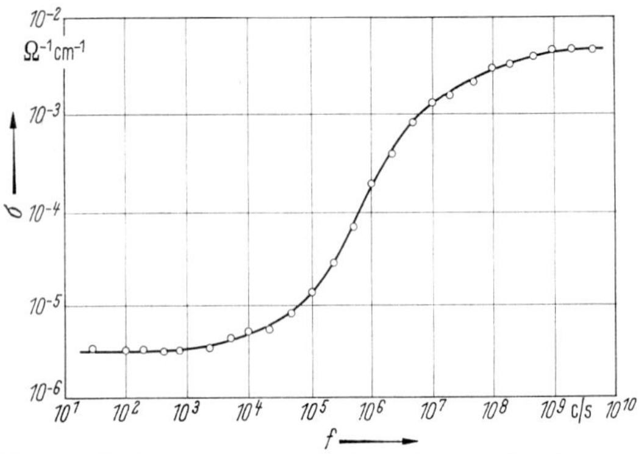

Fig. 30. $BaFe_{12}O_{19}$. Frequency dependence of the electrical conductivity at room temperature for a polycrystalline sample of $BaFe_{12}O_{19}$ with excess of iron. The ferrous content of the sample is 0.022% by weight [*68HW*].

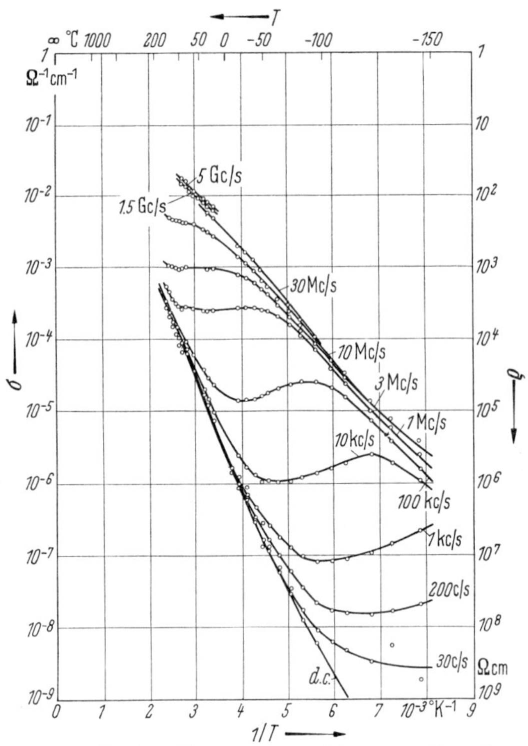

Fig. 31. $BaFe_{12}O_{19}$. Temperature and frequency dependence of the conductivity of a polycrystalline sample of $BaFe_{12}O_{19}$ with excess iron. Ferrous content is 0.022% by weight. Activation energy at room temperature about 0.4 eV for the d. c. conductivity and about 0.05 eV for the conductivity at a frequency of 5 Gc/s [*68HW*].

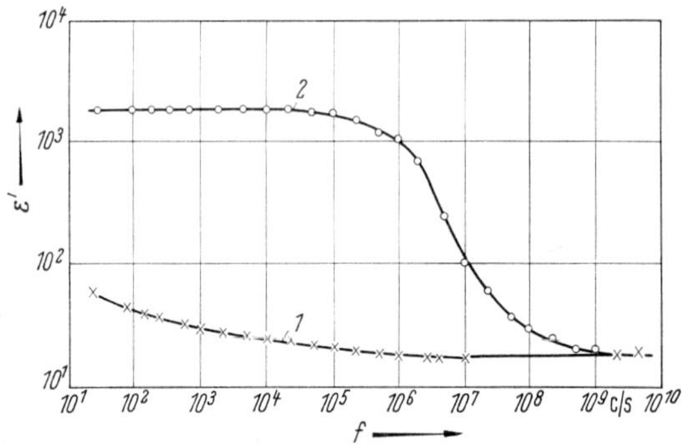

Fig. 32. $BaFe_{12}O_{19}$. Frequency dependence of the dielectric constant of polycrystalline samples. Sample *1* is practically stoichiometric, sample *2* has an excess of iron ions, or a deficiency of oxygen ions, it contains 0.022% by weight of Fe^{2+} ions [*68HW*].

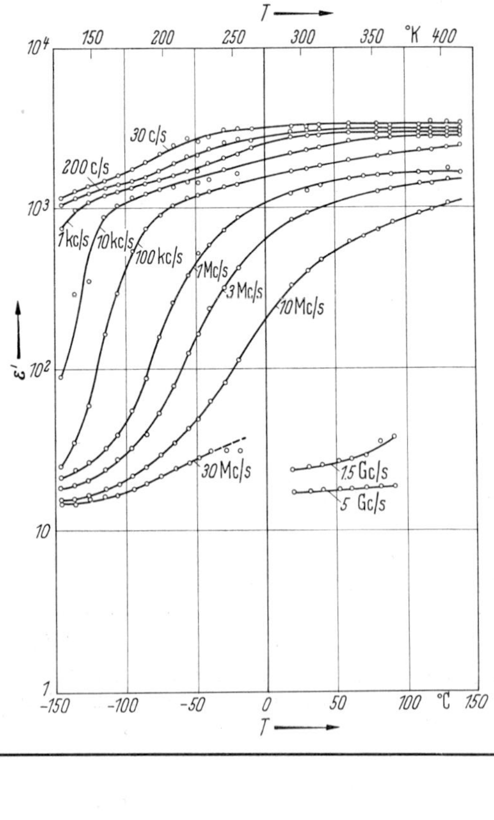

→

Fig. 33. $BaFe_{12}O_{19}$. Temperature and frequency dependence of the real part ε' of the dielectric constant of a polycrystalline sample of $BaFe_{12}O_{19}$ with excess iron. Ferrous content 0.022% by weight [*68HW*].

7.6.3 Mössbauer spectra, saturation magnetization and Curie temperature of ferrites with M structure

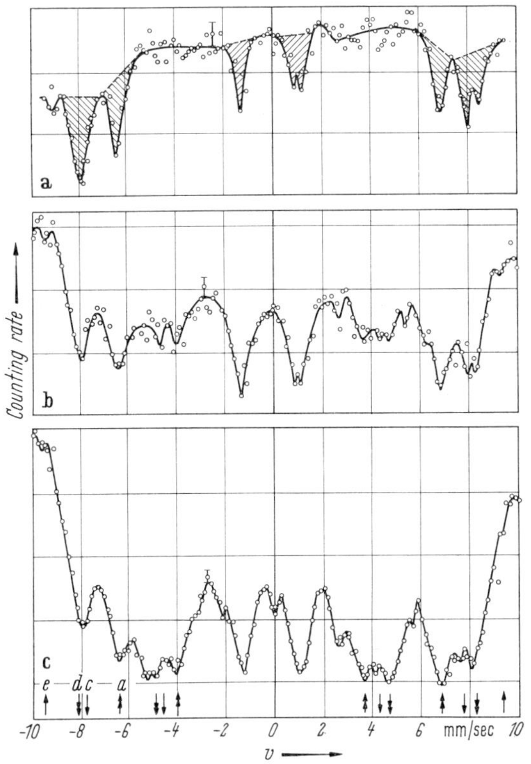

Fig. 34. Mössbauer spectra for several samples of $BaFe^{57}_{12}O_{19}$ at room temperature

a Single crystals (gamma rays ⊥ basal plane),
b Oriented crystals (gamma rays ⊥ basal plane),
c Oriented crystals (gamma rays ∥ basal plane), [*64LB*].

Sublattice	Number	Spin	H_{eff} ($T = 300$ °K) 10^5 Oe	ΔH_{dip} ($T = 300$ °K) 10^3 Oe
a octahedral	6	up	4.15	−9
b octahedral	1	up	4.55	+14
c tetrahedral	2	down	4.85	−2
d octahedral	2	down	5.15	+10
e 5-fold coordination	1	up	5.80	+9

H_{eff} is effective magnetic field at the nucleus.
ΔH_{dip} is difference in dipole field for spins parallel and perpendicular to the *c* axis of the crystal.
↓↑ means relative orientation of the magnetic moments of the ions on the lattice sides denoted by *a*, *b*, *c* etc.

Fig. 36b. →
Computed curves for the sublattice magnetization M_T relative to 0 °K values M^0_T, plotted against temperature relative to the Curie point. The correct algebraic sum of these curves gives saturation magnetization of the crystal as a function of temperature as given in Fig. 34 [*63LF*], [*64LB*].

	Spin	Site	Number
— — — — — — —	↑	5-fold	1
———————	↓	tetrahedral	2
×	↓	octahedral	2
- - - - - - - - - -	↑	octahedral	1
—·—·—·—·—	↑	octahedral	6

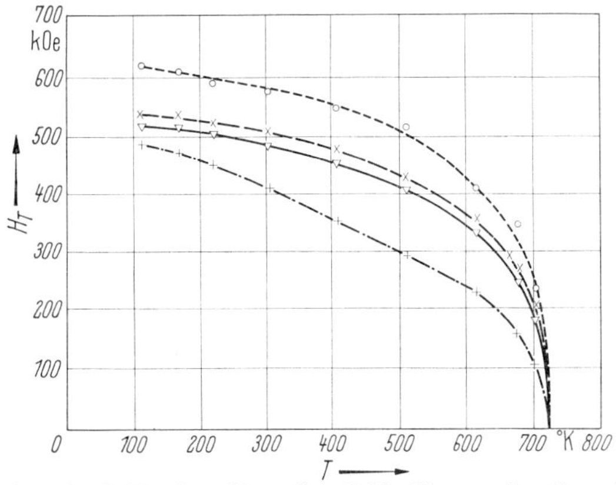

Fig. 35. $BaFe_{12}O_{19}$. Hyperfine fields H_T as a function of temperature for the iron ions at four different sites derived from the Mössbauer spectra of Fig. 35.

	Spin	Number of Fe^{3+} per unit cell
— — — ○ — — —	↑	1
——— × ———	↓	2
——— ▽ ———	↓	2
—·— + —·—	↑	6+1

Fig. 36a. $BaFe_{12}O_{19}$.
Hyperfine fields H_T from experiment relative to 0 °K values H^0_T, plotted against temperature relative to the Curie temperature Θ_C. The explanation of the signs is the same as in Fig. 35.
↓

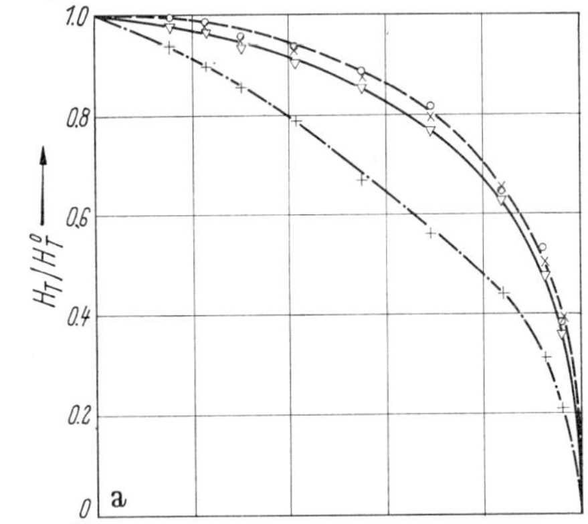

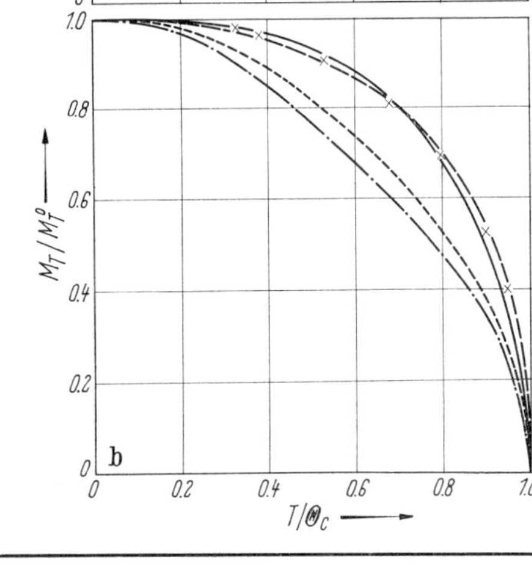

Tab. 6. Saturation magnetization of $BaFe_{12}O_{19}$

$4\pi M_s$ at 20 °C Gauss	Θ_C °K	Ref.
4630	723	*52WRGO*
4775	730	*59CSEFWGDFJ*
	725	*61AS*

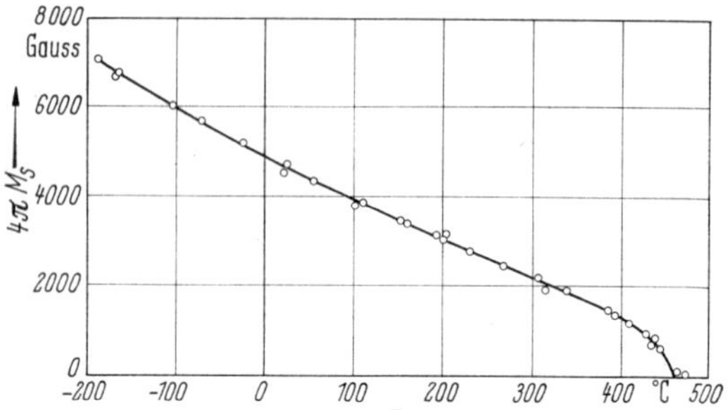

Fig. 37. $BaFe_{12}O_{19}$. The saturation magnetization for a single crystal measured along the *c* axis as a function of temperature [*52RSS*], [*53R*].

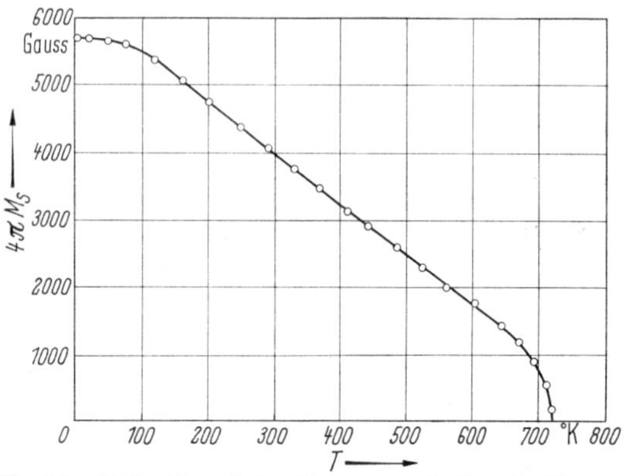

Fig. 38. $PbFe_{12}O_{19}$. Saturation magnetization $4\pi M_s$ as a function of temperature. $\Theta_C = 725$ °K [*59PR*].

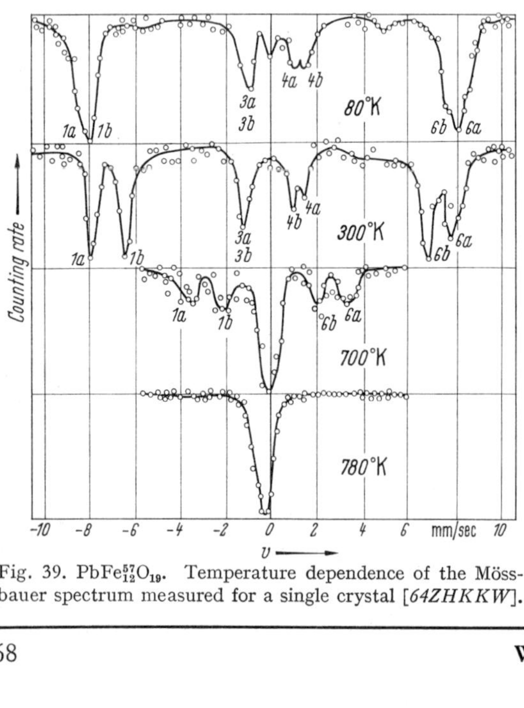

Fig. 39. $PbFe^{57}_{12}O_{19}$. Temperature dependence of the Mössbauer spectrum measured for a single crystal [*64ZHKKW*].

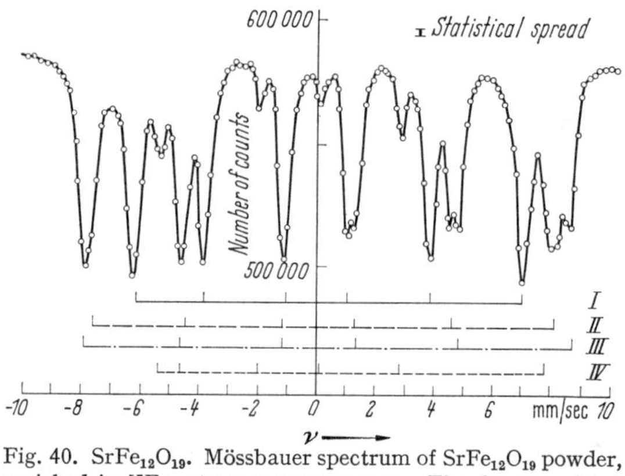

Fig. 40. $SrFe_{12}O_{19}$. Mössbauer spectrum of $SrFe_{12}O_{19}$ powder, enriched in ^{57}Fe, at room temperature. The decomposition of the spectrum into four subspectra I, II, III and IV is indicated [*66WR*].

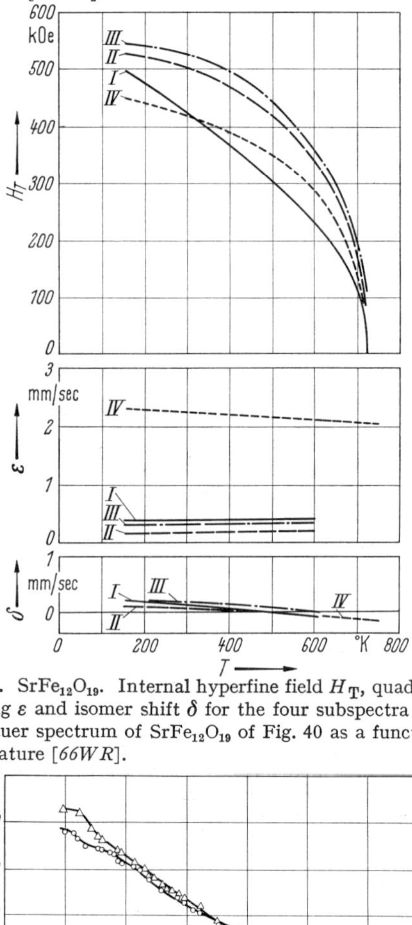

Fig. 41. $SrFe_{12}O_{19}$. Internal hyperfine field H_T, quadrupole splitting ε and isomer shift δ for the four subspectra of the Mössbauer spectrum of $SrFe_{12}O_{19}$ of Fig. 40 as a function of temperature [*66WR*].

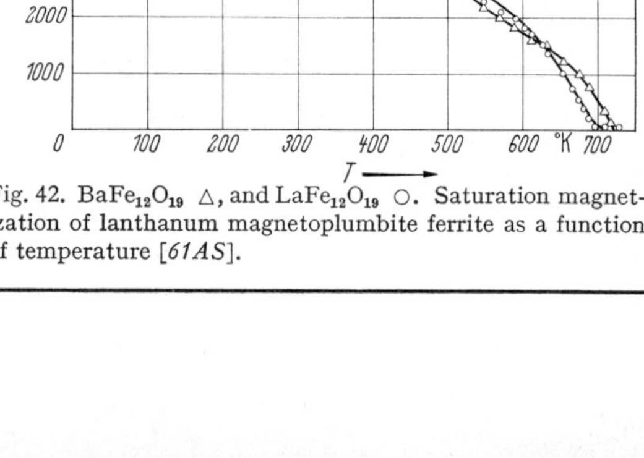

Fig. 42. $BaFe_{12}O_{19}$ △, and $LaFe_{12}O_{19}$ ○. Saturation magnetization of lanthanum magnetoplumbite ferrite as a function of temperature [*61AS*].

Tab. 7. Anisotropy of Curie temperatures.
For the spontaneous magnetization parallel to the *c* axis $\Theta_{C,2}$ is measured. For the spontaneous magnetization perpendicular to the *c* axis $\Theta_{C,1}$ is measured. [*65 BJKLP*].

Compound	Sample	$\Theta_{C,2}$ °K	$\Theta_{C,1}$ °K	$\Delta\Theta_C$ °K
$BaFe_{12}O_{19}$	single crystal	713.3	709.5	3.8
$BaFe_{12}O_{19}$	crystal-oriented sample	732.4	726.3	6.1
$SrCr_{3.2}Fe_{8.8}O_{19}$	crystal-oriented sample	584.4	577.0	7.4

Tab. 8. Curie temperature of the compounds $Ba_xLa_{1-x}Fe^{2+}_{1-x}Fe^{3+}_{11+x}O_{19}$ according to [*61AS*].

x	Θ_C °K
1	725
0.8	715
0.5	710
0	695

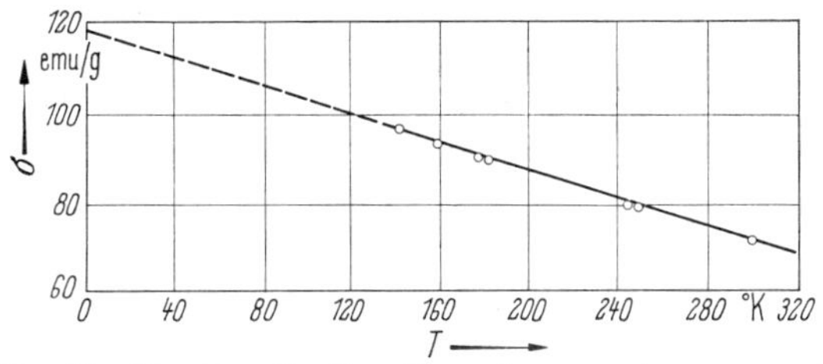

Fig. 43. $Na_{0.5}La_{0.5}Fe_{12}O_{19}$. Magnetic moment per gram, σ, vs. temperature for a single crystal, magnetized in the direction of easy magnetization (*c* axis) [*57SB*].

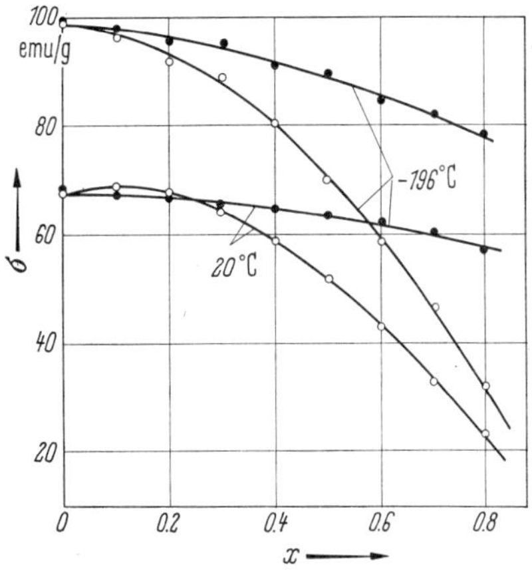

Fig. 44. Magnetic moment per gram, σ, as a function of composition [*61SA*].
$Ba_{1-x}La_xNi_xFe_{12-x}O_{19}$ ●, and $Ba_{1-x}Pr_xNi_xFe_{12-x}O_{12}$ ○.

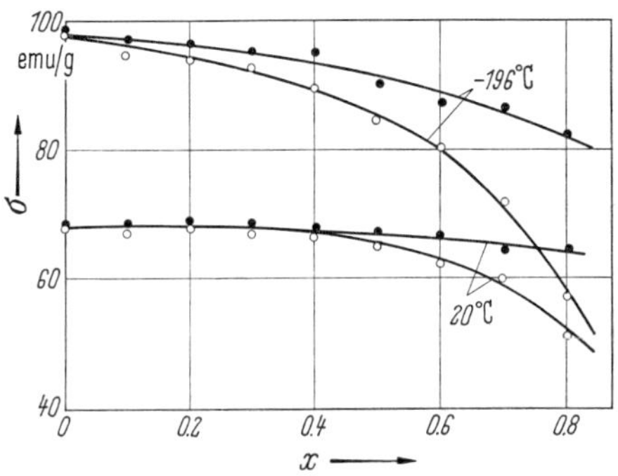

Fig. 45. Magnetic moment per gram, σ, as a function of composition [*61SA*].
$Ba_{1-x}La_xCoFe_{12-x}O_{19}$ ●, and $Ba_{1-x}Pr_xCoFe_{12-x}O_{19}$ ○.

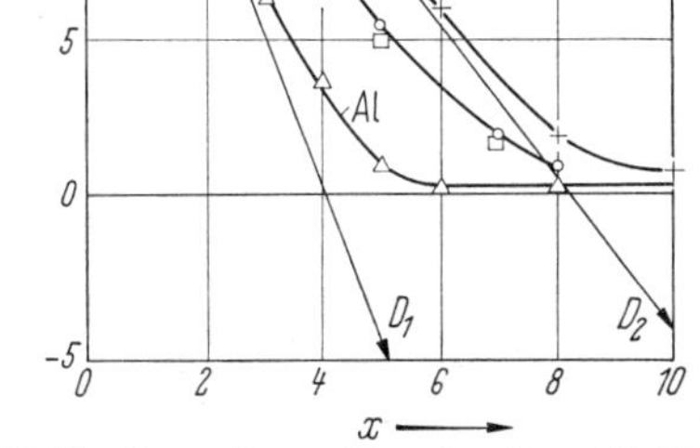

Fig. 46. $BaMe_xFe_{12-x}O_{19}$ systems for Me = Ga(+), Cr(○), and Al (△), and $SrCr_xFe_{12-x}O_{19}$ (□). Magnetic moment per molecule, p_m, in Bohr magnetons as a function of composition. D_1: theoretical curve for Al^{3+} on 2*a* and 12*k* sites; D_2: theoretical curve for Ga^{3+} on 4*a* and 12*k* sites [*58BDP*].

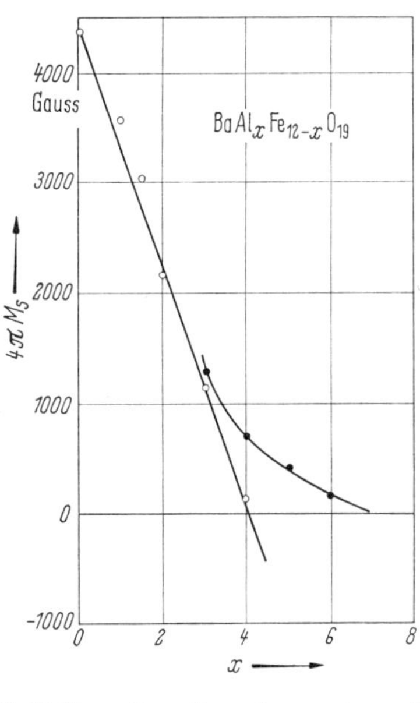

Fig. 47. $BaAl_xFe_{12-x}O_{19}$. Room temperature value of the saturation magnetization vs. aluminum content x for members of the series $BaAl_xFe_{12-x}O_{19}$ [*57vUS*].
Firing temperature: ○ 1400 °C and lower
● 1450 °C and higher

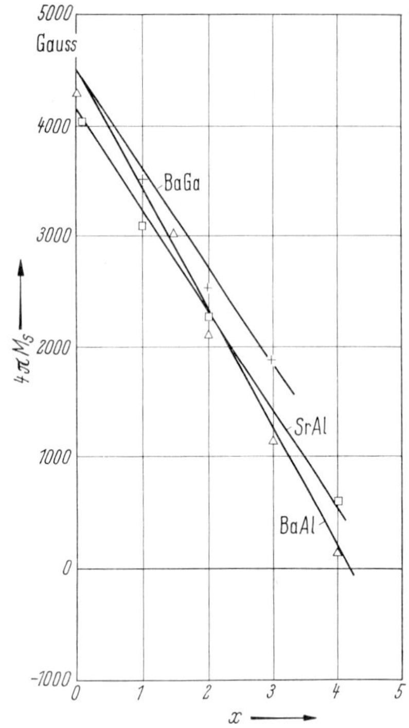

Fig. 48. Room-temperature value of the saturation magnetization of aluminum- and gallium-substituted **M**-type ferrites as a function of composition [*57vU*].
+ $BaGa_xFe_{12-x}O_{19}$, △ $BaAl_xFe_{12-x}O_{19}$, □ $SrAl_xFe_{12-x}O_{19}$.

→
Fig. 50. $BaGa_xFe_{12-x}O_{19}$ and $BaAl_xFe_{12-x}O_{19}$. Saturation magnetization of polycrystalline samples in dependence of the composition parameter x measured at liquid nitrogen temperature [*58MB*].

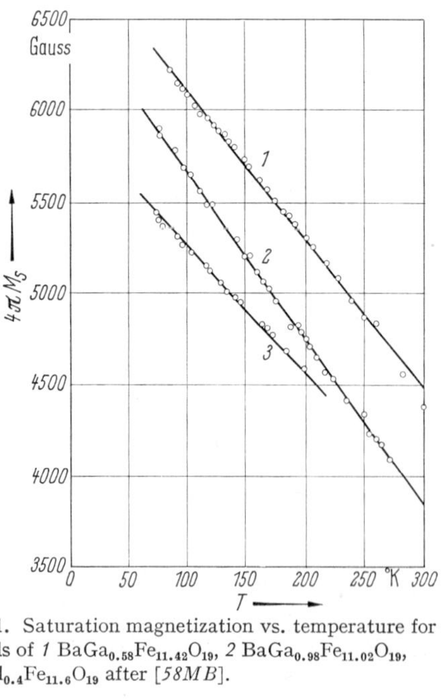

Fig. 51. Saturation magnetization vs. temperature for single crystals of *1* $BaGa_{0.58}Fe_{11.42}O_{19}$, *2* $BaGa_{0.98}Fe_{11.02}O_{19}$, *3* $BaAl_{0.4}Fe_{11.6}O_{19}$ after [*58MB*].

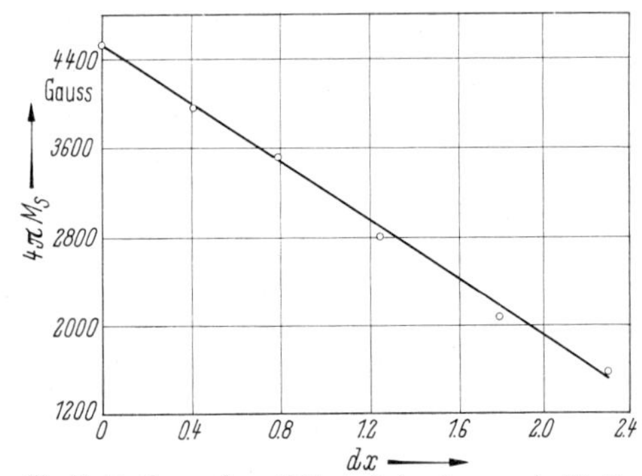

Fig. 49. $SrAl_xFe_{12-x}O_{19}$. Effect of aluminum substitution in Sr**M** materials on the room temperature saturation magnetization. The values of $4\pi M_s$ apply to X-ray density [*67GR*].

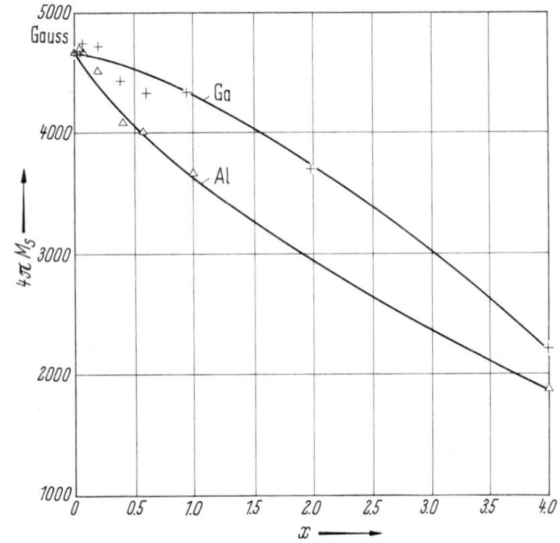

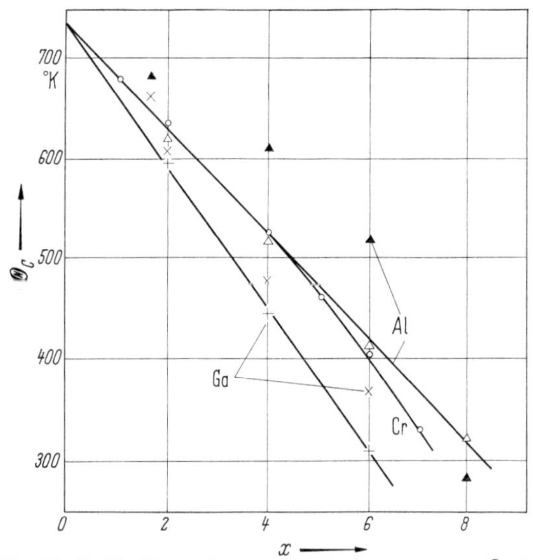

Fig. 52. $BaMe_xFe_{12-x}O_{19}$ system. Curie temperature Θ_C for the ferrites:
Me = Ga + [*59BDPP*]
× [*58MB*]
Me = Cr ○ [*59BDPP*]
Me = Al △ [*59BDPP*]
▲ [*58MB*]

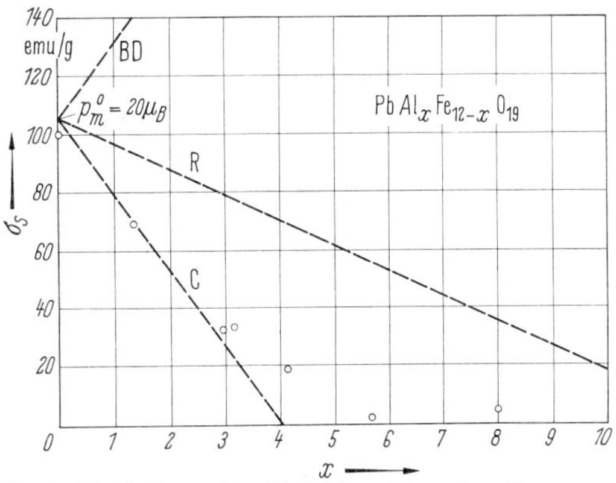

Fig. 53. $PbAl_xFe_{12-x}O_{19}$. Variation of the saturation magnetization of magneto-plumbite with Al content, measured for a crystal-oriented sample and extrapolated to 0 °K and $H=\infty$. For $PbFe_{12}O_{19}$ is found $p_m^0 = 20\ \mu_B$. Curve C theoretical values for Al substitutions on the octahedral sites in the S block of the **M** structure. Curves R and BD for Al substitution on other sites [*59BK2*].

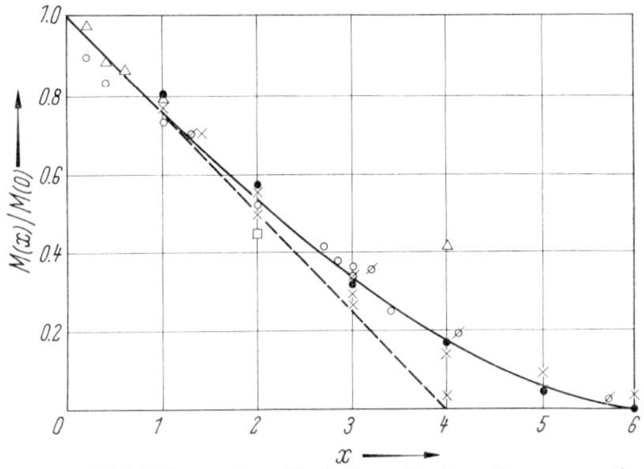

Fig. 54. $MeAl_xFe_{12-x}O_{19}$. Variation of saturation magnetization with composition. Me = Sr, Ba or Pb. After [*64dB*].

● [*59BDPP*]
△ [*58MB*]
× [*57vUS*]
ø [*58BK*]
□, ○ [*63dBdPB*]

——— $\frac{M(x)}{M(0)} = \frac{1}{2}\left[2 - \frac{x}{2} + \left(\frac{x}{6}\right)^{2.55}\right]$

---------- $\frac{M(x)}{M(0)} = 1 - \frac{x}{4}$

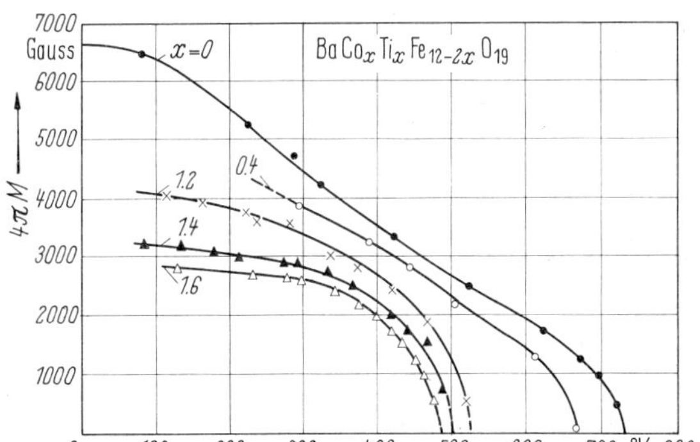

Fig. 55. $BaCo_xTi_xFe_{12-2x}O_{19}$. Magnetization as a function of temperature measured at a field strength of 6600 Oe [*59CSEFWGDFJ*].

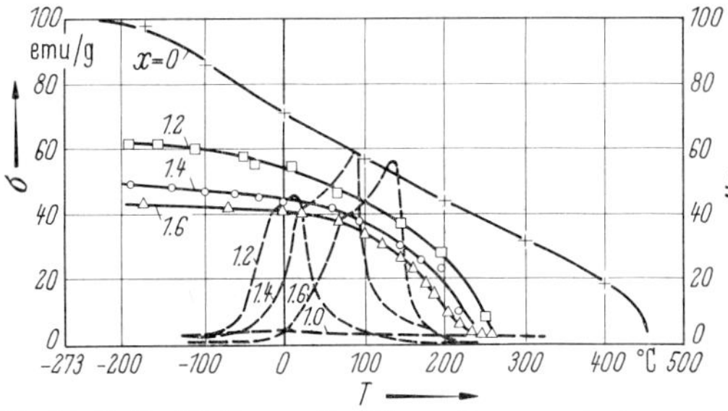

Fig. 56. $BaCo_xTi_xFe_{12-2x}O_{19}$. Magnetic moment, σ, measured at a field strength $H = 6600$ Oe and initial permeability, μ_i, (f=1600 c/s) for the compounds $BaCo_xTi_xFe_{12-2x}O_{19}$ as a function of temperature. Parameter x [*59CSEFWGDFJ*].

——— σ ---------- μ_i

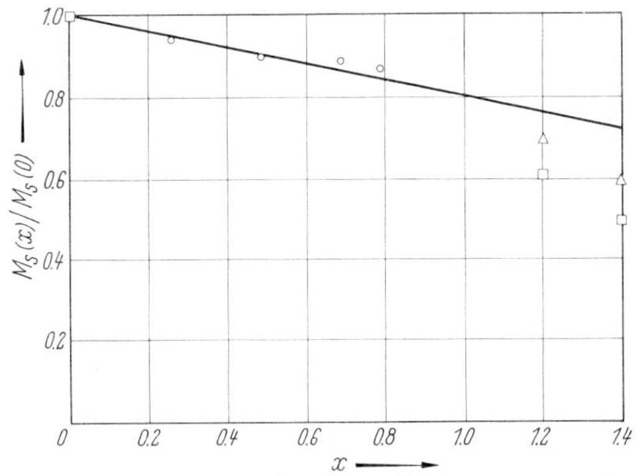

Fig. 57. $BaTi_xCo_xFe_{12-2x}O_{19}$. Normalized room temperature experimental values of saturation magnetization $M_s(x)/M_s(0)$ plotted against the composition parameter x after ○ [*63dBdPB*], □ [*59CSEFWGDFJ*], and △ [*59SW*].

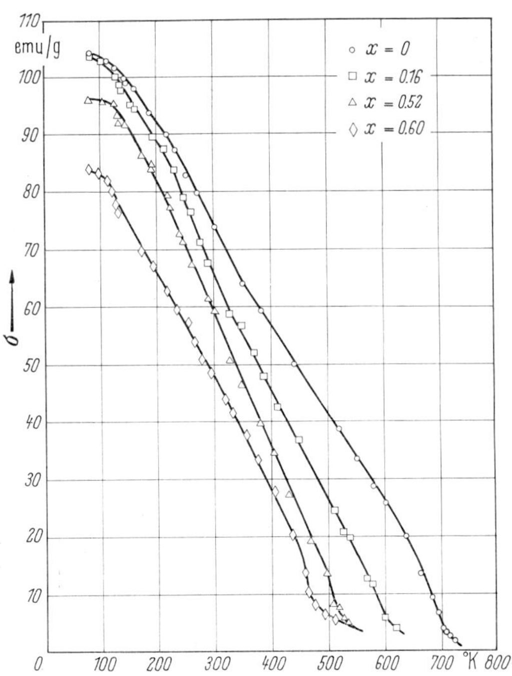

Fig. 58. $BaFe_{12-2x}Ir_xZn_xO_{19}$. Magnetic moment per gram, σ, vs. temperature at $H = 12\,000$ Oe for single crystals, after [*63TKS*].

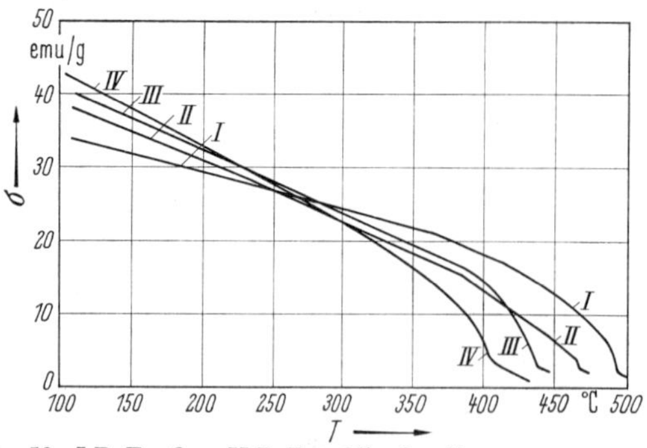

Fig. 59. I $BaFe_{12}O_{19}$; II $BaFe_{11.5}Ni_{0.5}O_{18.5}F_{0.5}$; III $BaFe_{11}Ni_1O_{18}F$; IV $BaFe_{10}Ni_2O_{17}F_2$.
Magnetic moment per gram, σ, of nickel-fluor-substituted **M** compounds as a function of temperature, measured on polycrystalline samples with a magnetic field of 6000 Oe [*62BRT*].

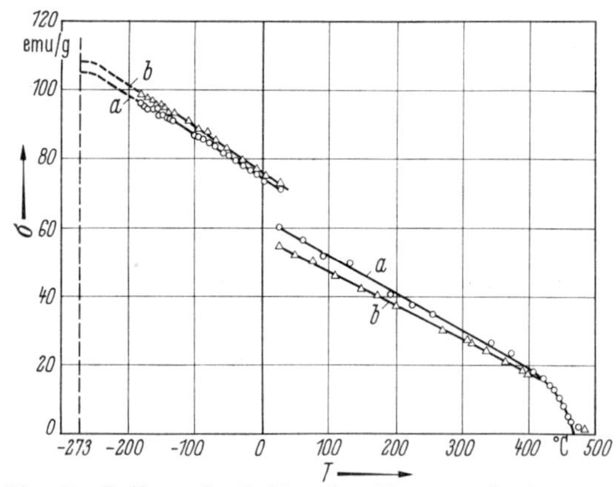

Fig. 60. $BaFe_{12-x}Co_xO_{19}F_x$. Specific magnetization, σ, vs. temperature T. *a)* $x = 0.3$; *b)* $x = 0.4$.
Left portion of curves is for single crystals with easy direction parallel to field; right portion on powders for Curie point determination [*63RLB*].

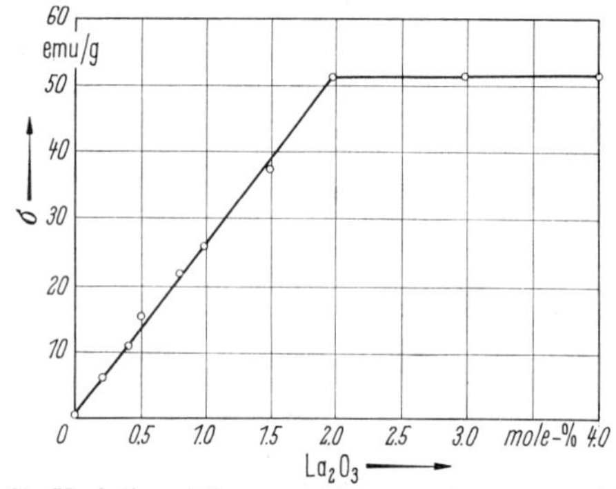

Fig. 61. Variation of the magnetic moment per gram with the content of La_2O_3 in calcium hexaferrite ($CaO \cdot 6Fe_2O_3$) at 296 °K [*63IK*].

Tab. 9. Curie temperature of some compounds of $BaIr_xZn_xFe_{12-2x}O_{19}$ according to [*63TKS*]

x	Θ_C °K
0	720
0.16	635
0.52	560
0.56	550
0.6	520

Tab. 11. Magnetic moment per gram, σ, at 78 °K for a number of substituted hexagonal compounds. (Formulas only represent compositions as initially mixed.) [*58MB*]

Compound	σ (78 °K) emu/g
$BaFe_{11.6}Zn_{0.6}O_{19}$	78.5
$BaFe_{11}ZnO_{18.5}$	82.1
$BaFe_{11}Zn_{1.5}O_{19}$	80.3
$BaFe_{11.6}Mg_{0.4}O_{19}$	65.4
$BaFe_{11}TiO_{19.5}$	58.0
$BaFe_{10}Cr_2O_{19}$	40.2

Tab. 10. Effect of fluoride-compensated cobalt substitutions in $BaFe_{12}O_{19}$ on the saturation magnetization. Measurements made on polycrystalline samples in a magnetic field of 6 kOe. [*63RB*]

Composition	σ Predicted emu/g	σ Observed emu/g	σ_s^0 (H = ∞) Predicted emu/g
$BaFe_{12}O_{19}$		76	100
$BaCo_{0.3}Fe_{11.7}O_{18.7}F_{0.3}$	78.3	76.5	103
$BaCo_{0.4}F_{11.6}O_{18.6}F_{0.4}$	79.0	85	104
$BaCo_{0.45}Fe_{11.55}O_{18.55}F_{0.45}$	79.4	95.2	104.5
$BaCo_{0.47}Fe_{11.53}O_{18.53}F_{0.47}$	79.8	100	105

Pauthenet [*52P*] has given a value of 76 Gauss cm³/g for the magnetization of a randomly oriented powder of $BaFe_{12}O_{19}$ in a field of 6 kOe, extrapolated to 0 °K. This field strength is well below the field required for saturation. Since this value of 76 Gauss cm³/g represented a material with a saturation magnetization of 20 Bohr magnetons per formula unit, for which σ_s^0 ($H = \infty$) is 100 emu/g the values of σ ($H = 6$ kOe) for the other powder samples were calculated on the basis of their calculated saturation magnetizations in Bohr magnetons, as follows: σ_s^0 (predicted) $= 76\ [n_B\,(\text{calc.})]/n_B\,(BaFe_{12}O_{19})$.
The saturation magnetizations n_B (calc.) were calculated assuming a tetrahedral site preference for the Co^{2+} ion, and random orientation with respect to the external field [*63RB*].

7.6.4 Effective spectroscopic splitting factor g_{eff} of ferrites with M structure

Tab. 12. Effective g-factors of compounds with **M** structure

Compound	f_{res} Gc/s	T °C	g_{eff}	Sample	Ref.
$BaFe_{12}O_{19}$	23.930	−196	2.00	single crystal disc	*55SB*
	23.980	20	1.98		
	23.980	112	2.00		
	23.980	155	2.01		
	23.980	200	2.02		
$BaFe_{12}O_{19}$	≅59	20	1.87	single crystal disc	*61WIS*
$BaFe_{12}O_{19}$	40 ··· 70	20	2.05	single crystal disc	*62B*
$BaFe_{12}O_{19}$	50 ··· 55	25	1.96	single crystal sphere	*62M*
$BaAl_xFe_{12-x}O_{19}$ $x < 1.0$	50 ··· 145	20	1.91	crystal-oriented disc	*58dPdBB, 64dB*
$SrAl_xFe_{12-x}O_{19}$ $x < 1.0$	50 ··· 145	20	1.91	crystal-oriented disc	*58dPdBB, 64dB*
$BaTi_xCo_xFe_{12-2x}O_{19}$ $x < 1.0$	23 ··· 145	20	1.91	crystal-oriented disc	*64dB*

7.6.5 Magneto-crystalline anisotropy of ferrites with M structure

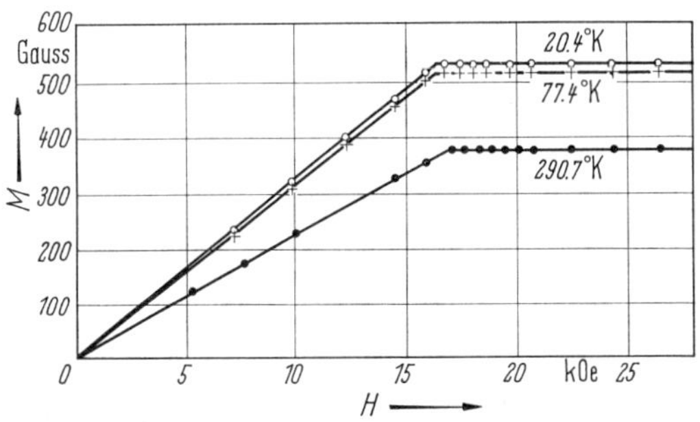

Fig. 62. $BaFe_{12}O_{19}$. Magnetization of a single crystal as a function of the magnetic field applied perpendicular to the c axis of the crystal [*59CSEFWGDFJ*].

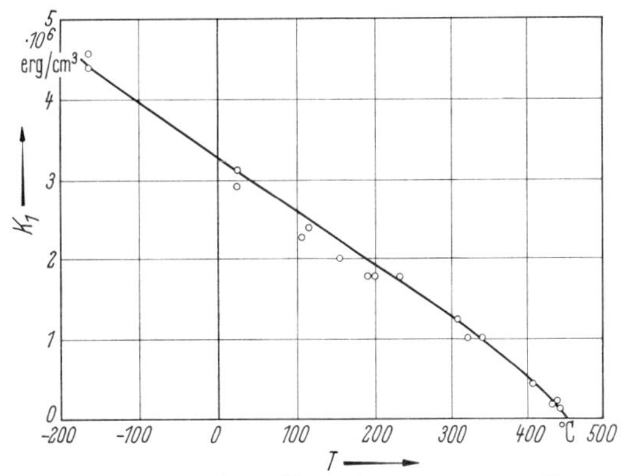

Fig. 63. $BaFe_{12}O_{19}$. Crystalline anisotropy constant K_1 measured as a function of temperature [*52RSS*], [*53R*]. For a theoretical discussion of the origin of the crystalline anisotropy see [*59SW*] and [*65F*].

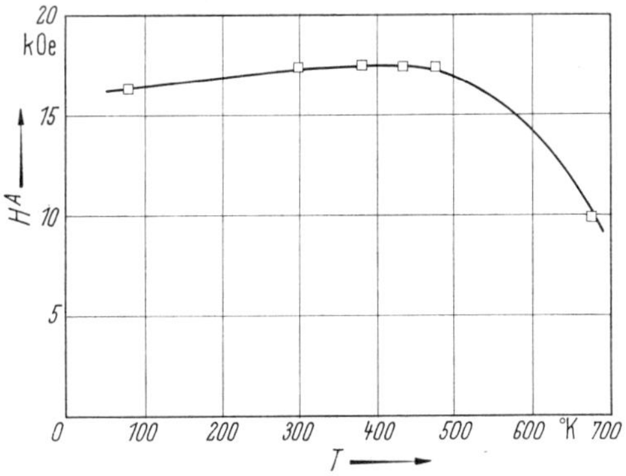

Fig. 64. $BaFe_{12}O_{19}$. Magneto-crystalline anisotropy field as a function of temperature [*59CSEFWGDFJ*].

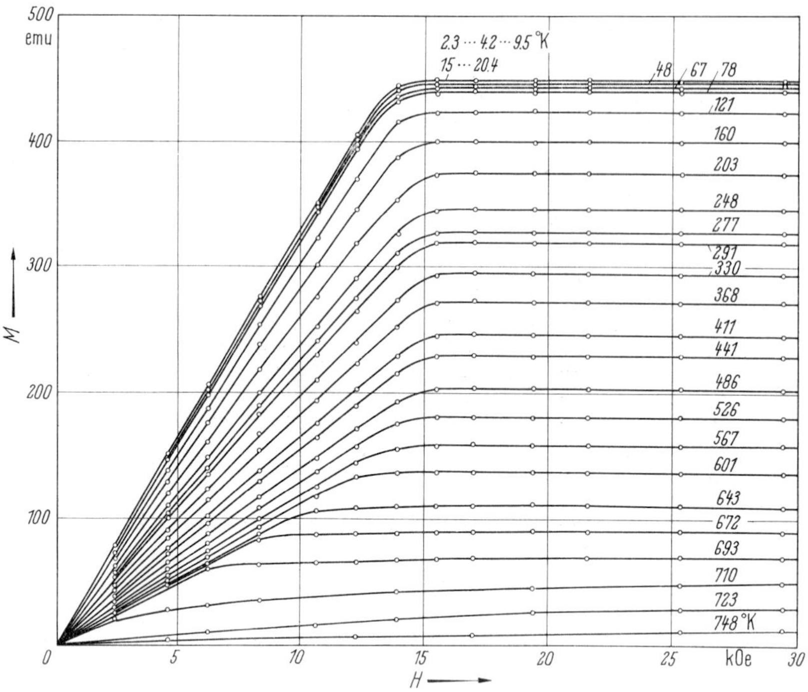

Fig. 65. $PbFe_{12}O_{19}$. Magnetization of a single crystal as a function of the magnetic field applied perpendicular to the c axis of the crystal. Parameter T in °K [59PR2].

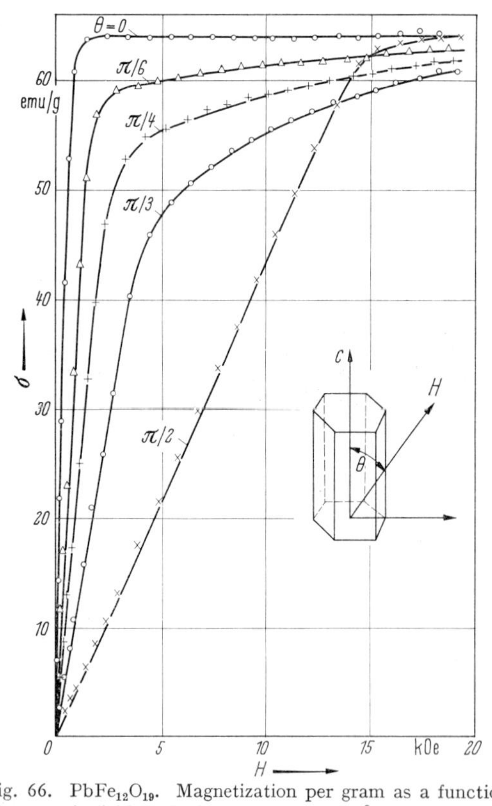

Fig. 66. $PbFe_{12}O_{19}$. Magnetization per gram as a function of magnetic field applied at various angles θ with respect to the c axis of a single crystal [59V].

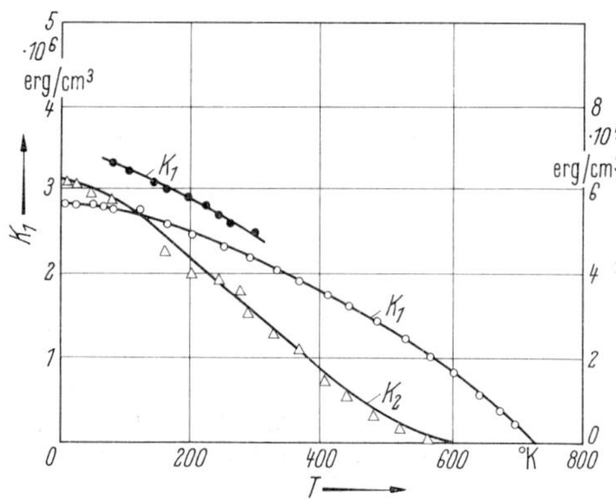

Fig. 67. $PbFe_{12}O_{19}$. Magneto-crystalline anisotropy coefficients K_1 and K_2 as functions of temperature. ○ and △ according to [59PR2], ● according to [59V1].

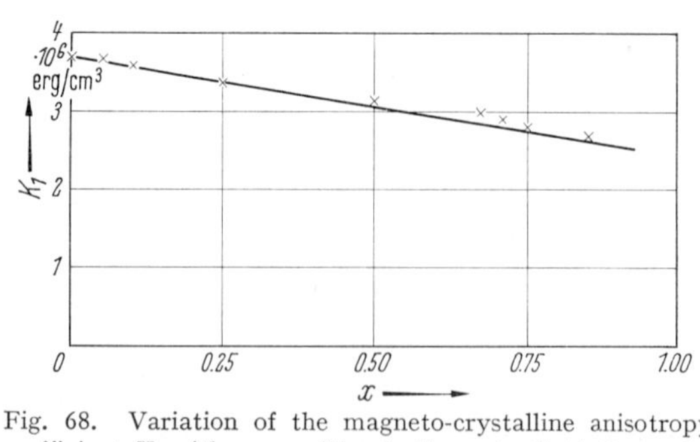

Fig. 68. Variation of the magneto-crystalline anisotropy coefficient K_1 with composition in the series $SrAl_xFe_{12-x}O_{19}$ at room temperature, after [64dB].

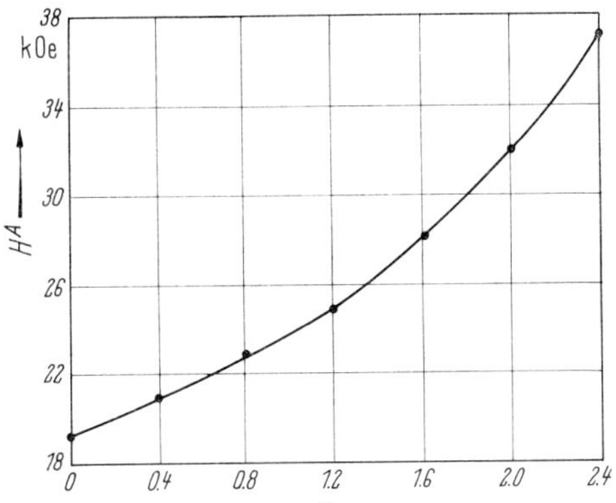

Fig. 69. $SrAl_xFe_{12-x}O_{19}$. Effect of aluminum substitution in Sr**M** materials on anisotropy field at room temperature [*67GR*].

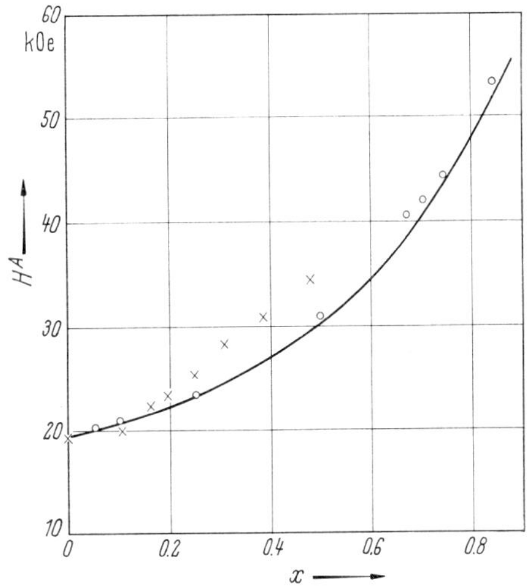

Fig. 70. $SrAl_xFe_{12-x}O_{19}$. Variation of anisotropy field H^A with composition at room temperature, after ○ [*64dB*] and × [*64T*].

Tab. 13. Magneto-crystalline anisotropy field H^A for $BaFe_{12}O_{19}$ as measured from ferromagnetic resonance experiments. For explanation see 7.1e. H_{res} = resonance field; h_{rf} = high frequency field

Sample	Directions of H_{res} and h_{rf}	T °C	f_{res} Gc/s	H_{res} in [kOe] for the component of h_{rf} parallel to H_{res}	H_{res} in [kOe] for the component of h_{rf} perpendicular to H_{res}	H^A kOe	Ref.
Single crystal disc c axis ⊥ disc	H_{res} in basal plane, h_{rf} in basal plane the angle between H_{res} and h_{rf} is 45°	−196	23.93	14.3	about 10	16.2	*55SB*
		20	23.98	15.05	17.15	17.0	
		112	23.98	15.40	17.55	17.3	
		155	23.98	15.40	17.15	17.3	
		200	23.98	15.35	16.85	17.3	
Single crystal disc c axis ∥ disc	H_{res} ∥ c axis, H_{res} ∥ disc, h_{rf} ⊥ H_{res}	20	58.2		3.844		
		20	58.4		3.886	18.4*)	*61WIT*
			58.6		3.964		
			58.8		4.042		
			59.0		4.140		
Single crystal sphere	H_{res} ∥ c axis, h_{rf} ⊥ H_{res}	25	52.7		1.900	17.55	*62M*
			53.42		2.021		
			53.5		2.100		
			55.5		2.850		
Single crystal discs		20	>40			17.0	*62B*
Crystal-oriented discs c axis ∥ disc	H_{res} ∥ c axis	20				17.5	*58dPdBB*
Single crystals and crystal-oriented samples		20	53.0			17.0	*63TKS*
Crystal-oriented very thin discs		20				17.5	*64dB*

*) Extrapolated value for H_{res} = 0, see Fig. 96.

Tab. 14. Room temperature magneto-crystalline anisotropy coefficient K_1 for Ga- and Al-substituted barium ferrites [58MB]

Compound	K_1 $10^6 \cdot$ erg/cm³
$BaFe_{12}O_{19}$	3.3
$BaGa_{0.4}Fe_{11.6}O_{19}$	3.5
$BaGa_{0.8}Fe_{11.2}O_{19}$	2.9
$BaAl_{0.2}Fe_{11.8}O_{19}$	3.3
$BaAl_{0.4}Fe_{11.6}O_{19}$	2.9

Tab. 15. Magnetic properties of crystals in the system $BaIr_xZn_xFe_{12-2x}O_{19}$, after [63TKS]

x	Magneto-crystalline anisotropy field H^A in [Oe] at room temperature		Θ_C °K
	from magnetization measurements	from ferromagnetic resonance	
0	17 000	17 000	720
0.16	9 600		635
0.52	15 000*)	16 000*)	560
0.56	16 000*)		550
0.60	20 000*)	22 000*)	520

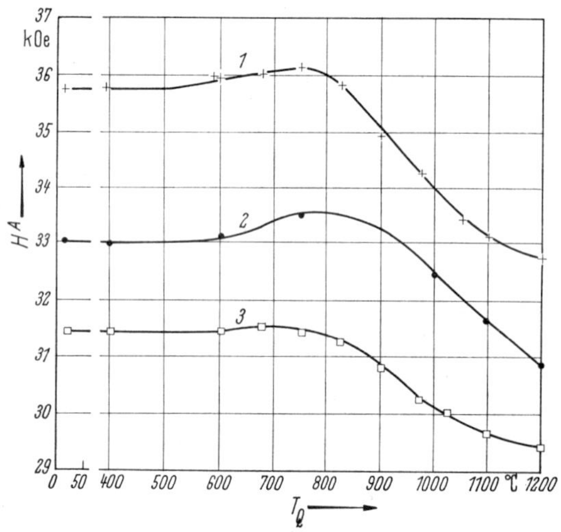

Fig. 71. Variation of anisotropy field with quenching temperature T_Q in aluminum substituted Sr**M** compounds [67GR]. *1* $SrAl_{2.25}Fe_{9.75}O_{19}$; *2* $SrAl_{2.10}Fe_{9.90}O_{19}$; *3* $SrAl_{1.92}Fe_{10.08}O_{19}$.

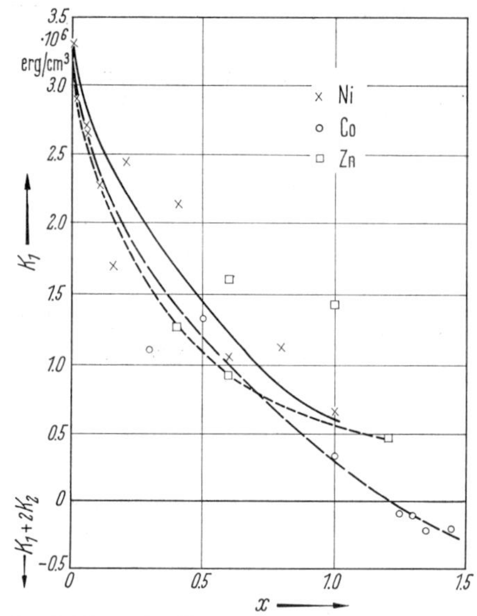

Fig. 72. $BaTi_xMe^{2+}_xFe^{3+}_{12-2x}O_{19}$. Room-temperature values of the magneto-crystalline anisotropy coefficients as functions of composition [59SW].
Me = Ni, Co, Zn.

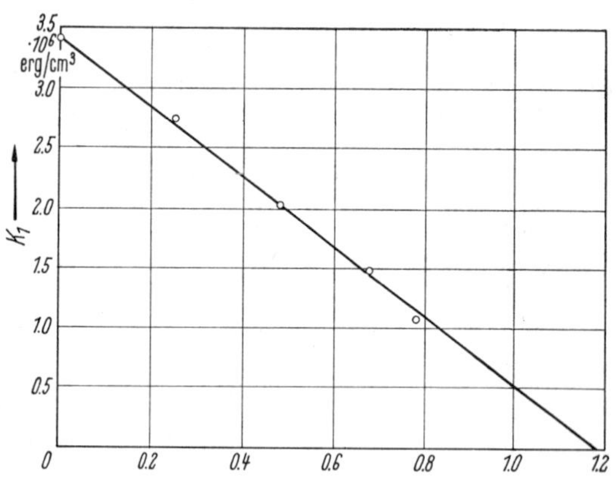

Fig. 73. $BaTi_xCo_xFe_{12-x}O_{19}$. Variation of the magneto-crystalline anisotropy coefficients K_1 and K_2 with composition at room temperature [64dB].

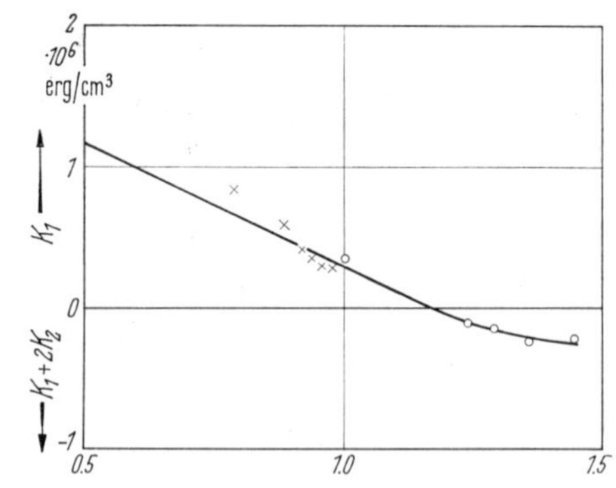

Fig. 74. $BaCo_xTi_xFe_{12-2x}O_{19}$. First order magneto-crystalline anisotropy coefficients K_1 and K_2 vs. composition.
× derived from the maxima of μ''-curves in Fig. 95,
○ magnetostatic measurements according to [59dBdPB].

*) Preferential plane for the magnetization.

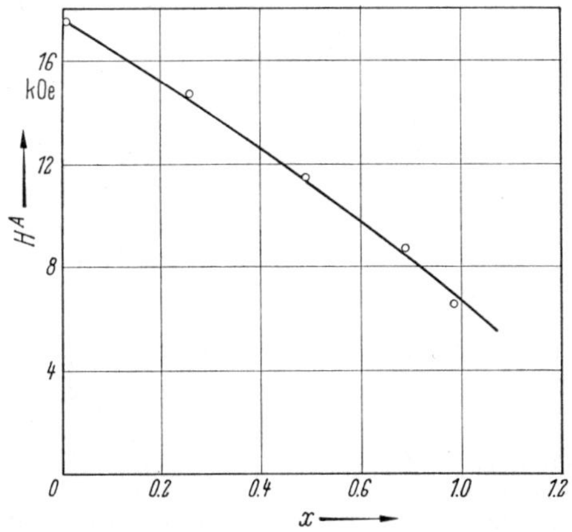

Fig. 75. $BaTi_xCo_xFe_{12-2x}O_{19}$. Variation of anisotropy field H with composition at room temperature [*64dB*].

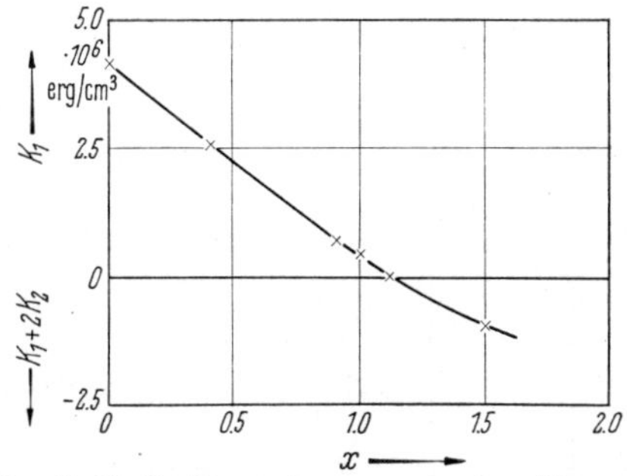

Fig. 76. $BaTi_xCo_xFe_{12-2x}O_{19}$ compounds. The magneto-crystalline anisotropy coefficients K_1 and K_2 as a function of cobalt content at 90 °K [*61LES*].

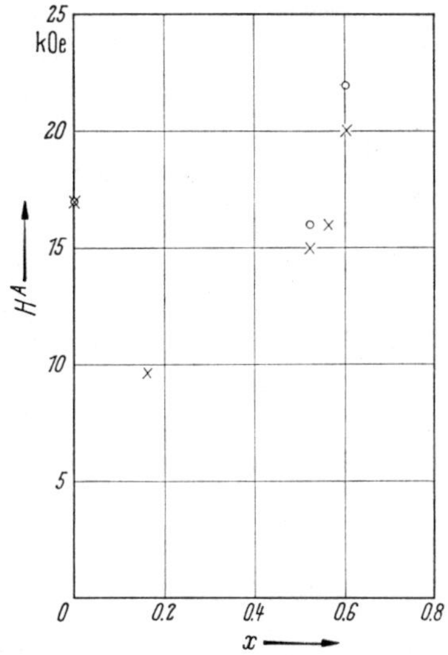

Fig. 77. $BaIr_xZn_xFe_{12-2x}O_{19}$. Magneto-crystalline anisotropy field H^A for compounds of the series Ir_xZn_x**M** [*63TKS*].
× derived from magnetization curves
○ derived from ferromagnetic resonance experiments.

7.6.6 Hysteresis properties of ferrites with M structure

Figs. 78 ··· 88 give static hysteresis properties of polycrystalline samples. For the dynamics of the magnetization processes in such samples see e. g. [*65ACGL*]. For a study of the magnetization reversal and the corresponding modifications of the domain structure see [*58WSR*], [*59BCGI*], [*60KE*], [*60P*], [*61KG*], [*62BC*], [*63SK*], [*63RTN*], [*63RT*], [*64RTR*], [*65KG*], [*65RT*], [*66SF*].

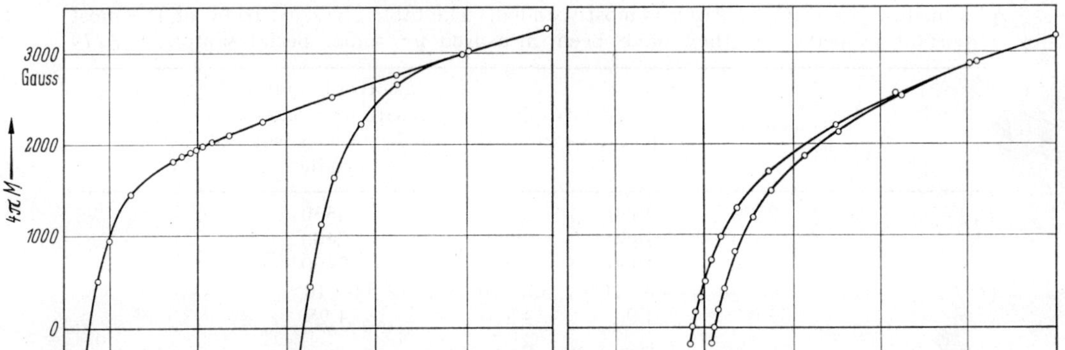

Fig. 78. $BaFe_{12}O_{19}$. Hysteresis loops at room temperature for ceramic samples with random crystal orientation and sintered at a) 1350 °C and b) 1400 °C, respectively [*52RSS*].

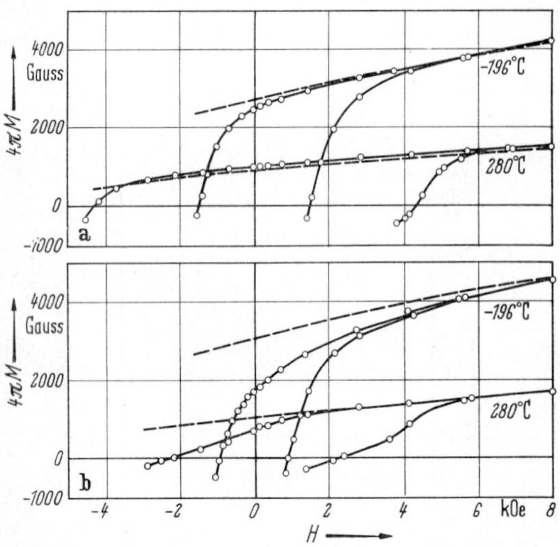

Fig. 79. $BaFe_{12}O_{19}$. Hysteresis loops for ceramic samples with random crystal orientation, measured at −196 °C and 280 °C, respectively. a) A material with small crystals, obtained by sintering at a low temperature, b) a material with larger crystals. Dashed lines: demagnetization curves calculated for pure rotation [52RSS]. See also [56GV].

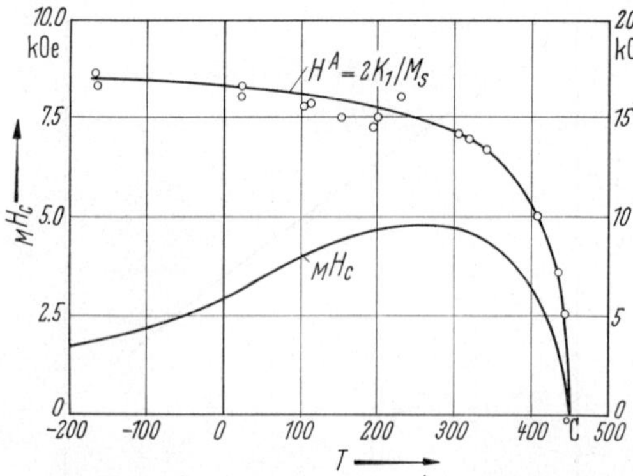

Fig. 80. $BaFe_{12}O_{19}$. The quantity $H^A = 2K_1/M_s$ compared with the coercive force $_MH_c$ for fine grained powder as a function of temperature. Since, for pure rotation, $_MH_c = 0.96\,K_1/M_s$, the difference between the two curves shows that wall formation and wall movement must take place [52RSS], [53R].

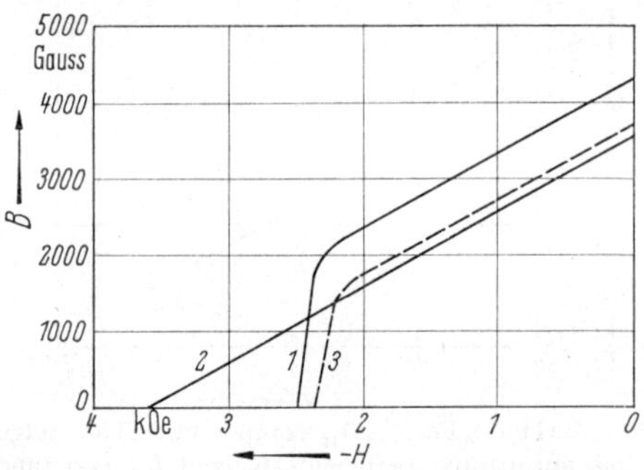

Fig. 81. *1* Sr-ferrite $SrFe_{12}O_{19}$, $(BH)_{max} = 4.3 \cdot 10^6$ Gauss Oe
2 Sr-ferrite $SrFe_{12}O_{19}$, $(BH)_{max} = 3.5 \cdot 10^6$ Gauss Oe
3 Ba-ferrite $BaFe_{12}O_{19}$, $(BH)_{max} = 3.7 \cdot 10^6$ Gauss Oe
Demagnetization curves of anisotropic oxide magnets, measured parallel to the preferential direction [66F].

Tab. 16. The general chemical formula for the ceramic permanent magnetic materials is $MeO \cdot nFe_2O_3$, where Me represents one of the metals Sr or Ba and n is a number which is in most cases somewhat smaller than 6 (e.g. 5.5) in order to obtain best permanent magnetic properties. For an improvement of the magnetic properties also a small amount of a modifier like SiO_2 or Al_2O_3 is mostly added. The table gives a survey of the most important properties as they have been measured for some special samples. [67T]

	isotropic	anisotropic (crystal-oriented)		anisotropic material with at moment highest energy product	
B_r	2200	3850	3650	4550	Gauss
$_BH_c$	1650	1850	3000	2400	Oe
$_MH_c$	2700	1900	3200	2460	Oe
$(BH)_{max}$	$0.95 \cdot 10^6$	$3.4 \cdot 10^6$	$3.2 \cdot 10^6$	$5.0 \cdot 10^6$	Gauss · Oe
ϱ	5.0	4.9	4.7	4.95	g/cm³
Me	Ba	Ba	Sr	Sr	
n	5.5	5.5	5.5	5.6	
modifier	SiO_2	SiO_2	SiO_2	SiO_2	
demagnetization curve in Fig.	82	83	84	85	

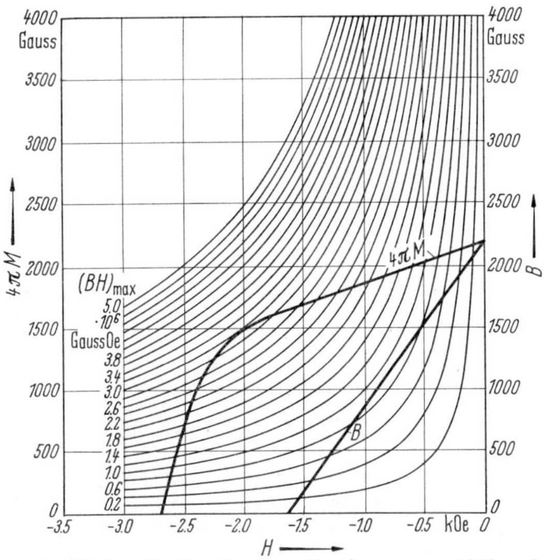

Fig. 82. MeO · nFe_2O_3. Demagnetization curves $B(H)$ and $4\pi M(H)$ of the isotropic ceramic permanent magnetic material mentioned in Tab. 16 [*67T*].

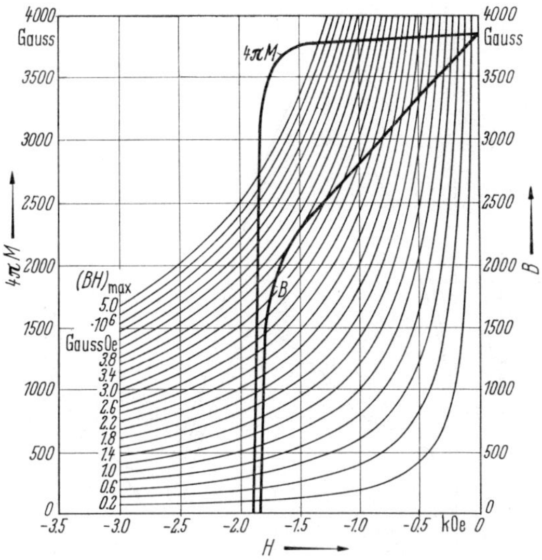

Fig. 83. MeO · nFe_2O_3. Demagnetization curves $B(H)$ and $4\pi M(H)$ of the anisotropic ceramic permanent magnetic material with high remanence mentioned in Tab. 16 [*67T*].

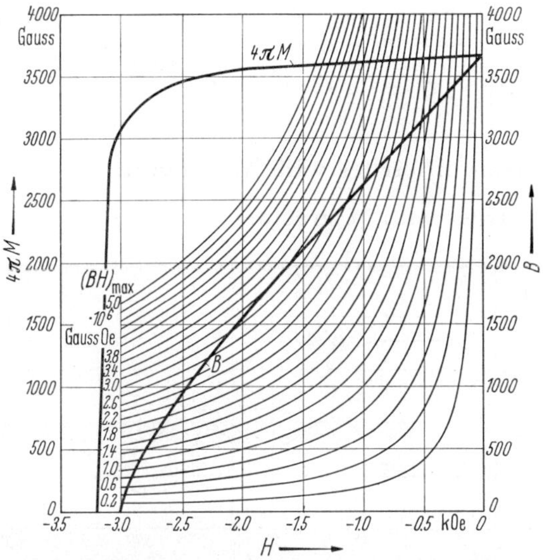

Fig. 84. MeO · nFe_2O_3. Demagnetization curves $B(H)$ and $4\pi M(H)$ of the anisotropic ceramic permanent magnetic material with high coercive force mentioned in Tab. 16 [*67T*].

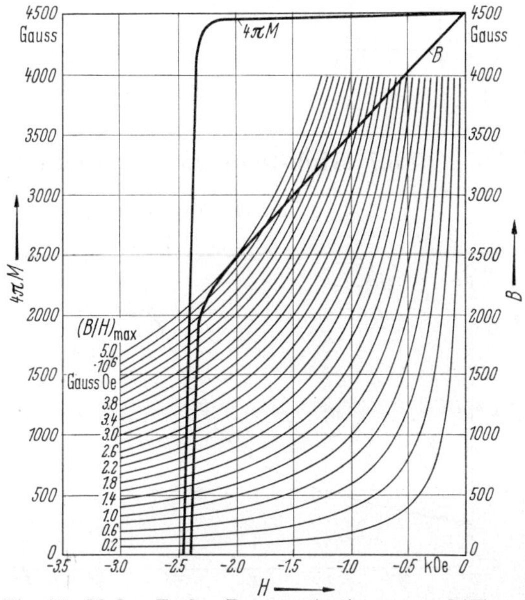

Fig. 85. MeO · nFe_2O_3. Demagnetization curves $B(H)$ and $4\pi M(H)$ of the anisotropic ceramic permanent magnetic material with an energy product $(BH)_{max}$ which is extremely high for this type of material [*67T*].

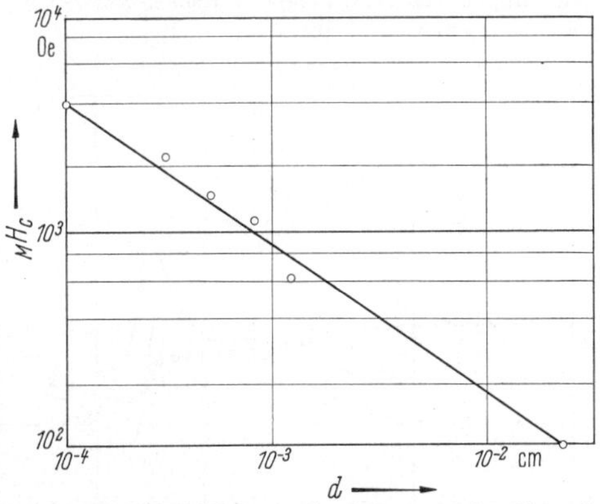

Fig. 86. $BaFe_{12}O_{19}$. The coercive force as a function of particle diameter d in sintered samples [*56SKT*].

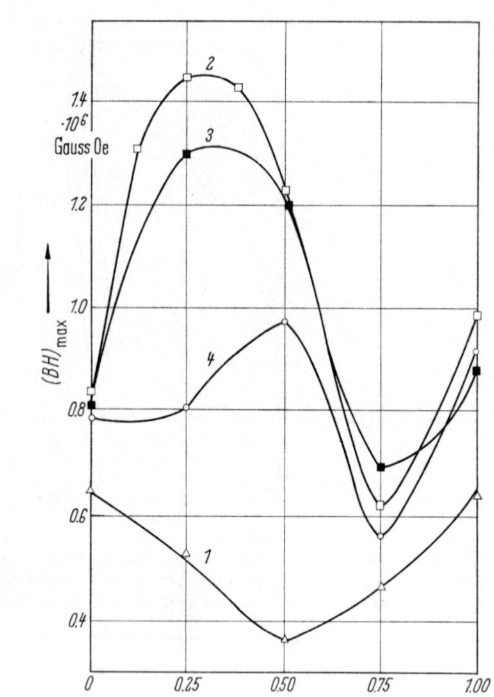

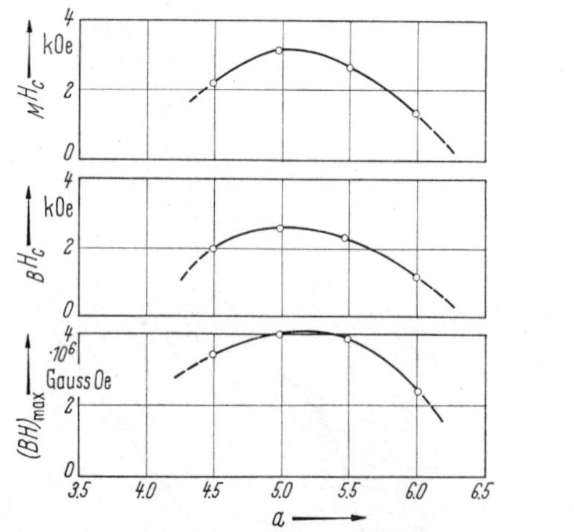

Fig. 88. $(SrO)_{1-x}(CaO)_x \cdot aFe_2O_3$. Hysteresis properties of crystal-oriented samples. The remanence induction is for all compositions equal to about 4000 Gauss [66F].

←

Fig. 87. $Ba_{1-x}Sr_xFe_{12}O_{19}$. Curves showing the composition dependence of $(BH)_{max}$ of mixed ferrites $Ba_{1-x}Sr_xFe_{12}O_{19}$ fired at various different temperatures [62BU].
1 1200 °C; *2* 1230 °C; *3* 1260 °C; *4* 1300 °C.

7.6.7 High-frequency magnetic properties of ferrites with M structure

7.6.7.1 Magnetic spectrum of ferrites with M structure

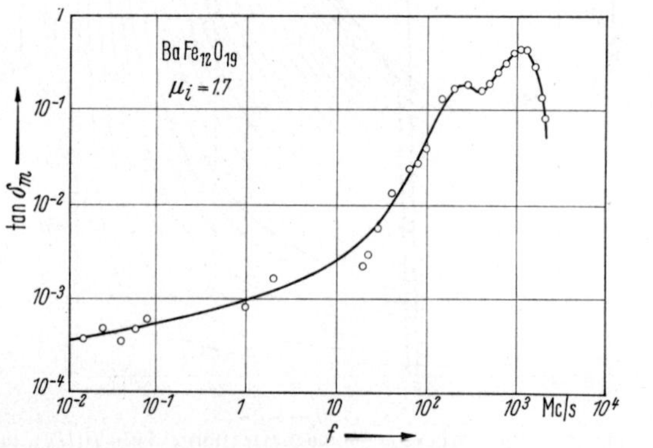

Fig. 89. $BaFe_{12}O_{19}$. Magnetic loss factor $\tan \delta_m$ at room temperature as a function of frequency for an isotropic polycrystalline sample with small crystals (average crystal dimension ≦ 1 μm) [59SW].

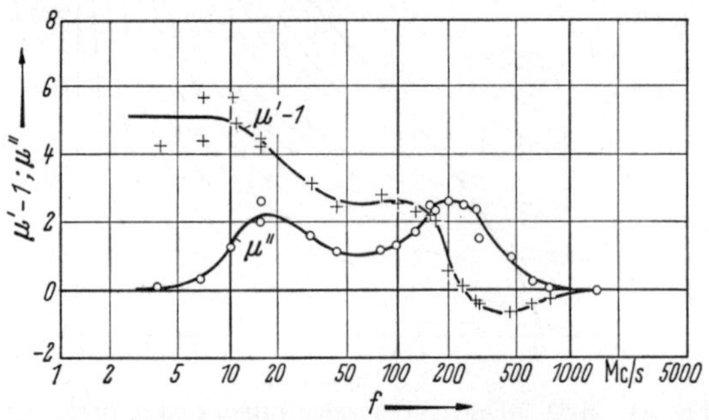

Fig. 90. $BaFe_{12}O_{19}$. The real and the imaginary part of the initial permeability as a function of frequency for a crystal-oriented sample. The r. f. magnetic field is parallel to the c axes of the crystallites [53W].

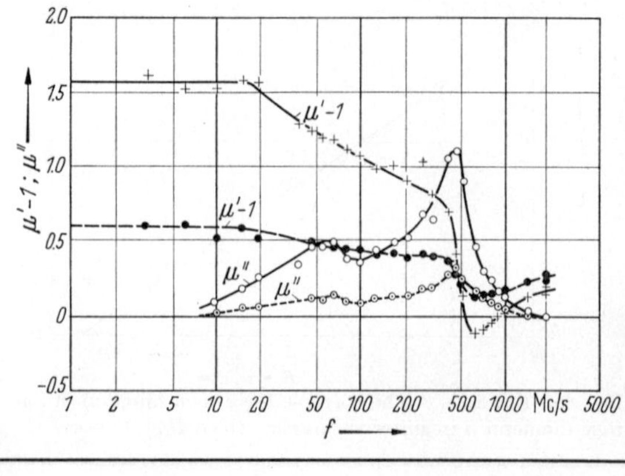

→

Fig. 91. $BaFe_{12}O_{19}$. The real and the imaginary part of the initial permeability as a function of frequency for two polycrystalline samples. In the crystal-oriented sample *1* the h. f. magnetic field is perpendicular to the c axes of the crystallites. In the isotropic polycrystalline sample *2* all directions of the c axis relative to the h. f. magnetic field occur [53W].
·········· crystal-oriented sample *1*
——— isotropic sample *2*

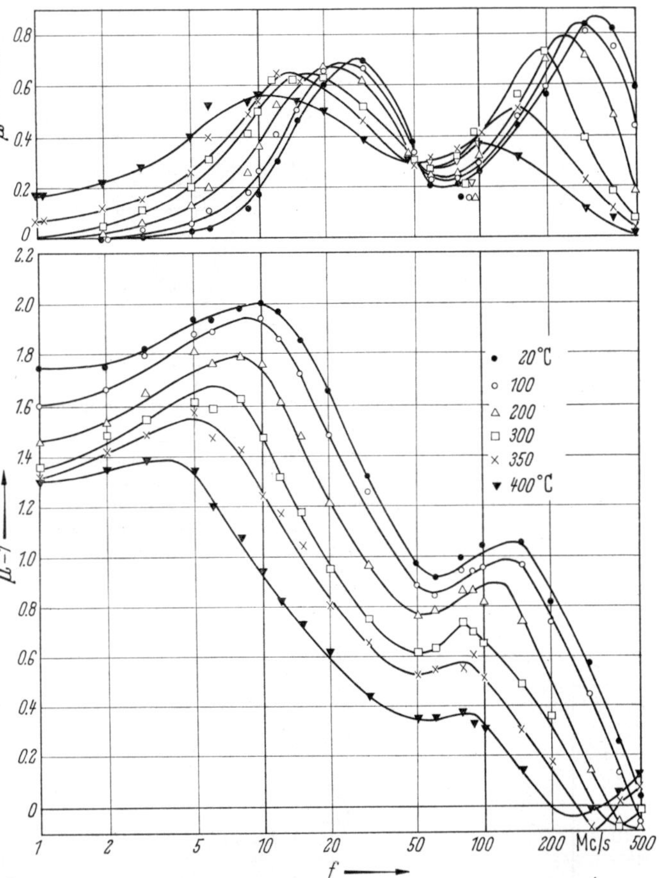

← Fig. 92. $BaFe_{12}O_{19}$. The real and the imaginary part of the initial permeability as a function of frequency for a polycrystalline sample with relative large crystals (average dimension about 20 μm) measured at various temperatures [67SW].

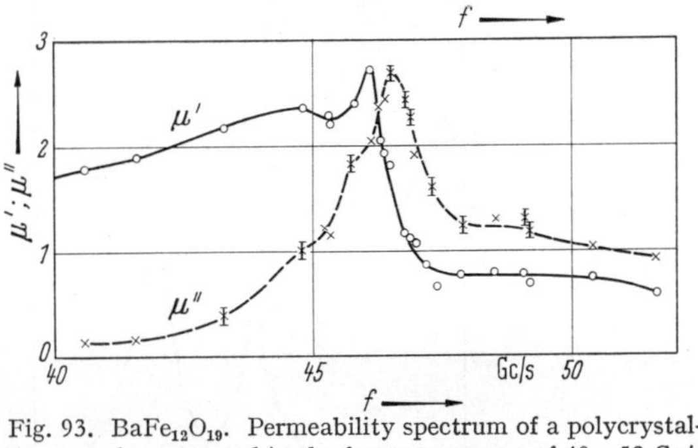

Fig. 93. $BaFe_{12}O_{19}$. Permeability spectrum of a polycrystalline sample, measured in the frequency range of 40 ··· 52 Gc/s on a rod with a cross section of 0.15 mm [67SS].

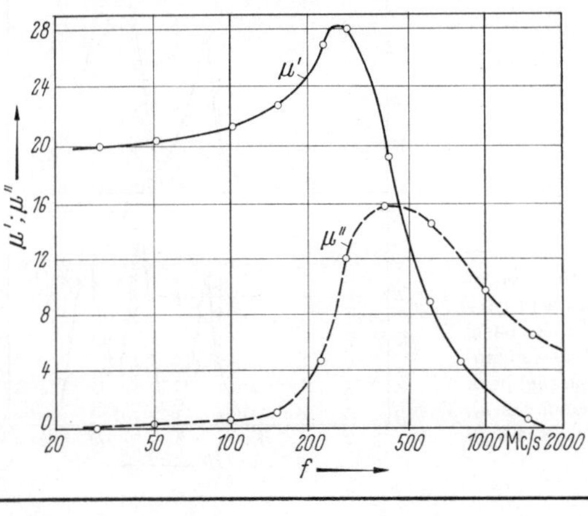

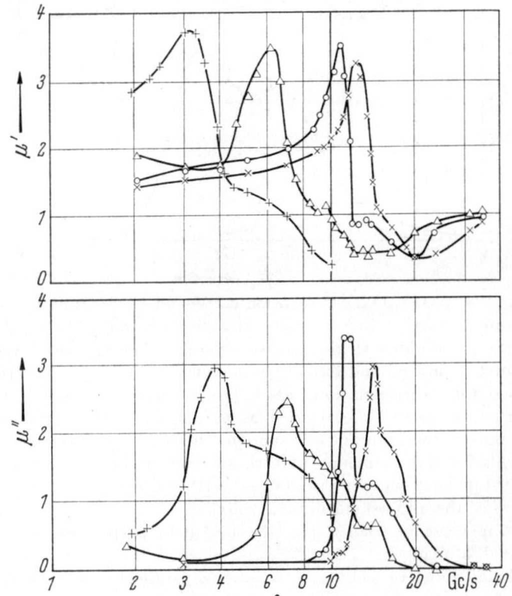

Fig. 95. $BaCo_x^{2+}Ti_x^{4+}Fe_{12-2x}O_{19}$. Permeability spectrum of polycrystalline samples [67SS].
+ $x = 0.98$, △ $x = 0.92$, ○ $x = 0.89$, × $x = 0.79$.

← Fig. 94. $BaCoTiFe_{10}O_{19}$. Magnetic spectrum of an isotropic polycrystalline sample [64NSVW].

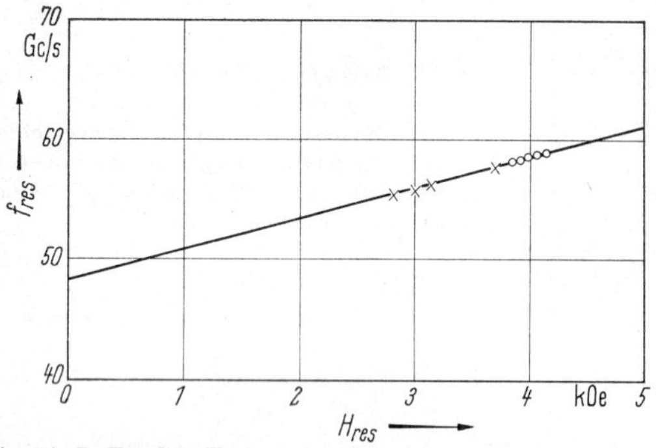

Fig. 96. $BaFe_{12}O_{19}$. Ferromagnetic resonance frequency, f_{res}, of a single crystal. The applied d. c. magnetic field, H_{res}, is parallel to the c axis of the crystal.
○ according to [*61WIT*]; × according to [*62M*].

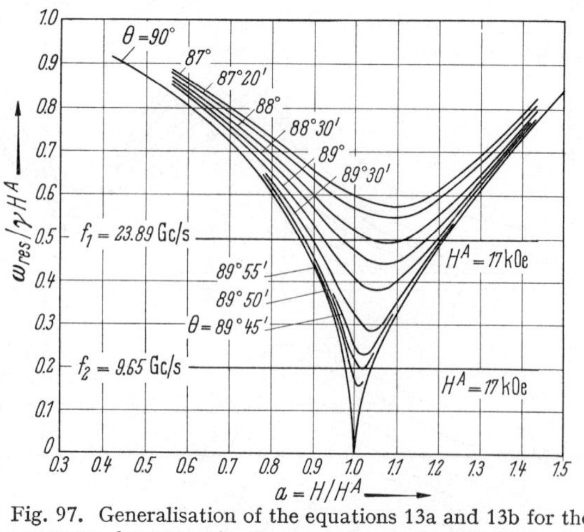

Fig. 97. Generalisation of the equations 13a and 13b for the resonance frequency in the case of a single-domain crystal for an arbitrary angle θ between the applied magnetic d. c. field and the c axis of the crystal. The ordinate is the resonance frequency, ω_{res}, relative to the "natural" resonance frequency, γH^A; the abscissa is the applied field relative to the anisotropy field, H^A, [*62H*].

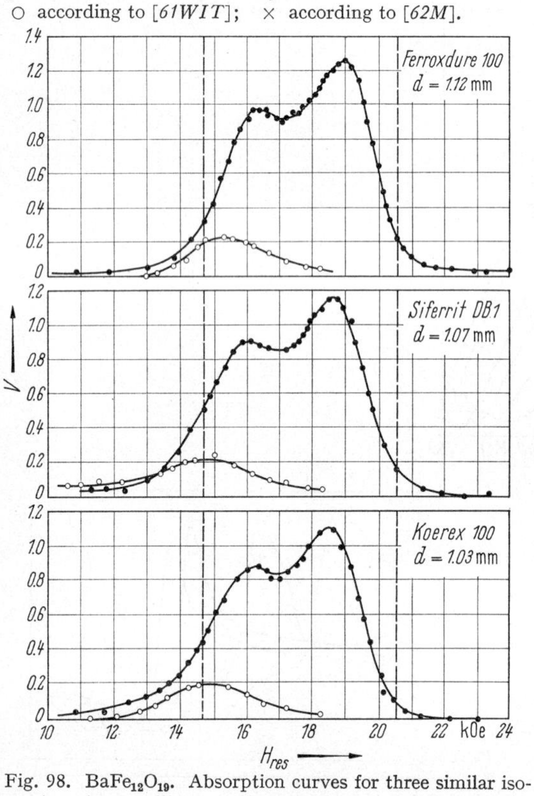

Fig. 98. $BaFe_{12}O_{19}$. Absorption curves for three similar isotropic polycrystalline samples at a frequency of 23.89 Gc/s with presumably single-domain crystals. The spheres (diameter d) are ground out of commercial materials which are used for permanent magnets. The absorption of the h. f. signal of 23.89 Gc/s is given as a function of the applied magnetic field strength. Compare the measured resonance fields for this frequency with those given in Fig. 97 for the crystals with various orientations with respect to the direction of the applied static field [*62H*].
• transverse resonance, i. e. the h. f. field perpendicular to the d. c. field,
○ longitudinal resonance, i. e. the h. f. field parallel to the d. c. field.

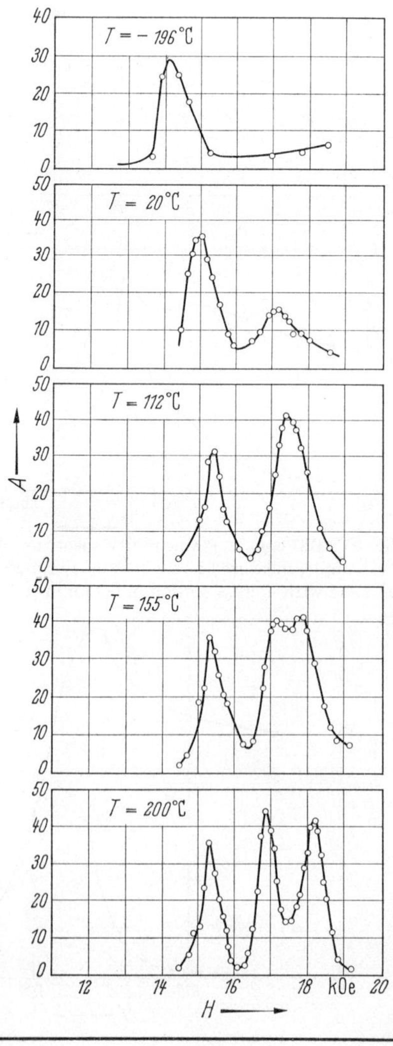

→

Fig. 100. $BaFe_{12}O_{19}$. Absorption peaks for a single crystal as a function of the static field, H, at five different temperatures (absorption is on an arbitrary scale). For an explanation see Fig. 99. The microwave field is in the basal plane at 45° with the static field, H, so that a component $h_{rf,y}$ and also a component $h_{rf,x}$ occurs. At low temperature ($T = -196$ °C) only the parallel peak is found (h_y-component), at higher temperature also the x component of h_{rf} gives rise to absorption since then the measuring frequency $f = 24$ Gc/s is high enough for these resonances also to occur [*55SB*].

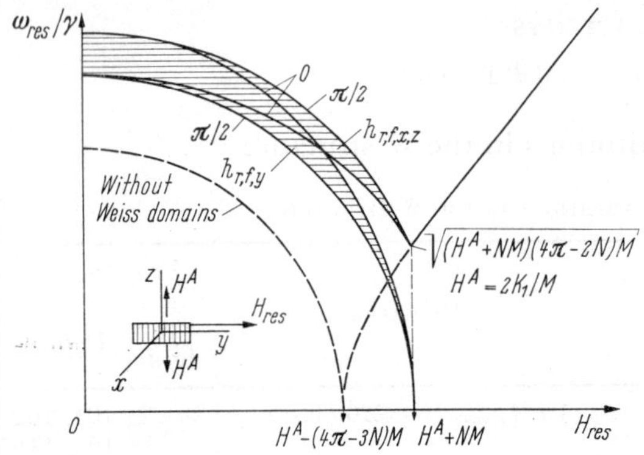

Fig. 99. $BaFe_{12}O_{19}$. Complication of Fig. 2 for the case of the presence of Bloch walls in the single crystal. Dependence of the resonance frequency on the strength of the applied magnetic field, H_{res}, perpendicular to the hexagonal axis for various orientations of the microwave field, h_{rf}, and of the Bloch walls with respect to the static field. The values of the angle between the Bloch walls, which are always parallel to the c axis, and H_{res} are indicated $(0, \pi/2)$. For $H_{res} > H^A + NM$ the specimen is saturated in the y direction. The ω_{res}/γ- and H_{res}-scales are the same. The graph applies approximately to $BaFe_{12}O_{19}$ at room temperature [*55SB*].

7.6.7.2 Line width of the ferromagnetic resonance of ferrites with M structure

Tab. 17. Room temperature measuring results for the line width ΔH of the ferromagnetic resonance of ferrites with **M** structure. The static magnetic resonance field H_{res} is parallel to the c axis of the crystal, i.e. parallel to the magneto-crystalline anisotropy field H^A. The h.f. measuring field is oriented perpendicular to the static field H_{res}

Compound	Sample	f_{res} Gc/s	H_{res} kOe	ΔH Oe	Ref.
$BaFe_{12}O_{19}$	single crystal disc, porosity about 3%	≅ 59	≅ 4	≅ 10	*61WIT*
	single crystal sphere	50···55		< 6	*62M*
	single crystal disc	55.5	7.6	50	*60BCdBdP*
	single crystal ellipsoid	57.7	5.0	53	*60BCdBdP*
	polycrystalline ellipsoid, crystal-oriented			1200	*60BCdBdP*
	isotropic polycrystalline sphere		9.65	2450	*62H*
	isotropic polycrystalline sphere		9.65	2600	*62H*
$BaAl_xFe_{12-x}O_{19}$ } $SrAl_xFe_{12-x}O_{19}$ }	polycrystalline, crystal-oriented	50···90		1500···2000	*58dPdBB*
$BaIr_xZn_xFe_{12-2x}O_{19}$	single crystal				
$x = 0$		55		53	*63TKS*
$x = 0.52$		17		1050	*63TKS*
$x = 0.60$		14		1175	*63TKS*
$BaTi_xCo_xFe_{12-2x}O_{19}$	polycrystalline, crystal-oriented	}			
$x = 0$		}		1600	*64dB*
$x = 0.25$		}		1850	*64dB*
$x = 0.48$		} 23···50		2500	*64dB*
$x = 0.68$		}		1700	*64dB*
$x = 0.78$		}		2000	*64dB*
$SrAl_xFe_{12-x}O_{19}$	polycrystalline, crystal-oriented				
$x = 0$				1600	*64dB*
$x = 0.1$				1800	*64dB*
$x = 0.2$				2000	*64dB*
$x = 0.5$				2500	*64dB*
$x = 1.0$				3300	*64dB*
$x = 1.35$				2500	*64dB*
$x = 1.42$				2300	*64dB*
$x = 1.50$				2300	*64dB*
$x = 1.70$				2700	*64dB*

7.7 W-type ferrites

(general chemical formula $BaMe_2^{2+}Fe_{16}O_{27}$)

7.7.1 Survey of chemical substitutions in the W structure

Tab. 18. Survey of chemical substitutions in the **W** structure

Compound	Symbol	Ref.	Remarks	Properties in	
				Tables	Figures
$BaFe_2^{2+}Fe_{16}^{3+}O_{27}$	Fe_2**W**	*52W, 52B, 62BRT*	$(4\pi M_s)_{20\,C°} = 5220$ Gauss $\Theta_C = 728$ °K, [*52WRGO*], [*62BRT*] $a = 5.88$ Å; $c = 32.845$ Å	20, 22	101, 102, 103, 116, 133
$BaMn_2Fe_{16}O_{27}$	Mn_2**W**	*59SW*	$(4\pi M_s)_{20\,°C} = 3900$ Gauss $\Theta_C = 688$ °K; $c = 32.94$ Å $\varrho_X = 5.30$ g/cm³ [*65V*]		
$BaNi_2Fe_{16}O_{27}$	Ni_2**W**	*59SW*	$c = 32.81$ Å; $\varrho_X = 5.34$ g/cm³ [*65V*]	22	104, 111, 115, 129
$BaCo_2Fe_{16}O_{27}$	CO_2**W**	*66SZZ*	$c = 32.84$ Å; $\varrho_X = 5.33$ g/cm³ [*65V*] For cobalt ion distribution [*62BI*], [*64PZ*], [*67SY*]	20, 22	102, 135
$BaNiFe^{2+}Fe_{16}O_{27}$	NiFe**W**	*59SW*	$(4\pi M_s)_{20\,°C} = 3450$ Gauss [*62 BRT*] $\Theta_C = 793$ °K		34
$BaZnFe^{2+}Fe_{16}O_{27}$	ZnFe**W**	*59SW*	$(4\pi M_s)_{20\,°C} = 4800$ Gauss $\Theta_C = 703$ °K	22	101, 104
$BaZn_{1.5}Fe_{0.5}^{2+}Fe_{16}O_{27}$	$Zn_{1.5}Fe_{0.5}$**W**	*59SW*		22	
$BaZn_2Fe_{16}O_{27}$	Zn_2**W**	*64ST*	$c = 32.85$ Å; $\varrho_X = 5.38$ g/cm³ [*65V*]	22	106, 107, 115, 129
$BaMg_2Fe_{16}O_{27}$	Mg_2**W**	*65V*	$c = 32.80$ Å; $\varrho_X = 5.12$ g/cm³ [*65V*]		
$BaNi_{0.5}Zn_{0.5}Fe^{2+}Fe_{16}O_{27}$	$Ni_{0.5}Zn_{0.5}Fe$**W**	*59SW*	$(4\pi M_s)_{20\,°C} = 4550$ Gauss $\Theta_C = 723$ °K	22	104
$BaNi_xZn_{1-x}Fe_{16}O_{27}$	Ni_xZn_{1-x}**W**	*64HH*	$(4\pi M_s)_{20\,°C} = 4500$ Gauss		
$BaCo_xFe_{1-x}Fe_{16}O_{27}$	Co_xFe_{1-x}**W**	*64PZ*	$\Theta_C = (775 \pm 5)$ °K from magnetic measurements; $\Theta_C = (758 \pm 5)$ °K from neutron diffraction; practically independent of x. [*66YSZP*]. For magnetic structure see also [*66SY*]	23	105, 117, 118, 119, 120, 121, 122, 123, 132
$BaCo_xZn_{2-x}Fe_{16}O_{27}$	Co_xZn_{2-x}**W**	*57vURS*	$0 \leq x \leq 2$		108, 124
$BaCo_xNi_{2-x}Fe_{16}O_{27}$	Co_xNi_{2-x}**W**	*62RPW, 64BM*	$0 \leq x \leq 2$	21	106, 115, 126, 127, 128, 129, 136
$BaCo_xCu_{2-x}Fe_{16}O_{27}$	Co_xCu_{2-x}**W**	*64BM*	$0 \leq x \leq 2$		
$BaCo_{0.75}Zn_{0.75}Fe_{0.5}^{2+}Fe_{16}O_{27}$	$Co_{0.75}Zn_{0.75}Fe_{0.5}$**W**	*59SW*		22	
$BaNi_2Al_xFe_{16-x}O_{27}$	Ni_2Al_x**W**	*57vUS, 57vURS*	$x \leq 7$	19, 21	109, 110, 128
$BaNi_2Ga_xFe_{16-x}O_{27}$	Ni_2Ga_x**W**	*57vUS 57vURS*	$x \leq 7$ $\Theta_C = 518$ °K for $x = 5$ $\Theta_C = 418$ °K for $x = 7$		
$BaNi_2In_xFe_{16-x}O_{27}$	Ni_2In_x**W**	*57vUS, 57vURS*	$x \leq 2.5$		
$BaNi_2Cr_xFe_{16-x}O_{27}$	Ni_2Cr_x**W**	*57vUS, 57vURS*	$x \leq 2.5$		

Tab. 18. (continued)

Compound	Symbol	Ref.	Remarks	Properties in Tables	Properties in Figures
$BaZn_2GaFe_{15}O_{27}$	$Zn_2Ga\mathbf{W}$	*57vUS, 57vURS*			
$BaZn_2Ga_3Fe_{13}O_{27}$	$Zn_2Ga_3\mathbf{W}$	*57vUS, 57vURS*	$\Theta_C = 508$ °K		
$BaZn_2Al_3Fe_{13}O_{27}$	$Zn_2Al_3\mathbf{W}$	*57vUS, 57vURS*	$\Theta_C = 558$ °K		
$BaNi_{0.012}Fe^{2+}_{1.98}Fe^{3+}_{15.417}Ni_{0.563}O_{26.417}F_{0.583}$		*62BRT*			112
$BaNi_{0.01}Fe^{2+}_{1.99}Fe^{3+}_{14.67}Ni_{1.33}O_{25.67}F_{1.33}$		*62BRT*			112
$BaNi_{0.86}Fe^{2+}_{1.14}Fe^{3+}_{13.67}Ni_{2.33}O_{24.67}F_{2.33}$		*62BRT*			112
$BaFe^{2+}_2Fe^{3+}_{14.65}Co_{1.35}O_{25.65}F_{1.35}$		*63RLB*	$\sigma^0_s = 109$ emu/g; $\Theta_C = 718$ °K		113
$BaFe^{2+}_2Fe^{3+}_{13.86}Co_{2.14}O_{24.86}F_{2.14}$		*63RLB*	$\sigma^0_s = 112$ emu/g; $\Theta_C = 701$ °K		113
$Ba_{0.27}Sr_{0.73}Co_{1.845}Fe^{2+}_{0.155}Fe_{16}O_{27}$		*60B*	$K_3 = -1.22 \cdot 10^4$ erg/cm³ at 300 °K $K_3 = +1.35 \cdot 10^5$ erg/cm³ at 77 °K		
$SrCo_xNi_{2-x}Fe_{16}O_{27}$	$SrCo_xNi_{2-x}\mathbf{W}$	*64BM*	$0 \leq x \leq 2$		
$SrCo_xCu_{2-x}Fe_{16}O_{27}$	$SrCo_xCu_{2-x}\mathbf{W}$	*64BM*	$0 \leq x \leq 2$		

7.7.2 Electric and dielectric properties of ferrites with W structure

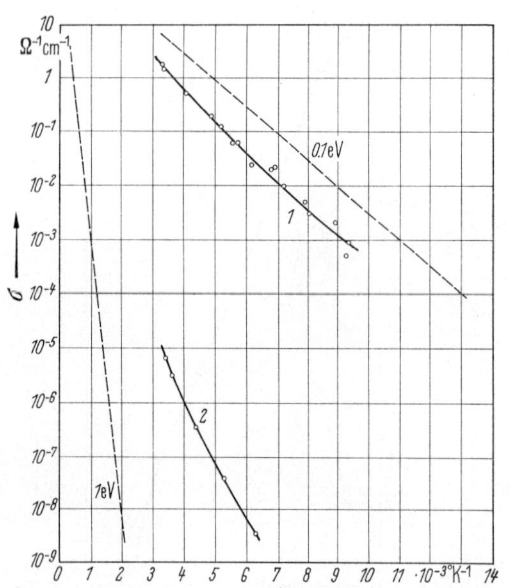

Fig. 101. *1* $BaFe^{2+}_2Fe^{3+}_{16}O_{27}$, ($Fe_2\mathbf{W}$) and *2* $BaZnFe^{2+}Fe^{3+}_{16}O_{27}$, ($ZnFe^{2+}\mathbf{W}$). Temperature dependence of the d. c. electrical conductivity of polycrystalline samples, after [*59SW*]. The dashed lines have been drawn with slopes corresponding to an activation energy of 0.1 and 1 eV.

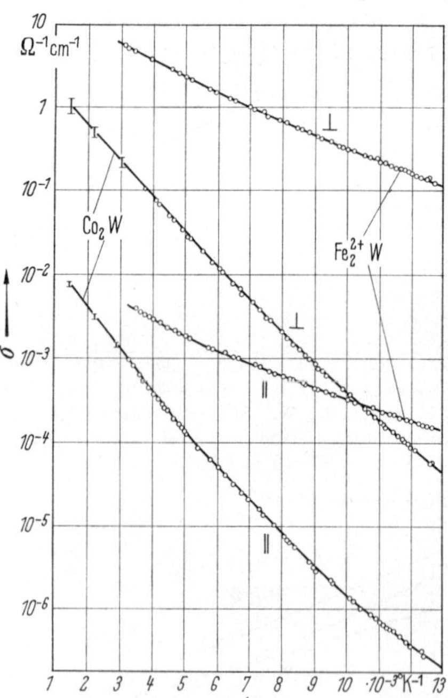

Fig. 102. $BaFe^{2+}_2Fe_{16}O_{27}$, ($Fe_2\mathbf{W}$) and $BaCo_2Fe_{16}O_{27}$, ($Co_2\mathbf{W}$). Temperature dependence of the electrical conductivity, $\sigma_{\parallel}$, (parallel to the *c* axis) and, $\sigma_{\perp}$, (perpendicular to the *c* axis) for single crystals. After [*66SZZ*].

Tab.19. Activation energies, q, in [eV] for the conductivity of the samples Fe_2**W** and Co_2**W** of Fig.102 [*66SZZ*]

Crystal	Direction of current	Temperature range [°K] 77…100	100…170	170…300	300…670
		q in [eV]			
Fe_2**W**	∥ c axis	0.025	0.029	0.041	—
	⊥ c axis	0.029	0.033	0.039	—
Co_2**W**	∥ c axis	0.059	0.079	0.095	0.095
	⊥ c axis	0.064	0.075	0.083	0.083

Tab. 20. Dielectric constant, ε, of polycrystalline isotropic samples of $BaNi_2Al_xFe_{16-x}O_{27}$ measured at a frequency of 9.5 Gc/s at room temperature [*64T*]

x	0.60	0.73	0.86	1.00
ε	15.4	14.9	15.1	14.5

7.7.3 Saturation magnetization and Curie temperature of ferrites with W structure

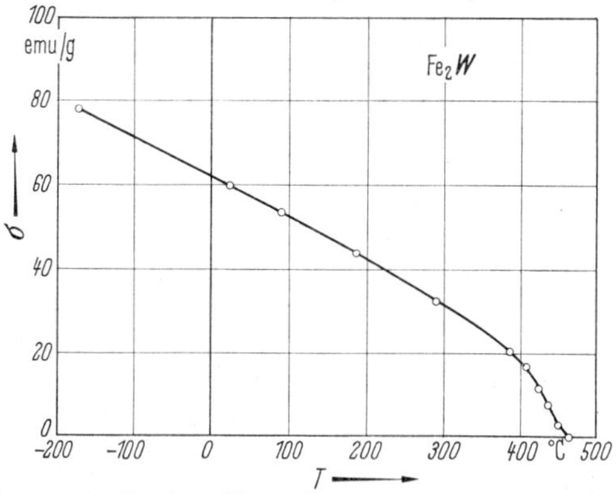

Fig. 103. $BaFe_{18}O_{27}$. Magnetic moment per gram as a function of temperature [*52WRGO*].

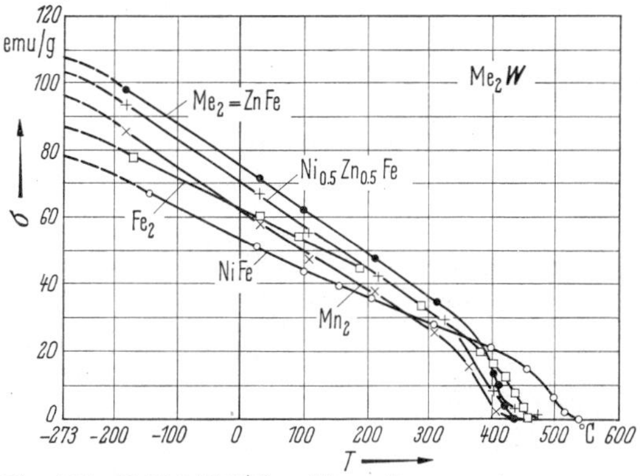

Fig. 104. $BaMe_2^{2+}Fe_{16}^{3+}O_{27}$. Magnetic moment per gram vs. temperature for a number of compounds with **W** structure, measured on polycrystalline specimens at a field strength of 6600 Oe [*59SW*].

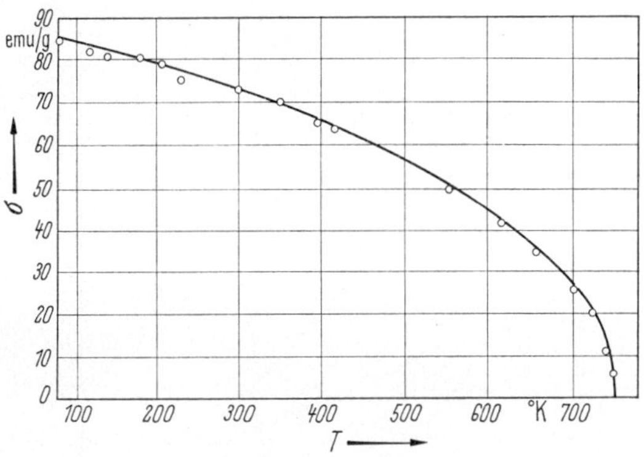

Fig. 105. $Co_{1.65}Fe_{0.35}$**W**. Magnetic moment per gram vs. temperature for a single crystal. Curie temperature $\Theta_C = (775 \pm 5)$ °K [*66YSZPZ*].

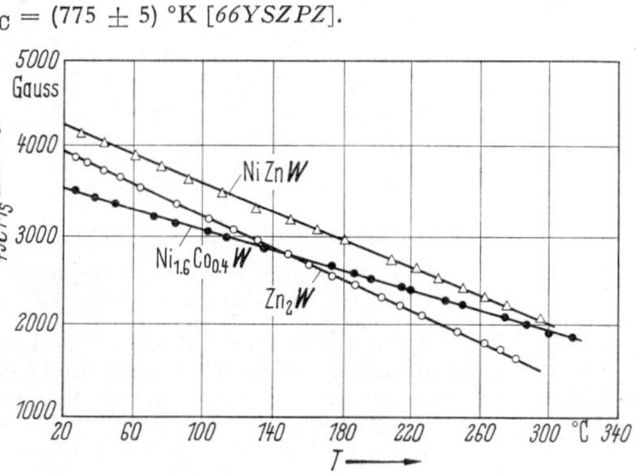

Fig. 106. Zn_2**W**, NiZn**W**, and $Ni_{1.6}Co_{0.4}$**W** compounds. Saturation magnetization vs. temperature [*64HH*].

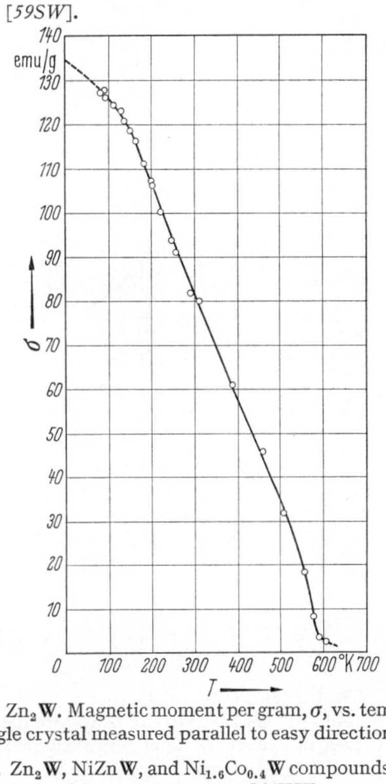

Fig.107. Zn_2**W**. Magnetic moment per gram, σ, vs. temperature for a single crystal measured parallel to easy direction [*64ST*].

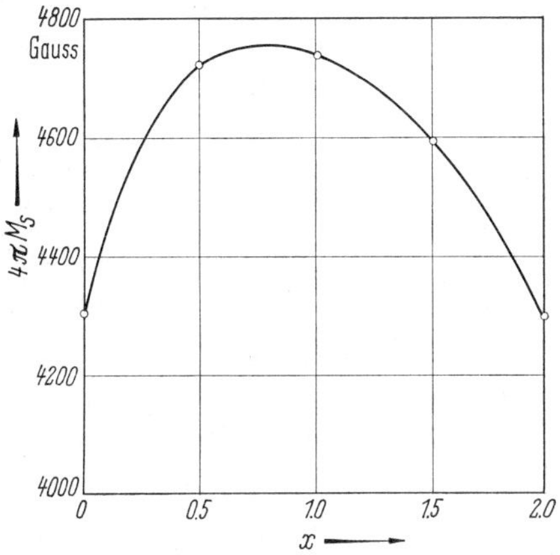

Fig. 108. $BaCo_{2-x}Zn_xFe_{16}O_{27}$. The dependence of the room-temperature saturation magnetization upon zinc content x [*57vURS*].

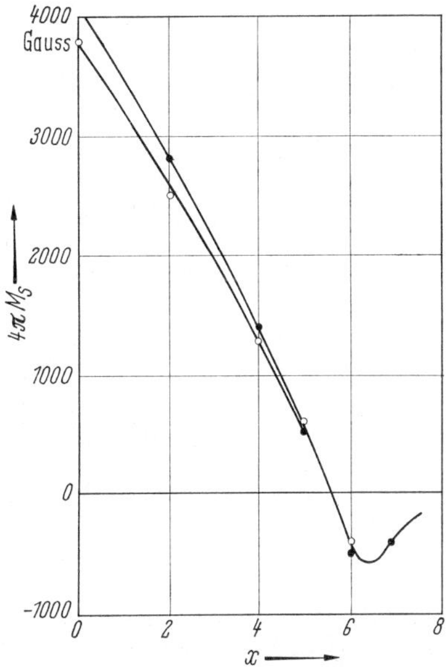

Fig. 110. A comparison of the dependencies of the room-temperature saturation magnetization upon aluminum content for the indicated series. For the significance of the sign of the magnetization, see subscript of Fig. 109 [*57vUS*].
● $Ba_{1.2}Ni_{1.5}Cu_{0.5}Al_xFe_{16-x}O_{27}$; ○ $Ba_{1.2}Ni_{2.0}Al_xFe_{16-x}O_{27}$.

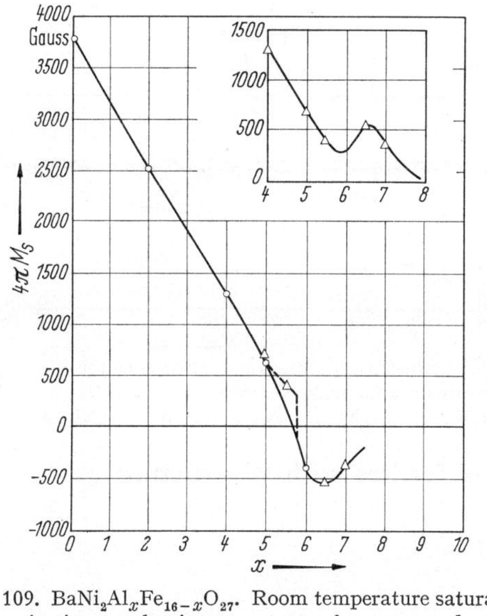

Fig. 109. $BaNi_2Al_xFe_{16-x}O_{27}$. Room temperature saturation magnetization vs. aluminum content x for compounds of the series $BaNi_2Al_xFe_{16-x}O_{27}$. Positive values of the saturation magnetization mean an octahedral domination of spins, negative values a tetrahedral domination [*57vUS*].

Firing temperature C°	Firing time h
○ 1400	10
△ 1400	20

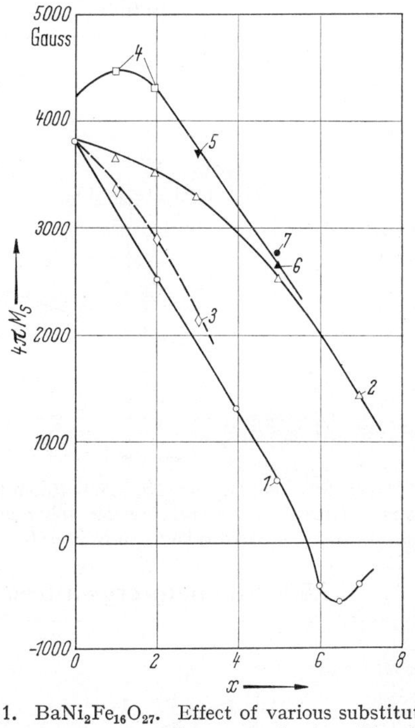

Fig. 111. $BaNi_2Fe_{16}O_{27}$. Effect of various substitutions on room-temperature saturation magnetization. For the significance of the sign of the magnetization, see subscript of Fig. 109 [*57vURS*].

1 Al for Fe
2 Ga for Fe
3 In or Cr for Fe
4 Zn for Ni

} x denotes the number of metal ions replaced per formula unit

5 $BaZn_2GaFe_{15}O_{27}$
6 $BaZn_2Ga_3Fe_{13}O_{27}$
7 $BaZn_2Al_3Fe_{13}O_{27}$

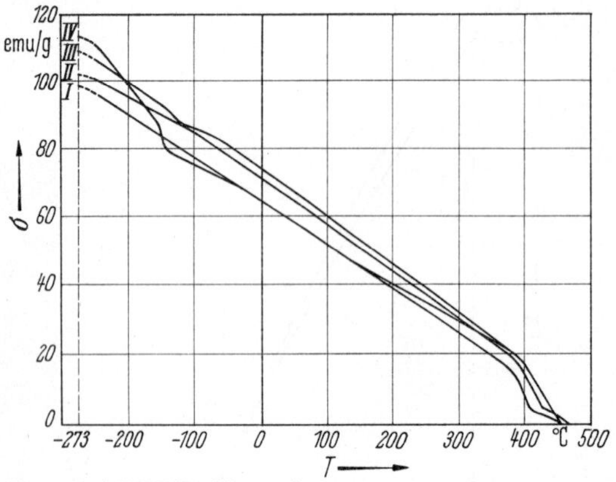

Fig. 112. [*62BRT*]. Magnetic moment per gram, σ, vs. temperature for a single crystal of:

		σ_s^0 [emu/g]	p_m^0 [μ_B]
I	$BaFe_2^{2+}Fe_{16}O_{27}$	98	28
II	$BaNi_{0.012}Fe^{2+}_{1.98}Fe^{3+}_{15.417}Ni_{0.563}O_{26.417}F_{0.583}$	101.8	29.17
III	$BaNi_{0.01}Fe^{2+}_{1.99}Fe^{3+}_{14.67}Ni_{1.33}O_{25.67}F_{1.33}$	108.5	31.12
IV	$BaNi_{0.86}Fe^{2+}_{1.14}Fe^{3+}_{13.67}Ni_{2.33}O_{24.67}F_{2.33}$	113.0	32.78

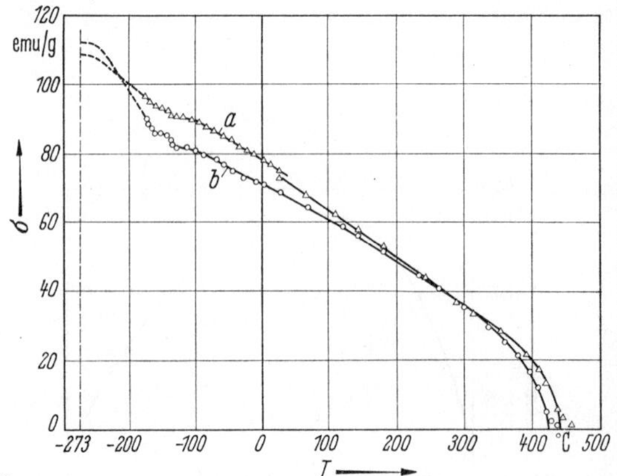

Fig. 113. $BaFe^{2+}Co_xFe^{3+}_{16-x}O_{27-x}F_x$ (**W** phase). Magnetic moment per gram, σ, vs. temperature. Curve *a* $x = 1.35$; Curve *b* $x = 2.14$ [*63RLB*].

7.7.4 Effective spectroscopic splitting factor g_{eff} of ferrites with W structure

Tab. 21. Room temperature values for the effective spectroscopic splitting factor g_{eff} of ferrites with **W** structure as measured on crystal-oriented spheres

Compound	x	f_{res} Gc/s	g_{eff}	Ref.
Co_xNi_{2-x}**W**	0…1	37.5	1.90 ± 0.10	*62RPW*
	0, 0.74, 0.8	37.56	1.85	*64HH*
Ni_xZn_{1-x}**W**	0…2	18…60	1.92…2.02	*64HH*
$BaNi_2Al_xFe_{16-x}O_{27}$	0…2	18…60	2.0	*64HH*

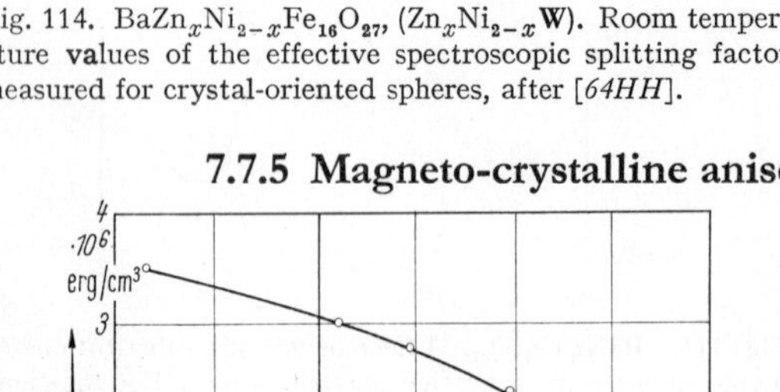

Fig. 114. $BaZn_xNi_{2-x}Fe_{16}O_{27}$, ($Zn_xNi_{2-x}$**W**). Room temperature values of the effective spectroscopic splitting factor measured for crystal-oriented spheres, after [*64HH*].

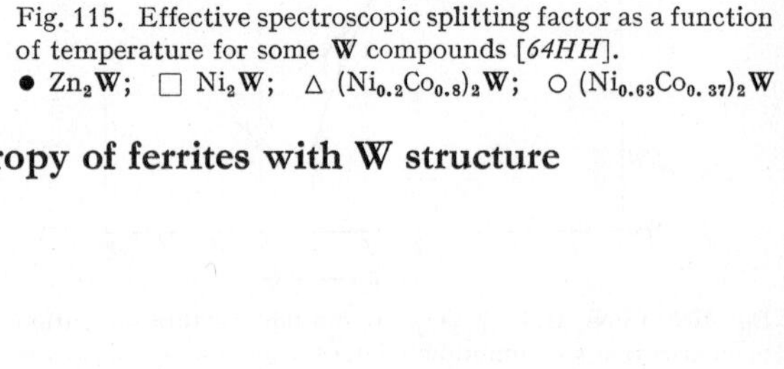

Fig. 115. Effective spectroscopic splitting factor as a function of temperature for some **W** compounds [*64HH*].
● Zn_2**W**; □ Ni_2**W**; △ $(Ni_{0.2}Co_{0.8})_2$**W**; ○ $(Ni_{0.63}Co_{0.37})_2$**W**

7.7.5 Magneto-crystalline anisotropy of ferrites with W structure

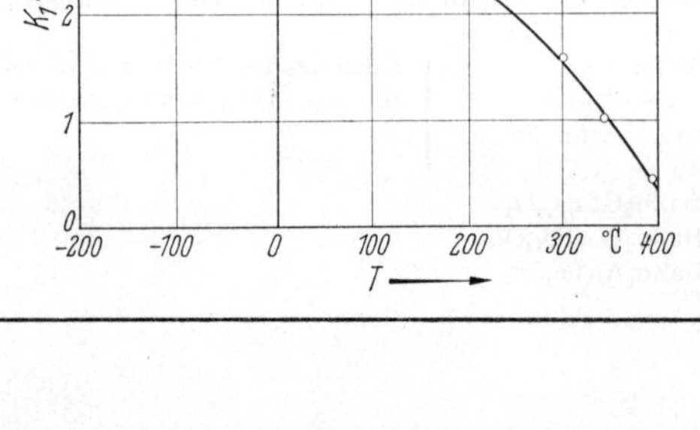

←

Fig. 116. $BaFe_{18}O_{27}$. The first-order magneto-crystalline anisotropy coefficient K_1 as a function of temperature [*52WRGO*].

Tab. 22. Magneto-crystalline anisotropy coefficient, saturation magnetization and anisotropy field derived from magnetization curves of crystal-oriented samples of some **W** compounds at 20 °C [59SW]

Compounds	Symbol	K_1 10^6 erg/cm³	$K_1 + 2K_2$ 10^6 erg/cm³	M_s Gauss	H^A Oe
$BaFe_2^{2+}Fe_{16}^{3+}O_{27}$*)	Fe_2**W**	+3.0		314	19 000*)
$BaZn_2Fe_{16}O_{27}$	Zn_2**W**	>0			9 800**)
$BaZnFe^{2+}Fe_{16}O_{27}$	ZnFe**W**	+2.4		380	12 500
$BaZn_{1.5}Fe_{0.5}^{2+}Fe_{16}O_{27}$	$Zn_{1.5}Fe_{0.5}$**W**	+2.1		380	11 100
$BaZnMnFe_{16}O_{27}$	ZnMn**W**	+1.9		370	10 200
$BaNi_2Fe_{16}O_{27}$	Ni_2**W**	+2.1		330	12 700
$BaZnNi_{0.5}Fe_{0.5}^{2+}Fe_{16}O_{27}$	$ZnNi_{0.5}Fe_{0.5}$**W**	+1.6		350	9 100
$BaZn_{0.75}Co_{0.75}Fe_{0.5}^{2+}Fe_{17}O_{27}$	$Zn_{0.75}Co_{0.75}Fe_{0.5}$**W**		−0.4	360	2 200
$BaCo_2Fe_{16}O_{27}$	Co_2**W**		negative		7 300**)

Tab. 23. Magneto-crystalline anisotropy coefficients and angle between the direction of the spontaneous magnetization and the c axis for Co_xFe_{2-x}**W** compounds. [64PZ]

Ferrite composition	290 °K				77 °K			
	K_1 erg/cm³	K_2 erg/cm³	$\|K_3\|$ erg/cm³	θ_0 deg	K_1 erg/cm³	K_2 erg/cm³	$\|K_3\|$ erg/cm³	θ_0 deg
Fe_2**W**	34 · 10⁵	0	≦ 10³	0	43 · 10⁵	0	≦ 10³	0
$Co_{0.3}Fe_{1.7}$**W**	28 · 10⁵	0	≦ 10³	0	25.4 · 10⁵	4.8 · 10⁵	0.6 · 10⁵	6
$Co_{0.5}Fe_{1.5}$**W**	23.8 · 10⁵	1 · 10⁵	≦ 10³	0	− 0.4 · 10⁵	21.4 · 10⁵	0.8 · 10⁵	10
$Co_{1.0}Fe_{1.0}$**W**	2.1 · 10⁵	5.6 · 10⁵	≦ 10³	0	−30 · 10⁵	30 · 10⁵	1.07 · 10⁵	45
$Co_{1.5}Fe_{0.5}$**W**	−6.6 · 10⁵	6.3 · 10⁵	≦ 10³	46	−49.8 · 10⁵	36 · 10⁵	0.8 · 10⁵	56

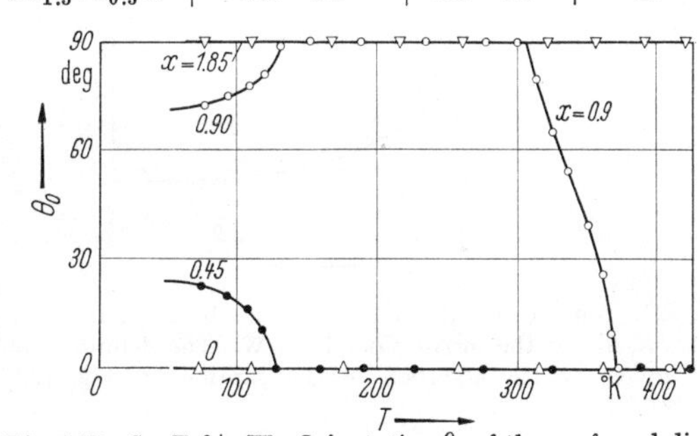

Fig. 117. $Co_xFe_{1-x}^{2+}$**W**. Orientation θ_0 of the preferred direction of magnetization (measured from the c axis) as a function of temperature for various samples, after [62B].

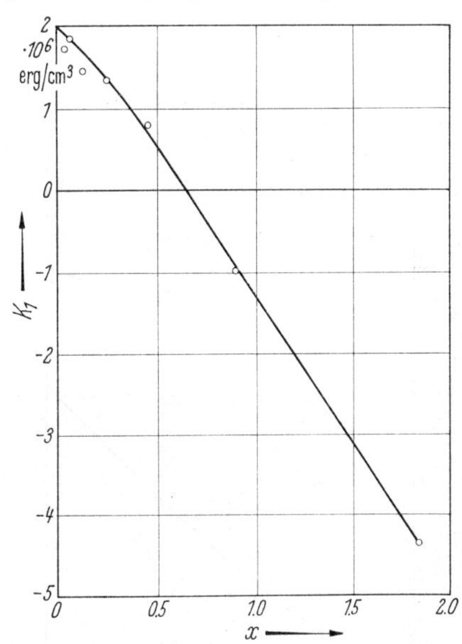

Fig. 119. Co_xFe_{2-x}**W**. Room-temperature values of first-order magneto-crystalline anisotropy coefficient, K_1, as a function of cobalt concentration, x, [60B2].

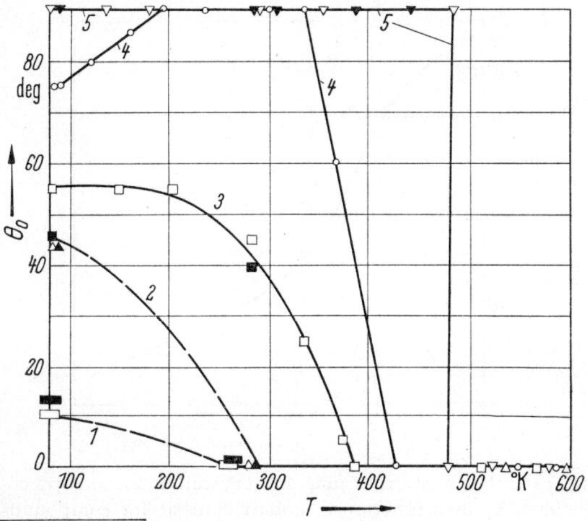

←

Fig. 118. $BaCo_xFe_{18-x}O_{27}$. Temperature dependence of the angle θ_0 between the spontaneous magnetization and the c axis for ferrites with the chemical composition $BaCo_xFe_{18-x}O_{27}$. *1* $x = 0.5$; *2* $x = 1.0$; *3* $x = 1.5$; *4* $x = 1.65$; *5* $x = 1.75$ [66YSZPZ].

*) Measurement on single crystals [52WRGO]. **) Result of resonance measurements on single crystals [64ST].

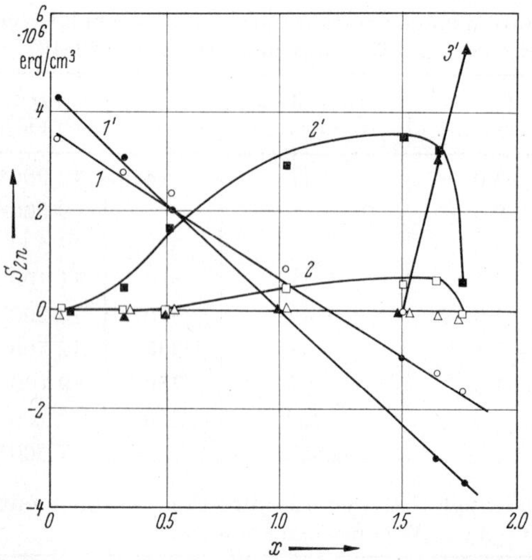

Fig. 120. $BaCo_xFe_{18-x}O_{27}$ ($Co_xFe_{2-x}W$). Dependence of the room-temperature magnetic anisotropy constants, S_{2n}, on the composition: *1* S_2 at 293°K; *2* S_4 at 293 °K; *1'* S_2 at 77 °K; *2'* S_4 at 77 °K; *3'* S_6 at 77 °K [*66YSZPZ*]. The anisotropy constants are defined according to the equation:

$$w_K = S_0 + S_2(\sin^2\theta - 2/3) + S_4(\sin^4\theta - 8/7\sin^2\theta + 8/35) + S_6(\sin^6\theta - 18/11\sin^4\theta + 8/11\sin^2\theta - 16/231) + S_6'\sin^6\theta\cos 6\Phi,$$

which expresses the anisotropy energy density in spherical harmonics [*62B1*].

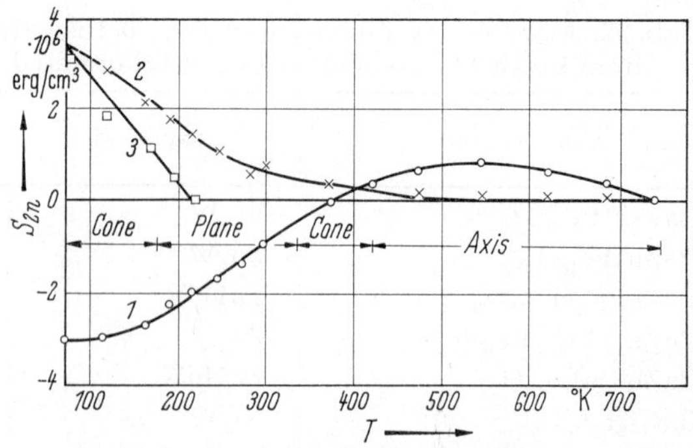

Fig. 121. Temperature dependence of the magnetic anisotropy constants, S_{2n}, for the ferrite $Co_{1.65}Fe_{0.35}W$: *1* $S_2(T)$; *2* $S_4(T)$; *3* $S_6(T)$. The definition of S_{2n} is given by the equation in the caption of Fig. 120. The temperature regions where the crystal shows a preferential axis, a preferential cone or a preferential plane for the magnetization are indicated [*66YSZPZ*].

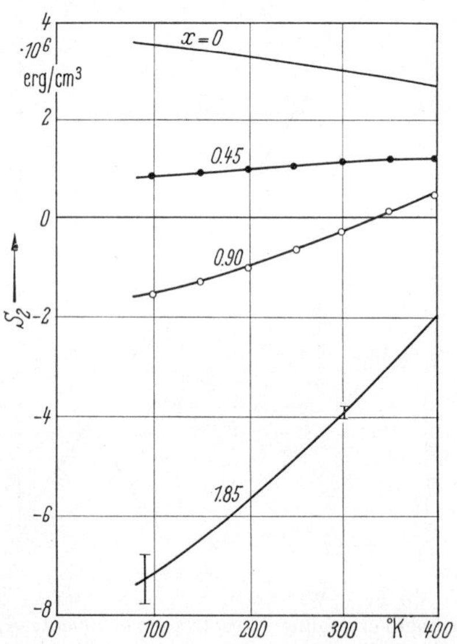

Fig. 122. $Co_xFe^{2+}_{1-x}W$. Experimental values of S_2 (for definition see caption of Fig. 120) as a function of temperature for three samples [*62B*]. The data for Fe^{2+}_2W were obtained from [*52WRGO*].

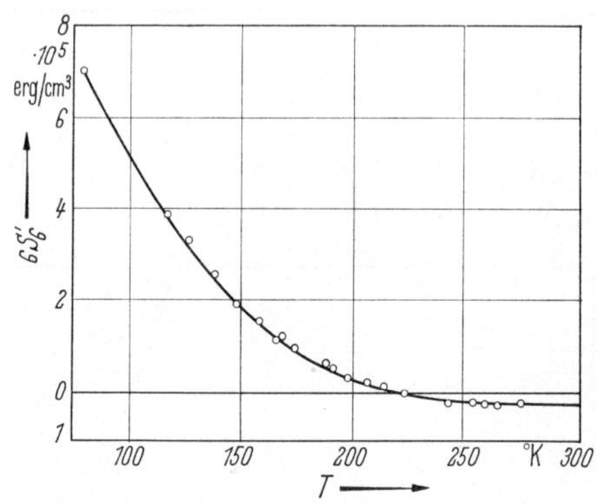

Fig. 123. Temperature dependence of the "basal" anisotropy constant, S_6', for the ferrite $Co_{1.65}Fe_{0.35}W$. The definition of S_6' is given by the equation in the caption of Fig. 120. [*66YSZPZ*].

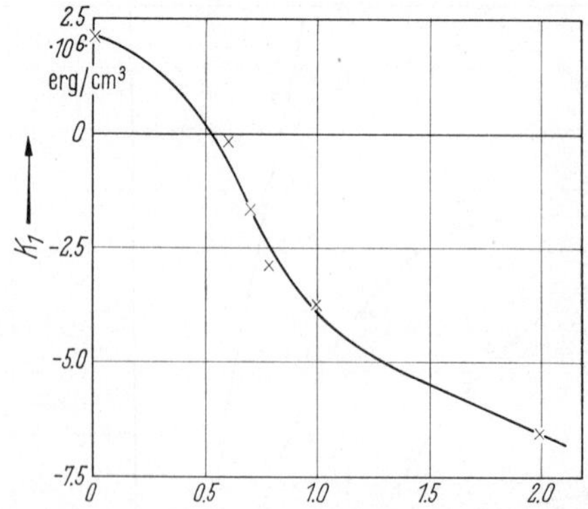

Fig. 124. The first-order magneto-crystalline anisotropy coefficient, K_1, as a function of cobalt content for compounds $Co_xZn_{2-x}W$ at 77 °K [*61LES*].

→

Fig. 125. $Co_{2-x}Zn_xW$. Temperature dependence of the initial permeability of polycrystalline samples of solid solutions [*67P*].

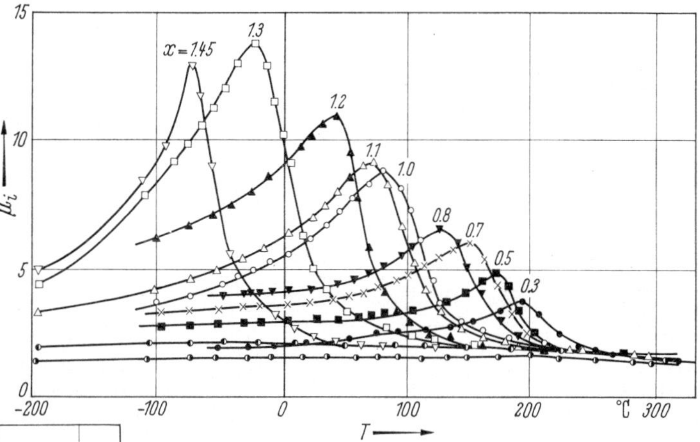

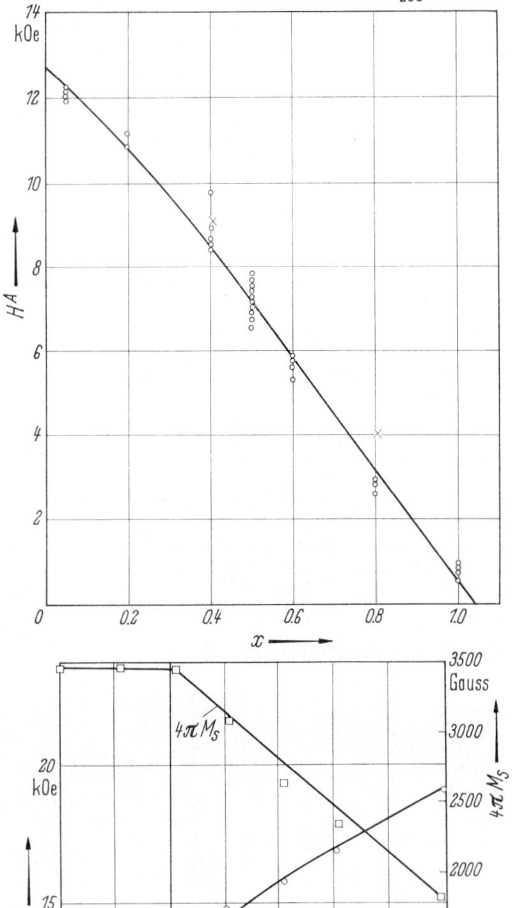

←

Fig. 126. $BaNi_{2-x}Co_xFe_{16}O_{27}$ compounds. Room temperature values of the uniaxial magneto-crystalline anisotropy field as a function of cobalt content measured for crystal-oriented samples of nickelcobalt **W** [*62RPW*].

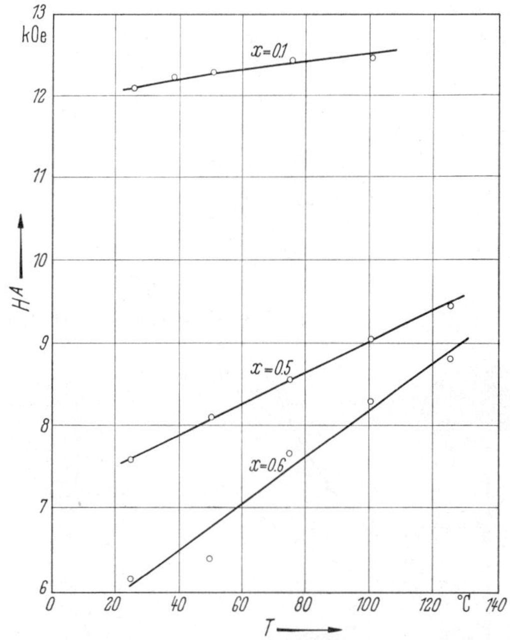

Fig. 127. $BaNi_{2-x}Co_xFe_{16}O_{27}$ compounds. Variation of the uniaxial magneto-crystalline anisotropy field with temperature in three samples of nickel cobalt **W**, after [*62RPW*].

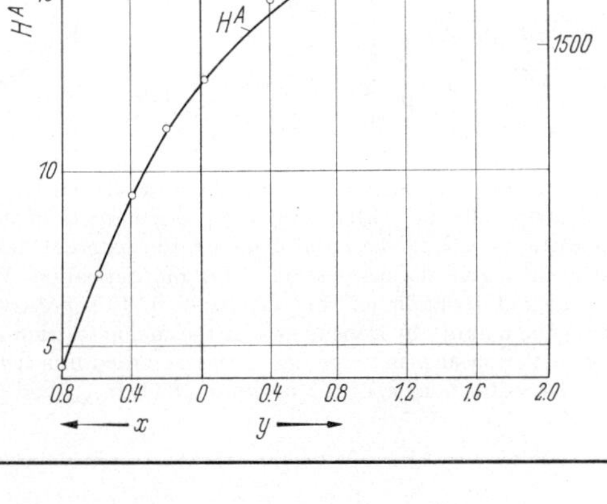

←

Fig. 128. $Co_xNi_{2-x}W$ and $BaNi_2Al_yFe_{16-y}O_{27}$, ($Ni_2Al_yW$). Room temperature values of the magneto-crystalline anisotropy field and magnetization as a function of cobalt and aluminum content, after [*64HH*]. See also [*64T*].

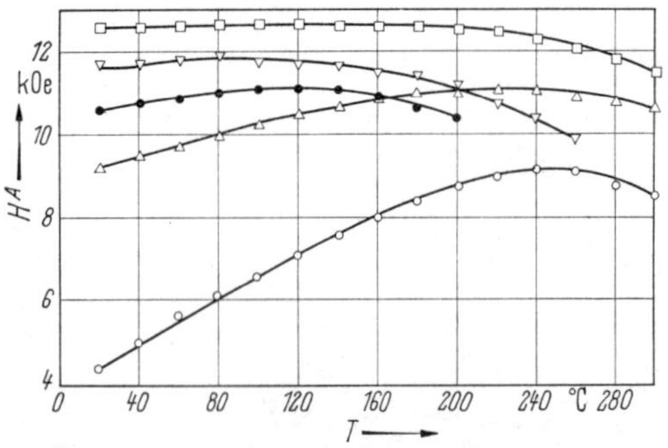

← Fig. 129. Magneto-crystalline anisotropy field as a function of temperature for various nickel-cobalt-zinc **W** compounds after [*64HH*]. □ Ni_2**W**, ▽ NiZn**W**, ● Zn_2**W**, △ $Ni_{1.6}Co_{0.4}$**W**, ○ $Ni_{1.26}Co_{0.74}$**W**

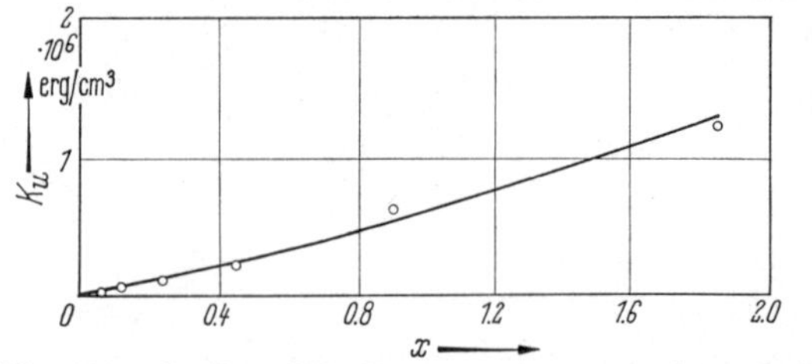

Fig. 132. Co_xFe_{2-x}**W**. Room temperature values of the anneal-induced uniaxial anisotropy, K_u, (for annealing at 100 °C) as a function of cobalt concentration, x. The anneal-induced anisotropy constant per cobalt ion is about $3 \cdot 10^{-16}$ erg. Activation energy for the annealing process 1.26 eV. Annealing time at 100 °C for $x \leq 0.12$ a few minutes, for $x = 0.9$ about 10^4 h. These data are comparable to those found in the case of the spinels $Co_xFe_{3-x}O_4$ [*60B2*]. For the anneal-induced anisotropy energy density in a crystal with hexagonal symmetry a phenomenological formula with four constants is given in [*65ZP*]:

$$w_U = U_1\alpha_3^2\beta_3^2 + U_2(\alpha_1\beta_1 + \alpha_2\beta_2)^2 + U_3(\alpha_2\beta_1 - \alpha_1\beta_2) + 2U_4\alpha_3\beta_3(\alpha_2\beta_2 + \alpha_1\beta_1)\,.$$

α_i direction cosines of the magnetization at measuring temperature,
β_i direction cosines of the magnetic field during annealing,
$x_3 \triangleq c$ axis.

For $BaCo_{1.5}Fe^{2+}_{0.5}Fe_{16}O_{27}$ is found [*65ZP*] after annealing for 15 minutes in a field of 15 kOe at 250 °C and afterwards cooling to room temperature at a rate of 3 degree/min:

$U_1 = -1.5 \cdot 10^5$ erg/cm³ $U_3 = 1.3 \cdot 10^5$ erg/cm³
$U_2 = -3.0 \cdot 10^5$ erg/cm³ $U_4 = 2.0 \cdot 10^5$ erg/cm³

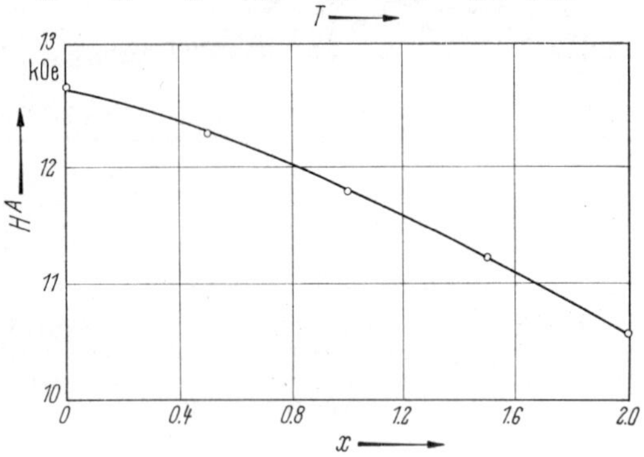

Fig. 130. $BaZn_xNi_{2-x}Fe_{16}O_{27}$, ($Zn_xNi_{2-x}$**W**). Room temperature values of the magneto-crystalline anisotropy field, H^A, as derived from ferromagnetic resonance experiments on crystal-oriented spheres after [*64HH*].

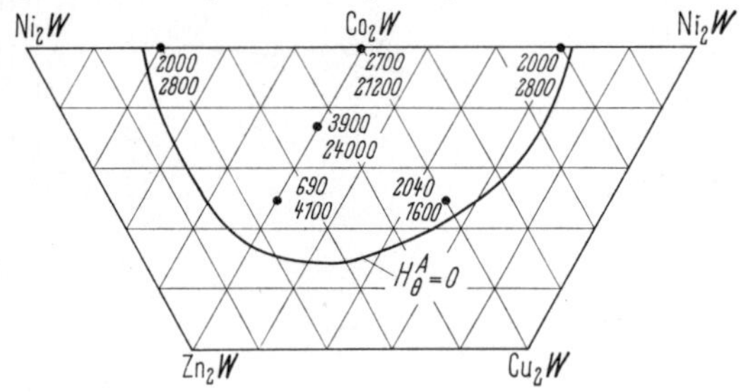

← Fig. 131. The curve represents the approximate location of zero anisotropy for **W**-type hexagonal ferrites at room temperature. The figures shown are the values of the line width, ΔH, and the anisotropy field, H_θ^A, in Oe measured at a frequency of 10···20 Gc/s for some polycrystalline compounds [*66BGH*].

7.7.6 Magnetostriction of ferrites with W structure

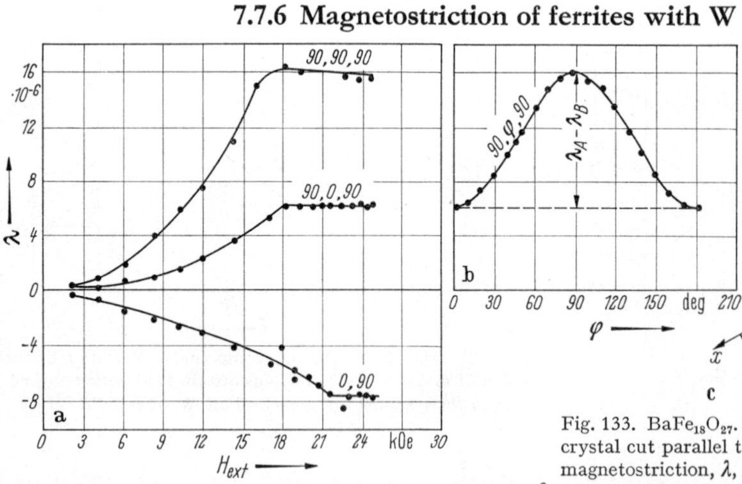

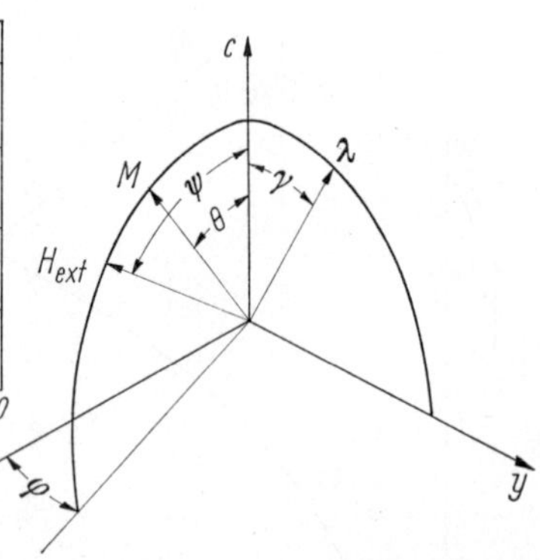

Fig. 133. $BaFe_{18}O_{27}$. Magnetostriction of a disc-shaped single crystal cut parallel to the basal plane: a) Dependence of the magnetostriction, λ, in the basal plane on the external field intensity, H_{ext}; the numbers on the curves denote the angles θ, φ and γ which give the orientation of the magnetization, M, in fields $H_{ext} > 19\,000$ Oe and the direction of measurement of the magnetostriction (cf. Fig. c); curve 0, 90 represents the magnetostriction along the direction $\gamma = 90°$ for magnetization along the c axis. b) Dependence of the magnetostriction along the direction y, on the orientation of the magnetization vector in the basal plane; the curve was recorded in a field of 21 000 Oe. c) Orientation of the direction of measurement of the magnetostriction, λ, of the magnetic field, H_{ext}, and of the magnetization vector, M, with respect to the coordinate axes [*64FZ*].

7.7.7 High-frequency magnetic properties of ferrites with W structure

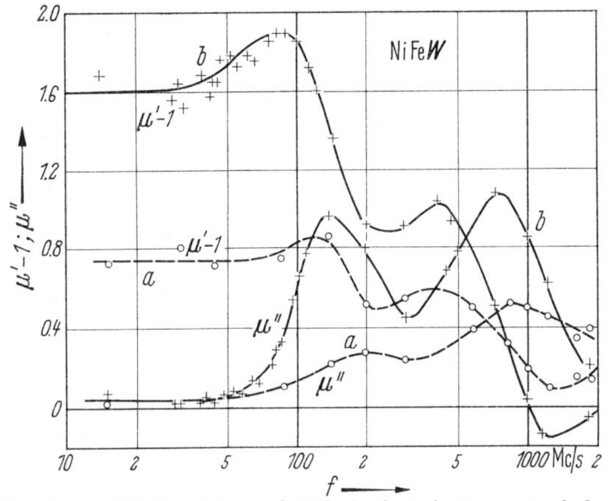

Fig. 134. NiFe**W**. The real and the imaginary part of the initial permeability of a polycrystalline sample. Sample *a* is isotropic, not crystal-oriented. Sample *b* is crystal-oriented and the h. f. measuring field is parallel to the *c* axes of the crystals [*59SW*].

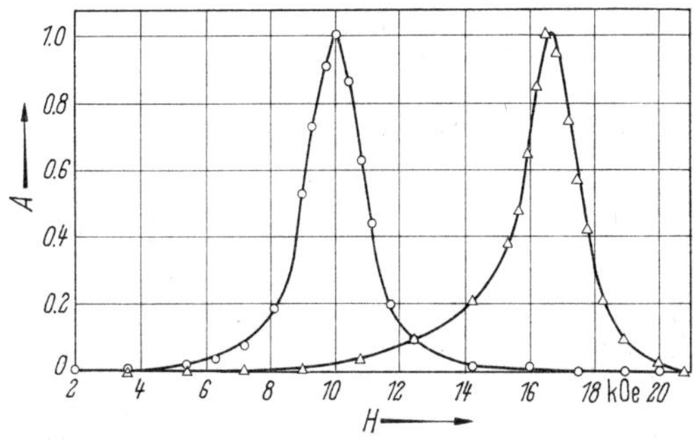

Fig. 136. $Ni_{1.26}Co_{0.74}$**W**. Resonance spectra of a crystal-oriented polycrystalline sphere at room temperature measured at a frequency of 37.56 Gc/s. The absorption, *A*, of the signal is given on an arbitrary scale as a function of the applied magnetic field, *H*. The anisotropy field strength, H^A, equals 4.5 kOe and $g_{eff} = 1.85$. After [*64HH*], see also [*62RPW*]. ○ $H_\parallel$ *c* axis, i. e. parallel to easy direction for magnetization; △ $H_\perp$ *c* axis, i. e. parallel to hard direction for magnetization.

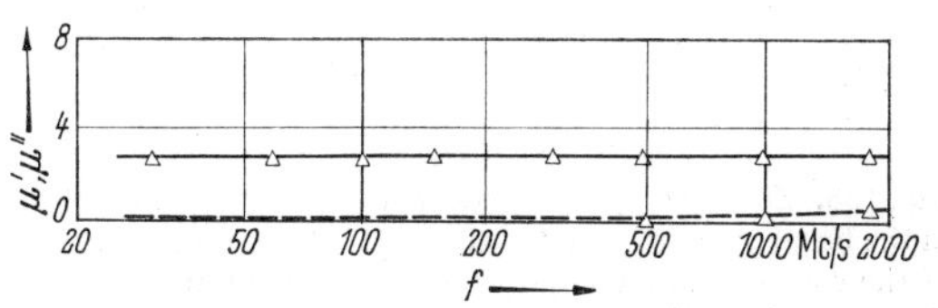

←

Fig. 135. Co_2**W**. Magnetic spectrum of an isotropic polycrystalline sample, after [*64NSVW*]. ———— μ' ; ------------ μ''.

7.8 Y-type ferrites

(general chemical formula $BaMe^{2+}Fe_6^{3+}O_{11}$)

7.8.1 Lattice parameters of ferrites with Y structure

Tab. 24. Lattice parameter *c* (with an accuracy of ±0.01 Å) and X-ray density ϱ_X of compounds of type Me_2**Y** of stoichiometric composition. The parameter *a* equals in all cases 5.88 Å [*65V*]

Me	*c* Å	ϱ_X g/cm³	Ref.
Fe	43.6	5.39	*59SW*
Mn	43.82	5.30	*65V*
Zn	43.60	5.41	*65V*
Co	43.56	5.35	*65V*
Mg	43.46	5.12	*65V*
Ni	43.42	5.38	*65V*

7.8.2 Electric and dielectric properties of ferrites with Y structure

Tab. 25. Room temperature electrical conductivities $\sigma_\perp$ in the basal plane and $\sigma_\parallel$ parallel to the *c* axis for single crystals of $BaZn_{1-x}Fe_x^{2+}Fe_6^{3+}O_{11}$, after [*67VI*]. For one sample also the corresponding activation energies $q_\perp$ and $q_\parallel$ are given

x	$\sigma_\perp$ $\Omega^{-1}cm^{-1}$	$\sigma_\parallel$ [$\Omega^{-1}cm^{-1}$]		ε'	$q_\perp$ eV	$q_\parallel$ eV
		d. c.	130 Mc/s			
0.01	260^{-1}	1300^{-1}	2800^{-1}	19		
0.03	58^{-1}	240^{-1}	1200^{-1}	16		
0.03	38^{-1}	460^{-1}			0.19	0.32
0.07	55^{-1}	1400^{-1}	1300^{-1}	18		
0.07	46^{-1}	460^{-1}				
0.09	21^{-1}	320^{-1}				

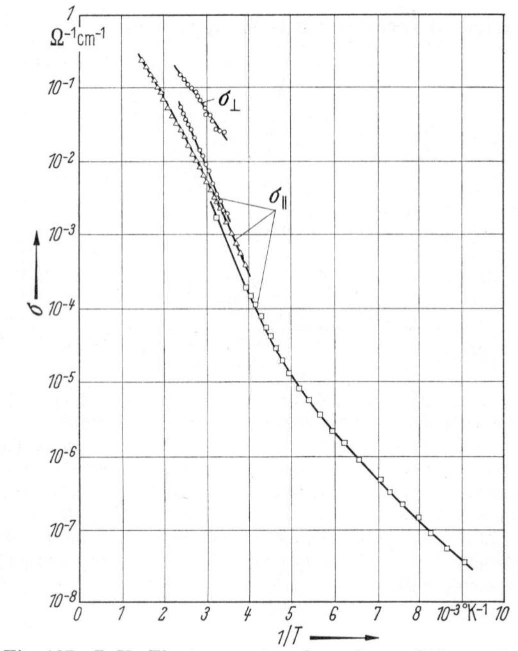

Fig. 137. Zn**Y**. The temperature dependence of the conductivity of single crystals. In the basal plane: $\sigma_\perp$; parallel to the c axis: $\sigma_\parallel$. ○ [*67V*]; △, □ [*67H*].

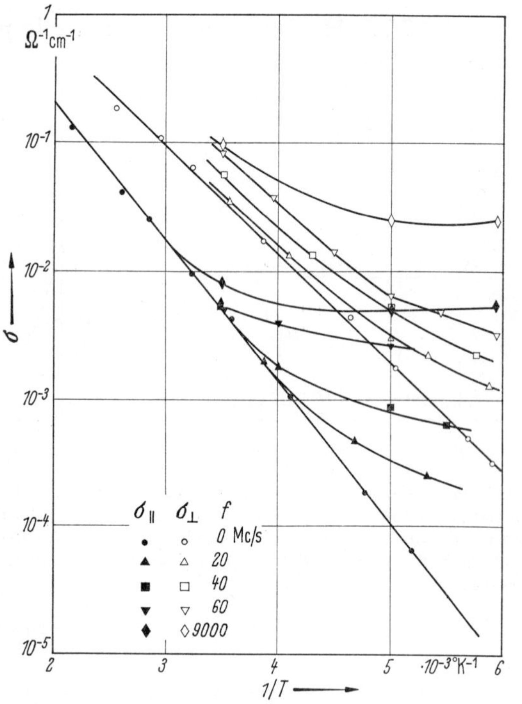

Fig. 138. Zn**Y**. Temperature dependence of the conductivity in the basal plane, $\sigma_\perp$, and of the conductivity in the direction of the c axis, $\sigma_\parallel$, of a single crystal at various frequencies of the measuring field. Direct current activation energies of the resistivity:

$q_\parallel = 0.24$ eV, $q_\perp = 0.18$ eV. After [*66KK*].

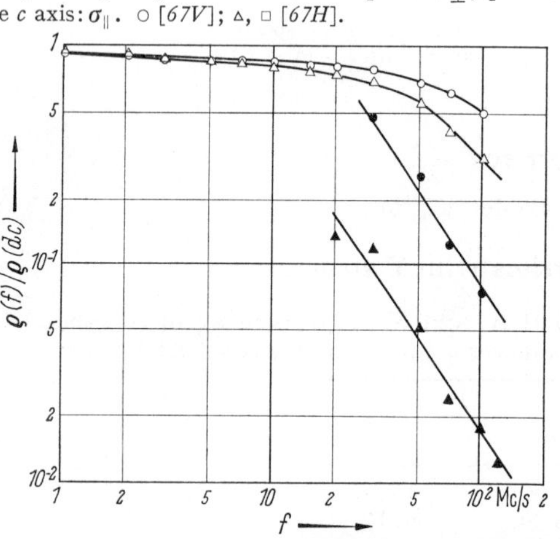

←

Fig. 139. Zn**Y**. Frequency dependence of $\varrho(f)/\varrho$(d. c.) at 300 °K and 77 °K for a single crystal. The resistivity at frequency f, $\varrho(f)$, is normalized with respect to the d. c. resistivity, ϱ(d. c.), of the crystal [*66KK*].

	300 °K left hand scale	77 °K right hand scale
$\varrho_\perp$	○	●
$\varrho_\parallel$	△	▲

7.8.3 Mössbauer spectra, saturation magnetization, spectroscopic splitting factor, and Curie temperature of ferrites with Y structure

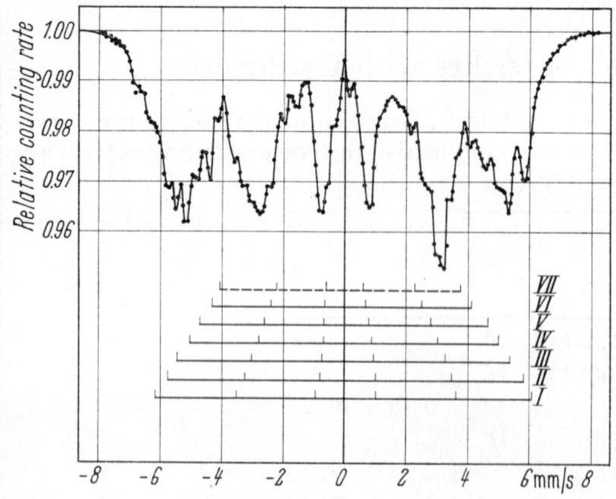

←

Fig. 141. $Mn_{0.31}Zn_{0.69}$**Y**. Mössbauer spectrum for $Mn_{0.31}Zn_{0.69}$**Y** absorber at 297 °K after [*67AAL*]. The hyperfine field is approximately equal to that of Zn**Y**, see Fig. 140. $\Theta_C = (143 \pm 5)$ °C.

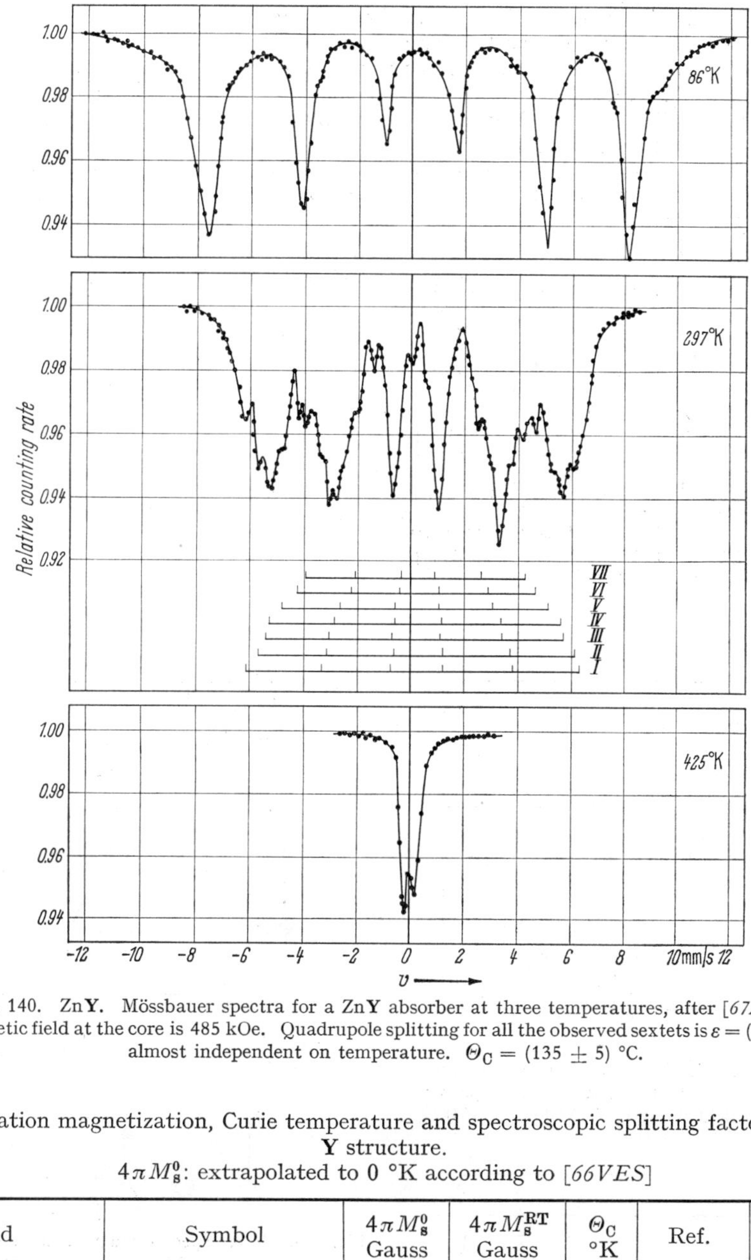

Fig. 140. Zn**Y**. Mössbauer spectra for a Zn**Y** absorber at three temperatures, after [*67AAL*]. At 86 °K the magnetic field at the core is 485 kOe. Quadrupole splitting for all the observed sextets is $\varepsilon = (0.09 \pm 0.04)$ mm/sec, almost independent on temperature. $\Theta_C = (135 \pm 5)$ °C.

Tab. 26. Saturation magnetization, Curie temperature and spectroscopic splitting factor of ferrites with **Y** structure.
$4\pi M_s^0$: extrapolated to 0 °K according to [*66VES*]

Compound	Symbol	$4\pi M_s^0$ Gauss	$4\pi M_s^{RT}$ Gauss	Θ_C °K	Ref.	g_{eff}	Ref.
$BaMgFe_6O_{11}$	Mg**Y**	2650*)	1500	550	*59SW*	≃ 2	*59SJ*
$BaNiFe_6O_{11}$	Ni**Y**	2230	1600	660	*59SW*		*62B2*
$BaZnFe_6O_{11}$	Zn**Y****)	7510	2550	400	*59SW*		
			≃ 2000		*62D*		
			3800		*62DKR*	1.87	*62B2*
$BaZn_{0.75}Fe_{0.25}Fe_6O_{11}$	$Fe_{0.25}Zn_{0.75}$**Y**		2400		*59SW*		
$BaCoFe_6O_{11}$	Co**Y**	3640	2300	610	*59SW*	3.70	*59BK1*
			5050		*59BK1*	2.68	*62B2*
$BaMnFe_6O_{11}$	Mn**Y**		2100	560	*59SW*		
$BaMn_{0.5}Zn_{0.5}Fe_6O_{11}$	$Mn_{0.5}Zn_{0.5}$**Y**		2300		*64G*	1.92	*64G*
	$Mn_{0.5}Zn_{0.5}$**Y** + 3.5% by weight Mn		2350		*64SDT*		

*) After slow cooling. For quenched samples $4\pi M_s^0 = 3320$ Gauss.
**) Lattice constants $a = 5.876$ Å, $c = 43.558$ Å according to [*57B*].

Tab. 27. Cation distribution and relative orientations of the magnetic moments of the magnetic ions (indicated by arrows) in ferrites with **Y** structure as derived from the low temperature saturation magnetization. Tetrahedrally surrounded ions are indicated by ( ), octahedrally surrounded ions are indicated by [ ], after [*66VES*]

Ferrite	Cation distribution	Remarks
Mg**Y**	$(Mg_{0.15}\overrightarrow{Fe}_{1.85})\ [Mg^{2+}_{0.85}\overrightarrow{Fe}_{0.83}\overleftarrow{Fe}_{3.32}]$	after slow cooling
	$(Mg_{0.2}\overrightarrow{Fe}_{1.8})\ [Mg_{0.8}\overrightarrow{Fe}_{0.84}\overleftarrow{Fe}_{3.36}]$*)	
	$(Mg_{0.25}\overrightarrow{Fe}_{1.75})\ [Mg_{0.75}\overrightarrow{Fe}_{0.85}\overleftarrow{Fe}_{3.4}]$	quenched
	$(Mg_{0.3}\overrightarrow{Fe}_{1.7})\ [Mg_{0.7}\overrightarrow{Fe}_{0.86}\overleftarrow{Fe}_{3.44}]$*)	
Zn**Y**	$(Zn_{0.85}\overrightarrow{Fe}_{1.05})\ [Zn_{0.05}\overrightarrow{Fe}_{0.99}\overleftarrow{Fe}_{3.96}]$	
Co**Y**	$(\overrightarrow{Co}_{0.18}\overrightarrow{Fe}_{1.82})\ [\overrightarrow{Co}_{0.16}\overrightarrow{Fe}_{0.84}\overleftarrow{Co}_{0.66}\overleftarrow{Fe}_{3.35}]$	
Cu**Y**	$(Cu_{0.19}\overrightarrow{Fe}_{1.81})\ [Cu_{0.81}\overrightarrow{Fe}_{0.84}\overleftarrow{Fe}_{3.35}]$	
Mn**Y**	$(\overrightarrow{Mn}_{0.85}\overrightarrow{Fe}_{1.15})\ [\overrightarrow{Mn}_{0.03}\overrightarrow{Fe}_{0.97}\overleftarrow{Mn}_{0.12}\overleftarrow{Fe}_{3.88}]$	
Ni**Y**	$(\overrightarrow{Ni}_{0.05}\overrightarrow{Fe}_{1.95})\ [\overrightarrow{Ni}_{0.19}\overrightarrow{Fe}_{0.81}\overleftarrow{Ni}_{0.76}\overleftarrow{Fe}_{3.24}]$	

Tab. 28. The ferrous content x and the Curie temperature for the initial permeability, T_C**), of $Zn_{1-x}Fe^{2+}_{x}$**Y** [*67V1*]

x	T_C [C°]
0.01[3]	88
0.01[9]	97
0.03	102.5

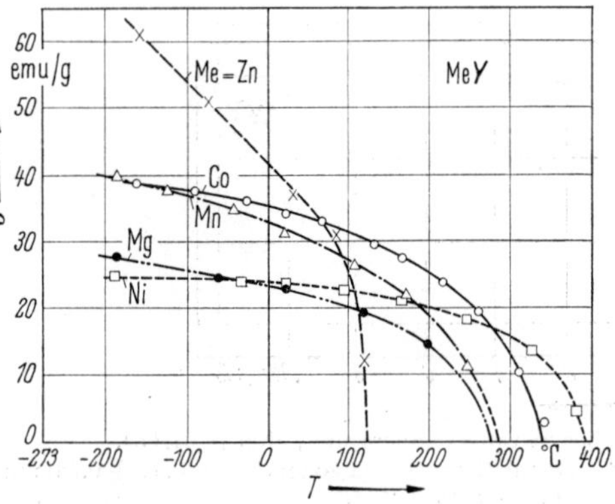

Fig. 142. Specific magnetization vs. temperature for a number of compounds with **Y** structure, measured on polycrystalline specimens at a field strength of 11 000 Oe [*59SW*].

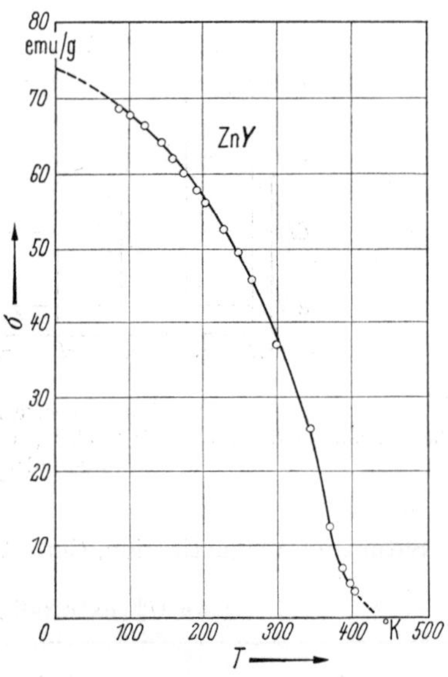

Fig. 143. Zn**Y**. Specific magnetization vs. temperature for a single crystal. Field of 10 500 Oe applied parallel to easy direction [*64ST*].

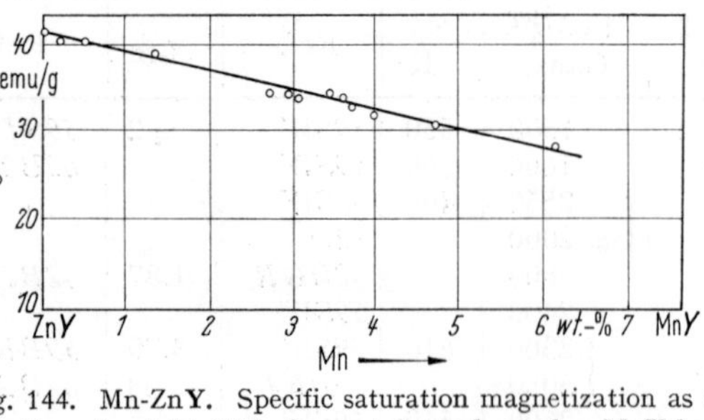

Fig. 144. Mn-Zn**Y**. Specific saturation magnetization as a function of weight %Mn in crystals; end member Mn**Y** has 7.81 weight% Mn, after [*64TDS*], [*65DTS*].

*) Formula obtained from X-ray analysis.

**) See list of symbols p. 553.

7.8.4 Magneto-crystalline anisotropy of ferrites with Y structure

7.8.4.1 Magneto-crystalline anisotropy derived from static measurements

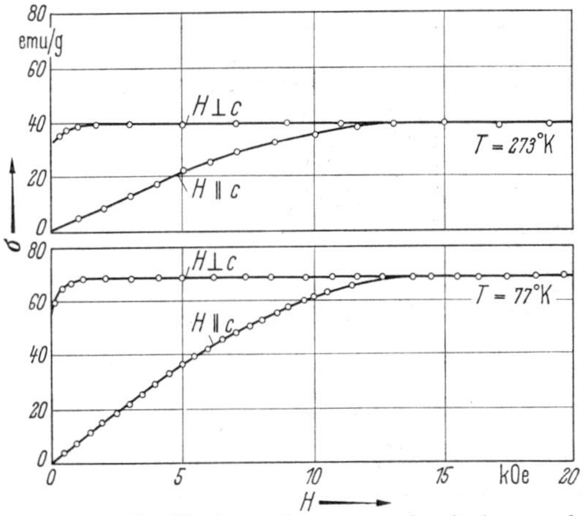

Fig. 145. ZnY. Magnetization curves of a single crystal as measured with the field parallel to the c axis (hard direction) and perpendicular to the c axis (preferential plane) [61E].

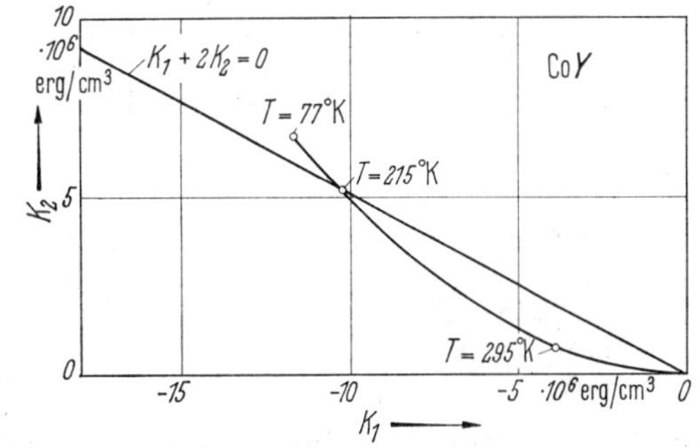

Fig. 146. CoY. The magneto-crystalline anisotropy coefficients, K_1 vs. K_2, as a function of temperature. The straight line represented by $K_1 + 2K_2 = 0$ separates the phases with preferential cone and preferential plane for the magnetization [66E], see also [60B].

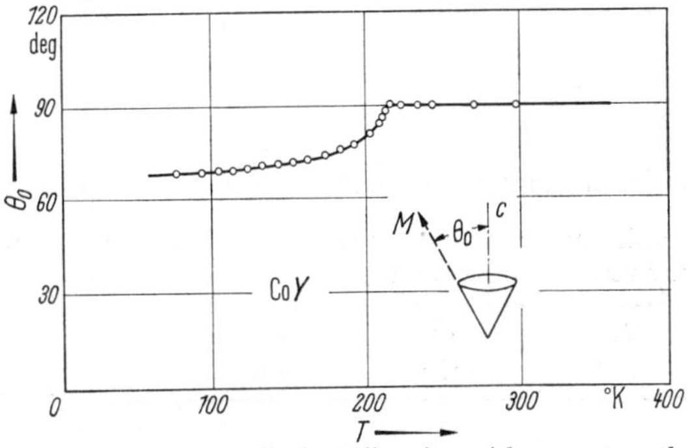

Fig. 147. CoY. Equilibrium direction with respect to the c axis of the spontaneous magnetization as a function of temperature [59SW].

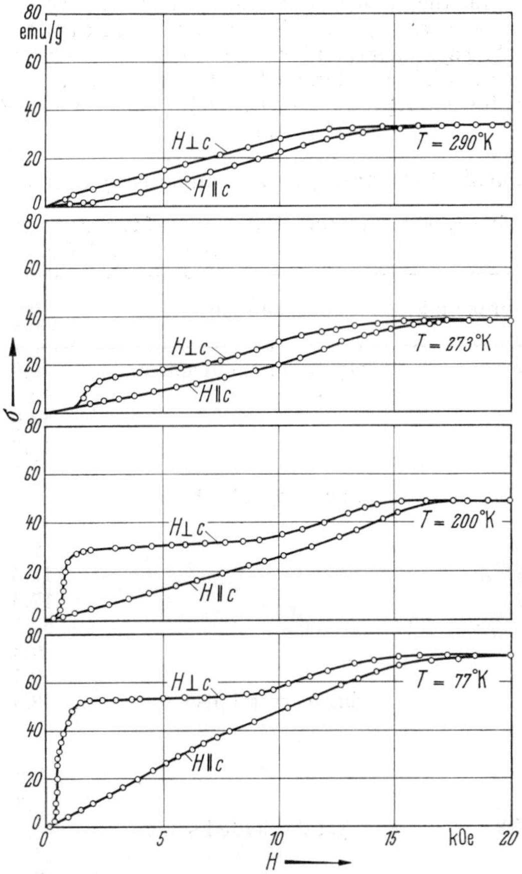

Fig. 148. $Ba_{0.05}Sr_{0.95}ZnY$. Magnetization curves of a single crystal. No spontaneous magnetization is observed at zero field strength [61E].

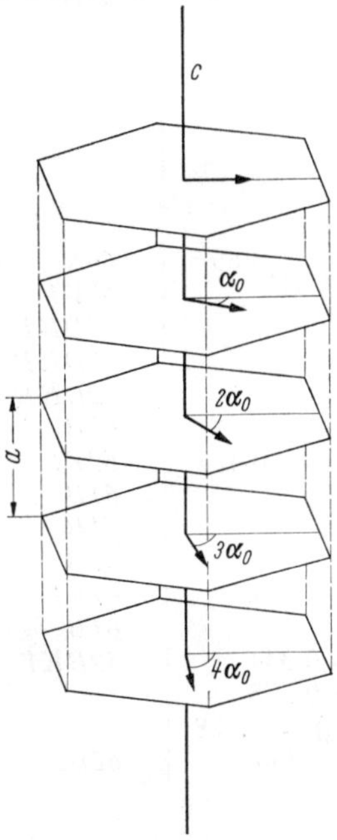

←
Fig. 149. SrZnY. Helical spin configuration in a hexagonal crystal with a preferential plane for the magnetization, at zero magnetic field strength, as proposed to explain the magnetization curves of Fig. 148 [61E].

Tab. 29. Room temperature values for the magneto-crystalline anisotropy of ferrites with **Y** structure

Compound	$K_1 + 2K_2$ 10^6 erg/cm³	H_θ^A Oe	H_ϕ^A Oe	Ref.	Remarks
Mg**Y**	−0.6	10 000		*59SW*	
Ni**Y**	−0.9	14 000		*59SW*	$T \leqq 50$ °K preferential cone, $\theta_0 = 74$ ° at 4 °K; for $T > 50$ °K preferential plane, [*66E*]
Zn**Y**	−1.0	9 000		*59SW*	$H_\phi^A < 1$ Oe ($K_3 < 6$ erg/cm³), [*67V1*]
		9 900		*64ST*	
$Fe_{0.25}Zn_{0.75}$**Y**	−0.9	9 500	< 1	*59SW*	
$Mn_{0.5}Zn_{0.5}$**Y**		9 560		*64G*	
Co**Y**	−2.6	28 000	155	*59SW*	see also Tab. 30 and Fig. 146, 147

Tab. 30. Magneto-crystalline anisotropy of Co**Y**. (See also Fig. 146, 147)

$T \leqq 215$ °K preferential cone for the magnetization $\theta_0 = 66°$ at 77 °K

$T > 215$ °K preferential plane, [*66E*].

Magnetic field-cooling at a temperature $T \leqq 215$ °K reveals the presence of a trigonal magneto-crystalline anisotropy, which is consistent with the crystalline symmetry, and which should be represented by an additional term $K_2^! \sin^3\theta \cos\theta \cos 3\phi$. The coefficient $K_2^!$ is found to be $K_2^! = 6 \cdot 10^5$ erg/cm³ at 117 °K, see [*60B2*]

T °K	$\lvert K_3 \rvert$ erg/cm³	H_ϕ^A Oe	Ref.
443	150	37	*59SW*
300	800	150	*59SW*
300	600		*60B2*
230	1890		*60B2*
77	2000	340	*59SW*

7.8.4.2 Magneto-crystalline anisotropy field derived from the ferromagnetic resonance frequency

Tab. 31. Room temperature values of the magneto-crystalline anisotropy fields of ferrites with **Y** structure derived from the ferromagnetic resonance frequency

Compound	Sample	f_{res} Gc/s	H_{res} kOe	H_θ^A kOe	H_ϕ^A Oe	Ref.
Zn**Y***)	crystal-oriented sphere	10.8		9.2		*61B*
		13.0		8.4		*61B*
		14.9		8.5		*61B*
		17.0		8.4		*61B*
				8.9···10		
	single crystal sphere	8.6	1.190	about 8.0 (for 8.6 … 37.7)		*62D*
		8.9	1.225			
		9.8	1.455			
		16.6	3.250			
		17.3	3.460			
		37.7	9.680			
		9.3		9.315	3.65	*66N*
		9.3	1.125	9.400		*67V1*
	single crystal disc	8.5	8.0	8.3		*62DKR*
			7.8	8.7		*62DKR*
			8.2	8.0		*62DKR*
		8···70		9.9		*62B2*
		9.3	0.950	9.5	0.4	*67V1*
Ni**Y**	crystal-oriented sphere	13.0		17.1		*61B*
		14.9		17.2		*61B*
		17.0		16.7		*61B*
$Ni_{0.4}Zn_{0.6}$**Y**	single crystal disc	9.3	0.825		0.6	*67V1*
Co**Y**	crystal-oriented disc	20.0		40.0		*61B*
	single crystal sphere	8···27	2.2	$K_1/M_s = 0.344$ $K_4/M_s = 0.275$ $(K_2 + \frac{3}{2} K_3) = 3.528$		*59BK1*
	single crystal disc	8···70		28	200	*62B2*

*) For substituted Zn**Y** ferrites see Tab. 32.

Tab. 32. Magneto-crystalline anisotropy fields for some substituted ZnY crystals*) measured at a frequency of 9.282 Gc/s at room temperature and at a frequency of 9.315 Gc/s at 77 °K. [67V1]

Compound	Sample	T °K	H_{res} Oe	H_θ^A Oe	H_ϕ^A Oe	ΔH Oe
ZnY	disc	295	950	9 500	0.4	13
$Zn_{0.97}Fe^{2+}_{0.03}Y$	disc	295	908	10 100	0.5	14
$Zn_{0.95}Fe^{2+}_{0.05}Y$	disc	295	803	12 000	0.1	14
$Zn_{0.92}Fe^{2+}_{0.03}Co_{0.05}Y$	disc	295	809	11 800	1.1	29
ZnY	disc	77	874	10 700	15.5	24
$Zn_{0.97}Fe^{2+}_{0.03}Y$	disc	77	815	11 600	18.8	60
$Zn_{0.92}Fe^{2+}_{0.03}Co_{0.05}Y$	disc	77	820		60.5	51
ZnY	sphere	295	1 125	9 400		50
$Zn_{0.96}Fe^{2+}_{0.03}Co_{0.01}Y$	sphere	295	1 040			47
$Zn_{0.94}Fe^{2+}_{0.03}Co_{0.03}Y$	sphere	295	970			38
$Zn_{0.92}Fe^{2+}_{0.03}Co_{0.05}Y$	sphere	295	929	11 900		45

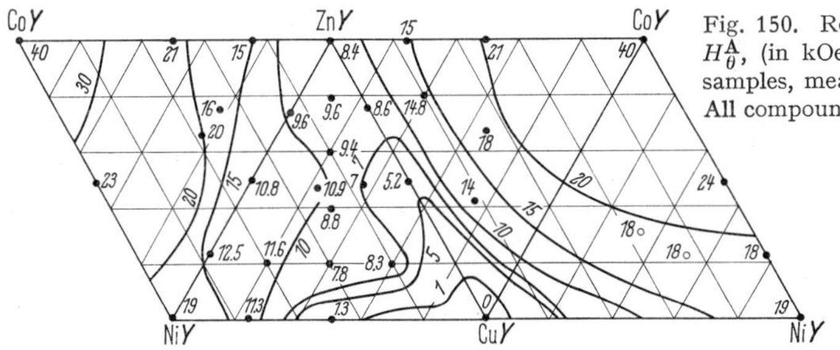

Fig. 150. Room temperature values of the anisotropy field, H_θ^A, (in kOe) of polycrystalline Y-type hexagonal ferrite samples, measured at a frequency between 10 and 20 Gc/s. All compounds have planar anisotropy [66BGH].

7.8.5 Hysteresis properties of ferrites with Y structure

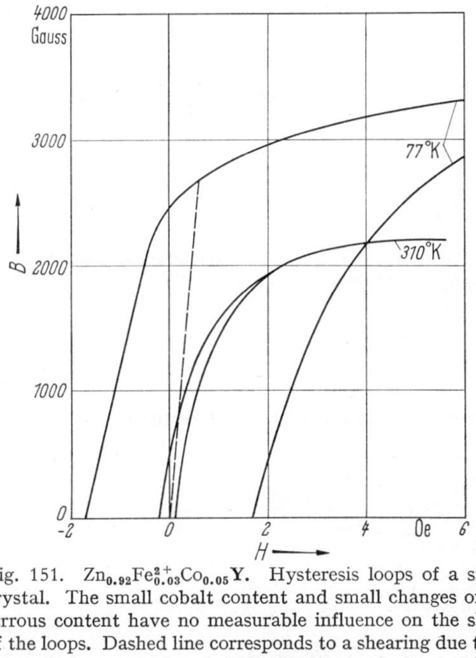

Fig. 151. $Zn_{0.92}Fe^{2+}_{0.03}Co_{0.05}Y$. Hysteresis loops of a single crystal. The small cobalt content and small changes of the ferrous content have no measurable influence on the shape of the loops. Dashed line corresponds to a shearing due to an internal demagnetizing factor, $N_{int}/4\pi$, of about 1/4000 [67V1].

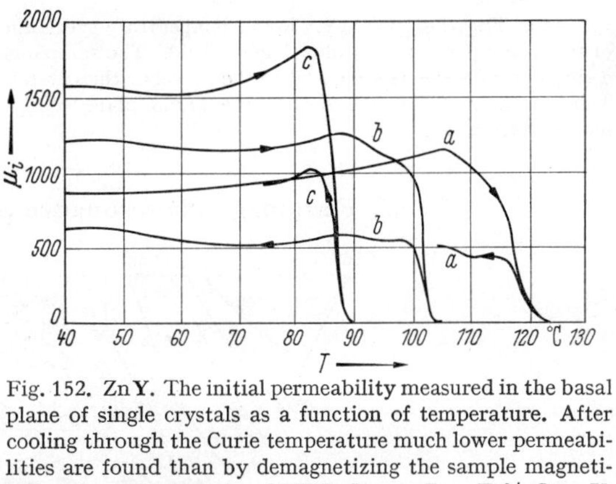

Fig. 152. ZnY. The initial permeability measured in the basal plane of single crystals as a function of temperature. After cooling through the Curie temperature much lower permeabilities are found than by demagnetizing the sample magnetically at lower temperature [67V1]. Curve *a* $Zn_{0.92}Fe^{2+}_{0.03}Co_{0.05}Y$, curve *b* $Zn_{0.97}Fe^{2+}_{0.03}Y$, curve *c* ZnY.

*) In the last column the measured line width ΔH of the sample is also quoted. They should be compared with those given in Tab. 33.

7.8.6 High-frequency magnetic properties of ferrites with Y structure

7.8.6.1 Magnetic spectrum of the initial permeability of ferrites with Y structure

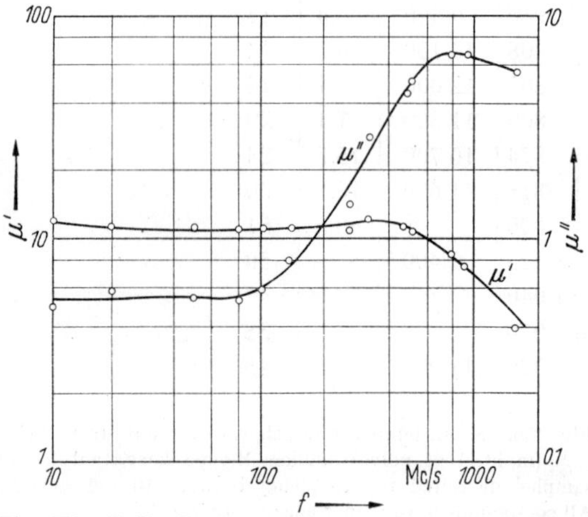

Fig. 153. Mg**Y**. Room temperature values of the real and the imaginary part μ' and μ'' of the initial permeability of an isotropic polycrystalline sample as a function of frequency [*59SW*]. For the corresponding properties of Ni**Y** and $Ni_{0.25}Zn_{0.75}$**Y** see [*61HGB*].

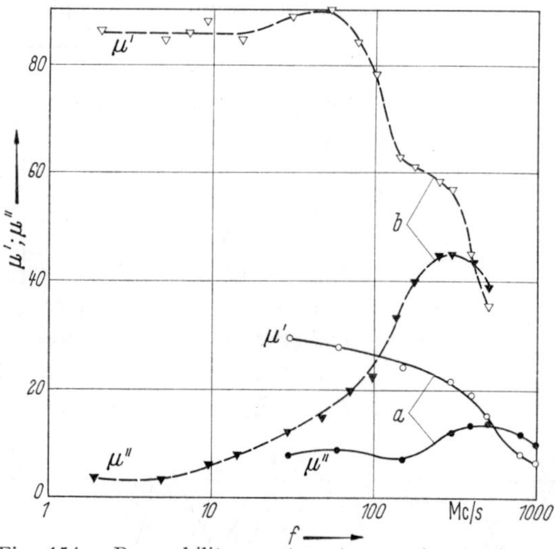

Fig. 154. Permeability spectra at room temperature of *a* a polycrystalline isotropic sample of $Zn_{0.97}Fe^{2+}_{0.03}$**Y**, *b* a crystal-oriented sample of $Zn_{0.8}Mn_{0.2}$**Y** where the h. f.-magnetic measuring field is parallel to the preferential plane for magnetization of the crystals [*67V1*].

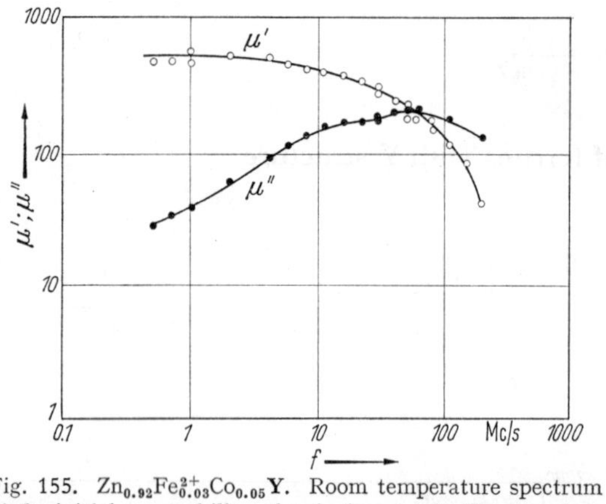

Fig. 155. $Zn_{0.92}Fe^{2+}_{0.03}Co_{0.05}$**Y**. Room temperature spectrum of the initial permeability of a single crystal. The sample is a ring cut with the axis parallel to the *c* axis of the crystal. The r. f. field is parallel to the preferential plane of the magnetization [*67V1*].

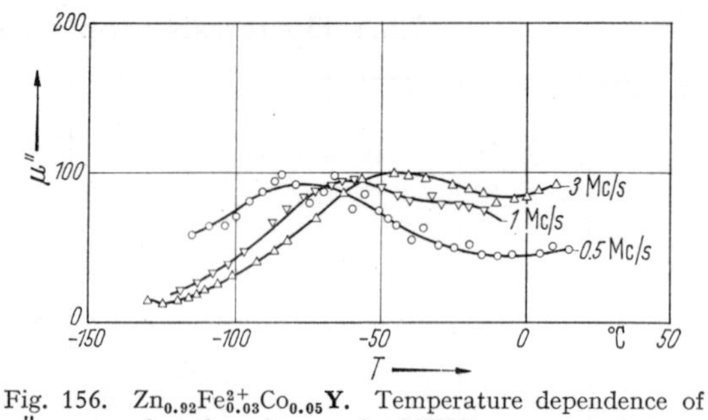

Fig. 156. $Zn_{0.92}Fe^{2+}_{0.03}Co_{0.05}$**Y**. Temperature dependence of μ'' measured at three frequencies [*67V1*].

7.8.6.2 Ferromagnetic resonance properties of ferrites with Y structure

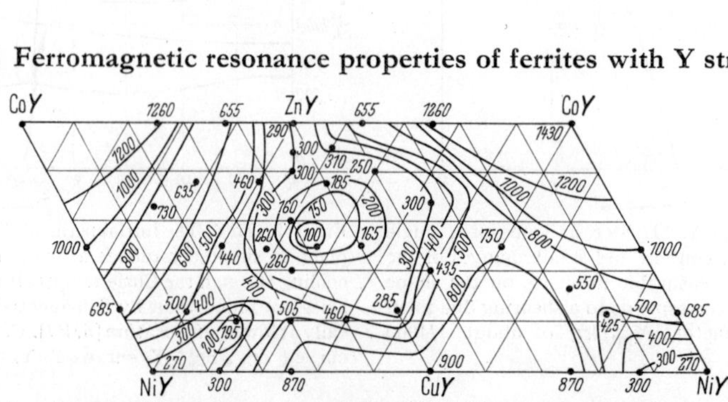

Fig. 157. Room temperature values of the microwave line width, ΔH, in Oe of polycrystalline samples of **Y**-type hexagonal ferrites, measured at frequencies of 10···20 Gc/s [*66BGH*].

Tab. 33. Room temperature measuring results for the line width ΔH of various samples of compounds with **Y** structure. The h.f. measuring field h_{rf} is oriented perpendicular to the static field H_{res}

Compound	Sample	Relative orientation of H_{res} and h_{rf} with respect to crystal axes	f_{res} Gc/s	ΔH Oe	Ref.
ZnY*)	spheres, crystal-oriented	$h_{rf} \perp c$ axis	10.8	530	*61B*
			13.0	570	*61B*
			14.9	600	*61B*
			17.0	600	*61B*
	disc $\perp$ *c* axis of the crystals	$h_{rf} \perp c$ axis	14.9	520	*61B*
	disc $\parallel$ *c* axis of the oriented crystals	$h_{rf} \perp c$ axis	14.9	430	*61B*
		$h_{rf} \parallel c$ axis	14.9	1000	*61B*
	single crystal sphere			200···300	*61B*
	single crystal disc			170···330	*61B*
	single crystal sphere	$H_{res} \perp c$ axis	8.6	60	*62D*
		$h_{rf} \perp c$ axis	8.9	70	*62D*
			9.8	70	*62D*
			16.6	75	*62D*
			17.3	75	*62D*
			37.7	100	*62D*
	single crystal sphere	$h_{rf} \perp c$ axis	8.9	13.8	*63D*
		$h_{rf} \perp c$ axis	9.0	8.0	*64TDS*
	single crystal disc $\parallel$ *c* axis of the crystal	$h_{rf} \perp c$ axis	8.5	12.8	*62DKR*
			8.5	14.4	*62DKR*
	single crystal disc $\parallel$ *c* axis of the crystal	$h_{rf} \perp c$ axis	8···12	20	*62B2*
			70	200	*62B2*
			8.26	300	*62B2*
			11.69	80	*62B2*
$Mn_{0.5}Zn_{0.5}$**Y**	single crystal sphere	$h_{rf} \perp c$ axis	9.0	3.8	*64SDT*
			17.3	4.9	
NiY	crystal-oriented sphere	$h_{rf} \perp c$ axis	13.0	500	*61B*
			14.9	580	*61B*
			17.0	640	*61B*
$Ni_{0.4}Zn_{0.6}$**Y**	single crystal sphere	H_{res}, $h_{rf} \perp c$ axis	9.3	18	*67V1*
CoY	single crystal sphere		8···27	1800	*59BK1*
	single crystal disc	$H_{res} \perp c$ axis	10	900	*62B2*
	disc $\parallel$ *c* axis	$h_{rf} \perp c$ axis	70	1300	*62B2*

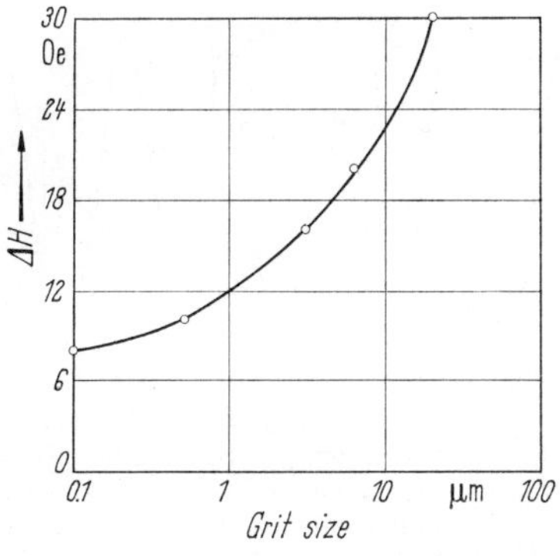

Fig. 158. **ZnY**. Ferromagnetic resonance line width, ΔH, measured at a frequency of 9.0 Gc/s for a single crystal sphere after various stages of the polishing of the sphere (diameter about 0.5 mm). The grit size of the polishing material ranges from 10 to 0.1 μm [*64TDS*].

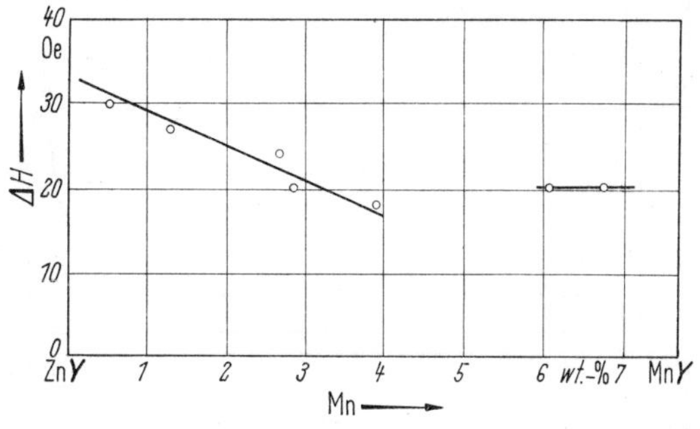

Fig. 159. Mn-Zn**Y**. Ferromagnetic resonance line width, ΔH, as a function of weight% Mn in unpolished single crystal spheres; end member Mn**Y** has 7.81 weight% Mn, after [*64TDS*], [*65TDS*].

*) For the line width of substituted Zn**Y** compounds see also Tab. 32.

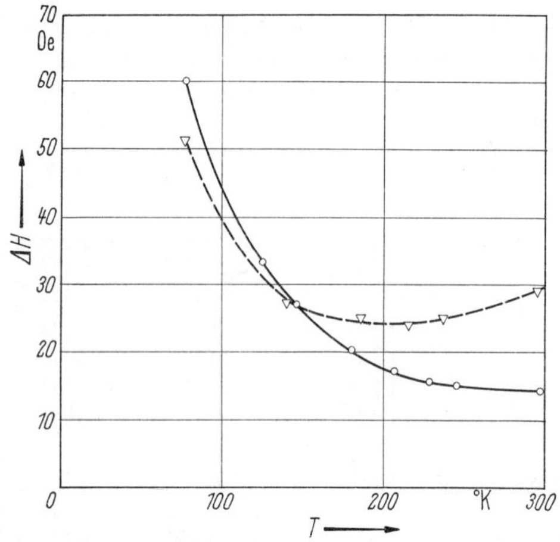

Fig. 160. Line width as a function of temperature for single crystal discs of $Zn_{0.92}Fe^{2+}_{0.03}Co_{0.05}Y$ (broken curve) and $Zn_{0.97}Fe^{2+}_{0.03}Y$ (solid line). See also Tab. 32 [*67V1*].

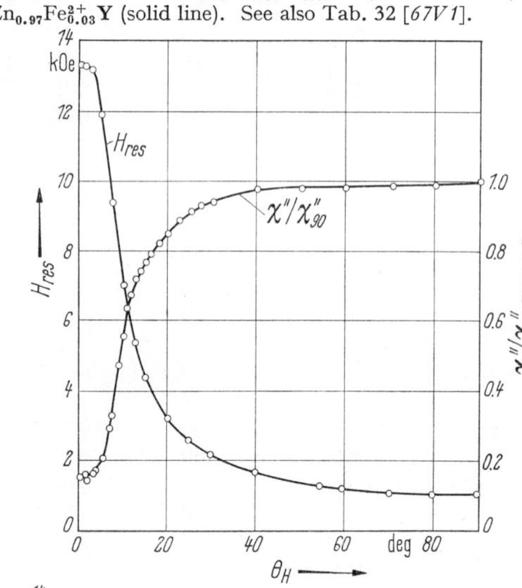

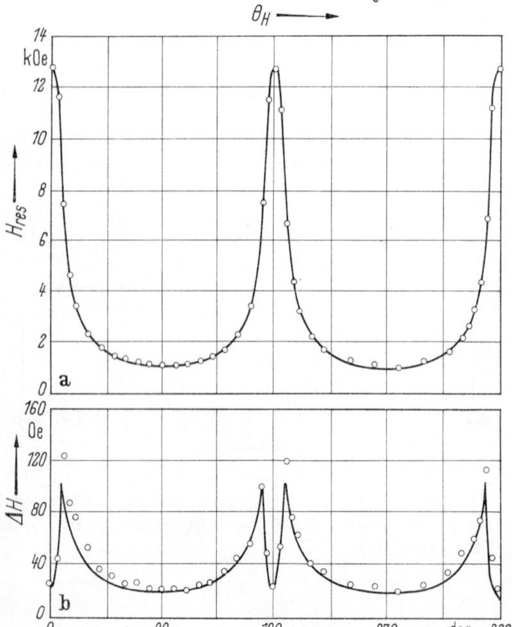

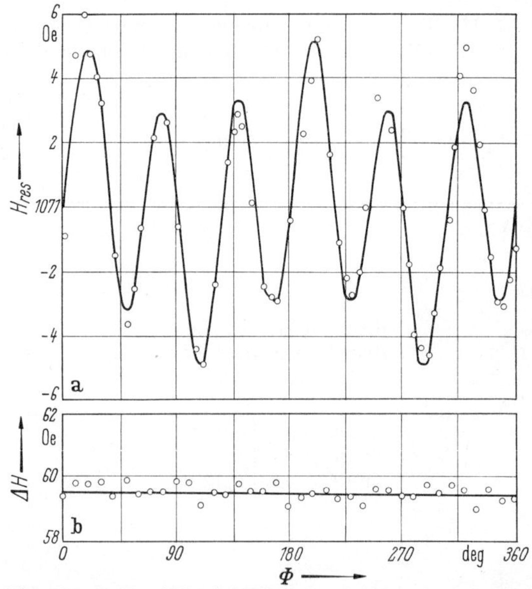

Fig. 163. Zn**Y**. a) The field, H_{res}, for the ferromagnetic resonance at 9.3 Gc/s of a single crystal sphere at room temperature. The field, H_{res}, is parallel to the basal plane, its direction in the basal plane is given by the angle Φ. b) The line width, ΔH, under the same conditions as in Fig. a) [*66N*].

← Fig. 161. Zn**Y**. Main resonance field, H_{res}, as dependent on the angle θ_H between this field and the c axis of a single crystal sphere. The high frequency field, h_{rf}, is oriented perpendicular to, H_{res}, and has a frequency of 9.0465 Gc/s. The imaginary part of the susceptibility at resonance is given relative to its value for $\theta_H = 0$ [*65M*].

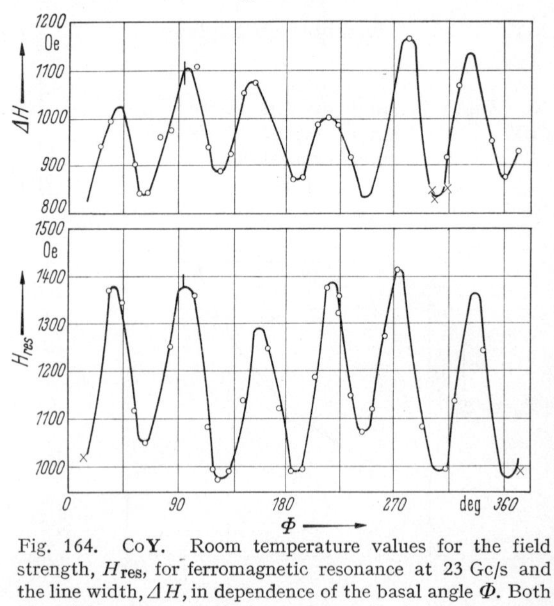

Fig. 164. Co**Y**. Room temperature values for the field strength, H_{res}, for ferromagnetic resonance at 23 Gc/s and the line width, ΔH, in dependence of the basal angle Φ. Both H_0 and h_{rf} are perpendicular to the c axis of a single crystal. The sample is a disc cut perpendicular to the c axis [*62B2*].

← Fig. 162. Zn**Y**. a) The field, H_{res}, for the ferromagnetic resonance at 9.3 Gc/s of a single crystal sphere at room temperature as a function of the angle, θ, between the direction of the static field, H_{res}, and the direction of the c axis of the crystal. b) The line width, ΔH, under the same conditions as in Fig. a) [*66N*].

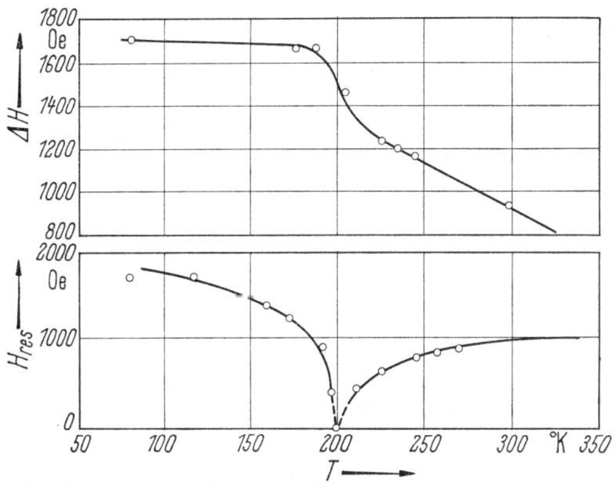

Fig. 165. CoY. Field for resonance, H_{res}, and line width, ΔH, of a single crystal as a function of temperature; for the experimental arrangement see caption of Fig. 164 [*62B2*].

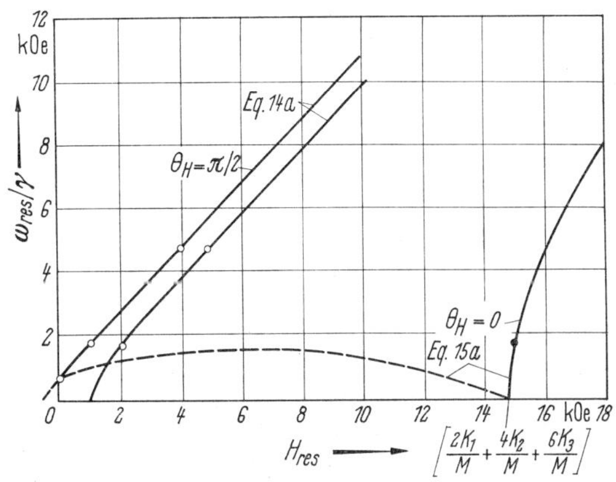

Fig. 166. CoY. The internal magnetic field, H_{res}, necessary for the ferromagnetic resonance in a single crystal. The curves have been calculated according to the theory of Belson and Kriessman (see 7.1e, Eq. (14a) and (15a)) with the following constants which were measured at X and K band frequencies:

$K_1/M = 344$ Oe, $K_4/M = 27.5$ Oe, $g_{eff} = 3.70$.

The angle between magnetic resonance field, H_{res}, and the c axis of the crystal is designated by θ_H [*59BK*].

7.8.6.3 Non-linear effects in the ferromagnetic resonance of ferrites with Y structure

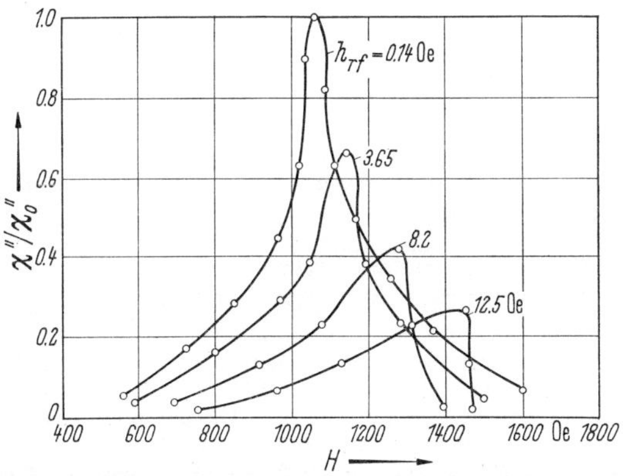

Fig. 167. ZnY single-crystal sphere. Variation of the normalized susceptibility and the field required for resonance with the incident r. f. power as a parameter: frequency = 8957 Mc/s, ΔH = 85 Oe. Magnetic field, H, parallel to c axis of the crystal. The r. f. field, h_{rf}, is oriented perpendicular to H. The imaginary part, χ'', of the susceptibility has been normalized by dividing through the corresponding value, χ_0'', measured at low amplitude of h_{rf} [*62D*].

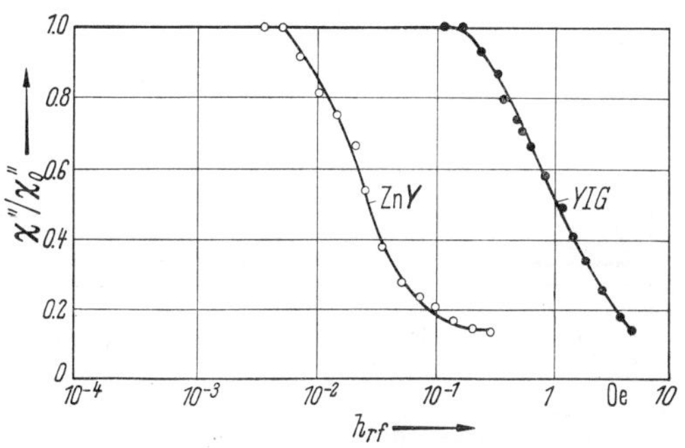

Fig. 168. ZnY. Normalized imaginary part of the magnetic susceptibility at resonance of a polished single-crystal sphere as a function of power, measured at room temperature at a frequency of 8.9 Gc/s. The static magnetic field is parallel to the c axis of the crystal, the r. f. magnetic measuring field is parallel to the basal plane. For comparison a curve measured for an yttrium iron garnet is also given. The low-power line widths measured for the ZnY and the YIG crystal are 13.8 and 6.5 Oe, respectively [*63D*].

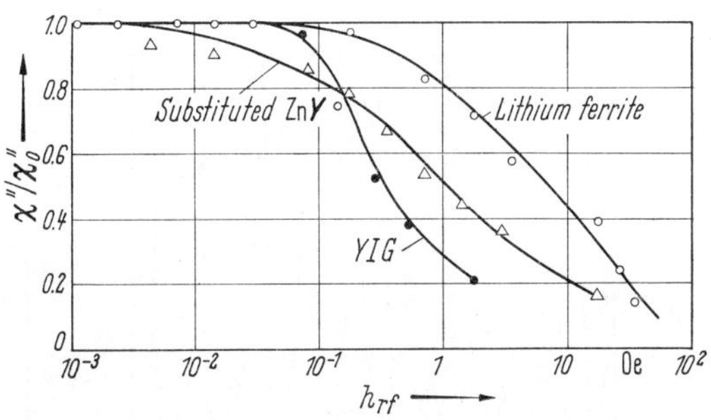

Fig. 169. Normalized imaginary part of the magnetic susceptibility at resonance as a function of incident power for a single crystal sphere of manganese-substituted ZnY (3.5 % by weight), measured at a frequency of 17.3 Gc/s at room temperature. The static magnetic field is parallel to the c axis of the crystal, the r. f. magnetic measuring field is parallel to the basal plane. The sharp knee in the curves for the ferrite of Fig. 168 is not found here. According to [*62LB*] this can be a consequence of inhomogeneities in the crystal. For comparison the corresponding property of an yttrium iron garnet crystal and that of the spinel lithium ferrite are given in the same figure [*64SDT*].

△ substituted ZnY;	ΔH = 4.9 Oe
● YIG;	ΔH = 2.35 Oe
○ Lithium ferrite;	ΔH = 5.1 Oe

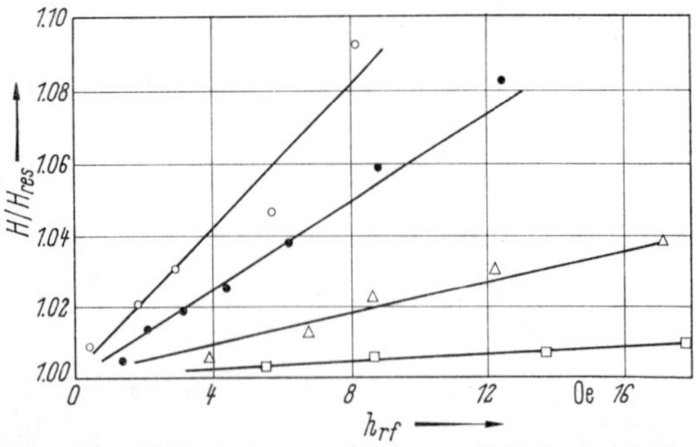

Fig. 170. Zn**Y** single-crystal sphere. Shift in the field required for resonance vs. the amplitude of the r. f. magnetic field – with frequency as a parameter. Magnetic field, H, parallel to the c axis of the crystal. The r. f. magnetic field is parallel to the basal plane. ΔH is the line width at low power [*62D*].

	Frequency f [Gc/s]	Low-power line width ΔH [Oe]
○	8.6	60
●	9.8	70
△	17.3	75
□	37.7	100

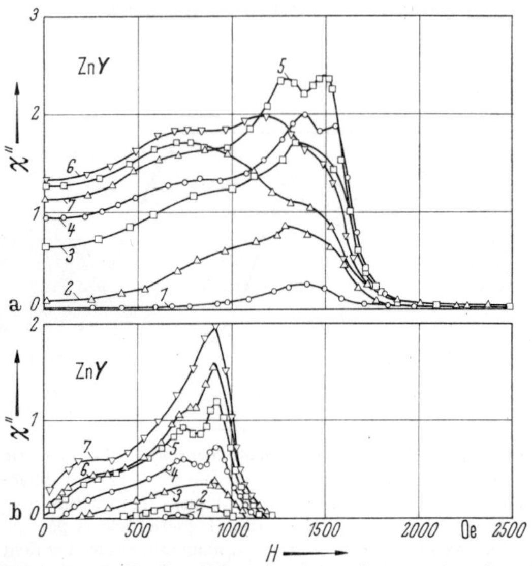

Fig. 172. Zn**Y**. Parallel pump absorption as a function of applied d. c. and r. f. magnetic fields. The pump frequency is 16.6 Gc/s and $T = 290$ °K. Magnetic field, H, parallel to the basal plane of the crystal, and $h_{rf} \parallel H$ [*63GH*].

Fig. a) Spherical sample;	Fig. b) Thin flake sample.
1 $h_{rf} = 3.8$ Oe	*1* $h_{rf} = 4.41$ Oe
2 $h_{rf} = 5.34$ Oe	*2* $h_{rf} = 4.95$ Oe
3 $h_{rf} = 9.54$ Oe	*3* $h_{rf} = 6.23$ Oe
4 $h_{rf} = 16.9$ Oe	*4* $h_{rf} = 7.85$ Oe
5 $h_{rf} = 30.2$ Oe	*5* $h_{rf} = 9.86$ Oe
6 $h_{rf} = 53.2$ Oe	*6* $h_{rf} = 12.5$ Oe
7 $h_{rf} = 86.7$ Oe	*7* $h_{rf} = 24.8$ Oe

→
Fig. 174. Zn**Y**. The absorption pattern of a single crystal sphere measured at a frequency of 9.0465 Gc/s and a high frequency field $h_{rf} = 0.69$ Oe. The r. f. field is perpendicular to the static field, H. The angle, θ_H [degree], between this static field and the c axis of the crystal is given as a parameter. The values of χ'' have been normalized to the highest peak that was found (i. e. at $\theta_H = 8.5°$). At higher microwave powers a second subsidiary maximum is found [*65M*].

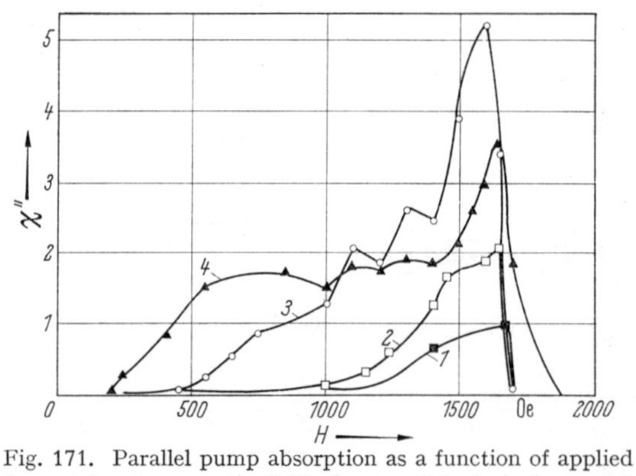

Fig. 171. Parallel pump absorption as a function of applied d. c. and r. f. magnetic fields for a polished single crystal sphere of manganese-substituted Zn**Y** with 3% by weight of manganese. The pump frequency is 17.2 Gc/s [*64TDS*], [*65DTS*].

1 $h_{rf} = 0.65$ Oe
2 $h_{rf} = 1.05$ Oe
3 $h_{rf} = 1.78$ Oe
4 $h_{rf} = 4.00$ Oe

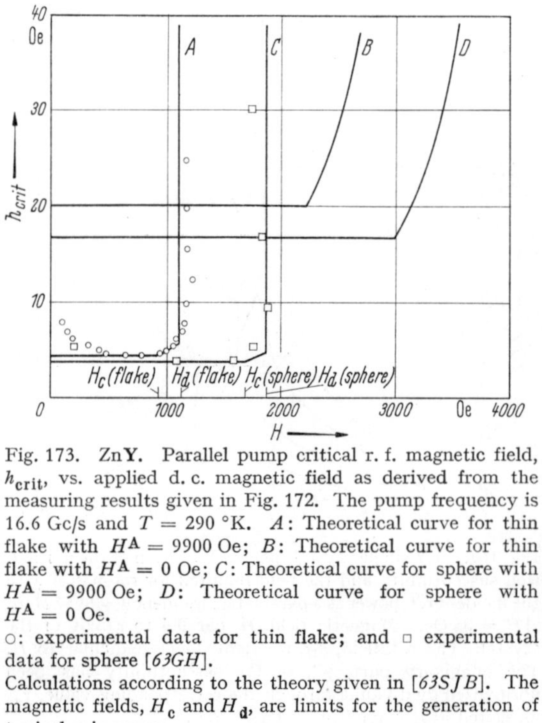

Fig. 173. Zn**Y**. Parallel pump critical r. f. magnetic field, h_{crit}, vs. applied d. c. magnetic field as derived from the measuring results given in Fig. 172. The pump frequency is 16.6 Gc/s and $T = 290$ °K. A: Theoretical curve for thin flake with $H^A = 9900$ Oe; B: Theoretical curve for thin flake with $H^A = 0$ Oe; C: Theoretical curve for sphere with $H^A = 9900$ Oe; D: Theoretical curve for sphere with $H^A = 0$ Oe.
○: experimental data for thin flake; and □ experimental data for sphere [*63GH*].
Calculations according to the theory given in [*63SJB*]. The magnetic fields, H_c and H_d, are limits for the generation of typical spin waves.

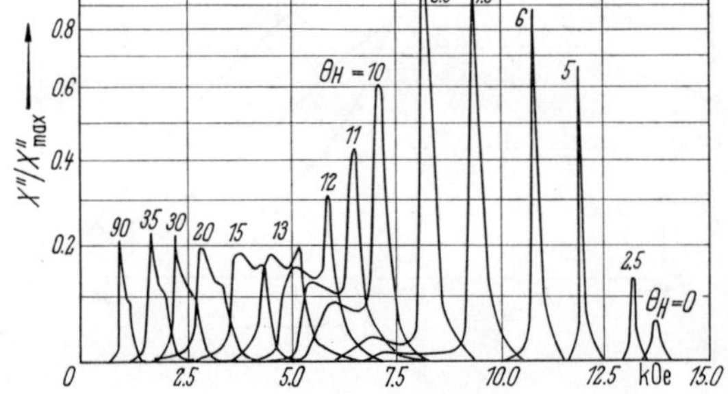

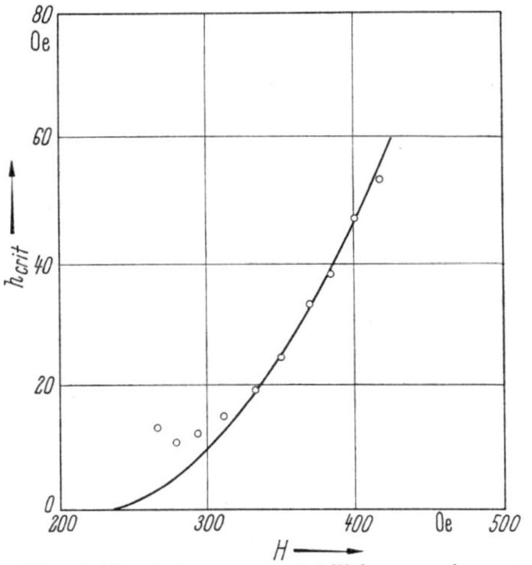

Fig. 175. ZnY. Spin wave instabilities occuring at pump frequencies far below the bottom of the spin-wave spectrum, the so-called magneto-elastic pump effect. Experimental points and theoretical curve for h_{crit} vs. H for a single crystal disc in a *transverse* pumping experiment. H and h_{crit} parallel to basal plane and $h_{crit} \perp H$. Pump frequency 1.3 Gc/s. T = 301 °K [*64TWA*]. For the theory of these sound wave instabilities or higher order magnon processes, see [*62M*], [*63WS*] and [*63AZW*].

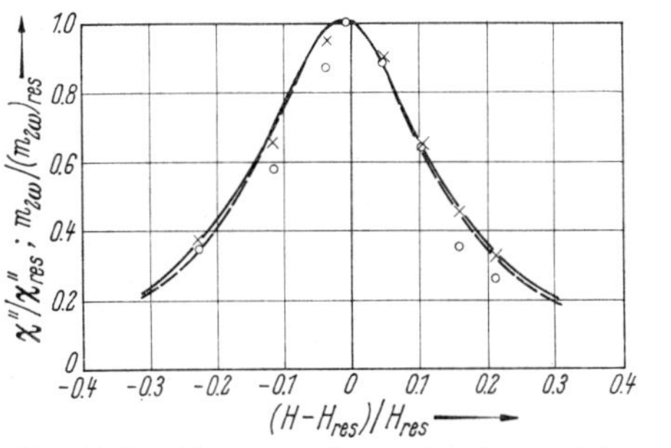

Fig. 176. ZnY. Absorption (crosses) and second harmonic magnetization (circles) in a single crystal disc at constant drive level of h_{rf} = 28 Amp/m, compared with theoretical curves. The d. c. field, H, and the r. f. field, h_{rf}, are parallel to the basal plane of the crystal, which itself is parallel to the plane of the disc. $h_{rf} \perp H$. The absorption, χ'', is given relative to its value at resonance, χ''_{res}, and also the amplitude of the second harmonic magnetization, $m_{2\omega}$, is given relative to its amplitude at resonance, $(m_{2\omega})_{res}$ [*62DKR*].

χ'' { ——— theor. × × × exp. $m_{2\omega}$ { — — — theor. ○ ○ ○ exp.

Tab. 34. Critical field strength, h_{crit}, for non-linear effects and the spin-wave line width ΔH_k in the case of *transverse* pumping at room temperature. The resonance field H_{res} is oriented parallel to the c axis of the crystal and the r.f. field parallel to the basal plane

Compound	f_{res} Gc/s	h_{crit} Oe	ΔH Oe	ΔH_k*) Oe	H_{res} kOe	Ref.
ZnY unpolished sphere	8.6	0.16	60		1.190	*62D*
	8.9	0.30	70		1.225	
	9.8	1.0	70		1.455	
	16.6	1.8	75		3.250	
	17.3	1.5	75		3.460	
	37.7	1.4	100		9.680	
ZnY polished sphere	8.9	0.01	13.8	6.6		*63D*
MnY-ZnY 3% by weight of Mn, polished sphere	17.3	0.06	4.9	0.74		*64SDT*

Tab. 35. Critical field strength h_{crit} for non-linear effects and the spin-wave line width ΔH_k in the case of *parallel* pumping at room temperature. The resonance field H_{res} and the r.f. magnetic measuring field are mutually perpendicular and they are both oriented parallel to the basal plane of the crystal

Compound	f_{res} Gc/s	h_{crit} Oe	ΔH Oe	ΔH_k*) Oe	Ref.
ZnY sphere	16.6	see Fig. 172, 173	—	8.0	*63GH*
disc perpendicular to c axis of crystal	16.6	see Fig. 172, 173	—	9.24	*63GH*
Manganese-substituted ZnY with 3% by weight of manganese polished sphere	17.2	0.65 see Fig. 171	3.8	1.24	*65DTS*

*) The line width for the spin waves with wave vector k, i. e. ΔH_k, is derived from the critical field strength h_{crit} according to the theory of Bady and Schlömann [*62BS*], [*63SJB*].

7.9 Z-type ferrites

(general chemical formula $Ba_3Me_2^{2+}Fe_{24}O_{41}$)

7.9.1 Lattice parameters of ferrites with Z structure

Tab. 36. Lattice parameter c (with an accuracy of ± 0.01 Å) and X-ray density ϱ_X of compounds of type Me_2**Z** of stoichiometric composition. The parameter a equals in all cases 5.88 Å [*65V*]

Me	c Å	ϱ_X g/cm³	Ref.
Fe	52.3	5.33	*59SW*
Mn	52.80	5.29	*65V*
Zn	52.37	5.38	*65V*
Co	52.31	5.35	*65V*
Mg	52.11	5.23	*65V*
Ni	52.10	5.38	*65V*

7.9.2 The resistivity of ferrites with Z structure

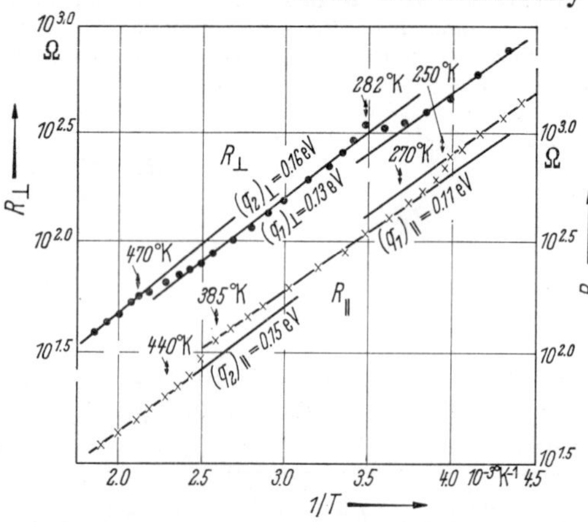

←

Fig. 177. Co_2**Z**($Ba_3Co_2Fe_{24}O_{41}$). The electrical resistance, R, of a single crystal as a function of temperature. The index ∥ denotes results obtained from measurements parallel to the c axis, the symbol ⊥ denotes perpendicular to this axis. The three parts of the curves as indicated by the straight lines $\log R$ vs. $1/T$ seem to correspond to the ranges of preferential cone, preferential plane, and preferential direction for the magnetization in the crystal, as is suggested by the authors. The activation energies calculated from the slopes of the straight lines drawn through the measuring points are indicated [*67BR*].

7.9.3 Saturation magnetization and Curie temperature of ferrites with Z structure

Tab. 37. Saturation magnetization and Curie temperature of some **Z** compounds [*59SW*]

Compound	Symbol	σ_s^0 emu/g	σ_s^{RT} emu/g	$4\pi M_s$ (20 °C) Gauss	Θ_C °C
$Ba_3Co_2Fe_{24}O_{41}$	Co_2**Z**	69	50	3 350	410
$Ba_3Cu_2Fe_{24}O_{41}$	Cu_2**Z**	60	46	3 100	440
$Ba_3Zn_2Fe_{24}O_{41}$	Zn_2**Z**	55	58	3 900	360
$Ba_3Ni_2Fe_{24}O_{41}$	Ni_2**Z**	54	—	—	—

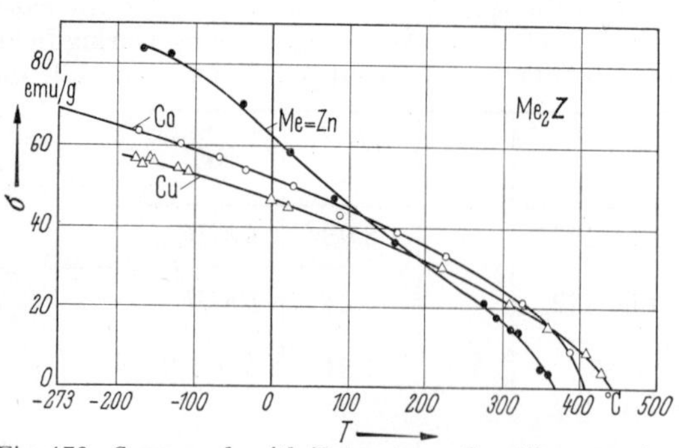

Fig. 178. Compounds with **Z** structure. Specific magnetization as a function of temperature for compounds with **Z** structure, measured on polycrystalline specimens at a field strength of 11 000 Oe for Co_2**Z** and Zn_2**Z** and at a field strength of 18 000 Oe for Cu_2**Z** [*59SW*].

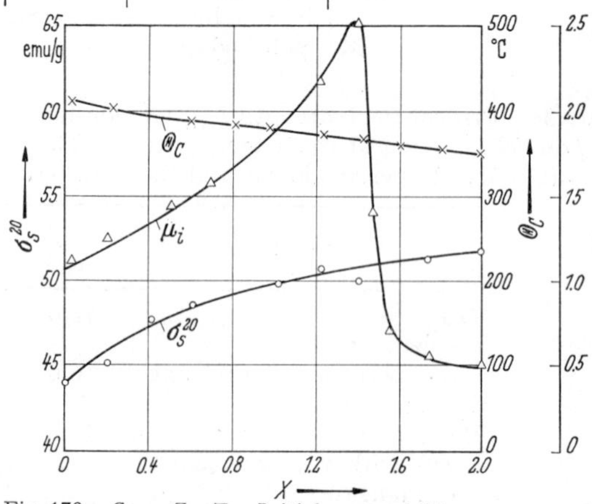

Fig. 179. $Co_{2-x}Zn_x$**Z**. Initial permeability, μ_i, saturation magnetization, σ_s^{20}, Curie temperature, Θ_C, measured on polycrystalline samples of solid solutions. The saturation magnetostriction, λ_s, measured in a field of 2000 Oe, is practically the same for all compositions: $(6\cdots7)\cdot10^{-6}$ [*67P*].

7.9.4 Magneto-crystalline anisotropy of ferrites with Z structure

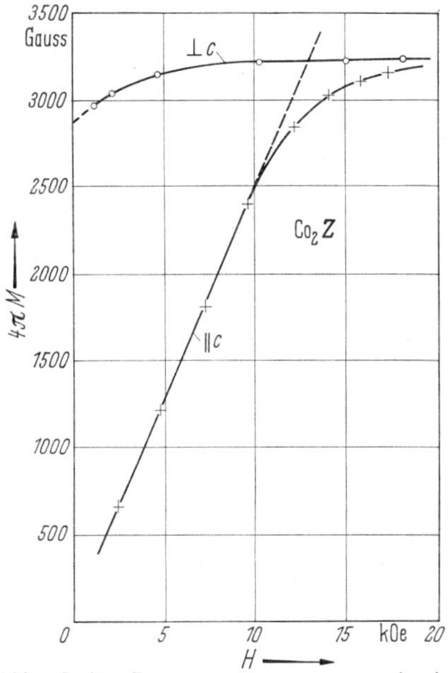

Fig. 180. Co_2Z. Room temperature magnetization curves of a single crystal for the cases of a magnetic field parallel and perpendicular to the c axis of the crystal [*59SW*].

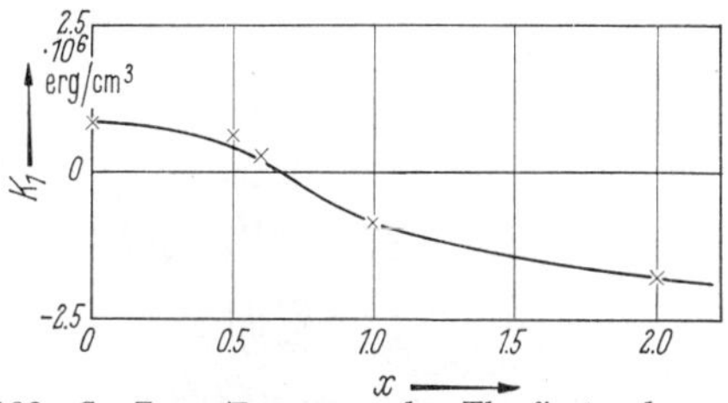

Fig. 182. $Co_xZn_{2-x}Z$ compounds. The first-order magneto-crystalline anisotropy coefficient, K_1, as a function of cobalt content for compounds $Co_xZn_{2-x}Z$ at 77 °K [*61LES*].

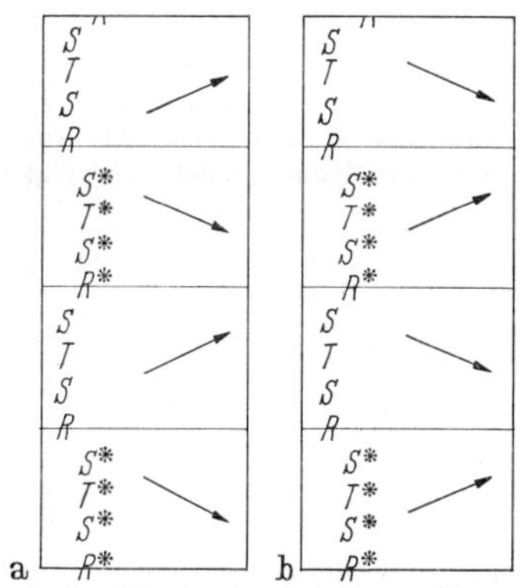

Fig. 183. Co_2Z. The ferrite Co_2Z has an intrinsic magneto-crystalline anisotropy with a symmetry lower than that of the crystal structure, but consistent with that of a subgroup of the structure. The existence of a trigonal anisotropy is demonstrated in the low temperature region where the crystal has a preferred cone for the magnetization. The figures represent puckered magnetization patterns within a unit cell of Co_2Z which result in trigonal anisotropy at 77 °K. The magnetized state is shown, with the net magnetization lying in the basal plane. R, S, T, etc., refer to the structural subgroups shown in Fig. 10 [*60B2*].

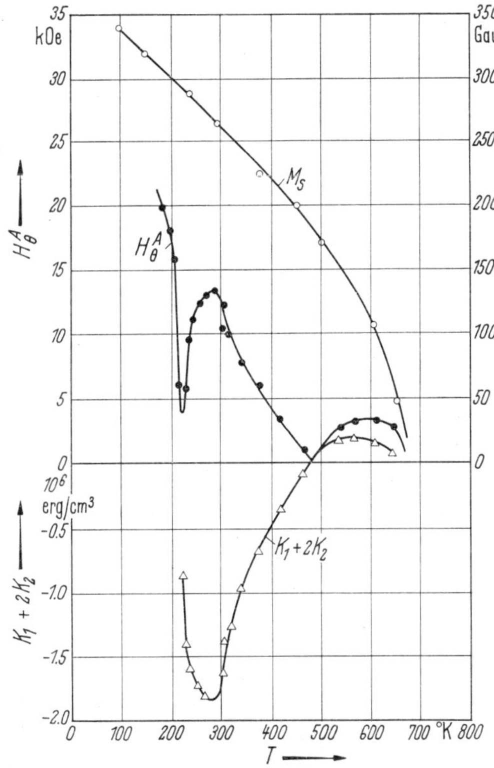

Fig. 181. Co_2Z. Magneto-crystalline anisotropy coefficients, $K_1 + 2K_2$, anisotropy field strength, H_θ^A, and saturation magnetization, M_S, as a function of temperature. For temperatures above 480 °K the triangles represent the (positive) values of K_1 [*59SW*].

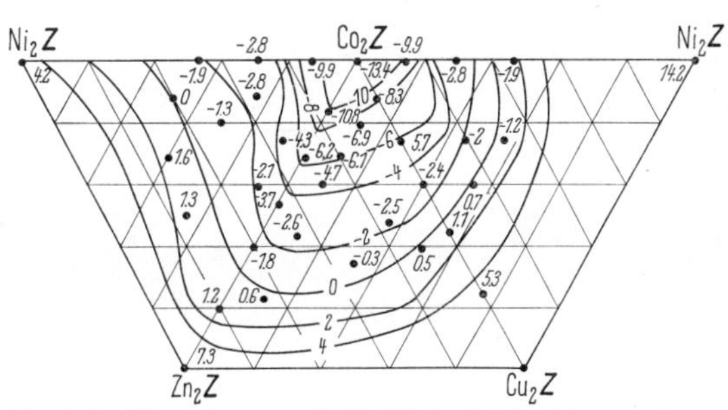

Fig. 184. The anisotropy field, H^A (in kOe), of Z-type hexagonal ferrites. Planar anisotropy is identified by a negative sign, and uniaxial anisotropy is indicated by a positive sign. Room temperature values measured at a frequency in the range from 10 ··· 35 Gc/s [*66BGH*].

Tab. 38. Planar magneto-crystalline anisotropy in $Co_xFe_{2-x}Z$ [*59SW*]

Compound	T °C	M_s Gauss	$\|K_3\|$ erg/cm³	H_ϕ^A Oe
$Co_{1.5}Fe^{2+}_{0.5}Z$	20	280	55	7
	158	220	≃9	≃2
$Co_{1.92}Fe^{2+}_{0.08}Z$	20	280	120	16
$CoZn_{0.5}Fe^{2+}_{0.5}Z$	20	290	25	3

7.9.5 High-frequency magnetic properties of ferrites with Z structure

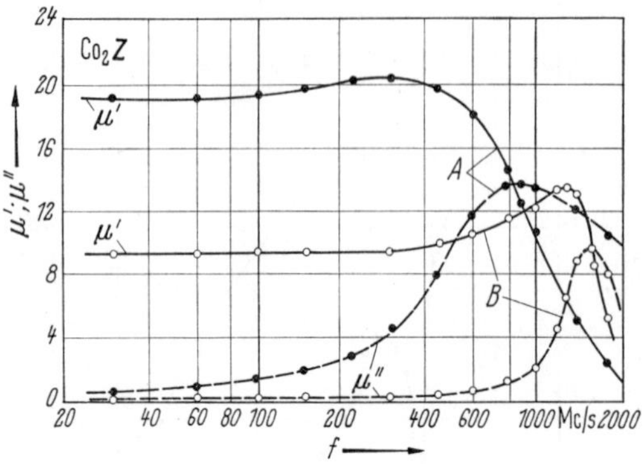

Fig. 185. Co_2**Z**. Magnetic spectra of two different isotropic polycrystalline samples:
A: Density 4.79 g/cm³; resistivity $2 \cdot 10^2\,\Omega$cm; the excess Fe_2O_3 is +1.0% by weight.
B: Density 4.60 g/cm³; resistivity $7 \cdot 10^7\,\Omega$cm; no excess or deficiency of Fe_2O_3 [*64NSVW*], see also [*61HGB*].

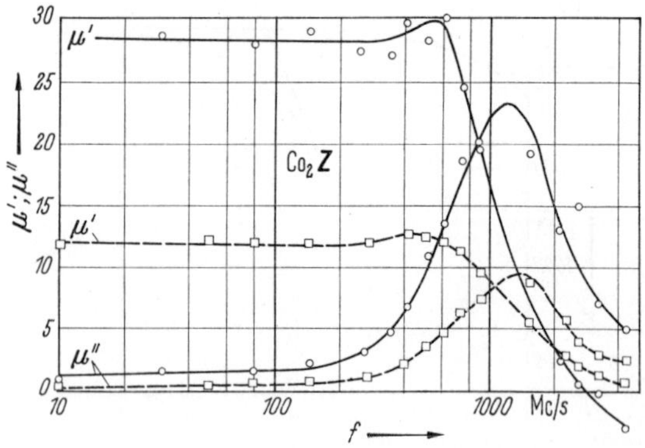

Fig. 186. Co_2**Z**. Magnetic spectra of two samples. The broken lines apply to an isotropic and the solid lines to a crystal-oriented sample. In the latter case the h. f. measuring field is oriented perpendicular to the *c* axis of the crystals [*59SW*].

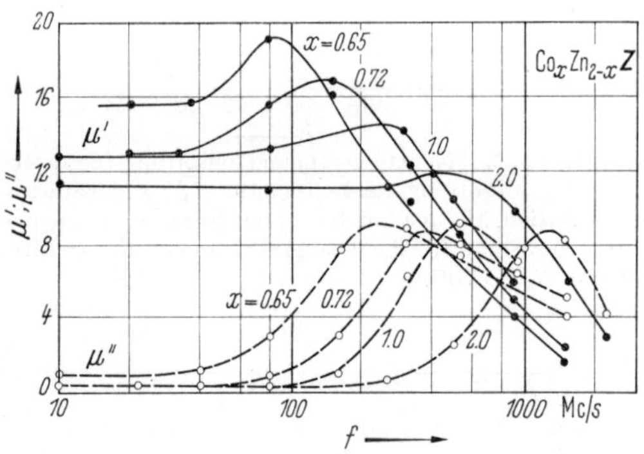

Fig. 187. Magnetic spectra of some isotropic polycrystalline samples of ferrites with chemical compositions in the series Co_xZn_{2-x}**Z** [*59SW*].

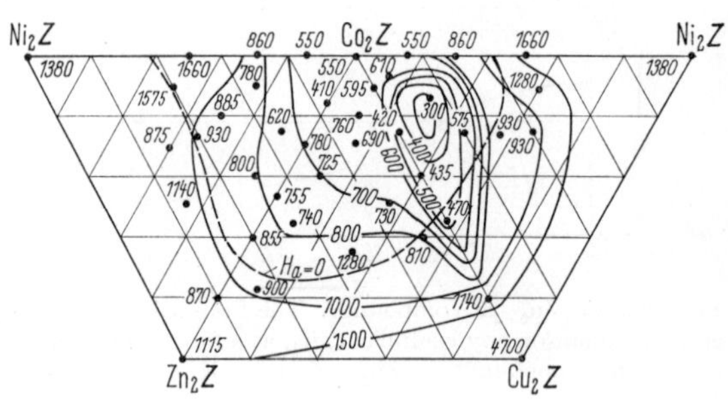

Fig. 188. Room temperature values of the microwave line width, ΔH, in Oe of polycrystalline samples of **Z**-type hexagonal ferrites measured at a frequency in the range of 10···35 Gc/s. The dotted line indicates the region of zero anisotropy [*66BGH*].

7.10 U-type ferrites

(general chemical formula $Ba_2Me^{2+}Fe_{18}O_{30}$)

Zn**U**: $c = 113.2$ Å; $a = 5.88$ Å [*68KTAS*]

$\varrho_X = 5.36$ g/cm³

Space group R3m, rhombohedral,

6 formula units to a triply-primitive hexagonal unit cell; $\Theta_C = (673 \pm 2)$ °K

Electrical conductivity $\sigma_\parallel = 10^{-7}\,(\Omega\text{cm})^{-1}$ and $\sigma_\perp = 2.5 \cdot 10^{-4}\,(\Omega\text{cm})^{-1}$

Ferrimagnetic resonance line width $\Delta H = 210$ Oe at $f = 30$ GHz

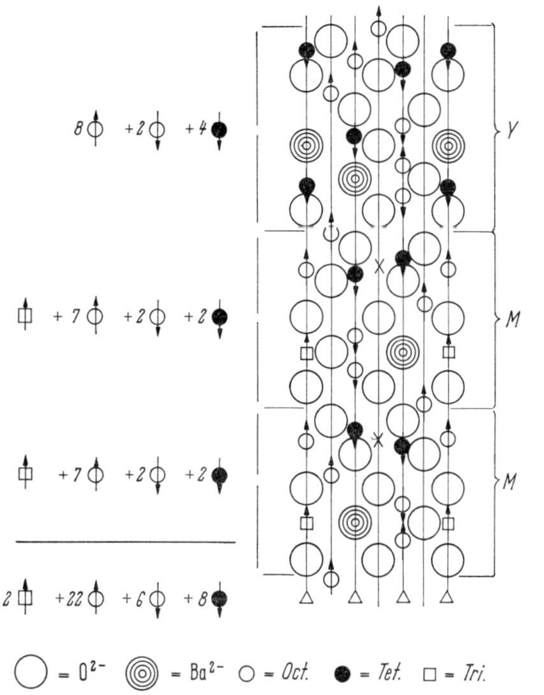

Fig. 189. ZnU. Collinear spin model; **M** = RS, **Y** = TS; according to [*68KTAS*].

↑ relative orientations of the magnetic moments of the metal ions.

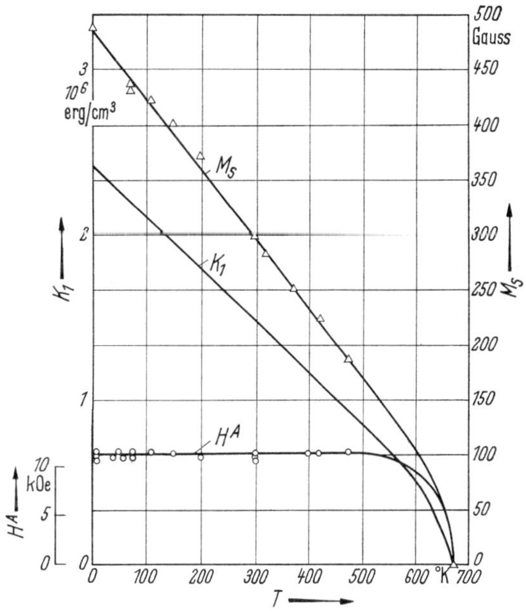

Fig. 190. $Ba_2ZnFe_{18}O_{30}$ (Zn**U**). Saturation magnetization, M_s, magneto-crystalline anisotropy coefficient, K_1, and anisotropy field strength, H^A, of a single crystal as a function of temperature [*68KTAS*].

7.11 Calcium ferrites

$Ca_2Fe_2O_5$, dicalciumferrite: Crystal structure orthorhombic [*57BBS*], the magnetic ordering is antiferromagnetic [*57DK*], [*66GGGWE*], [*66CHKB*], the preferential direction of the sublattice magnetization is the *c* axis [*66GGWG*]. Also the compound $Sr_2Fe_2O_5$ with the same structure exists [*66GGGWE*]. $Ca_2Fe_{2-x}Me^{3+}_xO_5$ with Me = Ga ($0 \leq x \leq 1.5$) or Me = Sc ($0 \leq x \leq 0.5$) have also $Ca_2Fe_2O_5$ structure; for the antiferromagnetic properties see [*67GGGWE*].

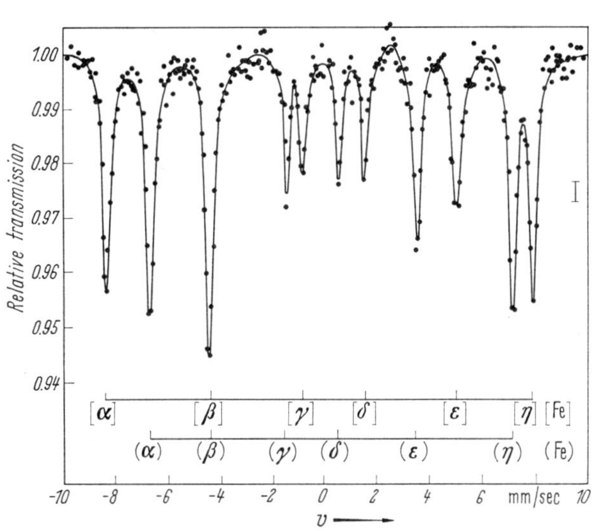

Fig. 191. Mössbauer spectrum obtained with a room temperature source of Co^{57} diffused into Pt and a polycrystalline absorber of $Ca_2Fe_2O_5$ at room temperature. Source approaching absorber is positive velocity [*66GWGG*].

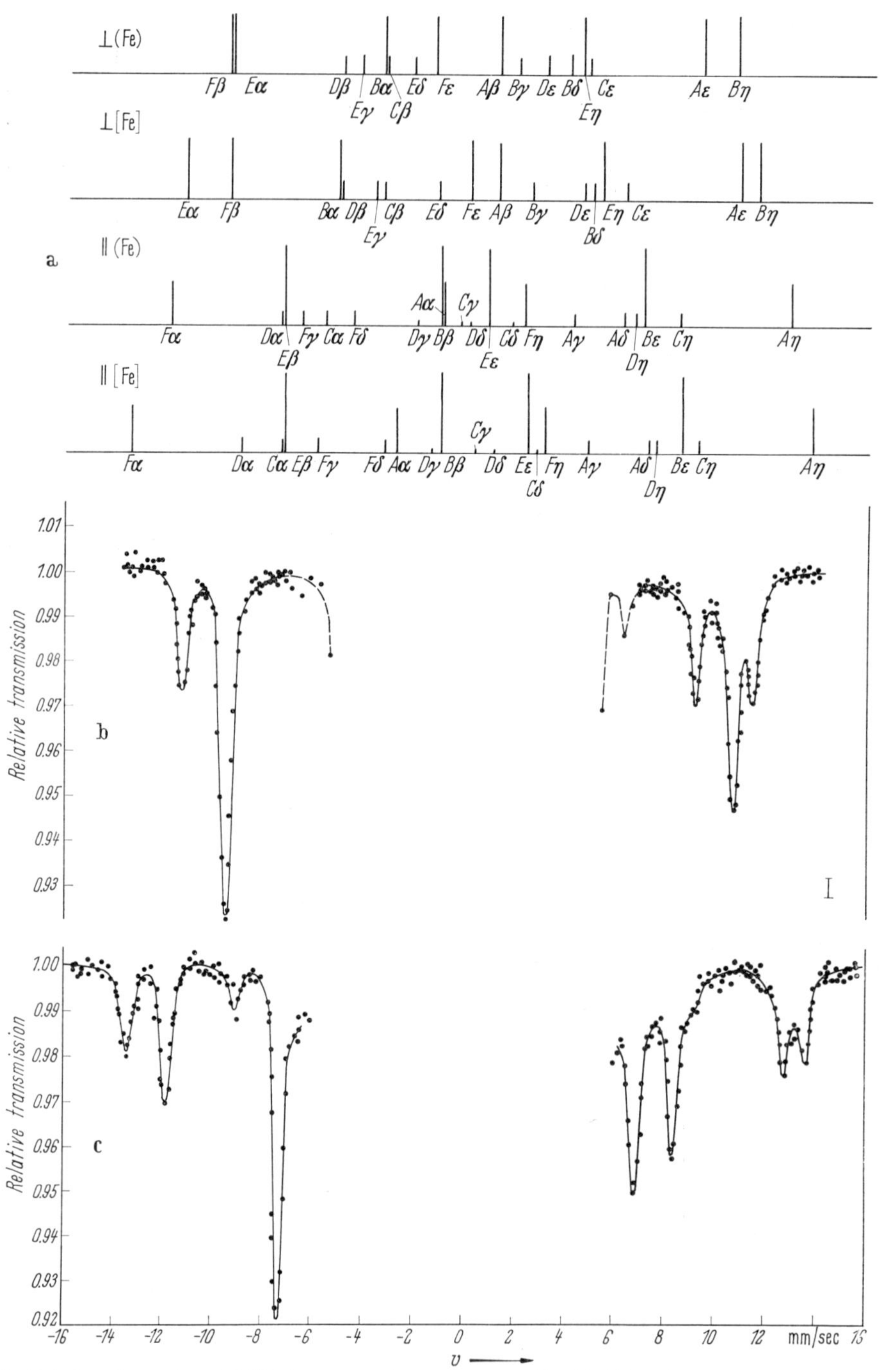

Fig. 192. Predicted and experimental Mössbauer spectra for a magnetized Co^{57} in α-iron source at room temperature and a single crystal of $Ca_2[Fe](Fe)O_5$. The crystallographic b axis of the $Ca_2[Fe](Fe)O_5$ and the magnetization direction of the source are parallel and perpendicular to the γ-ray propagation direction, respectively. (a) Expected locations and intensities for the octahedral [Fe] and tetrahedral (Fe) subspectra with the source magnetization direction perpendicular ($\perp$) and parallel or antiparallel ($\parallel$) to the spin orientation in the absorber. (b, c) Observed spectra with c axis perpendicular (b) and parallel (c) to the source magnetization direction [*66GWGG*].

$CaFe_2O_4$, monocalciumferrite: Crystal structure orthorhombic, space group Pnam — D_{2h}^{16}, [*55BBM*], [*56HPR*], [*57DK*] with antiferromagnetic ordering of the magnetic moments of the ions [*57DK*], [*66GGGWE*]. For Cr-substituted $CaFe_2O_4$ see [*67CHK1*], [*67CHK2*].

$CaFe_4O_7$, calciumdiferrite: Existence proved and interplanar spacings given in [*58PM*]. Hexagonal crystal structure, space group $R\bar{3}c$, with $a = (5.992 \pm 0.004)$ Å and $c = (31.121 \pm 0.008)$ Å, [*62CT*]; weakly ferromagnetic, [*62CT*], [*65H*].

$Ca_4Fe^{3+}_{14}O_{25}$, ($4\,CaO \cdot 7\,Fe_2O_3$): According to [*60BK*] called structure I, hexagonal R3c, $a = 6.0$ Å, $c = 95.0$ Å. This structure is stabilized by addition of small amounts of a third ion, trivalent or tetravalent, with a radius between 0.92 and 1.15 Å, e. g. $Ca_{3.5}Y_{0.7}Fe_{13.6}O_{25}$; other ions introduced Nd^{3+}, Gd^{3+}, Bi^{3+}, In^{3+}, Th^{4+} and Ge^{4+}.

$Ca_4Fe^{3+}_{14}O_{25}$, ($4\,CaO \cdot 7\,Fe_2O_3$): According to [*60BK*] called structure II and related to structure I, hexagonal P$\bar{3}$c, $a = 6.0$ Å, $c = 31.6$ Å. Some Ca^{2+} is replaced by Fe^{3+}. The Fe^{3+} is replaced by some divalent ions with a radius of about 0.8 Å, e. g. $Ca_{3.6}Fe^{II}_{0.4}Fe^{3+}_{14}O_{25}$.

$Ca_4Fe^{3+}_{20}O_{33}$, ($4\,CaO \cdot 10\,Fe_2O_3$): According to [*60 BK*] hexagonal R$\bar{3}$c, $a = 6.0$ Å, $c = 62.3$ Å. The Ca^{2+} is replaced by some trivalent or tetravalent ions similar to structure I. The Fe^{3+} is replaced by some Me^{2+} and Ca^{2+}, e. g. $Ca_{4.1}Y_{0.7}Mg_{1.6}Fe_{17.5}O_{33.0}$.

Tab. 39. Magnetic properties of CaO-Fe_2O_3 compounds [*60BK*]

	Structure I	Structure II	
saturation magnetization, σ_s^0 (extrapolated)	42	55	Gauss cm³/g
saturation magnetization σ_s^{20}	22	28	Gauss cm³/g
Curie temperature, Θ_C	150	175	°C
anisotropy field H_θ^A	2200	—	Oe
anisotropy field H_ϕ^A	≅50	—	Oe
permeability, μ_i (up to 1 Mc/s)	20	20	
(at 100 Mc/s)	10	10	
(at 300 Mc/s)	6	5	
resonance frequency, f_{res}	400	300	Mc/s

$CaO \cdot 6\,Fe_2O_3$, calciumhexaferrite: **M** structure obtained by adding a small amount of lanthanum oxide to calciumhexaferrite [*63IK*], see Tab. 4 and Fig. 61.

$CaFe_4O_7$, modified: [*66I*]. Calciumdiferrite modified by up to 2.0 mol-% of Me_2O_3, where Me = As, Sb, In, Bi, Y, La, Gd or other rare earth atoms, or up to 2.0 mol-% of MeO_2, where Me = Ge, Ti, Hf, Zr, Ce and Th. These metals stabilize the hexagonal crystal structure. Space group of the crystals R$\bar{3}$c with $a = (6.03 \pm 0.01)$ Å and $c = (31.5 \pm 0.1) \cdot 3$ Å; $7 \cdot 3$ formulae units $CaFe_4O_7$ per unit cell. Density, $\varrho = 4.60$ g/cm³. According to [*66I*] the crystal structure is similar to structure I of [*60BK*], there is however some difference in the interplanar d-spacings. Electric and magnetic properties are given in the Fig. 193 ··· 207.

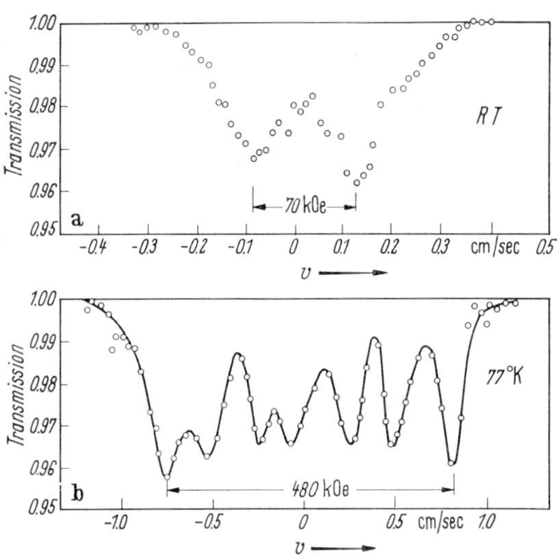

Fig. 193. Mössbauer transmission spectra of a polycrystalline sample of calcium diferrite at room temperature and at 77 °K. The internal magnetic fields are (70 ± 5) kOe and (480 ± 20) kOe at room temperature and at 77 °K, respectively [*66IY*].

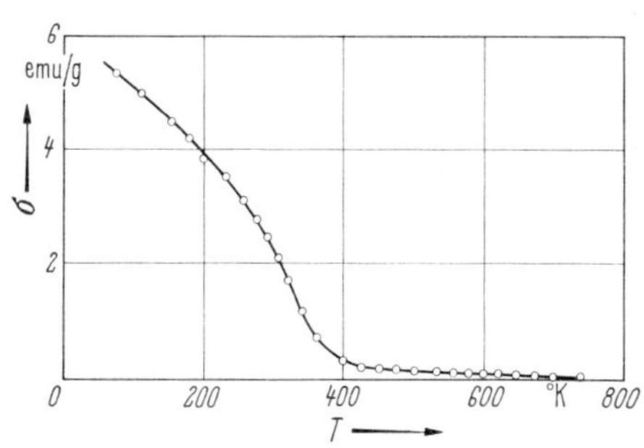

Fig. 194. Magnetic moment per gram vs. temperature for a polycrystalline sample of calcium diferrite at an applied field of 8 kOe. The Curie temperature is $\Theta_C = 130$ °C [*66IY*], see also [*62CT*].

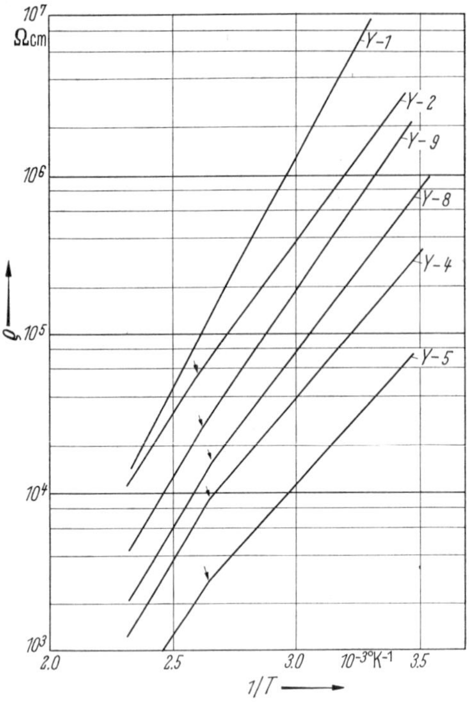

Fig. 195. The resistivity, ϱ, as a function of temperature for polycrystalline sintered samples of $CaFe_4O_7$ stabilized with various amounts of Y_2O_3 [*63IK*].

Specimen	Content of Y_2O_3 mole%	T_k*) °C	Θ_C**) °C	q***) eV
Y-1	0	–	–	0.59
Y-2	0.2	112	110	0.42
Y-4	0.6	105	103	0.36
Y-5	0.8	105	108	0.35
Y-8	1.5	103	107	0.42
Y-9	1.7	107	107	0.46

*) T_k: Temperature at the deflection point; (↓ in this figure).
**) Θ_C: The ferromagnetic Curie temperature.
***) q_ρ: The activation energy in lower temperature region.

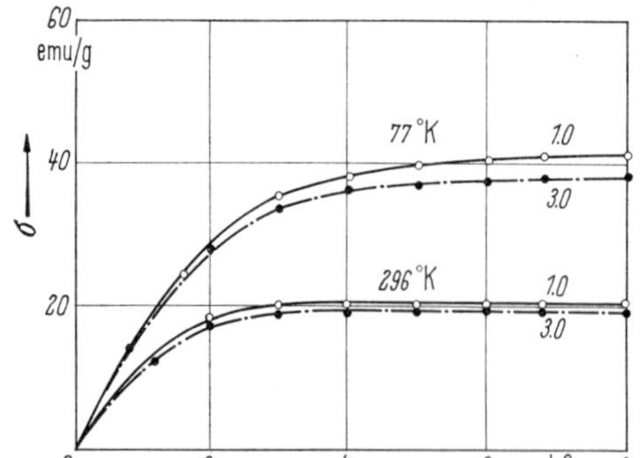

Fig. 197. Magnetic moment per gram vs. applied field at 77 °K and 296 °K for samples of stabilized $CaFe_4O_7$ containing 1.0 or 3.0 mole-% of CeO_2 [*66I*].

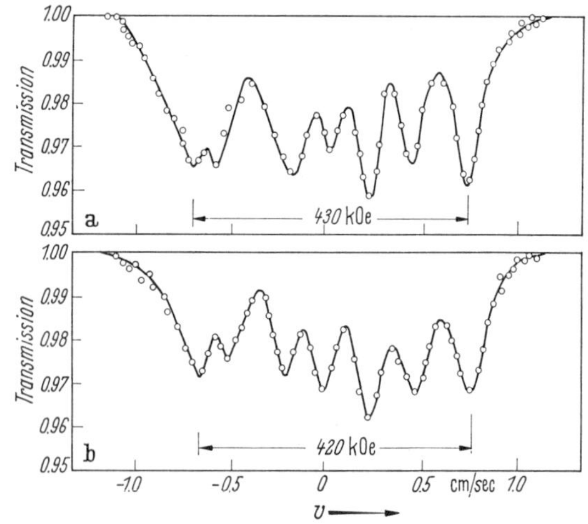

Fig. 196. Mössbauer transmission spectra at a temperature of 77 °K for samples of $CaFe_4O_7$ stabilized with *a* 0.4 mole-% and *b* 2.0 mole-% of Y_2O_3. The internal magnetic fields have been determined with an accuracy of ± 20 kOe [*66I*].

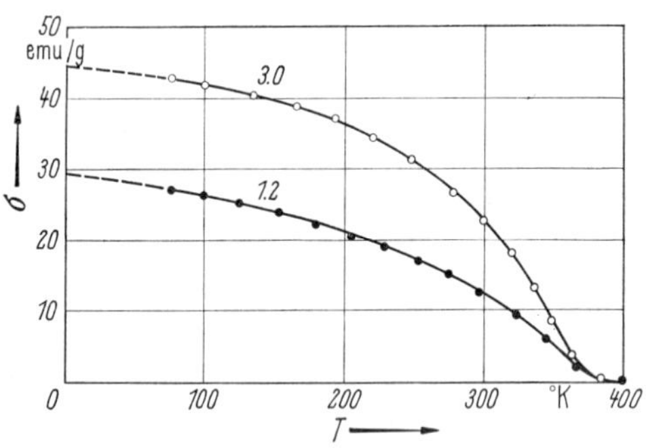

Fig. 198. Magnetic moment per gram vs. T for samples of stabilized $CaFe_4O_7$ containing 1.2 or 3.0 mole-% of Y_2O_3 [*66I*].

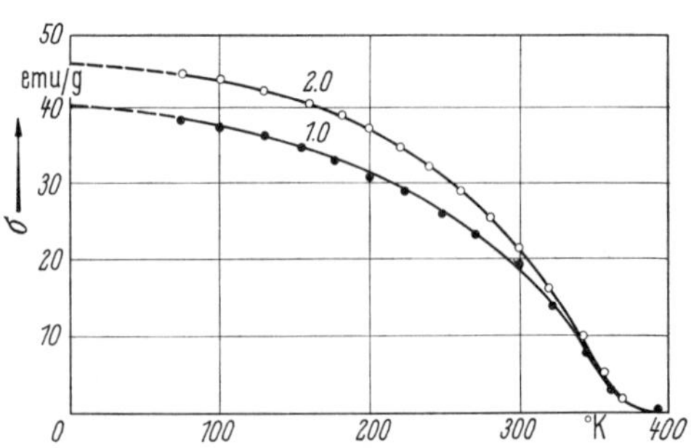

Fig. 199. Magnetic moment per gram vs. T for samples of stabilized $CaFe_4O_7$ containing 1.0 or 2.0 mole-% of CeO_2 [*66I*].

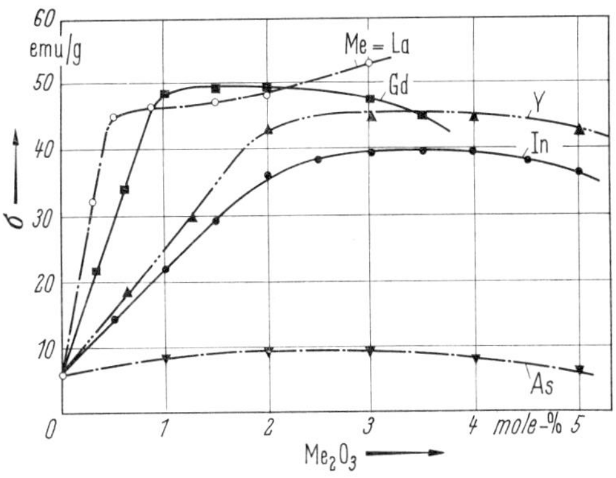

Fig. 200. Magnetic moment per gram extrapolated to 0 °K for samples of $CaFe_4O_7$ stabilized with various amounts of Me_2O_3 [*66I*].

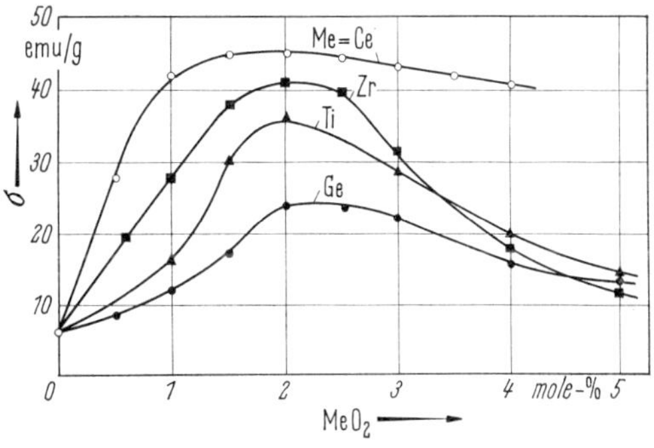

Fig. 201. Magnetic moment per gram extrapolated to 0 °K for samples of $CaFe_4O_7$ stabilized with various amounts of MeO_2 [*66I*].

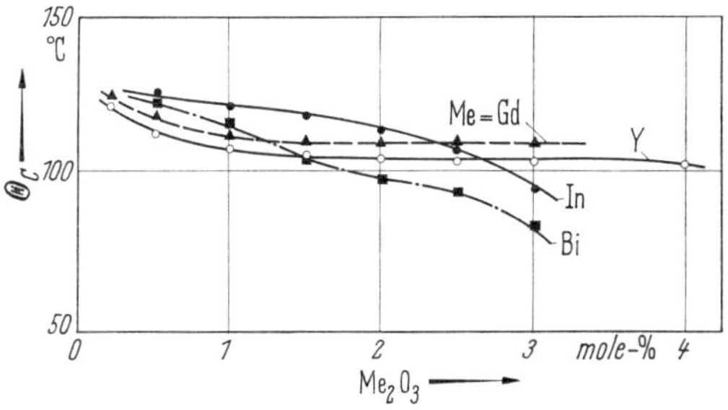

Fig. 202. Curie temperatures for samples of $CaFe_4O_7$ stabilized with various amounts of Me_2O_3 [*66I*].

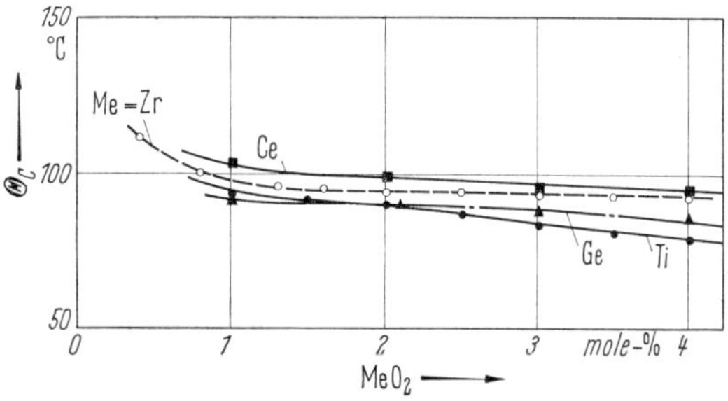

Fig. 203. Curie temperatures for samples of $CaFe_4O_7$ stabilized with various amounts of MeO_2 [*66I*].

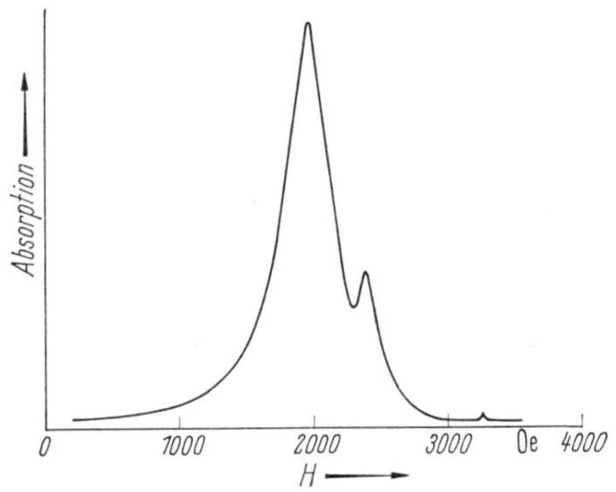

Fig. 204. Absorption curve at a frequency of 9.301 Gc/s as a function of the static field strength, H, for a $CaFe_4O_7$ single crystal platelet stabilized with 3.0 mole-% Gd_2O_3. Microwave field and static field are applied parallel and perpendicular to the direction of the c axis, respectively. The observed line width is 240 Oe. For calibration the electron spin resonance field strength for DPPH, i. e. diphenyl-picryl-hydrazine, is also given [*64ICK*].

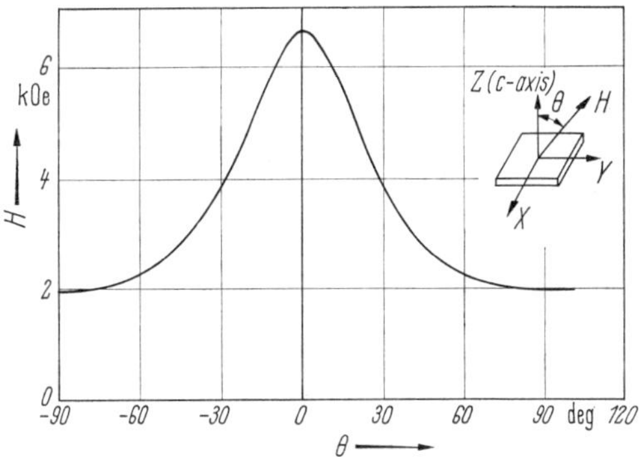

Fig. 205. Field for resonance as a function of polar angle, θ, for a single crystal platelet of $CaFe_4O_7$ stabilized with 3.0 mole-% Gd_2O_3 and measured at a frequency of 9.301 Gc/s. The basal plane of the hexagonal crystal is a preferential plane for the magnetization. $H_\theta^A = 2090$ Oe, $\gamma = 2.85$, $4\pi M_s = 1290$ Gauss. $H_\phi^A < 5$ Oe [*64ICK*].

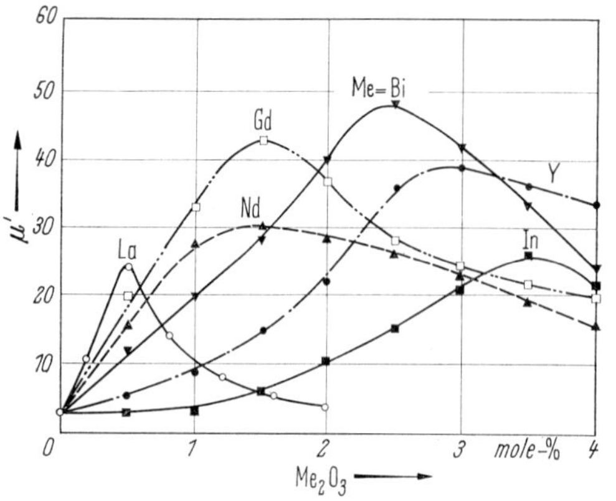

Fig. 206. The real part of the initial permeability at a frequency of 1 Mc/s for polycrystalline samples of $CaFe_4O_7$ with various amounts of additives Me_2O_3 [66I].

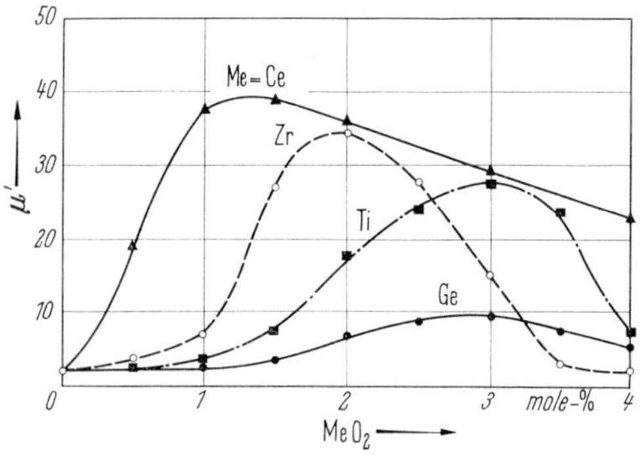

Fig. 207. The real part of the initial permeability at a frequency of 1 Mc/s for polycrystalline samples of $CaFe_4O_7$ with various amounts of additives MeO_2 [66I].

Tab. 40. Initial permeability and magnetic loss factor as a function of frequency for two samples of $CaFe_4O_7$ stabilized by 1.5 mole % of Gd_2O_3 and 2.5 mole % of ZrO_2, respectively [66I]

Sample	f[Mc/s]	0.5	1	50	100	150	230
1.5 Gd_2O_3	μ'	47.6	44.2	33.4	28.7	20.9	14.3
	$\tan \delta_m$	0.02	0.03	0.10	0.18	0.27	0.36
2.5 ZrO_2	μ'	31.9	29.0	22.2	16.3	13.8	10.4
	$\tan \delta_m$	0.02	0.04	0.11	0.23	0.38	0.42

7.12 References for 7

28J Johansson, K.: Z. Krist. **68** (1928) 87.

38A Adelsköld, V.: Arkiv Kemi. Mineral. Geol. **12A** (1938) 1.

48K Kittel, Ch.: Phys. Rev. **73** (1948) 155.

48N Néel, L.: Ann. Phys. (Paris) **3** (1948) 137.

50GD Gurry, R. W., and L. S. Darken: J. Am. Chem. Soc. **72** (1950) 3906.

51B Berry, L. G.: Am. Mineralogist **36** (1951) 512.

52B Braun, P. B.: Nature **170** (1952) 708.

52P Pauthenet, R.: Ann. Phys. (Paris) **7** (1952) 710.

52RSS Rathenau, G. W., J. Smit, and A. L. Stuijts: Z. Physik **133** (1952) 250.

52W Wijn, H. P. J.: Nature **170** (1952) 707.

52WRGO Went, J. J., G. W. Rathenau, E. W. Gorter, and G. W. van Oosterhout: Philips Tech. Rev. **13** (1951/52) 194.

53R Rathenau, G. W.: Rev. Mod. Phys. **25** (1953) 297.

53W Wijn, H. P. J.: Physica **19** (1953) 555.

54G Gorter, E. W.: Philips Res. Rept. **9** (1954) 295 and 403.

55BBM Bertaut, F., P. Blum, and G. Magnano: Compt. Rend. **241** (1955) 757.

55K1 Kojima, H.: Sci. Rept. Res. Inst. Tohoku Univ. **A7** (1955) 502.

55K2 Kojima, H.: Sci. Rept. Res. Inst. Tohoku Univ. **A7** (1955) 507.

55OKW Okamura, T., H. Kojima, and S. Watanabe: Sci. Rept. Res. Inst. Tohoku Univ. **A7** (1955) 411.

55S Suhl, H.: Phys. Rev. **97** (1955) 555.

55SB Smit, J., and H. Beljers: Philips Res. Rept. **10** (1955) 113.

55WA Weiss, M. T., and P. W. Anderson: Phys. Rev. **98** (1955) 925.

56GV Guillaud, C., and G. Villers: Compt. Rend. **242** (1956) 2867.

56HPR Hill, P. M., H. S. Paiser, and J. R. Rait: Acta Cryst. **9** (1956) 981.

56/57JWB Jonker, G. H., H. P. J. Wijn, and P. B. Braun: Philips Tech. Rev. **18** (1956/57) 145.

56K Kojima, H.: Sci. Rept. Res. Inst. Tohoku Univ. **A8** (1956) 540.

56SKT Sixtus, K. J., K. J. Kronenberg, and R. K. Tenzer: J. Appl. Phys. **27** (1956) 1051.

57A	Artman, J. O.: Phys. Rev. **105** (1957) 62.
57B	Braun, P. B.: Philips Res. Rept. **12** (1957) 491.
57BBS	Bertaut, F., P. Blum, and A. Sagnières: Compt. Rend. **244** (1957) 2944.
57BP	Berger, W., and F. Pawlek: Arch. Eisenhüttenw. **28** (1957) 101.
57DB	Deschamps, A., and F. Bertaut: Compt. Rend. **244** (1957) 3069.
57DK	Decker, B. F., and J. S. Kasper: Acta Cryst. **10** (1957) 332.
57/58K	Kooy, C.: Philips Tech. Rev. **19** (1957/58) 286.
57SB	Summergrad, R. N., and E. Banks: Phys. Chem. Solids **2** (1957) 312.
57/58SW	Stuijts, A. L., and H. P. J. Wijn: Philips Tech. Rev. **19** (1957/58) 209.
57vU	van Uitert, L. G.: J. Appl. Phys. **28** (1957) 317.
57vURS	van Uitert, L. G., M. H. Read, and F. J. Schnettler: J. Appl. Phys. **28** (1957) 280.
57vUS	van Uitert, L. G., and F. W. Swanekamp: J. Appl. Phys. **28** (1957) 482.
58BDP	Bertaut, F., A. Deschamps, and R. Pauthenet: Compt. Rend. **246** (1958) 2594.
58K	Kojima, H.: Sci. Rept. Res. Inst. Tohoku Univ. **A10** (1958) 175.
58MB	Mones, A. H., and E. Banks: Phys. Chem. Solids **4** (1958) 217.
58dPdBB	du Pré, F. K., D. H. de Bitetto, and F. G. Brockman: J. Appl. Phys. **29** (1958) 1127.
58PM	Philips, B., and A. Muan: J. Am. Ceram. Soc. **41** (1958) 445.
58WSR	Williams, H. J., R. C. Sherwood, and J. P. Remeika: J. Appl. Phys. **29** (1958) 1772.
59BBS	Bertaut, F., P. Blum, and A. Sagnières: Acta. Cryst. **12** (1959) 149.
59BCGI	Bates, L. F., D. J. Craik, P. M. Griffiths, and E. D. Isaac: Proc. Roy. Soc. London **253** (1959) 1.
59BDPP	Bertaut, F., A. Deschamps, R. Pauthenet, and S. Pickart: J. Phys. Radium, **20** (1959) 404.
59BK1	Belson, H. S., and C. J. Kriessman: J. Appl. Phys. **30** (1959) 175S.
59BK2	Bozorth, R. M., and V. Kramer: J. Phys. Radium **20** (1959) 393.
59CSEFW-GDFJ	Casimir, H. B. G., J. Smit, U. Enz, J. F. Fast, H. P. J. Wijn, E. W. Gorter, A. J. W. Duyvesteyn, J. D. Fast, and J. J. de Jong: J. Phys. Radium **20** (1959) 360.
59dBdPB	de Bitetto, D. J., F. K. du Pré, and F. G. Brockman: Proc. Symp. on Millimeter Waves, Polytechn. Inst. Brooklyn, M. R. I. Symposia Series **9** (1959) 95.
59PR1	Pauthenet, R., and G. Rimet: Compt. Rend. **249** (1959) 656.
59PR2	Pauthenet, R., and G. Rimet: Compt. Rend. **249** (1959) 1875.
59SJ	Schlömann, E., and R. V. Jones: J. Appl. Phys. Suppl. **30** (1959) 177.
59SW	Smit, J., and H. P. J. Wijn: Ferrites, New York and Eindhoven 1959.
59V1	Villers, G.: Compt. Rend. **249** (1959) 1337.
59V2	Villers, G.: Compt. Rend. **248** (1959) 2973.
60B1	Bickford Jr., L. R.: Phys. Rev. **119** (1960) 1000.
60B2	Bickford Jr., L. R.: J. Appl. Phys. **31** (1960)259S.
60BCdBdP	Bady, I., T. Collins, D. J. de Bitetto, and F. K. du Pré: Proc. IRE **48** (1960) 2033.
60BK	Braun, P. B., and W. Kwestroo: Philips Res. Rept. **15** (1960) 394.
60FSS	Frei, E. H., M. Schieber, and S. Shtrikman: Phys. Rev. **118** (1960) 657.
60GT	Goto, Ya., and T. Takada: J. Am. Ceram. Soc. **43** (1960) 150.
60H	Heimke, G.: J. Appl. Phys. **31** (1960) 271S.
60KE	Kooy, E., and U. Enz: Philips Res. Rept. **15** (1960) 7.
60KG	Kaczér, J., and R. Gemperic: Czech. J. Phys. **B10** (1960) 505.
60MK	Morrison, C. A., and N. Karayianis: Phys. Chem. Solids **4** (1958) 217.
60MS	Moruzzi, V. L., and M. W. Shafer: J. Am. Ceram. Soc. **43** (1960) 367.
60NPRG	Néel, L., R. Pauthenet, G. Rimet, and V. S. Giron: J. Appl. Phys. **31** (1960) 275S.
60P	Paulus, M.: Compt. Rend. **250** (1960) 2332.
60SLE	Smit, J., F. K. Lotgering, and U. Enz: J. Appl. Phys. Suppl. **31** (1960) 137S.
60ST	Sugimoto, M., and T. Takei: J. Sci. Res. Inst., Tokyo **54** (1960) 212.
61AS	Aharoni, A., and M. Schieber: Phys. Rev. **123** (1961) 807.
61B	Bady, I.: I.R.E. Trans. Microwave Theory Tech. **9** (1961) 59.
61BR	Banks, E., and M. Robbins: Nature **191** (1961) 1386.
61E	Enz, U.: J. Appl. Phys. Suppl. **32** (1961) 22S.
61G	Gersdorf, R.: Proefschrift Gem. Univers. Amsterdam 1961.
61HGB	Harvey, R. L., I. Gordon, and R. A. Braden: RCA Rev. **22** (1961) 648.
61KG	Kaczér, J., and R. Gemperle: Czech. J. Phys. **B11** (1961) 510.
61LES	Lotgering, F. K., U. Enz, and J. Smit: Philips Res. Rept. **16** (1961) 441.
61OMK	Okazaki, C., S. Mori, and F. Kanamaru: J. Phys. Soc. Japan **16** (1961) 119.
61SA	Smolenskii, G. A., and A. A. Andreev: Izv. Akad. Nauk SSSR, Ser. Fiz. **25** (1961) 1392; Bull. Acad. Sci. USSR, Phys. Ser. (English Transl.) **25** (1961) 1405.
61WIT	Wang, F. F. Y., K. Ishii, and J. B. Y. Tsui: J. Appl. Phys. **32** (1961) 1621.
62B1	Bickford Jr., L. R.: J. Phys. Soc. Japan Supp. B-1 **17** (1962) 272.

62B2	BUFFLER, C. R.: J. Appl. Phys. Suppl. **33** (1962) 1360.
62BC	BATES, L. F., and D. J. CRAIK: J. Phys. Soc. Japan, Suppl. B-1 **17** (1962) 535.
62BRT	BANKS, E., M. ROBBINS, and A. TAUBER: J. Phys. Soc. Japan, Suppl. B-1 **17** (1962) 196.
62BS	BADY, I., and E. SCHLÖMANN: J. Appl. Phys. **33** (1962) 1377.
62BU	BOROVIK, YE. S., and N. E. USIKOWA: Fiz. Metal. i Metalloved. **13** (1962) 470; Phys. Metals Metallog. (USSR) (English Transl.) **13** (1962) 150.
62CMP	COURTEL, R., H. MAKRAM, and G. PIGEAT: Compt. Rend. **254** (1962) 4447.
62CT	CHESSIAN, H., and E. T. TURKDOGAN: J. Am. Ceram. Soc. **45** (1962) 579.
62D	DIXON Jr., S.: J. Appl. Phys. Suppl. **33** (1962) 1368.
62DKR	DOUTHETT, D. D., I. KAUFMAN, and A. S. RISLEBY: J. Appl. Phys. **33** (1962) 1395.
62H	HEMPEL, K. A.: Z. Angew. Phys. **14** (1962) 488.
62KG	KOJIMA, H., and K. GOTO: J. Phys. Soc. Japan Suppl. B-1 **17** (1962) 201.
62LB	LAX, B., and K. J. BUTTON: Microwave ferrites and ferrimagnetics, McGraw-Hill N. Y. 1962.
62M1	MITA, M.: J. Phys. Soc. Japan **18** (1962) 155.
62M2	MORGENTHALER, F. R.: Proc. IRE **50** (1962) 2139.
62MR	MOUNTVALA, A. J., and S. F. RAVITZ: J. Am. Ceram. Soc. **45** (1962) 285.
62RPW	RODRIGUE, G. P., J. E. PIPPIN, and M. E. WALLACE: J. Appl. Phys. Suppl. **33** (1962) 1366.
62TSGW	TAUBER, A., R. O. SAVAGE Jr., R. J. GAMBINO, and C. G. WHINFREY: J. Appl. Phys. **33** (1962) 1381.
63AZW	AULD, B. A., R. E. ZOKHEIM, and D. K. WINSLOW: J. Appl. Phys. **34** (1963) 2281.
63BM	BOROVIK, YE. S., and YU. A. MAMALUY: Fiz. Metal. Metalloved. **15** (1963) 300; Phys. Metals Metallog. (USSR) (English Transl.) **15** (1963) 120.
63BY	BOROVIK, YE. S., and N. G. YAKOVLEVA: Fiz. Metal. Metalloved. **15** (1963) 511; Phys. Metals Metallog. (USSR) (English Transl.) **15** (1963) 28.
63D	DIXON Jr., S.: J. Appl. Phys. **34** (1963) 3441.
63dBdPB	DE BITETTO, D. J., F. K. DU PRÉ, and F. G. BROCKMAN: Final Report for USARDL Contract No. DA 36-039 SC-85 279.
63GH	GREEN, J. J., and B. J. HEALY: J. Appl. Phys. **34** (1963) 1285.
63H	HEMPEL, K. A.: Z. Angew. Phys. **16** (1963) 395.
63IK1	ICHINOSE, N., and K. KURIHARA: J. Phys. Soc. Japan **18** (1963) 170.
63IK2	ICHINOSE, N., and K. KURIHARA: Japan. J. Appl. Phys. **2** (1963) 735.
63L	LAROIA, K. K.: Indian J. Pure Appl. Phys. **1** (1963) 396.
63LF	LOEF, J. J. VAN, and P. J. M. FRANSSEN: Phys. Letters **7** (1963) 225.
63LS	LAROIA, K. K., and A. P. B. SINHA: Indian J. Pure Appl. Phys. **1** (1963) 215.
63RB	ROBBINS, M., and E. BANKS: J. Appl. Phys. **34** (1963) 1260.
63RLB	ROBBINS, M., S. LERNER, and E. BANKS: Phys. Chem. Solids **24** (1963) 759.
63RT	ROSENBERG, M., and C. TANASOIU: Rev. Phys. Akad. Rep. Populaire Roumaine **8** (1963) 383.
63RTN	ROSENBERG, M., C. TANASOIU, and I. NICOLAE: Studii Cercetari Fiz., **3** (1963) 249.
63SJB	SCHLÖMANN, E., R. JOSEPH, and I. BADY: J. Appl. Phys. **34** (1963) 672.
63SK	SHUR, I. S., and G. S. KANDAUROVA: Fiz. Metal. Metalloved. **16** (1963) 158.
63TKS	TAUBER, A., J. A. KOHN, and R. A. SAVAGE: J. Appl. Phys. Suppl. **34** (1963) 1265.
63WS	WHITE, R. M., and M. SPARKS: Phys. Rev. **130** (1963) 632.
63Z	ZÁVĚTA, K.: Phys. Status Solidi **3** (1963) 2111.
64BM	BOROVIK, YE. S., and YU. A. MAMALUY: Fiz. Metal. Metalloved. **18** (1964) 703; Phys. Metals Metallog. (USSR) (English Transl.) **18** (1964) 66.
64dB	DE BITETTO, D. J.: J. Appl. Phys. **35** (1964) 3482.
64FZ	FONTON, S. S., and A. V. ZALESSKII: Zh. Eksperim. i Teor. Fiz. (USSR) **47** (1964) 1693; Soviet Phys. JETP (English Transl.) **20** (1965) 1138.
64G	GREEN, M. A.: Thesis University of Arizona, p. 32, referred to in [*64TDS*].
64HG	HEMPEL, K. A., and P. GROSSER: Z. Angew. Phys. **17** (1964) 153.
64HH	HODGES, L. R., and G. R. HARRISON: J. Am. Ceram. Soc. **47** (1964) 601.
64ICK	ICHINOSE, N., S. CHIBA, and K. KURIHARA: Japan. J. Appl. Phys. **3** (1964) 235.
64KE	KOHN, J. A., and D. W. ECKART: Z. Krist. **119** (1964) 454.
64LB	LOEF, J. J. VAN, and A. BROESE VAN GROENOU: Proc. Intern. Conf. Magnetism Nottingham (1964) 646.
64NSVW	NECKENBÜRGER, E., H. SEVERIN, J. K. VOGEL, and G. WINKLER: Z. Angew. Phys. **18** (1964) 65.
64PZ	PEREKALINA, T. M., and A. V. ZALESSKII: Zh. Eksperim. i Teor. Fiz. (USSR) **46** (1964) 1985; Soviet Phys. JETP (English Transl.) **19** (1964) 1337.
64RL	ROTH, W. L., and F. E. LUBORSKY: J. Appl. Phys. **35** (1964) 966.
64RTR	ROSENBERG, M., C. TANASOIU, and C. RUSU: Phys. Status Solidi **6** (1964) 141.

64SDT	Savage Jr., R. O., S. Dixon Jr., and A. Tauber: J. Appl. Phys. **36** (1964) 873.
64ST	Savage, R. O., and A. Tauber: J. Am. Ceram. Soc. **47** (1964) 13.
64T	Taft, D. R.: J. Appl. Phys. **35** (1964) 776.
64TDS	Tauber, A., S. Dixon Jr., and R. O. Savage Jr.: J. Appl. Phys. **35** (1964) 1008.
64TWA	Tokheim, R. E., D. K. Winslow, and B. A. Auld: J. Appl. Phys. **35** (1964) 1010.
64vH	van Hook, H. J.: J. Am. Ceram. Soc. **47** (1964) 579.
64WT	Winkler, G., and W. Tolksdorf: „Unsere Forschung in Deutschland", Philips Hamburg, Germany 1964, p. 109.
64ZHKKW	Zinn, W., S. Hüfner, M. Kalvius, P. Kienle, and W. Wiedemann: Z. Angew. Phys. **17** (1964) 147.
65ACGL	Asti, G., M. Colombo, M. Giudici, and A. Levialdi: J. Appl. Phys. **36** (1965) 3581.
65BKLJP	Belov, K. P., L. I. Koroleva, R. Z. Levitin, Yu. V. Jergin, and A. V. Pedko: Phys. Status Solidi **12** (1965) 219.
65DTS	Dixon, Jr., S., A. Tauber, and R. O. Savage Jr.: J. Appl. Phys. **36** (1965) 1018.
65F	Fuchikami, N.: J. Phys. Soc. Japan **20** (1965) 760.
65GMCB	Gallagher, P. K., J. B. MacChesney, and D. N. E. Buchanan: J. Chem. Phys. **43** (1965) 516.
65H	Hirone, T.: J. Appl. Phys. **36** (1965) 988.
65KE1	Kohn, J. A., and D. W. Eckart: J. Appl. Phys. **36** (1965) 1171.
65KE2	Kohn, J. A., and D. W. Eckart: Am. Mineralogist **50** (1965) 1371.
65KG	Kojima, H., and K. Goto: J. Appl. Phys. **36** (1965) 538.
65M	Mita, M.: J. Phys. Soc. Japan **20** (1965) 1599.
65MCPSW	MacChesney, J. B., J. F. Potter, R. C. Sherwood, and H. J. Williams: J. Chem. Phys. **43** (1965) 3317.
65RT	Rosenberg, M., and C. Tanasoiu: Phys. Status Solidi **10** (1965) 613.
65SDT	Savage Jr., R. O., S. Dixon Jr., and A. Tauber: J. Appl. Phys. **36** (1965) 873.
65T	Tolksdorf, W.: Reactivity of Solids, Amsterdam: Elsevier 1965.
65V	Vinnik, M. A.: Zh. Neorgan. Khim. **10** (1965) 2137; Russ. J. Inorg. Chem. (English Transl.) **10** (1965) 1164.
65VAS	Vinnik, M. A., A. I. Agranovskaya, and N. N. Semenova: Izv. Akad. Nauk SSSR, Ser. Neorgan. Materialy **1** (1965) 1177.
65W	Winkler, G.: 5th Intern. Symposium Reactivity of Solids, Munich 1964, Elsevier, Amsterdam (1965) 572.
65ZP	Zalesskii, A. V., and T. M. Perekalina: Zh. Eksperim. i Teor. Fiz. (USSR) **48** (1965) 48; Soviet Phys. JETP (English Transl.) **21** (1965) 64.
66AST	Aucoin, T. R., R. O. Savage Jr., and A. Tauber: J. Appl. Phys. **37** (1966) 2908.
66BGH	Baden, R. A., I. Gordon, and R. L. Harvey: IEEE Trans. Magnetics **2** (1966) 43.
66BK	Belov, K. P., and L. I. Koroleva: Fiz. Tverd. Tela **8** (1966) 220; Soviet Phys.-Solid State (English Transl.) **8** (1966) 172.
66CHKB	Corliss, L. M., J. M. Hastings, W. Kunnmann, and E. Banks: Abstract paper 7th Int. Congress Int. Union Cryst. Moscow 1966.
66E	Enz, U.: Ned. Tijdsch. Natuurk. **32** (1966) 241.
66F	Friess, K.: Z. Angew. Phys. **21** (1966) 91.
66GGGWE	Geller, S., R. W. Grant, U. Gonser, H. Wiedersich, and G. P. Espinosa: Phys. Letters **20** (1966) 115.
66GGWG	Gonser, U., R. W. Grant, H. Wiedersich, and S. Geller: Appl. Phys. Letters **9** (1966) 18.
66I	Ichinose, N.: Japan. J. Appl. Phys. **5** (1966) 1140.
66IY	Ichinose, N., and T. Yoshioka: J. Phys. Soc. Japan **21** (1966) 1471.
66KK	Kasami, A., and S. Koide: J. Phys. Soc. Japan **21** (1966) 552.
66N	Neckenbürger, E.: IEEE Trans. Magnetics **2** (1966) 473.
66SF	Shtol'ts, E. V., and Ya. D. Fel'dman: Fiz. Tverd. Tela **8** (1966) 45; Soviet Phys.-Solid State (English Transl.) **8** (1966) 34.
66SY	Sizov, R. A., and I. I. Yamzin: Soviet Phys. "Doklady" (English Transl.) **11** (1966) 379.
66SZZ	Šimša, Z., A. V. Zalesskii, and K. Zvátěa: Phys. Status Solidi **14** (1966) 485.
66T	Tolksdorf, W.: IEEE Trans. Magnetics **2** (1966) 472.
66VES	Vinnik, M. A., A. P. Erastova, and Yu. G. Saksonov: Fiz. Tverd. Tela **8** (1966) 269; Soviet Phys.-Solid State (English Transl.) **8** (1966) 219.
66W	Winkler, G.: Z. Angew. Phys. **21** (1966) 282.
66vWR	van Wieringen, J. S., and J. G. Rensen: Z. Angew. Phys. **21** (1966) 69.
66YSZPZ	Yamzin, I. I., R. A. Sizov, I. S. Zheludov, T. M. Perekalina, and A. V. Zalesskii: Zh. Eksperim. i Teor. Fiz. (USSR) **50** (1966) 595; Soviet Phys. JETP (English Transl.) **23** (1966) 395.

67AAL	Albanese, G., G. Asti, and C. Lamborizio: Boston IUPAP Conference on Magnetism 1967.
67BR	Bungert, I., and M. Rosenberg: Phys. Status Solidi **20** (1967) K163.
67C	Cook Jr., C. F.: J. Appl. Phys. **38** (1967) 2488.
67CHK1	Corliss, L. M., J. M. Hastings, and W. Kunnmann: J. Appl. Phys. **38** (1967) 946.
67CHK2	Corliss, L. M., J. M. Hastings, and W. Kunnmann: Phys. Rev. **160** (1967) 408.
67GR	Greenwood, R. E., and G. P. Rodrigue: J. Appl. Phys. **38** (1967) 1401.
67GWGGE	Grant, R. W., H. Wiedersich, S. Geller, M. Gonser, and G. P. Espinosa: J. Appl. Phys. **38** (1967) 1455.
67KE	Kohn, J. A., and D. W. Eckart: Z. Krist. **124** (1967) 69.
67KEC	Kohn, J. A., D. W. Eckart, and C. F. Cook Jr.: Mat. Res. Bull. **2** (1967) 55.
67P	Petrova, I. I.: Fiz. Tverd. Tela **9** (1967) 944; Soviet Phys.-Solid State (English Transl.) **9** (1967) 738.
67SS	Severin, H., and J. P. Stoll: Z. Angew. Phys. **23** (1967) 209.
67ST	Savage Jr., R. O., and A. Tauber: Mat. Res. Bull **2** (1967) 469.
67SW	Schmitz, W., and H. P. J. Wijn: Z. Angew. Phys. **23** (1967) 204.
67SY	Sizov, R. A., and I. I. Yamzin: Soviet Phys. — Cryst. (English Transl.) **12** (1967) 111.
67T	Tomholt, F.: These data were kindly supplied by F. Tomholt, Philips Elcoma Division, Eindhoven, Netherlands.
67V1	Verweel, J.: Z. Angew. Phys. **23** (1967) 200.
67V2	Verweel, J.: J. Appl. Phys. **38** (1967) 1111.
67VAS	Vinnik, M. A., A. I. Agranovskaya, and N. N. Semenova: Zh. Neorgan. Khim. **10** (1967); Russ. J. Inorg. Chem. (English Transl.) **12** (1967) 18.
68HW	Haberey, F., and H. P. J. Wijn: Phys. Status Solidi **23** (1968) 231.
68KTAS	Kerecman, A. J., A. Tauber, T. R. Aucoin, and R. O. Savage Jr.: J. Appl. Phys. **39** (1968) 726.
68V	Verweel, J.: Philips Res. Rept. **23** (1968/69) not yet published.

6.2.2 Cr sulfide, selenide and telluride spinels

6.2.2.1 Comparison of the crystallographic, electrical, and magnetic properties of Cr spinels containing S, Se and Te

Tab. 1. Survey of binary sulfides, selenides and tellurides (ACr_2X_4) with spinel structure *[67RLW]*. See also *[68L]*, *[68MM]*, and references in the table

ACr_2X_4	Type of magnetic order	Θ_C °K	Θ_C Ref.	Type of conductivity
$A[Cr_2]S_4$; A = Fe^{2+}, Mn^{2+}, Co^{2+}	ferrimagnetic [1][2]			semiconducting [3]
$FeCr_2S_4$		≈180	*66K, 69GDP*	
$MnCr_2S_4$		67	*68DBL*	
$CoCr_2S_4$		220	*69GDP, 68L(2)*	
$Zn[Cr_2]X_4$; X = S, Se	antiferromagnetic [4][5]			semiconducting [3][5]
$A[Cr_2]X_4$; X = S, Se; A = Cd^{2+}, Hg^{2+}				semiconducting [6]
$CdCr_2S_4$	ferromagnetic	84.5	*66BWRL*	
$CdCr_2Se_4$	ferromagnetic	129	*66BWRL*	
$HgCr_2S_4$	metamagnetic	36	*66BWRL*	
$HgCr_2Se_4$	ferromagnetic	106	*66BWRL*	
$CuCr_2X_4$; X = S, Se, Te	ferromagnetic [5]			metallic [5]
$CuCr_2S_4$		420	*64L(1)* [7]	
$CuCr_2Se_4$		460	*64L(1)* [7]	
		(433)	*67RLW*	
$CuCr_2Te_4$		365	*64L(1)*	

Tab. 2. Crystallographic parameters for Cr spinels *[66RBW]*

Compound	*u* (X-ray diffraction)	*u* (neutron diffraction)	Debye-Waller factor	a_0 Å
$MnCr_2S_4$	0.3863	0.3876 [8]	0.18	10.11_0
$FeCr_2S_4$	0.3850	0.3840 [9]	0.59	9.99_5
$CoCr_2S_4$	0.3821	0.3830 [10]	0.35	9.92_3
$CuCr_2S_4$	0.3841		0.55	9.81_4
$ZnCr_2S_4$	0.3842		0.82	9.98_6
$ZnCr_2O_4$	0.3899		0.36	8.32_1
$ZnCr_2Se_4$	0.3849	0.3843 [11]	1.46	10.44_0
$MnCr_2O_4$	0.3890	0.3892 [10]	≈0	8.43_7
$CdCr_2O_4$	0.3896		0.33	8.59_6
$CdCr_2Se_4$	0.3894		1.25	10.74_0

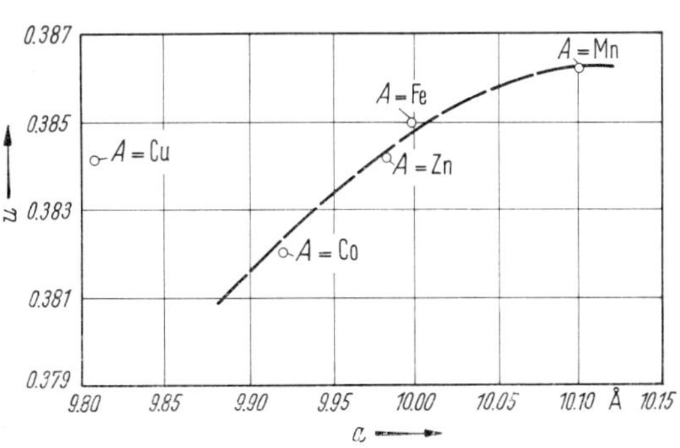

Fig. 1. ACr_2S_4 (A = Co, Cu, Fe, Mn, Zn). Crystallographic *u* parameter vs. cubic cell edge *a* for Cr spinels *[66RBW]*.

[1] *[56L(1)]*; [2] *[56L(2)]*; [3] *[65BRW]*; [4] *[56L(3)]*; [5] *[64L(1)]*; [6] *[66LR]*; [7] See Fig. 17 in 6.2.2.8 from *[68OTI]*. [8] *[65MD]*; [9] *[64SCP]*; [10] *[–DM]*; [11] *[65P]*.

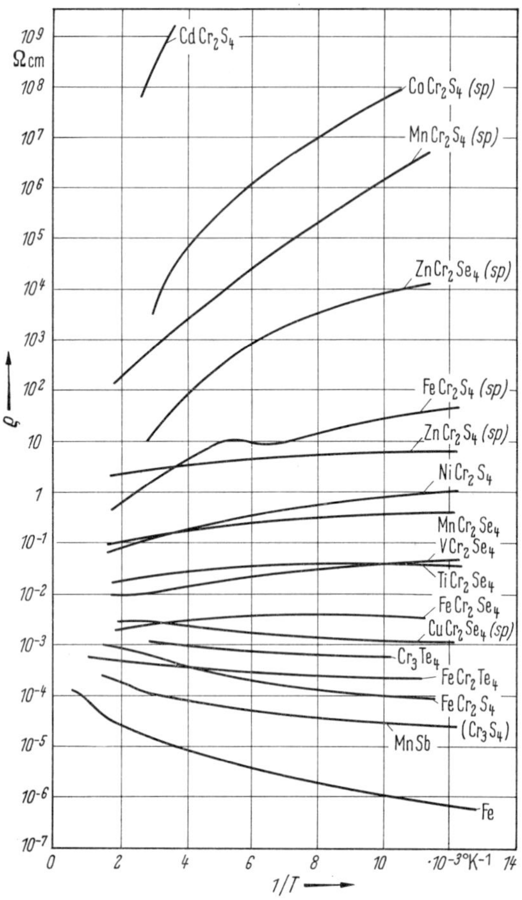

Fig. 2. ACr_2X_4 chalcogenides. Resistivity, ϱ, vs. reciprocal temperature. Compounds with normal spinel structure are marked by (sp). The other have the cubic Cr_3S_4 type structure which can be described as a defect-NiAs structure with ordered vacancies. Curves for Fe and MnSb are given for comparison [*68MM*]. [From W. Albers, G. van Aller and C. Haas, Coll. Intern. C.N.R.S. No. 157, Orsay, Oct. 1965 (Editions C.N.R.S. 1967)].

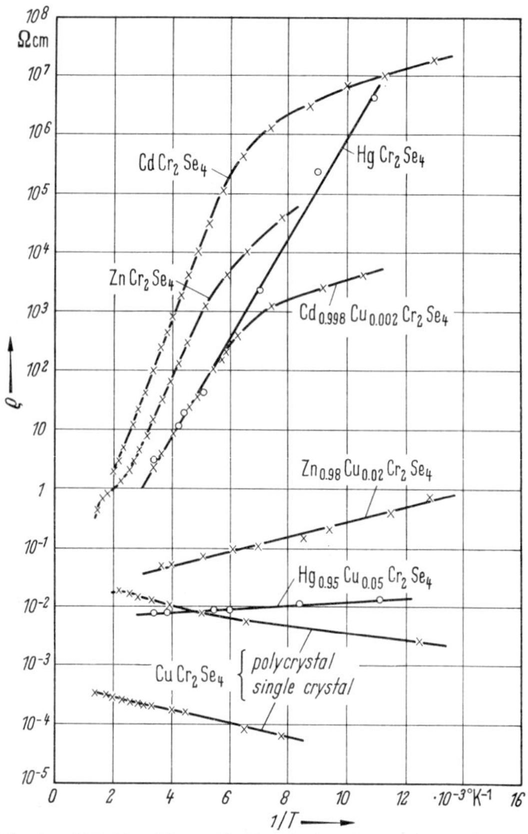

Fig. 3. XCr_2Se_4 (X = Cd, Hg, Zn). Comparison of resistivity vs. reciprocal temperature curves of pure compounds with those which have Cu substitutions [*68L*].

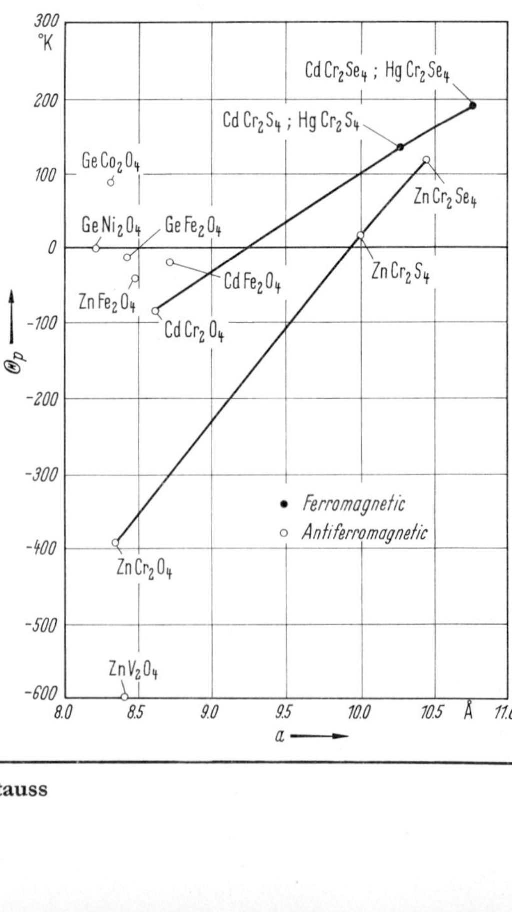

→

Fig. 4. Relationship between the paramagnetic Curie temperature and lattice constant for Cr-containing cubic spinels and related spinels [*66BWRL*]. For exact values, see the respective tables.

6.2.2.2 Comparative properties of $CdCr_2Se_4$, $CdCr_2S_4$, $HgCr_2Se_4$, and $HgCr_2S_4$

Tab. 1. Effect of pressure on the lattice constants and ferromagnetic Curie temperatures of Cr S and Se spinels [*69S(1)*]

Compound	a_0 Å	Θ_C °K	$\frac{d\Theta_C}{dp}$ °K/kbar	$\frac{d\Theta_C}{da_0}$ °K/Å
$CdCr_2Se_4$	10.755	128	−0.82	+66
$CdCr_2S_4$	10.244	84	−0.58	+46
$HgCr_2Se_4$	10.753	105.5	−0.95	+99
$HgCr_2S_4$	10.237	36.1	+0.14	−10

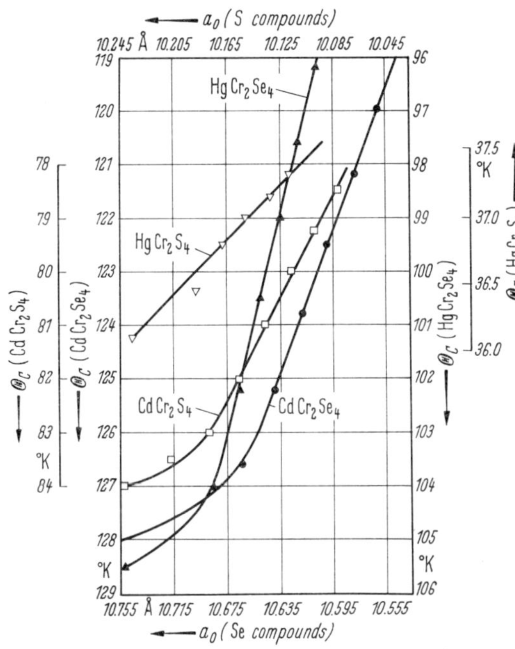

Fig. 1. Relationship between ferromagnetic Curie temperature Θ_C and lattice constant a_0 for Cr chalcogenide spinels. a_0 is varied by hydrostatic pressure [*69S(1)*].

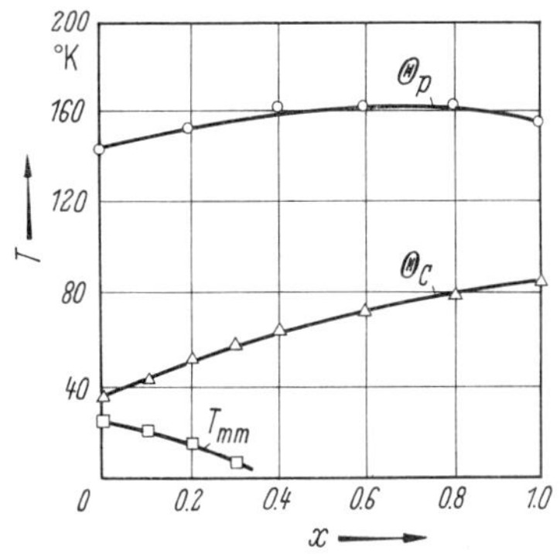

Fig. 2. $Hg_{1-x}Cd_xCr_2S_4$. Variation of the paramagnetic Curie temperature Θ_p, ferromagnetic Curie temperature Θ_C, and metamagnetic phase transformation temperature T_{mm} as functions of composition in the mixed system $CdCr_2S_4$-$HgCr_2S_4$. The metamagnetic transformation is to a non-colinear arrangement of magnetic moments at low temperatures which is characteristic of $HgCr_2S_4$ [*67BRW*].

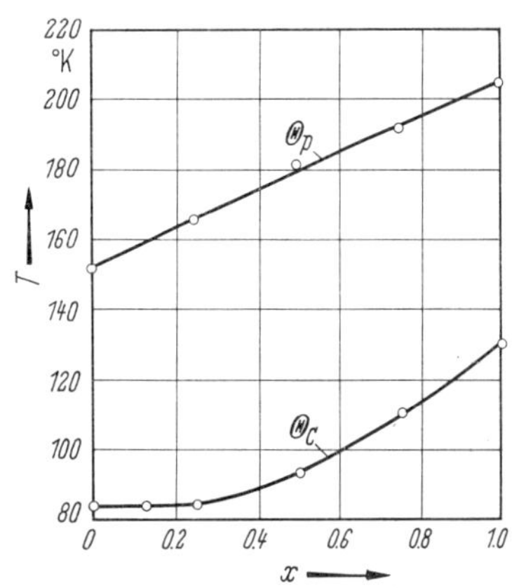

Fig. 3. $(1-x)CdCr_2S_4 \cdot xCdCr_2Se_4$. Dependence of the paramagnetic Curie temperature Θ_p and ferromagnetic Curie temperature Θ_C on composition in the mixed $CdCr_2Se_4$-$CdCr_2S_4$ system [*67WBR*].

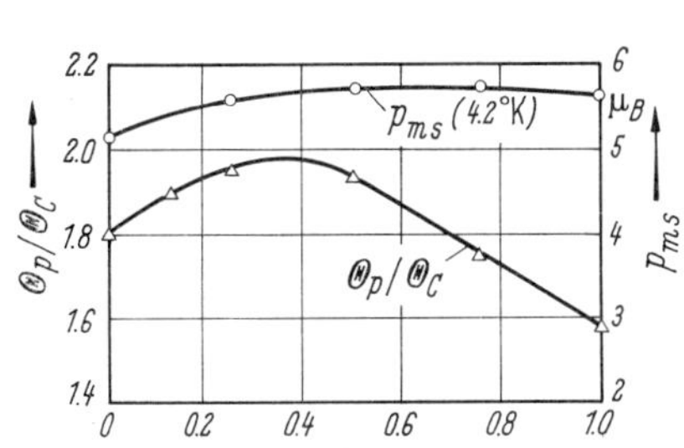

Fig. 4. $(1-x)CdCr_2S_4 \cdot xCdCr_2Se_4$. Variation of the saturation magnetic moment per molecule and of the ratio Θ_p/Θ_C (paramagnetic Curie to ferromagnetic Curie temperature) with composition of the mixed system $CdCr_2Se_4$-$CdCr_2S_4$ [*67WBR*]. For exact values, see the respective tables.

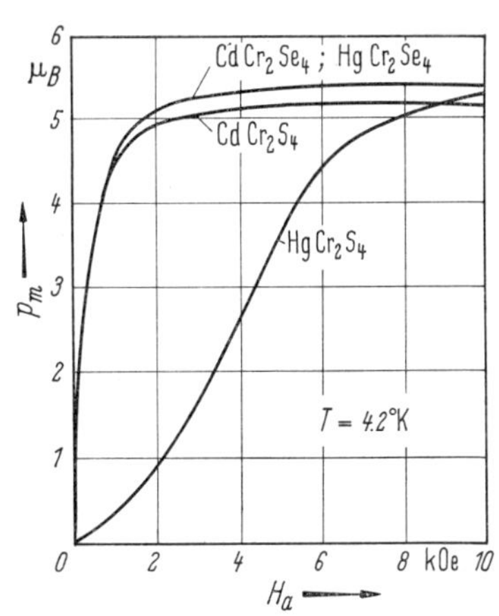

Fig. 5. $CdCr_2Se_4$, $CdCr_2S_4$, $HgCr_2Se_4$, and $HgCr_2S_4$. Magnetic moment per molecule p_m vs. applied field H_a for polycrystalline samples at 4.2 °K [*66BWRL*].

Tab. 2. Zero field nuclear magnetic resonance data (4.2 °K) for polycrystalline Cr S and Se spinels

Material	Cr			Se		Cd or Hg
	H_{eff} [kOe] isotropic	H_{eff} [kOe] anisotropic	quadrupole splitting [Mc/s]	H_{eff} [kOe] isotropic	H_{eff} [kOe] anisotropic	H_{eff} [kOe] isotropic
$CdCr_2Se_4$	−182.5 [1])	+2.30 [1])	0.90 [1])	−98.0 [1])	+9.2 [1])	+136.2 [1])
$CdCr_2S_4$	−191 [2])	+2.07 [2])	0.95 [2])			+167 [1])
$HgCr_2Se_4$	−197.4 [3])	+2.3 [2])	0.98 [2])	−91.7 [3])	+9.75 [2])	+446 [3])
$HgCr_2S_4$	−190 [3])	+2.3 [3])	0.95 [3])			+524 [3])

Tab. 3. $CdCr_2Se_4$; $CdCr_2S_4$. Effect of sample treatment on the magnetocrystalline anisotropy and FMR results at 4.2 °K [*68PB*] [4])

Sample treatment	K_1/M Oe	K_2/M Oe	FMR line width [5])	FMR angular variation [6])
$CdCr_2Se_4$, flux grown [7])	<25	<25	isotropic	(large demagnetizing field)
$CdCr_2Se_4$, annealed (reduced) in H_2	33	−48	isotropic	normal
$CdCr_2Se_4$, reannealed (oxidized) in Se	9	≈ 0	isotropic	normal
$CdCr_2S_4$, vapor transport grown [8])	75	160	anisotropic	anomalous
$CdCr_2S_4$, annealed (oxidized) in S	9	3	isotropic	normal
$CdCr_2S_4$, reannealed (reduced) in H_2	522	1771	anisotropic	anomalous
$CdCr_2S_4$, reannealed (oxidized) in S	⩽50	?	isotropic	(large demagnetizing field)

6.2.2.3 Cd-Cr selenide spinel

Tab. 1. General properties of $CdCr_2Se_4$ (room temperature unless noted)

Preparation is described in [*66BWRL*], [*66HP*], [*67HRBA*], [*67P*], [*68LS*], [*68PB*], [*69NS*]. At high temperatures and pressures $CdCr_2Se_4$ converts to a monoclinic defect-NiAs structure with weak magnetization and metallic eletrical properties. The high pressure phase can be retained at atmospheric pressure by quenching first to room temperature [*68BFLS*]. The cubic spinel phase is a semiconductor, *p*-type when pure or doped by replacement of Cd by Au or Ag, *n*-type when doped with Ga or In. It is a ferromagnet with very low anisotropy. The electrical and magnetic properties are very sensitive to stoichiometry [*68PB*]. Optical and electrical properties are strongly affected by magnetic ordering.

		Ref.
Lattice constant	$a_0 = (10.75 \pm 0.01)$ Å	*66BWRL* [9])
Se parameter	$u = 0.390$	*66BWRL*
Density	$\varrho = 5.675$ g/cm³ (high pressure phase 5.77)	*66LR* (*68BFLS*)
Compressibility	4.0 (300 °K), 3.6 (200 °K), 3.1 (80 °K) · 10^{-6} bar^{-1}	*69S(1)*
Paramagnetic Curie temperature	$\Theta_p = 204$ °K	*66BWRL*
Molar Curie constant	$C_m = 3.82$ cm³ °K mole^{-1}	*66BWRL*
Ferromagnetic Curie temperature	$\Theta_C = (129 \pm 1)$ °K	*66BWRL* [10])
Saturation moment per molecule	$p_{ms} = 6\ \mu_B$ (4.2 °K) (high pressure phase 0.035 μ_B)	*67LPS* (*68BFLS*)
Saturation magnetization	$4\pi M_s = 4480$ Gauss (4.2 °K)	*67LPS*
Resistivity	typical $\varrho = 10^2$ (300 °K), 10^5 (4.2 °K) Ω cm (high pressure phase $2 \cdots 5 \cdot 10^{-2}$)	[11]) (*68BFLS*)
Index of refraction	$n = 3.02$	*68BZ*

[1]) [*69S(2)*].
[2]) [*68BBB*].
[3]) [*69BBB*].
[4]) For other anisotropy and FMR data see also [*67BP*], [*67LPS*], [*68LS*], [*68VT*].
[5]) See Fig. 13 in 6.2.2.5.
[6]) See Fig. 12 in 6.2.2.5.
[7]) Average moment $p_{ms} = 5.94\ \mu_B$.
[8]) Average moment $p_{ms} = 5.55\ \mu_B$.
[9]) See also [*66RBW*], [*67P*], [*68BFLS*].
[10]) See also [*66MDAW*], [*68BSL*], [*69S(1)*], [*70M*].
[11]) See [*66LR*], [*67LPS*], [*68LS*].

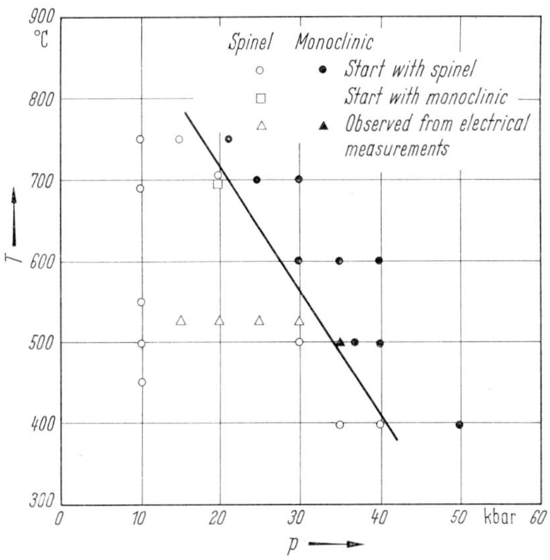

Fig. 1. $CdCr_2Se_4$. Pressure-temperature diagram for the phase change from cubic spinel to monoclinic defect-NiAs structure [*68BFLS*].

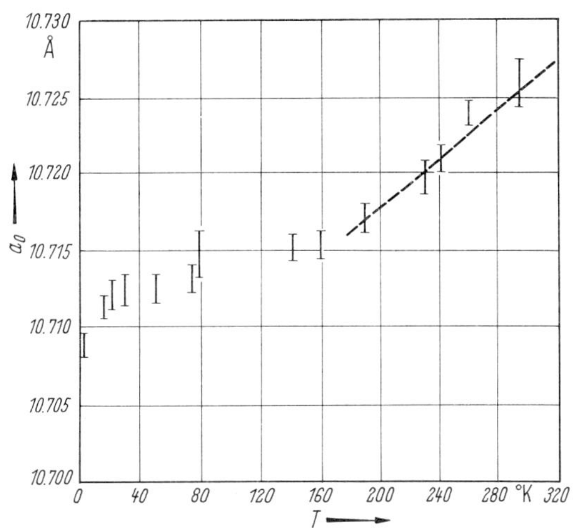

Fig. 2. $CdCr_2Se_4$. Variation of the lattice parameter a_0 with temperature for a single crystal [*69MKWWP*].

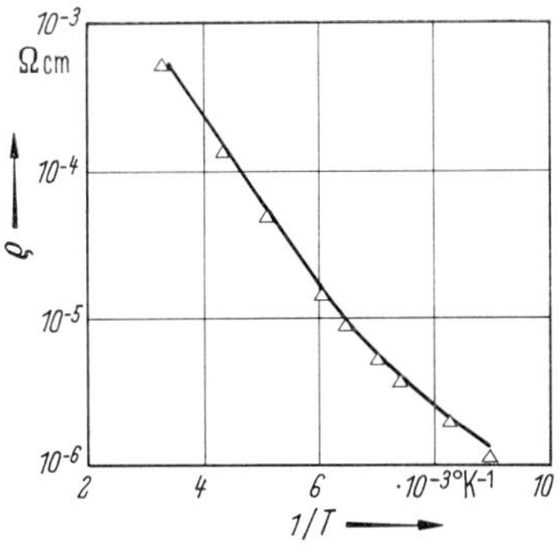

Fig. 3. $CdCr_2Se_4$. Electrical resistivity vs. reciprocal temperature for a pure *p*-type polycrystalline sample [*67L*]. See also [*66LR*], [*68LS*].

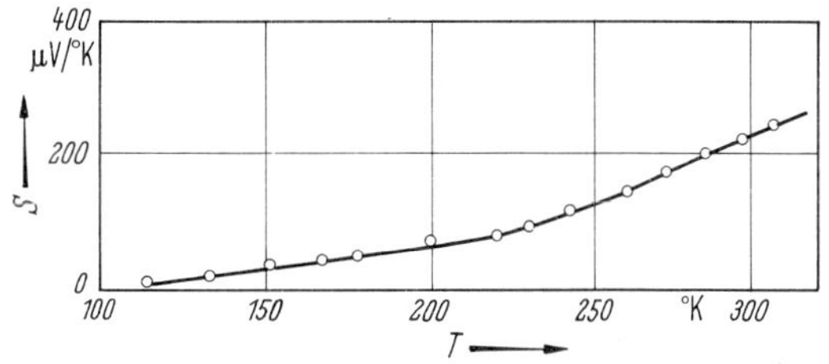

Fig. 4. $CdCr_2Se_4$. The Seebeck coefficient *S* vs. temperature for a pure polycrystalline sample [*67L*]. See also [*66LR*].

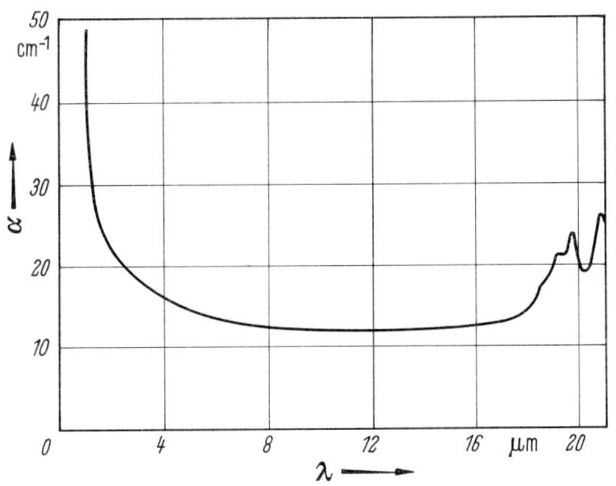

Fig. 5. $CdCr_2Se_4$. Infrared absorption spectrum of a single crystal in the [100] direction at 80 °K [*68BZ*].

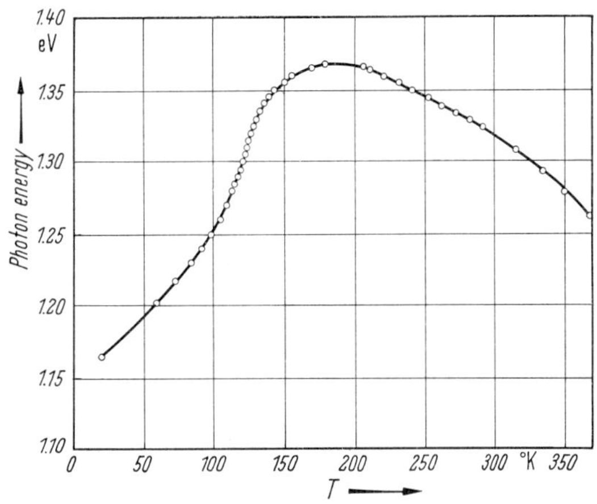

Fig. 6. $CdCr_2Se_4$. Optical energy gap ΔE vs. temperature in a single crystal. (Band edge position, taken arbitrarily at absorption coefficient $\alpha = 1500\ cm^{-1}$.) Application of a magnetic field produces an additional red shift [*66HP*]. See also [*68RBE*].

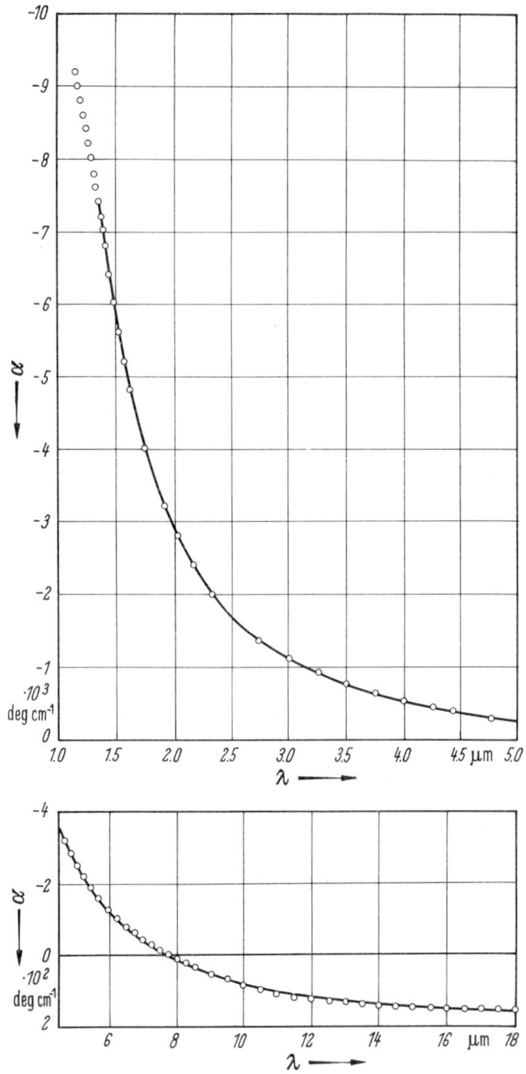

Fig. 7. $CdCr_2Se_4$. Faraday rotation vs. wavelength at 82 °K in a field of 15 kOe along the [100] direction for a single crystal. (Maximum figure of merit is 63 deg/cm dB at 1.3 μm). [*68BZ*].

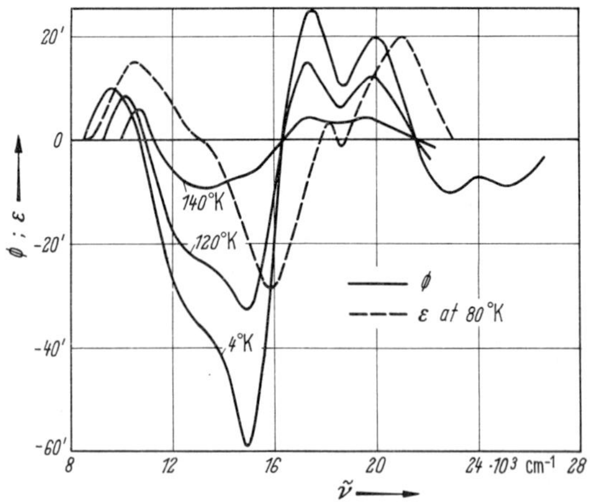

Fig. 8. $CdCr_2Se_4$. Magnetooptical Kerr effect vs. wave number (for normal reflection from a single crystal) for plane polarized light propagating parallel to the applied field: ϕ = rotation of polarization, ε = ellipticity [*69BHRZ*].

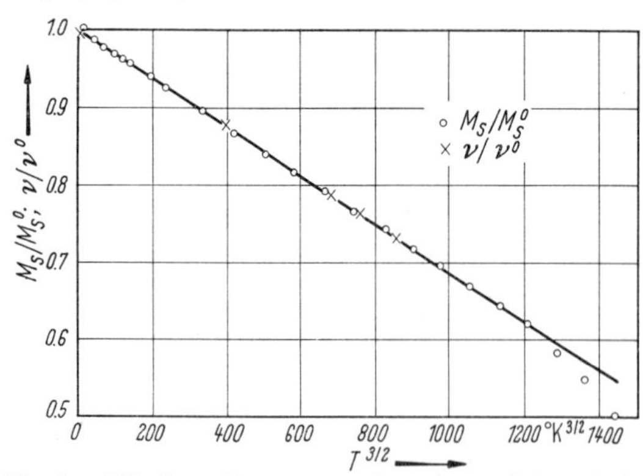

Fig. 9. $CdCr_2Se_4$. Temperature dependence of the relative saturation magnetization, M_S/M_S^0, and NMR frequency, ν/ν^0 of a polycrystalline sample, obtained by zero-field NMR (×) and magnetometer (in 10 kOe), plotted vs. $T^{3/2}$. Similar behavior is observed for $CdCr_2S_4$ [*68SRFDMW*].

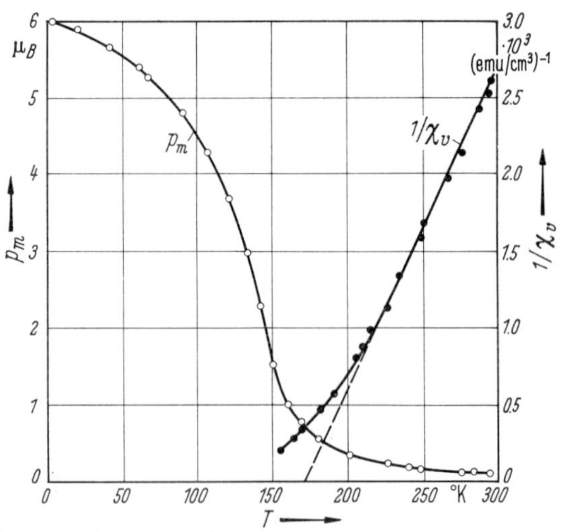

Fig. 10. $CdCr_2Se_4$. Temperature dependence of the magnetic moment per molecule and inverse susceptibility for a single crystal in 15.3 kOe field [*67LPS*].

Fig. 12. $CdCr_2Se_4$. Spin-wave (parallel pumped) line width ΔH_k vs. temperature at X-band [*68BSL*]. a) From 4.3 °K to within a few degrees of the Curie temperature (Θ_C = 130 °K). b) Low temperature behavior in greater detail.

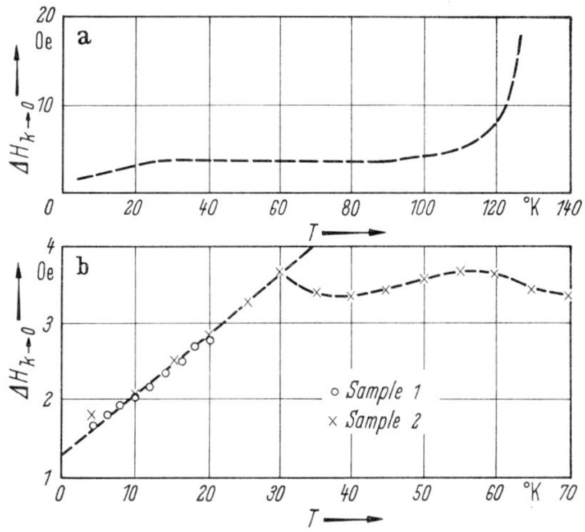

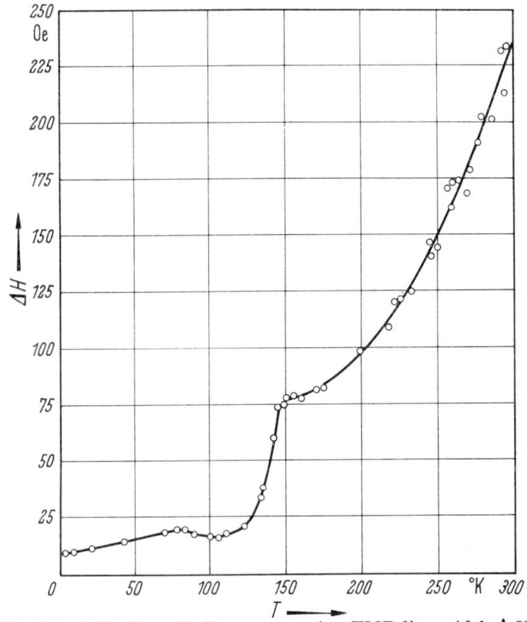

Fig. 11. $CdCr_2Se_4$. Uniform precession FMR line width ΔH vs. temperature at 13 Gc/s in the [100] direction [*67LPS*]. See also [*68LS*], [*68BSL*].

Fig. 13. $CdCr_2Se_4$. Divergence of spin-wave (parallel pumped) line width ΔH_k near the critical temperature T_0 [*68BSL*]. $(\Delta H_{k\to 0})^{-1} = (A/w)(T_0 - T)$; T_0 = 133 °K.

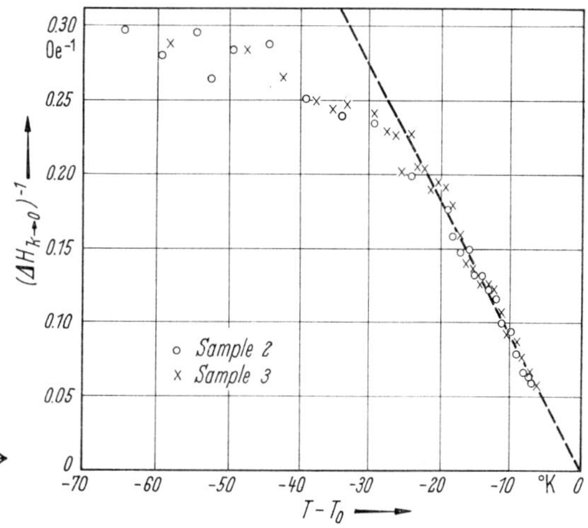

6.2.2.4 Cd-Cr selenide spinel with substitutions

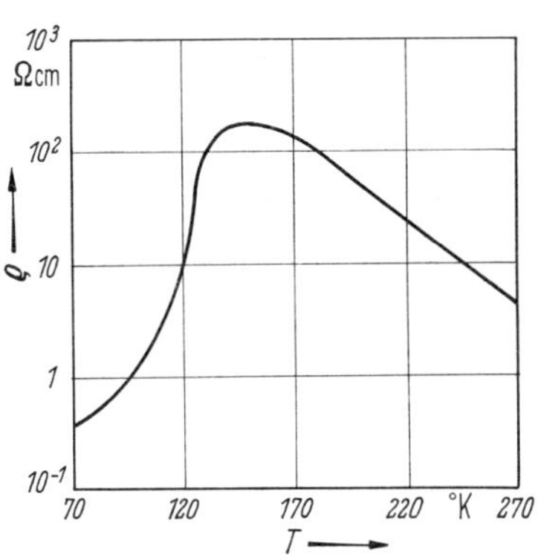

Fig. 1. *n*-type $CdCr_2Se_4$: 2% Ga. Resistivity ϱ vs. temperature for a polycrystalline sample [*67HRBA*].

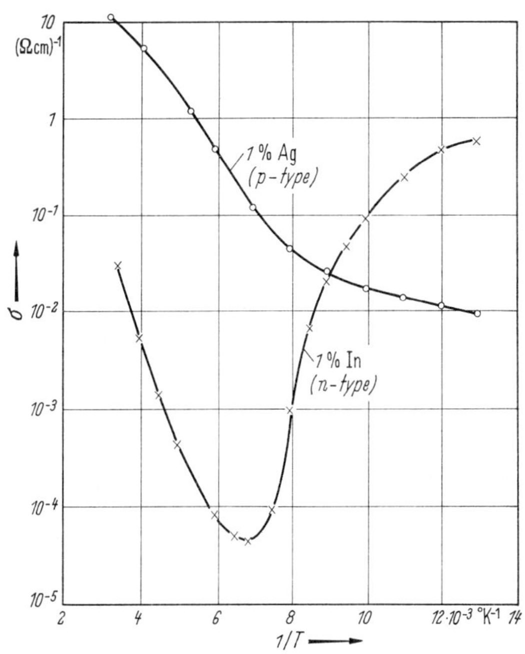

Fig. 2. $CdCr_2Se_4$ doped with 1% Ag (*p*-type) or 1% In (*n*-type). Conductivity σ vs. reciprocal temperature [*67L*]. See also [*68VT*].

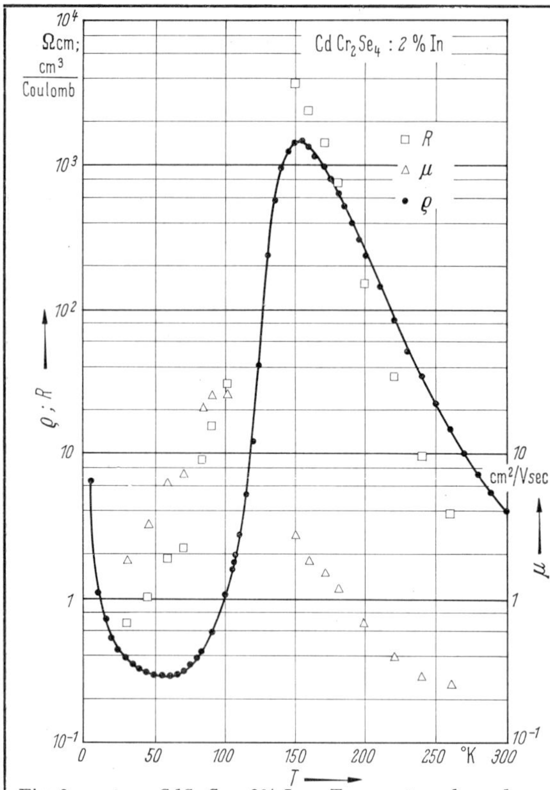

Fig. 3. *n*-type $CdCr_2Se_4$: 2% In. Temperature dependence of the resistivity ϱ, ordinary Hall coefficient R, and Hall mobility μ [*69AG*]. See also [*67L*], [*66LR*].

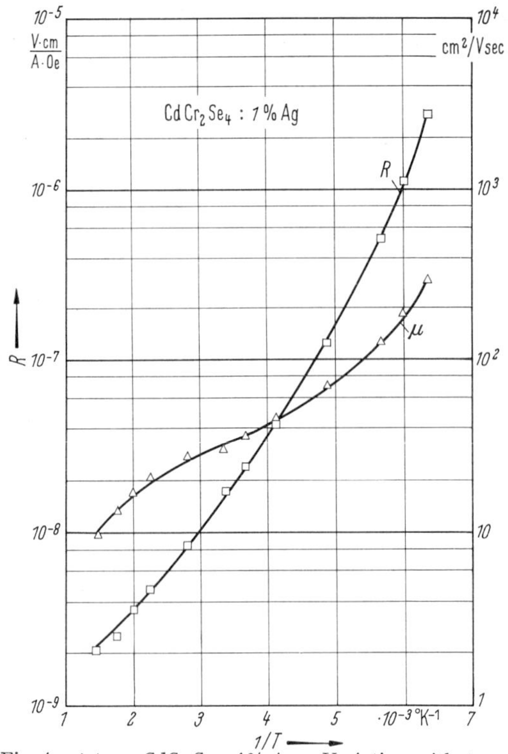

Fig. 4. *p*-type $CdCr_2Se_4$: 1% Ag. Variation with temperature of the ordinary Hall coefficient R and Hall mobility μ [*67L*]. See also [*68VT*].

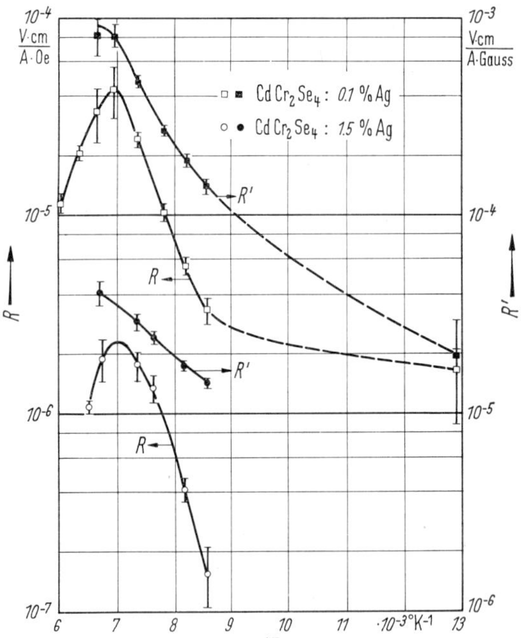

Fig. 5. *p*-type $CdCr_2Se_4$ doped with 0.1 or 1.5% Ag. Field independent ordinary (R) and anomalous ferromagnetic (R') Hall coefficients vs. reciprocal temperature [*67L*].

→

Fig. 6. *n*-type $CdCr_2Se_4$: 1% In. Field dependent Hall coefficient R' vs. field strength at various temperatures [*67L*].

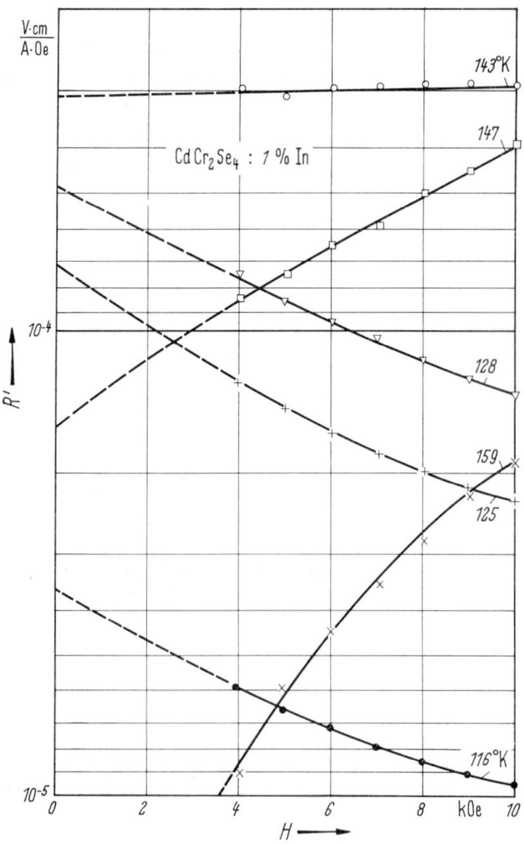

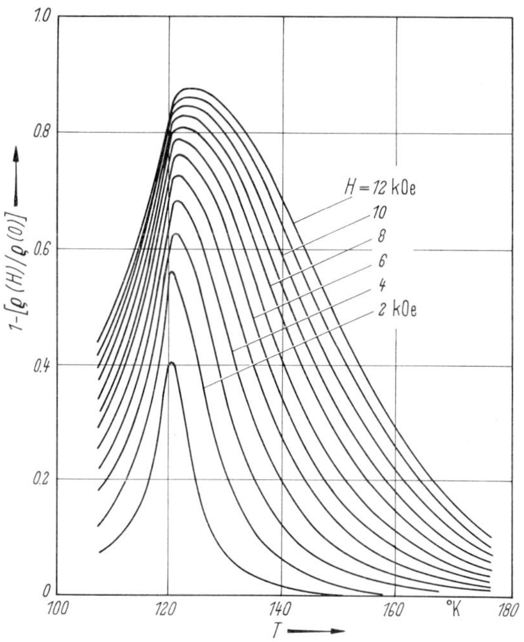

Fig. 7. *n*-type $CdCr_2Se_4$: 2% Ga. Relative magnetoresistance as a function of temperature in various magnetic fields [*67HRBA*].

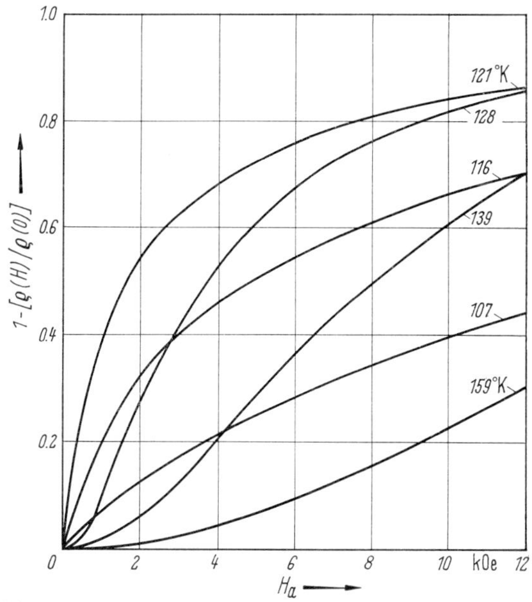

Fig. 8. *n*-type $CdCr_2Se_4$: 2% Ga. Relative magnetoresistance vs. field at various temperatures [*67HRBA*].

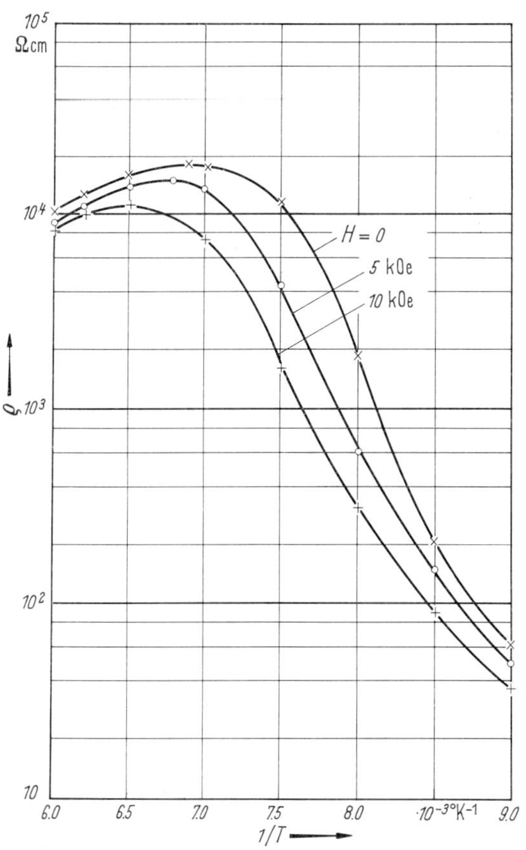

Fig. 9. *n*-type $CdCr_2Se_4$: 1% In. Transverse magnetoresistance vs. inverse temperature [*67L*].

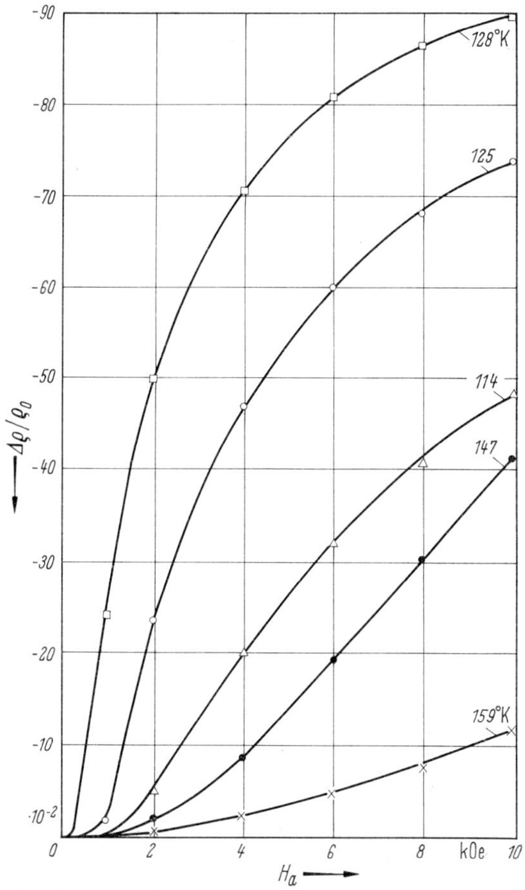

Fig. 10. *n*-type $CdCr_2Se_4$: 1% In. Relative magnetoresistance vs. applied field at various temperatures [*67L*].

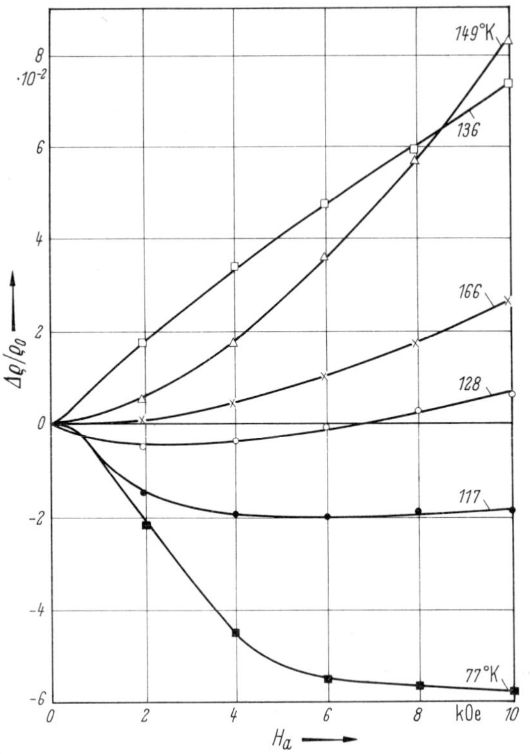

Fig. 11. p-type $CdCr_2Se_4$: 1% Ag. Relative magnetoresistance at different temperatures as a function of applied field [67L].

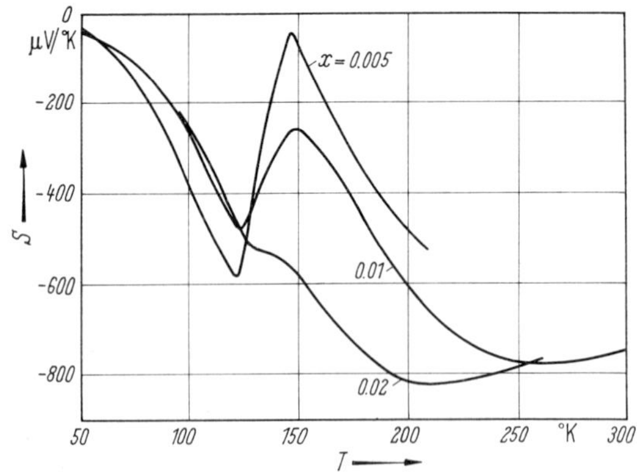

Fig. 13. $Cd_{1-x}In_xCr_2Se_4$. Seebeck coefficient S vs. temperature, in zero external field, for n-type $CdCr_2Se_4$ doped with 0.5, 1, or 2% In [69AG].

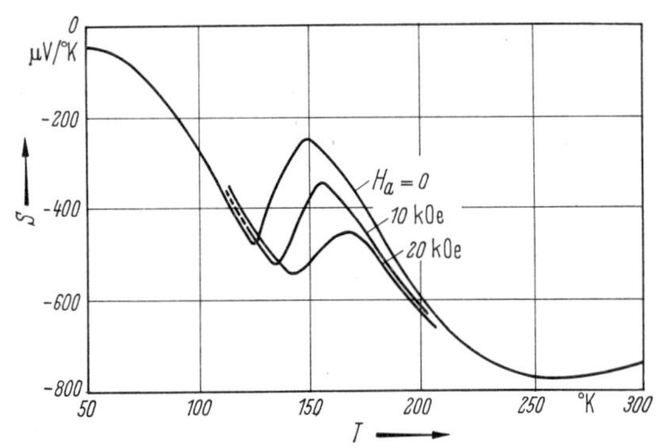

Fig. 14. $Cd_{1-x}In_xCr_2Se_4$ ($x = 0.01$). Effect of external magnetic fields on the temperature dependence of the Seebeck coefficient S of n-type $CdCr_2Se_4$: 1% In [69AG].

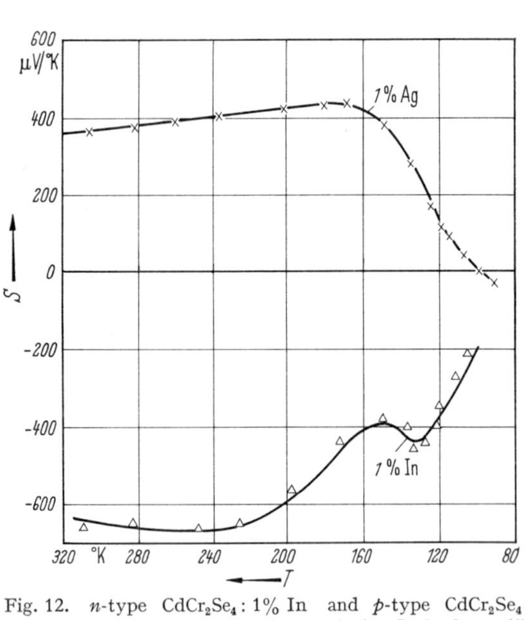

Fig. 12. n-type $CdCr_2Se_4$: 1% In and p-type $CdCr_2Se_4$: 1% Ag. Temperature dependence of the Seebeck coefficient S [67L].

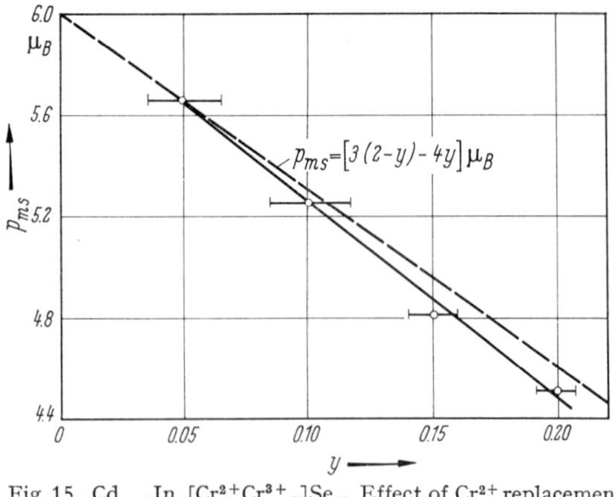

Fig. 15. $Cd_{1-y}In_y[Cr^{2+}_yCr^{3+}_{2-y}]Se_4$. Effect of Cr^{2+} replacement for Cr^{3+} on the saturation magnetic moment per molecule of $CdCr_2Se_4$ at 4.2 °K. Replacement occurs through substitution of In^{3+} for Cd^{2+}. Cr^{2+} sites are antiferromagnetically coupled to Cr^{3+} sites [68PB].

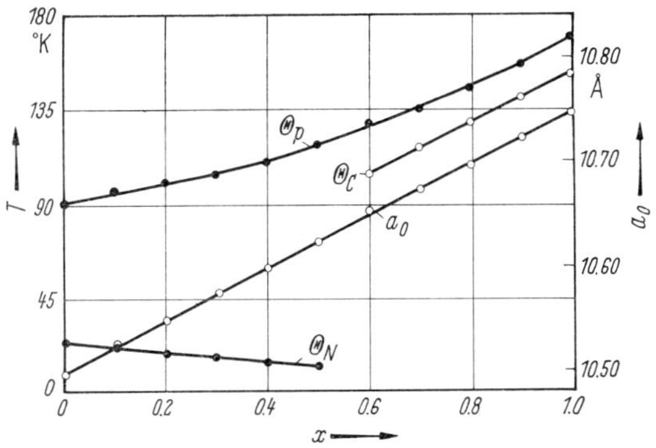

Fig. 16. $Zn_{1-x}Cd_xCr_2Se_4$. Variation of the paramagnetic Curie temperature Θ_p and of the critical temperature Θ_C or Θ_N on composition in the mixed system $CdCr_2Se_4$-$ZnCr_2Se_4$ [*69BMV*]. See also [*67BRW*].

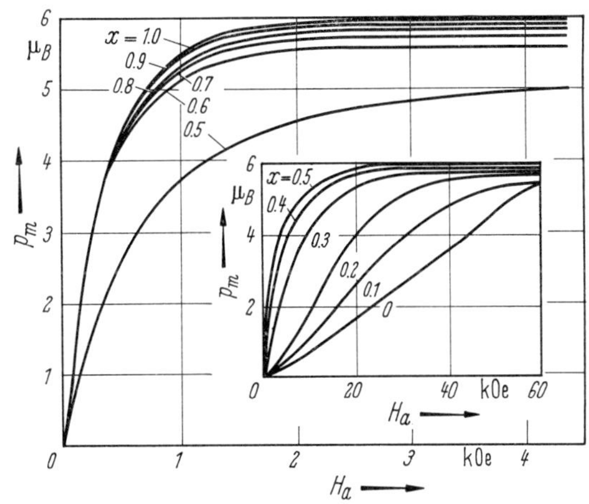

Fig. 17. $Zn_{1-x}Cd_xCr_2Se_4$. Variation of the magnetic moment per molecule with magnetic field at 4.2 °K in the mixed system $CdCr_2Se_4$-$ZnCr_2Se_4$ [*69BMV*].

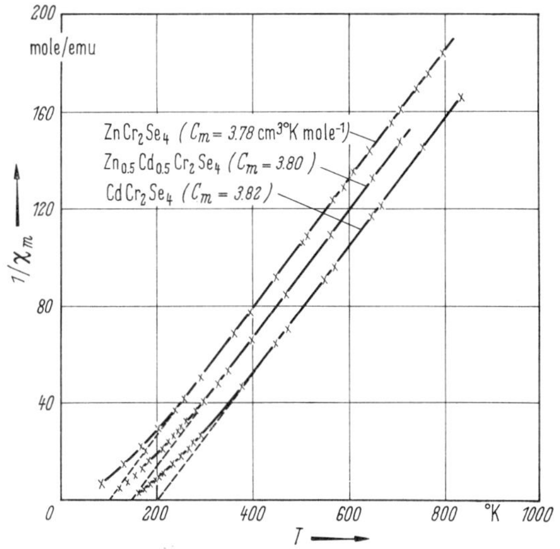

Fig. 18. $CdCr_2Se_4$-$ZnCr_2Se_4$. Dependence of the inverse magnetic susceptibility on temperature in the mixed system [*68L*].

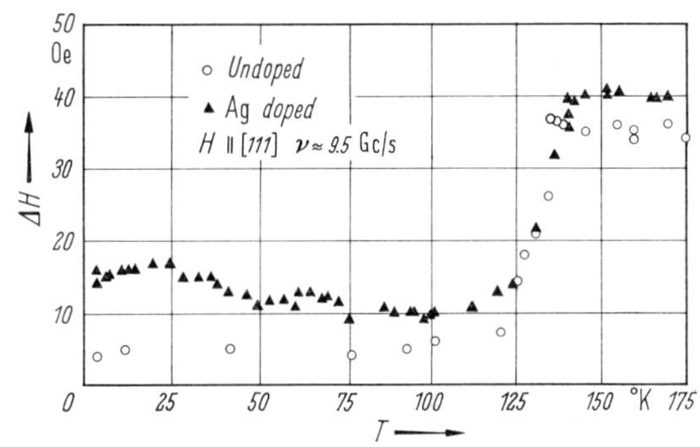

Fig. 19. $CdCr_2Se_4$: Ag. Effect of Ag doping on the uniform precession FMR line width ΔH [*68LS*].

6.2.2.5 Cd-Cr sulfide spinel and substituted compounds

Tab. 1. General properties of $CdCr_2S_4$ (room temperature unless noted)

Preparation is described in [*66BWRL*], [*66HP*], [*68PB*], [*69NS*]. $CdCr_2S_4$ is a semiconductor, *p*-type when pure or doped by Ag substitution for Cd, *n*-type when doped with In. It is a ferromagnet with low anisotropy. The electrical and magnetic properties are very susceptible to stoichiometry [*68PB*]. Optical and electrical properties are strongly affected by magnetic ordering.

		Ref.
Lattice constant	a_0 = 10.244 Å	*66BWRL*
S parameter	u = 0.390	*66BWRL*
Density	ϱ = 4.243 g/cm³	*66LR*
Compressibility	5.0 (300 °K), 4.2 (200 °K), 3.3 (80 °K) · 10^{-6} bar⁻¹	*69S(1)*
Paramagnetic Curie temperature	Θ_p = 152 °K	*66BWRL*
Molar Curie constant	C_m = 3.7 cm³ °K mole⁻¹	*67WBR*
Ferromagnetic Curie temperature	Θ_C = 84.5 °K	*66BWRL*
Saturation moment per molecule	p_{ms} = 6 μ_B	*69NS*
Resistivity	typical ϱ = 10^3 (300 °K), 10^6 (4.2 °K) Ω cm	[1])

[1]) See [*66LR*], [*69NS*].

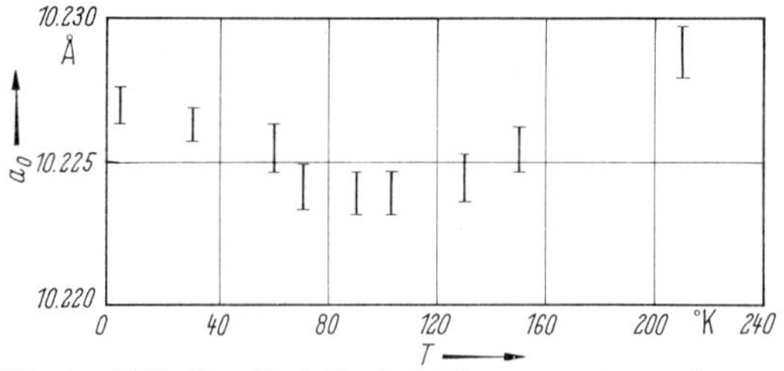

Fig. 1. $CdCr_2S_4$. Variation in lattice parameter a_0 of a pure single crystal as a function of temperature [*69MKWWP*].

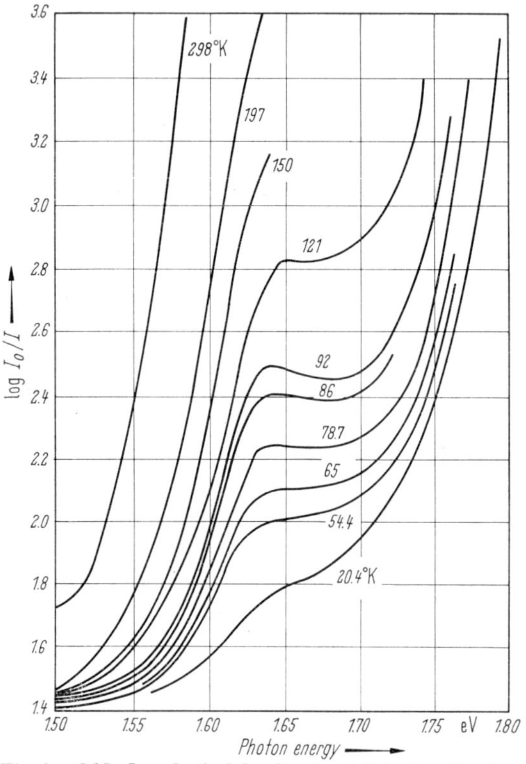

Fig. 3. $CdCr_2S_4$. Optical density, log I_0/I (ratio of incident to transmitted light), vs. photon energy for a 40 μm thick section of a single crystal at various temperatures. Between 1.0 and 1.5 eV several weaker absorption bands occur which shift very little with temperature [*66HP*].

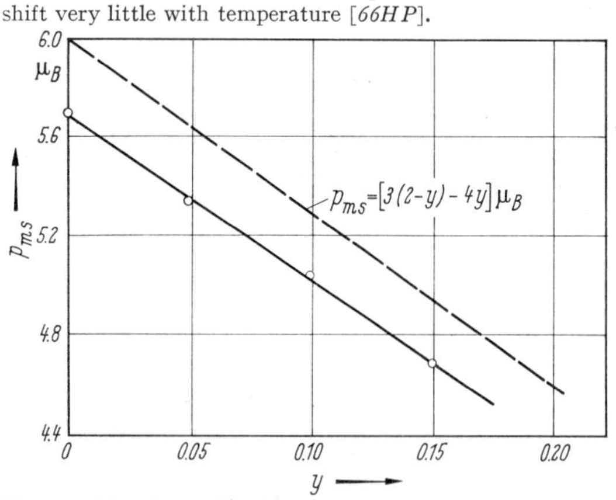

Fig. 4. $Cd_{1-y}In_y[Cr^{2+}_yCr^{3+}_{2-y}]S_4$. The effect of Cr^{2+} replacement for Cr^{3+} on the saturation magnetic moment per molecule of $CdCr_2S_4$ at 4.2 °K. Replacement is effected by substitution of In^{3+} for Cd^{2+}. Cr^{2+} couples antiferromagnetically to Cr^{3+} sites [*68PB*].

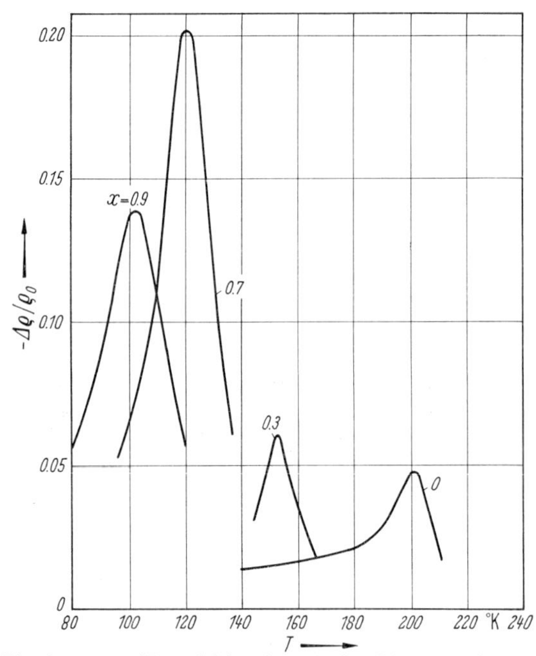

Fig. 2. p-type $(Fe_{1-x}Cd_x)_{0.98}Ag_{0.02}Cr_2S_4$. Magnetoresistance vs. temperature for several compositions in a field of 12 kOe [*69BHRZ*]. See also [*66LR*].

Tab. 2. Lattice constants and saturation moments of some substituted $CdCr_2S_4$ spinels [*69RMS*]

Composition	a_0 Å	p_{ms} μ_B
$Cd[V_{0.2}Cr_{1.8}]S_4$	10.251	5.30
$Cd[Ti_{0.2}Cr_{1.8}]S_4$	10.254	5.26
$Cd[Ti_{0.4}Cr_{1.6}]S_4$	10.257	4.58
$Cd_{0.8}In_{0.2}[Ni_{0.2}Cr_{1.8}]S_4$	10.251	4.59
$Cd_{0.8}In_{0.2}[Co_{0.2}Cr_{1.8}]S_4$	10.248	3.97
$Cd_{0.9}In_{0.1}[V_{0.1}Cr_{1.9}]S_4$	10.259	4.87
$Cd_{0.8}In_{0.2}[V_{0.2}Cr_{1.8}]S_4$	10.280	3.64
$Cd_{0.7}In_{0.3}[V_{0.3}Cr_{1.7}]S_4$	10.308	2.66

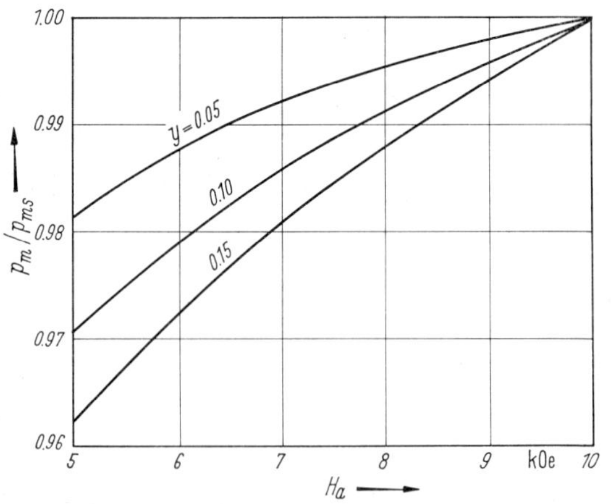

Fig. 5. $Cd_{1-y}In_y[Cr^{2+}_yCr^{3+}_{2-y}]S_4$. Relative magnetic moment p_m/p_{ms} vs. applied field for various Cr^{2+} (In^{3+} for Cd^{2+} substitution) concentrations in $CdCr_2S_4$. (Normalized at 10 kOe) [*68PB*].

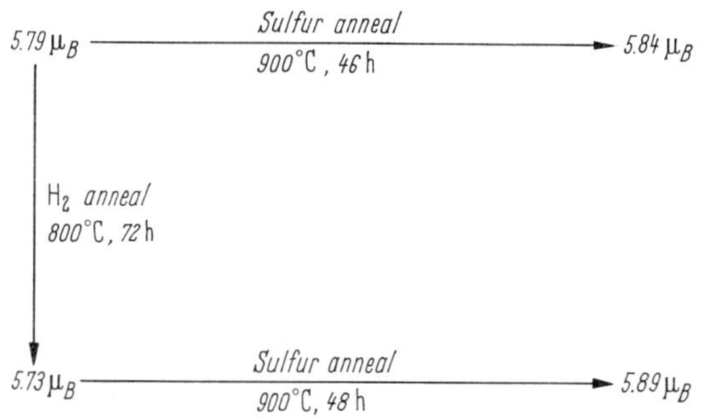

Fig. 6. $CdCr_2S_4$. Demonstration of the importance of sample stoichiometry to the magnetic properties of Cr spinels. The effect of annealing treatments on the saturation magnetic moment per molecule of a polycrystalline sample at 4.2 °K [*68PB*].

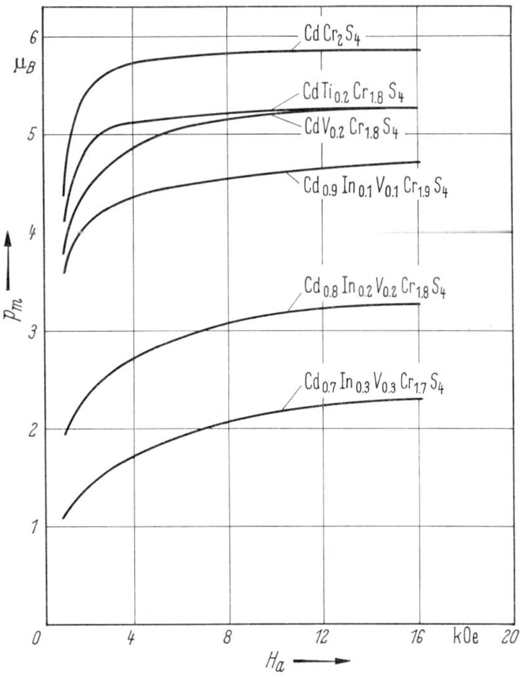

Fig. 7. Magnetic moment per molecule vs. applied field at 1.5 °K of pure $CdCr_2S_4$ and of $CdCr_2S_4$ with Ti or V substitutions for Cr (with and without simultaneous In substitutions for Cd). Ti^{3+} and V^{3+} show no local moments [*69RMS*].

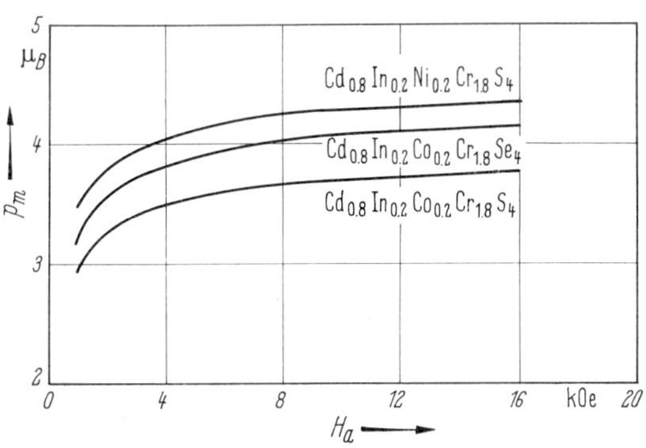

Fig. 8. $Cd_{0.8}In_{0.2}[X_{0.2}Cr_{1.8}]S_4$ (X = Ni, Co) and $Cd_{0.8}In_{0.2}Co_{0.2}Cr_{1.8}Se_4$. Magnetic moments per molecule vs. applied field at 1.5 °K. Ni^{2+} and Co^{2+} couple antiferromagnetically to Cr^{3+} sites [*69RMS*].

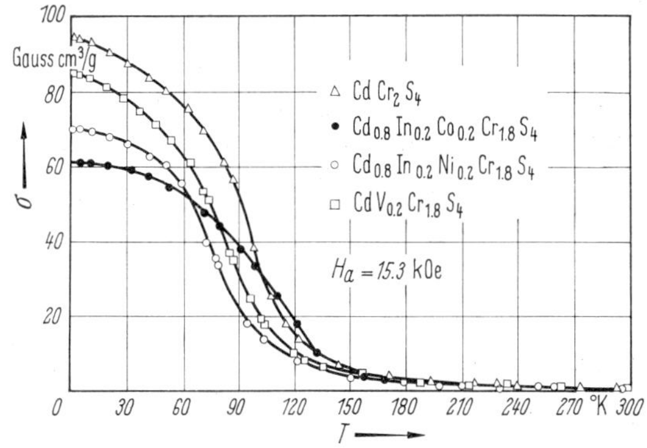

Fig. 9. $CdCr_2S_4$ and substituted compounds. Magnetisation σ vs. temperature in 15.3 kOe external field [*69RMS*].

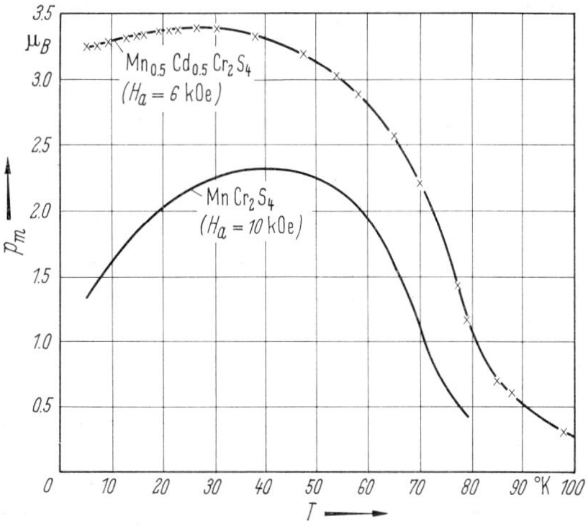

Fig. 10. The magnetic moment per molecule vs. temperature for the mixed spinel $Mn_{0.5}Cd_{0.5}Cr_2S_4$ in 6 kOe external field and for $MnCr_2S_4$ in 10 kOe field [*68L*].

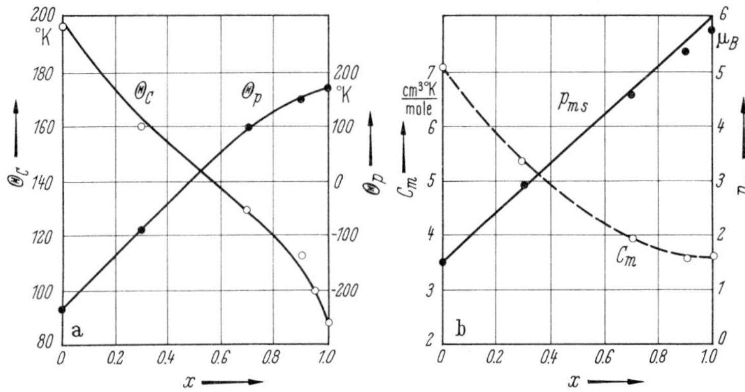

Fig. 11. $Fe_{1-x}Cd_xCr_2S_4$. (a) The paramagnetic Curie temperature Θ_p and ferromagnetic Curie temperature Θ_C and (b) the molar Curie constant C_m and saturation moment per molecule as functions of composition in the $CdCr_2S_4$-$FeCr_2S_4$ system [*69BHRZ*].

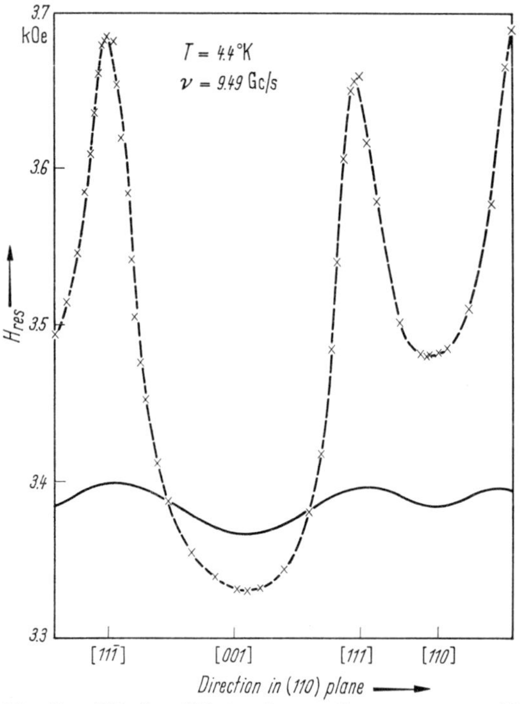

Fig. 12. $CdCr_2S_4$. Effect of annealing treatment: The angular variation in (110) plane of the ferromagnetic resonance field for single crystal spheres at 4.4 °K [*68PB*].

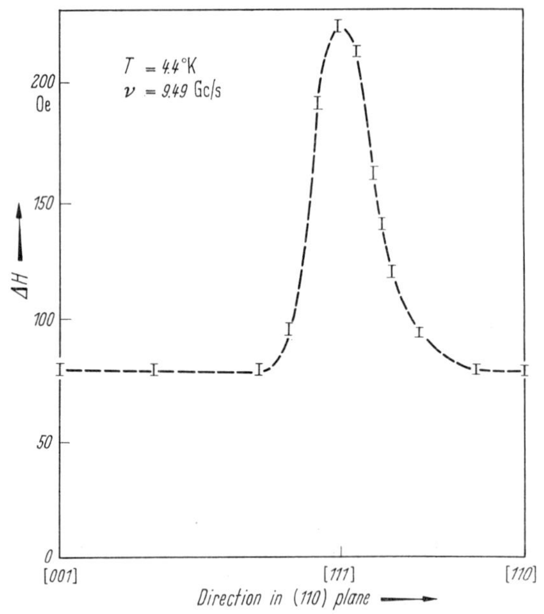

Fig. 13. $CdCr_2S_4$. Effect of annealing treatment: Anomalous angular dependence of the ferromagnetic resonance line width in the (110) plane for a single crystal sphere of low moment, as grown, at 4.4 °K [*68PB*].

6.2.2.6 Hg-Cr selenide, Hg-Cr sulfide spinels and their substituted compounds

Tab. 1. General properties of $HgCr_2Se_4$ (room temperature unless noted)

Preparation is described in [*66BWRL*], [*69LE*]. $HgCr_2Se_4$ is a semiconductor with complex behavior. Conductivity is of *p*-type above the magnetic ordering temperature, Θ_C, but becomes *n*-type at 77 °K [*69LE*]. Its electrical and ferromagnetic properties are expected to be seriously influenced by stoichiometry. See [*68PB*]. Optical and electrical properties are strongly affected by magnetic ordering.

		Ref.
Lattice constant	$a_0 = 10.753$ Å	*66BWRL*
Se parameter	$u = 0.390$	*66BWRL*
Compressibility	4.6 (296 °K), 3.1 (200 °K), 2.4 (80 °K) · 10^{-6} bar^{-1}	*69S(1)*
Paramagnetic Curie temperature	$\Theta_p = 200$ °K	*66BWRL*
Molar Curie constant	$C_m = 3.79$ cm³ °K mole^{-1}	*66BWRL*
Ferromagnetic Curie temperature	$\Theta_C = 106$ °K	*66BWRL*
Saturation moment per molecule	$p_{ms} = 5.64\ \mu_B$	*66BWRL*
Resistivity	$\varrho = 0.7\ \Omega$cm	*69LE*

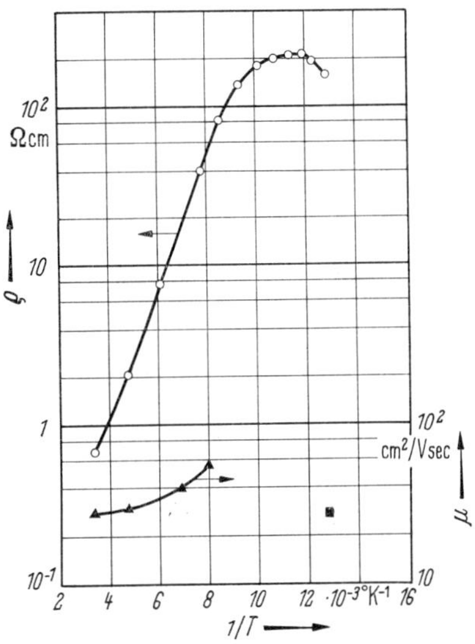

Fig. 1. $HgCr_2Se_4$. Electrical resistivity ϱ and Hall mobility μ for a pure single crystal vs. inverse temperature. Mobilities are shown only in the region of linear variation of Hall voltage with magnetic field. Conduction is p-type above Θ_C but n-type at 77 °K [*69LE*].

▲ Hall mobility; ■ electron mobility.

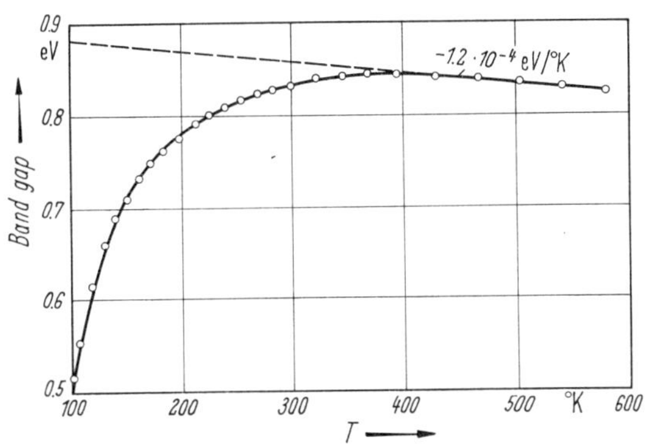

Fig. 2. $HgCr_2Se_4$. Optical band gap energy vs. temperature (Band edge position is taken arbitrarily as the point where the absorption coefficient reaches $\alpha = 2000\ cm^{-1}$). Magnetic fields produce additional red shifts [*69LE*].

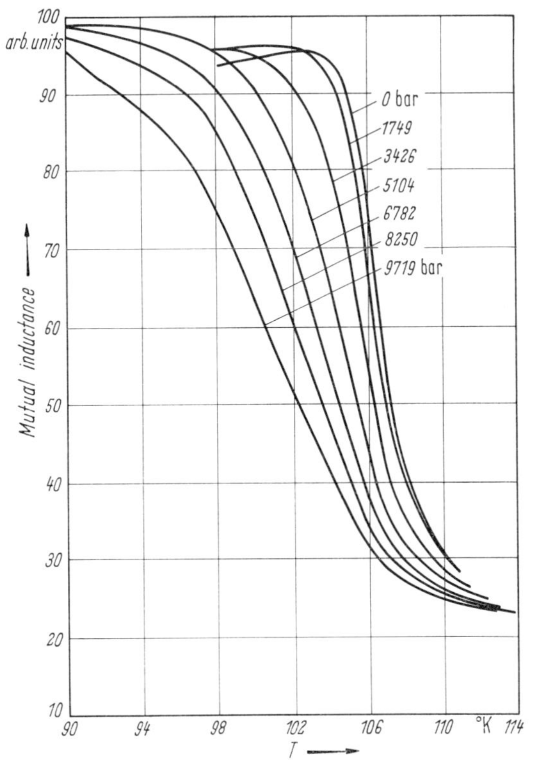

Fig. 3. $HgCr_2Se_4$. Variation with pressure of mutual inductance (arbitrary units) vs. temperature. The position of the upper knee determines the ferromagnetic Curie temperature Θ_C for each curve [*69S(1)*].

Tab. 2. General properties of $HgCr_2S_4$ (room temperature unless noted)

Preparation is described in [*66BWRL*], [*68HBE*]. $HgCr_2S_4$ is ferromagnetic above, metamagnetic below 25 °K and 7 kOe. Its moments, at low temperatures in zero field, are ordered in a noncolinear flat spiral having a propagation direction along a particular cube edge within each domain. This ordering is like that of $ZnCr_2Se_4$. Its electrical and magnetic properties are expected to be considerably affected by stoichiometry. See [*69PB*]. Optical properties are strongly affected by magnetic ordering.

		Ref.
Lattice constant	$a_0 = 10.273$ Å	*66BWRL*
S parameter	$u = 0.390$	*66BWRL*
Compressibility	4.1 (300 °K), 2.7 (77 °K), 2.2 (4.2 °K) · $10^{-6}\ bar^{-1}$	*69S(1)*
Paramagnetic Curie temperature	$\Theta_p = 142$ °K	*66BWRL*
Molar Curie constant	$C_m = 3.62\ cm^3$ °K $mole^{-1}$	*66BWRL*
Ferromagnetic Curie temperature	$\Theta_C = 36$ °K	*66BWRL*
Saturation moment per molecule	$p_{ms} = 5.46\ \mu_B$	*68HC*

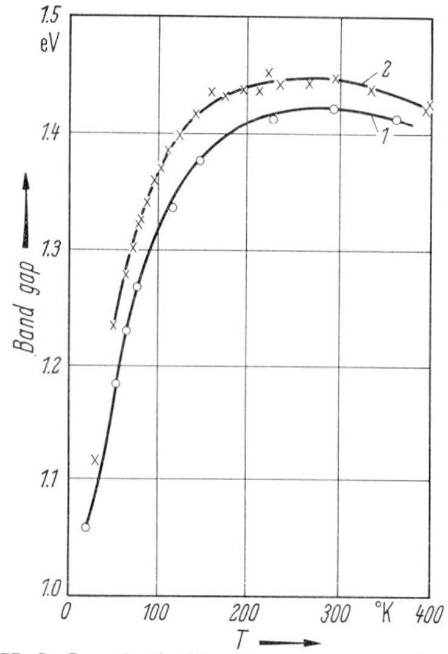

Fig. 4. $HgCr_2S_4$. Optical band gap energy vs. temperature. (1) Single crystal absorption measurements of the energy at which the absorption coefficient reaches $\alpha = 3000\ cm^{-1}$. (2) Polycrystalline diffuse reflection measurements of the energy for onset (on the high energy side) of a nonlinear change in reflection. Magnetic fields produce additional red shifts [*68HBE*].

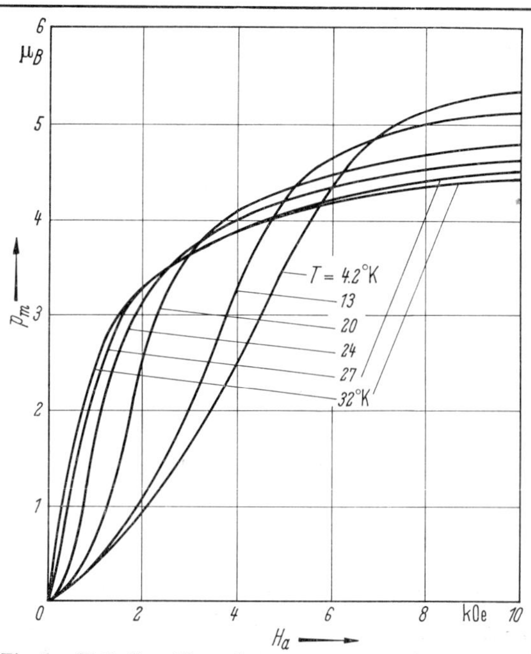

Fig. 5. $HgCr_2S_4$. Magnetic moment per molecule vs. applied field at various temperatures for a polycrystalline metamagnetic sample [*66BWRL*].

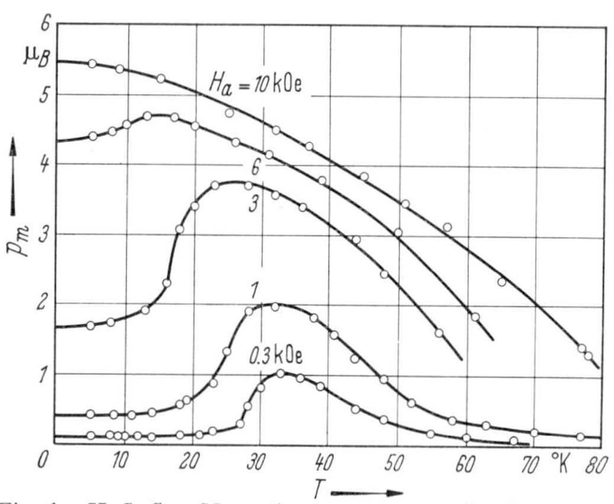

Fig. 6. $HgCr_2S_4$. Magnetic moment per molecule vs. temperature in different external fields for a polycrystalline sample [*66BWRL*].

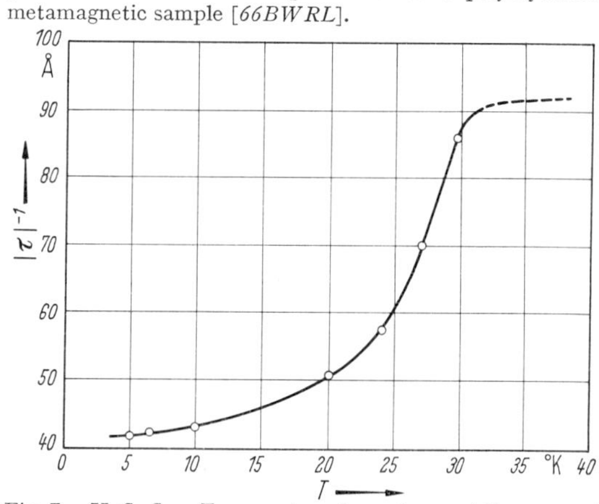

Fig. 7. $HgCr_2S_4$. Temperature dependence of the magnetization spiral wavelength τ in a powdered sample with zero applied field, obtained by diffraction of 1.7 Å neutrons. In contrast, $ZnCr_2Se_4$, which orders similarly, shows a temperature independent wavelength [*68HC*].

6.2.2.7 Cu-Cr S, Se and Te spinels

Tab. 1. $CuCr_2X_4$, with X = S, Se and Te. Preparation procedures, X-ray data, and magnetic properties [*64L(1)*]

Compound	Preparation	X-ray diagram	a_0 Å	p_{ms} at 4 °K μ_B	Θ_C *) °K	Θ_p °K	C_m cm³ °K mole⁻¹
$CuCr_2S_4$	300 h, 750 °C +100 h, 750 °C +300 h, 600 °C	spinel + 2 very weak reflections	9.822	4.58	420	390	2.40
$CuCr_2Se_4$	300 h, 750 °C +100 h, 750 °C +300 h, 600 °C +100 h, 900 °C	spinel + 1 hardly visible reflection	10.337	4.94	460	465	2.50
$CuCr_4Te_4$	160 h, 500 °C +100 h, 500 °C	spinel + 3 weak reflections	11.137	4.93	365	400	2.90

*) Determined from the σ^2-T curve.

Tab. 2. $CuCr_2Se_4$ and $CuCr_2Te_4$. Crystallographic parameter u and degree of inversion x at 508 °C, determined from X-ray and neutron diffraction. The parentheses contain the standard deviations [67C]

	$CuCr_2Se_4$			$CuCr_2Te_4$	
	X-ray	neutrons		X-ray	neutrons
x	0.065 (0.078)	0.023 (0.019)	x	−0.163 (0.256)	0.012 (0.018)
u	0.25756 (0.00034)	0.25738 (0.00021)	u	0.25595 (0.00099)	0.25628 (0.00038)

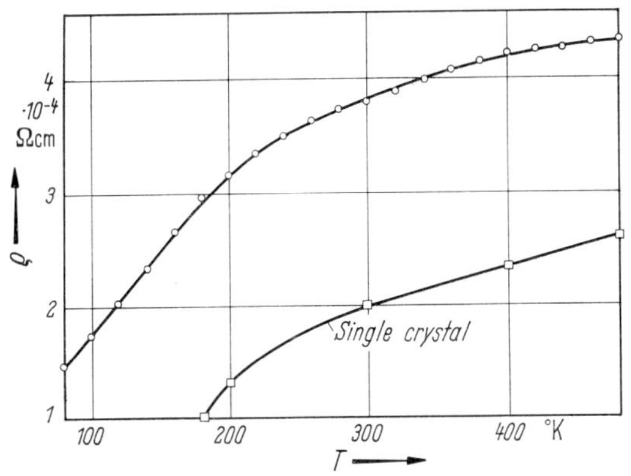

Fig. 1. $CuCr_2Se_4$. Resistivity of a polycrystalline sample as a function of temperature [67RLW]. Single crystal measurements [64F(1)].

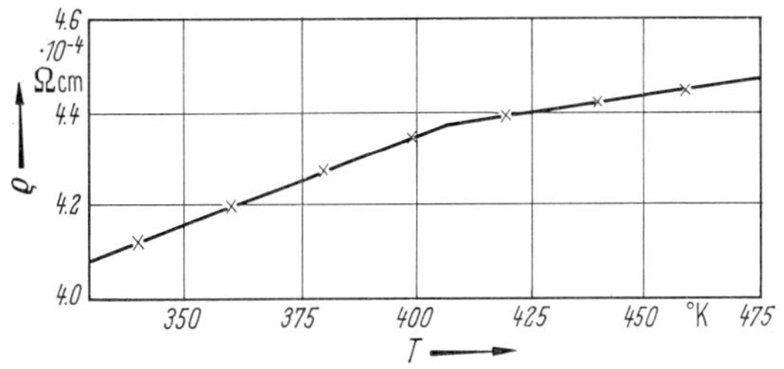

Fig. 2. $CuCr_2Se_4$. Resistivity of a polycrystalline sample as a function of temperature in the vicinity of the Curie temperature [67RLW].

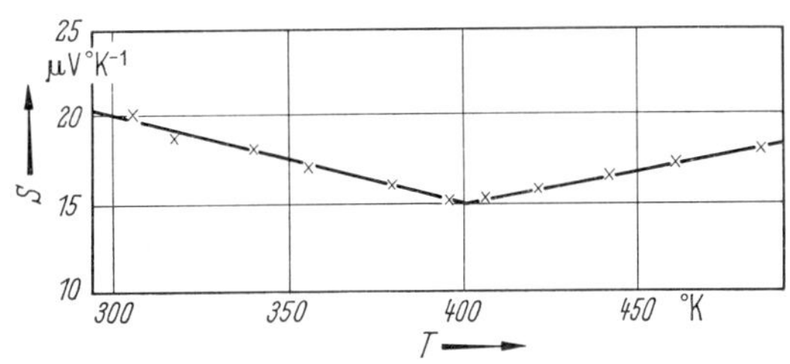

Fig. 3. $CuCr_2Se_4$. Seebeck coefficient of a polycrystalline sample as a function of temperature [67RLW].

Tab. 3. $CuCr_2Se_4$ and $CuCr_2Te_4$. Magnetic moments p_m on (B) and (A) sites, crystallographic u parameter, the overall isotropic temperature factor, B, and mean square deviation of the calculated X-ray intensities, R, at 298 °K and 4.2 °K [67C]

	$CuCr_2Se_4$		$CuCr_2Te_4$	
	T = 298 °K	T = 4.2 °K	T = 298 °K	T = 4.2 °K
p_m (B)	2.31787 (0.14420) μ_B	2.80788 (0.11231) μ_B	1.98204 (0.10940) μ_B	3.11183 (0.16388) μ_B
p_m (A)	−0.05454 (0.16281) μ_B	0.01421 (0.12817) μ_B	0.02611 (0.11773) μ_B	0.05322 (0.14094) μ_B
u	0.25761 (0.00034)	0.25907 (0.00024)	0.25606 (0.00027)	0.25610 (0.00049)
B	0.4332 (0.1913) Å²	−0.0755 (0.1259) Å²	0.7881 (0.2207) Å²	0.0775 (0.4103) Å²
R	1.266%	1.106%	1.092%	1.485%

Tab. 4. $CuCr_2Se_4$. X-ray density, preparation conditions, and electrical properties of a sintered (in vacuum under pressure) sample [67RLW]

	$CuCr_2Se_4$
Lattice parameter	a_0 = 10.334 Å
X-ray density	ϱ_X = 5.818 g/cm³
Pressure sintering	
Temperature	T = 750 °C
Pressure	p = 20 kpsi
Experimental density	ϱ_{exp} = 5.731 g/cm³
Experimental density/X-ray density	ϱ_{exp}/ϱ_X = 98.5%
Conductivity (298 °K)	$\sigma = 2.62 \cdot 10^3\ \Omega^{-1}\,cm^{-1}$
Seebeck effect (298 °K)	S = +20.5 μV °K⁻¹

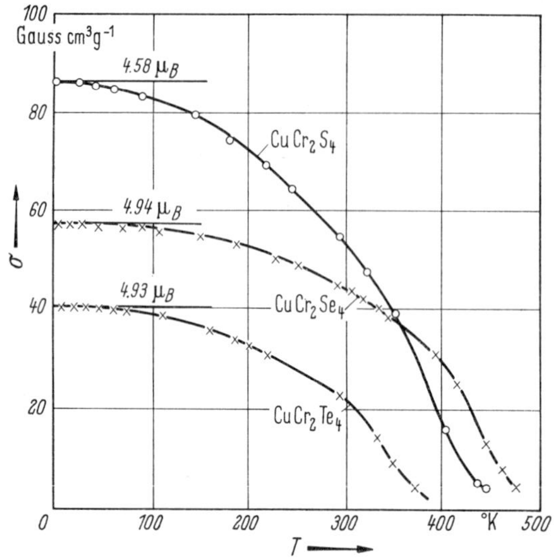

Fig. 4. $CuCr_2X_4$, X = S, Se, or Te. The magnetization per gram for H = 20 kOe as a function of temperature [*64L(1)*]. See Tab. 1.

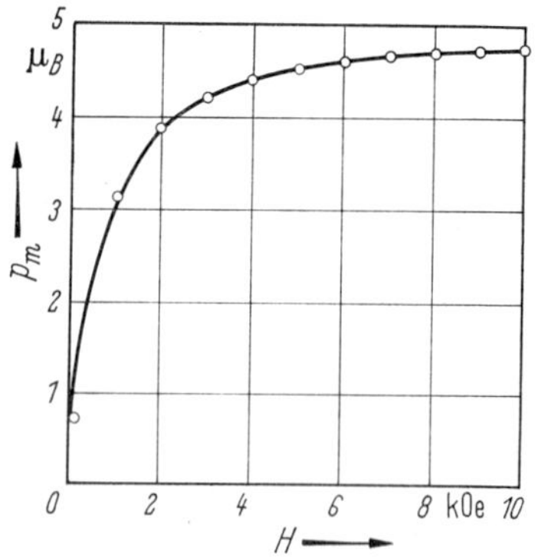

Fig. 5. $CuCr_2Se_4$. Magnetization as a function of magnetic field at 4.2 °K [*67RLW*]. See Tab. 4.

Tab. 5. Magnetic properties of $CuCr_2S_4$ and $CuCr_2Se_4$ [*64L(2)*]. Compare Tab. 1.

	$CuCr_2S_4$	$CuCr_2Se_4$	
Θ_C	425	460	°K
p_m (4 °K)	4.58	4.94	μ_B
C_m	2.36	2.50	cm³ °K mole⁻¹

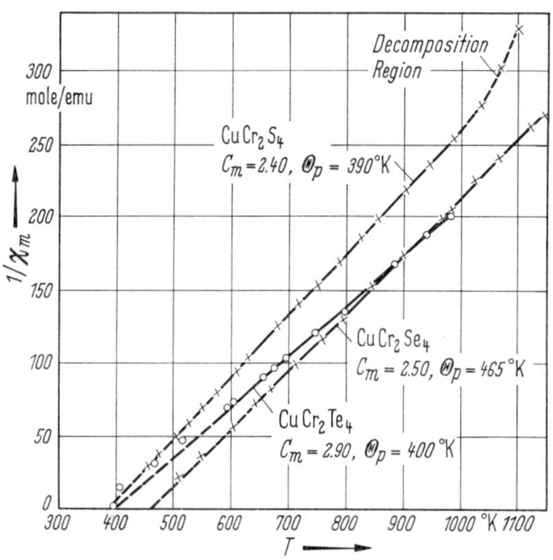

Fig. 6. $CuCr_2X_4$, X = S, Se, or Te. The reciprocal molar susceptibility vs. temperature [*64L(1)*]. C_m in [cm³ °K mole⁻¹]. See Tab. 1.

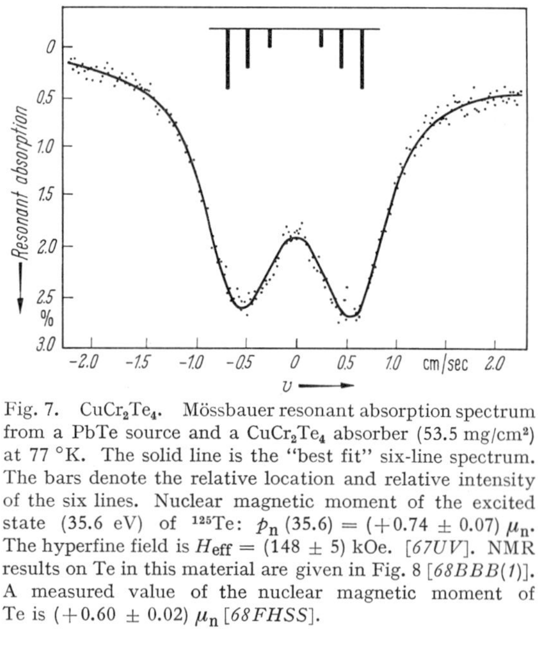

Fig. 7. $CuCr_2Te_4$. Mössbauer resonant absorption spectrum from a PbTe source and a $CuCr_2Te_4$ absorber (53.5 mg/cm²) at 77 °K. The solid line is the "best fit" six-line spectrum. The bars denote the relative location and relative intensity of the six lines. Nuclear magnetic moment of the excited state (35.6 eV) of ^{125}Te: p_n (35.6) = (+0.74 ± 0.07) μ_n. The hyperfine field is H_{eff} = (148 ± 5) kOe. [*67UV*]. NMR results on Te in this material are given in Fig. 8 [*68BBB(1)*]. A measured value of the nuclear magnetic moment of Te is (+0.60 ± 0.02) μ_n [*68FHSS*].

Tab. 6. $CuCr_2Se_4$ and $CuCr_2Te_4$. The hyperfine fields in kOe at various nuclei

	$CuCr_2Se_4$		$CuCr_2Te_4$		
	77 °K	0 °K	77 °K	0 °K	
^{63,65}Cu	+ 70.6 [1]	72.6 *) [1] 72.1 (4.2 °K) [2]	38.0 [1]	39.9 *) [1]	kOe
^{53}Cr	−159 [1]	−161 *) [1]	−145 [1]	−151 *) [1] (−)151.1 (1.4 °K) [3]	kOe
^{125}Te			−181 [3]	−187.5 (1.4 °K) [3]	kOe

*) Extrapolated values. [1]) [*67YWC*]; [2]) [*67L(1)*]; [3]) [*68BBB*].

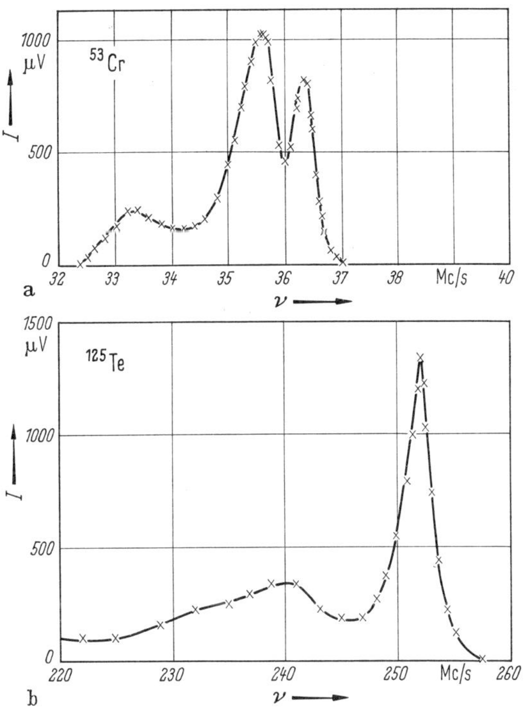

Fig. 8. $CuCr_2Te_4$. Spectra of ^{53}Cr (a) and ^{125}Te (b) taken using NMR spin-echo techniques at 1.40 °K and zero magnetic field. The highest frequency peak in the ^{53}Cr spectrum yields a hyperfine field of (−)151.1 kOe in agreement with the extrapolated value in Tab. 6. The hyperfine field at the Te nucleus (corresponding to the sharp spectral maximum) is −187.5 kOe at 1.4 and 4.2 °K, and −181 kOe at 77 °K [*68BBB(1)*].

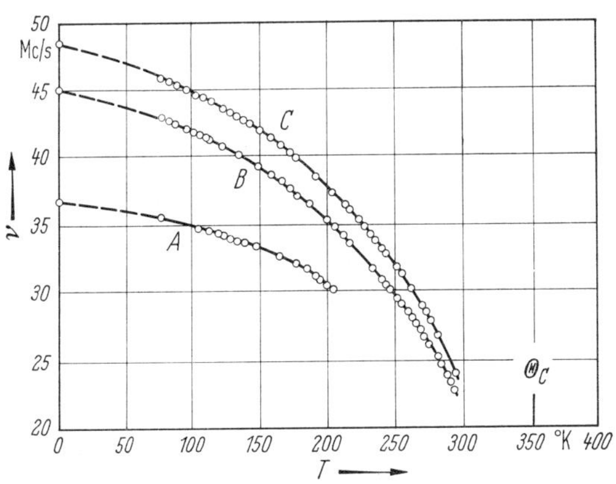

Fig. 10. $CuCr_2Te_4$. Temperature dependence of the NMR frequencies of ^{53}Cr (curve *A*), ^{63}Cu (curve *B*), and ^{65}Cu (curve *C*) in a powdered sample [*67YWC(1)*]. See also [*67YWC(2)*].

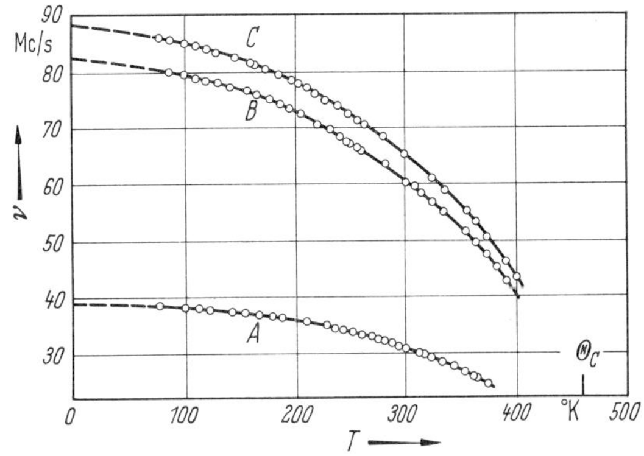

Fig. 9. $CuCr_2Se_4$. Temperature dependence of the NMR frequencies of ^{53}Cr (curve *A*), ^{63}Cu (curve *B*), and ^{65}Cu (curve *C*) in a powdered sample [*67YWC(1)*]. See also [*67YWC(2)*].

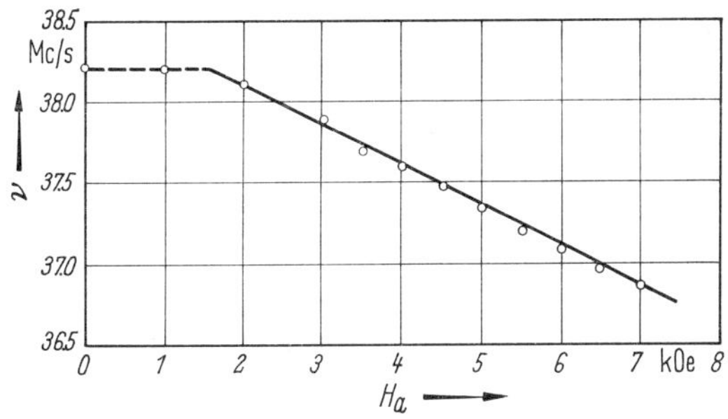

Fig. 11. $CuCr_2Se_4$. Dependence of the ^{53}Cr NMR frequency on external magnetic field.
$d\nu/dH = -0.24$ (Mc/s)/kOe for the linear portion of the curve, indicating that the hyperfine field for ^{53}Cr is antiparallel to the magnetization [*67YWC(2)*].

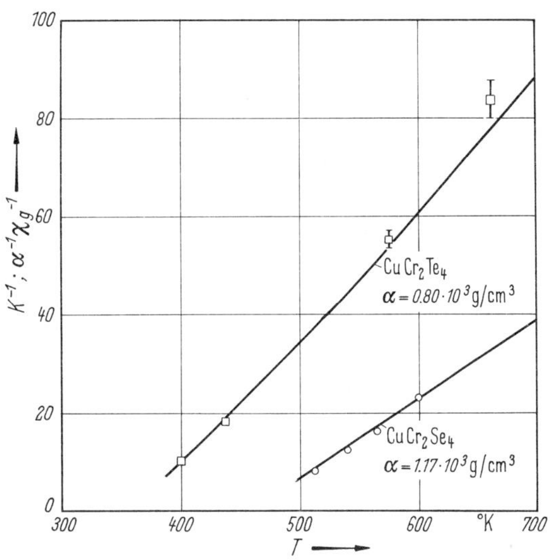

Fig. 12. $CuCr_2Se_4$ and $CuCr_2Te_4$. The reciprocal of the Knight shift, K^{-1}, (points) and the reciprocal of the gram susceptibility (solid lines) vs. temperature. The scale factors α are chosen so that the lines fit the circles [*67L(1)*].

Tab. 7. $CuCr_2S_4$. NMR frequencies, line widths ($\Delta\nu$), and relaxation times (T_1 and T_2) [*68K*]

Spinel	nucleus	T °K	ν Mc/s	$\Delta\nu$ Mc/s	T_1 msec	T_2 msec
$CoCr_2S_4$	^{53}Cr	77	47.4	2	0.3	$\approx T_1$
$CuCr_2S_4$	^{53}Cr	77	38.9	0.5	3.5	$\approx T_1$
		300	27	0.5	0.4	$\approx T_1$
	^{63}Cu	77	100.6 (^{63}Cu) 107.8 (^{65}Cu)	0.11	0.7	0.3
	^{65}Cu	300	65.6 (^{63}Cu) 70.35 (^{65}Cu)	0.11	0.1	0.08

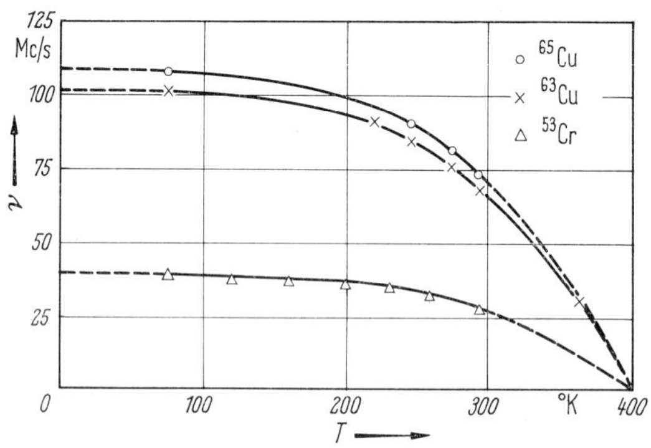

Fig. 13. $CuCr_2S_4$. Temperature dependence of the NMR frequencies for ^{65}Cu, ^{63}Cu, and ^{53}Cr nuclei [*68K*].

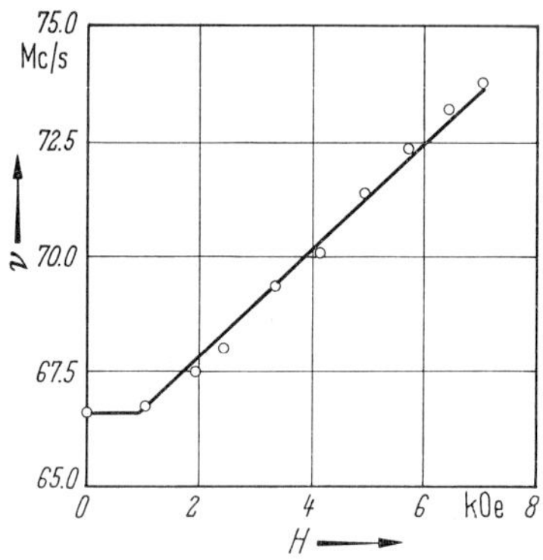

Fig. 14. $CuCr_2S_4$. Magnetic field dependence of the NMR frequency of ^{63}Cu at 300 °K [*68K*].
ν [Mc/s] $= 65.5 + 1.1285\,H$ (H in kOe). (In the original paper there is a — instead of the +.)

6.2.2.8 Cu-Cr S, Se and Te spinels with substitutions

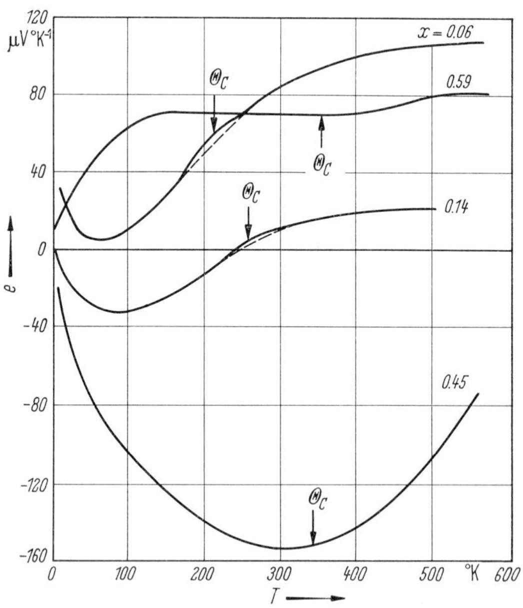

Fig. 1. $Fe_{1-x}Cu_xCr_2S_4$. Thermoelectric power e vs. temperature for various compositions [*68HB*].

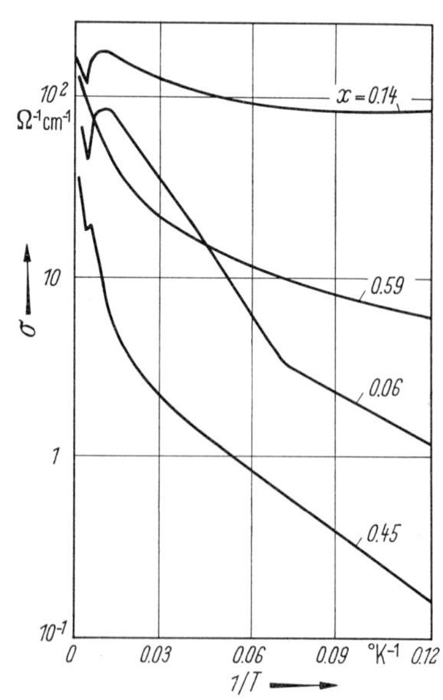

Fig. 2. $Fe_{1-x}Cu_xCr_2S_4$. Electrical conductivity vs. inverse temperature for various compositions [*68HB*].

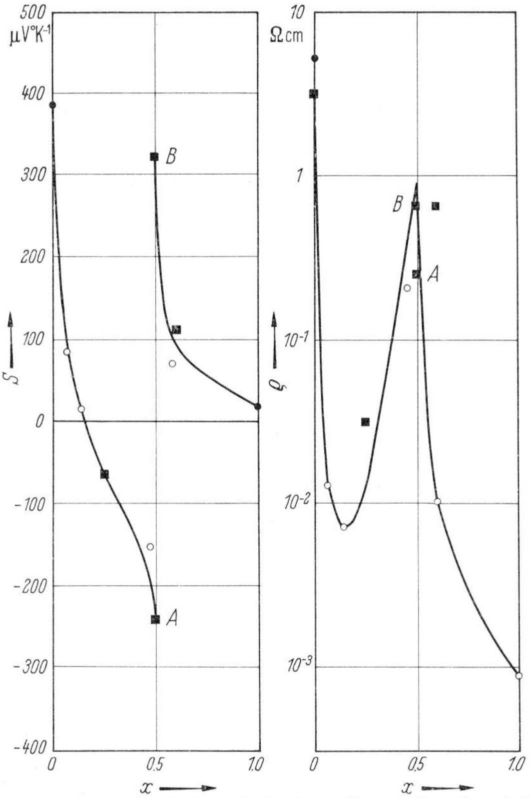

Fig. 3. $Fe_{1-x}Cu_xCr_2S_4$. Seebeck coefficient S and resistivity at RT as functions of x. Measurements obtained by others: Open circles (single crystal measurements) [*68HB*]; Closed circles [*65BRW*]. Data for curves corresponding to solid squares: Curve A: samples quenched from 700 °C; Curve B: samples slowly cooled with annealing at 500 °C and 100 °C [*69LVVV*].

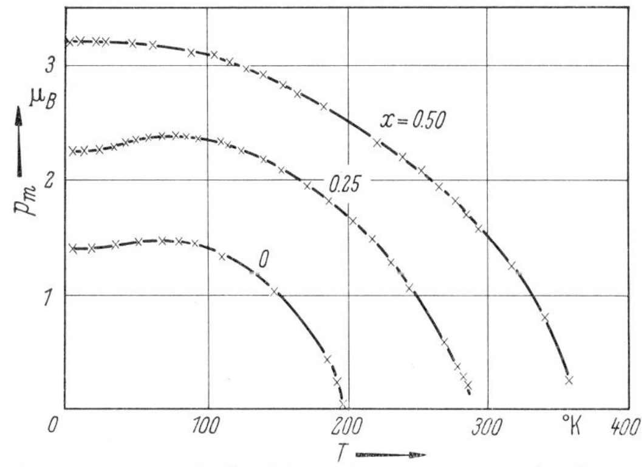

Fig. 4. $Fe_{1-x}Cu_xCr_2S_4$. Magnetic moment per molecule, p_m, vs. T curves measured with an applied field of 9 kOe for various compositions [*69LVVV*]. See Fig. 5 for additional compositions.

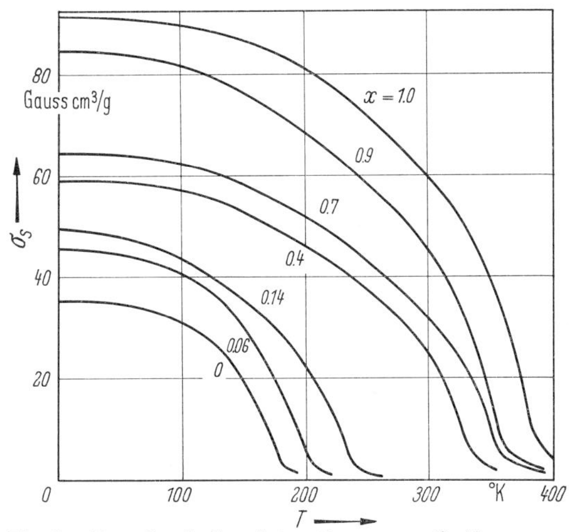

Fig. 5. $Fe_{1-x}Cu_xCr_2S_4$. Saturation magnetization σ_s vs. T. Compositions with $x = 0$, $x = 0.06$, and $x = 0.14$ were single crystal samples [*67HB*]. See also Fig. 4.

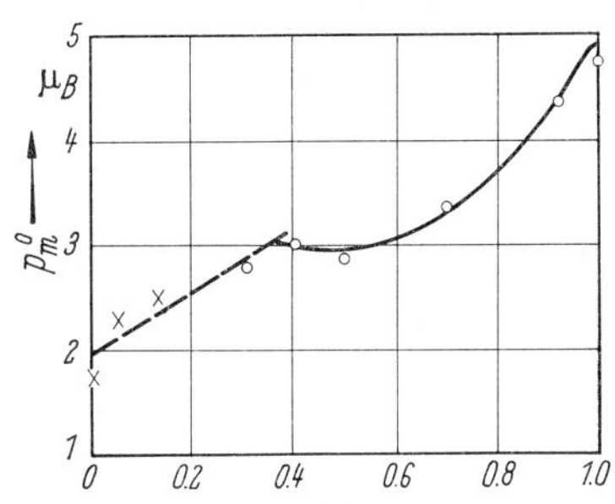

Fig. 6. $Fe_{1-x}Cu_xCr_2S_4$. Magnetic moment per molecule at 0 °K vs. x. × = single crystals, ● = powdered samples [*67HB*]. See also [*69LVVV*].

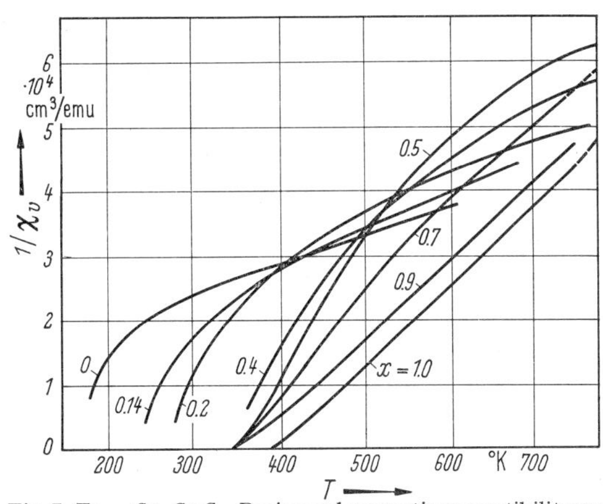

Fig. 7. $Fe_{1-x}Cu_xCr_2S_4$. Reciprocal magnetic susceptibility as a function of temperature for various compositions. For compositions with $x = 0$ and $x = 0.14$, single crystals were used [*67HB*].

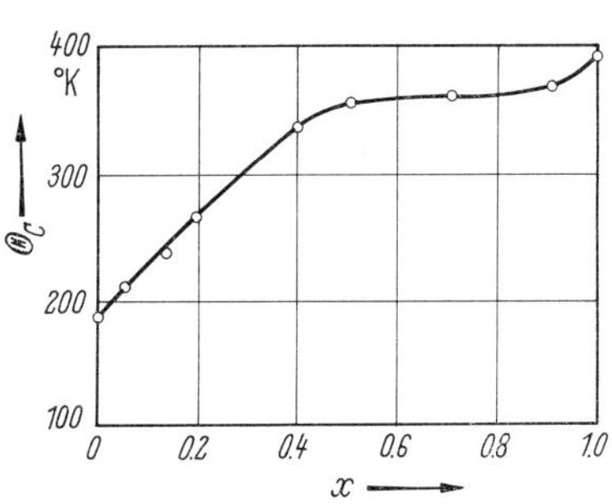

←

Fig. 8. $Fe_{1-x}Cu_xCr_2S_4$. Curie temperature vs. composition [*67HB*]. See also [*69LVVV*].

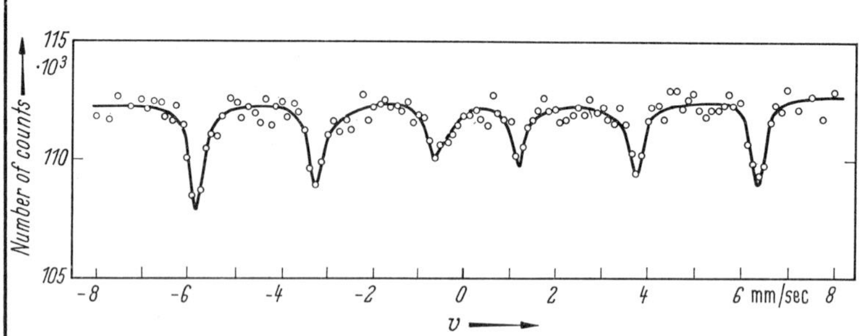

Fig. 9. $Fe_{0.5}Cu_{0.5}Cr_2S_4$. Mössbauer spectrum at liquid He temperature. 65 Mg/cm² sample. Source ^{57}Co in Pd [69LVVV].

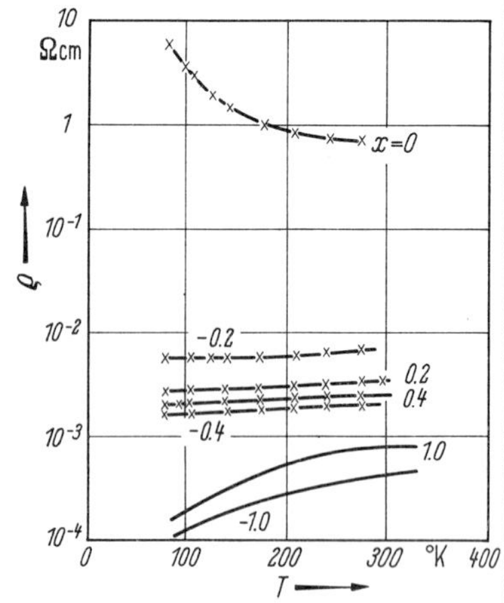

Fig. 11. $CuCr_{1+x}Ti_{1-x}S_4$. Logarithm of the resistivity vs. temperature for various compositions [68LV]. Curves for $x = 1.0$ and -1.0 from [65BRW].

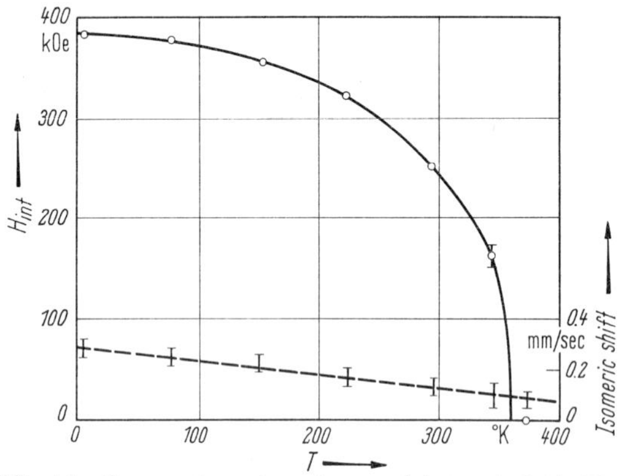

Fig. 10. Temperature dependence of internal field H_{int} (solid curve) and isomer shift (dashed curve) for Fe in $Fe_{0.5}Cu_{0.5}Cr_2S_4$ [69LVVV].

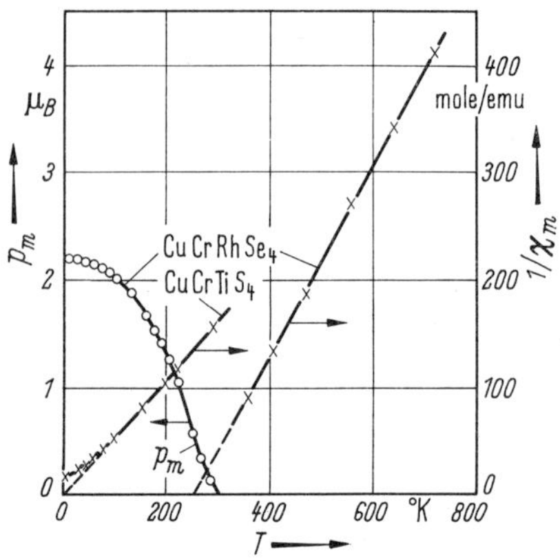

→

Fig. 12. $CuCrRhSe_4$ and $CuCrTiS_4$. Temperature dependence of the magnetization measured with a 9 kOe applied field and the reciprocal molar susceptibility [68LV].

Tab. 1. $CuCr_{1+x}Ti_{1-x}S_4$; $CuCr_{1+x}Rh_{1-x}Se_4$. Magnetic moment per molecule, p_m, ferromagnetic Curie temperature, Θ_C, molar Curie constant, C_m, asymptotic Curie temperature, Θ_a, Seebeck coefficient, S, and resistivity, ϱ [68LV]

Compound	x	p_m at 4.5 °K μ_B	Θ_C °K	C_m cm³ °K mole⁻¹	Θ_a °K	S at RT μV/deg	log ϱ (ϱ in Ωcm) at RT
$CuCr_{1+x}Ti_{1-x}S_4$	1.0	4.6	420	2.40	390	+16.0 [1]	−3.1 [1]
	0.4	3.2	210	2.01	195	+32	−2.6
	0.2	— [2]	100	1.98	105	+57	−2.5
	0	...	...	1.84	4	−250	−0.2
	−0.2	...	...	1.39	37	−58	−2.2
	−0.4	...	...	1.03	45	−30	−2.7
	−1.0					−11.8 [1]	−3.4 [1]
$CuCr_{1+x}Rh_{1-x}Se_4$	1.0	4.94	460	2.50	465	+40	−1.9
	0.75	4.0	425	2.16	425		
	0.10					+7.9	−3.0
	0	2.2	270	1.12	255	+6.3	−2.1
	−0.25	1.7	190	0.90	210	+7.4	−3.0
	−0.40	1.4	160	0.71	185		
	−0.55	0.88	115	0.45	168		
	−0.70	0.57	80	0.29	120		
	−1.0					+7.3	−2.9

[1] [65BRW].
[2] Measurements by H. Zijlstra; no saturation occurs at 4.5 °K in fields up to 90 kOe and $p_m = (1.6 + 6.2 \cdot 10^{-6} H)\ \mu_B$/mole for $10 < H < 90$ kOe.

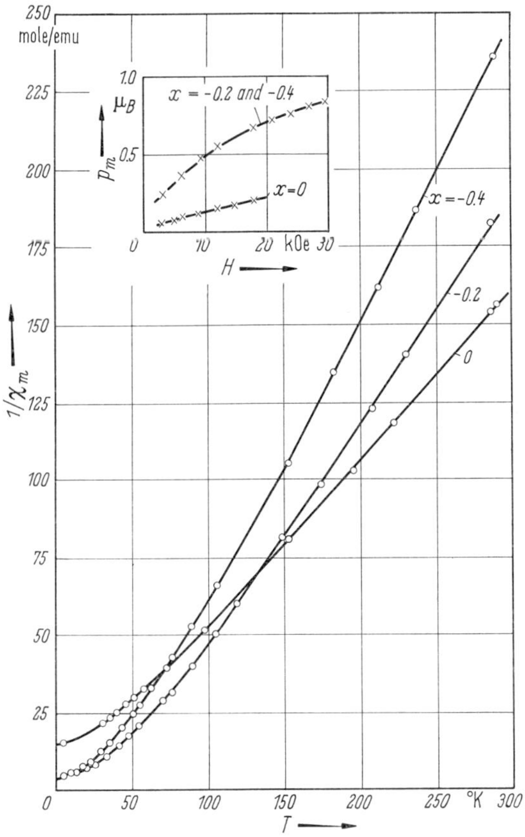

Fig. 13. $CuCr_{1+x}Ti_{1-x}S_4$. Temperature dependence of the reciprocal molar susceptibility. Inset: Magnetic field dependence of the magnetic moment p_m at 4.4 °K [68LV].

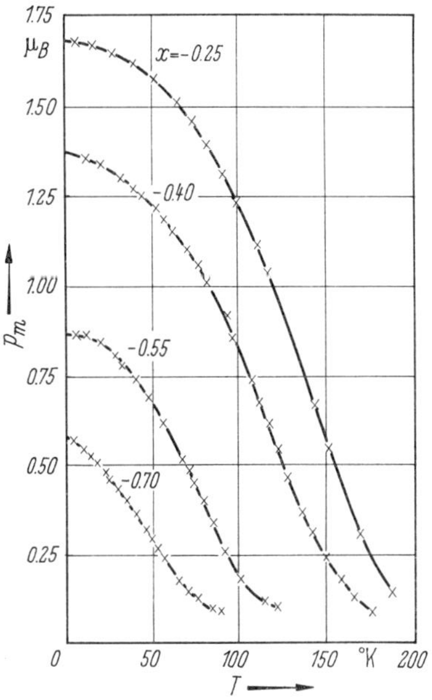

Fig. 14. $CuCr_{1+x}Rh_{1-x}Se_4$. Temperature dependence of the magnetization measured with a 9 kOe applied field [68LV].

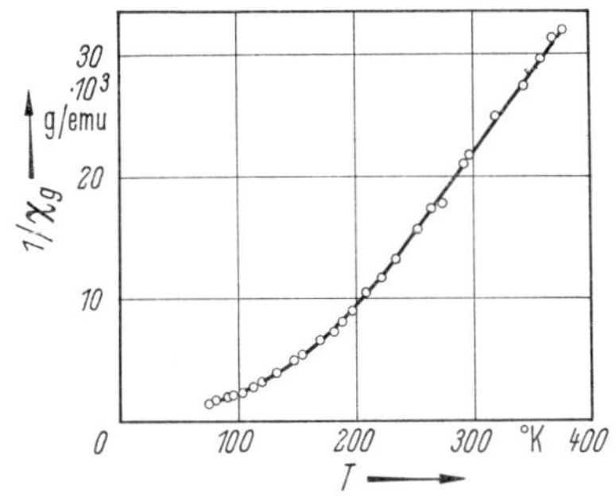

Fig. 16. $Cu_{0.5}In_{0.5}Cr_2Se_4$. Temperature dependence of the inverse susceptibility [69YC].

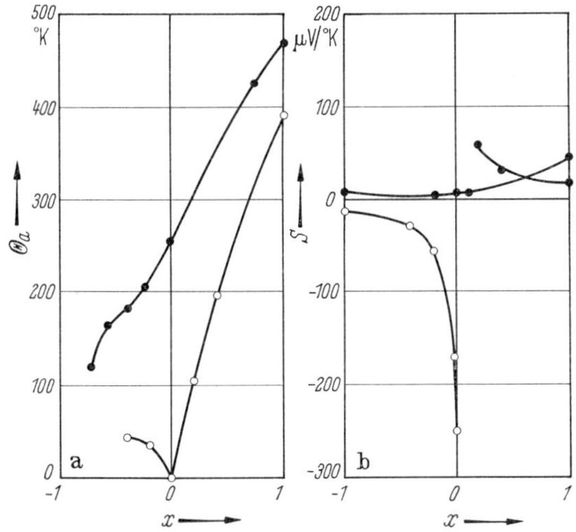

Fig. 15. $CuCr_{1+x}Rh_{1-x}Se_4$ (closed circles) and $CuCr_{1+x}Ti_{1-x}S_4$ (open circles). Asymptotic Curie temperature and Seebeck coefficient vs. composition [68LV].

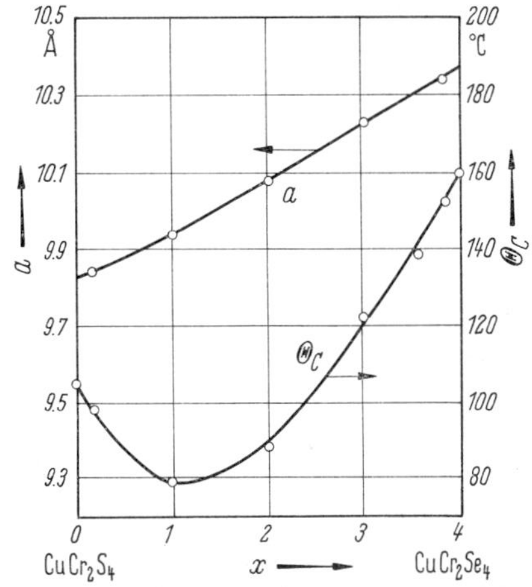

Fig. 17. $CuCr_2Se_xS_{4-x}$. Lattice parameters and Curie temperature vs. x [68OTI].

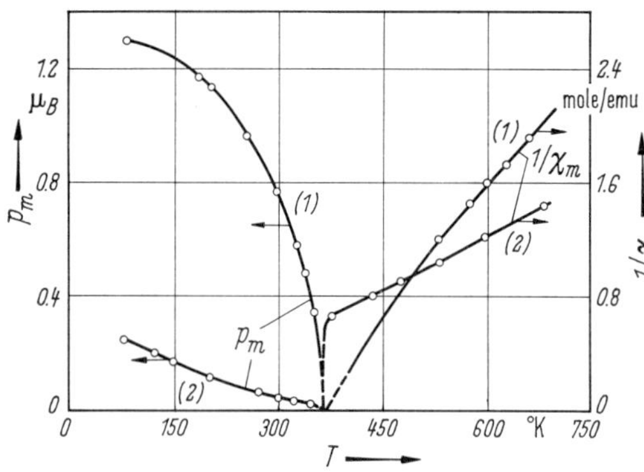

Fig. 18. $CuCr_2SeS_3$. Temperature dependence of the magnetization and susceptibility. Sample (1) and sample (2) were fired at 500 °C and 350 °C, respectively [68OTI].

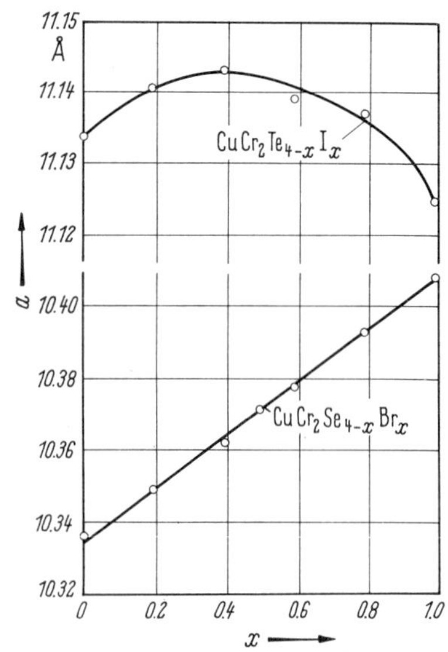

Fig. 19. $CuCr_2Se_{4-x}Br_x$; $CuCr_2Te_{4-x}I_x$. Variation of unit cell parameters with composition [68RBL].

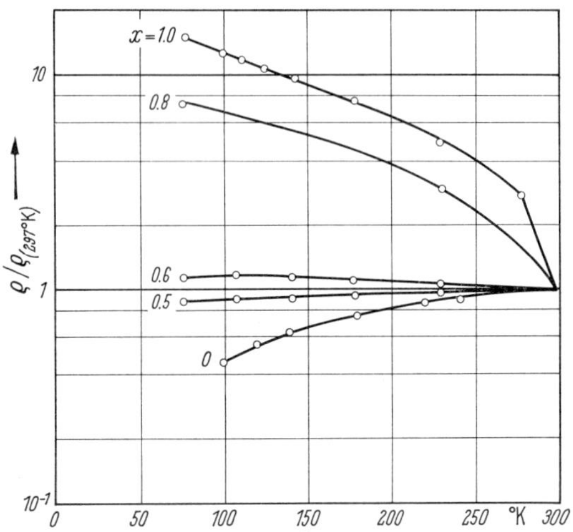

Fig. 20. $CuCr_2Se_{4-x}Br_x$. Variation of resistance ratio $\varrho/\varrho_{(297\,°K)}$ with temperature [68RBL].

Tab. 2. $CuCr_2Se_3Br$, $CuCr_2Te_3I$ and $CuCr_2Se_3Cl$. Magnetic properties and unit cell parameters [68RBL]

	a_0 Å	p_m μ_B (4°K, 10 kOe)	Θ_C °K	Θ_p °K
$CuCr_2Se_3Br$	10.416	5.25	274	345
$CuCr_2Te_3I$	11.125	4.10	294	...
$CuCr_2Se_3Cl$ [a])	10.377	4.40	...	...

Tab. 3. $CuCr_2Se_{4-x}Br_x$. Magnetic properties [68RBL]

$CuCr_2Se_{4-x}Br_x$	p_m μ_B (4 °K, 10 kOe)	Θ_C °K	Θ_p °K
$x = 0$	4.80	434	448
0.2	4.88	406	438
0.4	4.98	375	405
0.6	5.04	345	385
0.8	5.17	307	360
1	5.24	274	345

6.2.2.9 Fe-, Mn-, Co-, and Zn-Cr S and Se spinels and their substituted compounds

Fig. 1, see next page

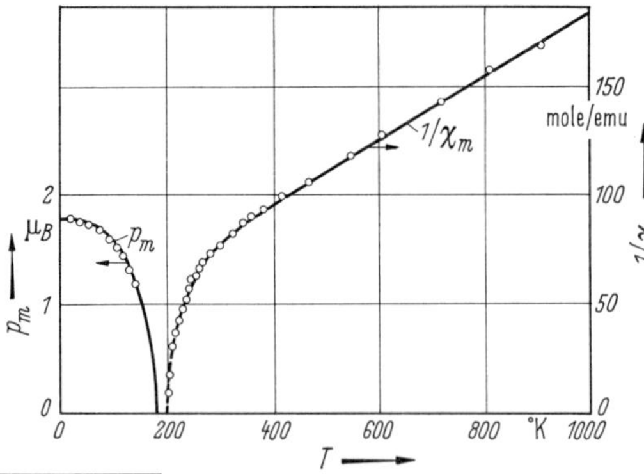

←

Fig. 2. $FeCr_2S_4$. Magnetization and reciprocal paramagnetic susceptibility vs. temperature [69GDP].

[a]) The compound $CuCr_2Se_3Cl$ is strongly attracted to a magnet at room temperature. This material is very hydroscopic, for which reason reliable magnetic measurements have not yet been made.

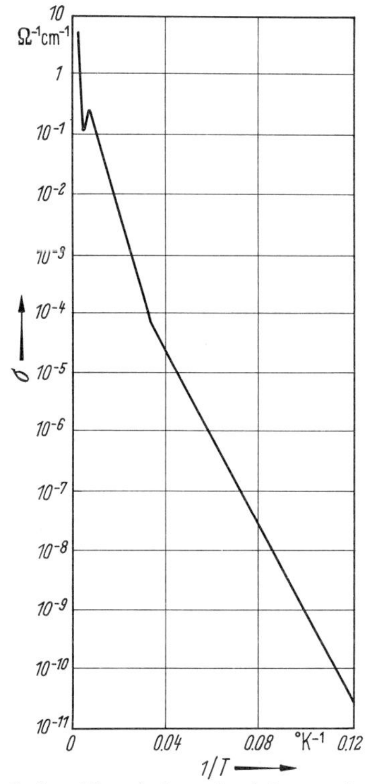

Fig. 1. $FeCr_2S_4$. Electrical conductivity vs. inverse temperature [*68HB*].

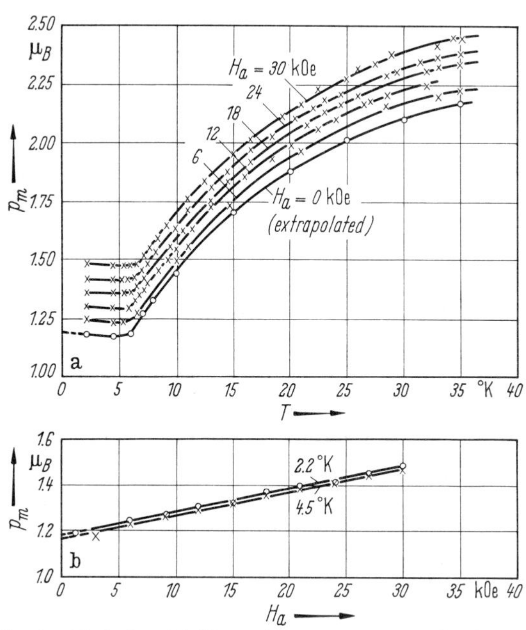

Fig. 4. $MnCr_2S_4$. (a) Magnetic moment per molecule vs. temperature for various applied fields. (b) Magnetic moment per molecule vs. applied field at 2.2 °K and 4.5 °K [*68L(1)*].

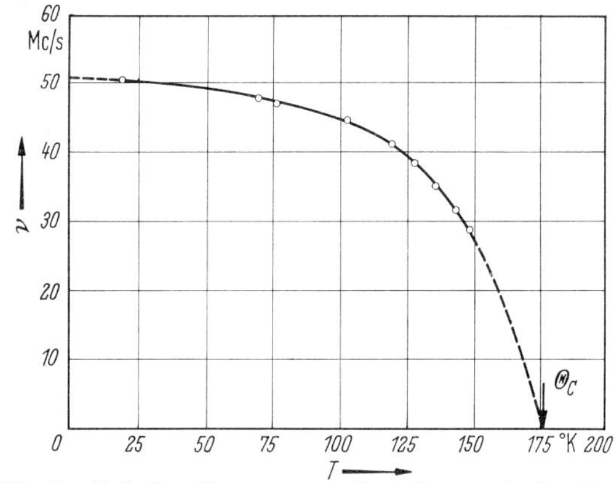

Fig. 3. $FeCr_2S_4$. Temperature dependence of the ^{53}Cr NMR frequencies [*66K*].

Tab. 1. Mössbauer data for $FeCr_2S_4$ [*67YM*]. See also [*68HS*] and [*67EST*] (Co^{57} source at 25 °C)

Spinel	*T* of the absorber °C	Isomer shift relative to stainless steel mm/sec	Line width mm/sec
$Fe[Cr_2]S_4$	25	0.72 ± 0.04	0.48 ± 0.04
	−80	0.80 ± 0.04	0.44 ± 0.04

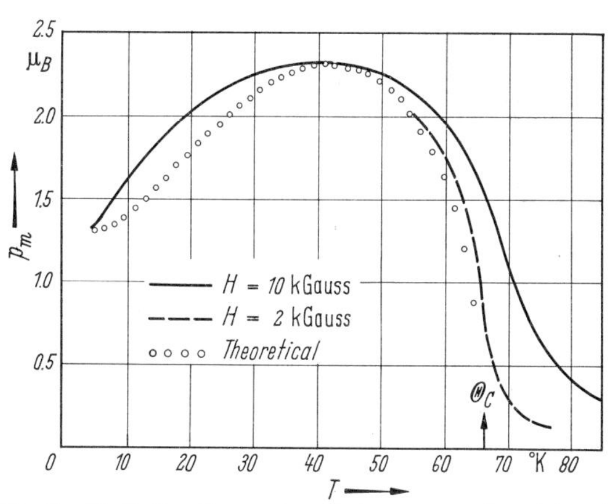

Fig. 5. $MnCr_2S_4$. Comparison between experimental and theoretical magnetization vs. temperature curves. The theoretical curve was computed for the molecular field approximation assuming collinear spin configuration with $p_m(A) = 4.7\,\mu_B$ and $p_m(B) = 3.0\,\mu_B$, and $J_{BB}/J_{AB} = -4$. The negative sign corresponds to a ferromagnetic *B-B* interaction [*65MD*].

Tab. 2. $MnIn_xCr_{2-x}S_4$. Crystallographic and magnetic parameters [*68DBL*]. See also [*65MD*] which gives 0.3876 as the neutron diffraction value for the *u* parameter of $MnCr_2S_4$

x	a_0 Å	p_m μ_B (4.2 °K, 10 kOe)	Θ_C °K	Θ_N °K	C_m cm³ °K mole⁻¹	Θ_p °K
0	10.108	1.27	67	...	7.25	+10 ± 3
0.2	10.168	1.15	48	...	6.98	− 3 ± 3
0.3	10.198	0.85	36	...	...	...
0.4	10.227	...	...	18	6.81	− 9 ± 3
0.6	10.298	...	...	17	6.33	−21 ± 3
0.8	10.352	...	...	16	6.27	−27 ± 3
1.0	10.418	...	...	13	5.94	−36 ± 3

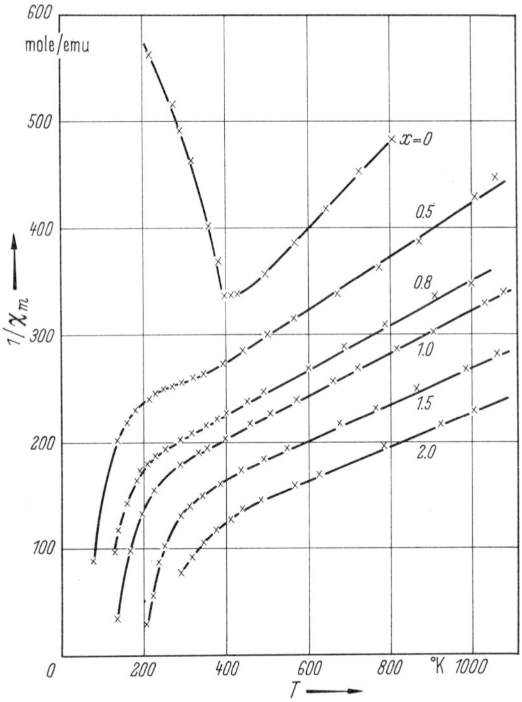

Fig. 7. $CoRh_{2-x}Cr_xS_4$. Reciprocal susceptibility vs. temperature [*68L*]. Curve for $x = 0$ after measurements by G. BLASSE.

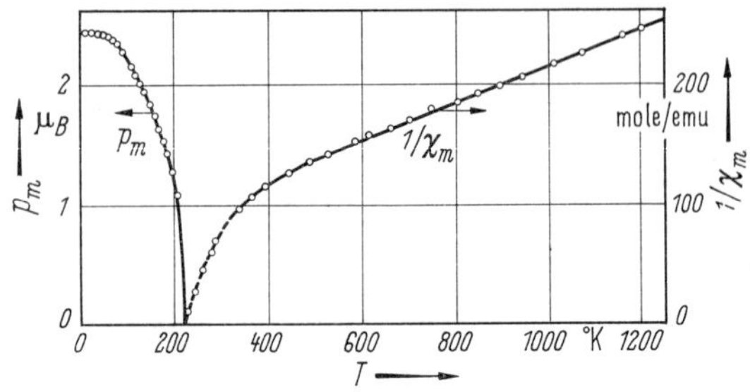

Fig. 6. $CoCr_2S_4$. Magnetization and reciprocal paramagnetic susceptibility vs. temperature [*69GDP*].

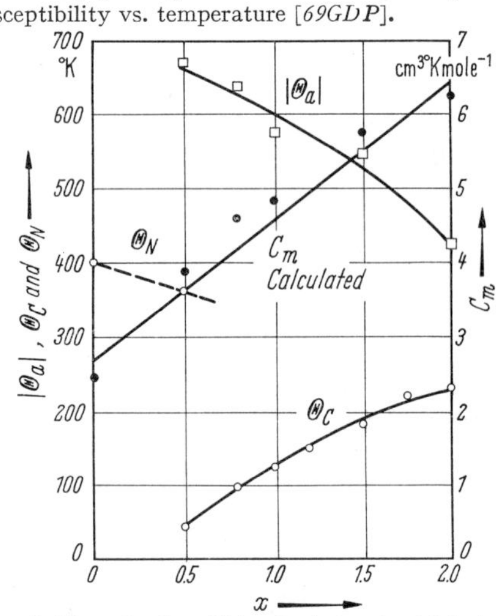

Fig. 8. $CoRh_{2-x}Cr_xS_4$. Molar Curie constant (black circles), asymptotic Curie temperature (squares), ferromagnetic Curie temperature and Néel temperature (open circles) vs. composition. The straight C_m vs. x curve is calculated using $g_{Cr} = 2$ and $g_{Co} = 2.4$ [*68L*].

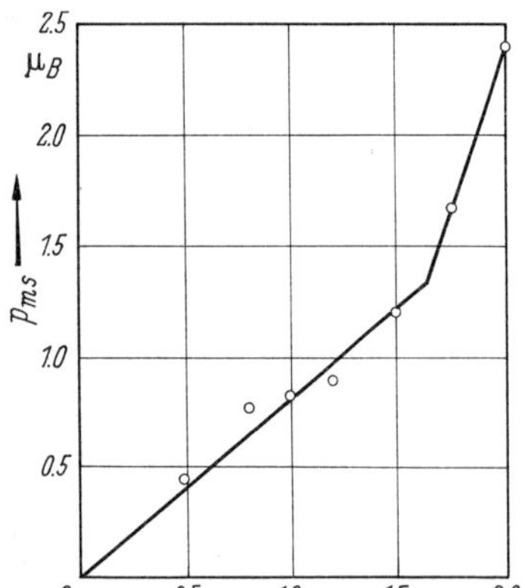

Fig. 9. $CoRh_{2-x}Cr_xS_4$. Saturation magnetic moment at 4.2 °K vs. composition [*68L*].

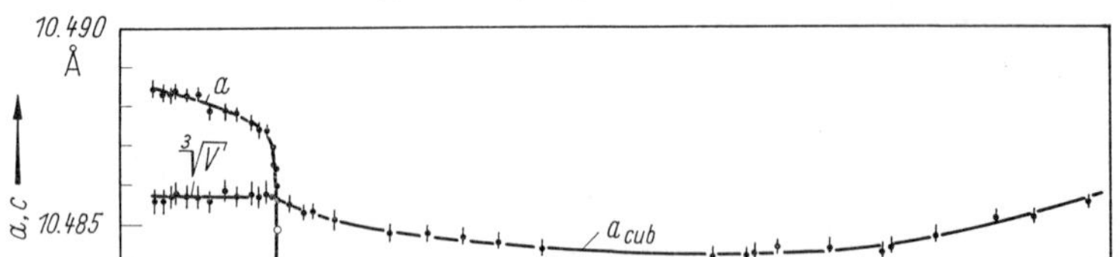

Fig. 10. $ZnCr_2Se_4$. Temperature dependence of the crystallographic parameters and the cube root of the unit cell volume. *a* and *c* are the tetragonal parameters and a_{cub} is the cubic (spinel) parameter [*66KK*].

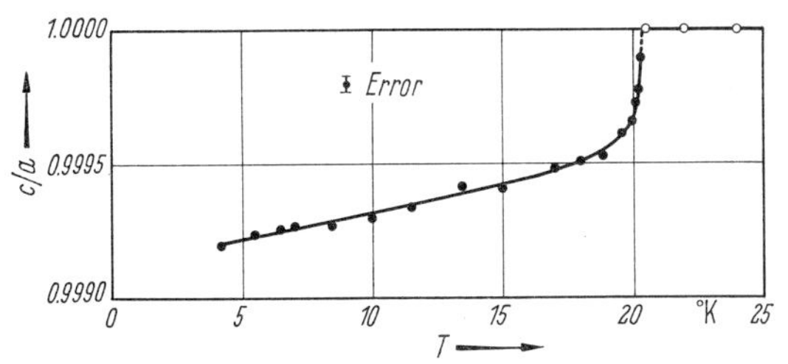

Fig. 11. $ZnCr_2Se_4$. Variation of the c/a ratio as a function of temperature [*66KK*].

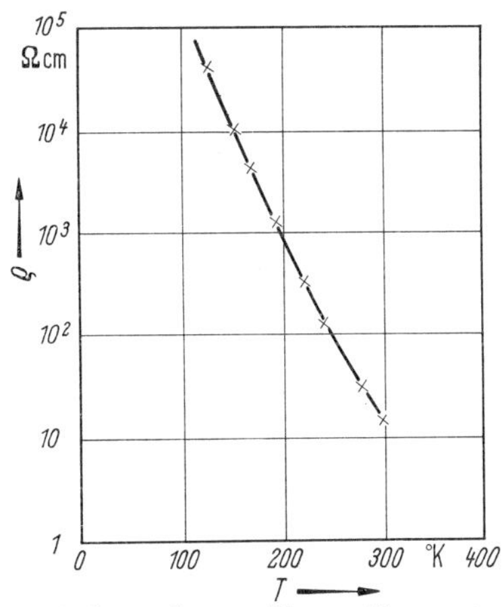

Fig. 12. $ZnCr_2Se_4$, polycrystalline. Temperature dependence of the logarithm of the resistivity [*64L(1)*].

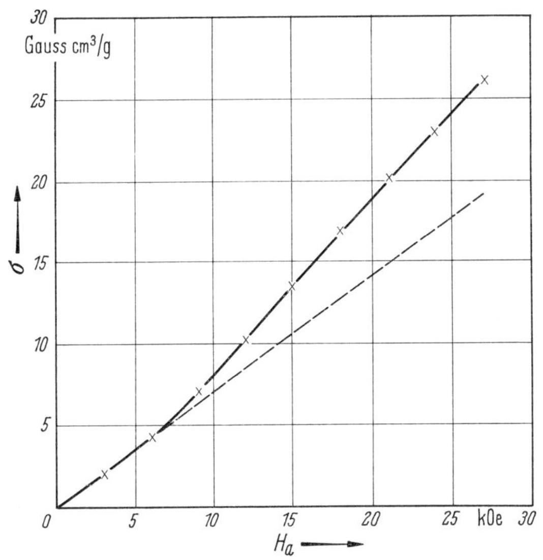

Fig. 13. $ZnCr_2Se_4$. Magnetization vs. external field at 4.2 °K. The slope of the dotted curve is equal to $7 \cdot 10^{-4}$ Gauss $cm^3\ g^{-1}\ Oe^{-1}$ [*65L*].

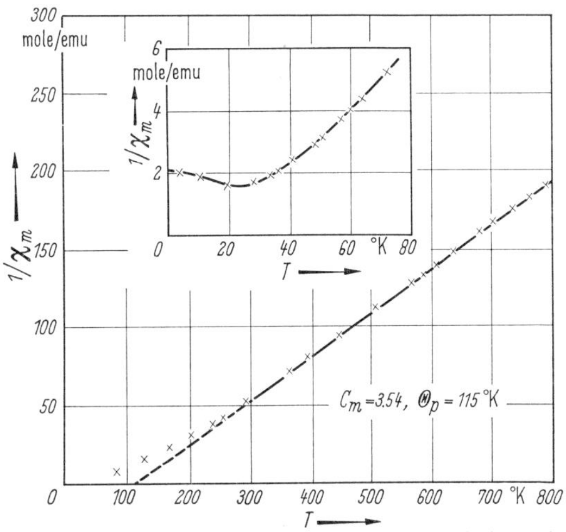

Fig. 14. $ZnCr_2Se_4$. Temperature dependence of the reciprocal molar susceptibility [*64L(1)*]. C_m in [cm^3 °K $mole^{-1}$].

6.2.2.10 References for 6.2.2.1 ··· 6.2.2.9

56L(1) LOTGERING, F. K.: Philips Res. Rept. **11** (1956) 218.
56L(2) LOTGERING, F. K.: Philips Res. Rept. **11** (1956) 337.
56L(3) LOTGERING, F. K.: Philips Res. Rept. **11** (1956) 190.
64L(1) LOTGERING, F. K.: Proc. International Conf. on Magnetism, p. 533 Nottingham 1964.
64L(2) LOTGERING, F. K.: Solid State Commun. **2** (1964) 55.
64SCP SHIRANE, G., D. E. COX, and S. J. PICKART: J. Appl. Phys. **35** (1964) 954.
65BLR BALTZER, P. K., H. W. LEHMANN, and M. ROBBINS: Phys. Rev. Letters **15** (1965) 493.
65BRW BOUCHARD, E. J., P. A. RUSSO, and A. WOLD: Inorg. Chem. **4** (1965) 685.
65L LOTGERING, F. K.: Solid State Commun. **3** (1965) 347.
65MD MENYUK, N., and K. DWIGHT: J. Appl. Phys. **36** (1965) 1088.
65P PLUMIER, R.: Compt. Rend. **260** (1965) 3348.
66BMW BUSCH, G., B. MAGYAR, and P. WACHTER: Phys. Letters **23** (1966) 438.
66BWRL BALTZER, P. K., P. J. WOJTOWICZ, M. ROBBINS, and E. LOPATIN: Phys. Rev. **151** (1966) 367.
66HP HARBEKE, G., and H. PINCH: Phys. Rev. Letters **17** (1966) 1090.
66K KHOI, D.: Compt. Rend. **262** (1966) 1555.
66KK KLEINBERGER, R., and R. KOUCHKOVSKY: Compt. Rend. **262** (1966) 628.
66LR LEHMANN, H. W., and M. ROBBINS: J. Appl. Phys. **37** (1966) 1389.
66MDAW MENYUK, N., K. DWIGHT, R. J. ARNOTT, and A. WOLD: J. Appl. Phys. **37** (1966) 1387.
66RBW RACCAH, P. M., R. J. BOUCHARD, and A. WOLD: J. Appl. Phys. **37** (1966) 1436.
67BP BERGER, S. B., and H. L. PINCH: J. Appl. Phys. **38** (1967) 949.
67BRW BALTZER, P. K., M. ROBBINS, and P. J. WOJTOWICZ: J. Appl. Phys. **38** (1967) 953.

67C	COLOMINAS, C.: Phys. Rev. **153** (1967) 558.
67EST	EIBSCHÜTZ, M., S. SHTRIKMAN, and Y. TENENBAUM: Phys. Letters **24A** (1967) 563.
67HB	HAACKE, G., and L. C. BEEGLE: J. Phys. Chem. Solids **28** (1967) 1699.
67HRBA	HAAS, C., A. M. J. G. VAN RUN, P. F. BONGERS, and W. ALBERS: Solid State Commun. **5** (1967) 657.
67L	LEHMANN, H. W.: Phys. Rev. **163** (1967) 488.
67L(1)	LOCHER, P. R.: Solid State Commun. **5** (1967) 185.
67LPS	LECRAW, R. C., H. VON PHILIPSBORN, and M. D. STURGE: J. Appl. Phys. **38** (1967) 965.
67P	PHILIPSBORN, H. VON: Helv. Phys. Acta **40** (1967) 810.
67RLW	ROBBINS, M., H. W. LEHMANN, and J. G. WHITE: J. Phys. Chem. Solids **28** (1967) 897.
67UV	ULLRICH, J. F., and D. H. VINCENT: Phys. Letters **25A** (1967) 731.
67WBR	WOJTOWICZ, P. J., P. K. BALTZER, and M. ROBBINS: J. Phys. Chem. Solids **28** (1967) 2423.
67YWC(1)	YOKOYAMA, H., R. WATANABE, and S. CHIBA: J. Phys. Soc. Japan **22** (1967) 659.
67YWC(2)	YOKOYAMA, H., R. WATANABE, and S. CHIBA: J. Phys. Soc. Japan **23** (1967) 450.
67YM	YAGNIK, C. M., and H. B. MUTHUR: Solid State Commun. **5** (1967) 841.
68BBB	BERGER, S. B., J. I. BUDNICK, and T. J. BURCH: J. Appl. Phys. **39** (1968) 658.
68BBB(1)	BERGER, S. B., J. I. BUDNICK, and T. J. BURCH: Phys. Letters **26A** (1968) 450.
68BFLS	BANUS, M. D., A. FERRETTI, M. C. LAVINE, and T. E. STACK: Solid State Res. Rep., Lincoln Laboratory, MIT (1968: 3) 28.
68BSL	BARTKOWSKI, R., J. P. SAGE, and R. C. LECRAW: J. Appl. Phys. **39** (1968) 1071.
68BZ	BONGERS, P. F., and G. ZANMARCHI: Solid State Commun. **6** (1968) 291.
68DBL	DARCY, L., P. K. BALTZER, and E. LOPATIN: J. Appl. Phys. **39** (1968) 898.
68FHSS	FRANKEL, R. B., J. J. HUNTZICKER, D. A. SHIRLEY, and N. J. STONE: Phys. Letters **26A** (1968) 452.
68HB	HAACKE, G., and L. C. BEEGLE: J. Appl. Phys. **39** (1968) 656.
68HBE	HARBEKE, G., S. B. BERGER, and F. P. EMMENEGGER: Solid State Commun. **6** (1968) 553.
68HC	HASTINGS, J. M., and L. M. CORLISS: J. Phys. Chem. Solids **29** (1968) 9.
68HS	HOY, G. R. and K. P. SINGH: Phys. Rev. **172** (1968) 514.
68K	KHOI, L. D.: Solid State Commun. **6** (1968) 203.
68L	LOTGERING, F. K.: J. Phys. Chem. Solids **29** (1968) 699.
68L(1)	LOTGERING, F. K.: J. Phys. Chem. Solids **29** (1968) 2193.
68LRBE	LEMS, W., P. J. RIJNIERSE, P. F. BONGERS, and U. ENZ: Phys. Rev. Letters **21** (1968) 1643.
68LS	LARSON, G. H., and A. W. SLEIGHT: Phys. Letters **28A** (1968) 203.
68LV	LOTGERING, F. K., and R. P. VAN STAPELE: J. Appl. Phys. **39** (1968) 417.
68MM	METHFESSEL, S., and D. C. MATTIS: Handbuch der Physik, Vol. XVIII/I, Magnetismus, p. 389, Springer-Verlag, Heidelberg/Berlin/New York 1968.
68OTI	OHBAYASHI, K., Y. TOMINAGA, and S. IIDA: J. Phys. Soc. Japan **24** (1968) 1173.
68PB	PINCH, H. L., and S. B. BERGER: J. Phys. Chem. Solids **29** (1968) 2091.
68RBL	ROBBINS, M., P. K. BALTZER, and E. LOPATIN: J. Appl. Phys. **39** (1968) 662.
68SRFDMW	STAUSS, G. H., M. RUBINSTEIN, J. FEINLEIB, K. DWIGHT, N. MENYUK, and A. WOLD: J. Appl. Phys. **39** (1968) 667.
68VT	VURAL, B., and E. E. THOMAS: Appl. Phys. Letters **12** (1968) 14.
69AG	AMITH, A., and G. L. GUNSALUS: J. Appl. Phys. **40** (1969) 1020.
69BBB	BERGER, S. B., J. I. BUDNICK, and T. J. BURCH: Phys. Rev. **179** (1969) 272.
69BHRZ	BONGERS, P. F., C. HAAS, A. M. J. G. VAN RUN, and G. ZANMARCHI: J. Appl. Phys. **40** (1969) 958.
69BMV	BUSCH, G., B. MAGYAR, and O. VOGT: Solid State Commun. **7** (1969) 509.
69GDP	GIBART, P., J. L. DORMANN, and Y. PELLERIN: Phys. Status Solidi **36** (1969) 187.
69LE	LEHMANN, H. W., and F. P. EMMENEGGER: Solid State Commun. **7** (1969) 965.
69LVVV	LOTGERING, F. K., R. P. VAN STAPELE, G. H. A. M. VAN DER STEEN, and J. S. VAN WIERINGEN: J. Phys. Chem. Solids **30** (1969) 799.
69MKWWP	MARTIN, G. W., A. T. KELLOGG, R. L. WHITE, R. M. WHITE, and H. PINCH: J. Appl. Phys. **40** (1969) 1015.
69NS	NEIDA, A. R. VON, and L. K. SHICK: J. Appl. Phys. **40** (1969) 1013.
69RMS	ROBBINS, M., M. A. MIKSOVSKY, and R. C. SHERWOOD: J. Appl. Phys. **40** (1969) 2466.
69S(1)	SRIVASTAVA, V. C.: J. Appl. Phys. **40** (1969) 1017.
69S(2)	STAUSS, G. H.: Phys. Rev. **181** (1969) 636.
69YC	YOKOYAMA, H., and S. CHIBA: J. Phys. Soc. Japan **27** (1969) 505.
70M	MIYATANI, K.: J. Phys. Soc. Japan **28** (1970) 259.
—DM	DWIGHT, K., and N. MENYUK (unpublished).

8 Index of substances for III/4a and III/4b — Substanzenverzeichnis für III/4a und III/4b

The index of substances includes the compounds from both part a and part b in Vol. III/4. The letters **a** and **b** before the page numbers refer to these parts.

The series is alphabetical:

The compounds with stoichiometric formulae are listed according to the alphabetically arranged gross formulae and ascending indices of each of the elements, for example, $LiTiAlO_4$ appears as $AlLiO_4Ti$. The systems with non-stoichiometric formulae (solid solutions, special systems, and series of solid solutions) are arranged simply according to the alphabetical order of the symbols of their elements, omitting the indices and connecting the symbols with hyphens, for example, $Ba_{1-x}Bi_xNi_xFe_{12-x}O_{19}$ is found as Ba–Bi–Fe–Ni–O, and $BiFeO_3$–$BaTiO_3$ as Ba–Bi–Fe–O–Ti.

These systems and also a large part of the gross formulae are identified in addition by an abbreviation: (P) = perovskite or perovskite related compound, (G) = garnet, (Sp) = spinel, and (HexF) = hexagonal ferrite.

Doped substances appear twice: first according to the formula of the host substance and secondly according to the symbol of the dopant, for example, $Al_5O_{12}Tm_3$: Fe appears also as: Fe in $Al_5O_{12}Tm_3$.

Numbers in italics refer to information given in the text.

Das Substanzenverzeichnis umfaßt die Substanzen beider Teilbände von Band III/4; die Buchstaben **a** und **b** vor den Seitenzahlen weisen auf die beiden Teilbände a und b hin.

Die Reihenfolge ist alphabetisch:

Die Verbindungen mit stoichiometrischen Formeln sind nach den alphabetisch geordneten Bruttoformeln und steigenden Indizes jedes Elementsymbols eingeordnet, z. B. $LiTiAlO_4$ bei $AlLiO_4Ti$. Systeme mit nichtstoichiometrischen Formeln (Mischkristalle und spezielle Systeme sowie Mischkristallreihen) sind, ohne Berücksichtigung der quantitativen Zusammensetzung, nach der alphabetisch geordneten Folge ihrer Elementsymbole, die einfach durch Bindestriche verbunden sind, eingeordnet, z. B. $Ba_{1-x}Bi_xNi_xFe_{12-x}O_{19}$ bei Ba–Bi–Fe–Ni–O, $BiFeO_3$–$BaTiO_3$ bei Ba–Bi–Fe–O–Ti.

Diese Systeme und auch ein großer Teil der Bruttoformeln sind zusätzlich durch eine in Klammern gesetzte Abkürzung gekennzeichnet: (P) = Perowskit oder perowskitähnliche Verbindung, (G) = Granat, (Sp) = Spinell und (HexF) = hexagonaler Ferrit.

Dotierte Substanzen erscheinen zweimal: erstens unter der Formel der Grundsubstanz und zweitens unter dem Symbol der Dotierung, z. B. $Al_5O_{12}Tm_3$: Fe auch unter: Fe in $Al_5O_{12}Tm_3$.

Kursiv gesetzte Seitenzahlen weisen auf eine Information innerhalb des Textes hin.

Mercedes-Druck, Berlin 61